2007 International Conference on Thermal, Mechanical & Multi-Physics Simulation and Experiments in Microelectronics & Micro System

London, United Kingdom
15-18 April 2007

Volume 1 of 2

IEEE Catalog Number: 07EX1736
ISBN: 1-4244-1105-X

**Copyright © 2007 by The Institute of Electrical and Electronics Engineers, Inc.
All Rights Reserved**

Copyright and Reprint Permissions: Abstracting is permitted with credit to the source. Libraries are permitted to photocopy beyond the limit of U.S. copyright law for private use of patrons those articles in this volume that carry a code at the bottom of the first page, provided the per-copy fee indicated in the code is paid through Copyright Clearance Center, 222 Rosewood Drive, Danvers, MA 01923.

For other copying, reprint or republications permission, write to IEEE Copyrights Manager, IEEE Operations Center, 445 Hoes Lane, Piscataway, New Jersey USA 08854. All rights reserved.

IEEE Catalog Number: 07EX1736

ISBN: 1-4244-1105-X

LOC: 2007922208

Additional Copies of This Publication Are Available from:

IEEE Service Center
445 Hoes Lane
Piscataway, NJ 08854
IEEE Service Center
445 Hoes Lane
Piscataway, NJ 08854
Phone: (800) 678-IEEE
                (732) 981-1393
Fax: (732) 981-9667
E-mail: customer-service@ieee.org

2007 International Conference on Thermal, Mechanical & Multi-Physics Simulation and Experiments in Microelectronics & Micro System

London, United Kingdom
15-18 April 2007

IEEE Catalog Number: CFP07566-POD
ISBN: 978-1-42441-105-4

Table of Contents

Thermal and hot spot evaluations on oil immersed power Transformers by FEMLAB and MATLAB software's 1
Kourosh Mousavi Takami, Hasan Gholnejad, Jafar Mahmoudi

Molecular Dynamics Simulation for the diffusion of water in amorphous polymers examined at different Temperatures 7
E. Dermitzaki, J. Bauer, H. Walter, B. Wunderle, B. Michel, H. Reichl

Mechanical Characterization and Viscoplastic-Damage Constitutive Model of SnAgCu Solder 16
Zhou Jun, Chai GuoZhong, Liu Yong, Liang LiHua, Hao WeiNa

Atom diffusion mechanism of thermo-sonic flip chip bonding interface 23
Fuliang WANG, Junhui LI, Lei HAN, Jue ZHONG

A coupled numerical and experimental study on thermo-mechanical fatigue failure in SnAgCu solder joints 27
M. Erinc, P.J.G. Schreurs, M.G.D. Geers

Magnetic Field Sensor Using a Physical Model to Pre-Calculate the Magnetic Field and to Remove Systematic Error due to Physical Parameters 33
Simon Hainz, Mario Jungwirth

Unsteady CFD Modelling for Electronics Cooling 39
James C. Tyacke, Paul G. Tucker

Combined Virtual Prototyping and Reliability Testing Based Design Rules for Stacked Die System in Packages 46
W.D. van Driel, R. A. Real, D.G. Yang, G.Q. Zhang, J. Pasion

Evaluation of Thermal Strains in BGA Packages Using Digital Speckle Correlation Technique and FEA 51
Adam Robert Zbrzezny, Vincent Chan, Hua Lu, Ming Zhou

Quantification of creep strain in small lead-free solder joints with the in-situ micro electronic-resistance measurement 56
Li Jiang, Keling Yang, Jiemin Zhou, Ke Xiang, Wenjie Wang

Computational Design and Optimisation of Mechanically Reinforced Masks for Stencil Lithography 62
Maryna Lishchynska, Marc A.F. van den Boogaart, Juergen Brugger, James C. Greer

Extended Finite Element for Electromechanical Coupling 69
Veronique Rochus, Daniel Rixen

Finite Element Modelling (FEM) of Green Electronics in Aeronautical and Military Communication Systems (GEAMCOS) 77
A. Chaillot, G. Massiot, C. Munier, I. Lombaërt-Valot, S. Bousquet, C. Chastanet, D. Plouseau, E. Munier, D. Maron, P. Raynal, S. Villard, R. Dumonteil

Thermomechanical Multiscale Modelling of Substrates 85
R.L.J.M. Ubachs, O. van der Sluis, W.D. van Driel, G.Q. Zhang

Creep Measurements of 200 µm - 400 µm Solder Joints 93
Mike Röllig, Steffen Wiese, Karsten Meier, Klaus-Jürgen Wolter

Finite Element Modeling of Thermoelastic Damping in Filleted Micro-Beams 102
Severine Lepage, Jean-Claude Golinval

Numerical Analysis of the Reliability of Cu/low-k Bond Pad Interconnections Under Wire Pull Test: Application of a 3D Energy Based Failure Criterion 109
Sébastien Gallois-Garreignot, Vincent Fiori, Stéphane Orain, Olaf van der Sluis

Thermal Simulation with Coupled Network Models on System Level 116
Adam Augustin, Bartosz Maj

A PERTURBATION METHOD FOR THE 3D FINITE ELEMENT MODELING OF ELECTROSTATICALLY DRIVEN MEMS 120
Mohamed Boutaayamou, Ruth V. Sabariego, Patrick Dular

iii

Table of Contents

Failure Analysis of Power Silicon Devices at Operation above 200 °C Junction Temperature 125
Vasile V.N. Obreja, Keith I. Nuttall, Octavian Buiu, Steve Hall

Simulation of Wafer Probing Process Considering Probe Needle Dynamics 131
Ilko Schmadlak, Torsten Hauck

Modeling of the mechanical behavior during programming of a non-volatile phase-change memory cell using a coupled electrical-thermal-mechanical finite-element simulator 136
Thomas Gille, Judit Lisoni, Ludovic Goux, Kristin De Meyer, Dirk J. Wouters

Surface Evolution of Strained Thin Solid Films: Stability Analysis and Time Evolution of Local Surface Perturbations 142
E. Dornel, J-C Barbé, J. Eymery, F. de Crécy

Gold Wire Bonding Induced Peeling in Cu/Low-k Interconnects: 3D Simulation and Correlations 150
Vincent Fiori, Lau Teck Beng, Susan Downey, Sebastien Gallois-Garreignotd, Stephane Orain

Reliability Based Design Optimisation for System-in-Package 159
S. Stoyanov, J. M. Yannou, C. Bailey, N. Strusevich

Multiphysic modelling of a microactuator based on the decomposition of an energetic material: application to microfluidics 167
Gustavo Adolfo Ardila Rodríguez, Carole Rossi

Crack Degradation Model Derived From Experimental Strain-Stress Data 175
Milos Dusek, Christopher Hunt

CFD Aided Reflow Oven Profiling for PCB Preheating in a Soldering Process 181
Ilja Belov, Mats Lindgren, Peter Leisner, Fredrik Bergner, Robin Bornoff

The Effect of Downscaling the Dimensions of Solder Interconnects on their Creep Properties 189
S. Wiese, M. Roellig, M. Mueller, K.-J. Wolter

The Dependence of Composition, Cooling Rate and Size on the Solidification Behaviour of SnAgCu Solders 197
Maik Mueller, Steffen Wiese, Mike Roellig, Klaus-Juergen Wolter

Electromigration in Large Volume Solder Joints 207
Karsten Meier, Mike Roellig, Steffen Wiese, Carsten Goette, Ulrich Deml, Klaus-Juergen Wolter

Multi-scale energy-based failure modeling of bond pad structures 214
O. van der Sluis, R.B.R. van Silfhout, R.A.B. Engelen, W.D. van Driel, G.Q. Zhang

Impact of Solder Overflow and ACLV Moisture Absorption of Mold Compound on Package Reliability 220
Richard Qian, Yong Liu, Scott Irving, Timwah Luk

Experimental Determination and Modification of Anand Model Constants for Pb-Free Material 95.5Sn4.0Ag0.5Cu 225
Wang Qiang, Liang Lihua, Chen Xuefan, Weng Xiaohong

Numerical Simulation and Thermal Failure Analysis of SOM Package 234
Jae Choon Kima, Jin Taek Chunga, Won Suk Lee, Gyoung Bum Kimb, Dong Jin Lee, Ji-Man Cho, J.-Hyuk Yu, Byeong-Kwon Ju, Sung Woo Hang, Heung-Woo Park, Sang-Kyeong Yun

A Mathematical Technique to estimate the High Frequency Current Inside the Silicon Die from the Noise Measurements 239
Bidyut K. Bhattacharyya, Gang Huo

Thermal-mechanical Modelling of Power Electronic Module Packaging 243
Hua Lu, Tim Tilford, Xiangdong Xue, Chris Bailey

Multi Physics Modelling of the Electrodeposition Process 249
M. Hughes, C. Bailey, K. McManus

Application of Higher Order Derivatives Method to Parametric Simulation of MEMS 257
Vladimir Kolchuzhin, Jan Mehner, Thomas Gessner, Wolfram Doetzel

Table of Contents

Multi-objective Parametric Approach to Numerical Optimization of Stacked Packages 263
Lukasz Dowhan, Artur Wymyslowski, Rainer Dudek

Research of Stacked VIA's Mechanical Stress 270
Tohru Nakanishi, Hideo Ohkuma, Hiroshi Ohira

Research of Material Properties Reliance for POP with Numerical Analysis 278
Tohru Nakanishi, Hiroshi Ohira, Hideo Ohkuma

Cure shrinkage characterization and its implementation into correlation of warpage between simulation and measurement 283
W.H. Zhu, Guang Li, Wei Sun, F.X. Che, Anthony Sun, C.K. Wang, H.B. Tan, B.Z. Zhao, N.H. Chin

Optimizing the Dynamic Response of RF MEMS Switches using Tailored Voltage Pulses 291
Vitaly Leus, Arnon Hirshberg, David Elata

The Effect of Visco-elasticity on the Result Accuracy of FEM Panel Warpage Simulations Supporting Industrial Microelectronics Packaging 295
Sven Rzepka, Axel Müller

Parametric Study of Fracture Properties in Polycrystalline MEMS 303
Fabrizio Cacchione, Alberto Corigliano, Sarah Zerbini

A Modelling Framework for the Reliability of Safety Critical Electronics 308
Diana Segura Velandia, Paul P. Conway, Antony Wilson, Andrew A. West, David C. Whalley

Development of an Internal Liquid Cooling System for CPU using CAE 314
Sebastian Flores, Song-Hao Wang, Chong-Qing Chang

Kinetic Characterisation of Molding Compounds 319
K. M. B. Jansen, C. Qian, L. J. Ernst, C. Bohm, A. Kessier, H. Preu, M. Stecher

A Study of Failure Mechanism and Reliability Assessment for the Panel Level Package (PLP) Technology 324
Ming-Chih Yew, Hsiu-Ping Wei, Ching-Shun Huang, Dyi-Chung Hu, Wen-Kung Yang, Kou-Ning Chiang

A Novel Micro/Nano Electromechanical Actuator with Integral Electrodes 332
Jofre Pallarès, Manuel Carmona, Marta Duch, Marta Gerbolés, Lluís Terés

Design of Metal Interconnects for Stretchable Electronic Circuits using Finite Element Analysis 340
Mario Gonzalez, Fabrice Axisa, Mathieu Vanden Bulcke, Dominique Brosteaux, Bart Vandevelde, Jan Vanfleteren

Numerical Modeling of Electrical Resistance of Interconnections in High-Tech Multilayer PCBs Manufactured by Magnetron Sputtering Deposition of Copper 346
Janusz Borecki, Artur Wymyslowski

Simulation of Failure Criteria in Dielectric Layers 352
Khalil Arshak, Ivor Guiney, Edward Forde

Dynamic mechanical behavior of SnAgCu BGA solder joints determined by fast shear tests and FEM simulations 357
Eberhard Kaulfersch, Sven Rzepka, Vijay Ganeshan, Axel Müller, Bernd Michel

Mechanical Characterization Analysis of flexible and stretchable Ultra-Thin Substrates by Experiment and FE Simulation 361
L. Wang, T. Zoumpoulidis, K.M.B. Jansen, M. Bartek, G.Q. Zhang, L.J. Ernst

Study on the Board-level SMT Assembly and Solder Joint Reliability of Different QFN Packages 366
Wei Sun, W.H. Zhu, Retuta Danny, F.X. Che, C.K. Wang, Anthony Y.S. Sun, H.B. Tan

Thin Film Interface Fracture Properties at Scales Relevant to Microelectronics 372
A. Xiao, L. G. Wang, W. D. van Driel, O. van der Sluis, D. G. Yang, L. J. Ernst, G. Q. Zhang

Design for Reliability with AuSn Interconnects 378
Rainer Dudek, Olaf Wittler, Wolfgang Faust, Birgit Brämer, Matthias Klein, Wei Jun, Bernd Michel

Table of Contents

Creep Analysis of a Lead-free Surface Mount Device 385
Pradeep Hegde, David Whalley, Vadim. V. Silberschmidt

Multi-Scale Modeling of Shock-Induced Failure of Polysilicon MEMS 392
Aldo Ghisi, Fabio Fachin, Stefano Mariani, Alberto Corigliano, Sarah Zerbini

Packaging Effects of Cu/Low-k Interconnect Structure 398
Ming-Che Hsieh

Validation of Dynamic Thermal Simulations of Power Assemblies Using a Thermal Test Chip 403
X. Jordà, M. Vellvehi, X. Perpinyà, J.L. Galvez, P. Godignon

Thermo-mechanical investigations on the effects of the solder meniscus design in solder joint lifetime for power electronic devices 409
M. Bouarroudj, Z. Khatir, S. Lefebvre, L. Dupont

Development and Assessment of Global-Local Modeling Technique Used in Advanced Microelectronic Packaging 416
F. X. Che, H.L.J. Pang, W. H. Zhu, Wei Sun, Anthony Y.S. Sun, C.K. Wang, H.B. Tan

Transport of Corrosive Constituents in Epoxy Moulding Compounds 423
M. van Soestbergen, L.J. Ernst, G.Q. Zhang, R.T.H. Rongen

Development of Reliability Verification System for Robust Package Design 428
Dongkil Shin, Hyunggil Baek, Joonyoung Oh, Dongok Kwak, Kuyoung Kim, Younghee Song, Jeongyeol Kim

The effect of board stiffness on the solder-joint reliability of HVQFN-packages 435
J. de Vries, M. Jansen, W. van Driel

Investigation of Mechanical Reliability of Cu/low-k Multi-layer Interconnects in Flip Chip Packages 442
Chihiro J. UCHIBORI, Xuefeng Zhang, Sehyuk Im, Paul S. Ho, Tomoji Nakamura

Packaging Design and Testing for High Temperature Applications > 150 °C 447
Thomas Schreier-Alt, Christian Rebholz, Frank Ansorge

An enriched cohesive zone model for numerical simulation of interfacial delamination in microsystems 454
M. Samimi, B.A.E. van Hal, R.H.J. Peerlings, J.A.W. van Dommelen, M.G.D. Geers

Underexpanded Micro-nozzle Flow Simulation with Coupled Thermal-Fluid Modeling 461
José Hermida Quesada, José A. Moríñigo, Francisco Caballero Requena

Optimization of Cu Low-k bond pad designs to improve mechanical robustness using the Area Release Energy method 468
R. A. B. Engelen, O. van der Sluis, R. B. R. van Silfhout, W.D. van Driel, V. Fiori

The chemical-mechanical relationship of the SiOC(H) dielectric film 472
Cadmus Yuan, O. van der Sluis, G. Q. Zhang, L. J. Ernst

Magnetically actuated microvalve for disposable drug infusor 478
M. Duch, J.Casals-Terré, J. A. Plaza, J. Esteve, R Pérez-Castillejos, E. Vallés, E. Gómez

Generating VHDL3AMS Models of Digital-to-Analogue Converters From MATLAB®/SIMULINK® 484
Alexandre Cesar Rodrigues da Silva, Ian Grout, Jeffrey Ryan, Thomas O'Shea

Prediction of the Influence of Induced Stresses in Silicon on CMOS Performance in a Cu-Through-Via Interconnect Technology 491
Chukwudi Okoro, Mario Gonzalez, Bart Vandevelde, Bart Swinnen, Geert Eneman, Peter Verheyen, Eric Beyne, Dirk Vandepitte

Effect of Processing Parameters and Hygro-thermo-mechanical Stresses on the Reliability of Flip Chip Bonding RFID Tags 498
D.G. Yang, E. de Bruin, B. Kasemset, W. D. van Driel

Efficient electrostatic-mechanical modeling of C-V curves of RF-MEMS switches 503
Jeroen Bielen, Jiri Stulemeijer

Table of Contents

Evaluation of Creep in RF MEMS Devices 509
Marcel van Gils, Jeroen Bielen, Gavin McDonald

Investigation of Carbon Nanotube Performance under External Mechanical Stresses and Moisture 515
Haibo Fan, Kai Zhang, Matthew M.F. Yuen

Wafer Probing Simulation for Copper Bond Pad Based BPOA Structure 519
Yumin Liu, Yong Liu, Scott Irving and Timwah Luk

Optimization and Robust Design of Electronics Assemblies under Fracture, Delamination and Fatigue Aspects 524
Jürgen Auersperg, Matthias Klein, Bernd Michel

Mechanical Characterization of III-V Nanowire Using Molecular Dynamics Simulation 532
Alex. W. Dawotola, C. A. Yuan, W. D. van Driel, E. P. A. M. Bakkers, G. Q. Zhang

Stress Relaxation Characterization of Hypoeutectic Sn3.0Ag0.5Cu Pb-free Solder: Experiment and Modeling 537
Gayatri Cuddalorepatta, Dominik Herkommer, Abhijit Dasgupta

Effect of Surrounding Air on Board Level Drop Tests of Flexible Printed Circuit Boards 545
Luciano Arruda, Germano Freitas

Numerical Simulation of Ion Drift within Ion Mobility Spectrometers in High Peclet Conditions using FEM Techniques 549
A. Kalms, M. Salleras, Z. Liu, J.G. Korvink, J. Goebel, G. Müller, J. Samitier, S. Marco

Electro-osmotic Flow Based Cooling System For Microprocessors 555
P. F. Eng, P. Nithiarasu, A. K. Arnold, P. Igic, O. J. Guy

Parameter Identification on Wafer Level of Membrane Structures 560
Steffen Michael, Siegfried Hering, Gisbert Hölzer, Tobias Polster, Martin Hoffmann, Arne Albrecht

A micromachined thermoelectric sensor for natural gas analysis: Thermal model and experimental results 565
G. Carles, S. Udina, M. Salleras, J. Santander, L. Fonseca, S. Marco

Behavioral Modelling of Vibrating Piezoelectric Micro-Gyro Sensor and Detection Electronics 571
S. Megherbi, R. Levy, F. Parrain, H. Mathias, O. Le Traon, D. Janiaud, J. P. Gilles

Validation of constitutive models for electrically conductive adhesives 575
Marcel Meuwissen, Monique van den Nieuwenhof, Henk Steijvers, Adri van der Waal, Tom Bots

Simultaneous Measurement of Young's Modulus and Damping Dependence on Magnetic Fields by Laser Interferometry 583
A. L. Morales, A. J. Nieto, R. Moreno, A. González, J. M. Chicharro, P. Pintado

Validation Of Simulation Platform By Comparing Results And Calculation Time Of Different Softwares. 588
Hikmat ACHKAR, Fabienne PENNEC, David PEYROU, Mahmoud AL AHMAD, Marc SARTOR, Robert PLANA, Patrick PONS

Multi-Physics Modelling for Microelectronics and Microsystems - Current Capabilities and Future Challenges. 593
C. Bailey, H. Lu, S. Stoyanov, M. Hughes, C. Yin, D. Gwyer

USING MOLECULAR MODELING TO UNDERSTAND CLEANER EFFICIENCY FOR BARC ("BOTTOM ANTI-REFLECTIVE COATING") AFTER PLASMA ETCH IN DUAL DAMASCENE STRUCTURES 601
Nancy Iwamoto, Deborah Yellowaga, Amy Larson, Ben Palmer, Teri Baldwin-Hendricks

Numerical Simulation of Creep Strain of PBGA Solders under Thermal Cycling 607
F. L. Sun, Y. Liu, L. F. Wang

Comparative Sensitivity Analysis for μBGA and QFN Reliability 611
Jürgen Wilde, Elena Zukowski

Table of Contents

Experimental Determination of Time-Independent Elastic-Plastic Behaviour of Solder Joints at High Strain Rates 618
S. Wiese, K. Meier, D. Scholz, A. Müller, M. Röllig, S. Rzepka, K. J. Wolter

The BGK kinetic model applied to the analysis of gas-structure interactions in MEMS 623
Attilio Frangi, Aldo Ghisi

Development of 2D Modeling Techniques for the Thermal Fatigue Analysis of Solder Joints of a Module Mounted in a 3D Cavity on a Printed Circuit Board 631
Y. S. Chan, S. W. Ricky Lee, Yuming Ye, Sang Liu

A Macro Model Based On Finite Element Method To Investigate Temperature And Residual Stress Effects On RF MEMS Switch Actuation 637
D. Peyrou, H. Achkar, F. Pennec, P. Pons, R. Plana

Atomistic Simulations of Interface Properties in Metals 641
Tomotsugu Shimokawa

Wire Bond Reliability for Power Electronic Modules - Effect of Bonding Temperature 647
Wei-Sun Loh, Martin Corfield, Hua Lu, Simon Hogg, Tim Tilford, C Mark Johnson

Reduced Order Electro-Thermal Models for Real-Time Health Management of Power Electronics 653
Mahera Musallam, Cyril Buttay, C Mark Johnson, Chris Bailey, Michael Whitehead

Reliability Test Method Overiew to Characterize Second Level Interconnects 659
M. Brizoux, H. Fremont, Y. Danto, W. C. Maia Filho

Kinetic Analysis of Electromigration Enhanced Intermetallic Growth and Void Formation in Pb-Free Solders 664
Brook Chao, Seung-Hyun Chae, Xuefeng Zhang, Paul S. Ho

Measurement of Dynamic Properties of MEMS and the Possibilities of Parameter Identification by Simulation 672
Matthias Ebert, Falk Naumann, Ronny Gerbach, Joerg Bagdahn

DESIGN-FOR-RELIABILITY TOOLS FOR HIGHLY INTEGRATED SYSTEM-ONPACKAGE TECHNOLOGY 678
R. V. Pucha, S. Hegde, M. Damani, K. J. Leee, K. Tunga, A. Perkins, S. Mahalingam, G. Lo, K. Klein, J. Ahmad, S. K. Sitaraman

Block-Diagram Based SIMULINK Analysis for the Drop Impact Response of a Mobile Electronic System 679
Jiang Zhou, Paul Corder, Ratna P. Niraula

Fluctuation Mechanism of Mechanical Properties of Electroplated-Copper Thin Films Used for Three Dimensional Electronic Modules 686
Hideo Miura, Ken Suzuki, Kinji Tamakawa

Computational Methods and High Speed Imaging Methodologies for Transient-Shock Reliability of Electronics 692
Pradeep Lall

Low-temperature and Pressureless Sintering Technology for High-performance and High-temperature Interconnection of Semiconductor Devices 705
Guo-Quan Lu, Jesus N. Calata, Guangyin Lei, Xu Chen

3D Modeling of Electromigration Combined with Thermal-Mechanical Effect for IC Device and Package 710
Yong Liu, Scott Irving, Timwah Luk, Lihua Liang, Shinan Wang

Fatigue Strength and Damage Behaviors of Multi-Scale Metallic Films and Multilayers 723
G. P. Zhang, X. F. Zhu, Y. P. Li, Z. G. Wang

Prognostics and Health Monitoring of Electronics 729
Michael Pecht, Brian Tuchband, Nikhil Vichare, Qu Jian Ying

The Changing Role of CFD in Electronics Thermal Design 737
Dr. John Parry

viii

Table of Contents

Digital and Continuous Liquid Cooling for Electronic Systems..738
M. Baelmans, H. Oprins, T. Stevens, F. Rogiers

Solid State Chemical Sensors: Technologies and Applications..745
Krishna C. Persaud, Anthony Flint, Robert W. Sneath,

Thermal and hot spot evaluations on oil immersed power Transformers by FEMLAB and MATLAB software's

Kourosh Mousavi Takami[1], Hasan Gholnejad[2] and Jafar Mahmoudi[3]

1,2:Ph.D. student in Mälardalen University, 3: professor in Mälardalen University

Box 883,721 23, IST Dep., Mälardalen University, Västerås, Sweden

1.kourosh.mousavi.takami@mdh.se, 2.hasan.gholinejad@mdh.se, 3.jafar.mahmoudi@mdh.se

Abstract

Transformers are important and expensive elements of a power system. Inordinate localized temperature rise, hottest spot temperature (HST), causes rapid thermal degradation of insulation and subsequent thermal breakdown. To prescribe the limits of short-term and long-term loading capability of a transformer, it is necessary to estimate the HST of transformer winding to as high a degree of accuracy as can possibly be made. These papers have now improved the accuracy of estimation of hottest spot temperature. Inordinate temperature rise in a power transformer due to load current is known to be the most important factor in causing rapid degradation of its insulation and decides the optimum load catering ability or the load ability of a transformer. The Top Oil Temperature (TOT) and Hottest Spot Temperature (HST) being natural outcome of this process, an accurate estimation of these parameters is of particular importance. IEEE / IEC among others have proposed procedure to estimate the temperatures, however, the accuracy of the predictions are not always as good as are desired. Unacceptable temperature rise may occur due to several fault conditions other than overloading, and hence warrant an online monitoring of the transformer.

Nomenclatures

A(t)	Time variation of heat generated.
C_p	Specific heat at constant pressure.
f	Friction factor.
g	Acceleration due to gravity.
h	Heat transfer coefficient.
K	Thermal conductivity.
l	Length.
p	Pressure.
P_b	Barometric pressure.
P^*	Estimated pressure value.
P'	Corrected pressure value.
q''	Heat flux at walls.
q'''	Heat generation per unit volume due to energy losses.
S_C, S_P	Source terms in the transport equation.
t	Time.
T	Temperature.
T_f	Surrounding temperature.
u,v	Velocity components.
x, y, z	Cartesian coordinates.

α	Relaxation parameter.
α_p	Relaxation parameter for pressure.
Γ	Diffusion coefficient .
Δt	Time difference.
ε	Emissivity.
ρ	Density.
μ	Dynamic viscosity.
σ	Stefan-Boltzmann coefficient .
σ_i	Surface tension .
Φ	Dependent variable .

Subscripts

i	In the position i.
t	At the time t.
u	For the u velocity component.
V	For the v velocity component .
w	At the tank wall.
x	in the x direction.
x_b	In the x direction at the bottom.
x_t	In the direction x at the top.
y	In the direction y.

1. Introduction

TRANSFORMERS are important components in a power system and if this device with every reason is gone out from network; very damages will be occurring in network. Localized temperature rise in the winding takes place due to load current coupled with short time over - loads, which the system is asked to cater to. Hot Spots, regions of inordinate temperature rise in the winding and top surface of insulating oil (Hottest spot temperature, HST, and Top Oil Temperature, TOT) causes rapid thermal degradation of insulation and subsequent thermal run away. In order to avoid catastrophic failure of such expensive equipment, monitoring of the condition of insulation there- in is undertaken on a regular basis. The genesis of the two zonal Temperatures TOT and HST is a complex process in itself, vitiated further by the thermal inhomogeniety of the winding [1] and dynamic nature of the load and ambient [2]. The thermal conductivity of the insulation part being largely different from that of the conductor and the fact that there is anisotropy in this quantity in both radial as well as axial directions complicates the estimation procedure. In the recent past, technical bodies, IEEE and IEC, have suggested empirical methods for estimating the temperature extremes. In view

1-4244-1105-X/07/$25.00 ©2007 IEEE

of the fact that the estimation of temperature using the formulations there in is not as accurate as are desired, new and improved methods need to be developed to mitigate the situation.

A perusal of current literature in this area indicates that in the recent past, conventional heat run tests performed after the transformer was manufactured were the only means of acquiring information on the magnitude of the possible HST, but the location of HST was not possible. The extrapolation of heat run test results to quantify the amount of insulation degradation was found to be highly inaccurate [2]–[5]. Recently, various methods have been suggested for direct measurements of HST in transformer windings using fiber-optic sensors and fluoro-optic thermometers and the optimal location of these sensors has also been demonstrated. It has been reported [6], [7] that optoelectronic modules have since been introduced in the full-size transformers.

Certain empirical formulae have been proposed [4], [8], [11] whereby transformer designers could estimate possible maximum winding temperature at the design stage. These formulae, although not very accurate, serve as broad guidelines for verifying the designs. It is therefore important to note that a worthwhile treatment of the thermal aspects of a transformer has been made only very recently [9], [10] wherein a semi theoretical model has been proposed to study certain aspects of heat transfer in cylindrical geometries. A comprehensive report generated by the CIGRE working group 12, examines, critically, all aspects of diagnosis of power transformers to show that the conventional factory heat run test and/or gas analysis fail to provide information on the magnitude of HST and overloading limits [2], [3]. In continuation to the IEEE Transformer loading guide [8], a recent addendum [12] to it gives further information on thermal modeling of a power transformer. An equivalent circuit analogy of transformer model considered by Swift *et al* [13]. Admittedly, these models are mathematically much simpler. The calculation of average bulk oil and winding temperature can be made, knowing the thermal capacity of the solid (winding) or liquid (oil) medium. Since the masses of winding and oil bulk are known from the design, the heat capacity of bulk oil and winding can be easily calculated. These models present difficulties when used to estimate temperature of the hottest spot, or top-oil temperature.

Earlier work in this area [1], [14] assumes a constant heat source operating over the entire winding structure, besides treating the winding to be thermally isotropic. Also, a single empirical heat-transfer formula suggested thus far for calculating the boundary layer temperature drop has been applied to all surfaces of the winding [14]. But this is found to be incorrect as the mechanism of natural heat convection in the axial direction is markedly different from that along the radial direction. Such an assumption holds true for layer-type winding only. The authors have now embarked on a technique for improving the accuracy of predicting the magnitude and the location

of the HST. This calls for a major revision of the thermal model considering the following changes.

➤ Oil is a Newtonian and incompressible fluid.
➤ Properties of solid materials and fluids change with the temperature.
➤ Fluid flow is laminar, unsteady and two-dimensional.
➤ There is internal heat generation.
➤ Effects of thermal radiation are included.
➤ The transformer winding shall be treated as a thermally anisotropic structure, which is actually true in practice.
➤ The heat source should be considered as a function of the local temperature (distributed heat source) in cylindrical or Cartesian characterizing.
➤ The initial temperature distribution function should be rewritten as a new function of the curvature of the transformer winding.

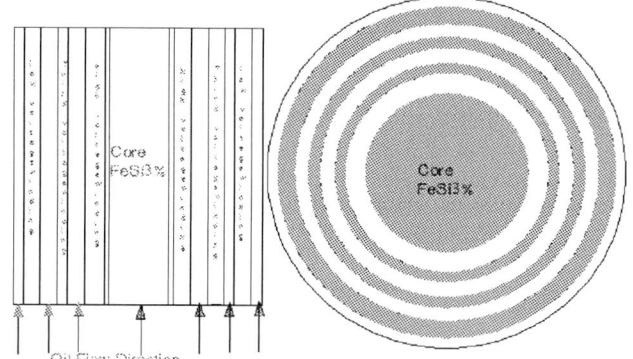

Fig.1. Simple model of transformer winding, HV = disc, LV = layer.

In most of work by another authorized, the thermal time constant of the winding was assumed to be a notional value of 5 min. In this paper, it has been found with calculation.

2. Experimental Data and *Governing Equations*

2.1. Simplified Transformer Winding and experimental data

Text The structure of a transformer winding is complex and does not conform to any known geometry in the strict sense. Under fairly general conditions, the transformer cores can be assumed rectangular in formation; hence, a layer or a disc winding is considered to be a finite annular cylinder. The thermal and physical properties of the system would be equivalent to a composite system of insulation and conductor. A simple geometry of a model transformer winding is given in Fig. 1. In the current paper, the heat generated in the body of the winding is taken to be a function of the local temperature [see (9)]. The heat is conducted away from here by the insulating oil in constant circulation in the vertical and horizontal ducts by a process of convection. However, in an actual transformer, the copper conductor is the heat source and is distributed over the volume of the cylinder defining the winding geometry. The heat generated in insulation due to dielectric loss is normally ignored.

The process of heat transfer in transformers is due both to conduction and to convection. The mixed boundary value problem describing the process automatically takes care of both the modes of heat transfer.

Figure 1 shows that:
- Oil only flows through ducts
- 10 degree winding rises are typical; makes two stage cooling almost impossible
- Number of ducts limited by short-circuit strength
- Demand a heat-run if two stage cooling is expected

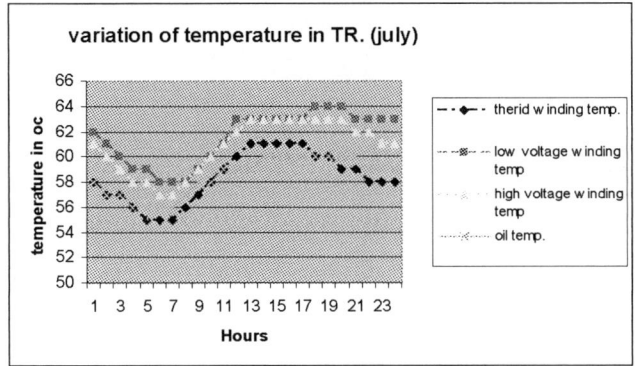

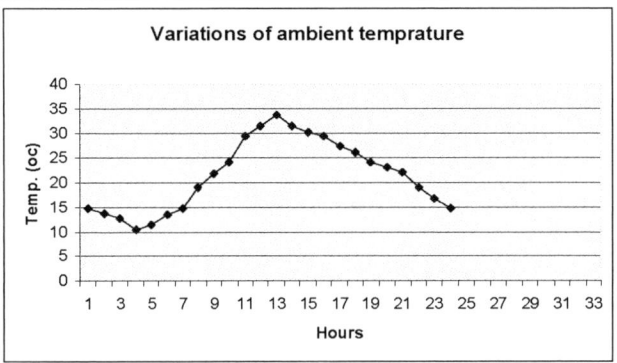

Fig.2. Variation of temperature in transformer and ambient for three winding 250MVA, 230/63/0.4KV transformer on July 2006 in Pounel 230/63/20KV substation – Iran

Figure 2 shows the variations of temperature in oil, LV winding, HV winding and third winding a 230/63/0.4 KV transformer and ambient temperature in July 2006.

According IEEE standard data's for 250 MVA Power transformer in IEEE loading guide 1995:[6]

Transformer Losses, W.

No Load	78100
Pdc losses ($I^2 R_{dc}$)	411780
Eddy losses	41200
Stray losses	31660
Nominal voltage	118 KV 230KV
Pdc at hot spot location	467 527
Eddy current losses at hot spot location	309 (0.65 pu) 157 (0.3 pu)
Per unit height to winding hot spot	1 1

Temperature Rise °C .

Rated top oil rise	38.3
Rated top duct oil rise	38.8
Rated hot spot rise	58.6 50.8
Rated average winding rise	41.7 39.7
Rated bottom oil rise	16
Initial top oil	38.3
Initial top duct oil	38.3
Initial average winding	33.2
Initial bottom oil	28
Initial hot spot	38.3

Transformer component weights, kg

Mass of core and coil assembly	172200
Mass of tank	39700
Mass of oil	37887

And *heat balances are:*
$$Q\,GC+ Q\,LW+ Q\,GSL= Q\,AO+ Q\,LO$$
Where,
$Q\,LW$ is the heat lost by the windings, W.
$Q\,GC$ is the heat generated by the core losses, W.
$Q\,GSL$ is the heat generated by stray losses, W.
$Q\,AO$ is the heat absorbed by the tank core oil, W.
$Q\,LO$ is the heat lost by the oil, W.
$Q\,GC=PCR$ are the rated core losses, W.

2.2. Governing *Equations* of *Heat Flow*

With the thermal conductivity, in rectangular coordinate system is written as

➤ Continuity equation

$$\frac{\partial \rho}{\partial t} + \frac{\partial}{\partial x}(\rho u) + \frac{\partial}{\partial y}(\rho v) = 0 \qquad (1)$$

➤ Linear momentum equation in the x direction

$$\frac{\partial(\rho u)}{\partial t} + u\frac{\partial}{\partial x}(\rho u) + v\frac{\partial}{\partial y}(\rho u) =$$
$$-\frac{\partial p}{\partial x} + \frac{\partial}{\partial x}(\mu\frac{\partial u}{\partial x}) + \frac{\partial}{\partial y}(\mu\frac{\partial u}{\partial y}) \qquad (2)$$

➤ Linear momentum equation in the y direction

$$\frac{\partial(\rho v)}{\partial t} + u\frac{\partial}{\partial x}(\rho v) + v\frac{\partial}{\partial y}(\rho v) =$$
$$-\frac{\partial p}{\partial y} + \frac{\partial}{\partial x}(\mu\frac{\partial v}{\partial x}) + \frac{\partial}{\partial y}(\mu\frac{\partial v}{\partial y}) - g(\rho - \rho_{ref}) \qquad (3)$$

➤ Energy equation

$$\frac{\partial(\rho C_p T)}{\partial t} + u\frac{\partial}{\partial x}(\rho C_p T) + v\frac{\partial}{\partial y}(\rho C_p T) =$$
$$\frac{\partial}{\partial x}(k\frac{\partial T}{\partial x}) + \frac{\partial}{\partial y}(k\frac{\partial T}{\partial y}) + A(t)q''' \qquad (4)$$

$$A(t) = \begin{cases} 0 \quad for \;\; fluids \\ -1.37\times10^{-9}\times t^2 + 0.805\times10^{-4}\times t + 0.1358 \\ for \;\; the \;\; compunents \;\; of \;\; the \;\; active \;\; unit \end{cases}$$

In the above equations, temperature T, is a function of space variables x, y and z and time variable t (i.e., $T = T(x, y, z, t)$), that z(thickness) is constant in this model. The term A (t) is coefficient of the heat source function, and has been modified here to take care of variation of resistivity of copper with temperature at time t. With this representation, the function A becomes time dependent, distributed, heat source.

2.2.1. Boundary condition

$x = 0, y = y, t = t, u = v = 0$

$$-k_w \frac{\partial T_w}{\partial x} = h_{yl} \left(\frac{T}{x=0} - T_f \right) \quad (5)$$

$$+ \varepsilon \sigma . \left(\left(\frac{T}{x=0} + 273 \right)^4 - (T_f + 273)^4 \right)$$

$\varepsilon = 0.2, \sigma = 5.67 \times 10^{-8} \, k/m^2 k^4$

$x = l_x, y = y, t = t, u = v = 0$

$$-k_w \frac{\partial T_w}{\partial x} = h_{yr} \left(\frac{T}{x=l_x} - T_f \right) \quad (6)$$

$$+ \varepsilon \sigma . \left(\left(\frac{T}{x=l_x} + 273 \right)^4 - (T_f + 273)^4 \right)$$

$y = 0, x = x, t = t, u = v = 0$

$$-k_w \frac{\partial T_w}{\partial y} = h_{yb} \left(\frac{T}{y=0} - T_f \right) \quad (7)$$

$$+ \varepsilon \sigma . \left(\left(\frac{T}{y=0} + 273 \right)^4 - (T_f + 273)^4 \right)$$

$y = l_y, x = x, t = t, u = v = 0$

$$-k_w \frac{\partial T_w}{\partial y} = h_{yt} \left(\frac{T}{y=l_y} - T_f \right) \quad (8)$$

$$+ \varepsilon \sigma . \left(\left(\frac{T}{y=l_y} + 273 \right)^4 - (T_f + 273)^4 \right)$$

$$W_c = 2.17 . f . \sqrt{p_b} (T_w - T_f)^{1.25} \quad (9)$$

$$q'' = h . (T_w - T_f) \quad (10)$$

$$h(x, y) = 2.17 . (T_w - T_f)^{0.25} \quad (11)$$

The boundary conditions and initial condition still remain the same as (5)–(8).

3. Results and discussions

3.1. Steady-State Thermal Performance

A software program in MATLAB using mathematical code has simulated the steady-state performance. While on this point, the authors wish to point out that the results obtained in this work seem to be in reasonable qualitative agreement with some of the published experimental results. However, actual calculations for comparison could not be performed for paucity of design details in these papers. The applicability of the model has also been verified on transformers of the following ratings:
• 250 MVA   OFAF

The cu-loss in the winding per disc/layer of the above transformers has been tabulated in appendix1. The total losses in transformer 250 MVA are 562,780 kW (0.23%).

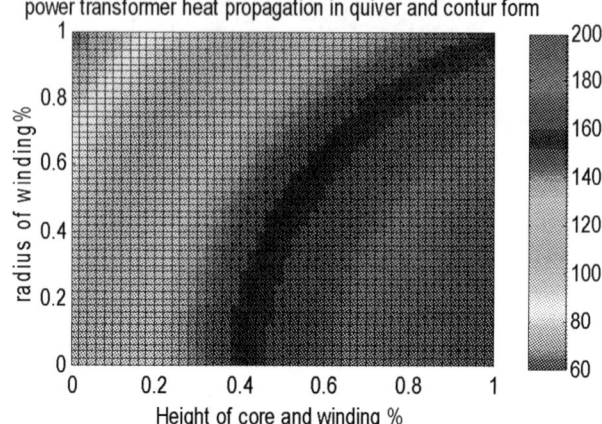

Fig. 3. Temperature distribution in the high voltage disc.

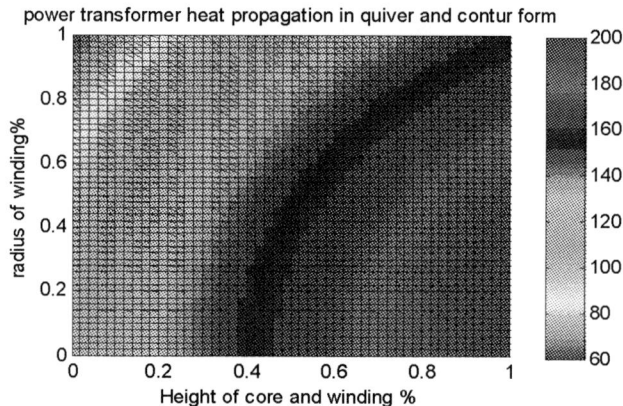

Fig. 4. Temperature distribution of low voltage (LV) winding layer.

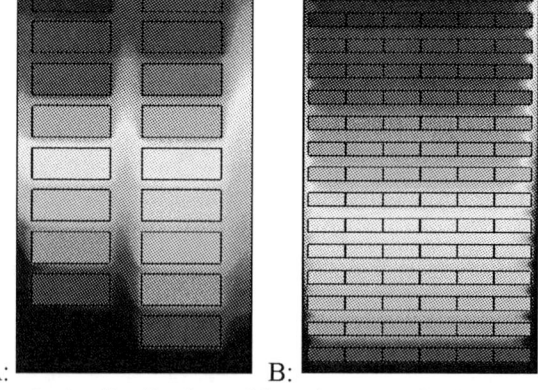

Figure 5: A- distribution of heat in core of winding with ONAN power transformer B- for OFAN (simulation by FEMLAB software)

While on this aspect, the authors wish to point out that the IEEE loading guide and other similar documents offer empirical relations for the calculation of the HST based on per-unit load. The formulations tend to ignore the possibilities of two transformers which are rating-wise identical but have a different winding structure and varying heat loss/unit volume. The method suggested by the authors gives due representation for this omission and, hence, is believed to give more accurate estimates.

It may be observed that the maximum temperature occurs in the neighborhood of 80–95% of the axial and 50% of the radial thickness of the disc. It is illustrated in table I. Fig. 3 and figures 5 shows the spatial temperature distribution in a disc-type Hv winding. It may be observed that the HST would occur around the location of the second or the third disc from the top as surmised. But this position would change if there is oil stagnancy in the cooling ducts. In a layer winding, the HST occurs at around 80 to 95% of winding from the bottom. The spatial temperature distribution for a layer winding in, plane has been shown in Fig. 4. Under OFAF modes of cooling, the magnitude and location of the HST at different loading have been shown in Tables I. The computed value of the HST, arrived at the estimated temperature of the hot spot and its location seem to be in line with the field measurements. However, there appears to be no published literature giving this information against which comparisons can be made.

However it is obviously necessary to limit the core temperature to values that cause no damage to the core itself, adjacent materials, or the oil. It has been shown in figures 3, 4, 5 and table 1 and 2that core temperatures as low as 110 C–120 C may degrade oil. This has led experts in the field to suggest that 130 C with 0.08m/s for oil speed would be a reasonable limit for the core temperature.

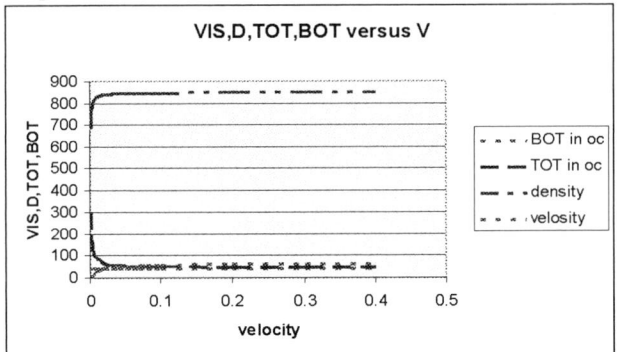

Figure6: viscosity, density, TOT and BOT versus velocity

3.2. **Unsteady-State Thermal Performance**

From a practice standpoint, the thermal performance in dynamic loading and dynamic ambient are of great interest. It has been assumed here that the boundary surface temperature as well as oil temperature gradient depends on time. Also, the methods of calculation of the thermal parameters are exactly same to that in the steady-state case. In this case (transient case), the heating process in a transformer winding has been simulated by introducing a transient thermal load.

1) Initial Function:
The initial function for transient heat-flow problem can be determined from the temperature distribution obtained from the steady-state solution.

2) Estimation of Winding Time Constant:

The time constant of the transformer winding can be determined. The thermal time constant of winding is around 4.5–5 min. These values have been expressly calculated (and not assumed as was done earlier) using the thermal parameters of the winding.

Table I: HST Magnitude and Location in ONAN

250KVA Transformer , Amb. Temp. =33 c , on humidity=60%					
	detail description			HST	HST cal
pu load	winding type	HST local	local R/L	IEC	M1
1	hvd	91	88/53	113	97
	lvd	94	85/43		
1.25	hvd	116	84/47	140	123
	lvd	120	82/57		
1.5	hvd	142	86/48	172	148
	lvd	147	81/44		
1.8	hvd	182	81/43	215	182
	lvd	185	81/43		
2*	hvd	205	89/39	245	221
	lvd	214	82/44		
2.2*	hvd	....	83/46	280	245
	lvd	244	85/48		

*There are instances [16] in which large power transformer are loaded to as high as 2.5 pu although the limit there-on is 1.5 pu as per the IEEE guide

Table II: variation of viscosity and density with temperature in oil of the transformer

Velocity m/s	Temperature in bottom (oc)	Temperature in top(oc)	Density	Viscosity
0.001	40.063	302.66	684.92	0.226
0.002	40.033	199.48	685	0.226
0.003	40.025	155.32	688	0.226
0.004	40.021	131.18	685.5	0.226
0.005	40.019	115.92	688.9	0.226
0.006	40.017	105.33	689	0.226
0.007	40.017	97.51	689.05	0.226
0.008	40.016	91.46	689.09	0.229
0.009	40.016	86.65	689.3	0.229
0.01	40.016	82.71	689.49	0.229
0.02	40.015	64.52	689.95	0.229
0.03	40.015	58.89	689.96	0.229
0.04	40.015	55.9	690	0.229
0.05	40.015	53.98	690.01	0.229
0.06	40.015	52.6	690.06	0.229
0.07	40.014	51.52	690.09	0.229
0.08	40.014	50.64	690.1	0.229
0.09	40.014	49.89	690.11	0.229
0.1	40.014	49.25	690.13	0.229
0.2	40.001	45.82	690.14	0.23
0.3	39.925	45.2	690.15	0.23
0.4	39.159	43.7	690.19	0.23

2) Velocity, density and viscosity

On the table II and figure 6 shows that with increasing of oil velocity will be decrease top oil temperature and density and viscosity will be increase in oil and so temperature will be constant in bottom oil. That means with increasing of velocity could improve the condition of cooling and decreasing the TOT and HST. Of course the medium of core and winding temperature would be more than TOT while oil velocity is over of 0.08m/s.

4. Conclusions

The heat source function has been taken as temperature dependent. The thermal model presented here can predict the hotspot location, with a higher degree of accuracy than was hither to possible. The result of the study on the rate of convergence indicates that the convergence is indeed quite rapid and nearly monotonic. Also, suggestions have been made as to the optimum numbers of terms to be considered to obtain sufficient accuracy. It has been shown that core temperatures as low as 110 C–120 C may degrade oil. This has led experts in the field to suggest that 130 C with 0.08m/s for oil speed would be a reasonable limit for the core temperature.

References

[1] M. K. Pradhan and T. S. Ramu, "Prediction of hottest spot temperature (HST) in power and station transformers," *IEEE Trans. Power Delivery*, vol. 18, pp. 1275–1283, Oct. 2003.

[2] R. Hurter and F. Viale, "Thermal aspects of large transformers, test procedures hot spot identification, permissible limits, their assessment in factory tests and service, overload limitations, effect of cooling system:- Presented in the name of study committee-12," in *Proc. CIGRE.12–13 30th Session*, vol. 1, 1984.

[3] J. Aubin, "Thermal aspects of transformers:-Presented in the name of study committee 12," in *Proc. CIGRE Paper, 12-1073 1990*, vol. 1, Aug./Sept. 1990.

[4] "Loading Guide for Mineral Oil-Immersed Power Transformer,", IEC- 354, 1991.

[5] M. V. Thaden, S. P. Meheta, S. C. Tuli, and R. L. Grubb, "Temperature rise test on a OFAF core-form transformer, including loading beyond name plate," *IEEE Trans. Power Delivery*, vol. 10, pp. 913–919, Apr. 1995.

[6] W. Lampe, L. Pettersson, C. Ovren, and B. Wahlstrom, "Hot spot measurement in power transformers," in *Proc. CIGRE Paper, 12—02 30th Session*, vol. 1, 1984.

[7] W. J. Mcnutt, J. C. McIver, R. V. Snow, and D. J. Fallon, "Transformer loading capability information derived from winding hot spot measurements," in *Proc. Int. Conf. Large High Voltage Electrical System, 30th Session*, vol. 1, CIGRE, Paris, France, 1984, Paper, 12-08.

[8] "IEEE Loading Guide for Mineral Oil Immersed Transformer,", C57.91.1995, 1996.

[9] L. W. Pierce, "An investigation of the thermal performance of an oil filled transformer winding," *IEEE Trans. Power Delivery*, vol. 7, pp. 1347–1358, July 1992.

[10] , "Predicting liquid filled transformer loading capability," *IEEE Trans. Ind. Applicat.*, vol. 30, pp. 170–178, Jan./Feb. 1994.

[11] "IEEE Loading Guide for Mineral Oil Immersed Transformer, C57.91/Corrigenda,", Draft Corrigenda to C57.91-1995, 2001.

[12] *IEEE Guide for Determination of Maximum Winding Temperature Rise in Liquid-Filled Transformers*, IEEE Std. 1538-2000, Aug. 2000.

[13] G. Swift, T. S. Molinski, and W. Lehn, "A fundamental approach to transformer thermal modeling. Part-I: Theory and equivalent circuit," *IEEE Trans. Power Delivery*, vol. 16, pp. 171–175, Apr. 2001.

[14] S. A. Ryder, "A simple method for calculating winding temperature gradient in power transformers," *IEEE Trans. Power Delivery*, vol. 17, pp. 977–982, Oct. 2002.

[15] M. N. Ozisik, *Boundary Value Problem of Heat conduction*. Scranton, PA: International Textbook, 1968, pp. 467–475.

[16] H. Nordman and M. Lahtinen, "Thermal overload tests on a 400 MVA power transformer with a special 2.5 p.u. short time loading capability," *IEEE Trans. Power Delivery*, vol. 18, pp. 107

Molecular Dynamics Simulation for the diffusion of water in amorphous polymers examined at different Temperatures

E.Dermitzaki, J.Bauer, H.Walter, B.Wunderle, B.Michel, H.Reichl

Fraunhofer Institute of Reliability and Microintegration (IZM), Dept.Mechanical Reliability and Micro Materials

Gustav-Meyer-Allee 25, D-13355 Berlin, Germany

Emmanouella.dermitzaki@izm.fraunhofer.de , Tel. +493046403200, Fax +493046403211

Abstract

The authors first investigate a polymeric composite material (epoxy resin with silicone nanoparticles) and determine its mechanical properties (Coefficient of thermal expansion-CTE and Young's Modulus-E) by a combined experimental and simulative (homogenisation) approach. The size distribution and the volume fraction percentage of the nanoparticles were determined as well as the variation of the mechanical properties of the pure epoxy resin compared to the composite material especially synthesized for this investigation. Then the diffusion coefficient of water in amorphous polymeric materials is measured (aromatic epoxy: 1,3-Bis-(2,3-epoxypropyl)-benzene and amine hardener:1,2-Diaminoethan). Mechanical characterization of the materials, diffusion experiments (gravimetric), and Molecular Dynamics (MD) simulations examine the temperature range of 300-400K, where diffusion coefficients will be calculated under thermodynamic boundary conditions by using a classical force-field MD. The diffusion mechanisms and the mechanical characterization are examined as a function of the stoichiometry and molecular weight of the polymers.

1. Introduction

Polymeric materials-epoxy based (with or without filler) are extensively used in microelectronic applications eg.packaging, due to their outstanding rheological and thermomechanical properties. Though these materials cause reliability problem when exposed to humid enviroment phenomena such as water absorption and diffusion take place. They may be further constrained by mechanical loading and a wide range of temperature variations. A combination of those parameters, temperature, humidity and stress, alter the mechanical properties of the materials at interfaces, lead to adhession loss, interface delamination and finally to crack propagation. These thermomechanical failure-mechanisms can cause complete failure of a microelectronic component and system [1].

Electronic packaging tends towards miniaturization and higher performance, forcing at the same time the development of new *reliable* nano-materials and the need for experimental and simulative methods on nano-scale for their characterization and formulation of lifetime models [2]. Therefore we need to establish a *structure property correlation*. This again requires *nano-analytical tools* in simulation and experiment, in order to study the material properties of epoxy resins with and without filler on nano-scale. We investigate the properties and the failure behavior of materials and material interfaces under the explicit consideration of their nano-structure and their induced effects *(nanoreliability)*. In order to make conclusive statements about the structure of the materials, we synthesize them ourselves, allowing at the same time a systematic variation of material parameters. In order to identify their properties we do mechanical characterisation and evaluate the coefficient of thermal expansion-CTE and the Youngs-Modulus-E, as those two are important quantities which describe the macroscopic response of epoxies. As nano-simulation methods we introduce first the method of homogenisation (FE-based) where we examine the epoxy resin with filler particles. Then molecular modeling will evaluate diffusion coefficients for epoxy resin under the external influence of temperature and moisture. These material parameters, partially ab-initio calculated, will provide us with material data, which can be input data to FE-analysis tools, and then be able to investigate the failure properties of the materials. In this paper we focus only on the material properties.

Nano-filled polymers display advantageous material properties with respect to the thermo-mechanical reliability of packaging and various investigations have been proposed describing composite polymer materials [3-6], by using homogenisation approaches in comparison also to experimental measurements. Also, according to [7-8] varying the weight percentage of filler content the CTE and E of the composite material do alter. The filler enhances the matrix properties and makes the composite depend on the volume fraction of the filler eg. Increasing CTE and decrease the E [7-8] with decreasing filler content. So the method of homogenisation is appropriate for our investigation using an epoxy matrix and soft silicone nano-particles. Having as input previous knowledge of the mechanical properties (CTE & E) of epoxy and filler we can thus calculate the mechanical properties of the composite material. These will then be compared with experimental values. Other variables in the study are filler content, filler size and filler size distribution. Then various microscopy techniques are used to measure the filler content [9-10].However homogenisation is a tool that uses finite element analysis-(FEA) and evaluates mechanical properties with the use of continuum theory. Thus making it inadequate to represent interactions on a molecular level and to observe size effects. Therefore we make use of molecular modelling. It was found that (main result briefly stated), no size effect can be reproduced from that. To include such phenomena simulation needs to include physics on

1-4244-1105-X/07/$25.00 ©2007 IEEE

the nanoscal. For ab-initio calculations of eg. a diffusion coefficient we have to make use of molecular modelling.

With molecular dynamics we can construct the explicit molecular structure of an epoxy resin in a representative volume element, describing the bulk of the material. Interatomic interactions such as bonded ,non-bonded and hydrogen bonding, between atoms are communicated by an empirical forcefield which allows modelling of the structure by finding the minimum energy. The forcefield is parameterized based on ab-initio calculations and by applying various thermodynamic constraints (pressure,temperature etc) we can demonstrate and calculate material properties such as diffusion coefficient [11-14] and also provide information and understanding of structural effects. Again the materials,pure epoxy resin are synthesized in our labs. Then we investigate the diffusion mechanisms and calculate the diffusion coefficient-D as a function of temperature, structure, volume, pressure and water concentration,formally represented in equation 1.

$$D = f(T, V, structure, P, C) \quad (1)$$

The materials we test are an aromatic epoxy & amine hardener, and four distinct combinations varying in stoichiometry of epoxy-hardener ratio and molecular weight of the hardener. Then the diffusion coeffiecient for water is calculated by MD and reproduced in experiments. Here we used gravimetric measurement to obtain the diffusion coefficient and standard methods to obtain the E (by DMA) and CTE(by TMA). Since this approach is a rather new discipline questions arise in terms of molecular modelling: How can the polymer microstrusture be represented correctly? In order to construct the polymer in a representative waya sensitivity study has been conducted with respect to chain interlinking, polarity of end-groups and geometrical configuration .

2. Homogenisation

To show the influence of the filler particles (silicone) on the macroscopic mechanical behaviour (E and CTE) of the epoxa matrix, we prepared samples of different bur well-known weight percentage of filler (0,5,10 and 15wt%). Experiments are to show the reproduction of that values by microscopical techniques,whereas FE-simulation and theoretical formulae given in literature [7] are to verify that structure-property correlation numerically. An overview of the procedure is given in Table 1.

Investigation	Experiment	Software
Identify size distribution of Filler	SEM experiments with FIB (Focused Ion Beam) material preparation	—
Calculate Volume Fraction of Filler	AFM (Atomic Force Microscopy) via SEM (Scanning Electron Microscopy)	digital image analysis & Theory
Measure Filler & Matrix properties	CTE (Coefficient of Thermal Expansion) & E(Young's Modulus)	FE

Table 1: Representation of the investigation

Theory: Coefficient of Thermal Expansion and Moduli for Polymeric Composites

Kerner developed a formulae for (infinitely) soft spherical particles describing the coefficient of volume expansion of a composite polymeric material, Equation 2[7]. Where α_m is the thermal expansion of the matrix, α_f the thermal expansion of the filler, K_m the bulk modulus of the matrix, K_f the bulk modulus of the filler and ϕ_m and ϕ_f are the volume fraction of the matrix and filler respectively.

$$a = \alpha_m \phi_m + \alpha_f \phi_f - (\alpha_m - \alpha_f)\phi_m \phi_f \frac{1/K_m - 1/K_f}{\phi_m / K_f + \varphi_f / K_m + 3/(4G_m)} \quad (2)$$

Figure 1 represents the relation of thermal expansion to the volume fraction by means of hard filler. Plot A is described by the 'rule of mixtures', plot B by the Kerner equation and plot C by Thomas Equation 3, where the filler particles are rodlike in shape with different orientations. [7]

$$\log a = \phi_m \log \alpha_m + \phi_f \log \alpha_f \quad (3)$$

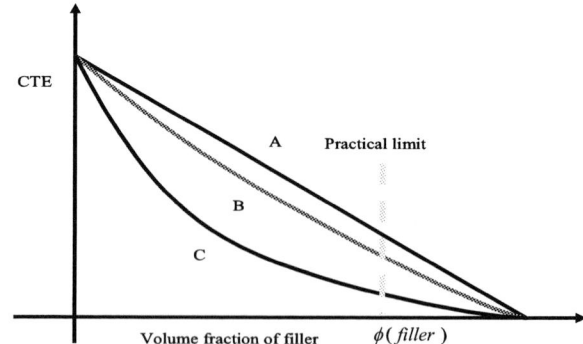

Figure 1: Coefficient of thermal expansion

The 'rule of mixtures' predicts the mixture of different components as a uniform blend (no structure) and the relation of thermal expansion with volume fraction is therefore a linear function of composition (no interaction between constituents). Equations 2 and 3 obey a non-linear relationship, due to mechanical restraints on the matrix by the filler particles. The curvature of graph C is larger than the one in B, since the interaction of rodlike particles with the matrix is larger. There is a 'break point' at maximum density, since there is always some maximum volume content where the filler particles touch one another. Equation 4 shows the Modulus of the composite material [7] where G is the shear modulus of the filled material, G_m the shear modulus of the unfilled matrix and v_1 the poisons ratio of the unfilled matrix.

$$\frac{1}{G} = \frac{1}{G_m}\left[1 + \frac{15(1 - v_1)}{7 - 5v_1} \times \frac{\phi_f}{\phi_m}\right] \quad (4)$$

The volume fraction of the filler ϕ_f was derived by the Composite Formula, Equation 5 [15] where ρ_f the density of the particles ρ_m the density of the matrix and w_f the weight percentage of the particles:

$$\phi_f = \frac{1}{1 + \frac{\rho_f}{\rho_m}\left(\frac{1}{w_f} - 1\right)} \quad (5)$$

Correlation of Area to Volume Content
The wt% of the materials is previously known and the volume fraction can be evaluated from equation 5. But how can we calculate the volume fraction of the specimens from images obtained from microscopy techniques? The idea is: Images depicted from microscopy techniques provide numerical information of the area of the specimen (cut through by a plane) and used as input to a contrast method (analySIS® software). The image filtering transforms the prototype image just by removing shades, sharpening edges, decreasing high illumination or enhancing local contrast around the region of the filler Figure 2. Then the software calculates the area content of the spheres.

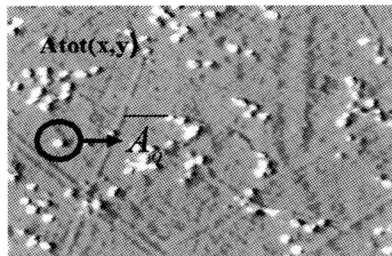

Figure 2: Filtered image from AFM

As can be shown (see appendix A), the volume content can be set equal to the area content for the case of spherical particles (equation 6) as a statistical mean value for a large number of spheres.

$$\frac{\overline{V}_o}{V_{Tot}} = \frac{\overline{A}_o}{A_{Tot}} \quad (6)$$

Microscopy Techniques
In this section first we compare results of AFM & SEM (Microscopy techniques) to our theory, equation 5. We get results of the area content of the dispersed nano-particles depicted from the image analysis (first sample: 5wt% filler). Then the average area is equalised, Equation 6, to the volume fraction of the spheres. AFM showed good correlation to theory, Figure 3 and was used further to obtain images from the 10 & 15wt% of filler. The discrepancy of the results can be explained by the surface scanning process and contrast generation, depicted in figure 4. The SEM graph slightly overestimates the area content, as sub-surface information still contributes to the image. This is due to secondary electrons coming from a distance beneath the surface. This effect is significant, as for given parameters of beam energy depths around 50 to 100nm are possible, which is of the order of the diameter of the particles [10]. On the other hand AFM works with a tip and a cantilever and generates a map of surface topography capturing particles lying only on the surface of the specimen [9].

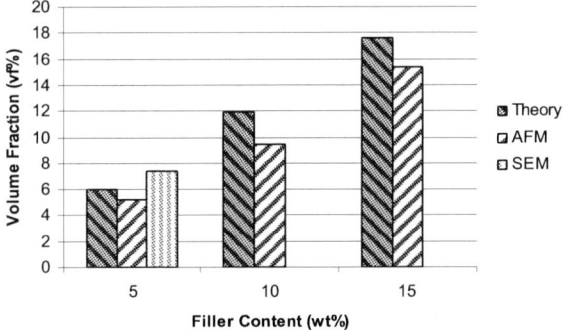

Figure 3: Statistics of image analysis & theory

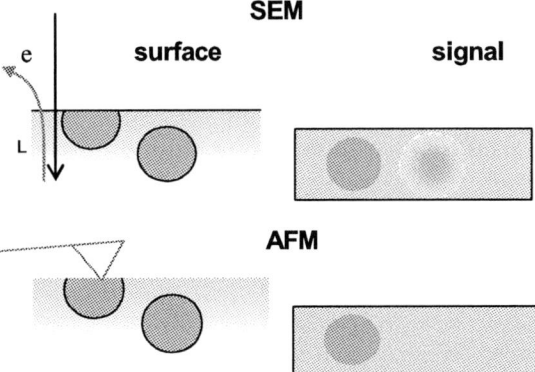

Figure 4: Illustration of signal from SEM and AFM

However the discrepancy between theory and AFM, depicted in figure 3, could be explained by:

- This is a statistical correlation on very few samples (<4) thus we need to examine more samples or larger area.
- The surface content measurement depends on manual adjustment of the contrast threshold value. This could also contribute to a slight over-or underestimation of the contents.

After correlation of AFM and SEM we employed SEM with FIB preparation technique to identify the size of the filler. Material was milled away and allowed the investigation of the inner part of the specimen, figure 5, avoiding artefacts from the polishing technique. The diameter of the nano-particles was estimated approximately to 170nm, which correlates well to the given values. Here SEM does not induce any error,als only lateral information is used for the measurement.

Figure 5: SEM image of FIB preparation

Homogenisation, Experiments & Sensitivity Analysis
The previously determined results of CTE and E for epoxy and filler were input into the homogenisation software. Numerical predictions of those properties were compared to the experimental data. This software uses an FE-method to determine properties of composite material such as E and CTE. It consists of a representative volume element and periodic boundary conditions and uses a Monte Carlo method to generate a geometry, where inclusions (i.e. particles) do not overlap. Before performing the simulations we did a sensitivity analysis. Employing a Gaussian distribution w.r.t particle size, at the same time keeping constant the total volume of all spheres in the cell (see figure 6) one can simulate a size distribution. With respect to a delta-distribution ($\sigma = 0$) we observed no discrepancy in the results. This means that in homogenisation we can not observe any size effects for spherical particles. Further analysis did even show, that it has no influence by how many spheres the volume is represented. This is to be expected, as a representative situation can already be modelled by a single sphere in a cell of equal filler to matrix ratio using periodic boundary condition. As in FE simulation is based on pure continuum´, no size-effect should be expected. Variations are only due to numerical, i.e. mesh-related issues.

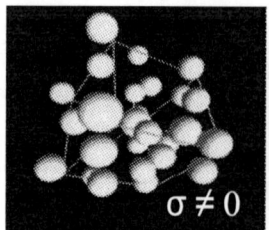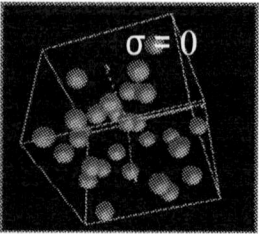

Figure 6: 30 Spheres with & without Gaussian distribution of 5wt%

Finally the diameter of the spheres was evaluated 170nm which is equivalent to 60 spheres for a cell of 1x1x1 µm for 5wt%. Table 2 shows the equivalence of volume fraction, as stated in equation 5, and the corresponding diameter of spheres. As no size effect was detected we may also keep constant the number of spheres for different wt% and use a different diameter for all spheres.

wt% of filler content	Volume fraction	Diameter of 60 Spheres (nm)
5	6,01	106,36
10	11,90	133,53
15	17,67	152,32

Table 2: different filler content for the use of 60 spheres

For the simulations the volume fraction of the filler was adjusted and the mechanical properties of CTE and E were obtained, Figure 7 & Figure 8, and compared to the pure epoxy resin.

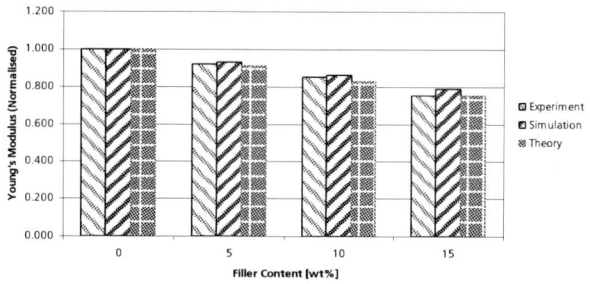

Figure 7: Young's Modulus of Experimental, Simulative and Theoretical Results normalised to pure epoxy

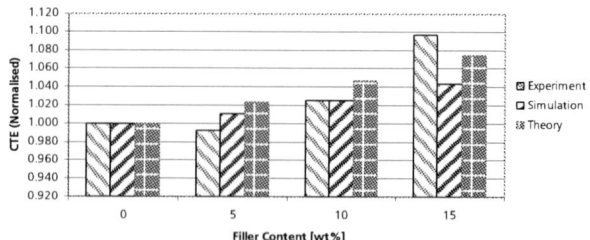

Figure 8: CTE of Experimental, Simulative and Theoretical Results, normalised to epoxy resin

For both graphs the theoretical tendency, calculated from Equations 1 & 4, is reproduced correctly and the discrepancy of the results is negligible considering the low number of samples the statistics is based on. Simulative results show only a difference of 2% when correlated with the experiment and theory. A further reason could be that the simulation assumes perfect conditions (ideal binding between the filler and the matrix, no gas inclusions) whereas in experiment the specimen might contain bubbles or local delamination.

Conclusions for Homogenisation

In the present study we were focused on the CTE and E-Modulus properties of the composite material compared to the pure epoxy resin with different filler content of silicone particles down into the nano-region. The size was determined approximately to 170nm with the aid of SEM with FIB material preparation and the volume fraction was acquired from AFM. The correlation of volume to area content of the particles satisfied our predictions considering the few samples we used. The correlation of experiment, simulation and theory of the Young's Modulus and CTE was good and the tendency was correctly predicted. Their negligible differences could be based on low statistics. Finally, sensitivity analysis proved that only the correct volume fraction is governing the mechanical quantities, thus not being able to detect any surface or size effect. The size of the particles and the variance in size do not enter in this scale yet and the physics governing is still classical. So as long as we consider particles with diameter around 100nm we may use the given theoretical formulae. This might change if we get down to smaller spheres (<10nm) as structural features (chain length, free volume) in the polymer might interact with the particles with then higher surface energy due to their small radius. For the classical regime, however, the homogenisation methods can quickly furnish accurate results, in particular for composites with less dissimilar properties as the ones considered here.

3 Molecular Dynamics

In homogenisation we could not model any size effect. Therefore we make use of MD and we model our structure, again an epoxy resin synthesized on a nano-meter scale. We try to establish a structure property correlation by bringing together experiments and simulations. As a test we measure and calculate the diffusion coefficient-D. For that purpose we set up a test matrix, which consists of 4 variations to the material's inner structure of the same epoxy and hardener. These materials differ in stoichiometry and molecular weight. So we keep stable the volume but change the polarity and the density branching of epoxy and hardener (see table 3). We identify these microscopic effects through D and mechanical properties (E & CTE). In molecular modelling first we need to represent correctly the microstructure. This is challenging, as it is very difficult and CPU time consuming to interlink the individual molecules adequately to form a 3D network. So another

solution needs to be found, relying on a masking of the un-reacted groups of high polarity to simulate non-interlinked chains. This first requires a sensitivity study as to polarity, chain length and molecular weight.

Materials Definition

The materials investigated are the aromatic epoxy RDGE Resorcindiglycidylether and the amine hardener DAE Diaminoethane, Figure 9. Four materials will be constructed; two will be 2/1 and 4/3 concerning their stoichiometric ratio (epoxy/hardener) and the other two having the above stoichiometry will differ in the molecular weight of their hardener. Table 3 shows the four materials.

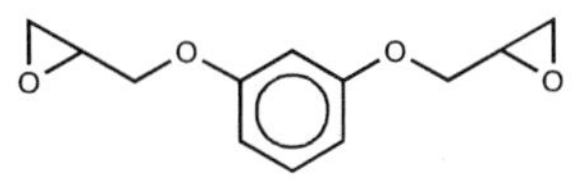

Epoxy: 1,3-Bis(2,3-epoxypropoxy)benzene (Resorcinol diglycidyl ether, RDGE)

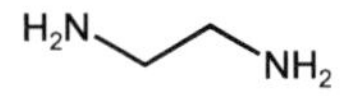

Hardener: 1,2-Diaminoethan

Figure 9: Structure of epoxy resin and hardener

Stoichiometry/ Epoxy hardener	No. of carbons in hardener chain
2/1	2
2/1	4
4/3	2
4/3	4

Table 3: Materials characterization

In molecular modelling we want first to interlink the epoxy with hardener and then calculate D. However the following limitations arise:

1. In diffusion mechanism within the presence of water and the un-reacted methyl or hydroxyl groups (in the structure as end groups which are highly polar), Figure 9, hydroging bonding takes place. Since this contributes to diffusivity we want to control the polarity of the end-groups.

2. There is a restriction of the possible number of atoms from 2000 to 10000 and we also want to keep down the CPU time, since the more atoms we use the more it requires large computetional resources .

3. If for example we want our structure to have 2,4nm dimensions, then we have to use 16 reactions of the Figure 9. This leads to an enormous number of different representations of our structure.

Therofore, limited from the above restrictions we perform a sensitivity analysis, where we systematically alter and control the geometry and chemical representation of the free end groups of the chains. So we gain knowledge of the contribution of polarity to diffusivity and apply a step by step interlink of the hardener with the epoxy in order to view some possible ways of interlinking (different geometrical presentations of the same substance). In sensitivity analysis *Material 2/1-2carbons* will be examined and in Figure 10 are examined different models of the reaction 2/1. By this approach we want to accomplish the most suitable packing of the microstructure for the molecular dynamics simulations. Model 1 and Model 2 have the same stoichiometry and molecular weight though differ in the polarity since in Model 2 the two epoxy resins react only with one of the amino-end groups of the hardener thus resulting in a different geometrical representation and in a more polar reaction. The contribution of the end group effects and the different geometrical representation of the chain will be derived from MD. In Model 2 we substitute two methyl groups (-CH3) in the un-reacted amino end group and we get Model 3,which is then reduced in polarity. By correlating them in MD we can view more explicitly the contribution of the amino group to diffusivity (end group effects and polarity). Finally Model 4 is a pseudo-cut of the epoxy end-groups placed on the other ends of each amine where additional H-atoms are placed at the ends (CH3 instead of CH2 and CH2OH instead of CHOH). This representation is neutral to polarity and this should evoke the full linking of the final structure. In chapter of sensitivity analysis the different diffusivity results will be presented. The step by step linking of Model 4 will follow the different conformations as suggested in Figure 11..

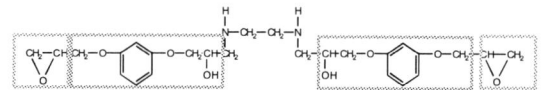

Model 1: simple addition of two resin molecules at both ends of hardener

Model 2: simple addition of two resin molecules at one end of hardener

Model 3: simple addition of two resin molecules at one end of hardener and addition of methyl groups to the other end of hardener

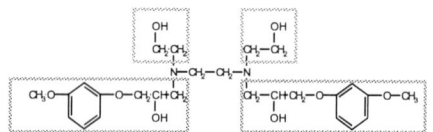

Model 4: a pseudo-cut of the structure to avoid un-reacted end-groups
based on Model 1 following correct stoichiometry

Figure 10: Different representations of reaction 2/1 epoxy-hardener

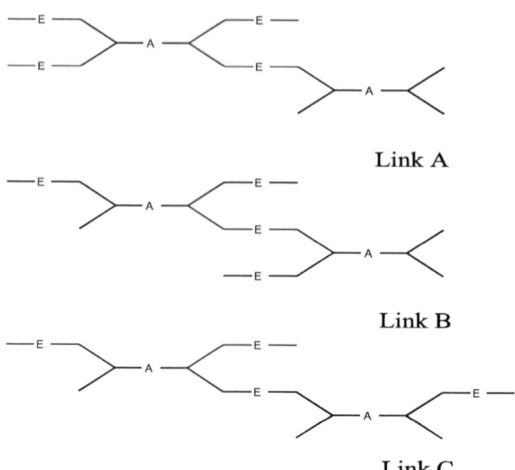

Link A

Link B

Link C

Figure 11: Stage 2 of linking where 2 chains of Model 4 react in different conformations.

The remaining unreacted end-groups of Figure 11 are:
- Link A: 1 primary amino group NH2, 1 secondary amino group NH3 and 3 epoxy groups
- Link B: 1 primary amino group NH2 1 secondary amino group NH and 3 epoxy groups of different shape
- Link C: 3 secondary amino groups NH and 3 epoxy groups of different shape

The number of isomeric structures increases as the number of monomers increases and the shape of the isomer will influence the simulation results due to different geometrical representation (the molecular weight is kept constant). These possible combinations of two molecules of amine-hardener (A) with four molecules of epoxy-monomer (E) will be checked in MD and the different percentage of their contribution to diffusivity. If they maintain no significant contribution to diffusivity then Link 3 will be the ultimate linkage of the microstructure, if not then each of the above link will give as different isomeric structures and so on till all the unlinked chains link and reach the final presentation of the microstructure.

Mechanical Characterisation

For the mechanical characterization of the materials the Elastic Modulus was derived from DMA -three point bending and the glass transition temperature-Tg and the CTE from TMA experiments. In Figure 12 and Figure 13 are presented the results-for investigation puproses we added a fifht material 2/1-6carbons.

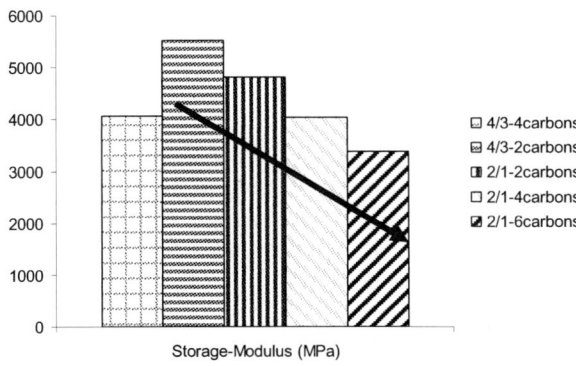

Figure 12: Experimental Modulus results at 20C

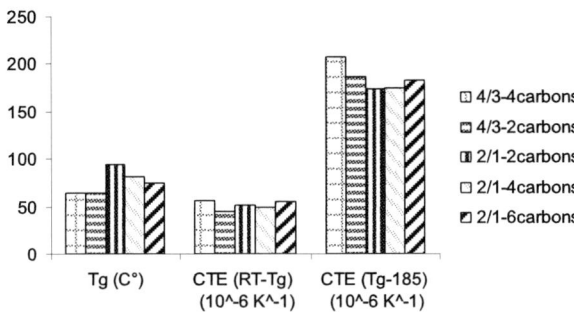

Figure 13: Experiment:Tg and CTE results

In figure 12: With respect to the 2/1 groups the 4/3 groups have unreacted end groups and are therefore more polar. This could induce a twofold effect: On one hand, the polar groups should stiffen the assembly (i.e. higher E-modulus) due to higher intermolecular forces. On the other hand should a lower branching density decrease the stiffness of the network. Though what is seen is the increase of E-modulus.Then as we add more carbons in the hardener chain (materials 2/1-4carbons, 2/1-6 carbons and 4/3-4carbons), we alter the molecular weight distribution and as a result the stiffness of the polymer. The E-modulus is decreased as we increase the chain length. In figure 13: The glass transition temperature devides the glassy from the rubbery state, where most of the pseudo-chemical bonds do not contribute any more to the stiffness of the material. The properties are determined by entropy-elasticity. So for the results depicted in figure 13 there could be the following interpretations: The longer the chains, the more free volume there is in the network and less non-covalent bonds need to be dissolved to reach the rubbery state. Tg is lowered. For the polar configurations the lower branching density seems to be the governing factor, lowering itself the Tg also.

Molecular Dynamics Simulation

The mechanism of water diffusion in polymers has been described to our previous work, see ref [16]. In sensitivity analysis 4 different models of material 2/1-

2carbons, figure 10, were generated with the aid of COMPASS forcefield. We used an NVT ensemble (control number of atoms, volume and temperature) at 320K and density of 1,1 g/cm³. Eight unlinked chains and 4 water molecules, which contribute to 1wt% of water absorption, were used in the simulation, figure 14, contributing to 18,36Å side length. The diffusion coefficient was determined from the self diffusion coefficient of the penetrant molecules. This can be evaluated from the slope of the mean square displacement as a function of time, where N_a is the number of atoms following with the aid of the simulation their trajectories, r_i id the position vector of the i-th atom in equation 7 [17]

$$D_a = \frac{1}{6N_a}\lim_{t\to\infty}\frac{d}{dt}\sum_{i=1}^{N_a}\left\langle\left[r_i(t)-r_i(0)\right]^2\right\rangle$$

$$MSD = <\left|r_i(t)-r_i(0)\right|^2>$$

$$D = \frac{a}{6}\;(7)$$

Where a is the slope of MSD as a function of time

The MSD is the averaged mean squared displacement over the number of atoms-water molecules in the polymer matrix.. To certify that the system has reached the normal diffusion regime, the slope of the log(MSD) as a function of log(t) must be close to unity (show linearity of the system).

Figure 14: Presentation of the 8 chains in a cubic cell of 18,36 Å side length.

Figure 15 shows the diffusion coefficient of the four models. Model 2 which is the most polar model, due to the unreacted amino group, has the highest diffusion coefficient and the contribution to polarity of the amino group is shown when compared to Model 3. The correlation of Model 1 and Model 2 suggests that geometry indeed plays a significant role to diffusion,while Model 4 proved its neutrality to polarity. Figure 16 correlates the 16 unlinked chains (Model 4) with the first linkage of those (LinkA). Simulations were carried out for 363K with cell dimensions for each siode length,24A°. 20 water molecules were used , relating to 5wt% absorption. The result shows that as the microstructure heads to the full linkage the diffusion coefficient decreases,as the mobility of the segmental chains decreases as well. In figure 17 we see the liniarity of the graph for the Model 4 at 363K.

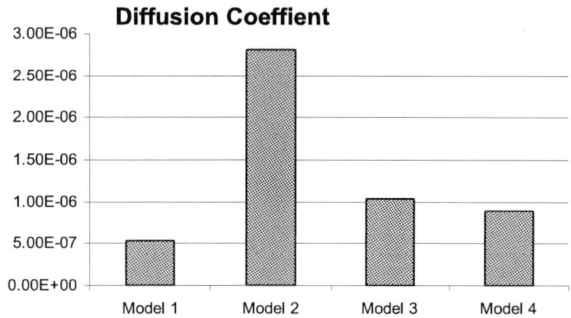

Figure 15: Determination of Diffusion Coefficient for all models at 320K

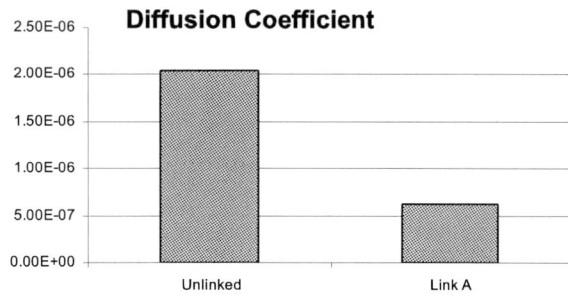

Figure 16: Comparison of diffusion coefficient before and after the first linkage of 16 chains according to Link A of Model 4 at 363K

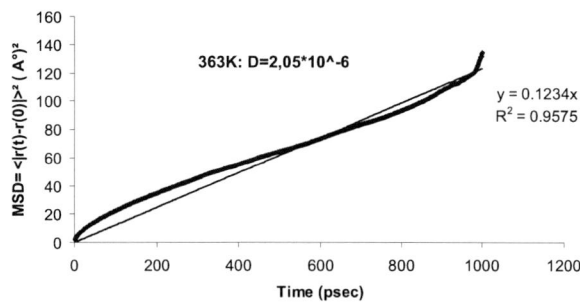

Figure 17 : Determination of Diffusion Coefficient of Model 4 at 363K

Conclusions

We have shown that changes in the chemical structure of an epoxy resin (stoichiometric ratio of epoxy/hardener and in the molecular weight of the hardener) have an effect on macroscale-mechanical characterization of the materials and what governs mostly the mechanisms is polarity. A sensitivity analysis of MD for the calculation of the diffusion coefficient, proved the "size effect" between various conformations of the same model-material: Different results were derived from models which maintained constant the molecular weight but had different geometry and models with the same geometry but had different polarity in their end-groups. The were

expected and therefore we continue our research with the full linkage of the microstructure.

References

1. O. Wittler, P. Sprafke, J. Auersperg, B. Michel, H. Reichl: *Fracture Mechanical Analysis of Cracks in Polymer Encapsulated Metal Structures*, Polytronic 2001, Potsdam, Germany, Oct. 21-24, 2001, proceedings, pp. 203-208
2. H. Reichl, A.Schubert, M.Töpper, "Reliability of flip chip and chip size packages", Microelectronics Reliability ,Vol 40 (2000), pp. 1243-1254
3. Hans Rudolf Lusti, Peter J.Hine, Andrei A.Gusev: Direct numerical predictions for the elastic and thermoplastic properties of short fibre composites submitted to Composites science and technology 2002
4. Andrei A.Gusev: "Numerical Identification of the Potential of Whisker-and Platelet-Filled Polymers" submitted to Macromolecules 2001
5. Michael Wissler, Hans Rudolf Lusti, Chantal Oberson,Albert H. Widmann-Schupak, Gianluca Zappini and Andrei A.Gusev: "Non-Additive effects in the elastic behaviour of dental composites" submitted to Advanced engineering materials 2003
6. Andrei Gusev, Hans Rudolf Lusti and Peter J.Hine "Stiffness and thermal expansion of short fiber composites with fully aligned fibers" submitted to advanced engineering materials 2002
7. E.Nielsen and Robert F.Landel. Mechanical Properties of Polymers and Composites, Marcel Dekker 1994
8. Gert Strobl, The physics of Polymers, Second Edition, Springer 1997
9. Linda C. Sawyer and David T. Grubb, Polymer Microscopy, Second Edition, Chapman & Hal, l 1996
10. S.K. Chapman, , S.K. Chapman, 1986 Working with Scanning Electron Microscope
11. N. Iwamoto, "Applying Polymer Process Studies Using Molecular Modelling ",*Adhesive Joining and Coating Technology in Electronics Manufacturing, 4th International Conference*, (2000), pp. 182-187
12. N.Iwamoto, "Advancing Polymer Process Understanding in package and Board Applications through Molecular Modelling", *50th Electron. Comp. Tech. Conf.*, USA, (2000), pp. 1354-1359
13. Lydia Fritz and Dieter Hofmann, " Behaviour of water/ethanol mixtures in the interfacial region of different polysiloxane membranes-a molecular dynamics simulation study" Polymer, Vol. 39, No. 12, (1998), pp. 2531-2536
14. Lydia Fritz and Dieter Hofmann, "Molecular Dynamics simulations of the transport of water-ethanol mixtures through polydimethylsolixane membranes", Polymer, Vo. 38, No. 5, (1997), 1035-1045
15. P.K Mallick, Composites Engineering Handbook, Marcel Dekker 1997

16. E.Dermitzaki et al, Diffusion of water in amorphous polymers at different temperatures using molecular Dynamics Simulation, *1st Electronics Systemintigration Technology Conference (ESTC)* (2006), September, 05-07.Dresden, pp.762-772

17. Dieter Hofmann et al, "Detailed Atomistic molecular modelling of small molecule dissufion and solution process in polymeric membrane materials", Moacromol. Theory Simul. , Vol 9, (2000), pp. 293-327

Appendix A

The surface of a sphere of radius R is: $A_{sp} = 4\pi R^2$

and its enclosed volume is: $V_{sp} = \dfrac{4}{3}\pi R^3$

Now we contemplate a volume of depth R of randomly distributed spheres of equal volume cut by a plane. Then the average volume of a cut sphere is:

$$\overline{V_o} = \frac{1}{2}V_{sp} = \frac{2}{3}\pi R^3$$

So, the ratio of the average volume of a sphere to the total volume is: $\dfrac{\overline{V_o}}{V_{tot}} = \dfrac{N\frac{2}{3}\pi R^3}{A_{tot}R} = \dfrac{N\frac{2}{3}\pi R^2}{A_{tot}}$

with N being the number of spheres cut.

The average area of a cut sphere is:

$$\overline{A_o} = \sum_i^N A_{cut} = \frac{N}{2R}\int_{-R}^{R} A_{cut}\,dl = \frac{N}{R}\int_0^R r^2\pi dl =$$

$$= \frac{1}{R}\int_0^R \pi\left(R^2 - l^2\right)dl =$$

$$= \frac{1}{R}\left(R^3\pi - \frac{1}{3}R^3\pi\right) = \frac{2}{3}R^2\pi N$$

where r=cut radius, l=varying depth

And the ratio is:

$$\frac{\overline{V_0}}{V_{tot}} = \frac{N\frac{2}{3}\pi R^2}{A_{tot}} = \frac{\overline{A_0}}{A_{tot}}$$

And therefore we generalize our theory for all dispersed spheres distributed in the matrix and cut by a plane.

Mechanical Characterization and Viscoplastic-Damage Constitutive Model of SnAgCu Solder

Zhou Jun[1,2], Chai GuoZhong[1], Liu Yong[2], Liang LiHua[1,2], Hao WeiNa[1]

The MOE Key Laboratory of Mechanical Manufacture and Automation,
Zhejiang University of Technology, Zhejiang, 310014 China
Fairchild-ZJUT Microelectronic Packaging Joint Lab,
Zhejiang University of Technology, Hangzhou 310032, China
Email:Zhouaanod@163.com

Abstract

Due to the toxicity of Pb present in Sn-Pb solders used in many electronic products, alternative solders need to be considered. Recently, there has foucused its interedt on Sn-Ag-Cu(SAC) alloy because of its comparative low melting temperature, the competitive price, and apparently good mechanical properties. The time- and temperature-dependent deformation behavior of Sn4.0Ag0.5Cu solder alloy was determined over strain rates ranging from 10-5s-1 to 10-2s-1 and temperatures ranging from 25℃ to 150℃. A viscoplastic-damage model incorrporating the effect of microstructural was developed for capturing the complex hierarchy of damage mechanisms, coupled with viscoplastic and stress state effects. The temperature and rate dependent flow properties of the matrix material have been obtained by inverse procedure. The model was also implemented into finite element program ABAQUS through its user defined material subroutine. The model is validated by comparing the predictions to the experimental data involving temperature, loading rate. The predictions have shown the ability of the modified viscoplastic-damage model coupled with microstructural damage to correctly describe the experimental observations: nonlinearity, strain rate sensitivity and damage evolution.

1. Introduction

Increasing environmental and health concerns about the toxicity of lead, as well as the possibility of legislation limiting the usage of lead-bearing solders, have stimulated substantial research and development efforts to discover substitute, lead-free solder alloys for electronic applications. SnAgCu solder has quite commonly been considered as the most viable replacement for SnPb solder. However, unlike lead solders, the recently employed lead-free solders do not have a long history and manufacturing process, and also engineering database have not been established[1]. Therefore, detailed studies of mechanical properties of SnAgCu solder have been quite limited. And most conventional mechanics approach assumes that the solder materials are perfect of defects-free, the influence of microstructure on deformation was not considered. However, a few literatures have shown that material deterioration and stiffness degradation are induced by the presence of micro-defects, especially initiate, coalescence and grow of voids under mechanical or thermal cycling load application[2-5]. Therefore, in order to model the lead-free solder material behavior

more precisely, it is indispensable to investigate mechanical damage behavior of lead-free solders induced by formation and growth of voids. Sung et al.[6] presented that a constitutive model for solder alloys should be able to describe both the steady and transient behaviors well, and it necessary to include the microstructure influence in a constitutive model.

Many constitutive models of lead-free or PbSn solder materials have been developed. Pao et al.[7]、Darveaux and Banerji[8] 、Sarihan[9] et al. presented some decoupled creep-plasticity models where the time-dependent and time-independent deforms are artificially separated. However, it is very difficult to separate the plastic strain from creep strain based on the mechanical experimental tests. So, some researchers(Busso et al.[10], McDowell et al.[11], Qian et al.[12], Skipor et al.[13] developed unified creep-plasticity models in order to describe solder material behavior more precisely. Most of them are phenomenological models. The influence of micro-structure on deformation was not considered. At the present time, several works(Tang and Basaran[14], Wei and Chow[15], Zhang et al.[16], Sharma and Dasgupta[17], Stolkarts et al.[18])concerned about damage coupled constitutive model had also been carried out.

The aim of this paper is to character mechanical properties of SnAgCu solder and determine more quantitatively their influence on the constitutive equation. Previously obtained experimental data for a lead-free solder Sn4.0Ag0.5Cu have been analyzed, some additional experiments were performed, and the previous studies complemented, as necessary. Young's moduli, plasticity, microcrack and damage, temperature effects are investigated. Then, modified viscoplastic-damage constitutive equations for SnAgCu solder was proposed to simulate the effect of voids on reliability and the macroscopic mechanical response. The present model also considers the influence of the grain sizes or phase sizes on deformation by incorporating the grain size in matrix viscoplastic flow rule. Based on Gurson-Tvergaard plastic potential equation and orthogonal rules, void volume fraction as damage variable was introduced into the material constitutive relation. The constitutive relation includes strain softening, strain hardening, strain rate sensitivity and void evolution. The parameters are identified by combined experimental and numerical techniques. A non-linear optimization method has been developed to optimize the micromechanical parameters in order to obtain the best agreement between the experimental data and the predicted curve. The model was also implemented into finite element program ABAQUS through its user defined material subroutine. The predictions have shown the ability of the modified viscoplastic-damage model coupled with damage to correctly describe the experimental observations: nonlinearity, strain rate sensitivity and damage evolution.

2. Experimental Results and Discussion

This paper has dealt with raw experimental data for the lead-free solder, Sn4.0Ag0.5Cu, collected from Fairchild-ZJUT Microelectronic Packaging Joint Lab. The specimen preparation, geometry, machining, pretesting treatment, testing procedures, and testing set-ups are reported in publications by wang et al.[19].

The SnAgCu solder material tensile behavior tests were carried out using a precision non-contact strain digital measurement technique under constant true strain rates from 1.0E-5/s up to 1.0E-3/s and at temperature of 25 ℃ ,75 ℃ ,150 ℃ . Strains, based on crosshead displacement measurements, were adjusted by removing machine compliance effects. Based on DSCM(Digital Speckle Correlation Method), a novel video control experimental technique was developed for measuring axial and transverse strains [20].A computer with data acquisition software was used to collect the data.The criterion for failure is at least 10% strain and failure within the gauge length of the material.

The specimen and its grips were heated in a chamber attached to the tensile equipment. The chamber can be heated by an electrical resistance heater. The chamber temperature can be controlled within ±1 °C. Prior to the tension test, each specimen was held at the test temperature in the chamber for about 30 min in order to achieve thermal equilibrium.

The microstructures of the solders of as-cast and after tensile tests were observed in a scanning electron microscopoe(SEM). Prior to SEM observation, these specimens were polished to a 0.25 μm diamond finish, followed by polishing on a microcloth with any abrasive. Etching was carried out using a 5% nitric acid and 95% alcohol solution. The SEM observations were carried out using an accelerating voltage of 15Kv.

As shown in Fig. 1, the background microstructure of the solder consists of globular Sn-rich phase matrix with dispersed about 15vol% Ag3Sn precipitates around the Sn-rich regions, and small amounts of globular Cu-Sn intermetallics were observed that were slightly larger tan the Ag3Sn precipitates. At the same time, a few micro-defects, such as voids, can be found.

Fig. 2 shows the fractography of the alloys after the SnAgCu solder solder alloy after tensile tests . Many fine crack could be observed at the section surface. The microscopic crack is initiated on a cavity or on a particle. Initiate, coalescence and grow of voids with propagation induced microcrack. As the crack grows, it becomes unstable so that the materials failed rapidly.

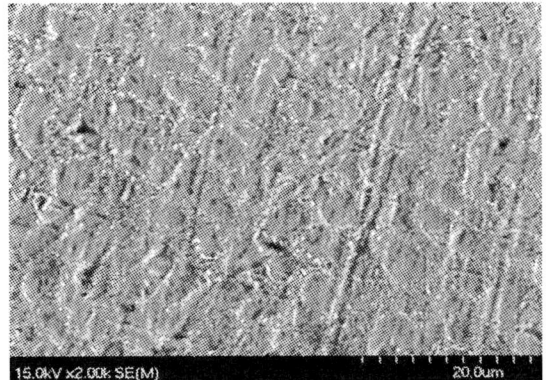

Fig. 1　optical micrograph of as-casted SnAgCu solder

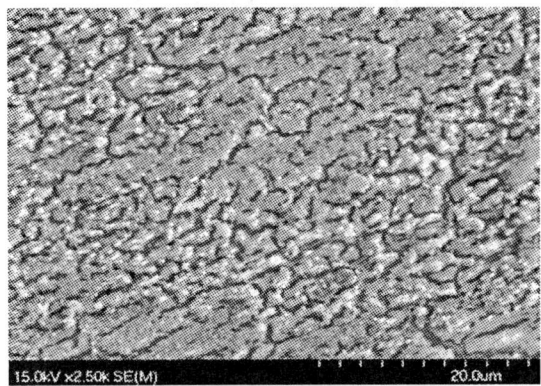

Fig. 2　optical micrograph of SnAgCu solder after tensile test

A typical tensile curve is shown in Fig. 3 where load are plotted versus displacement under various constant strain rates and at room temperature, exhibiting a strong influence of the strain rate on the non-linear behavior. The maximum load is increased by 12% for strain rates varying between 10E-5/S and 10E-3/S. This decrease is likely related to material softening. In general way, the strain softening is due to both the void growth and shear bands growth while the strain hardening is almost completely due to the plasticity in the matrix among a given strain value.

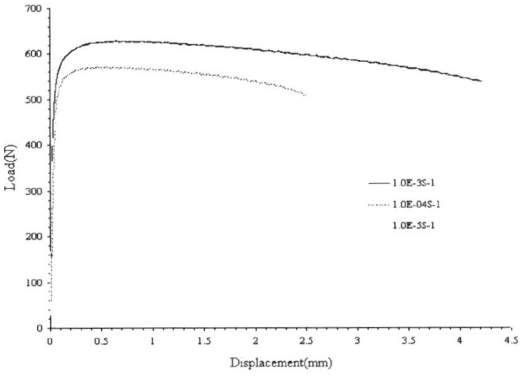

Fig. 3 Ture stress vs. true strain at room temperature

Fig. 7-9 show the typical true stress-strain vurves of the SnAgCu solter at different strain rates and temperatures. It is noted that the flow stresses of the Sn4.0Ag0.5Cu are strongly dependent on the test temperature and strain rate parameters. At a constant temperature, increasing the strain tate gave rise to a higher flow stress, while at a constant strain rate, raising the temperature resulted in lower flow stresses.

3. Constitutive Model

In non-linear solidmechanics, the material behaviour is often described by a rate-form(incremental) constitutive equation. In the case of elastic-viscoplasticity, the stress rate $\dot{\sigma}_{ij}$ depends linearly on the elastic strain rate tensor $\dot{\varepsilon}_{ij}^{e}$

$$\dot{\sigma}_{ij} = C_{ijkl}\dot{\varepsilon}_{kl}^{e} \qquad (1)$$

where C_{ijkl} is the isotropic elastic tangent stiffness tensor.

The macroscopic rate of strain tensor, $\dot{\varepsilon}_{ij}$, is written as the sum of an elastic part and a plastic part

$$\dot{\varepsilon}_{ij} = \dot{\varepsilon}_{ij}^{e} + \dot{\varepsilon}_{ij}^{p} \qquad (2)$$

The macroscopic viscoplastic part of the rate of deformation is derived from the normality rule and given by

$$\dot{\varepsilon}_{ij}^{p} = \dot{\lambda}\nu_{ij} \qquad (3)$$

where $\dot{\lambda}$ is the non-negative plastic multiplier. And

$$\nu_{ij} = \frac{\partial\phi}{\partial\sigma_{ij}} \qquad (4)$$

is the gradient of the plastic potential Φ. To determin the plastic multiplier, the loading-unloading conditions should be imposed in a Kuhn-Tucker form as

$$\dot{\lambda} \geq 0, \quad \phi \leq 0, \quad \dot{\lambda}\phi = 0 \qquad (5)$$

implying that during plastic loading, $\phi = 0$, $\dot{\lambda} > 0$, and $\dot{\phi} = 0$.

In this paper, the yield function to be considered is the Gurson yield condition[21], which was extended by Tvergaard and Needleman [22,23]. To account for the interaction between voids, they empirically introduced additional parameters (q$_1$, q$_2$, q$_3$) in the original Gurson model, as follows:

$$\phi(\sigma_{ij}, \bar{\sigma}, f) = \left(\frac{\sigma_e}{\bar{\sigma}}\right)^2 + 2q_1 f^* \cosh\left(\frac{3q_2\sigma_m}{2\bar{\sigma}}\right) - (1 + q_3^2 f^{*2}) \qquad (6)$$

where σ_{ij} is Cauchy stress tensor, σ_e is equivalent stress, $\bar{\sigma}$ is equivalent flow stress in the matrix material disregarding local stress variations. As follows

$$\sigma_m = \frac{1}{3}\sigma_{kk} \qquad (7)$$

$$\sigma_e = \sqrt{\frac{3}{2}\sigma'_{ij}\sigma'_{ij}} \qquad (8)$$

$$\sigma'_{ij} = \sigma_{ij} - \sigma_m\delta_{ij} \qquad (9)$$

The void fraction modification, f^*, was introduced by Tvergaard and Needleman, in order to simulate the rapid loss of strength accompanying void coalescence and is given by

$$f^* = f^*(f) = \begin{cases} f, & f \leq f_c, \\ f_c + \dfrac{q_1^{-1} - f_c}{f_F - f_c}(f - f_c), & f \geq f_c, \end{cases} \qquad (10)$$

where f_c is the critical void volume fraction, and f_F is the ultimate value of void volume fraction where the material completely fails.

The constitutive response of the matrix is considering as elastic-viscoplastic. A power law for the rate hardening including grain size effect, on the same format as Arrhenius relation and Dorn law, is used to describe the materials's viscous behavior. Thus

$$\dot{\bar{\varepsilon}}^{p} = \frac{ADEb}{kT}\left(\frac{b}{d}\right)^p \left(\frac{\bar{\sigma}}{\sigma_f(\bar{\varepsilon}^p)}\right)^m \exp(-\frac{Q}{RT}) \qquad (11)$$

where, A is a dimensionless material parameter;$D= D_0\exp(-Q/RT)$ is diffusion coefficient, Q is activation energy, k is Boltzmann's constant, b is burgers vector, d is average grain size, p is grain size exponent,m is strain rate hardening exponent, $\sigma_f(\bar{\varepsilon}^p)$ is yield stress related to equivalent plastic strain

$$\sigma_y(\bar{\varepsilon}^p) = \left[1 + \frac{\bar{\varepsilon}^p}{\varepsilon_0}\right]^n \qquad (12)$$

where n is strain harding exponent, ε_0 is reference strain.

The evolution of voids includes the growth of existing voids and the nucleation of new voids

$$\dot{f} = \dot{f}_{growth} + \dot{f}_{nucl} \qquad (13)$$

where the rate of the void fraction due to growth of existing voids is a function of the plastic volume rate

$$\dot{f}_{growth} = (1 - f)\dot{\varepsilon}_{kk} \qquad (14)$$

where $\dot{\varepsilon}_{kk}$ is the plastic strain rate tensor.

The nucleation of voids is a very complex physical process depending on the microstructure of the material. The commonly used strain controlled criterion proposed by Chu and Needleman [24] is expressed as

$$\dot{f}_{nucl} = \frac{f_N}{s_N\sqrt{2\pi}}\exp\left[-\frac{1}{2}\left(\frac{\bar{\varepsilon}^p - \varepsilon_N}{s_N}\right)\right]\dot{\bar{\varepsilon}}^p \qquad (15)$$

where f_N is the volume fraction of void nucleating particles, ε_N is the mean strain about which nucleation occurs, s_N is the standard deviation.

Following Ponthot[25], Carosio[26] has presented a "continuous" viscoplastic formulation. He introduced a yield function, $\overline{F}(\sigma_{ij}, \kappa_p, \dot{\lambda}) = 0$, from which a consistency condition can be derived under persistent viscoplastic flow when $\dot{\overline{F}} = 0$. Betegón[27] apply this procedure to the Gurson yield function to derive evolution of internal parameters as a function of the plastic multiplier.

According to Betegón's method, equating the plastic dissipation rate of a representative volume element with the plastic dissipation rate of the matrix material results in

$$(1-f)\overline{\sigma}\dot{\overline{\varepsilon}}^p = \sigma_{ij}\dot{\varepsilon}_{ij}^p \tag{16}$$

that allow expressing the evolution laws of the internal parameters in terms of the plastic multiplier by means of the plastic modulus

$$\dot{\overline{\sigma}} = h_1\dot{\lambda} + H_1\ddot{\lambda} \tag{17}$$

$$\dot{f} = h_2\dot{\lambda} \tag{18}$$

$$\dot{\overline{\varepsilon}}^p = h_3\dot{\lambda} \tag{19}$$

with plastic modulus

$$h_3 = \frac{\sigma_{ij}v_{ij}}{(1-f)\overline{\sigma}} \tag{20}$$

$$h_1 = \frac{\partial \sigma_f}{\partial \overline{\varepsilon}^p}\left(\frac{\dot{\overline{\varepsilon}}^p}{\varpi}\right)^{1/m} h_3 \tag{21}$$

$$H_1 = \frac{\overline{\sigma}}{m\dot{\varepsilon}^p}h_3 \tag{22}$$

$$h_2 = (1-f)v_{kk} + Ah_3 \tag{23}$$

where

$$A = \frac{f_N}{s_N\sqrt{2\pi}}\exp\left[-\frac{1}{2}\left(\frac{\overline{\varepsilon}^p - \varepsilon_N}{s_N}\right)\right] \tag{25}$$

$$\varpi = \frac{ADEb}{kT}\left(\frac{b}{d}\right)^p \exp(-\frac{Q}{kT}) \tag{26}$$

Then using these expressions and the viscoplastic consistency condition, the tangetial elasto-viscoplastic constitutive relations are obtained as

$$\dot{\sigma}_{ij} = C_{ijkl}^{vp}\dot{\varepsilon}_{kl} \tag{27}$$

with

$$C_{ijkl}^{vp} = C_{ijkl} - \frac{C_{ijmn}v_{mn}v_{pq}C_{pqkl}}{v_{mn}C_{mnpq}v_{pq} - \xi_1 h_1 - \xi_2 h_2}\left(1 - e^{-\frac{b}{c}t'}\right) \tag{28}$$

where

$$\xi_1 = \partial\Phi/\partial\overline{\sigma} \tag{29}$$

$$\xi_2 = \partial\Phi/\partial f \tag{30}$$

$$b = v_{ij}C_{ijkl}v_{kl} - \xi_1 h_1 - \xi_2 h_2 \tag{31}$$

$$c = \xi_1 H_1 \tag{32}$$

4. Model validation against experiments data and finite element calculations

In order to identify plastic deformation and failure properties of SnAgCu solder material, an iverse procedure to extract the flow properties was developed. The procedure is based on the comparison of the experimental tensile curves and curves generated by finite element calulations of the uniaxial tensile test. As Fig.4, The flow properties of non-damaged material up to the fracture strain is obtained after a couple of iterations. Finite element simulations of the uniaxial tensile test were carried out using a first extrapolation of the experimental flow curve beyond the necking start point. This method has been initially proposed by Norris et al.[28] and used later by several authors, e.g. Pardoen and Delannay[29], zhang et al.[30], Lassance et al.[31], Springmann and Kuna[32]. For the initial FE-run reliable starting values of solder materials can be guessed or extracted from literature. In subsequent runs the parameters are modified. The procedure is iterative by repeating the computation until the parameters provide the closest match to the experimental load-displacement curve.

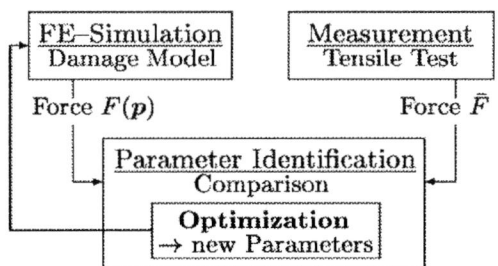

Figure 4. Identification method

First of all an overview is given about all the different geometric and material parameters that are involved in the finite element model of the tensile test. The geometry of the specimen and the values of the loading can be found in reference [19]. The elastic properties(E, v) are known for the materials investigated here, thus have not to be determined. Frequency factor D_0, Burgers vector b, and average grain size d can be gained by microstructural analysis. The initial void volume fraction f_0 can be determined in many cases by the chemical composition of the material. The void volume fraction at final failure f_F is material dependent and can be measured by microscopic investigations of the fracture surfaces. The load displacement curve is very insensitive to the

19

parameters ε_N and s_N, so they are fixed to 0.3 and 0.1. Under this assumptions there remain six parameters(A, p, m, Q, ε_0, n) for the hardening behavior of Eq.(11) and four (q_1, q_2, f_c, f_N) for the damage model to be identified.

The identification has been performed in two stages. In the first step, we assume that the material has only initial void growing without void nucleation, that is, f_N=0, f_c=1, f_F=1, q_1=1, q_2=1. That means, neither void nucleation nor void coalescence is modeled, only the assumed initial voids(f_0=0.002) can grow. In the second step, according to these viscoplastic parameters, the damage model parameters are calibrated by optimization routine to minimize the error between an experimentally results and one predicted by FE simulation. The calculated material parameters are summarized in Table 1.

Table 1. solder material properties

Material Parameters			
E(MPa)	44671-146.03T(°C)	v	0.3
Hardening Parameters			
A	5.48×10^9	$D_0(mm^2/s)$	57.6
$b(mm)$	4.08×10^{-7}	d	5.45×10^{-3}
p	2.76	m	12.5
n	0.12	$Q(KJ/mol)$	61
Damage Parameters			
q_1	1.397	q_2	1.254
q_3	1.9516	f_N	0.047
f_F	0.28	f_c	0.136

The numerical analysis was performed using the finite element code ABAQUS by user defined material subroutine. The axisymmetric FE mesh for this model is shown in Fig.4 (dimensions are indicated in Fig. 1). There are 330 elements (the axisymmetric FE model and 8-node biquadratic, reduced integration element:CAX8R) and 2480 nodes. The distribution of the elements is gradually fine toward X axis. The model is axisymmetric about the y-axis and a symmetry boundary condition is imposed on the bottom edge. Displacements are applied at the top edge and are ramped at a rate corresponding to the experimental crosshead displacement speed. Although this test should be carried out under constant strain rate conditions, the test specimen was not instrumented during testing: instead a constant crosshead displacement rate was applied.

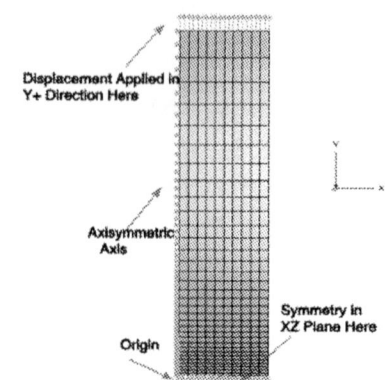

Fig. 5 Axisymmetric FE mesh for tensile specimen

Fig. 5 shows the distribution of Mises stress and damage at strain of 14%. It is noted that Mises stress isn't uniform distribution due to the presence of voids damage. Void damage concentrated on the middle of the specimen, then there were higher stress and firstly occurred material failure. So, the formation of the neck results at the center of the specimen can be observed. It reflected the experimental results correctly. The model does capture the deformation mechanisms accounting for voids damage.

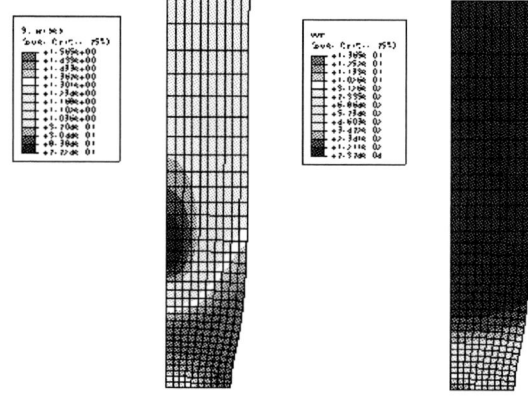

Fig. 6 Mises stress(L) and damage(R) distribution

In Fig. 7-9, numerical response compared with the experimental data for different temperatures and strain rates. A quite nice agreement is observed between nonlinear hardening experimental results and the predicted behaviour given the viscoplastic-damage model.

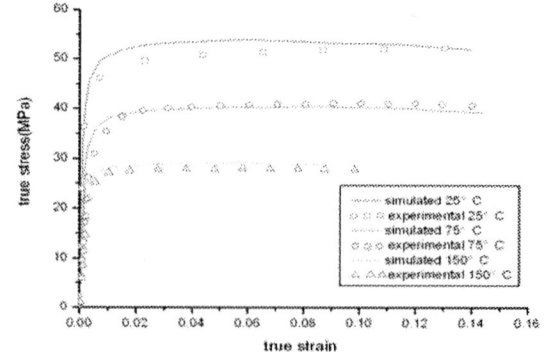

Fig. 7 True stress vs. true strain at 1.0E-3/s

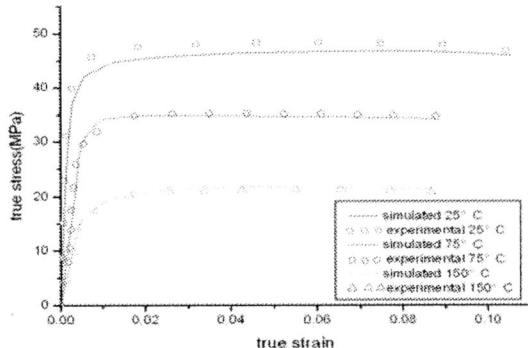

Fig. 8 True stress vs. true strain at 1.0E-4/s

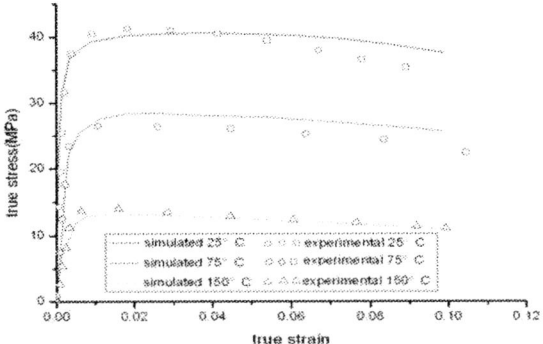

Fig. 9 True stress vs. true strain at 1.0E-5/s

5. Conclusions

A unified thermal viscoplastic-damage constitutive model coupled the effect of microstructure has been developed and implemented into ABAQUS6.5 through user defined material subroutine. The model allow structural analyses to be made of lead-free solder undergoing a deteriorating microstructure. Material parameters are determined from inverse procedure. Using computational simulations ,the predictions of the model verified against various tensile test data for lead-free solder. It is shown the ability of the viscoplastic model coupled with void damage to correctly describe the experimental observations: nonlinearity, strain rate sensitivity and void damage evolution are well captured.

However, several aspects need to be detailed and completed in the future. A future series of experiments on joint-scale test specimens will carry out to be better account for microstructure effects when it is applied predict the thermomechanical behavior and fatigue life of a particular packaging. In future work to develop a fatigue life model, efforts will be made to correlate a critical level of damage with fatigue life as a function of temperature.

Acknowledgments

This research project is supported by The MOE Key subject of Advanced Manufacturing Technology and Equipment of Zhejiang Province. The authors would also like to acknowledge funding support from Fairchild.

References

1. Amagai, M. et al., "Mechanical Characterization of Sn-Ag-Based lead-free Solders," Microelectronics Reliability, Vol. 42, No. 6(2002), pp.951-966

2. Yunus, M. et al, "Effect of voids on the reliability of BGA/CSP solder joints," Microelectronics Reliability, Vol. 43, No. 12(2003), pp. 2077-2086

3. Ladani, L. J., Dasgupta, A., "The Successive-Initiation Modeling Strategy for Modeling Damage Progression: Application to Voided Solder Interconnects," Proc 7th International Conference on Thermal, Mechanical and Multi-Physics Simulation and Experiments in Micro-Electronics and Micro-Systems, Como, Italy, April, 2006, pp. 570-575.

4. Wang D.J., Panton, R.L., "Experimental Study of Void Formation in eutectic and Lead-Free Solder Bumps of Flip-Chip Assemblies," J Electronic Packaging, Vol. 128, No. 3(2006), pp. 202-207

5. Kim, D.S., et al., "Effect of void formation on thermal fatigue reliability of lead-free solder joints," The 9th Intersociety Conference on Thermal and Thermo-mechanical Phenomena In Electronic Systems, Las Vegas, NY, USA, June. 2004, pp 325-329.

6. Yi, S., Luo G.X., Chian K.S., " A Viscoplastic Constitutive Model for 63Sn37Pb Eutectic Solders," Journal of Electronic packaging, Vol. 124, No 2(2002), pp. 91-96.

7. Pao, Y.H., Badgley, S., Jih, E., Govila, R., and Browning, J., "Constitutive Behavior and Low Cycle Thermal Fatigue of 97Sn3Cu Solder Joints," Journal of Electronic Packaging, Vol. 115, No. 6(1993), pp. 147-152

8. Darveaux, R., Banerji, K., "Constitutive Relations for Tin-Based Solder Joints," IEEE Trans. CHMT, Vol. 15, No. 6(1992), pp.1013-1024

9. Sarihan, V., "Temperature Dependent Viscoplastic Simulation of Controlled Collapse Solder Joint Under Thermal Cycling," Journal of Electronic Packaging, Vol. 115, No. 6(1993), pp.16-21.

10. Busso, E.P., Kitano, M., Kumazawa, T., "A Visco-plastic Constitutive Model for 60/40 Tin-Lead Solder Used in IC Package Joints," Journal of Engineering Materials and Technology, Vol. 114, No. 3(1992), pp.331-337.

11. McDowell, D.L., Miller, M.P., Brooks, D.C., "A Unified Creep-Plasticity Theory for Solder Alloys," Fatigue Testing of Electronic Materials, Atlanta, GA, USA, May, 1994, pp.42-59

12. Qian, Z., Liu, S., "A Unified Viscoplastic Constitutive Model for Tin-Lead Solder Joints," Advances in Electronic Packaging, Vol. 19, No.2(1997), pp. 1599-1604

13. Skipor, A.F., et al., "On the Constitutive Response of 63/37 Sn/Pb Eutectic Solder," Journal of engineering Materials and Technology, Vol. 118, No. 1(1996), pp. 1-11

14. Tang, H., and Basaran. C., " A Damage Mechanics-Based Fatigue Life Prediction Model for Solder joints," Journal of Electronic Packaging, Vol 125, No. 1(2003), pp.120-125

15. Wei Y., Chow, C.L., Lau, K.J., Vianco, P., Fang, H.E., "Behavior of lead-free Solder under Thermomechanical Loading," Journal of Electronic Packaging, Vol 126, No. 3(2004), pp.367-373

16. Zhang X.W., Lee, S-W. E., Pao, Y.H., "A Damage Evolution Model for Thermal Fatigue Analysis of Solder Joints," Journal of Electronic Packaging, Vol 122, No. 3(2000), pp.200-206.

17. Sharma P., and Dasgupta, A., "Micro-mechanics of Creep-fatigue Damage in Pb-Sn Solder due to Thermal Cycling-Part I:formulation," Journal of Electronic Packaging, Vol 124, No. 3(2002), pp.292-297

18. Stolkarts, V., Keer, L.M., and Fine, M.E., "Constitutive and cyclic damage model of 63Sn–37Pb Solder," Journal of Electronic Packaging, Vol. 123, No. 4(2001), pp. 351–355.

19. Wang Q., Liang L.H., and Liu Y., "Experimental Determination of Mechanical Properties for lead-free Material," 7th International Conference on Electronics Packaging Technology, Shanghai, china(2006), pp.276-279

20. Zhang, J.H., Feng, P., Cai, Z.S., Liang, L.H., "Application study on deformation measurement in mid-temperature environment for SnSb8.5 welding wire using DSCM," Journal of Zhejing University of Technology, Vol. 34, No. 5(2006), pp.567-570

21. Gurson, A.L., "Continuum theory of ductile rupture by void nucleation and growth: Part 1-yield criteria and flow rules for porous ductile media," Journal of Engineering Materials and Technology, Vol. 99, No, 1(1977),pp. 2-15

22. Tvergaard, V., "On Localization in Ductile Materials Containing Spherical Voids," International Journal of Fracture, Vol. 18, No. 4(1982),pp.237-252

23. Tvergaard, V., Needleman, A., "Analysis of the Cup-Cone Fracture in a Round Tensile Bar," Acta Metallurgica, Vol. 32, No. 1(1984), pp. 157-169

24. Chu, C.C., Needleman, A., "Void Nucleation Effects in Biaxially Stretched Sheets," Journal Engineering Materials and Technology, Vol.102, No. 3(1980), pp. 249-256

25. Ponthot, J.P., "An extension of the radial return algorithm to account f0 rate-dependent effects in frictional contact and viscoplasticity," Journal of Materials Process Technology, Vol. 80, No. (1998), pp.628-634

26. Carosio, A., William, K., Etse, G., "On the consistency of viscoplastic formulations," International Journal of Solids and Structures, Vol. 37, No. 37(2000), pp.7349-7369

27. Betegón, C., del Coz J.J., Peñuelas, I., "Implicit integration procedure for viscoplastic Gurson materials," Computer Methods in Applied Mechanics and Engineering, Vol. 195, No. 44-47(2006), pp.6146-6157

28. Norris D.M., Moran, B., Scudder J.K., Quinones D.F., "A computer simulation of the tension test," Journal of the Mechanics and Physics of Solids, Vol. 26, No. 1(1978), pp. 1-19

29. Pardoen, T., Delannay, F. "Assessment of void growth models from porosity measurements in cold drawn copper bars," Metallurgical and Materials Transactions A, Vol. 29A, No. 7(1998), pp.1895-1909

30. Zhang, Z.L., Odegard, J., Sovik, O.P., Thaulow, C.A., "A study on determining true stress-strain curve for anisotropic materials with rectangular tensile bars," International Journal of Solids and Structures, Vol. 38, No.6(2001), pp.4489-4505

31. Lassance, D., Fabregue, D., Delannay, F., Pardoen, T., "Micromechanics of room and high temperature fracture in 6xxx Al alloys," Progress in Materials Science, Vol. 52, No. 1(2007), pp.62-129

32. Springmann, M., Kuna, M., "Identification of material parameters of the Gurson-Tvergaard-Needleman model by combined experimental and numerical techniques," Computational Materials Science, Vol. 33, No. 4(2005), pp.501-509

Atom diffusion mechanism of thermo-sonic flip chip bonding interface

Fuliang WANG, Junhui LI, Lei HAN, Jue ZHONG
School of Mechanical and Electrical Engineering, Central South University
Changsha, HN province, 410083, China
Tel: +86-0139-7514-5436; E-mail: wangfuliang@mail.csu.edu.cn;

Abstract

The TSFC (thermosonic Flip chip) bonding was realized in a self- structured TSFC bonder. The atom inter-diffusion on Au-Ag bonding interface was characterized by TEM, and high-resolution TEM pictures reveales that the dislocation density in the bump increases after the acting of ultrasonic. And the interlaced dislocation slip lines were observed in the SEM pictures of bumps surface, which indicates the dislocations motion in the interior of bumps. A FEM model was used to simulate the stress on the bonding interface. It is noticed that the ultrasonic vibration causes high stress in the contact interface of bump and pad, which increases the dislocation density and provides short-circuit diffusion channel for Au and Ag atom inter-diffusion. Finally, a preliminary discussion about the atom diffusion, based on the atom diffusion theory, is also presented. Studies show that the stress is a significant component of atom diffusion driving force. The Gibbs free energy, chemical potential and acting force of Au and Ag atom on bonding interface are increased by the stress gradient. With the driving force caused by stress, the probability for atom to overcome the energy barrier increased, and the diffusion speed increased. And the atom diffusion depth on bonding interface is about 200-500nm in several hundred milliseconds, which forms good bonding strength.

1. Introduction

Thermo-sonic Flip chip (TSFC) bonding, an efficient area array microelectronic package technology, offers a snap, simpler assembly process in reducing the bond temperature, pressure and time, and it also provides strong metallurgical joining, which is thought to be more reliable than conductive adhesive and comparable to solder interconnection [1-4].

Although the thermosonic bonding technology has been applied in IC industry for many years, there is a lack of detailed understanding of bonding mechanism. Generally speaking, it is thought to contain several phases as follows: Firstly, the bumps of chip and pads of substrate are brought into contact under the tool force, producing some initial deformation but no adhesion due to the presence of surface oxides; And then, the ultrasonic vibration energy is propagated to the bonding interface through transducer and tool; Meanwhile, the surface oxides of bump and pads are broken up and fresh surfaces are exposed. At last, atom diffusion causes micro-welds in bonding interface and formed bonding strength [5].

However, people have no idea about how the atom diffusion happened on the bonding interface and how the bonding strength forms. In this work, the atom diffusion

mechanism of TSFC bonding interface was studied to gain a better understanding.

2. Experiments
2.1 Configuration of TSFC bonding system

The experiment table used in this work is a self-structured TSFC bonder, modified from the ultrasonic wedge bonder. The structure of bonding interface is shown as figure 1. There is no vacuum or groove on the tool, and ultrasonic vibration energy is propagated to chip by friction force between tool and chip which causes by bonding force. The bonding tool is a solid cylinder made by standard tungsten carbide steel (Young's modulus and density are $500GPa$ and $4900kg/m^3$ respectively), with a smooth end larger than chip surface.

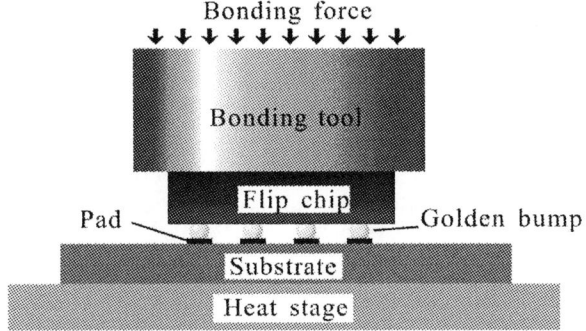

Fig.1 Structure of bonding interface

When ultrasonic energy is applied on tool, the chip and tool vibrate along a setting direction. Due to the bonding force, friction force appears on the tool and chip contact interfaces, and the ultrasonic energy is propagated from tool to chip and to the bumps/pads interface.

The $1\times1mm$ silicon Flip chip containing 8 peripheral Au bumps is employed in this study. Bumps are prepared by using a gold wire bonder with $2mil$ ($50\mu m$) gold wire, and resulted in approximately spherical bump of $80\mu m$ and $27\mu m$ in diameter and height respectively. The material of chip pad is aluminum. There are 28 pads on a chip, and 8 of them have Au bumps. The substrates are fabricated by Cu and coated with Ag ($10\mu m$ in thickness).

The bonding procedure began with the substrate sitting on a heated stage. The substrate was hold by a clamp. The temperature of substrate was maintained at 160℃. Then the flipped chip was transferred and brought into contact with substrate. The tool was brought into contact with the chip back in the center. At the instant the applied load reached predetermined value, 60 kHz ultrasonic vibration was applied through tool for 100ms to mount chip on substrate. Bonding conditions were set as table 1, and good bonding strength ($20g$/bump) were obtained.

Table 1 Conditions for TSFC bonding

Bonding force	1.76N
Substrate temperature	163℃
Ultrasonic power	2.8W
Bonding time	100ms

2.2 Atom diffusion on bonding interface

The interface of bump and pad was observed by TEM after bonding, and the inter-diffusion was found, as shown in figure 2, the left side is pad (made by Ag), and the right side is bump (made by gold). It can be seen that the atom diffusion truly happened on the bonding interface and formed bonding strength.

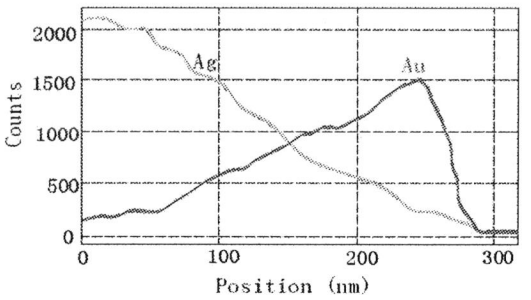

Fig.2 Atom inter-diffusion on bonding interface

2.3 FEM model of bonding interface

It is believed that the stress on bonding interface caused by ultrasonic vibration is the driver force of atom diffusion, so a 2-D FEM model was developed to simulate the stress distribution on bonding interface during bonding process [7].

The vibration amplitude and bonding force of chip in the FEM mode were set as $400nm$ (measured by a Laser Doppler Vibrometer) and $58.6MPa$ (equal to 30g/bump). And the simulation result is shown in figure 3, which indicates that the maximal von Misses stress located on one side of contact interface, and changed from one side to another with the cycled ultrasonic vibration. Figure 4 is the distribution of σ_y on bonding interface, and zero in the x-coordinate is the center of bump.

The simulation results show that the maximal stress on bump and pad are over $600MPa$ and $350MPa$ respectively.

Fig.3 von Misses stress on the bonding interface

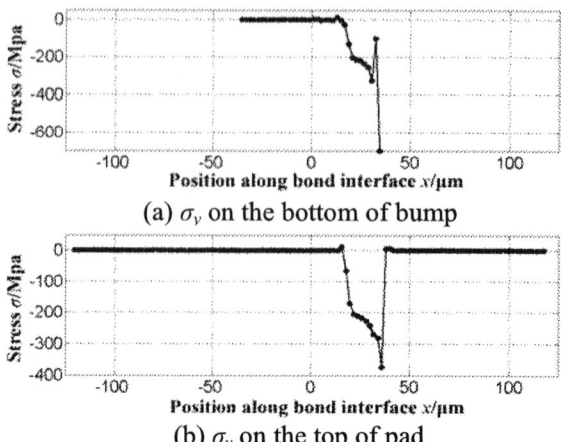

(a) σ_y on the bottom of bump

(b) σ_y on the top of pad

Fig.4 Distribution of σ_y on bonding interface.

3. Discussion

3.1 Effect of stress on dislocation multiplication

The great stress on bonding interface has affected the dislocation density and plastic deformation of bumps and pads. Normally, the typical dislocation density of metal in annealed state is $10^6 \sim 10^8/cm^2$. According to dislocation multiplication created by Frank-Read dislocation source, the dislocation lines on bumps and pads are elongated, deformed, moved and multiplicated when the greater stress act on them [8], as shown in figure 5.

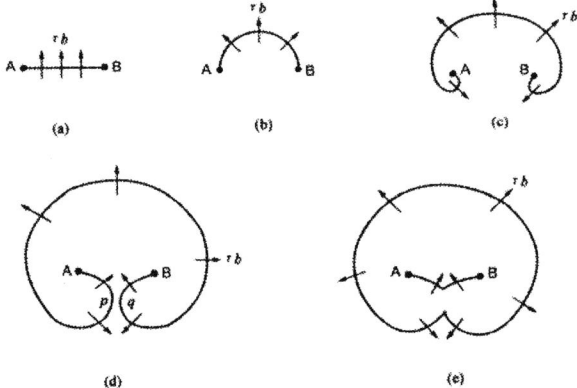

Fig. 5 Dislocation generation in a Frank-Read source

The critical stress to start the multiplication of Frank Read sources is:

$$\tau = \frac{Gb}{L} \qquad (1)$$

where G is the shear modulus (the shear modulus G of gold and silver is smaller than 300MPa), L is the distance between A and B, b is the absolute value of the Burgers vector and τ is the local component of the acting stress. Using characteristic values, L is in the range of about $100 \sim 200$ nm. G is in the range of $200 \sim 300 MPa$ [9].

With Eq.1, the critical stress of dislocation generation is about $0.3 MPa$. Actually, the critical stress of gold and silver is $0.79 MPa$ and $0.47 MPa$, which is much smaller than the simulated stress value on bonding interface [10].

Obviously, the stress on bonding area is high enough for dislocation multiplication. At the same time, the increase of dislocation density can decrease the critical stress for dislocation generation in a Frank-Read source. After the plastic deformation caused by dislocation movement, the dislocation density can be increased to $10^{10} \sim 10^{12}/cm^2$ and formed dislocation net [9].

With the TEM, the dislocation density of aluminum pad can be obtained as figure 6(a) and (b), which indicates that the dislocation density is sparse before ultrasonic acting and the density increased after the acting of ultrasonic vibration.

(a) before the acting of ultrasonic

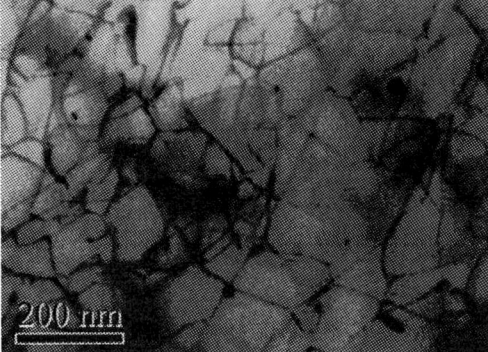

(b) after the acting of ultrasonic

Fig.6 Dislocation density of aluminum pad

When dislocations move to metal surface, parallel dislocation lines appeare. With the SEM, slip bands and crossed slip lines can be observed on the surface of golden bumps and inner of bumps, as shown in figure 7 and 8.

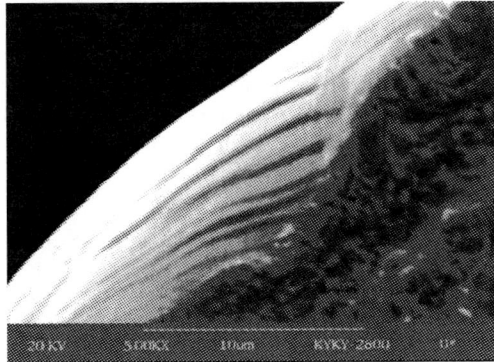

Fig.7 Slip bands on bump surface

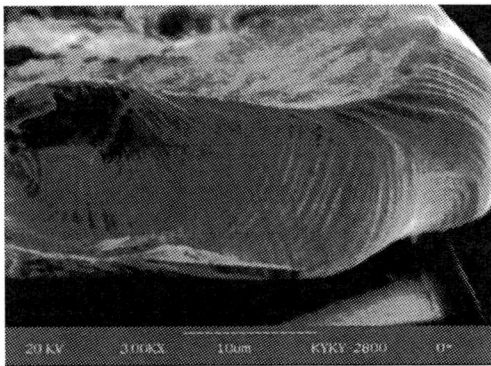

Fig.8 Crossed slip lines on the inner of bump

To draw a conclusion, with the effect of bonding force and ultrasonic vibration, the stress on the bump/pad causes dislocation multiplication and dislocation density incensement, and the dislocation motion is a main cause of plastic deformation of bump and pad.

Additionally, the strain rate has effects on dislocation multiplication. Under the bonding condition (163 ℃ in temperature and several MPa in stress), the strain rate is larger than $10/s$, which enhances the dislocation multiplication.

Except for increased dislocation density, the oxide layers of bonding surface are also broken by high stress on bonding interface and made ready for atom diffusion.

3.2 Atom diffusion channel and driving force

Usually, atom diffusion along dislocation line, free surface and grain boundary is much easy. The stress breaks the surface and increases the dislocation density, which provides short-circuit diffusion channels for Au and Ag atom inter-diffusion.

Atom diffusion along short-circuit diffusion channels need enough driving force. According to the materials physics theory [10], the driving force of atom diffusion is chemical potential gradient. The diffusion result is to decrease the Gibbs free energy of whole system, and the balance condition is that the Gibbs free energy everywhere is equal, i.e. gradient is 0.

Considering the bonding interface area which is composed by bumps and pads, the Gibbs free energy of the interface system is [10]:

$$dG = VdP - SdT + \sum_i \mu_i dn_i \qquad (2)$$

where VdP is the Gibbs free energy related to stress gradient, $\mu_i dn_i$ is the free energy of component i when stress is 0, SdT is the free energy related to temperature gradient.

As for the bonding interface system, components include Au and Ag, and the temperature of bumps and pads can be considered as constant during bonding, i.e. $dT=0$. So Eq.2 can be simplified as Eq.3:

$$dG = Vdp + \mu_{Au} dn_{Au} + \mu_{Ag} dn_{Ag} \qquad (3)$$

And the chemical potential of Au and Ag are [10]:

$$\mu_{Au} = \frac{dG}{dn_{Au}} = V_{m,Au}dp + \mu_{Au,\Delta p=0} \qquad (4)$$

$$\mu_{Ag} = \frac{dG}{dn_{Ag}} = V_{m,Ag}dp + \mu_{Ag,\Delta p=0} \qquad (5)$$

where $\mu_{Au,\Delta p=0}, \mu_{Ag,\Delta p=0}$ is the chemical potential of Au and Ag when stress gradient is 0, V_m is Molar Volume, unit is mol/m^3, and $V_m dp$ is the increase of Gibbs free energy caused by stress gradient.

So the force on y direction (perpendicular to the bonding interface) of Au and Ag atom on bonding interface is:

$$\begin{cases} F_{Au} = -\dfrac{\partial \mu_{Au}}{\partial y} \\[2mm] F_{Ag} = -\dfrac{\partial \mu_{Ag}}{\partial y} \end{cases} \qquad (6)$$

According to the diffusion equation [10], the diffusion flux on the y direction of Au and Ag atom is:

$$\begin{cases} J_{Au} = -B\dfrac{\partial \mu_{Au}}{\partial y}\rho_{Au} = -B\rho_{Au}F_{Au} \\[2mm] J_{Ag} = -B\dfrac{\partial \mu_{Ag}}{\partial y}\rho_{Ag} = -B\rho_{Ag}F_{Ag} \end{cases} \qquad (7)$$

where ρ is the density, unit is kg/m^3, B is the mobility of atom, both of them are related to crystal structure; and F is the driving force acting on the component, which is related to stress gradient.

So, the free energy, chemical potential and force of Au and Ag atom on bonding interface increases by the stress gradient caused by ultrasonic vibration and bonding force. With the increase of driving force, the probability for atom to overcome the energy barrier increases, and the diffusion speed increases. Finally, the atom diffusion depth on bonding interface is about 200-500nm in several hundred milliseconds, and formes good bonding strength.

Except for stress gradient, dislocations also decreases the energy barrier in a certain extent, and makes the atom diffusion easier.

Actually, the driving force on bonding interface is a much more complex physical chemistry process, and farther and deeper researches are expected.

4. Conclusions

The TSFC bonding was realized with the flip chip die constrained under the bonding force. With the SEM and TEM, the atom diffusion on bonding interface was observed, and a FEM model was used to simulate the stress on the bonding interface. Studies show that the stress caused by ultrasonic vibration and bonding force has effects on the atom diffusion:

1) The surface is broken by stress, and the the dislocation density is increased by stress, which provides short-circuit diffusion channel for Au and Ag atom inter-diffusion.

2) Stress is a significant component of atom diffusion driving force.

Increasing diffusion channel density and driving force can increase the atom diffusion area and depth; and increase the bonding strength and bonding reliability.

Acknowledgments

The authors are thankful to the The Natural Science Foundation of China (Contract No.: 50390064, 50575230, 50675227,50429501) and the China Department of Science & Technology Program 973 (Contract No.: 2003CB716202) for their supports, and thankful to ASMPT for their support.

References

1. Irwin, Ronda S.; Zhang, Wenge; Harsh, Kevin F.; Lee, Y.C. Quick prototyping of flip chip assembly with MEMS: Proceedings of the International Symposium, v 44, 1998, p 256-263

2. Yatsuda, H.; Iijima, H.; Yabe, K.; Iijima, O. Flip-chip STW filters in the range of 0.4 to 5 GHZ, In Proceedings: 2002 IEEE Ultrasonics Symposium, Oct 8-11 2002, Munich, Germany, v 1, 2002, p 11-18

3. Taizo T, Tomohiro I, Ikuo M，Thermosonic flip-chip bonding for SAW filter, Microelectronics Reliability, 2004,44(5):pp149-154

4. Pang, C.C.-H.; Sham, M.-L.; Hung, K.-Y. , Investigation on the high frequency thermosonic flip chip bonding under low temperature, Source: Proceedings of the 5th Electronics Packaging Technology Conference (EPTC 2003), 10-12 Dec. 2003 , Singapore, pp 376-379.

5. Schwizer, J., Mayer, M., Bolliger, D. et al, Thermosonic ball bonding: friction model based on integrated microsensor measurements, In proc, 24th IEEE/CPMT International Electronics Manufacturing Technology Symposium, Austin, TX, USA, 1999, pp.108-114.

6. Jun Qi, Ngar Chun Huang, Ming Li, Deming Liu, Mechanism analysis of process parameters effect on bondability in ultrasonic gold ball bonding, In Proceedings: HDP'07, June 27 -30, 2005, Shanghai, China, pp: 56-60.

7. Fuliang wang, Lei han, Jue zhong. Study on the Circle Band Interface of Thermosonic Flip Chip Bonding, Transactions of the China Welding Institution, 2006,27(11), 65-68.

8. Hirth JP, Lothe J. Theory of dislocations. New York: Wiley; 1982.

9. E. Orowan, in: Dislocations in Metals, American Institute of Mining and Metallurgy Engineering, New York, 1954, p. 103.

10. Shanglin Y., Tingdong X., Materials physics. Harbin: Press of HIT, China, 2004.

A coupled numerical and experimental study on thermo-mechanical fatigue failure in SnAgCu solder joints

M. Erinc, P.J.G.Schreurs, M.G.D.Geers
Eindhoven University of Technology
Materials Technology, WH 4.139, PO box 513, 5600MB Eindhoven, The Netherlands
m.e.erinc@tue.nl, +31 40 2472245

Abstract

In ball grid array (BGA) packages, solder balls are exposed to cyclic thermo-mechanical strains arising from the thermal mismatch between package components. Since fatigue cracks in solder balls are observed generally at the chip side junction, dedicated fatigue experiments are conducted using eutectic SnAgCu- Ni/Au specimens in order to mechanically characterize the bonding interface. Sn based solders are prone to thermal fatigue due to the intrinsic thermal anisotropy of the β-Sn phase. Bulk SnAgCu specimens are thermally cycled and mechanical tests are conducted to quantify the thermal fatigue damage. In both damage schemes a strong size effect is observed. Experimental results are used to develop a cohesive zone based fatigue damage evolution law. Fatigue crack propagation is predicted by an irreversible linear traction-separation cohesive zone law accompanied by a non-linear damage variable. Finally, bulk damage in SnAgCu due to thermal fatigue and the interfacial fatigue failure in BGA balls are combined to simulate a BGA solder ball exposed to thermo-mechanical fatigue in 2D. This combined approach gives a more realistic outcome when determining the overall mechanical response, since the microstructural entities and the solder ball itself are on the same size scale and thus the solder ball cannot be treated as a continuum.

1. Introduction

The ongoing miniaturization in microelectronics has led to the development of ball grid array (BGA) and flip-chip packages. As size gets smaller, local operating temperatures increase, hence the package components are exposed to higher mechanical strains. Another consequence of miniaturization is that in small parts mechanical properties become inhomogeneous due to size effects (Fig. 1). It is previously reported that these size effects cause microstructure driven damage localization in lead-free solders [1]. Therefore, an in-depth study relating microstructure and fatigue damage evolution is crucial for the physical understanding of the thermo-mechanical fatigue problem in solder joints. Driven by miniaturization, solder joints have become so small that the size of microstructural entities in the solder is comparable to that of the joint itself. In order to predict fatigue failure at these size scales, rather than scale independent continuum-based approaches (i.e. Coffin-Manson rule), appropriate damage formulations in the solder joint are needed.

Due to legislation and technological requirements, microelectronics industry has moved to eutectic SnAgCu solders as a replacement for the traditional SnPb solder. A broad review on solder materials, reliability issues and solder/substrate interactions can be found in the literature [2-9]. Solders are exposed to high homologous temperatures in service, therefore, in microstructural and mechanical characterization, evaluation of time-dependent properties becomes crucial, which can be found in [10-12]. A review on solder joint fatigue models is given in [13]. However, experimental validation of these models at the level of a single solder interconnect is hardly existing in the literature. A chip size experimental and numerical study of SnPb solder fatigue fracture is carried out by [14]. A solder ball sized coupled study on thermo-mechanical fatigue failure in SnAgCu is presented in [15].

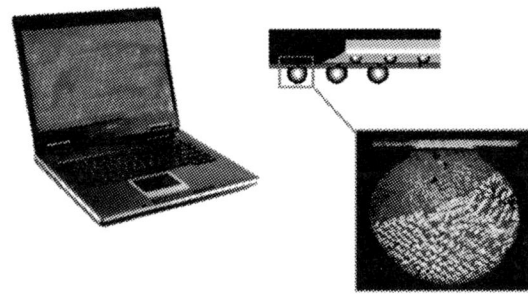

Fig. 1 Miniaturization amplifies the effect of microstructural discontinuities on the overall performance of the component

The objective of this study is: (1) to establish a link between microstructure and damage in lead-free solder joints under cyclic thermo-mechanical loading, (2) to characterize fatigue damage evolution for the damage mechanisms observed and (3) to simulate local deformations leading to fatigue crack propagation by separation of interfaces. The reliability of soldered joints are determined by the fatigue strength of the bump/pad junction. In order to evaluate this interface, soldered joint specimens are prepared and constant amplitude mechanical fatigue tests are conducted. Another important issue is that thermal fatigue loading causes very local deformations especially on Sn grain boundaries due to the thermal anisotropy of β-Sn; as reported many times in the literature [16-20]. In order to incorporate this intergranular fatigue damage, bulk SnAgCu specimens are thermally cycled. Experimental results are used to

1-4244-1105-X/07/$25.00 ©2007 IEEE   27

develop a cohesive zone based thermo-mechanical fatigue damage model for SnAgCu solder balls.

2. Experimental

To measure the mechanical strength of the bump/pad interface, solder joint specimens are prepared by reflowing SnAgCu paste on Ni/Au substrates for pure tensile and pure shear testing, as shown in Fig.1. Ni substrates are machined from 1mm thick 99.98% purity nickel plates followed by Au coating by ion sputtering to a thickness of 0.5-0.8μm. Ni/Au substrates are soldered by Sn3.8Ag0.5Cu solder flux on a hot plate at 250°C and air cooled. The solder thickness varied between 500±200 μm. For the tensile specimens, four different geometries are considered; the gage width, shown with an arrow in Fig.2, is taken as 10, 5, 3 and 1mm. Two 10mm, six 5mm, eight 3mm and eight 1mm specimens are produced. For the double joint shear specimens, the gage width is taken as 3 and 1mm and three specimens for each geometry are produced. For the single joint shear specimens, nine 3mm gage specimens are produced. On average, the solder volume in 10, 5, 3 and 1mm tensile samples was ~4, ~2, ~2 and ~0.4mm³, in 3mm and 1 mm double joint shear specimens ~1.2 and ~0.4mm³, in single joint shear specimens ~1.2mm³, respectively.

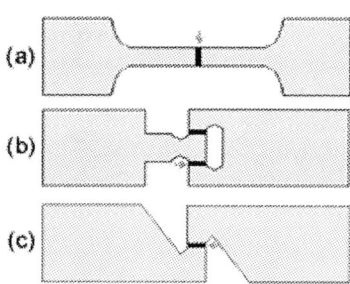

Fig. 2 Sn3.8Ag0.5Cu-Ni/Au soldered joints of shape: (a) tensile, (b) double joint shear, (c) single joint shear.

Fatigue tests were conducted in a computer controlled double spindle micro-tensile stage (Fig.3) where reaction force and elongation were recorded. An external extensometer and a 500N load cell was used and a triangular cyclic elongation was prescribed. Tests took place at constant room temperature. Ni/Au-SnAgCu fatigue specimens are cyclically loaded with constant amplitude. Fatigue life is determined by the decrease in the reaction force using the half life criterion. Strain applied to solder joints varied in the range -1%-3.5%.

For thermal cycling experiments, tensile specimens are machined from a cast Sn4Ag0.5Cu slab. Seven specimens each having 1, 2 and 3mm thickness are prepared. Thermal cycling was performed at Philips Applied Technologies, Eindhoven, between -40 and 125°C where the ramp rate was 11°C/min and the duration of a cycle was approximately 1 hour. In the furnace chamber, samples were allowed to expand and shrink without any mechanical constraints. Three samples with different thicknesses were removed from the furnace chamber at cycles 500, 1000, 1500, 2000, 3100 and 4000. No surface conditioning was applied to the

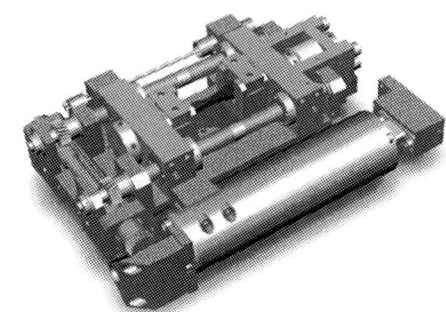

Fig. 3 Double spindle micro-tensile stage (Kammrath & Weiss).

samples after thermal cycling. The gage area of the tensile specimens were scanned with Orientation Imaging Microscopy (OIM) before and after thermal cycling. After thermal cycling, strain gages are glued to the samples with the intention of measuring a global elastic modulus which will be used later as a global damage indicator. Tensile tests were performed in the micro-tensile-stage mentioned above using a 20N loadcell and a 0.5μm/s clamp speed. An appropriate strain (in the order of e-5) was applied to produce a stress less than 2MPa in the cross-section of the specimen. A reference elastic modulus value of 64±1GPa is measured prior to thermal cycling.

3. Results

3.1 Interfacial fatigue failure in SnAgCu-Ni/Au

SnAgCu-Ni/Au soldered joints are mechanically cycled. For every fatigue specimen, stress vs. number of cycles is plotted. Results can be categorized in one of the three following schemes: a delamination regime (type A), an infinite life regime (type C), and in between type A and C, a gradual failure regime (type B). Microstructural observations suggest that there are two different failure mechanisms operative: (1) heterogeneous matrix deformation and (2) localization of deformation at the bonding interface. Selected stress-cycle curves for failure types A, B and C and the corresponding micrographs are shown in Fig.4 and Fig.5, respectively.

Type A failure can be examined in three stages: (1) incubation period in the beginning where stress reached at peak strain is steady, (2) fatigue failure stage where stress at peak strain decreases steeply and (3) the last stage where stress decreases gradually which is believed to be due to crack bridging and friction. Fractographs show highly localized damage at the bonding interface since the major crack propagates on the interface and there is almost no matrix deformation. If there are blowholes in the solder, main crack preferentially propagates on that weaker interface. Here the dominating failure mechanism is delamination. Strain life is between 1000-2000 cycles. Type A failure is observed in the fatigue specimens in which the initial stress was highest. In addition, solder volume being less than ~1mm^3 seemed to increase the tendency towards delamination.

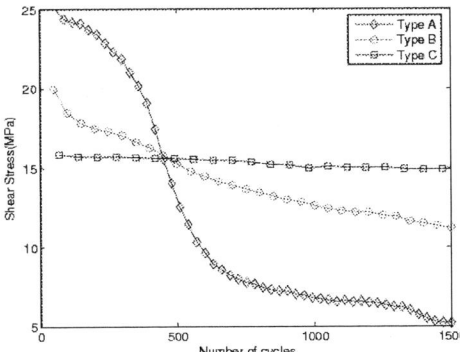

Fig. 4 Stress drop regimes observed in cyclic strain controlled experiments.

An infinite fatigue life (type C) is observed when the initial stress was less than 17MPa, and/or when solder volume exceeded 3mm^3. This suggests that strain life of SnAgCu solder joints is not only determined by the initial stress state but also depends on geometry. In the solder matrix many small cracks are observed. The pattern of these cracks coincides with the pattern observed in the polarized images taken before the fatigue tests. This suggests for the present case that fatigue cracks initiate on domain boundaries where there is a slight difference in the crystallographic orientation. Another interesting observation is that slip lines appeared in β-Sn dendrite arms. Slip lines are not observed in the eutectic matrix because the intermetallic components of the ternary eutectic matrix (i.e. eutectic Ag$_3$Sn and eutectic Cu$_6$Sn$_5$ are pinning points for dislocation movement. In type C specimens, a major fatigue crack was still not observed after 10,000 cycles. Our experiments suggest that there is a fatigue limit in both normal and tangential directions of the bonding interface, with a roughly estimated value of σ$_f$=17MPa. However, a separate testing procedure is necessary to determine the fatigue limit of the interface and its dependence on solder geometry more accurately.

Specimens in which fatigue failure proceeds gradually are referred as type B. Fractographs show heavy matrix deformation as well as an interfacial fatigue crack which has led to failure. Surface relief in the solder matrix consists of slip in β-Sn dendrites, many short cracks on domain boundaries in the eutectic matrix, parallel deformation bands 45° to the loading direction, and secondary fatigue crack propagation on grain boundaries. Strain life is between 2000-5000 cycles. Here, both deformation mechanisms mentioned above are operative.

Whether the fracture behavior will be abrupt (by delamination) or gradual depends on local soldering conditions. Existence of manufacturing defects on the surface, less wetting, less intermetallic growth will favor delamination leading to a sudden failure. If the fatigue crack propagation is taking place in the solder very close to the bump/pad, dimples and fatigue striations are observed in the fractographs and the expected failure type is gradual.

Shear specimens show more matrix deformation than tensile specimens. This suggests that the interfacial region is weaker in the normal direction than in the tangential direction. In the tangential direction, the scallop geometry of intermetallic compounds serve as a mechanical barrier and solder cannot delaminate easily. Therefore, energy dissipation due to plastic deformation in the matrix is promoted over damage localization at the bonding interface.

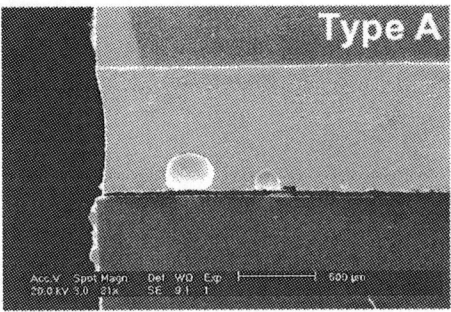

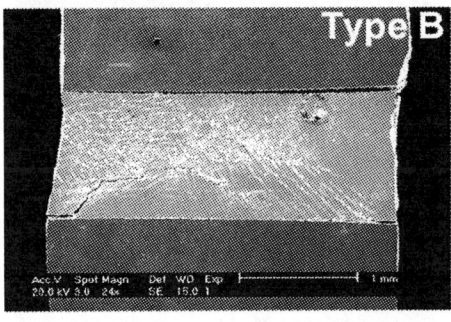

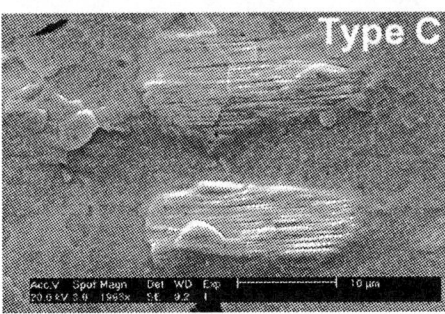

Fig. 5 Typical micrographs corresponding to A, B and C failure types.

3.2 Thermal fatigue damage in bulk SnAgCu

Bulk SnAgCu specimens were thermally cycled between -40 and 125°C. Fig. 6 shows the surface roughness images of a sample before thermal cycling and after 1000 cycles. It is important here to note that the z-axis on the top image is between ±2μm whereas the bottom image is between ±20µm. After 500 cycles, many small cracks appear on the grain boundaries. Further cycling to 1500 cycles results in heavy deformation patterns at the grain boundaries. After 2000 cycles, the damaged zones have grown to the interior of the grains. Especially around the triple points a large area appears to be affected.

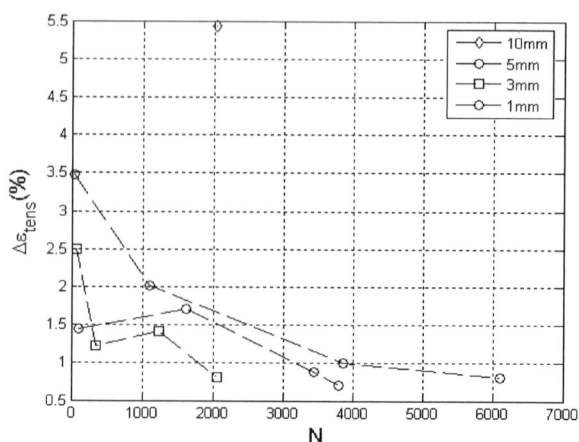

Fig. 7 Strain life versus strain on solder for SnAgCu-Ni/Au soldered joints.

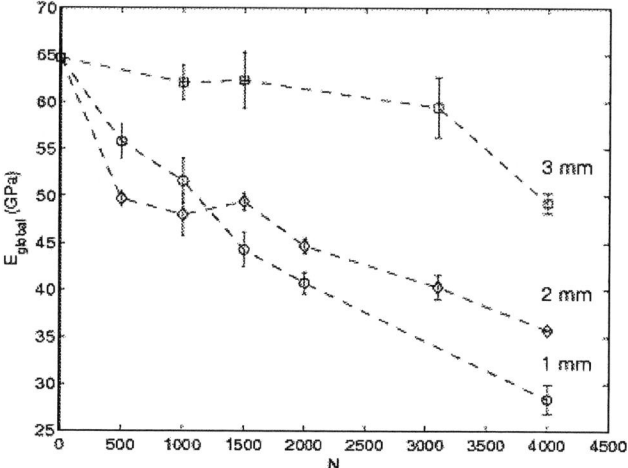

Fig. 8 Global elastic moduli measured after indicated number of thermal cycles

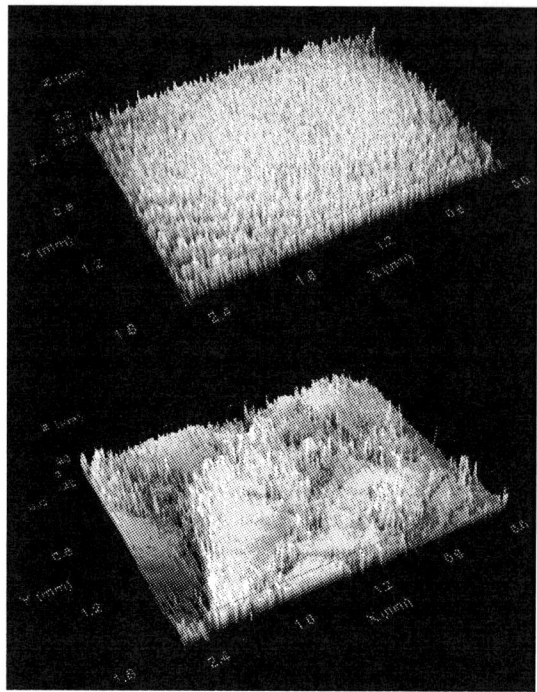

Fig. 6 Surface roughness patterns before thermal cycling (top) and after 1000 cycles (bottom).

3.3 Size effect

In both solder/pad interface and bulk solder studies, a strong size effect was seen on the results. Strain life versus applied strain for soldered joint specimens are shown in Fig. 7. Indicated strain values are corrected for the elongation in the substrate and thus correspond to the strain in the solder. As the gage width decreases, fatigue life under the same initial stress also decreases. Apparently, a smaller solder volume increases the tendency towards localized deformations at the bonding interfaces. In short, for SnAgCu-Ni/Au soldered joints, smaller joints are weaker than bigger joints upto 0.4mm³ solder. It can be suggested that solder joints less than 0.4mm³ probabilistically contain less defects on the surface, where fatigue cracks may initiate.

Global elastic moduli of bulk SnAgCu samples measured after various temperature cycles are plotted in Fig. 8. Although cyclic softening is observed in all 3 sets (1mm, 2mm and 3mm thick samples), the amount of deformation clearly shows a size-effect. Thinner samples are seen to suffer more from thermal fatigue than thicker samples.

4. Modeling

Based on previous cohesive zone formulations [21,22] an irreversible damage model for interface fatigue crack growth is described where the traction T is related linearly to the separation Δ for both normal (n) and tangential (t) directions. A damage variable D is introduced in the cohesive zone model which evolves from the undamaged state D=0, until complete failure D=1. The traction-separation law and the damage evolution law are shown below:

$$T_\alpha = k_\alpha (1 - D_\alpha) \Delta_\alpha$$

$$\dot{D}_\alpha = c_\alpha |\dot{\Delta}_\alpha| (1 - D_\alpha + r)^m \left\langle \frac{|T_\alpha|}{1 - D_\alpha} - \sigma_f \right\rangle$$

where k is the initial stiffness, σ_f is the fatigue limit, c, m and r are constants that control the decaying of the cohesive interaction and α refers to the normal and tangential directions of the cohesive zone. Test specimens are modeled using 2D plane stress elements in the commercial finite element package MSC Marc Mentat. Fatigue experiments are simulated by prescribing the experimental conditions at the specimen boundaries. In the simulations, damage evolution parameters are tuned to obtain the same global stress-strain response as measured in the experiments. Details of parameter determination in the damage evolution law is explained elsewhere [15,20]. By modeling the complete test-specimen, experimental stress-strain data is converted indirectly to the traction-displacement in the cohesive zones. For thermal fatigue experiments, the complete microstructure of the test specimens is modeled by incorporating the granular structure and the average orientations of the grains. The crystallographic orientations in OIM are given by Euler angles which are rotated to the FEM reference frame. The orthotropic thermal expansion coefficient of β-Sn is thereby incorporated correctly for every grain. The material parameters used in simulations are tabulated in Table 1. For SnAgCu, the hardening rule is based on experimental data points [15] and the creep model is implemented as given in [11].

Table 1. Linear material parameters used in parameter determination.

Material	model	E(GPa)	ν
SnAgCu	visco-el-pl	64.1	0.4
Cu	lin. el.	128	0.35
Ni	lin. el	198	0.31
Ni3Sn4	lin. el.	140.4[23]	0.2

The bulk SnAgCu damage model and the interfacial damage model addressing the SnAgCu/Ni-Au interface are combined to simulate the full microstructure of a BGA solder ball in 2D. The microstructure consists of a single crystal SnAgCu containing twins. The mismatch angle between the local β-Sn [001] directions of the solder bulk and the twins is 112.1° as measured by OIM. The solder ball is 500μm in diameter, the UBM consists of a 20μm Cu and 10μm Ni. There is a 2μm Ni$_3$Sn$_4$ IMC between the metallization and the solder. At the bump/pad interfaces and the twin boundaries cohesive zones with a 2nm thickness is placed. The cohesive law is described separately for the bump/pad interface and the bulk solder as explained in the previous section. A cyclic thermal load between -40 and 125°C is applied to all elements. At each cohesive zone an effective damage value D_{eff} is calculated from its normal and tangential components as follows:

$$D_{eff} = \left(D_t^2 + D_n^2 - D_t D_n\right)^{1/2}$$

In [22] the fatigue life of an interface is determined as the number of cycles where $D_{eff}=0.5$. The initial mesh and the contour plot of D_{eff} after 2000 temperature cycles is shown in Fig. 9. Nodal values are linearly averaged for

visualization. Damage evolution takes place both at the bonding interface and on the twin boundaries in the solder. The average D_{eff} after 2000 cycles in the top bump/pad interfacial cohesive element nodes is $D_{eff}=0.22$, whereas at the twin boundaries it is one scale lower being $D_{eff}=0.025$. In Fig.10 mean D_{eff} values of all interfacial cz elements are plotted against number of cycles. Extrapolating Fig.10, the solder ball in question will fail at its bond/pad interface approximately 9500 temperature cycles between -40 and 125°C according to the chosen failure criterion. The extrapolation follows a typical power law relation which is consistent with the solution of the governing equation at hand.

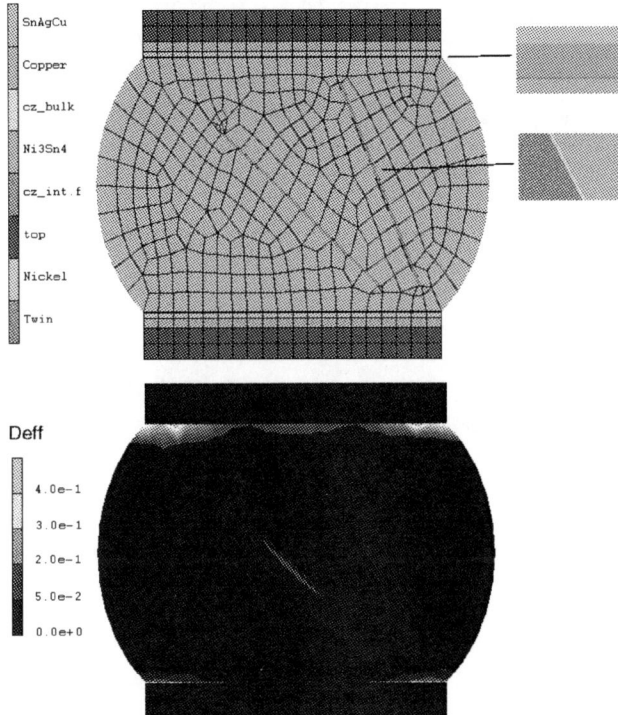

Fig. 9 Thermal cycling simulation of a SnAgCu solder ball, with an OIM scan microstrucutre, N=2000 cycles between -40 and 125°C.

As also predicted in the example above, the overall failure is generally observed at the bump/pad interface. There are a number of reasons for this preferential fatigue crack growth. First, the bump geometry of the solder ball localizes stresses at the bump/pad interfaces whereas for instance in an hourglass solder joint, the stress concentration is observed more to the middle of the joint. Second, as seen by fatigue experiments, damage evolution parameters of the Sn grain boundaries are lower than the ones for the SnAgCu-Ni/Au interface. It follows that the bump/pad interface is a weaker boundary and is prone to damage. Furthermore, for thermal fatigue damage to initiate on the grain boundaries, the so called 'mismatch factor' M plays an important role [24]. If a certain set of grain orientations and a grain boundary normal yields to

M=0, no fatigue damage is expected at that specific grain boundary.

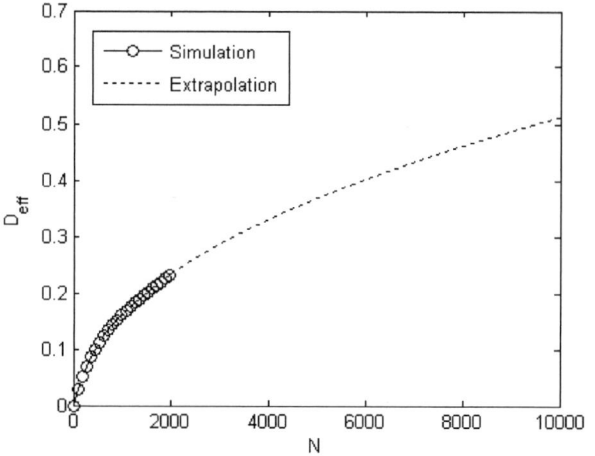

Fig. 10 Damage evolution at the bump/pad interface in a twinned SnAgCu solder bump under temperature cycling between -40 and 125°C.

4. Conclusions

- From the local strain fields observed in a soldered joint under tensile loading, strain localization is seen to be strongly microstructure dependent.

- Highest local strains are observed at the solder joint interface followed by grain boundaries. Thermal fatigue damage causes local deformations only at β-Sn grain boundaries. This issue is vital in solder balls and also in packages where solder is used in larger quantities (i.e. QFN packages) due to their polycrystalline matrix.

- SnAgCu reflects size dependent thermo-mechanical properties.

- Both the interfacial and intergranular damage mechanisms are characterized by dedicated fatigue experiments in inverse modelling.

- Using the data obtained, a microstructure-related fatigue life prediction tool for BGA solder balls in 2D is developed.

In this study, a thermo-mechanical fatigue damage model for 2D SnAgCu BGA solder ball is presented. Future study consists of making a database of SnAgCu solder ball microstructures in 3D, and expanding the model to a 3D chip size prediction tool.

Acknowledgments

This research is supported by the Technology Foundation STW, applied science division of NWO and the Technology programme of the Ministry of Economic Affairs. Special thanks to J.W.C. de Vries for his valuable contribution to the experimental part of this study.

References

1. Matin, M.A. Microstructure evolution and thermo-mechanical fatigue of solder materials. Ph.D Thesis, Technische Universiteit Eindhoven, ISBN 90-386-2887-0, 2005.
2. Abtew, M., Seldavuray, G., Mat.Sci.Eng.R 27:95—141, 2000.
3. Zeng, K., Tu, K.N. Mat.Sci.Eng.R 38:55—105, 2002.
4. Laurila, T., Vuorinen, V., Kivilahti, J.K. Mat.Sci.Eng.R 49:1—60, 2005.
5. Zeng, K., Vuorinen, V., Kivilahti, J.K. IEEE T Electron Pa M 25:162—167, 2002.
6. Ratchev, P., Vandevelde, B., deWolf, I.. IEEE T Device and Materials Reliability 4:5—10, 2004.
7. Sharif, A., Chan, Y.C. J Mater Sci Mater El 16:153—158, 2005.
8. Yoon, J., Kim, S., Jung, S. J Alloy Compd 392:247—252, 2005.
8. Erinc, M., Schreurs, P.J.G., Zhang, G.Q., Geers, M.G.D. J Mater Sci Mater El 16:93–700, 2005.
9. Erinc, M., Schreurs, P.J.G., Zhang, G.Q., Geers, M.G.D. Microelectron Reliab 44:1287—1292, 2004.
10. Wiese, S., Feustel, F., Meusel, E. Sensor Actuator A 99:188—193, 2002.
11. Wiese, S. Wolter, K.-J. Microelectron Reliab 44:1923—1931, 2004.
12. Lin, C.K., Chu, D.Y. J Mater Sci Mater Electron 16:355—365, 2005.
13. Lee, W.W., Nguyen, L.T., Selvaduray, G.S. Microelectron Reliab 40:231—244, 2000.
14. Towashiraporn, P., Subbarayan, G., Desai, C.S. Int J Solids Struct 42:4468—4483, 2005.
15. Erinc,M, Schreurs, P.J.G, Geers, M.G.D, Int J Solids Struct, In Press, 2007.
16. Matin, A.,Coenen, E.W.C., Vellinga, W.P., Geers, M.G.D. Scripta Mater, 53:8, 927—932, 2005.
17. Subramanian, K.N., Lee, J.G. J Mater Sci Mater El 15:235—240, 2004.
18. Telang, A.U., Bieler, T.R., Lucas, J.P., Subramanian, K.N., Lehman, L.P., Xing, Y., Cotts, E.J., 2004. J Electron Mater 33(12), 1412--1423.
19. Vianco, P.T., Rejent, J.A., Kilgo, A.C. J Electron Mater, 33(12), 1473--1484, 2004.
20. Erinc,M, Schreurs, P.J.G, Geers, M.G.D. Proc. 7th World Congress on Computational Mechanics, Los Angeles, 2006.
21. Roe, K.L., Siegmund,T., Eng Fract Mech 70:209—232, 2003.
22. Abdul-Baqi, A., Schreurs, P.J.G., Geers, M.G.D. Int J Solids Struct 42:927—942, 2005.
23. Jang, G.Y., Lee, J.W., Duh, J.G., J Electron Mater 33:1103—1110, 2004.
24. Ubachs, R.L.J.M. Thermomechanical modelling of microstructure evolution in solder alloys. Ph.D Thesis, Technische Universiteit Eindhoven, ISBN 90-386-2967-2, 2005.

Magnetic Field Sensor Using a Physical Model to Pre-Calculate the Magnetic Field and to Remove Systematic Error due to Physical Parameters

Simon Hainz
KAI Kompetenzzentrum
Automobil- und Industrie- Elektronik
Europastrasse 8, 9524 Villach, Austria
simon.hainz@k-ai.at

Mario Jungwirth
University of Applied Sciences Upper Austria
Campus Wels
Stelzhamerstrasse 23, 4600 Wels, Austria
m.jungwirth@fh-wels.at

Abstract

Speed and angle measurements of rotating shafts are very important in automotive applications. Typical sensing arrangements for angular measurements using the magnetic principle are analyzed in this paper [1].

It is shown that such sensor arrangements are prone to phase errors. The phase error mainly depends on the distance between sensor element and rotating shaft. By employing finite element simulations, a variety of frequently used magnetic field sensor configurations are investigated. Measurements complement the simulations and confirm correct simulation results.

Due to mounting tolerances and mechanical vibrations in automotive applications the sensor distance is varying and dynamic phase errors appear. The accuracy of measurement can only be improved if these errors get compensated. This can be done with the help of digital signal processing.

1. Introduction

Speed and position measurements of rotating shafts are very important in the field of mechanical engineering. In automotive applications sensors for such measurements have the largest market share by unit of all sensor types [1].

Common principles for angular measurements use the optic or magnetic principle. In automotive applications sensors have to work in harsh environments (temperature variations, mechanical vibration, oil contamination) and must be cheap in production. Optical sensors are temperature dependent and not resistant to oil contamination. Magnetic field sensors (e.g. Hall Sensors) are immune to oil contamination and cheap in production. This makes them very suitable for automotive applications.

Figure 1 shows a typical sensing arrangement for angular position detection using magnetic sensors. A magnetic pole wheel is mounted on a shaft. The magnetic regions on the wheel pass the sensor if the wheel rotates.

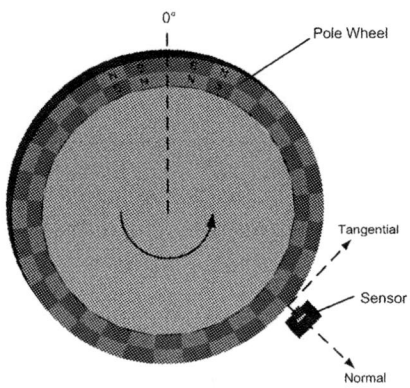

Figure 1: Pole Wheel with Alternating Coding

Using a Hall element as magnetic field sensor, each magnetic north pole on the wheel causes a positive voltage and each south pole a negative voltage at the output of the sensor. Zero crossing detection algorithms are commonly used to digitize the analogue Hall voltage. The output is an electric reproduction of the magnetization pattern (coding) of the wheel. Knowing the coding the rotational speed can be determined from the output signal.

There are different demands on the sensor system: In ABS (Antilock Braking System) applications the speed of the wheel must be measured in order to detect blocked wheels; in steering wheels the important information is the angle of the shaft; in camshaft applications the angle of the shaft is important for the timing of the fuel injection.

These different requirements call for different magnetic coding. For speed measurements a regular coding with alternating north and south poles is used. With regular coding, the output frequency of the sensor is directly proportional to the revolution speed of the shaft.

If both speed and angle has to be measured, the coding of the wheel must be irregular. Figure 1 shows an example of such an irregular coding where the longer south pole defines the point of origin. Wheels with more complex coding as shown in Figure 2 can also be used. Especially in camshaft applications such irregular coding is common.

---

[1] This work was jointly funded by the Federal Ministry of Economics and Labour of the Republic of Austria and the Carinthian Economic Promotion Fund (KWF).

1-4244-1105-X/07/$25.00 ©2007 IEEE

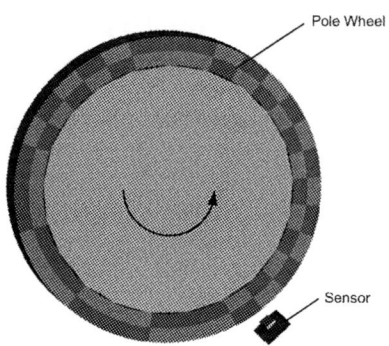

Figure 2: Pole Wheel with Irregular Coding

Another sensing arrangement to measure the angular position is shown in Figure 3. Instead of the active pole wheel a cogwheel is used. A permanent magnet (back bias magnet) placed behind the sensor generates a constant magnetic field. This field gets shaped by the rotating cogwheel. If a cog is in front of the sensor, the magnetic field strength at the sensors position is high. If a gap is in front of the sensor, the field strength is low. The variation of the field strength can be measured using a magnetic field sensor.

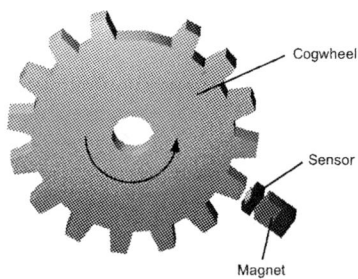

Figure 3: Cogwheel with Back Bias Magnet

Crankshaft coding typically uses 60 cogs with one 'missing' cog. Cogwheels with this coding can still be used to transmit mechanical forces. Wheels with more complex coding (more 'missing' cogs) are also used but they are not able to transmit mechanical forces.

The functionality of such a sensing arrangement can be understood easily, but to determine the angular accuracy magnetic field calculations have to be carried out.

In Section 2 an introduction to the Finite-Element-Method (FEM) and usage within electromagnetic problems is given and simulation results for realistic configurations are presented in Section 3. These results are verified by measurements in Section 4. Section 5 describes the systematic phase error compensation using digital signal processing [2]. Section 6 concludes with a summary of the work.

2. Methods for Solving Electromagnetic Problems

Analytical methods can rarely be used as solving the partial differential equations describing the magnetic field (Maxwell's equations) is either not possible or only with major simplifications [3].

Optimized designs require rather complex geometries and one must apply numerical methods to solve the field equations [4]. The Finite-Element-Method (FEM) is one of the most frequently used numerical methods today [5]. FEM uses locally defined weighting- and approximation-functions (local Galerkin method) to solve second-order-partial differential-equations (our Maxwell's equations). A further strength of FEM is the use of simple but flexible (finite) elements like triangles and tetrahedrons in order to generate very flexible grids. Automatically generated grids, adaptive meshing and algorithms for optimization in 3D space are common and state-of-the-art.

Today there are a lot of commercial tools based on the FEM, which are also applicable for electromagnetic field simulation. A comprehensive overview on international activities in electromagnetic simulations can be found in [6].

3. FEM Simulations

Static and dynamic FE-simulations of cog- and pole wheel configurations (see Figure 6 and Figure 1) where performed with COMSOL Multiphysics®, Version 3.3 [7].

The simulated sensor has a Hall element which is only sensitive to the normal (radial) component of the magnetic flux density. In order to compare FE-simulations with measurements, the normal component of the flux density across the Hall element has been calculated using the simulation-parameters shown in Table 1.

Table 1: Parameters for Static and Dynamic FE-Simulations

	Relative Permeability	Electric Conductivity [S]	Remanent Flux Density [T]
Air	1	0	0
IC Chasing	1	0	0
Sensor	1	1.00E-12	0
Lead Frame	1	6.00E+07	0
Magnet	1	0	1
Magnet Chasing	4000	1.12E+07	0
Cogwheel	5000	4.03E+06	0

3.1. Pole Wheel Simulations

Using the »Magnetostatics, No Currents Mode« for static simulations, the simulation of a pole wheel with coding as shown in Figure 1 was carried out.

Figure 4 shows the simulation result with the distance d between Hall element and surface of the wheel as parameter. The X-Axis shows the position on the circumference of the magnetic wheel. The length of each magnetic pole is 2.5 mm.

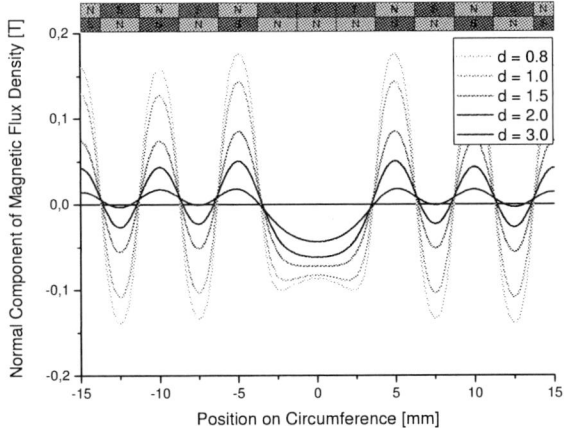

Figure 4: FEM Simulation of the Normal Component of a
Pole Wheel with Variation in Distance d

The output voltage of the Hall element is directly proportional to the normal component of the magnetic flux through the element (see Figure 4). In a first approximation the output voltage is an electrical reproduction of the magnetization pattern. By using a Schmitt trigger (with small hysteresis) a digital output signal can be generated from the analog sensor signal.

Assuming a noiseless setup the ideal switching point would be at 0 V (the zero crossing point).

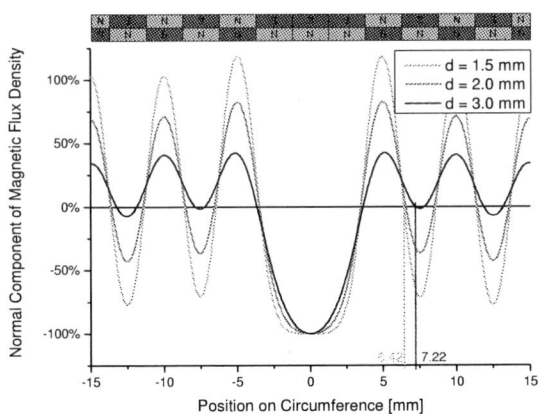

Figure 5: FEM Simulation of the Normal Component of a
Pole Wheel with Variation in Distance d, Normalized

Normalizing the magnetic flux shown in Figure 4 shows the position of the zero crossing more clearly (see Figure 5). The position of zero crossing depends on the distance between sensor and pole wheel. Varying the distance causes a displacement of the zero crossing point resulting in a systematic phase error.

Increasing the sensor distance from 1.5 mm to 3 mm causes the zero crossing point to move from 6.42 mm to 7.22 mm, which corresponds to 32 % of a 2.5 mm long magnetic region.

3.2. Static Cogwheel Simulations

Transient simulations of the cogwheel were done using the »Perpendicular Induction Currents, Vector Potential Application Mode« with moving mesh (ALE – Arbitrary Langrangian-Eulerian). Figure 6 shows the simulated magnetic circuit. The iron casing of the permanent magnet is used to shape the magnetic field and maximize the field variations of the normal component at the sensor position.

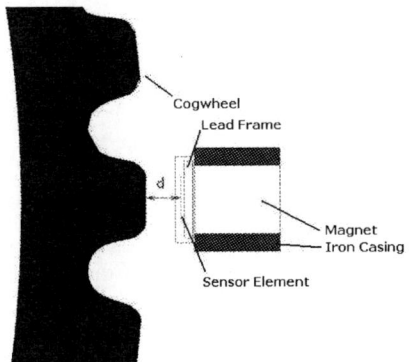

Figure 6: Cogwheel Configuration
with Hall Sensor and Back-Bias Magnet

The simulation results are shown in Figure 7. The amplitude and shape of the flux density changes with the distance d between sensor and wheel (see Figure 6).

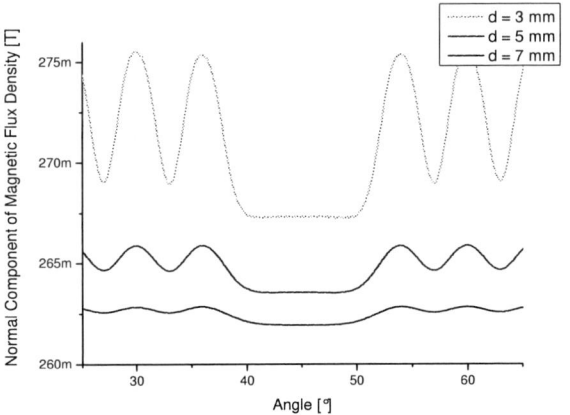

Figure 7: FEM Simulation of the Normal Component of a
Cogwheel with Variation in Distance d

The Hall element converts the flux to a proportional electrical Hall voltage. Due to the offset the digitization of the voltage is more complicated than in the case of a pole wheel. A digital signal post-processing unit calculates and subtracts the offset of the Hall voltage as shown in Figure 8. The resulting signal can be digitized with a simple Schmitt trigger.

Again a displacement of zero crossing points as a function of the sensor distance d resulting in a systematic phase error can be seen in Figure 8.

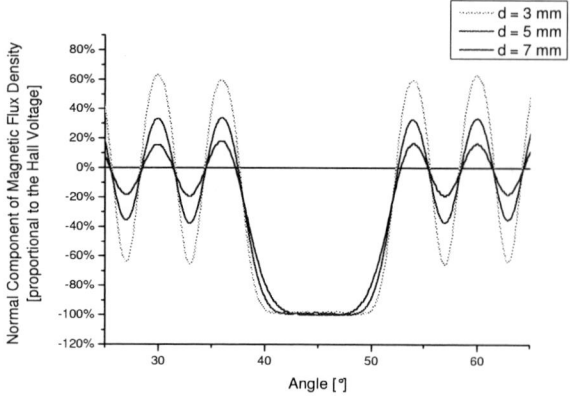

Figure 8: FEM Simulation of the Normal Component of a Cogwheel with Variation in Distance *d*, Without Offset, Normalized

This solution is implemented in the TLE4982C from Infineon, an integrated Hall sensor with variable offset cancellation [10].

3.3. Dynamic Cogwheel Simulations

Results of dynamic simulations carried out with different rotational speeds are shown in Figure 9. The rotating wheel causes variations of the magnetic flux. Flux changes induce eddy currents in the lead frame of the sensor chip which depend on the rate of flux variation. These eddy currents generate their own magnetic field which acts against the original field (Lenz's law). The rotational speed increases from 200 rpm (curve with highest amplitude) to 5000 rpm and the resulting flux densities decrease correspondingly.

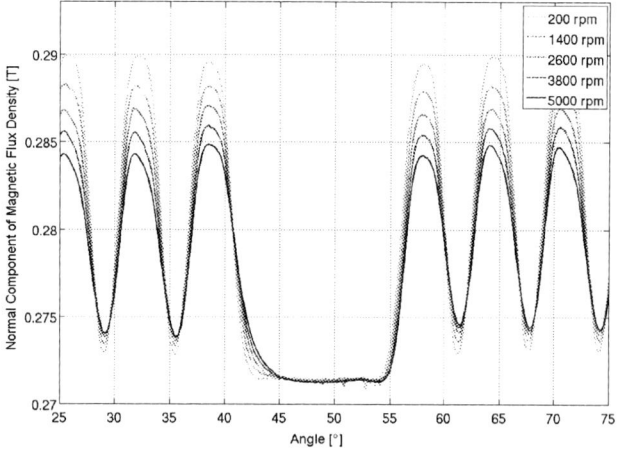

Figure 9: FEM Simulation of the Normal Component of a Cogwheel with Variation in Rotational Speed (200 rpm to 5000 rpm)

In Figure 10 a snap shot of the rotating cogwheel with the sensor is shown. The normalized distribution of the induced eddy current density in the lead frame can be seen. Due to the high electrical conductivity of the lead frame material the eddy currents will decay rather slowly. This effect can also be observed as a time delay in the signal, see Figure 9.

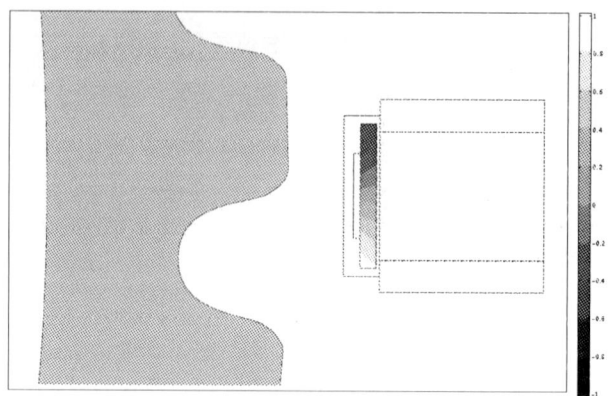

Figure 10: Induced Eddy Currents in Lead Frame due to Cogwheel Rotation, Normalized

4. Measurement Results

A test bench was built in order to verify the FEM simulation results. The setup consists of a DC motor to turn the pole/cog wheel. A stepper motor varies the distance between sensor and rotating wheel, the angle of the wheel is measured by an incremental encoder. The rotation speed of the wheel is controlled by a microcontroller. A LabVIEW program controls the test bench. Figure 11 shows the block diagram of the setup.

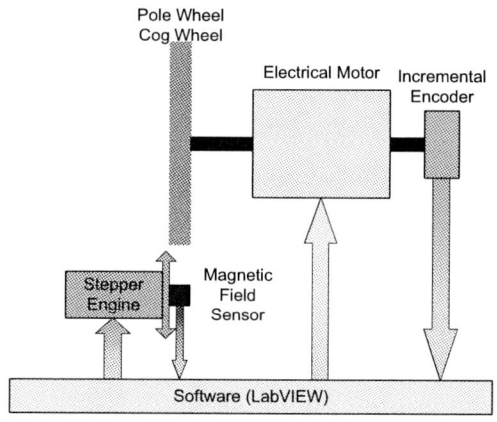

Figure 11: Block Diagram of the Test Bench

The incremental encoder generates 5000 pulses per revolution and thus allows an angular resolution of 0.072°. The rotational speed of the motor can be varied from 100 to 3000 rpm. The test bench allows to measure pole wheels and cog wheels with back bias magnets.

For measuring the magnetic field the mono cell Hall sensor »Honeywell SS495A1« with a sensitivity of 3.125 G ± 0.094 G was used. In order to reduce noise measurements shown in Figure 13 are averaged over three revolutions.

36

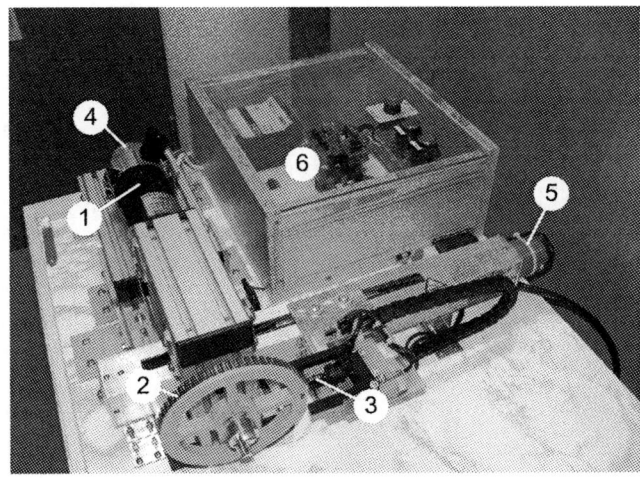

Figure 12: Measuring station

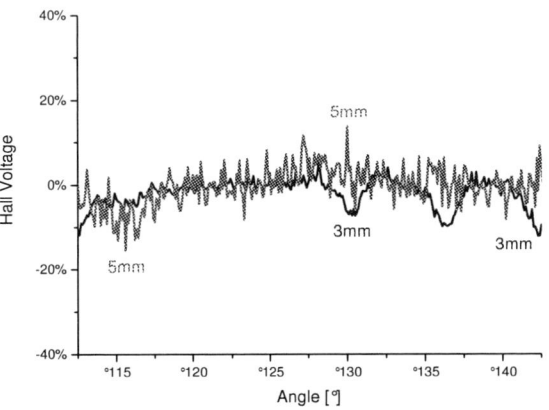

Figure 14: Difference between FEM Simulation and Measurements at a Sensor Distance of 3 and 5mm, Normalized

Figure 12 shows the test bench with DC motor (1), cogwheel (2), sensor element (3), incremental encoder (4), stepper motor for sensor distance variations (5) and microcontroller (6).

Figure 13 compares measurements and simulations of distances of 3 and 5 mm in normalized representation. It can be seen that the measurement results are in good agreement with FEM simulations.

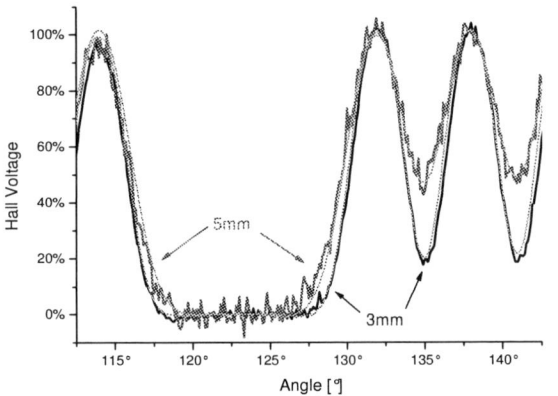

Figure 13: Measurements vs. FEM Simulation, Normalized
dark lines: Measurements; Sensor Distance d = 3 e.g. 5 mm
light lines: FEM Simulation Results d = 3 e.g. 5 mm

The deviation between measurements and simulations is shown in Figure 14. It can be seen that the match is quite acceptable and the deviation is below 15 %. The main fraction of the error is caused by noise of the measured signal.

5. Compensation of the Angular Phase Error

State of the art sensor concepts use zero detection algorithms to digitize the analog output voltage of the sensor element. The sensors concepts are optimized to detect the zero crossing point of the signal and achieve an angular accuracy below $0.1°$, but they do not take into account the systematic phase error (Section 3) as a result of distance variations [2].

Compensating the systematic phase error improves the angular accuracy. The »Data Predictive Decision Feedback Equalizer« (pDFE) presented in [2] uses this approach and compensates the dependence of the Hall signal on distance variations. An adaptive algorithm is used to estimate the physical parameters (distance, magnetization strength,...) which allows the pDFE to calculate the magnetic field and the systematic phase error.

As shown in Section 2 magnetic field calculations are complex and require high processing power. The performance of integrated sensors is limited and therefore a simplified model is used for calculations.

The normal component of the magnetic flux density of a single permanent magnet is depicted in Figure 15. Assuming that each magnetic region on the pole wheel generates such a flux distribution and the different regions do not influence each other, the Hall voltage of the wheel can be calculated using linear superposition. Although permanent magnets are nonlinear and superposition is not allowed the simplified linear model has been used because magnetization strength and therefore nonlinearities are small in this application. FEM simulations show that the linear model produces results which are in good agreement with FEM simulations.

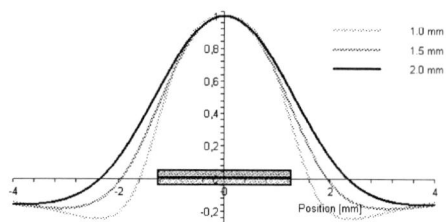

Figure 15: Normal Component of Magnetic Flux Density
of a Permanent Magnet

Figure 16 shows the FEM simulation of a weakly magnetized pole wheel and calculation results using the linear model with a sensor distance $d = 1$ mm. For better representation the difference between FEM and linear model is scaled by a factor of 10 and it can be seen that the deviation is less than 2 %.

Further FEM simulations have shown that the linear model provides good results if the sensor distance is between 1 mm and 7 mm.

Packaging and mounting of the integrated circuit causes distances greater than 1 mm. The maximal distance limited by the sensitivity of the Hall element is approximately 6 mm. This means that the linear model can be used in this application.

Compensating the systematic phase error based on the linear model is computationally easier than a FEM model and the remaining error is merely 2 % for distances greater than 1 mm. Finite bit width, noise, etc. in the post processing DSP cause additional errors. The 2 % error is the theoretical optimum if the linear model is used for calculations.

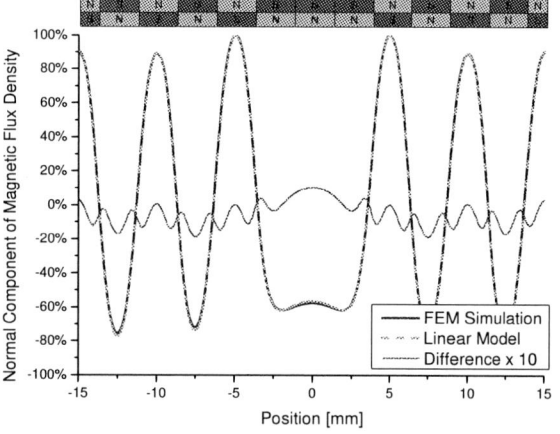

Figure 16: FEM Simulation vs. Simplified Physical Model
of a Pole Wheel at a Sensor Distance d = 1 mm;
The Difference is multiplied by a factor of 10

6. Conclusions

Magnetic sensing arrangements were analyzed using static and dynamic finite element simulations. Current integrated sensors detect zero crossings within 0.1° accuracy but ignore the systematic phase error.

For distances in the range of 1 – 2 mm the phase error is negligible, in arrangements with fixed distance the phase error can be compensated statically but in arrangements with varying distances dynamic phase compensation is required.

The systematic phase error was simulated and verified by measurements. It could be shown that the deviation is below 15 %, noise contributing a considerable fraction of the deviation.

Acknowledgments

The authors acknowledge the valuable contributions of Dirk Hammerschmidt and Tobias Werth from Infineon Technologies Austria, Erwin Ofner from Carinthia University of Applied Sciences and Daniel Hofinger from University of Applied Sciences Upper Austria, Campus Wels.

References

1　J. Marek, H.-P. Trah, Y. Suzuki, I. Yokomori, Sensors for Automotive Technology, Wiley VCH (2003)

2　S. Hainz, H. Grünbacher, D. Hammerschmidt, E. Ofner, T. Werth, „Position Detection in Automotive Application by Adaptive Inter Symbol Interference Removal," IEEE Sensor Conference, Daegu, South Korea, October 2006

3　S. Eidelmann et.al. Analytic Methods in the Theory of Differential and Pseudo-Differential Equations of Parabolic Type, Birkhäuser, Vol. 152 (2004)

4　T.J.R. Hughes, The Finite Element Method, Dover Publications Inc. (2000)

5　K.J. Bathe, Finite Element Procedures, Springer (2002)

6　www.compumag.co.uk

7　www.comsol.com

8　E. Ramsden, Hall-Effect Sensors, Theory and Application, Elsevier (Oxford, 2006)

9　"Differential Hall sensor - TLE4925 / TLE4925C," Data Sheet, Infineon Technologies (2003)

10　"Programmable True Power On Hall Sensor – TLE4982C," Data Sheet, Infineon Technologies (2005)

11　Edward A. Lee, David G. Messerschmitt, Digital communication, second Edition, Kluwer Academics Publishers (1994)

12　J. M .Cioffi, W. L. Abbott, H. K. Thapar, C. Mi. Melas, K. D. Fisher, "Adaptive equalization in magnetic-disk storage channels," IEEE Communications Magazine (1990)

13　J. E. C. Brown, P. J. Hurst, B. C. Rothenberg, S. H. Lewis, "A CMOS adaptive continuous-time forward equalizer, LPF, and RAM-DFE for magnetic recording," IEEE Journal of Solid-State Circuits, Vol. 34, No. 2 (1999)

14　J. R. Treichler, C. R. Johnson, M. G. Larimore, Theory and Design of Adaptive Filters, Prentice Hall (New Jersey 2001)

15　B. Widrow, S. D. Stearns, Adaptive Signal Processing, Prentica Hall (New Jersey 1985)

16　S. Haykin, Adaptive filter theory, Fourth Edition, Prentice Hall (New Jersey 2002)

Unsteady CFD Modelling for Electronics Cooling

James C. Tyacke and Paul G. Tucker
University of Wales, Swansea
Singleton Park, Swansea, SA2 8PP, UK
191271@swansea.ac.uk

Abstract

Numerical simulations of the relatively low Reynolds number flow (Re=14,200, based on channel height) in a ribbed channel are made using hybrid methods originating from the aerospace industry. Based on Reynolds-averaged Navier Stokes (RANS) and Large Eddy Simulation (LES), hybrid methods are explored using RANS-LES and RANS-Implicit-LES (RANS-ILES) models. Using a hybrid method with simple models we see how coarse we can make the grid and obtain reasonable heat transfer predictions for this particular case. Results from using different LES and hybrid RANS-(I)LES approaches are in better agreement than those using different RANS models..

1. Introduction

With shrinking systems and increasing power densities, accurate numerical modelling of system flows is becoming ever more important. Inadequate thermal analysis can lead to an electronic component or system failing prematurely due to increased temperatures. Bailey [1] highlights that many reliability assessments are based on old statistical data and products are often designed and tested afterwards. There is therefore, a growing need for more accurate reliability assessments based on virtual prototyping and CFD. Lasance [2] and Joshi et al. [3] highlight various aspects still adversely affecting the modelling of electronics systems such as transitional flow regimes and the CFD models constants being matched to experimental data to gain better results. This model calibration is meaningless when applying a CFD model to a different system and may result in a problematic design process. Simulation is made more difficult by complex geometry, strong streamline curvature and thermally induced instabilities.

Flows may be resolved precisely using Direct Numerical Simulation (DNS) but this is very computationally expensive. With LES, only eddies smaller than the grid width are modelled using a sub grid scale (SGS) model whereas RANS uses a model for the entire turbulent field, losing any time dependant flow features in the process. Implicit LES (ILES) does not make use of a model for sub grid stresses, but relies on errors arising from the discretisation process to, for example, dissipate energy from the flow. The use of (I)LES and hybrid RANS-(I)LES models to capture key flow characteristics has become an area of interest for problem aerospace flows where time dependent separated flows are of importance. Hybrid RANS-(I)LES models make use of economical RANS near walls and (I)LES elsewhere to resolve time dependent features. One

specific hybrid technique is known as DES, where, for a wall normal distance less than $C_{DES}\Delta$ (C_{DES} =0.65, Δ=max. grid spacing), RANS is used, otherwise, LES is used. An overview of RANS, LES and hybrid turbulence modelling approaches is given by Spalart [4].

Channel flows with ribs or cubes mounted on walls may be used to represent idealised circuit boards. Both these types of flow have been studied by many. Acharya [5] compare experimental data numerical heat transfer predictions for the standard and nonlinear k-ε model of Speziale [6]. These are based on the transport of kinetic energy and dissipation. Results are generally similar with the non-linear model improving streamwise turbulence intensities. This ribbed channel is also studied here with the case details shown in Figure 1 and Table 1. Iacovides and Raisee [7] apply low Re eddy viscosity models and a Reynolds stress model to 2D and 3D ribbed channel flows. It was found that low Re models are necessary to predict heat transfer correctly, the Reynolds Stress model giving superior predictions. Tafti [8] compares quasi-DNS (ILES grid substantially to coarse for a true DNS) and LES simulations on different grids for a ribbed duct at a bulk Reynolds number of 20,000. Heat transfer predictions using LES and the dynamic Smagorinsky model seemed insensitive to grid resolution whereas clear differences could be seen between quasi-DNS on different grids. Viswanathan and Tafti [9] revisit this case using the Detached Eddy Simulation (DES) hybrid method with different grids and a RANS simulation. DES results compare well with previous LES results and experimental data whereas the RANS model failed to capture key features of the flow. DES was sensitive to grid resolution as the interface is based on the grid spacing. For wall mounted cubes the mean flow is much more 3D and is not well represented using RANS models. A case such as this including conjugate heat transfer is visited by Zhong and Tucker [10]. RANS, LES and hybrid predictions revealed that LES and hybrid heat transfer predictions were more consistant, where as RANS results were unpredictable showing no particular trend. This case has also been studied by many at the 8th ERCOFTAC/IAHR/COST workshop on refined turbulence modelling (see Hellsten and Rautaheimo [11]), where similar results were observed. Tucker and Liu [12] investigate a CPU case with a heater element to study heat transfer with several LES and hybrid models, no one model was found to be ultimately better. Eveloy et al. [13] study single and multi-component PCBs using both a zero-(LVEL) and two-equation (k-ε) turbulence models. LVEL is based on differentiation of a Taylor series based law of the wall (see Spalding [14]). Errors were found downstream of components around regions of complex geometry and flow separation. This indicates the reliability of various RANS turbulence models when applied to complex geometry flows is questionable and that RANS models may not produce important flow features.

A good prototype for assessing the performance of different modelling techniques applied to electronics is a ribbed channel. Therefore, the relatively low Reynolds number, non-isothermal ribbed channel flow of Acharya et al. [5] is studied as a continuation of work by Liu et al. [15]. To show the wide range of heat transfer results possible, RANS heat transfer results are also presented as a comparison. Low Re k-ε models have very high (near DNS) near wall normal grid demands so a k-l model becomes a better choice near walls. This also overcomes streak resolution issues with LES, hence the simple Wolfshtein [16] k-l RANS model is applied near walls. LES and RANS-(I)LES methods are explored, the main attraction being accurate heat transfer prediction.

2. Modelling Approaches

The governing equations for RANS and (I)LES for incompressible flows may be written in the same Cartesian weakly conservative tensor form:

$$\frac{\partial \tilde{u}_j}{\partial x_j} = 0 \tag{1}$$

$$\rho \frac{\partial \tilde{u}_i}{\partial t} + \rho \frac{\partial (\tilde{u}_i \tilde{u}_j)}{\partial x_j} = \delta_{1j}\beta - \frac{\partial \tilde{p}}{\partial x_i} + \frac{\partial}{\partial x_j}\left[(\mu + \mu_T)\frac{\partial \tilde{u}_i}{\partial x_j} \right] - \frac{\partial \tau_{ij}}{\partial x_j} \tag{2}$$

$$\rho \frac{\partial \tilde{T}}{\partial t} + \rho \frac{\partial (\tilde{u}_j \tilde{T})}{\partial x_j} = -\rho \delta_{1j}\alpha \tilde{u}_j + \frac{\partial}{\partial x_j}\left[\frac{\mu}{\mathrm{Pr}}\frac{\partial \tilde{T}}{\partial x_j} \right] - \frac{\partial h_j}{\partial x_j} \tag{3}$$

In the above equations, $\tilde{u}_i$ is a fluid velocity component (i=1-3 representing streamwise, wall-normal and spanwise directions respectively), ρ the fluid density, μ dynamic viscosity, $\tilde{p}$ the periodically reduced static pressure, $\tilde{T}$ periodic temperature, t time and x_j (j=1-3) the spatial coordinate. The tilde represents ensemble averaging and spatial filtering for RANS and LES respectively. The subscript 'T' denotes whether a RANS model or LES subgrid scale (SGS) model is to be used. The α and β symbols represent mean temperature and pressure gradients in the periodic streamwise direction (see Patankar et al. [17]) and are calculated using Equations 19 and 20. The Reynolds stress tensor found in the RANS formulation and the SGS stress are denoted by τ_{ij}, whereas the turbulent heat flux tensors are marked by h_j. The unknowns τ_{ij} and h_j must be approximated to allow a solution to be found.

LES

Essentially the Smagorinsky LES model (Smagorinsky [18]) is analogous to Prandtl's mixing length turbulence model. The SGS viscosity for the Smagorinsky model may be written as:

$$\mu_{SGS} = \rho(l)^2 \left| \bar{S} \right| \tag{4}$$

where $\left| \bar{S} \right| = \sqrt{2 S_{ij} S_{ij}}$ and the mean strain rate, $S_{ij} = (\partial \tilde{u}_i / \partial x_j + \partial \tilde{u}_j / \partial x_i)/2$.

Prandtls assumption that the turbulence length scale may be written as κy where κ is the Karmen constant allows us to take the minimum of the mixing length and the LES length scale, $\Delta = (\Delta_x \Delta_y \Delta_z)^{1/3}$ to model near wall behaviour. i.e. $l = \min(\kappa y, C_s \Delta)$ in the above. The Smagorinsky constant, $C_s = 0.1$.

k-l based RANS-LES

To calculate the eddy viscosity the k-l based hybrid approach involves the use of the Wolfshtein [16] RANS model near the wall and the Yoshizawa model (Yoshizawa [19]) for the LES region. Only constants and length scales need to be changed between the RANS and LES zones, the transport equation for kinetic energy is computed as below.

$$\frac{\partial k_T}{\partial t} + \frac{\partial \tilde{u}_j k_T}{\partial x_j} = \frac{1}{\rho}\frac{\partial}{\partial x_j}\left[\left(\mu + \frac{\mu_T}{\sigma_k} \right)\left(\frac{\partial k_T}{\partial x_j} \right) \right] + P_{K_T} - \varepsilon_T \tag{5}$$

The Schmidt number for k is $\sigma_k = 1$ and the turbulence production and dissipation terms are given by P_{KT} and ε_T:

$$P_{k_T} = 2\mu_T S_{ij} S_{ij} \tag{6}$$

$$\varepsilon_T = C_\varepsilon k_T^{3/2} / l_\varepsilon \tag{7}$$

also, $\mu_T = \rho C_\mu l_\mu k^{1/2}$ (8)

The length scales are defined as:

$$l_{\varepsilon, RANS} = 2.4 y (1 - e^{-0.263 y^*}), \tag{9}$$

$$l_{\mu, RANS} = 2.4 y (1 - e^{-0.016 y^*}), \tag{10}$$

$$l_{\varepsilon, LES} = l_{\mu, LES} = (\Delta_x \Delta_y \Delta_z)^{1/3} \tag{11}$$

where $y^* = y \rho k_T^{1/2} / \mu$. Constants in the RANS region are defined as $C_\varepsilon = 1$ and $C_\mu = 0.09$ and for the LES region, $C_\varepsilon = 1.05$ and $C_\mu = 0.07$ are used. For the hybrid RANS-LES model, the interface between the two modelling approaches is taken as 10% of the rib height. This is chosen due to the highly separated nature of the flow and based on previous work gives better predictions than using 20% of the rib height as the interface. For k-l based

RANS-LES, length scales between the RANS and LES regions differ greatly. For this reason, length scales are smoothed using Equation 12.

$$l_s^{new} = \frac{l_{j+1}^{old}\delta y_{j+1} + l_j^{old}(\delta y_{j+1} + \delta y_{j-1}) + l_{j-1}^{old}\delta y_{j+1}}{2(\delta y_{j+1} + \delta y_{j-1})} \qquad (12)$$

k-l RANS-ILES

Again the Wolfshtein k-l model is used near walls, however this time a wall distance function is used based on a Hamilton-Jacobi (HJ) equation. This smoothly blends the length scale near the wall (based on minimum normal wall distance) with a length scale in the ILES zone which is essentially equal to zero. The solution is completely ILES by 10% of the rib height, the aim being to model near wall dynamics well enough and allowing numerical dissipation to remove energy from the rest of the flow (see later). Numerical dissipation stems directly from discretisation of spatial and temporal terms as well as via low-level terms left fully implicit (see Tucker and Liu [12]). The HJ to solve for the wall distance function, $\tilde{d}$ is shown below.

$$\left|\nabla\tilde{d}\right| = 1 + f(\tilde{d})\nabla^2\tilde{d} + g(d) \qquad (13)$$

$$f(d) = \varepsilon_0\tilde{d} \text{ and } g(d) = \varepsilon_1(d/L)^n \qquad (14)$$

The length scale L is the distance from walls to the ILES region with n being a positive integer. The functions $f(d)$ and $g(d)$ allows smooth blending between the RANS and ILES zones. For further details see Tucker [20].

Thermal modelling

For all computations the eddy diffusivity model below is used:

$$h_j = -\frac{\mu_T}{\mathrm{Pr}_T}\frac{\partial\tilde{T}}{\partial x_j} \qquad (15)$$

Again μ_T may represent μ_t or μ_{SGS} for RANS or LES zones respectively.

For the *k-l* based RANS-(I)LES, the value of the turbulent Prandtl number varies dramatically between the RANS and LES zones. Therefore the harmonic mean is taken for Pr_T. At control volume faces which are located halfway between node points the diffusion coefficient between node points i,j,k and i,j+1,k takes the form:

$$\Gamma = \frac{2\Gamma_{i,j,k}\Gamma_{i,j+1,k}}{\Gamma_{i,j,k} + \Gamma_{i,j+1,k}} \qquad (16)$$

Similar expressions can be formulated for other cell faces. For RANS regions $\mathrm{Pr}_T = 0.9$, for LES, $\mathrm{Pr}_T = 0.4$. For the Smagorinsky LES $\mathrm{Pr}_T = 0.4$ for the whole flow.

The thermal diffusivity can be obtained by substituting Equation 15 into Equation 3, shown here.

$$\Gamma = \frac{\mu}{\mathrm{Pr}} + \frac{\mu_T}{\mathrm{Pr}_T} \qquad (17)$$

To calculate the Nusselt number from along the wall with a constant heat flux $q_w^{"}$ the following is used:

$$Nu = \frac{q_w^{"}D_h}{k(T_w(x) - T_b(x))} \qquad (18)$$

$T_w(x)$ and $T_b(x)$ are the wall and bulk temperatures along the x direction. The thermal conductivity k, is given by $\mu c_p / \mathrm{Pr}$ and D_h is the hydraulic diameter.

Configuration, boundary conditions and numerical details

A finite volume solver (Tucker [21]) is employed. The code uses staggered grids between the pressure and velocity nodes to increase energy conservation properties. This is important for LES simulations. For the LES and RANS-LES models, second-order central differencing was used for spatial terms and the Crank-Nicolson scheme for temporal discretisation. For the RANS-ILES cases, the discretisation schemes are varied to dissipate energy. It was found that changing from a second order central difference spatial scheme to the CONDIF scheme by Runchal [22] introduced too much dissipation and led to velocity under-predictions and poor heat transfer results. Therefore only the Crank Nicolson and Galerkin temporal schemes are used to improve the solution.

The ribbed channel being considered is shown in Figure 1. The flow parameters are shown in Table 1 and are the same as those used by Acharya et al. [5]. The Reynolds number is based on the channel height, H and the bulk velocity, U_0. In the streamwise and span wise directions (x and z respectively), periodic boundary conditions are applied for both the flow and temperature fields. For the z-direction, a cyclic TDMA algorithm (Patankar et al. [17]) was used, the periodic Gauss-Siedel algorithm being used in the x-direction. Impermeability and no slip conditions were applied at walls with a constant heat flux along the bottom wall either side of the rib, which had adiabatic conditions imposed on its surfaces.

The mean pressure and temperature gradients (β and α respectively) were calculated using the following:

$$\beta_{new} = \beta_{old} - \rho\left(\frac{(Q_{new} - Q_0) - 0.5(Q_{old} - Q_0)}{0.5\Delta t H z_{max}}\right) \qquad (19)$$

$$\alpha = \frac{q_w''}{\rho c_p H U_0} \tag{20}$$

Q_0 is the volume flow rate, with Hz_{max} giving the cross-sectional area of the channel. The subscripts 'new' and 'old' denote new and old time levels. The channel width is $9.5e$. Grids were stretched towards walls to capture the boundary layer region, attention also paid to the separated region downstream of the rib. In the spanwise direction the grid is uniform for 3D models. For 2D RANS simulations a 199x142 grid was used. For all other cases a coarse, $62{\times}57{\times}17$ and a finer, $121{\times}112{\times}33$ grid have been applied. Further grid details can be found in Liu et al. [15].

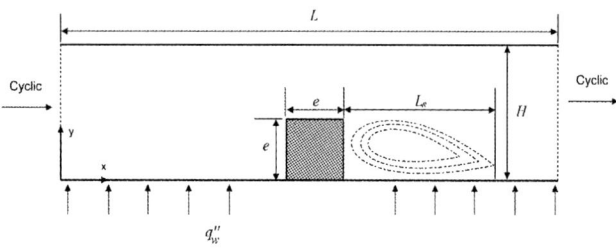

Figure 1. Schematic of the ribbed channel configuration

Table 1. Flow parameters for the ribbed channel flow

Re	U_0	L	H
14200	3.6m/s	0.127m	0.061m

e		D_h	q_w''
0.00635m		0.1016	280W/m²

3. Results and discussion

Tables 2 and 3 show the average percentage errors for the normalised mean velocity. The following equation is used to calculate the percentage errors:

$$\text{Error}_\phi = \frac{\sum_{i=1}^m \left| \phi_{exp} - \phi_{num} \right|}{\sum_{i=1}^m \left| \phi_{exp} \right|} \tag{21}$$

where ϕ may be the mean velocity or other statistic, m the total number of experimental data points and the subscripts 'exp' and 'num' representing experimental and numerical values respectively. Where numerical and experimental data points do not exactly coincide, a stiff quadratic spline interpolation is used.

Models A and B refer to the k-l RANS-ILES models using central differencing and Galerkin time schemes respectively. Models C and D mark the k-l based RANS-LES and LES models respectively. Models E-H represent the same models in the same order for the coarse grid.

Profiles are shown for several locations at $x/e = 2.3$, 10, 10.5, 11.1, 13.6, 16.2 and 17.6. Locations 10 and 10.5 correspond to the center and right corner of the rib. Other locations except for $x/e = 2.3$ are downstream of the rib.

Models E and F (k-l RANS-ILES with central differencing and Galerkin time schemes) fail to accurately capture the mean velocities. Model H (Smagorinsky LES) shows some improvement with model G (k-l RANS-LES) giving the best results for the coarse grid. Table 3 shows mean velocity errors for V. Errors are greatest in the recirculation region just downstream from the rib, with improvement on top of the rib.

Table 2. Mean streamwise velocity (U/U_0) errors for models A-H

Model:	A	B	C	D	E	F	G	H
x/h								
10	25	23	19	17	46	44	23	33
10.5	28	26	22	20	47	46	26	36
11.1	33	31	28	27	49	48	33	42
13.6	29	26	27	27	47	45	30	38
16.2	30	27	26	26	49	47	27	38
17.6	26	22	20	21	47	44	21	33
2.3	21	17	12	13	44	41	16	28

Table 3. Mean cross-stream velocity (V/V_0) errors for models A-H

Model:	A	B	C	D	E	F	G	H
x/h								
10	15	21	18	16	34	29	23	12
10.5	33	31	31	37	54	50	32	29
11.1	60	59	55	55	55	57	74	71
13.6	63	51	59	68	72	71	61	70
16.2	57	50	48	50	73	67	30	55
17.6	45	43	40	44	76	66	28	53
2.3	61	62	32	48	83	82	54	83

Tables 4 to 6 display the resolved normal ($\overline{u'u'}^{1/2}/U_0$ and $\overline{v'v'}^{1/2}/U_0$) and shear-stress ($-\overline{u'v'}/U_0^2 \times 1000$) errors at different locations. Due to a shear layer from the top rib surface, peaks appear in the data which decrease downstream of the rib. Near the rib these peaks are underpredicted by all modelling methods. The finer grid performs much better in predicting the streamwise normal stresses. The same trend can be found in the $\overline{v'v'}^{1/2}$ predictions although the difference between grid resolutions is not so pronounced. The normalised shear stress errors in Table 6 show the stresses are not captured well. Errors are greatly reduced moving to a finer grid especially on the rib ($x/h = 10\text{-}10.5$). The different models produce similar average errors using the finer grid.

Table 4. Normal stress ($\overline{u'u'}^{1/2}/U_0$) errors for models A-H

Model:	A	B	C	D	E	F	G	H
x/h								
10	39	43	41	40	94	99	39	93
10.5	36	41	40	40	88	93	41	89
11.1	35	35	37	38	70	70	41	73
13.6	17	14	15	17	48	48	16	46
16.2	13	8.8	9.8	13	46	43	6.5	39
17.6	12	7.3	7.9	12	43	40	4.4	35
2.3	12	8	8.2	8.4	52	53	6.1	53

Table 5. Normal stress ($\overline{v'v'}^{1/2}/U_0$) errors for models A-H

Model:	A	B	C	D	E	F	G	H
x/h								
10	23	24	26	29	26	28	36	35
10.5	18	19	24	26	22	25	36	38
11.1	32	31	35	35	31	32	41	43
13.6	21	16	20	22	29	26	24	26
16.2	21	15	19	18	36	31	21	24
17.6	20	14	15	16	36	32	20	24

Table 6. Shear stress ($-\overline{u'v'}/U_0^2 \times 1000$) errors for models A-H

Model:	A	B	C	D	E	F	G	H
x/h								
10	54	60	48	50	165	179	64	174
10.5	53	53	48	52	84	84	63	100
11.1	65	70	70	66	75	77	73	78
17.6	65	55	55	55	70	68	54	53

Heat transfer predictions for the models in this study are shown by the Nusselt number in Figures 2 and 3 for the relatively fine and coarse grids respectively. The two peaks in the experimental and modelled data are due to flow impingement and reattachment after the rib. The exact location of these is dependant upon the model used. All models under predict heat transfer, giving better results upstream of the rib. This is not suprising as there is flow seperation and more recirculation downstream of the rib. Figure 4 shows the Nusselt number profile for various RANS models as mentioned earlier, for details see Liu [23]. The range of results is vast in comparison to the LES and hybrid approaches. Although there are definitely differences between the LES and hybrid approaches, there is a strong agreement with the experimental data.

Table 7. Nusselt number error for models A-H

Model:	A	B	C	D	E	F	G	H
Nu	16	5	11	11	7	6	17	18

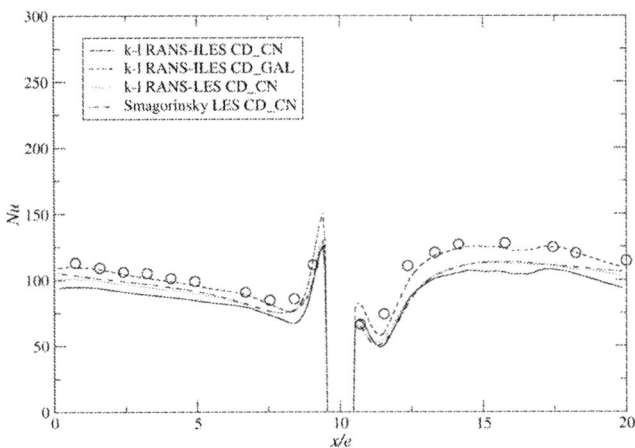

Figure 2. Local Nusselt number distributions for finer grid (models A-D)

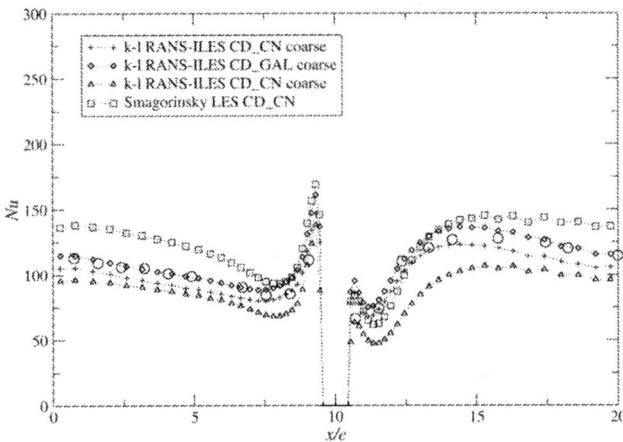

Figure 3. Local Nusselt number distributions for coarse grid (models E-H)

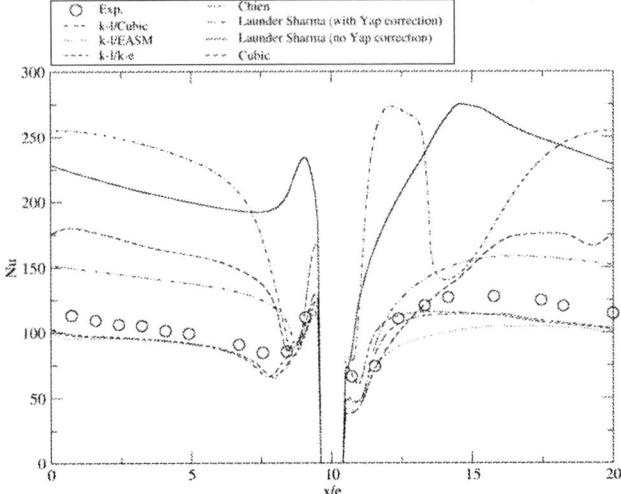

Figure 4. Local Nusselt number distributions for 2D RANS models

Table 7 shows the average error in the Nusselt number for all the models. Models B,E and F give the best results. Although the velocities were not predicted as well using the coarse grid, the average errors in the heat transfer predictions were not affected too much. Changing the time scheme from the central difference to the Galerkin gave a relatively large change in the Nusselt number, the average Courant number being just 0.16. There is greater variation in the Nusselt number profiles using the coarse grid.

4. Conclusions

LES and RANS-(I)LES models were used to simulate the flow over a ribbed channel with heat transfer. Model G (k-l RANS-LES) predicted the flow well on a coarse grid, although heat transfer results were not as good as the other models. Accurate flow prediction on a coarse grid indicates it may be worthwhile obtaining an intitial solution on a coarse grid and interpolating onto a finer grid for final accuracy improvement to reduce computation time. The RANS-ILES models gave good heat transfer results but failed to capture the flow well using a coarse grid which is of concern. On average, flow statistic errors are nearly halved using a finer grid. Using an explicit model seems to help predict the flow better, but did not result in more accurate heat transfer predictions. Generally results were worst just downstream of the rib where flow seperation, recirculation and reattachment occur. Largest errors occurred near surfaces such as the top of the rib and the channel floor. Only linear LES models are considered here. Future work using non-linear models may predict near wall aniostropy and improve thermal predictive accuracy. An important point is that the LES and hybrid RANS-(I)LES models showed the same trends whereas RANS model choice seems to be a more crucial factor.

Acknowledgments

This work was funded under an Industrial Prime Faraday CASE Award. The funding of Flomerics PLC is gratefully acknowledged.

References

1. Bailey, C, "Modelling the Effect of Temperature on Product Reliability," Proc19[th] IEEE SEMI-THERM Symposium, 2003, pp. 324-331.
2. Lasance, C. J. M., "The Conceivable Accuracy of Experimental and Numerical Thermal Analysis of Electronic Systems," *Seventeenth IEEE SEMI-THERM Symposium*, 2001, pp. 180-198.
3. Joshi, Y., Baelmans, M., Copeland, D., Lasance, C. J.M., Parry, J., Rantala, J., "Challenges in Thermal Modeling of Electronics at the System Level: Summary of Panel Held at the Therminic 2000," IEEE transactions on components and packaging technologies, Vol. 24, No. 4 (2001).
4. Spalart, P. R., "Strategies for turbulence modelling and simulations," International Journal of Heat and Fluid Flow, Vol. 21, 2000, pp. 252-263.
5. Acharya S., Dutta, S., Myrum, T. A., Baker, R. S., "Periodically developed flow and heat transfer in a ribbed duct," Int. J. Heat Mass Transfer, Vol. 36, No. 8 (1993), pp. 2069-2082.

6. Speziale, C. G., "On non-linear k-l and k-ε models of turbulence," Journal of Fluid Mechanics, Vol. 178, 1987, pp. 459-475.
7. Iacovides, H., Raisee, M., "Recent progress in the computation of flow and heat transfer in internal cooling passages of turbine blades," International Journal of Heat and Fluid Flow, Vol. 20, (1999), pp. 320-328.
8. Tafti, D. K., "Evaluating the role of subgrid stress modeling in a ribbed duct for cooling of turbine blades," International Journal of Heat and Fluid Flow, Vol. 26, 2005, pp. 92-104.
9. Viswanathan, A. K., Tafti, D. K., "Detached Eddy Simulation of Turbulent Flow and Heat Transfer in a Ribbed Duct," Transactions of the ASME, Vol. 127, 2005, pp. 888-896.
10. Zhong, B., Tucker, P. G., "k-l based hybrid LES/RANS approach and its application to heat transfer simulation," International Journal for Numerical Methods in Fluids, Vol. 46, 2004, pp. 983-1005.
11. Hellsten, A., Rautaheimo, P., "Proceedings of the 8th ERCOFTAC/IAHR/COST Workshop on Refined Turbulence Modelling", 1999.
12. Tucker, P. G., Liu, Y., "Contrasting a novel temporally oriented Hamilton-Jacobi-equation-based ILES method with other approaches for a complex geometry flow," International Journal for Numerical Methods in Fluids, Vol. 48, 2005, pp. 1241-1257.
13. Eveloy, V., Lohan, J., Rodgers, P., "A Benchmark Study of Computational Fluid Dynamics Predictive Accuracy for Component-Printed Circuit Board Heat Transfer," IEEE Transactions on components and packaging technologies, Vol. 23, No. 3 (2000).
14. Spalding, D. B., "A single formula for the law of the wall," Transactions of the ASME Series A, Journal of Applied Mechanics, Vol. 28, No. 3 (1961), pp. 444-458.
15. Liu, Y., Tucker, P. G., Lo Iacono, G., "Comparison of zonal RANS and LES for a non-isothermal ribbed channel flow," International Journal of Heat and Fluid Flow, Vol. 27, 2006, pp. 391-401.
16. Wolfshtein, M., "The velocity and temperature distribution in one-dimensional flow with turbulence augmentation and pressure gradient," International Journal of Heat and Mass Transfer, Vol. 12, 1969, pp. 301-318.
17. Patankar, S. V., Liu, C. H., Sparrow, E. M., "Fully developed flow and heat transfer in ducts having streamwise-periodic variations of cross-sectional area," ASME J. Heat Transfer Vol. 99, 1997, pp. 180-186.
18. Smagorinsky, J., "General circulation experiments with the primitive equations. I. The basic experiment," Monthly Weather Review, Vol. 91, 1963, pp. 99-165.
19. Yoshizawa, A., "Bridging between eddy-viscosity-type and second-order models using a two-scale DIA," In: *Proceedings of the 9th International*

Symposium on Turbulent Shear Flow, Kyoto, Vol. 23, No.1 (1993), pp.1-6.

20. Tucker, P.G., "Novel MILES computations for jet flows and noise," International Journal of Heat and Fluid Flow, Vol. 25, 2004, pp. 625-635.

21. Tucker, P. G., Computation of Unsteady Internal Flows, Kluwer Academic Publishers, (Dordrecht, 2001).

22. Runchal, A. K., "CONDIF: A modified central-difference scheme for convective flows," International Journal for Numerical Methods in Engineering, Vol. 24, 1987, pp. 1593-1608.

23. Liu, Y., "Numerical simulations of unsteady complex geometry flows," Ph.D. thesis. School of engineering, University of Warwick, UK, 2004.

Combined Virtual Prototyping and Reliability Testing Based Design Rules for Stacked Die System in Packages

W.D. van Driel[1,2], R. A. Real[3], D.G. Yang[1], G.Q. Zhang[1,2], J. Pasion[3]

[1] NXP Semiconductors, Nijmegen, The Netherlands
[2] Delft University of Technology, Delft, The Netherlands
[3] NXP Semiconductors, Cabuyao, Philippines

Abstract

Since the last 2-4 years, the focus in microelectronics is gradually changing from front-end to packaging. More added values are put into packages, where System in Packages (SiP) is an answer for the ongoing function integration trend. In SiP several dies are placed into one package, either side-by-side or on top of each other. The miniaturization trend more or less forbids placing dies side-by-side, since it will make the package larger. Several stacking die concepts exist, in this paper we have investigated two different ones: silicon spacer versus ball spacer. In the silicon-spacer concept, a thin piece of silicon is used to separate the actives dies in the stack. In the glue-spacer concept this is accomplished with a filler-filled die-attach. Virtual prototyping techniques are used to explore the stress/strain hotspots for different package types, being QFN, BGA, QFP, and LQFP using both stacking concepts. It is found that the QFN package type has the highest stress levels compared to BGA and QFP. Optimization techniques are used to explore the design space of the worst-case package type. For example, it is found that the spacer thickness should be equal or thinner than the die stacked on top of it to prevent the occurrence of die crack. Standard qualification experiments on specific worst-case design will be conducted in future to verify the calculated responses. By combining virtual prototyping techniques with smartly chosen reliability tests allows that possible failure mechanisms within stacked die SiP packages to be better understood and thus prevented.

1. Introduction

Packaging has evolved during the last 3 decades, starting with 2 pins transistors (TO) in the late 1960s, wire bonding technologies in the mid 1970s, surface mount technologies (SMT) in the 1980s, and Flip Chip (FC) technologies in the 1990s. To replace the wire bonding technology, per today, the major technology trend for microelectronic packages is focused on function integration. Function integration into one package is covered by the development of the System in Package (SiP) in which several dies are placed into one package. Figure 1 indicates the packaging development trend over the years. As years have progressed, the number of package styles exploded rather than evolved. In an evolving mode, package styles are replaced but this is not the case: old package styles like SDIP are still manufactured and sold at rather high volumes. Instead, new package styles are continuously added to the market and SiP is one of the latest.

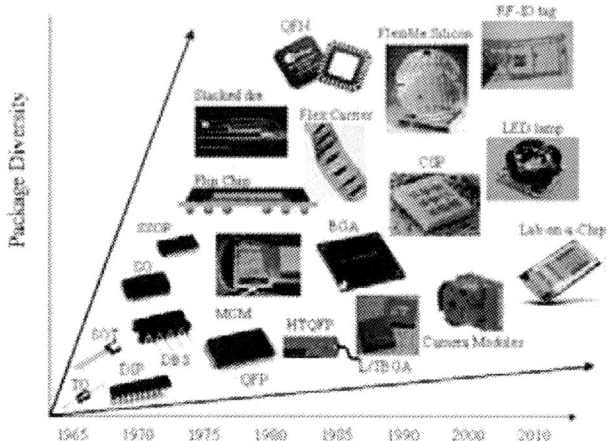

Figure 1: Packaging development trend.

The technical feasibility and unique potential of SiP offer better performance due to reduced interconnect length and power consumption, smaller form factor, higher device density, integration of devices from heterogeneous substrates, and the capability to process different functional entities on different wafers, in different fabs and by different manufacturers; opening new possibilities for future devices. The focus now lies on innovative manufacturing and integration schemes, which meet both economic and technical demands. Vertical chip stacking can be performed as chip-to-chip, chip-to-wafer, or wafer-to-wafer processes. Even though the fundamental principles of SiPs and single die packages are similar, the range of applications requires a variety of different manufacturing processes. Besides this, SiP introduction leads to increased chances and consequences of failures, increased design complexity, dramatically decreased design margins and increased difficulty to meet quality, robustness, and reliability requirements. This paper highlights our major research and development results for state-of-the-art virtual prototyping and virtual reliability qualification of SiP with stacked die technology. Thermo-mechanical reliability issues have been identified as major bottlenecks in the development of future microelectronic components [1]. In this paper, we focus on the virtual thermo-mechanical prototyping and qualification ranking for different packaging families, including QFN, BGA, QFP, and LQFP. Combined with a specific experimental matrix, the results of our research project are used to predict, qualify, and optimize the

1-4244-1105-X/07/$25.00 ©2007 IEEE   46

thermo-mechanical behavior of the SiP package reliability, against the actual requirements prior to major physical prototyping, manufacturing investments and reliability qualification tests.

2. Stacking Dies

Two stacked die concepts are evaluated, being silicon-spacer and glue-spacer. In the silicon-spacer concept, a thin piece of silicon is used to separate the active dies in the stack. In the glue-spacer concept this is accomplished with a filler-filled die-attach. Figure 2 shows schematically both stacking concepts for two dies. Of course, both concepts can be used for stacking more dies as well. Introducing stacks of such a stiff material, Silicon, into the package increases the bending resistance. Associated with that is the increased risk and/or vulnerability for cracks during assembly and/or reliability testing, either in the package body (moulding compound) or in the die itself. Figure 3 gives examples of such failures. Important design issues to prevent these failures in stacked SiP are:

- What properties are optimal for the ball spacer material?
- What thickness is optimal for the silicon spacer?
- What is the maximum overhang of the daughter die?
- What are the limits of both concepts, in terms of number of dies to be stacked?
- Which package family is more suited, which one less [2]?

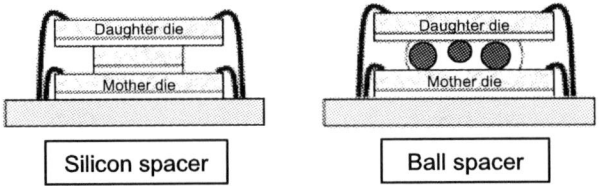

Figure 2: Investigated stacking concepts.

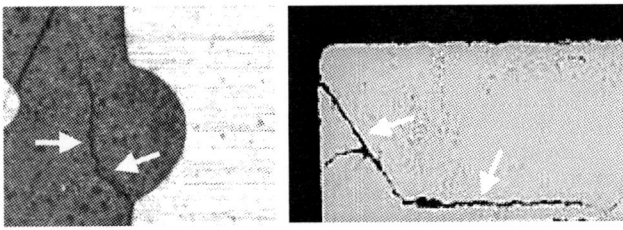

Figure 3: Fractures within package body (left) and silicon (right).

3. Finite Element Modeling

Parametric models are developed to obtain 'package stress/strain hotspots' using the state-of-the-art virtual prototyping / qualification techniques. FE models are constructed for both stacking concepts in QFN, BGA, QFP, and LQFP packages, see Figure 4 for examples. The nonlinear FE models include isotropy for silicon, visco-elasticity for moulding compound and die-attach, elasto-plasticity for copper, and orthotropic visco-elasticitity for FR4 [3]. Calculated response are moisture intake, package warpage, die stress values, and interface stress levels as a concequence of the thermal and hygro-swelling changes during manufacturing and testing. We have used the 'wetness' approach [4, 5, 6], which assumes continuity of the weighted moisture concentration across interfaces of different materials. The wetness is defined as $W = C/C_{sat}$. Using the wetness approach, the moisture diffusion implementation in commercial available FE software codes becomes straightforward with the help of appropriate user subroutines. All the appropriate materials (moulding compound, substrate, die-attach) have been characterized with regard to their moisture behavior under MSL conditions. It is assumed that the moisture uptake in the polymer materials can be described with Fick's Law of diffusion. Note that for different package families there are different material types.

The prediction models for the different package types are combined with advanced simulation-based optimization methods, such as sequence DOE and stochastic RSM techniques, to evaluate the design space of the different concepts [7, 8].

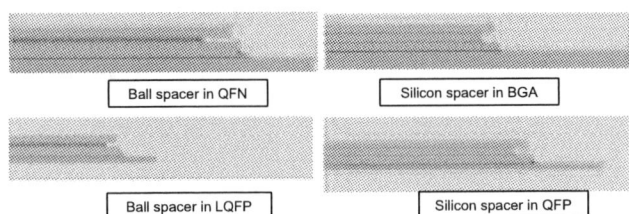

Figure 4: Example parameteric models of different package types.

4. Results

Looking at the calculated stress/strain responses, the FE results identify the hotspots for the different stacking concepts. These hotspots are:

1. Package warpage
2. Die-crack for mother, daughter, and spacer die
3. Die-to-spacer delamination
4. Die-to-compound delamination
5. Compound-to-frame delamination
6. Body crack

Figure 5 shows the moisture concentration after MSL1 conditions for a QFN package and compares the silicon and ball spacer concept. Clearly, the package is not fully saturated after MSL1. This in contrast to a single die QFN, which will be fully saturated after MSL1. The moisture gradients are input for the further hygro-thermo-mechanical stress calculations.

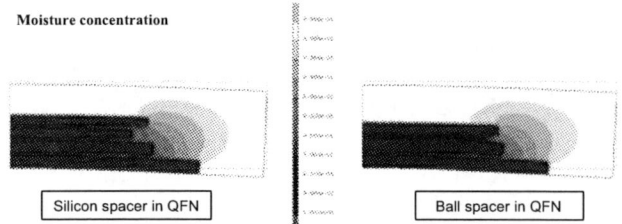

Figure 5: Moisture concentration comparing both stacking concepts in QFN.

Figure 6 shows a typical result of stress responses. The figure shows the maximum stresses in the dies for the silicon spacer concept in QFN. Hotspots in terms of stresses can be found at the center of the daughter and spacer die as well as the point where the daughter dies hang over the spacer. These points are used in the optimisation part.

Figure 6: Stress response for the silicon spacer concept in QFN.

Comparing the different package styles the stress response can be totally different. Figure 7 shows the deformed structure for the silicon spacer concept for QFN and BGA. In a BGA, the stiffness of the substrate is about 15GPa, which is a bit lower than the one for the compound, approximately 20GPa at 25degC. In the QFN package, the frame is made of copper, with a typical stiffness of 123GPa, which is close to values for Silicon. As a consequence of another stiffness distribution in the package, the resulting deformations are totally different, as Figure 7 shows. In QFN, the frame is able to pull the stack above, but in BGA, it is the compound that opens the stack. This has significant implications for the reliability, in terms of outer design boundaries, for the stacking concepts.

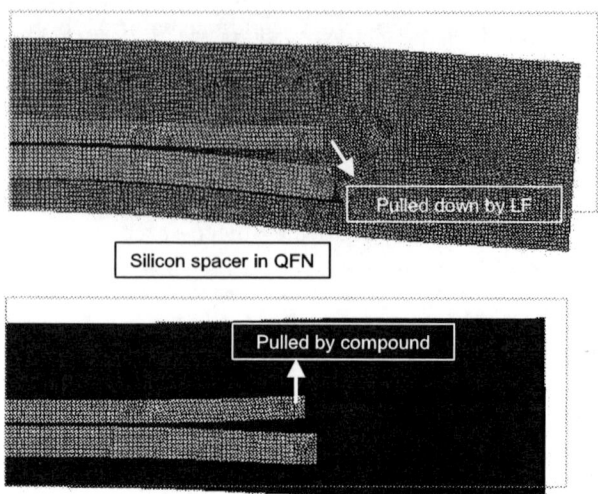

Figure 7: Deformed structure for silicon spacer concept in QFN versus BGA.

Given the six hotspots mentioned above, a ranking is made for each concept in the selected package types. Table 1 lists the worst-case package type for each of the six hotspots. Of course, these hotspots are related with possible failure modes. Looking at Table 1, the stress response of the QFN package family is found to be most critical. Therefore, this package is chosen to perform the optimization step.

Table 1: Hotspot ranking for the different package styles.

	Worst-case Package	
Hotspot	Silicon spacer	Ball spacer
Package warpage	QFN	QFN
Die-crack	QFN	BGA
Die-to-spacer delamination	QFN	QFN
Die-to-compound delamination	QFP	QFP
Compound-to-frame delamination	QFP	QFP
Body crack	QFN	QFN

A space-filling Latin-Hypercube DOE consisting of 9 input parameters and over 100 calculations is constructed. Using the parametric non-linear 3D FEM models, FEM simulations are carried out for all the 100 numerical experiments, and the earlier mentioned output variables (hotspots) are used as the response parameter.

For all response parameters quadratic models with interactions are used for RSM generation. Using OPTIMUS [9] automatic running procedure based on

48

cross-validation, the unimportant model terms were deleted. The regression statistics are indicated that for all quadratic models the accuracy requirements are satisfied (in all cases: $R^2 > 0.9$).

Figure 8 shows a typical result of the optimization process. The figure shows the stress in the daughter die as function of its thickness and the thickness of the spacer die for a 12x12 mm^2 body size; a 75% pad-body-ratio; a 90% die-pad-ratio, and a 200μm leadframe thickness. Assuming an allowable stress level of 150MPa for Silicon [10], a clear risk area can be identified: a thick daughter die with a thick spacer in between. This will need an experimental verification. A design rule that can be deducted from Figure 8, is the fact that the silicon spacer should always be thinner than the thickness of the die on top of it. This is independent on the amount of stacks created. Other significant parameters that play a role for this stress reponse are:
- Body size: larger is worse
- Die-to-pad ratio: larger is worse
- Leadframe thickness: thinner is better

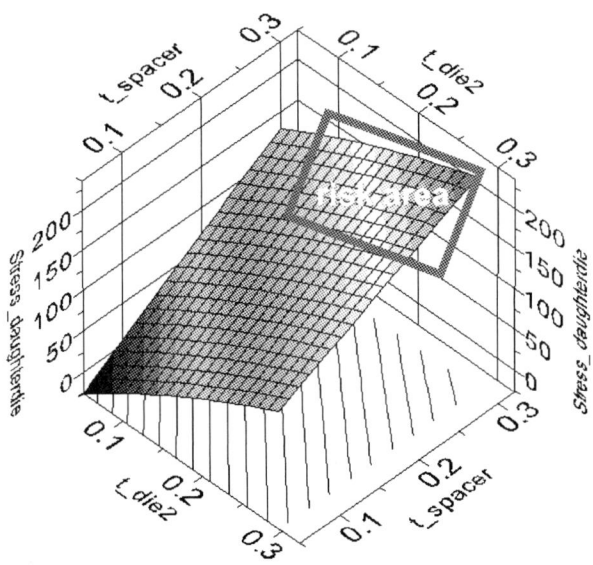

Figure 8: Response surface for daughter die crack in QFN: daughter die versus spacer die thickness.

Figure 9 shows the response surface for the stress in the compound as a function of pad-to-body and die-to-pad ratio. Other parameters are fixed to: 12x12 mm^2 body size; 150μm stacked die thickness, 150μm spacer die thickness, and 200μm leadframe thickness. For compounds, typical strength is in the order of 80-100MPa [1], in Figure 9 this value is not reached. But the figure indicates that increasing the number of stacks in the package, this risk will be a realistic one. The figure also indicates the effect of the die-to-pad ratio on the stress response in the compound. While increasing the ratio, the risk for body crack is higher. This needs an experimental

verification. A design rule that can be deducted should relate the number of stacks with the maximum allowable. Other significant parameters that play a role for this stress reponse are:
- Body size: larger is worse
- Leadframe thickness: thinner is better

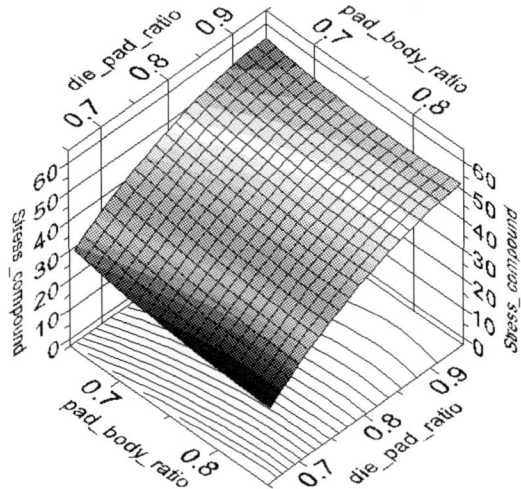

Figure 9: Response surface for compound crack in QFN: pad-to-body versus die-to-pad ratio.

5. Experimental Verifications

Standard qualification experiments will be conducted to verify the above mentioned calculated responses. Based on the optimization results, a smartly selected number of samples are created for which reliability qualification tests are performed. These samples are at the boundaries of possible failure modes as die-crack, die-to-die and die-to-compound delamination. For example, to investigate possible fracture in the daughter die a sample will be built with a 7x7mm2 body size, fixed mother and daughter die thickness and several spacer thicknesses, e.g. 150μm versus 300μm. The simulation results predict that the version with the 150μm thickness will survive (stress level is 85MPa) and the 300μm version not (170MPa). The samples will be subjected to TMCL testing, test until failure, and monitor failures at given cycles. The reliability tests are performed until failures to obtain acceleration factors for the specific failure mode.

For those legs that fail for the worst-case package style, see Table 1, identical samples will be built for the other package types. In this way, the ranking over the different package styles can be verified. At present, the sample building is in progress.

6. Conclusions

In this paper we have investigated two different concepts of die stacking in SiP packages: silicon spacer versus ball spacer. Virtual prototyping techniques are used to explore the stress/strain hotspots for different package types, being QFN, BGA, QFP, and LQFP. In general, QFN has the highest stress levels compared to laminate based (less stiff carrier) and QFP (has a bottom plastic balancer).

Optimization techniques are used to explore the design space of the worst-case package type. For example, it is found that the spacer thickness should be equal or thinner than the die stacked on top of it to prevent the occurrence of die crack. Standard qualification experiments on specific worst-case design will be conducted in future to verify the calculated responses.

By combining virtual prototyping techniques with smartly chosen reliability tests allows that possible failure mechanisms within stacked die SiP packages to be better understood and thus prevented.

Acknowledgments

The authors acknowledge the fruitful discussions with Rik Bressers and Marc Donker from NXP Nijmegen, Frederick P Arellano, Jerry Tan, Jomar Amistoso, Katherine V Martinez from NXP Philippines and Y.S. Chou from NXP Koashung.

References

1. G.Q. Zhang, W.D. van Driel, X.J. Fan (editors), Mechanics of Microelectronics, Series: Solid Mechanics and Its Applications, Vol. 141, ISBN: 1-4020-4934-X, 2006.
2. W.D. van Driel et al., Virtual Prototyping based Design Optimization of the Substrate, Leadframe, and Flip Chip Package Families with Low-k Technology, Proc EuroSimE2006, pp. 583 - 588.
3. W.D. van Driel et al, "Packaging Induced Die Stresses - Effect of Chip Anisotropy and Time-dependent Behavior of a Moulding Compound", Journal of Electronic Packaging 125 (4), 2003, pp. 520-526.
4. M.A.J. van Gils, et al, "Characterisation and Modelling of Moistures Driven Interface Failures", Microelectronics Reliability 44 (11), 2004, pp. 1317 - 1322.
5. E.H. Wong et al, "Moisture Diffusion and Vapour Pressure Modeling of IC Packaging", Proc. ECTC 1998, pp 1372-1378.
6. R. Dudek et al, "Studies on Moisture Diffusion and Popcorn Cracking", Proc. EuroSimE 2002, pp. 225-232.
7. W. van Driel et al, "Response Surface Modelling for Non-linear Packaging stresses", Journal of Electronic Packaging 125 (4), 2003, pp. 490-497.
8. A. Wymysłowski, G.Q.Zhang, W.D. van Driel, L. J. Ernst, Virtual Thermo-Mechanical Prototyping of Microelectronics and Microsystems, In: Micro- and Opto- Electronic Materials and Structures: Physics, Mechanics, Design, Reliability, Packaging Volume 1 Materials Physics / Materials Mechanics. Volume 2

Physical Design / Reliability and Packaging, Suhir, Ephraim; Lee, Y.C.; Wong, C.P. (Eds.), ISBN-10: 0-387-27974-1/ ISBN-13: 978-0-387-27974-9, 2006.
9. OPTIMUS Version 5.0, optimisation software tool, Noesis Inc., Manual, 2005.
10. Wu, J.D., C.Y. Huang, C.C. Liao, Fracture strength characterization and failure analysis of silicon dies, Microelectronics Reliability 43, pp. 269–277, 2003.

Evaluation of Thermal Strains in BGA Packages Using Digital Speckle Correlation Technique and FEA

Adam Robert Zbrzezny[1], Vincent Chan[1], Hua Lu[2], Ming Zhou[2]

[1]Advanced Micro Devices
Package Development Engineering, 105 Commerce Valley Drive West
Markham, Ontario, Canada, L3T 7W3
[2]Ryerson University
Department of Mechanical Engineering, 350 Victoria Street
Toronto, Ontario, Canada, M5B 2K3
adam.zbrzezny@amd.com

Abstract

A digital speckle correlation (DSC) technique was applied to second level interconnects of flip chip and wire bonded packages in order to measure in-situ deformations induced by temperature cycling. The measurements allowed for the total normal and shear strains to be evaluated at various temperatures for different package types having lead-free and Sn-Pb metallurgies. It was observed that among the flip chip packages the shear strains varied with temperature and were greater for the Sn-Pb package than for the lead-free package. As expected, the BGA package with a metal ring stiffener exhibited the lowest strains. The results from the FEA model of the wire bonded PBGA correlated well with the experimental data obtained by DSC.

1. Introduction

The reliability of a microelectronic package is largely dependent upon the state of stress and strain in solder interconnects, pad interfaces and nearby surrounding materials. As the temperature of a package varies periodically, the stress and strain in those areas fluctuate as well, leading to a development of a looped stress-strain curve. The area of the hysteresis loop can be quantatevely used to determine material degradation caused by cycling stress. As damage accumulates, the fatigue cracks will eventually develop.

The physics-based material and structure modeling should play a leading role in the reliability research given its enormous cost saving potential. Yet the industry overwhelmingly relies on a practice of accelerated life testing in conjunction with the statistic based data analysis.

This study used an experimental mechanics-based approach, the key of which was to obtain the interconnect strains directly from the solder-joints in real packages under non-accelerated temperature variation. Given that the scale of the damage sensitive site is in the range of microns, a high-resolution strain measurement technique, Digital Speckle Correlation, was employed. A numerical approach was also used in data post-processing so that the measured "total strain" was analyzed to yield stress, creep strain-rate and strain energy density. Once stress was solved, the elastic, plastic and creep components that constituted the total strain could be separately obtained.

The objectives of this study were to experimentally determine the solder joint strains in response to temperature variation in BGA assemblies, and to analyze the strain data to obtain stress response, thus to construct the stress-strain hysteresis curves. The measured and calculated data were to be used to make a comparison with the modeling results obtained by finite-element analysis (FEA), where possible.

2. Digital Speckle Correlation Measurement Technique

With adjustable macro to micro spatial resolution, the DSC method is capable of extracting whole-field in-plane deformation parameters of a surface. The methodology is based on the concept that originates from digital image processing and pattern recognition [1-3]. A pair of images is processed by evaluating the correlation between the reference and deformed images. A numerical procedure is established to converge the actual values of deformation parameters at a point by point manner. The method requires that a surface under the probe exhibit a random (speckle) pattern, or essentially a variable light reflectivity. Surfaces with natural texture showing sufficient variation in reflectivity may be exempted from coating with artificial speckles. For a surface experiencing in-plane deformation, the motion of the points at the pixel locations of a digital image is regulated by the two-dimensional deformation kinematics. Assuming that the de-correlation between a pair of the image subsets is only attributable to the surface in-plane deformation, the degree of the de-correlation (or the dissimilarity) between the image subsets can be quantified by a factor S defined as:

$$S(u,v,\partial u/\partial x,\partial u/\partial y,\partial v/\partial x,\partial v/\partial y)=1-\frac{\sum f(x,y)g(x^*,y^*)}{\sqrt{\sum f^2(x,y)\sum g^2(x^*,y^*)}} \quad (1)$$

The S function has six independent variables, namely the two displacement components u and v, and the four partial derivatives $\partial u/\partial x$, $\partial u/\partial y$, $\partial v/\partial x$ and $\partial v/\partial y$. The six deformation parameters are defined at a point of interest, usually the center point of an image subset. The discrete function $f(x, y)$ is a subset of the reference image, that is the image recorded before deformation. The $g(x^*, y^*)$ is an artificially constructed image subset that is converted from the originally recorded deformed image $g(x, y)$. The values for $g(x^*, y^*)$ are usually not the originally recorded pixel readings but rather interpolations of the regular pixel readings of $g(x, y)$. Let (x, y) be the coordinates of a point before deformation and (x^*, y^*) be the new coordinates of the same point after deformation. The following kinematic relation prevails:

$$x^* = x + u + \frac{\partial u}{\partial x}\Delta x + \frac{\partial u}{\partial y}\Delta y, \qquad y^* = y + v + \frac{\partial v}{\partial x}\Delta x + \frac{\partial v}{\partial y}\Delta y \qquad (2)$$

Newton-Raphson algorithm is used in searching for the root of the S function in an iterative way. As S is minimized, the deformation parameters are considered to have simultaneously converged to their actual values. The strain components are then obtained from the displacement gradients with the following relations:

$$\varepsilon_x = \sqrt{1 + 2\frac{\partial u}{\partial x} + (\frac{\partial u}{\partial x})^2 + (\frac{\partial v}{\partial x})^2} - 1$$

$$\varepsilon_y = \sqrt{1 + 2\frac{\partial v}{\partial y} + (\frac{\partial v}{\partial y})^2 + (\frac{\partial u}{\partial y})^2} - 1 \qquad (3)$$

$$\gamma_{xy} = \arcsin \frac{\frac{\partial u}{\partial y} + \frac{\partial v}{\partial x} + \frac{\partial u}{\partial x}\frac{\partial u}{\partial y} + \frac{\partial v}{\partial x}\frac{\partial v}{\partial y}}{(1 + \varepsilon_x)(1 + \varepsilon_y)}$$

For small deformation, the equations (2) are approximated by simpler forms given below:

$$\varepsilon_x = \frac{\partial u}{\partial x}, \quad \varepsilon_y = \frac{\partial v}{\partial y}, \quad \gamma_{xy} = \frac{\partial u}{\partial y} + \frac{\partial v}{\partial x} \qquad (4)$$

For problems that involve temperature changes, the measured strains are the sums of the mechanical strains and thermal strains. For thermally isotropic materials that respond linearly to the temperature change ΔT with a constant coefficient of thermal expansion α, equations in (2) and (3) become:

$$\varepsilon_x = \sqrt{1 + 2\frac{\partial u}{\partial x} + (\frac{\partial u}{\partial x})^2 + (\frac{\partial v}{\partial x})^2} - 1 + \alpha \cdot \Delta T$$

$$\varepsilon_y = \sqrt{1 + 2\frac{\partial v}{\partial y} + (\frac{\partial v}{\partial y})^2 + (\frac{\partial u}{\partial y})^2} - 1 + \alpha \cdot \Delta T \qquad (2')$$

$$\gamma_{xy} = \arcsin \frac{\frac{\partial u}{\partial y} + \frac{\partial v}{\partial x} + \frac{\partial u}{\partial x}\frac{\partial u}{\partial y} + \frac{\partial v}{\partial x}\frac{\partial v}{\partial y}}{(1 + \varepsilon_x)(1 + \varepsilon_y)}$$

and

$$\varepsilon_x = \frac{\partial u}{\partial x} + \alpha\Delta T, \quad \varepsilon_y = \frac{\partial v}{\partial y} + \alpha\Delta T, \quad \gamma_{xy} = \frac{\partial u}{\partial y} + \frac{\partial v}{\partial x} \qquad (3')$$

DSC obtained in-plane strains are independent of prior knowledge of the material's constitutive laws, and do not contain any error due to numerical differentiation of

discrete displacement data. The accuracy of the strain measurement is around 200 $\mu\varepsilon$ under normal test conditions.

3. Digital Speckle Correlation Experiments

DSC experiments were carried out on four different components. The experimental matrix, given in Table 1, consisted of four cells: a) Cell 1 was a 35 x 35 mm FCBGA with Sn-Pb balls assembled with lead-free paste, b) Cell 2 was a 35 x 35 mm FCBGA with lead-free balls assembled with lead-free paste, c) Cell 3 was a 45 x 45 mm metal ring reinforced FCBGA with Sn-Pb balls assembled with Sn-Pb paste, and d) Cell 4 was a 31 x 31 mm WBBGA with lead-free balls assembled with lead-free paste.

Table 1: Experimental matrix.

Cell	Package Style/Size	Interconnect Metallurgy	Ball Count
1	FC/35x35 mm	Sn-Pb balls/SAC paste	1201
2	FC/35x35mm	SAC balls/SAC paste	1201
3	FC with metal ring/45x45mm	Sn-Pb balls/Sn-Pb paste	2140
4	Overmolded WB/31x31mm	SAC balls/SAC paste	564

Previous failure statistics indicated that for that type of BGA assemblies the solder joint failure usually occurred right underneath the die edge or in the outermost corner of the package. In this investigation, the analyses were performed on the solder joints underneath the die edge, as indicated in Figure 1.

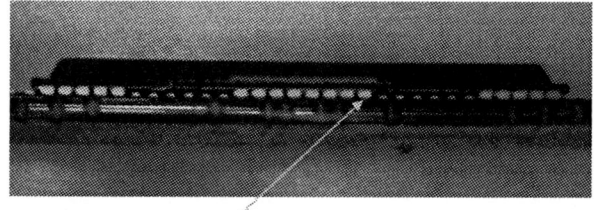

Center-right solder ball

Figure 1: Cross-sectioned PBGA sample; indicated is a solder joint for DSC analysis.

The following procedure was used to obtain the measurements. First, the sample was cut with a programmable diamond saw to expose the solder joints of interest. Next, the sample was placed in a vacuum thermal chamber that was specially designed to restrict the sample motion during testing. In this study, a three–cycle thermal profile with a starting point at room temperature was followed. The temperature extremes were 0°C and 115°C. The actual ramp rate realized was around 4°C /min and dwell time was set at 15 minutes, see Figure 2 for details.

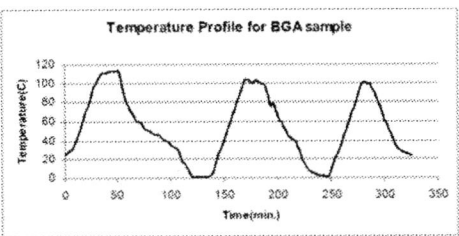

Figure 2: Thermal profile used in DSC measurements.

An optical imaging system was used to record the solder joint images during the test. The system consisted of a CCD (charge-coupled devise) camera mounted on a three-dimensional motion stage. A micro-zoom lens attached to the camera could be adjusted to fill a view field with a dimension around 1 mm square. In this study of the 0.65 mm diameter solder joint, a medium optical magnification was used. This arrangement resulted in a view area covering the whole solder joint as well as a part of the PCB below the pad and another area of the component substrate above the pad. With a CCD pixel resolution of 525x640, the system had a capacity of laying measurement with a spatial resolution at sub-micron scale. Typically, in solder joint applications, an average strain in a targeted area of 0.1x0.1 mm can be obtained by laying about 100 data points in that area. However, at such large optical magnification, the measurements were limited to only one solder joint per test. Figure 3 shows the cross-section of the solder joint before and after coating with speckles.

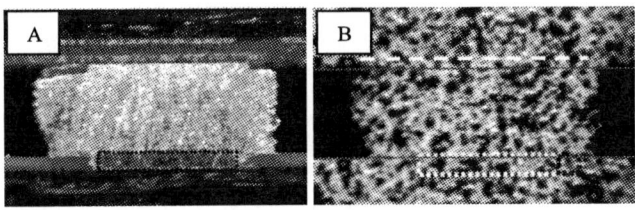

Figure 3: Cross-section of the joint before (A) and after (B) coating with speckles.

4. Digital Speckle Correlation Experimental Results

The recorded image series were processed in two different ways. First, the images were processed to obtain deformation in a large area that covered the solder ball and the pad areas together. Some typical results are shown in Figure 4, in which the displacement vector and strain tensor are given at 95 °C of the first cycle. The whole-field information helped to identify locations of stress concentration across the entire area. Also, the evolution of each component of the displacement vector and strain tensor could be tracked for assessing the trend of change with the temperature variation.

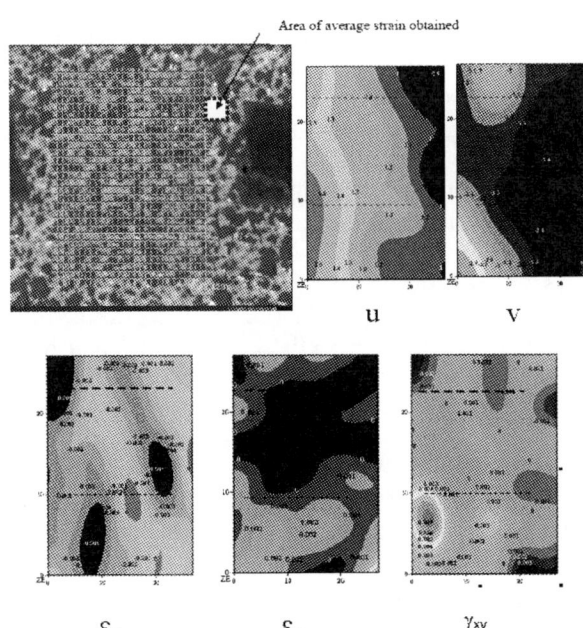

Figure 4: Sample results showing measurements at 95°C.

Secondly, based on the whole field information, the specific values of these deformation components were obtained at the identified local sites. That was usually done at selected points to facilitate the detailed physics-based analysis. The so-called point is a small area that actually contains many pixels. In this study, the quantitative analysis was based on the raw data obtained in a small area located at the upper right corner of the solder ball. The area was of 30 μm × 30 μm and had 10 × data points. The analysis used the average of the shear stain γ_{xy} data obtained in the area. Random noise in the digitized image pixel readings is inherent to the CCD technique, especially when the temperature of the CCD processor gets high with time. There are also other factors that contribute to data scattering, namely temperature induced air current, light illumination instability or change in surface reflectivity. While the use of average strain in the small area alleviates such problems, such data still scatters as shown in Figure 5.

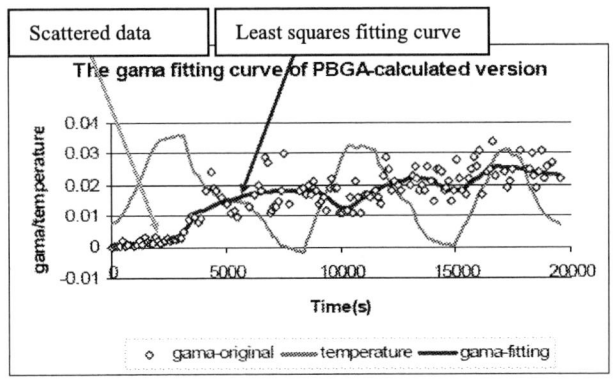

Figure 5: Measured shear strain and the fitting curve for Cell 4.

Further treatment of the raw data involved fitting of a least-squares regression curve to the discrete strain data. The final analysis was based on the strain data at individual time/temperature points which were obtained from the fitting curves. In this particular case, the highest shear strain obtained reached about 2.6 %.

Using the procedure outlined above, it was observed that among the flip chip packages the shear strains varied with temperature and were the highest for Cell 1 followed by Cell 2 and 3, Figure 6. The results confirmed that the Sn-Pb package from Cell 1 experienced higher strains than the lead-free package from Cell 2. Since the lead-free solder possesses much higher elastic modules and is more creep resistant than the Sn-Pb solder, the results were in-line with the expectations. The package from Cell 3 saw the lowest strains and it was attributed to the metal ring stiffener attached to the flip chip substrate.

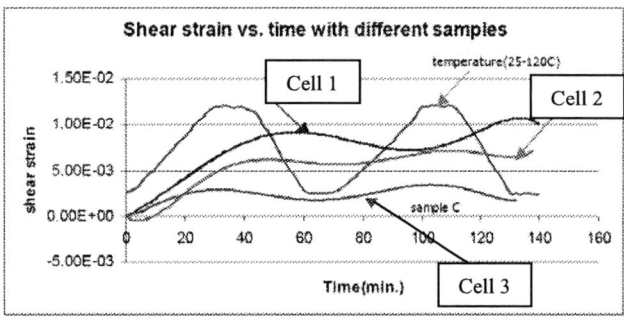

Figure 6: Measured shear strains for different samples.

5. FEA Simulation of PBGA under Thermal Loading

In parallel with the DSC measurements, a 3-D FEA model of the wire bonded package was created in ANSYS v.10. Thermal simulation was run with the same temperature profile as used for the DSC experiments. An octant-symmetry non-linear global model was used to determine the highest stressed joints in the package, Figures 7-8.

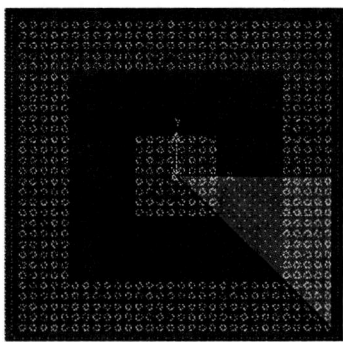

Figure 7: Solder ball layout of the package. Shadowed portion indicates 1/8th symmetry used to create the model.

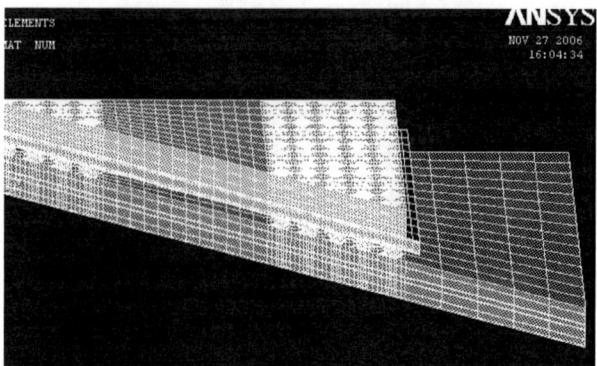

Figure 8: FEA global model of the package.

Subsequently, a submodel was built and the strains were evaluated at different temperatures. A commercially available package, ReliANS [4], was used to create the model. Lead-free solder (SAC305) was modeled as a viscoplastic material using the Anand model, the constants for which were taken from Rodgers et al. [5] and are given in Table 2 for reference.

Table 2: Fitted Anand model constants for SnAgCu.

Description	Symbol	Units	SnAgCu
Pre-exponential factor	A	1/s	107.65
Activation energy	Q/R	K	7,619
Stress multiplier	ξ	–	59.36
Strain rate senisitivity of stress	m	–	4.03
Coefficient for deformation resistance saturation value	$\hat{S}$	MPa	86.28
Strain rate sensitivity of saturation value	n	–	0.0046
Hardening coefficient	h_0	MPa	9,002
Strain rate sensitivity of hardening coefficient	a	–	1.30
Initial value of S	S_0	MPa	22.64

The PWB and BGA substrates had four copper layers and were made of FR4 and BT resin respectively. The silicon die was attached to the substrate with a die attach and was overmolded. The properties of the materials used in the package were taken from Madenci et al. [4].

The maximum strains were observed in the outermost corner solder joints. Figure 9 shows the total shear strains for the solder joints in the corner vicinity.

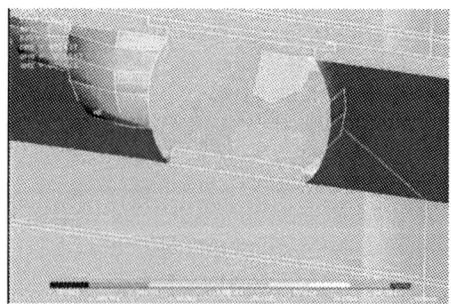

Figure 9: Maximum shear strains in package from Cell 4.

The shear strains were also calculated for the solder joints directly under the die edge. The total strains for those joints are given in Figure 10.

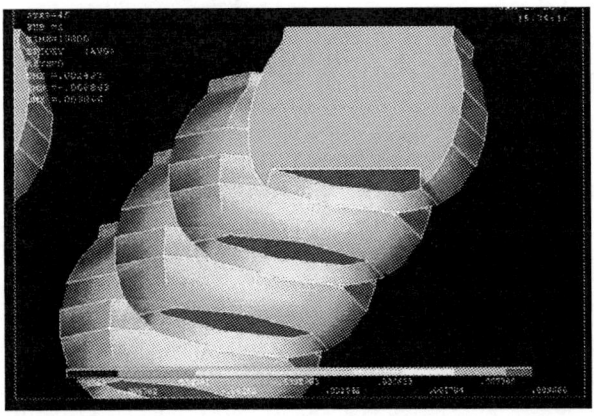

Figure 10: Shear strains in package from Cell 4 in the solder joints under die edge; for clarity, only the solder material is shown.

The maximum shear strains were found near the PWB pad and their magnitude was around 0.8%. For comparison, Figure 11 shows the DSC measured shear strains for the same package type for the same joint location.

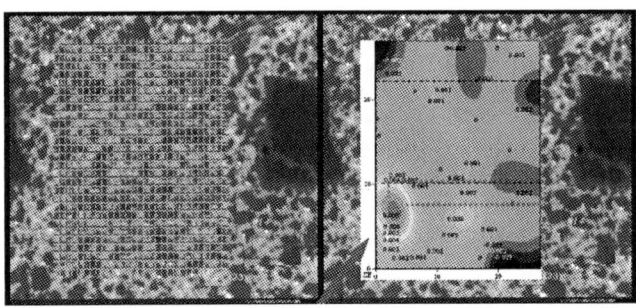

Figure 11: Measured shear strains in package from Cell 4.

As can be seen, the maximum shear strains were also 0.8% and they were located in the same area as those calculated with ASNYS. This excellent correlation suggests that the DSC technique coupled with FEA constitute viable options to expensive thermal cycling, at least at a development stage. Once the maximum shear strains are obtained, one can use existing models [6-8] to calculate the expected characteristic lives of the packages.

6. Conclusions

The current work concentrated on twofold objectives: (1) to measure total strains in BGA packages of different construction using DSC, and (2) to compare the experimental results with the results obtained with FEA. It was shown, that the DSC method, despite some data

scattering issues, could be reliably used for packaging benchmarking. In particular, it was shown that Cell 3 package, with a stiffener ring, experienced the lowest shear stresses from among Cell 1, 2 and 3. In addition, the FEA results for Cell 4 package correlated very well with the DSC measurements. Consequently, it is felt that the DSC and FEA together are powefull tools in assessing the state of stress in electronic packages.

References

1. Hua Lu, "Application of Digital Speckle Correlation to Microscopic Strain Measurement and Materials' Property Characterization", *Transaction of the ASME Journal of Electronic Packaging,* Vol. 120, September 1998, pp. 275-279.
2. Hua Lu *et al.*, "Hybrid Reliability Assessment For Packaging Prototyping," *Journal of Microelectronics Reliability*, Vol. 45, March 2005, pp. 597-609.
3. Pang, J. H. L. *et al.*, "Application of Digital Speckle Correlation to Micro-deformation Measurements of A Flip Chip Assembly", *Proc 53rd Electronic Components and Technology Conf*, New Orleans, LA, May. 2003, pp. 926-932.
4. Madenci, E. *et al.*, <u>Fatigue Life Prediction of Solder Joints in Electronic Packages with ANSYS</u>, Kluwer Academic Publishers, (Boston, 2003).
5. Rodgers B. *et al.*, "Experimental Determination and Finite Element Model Validation of the Anand Viscoplasticity Model Constants for SnAgCu, *Proc 6th Int. Conf. on Thermal, Mechanical and Multiphysics Simulation and Experiments in Micro-Electronics and Micro-Systems, EuroSimE 2005*, Berlin, Germany, April. 2005, pp. 490-496.
6. Schubert, A. *et al*, "Fatigue Life Models for SnAgCu and SnPb Solder Joints Evaluated by Experiments and Simulation", *Proc 53rd Electronic Components and Technology Conf*, New Orleans, LA, May. 2003, pp. 603-610.
7. Syed, A., "Updated Life Prediction Models for Solder Joints with Removal of Modeling Assumptions and Effect of Constitutive Equations", *Proc 7th Int. Conf. on Thermal, Mechanical and Multiphysics Simulation and Experiments in Micro-Electronics and Micro-Systems, EuroSimE 2006*, Como, Italy, April. 2006, pp. 1-9.
8. Wiese, S. *et al*, "Microstructural Dependence of Constitutive Properties of Eutectic SnAg and SnAgCu Solders", *Proc 53rd Electronic Components and Technology Conf*, New Orleans, LA, May. 2003, pp. 197-206.

Quantification of creep strain in small lead-free solder joints with the *in-situ* micro electronic-resistance measurement

Li Jiang[1], Keling Yang[1], Jiemin Zhou[2], Ke Xiang[1], Wenjie Wang[1]

1.School of Physics Science and Technology, 2. School of Energy and Power Engineering

Central South University, Changsha 410083, PR China

jl806.student@sina.com, yycolin@hotmail.com, +86-731-8836806

Abstract

Single shear lap creep specimens with a 1 mm^2 cross sectional area (similar in size to small lead-free solder joints used in electronic packaging and jointing) between thin copper strips were developed and fabricated using lead-free solder (Sn-3.5Ag) to quantify their creep strain with *in situ* micro electronic-resistance measurement. Where the solder joints' micro electronic-resistance is *in situ* measured by an electronic testing system (tailor-made for the micro electronic-resistance and stress measurement) and recorded by a PC via serial port, then all data of micro electronic-resistance and elapsing time are formed in curves. They are used to describe the solder joints' micro electronic-resistance and electronic-resistance strain varied with time. Most of curves can reveal the continual development of damage and fracture mechanisms which are consistent with observations generated by literatures. The quantitative relationship between electronic-resistance strain and mechanical-creep strain was proved theoretically using a mathematic model. These mean that the *in-situ* micro electronic-resistance measurement can be used as an alternative quantification of creep strain in small lead-free solder joints. Thus, provide an alternative and simplified evaluation method about the reliability of a solder joint.

1. Introduction

Mechanical fracture of solder joints in electronic packages is a major cause of failure and decreased reliability in electronic application. And the creep strain is an important character of mechanical properties. Thus, numerous investigations of the creep strain in lead-free solder joints can be found in literatures. However, most of them are developed by using optical method and/or *in-situ* SEM method. For example, J.McDougall measured the creep strain distribution in small crept lead-free solder joints optically by following the change in shape of a scratch, enhanced the digital images using Adobe Photoshop® and then extracted data points from these images using 'Datathief' data digitizing software[1]. K.J.LAU investigated shear failure of solder joints using an *in-situ* SEM method. SEM micrographs were captured and the load-displacement data were acquired during the shear test. Simultaneously, strains were measured with a view to compare microscopic observations of a laser grid on the copper strip and finite element analysis results[2]. Some literatures[3-9] have investigated and/or quantified the creep strain in small lead-free solder joints microcosmically. Since the solder joints are used in electronic application, their mechanical character should be correlative with electronic character in theory and fact.

Thus, the creep strain in small lead-free solder joints can be quantified following their electronic characters change, such as resistance.

The main purpose of current work is to quantify the basic characteristics of creep strain in small Sn-3.5Ag solder joints with an *in-situ* micro electronic-resistance measurement. A creep strain test was carried out to investigate the basic quantitative relationship between the electronic-resistance of solder joint and creep strain in it. The solder joint's micro electronic-resistance is *in-situ* measured by an electronic testing system (tailor-made for the micro electronic-resistance and stress measurement) and recorded by a PC via serial port, then all the data of micro electronic-resistance and elapsing time are formed in curves during the creep strain test. In order to lucubrate and discuss the above quantitative relationship in theory, a new mathematical model has been developed.

2. Specimens fabrication and Experiments

2.1 Specimens fabrication

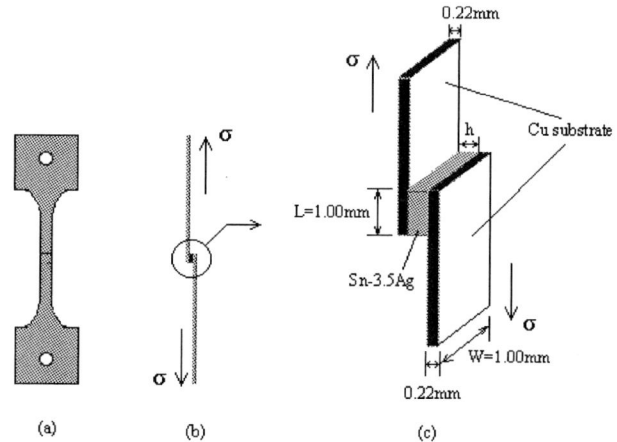

Fig.1. The front elevation (a) and profile (b) of a shear lap creep specimen, and the megascopic stereograph(c) of solder joint.

The solder alloys used in this investigation was Sn-3.5Ag. The specimen's dimensions are shown in Fig.1(c). Before solder joints were fabricated, the Cu substrates were polished with emery cloth to smooth their surfaces and eliminate the external oxides. A solder mask was applied near the narrow end of the Cu substrate to achieve a cross-sectional area of ≈1mm^2[1],similar in size to small solder joints used in electronic packing and jointing. Then the Cu substrates were chemically cleaned with a dilute solution of alcohol. After drying, flux was applied to the narrow end of Cu substrate in order to facilitate the flowing of solder and prevent formation of oxides. Two

Cu substrates were then clamped and soldered with lead-free solder, approximately 1 mm^2 in area. The solder joint were sandwiched between the two Cu substrates as shown in Fig.1(C).

2.2 Experimental procedures

In order to capture the micro electronic-resistance of solder joints during creep strain testing, a tailor-made test system (shown in Fig.2) is developed and fabricated.

After the specimen and micro electronic-resistance test probe were prepared and fixed at the test position, then run PC software programmed for quantifying the creep strain in solder joints. And then power the solder joints test instrument, shown in Fig.2(a). After the instrument was calibrated (if it is needed), test load is then placed. After all the needed operations were finished, the test for quantification of creep strain in small lead-free solder joints can start. During the testing, all real-time data were captured by solder joints test instrument, and via a RS-232 serial port sent to a personal computer, in where the data of micro electronic-resistance were processed and formed in curves which were displayed on PC screen. These curves can reveal the continual development of damage and fracture mechanisms consistent with observations generated by literatures. When the specimen fails, the test process is ended automatically.

3. Results

During shear creep testing, the micro electronic-resistance of solder joint was measured and plotted in PC screen. Fig.3 and Fig.4 are the results of the A specimen and B specimen, respectively. Fig.3(a) and Fig.4(a) show the first-hand data of the micro electronic-resistance of solder joint throughout an experiment. In order to be easy to investigate the relationship between electronic-resistance change and mechanical creep strain, a new phrase "electronic-resistance strain" was referred here. The electronic-resistance strain was defined as equation (1).

$$\varepsilon_R = (R_t - R_0)/R_0 \qquad (1)$$

Where R_t is the electronic-resistance value of solder joint when shear testing elapsed t second and R_0 is the initial electronic-resistance value of solder joint.

From the micro electronic-resistance strain data captured by the tailor-made test system, the electronic-resistance strain ε_R was computed as follows: Firstly, use the smooth option to smooth noisy data and/or to create a smooth derivative directly from the raw data, using OriginLab$^®$ software. Considering the *in-situ* micro electronic-resistance measurement has intrinsical noise, the numerical point FFT smoothing and/or Savitzky-Golay smoothing are applied necessarily. The raw and

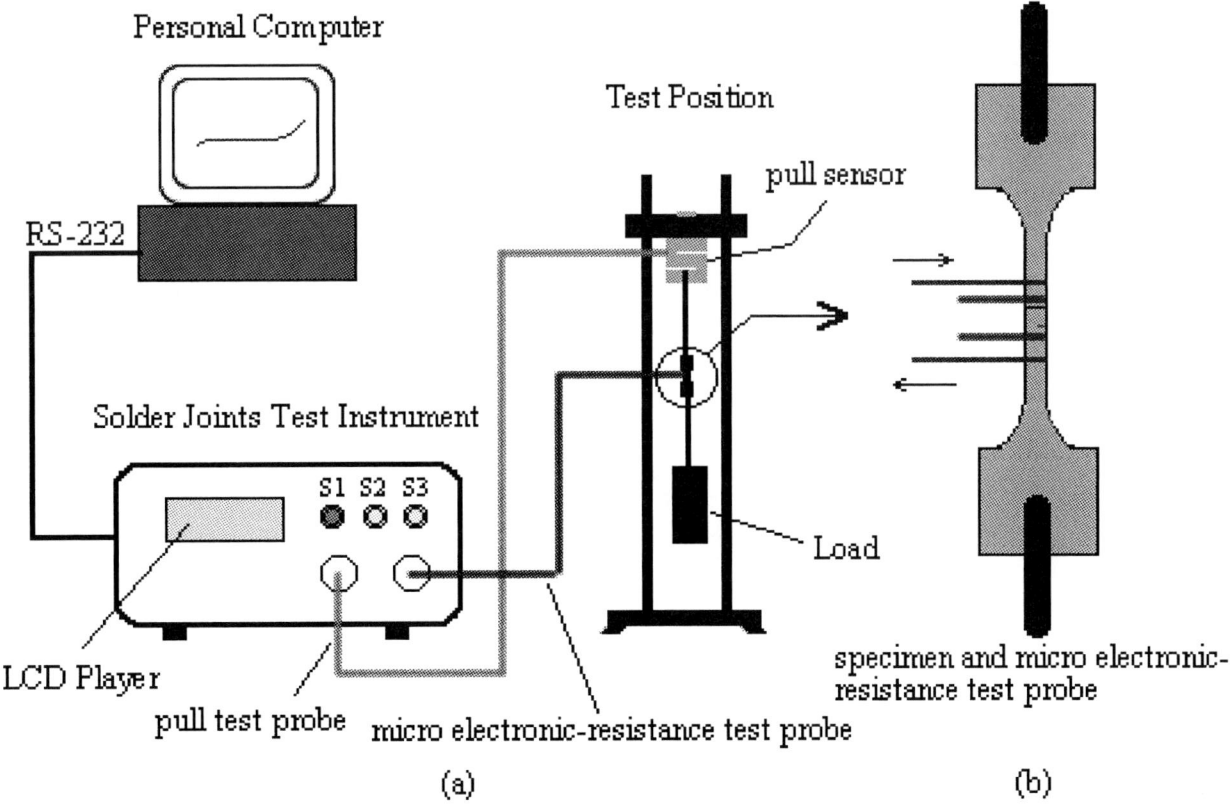

Fig.2. The whole test system (a) with *in-situ* micro electronic-resistance measurement, consisted of a personal computer, a solder joints test instrument and a test position, with the specimen and micro electronic-resistance test probe (b).

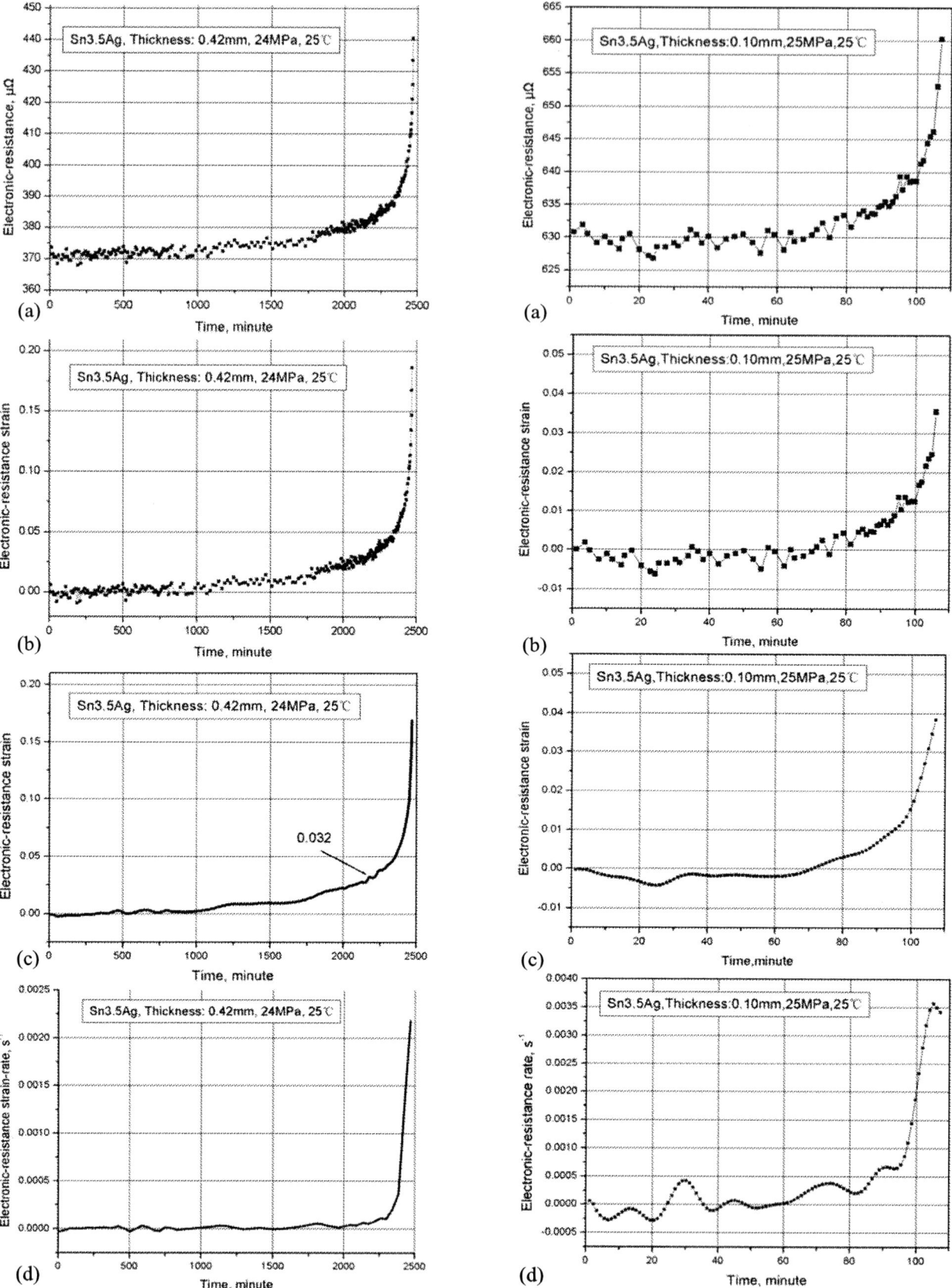

Fig.3. A specimen of Sn-3.5Ag solder joint, 0.42mm thick, deformed at 25℃ at 24MPa (a) Electronic-resistance of the solder joint during shear testing, (b)electronic-resistance strain history of the solder joint, (c)electronic-resistance strain curves smoothed by OriginLab® software, (d)electronic-resistance strain-rate varied with time .

Fig.4. B specimen of Sn-3.5Ag solder joint, 0.10mm thick, deformed at 25℃ at 25MPa (a) Electronic-resistance of the solder joint during shear testing, (b)electronic-resistance strain history of the solder joint, (c)electronic-resistance strain curves smoothed by OriginLab® software, (d)electronic-resistance strain-rate varied with time .

smoothed curves are shown in Fig.3(b) and Fig.3(c), Fig.4(b) and Fig.4(c). Secondly, in order to examine how the electronic-resistance strain-rate varied with time, polynomial equations of order 3-9 were used to obtain a good fit to each set of micro electronic-resistance data using OriginLab® software. The derivative was evaluated to obtain the electronic-resistance strain-rate, and plotted in Fig.3 (d), Fig.4(d).

Most of the electronic-resistance strain-time curves can reveal the continual development of damage and fracture mechanisms which are consistent with observations generated by literatures. And when the thickness of solder joint was sufficient, its electronic-resistance strain history approximately exhibited traditional primary, secondary, and tertiary strain behaviors, such as Fig.3(c). If the solder joint was too thick, its primary strain behavior was not easy to observe obviously, such as Fig.4(c). However, the propensity to creep and develop fracture was obvious.

Fig.3 illustrates the creep history of a lead-free solder joint, A specimen, which thickness is 0.42mm, subjected to shear-creep deformation under a shear stress of 24MPa for about 2466.4minutes. The initial and final electronic-resistances of A specimen are 371μΩ and 440μΩ. The final electronic-resistance should be infinite, when the solder joint was fractured completely. Since the specimen was fractured completely, it was not necessary to continue the measurement. Thus, the final electronic-resistance was captured at 10 seconds before test process was ended automatically. The maximum value of electronic-resistance strain(ε_R) of A specimen is equal to 0.18. The initial value of ε_R of tertiary creep is near equal to 0.032. The maximum electronic-resistance strain-rate was measured to be $2.2 \times 10^{-3} s^{-1}$. And the electronic-resistance strain-rate plotted in Fig.3(d) indicates that the ε_R accumulates very slowly before entering of tertiary creep, but very quickly after entering of tertiary creep. The electronic-resistance strain-time history exhibited the secondary and tertiary strain behavior distinctly shown in Fig.3(c), but the primary strain behavior can't be exhibited clearly.

Fig.4 shows the creep history of a lead-free solder joint, B specimen, which thickness is 0.10mm, subjected to shear-creep deformation under a shear stress of 25MPa for about 107.1minutes. The initial and final electronic-resistances of B specimen are 630μΩ and 660μΩ, respectively. Unlike A specimen in Fig.3, the initial and final electronic-resistances of B specimen are larger than A specimen, and the creep time of B specimen is shorter than A specimen. Compared with in Fig.3, the highest electronic-resistance strain value of B specimen is lower, which is 0.036. The maximum rate of electronic-resistance strain was measured to be $3.5 \times 10^{-3} s^{-1}$. The electronic-resistance strain-time history exhibited the secondary and tertiary strain behavior distinctly, but the primary strain behavior missed. The negative electronic-resistance strains indicated in Fig.4(b) and Fig.4(c) are probably resulted of system errors and stochastic errors.

4. Discussion

4.1. Feasibility of testing

In order to lucubrate and discuss the quantitative relationship between electronic-resistance strain and mechanical creep strain, a mathematical model was developed, in view of Griffth's theory of the fracture [10, 11]. Fig.5 shows the dimensions of a solder joint, which there is an equivalent crack in its centre. During shear creep was testing, the crack was being developed under the shear stress of σ, and the electronic-resistance of whole solder joint changed concomitantly.

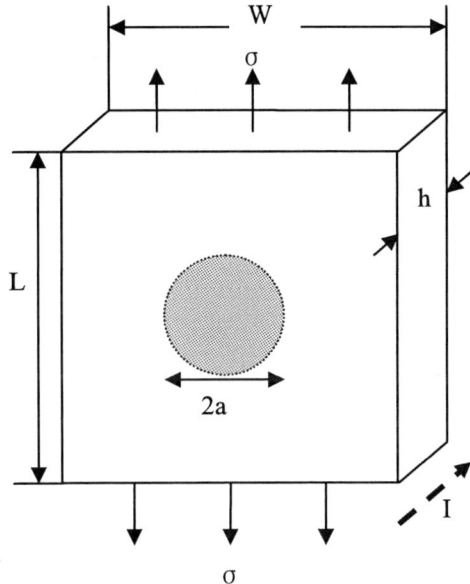

Fig.5. Solder joint with an equivalent crack developed under stress of σ.

When there is no crack in the solder joint, its electronic-resistance R_0 is computed as equation (2).

$$R_0 = \rho \frac{h}{S} \qquad (2)$$

Where, ρ is the resistivity of Sn-3.5Ag and S is the equivalent section area of solder joint at vertical position of electric current. $S = WL$, W is the width and L is the length of solder joint. Since the crack in solder joint develops under the stress of σ, the electronic-resistance R is computed as equation (3).

$$R = \rho \frac{h}{WL - S_c} \qquad (3)$$

Where S_c is the equivalent area of the crack in solder joint, given by equation (4), in which a is the equivalent width of the crack.

$$S_c = f(a). \qquad (4)$$

Equation (3) assumes that the change of L is neglected, i.e. S does not change markedly comparing with the development of S_c. Thus, the electronic-resistance of solder joint during shear creep test is computed as

$$R = \rho \frac{h}{WL - f(a)} \qquad (5)$$

Equation (5) indicates that the development of crack in solder joint can be captured by an *in-situ* electronic-resistance measurement, during solder joint crept under a shear stress. The electronic-resistance of solder joint is equal to R_0, When there is no crack in the solder joint, i.e. $f(a)$ is equal to zero. The $f(a)$ is weensy in the initial stage of creep, thus R is equal to R_0 approximately. During the solder joint crept under a stress of σ, the equivalent crack in solder joint was developed continuously. Synchronously, R increased from even nearly equating R_0 to a biggish value. And the relationship between R and R_0 was linear approximately. This linear relationship was remained until the crack of solder joint was prodigious and solder joint was fractured, here $f(a) \approx WL$ and R tended to infinity. The above creep behavior which is described as equation (5) was consistent with results of experiments.

In order to estimate the electronic-resistance strain varied with the width of the crack, equation (5) is differenced as:

$$dR = \frac{d\rho h}{WL - f(a)} + \rho \frac{dh}{WL - f(a)}$$
$$+ \rho \frac{hf(a)'da}{[WL - f(a)]^2} - \frac{Wh\rho dL}{[WL - f(a)]^2} \qquad (6)$$

$$\frac{dR}{R} = \frac{d\rho}{\rho} + \frac{dh}{h} + \frac{f(a)'da}{[WL - f(a)]} + \frac{WdL}{[WL - f(a)]} \qquad (7)$$

Because the temperature was constant during testing, ρ of solder joint was constant also. Thus, the electronic-resistance strain results from height strain and the development of crack. On the assumption that electronic-resistance strain which was due to height-strain of solder joint is not obvious, i.e. $\dfrac{dh}{h} \ll \dfrac{f(a)'da}{[WL - f(a)]}$. And because of the geometric relationship in the shear deformation, given $dL \approx \sqrt{2hdh}$, dL can be neglected. Thus, the electronic-resistance strain is computed as

$$\varepsilon_R = \frac{\Delta R}{R} \approx \frac{f(a)'\Delta a}{[WL - f(a)]} \qquad (8)$$

For example, assume that the equivalent crack is round, where $f(a) = a^2\pi$ and $f(a)' = 2a\pi$. The electronic-resistance strain is computed as equation (9), in which $a < W/2$.

$$\varepsilon_R = \frac{\Delta R}{R} \approx \frac{2a\pi\Delta a}{[WL - a^2\pi]} \qquad (9)$$

The equation (9) was plotted in Fig.6.

4.2. Electronic-resistance strain behavior

The electronic-resistance strain creep history shown in Fig.3(c) and Fig.4(c) indicates that the secondary creep behavior persisted for a long time. The creep strain in small lead-free solder joints accumulates slowly. The tertiary creep, such as creep fracture, is clear and obvious. It is probable that the model of development of crack in solder joint on secondary and tertiary creep stages was hypothesized. The primary creep is not obvious, because of several reasons. The precision of the test system is not enough, and there is noise in the test system during testing. The strain of primary creep is very small, because the dimension of the solder joint is smaller with respect to [12]. On the other hand, the strain that the crack generated initially is smaller than that the crack developed. If there are the voids in the solder joint, the primary creep can not be observed obviously.

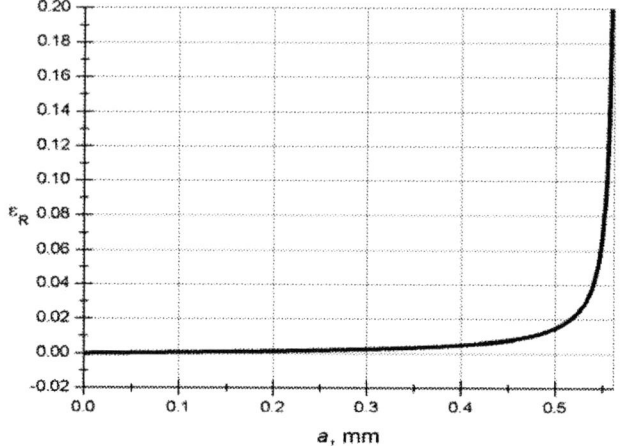

Fig.6. Electronic-resistance strain varied with width of crack. W=L=1mm, $\triangle$a=0.001mm

5. Conclusions

An *in-situ* creep strain measurement method has been developed to measure the creep strain and capture the creep behavior in the small lead-free solder joint. The results of this method can reveal the quantitative relation between mechanical shear-creep strains and electronic-resistance strains approximately. The method can be further developed to study their quantitative relationship. The following summarizes specific findings in this study:

(1)Where the *in-situ* micro electronic-resistance measurements are recorded they can reveal the continual development of creep and fracture mechanisms which are consistent with observations generated by literatures. And a rough mathematical model has developed, which has proved the quantitative relationship between electronic-resistance strain and development of crack in solder joint. These mean that this method provides an alternative and simplified evaluation method about the reliability of a solder joint.

(2)Most of the electronic-resistance strain-time curves can indicate the second and tertiary behavior clearly. And the tertiary behavior is exhibited very markedly. By comparison, the primary behavior can't be captured and exhibited as clearly as second and tertiary behavior. Thus, the precision of the electronic testing system (tailor-made for the *in-situ* micro electronic-resistance and stress measurement) is needed to be improved.

(3)The mathematical model referred in this article is too rough to describe the tertiary behavior accurately. It is necessary to research further in primary behavior of shear creep using *in-situ* electronic-resistance measurements and excogitate a better mathematical model.

Acknowledgments

We are grateful to the National Natural Science Foundation of China (Grant No:5957081), because this work was supported by the National Natural Science Foundation of China.

References

1. J.McDougall, *et al*, "Quantification of creep strain distribution in small crept lead-free in-situ composite and non composite solder joints," Materials Science and Engineering, A285 (2000), pp.25-34.

2. K.J.LAU, C.Y.TANG, *et al*, "Microscopic experimental investigation on shear failure of solder joints," International Journal of Fracture,130(2004), pp.617-634.

3. J.Zhao, *et al*, "Fratigue crack growth behavior of 96.5Sn-3.5Ag lead-free solder," International Journal of Fatigue,23(2001), pp.723-731.

4. Jie Zhao, *et al*, "Fatigue crack growth behavior of Sn-Pb and Sn-based lead-free solders," Engineering Fracture Mechanics, 70(2003), pp.2187-2197.

5. S.C.Chen, *et al*, "The numerical analysis of strain behavior at the solder joint and interface in a flip chip package," Journal of Materials Processing Technology, 171(2006), pp.125-131.

6. Chih-Kuang Lin, *et al*, "Creep properties of Sn-3.5Ag-0.5Cu lead-free solder under step-loading," J Mater Sci: Mater Electron, 17(2006), pp.577-586.

7. M.Kerr, N.Chawla, "Creep deformation behavior of Sn-3.5Ag solder/Cu couple at small length scales," Acta Materialia, 52(2004), pp.4527-4535.

8. Jicheng Gong, *et al*, "Modelling of Ag_3Sn coarsening and its effect on creep of Sn-Ag eutectics," Materials Science and Engineering, A427(2006), pp.60-68.

9. S.Wiese, F.Feustel, E.Meusel, "Characterisation of constitutive behaviour of SnAg, SnAgCu, and SnPb solder in flip chip joints," Sensors and Actuators, A99 (2002), pp.188-193.

10. Guangzhong Wang, <u>Fatigue of Materials</u>. National Defence Industry Press (Beijing, 1991), pp.144-145. (in Chinese).

11. Richard G. budynas, <u>Advanced Strength and Applied Stress Analysis(Second Edition)</u>, McGraw-Hill Companies,Inc (1999), pp.520-523.

12. Masazumi Amagai, *et al*, "Mechanical characterization of Sn-Ag-based lead-free solders," Microelectronics Reliability, 42(2002), pp.951-966.

Computational Design and Optimisation of Mechanically Reinforced Masks for Stencil Lithography

Maryna Lishchynska[1], Marc A.F. van den Boogaart[2], Juergen Brugger[2], James C. Greer[1]

[1] Tyndall National Institute, Cork, Ireland; email: marynal@tyndall.ie

[2] EPFL, Laboratoire des Microsystèmes, CH-1015 Lausanne, Switzerland

Abstract

Identifying, predicting and optimising stencil lithography is critical to the successful and timely development of this technique with a wide range of potential applications such as deposition on non-conventional and unstable materials (i.e. bio-chemical, hydrophobic), patterning heterostructures (epitaxial, magnetic, complex oxides, piezoelectric materials) and deposition of nanodevices onto CMOS. Previously confirmed for cantilever-like stencils is the thesis that degrading effects of stress-induced deformation of stencils can be overcome by strategic placement of corrugating structures. This approach is further exploited in this work to mechanically stabilise complex stencil designs. This involved studying the evolution of stencil deformation due to deposition induced stress and iterative design of optimal corrugation structures to be incorporated into the stencils. It is shown that degrading effects of stress-induced deformation of stencils can be significantly reduced which subsequently improves pattern definition. Reduction in deformation and in pattern distortion in the range of 50% to 96% was achieved.

1. Introduction

Stencil lithography is emerging as an alternative to traditional photolithography techniques permitting fabrication of nanometer scale surface topologies. Stencilling is a direct or additive patterning technology where a controlled amount of a material is directly deposited through the stencil apertures onto the substrate [1]. The inevitable accumulation of the deposited material on the mask itself and subsequent out-of-plane stencil deformation due to the deposition-induced stress [2] pose serious limitations to stencil applications. These phenomena result in an increased gap between the substrate and the stencil, and it is important to reduce this distance to achieve tight tolerance for pattern definition. Due to the geometry of the deposition system (finite size material source, finite source-stencil distance and resulting angular deposition flux), too large stencil-substrate gap causes excessive pattern distortion and blurring [3].

Modelling and simulation techniques are of great assistance for assessing the stress-induced deformation of stencil membranes, predicting pattern distortion and aiding stencil design optimisation. Moreover, for correct process window definition, fabrication of efficient MEMS tools and consequently improved pattern resolution, an accurate modelling of the membranes and a good understanding of related phenomena/effects is needed.

State-of-the-art modelling and simulation for nanolithography pattern transfer includes modelling the deformation of the stencil mask as well as prediction of pattern distortions. Nevertheless, there appears to be no significant investigations into reducing out-of-plane deformations of the mask, which is vital for pattern definition by nanostencil lithography. Our previous studies on simple cantilever-like masks [3, 4] confirmed that the above-mentioned limitation can be overcome by corrugating critical areas on the stencils. Such corrugation structures mechanically reinforce masks substantially reducing their deformability. Ref. [4] describes the basic flow of the fabrication process in detail and provides a computational and experimental confirmation of the stabilizing effect of corrugation on membrane deformation. One of the most important conclusions drawn from our previous studies with cantilever-like membranes is that the largest reduction in deformation, and consequent improvement in pattern definition are achieved with corrugation rims located perpendicular to the direction of increasing deformation or "deformation isolines" (fig. 1a). This guideline is exploited in present work as a practical design rule for stabilisation of complex apertured membranes (fig. 1b) fabricated for specific engineering applications.

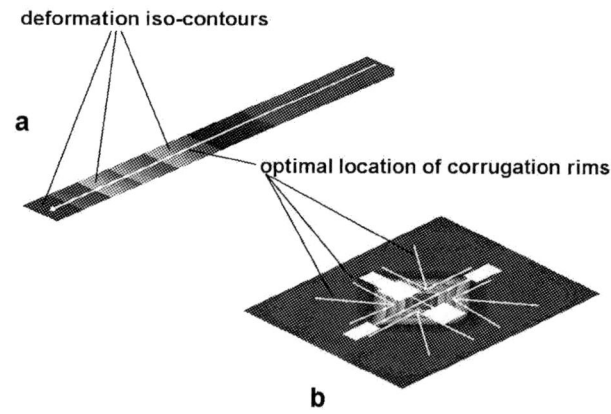

Fig. 1. Deformation iso-contours in plain cantilever membrane (a) and a fragment of apertured membrane/stencil (b); bold white lines indicate the direction where most effective corrugation structures should be located.

This paper aims to demonstrate the efficiency of the finite element method (FEM) applied to the design and optimisation of stabilisation structures incorporated in lithography stencils as well as prediction of pattern distortion/blurring due to the stencil deformation. Various stencil designs devised for patterning of nanoresonators interconnections and other structures were modelled and their deformation due to the deposition induced stress was simulated. It was found that by analysing the "deformation iso-contours map" it is possible to make effective predictions for the stabilisation of the structures. As this "deformation map" evolves with incorporation of reinforcing corrugation rims, these designs can be further optimised in order to achieve a required reduction in stencil deformation allowing a specified tolerance to pattern blurring to be targeted.

2. Modelling optimal geometry of stabilization structures for stencils with elementary shape apertures

As a first approach, in this section we consider stencils membranes with apertures with simple (canonical) shapes. The intention is to test the "deformation isolines map" technique on simpler structures as well as applying the methodology to stabilisation of more complex stencil geometries. Thus, membranes with a round aperture, one slit or two slits (table 1), single-corner (fig. 2) and double-corner (fig. 3) were considered.

All these membranes were modelled with the FEM software tool Coventor [5] and all simulations geometries are based on a 500nm thick SiN membrane. Thin-film properties of the materials used in simulations are presented in table 2. It was found that fabrication residual stress in SiN layer induces very little out-of-plane deformation of the membrane (table 1). Thus, experimentally measured SiN residual stress of 200 MPa incorporated into simulations results in sub-nanometre range of deformation. On the other hand, deposition of highly stressed (1230 MPa) 50-nm thick Cr results in an increase in the deformation by four orders of magnitude (table 1). Analysis of resulting deformation iso-contours helped to establish optimal geometry of stabilisation rims for the considered canonical cases. Introduction of corrugation rims shown in table 1 reduces the deformation by up to 90%. Two more membranes with a single- and double-corner aperture geometries were studied and suggested geometries of stabilisation rims were found to reduce the deformation by 77% and 64%, respectively (fig. 2, 3).

Table 1. Pictorial summary of results on stabilisation of membranes with simple shape apertures

Stencil design	Effect of fabrication residual stress in SiN	Evolution of deformation after Cr deposition	Suggested geometries of stabilisation structures
Round aperture	Max vertical deformation 0.0026nm	Max vertical deformation 0.11µm	
One slit	Max vertical deformation 0.21nm	Max vertical deformation 0.46µm	
Two slits	Max vertical deformation 0.13nm	Max vertical deformation 0.47µm	

Table 2. Material properties used in simulation study

Material	Layer thickness, nm	Young's modulus, GPa	Poisson's ratio	Intrinsic stress, MPa	Reference
SiN	500	276	0.27	200	[6]
Cr	50	277	0.3	1230	[7]

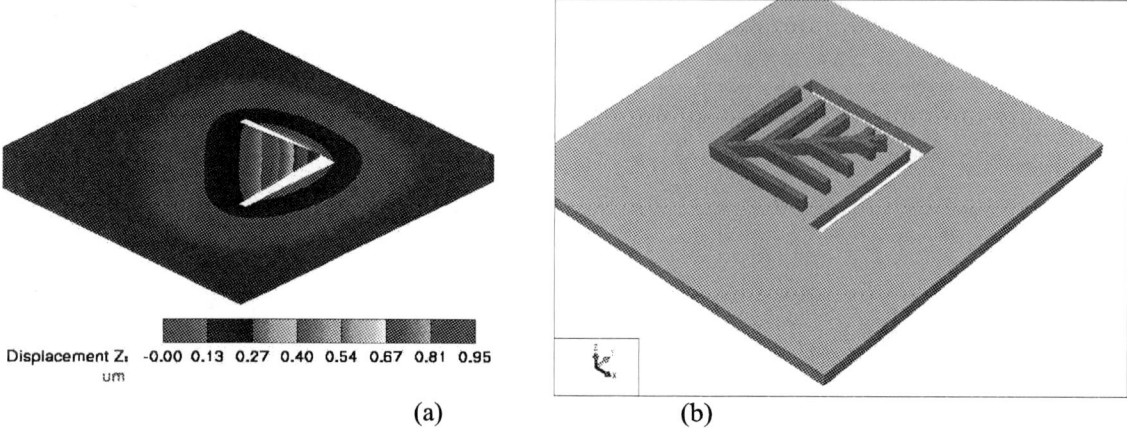

(a)   (b)

Fig. 2. Simulated deformation (a) and suggested geometry of corrugation (b) for membrane with single corner aperture. White areas are the apertures, dark grey - the corrugation rims, light grey - the stencil membrane.

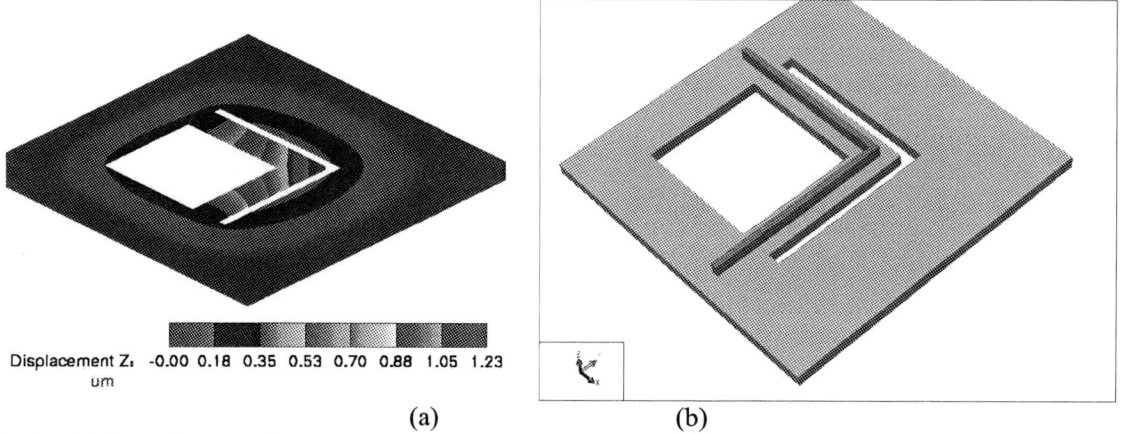

(a)   (b)

Fig. 3. Simulated deformation (a) and suggested geometry of corrugation (b) for membrane with double corner aperture.

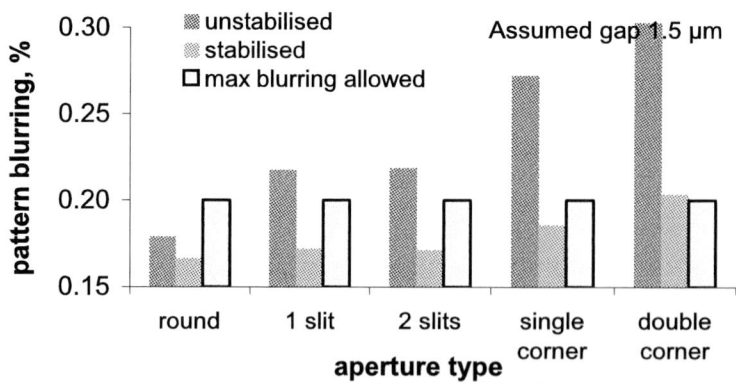

Fig. 4. Effect of stencil stabilisation translated in terms of pattern definition.

Due to the physics of the deposition process and geometry of the deposition setup [4] the stencil deformation leads to an increased gap between the stencil and the substrate which causes significant pattern distortion such as blurring. From purely geometrical considerations [3] vertical stencil deformation δ translates into broadening of the pattern b according to the following formula:

$$b \approx \frac{G_{effective}S}{D} = \frac{(G+\delta)S}{D}, \qquad (1)$$

where G is the gap between the stencil and the substrate, S is source size, D is the distance from the source to the stencil, $b=B-W$, W is aperture width (corresponds to the ideal pattern width). Graph presented in figure 4 summarises results of this section's membrane stabilisation study in terms of reduced pattern blurring with respect to the maximum allowed blurring value set by fabrication requirements. Thus, suggested corrugation geometries achieve the target for reduction in deformation (i.e. keep pattern blurring within acceptable range). Figure 4 also shows that maximum allowed gap with chosen corrugation geometries is 1.5μm. Unstabilised membranes can only tolerate much smaller gap which reduces the process window.

These results were used as guidelines for the work on real stencils presented in the following section.

3. Stabilisation of experimental complex stencil designs

This section presents results on FEM modelling deposition induced deformation in experimental stencils designed by EPFL and CNM. In the first subsection we show in detail the evolution of membrane deformation and process of establishing the optimal geometry of stabilization structures for one stencil. Various other stencil designs and results on their optimization are summarized in the second subsection.

3.1. Full design flow of corrugation geometry for nanoresonator stencil and evaluation of stabilization effect on pattern distortion: stencil design A

Consider a whole stencil designed for patterning nanoresonating devices (design A). The stencil area is 2340x1036 μm². Figure 5 shows a simulated deformation of the entire stencil. Figures 6 present "deformation isoline maps" of 4-component and 1-component fragments of the stencil. Analysis of figures 6 and 7 yields the conclusion that similar deformation patterns, at least

in qualitative terms, are obtained when considering different fragments of the stencil.

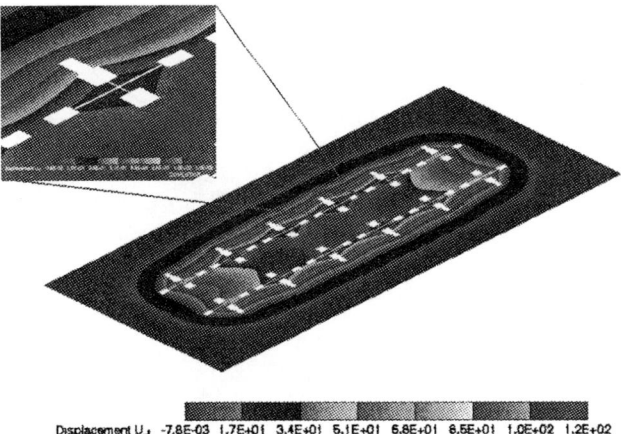

Fig. 5. Simulated "deformation isolines map" of the whole stencil with enlarged fragment (component) of the membrane (displacement shown in μm).

Accurate simulation of the entire unstabilised stencil proved to be extremely time consuming computationally. Simulation of a corrugated membrane would require even more resources as a very fine mesh is vital for accurate modelling of the increased number of structural features (corrugation rims). Since simulations of the entire membrane and its one component qualitatively result in a very similar deformation patterns, for stabilisation purposes, only one component of the entire stencil is used in the design of corrugation patterns in this study. The resulting optimal corrugation geometry can then be extrapolated across the full stencil. Although such an approach may not produce the optimal corrugation geometry due to neglected interactions between fragments, as long as the spacing between fragments is sufficiently large, our approach is a viable approximation for stabilising the stencil locally as the isocontours, not the magnitude of the displacement, determine the stabilisation structures.

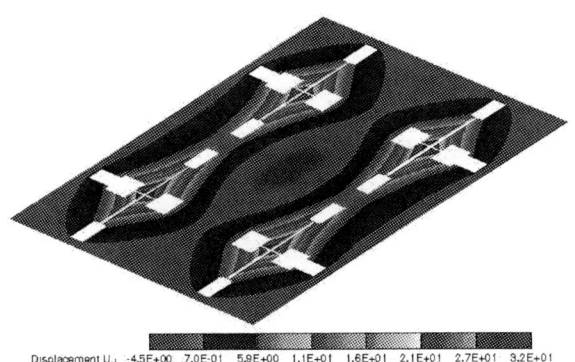

(a)

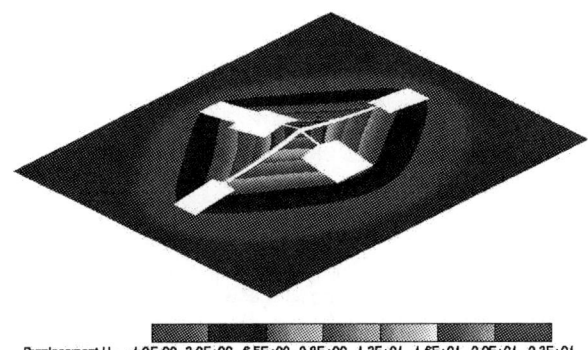

(b)

Fig. 6. Simulated deformation (μm) of four-component (a) and one-component (b) stencils.

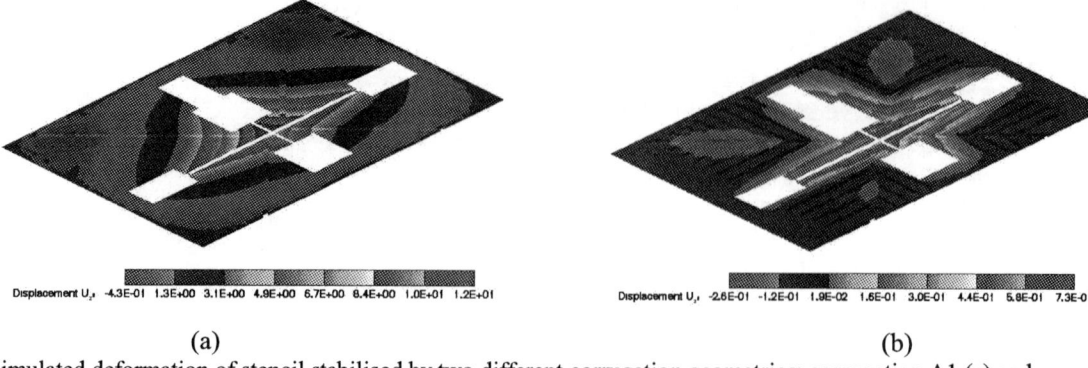

(a)   (b)

Fig. 7. Simulated deformation of stencil stabilised by two different corrugation geometries: corrugation A1 (a) and corrugation A2 (b).

Two different geometries of stabilisation structures for the stencil were suggested and numerically tested. Firstly, corrugation A1 was introduced as the most logic stabilisation type based on analysis of stencil "deformation isolines map" in figure 6b. The results (fig. 7a) show a reduced degree of stencil deformation but not sufficient to achieve the pattern resolution desired.

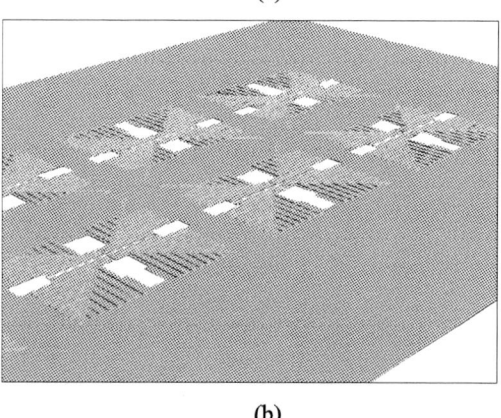

(a)

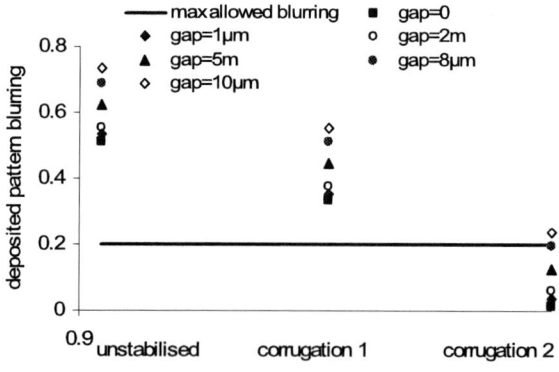

(b)

Fig. 8. Detailed schematic of one-component stencil with corrugation rims (a) and 3D model of the corresponding whole stencil incorporating stabilisation structures (b).

Analysis of the image suggests further adjustments to the corrugation scheme (fig. 7b). This second revision was tested and simulation results revealed a significant

reduction in membrane deformation of 96%. A 2D layout of the one-component stencil incorporating stabilisation rims together with 3D model of the whole stabilised membrane are presented in figure 8. The evolution of reduction in stencil deformation for considered corrugation geometries is plotted in figure 9.

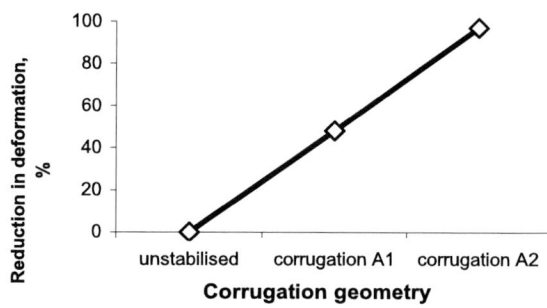

Fig. 9. Plot of percent values of reduction in deformation for various corrugation geometries.

Fig. 10. Values of calculated blurring (μm) of deposited pattern for various corrugation geometries and stencil-substrate gaps.

Formula (1) was applied here in order to access blurring of deposited pattern before and after incorporation of stabilisation structures. Various gap values in the realistic range between 0 and 10 μm were

Table 3. Summary of results on stencils stabilisation

Stencil design	Unstabilised stencil geometry (1 component) and deformation isolines	Deduced optimal geometry of stabilised stencil (stencil in light grey, apertures in white, corrugation rims in dark grey)	Reduction in stencil deformation, %
B	max deformation = 4μm		85.5
C	max deformation = 8.7μm		80.5
D	max deformation = 37μm		51
E	max deformation = 15μm		50
F	max deformation = 1.5μm		65

used in calculations. Figure 10 shows that there is no gap tolerance for unstabilised stencil at all. In fact, pattern distortion exceeds maximum allowed value of blurring even for a zero gap, an ideal case that is never achieved with the current fabrication process. The plot also shows that corrugation geometry A2 allows for a gap tolerance up to 8μm. This implies an increase of the process window allowing for greater control of the pattern during fabrication.

3.2. Stabilisation of other stencil designs

This subsection presents results on simulation of deposition stress induced deformation of MEMS stencils B, C, D, E and F. Stencils B, C, D, E are designed for deposition of nanoresonators, stencil F is intended for fabrication of three levels of interconnects to realise electrical measurement on a single molecule or nanowire. Simulation results on stencil deformation and final suggestions for optimal corrugation geometry based on analysis of deformation profiles are presented in table 3.

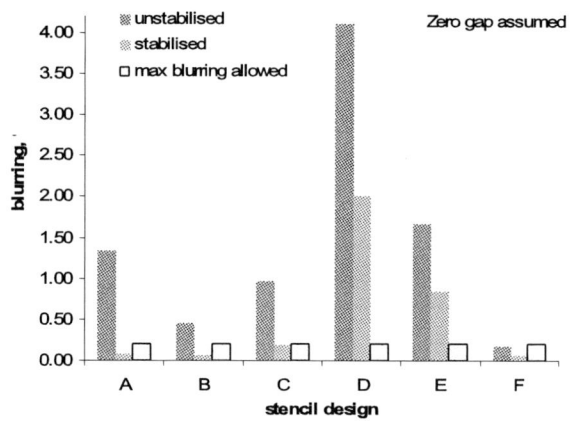

Fig. 11. Values of calculated blurring (μm) of deposited pattern for various stencil designs.

Figure 11 presents values of pattern broadening estimated via formula (1) for unstabilised and stabilised stencils considered in this study. Visibly, for majority of stencil designs pattern distortion is reduced to below maximum allowed value, though for some designs it is still above the limit. Nevertheless, corrugation significantly reduces pattern broadening in all studied cases which improves stencil usability. Achieving the blurring targets for the structures D and E may require optimization of additional processing conditions.

4. Conclusions

Various stencil designs aimed for deposition of nanoresonators and interconnects were modelled and their deformation due to the deposition induced stress was simulated. Analysis of "deformation isolines map" based on previous experience resulted in suggestions on optimal geometries of stabilisation structures. These were tested and are showed to achieve a significant reduction in deformation for all stencils.

Studied cases resulted in deformation reduced by 50% to 96% which translates into the pattern blurring level below the maximum allowed value in most cases. It was also shown that there is no gap tolerance for unstabilised stencils. In fact, pattern distortion exceeds maximum allowed value of blurring even for a zero gap between the substrate and the stencil, an ideal case that is never achieved for the current fabrication process. For some stencil designs the suggested corrugation geometries allow for a gap tolerance up to 8μm. This implies an increase of the process window which is an important advantage from practical viewpoint.

Results presented here are based on the assumption of Cr as a deposited material which is the "worst case scenario" because of high deposition stresses inherent to Cr films. Therefore, for "better cases" with lower deposition stress materials the probability of achieving the target reduction in deformation (pattern blurring below maximum allowed value) is secure.

Acknowledgments

This project is funded by EC-funded project NaPa (contract no. NMP4-CT-2003-500120).

References

1. M. A.F. Van den Boogaart, et al, "DUV-MEMS stencils for high-throughput resistless patterning of mesoscopic structures", J. Vac. Sci. Technol. B, **22**(6), pp. 3174 – 3177, 2004.
2. M. Ohring, The materials science of thin films, London, Academic press, 1992.
3. M. Lishchynska, V. Bourenkov, M. A. F. van den Boogaart, L. Doeswijk, J. Brugger, J.C. Greer. Predicting mask distortion, clogging and pattern transfer for stencil lithography. *Microelectronics Engineering*, 84, 2007, 42-53.
4. M.A.F. van den Boogaart, M.Lishchynska, L.M. Doeswijk, J.C. Greer, J. Brugger. Corrugated membranes for improved pattern definition with micro/nanostencil lithography. Sensors & Actuators A 130-131 (2006) 568-574.
5. CoventorWare version 2006. Reference Guides and Tutorials.
6. H.Baltes, et al, CMOC-MEMS, Weinhem, Wiley-VCH, 2005.
7. R.Whiting & M.A.Angadi, Young's modulus of thin films using a simplified vibrating reed method, Meas. Sci. Technol., 1, pp. 662-664, 1990.

Extended Finite Element for Electromechanical Coupling

Véronique Rochus and Daniel Rixen
Delft University of Technology, faculty 3mE
Dpt. of Precision and Microsystems Engineering, Engineering Dynamics
Mekelweg 2, 2628 CD Delft, The Netherlands
V.Rochus@tudelft.nl, D.J.Rixen@tudelft.nl
Phone: +31-(0) 15-278 18 81

Abstract

In MEMS modelling the electro-mechanical coupling takes an important place. Indeed many devices use electrostatic forces as actuator. The numerical modelling of this type of problem needs a strong coupling between the mechanics and the electrostatic field. In fact when the structure is moving, the electrostatic field around it has to be modified in consequence. The first solution is to use finite element method to model the electrostatic field. In this case the mesh has to be updated depending on the displacement of the structure. Many researches have been performed to deform properly the electrostatic mesh, but when large displacement are taken into account, the elements become distorted. Furthermore, when the pull-in is achieved, the electrodes are in contact and the layer of electrostatic elements is totally squeezed. The second usual solution is to use the boundary element method to model the electrostatic field. In this case, there are no more remeshing problem, but the computational time is larger and singularity problems appears when the electrodes become in contact.

One solution for this remeshing problem is to use extended finite elements (X-FEM) which are a new type of elements tailored to simulate problems involving discontinuities and moving boundaries. Initially this methodology was created for crack propagation problems [1, 3, 5], but its application has been extended to several other problems such as elastic problem involving inclusions, flow-structure interaction and solidification problems. In this paper the concept of extended finite elements is applied to develop modelling approaches for the electro-mechanical coupling. The method will be illustrated here for a one-dimensional problem and implementation issues relative to the two-dimensional case are discussed.

1. Introduction

This research aims at modelling of electro-mechanical coupling that normally takes place in some micro-electro-mechanical systems (MEMS) like micro-resonator or RF switches. The problem can be described as a conducting mechanical structure with applied voltage, which generates a surrounding electrostatic field. The electrostatic field, in its turn, causes an appearance of electrostatic force, applied to the structure. This type of problem is a strongly non-linear problem since the electric domain changes with the deformation of the structure. The usual numerical techniques to model this type of electro-mechanical problem are the finite element method and the boundary element method. The mechanical structure is usually simulated by a finite element model and the electrostatic domain is solved by either finite element method or boundary element method. For both cases some problems appear when the structure undergoes large displacements and when the electrodes come into contact. Indeed, the electrostatic finite element mesh has to be modified as the structure moves. Moreover, the electrostatic mesh can be severely deformed if the structure undergoes large displacements. Furthermore, when the electrodes come into contact, the elements between the electrodes have to be deleted. The boundary element method proposes a partial solution to this problem: the electrostatic domain is meshed only on the boundary hence allowing large displacements to the structure. However this increases the computation time. Moreover, when the structure comes into contact, the boundary element can no longer be applied since it requires that a gap exists between the electrodes. In order to simplify and improve modelling of structures moving in an electric field, we propose to make use of the concept of eXtended Finite Elements. They are a new type of elements tailored to simulate problems involving discontinuities and moving boundaries.

The basic idea is to have an electrostatic mesh covering the entire domain and that does not change while the structure part is moved within the field. The electro-mechanical problem is considered as a bi-material problem where the mechanics and the electricity are computed and coupled on a single element. Following the variational approach developed in [4], electrostatic forces may be derived and applied at the interface of the element. The electromechanical problem may then be solved and the results correspond very well to the analytical solution for a one dimensional problem. A short discussion will be also given about implementation issues of this techniques in 2 dimensions.

2. eXtended Finite Element Theory

The extended finite element method consists in discretising the entire electro-mechanical problem with a fixed mesh and in following the interface between two domains through this mesh. At the interface the physical field or its gradient are no more continuous. To model this discontinuity, special shape functions are used to enrich the usual discretisation. For instance, the mechanical displacement u is enhanced by discontinuous shape functions M_i such as:

$$u(x,t) = \sum_i N_i(x)U_i(t) + \sum_j M_j(x,t)A_j(t) \qquad (1)$$

1-4244-1105-X/07/$25.00 ©2007 IEEE

where N_i are the standard shape functions and M_j are the enriched shape functions taking the discontinuity into account. New unknowns A_j are introduced to model the discontinuity.

There are different ways to create these additional shape functions. Moes [2] proposes to define these functions based on the standard shape functions by the relation:

$$M_j(x,t) = N_j(x)\Theta(x,t) \tag{2}$$

where $\Theta(x,t)$ is defined by:

$$\Theta(x,t) = \sum_i |\psi_i| N_i - \left| \sum_i \psi_i N_i \right| \tag{3}$$

where $\psi(x,t)$ is the level set field describing the location of the interface, and ψ_i is the value of the level set of the node i. The level set is described by a surface function intersecting the problem plan at the interface.

First this methodology will be applied in a one-dimensional electro-mechanical example and the electro-mechanical problem will be solved by computing electrostatic forces at the interface. Then this methodology will be adapted to the two dimension problem.

3. Implementation in One Dimension

3.1 Shape Functions in 1D

In this section the shape functions for a pure mechanical bi-material element are presented in one dimension. In that case the usual shape functions N_i on a reference element are the following:

$$\begin{cases} N_1 = 1 - \eta \\ N_2 = \eta \end{cases} \tag{4}$$

These shape functions are represented in Figure 1. They allows to model linear behaviour in the element.

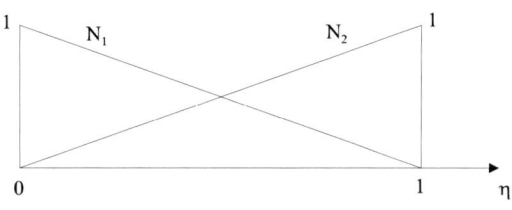

Figure 1: Linear shape functions in 1D.

Then the enriched shape functions are added to model the discontinuity. The enriched shape functions, following the approach of Moes [2], are given by (2) where Θ is defined by the level set Ψ equal to the signed distance between the point x and the discontinuity which is located at a distance Γ from the first node. In the 1D model, this

enrichment Θ becomes:

$$\begin{cases} \Theta_a = \dfrac{\eta}{\Gamma} & \text{for } 0 \le \eta \le \Gamma \\ \Theta_b = \dfrac{1-\eta}{1-\Gamma} & \text{for } \Gamma \le \eta \le 1 \end{cases} \tag{5}$$

The plot of these functions is illustrated in Figure 2 and

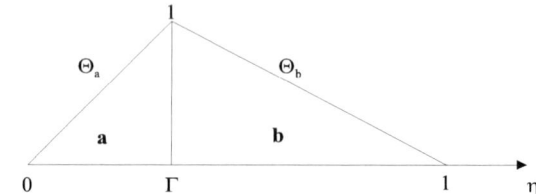

Figure 2: Extended finite element.

the enriched shape functions are:

$$\begin{cases} M_{1a} = (1-\eta)\dfrac{\eta}{\Gamma} \\ M_{2a} = \eta\dfrac{\eta}{\Gamma} \\ M_{1b} = (1-\eta)\dfrac{1-\eta}{1-\Gamma} \\ M_{2b} = \eta\dfrac{1-\eta}{1-\Gamma} \end{cases} \tag{6}$$

The sum of these two shape functions corresponds to the expression of Θ (see equation (5)).

$$\begin{cases} M_{1a} + M_{2a} = \dfrac{\eta}{\Gamma} \\ M_{1b} + M_{2b} = \dfrac{1-\eta}{1-\Gamma} \end{cases} \tag{7}$$

which correspond of the plot in Figure 2

3.2 Application to Electro-mechanical Coupling

The extended finite elements methodology is now applied to electro-mechanical coupling in one dimension. The element is divided in two sub-domains: the mechanical domain called "a" (in grey) which represents a conducting material and the electrostatic domain "b" (in white) representing a non-conducting medium. The part "b" represents, for instance, the air in which the structure moves (see Figure 3).

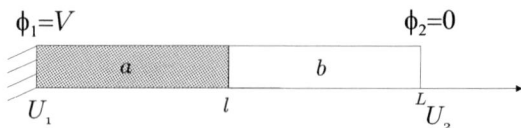

Figure 3: Extended finite element.

In electro-mechanical problems, the physical unknowns are the displacement u and the voltage ϕ. Both fields will be discretised on domain a and b. Hence we assume that there is a mechanical field and an electrostatic field in both parts and define:

$$\begin{cases} u_a = N_1 U_1 + N_2 U_2 + M_{1a} A_1 + M_{2a} A_2 \\ u_b = N_1 U_1 + N_2 U_2 + M_{1b} A_1 + M_{2b} A_2 \\ \phi_a = N_1 \Phi_1 + N_2 \Phi_2 + M_{1a} B_1 + M_{2a} B_2 \\ \phi_b = N_1 \Phi_1 + N_2 \Phi_2 + M_{1b} B_1 + M_{2b} B_2 \end{cases} \quad (8)$$

where N_i are the standard shape functions, and M_{ia} and M_{ib} are the enriched shape functions for each domains:

$$\begin{cases} N_1 = 1 - \frac{x}{L} \\ N_2 = \frac{x}{L} \end{cases} \quad \begin{cases} M_{1a} = \left(1 - \frac{x}{L}\right) \frac{x}{l} \\ M_{2a} = \frac{x}{L} \frac{x}{l} \\ M_{1b} = \left(1 - \frac{x}{L}\right) \frac{L-x}{L-l} \\ M_{2b} = \frac{x}{L} \frac{L-x}{L-l} \end{cases} \quad (9)$$

where l is the position of the interface.

The unknowns of this problem are U_1, U_2, Φ_1 and Φ_2, the displacement and the electric potential at the extremities of the element. and the new unknowns A_1, A_2, B_1 and B_2 used to model the discontinuity of the mechanical field and the electric potential.

3.2.1 Electrostatic Potential

Along the extended element including the conductor structure and the air gap between the electrodes, the electric field is discontinuous. Indeed the voltage is contant on the conductor (mechanical part a) and decreases linearly on the electric domain b.

A potential difference is applied between the extremities of the element ($\Phi_1 = V$ and $\Phi_2 = 0$) as shown in Figure 3. The stiffness matrix associated to the electrostatic field may be computed by integration over both parts of the element:

$$\delta \Phi^T K_{\phi\phi} \delta \Phi = \frac{1}{2} \int_0^l \frac{\partial \delta \phi}{\partial x} \varepsilon_a \frac{\partial \delta \phi}{\partial x} dx + \frac{1}{2} \int_l^L \frac{\partial \delta \phi}{\partial x} \varepsilon_b \frac{\partial \delta \phi}{\partial x} dx \quad (10)$$

where ε_a and ε_b are the permittivity of domain a and b, respectively. Considering the structure as a perfect conductor, the voltage on this part has to be constant. To keep the voltage constant on the mechanical part, the permittivity of this domain is imposed to be a very large number compared to the void permittivity. In this case we will take $\varepsilon_a = 1$ and $\varepsilon_b = \varepsilon_0$.

Solving the pure electrostatic problem, the obtained potential along the element is constant on the mechanical part and decreases linearly between the electrodes as plotted in Figure 4.

3.2.2 Electrostatic Forces at the Interface

To compute the electrostatic forces applied at the interface between the mechanical structure and the electrostatic domain, an energetic approach has been chosen as proposed by the author in paper [4]. This method consists in determining the electrostatic forces at the nodes of a finite element by an integration on its volume. The finite element formulation is:

$$f_{elec}^T \delta u = \frac{1}{2} \int_\Omega D^T F \left(\mathbf{grad} \delta u\right) d\Omega \quad (11)$$

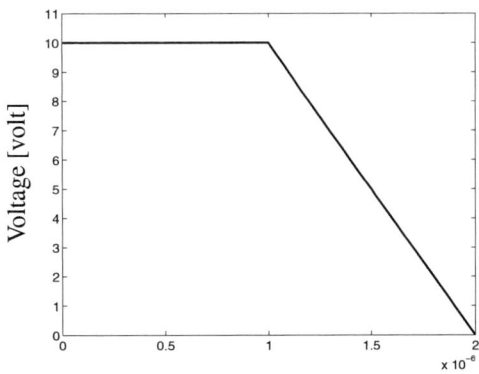

Figure 4: Electrostatic potential in one extended finite element.

with

$$F = \begin{pmatrix} \frac{\partial \phi}{\partial \xi} & 0 & 2\frac{\partial \phi}{\partial \eta} & -\frac{\partial \phi}{\partial \xi} \\ -\frac{\partial \phi}{\partial \eta} & 2\frac{\partial \phi}{\partial \xi} & 0 & \frac{\partial \phi}{\partial \eta} \end{pmatrix} \quad (12)$$

where D is the electrostatic displacement. In one dimension this expression is reduced to:

$$f_{elec}^T \delta u = -\frac{1}{2} \int_0^L \left(\frac{\partial \phi}{\partial x}\right)^2 \varepsilon \frac{\partial \delta u}{\partial x} dx \quad (13)$$

The same method is now applied to the extended finite element. The electrostatic forces is computed on each subdomain of the element by:

$$f_{elec}^T \delta u = -\frac{1}{2} \int_0^l \left(\frac{\partial \phi_a}{\partial x}\right)^2 \varepsilon_a \frac{\partial \delta u_a}{\partial x} dx - \frac{1}{2} \int_l^L \left(\frac{\partial \phi_b}{\partial x}\right)^2 \varepsilon_b \frac{\partial \delta u_b}{\partial x} dx \quad (14)$$

The electrostatic potential being constant on the conductor structure, the first term disappears and the electrostatic forces are computed only by the integration on domain b.

3.2.3 Electro-mechanical Coupling

Now the complete electro-mechanical problem will be considered. The mechanical stiffness of an extended one-dimensional element is obtained by:

$$\delta u^T K_{uu} \delta u = \frac{1}{2} \int_0^l \frac{\partial \delta u}{\partial x} E_a \frac{\partial \delta u}{\partial x} dx + \frac{1}{2} \int_l^L \frac{\partial \delta u}{\partial x} E_b \frac{\partial \delta u}{\partial x} dx \quad (15)$$

where E_a and E_b are the Young's modulus of the domain a and b, respectively. In the present case, a mechanical behaviour exists only on the domain a and E_b will be set to zero.

The equilibrium position of the electro-mechanical problem may be obtained solving the system:

$$\begin{cases} K_{uu} u = f_{elec} \\ K_{\phi\phi} \Phi = q_{elec} \end{cases} \quad (16)$$

where the array u contains the mechanical degrees of freedom u_i and A_i and Φ contains Φ_i and B_i. q_{elec} are the charges on the nodes. This system of equation is non-linear since the electrostatic force f_{elec} is a non-linear function of the electric potential (see equation (14)) and because the position of the interface l after deformation has to be taken into account in the computation of the electrostatic stiffness $K_{\phi\phi}$ and force f_{elec}.

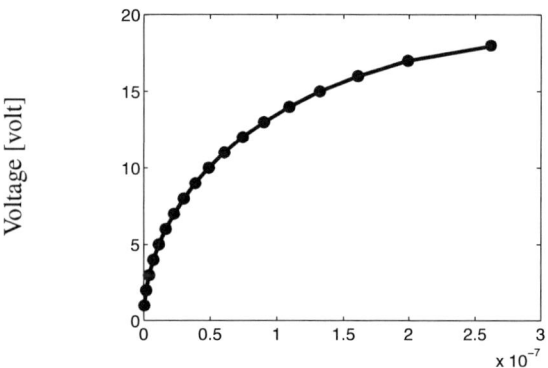

Displacement of the interface [m]

Figure 5: Displacement of the interface (dots) and analytical solution (plain line).

In Figure 5 the displacement obtained with the extended finite element method (represented with dots) is compared to the analytical solution (plain line). These two results fit very well for the stable part of the curve.

4. 2D Electrostatic Problem

We will now investigate possible shape functions for extending the triangular linear elements in two dimensions. Triangular elements are often used in practice in automatic meshing tools and are thus very common in practical models.

4.1 Moes' Shape Functions

For a triangular form Moes [2] proposes for θ the following function:

$$\Theta = \sum_i |\psi_i| N_i(x,y) + \left|\sum_i \psi_i N_i(x,y)\right| \qquad (17)$$

where ψ_i is the value of the level set of the node i. Let us assume that, in the reference coordinate space, the interface passes through the nodes (C_1,C_2) and (D_1,D_2) where $C_1 = 0$ and $D_1 = 1 - D_2$ as shown in figure 6. The level set may be chosen as:

$$\psi(\xi,\eta) = a\xi + b\eta + c \qquad (18)$$

where $a = (D_2 - C_2)$, $b = -D_1$ and $c = C_2 D_1$ so that $\psi = 0$ corresponds to the equation of the interface between the two domains. Note that this choice is not unique for a,b,c.

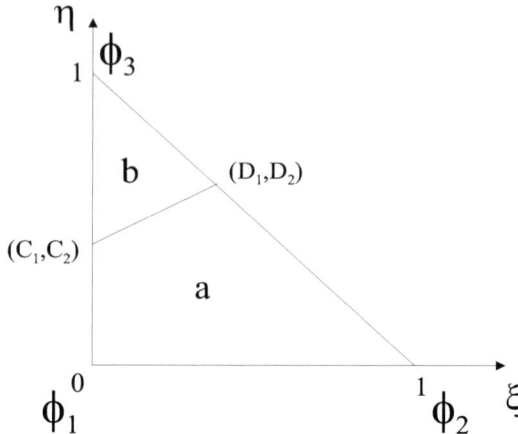

Figure 6: Extended finite element in 2D.

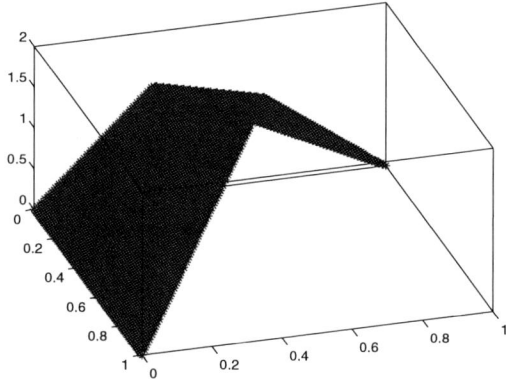

Figure 7: Enrichment hat function Θ.

The enriched shape functions and the linear shape functions are:

$$\begin{cases} N_1 = 1 - \xi - \eta \\ N_2 = \xi \\ N_3 = \eta \end{cases} \qquad \begin{cases} M_{a1} = -2(b+c)\eta(1-\xi-\eta) \\ M_{a2} = -2(b+c)\eta\xi \\ M_{a3} = -2(b+c)\eta\eta \\ M_{b1} = 2(c+a\xi-c\eta)(1-\xi-\eta) \\ M_{b2} = 2(c+a\xi-c\eta)\xi \\ M_{b3} = 2(c+a\xi-c\eta)\eta \end{cases}$$
$$(19)$$

The hat function Θ modelling the discontinuity is plotted in figure 7.

$$\Theta_a = \sum_i M_{ai} \qquad \Theta_b = \sum_i M_{bi} \qquad (20)$$

We can observe that the line of discontinuity is not horizontal, which seems to indicate that these extended formulation might not be suitable to impose a constant potential (as needed for a conductor inside for instance). Let us check if this extended shape functions are suitable for the electrostatic problem

72

So let us investigate if the extended field above can be used to model an electric potential constant on one domain and varying on the other. The potential is discretised by:

$$\begin{cases} \phi_a = N_1\Phi_1 + N_2\Phi_2 + N_3\Phi_3 + M_{1a}B_1 + M_{2a}B_2 + M_{3a}B_3 \\ \phi_b = N_1\Phi_1 + N_2\Phi_2 + N_3\Phi_3 + M_{1b}B_1 + M_{2b}B_2 + M_{3b}B_3 \end{cases}$$ (21)

Considering that the potential is constant on the quadrangular domain a (see Figure 7), we impose $\frac{\partial \phi_a}{\partial \xi} = 0$ and $\frac{\partial \phi_a}{\partial \eta} = 0$ for all values of ξ and η on the domain a. After development we obtain the following constraints:

$$\Phi_1 = \Phi_2 \quad \text{and} \quad B_1 = B_2 = B_3 = -\frac{(\Phi_1 - \Phi_3)}{2(b+c)}$$ (22)

Only two unknowns are left, as expected, the potential Φ_3 and $\Phi_1 = \Phi_2$.

We will now consider that the potential is constant on the triangular domain b so that $\frac{\partial \phi_b}{\partial \xi} = 0$ and $\frac{\partial \phi_b}{\partial \eta} = 0$ for all values of ξ and η on the domain b. The conditions to have a constant potential on this domain are

$$B_2 = B_1 = B_3 \quad \text{and} \quad \begin{cases} B_1 = \dfrac{\Phi_1 - \Phi_2}{2a} \\ B_1 = \dfrac{\Phi_3 - \Phi_1}{2c} \end{cases}$$ (23)

In this case, setting the triangular domain b to a constant potential Φ_3 implies a relation between Φ_1 and Φ_2. This can be understood as follows.

The condition $B_2 = B_1 = B_3$ enforces the linearity of the discretisation field on domain b and as a consequence, also the linearity of the field on the quadrangular domain a. The relation between Φ_1 and Φ_2 implied by the second set of constraints in (23) imposes that the values of the electric potential at the nodes "1", "2", "C" and "D" are coplanar. So the situation where Φ_3 is set at a potential V (higher electrode) and Φ_1 and Φ_2 are on the grounded lower electrode, is possible only if the interface is parallel to the lower electrode.

Thus it is impossible to impose a constant potential inside the triangular part of the extended element and a linear field on the quadrangular part. We must therefore conclude that the shape functions derived from the approach described in [2] not suitable for electro-mechanical modelling in the vicinity of conductors. Hence in the following section, we build a different extended field to circumvent this shortcoming.

4.2 Quadratic Enriched Shape Functions

The underlying idea of this new approach can be best understood by observing that if the extended element was build out of two finite elements (one for the conductor and one for the non-conducting part), it would straightforward to simulate the behaviour of electro-mechanical problem. So, we will try to use quadrangular shape functions for the trapezoidal part and triangular shape functions for the triangular part.

Therefore we will introduce two successive changes of variables corresponding to two isoparametric transformation of the physical space: a first between (X, Y) and (ξ, η), and a second between (ξ, η) and (s, t), as presented in Figure 8. Note that the second transformation is defined separately for the triangular and the quadrangular part if the partitioned element.

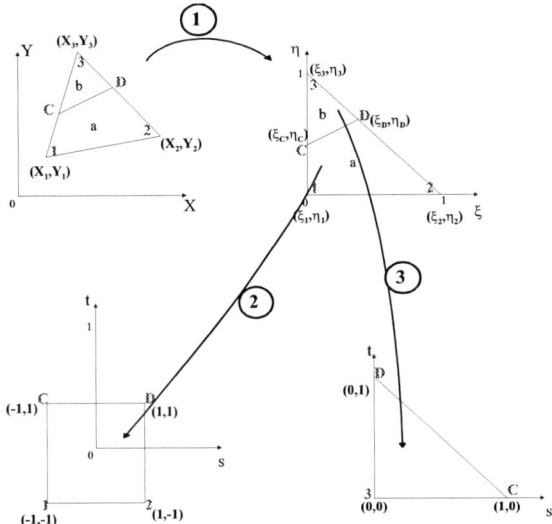

Figure 8: Successive transformations for an extended triangular element.

4.2.1 First change of variables

The first change of variables is identical for both domains a and b and may be expressed by the relation:

$$\begin{cases} X = N_1X_1 + N_2X_2 + N_3X_3 \\ Y = N_1Y_1 + N_2Y_2 + N_3Y_3 \end{cases} \qquad \begin{cases} N_1 = (1-\xi-\eta) \\ N_2 = \xi \\ N_3 = \eta \end{cases}$$ (24)

The coordinates of the point C and D is obtain by computing the intersection between the level set boundary and the edge of the triangle. In the second space, the position of these points are:

$$\begin{cases} \xi_C = 0 \\ \eta_C = \dfrac{(X_C - X_1)}{(X_3 - X_1)} \end{cases} \qquad \begin{cases} \xi_D = \dfrac{(X_3 - X_D)}{(X_3 - X_2)} \\ \eta_D = \dfrac{(X_D - X_2)}{(X_3 - X_2)} \end{cases}$$ (25)

4.2.2 Second Transformation - Quadrangular Part

The usual shape functions for a quadrangle are:

$$\begin{cases} M_1 = (1-s)(1-t)/4 \\ M_2 = (1+s)(1-t)/4 \\ M_3 = (1+s)(1+t)/4 \\ M_4 = (1-s)(1+t)/4 \end{cases}$$ (26)

The relation between the reference space (s,t) and the intermediate space (ξ, η) is given by the following isoparametric transformation:

$$\begin{cases} \xi(s,t) = \xi_1 M_1 + \xi_2 M_2 + \xi_D M_3 + \xi_C M_4 \\ \eta(s,t) = \eta_1 M_1 + \eta_2 M_2 + \eta_D M_3 + \eta_C M_4 \end{cases} \quad (27)$$

The last two shape functions will be used to enhance the solution field since the shape functions M_1 and M_2 are associated to the nodes 1 and 2 that already define the amplitude of the shape functions N_1 and N_2 in the basic element. The total discretisation of the extended field is thus:

$$\begin{aligned} \phi_a = \ & N_1\left(\xi(s,t), \eta(s,t)\right)\Phi_1 + N_2\left(\xi(s,t), \eta(s,t)\right)\Phi_2 \\ & + N_3\left(\xi(s,t), \eta(s,t)\right)\Phi_3 + M_4(s,t)B_C + M_3(s,t)B_D \end{aligned} \quad (28)$$

It is easy to verify that the part of the hat function θ in domain a can be found by setting $B_C = B_D$:

$$\Phi_a = M_3 + M_4 \quad (29)$$

This is depicted in figure 9.

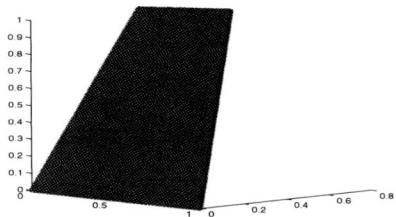

Figure 9: Enrichment of domain a.

Let us now compute the electric stiffness matrix. The electric stiffness matrix is computed by the relation:

$$K_{\phi\phi_a} = \int_{S_a} B_a^T(X,Y)\,\varepsilon_a\,B_a(X,Y)\,dX\,dY \quad (30)$$

where S_a is the surface of integration in the space (X,Y) and where $B_a(X,Y)$ is the matrix of electric field shape functions obtained from the derivatives of the potential shape functions. The change of variables between the space (X,Y) and the reference space (s,t) may be obtained by the relation (24) and (27) which allow us to write $X(s,t)$, $Y(s,t)$ and the Jacobian J of this transformation. The electric stiffness matrix may then be computed on the reference space by:

$$K_{\phi,\phi_a} = \int_{\Omega_a} B_a^T(s,t)\,J^{-T}\,\varepsilon_a\,J^{-1}\,B_a(s,t)\det(J)\,ds\,dt \quad (31)$$

where

$$B_a(s,t) = \begin{pmatrix} \frac{\partial N_1}{\partial s} & \frac{\partial N_2}{\partial s} & \frac{\partial N_3}{\partial s} & \frac{\partial M_4}{\partial s} & \frac{\partial M_3}{\partial s} \\ \frac{\partial N_1}{\partial t} & \frac{\partial N_2}{\partial t} & \frac{\partial N_3}{\partial t} & \frac{\partial M_4}{\partial t} & \frac{\partial M_3}{\partial t} \end{pmatrix} \quad (32)$$

4.2.3 Second Transformation - Triangular Part

The shape functions for the triangular domain are the same as the initial change of variables:

$$\begin{cases} G_1 = (1 - s - t) \\ G_2 = s \\ G_3 = t \end{cases} \quad (33)$$

and the isoparametric transformation becomes:

$$\begin{cases} \xi(s,t) = \xi_3 G_1 + \xi_C G_2 + \xi_D G_3 \\ \eta(s,t) = \eta_3 G_1 + \eta_C G_2 + \eta_D G_3 \end{cases} \quad (34)$$

As for the quadrangular part only the shape function not associated to the basic node (here G_2 and G_3) will define the enrichment field. The total enhanced shape functions are thus, for domain b

$$\begin{aligned} \phi_b = \ & N_1\left(\xi(s,t), \eta(s,t)\right)\Phi_1 + N_2\left(\xi(s,t), \eta(s,t)\right)\Phi_2 \\ & + N_3\left(\xi(s,t), \eta(s,t)\right)\Phi_3 + G_2(s,t)B_C + G_3(s,t)B_D \end{aligned} \quad (35)$$

Obviously since the added degrees of freedom B_C and B_D used here and used in the shape functions (28) in domain a are identical, the continuity of the potential field is guarantied while the electric field, gradient of the potential, can be discontinuous due to the enrichment field. Again we have that the hat function θ in the extended theory is

$$\Phi_b = G_2 + G_3 \quad (36)$$

The electric stiffness matrix is computed by the relation

$$K_{\phi,\phi_b} = \int_{S_b} B_b^T(X,Y)\,\varepsilon_b\,B_b(X,Y)\,dX\,dY \quad (37)$$

where S_b is the surface of integration in the space (X,Y). Using (24) and (34) the electric stiffness matrix may be computed on the reference space by

$$K_{\phi,\phi_b} = \int_{\Omega_b} B_b^T(s,t)\,J^{-T}\,\varepsilon_b\,J^{-1}\,B_b(s,t)\det(J)\,ds\,dt \quad (38)$$

where

$$B_b(s,t) = \begin{pmatrix} \frac{\partial N_1}{\partial s} & \frac{\partial N_2}{\partial s} & \frac{\partial N_3}{\partial s} & \frac{\partial G_2}{\partial s} & \frac{\partial G_3}{\partial s} \\ \frac{\partial N_1}{\partial t} & \frac{\partial N_2}{\partial t} & \frac{\partial N_3}{\partial t} & \frac{\partial G_2}{\partial t} & \frac{\partial G_3}{\partial t} \end{pmatrix} \quad (39)$$

4.2.4 Simple Verification of the Element

We will consider the very simple 2D case of a unit square domain where a voltage of 1V is imposed on the top and where the lower edge is grounded. Half of the domain is conducting and the domain is modelled with 2 triangular extended elements as built in the previous section (See figure 10).

First the interface between conductor and vacuum is taken parallel to the electrodes. The computed electric potential is plotted in figure 11: it is observed that the potential is constant on the conductor and decreases linearly between the electrodes as expected.

74

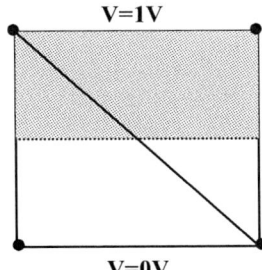

Figure 10: Simple 2D model.

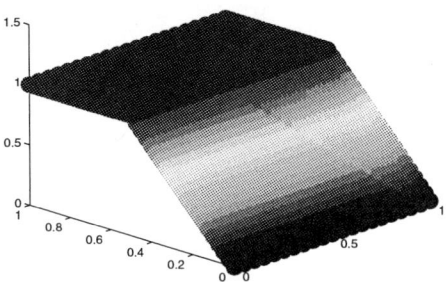

Figure 11: Electrostatic potential computed by two eXtended Elements.

Next the electric potential is also computed for the case where the interface is not parallel to the extremities as shown in figures 10. The computed potential is shown in 13. The potential is again piecewise linear and clearly the new shape functions of the eXtended Finite Elements can properly handle the computation of the electric potential even if the interface is oblique.

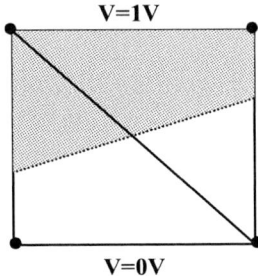

Figure 12: Simple 2D model.

5. Conclusions

The issue of mesh moving is a real challenge when modelling electro-mechanical devices with finite elements. This paper investigates the application of the Extended Finite Element approach to model the motion of a structure in

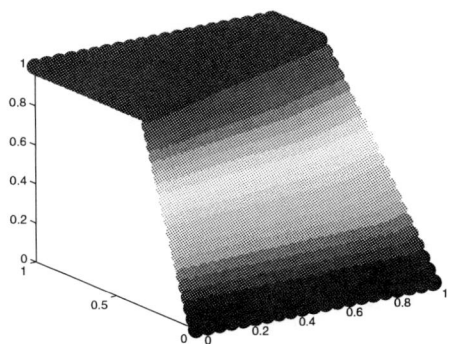

Figure 13: Electrostatic potential computed by two eXtended Elements.

an electrostatic field. The electro-mechanical forces are derived from the variational methodology proposed in paper [4].

When applied to a simple one dimensional problem, the eXtended Finite Element approach finds the exact solution for the strongly coupled electro-mechanical problem: the exact electrostatic potential along the element is retrieved and, under the action of the electrostatic forces on the interface, the correct relation between deformation and applied voltage is found.

For the two dimensional case, we have discussed why the enhancement strategy proposed in [3] is not suitable for modelling the electric field jump in practical problems. We propose in this paper different enhancement shape functions that guaranty that the potential field across a conductor/vacuum interface can be properly approximated. It is expected that the same approach can be applied in three dimensions. This will be investigate in the future together with the global efficiency of the new elements in solving the electromechanical coupling of real microsystems.

Acknowledgments

The authors want to acknowledge the support of the Koiter Institute of the Delft University of Technology and of the MicroNed program financed by the ministry of economical affairs of The Netherlands. The first author acknowledges the financial support of the Belgian National Fund for Scientific Research.

References

[1] T. BELYTSCHKO AND T. BLACK, *Elastic Crack Growth in Finite Elements with Minimal Remeshing*, International Journal of Numerical Methods in Engineering, Vol. 45, No. 5 (1999), pp. 601–620.

[2] N. MOËS, M. CLOIREC, P. CARTRAUD, AND J. F. REMACLE, *A computational approach to handle complex microstructure geometries*, Computer Methods in Applied Mechanics and Engineering, Vol. 192 (2003), pp. 3163–3177.

[3] N. MOES, J. DOLBOW, AND T. BELYTSCHKO, *A Finite Element Method for Crack Growth without Remeshing*, International Journal of Numerical Methods in Engineering, Vol. 46 (1999), pp. 131–150.

[4] V. ROCHUS, D. J. RIXEN, AND J.-C. GOLINVAL, *Monolithical Modeling of Electro-Mechanical Coupling in Micro-Structures*, International Journal for Numerical Methods in Engineering, Vol. 65, No. 4 (2006), pp. 461–493.

[5] N. SUKUMAR, T. BELYTSCHKO, C. PARIMI, N. MOËS, AND U. SHUJI, *Modeling Holes and Inclusions by Level Sets in the Extended Finite Element Method*, Computer Methods in Applied Mechanics and Engineering, Vol. 190 (2001), pp. 6183–6200.

Finite Element Modelling (FEM) of Green Electronics in Aeronautical and Military Communication Systems (GEAMCOS)

A. Chaillot[1] and G. Massiot[1], C. Munier[1], I. Lombaërt-Valot[1], S. Bousquet[2], C. Chastanet[2],
D. Plouseau[3], E. Munier[3], D. Maron[4], P. Raynal[4], S. Villard[5], R. Dumonteil[5]
[1] EADS France, 12 rue Pasteur BP 76, 92152 SURESNES Cedex, France
agnes.chaillot@eads.net, tel: +33 (0)1 46 97 30 96
[2] AIRBUS France, [3] EADS Secure Networks, [4] ACTIA, [5] TECHCI

Abstract

As part of the European LIFE project GEAMCOS, this paper presents the simulation evaluation of lead-free electronic assemblies submitted to harsh environments encountered in aeronautical and military communication applications. It addresses the following questions:

— the determination of an optimal accelerated test condition in the temperature range of -40°C/ 100°C for assemblies with SnAgCu alloy,

— the evaluation of the life-time of five assemblies with SnAgCu and SnPb,

— a sensitivity analysis on design and serigraphy parameter variations,

— the determination of a test protocol in vibration environments.

All the results give relevant life-time predictions. The CSP and PQFP components should have longer time-to-failure with SnAgCu solder joints than with SnPb joints. The contrary is expected for 1st and 2nd level interconnects of large BGA packages. In any case, the factor is found to remain less than about 2. The PCB thermal expansion coefficient has the highest impact, followed by the solder joint height. The first level interconnect reliability is not impacted by any of the parameter changes.

Simulation work in GEAMCOS project shall go on in 2007-2008 with an experimental validation of the thermo-mechanical simulations. This will require to identify a new constitutive material model for SnAgCu and to calibrate the simulations with experimental test results.

1. Acronyms and symbols

BGA   Ball Grid Array package
CSP   Chip Scale Package
CTE   Coefficient of linear Thermal Expansion
FEM   Finite Element Modelling
PBGA Plastic Ball Grid Array
PCB   Printed Circuit Board
PQFP Plastic Quad Flat Pack
α Coefficient of linear Thermal Expansion [ppm.K^{-1}]
$\Delta\varepsilon$ Strain range
$\Delta\varepsilon_{eq}$ Equivalent strain range
$\Delta\gamma$ Shear strain range
$\Delta\gamma_{eq}$ Equivalent shear strain range
E Young's modulus [GPa]
E_p Tangent (plastic) modulus [MPa]
ε Strain
ε_{therm} Thermal expansion

ε_{zz} Tensile strain
g Normal gravity (9.81 m.s^{-2})
k Boltzmann constant (8.616·10^{-5} eV.K^{-1})
N_f Median number of cycles to failure
f Cycle frequency [Hz]
ν Poisson's ratio
σ Stress [MPa]
σ_y Yield stress [MPa]
r.m.s   Root mean square
T Temperature [°C]
T_g Glass transition temperature [°C]

2. Introduction

Concerning lead-free electronics, specifications requested for aeronautical and military communication sectors (reliability, life-time, security, maintainability) don't permit to use simply alternative solutions already proposed for consumer electronics without some changes [1]. The main tasks of this three and a half year project are thus to evaluate lead-free technologies in harsh environments and to transfer these new technologies to other sectors (space, military systems…) with the same reliability needs.

After a brief review of the different steps of the GEAMCOS project, this paper gives the simulation results obtained on lead-free assemblies submitted to harsh environments (under temperature cycling and under vibration environments). The scope of these simulations is firstly to determine optimal accelerated test conditions for civil aircraft. It is then to evaluate the reliability of the assemblies in terms of mean life-time in thermal accelerated conditions, according to design and serigraphy parameter variations.

Some innovative techniques in modelling of electronic assemblies are developped here in order to be able to run whole series of large Finite Element Models (FEM) like BGA 1508 assemblies (Ball Grid Array), and also to apply fatigue laws in a correct and legible manner. A specific fatigue law for the lead-free alloy is proposed, based on published fatigue test results.

3. Presentation of GEAMCOS project

The present work is part of the European LIFE project GEAMCOS supported by European commission. This acronym stands for Green Electonics in Aeronautical and Military Communication Systems.

Its first objective is to qualify mixed assemblies (lead-free alloy and leaded components) used in order to repair boards during the forward phase transition. For that

1-4244-1105-X/07/$25.00 ©2007 IEEE

purpose, a test board (cf. Figure 1) was constructed with the characteristics specified in Table 1.

It was then submitted to temperature cycles and vibration loads defined by FEM studies presented here.

Figure 1: GEAMCOS forward test board

Components	BGA, CICGA, SQFP, PQFP, TQFP, PLCC, LCC, SSOP, C0709, DO27A, DO213, DIL	
PCB	Laminate material	- high T_g and low CTE FR4 - current FR4 (medium T_g)
	Finish	NiAu, Sn, SnPb
Solder material for reflow and wave	Sn3Ag0.5Cu	
No clean flux cleaned		

Table 1: Main characteristics of the forward test board

The project then focuses on full lead-free technology (lead-free alloy and components). It includes an experimental validation of the processes on another test board specified in Table 2, as well as a numerical estimation of its reliability with FEM simulations (according to design and production parameters).

Components	PBGA, CBGA or CCGA, Passive, LLP, D2PACK, ceramic CSP , QFP, Mini-melf...	
PCB	Laminate material	high T_g and low CTE FR4 (halogen free)
	Finish	NiAu, Sn, SnPb
Solder material for reflow and wave	Sn3Ag0.5Cu	
No clean flux cleaned		

Table 2: Main characteristics of the full lead-free test board

In a final step, a functional full lead-free demonstrator will be developed and tested in harsh environments.

Note that during temperature cycles, the reliability of daisy-chain and passive packages is tested with electrical monitoring. It will then be possible to correlate thermo-mechanical FEM results with experimental data as soon as full lead-free vehicles will have been tested.

4. Construction of thermo-mechanical FEM models

To be representative of a large number of package types used in on-board electronic equipments, five component packages were modelled under accelerated test conditions in temperature:

— a BGA 84 Chip Scale Package (CSP) with 0.5-mm pitch,

— a PBGA 400 with 0.8-mm pitch,

— a PBGA 896 with 1-mm pitch,

— a PBGA 1508 with 1-mm pitch,

— a PQFP 208 (Plastic Quad Flat Pack) with 0.5-mm pitch.

All the input data required for the component package designs are based on optical observations on the GEAMCOS forward board and on datasheets provided by component suppliers. Models are fully parametrised so that any geometric and material variations can easily be input in this study.

Concerning material properties, for most materials, a thermo-elastic model is sufficient as long as the temperature remains below glass transition and stresses stay below the yield stress. This is the case here for all materials except copper pads, SnPb and SnAgCu. The copper pads and pins are subjected to noticeable stress and thus require an elastoplastic model in order to account for plastification when the yield stress is reached (for Cu: elastoplastic yield stress σ_y=70 MPa, tangent modulus E_p=2000 MPa, kinematic hardening assumed [2]). All material properties are documented in the Table 3.

	E [GPa]	ν	α [ppm.°C^{-1}]	Ref.
Si	140	0.22	2.5	[3]
Mold compound	30	0.24	8	[4]
Cu	120	0.34	17	[3]
BT	22	0.2	x: 13, y:16, z: 57	[5, 6]
FR4 IS400	19	0.15	x: 15, y:12, z: 40	[7]
Underfill	8	0.33	32	[6]
SnAgCu	$f(T)$	0.36	17.6	[8,9]
SnPb	$f(T)$	0.4	23.9	[10,11]

Table 3: Material properties

for SnAgCu [8, 9]:

E [MPa] = 50200 [MPa] × (1 – T [°C] / 200)

σ_y [MPa] = 57.0 at -55 °C

σ_y [MPa] = 35.1 at 25 °C

σ_y [MPa] = 26.1 at 70 °C

σ_y [MPa] = 21.4 at 100 °C

and the tangent modulus E_p sets to zero.

and for SnPb [10, 11]:

E [MPa] = 32400 [MPa] × (1 –T [°C] / 370)

σ_y [MPa] = 23.9 at -20 °C

σ_y [MPa] = 22.0 at 0 °C

σ_y [MPa] = 18.2 at 50 °C

σ_y [MPa] = 8.3 at 100 °C

E_p [MPa] = 1780 [MPa] × (1 – T [°C] / 136)

The relevant characteristics of the PCB are given by: the base material properties (FR4), the thickness of each layer and the percentage of surface covered with copper in each layer.

Basic simulations (tractions by a strain of 10^{-2} in the 3 directions and free thermal expansion of 1°C) were performed on a PCB sample in order to determine its equivalent thermo-elastic properties, given in Table 4.

	Symbols	Units	Values	Ref.
Young's modulus	E	[GPa]	29	
Poisson's ratio	ν	-	0.29	
In-plane CTE	α_x	[ppm.°C⁻¹]	16	[3, 7]
	α_y	[ppm.°C⁻¹]	14.5	
Out-of-plane CTE	α_z	[ppm.°C⁻¹]	36	

Table 4: Equivalent properties of the PCB

The viscoplastic model used for SnAgCu applies to the eutectic alloy Sn3.8Ag0.7Cu published in [8] (as a first approximation and due to the lack of data on Sn3Ag0.5Cu, the alloy actually used in GEAMCOS). It is based on test results made with normalised bulk samples. Two flow regimes are taken into account at low and high creep rates [12].

For SnPb, the creep rate is quite higher than for SnAgCu. This has the consequence that SnPb solder joints have higher deformations and lower stresses than SnAgCu in a given assembly (cf. Table 5).

$$\dot{\varepsilon}_{creep} = B \sinh^n \left(\frac{\sigma}{\sigma_0} \right) \exp \left(-\frac{Q}{kT} \right)$$

	B [s⁻¹]	n	σ_0 [MPa]	Q [eV]	Ref.
Sn3.8Ag0.7Cu	501.3	4.96	31.6	0.47	[8]
Sn40Pb	4.10⁶	3.3	8.27	0.81	[12]

Table 5: Constitutive equation parameters for solder materials

As an example, a mesh overview of the nominal model of the BGA 1508 is given in Figure 2. Only 1/8th of the structure is represented, because of the plane symmetries. Closer views are given in Figure 3. The type of element is 8-node hexaedron. Contact elements are used in transitional areas between critical domains like bumps and solder balls, and non critical domains like the mold compound, the PCB etc. These elements are considered perfectly glued. This feature enables to model in a single computation the first level (bumps) and second level (balls) interconnects with full visco-plastic behaviour inside large BGA packages.

The model is submitted to a static non linear analysis in plasticity and creep with small displacements. The loading consists of a succession of three thermo-mechanical cycles in uniform temperature as defined below:

— Initial temperature: 20°C,
— Maximum temperature: 100°C,
— Minimum temperature: –40°C,
— Ramp rate: 5°C.min⁻¹ (or 10°C.min⁻¹),
— Dwell time: 15 min (or 45 min),
— Number of cycles used for FEM evaluation: 3.

Bibliography doesn't propose relevant fatigue models for SnAgCu, whereas fatigue models are really the key of the fatigue life-time assessment by simulation. Therefore, a new model is proposed here to overcome the difficulty.

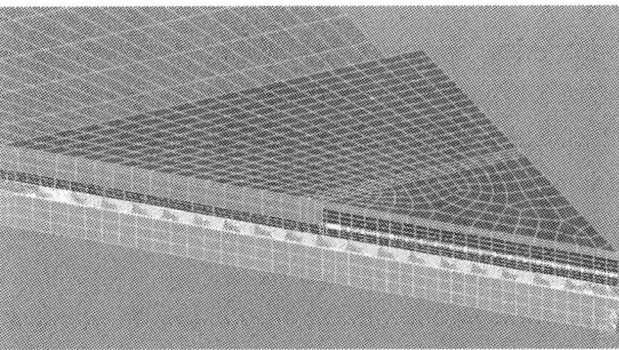

Figure 2: Mesh overview of the BGA 1508, 1-mm pitch

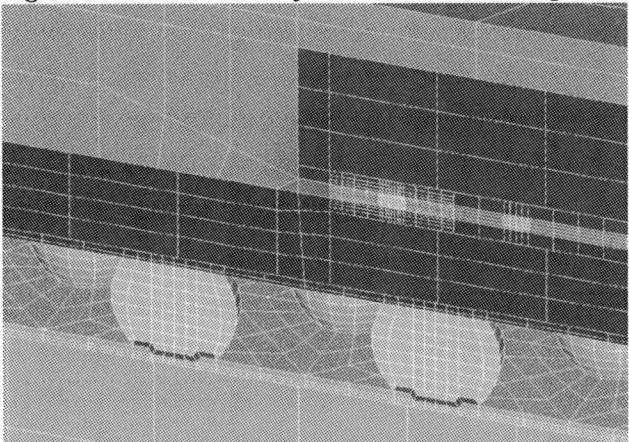

Figure 3: Closer view on the interconnects of the BGA 1508 (balls and bumps)

The assessment of the interconnect fatigue is made following a version of Coffin Manson's law modified by a correction factor in frequency. The law that is proposed for SnAgCu is the following (frequency f in Hz):

$$N_f(50\%) = \frac{1}{2} \left(\frac{f}{10^{-3}} \right)^{0.14} \left(\frac{\Delta\varepsilon_{eq}}{2.7} \right)^{-1/0.75}$$

It is based on raw fatigue test results published in [13], made with normalised test samples. Note that the tests have been made at 25°C for Sn3.9Ag0.6Cu (due once again to the lack of data on Sn3Ag0.5Cu). The modified Coffin Manson's law is plotted in the Figure 4, together with the experimental data from [13].

As for Sn3.9Ag0.6Cu, a frequency-modified Coffin Manson's law is used for Sn40Pb at 35°C. It has been published in [14]:

$$N_f(50\%) = \frac{1}{2} \left(\frac{f}{10^{-3}} \right) \left(\frac{\Delta\gamma_{eq}}{2 \times 0.945} \right)^{-1/0.52}$$

The strain range in both reliability laws will be obtained with the help of an automated post-treatment, by calculating the equivalent strain range $\Delta\varepsilon_{eq}$ or the shear strain range $\Delta\gamma_{eq}$ according to [15]. They are given by the following expression:

$$\Delta\gamma_{eq} = \sqrt{3}\Delta\varepsilon_{eq} = \left(3 \left(\Delta_{cycle} (\varepsilon_{zz} - \varepsilon_{therm}) \right)^2 + \left(\Delta_{cycle}\gamma \right)^2 \right)^{1/2}$$

where $\Delta_{cycle}x$ designates the range of a value x during a complete cycle.

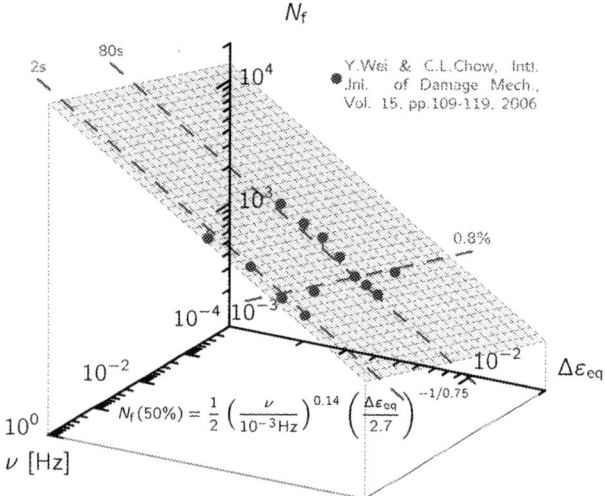

$$N_f(50\%) = \frac{1}{2}\left(\frac{\nu}{10^{-3}\,\mathrm{Hz}}\right)^{0.14}\left(\frac{\Delta\varepsilon_{eq}}{2.7}\right)^{-1/0.75}$$

Figure 4: Reliability law for Sn3.9Ag0.6Cu at 25°C

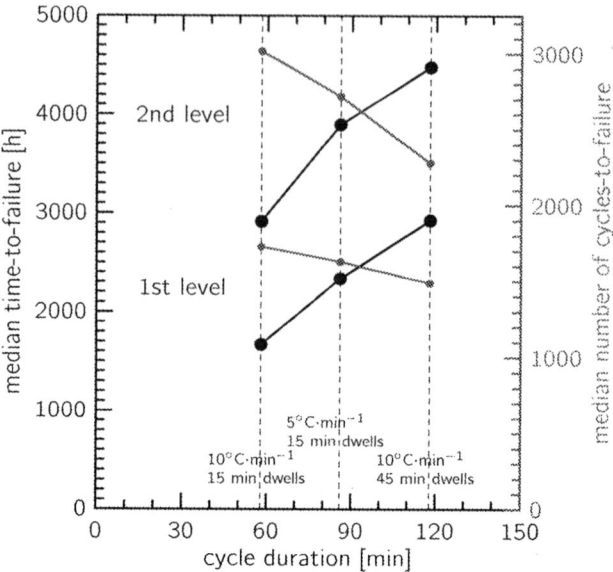

Figure 5: Time to failure versus cycles to failure with different accelerated test conditions at –40°C/+100°C for the BGA 400

5. Optimal accelerated test cycle in temperature

The objective here was to fix an optimal accelerated test cycle in temperature for the GEAMCOS project. It focuses on one package type: a BGA 400 and two temperature ranges (–40°C/+100°C and –30°C/+70°C, according respectively to aeronautic and military communication systems). In order to understand what happens to a solder joint, we consider the median time and cycles to failure of the critical (outer) solder ball/bump of the PBGA.

In the range –40°C/+100°C, three test conditions were computed (cf. Figure 5):

a. a rapid ramp of 10°C.min⁻¹ followed by a short dwell of 15 min,

b. a rapid ramp of 10°C.min⁻¹ followed by a long dwell of 45 min,

c. a slow ramp of 5°C.min⁻¹ followed by a short dwell of 15 min.

It results that the time to failure is the shortest with test condition (a), which is the most accelerated. The reason is that the decreased number of cycles to failures of (b & c) with respect to (a) is more than compensated by the long duration of a single cycle. The same consideration applies directly with the temperature range –30°C/+70°C. Therefore, experimental thermal tests of GEAMCOS will be done according to conditions (a).

6. Example of thermo-mechanical simulations to assess the reliability of PBGA 1508

Using the methodology described in paragraph 4, results are given here, as an example, for the PBGA 1508. The simulations results give first the location of the weakest interconnects. The weakest ball is at the outest corner of the package and the weakest bump is at the die corner (cf. Figure 6). This is somewhat surprising, as plastic BGA packages are known to have failures at the ball located vertically underneath the die corner. This may be explained by the low coefficient of thermal expansion of the overmold (Sumikon) given at 8 ppm·K⁻¹. The value is unusually low and it will be required to proceed to further measurements by Thermo-Mechanical Analysis.

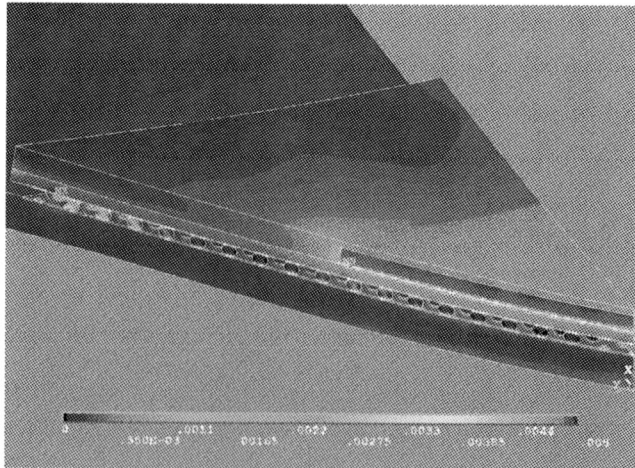

Figure 6: Total mechanical equivalent strain at the end of the 100°C dwell (displacements × 25) for the BGA 1508

In order to understand what happens to a solder joint, let us consider the stress/strain history in shear of the critical ball of the PBGA 1508 (cf. Figure 7). It appears that, for the second level of interconnects, the CTE of the PCB and the nature of the solder material (SnAgCu or SnPb) have a great impact on the mechanical behaviour and thus on the reliability of the assembly.

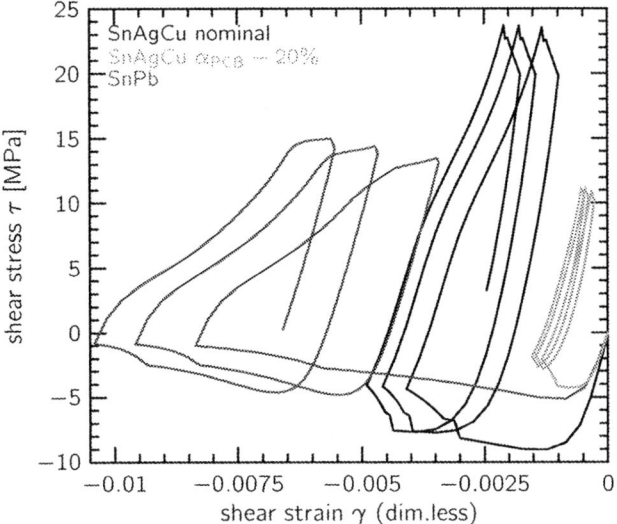

Figure 7: Critical ball of the PBGA 1508 in shear direction (–40°C/100°C)

More data are given in Table 6, which indicates the quantitative impact predicted when design or manufacturing parameters are varied. The parameter variation is as follows:

— coefficient of linear thermal expansion of the PCB increased or decreased by 20 %,
— pad area decreased or increased by 20 % (variations in diameter are 10 %), resulting in a change of the standoff height,
— standoff height decreased or increased by 20 %,
— change of alloy with SnPb instead of SnAgCu.

	N_f (50 %) [cycles] - SnAgCu			
	2nd level		1st level	
Nominal	4800		1600	
Parameter/Variation	*–20 %*	*+20 %*	*–20 %*	*+20 %*
Pad area (Ø ± 10 %)	4700	4900	1600	1600
Ball height	3800	5600	1600	1600
CTE of the PCB	12000	2200	1600	1600
	N_f (50 %) [cycles] - SnPb			
	2nd level		1st level	
Nominal	7200		3500	

Table 6: Median fatigue life-time of the PBGA 1508 (-40°C/100°C, 10°C.min^{-1}, 15 min dwell)

The comparison between SnPb and SnAgCu shows that the SnAgCu is predicted to be less reliable than the classical SnPb alloy with this component package.

The bumps are not affected at all by the changes of parameters: their fatigue life-time is fully determined by the interface between the die and the substrate. The role of the underfill is probably determinant.

For the balls, the most dominant impact is the CTE of the PCB. The mismatch between the CTE of the PCB and the CTE of the PBGA substrate is an important point to check but it may be quite difficult because the design of the PCB is purely determined by routing. The next important parameter is the standoff height. The higher the standoff, the longer the life-time. A change of the interconnect design (columns instead of balls for instance) may then bring a benefit to the reliability of the assembly.

7. Example of thermo-mechanical simulations to assess the reliability of a PQFP 208

The same FEM methodology was applied to a PQFP 208 component with 0.5-mm pitch. In this case, the largest displacements are also observed at 100°C at the end of the dwell. They are about 26 μm at the solder joints located in the corner of the package (cf. Figure 8).

A series of simulations were made on the PQFP 208 with modification of the CTE of the PCB and of the solder thickness.

For the postprocessing, views of stresses and strains were not sufficient to determine the position where the rupture is initiated. Reliability evaluations were then conducted in different areas of the corner solder joint (cf. red, yellow, green, blue and purple areas in Figure 9). The less reliable area was always located at the top of the joint. Therefore, all the reliability results (strain range, inelastic work density and median cyclic fatigue life) were extracted in this area (cf. Table 7). It means for example that the given number of cycles indicates a complete rupture in this area. The main conclusions of these results are the following:

— PQFP208 solder joints are very reliable (> 1000 cycles in the red area),
— the CTE of the PCB has a major impact on the reliability of this component and PCB must be carefully chosen,
— it is necessary to control the thickness of the solder joints,
— the predicted thermo-mechanical reliability with SnAgCu is higher than with SnPb in the temperature range –40 to +100°C.

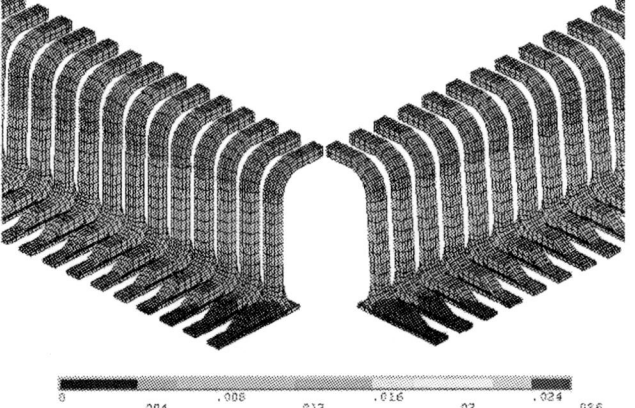

Figure 8: Isovalues of the displacement sums in mm at the end of the last dwell at 100°C

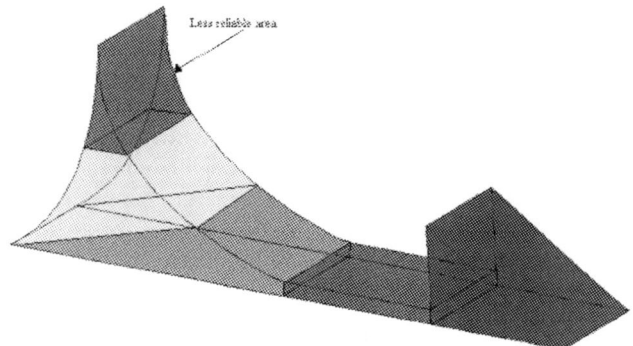

Figure 9: Area in red in which reliability results are given

	N_f (50%) [cycles] - SnAgCu	
Nominal	2300	
Parameter / Variation	*−20%*	*+20%*
CTE of the PCB	5100	1200
Solder thickness	1200	3700
	N_f (50%) [cycles] - SnPb	
Nominal	1500	

Table 7: Median fatigue life-time of the PQFP 208 (−40°C/100°C, 10°C.min⁻¹, 15 min dwell)

8. Synthesis of parametric thermo-mechanical FEM

As mentionned in the introduction, this FEM model was also applied to other BGA packages using the same methodology as for the PBGA 1508:

— a BGA CSP 84 with 0.5-mm pitch,
— a PBGA 400 with 0.8-mm pitch,
— a PBGA 896 with 1-mm pitch.

The same tendencies as with PBGA 1508 were obtained in terms of parameter variations for the PBGA 896, as illustrated in Figure 10.

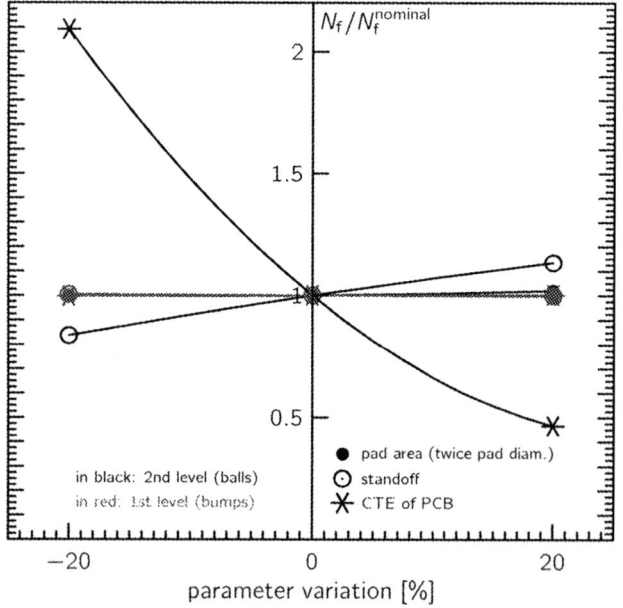

Figure 10: Sensitivity study of the critical ball of the PBGA 896 in shear directions (−40°C/100°C)

For the first level of interconnects (bumps), the Figure 11 shows a clear advantage of SnPb bumps. SnAgCu bumps are predicted to fail in about twice less time than SnPb bumps. The interconnect numbers as well as the parameters variations have no influence on the behaviour of the bumps. The reliability of the bumps is purely a property of the interface substrate/die/underfill.

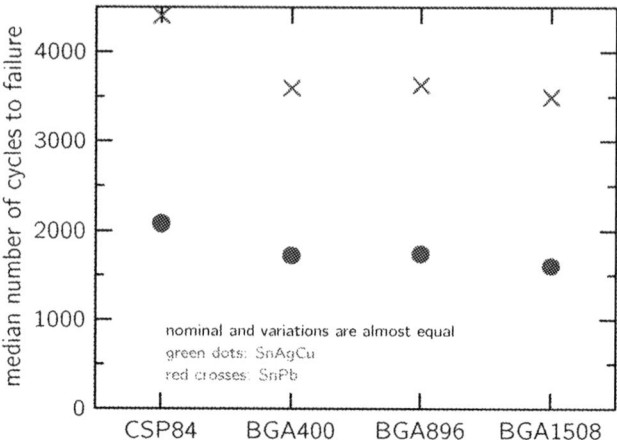

Figure 11: Calculated reliability of the first level of interconnects - SnAgCu versus SnPb (−40°C/100°C)

The analysis is more complex for the second level of interconnects (balls) due to pitch size and to the CTE mismatch (the CTE mismatch is increased for the CSP for which die and substrate have the same size). We can notice on Figure 12 that:

— the PBGA 1508 is predicted to fail earlier than the PBGA 896, both having 1-mm pitch.
— the component with the smallest pitch (CSP 84) is the weakest package.

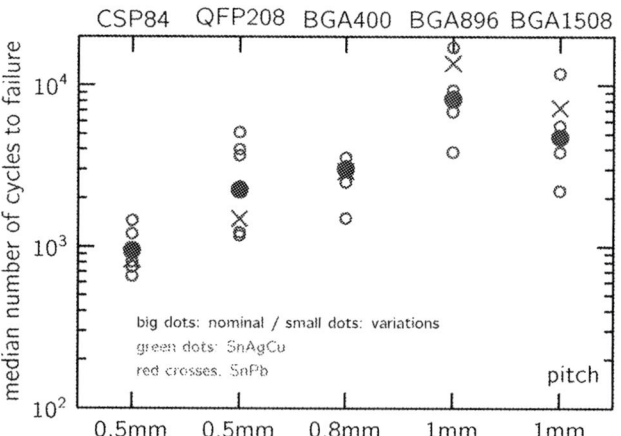

Figure 12: Calculated reliability of the second level of interconnects - SnAgCu versus SnPb (−40°C/100°C)

9. Construction of vibration FEM models

For vibration environments, another FEM model was constructed, taking into account component locations and weights, as well as the board fixture. As a first approximation, the global test board of the Figure 1 was simulated as explained below (cf. Figure 13):

— the PCB is modelled by its neutral fibre and by shell elements,
— components are modelled by punctual masses linked to the PCB with beams,

— the screws are located at the center and at the four corners of the PCB.

In this model, the density of the PCB was calculated considering the weight of the PCB without any components (2260 kg.m^{-3}). Only linear properties of materials are considered in such dynamic simulations and the damping coefficient was arbitrarily chosen equal to 2 %. Inputs like the PCB Young's modulus and the damping coefficient need to be calibrated with experimental data. Nevertheless, they already help to define an adequate test protocol for GEAMCOS.

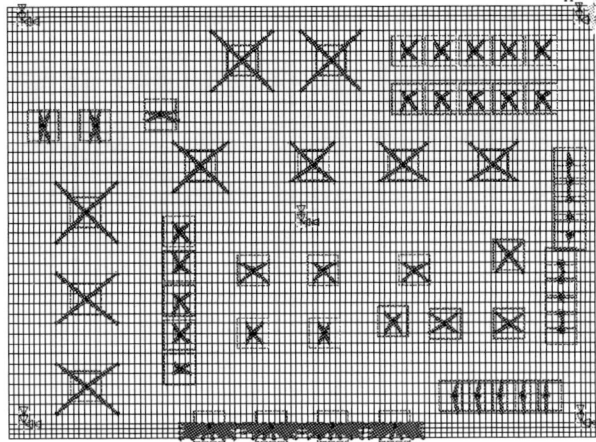

Figure 13: Model for vibration simulations

10. Optimal test protocol in vibration

The experimental tests in vibration were also directed by FEM results. The determination of harmonic modes and the response of the board to a sine excitation of 1 g determine:

— the most strained components (to follow with laser velocimetry during robustness and fatigue vibration tests),

— their response in g to such an excitation (which indicates the level of acceleration in g r.m.s (root mean square) to apply for rigidity tests to obtain a critical level of 80 g on the board).

In our case with FEM, the most strained components are observed for the harmonic mode n°2 (90 Hz) (cf. Figure 14) or for the mode n°4 (100 Hz).

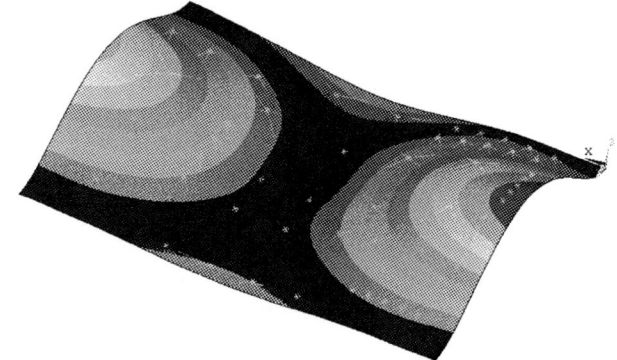

Figure 14: Shape of the harmonic mode n°2

Their response to 1 g in sine is about 20 g and indicates that the critical level of 80 g will be obtained with 4 g in input (with the hypothesis of the linearity of the dynamic behaviour).

Considering these results, we have decided to apply the protocol of test described in Table 8 for GEAMCOS and to point the laser beam on the 4 points specified by a white square in Figure 15.

	Modal analysis	Robustness	Fatigue
Excitation type and level	Sine 1 g	Sine 4 g	Random vibration with aeronautic profile
Frequency range [Hz]	50 to 500	50 to 500	To be defined
Scan duration	10 min	40 min	5 hours
Boards and their finishes (excitation axis)	1 in NiAu (x, y, z)	3 in NiAu (x, y, z axis) 1 in Sn (z axis)	3 in Sn (x, y, z axis) 1 in NiAu (z axis)

Table 8: Protocol of experimental tests in vibration

Figure 15: Test bench for vibration analysis (at Emitech)

11. Conclusions and perspectives

The large panel of simulations presented in the report gives a first insight of the thermo-mechanical life-time reliability of electronic assemblies with lead-free solder joints. Notwithstanding that material data for SnAgCu published so far is quite limited, the report demonstrates that these data lead to quite satisfactory results, when compared with what is experimentally known from lead-free assemblies. All BGA, CSP and PQFP models give relevant life-time predictions, that will be confronted with future test results.

The CSP and the PQFP packages should have a longer time-to-failure with SnAgCu solder joints than with classical SnPb joints. The contrary is expected for the 1st and 2nd level interconnects of large BGA packages, where SnPb should still be more reliable. But in any case, the factor in life-time is found to remain less than about 2.

The parameter variations show without surprise that the 2nd level reliability is strongly impacted by the relative coefficient of thermal expansion of the PCB to the component. The solder joint height has also a noticeable impact. Yet the first level reliability is not impacted by any of the parameter changes.

Simulations with various temperature ramp rates and dwell times have led to the choice of an optimal temperature cycle for accelerated testing. It results from a compromise between a high acceleration factor, a correct level of stress relaxation at high temperature and a good heat transfer until temperature dwells are effectively reached. In fact, the cycle chosen is a classical cycle for SnPb assemblies (15-min dwell time, $10°C.min^{-1}$ ramp rate) and thus, allows comparisons.

Dynamic models were also constructed to determine an adequate protocol of test in vibration. It has permitted to select an adequate excitation level for robustness and fatigue tests. The most strained components of the forward board were identified and followed by laser velocimetry during experimental tests.

This part of GEAMCOS project shall go on in 2007-2008 with an experimental validation of these simulations. This will require to identify a new constitutive material model for Sn3.0Ag0.5Cu and subsequently to calibrate the simulations with all experimental test results. All the analyses will be reviewed then.

Acknowledgments

This work was supported by European commision. The authors also wish to thank Marc Grieu, PhD thesis student at EADS France involved in vibration simulations.

References

1. JCAA/JG-PP Lead-Free Solder Project Joint Test Report, http://acqp2.nasa.gov/JTR.htm, downloaded on January 2007.

2. Jeomshik Yang, Se-Hyeong Lee, Dong Nyung Lee, "Analysis of bending of cross sections of tensile tested electrodeposited copper foils", *Material Science & Engineering A*, Vol. 234–236, pp. 149–153, 1997.

3. J.A. King, Materials Handbook for Hybrid Electronics, J.A. King Editor, Artech House, Norwood (USA), ISBN 0 89006 325 7, 610 pages, 1988.

4. Sumitomo Bakelite Co. Ltd, "Sumikon, EME–G770", *product datasheet* on multi-aromatic resin, Br/Sb free pkg, low warpage, downloaded on March 2006, www.idt.com/products/files/8773/A-0305-02.pdf .

5. Motorola Semiconductor, "Plastic Ball Grid Array (PBGA)", *application note, ref. AN1231/D,* review 2, April 1996.

6. Xilinx Inc., "Package Datasheet - FF1152", document without reference, 3 pages, not dated.

7. Isola Group, "BIS400 IS420/1, Temperature resistant middle Tg and high Tg base materials with low z-axis expansion", *datasheet reference* 800384, 6 pages, November 2005.

8. John H.L. Pang, and B.S. Xiong, "Mechanical Properties for 95.5Sn3.8Ag0.7Cu Lead-Free Solder Alloy", *IEEE Trans. on Components and Packaging Technologies*, Vol. 28., No. 4, pp. 830–840, December 2005.

9. A. Schubert, R. Dudek, E. Auerswald, A. Gollhardt, B. Michel, H. Reichl, "Fatigue Life Models for SnAgCu and SnPb Solder Joints Evaluated by Experiments and Simulation", *53rd ECTC conference*, pp. 603-610, 2003.

10. Veronique Audigier, "Étude du comportement en fatigue des joints brasés de composants électroniques montés en surface - Impact sur la fiabilité des assemblages", PhD thesis, ENSAM, No d'ordre 96.17, July, 27th, 1996.

11. International Tin Research Institute, "Solder Alloy Data - Mechanical Properties of Solder and soldered Joints", ITRI publication N. 656.

12. Robert Darveaux, "Effect of Simulation Methodology on Solder Joint Crack Growth Correlation", *50rd ECTC conference*, pp. 1048–1058, May 2000.

13. Y. Wei, C.L. Chow, "Isothermal Fatigue Damage Model for Lead-Free Solder", *International Journal of Damage Mechanics*, Sage Publications, Vol. 15, pp. 109-119, April 2006.

14. Harvey D. Solomon, "Fatigue of 60/40 solder", *IEEE Transactions on Components, Hybrids, and Manufacturing Technology,* Vol. CHMT–9, No.4, pp. 423–432, December 1986.

15. Julie A. Bannantine, Jess J. Comer, and James L. Handrock, Fundamentals of Metal Fatigue Analysis, Prentice-Hall, 273 pages, 1990.

Thermomechanical Multiscale Modelling of Substrates

R.L.J.M. Ubachs[a,b], O. van der Sluis[a,b], W.D. van Driel[a,c], and G.Q. Zhang[a,c]

[a]Department of Precision and Microsystem Engineering, Delft University of Technology, the Netherlands
[b]Philips Applied Technologies, High Tech Campus 7, 5656 AE Eindhoven, The Netherlands
[c]NXP, IMO-BE Innovation BY1.055 Gerstweg 2, 6534 AE Nijmegen, The Netherlands
Corresponding author: O. van der Sluis, olaf.van.der.sluis@philips.com

Abstract

For the thermomechanical analysis of mutlilayer substrates, detailed FE is not feasible. To capture all features, the number of elements necessary would lead to extreme computational loads. To overcome this, an elementwise homogenization method utilizing multiple representative volume elements is presented. Employing a global-local step it is still possible to obtain accurate values for local quantities of interest. The approach is demonstrated for thermal and mechanical problems. For the mechanical case the procedure is validated by a confrontation with experimental results showing the ability of the technique to indicate potential problem areas using the global/homogenized step and predict failure sites using the local step.

1 Introduction

The ongoing miniaturization in the microelectronics industry also has its implications for the substrates. Currently, multilayer substrates are in use that consist of alternating dielectric and metallization layers. Due to the reduction in thickness of these layers and the decrease of track pitch, track width, and feature size, reliability has become more of an issue. Differences in coefficients of thermal expansion of the various materials used can lead to warpage of a substrate. Not only can warpage of the substrates adversely affect chip adhesion, substrate cracks have been observed as a result of thermomechanical cycling, occasionally propagating through the copper interconnecting lines, see Fig. 1.

Performing accelerated tests on substrates to identify possible failure sites is a time consuming and expensive process. Here, numerical tools would be able to provide a significant benefit to the substrate design and analysis process. Not only can these methods be used to predict the thermal and mechanical behaviour, they may also serve to provide a better understanding of the mechanisms involved. However, using detailed finite element (FE) models results in extreme computational loads due to the amount of detail that needs to be modelled, making this approach unfeasible.

To overcome this, instead of detailed modelling, homogenization techniques can be used. These methods reduce both the computational costs and the hardware demands. A drawback, however, is that local failure can not be captured straightforwardly using homogenized FE calculations, necessitating a multiscale approach.

Multiscale simulations can be done either sequentially or simultaneously. In the sequential approach (global-local) a coarse/homogenized FE calculation is performed of the entire board using apparent material properties. Potential problem areas are then located and detailed FE models of only those regions are used to calculate the local field variables, using boundary conditions derived from the global simulation.

Examples of simultaneous approaches are adaptive remeshing, element overlay techniques, and generalized FEM. With adaptive meshing the mesh is locally refined every increment based on some criterion. For overlay techniques, locally a fine mesh is superimposed on top of the coarse global mesh [2].In generalized FEM, locally elements are enriched with additional degrees of freedom to capture the local field variables more accurately.

In this contribution a global-local FE modelling technique is demonstrated that is able to evaluate the thermal and mechanical behaviour of an entire substrate and yield accurate values for local quantities, such as stresses and strains, when necessary. The proposed multilevel approach comprises of a global homogenized calculation, followed by a localization step. The global calculation allows for the determination of critical regions. These regions are subsequently modelled in detail to evaluate the local quantities of interest.

2 Modelling approach

Standard homogenization techniques deal with two length scales: the macro and the micro scale. The material supposedly behaves homogeneous on the macro scale, but heterogeneously on the micro scale. It is assumed that a representative unit cell can be defined whose response will equal the response of the macro scale. This approach works well for periodic and statistically homogeneous materials. However, the substrates in question are neither. Every layer has different average orientations and volume fractions of composing materials, and even in one layer the heterogeneity is great.

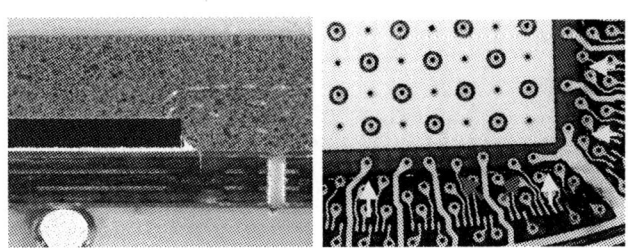

Figure 1: Cracks in substrates.

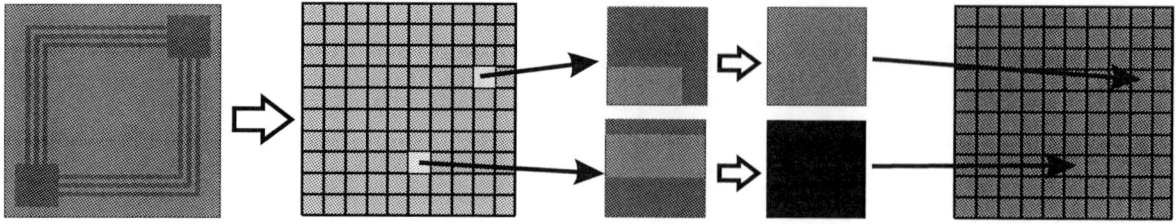

Figure 2: Schematic of the modelling procedure.

The problem of spatial variations of the volume fraction is also encountered in functionally graded materials (FGMs) [1, 7]. Homogenization of these types of materials are performed by assigning different effective properties to regions whose compositions deviate too much from each other. FGMs usually deal with a matrix material and a some kind of filler, eg. particles or fibres, whose volume fraction varies throughout the product. A series of similar unit cells can be defined that are statistically similar and only differ in volume fraction of the filler material/particles. For a substrate –viewing the glass fibre laminate as the matrix material and the metal circuit as the inclusions– the issue is more complex. There are various types of inclusions, and their relative size, compared to macroscopic dimensions– is typically larger then in the case of FGMs. This necessitates the choice of multiple unit cells with significantly different statistical properties.

In [3] and [4] this problem is tackled by defining unit cells for each feature present in the substrate. They then proceed to obtain effective properties for each feature. Subsequently, the substrate is divided into regions, each containing several features. Next, an inventory is taken of the features per region, taking into account orientation and location of the features, and the effective properties of the regions is determined. Finally, the homogenized regions are assembled back into the substrate and a macro simulation can be preformed.

The homogenization approach adopted here, uses not a single, but several unit cells that are characteristic for various features/structures found in the substrate [6].

2.1 Apparent properties

The homogenization procedure is implemented for solving mechanical and steady state thermal problems. The governing equations read

$$\vec{\nabla} \cdot \boldsymbol{\sigma} = \vec{\nabla} \cdot (\mathbb{C} : \boldsymbol{\varepsilon}) = 0, \tag{1}$$

$$\rho C_p \frac{\partial T}{\partial t} = -\vec{\nabla} \cdot \vec{q} = \vec{\nabla} \cdot \left(\boldsymbol{k} \cdot \vec{\nabla} T \right) = 0, \tag{2}$$

where $\vec{\nabla}$ is the gradient operator, $\boldsymbol{\sigma}$ is the Cauchy stress tensor, $\mathbb{C}$ the fourth order stiffness tensor, $\boldsymbol{\varepsilon}$ the strain tensor, ρ density, C_p heat capacity, $\vec{q}$ heat flux, T temperature, t time, and $\boldsymbol{k}$ thermal conductivity.

The required apparent properties then are $\bar{\mathbb{C}}$ and $\bar{\boldsymbol{k}}$, which are defined as follows:

$$\langle \boldsymbol{\sigma} \rangle = \int_\Omega \mathbb{C} : \varepsilon \, \mathrm{d}\vec{x} = \bar{\mathbb{C}} : \langle \varepsilon \rangle, \text{ with } \langle \varepsilon \rangle = \int_\Omega \varepsilon \, \mathrm{d}\vec{x} \tag{3}$$

$$\vec{\nabla} \cdot \left(\bar{\boldsymbol{k}} \cdot \langle \vec{\nabla} T \rangle \right) = 0, \text{ with } \langle \vec{\nabla} T \rangle = \int_\Omega \vec{\nabla} T \, \mathrm{d}\vec{x} \tag{4}$$

To calculate the apparent properties of a unit cell, several FE simulations are performed, subjecting the cell to various loading conditions. Using Eq.(3) the apparent properties can then be calculated. For the simulations periodic boundary conditions are imposed.

Also, for the thermal apparent properties, the thermal circuit model has been used to obtain them analytically. In the thermal circuit model each component (region of material) is replaced by an equivalent resistor R. The total resistance of components in series can be found by summation, $R = \sum R_i$, and for components in parallel by $R = \left(\sum 1/R_i \right)^{-1}$. Acquiring the effective properties analytically has the advantage that if some material parameter changes only the expressions need to be reevaluated. If FE calculations are used, all unit cell simulations have to be repeated. It was found that the results obtained employing FEM and those employing the circuit model were comparable.

2.2 Procedure

For the homogenization step, schematically depicted in Fig. 2, each of the layers is considered separately. The layers are meshed without accounting for the details of the circuitry. Then, each element is analysed, extracting relevant structure information, i.e. for a metallization layer, the volume fraction of copper, the anisotropy ratio and the averaged orientation of the copper circuit. Next, the appropriate type of unit cell is identified for the element and the apparent properties are assigned taking into account the orientation and the volume fraction. In this way a elementwise homogenized model for the entire substrate is build. For more information on the procedure see [6].

3 Thermal homogenization

To validate the thermal homogenization model, results obtained using the homogenization procedure are compared with a reference calculation. Two square unit cell types are defined, one containing a copper trace and one containing a square copper pad embedded in epoxy (Fig. 3). Only the diagonal components of the apparent conductivity tensor are assumed to be nonzero. Employing the thermal circuit model to obtain the apparent conductivity yields for the 'trace':

$$\bar{k}_{11} = [k_c h_t + k_e (h - h_t)] \, h^{-1}, \tag{5a}$$

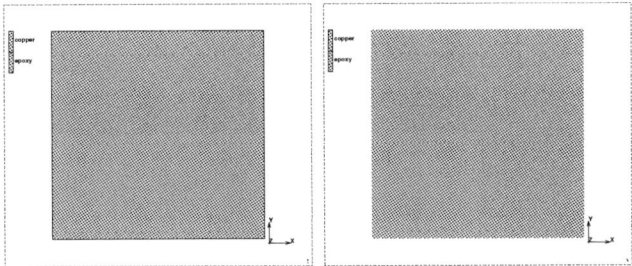

Figure 3: Unit cells for thermal homogenization, orange denotes copper.

$$\bar{k}_{22} = \frac{k_c k_e h}{k_e h_t + k_c (h - h_t)}, \tag{5b}$$

$$\bar{k}_{33} = \left[k_c h_t + k_e (h - h_t) \right] h^{-1}, \tag{5c}$$

and for the 'pad':

$$\bar{k}_{11} = w \left[\frac{h w_t}{k_c h_t + k_e (h - h_t)} + \frac{(w - w_t)}{k_e} \right]^{-1}, \tag{6a}$$

$$\bar{k}_{22} = h \left[\frac{w h_t}{k_c w_t + k_e (w - w_t)} + \frac{(h - h_t)}{k_e} \right]^{-1}, \tag{6b}$$

$$\bar{k}_{33} = \left(w_t k_c h_t - w_t k_e h_t + k_e w h \right) (w h)^{-1}. \tag{6c}$$

Here h and w are the heights and widths of the unit cell and the feature (subscript t) and the subscripts e and c denote values for epoxy and copper respectively.

The reference calculation is performed using a detailed model of a simple substrate containing only one signal layer, Fig. 4. The substrate is loaded by applying 1 W of power to the lower left copper pad and 0.2 W to the pad in the center. A natural convection of 10 Wm^{-2}K^{-1} is assumed on the top and bottom surface, with the surroundings at 0°C. The thermal conductivity of Cu and epoxy have been taken to be 400 W(mK)$^{-1}$ and 0.2 W(mK)$^{-1}$ respectively. Next, homogenized calculations are performed with equal boundary conditions.

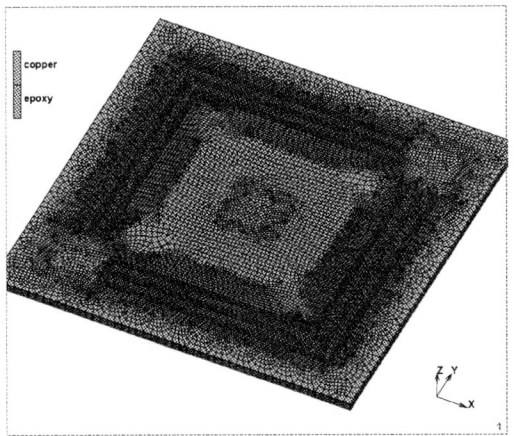

Figure 4: Detailed finite element model, using 145,194 elements.

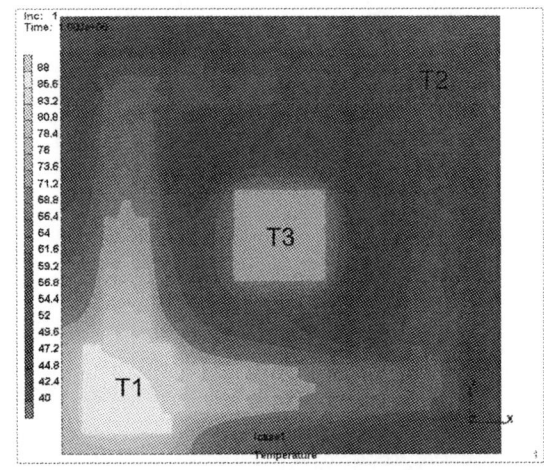

(a) CPU time: 46.81 s, $T1 = 86.3$°C, $T2 = 53.8$°C, $T3 = 73.8$°C.

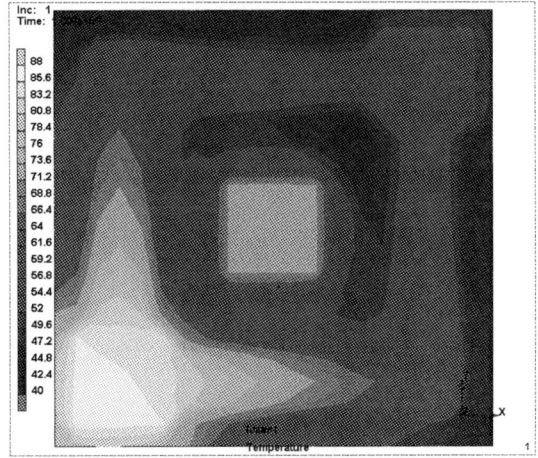

(b) 800 elements, CPU time: 0.42 s, $T1 = 86.0$°C, $T2 = 52.7$°C, $T3 = 71.5$°C.

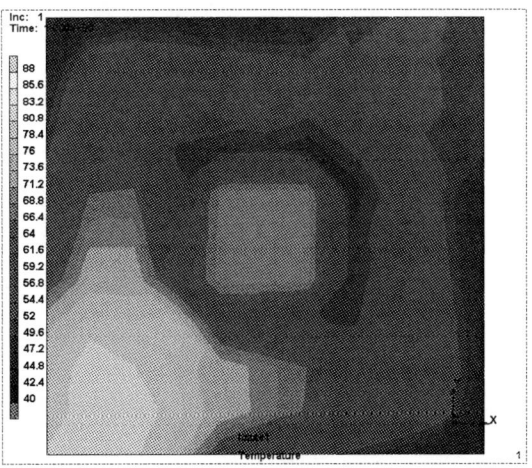

(c) 200 elements, CPU time: 0.14 s, $T1 = 82.5$°C, $T2 = 52.0$°C, $T3 = 69.9$°C.

Figure 5: Temperature distributions obtained by detailed FE modelling (a) and the homogenization procedure (b) and (c).

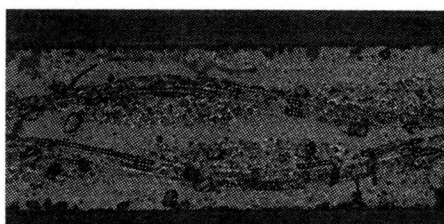

Figure 6: Side view in 11-direction of the substrate showing the glass fibres.

Fig. 5 shows the steady state temperature profiles obtained with the reference calculation and with two homogenized calculations. Since for the homogenized calculations the traces and pads do not need to be captured in detail, the number of necessary elements reduces drastically. This then leads to calculation times which are orders of magnitude lower in comparison. Furthermore, comparing local values indicates that the homogenized solution yields accurate results. The homogenized substrate containing 800 elements yields a maximum error for the three sample points, chosen in the center of the copper pads, of only 3%, while that containing just 200 elements introduces an error of 5%.

4 Mechanical homogenization

In this section results from the homogenization procedure are discussed for mechanical problems. For validation of the approach four point bending experiments have been performed on a number of substrates. The tests are

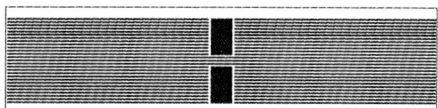

Figure 7: Circuit layout of both signal layers of sample A. No vias in the dielectric layer.

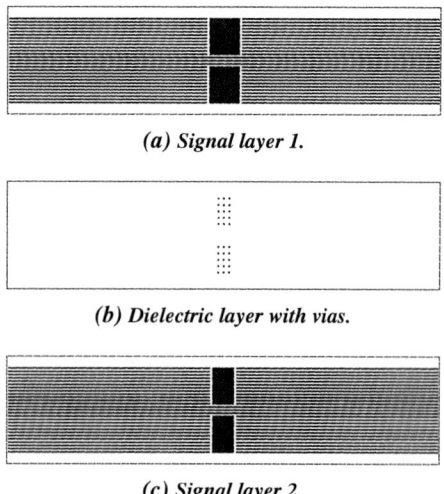

(a) Signal layer 1.

(b) Dielectric layer with vias.

(c) Signal layer 2.

Figure 8: Circuit layout of layers of sample B.

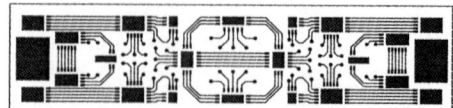

(a) Signal layers 1 and 2.

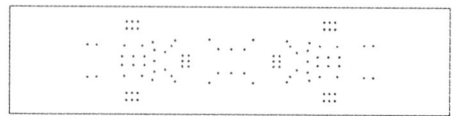

(b) Dielectric layer with vias.

Figure 9: Circuit layout of layers of sample C.

performed using a tensile stage with a 20 N load cell. The outer span of the bending setup is 20 mm, while the inner span is 10 mm.

The specimens consist of two Cu signal layers, separated by a dielectric core layer. The core material is FR4 with two woven glass fibre mats, see Fig. 6. The glass fibres are directed along the length, the 11-direction, of the specimen and perpendicular to it, the 22-direction. The core material will therefore exhibit orthotropic behaviour. The thickness of the dielectric layer was measured to be 210 μm and of the Cu plating 474 μm. The length and width of the specimens are 40 mm and 10 mm, respectively.

Three sample substrates with varying circuit layouts have been used, see Figs. 7–9. They were all subjected to bending deformation until failure occurred. For each of the samples a typical load-deflection curve is shown in Fig. 10. Failure occurred at identical locations for all specimens of a sample, see Fig. 11.

Tensile tests have not been used as the clamping of the specimens was expected to influence the measurements.

4.1 Parameter identification

First, the material parameters necessary for the simulations were determined. For this purpose four point bending

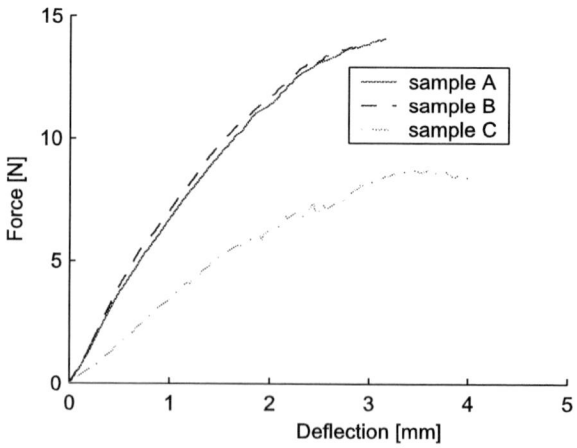

Figure 10: Experimental force-deflection curves for the three samples.

88

Figure 11: Typical fractures of the test samples.

tests are performed on bare substrates, without Cu, and full layered substrates, with Cu plating on both sides of the dielectric layer. Fig. 12 shows the resulting load-deflection curves for the two types of specimens.

From these measurements the Young's moduli of the plated Cu and the FR4 material are determined. This is done by solving the the four point bending problem analytically. First however, FE calculations –with estimated parameters– where performed to determine if either the beam bending or plate bending theory are applicable in this case. It was found that for the thickness-width ratio of the test samples used, the FE solution approached the plate bending theory solution closest, indicating its use to be appropriate.

The classical plate equation reads

$$\nabla^2 \left[D(x,y)\nabla^2 w(x,y) \right] = f(x,y), \tag{7}$$

$$\text{with } D = \int_{-t/2}^{t/2} \frac{Ez^2}{(1-\nu^2)} \, \mathrm{d}z.$$

Here E denotes the Young's modulus, ν the Poisson's ratio, I the momentum of inertia, w the deflection, f the applied

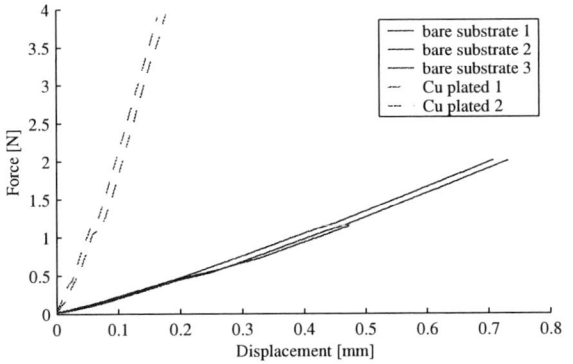

Figure 12: Load-displacement (of the upper span) curves for the 4 point bending tests of the bare substrate and the Cu plated substrate.

Table 1: Confrontation between experimental and simulation results.

| Sample | Bending stiffness [N] | | Error |
	Experiment	Simulation	[%]
A	7.92211	8.46184	6.81
E	8.42628	9.75263	15.7
F	3.75352	3.51184	6.44

force, and D the flexural rigidity of the plate.

Integration and application of the relevant boundary conditions then yields –for the case of the Cu plated specimens– the displacement of the upper span as

$$w|_{x=a} = \frac{P(4a^3 - 3a^2L)}{12 \left(\frac{2E_cI_c}{1-\nu_c^2} + \frac{E_eI_e}{1-\nu_e^2} \right)}. \tag{8}$$

Here P is the total applied force, L is the outer span width, and a the position where the inner span contacts the specimen. The subscripts c and f denote the Cu and the FR4, respectively.

For the Poisson's ratios values from the literature have been taken, 0.3 for Cu and 0.4 for FR4. The resulting values for the Young's moduli are 20.44 GPa for FR4 and 106.5 GPa.

4.2 Simulation

Using the above values, unit cell calculations have been performed to obtain apparent properties necessary for the homogenized models. The same unit cells, with varying volume fraction, as for the thermal simulations have been used, see Fig. 3.

After analysis of the samples to obtain the geometry parameters (volume fraction, orientation, and anisotropy) the homogenized models can be assembled. Because of symmetry conditions only one quarter of the samples needs to be modelled. Table 1 shows a comparison between bending stiffnesses obtained experimentally and using simulations. With errors under 10% for samples A and C, the homogenization procedure is considered to perform very well. For sample B a larger error is found, however, the method does predict the correct trend yielding the highest bending stiffness for this sample. Furthermore, the solution might still be acceptable, depending on the requirements.

Resulting strain energy and major stress distributions in the top right quarter of the samples are shown in Figs. 13–15. As a result of the homogenization, the field variables are averaged over each element. Although the Cu circuit can be recognized from the stress field, no conclusion can be made about the local values. However, the homogenized solution can be used to indicate possible problem areas in the substrate. The stress distribution is not considered a suitable measure for this because of the large difference in elastic properties between Cu and FR4/epoxy. The strain energy would be a better indicator. Comparison with experimental results shows that the failure sites coincide with regions exhibiting elevated strain energy values.

4.3 Local solution

It was shown above that the homogenization procedure can be used to indicate potential problem areas in the substrates. However, no quantitative data on local field variables is obtained. Employing a global-local approach this data can be obtained. After a potential problem area has been identified with the homogenized model, a detailed FE model of the region is build and a local simulation is performed. The boundary conditions of this local simulation are taken from the global/homogenized calculation. Fig. 16 shows the stress field for the two regions denoted in Fig. 13. The displacements of the outer edges are prescribed by interpolation of the displacements found using the homogenized calculation. Because of the required interpolation, results at the prescribed boundaries are unreliable. However, at some distance from these boundaries, this effect is not noticeable any more, as becomes clear from Fig. 16, away from the edges the solutions of both regions match. Furthermore, the position with the highest stress predicted by the local simulations coincides with the experimentally found fracture site.

4.4 Failure simulation

Simulations have also been performed trying to capture the substrates failure. For this purpose sample A is modelled again using the homogenization approach, but the region where fracture occurred has been modelled in detail. The two nonconforming meshes have been coupled using the 'insert' option in MSC.Marc-Mentat, see Fig. 17. Failure is assumed to occur when a maximum stress limit is reached. If this happens the elasticity modulus at that point is taken as 10% of the initial value. For Cu the maximum tensile stress is taken at 150 MPa, and for FR4 30 MPa. Fig. 18 shows the result from this simulation. Initial failure occurs at the traces spanning the entire length of the specimen after which the crack goes through the FR4 material between the Cu pad and the traces. This is consistent with the experimentally observed crack path.

5 Conclusion

An elementwise homogenization technique has been demonstrated to evaluate the thermal and mechanical behaviour of multilayer substrates. The approach uses mul-

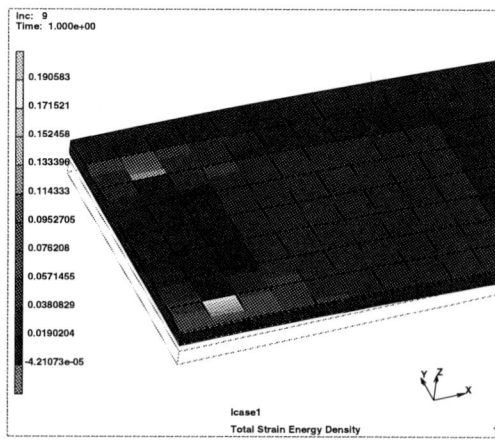

(a) Major stresses [MPa].

(b) Strain energy [MJ/m^3]

Figure 13: Sample A under 4 point bending, inner span displacement 0.3 mm.

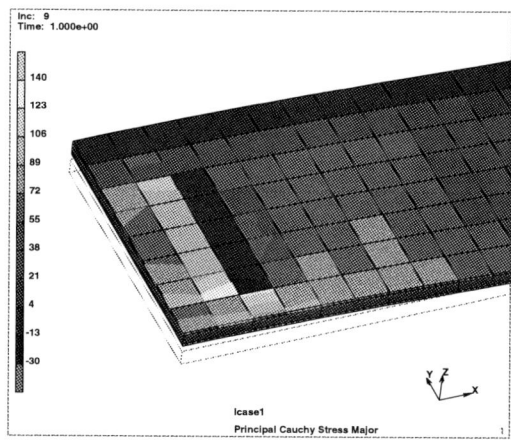

(a) Major stresses [MPa].

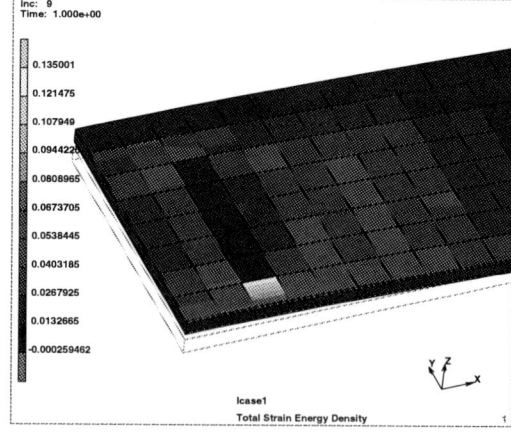

(b) Strain energy [MJ/m^3]

Figure 14: Sample B under 4 point bending, inner span displacement 0.3 mm.

tiple unit cells to capture the apparent properties of different regions of the substrate, accounting for the orientation, anisotropy, and volume fraction of Cu. It is shown that the approach significantly reduces the computational effort necessary to evaluate the behaviour of a substrate.

A comparison of global mechanical properties, e.g. bending stiffness of the substrates, with experimentally obtained results showed a good agreement. For typically local properties, e.g. stress, satisfactory qualitative results are found, allowing the homogenized results to be used to identify potential problem areas. These ares are then further investigated employing a global-local step to obtain accurate local quantities of interest. The procedure accurately predicts the critical region in the substrate; the highest local stresses are found at the positions where fracture is seen to occur experimentally. Finally, employing a mixed homogenized-detailed FE model and applying a maximum stress failure criterion, also the correct crack path is predicted.

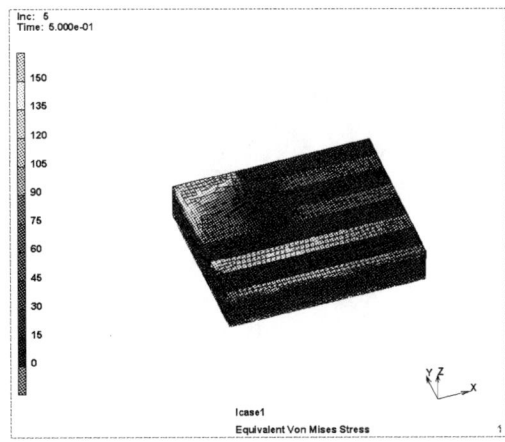

(a) Region I

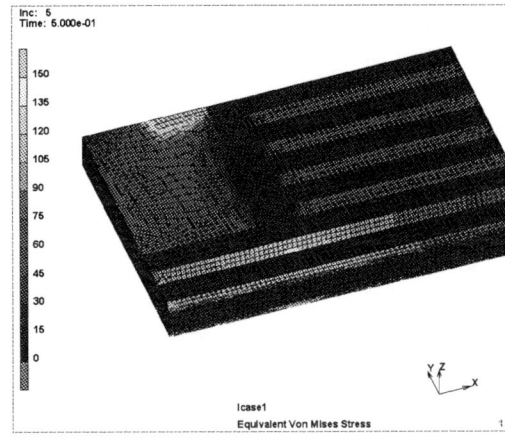

(b) Region II

Figure 16: Local calculations of the critical region of sample A.

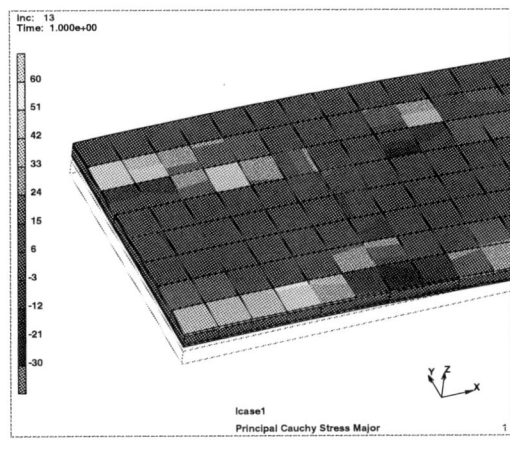

(a) Major stresses [MPa].

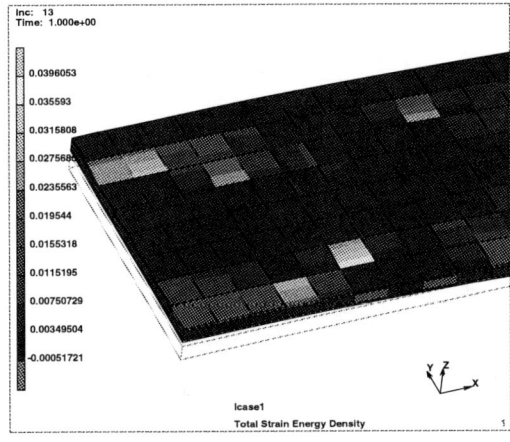

(b) Strain energy [MJ/m³]

Figure 15: Sample C under 4 point bending, inner span displacement 0.3 mm.

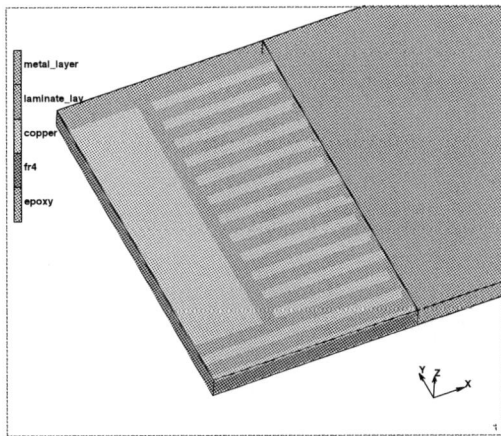

Figure 17: Combined detailed-homogenized mesh for sample A.

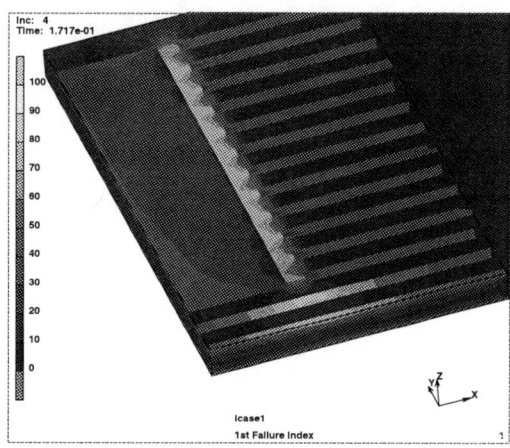

Figure 18: Failure index [%] after a four point bending simulation of sample A. A value of 100% means the failure criteria has has been reached.

References

[1] J.R. Cho and D.Y. Ha. Averaging and finite-element discretization approaches in the numerical analysis of functionally graded materials. *Materials Science snd Engineering A*, 302:187–196, 2001.

[2] J. Fish. The *s*-version of the finite element method. *Computers and Structures*, 43(3):539–547, 1992.

[3] J.L. Grenestedt and P. Hutapea. Influence of electric artwork on thermomechanical properties and warpage of printed circuit boards. *J. of Applied Physics*, 94(1): 686–696, 2003.

[4] P. Hutapea, J.L. Grenestedt, M. Modi, M. Mello, and K. Frutschy. Prediction of microelectronic substrate warpage using homogenized finite element models. *Microelectronic Engineering*, in press, 2006.

[5] S.H. Lee, J.H. Song, Y.C. Yoon, G. Zi, and T. Belytschko. combined extended and superimposed finite element method for cracks. *Int. J. for Numerical Methods in Engineering*, 59:1119–1136, 2004.

[6] R.L.J.M. Ubachs, O. van der Sluis, W.D. van Driel, and G.A. Zhang. Multiscale modelling of multilayer substrates. *Microelectronics and Reliability*, 46(9–11): 1472–1477, 2006.

[7] K. Vemaganti and P. Deshmukh. An adaptive global–local approach to modeling functionally graded materials. *Computer Methods in Applied Mechanics and Engineering*, 195:4230–4243, 2006.

Creep Measurements of 200 μm – 400 μm Solder Joints

Mike Röllig, Steffen Wiese, Karsten Meier, Klaus-Jürgen Wolter
Dresden University of Technology
Electronics Packaging Laboratory, IAVT
Email: roellig@avt.et.tu-dresden.de, phone: +49 351 463 36594

Abstract

For the last decades, many mechanical measurements on solder alloys were carried out. As a matter of fact, the microstructure of the solder materials is affected by their compositions. In addition, external variables like the reflow cooling rate, solder volume, thermal mass of the package and pad metallization may have an influence. For those reasons the discrepancies of creep measurements on solder contact specimen are larger than on tensile samples.

A motivation for the creep measurement activities is the lifetime prediction of electronic components, which have solder joints for electric-mechanical connection on their interposer or printed circuit board. Structure-mechanic simulation tools like the FEM can calculate the mechanical interactions between the assembled materials of such complete packages. Often, the solder joints are the weakest participants in the whole assembly and determine the total lifetime. Nevertheless, every simulation is highly dependent on the material laws. Therefore, the FEM needs an accurate fatigue model and a precise material model for the lifetime prediction of this solder.

The paper presents a new experimental design for measuring the creep behaviour of area arrayed solder bumps in different sizes of various packages. It focuses on the feasibility of the measurement of industrial manufactured FC, CSP or BGA packages. First measurements were accomplished on solder bumps with 200μm and 400μm diameter.

The test setup works by cyclic reversible shear force initiation into solder joints. It operates in the temperature range between T = [20...125] °C. High-resolution force adjustment and displacement measurement enables a steady state strain rate measurement range of $[10^{-2}...10^{-8}]$ 1/sec. Industrial demands for introducing the new SnAgCu base solders required a concentration on various high Sn-based alloys.

Introduction

One tendency in lifetime-prediction of electronic packages under thermo-mechanical load is the implementation of FEM-Simulation. The use of simulations has accelerated the ability to generate conclusion about acceptability of new package concepts and about reliability of electronic components. However, the accuracy of simulated results strongly depends on the material models. The introduction of lead free solders in electronic products to replace lead-containing solders comes along with the need of new material models. The typically used tin-rich solders, such as SnAg or SnAgCu mixture systems, operate usually at high homologous temperatures $T_{hom} > 0.5$ [10], [6]. Therefore, the time-dependent creep deformation inside the solder alloy dominates over the instantaneous plastic deformation. During the service of electronic packages, the thermo-mechanical load stresses the solder joints during the service or accelerated tests. The main contribution of plastic strain in the solder joints comes along with the creep strain [4].

Accordingly, the investigation in solder creep behaviour becomes crucial. Many authors have published creep measurements and material functions for different solder alloys [7], [8], [9], [1], [6]. However, even for uniform solder composition the creep data show a wide spread. Steady state strain rates at a given degree of stress and temperature vary over a range of factor 10^2. The differences of the solder specimen may have strong influences and the creep strain sensitivity on stress and temperature depends more or less on the final material and its condition.

For that reason, a creep measurement unit has been constructed to measure material behaviour on real electronic components with representative manufacturing conditions.

Creep behaviour of tin-rich solders

The creep of solder alloys passes through different phases over time. Firstly, after application of a mechanical load on the solder material, the creep deformation runs through a transient phase. The growing dislocation density leads to an increase of material hardening and decreasing strain velocity. Simultaneously a softening mechanism works. At the point when hardening and softening deformation mechanism acquire an equal absolute velocity, the solder strain rate has reached its steady state. The tertiary phase of creep starts with the onset of the material degradation. This is achieved by an increase of the strain rate. For many technological applications, the description of steady state creep is sufficient.

The physical-state variables mechanical stress σ and the ambient temperature T affect the steady state strain rate from the outside.

$$\dot{\varepsilon}_{ss} = f(\sigma, T) \qquad (1)$$

However, the steady state strain rate is also influenced by the condition of the microstructure. The mixture of Sn,

Ag and copper lead to the formation of various intermetallic-compounds (IMC), such as Sn_3Cu and Cu_6Sn_5. These IMCs are much harder, than the ß-Sn-regions. The little particles are spread relatively homogenously over the whole solder joint interconnection. However, around the ß-Sn-dendrites there are local conglomerations. The number of IMC depends on the Ag and Cu concentration inside the solder. It is characterized by the solder paste or by solder ball, which is applied on the component as well as the chemical reaction between the liquid solder and the connection pad surface metallurgy, takes place during reflow soldering process. The consequence is a different microstructure depended on the amount of dissolved material. The movement and mobility of dislocations inside the Sn-matrix strongly depends on the number of the enclosed hard particles.

Even if the final composition of solder alloy is known, many other parameters may influence the microstructure of solder joints. Such parameters are:

- solder cooling rate
- solder joint size
- thermal mass of PCB and component
- thermal aging condition
- thermo-mechanical aging condition

Accordingly, the creep deformation behaviour is a complex function of the material and its condition. All specimen types, which are different to real solder joints in electronic packages and its material conditions, may generate non-representative creep data. For industrial use, it is necessary to generate creep data, which have been measured on product adequate solder joints. There is a need in creep measurements of FC, CSP and BGA packages.

Experimental setup

Based on the requirements on the creep specimen an adapted measurement unit has been constructed (Fig. 1). The principle method of operation is shown in figure 2 and the functional realization is depicted by figure 1. The creep specimen is located between two equal heater ceramics (A). The lower ceramic sheet is clamped on the X-Y-adjustment table (B) represented by node 1. A controlled force load inducts a linear guided slider (E), which is driven by a step motor (F) (node 3, fig. 2). Further, a S-shaped force sensor (D) transmits the force from the slider to one of the two ceramic heaters. There, the force loads the creep specimen. A LVDT displacement sensor (C) records the deformation of the solder joints with respect to time. This measurement representative takes place on node 2.

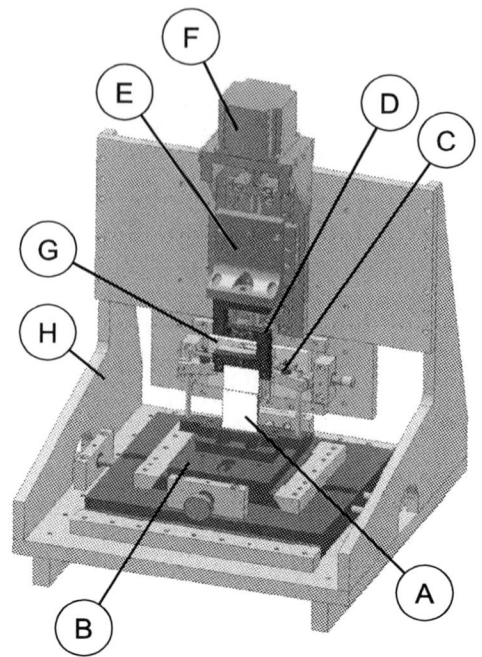

Fig. 1 CAD-construction of the creep measurement unit with its main attachment parts

The slider's minimum linear displacement increment of 156 [nm/step] has been reached. In dependence on the used force sensor, a certain steady state force control accuracy can be realized.

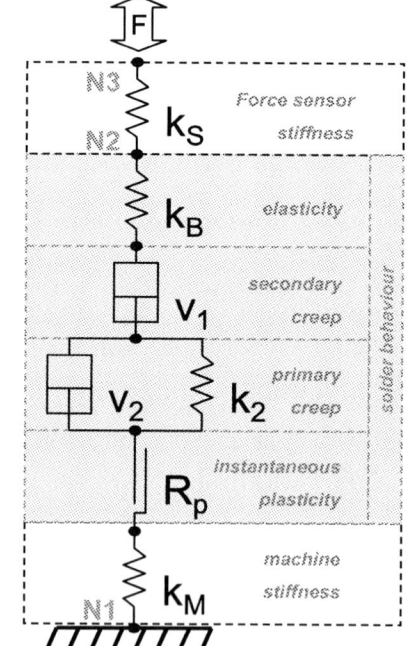

Fig. 2 principle method of operation of the creep measurement unit

measurement region [N]	typical bump area [mm²]	force accuracy [mN/step]	typical components
± 10	0.44	7.2	FC
± 20	0.87	14.4	FC, CSP,
± 50	2.17	36.0	CSP, BGA
± 100	4.35	72	BGA

Tab. 1 guidelines of force sensor selection in dependence on the solder joint connection area

The number and sizes of solder joints attached to the component result in a certain total joint area. Consequently, the force sensor stiffness must always be adjusted. On one hand, the sensor working range should be completely used within a safety margin, and on the other hand, the elastic stress inside the sensor material may not reach the yield stress. Therefore, a series of S-shaped force sensors with variations in stiffness k_S has been designed by FEM. For its realization, the very stiff temperature hardening Aluminum alloy 2017 has been chosen.

material	E-Modul [GPa]	$R_{P0.2}$ [MPa]
Al 99.5%	70	41
Al-4.0Cu-0.7Mg-0.7Mn (Al 2017 T451)	72.5	275

Tab. 2 relevant force-sensor material data in comparison to pure Aluminum

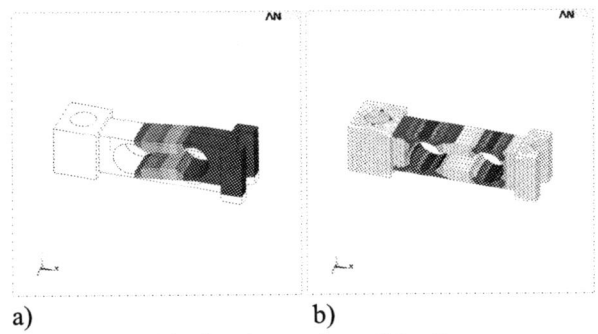

a)   b)

Fig. 3 FEM-Model of active region of the force sensor, a) displacement under force load, b) elastic strain distribution

For analysis, an electronic component is fixed between two heat able ceramic sheets. Each ceramic consist of the very stiff Al_2O_3 with a nominal thickness of 0.635 mm. The outer dimensions are 50 x 30 mm². One surface carries a thick-film resistor (max. 15 W) with printed AgPd-connection pads to enable a local heat. Because of its very good thermal conductivity, the ceramic transmits the heat directly into the electronic component.

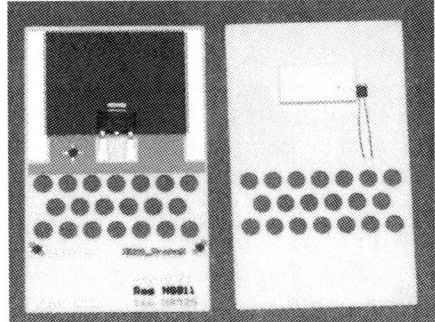

Fig. 4 front (left) and back (right) side of the ceramic heater for creep specimen, left: thick-film resistor with PT1000 temperature sensor for heat regulation, right: example of a creep specimen inclusive PT100 temperature sensor for observation measurement

E-Modul [GPa]	therm. cond. [W m⁻¹ K⁻¹]	therm. expansion [10⁻⁶ K⁻¹]
320	24	7.2

Tab. 3 selective properties of Al_2O_3-ceramic [11]

The heat distribution over the resistor surface is uniform (see fig. 5). Using an infrared camera system the induced heat has been observed. For this application, the calibration of the camera system was performed according to the emission ratio of the resistor surface.

Additionaly, the ceramic has three rows of holes to lower a parasitic heat flow toward the ceramic clamps. A temperature controller within an accuracy of ± 0.5 K regulates the induced heat at the hot spot.

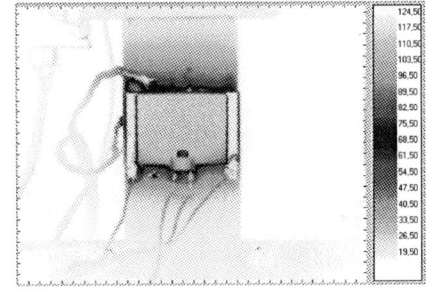

Fig. 5 infrared image of the locally heated region at the ceramics carrying the creep specimen, component temperature 75°C

The assembling of the creep specimen between the heat ceramics allows the global induced force to works as shear force loads on the solder joints. The fixture realizes a two-component epoxy resin adhesive with a temperature resistance up to 200 °C.

As shown in figure 2 the force sensor spring k_S and the elements of solder material behaviour belong to one functional chain. The spring k_B represents the elasticity of the solder material. The Newton element v_1 has a time dependent continuous working behaviour without damping. It represents the secondary creep behaviour, when the hardening and the softening creep mechanism are in steady state. The primary creep is modeled by parallel connection of the damper v_2 and a spring k_2. This models damped strain behaviour continuously. The displacement velocity decreases in dependence on the spring stiffness k_2. Hardening mechanisms lead to a decrease of the creep strain rate during the primary creep phase.

The Masing element R_P stands for the instantaneous plastic behaviour. This element is inactive up to a certain yield load. After exceeding a limit the slider becomes active and a time independent deformation occurs. This element will never be activated by a correctly arranged creep measurement routine.

The machine stiffness k_M is considered at the end of the functional chain. The elasticity value includes the bending stiffness of the machine, the stiffness of the creep specimen clamping as well as the stiffness of the solder joint connection area on the substrate. In all creep measurement units the time dependent contribution of any material around the solder must be avoided. Furthermore, the stiffness of the whole machine is supposed to be as high as possible to decrease the elastic influence in time independent behaviour measurements.

Determination of mechanical data from solder joints

The preferred creep measuring method is the force driven test. Using a closed loop force control a constant shear force loads the connection pads of the solder joints. Two independent working measuring LVDT-sensors observe the displacement of the connection pads, within an accuracy of ±100 nm. The absolute deformation and the velocity can be determined. The shear force loads the solder joints reversibly without causing hard damage to the alloy. This method allows repeating measurements on one creep sample. Abnormalities caused by sample variation can be avoided.

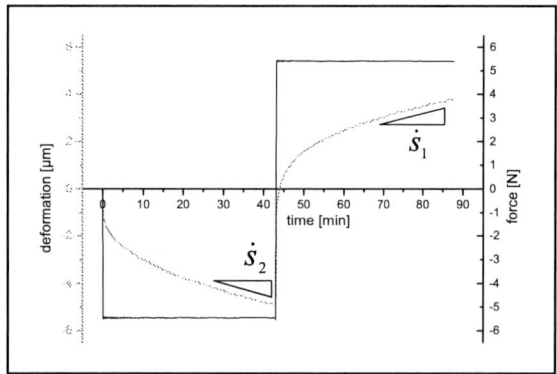

Fig. 6 example of one full cycle creep measurement, left scale: displacement of solder joint connection pads, right scale: applied shear force load

The determined force loads and displacement rates must be converted into equivalent values for stress and steady state strain rate. However, the measurement of real solder joints comes along with a multi-axial stress

condition inside the solder joint (fig. 7). Because of the outer shape of real solder joints a strong inhomogeneous distribution of stress occurs. Consequently, the creep strain is concentrated along the highest stress regions.

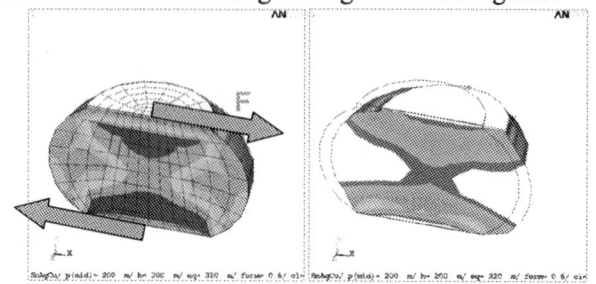

a) equiv. stress   b) equiv. creep strain

Fig. 7 FEM-calculation for observation of the stress strain behaviour of real solder joints, example: PadØ=200µm, a) stress distribution inside a solder joint during shear force loading, b) creep strain distribution after 5µm displacement

The determination of stress and strain based on the simplified equations (2 & 3) may not be applicable for this strong inhomogeneous distribution.

$$\sigma_M = \tau \cdot \sqrt{3} \qquad (2)$$

$$\varepsilon_M = \frac{\gamma}{\sqrt{3}} \qquad (3)$$

These equations are derivations of the von Mises equivalent equations (4) and (5) with the assumption of uni-axial stress conditions.

$$\sigma_M = \sqrt{\frac{1}{2} \cdot \left[(\sigma_x - \sigma_y)^2 + (\sigma_y - \sigma_z)^2 + (\sigma_z - \sigma_x)^2 + 6 \cdot (\tau_{xy}^2 + \tau_{yz}^2 + \tau_{xz}^2) \right]}$$

(4)

$$\varepsilon_M = \frac{\sqrt{2}}{3} \cdot \sqrt{\left[(\varepsilon_x - \varepsilon_y)^2 + (\varepsilon_y - \varepsilon_z)^2 + (\varepsilon_z - \varepsilon_x)^2 + \frac{3}{2} \cdot (\gamma_{xy}^2 + \gamma_{yz}^2 + \gamma_{xz}^2) \right]}$$

(5)

In [2] one suggestion has been published to avoid a systematical error by converting primary data into secondary data. The derived stress condition must be corrected in dependence on the outer shape of the solder joint. All following creep measurements have been corrected.

Preparation of Creep samples

As already mentioned the creep unit is designed for a wide range of sample packages. However, the requirements for well-prepared creep samples are the same. To study effects on solder joints it is necessary to observe the microstructure of the joints of interest. It is recommended to perform a cross section of the solder joint after the measurements. Full-populated area array packages come along with many solder joints. This may result in errors of measurement, because individual or groups of damaged solder joint might not be detected.

A technical reason for decreasing the total number of solder joints is the need of lowering the force load. Hence, the ratio of machine stiffness to specimen stiffness increases.

The minimum number of solder joints for a stabilized construction of components and the printed circuit board (PCB) is four. These bumps are usually not able to carry the whole component, so that they would collapse under the weight. The use of spacer sheets avoids this phenomenon. Of course, the spacer thickness should be in a representative range to ensure a very similar solder joint geometry. Appropriate spacer materials are ceramics and glasses because of their smooth surfaces, which reduce physical attachment, and their availability in different thicknesses.

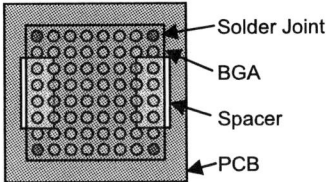

Fig. 8 scheme of the area-array component assembly with application of spacers

The best creep specimen quality can be reached if only the wanted solder joint positions are attached with solder bumps on the component and the solder paste is only printed on the complement connection pad.

The package materials, which surround the solder joints, also influence the creep measurement results. So the visco-elastic or even visco-plastic deformation behaviour of the organic material of the PCB or the interposer has a contribution to the time dependent displacement measurement. Typically, FR4 and High-Tg FR4 show glass transition temperatures around 140°C to 180°C. The material flow can start below the glass transition temperature, because of their amorphous softening behaviour.

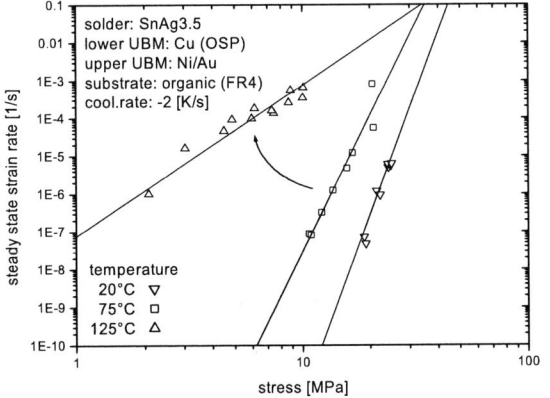

Fig. 9 creep measurement of SnAg3.5 solder alloy on a BGA-Package, upper UBM: Ni/Au, lower UBM: Cu/OSP, solder volume: approx. 3 x 10^{-2} [mm^3]

The figure 9 shows creep measurement of SnAg3.5 solders on Cu as well as on Ni/Au UBM. The component substrate and also the PCB substrate consists of organics. The measurements at a temperature of 125°C show atypical tendencies. The absolute steady state strain rates appear to be very high at low stresses. Additionally, the stress sensitivity decreases extremely (n = 4) in comparison with values at 75°C and 20 °C. This suggests very high activation energy. Here the visco-elastic strain contributes to the creep strain measurements. The creep measurements on real packages are only valid for temperatures below 100°C.

The softening property of periphery materials hinders the determination of creep properties at high temperatures. For that reason, a further creep specimen has been created, which finally reaches adequate surface metallization properties. The main improvement on that is the substitution of the organic substrate material by Al$_2$O$_3$ ceramic. A strong requirement on the production of adequate creep samples are the application of very similar substrate metallizations, such as copper and tin (1 µm) or nickel (5µm) and gold (300nm). The surface metallization strongly influences the final solder composition inside the solder joint. For that reason, a universal ceramic chip has been designed and with it, the technological configuration was developed.

The first steps are the physical vapor deposition of the seed layers titanium and copper based on thin film technology. After an electro-plating process under PCB-manufacturing conditions, the copper thickness is increased to up to 15 µm. Industrial manufactured PCBs typical carry electroplated copper of approx. 10-20 µm. The metallization structure is given by a photolithography and back-etching process. The thin copper metallization allows very hard shaped structures, because of the low under-etching. The solder pad openings are defined by a solder resist.

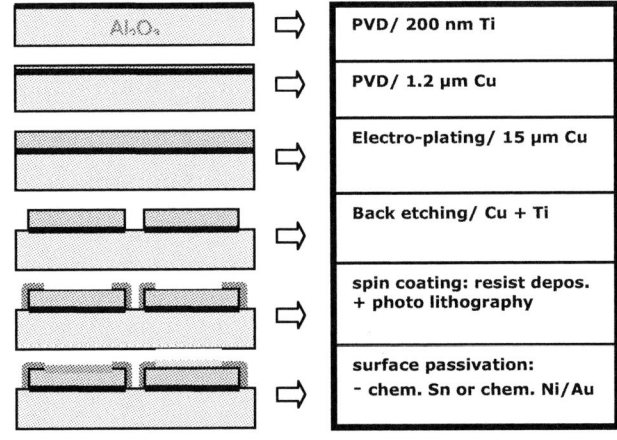

Fig. 10, schematic process flow of manufacturing the base of creep measuring specimen

The openings of the solder connection pads must be in a very good quality, since no soldering defects may occur and the solder joint shape must be precisely repeatable. The thickness of solder resist should be as thin as possible to avoid an influence of resist neck in the material measurements. At the end, a certain connection-pad surface metallization finishes the universal chip.

The universal chip carries structures from 100μm, 200μm, 300μm and 400μm diameter openings. In dependence of chosen structure sizes the dicing process manufactures single chips. Additionally, the different pad openings allow the investigation of material behaviour in different solder volumes (approximately two decades).

padØ [μm]	400	300	200	100
approx. volume [mm³]	3×10^{-2}	1×10^{-2}	4×10^{-3}	5×10^{-4}

Tab. 4 overview of the possible investigated solder volumes by using the universal ceramic chip structures

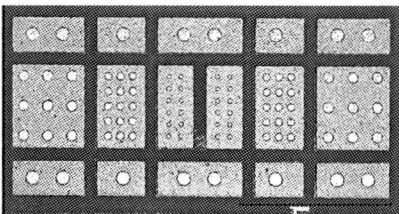

Fig. 11 universal ceramic chip layout with 100μm – 400μm pad diameter

Usually the solder bumps are applied by printing solder paste or by application of single solder balls. The availability of solder in these kinds is limited to common alloys, such as SnAg3.5, SnAg3.0Cu0.5, SnAg3.8Cu0.7. Providing non-typical solders as pastes or miniaturized balls are quite difficult. However, these alloys are necessary for study purposes. Therefore, a manufacturing method was created, where a micro punching tool allows the separation of small pre-forms out of a solder foil.

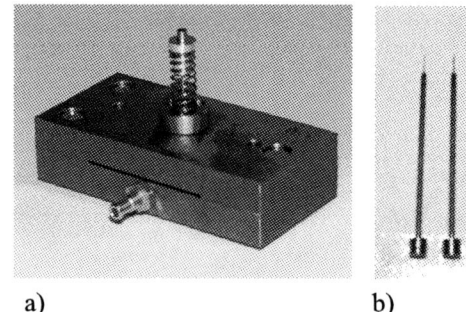

Fig. 12 a) micro punching tool with precisely guided punch and vacuum exhaust for foil fixation, b) punches in different sizes form Ø100 μm – Ø2500 μm

The providing of cast bar material in different alloy compositions is the easier part. However, the production of powder in grain size classes 4 – 6 comes along with high efforts. For this reason, roller systems are used to compress the solder bars to thin solder foils. Up to now the limitation of thickness is 100 μm.

 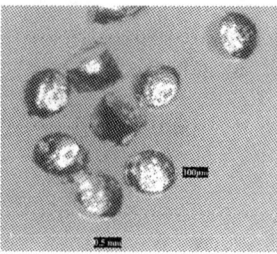

a)   b)

Fig. 13 a) rolled solder foils b) punched solder pre-forms

The following summarized developments give an effective degree of freedom for wide creep sample preparation:

- spacer application for reduction of solder joint number on test specimen
- the development of the ceramic based universal chip with adequate PCB pad metallization
- the development of micro punching tool for solder pre-forms from untypical solder alloys

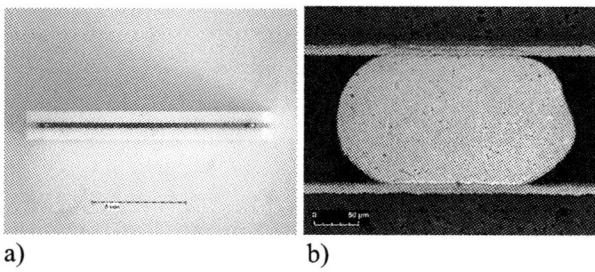

a)   b)

Fig. 14 a) face to face soldered universal chip with four joints, b) reflow soldered joint

As mentioned above the material conditions caused by manufacturing process should be regarded with care. The overall conditions of the experimental specimen were:

- cooling rate –2K/sec
- reflow peak temperature 240°C, liquidus duration approx. 45 sec
- thermal storage at 110 °C for 24 h

Experimental results
Measurements of SnPb36Ag2 solder joints:
Several authors have published secondary creep data of the common used near eutectic SnPb solders. It has been described that the differences in measurements of lead-containing solders are not as high as for lead-free solders. The developed creep machine and the novel specimen have been used for the common available SnPb36Ag2 solder.

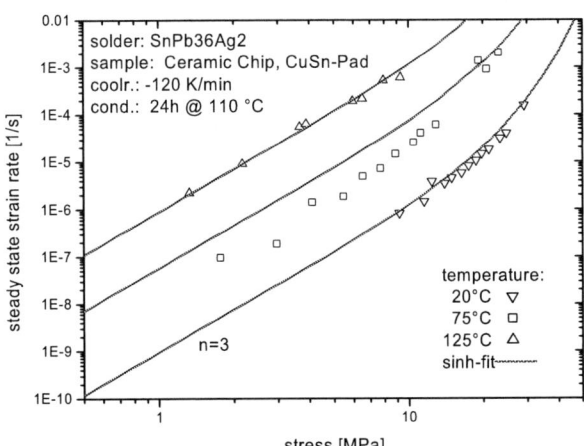

Fig. 15 steady state creep behaviour of SnPb36Ag2 solder joints, connection pad Ø400 µm

The material function fit was done by using the equation (6). Darveaux suggests the Garofalo-approach for the mathematical description of steady state strain rate $\dot{\varepsilon}_{ss}$ of solder alloys [6]. The stress exponent n gives the sensitivity of stress dependence. With σ_N the stress region of the power law breakdown is adjusted and the Q' is the activation energy.

$$\dot{\varepsilon}_{ss} = A \cdot \sinh\left(\frac{\sigma}{\sigma_N}\right)^n \cdot \exp\left(\frac{-Q'}{R \cdot T[K]}\right) \qquad (6)$$

	solder	A [1/s]	σ_N [MPa]	n	Q' [kJ/mol]
own m.	SnPb36Ag2	4e+5	13.5	3.0	63
Schubert [5]	SnPb40Ag1	1.5e+3	10	2.76	53.5

Tab. 5 coefficients for eq. (6), comparison of own results vs. Schubert et al.

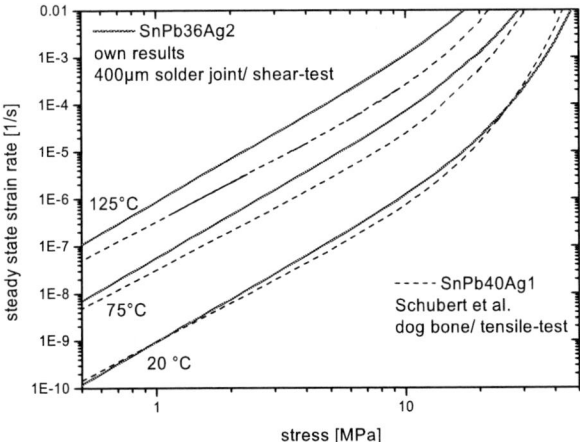

Fig. 16 comparison of lead-containing solders, own measurements and from Schubert et al.

Schubert et al. [5] measured the SnPb40Ag1 solder dog-bone specimen under tensile test conditions at different temperature niveaus form 20°C up to 150°C.

At 20°C the measured steady state strain rate of both independent measurements are almost equal. Furthermore, the stress sensitivity of steady state strain rate at the lower stress region is very similar, expressed by the stress exponent n ≈ 3. The activation energy of our own measurement is determined by the lowest and highest temperature (20°C / 125°C). The determined activation energy is higher than the suggested activation energy of Schubert et al.

Measurements of Sn-rich solder joints:

As mentioned before the developed technology of sample preparation easily allows the investigation of non-typical solder alloys. The measurement of the solder SnAg1.3Cu0.5Ni0.05 (A) in comparison with SnAg1.3Cu0.2Ni0.05 (B) is shown in figure 17. The pre-forms were soldered on 200µm Cu/Sn pads. Both solder alloys show almost the same creep behaviour, which is marginal influenced by the slight copper difference. It is assumed that during the liquid phase of soldering the copper diffuses into the solder alloy, which leads to a general increase in its concentration. The final difference of copper contingent in both solder joints is low.

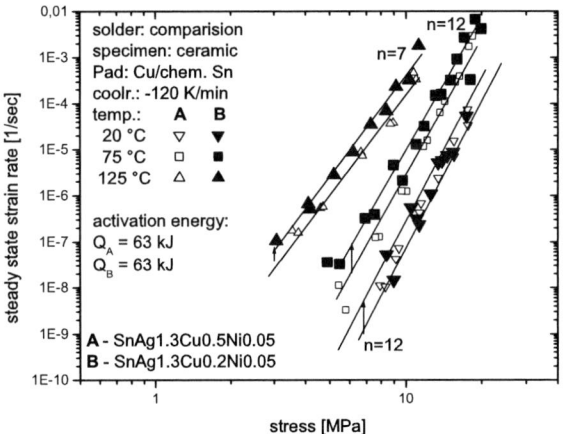

Fig. 17 creep data of SnAg1.3CuXNi0.05 Solder joints on ceramic chip with Cu/Sn UBM, padØ = 200µm

The measurement of the SnAg2.7Cu0.4Ni0.5 alloy is shown in figure 18. The pad metallization on this creep specimen was again Cu/Sn. Therefore, the final copper concentration is affected by the diffusion and it is supposed to be higher as given. At low temperatures the stress exponent is highest at n = 14. It decreases if the temperature rises. This alloy shows a stronger temperature dependency than the solder with a significant lower silver content.

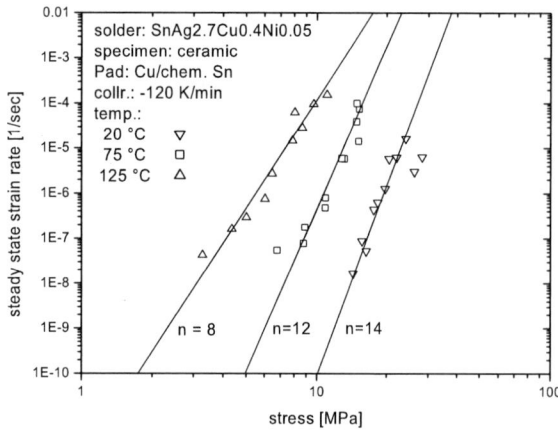

Fig. 18 creep data of SnAg2.7Cu0.4Ni0.05 Solder joints on ceramic chip with Cu/Sn UBM, padØ = 200μm

solder	A [1/sec]	σ_N [MPa]	n	Q' [kJ/mol]
SnAg1.3Cu0.2Ni0.05	8E+5	10	7	75
SnAg1.3Cu0.5Ni0.05	6E+4	9	7	70
SnAg2.7Cu0.4Ni0.05	2E+7	7	5	100

Tab. 6 coefficients for eq. (6) of non-typical solder alloys

Discussion

The material function fit was done by the sinh-law, because it was determined that the stress exponent has strong temperature dependences. For that reason, the power law break down section of this function is used to fit the measurements. Hence, the extrapolation into low strain rate regions (below: 10^{-8} [1/sec]) could lead to wrong tendencies.

The main creep strain contribution in solder joints in electronic packages occurs at the high temperatures. Therefore, the measurement of creep at high temperatures is necessary and should be respected by the determination of material functions.

The solder increases its stiffness at low temperatures. An extrapolation into temperatures far below 20 °C might lead to very high stresses. Only few publications describe measurements at temperatures lower than room temperature. The technical efforts are very high for such measurements and need further investigations.

Conclusions

In general, the investigation of solder joints on real electronic packages can be done with the presented creep measurement unit. In account of this, solder alloy impurities caused by connection pad metallization will be considered more properly in future. Furthermore

- creep measurements on real Packages are possible, within a temperature safety margin dependent on the glass transition temperature of the organic substrates (T < 100 °C)

- the presented universal ceramic chip eliminates every time dependent strain influence of substrates, but carries typical PCB metallization Cu/Sn or Ni/Au
- the development of the micro-punch allows to manufacture solder pre-forms to fasten the investigation of non-typical solders
- creep measurements on 200μm solder joints with non-typical alloys were presented
- variations of copper contents inside the final alloy do not change the stress exponent and the absolute steady state strain rates are almost equal
- an increase of silver content in tin-rich alloys leads to an absolute arising of the creep resistance
- the stress exponent decreases by an increase of measurement temperature.

Acknowledgments

The authors would like to thank the German Federal Ministry of Education and Research (BMBF) and the Qimonda Dresden GmbH & Co. OHG for their financial support.

References

[1] S. Wiese, A. Schubert, H. Walter, R. Dudek, F. Feustel, E. Meusel, B. Michel, „Constitutive behaviour of lead-free solder vs. lead-containing solders – experiments on bulk specimen and flip-chip joints", Electronic Components and Technology Conference, 2001

[2] M. Röllig, S. Wiese, K.-J. Wolter, "Extraction of material parameters for creep experiments on real solder joints by FE analysis", EuroSimE, 2006

[3] S. Wiese, M. Röllig, M. Müller, S. Rzepka, K. Nocke, C. Luhmann, F. Krämer, K.Meier, K.-J. Wolter „The influence of Size and Composition on the Creep of SnAgCu Solder Joints", 1st Eelectronics Systemintegration Technology Conference, Dreden, Germany, 2006

[4] A. Schubert, R. Dudek, H. Walter, E. Jung. A. Gollhardt, B. Michel, H. Reichelt, „Reliability Assessment of Flip-Chip Assemblies with lead-free Solder Joints", Electronic Components and Technology Conference, 2002

[5] A. Schubert, H. Walter, R. Dudek, B. Michel, G. Lefranc, J. Otto, G. Mitic, „Thermo-mechanical Properties and Creep Deformationof Lead-Containing and Lead-free Solders", International Symposium on Advanced Packaging Materials, 2001

[6] R. Darveaux, K. Banerji, "Constitutive Relation for Tin-Based Solder Joints", Components, Hybrids and Manufacturing Technology, Vol. 15, No. 6, 1992

[7] R. Darveaux, "Shear deformation of lead free solder joints", Electronic Components and Technology Conference, 2005

[8] S. Déplanque, W. Nüchter, M. Spraul, B. Wunderle, R. Dudek, B. Michel, „Relevance of primary creep in thermo-mechanical cycling for life-time prediction in Sn-based solders", 6th. EuroSimE conference proceedings, 2005

[9] J. H. L. Pang, B. S. Xiong, T.H. Low, "Creep and fatigue Charcterization of lead free 95.5Sn-3.8Ag-0.7Cu solder", Electronic Components and Technology Conference, 2004

[10] H. Berek, "Degratationsmechanismen: Experimentelle Befunde an Flachbaugruppen", Institutskolloquium: „Ausfälle genau vorhersagen?" – Grüne Elektronik und ihre Folgen, Dresden, 2004

[11] Kyocera, Data sheet for industrial ceramics, "Electronic Fine Ceramics", 2006

Finite Element Modeling of Thermoelastic Damping in Filleted Micro-Beams

Séverine Lepage and Jean-Claude Golinval
University of Liege
Aerospace and Mechanical Engineering Department
Chemin des chevreuils, 1, B-4000 Liege, Belgium
SLepage@ulg.ac.be
Phone: +32-(0) 4-366 48 52

Abstract

In the design of micro-electromechanical systems (MEMS) such as micro-resonators, dissipation mechanisms may have detrimental effects on the quality factor (Q), which is directly related to the response amplitude of the system excited at its natural frequency. One of the major dissipation phenomena to consider in such high-Q micro-systems is thermoelastic damping. Hence, the performance of such MEMS is directly related to their thermoelastic quality factor which has to be predicted accurately. Analytical models have been developed in the literature to assess thermoelastic damping. However, these models are based on very restrictive assumptions and can only be used for simple configurations such as bending beams or bars in extension. In order to investigate structures of complex geometry, a numerical approach is required. In this paper, the finite element method is used to study the effect of the anchor on the thermoelastic quality factor of clamped-clamped silicon beams. Moreover, the effect of fillet at the anchor/beam interface is investigated. It is shown that smoothing the junction between the beam and the anchor can be benefic for some configurations while it can be detrimental in others. This study highlights the importance to have tied manufacturing tolerances for the purpose of high-Q micro-resonators fabrication.

1. Introduction

Thermoelastic damping has been identified as an important loss mechanism in numerous high-Q micro-resonators, see for example Refs. [1, 2, 3, 4, 5]. The ability to accurately model and predict energy loss due to the thermoelastic effects is therefore a key requirement in order to improve the performance of high-Q resonators. Although most studies of thermoelastic quality factor till date have been based on analytical models, which are subject to very restrictive assumptions so that they are not sufficiently accurate to predict the behavior of complex 3-D structures. In this paper, a finite element formulation has been developed in order to analyze the behavior of systems that are not analytically tractable.

Micromachining processes yield shapes that are not geometrically perfect. Figure 1 shows an example of geometrical imprecisions induced by MUMPS fabrication process [6]. This shows that the process yields rounded corners instead of sharp 90° corners and a beam that is narrower by 20 % than the layout width. Therefore, when designing a high-Q micro-resonator, it is important to study the influ-

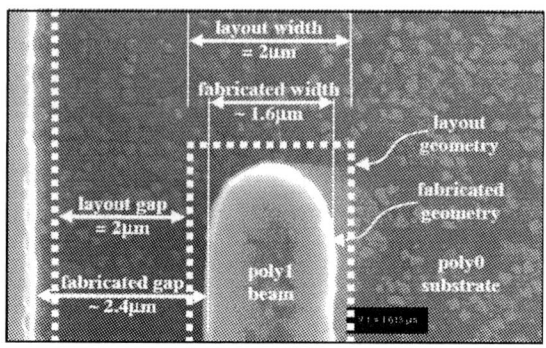

Figure 1: Geometric errors induced by MUMPS fabrication process [6].

ence of shape variation on its quality factor. This paper focuses on the analysis of clamped-clamped silicon beams. Due to manufacturing processes, the beam to anchor interface may be filleted instead of being sharp cornered and the beam width may also vary. The influence of such shape variation has therefore to be studied on the thermoelastic quality factor.

Firstly, thermoelastic damping in beam resonators is briefly reviewed and the thermoelastic finite element formulation is derived. Finally, finite element analyses are carried out to study the effect of the anchor on the thermoelastic quality factor of clamped-clamped silicon beams. Moreover, the effect of fillet at the anchor/beam interface is investigated.

2. Thermoelastic Damping

The basic notions of thermoelasticity are well known [7]. In isotropic solids with a positive thermal expansion coefficient, an increase of temperature creates an expansion and inversely, a decrease of temperature produces a compression. Similarly, an expansion lowers the temperature and a compression raises the temperature. Therefore, when a thermoelastic solid is set in motion, it is taken out of equilibrium, having an excess of kinetic and potential energy. The coupling between the strain and the temperature fields induces an energy dissipation mechanism which causes the system to return to its static equilibrium. The relaxation of the thermoelastic solid is achieved through the irreversible flow of heat driven by local temperature gradients that are generated by the strain field. Thermoelastic damping results from this dissipation which is not always measurable. When the vibration frequency is much lower than the relaxation rate, the solid is always in thermal equilibrium and the vibrations are isothermal. On the other hand, when the vibration frequency is much higher than the relaxation rate, the system has no time to relax and the vibrations are adiabatic. Hence, it is only when the vibration frequency is of the order of the relaxation rate that the energy loss becomes appreciable.

2.1 Analytical Models

Zener [8] was the first to develop expressions to approximate the thermoelastic damping. His theory is based on an extension of Hooke's law involving stress σ, strain ε as well as their first time derivatives $\dot{\sigma}$ and $\dot{\varepsilon}$ [8]:

$$\sigma + \tau_\varepsilon \dot{\sigma} = E_R(\varepsilon + \tau_\sigma \dot{\varepsilon}). \qquad (1)$$

This model is called the "Standard Anelastic Solid" model. The three parameters τ_ε, τ_σ and E_R have the following physical interpretation:

- τ_ε is the relaxation time at which the stress relaxes exponentially when the strain is kept constant.

- τ_σ is the relaxation time at which the strain relaxes exponentially when the stress is kept constant.

- E_R is the elastic modulus after all relaxations have occurred.

The unrelaxed value of the elastic modulus E_U can be defined using the three previous parameters:

$$E_U = E_R \frac{\tau_\sigma}{\tau_\varepsilon}. \qquad (2)$$

In order to analyze the characteristics of the solid vibrations, the stress and the strain are considered to vary harmonically at the natural pulsation ω_n. The dissipation in the solid can be measured by Q^{-1}, the inverse of the quality factor of the resonating structure, which is defined as the fraction of energy lost per cycle:

$$Q^{-1} = \Delta_E \frac{\omega_n \tau}{1 + (\omega_n \tau)^2}, \qquad (3)$$

where $\tau = \sqrt{\tau_\sigma \tau_\varepsilon}$ is the effective relaxation time and $\Delta_E = \sqrt{\frac{\tau_\sigma}{\tau_\varepsilon}} - \sqrt{\frac{\tau_\varepsilon}{\tau_\sigma}} = \frac{E_U - E_R}{\sqrt{E_R E_U}}$ is the relaxation strength.

Thus, the dissipation exhibits a Lorentzian behavior as a function of $\omega_n \tau$ with a maximum value of $\Delta_E/2$ when $\omega_n \tau = 1$. This agrees with the previous qualitative explanation. When the frequency is small compared to the relaxation rate, i.e. $\omega_n \tau << 1$, the thermoelastic dissipation is negligible and the oscillations are isothermal. On the other hand, when the frequency is large compared to the relaxation rate, i.e. $\omega_n \tau >> 1$, the oscillations are adiabatic. Therefore, it is only when the frequency is of the order of the relaxation rate, i.e. $\omega_n \tau \approx 1$, that the thermoelastic dissipation takes importance.

For a beam in flexion, assuming that the relaxation occurs only through the first transverse conduction mode and

that the thermoelastic natural frequency ω_n can be approximated by the isothermal frequency $\omega_{o,n}$, the inverse of the quality factor for a thermoelastic flexural beam resonator can be expressed as follows

$$Q^{-1} = \frac{E\alpha^2 T_o}{C_v} \frac{2\zeta^2/\pi^2}{1 + (2\zeta^2/\pi^2)^2}, \qquad (4)$$

where E is the Young modulus, α is the heat expansion coefficient, C_v is the heat capacity at constant volume, T_o is the reference temperature and ζ is a dimensionless parameter which depends on the thermal diffusivity $\chi = \kappa/C_v$ where κ is the thermal conductivity, the beam thickness b and the isothermal frequency $\omega_{o,n}$: $\zeta = b\sqrt{\frac{\omega_{o,n}}{2\chi}}$. Lifshitz and Roukes (LR) [9] proposed an analysis based on the same fundamental physics but in which the transverse temperature profile is more accurately modeled. Their model gives the following expression for the inverse of the quality factor:

$$Q^{-1} = \frac{E\alpha^2 T_o}{C_v} \left(\frac{6}{\zeta^2} - \frac{6}{\zeta^3} \frac{\sinh\zeta + \sin\zeta}{\cosh\zeta + \cos\zeta} \right). \qquad (5)$$

The quality factors predicted by LR model differ from Zener's ones by between 2 % and 20 % depending on the value of the dimensionless parameter ζ. Indeed, it can be showed that the quality factor given by equation (5) is bounded between two Lorentzians:

$$\Delta_E \frac{2\sqrt{6}}{5} L\left(\frac{\zeta^2}{\sqrt{24}}\right) \leq Q^{-1} \leq \Delta_E \frac{\sqrt{6}}{2} L\left(\frac{\zeta^2}{\sqrt{24}}\right), \qquad (6)$$

where the Lorentzian L is defined as

$$L(\eta) = \frac{\eta}{1 + \eta^2}. \qquad (7)$$

For small values of ζ, the quality factor tends to its lower Lorentzian bound. While for large values of ζ, it tends to its upper Lorentzian bound. Zener's solution corresponds to the following Lorentzian:

$$Q^{-1} = L\left(\frac{2\zeta^2}{\pi^2}\right). \qquad (8)$$

It results from this comparison that expressions (4) and (5) differ by less than 2% on the isothermal side of the peak (low ζ). While on the adiabatic side of the peak (high ζ), the difference can reach 20 %. Hence, when considering configurations located on the adiabatic side of the peak, it is better to use LR model than Zener's approximation.

2.2 Finite Element Formulation

The analytical models are based on very restrictive assumptions and can only be used for simple beam-like configurations. Even if some work have been carried out to extend the analytical models to polycrystalline beams [10], laminated beams [11] or uniform rings of rectangular cross-section [12], in order to investigate complex structures (i.e. non rectangular geometry, anisotropic material,...), a numerical approach is required. The finite element method can be used to solve the dynamics of thermoelastic structures [13, 14].

The thermoelastic finite element formulation can be derived from Hamilton's variational principle in which mechanical and thermal degrees of freedom are considered simultaneously. The displacement field $\mathbf{u}$ and the temperature increment θ are related to the corresponding node values $\mathbf{u_u}$ and $\mathbf{u_\theta}$ by the mean of the shape function matrices $\mathbf{N_u}$ and $\mathbf{N_\theta}$:

$$\mathbf{u} = \mathbf{N_u u_u}, \tag{9}$$

$$\theta = \mathbf{N_\theta u_\theta}. \tag{10}$$

Therefore, the strain field ε and the thermal field $\mathbf{e}$ are related to this nodal values by the shape function derivative matrices $\mathbf{B_u}$ and $\mathbf{B_\theta}$

$$\varepsilon = \mathcal{D}\mathbf{N_u u_u} = \mathbf{B_u u_u}, \tag{11}$$

$$\mathbf{e} = -\nabla\mathbf{N_\theta u_\theta} = \mathbf{B_\theta u_\theta}, \tag{12}$$

where ∇ is the gradient operator and $\mathcal{D}$ is the derivation operator defined so that $\varepsilon = \mathcal{D}\mathbf{u}$ according to the displacement compatibility equation. The finite element discretisation leads to the following dynamic equilibrium equation that governs the thermoelastic behavior of the system :

$$\begin{pmatrix} \mathbf{M_{uu}} & 0 \\ 0 & 0 \end{pmatrix}\begin{pmatrix} \ddot{\mathbf{u}}_u \\ \ddot{\mathbf{u}}_\theta \end{pmatrix} + \begin{pmatrix} 0 & 0 \\ \mathbf{C_{\theta u}} & \mathbf{C_{\theta\theta}} \end{pmatrix}\begin{pmatrix} \dot{\mathbf{u}}_u \\ \dot{\mathbf{u}}_\theta \end{pmatrix}$$
$$+ \begin{pmatrix} \mathbf{K_{uu}} & \mathbf{K_{u\theta}} \\ 0 & \mathbf{K_{\theta\theta}} \end{pmatrix}\begin{pmatrix} \mathbf{u}_u \\ \mathbf{u}_\theta \end{pmatrix} = \begin{pmatrix} \mathbf{F}_u \\ \mathbf{F}_\theta \end{pmatrix}, \tag{13}$$

where $\mathbf{M_{uu}}$ is the mass matrix, $\mathbf{C_{\theta u}}$ is the damping matrix due to thermo-mechanical coupling effect and $\mathbf{C_{\theta\theta}}$ is the damping matrix due to the thermal field. Matrix $\mathbf{K_{u\theta}}$ is the stiffness matrix due to thermo-mechanical coupling. Matrices $\mathbf{K_{uu}}$ and $\mathbf{K_{\theta\theta}}$ are the stiffness matrices due to mechanical and thermal fields, respectively. Vectors $\mathbf{F}_u$ and $\mathbf{F}_\theta$ are the force vectors due to mechanical and thermal fields, respectively.

Thermoelastic effects modify the quality factor of the response, inducing both damping and resonance frequency shift. In order to quantify the quality factor of a structure, a modal analysis has to be carried out. Equation (13) takes the general form

$$\mathbf{M\ddot{q}} + \mathbf{C\dot{q}} + \mathbf{Kq} = 0, \tag{14}$$

where $\mathbf{C}$ and $\mathbf{K}$ are non-symmetric matrices. This problem may be transformed into a linear problem of twice the size through a linearization procedure. Partitioning the eigenvectors into thermal and mechanical degrees of freedom and substituting the time derivative of the thermal degrees of freedom by their values, the eigenvalue problem to solve may be rewritten in the form

$$\begin{pmatrix} -\mathbf{K}_{uu} & -\mathbf{K}_{u\theta} & 0 \\ 0 & -\mathbf{K}_{\theta\theta} & 0 \\ 0 & 0 & \mathbf{M}_{uu} \end{pmatrix}\begin{pmatrix} \mathbf{x}_u \\ \mathbf{x}_\theta \\ \dot{\mathbf{x}}_u \end{pmatrix}$$

$$= \lambda \begin{pmatrix} 0 & 0 & \mathbf{M}_{uu} \\ \mathbf{C}_{\theta u} & \mathbf{C}_{\theta\theta} & 0 \\ \mathbf{M}_{uu} & 0 & 0 \end{pmatrix}\begin{pmatrix} \mathbf{x}_u \\ \mathbf{x}_\theta \\ \dot{\mathbf{x}}_u \end{pmatrix}. \tag{15}$$

If the number of mechanical and thermal degrees of freedom is denoted n_u and n_θ, respectively, the eigenvalue problem (Eq. 15) has $2n_u$ conjugate complex eigenvalues and n_θ real eigenvalues. The $2n_u$ eigenvalues correspond to the mechanical eigenfrequencies and the n_θ ones to the thermal eigenfrequencies.
Solving the thermoelastic eigenvalue problem with a nonsymmetric block Lanczos algorithm allows to calculate the complex eigenvalues of the thermoelastic structure and hence, determine the quality factor of the corresponding mode. The quality factor of the nth mode is given by

$$Q = \frac{\omega_i}{2\omega_r}, \tag{16}$$

where ω_r and ω_i are the real and imaginary parts of the nth conjugate complex eigenvalue of Eq. (15). Note that another way to determine the quality factor is to carry out a harmonic analysis and to derive the value of the quality factor from the frequency response function of the structure.

All these finite element developments are implemented in a software called Oofelie ("Object Oriented Finite Element Led by Interactive Executer"). This software is written in C++ language so that it allows to solve multiphysic problems with strong coupling [15, 16].

3. Clamped-clamped Silicon Beam

In numerous micro-resonators, the vibrating part consists in a clamped-clamped silicon beam. In this section, the test case beam has the following dimensions: a length L of 90 μm, a height h of 4.5 μm and a width w of 4.5 μm (Figure 2). The thermal and mechanical properties of silicon at $T_o = 298\ K$ are: $E = 1.58\ 10^{11}\ N/m^2$, $\rho = 2300$ kg/m^3, $\nu = 0.2$, $c_v = 711\ J/kgK$, $\alpha = 2.5 10^{-6}\ K^{-1}$ and $k = 170\ Wm^{-1}K^{-1}$. The thermoelastic quality factor is determined for the first bending mode in plane OYZ.

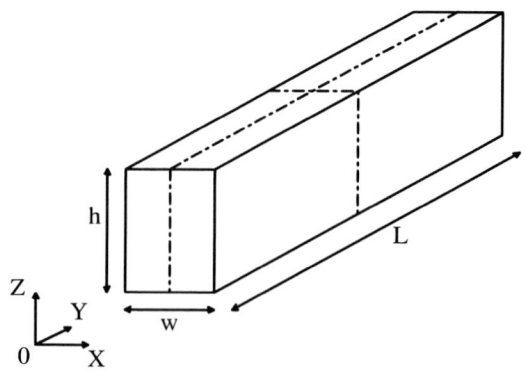

Figure 2: Beam geometry.

In this analysis, 8-node quadrilateral plane stress elements are used. Each node i has two mechanical degrees of freedom (v_i, w_i), the displacements of node i along the two directions (Y, Z). At each node, a thermal degree of freedom represents the temperature increment θ_i. The displacement field and the thermal field use bi-quadratic interpolation functions as shape functions.

Making use of the symmetry of the problem, only one half of the beam is modeled. The mesh comprises 80 elements along the half length and 8 elements along the vibrating height. The boundary conditions are such that one of the extremity is clamped and the symmetry conditions are imposed on the symmetry plane. The temperature increment of the central node of the clamped extremity is fixed to zero.

Figure 3 represents the thermoelastic mode. The temperature increment field is coherent with the strain field:

- The maxima of the temperature decrease and increase are located at the clamped extremity where the strain maxima are located.

- The temperature of the neutral fiber does not vary.

- The middle of the beam shows local maxima of the temperature decrease and increase.

The corresponding quality factor is equal to 13 258. The frequency shift due to the thermoelastic damping is equal to 2.2169e-5 and the attenuation is equal to 3.7713e-5.

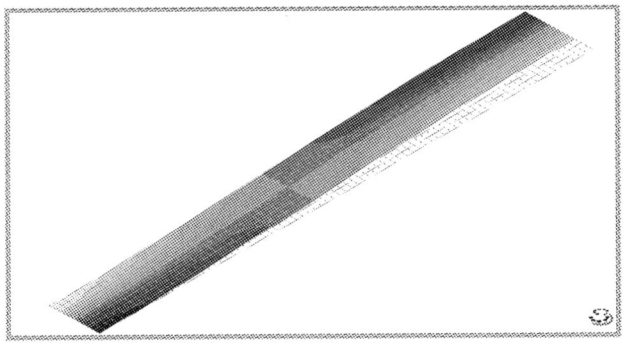

Figure 3: Thermoelastic mode of the quadrilateral finite element model (White: temperature increase, Black: temperature decrease).

3.1 Effect of Beam Height

As predicted by the analytical models, the thermoelastic effects on the behavior of an oscillating beam depend on the aspect ratio of the beam. For a beam of a fixed length of 90 μm, the first natural frequency is calculated using the thermoelastic models as a function of its height. Figure 4 represents the frequency shift, which is given by $\Re(\omega_n)/\omega_{o,n} - 1$, and the attenuation, $\Im(\omega_n)/\omega_{o,n}$, where $\omega_{o,n}$ is the isothermal natural frequency. The frequency shift increases with the height of the beam, but the slope decreases so that the frequency shift tends to reach an upper limit. The frequency shift varies from zero for small values

of h, for which the beam can be considered as isothermal, to the adiabatic threshold value for large values of h. The attenuation exhibits a maximum for a beam height of 5.3 μm. The analytical and finite element models give similar results. As the beam height to length ratio increases, Poisson's effect is reinforced, leading to larger errors in LR model. Indeed, when the beam is not slender enough, Euler-Bernoulli assumption becomes too restrictive. Figure 5 shows the variation of the quality factor with the height of the beam. The quality factor reaches its minimum for a beam height of 5.3 μm.

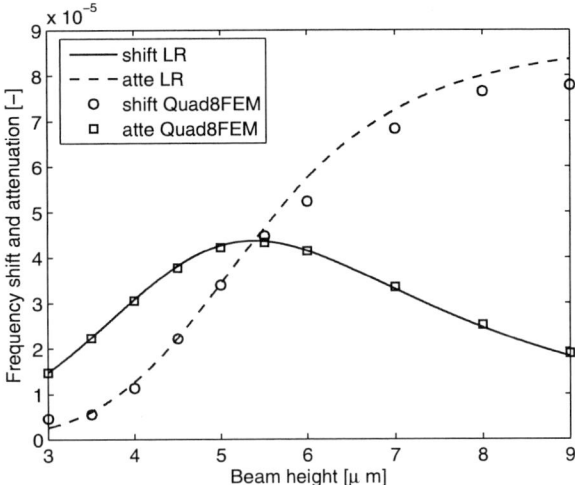

Figure 4: Variation of the frequency shift and attenuation versus the beam height (beam length: 90 μm).

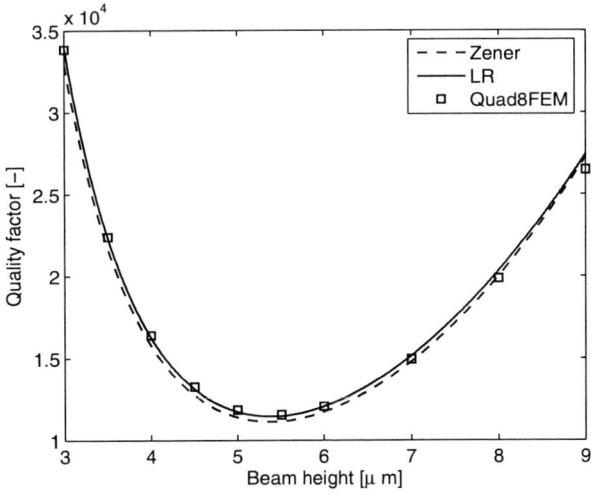

Figure 5: Variation of the quality factor versus the beam height (beam length: 90 μm).

3.2 Influence of Anchor

Anchors are added to the clamped-clamped beam in order to understand their influence on the thermal and mechanical fields. The anchor dimensions are such that they include most of the stress and temperature distribution that extend into the anchor structure [17]. The width w of the anchor is the same as the beam one, its height is equal to three times the height h of the beam and its length is equal to two beam heights (Figure 6). The three sides of the anchor rectangle which are not attached to the beam (delimited by bold lines in Figure 6) are clamped. Due to the symmetry plane (represented by dotted lines in Figure 6), one half of the structure is modeled using quadrilateral finite elements.

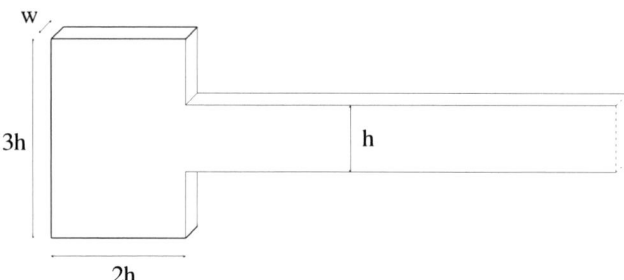

Figure 6: Geometry of the half beam with anchor.

The effects of anchor are studied on two test cases. One is located on the isothermal side of the quality factor curve. The beam height is set to 4.5 μm. The other one is located on the adiabatic side of the quality factor curve. The beam height is equal to 7 μm. For both cases, the beam width is set to 4.5 μm and the beam length is equal to 45 μm. Table 1 compares the quality factor, the damping and the resonant frequency with and without anchor for both cases. In both cases, the addition of the anchor decreases the resonant frequency due to an effective lengthening of the beam and decreases the damping due to the reduction of strain and temperature gradient near the anchor to beam interface. Relaxing the strain near the ends of the beam increases the performance of the resonator. In the isothermal case, the decrease in damping is more important than the decrease in resonant frequency, ultimately raising the quality factor. On the other hand, the decrease in damping does not overcome the decrease in the resonant frequency for the adiabatic case, leading to a smaller quality factor.

Fillets are added to the geometry to smooth out the intersection between the beam and the anchors. To study their effect on thermoelastic damping, the quality factor is calculated for two fillet sizes; the radius R is set to 10 % and 20 % of the beam height. Figures 7 and 8 compare the temperature distribution for the isothermal configuration with a fillet of 0.45 μm and 0.9 μm. The configuration with the larger fillet presents higher temperature gradient at the beam to anchor junction. Table 2 shows the influence

Table 1: Effect of anchor on the quality factor of C-C silicon beams.

Height [μm]	Anchor	Q [-]	ω_r [rad/s]	ω_i [rad/s]
4.5	Yes	14553	939	2.7317e7
4.5	No	13258	1104	2.9296e7
7	Yes	14296	1406	4.0210e7
7	No	14942	1493	4.4629e7

of the fillet radius on the quality factor, damping and resonant frequency for both the isothermal and adiabatic cases. A larger fillet induces an increase in the resonant frequency as the beam becomes effectively shorter. As the effective beam height increases near the beam to anchor junction, the effective relaxation time for damping in that region increases. In the isothermal case, the increase in the damping outpaces the increase in the resonant frequency, decreasing the quality factor. Conversely, in the adiabatic case, the increase in the resonant frequency outpaces the increase in the damping, giving a higher quality factor. Globally, increasing the fillet has similar consequences than increasing the effective height of the beam. In the adiabatic side, an increase in the height yields a larger quality factor, whereas in the isothermal side, it yields a lower quality factor.

Table 2: Effect of fillet on the quality factor of C-C silicon beams.

h [μm]	R [μm]	Q [-]	ω_r [rad/s]	ω_i [rad/s]
4.5	0	14553	939	2.7317e7
4.5	0.45	14351	960	2.7569e7
4.5	0.9	14166	984	2.7879e7
7	0	14296	1406	4.0210e7
7	0.7	14312	1425	4.0786e7
7	1.4	14417	1438	4.1463e7

The finite element simulations show that the isothermal and adiabatic configurations have completely opposite behaviors. The presence of the anchor decreases the quality factor of the isothermal configuration while it increases the quality factor of the adiabatic configuration. Two radii of the fillet are considered. The simulations shows that the larger the fillet, the higher the quality factor of the isothermal configuration and the lower the quality factor of the adiabatic configuration. The thermoelastic finite element method allows to model complex structures that are not analytically tractable and to study the effect of features such as anchor or fillet which cannot be taken into account in analytical models. It is shown that smoothing the junction between the beam and the anchor can be benefic for some configurations while it can be detrimental in others.

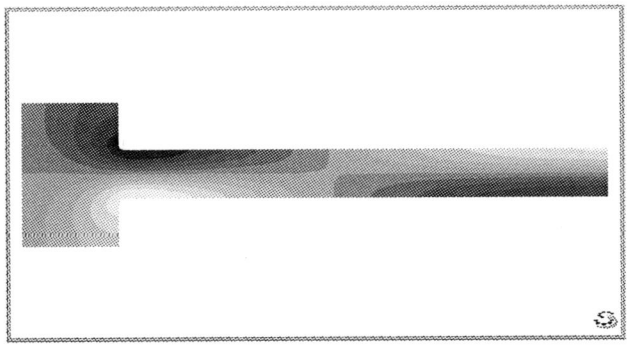

Figure 7: Temperature distribution for isothermal case with anchor and fillet (radius=0.45 μm).

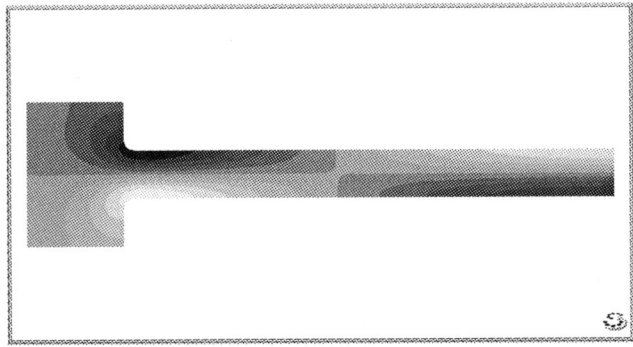

Figure 8: Temperature distribution for isothermal case with anchor and fillet (radius=0.9 μm).

4. Conclusions

The application of the finite element method makes possible the determination of the quality factor of complex structures, and it allows a better understanding of the phenomena occurring in thermoelastic vibrations. Due to the strong interaction between the thermal and mechanical fields, the influence of a parameter such as anchor or fillets on the multiphysic behavior of a MEMS is not straightforward. Thanks to the developed finite element approach, the different parameters that influence the behavior of the micro-resonator can be identified, so that the physical phenomena are better understood and the design of the micro-resonator can be modified in order to improve its quality factor and hence, its performances. On top of that, this study highlights the importance to have tied manufacturing tolerances for the purpose of high-Q micro-resonators fabrication.

Acknowledgments

The author S. Lepage is supported by the Belgian National Fund for Scientific Research (FNRS), which is gratefully acknowledged. This work is also supported by the Communauté Française de Belgique - Direction Générale de la Recherche Scientifique in the framework Actions de Recherche Concertéées (convention ARC 03/08-298).

References

[1] A. Duwel, J. Gorman, M. Weinstein, J. Borenstein, and P. Ward, "Quality Factors of MEMS Gyros and the Role of Thermoelastic Damping," in *Proceedings of the 15th International Conference on Microelectromechanical Systems (MEMS)*, (Las Vegas, NV), pp. 214–219, January 20-25 2002.

[2] A. Duwel, J. Gorman, M. Weinstein, J. Borenstein, and P. Ward, "Experimental Study of Thermoelastic Damping in MEMS Gyros," *Sensors and Actuators A*, no. 103, pp. 70–75, 2003.

[3] B. Houston, D. Photiadis, M. Marcus, J. Bucaro, X. Liu, and J. Vignola, "Thermoelastic Loss in Microscale Oscillators," *Applied Physics Letter*, vol. 80, pp. 1300–1302, February 2002.

[4] B. Houston, D. Photiadis, J. Vignola, M. Marcus, X. Liu, D. Czaplewski, L. Sekaric, J. Butler, P. Pehrsson, and J. Bucaro, "Loss due to Transverse Thermoelastic Currents in Microscale Resonators," *Materials Science and Engineering A*, no. 370, pp. 407–411, 2004.

[5] R. Abdolvand, G. Ho, A. Erbil, and F. Ayazi, "Thermoelastic Damping in Trench-refilled Polysilicon Resonators," in *Transducers'03, the 12th International Conference on Solid State Sensors, Actuators and Microsystems*, (Boston), pp. 324–327, June 8-12, 2003.

[6] J. Clark, D. Garmire, M. Last, and J. Demmel, "Practical Techniques for Measuring MEMS Properties," *Technical Proceedings of the 2004 NSTI Nanotechnology Conference and Trade Show*, vol. 1, 2001.

[7] W. Nowacki, *Thermoelasticity*. Oxford: Pergamon Press, 1986.

[8] C. Zener, "Internal Friction in Solids," *Physical Review*, vol. 52, pp. 230–235, August 1937.

[9] R. Lifshitz and M. Roukes, "Thermoelastic Damping in Micro-and Nano-mechanical Systems," *Physical Review B*, vol. 61, pp. 5600–5609, February 2000.

[10] V. Srikar and S. Senturia, "Thermoelastic Damping in Fine-grained Polysilicon Flexural Beam Resonators," *Journal of Microelectromechanical Systems*, vol. 11, no. 5, pp. 499–504, 2002.

[11] J. Bishop and V. Kinra, "Thermoelastic Damping of a Laminated Beam in Flexure and Extension," *Journal of Reinforced Plastics and Composites*, vol. 12, pp. 210–226, February 1993.

[12] S. Wong, C. Fox, and S. McWilliam, "Thermoelastic Damping of the In-plane Vibration of Thin Silicon Rings," *Journal of Sound and Vibration*, vol. 293, no. 1.

[13] S. Lepage and Golinval, J.-C., "Finite Element Modeling of the Thermoelastic Damping in Micro-Electromechanical Systems," in *Acomen 2005, Third International Conference on Advanced Computational Methods in Engineering*, (Ghent, Belgium), 30 May-2 June 2005.

[14] S. Lepage, O. Le Traon, I. Klapka, S. Masson, and Golinval, J.-C., "Thermoelastic Damping in Vibrating Beam Accelerometer: A New Thermoelastic Finite Element Approach," in *Caneus 2006*, (Toulouse, France), August 27- September 01 2006.

[15] I. Klapka, A. Cardona, and M. Géradin, "An Object-Oriented Implementation of the Finite Element Method for Coupled Problems," *Revue Européenne des Eléments Finis*, August 1998.

[16] I. Klapka, A. Cardona, and M. Géradin, "Interpreter OOFELIE for PDEs," *European Congress on Computational Methods in Applied Sciences and Engineering, ECCOMAS 2000, Barcelona, 11-14 September*, 2000.

[17] J. Gorman, "Finite Element Model of Thermoelastic Damping in MEMS," Master's thesis, Massachusetts Institute of Technology, June 2002.

Numerical Analysis of the Reliability of
Cu/low-k Bond Pad Interconnections Under Wire Pull Test:
Application of a 3D Energy Based Failure Criterion

Sébastien Gallois-Garreignot[a], Vincent Fiori[b]*, Stéphane Orain[c], Olaf van der Sluis[d]

[a] Freescale, 850 rue Jean Monnet, 38920 Crolles, France
[b] STMicroelectronics, 850 rue Jean Monnet, F-38926 Crolles, France
[c] NXP, 860 rue Jean Monnet, 38920 Crolles, France
[d] Philips Applied Technologies, 5656 AE Eindhoven, The Netherlands
* Phone: (33) 4.38.92.25.58   Email: vincent.fiori@st.com

Abstract

Due to size reduction in die manufacturing and introduction of brittle dielectric materials, crack related failures occur currently, mainly in interconnect levels. By means of Finite Element (FE) simulations, an energy based failure criterion named Nodal Release Energy (NRE) Method, inspired by the so-called Area Release Energy (ARE) one developed by Philips Applied Technologies, is used to numerically predict the mechanical related failures. More precisely, the failure index is applied to investigate wire bonding induced peeling. In this paper, the NRE method is presented and its added value to forecast delamination failures in a typical microelectronics stack is demonstrated.

The NRE method is related to fracture mechanics and founded on propagation approach. Two FE calculations are used to evaluate the energy quantity: an uncracked and cracked one. In the latter model, a virtual crack is inserted. Aiming to compare the NRE values with known physical quantities experimentally measured such as critical adhesion energy, a relation bridging the gap from NRE to the Energy Release Rate is given. This relation is based on the crack extension method and relates to Griffith theory. The accuracy of the NRE method is investigated through comparisons with 2D and 3D analytical cases. Results show that the method provides a good approximation. The NRE behaviour with respect to key numerical parameters will be studied.

At last, a typical bond pad structure under a wire pull test is simulated. Both stress and energy based analyses are carried out. The critical interface is investigated with both post processing methods. Results based on the energy criterion show that delamination interface is in agreement with experimental observations, in contrast to stress based values. However, it is also shown that simulation results can depend on the prescribed crack length, suggesting a accurate definition of the cracked model. The main assumptions done in this study are discussed, trying to define the associated uncertainties, particularly residual stress and crack morphology features. Finally, the added insights provided by NRE method and its ability to help in design and process development for advanced IC technologies are demonstrated.

I-Introduction

According to Moore's law established in the sixties, the number of transistors in a die is multiplied by four every three years. Added to the necessity of always smaller devices for integration purpose in products, this leads to a reduction of the critical dimensions. To face this reduction and ensure higher performance, a new kind of dielectric materials has been introduced recently in integrated circuit (IC). Unfortunately, these materials, called low-k or ultra low-k due to their low dielectric constant (k), have poor mechanical and adhesion properties [1-2]. These issues, added to the stress induced by front end (FEoL) processes (thermal mismatch, intrinsic stresses, etc.) assembly and packaging steps (wire bonding, flip chip, molding, etc.), affects die reliability and cracks are commonly observed at the interconnect levels [3-4].

Mechanical modelling appears to be an important tool to forecast these failures and to speed up IC development. Brittle delamination is suspected to be the main failure mode, thus fracture mechanics is the best approach to address it. Several numerical methods have been already proposed to describe fracture phenomena within a FE framework [5-6-7-8]. However, due particularly to a higher complexity of IC structures, 3D simulations are needed to provide reliable results, since most of the commonly used interconnect architectures can not be described by 2D models. By consequence, most of the existing methods are not suitable in an industrial use. To reach the aforementioned requirements, a novel method has been developed by Philips Applied Technologies, more precisely the so-called Area Release Energy (ARE) demonstrated its performance [9-10].

In the present article, the Nodal Release Energy (NRE) method, based on ARE, is detailed, and its numerical sensitivity is investigated. In addition to this failure index, a relation between NRE and the Energy Release Rate is given. As a result, the calculated values can be linked directly to physical quantities and the interface toughness. Then, this method is validated by a comparison with analytical cases. Finally, a typical bond pad structure submitted to wire pull test is simulated as an application case. By an investigation of NRE values, the layer most likely to delaminate is identified. The impact

1-4244-1105-X/07/$25.00 ©2007 IEEE

of the assumptions made during the study on the provided insights and results is discussed.

II-Nodal Release Energy Methodology

Formulation

Based on fracture mechanics (which is known to be more accurate to forecast delamination failures in brittle materials than stress based approach), the propagation approach is the study of the growth of a pre-existing crack of which the location and geometry must be defined a priori. As a result, the Energy Release Rate (ERR or G) is evaluated and compared with the considered critical value, the interface toughness in our case (G_c), from which the crack propagates. In our paper, an alternative energy based method is formulated, called the NRE method. It is based on two numeric models: an uncracked and cracked one in which a virtual crack is inserted.

-Firstly, the undamaged model without crack is solved. Nodal forces are picked at the future location of the crack (figure 1-a). It should be noticed that both crack morphology and crack orientation have to be assumed at this step.

-Secondly, the crack is inserted by releasing nodes within the initial model after which the damaged model is solved. Displacements are picked on nodes of the two opposite lips of the crack (figure 1-b: u_i^+ and u_i^-).

The formulation of NRE is a multiplication of forces (F_i), acting on the undamaged model, and displacements (u_i), induced by the crack opening resulting in an energy quantity (*n* represents the number of nodes that belong to the released area). This value is divided by the crack area (*A*) and leads to an energy rate:

$$NRE_{tot} = \frac{1}{A} \cdot \sum_{i=1..n} F_i \cdot \left(u_i^+ - u_i^- \right) \quad (I)$$

The reader will note that NRE is an averaged energy. Numerical instabilities around the crack tip which could appear are then get round, contrary to some others which use only nodal values at the vicinity of the crack tip [7].

However, it must be kept in mind that NRE has not the same meaning that the Energy Release Rate. Indeed, according to Griffith theory, G is defined as the energy released by a small crack increase from *a* to *a+Δa*, contrary to NRE which represents an increase from 0 to *a*. So, NRE cannot be directly compared with the experimentally measured physical quantity G. Therefore, the following relation between NRE and G is introduced, based on the crack extension method:

$$G = \frac{NRE_{a+da} \times (a+da) - NRE_a \times a}{2da} \quad (II)$$

Where *da* is the crack extension, NRE_{a+da} and NRE_a is the energy computed for a semi-crack length of, respectively, *a+da* and *a*. In practice, the crack extension

value is defined by the finite element size. In order to study the accuracy of this method, analytical and numerical results are compared in the next section.

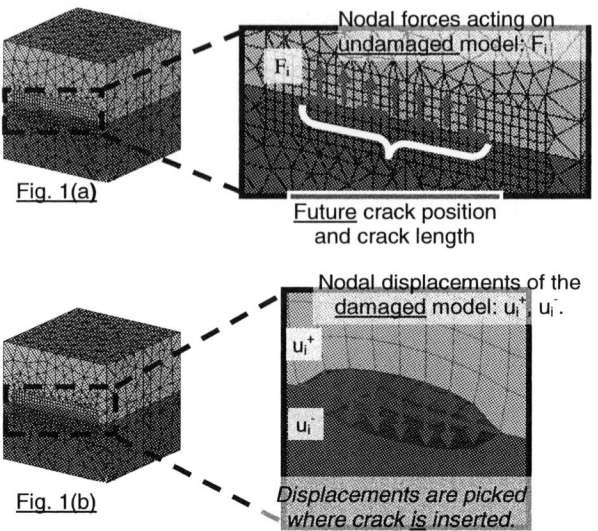

Fig. 1(a)

Fig. 1(b)

Figure 1: NRE Method methodology. Fig. 1(a): uncracked model, Fig. 1(b): cracked model

III- Numerical sensitivity and accuracy investigations

Aiming to study numerical sensitivity, a 2D infinite plate with a straight crack and a 3D cylindrical bar with a penny shaped crack problems [11, 12] are chosen as the validation models. Mesh density and results consistency from 2D to 3D are studied by comparing the NRE quantity with the analytical one.

Validation model description

* 2D analytical solution: the 2D validation case is a infinite homogeneous plate under remote tensile loading (fig.2), modelled as a 2D plane strain problem. The FE model should be large enough to ensure that boundary stress fields are not affected by the crack presence.

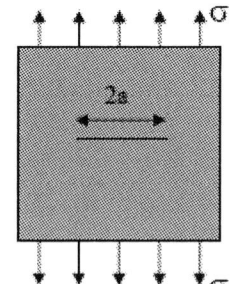

Figure 2: 2D Plane strain analytical case (a= semi crack length, σ= load applied)

The stress intensity factor (*K*) is given by the following relation [11]:

$$K = \sigma \sqrt{\pi a} \quad (III)$$

Then, the energy released rate (G) is expressed as:

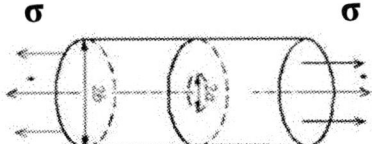

$$G = \frac{\kappa+1}{8\mu} K^2 \; (IV) \quad \text{where} \begin{cases} \mu = \dfrac{E}{2(1+\nu)} \\ \text{E = Young's Modulus} \\ \nu = \text{poisson's ratio} \\ \kappa = \text{3-4v } (plane\ strain) \\ a = \text{half crack length} \end{cases}$$

* <u>3D analytical solution</u>: The 3D validation model is in reality a pseudo 3D solution. Indeed, the cylindrical bar (*b*: bar radius) with a penny crack shape (*a*: crack radius) under tension with homogeneous material (fig.3) could be described by a 2D axisymmetric model

<u>Figure 3:</u> 3D/2D axisymmetric analytical case

The stress intensity factor is [12]:

$$K = \sigma \frac{F b^{3/2}}{\sqrt{(b-a)(b+a)}} \sqrt{\pi a} \quad (V)$$

where $\quad F = \dfrac{2}{\pi}\left(1 + \dfrac{1}{2}\dfrac{a}{b} - \dfrac{5}{8}\dfrac{a^2}{b^2}\right) + 0.268\dfrac{a^3}{b^3}$

G is linked with *K* by:

$$G = (1-\nu)\frac{K^2}{2\mu} \quad (VI)$$

Mesh sensitivity investigations

The accuracy of the NRE according to the mesh refinement is hereafter studied. Figure 4 shows the variations of NRE for different crack discretizations. It is found that, with both 2D and 3D models, a convergent behaviour of NRE is observed according to the size of the finite element in the crack. More precisely, with a half crack length defined by only 5 elements, the NRE procedure provides less than 5% error compared to NRE asymptotic value (considered to be the value calculated with a crack discretization above 25 elements).

Furthermore, additional works demonstrated that the accuracy is correlated with the features of the stress field at the vicinity of the crack area. More precisely, for a given discretization level, increasing the stress field unsteadiness along the crack would slightly deteriorate the precision. However, the convergence ability of the NRE remains whatever the model.

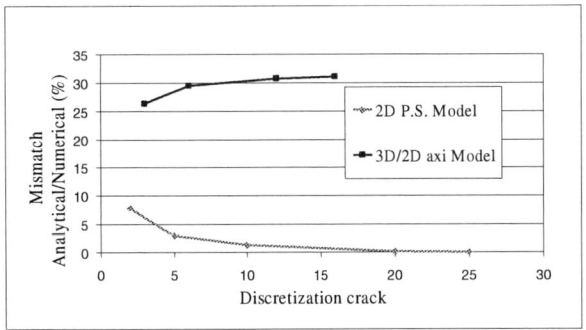

Figure 4: Comparison between the NRE values and G analytical ones according to the 2D Plane strain (P.S.) and 3D/2D axisymmetric models

Comparison NRE vs. G

Even if G and NRE have not the same formulation, the 2D model results are close to the G theoretical value in some particular cases, as shown in figure 4. Indeed, with a crack defined by 5 elements, the mismatch of NRE with G analytical values is less than 5 % in the case of the 2D model. However, in the 3D model, the results show an overestimation of about 30 %. Thus, in some particular cases where the G evolution as a function of the crack length is linear, NRE Method gives G values. However, to match ERR definition in all situations, a relation between G and NRE is proposed (see §II).

Figure 5 plots the error between $G_{analytical}$ and $G_{numerical}$, estimated from NRE, according to the crack extension value *da*. As described in equation *(II)*, a crack extension needs to be assumed from the suspected crack propagation path (here assumptions on crack morphology are imposed by the analytical case parameters).

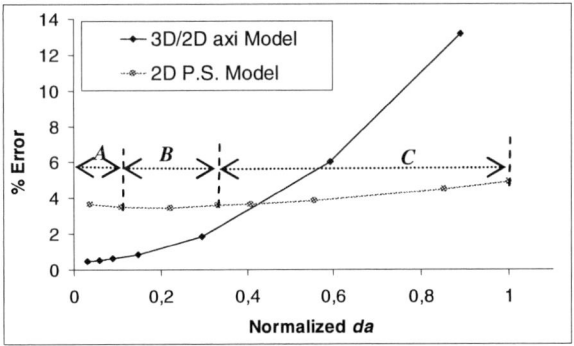

<u>Figure 5:</u> Comparison between the G numerical values and analytical ones according to the normalized crack extension

Results show that, generally speaking, the smaller the *da*, the lower the error since *da* is close to the infinitesimal crack extension as defined by the Griffith theory.

The 2D model has a lower dependency with respect to *da* and a steady error of about 4% whatever the *da* values is observed. For the low da values (fig. 5 zone *B*), this result comes from the error induced by the energy estimation accuracy related to the whole crack discretization. As for the high *da* values (zone *C*), the error from NRE to G remains limited due to the previously explained linearity behaviour of G(a) in the

111

considered model. Furthermore, the use of a too small crack extension *da* slightly increases the error (fig.5 zone A). Indeed, in that case the difference between the NRE_{a+da} and NRE_a is not substantial enough and leads to an unstable operation. It is also noted that, by consequence, a higher stress value would imply a lower necessary crack extension.

So, in order to obtain a good approximation of G and to set a suitable value of *da*, attention must be paid to the variation from NRE_a to NRE_{a+da} too. By consequence, the behaviour of NRE according to the crack length plays a role in the frame of a G evaluation.

Finally, the behaviour of NRE Method has been studied. Through a benchmark between analytical and numerical results, the accuracy of NRE and its ability to evaluate G has been shown. However, results underline that some of the numerical parameters such as the crack extension have to be chosen carefully. Hence, aiming to perfectly manage the G precision, a study of the NRE variations as a function of the crack length would be required.

IV- Application case

Model description

As introduced previously, the use of low-k materials with poor mechanical properties leads to an increase of delamination occurrences in interconnects. The aim of this part is to apply the NRE Method in order to forecast failures in Cu/low-k stacks.

With wire bonded products, one of the critical qualification steps for the die reliability is the so-called wire pull test. This is a destructive test while the wire bonding process is checked (fig.6). It consists in pulling until failure the bonded gold wire with a hook.

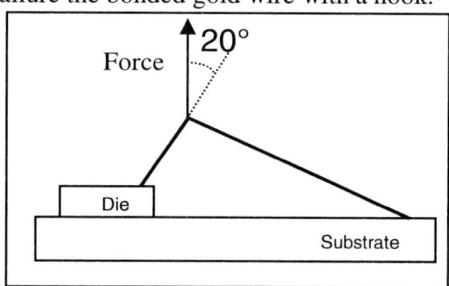

Figure 6: Schematic view of the simulated wire pull test.

The force prescribed by the hook is recorded and the failure modes are observed to achieve bonding process qualification. One of the following failure modes are observed: (i) the wire breaks in the capillary's neck region, the related failure mode is called "neck break" (fig.7-a), this result is a "pass" test. (ii) Poor intermetallic formation leads to adhesion problem between the wire and the aluminium pad. These kinds of failures are mostly related to chemical issues and are not addressed in this work. (iii) Due to the low structural strength of the porous low-k materials and associated interfaces, cracks can also appear in interconnect levels during this destructive test (fig.7-b-c). The associated failure mode, called "pad

peeling" is closely investigated here. Thus, the loading conditions represent a wire pull test with 20° angle.

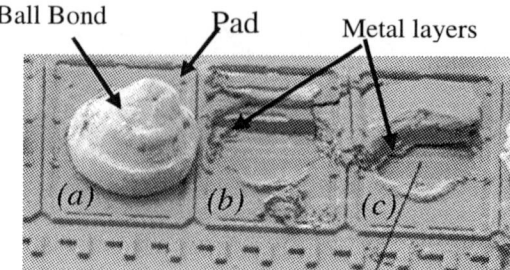

Figure 7: Failures observed by SEM after a wire pull test (*courtesy of C2A Charac. Team*): (*a*) Ball neck break (b) (c) Pad peeling.

A 2D plane strain model is used. The effect of the number of levels in the IC stack is particularly studied. So, to prevent local and patterning effect, metal lines and vias are not modelled.

The interconnect stack is made with full metal (Cu) and full dielectrics (low-k and SiO_2) materials. The thin TaN/Ta barrier properties are included in the copper material. However, as suggested by the observed failure location, the nitride barriers (SiCN) are numerically described. A typical stack in microelectronic devices (fig.8), composed by 7 levels of metal is built. 5 over the 7 levels are made of thin metal (X Levels) and the 2 others are thick (Z levels). The thin layers are surrounded by low-k dielectric whereas the Z levels are surrounded by SiO_2.

A linear elastic behaviour is used for all materials, and residual stress induced by FEoL steps is not simulated.

The gold wire and the front-end models use a free meshing with triangular finite elements. The back-end (BEoL) mesh is mapped with rectangular elements.

Despite the fact that all BEoL interfaces (Cu/SiCN - SiCN/Low-k - Low-k/Cu - SiO_2/Cu) are tested, only results concerning the most critical ones, which are the Dielectric/Copper pair, are reported here.

For comparative purpose between the distinct material pairs involved, G needs to be related with the corresponding critical adhesion energy Gc (see table 1). Values have been determined experimentally (*e.g.*, by a four points bending test [2]) and literature inputs are used in this work. It should be noticed that a large range of values is found according to the source [13-14] which underlines both the property dependence to process conditions and the difficulty to measure precisely such data. Adhesion property of TaN-Ta is supposed to be close to copper. In consequence, interface *X/TaN*-Ta is considered equal to *X/Cu*.

Interface	G_c [J/m²]
Low-k/TaN-Ta	$4.5^{\pm 1.5}$
SiO$_2$/TaN-Ta	15

Table 1: Critical energy release rate for interface Low-k/TaN-Ta and SiO$_2$/TaN-Ta.

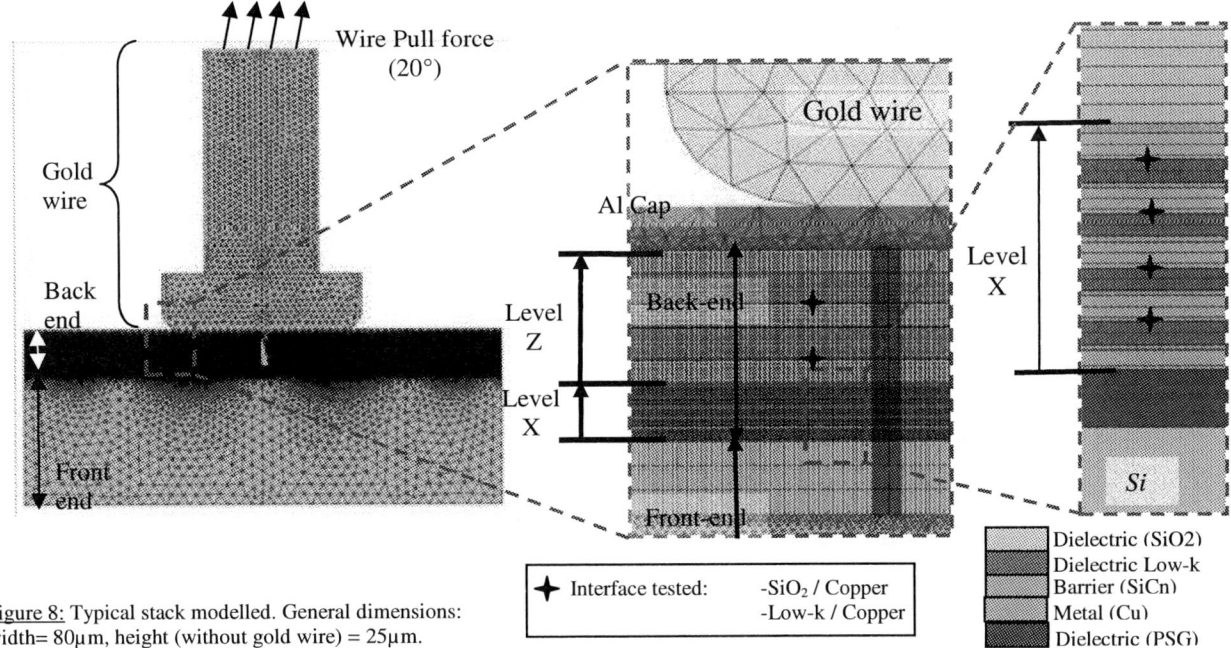

Figure 8: Typical stack modelled. General dimensions: width= 80µm, height (without gold wire) = 25µm.

Interface tested: -SiO₂ / Copper
-Low-k / Copper

Dielectric (SiO2)
Dielectric Low-k
Barrier (SiCn)
Metal (Cu)
Dielectric (PSG)

As for the crack morphology, an adhesive one (or interface crack) is assumed. It will be inserted below the junction of the pad and the stack (figure 8).

Simulation results

The stress induced by the wire pull test within the pad increases from the die bottom to the junction of the aluminium cap and the gold wire (fig.9). It is interesting to note that with a stress based approach, the critical interface is the top low-k interface. In addition to that, the discrimination between each low-k interfaces (3.5% difference between Maximum and Minimum stress) is not obvious.

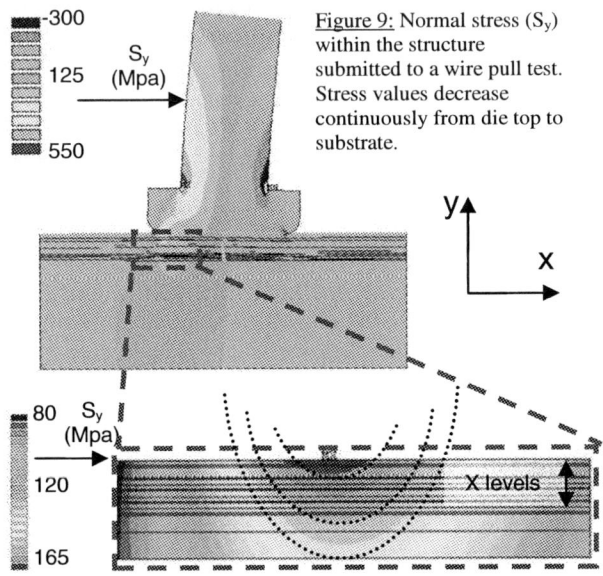

Figure 9: Normal stress (S_y) within the structure submitted to a wire pull test. Stress values decrease continuously from die top to substrate.

On the contrary, with the previously described energy based approach, clearer results are found. Indeed, the X Levels are found to be more susceptible to failure than Z ones (fig.10). This result confirms the general observation

that the use of low-k degrades the overall die reliability. Moreover, a closer look at the results in the thin layer region suggests that the middle of the X stack is the most critical part. Indeed, the highest NRE values are evaluated at V2/M3 and V3/M4 interfaces (Vx represents the dielectric layer number and Mx the copper layer number, so V2/M3 is the interface between the second dielectric layer from the bottom and the third copper layer). More precisely, these two interfaces have energy of about 17% higher than those at the bottom levels. Hence, the discrimination between interfaces is clear.

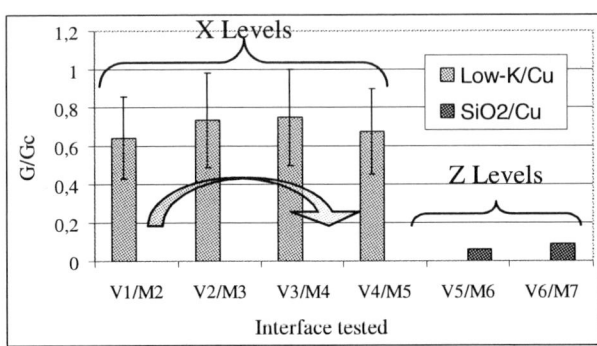

Figure 10: Energy release rate normalized and related with G_c for dielectric/copper interface at X and Z levels.

Trying to understand these results, assumption is made that the stiff layers above and below the X levels limit the deformation and thus, the energy released at the levels in the centre of the structure (less influenced by the stiff layers) could lead to crack propagation.

Thus, NRE method clearly provides an easier discrimination between interfaces than stress based approach. However, the two approaches give dissimilar trends. So, a benchmark between numerical and experimental results has been done.

113

With similar stacks and loading conditions, experimental failure analyses report that delaminations occur mainly in X levels and particularly, near the middle of the stack, see figure 11.

Figure 11: FIB cuts and SEM observations after a wire pull test (courtesy of C2A Charac. Team). Failure occurs at metal 4

The NRE results are in agreement with these observations which underlines the consistency of the failure index for this problematic. Note that similar insights were found in [10].

Effect of the initial crack length

The added insights of the NRE Method compared to stress approach have been shown. However, even if the previous statement is valid in the frame of a propagation approach, the morphology of the virtual crack needs to be assumed, in particular the crack length. To investigate the sensitivity of this choice, different crack lengths are simulated. Only the most critical interfaces, i.e. the low-k/Cu ones (X levels) are considered.

As described in figure 12, an increase of the crack length leads to dissimilar results. More precisely, with a half crack length from 0 to 0.7μm, the critical interface is found in the middle of the stack. However, for higher crack length values, it moves to the top of the X levels and V4/M5 interface is found to be the most critical one. In our case, G_c is considered to be equal to $4.5^{+/-1.5}$ J/m². So, to reach this value, a crack length of 1.8μm would be needed and the corresponding critical interface would be V4/M5 (fig.12).

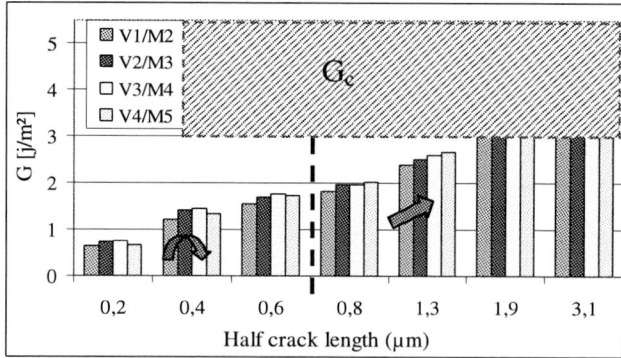

Figure 12: Energy release rate at Low-K/copper interface for different crack lengths.

As a result, it is found that the predicted failed interface depends on the crack length choice. This can be explained by the fact that energy is influenced by either the stress or the strain depending on the considered application. On the other hand, in our model, the top of the stack is the higher constrained part of the device (figure 9). While increasing the crack length, it seems that the influence of the stiffer layers nearby the X levels decreases and the effect of this stress becomes preponderant. In others words, with thicker X levels, the influence of the strain components induced by the stiff layers would be more important than the stress effect. Thus, even with cracks large enough to reach the threshold value, the middle of the stack would remain the most critical region.

Discussion

The propagation approach suggests that the suitable crack length to be used should be the one which leads to evaluate the closest G values to Gc. This underlines the need for accurate values of both measured and evaluated Gc. Furthermore, results show that the simulated model, loading conditions, and more generally speaking the FE strategy must be the most realistic as possible. Indeed, even qualitative insights are affected by both intrinsic model properties and FE results.

More precisely, it is well-known that during the process steps, residual stress is induced in the back-end. The crack propagation would be strongly influenced by the stress values within the pad. Hence it is expected that initial stress conditions should be included over the height of the pad structure. Obviously, the loading conditions are also important for the stress results. According to its modelling parameters, it would be more and less aggressive for the pad and would affect G (i.e. G/Gc ratio) values.

Moreover, geometry singularities and local defaults within a patterned structure could promote and weaken crack nucleation and propagation. The present model, with full sheets and without bonding simulation, cannot address these phenomena. Same consequences are observed during the bonding process during which the pressure and ultrasonic displacements applied between the pad. Indeed, the wire bonding process itself cause micro-cracks [15] and to weaken the Cu/Low-k stack.

These uncertainties and their related potential modelling improvements have to be kept in mind aiming to refine the analysis in future investigations.

In the present study, the added insights of the NRE method, compared to stress based approach, has been shown. Indeed, with carefully choosen crack parameters, numerical results are able to forecast the experimentally observed delaminated interface. Whereas with a stress based analysis, delamination location does not match reality and the stress range within the stack does not provide a clear discrimination.

V- Conclusion

In this paper, an energy-based methodology aiming to forecast peeling failure is proposed, validated and applied. NRE method is an energy method inspired on ARE and based on a fracture mechanics approach. Its formulation is described, and a relation between G and NRE is given.

Then, the accuracy and the numerical sensitivity are benchmarked on two analytical cases. Moreover, generic guidelines concerning the FE parameters to be used to ensure reliable results, in particular the crack discretization and crack extension values are provided.

In order to investigate the failure region within IC, the NRE method is applied on a typical die stack under wire pull test in 2D. The predicted interface peeling shows good agreement with observation, contrary to stress based prediction.

A critical assessment of the NRE method underlines some uncertainties and key factors. Both modelling and experimental topics are discussed and particular attention is needed on the residual stress evaluation and on the G_c measurement. Indeed, it has been shown that the delamination location is dependent on the prescribed crack length and by extension on the threshold value G_c. This provides further work suggestions.

However, it has been also shown that NRE method gives added insights compared to stress analysis. Indeed, with reasonable crack parameters, the experimental observations are confirmed by a clear prediction of the delaminated region.

Finally, the proposed method could help in many application cases [15]. The described simulation method is useful for different problematics such design optimization and process development.

Acknowledgments

Thanks are particularly due to the following contributors to this work: Crolles2 Alliance (C2A) physical characterization and mechanical modeling teams especially Torsten Hauck of Freescale, Munich for his valuable inputs concerning the analytical cases.

References

1. T.M. Moore et al., Characterization and Metrology for ULSI Technology 550, (2000), pp.431-439
2. M. Damayanti et al., "Adhesion study of low-k/Si system using 4-point bending and nanoscratch test", *Materials Science and Engineering* B 121, (2005), pp. 193-198
3. S. Balakumar et al., "Peeling and delamination in Cu/SiLK process during Cu-CMP", *Thin Solid Films* 462–463, (2004), pp. 161-167
4. Ikegami, "Mechanical Problems in the Production Process of Semiconductor Devices", *JSME International Journal*, vol.33 ,n°11990
5. Raju Is, Newman Jc., "Three-Dimensional Finite-Element analysis of finite-thickness fracture specimens", Report NASA TN D 8414, (1977), 40 p.
6. Shivakumar Raju, "An equivalent domain integral method for three-dimensional mixed-mode fracture problems", *Engineering Fracture Mechanics*, vol.42 , N°6, (1992), pp. 935-959
7. Krueger, R., "The Virtual Crack Closure Technique: History, Approach and Applications." NASA/CR-2002-211628 ICASE Report No. 2002-10.
8. T. Elguedj, "Appropriate extended functions for X-FEM simulation of plastic fracture mechanics". *Computer Methods in Applied Mechanics and Engineering.* vol.195, Issues 7-8 ,2006, p.501-515
9. M.A.J. van Gils et al, "Analysis of Cu/low-k Bond Pad Delamination by using a Novel Failure Index,", accepted in *Microelectronics Reliability*.
10. O. van der Sluis et al, "Efficient damage sensitivity analysis of advanced Cu Low-k bond pad structures using the Area Release Energy criterion. Submitted to *Microelectronics Reliability*.
11. C.T.Sun, W.Quian. "The use of finite extension strain energy release rates in fracture of interfacial crack". School of Aeronautics and Astronautics, Purdue University. West Lafayette.
12. Y. Murakami. "Stress Intensity Factors Handbook". Volume 1 and 2, Pergamon Press, Oxford, (1990)
13. N. Cherault (Thesis) "Caractérisation et modélisation thermomécanique des couches d'interconnexions dans les circuits sub-microélectroniques". Alliance. 2006
14. T. Sherban et al., International Interconnect Technology Conferences, p.257-259, 2001
15. Vincent Fiori et al. "3D Multi Scale Modeling of Wire Bonding Induced Peeling in Cu/Low-k Interconnects: Application of an Energy Based Criteria and Correlations with Experiments." submitted ECTC 2007.

Thermal Simulation with Coupled Network Models on System Level

Adam Augustin
Freescale Semiconductor
Schatzbogen 7, D-81829 Munich, Germany
Email: adam.augustin@freescale.com

Bartosz Maj
Continental Automotive Systems AG
P.O. Box 90 01 20, D-60441 Frankfurt, Germany
Email: bartosz.maj@contiautomotive.com

Abstract— The paper describes thermal modeling with coupled network models by means of a typical optimization process for semiconductor products. This becomes especially important in automotive applications with small thermal budgets. Using two different module configurations, the entire modeling process is discussed in detail. The approach with boundary condition independent compact models provides a much faster simulation, enabling execution of thermal case studies for different mechanical designs, considering diversity of transient loading conditions.

I. INTRODUCTION

Modern design flows for electronic components and devices as well as virtual prototyping environments rely heavily on simulation models, not only in the electric and mechanic domains, but more and more in the thermal domain as well. The additional focus on thermal investigations is driven through multiple factors such as the increasing packing density, integration of power and logic functions on the same die or module and installation space limitations. Additional optimization of the mechanical build up of whole modules also becomes a high priority task during the development. To fulfill these needs, different approaches may be used for temperature prediction in early development stages.

For simple problems, where idealized boundary conditions can be used without compromising the results, the use of analytical solutions of the heat transfer equation is very common [1]. Limitations of this kind of solution lie within the complexity of the examined devices, rendering the approach unusable for devices with many heat sources and extensive control cycles. In systems with an extensive number of degrees of freedom, only numerical finite element methods (FE) lead to reliable results [2]. The shortcoming of this method is the excessive time consumption of such approaches and the complexity of operating the software. On the other hand FE simulation with ANSYS [3] or FLOTHERM [4] is an industry standard and already present in the design flows of many companies.

The most promising approach is the usage of methods to reduce the system order of models [5]–[7], without sacrificing the accuracy needed. Additionally, this approach allows incorporating the results into a simulation environment, familiar to the most concerned users, because of the possibility of the full integration into a standard network simulator. Boundary

condition independent modeling is needed in order to have the full options for model coupling [8], [9]. In this work we apply a network model, introduced in [10], which is based on model order reduction. Deriving from this fact, electro-thermal simulations are possible. These are also manifoldly developed by different research groups [11]–[14].

The paper is structured as follows. In the next section a brief description of the used modeling method is given, describing the generation of the models. Section III delineates in detail the properties of the utilized models. In section IV some results will be discussed, focusing on the comparison between the network modeling approach and the common model order reduction. At the end of the paper some conclusions are given.

II. MODELING METHOD

In order to calculate different module configurations our approach is based on boundary condition independent models presented in [10]. The method provides compact models which can be used in every SPICE-similar network simulator.

In [10], the generation of compact models starts with an FE-model which is build using ANSYS. Here, an accurate and global model can be build considering complex geometrical structures as well as each particular material. By means of model order reduction methods (Krylov subspaces via the block Arnoldi algorithm [15], [16]) a small model in state space representation

$$
\begin{aligned}
\mathsf{V}^\mathsf{T}\mathsf{M}\mathsf{V}\,\dot{z}(t) + \mathsf{V}^\mathsf{T}\mathsf{A}\mathsf{V}\,z(t) &= \mathsf{V}^\mathsf{T}\mathsf{B}\,u(t) \\
y(t) &= \mathsf{C}\mathsf{V}\,z(t)
\end{aligned}
\tag{1}
$$

with the state vector $z(t)$ is generated using the command-line tool *MOR for ANSYS* developed on IMTEK, the University of Freiburg in Germany [17]. The matrices M and A are the system matrices, B the input and C the output matrix, respectively. Vector $u(t)$ describes the input (thermal flow) and vector $y(t)$ the output (temperatures). The Krylov subspace projection is represented by matrix V. Now, the reduced model is transformed into the following compact dynamic thermal model (Kirchhoffian network)

$$
\mathsf{H}\,\dot{T}(t) + \mathsf{K}\,T(t) = P(t)
\tag{2}
$$

with the capacitance matrix H and the conductance matrix K. The elements in vector $T(t)$ stand for potentials and the

1-4244-1105-X/07/$25.00 ©2007 IEEE

elements in $P(t)$ for electrical fluxes, respectively. Eq. (2) is a set or ordinary differential equations and can be implemented in any SPICE-similar network simulator by means of, e.g. the behavioral modeling language Verilog-A.

III. APPLICATION AND MODELING

The presented approach can be used during the early phase of development of a new system, giving the possibility to extensively test the new design. Thermal simulations are done in the network simulator environment, allowing the design engineer to test different load scenarios as well as different PCB (printed circuit board) heat sink options. Thanks to the time savings given by this approach, many more options can be tested for a successful system optimization.

A. Finite Element Model Preparation

First, for all components separate network models have to be build. In our case it is a semiconductor device (see Fig. 1), a PCB and two heat sinks. Since the mounting position of

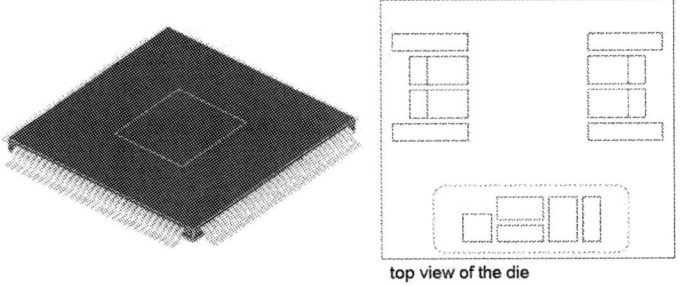

Fig. 1. Model of the investigated smart power device in an exposed pad package with 128 leads. The marked heat sources in the right picture are permanently active

the device varies over the used boards, as shown in Fig. 2, it is not possible to generate only one compact model for the PCB. Therefore, the boards were modeled conjoined with their appropriate heat sink. Hence, three boundary condition independent compact models have been generated: one package model and two different PCB/heat sink combinations.

B. Network Model Preparation

The right picture in Fig. 1 shows the top view of the die with a schematic floor plan of active regions. As can be seen, there are 17 active heat sources whereof the lower five are permanently active (each source with different dissipated power) and the remaining 12 sources are driven dynamically. The load profiles for these heat sources are pulses with certain amplitudes and different duty cycles. All these heat sources are modeled as separate I/O-ports (input/output) in the network model. Additionally, there are several thermal sensors distributed over the entire die surface. That positions have been modeled as single I/O-ports, as well. In the end, the final network model of the device has 41 ports including the top and bottom of the package, too.

For the network models of the PCB/heat sink combinations, only three ports have been defined. The first one is the

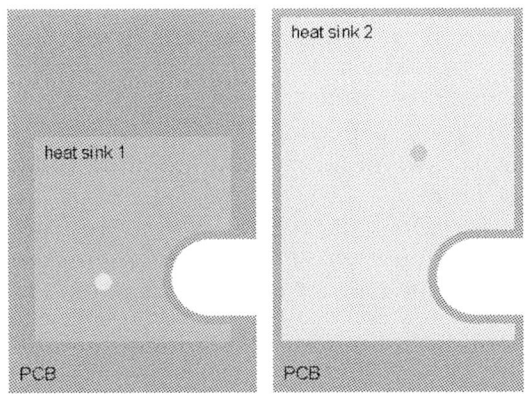

Fig. 2. View on the back side of the PCB with two different heat sink options for the same application. The point within the heat sink indicates the position of the smart power device placed on the top side of the PCB

connection to the exposed pad of the package, followed by the bottom of the heat sink in order to connect the convection model to it. Finally, the position of a thermocouple under the board has been modeled as one I/O-port. In Fig. 3 the resulting coupling between all models is shown. Since the particular

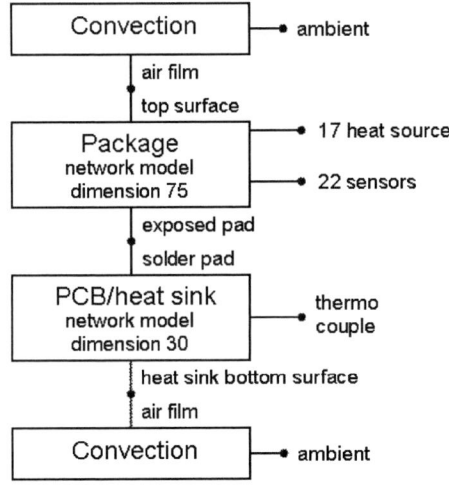

Fig. 3. Block diagram showing the coupling of compact thermal network models

models of the module are boundary condition independent on top of the package as well as on the bottom of the heat sink, two simple convection models have been attached.

IV. RESULTS

Looking at both setups in Fig. 2 the comparison of thermal performance seems to be clear. Left combination (combi1) has a much smaller heat sink, therefore the performance should be worse compared to the right combination (combi2). However, considering material properties of both heat sinks, the situation becomes much more ambiguous. Thermal conductivity of the alloy in heat sink of combi1 is about 390 W/mK compared to the material of heat sink of combi2 with only 130 W/mK. Obviously,

117

the geometrical properties can not be the sole indicator to contribute to a decision.

It must be stated that the modeling of convection is slightly bastardized using only one I/O-port, since the areas with this condition (top of package and bottom of heat sink) are relatively large and show a non-homogeneous temperature distribution, generally. In [10], a much more accurate approach for convection is introduced. This is achieved using Lagrange polynomials with arbitrary order covering even several hot spots on a surface.

For comparison we choose a worst case scenario, which means that all dynamic heat sources (the upper ones in right picture of Fig. 1) are active for a period of six seconds. The basis for our network models are order reduced models generated using multiple expansion points in order to consider aimed switching conditions. This also leaves all options open for a wide range of input functions. For instance, in our case, the frequency of the applied input functions varies in the range of 15 Hz with different duty cycles and amplitudes. Due to some necessary constrains (see [10]) for the reduced model which is the starting point for the network model (e.g. the same number of inputs and outputs), we compare the network results also with another reduced model which was generated without any constrains. Since the pure reduced models can not be coupled, the entire modules (device, PCB and heat sink) are modeled as one "large" compact model. For the order of these models we choose 200 with 17 inputs and 27 outputs, respectively. For this dimension, the error to the full FE-model stays under 1 %.

Fig. 4 shows a comparison between both combinations at the I/O-port of thermo couple on the one hand and the differences between the reduced and the network model on the other hand. The time range is showing only the interesting part where the

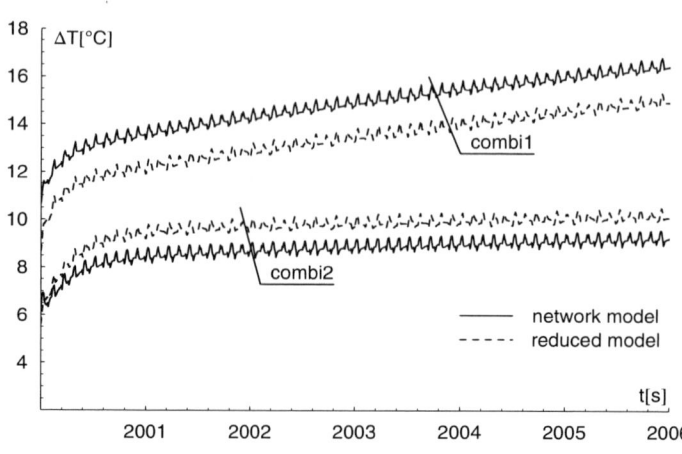

Fig. 5. Temperature response in one of the build in sensors on the die. ΔT is the difference between the resulting temperature and ambient temperature

by the additional transformation (from compact to network model). The reason for the difference between network and reduced model will be explained later in this section, even though they would be neglected in our application. In Fig. 5 the results at one of the build-in sensors on the die are shown. The sensor is placed in the vicinity of the upper edge of the die (see Fig. 1). The behavior is similar to Fig. 4, but in combi2 the temperature of the pure reduced model is higher than in network model.

The differences between both modeling methods can be explained in a slightly different treatment of the boundary condition on top of the package. For the pure reduced model the convection in ANSYS is applied as a surface load, already. The network model is boundary condition independent, so that the condition is applied within the network simulator on the particular I/O-port describing the top of the package by one port only. Due to the required modeling variation, the resulting differences are acceptable and negligible, considering the problems and possible sources of error in thermal measurement setups. Since the die surface is near to the top surface of the package, the applied boundary condition affects all I/O-ports defined at die surface. In the network considering combi1 the influence seams to be not as high as in the network with combi2, since the ratios for all ports are the same. This is why the coupled network including combi1 has always higher temperatures then the reduced model.

Using the compact network modeling approach, the accuracy is not the only quality indicator, but also the time consumption. For the compact model an up-to-date workstation needs 29.1 s for the calculation of 6300 time steps. For the same number of times steps, the coupled network simulations needs about 83 minutes. However, the here used UNIX workstation is much slower than the up-to-date workstation. The pure processor speed is 6.8 times slower. Finally, comparing these numbers to a full FE-model calculation, which would take about 40.3 hours (on the up-to-date workstation) for the same number of time steps, even the coupled network simulation is very fast. The latter number is projected form

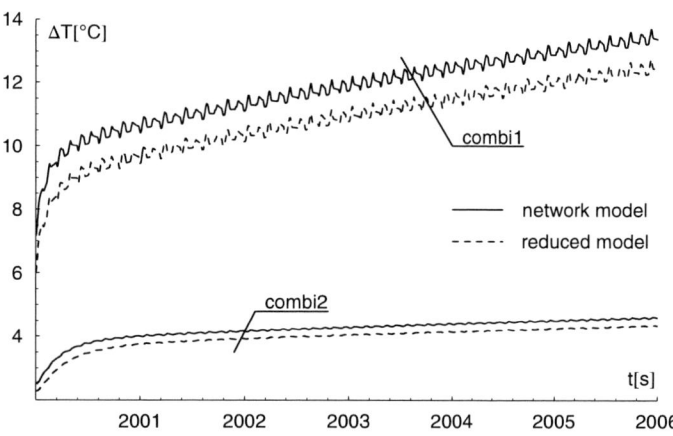

Fig. 4. Temperature response on the I/O-thermocouple at the PCB. Comparison between both combinations as wells as between two modeling methods. ΔT is the difference between the resulting temperature and ambient temperature

dynamic functions are active. There is a significant difference between both heat sinks showing that combi2 is much better. Both modeling approaches show the same transient behavior, which means, that the dynamic character does not change a simulation with less time steps, of course.

The shift on the x-axis in the above figures is necessary due to the permanently active heat sources. Considering these heat sources the shift by 2000 seconds guarantees a steady state allowing to start the dynamic input functions. This is also the reason why the plotted curves don't start at the ambient temperature ($0\,°C$ in the plots).

V. CONCLUSIONS

The work presented in this paper focuses on the use of the model reduction method in thermal modeling. Boundary condition network modeling, based on this method, is applied on a typical semiconductor application. The use of such an approach for thermal simulations in early development stages and additional optimizations of the thermal design is shown in the extract of the examination. Beyond the use in the early development stages, the compact network models can also be used for electro-thermal simulations, without need of creating an additional model for this task. This is possible as the models are calculated directly within a network simulator environment. The accuracy of the results, considering the constrains in the utilized models, can be accepted. However, by a better modeling of the convection more exact results can be expected.

ACKNOWLEDGMENT

The authors would like to thank Julia Czernohorsky and Matthias Viering for their technical support and help on the model preparation.

REFERENCES

[1] H. S. Carslaw, J. C. Jaeger: *Conduction of Heat in Solids*, Oxford University Press, 1959
[2] A. Augustin, T. Hauck, A. Ghazinour: "Thermo-Electric Simulation of a 77GHz Radar Transmitter Chip for Automotive Applications", *EuroSimE Proceedings*, 2006, pp. 476-479

[3] ANSYS Inc., http://www.ansys.com
[4] Flomerics Inc., http://www.flomerics.com/flotherm/
[5] A. C. Antoulas, D. C. Sorensen: "Approximation of large-scale dynamical systems: An overview", *Applied Mathematics & Computer Science 11*, (5) 2001, pp. 1093-1121
[6] T. Bechtold, E. B. Rudnyi and J. G. Korvink: "Error indicators for fully automatic extraction of heat-transfer macromodels for MEMS", *Journal of Micromechanics and Microengineering*, v. 15, No. 3, 2005, pp. 430-440
[7] A. Augustin, T. Hauck, B. Maj, J. Czernohorsky, E. B. Rudnyi and J. G. Korvink: "Model Reduction for Power Electronics Systems with Multiple Heat Sources", *THERMINIC Proceedings*, 2006, pp. 113-117
[8] L. Codecasa: "A Novel Approach for Generating Boundary Condition Independent Compact Dynamic Thermal Networks of Packages", *IEEE Transactions on Components and Packaging Technologies*, vol. 28, 2005, pp. 593-604
[9] H. Vinke and C. J. M. Lasance: "Compact models for accurate thermal characterization of electronic parts", *IEEE Transactions on Components Packaging and Manufacturing Technology Part A*, vol. 20, 1997, pp. 411-419
[10] A. Augustin and T. Hauck: "A New Approach to Boundary Condition Independent Compact Dynamic Thermal Models", *23rd IEEE SEMI-THERM Symposium*, 2007
[11] A. Ammous, S. Ghedira, B. Allard, H. Morel, and D. Renault: "Choosing a thermal model for electrothermal simulation of power semiconductor devices", *IEEE Transactions on Power Electronics*, vol. 14, 1999, pp. 300-307
[12] L. Codecasa, D. D'Amore, and P. Maffezzoni: "An Arnoldi based thermal network reduction method for electro-thermal analysis", *IEEE Transactions on Components and Packaging Technologies*, vol. 26, 2003, pp. 186-192
[13] Y.C. Gerstenmaier, A. Castellazzi, G. Wachutka: "Electro-thermal simulation of multi-chip-modules with novel transient thermal model", *THERMINIC Proceedings*, 2004, pp. 329-334
[14] T. Bechtold, E. B. Rudnyi, and J. G. Korvink: "Automatic generation of compact electro-thermal models for semiconductor devices", *IEICE Transactions on Electronics*, vol. E86C, 2003, pp. 459-465
[15] R. W. Freund: "Krylov-subspace methods for reduced-order modeling in circuit simulation", *Journal of Computational and Applied Mathematics*, Vol. 123, pp. 395-421, 2000
[16] P.J. Heres: "Robust and efficient Krylov subspace methods for model order reduction", *PhD Thesis*, 2005, Eindhoven University of Technology
[17] http://www.imtek.uni-freiburg.de/simulation/mor4ansys/

A PERTURBATION METHOD FOR THE 3D FINITE ELEMENT MODELING OF ELECTROSTATICALLY DRIVEN MEMS

Mohamed Boutaayamou, Ruth V. Sabariego and Patrick Dular

Dept. of Electrical Engineering and Computer Science (ELAP)

University of Liège Sart Tilman Campus, Building B28 , B-4000, Liège, Belgium

mboutaayamou@ulg.ac.be, +32 (0)4 366 37 10

Abstract

In this paper, a finite element (FE) procedure for modeling electrostatically actuated MEMS is presented. It concerns a perturbation method for computing electrostatic field distortions due to moving conductors. The computation is split in two steps. First, an unperturbed problem (in the absence of certain conductors) is solved with the conventional FE method in the complete domain. Second, a perturbation problem is solved in a reduced region with an additional conductor using the solution of the unperturbed problem as a source. When the perturbing regions are close to the original source field, an iterative computation may be required. The developed procedure offers the advantage of solving sub-problems in reduced domains and consequently of benefiting from different problem-adapted meshes. This approach allows for computational efficiency by decreasing the size of the problem.

1. Introduction

Increased functionality of MEMS has lead to the development of micro-structures that are more and more complex. Besides, modeling tools have not kept the pace with this growth. Indeed, the simulation of a device allows to optimize its design, to improve its performance, and to minimize development time and cost by avoiding unnecessary design cycles and foundry runs. To achieve these objectives, the development of new and more efficient modeling techniques adapted to the requirements of MEMS, has to be carried out.

Several numerical methods have been proposed for the simulation of MEMS. Lumped or reduced order models and semi-analytical [1][2] methods allow to predict the behaviour of simple micro-structures. However they are no longer applicable for devices, such as combdrives, electrostatic motors or deflectable 3D micromirrors, where fringing electrostatic fields are dominant [3]. The FE method [4] can accurately compute these fringing effects at the expense of a dense discretization near the corners of the device.

The scope of this work is to introduce a perturbation method for the FE modeling of electrostatically actuated MEMS. An unperturbed problem is first solved in the complete domain taking advantage of any symmetry and excluding additional conductive regions and thus avoiding their mesh [5][6]. Its solution is applied as a source to the further computations of the perturbation problems when conductive regions are added [5]. If the coupling between regions is significant, an iterative procedure is used to obtain an accurate solution. Successive perturbations in each region are thus calculated not only from the original source region to the added conductor but also from the latter to the former.

The perturbation method benefits from the use of different subproblem-adapted meshes, this way computational efficiency increases.

As test case, we consider a micro-beam subjected to an electrostatic field created by a micro-capacitor. The micro-beam is meshed independently of the complete domain between the two electrodes of the device. The electrostatic field is computed in the vicinity of the corners of the micro-beam by means of the perturbation method. For the sake of validation, results are compared to those calculated by the conventional FE approach. Furthermore, the accuracy of the perturbation method is discussed as a function of the extension of the reduced domain.

2. Electric Scalar Potential Weak Formulation

We consider an electrostatic problem in a domain Ω, with boundary $\partial\Omega$, of 2D or 3D Euclidean space. The conducting parts of Ω are denoted Ω_c, with boundary $\partial\Omega_c$, and dielectric ones Ω_d, with $\Omega = \Omega_c \cup \Omega_d$. The governing differential equations and constitutive law of the electrostatic problem in Ω are

$$\mathbf{curl}\,e = 0\,,\quad \mathrm{div}\,d = q\,,\quad d = \varepsilon\,e\,, \qquad \text{(1a-b-c)}$$

where e is the electrostatic field, d is the electric flux density, q is the electric charge density and ε is the electric permittivity. In charge free regions, we obtain from (1a-b-c) the following equation in terms of the electric scalar potential v

$$\mathrm{div}\,(-\varepsilon\,\mathbf{grad}\,v) = 0\,. \qquad (2)$$

The electrostatic problem can be calculated as a solution of the electric scalar potential formulation obtained from the weak form of the Laplace equation (2) as

$$(-\varepsilon\,\mathbf{grad}\,v\,,\,\mathbf{grad}\,v')_\Omega - <\mathbf{n}\cdot\mathbf{d}\,,\,v'>_{\Gamma_d} = 0,$$

$$\forall\,v' \in F(\Omega), \qquad (3)$$

1-4244-1105-X/07/$25.00 ©2007 IEEE

where $F(\Omega)$ is the function space defined on Ω containing the basis functions for v as well as for the test function v'; $(\,\cdot\,,\,\cdot\,)_\Omega$ and $<\cdot\,,\,\cdot>_\Gamma$ respectively denote a volume integral in Ω and a surface integral on Γ of products of scalar or vector fields. The surface integral term in (3) is used for fixing a natural boundary condition (usually homogeneous for a tangent field constraint) on a portion Γ of the boundary of Ω; the normal $\mathbf{n}$ is exterior to Ω.

3. Perturbation Method

An unperturbed problem is first defined in Ω without considering the properties of a so-called perturbing region $\Omega_{c,p}$ which will further lead to field distortions. At the discrete level, this region is not described in the mesh of Ω. The perturbed problem focuses thus on $\Omega_{c,p}$ and its neighborhood, their union Ω_p being adequately defined and meshed will serve as the studied domain. Electric field distortions appear when a perturbing conducting region $\Omega_{c,p}$ is added to the initial configuration. The perturbed problem is defined as an electrostatic problem in Ω_p.

Particularising (1a-b-c) for both the unperturbed and perturbed problems, we obtain

$$\mathbf{curl}\,e_u = 0 \,, \ \mathrm{div}\,d_u = 0 \,, \ d_u = \varepsilon_u\,e_u \,, \qquad (4\text{a-b-c})$$

$$\mathbf{curl}\,e_p = 0 \,, \ \mathrm{div}\,d_p = 0 \,, \ d_p = \varepsilon_p\,e_p \,, \qquad (5\text{a-b-c})$$

where the subscripts u and p refer to the unperturbed and perturbed quantities, respectively. Equations (4b) and (5b) assume that no charge density exists in the considered regions. The source term of the perturbed formulation is determined from the electric potential distribution in the unperturbed domain.

Subtracting the unperturbed equations from the perturbed ones, one gets

$$\mathbf{curl}\,e = 0 \,, \ \mathrm{div}\,d = 0 \,, \ d = \varepsilon_p\,e + (\varepsilon_p - \varepsilon_u)\,e_u \,, \ (6\text{a-b-c})$$

with the field distortions: $e = e_p\text{-}e_u$ and $d = d_p - d_u$. Note that if $\varepsilon_p \neq \varepsilon_u$, an additional source term given by the unperturbed solution $(\varepsilon_p - \varepsilon_u)\,e_u$ is considered in (6c). For the whole of this study, let us suppose that ε_p is the same as ε_u. The zero tangential electric field on the boundary $\partial\Omega_{c,p}$ of the so-considered perfect conductors leads to a condition on the electric scalar potential as follows

$$v = -v_u\big|_{\partial\Omega_{c,p}} . \qquad (7)$$

This way, v_u acts as a source for the perturbation problem.

Two independent meshes are used. A mesh of the whole domain without the additional conductive regions, and a mesh of the perturbing region. A projection of the results between one mesh and the other is then required.

A. Unperturbed electric scalar potential formulation

The unperturbed problem is obtained by particularising ($v = v_u$) and solving (3) as follows

$$(-\varepsilon\,\mathbf{grad}\,v_u\,,\,\mathbf{grad}\,v')_\Omega - <\mathbf{n}\cdot\mathbf{d}_u\,,\,v'>_{\Gamma_u} = 0,$$

$$\forall\,v' \in F(\Omega). \qquad (8)$$

B. Perturbed electric scalar potential formulation

The source of the perturbation problem, v_s, is determined in the new added perturbing conductive region $\Omega_{c,p}$ through a projection method [7]. Given the conductive nature of the perturbing region, the projection of v_u from its original mesh to that of $\Omega_{c,p}$ is limited to $\partial\Omega_{c,p}$. It reads

$$<\mathbf{grad}\,v_s\,,\,\mathbf{grad}\,v'>_{\partial\Omega_{c,p}} - <\mathbf{grad}\,v_u\,,\,\mathbf{grad}\,v'>_{\partial\Omega_{c,p}} = 0,$$

$$\forall\,v' \in F(\partial\Omega_{c,p}). \ (9)$$

In case of a dielectric perturbing region, the projection should be extended to the whole domain $\Omega_{c,p}$. Besides, we choose to directly project $\mathbf{grad}\,v_u$ in order to assure a better numerical behaviour in the ensuing equations where the involved quantities are also gradients.

The perturbation problem is completely characterised by (3) applied to the perturbation potential v_p as follows

$$(-\varepsilon\,\mathbf{grad}\,v_p\,,\,\mathbf{grad}\,v')_{\Omega_{c,p}} - <\mathbf{n}\cdot\mathbf{d}_p\,,\,v'>_{\Gamma_p} = 0,$$

$$\forall\,v' \in F(\Omega_{c,p}), \qquad (10)$$

with a Dirichlet boundary condition defined as $v_p = -v_s$. The process of the resolution and the projection of the electric scalar potential from one mesh to another are represented in Fig. 1.

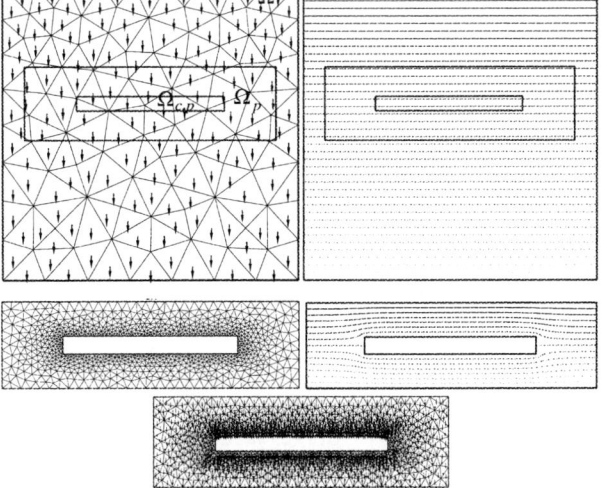

Fig. 1. Unperturbed (mesh of Ω and distribution of the unperturbed electric field e_u (*top left*) and the unperturbed electric potential v_u whose gradient has to be projected on $\partial\Omega_{c,p}$ (*top right*)) and perturbation problems (*middle left*: adapted mesh of Ω_p; *middle right*: distribution of v in Ω_p; *bottom*: distribution of e in Ω_p.

4. Iterative Sequences of the Perturbation Electric Scalar Potential Problem

When the perturbing region $\Omega_{c,p}$ is closed to the original source field, iterative sequences have to be carried out.

The starting point of the iterative process is to determine the unperturbed electric scalar pontential v_u in Ω as shown previously. This solution is then projected from its original mesh to that of the added conductor $\Omega_{c,p}$ and used as a source for the so-called perturbation problem. This way, we obtain a potential on $\partial\Omega_{c,p}$ that will counterbalance the potential on $\partial\Omega_c$. A new source, $v_{s,i}$, for the initial configuration has to be then calculated, with i (i = 0, 2, ...) refers to the iteration number. This is done by projecting v_p from its mesh to that of the unperturbed problem defined in Ω as follows

$$<\mathbf{grad}\ v_{s,i}\ ,\ \mathbf{grad}\ v'>_{\partial\Omega_c} - <\mathbf{grad}\ v_p\ ,\ \mathbf{grad}\ v'>_{\partial\Omega_c} = 0,$$

$$\forall\ v' \in F(\partial\Omega_c). \quad (11)$$

We obtain a new unperturbed electric potential $v_{u,i}$ in Ω as

$$(-\varepsilon\ \mathbf{grad}\ v_{u,i}\ ,\ \mathbf{grad}\ v')_{\Omega} - <\mathbf{n}\cdot\mathbf{d}_{u,i}\ ,\ v'>_{\Gamma_{u,i}} = 0,$$

$$\forall\ v'\ F(\Omega), \quad (12)$$

with a Dirichlet boundary condition defined as $v_{u,i} = -v_{s,i}$.

This sequence refers to the first iteration. Next iteration, a new perturbation problem is solved as described in section 3.B.

This iterative process is repeated until convergence, i.e. the relative difference of the electric field with respect to the conventional FE solution has to be smaller than a given tolerance.

5. Application

A parallel-plate capacitor (Fig. 1) is considered as a 2D FE test case to illustrate and validate the perturbation method for electrostatic field distortions (length of plates = 200 μm, distance between plates: $d = 200$ μm). The conducting parts Ω_c of the capacitor are two electrodes between which the difference of electric potential is $\Delta V = 1V$ (upper electrode fixed to 1V). The perturbing conductive region $\Omega_{c,p}$ is a micro-beam (length = 100 μm, width = 10μm). This perturbing region is placed at a distance d_l of the electrode at 1V.

First, we study the accuracy of the perturbation method as a function of the size of the perturbing domain. In this case, $d_l = 95$ μm. Fig. 2 shows examples of meshes for the perturbing problems. An adapted mesh, specially fine in the vicinity of the corners of the micro-beam is used. Note that any intersection of perturbation problem boundaries with the unperturbed problem material regions is allowed.

The electrostatic field between the plates of the capacitor is first caculated in the absence of the micro-beam. The solution of this problem is then evaluated on the added micro-beam and used as a source for the so-called perturbation problem.

In Fig. 3 (a), the local electric field is depicted for different sizes of the perturbation domain. The first one is a rectangular bounded perturbation region (length = 170 μm, width = 50μm). The second one is a rectangular perturbation domain as well (length = 180 μm, width = 150μm). The third one is an extended perturbation region to infinity through a shell transformation [8].

Comparing with the conventional FE solution, we can observe that the relative error of the local electric field is under 1.5% when the perturbation domain is extended to infinity through a shell transformation (Fig. 3 (b)). This justifies our choice for this kind of perturbation region for the whole of our study.

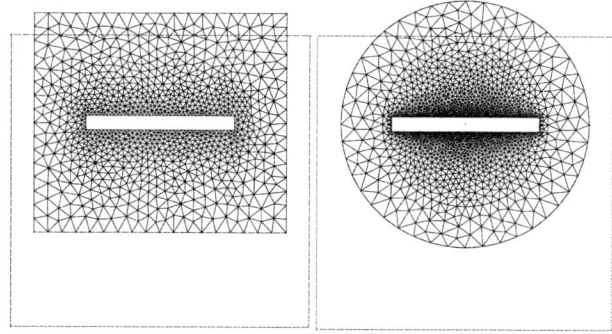

Fig. 2. Meshes for the perturbation problems without (*left*) and with a shell for transformation to infinity (*right*).

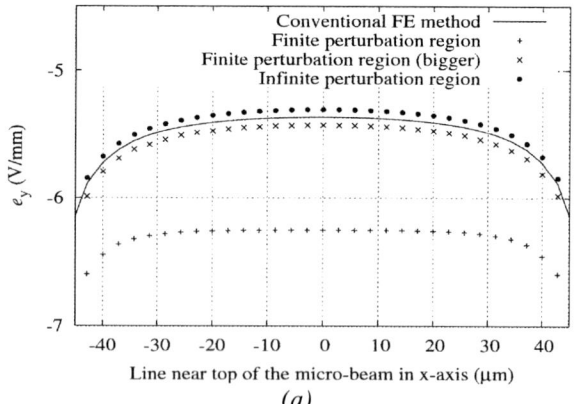

(a)

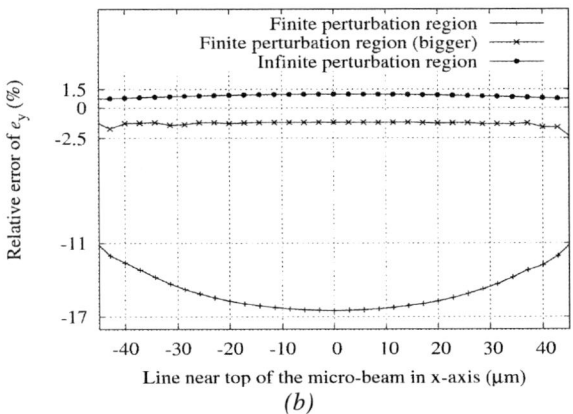

(b)

Fig. 3. Electric field (*y*-component) computed near the micro-beam for different perturbing regions *(a)*. Relative error of the electric field (*y*-component) with respect to the FE solution in each perturbing region *(b)*.

The relative error of the electric potential and the electric field near electrode at 1V increases when the micro-beam is close to the latter (Fig. 4) what highlights a significant coupling of these regions. A more accurate solution for close positions needs an iterative process to calculate successive perturbations in each region.

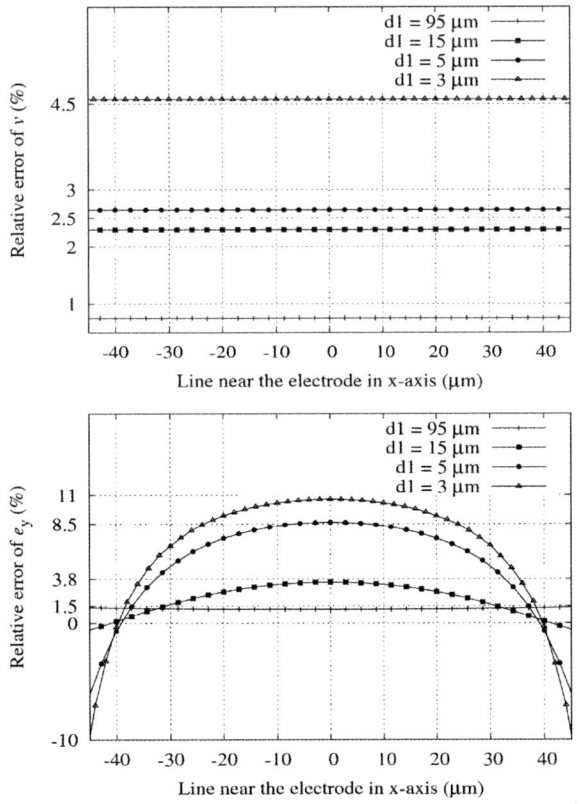

Fig. 4. Relative error of the electric potential *(top)* and the electric field (*y*-component) *(bottom)* computed near electrode at 1V for several distances separating electrode at 1V and the micro-beam.

To illustrate the iterative perturbation process, the distance $d_1 = 3$ μm is chosen as an example (Fig. 5 and Fig. 6). Successive perturbation problems defined in each region are solved.

At iteration 0, the unperturbed electric potential scalar is computed in the whole domain Ω. Projecting this quantity in the domain Ω_p leads to a perturbed electric potential scalar v_p ensuing the electric field perturbation e_p. At this iteration, the relative error of the electric scalar potential computed near the micro-beam and the electrode at 1V with respect to the conventional FE technique is bigger than 2.5% (Fig. 5). Besides, the difference between the *y*-components of e_p and the reference solution (FE) is considerable (relative error up to 14.1 %) which is due to a significant coupling between these regions (Fig. 6). At iteration 1, v_p is projected from its mesh to that of Ω where a new perturbation problem is solved and its solution is projected again in Ω_p (at iteration 2). The relative error of the local electric field at iteration 29 is reduced to 1.1 %.

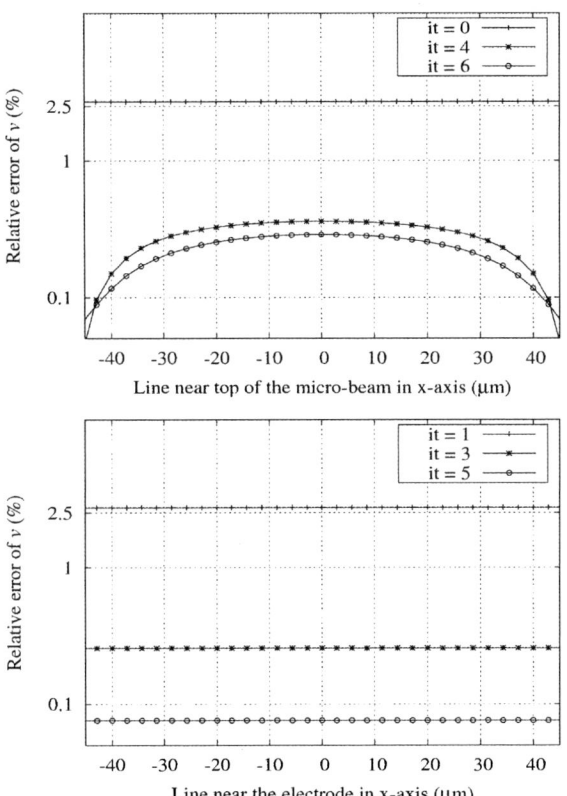

Fig. 5. Relative error of the electric scalar potential computed near the micro-beam *(top)* and electrode at 1V *(bottom)* in some iterations.

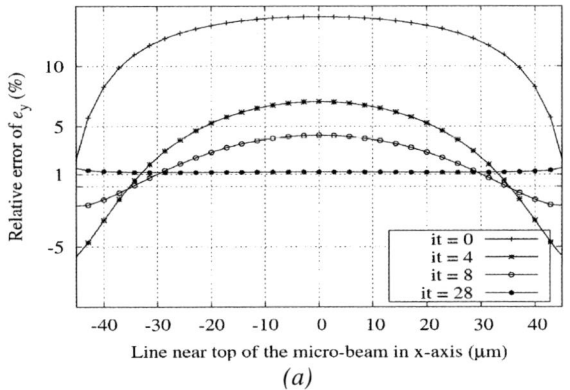

(a)

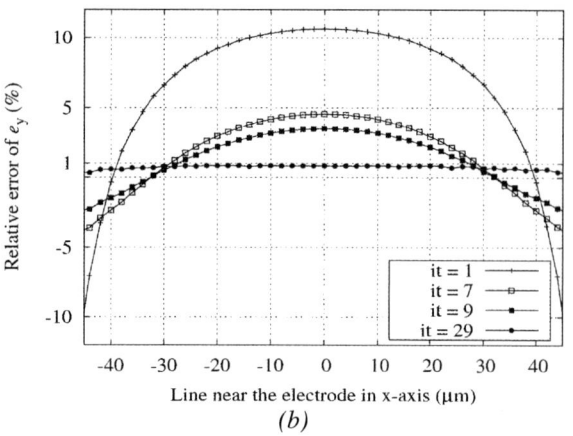

(b)

Fig. 6. Relative error of the electric field (*y*-component) computed near the micro-beam *(a)* and electrode at 1V *(b)* in some iterations.

In order to highlight the relationship between the distance separating the micro-beam and electrode at 1V and the number of iterations required to achieve the convergence, several positions d_1 of the micro-device are considered (Fig. 7). For each of them, the perturbation problem is solved and an iterative process is carried out till the relative error of the local electric field is smaller than 1%.

As expected, several iterations are needed to have an accurate solution when the micro-beam is closed to the considered electrode.

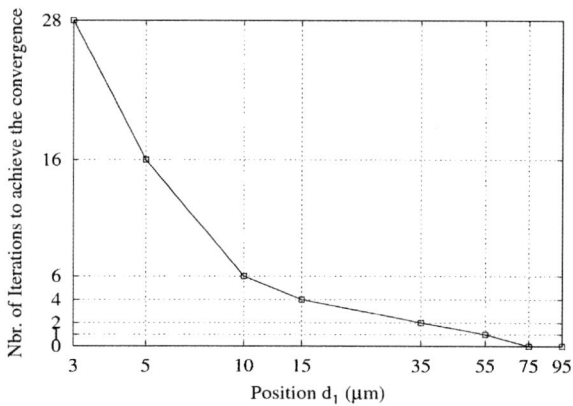

Fig. 7. The number of iterations to achieve the convergence versus the distance separating electrode at 1V and the micro-beam.

Conclusion

A perturbation method for computing electrostatic field distortions due to the presence of conductive microstructure has been presented. First, an unperturbed problem (in the absence of certain conductors) is solved with the conventional FE method in the complete domain. Second, a perturbation problem is solved in a reduced region with an additional conductor using the solution of the unperturbed problem as a source.

In order to illustrate and validate this method, we considered a 2D FE model of a capacitor and a moving micro-beam. Results are compared to those obtained by the conventional FE method. When the moving region is closed to the electrostatic field source, several iterations are required to obtain an accurate solution. Successive perturbations in each region are thus calculated not only from the original source region to the added conductive perturbing domain, but also from the latter to the former.

The convergence acceleration of iterative sequences of the perturbation method will be treated in a future work.

Acknowledgments

This work is supported by the Belgian Science Policy (IAP P6/21) and the Belgian French Community (ARC 03/08-298). P. Dular is a Research Associate with the Belgian National Fund for Scientific Research (F.N.R.S.).

References

1. Younis, M. I. *et al*, "A reduced-order model for electrically actuated microbeam-based MEMS," *Journal Microelectromechanical Systems*, Vol. 12, No 5, (2003), pp. 672-680.
2. Osterberg, P.M., Senturia, S.D., "M-TEST: A Test Chip for MEMS Material Property Measurement Using Electrostatically Actuated Test Structures," *Journal Microelectromechanical Systems*, Vol. 6, No. 2 (1997), pp. 107-118.
3. Boutaayamou, M., Nair, K. H., Sabariego, R. V., Dular, P., "Finite Element Modeling of Electrostatic MEMS Including the impact of Fringing Field Effects on Forces," Proc 7th EuroSimE 2006, Como, Italy, April, pp. 417-421.
4. Rochus, V., Rixen, D., Golinval, J-C., "Modeling of Electro-Mechanical Coupling Problem using the Finite Element Formulation," Proc 10th SPIE Vol. 5049, 2003, pp. 349-360.
5. Dular, P., Sabariego, R. V., "A perturbation method for computing field distortions due to conductive regions with h-conform magnetodynamic finite element formulations," to be published in IEEE Trans. Mag., 2007.
6. Sabariego, R. V., Dular, P., "A perturbation technique for the finite element modelling of nondestructive eddy current testing," Electromagnetic Fields in Mechatronics, Electrical and Electronic Eng. Vol. 27. IOS Press, 2006.
7. Geuzaine, C., Bmeys, B., Henrotte, F., Dular, P., Legros, W., "A Galerkin projection method for mixed finite elements", IEEE Trans. Mag., Vol. 35, No. 3, 1999, pp. 1438-1441.
8. Imhoff, J.F., Meunier, G.,Brunotte, X., Sabonnadière, J.C., "An original solution for unbounded electromagnetic 2D and 3D problems throughout the finite element method", IEEE Trans. Mag., Vol. 26, No. 5, 1990, pp. 1629-1631.

Failure Analysis of Power Silicon Devices at Operation above 200 °C Junction Temperature

Vasile V.N. Obreja [1], Keith I. Nuttall [2], Octavian Buiu [2], and
Steve Hall [2]

[1]National R&D Institute for Microtechnology (IMT), Str. Erou Iancu Nicolae, 077190, Bucharest, Romania
Tel: +4021-490 8212, Fax: +4021- 490 8238, E-mail:vasileo@imt.ro
[2]Department of Electrical Engineering and Electronics, The Liverpool University, *L69 3BX*, Liverpool, UK
Tel: +44-151 –794-4508, Fax: +44-151-794-4540 E-mail: ee16@liv.ac.uk, O.Buiu@liv.ac.uk

Abstract

A temperature of 200 °C for the PN junction of power silicon devices (diodes, thyristors, transistors) is known as a limit for their reliable performance. PN junction failure after operation above this temperature consists in excessive high current or even electrical shortcircuit when reverse bias voltage is applied. Enough information in the published literature on this subject does not exist at this time. PN junction devices available at this time on the market were measured and placed in a hot chamber at constant ambient temperature higher than 200 °C. Junction blocking voltage was applied and the level of leakage current was monitored. The silicon die after de-capsulation of failed devices exhibits a small region of material degradation located at the junction periphery, causing excessive high leakage or short-circuit of the junction. The leakage current flowing at the interface between semiconductor and the passivating dielectric layer from the junction edge is a key factor involved in device failure. Lower level and uniform distribution of this current around the junction periphery can enable reliable operation above 200 °C. Experimental results are presented and analyzed .

1. Introduction

In the data sheets of commercial power silicon devices manufactured at this time, specified values of the maximum permissible junction temperature, T_{JM}, during device operation are limited to 200 °C. Operation above this value may lead to device failure. As a consequence, for reliable operation at higher temperature towards 300°C, devices based on wide band-gap semiconductor materials have been developed. Nevertheless commercial available devices based on such materials have a value of T_{JM} also limited to 200 °C. High level of the leakage current for silicon devices due to junction edge has been known as major limitation for reliable operation even below 200 °C, [1-3]. In [4 -5] it is shown that the junction edge current may be dominant from room temperature up to 300 °C. Preliminary data about failure mechanism related to the junction edge current were reported in [6]. Less information there is in the literature regarding the unreliable operation and failure mechanisms of power silicon devices above 200°C.

The purpose of this work is to provide new experimental results and failure analysis from an investigation on operation of commercial power silicon devices at junction temperature, T_J, in a range of 225 – 275°C.

2. Experimental procedure

Operation tests of power silicon diodes, thyristors and even power MOSFETs at junction temperature higher than 200 °C have been carried out. The devices mounted on large heatsinks were placed in a hot chamber at constant ambient temperature. In the device blocking state by means of a suitable circuit, the desired voltage was applied, so that, the level of leakage current to be monitored. Device failures under blocking bias voltage have appeared starting from 200 °C in the hot chamber. After device failure, excessive high current in the blocking state or electrical short-circuit was manifested. The failed devices have been de-capsulated and the silicon die was released without any damage. Glass passivated junctions as shown in Fig.1 or planar oxide passivated junctions as shown in Fig.2 were exhibited by the silicon dice from the de-capsulated failed devices. Small spots of degraded passivation material (glass or oxide) located at the junction (die) corner or outside a corner, as shown in Figs.1-2 have been found by microscope inspection. The local spot or defect causing excessive high current (electrical short-circuit) does not represent more than a few percent of the junction area or its perimeter. By suitable removal of this defect (for example material etching), partial or total recovery of the junction current capability takes place.

It has been observed that before failure, increase in the level of leakage current with the time takes place. By means of an I-V curve tracer, reverse (blocking) electrical characteristics have been displayed at junction temperature in a range of 200 – 275 °C. A hysteresis effect is manifested by the displayed unstable current-voltage characteristic before failure, [7-8]. This effect consists in the fact that two different values of the blocking leakage current can be seen for the same value of the applied sine wave voltage. Consequently, fluctuation of the junction instantaneous temperature takes place when the sine wave voltage is applied. Such behavior is caused by non-uniform distribution of the leakage current at the junction periphery. For a small portion of the junction periphery, higher leakage current flow is possible than for another portion of the same length taken on the periphery. In such a portion higher local temperature is possible, [9], and finally, it may

1-4244-1105-X/07/$25.00 ©2007 IEEE

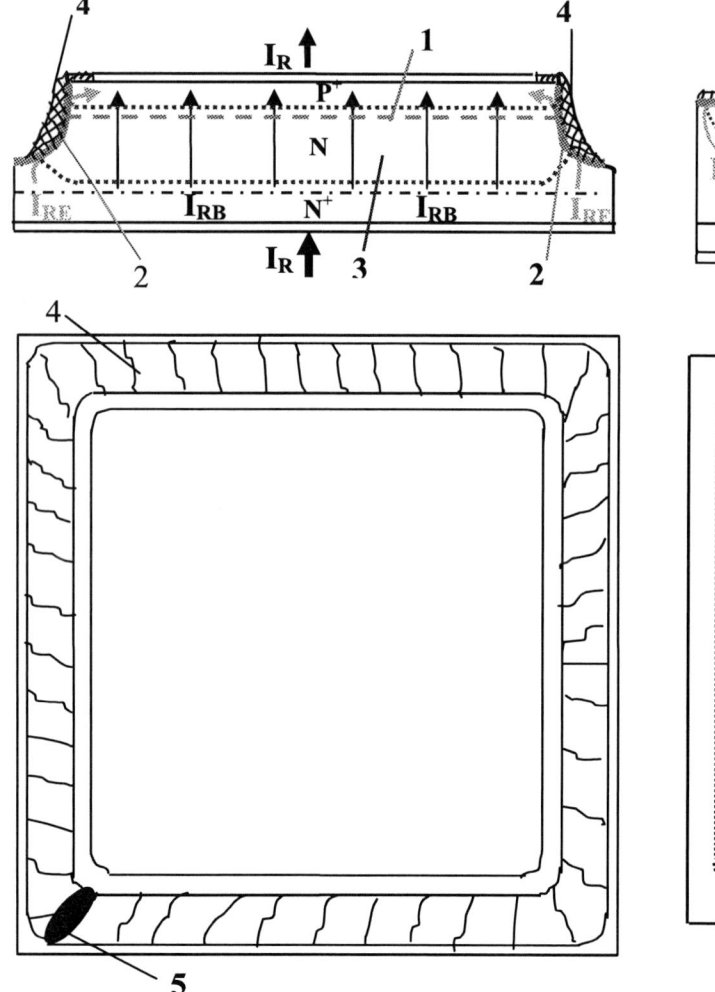

Fig.1. – Silicon diode die with glass passivated junction (cross section and view from the front side); 1- PN junction; 2 – dielectric–semiconductor interface at junction periphery; 3 – charge depletion layer in the junction bulk; 4- glass passivation layer; 5- spot of material degradation at the junction edge;

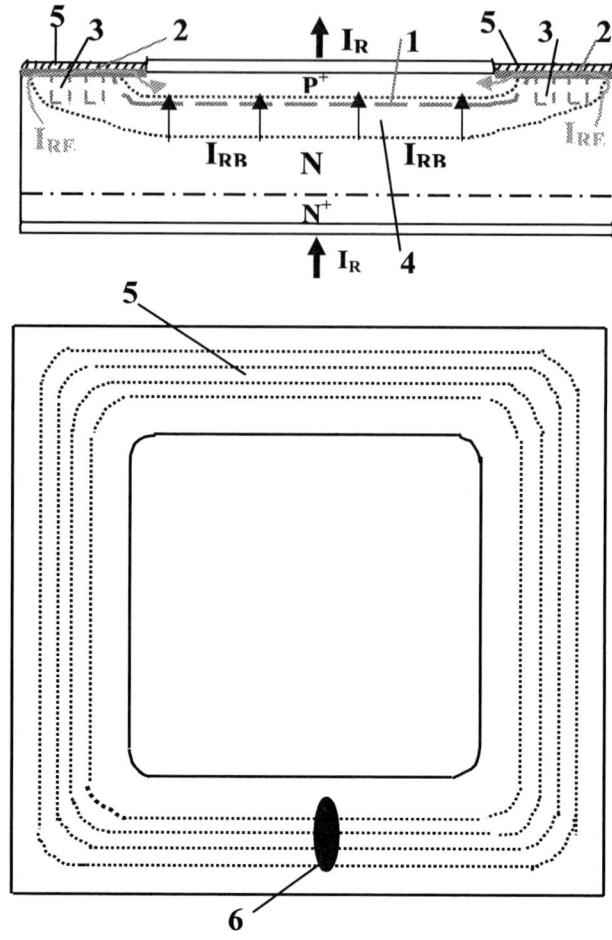

Fig.2. – Planar silicon diode die with guard rings and oxide passivated junction (cross section and view from the front side); 1- PN junction; 2 – dielectric–semiconductor interface at junction periphery; 3- P^+ diffused guard rings   4– charge depletion layer in the juction bulk;   5- oxide passivation layer; 6- spot of material degradation at the junction edge;

become the location of the spots of material degradation shown in Figs. 1-2.

3. Experimental results, analysis and discussion

Typical reverse (blocking) electrical characteristics plotted in linear scales for encapsulated diodes are shown in Figs. 3- 4. These characteristics have been measured by applying reverse bias voltage for short time to avoid device selfheating. The characteristics in Fig.3 correspond to two samples of controlled avalanche diode. The main ratings from the data sheet are the maximum rectified current of 40A, and the maximum permissible repetitive reverse voltage, V_{RRM}, specified at 1200 V for a maximum junction temperature, T_{JM} =160 ^oC. At this temperature, short duration (10 µs) applied voltage is allowed to reach the breakdown voltage value of about 2000 V. Operation for 10 µs in the breakdown region is

possible only for a limited value of the avalanche current. The maximum value of the reverse current, I_{RRM}, indicated in the data sheet at V_{RRM} =1200 V and T_J = 160^oC is 4.5 mA. For the two samples in Fig.3 one can see that the level of I_{RRM} is significantly lower than the specified value of 4.5 mA even at higher temperature of 200 ^oC. In the device data sheets, usually, higher specified values are given than the actual measured ones, to cover all the manufactured devices and possible variations. Specification of 4.5 mA in the data sheet and the measured results from Fig.3 indicate that large fluctuations of the I_R level are possible from sample to sample, on behalf of the junction edge current component, I_{RE}, shown in Figs. 1-2. The bulk component, I_{RB}, can be significantly lower than I_{RE} ($I_R = I_{RE} + I_{RB}$).

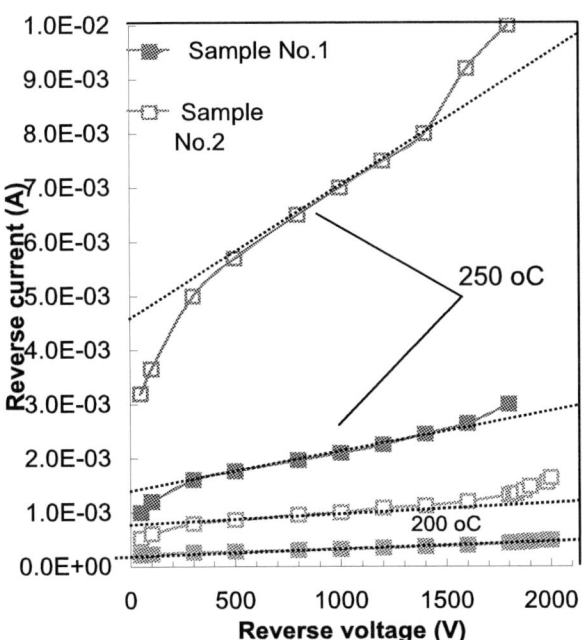

Fig.3. – Current – voltage characteristics at 200 °C and 250 °C for two samples of commercial controlled – avalanche rectifier diode (40A, 1200 V)

Linear variation of current with the voltage for a portion of the current-voltage characteristic is visible in Fig.3. This resistive behavior could be attributed to the flow of leakage current, I_R, through the interfacial layer from the junction periphery (Figs. 1-2). The resistance of this very

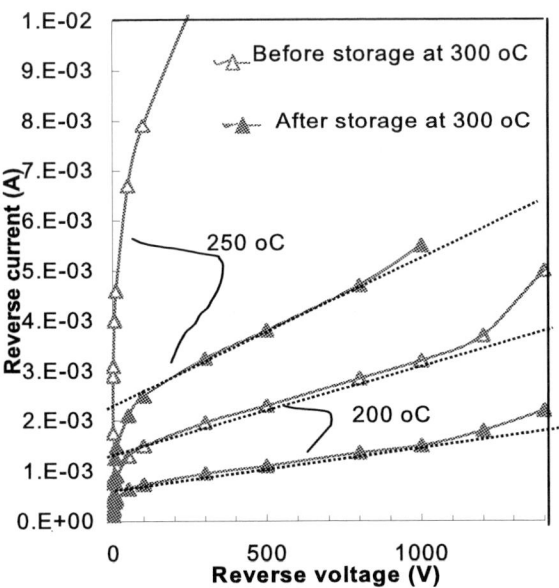

Fig.4. – Current – voltage characteristics at 200 °C and 250 °C for a sample of ordinary commercial rectifier diode (25A, 1200 V)

thin layer interfacing the semiconductor junction periphery and the dielectric passivating material layer may determine the electrical characteristic of the junction with its partial linearity. For the samples used in Fig.3 no failure occurred at 200 °C for V_{RRM} =1200 V. Nonetheless failure occurred at 250 °C for the sample No.2 but not yet for the sample No.1. Failure of the sample No.1 occurred at 275°C when the current level was higher. The defects after failure were found at the junction corners (Fig.1).

Other results shown in Fig.4 for an ordinary rectifier diode of 25 A and V_{RRM} =1200 V from the same manufacturer as for the device samples used in Fig.3 indicate that significant reduction in the level of I_R is possible after device storage at 300 °C. Such behavior is possible due to the changes in the resistive interfacial layer located at the junction periphery between semiconductor and the passivation material (Fig.1). For this diode sample, failure occurred at 250 °C and V_{RRM} =1200 V, although significant reduction in the I_R level took place.

The non-uniform distribution of the junction edge current component (I_{RE}) at the junction periphery is a major problem. Suitable devised experiment for a diode die placed on hot plate, [9 -10], and use of an infrared imaging microscope, revealed non-uniform distribution of I_R at T_J = 220 °C, as shown in Fig.5. The applied voltage was only 500 V. Higher temperature regions of about 228°C at the two corners from the left side, in comparison with 220 °C for a good part of the die area indicate that most of the I_R leakage current is concentrated at the two corners. The square shape silicon die (glass passivation) used in Fig.5 has a size of about 14.5 mm. The size of silicon die in the case of Fig.3 is about 5.5 mm and in the case of Fig.4 is about 4 mm. Electrical characteristics for the diode die used in Fig. 5 are shown in Fig.6. It is seen from Fig.6 that visible linear electrical characteristics are exhibited up to 1000 V at temperature of 200 and 225 °C. Although linearity is manifested up to 1000V in Fig.6 at 225 °C, non-uniform distribution of I_{RE} on the junction perimeter is possible at lower voltage of 500 V as shown in Fig. 5. From Fig. 6 it is seen that at 250 °C and above 300 V, deviation from linear variation takes place due to further accentuation in the non-uniform distribution of I_{RE} around the junction edge. Attempt of operation for longer time at this temperature and applied voltage of 1000 V will result in device failure with a defect in the corner (Fig.1).

As it is known, the value of T_{JM} specified for commercial thyristor devices available at this time is limited to 125- 150 °C, [11-12]. Despite the diode case (Fig.1), the silicon die of thyristor devices has two identical blocking PN junctions, one at the front side and the other at the back side. Reliable operation of thyristors at T_{JM} = 175 - 200 °C would be beneficial in applications. Nonetheless due to the I_{RE} component, visible split is possible for the blocking electrical characteristics correspondig to the two identical junctions, starting from

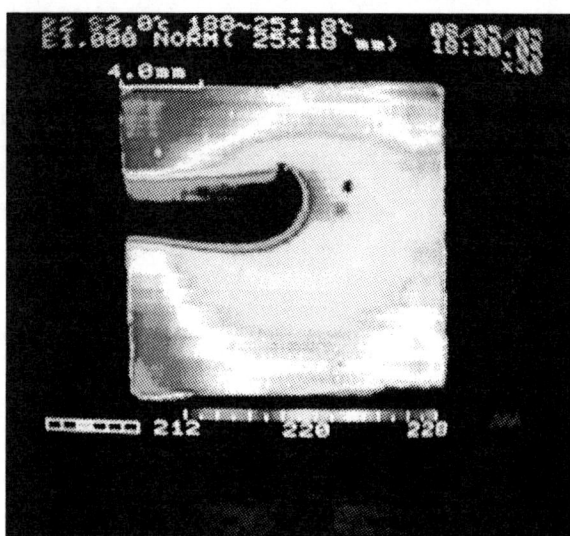

Fig.5. – Infrared microscope image of a glass passivated die of silicon rectifier diode on hot plate at 220 °C and applied DC voltage V_R =500 V (device ratings 250A, 1600 V)

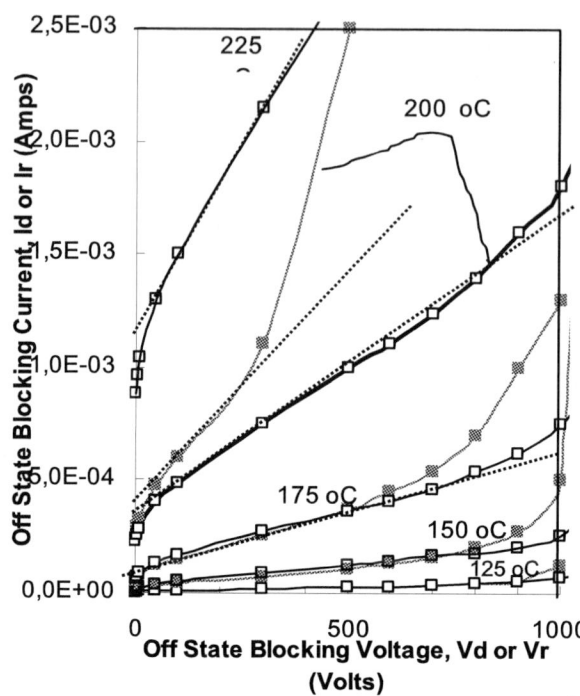

Fig.7. - Off - state direct and reverse blocking current – voltage characteristics in the temperature range 125 – 225°C for a 16A and 600 V thyristor

limited to 150 –175 °C, [13]. The silicon die is different from a bipolar transistor because instead of a single blocking junction, many cell junctions connected in parallel can operate in the blocking state. Typical electrical characteristics above 200 °C, for a 7A and 600V commercial power MOS transistor are shown in Fig.8. One can see that above 400 V, visible deviation from linear dependence takes place. Consequently, unreliable operation at 250 °C and 600 V is possible.

In the above, results for PN junction devices with high charge carrier lifetime were analyzed. For fast switching diodes based on low lifetime in the junction base, reliable operation even at 150 °C junction temperature may be a problem. The I_R level for such diodes is significantly higher than for the similar standard recovery ones (high lifetime in the junction base), [8]. This time, the I_{RB} component (Figs. 1-2) is without doubt the primary component of the I_R current. Nonetheless the junction edge component (I_{RE}) is not negligible. In Figs. 9-10 electrical characteristics at 150 °C and 200 °C are shown. The sample No.1 corresponds to junction termination shown in Fig.1 and the sample No.2 corresponds to the termination shown in Fig.2. In Figs. 9-10, the same linear voltage dependence is exhibited as in the case of standard recovery diodes (Figs. 3-4). At 150°C and above 1000 V, visible deviation from linear variation takes place. Operation at 150 °C and 1200 V (the maximum value of V_{RRM} specified in the data sheet for this temperature) resulted in device failure with spots of material degradation located at the junction edge .

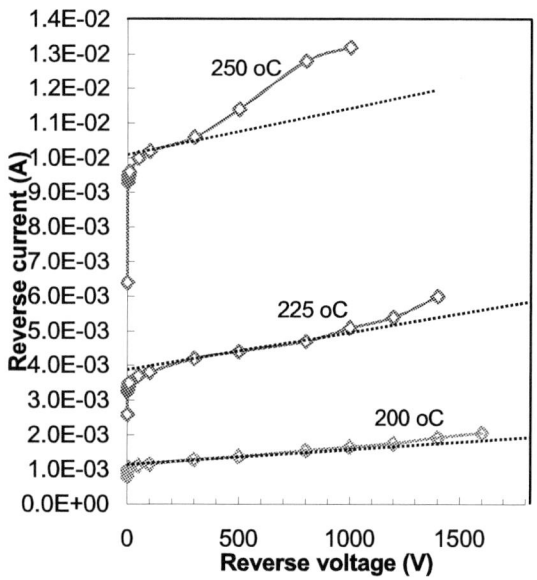

Fig.6. – Current – voltage characteristics at 225 °C for the silicon die of rectifier diode sample used in Fig.5 (device ratings 250A, 1600 V)

150 °C. Typical results are shown in Fig.7 for a 16 A and 600 V thyristor. Device failure has occurred for the thyristor sample used in Fig.7 when direct bias voltage of 600 V was applied for longer time at 200 °C. The defect was found near the die corner.

In the case of power MOS transistors failures have occurred above 200 °C. In the data sheets, specified values for the maximum operation temperature (T_{JM}) are

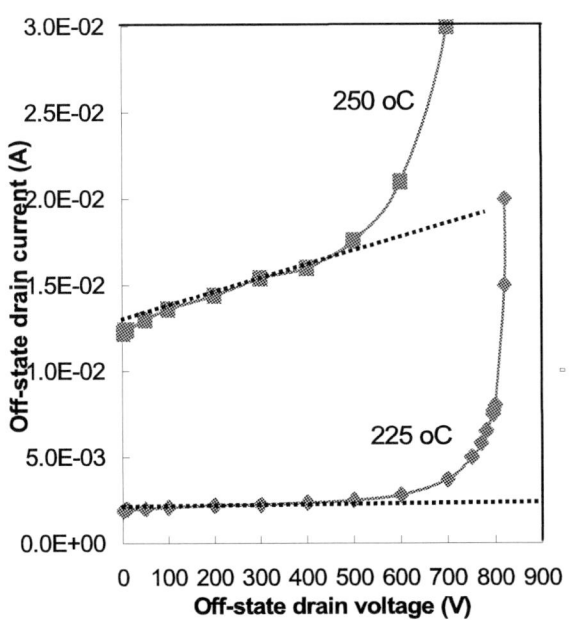

Fig.8. - Off - state drain electrical characteristics at 225 and 250 °C for a 7A and 600 V power MOS transistor

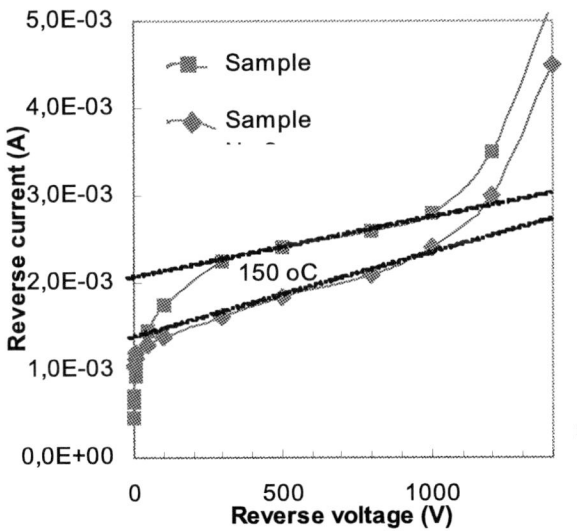

Fig.9.- Electrical characteristics at 150°C for 20A and 1200 V fast switching diodes (low lifetime in the junction base)

4. Conclusions

Above 200 °C the junction edge leakage current of commercial high voltage devices can cause device definitive failure, manifested as excessive high current or electrical short-circuit when reverse voltage is applied. A spot of material degradation located at the junction periphery has been found for the failed devices. This small region of degraded material causes electrical short-

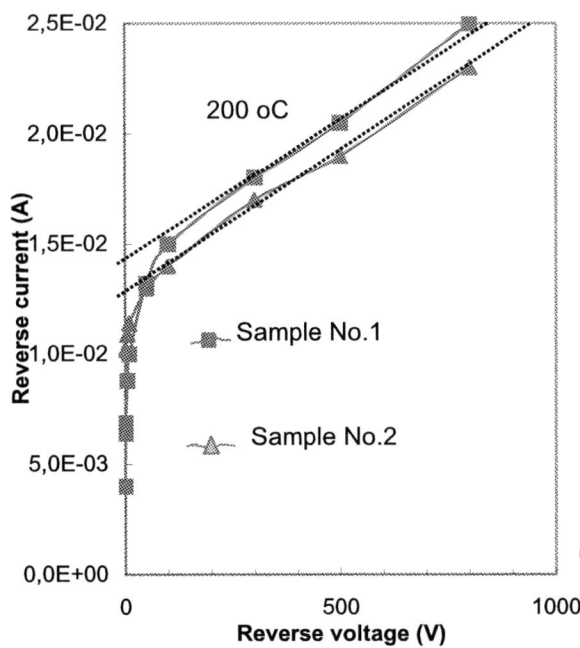

Fig.10.- Electrical characteristics at 200°C for 20A and 1200 V fast switching diodes (the same samples as in Fig.9)

circuit or very high current in the blocking state. Non-uniform distribution of the leakage current around the junction perimeter favors electrical characteristic instability followed by device failure. The PN junction current-voltage characteristics plotted in linear scales exhibit a portion of linear variation. This behavior could be attributed to the resistance of the interfacial layer situated between the semiconductor and the passivating dielectric layer from the junction periphery. For commercial available devices, this layer has significant influence on the level of reverse current, and on device reliability at junction temperature higher than 200°C.

Acknowledgments

Partial support for this work from the "CEEX" Romanian R&D Program by means of the contract No. 310/13.09.2006-AMCSIT-Politehnica is gratefully acknowledged .

References

1. R.R. Verdeber et al, " SiO2/Si3N4 passivation of high power rectifiers" *IEEE Trans-Electron Devices,* vol. 17, 1970, pp. 797 -800
2. S. Salkalachen, N. Krishnan, S. Krishnan, H. Satyamurthy and K. Srinivas, "Edge passivation and related electrical stability in silicon power devices", *IEEE Trans. Semiconductor Manufacturi*ng, vol.3, 1990, pp. 12-17
3. J.R. Trost, R.S. Ridley, M. Kamal Khan and T.Grebs, "The effect of charge in junction termination extension passivation dielectrics," *Proc.11*[th]

International Symposium on Power Devices and ICs, Switzerland, May 1999, pp. 189 -192

4. V.V.N. Obreja, " On the leakage current of present-day manufactured semiconductor junctions" *Solid-State Electronics,* vol.49, no.1, 2000 , pp. 49-57

5. V.V.N. Obreja, "An experimental investigation on the nature of reverse current of power silicon PN junctions", *IEEE Trans. Electron Devices,* vol.49, No.1, 2002, pp.155-163

6. K.I. Nuttall, O. Buiu and V.V.N. Obreja, "Surface leakage current related failure of power silicon devices operated at high junction temperature", *Microelectronics Reliability,* vol. 43, Sept-Nov.2003, pp. 1913 –1918

7. V.V.N. Obreja, C. Codreanu, K.I. Nuttall, O. Buiu, "Reverse current instability of power silicon diodes (thyristors) at high temperature and the junction surface leakage current", *Proceedings IEEE International Symposium on Industrial Electronics (ISIE2005),* June 2005, Dubrovnik, Croatia, pp. 417 – 422

8. V.V.N. Obreja, C. Codreanu, K.I. Nuttall, "Reverse leakage current instability of power fast switching diodes operating at high junction temperature" in *Proceedings 36th IEEE Power Electronics Specialists Conference(PESC2005),* June 2005, Recife, Brazil, pp. 537-540

9. V.V.N. Obreja, C. Codreanu, K. Nuttall, O. Buiu, "Experiments on behaviour of power silicon PN junctions under reverse bias voltage at high temperature ", *Proceedings EuroSIME2004 – Thermal &Mechanical Simulation and Experiments Microelectronics and Microsystems,* May 2004, Brussels, Belgium, pp. 185 –190

10. V.V.N. Obreja, C. Codreanu, K.I. Nuttall, I. Codreanu, "Peaks in temperature distribution over the area of operating power semiconductor junctions related to the surface leakage current" *Proceedings 6th International Conference on Thermal, Mechanical and Multiphysics Simulation and Experiments in Micro-electronics and Micro-systems (EuroSIME2005),* April 2005, Berlin, pp. 584 -589

11. V.V.N. Obreja, C. Codreanu, K. Nuttall, O. Buiu, C. Podaru, "The operation temperature of silicon power thyristors and the blocking leakage current", *Proceedings 35th IEEE Annual Power Electronics Specialists Conference (PESC04),* June 2004, Aachen, Germany, pp. 2990 – 2993

12. V.V.N. Obreja, E. Manea, C. Codreanu, M. Avram, C. Podaru, "The junction edge leakage current and the blocking I-V characteristics of commercial glass passivated thyristor devices", *Proceedings 2005 International Semiconductor Conference (CAS2005),* Oct. 3-5, 2005, Sinaia, Romania, pp. 447 – 450

13. V.V.N. Obreja, C. Podaru, E. Manea, A. Coraci, C. Codreanu, "Experimental drain I-V characteristics of power MOS transistors and the nature of reverse leakage current of oxide passivated PN junctions", *Proceedings 2005 International Semiconductor Conference (CAS2006),* Sept., 2006, Sinaia, Romania, pp. 305 – 308

Simulation of Wafer Probing Process Considering Probe Needle Dynamics

Ilko Schmadlak, Torsten Hauck
Freescale Halbleiter Deutschland GmbH
Schatzbogen 7; D-81829 München; Germany
ilko.schmadlak@freescale.com, +49 89 92103685

Abstract

One key failure root cause during wafer production is the electrical test by probing all interconnects of each die of the wafer. The probing process can result in excessive damage of the Back End of Line (BEOL) wafer layers underneath the probe pad, especially if brittle Low-k dielectrics are used. The industry is trying to reduce design limitations for these under pad areas.

A 3D sub-modeling simulation approach has been used to investigate stress states in layers below the probe pad. The dynamic behavior of the needle is determined by modal analysis. Its representation and calculation is done by an analytical model using the modal superposition method. These small, specialized simulation models help to distribute the simulation effort, by handling specific aspects of the real process, such as the contact problem, needle dynamics and homogenization of fine structures. The results showed that the needle dynamics can be neglected during further studies. The static load due to the maximum displacement of the needle tip during the probe event is one magnitude higher then any dynamic driven load.

1. Background

The electrical test of wafers is established through mechanical contact between the probe machine and the wafer. The contact is made by a defined movement of the wafer towards the fixed probe card, with the required set of probe needles. These needles test every pad on a die during one event. The contact needs to satisfy a certain quality in order to avoid false fails during the test. The current technology has a limit of around 1000 pads per die or needles per probe card. In case of a dual probe card, even two dies can be tested in one go. Due to the size of the probe pads and the described density at die level, the probe event needs to be adjusted very tightly [1]. The reason for this is a minimum contact force that needs to be achieved for all probe pads in order to penetrate through the native aluminum oxide layer. If the load is too high the sensitive structures underneath will fail. In opposition to this, the probe process should also be fast. This maximum load is relatively high for the probing on bond pads since the structures beneath the bond pads are supported by rigidizing features. This means they are especially designed to withstand the forces that act during wire bonding and therefore probing. However, this probe location has the disadvantage that the probing destroys the surface of the pad and can expose the copper underneath. To achieve a valid probe result, the process needs to be repeated. Usually each pad is tested 6-8 times. After that, the pads often do not meet the required surface quality, needed for a reliable wire bond process (Fig. 1). As a result the probability for the wire bonds to fail, rises. In order to avoid this issue, the probe over passivation design was developed in which the bond pads were extended to allocate a special probe area on the pad and eliminate probe wire bonding interferences. The longer pads do not increase die size and require only one mask change. In order to obtain the maximum pay-off from this design, the design freedom underneath the extended probe area should not be compromised. Nevertheless, the probe over active process is very sensitive and a number of different failure types occurred in the under pad structures.

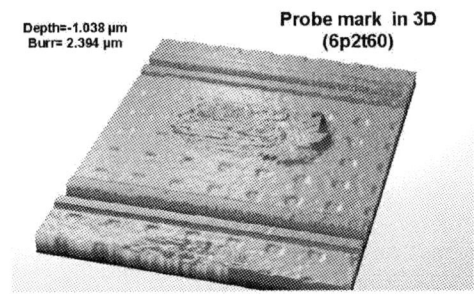

Fig.1: 3D probe mark relief

2. General Modeling Approach

Finite Element (FE) modeling and simulation was performed in order to assess the probe process and its impact on the structural integrity of the BEOL-layers underneath. The approach considered a detailed FE-model of a test vehicle probe pad, in order to allow a good comparison with results of planned nano-indentation and probe tests on the same pad. Due to the complexity of the design, other simplifications had to be made. A lot of the very small repetitive patterns in the layers were substituted by volumes with equivalent material properties to guarantee the same global stiffness.

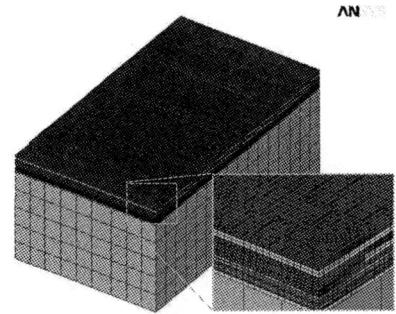

Fig.2: FE-pad model with detailed BEOL structure

1-4244-1105-X/07/$25.00 ©2007 IEEE

The homogenization procedure which was used to derive the orthotropic material properties requires the FE-modeling of the patterns smallest repetitive unit cell (RUC). This model is then simulated under 7 different loading and boundary conditions. Three simulations cover all three axial stiffnesses of the RUC, and allow the calculation of the three Young's moduli plus Poisson's ratios. The next three simulations test the structures resistance against the three shear modes from which all shear moduli can be derived. The last simulation calculates the global coefficient of thermal expansion of the structure [2]. The displacements of the RUC's sides have to be restricted by the right boundary conditions in order to fulfill the required symmetries. Its shape always needs to satisfy the definition of a parallelepiped, when subjected to the test load.

As a result the model size was decreased considerably by keeping the needed resolution in structure details. In order to allow the simulations to run through in a reasonable amount of time, the probe process as a whole had to be simplified accordingly. In order to transfer the pressure distribution into the correct nodal forces for an arbitrary mesh, the principle of virtual work was used [3]. The approach requires that the pressure distribution over the probe pad is known. It is therefore either simulated outside, in a small contact model or can be defined as a function. Eq. (1) allows the calculation of the in plane and out of plane nodal forces for every finite element in the defined touch down area, knowing the probing pressure distribution. The equation needs to be satisfied for every arbitrary virtual displacement δu.

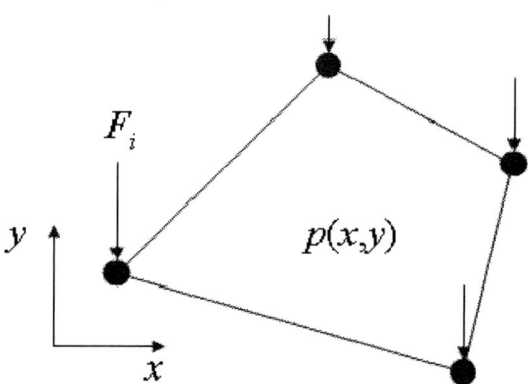

Fig.3: Nodal forces need to be calculated for every arbitrary shaped element with known pressure distribution

$$\sum_{i=1}^{4} F_i \delta u_i = \iint_{A_e} p(x, y)\delta u(x, y)\mathrm{d}A \qquad (1)$$

This method decouples the contact problem and the calculation of the loading condition from the pad model and enables its efficient usage. Another aspect of the probe process, the effect of needle dynamics, had to be investigated further in order to decide whether those

effects can be neglected or need to be considered in the analysis. The background is the dynamic behavior of the needle, which leads to high contact forces shortly after the first contact between pad and needle has been established. The moving pad, making contact with the needle, results in a sudden acceleration peak for the needle tip. The following vibration is the result of that excitation and is damped quickly which leaves only the static probe force as a load. Fig. 4a-c show the needle tip and probe pad displacements as well as their speeds and accalerations respectively during a probe event. The charts in Fig.4c show that the needle is heavily excited in the instant of contact. It is essential to know what effect that acceleration peak of the needle tip has on the contact forces that occur due to that.

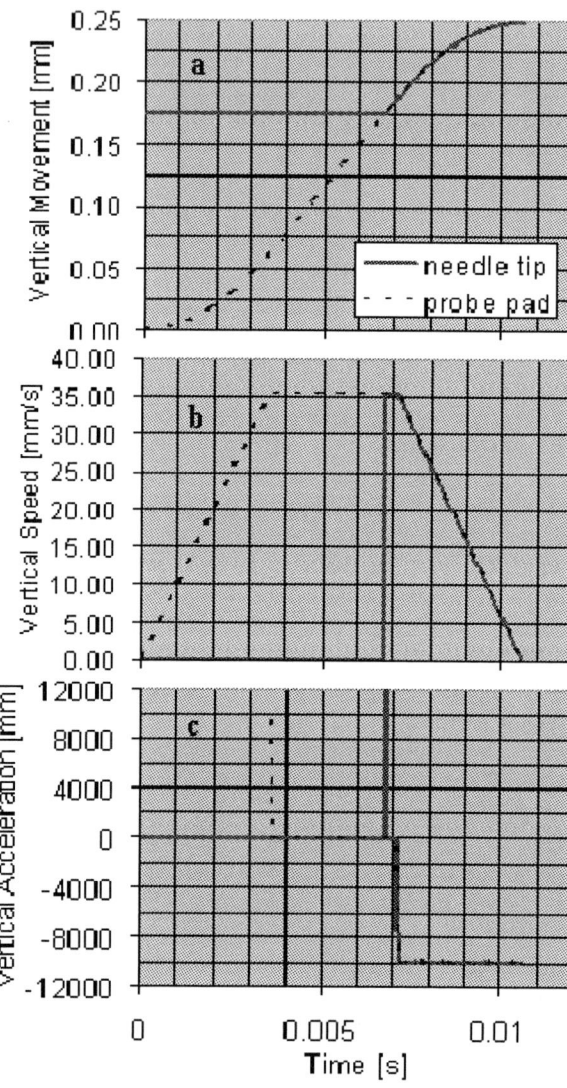

Fig.4a-c: Probe pad and needle tip displacements (4a), speeds (4b) and accelerations (4c) for typical probe event

3. Effects of Needle Dynamics

In order to describe the dynamic behavior of the needle tip, a parametric Finite Element model was developed, and a modal analysis was performed.

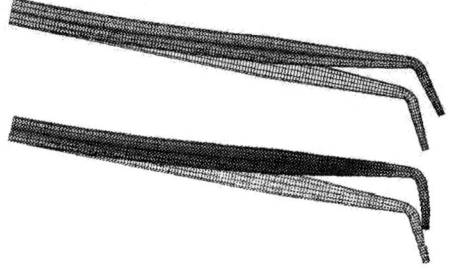

Fig.5a-b: Needle deformation for scrubbing (5a) and stubbing mode (5b)

Fig.6a-c: First three mode shapes of needle under stubbing conditions

It was assumed that the highest contact forces due to needle dynamics would appear in the case of pure stubbing. This means that no scrub is achieved during probing. This results in higher static probe forces but decreased quality of the established contact. It is therefore desirable to avoid stubbing completely. In this analysis, stubbing was established by simply locking the horizontal displacement u_x of the needle tip. In order to describe the dynamic behavior of the needle during the probe event in a simplified way, the modal superposition method was used. Therefore, only the vertical displacements u_y of the first 10 modes of the needle tip had to be extracted since they carry all the needed information of the dynamic needle behavior inside them. The needle tip is as well the only node, at which a load through contact is applied. As mentioned, the system is reduced to an order of 10, since not all of the calculated natural modes are needed for an accurate description. The higher modes are considered nonsignificant due to their small amplitudes. The solutions for the decoupled equations are calculated and superposed in order to derive the global solution. The following equations describe the different steps.

Prescribed Motion of the Wafer

$$w = v_0 t + \frac{(v_0 t)^2}{4\delta} \qquad (2)$$

Vertical Displacement of the needle tip

$$u_y = \sum_{i=1}^{10} a_i q_i \qquad (3)$$

Equations of Motion

$$\ddot{q}_i + 2\omega_i \xi \dot{q}_i + \omega_i^2 q_i - a_i f = 0, \quad i = 1, \cdots, 10 \qquad (4)$$

Contact Condition

$$\sum_{i=1}^{10} a_i q_i + \frac{f}{c} = w \qquad (5)$$

$$c = \begin{cases} \text{contact stiffness } c_0 & \text{penetration} \\ \text{no stiffness} & \text{no penetration} \end{cases} \qquad (6)$$

Initial Conditions

$$q_i(0) = 0, \dot{q}_i(0) = 0 \qquad (7)$$

Once the dynamic behavior is described by such a decoupled system, its superposed solution can be derived with any mathematical software.

For the following calculations the initial time $t=0$ was defined to be the instant of first contact between moving wafer and the fixed probe needle. Eq. (2) defines the motion of the wafer, which reflects the process sequence shown in Fig.3a. The movement of the wafer after contacting the needle, is called overdrive. It is one key parameter for the definition of the process window [4]. The displacement of the needle is described by Eq. (3) using the first 10 natural modes. Eq. (4) is the decoupled system of differential equations that describes the needle motion after excitation. The results of the modal analysis from the FE simulation consider a normalization of the mass matrix. Hence, the mass matrix does not need to be considered in Eq.(4). In Eq.(5) the conditions for the contact are defined. The contact is realized by the parameter c which can be imagined like a spring with variable stiffness. For the case that the contact force on the probe pad is positive, c equals the contact stiffness $c_0=15000$ mN/μm. In case it is negative c will be set to a very small value, close to 0. The value for c_0 was motivated by the stiffness of the probe pad and reflects its elastic deformation during contact. The inertia of the probe pad was neglected in this analysis due to its nonsignificant effect [4].

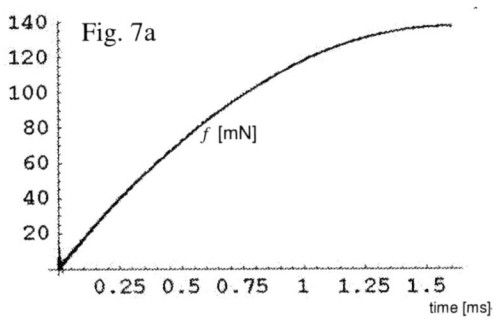

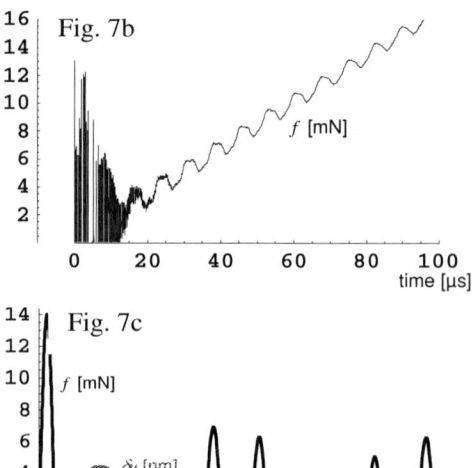

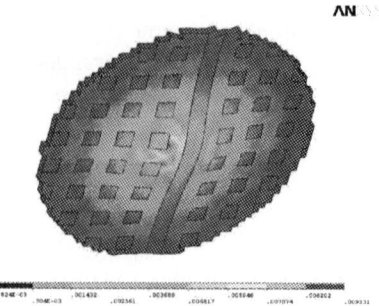

Fig.8: First principle stress contour plot of rigidizing layer (donut pattern) underneath probe event)

Fig.7a-c: Probe force during probe event over time (7a), dynamic effects during first 100µs (7b), contact force and relative displacement between pad and needle (7c)

In Fig. 7a to 7c the magnitude of the dynamic contact forces can be compared to the static contact forces that are reached during pure stubbing. The highest dynamic contact force appears during the first 0.05µs after contact between pad and needle is established. In Fig. 7c the contact force f and the relative displacement δu between needle tip and probe pad surface are shown. The two graphs demonstrate the work principle of the contact constrain. During penetration, high contact force peaks appear. In case of lift off the contact force is 0. The chart in Fig.76a shows that the static force reached at the full wafer overdrive is of a factor of 10 larger than the dynamic contact forces from the needle excitation shortly after first contact was made. The conclusion was that the needle dynamics did not have to be considered for the case that was simulated. Only the static force at maximum overdrive needs to be simulated.

4. Discussion

A number of static simulations with the probe pad model had been done afterwards. Stubbing and scrubbing was considered. Unfortunately the model and the simulation results have never been validated or verified by single, instrumented probe tests and nano-indentation tests on the modeled prototype test pads, as planned. Only a general comparison between failure analysis findings and stress contour plots from simulation indicated the value of the model and the correctness of the approach. In Fig. 8 high stress level discontinuities appear at copper-dielectric material edges. Similar to that a cross section of a structure that failed due to probing in Fig. 9.

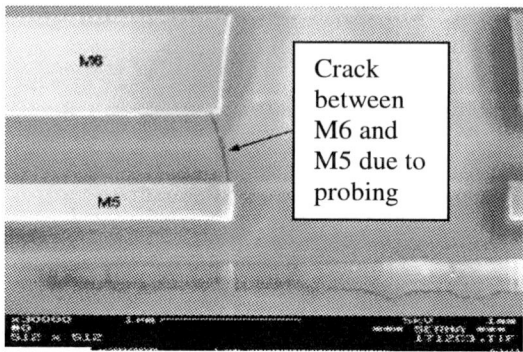

Fig.9: Crack that propagated from copper edge in M6 to copper edge in M5 due to probing

Nevertheless, the model needs to be verified and validated by a number of well defined tests. The simulation considers a lot of unknowns that can only be derived from such tests. The friction between needle and pad during scrubbing as well as the unknown influence of the native aluminum oxide layer are just two examples. The aluminum oxide builds up very quickly and has to be penetrated during the probe event while showing very different material properties compared to aluminum.

5. Conclusion

The main result of the analysis was that the failures found after probing, considering the given needle parameters used in this simulation, had been caused by the static and not the dynamic loads. This leads to the statement that the optimization of the probe process can focus on the parameters influencing the maximum static contact force like wafer overdrive or the tolerance for coplanarity of needle tips in the probe card.

The results of the probe simulations considering the static load only, are yet to be verified and adjusted by measurements, to allow statements about design and process optimizations. Nevertheless the first simulations have already shown their ability to identify locations of highest concern in the BEOL stack underneath the probe pad.

Acknowledgments

This work arose in the Crolles Modeling Cooperation between ST Microelectronics, NXP and Freescale Semiconductor. We would therefore like to thank all our supporters from the CMC team. In addition to that we would like to thank Nadine Aldahan from the Freescale Semiconductor Probing Organization and her group for the continuous support and advice.

References

1. Chiang C.-L., Hurley D.T., "Dynamics of Backside Wafer Level Microprobing", *IEEE* 98CH36173 36th Annual International Reliability Physics Symposium, Reno, Nevada,1998
2. Fiori V., Orain S., "A MultiScale Finite Element Methodology to Evaluate Wire Bond Pad Architectures", EuroSimE 2005 Proceedings pp, 648 - 655
3. Ziegler F. "Technische Mechanik der festen und flüssigen Körper", Springer-Verlag Wien 1985
4. Lui Y., Desbiens D., Irving S., Luk T., Edborg S., Hahn D., Park S., "Probe Test Failure Analysis of Bond Pad Over Active Structure by Modeling and Experiment", ECTC 2005, Proceedings pp, 861 - 866

Modeling of the mechanical behavior during programming of a non-volatile phase-change memory cell using a coupled electrical-thermal-mechanical finite-element simulator

Thomas Gille [1,2], Judit Lisoni[1], Ludovic Goux[1], Kristin De Meyer[1,2], and Dirk J. Wouters[1]

[1]IMEC, Kapeldreef 75, 3001 Leuven, Belgium

[2]ESAT/INSYS, KU Leuven, Kasteelpark Arenberg 10, 3001 Leuven, Belgium

Abstract

During programming of a phase-change memory cell, the material is locally heated up to high temperatures (>600°C), to induce phase transitions as crystallization (SET) or amorphization by quenched cooling (RESET). In this work, the thermo-mechanical stresses induced in a line-type phase-change memory cell were examined using electrical-thermal-mechanical coupled finite-element simulation. Specific procedures are described, for implementing particular mechanical characteristics of the phase-change material, i.e. the volume change on crystallization, the relaxation and the phase-dependent Young moduli. Their respective effects during SET and RESET programming are investigated.

1. Introduction

Phase-change random-access memory (PCRAM) is currently one of the most widely investigated alternative non-volatile memory technologies. Main merit of the phase-change cell is its excellent scalability, making it a serious candidate for replacement of (NOR) Flash technology, with working demonstrators already up to 512Mb in 90nm technology [1]. However, while endurance is higher than Flash, it is still limited to $<10^9$ cycles on the array level, and found to be strongly depending on program conditions. One of the defectivities observed is stuck in the high-resistive state (RESET), presumably due to detaching of the phase-change material from the contacting electrode.

In PCRAM, the memory element is a resistor made of chalcogenide phase-change material (typically a Ge-Sb-Te compound), where the material phase constitutes the information bit (low-resistive crystalline or high-resistive amorphous). The programming to the high-resistive state occurs by current-induced Joule heating above the melting temperature (>600°C) followed by fast quenching in the amorphous phase (RESET), while the programming to the low-resistive state again is achieved by heating to slightly below the melting temperature resulting in fast solid-state crystallization (SET). It is obvious that the combined effect of such a high temperature rise locally in the memory cell structure, together with the volume change on melting, resp. crystallization, may result in high thermo-mechanical stresses, possibly leading to cell defectivity.

Electro-thermal combined simulations have been reported by different groups, and proved very useful to understand the electronic-switching mechanism [2] as well as the phase-change mechanism (using appropriate nucleation and crystallization models) in different cell configurations and programming conditions [3].

Likewise, mechanical simulation would greatly help understanding the mechanical behavior of the PCRAM cell during SET and RESET programming. However, mechanical simulation of PCRAM is scarce in the literature, and its coupling with electro-thermal simulation is not yet reported.

Extensive mechanical characterizations of phase-change material (PCM) were investigated by the group of Prof. Wuttig, having reported on the existence of different biaxial elastic moduli for the crystalline and amorphous states of PCM [4], on the stress relaxation and related viscosities [4] and on the volume change of PCM during crystallization [5].

In this work, we studied a line-type PCRAM cell, using a growth-dominated doped-Sb-Te PCM material. This line-type PCRAM cell was selected because of low-voltage and fast pulse-width (<50nsec.) for RESET and SET programming [6]. A coupled electro-thermal-mechanical-simulation is performed using finite-element modeling. The implementation and effects of the volume change during crystallization, of the relaxation, and of the different elastic moduli are studied.

2. Finite element modeling method

For the finite-element simulation (modeling, solving and post-processing), we used the commercial program MSC.MARC MENTAT (64bit version), which is described with more details in [7]. This program can handle electro-thermo-mechanical coupled simulations. Non-default material properties, such as volume changes, can be built in by programming user-defined subroutines. For each of the phenomena, a different subroutine had to be developed and was thoroughly tested to ensure its accuracy.

The line cell consists of a PCM bridge connecting two electrode contact-pads. As shown in Fig. 1, the active (= phase-switching) area is situated in the middle of the bridge. The bridge dimensions are 400 nm long, 100 nm wide and 20nm thick. The volume of a unit-cell in the bridge is 5x5x5 nm^3. Due to symmetry, only one quarter of the line is considered in simulation (defined by the dashed line in Fig. 1). Note that for sake of clarity the additional SiO$_2$ isolation layer on top of the cell is not shown in Fig. 1.

The electrical program conditions used for the simulation are a 50ns-long voltage pulse of 1.4V amplitude for the RESET signal, and a 150ns-long pulse of 1.1V for the SET signal. A trailing edge of 20ns is used for both RESET and SET.

The thermal boundary conditions are defined by fixing the temperature at the outer surface of the SiO$_2$ insulating layer on top of the cell at 300K. To set the mechanical boundary conditions, the nodes of the outside faces of the memory cell are mechanically locked in the direction perpendicular to the free faces. To ensure a good accuracy of the simulations, the time-step is forced to be smaller when the temperature rate is higher, e.g. at the rising and trailing edges of each signal, so that the time-step varies between 5ns and 0.5ns during the signals. When the bridge has cooled down sufficiently the time-step increases to 65ns.

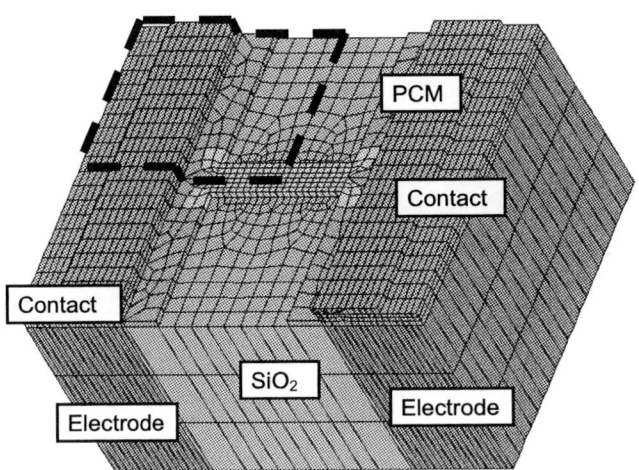

Figure 1: 3-D view of the line-type PCRAM cell structure and simulation grid. Because of symmetry, only a quarter of the cell is simulated (defined by the dashed line).

3. Implementation and effect of volume change

The antimony-telluride compounds experience a large (>5%) thickness decrease upon crystallization [5], as shown in Fig. 2, which should theoretically introduce a stress of 1.5GPa [5]. Conversely this volume decrease is balanced by an increase during melting. The multiple volume increases and decreases present a heavy cyclic load on the materials during program cycling. Therefore it is necessary to accurately simulate this behavior.

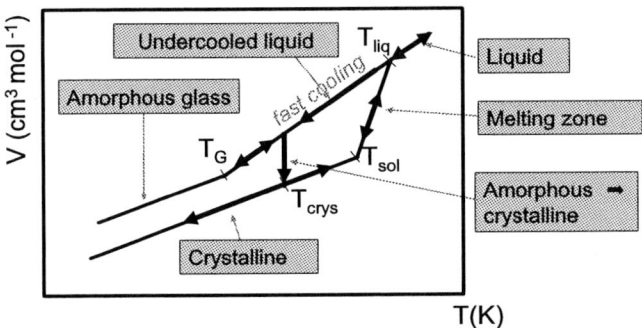

Figure 2: Molar volume vs. temperature during thermal cycling of the phase-change material (schematic).

The volume increase upon melting has been modeled as a modification of the coefficient of thermal expansion (CTE). Assuming the relative volume change X and the temperature region during melting ΔT, the one-dimensional volume strain component, ε_x^{vol}, would be expressed as follows:

$$\varepsilon_x^{vol} = \frac{\sqrt[3]{(1+X)}-1}{\Delta T} \qquad (1)$$

The volume change is obtained after applying Eq. 1 over three dimensions. In the melting region, i.e. between solidus and liquidus temperature (see Fig 2), a linear interpolation with respect to the temperature is calculated.

Using the above-mentioned implementation, Fig. 3 shows the simulated thermal strain as a function of the temperature for several nodes during RESET programming. Node 1 is situated close to the center of the bridge, node 3 is at the end of the bridge, and node 2 is in-between (see Fig 3). During heating, at the solidus temperature, the state of the PCM changes from crystalline to melting and the volume increases. Since this is modeled as an addition to the CTE, the slope of the thermal strain curve vs. temperature increases. From Fig. 3, full melting occurs only for nodes 1 and 2. During cooling, node 1 in the center gets fully amorphized and has high strain, while node 3 at the edge gets fully crystallized and is back to initial low-strain. As for the final strain of intermediaste node 2, which is shared by 6 amorphized elements and 2 crystalline elements, it is inbetween that of fully amorphized and fully crystallized nodes.

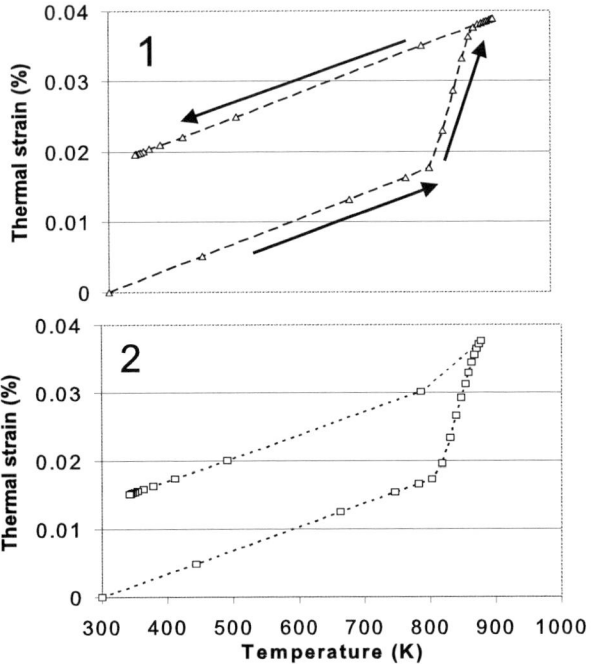

137

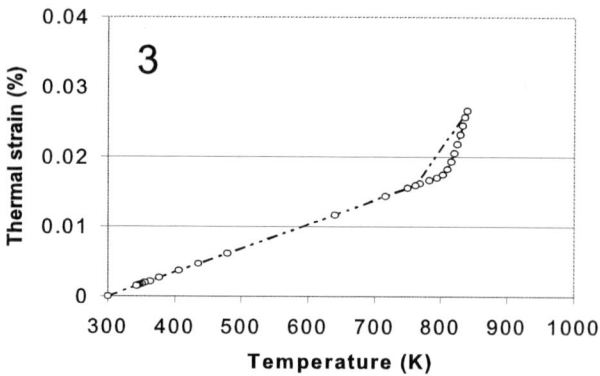

Figure 3: Evolution of the thermal strain vs. temperature during RESET programming for three nodes.

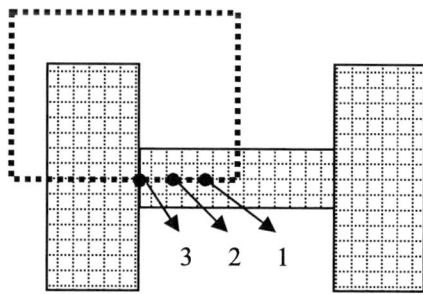

Figure 4: location of the three nodes

Figure 5 shows the simulated thermal strain as a function of the temperature for the three nodes during SET programming. For the growth dominated doped-Sb-Te material under study, the crystallization of the amorphized material is dominated by the growth of the crystal into the undercooled material, which means that the outer elements of the amorphous zone are crystallized first. In the simulation, the volume decrease upon crystallization has not been interpolated, as the crystallization of an amorphized element occurs in one time-step. Because node 2 is shared by both amorphous and crystallized elements, it is first to crystallize. Node 1 crystallizes at a higher temperature, firstly because it is closer to the center and thus hotter (i.e., with a lower undercooling that drives the crystal growth), and secondly because it cannot crystallize before one of its neighbors has crystallized (untill then, the temperature further increases). To verify the correctness of the code, a comparison between two different RESER/SET cycles has been made, i.e. with and without incorporating the volume change effect. We observed that the final stress states after RESET/SET cycle with and without the volume change effect are identical, which is consistent because the implementation only involves elastic effects.

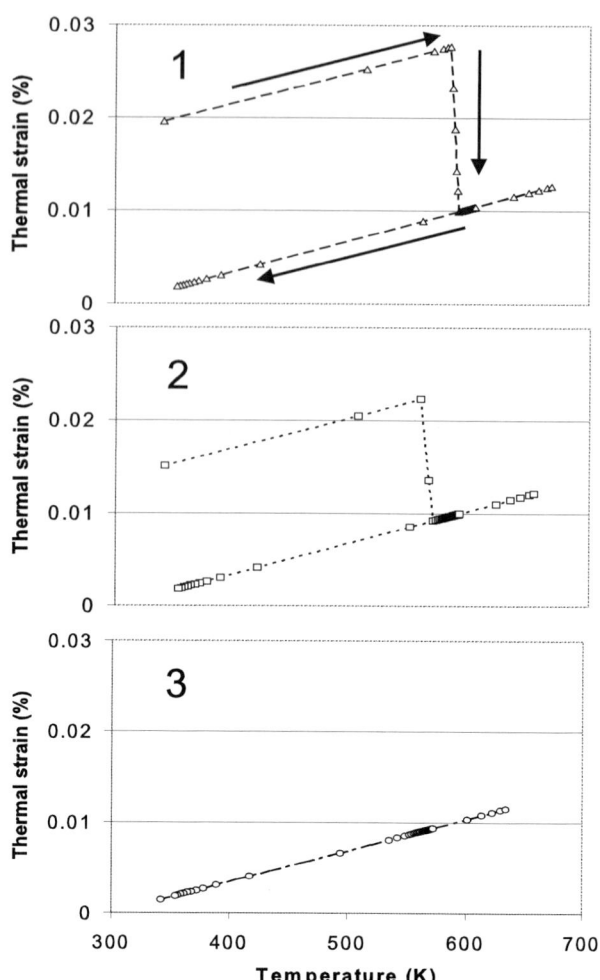

Figure 5: Evolution of the thermal strain vs. temperature during SET programming for three nodes.

4. Implementation and effect of viscous flow

It has been reported that the measurement of the stress induced by crystallization gives only 10% of the predicted value [4], and it was argued that the other 90% was released by plastic flow in the amorphous phase [4]. This result suggests that relaxation can play a great role in the stress evolution of the PCRAM cell.

We implemented the relaxation according to the Maxwell model [8], whereby the total strain rate is expressed as the sum of the elastic strain rate and the plastic strain rate:

$$\dot{\varepsilon} = \frac{\dot{\sigma}}{E} + \frac{\sigma}{\eta} \qquad (2)$$

The first part represents the elastic strain rate, characterized by the Young modulus, and the second part represents the creep strain rate, characterized by the viscosity. From Eq. 2, the equivalent creep strain can be easily deduced and implemented in the user-defined subroutine as follows:

$$\Delta \varepsilon_{creep} = \frac{\sigma_{eq}}{\eta} \Delta t \qquad (3)$$

To test the accuracy of the subroutine, we tested our simulation by comparing results to the experimental measurements obtained in Ref. 4. In this latter reference, stress relaxation curves are measured at different temperatures (see inset of Fig. 6), and the stress decreases exponentially consistently with the integration of the following equation:

$$\frac{\dot{\sigma}}{Y_f} + \frac{\sigma}{6\eta} = 0 \qquad (4)$$

This equation is derived from the general Maxwell model [4]. It predicts the exponential decrease of the stress of a blanket phase-change film (~1μm thick) on a thick substrate, as a function of the time. In our simulation, the viscosity was implemented as a fitting parameter, and the equivalent creep strain increment was slightly altered as follows in order to fully obey Eq. 4:

$$\Delta \varepsilon_{creep} = \frac{\sigma_{eq}}{6\eta} \Delta t \qquad (5)$$

Fig. 6 shows the good match between the experimental data and our simulation. However, since Eq. 4 has been derived for a ~1μm thick layer with a free surface, while in our simulation the material is very thin (~20nm) and constrained in all three dimensions, the original equation (without '6' in the denominator) has been used in the effective simulation.

Simulations of the effect of viscous flow during the RESET and SET programming however indicate that creep strain is negligible due to the short time frames involved (<100nsec.). The creep strain is 7 orders of magnitude lower than the elastic and thermal strain. However, on larger timescales (a few hours), and at higher temperatures (85°C), a significant relaxation could be observed (see Fig. 7).

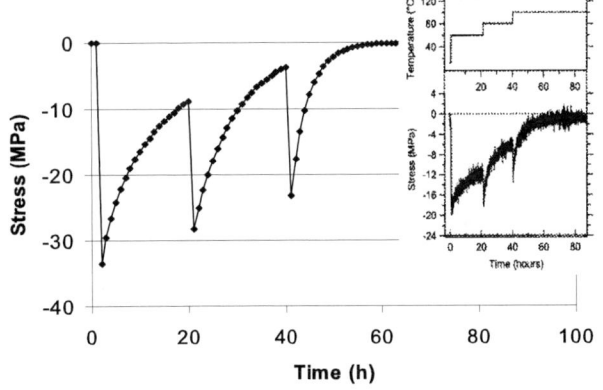

Figure 6: Simulated stress evolution with time, taking the experimental conditions of Ref. 4. The applied tim-temperature profile and experimental results are shown in the inset.

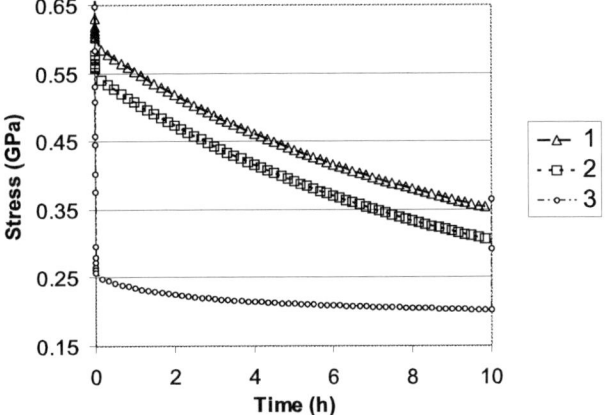

Figure 7: Long-term stress relaxation after RESET at 85°C

5. Effect of different Young moduli

Different Young moduli for the amorphous and crystalline phases were measured in Ref. 4. They are 37.1GPa and 15.8GPa for the crystalline phase and the amorphous phase respectively. This different Young moduli may have important effects on possible stress build-up during cycling. Indeed, while it was shown above that, in case of a single Young modulus, the volume change upon crystallization has no net effect during a complete RESET/SET cycle, a residual effect can be expected in case of different Young moduli in the different states.

Only data for the Young modulus in the crystalline and amorphous state are available. Since the Young modulus is related to the density [9] and as the density in both the undercooled and liquid state is lower than the density in the amorphous state (see Fig. 2), we extrapolated the Young modulus in the amorphous phase to the liquid and undercooled states. Taking the smaller amorphous value for the liquid state may be further validated by the fact that the measured bulk moduli of liquefied metals are smaller than their solid counterparts

139

[10, 11]. When the temperature of an element rises above the liquidus temperature, a linear interpolation with respect to the temperature between the amorphous and crystalline Young modulus is made.

To test the correct implementation of this effect, an initial crystalline cell (state C, Fig. 8) was RESET and quenched to the amorphous phase (state A, following the trajectory n°1 in Fig. 8). After that, the cell was SET through two different paths. The cell was either allowed to crystallize at a temperature Tc (<Tsol), following the trajectory n°2 ("normal operation"), or it was "forced" to follow trajectory n°3, i.e. was heated up to the melting, following the reverse of the trajectory n°1.

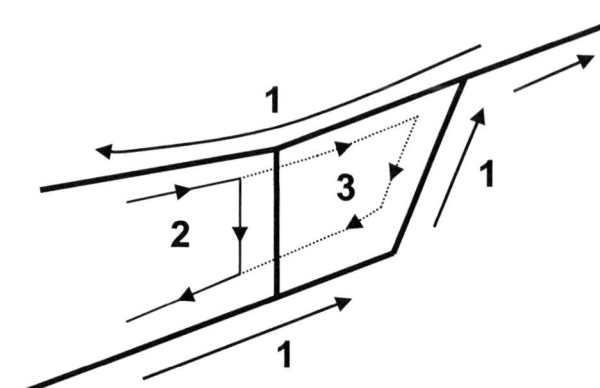

Figure 8: Trajectories denoting RESET (1) , SET (2) and 'forced' SET (3) used during simulation. "C" is denotes the crystalline (SET) state, an A the amorphous (RESET) state

Figure 9(a) shows the Von Mises stresses in the amorphized "RESET" state, indicating that during switching very high stresses (up to 1.36GPa) can be induced. Figure 9(b) shows the stresses in the cell after the complete RESET/SET cycle following trajectory n°3. After fine-tuning of the simulation time-step (most important is to have a small time-step close to the transition region where the Young modulus changes from the amorphous to the crystalline phase), a negligible residual stress (<70MPa) is obtained. Note that the stress profile does not return to the start state because of small calculation errors during transitioning between different Young moduli. The aforementioned measures serve to reduce the error: the maximum Von Mises stress of 70 MPa, is considered to be negligible (The error is further cumulative over multiple reset-set cycles).

However, when following the SET trajectory n°2, a large residual stress of 643 MPa is obtained, as can be seen in Fig 9 (c). Furthermore, Fig. 9 (d) shows the residual stresses after following the same SET trajectory as in case (c), however implementing a single value of the Young Modulus. In this case, no significant stress build-up is found, clearly illustrating the effect of the different Young moduli.

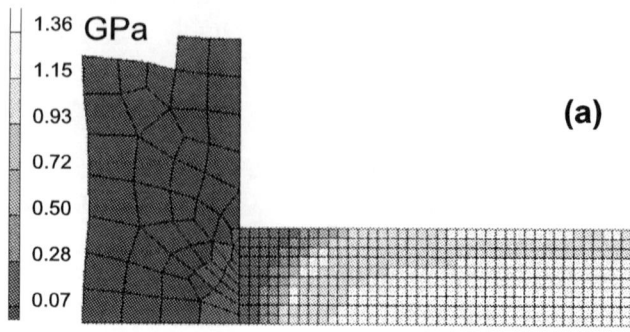

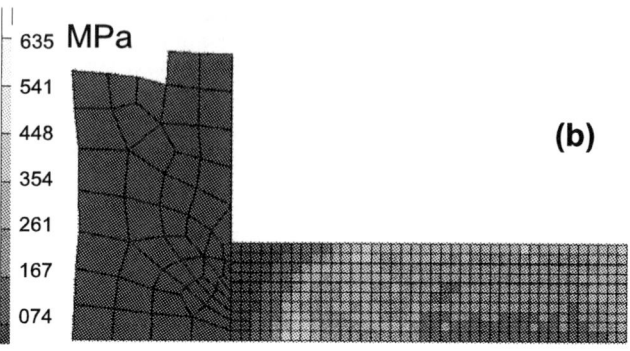

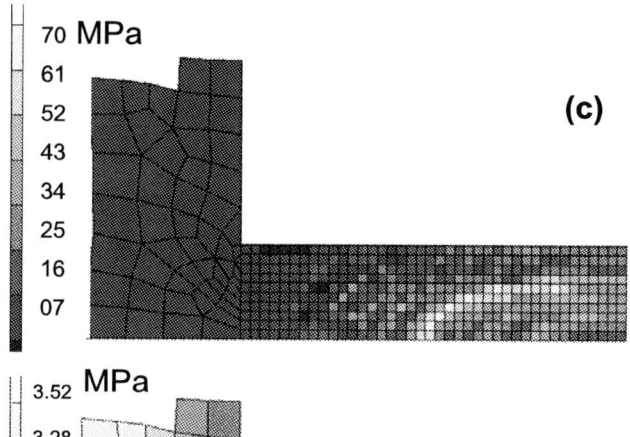

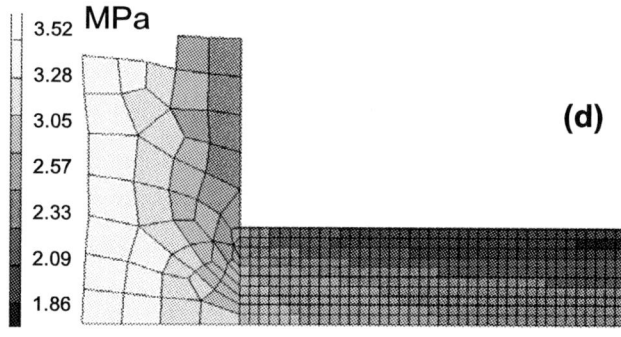

Fig 9: Von Mises stress profiles in the PCRAM cell (a) after RESET (in state A), and (b-d) in state C after a complete RESET-SET cycle : (b) following normal SET trajectory no.3, (c) following "forced" SET trajectory no.2, and (d) following normal SET trajectory no.3 but using only a single value for the Young modulus.

6. Conclusions

The mechanical behavior of the line-phase cell was explored using coupled electro-thermo-mechanical finite element simulation. It is shown that high mechanical stresses are induced by the RESET pulse, while the combined effect of volume decrease on crystallization and the difference in Young moduli in the crystalline and amorphous state results in a netto stress build-up after a complete RESET/SET cycle.

References

1. Song Y. J. et al., "Highly Reliable 256Mb PRAM with Advanced Ring Contact Technology and Novel Encapsulating Technology," *Symposium on VLSI Technology Digest of Technical Papers,* Honolulu, Hawaii, June 2006

2. Pirovano A. et al, "Electronic Switching in Phase - Change Memories," *IEEE Trans. Elec. Dev.,* Vol. 51, No. 3 (2004), pp 452-459.

3. Gille T. et al, "Impact of Material Crystallization Characteristics on the Switching Behavior of the Phase Change Memory Cell," *Mater. Res. Soc. Symp. Proc. 918E,* 2006, pp 53-58

4. Kalb I. J. et al, "Viscosity and Elastic Constants of Thin Films of Amorphous TE Alloys used for Optical Data Storage,"*J. Appl. Phys.*, Vol. 94, No. 8 (2003), pp. 4908- 4912

5. Leervad Pedersen T. P. et al, "Mechanical Stresses upon Crystallization in Phase Change Materials," *J. Appl. Phys.*, Vol. 79, No. 22 (2001), pp. 3597-3599

6. Lankhorst M. H. R. et al, "Low-cost and Nanoscale Non-Volatile Memory concept for Future Silicon Chips," *Nature Materials*, Vol. 4, No. 4 (2005), pp. 347-352

7. MSC.Marc, version 2005 r3, Juli 2005, MSC.Software Corp

8. Maxwell J. C., "On the Viscosity or Internal Friction of Air and Other Gases," *Roy. Soc. Proc.*, XV, 1867, pp 14-17

9. Omeltchenko A. et al, "Structure, Mechanical Properties, and Thermal Transport in Microporous Silicon Nitride – Molecular Dynamics Simulations on a Parallel Machine," *Europhys. Lett*, Vol. 33, No. 9 (1996), pp. 667- 672

10. Hixon R. S. et al, "Thermophysical Properties of Solid and Liquid Tungsten," *Int. J. Thermophys.*, Vol. 11, No. 4 (1990), pp. 709- 718

11. Hasegawa M. et al, "Bulk Moduli of Solid and Liquid Metals," *J. Phys. F : Met Phys*, Vol. 11 (1981), pp. 977- 994

Surface Evolution of Strained Thin Solid Films:
Stability Analysis and Time Evolution of Local Surface Perturbations

E. Dornel[1], J-C. Barbé[1,*], J. Eymery[2], and F. de Crécy[1]

[1] Léti-MINATEC D2NT/LSCDP - 17, rue des Martyrs 38 054 GRENOBLE CEDEX 9, France
[2] Équipe mixte CEA-CNRS-UJF « Nanophysique et Semiconducteurs », CEA/DRFMC/SP2M, 17, rue des Martyrs, 38054 Grenoble cedex 9, France.
* Corresponding author: jean-charles.barbe@cea.fr

Abstract

A discrete formulation of the surface diffusion potential is developed taking into account both the capillarity and the elasticity contributions. This potential, also valid in the large curvature regime, is applied to study the surface evolution of strained thin solid films. The home made numerical tool MoveFilm, used to track the surface evolution due to surface diffusion, is coupled to a finite element solver Cast3M to calculate the mechanical problem. The numerical results are consistent with known analytic results in the small perturbation approximation. This validates the code and our formulation of the diffusion potential. To go further, some non-linear strain effects are analysed concerning the destabilization dynamics imposed by an initial local surface perturbation. For low biaxial strain (1%), the perturbation amplitude grows and reaches a maximum before a lateral expansion of the perturbation. During the evolution, the profile remains quasi-sinusoidal and selects a specific wavelength. For larger strain, a specific wavelength is also selected but the profile shape is highly asymmetric and exhibits sharp grooves similar to cracks.

1. Introduction

Thin solid films are basic building blocks in microelectronics and optoelectronics. However, due to the shrinking of the layer thickness imposed by technology requirements, the surface to volume ratio increase may lead to morphological instabilities. Although a nearly plane solid surface flattens due to capillarity [1], the surface of a stressed solids may be unstable [2-4]. For a given transport mechanism, such instability results from the competition between stabilizing effect of the surface tension which tends to flatten the surface and destabilizing effect of the elastic strain. In this study, we consider the surface diffusion as the predominant mass transport process. It is described by the surface diffusion potential μ.

The following section presents the main assumptions and analytical developments used to solve the problem of surface diffusion taking into account elastic effects (the numerical method was previously described in reference [5]). The validation of the formulation of the diffusion potential is detailed at the end of section 2. Then, the dynamics of the evolution will be discussed in section 3, and the conclusion is given in section 4.

2. Surface diffusion phenomenology and numerical approach

In this paper, the notion of potential is enlarged to a diffusion potential μ to take into account both capillarity and elasticity. μ is defined as the derivative of the Helmholtz free energy F with respect to the number of atoms N at constant temperature T and strain [5, 6]:

$$\mu = (\partial F / \partial N)_{T,strain} \qquad (Eq.1)$$

The Helmholtz energy variation is composed of two terms corresponding to a surface contribution (reduced to the capillary work neglecting the surface stress [7]) and to a volume contribution (reduced to the volumic strain energy work):

$$\delta F = \delta \int_{surface} \gamma \, dS + \delta \int_{volume} \omega \, dV \qquad (Eq.2)$$

where γ is the surface energy and ω is the elastic energy density calculated with the elastic strain tensor $\boldsymbol{\varepsilon}$ and the volumic elastic stiffness tensor $\boldsymbol{C_v}$:

$$\omega = \frac{1}{2} \boldsymbol{\varepsilon}^t : \boldsymbol{C_v} \cdot \boldsymbol{\varepsilon} \, dV \qquad (Eq.3)$$

According to the Nernst-Einstein equation [1], the surface potential gradient will produce a drift of surface atoms with the following flux:

$$\vec{j} = -\left(\frac{D_s n}{k_B T}\right) \cdot \vec{\nabla}_S \mu \qquad (Eq. 4)$$

where $\vec{\nabla}_S$ is the surface gradient operator, D_s is the surface diffusivity, n is the number of atoms per unit area, and $k_B T$ is the thermal energy. The surface evolution is conditioned by the local flux balance [5].

According to [5], the capillary contribution to the diffusion potential on a 1D surface can be discretized as:

$$\mu_{cap} = \mu_{cap}^0 + 2\gamma \Omega_0 \times \frac{\sin(\theta_l) - \sin(\theta_r)}{\dfrac{b_l}{1 + Tr[\varepsilon_l]} + \dfrac{b_r}{1 + Tr[\varepsilon_r]}} \qquad (Eq. 5)$$

where Ω_0 is the unstrained atomic volume, μ_{cap}^0 is a constant and the other terms are defined on figure 1. Equation 5 accounts for isotropic surface energy and slightly differs from reference [5] by the terms $Tr[\varepsilon_i]$, the trace of the volumic strain tensor at the surface. These terms are necessary to take into account the change of volume due to strain.

1-4244-1105-X/07/$25.00 ©2007 IEEE

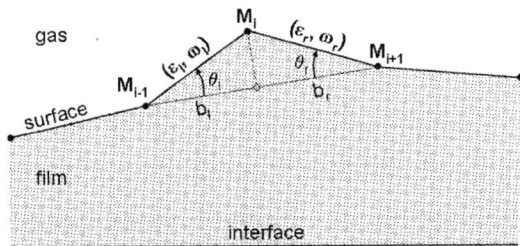

Figure 1: Sketch of surface potential calculation according to Eq. 5 and 6 (see text). b_l (resp. b_r) is the distance between the left point M_{i-1} (resp. the right point M_{i+1}) to the orthogonal projection of the central point M_i on (M_{i-1}, M_{i+1}). θ_l (resp. θ_r) is the angle between $M_{i-1}M_{i+1}$ and $M_{i-1}M_i$ (resp. $M_{i-1}M_{i+1}$ and $M_{i+1}M_i$).

The diffusion potential is calculated considering an infinitesimal increment of matter on the surface [5]. For the elastic contribution to the diffusion potential, we suppose that the matter accretion is performed in coherent homoepitaxial conditions. In consequence, the strain of the added matter is equal to the strain of the underneath matter.

Our discrete approach leads to the following formulation of the elastic part of the diffusion potential:

$$\mu_{el} = \mu_{el}^0 + 2\Omega_0 \frac{\omega_l \cdot b_l + \omega_r \cdot b_r}{\dfrac{b_l}{1 + Tr[\varepsilon_l]} + \dfrac{b_r}{1 + Tr[\varepsilon_r]}} \qquad (Eq .6)$$

where ω_i is the strain energy density taken at the surface, and μ_{el}^0 is a constant.

The evolution of the 2D film is described by a 1D line discretized into points. However, the complete 2D structure is needed to solve the mechanical problem and to calculate energy densities and strains at the surface. As illustrated in figure 2, our strategy is to alternate the surface evolution steps through our surface diffusion tool MoveFilm and the resolution of the mechanical problem using an internal numerical code Cast3M [8] that uses the finite element method.

For the mechanical part of the simulation, the inputs are the coordinates of the points at the surface. From this surface discretization a 2D mesh is built under Cast3m and the loading is applied as imposed displacements at the interface between the film and its substrate. Along the interface, the displacements perpendicular to the interface are imposed to be null: this condition is equivalent to an infinitely rigid substrate. The component of the displacement parallel to the interface linearly increases from the left hand side of the simulation box to the right hand side: this condition is equivalent to an imposed deformation at the interface. On the left hand side (respectively right hand side) of the simulation box, only the component of the displacements parallel to the interface is constraint and is equal to the imposed displacement on the node at the left hand side (resp. right

hand side) of the interface. These conditions are illustrated on figure 3.

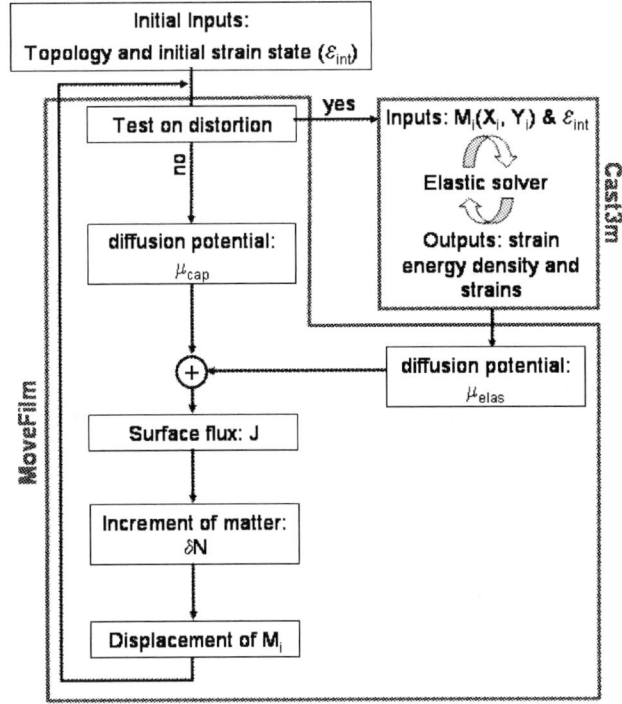

Figure 2: Schematic of the strategy of simulation.

Calculations have been performed with silicon characteristic values, namely a Young modulus of 170 GPa, a Poisson ratio of 0.28 and a surface energy of 1 J.m^{-2} [9-11]. The initial film thickness h_0 is 10nm, and the time scale has been dimensioned with a factor B:

$$B = \frac{D_s \cdot \gamma \cdot n \cdot \Omega_0^{\ 2}}{k_B T} \qquad (Eq. 7)$$

B has been measured experimentally thanks to annealing experiments of Si film at 1100°C under H$_2$ atmosphere at 10 Torr (not shown here). It is found that $B = 2.10^{-4}$ µm^4.s^{-1}. This value has the same order of magnitude than similar estimations found in literature [12]. To use the numerical result shown in this paper, with a different length scale H and/or at a different temperature or pressure, the time scale has to be rescaled by the factor H^4/B. For example, if half of the initial film thickness is used, then the evolution speed is increased by a factor 8. In addition, due to the exponential dependence of the coefficient diffusion D$_s$ on the temperature, a 40° temperature increase makes the B parameter and the diffusion speed roughly twice higher (it is evaluated considering an Arrhenius law with an activation energy of 3.5eV [13]).

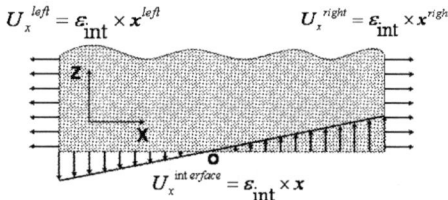

Figure 3: *Sketch of the boundary conditions imposed on the displacement component parallel to the interface. The component of the displacements perpendicular to the interface is free on the vertical sides and null on the interface.*

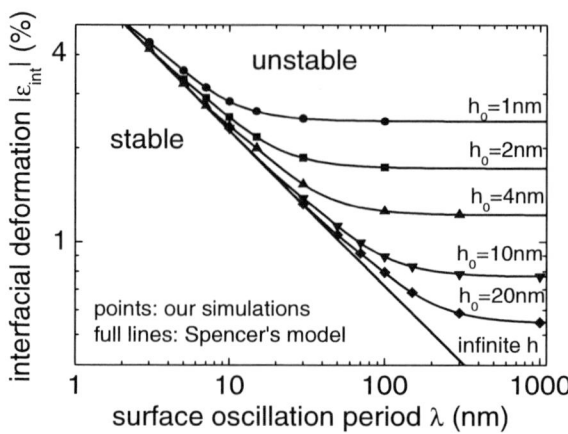

Figure 4: *Domains of stability of a thin silicon film on an infinitely rigid substrate as a function of the thickness h_0 of the film, the wavelength λ of a small sinusoidal perturbation, and the interfacial deformation ε_{int} at the film/substrate interface.*

Validation from the small perturbation theory:

The validity of this formulation of the diffusion potential has been checked through the stability analysis of strained thin film. Our procedure was to impose first a small sinusoidal perturbation of wavelength λ on the flat free surface of a film on a rigid substrate, then to calculate the sign of the perturbation growth rate to determine the stability domain. This elasticity problem, known as the Asaro-Tiller-Grinfeld instability [2, 4], has been studied more recently by Spencer [14]. The numerical results (symbols) are presented in figure 4 where they are compared to the analytical results of Spencer (full lines). This figure presents the critical wavelength λ_c above which the film is unstable for a given interfacial strain ε_{int} and a given thickness h_0 (see inset of figure 6 showing the variation of the growth rate with the wavelength of the sinusoidal perturbation). In a practical way, this critical value was obtained by dichotomy at fixed λ.

Figure 4 illustrates the possibility, at a given strain, to find a critical thickness h_c below which the film is always stable: this critical thickness decreases when the imposed strain increases. For example, a strained thin silicon film with 1% deformation leads to a critical thickness of order 6 nm. This observation is of critical interest for microelectonics applications where strained substrates, for example strained Silicon On Insulator (sSOI) substrates, are currently used in order to improve the mobility of carriers. It can be underlined that the destabilization scheme of the strained films does not depend on the sign of the strain (compression or tension), but only on the strain energy.

The very good agreement between our numerical results and the analytical model of Spencer validates our formulation of the elastic part of the diffusion potential and allows studying the dynamics of the morphological evolution of strained films in the next section.

3. Dynamics of the evolution from a local surface perturbation

One important point is to determine the natural wavelength selected by the system. The symmetry imposed by the simulated domain may disrupt the physical symmetry (see [15-16]). To release this numerical constraint, we choose to investigate the evolution of a very small local perturbation (Dirac-like perturbation) on a completely flat surface. As the perturbation grows laterally, the domain is enlarged during the simulation to get rid of the edge effects. Therefore, the developed mode is not disrupted by the boundary conditions and no CPU time is wasted.

In order to perform dynamic studies including elasticity in a reasonable CPU time, we calculate with each iteration, the distortions of the triangles (M_{i-1}, M_i, M_{i+1}) that combine both the rotation and the stretching of the triangles. An empirical threshold value of the distortion was then determined in order to reach the numerical convergence. The elastic energy ω and the deformation $Tr[\varepsilon]$ are only refreshed when the threshold value is reached.

The dynamic scheme observed is the following. At the beginning of the simulation, the Dirac-like perturbation smoothens and the surface oscillates around the initial thickness. The pseudo-wavelength of the oscillation increases until it reaches the boundary of the unstable domain given in Fig. 4. At this time, the perturbation starts to grow in amplitude and a wavelength λ_{instab} is selected. Figure 5 illustrates the profile for a 1% interfacial strain with a selected wavelength λ_{instab}. It is observed that, the profile is pseudo-sinusoidal (see the inset of Fig. 5) and remains symmetric around the initial film thickness. An example of the evolution of the pseudo-wavelength versus time is given in Fig. 6 for a 1%

144

interfacial strain. It can be underlined, for the same imposed deformation ($\varepsilon_{int} = 1\%$, see Fig. 5), the local deformation is about 1.2% in the valleys.

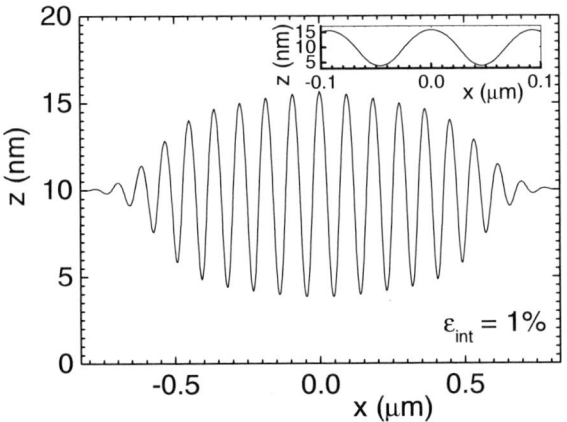

Figure 5: *Surface profile of a 10nm thick silicon film with a 1% interfacial strain at t = 15ms. The centre of the perturbation (given in the inset) remains almost sinusoidal and symmetric compared to the initial film thickness.*

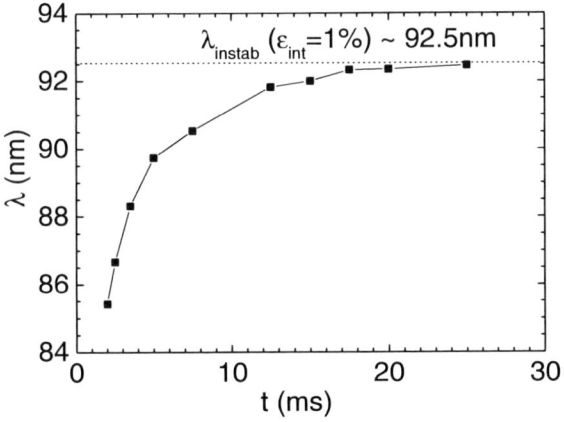

Figure 6: *Wavelength λ developed from a Dirac-like surface perturbation as a function of time for a 1% deformation. The wavelength is calculated as the average on the 4.5 central periods.*

Figure 7 presents the natural wavelength computed (symbols) as a function of the imposed strain for a 10nm thick silicon film. The lines are the analytical results of Spencer λ_c (dashed line) and λ_m (full line) [14] corresponding respectively to the critical wavelength and to the maximal growth rate wavelength (see inset in figure 7). Our numerical results demonstrate that the natural selected wavelength is the wavelength for which the growth rate is maximum λ_m.

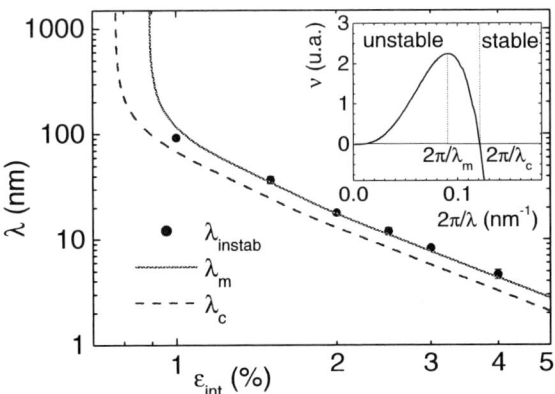

Figure 7: *Wavelength λ_{instab} developed from a Dirac-like surface perturbation for $h_0 = 10nm$ (see Fig. 5) as a function of the deformation. λ_{instab} is compared to the maximal growth rate wavelength λ_m predicted by the small perturbation theory [14] and to the critical wavelength λ_c under which no perturbation is developed (see the stable domain shown in Fig. 4). The inset shows the growth rate v of a sinusoidal perturbation defined by Spencer [14]. v is maximal for λ_m and null for λ_c.*

This figure shows a small discrepancy from the theoretical prediction for the smallest strain value (1%). This can be explained by the time evolution of the wavelength shown in Fig. 6. As said before, the local initial perturbation first smoothes and expands laterally with a pseudo-periodic oscillation. For $\varepsilon_{int} = 1\%$, λ_{instab} reaches only about 3/4 of λ_m whereas the perturbation amplitude is already half of the initial thickness (see Figures 6 and 7). Therefore, the small perturbation theory cannot be rigorously applied in this case. The wavelength of the perturbation converges to $\lambda_{instab} = 92.5$ nm whereas the theoretical value is $\lambda_m = 117$ nm [14].

Figure 8 shows the evolution of the perturbation envelop for a 1% interfacial strain. It is observed that after the growing amplitude regime, the perturbation amplitude is stabilized. For example, the minimum thickness does not go below about 4nm for a 1% interfacial strain. Once the maximum perturbation amplitude is reached, the film only exhibits a linear lateral expansion.

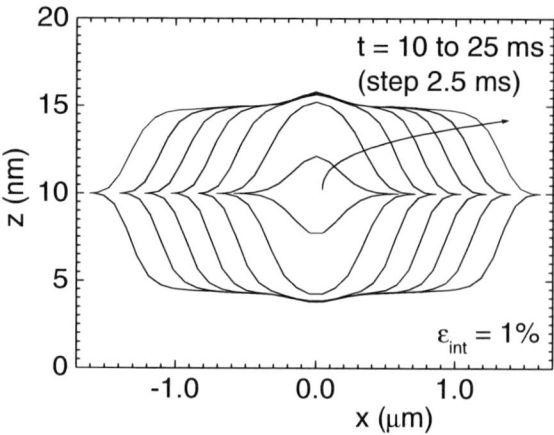

Figure 8: *Perturbation envelop for different time for a 10nm thick silicon film with an interfacial strain of 1%. It illustrates first the amplitude expansion and then the lateral expansion of the perturbation.*

Figure 9 shows the variation of the minima and maxima of the perturbation as a function of time for a 1% deformation. The perturbation first exhibits an exponential amplitude growth as predicted by Spencer [14]. Then, its amplitude is stabilized to about 3.8nm. It is observed that the minima and the maxima are roughly equal during the whole simulation, so for a low deformation, the perturbation always remains symmetric around the initial thickness (see Fig. 5).

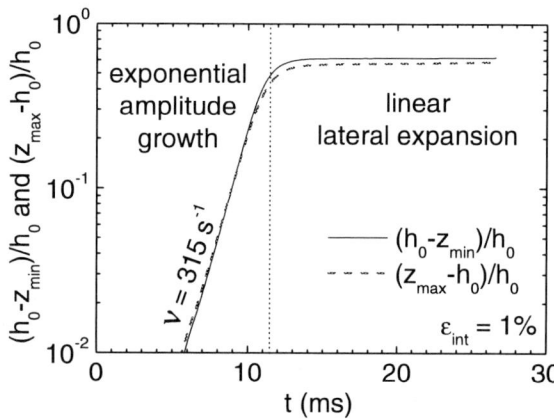

Figure 9: *Variation of the minimum and maximum of the perturbation as a function of time for a 1% deformation. The perturbation has an exponential amplitude growth before 11.5ms and a linear lateral expansion after 11.5ms.*

The growth-rate dependence of a sinusoidal perturbation (called ν in the inset of Fig. 7) with the deformation ε_{int} is given by the slope of the lines obtained in the exponential growth regime (see for example Fig. 9). This growth rate has been calculated at the early stage of

the growing perturbation (i.e. when the small perturbation hypothesis is still valid) and drawn in figure 10 for several deformations. In the approximation of infinite thick film, ν is proportional to ε_{int}^{k} with $k = 8$ [14]. From Fig. 10, k is found to be equal to 8.4. At low deformation (for example 1% in this curve), the perturbation wavelength is not yet stabilized at this early stage/time of the simulation (see Figs. 6 and 9) and the approximation of infinite thickness is not valid anymore, so the expected growth rate is lower than the $\nu \propto \varepsilon_{int}^{k}$ curve as shown in Fig. 10.

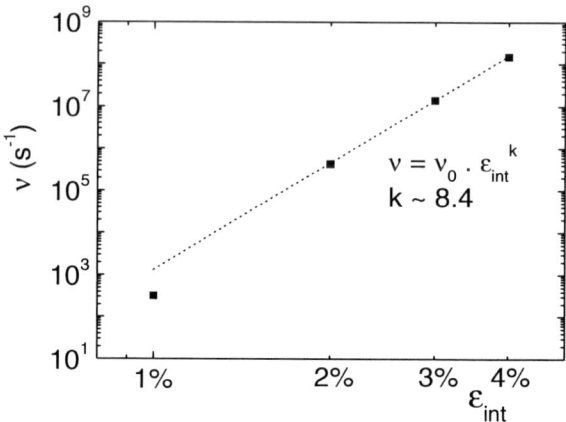

Figure 10: *Growth rate ν of the perturbation as a function of the deformation ε in logarithm scale. For large deformation, ν is proportional to ε_{int}^{k}. k is found to be equal to 8.4.*

We then have investigated why the minimum thickness is anchored at 3.8nm for $\varepsilon_{int} = 1\%$ (see Figs. 8 and 9). As the perturbation stays barely sinusoidal (see Fig. 5 and 9), we have computed the potential on the surface of a sinusoidal perturbation (static calculations) with $\lambda = \lambda_m (\varepsilon_{int} = 1\%)$. The results are shown in Fig. 11. It is observed that for a large perturbation amplitude a ($a > 0.6\ h_0$), the maximum of the diffusion potential is shifted away from the valley of the sinusoidal surface. That means the valleys of a sinusoidal perturbation with $a > 0.6\ h_0$ and $\varepsilon_{int} = 1\%$ will be filled. Therefore, the amplitude of a perturbation will only grow until reaching $0.6\ h_0$ and then will stop to grow. For $\varepsilon_{int} = 1\%$ and $h_0 = 10$nm, this leads to a minimum thickness of 4nm, which is in good agreement with the maximum perturbation amplitude reached in the lateral expansion regime (see Fig. 9). It is worth noting that the shift of the maximum of the potential is due to the elastic part: the maximum strain does not correspond to the minimum thickness for a large perturbation.

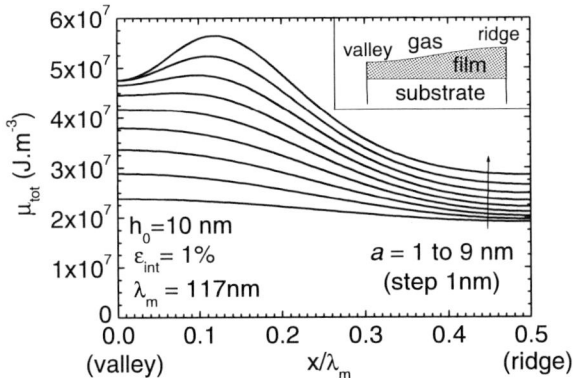

Figure 11: Total diffusion potential (including capillarity and elasticity) on a sinusoidal perturbed surface with different amplitudes a as a function of position x. $x/\lambda_m = 0$ (resp. 0.5) corresponds to the valley (resp. ridge) of the sinusoidal perturbation. For large perturbation ($a > 0.6 h_0$), the minimum of the potential does not correspond to the valley of the perturbation, so that the valley must fill. This phenomenon explains why, for $\varepsilon_{int} = 1\%$, the minimum of thickness does not reach the substrate and is stuck at 3.8nm.

Figure 12 gives an example of a surface profile for a high interfacial strain ($\varepsilon_{int} = 2\%$). It is observed that the surface profile is strongly asymmetric even at the beginning of the perturbation growth of 1nm amplitude. At this time, most of the evolution is observed on the minima that become sharp and deepen quicker than the ridges growth. This behaviour is very different to the surface profile evolution of a low interfacial strain film (see the example given in Fig. 5). Figure 13 illustrates the elastic energy density for a high interfacial strain ($\varepsilon_{int} = 2\%$). It is observed that the elastic energy is concentrated on the first minima, due to the high curvature in the minima. The strain in the two principal minima of the profile of Fig. 12 reaches 6% i.e. three times the imposed interfacial deformation. We can also observe in Fig. 14 showing the minimum and maximum of the height evolution that, for $t > 5.6\mu s$, the perturbation growth is more than exponential for these asymmetric perturbations and that the $v = v_0 \cdot \varepsilon_{int}^{\,k}$ law (with $k \sim 8$) is not valid anymore. At this time, the topology looks like a crack. Moreover, it must be pointed at that the perturbation evolution speed is $2^k \sim 300$ times larger when the interfacial strain is doubled. For example, it takes about two thousand less time to develop the profile on Fig. 12 compared to the one corresponding to Fig. 5.

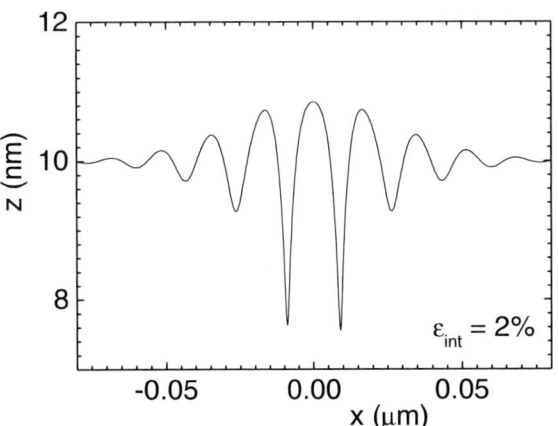

Figure 12 Surface profile at t = 8µs corresponding to the evolution of a Dirac-peak on a 10nm thick silicon film with an interfacial strain of 2%. It illustrates the highly asymmetric profile obtained for a high interfacial strain. The thickness minima become sharp and deepen quicker than the maxima growth.

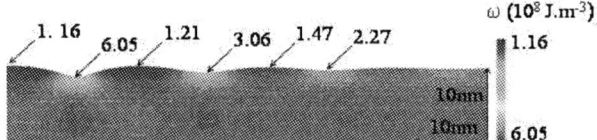

Figure 13: Elastic energy density ω of the right part of the simulated structure at t = 7µs corresponding to the evolution of a Dirac-peak on a 10nm thick silicon film with an interfacial strain of 2%. The horizontally and vertically scales are respected. The energy is concentrated on the first minima.

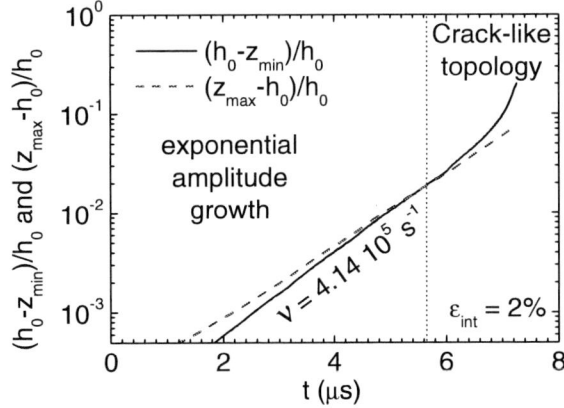

Figure 14: Variation of the minimum and maximum of the perturbation as function of time for a 2% interfacial deformation.

To quantify the evolution of the asymmetry on the profile when ε_{int} is high, the ratio of Δz_{min} over Δz_{max} is plotted in Fig. 15, where $\Delta z_{min} = h_0 - z_{min}$ (resp. $\Delta z_{max} = z_{min} - h_0$) is the difference between the minimum (resp. maximum) film thickness z_{min} (resp. z_{max}) and the initial film thickness h_0. It is observed that for $\varepsilon_{int} = 1\%$, the asymmetry $\Delta z_{min}/\Delta z_{max}$ is close to 1 within 10%. On contrary, for $\varepsilon_{int} \geq 2\%$, the asymmetry is continuously increased with ε. For example, the asymmetry ratio for $\varepsilon_{int} = 2\%$ reaches two when the film perturbation is still less than 15% of the film thickness.

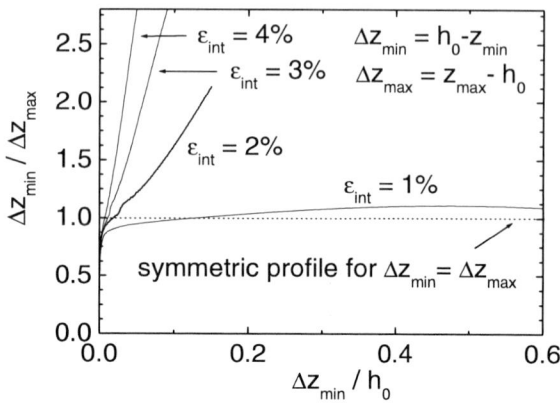

Figure 15: Quantification of the asymmetry as a function of the amplitude of the perturbation, for different imposed interfacial strain.

4. Conclusions

A numerical method for the calculation of the surface diffusion potential has been developed. The method presented in reference [5] has been extended to take into account the strain effect. As in [5], this method can be applied to highly curved and to non-flat film with a large substrate interfacial strain.

This method has been applied to strained thin film on a rigid substrate and goes beyond the small perturbation theory developed by Spencer [14]. Simulations have been carried out to investigate the non-linear dynamics effects. As the small perturbation hypothesis is not valid anymore when the perturbation is growing, we quantify the discrepancies from the predicted small perturbation theory values on the spontaneous selected wavelength λ_{instab} and on the growth rate v.

The scheme of a Dirac-perturbation evolution is the following. First, the film surface oscillates until a wavelength is selected and the perturbation amplitude grows exponentially. Then, for a low interfacial deformation, the amplitude of the film surface oscillation reaches a maximum. After that, only a lateral expansion of the perturbation is observed. The scheme of the perturbation evolution for films with high interfacial strain is different. Indeed, the oscillations are rapidly asymmetric and the minima sharpen. The growths of the profile extrema are more than exponential and the profile minima are similar to a crack-like topology.

The surface profile evolution of large perturbation and large strain is difficult to track. Indeed, the mesh must be decreased for large curvature leading to a large increase of the CPU time for the mechanical calculation. Finally, 3D simulations could be performed with the same numerical method. Different perturbation evolutions could be expected, as the second dimension curvature should change the diffusion potential value.

Acknowledgments

We wish to acknowledge O. Cueto, D. Jamet, and B. Mathieu for advices in numerical computation. Our gratitude is also extended to P. Müller, J. J. Métois and A. Saùl for scientific discussions on surface effect on strained films. We are grateful to F. Lançon and J. Villain for many useful discussions.

References

1. Mullins, W. W., "Flattening of a Nearly Plane Solid Surface due to Capillarity", *J. Appl. Phys.* Vol. 30, No. 1 (1957), pp. 77-83

2. Asaro, R. J., and Tiller, W. A., "Interface Morphology Development during Stress Corrosion Cracking: Part I, Via Surface Diffusion", *Metall. Trans. A*, Vol. 3 (1972), pp. 1789-1796.

3. Grinfeld, M. A., "Instability of the Separation Boundary between a Non-hydrostatically Stressed Elastic Body and a Melt", *Sov. Phys. Dokl.* Vol. 31 (1986) pp. 831-834.

4. Grinfeld, M. A., "The Stress Driven Instability in Elastic Crystals: Mathematical Models and Physical Manifestations", *J. Nonlinear Sci.*, Vol. 3 (1993), pp. 35-83.

5. Dornel, E. *et al*, "Surface Diffusion Dewetting of Thin Solid Film : Numerical Method and Application to Si/SiO$_2$", *Phys. Rev. B*, Vol. 73 (2006), No. 11, pp. 115427-115437.

6. Herring, C., Physics of powder Metallurgy, McGraw-hill Book (New York, 1951), edited by Kingston, W. E.

7. Müller, P., and Saùl, A., "Elastic effects on surface physics", *Surf. Sci. Reports*, Vol. 54 (2004), pp. 157-258.

8. http://www-cast3m.cea.fr/cast3m/index.jsp

9. Wortman, J. J., and Evans, R. A., "Young's Modulus, Shear Modulus, and Poisson's Ratio in Silicon and Germanium", *J. Appl. Phys.*, Vol. 36 (1965), No. 1, pp153-156.

10. Métois, J. J., and Müller, P., "Absolute surface energy determination", *Surf. Sci.*, Vol. 548 (2004) pp. 13-21.

11. Eaglesham, D. J. , "Equilibrium Shape of Si", *Phys. Rev. Let.*, Vol. 70 (1993), No. 11, pp. 1643-1646.

12. Sudoh, K., et al., "Numerical Study on Shape Transformation of Silicon Trenches by High-Temperature Hydrogen Annealing", *Jap. J. of Appl. Phys.*, Vol. 43, No. 9A (2004), pp. 5937-5941.

13. Kuribayashi, H., et al., "Shape transformation of silicon trenches during hydrogen annealing", *J. Vac. Sci. Technol. A*, Vol. 21 (2003), No. 4, pp. 1279-1283.

14. Spencer, B. J. *et al*, "Morphological instability in epitaxial strained dislocation-free solid films: Linear stability theory", J. Appl. Phys., Vol. 73 (1993), No. 10, pp. 4955-4970.

15. Yang, W. H, and Srolovitz, D. J., *J. Mech. Phys. Solids*, Vol. 42 (1994), No. 10, pp. 1551-1574.

16. Kassner, K. *et al*, "Phase-field modeling of stress-induced instabilities", *Phys. Rev. E*, Vol. 63 (2001), No. 036117.

Gold Wire Bonding Induced Peeling in Cu/Low-k Interconnects: 3D Simulation and Correlations.

Vincent Fiori [a,*], Lau Teck Beng [b], Susan Downey[c], Sebastien Gallois-Garreignot[d], Stephane Orain [e].

[a]STMicroelectronics, 850 rue Jean Monnet, F-38926 Crolles, France.
[b]Freescale Semiconductor, Malaysia Sdn Bhd.
[c]Freescale Semiconductor Inc., Austin, TX 78735.
[d]Freescale Semiconductor, 850 rue Jean Monnet, F-38926 Crolles, France.
[e]NXP, 860 rue Jean Monnet, 38920 Crolles, France.
[*]Phone: (33) 4.38.92.25.58 Email: vincent.fiori@st.com

Abstract

Amongst solutions to connect the die to the package, thermosonic wire bonding process remains widely used. However, the introduction of low-k dielectric materials, and the feature size decrease of IC chips to follow Moore's law, pose great integration challenge.

This paper aims to demonstrate the compliance of the proposed modeling approach with the aids of experimental validations. 3D multi scale simulation of both bonding process and wire pull test is carried out. Using a previously validated homogenization procedure to include pad structure description even at the global scale, stress fields acting in both the gold wire and the copper/low-k stack have been evaluated and discussed. The modeling strategy also includes an in-house developed energy based analysis.

For the experimental part, a wide range of wire bond trials have been performed in order to qualify the 65-nm technology node. On behalf of that, the effect of the bonding conditions has been studied. More precisely, it was found that the peeling failure rates are significantly dependant on the used wire types and their respective bonding parameters.

In this paper, some numerical parameters are firstly discussed and the most suitable modeling strategy is proposed. Hence, typical results are presented, and the comparison of the peeling hazard induced by distinct bonding conditions is carried out. Simulations are then faced to experimental results and a good agreement is found. In addition to that, the complementary nature of the energy based failure criteria is highlighted through a clearer determination of the forecasted location of the failed interface in the IC stack.

Finally, the simulation procedure with confirmed experimental results demonstrates its ability in design and process optimizations by providing a better understanding of pad peeling failure mechanisms.

1. Introduction

Amongst solutions to connect the die to the package, thermosonic wire bonding process remains widely used. However, the introduction of low-k dielectric materials, and the feature size reduction on IC chips following Moore's law, pose great integration challenge [1]. On one hand, as the k value is reduced by introducing porosity into low-k materials, their mechanical behavior becomes less robust [2]. On the other hand, with the critical dimension reduction, stresses induced by both front-end of line (FEOL) and back-end processes like wire bonding tend to increase [3] [4]. Delamination failure modes are now commonly observed and mechanical integrity of IC has became a key integration issue [5].

Modeling is one of the major tools used to understand and reduce mechanical related defects. Various design concepts can be evaluated early in development, without actual silicon. More precisely, it is observed that the interconnect pad architecture plays a great role in peeling failure rate [6], and finite element modeling has demonstrated its ability to provide insights for bond pad design optimization [7]. However, due to the complexity of the physical phenomena involved, added to very detrimental aspect ratio from die level to interconnect one, specific modeling methodologies need to be developed.

2. Modeling Methodology

2.1 Multi scale and homogenization procedure

Typical dimensions involved in wire bonding induced peeling failures have a very wide range: During the wire bonding process, macroscopic and uneven loads are imposed by the gold wire, with bonded ball diameter (BBD) of about 32um and 36um for 40um and 50um bond pad pitch respectively. On the other hand, in the peeling region, i.e. the interconnect levels, very thin layers such as the hard masks with a typical thickness of 30nm are concerned. Moreover, due to the bond pad architecture, the use of a three dimensional (3D) analysis is mandatory and a common plane strain assumption is not suitable. Indeed, a two dimensional approach has many drawbacks: For example, the distinction between via and metal line is not possible, and interconnect patterns such as dielectric enclosure can not be described. Hence, to consider the very detrimental aspect ratio would lead to a huge amount of elements with a single 3D FE approach and multi-level modeling need to be performed.

The main steps of multi level modeling are the following (Figure 1):

Compute a first simulation at global level to get displacements field.

At the global level, locate the maximum strained area, i.e the future location of the micro model.

Apply the displacement field calculated at the macro level as boundary conditions of the micro model in order to reach the local stress field.

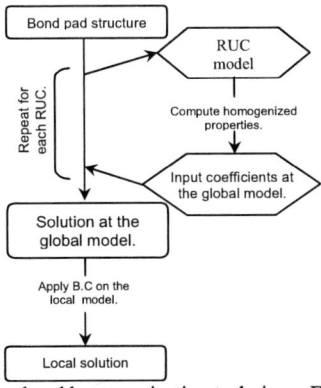

Figure 1 Multi Level and homogenization technique: Flow chart.

The key point of the multi scale method is to be able to describe the materials and the bond pad layout which compose the macro model with a sufficient amount of details in order to calculate properly the displacement field. At this stage, a homogenization procedure needs to be defined. This accounts for the geometrical details by modifying the mechanical properties without meshing all the interconnect layer patterning.

Several ways to perform this step can be used. However, it has been shown that precautions must be taken, in particular concerning the choice of the representative unit cell (RUC) and the rheological model to be applied [8]. Thus, the following method is carried out in this work:

One periodic RUC for each interconnect level is chosen. More precisely, the Inter Metal Dielectric (IMD) architecture and the Metal geometry are considered separately and thus for each interconnect level. This leads to a dozen of equivalent property sets (e.g of a six metal levels (ML) interconnect stack). As for the equivalent behavior of the RUCs, considering both the pad structure designs and the material properties to be homogenized [11], a linear orthotropic model ensures both reliable results and computation efficiency. Hence, the set of equivalent properties for each cell is made up of nine coefficients: (i) Young's moduli E_x, E_y, E_z, (ii) Poisson's ratio ν_{xy}, ν_{xz}, ν_{yz}, and (iii) shear moduli G_{xy}, G_{xz}, G_{yz}. The coefficients are calculated by applying simple loading cases on the RUC and then by extracting reaction forces acting on its boundaries. Consequently, the whole IC stack at the global level will be described by hundreds of parameters.

Then, the stand alone evaluated coefficients are included in the global model and the solution is calculated. Finally, degrees of freedom (DOF) are interpolated and applied to the local model, bridging the gap between the scales and providing stress field at the IC level.

2.2 Loading conditions

Once the finite elements models are built, suitable loads must be applied. Even though the peeling failures are mostly observed during the wire pull test, it is already known that such failures are inherently caused by the wire bonding process itself, especially with an unoptimized bonding process. It is likely that unoptimized bonding parameters will cause micro-cracks in the bond pad structure during bonding [12]. Hence, in a comprehensive modeling analysis, both the bonding and pull test must be considered.

However, thermosonic bonding mechanism is complicated and the physical motion of the capillary and free air ball (FAB) is difficult to be modeled precisely. Some assumptions need to be made, particularly in the aspect of Au-Al intermetallic formation [13][14], the interaction of this with the bonding parameters [15][16] and its effects on pad peeling failure rate. Despite the fact that some technological barriers remain in analyzing the whole mechanism by simulation, some promising works have been published recently proposing various approaches for investigating, although often separately, the various factors of wire bonding physics [3][17][18][19].

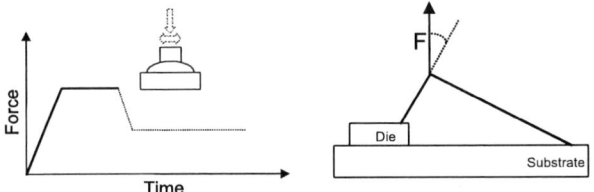

Figure 2 Schematic of the applied loads: Stage 1 (left): Simplified bonding process, in solid lines the simulated steps. Stage 2 (right): Wire pull test.

In this paper, considering both the requirements for 3D bond pad architecture considerations and the aforementioned limitations, a simplified modeling of the bonding process is proposed, followed by the standard wire pull test simulation (Figure 2): Starting from the bonded ball which has non linear material behavior, both contact and ultrasonic steps are reproduced in a 3D model. The model considers that a rigid capillary carries the wire with a normal force and horizontal displacement resulting from the wire bonding parameters. Whereas the applied pressure value is provided directly by the parameter of bonding machine, the determination of the ultrasonic displacement value is rather subjective as it is not quantitatively controlled once the contact is established with the bond pad surface. The tangible displacement of the capillary can be measured as a function of the 'power' parameter in the wire bonder. In this modeling work, despite the decrease of ultrasonic displacement during the contact stage, it is assumed that the capillary and FAB displacement is consistently following the 'power' value used in the model. In later stage, the capillary is removed from the model and wire pull test at a given angle is performed. The force acting during this load step is extracted from the measured wire strain-stress curves provided by the wire suppliers.

In this paper, two sets of bonding conditions, wire types, and corresponding bonding capillaries are used. and are designated Set A and Set B as shown in Table 1. More precisely, each set has its own wire and capillary types, and the associated bonding machine parameters. Hence, the distinction according to the used bonding set will be included through both the related strain/stress curves and a couple bonding parameters. Note that due to confidential concerns, quantitative values are normalized by set A ones. However, the following description of the sets is provided: loading condition sets will be used for 2N and 3N gold wires:

TABLE 1: LOADING PARAMETERS

	Set A	Set B
Gold bonding wire		
Type	2N	3N
Gold purity (%)	99	99.9
Bonding capillary	A	B
Bonding parameters		
Contact force	1	1.1
Bonding power	1	1
Wire pull process		
Maximum load	1	0.78

It should be noticed that the differences induced by the two sets in term of peeling failures can not directly be guessed. Indeed, on one hand the wire used in set A is stiffer and shows higher fracture threshold, on the other hand the contact force is lower for this set. Thus, since the corresponding failure hazard would be affected by the whole parameters, the trend from set A to B is not obvious and modeling would bring insights on that.

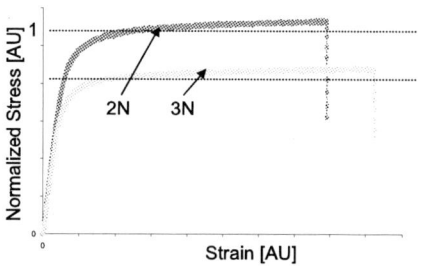

Figure 3 Stress strain curves of wires used in simulated bonding sets.

2.3 Failure criteria

Finally, suitable criteria must be defined according to the observed failure modes and by taking into consideration of the material set used. Despite the fact that in-situ investigations of the failure modes are tricky, the most likely scenario in the case of peeling failures is brittle fractures. Furthermore, micro-crack is highly suspected at the early stage of peeling failure, which is initiated at the many interfaces of metal layer and low-k

dielectric. Hence, the targeted failure mode, which is brittle delamination, is related to fracture mechanics and specific energy based numeric tools are developed [9][10]. The so-called Nodal Released Energy (NRE) post processing procedure is based on the computation of energetic quantity from the nodal solution. To evaluate the NRE value, two simulations are required: One with an undamaged model, the other with the damaged one where a virtual crack has been inserted. At this step the crack properties need to be defined: The out of plane locations of the crack is at the weakest interfaces of the model, and the peak stress values are analyzed to put the crack in the suitable region. Figure 4 shows the post processing flow chart and formula.

Note that the numerical parameters to be used, the sensitivity and the ability to link this value with the energy release rate has been previously validated; and additional insights compared to a standard stress based analysis have been demonstrated [9].

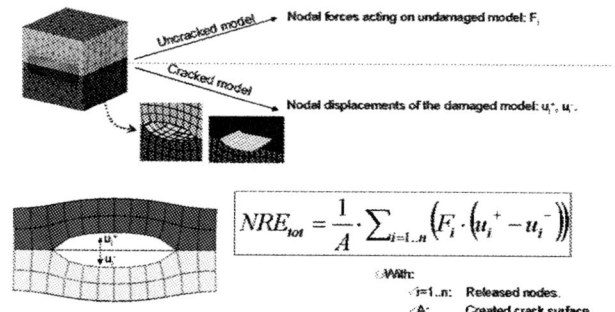

Figure 4 Energy based failure criteria: Nodal Release Energy (NRE), flow chart and formula.

3. Finite Element Models and modeling parameter settings

3.1 Finite Element (FE) Models

The three dimensional bonding model consists of a silicon die containing the interconnect layers, the gold wire and ball, and the capillary. The simulated die stack is seven copper metal levels (7ML) including five low-k levels and the two others containing classical silicon dioxide dielectric. Above the copper/low-k stack, the oxide passivation and aluminum are modeled. Dimensions and material properties are those used in advanced products of 65-nm technology, in which ultra porous low-k dielectric is integrated. Figure 5 depicts the two models used in that multi scale simulation, a meshed interconnect unit cell is provided as an example. On the local model, particular care must be taken to prevent (i) mesh dependency, a Manhattan type mesh with a constant element size is used for all the simulated interconnect pad structures; (ii) interpolation error when transferring boundary conditions from global to local where dimensions of the local model must kept constant in all simulations, disregard the periodicity of the pad design.

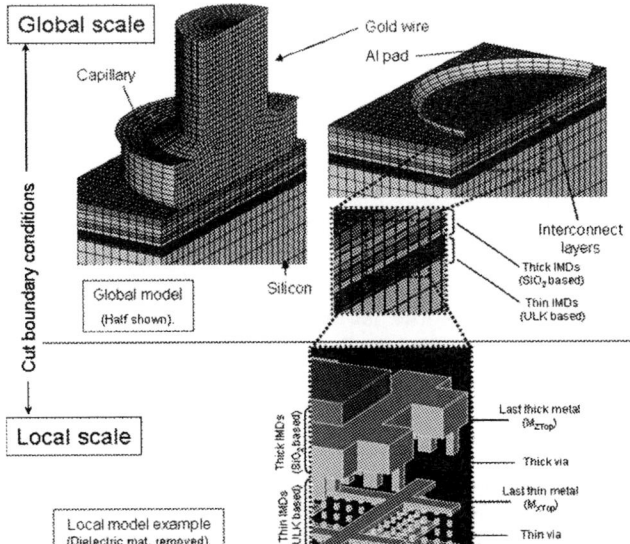

Figure 5 Finite Element Models: Bond pad & wire assembly at the global scale (top) and interconnect structure at the local scale (bottom).

The simulated bond pad layout is made of crossed copper lines at all metal levels. Keeping the layout unchanged, various pads having distinct metal densities are modeled and results are presented in the first part of this paper.

Despite this work is not dedicated on pad structure dependency study and optimization, it must be noted that, for a given metal layout, the equivalent stiffnesses may not change even if the local structure are modified. As an example, comparing pad layouts having just a proportional in plane change would not provide any input. By consequences, in such structures, the pad strength evaluation would be only affected by local considerations that could not be investigated in the global model. Furthermore, it has been demonstrated that only an energy based failure criteria is able to fully discriminate structures having such properties [20].

In the frame of the multi level approach, the equivalent properties of the pad structure need to be computed as defined in 0. Figure 6 shows the effective Young's modulus of a typical simulated structures, which is one of the main properties that affect pad strength.

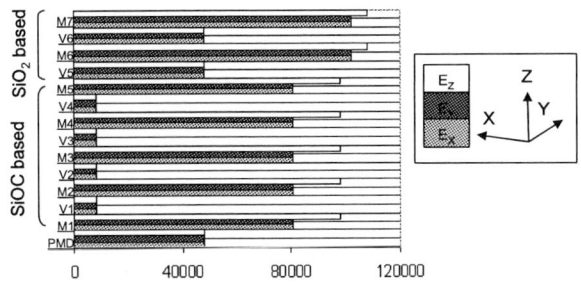

Figure 6 Homogenized properties for a typical bond pad layout. Effective Young's moduli in the three directions and per interconnect layer.

Preliminary modeling investigations

In this section, discussion on simulation methodology is proposed through two issues: The one, caused by a possible mean to achieve CPU resource reduction, is related to the suitable die model to be considered. The other will be more subjected to discussion and concerns the bonding interface itself. More precisely it deals with the management of the contact behavior between the ball bottom part and the die top surface.

Ball/Die interactions

In order to reduce CPU time and finite elements amount, an option exists to perform the simulation in two steps: (1) A first model with the wire & capillary assembly and a simplified die model to evaluate the loading conditions at the top die surface. (2) Then, a second model with the die & interconnect parts to reach the stress fields acting at the copper/low-k layers. Unfortunately, this approach would not be accurate due to the presence of some interactions of the ball with the pad structure. More precisely, Figure 7 shows that the normal stress acting on the die surface depends on the metal density of the underneath layers. Hence, considering either rigid substrate or a given pad stiffness would not be accurate to precisely evaluate stress induced in distinct pad architecture options. This justifies the need to include an accurate description of the IC layers to evaluate the solution in the ball region and at its interface with the die.

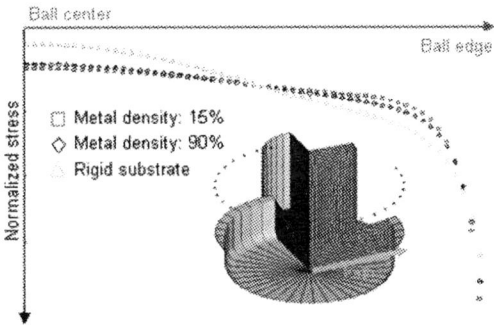

Figure 7 Effect of the pad stiffness on the normal stress induced at the ball/die interface.

Ball/Die contact behavior definition

Two contact element pairs at interfaces are being considered: (i) between the capillary and the wire, (ii) between the ball and the pad surface.

Since adhesion forces between capillary and wire

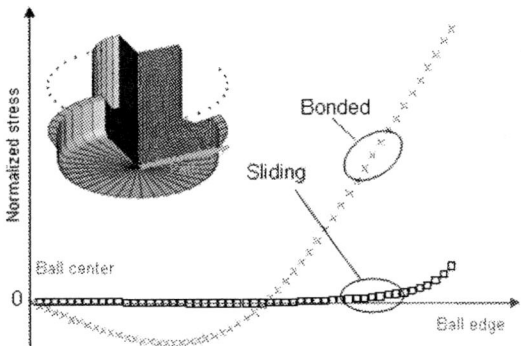

Figure 8 Effect of contact behavior at ball/pad interface on normalized shear stress.

remain limited, a perfect sliding contact should be the suitable law. However, the question is trickier for the other interface: Indeed, the adhesion forces are gradually created during the bonding stage, the friction law acting between the wire and the bond pad ranges from a sliding contact to a totally bonded one. To investigate friction effects, a vertical force is applied to the capillary. Figure 8 shows the effect of the contact behavior at the ball/die interface on the shear stress distribution. As expected, higher shear stresses are induced in the case of a bonded contact compared to the pure sliding one. Hence, in this paper and for the worst case, a bonded contact law will be applied.

4. Simulation results

4.1 Typical results at the wire level

In this section, typical results at the wire level are provided for a fixed wire type and pad structure. This preliminary analysis is carried out in order to capture general considerations from the before described simulation method.

<u>Bonding process</u>

As exposed in 0, the bonding process itself is performed into two steps: The one related to the contact stage, while normal pressure is applied, then, tangential displacement corresponding to the ultrasonic vibration is imposed. Figure 9 depicts the sequential equivalent plastic stress induced.

Finally, the analysis of the stress and strain fields in the wire and at the end of the bonding stage (Figure 10) can be summarized as follows: The plastic deformations in the wire are concentrated at 2 locations:

- An annular shaped area at ball edge and at the pad surface vicinity.
- In the capillary region of the wire neck.

It should also be noticed that the maximum compressive stress is found below the capillary foot print.

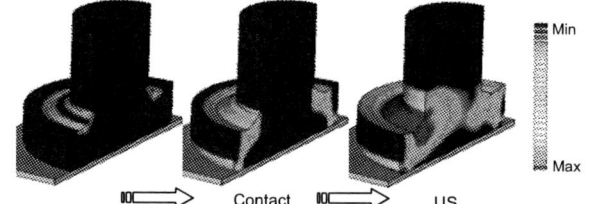

Figure 9 Equivalent plastic stress during bonding in the wire.

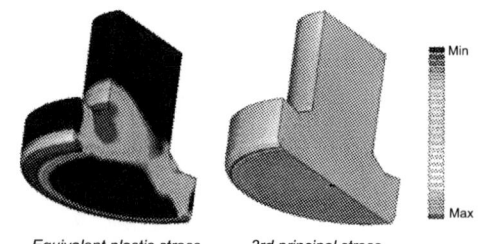

Figure 10 Equivalent plastic and principal compressive stress components at the end of the bonding stage.

<u>Wire pull test</u>

Similar analysis can be done for the wire pull test. According to the pulling angle, the applied loads are a combination of pure pull and tangential solicitations through the wire top. Figure 11 depicts the plastic and Von Mises stresses induced. Results are the following:

The main plastic deformation is concentrated at the wire neck.

Due to the pull angle, both tensile and compressive areas are found at the pad surface.

In addition to that, strong similarities with some observed location and fracture path of a wire neck break failure are pointed out.

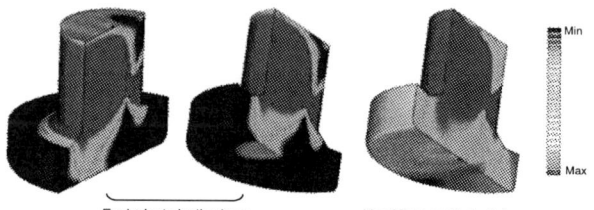

Figure 11 Equivalent plastic and Von Mises equivalent stress components in the wire after wire pull test.

4.2 Typical results at the die level

<u>Stress based analysis</u>

Using the described multi scale methodology, simulation is carried out at the global scale in order to provide general considerations on the stress fields induced by both considered load steps. Typical results are presented for given sets of bond pad (i.e. homogenized properties) and bonding conditions (i.e. contact force, US power and wire/capillary assembly). Despite the fact that local IC investigations are not the main purpose of this work, it must be noticed that, in the frame of a whole multi scale procedure, this preliminary analysis at the global scale is mandatory to locate the micro model with respect to the macro one.

The bonding process and the wire pull test are being considered separately, and stress maps are provided for a dielectric layer situated in the middle of the interconnect stack. Significant stress components are illustrated for each load case.

Bonding process

Figure 12 shows the typical stress figures induced by the bonding stage at an IMD level. Both normal and shear loadings are applied according to the bonding parameters as described in 0. Furthermore, to prevent from architecture orientation effect, shear load is applied diagonally through the bond pad. From these plots, results are as follows:

The ultrasonics (US) induce both tensile and compressive areas in the pad structure. The compressive one is located in the front side of the US displacement, whereas the tensile one is at the opposite side.

The maximum stressed area is located just below the capillary tip near the edge of the ball.

According to these results, the position of the next modeling level can be defined. It can also be assumed that it corresponds to the likely failure region in the pad structure during the bonding process.

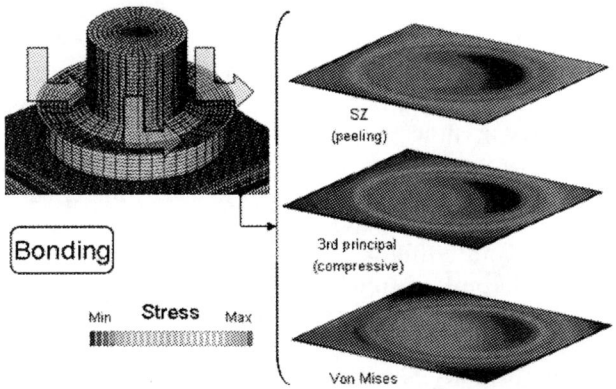

Figure 12: Typical stress figures induced by the bonding stage at an IMD level: Extracted fields at the middle on the copper/low-k stack. Peel (top), compressive (center) and Von Mises (bottom) stress components shown.

Wire pull test

Figure 13 shows the typical stress figures induced by the wire pull test at an IMD level. A 20° angle of traction load is applied at the highest point of wire cross section. Note that the stress fields at this stage differ radically from those calculated during the bonding process. More precisely:

A tensile peeling stress is observed on one side of the wire and due to the pulling angle, a compressive stress also presents at the opposite side.

The maximum stressed area is this time located below the wire at a central region of the bond pad.

Again, this defines the next modeling level location and also corresponds to the most likely failure region in the pad structure during this load.

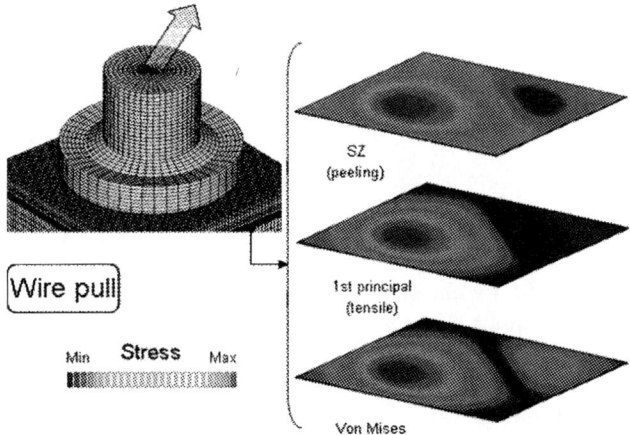

Figure 13: Typical stress figures induced by the wire pull test at an IMD level: Extracted fields at the middle on the copper/low-k stack. Peel (top), tensile (center) and Von Mises (bottom) stress components shown.

Energy based analysis

Prior to any bonding set usage related dependence, a deeper analysis is carried on at the die level. Both loading steps are performed on standard bond pad architecture and maximum stress locations are determined on the top die surface. More precisely, distinct critical sites are found with respect to the considered step: During the bonding, the maximum stressed area is located just below the capillary tip near the edge of the ball. Whereas, after the wire pull test, peak stress is located below the wire, in a rather central region of the bond pad. These considerations will be used in the energy based analysis. Indeed, on one hand the in plane locations of the likely crack nucleation sites are defined and NRE can be computed at the global scale. On the other hand, the degrees of freedom (DOF) calculated at the boundaries of these regions can be interpolated to be applied in the local interconnect models.

Thus, the NRE is evaluated at each low-k interface at the vertical points of the stress extrema. The corresponding NRE values are plotted on Figure 14 separately for bonding and wire pull processes. Opening modes are split into a peeling or crushing component (mode I), and a shear one (mode I/II). Hence, for the considered bonding parameters, pull test angle and force magnitude, results are analyzed as follows:

The opening mode I is dominant for both loading conditions, even during bonding. Furthermore, the mode mixity is found roughly constant as a function of the interconnect depth.

The maximum NRE value is found in the top region of the interconnect, but not exactly at the uppermost interface. It must be noticed that, contrary to stress values which show a continuous decrease from die top to substrate [9], the maximum energy quantity is found in

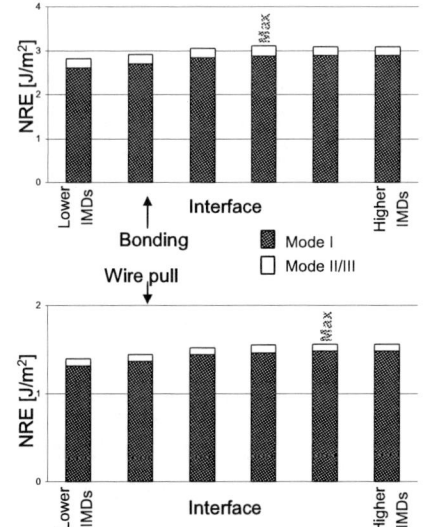

Figure 14 NRE results at interfaces for bonding process (top) and wire pull test (bottom).

155

the inside of the stack. This suggests that the main delamination path would not be located at the highest low-k interface, but slightly beneath. This feature is in good accordance with experimental observations (Figure 15) and underlines the relevance of the NRE procedure.

NRE values computed for bonding process are about two times higher than those for pull test. This confirms engineering opinion suggesting that, despite the fact that most of the peeling failures are observed during wire pull test, the bonding process plays an important role in interconnect damaging.

From these results, assessment for the peeling failure is proposed: During bonding, the delamination would nucleate first in a circular pattern around the edge of the bonded ball. Since tensile stress state remains limited during this step, the integrity of the pad is not altered, and cracks are only initiated. Then, the traction load induced during the wire pull test would create another nucleation site concentrated in the centre of the bond pad. Finally, both nucleation sites would merge and lead to the complete delamination of the interconnect levels. In addition to the stress and energy based analyzes, the experimental horseshoe shaped of the fracture path would support this scenario (Figure 16).

Figure 15 FIB cross section after pull test: Typical fracture path for peeling .failure, delamination located at the upper part of the low-k stack (courtesy of C2A Charac. team).

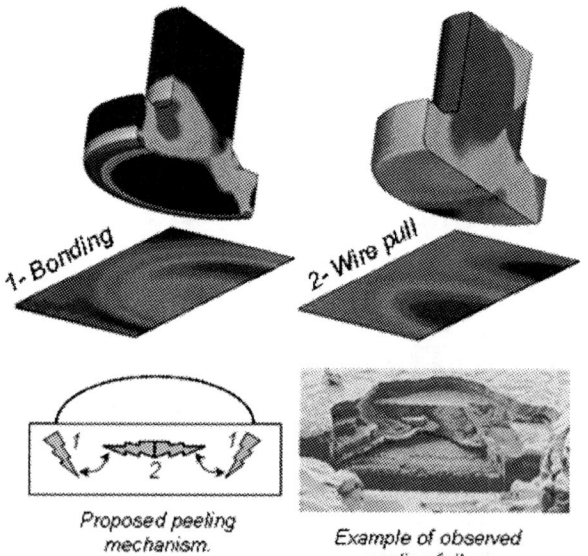

Figure 16: Peeling mechanism assessment (bottom left) from stress analysis during bonding (top left) and wire pull (top right). Bottom right: Post mortem picture of a peeling failure example observed after pull test.

4.3 Effect of used wire type on the die reliability

In this section, the proposed modeling methodology is applied and peeling hazard for the two kinds of bonding sets a presented in 0 is compared. Figure 17 shows the maximum peel stress calculated in the middle of the IC stack. Obviously, results underline that one of the bonding set induces significant higher stress compared to the other one. More precisely, for both simulated stages (i.e. bonding process and wire pull), similar trend is observed: Bonding set A induces additional stress of about 20%. The result also enables to clarify the effect of the simultaneously changes from Set A to B concerning both the wire stiffness and bonding parameters on die peeling hazard. More precisely, modeling outputs underline that for the two considered bonding sets, the dominant factor is the wire type, contrary to the used machine parameters which play role at a lower order of magnitude.

In the next section, the latter comparative result will be faced with experimental tests.

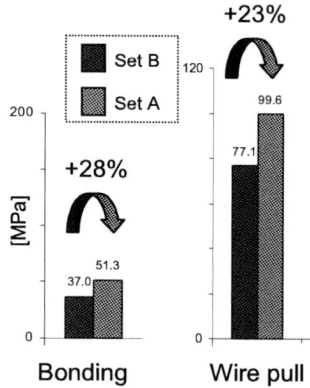

Figure 17 Maximum peel stresses in the IC stack evaluated at bonding stage and wire pull test for 2 wire types.

5. Experiments and Correlations

5.1 Test procedure

In order to qualify a technology node, a wide range of wire bond trials are performed and various wire bond responses are collected. One of the important responses in wire bond is the wire pull test. The test procedure consists of the following steps: First wire bonding operation is tried, followed by a wire pull test, then a visual observation is performed to determine the related failure mode and to check the integrity of the bond. Finally, categorization of the results in order to provide a reliable statistical data and rates are obtained with respect to each failure mode: Wire neck break, non-sticking corresponding to intermetallic issue, or peeling with delamination at die level. Note that in terms of product qualification, only the neck break failure mode is accepted.

156

5.2 Experimental DOE results and correlations

In the frame of 65-nm node qualification, several factors are studied on a test vehicle aiming to set the bonding recipes and pad architectures. Amongst others, these include IC material changes, wire/capillary types and a large range of bonding parameters. This paper is focused on the effect of the used bonding conditions, and results obtained at the early stage of the process optimization are used. Indeed, a significant rate of peeling failures was encountered with unoptimized parameter sets. It has been also found that peeling amount was highly dependent on the bonding set used. Peeling rates obtained with set A and set B are presented in Figure 18. From these results, the set A obviously induces a higher peeling rate.

Furthermore, the numerical study previously presented is in good accordance with the latter experimental result.

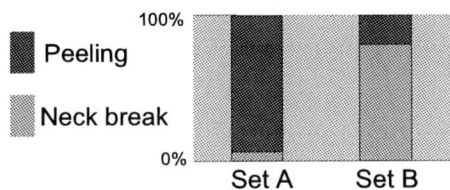

Figure 18:.Experimental peeling rates after wire pull test for the two wire types and bonding conditions.

6. Conclusion

In the frame of the wire bonding process development of advanced technologies, mechanical modeling has been employed to particularly address peeling related failures. More precisely, 3D multi scale simulation of both bonding process and wire pull test has been completed. Using a previously validated homogenization procedure to include pad structure description even at the global scale, stress fields acting in both the gold wire and the copper/low-k stack have been evaluated and discussed. The modeling strategy includes two kinds of analysis, a stress based one and an energy based one. The latter is founded on a propagation approach, as suggested by the observed failure mode. Stress based analyses at both the wire and die levels provide deeper understanding of the failure mechanisms. Furthermore, the complementary nature of the energy based failure criteria has been highlighted through a clearer determination of the forecasted location of the failed interface in the IC stack.

In addition to the numerical work, correlations with experiments have been achieved. Indeed, during the 65-nm technology node development, a wide experimental DOE provided reliable statistical data, which has been used here. The effect of the bonding conditions, through variation of the wire type and the parameters of the bonding machine on the peeling failure rate has been particularly studied. Modeling results have been found to be in agreement with experiments. In this work, it has been numerically validated that the wire stiffness and strength, combined with their respective bonding parameters, affects greatly the die reliability.

Finally, the simulation procedure with confirmed experimental results demonstrates its ability in design and process optimizations by providing a better understanding of pad peeling failure mechanisms.

Acknowledgements

Thanks are particularly due to the following contributors to this work: ASM Singapore for developing bonding process parameters to eliminate all pad peeling, K&S for providing the wire curves and FAB displacement measurements, the whole Crolles Alliance Assembly (CAA) team at Grenoble and Crolles2 Alliance (C2A) common IO and mechanical modeling teams for their valuable inputs.

References

[1] Goldberg C., Integration of a Mechanically Reliable 65-nm node technology for low-k and ULK Interconnects with various Substrate and Package Types, IEEE, 2005.

[2] Dauskardt, R.H., Stroband, S., Reliability of Advanced Interconnect Structures: New Materials and Length Scale Challenges, Stanford Material science and Engineering report, 2002.

[3] Viswanath A., Numerical Analysis by 3D Finite Element Wire Bond Simulation on Cu/Low-K Structures, EPTC, 2005.

[4] Hoofman R.J.O.M, Reliability Challenges accompagnied with interconnect downscaling and ultra low-k dielectrics, IEEE, 2005.

[5] Don Scansen, Impact of Low-k dielectrics on microelectronics reliability, IEEE CCECE/CCGEI, Saskatoon, May 2005.

[6] Ming-Dou Ker, Fully Process-Compatible Layout Design on Bond Pad to Improve Wire Bond Reliability in CMOS ICs, IEEE Transactions Components and Packaging Technologies, VOL. 25, NO. 2, JUNE 2002.

[7] Mercado L., Goldberg C., Kuo S-M., A simulation method for predicting mechanical reliability with low-k dielectrics, ECTC proceedings, 2003.

[8] Fiori V., Orain, S., A Multi-scale Finite Element Methodology to Evaluate Bond Pad Architectures, 6th Eurosime Conference Proc., Germany, 2005, pp. 648-655.

[9] Gallois-Garreignot S., Numerical Analysis of the Reliability of Cu/low-k Bond Pad Interconnections Under Wire Pull Test Application of a 3D Energy Based Failure Criteria, submitted Eurosime 2007.

[10] M.A.J. van Gils et al, Analysis of Cu/low-k Bond Pad Delamination by using a Novel Failure Index, 6th Eurosime Conference Proc., Germany, (2005), pp. 190-196.

[11] A. Baldacci, C. Rivero, P. Gergaud, M. Grégoire, O. Sicardy, O. Bostrom, P. Boivin, J.S. Micha, O. Thomas, Stresses in copper blanket films and damascene lines : measurements and finite element analysis, ESSDERC 2004, September 20-24 2004, Leuven.

[12] Jonathan Tan & Al., Wire-bonding process development for low-k materials, Microelectronic Engineering 81 (2005).

[13] Tumala, R.R., Fundamental of Microsystems Packaging. McGraw-Hill, NY 2005.

[14] Harmann G.G., Wire Bonding in Microelectronics, McGraw-Hill, NY 1997.

[15] Chang W.R., Kim J.K, Liu P.C.K, Process Window for low temperature Au wire bonding, J.Electron. Mater. 33, 146-155, 2004.

[16] Jeng Y.R., Hong J.H., A microcontact approach for ultrasonic bonding in microelectronics, J. Tribol., 3515-3521, 2001.

[17] Jeon I. & Al., The Effect of Ultrasonic Power on Bonding Pad and IMD Layers in Ultrasonic Wire Bonding, ISEMP, 2001.

[18] Ding Y., Numerical analysis of ultrasonic wire bonding: Effects of bonding parameters on contact pressure and frictional energy, Mechanics of Materials, 2005, under review.

[19] Liu Y., Thermosonic Wire Bonding Process Simulation and Bond Pad Over Active Stress Analysis, ECTC, 2004.

[20] Vincent Fiori, Lau Teck Beng, Susan Downey, Sebastien Gallois-Garreignot, Stephane Orain; 3D Multi Scale Modeling of Wire Bonding Induced Peeling in Cu/Low-k Interconnects: Methodology, Energy Based Analysis and Correlations with Experiments. Accepted in 57th ECTC conference, 2007.

Reliability Based Design Optimisation for System-in-Package

S. Stoyanov [1], J. M. Yannou [2], C. Bailey [1], N. Strusevich [1]

[1] Centre for Numerical Modelling and Process Analysis, University of Greenwich
30 Park Row, Greenwich, London SE10 9LS, UK
[2] NXP, 2 Rue de la Girafe, BP 5120, 14079 Caen, Cedex 5, France
E-mail: s.stoyanov@gre.ac.uk   Phone: +44 (0)20 8331 8520

Abstract

This paper discusses a reliability based optimisation modelling approach demonstrated for the design of a SiP structure integrated by stacking dies one upon the other. In this investigation the focus is on the strategy for handling the uncertainties in the package design inputs and their implementation into the design optimisation modelling framework. The analysis of thermo-mechanical behaviour of the package is utilised to predict the fatigue life-time of the lead-free board level solder interconnects and warpage of the package under thermal cycling. The SiP characterisation is obtained through the exploitation of Reduced Order Models (ROM) constructed using high fidelity analysis and Design of Experiments (DoE) methods. The design task is to identify the optimal SiP design specification by varying several package input parameters so that a specified target reliability of the solder joints is achieved and in the same time design requirements and package performance criteria are met.

1. Introduction

System-in-Package (SiP) technology was developed to provide fully functional electronic systems and sub-systems that integrate several functionally different devices (like IC- and RF-chips) and optical, MEMS, sensor and other components into a single package. The 3D micro integration design concept of the SiP structures and the increased package complexity/functionality combined with shorter times allowed for the design cycles is resulting in a decreased knowledge about the performance and the reliability of these electronic modules [1]. A major challenge is to understand the risks for failure and the associated failure modes, and to qualify the package from a reliability stand of point. Another aspect of the SiP design challenge is to assess and take into account the thermal, mechanical and electrical behaviour of a particular SiP module so that the structure can be optimised before it is actually manufactured. Unfortunately, there is little design knowledge and experience about SiP, with the options for real testing that can aid the design for manufacture being also limited.

Simulation based optimisation for virtual design prototyping of various electronic packages and manufacturing processes has proven as an effective approach for process characterisation and product development at the early design stages [2-4]. Computational modelling and simulation approach that exploits numerical analysis tools and methods (such as Finite Element Analysis or reduced order models) integrated with optimisation techniques can aid the identification of the optimal design/process specification and the formulation of design rules for optimal performance/reliability of the developed SiP structures. The virtual design optimisation approach is a strategy that can deliver the deterministic optimal package design based on the variation of a number of input parameters so that imposed constraints and design requirements are satisfied.

However, in reality such optimal package design, from deterministic point of view, may be far from a reliable and safe design solution. The reason for this is that the design of a real system, including the design of a SiP structure, often includes parameters that have uncertainties associated with them. This is a result of the natural variations that exist in the manufacturing and/or operational process parameters (e.g. operational temperature, humidity, etc), the tolerances in the dimensions of the manufactured structures, the physical properties of the materials, etc. It is very difficult and often impossible to control such existing variations. These tolerances and variations of the input design parameters may have significant impact on the system behaviour and can lead to variations and scatter of the response parameters that define the target requirements for performance and reliability. Therefore, the uncertainties in the responses/behaviour of the deterministic optimal design can result in performance that violates the specified requirements and reliability criteria. In order to ensure reliability of the designed system the uncertainties associated with the input parameters must be taken into account and brought into the modelling framework so that the optimal solution always meets the design constraints despite of the existing variations in the system/process response parameters.

Three key aspects are emphasised and discussed in the paper: (1) thermo-mechanical life-time assessment of the lead-free (SAC) SiP interconnects and warpage of the package using finite element analysis, (2) modelling of the uncertainties of the SiP design inputs and responses, and (3) optimal SiP design identification through reliability-driven numerical optimisation. The optimisation modelling incorporates the development of Reduced Order Models (ROM) for fast evaluation and assessment of the SiP thermo-mechanical response parameters. The ROM are developed using the results from high fidelity analysis (Finite Element Analysis) conducted for limited number of experimental SiP design configurations and the relevant response surface modelling.

1-4244-1105-X/07/$25.00 ©2007 IEEE

2. SiP Structure and Design Parameters

2.1. Geometric Details

The structure under investigation is a stacked dies SiP. The active die is flipped onto the passive die. The board level solder joints are designed in two peripheral rows along each side of the passive die and are located on the same side of the passive die where the active die is placed. The external row of joints is 11x11 and the second row had 9x9 configuration pattern. The pitch size used to distribute the solder joints is 0.5 mm. Figure 1 illustrates the SiP.

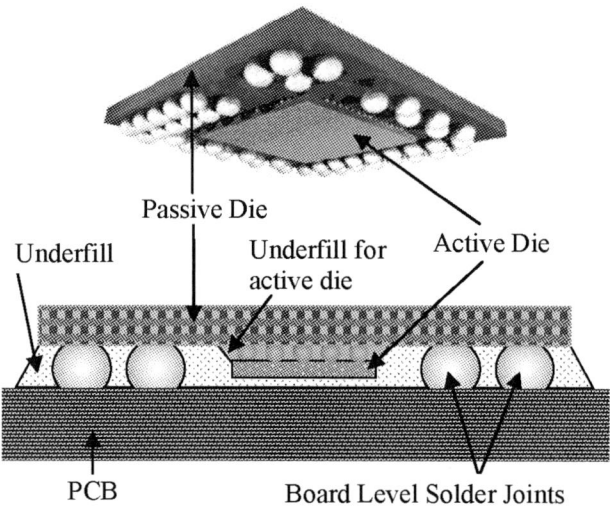

Figure 1: SiP structure.

This SiP component is then placed on a printed circuit board (PCB). To improve the thermo-mechanical reliability of the board level solder joints, underfill material is used to fill the gap between the PCB and the passive die.

Table I details some of the dimensions of the SiP assembly.

Table I. SiP nominal design and parameter variations.

System in Package Structure	Nominal (Initial) Design	Design Parameter Variation
Active Die Dimensions [mm]	Area: 2.26 x 2.26	none
	Thickness: 0.15	none
Passive Die Dimensions [mm]	Area: 5.7 x 5.7	none
	Thickness: 0.15	0.15 - 0.25
PCB Thickness [mm]	0.8	0.8-1.2
Stand-off Height of Board Level Solder Joints [μm]	210	210 - 260
I/O pitch [mm]	0.5	none
Number I/O (two rows, peripherally)	11x11 (1st row)	none
	9x9 (2nd row)	none

In Table I, the second column specifies the geometry of the nominal (or initial) design of the SiP while the third column of the table provides details on some possible design variations of the SiP assembly parameters that are feasible to implement.

As detailed in Table I, we will consider the following SiP design parameters with potential to vary from their nominal values:

1. PCB thickness (HPCB);
2. Board level solder joints stand-off-height (SOH);
3. Passive die thickness (HDIE).

Because any of these design parameters can be changed within the specified lower and upper bounds, we call them *design variables*. The bounds which define the variation limits for each of the three design variables are specified in the third column of Table I. Note that the term variation here is not related to the uncertainty qualification but has the meaning of limits for changing the value of the variable. By changing the value of any of these design variables, design modifications of the SiP structure can be generated. A set of values for the specified design variables that define a certain design is named shortly as a *design point*.

Due to the existing symmetry in the SiP structure, it is sufficient to represent in the computer model only one-eight part of the assembly. The one-eight symmetric models lower significantly the compute times for the undertaken simulations. The symmetry planes are taken into account in the 1/8 modelled part by using the appropriate boundary conditions set up for the analysis.

All solder interconnects inside the modelled one-eight part of the SiP are taken into account and are included into the finite element model representation. Once the geometry and all model components are defined, a finite element mesh grid is generated throughout the whole domain of the model. The finite element model (1/8 of package) is shown in Figure 2.

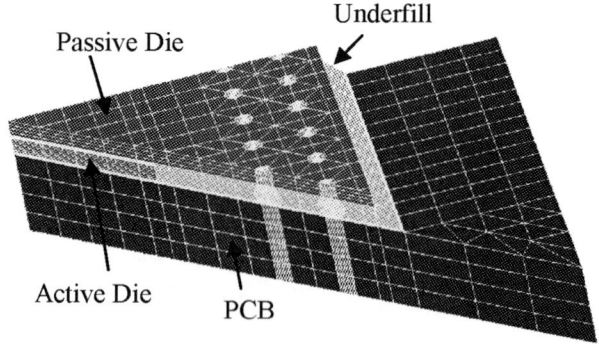

Figure 2: 1/8 Finite element model of SiP.

2.2. Material Models

In this modelling study, non-linear material behaviour is considered for the solder material; the rest of the materials are assumed to behave in an elastic manner.

Temperature dependent properties of the SiP materials are also incorporated in this analysis. Orthotropic coefficient of thermal expansion is implemented for the organic composite package carrier (the PCB). The detailed properties of all materials are reported in Table II.

Table II. SiP materials and their properties.

Materials	CTE ($10^{-6}/$°C)	Young's Modulus E (MPa)	Poisson ratio ν
PCB	16.0 (xy) 52.4 (z)	25,000	0.3
Solder : SnAgCu	20.0	61251-58.5T(°K)	0.36
Die: Silica	2.3	169,000	0.26
Underfill Epoxy	28.0	3,500	0.3

Under temperature cycling the solder materials experience deformations due to time-dependent plasticity and creep accompanied by stress relaxation [5, 6]. The cycling stresses are result of the coefficient of thermal expansion miss-match between the materials used in the package. The inelastic strain rate of the solder material is modelled in this study using the widely accepted *sinh* constitutive law. The inelastic strain rate $\dot{\varepsilon}_{ij}^{cr}$ using *sinh* expression is given as

$$\dot{\varepsilon}_{ij}^{cr} = A(\sinh(\alpha\,\sigma_{eff}))^n \exp\left(-\frac{Q}{RT}\right) \quad (1)$$

where R is the gas constant, T is the temperature in Kelvin, σ_{eff} is the effective (Von Mises) stress, and all other symbols represent material properties. For Sn3.9Ag0.6Cu solder alloy the creep constants are taken from Schubert et al [7] and have the following values: $A = 277984\,s^{-1}$, $\alpha = 0.02447\,MPa^{-1}$, $n = 6.41$ and $Q = 6500R$.

To enhance the overall reliability of the SiP structure, an underfill material is utilized. The underfill selected in this investigation has coefficient of thermal expansion $CTE=28$ ppm/°C and Young's Modulus $E=3.5$GPa.

2.3. Computational Analysis of SiP Thermo-Mechanical Behaviour and Life-time Modelling

The thermo-mechanical response of the SiP structure is analysed under accelerated thermal cycling. A cycle lasts for 2 hours and consists of four stages: ramp up from -40 °C to 125 °C for 45 minutes; hold at the higher temperature for 15 minutes; ramp down to -40 °C for 45 minutes; and finally hold at -40 °C for 15 minutes. In this study we assume a stress-free state for the SiP at 20°C; this is also the starting temperature for the thermal cycling.

The time dependent inelastic deformations are highly non-linear; hence transient analysis is required to simulate the solder behaviour. In such a transient analysis, the time domain of the thermal cycle is divided into time steps with thermal load at any time step the temperature change between this step and the previous step.

Inelastic strain and stress in Sn3.9Ag0.6Cu solder joints and package deformations are predicted from the non-linear FEA. These response values are used to calculate the damage in solder joints and to judge the thermal fatigue reliability of the package interconnects. In this study, the volume weighted average value, Wp, of the accumulated creep energy density per thermal cycle in the solder joints is considered. The Wp is also known as the inelastic work density and will be referred shortly as the *damage*. This quantity is used subsequently in a life-time model to make prediction for the solder joints mean cycles to failure.

The damage parameter Wp is calculated from

$$Wp = \sum_{\Delta t} \frac{\sum\limits_{i=1}^{IP} \int\limits_{V_i} \sigma^T (\Delta\varepsilon^{vp}) dV}{\sum\limits_{i=1}^{IP} \int\limits_{V_i} dV} \quad (2)$$

where the outer sum is over the time steps Δt that cover a full thermal cycle, "IP" is the number of the integration points used to calculate the inelastic work density, V_i is the volume associated with the integration point with index "i", σ is the stress tensor, $\Delta\varepsilon^{vp}$ is the tensor of the visco-plastic strain increment for Δt. Accumulated inelastic energy density per cycle in the most critical solder joint is calculated over a thin layer of solder mesh elements at the passive die interface (the volume V in Eq. 2). This is the critical location within the solder joint where it was found the crack will initiate and propagate.

A number of simplifications and assumptions are made in the analysis. All initial stresses in the package are neglected. The damage parameter Wp is calculated from the results associated with the second thermal cycle that is found to provide a stabilized hysteresis loop. Modelling the temperature cycling regime assumes an isothermal loading throughout the package. Finally, perfect adhesion between all materials is assumed.

The SnAgCu solder joint life prediction model [8] used to correlate the damage Wp to life-time in terms of cycles to failure is:

$$N_f = (0.0014\,Wp)^{-1} \quad (3)$$

The above life prediction model is function of the accumulated creep energy density per cycle Wp (in MPa) and predicts the mean life N_f of SnAgCu solder joints in terms of number of cycles to failure. The constants in Equation 3 are obtained by fitting a linear regression model to sets of experimental data assuming the hyperbolic sine constitutive equation for the solder creep behaviour [8].

The software package PHYSICA [9] is used to model and predict the evolution of thermal stresses and strains in this SiP package during a thermal cycle. Figure 3 illustrates the contours of the inelastic work density across the solder joints associated with the 1/8 part of the constructed finite element model of the SiP with the initial design specification. The results from this finite element simulation show that the most critical solder joint (i.e. likely to fails first) is the one located at the corner of the package. It was also found that variations in the values of the analyzed design parameters do not affect the location of the most exposed to the creep damage joint and it always stays at the corner of the SiP structure.

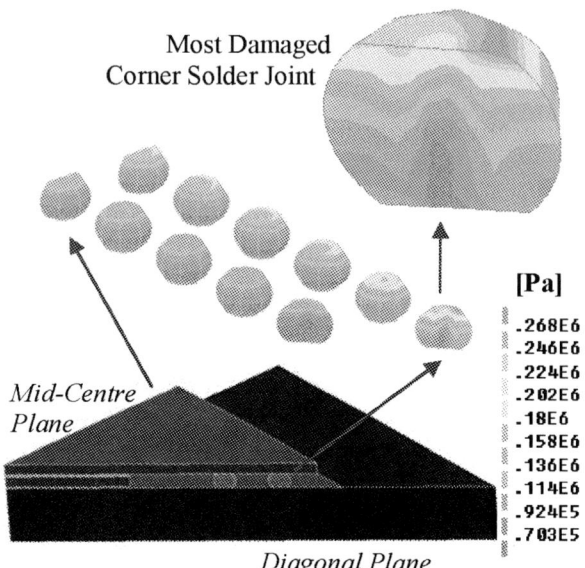

Figure 3: Contour levels for inelastic work density across SiP solder joints (initial design) at the end of a thermal cycle.

The non-linear FEA provides us also with predictions for the deformations across the SiP assembly. A response of interest is the maximum warpage of the SiP during the thermal cycling. This quantity is defined as the difference between the minimum and maximum out-of-plane deflection of the package and is denoted as D_w. It is found that the minimum and maximum out-of-plane displacements are occurring at centre and at the corner of the SiP passive die respectively. The maximum warpage occurs at the highest temperature during the thermal cycle (125°C). Figure 4 illustrates the warpage of the SiP for the initial design (deformation is magnified by factor 50).

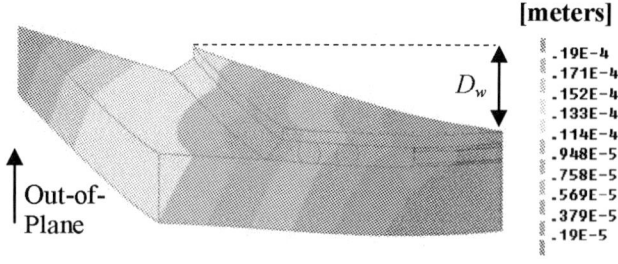

Figure 4: Contour levels of the out-of plane deformations across SiP assembly at 125°C during a thermal cycle.

3. Design of Experiments and Response Surface Modelling for Fast Design Evaluations

The non-linear finite element analysis outlined in the previous section is capable of predicting the SiP responses of interest. This is a compute intense method and often is not suitable for design purposes where many design evaluations will be required during the iterative design optimisation process. However, as it was demonstrated, a finite element analysis has the advantage of predicting with great accuracy the behaviour of the analysed system.

In order to benefit from finite element analysis capabilities in the design process, reduced order models based on Design of Experiments and Response Surfaces can be constructed. These models allow us to undertake fast evaluations of the response of interest for different design specifications [4].

The optimisation modelling for the SiP structure involves the following steps:

1. Identify the experimental design points in the three-dimensional design variables space (HPCB, SOH, HDIE).
2. Undertake finite element analysis at each design point (*PHYSICA* simulation). Obtain data for solder joint cycles to failure N_f (using Eq.2 and the predicted solder damage Wp) and the maximum package warpage D_w.
3. Use the above SiP response data to construct Response Surface (RS) approximation for N_f and D_w by fitting functions to the data points (second order polynomials are demonstrated).

3.1. Design of Experiments (DoE) Method

The DoE is performed to identify the set of design points at which the finite element analysis will be undertaken to provide predictions for the SiP response under thermal cycling. The focus here is on the cycles to failure of solder joints N_f and the maximum warpage of the package D_w. The DoE method decided in this study is the 15 points Central Composite Design (CCD). It is a combination of the factorial, axial and the central points of the 3-dimensional design space cube defined by the limits of the three design variables.

The way the DoE points are dealt with is to use normalised (scaled) values of the design variables in the interval [-1, 1]. In this, -1 corresponds to the lower limit of the design variable and 1 corresponds to the upper limit of the design variable. This transformation is detailed in Table III.

The design point number is given in the first column of Table IV and the design variable scaled values for each of the experimental points of the CCD are given in the next three columns of the table. After running the 15 DoE simulations, numerical predictions for the solder joints cycles to failure N_f and the maximum warpage of the package D_w become available. These predictions are detailed in the fifth and the sixth column of Table IV respectively.

Table III. Design Variable scaling.

	PCB thickness HPCB	Stand-off Height SOH	Passive Die Thickness HDIE
Un-scaled Limits [mm]	0.8 to 1.2	0.21 to 0.26	0.15 to 0.25
Scaled Limits (dimensionless)	-1 to 1	-1 to 1	-1 to 1

Table IV: The 15 scaled points of CCD and SiP response.

Design Point	HPCB	SOH	HDIE	Cycles to Failure N_f	SiP Warpage D_w [μm]
1	-1	-1	-1	2 990	11.92
2	1	1	1	2 255	7.52
3	-1	1	1	2 780	10.75
4	1	-1	-1	2 409	7.93
5	-1	-1	1	2 809	11.55
6	1	1	-1	2 437	7.35
7	-1	1	-1	2 973	11.05
8	1	-1	1	2 232	8.10
9	0	0	-1	2 659	9.23
10	0	0	1	2 480	9.24
11	0	-1	0	2 537	9.68
12	0	1	0	2 542	8.99
13	-1	0	0	2 877	11.38
14	1	0	0	2 319	7.77
15	0	0	0	2 548	9.31

3.2 Response Surface (RS) Modelling

After obtaining the SiP responses at the experimental design points as detailed above, the next stage in the modelling procedure is to construct approximation models to the solder joint cycles to failure (life-time) and SiP warpage. Second order polynomials are used to fit the data in Table IV by conducting least square techniques. The coefficients of the constructed RS polynomial approximations are detailed in Table V. The polynomials are based on the inputs of the scaled design variables.

The RS models can now be used to evaluate approximately the solder joints life-time and the package warpage for any design point (i.e. for any combination of values for PCB thickness, solder joint stand-off-height and passive die thickness) without running any finite element simulations. These RS models are reduced order

models and can substitute the compute intensive finite element analysis as an approach for design evaluation.

Table V: RS polynomial coefficients for SiP responses (based on scaled input design variables).

RS Polynomial Term	Warpage D_w RS Model Coefficient	Life-time N_f RS Model Coefficient
Constant	9.31213	2548.13333
HPCB	-1.79940	-277.70000
SOH	-0.35180	1.00000
HDIE	-0.03290	-91.20000
HPCB*SOH	0.06263	12.12500
HPCB*HDIE	0.12563	1.87500
SOH*HDIE	0.00713	-2.12500
HPCB**2	0.26233	49.83333
SOH**2	0.02333	-8.66667
HDIE**2	-0.07717	21.33333

The quality of the RS polynomial approximation is evaluated by number of techniques for estimating its predictive power and accuracy. These include analysis of the calculated efficiency measures, analysis of variance (ANOVA) and analysis of the residuals. For example, for both RS models we have found a very good predictive capability indicated by the coefficient of multiple determination, $R_{adj}^2 = 99.96$ to 99.99%.

4. SiP Design Optimisation

If we come back to the SiP structure and the specified design variables which provide flexibility to obtain different specification for the package and the assembly, the question now is what would the optimal design be? In order to identify the optimal design specification, we need to know:

1. What are the design variables we can change in order to have different designs that then can be evaluated. In this study we have already defined the three design variables of interest (HPCB, SOH and HDIE).

2. With respect what aspect/criterion we would like to have an optimal design specification (i.e. which aspect of the SiP we want to optimise).

3. What are the requirements this optimal design must satisfy (e.g. reliability, or any other).

Once the above questions are answered, we can formulate the design task as a mathematical problem and solve it using optimisation techniques.

4.1 Deterministic Design Optimisation

For our problem, the following formulation of the design task is given:

Find values of the design variables HPCB, SOH and HDIE that

Minimise   Warpage of SiP, D_w (4.0)

Subject to:

(1) Life-time $N_f \geq 2\ 700$ (4.1)

(2) SOH + HDIE ≤ 0.40 mm   (4.2)

(3) $0.8 \leq$ HPCB ≤ 1.2 mm   (4.3)

(4) $0.21 \leq$ SOH ≤ 0.26 mm   (4.4)

(5) $0.15 \leq$ HDIE ≤ 0.25 mm   (4.5)

The design task (4) requires a solution for which the warpage of the package is minimised (4.0) while satisfying the life-time constraint (4.1). The constraint (4.1) states a requirement for the solder joints fatigue mean life to be no less than 2700 cycles. An additional constraint (4.2) is included in the design formulation. It requires the total thickness of the SiP package to be less than or equal to 400 microns. Constraints (4.3)-(4.5) account for the design variable limits.

In the above optimisation task the warpage and life-time evaluation of different designs during the iterative solution procedure exploits the representative RS models developed for the two responses. No calls to finite element analysis software are performed at this stage. The above optimisation problem is defined and solved using VisualDOC [10]. The optimal solution has been found first by using gradient optimisation numerical techniques [10]. To verify that the found optimal design is the true global optimum, a non-gradient optimisation of the same design task was performed. It confirmed the already identified optimal solution.

Based on the solution of the design task (4), a deterministic optimal solution for the design of the SiP structure has been found. The optimal design results are reported in Table VI. The optimal passive die thickness is 150 μm (value of the lower bound) and the optimal values for PCB thickness and solder joint stand-off- height are respectively 0.971 mm and 250 μm. Note, at this optimal design specification for the SiP assembly the life-time constraint (4.1) and the SiP thickness constraint (4.2) become both *active* (i.e. have values equal to the imposed limits and satisfy the constraints as equality). Any effort for further improvement of the objective (4.0), i.e. D_w minimisation, will cause one or both of the constraints to become violated. This would result in a design specification outside the feasible domain of (4.1)-(4.5).

Table VI: Deterministic optimum.

	Initial	Optimal
HPCB [mm]	0.800	0.971
SOH [mm]	0.210	0.250
HDIE [mm]	0.150	0.150
Warpage D_w [μm]	11.90	9.341
Life-time N_f [cycles]	2990	2700
SiP thickness, SOH+HDIE [mm]	0.360	0.400

4.2 Uncertainty Modelling and its Effect on Reliability

A deterministic optimum specification is not necessary a reliable optimal solution. The reason for this is in the uncertainty which normally is associated with the design variables. Such variations from the deterministic optimal values can lead to SiP structures that fall outside the failure free design domain. In this study, a design for the SiP is defined as *reliable* if it satisfies the constraints (4.1)-(4.2) given in the formulation of the design problem (4). Here we will not be concerned if the limit constraints (4.3)-(4.5) are violated as a result of the design variable uncertainty. We are going to assume feasible SiP design specifications at and near the design variable limits. However, there is no limitation to consider and include in the reliable domain formulation all or any of the design variable limit constraints.

In this study we consider variations (uncertainty related) for the design variables which follow and can be described using normal (Gaussian) distribution. Normally the distribution of the probabilistic input design variables is known and can be specified through certain distribution parameters. In this study the distribution is defined by two parameters, the mean value and the standard deviation. The following standard deviations define the distributions that account for the variable uncertainty:

a) HPCB: standard deviation $\sigma_{HPCB} = 16$ μm;

b) SOH: standard deviation $\sigma_{SOH} = 2$ μm;

c) HDIE: standard deviation $\sigma_{HDIE} = 2.5$ μm;

Figure 5 shows the probability density function (PDF) for the HPCB scaled design variable with mean value 0 (un-scaled value of 1 mm). It also shows the cumulative density function (CDF) for the same scaled variable. In a similar manner, PDFs and CDFs functions can be considered for the other two design variables.

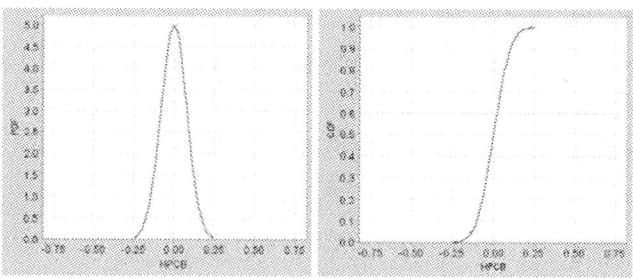

Figure 5: Probability and cumulative density functions for mean scaled value 0 and scaled standard deviation 0.08.

The uncertainties of the input parameters will affect the responses of the SiP assembly and will lead to variations in the values of the responses of interest. In particular, the life-time N_f and the thickness of the package values will follow a certain distribution profile as a result of the variation. In general, the uncertainty properties of the responses are unknown. Therefore, when uncertainties are included in the design optimisation task, we need to estimate the random properties of the responses. Different methods can be used to obtain this information. One way is to calculate the response mean value and standard deviation and to use this information to judge the probability of failure with respect to that response.

The most common simulation method is to run a Monte Carlo Simulation [10]. This is the technique used in this study. The basic idea is, for fixed nominal values of the design input variables, to generate large number of random design points according to the distributions of the probabilistic input variables. The values of the design input variables of the generated points in the set represent the variations due to uncertainty from the fixed nominal values. For each of these points the values of the responses of interest are calculated. The values of the fatigue life and warpage response in the SiP design task are calculated using the Response Surface models.

4.3 Design Optimisation with Uncertainties (Probabilistic Optimisation)

In reliability based optimisation the aim is to account for the variations of the responses that define the reliable design domain and to ensure that the deterministic optimal solution is moved from the boundary of the active constraints inside the feasible domain. Therefore, the aim is to minimise or satisfy constraints that involve system responses and related probability of failure. This reliable optimum design is called a probabilistic optimum. To define the probabilistic optimum one must specify what probability of failure will be acceptable.

To demonstrate the reliability based design optimisation strategy, the following re-formulation of the design task (4) is given:

Find values of the design variables HPCB, SOH and HDIE that

Minimise	Warpage of SiP, D_w	(5.0)
Subject to:		
	(1) P(Life-time $N_f \leq 2\ 700$) ≤ 0.05	(5.1)
	(2) P(SOH + HDIE ≥ 0.40 mm) ≤ 0.05	(5.2)
	(3) $0.8 \leq$ HPCB≤ 1.2 mm	(5.3a)
	Standard deviation $\sigma_{HPCB} = 16\ \mu m$	(5.3b)
	(4) $0.21 \leq$ SOH ≤ 0.26 mm	(5.4a)
	Standard deviation $\sigma_{SOH} = 2\ \mu m$	(5.4b)
	(5) $0.15 \leq$ HDIE ≤ 0.25 mm	(5.5a)
	Standard deviation $\sigma_{HDIE} = 2.5\ \mu m$	(5.5b)

As evident from the above formulation, the solution of this optimisation problem will account for the variation of the input design variables (the constraints (5.3a)-(5.5b)). The constraint (5.1) states that the probability of the fatigue life being less than or equal to 2700 cycles to failure must be no greater than 0.05 (i.e. 5 % probability of failure limit with respect to the life-time requirement). Similarly, the constraint (5.2) is re-formulated to represent a reliability requirement on the package

thickness, i.e. the probability of SiP thickness (SOH+HDIE) becoming great than or equal to 400 microns must be no greater than 0.05. By solving this problem we can find a solution (the probabilistic optimum) which, despite the uncertainty of the input parameters, will be always 95 % reliable. This reliability is with respect to design constraints (4.1) and (4.2).

VisualDOC software package has incorporated features for probabilistic design optimisation and therefore is used once again to specify and solve the design task (5). Note, the same optimisation techniques as applied in deterministic optimisation can be used to solve the probabilistic design task. However, there is some extra calculation efforts associated with running the Monte Carlo simulation at each of the design optimisation iterations in order to evaluate the probabilities of failure as defined in (5.1) and (5.2).

The solution of the design task (5) is reported in Table VII. The previously found deterministic optimum is also included in the table.

Table VII: Probabilistic optimum.

	Initial	Determ. Optimum	Probab. Optimum
HPCB [mm]	0.800	0.971	0.945
SOH [mm]	0.210	0.250	0.242
HDIE [mm]	0.150	0.150	0.150
Warpage D_w [μm]	11.90	9.341	9.721
Life-time N_f [cycles]	2990	2700	2 741
P(Life-time $N_f \leq 2\ 700$)	-	0.5	0.05
SiP thickness SOH+HDIE [mm]	0.360	0.400	0.392
P(SOH+HDIE ≥ 0.40 mm)	-	0.5	0.00

The last two columns of the table compare the deterministic and probabilistic solution. It is clear that by moving the deterministic optimum from the active constraints boundary inside the feasible domain, we have compromised on the level to which our objective is minimised, the SiP warpage (from 9.341 up to 9.721 μm). However, what we have gained by doing this is that our probabilistic optimum is now 95 % reliable. This compares with 50 % reliability of the deterministic optimum. In particular, at the probabilistic optimum the cycles to failure in terms of mean value are 2741 and only 5% of the SiP structures will have life-time less than 2700 as a result of the uncertainties of the input design variables. Figure 6 shows the Monte Carlo simulation output for the life-time response at the probabilistic optimum. This run uses 3000 points to compute the life-time response standard deviation from the mean value (2741) and to estimate the probability of failure (with respect to the 2700 cycles limit).

Note that at the probabilistic optimum the probability of failure for SiP thickness constraint (5.2) becomes 0.

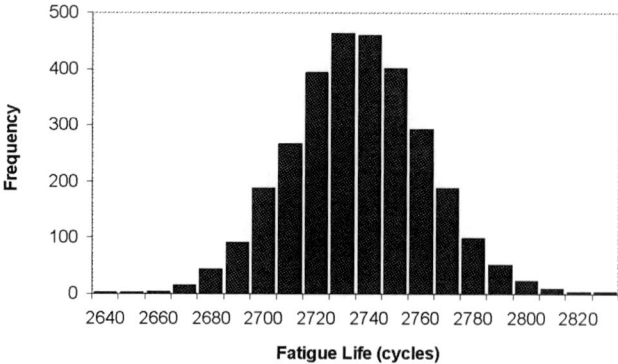

Figure 6: Monte Carlo simulations (3000 points) for fatigue life at the probabilistic optimum.

4. Conclusions

This paper has demonstrated a modelling framework for reliability based design optimisation. It is shown that the deterministic optimum might not be a reliable solution. It is important to bring into the design problem formulation probability of failure constraints. The advantage of such design approach is that it can deliver a more realistic design solution and provides the opportunity to account for the variations of the input design parameters.

A SiP structure has been optimised under a set of design constraints. A reliable design which satisfies a requirement for 95% reliability with respect to the system life-time and package thickness constraints is demonstrated. The optimal solution has been found in a very efficient and automated way. The concept of exploiting reduced order models based on response surface modelling and design of experiments techniques has been also incorporated in the calculation procedure. It was shown that the usage of reduced order models is extremely critical aspect in the implementation of the design optimisation with uncertainties approach.

Acknowledgments

This work was financially supported by the UK Engineering and Physical Sciences Research Council (EPSRC) and the Innovative electronics Manufacturing Research Centre (IeMRC) through Design for Manufacture for SiP project. Special thanks are extended to our academic parthner Lancaster University and our industrial collaborators: NXP Semiconductors, Flomerics Ltd, Coventor and SELEX Sensors and Airborne Systems.

References

1. Tai, K.L, "System-In-Package (SIP): Challenges and Opportunities," *Proceedings of Asia and South Pacific Design Automation Conference (ASP-DAC'00)*, Yokohama, Japan, 2000, pp. 191-196.
2. Vanderplaats, G. N., Numerical Optimisation Techniques for Engineering Design: with Applications, VR&D, Colorado Springs (1999).

3. Zhang, G Q, Maessen, P, Bisschop, J, Janssen, J, Kuper, F and Ernst, L, "Virtual Thermo-Mechanical prototyping of Microelectronics – the Challenges for Mechanics Professionals", *Proceedings of EuroSIME*, 2001, pp. 21-24.
4. Stoyanov, S., Optimisation Modelling for Microelectronics Packaging and Product Design, PhD Thesis, University of Greenwich, London, UK, 2004.
5. Hua, F., "Pb-free Solder Challenges in Electronic Packaging and Assembly", *The 53-rd IEEE Electronic Components and Technology Conference Proceedings*, New Orleans, Louisiana, USA, June 2003, pp. 58-63.
6. Lau, J. H., "Design, Materials, Process and Reliability of Lead-free Solders for Robust IC Electronic and Optoelectronic Packaging", *Short Course Notes of the 5-th Electronics Packaging Technology Conference*, Singapore, December 2003.
7. Schubert, A., *et al*, "Reliability Assessment of Flip-Chip Assemblies with Lead-Free Solder Alloys", *The 52-th IEEE Electronic Components and Technology Conference Proceedings*, San Diego, CA, USA, May 2002, pp. 1246-1255.
8. Syed, A, "Accumulated Creep Strain and Energy Density Based Thermal Fatigue Life Prediction Models for SnAgCu Solder Joints", *Proceedings of the 54-th Electronic Components and Technology Conference Proceedings*, Las Vegas, Nevada, USA, June 2004, pp. 737-746.
9. PHYSICA, http://www.physica.co.uk, Multi-physics Software Ltd
10. VR&D VisualDOC (Version 5.1), http://www.vrand.com

Multiphysic modelling of a microactuator based on the decomposition of an energetic material: application to microfluidics

Gustavo Adolfo Ardila Rodríguez, Carole Rossi

LAAS-CNRS, 7 avenue du colonel Roche, 31077 Toulouse Cedex 04, France

Email : gaardila@laas.fr

Abstract

We present the conception of a micro pressure source for microfluidics applications. It consists in 3 main parts: (i) a heating resistance built on a dielectric membrane, (ii) an energetic material which decomposes generating a high volume of biocompatible gas, (iii) an elastomer membrane with high elastic properties. PDMS is chosen as the membrane material for its elastic properties. A bimetallic energetic material composed of Mn and Co is chosen because of the high amount of biocompatible gas liberated from its decomposition [1].

When the actuation is required, the energetic material is heated beyond the ignition temperature to generate a high concentration of gas; the gain in pressure produces the deformation of the elastic membrane. Individual models have been developed for each physical phenomenon: (i) the heating of the resistance by an electro-thermal model, (ii) the gas generation from the decomposition of the energetic material based on an ideal gas model and a mass transfert model and coupled with the elastic membrane deformation with a mechanical deformation model. This paper presents each individual model under COMSOL Multiphysics that has been correlated with experimental results. Then, the implementation of all models into one global model allow us to predict the actuator performance as a function of input electrical signal.

1. Introduction

MEMS technologies in general and Microfluidics in particular are one of the most developed technologies in the last years. Further miniaturization and increase level of systems complexity are the current trend in Microsystems. In that context, new tools for the conception steps are essential to optimize the performance of the system, to save time and money during the development phase.

With the increased complexity level, multiphysics approaches taking into account the interactions between different physical phenomena, are necessary for the systems conception.

Some works have been recently reported on the multiphysics modeling of micro actuators [2-7]. This bibliography shows the difficulty in validating complete 3D models experimentally; often simplified models are validated either by analytical models or experimentally. Depending on the complexity of the structure or the multiphysic coupling, simplified models often leading to linearization of highly non-linear expressions are preferred.

In Microfluidics, an important issue is the integration of reliable, small and easy to use micro actuation parts for pumping and valving of liquids. Over the past decade a large variety of micro valves and pumps based on several effects including piezoelectric [8], electrostatic [9] and (thermo) pneumatic have been proposed [10, 11]. These are often expensive and difficult in design and fabrication. In addition, their deflection is too small to pump or completely close a microfluidic channel necessary for a good valve. An exception is based on pneumatic actuation. However, these actuators are usually too large to be fabricated in a dense array format and to be fully integrated into the microfluidic lab-on-a chip [12]. An emerging concept is to integrate pressure sources directly in the microfluidic platform by integrating energetic materials on a micro heater fabricated into the microchannel [13]. When the actuation is required, the energetic material is heated up and generates a high volume of gas which therefore locally increases the pressure. The micro/nanoactuator based on these mechanisms can be fairly simply fabricated, low cost, very compact (<1mm²), fast, biocompatible, integratable with different technologies (silicon, glass, plastics), adjustable and could be integrated on large scale.

The reliable integration in submillimeter scale, of, a microheater, a solid energetic material and an elastic membrane into a microchannel raises several modelling and technological challenges. The first one is to clearly understand the physics on which these actuators operate through multiphysics models, building up individual models for each physical phenomenon and then integrating them into a full package, enabling to make virtual prototyping. The second challenge is the elaboration of energetic materials that keeps mechanical actuation capabilities when it is deposited in a thin layer (thickness <100μm). The third one is to set up the MEMS technologies for the integration and interfacing of all materials (energetic, substrate, heater, polymeric membrane and fluid) at submillimeter scale.

In this paper we focus on the modelling issues. The first part of the paper is the presentation of individual multiphysics model built up using COMSOL Multiphysics. The numerical modelling issues will be discussed and compared with experimentation. In a second part, the integration of all individual models into a general model will be presented and multiphysics interfacing issues will be discussed.

2. Description of the microactuator

The proposed micro actuator made on silicon consists of 3 main parts as illustrated in Figure 2:

- One heating platform: it is made of one thin dielectric membrane (SiO₂/SiNₓ) on which is deposited a polysilicon resistor [14].

- One elastic membrane which encapsulates energetic material: PDMS (PolyDiMéthylSiloxane) has been chosen for the demonstrator for its easy prototyping. [15]

- One energetic material [1]

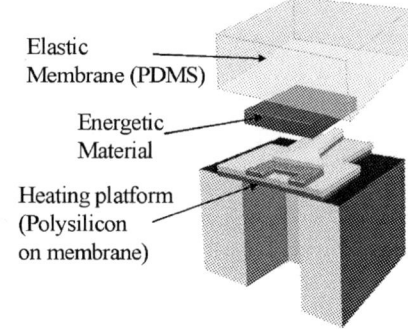

Figure 1. Micro actuator structure 3D view

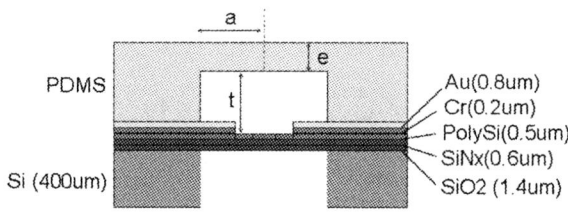

Figure 2. Micro actuator structure cut view characteristics

The micro heater and membrane structures are shown in Figure 2 and Figure 3. The geometrical parameters taken for the simulation are given in the Table 1..

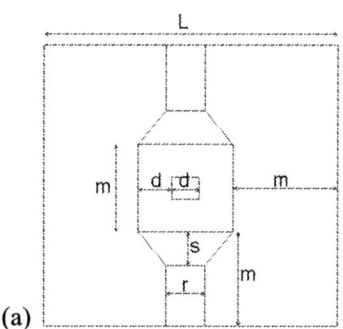

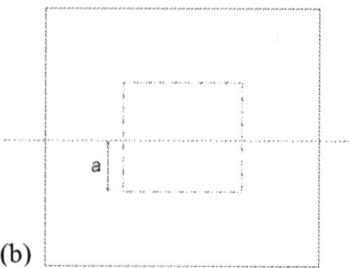

Figure 3. Membrane (a) and heater geometrical (b) characteristics.

Table 1. Micro actuator geometrical parameters.

Geometrical parameter	Value (µm)
L	500
m	166
r	92
d	55
s	72
a	250
e	30
t	100

When the actuation is required, a current flows in the polysilicon resistor heating up by the Joule effect the polysilicon and by thermal conduction the energetic material in contact with the heater. When the ignition temperature of the energetic material is reached, a high volume of air is released. Then, the pressure in the cavity increases and deforms the elastic membrane.

3 models describe the microactuation:

- the electro-thermal model to simulate the heating by Joule effect,

- the gas generation from the decomposition of the energetic material, once the ignition temperature (250°C) [1] is reached,

- a mechanical model to predict the membrane deformation as a function of gas pressure.

3. Mathematical models

3.1 Electro thermal model

The electro-thermal model includes the well known heat flow equation and the electric potential equation. The link between the two equations is done by the Joule effect ($Q = \sigma |\nabla V_e|^2$).

Electric potential

$$-\nabla.(\sigma \nabla V_e) = 0 \qquad (1)$$

Heat flow

$$\rho.C_p \frac{\partial T}{\partial t} - \nabla.(k \nabla T) = Q \qquad (2)$$

3.2 Gas evolution model

The evolution of the enclosed gas in the cavity follows the Fick's law of diffusion for the concentration of species (c): [16]

$$\frac{\partial c}{\partial t} - \nabla \cdot (D \nabla c) = R \quad (3)$$

Where D is the diffusion coefficient and R the reaction term. In this approach no convection is considered. The pressure in the cavity could be directly related to the gas concentration. Considering the gas as ideal we obtain:

$$P = cMT \quad (4)$$

where M is the air molar mass.

3.3 Membrane deformation model

The classical mechanical model is based on the small deformation approximation: the membrane deformation-to-thickness ration is very small ($w/e \ll 1$) and the membrane deformation-to-lateral dimension ($w/2a \ll 1$) is low [17]. In our case, both assumption are not true ($w/e \sim 3\text{-}5$ and $w/2a \sim 0.2\text{-}0.4$), therefore, the PDMS membrane deformation is simulated using a large deformation model.

The strain tensor is expressed as follow [17]:

$$\varepsilon_{ij} = \frac{1}{2}\left(\frac{\partial u_i}{\partial x_j} + \frac{\partial u_j}{\partial x_i} + \frac{\partial u_k}{\partial x_i} \cdot \frac{\partial u_k}{\partial x_j}\right) \quad (5)$$

where u represents the deformation in a given direction, x is the spatial coordinate and the sub indices i,j,k represent each component in x, y, z direction. The non-linear term $\partial u_k / \partial x_i \cdot \partial u_k / \partial x_j$ characterizes the large deformation behavior.

The membrane stress can be calculated from the First Piola-Kirchhoff stress p 18] as:

$$p = \frac{\partial Whyp}{\partial \nabla \vec{u}} \quad (6)$$

where $\nabla \vec{u}$ is the displacement gradient and $Whyp$ is a strain energy function which characterizes the hyper elastic properties of the PDMS.

The Neo-Hookean Model proposed by COMSOL Multi-physics [19] considers the strain energy function as follow:

$$Whyp = \frac{1}{2}\mu\left(I_1' - 3\right) + \frac{1}{2}\kappa\left(J_{el} - 1\right)^2 \quad (7)$$

where μ and κ are the shear and bulk modulus of the PDMS, respectively. J_{el} is the elastic part of the total volume change of the PDMS:

$$J_{el} = \frac{J}{J_{th}} = \frac{J}{(1 + \varepsilon_{th})^3} \quad (8)$$

Here, J represents the ratio between the current volume and the original volume and J_{th} is the thermal expansion expressed in terms of the thermal strain ε_{th}.

I_1' is the first modified strain invariant $: I_1' = (C_{11} + C_{22} + C_{33})J_{el}^{-\frac{2}{3}}$ where C_{11}, C_{22} and C_{33} are the matrix components of the Cauchy-Green tensor $C = F^T F$, where F is the deformation gradient matrix containing information about deformation and rotation.

4. Model implementation

The equations (1-4) and (5-8) have been solved numerically using the Finite Element Method (FEM). Several material properties are required to solve the different mathematical equations presented in the section 2:

- The electrical conductivity (σ), thermal conductivity (k) and the volumic mass (ρ) of all the structural materials and air: see Table 2.
- The Young modulus (E), the Poisson coefficient (v) and the residual stress (σ_o) of the PDMS.
- The air characteristics: volumic mass (ρ_{air}) and molar mass (M) depend on P, T and initial volume, respectively.

Table 2. Properties of the materials used for the fabrication of the micro actuator.

Material	σ (S/m)	k (W/m.K)	ρ (kg/m³)	Cp (J/Kg.K)
Silicon	Not used	100	2328	702
SiO2	Not used	1.4	2270	1000
SiNx	Not used	49	2270	1000
Polysilicon (Bore implanted)	$\sigma_{poly}(T)$ 5146.3 (23°C)	100	2328	700
Gold	4.4×10^7	297	19300	128.7
Chromium	7.3×10^6	93.7	7150	449
PDMS	Not used	0.15	965	1200
Air	Not used	$k_{air}(T)$	$\rho_{air}(T)$	$Cp_{air}(T)$

4.1 Thermal properties of the gas produced by the energetic material

As discused in section 4.1 (equation 9) the gas liberated by the energetic material is composed of N_2, O_2 and H_2O vapor, they are present in the air but not in the same proportion. Nevertheless as a first approach in the gas generation from the energetic material, we consider the gas generated as air.

The Nusselt (Nu) number has been used to determine the convection coefficient (h), given by:

$$h = Nu \frac{t}{2a} \quad (10)$$

$$Nu = A(Gr.\mathrm{Pr})^n \left(\frac{t}{2a}\right)^m \quad (11)$$

Where A, n and m are constant depending on the value of the product $Gr \times Pr$. The Grashof (Gr) and Prandtl (Pr) numbers of air contained in the micro cavity are written as follows :

$$Gr = \frac{g \cdot \beta_{air}(T_{max} - T)t^3 \rho_{air}^2}{\mu_{air}^2} \quad (12)$$

Where g is the gravity with a value of 9.8m/s², β_{air}, ρ_{air}, μ_{air} are the air expansion coefficient, density, dynamic viscosity respectively.

They are temperature dependant and the values can be found at atmospheric pressure in the literature [16].

The air is contained in a square cavity. Its lateral dimension is *2a* and height is *t* (see Figure 2 and). We calculate the Grashof number for the highest conditions in pressure and temperature in the cavity as function of time, when the maximal electrical power studied is applied (100mW). We obtained a peak of pressure of *P*=466.7MPa and a temperature at the heating platform *(T$_{max}$)* of 342°C, the Grashof number is equal to 606 and the air Prandtl number is equal to 0.7 [16].

Gr×Pr = 424.2 <<1700 that is the limit of the conduction regime [16]. Therefore the air contained in the cavity is in a purely conduction regime and convection coefficient is set to 0.

4.2 Gas diffusion

The diffusion coeffcient can be calculated as function of the nature of the gas, the pressure and the temperature in the cavity as: [16]

$$D = 435.7 \frac{T^{3/2}}{p\left(V_A^{1/3} + V_B^{1/3}\right)^2} \sqrt{\frac{1}{M_A} + \frac{1}{M_B}} \quad (13)$$

V indicating the molecular volume of the constituents *A* and *B* and *M* as the molecular weight, for the air *V*=29.9 and *M*=28.9. In our model the diffusion coefficent can not be considered constant as the temperature and the pressure in the cavity changes in time and are not homogeneus in space.

4.3 Meshing optimization

The electro-thermal, membrane deformation and gas generation models are implemented under COMSOL Multiphysics. The optimized meshing for the simulation is determined by performing an independent grid study to minimize the modeling error. When the change in the solution between subsequent stages of meshing refinement is considered to be negligible, the lower, but still sufficient, mesh resolution is kept. Figure 4 illustrates the procedure for the heating platform: the polysilicon microheater maximum temperature and time of simulation is compared among different stages of meshing refinement: the compromise between simulation time and accuracy is reached for the rounded point and gives the good optimal meshing.

For the heater platform, the optimized mesh consists of 7707 tetrahedral and prism elements on one quarter of the microheater structure. Figure 5 shows the meshing of the heater platform separated in the substrate and the dielectric membrane.

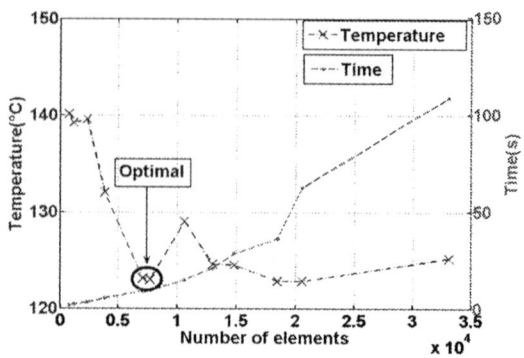

Figure 4. Example of mesh optimization procedure for the heating platform model

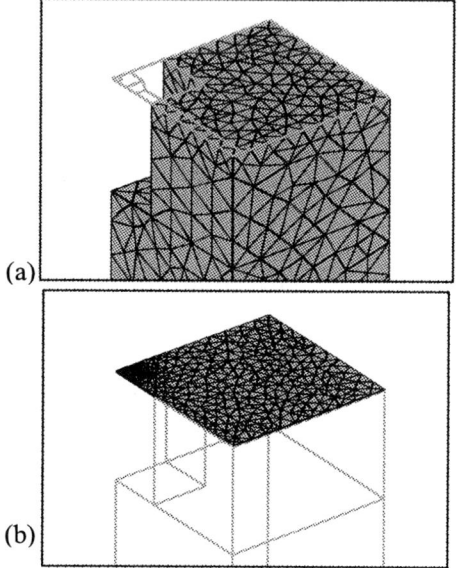

Figure 5. Heating platform. Substrate (a) and dielectric membrane (b) meshing

The optimized meshing for the PDMS membrane contains 5104 tetrahedral elements (see Figure 6).

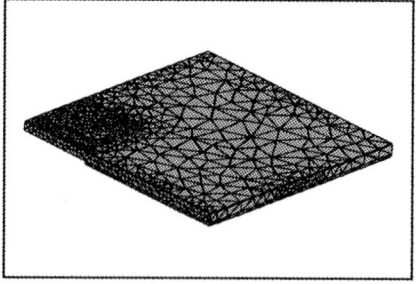

Figure 6. Meshing of the elastic membrane

The optimized meshing for the 2D axisymetric gas generation model contains 12103 triangular elements (see Figure 7).

5. Individual model implementation and experimental validation

5.1 Heating platform

The results for the microheater simulation are illustrated in Figure 8. Here the temperature of the heater dielectric membrane is given at the steady state and for the following conditions: Applied Voltage: 4V; electrical power: 29.8mW; convection coeffcient h=0.

From the simulation, the required power and the resulting equivalent temperature distribution are obtained.

The graph of Figure 9 gives the heater maximal temperature as a function of the electrical power with and without taking into account the temperature effect on σ_{poly}. A comparison with Infrared (IR) thermal characterization, made with a CEDIP JADE III MW camera, enables to evaluate the modeling error. From this graph, below a maximal temperature of 116°C the maximal error in the measurement is 33%, above this temperature the maximal error reduces to 16%. The difference between the model and the measurements could be explained by the uncertainty to evaluate the value of the emissivity coefficient (ε) of the different materials using the IR measurement.

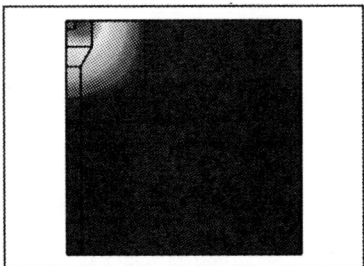

Figure 8. Example of temperature distribution for a 500μm×500μm dielectric membrane heating platform.

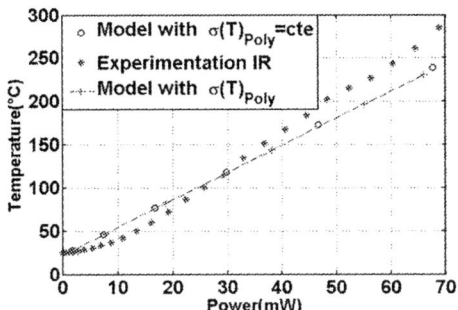

Figure 9. Heater maximal temperature vs. electrical power

Figure 11 gives the theoretical and experimental maximum membrane deformation as a function of the pressure inside the cavity. The curves correspond to a 30μm thick PDMS membrane fabricated at LAAS and having a Young Modulus of 4MPa measured by the "Bulge test" [20] and nanoindentation [21]. The comparison shows a difference in the non-linearity of the curves, experimental curves are almost linear because of the separation effect of the membrane from the substrate as pressure increases.

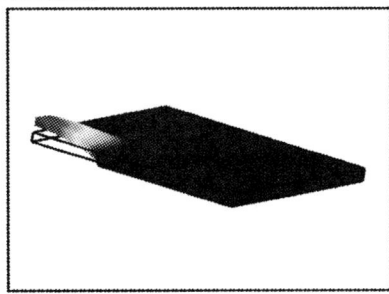

Figure 10. Membrane deformation for a 500μm×500μm membrane being 30μm thick.

5.3 The gas generation and its evolution

As a first approach in the implementation of the model describing the gas generation, some simplifations were made:

- A 2D axisymmetric model is implemented to validate the concept before implementing a full 3D model,

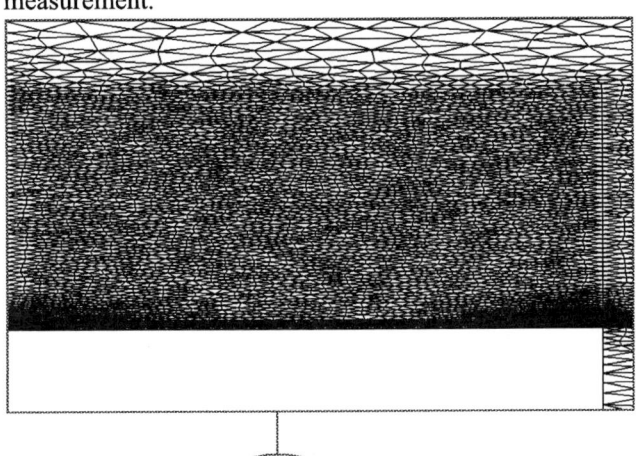

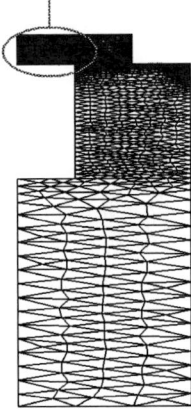

Figure 7. Meshing of the 2D axisymmetric model for the gas generation including the electro-thermal model

5.2 Membrane deformation

The results for the elastic membrane deformation simulation are illustrated in Figure 10. Here the deformation (w) of the PDMS membrane with parameters E=4MPa; v=0.48; σ_o=0 is given at the steady state and for the following conditions: Applied pressure: 0.01MPa; boundary: clamped all around the membrane.

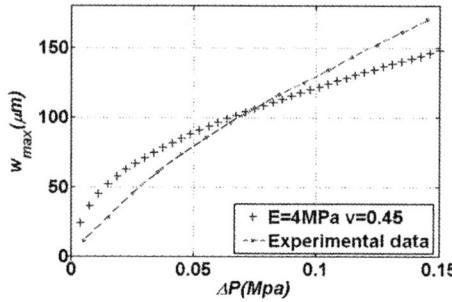

Figure 11. Comparison between experimental and simulated results for the deformation of a 500μmx500μm membrane.

- As the geometry of the heating resistance is different from the 3D model, a calibration of the 2D model in terms of applied tension and obtained heating temperature was made by modifying the electrical conductivity of the heating material,

- The amount of energetic material deposited on the heating platform is calculated from the profilometer measurement (see section 4.2), the mean thickness of the deposited material is 1.4μm. We use this value to consider an uniform desposit of the material,

- The gas liberated from the energetic material is composed of more than one species (see section 4.2), in this first approach the gas is considered as air, even if the proportion of the individual species are not exacly the same,

- The gas generation is considered to happen instantly once the ignition temperature is reached, in that way the simulation is made in two times: First; an stationary simulation considering the energetic material deposited over the heating platform. In this simulation the temperature distribution is obtained for a given applied voltage, in this case 6.3V. Then, the assessment of the quantity of material that reaches the ignition temperature is made as shown in Figure 12. This is taken as the initial condition for the next step; Secondly, once the quantity of the material that decomposes in gas is known, the transient simulation of the gas evolution can be calculated as a function of the electrical applied power. The observation point where the transient measurements where made is shown in Figure 12.

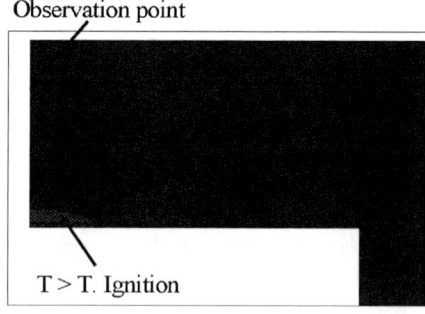

Figure 12. Stationary electro-thermal simulation. Evaluation of the energetic material with a temperature higher than the ignition temperature. Observation point for the transient simulation

Figure 13 shows the fast concentration evolution (2ms) compared to the thermal evolution of the system, starting from the initial air concentration (41.44mol/m³) calculated at ambient temperature and atmospheric pressure to the final concentration of the cavity (51.34 mol/m³).

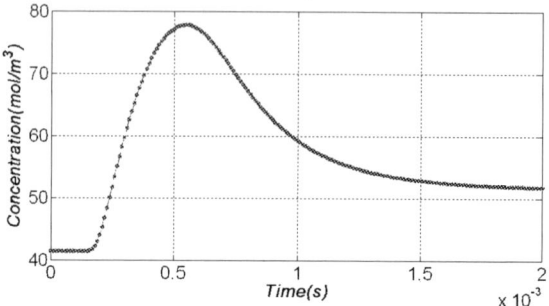

Figure 13. Gas concentration evolution on the observation point near the PDMS membrane

Figure 14 shows the evolution of the temperature in the cavity as the actuator is turned off, it takes 3.6s to reach the stationary state at ambient temperature.

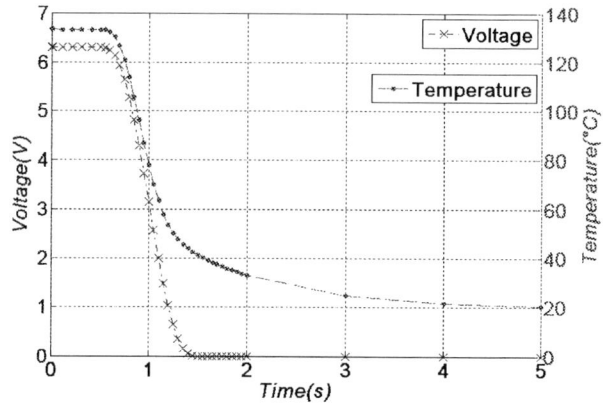

Figure 14. Temperature evolution on the observation point under the PDMS membrane

Figure 15 shows the evolution of the pressure in the cavity as the actuator is turned off. After 1ms from the decomposition of the gas, a high peak of pressure is detected (see the zoom in Figure 15). The assessment of the effect of this pressure peak on the global performance of the system is under study. A first stationary state is reached at 0.073MPa once the actuator is still on, then the pressure follows the temperature evolution in the cavity, reaching the final pressure 0.025MPa.

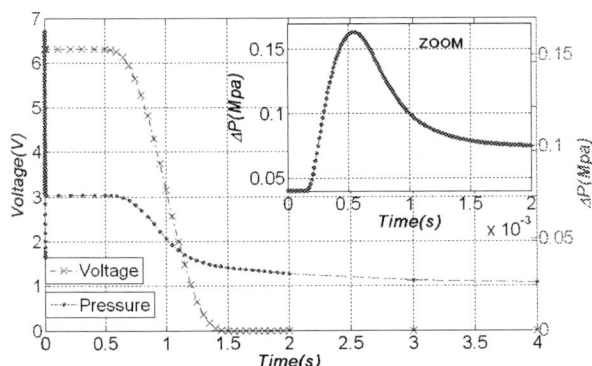

Figure 15. Pressure evolution on the observation point under the PDMS membrane

6. Global model assessment

The goal of the final system model is to predict the amplitude of the actuation as a function of the input signal, that is, the maximal deformation of the membrane as a function of the electrical applied power (w_{max}=f($I{\times}Ve$)).

From the axisymmetric model, a pressure is obtained as a function of the applied power, Figure 16 shows that the minimal electrical power required to obtain the actuation is 68.7mW. From this pressure data, we can obtain the deformation of the hyperelastic membrane with the 3D model (see section 5.2).

Figure 16 gives also the maximal deformation of the elastic membrane as function of the electrical power.

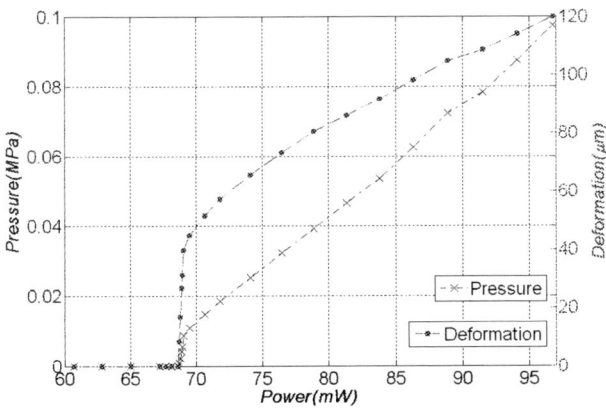

Figure 16. Global model assessment in stationary state. Cavity pressure and maximal deformation of the elastic membrane as a function of the applied electrical power

7. Conclusions

In this article we have presented a micro pressure source based on the decomposition of an energetic material to illustrate a multiphysics conception approach based on the COMSOL Multiphysics software.

In this actuator, an electrical power of 86.3mW is applied to the heating platform to obtain a membrane maximal deformation of 100μm.

First an electro-thermal model of a heating platform has been implemented and validated by IR measurements.

Simulations results on the micro balloon blowing show a reasonable agreement with experimental data. Still, some discrepancies are found in the non-linearity of the curves. Possible modifications on the model should be considered. At the same time, technical aspects, as the adherence between the substrate and the elastic membrane will be optimized in order to characterize the membrane properties and increase the system's performance.

A first approach of a model for the gas generation from the decomposition of the energetic material is presented, results show quite fast transient behavior in the gas concentration dynamics inducing a fast peak in the cavity pressure, a dynamic evaluation of the membrane response to this peak must be considered, this will lead us to a full multiphysics coupling with the membrane deformation as a fluid-structure interaction.

The advantage in actuating with an energetic material is exposed as the efficiency in the level of actuation increases reducing the electrical power needs. The reduced power consumption of this actuator makes it quite appropiated for microfluidic portable applications or applications where nowadays high integrated pressure sources are required [22].

Next a global multiphysics model including the interaction of the acturator with the actuated fluid in a microchanel is envisaged.

Acknowledgments

This research is fully supported by a French National Research Agency ANR grant. The authors gratefully acknowledge the services TEAM and 2I from the LAAS-CNRS for their competences and skills that made possible this work. The authors also would like to thank B. Chaudret, S. Sabo-Etienne and P. fau from the LCC CNRS and A. F. Mingotaud, M. Mauzac and J. Fitremann from the IMRCP for their colaboration in the project. The authors also wish to thank the COMSOL support team for their services.

References

1. C. Pradère, S. Suhard, L. Vendier, G. Jacob, B. Chaudret, S. Sabo-Etienne, to be published.
2. Cabal, A. *et al*, "Thermal actuator with optimized heater for liquid drop ejectors", *Sensors and Actuators A* Vol. 123-124 (2005) pp. 531-539.
3. Harouche, I. P. F. and Shafai, C. "Simulation of shaped comb drive as a stepped actuator for microtweezers application", *Sensors and Actuators A,* Vol. 123–124 (2005), pp. 540–546.
4. Park, Y. H. and Park, K. C. "High-fidelity modeling of MEMS resonators—Part II: Coupled beam-substrate dynamics and validation", *J. Micromech. Syst.* Vol. 13 (2).
5. Pandiyan, J. *et al*, "Modelling & simulation of novel three arm MEMS actuators & its application", *International MEMS Conference 2006, Journal of Physics: Conference Series,* Vol. 34 (2006), pp. 436–441.

6. Zhu, Y. *et al*, "A thermal actuator for nanoscale *in situ* microscopy testing: design and characterization", *J. Micromech. Microeng.* Vol. 16 (2006), pp. 242–253.

7. Atre, A. "Analysis of out-of-plane thermal microactuators" *J. Micromech. Microeng.* Vol. 16 (2006), pp. 205–213.

8. Truong, Thai-Quang and Nguyen, Nam-Trung, "A polymeric piezoelectric micropump based on lamination technology", *J. Micromech. Microeng.* Vol. 14 (2004), pp. 632–638.

9. Bourouina, T. *et al*, "Design and simulation of an electrostatic micropump for drug-delivery applications", *J. Micromech. Microeng.* Vol. 7 (1997), pp. 186–188.

10. Baek, J. *et al*, "A pneumatically controllable flexible and polymeric microfluidic valve fabricated via in situ development", *J. Micromech. Microeng.,* Vol. 15 (2005), pp. 1015-1020.

11. Chien-Chong, H. *et al*, "An on-chip air-bursting detonator for driving fluids on disposable lab-on-a-chip systems", J. Micromech. Microeng. Vol. 17 (2007) 410–417.

12. Song, W. H. and Lichtenberg, J., "Thermo-pneumatic, single-stroke micropump », *J. Micromech. Microeng.,* Vol. 15 (2005), pp. 1425-1432.

13. Rossi, C. and Estève, D., "Micropyrotechnics, a new technology for making energetic microsystems: review and prospective", Sensors and Actuators A, Vol. 120 (2005), pp. 297–310.

14. Rossi, C. *et al*, "Realization and performance of thin SiO2/SiNx membrane for micro-heater applications", *Sensors and Actuators A* Vol. 64 (1998), pp. 241-245.

15. Armani, D. K. *et al*, "Re-configurable fluid circuits by PDMS elastomer micromachining", *12th Int. Conf. on MEMS (MEMS 99) (Orlando, FL) pp.* 222–227.

16. Holman, J. P., <u>Heat Transfer,</u> Mc Graw Hill (1997)

17. Timoshenko, S. P. and Woinowsky-Krieger, S., <u>Théorie des plaques et coques,</u> Béranger (Paris, 1961).

18. Tadmor, E. B. *et al*, "Mixed finite element and atomistic formulation for complex crystals", *Phys. Rev. B*, Vol. 59, No. 1 (1999), pp. 235-245.

19. Hoel, A. and Jullien, M. C., "Modelisation of a hyperelastic polymer membrane deformation", *COMSOL Multiphysics Conference 2005, (Paris),* pp. 37-39.

20. Poilane, C. *et al*, "Analysis of the mechanical behaviour of shape memory polymer membranes by nanoindentation, bulging and point membrane deflection tests", *Thin Solids Film* Vol. 379 (2000), pp.156-165.

21. Zhang, W. *et al*, "Novel room-temperature first-level packaging process for microscale devices", *Sensors and Actuators A* Vol. 123-124 (2005), pp. 646-654.

22. Thorsen, T. *et al*, "Microfluidic large-scale integration" , *Science,* Vol 298 (2002), pp 580.

Crack Degradation Model Derived From Experimental Strain-Stress Data

Milos Dusek and Christopher Hunt
National Physical Laboratory
Queens road, Teddington, UK
milos.dusek@npl.co.uk, chris.hunt@npl.co.uk, tel: +44208977 3222

Abstract

A new concept of lifetime prediction of solder joints is presented, based on a fatigue cycling experiment under isothermal conditions. A common approach of lifetime prediction of solder joints is based on damage model based on the Coffin-Manson equation, which uses the plastic strain range from a single hysteresis loop to calculate a number of cycles to failure. This was later modified by Engelmaier.

Then Morrow developed a damage model based on strain energy density, which takes the area of a single hysteresis loop. These stress-strain hysteresis loops are calculated in FEA modelling, but with the assumption that there is no crack. Hence only engineering values of strain and stress are estimated, and the load-bearing area of a solder joint is assumed to be constant and not cracked. This is clearly an over simplification as solder will crack under fatigue conditions during thermal cycling. A further simplification is that the stress is calculated from an inverse solution of secondary creep rate equation.

The correct approach should reflect the true stress and strain (measurement of force and actual load bearing area) developed inside a solder joint as a function of time. A clear correlation of solder joint area with electrical resistance can be shown to exist, hence the load bearing area can be obtained by monitoring of solder joint electrical resistance, and therefore the true stress can be calculated. Measurements following this approach show that the strain-stress hysteresis loop area does not collapse (using a constant displacement controlled profile of the solder joint), but that the hysteresis loop expands on the stress axis. Since, an expansion of a hysteresis loop on the stress axis would cause a divergence of stress energy density, the true shear strain is shown to be decreasing under the same displacement in each cycle. A crack degradation model will be presented using the measured true stress strain hysteresis loop area through out the fatigue life of a solder joint.

1. Introduction

Although the principal aim of a solder joint is to achieve both an electrical connection and a mechanical attachment, it is the latter that generally gives cause for concern. In terms of the bond strength, or the force needed to break a solder joint by push off or mechanical shock, the solder joint is considerably over-engineered. However, fatigue resistance poses a very different situation, and the ability of solder to withstand considerable damage, while retaining the component connection, is critical. Stronger materials that could resist the fatigue damage, would result in an unacceptable

change in failure mode, with failures occurring in the component body or in the substrate rather than in the solder.

A comprehensive understanding of the mechanical behaviour of SnPb solder has been built up over more than 50 years. However, with the implementation of the European legislation, lead-containing solders have now been banned in certain applications from 1 July 2006, and there is a dearth of information on the mechanical behaviour of their lead-free replacements. Hence there is a critical requirement to acquire materials data for these new lead-free solders, both to gain confidence in their performance and to supply the performance information for reliability modelling. At the same time there is a realisation that acquiring data in tension from typical materials laboratory instruments is not appropriate and that any new materials data should be obtained from solder samples having volumes similar to those of solder joints. Moreover the joints should be loaded in shear (not tension), mirroring the practical situation in the field. Consequently, there is a requirement to generate new data on the new materials, and to acquire them in a shear mode. This paper describes a new approach developed at NPL to overcome these constraints, resulting in a new instrument [Ref 1 pp.468] with a sample geometry that permits small solder volumes to be studied in shear.

Creep, relaxation and fatigue are the key parameters that need to be measured. These mechanical measurements must be made at various temperatures. Since small samples are to be studied the displacement measurement accuracy required needs to be better than $0.1 \ \mu m$. These requirements are quite demanding and as a consequence a number of novel solutions have been generated [Ref 2, pp. 67].

A solder sample undergoing repetitive cyclic strain fatigue will fail gradually. The resistive force to the applied strain will decrease as a crack grows and microstructural changes occur. In each repeating strain cycle the maximum and minimum force will reduce as the crack grows. The rate of degradation (the difference between the maximum and minimum forces per cycle) is a measure of the solder performance. The point used to define the number of cycles to failure does not necessarily have to coincide with a 100% crack length. Hence a load drop parameter is defined as the reduction in the ratio of the difference between the maximum and minimum load at n cycles, compared to the difference between the maximum and minimum load at zero cycles. Whilst a proposed Japanese standard for tensile solder specimens recommends a 20% drop of load to define life-time [Ref 3], a similar US standard [Ref 4, pp 57] recommends a

50% load decrease. Alternatively, another definition of life-time can be derived from the crack length in the test piece, or an electrical resistance increase of tested specimen, as a direct cause of fatigue cracking.

2. Model Solder Test Specimen

Solder used in electronics is a relatively ductile material and is used predominantly at very high homologous temperatures 0.6Tm - 0.8Tm (Tm is melting temperature in K). Creep rates in any materials at these temperatures are high, and hence creep tensile testing has traditionally been performed in the tensile mode. But as mentioned above there is a requirement to obtain the new materials data from small solder joints in shear. Since testing in shear using a traditional test instrument is challenging, it is better to use an instrument that operates along a single axis, and have this axial loading converted into a shear action within the specimen. In Figure 1 a specimen format is proposed that allows the central solder section to be loaded in plain shear, while applying an axial load overall to the sample [Ref 5].

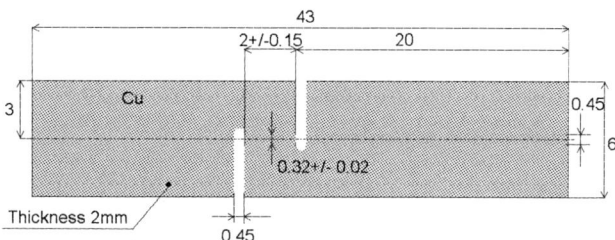

Figure 1. Drawing of the test piece (all dimensions are in mm)

This design has the advantages that it can accommodate any solder, and that the copper bar can be plated using conventional PCB fabrication finishes (e.g. ENIG, immersion tin, silver) prior to soldering. A typical joint specimen is illustrated in Figure 2.

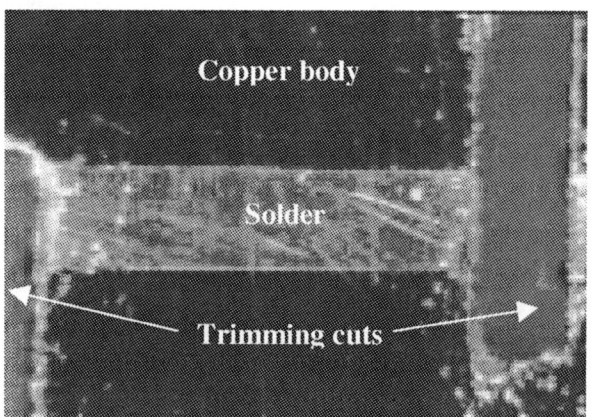

Figure 2. Solder joint formed inside the copper body of a test piece

3. Test Measurement Method

In order to determine the displacement profiles to be applied to the copper samples, strains experienced in surface mount components have to be estimated. In Table 1 and Table 2 typical parameters are given for ceramic chip resistor and relevant materials. To calculate the strain range experienced by a chip resistor the following assumptions were made: the alumina in a 2512-chip resistor has a CTE = 4.5 ppm/°C, and is mounted on a FR4 substrate with a CTExy = 18 ppm/°C; the assembly is subjected to thermal cycling between –55 and +125 °C, i.e. a temperature span (ΔT) of 180 °C. In Table 2, column 2, the resistor body length (L) is listed. In column 3, ΔL is the difference in expansion between the resistor body and the substrate when measured from the centre point over a temperature change of 180 °C. For a chip resistor the average stand-off height is 80 μm, and the calculated approximate average shear strain range (γ) is recorded in the fourth column. The displacement applied to the model joint to simulate the same shear strain as experienced by a single resistor joint, is given in column 5. The value is 4 times that of column 3 as the joint height is four times greater, i.e. 320 μm as opposed to 80 μm (stand-off height).

Table 1. Experimental variables used to calculate

ΔT	180	°C
SoH	80	μm
CTE pcb	18	ppm/°C
CTE Al$_2$O$_3$	4.5	ppm/°C
Gap	320	μm

Table 2. Experimental total strain range

Resistor type	L [mm]	ΔL [μm]	γ	Δl (μm)
2512	6.2	7.5	0.094	30

A displacement profile used to achieve an equivalent deformation to that experienced by a type 2512-resistor in a –55 to +125 °C thermal cycle, is shown in Figure 3. There is a difference between thermal cycling the PCB assembly and the copper-solder test specimen tested by the instrument. However the isothermal strain cycling in the instrument is a first approximation against which various solders can be assessed. In a full thermo-mechanical fatigue test the displacement profile would be synchronised with a temperature profile, simulating a thermal cycling experiment. The isothermal experiment may actually be a more severe test on a material. This is because the strain range is constant here, irrespective of temperature, but the creep rates will differ.

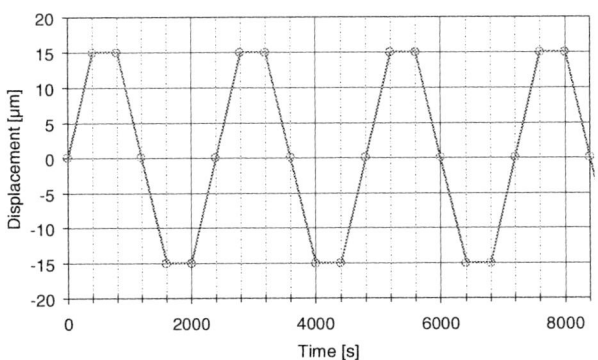

Figure 3. Displacement profile (equivalent to 2512-type resistor)

Failure can be defined by a load drop or by an increase in joint resistance during the strain cycling. Changes from 80%-50% of the initial value are usually taken as performance criteria for a number of life-cycles. In addition, the resistance across the solder joint can be monitored using a 4-point measurement method; as the crack area increases the resistance through the solder joint increases. It is assumed that there is no change in resistance due to microstructural changes, and hence the resistance increase (in $\mu\Omega$) can be correlated with crack growth. This method can be used to augment monitoring of the degradation of the solder joint and of the variation in the load drop. Since the method is very sensitive and 0.1 $\mu\Omega$ increases in resistance can be detected, it is important to monitor specimen temperature and correct for thermal fluctuations.

A disadvantage of the specimen design is the elastic compliance in the copper body of the specimen. The specimen displacement measured is a combination of the deformation in the solder and the extension of the copper. This elastic extension in the copper can be added to the measured deformation for solder (displacement) as expressed in Equation 1.

$$L_{total} = L_{compl} + L_{solder} \qquad \text{Equation 1}$$

Where:

L_{total} - total measured displacement (deformation)

L_{compl} - the displacement correcting for elastic deformation in the copper body

L_{solder} - the displacement of the solder following the applied displacement profile

The maximum and minimum nominal stresses from each cycle are plotted in Figure 4. One failure criterion in measuring the life time of solder joints can be a relative load (or nominal stress) decrease by 50%. Other criteria can be defined as a resistance increase (Figure 4, green line), which can be correlated with crack length. In Figure 4 the maximum and minimum nominal stresses (assuming the cross-sectional area is constant) are plotted for each strain cycle (SnAgCu – SAC - solder). There is an initial strain hardening as the solder joint reaches the maximum shear stress range of 54 MPa (-27 MPa to + 27 MPa), which is reduced to 27 MPa (-13.5 MPa to +13.5 MPa; ~50%) after 80 cycles. The lifetime of the solder joint is then defined as the point of 50% load (nominal stress) reduction.

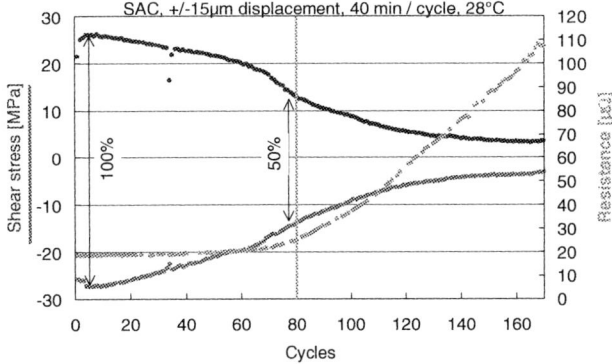

Figure 4. Maximum (blue) and minimum (red) nominal stresses, and resistance (green) across a solder joint during strain cycling

In Figure 4 the resistance across the solder joint, monitored in every cycle (green symbol), is also shown. The initial resistance of about 18.5 $\mu\Omega$ increases steadily and slowly up to 70 cycles. Assuming the resistivity of the solder joint and the solder joint gap remain constant, the only variable is the area of the solder joint perpendicular to the measurement current (~10A), which is related directly to the presence of a crack.. Hence the resistance measurement is directly related to the development of a crack.

This correlation is demonstrated by re-plotting the data in Figure 5, in which values have been normalised at zero cycles. It can be seen that the more rapid increase in resistance, and the 50% increase in resistance value correspond with the point of the 50% load drop. Therefore the measurement of resistance can be used as a complementary technique to the measurements of load decrease. It is assumed the resistance value changes are purely a geometric effect, whereas the load drop may be influenced by other materials changes, such as strain hardening, and property changes due to the presence of any intermetallics. Hence the resistance data can be used to predict crack growth rates.

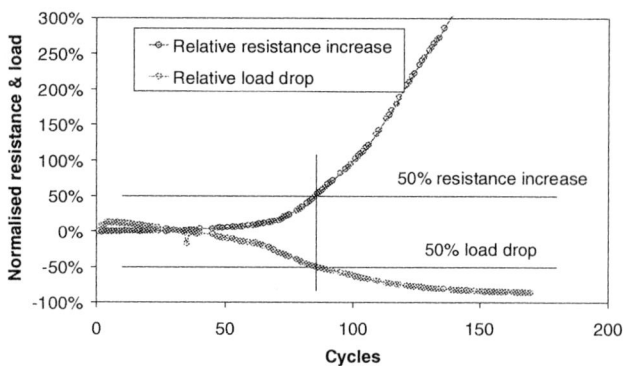

Figure 5. Normalised plot of load drop (red) and resistance increase (blue)

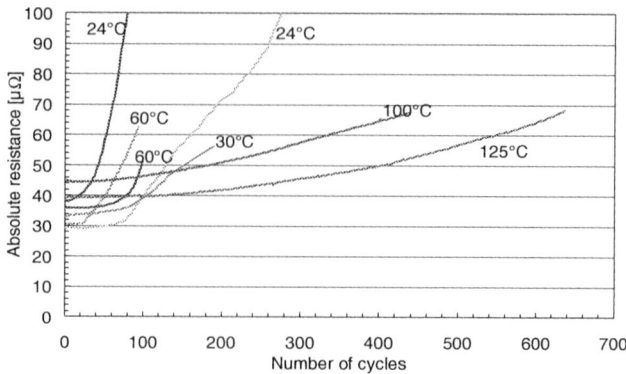

Figure 6. Absolute resistance as a function of number of cycles

In Figure 6 absolute resistance is plotted for various solder joints at various temperatures, the resistance being measured at the end of each cycle. The initial resistance should be, for identical geometries, only a function of material and temperature. It can be seen that there is a variation in the solder joint size (length) since highest initial resistance (at 0 cycles) was recorded at 100°C and not 125°C. By normalising the relative resistance, the geometry and temperature variations can be eliminated as demonstrated in Equation 2. The relative resistance increases are plotted in Figure 7.

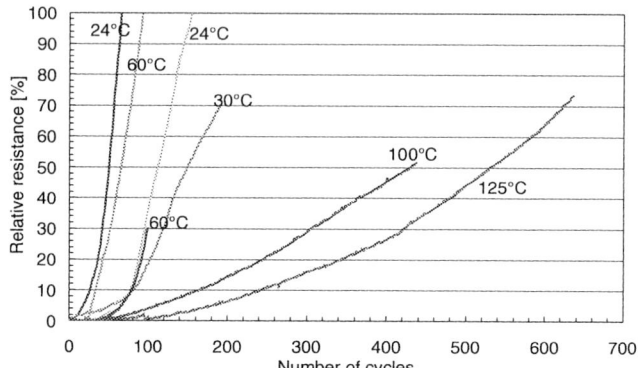

Figure 7. Relative resistance as a function of number of cycles

$$R_r = \frac{R_i - R_0}{R_0} \cdot 100\% \qquad \text{Equation 2}$$

Where:
R_r – Relative resistance increase
R_i – resistance measured at the end of each cycle
R_0 – nominal resistance (before 1st cycle)

Crack growth can be estimated using the relative resistance of a solder joint. An electrical resistance model of solder joint is shown in Figure 8, and comprises a number of equally thick layers. In this analysis one of the layers contains the crack, and is being shorted, with a resultant resistance increase (number of cycles). The resistance measured across the solder joint does include a resistance in the copper body (Rc) and the solder joint (Rs) as expressed in Equation 3. A single crack model is used due to equivalence in electronic resistance circuitry i.e. two opposite cracks behave like one of double thickness.

The resistance across the copper part is very low (1.2 µΩ), due to larger geometry, and is in fact much lower than the resistance through the solder layers. For simplicity the crack length is assumed to progress along one interface only - see in Equation 4.

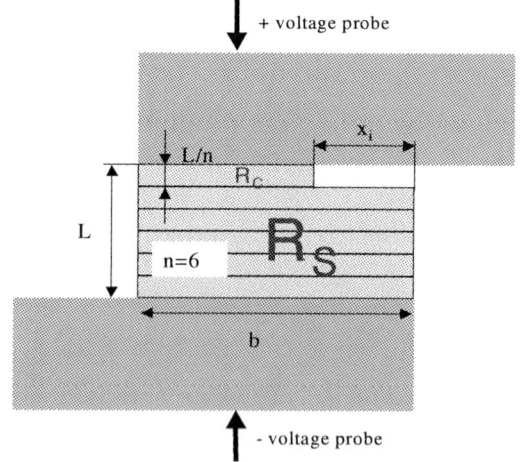

Figure 8. Model of solder joint for calculation of a crack length

$$R_0 = R_S + R_C \qquad \text{Equation 3}$$

where: Rc << Rs

$$x_i = \frac{b \cdot n}{n + \dfrac{R_0}{\Delta R_i}} \qquad \text{Equation 4}$$

where:

x_i - crack length

b   - solder joint length (in direction of crack propagation)

n   - number of model layers (n is set to 6 in this experiment)

R_0 - nominal resistance of solder joint

ΔR_i - absolute increase of resistance after i-th cycle ($\Delta R_i = R_i - R_0$)

The only unknown variable in Equation 4 is n the number of layers in the model of the solder joint. The thickness of a crack is hence the gap (L) divided by number of layers (n). To estimate n, simulations was undertaken on one measurement in which n was varied from 5 to 100. Results of this simulation are shown in Figure 9. It can be seen that a small noise at the beginning of experiment contributed largely to a scattering of results of crack length for large values of n (small crack height). After the analysis of noise the value of n was arbitrary set to 6. According to the model larger values of n would significantly open and close the crack, which does not accord with observations using a microscope.

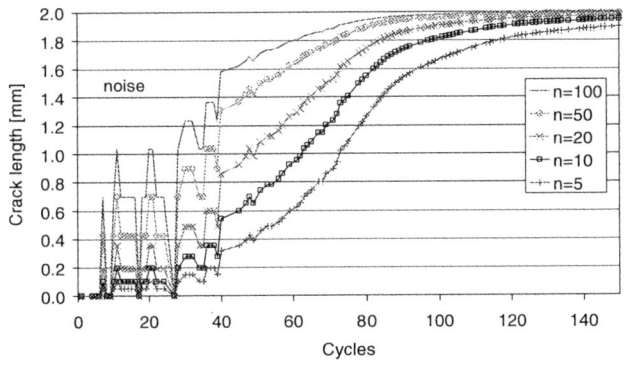

Figure 9. Simulation of crack length as a function of the number of layers n

In Figure 10 the relative resistance measurements have been converted to crack length according to the model with n=6 (Equation 4). In general, crack growth is slower at higher temperatures with slow mechanical cycling where creep and stress relaxation dominate the deformation mode.

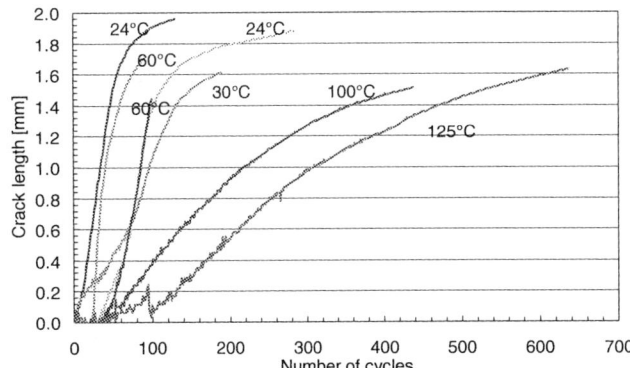

Figure 10. Crack length as a function of the number of cycles for n=6 and various solder joint temperatures

Figure 11 presents a comparison of the engineering and true shear stresses present for measurement taken at 24°C. Usually, the true stress cannot be calculated, since the area over which the load is acting is unknown, but in this case the crack area can be predicted from the relative change in resistance. The calculation indicates that, unlike the engineering stress, the true stress increases as the crack grows. These results show that the final true stress achieves values close to the yield point [Ref 7]. These higher stresses are more consistent with failure of the joint than are the engineering values. Comparisons with literature values are limited by the low strain rates and relaxation that were present in these experiments.

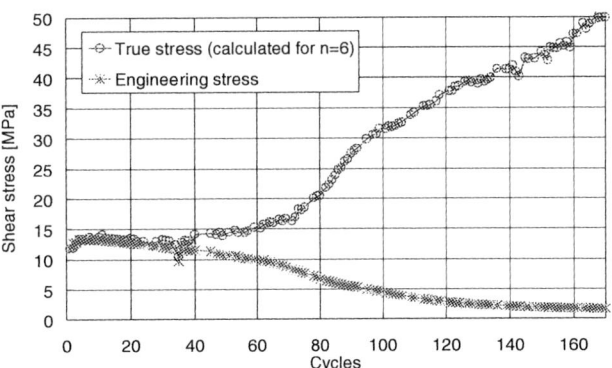

Figure 11. Comparison of true (calculated) and nominal (measured) stresses at 24°C

It is useful to be able to relate crack length and joint resistance – see Figure 12. For example, a 50% increase in resistance corresponds to a crack growth of 75% relative to the solder joint size, for n=6 (Equation 4).

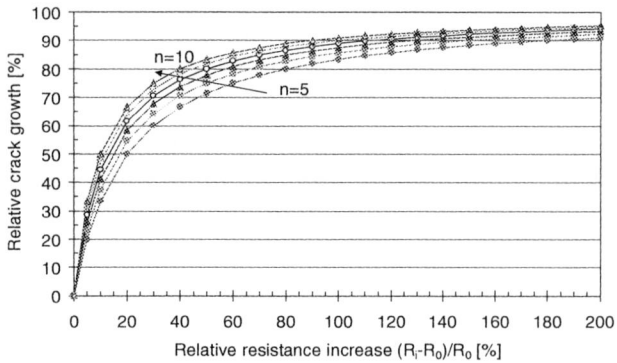

Figure 12: A model relation between relative resistance increase and crack growth for number of layers n=5 to 10

4. Conclusions

A new approach for measuring solder fatigue is described. This is based on the construction of an instrument to apply defined loading under precision (displacement and force) control to measure the materials properties. The experimental design includes a new sample geometry that permits small solder volumes to be studied in pure shear. These two elements permit the necessary solder properties to be collected for new solder alloys under geometric conditions and temperature that reflect common usage. Preliminary data thus acquired are presented, which can be utilised by FEA modelling of solder performance.

A 4-point measurement system for resistance monitoring has been evaluated and the results found to correlate well with load decrease during isothermal fatigue testing of solders.

A route to predict a fundamental reliability variable of solder has been identified.

Acknowledgments

The work was carried out as part of a project in the Materials Processing Metrology Programme of the UK Department of Trade and Industry.

References

1. Wassink, K. R. J.: Soldering in electronics, 2nd edition, Electrochemical Publications, 1989, pp. 468
2. Plumbridge, W., Matela, R. J., Westwater, A. : Structural integrity and reliability in electronics, Kluwer Academic Publishers, 2003, ISBN: 1-4020-1765-0
3. IEC 62137-1-5 standard: Environmental and endurance test methods for surface mount solder joints - Part 5 Mechanical shear fatigue test - draft
4. Clech, J. P.: Acceleration factors and thermal cycling test efficiency for lead-free Sn-Ag-Cu assemblies, SMTA International Chicago, 2005, pp. 902
5. Dusek, M., Hunt, C.: Test approach to isothermal fatigue measurements for lead-free solders, NPL report DEPC-MPR 048, March 2006
6. Dusek, M, Hunt, C.: The Measurement of Creep Rates and Stress Relaxation for Micro Sized Lead-free

Solder Joints, NPL report DEPC-MPR 021, April 2005
7. Pang, J.H.L.: Bulk solder and solder joint properties for lead-free 95.5Sn-3.8Ag-0.7Cu Solder Alloy, Conference proceedings ECTC 2003 International, Chicago
8. Astrom, K. J., Hagglund, T.: PID Controllers: Theory, Design, and Tuning, International Society for Measurement and Control, ISBN-10: 1556175167, 2nd edition (January 1, 1995)

CFD Aided Reflow Oven Profiling for PCB Preheating in a Soldering Process

Ilja Belov[1], Mats Lindgren[2], Peter Leisner[1,3], Fredrik Bergner[4], Robin Bornoff[5]

[1]School of Engineering, Jönköping University
Box 1026, SE-551 11 Jönköping, Sweden
E-mail: ilja.belov@jth.hj.se, phone: +46 36 10 16 86

[2]Kitron AB
Box 1053, SE-551 10 Jönköping, Sweden
E-mail: mats.lindgren@kitron.com, phone: +46 36 559 40 00

[3]SP Technical Research Institute of Sweden
Box 857 SE-501 15 Borås, Sweden
E-mail: peter.leisner@sp.se, phone: +46 33 16 54 45

[4]Flomerics Nordic AB
Romansvägen 6, SE-131 40 Nacka, Sweden
E-mail: fredrik.bergner@flomerics.se, phone: +46 8 601 04 62

[5]Flomerics Group PLC
81 Bridge Road Hampton Court Surrey KT8 9HH UK
E-mail: robin.bornoff@flomerics.co.uk, phone: +44 20 8487 3000

Abstract

A CFD-aided reflow oven profile prediction algorithm has been developed and applied to modelling of preheating of a PCB with non-uniform distribution of component thermal mass in a forced air convection solder reflow oven. The iterative algorithm combines an analytic approach with CFD modelling. It requires an experimentally validated CFD model of the solder reflow oven and a CFD model of the PCB as main inputs. Results of computational experiments have been presented to reveal good agreement between predicted PCB profiles and corresponding CFD calculations. Application guidelines contained in the description of the algorithm will assist potential users both during the virtual prototyping phase of a PCB including designing for assembly and in the phase of reflow oven profiling.

1. Introduction

It has already been recognized that thermal analysis of a PCB in the early design phase should include reflow soldering process considerations that can affect the PCB layout and choice of IC packages [1, 2]. A well designed reflow soldering process for a particular PCB layout ensures robust production process, and forms the basis for long service life of the product.

Forced air convection solder reflow ovens designed for reliable reflow soldering are conveyorized in order to gradually pass a PCB through the oven zones, thus enabling preheating (often including ramp and soak), reflow soldering and cooling. Component density in a particular PCB region is an important factor that affects the rate of heat transfer from the oven to the PCB. Being successfully solved for eutectic Sn-Pb solders in the convection solder reflow ovens with high heat transfer rate, this problem arose again as a consequence of the RoHS directive. Since the most common lead-free solders have higher melting points, the process window becomes narrow, and the sensitivity of the process parameters to non-uniform component thermal mass distribution on a

PCB becomes very high. The temperature difference between the hottest component (maximum case temperature) and the coldest solder joint (minimum reflow temperature) should not exceed 10-20°C (depending on the package size). Such a temperature variation on the PCB forms a so called ΔT problem. It is partly solved by introducing a soak zone in the Ramp-Soak-Spike (RSS) process, serving to bring the temperature of all the PCB regions to approximately the same level, with subsequent simultanious reflow for all the assembly parts [3]. On PCBs with uniform component thermal mass distribution the ramp-to-spike (RTS) process can be used for lead-free soldering to exclude the soak zone [3]. In order to solve the ΔT problem for the boards with non-uniform distribution of component thermal mass and at a narrow process window the soak zone is still needed as well as very carefully chosen oven temperature settings.

A number of additional requirments have to be met by the reflow soldering process. E.g. during preheating they include limitations on minimum and maximum preheat temperatures, preheat time, and the maximum slope or ramp rate for the PCB temperature (see table 1), [4].

Table 1. Lead-free reflow profile recommendation [4]

Parameter	Value
Minimum preheat temperature, $T_{p,min}$	150 °C
Maximum preheat temperature, $T_{p,max}$	200 °C
Preheat time	60-180 s
Maximum ramp rate	3 °C/s

In the phase of physical prototyping of a PCB, the oven profile is experimentally obtained by advanced thermal profiling equipment. A number of thermocouples are placed on a test PCB and along the oven conveyor. Special software supplied with the equipment predicts the oven profile based on measured temperatures.

1-4244-1105-X/07/$25.00 ©2007 IEEE

Disadvantages of this approach include the need for a physical prototype of the PCB. Several process runs are required to attain the desired oven profile. Practically, it can take up to one working day for one or two engineering staff for reaching the desired oven profile for a particular populated PCB. In addition, there is no guarantee that the actually hottest and coldest spots are captured on a PCB by temperature measurements. This conventional profile prediction procedure is fairly reliable for PCBs balanced in terms of component thermal mass, and eutectic Sn-Pb solders. This procedure, however, might require further modification to improve oven profile prediction for PCBs with non-uniform component thermal mass distribution and lead-free solders. Improvements to the oven profiling procedure in this case can include e.g. using analytic methods and/or numerical modelling tools along with temperature measurements. A number of works were published on the methods of modelling of the reflow soldering process [1, 2, 5].

The proposed CFD-driven oven profile prediction algorithm or the oven profile generator is integrated into the PCB and reflow soldering process design, as shown in figure 1. This algorithm is applied to modelling of populated PCB preheating in an air convection solder reflow oven, and does not require a physical prototype of the PCB. It relies only on a validated CFD model of the solder reflow oven and a model of the PCB.

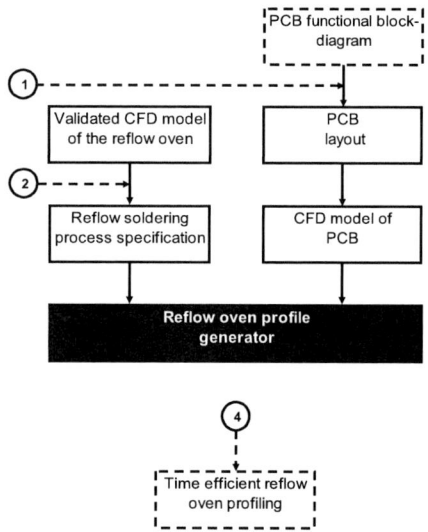

Figure 1: Reflow oven profile generator as a part of the PCB and reflow soldering process design

Being used together with the presented algorithm, these models should result in starting points for real oven profiling thereby minimizing the amount of profiling runs required. The iterative nature of the algorithm should enable both the process and the PCB layout modifications in the phase of virtual prototyping of PCBs.

2 Description of the algorithm

A validated CFD model of the solder reflow oven and the CFD model of PCB are the key inputs to the algorithm. There exist a number of commercially available simulation tools, enabling user friendly, fast and reliable PCB and solder reflow oven models construction, based on standard smart parts, component libraries and material data bases [6, 7]. For now, let us assume that a validated CFD model of the solder reflow oven is available as an input to the algorithm, as well as the CFD model of the PCB. Reflow soldering process specification should also be provided as an input. In addition to the data provided in table 1 it includes initial temperature of the assembly, a preferable conveyor speed, the maximum and minimum process temperatures, as well as target ΔT_{PCB} at the end of preheating. A number of tolerance values on the process parameters will be introduced later. The temperature settings for each oven zone form the main output of the algorithm. The block-diagram of the reflow oven profile generator is presented in figure 2, and each step of the algorithm is explained below.

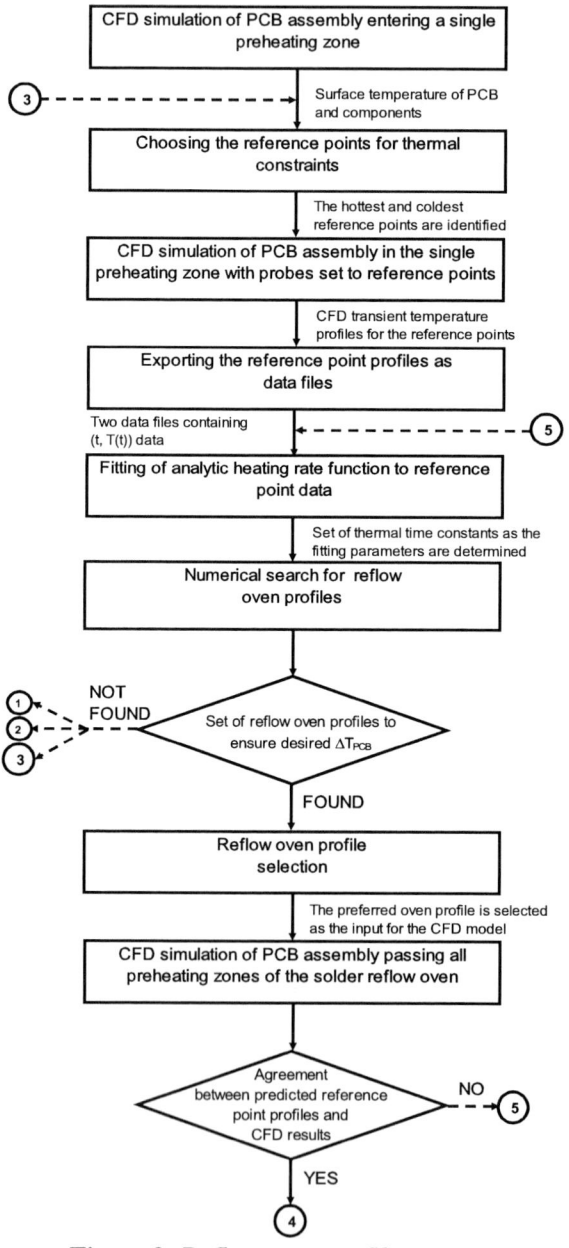

Figure 2: Reflow oven profile generator

1. CFD simulation of PCB assembly entering a single preheating zone. The purpose of this step is to get complete surface temperature contour plots for the PCB at different times, during the heating of the PCB assembly from the initial temperature to the steady state or equilibrium temperature determined by the temperature supported in the preheating zone. It is important here that the reflow oven model reflects the actual conveyor dynamics since the transient temperature profile of the PCB is affected by gradual entering the PCB to the preheating zone. Different regions on the PCB have different heating rate, depending on the local thermal mass. Few snapshots of PCB surface temperature distribution are sufficient for identification of the hottest and the coldest PCB regions during preheating.

2. Choosing reference points for thermal constraints. Based on the results from step 1, the hottest and coldest reference points are identified, to be used in the numerical search for reflow oven profiles. In order to expedite the reference point identification, it is reasonable to point out the solder paste locations on the PCB, which, as a rule, are beside (e.g. QFP) or under (e.g. BGA) components. Computational experiments showed that choosing the absolute coldest and hottest spots on the overall PCB surface as reference points can cause problems in bringing both of them to equal temperature during preheating. Rather, the coldest point located on the solder paste can be chosen as one reference point, and the hottest point on the solder paste can be chosen as the other reference point, see figure 3. The main purpose of PCB preheating is to bring the selected reference points, and thereby all the points of the solder paste, to the maximum preheat temperature specified in table 1.

It was confirmed by CFD simulations that during preheating the hottest point of the solder paste is warmer than the hottest component case. With the view of the following reflow soldering such a choise of the hottest reference point makes the thermal constraint for component maximum case temperature, $T_{comp,max} < T_{spec,case}$, automatically satisfied. On the other hand, the choise of the coldest point of the solder paste as the coldest reference point leads to $T_{solder,min} > T_{reflow}$, i.e. the temperature at all solder paste locations is above the solder reflow temperature.

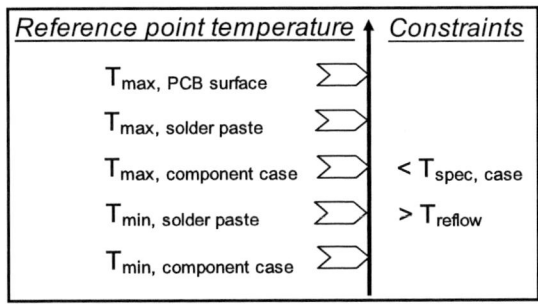

Figure 3: Reference points and thermal constraints

It is worth noting that too large difference in temperature between the hottest and coldest reference

points along with small ΔT_{PCB} might result in failure to find the adequate oven profile for PCB preheating. In this case the thermal constraints must be relaxed by choice of new reference points (see label 3 in the block-diagrams of figure 2 and figure 3). E.g. the component maximum case temperature can be chosen as the hottest reference rather than the warmer point in the solder paste region, figure 3. Another solution to obtain the the adequate oven profile for populated PCBs is of course decreasing the conveyor speed (see label 2 in the figures) or at extreme case, changing the PCB layout (see label 1).

3. CFD simulation of PCB assembly in the single preheating zone with probes set to reference points. Once the reference points are chosen, the single preheating zone simulation can be repeated to monitor the transient temperature exactly in these points. Thus two time-domain temperature profiles are obtained by this step. Due to the reference point selection the temperatures of all the other points of the same type on the PCB (in this case, all located on the solder paste regions) should fall between the temperatures of the reference points at each moment of time during preheating.

4. Exporting the reference point profiles as data files. By this step, two data files containing *(Time, Temperature (Time))* data are created to correspond to the temperature profiles for respective reference points. These files will then be used as the input for a curve fitting routine in the next step.

5. Fitting of analytic heating rate function to reference point data. The temperature rise of a material from the initial temperature $T_{initial}$ that occurs during the heat cycle in a transient condition is determined by Equation (1) [8]:

$$T_{end} - T_{start} = (T_{oven} - T_{start})(1 - e^{-t/\tau}) \qquad (1)$$

In this one-dimensional model T_{start} is the initial temperature of the material at the moment of time $t = 0$, T_{end} is the temperature of the material at the moment of time t, T_{oven} is the temperature of the air in the reflow oven, used in the single preheating zone simulation, $\tau = RC$ is the time constant with R as the thermal resistance from the air to the measurement point on the PCB, C as the thermal capacitance defined by $C = c_p \rho V$. Here c_p is the specific heat of the material, ρ is the material density, and V is the material volume.

A common assumption when using model (1) with analytically determined C and R, for PCB profile prediction in a solder reflow oven includes no heat spreading in the plane of the PCB [2] at a large uniform heat transfer by convection in the direction out-of-plane of the PCB. However, in the conveyorized reflow ovens where the PCB moves slowly from one oven zone to another this assumption is corrupted. At the transition between the oven zones, some parts of the PCB surface are located in one temperature environment whereas the other parts are found in another temperature environment at the same moment of time. This creates temperature gradients on the PCB and makes it quite difficult to predict analytically the values of the thermal resistance

183

and thermal capacitance for using them in Equation (1). The described drawbacks become especially critical for the narrow process windows (as in lead-free reflow soldering) where errors caused by the model simplification can make adequate prediction of a feasible reflow oven profile impossible.

The main idea of the proposed algorithm is to use the so called local time constant $\tau = RC$ as the fitting parameter without attempting to analytically evaluate R and C, separately. Such an approach implicitly takes into account three-dimensional heat transfer at different modes (conduction, convection, radiation) in the PCB regions around the reference points. Moreover, the proposed approach takes into account the heating rate of the PCB assembly, when it is placed into the changing reflow oven environment. At large heat transfer coefficients or large Biot numbers the temperature gradient in the in-plane direction of the PCB assembly is influenced by the heating rate [2]. Therefore, the time constant would in this case be dependent on the heating rate too. In order to obtain a good fit for the transient temperature data in the reference points it might be beneficial to split the heating rate function curve into several profiles according to heating rate, r (e.g. $3\,^{\circ}\text{C/s} \ge r > 1.5\,^{\circ}\text{C/s}$, $r \le 1.5\,^{\circ}\text{C/s}$) extracted from CFD calculated data. Curve fitting is thus made for each data set with its corresponding heating rate function curve determined by its respective time constant.

In general, by step 5, two sets of time constants are identified, $\tau_{cold} = \{\tau_{1,cold}, \tau_{2,cold}, ... \tau_{n,cold}\}$ and $\tau_{hot} = \{\tau_{1,hot}, \tau_{2,hot}, ... \tau_{k,hot}\}$. Each of them approximates the time dependent effective thermal mass of the material at the reference points of the PCB. Two profiles are constructed according to Equation (1), both for the coldest and for the hottest reference points.

6. Numerical search for reflow oven profiles. The implemented software program searching for reflow oven profiles requires the data obtained in steps 1-5 of the algorithm, as well as additional information regarding the process specification, which should well correspond to the CFD model of the solder reflow oven. Though elaboration of the numerical routine searching for reflow oven profiles is out of scope of the present paper, it is expedient to explain in more detail its main inputs and output. An example of the input data set is provided in table 2. The output includes a set of piecewise reflow oven profiles $T_{oven}(t)$ satisfying the process specifications to bring piecewise functions $T_{hot}(t)$ and $T_{cold}(t)$ for the instant temperatures at the reference points to the maximum preheat temperature (200°C). The maximum preheat temperature is specified through the preheat target temperature, 197°C, and the preheat temperature tolerance, $\pm 2.5\,^{\circ}\text{C}$. It means that in the end of preheating or in the end of the fifth zone, the temperature difference ΔT_{PCB} between the two reference points should not exceed $5\,^{\circ}\text{C}$, which is quite common to have in a real reflow soldering process, see figure 4. Piecewise functions $T_{hot}(t)$ and $T_{cold}(t)$ are defined by Equation (1) for each reflow

oven zone, with the time constants taken from sets τ_{hot} and τ_{cold} depending on the heating/cooling rate of the reference points and the time constant transition slope, which in this study was taken to be 1.5 °C/s.

An initial temperature tolerance is applied in a similar way as the preheat temperature tolerance, to set the maximum allowed deviation of the reference point temperatures from the initial temperature of the assembly, in the beginning of the first preheating zone. In order to reduce the number of analysed reflow oven profiles in the software routine, and thereby to save computational time, the maximum temperature difference between the coldest and hottest reference point profiles is introduced. This parameter specifies the allowed temperature difference between the reference points in the beginning of each preheating zone, to let the profiles be, in a sence, linked to each other during PCB preheating.

Table 2. Input data for the oven profile search routine

Parameter	Value
Number of preheating zones	5
Length of preheating zones	330 mm
Conveyor speed	9.18 mm/s
Preheat target temperature, T_p	197 °C
Preheat temperature tolerance,	$\pm 2.5\,^{\circ}\text{C}$
Time delay for coldest point	14.3 s
Time delay for hottest point	9.8 s
Maximum temperature difference between coldest and hottest profiles	40 °C
Initial temperature tolerance	$\pm 1\,^{\circ}\text{C}$
Initial temperature of the assembly	25 °C
Maximum ramp rate	3 °C/s
Time constant transition slope	1.5 °C/s

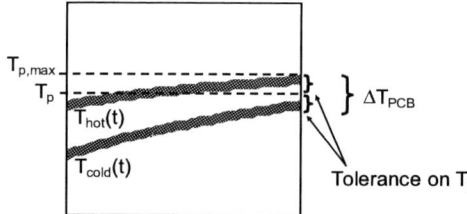

Figure 4: Tolerance on preheat target temperature

Finally, time delay parameters for the coldest and the hottest reference points reflect the fact that heating of the reference points does not occur immidiately when the PCB enters the solder reflow oven. Moreover, since the reference points are located in different parts of the PCB and are characterized by different time constants their respective time delays differ. The time delay parameters can be easily obtained from the simulation performed in step 3 of the algorithm. Time delays are actually not considered in the oven profile search routine, where it is assumed that heating of reference points occur synchronously. They will be taken into account when plotting the reference point temperature profiles to verify them with respective CFD calculated profiles.

When a PCB under test is highly unbalanced in terms of component thermal mass step 6 can result in no feasible reflow oven profiles. In this case, the iterative study of the reflow soldering process and the PCB assembly has to be performed. First of all, changing the type of reference points has to be tested, according to guidelines provided in step 2. As the second remedy, the process specification can be adjusted, including e.g. the conveyor speed or the solder paste, which would affect the ramp rate. Finally, the PCB layout can be modified together with choice of components in order to balance the PCB in terms of more uniform distribution of the component thermal mass, see figure 1 and figure 2.

7. Reflow oven profile selection. Having obtained a set of reflow oven profiles, the preferred oven profile is selected for use as the input for the CFD simulation in the next step. Selection criteria can differ and rely upon the expert knowledge e.g. about particular solder paste, etc. For PCBs highly unbalanced in terms of component thermal mass only a limited number of reflow oven profiles will be supplied by step 6, so that selection procedure for the preferred oven profile in this case is not a difficult problem, which is therefore omitted in the present study.

8. CFD simulation of PCB assembly passing all preheating zones of the solder reflow oven. This step is necessary for checking agreement between the predicted reference point temperature profiles and the respective CFD results. Having a well validated model of the solder reflow oven and a high-quality CFD model of the PCB increases confidence in the feasibility of the selected reflow oven profile prior to testing in the real solder reflow oven. If the discrepancy between CFD calculated and predicted results is too large, a better curve fitting must be performed in step 5. In fact, sets $\tau_{cold} = \{\tau_{1,cold}, \tau_{2,cold}, ... \tau_{n,cold}\}$ and $\tau_{hot} = \{\tau_{1,hot}, \tau_{2,hot}, ... \tau_{k,hot}\}$ can be supplemented with more elements to better approximate the CFD calculated data obtained in step 3.

3. Reflow oven and PCB models

Transient CFD simulations have been performed with a commercial software program (Flotherm 6.1 from Flomerics), which is based on the finite-volume computation method. In the finite volume method the solution domain is subdivided into non-overlapping finite volumes represented by cuboid grid cells, with an option of localized gridding. The turbulence has been approximated by the automatc algebraic model. The x-high and x-low faces of the computational domain are set open, meaning the free boundary of constant pressure

through which air can flow, figure 5. The other faces of the domain have symmetry boundary conditions, i.e. a frictionless, impermeable and adiabatic planar surfaces through which neither air nor heat can flow. The volume bounded by the domain faces represents the space inside the solder reflow oven just above and below the PCB. The preheating zones of the air convection reflow oven are modelled by a set of sources applied to Z-velocity, pressure and transient temperature, and placed directly on the z-high and z-low domain faces. In total, twenty sources have been introduced, ten two dimensional source objects of width $w = 25.7$ mm facing each side of the PCB. The source for pressure is activated as the source of mass flow ($6.2678 \cdot 10^{-3}$ kg/s), to represent the 0.7 m/s air flow impressed from the top and from the bottom sides of the reflow oven. In order to simulate the conveyor of the oven moving with speed v the temperature source has a transient attribute, as shown in figure 6. Five temprerature levels are found in the figure, corresponding to the five preheating zones. Each of ten temperature sources on the top and on the bottom of the oven except for the first source (see numbering of sources and respective profiles in figure 5 and figure 6), have activation time delay t_d, depending on the source number n ($n = 1,...,10$):

$$t_d = w \cdot (n-1) / v \qquad (2)$$

Switching of the temperature sources to another temperature level occurs gradually, so that each source keeps a particular temperature level during time $t_{zone} = L / v$, where L is the preheating zone length. Figure 5 exhibits two temperature visualization planes at time $t = 10.3$ s after the first four temperature sources have been activated, to maintain temperature $T = 149$ °C. The process of PCB entering the preheating zone of the oven can be viewed in the figure.

The constructed oven model corresponds to the reflow oven *Heller 1800EXL*, and is validated within a commercial project at company *Kitron AB*. For the present study the validated reflow oven model is one of the available inputs to the algorithm, and the model validation procedure is therefore omitted.

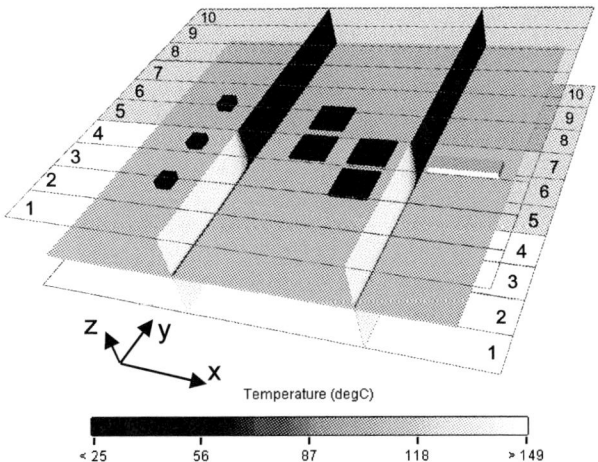

Figure 5: CFD model of the PCB in the air convection reflow oven with ten-source zone representation

The tested PCB consists of a 262.7×257×1.6 mm³ FR4 substrate with six 35 μm layers of copper. Four plastic surface mounted QFP packages are placed in the middle of the PCB which are approximated by four 25×25×2 mm³ cuboid modelling objects. Three surface mounted ferrite beads on the left-hand side of the PCB (figure 5) are simulated as 11×12×6 mm³ cuboid modelling objects. A 41.8×11.8×4 mm³ plastic connector is found on the right-hand side of the PCB. All the PCB components were assigned an effective material with thermal conductivity of 15 W/mK, which is a reasonable approximation for plastic packages. The thermal conductivity of the materials is provided in table 3. Localized gridding have been used in order to better mesh the electronic components and the area around them, where the solder paste location is assumed.

Table 3. Thermal conductivities of materials used in simulations

Material (object)	Thermal conductivity, W/mK
FR4 (PCB)	0.3
Copper (PCB metal layers)	385
Plastic package material (components, connector)	15

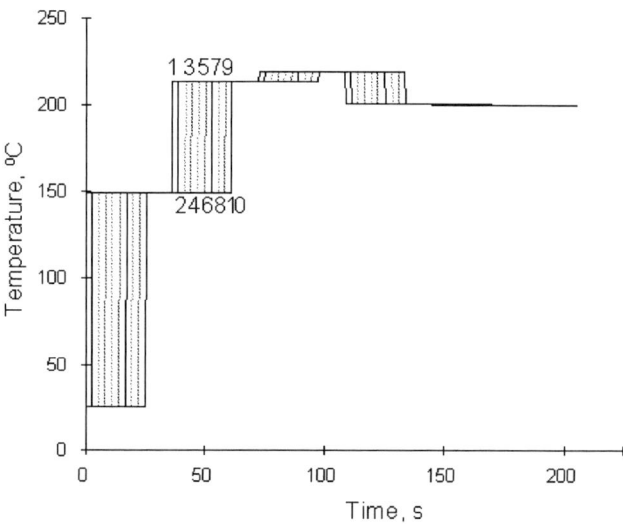

Figure 6: Transient functions for ten thermal sources to simulate a moving reflow oven conveyor

4. Computational experiment

Both the validated solder reflow oven model and the PCB model constructed in the previous section enable demonstration of the CFD driven reflow oven profiling. The results from each step of the profiling algorithm are presented here.

In the first step, a CFD transient simulation is performed for the PCB entering a single preheating zone of the solder reflow oven, with the temperature set to 200 °C. A few snapshots of the PCB surface temperature

are made in order to determine the hottest and the coldest parts of the PCB and particularly in the solder paste regions. The snapshot of surface temperature taken at $t = 89$ s is shown in figure 7.

The second step of the algorithm consists in selection of reference points. Dash-pattern circles in figure 7 indicate the hottest regions of the PCB with the solder paste (close to the connector). Solid-pattern circles reveal the coldest spots in the solder paste (near one of the ferrite beads and beside the two plastic packages).

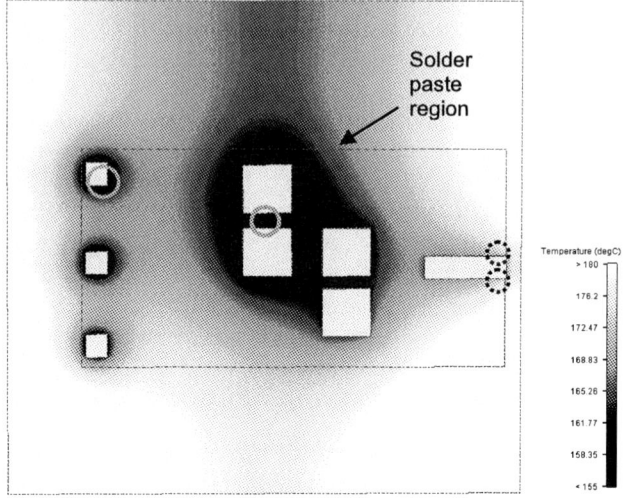

Figure 7: PCB surface temperature (t = 89 s) during preheating up to 200°C

Temperature probes 1-6 are thus set to the points on the PCB at solder paste locations, which have been revealed to be the hottest and the coldest points, see figure 8. Probe 7 in figure 8 is set on the PCB surface for comparison purpuses.

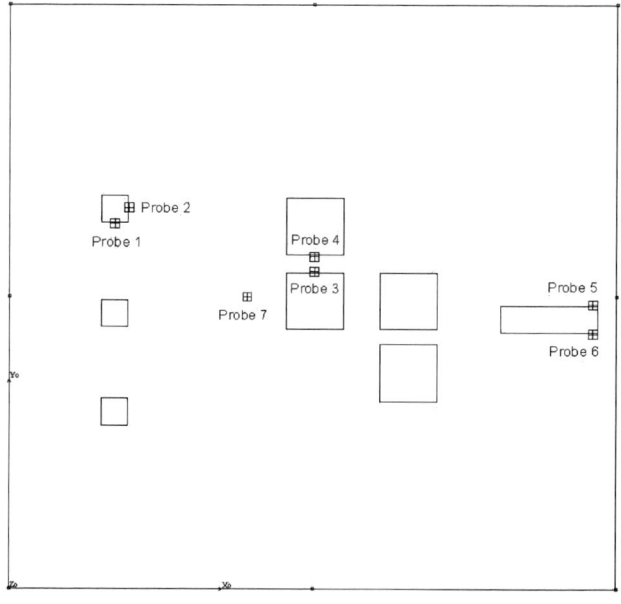

Figure 8: Temperature probes set in the coldest and in the hottest PCB regions with solder paste

In the third step, a single preheating zone CFD simulation is performed as in the first step. The temperature profiles displayed in figure 9 confirm the choice of Probe 4 and Probe 6 as being the coldest and the hottest reference points, respectively. The temperature profiles of the other probes reside, as expected, between the reference point profiles.

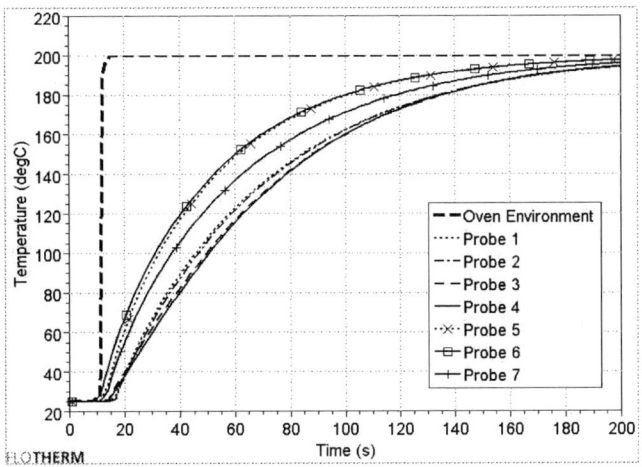

Figure 9: Temperature profiles of the reference points (Probes 4 and 6) during single-zone PCB preheating in comparison with the other temperature probes

The reference point profiles are exported as data files in the fourth step of the algorithm. An example of thermal data for the coldest reference point is provided in table 4.

Table 4. Data file structure corresponding to the coldest reference point

t, s	T, °C	Comment
0.93333334 1.8666667 2.8	25.000029 25.00003 25.004951	Omitted data due to poor defined profile in the initial part of preheating
. . .	. . .	
19.6 20.533333 21.466665	37.167603 39.188446 41.204613	Set of data used for curve fitting
. . .	. . .	
60.18847 61.1218, *62.31311* 63.50442, 64.69573	115.468475 116.87859 *118.6447* 120.37944 122.08258	◀ RC-transition point, where the temperature profile slope becomes $r < 1.5$ °C/s
. . .	. . .	
198.27567 199.209 200.14233	194.1294 194.22948 194.3279	

Data points in initial part of the preheating have been excluded from the subsequent curve fitting procedure, since they cannot be well fitted with the function given by Equation (1). The initial part of preheating cannot be

described by trivial models. Instead, it is decided to model it by introducing time delays in step 6, which supply satisfactory results.

Heating rate function fitting results obtained in the fifth step of the algorithm, are provided in figure 10. Numerical results of the curve fitting are reported in table 5.

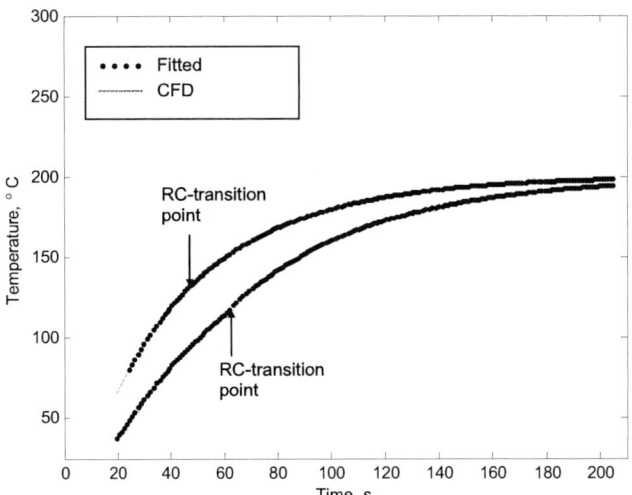

Figure 10: Curve fitting for CFD calculated data related to the coldest (lower graph) and to the hottest (upper graph) reference points

Table 5. Time constants and the maximum curve fitting errors for RC-transition slope r = 1.5 °C/s

Coldest reference	Hottest reference
$\tau_{1,cold} = 63.6$ s	$\tau_{1,hot} = 40.4$ s
$\Delta_{max} = 3.7\%$	$\Delta_{max} = 1\ \%$
$\tau_{2,cold} = 53.0$ s	$\tau_{2,hot} = 43.5$ s
$\Delta_{max} = 0.3\%$	$\Delta_{max} = 0.2\%$

For the hottest reference point the curve fitting actually starts from the moment of time $t = 24.3$ s. Before that time the slope extracted from the CFD calculated data exeeds the maximum ramp rate specified in table 2 as 3 °C/s, and therefore there is no need of finding an analitic function for heating rates higher than 3 °C/s, thereby worthenning the maximum curve fitting error.

Numerical search for reflow oven profiles is performed by the sixth step, taking into account the data from table 2 and the time constants from table 5. The output of the Matlab routine is formed as a text file containing a number of oven profiles.

The selected oven profile (the seventh step of the algorithm) is presented in table 6. The selection has been based on the data from table 1 regarding the preheat time from the minimum to the maximum preheat temperatures, applied to the predicted temperature profiles of the reference points (the solid-pattern profiles in figure 11).

Concluding the algorithm realization, a CFD simulation is performed for the PCB passing all five preheating zones of the solder reflow oven, set to the temperatures provided by table 6. The CFD thermal transient output in the reference points (dot-pattern curves in figure 11) should be compared to the respective predicted profiles (solid-pattern curves). The results of comparison reveal the maximum difference between the CFD calculated and predicted data being not larger than 2 °C, in spite of only two time constants per reference point used in curve fitting. The reference points are brought to the preheat target temperature with the demanded tolerance ± 2.5°C, meaning that the temperatures of all the other points on the solder paste regions are within the ± 2.5°C tolerance region around the preheat target temperature.

Table 6. Selected oven profile with respective zone-related time constants

Zone nr.	1	2	3	4	5
Temperature setting, °C	149	214	219	201	200
Hottest reference time constant, s	40.4	40.4	43.5	43.5	43.5
Coldest reference time constant, s	63.6	63.6	53.0	53.0	53.0

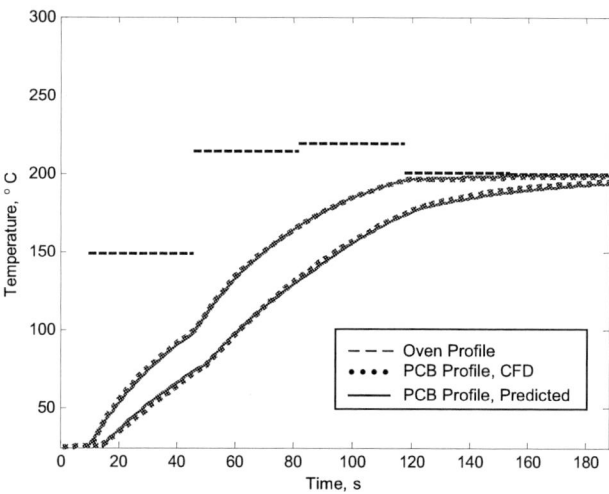

Figure 11: Modelling results for the coldest (lower) and hottest (upper) reference points during PCB preheating in the solder reflow oven with five preheating zones

5. Conclusion

The presented study was focused on the development and application of a reflow oven profile prediction algorithm, which combines both the analytic approach and computational fluid dynamics. Modelling results for preheating of a populated PCB with non-uniform distribution of component thermal mass in an air convection reflow oven led to the following conclusions.

The algorithm provides reliable prediction of both the reflow oven profile and the PCB preheating profile,

provided that the validated CFD model of the solder reflow oven and a high-quality PCB model are availabe as the input. In future, the reflow and the cooling phases will be included in the algorithm.

The main idea of the algorithm is to fit the one-dimensional function for the temperature rise in a material during the heat cycle to the CFD calculated temperature data for the coldest and the hottest reference points on the PCB, with the local time constant as the curve fitting parameter. The number and the value of time constants used for the particular point of the PCB depend on the heating/cooling rate.

Three-dimensional local heat transfer in all modes is taken into account by simple calculating the effective time constant related to a particular location on the PCB, instead of evaluating separately a thermal resistance and a thermal capacitance. Good agreement between predicted and CFD temperature results for PCB preheating demonstrated feasibility of the algorithm.

The algorithm application is reccommended to obtain a starting point in profiling of a real forced air convection reflow oven, for soldering processes on PCBs with non-uniform component thermal mass distribution, characterized by a narrow process window, like in lead-free reflow soldering. On the other hand, the iterative nature of the algorithm enables integration in the PCB and assembly process design phase where a physical prototype of the board is not available.

References

1. Sarvar, F. et al, "Effective Transient Process Modeling of the Reflow Soldering Process: Use of a Modeling Tool for Product and Process Design", *IEEE Trans CPMT–C*, Vol. 21, No.3 (1998), pp.165-171.
2. Van Steenberge, N., Vandevelde, B., Schildermans, I., Willems, G., "Analytical and Finite Element Models of the Thermal Behavior for Lead-free Soldering Processes in Electronic Assembly," *Proc. EuroSimE 2005*, April 2005, pp. 675-680.
3. Suraski, D., "The Benefits of a Ramp-to-Spike Reflow Profile," *Surface Mount Technology (SMT)*, April 1, 2000, http://smt.pennnet.com.
4. "Reflow Soldering Guidelines for Lead-Free Packages," *Application Note 353-1.0*, Altera Corp., Jul 2004, http://www.altera.com.
5. Lempinen, J, Tuominen, A, Hämeenoja, O, Tuokko, I, Kerminen, T, "Use of CCD for Reflow Process Design," *Proc. IMAPS Nordic Annual Conference*, Gothenburg, Sweden, September, 2006, pp. 91-93.
6. "Flotherm v.6.1 User Manual," Flomerics PLC. (2005).
7. "FloPCB User Manual," Flomerics PLC. (2006).
8. Steinberg, D. S., Cooling techniques for electronic equipment, John Wiley & Sons, NY, 1991.

The Effect of Downscaling the Dimensions of Solder Interconnects on their Creep Properties

S. Wiese, M. Roellig, M. Mueller, K.-J. Wolter
Dresden University of Technology, Electronic Packaging Laboratory, Germany
TU Dresden, Fak. ETIT, IAVT, D-01062 Dresden
wiese@avt.et.tu-dresden.de, +49 351 463 3317

Abstract

The creep behaviour of solders is an important input for accurate material models for FE-analysis of electronic assemblies. Usually the mechanical behaviour of solders, is been determined by tensile tests on bulk solder specimens. Although performing these tests is not complicated and the results are easy to interpret, one of the key problems lies in the fact that solder joints are very small and therefore cannot be represented by large tensile specimens. The paper describes the attempts to gain deformation data on ultra small solder joints. It compares creep data that was experimentally gained on bulky samples and on small solder joints.

1. Introduction

In order to achieve improved packaging density, electrical performance and manufacturability, electronics package architectures are changed steadily. Such design changes may lead to significant changes in manufacturing processes or employed materials. All these factors together may create unknown reliability risks. Therefore new developments in electronics packaging physical design are usually accompanied by a number of different reliability considerations. While thermal cycling tests are used to assure the practical reliability of a design and to detect potential new failure mechanisms, theoretical analysis like FE-analysis are a most powerful tool to find the most important root causes for insufficient reliability figures.

While FE-analysis may provide useful information for improvements in the package physical design, the best material selection or optimization of manufacturing processes; its accuracy is dependent upon the precision of the description of deformation behaviour of the employed materials. Among these materials there is particular interest in polymers and solders. Both show most complex temperature- und time-dependent deformation behaviour. In addition to that solder interconnect failure is also a major concern for package-level-reliability. In the past, the major part of failures at package- and board-level was due to fatigue of solder joints.

Therefore, the adequate characterization of the mechanical behaviour of solder is one of the key issues to address, in order to introduce the advantageous FEM simulations as a standard practice in the reliability analysis of electronic packages and assemblies. Incorrect assumptions about the constitutive behaviour, which might be causally connected to unsuited experimental methods, can cause misleading FEM simulation results.

2. Methodology of Characterization

Characterization of constitutive material behaviour is usually done through tensile tests using standardized samples. Such samples are typically simple bars or rods with dimensions of at least a few millimeters. Figure 1 shows typical solder bulk specimens. Such specimens are usually fabricated from standard bar-solder which is melted and solidified in a aluminium mould. Alternatively mechanical preparation methods can be used to prepare bulk solder specimens.

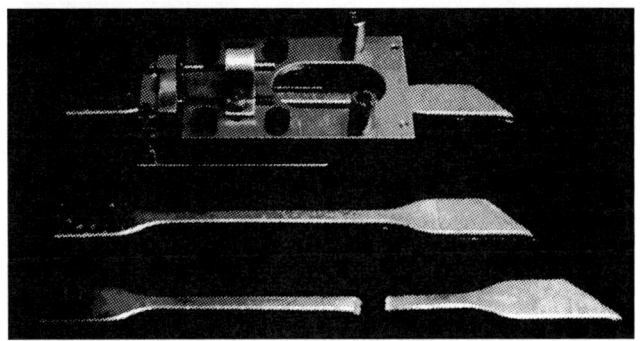

Figure 1: Bulk specimens: after test (foreground), before test (centre), with clamped strain sensor (background)

The idea behind the typical dimension of bulk samples is that their volume is large enough to represent the material behaviour of large structures in civil engineering applications. However the situation in the small structures of electronic assemblies is quite different. These structures are often so small (a few to a hundred micrometers), that structural members as single grains may significantly influence their mechanical behaviour. Therefore the characterization of material behaviour for electronic packaging cannot be based on the experimental methods that have been established for large structures. New experiments need to be created in order to meet the needs of micrometer-sized structures.

In order to obtain relevant constitutive data for real interconnections, specimens had to be scaled down to a typical solder joint size. The simplest idea is to use solder joints as specimens. One possible way to determine the behaviour of such joints is to measure the shear force vs. shear angle in the course of simple shear of a joint. The primary task consisted of designing a suitable test setup for such an experiment.

The loading condition in the test setup is a simple reversal shear under isothermal conditions with appropriate cyclic strain rates and amplitudes. The first designs were presented by Mei [1], Subrahmanyan [2], and Nir [3]. While the investigations of Mei and coworkers were performed on a microactuated but still large-scale setup, Subrahmanyan made the first attempts to significantly reduce the dimensions of the setup. Small setups suit the small specimens best, as errors due to temperature fluctuation and vibration are reduced. The setup introduced by Nir and co-workers had other innovations. It used linear microactuators (antiparallel coil system), which provide a zero backlash when the direction is reversed. The specimens were held in place by epoxy glue, minimizing the damage on the specimen before the test. A test setup for ultra small flip chip joints was created based on these ideas [4].

The specimen consisted of two silicon chips that were bonded together in an arrangement of 4 corner solder joints [4]. The mechanical characterization of these solder joints required well-controlled movement of the two silicon chips relative to each other, in such a way that all bending moments and out-of-plane forces were eliminated or at least minimized. This was achieved through a symmetric grip configuration of two identical specimens (Fig. 2). The load was generated by a piezoelectric translator, which provides smooth movement with subnanometer resolution over a wide range of velocities. A force sensor was mounted between the inner grip and the piezoelectric translator. The translation direction was in line with the gravity axis, eliminating the need for bearings which would otherwise be required to compensate for weight. Bearings cause frictional forces of at least 150 mN, which make high resolution force measurements extremely difficult.

A high resolution displacement measurement was achieved using a dual laser beam interferometer. Each laser beam was reflected off one of two neighboring edges of the specimens' silicon chips. In order to minimize errors due to temperature changes or vibrations, the laser interferometer heads were borne symmetrically to the specimen.

Although the test setup for shear experiments on flip chip joints is able to record very precisely the force displacement curves of ultra small flip chip solder joints, it suffers from certain limitations in the degree of experimental freedom. One of these limitations is the specimen design. When the flip chip test setup was created [4, 5], most of the interest in the area of mechanical behaviour focused on flip chip joints and on SnPb solders. With the development of new area-array-packages and the switch to lead free solders, the focus of interest also changed. The most critical parts in modern packages are solder balls having a diameter of 300...800 µm and consisting of various SnAgCu-base alloys. Therefore a new test setup [6] was designed, that was able to perform shear experiments on solder balls (Fig. 3). The targeted specimens have been BGA-, FBGA- and CSP-packages.

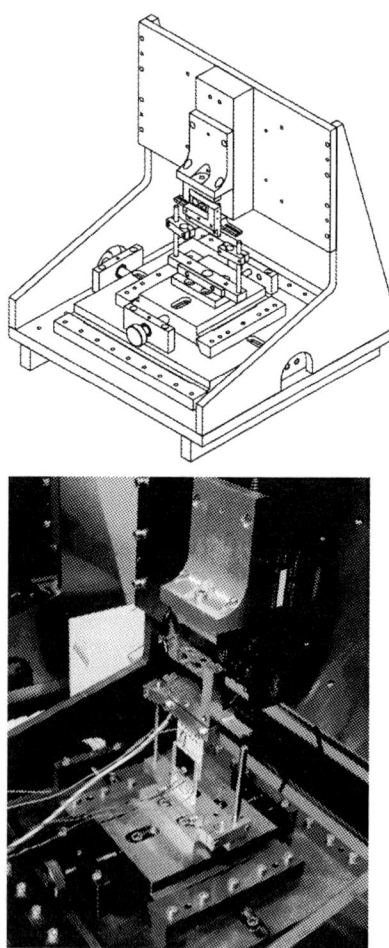

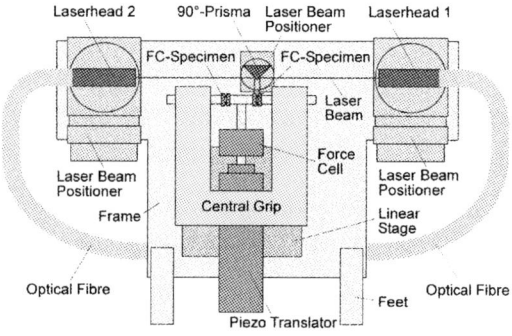

Figure 2: Test setup for flip chip specimens

Figure 3: Test-Setup for Solder Ball specimens

The test setup solder ball specimens (Fig. 3) consisted on a vertical directed stepper motor translation stage, that provides submicrometer movements in z-axis. A specially designed force sensor is mounted on the stepper motor translation stage. The force sensor is characterized by compliance in z-axis and high stiffness in x, y-directions. The z-axis compliance is critical for adjusting the load force in creep testing. The x, y-stiffness provides guidance for the relative movement of an area array package versus the board. Such movement is needed to create a well defined shear deformation in the solder balls of the area array package

In order to grip the area array package specimen in the test setup, it is glued on between two alumina sheets by an epoxy adhesive (Fig. 4). The upper sheet is mounted to the force sensor. The lower sheet is hold by the x, y-table, that is situated on the opposite side of the force sensor. In order to fix the specimen into the test setup it is first glued to the upper sheet, before it is attached to the lower sheet. On the backsides of two alumina sheets there are thick film heaters on to create the desired testing temperatures.

Figure 4: Mounting of an area array specimen via alumina sheets, which provide different test temperatures by integrated thick film heaters on their backside

However there can be some shortcomings in the experimental methodology. Since the highest test temperature (125°C) can be near the glass transition of the organic board, there can be some unexpected compliance of the organic base material underneath the Cu pad on the board.

In order to surely eliminate this effect, new specimens were created on a ridged alumina substrate. The specimens used a five bump design, which is detailed described in [7]. Briefly the specimen consists on two alumina chips (3,2 mm X 1,2 mm) soldered against each other by 4 corner balls and one centre ball (Fig. 5). The solder balls were manually made in the lab by micro-punching circular platelets (diameter = 300 μm, thickness = 140 μm) out of thin solder sheets. The sheets were rolled from solder ingots. The balls were soldered on a Cu-metallization, that was electroplated on the alumina chips, to provide circular landing pads with a diameter of 200 μm.

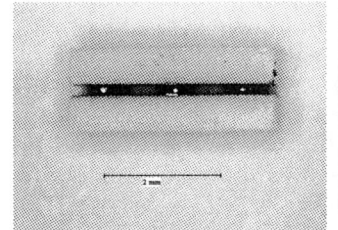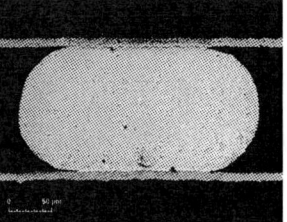

Figure 5: Specimens for Micro Solder Ball (footprint 200μm). Solder balls are soldered on electroplated Cu-metallization on alumina.

3. Experimental Findings

In the last years there have been a myriad of creep tests conducted on bulk solder specimens on the eutectic SnAg and SnAgCu systems. Although performing these tests is not complicated and the results should be easy to interpret, there is a remarkable scatter in the received creep data. This scatter might be caused by two factors. The first one is material related. Since different studies used different preparation procedures to fabricate specimens, it is most likely that huge variety of microstructures has been examined by the various investigations. The second factor is methodology related. The experimental methods that were used in order to determine the creep properties vary strongly between different investigations. These variations include the type of test (constant-load test, load-stepping test, constant-strain-rate tests, strain-rate-change test), the choosen stress or strain rate range, the test temperatures and the method of determination of steady state strain rate.

In order to compare creep data between bulk specimens and micro solder balls one should choose parameters that are relatively independent from test conditions. Such parameters are above all the stress exponents (n) and the activation energy (Q). However in some cases there is a significant dependence of Q and the test temperature.

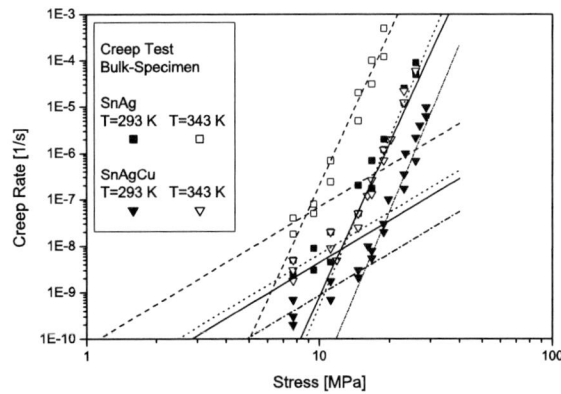

Figure 6: Results from creep tests on bulk SnAg3.5 and SnAg3.8Cu0.7 samples at test temperatures of T = 20 °C, T = 70 °C from [8]

191

The typical creep behaviour of SnAg and SnAgCu bulk material is depicted in figure 6. The behaviour is characterized by a stress exponent of $n = 11$ for SnAg3.5 and $n = 12$ for SnAg3.8Cu0.7. The activation energy for both alloys is about $Q = 61$ kJ/mol. Apart from a few exceptions the results for the stress exponent are confirmed by a large number of other studies [9-21]. However the values for activation energy vary strongly ($Q = 46.6 \ldots 108.5$ kJ/mol), which might by due to the different test temperatures used by the different investigations.

use of different under bump metallizations (UBM) led to very different creep properties. When SnAg solder was deposited onto a Cu UBM, it showed a much lower creep resistance and a lower stress exponent (n = 11) than the same solder deposited onto a NiAu UBM (n = 20).

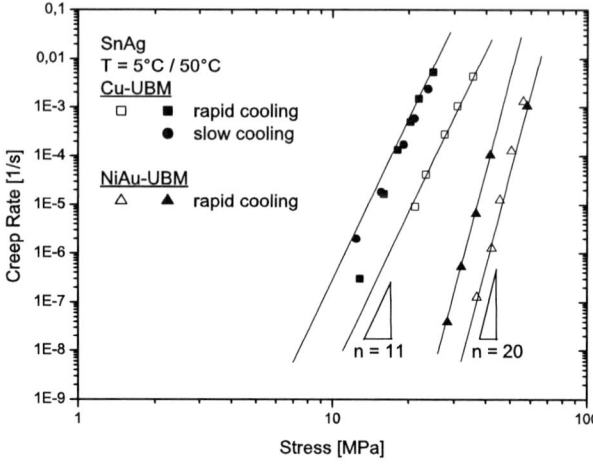

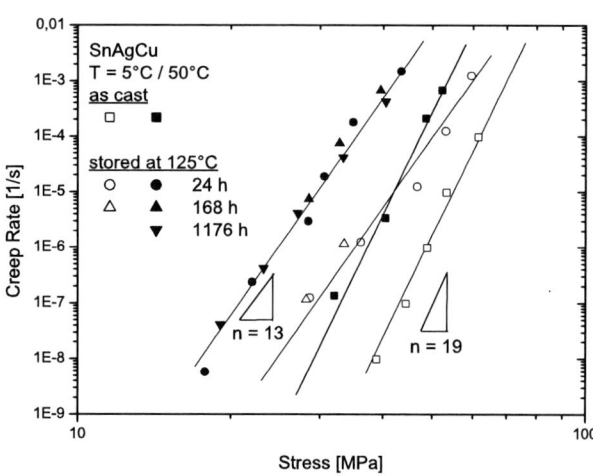

Figure 7: Results from creep tests on flip chip SnAg3.5 and SnAg3.8Cu0.7 solder joints at test temperatures of T = 5 °C, T = 50 °C from [8]

The results of the creep experiments on flip chip joints are depicted in the diagrams in figure 7. The steady state creep rate is plotted versus the applied stress. The creep data was fitted using a simple power law (Eq. 1). The parameters are summarized in table 1. The upper diagram in figure 7 shows the results for SnAg3.5 solder. It illustrates that different cooling rates have not led to significant differences in creep behaviour. However, the

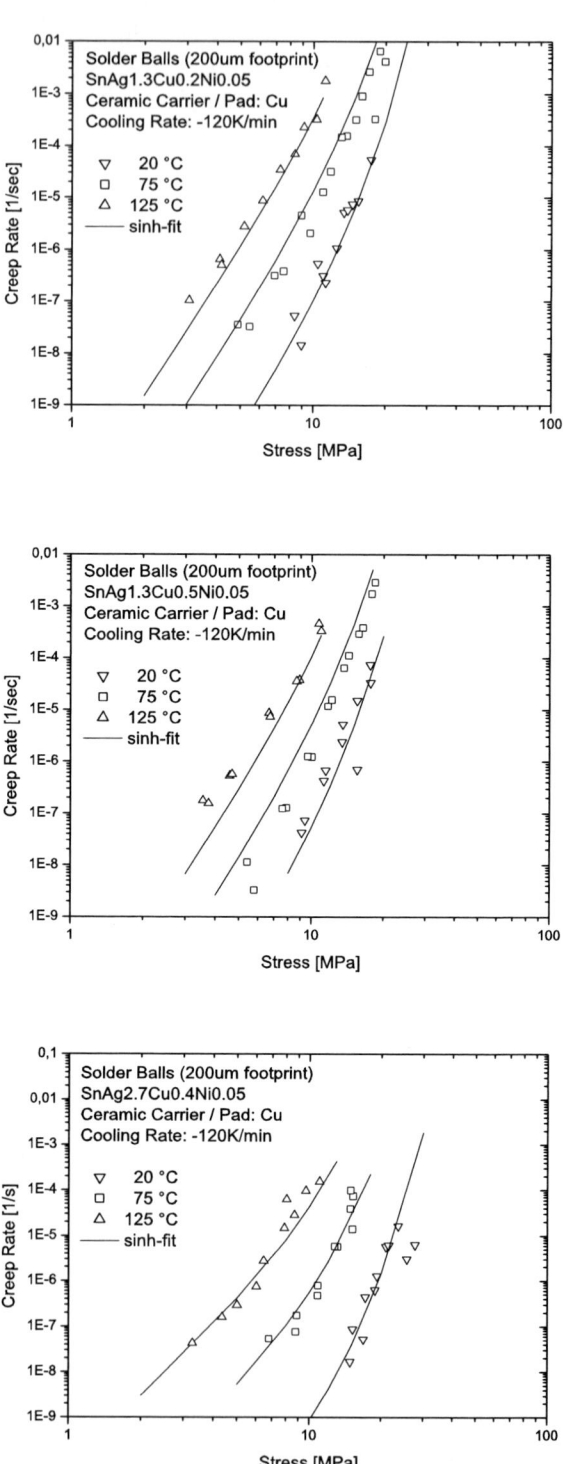

Figure 8: Creep data for lab made micro solder balls made from solder ingots rolled into sheets. The compositions of the solder ingots have been SnAg1.3Cu0.2Ni0.05; SnAg1.3Cu0.5Ni0.05 and SnAg2.7Ag0.4Ni0.05.

$$\dot{\varepsilon} = A \left(\frac{\sigma}{\sigma_N} \right)^n \exp\left(-\frac{Q}{RT} \right) \qquad (1)$$

Table 1: Parameters for power law creep model (Eq. 1) for eutectic SnAg solder when $\sigma_N = 1$ MPa

Flip-Chip Joints	A [s^{-1}]	n	Q [kJ/mol]
SnAg[1]	2E-6	11	73,2
SnAg[2]	6E-25	20	74,9
SnAgCu[3]	6E-23	19	84,2
SnAgCu[4]	1E-12	13	75,2

[1] Flip chip joints, Cu UBM, Umnicore Microbond SnAg3.5, as-cast
[2] Flip chip joints, NiAu UBM, Umnicore Micro-bond SnAg3.5, as-cast
[3] Flip chip joints, Cu UBM, Fremat Sn95.5Ag3.8Cu0.7, as-cast
[4] Flip chip joints, Cu UBM, Fremat Sn95.5Ag3.8Cu0.7, 24h … 1176h/125°C

The bottom diagram in figure 7 shows the results for SnAg3.8Cu0.7 solder. As cast SnAgCu-solder shows a significant higher stress exponent (n=20) than as cast SnAg solder. After thermal storage the stress exponent declines to a value of n = 13.

The results of the creep experiments on micro solder balls (200 µm diameter) are depicted in the diagrams in Figure 8. The compositions of the solder ingots, that were manufactured and analysed by JL Goslar, have been SnAg1.3Cu0.2Ni0.05; SnAg1.3Cu0.5Ni0.05 and SnAg2.7Ag0.4Ni0.05. It is believed that due to the manufacturing process, the solder balls will contain solder composition very similar to the ingots. The diagrams in Figure 8 show the steady state creep rate plotted versus the applied stress. Creep rate and stress values have been calculated using the methodology described in [22]. The creep behaviour was modelled using a sinh-law (fit equation (2)). The fit parameters for the lab made micro-solder balls are given in Table 2.

$$\dot{\varepsilon} = A \sinh\left(\frac{\sigma}{\sigma_N} \right)^n \exp\left(-\frac{Q}{RT} \right) \qquad (2)$$

Table 2: Parameters for sinh-law creep model (Eq. 2) for SnAgCu base alloys in lab made micro solder balls (footprint 200 µm)

Micro Solder Ball (200 µm footprint)	A [s^{-1}]	σ_N [MPa]	n	Q [kJ/mol]
SnAg1.3Cu0.2Ni0.05	8E+5	10	7	75
SnAg1.3Cu0.5Ni0.05	6E+4	9	7	70
SnAg2.7Cu0.4Ni0.05	2E+7	7	5	100

The micro solder joints of SnAg1.3Cu0.2Ni0.05 and SnAg1.3Cu0.5Ni0.05 show a stress exponent of n = 12 at temperatures of T = 20 °C and T = 75 °C. The activation energy can be estimated with 63 kJ/mol. The micro solder joints of SnAg2.7Ag0.4Ni0.05 shows a stress exponent of

n = 14 at a temperature of T = 20 °C and n = 12 at a temperature of T = 75 °C. The stress exponent at a temperature of T = 125°C is significantly lower for all alloys and has a value of n = 7 … 8.

4. Microstructural Considerations

The microstructure of the specimens was analyzed using standard metallographic preparations. In the final step ion beam polishing or chemical etching techniques were used to create a topography between the and the intermetallic particles. The goal of the microstructural analysis is to receive information about the spatial arrangement of intermetallic particles.

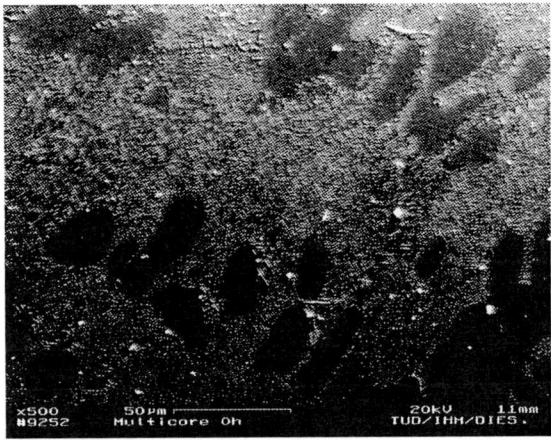

Figure 9: Microstructure of SnAgCu bulk specimen (scanning electron microscopy).

The typical microstructure of a SnAgCu-bulk specimen is shown in figure 9. The microstructure is characterized by a few β-Sn-dendrits that are surrounded by a relatively large area of the quasieutectic, that consists small intermetallic particles that are finely dispersed into the β-Sn-matrix.

Figure 10: Microstructure (light microscopy) of flip chip specimen joint (printed solder paste), SnAg3.5, NiAu-metallization, after soldering

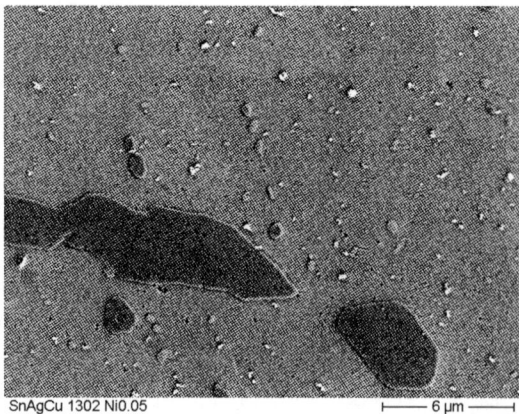

Figure 11: Microstructure (scanning electron microscopy) of micro solder ball, SnAg1.3Cu0.2Ni0.05 (upper image), SnAg1.3Cu0.5Ni0.5 (centre imag.) SnAg2.7Cu0.4Ni0.05 (bottom image), Cu-Metallization (200 μm footprint), after soldering

In contrast to the bulk microstructure the SnAg3.5 flip chip joints (Fig. 10) show a columnar structure. Therefore it can be concluded, that nucleation of primary Sn crystals took place at the interface to the UBM. Subsequently the solidification was directed from one metallization side to the opposite metallization side.

A very similar microstructural appearance was found on SnAgCuNi micro solder joints (Fig. 11). These joints

do not show the well known dendritic or columnar microstructures. Comparing the different compositions of the SnAgCuNi micro solder joints, it shows that a higher Ag and Cu content leads to a larger number intermetallic particles within the β-Sn-matrix.

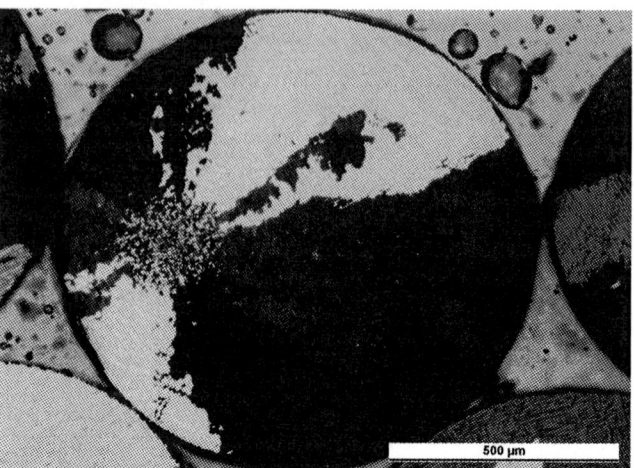

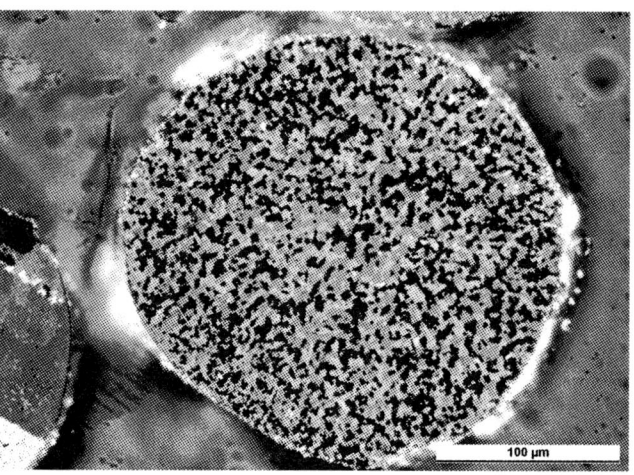

Figure 12: Microstructure (polarized light microscopy) of a SnAg3Cu0.5 solder balls. The diameters of the balls are d = 1100 μm (upper picture) and d = 270 μm (bottom picture). The cooling rate for both solder balls has been 1.1 K/s [23]

Experiments from [23] give a reasonable explanation, why the microstructures between bulk specimens, medium sized solder balls and tiny flip chip joints are different. As can be seen from the polarized light images in figure 12, in large solder balls there forms a small centre area with many randomly orientated grains. This is where the nucleation process supposedly starts. This area is surrounded by a ring that contains only a small number of differently orientated grains. It can be assumed that this is caused by a rapid crystal growth that starts from the centre area to the outside. Very small solder balls (d < 300 μm) consist only on large number of randomly orientated grains. Therefore they have a large number of grain boundaries compared with their volume. Moreover mechanisms of dendritic growth that can be observed in large solder volumes, may not occur in such small solder joints.

5. Discussion

Because of the different approaches in specimen design, test setups and experimental methodology, that were necessary to investigate to solder creep behaviour at different size levels, the results of the study are not free from ambiguity. However, new findings resulted from the conducted experiments which allow a more complete understanding of the complex nature of creep in SnAg/SnAgCu alloys.

At ultra small volumes (flip chip joints) the material of the metallization has a much stronger impact on the creep properties. The metallization material changes not only the absolute creep strength but also the stress sensitivity (stress exponent) of the creep behaviour. Such effects could not have found on bulk level.

Although the activation energy seem to change with size, the results on the micro solder joints show, that this must not be necessarily the case. Earlier assumptions [24], that the smaller size of precipitates affects the higher activation energy are not longer hold.

The differences creep behaviour between bulk solder specimens and tiny solder joints can be well understood by the differences in microstructure. The absence of dendritic growth result in a large number of grain boundaries. Since certain intermetallics such as $AuSn_4$ accumulate preferable at grain boundary triples. At such locations they form effective barriers for dislocation movement. This may form an additional strengthening mechanism to the intrinsic strengthening by Ag_3Sn-, Cu_5Sn_6-intermetallic particles within the β-Sn-matrix.

6. Conclusions

Creep experiments on SnAgCu-base solders have been conducted on several types of specimens: flip chip, micro solder ball and bulk specimens. Various test setups have been developed, that are specially adapted to these kinds of specimens.

The results of the experiments show that the influence of size and composition of SnAgCu-base solders is very complex. Size and composition are no independent factors. On bulk specimens the stress exponent was neither influenced by temperature nor by alloy type. Generally stress exponents between n = 11 … 12 have been found. No correlation between the stress exponent and the alloy type could be found. The creep behaviour follows a simple power law.

On micro solder ball specimens the stress exponent is influenced by temperature. Stress exponents at T = 125 °C (n = 7…8) are typically lower than at room temperature (n = 12 …14). Therefore the best fit was achieved using a sinh-law.

On flip chip specimens very high stress exponents (n = 11 … 20) were detected. Due to small range of test temperatures used in these investigations (5 °C … 50 °C), no fit-function could have been concluded.

Microstructural analysis point out, that bulk solder has typically a dendritic microstructure. At very small solder joints no dendritic growth was detected. The different nucleation behaviour is supposed to be the reason for the differences in creep properties.

Acknowledgments

The work for this paper was supported within the scope of technology development by the EFRE fund of the European Community, funding of the State Saxony of the Federal Republic of Germany, funding by the German Research Council (DFG) and funding of the German Federal Ministry of Education and Research (BMBF). The authors would like to thank Dr. Gunter Hagen from KMS Kemmer, Dresden for providing the micro punching tool. A particular thank is given to Ms. Sandy Bennemann from Fraunhofer IWM, Halle and Ms. Anke Schoene from TU Dresden for her support in microstructural analysis. Especially acknowledged is the support and the scientific advice by Dr. Sven Rzepka of Qimonda, Dresden.

References

1. Mei, Z., Morris, Jr.; Shine, M. C., Summers, T. S. E.: "Effects of Cooling Rate on Mechanical Properties of Near-Eutectic Tin-Lead Solder Joints," Journal of Electronic Materials, vol. 20 (1991), pp. 599-608
2. Subrahmanyan, R.: "A damage integral approach for low cycle and thermal fatigue," Diss., Cornell University, Ithaka 1990.
3. Nir, N., Dudderar, T. D., Wong C. C., Storm, A.R.: "Fatigue Properties of Microelectronics Solder Joints," Trans. of the ASME, vol. 113 (1991), pp. 92-101
4. Wiese, S.; Feustel, F.; Rzepka, S.; Meusel, E.: Experimental Characterization of Material Properties of 63Sn37Pb Flip Chip Solder Joints. Electronic Packaging Materials Science X - MRS Symposium Proceedings, April 14-16 (1998), San Francisco, Belton, D. J.; Gaynes, M.; Jacobs, E. G.; Pearson, R.; Wu, T. (Ed.), Materials Research Society, Warrendale, Bd. 515 (1998), pp. 233-238
5. Wiese, S.; Feustel, F.; Meusel, E.: Influence of loading history on cyclic mechanical properties of Flip Chip Joints: Experiments and Simulation. Proceedings of the Micro Materials '97 Conference, April 16-18 (1997), Berlin, ddp Goldenbogen, Dresden, pp. 279-282
6. Röllig, M.; Wiese, S.; Meier, K.; Wolter, K.-J.: Creep Measurements of 200 µm – 400 µm Solder Joints. see these Proceedings.
7. Röllig, M.; Dudek, R.; Wiese, S.; Wunderle, B.; Wolter, K.-J.; Michel, B.: Novell test concept for experimental lifetime prediction of miniaturized lead-free solder contacts. Proceedings of EuroSIME 2005, Berlin, April 18 -20 (2005), pp. 86 – 90

8. Wiese, S.; Wolter, K.-J.: Microstructure and creep behaviour of eutectic SnAg and SnAgCu solders. Microelectronics Reliability. Vol. 44 (2004), pp. 1923-1931

9. Mavoori, H et al.: "Creep, Stress Relaxation, and Plastic Deformation in Sn-Ag and Sn-Zn Eutectic Solders". Journal of Electronic Materials, Vol. 26 (1997), pp.783-790.

10. Neu, R. W. et al: "Thermomechanical Behaviour of 96Sn-4Ag and Castin Alloy", Journal of Electronic Packaging, Vol. 123 (2001), pp. 238-246.

11. Raeder, C. H.; Schmeelk, G. D.; Mitlin, D.; Barbieri, T.; Yang, W.; Felton, L. F.; Messler, R. W.; Knorr, D. B.; Lee, D: Isothermal Creep of Eutectic SnBi and SnAg Solder and Solder Joints. Proceedings of 1994 IEEE/CPMT Int'l. Electronics Manufacturing Technology Symposium (1994), pp. 1-6.

12. Wu, K.; Wade, N.; Cui, J.; Miyahara, K.: Microstructural Effect on the Creep Strength of a Sn-3.5%Ag Solder Alloy. Journal of Electronic Materials, Vol. 32 (2003), pp. 5-8.

13. Lang, F.; Tanaka, H., Munegata, O.; Taguchi, T.; Narita, T.: The effect of strain rate and temperature on the tensile properties of Sn-3.5Ag solder. Materials Characterizations. Vol. 54 (2005), pp. 223-229

14. Shohji, I.; Yoshida, T.; Takahashi, T.; Hioki, S.: Tensile properties of Sn-Ag based lead free solders and strain rate sensitivity. Material Science and Engineering, Vol. A366 (2004), pp. 50-55

15. Plumbridge, W. J., Gagg, C. R.: Effects of strain rate and temperature on the stress-strain response of solder alloys. Journal of Material Science: Materials in Electronics, Vol. 10 (1999), pp. 461-468

16. Guo, Z.; Pao, Y.-H.; Conrad, H.: Plastic Deformation Kinetics of 95.5Sn4Cu0.5Ag Solder Joints. Journal of Electronic Packaging, Vol. 117 (1995), pp. 100 – 104

17. Kerr, M.; Chawala, N.: Creep deformation of Sn-3.5Ag solder/Cu couple at small length scales. Acta Materialia, Vol. 52 (2004), pp. 4527-4535

18. Morris Jr., J.W.; Song, H.G.; Hua, F.: Creep Properties of Sn-rich Solder Joints. Proc. IEEE 53th Electronics Components and Technology Conference, New Orleans, May 27 - 30, pp. 54 – 57

19. Ochoa, F.; Deng, X.; Chawala, N.: Effects of Cooling Rate on Creep Behavior of a Sn-3.5Ag Alloy. Journal of Electronic Materials, Vol. 33 (2004), No. 12, pp. 1596 – 1607

20. Dutta, I.; Park, C.; Choi, S.: Impression creep characterization of rapidly cooled Sn-3.5Ag solders. Materials Science and Engineering, A 379 (2004), pp. 401 – 410

21. Whitelaw, R.S.; Neu, R.W.; Scott, D.T.: Deformation Behavior of Two Lead-Free Solders: Indalloy 227 and Castin Alloy. Journal of Electronic Packaging, Vol.121 (1999), pp. 99 – 107

22. Röllig, M.; Wiese, S.; Wolter, K.-J.: Extraction of material parameters for creep experiments on real solder joints by FE-analysis. Proceedings of the Eurosime Confernce 2006, Como (I), April 24 -26 (2006), pp. 281-289

23 Mueller, M.; Wiese, S.; Roellig, M; Wolter, K.-J.: The Dependence of Composition, Cooling Rate and Size on the Solidification Behaviour of SnAgCu Solders. see these Proceedings.

24. Wiese, S.; Schubert, A.; Walter, H.; Dudek, R.; Feustel, F.; Meusel, E.; Michel, B.: Constitutive Behaviour of Lead-free Solders vs. Lead-containing Solders -Experiments on Bulk Specimens and Flip-Chip Joints. Proceedings IEEE 51th Electronic Components and Technology Conference, Orlando, 29.05.-01.06.2001, pp. 890-902

The Dependence of Composition, Cooling Rate and Size on the Solidification Behaviour of SnAgCu Solders

Maik Mueller, Steffen Wiese, Mike Roellig, Klaus-Juergen Wolter
Dresden University of Technology, Electronics Packaging Laboratory, Germany
TU Dresden, Institut für Aufbau- und Verbindungstechnik der Elektronik (IAVT), D-01062 Dresden, Germany
maik.mueller@avt.et.tu-dresden.de, +49 351 463 33172

Abstract

The scope of this study is to investigate the influences on the solidification of the microstructure of SnAgCu solders. It will be shown that the solidification process depends on solder composition, specimen size and manufacturing conditions. The influence of solder composition has been investigated on bulk solder ingots by varying the Ag content from 3.0 wt% to 3.8 wt% and the Cu content from 0.4 wt% to 1.5 wt%. The influence of an additional Au content was investigated on a SnAg3.0Cu0.5Au0.14 solder. Solidification experiments with different cooling rates from 0.006 K/s to 0.6 K/s have been carried on bulk solder ingots (length 23 mm; Ø 7 mm) of the alloys SnAg3.0Cu0.5, SnAg3.8Cu0.7 and SnAg3.5Cu0.4. The results will point out the influence of cooling gradients on the microstructure.

In order to compare the microstructure of these large specimen with real solder joints, solidification experiments with cooling rates from 0.33 K/s to 10.9 K/s have been carried out on SnAg3.0Cu0.5 solder balls with four different sizes (Ø 130 µm, Ø 270 µm, Ø 590 µm and Ø 1100 µm). Phase sizes and shapes as well as grain orientations have been investigated and compared. Experiments on directed solidification have been carried out in order to investigate the dendritic growth of β-Sn dendrites. For that purpose a temperature gradient was generated inside the solder during solidification.

1. Introduction

During the last years many studies and investigations on the solidification behaviour of lead free solders were published in order to improve the reliability of lead free interconnects [1-4]. A mayor influence on material behaviour is the microstructure that is formed during the solidification process. This microstructure is responsible for the mechanical properties of the interconnection and depends on various influences. Therefore it is necessary to understand the solidification process and its influences in order to interpret differences in material behaviour.

The microstructure of SnAgCu consists of three phases: β-Sn and the intermetallic compounds Ag_3Sn and Cu_6Sn_5. According to the phase diagram the eutectic composition with the melting point at 217 °C is around Sn95.5Ag3.8Cu0.7 [5]. It is known that changes in composition can change size, shape and number of those phases [1, 5, 6]. The determining process for the formation of the microstructure is the reflow process. This dynamic process is not comparable with the state of equilibrium in the phase diagram [2]. That means that the parameter cooling rate influences the phase growth of intermetallics and changes the phase diagram [3, 4].

Apart from the choice of the Ag and Cu content in the solder paste, it has to be considered that the composition also changes during the reflow process. This is caused by the dissolution of substrate metallisation [7] such as Ni/Au, Ag and Cu. This depends on peak temperature and time above liquidus of the reflow profile. That means that soldering on a Cu metallisation will cause an increase of the Cu content and a decrease of the thickness of the metallisation. In [7] is shown that the grade of dissolution also depends on the deposition process of the metallisation material.

Another influence is the temperature difference caused by different thermal masses of the PCB and the component. This causes a temperature gradient in the solder bump during solidification and may have influence on the growth of dendritic structures like β-Sn according to [8, 9].

Another challenge for the use of lead free solders in interconnects is the miniaturisation. At the moment Flip-Chip bumps with heights of less than 90 µm are already in use [10]. As explained in [11, 12] it is planed to increase the pitch sizes down to 20 µm in the next years. FE simulations were already carried out in order to prove the concepts of these interconnections with bump diameters of 50 µm to 20 µm [13, 14]. In [15] it was already explained that material behaviour changes with size and possible reasons for that are explained in [16]. Lehman et al. describes that microstructure changes with bump size and the appearance of "interlaced" grain structures (areas with a large number of small, randomly oriented grains) appear more frequent with decreasing solder volume. In [17] it is described that the material behaviour of Sn is anisotropic. That means that material behaviour depends on grain orientation. This is a possible reason why solder balls with less assumed stress sometimes fail earlier than those with higher assumed stress.

The mentioned influences allow a large variety of different microstructures. The research presented in this study points out how the microstructure can be changed by process parameters and what differences between large bulk specimen and small solder bumps can be found. This fundamental research is necessary to understand the solidification process and may allow interpretations on differences found in the measured material data of specimen with different compositions, cooling rates and solder volumina.

1-4244-1105-X/07/$25.00 ©2007 IEEE

2. Experiments on Bulk Solder Ingots

Experimental Procedure

For the experiments on bulk solder ingots a test setup was developed consisting of a heating block, a controller unit and a software interface (Fig. 1a). The heating block (Fig. 1b) was made of Al. It has eight, symmetric aligned drills for test tubes (inner Ø 7 mm). The solder is melted in these test tubes to prevent interfacial reactions, which cause changes in solder composition. The temperature management is realised by power resistors for heating and fans for cooling (Fig. 1b). The whole setup is controlled by a programmable PID controller. The software interface documents the experimental data and transmits the necessary parameters to the controller, in order to realise a desired temperature profile. This setup allows automated experiments with accurate cooling rates from 0.35 K/min to 37.4 K/min (0.006 K/s to 0.6 K/s) on eight samples at a time.

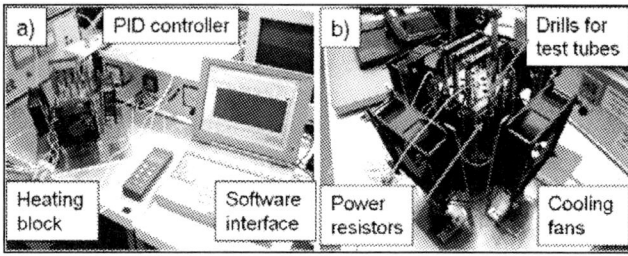

Figure 1: a) Test setup consisting of heating block (left), controller unit (back), software interface (right); b) detail of heating block

For the experiments three different alloy compositions were used: near eutectic SnAg3.8Cu0.7 (Multicore), SnAg3.0Cu0.5 (Alpha Metals) and SnAg3.5Cu0.4 (lab fabrication). The raw solder bars were melted and filled in test tubes. After the solidification, the ingots were cut to a final length of approx. 23 mm. In the beginning of the experiment, the samples were held at 250 °C for 15 min, to solute all intermetallics and allow stable start conditions for the subsequent cooling process. Finally cross sections were made to analyse the microstructure.

Results on Solder Composition and Cooling rate

Fig. 2 and Fig. 3 compare the microstructure of the SnAg3.8Cu0.7 and SnAg3.0Cu0.5 solders. The near eutectic SnAg3.8Cu0.7 solder has a very heterogeneous microstructure (Fig. 2a). In the centre regions eutectic solidified solder can be found with very small (< 2 μm), disperse arranged, globular intermetallics (Fig. 2c, 2d). The outer regions have a coarse microstructure with large Ag_3Sn plates covered in β-Sn (Fig. 2b). These plates reach lengths up to 2 mm and a thickness up to 15 μm. Small Ag_3Sn needles and Cu_6Sn_5 intermetallics can be found between those large structures (Fig. 2b). There are also β-Sn dendrites in the microstructure.

The dominant phase in the microstructure of SnAg3.0Cu0.5 is β-Sn (Fig. 3). This phase grows dendritic into the solder and shows different growth directions. Between the dendritic structures, intermetallics of different sizes have been formed (Fig. 3c, 3d), which mark the eutectic.

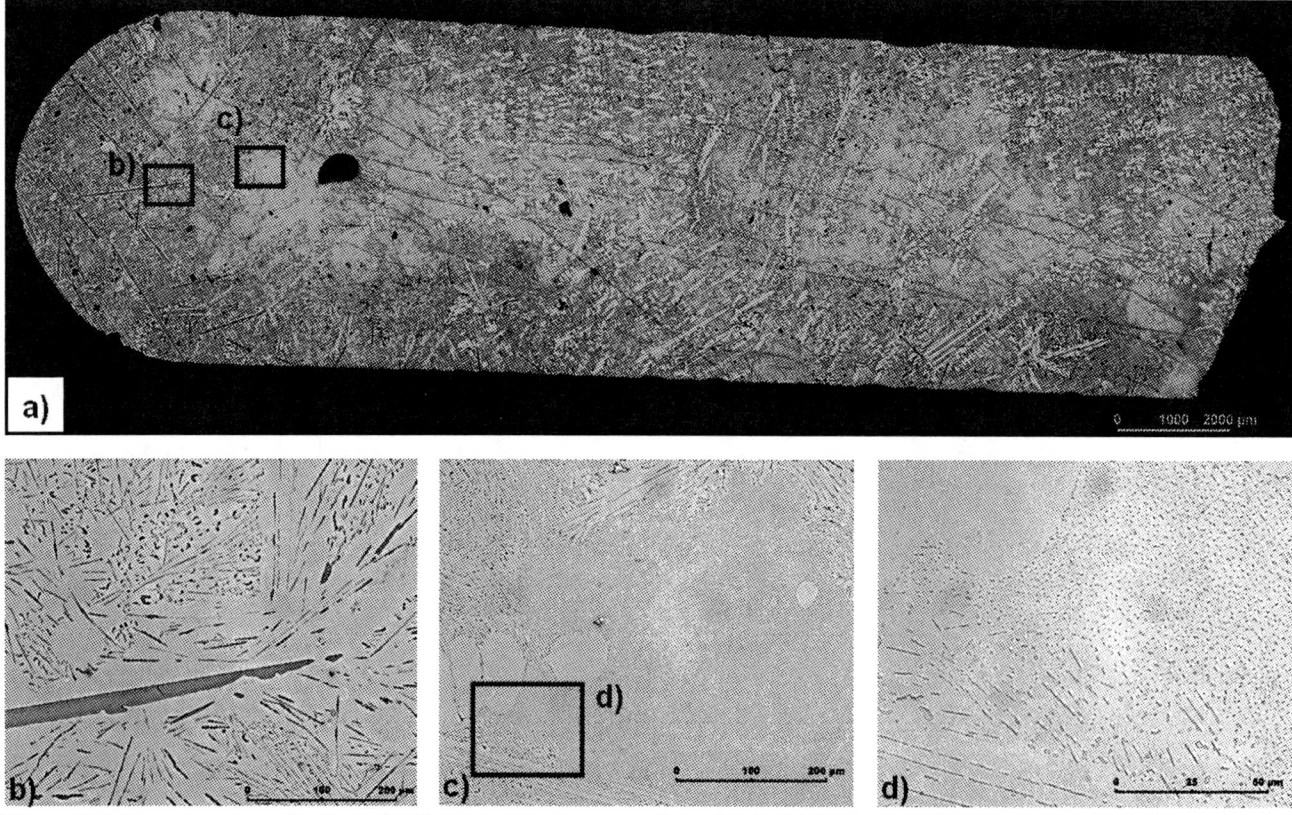

Figure 2: Microstructure of a SnAg3.8Cu0.7 solder ingot cooled with 0.0083 K/s (0.5 K/min)

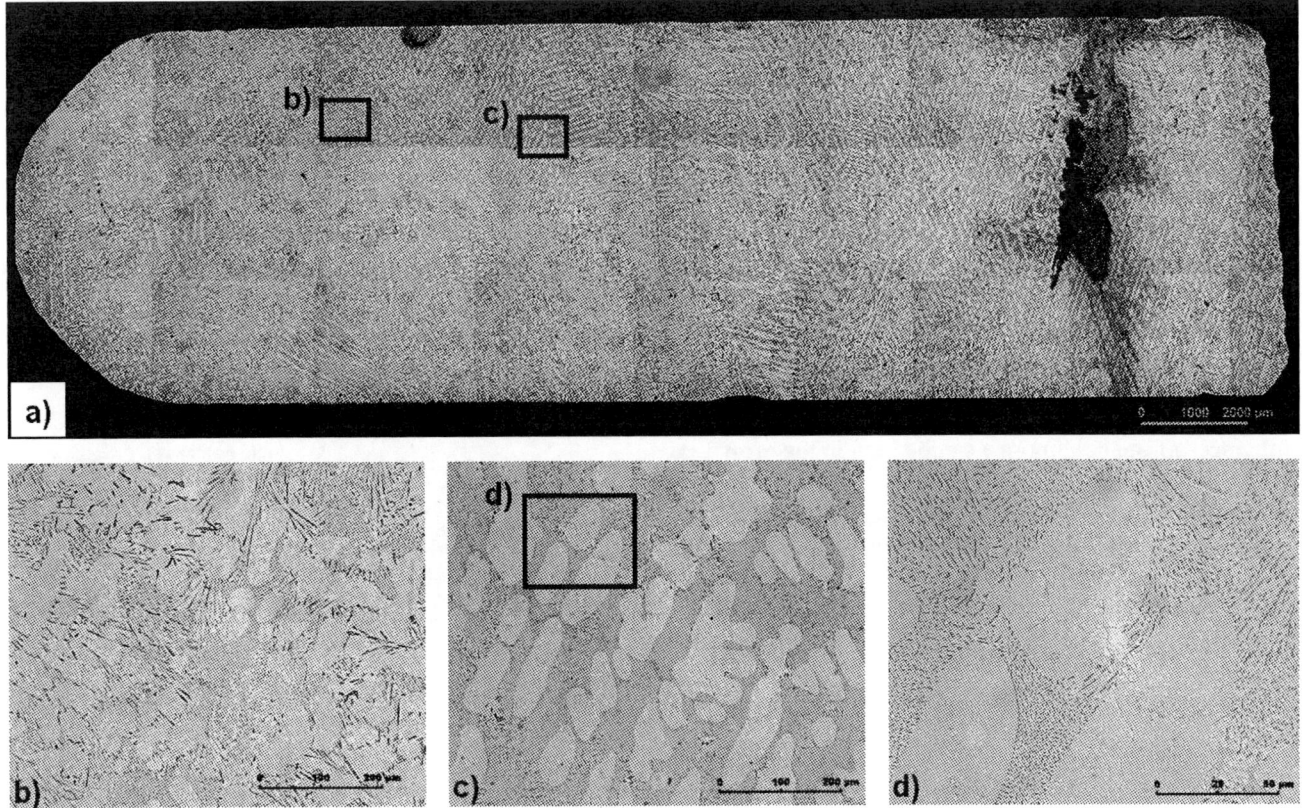

Figure 3: Microstructure of a SnAg3.0Cu0.5 solder ingot cooled with 0.1683 K/s (10.1 K/min)

SnAg3.8Cu0.7	SnAg3.5Cu0.4	SnAg3.0Cu0.5
-0.0083 K/s (-0.5 K/min)	-0.0058 K/s (-0.35 K/min)	-0.0056 K/s (-0.34 K/min)
-0.585 K/s (-35.1 K/min)	-0.55 K/s (-33 K/min)	-0.623 K/s (-37.4 K/min)

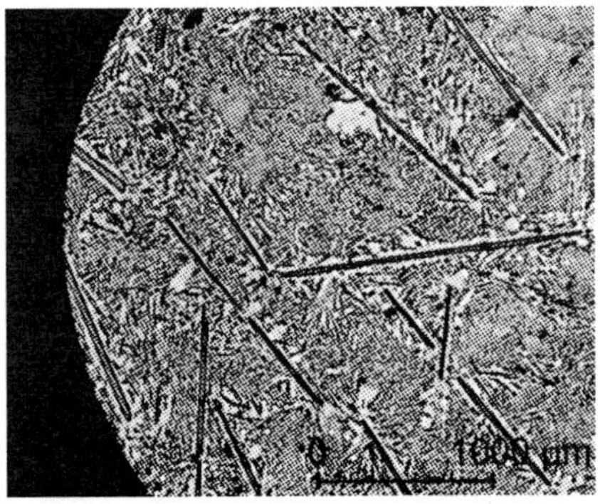

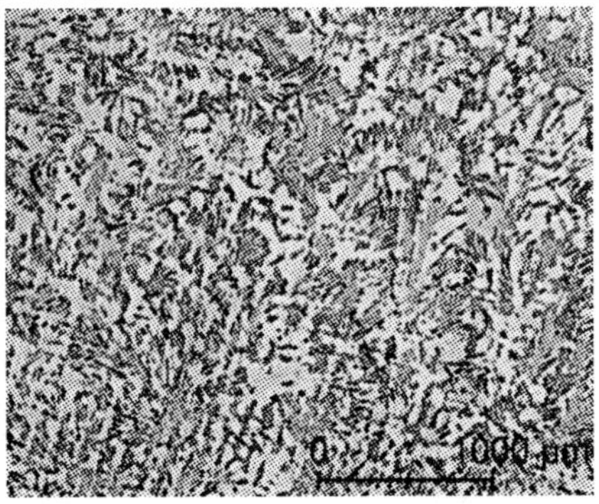

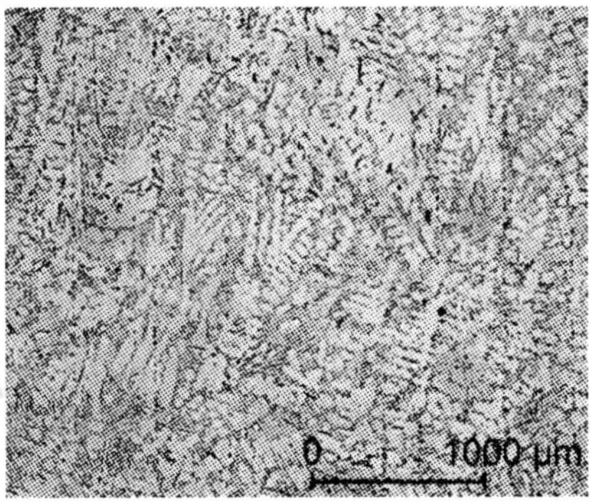

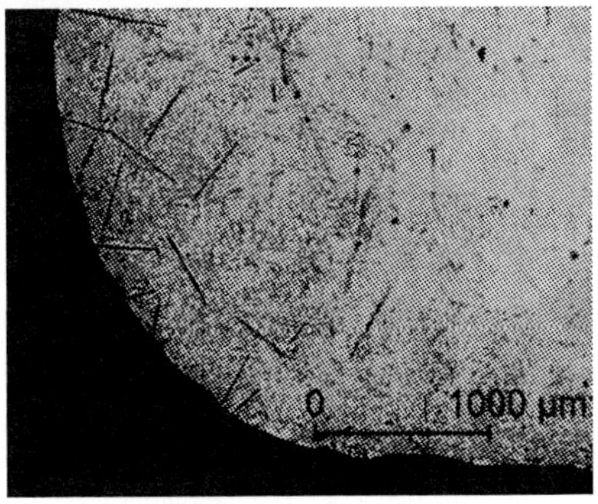

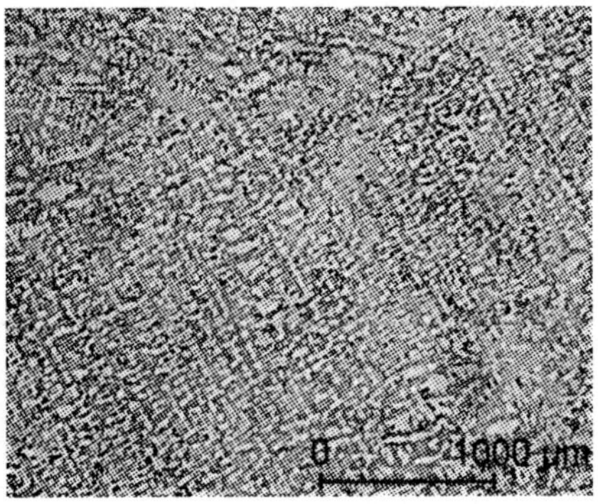

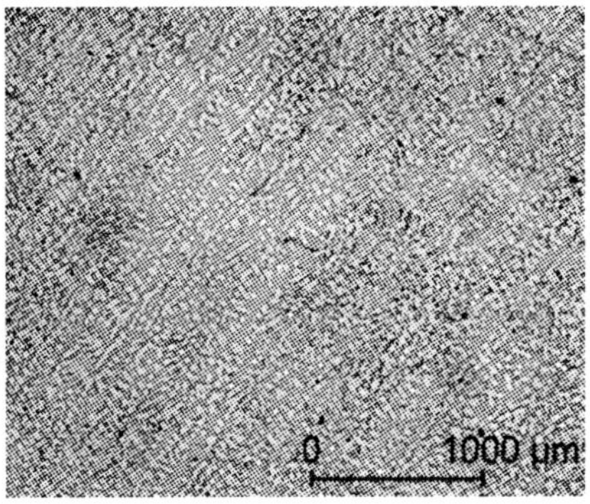

Figure 4: Influence of cooling rate and solder composition on the microstructure of SnAgCu solder ingots

The presence of primary solidified Ag₃Sn intermetallics can not be excluded. In Fig. 3b larger intermetallics are shown that may have nucleated as primary phases. But these are comparatively small to the large plates found in the SnAg3.8Cu0.7 solder.

The microstructure of the SnAg3.5Cu0.4 solder looks similar to the SnAg3.0Cu0.5 alloy (Fig. 4). Only the size of Ag₃Sn intermetallics is slightly larger.

The results of the experiments on solder composition and cooling rate are summarized in Fig. 4. By increasing the cooling rate, large Ag₃Sn intermetallics in the SnAg3.8Cu0.7 solder become smaller and the eutectic region in the centre of the specimen enlarges. In the SnAg3.0cu0.5 and the SnAg3.5Cu0.4 solders the influences of the cooling rate can be seen in the growth of the β-Sn dendrites. With faster cooling rates the thickness of the β-Sn decreases and the dendrites become longer. Faster cooling also leads to a more directed growth of the dendritic structures and decreases the size of intermetallics found between the dendrites.

3. Experiments on Solidification Front

Experimental Procedure

For experiments on directional solidification a test setup was developed that is able to generate a temperature difference inside a solder sample. The heating tool shown in Fig. 5 consists of five separate segments that can be heated independently. The segments are isolated from each other by a small gap of air (approx. 1 mm). This test setup can solidify a solder ingot with a length of 100 mm in a test tube (inner Ø 11 mm).

The experimental procedure to generate solidification from inside out is shown in Fig 6. In step 1, the liquid solder was filled in the preheated test setup with a temperature of 250 °C. After that, two thermocouples (K-type) were fixed inside the liquid solder, to measure the temperature gradient (step 2). The thermocouples are placed in the middle (segment 3) and at the top of the specimen (segment 1). At a stable temperature of 230 °C the heaters of segment 3 are turned off, causing a steady gradient of 2 K (step 3) between the middle and the outer segments. In step 4 the heating segments 2 and 4 are

turned off. The maximum stable gradient is about 5 K. In step 5 the segments 1 and 5 are turned off and the whole test setup cools down with a cooling rate of approx. 0.217 K/s (13 K/min). Finally the sample was cut in five pieces representing each segment.

Experiments on solidification front were carried out on three different alloys: SnAg3.0Cu0.5, SnAg3.0Cu0.5Au0.14 and SnAg3.0Cu1.5. The solders, which contain additional impurities of Au and Cu, were lab fabricated.

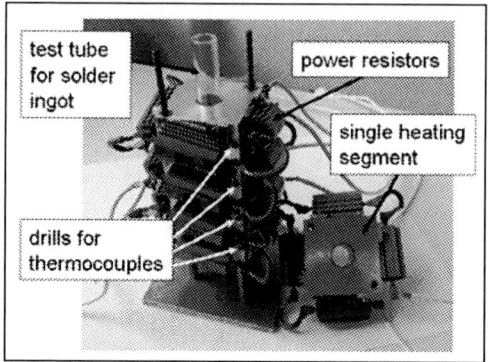

Figure 5: Five segment heating tool

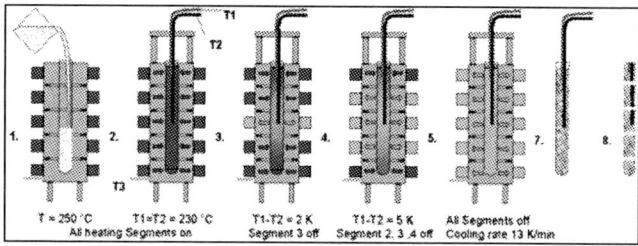

Figure 6: Experimental procedure for experiments on solidification front

Results

The result of the experiment on solidification front is shown in Fig. 7. The typical microstructure of the three interesting segments 1, 3 and 5 is shown for SnAg3.0Cu0.5. At the starting point of the solidification (segment 3 – Fig. 7b), a chaotic arrangement of β-Sn dendrites with no main growth direction can be found.

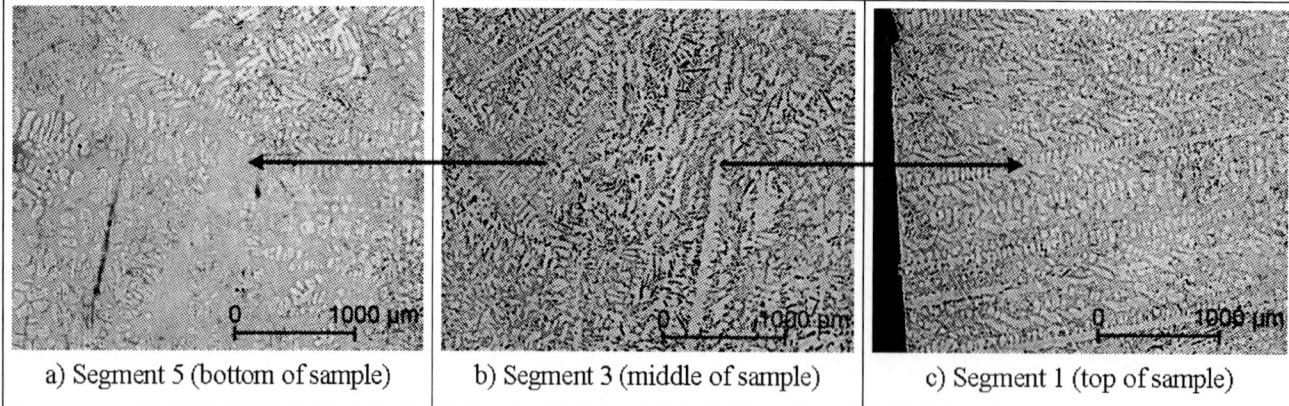

| a) Segment 5 (bottom of sample) | b) Segment 3 (middle of sample) | c) Segment 1 (top of sample) |

Figure 7: Microstructure of directional solidified SnAg3.0Cu0.5 (arrows indicate the planed direction of the solidification front)

200

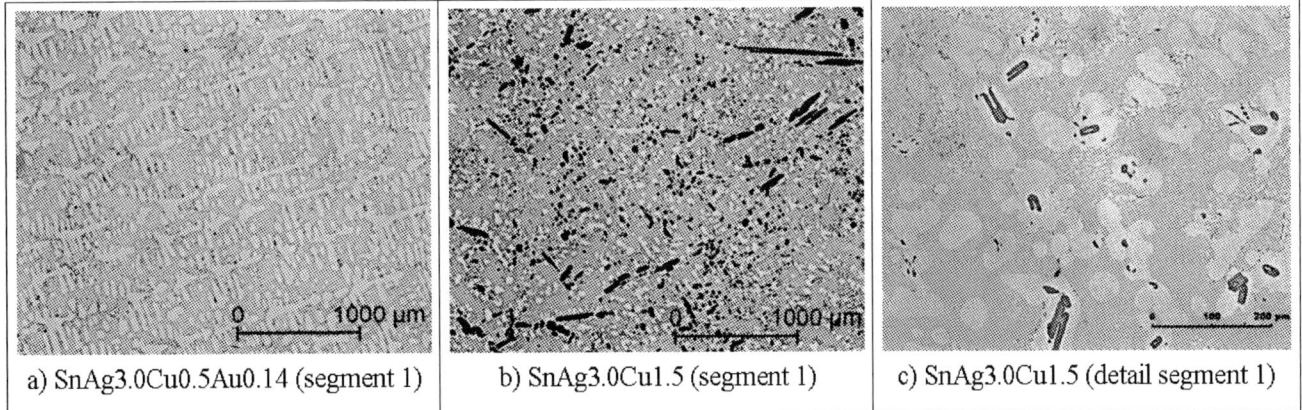

| a) SnAg3.0Cu0.5Au0.14 (segment 1) | b) SnAg3.0Cu1.5 (segment 1) | c) SnAg3.0Cu1.5 (detail segment 1) |

Figure 8: Influence of additional Au and Cu on the growth of dendritic β-Sn

The top of the sample (segment 1 – Fig. 7c) looks completely different. Large β-Sn dendrites are found aligned by the propagation direction of the solidification front. In segment 5 (bottom – Fig. 7a) a different microstructure was formed. It consists of smaller dendrites and large eutectic solidified regions.

The influence of additional Au and Cu on microstructure and dendritic growth is shown in Fig. 8. The SnAg3.0Cu0.5Au0.14 solder shows no significant differences in microstructure, compared to pure SnAg3.0Cu0.5. The formation of $AuSn_x$ intermetallics has no influence on dendritic growth. The solution of additional Cu into the solder causes the formation of large Cu_6Sn_5 intermetallics (Fig. 8b and 8c). Those structures are surrounded by pure β-Sn (Fig. 8c). The dendritic growth is disturbed by those primary intermetallics. There is also no solidification direction visible in the microstructure of the SnAg3.0Cu1.5 solder sample (Fig. 8b).

4. Experiments on SnAg3.0Cu0.5 Solder Balls

Experimental Procedure

The following experiments were carried out to investigate the solidification behaviour of small solder joints and its dependence on size and cooling rate. The joints were made of a SnAg3.0Cu0.5 solder bar, which was rolled to a foil with a height of 140 μm in order to punch preforms with different sizes out of it (Tab. 1).

Table 1: Size of the punched preforms and the investigated solder joints made of SnAg3.0Cu0.5

Preforms			Solder joint	
Ø [μm]	Heigth [μm]	Preforms per joint	Volume [μm³]	Ø [μm]
2500	140	1	6,87E+08	1100
700	140	2	1,08E+08	590
300	140	1	9,90E+06	270
100	140	1	1,10E+06	130

In order to get spherical joints with a defined cooling rate the test setup shown in Fig. 9 was used. The preforms were fluxed (WS600 - Alpha Metals) and placed in the holes of the ceramic. This setup was placed on the preheated heating plate at 250°C. The temperatures of the

heater and the ceramic were measured by K-type Thermocouples. After 5 min at 250°C the ceramic was removed and cooled down naturally by using different thermal masses and cooling methods (Fig. 10).

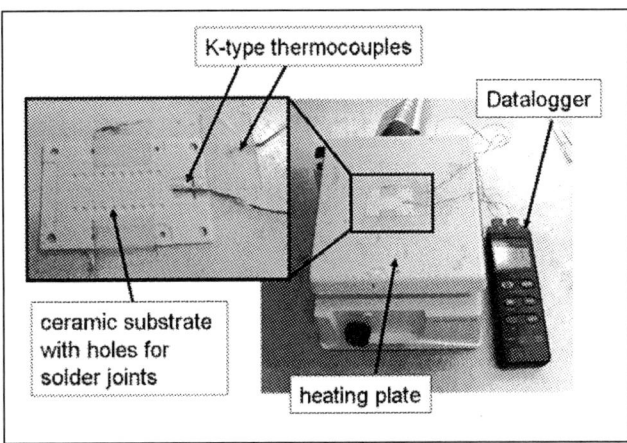

Figure 9: Test setup for experiments on solder balls

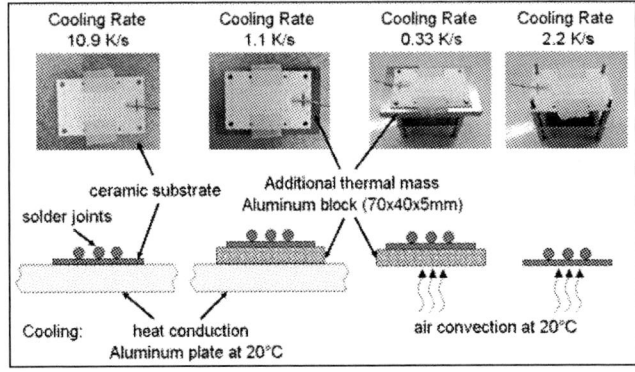

Figure 10: Realisation of different cooling rates

A comparison of the different ball sizes is shown in Fig. 11. The possible cooling profiles with rates from 0.33 K/s to 10 K/s (mean value between 240 °C and 140 °C) are shown in Fig. 12.

Cross sections of six joints per size and cooling rate were prepared and investigated by polarized light microscopy in order to get information on grain orientation.

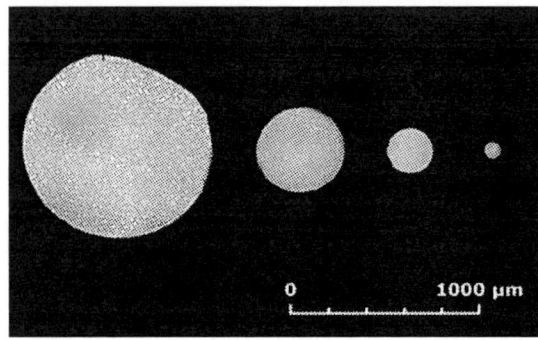

Figure 11: Comparison of the different ball sizes

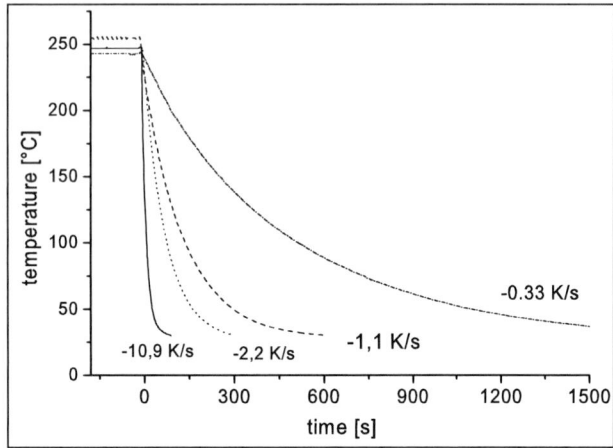

Figure 12: Cooling profiles of solder ball experiments

Results

The influence of cooling rate on the microstructure is shown in Fig. 13. A change in cooling rate does not cause a significant difference on phase sizes and dendritic growth in the microstructure of the investigated balls. The only difference caused by cooling rate was investigated for small balls at slow cooling. The microstructure of

these balls contains sometimes large intermetallics, which have formed as primary phases. The appearance of those intermetallics has been investigated on small solder balls (590 μm, 260μm and 130 μm), which were cooled down with rates of 0.33 K/s and 1.1 K/s.

The influence of cooling rate on grain orientation is shown in Fig. 14 with two solder balls per size and cooling rate. It is remarkable that balls which were solidified under similar conditions sometimes look completely different (Fig. 12 – e.g. 1100 μm balls at 0.33 K/s or 130 μm balls at 1.1 K/s).

Sometimes those balls seem to solidify with large areas with different grain orientations and sometimes with small randomly, orientated grains. It is also possible that a solder bump consists of both types of grain structures. A change in cooling rate seems to have no significant influence on that (Fig. 14).

Fig. 13 and Fig. 15 show a comparison of the microstructure of the different ball sizes. The shown pictures represent the general microstructure that is found in the balls of the accordant size. The dominant phases in the larger balls (Ø 1100 μm and Ø 590 μm) are β-Sn dendrites. The smaller balls show a finer microstructure with small precipitations and less Sn dendrites. Fig. 16 shows that those small randomly, orientated grains mark areas with a very fine microstructure. Regions were β-Sn Dendrites are formed show major grain orientations.

Fig. 14 and Fig. 15 show that the areas of small randomly, orientated grains become more and more dominant by decreasing the solder volume. The results on the smallest investigated ball size indicate that 3 out of 4 balls completely consist of small randomly, orientated grains. Fig. 13 and Fig. 15 also show that microstructure becomes finer by decreasing the ball volume. This effect is not influenced by cooling rate.

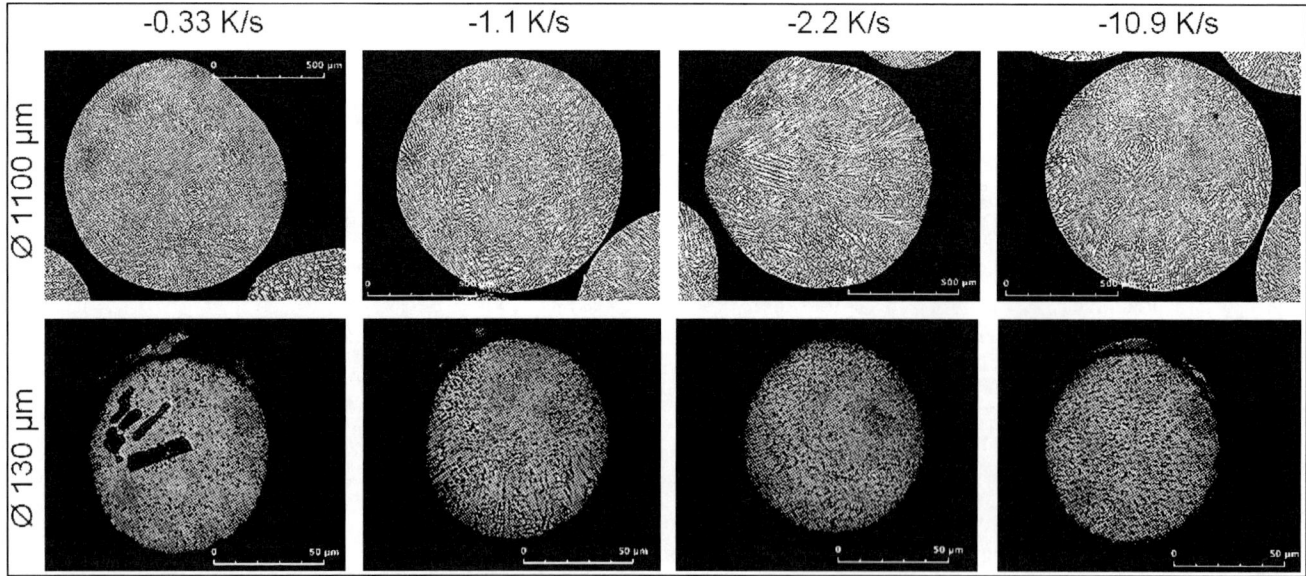

Figure 13: The influence of cooling rate on the microstructure of SnAg3.0Cu0.5 solder balls with different sizes (upper line: Ø 1100μm; lower line: Ø 130μm)

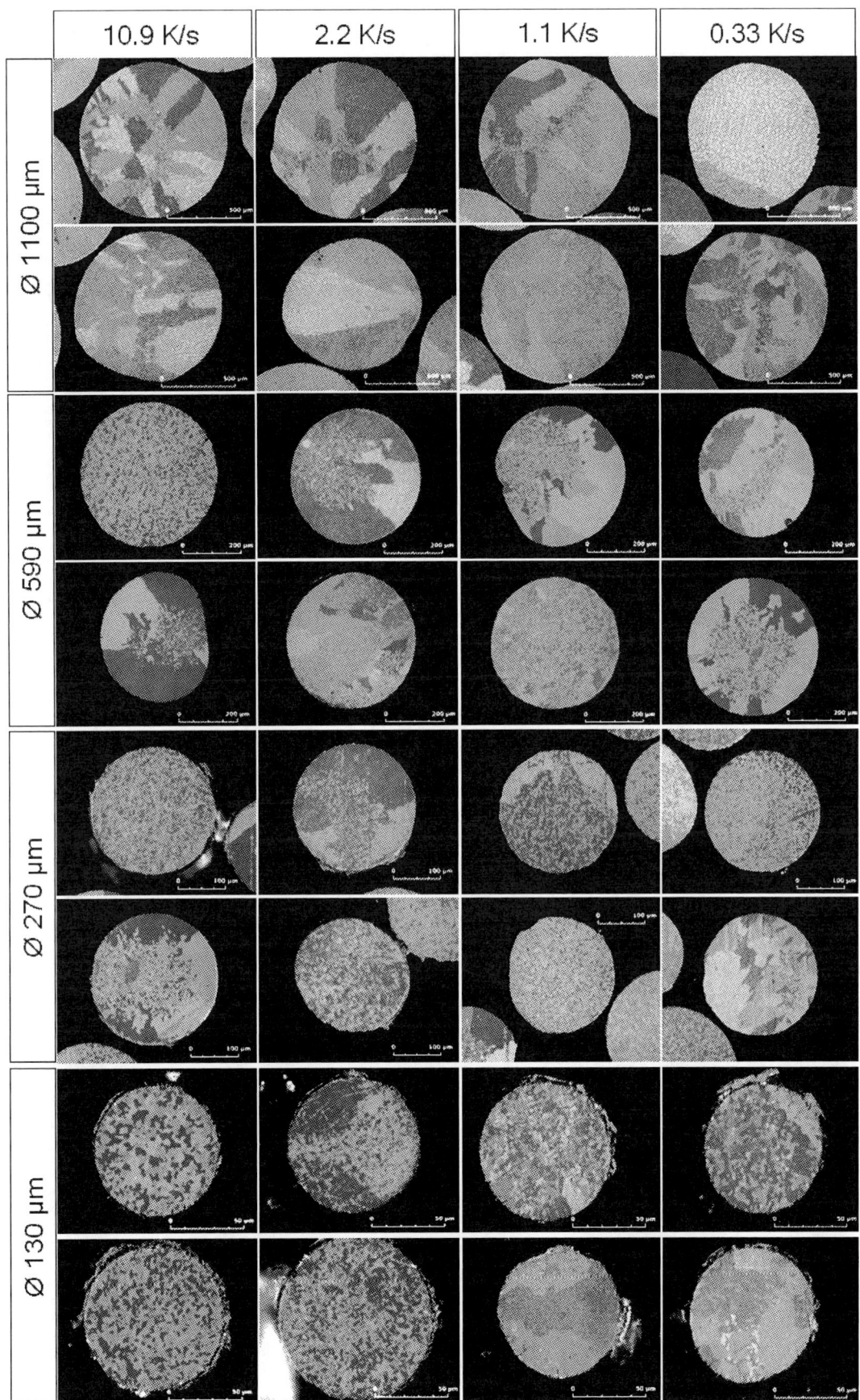

Figure 14: Dependence of cooling rate and solder volume on grain orientation of SnAg3.0Cu0.5 solder balls (polarized light microscopy)

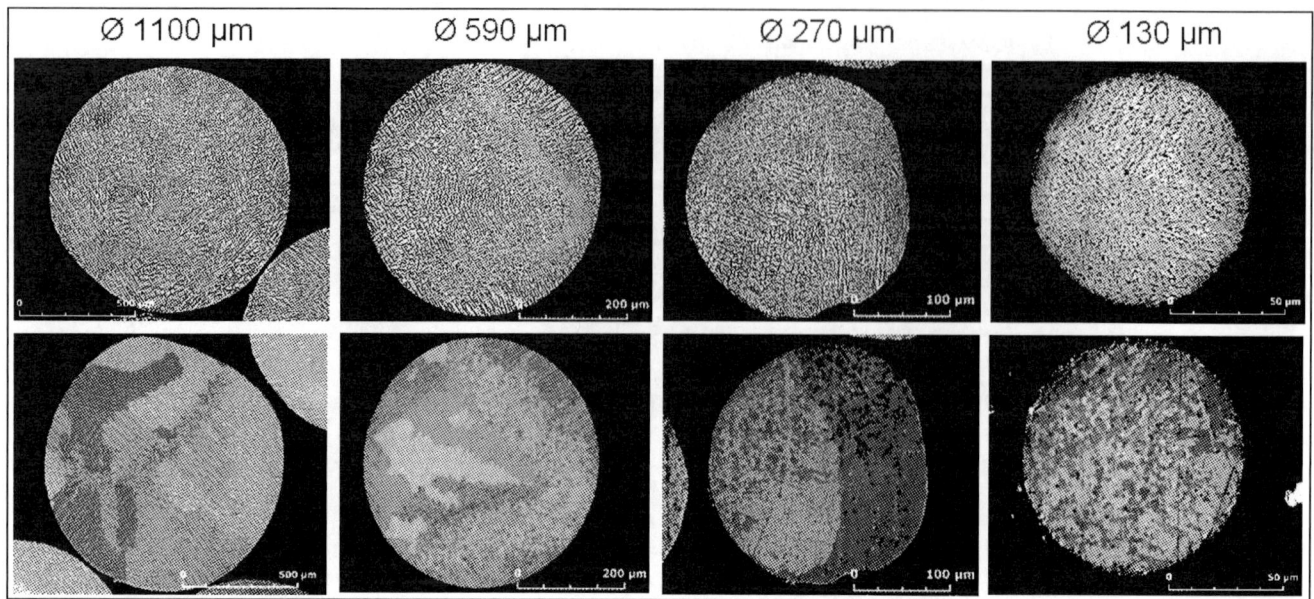

Figure 15: Influence of the solder volume on the microstructure (upper line) and the grain orientation (lower line) of SnAg3.0Cu0.5 solder balls cooled with 1.1 K/s

In Fig. 16 the microstructure of an Ø 1100 µm ball cooled with 1.1 K/s is shown. Fig. 16a-c show the grain orientation and Fig. 16d-f show the corresponding details on microstructure of the ball. In Fig. 16a two completely different areas can be seen. There is a fine structured region with small randomly, orientated grains (Fig. 16b). This region is surrounded by six large, coarse areas characterized by three major grain orientations, which are represented by the different greyscales (black, grey, white - Fig. 16a, Fig. 16c). It is remarkable that those major grain orientations correspond with those in the fine structured region (Fig. 16b). It has also been investigated that a large number of the investigated balls showed only three mayor grain orientations (Fig. 14). Fig 16b and Fig 16e show details of the fine structured area. It can be seen, that the microstructure coarsens from inside out, i.e. the grain size and the size of the intermetallics increase from the inside of this region to the outside. At the edge of this area β-Sn dentrites can be found and the fine microstructure suddenly turns into coarse and conical growing areas with major grain orientations. Fig. 16a also shows that those coarse areas expand with increasing

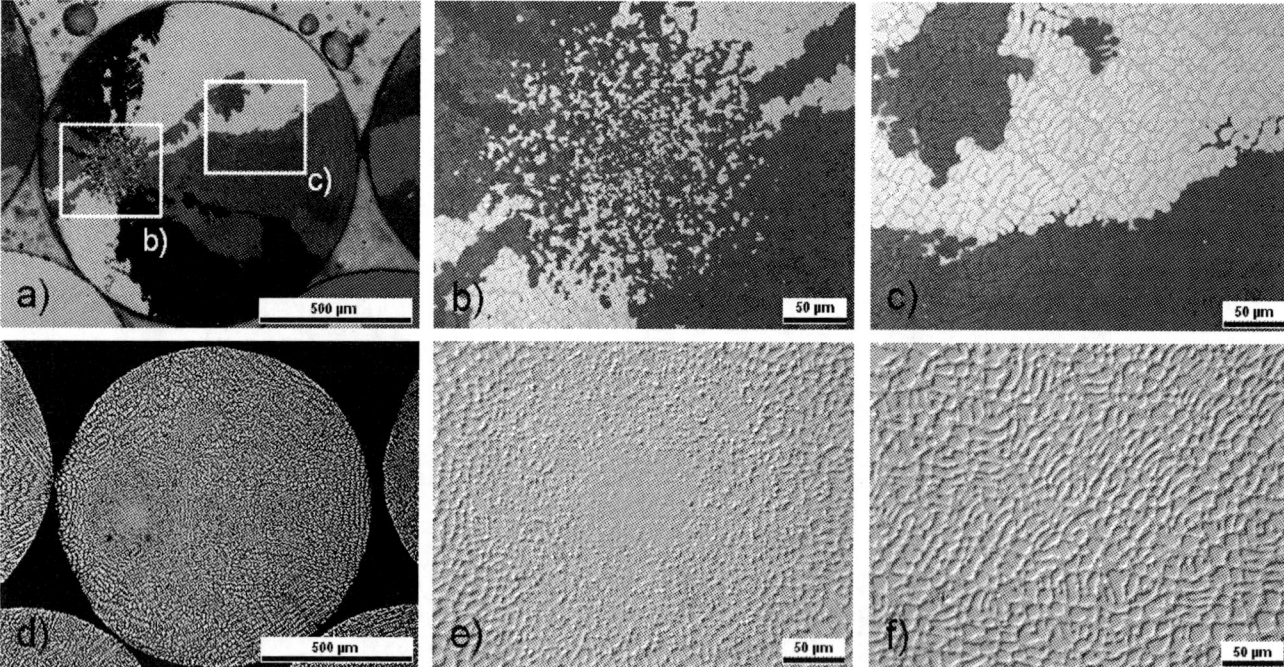

Figure 16: Grain orientation (a, b, c: polarized light microscopy) and corresponding microstructure (d: bright field microscopy; e, f: Differential Interference Contrast) of an as cast SnAg3.0Cu0.5 solder ball with Ø 1100µm, cooled with 1.1 K/s;

distance from the fine structured region. Fig. 16c and Fig. 16f show details on the microstructure of the coarse area. It consists of β-Sn dendrites with eutectic solidified regions in between. Fig. 16a also shows that the grain orientation of the coarse, dendritic areas is mirrored by the fine structured region. That means that opposite dendrites have the same grain orientation.

The fine structured region has not to be spherical as Fig. 16b indicates. Other investigated balls showed ring shaped areas with the same type of fine microstructure (see Fig. 14 and Fig. 15).

5. Discussion

A comparison between the microstructure of the large SnAg3.0Cu0.5 solder ingots (V≈8.85E+11 μm³) in Fig. 3 and the small SnAg3.0Cu0.5 solder balls (V<6.87E+08 μm³) in Fig. 13 and Fig. 15 reveals strong differences even at similar cooling rates. Indeed the dominating phase in these microstructures is dendritic β-Sn but the size of the dendrites and the intermetallics are much smaller in the solder balls compared with those found in the ingots. This indicates a scaling effect of the microstructure depending on solder volume. Another significant difference, that was not expected, is the size of the eutectic solidified regions between the β-Sn dendrites. The analysis on the microstructure of the solder ingots showed that these regions become larger with higher cooling rates. The eutectic solidified regions in the solder balls are much smaller although the cooling rate was much faster. Furthermore the microstructure of the solder ingots is highly depending on cooling rate, while the solder balls showed no significant difference in phase size and distribution at different cooling gradients. Another discrepancy is the appearance of large intermetallics in the microstructure of slow cooled solder balls. The growth of primary solidified intermetallics of that size (compared to the solder volume) has not been investigated in the microstructure of the SnAg3.0Cu0.5 solder ingots at all cooling rates.

These differences indicate a different solidification behaviour that is not explainable by a scaling effect. A possible reason for the unequal behaviour depending on solder volume might be the difference in undercooling, as mentioned in [16]. Lehman et al. explained that undercooling is proportional to the inverse ball size. This suggests that large solder volumina behave more similar to the phase diagram than small solder volumina. That means that the process of solidification is more dynamic in small solder balls and the influence of undercooling becomes more and more important while the dependence on cooling rate decreases.

The appearance of primary solidified intermetallics in solder balls and its dependence on cooling rate has also been investigated in [1, 2]. Those studies come to the result that at slow cooling gradients the formation of primary intermetallics becomes more likely. The reason for that phenomenon is the elapsed time for cooling down from melting point to the necessary temperature for the nucleation of β-Sn [2]. This time is longer for slower cooling rates and therefore the intermetallics have more time to grow and become larger.

The strong difference in microstructure of large solder ingots and small solder balls leads to the conclusion that this might cause differences in material behaviour. In [16] material data was measured on large bulk specimen and small Flip-Chip bumps. The comparison of the data showed strong differences. The results in the present paper indicate that this may be caused by the different microstructure. This means large solder volumina do not reflect the microstructure and material behaviour of small solder volumina and the specimen size for creep measurements has to be chosen for the accordant case in the simulation. This makes the acquisition of material data much more complicated.

Interpretations of the results, which are shown in Fig. 16, allow a possible reproduction of the solidification process of the solder ball. The fact that the coarse areas (Fig. 16a) grow conical and expand with increasing distance from the fine structured region allows the conclusion that solidification starts in the fine structured areas. The subsequent dendritic growth depends on conditions that have to be fulfilled for the surrounding melt. These conditions for dendritic growth seem to be less often reached in small solder volumina.

It is assumable that every solder ball has one or more fine structured regions, where the solidification process starts. This area is supposed to be the best cooled inside the ball, referring to the experiment on solidification front. This position could be the contact zone between solder and ceramic substrate for the experiments presented in this study. In real interconnects this zone is supposed to be on the interface. Cross section pictures in [16] show this effect.

The dendritic growth around the solidification centre is also characterized by a symmetric arrangement of the grain orientations (Fig. 16a and Fig 14). This phenomenon is described as cyclic twinning [17]. Bieler et al. showed that the random grain orientations form twins that are arranged in a nominal angle of 60°.

6. Conclusions

The influences of cooling rate and composition on the microstructure of SnAgCu solders have been investigated by solidification experiments on bulk solder ingots (V≈8.85E+11 μm³) at cooling rates from -0.0056 K/s to -0.623 K/s.

Faster cooling decreases the size of Ag_3Sn intermetallics and β-Sn dendrites.

A silver content above 3.5 wt% causes the formation of large primary Ag_3Sn intermetallics. SnAgCu solders with more than 1 wt% Cu solidify with large primary Cu_6Sn_5 intermetallics in the microstructure. The formation of primary intermetallics slows down the dendritic growth of β-Sn.

Experiments on directional solidification have been carried out in order to investigate the growth of β-Sn dendrites. The formation of this phase can be directed by temperature gradients inside the solder.

In order to compare these results on solder ingots with the microstructure in small SnAg3.0Cu0.5 solder balls, solidification experiments with ball sizes of Ø 1100 μm, Ø 590 μm, Ø 270 μm and Ø 130 μm have been carried out at different cooling rates from 0.33 K/s to 10.9 K/s.

The analysis of the microstructure showed that the formation of large primary intermetallics increases with slower cooling rates and smaller ball sizes.

There is no significant influence of the cooling rate on the grain orientation and dendritic growth of β-Sn found in these small solder bumps.

The experiments show that the microstructure becomes finer and the formation of small, randomly oriented grains becomes more frequent with smaller solder volume.

References

1. Kang, S.K.; Lauro, P.A.; Shi, D.-Y.; Henderson, D.W.; Gosselin, T.; Sarkhel, A.; Goldsmith, C.; Puttlitz, K.J.: "Ag₃Sn Plate Formation in the Solidification of Near-Ternary Eutectic Sn-Ag-Cu" *Journal of Metals*, (6/2006), pp. 61-65.
2. Lehman, L.P.; Kinyanjui, R.K.; Zavalij, L.; Zribi, A.; Cotts, E.J.: "Growth and Selection of Intermetallic Species in Sn-Ag-Cu No-Pb Solder Systems based on Pad Metallurgies and Thermal Histories" *Proc IEEE 53th Electronic Components and Technology Conference,* New Orleans, Louisiana, 2003, pp. 1215-1221.
3. Ochoa, F.; Williams, J.J.; Chawla, N.: "Effects of Cooling Rate on the Microstructure and Tensile Behavior of a Sn-3.5wt.% Solder" *Journal of Electronic Materials*, Vol. 32 (12/2003), pp. 1414-1420.
4. Kang, S.K.; Choi, W.K.; Shi, D.-Y.; Henderson, D.W.; Puttlitz, K.J.: "Microstructure and mechanical properties of lead free solders and solder joints used in microelectronic applications" *Journal of Research and Development*, Vol. 49 (2005), pp. 607-619.
5. Metallurgy division of the Material Science and Engineering Laboratory, "Ag-Cu-Sn Phase Diagram & Computational Thermodynamics", *http://www.metallurgy.nist.gov/phase/solder/agcusn.html,* Date: 2006-04-11.
6. Lu, H.Y.; Balkan, H.; Ng, K.Y.S.: "Effect of Ag content on the microstructure development of Sn-Ag-Cu interconnects" *Journal of Material Science: Mater Electron,* Vol. 17 (2006), pp. 171-188.
7. Kang, S. K.; Shi, D.Y.; Fogel, K.; Lauro, P.; Yim, M.-J.; Advocate, G.G. Jr.; Griffin, M.; Goldsmith, C.; Henderson, D.W.; Gosselin, T.A.; King, D.E.; Konrad, J.J.; Sarkhel, A.; Puttlitz, K.J.: "Interfacial Reaction Studies on Lead (Pb)-Free Solder Alloys" *IEEE Transactions on Electronics Packaging Manufacturing*, Vol. 25 (2002), pp. 155-161.
8. Haasen, P. *Physikalische Metallkunde 3., neubearbeitete und erweiterte Auflage*, Springer (Berlin, Heidelberg, 1994), pp. 188-212.
9. Yoshioka, H.; Tada, Y; Kunimine, K.; Furuichi, T.; Hayashi, Y.: "Heat transfer and solidification processes of alloy melt with undercooling: I. Experimental Results" *Acta Materialia*, Vol. 54 (2005), pp. 757-763.
10. Ebersberger, B.; Bauer, R.; Alexa, L.: "Qualification of SnAg Solder Bumps for Lead-free Flip Chip Applications" *Proc IEEE 54th Electronic Components and Technology Conference*, Las Vegas, NV, 2004, pp. 683-691.
11. Morris, James E.: "Nanopackaging and Electronics Packaging" *Proc IEEE 1st Electronic Systemintegration Technology Conference,* Dresden, 2006, pp. 873-880.
12. Albrecht, H.-J.: "Advanced Packages and Board Level Reliability" *Proc 1st Electronic Systemintegration Technology Conference,* Dresden, NV, 2006, pp. 1108-1117.
13. Stoyanov, S.; Bailey, C.: "Optimisation Modelling for Design of Advanced Interconnects" *Proc 1st Electronic Systemintegration Technology Conference,* Dresden, NV, 2006, pp. 203-208.
14. Aggarwal, Ankur O.; Raj, P. Markondeya; Sundaram, Venky; Ravi, D.; Koh, Sauwee; Mullapudi, Ravi; Tummala, Rao R.: "50 Micron Pitch Wafer Level Packaging Testbed with Reworkable IC-Package Nano Interconnects" *Proc IEEE 55th Electronic Components and Technology Conference,* Orlando, FL, 2005, pp. 1139-1146.
15. Wiese, S.; Roellig M.; Wolter, K.-J.: "Creep of Thermally Aged SnAgCu-Solder Joints" *Proc IEEE 54th EuroSimE 2005 Conference,* Berlin, 2005, pp. 79-85.
16. Lehman, L.P.; Kinyanjui, R.K.; Wang, Y.; Xing, L.; Zavalij, P.; Borgesen, P.; Cotts, E.J.: "Microstructure and Damage Evolution in Sn-Ag-Cu Solder Joints" *Proc IEEE 55th Electronic Components and Technology Conference,* Orlando, FL, 2005, pp. 674-681.
17. Bieler, T.R.; Jiang, H.; Lehman, L.P.; Krikpatrick, T.; Cotts, E.J.: "Influence of Sn Grain Size and Orientation on the Thermodynamical Response and Reliability of Pb-free Solder Joints" *Proc IEEE 56th Electronic Components and Technology Conference,* San Diego, CA, 2006, pp. 1462-1467.

Electromigration in Large Volume Solder Joints

Karsten Meier[1], Mike Roellig[1], Steffen Wiese[1], Carsten Goette[2], Ulrich Deml[2] and Klaus-Juergen Wolter[1]

[1] Dresden University of Technology, Electronics Packaging Laboratory, Germany

[2] Siemens AG, Siemens VDO Automotive, Regensburg

TU Dresden, IAVT, D-01062 Dresden, Germany

meier@avt.et.tu-dresden.de, phone: +49 351 463 35291, fax: +49 351 463 37035

Abstract

Among the various reliability related concerns in automotive and power electronics, electromigration in solder joints might be one of the least investigated degradation effects. It requires high current densities that usually only occur in small flip chip solder joints. Even if the current densities in power electronic components, like IGBTs (TO 263), cannot be compared with that in ultra small flip chip joints, there is an enormous thermal factor that helps to promote electromigration processes even in the larger volume solder joints of power electronic components.

The scope of the paper is to present the results of electromigration and thermomigration experiments on large volume solder joints of power electronic components (TO packages). Two solder alloys – SnAgCu and SnPb – were tested for electromigration behaviour. An experimental setup will be described that allows electromigration testing at very high absolute currents (up to $I = 100$ A) which are necessary for electromigration testing of these large volume solder joints. The numerous difficulties of transmitting such high currents to the solder joints of the relevant TO-components will be discussed. Thermographic analysis of the test structures will be presented that show accurate adjustment of test temperatures at the solder joints of the TO-components.

Metallographic analysis show the migration of lead metallisation into the solder joints and therefore grow of intermetallic phases. However, the degradation of the large volume joints is different from that of ultra small flip chip. Instead of severe Kirkendall voiding there is rather changes in the microstructure of the solder. The most important degradation mechanisms are the formation of intermetallic phases in the solder and phase separation.

The paper will describe the electromigation processes in the tested large volume joints. The degradation processes will be compared to that of degradation by thermomigration, in order to separate the electrical driven effects from the thermal effects. The results on large volume joints will be compared to that of small volume joints. Also a comparison between SnAgCu and SnPb solder will be presented.

1. Introduction

Electromigration became a reliability issue in microelectronic devices first. Due to the very small cross sectional area of a interconnection line in a semiconductor component very high current densities are reached at low currents even without taking current crowding into consideration [1, 2]. However, in a microelectronic device a interconnection line in general is a encapsuled monolythic system made of Aluminium or Copper. In numbers of research activities constitutive equations and effects of electromigration where studied and published for this scenario [3, 4]. In these works the electron wind is pointed out to be the main driver for migration processes concluding that there is a dependence of electromigration effects on the direction and density of the electron flow. In addition the temperature and the mechanical stress are known to have an influence too. Kirkendall voiding and mechanical stress are the main failure sources for interconnection lines.

Due to the ongoing scale down of solder joint size in FBGA, CSP and Flip Chip components, for example, electromigration became a reliability risk for the component level too. Geometrical circumstances at the connection of trace and solder joint lead to local current density concentrations known as current crowding (Fig. 1) [5]. In contrast to a monolythic conductor in a solder joint a material composition is treated by the electrical current and accelerated growth and seperation of intermetallic phases are reliability concerns as well as Kirkendall voiding.

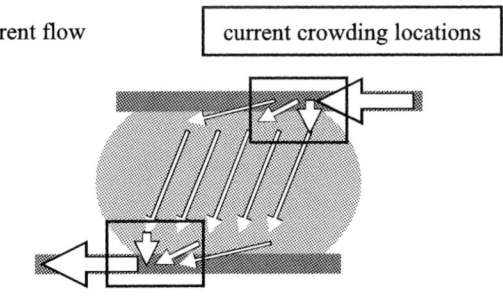

Figure 1 Current crowding locations in a Flip Chip solder joint

In a large volume solder joint much lower current densities are assumed due to the bigger geometrical dimensions. Thus very high currents are necessary to reach high current densities at local maxima. Components in power electronic or automobile applications for example are loaded with extreme high currents [6]. Nevertheless, the current densities in component solder joints of such devices still do not match the current densities in small Flip Chip joints or even monolythic interconnection lines. High temperatures can activate or accelerate electromigration processes or enable thermomigration though. Joule heating or aggravate

1-4244-1105-X/07/$25.00 ©2007 IEEE

ambient conditions are major heat sources (e. g. in so called under-the-hood-applications).

Whereas Kirkendall voiding, as an effect of electromigration, is a serious problem in microelectronic devices and very small solder joints, the growth and seperation of intermetallic phases needs to be observed in large volume solder joints. Intermetallic phases have much different mechanical properties to those of the solder and so they are expected to impair the reliability of the joint.

Apparently there is a need for the investigation of electromigration and its effects on large volume solder joints at least for applications in the power electronics and automotive field.

2. Specimen Design and Experimental Setup

To investigate electromigration effects on real TO-components a FR4-PCB test board has been designed to enable a high current load on a number of components at a time as well as easy handling and seperation for metallographic analysis (Fig. 2). Thick Cu traces on the subsrate were choosen to keep the current density in the trace low. The lead material of the TO-components was Copper/Nickel/Tin for SnAgCu solder or Copper/Tin for SnPb solder metallisation. Each test board was assembled with 7 TO-components in a serial chain. Therefore, one lead was connected to the plus the other to the minus terminal of the current source. Because of this configuration it is possible to investigate both current flow directions on each component.

Figure 2 FR4-test board for electromigration experiments on real TO-components

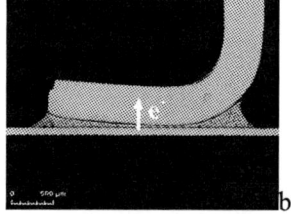

Figure 3 TO263-component with Cu-bar bypass (a) and electron flow through the joint connected to the current source minus terminal (b)

The solder joint connected to the current source minus terminal is stressed with a current flowing from the TO-component lead through the solder joint into substrate trace. The actual electron flow has the opposite direction

(Fig. 3-b). The solder joint connected to the current source plus terminal is stressed with a contrary current and electron flow respectively. To prevent the actual semiconductor device from damage the TO-component leads were shortened by a Copper bar (Fig. 3-a).

Two solder compositions were selected to be tested in all experiments. A SnAgCu solder composition as a common used lead-free solder was selected to investigate the effects of electromigration on the growth and seperation of intermetallic phases. To investigate these effects on a SnPb solder specimen with SnPb solder were prepared too. The specimen assembling was done on an industrial reflow line to get solder joints equal to those that are commercially produced.

The specimen were mounted on encased heat stages to ensure a homogenous temperature distribution. The use of the heat stages allowed experiments at controlled elevated temperatures. Two specimen can be mounted on one heat stage to achieve equal experimental conditions and to extend the capacity of the setup. The experimental setup is equiped with DC current sources providing currents up to $I = 60$ A at up to $V = 15$ V.

In order to avoid undesirable Joule heating solid state relays with a switching capability of $I = 100$ A have been integrated in the test system. Triggered by a pulse width modulated signal they are used to reduce the duty cycle T.

$$T = \frac{t_{ON}}{t_{ON} + t_{OFF}} \qquad (1)$$

The complete experimental setup including current sources, heat stages and temperature and duty cycle control units is shown in figure 4.

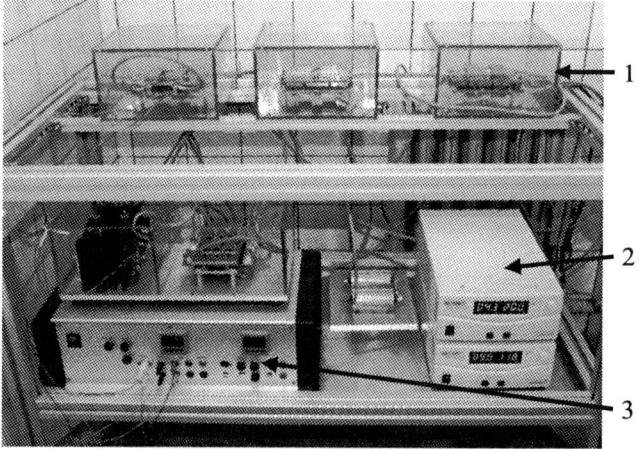

Figure 4 Setup for electromigration experiments: 1 - heat stages, 2 - current sources, 3 - temperature and duty cycle control units

Using a infrared camera system the temperature of the specimen, caused by Joule heating, was observed while the duty cycle was varied. In figure 5-a and 5-b the temperature distribution on a specimen for duty cycles of $T = 1$ and $T = 0.1$ at a certain current are shown. It can be

208

seen that there is a homogeneous temperature distribution over the whole specimen. Molten solder indicated that in contrast to the maximum temperature, pictured in figure 5-a, the specimen temperature exceeds the liqiudus temperature of the solder without reducing the duty cycle. Displaying a temperature of about $\vartheta = 80\ °C$ is caused by the saturation effect of the camera sensor. Hence reducing the duty cycle gets essential. Using a duty cycle of $T = 0.1$ the maximum specimen temperature could have been decreased to $\vartheta = 42\ °C$.

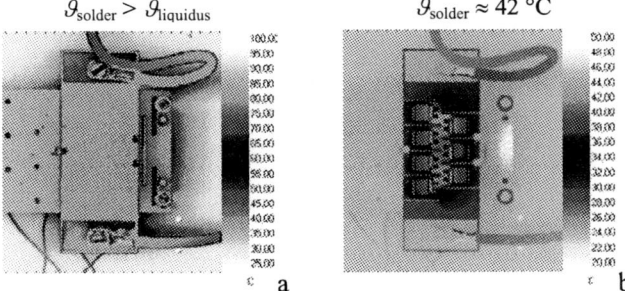

$\vartheta_{solder} > \vartheta_{liquidus}$ $\vartheta_{solder} \approx 42\ °C$

a   b

Figure 5 Temperature distribution over a electro-migration specimen for a certain current with T = 1 (a) and 0.1 (b)

Shorten the duty cycle reduces Joule heating essentially but increases the duration of a experiment by the factor f_T.

$$f_T = \frac{1}{T} \qquad (2).$$

Therefore, the minimum temperature for the experiment was specified to be $\vartheta = \vartheta_1$ inducing the shortest duty cycle of the highest current of $I = I_3$ to be $T = 0.2$ and so $f_T = 5$.

3. High Temperature Storage

High Temperature Storage tests were performed on SnAgCu and SnPb specimen in addition to electromigration experiments. Their results were used to specify temperature driven degradation processes and separate these from electrical driven degradation processes. High Temperature Storage was carried out at a temperature ϑ_{HTS} over a time of $t_{HTS} = t_{HTS-1}$, t_{HTS-2} and t_{HTS-3} [6]. Metallografic methods were used to analyse the high temperature stressed specimen.

4. Experiments on Electromigration

Electromigration experiments were performed under several current and temperature conditions. The current was varied between $I = I_1$, I_2 and I_3 ($I_1 < I_2 < I_3$) to achieve different current densities. For calculating the current densities the bottom area of a TO-component (0.8×1.6 mm²) lead was used. A current density of $S = 10$ kA/cm² was found to be the critical value to activate electromigration processes in interconnect lines

in [2]. By the combined usage of duty cycle controlled Joule heating and heat stages, the specimen temperature was constantly kept at $\vartheta_{EM} = \vartheta_{EM-1}$, ϑ_{EM-2} and ϑ_{EM-3} ($\vartheta_{EM-1} < \vartheta_{EM-2} < \vartheta_{EM-3}$). Experiments at ϑ_{EM-1} are supposed to show the electrical driven effects. Increasing the temperature to $\vartheta_{EM} = \vartheta_{EM-2}$ and ϑ_{EM-3} has the aim to accelerate electrical driven effects not detectable at ϑ_{EM-1}. Comparing the results of the experiments at $\vartheta_{EM} = \vartheta_{EM-2}$ and ϑ_{EM-3} to the results of the High Temperature Storage should enable detecting electrical driven effects.

The duration of an electromigration experiment at a specific condition (I_x, ϑ_y) was determined to be t_{EM}. The time t_{EM} means the time under current flow. The actual duration of an electromigration experiment t_{EM}' is calculated by

$$t_{EM}' = f_T \cdot t_{EM} \qquad (3).$$

An overview of the electromigration experiment condition is given in Table 1. Again metallografic methods are used for analysing the current and temperature stressed specimen too.

T [-]		current I		
		I_1	I_2	I_3
temperature ϑ	ϑ_1	1	0.31	0.2
	ϑ_2	1	0.62	0.3
	ϑ_3	1	0.62	0.52

Table 1 Current, temperature and duty cycle conditions for electromigration experiments

5. Experimental Results
High Temperature Storage

Specimen with SnAgCu solder joints show coarsening of their microstructure due to the storage at ϑ_{HTS} for $t_{HTS} = t_{HTS-1}$, t_{HTS-2} and t_{HTS-3}. As cast solder joints (Fig. 6 I-a) show numbers of small globular Cu_xSn_y and Ag_3Sn intermetallic particles surrounding bigger Tin dendrites. After a storage for up to t_{HTS-3} the small intermetallic particles coalseced to bigger ones (Fig. 6 I-a to I-d). As can be seen in figure 6 II-a to II-d specimen with SnPb solder joints show similar processes. As cast solder joints show a fine dispersed Tin an Lead rich phases. During a storage for up to t_{HTS-2} the solder microstructure coarsens likewise. The microstructure now consists of coarse Tin and Lead rich phases. A further storage for up to t_{HTS-3} does not further change the microstructure significantly so a equilibrium state seems to be reached.

At the as cast condition solder joints of both compositions show a hillock shaped Cu_6Sn_5 phase with a thickness of about 1 to 2 µm at the intereface of solder and substrate copper. A Cu_3Sn phase is propably present but not detectable yet. During storage for up to t_{HTS-3} the Cu_6Sn_5 phase grows to a thickness of about 5 µm while it gets a more planar shape. Again the thickness of the Cu_6Sn_5 phase in SnPb solder joints seems to reach that value already after t_{HTS-2}. For t_{HTS-3} between the substrate copper and the Cu_6Sn_5 phase a planar shaped Cu_3Sn phase grows to a thickness of about 2 μm (SnAgCu) and 4 μm (SnPb) respectively.

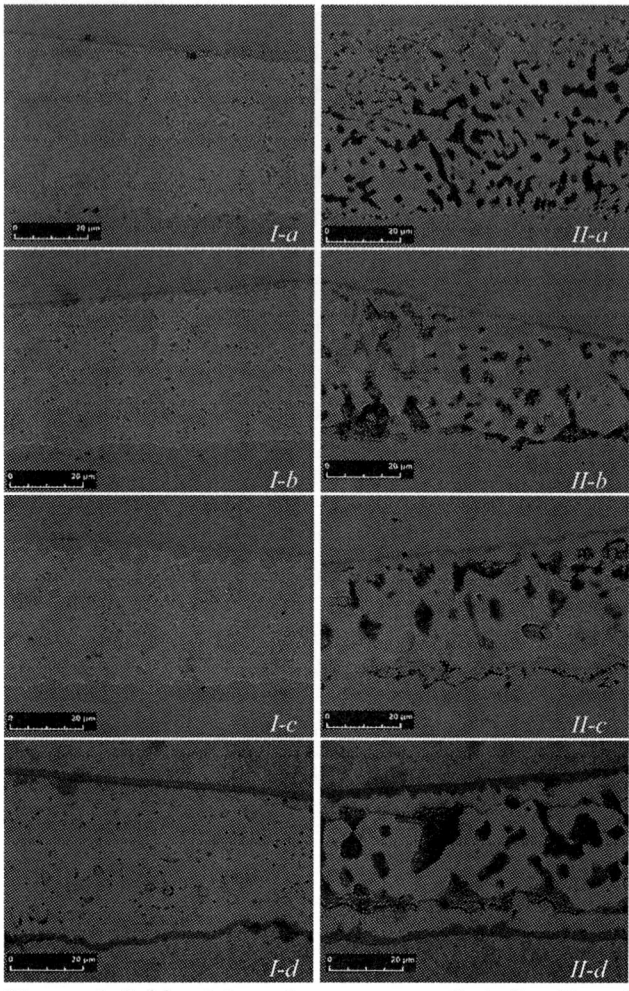

Figure 6 Solder joint microstructure: SnAgCu (I) and SnPb (II) in the condition as cast (a) and after storage for $t_{HTS} = t_{HTS-1}$ (b), t_{HTS-2} (c) and t_{HTS-3} (d) at ϑ_{HTS}

The Nickel/Tin surface finish of the TO-component leads for SnAgCu solder prevents the growth of Cu_xSn_y phases at this interface during the reflow process. A thin planar as well as needle-shaped Ni_3Sn_4 phase was detected instead of the Cu_xSn_y phases. The tickness of this Ni_3Sn_4 phase varies between less then 1 to 1.5 μm. Also some phase seperation has been indicated. During thermal storage the thin Ni_3Sn_4 phase changes to a planar shaped $(Ni_x,Cu_y)_3Sn_4$ phase and a hillock shaped $(Cu_x,Ni_y)_6Sn_5$ phase grows on top. This $(Cu_x,Ni_y)_6Sn_5$ phase grows to a thickness between 1 to 8 μm.

At the interface of the SnPb solder and the TO-component lead Cu_6Sn_5 and Cu_3Sn phases were observed with a hillock as well as a fingerlike shape in the as cast conditon. The thickness of the planar Cu_6Sn_5 phase was

about 1 to 2 μm and the thickness of the planar Cu_3Sn phase was about 1 μm. Some phase seperation has been indicated at the finger-shaped areas. Cu_3Sn phases arc propably present at this state due to the chemical deposition of Tin as the lead surface finish. Caused by the thermal storage the thickness of the planar Cu_6Sn_5 phase increases up to about 3 to 8 μm after t_{HTS-2} and the thickness of the planar Cu_3Sn phase increases up to about 1 to 5 μm respectively.

In SnPb solder joints with a very low stand-off a bonding of the Cu_6Sn_5 phases have been obeserved between TO-component lead and substrate after storage for t_{HTS-2} (Fig. 7).

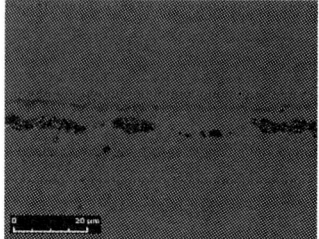

Figure 7 Cu_6Sn_5 bond between TO-component lead and substrate after storage for t_{HTS-2} at ϑ_{HTS}

Experiments on Electromigration

Looking at the microstructure of the SnAgCu and SnPb solder no signficant influence of the current load has been detected (Fig. 8). Solder joints stressed with currents of $I = I_1, I_2$ or I_3 at a temperature of ϑ_{EM-1} show almost no coarsening independent from the direction of the current flow. Figure 9 shows SnAgCu solder joints after loading with currents of $I = I_1$ (a) and $I = I_3$ (b) at a temperature of ϑ_{EM-1}. As can be seen both solder joints show similar coarse microstructure. Experiments with a current load of I_1 at ϑ_{EM-2} show a stronger but still weak coarsening compared to specimen exposed to a thermal storage for t_{HTS-2} at ϑ_{HTS} (Fig. 8 I-b and II-b). Specimen loaded with a current of $I = I_1$ and I_2 at a temperature of ϑ_{EM-3} show a coarse microstructure comparable to specimen exposed to a thermal storage for t_{HTS-2} at ϑ_{HTS}.

No electromigration effect has been recognised on the formation of intermetallic phases at the interface substrate copper and solder for both solder compositions. Depending on the temperature load Cu_3Sn and Cu_6Sn_5 phases at the interface of substrate copper and SnAgCu (Fig. 8 I-a to I-c) as well as SnPb (Fig. 8 II-a to II-c) solder grow independently from the current density and the direction of current flow. The same observation has been made for the interface of TO-component lead and SnPb solder.

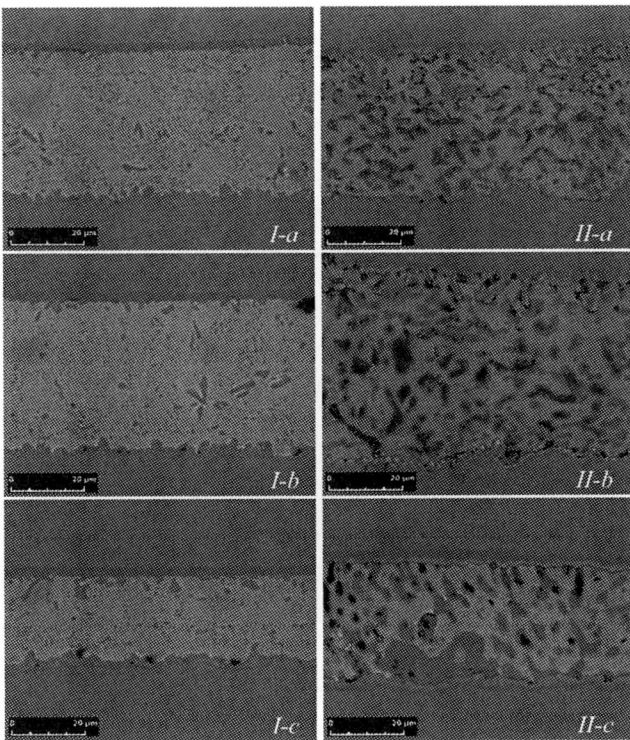

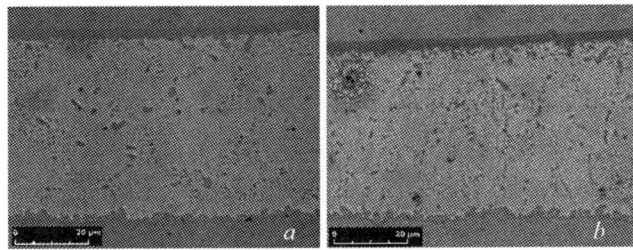

Figure 8 Coarsened solder joint microstructure: SnAgCu (I) and SnPb (II) after exposure to a current of $I = I_1$ at $\vartheta_{EM} = \vartheta_{EM-1}$ (a), ϑ_{EM-2} (b) and ϑ_{EM-3} (c)

Figure 9 Coarsened SnAgCu solder joint microstructure after exposure to currents of $I = I_1$ (a) and $I = I_3$ (b) at ϑ_{EM-1}

At the interface of TO-component lead and SnAgCu solder the formation of a homogenous planar shaped $(Cu_x,Ni_y)_6Sn_5$ phase has not been recognised. The initial Ni_3Sn_4 phase changes to a planar shaped $(Ni_x,Cu_y)_3Sn_4$ phase. Only under high temperature conditions at $\vartheta_{EM} = \vartheta_{EM-2}$ and ϑ_{EM-3} a partial growth of hillock shaped $(Cu_x,Ni_y)_6Sn_5$ phases has been seen (Fig. 10). Under a certain current load the partial growth of the $(Cu_x,Ni_y)_6Sn_5$ phase seems to depend on the temperature load. The higher the temperature ϑ_{EM} the more partial $(Cu_x,Ni_y)_6Sn_5$ phase growth can be observed. At a certain temperature condition no current dependence could be established. So there was no $(Cu_x,Ni_y)_6Sn_5$ phase growth at ϑ_{EM-1} under any investigated current condition and at $\vartheta_{EM} = \vartheta_{EM-2}$ and ϑ_{EM-3} similar phase growth under any investigated current condition has been found. However, comparing solder joints stored at ϑ_{HTS} for t_{HTS-2} (Fig. 6 I-c) and solder joints stressed with $I = I_1$ at ϑ_{EM-3}

(Fig. 8 I-c) a significant difference in the phase formation can be observed.

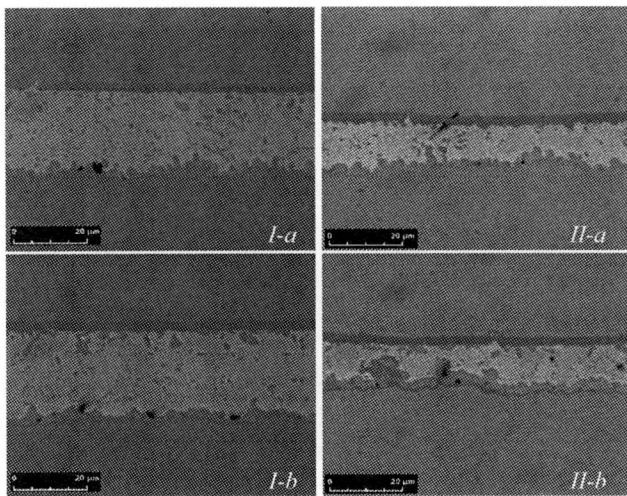

Figure 10 Partial growth of hillock shaped $(Cu_x,Ni_y)_6Sn_5$ phases under current loads of $I = I_1$ (I) and I_2 (II) at temperatures of $\vartheta_{EM} = \vartheta_{EM-2}$ (a) and ϑ_{EM-3} (b)

In SnAgCu solder joints connected to the plus terminal of the current source a intermetallic phase growth at the interface solder and component has been detected (Fig. 11). Regarding to the site on top of the Nickel surface finish, the electron flow direction and the colouring in the microsections it is assumed to be a $(Ni_x,Cu_y)_3Sn_4$ phase. Looking at the Nickel surface finish thickness at the phase growth locations a decrease from $d_m \approx 2$ µm (no phase growth) to $d_p \approx 1$ µm (phase growth) can be documented (Fig. 12). As exemplified pictured in Figure 11 the finger- partially needle-shaped phase grows at locations with a lower stand-off. Beneath the lead bend as well as the lowest stand-off no phase growth has been seen.

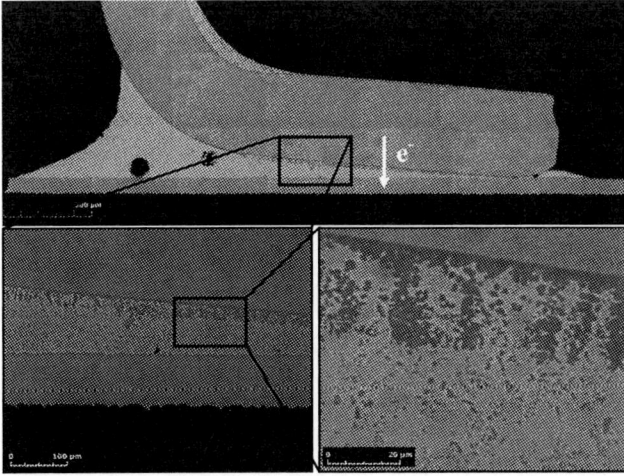

Figure 11 Intermetallic phase growth in SnAgCu solder joints at the component interface ($I = I_2$ @ $\vartheta_{EM} = \vartheta_{EM-2}$)

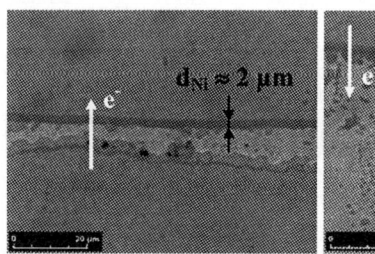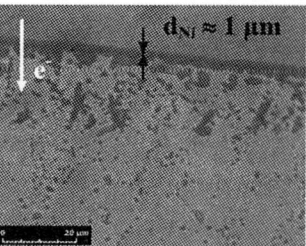

Figure 12 Consumption of surface finish Nickel for the growth of $(Ni_x,Cu_y)_3Sn_4$ phases in a SnAgCu solder joint stressed with $I = I_2$ at $\vartheta_{EM} = \vartheta_{EM-3}$: lead connected to minus (a) and plus (b) terminal

6. Discussion

No microstructure change or coarsening was observed in SnAgCu as well as SnPb solder joints driven by the electrical current. Recognised microstructure coarsening increased with higher experiment temperatures and is independent from the applied current. So it can be assigned to be only thermally driven. This is encouraged by the fact that even at a current of $I = I_3$ only a low current density is achieved in the TO-component solder joints compared to current densities applied in experiments on Flip Chip joints or even interconnection lines [2]. Hence no additional affect on solder joint degradation and therefore reliability by microstructure coarsening due to electromigration can be assessed.

Regarding the growth of intermetallic phases at the interface of substrate and component respectively and solder no electromigration effect has been found except the interface component and SnAgCu solder. During High Temperature Storage a homogenous $(Cu_x,Ni_y)_6Sn_5$ phase grows at this interface. The growth of this $(Cu_x,Ni_y)_6Sn_5$ phase is affected by the exist of a current flow $(I \geq I_1)$ independent from its direction. Then the phase grows only partially in a hillock shape. For its growth the $(Cu_x,Ni_y)_6Sn_5$ phase needs Copper diffusing from the substrate pad towards the component lead, because the Nickel metallisation prevents Copper diffusion coming from the component lead. This mechanism is probably interrupted or at least interfered in solder joints with an electron flow directed to the substrate pad. The reason for impairing the phase growth in solder joints with the opposite electron flow needs to be studied further.

The lack of a Nickel metallisation and the proximity to substrate or component copper prevent an electromigration effect on the growth of intermetallic phases at the other investigated interfaces. So they are growing likewise they do in High Temperature Storage experiments.

The growth of the $(Ni_x,Cu_y)_3Sn_4$ phase at the interface component and SnAgCu solder connected to the plus terminal of the current source is probably caused due to the electron flow coming from the component lead at this sites. The interrupt or at least interference of the Copper diffusion from the substrate pad is supporting this process. There is no or to less Copper to replace the Nickel in the growing intermetallic phase and to change it

to a $(Cu_x,Ni_y)_6Sn_5$ phase. Also the reduced thickness of the Nickel surface finish underneath the growing $(Ni_x,Cu_y)_3Sn_4$ phase points to a growth of that phase type and therefore a consumption of the Nickel metallisation. The growth of the $(Ni_x,Cu_y)_3Sn_4$ phase at the lower stand-off regions might be caused by current density distribution within the solder joint but needs to be studied further.

Dramatic electromigration effects on the growth of intermetallic phases as they are found in experiments on small Flip Chip solder joints were not observed due to the different geometrical dimensions and thus much lower current densities. Also intense current crowding like it takes place in Flip Chip solder joints is not expected in a TO-component solder joint.

7. Conclusions

The following effects have been observed in the investigated large volume solder joints: In SnAgCu and SnPb solder joints strong microstructure coarsening and growth of intermetallic phases has been observed because of High Temperature Storage at ϑ_{HTS} for up to t_{HTS-3}. No microstructure coarsening has been detected due to electrical currents of $I = I_1$, I_2 and I_3 for SnAgCu as well as for SnPb solder. Microstructure coarsening seen in accomplished electromigration experiments was assigned to be driven by temperature conditions. Formation of Cu_3Sn and Cu_6Sn_5 intermetallic phases at the interface substrate copper and SnAgCu or SnPb solder was not effected by the flow of an electrical current. Also intermetallic phase formation at the interface TO-component lead and SnPb solder was also not influenced by the flow of an electrical current. At the interface of component and SnAgCu solder the growth of a homogenous $(Cu_x,Ni_y)_6Sn_5$ phase is affected by a current flow independently from its direction. The growth of a finger-shaped $(Ni_x,Cu_y)_3Sn_4$ phase at the interface component and SnAgCu solder is caused by an electron flow coming from the component and supported by the interrupt or interference of the Copper diffusion coming from the substrate pad.

Acknowledgments

The authors would like to gratefully thank Mr. U. Deml and Dr. C. Goette from Siemens VDO Automotive Regensburg for their support in materials and specimen preparation as well as for many productive discussions.

References

1. Black, James R., "Electromigration Failure Modes in Aluminum Metallization for Semiconductor Devices", Proceedings of the IEEE, Vol. 57, No. 9 (1969)
2. Blech, I. A., "Electromigration inthin aluminum films on titanium nitride", Journal of Applied Physics, Vol. 47, No. 4 (1976)
3. Liang, L. H., "Electro-Migration Study in Solder Joint and Interconnects of IC packages", *Proceedings 7th EuroSimE Conference*, 2006

4. Ye, Hua, "Numerical Simulation of Stress Evolution During Electromigration in IC Interconnect Lines", IEEE Trans. on Components and Packaging Technologies, Vol. 26, No. 3 (2003)

5. Yeh, Everett C. C., "Current-crowding-induced electromigration failure in flip chip solder joints", Applied Physics Letters, Vol. 80, No. 4 (2002)

6. Auerbach, F., "Power-Cycling-Stability of IGBT-Modules", IEEE Industry Applications Society, Annual Meeting, New Orleans, 1997

Multi-scale energy-based failure modeling of bond pad structures

O. van der Sluis[1,2,*] R.B.R. van Silfhout[1], R.A.B. Engelen[1], W.D. van Driel[2,3], and G.Q. Zhang[2,4]

[1]Philips Applied Technologies, High Tech Campus 7, 5656 AE Eindhoven, The Netherlands
[2]Department of Precision and Microsystem Engineering, Delft University of Technology, the Netherlands
[3]NXP Semiconductors, IMO-BE Innovation BY1.055 Gerstweg 2, 6534 AE Nijmegen, The Netherlands
[4]NXP Semiconductors, High Tech Campus 60, 5656 AE Nijmegen, The Netherlands

Abstract

Thermo-mechanical reliability issues have been identified as major bottlenecks in the development of future microelectronic components. This is caused by the following technology and business trends: (1) increasing miniaturisation, (2) introduction of new materials, (3) shorter time-to-market, (4) increasing design complexity and decreasing design margins, (5) shortened development and qualification times, (5) gap between technology and fundamental knowledge development [22]. It is now well established that for future CMOS-technologies (CMOS065 and beyond), low-k dielectric materials will be integrated in the back-end structures [8]. However, bad mechanical integrity as well as weak interfacial adhesion result in major thermo-mechanical reliability issues. Especially the forces resulting from packaging related processes such as dicing, wire bonding, bumping and molding are critical and can easily induce cracking, delamination and chipping of the IC back end structure when no appropriate development is performed [4]. The scope of this paper is on the development of numerical models that are able to predict the failure sensitivity of complex three-dimensional multi-layered structures while taking into account the details at the local scale of the microelectronic components by means of a multi-scale method. The damage sensitivity is calculated by means of an enhanced version of the previously introduced Area Release Energy (ARE) criterion. This enhancement results in an efficient and accurate prediction of the energy release rate (ERR) at a selected bimaterial interface in any location. Moreover, due to the two-scale approach, local details of the structure are readily taken into account. In order to evaluate the efficiency and accuracy of the proposed method, several two-dimensional and three-dimensional benchmarks will be simulated. The paper focusses on the enhanced ARE method, including several two- and three-dimensional benchmarks.

1 Introduction

The introduction of new low-k materials, such as Black Diamond-I (CMOS090) and Black Diamond-II(x) (for CMOS065), results in major reliability issues. Especially the latter material, being porous, will drastically reduce the thermo-mechanical performance of the IC stack [4]. Indeed, Liu *et al.* [14] prove in their numerical analyses that the crack driving force increases with decreasing stiffness of the low-k material. Finite element modeling combined with experimental observations and validations, pro-

vides a way to gain more fundamental knowledge and ultimately, to understand, predict and prevent reliability issues. However, numerical methods and models that are readily available in commercial finite element packages, are currently not sufficient to model the mentioned phenomena accurately and efficiently. First, due to the inherent scale-difference between application ([cm] to [mm]) and smallest geometry detail ([μm] to [nm]), a multi-scale method should be used to cover these length-scale differences in an appropriate way. Second, delamination of three-dimensional multi-layered structures should be taken into account. Third, the three-dimensional geometry of the back-end is complex and the individual material and interface properties should be measured accurately.

For this purpose, we have developed a numerical framework that takes into account (i) the scale difference by means of a homogenization step, (ii) delamination sensitivity between the different materials, and (iii) the complex three-dimensional geometry of the bond pads [6, 18]. Especially the introduction of an energy-based failure criterion, the Area Release Energy (ARE) method has proven to be a rather efficient way to evaluate the damage sensitivity of complex multi-material structures in three dimensions. It has also been proven that interface stresses do not provide meaningfull results with respect to the delamination sensitivity thereby confirming the applicability of energy-based criteria, based on principles of fracture mechanics. However, the original ARE method does not calculate the value for the energy release rate, even though it is an energy-based value. The calculated values can therefore not be compared with measured interface strengths. To this end, numerical fracture mechanics methods, like the J-integral method [11] or the virtual crack closure technique (VCCT) [17] can be employed. However, these methods only provide accurate values for the energy release rate when using proper crack tip meshes, proposed by Barsoum [3] for the J-integral method and by Smith and Raju [19] for the VCCT method. As a result, generation of three-dimensional models including these typical crack tip meshes is a cumbersome task to evaluate even one single crack with given location, size and geometry in a structure, as is performed by Wang *et al.* [21] and Lui *et al.* [14]. These shortcomings have motivated the development of the enhanced ARE method. This method, based on a two-scale scheme, facilitates accurate calculation of ERR values at an arbitrary location while preserving the main advantage of the original ARE method: efficient damage sensitivity analysis of complete interfaces within gen-

---

*corresponding author: olaf.van.der.sluis@philips.com

1-4244-1105-X/07/$25.00 ©2007 IEEE

eral three-dimensional structures. The method will be explained in more detail in the next section, after which several analytical and numerical benchmarks will be solved in order to assess the proposed method.

2 Interface fracture mechanics

In order to give a more fundamental basis to the concepts that will be discussed in this paper, a short outline of interfacial fracture mechanics will be given. For more detail, the reader is referred to [10, 15].

Although the strength of the elastic singular stress field near the tip of an interface crack has the usual $\sqrt{r}$-singularity, it also exhibits an oscillatory behavior near the crack tip region. This behavior is a characteristic feature of interface cracks and results in a coupling between the stress intensity factors K_1 and K_2:

$$\sigma_{ij} = \frac{\Re(\mathbf{K}r^{i\varepsilon})}{\sqrt{2\pi r}}\tilde{\sigma}_{ij}^{\mathrm{I}}(\theta, \varepsilon) + \frac{\Im(\mathbf{K}r^{i\varepsilon})}{\sqrt{2\pi r}}\tilde{\sigma}_{ij}^{\mathrm{II}}(\theta, \varepsilon) \quad (1)$$

where the complex stress intensity factor $\mathbf{K} = K_1 + iK_2$, which uniquely characterizes the singular stress field [15]. The local polar coordinates for a coordinate system located on the crack front in planes perpendicular to the crack front are represented by r and θ. The functions $\tilde{\sigma}_{ij}^{\mathrm{I}}(\theta, \varepsilon)$ and $\tilde{\sigma}_{ij}^{\mathrm{II}}(\theta, \varepsilon)$ describe the angular distribution of the stresses around the crack tip and are given in Rice and Shih [16]. The bimaterial index ε is defined by

$$\varepsilon = \frac{1}{2\pi}\ln\left(\frac{1-\beta}{1+\beta}\right) \quad (2)$$

where β is one of the Dundur's elastic mismatch parameters and is related to the shear moduli μ_i

$$\beta = \frac{\mu_1(\kappa_2 - 1) - \mu_2(\kappa_1 - 1)}{\mu_1(\kappa_2 + 1) + \mu_2(\kappa_1 + 1)} \quad (3)$$

where $\kappa_i = 3 - 4\nu_i$ for plane strain problems. The subscripts 1 and 2 are associated with the different materials on either side of the interface. The mode mixity is defined as

$$\tan\psi = \frac{\Im(\mathbf{K}\ell^{i\varepsilon})}{\Re(\mathbf{K}\ell^{i\varepsilon})} \quad (4)$$

where ℓ is a reference length which has to be provided for each calculated or measured ψ-value. The energy release rate (ERR) for an interface is given by

$$G = \frac{1 - \beta^2}{E_*}(K_1^2 + K_2^2) \quad \text{with} \quad \frac{1}{E_*} = \frac{1}{2}\left(\frac{1}{\bar{E}_1} + \frac{1}{\bar{E}_2}\right) \quad (5)$$

where $\bar{E}_i = E_i/(1 - \nu_i^2)$. The crack tip opening displacements (CTOD) are asymptotically specified by $\mathbf{K}$ according to

$$\delta_2 + i\delta_1 = \frac{8}{(1 + 2i\varepsilon)\cosh\pi\varepsilon}\frac{\mathbf{K}r^{i\varepsilon}}{E_*}\sqrt{\frac{r}{2\pi}} \quad (6)$$

In our finite element analyses, the ERR has been chosen as crack driving force parameter, which will be calculated

by the J-integral method [11]. For accurate calculation of the J-integral value, the singularity, denoted by r^λ, where r is the distance from the crack tip and λ is the order of the singularity, should be captured properly. For homogeneous materials, for which $\lambda = 0.5$, Barsoum [3] has shown that the singularity can be described exactly when using so-called quarter point elements. However, as pointed out by Abdel-Wahab and de Roeck [1], these elements cannot be used for λ-values other than 0.5, unless a fine mesh is used. This is confirmed by He et $al.$ [9], who show that for interface cracks, convergence upon mesh refinement is obtained, however, with extremely fine meshes for high elastic mismatch values.

3 Enhanced ARE approach

As was already mentioned in the Introduction, one of the major disadvantages of fracture mechanics analysis is the necessary assumption of the location, shape and size of an initial defect [5]. Especially in the numerical analysis of complex three-dimensional geometries, inserting these defects into the model is quite a cumbersome task. The reason for this is the crack tip mesh requirement, which, particularly in three dimensions, is not straightforward to insert in these models. Consequently, most of the analyses are restriced to one or several crack location instead of complete interfaces [14, 21]. This has motivated the development of the so-called Area Release Energy (ARE) method [6, 18]. The ARE value predicts the delamination sensitivity of interfaces without knowing a priori the exact location of the delamination. Instead, the amount of energy is calculated that is released upon delamination for any position along any interface. As a result, an instant overview of the critical areas within any interface through a contour map is given thereby providing a direct comparison between different bond pad designs. The ARE value is calculated for each node i in an interface according to

$$G_i^{\mathrm{ARE}} = \frac{1}{2A_i}\sum_{j=1}^{n}\mathbf{F}_j^T[\mathbf{u}_j] \quad (7)$$

in which n is the number of nodes that are released within the area A_i around node i (in 2D: $A_i = 2\ell t$; in 3D: $A_i = \pi\ell^2$, where 2ℓ is the size of the defect), $\mathbf{F}_i$ is the force vector acting on the nodes before release, $[\mathbf{u}_i]$ is the crack opening displacement vector and t the thickness. For more information about the ARE method, the reader is referred to [6, 18]. Although this method provides a flexible way to capture the failure sensitivity of complex three-dimensional multi-material structures in the sense that critical locations are automatically identified without assuming a predefined location, several limitations have been identified: (i) due to the fact that the energy is calculated from a crack length increase from $0 \rightarrow 2\ell$, the method gives a total energy value. Recalling that the ERR is defined as the energy that is released from growing an existing crack from a to $a + da$, the ARE values cannot be linked directly to interface strength values. A rather straightforward extension would be to perform an additional VCCT analysis [13, 17] which would then result in the desired ERR values. However, as explained by [13], the VCCT step in 3D requires symmetric orthonormal meshes around the crack front to get accurate results. For non-orthonormal meshes, corrections are proposed by Smith and Raju [19]. These corrections still require a mesh that should incorporate the shape of the crack front in an approximate way, i.e., the nodes should still be located on the (possibly curved) crack front. Clearly, for typical ARE calculations, i.e. energy calculations at *each* node within an interface, this is not feasible; (ii) although a length scale ℓ is present in the calculations (recall that an *area* is released instead of only one node), the mesh density should still be large to get converged results. This is caused by the fact that in three dimensions, a circular area should be released whereas a circular geometry cannot be captured accurately in regular grids. As a result, mesh insensitivity is in some cases only obtained with prohibitively large models.

Motivated by these issues, an enhanced ARE approach is proposed in this paper: calculate an accurate value for the ERR (which can be linked directly by measured interface strength values) using a two-step approach: (1) perform a global simulation in which the global degrees of freedom are calculated, without the presence of a crack; (2) perform a local simulation in which the (interpolated) global degrees of freedom are prescribed at the boundaries of the local model for any location $\mathbf{x}$ in a pre-defined interface $\mathcal{I}(\mathbf{x})$. This local model contains a proper three-dimensional crack tip mesh thereby assuring reliable ERR values. This approach is depicted in Fig.1.

The model on the left corresponds to the 'scale 1' model while the local crack tip mesh model is depicted on the right. Notice that in the crack tip model, the crack shape and size can be modified easily, while also different material interfaces and possibly other geometries (*e.g.*, including a via) can be taken into account. Advantages of this two-step method are: (a) the calculated energy level is indeed the ERR value which can directly be linked to measured interface strength values; (b) the geometry of the local crack tip model is flexible: both the geometry of the crack front as well as the material geometry (in fact, the local crack tip mesh depicted in Fig.1 is a very simple example). In the original ARE method, only penny-shaped

cracks could be used; (c) the main advantage of the original ARE method is preserved: flexible energy release calculation in any location at a pre-selected interface $\mathcal{I}(\mathbf{x})$ without modifying the original three-dimensional model.

In the next section, several analytical and numerical benchmarks will be discussed.

4 Benchmarks

In this section, several benchmark problems will be solved using the proposed enhanced ARE method as described in the previous section. The first three benchmarks are analytical benchmarks. In addition, two numerical benchmarks will be discussed. From a computational time point of view, the local model should be as small as possible, whereas from an accuracy point of view, these dimensions should be as large as possible. This trade-off will be studied in these benchmarks, in which the influence of the size of the local models on the resulting ERR values will be studied which provides some guidelines for the accuracy and application of the method.

4.1 Penny-shaped crack in a uniform infinite medium

As a first benchmark, a penny-shaped crack in a uniform infinite medium under remote tensile stress will be discussed. The geometry and boundary conditions are given in Fig.2(a). The analytical expression for the stress intensity factor is [20]

$$K_I = \frac{2}{\pi}\sigma_{zz}^\infty\sqrt{\pi a} \tag{8}$$

The following values have been assumed: $E = 70000$ MPa, $\nu = 0.3$, $a = 1$ mm and $\sigma_{yy}^\infty = 100$ MPa. This results in $K_I = 112.8$ MPa$\sqrt{\text{mm}}$ and $G = 0.166$ N/mm. The local crack model used in the local ARE simulations is depicted in Fig.2(b). Notice that the penny-shaped crack front has been modeled using an appropriate three-dimensional crack tip mesh.

To study the dependency of the size of the local crack model, the dimensions $x_1 \times x_2 \times x_3$ have been varied: $6 \times 6 \times 4$, $8 \times 8 \times 6$, and $10 \times 10 \times 8$. The results are given in Table 1. It can be concluded that these results are quite accurate, even though the global displacement solution vector, applied on the local model, does not correspond

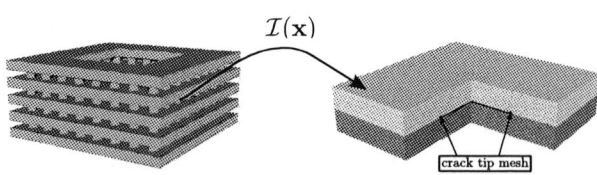

Figure 1: The proposed framework for the new ARE calculation: on the left, an example of the global model; on the right, an example of the local crack tip mesh model for any position x in a selected interface $\mathcal{I}(\mathbf{x})$

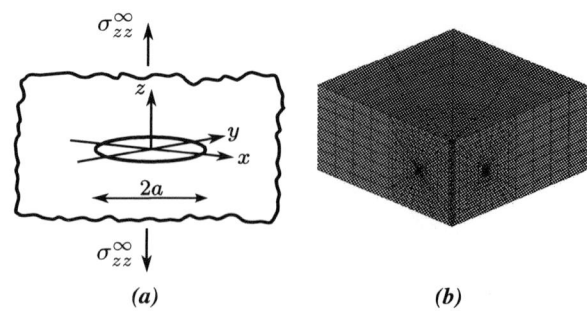

Figure 2: (a) Geometry and boundary conditions for the uniform penny-shaped crack benchmark, (b) the local crack model

to a cracked geometry which confirms the potential of the proposed method. It is remarked that all results have been checked on mesh insensitivity.

Table 1: Results of the ERR of the uniform penny-shaped crack benchmark

dimensions [mm³]	ERR value	error [%]
$6 \times 6 \times 4$	0.1454	12.4
$8 \times 8 \times 6$	0.1592	4.0
$10 \times 10 \times 8$	0.1640	1.0

The main application of the ARE method is the failure sensitivity of multi-layered complex three-dimensional structures in which the focus is on interface fracture. Therefore, the following benchmarks deal with the ERR calculation at bimaterial interfaces.

4.2 Straight interface crack in an infinite plate

The first benchmark of a bimaterial interface is the two-dimensional one, introduced and solved by Rice [15], see Fig.3. The stress intensity factor is expressed by

$$\mathbf{K} = (\sigma_{yy}^\infty + i\sigma_{xy}^\infty)(1 + 2i\varepsilon)\sqrt{\pi a}(2a)^{-i\varepsilon} \qquad (9)$$

The calculations have been performed with $E_1 = 70000$ MPa, $\nu_1 = 0.2$, $E_2 = 35000$ MPa and $\nu_2 = 0.2$. For this material combination, $\beta = -0.125$ and $\varepsilon = 0.04$. Again, $a = 1$ mm and $\sigma_{yy}^\infty = 100$ MPa. This results in $\mathbf{K} = 177.57 - 9.26i$ MPa$\sqrt{\text{mm}}(\text{mm})^{i\varepsilon}$ and $G = 0.64$ N/mm. For this calculation, the local crack model has been generated and is illustrated in Fig.3(b). It should be noted that the analytical solution requires continuity of the normal strain component ε_{xx} at the boundaries at infinity. This can be either taken into account by prescribing additional stresses at the edges, or to tie the nodal displacements at the edges in x-direction resulting in freely contracting, but straight, edges. The latter option has been used in the presented simulations. Also for this benchmark, the dimensions of the local model have been varied: 5×5, 10×10 and 14×14. The ERR is calculated from (9) by using (5). The results are given in Table 2. The simulation results converge to the analytical solution which indicates that the method is quite accurate. This benchmark also illustrates that the method is not restricted to only one

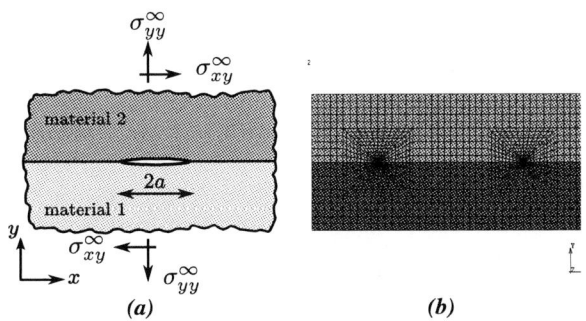

Figure 3: (a) Geometry and loading conditions of the Rice benchmark; (b) zoomed-in part of the local crack model

Table 2: Results of the ERR of the Rice benchmark

dimensions	ERR value	error [%]
5×5	0.567	11.4
10×10	0.605	5.4
14×14	0.622	2.8

kind of crack geometry, as was indeed the case in the previous ARE method where only penny-shaped cracks could be used.

4.3 Penny-shaped crack in an infinite bimaterial

The final analytical benchmark is a penny-shaped crack in a bimaterial interface. The geometry and boundary conditions are given in Fig.4. The analytical solution for the stress intensity factor $\mathbf{K}$ to this problem has been provided by Kassir and Bregman [12]:

$$K_1 + iK_2 = 2\sigma_{zz}^\infty \frac{\Gamma(2 + i\varepsilon)}{\Gamma(\frac{1}{2} + i\varepsilon)} \sqrt{a}(2a)^{-i\varepsilon} \qquad (10)$$

where Γ the Gamma function with complex argument.

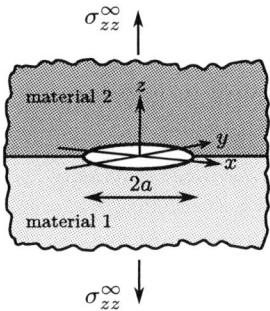

Figure 4: Geometry and boundary conditions for the bimaterial penny-shaped crack benchmark

The calculations have been performed with $E_1 = 70000$ MPa, $\nu_1 = 0.2$, $E_2 = 35000$ MPa and $\nu_2 = 0.2$. Again, $a = 1$ mm and $\sigma_{yy}^\infty = 100$ MPa. This results in $\mathbf{K} = 112.97 - 7.64i$ MPa$\sqrt{\text{mm}}(\text{mm})^{i\varepsilon}$ and $G = 0.260$ N/mm. For this calculation, the local crack model essentially equals the local model for the uniform penny-shaped crack, see Fig.2(b), except for the fact that now two materials are present in the model (see also the right picture in Fig.1). Also for this case, the analytical solution requires continuity of ε_{rr} at the interface boundary at $r \to \infty$. This has been modeled with nodal tyings at the outer edges, such that (radial) contraction is allowed at the outer edges while the edges remain straight. Another way of achieving this, would be to prescribe additional stresses at the outer edges, as explained by Ayhan et al. [2]. The results of varying local model dimensions are given in Table 3. It can be seen that the ERR is calculated very accurately and converges to the analytical solution.

Indeed, the three benchmark cases confirm that the proposed two-level ARE method indeed captures the local stress field around any crack front thereby providing ac-

217

Table 3: Results of the ERR of the bimaterial penny-shaped crack benchmark

dimensions	ERR value	error [%]
$6 \times 6 \times 4$	0.233	10.2
$8 \times 8 \times 6$	0.252	2.9
$10 \times 10 \times 8$	0.258	0.6

curate ERR values. In the next section, a two- and a three-dimensional application of the method will be illustrated.

5 Application

5.1 Two-dimensional bond pad structure

This application serves to illustrate the applicability in a more complex geometry: a two-dimensional cross-section of a characteristic part of a typical bond pad design, see Fig.5. The width of the model is 16 μm and the height is 4.71 μm. A vertical displacement of 0.1 μm is prescribed on the top whereas the displacements at the bottom and left side are suppressed. All materials are assumed linear elastic for this benchmark.

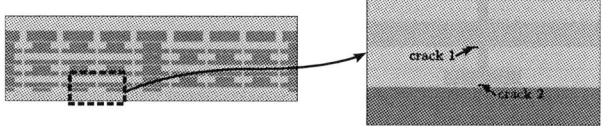

Figure 5: The two-dimensional bond pad model including a zoom of the two crack locations crack1 and crack2

Clearly, for this case, no analytical solution exists. Therefore, two models have been made in which a crack with $a = 0.13$ [μm] (which corresponds to the diameter of a via) is inserted at two different locations, indicated in the right picture of Fig.5. Crack 1 is located at the metal/dielectricum interface, whereas crack 2 is inserted at the metal/hardmask interface. This approach is also used by Wang et al. [21] and Liu et al. [14] which clearly illustrates the disadvantage of the application of ERR calculations in the conventional way: a crack location (and geometry in 3D) has to be chosen and for each location, different models have to be generated. The thus calculated reference solution for crack 1 is $G = 2.55$ N/m and for crack 2 is $G = 5.76$ N/m. These solutions are the converged values for several mesh density values. For the ARE approach, several local models have been generated with varying dimensions, indicated by Fig.6. Notice that here, only the location (e.g., an interface or as in this case, a single point) of the local model has to be provided in the global model. The

Table 4: Results of the errors in the ERR of the two-dimensional bond pad structure benchmark

dimensions	crack 1 [%]	crack 2 [%]
small	6.3	12.6
medium	3.9	4.5
large	1.1	1.5

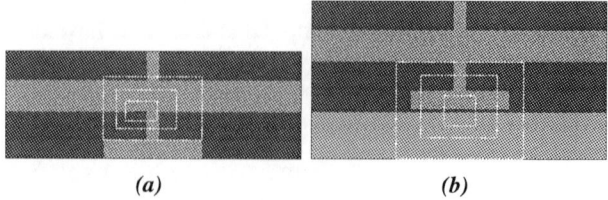

Figure 6: Different model sizes for the the two-dimensional bond pad benchmark (a) for crack 1 and (b) for crack 2

crack tip meshes are equal to the mesh shown in Fig.3(b), only with different dimensions and materials. The results for both crack location are given in Table 4. The regions illustrated in Fig.6 are denoted in the table by small, medium and large for convenience. From the results, it can be concluded that even for more complex geometries, the method is rather accurate, even for small local models.

5.2 Three-dimensional bond pad structure

An important application of the ARE method is the evaluation of the failure sensitivity of three-dimensional bond pad structures, which have already been evaluated with the existing ARE approach [6, 18], however, with the aforementioned shortcomings. As a benchmark example, the pad design as illustrated in Fig.7 will be used. Notice that the two-dimensional geometry from Section 5.1 is a cross-section of the three-dimensional geometry from Fig.7. For this benchmark, a penny-shaped crack of $2a = 0.13$ μm is inserted as depicted in the right picture of Fig.7. Analogous to the two-dimensional benchmark, a vertical displacement of 0.1 μm is prescribed on top of the structure. Again, a reference solution is obtained by fully meshing the structure including an appropriate crack tip mesh. The converged ERR value is $G = 0.707$ N/m.

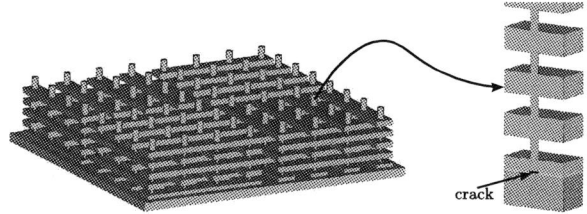

Figure 7: The three-dimensional bond pad model including a zoom of the crack location. For visualization purposes, the dielectrium and hardmask materials are removed from the pictures

For the ARE approach, three different local models have been generated with equal dimensions as the ones for the two-dimensional benchmark (crack 2) and indicated in Fig.6(b). The calculated ERR values are given in Table 5. It can be concluded that the results for the three-dimensional benchmark are even more accurate than the two-dimensional benchmark.

218

Table 5: Results of the ERR of the three-dimensional bond pad structure

dimensions	ERR value	error [%]
small	0.737	4.2
medium	0.720	1.8
large	0.711	0.6

6   Conclusions

In this paper, an enhanced version of the ARE method is proposed. In addition to the already efficient way of calculating energy values in an interface, already possible with the original ARE method, the calculated energy values are now accurate values for the ERR. Therefore, the values can directly be combined with interface strength values (or bulk strength values in the case for uniform materials). The method is based on a two-scale approach: at global level, the degrees of freedom are calculated without taking into account the cracked state whereas at local level, an arbitrary crack geometry is used for the calculation of the ERR. The benchmarks indicate that even for small local models, accurate ERR values are obtained. In this way, the geometry of the considered structures as well as the crack geometry and corresponding materials can be taken into account flexibly.

Future work will firstly focus on ARE computations within selected interfaces combined with already measured interface strength values by four-point bending tests and on the optimisation of the CPU time of the local models by means of numerical condensation. Secondly, the method can be combined with the dedicated cohesive zone model implementation by Van Hal *et al.* [7]: the ARE calculation will provide critical locations, whereas the cohesive zone models will subsequently be applied to model to transient crack propagation.

References

[1] Abdel Wahab M, de Roeck G (1995). "A 2-D five-noded finite element to model power singularity." *International Journal of Fracture*, 74:89–97.

[2] Ayhan A, Kaya A, Nied H (2006). "Analysis of three-dimensional interface cracks using enriched finite elements." *International Journal of Fracture*, 142:255–276.

[3] Barsoum R (1976). "On the use of isoparametric finite elements in linear fracture mechanics." *International Journal for Numerical Methods in Engineering*, 10:23–37.

[4] van Driel W (2007). "Facing the challenge of designing for Cu/Low-k reliability." *Microelectronics Reliability*, accepted for publication.

[5] Ernst L, van Driel W, van der Sluis O, Corigliano A, Tay A, Iwamoto N, Yuen M (2007). "Fracture and delamination in microelectronics." *Key Engineering Materials*, accepted for publication.

[6] van Gils M, van der Sluis O, Zhang G, Janssen J, Voncken R (2007). "Analysis of Cu/low-k bond pad delamination by using a novel failure index." *Microelectronics Reliability*, 47:179–186.

[7] van Hal B, Peerlings R, Geers M, van der Sluis O (2007). "Cohesive zone modeling for structural integrity analysis of IC interconnects." *Microelectronics Reliability*, accepted for publication.

[8] Hartfield C, Ogawa E, Park YJ, Chiu TC (2004). "Interface reliability assessments for Copper/low-k products." *IEEE Transactions on Device and Materials Reliability*, 4:129–141.

[9] He M, Evans A, Hutchinson J (1994). "Crack deflection at an interface between dissimilar elastic materials: Role of residual stresses." *International Journal of Solids and Structures*, 31:3443–3455.

[10] Hutchinson J, Suo Z (1992). "Mixed mode cracking in layered materials." *Advances in Applied Mechanics*, 29:63–191.

[11] Kanninen M, Popelar C (1985). *Advanced Fracture Mechanics*. Oxford University Press, New York.

[12] Kassir M, Bregman A (1972). "The stress-intensity factor for a penny-shaped crack between two dissimilar materials." *Journal of Applied Mechanics*, 39:308–310.

[13] Krueger R (2002). *The virtual crack closure technique: history, approach and applications*. NASA Report CR-2002-211628.

[14] Liu X, Lane M, Shaw T, Simonyi E (2007). "Delamination in patterned films." *International Journal of Solids and Structures*, 44:1706–1718.

[15] Rice J (1988). "Elastic fracture mechanics concepts for interfacial cracks." *Journal of Applied Mechanics*, 55:98–103.

[16] Rice J, Sih G (1965). "Plane problems of cracks in dissimilar media." *Journal of Applied Mechanics*, 32:418–423.

[17] Rybicki E, Kanninen M (1977). "A finite element calculation of stress intensity factors by a modified crack closure integral." *Engineering Fracture Mechanics*, 9:931–938.

[18] van der Sluis O, Engelen R, van Driel W, van Silfhout R, van Gils M (2007). "Efficient damage sensitivity analysis of advanced Cu Low-k bond pad structures using Area Release Energy." *Microelectronics Reliability*, accepted for publication.

[19] Smith S, Raju I (1998). "Evaluation of stress-intensity factors using general finite-element models." In *Fatigue and Fracture Mechanics: 29th Volume, ASTM STP 1321*, edited by T Panontin, S Sheppard. American Society for Testing and Materials.

[20] Tada H, Paris P, Irwin G (1985). *The stress analysis of cracks handbook*. Paris Productions, St. Louis.

[21] Wang G, Ho P, Groothuis S (2005). "Chip-packaging interaction: a critical concern for Cu/low k packaging." *Microelectronics Reliability*, 45:1079–1093.

[22] Zhang G, van Driel W, Fan X (2006). *Mechanics of Microelectronics*. Springer, Dordrecht, The Netherlands, ISBN 1-4020-4934-X.

Impact of Solder Overflow and ACLV Moisture Absorption of Mold Compound on Package Reliability

Richard Qian
Fairchild Semiconductor (Suzhou) Corp., 1 Sutong Road, Suzhou, China

Yong Liu, Scott Irving and Timwah Luk
Fairchild Semiconductor Corp., 82 Running Hill Road, South Portland, ME 04106, USA
Email: yliu@fairchildsemi.com; Phone: (207) 761-3155; Fax: (207) 761-6339

Abstract

Moisture induced die surface metal layer corrosion is a critical reliability problem which may cause eletrical failures. In this paper, the impact of solder overflow and moisture absorption of mold compound on the die surface metal layer are investigated through modeling work. Comprehensive analysis of the modeling and simulation results for die surface metal layer moisture concentration is presented. The assembly process is optimized to reduce solder overflow after the modeling. Practical tests of the solder overflow and moisture absorption factors of mold compound are carried out. Both modeling and testing have confirmed that solder overflow and mold compound's moisture absorption have a significant impact to die surface metal layer's moisture concentration.

1. Introduction

Assembly process induced faults (such as solder overflow) and moisture absorption in mold compound could generate high moisture concentration at the die surface metal layer [1~2]. It would induce metal corrosion. Delamination between soft solder and mold compound would appear if the adhesive strength becomes low at high moisture absorption levels. While solder overflow and moisture absorption will speed up the delamination at the die surface metal layer during ACLV test. This issue has been existed in the industry for a long time, but there are few papers that have shown effective methodology how to resolve it.

Modeling and simulation with a smaller amount of empirical testing is an effective way to evaluate solder overflow and the mold compound's moisture abstoption impact to moisture concentration at the die surface metal layer [3]. This allows us to reduce possible small defects in package , improve final product quality and reliability by minimizing the possibility of package failure. In this paper, a transient non-linear dynamic finite element framework is built for moisture diffusion during the ACLV test. The simulation work has two parts: The first part is to check the impact of solder overflow, to compare die surface metal layer's moisture concentration between solder overflow model and solder without overflow model. The initial delamination between the overflowed solder and mold compound is considered with the solder overflow model due to the low adhesive strength between mold compound and solder. The second part is to investigate the impact of the mold compound's moisture absorption. Both with solder overflow model and without solder overflow models are investigated. Moisture concentration level on the die metal layer is simulated with different mold compounds under the ACLV test conditions (96 hours 121°C/100%RH). Finally comprehensive analysis of the modeling and simulation results for die surface metal layer moisture concentration is presented. The assembly process is optimized to reduce solder overflow after the modeling. Practical tests of the solder overflow and moisture absorption factors of mold compound are performed by Fairchild Suzhou site. Through this study, excellent reliability performance is expected to be achieved.

2. Solder Overflow in Die Attach Process and Finite Element Models description

During the eutectic die attach process, a lead frame is held on the surface of heater block. The soft solder wire is then placed on the lead frame. The temperature of the heater block ramps up to the melting temperature of solder which makes the wire fuse [4]. The melt solder is not flat due to surface tension (as shown in Fig.1). It will be flattened to reduce die tilt and increase solder coverage. A typical melt solder flattening process is shown in Fig. 2. A smoothing tool in the solder attach equipment moves down and flattens the melt solder and changes the round solder bump into square shape. Finally the cavity collet head picks the die and moves the die to the flattened melt solder on the lead frame.

(a) melt solder on lead frame

Fig. 1. Melt solder on DAP.

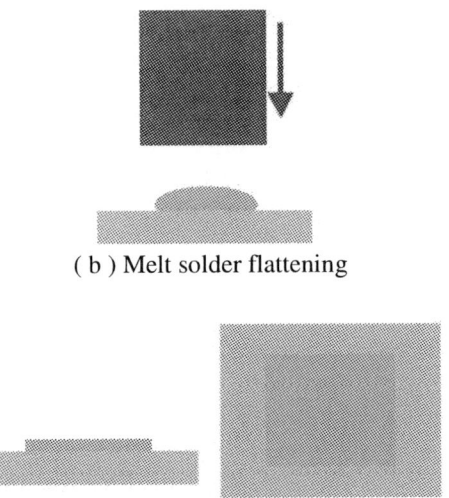

(b) Melt solder flattening

(c) Flattened melt solder.
Fig. 2. Melt solder flattening

Demands for larger die sizes have driven an increasement of solder size in package, which may induce solder overflow in the melt solder flattening process. Figure 3 shows a solder overflow example due to solder attach process. Solder overflow appears beside the middle lead. A solder overflow finite element model is generated based on this real example. Fig. 4 shows the 3D finite element mesh of a typical TO220 package and solder overflow from lead frame. Lead frame, solder, die, wire bonding and epoxy mold compound are considered in this model. The size of die is 4.7 mm x 6 mm with 0.3 mm thickness. With solder overflow model, solder size (5.2 mm x 6.2 mm with 0.05mm thickness) is a little larger than die size. For the model without solder overflow, solder size is same as die size.

(a) Solder overflow due to die attach process

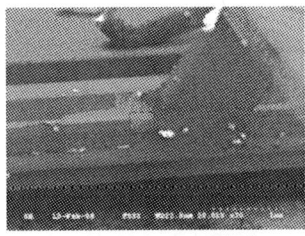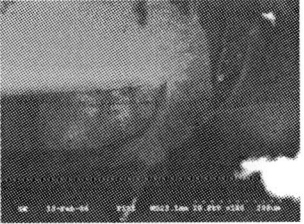

(b) Actual picture of solder overflow
Fig. 3. Solder overflow from lead frame

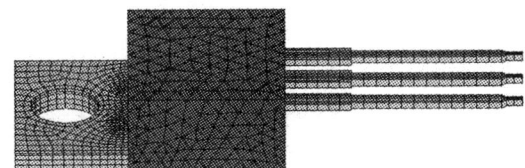

(a) mesh of the 3D model

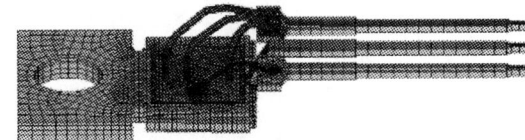

(b) Mesh of 3D model (hiding the EMC)

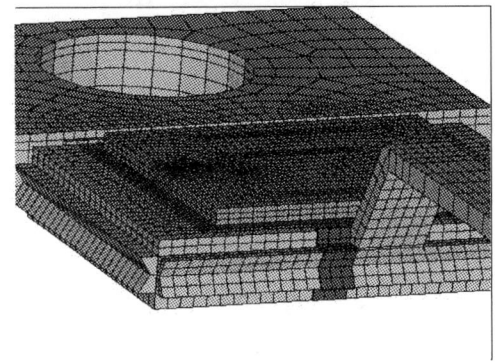

(c) Mesh of solder overflow from lead frame
Fig. 4. Solder overflow from lead frame DAP

Ideal model is used in the case without solder overflow simulation. For the solder overflow model, the initial delamination between the overflowed solder and mold compound due to the low adhesive strength between solder and mold compound is introduced. The initial delamination may propagate during ACLV test due to CTE mismatch of different material in package. In order to solve the problem effectively, only the initial delamination is considered in simulation.

The saturated moisture concentration of mold compound is 1.14E-2 mg/mm^3. Moisture diffusivity is 3.47E-06 mm^2/s. In the second part of the simulation work, the new mold compound's saturated moisture level is 0.84E-2 mg/mm^3, moisture diffusivity is 2.41E-06 mm^2/s.

3. Effect of solder overflow

Die size is relative large. Moisture concentration at die surface metal layer may be quite different after the ACLV test. Six different positions at die surface metal layer are compared throughout the ACLV test. Since the gate area is much smaller than source area, one position is selected at gate metal area and the other five positions are selected at source metal area (as shown in Fig. 5) in the models..

The moisture cannot penetrate the lead frame, therefore, the diffusion path is through the mold compound.

Moisture diffuses into the package through the top surface, the vertical surface and bottom surface of mold compound, as shown in Fig. 6.

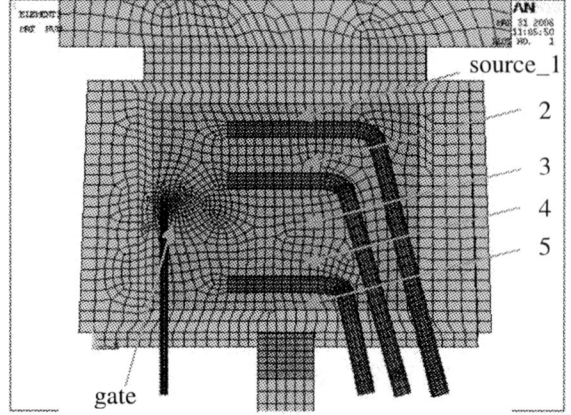

(a) Without solder overflow

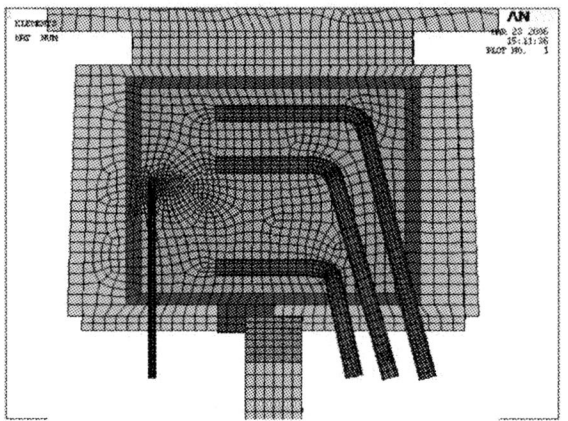

(b) With solder overflow

Fig. 5. Mesh of lead frame, solder, die and wire bonding.

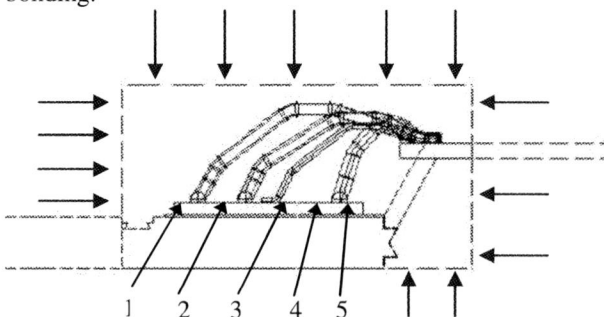

Fig. 6 Moisture diffuses into the package through mold compound surface.

Moisture soak simulation results are shown in Fig. 7 to Fig. 9 and Table 1. Figure 7 shows moisture concentration history at die surface metal layer's six positions in the whole ACLV test without solder overflow. Moisture concentration at position source 1 is much higher than that of the other five positions. The reason is that position of source 1 is near the left surface of mold compound.

Moisture can diffuse through mold compound and reach position source 1 relative easy. Figure 8 shows moisture concentration history at the six positions in ACLV test with solder overflow. Compared with no solder overflow model's result, moisture concentration at all of the six positions is increased. Moisture concentration at position of source 5 and gate, which is near the delamination area between mold compound and solder, is increased dramatically.

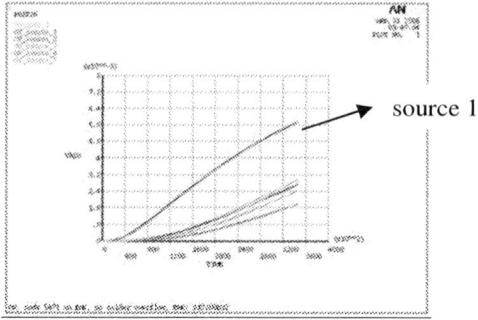

Fig. 7. Moisture concentration history of mold compound at die surface without solder flow.

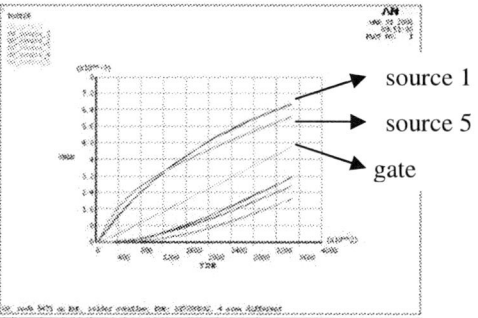

Fig. 8. Moisture concentration history of mold compound at die surface with solder flow.

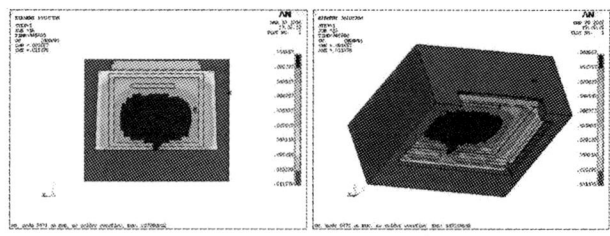

(a) Without solder overflow

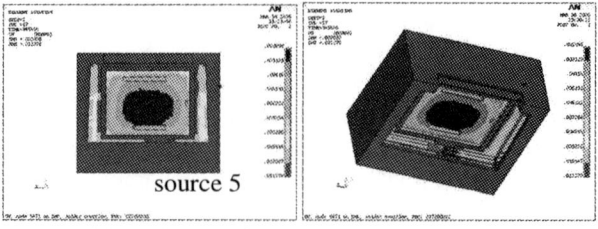

(b) With solder overflow

Fig. 9. Moisture concentrate contour of EMC.

222

Fig. 9 gives simulation results of moisture soak after 96 hours at 121ºC/100%RH. The outer boundary and delamination area of mold compound has the highest moisture, and the inner boundary at the internal surface of die has much lower moisture levels. Table 1. gives the moisture concentration of six positions with and without solder overflow after 96 hours with 121ºC/100%RH.

Table 1 Moisture concentration at die surface metal layer after ACLV test.

	Moisture concentration (1E-3 mg/mm^3)					
	gate	source				
	1	1	2	3	4	5
no solder overflow	2.95	5.74	2.74	1.77	1.77	2.42
solder overflow	4.84	7.37	3.27	2.12	2.72	6.06

4. Effect of mold compound

The effect of the mold compound's moisture absorption is investigated in this section. The results of a new mold compound are shown in Fig.10–Fig.12. Comparison of the two mold compounds is listed in Table. 2 and Table. 3. Fig. 10 and Fig. 11 show moisture concentration history at die surface metal layer's six position in the whole ACLV test. Fig. 12 gives simulation results of moisture soak after 96 hours with 121ºC/100%RH. The results show the similar trends of using previous mold compound but have a lower moisture concentration at the metal layer.

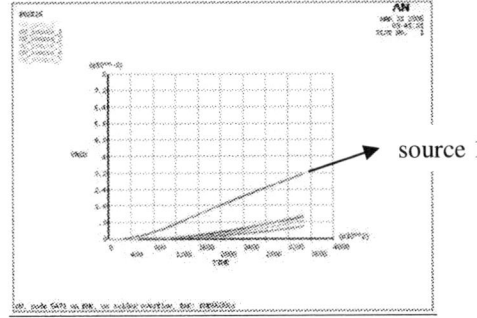

Fig. 10 Moisture concentration history of mold compound at die surface without solder flow (new EMC).

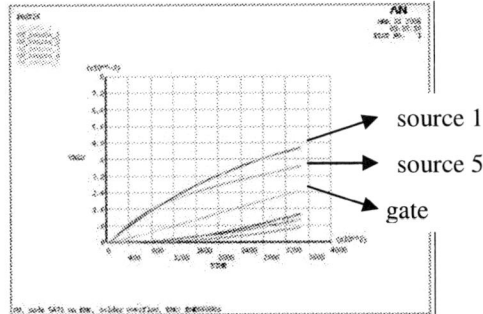

Fig. 11 Moisture concentration history of mold compound at die surface with solder flow (new EMC).

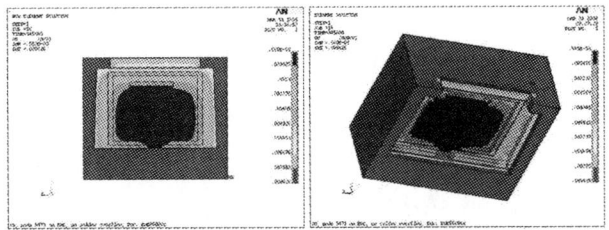

(a) No solder overflow

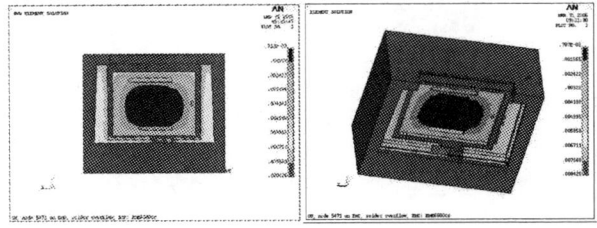

(b) With solder overflow

Fig. 12. Moisture concentration contour of new EMC.

Table 2 gives metal layer's moisture concentration comparison of the two mold compounds without solder overflow. Table 3 gives comparison of two mold compounds with solder overflow. The results show that the low moisture absorption mold compound induced lower moisture concentration at die's metal layer. The results also show that the low moisture absorption mold compound and without solder overflow model induces the lowest moisture concentration at the die surface metal layer.

Table 2. Moisture concentration of die surface metal layer after ACLV test (no solder overflow).

	Moisture concentration (1E-3 mg/mm^3)					
	gate	source				
	1	1	2	3	4	5
initial EMC	2.95	5.74	2.74	1.77	1.77	2.42
new EMC	1.14	3.13	1.05	0.59	0.58	0.86

Table 3. Moisture concentration of die surface metal layer after ACLV test (with solder overflow from lead frame).

	Moisture concentration (1E-3 mg/mm^3)					
	gate	source				
	1	1	2	3	4	5
initial EMC	4.84	7.37	3.27	2.12	2.72	6.06
new EMC	2.48	4.57	1.35	0.71	1.1 0	3.65

5. Process improvement and Experimental data

A solder rim is introduced around the melt solder during the solder flattening process (see Fig. 13). The solder rim will clamp the lead frame DAP and prevent solder overflow when the solder is flattened.

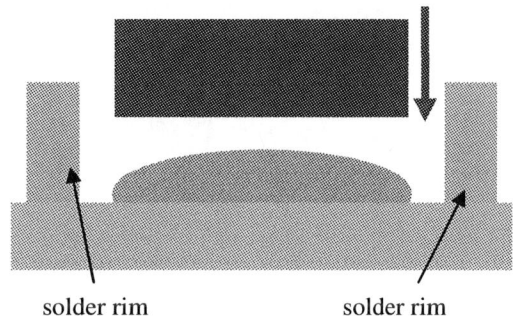

Fig. 13 Solder flattening process improvement.

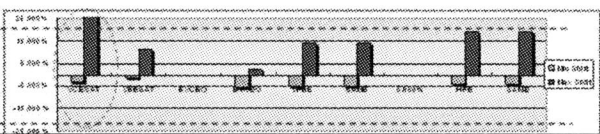

(a) Without solder overflow, high moisture absorption of mold compound (initial mold compound)

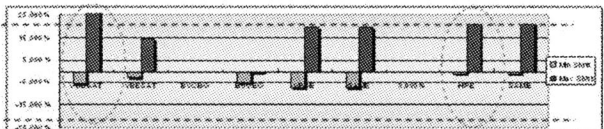

(b) With solder overflow, high moisture absorption of mold compound (initial mold compound)

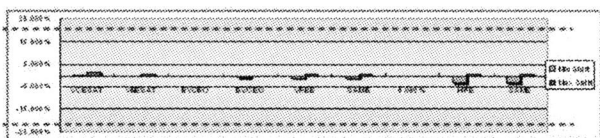

(c) Without solder overflow, low moisture absorption of mold compound (new mold compound)

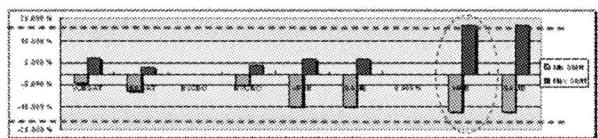

(d) With solder overflow, low moisture absorption of mold compound (new mold compound)
Fig. 14 Electrical test results after ACLV test.

Different mold compound, with and without solder overflow packages were run to determine the efficiency of process and material change. The sample size consists of 308 units. Fig. 14 shows the electrical test results after ACLV (96 hours, 121°C/100%RH). For the initial mold compound, there are 5 VCESAT failures in without solder overflow units and 7 VCESAT & 1 HFE failures in with solder overflow units. For the new mold compound, there is 1 HFE failure in with solder overflow units and no failure in without solder overflow units.

6. Conclusion

Moisture induced die surface metal layer corrosion is investigated in this paper. Both the effect of assembly process induced faults (solder overflow) and moisture absorption in mold compound are simulated. The work has allowed us to reach the following conclusions:
1. Both solder overflow and mold compound's moisture absorption have a larger impact on die surface metal layer's moisture after ACLV test.
2. Experimental data shows a significant improvement in reliability for package by using a solder rim during the die attach process and using low moisture absorption mold compound, as predicted by the simulations.

Acknowledgements
The authors wish to thank the support from Automation Development in Maine and Assembly Engineering Suzhou, Fairchild Semiconductor Corp.

References
1. Lei L. Mercado and Brian Chavez, "Impact of JEDEC Test Conditions on New-Generation Package Reliability", *Proc 51st Electronic Components and Technology Conf, Orlando, Florida, May 2001*.
2. Xuejun Fan, G.Q.Zhang, W.van Driel & L. J. Ernst, "Analytical Solution for Moisture-Induced Interface Delamination in Electronic Packaging", *Proc 53rd Electronic Components and Technology Conf, New Orleans,, May 2003*.
3. Yong Liu, Scott Irving, Mark Rioux, Andrew J. Schoenberg and David Chong, "Die Attach Delamination Characterization Modeling for SOIC Package ", *Proc 52nd Electronic Components and Technology Conf, San Diego, CA, May 2002*.
4. Yong Liu and Scott Irving, "Impact of the Die Attach Process on Power & Thermal Cycling For a Discrete Style Semiconductor Package", *EuroSimE*, 2004.

Experimental Determination and Modification of Anand Model Constants for Pb-Free Material 95.5Sn4.0Ag0.5Cu

Wang Qiang, Liang Lihua, Chen Xuefan, Weng Xiaohong
Fairchild-ZJUT Microelectronic Packaging Joint Lab, Zhejiang University of Technology, Hangzhou 310032, China

Yong Liu, Scott Irving and Timwah Luk
Fairchild Semiconductor Corp, South Portland, Maine 04106, USA

Abstract

A series of tensile tests for pb-free solder material 95.5Sn4.0Ag0.5Cu are conducted under a wide range of temperatures and constant strain rates to obtain the required data for fitting the material parameters of the Anand model. Based on these test results, empirical equations of the tensile strength, elastic modulus and yield stress are fitted as a functions of temperature. It is found that the temperature and strain rate have demonstrated crucial effects on tensile and creep properties of SnAgCu solder material. The test results have also displayed certain viscoplastic behavior, temperature dependence, strain rate sensitivity and creep resistance. A procedure for the determination of Anand material parameters through data fitting is proposed. In order to capture experimental data accurately, a modified Anand model with certain parameters which are the functions of the temperature and strain rate with the quadratic polynomial formula is discussed. Good agreements between the modified model prediction and experimental data have been obtained.

Key words: Lead-free material; Anand model; tensile tests

1. Introduction

With the development of microelectronics and surface mounted technologies, there are a lot of researchers whose studies focus on studying the solder joint reliability of various Pb-solders. But Pb compound has been cited by International Environmental Protection Agency as one of the top 17 chemicals posing the greatest threat to human beings and the environment. As a result, developing feasible alternative lead-free solders for electronic assemblies is a necessity. Most fatigue life modeling methodologies for solder joints require the estimation of stress, strain and plastic work in a finite element model. The finite element technique requires the user to provide true information of the material properties and boundary conditions for the model before it can be implemented successfully. There are many experimental data and constitutive models for SnPb based solder material[1], examples are 60Sn40Pb, 63Sn37Pb, and 62Sn36Pb2Ag[2-7]. At the same time, a variety of models have been proposed for simulating the reliability of eutectic SnPb solder joints[2-7]. While in recent years, some studies have been reported in extending these models to lead-free soldered assemblies. The Anand model which has been used for SnPb solder material, is now applied to represent the inelastic behavior for lead-free solder material. ANSYS® finite element software offers the Anand model as a standard option with 9 constants. Amagai et al. [8] have presented the Anand constants by tests for 95.75Sn3.5Ag0.75Cu and 98.5Sn1.0Ag0.5Cu, but it neglected to include one of the nine required constants, s_0, the initial value of deformation resistance. Kim et al. [9], published all nine constants, which has indeed updated the work by Amagai et al for 98.5Sn1.0Ag0.5Cu. Rodgers et al [10] have presented the test results for 95.5Sn3.8Ag0.7Cu, but their curves of Anand are quite different to those obtained by Reinikainen et al [11] for 95.5Sn4.0Ag0.5Cu. Chen et al. [12] introduced a modified Anand model with h_0, the hardening coefficient, which is the function of temperature and strain rate. But this work just modifies the h_0 which might not be able to solve the entire problem with 9 parameters. More work is needed in this research field.

In this paper, the material properties of 95.5Sn4.0Ag0.5Cu are investigated with different temperatures and strain rates. The empirical equation for elastic modulus is fitted as a functions of temperature. The data of yield stress and tensile strength are determined as the function of temperature as well. Nine Anand constants are obtained using a nonlinear fitting procedure. To capture experimental curves more accurately, three parameters, s_0, $\hat{s}$ and ξ are modified to be quadratic function of temperature and strain rate. The modified model has substantially improved the data fitting quality.

2. Anand Model

Anand [13] developed a constitutive equation for the rate-dependent deformation of metals at high temperatures (i.e. in excess of the homologous temperature of $0.5T_m$). Although it aimed at hotworking of steels and other structural metals, it has been adopted successfully to represent the viscoplastic behaviour of solder materials. The model was initial proposed by Anand and it was subsequently developed by Brown et al. [14]. It unifies the creep and rate-independent plastic behaviour of the solder by making use of a flow equation and an evolution equation [16]. The basic equations are listed as following.

1-4244-1105-X/07/$25.00 ©2007 IEEE

Flow Equation:

$$\dot{\varepsilon}_p = A \exp\left(-\frac{Q}{RT}\right)\left[\sinh\left(\xi\frac{\sigma}{s}\right)\right]^{1/m} \tag{1}$$

where $\dot{\varepsilon}_p$ is inelastic strain rate, A is preexponential factor, Q is activation energy, R is gas constant, T is absolute temperature, ξ is multiplier of stress, σ is equivalent stress, m is strain rate sensitivity.

evolution Equation:

$$\dot{s} = \left\{h_0\left(|B|\right)^a \frac{B}{|B|}\right\}\dot{\varepsilon}_p \tag{2}$$

where

$$B = 1 - \frac{s}{s^*} \tag{3}$$

and

$$s^* = \hat{s}\left[\frac{\dot{\varepsilon}_p}{A}\exp\left(\frac{Q}{RT}\right)\right]^n \tag{4}$$

where h_0 is the hardening/softening constant, a is the strain rate sensitivity of hardening/softening. The quantity s^* represents a saturation value of deformation resistance s, associated with a set of given temperature and strain rate as shown in (4). $\hat{s}$ is a coefficient, and n is the strain rate sensitivity for the saturation value of deformation resistance, respectively. The set of Anand constitutive equations can account for the physical phenomena of strain-rate and temperature sensitivity, strain rate history effects, isotropic strain-hardening and the restoration process of dynamic recovery.

There are nine material parameters of the unified rate-dependent Anand model A, Q/R, ξ, m, h_0, $\hat{s}$, n, a, and s_0. The last one is the initial value of the deformation resistances, which is needed in solution of the evolution of deformation resistance in (2).

The flow equation is similar to that used for describing the steady-state secondary creep but with the addition of an internal state variable defined as the deformation resistance, s, which depends on the temperature- and rate-dependent strain history of the material. The assumed relationship between the stress of the model and the deformation resistance variable is

$$\sigma = cs, \quad c = \left(T, \dot{\varepsilon}_P\right) \tag{5}$$

where, by definition,

$$c = \frac{1}{\xi}\sinh^{-1}\left(\frac{\dot{\varepsilon}_p}{A}\exp\left(\frac{Q}{RT}\right)\right)^m \tag{6}$$

3. Experiment Procedure

To obtain the data for the fitting parameters of lead-free material 95.5Sn4.0Ag0.5Cu, a series of tensile tests are carried out. The test specimens are made in a casting dog bone shape based on the ASTM E 8M-04 standard, with a diameter of 4mm and a gauge length of 20mm as shown in figure 1. All the tests are performed using a Reger screw-driven universal test machine with a 3,000N load cell and an integrated thermal chamber. The temperature is kept constant during testing.

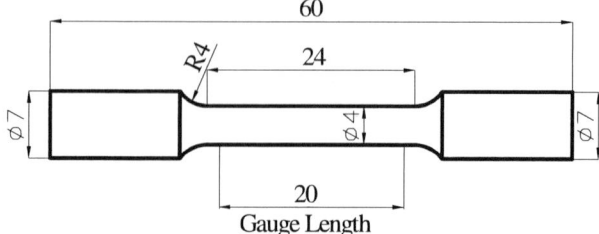

Figure 1: Tensile test specimen

The axial strain is obtained from an extensometer which is fixed on the specimen at room temperature. At high temperatures, a precision non-contact strain digital measurement technique, DSCM (digital speckle correlation method), is adopted to obtain the true strain of gauge length. Two points are marked on specimen for gage length. When the test is conducted, data of two points are collected and analyzed through CCD camera, video data acquisition system. figure 2 shows the the non-contact digital strain measurement system.

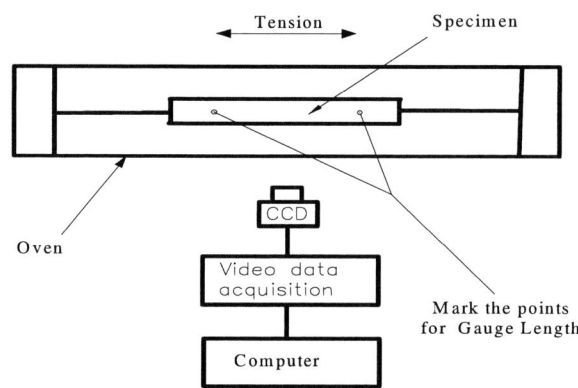

Figure 2: Digital strain measurement system

Note that tests should be carried out under constant strain rate conditions. However due to the different function of our test equipment, the constant crosshead displacement rate is applied. The crosshead displacement rate is calculated from the strain rate according to the scale of the specimen. Each test must be repeated 6 times with the same condition, the same strain rate and the same temperature, based on the ASTM E 8M-04 standard. The test results are calculated by averaging all data in the 6 time tests.

Finally, the temperature dependent Young's modulus of this lead-free material can be obtained from the stress-strain curves at the range of temperature from 25°C to 200°C. The data of load and displacement collected in real time by the software of test machine are converted to true stress and true strain.

4. Test Results and Discussion

Figures 3-5 show the true stress versus true strain curve of 95.5Sn4.0Ag0.5Cu at three temperatures with a range of strain rates. It can be seen from these figures that there appear a general feature of strain hardening behavior at the beginning in the stress-strain curves, and the deformation behavior of 95.5Sn4.0Ag0.5Cu is highly rate dependent. After that, the plastic flow reaches the steady state. It can be seen that the tensile strength and yield stress increase with the strain rate increases.

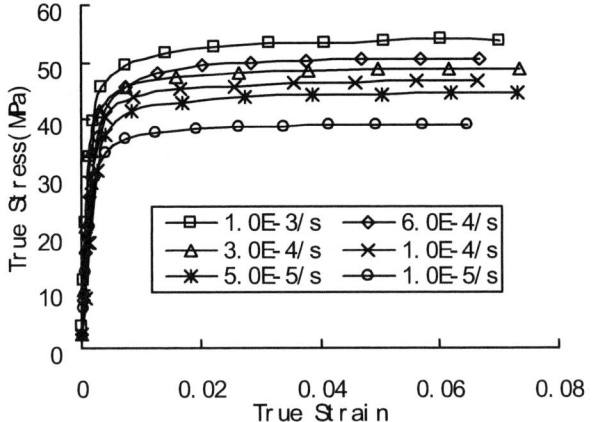

Figure 3: True stress versus true strain for 95.5Sn4.0Ag0.5Cu at 25 ℃ and different strain rates

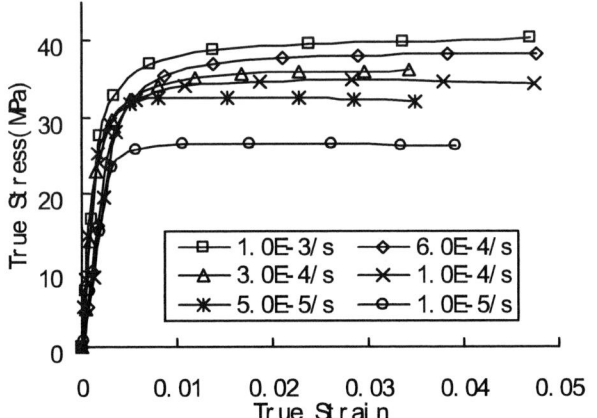

Figure 4: True stress versus true strain for 95.5Sn4.0Ag0.5Cu at 75 ℃ and different strain rates

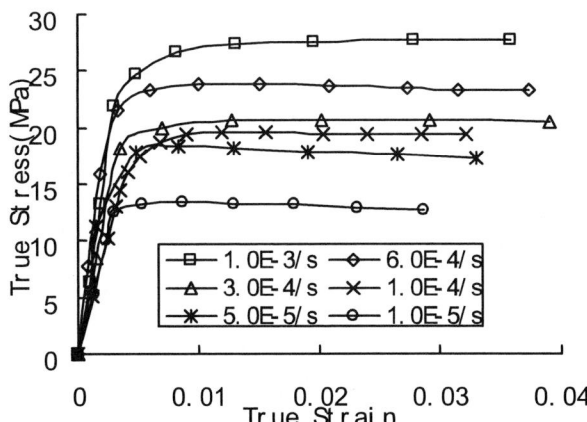

Figure 5: True stress versus true strain for 95.5Sn4.0Ag0.5Cu at 150 ℃ and different strain rates

Figures 6-8 summarize the true stress versus true strain for 95.5Sn4.0Ag0.5Cu at strain rate 1.0E-3/s, 1.0E-4/s and 1.0E-5/s with three temperatures. It can be seen from these figures that the deformation behavior of the solder is highly temperature dependent. Here, the value of saturation stresses can be obtained from constant displacement rate tests when the test has reached steady state. For example, with a strain rate of 1.0E-4/s at 25°C, the stress saturates to about 45.2 MPa, as shown in figure 7. From those figures, one can get the saturation stresses at different strain rates and different temperatures. With these test data, Anand parameters can be obtained based on a fitting algorithm. It can also be seen that the tensile strength and yield stress increase while the temperature decreases.

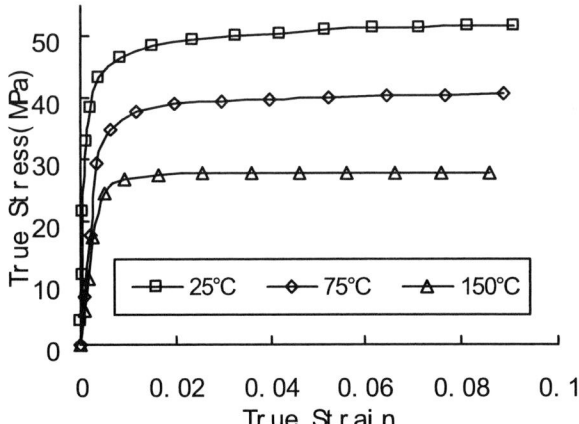

Figure 6: True stress versus true strain for 95.5Sn4.0Ag0.5Cu at strain rate 1.0E-3/s and three temperatures

Figure 9 shows the variation of the tensile strength and yield stress with temperature at the strain rate of 7.5E-3/s. It can be clearly seen from figure 9 that the tensile strength and yield stress are directly related to temperature.

227

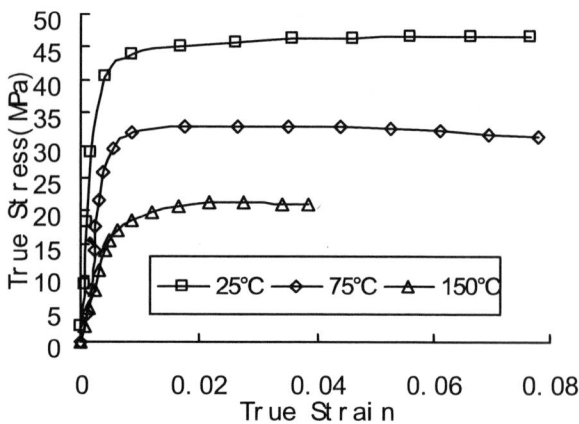

Figure 7: True stress versus true strain for 95.5Sn4.0Ag0.5Cu at strain rate 1.0E-4/s and three temperatures

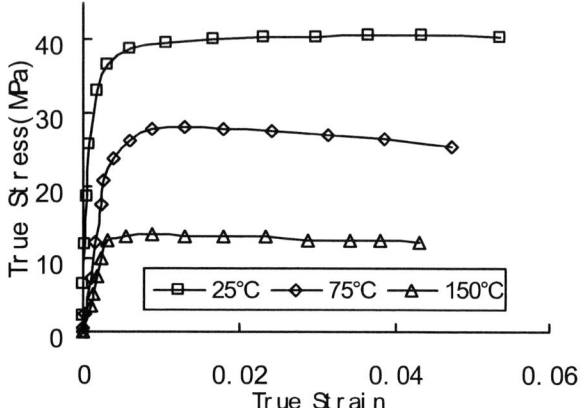

Figure 8: True stress versus true strain for 95.5Sn4.0Ag0.5Cu at strain rate 1.0E-5/s and three temperatures

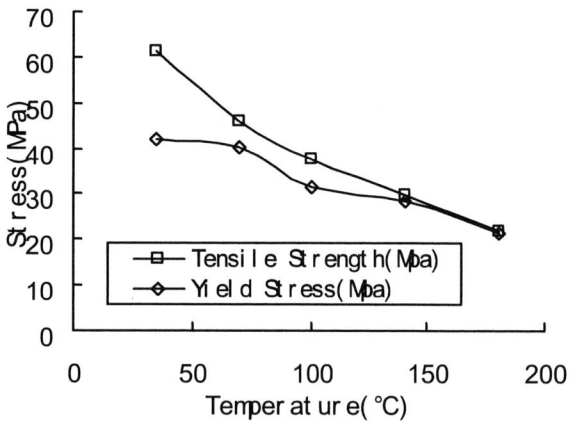

Figure 9: Tensile strength and yield stress versus temperature at strain rate 7.5E-3/s

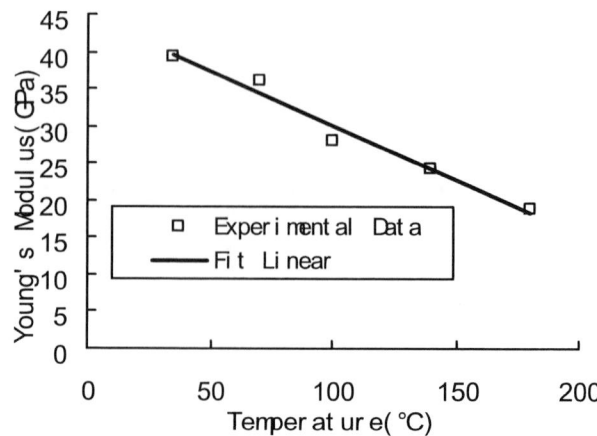

Figure 10: Temperature dependent Young's modulus at strain rate 7.5E-3/s

The temperature dependent Young's modulus for this lead-free material is also obtained from the stress-strain curves in the range of temperature from 25°C to 200°C at 7.5E-3/s.

The results are plotted in figure 10. It has observed that the Young's modulus is approximately a linear function of temperature at strain rate 7.5E-3/s:

E(GPa)=44.7-0.146T(°C).

We should indicate that, it is likely that because the tests are carried out on the cast test specimens, the results may not represent the true behavior of actual solder joint. In addition, the scale of the microstructure and the presence of intermetallics at the specimen surface might have an effect on the viscoplastic behavior. Finally, the annealing and aging of the specimens prior to testing could also affect the material properties. These factors will be considered in our future studies.

5. Anand Model Parameter Estimation

In order to apply the Anand model to simulate the thermomechanical responses of solder joints in electronic packaging and surface mount technology, the material parameters of the constitutive relations must be determined first. The data of load and displacement at real time by the software of test machine are converted to true stress and true strain. Then they are fitted into Anand model parameters by First Optimization®, an effective non-linear fitting software.

When the steady state plastic flow occurs, the saturation stress appears. For viscoplastic deformation behavior, this steady state plastic flow happens when the plastic flow has been fully developed and equals to the applied strain rate at a fixed temperature and a fixed strain rate. On the other hand, at a given stress and temperature, one can gain the steady state creep rate. This steady state creep rate is equivalent to the steady state plastic flow in constant strain rate testing when the saturation stress

228

equals the given stress in steady state creep testing. The saturation stress can be obtained as follows:

$$\sigma^* = \frac{\hat{s}}{\xi}\left[\frac{\dot{\varepsilon}_p}{A}\exp\left(\frac{Q}{RT}\right)\right]^n \sinh^{-1}\left[\left(\frac{\dot{\varepsilon}_p}{A}\exp\left(\frac{Q}{RT}\right)\right)^m\right] \quad (7)$$

Where the material parameters A, Q/R, m, n, and $\hat{s}/\xi$ in (7) can be fitted to the $\dot{\varepsilon}_p$-σ^* pair data obtained from constant strain rate tests.

Values of the nine material parameters A, Q/R, ξ, m, h_0, $\hat{s}$, n, a, and s_0 of Anand model for 95.5Sn4.0Ag0.5Cu solder are determined by the procedures outlined as follows.

a) Determination of the saturation stresses σ^* and steady state plastic strain rates from all the tests.

b) Nonlinear fitting of A, Q/R, m, $\hat{s}/\xi$, and n in (7).

c) Determination of ξ and $\hat{s}$. Using the values determined in step b), the parameter ξ is selected such that the ratio of σ/s is less than unity, and $\hat{s}$ is then determined from the combined term $\hat{s}/\xi$.

d) Nonlinear fitting of h_0, a, and s_0. The stress of Anand model can be expressed as

$$\sigma = \sigma^* - \left[\left(\sigma^* - cs_0\right)^{(1-a)} + (a-1)\left\{(ch_0)\left(\sigma^*\right)^{-a}\right\}\varepsilon_p\right]^{\frac{1}{(1-a)}} \quad (8)$$

with

$$c = \frac{1}{\xi}\sinh^{-1}\left(\frac{\dot{\varepsilon}_p}{A}\exp\left(\frac{Q}{RT}\right)\right)^m \quad (9)$$

The values of material parameters h_0, a, and s_0 are obtained from the σ-ε_p curves (especially the transient state) with various temperatures and strain rates. In the fitting, the saturation stresses given in step a) are also used. The material parameters for 95.5Sn4.0Ag0.5Cu solder determined by following this procedure are listed in Table 1. It can be clearly seen from figure11 that the saturation stress is temperature and strain rate dependent. Comparisons of the model simulation with test curves of 95.5Sn4.0Ag0.5Cu solder are shown in figures 12-17 respectively.

Table 1: Fitted Anand model constants for SnAgCu

Description	Symbol	Units	SnAgCu
Initial value of s	s_0	MPa	20
Activation energy	Q/R	K	10561
Pre-exponential factor	A	1/s	325
Stress multiplier	ξ	-	10
Strain rate sensitivity of stress	m	-	0.32
Hardening coefficient	h_0	MPa	8.0E5
Coefficient for deformation	$\hat{s}$	MPa	42.1

resistance saturation value			
Strain rate sensitivity of saturation value	n	-	0.02
Strain rate sensitivity of hardening coefficient	a	-	2.57

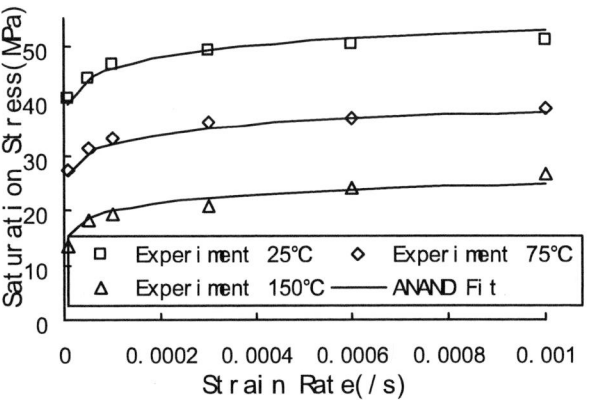

Figure 11: Saturation stress from experiment and Anand fit at different temperatures

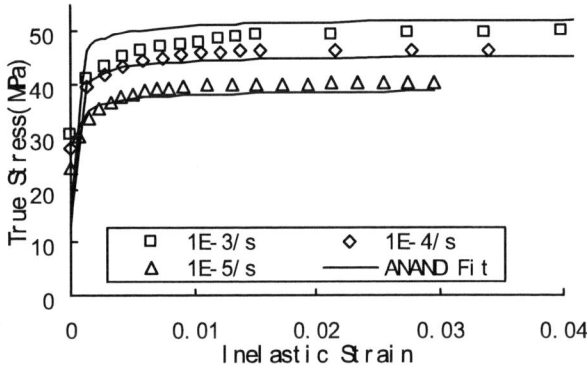

Figure 12: Experiment data and Anand fit curve at 25 °C

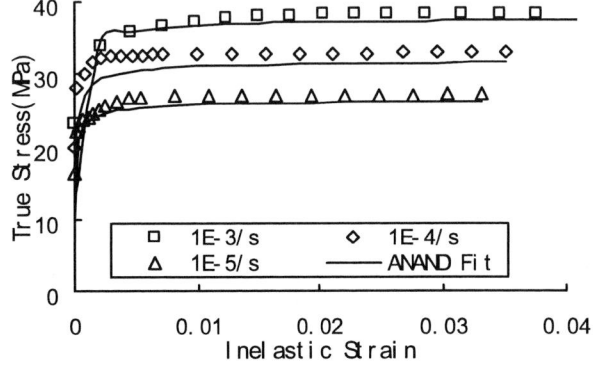

Figure 13: Experiment data and Anand fit curve at 75 °C

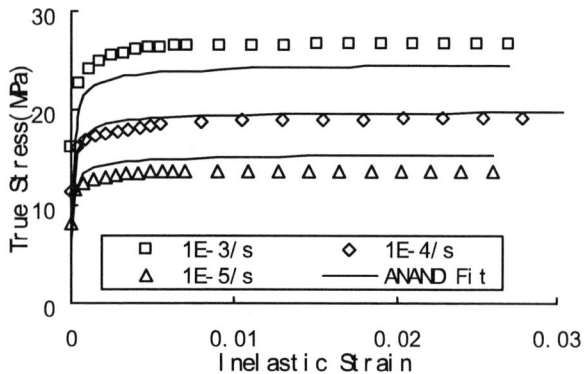

Figure 14: Experiment data and Anand fit curve at 150 ℃

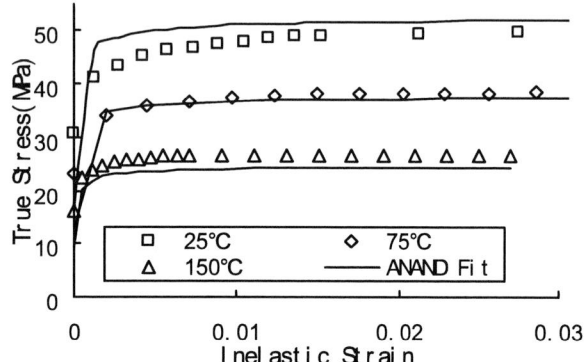

Figure 15: Experiment data and Anand fit curve at 1E-3/s

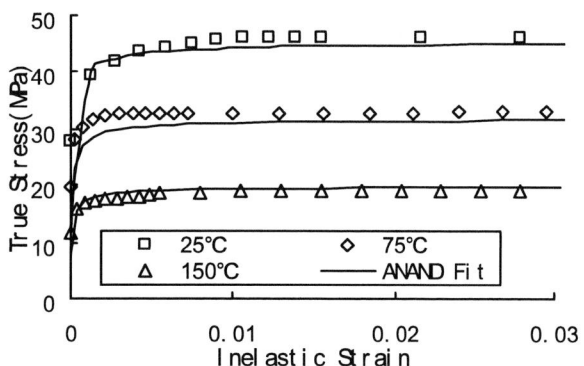

Figure 16: Experiment data and Anand fit curve at 1E-4/s

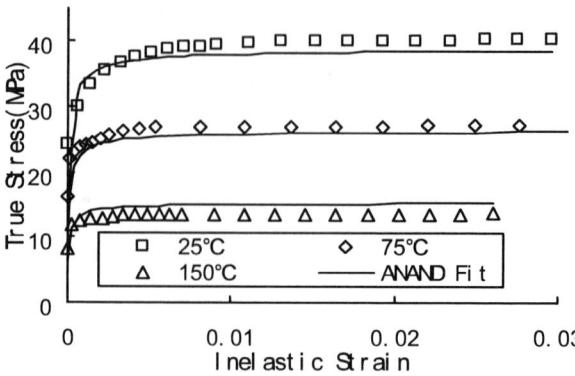

Figure 17: Experiment data and Anand fit curve at 1E-5/s

Above figures show that the Anand model can represent the non-linear deformation behavior at different strain rate and temperatures. Nevertheless, the result of Anand model fitting is not perfect to us. One can see there are some deviations from the experimental data and some errors may also come from the experiments. The result shows that Anand model has limited ability to describe the temperature/strain rate cross dependent relationship, especially at high homologues temperature. A modified Anand model is proposed in the next section.

6. Modified Anand Model Parameter

As mentioned in previous section, Anand model needs more temperature sensitivity and strain rate sensitivity to fit lead free material better. Some researcher suggested modifying the s_0 as linearly temperature dependent [10]. Pei *et al.* [15] modified all nine parameters as linearly temperature dependent, this requires a large amount of work. However, after observation through test curves, it has found that some parameters are only slightly dependent on temperature. So those parameters could be treated as constants. In this study, three parameters s_0, $\hat{s}$ and ξ will be further investigated based on our observation and study of test data, they are considered as the functions of both temperature and rate dependent:

$$s_0 = a_0 + a_1 T + a_2 T^2 + a_3 \dot{\varepsilon}_p + a_4 \dot{\varepsilon}_p^2$$
$$\hat{s} = a_0 + a_1 T + a_2 T^2$$
$$\xi = a_0 + a_1 T + a_2 T^2 \qquad (10)$$

Non-linear optimization fitting software is used for the model fitting. The constants of this lead free solders are listed in Table 2 and Table 3. The comparisons of the modified Anand model simulation with test curves of 95.5Sn4.0Ag0.5Cu are shown in figures 18-24 respectively.

Table 2: Anand constants used for equation (10)

Description	Symbol	Units	SnAgCu
Initial value of s	s_0	MPa	-
Activation energy	Q/R	K	12480
Pre-exponential factor	A	1/s	4223
Stress multiplier	ξ	-	-
Strain rate sensitivity of stress	m	-	0.34
Hardening coefficient	h_0	MPa	2.98E5
Coefficient for deformation resistance saturation value	$\hat{s}$	MPa	
Strain rate sensitivity of saturation value	n	-	0.03
Strain rate sensitivity of hardening coefficient	a	-	1.98

Table 3: Three modified coefficients for equation (10)

Symbol	a_0	a_1	a_2	a_3	a_4
s_0	18	0.24967	-6.25E-4	5925	-3.63E6
$\hat{s}$	-7	0.33	-6.48E-4	-	-
ξ	18.7	0.016	-1.1E-4	-	-

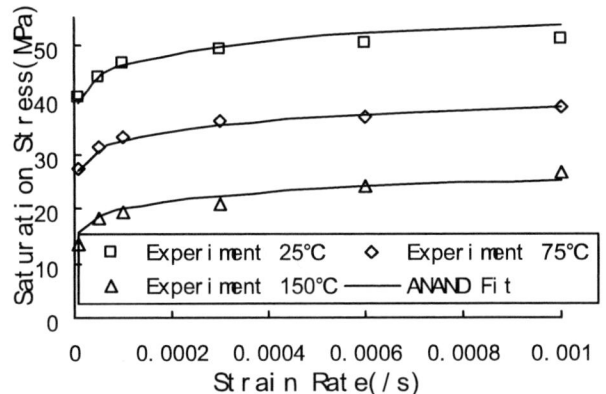

Figure 18: Saturation stress from experiment and modified Anand fit at different temperatures

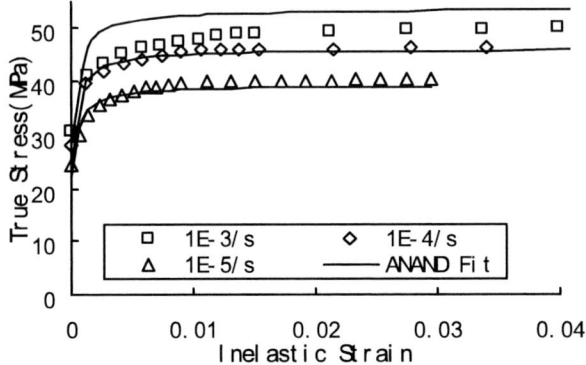

Figure 19: Experiment data and modified Anand fit curve at 25 °C

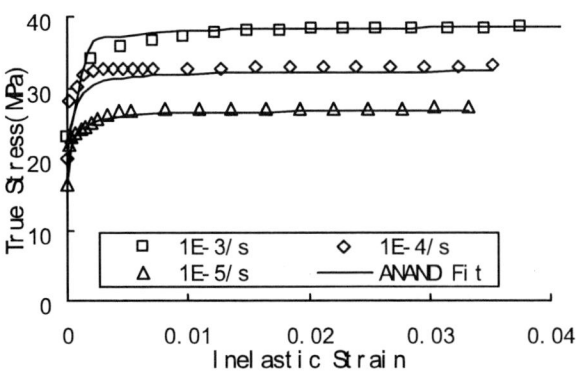

Figure 20: Experiment data and modified Anand fit curve at 75 °C

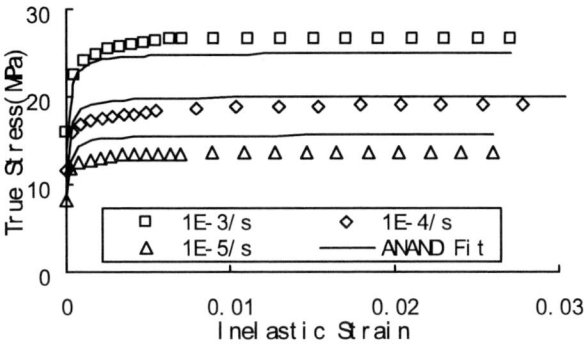

Figure 21: Experiment data and modified Anand fit curve at 150 °C

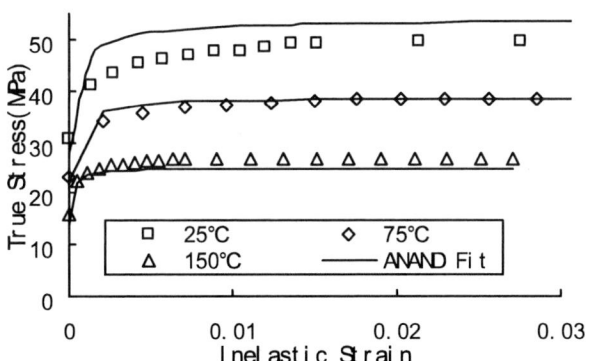

Figure 22: Experiment data and modified Anand fit curve at 1E-3/s

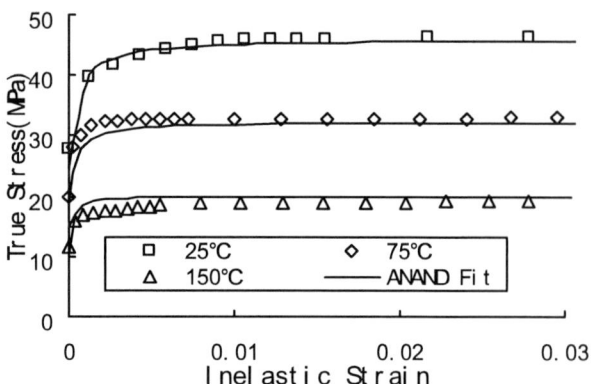

Figure 23: Experiment data and modified Anand fit curve at 1E-4/s

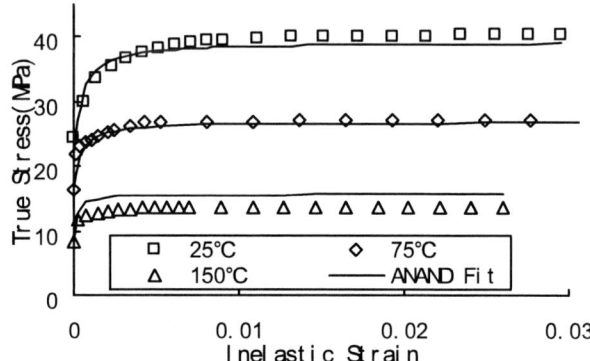

Figure 24: Experiment data and modified Anand fit curve at 1E-5/s

Compare figures 11-17 with figures 18-24, it is obvious that the temperature and strain rate dependent Anand model fits much better than the initial version. More parameters are an additional to this improvement. Even Chen *et al.* [12] modified h_0, the hardening co-efficient, as both temperature and strain rate dependent. Nevertheless, just to modify the h_0 may not be able to effectivly solve the entire 9 parameter system. The temperature dependence of other parameters in the model suggest the form of Anand model has its limitation on solder materials. Addidional models need to be explored to describe the deformation of solder material more accurately and more efficiently.

7. Conclusions

In this paper, tensile tests for Pb-free material 95.5Sn4.0Ag0.5Cu with different temperatures and strain rates have been conducted. It is found that this type of lead-free material has strong dependence on both test temperature and strain rate. The values of tensile strength and yield stress have directly related to temperature. The Young's modulus may approximately be expressed as a linear function of temperature.

The Anand model is applied to represent the inelastic deformation behavior of this pb-free solder

95.5Sn4.0Ag0.5Cu. A fitting procedure is presented based on the non-linear data fitting software First Optimization®. To fit stress-strain curves more perfectly, a modified method is also discussed in this work. The materials parameters of Anand model obtained through the modified method are verified to see if it can predict the viscoplastic deformation behavior. As a result, modified Anand model can indeed fit much better than the initial Anand model.

Since the input Anand parameters in ANSYS have no the function of both temperature and strain rate for modified Anand model, a user-defined subroutine code may be developed.

Acknowledgments

The authors gratefully acknowledge support from the Chinese National Science Foundation (NSF) No. 10372093 and Fairchild Semiconductor Corp. Thank Fairchild, Suzhou for their help to supply the test materials.

References

1. Wilde, J. "Rate Dependent Constitutive Relations Based on Anand Model for 92.5Pb5Sn2.5Ag Solder", *IEEE TRANSACTIONS ON ADVANCED PACKAGING*, August 2000, Vol. 23, No. 3.
2. Darveaux, R. and Banerji, K., "Constitutive Relations for Tin-Based Solder Joints", *IEEE Trans. CHMT*, 15, No. 6, (1992), pp. 1013–1024.
3. Weinbel, R. C. *et al*, "Creep Fatigue Interaction in Eutectic Lead-Tin Solder Alloys", *J. Mater. Sci. Lett.*, 6, (1987), pp. 3091–3096.
4. Lau, J. and Rice, J. R. "Thermal Stress/Strain Analyses of Ceramic Quad Flat Pack Packages and Interconnections", *Proc. Elect. Components and Technology 40th Conf.*, Las Vegas, NV, 1990, Vol. 1, pp. 824–834.
5. Sarihan, V. "Temperature Dependent Viscoplastic Simulation of Controlled Collapse Solder Joint Under Thermal Cycling", *ASME J. Electron. Packag.* 115, (1993), pp. 16–21.
6. Pao, Y. H. *et al*, "Constitutive Behavior and Low Cycle Thermal Fatigue of 97Sn3Cu Solder Joints", *ASME J. Electron. Packag.*, 115, No. 6, (1993) pp. 147–152.
7. Knecht, S. and Fox, L. R. "Constitutive Relation and Creep-Fatigue Life Model for Tin-lead Solder", *IEEE Trans. CHMT*, 13, No. 2, (1990), pp. 424–433.
8. Amagai, M. *et al*, "Mechanical Characterization of Sn-Ag-Based Lead-Free Solders", *Microelectronics Reliability*, Vol. 42, No. 6, (2002), pp. 951-966.
9. Kim, Y. *et al*, "Vibration Fatigue Reliability of BGA-IC Package With Pb-Free Solder and Pb-Sn Solder", *Proceedings of 53rd Electronic Components and Technology Conference*, New Orleans, LA, May. 2003, pp. 891-897.

10. Rodgers, B. *et al*, "Experimental Determination and Finite Element Model Validation of the Anand Viscoplasticity Model Constants for SnAgCu", *EuroSimE 2005*, Berlin, Germany, April. 2005, pp. 490-496.

11. Reinikainen, T. *et al*, "Deformation Characteristics and Microstructural Evolution of SnAgCu Solder Joints", *EuroSimE 2005*, Berlin, Germany, April. 2005, pp. 91-98.

12. Chen. *et al*, "Modified Anand Constitutive Model for Lead-Free Solder Sn-3.5Ag", *The Ninth Intersociety Conference on Thermal and Thermomechanical Phenomena in Electronic Systems.*, *IEEE*, 2004, pp. 447-452.

13. Anand, L. "Constitutive Equations for the Rate-Dependent Deformation of Metals at Elevated Temperatures", *Journal of Engineering Materials and Technology,* Transactions of the ASME, vol.104, No. 1, (1982), pp. 12-17.

14. Brown, S. *et al*, "An Internal Variable Constitutive Model for Hot Working of Metals", *International Journal of Plasticity*, vol. 5, No. 2, (1989), pp. 95-130.

15. Pei, M. *et al*, "Constitutive Modeling of Lead-Free Solders", *Advances in Electronic Packaging 2005, IEEE*, San Francisco, CA, Jul. 2005, pp. 1307-1311.

16. Zhang, Q. *et al*, "Constitutive Properties and Durability of Lead-Free Solders", *CALCE EPSC Press*, Maryland, 2003, pp. 65-137.

Numerical Simulation and Thermal Failure Analysis of SOM Package

Jae Choon Kim[a], Jin Taek Chung[a], Won Suk Lee[a], Gyoung Bum Kim[b], Dong Jin Lee[b], Ji-Man Cho[c], Ji-Hyuk Yu[c]
Byeong-Kwon Ju[c], SungWoo Hang[b], Heung-Woo Park[d], Sang-Kyeong Yun[d]
Department of Mechanical Engineering, Korea University[a]
Department of Electrical Engineering, Korea University[b]
Display and Nanosystem Lab, College of Engineering, Korea University[c]
SOM Group, OS Division, Samsung Electro-Mechanics[d]
Email: jchung@korea.ac.kr

Abstract

This paper presents a numerical analysis and experimental study on failure of the spatial optical modulators (SOM) package and its preventing from thermal damage. The SOM package has been developed for projection displays. Most heat sources of the SOM package were mounted on a glass plane that was coated with SiO2 and TiO2 for optical performance. The heat sources were distributed by entering the laser beam to the operating mirror, heating of IC chip and ohmic heating in the circuit of glass plane. In designing the SOM package, it is important to consider the thermal influence to the SOM package's performance and reliability. In this study, the electrical and thermal simulation by CFD and experiments using infrared camera are carried out for the enhancement of thermal reliability. The results are in good agreement with the experiments results and the maximum temperature of glass plane was decreased by the change of glass plane structure change.

1. Introduction

The issue of temperature is considered of a critical factor that impacts the performance of the product in designing electronic devices. The SOM package which is an optical module that drives mirrors in nano scale movement also responds sensitively to temperature. The piezoelectric element drives the mirror of the nano scale to shift vertically and the light reflected from the transmitted laser creates the image. [1] The SOM Mirror is controlled by the input signal of the IC and the IC chip generates heat during the process. The heat impacts the drives of the SOM. The Pattern Line that transmits input and output power and signal to the IC Chip and SOM is composed of layers of Cr and Gold with another layer of Cr. It generates immense heat according to the electric current and voltage, and the failure of line occures by more heat and structural stress.

N. Cordero[2] conducted a study on analysis the Joule-heating using CFD. He studied the output on the Copper surrounded by Glass, SU8, and SiO2, and observed the heat generating by using CFD. As a result, he proved that CFD modeling could also calculate joule-heating of micro size conductors.

The simultaneous analysis of EM(Electro-Migration) in its electronic, mechanical, and thermal aspects can be conducted utilizing multi-physics. L.H.Ling[3] analyzed solder joint used in IC Package and the electronic, mechanical, and thermal aspects of the EM(Electro-Migration) phenomenon of interconnects. He also looked into the location and temperature where EM is generated and its structural development.

In order to, however, more clearly define the failure of package, we must conduct joint research of CFD, and thermal experiments using infrared cameras. This paper presents the failure phenomenon by analyzing thermal and structural elements of SOM Package and thermal measurement experiments.

2. Thermal analysis of the Package

A thermal analysis was performed by modeling the package and using the Computational Fluid Dynamics (CFD) in order to confirm the internal thermal distribution. The Flotherm which is a commercial code was used to perform the computation. The results were also verified through series of experiments. The structure of the SOM Package is composed of IC that controls the input and output on the glass pane, the SOM that controls laser reflections, and a passivation by epoxy on the top as shown in Figure 1.

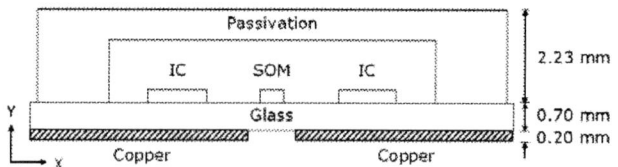

Figure 1: Schematic of SOM package

It is the IC Chip that mainly emits heat in the package. The heat source of the IC Chip is approximately 80mW and around 15-20% of the total input is generating heat. The Figure 2 represents the computational results showing the distribution of temperature. The terms in Table 1 was used as boundary and test conditions. The Figure 2 (a) presents the temperature distribution on the cross section of the package. The maximum temperature of the IC chip was 57℃ and the comparable figure

Table 1. Material Property

Material	Conductivity
Glass	0.89W/mK
SOM & IC	110W/mK
Au	301W/mK
Cr	69.1W/mK
Cu	385W/mK
Passivation silicone	0.35W/mK
Ambient Temperature	20℃

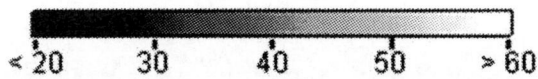

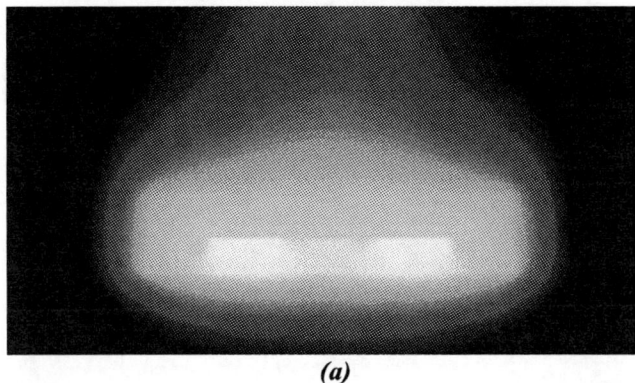

(a)

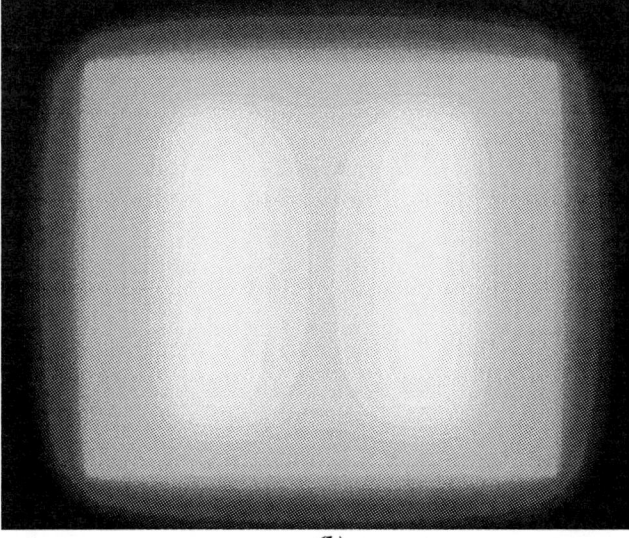

(b)

Figure 2: Temperature distribution of (a) across a section, (b) a glass plate obtained by CFD

for the Glass Plate was 53℃. The Figure 2 (b) is the simulation results of the temperature distribution of the glass plate. Compared to the Figure 3 which is the result of the experiments, the error range is within 5%. The following are ways to increase the cooling performance of the package. The structure of the package can be

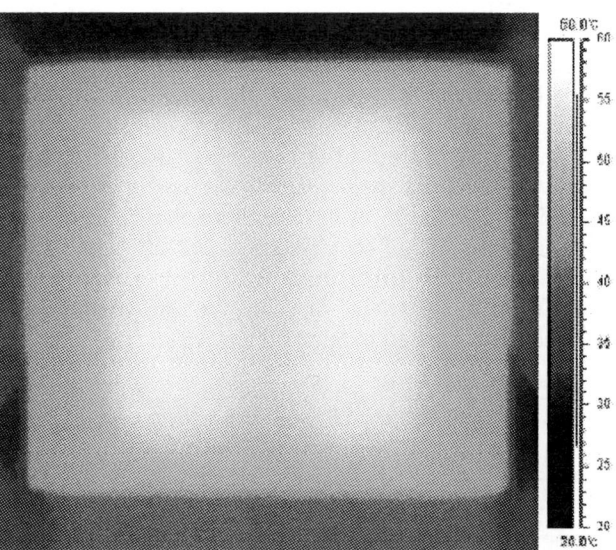

Figure 3: Temperature distribution of a glass plate obtaind by experiment

Table 2. Temperature comparision

	Maximum Temperature
Case 1	57.4℃
Case 2	56.1℃
Case 3	54.7℃

altered; a metal layer can be added to the glass plane; a heat sink or a fan can be loaded. In the case of SOM package, it is difficult to go with the aggressive methods such as loading a fan or a heat sink. Therefore, we could consider adding a metal layer on the glass plane[4] or enhancing the convection heat transfer coefficient by altering the direction of gravity on the package. Table 2 displays the maximum temperature of three cases of applied to increase cooling performance. The case 1 is for normal condition and case 2 differs from case 1's gravity direction. The case 3 has a copper layer attached on glass. Comparing the simulations in which the package was standing both vertically and horizontally, the thermal difference is around 1℃ which translates into a 2% cooling effect and the maximum temperature is lower around 3℃ than the case 1 when the package has a copper layer.

3. Thermal analysis of the Glass Line using CFD

The heat generated along the pattern line of the nano scale used in the package mainly causes the failure in the line. Therefore, in order to analyze the line failure, a thermal analysis is required. The pattern line on the glass that transmits the input and output power and signal from the SOM package to the IC generates intense heat even at low levels of electric current and voltage. A thermal model for CFD was made in real conditions in which the line failure occurs, and a transient analysis was conducted. The result was then compared with the test results.

The thermal model used in the computations is presented in Figure 4. A pattern was created using gold and Chromium on the glass plane of 0.7mm thick. Chromium was used because the low adhesion ratio of gold makes patterning difficult if used solely. In order to conduct a Joule-heating analysis on the triple layered patter line, the three layers were formatted into a single layer and the value of each electric resistively and thermal conductivity was substituted for a single value. The location of the pattern line on the glass was made identical with the conditions of the experiment.

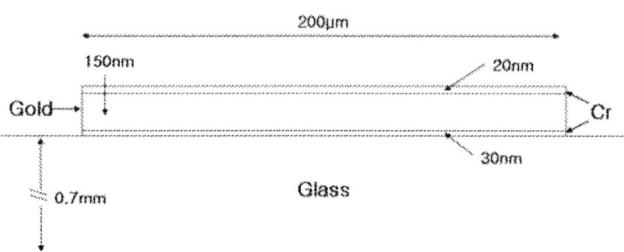

Figure 4: Cross section of Omhic heating Model

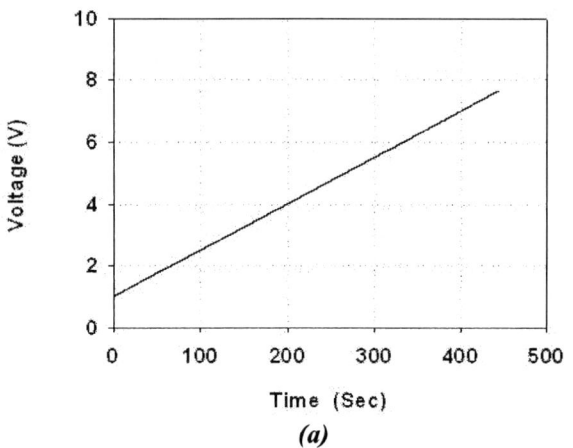

(a)

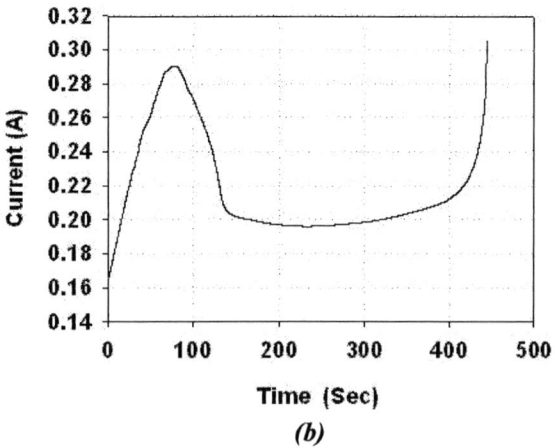

(b)

Figure 5: Loaded values (a) Voltage, (b) Current

Identical boundary conditions were used to compare the values of the experiment results. The Figure 5 (a)

displays increasing voltage of the pattern line by 30mV per every two seconds starting at 1 volt and presents the differing value until the failure occurs in the line. The Figure 5 (b) is the value of electric current that is confirmed along the pattern line according to the time and voltage change and shows the two peak values around 70 seconds and 450 seconds. The value does not change greatly in between the 150 seconds and 400 seconds range.

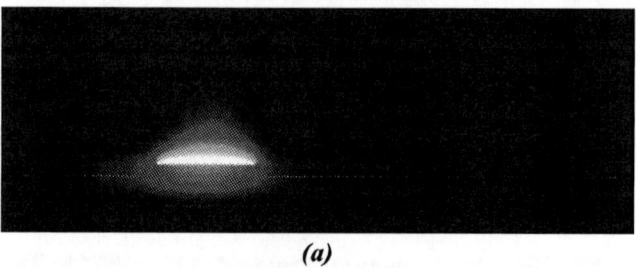

(a)

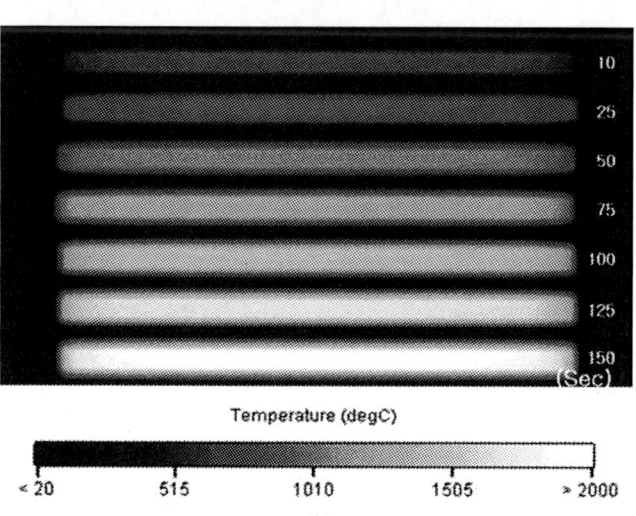

(b)

Figure 6: Temperature distribution at line according to time level

Figure 6 (a) represents the sectional diagram of the heat generated from the line after 150 seconds have past and the Figure 6 (b) displays of the temperature change in the pattern line associated with the time and voltage changing. After 75 seconds, it surpassed 1064℃ which is the melting point of gold. Exceeding the 150 seconds mark, it went over 1907℃ which is the point Chromium melts. The maximum temperature was recorded in the mid point of the line. The time of line failure deducted by the results according to the conditions of Figure 4 is around the 70 seconds mark. The voltage and current at this point are 2.0645V and 0.2876A, respectively.

4. Temperature measurement of Glass Line

A thermal measurement on the pattern line using an infrared camera was also conducted to verify the thermal analysis and more accurately observe the failures. The pattern lines used in the experiment were identical to that of the Figure 3 which is the structure of the SOM package pattern line. The specification of the infrared camera is given on Table 3.

Figure 5 (a) shows of the changes in temperature after putting consistent voltage on both ends of the line. The maximum change of the line according to the given timeframe is presented in Figure 7.

Table 3. Specification of Infrared Camera

Pixel number	320*240pixel
Spetial resolution	1.3mrad
Spectral range	7.5 to 13μ
1Pixel area	18μ m²

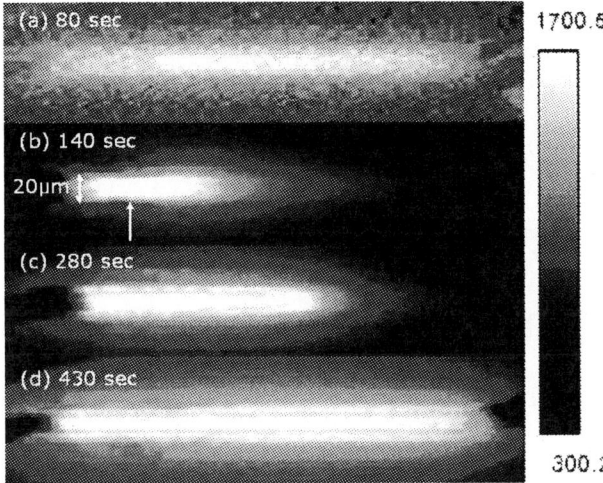

Figure 7: Temperature distribution after (a)80 seconds, (b)140 seconds, (c)280 seconds, (d)430 seconds at line

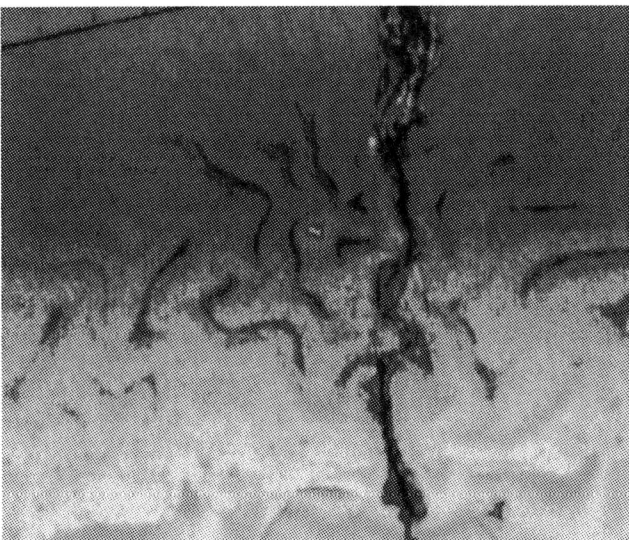

Figure 8: 1000times scale up of failure spot

Figure 7 (a) displays failure spot of the line at 80 seconds. The maximum temperature is around 1100 ℃ at 80 seconds. Figure 7 (b) presents the location that located at the three-quarters point of the line and occurs inclined toward the cathode. The maximum temperature of the line is around 1700 ℃ at the location. Gold melts at 1064 ℃, and it is concluded that the layer of gold between the two Chromium layers has already melted. Figure 7 (c) shows the heat transmit to the anode. Figure 7 (d) is the temperature distribution right before the line is cut. The maximum temperature is over the 1700 ℃. The melting point of Chromium is also 1907 ℃, which means that the structural failure could occur as the temperature of the line reaches the melting point. The observations with an optical microscope revealed that the most massive destruction occurred at the point of maximum temperature and in other areas as well. The results of experiment are simular to simulation's until 80 seconds where temperature records around melting point of gold 1064 ℃, however the error occur after 80seconds. The gold layer between the two Chromium layers melts already at 1064 ℃ before Chromium layers melt. The simulation did not consider the phase changing. That can be the reasons for difference between simulation and experiment's results.

Figure 8 shows results at spot of pointed in Figure 7 (b) that were observed by microscope. The wrinkles of occupied beside the failure line exist around pattern line edge mainly. It seems to make on process of solidification after fusion.

5. Conclusions

A predictive thermal analysis model was developed and applied for analzing SOM package temperature and was verificated by experiments. One package structure was proposed for good thermal performance and compared maximum temperature with original package. The maximum temperature of proposed package is lower 2.7 ℃ than original package.

The package install direction is important in electric devises decision because the gravity could affect the heat transfer coefficient. The simulation conducted according to gravity direction using the model. The maximum temperature of package was laid horizontally with gravity direction is lower 1.3 ℃ than the package was laid vertically with gravity direction.

A transient thermal-electric analysis model was also developed and applied for analysis about package pattern line failure. It was verificated by experiments. Compared with previous studies of failure, it can obtain the temperature change of the micro scale line with respect to time and voltage change are obtained using infrared thermal camera.

Since the temperature and time when the failure occurs at package under high current density stressing are defined the results of this study can help to study about electromigration.

Then, the analysis model proceeded and developed to provide a design gideline of the package to avoid thermal problem.

Acknowledgments

This was supported by the Brain Korea 21 Project in 2006. This work was also supported by IITA in Korea MIC, Leading edge R&D Program.

References

1. Sang Kyeong Yun. *et al*, "Spetial Optical Modulator(SOM): Samsung's Light Modulator for the Next Generation Laser Diaplay" *The 6th International Meeting on Information Display and The International Display Manufacturing Conf*, Daegu, Korea, August. 2006, pp. 551-555.
2. N. Cordero, J. West and H. Berney, "Thermal Management of a Joule-heating Microreactor using Modelling Tools" *The 4th Int. Conf. on Thermal, Mechanical and Multiphysics Simulation and Experiments in Micro-Electronics and Micro-Systems, EuroSimE,* Mar. 2003, pp. 85-88.
3. L.H.Liang, Y.J, Xu and Y. Liu "Electro-Migration Study in Solder Joint and Interconnects of IC packages" *The 7th Int. Conf. on Thermal, Mechanical and Multiphysics Simulation and Experiments in Micro-Electronics and Micro-Systems, EuroSimE,* April. 2006, pp. 1-7.
4. John Lohan. *et al* "Experimental and Numerical Investigation into the Influence of Printed Circuit Board Construction on Component Operating Temperature in Natural Convection" *IEEE Transsactions on Components and Packaging Technologies,* Vol.23 No 3, Semtember. 2000, pp.578-586

A Mathematical Technique to estimate the High Frequency Current Inside the Silicon Die from the Noise Measurements

Bidyut K. Bhattacharyya (IEEE Fellow) and Gang Huo
Intel Corporation, Oregon, USA
Email: Bidyut.K.Bhattacharyya@intel.com

Abstract

In this paper, we have shown a method to determine the current drawn by the device as a function of time. It is normally difficult to determine the switching current drawn by the device as a function of time. In this paper, the current drawn by the device is determined by measuring the voltage (V_X) at some desire location on the top of the device and then by simulating the impedance profile from the basic topology of the power delivery network. The L, C and R of the power deliver network were assumed as the fundamental parameters of our methodology. In our methodology, some of the AC parameters of power delivery network including the average current drawn by the device, the average voltage at the measuring node and the steady power supply voltage, were assumed to be known.

Introduction

There are various articles in the literatures, which describe how to measure the power and ground noise generated by the devices. There are papers (reference-1, 3) which describe how to generate the desire currents drawn by the device, in frequency domain. In reference-1, it was shown that current waveform, generated by a CMOS device always leads to a constant $I(\omega)$ up to 800MHz. Thus as long as the Power Delivery Network (PDN) has fundamental frequencies less than 800MHz one can assume constant current in frequency domain. Normally the fundamental resonance frequencies of the Power Delivery Network are less than 300MHz (for a very good PDN design). In time domain, constant $I(\omega)$, translate to a Dirac Delta function waveform, $I_0\delta(t)$. This is quite different than, what is described in reference-2, 3 by Alex Waizman et al. In Reference-3, one was trying to achieve a periodic square wave current $I(t)$. This periodic wave expanded in terms of sine and cosine term with decaying amplitude ($\sim 1/n$) to determine $I(\omega)$. This kind of square wave current was achieved by turning the device clock on for N cycles and then turning the device clock off for N cycles. In this way, one forms a square wave current waveform with period NT_{CLOCK}. Where, T_{CLOCK} is the internal clock period used to drive the device. It is important to mention that most of the articles in the literatures are to define a set of experimentation to generate the desire current drawn by the device [reference-(1,3)] and then determine the value of $Z(\omega)$ after measuring the noise voltages. There are practically no articles, which describe the current drawn by the device while running an application.

In real life, the current drawn by the device is a constant current [$<I(t)>$] superimposed by a throttling current having various amplitudes in various time. This can be expressed using equation-(1) as shown below,

$$I(t) = <I(t)> + I_{AC}(t) \ldots\ldots\ldots\ldots\ldots\ldots(1)$$

The second term of the above equation is throttling term. In real life one need to understand how the device will generate the throttling term to produce the AC noise. This noise practically determines the device performance. The noise generated by the throttling term is normally less than if one assumes the step current where current will ramp from zero amperes to maximum I_{DC} in short time. People believe this current ramp occurs in few clock cycles. The few clock cycles are less than the response time of the power delivery network (PDN) or the inverse of the resonance frequencies of PDN. Normally the resonance frequencies of the PDN are less than 300MHz.

Modeling Methodology and Calculation of Current Drawn by the Device

In figure-(1) we show the schematic diagram of the power deliver network. R_{DIE} and C_{DIE} determine the die decoupling resistance and die decoupling capacitance. This capacitance C_{DIE} and resistance R_{DIE} deliver the instantaneous power required by the device. These two parameters, together with the absolute value of L determine the instantaneous voltage drop (ΔV_x) on Power Delivery Network. If package inductance (L) and resistance (R) increases then the voltage drop (ΔV_x) will also increases for a given time dependent current. In our present methodology we measure the voltage V_x as seen by the device at C4 bumps (area array solder balls on the

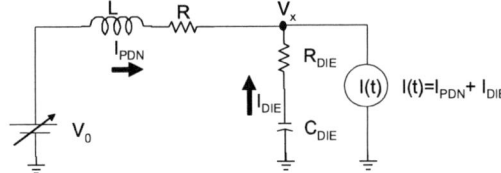

Figure-(1): Schematic Diagram of the Power Delivery Network. In real life the voltage V_0 is variable due to the complex network. After package inductance, L, one encounters mother board resistance, inductance and discrete capacitance components placed in package.

top of the die) and then estimate the current drawn by the device using the values of L, R, C_{DIE} and R_{DIE} of the PDN.

In this section we will generate an expression for the current drawn by the device, $I(t)$, as a function of time. The voltage measures at C4 bump is assumed to be V_X.

1-4244-1105-X/07/$25.00 ©2007 IEEE   239

This is shown in figure-(1). Normally this voltage is different at different C4 bumps. There are many C4 bumps in a device. Simulation, using 30amp of current ramp in short time, shows that the maximum variation in voltage (due to finite impedance of power bus inside the devive) bewteen any two C4 ground (or power) bumps on the top of a given device is less than 15mV. This current ramp normally generates noise somewhere between 100-200mVolt, when the supply voltage is about 1.2Volt. Since the variation of voltages between any two C4 bumps is small and hence for all practical purpose we will assume the voltage, $V_X(t)$, at every power and ground bumps on the device is same at any given time. That is represented by $V_X(t)$ in figure-(1) & (2).

Given above assumption, let us first calculate the current delivered by C_{DIE} and R_{DIE}. If I^1_{DIE} and I^2_{DIE} are the current delivered by the C_{DIE}, at time $t = t_1$ and also at time $t = t_2$ then one can write the following expression to determine the current through the C_{DIE} . V^1_X and V^2_X are the measured voltages at time $t = t_1$ and at time $t = t_2$ respectively.

$$A_{DIE} I^2_{DIE} = B_{DIE} I^1_{DIE} - (V^1_X - V^2_X)/R_{DIE} \ldots\ldots\ldots(2)$$
Where,
$$A_{DIE} = [1-(t_1-t_2)/(2\Gamma)], \quad B_{DIE} = [1+(t_1-t_2)/(2\Gamma)] \ldots\ldots(3)$$
and $\Gamma = R_{DIE} C_{DIE}$.

Equation-(2) can be solved iteratively to determine the current I_{DIE} delivered by the die decoupling capacitance at various time. The average value of this current over long time should be zero.

The current delivered by the PDN can also be written by the following expression. This is also an iterative equation which determines the current at time $t=t_2$ knowing the current drawn at time $t = t_1$ ($<t_2$). At time $t=0$ I_{PDN} is assumed to be zero.

$$A_{PDN} I^2_{PDN} = -B_{PDN} I^1_{PDN} + [V_0 - 0.5(V^1_X + V^2_X)]/R \ldots\ldots(4)$$
Where,
$$A_{PDN} = [0.5 - L/\{R(t_1-t_2)\}] \text{ and } B_{PDN} = [0.5 + L/\{R(t_1-t_2)\}] \ldots\ldots(5)$$

While deriving the equation-(2) and equation-(4) we assumed the average voltage and average current between any two consecutive times (linear extrapolation). Equations-(2) & (3) also determine the total current, $I(t)$. This current, $I(t)$, is shown in equation-(7) and (8).

In real life the actual interconnect is very complex. The complex interconnect is shown in figure-(2). Where the voltage V_0, defined in figure-(1) is also variables due to the inductance and resistance of the mother board, and also due to the added package decoupling capacitance.

The recursive equation-(2) and equation-(4) is vaild even though the voltage V_0 is a function of time. If V_0 is a function of time then one needs to replace V_0 in equation-(4) by average value of voltage at time $t=t_1$ and at time $t=t_2$. Thus V_0 can be represented by the following equation:-

$$V_0 = (V^1_0 + V^2_0)/2 \ldots\ldots\ldots\ldots(6)$$

Expression for the final current and an expression for the goodness parameter $|\eta|$ of the data fit

Figure-(1) and (2) shows the current drawn by the device, $I(t)$. At any instant of time this current can be written in terms of the following equation:-
$$I(t) = I_{PDN}(t) + I_{DIE}(t) \ldots\ldots\ldots\ldots(7)$$

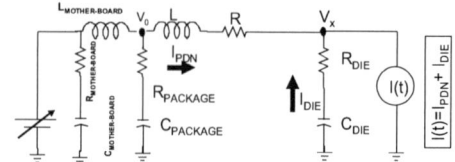

Figure-(2): This figure shows that actual PDN network has more variables that what is shown in figure-(1). This figure also shows that V_0 is also a function of time.

The instantaneous current $I^1(t_1)$ drawn by the device (required to solve the recursive equation) can also be written by the following equation:-
$$I^1(t_1) = I^1_{PDN}(t_1) + I^1_{DIE}(t_1) \ldots\ldots\ldots\ldots(8)$$

While determinining all or part of the L, C and R parameters, as shown in figure-(1) and figure-(2), we have used a parameter η^2, which we call the goodness of the data fit. For a given set of L, C and R parameters [described in figure-(1,2)], the goodness of the fit is defined by the following mathematical expression:-

$$\eta^2 = [<I(t)> - \sum_{i=0}^{k} I(t_i)/k]^2 \ldots\ldots\ldots\ldots\ldots(9)$$

The objective of the equation-(9) is to minimize the value of η for a given values of L, C and R while perfoming the data fit. The $<I(t)>$ is the average current and k is the number of measured data points. $<I(t)>$ is a measured parameter in our methodology. It is importamt to mention that for a given V_X, and given L,C and R values one can always generate an expression for current using equation-(2) and (4). Thus one will have redundancy in the determination of the current drawn by the device. In order to eliminate that redundancy one needs to input some other known AC parameters, like decoupling capacitance or the value of resonance frequency etc. from some other measurements as described in reference-(1,3).

Verification of the Equation-(2) and (4) using square wave current signals through circuit simulations

In this section we have generated a simulated noise voltage, V_X, given a known current $I(t)$ drawn by a device with $V_0 = 1$ Volt. In this case we have kept the voltage V_0 constant and used figure-(1) as our basic model. In order to determine the current drawn by the device as a function of time we used equation-(2) and (4) and also the simulated outout voltage, V_X. The currents which are determined using equation-(2) and (4) is then used to determine the total current $I(t)$ drawn by the device [equation-(7)]. This current is then compared with the simulated injected known current to determine how accurately we can estimate the current using equation-(2) and (4) and for a given goodness of the fit $|\eta|$. In figure-(3) we show the known current that we have used to simulate the voltage V_X [see figure-(1)] on the C4 bumps. The injected currents are square wave currents of magnitude 50amp with variable time widths. The PDN parameters or L,C and R values used to simulate the noise voltages, V_X, are shown below.

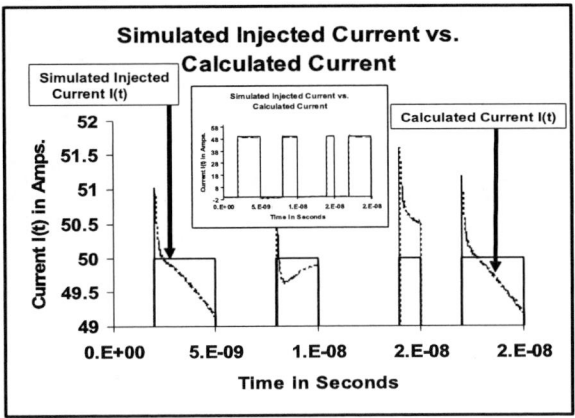

Figure-(3): This figure shows the simulated injected known current waveform vs. the calculated current using the simulated voltages. This plot is achieved with L, C and R-values, which produced minimum η^2.

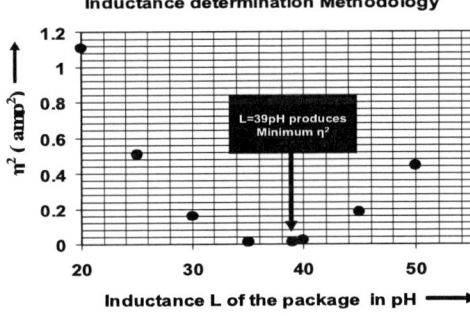

Figure-(4): This figure shows that when the package inductance L= 39pH the value of goodness of the fit is minimum. In this picture, we have assumed that C_{DIE}, R_{DIE} and R are measurable parameters. L is the only parameters we have changed in our present model.

R=5mOhm, C_{DIE} =120nF, R_{DIE} = 1mOhm, L =38pH. In the same figure-(3) we show the current I(t) drawn by the device when calculated using the simulated values of the voltages V_X. The best data fit was achieved with L=39pH, which is consistant with the simulated input values of these PDN parameters. While determining the current drawn by the device we assumed fixed value of C_{DIE} =120nF. Figure-(4) shows the goddness of the data fit. The value of η is minimum when L assumes 39pH. This is demonstrated in figure-(4). Even though the data has the best fit with L=39pH, but the calculated device current values, I(t), has an overshoot and undershoot. The

magnitude of this overshoot and undershoot is about 1.5 amp. This is about 2-2.5% of the maximum current swing 50 Amp. This overshoot and undershoot is also shown in figure-(3). One of the reason for this overshoot and undershoot is not having sufficient data points around the transition point, this is shown in figure-(5).

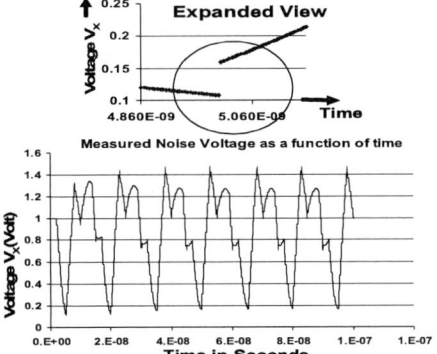

Figure-(5): The simulated voltage V_X as a function of time for square wave current waveform as shown in fig-(3). The top figure shows the expanded view in minor time scale to demonstrate discontinuities.

In figure-(5) we show the expanded view of the simulated noise voltage. This figure-(5) shows that there are sudden discontinuities in V_X around the transition point. This happens when currents [see figure-(3)] are switching from high to low or from low to high. When calculating I (t), around the transition point, we used the linear extrapolation of voltage and current between two consecutive times, that linear extrapolation is already reflected in equation-(2) and (4). That extrapolation causes the overshoot and undershoots of our present data fit methodology. In future one needs to develop some other extrapolation technique, which will help us to reduce the calculated error(<1%) while calculating the current drawn by the device.

Determination of current drawn by a real device while the device is in operation and running an application.

In this section we will discuss a real life case where we have collected noise data from a real device. The data was collected when the device was running an applications. The noise was measured beween the power and ground bumps using two sense lines [50 Ohm each see figure-(6)]. These sense lines are connected to two C4 bumps inside the device, one of the bump is power bump and the other one is ground bump. These two lines are also terminated at the measurement points using a 100 Ohm resistors. Figure-(7) shows the measured voltage for a real life device. When the particular application was running the average measured current delivered by the battery at 1.2 volt was about 13.7 Amp. Thus 13.7amp is the expected <I(t)> for this real life case. In this case, the part was running at 2.9GHz. The figure-(1) is used as our base model. Actually the figure-(2) will be much better model for this real life case, with V_0 as another set of measured data points near package decoupling capacitance. However, in this paper we have estimated an average V_0 using equation-(10) together with figure-(1) as our base operating model. Thus, V_0 will be an average driving voltage required to generate the desired current.

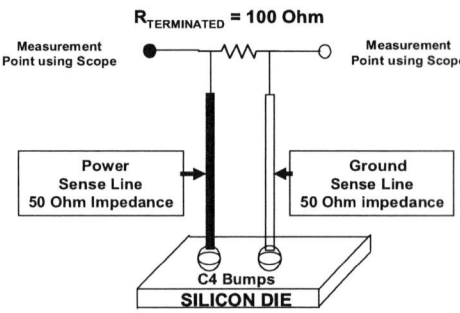

Figure-(6): This figure shows the schematic diagram of the measurement technique while one running an applications in real system.

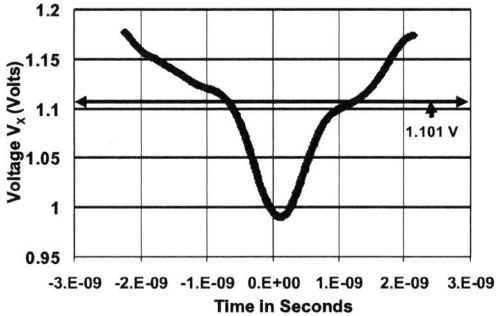

Figure-(7): This figure shows a section of the measured voltage, Vx, as a function of time while running an application. The average current delivered by the Voltage Regulator at 1.2 volt was 13.7 Amp.

$$V_0 = <I(t)>R + <V> \dots\dots\dots\dots\dots(10)$$

Where, $<V>$, is the average voltage (1.101V) the part sees at the measurement point. This is shown in figure-(7) using solid dark line. In this case the measured value of the die decoupling, C_{DIE} was 92nF. Equation-(10) and from the known value of C_{DIE}, we have perform the data fit to determine L, R and R_{DIE}. The result is shown in table-I below. We have observed that the sensitivity of the R_{DIE} to the actual average current or to the goodness of the fit η is negligible. As R_{DIE} decreases, the current drawn by the device looks much nosier than what we have shown in figure-(8). The table-I shows that the average driving voltage, V_0, is in the range of 1.18 to 1.17 Volts. This is due to the assumption that the maximum resonance frequency of the PDN in our present case is in the range of 250-300MHz. This table-I also shows that the effective DC resistance R is in the range of 5-6 mOhm. In addition, the effective package inductance was observed to estimated to be in the range of 3.4-3pH. The simulated package inductance data from physical geometries was 3.8pH. Thus, this is a close agreement.

In figure-(8) we also show the various components of current drawn by the device while running an application.

The data shows that approximately 20amp current was supplied by die decouple capacitors when the total current required by the device was 35Amp. The current supplied by PDN is about 15Amp. It may be worth mentioning that the worst dI(t)/dt for this real life case are about 31amp/nsec, this is calculated using figure-(8). Thus 20 amps current step requires 2.3 clock periods.

Table-I: Showing the data fit result of real life case. Estimated resonance frequency of PDN was somewhere between 250-300 MHz in this case. C_{DIE}=92nF and V_0 maximum can never be more than 1.2 Volt.

$<Vo>$ (Volt)	R (m-Ohm)	L (pH)	$1/[2\pi(LC_{DIE})^{1/2}]$ (MHz)	$f_{Resonance}$ of PDN (MHz)
1.199	7.12	3.67	274.039654	226.3601839
1.1837	6	3.34	287.258712	249.1206872
1.17	5	3.02	302.094606	271.8184498
1.151	3.6	2.52	330.70912	310.5346817
1.1426	3	2.28	347.679448	331.5211549
1.1262	1.8	1.6874	404.145195	395.1203757
1.1152	1	1.1676	485.846859	481.0378249
1.108	0.5	0.7284	615.122303	612.6896006

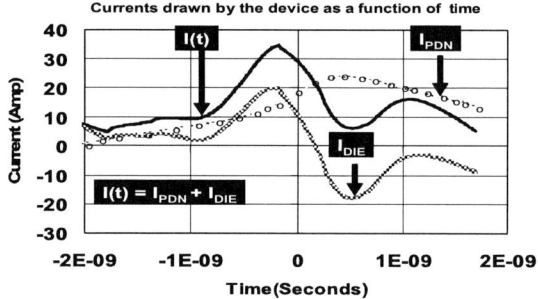

Figure-(8): This figure shows the various components of current I (t) drawn by the real device.

Conclusions

In this paper, we have shown a methodology to determine the current drawn by the device as a function of time. In our method, we measure the voltage close to the die and then use either figure-(1) as our base model with L, C and R are the parameters of the problems. Some of the parameters are estimated, calculated, or measured using various techniques available in literatures [reference-(1, 3)]. This is required in order to remove the redundancy of the simulated current drawn by the device. The remaining parameters were obtained by data fit.

References

1. "A Method to measure impedance of Chip/Package/Board Power Supply system ...." by Yaping Zhou et al; IEEE-EPEP conference proceedings 2006, Page-33, 36.
2. "Extended adaptive voltage positioning"; by Alex Waizman and Chee-Yee Chung, IEEE-EPEP conference proceedings, 2002.
3. "CPU Power Supply Impedance Profile Measurements using FFT ..." by Alex Waizman, IEEE-EPEP conference proceedings, page-29-32, Oct. 27-29, 2003.

Thermal-mechanical Modelling of Power Electronic Module Packaging

Hua Lu[†], Tim Tilford, Xiangdong Xue and Chris Bailey

School of Computing and Mathematical Sciences, University of Greenwich, 30 Park Row, London SE10 9LS, UK

[†]Email: h.lu@gre.ac.uk

Abstract

In this paper the reliability of the isolation substrate and chip mountdown solder interconnect of power modules under thermal-mechanical loading has been analysed using a numerical modelling approach. The damage indicators such as the peel stress and the accumulated plastic work density in solder interconnect are calculated for a range of geometrical design parameters, and the effects of these parameters on the reliability are studied by using a combination of the finite element analysis (FEA) method and optimisation techniques. The sensitivities of the reliability of the isolation substrate and solder interconnect to the changes of the design parameters are obtained and optimal designs are studied using response surface approximation and gradient optimization method.

1. Introduction

Power electronic modules(PEMs) are widely used in aerospace and automotive applications for the conversion and control of electrical power. A power module consists of several layers of insulator such as ceramic, conductor, and semiconductor, some metal wires, encapsulations, metal bars and the casing [1]. They are assembled together in the packaging process to form the power electronic circuit and the mechanical structure. Power electronic modules usually dissipate large amount of heat and operate in harsh environments. The stresses caused by the thermal, mechanical and electric voltage/currents greatly affect the long term reliability of these devices. The reliability of a power module is in general determined by the geometric design, choice of material and manufacturing technologies used. Since there are many different materials and joining technologies involved in a power module, it is a great challenge to the electronics packaging industry to address the power module reliability issues from the very early design stages of the product manufacturing cycle.

Two component structures are discussed in this paper: the isolation substrate and the chip mount-down solder interconnect. Of all the components in a power module, the alumina or AlN isolation substrates and the bonded copper conductor layer are the basic structures on which all other components are built on. The delamination of the copper tracks is an important reliability issue [2] whereas the solder interconnect of the chip mount-down is one of the major failure mechanisms [3]. Therefore the reliability of these two structures of the power module are fundamental to the reliability of the whole module and have attracted much research interest [4, 5, 6,7]. In this paper direct copper bonding (DCB) substrate and chip mount-down solder interconnect under thermal-mechanical loading conditions are modelled in order to understand the stress state in varying designs so that a physics-of-failure method can be used to predict the substrate's sensitivity to design parameters and the optimal design under certain constraints.

2. Isolation Substrate

The isolation substrate is a critical component in a PEM. Its functions are to provide mechanical support for other component and to electrically isolate the mounting plates and heat sink from the conductors wirebonds and semiconductor components. The isolation substrate is formed from a ceramic plate – usually alumina or Aluminium Nitride (AlN). A thin layer of copper (metallisation) is directly bonded to both sides of the ceramic. The copper layer on one side of the ceramic is etched to form a number of conductors.

To assess the substrate design the magnitude of the direction perpendicular to the substrate plane, i.e. the peel stress, induced by a thermal load has been selected as a performance metric. A response surface optimisation approach [8], has been used to evaluate optimal design and parameter sensitivity. The response surface has additionally been used in conjunction with parameter uncertainty data and a Monte-Carlo algorithm [9,10] to produce peel stress distribution data, an indicator toward component/product lifetime.

In order to determine the stresses induced by thermal load a numerical model has been developed. The Finite Element software package ANSYS [www.ansys.com] has been used to model the stresses and deformation of a simplified isolation substrate. The substrate geometry is symmetric about two axes. This symmetry is exploited, thus only a quarter of the isolation substrate tile structure has been modelled.

In order to reduce the number of degrees of freedom of the problem, shell elements were used. Two layers of shell elements were used for the ceramic layer to capture the behaviour in the ceramic close to the interface between the etched copper conductor and the ceramic more accurately. The layer close to that interface has a fixed thickness of 0.04mm regardless of the total thickness of the ceramic layer. This layer will be referred to as the top ceramic layer in this paper. Figures 1 and 2 shows a typical mesh and the normal stress distribution in the top ceramic layer. The FEA model geometry consists of two materials. The copper metallisation is subject to stresses near or in excess of the yield stress. It has therefore been modelled as an elastic-plastic material. The ceramic substrate is a brittle material which exhibits only limited plasticity at very high temperature and is therefore modelled as an elastic material. The material properties used in the simulation are listed in Table 1, in which E,

1-4244-1105-X/07/$25.00 ©2007 IEEE

ν, α, σ_y, and ε are the Young's modulus, the Poisson's ratio, the coefficient of thermal expansion (CTE) and the yield stress, respectively.

The damage indicator used in this work is the stress but at the bi-material interface the extrema are expected to be very mesh-dependent. Therefore, the elements with the highest stress magnitudes are identified, the stress values for these elements are volume averaged and the results are used as the damage indicator.

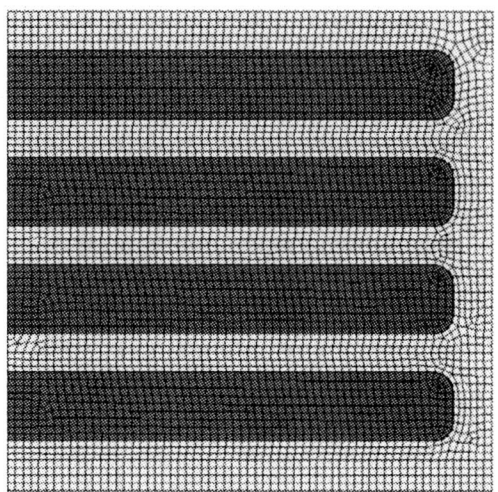

Figure 1: Typical FE model of the isolation substrate model.

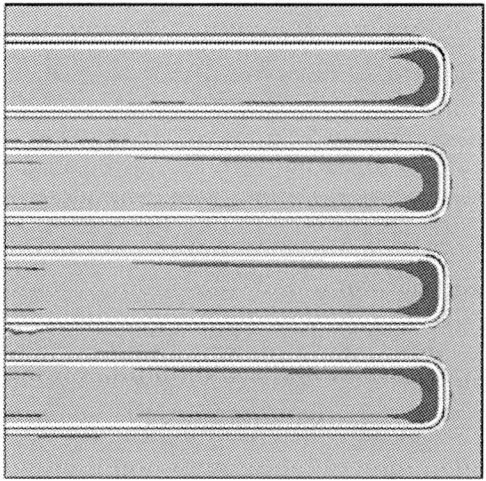

Figure 2: Out-of-plane normal stress distribution at the top ceramic layer.

Table 1: Material properties in the isolation substrate model.

	E(GPa)	ν	CTE(ppm/K)	σ_y(MPa)	ε (MPa)
AlN	310	0.24	5.6		
Cu	103.42	0.3	17	172	425

The design of experiment (DOE) [11,12] and the least squares methods [13] were used in order to form a response surface function. The substrate tile design has been described using six parameters. A random Latin Hypercube[14,15] method was used to select 28 design points within the predefined permissible range of each of the 6 parameters. The ANSYS FEA model was used to determine the peak peel stress within the substrate geometry in response to each of the discrete designs. The least square method, as implemented in the VisualDoc® package [16] fits a polynomial function consisting of linear, interaction and quadratic terms to the simulation data. The response surface is a function giving a value for the response variable (stress/lifetime/etc.) in reaction to a set of design variables. The general form of the response surface equation with six scaled design variables (labelled x_1 to x_6) is given in Equation 1. The design variables are scaled so that they take values between -1.0 and +1.0 as they vary over their permissible range. With this scaling the relative influence of the variables can be determined from the relative magnitude of the coefficients (a_0 to a_{27} in this case). These sensitivity results are shown in Figures 3-5. It can be concluded from the results that the thickness of the ceramic substrate, conductor spacing and conductor corner radius are the most important design parameters. The optimal design for the substrate can be determined from finding the minima of the response surface.

$$F(x) =$$
$$a_0 + a_1x_1 + a_2x_2 + a_3x_3 + a_4x_4 + a_5x_5 + a_6x_6 +$$
$$a_7x_1x_2 + a_8x_1x_3 + a_9x_1x_4 + a_{10}x_1x_5 + a_{11}x_1x_6 + \quad (1)$$
$$a_{12}x_2x_3 + a_{13}x_2x_4 + a_{14}x_2x_5 + a_{15}x_2x_6 + a_{16}x_3x_4 +$$
$$a_{17}x_3x_5 + a_{18}x_3x_6 + a_{19}x_4x_5 + a_{20}x_4x_6 + a_{21}x_5x_6 +$$
$$a_{22}x_1^2 + a_{23}x_2^2 + a_{24}x_3^2 + a_{25}x_4^2 + a_{26}x_5^2 + a_{27}x_6^2$$

Once the design engineer has determined the optimal design the component can progress to the manufacturing stage. The manufacturing process is imperfect and the final product will have small deviations from the design specification. These deviations or uncertainties have an impact on the performance and reliability of the product. In order to demonstrate how the manufacturing uncertainties impact on product reliability a Monte-Carlo method has been used in conjunction with the response surface function to evaluate the distribution in peel stress magnitude in a sample of components.

The Monte-Carlo method utilises the response surface to determine the stress in response to a set of design parameters. These design parameters are obtained by combining the optimal design values and an uncertainty value. In this work the uncertainty values have been generated by a Box-Muller transform [17]. The Box-Muller transform generates these values in a normal distribution around a mean (the optimum value). The magnitude of the standard deviation can be determined from manufacturing quality control processes. However, in this work the simplification of setting the standard deviation to 0.5% of the optimal value has been made. The algorithm was used to produce a total of 10 million sample points. The resulting peel stress was evaluated for each. The distribution of peel stress is shown in Figure 6. This distribution shows that, with a standard deviation of 0.5% of the optimum value, the peel stress results vary over a substantial range, impacting significantly on the overall component/product lifetime. This result shows that if the stress for the deterministic optimal design satisfies the design requirement there is still a possibility that the product would fail because of the manufacturing uncertainties.

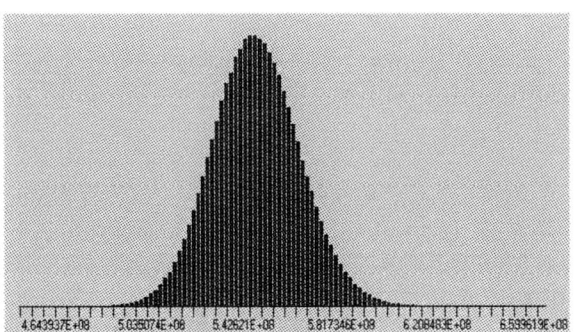

Figure 6. Peel Stress Distribution -10 million samples.

3.   Chip Mount-down Solder Interconnect

Chip mount-down solder interconnects provide mechanical support to the chip and electrical connection from the chip to other components in the electric circuit. Figure 7 shows a simplified 2D chip mount-down interconnect model. Only one half of the device needs to be modelled because of the symmetry. The elastic material properties used in the modelling are listed in Table 2.

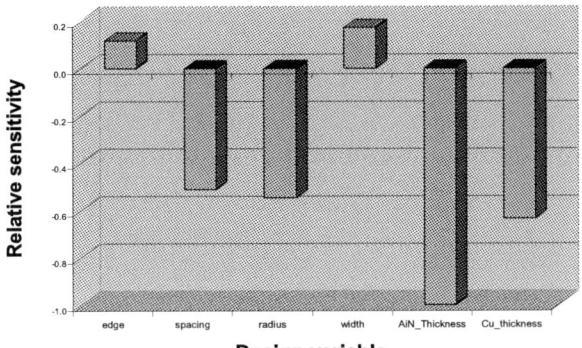

Figure 3. Sensitivity analysis – Linear terms.

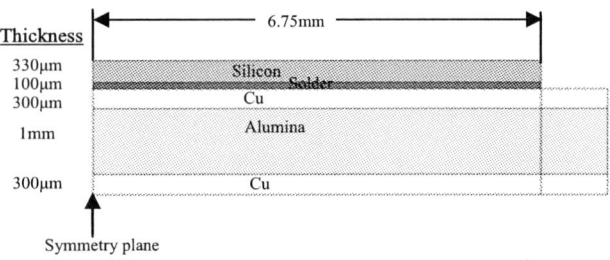

Figure 7. The basic 2D chip mount-down model.

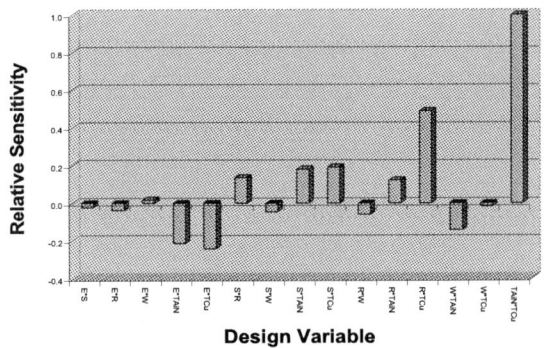

Figure 4. Sensitivity analysis – Interaction terms.

Table 2: Material properties used in the modelling.

	E (GPa)	ν	CTE (ppm/K)
Sn3.5Ag	54.05-0.193T	0.4	21.85+0.02039T
63Sn37Pb	36.68-0.56T	0.4	24
Silicon	113	0.29	3
Cu	115	0.31	17.3
Alumina	370	0.22	7.4

Solder fatigue fracture is assumed to be the failure mechanism. Two solder materials are considered in this work: the eutectic SnPb solder and the Sn3.5Ag lead-free solder. Because of the high homologous temperature of solder alloys, the deformation of these solders is modelled using a creep law. The strain rate is represented by Equation 2.

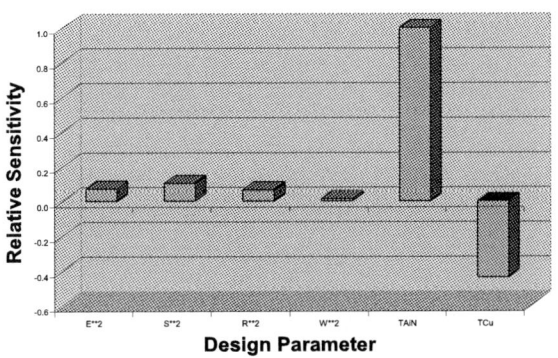

Figure 5. Sensitivity analysis – Quadratic terms.

245

$$\dot{\varepsilon}_{cr} = A \times \sinh^{n}(\alpha\sigma_{e}) \exp(\frac{-Q}{RT}) \qquad (2)$$

where R is the gas constant, T is the temperature in Kelvin, σ_{e} is the von Mises equivalent stress, A, n, α, Q are material constants and the values are listed in Table 3 [18].

Table 3: Creep parameters for solder materials.

	A(s)	n	α (1/MPa)	Q/R
SnPb	9.60E+04	3.3	0.087	8110
SnAg	9.00E+05	5.5	0.06527	8690

The first part of the work on solder interconnect is to obtain the damage indicators and evaluate effects of temperature range and median temperature on the fatigue life of chip mountdown solder interconnect. Four temperature profiles have been used in the modelling. The ramp and dwell times of the profiles are all 15 minutes and the temperature extremes are listed in Table 4.

Table 4: Thermal cycling temperature profiles. The unit in Celsius °C.

Cycle	Tmin	Tmax	Tmed	ΔT
1	-55	125	35	180
2	-25	155	65	180
3	-40	110	35	150
4	-10	140	65	150

The plastic work density per temperature cycle, ΔW, has been used as the damage indicator. Approximately, the fatigue lifetime is inversely proportionally to ΔW [18].

The multiphysics software package PHYSICA [19] has been used to carry out the modelling. Figure 8 shows a typical distribution of the accumulated plastic work density in the solder interconnect. The maximum ΔW value is found at the chip-solder interface at the edge of the solder layer, and this is the location where cracks initiate. The value of ΔW around the most damaged location has been used as an indicator of the reliability of the chip mountdown interconnect.

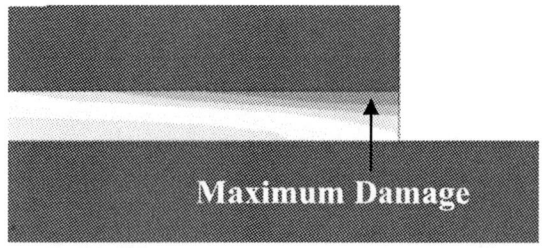

Figure 8: Typical distribution of accumulated plastic work density in solder joint.

The results are listed in Table 5. For SnAg solder interconnect, the ΔW value for Cycle 1 is about 27% higher than cycle 4 making it the most damaging temperature profile. The results also suggest that for temperature profiles with the same median temperature, the temperature range determines lifetime while for temperature profiles with the same temperature range the lower the median temperature the lower the lifetime.

Table 5: Plastic work per cycle.

Test Cycle	ΔW (MPa)	
	SnAg	SnPb
1	0.37	0.43
2	0.34	0.36
3	0.29	0.33
4	0.27	0.28

For the SnPb solder, a model relating ΔW to crack initiation and propagation rate can be expressed as:

$$N_{0} = 5.42 \times 10^{7} / \Delta W \qquad (3)$$

$$\frac{dl}{dN} = 5.792 \times 10^{-14} \Delta W^{1.13} \qquad (4)$$

where N_{0} is the number of cycles to crack initiation and l is the crack length and N is the number of cycles. The results are listed in Table 6.

Table 6: Number of cycles to crack initiation and crack propagation rate for SnPb solder interconnect.

Cycle	N_{0}	dl/dN (μm/cycle)
1	126	0.135
2	150	0.111
3	162	0.101
4	196	0.081

The second part of the work on solder interconnect is to analyse how reliability is affected by design variables. Design of experiment (DOE) and response surface methods (RSM) have been used to calculate the reliability sensitivity parameters for the changes in the die width, solder and substrate thickness, and gradient optimization method has been used to find the optimal design parameters [8]. The design space is defined in Table 7.

Table 7: Design space for the optimization of solder interconnect.

	Variable name	min	max	median
Die width	x_1	47.25	87.75	67.5
substrate thickness	x_2	1.4	2.6	2
solder thickness	x_3	0.7	1.3	1

A 27 points DOE was used. The plastic work density ΔW was defined as the objective function. The maximum von Mises stress in the die is an important response function to monitor because as the solder joint becomes more reliable the stress in the die may exceed the die strength. Figure 9 shows the location where these two functions are defined.

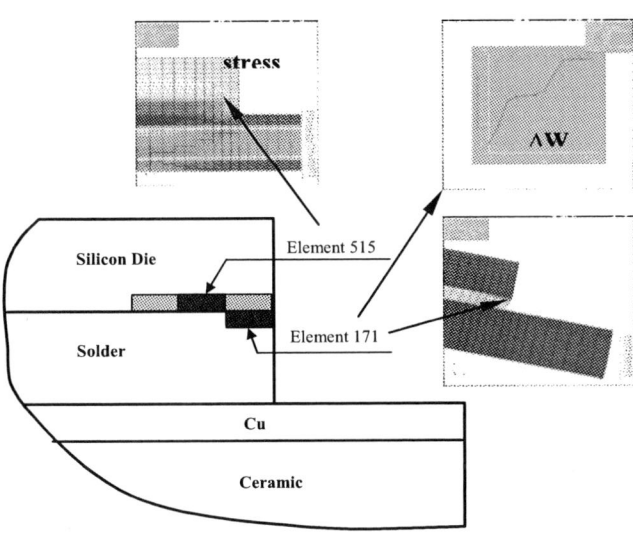

Figure 9: Locations where the stress and ΔW objective functions are defined.

For each design point in the DOE a FEA simulation is carried out and the results are then fitted to a quadratic response surface using the general optimization software VisualDoc®. In this analysis the temperature profile has a minimum temperature of -25°C, a maximum temperature of 125°C, a dwell time of 15 minutes, and a ramp time of 15 minutes. The resulting ΔW response surface has been expressed in Equation 5.

$$\Delta W(x_1, x_2, x_3) =$$
$$798700 + 36580 x_1 - 3.2654 \times 10^7 x_2 - 2.7605 \times 10^8 x_3 +$$
$$6.8587 \times 10^6 x_1 x_2 - 2.7435 \times 10^7 x_1 x_3 + 5.5556 \times 10^8 x_2 x_3 +$$
$$1.3548 \times 10^5 x_1^2 + 8.4877 \times 10^9 x_2^2 + 8.0247 \times 10^{10} x_2^2$$

$$(5)$$

This analytical response surface equation can then be used to find the optimal design and calculate the sensitivity with regard to any objective function and any location in the design space. The first sign of the relative importance of the variables can be found in the coefficient of the linear terms. Obviously, solder thickness has the greatest absolute linear coefficient value and therefore is therefore expected to be the most important variable. The fact that its value is negative means that ΔW decreases as solder thickness increases.

The sensitivity with regard to each design variable can be defined as the derivative of Equation 5. Because the equation is quadratic sensitivity is location dependent. For example at the centre of the design space, the sensitivity values with regard to the three design variables are 8.9×10^4, 2.3×10^6 and -1.16×10^8 respectively. This again shows that solder thickness has the greatest impact on the fatigue life of the chip mountdown solder interconnect.

The response surface has also been used to find the best design, or the design with the lowest ΔW, in the specified design space. Instead of using a direct method, the optimization was carried out using the response surface in order to save computing efforts. The optimal design for minimum ΔW has been detailed in Table 8.

Table 8: Optimal design of the chip mountdown.

	Initial	Optimal
ΔW (kJ)	585	557.7
Die stress (MPa)	80.3	86
x_1 (m)	0.0675	0.0668
x_2 (m)	0.002	0.00147
x_3 (m)	0.001	0.0013

Compared with the initial design, the value of the objective function ΔW has an improved by about 5%. It is important to note that the maximum stress in the die has increased by about 7%. This stress is still low compared to the strength of the die but if die cracking is possible the optimization process should include a constraint on the maximum stress in the die.

Following the deterministic optimization analysis, the optimisation with uncertainties is now being carried out in which the stress in the die is constrained to the die fracture strength and the design parameters were all assumed to have a normal distribution. The results which include the optimal design results and n-Sigma design results will be published elsewhere.

4. Conclusions

Thermal-mechanical computer modelling techniques have been used in conjunction with numerical optimization tools to evaluate the reliability of isolation substrate and chip mountdown solder interconnects. For the isolation substrate the conductor edge distance, the conductor corner radius and the substrate thickness have been found to be the most important factor affecting the reliability. For the chip mountdown the thickness of the solder interconnect that has the strongest effect on lifetime. For the eutectic SnPb solder the number of cycles to initiation and crack propagation rate have been obtained for four temperature profiles.

Acknowledgments

The authors wish to acknowledge the support of the Innovative electronics Manufacturing Research Centre (IeMRC) and the United Kingdom Department of Trade and Industry for the project 'Modelling of Power Modules for Lifetime, Accelerated Testing, Reliability and Risk'. The authors would like to thank project partners Semelab Ltd, Dynex Semiconductor Ltd., Goodrich Engine Control Raytheon Systems Ltd., SR Drives Ltd., and Areva T&D Ltd. for their support. The authors would also like to thank Prof. Mark Johnson and his colleagues at the University of Sheffield and University of Nottingham.

References

1. Sheng, W.W. and Colino, R.P., Power Electronic Modules, CRC Press (2005)

2. Shammas, N.Y.A., "Present problems of power module packaging technology", *Microelectronics Reliability*, Vol. 43, Issue 4 (2003), pp. 519-527

3. Pooch, M.-H., Dittmer, K.J., Gabisch, D., "Investigations on the damage mechanism of aluminum wire bonds used for high power applications", *Proc. EUPAC 96*, (1996) pp. 128-131

4. Günther, M., Wolter, K, Rittner, M, Nüthter, "Failure Mechanisms of Direct Copper Bonding Substrates", *Proceedings of Electronics Systemintegration Technology Conference* (ESTC), Dresden, Germany, 2006, pp.714-718

5. Dupont, L., Khatir, Z., Lefebvre, S., and Bontemps, S., "Effects of metallization thickness of ceramic substrates on the reliability of power assemblies under high temperature cycling", *Microelectronics Reliability*, vol. 46 (2006) pp.1766–1771

6. Yoshiyuki Nagatomo and Toshiyuki Nagase, "The study of the power Modules with High Reliability for EV Use", *Proceedings of the 17th International Electric Vehicle Symposium* (2000).

7. M. H. Poech and R. Eisele, A modelling approach to assess the creep behaviour of large-area solder joints, *Microelectronics Reliability*, Vol. 40, Issues 8-10, (2000), pp.1653-1658

8. Box, G. E. P. and Wilson, K.B. (1951) On the Experimental Attainment of Optimum Conditions (with discussion). Journal of the Royal Statistical Society Series B 13(1):1-45.

9. N. Metropolis and S. Ulam, "The Monte Carlo Method", *Journal of the American Statistical Association*, vol. 44, p335 (1949)

10. Nicholas Metropolis, Arianna W. Rosenbluth, Marshall N. Rosenbluth, Augusta H. Teller and Edward Teller, "Equation of State Calculations by Fast Computing Machines", *Journal of Chemical Physics*, vol. 21, p. 1087 (1953)

11. Cochran, W.G. and G.M. Cox. 1950. Experimental Design. John Wiley, New York

12. Jiju Anthony, Design of Experiments for Engineers and Scientists, (Elsevier 2003)

13. C.F. Gauss, "Theoria motus corporum coelestium in sectionibus conicis Solem ambientiu", (1809)

14. Iman R.L., Helton J.C. and Campbell J.E. "An approach to sensitivity analysis of computer models", *Journal of Quality Technology*, Parts I and II. 13(3,4), pp174-183 and pp232-240 (1981).

15. McKay, M. D., W. J. Conover and R. J. Beckman, "A Comparison of Three Methods for Selecting Values of Input Variables in the Analysis of Output from a Computer Code", *Technometrics* vol. 21, pp 239-245(1979).

16. VisualDoc is a product of VR&D: www.vrand.com

17. Box, G.E.P, M.E. Muller, "A note on the generation of random normal deviates", *Annals Math. Stat*, V. 29, pp. 610-611, 1958.

18. Lau, J.H. (editor), Ball Grid Array Technology, McGraw-Hill (1995), p.396

19. PHYSICA is a product of Physica Ltd: www.physica.co.uk

Multi Physics Modelling of the Electrodeposition Process

M.Hughes, C.Bailey, K.McManus
School of Computing and Mathematical Sciences
University of Greenwich
Park Row, Greenwich, London SE10 9LS
m.s.hughes@gre.ac.uk, c.bailey@gre.ac.uk, k.mcmanus@gre.ac.uk

Abstract

This paper describes ongoing research into the development of multi-physics models of the electrodeposition process. This is part of the collaborative project – MEMSA (Modelling the Electrodeposition process for Microsystems Applications) - between the universities of Greenwich and Heriott-Watt, and our industrial partners: Merlin Circuits and Raytheon Systems. The aim of this research is to build numerical models that can predict all aspects of the electrodeposition process, and to verify these models against experimental data gathered.

This paper focuses on model development for (*i*) the representation of the moving interface through a level-set technique, and (*ii*) the implementation of the associated moving boundary conditions and source terms together with considerations regarding the electrode kinetics boundary condition. Accurate modelling of the electrode kinetics is crucial to any electrodeposition model as it drives the deposition process and influences the distribution of the solved variables of which it is itself a non linear function.

The multi-physics code PHYSICA provides the framework in which the electrodeposition models will be built. This paper will be of particular interest to applied modellers wanting to modify Computational Fluid Dynamics (CFD) codes to simulate the electrodeposition process.

1. Introduction

Electrodeposition is a process that is truly multi-physics in its nature and of considerable importance to the microsystems and semiconductor industries. The reduction in length scales and the replacement of aluminium interconnects and trenches with copper has increased the operational speed of CMOS devices, largely due to copper's higher conductivity and reduced metallization capacitance. Much important modeling work has been applied to the simulation of these types of trenches and investigating the feature filling with respect to void formation. In particular Wheeler et al. [1,2], have published interesting results from modeling at the sub-micron scale where chemical additives have been used to produce a superconformal, 'bottom-up' filling of high-aspect ratio features for the deposition of Damescene copper. This process is known as CEAC (Curvature Enhanced Accelerator Coverage), However at larger length scales, (mm), this process does not necessarily scale up and the problem of ion depletion within the high-aspect ratio features can cause problems of void formation, through effects such as 'current-crowding',(Figure 1). This situation may be ameliorated by using forced convection to replenish the supply of reacting ions to the electrode surface. However because of the possibility of flow dead-zones in these trenches forced convection may not be sufficient to improve matters. The application of pulse reversed waveforms may diminish void formation [3] by improving the distribution of the time averaged deposition rate along the trench side walls, possibly because the concentration of reacting ions has sufficient time to recover during the plating off-time

Attempts to numerically model the electrodeposition process are challenging as they must solve a system of coupled non-linear equations with the added complication that the governing equation set changes under different physical situations; for example as the deposition current varies from primary to secondary, tertiary or diffusion limited regimes [4,5] Additionally the representation of electrode kinetics, the driving force for deposition, is of key importance and is complicated by its influence from the electrode surface overpotential and the concentration of reacting ions in the immediate vicinity of the depositing interface. These factors can in turn be influenced by effects such as forced convection of the electrolyte replenishing the ion supply to the deposition interface, the electrode potential difference or total current applied to the electrolytic cell. The governing equations may therefore include all or a combination of the momentum, heat, concentration and electric potential equations with various degrees of intercoupling by electromigration, convection and importantly through the reaction rate boundary condition at the electrode surface.

Much of the modeling work to date has focused on deposition within particular current distribution regimes where assumptions can be made about the deposition process and some simplification of the equations may therefore be possible, for example electromigration if an excessive supporting electrolyte is used [4,6], or electric field if the deposition rate is diffusion controlled and the surface overpotential can be provided from experimental voltammetry [1]. It is clear that developing a model to solve the full equation set with the electrode overpotential being implicitly calculated is a challenging task which must consider the underlying physics carefully, ideally supported by experimental results.

A suitable technique must be chosen to represent the moving interface, here the Level Set Method [1,2,7,10] has been chosen. This paper considers these issues from the viewpoint of developing a model from scratch within a CFD framework. This framework aims to provide a good starting point for model development as momentum and heat equations are then implicitly handled. Further to this a sensible first step is to build a model for the simplest electrodeposition scenario, namely, that of the primary current distribution regime with a single ionic species.

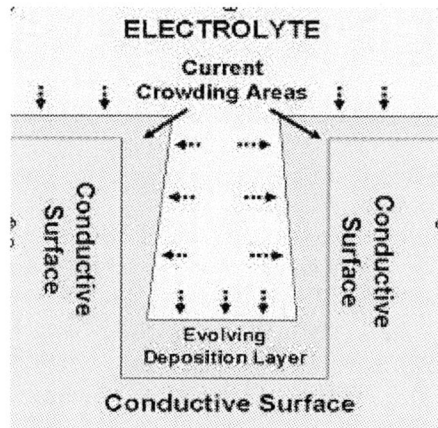

Figure 1: Current Crowding effect

In this scenario a DC current drives the process and the deposition rate is governed by Ohm's law this is discussed later in Section 5. Further to this, progression of the model into secondary and tertiary current distributions is considered in Section 6, these deposition regimes introduce more numerical complexity because of the non-linear reaction rate and boundary constraints at the deposition surface. In Section 7 future work towards the goals of the MEMSA project are considered together with some thoughts on how to address them. Implementation of the model framework at this stage of the project aims to establish numerical stability and reflect qualitative experimental behaviour.

We begin with an overview of the governing equations and a brief discussion of the current deposition regimes and subsequent boundary conditions.

2. Governing equations and deposition current deposition regimes.

The governing equations for the electrodeposition process are:

The Navier-Stokes equation if the electrolyte is under the influence of forced convection:

$$\rho \frac{du}{dt} + \rho u \nabla u = -\nabla P + \mu \nabla^2 u + S_u \qquad 1$$

where S_u represents momentum source for forced convection such as electroyte stirring. Together with the continuity equation:

$$\rho (\nabla . u) = 0 \qquad 2$$

and the temperature equation with external heating S_T

$$\rho C_p \frac{dT}{dt} + \rho C_p u \nabla T = k \nabla^2 T + S_T \qquad 3$$

The flux of ionic species is given by Paunovic and Schlesinger [8]:

$$N_i = -z_i e \upsilon_i c_i \nabla \phi - D_i \nabla c_i + u c_i \qquad 4$$

where $\phi, c_i, D_i, e, \upsilon_i, z_i$ are respectively; electrolyte electric potential, concentration and diffusion coefficent of the ith ionic species, elementary charge, ion mobility.and ion species valency. The first term on the RHS represents ion drift due to the electric field, the second diffusion of ions and the last term movement by convection. Ionic mobility is given by

$$\upsilon_i = \frac{D_i}{kT} ; \quad k \text{ is the Boltzman constant} \qquad 5$$

Migration is essentially an electrostatic effect that arises due the application of a voltage on the electrodes if there is a large quantity of the electrolyte (relative to the reactants) it is possible to ensure that the electrolysis reaction is shielded and not significantly affected by migration, in such circumstances the first term on the RHS can be neglected [4].

Concentration of ionic species can be represented by taking the divergence of the above term and expressing this in the total derivative for concentration of species to give the equation:

$$\frac{\partial c_i}{\partial t} + \nabla \cdot (u c_i) = \nabla \cdot (D_i c_i) + e z_i \nabla \cdot (\upsilon_i c_i \nabla \phi)$$

$$\updownarrow \qquad \updownarrow \qquad \updownarrow \qquad 6$$

convection   diffusion   migration

The equation set is closed with the electric potential equation togther with suitable boundary conditions for the complete equation set. The time scale for establishing a DC field is much faster than for establishing concentration gradients so under DC conditions the electric field can be expressed through electric potential as a Poisson equation without time influence:

$$\nabla^2 \phi = -\frac{4\pi}{\varepsilon} \sum e z_i c_i ; \qquad 7$$

where ε is the dielectric constant. An alternative to solving equation 7 as given by Griffiths [9] is to enforce electroneutrality in the bulk electrolyte, in which case the electric field becomes an unknown constant which is determined as part of the overall solution from the governing condition:

$$\sum_1^n z_i c_i = 0 \qquad 8$$

As with equation 7 this condition applies at every point in the solution domain, except at the thin layers adjacent to the electrode boundaries, the electrical double layer [9] which is of the order of $<\sim 1000$ Angstrons in width. In these thin layers the deposition current is accounted for by an electrode kinetic function, typically the Butler-Volmer equation [1,4,5,9]. In this electrical double layer the electroneutrality condition breaks down and a spacial charge exists [8]. This charge is referred to

250

as the surface overpotential and its value is one of the parameters that drive the reaction rate through the Butler-Volmer equation (Figure 2). In line with this authors present knowledge and literature read to date, the double layer is not explicitly taken account of with DC conditions. Instead the overpotential is either specified [1] or details of its explicit calculation are not given special attention [4,5]. However this region will effectively present a discontinuity to the electric potential distribution and therefore some thought is necessary towards the application of the boundary conditions for equation 7, this is discussed later in Section 6.

For AC conditions at low frequency it is likely that the above equations (6,7) can still be utilised. However at higher frequencies and if the numerical model is to implicitly calculate the overpotential it may be necessary to introduce a sub-model to calculate the overpotential which approximates the layer as a plate capacitor. This complication may be bypassed if sufficient overpotential vs applied voltage or current data is available.

3. Boundary Conditions

Ritter et al. [4] and Drese [5] give concise descriptions of four deposition regimes, the relevant equations and boundary conditions are listed here, the heavy line in Figure 2 below being the cathode-electrolyte interface

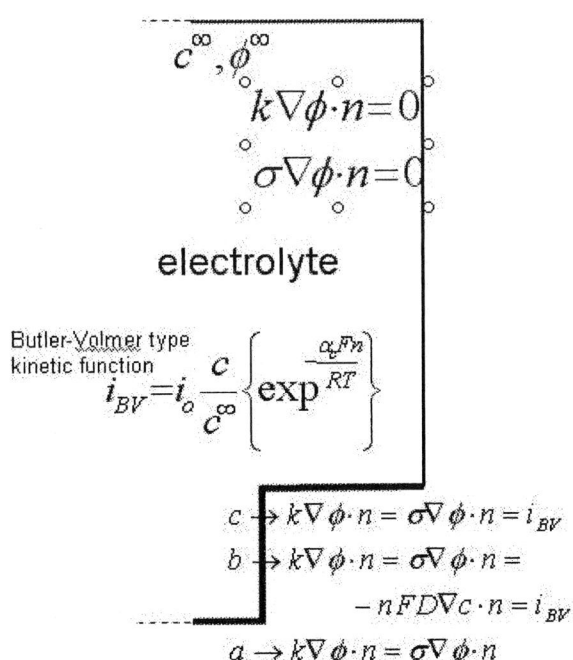

Figure 2: Boundary Condition schematic

- Tertiary current distribution

The deposition current, i_{bv}, at the cathode is given by the Butler-Volmer equation and is a function of the local interface concentration to bulk concentration of reacting ions C^{Intfce} / C^{∞} and electrode overpotential, η. At the

electrolyte-cathode interface condition b in Figure 2 needs to be enforced.

- Secondary current distribution

If concentration gradients can be ignored because the concentration of ions is very high then the electric potential equation is solved with condition c in Figure 2.

- Primary current distribution

If the resistance of the electrolyte is much higher then that of the interface then the Current density passing through the electrode is given by Ohms law, condition a in Figure 2 is applied.

- Diffusion limited current distribution

At sufficiently high overpotentials, a limiting current is reached as the ionic concentration at the interface approaches zero and electric potential equation can be ignored. At the interface, c = 0, and the deposition current is calculated as $I_{dep} = nFD \dfrac{dc}{dn}$

4. Moving the interface

To advect the deposition interface the level set method of Osher and Sethian [10] was chosen.. This is a numerical method for tracking interfaces and shapes that has been sucessfully applied to the electrodeposition process [1, 7]. It has the advantage of sucessfully handling surfaces that have sharp cusps or corners, without smearing, through a fixed mesh and has the advantages of a Eularian approach. Within the CFD code Physica the existing Level Set algorithm can be readily modified by decoupling its propagation from the momentum velocity and replacing this with a deposition velocity, v_{dep} which is calculated as below:

$$v_{dep} = \frac{i_{BV}\Omega}{nF} \; ; \text{ where } \Omega \text{ is molecular volume,}$$

n is charge number and F, Faradays constant, the units of v_{dep} are metres/sec.

In the level set method a variable φ is used to keep track of the moving interface. This variable is initialised to zero along the interface at the start of a simulation and at all other places in the computational domain stores a value representing the shortest distance to the interface with positive values in front and negative values behind. The calculated deposition velocity, v_{dep}, is then used to advect the interface and the new interfacial distances are updated as φ is reinitialised. Because the distance function, φ, is updated at computational nodes diffusional smearing from the propagation of the front can be kept to a minimum. The procedure is as follows:

- Initialise the level set function, $\varphi(x,y,z)=0$
- Update material properties
- Solve $\dfrac{\partial \varphi}{dt} + v_{dep}\nabla\varphi$, this updates the interface position only.
- At the end of the timestep reinitialise φ in locations other then the interface to update distance from the interface by iterating:

- $\varphi_{i+1} = \varphi_i + \Delta\tau\, S(\varphi_o)(1 - \nabla\varphi)$ where φ_o is the value of the variable at the start of the reinitialisation, $\Delta\tau$ is a pseudo time step that is set to be $\frac{1}{10}th$ of the minimum distance between the current computational cell centre and the centre of the adjacent cell d_{ap}^{Max}, which is closest to the zero level set.

$S(\varphi_o)$ is a sign function calculated by

$$S(\varphi) = \frac{\varphi_0}{\sqrt{(\varphi_0)^2 + (d_{ap}^{max})^2}}.$$ Futher details of the scheme can be found in [1,7].

5. Simulating the Primary current Regime

The primary current distribution provides a good first target for model development, because of the simpler governing equation set. Under these conditions the deposition current can be modeled using Ohm's law. If we make the assumption that the concentration of reacting ions is sufficiently high then we can ignore the influence of the ion concentration, equation 6, which has the advantages of reducing the equation set to that of solving a Laplace equation for electric potential using equation 7 with the RHS reduced to zero. At the deposition interface the current normal to the surface is given by boundary condition a in Figure 2.

The computational grid is shown in Figure 3 below, together with the boundary conditions for the electric potential equation. In this instance potentials are fixed at either ends of the domain. An alternative is to replace the fixed potential boundary condition, $\phi = 1.0$, by specifying a current boundary condition, i.e.

$k\dfrac{\partial\phi}{\partial n} = I_{anode}$.

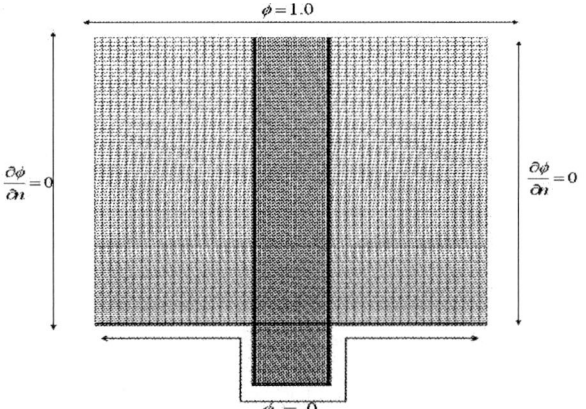

Figure 3: Computational grid and wall boundary conditions for DC conditions

At trench interfaces 'current crowding' effects occur because of a pinching effect on the electric field from sharp corners. In these instances voids may be formed as the current and hence deposition rate is higher at these edges. The results of this phenomenon on deposition can

be seen in Figure 4 below as time increases a void in enclosed in the trench as seen in the RHS picture.

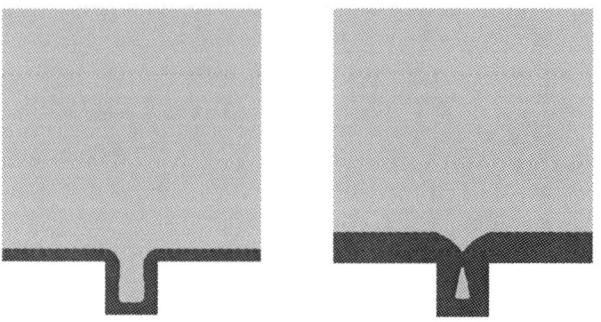

Figure 4: deposition through time

Under the primary regime, no special attention needs to be made at the interface boundary in terms of the conditions except to ensure that the electric conductivity across the interface is calculated using a harmonic average. If this is applied then the electric potential across the interface is rendered numerically continuous and the interface condition becomes similar to conjugate heat transfer in standard CFD simulations in that the condition $k_{electrolye}\dfrac{\partial\phi}{\partial n} = K_{metal}\dfrac{\partial\phi}{\partial n}$ is automatically satisfied.

Additionally it is convenient to note that under the circumstances where $K_{metal} \gg K_{electrolyte}$, and K_{metal} is large as is the case with metals and given that the currents imvolved are small, the cathode boundary condition ($\phi = 0$) will permeate through the metallic deposited layer and anchor the interface electric potential on the metal side to the applied boundary condition.

Calculating the electric potential gradients and hence current from Ohm's law can be achieved in an unstructured discretisation scheme by utilising Gauss's divergence formula as shown below where k is the electrical conductivity and norx, nory, norz are the Cartesian components of the face normal vectors. So for example when calculating the x-direction gradient, then only norx contributions are used.

$$k\times\frac{\sum^{faces}\phi_{face}\cdot Area_{face}\cdot(norx_{face}\,|\,nory_{face}\,|\,norz_{face})}{Cell\_volume}$$

The gradients are taken from both sides of the interface separately, with the value at the interfacial cell face ϕ_{face} , calculated by extrapolation along the gradient of ϕ as shown in Figure 5 overleaf.

This complication is necessary for taking gradients across a region that encloses materials with different electrical conductivities. It may not be necessary for the calculation of the deposition driving current as the electric current is only required in the electrolyte region of the computational domain and in particular in the cells adjacent to the interface on the electrolyte side

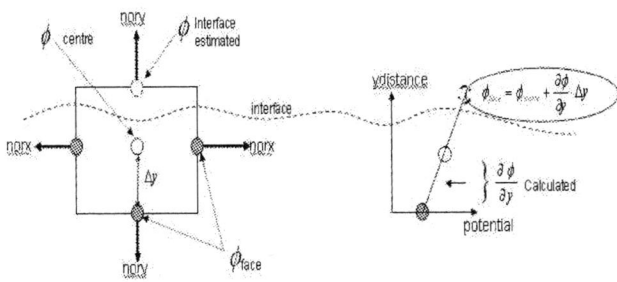

Figure 5: Calculating gradients across an interface

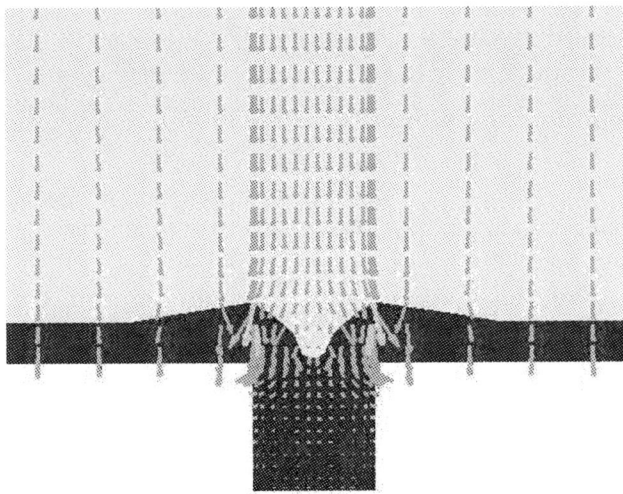

Figure 6: Current distribution through interface

6. Simulating the Tertiary Regime

Advancing the model to introduce ionic concentration involves greater restrictions at the deposition interface. This is now considered under DC conditions with a single ionic reacting species and an assumed constant overpotential.

In this scenario, equation (6) is solved for bulk concentration $\bar{c} = \dfrac{c}{c^\infty}$ together with the equation for electric potential (7) with the RHS again equated to zero as only one species is considered. At the boundary between the metal-electrolyte interface, condition b in Figure 2 needs to be satisfied and hence the position of electrolyte side interface cells must be tracked throughout the simulation so that the boundary source terms for equations 6 and 7 can be applied.

A schematic of the solution domain is shown in Figure 7 above where BV stands for the deposition current as given by a Butler-Volmer equation.

A complication with these boundary conditions is that the current passing through the interface is goverened by the surface kinetics and is defined by the Butler-Volmer equation. If we now consider the solution of equation 7.

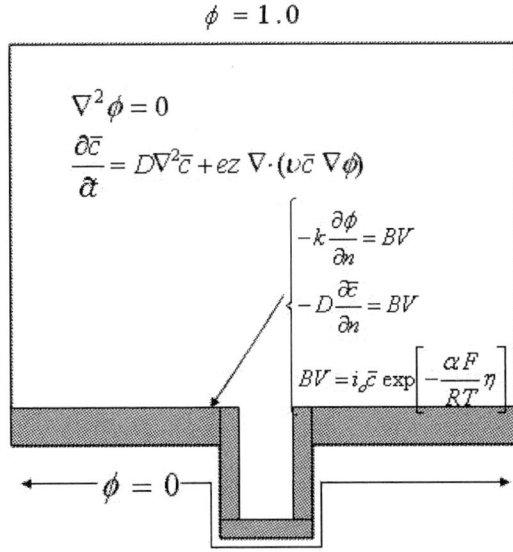

Figure 7: The solution domain

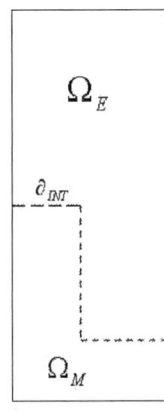

over the entire solution domain $\Omega_M + \Omega_E$, as was the case in section 5, then the current, I_{dep}, passing across the interface ∂_{INT} is governed by conduction alone and of course influenced by the relative values of the applied boundary conditions.

In the tertiary and secondary current regimes the current crossing the interface is governed by the surface kinetic function and therefore an appropriate 'sink' boundary condition must be applied to the electrolyte-side computational cells that are adjacent to the interface. Assuming that an appropriate boundary condition is applied here and equation 7 is computed over the entire domain, $\Omega_M + \Omega_E$, then the current passing across the interface will be incorrect as in addition to the applied sink it will contain a conduction contribution. This can be avoided by splitting the computation domain into two sides, Ω_M and Ω_E and linking these regions by appropriate sink/source type boundary conditions; the current leaving the electrolyte should be equal to the current entering the deposited metal. To recover current from this type of calculation the technique discussed section 5 is used. Figure 8 below explains this idea showing the results from a 1D test case in which the unequal spacing of the grid cells around the interface area is a way of testing the current calculation within the model (otherwise not a sensible grid arrangement). Current passing across the interface is continuous and is of equal magnitude to the computed deposition current from the Butler-Volmer equation.

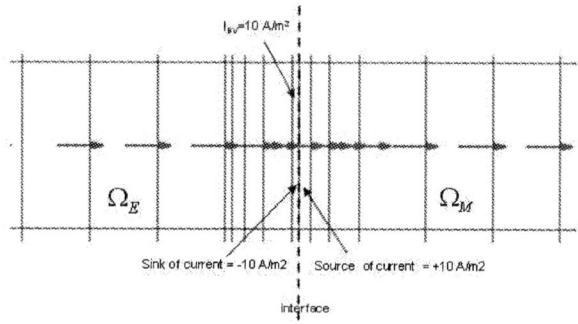

Figure 8: Current across an interface; splitting the domain.

This situation requires careful application of the current source terms in the Ω_M side when handling corners. Figure 9 below shows the recovered current from such a situation, here the required source for the electric potential equation in Ω_M at the corner cell, A, is calculated as the sum of the sinks at B and C.

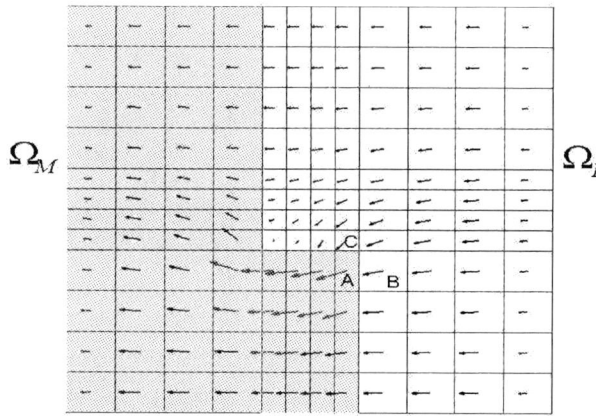

Figure 9: Current across an interface

These sources and sinks are calculated using the 'up-to- date' iteration values of deposition current as returned by the Butler-Volmer equation and are applied as flux type boundary conditions either side of the interface as shown below

$$S_\phi = A_{face} \times Coeff \times (\frac{\pm I_{BV}}{Coeff} - \phi_{cell}) ; \qquad 9$$

where I_{BV} is the deposition current, ϕ_{cell} is the value of electric potential at the cell center and a small value is used for the $Coeff$ (i.e. 1E-10) and the cell face area, A_{face} is estimated in the computational cell as:

$$\sum_{i=1}^{all\_cell\_faces} Area_{face} \cdot \frac{\nabla \varphi}{|\nabla \varphi|} \cdot n_{face} \cdot d_{face}$$

where $\nabla \varphi$ is the gradient of the level set function and $Area, n$ are the cell face areas and normals repectively and

d has the value 1 if the cell face is on the interface and zero elsewhere. It is updated at the end of a time step in line with the level set distance variable φ and gradient $\nabla \varphi$.

For simulations in which surface overpotential data is known *a priori* the distribution of electric potential is only required in Ω_E because there is no need to calculate the overpotential from the distribution of electric potential across the interface. Under these circumstances the solution of the electric potential equation may be limited to the electrolyte region, Ω_E, together with appropriate current sink at the interface and the deposition current estimated from the current that is recovered from the electric potential in the cells adjacent to the interface or from Butler-Volmer equation itself. The former of these being more likely to give a smoother deposition profile at the surface as it is a solved variable and will consequently somewhat numerically smooth the deposition current as calculated from the Butler-Volmer equation.

The Boundary condition for the solution of reacting ion concentration may also be applied in a similar manner to the current, as a flux loss at the interface,

$$D\frac{\partial \overline{c}}{dn} = -I_{BV}$$ or by fixing the interface concentration to zero if the concentration drops below a tolerance and the deposition current enters the diffusion limited regime. The application of the latter of these conditions is similar to equation 9:

$$S_\phi = A_{face} \times Coeff \times (0 - \phi_{cell}) ;$$

Coeff is calculated from $D \times Area / dn$ where dn is the distance between the computational cell centre and the interface. The larger the value of Coeff then the stronger is the tie of the computational cell centre to the applied surface zero value. If the magnitude of the diffusional coefficient, D, is extremely small (i.e. 10^{-9}-10^{-10}) , then in practice it may be necessary to increase the value of *Coeff* by a possible order of magnitude so that the influence of this source term is more dominantly felt.

The solution procedure for the simulation is as follows:

Start of time step-
- Store interface position in variable $d \begin{cases} 1 - int \ erface \\ 0 - otherwise \end{cases}$

Start of iterations
- Calculate I_{dep} from Butler-Volmer function *
- Solve for ϕ in Ω_E ; applying $-k\frac{\partial \phi}{\partial n} = I_{BV}$ at d=1
- Solve for $\overline{c}$ in Ω_E applying $-D\frac{\partial \overline{c}}{\partial n} = I_{BV}$ at d=1

End of iterations
- Update surface level set variable and associated parameters

End of time step

To improve numerical stability the calulation of I_{dep} can be moved from the iteration loop to the start of timestep.

Figure 10 (a-d) shows the deposition profile through time, showing a tendency for the deposition to be more concentrated at the top corners of the trench. This is highlighted in Figure 11 which shows the profile of I_{dep}, it has higher magnitude at the top trench edges in line with the deposition rate. Figure 12 shows vectors of the current profile in the electrolyte side of the trench region and also highlights a higher concentration of electric current at the trench corners.

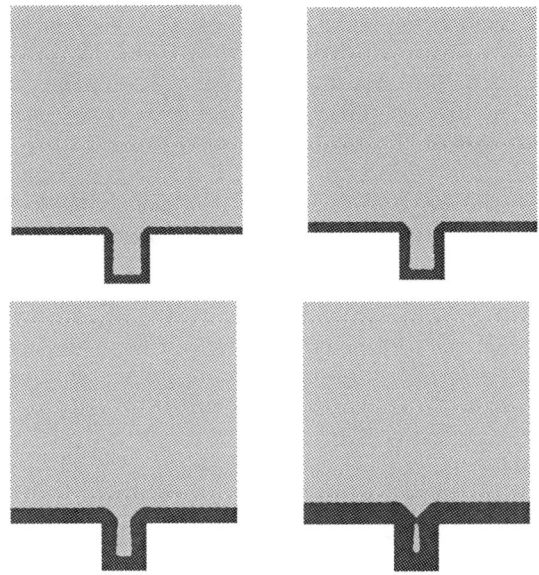

Figures 10 (a-d): deposition profile through time.

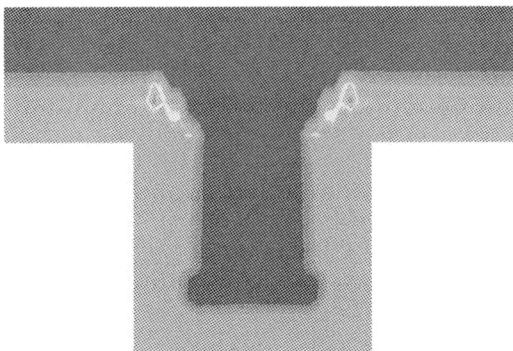

Figure 11: Surface Current from Butler-Volmer equation

7. Consideration of the overpotential calculation

The overpotential which acts over a tiny 'double-layer' effectively presenting a discontinuity to the electric potential equation. Its influence also feeds back into the electric field and ionic concentration equations by means of the electrode kinetics and hence errors in its calculation can result in numerical instability.

Many publications concerning the numerical simulation of the electrodeposition process do not give special attention the calculation of the overpotential and details seem to be hidden in the discretisation scheme or implied to be calculated as the potential difference

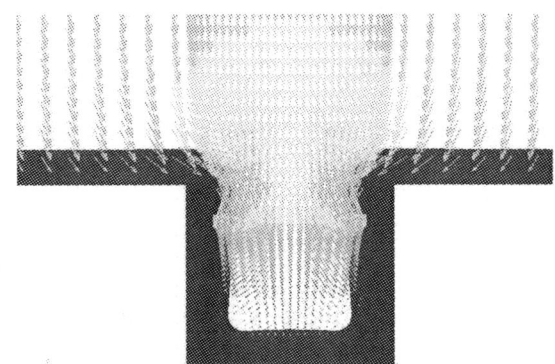

Figure 12 Current vectors in electrolyte region

$\left(\phi_{metal} - \phi_{electrolyte}\right)$ across the interface. Consideration of the boundary conditions gives some insight into the behaviour of these equations at the interface. The boundary conditions b in Figure 2 form a closed set of equations that can be iterated to a steady state solution by assuming that the gradients are to be taken over a layer that spans the interface boundary into the bulk region, Δn_{ϕ} for electric potential and Δn_c for concentration. Known constant values are applied for the cathode transfer coefficient, α_c, Temperature, T and exchange current density, I_o. The suffix i denotes interface values.

$$I_{dep} = I_o \frac{c^i}{C_{bulk}} \exp\left\{-\frac{\alpha_c F \eta}{RT}\right\}$$

$$\phi^{i+1} = \frac{I_{dep}}{k} \Delta n_{\phi} + \phi_{bulk} \qquad\qquad 10$$

$$c^{i+1} = \frac{I_{dep}}{zDF} \Delta n_c + c_{bulk}$$

If we couple into the above equation an estimation of the overpotential as the difference in potential across the boundary Δn_{ϕ}, i.e. $\eta = \phi^i - \phi_{bulk}$ we find that the final converged values are srongly dependent on the distance Δn_{ϕ} as shown in Figure 13 below. An obvious high level concluson might be drawn: It seems likely that any numerical model which attempts to implicitly calculate η will encounter difficulty in resolving the grid spacing around the interface region to a sufficient level and that the calculated deposition current will therefore be mesh dependent. Future work in this area will be essential in moving the model forwards into AC waveforms. Two lines of investigation are presently envisaged to introduce a sub-model from which the overpotential can be calculated namely:

i) Solve the equation set (10) in computational cells immediately adjacent to the interface in the electrolyte region. This method would use the cell values of concentration and electric potential within these interface cells as the bulk values and attempt to provide a more accurate estimation of the interface concentration, potential and deposition current by iterating the equation set.

ii) To estimate capacitance and charge based on parallel plate capacitor approaches such as the Stern model [8].

Both approaches will be investigated during the course of this project ideally with aid of accompanying experimental verification.

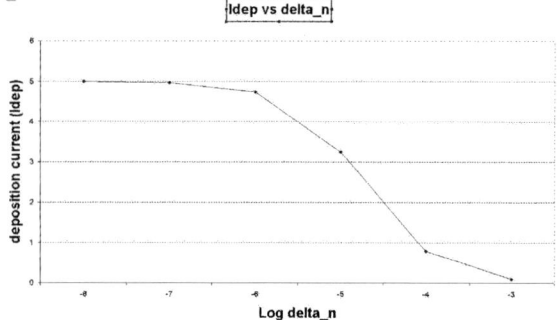

Figure 13: Calculated Deposition current vs log Δn

8   Conclusions and future work

The code framework for modelling the electrodeposition process has been implemented within the multi-physics code PHYSICA. Issues concerning the application of surface source terms and movement have been considered. The model shows numerical stability and exhibits qualitative behaviour for primary and tertiary current regimes with known overpotential.

Future work will address:

i) Testing the model against experimental data and subsequent tuning to give quantitatively correct behaviour.

ii) Investigating the calculation of surface overpotential through sub-models or relationships that may link the overpotential to concentration distribution.

iii) Advancing the model towards the simulation of pulse reverse plating and AC waveforms.

iv) Application of megasonic vibration within the model to encourage mixing and the replenishment of ions within trenches of high aspect ratio.

Glossary

C – concentartion of reacting ions – *moles/m³*.

C^i – interface concentration of reacting ions. – *moles/m³*.

C^∞ – interface ionic concentration – *moles/m³*

$\bar{c}$ –Dimensionless concentration ratio [0,1]

D –Diffusion coefficient – m²/s.

e – Elementary charge – $1.602.10^{-19}$.

F – Faradays constant – *C/mole*.

I_{BV} – Deposition current from But-Vol equation – *A/m²*.

I_{dep} –Deposition current – *A/m²*.

I_o –Exchange current density – *A/m²*.

K –electrical conductivity – $(\Omega m)^{-1}$.

$n|Z$ –ion valency

R –Universal gas constant – 8.314 *J/mole k*

T –temperature – K

α_c –Cathode transfer coefficient.

K – Boltzman constant – $1.38^{-23} m^2 kg/s^2 k$.

ϕ – electric potential – volts.

φ – Level Set distance variable – *m*.

σ – metal electric conductivity – *volts*.

Acknowledgments

The authors acknowledge the financial support of the Engineering and Physical Sciences Research Council (EPSRC) under Grant EP/C513061/1, as well as our industrial partners Merlin Ciurcuits and Raytheon Systems. We also acknowldge Jens Kaufmann and Professor Marc Desmuilliez from Heriot-Watt for useful discussions about the electrodeposition process. Special thanks are also given to Dr Nick Croft of Swansea University and Dr Dan Wheeler of National Institute for Standards and Technology (NIST) for helpful discussions on the modelling aspects of this research.

References

1. Wheeler D, Josell D, Moffat T.P, "Modeling Superconformal Electrodeposition Using the Level Set Method." *J.Electrochem. Soc.* Vol 150, No 5, (2003), ppC302-C310.

2. Wheeler D, Guyer J.E, Warren J.A, "A Finite Volume PDE Solver Using Phython". URL http://www.ctcms.nist.gov/fipy/download/fipy.pdf

3. West A.C, Cheng CC, Baker B."Pulse Reverse Copper Electrodeposition in High Aspect Ratio trenches and Vias". *J.Electrochem. Soc.* Vol 145, No 9, (1998), pp 3070-3074

4. Ritter G, McHugh P, Wilson G, Ritzdorf T, "Three dimensional numerical modelling of copper electroplating for advanced ULSI metallisation" *Solid-State Electronics*, 44, (2000), pp 797-807.

5. Drese K.S, "Design Rules for Electroforming in LIGA Process", *J.Electrochem. Soc*, Vol 151, No 6, (2004), D39-D45.

6. Drews T,O et al, "Multiscale simulations of copper electrodeposition onto a restistive substrate", *IBM J. RES & DEV*, Vol 49, No 1, (2005), pp49-63.

7. Adalsteinsson D, Sethian J.A, " A Level-Set Approach to a Unified Model for Etching, Deposition and Lithography I: Algorithms and Two-Dimesional Simulations", *J. Comput. Phys*, Vol 120, (1995), pp128-144

8. Paunovic M, Schlesinger M, "Fundamentals of Electrochemical Deposition", *Wiley-Interscience*, ISMN 0-471-16820-3.

9. Griffiths S.K et al, " Modeling Electrodeposition for LIGA Microdevice fabrication", *SAND98-8231*, (1998), Distribution Category UC-411.

10. Osher, S., Sethian, J.A, "Fronts propagating with curvature-dependent speed: Algorithms based on Hamilton-Jacobi formulations." *J. Comput. Physics*, vol. 79 (1988) pp 12–49

Application of Higher Order Derivatives Method to Parametric Simulation of MEMS

Vladimir Kolchuzhin[1], Jan Mehner[2], Thomas Gessner[2] and Wolfram Doetzel[1]

[1]Chemnitz University of Technology, Department of Microsystems and Precision Engineering, Germany
[2]Fraunhofer Institute for Reliability and Microintegration, Department Multi Device Integration, Germany
Email: vladimir.kolchuzhin@etit.tu-chemnitz.de

Abstract

The paper demonstrates the advanced simulation methodology based on differentiation of the discretized Finite Element (FE) equations to parametric simulation of Micro-Electro-Mechanical-Systems (MEMS). The idea of the approach is to compute not only the governing system matrices but also high order derivatives (HOD) with regard to design parameters by means of Automatic Differentiation (AD). As result, Taylor vectors of the model response can be expanded in the vicinity of the initial position with regard to dimensional and physical parameters. The objective of this presentation is to demonstrate the viability of HOD methods to parametric simulation of MEMS in the static, modal, frequency response domains on the basis of the structural analysis and macromodeling.

1. Introduction

Modeling and simulation of MEMS is of vital importance to develop innovative products and to reduce time-to-market at lower total costs. Advanced design methodologies and a variety of software tools are utilized in order to analyze complex geometrical structures, to account for interactions among different physical domains. Computer simulations provide a deep understanding of the device behaviour and lead to systems with optimized performance parameters.

MEMS components are commonly analyzed by FE techniques [1, 2]. But coupled electrostatic-structural analysis of a typical MEMS device can be quite time consuming using the traditional finite element approach, Fig. 1a. For example, full-mesh transient self-consistent coupled electro-structural simulation of MEMS devices still take a few hours using commercial finite element solvers [2, 3].

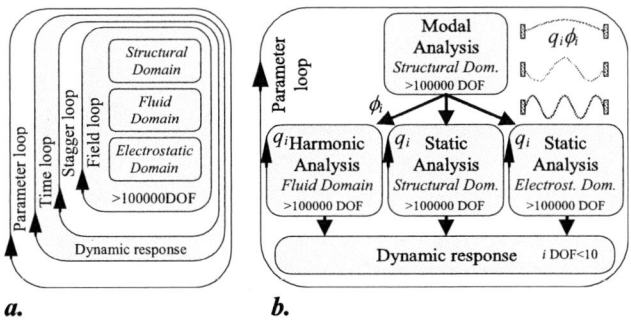

a.   *b.*

Figure 1: a. Full-mesh simulation and b. Mode Superposition Method

Currently, parametric models of complex devices are extracted by numerical data sampling and subsequent function fit algorithms. Each sample point must be ob-

tained by a separate FE run, Fig. 2. Usually, one needs between several ten and some hundred of sample data to capture the influence of design parameters accurately, which is cumbersome for practical use [2].

On the other hand, there exist numerous sophisticated tools, established for electronic design automation, which allow circuit and control system virtual prototyping. The problem is to obtain a macromodel from FE results, retaining accuracy, that exhibits just input/output interface terminals of a device by means of internal state variables and can be directly linked into the circuit and system-level schematic [1, 2]. Mode superposition method using the natural frequencies and mode shapes ϕ_i to characterize the dynamic response of a flexible structure has become an attractive alternative for fast simulations [4]. Recent contributions have shown that the deformation state and dynamics of flexible electro-mechanical systems can be described efficiently by modal superposition [4-9]. Essential speed up for the structural analysis is achieved since the deformation state of the mechanical system is represented by a weighted combination of a few eigenmodes, Fig. 1b. Capacitance-stroke functions provide non-linear coupling between the modes and the electrical quantities, such us electrostatic forces and electrical current, if stroke is understood as modal amplitude q_i [4]. In [5] was demonstrated an approach for adding dissipative effects into existing macromodels for damped harmonic and transient analyses. This method has been successfully applied to squeeze and slide film problems with nontrivial plate shapes and arbitrary motion [5, 3].

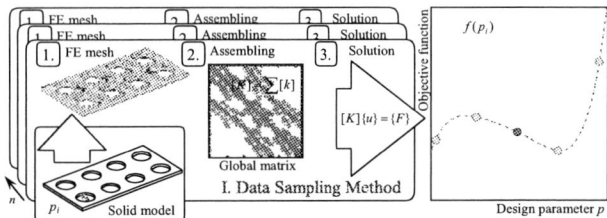

Figure 2: Flow chart of data sampling FE techniques

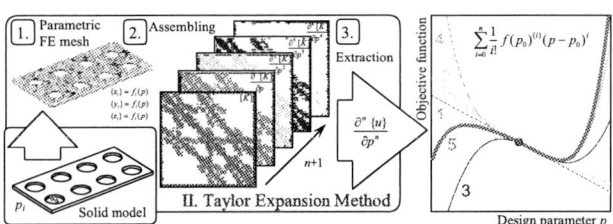

Figure 3: Flow chart of parametric FE techniques

The paper presents an advanced simulation methodology based on differentiation of the discretized FE equa-

1-4244-1105-X/07/$25.00 ©2007 IEEE   257

tions in the static, modal, frequency response domains with regard to set of design parameters p, Fig. 3. This approach takes to account parameter variations in a *single* FE run. The key idea of the new approach is to compute not only the governing system matrices of the FE problem but also n high order partial derivatives with regard to design parameters by means of Automatic Differentiation (AD). As result, Taylor vectors of the systems response can be expanded in the vicinity of the initial position capturing dimensions and physical parameters. Short review of the applications of Higher Order Derivatives Method for different types of problems will be presented. The objective of this presentation is to demonstrate the viability of HOD methods for the parametric FE simulation of micro-electro-mechanical-systems in the static, modal, frequency response domains on the basis of the structural analysis. Numerical details, accuracy and observed problems will be discussed on several examples.

2. Theory

Taylor series expansion is a common engineering approach to estimate the structural response versus parameter variation. The coefficients of the Taylor series can be expressed in terms of derivatives of the goal function at the initial point p_0:

$$f(p) \approx \sum_{i=0}^{n} \frac{1}{i!} f(p_0)^{(i)} (p - p_0)^i .$$
(1)

The applications of higher-order sensitivity analysis to FE problem as a way to increase the efficiency, accuracy and robustness was started in the 1990s [10-12]. Paper [13] reviews different techniques for structural design sensitivity analysis, including global finite differences, continuum derivatives, discrete derivatives, and automatic differentiation. Parametric technologies have been applied for CFD problem [10] and thickness optimization of shell structures in mechanical engineering [11], in electromagnetism [14-16]. Later this approach was used for microwave device design [18] and parametric model extraction for MEMS [20-23]. First commercially available Variational Technologies software was developed by the French company CADOE S.A. [24].

2.1. Taylor expansion of FE solutions

Taylor expansion of the FE solution can be obtained by differentiation corresponding ordinary FE problems, e.g. linear static problem:

$$[K]\{u\} = \{F\} .$$
(2)

The equations of FE problems are expressed in matrix notation, where $\{u\}$ is the unknown solution vector, $[K]$ is the governing FE matrix (e.g. stiffness matrix for mechanical problems) and $\{F\}$ is the load vector.

Higher order derivatives of FE solutions $\{u\}^{(n)}$ are computed recursively by:

$$[K]\{u\}^{(n)} = \{F\}^{(n)} - \sum_{i=1}^{n} C_n^i [K]^{(i)} \{u\}^{(n-1)} .$$
(3)

The matrix $[K]$ in (3) already factorized at the p_0.

Displacement derivatives $\{u\}^{(n)}$ have to be calculated first. Secondary derived variables like stresses or strain energy can be obtained from displacement derivatives [16, 20].

The theoretical aspects of the higher order derivatives method and Taylors expansions of FE solutions are presented in [11].

2.2. Derivatives of FE matrices

Different types of design parameters can be handled by this method. The most obvious ones are continuous parameters such as geometrical dimensions, material properties, etc. The methods can also deal with discrete parameters such as boundary conditions or loads. The general algorithm to extract a global finite element matrix $[K]$ and its n high order derivatives, based on the parametric finite elements and automatic differentiation, is presented more detailed in [16, 19, 20]. The derivatives of a global FE matrix $[K]$ can be obtained from a superposition of derivatives of elementary matrices $[k]$:

$$\frac{\partial^n}{\partial p^n}[K] = \sum_{el} \frac{\partial^n}{\partial p^n}[k]$$
(4)

The derivatives of an elementary stiffness matrix $[k]$ with regard to geometrical parameters p is obtained by equation:

$$\frac{\partial^n}{\partial p^n}[k] = \sum_{i}^{ngp} \frac{\partial^n}{\partial p^n}[B_i(p)]^T [D][B_i(p)] |J_i(p)| w_i$$
(5)

where $[B]$ is the strain-displacement matrix, $[D]$ is the material matrix, $[J]$ is the Jacobian matrix and w is the weighting factors.

The Jacobian matrix is a function of node coordinates of the element $[J(p)] = f(x(p), y(p), z(p))$. The $[B(p)]$ matrix is a function of the inverse Jacobian matrix $[B(p)] = f([J(p)]^{-1})$. Thus, the derivatives of elementary stiffness matrix (5) can be expressed from analytical functions $x(p)$, $y(p)$, $z(p)$, which connect node coordinates of the element with involved geometrical parameters p.

To get derivatives of the inverse Jacobian matrix $[J]$, the determinant of the matrix $|J|$ and the product of $[B]^T[D][B]|J|$ we use Automatic Differentiation [25].

2.3. Automatic Differentiation

In this section, we briefly review Automatic Differentiation techniques and describe our tool used for differentiation of the stiffness matrix. Difficulties arose mainly from the fact that extraction of high order partial derivatives becomes numerically unstable and time consuming. In contrast to symbolic differentiation, which propagates mathematical functions, AD algorithms calculate numerical values. Efficiency of AD relies on the fact that every function is nothing else than a sequence of arithmetic operations and elementary functions. The beauty of the approach is that higher order derivatives follow from a recursive combination of existing derivatives with binomial coefficients. Chain rules of differential calculus describe how to combine partial derivatives and binomial coefficients in order to form elementary mathematical operations and where to store results in arrays. What is remarkable is that AD gives an exact (up to the machine precision) representation of high order derivatives.

There are mainly two ways to implement an automatic differentiation tool. One is source-to-source transformation tool, and the other are object oriented AD tools. Several tools have been developed to handle the automatic differentiation process. The characteristics are summarized in the following table and details are available in [25-30].

Tools	Language	Technique	Comment
ADIC	C/C++	source transformation	up to 2nd order , 1997
ADIFOR	Fortran77	source transformation	1st order, 1991
ADiMat	Matlab	operator overloading	up to 2nd order (Jacob. and Hessian) 2002
ADOL-C	C/C++	operator overloading	1st and higher der. of vector functions, 1992
FFADLib	C/C++	operator overloading	derivatives up to n order, 2000 [26]
ADOGEN	C/C++	automatically generated the source code	patented, proprietary tools for derivate matrices [23]

While various AD tools are currently under academic research and development, we have prototyped the tool based on [28] implemented as a subroutine call in MATLAB. The tool supports the differentiation of matrices (including inverse matrix and determinant) and matrix-vector operations. Partial derivatives of matrices and vectors are arranged and stored in successive planes in a third dimension whereby the multi-index $\mu = \{p_1, \ldots, p_n\}$, which defines the partial derivatives in multivariable case (set of n parameters), corresponds to the plane location index R referred as 'tuple structure' in [28]:

$$R(\mu) = \sum_{j=1}^{n} \binom{K_j + j}{j}, \text{ where } K_j = \left(\sum_{i=0}^{j-1} \mu_{n-i} \right) - 1 \cdot$$

2.4. Parametric FE Mesh

In contrast to ordinary FEM, one needs a special parametric FE model what captures the variation of geometrical design variables without a re-mesh procedure. Parametric means that the node table contains not only a single numerical value for each spatial direction as it's supported by most CAD tools, rather each node i will be described by analytical functions $\{x_i(p), y_i(p), z_i(p)\}$, usually polynomials with respect to geometrical parameters p.

In order to automate the parametric mesh-morphing procedure, it is necessary to describe the shape change at the outer boundaries. Any three-dimensional linear transformation (rotation, translation, skew) can be represented by a coordinate transformation matrix. Partial derivatives of the transformation matrix with regard to design parameters p also can be compute by means of AD.

In a next step, the perturbation is extended to the interior domain. The accuracy of parametric results depends mainly on the quality of mesh perturbations caused by mapping of global parameters p to the nodal table. Generally, internal finite element nodes must move smoothly with respect to dimensional modifications, especially in case of large displacements or complicated shape (e.g. perforation holes, sharp notches). Similar problems are widely known from mesh morphing of coupled domain analyses [3, 31]. A convenient way to transform global parameters to internal nodal coordinates is a Laplacian algorithm used for mesh-morphing of FEM [31]. If the resulting mesh quality is not acceptable, one must utilize nonlinear algorithms for mesh perturbations [17].

3. Examples and Discussion

Parametric finite element algorithms have been prototyped in MATLAB [20] and tested by several examples (micromirror cell, comb drive, fixed-fixed beam). Solid modeling and mesh generation has been realized within ANSYS/Multiphysics [3]. The perturbation of internal nodes with respect to parameter can be obtained by solving Laplace's equation with Dirichlet boundary conditions [22].

3.1. Static analysis

Strain energy of a fixed-fixed beam (125x30x2 μm) under mechanical load is shown in Fig. 4.

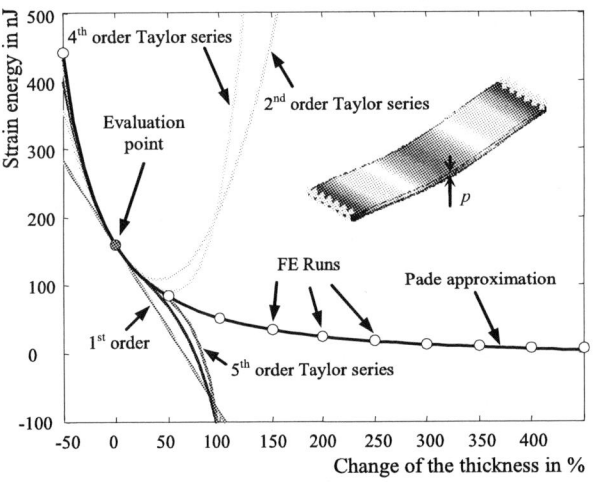

Figure 4: Static response: Strain energy vs. thickness

The thickness of the beam was used as design parameter. Performing a parametric structural static analysis we obtain all derivatives of the global matrix $[K]^{(n)}$ and the solution vector $\{u\}^{(n)}$. The derivatives of strain energy $W_{\text{SENE}}^{(n)}$ are extracted from derivatives of $[K]^{(n)}$ and $\{u\}^{(n)}$ at the p_0 by means of AD for the product $\{u\}^T [K] \{u\}$:

$$\frac{\partial^n}{\partial p^n} W_{SENE}(p) = \frac{1}{2} \frac{\partial^n}{\partial p^n} \{u(p)\}^T [K(p)] \{u(p)\} \cdot$$

Strain energy have been computed up to 5th order. Basic transformation of a Taylor series expansion to a Pade approximation [32] allows extending of the acceptable parameter range to about 400 percent!

Silicon is an anisotropic material. Dependence of the modulus of elasticity E and Poissons ratio v of Si on orientation in a (100)-plane can be approximated by equations:

$$E_{(100)}(\theta) = \frac{1}{0.768 - 0.704 \cdot \cos^2(\theta) \cdot \sin^2(\theta)} \quad (6)$$

$$v_{(100)}(\theta) = \frac{0.214 - 0.708 \cdot \cos^2(\theta) \cdot \sin^2(\theta)}{0.768 - 0.708 \cdot \cos^2(\theta) \cdot \sin^2(\theta)} \quad (7)$$

It is possible to involve material properties as design parameters taking into account the dependence of a matrix $[D(E,v)]$ on the parameter θ. Dependence of strain energy of a displaced Si-beam on crystalline orientation in a (100)-plane is shown in Fig. 5.

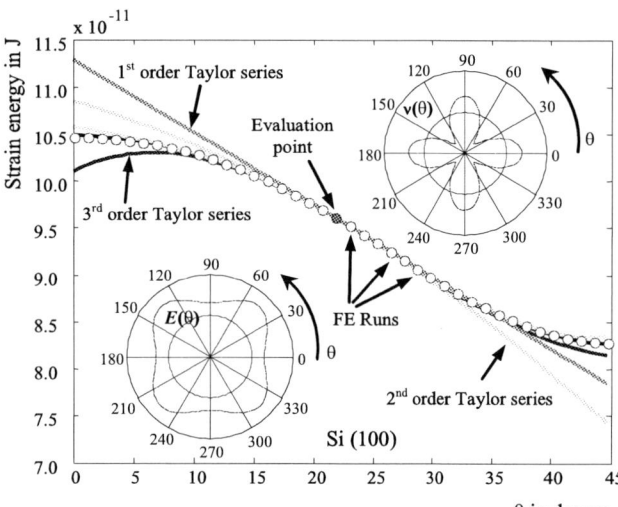

Figure 5: Static response: Strain energy vs. orientation

A next benchmark for estimating the HOD method accuracy is a micromirror cell with two degrees of freedom, the transversal shift and tilt angle [3], Fig. 6. The meshed model with 8-node brick elements has about 5,000 DOF.

The macromodel generation pass requires a series finite element solutions of the structural and electrostatic domains [3, 4]. At each sample point, the microstructure is displaced to a linear combination of mode shapes in order to extract the strain energy and mutual capacitances [4, 7, 8].

With the HOD method a time consuming FE data sampling process can be replaced by a single parametric FE run. Strain energy has been computed with regard to modal amplitude q_3. Experimental results demonstrate that the generated models accurately predict effects over a wide range of modal amplitudes. Fig. 6 shows that the response functions could be captured correctly by 2nd order polynomials in the linear case, since the strain energy function is a very smooth one.

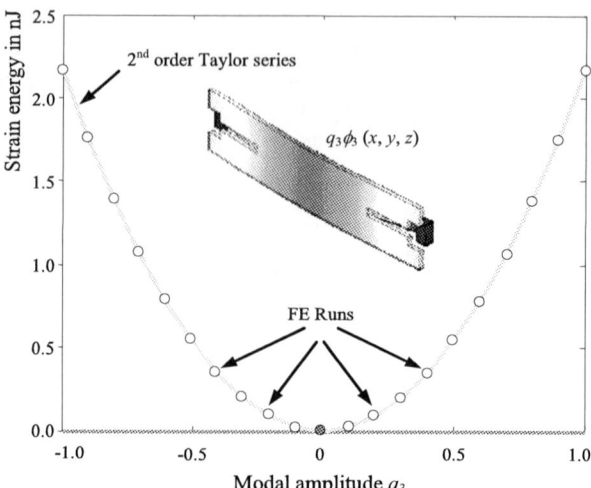

Figure 6: Static response: Strain energy vs. modal amplitude

3.2. Modal analysis

Parametric modal analysis is important for identification of dimensional parameters by vibration analyses and parametric macromodeling of MEMS [9].

As matrix $[K - \lambda_i M]$ is singular one especially algorithm (known as Nelson's method for first eigenvector derivative [33]) is applied for parametric modal analysis [34]. Exemplarily, dependence of frequency on length of comb drive fingers is shown in Fig. 7.

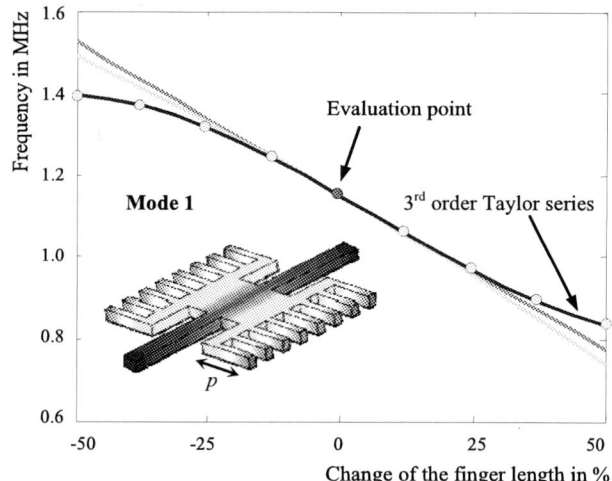

Figure 7: Modal response: Frequency vs. length of comb drive finger

3.3. Harmonic analysis

Squeeze film analysis simulates the effects of fluid damping of structure moving in small gaps between fixed walls. Reynolds squeeze film equation known from Lubrication theory is widely used in MEMS design to estimate the damping [5, 3]. The damping c and stiffness k coefficients are frequency-dependent:

$$c = \frac{F^{Re}}{v_z}, \quad k = \frac{F^{Im}\omega}{v_z}, \quad (8)$$

where F^{Re} is the real part of the pressure force. F^{Im} is the imaginary part of the pressure force, ω is the circular frequency and v_z is the normal velocity component of the moving structure.

A harmonic analysis is used to determine the stiffening and damping coefficients with regard to frequencies [5, 3]. At each frequency step ω_i, we factorize and solve a linear system to compute these coefficients:

$$\begin{bmatrix} K & -\omega C \\ \omega C & K \end{bmatrix} \begin{Bmatrix} p_1 \\ p_2 \end{Bmatrix} = \begin{Bmatrix} v_z \\ 0 \end{Bmatrix}, \qquad (9)$$

where $p = p_1 + ip_2$ is the pressure.

The HOD approach correctly takes into account the frequency-dependence in a single FE run. A rectangular plate under transverse motion is modeled exemplarily to compute the effective damping and squeeze stiffness coefficients, Fig. 8.

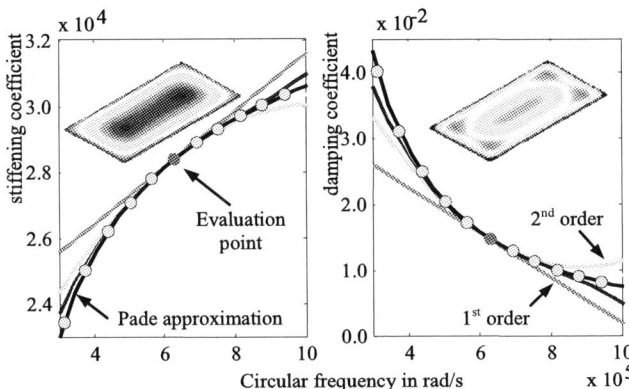

Figure 8: Frequency response: stiffening and damping coefficients

3.4. Discussion

The results from numerical case studies of test problems in linear static, modal and harmonic structural domains have been presented in Fig. 4-8. Comparisons with ordinary FE results have been used to validate the accuracy of the parameterized solutions, where up to fifth orders polynomials have been computed. The number of differential planes is comparable with the number of samples required for polynomial fit methods. Fortunately, Automatic Differentiation of additional planes is usually less expensive compared to FE solution runs needed for sampling. Therefore, the extra time taken for parametric modeling and mesh at the beginning of variational techniques disappears rapidly. Benefits of the novel approach compared to ordinary data sampling procedures needed for parametric modeling of MEMS become obvious for large and multi-parameter problems. The data sampling process is expected to grow as n^k (where n is the number of design parameters and k is the number of states for each design parameter). The method is also very efficient for large matrices, because the factorization of the system increases with the matrix dimensions. Essential speed-up can be achieved for sensitivity analyses and data sampling

needed for Reduced Order Macromodeling of MEMS [22].

By using the innovative HOD method, the simulation results become directly polynomial functions in terms of design parameters. It is necessary to point out the necessity for additional memory, parametric mesh-morphing procedures and to have access to the source code.

4. Conclusions

As illustrated in the paper, the HOD parametric FE technologies are a promising alternative to existing data sampling and function fit procedures utilized for MEMS parametric model extraction from ordinary FE analyses. The algorithms support static, modal and harmonic analyses of structural, electrostatic, thermal and fluidic domains.

In the near future, this key technology will be state of the art for MEMS modeling and macromodel generation. Further work will be focused on automated generation of MEMS *parametric* Reduced Order Models capturing inertial and damping effects.

Acknowledgments

The German Research Foundation through the projects SFB379 "Arrays of micromechanical sensors and actuators" has supported this work.

References

1. Senturia, S.D., <u>Microsystem Design</u>, Kluwer Academic Publishers, (Boston, 2001).
2. Mehner, J., Bennini, F., Dötzel, W., <u>CAD for Microelectromechanical Systems. System Design Automation - Fundamentals, Principles, Methods, Examples,</u> Kluwer Academic Publishers, (Boston, 2000), pp. 111-132.
3. ANSYS Inc.: "ANSYS Users Guide, Theory Reference manual," Canonsburg, USA.
4. Bennini, F., Mehner, J., Dötzel, W., "System Level Simulations of MEMS Based on Reduced Order Finite Element Models," *Int. Journal of Computational Engineering Science*, Vol. 2, No. 2 (2003), pp. 385-388.
5. Mehner, J., Dötzel, W., Schauwecker, B., Ostergaard, D., "Reduced Order of Fluid Structural Interactions in MEMS Based on Modal Projection Techniques," *The 12th Int. Conf. on Solid State Sensors, Actuators and Microsystems*, Boston, USA, June. 2003, pp. 1840-1843.
6. Nayfeh, A. H., Younis, M. I. and Abdel-Rahman, E. M., "Reduced-Order Models for MEMS Applications," *Nonlinear Dynamics*, 41 (2005), pp. 211-236.
7. Mähne, T., Kehr, K., Franke, A., Hauer, J. and Schmidt, B., "Creating Virtual Prototypes of Complex Micro-Electro-Mechanical Transducers Using Reduced-Order Modelling Methods and VHDL-AMS," *In Forum on Specification and Design Languages*, 2005, pp. 209-221.
8. Schlegel, M., Bennini, F., Mehner, J., Herrmann, G., Mueller D., Doetzel, W., "Analyzing and Simulation of MEMS in VHDL-AMS Based on Reduced Order

FE-Models," *IEEE Sensors Journal*, Vol. 5, No. 5 (2005), pp. 1019-1026.

9. Gugel, D., Doetzel, W., "Parametric reduced order modelling for industrial MEMS design processes," *1st European Conference & Exhibition on integration issues of miniaturized systems - MEMS, MOEMS, ICs and electronic components*," Paris, France, March. 2007. (Accepted for publication)

10. Bischof, C., Corliss, G., Green, L., Griewank, A., Haigler, K., Newman, P., "Automatic differentiation of advanced CFD codes for multidisciplinary design," *Journal on Computing Systems in Engineering*, (1992), pp. 625-638.

11. Guillaume, Ph., Masmoudi, M., "Computation of high order derivatives in optimal shape design," *Numerishe Mathematik*, Vol. 67, No. 2 (1994), pp. 231-250.

12. Ozaki, I., Kimura, F., Berz, M., "Higher-order sensitivity analysis of finite element method by automatic differentiation," *Computational Mechanics*, Vol. 16, Is. 4 (1995), pp. 223-234.

13. F. van Keulen, Haftka, R.T., Kim, N.H., "Review of options for structural design sensitivity analysis. Part 1: Linear systems," *Comput. Methods Appl. Mech. Eng.*, Vol. 194 (2005), pp. 3213-3243.

14. Petin, P., Coulomb, J.L., Conraux, Ph., "High Derivatives for Fast Sensitivity Analysis in Linear Magnetodynamics," *IEEE Trans. on magnetics*, Vol. 33, No. 2 (1997).

15. Lebensztajn, L., "Sensitivity Analysis and High Order Derivatives Applied to the Modeling of Permanent Magnets," *IEEE Trans. on magnetics*, Vol. 34, No. 5 (1998).

16. Nguyen, T. N., Coulomb, J.-L., "High Order FE Derivatives versus Geometric Parameters. Implantation on an Existing Code," *IEEE Trans. on magnetics*, Vol. 35, No. 3 (1999).

17. Perrin, S., Beley, J-D., Bao, V., "Parametric and Topologic Optimization Applied To Powertrain Behaviour," *World Automotive Congress*, Helsinki, Finland, June. 2002.

18. Thon, B., Bariant, D., Bila, S., Baillargeat, D., Aubourg, M., Verdeyme, S., Guillon, P., Thevenon, F., Rochette, M., Puech, J., Lapierre, L., Sombrin, J., "Coupled Padé approximation-finite element method applied to microwave device design," *Microwave Symposium Digest IEEE MTT-S International*, Vol. 3 (2002), pp. 1889-1892.

19. Garreau, S., Perrin, S., Rochette, M., "Application of high order derivatives to structural probabilistic analysis," *CADOE / European Space Agency Project*, Final Report, Jan. 2005.

20. Mehner, J. E., *et al*, "Parametric Model Extraction for MEMS Based on Variational Finite Element Techniques," *Proc. 13th Int. Conf. on Solid State Sensors, Actuators and Microsystems*, Seoul, Korea, June. 2005, pp. 776-779.

21. Kolchuzhin, V., Mehner, J., Dötzel, W., "Geometrically Parameterized Finite Element Model of the Silicon Strain Gauge," *7. Chemnitzer Fachtagung Mikrosystemtechnik- Mikromechanik & Mikroelektronik*, Chemnitz, Germany, Oct. 2005, pp. 190-195.

22. Kolchuzhin, V., Mehner, J., Gessner, T., Dötzel, W., "Parametric Finite Element Analysis for Reduced Order Modeling of MEMS," *Proc. 7th Int. Conf. on thermal, mechanical and multiphysics simulation and experiments in microelectronics and microsystems*, Como, Italy, Apr. 2006, pp. 220-225.

23. Kolchuzhin, V., Mehner, J., Gessner, T., Dötzel, W., "Parametric Simulation of MEMS Based on Automatic Differentiation of Finite Element Codes," *Technical Proc. of the 2006 NSTI Nanotechnology Conf. and Trade Show* Boston, USA, May. 2006, Vol. 3, pp. 507-510.

24. http://www.cadoe.com/

25. http://www.autodiff.org/

26. Corliss, G., Faure, C., Griewank, A., Hascoet, L., Naumann U., Automatic differentiation of algorithms: from simulation to optimization, Springer, (Boston, 2002).

27. http://www-unix.mcs.anl.gov/autodiff/adtools/

28. Tsukanov, I., Hall, M., "Data Structure and Algorithms for Fast Automatic Differentiation," *Int. Journal for Numerical Methods in Engineering*, Vol. 56, No. 13 (2003), pp. 1949-1972.

29. Griewank, A., Utke, J., "Evaluating higher derivative tensors by forward propagation of univariate Taylor series," Technical Report, TU Dresden, Institut fur Wissenschaftliches Rechnen, Dresden, 1995.

30. Bischof, C. H., H. M. Bücker, B. Lang, A. Rasch, and Slusanschi E., "Automatic Differentiation of Large-Scale Simulation Codes is no Illusion," *Preprint of the Institute for Scientific Computing RWTH-CS-SC-02-12*, Aachen University, Aachen, 2002.

31. Zhulin, V.I., Owen, S.J., and Ostergaard, D.F., "Finite Element Based Electrostatic-Structural Coupled Analysis with Automated Mesh Morphing," *Technical Proc. of the 2000 Int. Conf. on Modeling and Simulation of Microsystems*, San Diego, USA, March. 2000, pp. 501-504.

32. Chaffy, C., "How to compute multivariate Pade approximants," *Proceedings of the 5th ACM symposium on Symbolic and algebraic computation*. Waterloo, Canada. 1986, pp. 56-58

33. Nelson, R. B., "Simplified calculations of eigenvector derivatives," *AIAA Journal*, Vol. 14, No. 9 (1976), pp.1201-1205.

34. Liu, Z.-S., Chen, S.-H., Zhao, Y.-Q. and Shao, C.-S., "Computing eigenvector derivatives in structural dynamics," *Acta Mechanical Solida Siniea*, Vol. 6, No. 3 (1993), pp.291-299.

Multi-objective Parametric Approach to Numerical Optimization of Stacked Packages

Łukasz Dowhań[1], Artur Wymysłowski[1], Rainer Dudek[2]

[1]Wrocław University of Technology, Faculty of Microsystem Electronics and Photonics,
ul. Grabiszyńska 97, 53-439 Wrocław; Poland
[2]Fraunhofer Institute for Reliability and Microintegration (IZM),
Micro Materials Center Berlin and Chemnitz,
Gustav-Meyer-Allee 25, D-13355 Berlin, Germany
e-mail: Artur.Wymyslowski@pwr.wroc.pl, Lukasz.Dowhan@pwr.wroc.pl

Abstract

The main aim for development of small electronic packages is supported by an ongoing development of portable communications devices. Thin silicon dies are belived to improve as device performance as its reliability. Additionally, novel packaging techniques as stacked packaging reduce packaging cost, size, improve functinality and relibility. In case of the stacked packages, wafers are stacked to form 3D multi-chip package. On the other hand, the electronic market requires novel and efficient numerical designing tools to deal properly with the optimization. The goal of the current work was to design a reliable numerical model of the stacked package and afterwards perform numerical multi-objective optimization in reference to a number of variables, which influence the stacked package reliability.

1. Introduction

The progress in microprocessor's technology and computer science is a driving force for application of advanced numerical tools in daily engineer practice. A vast number of the engineering problems can be properly addressed by numerical simulations methods and efficient optimization algorithms, which in total is referenced as numerical design for optimization or numerical prototyping. This technique has rapidly evolved because of the development of faster microprocessors and RAM modules which improve and speed up the calculations and optimization algorithms. The current paper focuses on the multi-objective optimization methods adopted to novel and promising 3D stacked packaging.

Nowadays, the small, light and relible devices with an advanced functionality are demanded. One of the possibilities to achieve such features is the application of 3D stacked packaging. It is an electronic packaging in which the third (Z) dimension is used to stacked the chips or packages on top of each other (Figure 1). In that way a surface on PCB can be saved and the memory capacity can be enhanced at the same time. Stacked packages may incorporate a standard flip-chip wafer that is bonded upside-down to a bottom silicon wafer prepared with through-silicon vias and/or standard wire bonding.

In the paper a parametric numerical model of a 3D stacked package has been developed and optimized for a selected reliability multi-objective. A numerical model was created using the finite element method and ABAQUS v.6.5 software package. On of the reasons for using ABAQUS it was the embedded Python scripting language. Thanks to Python, the FEM model was parameterized in order to find during the optimization stage the essential design parameters (material properties) and their influence on selected reliability multi-objective. In this case the S11 stresses were selected placed in two different positions over the defined thermal cycle.

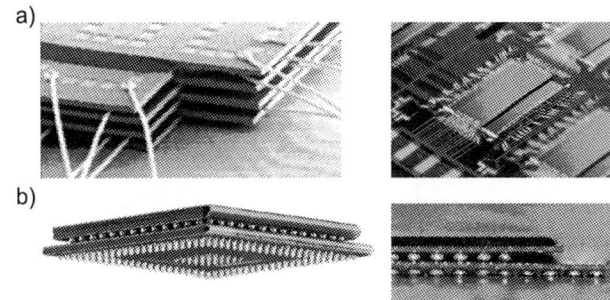

Figure 1. a) die stacking, b) package stacking

2. Single vs. Multi-objective Optimization

The problem of the packaging optimization seems to be one of the most important challenge in the area of numerical prototyping in microelectronics. For the recent years there has been an outstanding progress in many microelectronic fields and stacked packages are one of the vivid examples. In the previous research studies done by authors in reference to numerical prototyping of stacked packages both direct and indirect optimization procedures were implemented, which allowed for single-objective optimization.The current paper focuses on continuation of the previous research concerning stacked packages but towards multi-objective optimization. According to some researchers the multi-objective optimization should be the firsthand process, while a single-objective optimization process should be only an exception.

Optimization is in fact a separate branch of knowledge which focuses on development an application of different algorithms in order to find out an optimal solution due to a selected criteria and defined constraints. One of the crucial aspects of optimization is a vast number of optimization algorithms which are clasified according to the type of a problem. This means that certain optimization problems are preferred or even unique for a chosen type of a problem and almost useless in another type of problems. In engineering applications and

1-4244-1105-X/07/$25.00 ©2007 IEEE   263

especially microelectronic packaging there is a whole spectrum of optimization tasks which are defined either by mathematical formula or described by numerical model. Nevertheless, there seem to be one non-disputable aspect that is to be addressed properly in the near future which is multi-objective versus single-objective optimization. From that perspective a multi-objective optimization is highly complicated and requires an expert involvement for e.g. proper definition of the so-called objective function and appropriate mathematical algorithms.

Advanced optimization problems arise most often in case of a large number of input variables, nonlinear output behavior and multi-extreme responses with local and global aspects. The latter is the key selection criteria of an optimization algorithms as the global extreme is most often a final expectation. No matter of a chosen optimization algorithm, the main goal of the optimization can be formulated as follows: "Finding out such values of the input variables in the design space that would satisfy the optimization criteria of the selected objective function under defined constraints". From the mathematical point of view the optimization problem can be defined as follows:

$$
\begin{aligned}
min \; & f(\boldsymbol{X}), \; for \; \boldsymbol{X} = (x_{1,}x_{2,}\ldots,x_n)^T \\
& for \\
& c_i(\boldsymbol{X}) = 0, \; i = 1,2,\ldots,m \\
& c_i(\boldsymbol{X}) \geqslant 0, \; i = m+1,\ldots,n
\end{aligned}
\tag{1}
$$

where $f(X)$ is the objective function, X is the column vector of n independent input variables while $c_i(X)$ are the defined constraints of the input variables.

As mentioned earlier, due to a vast number of the optimization algorithms and different problems there is a need for a proper selection of an optimization algorithm according to the optimization problem, which would be capable of reducing e.g. total number of experiments and differentiate between local and global extreme. Additionally the appealing problem is connected with definition of the response function either it is defined by a mathematical or numerical model. In case of mathematical models an iterative algorithms would preferable while in case of numerical models a combined DOE/RSM approach followed up by an appropriate optimization algorithm would be the most convinient .

Probably the most appealing problems of optimization are due to the linear or non-linear character of the objective function, ability of differentiating between local and global extreme and single or multi-objective capabilities. In case of electronic packaging and single-objective optimization problems the most often applied optimization algorithms are based on gradient and simpleks methods, genetic algorithms and simulated annealing. In case of a multi-objective optimization problems there is a need for finding a compromise between different objective functions in order to find the optimal solution in reference to all the selected objective functions. In contrast to the single objective optimization,

the multi-objective optimization can be defined as follows:

$$
\begin{aligned}
min \; & F(\boldsymbol{X}) = \left[f_1(\boldsymbol{X}), f_2(\boldsymbol{X}), \ldots, f_m(\boldsymbol{X}) \right] \\
& for \; \boldsymbol{X} = (x_{1,}x_{2,}\ldots,x_n)^T \\
& c_i(\boldsymbol{X}) = 0, \; i = 1,2,\ldots,m \\
& c_i(\boldsymbol{X}) \geqslant 0, \; i = m+1,\ldots,n
\end{aligned}
\tag{2}
$$

where $f_i(X)$ are the objective function and the $F(X)$ is the optimization criteria based on the set of m objective functions.

In engineering applications the multi-objective optimal solution can formulated in a form of a set, which is optimal in the Pareto sense. The set means that there is not only one optimal solution but a whole set of solutions, which are equally acceptable from an engineering point of view and equivalent from the mathematical point of view. The solution is optimal in the Pareto sense, if there is not a better solution in reference of at least one criteria without worsening the solution in reference to all the other, which is presented in the figure 2.

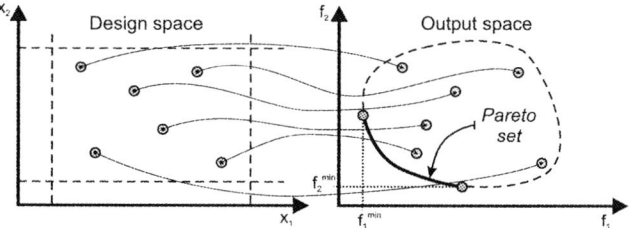

Figure 2. The Pareto set definition

Apart from the Pareto set approach, there are other multi-objective methods of which the most important ones in reference to practical and engineering applications are:
• Weighted Objective Method; the above method is based on transforming the multi-objective optimization into a single-objective optimization by defining a new criteria in a form of the objective function $z(X)$, which consist of a weighted sum of all the objective functions $f_i(X)$:

$$
z(\boldsymbol{X}) = \sum_{i=1}^{N} \left(w_i \cdot f_i(\boldsymbol{X}) \right)
\tag{3}
$$

where the weight coefficients w_i are to fulfill the following criteria:

$$
0 \leqslant w_i \leqslant 1 \; , \; \sum_{i=1}^{N} w_i = 1
\tag{4}
$$

In fact, the final solution in case of the weighted objective function is equivalent to a subset of the Pareto set, which can be graphically represented as an intersection point of a Pareto set with a line defined by the objective function $z(X)$. Apparently the intersection point is depended on the actual values of weight coefficients w_i, which is shown graphically in the figure 3.

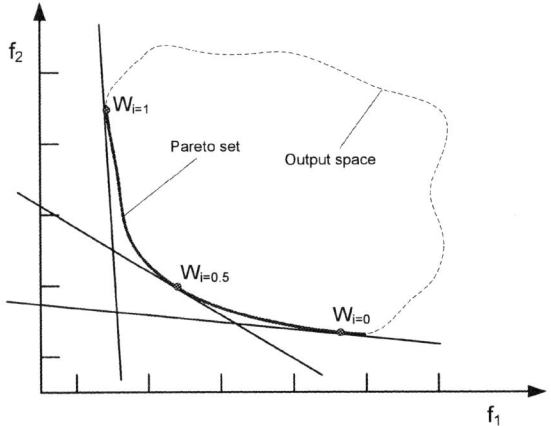

Figure 3. Intersection of the objective function $z(X)$ with the Pareto set.

- Hierarchical optimization method; similarly to the previous case, the above method is based on transforming the multi-objective optimization into a single-objective optimization by performing sequentially and iteratively optimization for separate objective functions $f_i(X)$ by a step procedure starting from the most important to less important by fulfilling the following criteria:

$$f_{i+1}(X) \leqslant \left(1 \pm \frac{\varepsilon_i}{100}\right) \cdot f_i(X) \quad (5)$$

where ε_i is an acceptable percent difference from the optimal value of the objective function $f_i(X)$.

- Evolutionary algorithms method; the above method is based on selected genetic algorithms and is most often implemented in case of a large number of input variables, when the Pareto set is to big for a classical methods mentioned above.

In the current research the multi-objective optimization was implemented in order to minimize the probability of failure due to selected failure criterias . In the previous research authors perforemd single-objective optimization taking into account warpage of the whole packge as a failure criteria, which is in fact a common approach in most of the applications. Unfortunatelly, there are much more failure criteria that may be equally important for failure as: bending, shear stresses, delamination, fatigue, etc. In working conditions some of them may be independent on the warpage failure criteria and require separate considerations and attention. Therefore as an example of multi-objective optimization, authors desided to find out the Pareto set for two stress components. In order to define the Pareto set, the weighted objective method was selected with an iterative change of w_i weight coefficients as shown in the figure 3.

3. Package Description

As an optimization carrier a stacked memory package was selected. The package consisted of three silicon chips which were separated by two silicon spacers and thermally conductive layers of wet adhesive material. The silicon dies were molded in molding compound material.

On the lower part of the package, besides the substrate, there are thin layers of copper film/plating and nickel plating. Figure 4 shows the symmetric 3D quarter of the module and the cross section of the selected area.

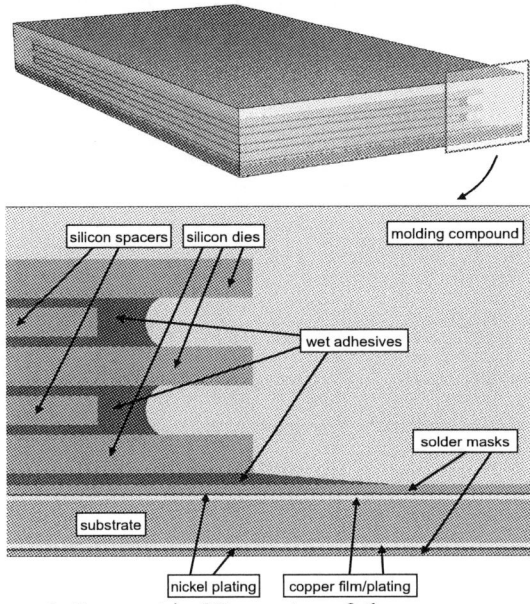

Figure 4. Symmetric 3D quarter of the memory module and detailed cross section of the selected area

Currently, the cell phones applications and digital cameras operate on memory modules based on two stacked dies – one for the logic circuits and second for the memory architecture. High capacity DRAM and flash memory systems consist of four or even more stacked modules.

3.1. Material Properties

An exact definition of material properties is one of the crucial aspects in FE-modeling. The selected optimization device can be properly described by elastic, plastic and viscoelastic material models, depending on the layer. Table 1 shows the main material models used in order to define the numerical model.

Table 1. Material properties used to FE-analysis

layer / material	elastic	plastic	viscoelastic
molding compound	X		X
chip/spacer	X		
adhesives	X		X
solder mask	X		X
nickel plating	X	X	
copper film/plating	X		
substrate	X (orthotropic)		

3.2. Loads – Thermal Cycle

As a load, one thermal cycle was defined. Because the maximum stresses and warpage occur at the lower dwell of the cycle, a 5 minute cooling step from 448 K to 293 K was chosen for the analysis, which is presented in figure 5.

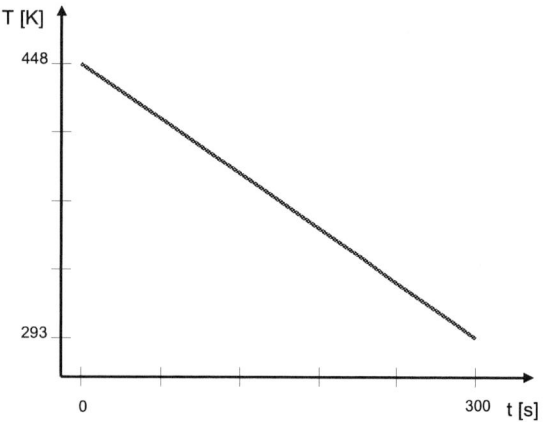

Figure 5. Defined temperature profile of the thermal cycle

4. Numerical FE-model

As a first approach to design of optimization of the stacked package a 2D numerical model was created. The model was build as a cross section through the shorter edge of the 3D model shown in Fig. 2. This model was created using the ABAQUS v.6.5 software.

The optimization process is relatively time-consuming because some of the optimization algorithms need to carry out lots of iterations. That was the reason to create 2D model (which consisted of much less elements than the 3D model). This model was based on "generalized plane strain" elements, which assume the final response to reflect the real behavior of 3D model in a cross section. In the figure 6 there is presented the created 2D FE-model of the stacked package.

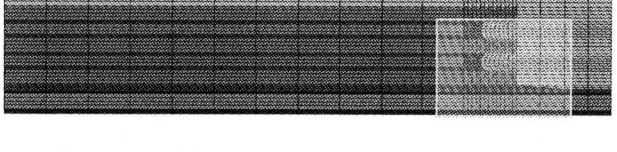

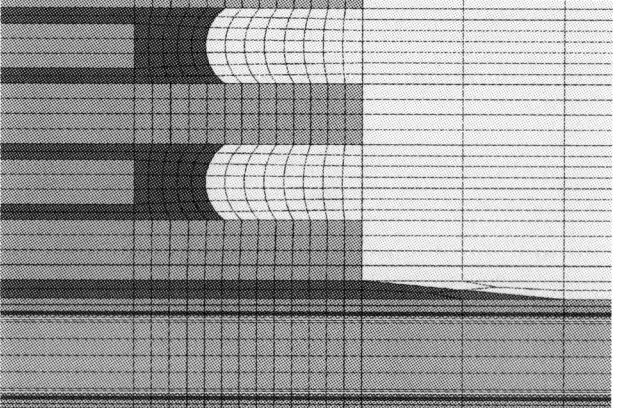

Figure 6. Meshed 2D model and the detailed area

In order to create 2D parametric model of the device, the Python scripting language incorporated in ABAQUS was used. Knowing the basics of Python it is relatively easy to create a model in which it is possible to parameterize the geometrical dimensions, material properties or loads and boundary conditions. Parametric studies are only supported in ABAQUS Standard/Explicit. However, using Python, it is also possible in the ABAQUS/CAE graphical environment.

5. Summary of the single-objective Optimization

The single-objective optimization of the presented memory module was described in detail in [1,2]. In the single-objective optimization two processes were performed: direct optimization which adopts evolutionary algorithm and indirect approach based on DOE/RSM methods. The optimization was done with the help of NOESIS-OPTIMUS and WRUT-VPT software packages. For direct optimization the differential evolution (DE) algorithm was adopted. DE is a simple, relatively easy to implement, evolutionary algorithm based on the idea of population (likewise genetic algorithm). The indirect optimization approach was based on DOE/RSM methods and VPT software package. It consisted of two stages:

1. DOE → orthogonal L16 scheme + ANOVA
2. DOE → Latin Hypercube + RSM → Kriging

This solution allows decreasing the total number of experiments needed to obtain reliable results. The advantage of this optimization method is the possibility of improving the quality of responses with the reduced number of experiments, compared to traditional methods as e.g. direct approach. One of the additional benefits is the possibility of application of the advanced optimization including sensitivity and tolerance design [3,4].

The optimization process based on DE was very time-consuming. The process consisted of 510 experiments. The most significant impact on the response (warpage) was noticed during the changes of material properties: CTE_molding and also CTE_adhesive and CTE_substrate. It was also noticed that the optimal value of warpage was achieved with the boundary input parameters. The indirect DOE/RSM optimization consisted of 36 numerical experiments (16 – stage 1, 20 – stage 2). The optimal response (warpage) of the investigated package was obtained after stage 1 of the optimization using the L16 orthogonal table experiment. In this case, the optimal warpage value was again calculated at the space boundary. Both of the presented optimization processes brought very similar results. The highest impact on the warpage was observed by changes of the three material parameters (CTE_molding, CTE_bstage, CTE_substrate). It was confirmed by the ANOVA analysis. CTE_molding influenced the warpage the most (67,67% - see table 2). The changes of geometrical dimensions (adhesive_thick, chip_thick) influenced the result insignificantly (4% - adhesive_thick; 2,45% - chip_thick).

Table 2. The summary of the single-objective optimization

Parameters	Nominal	Low	High	Optimization	
				Direct (DE)	Indirect (DOE/RSM)
CTE_molding	10	8	22	21.991	22
CTE_bstage	110	70	130	70.066	70
CTE_substrate	14.5	12	20	12.001	12
chip_thick	0.135	0.125	0.145	0.145	0.145
adhesive_thick	0.035	0.025	0.045	0.025	0.025
number of experiments				510	36
GOAL (warpage)	0.02728			**0.01155**	**0.01153**

6. Multi-objective Optimization

6.1. Assumptions and goal of the multi-objective optimization

In fact, the multi-objective optimization was prepared on the reults of the single-objective optimization basing on the knowledge of the most essential input parameters in reference to the warpage. As a design space, two most significant material parameters were chosen – CTE values of molding compound and adhesive layers. Figure 7 shows the design space. As the output, there were two stress S11 components selected as shown in the figure 8 - at Point 1, at half of the package and at Point 2 on the interface between chip and molding compound. These two criteria were chosen for the multi-objective optimization.

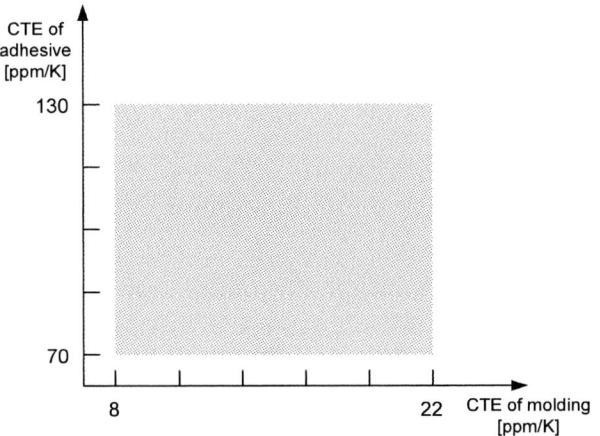

Figure 7. The selected design space for the multi-objective optimization

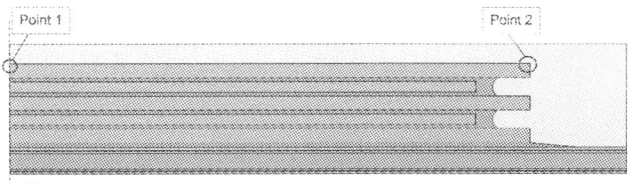

Figure 8. The two criteria chosen for the multi-objective optimization

6.2. Optimization Procedure

As mentioned earlier, the goal of the multi-objective optimization was to obtain the Pareto set for chosen failure criteria through the Weighted Objective Method. Authors decided, that the Pareto set for two stress components S11 can be defined through the weighted objective method by an iterative change of w_i weight coefficients as shown in the figure 3. Thus, the weights in the multi-objective problem in that way would be reduced to simple single-objective. The whole optimization procedure was implemented in the self developed WRUT-VPT software package. The multi-objective procedure was developed in Python scripting language. Thanks to the reduction of the multi-objective to single-objective problem the optimization was carried out using the TNC (Truncated Newton Code) non-linear optimization function available in the SciPy module of Python.

As it was mentioned earlier, Weighted Objective Method provides a weighted objective function $z(x)$, which is defined by formula 3 and 4 as a linear sum of objective function $f_i(X)$ and weight coeeficients w_i. The slope of the function is dependent on the values of weight coefficients but always be tangent to the Pareto set as shown in the figure 3 . The above assumption was directly implement to define the Pareto set. This was acheived by iterative change of the weight values thus leading to several different intersection (tangential) points, which through the followed up interpolation lead to represenation of the Pareto set. In current analysis in order to draw the Paraeto set, there were 5 different weight coeffcient values assumed changing in a range from 0 to 1.

6.3. Optimization results

Using the multi-objective optimization discussed in this paper it was possible to obtain the Pareto set (the set of equvalent optimal solutions) and also the parameters from the design space which optimize the output values. Fig. 9 shows the obtained Pareto set and the points from the design space which correspond to the optimal solutions.

As it is seen on the response diagram, some of the optimal values are placed inside the output set. It is caused by the single-objective algorithm (TNC) which is used after weighting the multiple criteria. This algorithm is not fully aderquate for optimizing the numerical problems and sometimes it can be trapped in local extrema.

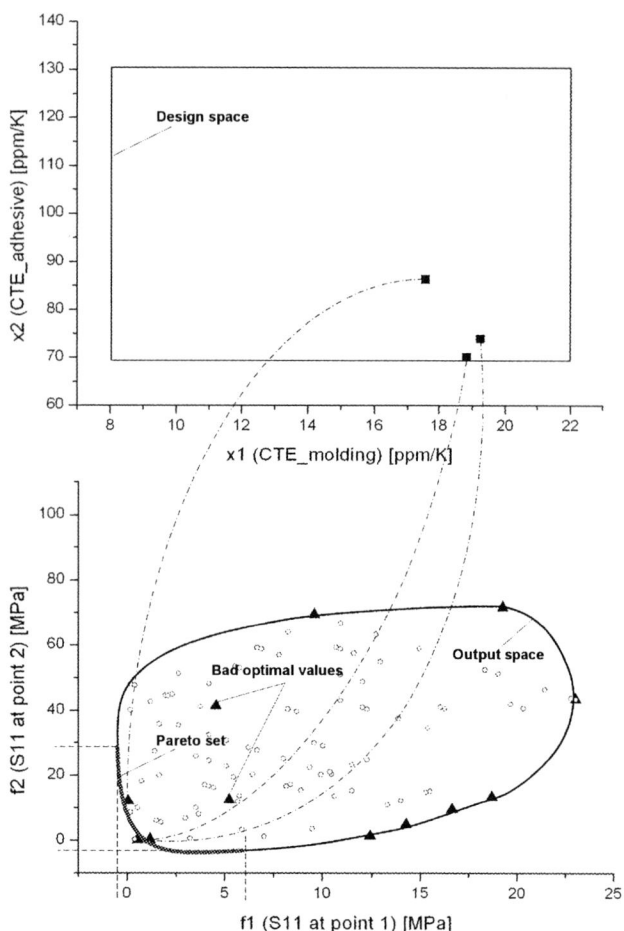

Figure 9. The obtained Pareto set and the corresponding values of the input parameters

The parameters from the design space which give the optimal solution ale placed in lower range of CTE adhesive values and in higher range of CTE molding compound values. This results correspond with the previous single-optimization [1,2]. However, in thisd approach the resulted input parameters differ from the prevoius ones becouse two criteria were taken in the consideration.

6.4. Verification Experiment

In order to verify the results, the verification experiment was carried out. This experiment was based on carrying out the simulations using the 100 randomly chosen points from the design space and to evaluate the response values for selected stress components criteria. The purpose of this experiment was to explore the whole design space and to find the potential extrems thus confirming the applied multi-objective optimization procedure based on weighted objective method. Figure 9 and 10 illustrate the achieved results.

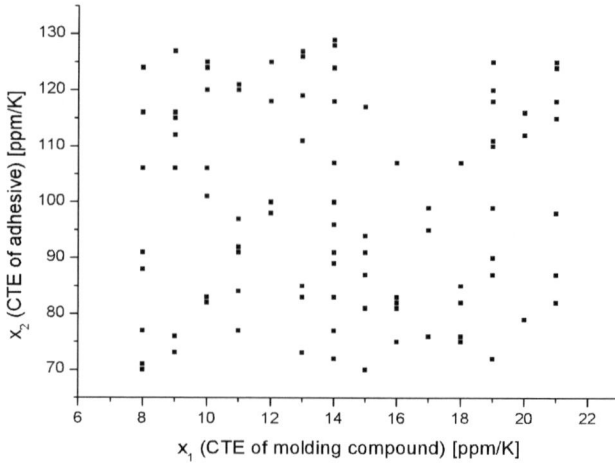

Figure 10. Random distribution of the input parameters in design space

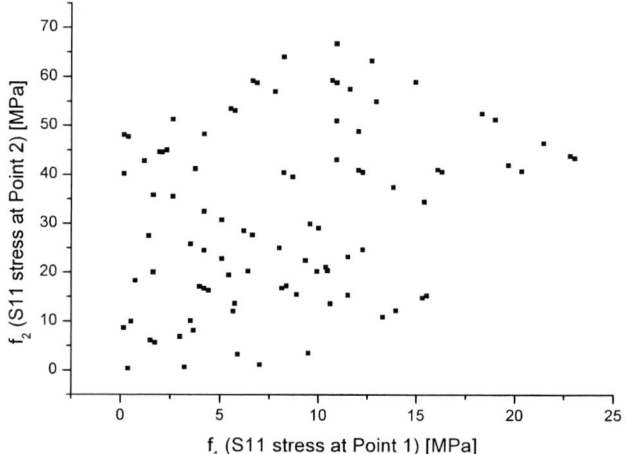

Figure 11. Evaluated distribution of the output parameters (dependent on the randomly chosen input parameters) and the coresssponding Pareto set.

7. Conclusions

The presented paper describes a parametric approach to multi-objective optimization method of stacked packages based on a memory module. The proposed optimization process included the parametric numerical model elaborated in the ABAQUS and Python scripting language. The optimization was done thanks to the self designed WRUT-VPT tool. The objective of the optimization was to minimize the selected stress components S11 at two different points of the investigated package.

However, for both as the parametric model as optimization procedure two simplifications were made. Namely, the material dependences were not completely defined, i.e. the dependence of viscoelastic properties on changes of CTE. This simplification was made because the mathematical description of this problem is not well known yet. The second problem is due to the single-objective TNC optimization algorithm. The optimization procedure works fine generally with the analitical functions. In case of numerical models where results are error biased this algorithm may not be sufficient enough. Mainly, the greates danger is that it can be easily trapped into local extrema. To avoid such situation, the global optimization algorithms should be used instead (genetic, simulated annealing, tabu search) or, which is more suitable, the DOE/RSM techniques. Unfortunatelly the global optimization algorithms need to carry out lots of iterations which is very time consuming, which is the key factor in case of numerical prototyping. Instead, as an valuable alternative the DOE/RSM techniques should be rather used [1,2].

In the future, the research work will continue on defining materials behavior and will focus on bounding the multi-objective optimization with the DOE/RSM techniques to obtain both fast and reliable numerical optimization process.

References

1. Łukasz Dowhań, Artur Wymysłowski, Rainer Dudek, Jürgen Auersperg, "Parametric Approach to Numerical Design for Optimization of Stacked Packages", *ESTC Conf*, Dresden, Germany, 2006,

2. Łukasz Dowhań, Artur Wymysłowski, Rainer Dudek, Jürgen Auersperg, "Numerical Approach to Optimization of Stacked Packages", *IMAPS Conf,* Kraków, Poland, 2006,

3. A. Wymysłowski, W.D. van Driel, G.Q. Zhang, J. van de Peer, N. Tzannetakis, "Smart and sequential approach to numerical prototyping in micro-electronic applications", *JMEP 2(1)*, 2005,

4. W.D. van Driel, J. van de Peer, N. Tzannetakis, A. Wymysłowski, Q. Zhang, "Advanced Numerical Prototyping Methods in Modern Engineering Applications", *Proc 5th EuroSimE Conf*, 2004,

5. Jürgen Auersperg, Rainer Dudek, Bernd Michel, "Combined Fracture, Delamination Risk and Fatigue Evaluation of Advanced Microelectronics Applications towards RSM/DOE Concepts", *Proc 7th EuroSimE Conf*, 2006

Research of Stacked VIA's Mechanical Stress

Tohru Nakanishi[1],
Nakanishi Professional Engineering Office[1]
Moriyama Shiga Japan
nakanishi-proeng@hotmail.co.jp

Hideo Ohkuma[2]
HTO Inc.[2]
Yasu Shiga Japan
ohkuma@athena.ocn.ne.jp

Hiroshi Ohira[3],
Flextronics Aichi K.K.[3]
Kamizue Komaki Aichi Japan
hiroshi.ohira@jp.flextronics.com

Abstract

The high density packaging has been focused with the Stacked via technologies. (The via is lettered as VIA hereafter because this is focused in this paper.) The stacked VIA technology becomes to be used for high density packaging as of today, however, it could not be said that the influence of the VIA stacking has been understood sufficiently.

In this paper, the influence of one to five stacked VIA technologies are studied with the parameter of stress and strain on the view point of reliability, comparing the condition that these VIAs are located directly on the RFP (Resin Filled PTH (Pin Through Hole)), and the other condition that these are located left and right with some distance from RFP.

The guideline as to the optimized design of the substrate that has the stacked VIA is provided.

1. Introduction

For the purpose of down-sizing consumer electric products, such as handy phone, DVC (Digital Video Camera), DSC (Digital Still Camera), notebook PC. CSP (Chip Scale Package) and the stacked VIA technology are becoming the important new high density packaging technology, instead of the fine pitch QFP (Quad Flat Package) mounted using perimeter leads, with being able to utilize the conventional surface mounting technology and to treat them on the same manner as the common package. They let the area of surface mounting be small. Besides, the use of build-up boards has been tried. However, the reliability of the the stacked via depends on the structure of the package, the motherboard, and the material properties.

In this paper, the numerical analysis of the stressed copper with material nonlinear FEM (Finite Element Method), is performed to evaluate the influence of the stacked VIAs controlling the reliability of the package and the build-up board for the fatigue fracture by the temperature cycle stress.

In this assessment, the compact modeling and analysis method on the view point of the reduction of the time needed for the nonlinear numerical analysis, are applied[1-3], for VIA joint which are very small, comparing to the whole structure.

2. Parameters Studied for Stacked VIA's stress and Strain

It can be said that the build-up boards improve the design capability because they increase the wiring density by the placement of the VIAs on the pads for mounting the package. And, it is also said the placement of the VIA on the RFP makes it easier to do the In/Out fan-out on a PWB (Printed Wiring Board). However, the more fine the pitch of the package, the more difficult it is to maintain and/or improve the reliability such as the solder connection.

Especially, the difference of the coefficient of thermal expansion between the package and the build-up board, causes the large strain into the solder, and repeats the plastic displacement by the thermal cycle stress. The accumulation of this strain causes the fatigue fracture, and finally crack of the solder.

Via on RFP structure is indispensable to improve the wiring density of build-up board and mount the fine pitch package. On the other hand, the base FR4 PWB structure and the board thickness depend on the specification of the final product needs. Therefore, the specification of the board on which package is mounted, especially the part of solder connection and the stacked VIA, should be studied in detail. On this purpose, in order to obtain the most reliable design, the influence of the stacked VIAs is studied this time in the actually applicable range.

The popular high density PCB is subjected for trying to get the technical discussion wide. That is the PCB which has 5 build up and copper layers both top and bottom side, shown in Figure-1. The horizontal size of this PCB is 49.0 (mm) and the 11.6 (mm) square silicon chip is mounted at the center on it. The chip thickness (height) is 0.5 (mm). The thickness of copper layer is 21.0 (μm) for only two copper layers located out of core layer. The other 4+4 copper layers located out of it have 15.0 (μm) thickness. The outermost insulation layers located both side have 30.0 (μm) thickness and the other inner insulation layers have 33.0 (μm) thickness.

The modeling is created with two dimensions. The center of stacked VIAs is located with the distance of 20.0 (mm) horizontally from the center of this PCB. The thickness of core is 800.0 (μm) and the outermost solder resist layers on both side have 15.0 (μm). The silicon chip is mounted with one such as solder, however, the stress and the strain of the solder are not discussion point this time. Therefore, the solder balls are not modeled, and silicon chip are mounted on the resist layers directly in modeling.

The center of RFP in core is located with the distance of 20.24 (mm) from the center of the PCB, and the diameter of resin filled portion is 160.0 (μm), at the outer side of it, there is copper wall with 20.0 (μm) thickness.

The number indicating in Figure-1 except the dimensions are corresponding to the material numbers, given in Table-1 and Table-2 described after.

1-4244-1105-X/07/$25.00 ©2007 IEEE

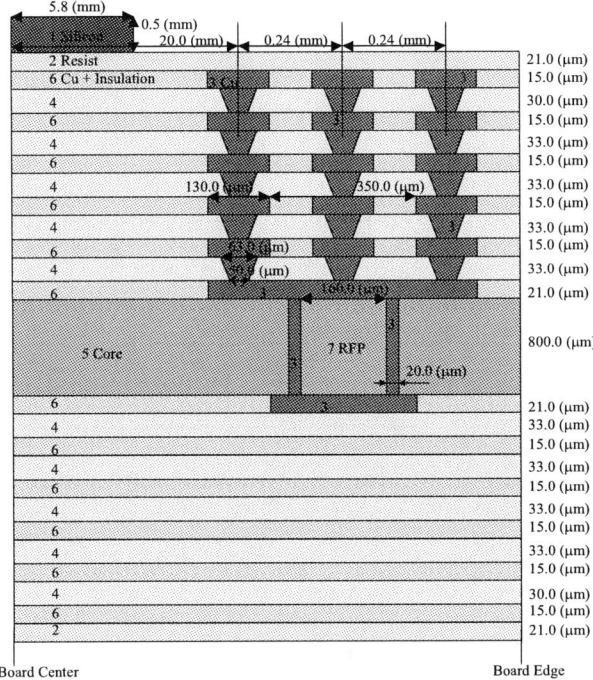

Figure-1. Subjected PCB inside structure

3. Modeling

The modeling whole shape is shown in Figure-2. The part magnification of Figure-2 is the Figure-3 that shows the area of build up portion (chip neighbor area). And the stacked VIA modeling is shown in Figure-4. There are three cases. The first is the case that the stacked VIAs are located directly upper side on RFP. And the second one is that the center of these VIAs is located left from the center of RFP with the distance of 0.24 (μm). And the third one is that the center of these VIAs is located right from the RFP with the same distance as second case. The first case is named as VIA-on-RFP case, the second case is named as Left-VIA case, and the third case is done as Right-VIA case hereafter.

In the calculation of VIA-on-RFP case, 10 VIAs located right and left of RFP are modeled as the insulation property. In the Left-VIA and Right-VIA cases, the same modeling method is applied. The meshing is shown in Figure-5. Figure-6 shows the meshing of the VIA-on-RFP case. And the Figure-7 shows the Left-VIA case.

The material properties used in the analysis are given in Table-1 and Table-2. Y is the vertical gravity direction.

As the boundary condition, the board center and the bottom of the board is cramped. The temperature stress to be applied in this model is 140 to -40 degree C.

Figure-2. Modeling whole shape

Figure-3. Chip Neighbor Area (Magnification of Fig.2)

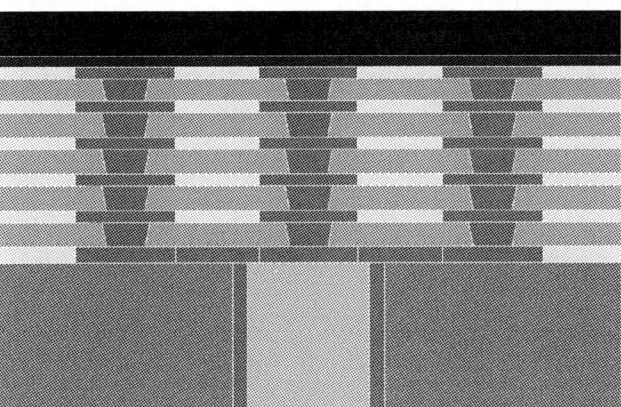

Figure-4. Stacked VIA Modeling (Magnification of Fig.2)

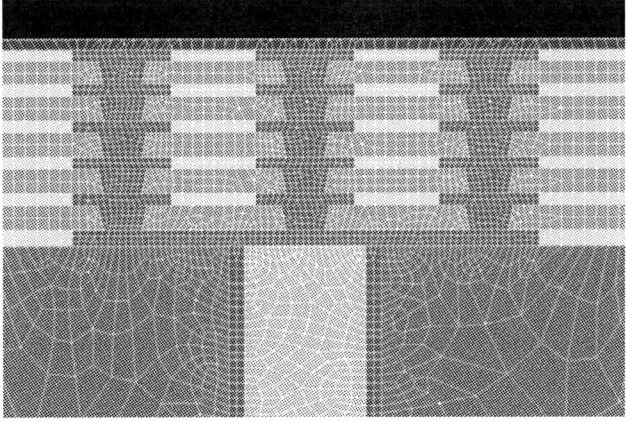

Figure-5. Meshing sample neighbor of Stacked VIA

4. Numerical Analysis Result

As the result of numerical analysis, the displacement and stress field calculated with the VIA-on-RFP modeling is shown in Figure-8 with Mises equivalent. The display bar under the deformed shape shows the stress Mises

271

scalar. The Figure-9 is the magnification of the 5 VIAs part. The Figure-10 is the sample of numerical analysis result as deformed stress field with 4 stacked VIAs model at Right-VIA case.

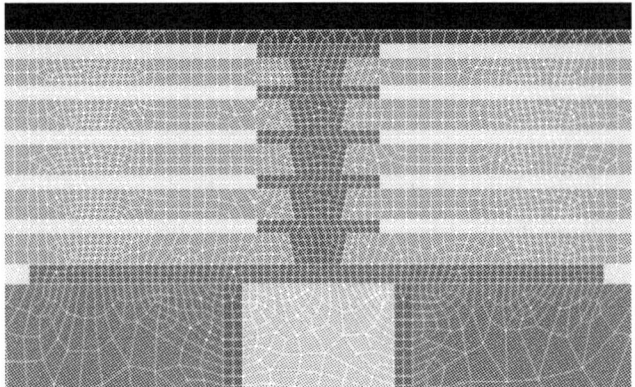

Figure-6. Meshing sample of Stacked VIA on RFP case

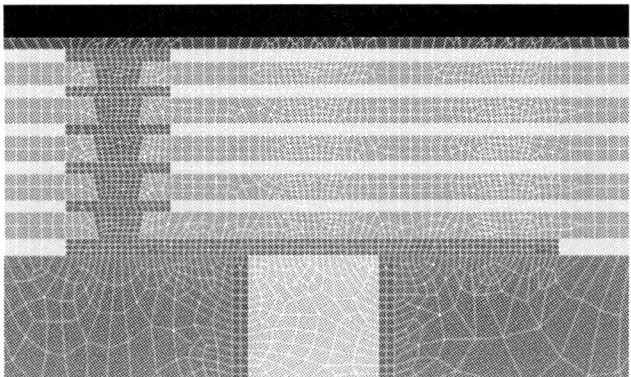

Figure-7. Meshing sample of left Stacked VIA case

Table-1. Material Properties applied into Analysis

No.	Material	Elasticity (kgf/mm²)		Poisson Ratio (-)	
		< Tg	> Tg	< Tg	> Tg
1	Chip Silicon	15300.0		0.17	
2	Solder Resist	275.5	6.83	0.31	
3	Copper	12000.0		0.35	
4	Build-up Insulation	408.2	4.08	0.30	
5	Core	x=2612.0 y=816.3 x=2612.0	x=1234.7 y=384.7 x=1234.7	xy=0.4 yz=0.4 zx=0.17	xy=0.23 yz=0.23 zx=0.32
6	Copper 75% + Insulation 25%	9102.05	9001.02	0.3375	
7	RFP	584.0	5.48	0.30	

Table-2. Material Properties applied into Analysis

No.	Material	Thermal Expansion Coefficient (10^{-6} 1/K)		Tg
		< Tg	> Tg	deg C
1	Chip Silicon	3.5		
2	Solder Resist	70.0	150.0	100.0
3	Copper	17.0		
4	Build-up Insulation	46.5	137.5	156.0
5	Core	x=14.8 y=35.0 z=14.8	x=9.6 y=279.6 z=9.6	168.0
6	Copper 75% + Insulation 25%	20.2	21.77	
7	RFP	35.0	80.0	135.0

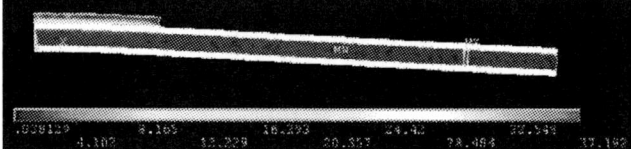

Figure-8. Deformed Stress Field of VIA-on-RFP case

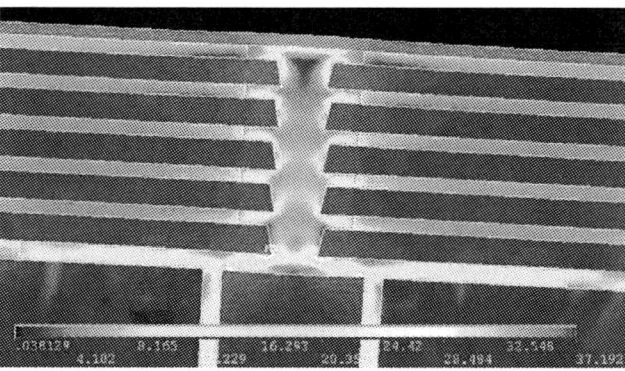

Figure-9. Magnification of Figure-8, 5 VIAs-on-RFP case

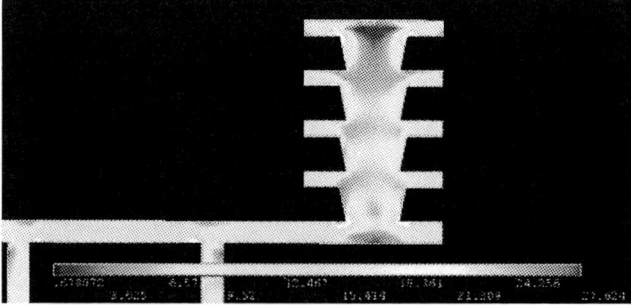

Figure-10. Deformed Stress Field of 4 Right-VIAs case

The maximum copper stress and strain of VIA-on-RFP case is summarized in Table-3 and Figure-11. The ones of Left-VIA case are also summarized in Table-4 and Figure-12, and the ones of Right-VIA case is done in Table-5 and Figure-13.

5. Discussion

As the overall observation, increasing the number of VIA stack, the maximum stress to be happened in the copper line becomes bigger. It could be said that the fewer stacks are stronger against the stress. However, the variation (increasing value) of the maximum stress by increasing the number of stack becomes small. If the number of stacks is increased more, the slope of the tangent might reach the area to be able to recognize as zero. On the other hand, in these area, some the other concerns would be happened. The stress value under 20.0 (kgf/mm2), which is recognized as the safe level for copper, is not observed in all cases. That means even if the number of stack is one, the copper line has the risk causing the crack.

From the Table-3 and Figure-11, it could be said that the location to be stressed into copper line shows the same manner in VIA-on-RFP case.

As to the Left-VIA case, in only 1 stack case, the location to happen the maximum stress moves from VIA bottom connection to the location of capped copper right edge over RFP.

Right-VIA case is more complicated and some other maximum stressed locations are observed. Comparing to the Left-VIA case and VIA-on-RFP case, the distance from the board center to the stacked VIA is slightly bigger than others. That means the displacement of VIA area becomes bigger and the influence of thermal expansion and/or shrink makes bigger. On the other hand, the board edge is free end in this modeling. This means the stress is releasable. It is thought that these complications show the lowest stress value, in parallel with the some maximum stressed locations.

5. Conclusions

The stacked VIA method expand the capability of high density packaging. On the other hand, it could not be said yet that the physical phenomenon and the influence of stack are understood sufficiently.

This time focused on 5 stacked VIAs and 3 stacked patterns are evaluated. One is the case that the VIA is located on RFP, another is that the VIA is located left with slight distance from the RFP, and the last one is the case of locating right. From the modeling and the numerical analysis result, the case which the stacked VIA locates on the RFP shows the worst data. And the case that the stacked VIA locates outer side with slight distance from RFP shows the lowest stress data. The complicated maximum stressed location is observed and the approximate equation is obtained to be able to assess the stress and strain value.

The stress and the strain were evaluated with the temperature stress only from 140 degree C to minus 40 degree C, in this paper. Some plastic materials have the grass transportation degree over 140. Therefore, if the condition is over this Tg, it is thought that the variation of these evaluated data becomes bigger.

For leading the high density packaging, the further advanced analysis of the complicated phenomenon is expected.

References

1. Tohru Nakanishi, Tomohiro Hase, "Optimal Thermo-Structural Analysis for High Density Package Mounting on Build-up Board", Japan Society of Electronics Engineering, Vol.127, No.11 (2003), pp.504-509

2. T.Nakanishi, et al, "Fatigue Life Analysis by Finite Element Method for CSP Mounting on Build-up Board", Proceedings of the International Conference on High Density Interconnect and Systems Packaging, Denver, Colorado, April 26-28 (2000), pp.283-288

3. T.Nakanishi, et al, "Optimal Thermo-Structural Analysis for CSP/FCA/MCM Mounting on Build-up Board", Proceedings of the 3rd International Conference on Benefiting from Thermal and Mechanical Simulation in (Micro)-Electronics EuroSIME 2002, Paris, France, April 15-17 (2002), pp.369-376

Table-3. Cu Max Stress & Strain of VIA-on-RFP case

Number of Cu Stack	1	4	3	2	5
von Mises Stress (kgf/mm²)	29.737	33.327	35.631	36.930	37.192
von Mises Total Strain (%)	0.3345	0.3749	0.4008	0.4155	0.4184
Location Type A	VIA bottom left connection between VIA and RFP Type-A	same as ←	same as ←	same as ←	same as ←

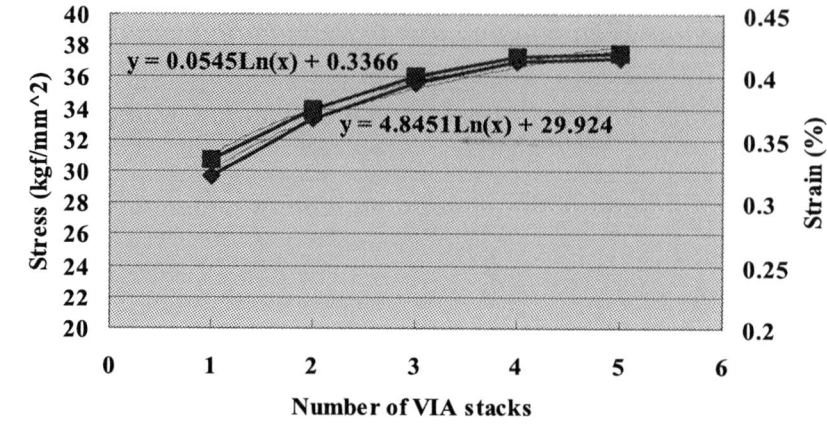

Figure-11. Max Stress and Strain VIA-on-RFP case

Table-4. Cu Max Stress & Strain of Left-VIA case

Number of Cu Stack	5	4	3	2	1
von Mises Stress (kgf/mm^2)	22.655	26.007	27.928	28.969	29.192
von Mises Total Strain (%)	0.2549	0.2926	0.3142	0.3259	0.3284
Location Type-B	VIA bottom right connection between VIA and RFP Type-B	same as ←	same as ←	same as ←	same as ←
von Mises Stress (kgf/mm^2)	24.723				
von Mises Total Strain (%)	0.2781				
Location Type-C	Capped copper right edge over RFP Type-C				

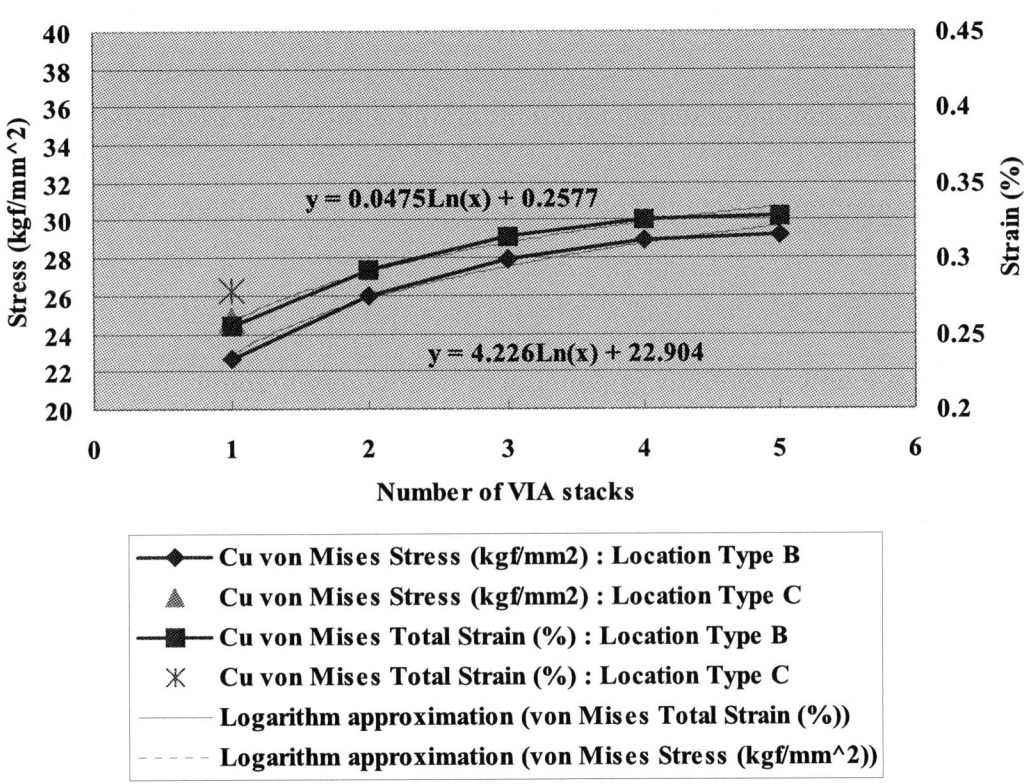

Figure-12. Max Stress and Strain Left-VIA case

Table-5. Cu Max Stress & Strain of Right-VIA case

Number of Cu Stack	1	2	3	4	5
von Mises Stress (kgf/mm^2)	21.139	24.308	26.214	27.274	27.512
von Mises Total Strain (%)	0.2378	0.2735	0.2949	0.3068	0.3095
Location Type-D	VIA bottom left connection between VIA and RFP Type-D	same as ←	same as ←	same as ←	same as ←
von Mises Stress (kgf/mm^2)	24.656				
von Mises Total Strain (%)	0.2774				
Location Type-E	Capped copper left edge over RFP Type-E				
von Mises Stress (kgf/mm^2)		24.470			
von Mises Total Strain (%)		0.2753			
Location Type-F		Capped copper right edge over RFP Type-F			
von Mises Stress (kgf/mm^2)	21.387				
von Mises Total Strain (%)	0.2461				
Location Type-G	VIA top right Type-G				
von Mises Stress (kgf/mm^2)				27.624	
von Mises Total Strain (%)				0.3108	
Location Type-H				VIA top left Type-H	

Max Stress & Strain Right-VIA case

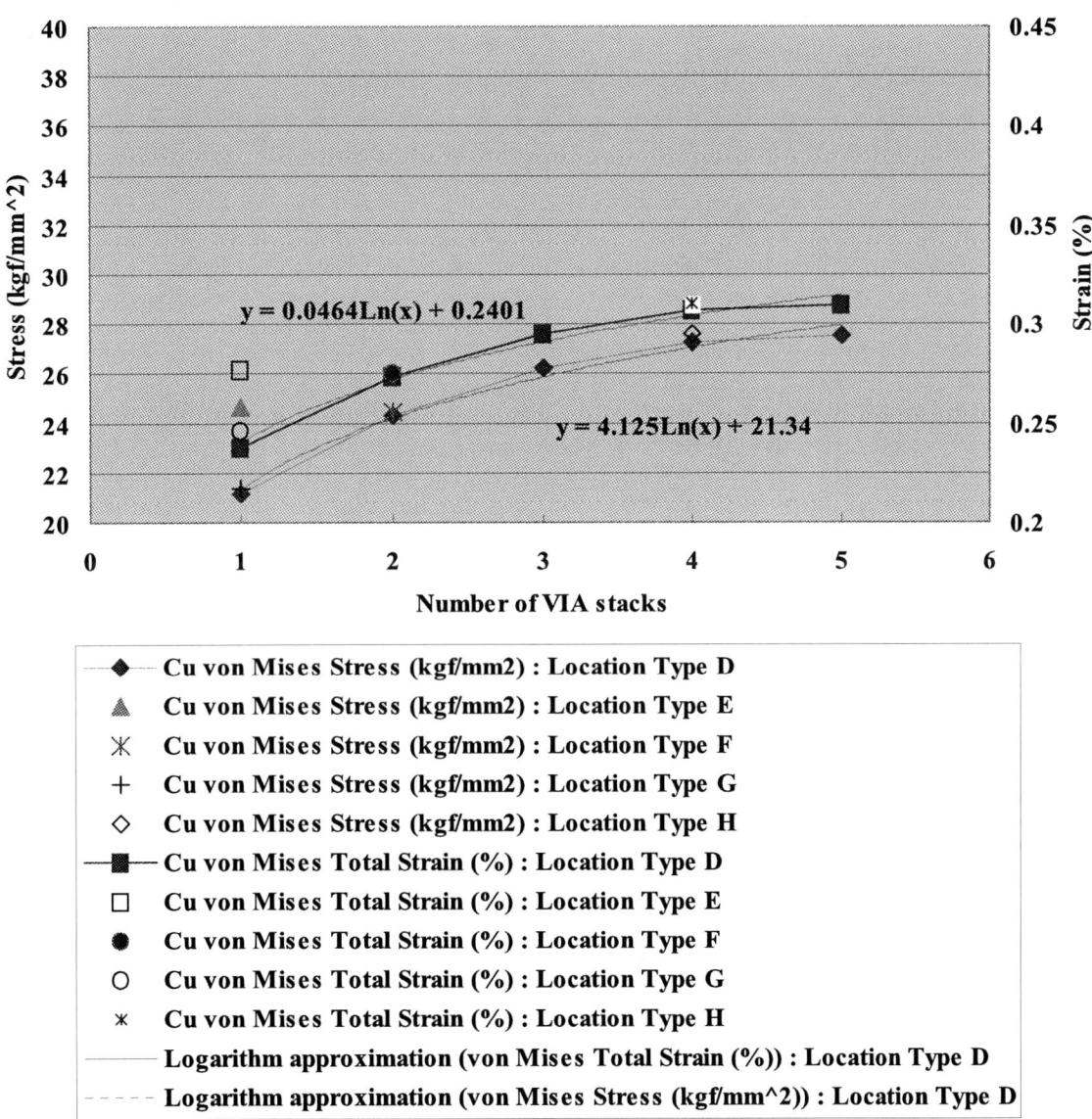

Figure-13. Max Stress and Strain Right-VIA case

Research of Material Properties Reliance for POP with Numerical Analysis

Tohru Nakanishi[1],
Nakanishi Professional Engineering Office[1]
Moriyama Shiga Japan
nakanishi-proeng@hotmail.co.jp

Hiroshi Ohira[2],
Flextronics Aichi K.K.[2]
Kamizue Komaki Aichi Japan
hiroshi.ohira@jp.flextronics.com

Hideo Ohkuma[3]
HTO Inc.[3]
Yasu Shiga Japan
ohkuma@athena.ocn.ne.jp

Abstract

The POP (Package on Package) has been the core technology for high density package in the area of electronics devices, such as Mobile Phone, Digital Camera, Digital Video Camera and Notebook PC. And its utilization has been increased rapidly. The development of the new technology POP has been expected and the most focused concern is the reliability of it.

In this paper, the reliance of the material properties on the numerical analysis for the reliability assessment of POP is studied, especially focusing the encapsulant. The result of the numerical analysis with the material properties, that the material supplier announced, did not support the result of the actual hardware testing. Then through the research process, we had focused the material properties, and then we actually measured the material properties with the piece of the actual hardware parts which are constructing this product. As the result, we found the order difference between the material supplier announced data and our measured data.

It is recognized strongly that the reliance of the material properties is extremely important. With this established numerical model, we had studied the optimized parameter of the encapsulant.

1. Introduction

The technology revolution for achievement of high density packaging has been strongly required for the mobile electrical products, such as mobile phone, notebook computer, digital camera, DVC (Digital Video Camera), DSC (Digital Still Camera). The POP (Package on Package) technology becomes one of the solutions for this achievement. However, the experience of industry is not enough, and the competition of its development has been accordingly heated up.

On the other hand, the numerical analysis is one of the beneficial tools for supporting these developments. That brings the sufficient data to be able to assess the parameters to control the design feasibility and the reliability.[1-3]

In this paper, we report the importance and the reliance of material properties, in which are used in the numerical analysis. In the development activity, the numerical analysis was performed on the view point of the reliability assessment. In first numerical analysis, the material properties information that had been obtained by each material supplier was applied into the modeling. In parallel with this numerical analysis, the actual hardware testing was performed as to some cases. As the result, we had faced the problem that the numerical analysis result did not support the result of actual hardware testing.

In the process of root cause study, we had focused the material properties of encapsulation (plastic), and we measured them by ourselves.

As the result, our measurement showed the big difference from the information that the material supplier announced. Therefore, another numerical analysis was performed with our measured material properties, and then, the result supported the actual hardware testing data.

2. POP Structure

The popular POP structure is shown in Figure-1. This is the cross-section figure that the industry is focusing as the key technology. In order to minimize the packaging size of horizontal direction, the method that two (or more) PCBs (Printed Circuit Board) are stacked is introduced. On the other hand, the hypertrophy of the vertical direction is not accepted. Therefore, the thinnest substrate (carrier board) is designed, and the connection between the chip and substrate is getting smaller. The substrate which is subjected this time has 150.0 (μm) thickness for upper one (The upper stacked printed circuit board which mounts the silicon chip is lettered as the upper substrate. And the lower staked printed circuit board that also mounts the chip is lettered as the lower substrate hereafter.) and 320.0 (μm) thickness for lower one.

The motherboard has the 20.0 (mm) horizontal length square and the 0.7 (mm) thickness. The horizontal size of the substrate is 15.0 (mm) square for both the upper and the lower substrate. Some silicon chips are stacked horizontally in the upper and lower plastic package. The solder balls, that connect the upper substrate and the lower substrate, have 500.0 (μm) diameter and are located with the pitch of 650.0 (μm). The lower substrate has the 300.0 (μm) diameter's solder balls that are located between the lower substrate and the mother board with the pitch of 500.0 (μm).

3. Numerical Analysis and Actual Hardware Testing

The first modeling for the structural numerical analysis of Figure-1 is established with full modeling, shown in Figure-2. This shows the encapsulation modeling case. The encapsulant is shown in red. Four outer-most lead free solder balls are modeled in detail, and the remains are modeled with rough elements[1-3], shown in Figure-3. The meshing sample is shown in Figure-4.

After the correlation is evaluated with the displacement of the actual hardware testing by temperature stress, the model is revised with one-quarter model, shown in Figure-5 and Figure-6. Figure-5 shows the modeling case that the encapsulation is not filled in the part of the solder connection of Figure-1. Figure-6 is the modeling that the encapsulation fills the surroundings of the solder balls. The center cross-section shape of Figure-6 is shown in Figure-7.

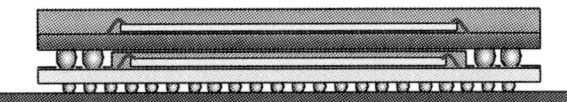

Fig-1. POP (Package on Package) Cross-section sample

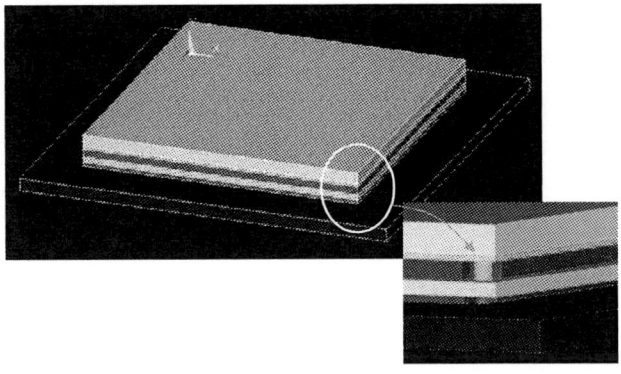

Fig-2. First Full Model Whole Shape

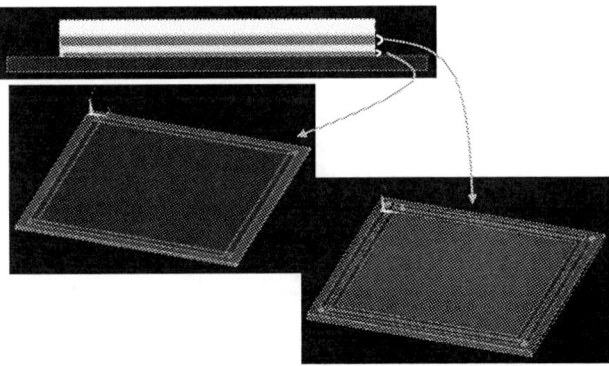

Fig-3. First Full Model : Lead Free Solder Ball & Encapsulation Modeling

The material properties are obtained from the material supplier. These are given in Table-1 and Table-2 except encalsulant. Y is the vertical gravity direction. The data of encapsulant is given in Table-3. Three supplier's materials are subjected this time.

The solder strain as the numerical analysis result with the temperature stress, that is from 115 degree C to minus 40 degree C, is given in Table-4. As a sample of the numerical analysis result, the solder strain in Encapsulant A case is shown in Figure-8. This solder is the outermost

one located under the upper substrate. And the deformation shape are shown in Figure-9 and Figure-10.

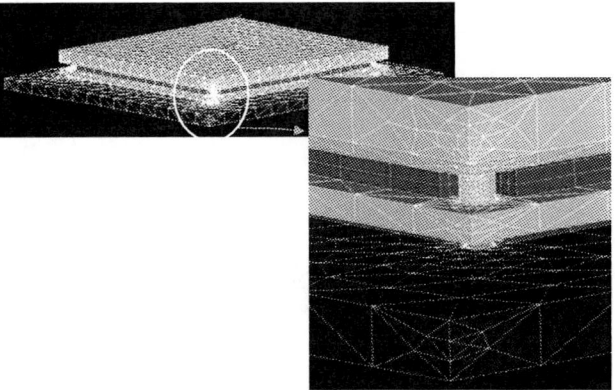

Fig-4. First Full Model : Meshing Sample

Table-1. Material Properties of except Encapsulant

No.	Material	Elasticity (kgf/mm^2)		Poisson Ratio (-)	
		< Tg	> Tg	< Tg	> Tg
1	Chip Silicon	15300.0		0.17	
2	Solder	4908.2		0.20	
3	Mold	1200.0		0.38	
4	Substrate	x=3000.0 y=1000.0 x=3000.0	x=1400.0 y=450.0 x=1400.0	xy=0.4 yz=0.4 zx=0.17	xy=0.23 yz=0.23 zx=0.32
5	Mother PCB	x=3000.0 y=1000.0 x=3000.0	x=1400.0 y=450.0 x=1400.0	xy=0.4 yz=0.4 zx=0.17	xy=0.23 yz=0.23 zx=0.32

Table-2. Material Properties of except Encapsulant

No.	Material	Thermal Expansion Coefficient (10^{-6} 1/K)		Tg
		< Tg	> Tg	deg C
1	Chip Silicon	3.5		
2	Solder	20.04		
3	Mold	2.8	24.1	100.0
4	Substrate	x=15.0 y=33.0 z=15.0	x=12.0 y=250.0 z=12.0	168.0
5	Mother PCB	x=16.4 y=33.0 z=16.4	x=4.31 y=250.0 z=4.31	125.0

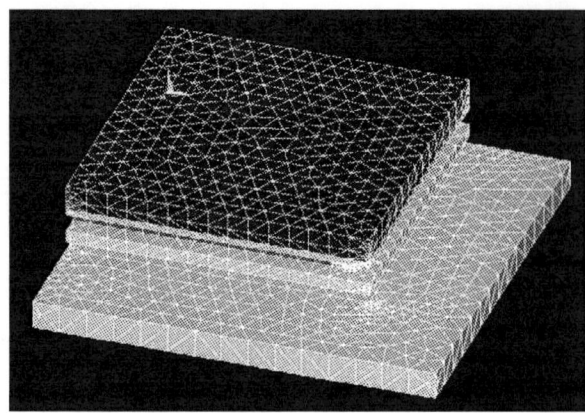

Fig-5. 1/4 Modeling without Encapsulant

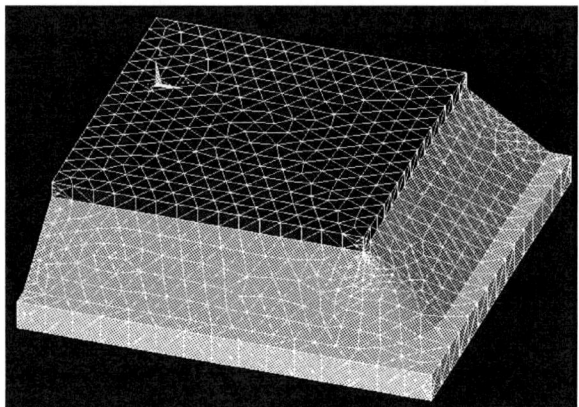

Fig-6. 1/4 Modeling with Encapsulant

Fig-7. Cross-section sample of Fig.6

Fig-8. Cross-section sample of Fig.2

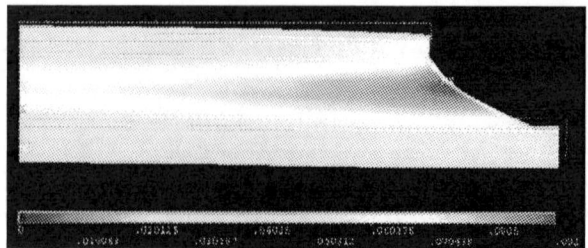

Fig-9. Cross-section sample of Fig.2

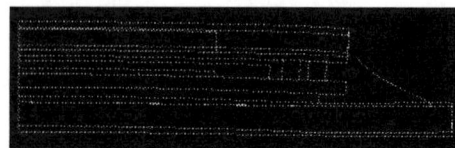

Fig-10. Cross-section sample of Fig.2

Table-3. Material Properties of Encapsulant announced by Supplier

	Elasticity (kgf/mm^2)		C.T.E. (kgf/mm^2) x 10^{-6}	
	< Tg	> Tg	< Tg	> Tg
A	5.1	No Data	80.0	210.0
B	755.1	91.84	37.0	138.0
C	377.55	19.39	47.0	159.0

Table-4. Solder Max Strain with Table-1 properties

	No Encap	Encap A	Encap B	Encap C
Solder Strain (%)	0.428	0.331	0.965	1.054

On the other hand, the actual hardware temperature cycle testing result is given in Table-5. Ten POP actual hardware were stressed with the temperature 115 to minus 40 degree C. With comparing Table-4 and Table-5, it is obvious that the result of numerical analysis does not support the actual hardware testing result.

Here, we made the test piece sample to get the data of each part's material properties, such as elasticity which was depending on the temperature, and then, measured them by ourselves. As to the encapsulant, the measured data is given in Table-6, that is corresponding to Table-3. After the measurement, it was cleared that our obtained

280

data showed the order difference from the supplier's catalogue value.

Table-5. Reliability Cycles tested with actual hardware

	No Encap	Encap A	Encap B	Encap C
Reliability (Cycles)	3100	500	2250	750

Table-6. Material Properties tested with actual hardware

	Elasticity < Tg (kgf/mm^2)	Elasticity > Tg (kgf/mm^2)	C.T.E. < Tg (kgf/mm^2) x 10^{-6}	C.T.E. > Tg (kgf/mm^2) x 10^{-6}
A	25.0	2.5	32.9	197.0
B	68.02	8.27	38.0	1115.0
C	70.318	3.61	55.0	150.0

Table-7. Solder Max Strain with Table-4 properties

	No Encap	Encap A	Encap B	Encap C
Solder Strain (%)	0.215	0.673	0.306	0.478

Table-8. Estimated Reliability Cycles on numerical analysis

	No Encap	Encap A	Encap B	Encap C
Reliability (Cycles)	3100	482	1743	842

4. Numerical Analysis with Measured Material Properties

The numerical analysis is re-performed with Table-6. As that result, the solder strain is given in Table-7. The each solder strain field is shown in Figure-11 to Figure-14 as a sample with no-encapsulation case.

From these results, the reliability cycles of solder connection are estimated on the condition that the power number n is set to 1.63 as describing the equation number 1 to 4 below. This n number 1.63 is the experimental one. In equation 1, N_f means the cycle quantity to reach the fatigue fracture, C means that metal's unique constant, ε means the plastic strain to be taken by one cycle, n means the constant, that means the relationship between the

plastic strain and the reliability. The calculation result is summarized in Table-8.

$$N_f = C(\varepsilon)^{-n} \quad \quad ...(1)$$

Encapsulatin A

$$N_{fA} = 3100 \times \left(\frac{0.00673}{0.00215}\right)^{-1.63} = 482.6 \quad ...(2)$$

Encapsulatin B

$$N_{fA} = 3100 \times \left(\frac{0.00306}{0.00215}\right)^{-1.63} = 1743.9 \quad ...(3)$$

Encapsulatin C

$$N_{fA} = 3100 \times \left(\frac{0.00478}{0.00215}\right)^{-1.63} = 842.9 \quad ...(4)$$

The Table-8, that is the solder connection reliability cycles obtained as the numerical analysis result, could support tightly the Table-5, that is obtained by the actual hardware testing.

On the other hand, it was found that the part which shows the maximum stress and strain in the numerical analysis, such as the solder balls, is depending what encapsulant was applied and/or whether the encapsulation is applied or not. And from the actual hardware testing result, it is found that the defected point is different among the Encapsulant A, B, C and no encapsulation case. The location that shows the maximum stress and strain in the numerical analysis result has closely reproduces the defected location of actual hardware testing.

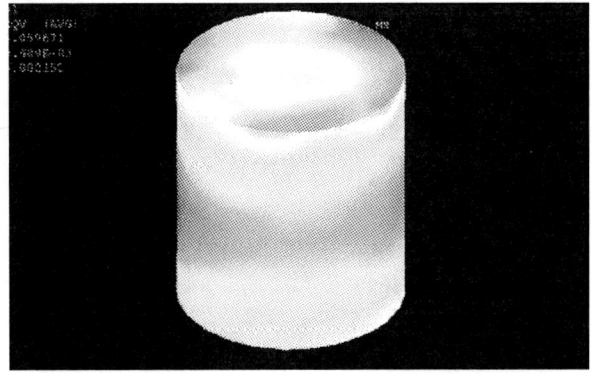

Fig-14. No encapsulation model Strain Filed of outer-most Solder Ball located under Upper Substrate

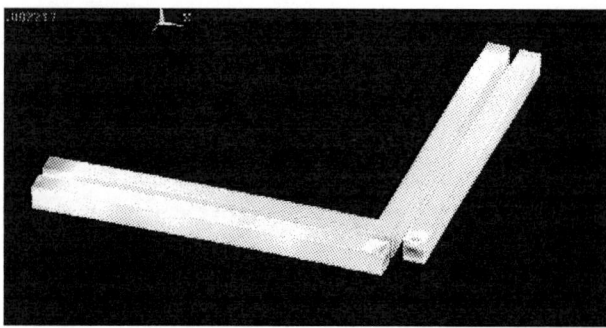

Fig-15. No encapsulation Strain Field of Solder balls except outer-most Solder Ball located under Upper Substrate

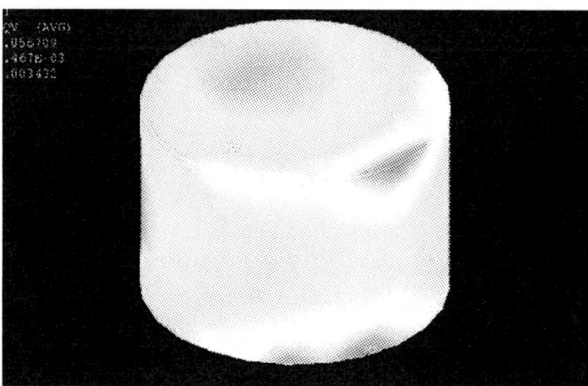

Fig-16. No encapsulation model Strain Filed of outer-most Solder Ball located under Lower Substrate

Fig-17. No encapsulation model Strain Field of Solder balls except outer-most Solder Ball located under Upper Substrate

5. Conclusions

In this paper, the material properties, that are applied into the numerical analysis, are focused, and it is shown that there is the risk, that the variation from the proper material properties does not support the actual physical phenomenon.

If the analysis goes to the parameter study phase with the model which does not reproduce the actual physical phenomenon, the primal and essential objective could never be achieved.

When the data, obtained by ourselves, was shown to the material supplier, and the difference from the catalogue data was questioned, it was the answer that the data was varied by the measurement condition and the measurement was basically difficult.

It is shown that it is very important the material properties even given by the material supplier is treated carefully.

References

1. Tohru Nakanishi, Tomohiro Hase, "Optimal Thermo-Structural Analysis for High Density Package Mounting on Build-up Board", Japan Society of Electronics Engineering, Vol.127, No.11 (2003), pp.504-509
2. T.Nakanishi, *et al*, "Fatigue Life Analysis by Finite Element Method for CSP Mounting on Build-up Board", Proceedings of the International Conference on High Density Interconnect and Systems Packaging, Denver, Colorado, April 26-28 (2000), pp.283-288
3. T.Nakanishi, *et al*, "Optimal Thermo-Structural Analysis for CSP/FCA/MCM Mounting on Build-up Board", Proceedings of the 3rd International Conference on Benefiting from Thermal and Mechanical Simulation in (Micro)-Electronics EuroSIME 2002, Paris, France, April 15-17 (2002), pp.369-376

Cure shrinkage characterization and its implementation into correlation of warpage between simulation and measurement

W.H. Zhu[1], Guang Li[2], Wei Sun[1], F.X. Che[1], Anthony Sun[1], C.K. Wang[1], H.B. Tan[1], B.Z. Zhao[2] and N.H. Chin[2]

[1] Packaging Design and Analysis Center, United Test and Assembly Center Limited,
5 Serangoon North Ave 5, Singapore 554916; E-mail: zhuwenhui0112@yahoo.com.sgT

[2] Cookson Semiconductor Packaging Materials,
12 Joo Koon Road, Singapore 628975

Abstract

In this work, a new approach was proposed to characterize the cure shrinkage of EMC by using the EMC/Cu bi-layer strip specimens. The warpage of bi-layer strip was measured at different temperature using Shadow Moiré. The results show that warpage at molding temperature was non-zero and zero-warpage temperature shifted from molding temperature (175degc) to higher temperature due to cure shrinkage effect. From Timoshenko's beam theory, the cure shrinkage was calculated as 1st order approximation theoretically either from the warpage at molding temperature or from zero-warpage temperature.

The determined cure shrinkage together with thermal shrinkage obtained from TMA tests was used to predict the warpage of the different EMC/Cu strips. Good correlation was observed in the wide temperature range.

As comparison, direct measurement of the cure shrinkage was also done using long rectangular bar specimens. Cure shrinkage was determined by extracting thermal shrinkage from total shrinkage.

Cure shrinkage of 2 EMCs were characterized and then applied to PBGA matrix. Warpage of the PBGA EMC/substrate maps was measured using Shadow Moiré and simulated as well for the molding compounds (EMCs) after 3 different processes, i.e. after transfer molding (TM), post mold cure (PMC) and PMC + Reflow at 260degc for 3 times (RF260X3). Consistence between simulation and experiments was found when cure shrinkage was considered. The presented data show the necessity and importance of cure shrinkage in warpage prediction simulation.

1. Introduction

Epoxy molding compound (EMC) is one of the key packaging materials. Its moldability and physical properties dominate the package reliability and warpage. One of the important features of the EMC is its shrinkage when reaction between the resin and hardener takes place during molding. Such shrinkage is commonly termed as cure shrinkage or chemical shrinkage originating from cross-linking, and causes more warpage in the packages in addition to that induced by CTE mismatch among packaging materials. With strict and stringent warpage requirement of packages, such as PoP, SD and memory cards, Quad Flat Non-lead (QFN) packages and various 3D packages, accurate prediction of warpage becomes very critical, and consequently cure shrinkage comes as very important player in warpage characterization. On one hand, miniaturization and low profile of packages results in less flexibility of geometry in package design and thus more challenges in development of epoxy molding compound to meet the low warpage requirements. On the other hand, increasing mould map / panel size is used in manufacturing to improve the production efficiency and to cut costs, leading to higher panel warpage and co-planarity problems in surface mounting of packages. To conquer such situation, we are urged to develop an approach to predict the warpage precisely and the way to optimization.

Shrinkage is composed of both thermal and chemical parts, as shown in Fig. 1. The chemical shrinkage or cure shrinkage, unfortunately ignored in the most of past simulation and analysis, results in additional residual stresses in the package and consequently contributes to the warpage, therefore becomes the key for accurate simulation of warpage as addressed in refs. [1,2].

Cure shrinkage was determined by researchers, mostly from molding compound (EMC) material manufacturers [3,4], by measuring the dimensional change of disc EMC specimens. The challenges for such direct measurement come from two aspects. Firstly, accurate measurement of dimensional change is difficult as cure shrinkage is normally small; Secondly, cure shrinkage such measured, even if it is accurate, is composed of both elastic and plastic parts in which only elastic parts contributes to warpage [5]. In fact, it is very hard to distinguish elastic cure shrinkage from total cure shrinkage, and this part really depends on a lot of factors such as composition of EMC. In this circumstance, we propose a new approach to characterize cure shrinkage in this work.

In the new method, EMC/Cu bi-layer strip specimens were used to measure warpage at different temperatures. After molding EMC onto Cu substrate, zero-warpage temperature, initially at molding temperature, shifted to a higher temperature due to cure shrinkage, while the warpage at molding temperature is no longer zero. Then the cure shrinkage can be calculated either from the warpage at molding temperature or from zero-warpage temperature by using Timoshenko's bi-layer beam model.

As comparison, direct measurement of the cure shrinkage was also illustrated using long rectangular bar specimens. This will be discussed in section 4. For the case studied here, a correction factor of about 0.6 must be introduced so as to extract elastic cure shrinkage from total cure shrinkage.

To further illustrate the effect of cure shrinkage on warpage of general packages, a series of tests were conducted for PBGA maps molded using 2 EMC materials onto substrate without inclusion of die. Warpage was measured using Shadow Moiré and simulated at 3 different conditions, namely after transfer molding (TM), post mold cure (PMC) and PMC + Reflow at 260degc for 3 times (RF260X3). The FEA simulation results corresponding to different processes with or without consideration of cure shrinkage were compared to demonstrate the cure shrinkage effect. Results show that a good and consistent agreement between simulation and experiments was achieved when cure shrinkage was considered in simulation.

2. New approach for cure shrinkage characterization

Cure shrinkage refers to volumetric reduction of the EMC before and after curing. The total shrinkage of EMC follows the profile in Fig.1.

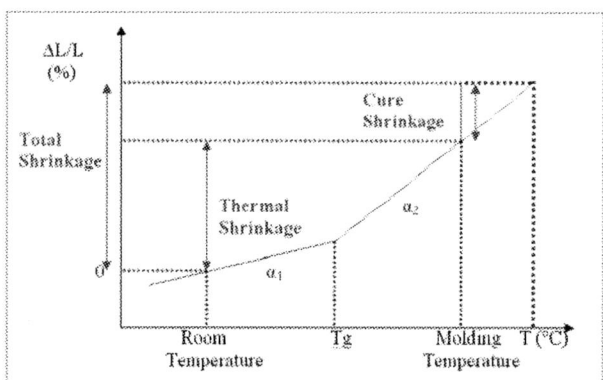

Fig. 1: Shrinkage in EMC at different temperatures

It can be seen from Fig.1 that total shrinkage includes two parts, thermal shrinkage due to temperature change and cure shrinkage due to cross-linking, i.e.:

Cure shrinkage (CS)
= Total shrinkage (TS) – Thermal shrinkage (ThS) (1)

Different ways have been proposed to determine cure shrinkage. Ken Oota [3] from Sumitomo studied various Epoxy molding compounds (EMCs), and found a very strong correlation between total cure shrinkage and free volume of materials, and a linear relationship between cure shrinkage and free volume was obtained. Similar relationship was also validated in M. Ogata's work [4]. Cookson also investigated the dependence of cure shrinkage upon Tg and other parameters. In general, cure shrinkage was in a range of 0.06% to 0.35% for most

green molding compound. This extra shrinkage results in additional warpage esp. at low temperature including room temperature and must be quantified so as to understand warpage extensively.

Experimental determination of cure shrinkage is quite challenging since the length change induced by cure shrinkage is of small percentage. Direct measurement of specimen's dimension change faces 2 bottlenecks: (1) accuracy and (2) the fact that, of the total cure shrinkage, only elastic cure shrinkage which is retained after solidification of EMC, contributes to warpage.

Here we propose an indirect approach to determine cure shrinkgae by using a bi-layer specimen. The specimen is prepared by molding the EMC onto a substrate made of a known material. Since cure shrinkage comes to play after molding, the bi-layer strip specimen will warp when cooling down from molding temperature to room temperatue. Practically, cure shrinkage after being fully cured is of more interest as this should be consistent parameter of any given EMC. Theoretically, measuring warpage at room temperature is already able to chracterize the cure shrinkage, however, to ensure accuracy and provide consistent data, cure shrinkage can be determined based on the warpage data in a wide range of temperatue using Shadow Moire. Since cure shrinkage is extracted by substracting thermal shrinkage from total shrinkage, accuracy of cure shrinkage is affected by thermal property data, and a group of warpage data at various temoperature points will help minimize the errors and improve accuracy. A substrate material with significant difference CTE from EMC is preferred so that the warpage can be measured with high accuracy.

As 1st order approaximation of warpage prediction of bi-layer strips, Timoshenko's bi-layer beam model can be applied assuming the strip is a thin beam. When temperature changes from T_o to T, it will deform due to different expansion of two layers, causing warpage w as in Fig. 2. The bowing curvature can be expressed as follows [7,8] when the deflection is not large:

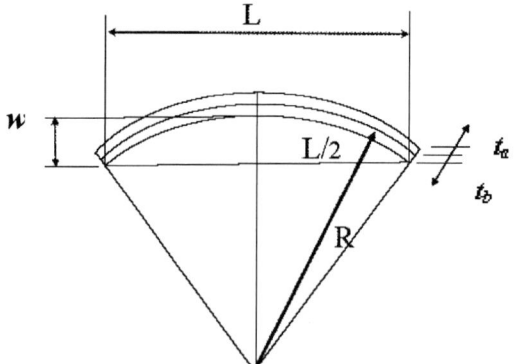

Fig.2: Geometric analysis of a warping bi-layer beam

$$\frac{1}{R} - \frac{1}{R_o} = \frac{6(1+p)^2 (\Delta L_b - \Delta L_a)}{t \left[3(1+p)^2 + (1+pq)\left(p^2 + \frac{1}{pq}\right) \right]} \quad (2)$$

284

where $\Delta La = \alpha_a$ (T-To), and $\Delta L_b = \alpha_b$ (T-To) are total shrinkage of layer a and layer b, respectively. $1/R_o$ is initial curvature of the strip at T_o (μm^{-1}), $1/R$ is Curvature of the strip at T, α_a and α_b are the effective CTE of low and high expansion materials (ppm/°C), $q = E_a/E_b$, where E_a and E_b are the Young's Modulii of the two materials, $p = t_a/t_b$, with t_a and t_b the thickness of the bimorph layers and $t = t_a + t_b$, the total thickness of the strip.

Please note the above equation is for an ideal bi-layer beam simply supported at both ends, while deflection is small without external mechanical force applied. Assuming initial curvature is zero ($1/R = 0$), the deflection at the mid point can be given as [7]:

$$(R - t_2)^2 = (R - w - t_2)^2 + \left(\frac{L}{2}\right)^2 \quad (3)$$

The radius of curvature after deflection is:

$$\frac{1}{R} = \frac{8w}{L^2 + 4w^2 + 8wt_2} \quad (4)$$

Provided that the deflection and the thickness are less than 10% of its length, $8wt_2$ and $4wd^2$ can then be neglected, therefore:

$$w = \frac{L^2}{8R} \quad (5)$$

Combining Eq. (2) with Eq. (5), we have:

$$w = \frac{3(1+p)^2 L^2 (\Delta L_b - \Delta L_a)}{4t\left[3(1+p)^2 + (1+pq)\left(p^2 + \frac{1}{pq}\right)\right]} \quad (6)$$

From Eq. (6), total shrinkage of the EMC (material b) can be derived from warpage, when Young's modulus, CTE and the thickness of substrate and EMC are given. Especially at molding temperature we have:

$$CS = \frac{4t\left[3(1+p)^2 + (1+pq)\left(p^2 + \frac{1}{pq}\right)\right]w_m}{3(1+p)^2 L^2} \quad (7)$$

Where W_m is the warpage at molding temperature, CS refers to cure shrinkage.

On the other hand, zero warpage temperature, Tz, will be shifted to a higher temperature due to cure shrinkage. Theoretically, cure shrinkage can also be determined from zero warpage temperature shifted from 175degc to Tz as follows:

CS= (Effective CTE of EMC – CTE of substrate) *

(Tz – molding temperature)

= Difference of expansion between EMC and substrate (Cu here)

$\approx$ ($\alpha_2 - \alpha_{Cu}$) * (Tz – Molding T), if Tg< molding T.

$$(8)$$

3. Experimental set-up of bi-layer beam system

Cure shrinkage causes additional warpage of the bi-layer system. Warpage should be zero at molding temperature, usually 175degc, provided there is no cure shrinkage. However, zero warpage temperature will be shifted to a higher temperature due to cure shrinkge. This can be addressed in Fig. 3.

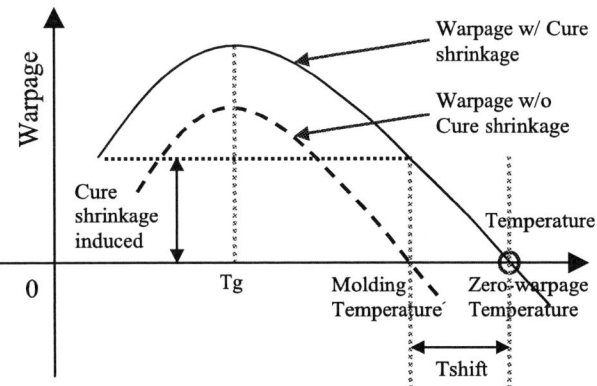

Fig. 3: Warpage of the bi-layer beam with and without cure shrinkage

It can be seen that warapge at molding temperature and zero warpage temperature are 2 characteristic data for cure shrinkage determination. A more elaborated way to cure shrinkage calculation is to use warpage in a wide range of temperature and best fitting the data to derive cure shrinkage. For this purpose, a series of tests were deisgned to measure cure shrinkage.

3.1 Experimental design of bi-layer beam system

To charactrerize cure shrinkage of EMCs, we designed EMC/Cu bi-material strips by molding 2 selected molding compounds onto copper strips, then the warpage of the bi-layer stripe was measured at different temperature. The specimen diemsions are shown in Fig.4.

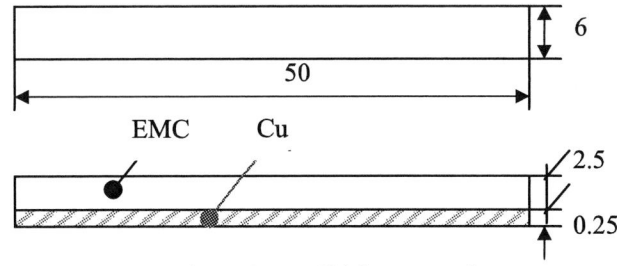

Fig. 4 Dimensions of bi-layer specimen

3.2 Warpage of bi-layer strips at different temperatures

Warpage in a temperature range from room temperature to 260degc was measured using Shadow Moiré. The EMC was post mold cured for 4 hours before tests. Fig. 5 shows a typical 3D contour obtained from Shadow Moiré. 4 or more samples in each case were tested to reproduce data. The detailed results are listed in Table 1 and plotted in Fig.6.

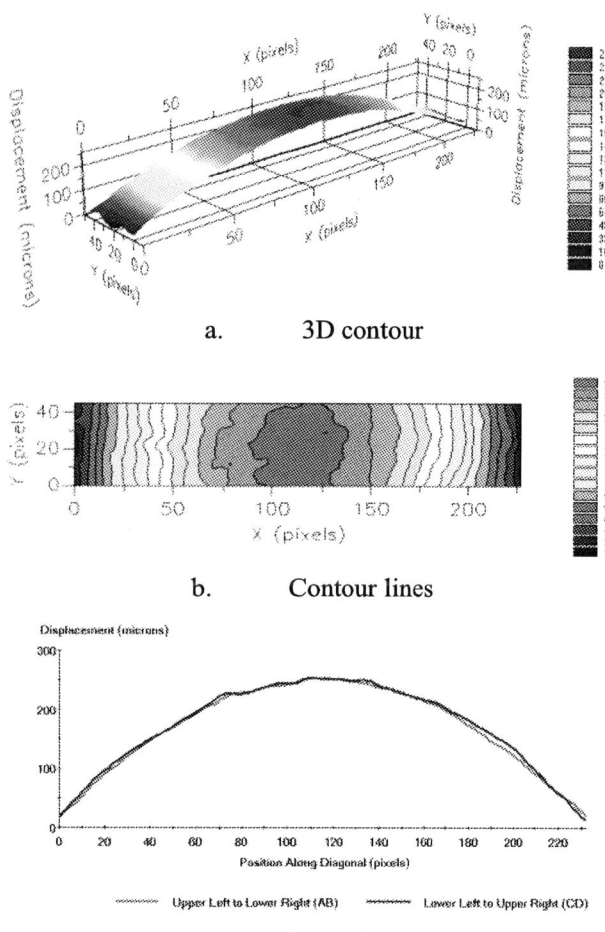

a. 3D contour

b. Contour lines

c. 2D diagonal extraction

Fig.5. Typical Shadow Moiré warpage plots @ 28degc

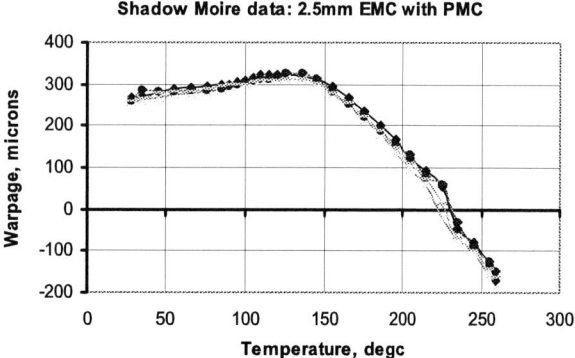

Fig. 6. Warpage versus temperature measured for 2.5mm EMC

4. Correlation with Timoshenko's bi-layer model

From the experimental results obtained in a temperature range from room temperature to 260degc, the cure shrinkgae can be determined using Timoshenko's beam theory as first order approximation following Eqs. (7) and (8). That is 0.135% for the selected EMC. The cure shrinkage can also be obtained by best-fitting the experimental data in the temperature

Table 1: Warpage results at selected temperature

T, degc	S1	S2	S3	S4	Average
28	268	259	258	257	261
35	273	262	287	264	272
45	278	267	282	277	276
55	289	274	281	281	281
65	291	279	285	282	284
75	294	283	287	288	288
85	299	289	290	296	294
90	300	296	295	299	298
95	305	296	300	300	300
100	310	302	304	300	304
105	317	307	308	304	309
110	321	311	312	308	313
115	324	314	314	309	315
120	323	312	318	309	316
125	323	314	326	312	319
135	320	312	326	314	318
145	314	304	313	302	308
155	296	280	283	279	285
165	269	250	252	254	256
175	236	215	221	220	223
185	201	182	187	186	189
195	167	137	154	146	151
205	124	95	129	110	115
215	93	58	78	73	76
225	57	-22	59	14	27
235	-46	-68	-31	-67	-53
245	-80	-107	-88	-100	-94
255	-134	-157	-126	-144	-140
260	-171	-185	-149	-162	-167

range selected (i.e. from room to high temperature). In implementation of cure shrinkage into warpage prediction, there are two alternative ways. Firstly, we can convert cure shrinkage into additional coefficient of thermal expansion (CTE), and combine it with thermal shrinkage to form an effective CTE which is temperature dependent; secondly, we can simply shift the reference temperature in ANSYS from molding temperature to zero-warpage temperature for warpage calculation. While temperature shift due to cure shrinkage actually can be obtained in two ways, one from the cure shrinkage at molding temperature and the other directly from T_0 where zero-warpage occurs as in experiemnts. The two methods should give the same results. However, since Shadow Moire will face accuracy challenges at T_0, accurate zero-warpage temperature determination will inevitably contain sometimes big errors of +/- 10degc. To improve this it is suggested to use Timosheno's beam model, as in Eq. (7), for the first order approximation to calculate cure shrinkage from the warpage at molding temperature, and further based on the CTE over molding temperature to determine the temperature shift.

$$Tshift = CS / (\alpha_{EMC} - \alpha_{Cu}) \qquad (9)$$

Expansion of the EMC is the addition of thermal expansion plus cure shrinkgae by taking 175degc as reference temperature, or equivalently the expansion with starting temperature of 175degc (molding temperature) + Tshift in Eq. (9) due to cure shrinkage. With that consideration, the correlation between beam model prediction and actual measurement was done and shown in Fig.7.

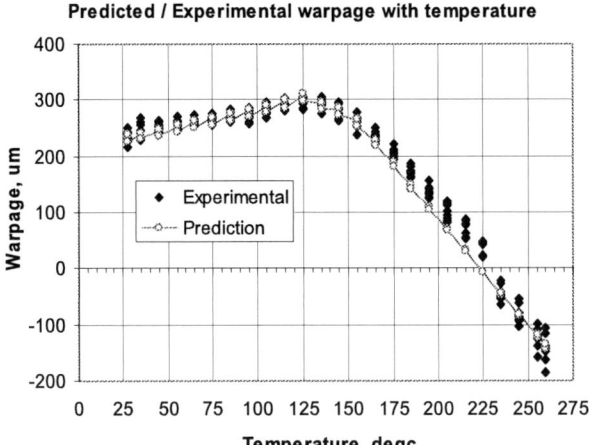

Fig.7. Correlation based on Timoshenko's beam model

It is notable that thermal shrinkage /avergae CTE of EMC is very critical in the determination of the cure shrinkage induced temperature shift. The commonly used $\alpha1$, $\alpha2$ and Tg to calculate thermal shrinkage causes errors, especially around Tg, and therefore deviation of prediction and measurement. What is even worse is that the errors are dependent upon testing method set-up. To ensure consistent correlation and accurate cure shrinkage characterization, it is recommeded to use complete TMA curve of the EMC to calculate shrinkage.

5. Direct measurement of cure shrinkage

Other methods for cure shrinkage characterization have been used before. Ken Oota [3] used a thin molded disc to determine cure shrinkage by measuring its dimension change before and after cooling down, which is total shrinkage, and the calculated thermal shrinkage using TMA data. The difference between total and thermal shrinkage is considered as cure shrinkage. As comparison, similar measurement was done, however, using rectangular stripes. Due to geometric effect of the bar and Poisson effect and constrained by the precision of the measurement tools, direct measurement using rectangular bar was not accurate. The cure shrinkage such determined was used as reference only.

Considering conversion of EMC from liquid to solid state, a correction factor of about 0.6 was introduced to

consider the contribution of elastic cure shrinkage to warpage [6]. Taking this into consideration, the determined cure shrinkage by such way was 0.15%, fairly agreeable with that obtained from our new method.

6. 3D FEA analsyis on PBGA maps and correaltion

To further illustrate the necessity of cure shrinkage in simulation and analysis, we applied the cure shrinkage data obtained for other 2 different EMCs following above methodology to warpage prediction of actual PBGA matrix. The objective is to extrapolate the cure shrinkage results to actual product to verify its effectiveness.

The test matrix was prepared by molding EMC onto substrate of 35mmX35mm without die attached. 2 EMC materials were selected. The reason to use such EMC / substrate matrix is to eliminate the effect of geometrical and material effect of other components on warpage. Therefore, the results were expected to be more applicable. Warpage of the samples after 3 different processes, namely after transfer molding (TM), post mold cure (PMC) and PMC+ReFlow at 260degc for 3 times (RF260X3) were measured using Shadow Moiré facility at room temperature.

The thermal and mechanical properties of the selected EMCs were characterized on TMA and DMA facilities after above mentioned 3 different process conditions, i.e. after transfer molding (TM), post mold cure (PMC) and PMC+ReFlow at 260degc for 3 times (RF260X3). The data were then used as inputs for simulations to investigate how the material properties, e.g. CTE and modulus as well as chemical shrinkage, affect the warpage behavior. Octant models as shown in Fig. 8 were applied. The substrate with effective molding area of 31mmX31mm was modeled as Solder mask + core BT-Cu + solder mask tri-layer structure. In one part, the total shrinkage was applied to FEA analysis of the EMC/substrate system, i.e. cure shrinkage was considered; while in another part the test platform was simulated using only thermal shrinkage data without consideration of cure shrinkage. The obtained results were then compared to demonstrate the contribution of cure shrinkage to warpage.

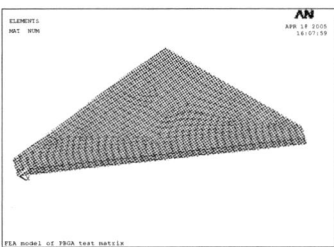

Fig.8. FEA model of test PBGA matrix

The simulation results after 3 different processes for the 2 selected EMCs were shown in Figs. 9-11, where in each of the Figures, the upper 3 plots are for EMC-1, while the lower 3 plots for EMC-2. Fig.9 shows the warpage contours when cure shrinkage was not considered and thermal shrinkage was deduced using Tg, α1 & α2; whereas for Fig.11 the results taking cure shrinkage into account and using thermal shrinkage data directly derived from TMA measurement. The difference between Fig.9 and Fig.10 is that the latter using thermal shrinkage directly from TMA curve.

Figures. 12 to 14 compare Shadow Moiré results and FEA simulation data for the cases in Figs. 9-11. It is clearly seen that:

(a) When no cure shrinkage was considered, and traditional Tg, α1 & α2 values used in simulation, the correlation is very poor, as shown in Fig. 12, the trend and warping pattern after different process can not be predicted;

(b) If thermal shrinkage data were derived directly from TMA test, the trend between simulation and measurement could be greatly improved, however, a tremendous discrepancy of warpage values was seen, as in Fig. 13;

(c) With cure shrinkage inclusion and application of TMA thermal shrinkage data, the correlation was perfect for both EMCs and after all various process conditions, as in Fig. 14.

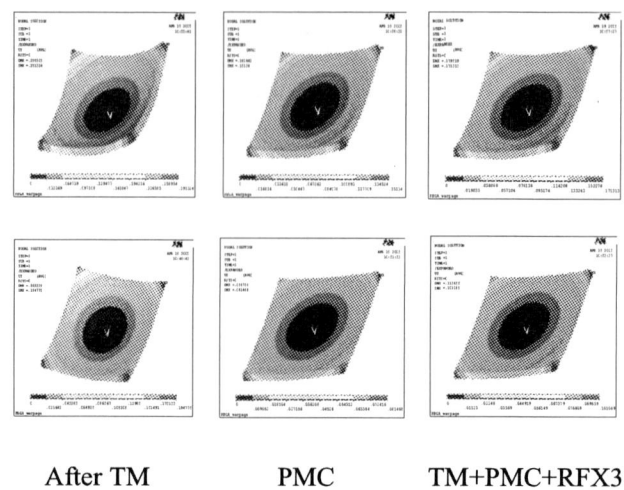

After TM   PMC   TM+PMC+RFX3

Fig.11 Warpage contour in PBGA - CS considered, using thermal shrinkage from TMA curve

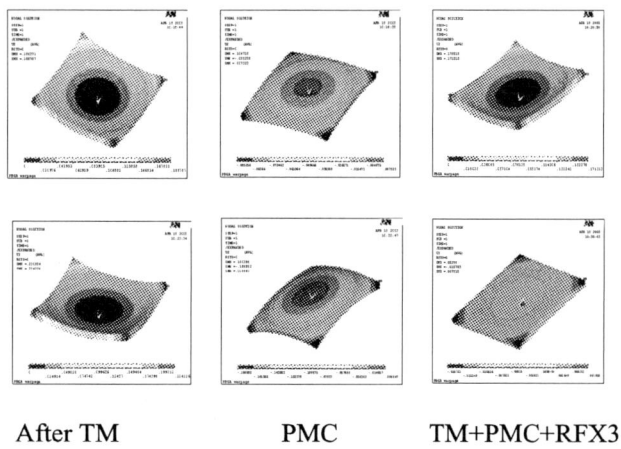

After TM   PMC   TM+PMC+RFX3

Fig.9. Warpage contour in PBGA - No CS, Tg, α1 & α2 used as simulation inputs

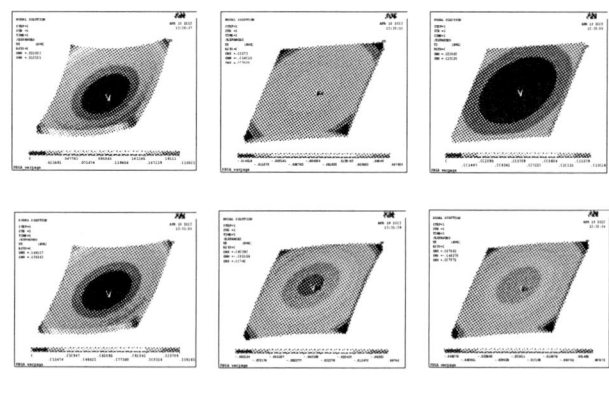

After TM   PMC   TM+PMC+RFX3

Fig.10 Warpage contour in PBGA - No CS, using thermal shrinkage from TMA curve

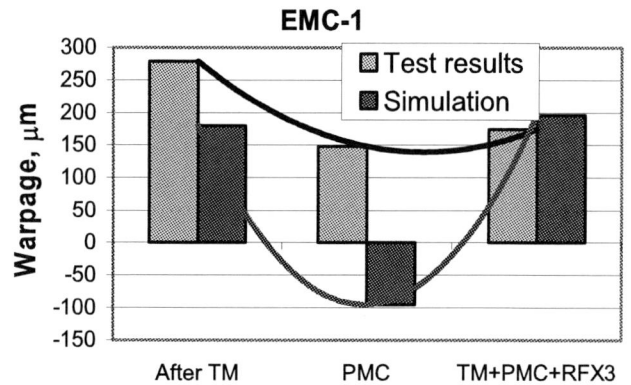

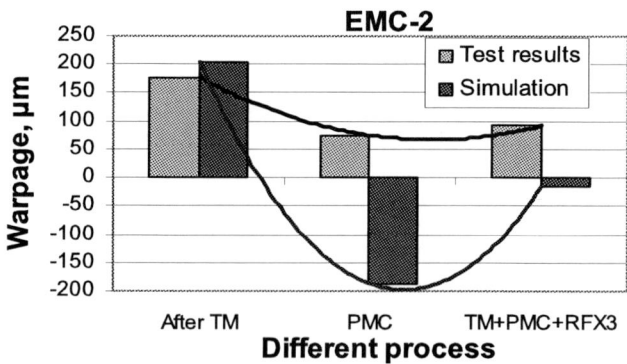

Fig. 12. Correlation between simulation and test - No CS, Tg, α1 & α2 used as simulation inputs

Modulus effect on warpage was also verified by using either DMA data or derived modulus at low and

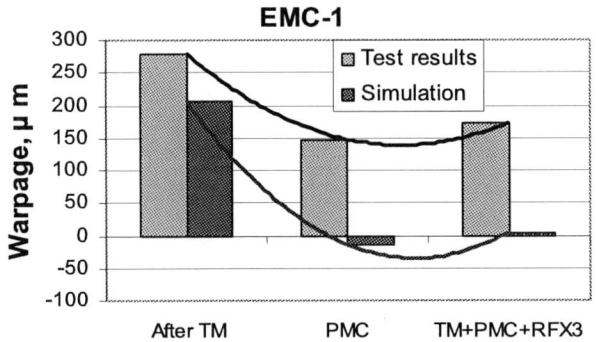

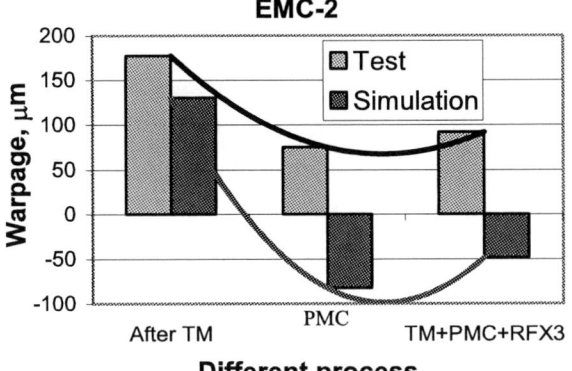

Fig. 13. Correlation between simulation and test - No CS, using thermal shrinkage from TMA

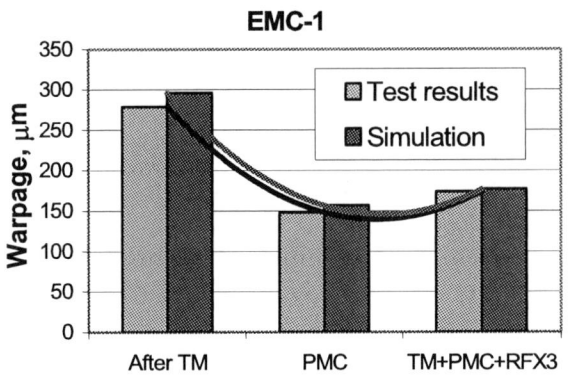

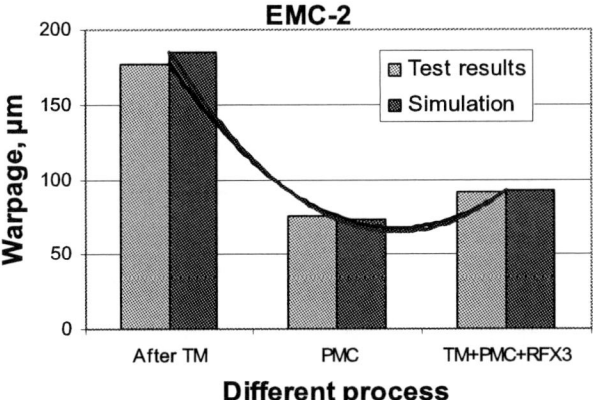

Fig. 14. Correlation between simulation and test – CS considered, using TMA from thermal shrinkage

high temperature. It was found that at low temperature it is only a second level effect. However, use of DMA data but it is recommended, especially at high temperature, where modulus is very low and maximum warpage is achieved. This is critical when high temperature warpage, for example, of PoP packages, is urging accurate characterization.

Good agreement of simulation with measurement under conditions of considering cure shrinkage and using TMA thermal shrinkage and DMA modulus data suggested the inclusion of cure shrinkage and identified the significance of correct application of TMA and DMA data. It demonstrates that accurate characterization of warpage is feasible but cure shrinkage must be considered, and material data must be reliability and properly used.

7. Discussion

In our new approach to characterize cure shrinkage, a simple EMC/Cu or a known material as substrate was used. Through measurement of warpage of the bi-layer beam system in a wide temperature range using Shadow Moiré, the elastic cure shrinkage of the EMC was derived by best fitting the experimental data. Material characterization data including thermal expansion (from TMA) and mechanical property (from DMA) were recommended to use to ensure accuracy of cure shrinkage determination. As 1st order approximation, elastic cure shrinkage can be obtained either from the warpage at molding temperature or from the Tshift, a temperature shift due to cure shrinkage. In summary, the procedure of cure shrinkage characterization is as follows:

a. Warpage measurement of EMC/Cu bi-layer system in a wide temperature range;
b. TMA and DMA characterization of the testing EMC;
c. Best-fitting warpage data to obtain cure shrinkage;
d. 1st order estimation of cure shrinkage using Wm, warpage at molding temperature; and Tshift, temperature shift from molding temperature to new zero-warpage temperature;
e. Verification using Timoshenko's beam model.

Compared with traditional direct measurement approach for cure shrinkage determination, our new method has high sensitivity and thus high accuracy of measurement, applicable to the whole temperature range. Besides, elastic cure shrinkage can be derived directly, whereas the plastic cure shrinkage that actually does not contribute to warpage is excluded.

Obviously, final cure shrinkage value varies with TMA and DMA data. This fact urges proper characterization of thermal and mechanical property of EMC and correct application. For TMA data, using $\alpha 1$, $\alpha 2$ and Tg will cause extra errors and should be abandoned; instead thermal shrinkage data should be applied directly to capture precisely the warpage trend.

More errors can be seen at high temperature. This partially is originated from modulus errors in DMA data. In rubber stage, modulus of EMC is very low, warpage becomes more sensitive to the modulus variation. Considering that zero-warpage temperature is at high temperature range, and moreover Shadow Moiré is not capable enough to capture the zero-warpage temperature, it is suggested to use both the warpage at molding temperature and Tshift to estimate the elastic cure shrinkage as first order approximation.

FEA analysis of the bi-layer could help clarify the accuracy of Timoshenko's beam model. That part of work can provide comprehensive comparison using 2D and 3D models and calibrate the effectiveness of beam model. The results will be presented separately in another paper due to length constrain of this paper.

For practical purpose, correlation for real packages with actual die will be significant. Theoretically, with or without die has no detriment for our new method. However, application of cure shrinkage data to real packages will be very demonstrative, and should be part of future works to further validate our method.

In our work, cure shrinkage is considered for fully cured EMC. No efforts have been put into shrinkage at different curing stages that were studied, e.g. in Ref. [9, 10]. This is conforming to actual industrial practice where post molding cure is usually one of the processes to ensure full-cross-linking in EMC and thus stabilized property of EMC.

8. Conclusions

1) New method to characterize cure shrinkage / total shrinkage of EMC was presented by measurement of warpage of bi-layer beam in a wide temperature range. Simple estimate of the cure shrinkage was initiated. As first order approximation, Timoshenko's beam model provides a good correlation.

2) Advantages of this new method are proposed. Firstly, it characterizes directly the elastic cure shrinkage; and secondly, it has higher accuracy than other direct measurement methods.

3) Cure shrinkage of 3 molding compounds was characterized. The cure shrinkage data of 2 EMCs were applied to PBGA matrix. It was illustrated clearly that accurate warpage prediction could be achieved when cure shrinkage data was applied and both TMA and DMA data were properly adopted.

References

1. Kiyoshi Miyaki, " Thermal-viscoelastic analysis for warpage of ball grid arrange packages taking into consideration of chemical shrinkage of molding compound", Journal of Japanese Electronics Society, Vol.7 (1), 2004, pp.54-61.

2. Gerard Kelly, Colin Lyden,et al, "Importance of molding compound cure shrinkage in the stress and warpage analysis of PQFP's", IEEE Transactions on Components, Packaging, and Manufacturing Technology, - Part B, Vol. 19(2), 1996, pp.296-300.

3. Ken Oota and Masumi Saka, "Cure Shrinkage Analysis of Epoxy Molding Compound," *Polymer Engineering and Science,* August 2001, Vol. 41, pp. 1373-1379.

4. M. Ogata, Noriyuki Kinjo, et al, " Effect of curing acceleratiors on Physical Propertiesof Expoxy Molding Compound", Journal of Applied Polymer Science, Vol. 44, 1992, pp.1795-1805.

5. M. Ogata, S.Eguich, T. Ishi and T. Kawada, " Cross-linking effects to cured properties of phenol novolac epoxy resin", Journal of Thermosettiing Plastics, 1999.

6. Sindee L Simon, Gregory B. Mckenna, Oliver Sindi, " Modeling the evolution of the dynamic mechanical properties of a commercial epoxy during cure after gelation".

7. Timoshenko, S.P., "Strength of materials", 1955, Pt. 1, Third Edition, Mc Graw Hill Publishers, New York

8. Timoshenko, S.P., "Theory of elasticity", 1951, Mc Graw Hill Publishers, New York.

9. Ernst, L.J., *et al.* (2006) "Fully Cure-Dependent Modelling and Characterization of EMC's with Application to Package Warpage Simulation", in: Proc. Of IEEE CPMT Int. Symp. on Advance Packaging Materials 2006, Atlanta, March 2006.

10. Yang, D. G., Jansen, K. M.B., Wang, L.G., Ernst, L.J., et al, IEEE Transactions on Components, Packaging, and Manufacturing Technology, Vol. 27, 2004, 676-683.

Optimizing the Dynamic Response of RF MEMS Switches using Tailored Voltage Pulses

Vitaly Leus, Arnon Hirshberg, and David Elata
Technion – Israel Institute of Technology
Haifa 32000, Israel
Email: elata@tx.technion.ac.il , Phone: +972 4 8292617.

Abstract

This paper presents analytic modeling, simulations, and experimental validation of RF MEMS switch actuation using tailored voltage waveforms. This actuation scheme can eliminate impact bouncing during switch activation and can significantly reduce harmonic oscillations during switch release.

1. Introduction

Electrostatic RF MEMS switches are a potential alternative to current IC switches because of their low power consumption, high isolation, and low insertion loss [1]. RF MEMS switches may be used in a wide range of communications, aerospace and other applications [1, 2].

Proper operation of switches requires that their electromechanical dynamic response meets specific parameters. In many electrostatic switches a short response time is desirable. However, a short switching time often means that the movable electrode impacts the fixed electrode with a high velocity. This impact may lead to a rebound of the movable electrode (i.e. bouncing [3]), and to accumulation of damage (e.g. pitting in Ohmic switches [1]). One way of reducing impact bouncing is packaging the switch with an ambient pressure that is sufficient to dampen its dynamic response. An ambient pressure can also damp harmonic oscillations during switch release [3]. However, the damping which is induced by the ambient pressure increases the response time of the switch.

This work considers an alternative way of eliminating the impact bouncing and harmonic oscillations, by tailoring the actuation voltage. Since no ambient pressure is required, the response time remains as short as possible for a given value of applied driving voltage.

2. Modeling the dynamic response of MEMS switches

In this work the switch is modeled as a 1D parallel-plates actuator, schematically illustrated in Fig. 1.

The actuator is constructed from a top electrode of mass m and area A that is suspended on a linear elastic spring with stiffness k, above a fixed bottom electrode. The bottom electrode is coated with a dielectric layer of thickness d, and the initial gap between the top electrode and the dielectric is g. The fixed bottom electrode is electrically grounded and a voltage V is applied to the top electrode.

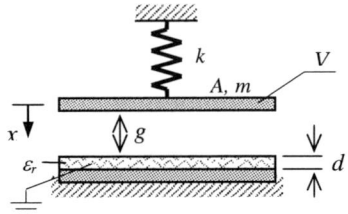

Figure 1. The parallel-plates actuator with a dielectric layer coating the bottom electrode.

The dynamic response of the parallel-plates actuator when it is subjected to a step-function voltage, is derived from the Hamiltonian of the system given by [4]

$$H = \frac{1}{2}m\dot{x}^2 + \frac{1}{2}kx^2 - \frac{\varepsilon_0 A}{2(d_0/\varepsilon_r + g - x)}V^2 \quad (1)$$

where ε_0 is the permittivity of free-space, ε_r is the relative permittivity of the dielectric, and $\dot{x}$ is the velocity of the movable electrode. The Hamiltonian is the sum of the kinetic energy, the elastic potential, and the electrostatic potential of the deformable capacitor and of the voltage source. The effect of fringing fields is not considered in this analysis.

The Hamiltonian may be rewritten in the normalized form

$$\tilde{H} = \frac{1}{2}\dot{\tilde{x}}^2 + \frac{1}{2}\tilde{x}^2 - \frac{1}{2}\frac{\tilde{V}^2}{1+\xi-\tilde{x}} \quad (2)$$

where

$$\tilde{H} = \frac{H}{kg^2}, \quad \tilde{x} = \frac{x}{g}, \quad \tilde{t} = \sqrt{\frac{k}{m}}t, \quad \dot{\tilde{x}} = \frac{d\tilde{x}}{d\tilde{t}}$$

$$\xi = \frac{d}{\varepsilon_r g}, \quad \tilde{V}^2 = \frac{\varepsilon_0 A}{kg^3}V^2 \quad (3)$$

and ξ is a dimensionless number that measures the properties of the dielectric layer.

The location of the movable electrode as function of time can be computed from the momentum equation that is derived from the Hamiltonian in the form

$$\ddot{\tilde{x}} = \frac{\partial^2 \tilde{x}}{\partial \tilde{t}^2} = -\frac{\partial \tilde{H}}{\partial \tilde{x}} = -\tilde{x} + \frac{1}{2}\frac{\tilde{V}^2}{(1+\xi-\tilde{x})^2} \quad (4)$$

The momentum equation (4) can be integrated twice in time to find the displacement of the electrode as function of time $\tilde{x}(\tilde{t})$. Figure 2 illustrates the trajectory of the top electrode for $\xi = 0$ and several values of applied step-function voltage. These simulation results were obtained by numerical time-integration of (4) using Matlab.

1-4244-1105-X/07/$25.00 ©2007 IEEE

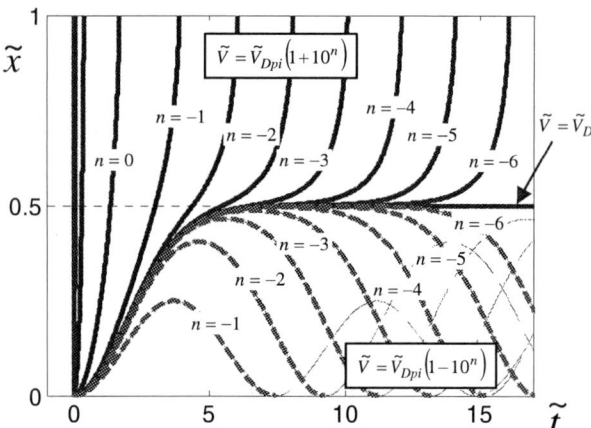

Figure 2. Dynamic response of the parallel-plates actuator for various values of applied voltage. For voltages below the dynamic pull-in voltage the response is periodic (dashed lines) and for voltages above this critical value the response is non-periodic (solid). These simulations are for the case of $\xi=0$.

Below a critical value of the applied voltage, the dynamic response of the actuator is periodic. For voltages above this critical value, the response is non-periodic, the velocity of the movable electrode is always positive, and it eventually collapses into contact with the fixed electrode. When the critical voltage is applied to the system, the movable electrode converges to an unstable equilibrium state [4]. This critical dynamic state is the *Dynamic Pull-In* state of the system that is found to be

$$\tilde{x}_{DPI} = \frac{1}{2}(1+\xi), \quad \tilde{V}_{DPI} = \frac{1}{2}(1+\xi)^{3/2} \quad (5)$$

In this work damping is neglected and accordingly, once the voltage is applied (at $\tilde{t}=0$ when $\tilde{x}=0$ and $\dot{\tilde{x}}=0$), the Hamiltonian $\tilde{H}_0 = \tilde{H}_{(\tilde{t}=0)}$ is unchanged. Setting $\tilde{H}_0 = -\tilde{V}^2/(1+\xi)/2$ it follows that

$$\tilde{H} = \frac{1}{2}\dot{\tilde{x}}^2 + \frac{1}{2}\tilde{x}^2 - \frac{1}{2}\frac{\tilde{V}^2}{1+\xi-\tilde{x}} = -\frac{1}{2}\frac{\tilde{V}^2}{1+\xi} \quad (6)$$

From (6), considering only positive velocities, the velocity of the movable electrode is found to be

$$\dot{\tilde{x}} = \sqrt{\frac{\tilde{x}\tilde{V}^2}{(1+\xi-\tilde{x})(1+\xi)} - \tilde{x}^2} \quad (7)$$

The impact velocity can be computed from (7) by setting $\tilde{x}=1$

$$\dot{\tilde{x}}_{(\tilde{x}=1)} = \sqrt{\frac{\tilde{V}^2}{\xi(1+\xi)} - 1} \quad (8)$$

For ordinary RF MEMS switches ξ is of the order of 1% or less, consequently the impact velocity is considerable. Such a high impact velocity leads to a rebound of the movable electrode extending the effective switching time of the actuator.

When the voltage is reduced below the hold-down voltage of the switch, the movable part is released from the down position. This spontaneous release inevitably leads to the harmonic oscillations of the movable electrode for some duration of time that depends on amount of damping in the system.

These unwanted effects can be alleviated by packaging the switch in ambient pressure. However, the damping which is induced by the ambient pressure increases the response time of the switch.

In this work an alternative way for reducing the high impact velocity and release oscillations by application of short voltage pulses is proposed.

3. Short-pulse voltage actuation

The unloaded electrostatic actuator is assumed to be at rest, and is instantaneously subjected to a constant voltage for a limited time interval. To simplify the analysis, it will be assumed that the voltage pulse is applied in a displacement interval rather than in a time interval (though the two intervals are uniquely related as will be shown later). In other words, we assume that the applied voltage is defined by

$$\tilde{V}^2 = \begin{cases} \tilde{V}^2 & 0 \le \tilde{x} < \tilde{x}_0 \\ 0 & \tilde{x}_0 \le \tilde{x} \le 1 \end{cases} \quad (9)$$

The purpose of this actuation strategy is to switch the actuator but ensure that the contact between the two electrodes occurs with zero impact velocity.

During the application of the voltage pulse $0 \le \tilde{x} \le \tilde{x}_0$, the Hamiltonian is equal to its value immediately after application of the voltage pulse, and (6) may be rewritten in the form

$$\frac{1}{2}\dot{\tilde{x}}^2 + \frac{1}{2}\tilde{x}^2 = \frac{1}{2}\tilde{V}^2\left(\frac{1}{1+\xi-\tilde{x}} - \frac{1}{1+\xi}\right) \quad (10)$$

This means that the work done by the voltage source is entirely invested in the kinetic energy and elastic potential of the movable electrode.

After the end of the pulse ($\tilde{x}_0 \le \tilde{x} \le 1$) no additional work is done by the voltage source and therefore

$$\frac{1}{2}\dot{\tilde{x}}^2 + \frac{1}{2}\tilde{x}^2 = \frac{1}{2}\tilde{V}^2\left(\frac{1}{1+\xi-\tilde{x}_0} - \frac{1}{1+\xi}\right) \quad (11)$$

From this moment until contact, the movable electrode is in a free-flight. Nullifying the velocity at contact, i.e. setting $\dot{\tilde{x}}=0$ at $\tilde{x}=1$, and solving (11) for $\tilde{x}_0$ yields

$$\tilde{x}_0 = \frac{(1+\xi)^2}{1+\xi+\tilde{V}^2} \quad (12)$$

From the previous section (Fig. 2) it is clear that the applied voltage must not be smaller than the dynamic pull-in voltage (or else no contact will be achieved at all). For the minimal value of $\tilde{V}=\tilde{V}_{DPI}$ (see (5)) it is found from (12) that the voltage-pulse interval extension is bounded from above by

$$\tilde{x}_0 \leq \frac{4(1+\xi)}{4+(1+\xi)^2} \qquad (13)$$

It can be deduced from (12) that as the applied voltage increases, the voltage pulse interval extension decreases. The reason is that at high actuation voltages, a shorter interval is sufficient to achieve the necessary work required for a longer free-flight.

The dynamic time-response of the system was simulated using Matlab. Figures 5a,b present the displacement and the velocity of the top electrode, respectively, as function of time for three different voltages. The impact velocity of the movable electrode is extremely high when the switch is actuated with step-function. But using short pulse voltage actuation, the impact velocity is drastically reduced.

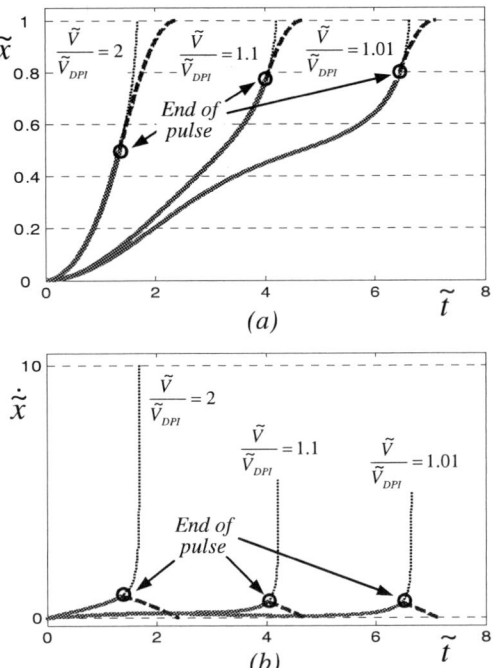

(a)

(b)

Figure 5. Simulated time-response of the parallel-plates actuator for three different applied voltages (for $\xi=0.01$): (a) displacement as function of time, (b) velocity as function of time. The solid lines are the trajectory during the pulse and the dashed lines are the trajectory of the free-flight. The dotted line is the trajectory associated with a step-function voltage.

The same actuation strategy can be performed when the actuator is released from the bottom position in order to reduce the harmonic oscillations of the movable electrode. Figure 6 presents the simulated dynamic response of the actuator while short voltage pulses are applied for both intervals: switching interval and release interval. It is shown that by applying an appropriate short voltage pulse, the impact velocity of the movable plate may be drastically reduced such that contact bouncing is effectively eliminated. It is also shown that by applying the similar pulse in release, the oscillations of the movable plate can also be eliminated.

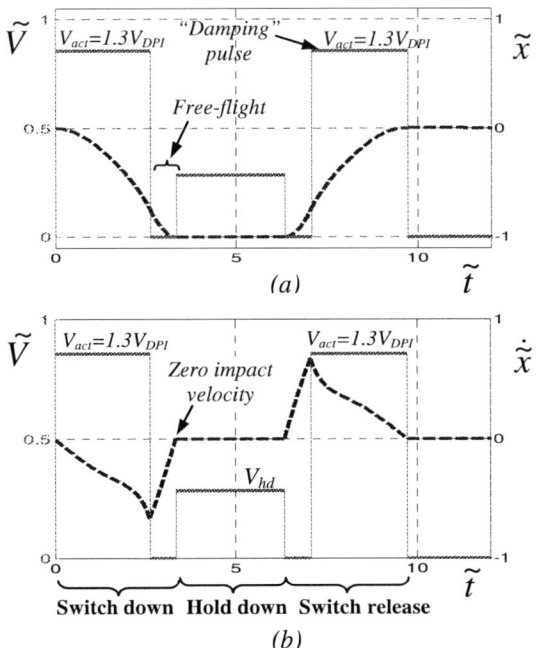

(a)

(b)

Figure 6. Simulated response of the parallel-plates actuator (both the switching and release responses) when the structure is actuated with the proposed short voltage-pulses: (a) displacement and (b) velocity (dashed lines).

4. Experimental results

To validate the analysis, a parallel-plates test structure was designed and fabricated using SOI technology (Fig. 7).

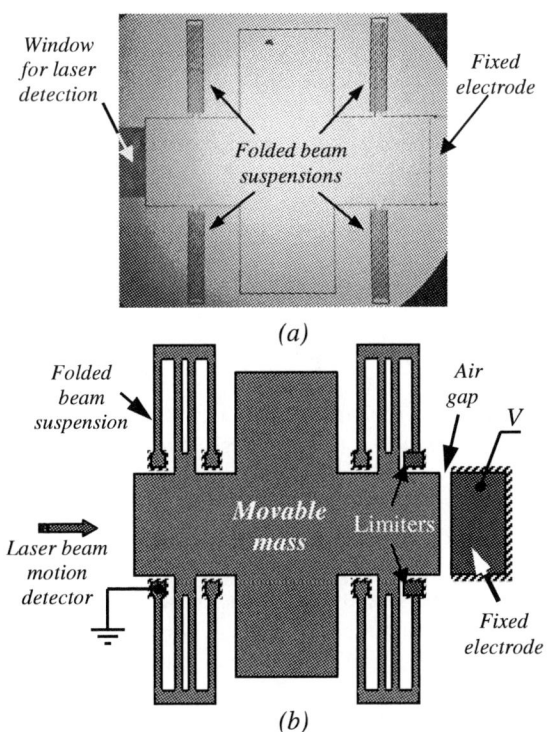

(a)

(b)

Figure 7. (a) Micro photo and (b) schematic top view of the parallel-plates test structure.

The large mass of the movable part and a relatively low stiffness of the support beams provide a low natural frequency. This enables to detect the mass motion with a high time-resolution at different locations on the device. This gives a more detailed picture of the dynamic response relative to what can be achieved be measuring the capacitance of the system (i.e. by measuring the velocity in different locations on the mass, rotation can be detected as well). The movable mass is symmetrically supported by four folded beams.

The initial gap is 6μm, and mechanical limiters stop the movable mass at a minimal gap of 1μm. The motion detection (both displacement and velocity) is performed using a laser vibrometer. As the motion of the movable mass occurs in the in-plane direction, the perpendicular laser beam was reflected to the horizontal direction using a 45° tilted mirror.

Figure 8 presents the measured response of the test structure (both displacement and velocity) at a low pressure of $5 \cdot 10^{-3}$ torr when a step-function voltage is applied to the fixed electrode. The response includes bouncing due to the high impact velocity, and release oscillations.

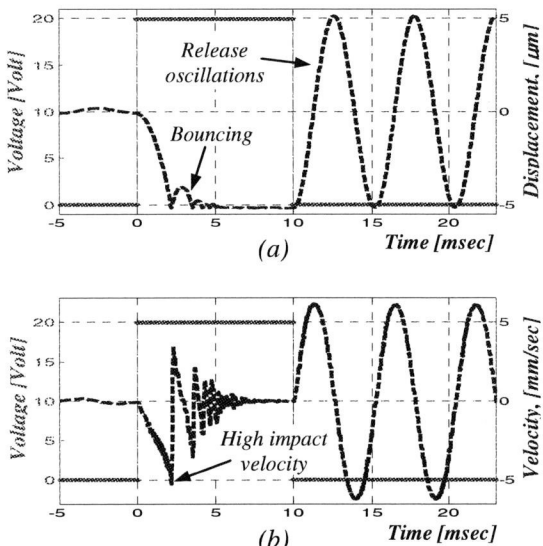

Figure 8. The measurement results (in vacuum) for the movable mass (a) displacement and (b) velocity (dashed lines), when the structure is actuated with the regular step-function voltage (solid line).

Figure 9 presents the measured response of the test structure (both displacement and velocity) at the same pressure when the proposed short voltage-pulses are applied to the fixed electrode. The measured results resemble the simulations: the velocity at contact is significantly reduced, and the release oscillations are considerably decreased.

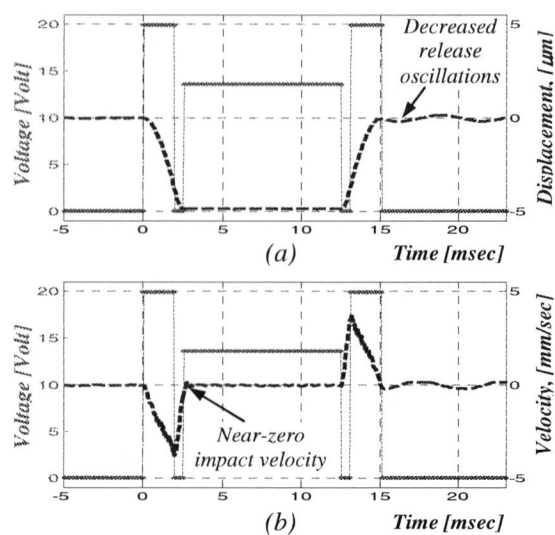

Figure 9. The measurement results (in vacuum) for the movable mass (a) displacement and (b) velocity (dashed lines), when the structure is actuated with the proposed short voltage-pulses (solid line).

5. Conclusions

A novel voltage waveform for optimal actuation of electrostatic RF MEMS switches was presented. The proposed actuation scheme was modeled analytically, numerically, and validated experimentally. It was shown that by applying an appropriate short voltage pulse, the switch can be closed with near-zero impact velocity. Similarly, by application of the same voltage pulse during the switch release, harmonic oscillations can be considerably reduced.

Similar work [5] was published after this abstract was submitted. However in our work we derive analytic expressions for impact velocity and pulse duration (e.g. (8) and (12)) which may be used as design rules.

Acknowledgments

This work was preformed within the context of AMICOM - The European Network of Excellence on RF MEMS and RF Microsystems.

References

1. Rebeiz, G. M., RF MEMS: theory, design, and technology. (Hoboken, N.J., 2003).
2. J. J. Yao, "RF MEMS from a device perspective," *JMM*, Vol. 10 (2000), pp. R9-R38.
3. Steeneken, P. G., *et al*, "Dynamics and squeeze film gas damping of a capacitive RF MEMS switch," *JMM*, Vol. 15 (2005), pp. 176-184.
4. Elata, D., *et al*, "On the dynamic pull-in of electrostatic actuators with multiple degrees of freedom and multiple voltage sources," *JMEMS*, Vol. 15 (2006), pp. 131-140.
5. Czaplewski, D. A., *et al*, "A soft-landing waveform for actuation of a single-pole single-throw ohmic RF MEMS switch," *JMEMS*, Vol. 15 (2006), pp. 1586-1594.

The Effect of Visco-elasticity on the Result Accuracy of FEM Panel Warpage Simulations Supporting Industrial Microelectronics Packaging

Sven Rzepka and Axel Müller
Qimonda Dresden GmbH & Co OHG, Dept. QD BET CMI
Königsbrücker Straße 180; D-01099 Dresden; Germany
sven.rzepka@qimonda.com

Abstract

Based on measurement results of BGA substrate panel warpage, the accuracy of two FEM simulation approaches has been assessed. Relying on elastic/plastic material models, the first approach has only been capable of following the most fundamental trend qualitatively. In terms of magnitude, these simulations already exaggerated the dependency of the warpage on the die thickness by a factor of two. Even worse, this approach wrongly estimated the second basic trend. Reducing the area of the dies without changing their arrangement, the real substrate panel warpage changes towards negative magnitudes while the simulation results directly pointed in the opposite direction. In the alternative FEM approach, visco-elastic behavior of the organic materials and chemical shrinkage of the mold compound have been added to the models. Afterwards, the experimental results could all be matched in a wide range of die sizes. The residual root mean square error was as small as the measurement inaccuracy. Sensitivity analyses clearly showed the root cause for why the first approach must have failed. They also optimized the simulation effectiveness by minimizing the effort in simulation and material characterization for the accuracy required. The paper details the developed scheme for trustworthy FEM substrate panel warpage estimations.

1. Introduction

During the past years, extensive work has been done in characterizing the behavior of the organic materials used in microelectronics packaging like mold compounds, die adhesives, substrates and solder masks [1-3]. This work has mainly focused on the visco-elasticity as substantial stress relaxation and creep has been found to occur in these materials under test and service conditions.

Nevertheless, many thermo-mechanical FEM simulations performed in microelectronics industry still rely on elastic models only. Obviously, people assume visco-elastic properties can be neglected because they would alter the results only marginally while they increase the execution time dramatically. Even stronger they believe that trends are not changed by the visco-elastic effects.

The following paper assesses the results of a typical thermo-mechanical simulation done in packaging industry. It concerns the panel warpage, which is a major risk factor for manufacturability and packaging yield of future products. In parametric studies based on elastic/plastic and on visco-elastic material models, respectively, the warpage of substrate panels is predicted and compared to experimental measurements.

2. First Level Packaging

In BGA technology, the dies cut off the thinned wafer are mounted to the substrate panel by die attach and wire bonding or by flip chip soldering, respectively. Subsequently, these arrays of dies and their interconnects are encapsulated by an organic material. Usually, this is done by over-molding the substrate panel without covering the substrate pads for the BGA solder balls (Fig. 1). Molding and curing the organic compound takes place at elevated temperatures. Afterwards, the panel is cooled down to room temperature at which flux and BGA balls are applied to the substrate pads. Then, the panel is reheated to activate the flux and to solder the balls while the final step, in which the individual components are separated, is carried out at room temperature again. Thus, the manufacturing process involves a number of temperature changes during which the panel may be deformed because of the differences in coefficient of thermal expansion (CTE) within the BGA system. Excessive panel warpage may trigger malfunctions of the automated handling system (e.g., failures in vacuum sucking, blocking in front of narrow slots) as well as failures during solder ball placement and component separation. Thus, panel warpage at room temperature is a serious risk affecting manufacturability and packaging yield.

3. Nature of Panel Warpage

Studying the nature of panel warpage, mold compound and substrate material are found to compete with each other. Both elements are attached to the silicon die,

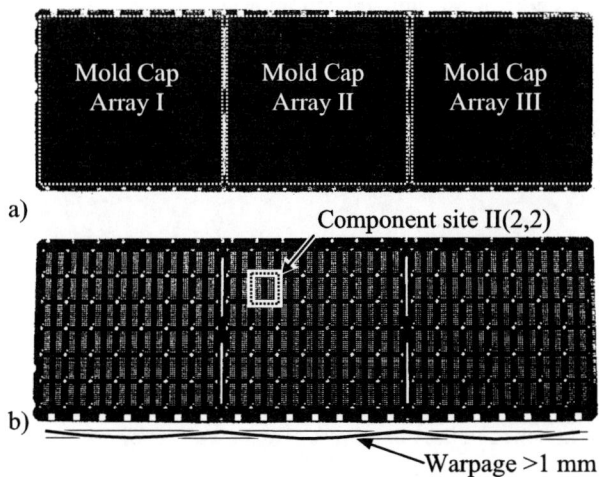

Fig. 1: BGA substrate panel showing excessive warpage
a) mold cap side, b) BGA ball pad side

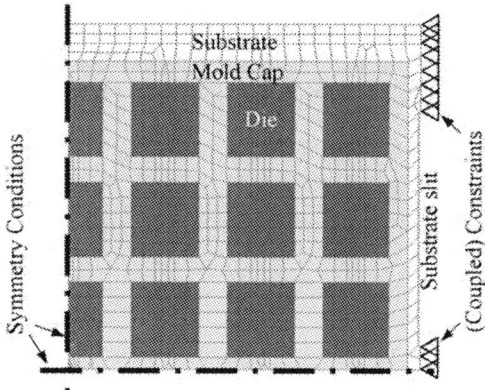

Fig. 2: FEM mesh a substrate panel array (one quarter)

Element / Material	Temperature	Young's Modulus [GPa]	Poisson's Ratio [1]	CTE [10^{-6}/K]	Comment
Die / Silicon	all	160	0.22	2.4	-
Die Adhesive / Epoxy	$< T_G$	1.0	0.35	30	$T_G = 40°C$
	$> T_G$	0.01	0.45	200	
Substrate / Epoxy-Glass Laminate	$< T_G$	X,Y: 13 Z: 4.5	XZ: 0.39 XZ,YZ: 0.25	X,Y: 18 Z: 56	$T_G = 190°C$
	$> T_G$	XY: 8.0 Z: 3.0	XZ: 0.39 XZ,YZ: 0.28	XY: 10 Z: 177	
Mold Cap / Epoxy (filled)	$< T_G$	26.8	0.25	7.1	$T_G = 114°C$
	$> T_G$	0.8	0.40	38	
Pad / Copper	all	120	0.34	16.5	$\sigma_Y = 50$ MPa
Solder Mask / Epoxy	$< T_G$	4.9	0.25	60	$T_G = 100°C$
	$> T_G$	0.1	0.35	130	

Table 1: Elastic /Plastic Material Properties (Plasticity is considered for copper by a bi-linear model)

which is very stiff and has the minimum CTE of the entire system. Thus, thermal expansion of both of its organic neighbors is restricted the same way: The interface to the die is constraint most. Consequently, both elements would warp in opposite direction when temperature is changed. However, they are directly bonded to each other outside the die area. Hence, the panel warpage seen in reality actually is an effective one. To a large extent, it is determined by the imbalance in stiffness and expansion between these two organic elements. If the mold cap is dominating, the panel with the dies on its topside shows a smiling warpage at room temperature. Increasing the thickness of the dies weakens the mold cap side and supports the substrate side instead. So, warpage changes towards crying. This way, panel warpage has been found to be a deterministic phenomenon, which can be controlled by a limited number of design parameters. Besides the material properties, which can be changed in discrete sets only, geometric dimensions allow a continuous adjustment for the effective warpage being kept within the acceptable bounds avoiding malfunctions and failures as mentioned above.

4. Thermo-mechanical Modeling

Although the number of design parameters is limited, predicting the magnitude of panel warpage precisely still is not trivial as numerous co-relations and interactions between the parameters as well as second order effects exist. Hence, predicting the magnitude of panel warpage is a typical task for thermo-mechanical simulation applying FEM.

A) Geometric Model and Boundary Conditions

A 3-D geometric model of the panel is shown in figure 2 also indicating the boundary conditions. Applying twofold symmetry conditions, only one quarter of one array of components needed to be meshed. The top edge of the panel is always free while the right edge is either free or attached to the next array. Thus, two sets of constraints were considered in the FEM model. The first set simply puts no constraints on the right edge as well while the second set forces the cross section areas outside the slit

(see Fig. 2) to stay straight, as they are part of a cutting plane. The latter case is modeled by a rigid constraint surface controlled by a pilot node [4]. Applying the scripting language of the FEM code, the dimensions of all elements are input as parameters and the mesh size is controlled based on correlations between them. This way, fully automated mesh generation was provided for the parametric study and poorly shaped elements (e.g., aspect ratios smaller than 1:20 or angles below 20°) were still avoided.

B) Material Models

Table 1 surveys the elastic and the plastic material properties considered in modeling the elements of the BGA panel. The data of the organic material has been measured in-house by dynamic mechanical analyses (DMA) while that of silicon and copper has been taken from literature [5]. In the actual FEM material models, more data points are taken from DMA and included as the temperature dependent properties. Thus, the set of material data applied has become quite comprehensive and seems as adequate as most of the other data sets typically used in industry these days for this kind of simulations.

Nevertheless, the data listed in table 1 does not account for any time dependent behavior of the organic materials. However, a 1 Hz DMA has shown substantial visco-elastic effects to occur. For example, the mold compound studied here has a loss modulus at the glass transition temperature T_G that reached more than 25% of the storage modulus' magnitude. That means, the force response of the sample to an elongation would be delayed considerably so that this effect should perhaps not be neglected. This observation finally triggered measurements of all organic materials included in the BGA system in order to characterize their visco-elasticity based on the concept of thermorheological simplicity as represented by equation (1) and (2),

$$G(t') = G_0\left[\alpha_\infty^G + \sum_{i=1}^{n} \alpha_i^G \exp\left(-\frac{t'}{\tau_i}\right)\right]$$

$$K(t') = K_0\left[\alpha_\infty^K + \sum_{i=1}^{n} \alpha_i^K \exp\left(-\frac{t'}{\tau_i}\right)\right] \quad (1)$$

$$\log_{10}(A) = \frac{C_1(T - T_{Ref})}{C_2 + (T - T_{Ref})} \quad (2)$$

in which G and K are the shear modulus and the bulk modulus, respectively, with the indices 0 and ∞ denoting their instantaneous ($t' = 0$) and the ultimate ($t' = \infty$) values while i is the number of the Prony term running from 1 to n. The factors α_i stand for the relative modulus with $\alpha_i^G = G_i/G_0$ and $\alpha_i^K = K_i/K_0$, the symbols τ_i denote the relaxation time of each Prony term and t' is the effective time computed as $t' = t \cdot A(T, T_{Ref})$ with t being the actual time and A the shift factor, which depends on the actual temperature T and the reference temperature T_{Ref}. Finally, C_1 and C_2 are the coefficients of the William-Landel-Ferry (WLF) shift function. The concept of thermorheological simplicity allows the substitution of temperature by time as it assumes the visco-elastic relaxation of the deviatoric (G) and the volumetric (K) stresses, eq. (1), at high temperatures being identical to that at low temperatures if the time is properly scaled such as by the WLF shift function, eq. (2). This way, the complete relaxation curve, which may easily span more than a dozen decades in time, can be obtained by combining several measurements each spanning a few decades only but being performed at different temperatures ([1]). The resulting so-called master curves then allow predicting the relaxation times at any temperature as well as the mechanical response under prescribed temperature histories. Figure 3 shows the master curves obtained for the Young's modulus (E) and the Poisson's ratio of the four organic materials comprised in the BGA panel. Table 2 lists the coefficients of the Prony series and the WLF shift function constants as they are included in the material models for two of these materials. A degenerated WLF shift function is used for the mold compound as approximation to $\log_{10}(A) = 0.1265 (T - T_{Ref})$, which was the best fit to the shift function but is not readily supported by the FEM code [4].

In addition to visco-elasticity, the chemical shrinkage during the cure of the mold compound is omitted in the data set listed in table 1. In reality, however, the magnitude in chemical shrinkage often is the key parameter for choosing a particular material. For example, if a new product shows too much crying type panel warpage after post mold cure, replacing the mold compound by a material with very similar elastic properties but larger chemical shrinkage has often solved the problem without generating new ones. Unfortunately, accounting for this

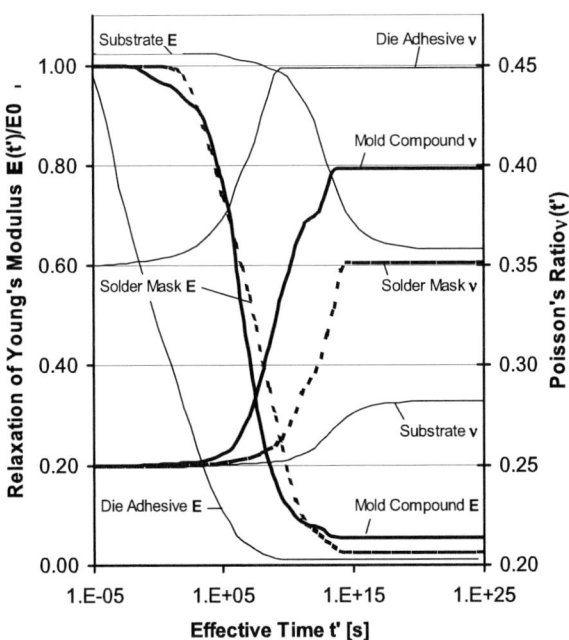

Fig. 3: Master curves of the organic materials Young's modulus has been normalized by the instantaneous modulus $E_0 = E(t' = 0)$

shrinkage is not possible in simulations based on elastic material models only. The extra compressive strain introduced would permanently exaggerate the deformation of

i	Mold Compound			Solder Mask		
	τ_i	α_i^G	α_i^K	τ_i	α_i^G	α_i^K
1	1e-01s	0.0193	0.0176	3.7e01 s	0.0060	0.0058
2	1e00 s	0.0153	0.0139	3.7e02 s	0.0596	0.0584
3	1e01 s	0.0060	0.0054	3.7e03 s	0.0865	0.0847
4	1e02 s	0.0349	0.0317	3.7e04 s	0.0984	0.0963
5	1e04 s	0.0951	0.0865	3.7e05 s	0.0895	0.0876
6	1e05 s	0.0625	0.0568	3.7e06 s	0.0924	0.0905
7	1e06 s	0.2137	0.1942	3.7e07 s	0.1491	0.1460
8	1e07 s	0.1984	0.1804	3.7e08 s	0.0984	0.0963
9	1e08 s	0.1454	0.1322	3.7e09 s	0.1431	0.1401
10	1e09 s	0.0728	0.0622	3.7e10 s	0.0835	0.0817
11	1e10 s	0.0423	0.0385	3.7e11 s	0.0179	0.0175
12	1e11 s	0.0213	0.0193	3.7e12 s	0.0239	0.0234
13	1e13 s	0.0237	0.0216	3.7e13 s	0.0268	0.0263

Mold Compound			Solder Mask		
C_1	C_2	T_{Ref}	C_1	C_2	T_{Ref}
1.265e9	1.0e10	25 °C	36.03	165.1	23 °C

Table 2: Coefficients of Prony series and WLF shift functions for mold compound and solder mask

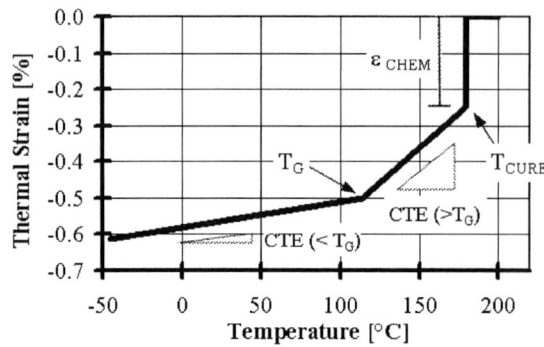

Fig. 4: Modeling the chemical shrinkage of the mold material by an extra strain at the curing temperature

the mold cap side so that the panel warpage could not be predicted realistically. Hence, chemical shrinkage can only be added to material models that consider visco-elasticity as well.

In this study, the chemical shrinkage has been added to the visco-elastic model of the mold compound material as effective value. The magnitude of 0.25% is about two third the magnitude obtained by the test commonly used in industry for quantifying the shrinkage. In those tests, the relative difference between the volumes of the molding tool cavity and the fully cured mold cap at room temperature is called chemical shrinkage. Actually, this is not correct as this difference is even dominated by thermal expansion effects [6]. Nevertheless, this combined value called chemical shrinkage is quite generally preferred over the academically exact set of several individual values because it is easy to obtain, it provides a single measure for practical material assessments and it has shown sufficient stability (i.e., constancy) across the molding and curing process windows. Of course, even this combined value cannot directly be included in the material model as long as the FEM simulation does not explicitly include the curing process itself, during which the chemical shrinkage occurs simultaneously to the growth in stiffness of the mold compound. In order to avoid the extra simulation effort, the effective value has been determined in a separate calibration step. Being added to the thermal strain at curing temperature T_{CURE} as shown in figure 4, the effect of chemical shrinkage is modeled accurately although the stiffness is already at the level of the fully cured material.

C) Load Model

The stress-free state is assumed at curing temperature. During the free cool down to room temperature, panel warpage occurs. The cool down follows an exponential decay function with 90% of the thermal change passed in the first 4 minutes after the start. This process is approximated by 9 load steps, in which the temperature is ramped down by 20 K or less. The temperature / time history is needed in the visco-elastic simulations only while time has not meaning in simulations just based on elastic / plastic models for the organic materials.

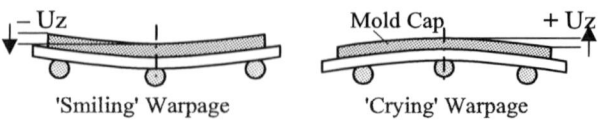

Fig. 5: Warpage criterion Uz and sign convention

5. Panel Warpage Measurements

Experimental data were available for BGA panels of several 512Mb memory products. The package size is 10.0x10.5 mm² for all these products. Consequently, the arrangement of the dies on the substrate panel was also identical. In each array, the same number of dies was placed along length and width of the panel. Despite the identical functionality, the size of the dies was different because 110 nm, 80 nm, or 70 nm technologies have been applied. In addition, the thickness of the dies was varied for this warpage study.

Panel warpage is quantified as the difference between highest and lowest point of the mold cap top surface within each substrate array (Fig. 5). A negative magnitude means the mold cap array forms a valley. So, its center is lower than the edges, which also is called 'smiling' warpage. Consequently, a positive magnitude means 'crying' warpage as the mold cap array forms a hill indicating that the thermal shrinkage of the substrate side is dominant. Figure 6 plots the warpage measurement results obtained at room temperature. All three series show the trend explained in section 3: Increasing the die thickness changes the warpage towards (stronger) crying, i.e., towards positive Uz values. A second trend has become clear as well: A decrease in die size leads to more smiling warpage, i.e., to more negative Uz. Obviously, smaller dies support the mold cap side. Thirdly, Fig. 6 indicates by means of the two curves for the 70 nm technology that the measurement accuracy is on the order of ±0.1 mm.

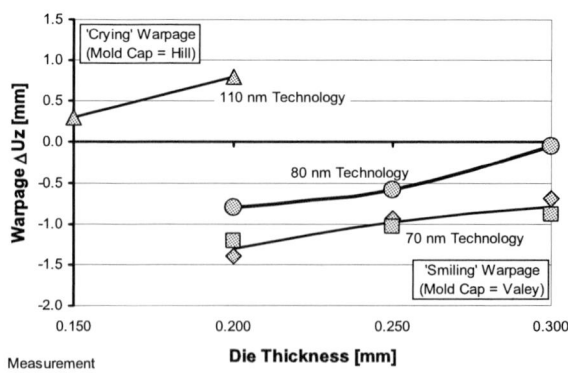

Fig. 6: Measurement results
Mold cap array warpage of 512M substrate panels with the same number and arrangement of the dies but manufactured at different technology nodes

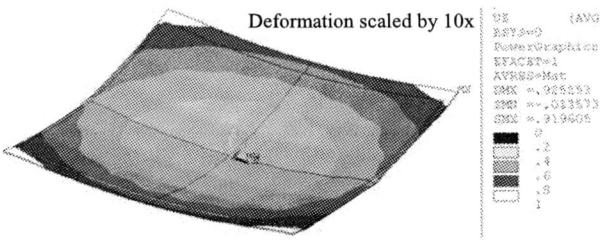

Fig. 7: Uz contour plot showing the warpage of one mold cap array after symmetric expansion from the one quarter meshed (80 nm technology, 200 μm dies)

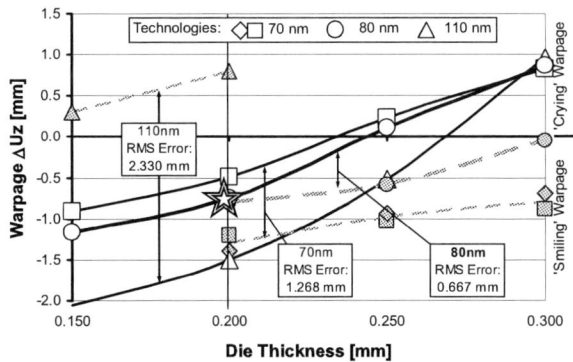

Fig. 8: Results of the elastic/plastic simulation (solid lines) as compared to the measurement results (dashed lines) – RMS … root mean square

6. Results of Elastic/Plastic Simulations

In warpage estimation, the deformation result is directly obtained as primary solution of the FEM equation. Figure 7 shows the Uz contour plot for the case of 80 nm technology and 200 μm thick dies. The mold cap forms a valley, which means the warpage is of 'smiling' type. The warpage magnitude, i.e., the maximum difference in Uz across the mold cap, is about –0.8 mm.

In figure 8, the results of all elastic/plastic simulations are plotted together with the warpage measurement readings. In the case of 80 nm technology, the simulation prediction matches the experimental reading very closely for the typical die thickness of 200 μm (as indicated by the star). Increasing the die thickness, the simulation result also predicts the warpage values to grow. This way, the qualitative trend, which was explained in section 3 and that was found in the measurements (section 5), can also be observed in the simulation results for all three technologies (70 nm, 80 nm, and 110 nm): The thicker the die the stronger the warpage is changed towards 'crying'.

Aside this fundamental trend, however, large discrepancies exist between the simulation and the experimental results. First, there is a quantitative difference with respect to this trend already. In the case of 80 nm technology, the average growth in warpage is simulated as 13.9 μm per 1.0 μm increase in die thickness while the measured data rather show 7.5 μm per 1.0 μm. In the case of 70 nm technology, these average slope values are 11.9 μm / 1.0 μm (simulation) vs. 5.2 μm / 1.0 μm (experiment) while they are 20.0 μm / 1.0 μm (simulation) vs. 10.0 μm / 1.0 μm (experiment) for 110 nm technology. Hence, the simulation exaggerates the dependency of warpage on the die thickness by a factor of two.

Second, the simulation results are misleading even with respect to trends. A typical die thickness is 200 μm. For dies of this thickness, simulation and experiment likewise result in 'smiling' mold cap warpage of about -0.8 mm in the case of 80 nm technology. The results for 70 nm technology, however, show simulated warpage magnitudes of -0.5 mm while the measured values are -1.4 mm …-1.2 mm. The 110 nm results are even more different. The simulation predicts -1.5 mm, which is strongly 'smiling', while the real samples clearly show

'crying' mold cap warpage of +0.8 mm in magnitude. That means, the simulations wrongly estimates the second trend found in the experiments (sec. 5) by pointing into the opposite direction. Unfortunately, this trend reversal is not limited to 200 μm thick dies but applies to almost any die thicknesses studied.

7. Assessment of the Result Mismatch

The extrapolation of the experimental and simulation results, as shown in figure 9, reveals that a specific die thickness can be found for each technology, at which the warpage results match. In the case of 70 nm technology, this point is at about 75 μm die thickness while the 80 nm results match at 200 μm and the 110 nm results at about 310 μm die thickness. Thus, there is a systematic behind the differences discussed so far: i) The die thickness, at which the results match, monotonously grows with the die size. ii) Outside this point, the simulation results exaggerate the dependency of warpage on the die thickness.

These two findings suggest that the models applied to the simulations are too rigid. They overreact to any small change in the system. The real samples show a softer and a more forgiving behavior.

In order to validate this root cause, the effect of some geometric and elastic material parameters was assessed. The substrate thickness was varied by ±20 μm; the Young's moduli and the CTE of substrate and mold compound were all scaled by factors 1.2 and 0.8 each. In addition, the thickness of the die adhesive film was varied between 30 μm and 70 mm. All these changes as well as the three different constraint conditions (sect. 4C) intent to check for possible deviations of the real situation from the nominal input data due to the range specified as tolerance band or process window for those parameters.

Taking these geometric and elastic parameters as additional factors of a DoE plan, the full factorial scheme would end up with $3^8 \cdot 4 = 26,224$ legs. The simulation effort would be as big as three full years if the effective run time would just be one hour per leg (where 'effective run time' is the 'actual run time' divided by the 'number of

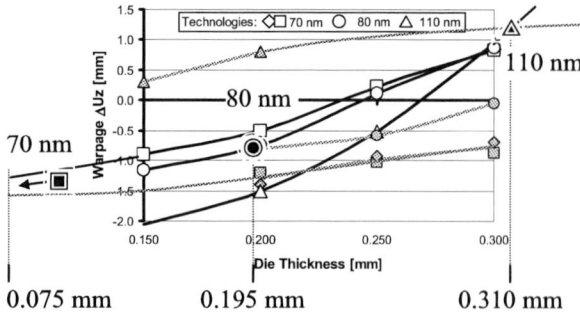

Fig. 9: Extrapolation of the results to find the die thicknesses where simulations and experiments match

jobs executed in parallel'). Reducing the scheme to 'one factor at the time', 648 legs would still remain when the additional factors would be applied to each of 12 cases (3 technologies, 4 die thicknesses) of the basic set. Even simulation studies can hardly handle such a large number of legs – and still, the results would be limited to one stripe per factor within the design space. Only sophisticated DoE schemes like the Latin hypercube method [7] or the D-optimal plan [8, 9] allow covering the design space of this sensitivity analysis fairly comprehensively with a feasible number of legs. In this study, a D-optimal plan with 69 legs has been applied.

The additional factors were found to have the following effects: Warpage is moved in the direction of (more) positive magnitudes, i.e., towards (stronger) 'crying' mold cap warpage, when

- the thickness, the CTE or the Young's modulus of the substrate is increased, when
- the CTE or the Young's modulus of the mold compound is reduced, or when
- the thickness of the die adhesive film is increased.

The warpage is moved into the opposite direction when the values of these factors are changed the other way round. The effects of all these factors are widely independent of each other. Thus, the total effect of any combination of these factors is just a superposition of the individual effects.

All these factors have only moved the warpage curves parallel, i.e., by the same increment for all die thicknesses within a particular technology. In other words, none of these factors alone and no combination of them has been capable of modifying the slope of the result plot (that was seen in fig. 8). The predicted dependency of warpage on the die thickness remained at its exaggerated level. That means, neither geometric tolerances nor any elastic material parameter can change the rigidity of the system – not in the simulation and not in reality, too.

In addition, the DoE results showed the constraint boundary conditions to have no significant effect at all. Obviously, the slits in the substrate decouple the three arrays from each other close to perfectly.

Hence, the DoE study validated the root cause: Elastic / plastic material data alone is not sufficient. It does not allow achieving correct warpage results throughout the usual range of die thicknesses. The models applied in the simulation need to account for an additional phenomenon, which reduces their rigidity and leads to more compliance.

8. Results of Visco-elastic Simulation

Starting with the 80 nm technology, the results of the visco-elastic simulations have been compared to the measurement readings of mold cap warpage. As seen in figure 10, the experimental results could be matched very closely. In the case of die thicknesses 200 μm and 250 μm, the simulation result deviated from the measurement by just -0.023 mm and 0.044 mm, respectively. Although the difference was larger for the 300 μm thick die (-0.392 mm), the root-mean-square (RMS) error of the simulation results sank to 0.279 mm from 0.667 mm in the case of time independent simulations.

The main goal of reducing the rigidity of the FEM models has been reached. The average slope of the computed warpage vs. die thickness curve is now just 6.3 μm per 1.0 μm die thickness instead of 13.9 μm per 1.0 μm before. That means, the fundamental shortcoming of the elastic / plastic simulations has been overcome.

Widening the field to 110 nm and 70 nm technologies, figure 10 shows the simulation results to match the experimental readings very closely as well. The trend reversal, which was discussed in section 6 for 200 μm thick dies, has completely disappeared. Moreover, the quantitative accuracy has reached a very high level. The remaining RMS error of the simulation results is just 0.061 mm and 0.181 mm for large dies (110 nm technology) and small dies (70 nm), respectively. In total across the study, the RMS error is 0.19 mm, which is quite comparable to the measurement accuracy that was found to be ±0.1 mm (see section 5), while the RMS error was as big as 1.5 mm in the elastic / plastic simulation study.

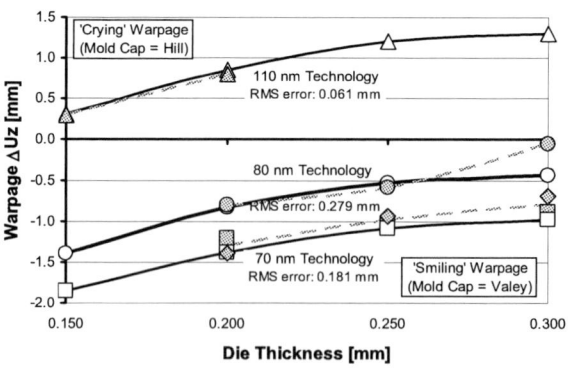

Fig. 10: Results of visco-elastic simulation (solid lines) matching the measurement results (dashed lines) very closely in all three technologies

9. Efficiency of Visco-elastic Simulations

Expanding the models of the organic materials mold compound, solder mask, die adhesive, and substrate to account for visco-elasticity, the simulation run time is increased by a factor of 40. Instead of directly solving the system of FEM equations in one step, many time steps with numerous iterations were needed. Thus, a second sensitivity analysis was carried out. It targeted on the improvement of simulation efficiency by identifying many small features like interconnect lines on the substrate, die adhesive squeeze-outs around the die, small radii at the corners of the die adhesive film and at some slits in the substrate panel that do not have a significant effect on the warpage. In addition, the geometric nonlinearities (i.e., large deformations) have also been found to have a negligible effect only. Furthermore, thin films like solder mask and die attach were combined with parts of the adjacent parts of the construction (which are substrate and die, respectively) into multilayer elements. Finally, plasticity of the remaining copper films was also found being ballast only. Omitting all these superfluous details, the number of elements needed in the FEM mesh (shown in Fig. 2) could be reduced to one tenth of its initial value while the warpage results obtained changed by 0.002 mm only. Compared to the average RMS error of about 0.2 mm, this additional deviation is insignificant indeed. At the same time, however, the simulation speed increased by more than factor 200 so that the visco-elastic simulation runs finally ended up being even faster (five times faster) than the elastic/plastic simulations done before. That means, the revision by rigorous de-featuring has really boosted the simulation efficiency: Accurate results are now achieved in a much shorter time.

Despite the substantial gain in simulations efficiency, the total effort remains larger when visco-elastic effects are added to the material models since the determination of all the extra material parameters is very time consuming as well. Therefore, one more sensitivity study was conducted to critically assess the contribution of each organic material on the total warpage magnitude.

Table 3 compiles the total RMS errors obtained of seven different sets of organic material models applied to four die thicknesses times three die sizes each. The result clearly shows that chemical shrink of mold compound and visco-elasticity of mold compound and solder mask are very important. Neglecting them, the RMS error is increased by factor 1.8 to 3.8. On the other hand, the RMS error increases by maximum 10% only if visco-elasticity of die adhesive and substrate materials is neglected. This means, the time-consuming sample preparation and determination of visco-elastic parameters can be omitted for these two materials.

The results of this sensitivity study is very reasonable. The mold cap is the one of the key-elements determining the warpage behavior (sect. 3). Chemical shrinkage and visco-elasticity of its material must undoubtedly have a great impact. In contrast to this, the solder mask may appear being of minor importance. However, it should be

Case No.	Visco-Elasticity				ε_{CHEM}	Total RMS		Remark
	Sub	SM	DA	MC	MC	[mm]	[%]	
1	X	X	X	X	X	0.188	100	Ref.
2	X	X		X	X	0.194	103	
3		X	X	X	X	0.202	107	
4		X		X	X	0.207	110	Opt.
5				X	X	0.325	173	
6	X	X	X	X		0.687	365	
7		X		X		0.710	378	

Table 3: Root Mean Square (RMS) error of simulation warpage results obtained across the samples of 3 technologies (70 nm, 80 nm, 110 nm) with 4 die thicknesses each (150 µm … 300 µm) depending on the features included in the models of the organic materials (BM… benchmark; Opt…Effort/Accuracy optimum)

mentioned that this rather thin film has the maximum distance to the neutral plane as it is placed at the surface opposite to the top of the mold cap. Based on fundamental mechanics theories, the effect of a film on the total stiffness of the system scales with the square of its distance to the neutral plane. In addition, the loading effect also grows with this distance as the leverage increases. Thus, the fairly large effect of the solder mask is justified as well.

The die adhesive film is located just in the vicinity of the neutral plane. So, it can not affect the warpage much. The epoxy of the laminate material used as substrate in this study has a glass transition temperature T_G of about 190 °C. According to the theory of polymer mechanics [10], visco-elastic effects that are relevant to technical processes are more or less restricted to a temperature range of 20…30 K around the glass transition temperature. Outside this range, the time constants of the transient processes are either too large (like years and more) or too small (like microseconds). Thus, elastic models can model the behavior of polymers very precisely for all temperatures outside this range around T_G. In the warpage situation under investigation, substantial stress is built up in the panel only after cooling to 150 °C and below, which means, after leaving the temperature range, in which the resin of the substrate could show significant visco-elastic effects. The mold compound, however, has a T_G of 114 °C. So, its visco-elasticity is very well relevant.

10. Summary and Conclusion

Substrate panel warpage is an important manufacturing risk and a yield detractor in the fabrication process of BGA components. It occurs after post mold cure. Its magnitude is determined by details of geometry and material within this substrate panel system. FEM simulation analysis is involved to pre-optimize the design. In this study, the result accuracy of two different FEM approaches has been assessed, by comparing to experimental warpage results.

The first FEM approach was based on 3-D elastic / plastic models. It was able to follow the quality of the most fundamental trend of the experimental results only. The warpage magnitude increased when the mold cap clearance (i.e., the mold cap thickness above the die) was reduced by an increased die thickness. However, this FEM approach already failed matching the quantity of this trend as it exaggerated the dependency of the warpage magnitude on the die thickness by a factor of two. Moreover, this elastic / plastic approach has lead to a completely wrong prediction towards a second basic trend, which is the effect of the die size (i.e., the area covered by silicon under the mold cap) on the warpage. For 200 µm thick dies, the elastic / plastic simulations misestimated a change towards positive warpage magnitudes (i.e., towards 'crying' warpage) when the die area is reduced while the experiments clearly showed the opposite trend being true: Warpage changes towards 'smiling' when the die size is reduced.

A comprehensive sensitivity analysis applying D-optimal DoE schemes and response surface evaluation methods has clearly made sure, the shortcomings of the elastic / plastic FEM approach are neither caused by any tolerances in geometric or elastic material parameters nor by the constraint boundary conditions applied. These details are able to shift the result curves but cannot change the slope because they do not remove the root cause of the result mismatch, which is the excessive rigidity the FEM simulations showed.

In the second FEM approach, visco-elasticity has been added to the models of all organic material and chemical shrink to the mold compound. Based on this advanced set of models, the shortcomings of the first approach have been overcome completely. The RMS error of the simulation results, which was as big as 1.5 mm when sticking to approach one, declined to 0.2 mm across the full study, which included three different die sizes (areas) with four different die thicknesses each. This remaining RMS is of the same order as the experimental uncertainties, which were found to be about ±0.1 mm in size.

A further sensitivity analysis has focused on the simulation efficiency. It allowed identifying the geometric details, which only had insignificant effect on the warpage result but substantially increased the computational time. After removing these superfluous details from the models, the run time of visco-elastic simulations even was factor five smaller than that of the initial simulations based on elastic / plastic models while the result accuracy was practically kept constant.

The third sensitivity analysis finally revealed in addition that mold compound and solder mask are the only parts of the substrate panel system, whose visco-elastic behaviors significantly influence the magnitude of the mold cap warpage. Neglecting the visco-elastic data of substrate and die adhesive allowed cutting the effort of material characterization into half while the simulation result accuracy was changed by less than 10% only. Hence, the optimum between effort and simulation result accuracy was found.

The set of models presented and discussed here provides for effective and accurate FEM estimation of substrate panel warpage and has successfully been introduced to industrial practice.

Acknowledgments

The authors would like to thank Dr. Claudia Luhmann for providing the visco-elastic material data.

References

1. Kiasat, M.S. *et al*, Time and Temperature dependent Thermo-Mechanical Characterization and Modeling of a Packaging Molding Compound, Proc 2nd *Thermal and Mechanical Simulation and Experiments in Microelectronics and Microsystems Conf. (EuroSimE 2001)*, Paris, France, April 2001, pp. 155-62
2. Bressers, H.J.L., van-Driel, W.D., Jansen, K.M.B., Ernst, L.J., Zhang, G.Q., From chemical building blocks of polymers to microelectronics reliability, *Proc. 5th Thermal and Mechanical Simulation and Experiments in Microelectronics and Microsystems Conf. (EuroSimE 2004)*, Brussels, Belgium, May 2004, pp.621-5
3. Jansen, K.M.B., Ernst, L.J., Bressers, H.J.L., Effect of Chemistry on Viscoelastic Properties of Moulding Compounds, *Proc. 7th Thermal, Mechanical and Multi-Physics Simulation and Experiments in Microelectronics and Microsystems Conf. (EuroSimE 2006)*, Como, Italy, April 2006, pp. 605-10
4. ANSYS Inc., ANSYS Release 10.0, Finite Element Analyses Software, Canonsburg 2006
5. MatWeb, Material property data base; http://www.matweb .com
6. Saraswat, M.K., Jansen, K.M.B., Ernst, L.J., Cure shrinkage and bulk modulus determination for moulding compounds, *Proc. 1st Electronic System-Integration Technology Conf.*, Dresden, Germany Sept. 2006, pp. 782-7
7. Wyss, G.D., Jorgensen, K.H., Latin Hypercube Sampling, http://www.prod.sandia.gov/cgi-bin/techlib/ access- control.pl/1998/980210.pdf, pp. 3-14
8. Montgomery, D.C., <u>Designs and Analysis of experiments</u>, John Wiley & Sons Inc. (Hoboken, NJ, 2005), pp. 136, 441
9. Wilhelm Kleppmann, <u>Taschenbuch Versuchsplanung</u>, Hanser Verlag (München, 2003), p. 213
10. Schwarzl, F.R., <u>Polymermechanik – Struktur und mechanisches Verhalten von Polymeren</u>, Springer Verlag (Berlin Heidelberg New York, 1990), p. 88

Parametric Study of Fracture Properties in Polycrystalline MEMS

Fabrizio Cacchione*, Alberto Corigliano*, Sarah Zerbini**

* Department of Structural Engineering. Politecnico di Milano. Piazza L. da Vinci, 32. 20133 Milano, Italy
** MEMS Product Division. STMicroelectronics, via Tolomeo 1. 20010 Cornaredo, (Milano), Italy

Abstract

A parametric numerical study aimed at understanding the influence of the average grain size on the mechanical behaviour of polysilicon films was addressed in this paper. A 2D geometrical model of the polycrystal coupled with a Finite Element (FE) procedure was employed. To simulate inter-granular and trans-granular crack propagation a cohesive crack model was used. The results were finally analyzed using the Weibull weakest link approach.

1. Introduction

The majority of MEMS used in research and market applications are made of polycrystalline materials. Among them, the most popular and used one is polysilicon, both for its well established technology, that comes from the IC world, and for its outstanding mechanical properties.

After almost two decades of researches from many different scientific groups worldwide [1-5], if one tries to collect all the data regarding mechanical properties of polysilicon at the microscale, he still will notice that the results are very dispersed when looking for instance to Young's modulus or fracture strength.

Many round robin tests [1, 2] were carried out during the years in order to understand the source of this scatter. A partial conclusion is that there are two main reasons which can justify the large experimental scatter: the polysilicon production process (i.e. the entire sequence of growth, masking, etching and oxidation; process parameters...) and the way in which each laboratory performs the mechanical characterization (i.e. the equipment, the size and shape of the specimen, the environmental conditions...).

A complete characterization of the material must take into account the above aspects and the role they play on its final mechanical properties.

This paper presents a numerical study concerning the influence of the average grain size on the fracture properties of polysilicon, here defined as the radius of a circle having area equal to the average area of the grains.

The paper is organized as follows: in Section 2 the main ingredients of the geometrical representation of a numerical polycristal are discussed, together with the material models used in FE simulations. Section 3 is dedicated to a brief description of the adopted step by step numerical strategy. Sections 4 and 5 contain a description of the performed numerical experiments and a discussion on the main results obtained.

2. Geometrical representation and material models

The numerical study was performed in a 2D Finite Element framework. Both the topological description of the polycrystal and its spatial discretization into finite elements are needed (Fig. 1).

The bi-dimensionality of the model can be mainly justified by the following hypotheses:

- a large number of deposition processes gives rise to a material with a columnar structure;
- the thickness of the deposited layer is small if compared to the in-plane dimensions of the structure;
- in a simple tensile test, the loads act only in the plane of the wafer.

2.1 Geometrical representation

A Voronoi tessellation technique (see e.g. [6]) was used to divide the specimen or structure into an aggregate of grains. Every grain represents a silicon crystallite. Various silicon crystallites are connected by a grain boundary. In this study the grain boundary was modeled as a zero-thickness line, since usually polysilicon does not contain a big amount of amorphous phase between two grains.

The specimen was then spatially discretized using 6-noded triangular FE. The size of the elements was designed in order to avoid size-depending solutions (Fig. 1).

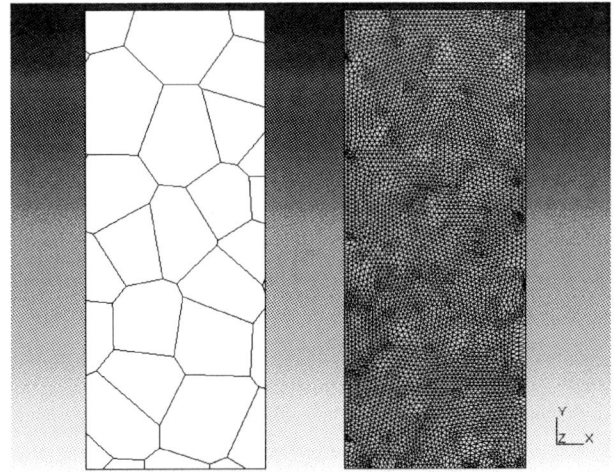

Figure 1. A tesseled specimen and its FE mesh

When cohesive crack models are used, the side length of an element involved in the fracture process must be much smaller than the cohesive zone radius, which can be computed by means of the following equation [7]:

$$R_c \cong \frac{\pi}{2}\left(\frac{K_{IC}}{t_{max}}\right)^2 = \frac{\pi}{2}\left(\frac{1.1}{2000}\right)^2 \frac{MPa^2 \, m}{MPa^2} = 0.47 \, \mu m \qquad (1)$$

where K_{Ic} is the fracture toughness in mode I and t_{max} is the maximum stress admissible in the cohesive law.

On the basis of Eq. (1) it was decided to limit the side length of FE in the spatial discretization to 0.1 μm.

2.2 Constitutive models

Every crystalline grain was considered to be linear elastic up to rupture. Since the crystalline structure of silicon is cubic with centered faces, the elastic stiffness matrix in the local frame has the structure given in (Eq. 2). The elastic stiffness values were found in the literature [8] and were used for the parametric study discussed in Sections 4 and 5.

$$\mathbf{S}^{loc} = \begin{bmatrix} 165.7 & 63.9 & 63.9 & 0 & 0 & 0 \\ 63.9 & 165.7 & 63.9 & 0 & 0 & 0 \\ 63.9 & 63.9 & 165.7 & 0 & 0 & 0 \\ 0 & 0 & 0 & 79.6 & 0 & 0 \\ 0 & 0 & 0 & 0 & 79.6 & 0 \\ 0 & 0 & 0 & 0 & 0 & 79.6 \end{bmatrix} GPa \quad (2)$$

Fracture initiation was governed by a stress norm involving the traction and shear forces acting on the edge of a FE. A new interface was dynamically inserted into the model whenever the stress norm exceeds a pre-determined value on at least one edge (a common border of two adjacent FE). The interface behavior was governed by a Camacho-Ortiz model [7] (Fig. 2). The two parameters of the model for a mode-I fracture are the energy release rate and the maximum traction that the material can sustain once the fracture process is activated. In this work the first interface parameter was chosen from [9], while the second one was chosen considering that the typical critical stress for polysilicon is around 1.5 – 2.0 GPa [1-5].

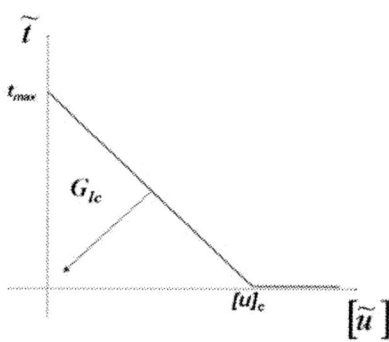

Figure 2. Mode I Chamacho-Ortiz model

3. Static – Relaxed dynamic algorithm

The algorithm used for the solution of the fracture problem belongs to the category of dynamic analyses with static-relaxation (see e.g. [10]). The first phase of the analysis is accomplished by a sparse direct frontal solver that solves the linear static problem assigning a fixed value of displacement on the constrained boundary. A routine computes the value of the prescribed displacement that causes the first crack initiation and stores the value of the corresponding reaction force. Once the fracture starts propagating, the problem becomes non linear and the code switches from the static solver to a statically-relaxed dynamic one. This algorithm is based on the central difference method for the numerical time integration of the equations of motion. The equivalent static solution is obtained by simulating a dynamic event and properly manipulating the values of the mass and damping matrices of the system in order to converge to the static equilibrium as fast as possible.

The choice of the dynamic relaxation algorithm instead of a static solver was motivated by the brittle behavior of polysilicon. In a typical force versus displacement curve of a pure tensile specimen (Fig. 3), it is possible to note a very sharp decrease in the response of the system after the peak. It is well known that a static solver with an iterative Newton-Raphson procedure converges with difficulty to the solution in a problem of this kind, thus justifying the choice of a dynamic non iterative solver.

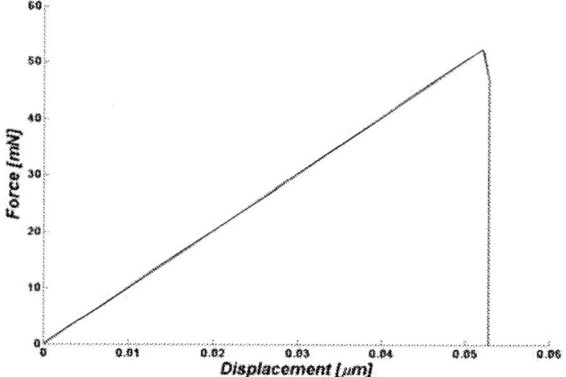

Figure 3. A typical force vs. displacement plot

For every assigned displacement, the algorithm looks for a static solution. Once converged to a steady-state, fracture activation is checked on every FE edge. If the norm of the stress is below the fixed threshold on every edge, the solution computed is stored and a new value of displacement is imposed. Otherwise new cracks are opened and a new static solution is sought. The algorithm goes on incrementing the applied displacements until the reaction forces approaches zero. A flow-chart of the used algorithm is shown in Fig. 4.

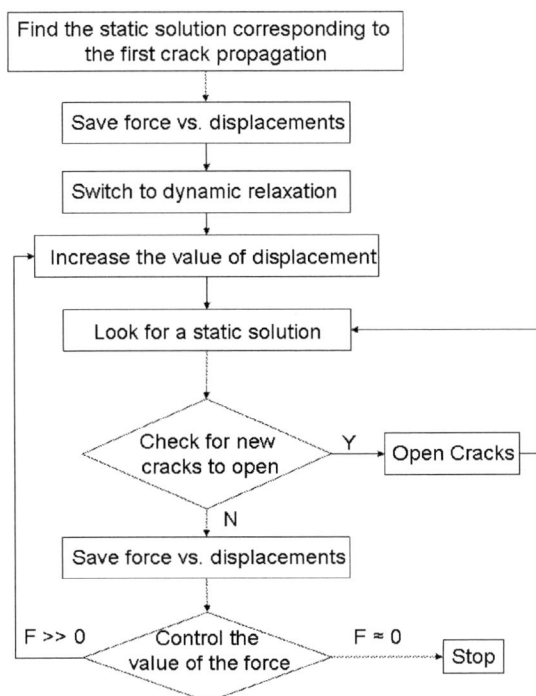

Figure 4. Flow-chart of the algorithm used

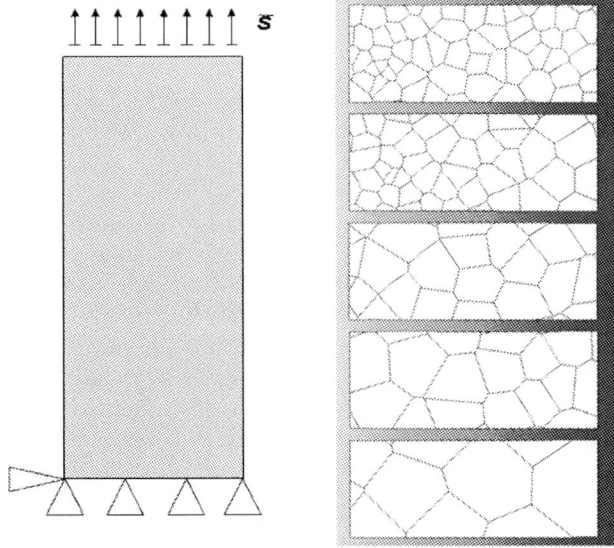

Figure 5. Static scheme of the specimen and examples of specimens with the five different grain sizes

4. Numerical tensile tests

Since polysilicon is a very brittle material, its strength properties are often evaluated in the framework of Weibull statistics [5, 11]. In order to obtain a statistically meaningful number of numerically simulated tensile tests, it is necessary to simulate a large number of nominally identical specimens for every assigned grain size. The results can then be considered as coming from different tests performed on a large number of identical structures and interpreted by means of the Weibull approach.

Numerical tensile tests were realized on a specimen with a 5 by 2 μm rectangular shape (Fig. 5) in plane stress conditions. The influence of the grain size was investigated on three different categories of polysilicon:

- *I.* with grains and grain boundaries having the same strength properties;
- *II.* with grains weaker than grains boundaries;
- *III.* with grain boundaries weaker than grains;

For every set of materials *I.*, *II.*, *III.*, five subsets with different average grain sizes was created. The grain sizes investigated were 0.2 / 0.3 / 0.4 / 0.55 / 0.75 μm (Fig. 5). Each subsets consisted of 50 different specimens. Each specimen in a subset differing from the other for the disposition and shape of the grains and for a randomly generated orientation of every grain.

5. Results and discussion

From thee force versus displacement plot obtained in every simulation, it was possible to compute the maximum nominal stress that acted during the test. This last can be interpreted as the nominal strength obtained in the virtual test. All the data of a specific subset were then used to compute the experimental cumulative distribution function. The data were then fitted using the Weibull distribution. The parameters obtained from the fitting procedure were Weibull's module m, that can be interpreted as a measure of the spread of the data, and the threshold value σ_0, i.e. the value of stress that causes the failure for the 63% of the entire set of structures.

5.1 Threshold stress

In Table 1 the nominal strength measured on the numerical tensile tests are shown at varying average grain size and fracture activation properties of polysilicon.

From table 1 it is possible to notice that for every material the value of the threshold stress remains almost constant (with variations of less than 1% of the mean value going from one grain size to another) showing an independence from the grain size. Moreover, the results of the simulations confirm the brittleness of the material. In fact the values of the threshold stress are close to the opening stress of the weakest interface. This result is in agreement with the use of Weibull theory for the evaluation of material's strength, also called the *weakest link approach*.

Another interesting remark can be done concerning the ratio of the average value of σ_0 over $t_{max\ lower}$, where $t_{max\ lower}$ is the lower value of interface resistance in the model. For the material *I.* (grain and grain boundary with the same strength) this value is 0.9333; for the material *II.* with the grains weaker than the grain boundaries is

0.9485 and for the last one *III.* is 0.9749. Interpreting this ratio as a performance index of the material, we should conclude that the material with weak grain boundaries appears to be the best one. A partial, and by no means exhaustive, explanation of this result can be the following. When the grains are weaker than the grain boundaries, the crack initiates from an edge located inside the grain; vice-versa, when the grain boundary is weaker than the grain, the first opened interface is to be sought onto the grain boundary. Since the total number of edges on the grain boundaries is small compared to the total number of edges into the grains, the probability to find a broken interface, and thus a crack, into the grain is bigger than the one of finding it onto the grain boundary. It turns out that it will be easier for a material with grains weaker than grain boundaries to break when the mean stress is close to the critical value for crack propagation.

σ_0	Interface resistance [GPa]		
Average Grain Size [μm]	Mat I : 2.0 Grain 2.0 Grain B.	Mat II : 1.5 Grain 2.0 Grain B	Mat III : 2.0 Grain 1.5 Grain B
250	1.8663	1.4183	1.4645
300	1.8652	1.4183	1.4581
400	1.8567	1.4273	1.4545
550	1.8644	1.4269	1.4670
750	1.8801	1.4246	1.4674

Table 1. Numerically obtained threshold stress at varying average grain size and kind of material

5.2 Weibull modulus

Table 2 shows the variation of the Weibull modulus *m* with respect to the grain size. In this case the link between the grain size and the modulus appears to be evident. For every material, as the grains become smaller, there is a trend toward an increase of modulus *m*, i.e. toward a reduction of the spread of the interval in which there is a sharp variation in the fracture probability. This effect is intuitively understandable. In a specimen with a small number of grains the overall mechanical behavior is strongly influenced by the shape and orientation of its grains. As the number of grains increases, the influence of a single grain onto the global behavior becomes less important and, if the number of grains is very large, the global response of the system should always converge to a mean value.

The second aspect that is worth noticing is the comparison of the modulus for the three material models analyzed. The material with the highest values of the modulus *m* is the one with trans-granular privileged crack propagation (*II.*), while the material with an inter-granular crack path (*III.*) appears to be the *less* reliable, because its modulus is always smaller than the one obtained in the

previous case. Finally, the material *I.*, without any privileged path seems to exhibit an intermediate behavior.

m	Interface resistance [GPa]		
Average Grain Size [μm]	Mat I : 2.0 Grain 2.0 Grain B.	Mat II : 1.5 Grain 2.0 Grain B	Mat III : 2.0 Grain 1.5 Grain B
250	72.0554	76.3107	54.1231
300	61.4907	74.8009	53.2615
400	62.2375	74.0283	35.2616
550	53.0305	63.5268	31.1036
750	51.5074	50.1098	36.1985

Table 2. Numerically obtained Weibull modulus at varying average grain size and kind of material.

6. Conclusions

The main purpose of this paper was to discuss the effect of the grain size on the global mechanical behavior of polysilicon films through the results of numerical simulations.

To reach this goal, a 2D FE procedure was developed which makes use of Voronoi tessellation, statically–relaxed dynamic algorithms and cohesive softening interfaces.

The simulations were carried out on sets of nominally identical specimens varying both the grain size and the material fracture properties.

The results show that the weakest link approach can be successfully used for this kind of materials. The numerical data were interpreted making use of the Weibull distribution; it was then possible to study the effect of the granulometry on the threshold stress and the modulus. The first one is not greatly influenced by the grain size, while the second one appears to grow with a reduction in the grain size.

7. Acknowledgments

The contribution of EU NoE Design for Micro & Nano Manufacture (PATENT-DfMM), contract n°: 507255 is gratefully acknowledged.

8. References

1. Sharpe, W.N., Brown, J.S., Johnson, G.C., Knauss W.G., "Round-robin tests of modulus and strength of polysilicon", *Materials Research Society Proceedings*, San Francisco, CA, 1998, vol. 518, pp. 57–65.
2. La Van, D.A., Tsuchiya, T., Coles, G., Knauss, W.G., Chasiotis, I., Read, D., "Cross Comparison of Direct Strength Testing Techniques on Polysilicon Films", In: Muhlstein, C., Brown, S.B. (Eds.), Mechanical Properties of Structural Films, ASTM STP 1413, American Society for Testing and Materials, West Conshohocken, PA, pp. 1-12.

3. Corigliano A., De Masi B., Frangi A., Comi C., Villa A., Marchi M., 2004. "Mechanical characterization of polysilicon through on chip tensile test", *J. of Microelectromechanical Systems*, vol. 13, pp. 200-219.

4. Bagdahn, J., Sharpe W.N., Jadaan, O., 2003. Fracture strength of polysilicon at stress concentrations, *J. of Microelectromechanical systems*, vol. 12, issue 3 , pp. 302-312.

5. Cacchione, F., Corigliano, A., De Masi, B., Riva. C., 2005. "Out of plane vs. in plane flexural behaviour of thin polysilicon films: mechanical characterization and application of the Weibull approach". *Microelectronics Reliability*, vol. 45, pp. 1758-1763.

6. Mullen R.L., Ballarini R., Yin L., Heue A. H. "Monte Carlo simulation of effective elastic constants of polycrystalline thin films". *Acta Materialia,* Vol. 45, (1997), pp. 2247-2255.

7. Camacho G. T., Ortiz M., "Computational Modelling of Impact Damage in Brittle Materials", *Int. J. of Solid Structures*, vol. 33, n. 20-22, (1996), pp. 2899-2938.

8. Brantley. W.A., " Calculated elastic constants for stress problems associated with semiconductor devices", *J. App. Phys.*, vol. 44, pp. 534-535, 1973

9. Chasiotis, I, 2006. Fracture Toughness and subcritical crack growth in polycrystalline silicon, *Journal of Applied Mechanics*, vol.73, pp. 714-722.

10. Oakley, D.R., Knight, N.F. Jr., 1995. Adaptive Dynamic Relaxation algorithm for non-linear hyperelastic structures. Part I. Formulation, *Comp. App. Mech. Eng.*, 126, pp. 67-89.

11. Weibull, W. 1951. "A statistical distribution of wide applicability", *J. Appl. Mech.*, vol. 18, pp. 293-297.

A Modelling Framework for the Reliability of Safety Critical Electronics

Diana Segura Velandia, Paul P. Conway, Antony Wilson, Andrew A. West, David C. Whalley
IeMRC, Innovative Manufacturing Research Centre, Loughborough University
Loughborough, Leicestershire, LE11 3TU, UK
A.A.West@lboro.ac.uk
Tel: +44 (0) 1509 227 677, Fax: +44 (0) 1509 227 648

Abstract

With the application of a methodical modelling architecture, analysis of the effects of both design and manufacturing processes on life cycle expectancy and manufacturing yields can provide a platform for systems level reliability, enabling the prediction of product failure. However, the calculation reliability has been historically centred on physics-of-failure approaches (i.e. at the 'component level') that cannot address growing system complexity, component miniaturisation and the coupling of cause and effects throughout the manufacturing value chain. In part to understand the effects of this shift to 'system-level' failure causes (e.g. design, manufacturing processes, software), the usability of an integrated modelling framework to methodically understand and characterise both the design and manufacturing operations is being assessed in this study. Application of this modelling framework will enable the simulation, evaluation and prediction of product reliability, including system-level effects.

Reliability Models for Safety Critical Electronics

Electronics manufacturers benefit from integrating reliability estimates at an early stage of the product design process. The focus of reliability prediction at this stage will indicate if the proposed design fulfils the reliability design requirements. A designer, might for instance, require information to assess the impact of design changes on the reliability of a new product that is a variant of an existing product. To achieve this, a reliability model is often needed to provide a clear picture of interdependencies within a framework for developing quantitative and qualitative reliability estimates to guide the design trade-off analyses.

The selection of the model is usually made by the design engineers, but it also depends upon company policies based upon the application in consideration. It also varies according to the product life cycle and is driven by the critical parts in the system. Ideally, a design for reliability (DfR) process should enable some form of prediction for product failure rates. System-level predictions provide a means of evaluating design alternatives in terms of their reliability estimates. However, these reliability calculations require knowledge not only about the components and the operating conditions expected, but also of the design and the manufacturing processes.

Reliability Prediction

Currently available reliability prediction technologies and techniques comprise 4 main categories of approach:

Traditional reliability models include (i) empirical and (ii) physics of failure models; (iii) 'enhanced' reliability models or more recent models and (iv) meta-models or frameworks that seek integration among existing models, data and human expertise as it is shown in *Figure 1*.

Empirical based models and physics-of-failure are data-driven models require[1]:

- In-house test data or operational field data;
- System reliability (consolidated assessment technique);
- Comparative-item data (data on similar products operating in similar environments);
- Translation (accounts for factors affecting field reliability);
- Empirical (observed field failure data), and
- Physics-of-failure (models each failure mechanism for each component)

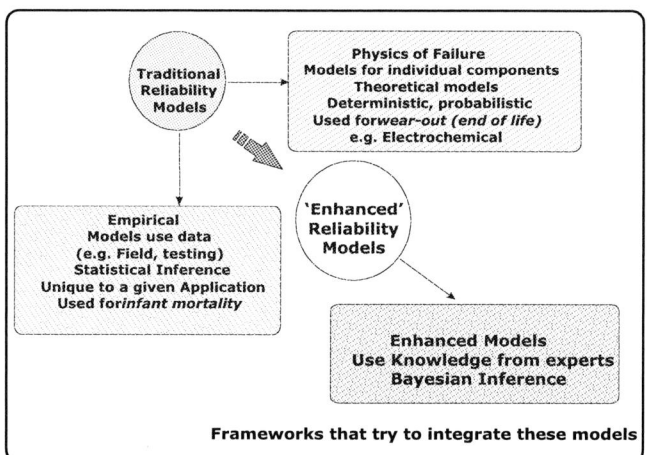

Figure 1 . Classification of the Literature in Reliability Models

Physics-of-failure models have been used as a bottom-up approach for assessing time to failure due to known component failure mechanisms [2, 3]. However, they are difficult to apply to the whole system because they are dependent upon identifying a validated model for all potential failure mechanisms. Similarly, empirical models relying on good quality observed failure data, are difficult to apply in applications other than the particular application they were originally developed from. Nevertheless, they are valuable when performing early trade-off analyses of competing designs.

Traditional 'component level' reliability prediction models cannot cope with the growing system complexity and component miniaturisation that has resulted in a shift

1-4244-1105-X/07/$25.00 ©2007 IEEE

to 'system level' failure causes, including manufacturing, design, and systems management. Therefore, the need to develop models that can address these system-failure effects on reliability estimates has been recognised [2, 4] 'Enhanced models' have tried to integrate human knowledge from design and manufacturing experts [5, 6]. In contrast to traditional models, these enhanced models do not use data, rather, knowledge elicited from experts at various stages in the product life cycle, i.e. design, manufacture, testing. These models draw upon past knowledge that has not been formalised.

'Enhanced models' integrate knowledge elicited from experts based on their lessons learned in two ways: by using expert's judgment as grading factors that can be applied to the design or manufacturing process e.g. [7], or by using Bayesian inference e.g. [8, 9].

A framework is required that provides replicable reliability estimate results while benefiting from the integration of the traditional 'component level' reliability prediction models, the most recent 'enhanced models' and in-house or trustworthy reliability databases. Few researchers have proposed frameworks that aim to integrate some of these data-knowledge components [4, 10].

For reliability prediction of aircraft electronics, Marshal et al. [10, 11] proposed the REMM (Reliability Enhancement Methodology and Modelling) methodology. The REMM framework makes use of various reliability techniques and tools at different stages of the product life cycle, for example:

Table 1. Reliability tools and techniques for integration in reliability frameworks

Stage	Reliability tools and techniques
Concept stage	Human Expert Knowledge
Design	FMCEA (Failure Mode Cause & Effect Analysis), FTA (Failure Tree Analysis), PoF (Physics of Failure)
Development	Growth tests, FRACAS (Failure Reporting, Analysis, and Corrective Action System)
Manufacture and test	SPC (Statistical Process Control charts, Burn in
Operation and maintenance	Field data

Due to a lack of understanding of the processes (business, management, design and manufacturing) within an organisation that can impact reliability, the models and frameworks for reliability prediction proposed in the literature are difficult to deploy in real practice. Therefore, a modelling framework that can address this gap is further discussed in this paper.

2. Methodology-Enterprise Modelling

For this investigation, a group of three electronic manufacturers was studied. Each company produces safety critical, harsh environment, aero-engine applications. These companies use quality components, employ mass solder dispensing systems, automated placement, and mass reflow soldering ovens. However, they still require a substantial amount of manual intervention including inter-process transportation, soldering of special components, wiring and fitting of connectors and rework operations. Production is also characterised by a broad product mix and typically very low volume batch manufacture.

The general experience of these manufacturers is that their design and manufacturing setup processes are currently time-consuming, result in poor first-time yields and produce a surprisingly high low-hour service return rate. The reasons for this situation are not fully understood, but are believed to be the result of:

- defects generated during the manufacturing flow,
- lack of formalised process knowledge,
- lack of knowledge concerning the implications of design features on manufacturing performance,
- lack of in-process performance monitoring and analysis,
- the impact of new materials and components due to environmental legislation (e.g. WEEE, RoHS)
- lack of knowledge of the impact of available enhancing processing capabilities
- difficulties in dealing with component obsolescence
- inability of current data collection tools to support the needs of the plant personnel, such as knowledge sharing, data mining

Designers associated with these companies generate reliability prediction models in two ways. First, if the designer is associated with the organisation which actually use the system, quantitative reliability analysis tends to be procedural. If the designers are associated with the organisation, which is providing items to customer requirements, then the quantitative reliability requirement is less often a specified procedure. Therefore, even though three electronic manufacturers participating in this study, they are categorised into two camps, i.e. design-in-house and design-to-requirements.

The modelling approach used in this research to understand the effects on these problems and to include system-failure causes in a reliability model is CIMOSA (open system architecture for computer integrated manufacturing systems) [12]. The tools used to support the modelling representation include Microsoft Visio [13] and iGrafx [14]. Additionally, various artificial intelligence (AI) techniques shall be used to cover the requirements of the enterprise modelling [15]. The AI techniques used range from knowledge representation, rule-based systems and case-based reasoning [16].

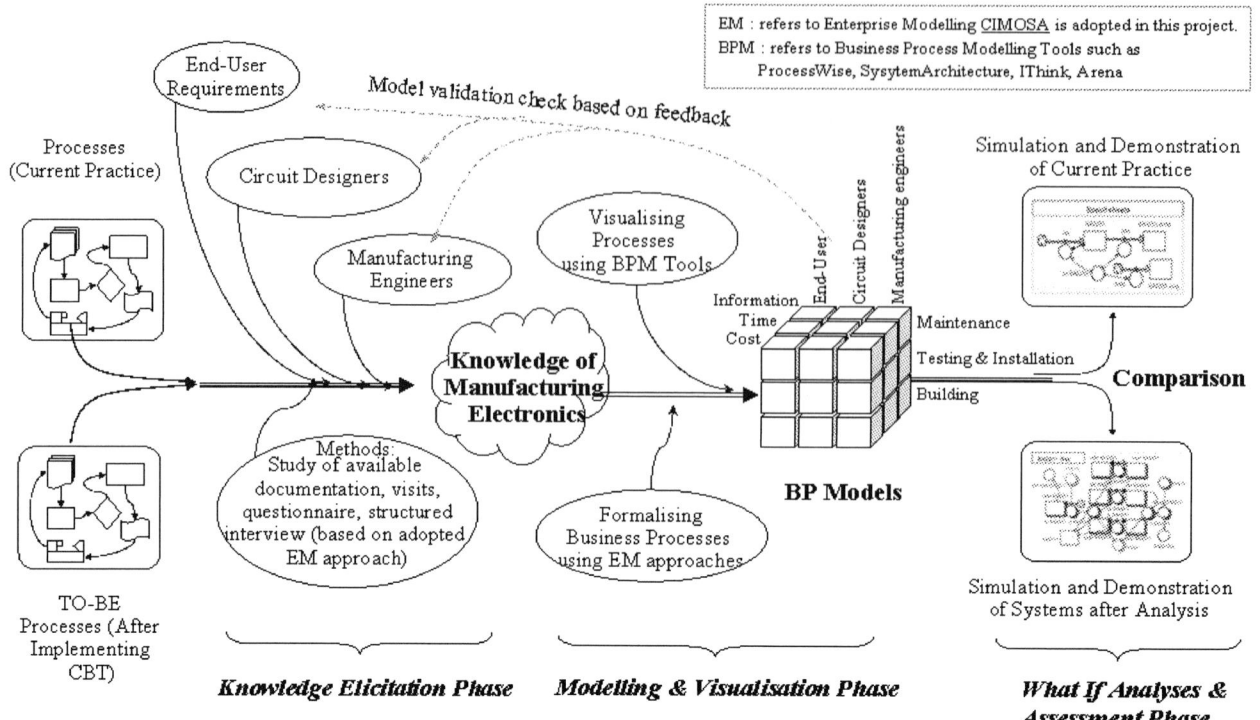

Figure 2. Requested layout of text area and margins. Source: Adapted from [17]

Essentially, CIMOSA allows organisations to integrate all the information available about the organisation to provide better decision-taking procedures. Its integrated architecture contributes to the development of models at levels of abstraction. It also allows to link different resource types such as shop-floor personnel, machines, software applications, databases) in different models.

The four views are established in a hierarchical arrangement of integrated domains (selected groups of study e.g. manufacturing operations) as shown in *Figure 3*. Only the 'function view' needs state (or event) relationships and hence is the most complex and most supported. The other views need less information about the state in the integrated domains. This arrangement permits the modelling process to be separated from the information knowledge, which is relatively static, hence, facilitating and making the model execution more independent.

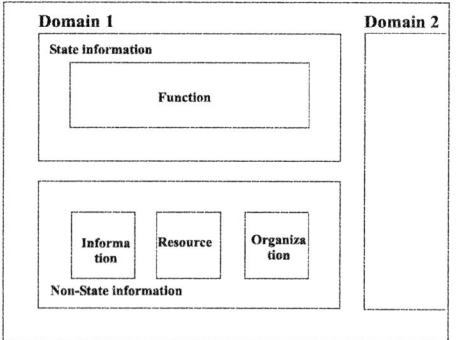

Figure 3. Hierarchical CIMOSA views. Source: [18]

The use of the CIMOSA modelling approach in this project comprises of three phases, namely *knowledge elicitation phase*, *modelling and visualisation phase* and *what-if analysis & assessment phase* (Refer to *Figure 2*). Initially, the current practice needs to be identified as well as all external interaction. At this stage, sufficient information about the way the existing "AS-IS" system behaves and interacts with other internal and external entities is critical. Several knowledge elicitation techniques can be implemented. The decision about what techniques to use depends on the domains to be modelled and compliance with the modelling method. For the purpose of this project, the following areas were selected for elicitation:

- *End User requirements*: definition of the end-user requirements and objectives.
- *Circuit Designers*: capture of the design knowledge including design rules, and design expertise and experiential design knowledge.
- *Manufacturing Engineers*: detailed mapping of manufacturing processes.
- *Study of available documentation*: elicitation from written documents, past failure reports, current practices and approaches to reliability, historical cases, databases, yield data, in-service reliability data.

After the knowledge is elicited from the selected domain areas, the knowledge of the AS-IS system can be modelled for various purposes, for example, (i) to enable a comparison of system entities when system changes are put into practice and, (ii) to perform "what-if" scenario studies to assess the prediction of the system and its

310

behaviour. In addition, the defects that cause yield failures were individually classified and logged at all official stages of inspection and test by all three manufacturers involved in the project.

3. Application of the CIMOSA Modelling Framework in Electronics Manufacturing

Process Mapping and Models

Various areas of the manufacturing of safety critical electronics were elicited using CIMOSA for each of the views previously explained. The models generated include different aspects of what an organisation needs to support their activities. Each of the models consists of business processes supported by events that control the flow of action and constructs representatives of each of the views (i.e. function, organisation, resource and information). This means that different aspects of the organisation were made explicit, i.e. 'what' is done, 'how' it is done, 'who' does it and 'when' it happens. Other aspects that can be included are the 'how much' and 'where'.

An example of a process model is shown in *Figure 4*. A 'CIMOSA function view' diagram is shown, modelling the process of the printed circuit assembly (PCA) storage and moisture control process within the assembly of PCA. The example model shows some CIMOSA constructs. Trapezoids were used to represent or describe actions at the start or end of the enterprise activities (EA). Enterprise activity rectangles containing the name and type of action can be decomposed into sub-processes. The procedural rules are not represented in the drawing but they were elicited in the form of first order predicate rules to be used by the model to describe synchronisation of concurrent steps and alternatives.

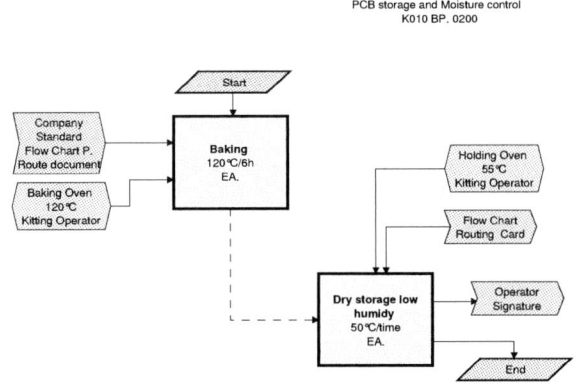

Figure 4. Extract from CIMOSA model of 'PCB storage and moisture control processes'

In addition, *Figure 4* also shows how the 'function view model' is linked to other CIMOSA views, e.g. 'information', 'resource' and 'organisation'. That is, various inputs and outputs are used to describe the functionality of the enterprise activities. For instance, the 'resource input and output' show the machines, equipment and operators required to accomplish the baking process.

Requirements for modelling

One of the most important requirements for any modelling technique is knowledge capture and representation i.e. the way in which information is captured, expressed and further used. An area of concern that arose in this investigation relates to how the same information about manufacturing processes is expressed differently among the organisations. Therefore, an agreed common ontology is being developed to support the enactment and the comparisons between models across the participating companies. However, the adoption of this terminology might not be easily adopted by all industrial partners because they are accustomed to 'reason' with that terminology and removing those terms may lead to difficulties in expressing information within the organisation. At the present time, when a new term is agreed upon, the differences are specified and when the new term is used (e.g. in comparison exercises), it is possible to translate the term and support communication.

Defect Modelling

On commencement of the defect modelling project activities, the initial data mining and process mapping exercises allowed to identify that the defect classifications were not consistent across all three electronics manufacturers. Also, in all but the most obvious cases, the three data sets were found to be insufficiently concise to provide neither, an apparent evidence trail to the defect's root cause, nor any such backup data that might at least facilitate the retrospective traceability to the root cause.

This is best illustrated by an example of previous misdiagnosis that had actually been uncovered by one of the manufacturers in its own defect classification system. This manufacturer has adopted the defect codification system proposed by the US Military in its specification MIL-STD-2000 Rev A. Condition A8 – Non Soldered Connections, would appear relatively unambiguous but, it is commonly perceived to suggest that the solder needed to form the solder joint is absent. But a misconception that proved to be common with assembly line operators across all three manufacturers interprets the meaning as the defect was due to a lack of solder being present and that the root cause must reside in the solder supply process – in this case the solder paste printing process. But on this particular occasion, a data trend was observed that suggested that this lack of solder was effecting one particular surface mount device (SMD). Without this trend being observed any corrective action may well have been incorrectly targeted. Further examination of residual unused stocks of said component determined that it required special processing to cut, form and pre-tin its leads and that the current assembly sequence employed often resulted in co-planarity anomalies in the z-axis of its fine pitch leads. Once discovered, the sequence was rearranged and the root cause removed. But the exercise had illustrated that the data logged under A8 – Non Soldered Connections (by far the highest frequency defect category) could not be relied on for even the most superficial levels of root cause analysis and defect modelling i.e. the identification of the process host of the defect opportunity.

This newly acquired knowledge has now been captured by CIMOSA modelling and the potential root causes mapped to the multiple process possibilities and shared amongst the manufacturers involved. As this scenario of misclassification and misdiagnosis is equally possible within the codifications used by all three of the manufacturers, the data from all three must also be regarded with the same uncertainty - thus suggesting that data alone will not provide root cause analysis, but data trends and CIMOSA models can highlight a previously unconsidered area for process improvement in the enhancement of soldering process yields i.e. materials handling and component pre-treatments prior to automated placement processes.

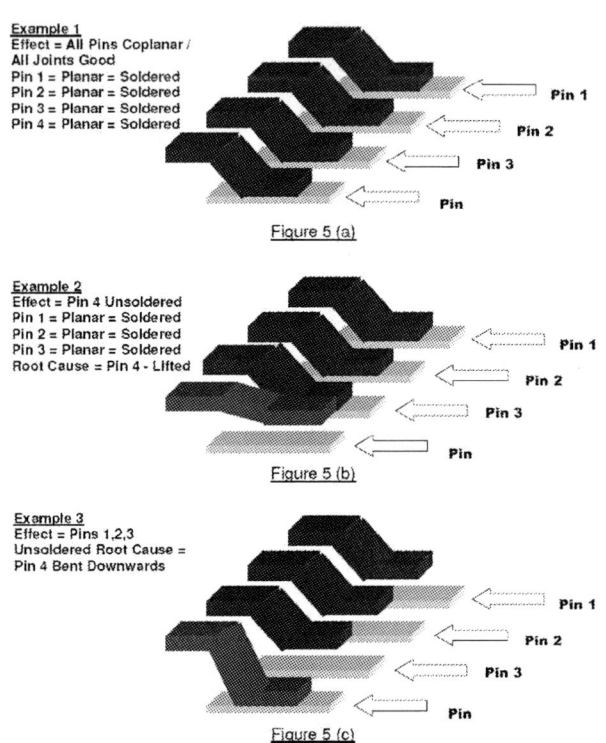

Figure 5. Lead Co-planarity root cause of assumed soldering defects

4. Conclusions

A methodology for capturing and describing the linkages between design and manufacturing process and other variables (e.g. yield, product reliability, cost and quality) has enabled the formalisation of the components of the current system and their interactions into a structured and standard format. At its present level of detail, the model does not support the prediction and simulation of process steps and their effects on the reliability. However, the CIMOSA component based approach will be further supplemented by dynamic modelling. The models obtained have enhanced the company procedures and are more effectively supporting manufacturing personnel.

Acknowledgments

The authors wish to express their gratitude to the industrial collaborators. This work was supported financially by the Innovative Electronics Manufacturing Research Centre (IeMRC).

References

1. Goel, A. and R.J. Graves, Electronic system reliability: collating prediction models. IEEE Transactions on Device and Materials Reliability, 2006, 6(2): p. 258- 265.
2. Denson, W., The History of Reliability Prediction. IEEE Transactions on Reliability, 1998, 47(3).
3. Jones, J. and J. Hayes, A comparison of electronic-reliability prediction models. Reliability, IEEE Transactions on, 1999, 48(2): p. 127-134.
4. Wong, K.L. A new framework for electronic assembly/system reliability prediction. in Reliability and Maintainability Symposium, 1996 Proceedings. 'International Symposium on Product Quality and Integrity'., Annual. 1996.
5. James, I., et al. Reliability metrics and the REMM model. in Reliability and Maintainability, 2004 Annual Symposium - RAMS. 2004.
6. Marshall, J., L. Walls, and J. Jones. Enhancing Product Reliability using REMM. in Reliability and Maintainability Symposium. 2002.
7. Denson, W.K. and S. Keene. A new reliability prediction tool. in Reliability and Maintainability Symposium, 1998. Proceedings., Annual. 1998.
8. Russell, H., et al., Eliciting Engineering Knowledge about Reliability during Design – Lessons Learnt From Implementation. Quality and Reliability Engineering International, 2001, 17: p. 169-179.
9. Lesley, W. and Q. John, Eliciting Prior Distributions from Engineering Experts to Support Bayesian Reliability Growth Modelling. Management Science, 1998, 14.
10. Marshall, J., L. Walls, and J. Jones, Reliability enhancement methodology and modelling - the REMM project. The Aeronautical journal., 2002, 106: p. 195-202.
11. Marshall, J., L. Walls, and J.A. Jones, Enhancing Product Reliability Using REMM. Proceedings, annual Reliability and Maintainability Symposium., 2002(2002): p. 372-378.
12. Vernadat, F., Enterprise Modeling and Integration. First ed. 1996: Chapman and Hall.
13. Microsoft® Office Visio® Professional 2003 SP2. 2003, Microsoft Corporation.
14. iGrafx Process 2006 for Six Sigma. 2005, Corel corporation.
15. Stader, J. and P. Jarvis. Intelligent Support for Enterprise Modelling. in 13th biennial European Conference on Artificial Intelligence (ECAI-98). 23-28 August 1998. Brighton Centre, Brighton, UK.
16. Segura-Velandia, D.M., et al. A Case-Based Reasoning Approach for Low Volume, High Added

Value Electronics. in 8th Electronics Packaging Technology Conference,

EPTC 2006. 6- 8 December 2006. Pan Pacific Hotel, Singapore.

17. Monfared, R.P., Practical Application of Enterprise Modelling. 2003: Wolfson School of Mechanical and Manufacturing Engineering, Loughborough University.

18. Goranson, H.T. The CIMOSA Approach as an Enterprise Integration Strategy. in First International Conference in Enterprise Integration Modeling. 1992.

Development of an Internal Liquid Cooling System for CPU using CAE

Sebastian Flores[1], Song-Hao Wang[2], Chong-Qing Chang[3]
1, 2: Mechanical Engineering Department, Kun-Shan University, Taiwan, ROC
Tel: 886-06-2050000-267
3: System Engineering Department, National Tainan University, Taiwan, ROC
Tel: 886-06-2062116

Abstract

Liquid cooling is currently one of the best and most immediate options for non uniform heat sources, high performance CPU/GPU. One of the main drawbacks of this alternative is that many customers may not be very enthusiastic to install a device that could some how leak into their computers. This project intends to develop a new generation liquid cooling system that is leakage-free.

This paper presents the development process of an internal liquid cooling system for Center Process Unit (CPU) or Graphics Process Unit (GPU), which can be installed inside the computer case. Commercial available computer aided design (CAD) software; Computational Fluid Dynamics (CFD), rapid prototyping (RP) technology, Computer Aided Manufacturing (CAM) and computer aided engineering (CAE) were used in this forward engineering process for the design and fabrication of the critical parts. Previous work from this project emphasized on the development of an internal liquid pump, mainly by using CAD and PR technology [1]. At current stage the project is emphasized on the design, simulation and testing of different water-blocks, that work best with the flow rate of the internal pump.

The optimum water-block design involves major dimensions as well as different internal geometries in order to direct the inlet fluid to a certain specific area and also various structures that can increase the heat extraction process from the CPU/GPU to the exterior of the system. The heat exchange process presents different scenarios; such as outlet fluid temperature, uniformity of temperature distribution, hot spots, pressure drop, etc. The criteria used for determining the optimum design will be discussed using CFD data, mathematical models and experimental results.

Finally the results obtained with adequate working conditions as well as the system performance trends are shown graphically.

1. Introduction

With the increasingly tendency of higher density microprocessors and new software requirements for example; Basic Microsoft's Vista requires at least 800MHz processor, 512 MB storage and a video card supporting DirectX 9. For upgraded Vista, 1GHz processor, 1GB storage and 128MB video card is needed.

Heat dissipation has become more of an issue since current microprocessors are reaching a point where they can no longer be cooled by traditional air-cooling systems. Another factor adding more complexity to the cooling process is that new dual-core chips have a greater non- uniform heat distribution, two "hot spots" will require a different cooling approach.

Current microprocessors consume up to more than 100 W/cm^2, and the heat generation of the CPUs also follows Moore's Law. With the amount of transistors doubling-up every 18 months so does in very similar pattern heat generation.

Common CPU failure mechanisms tend to be mechanical (solder bump bridging, die fracture, corrosion) and electrical (overstress, migration and diffusion, gate oxide breakdown) following the Arrhenius equation, which states that (for die temperatures operating in the range of -20°C to 140°C) every 10°C decrease in temperature reduces the failure rate by approximately a factor of 2. Therefore a reduction in chip failure rates with lower operating temperatures can be expected.

Traditional cooling of microelectronics makes use of air in natural convection or forced convection systems, it is found that these systems are insufficient in an increasing number of applications.

Liquid cooling can be directed to the exact place where the heat is being generated. This is something that can only be achieved with this technique. The liquid cooling system requires an active mechanism to move the fluid from the liquid block, where it absorbs the heat generated by the CPU/GPU, into the radiator where the heat transferred trough air cooled fins into the ambient, where the heat is transferred trough air cooled fins into the ambient.

Previous work of this project emphasized on the development of a liquid cooling system that can be installed in the interior of the computer case, following the concepts of forward engineering and using RP Technology to generate parts for the device [1].

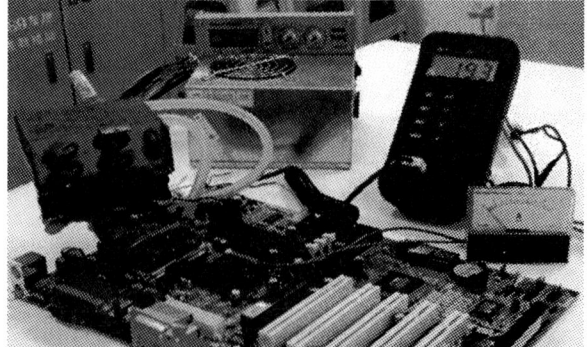

Fig. 1 Internal Liquid CPU Cooler with Testing Device

1-4244-1105-X/07/$25.00 ©2007 IEEE

This paper presents the study of different water-block designs to best fit the internal liquid pump, using commercial available CFD software (STAR-CD), targeting the understanding of the flow maldistribution and its effects on the heat performance of the water-block.

2. Assumptions and Simulation Method

The study of flow field and heat transfer was performed using commercially available CFD software, STAR-CD. The governing equations were solved using PRO-STAR which uses the finite volume method.

There are 12 different cases (3 major designs) analyzed in this paper, each case was simulated for different operating conditions such as, inlet geometry and flow rate, liquid path and input heat.

The heat source is located on the opposite side of the inlet and outlet of the water-block, which can be referred as the "back side" of the devise, the remaining walls are considered to be adiabatic.

The water-block is made of bronze and the wall thickness connected to the heat source is 2 mm. The heat source generates 50 W or 100 W at a constant power. The thickness of the channels inside the water-block varies from 1mm or 7mm according to the design.

Four different internal structures are presented; the first is a hollow block with no internal configurations to force the flow. The second is a hollow block with grids (2mm by 2mm) evenly distributed on the bottom as shown in Fig. 2. The third is a labyrinth style design, with the inlet on one end of the water-block and the outlet on the opposite side (labyrinth I) as shown in Fig. 3. To reduce pressure drop, a labyrinth with double passage (labyrinth II) was created Fig. 4. For labyrinth design I the cooling channels have a 4 mm width, and 8 mm height. For design II the passage has cooling channels of 7 mm wide. For the hollow block design the inlet is located on one corner of the water-block and the outlet is placed on the opposite corner, letting the fluid flow freely inside the water-block.

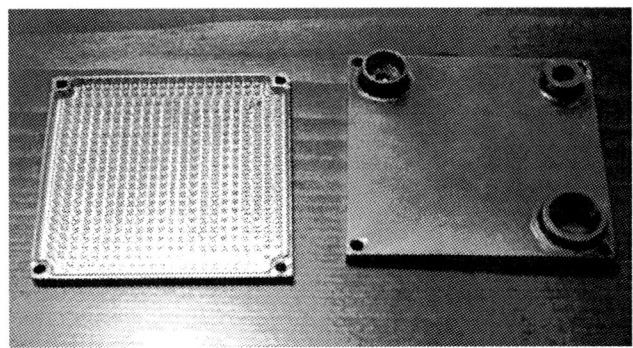

Fig. 2 Hollow water block with grids

Fig.3 Water block with Labyrinth I

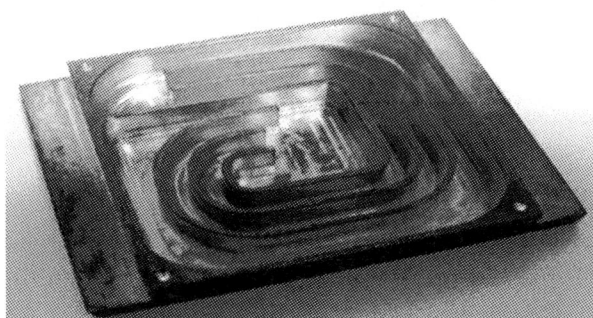

Fig.4 Water block with Labyrinth II design

The major assumptions for the simulations are:

1) The inlet liquid temperature T_{in} is a constant, without consideration of effect of the radiator;

2) Thermal energy is 100% carried out by the liquid flow; therefore all surfaces of the water block are defined as adiabatic, except for the bottom surface which is in contact with a uniform heat source. This is a more conservative assumption because heat transformed through convection and radiation is neglected in the simulation.

3) The fluid is incompressible;

4) Three-dimensional steady state flow and heat transfer;

5) Solid and fluid properties do not vary with temperature.

Therefore, major input values for this computer aided engineering analysis are heat flux q, liquid inlet temperature T_{in}, and liquid inlet flow rate Q_{liq} plus Water-block material properties:

o Aluminum: k=237 w/mok; Cv=903 j/kg ok; density = 2702 kg/m^3

o Bronze: k=401 w/mok; Cv =385 j/kg ok; density = 8933 kg/m^3;

o Water: k=0.62 w/mok; Cv =4181 j/kg ok; density = 998 kg/m^3;

o Major dimension for the water-block:

o Width: 50 mm x 50 mm; Height: 10 mm;

o Wall thickness: 1 mm or 2 mm;

o Working conditions;

o Q_{liq} = 22(cm^3/s) or 5.5(cm^3/s) as specified; W_{cpu}=57.2(Watts); T_{in}=298 (^{o}k)

Inlet pressure of the liquid flow was set to zero. The initial temperature of the liquid and the water-block material were set at 298°k which is the average inlet temperature registered in the testing facilities shown in

Fig. 1. Fluid inlet and outlet area equal. The liquid flow rate set at 22 (cm³/s) is measured from the liquid pump mentioned in the authors' previous work [1]. It is known from reference [16] work that the results of the simulation are independent from the mesh size, so the grid dependence test was omitted.

2. Results and Discussions

Table 1, 2, 3 represent maximum fluid velocity, pressure drop and maximum temperature under different water-block structures and working conditions. It is obvious that under same structure and inlet flow, the pressure drop and the maximum fluid speed are almost the same, no mater how the heat source varies. This is because of the assumptions of incompressible flow and constant liquid properties under different temperature. This confirms the results of the simulation.

The advantage of the grids design is very obvious from the CAE analysis, because with moderate increase of pressure-drop, the temperature is reduced quite significantly.

Labyrinth I design, leads to the lowest temperature. However, the pressure-drop increased dramatically to 6204 Pa, compared with 462 Pa from hollow block design. To compensate this drawback, labyrinth II design which has two parallel paths was considered. Through the labyrinth II design, a happy medium seems to be found.

Table 1. Maximum fluid velocity (m/s)

Design	57.2W, 5.5 cm³/s	57.2W, 22 cm³/s	100W, 22 cm³/s
Hollow	0.6	0.7	0.7
Grids	0.6	0.7	0.7
Lab I	0.6	1.23	1.23
Lab II	0.6	0.8	0.8

Table 2. △ Pressure drop (Pa)

Design	57.2W, 5.5 cm³/s	57.2W, 22 cm³/s	100W, 22 cm³/s
Hollow	173	462	463
Grids	299	625	625
Lab I	637	6204	6206
Lab II	223	994	995

Table 3. △ Maximum temperatures raise (°C)

Design	57.2W, 5.5 cm3/s	57.2W, 22 cm3/s	100W, 22 cm3/s
Hollow	18	10.7	19
Grids	12	7.5	13
Lab I	8	3	5
Lab II	19	6	10

A picture of the temperature distribution of the water-block viewed from the bottom is shown in Fig.5.

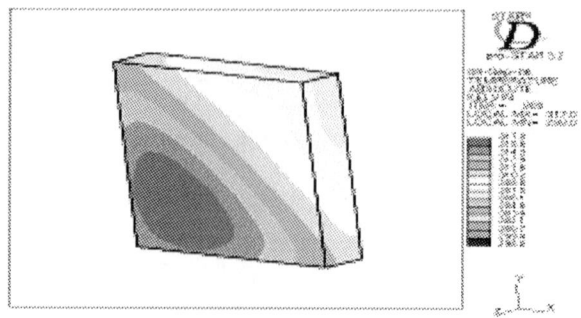

Fig. 5 Temperature distribution of hollow block viewed from bottom

Fig. 6 shows the temperature distribution of labyrinth design I viewed from the top. One of the most interesting conclusions from the CAE simulation is that with this technology, the location of the hot-spots can be controlled easily by properly locating the incoming cooling liquid.

Fig. 6 Temperature distribution of Labyrinth design I viewed from top

Fig. 7 through 8 represents the vector fields of liquid velocities from labyrinth I, II and hollow-block designs. For the labyrinth I and II designs, the liquid velocities are much higher and distributed more evenly throughout the flow paths. As a result, the forced convection plays the major role in the heat transfer process.

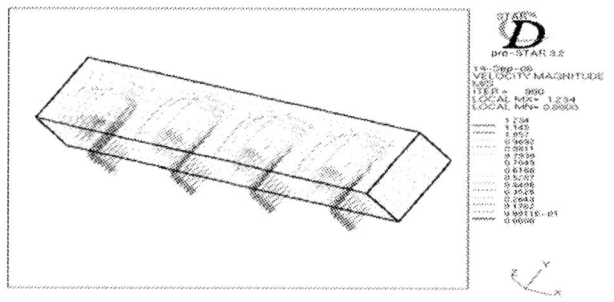

Fig. 7 Labyrinth I vector fields of liquid velocities viewed from a cross section

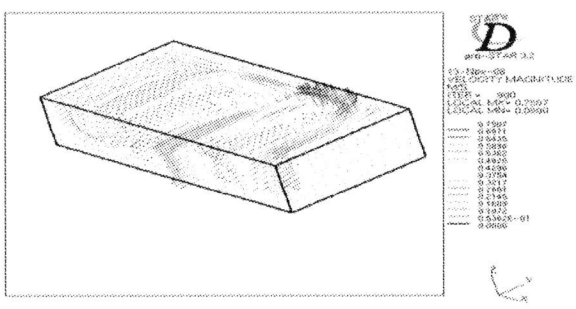

Fig. 8 Labyrinth II vector fields of liquid velocities viewed from a cross section. Refer to Fig. 4 for geometry.

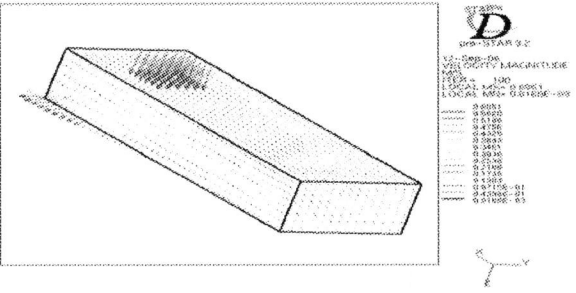

Fig. 9 Hollow-block vector fields of liquid velocities viewed from a cross section

Fig.10 presents the pressure drop vs. temperature rise for different water-block designs. It is seen in the hollow block design that liquid free flow has a modest cooling effect on the water-block.

The presence of grids enhances the cooling over 31% while increasing the pressure drop by 35%

Although Lab II design reduces the temperature by 47% compared with hollow block, it increases pressure drop by 115%. Since this project is targeting a cooling system that can be fitted inside a PC case, size and power of the components become constrain, so pressure can not be unlimited.

Lab I design is excluded from Fig. 10 due to its extremely high pressure drop value.

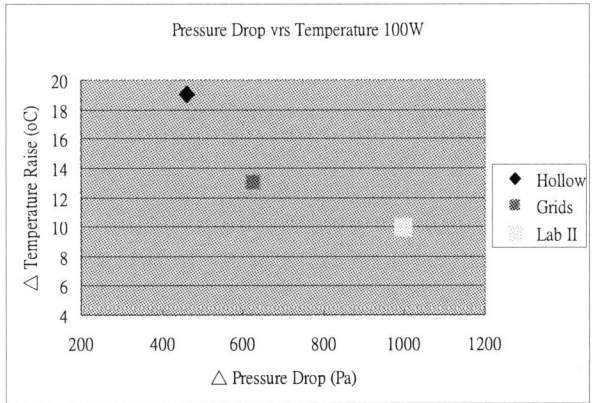

Fig. 10 Pressure drop vs. temperature rise for different water-block designs

3. Conclusions and outlook

It can be concluded that from the CAE analysis, designs of labyrinth (parallel channels) as well as the grids on the wall are very effective to increase the efficiency of water-blocks by reducing the working temperature.

It is seen that longer labyrinth length leads to lower temperature but higher pressure drop. Therefore, a suitable combination of thermal performance and pressure drop should be carefully considered. Because both parameters are closely linked, and pressure drop of a pump system can not be unlimited.

It can be concluded from the analysis that, designs of multiple labyrinths (parallel channels) are good candidates for water-block designs. The combination of the multiple labyrinth and the grid pattern seems to be an effective way to increase the heat transfer and so maintain low temperatures and low pressure drops at the junction interface which reduces the risk of premature failures on the CPU.

Further CAE analysis and future experimental verification is planed regarding the device performance tested using standard CPU cooling performance testing device.

4. Acknowledgement

The financial support through Taiwan's National Science Foundation (NSC 95-2221-E-168 -001) is greatly acknowledged. The authors would also express our appreciation to Dr. S.M. Lin and Dr. C.W. Lee of Kun-Shan University for their great advises and discussions through out the CAE analysis.

References

1. Wang, S.H., Flores, S., Chang, C.C. "Development of an Internal Liquid Cooling System for CPU using RP Technology", ICEPT2006 (IEEE) Proceedings, Shanghai, 2006, pp744

2. Culham, Y.S. Muzychka, "Optimization of Plate Fin Heat Sinks Using Entropy Generation Minimization", Microelectronics Heat Transfer Laboratory, Department of Mechanical Engineering University of Waterloo, 2000.

3. Butterbaugh, M. A. and Kang, S. S., "Effect of Airflow Bypass on the Performance of Heat Sinks in Electronic Cooling," ASME, Adv. Electronic Packaging, EEP-Vol. 10-2, pp. 843-848, 1995.

4. Luo, J. W., "Thermal Analysis of the CPU Heat Sink with Dual Fan Assemblies for a Personal Computer," Thesis for Master of Science, Department of Mechanical Engineering Ta-Tung University, 2000.

5. Maveety, J. G., and Jung, H. H., "Design of an Optimal Pin-Fin Heat Sink with Air Impingement," Int. Comm. Heat Mass Transfer, vol. 27, no. 2, pp. 229240, 2000.

6. Martin, H., "Heat and Mass Transfer between Impinging Gas Jets and Solid Surfaces", Advanced Heat Transfer 16, pp. 1-60, 1977.

7. Marto, P. J. and Peterson, G. P., "Application of Heat Pipes to Electronics Cooling," Advances in Thermal Modeling of Electronic Components and Systems, pp. 283-336, Hemisphere, Washington, D. C., 1988.

8. Petterson, G. P., "Heat Pipes in the Thermal Control of Electronic Components," Proc. Int. Heat Pipe Symp. 3rd, Tsukuba, Jpn., pp. 2-12, 1988.

9. Chi, S. W., "Heat Pipe Theory and Practice," McGraw-Hill, New York, 1976.

10. A. Bar-Cohen, "Optimization of vertical pin-fin heat sinks in natural convective heat transfer," in Proc. 11th Int. Heat Transfer Conf., vol. Kyonju, Korea, Aug. 23–28, 1998, pp. 501–506.

11. A. D. Kraus and A. Bar-Cohen, Thermal Analysis and Control of Electronic Equipment. Washington, DC: Hemisphere, 1983.

12. C. Y. Liu and Y. H. Hung, "Heat transfer and flow friction characteristics for compact cold plates," J. Electron. Packag., vol. 125, pp. 104–113, 2003.

13. G. Hetsroni, A. Mosyak, and Z. Segal, "Nonuniform temperature distribution in electronic devices cooled by flow in parallel microchannels," IEEE Trans. Compon. Packag. Technol., vol. 24, no. 1, pp. 16–22, 2001.

14. J. T. Teng, J. C. Chu, M. S. Liu, C. C.Wang, and R. Greif, "Investigation of the flow mal-distribution in microchannels," presented at the ASME Congress 2003, paper no. IMECE2003-41 323.

15. A. B. Datta and A. K. Majumdar, "Flow distribution in parallel and reverse flowmanifolds," Int. J. Heat Fluid Flow, vol. 2, no. 4, pp. 253–262, 1980.

16. 16. Lu, M.C. and Wang, C.C. "Effect of the Inlet Location on the Performance of Parallel-Channel Cold-Plate", IEEE Transactions on Components and Packaging Tech., vol. 29, no. 1, March 2006

Kinetic Characterisation of Molding Compounds

K.M.B. Jansen[1], C. Qian[1], L.J. Ernst[1], C. Bohm[2], A. Kessler[2], H. Preu[2], M. Stecher[2]
[1]Delft University of Technology, Mekelweg 2, 2628 CD Delft, The Netherlands
[2]Infineon Technologies AG (AIM AP), 81726 Munich, Germany
k.m.b.jansen@tudelft.nl

Abstract

During the packaging of electronic components stresses are generated due to curing effects and the difference in thermal shrinkage between molding compound and die. For a reliable simulation of the stresses generated in a package during cure and subsequent cooling it is essential to have accurate data for the thermal and mechanical properties of the molding compound, die, solder and substrate materials. Of these materials the molding compound is by far the most difficult one to model since these properties vary widely and are time, conversion and temperature dependent.

The present paper consists of an extensive study on serveral commercial molding compounds of which the kinetic parameters we obtained by analysing both isothermal and non-isothermal Differential Scanning Calorimetry (DSC) data.

For the modelling of the kinetics data it turned out to be necessary to take the diffusion limitation effect (incomplete cure at lower temperatures) into account. The glass transition versus conversion could be modelled satisfactory using the well known DiBenedetto equation.

1. Introduction

Experiments with different loading conditions showed that time and temperature dependent effects have significant impact on the type and extent of observed failure modes like cracks, interconnect reliability etc. Especially the molding compound is responsible for high stress and strain levels on the chip and interconnects. Depending on the selected material set, these results vary greatly. Reliable material data is thus essential for obtaining accurate stress predictions.

Material selection for future products and transfer of results from accelerated testing to product assembly and application conditions is only possible by understanding and optimizing relevant material properties. To be able to consider these in simulation, detailed thermal-mechanical properties of the molding compounds are crucial.

For a full characterisation of the molding compound properties it is required to know the

- Conversion as a function of time and temperature (the so-called cure kinetics)
- Shear modulus as a function of time, temperature and conversion
- Bulk modulus as a function of time, temperature and conversion

- Coefficient of thermal expansion as a function of temperature and conversion
- Cure shrinkage as a function of conversion

In a joint project between TU Delft and Infineon Technologies we therefore aim to fully characterise a set of six commercially available molding compounds. This is a rather involved task since most of the standard measurement techniques cannot be used for materials which react during characterisation. In the past we therefore developped experimental methods to characterise the change in mechanical properties during cure [1,2]. The material used was a model epoxy novolac resin with relatively simple chemistry. In the present project we will use these techniques on commerial molding compounds, which are known to have a complex chemistry and consists not only of the resin, hardener and filler but also of accelerator, flame retarder, mold-release agent, tackifyers and colorant.

The present paper is the first in a series of papers and will focus on the kinetics part only. The other topics will be published in subsequent papers.

The modelling of the conversion as a function of time and temperature is usually referred to a the cure kinetics. Reliable cure kinetics models are important for two reasons. First of all, they are needed to predict cure shrinkage and stress development during the actual molding process and are thus indispensible if curing is to be taken into account in process simulations and optimisations. This requires a good description for the kinetics at a higher temperature range (molding and post curing are usually done at 175-200 °C). The second reason for accurate kinetic modelling is that most characterisation experiments are performed at lower temperatures where the reaction is slower such that there is enough time to do the measurements. Therefore, the kinetic modelling should be accurate at the lower temperature region as well.

2. Kinetic model

The conversion α in the molding compound can be defined as the number of reacting epoxy groups divided by the total number of epoxy groups and varies between 0 and 100%. The reaction rate at a certain instant depends on how much of the unreacted epoxy and hardener is present and thus depends on the degree of cure itself. A convenient model which is relatively simple and sufficiently accurate is the Kamal-Sourour equation:

1-4244-1105-X/07/$25.00 ©2007 IEEE   319

$$\frac{d\alpha}{dt} = k_T(T)\,\alpha^m(1-\alpha)^n,\qquad\qquad \text{Eq.1}$$

where m and n are constants of order unity. The above model is therefore able to capture the common n^{th} order kinetics (if $m=0$) as well as autocatalytic reactions ($n=0$). k_T is the temperature dependent rate constant with units [1/s]. In literature it is often assumed that the reaction rate follows an Arrhenius temperature dependency

$$k_T(T) = k_0 \exp\left[\frac{-E_0}{RT}\right].\qquad\qquad \text{Eq.2}$$

Here R denotes the gas constant (8.314 J/mol/K), E_0 is the activation energy and T is the absolute temperature (in degrees Kelvin). Note that in principle also other alternatives for $k_T(T)$ are possible.

The parameter needed in all subsequent modelling is, however, not the reaction *rate* but the degree of cure α. In order to find this we must rewrite and integrate Eq.1

$$\int_0^\alpha \frac{d\alpha}{\alpha^m(1-\alpha)^n} = \int_0^t k_T(T)dt \equiv \tilde{t}_c,\qquad \text{Eq.3}$$

The left hand side is a function of α only whereas the right hand side can be interpreted as the dimensionless cure time $\tilde{t}_c$. Therefore the above equation can be rewritten as

$$F(\alpha) = \tilde{t}_c$$

and the desired conversion is obtained by inversion

$$\alpha(T,t_c) = F^{-1}(\tilde{t}_c).\qquad\qquad \text{Eq.4}$$

3. Diffusion limitation

The above kinetic model is not yet complete since it does not account for the fact that at relatively low temperatures the reaction stops at a certain maximum conversion level. A molding compound stored for one week at room temperature for example will react not further than about 10 to 20% conversion whereas the equations above predict 100% conversion after an infinitely long time. For higher temperatures these maximum conversion levels are higher (e.g. about 90% for 90 C). The reason for this phenomena is that during isothermal cure the glass transition temperature rises to a value above the cure temperature. That means that the material vitrifies into a glassy state which hinders the diffusion of the molecular segments. Therefore reactive sites will have more and more difficulty to meet until, eventually, the reaction stops completely. A simple way to incorporate this so-called diffusion limitation effect is by modifying Eq.1 such that the maximum conversion level is incorporated explicitly:

$$\frac{d\alpha}{dt} = k_T\,\alpha^m(\alpha_{max}-\alpha)^n.$$

In that way the condition of zero reaction rate at the maximum conversion is automatically satisfied. For convenience this can be written in the form of Eq.1 by normalising with respect to α_{max}:

$$\frac{d\hat{\alpha}}{dt} = k_T'\,\hat{\alpha}^m(1-\hat{\alpha})^n,\qquad \hat{\alpha} = \frac{\alpha}{\alpha_{max}}.\qquad \text{Eq.5}$$

The temperature dependency of α_{max} decreases inversely with decreasing (absolute) temperature [4]

$$\alpha_{max}(T_c) = a_0 - \frac{a_1}{T},\qquad 0 \le \alpha_{max} \le 1\qquad \text{Eq.6}$$

where T_{g1} denotes the final (fully converted) glass transition temperature and T is the applied cure temperature in degrees Kelvin. Notice that his maximum conversion α_{max} is bounded between 0 and unity. Direct estimates of a_0 and a_1 can be obtained by assuming that the limits for α_{max} (conversion levels of 0 and 1) occur at the lowest and highest glass transition temperatures (T_{g0} and T_{g1}, respectively). In this study, however, we will treat a_0 and a_1 as fit parameters.

The diffusion limitation can only be observed from the isothermal cure experiments since in the non-isothermal heat rate tests the temperature will always increase until curing is complete. An example of the diffusion limitation effect can be seen in Figure 1 for the 90 C cure experiments. This experiment clearly show that the conversion level does not increase beyond 90%, even for prolonged curing times of 10 hours.

4. Experimental

The molding compounds considered here are commercially availble and are refered to a materials A, C, E, F, G and H. They all have an ash content (filler concentration) in the range of 87 to 90%. The reaction kinetics of these compounds were analysed with a TA-Instruments DSC 2980 Differential Scanning Calorimeter. This device accurately measures the heat released during reaction, H, per unit of time. The total reaction heat is obtained by integrating the recorded heat flow dH/dt over the time needed to complete the reaction. The fact that each time when an epoxy group reacts a fixed amount of heat is released, means that the reaction rate and heat flow must be proportional. Therefore the experimentally obtained reaction rate becomes

$$\frac{d\alpha}{dt} = \frac{1}{H_{tot}}\frac{dH}{dt}.\qquad\qquad \text{Eq.7}$$

The experimental conversion level then simply follows by integration of the above expression.

Two sorts of experiments were performed to determine the kinetic parameters. Firstly uncured samples were subjected to a series of different heating rate tests (1, 2, 5, 10 and 15 °C/min) and the released reaction heat was recorded as a function of temperature. Secondly a series of isothermal cure experiments was performed. In these experiments the molding compound samples were cured at a specific temperature and for a specific time, after which the conversion level is obtained from a second DSC scan (10 °C/min) as $\alpha=1-H_{res}/H_{tot}$, where H_{res} is the measured residual heat of reaction. The conditions of these isothermal cure experiments are close to those expected in the characterisation tests and the isothermal expriments are therefore important to include in the kinetic modelling. This second set of experiments also provides the relation between conversion level and glass transition temperature.

5. Isothermal cure experiments

Samples of about 10 mg were cured at temperatures of 90 to 120 C for 10 to 300 minutes (and more if needed). Curing for shorter times resulted in an undesirable large contribution of the reaction progress during the heating period. That means that for this type of experiments it is not useful to apply curing temperatures above T_g (typically 120 C for these materials). A typical plot of the data obtained with isothermal cure experiments is shown in Figure 1. This shows for example that it takes about 80 minutes for the reaction to complete at 120 C and 240 minutes at 110 C.

An important reason to perform isothermal cure experiments is that together with the conversion level also a glass transition temperature can be extracted for each of the experiments. If both the T_g and the conversion are collected in a single plot we obtain for compound G a plot as shown in Figure 2. The full line is the curve according to the so-called DiBenedetto equation

$$T_g(\alpha) = T_{g0} + \frac{(T_{g1} - T_{g0})\lambda\alpha}{1 - (1-\lambda)\alpha}, \qquad \text{Eq.8}$$

where T_{g0} and T_{g1} denote the glass transition temperature at $\alpha = 0$ and 1, respectively and λ is a shape parameter with a value between 0.2 and 0.6. Since the glass transition temperature can usually be determined with more accuracy than the heat of reaction, from now on the reported conversion levels from isothermal cure experiments will be determined from in Eq.8 instead of using the measured residual heat.

Isothermal cure experiments as described above were performed for all molding compounds under study. This resulted in plots similar to Figures 1 and 2. The corresponding DiBenedetto parameters are listed in Table 1. Please note that these parameters for the glass transition temperature only apply for T_g's determined from DSC experiments at 10 C/min heating. Glass transition temperatures determined from DMA experiments are typically 10-15 C higher and depend on the heating rate as well.

In order to investigate the diffusion limitation effect additional isothermal experiments were performed at room temperature and at 70 C. An example of a plot of the thus determined maximum conversion levels versus the reciprocal temperature is shown in Figure 3. The full line is the fit according to the Gonzalez equation (Eq.6). The corresponding parameters a_0 and a_1 are also included in Table 1.

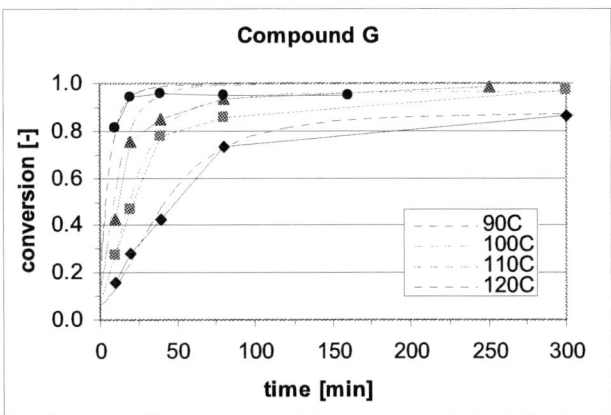

Fig.1: Isothermal cure experiments of compound G. Full lines are predictions (see text).

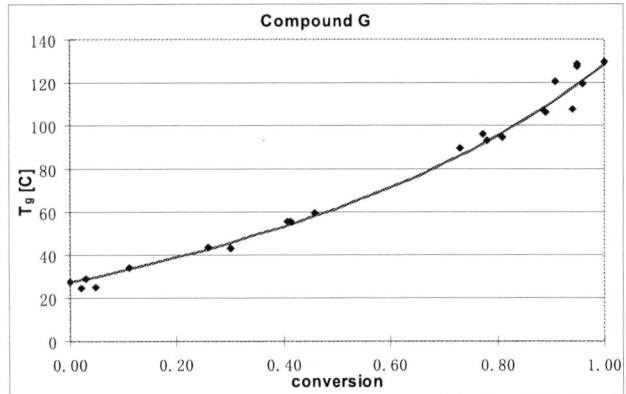

Fig.2: Glass transition versus conversion obtained from isothermal cure experiments. Full line: DiBenedetto fit.

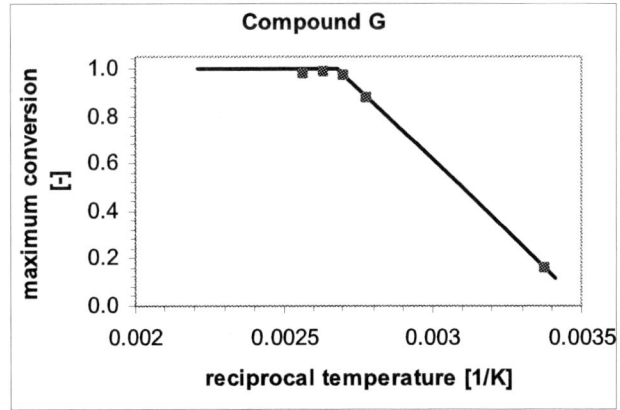

Fig.3: Measured maximum conversion levels (symbols) and fit to Equation 6 (full line)

	A	C	E	F	G	H
T_{g0}[°C]	29.7	20.6	32.4	17.8	23.7	17.8
T_{g1}[°C]	136.6	114.4	124.6	103.9	132.8	81.9
λ [-]	0.51	0.251	0.357	0.566	0.559	0.608
a_0 [-]	4.446	4.469	4.130	4.408	4.226	4.876
a_1 [K]	1289	1293	1181	1231	1204	1382

Table 1: Parameters for diffusion limitation and the DiBenedetto equation

5. Heat rate tests

The DSC tests at different heating rates were used to obtain first estimates of the parameters in the (modified) Kamal-Sourour equation (Eqs.1 and 5). A typical result is shown in Figure 4 (full lines).

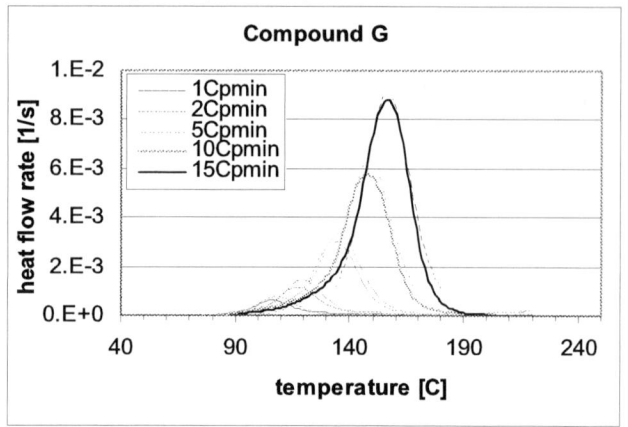

Fig.4: DSC results of heat rate tests (full lines) and final model predictions (dashed lines).

This shows a collection of heat flow curves with increasingly larger peaks and a maximum which shifts to higher temperatures with increasing heating rate. Notice that when these peaks are integrated with respect to time the total heat of reaction turns out to be approximately independent of the heating rate. For the molding compounds in this study these total heat of reaction varies between 12 and 30 J/g. The procedure for evaluating the kinetic constants is as follows:

o First determine the total heat of reaction H_{tot} and the initial glass transition temperature T_{g0} from the constant heat rate tests on uncured samples. These data are never exactly equal since part of the heat may remain unrecorded because of the error of DSC machine and H_{tot} also depends on the choice of the baseline.;

o Then calculate the activation energy E_a and the front factor k_0 from the slope of the reciprocal of the peak temperature versus the logarithm of the applied heating rate (Kissinger method [5,6]);

o Next divide the experimental reaction rate (Eq.7) by k_T using E_0 and k_0. In a plot of the thus obtained normalized reaction rate versus conversion all heat rate date then should overlap (Figure 5). The constants m and n are then obtained by fitting to the remaining part of Eq.5.

o From the isothermal experiments we first determine the DiBenedetto parameters and the diffusion limitation constants a_0 and a_1 (see section 5);

We then have a complete set of kinetic parameters which can be used to predict conversion levels at any desired temperature and curing time. As a check for internal consistency we therefore used the thus determined parameter set to make predictions for the isothermal cure data. For some materials (like compound F) this turned

out to work quite well, but for others there appeared to be a systematic under prediction of the conversion levels at lower curing times. There can be many reasons for this and below we will discuss this in more detail and propose a way to explain the difference.

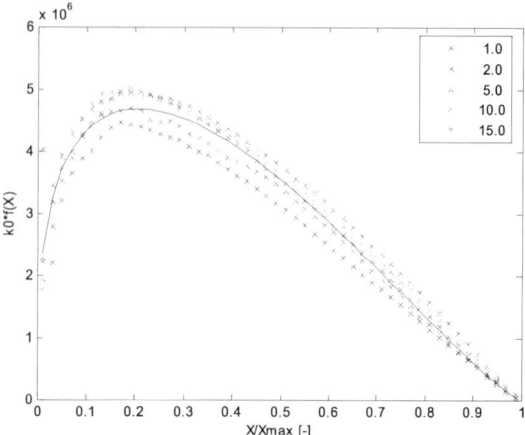

Fig.5: Normalized reaction rate plot (symbols) to determine constants m and n; Full line: fit.

The first explanation for the apparent inability to predict both heat rate and isothermal data with the same model is that the reaction kinetics is more complex than can be described with the present model. In literature a more general version of Eq.2 is proposed which contains the additional term $k_1 exp(-E_1/RT)[1-\alpha]^p$ (Prime, [5]). Alternatively, the experimental data can be tried to model using the so-called "model free kinetics" method [7]. In this method the activation energy is not a constant but is allowed to vary with conversion. Both methods are reported to be capable of reproducing experimental data in a satisfactory way. The problem is, however, that it is then necessary to introduce (many) extra fit parameters at the cost of a loss of simplicity relative to the original kinetic model. Moreover, the problem may not even be the lack of fit parameters since good fits for the heat rate data can be obtained with the present model for *several*, distinct, sets of parameters (k_0, E_0, m, n). A small change in the activation energy is for example compensated by a change in k_0 [7,8]. Furthermore also simultaneous changes in m and n (and k_0) are possible which hardly affect the predictions for the peaks in the heat rate experiments but have a rather large effect on the isothermal predictions. It should therefore be possible to optimise the kinetic parameter set for the present model such that it fits both sets of experiments. The predictions then will no longer result in perfect agreement with the experimental data but this can be justified by considering the fact that for the present materials there is always some uncertainty in the experimental data itself.

The explanation reason has to do with the assumed temperature dependency (Eq.2). As already mentioned, the isothermal cure tests are restricted to the temperature region below the glass transition whereas the heat rate tests are more sensitive to the region above T_g. This suggests that there can be a difference in activation energy below and above T_g. This was tested by shifting the isothermal cure curves along the log(t) axis until they overlapped (notice that this in fact uses the dimensionless cure time concept as mentioned in section 2). The temperature dependency of the shift factors, however, revealed that the change in activation energy was small.

The last explanation for the discrepancy between heat rate and isothermal data is an initial conversion level which appears to be present for the isothermal data but not for the heat rate test data. This effect can be seen by focusing on the initial part of the measured isothermal data (Figure 1). It appears then that extrapolation to zero curing time does not result in a zero conversion level but in an apparent initial cure of about 5-15%. Such initial conversion levels may have resulted for several reasons. First of all, some cure during storage and sample preparation may have occurred. The second reason is the conversion during heating up. In order to minimize the cure during heating, a heat ramp of 20 °C/min was programmed. Even with this fast heating rate, it takes about 4 or 5 minutes to reach the required isothermal temperature. In practice, it takes even longer since the DSC software slows down the heating rate when approaching the desired temperature. The effect of initial conversion can be taken into account by modifying the lower integration limit to a non-zero value α_0 in equation 3. It then turned out that with initial cure levels of 0.007, 0.15, 0.095, 0.00, 0.05 and 0.03 for compounds A, C, E, F, G and H respectively a good agreement between measured isothermal conversions and predictions from the heat rate tests could be obtained. An example is shown for material G in Figure 1 (dashed lines are predictions, symbols with full lines are measured data). The predictions for the heat rate test are shown in Figure 4 as the dashed lines.

The recommended kinetic parameter sets for all materials evaluated in this study are listed in Table 2.

	A	C	E	F	G	H
k_0 [10^6 s^{-1}]	3.10	7.98	11.1	1.63	12.9	8.53
E_0 [kJ/mol]	66.1	71.3	72.7	64.7	70.2	71.3
m [-]	0.71	0.272	0.168	0.344	0.671	0.313
n [-]	1.60	1.13	1.07	0.98	1.35	1.23
H_{tot} [J/g]	21.7	13.85	21.73	12.18	30.1	15.5

Table 2: Kinetic parameters for Equations 2 and 5

6. Conclusions

The kinetic parameters for six commercial molding compounds were carefully evaluated from a set of heat rate tests and isothermal cure tests. In particular care was taken to also have reliable predictions in the lower temperature region. For that purpose the kinetic model was modified to include the diffusion limitation effect (maximum cure levels at temperatures below the glass transition). These maximum cure levels were seen to be proportional to the reciprocal of the absolute cure temperature. The conversion dependency of the glass transition temperature was modelled using the well known DiBenedetto equation.

Acknowledgments

We kindly acknowledge the material suppliers Hitachi, Shin-Etsu and Sumitomo for supplying the molding compounds used in this study.

References

1. Jansen, K.M.B., Wang, L., van 't Hof, C., Ernst, L.J., Bressers, H.J.L., Zhang, G.Q., "Cure, Temperature and time dependent constitutive modelling of moulding compounds", EuroSimE Conference, Brussels, May 2004, p.581-585

2. Yang, D.G., Jansen, K.M.B., Ernst, L.J., Zhang, G.Q., Bressers, H.J.L., Janssen, J.H.J., Measuring and modelling the cure dependent rubbery moduli of epoxy resin, EurosimE Conference, Berlin, April 2005

3. Yang, D.G., Jansen, K.M.B., Ernst, L.J., Zhang, G.Q., van Driel, W.D., Bressers, H.J.L., "Modelling of cure-induced warpage in plastic IC packages", EuroSimE Conference, Brussels, May 2004, p.33-40

4. Gonzalez, V.M., Casillas, N., "Isothermal and temperature programmed kinetic studies of thermosets", *Polym Eng Sci*, 29 (1989), pp. 295-301.

5. Prime, R.B., "Thermosets", in "Thermal Characterization of Polymeric Materials", E.A. Turi, Ed., Academic Press, New York (1997), pp.1380-1766

6. Starink, M.J., "The determination of activation energy from linear heating experiments", *Thermochimica Acta*, 404 (2003), pp.163-176

7. Salla, J.M. et al., "Isoconversional kinetic analysis of a carboxyl terminated polyester resin", *Thermochimica Acta*, 388 (2002), pp.355-370

8. Malek, J., "The kinetic analysis of non-isothermal data", *Thermochimica Acta*, 200 (1992), pp.257-269

A Study of Failure Mechanism and Reliability Assessment for the Panel Level Package (PLP) Technology

Ming-Chih Yew[1], Hsiu-Ping Wei[1], Ching-Shun Huang[2], Dyi-Chung Hu[2], Wen-Kung Yang[2] and Kou-Ning Chiang[1]

[1]Advanced Microsystem Packaging and Nano-Mechanics Research Lab., Dept. of Power Mech. Eng.
Advanced Packaging Research Center, National Tsing Hua University, HsinChu, Taiwan.
Phone: 886-3-5742925   Fax: 886-3-5745377   E-mail: knchiang@pme.nthu.edu.tw

[2]Advanced Chip Engineering Technology Inc.
No. 65, Kuang-Fu North Rd., Hsin-Chu Industrial Park, Hu-Kou, HsinChu, Taiwan

Abstract

In this study, a new packaging technology, chip-on-metal (COM) panel level package (PLP), is proposed to resolve the problem of assembling a fine-pitched chip to a coarse-pitched substrate. During the manufacturing process, the filler polymer material is selected to fill the trench around the chip and provide a smooth surface for the redistribution lines. Therefore, the solder bumps could be located on both the filler polymer and the chip surface, and the pitch of the chip side is fanned-out. In our previous research, it was shown that the thermo-mechanical behavior of the COM PLP is different from the convention wafer level package (WLP) because of the designed packaging structure. In this study, the reliability characteristic of the proposed PLP technology is investigated and discussed through finite element analysis (FEA). The macro-micro modeling methodology is applied to assist in the reliability assessment of the trace/pad junction. From the simulated results, the mean cycle to failure of the solder joints can be highly increased by the proposed packaging technology. However, the new failure mode may happen at the metallic redistribution layer. The reliability of the signal trace in the COM PLP can be improved by an experienced design of the trace layout. Thus, the proposed PLP technology will have a high potential for various applications in the near future.

1. Introduction

Wafer level package (WLP) is a cost-effective solution for electronic packaging, which has been increasingly applied during recent years. The requirements to be fulfilled by applying this process are easy and stable manufacturing, low cost, good testability and sufficient electrical plus thermal performance. Besides, the WLP is preferred because of its re-workability which could increase the yield of product in multi-chip module and integrated circuit (IC) stacking module. However, because of the direct mounting of electronic components on the organic board, the coefficient of thermal expansion (CTE) mismatch in the packaging usually needs to be concerned and the reliability issues of the WLP with a large die size remain a problem.

To resolve the reliability problems produced by the CTE mismatch between the chip and organic substrate, many improvements have been, and are being developed. One of the improvements to the WLP is the soft stress-buffer-layer (SBL) structure, which is formed under the solder bumps to release the stresses in the solder joints [1]-[3]. This thick SBL coated on the die side can increase packaging reliability, but the yielding rate of this process is low. Underfilling (or undermolding) is another solution to the reliability problem, however the re-workability of WLP gets worse, and other solutions are sought to avoid excessive solder straining [4]. In 2000, Kawahara [5] proposed the structure of high copper posts to increase the gap between the chip and PCB, so as to reduce the stresses in the solder joints. On the other hand, Yew et al. [6] found that the stresses from CTE mismatch can be released by using a delaminating layer underneath the solder joints. Although these proposals provide effective modifications of the traditional WLP and enhance its reliability, less of them discussed about the assembly problem of WLP which is becoming serious nowadays.

As the improvement of the IC manufacturing process, the density of the transistor as well as the metallization in each die increase progressively. The difference of pad pitch between the die side and the substrate side becomes more and more obvious because of the limitation of fabrication technology and cost in the substrate. To fan-out the interconnection from the chip side, the chip carriers and lead frames are commonly chosen as the intermedium. In 2005, Yuan et al. [7] proposed a glass WLCSP to resolve the challenge faced by packaging house by transferring the twelve-inch semiconductor wafer into eight-inch packaging equipment. However, the result still had reliability problems and needed to be further improved. In this study, the chip-on-metal (COM) panel level package (PLP) technology, which is partly based on preceding technology, was proposed. The chip in the proposed package is first attached to a specific panel used as the chip carrier, and then the trench between the chips is filled with the selected material. As shown in Fig. 1, the solder bumps could be located on both the filler polymer and the chip surface, and the pitch in the chip side can be extended through this manner.

In the COM PLP, the solder on polymer (SOP) structure was developed to expand the chip area and also provide a buffer layer for the deformation energy from the

CTE mismatch. From the investigation [8]-[9], it is shown that the thermo-mechanical behavior of the COM PLP is different from the conventional WLP due to the application of new materials and structure. In this study, the failure mechanism and reliability assessment for COM PLP in board level will be discussed by using finite element analysis (FEA). Both the reliability analyses of solder joint and metallic trace are considered, and the suggested design rules will be provided.

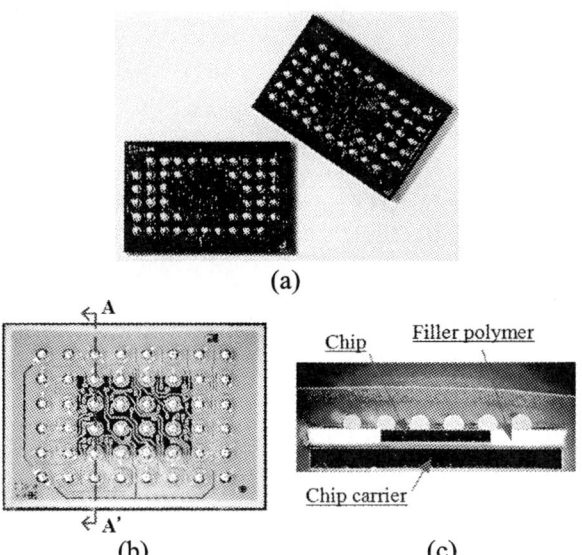

(a)

(b)   (c)

Figure x. (a) The focused testing sample in this study; (b) The IC device packaged using COM PLP technology; (c) The cross-sectional view of A-A'.

2. Fundamental theory

2.1 Solder joint thermal fatigue life prediction

Early solder joint fatigue models were developed based on experimental thermal cycling tests. Most models that address fatigue require the stress/strain data in order to predict service life. However, with the decreasing size of the solder joint, the experimental collection of stress/strain data is becoming increasingly difficult, and FEA is becoming the more practical route for obtaining stress/strain relationships [10]. In this study, the three-dimentional finite element (FE) model is established to assist the prediction of solder joint fatigue life in the COM PLP. The maximum equivalent plastic strain after one temperature cycle is calculated and applied to the Coffin-Manson equation (Eq. 1) to estimate the number of cycles to failure of the solder ball under temperature cycling. The Coffin-Manson equation [11] is widely adopted by electronic packaging researchers to predict the fatigue life of packages, and the equation for the solder modelled here can be expressed as:

$$N_f = 0.4405 \left(\Delta \varepsilon_{eq}^{pl} \right)^{-1.96} \tag{1}$$

where N_f denotes the mean cycles to failure and $\Delta \varepsilon_{eq}^{pl}$ represents the incremental equivalent plastic strain per

cycle loading. The incremental equivalent plastic strain per cycle is defined as:

$$\Delta \varepsilon_{eq}^{pl} = \sum d\varepsilon_{eq}^{pl} \tag{2}$$

where $d\varepsilon_{eq}^{pl}$ represents the incremental equivalent plastic strain in each of the sub-steps within one cycle. The incremental equivalent plastic strain of each sub-step is defined as:

$$d\varepsilon_{eq}^{pl} = \frac{\sqrt{2}}{3} \sqrt{ \begin{array}{l} (d\varepsilon_x^{pl} - d\varepsilon_y^{pl})^2 + (d\varepsilon_y^{pl} - d\varepsilon_z^{pl})^2 \\ + (d\varepsilon_z^{pl} - d\varepsilon_x^{pl})^2 + \frac{3}{2}(d\gamma^{pl})^2 \end{array} } \tag{3}$$

where $d\gamma^{pl2} = (d\gamma_{xy}^{pl})^2 + (d\gamma_{yz}^{pl})^2 + (d\gamma_{zx}^{pl})^2$ and $d\varepsilon_x^{pl}$, $d\varepsilon_y^{pl}$, $d\varepsilon_z^{pl}$, $d\gamma_{xy}^{pl}$, $d\gamma_{yz}^{pl}$ and $d\gamma_{zx}^{pl}$ are the incremental equivalent plastic strain components of each sub-step acting on the solder joint.

2.2 Principle of the macro-micro modeling

In conventional WLP, the solder bumps are directly formed on the pad which is defined on the surface of the chip. Generally speaking, the solder joints in WLCSP are the weakest portion because the most energy induced by CTE mismatch is absorbed by the junction. Nevertheless, the SOP structure in the COM PLP makes the connection between the chip and the substrate soft such that, and the reliability of the signal trace should also be concerned. Due to the relatively small dimension and the complicated variance of trace structure in the real testing sample, it is difficult and time-consuming to use a single finite element model to simulate the details of the interconnection features. Therefore, the macro-micro (global-local) modeling methodology is performed in this study and the selected portion of copper trace is modeled in the micro model. Fig. 2 shows the main steps of macro-micro modeling.

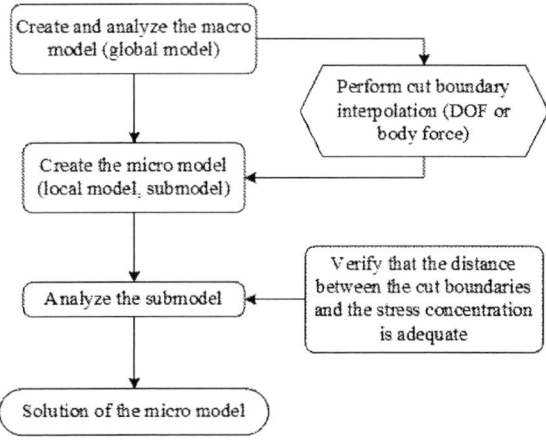

Figure 2. The flow chart of the macro-micro methodology in finite element analysis.

The concept of macro-micro modeling is to build two models. The first one is a macro (global) model with a relatively coarse mesh pertaining to the global boundary conditions (BC). The second one is called a micro (local) model, containing a detailed mesh for the specific area of interest. The micro model's BC is inherited from the global model by interpolating the nodal displacements and body forces from the global model. The accuracy of the micro model is dependent on the coarseness of the macro model and the details of the micro model [12]-[13]. It should be noted that the effect of the representative unit cell (RUC) in the macro model is assumed to be neglected, and it will be considered in the near future.

3. Design specifications of the COM PLP

The proposed COM PLP applies the filler polymer to broaden the area of chip, thus providing the base material when proceeding to the interconnect redistribution. As shown in Fig. 3, the manufacture of the PLP can be accomplished by a semiconductor processing compatible technology. First, the chips are diced from the wafer and back-sided attached to the chip carrier which has a bigger area. The selected material of chip carrier is metal or metallic alloy in order to enhance the thermal performance of the PLP. Therefore, the accumulated heat from the IC device can be effectively dissipated through the back-sided carrier. The trenches between the chips are filled with a special filler polymer and supply good coplanarity at the edge. Then the photo-pattern-able lamination, which was used as the bottom insulating layer (BIL) of redistribution metal traces, is spin coated on the surface and defined by lithography process. In the next step, as shown in Fig. 3(c), the Cu redistribution layer is deposited by electroplating after sputtering. Subsequently, the second lamination, which was adopted as the protection layer, is spin coated and defined to form the solder pads. Finally, a standard solder ball attachment process, which includes UBM depositing, ball placement and reflow, is used to accomplish the COM PLP.

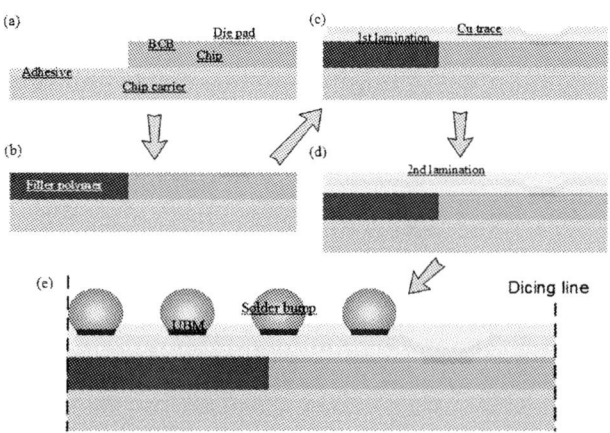

Figure 3. Schematic illustration of the manufacturing process of the COM PLP.

In this study, the IC device with a chip size of 2.5mm × 3.5mm was selected to conduct the developed PLP technology (Fig 1(a)). The filler polymer and lamination material are chosen to have a small Young's modulus, i.e. 50MPa and 90 MPa, thus further providing the stress buffer layer for the solder joints. The exterior dimension after packaging is 4mm × 6mm, and the pitch of the contact solder pads is 0.5mm. The solder pads on the chip are designed to be solder mask defined with a diameter of 0.29mm. Besides, the diameter of the 63Sn/37Pb solder bump is 0.3mm. It should be noted that the original pitch size of the chip is about 0.1 mm and has been successfully extended to 0.5mm by the proposed PLP technology without any intermediate.

Table 1. The material properties of the chip-on-metal (COM) PLP in the finite element model.

	Young's modules (GPa)	Poisson ratio	CTE (ppm/°C)
Chip carrier	148	0.3	5
Solder mask	3.4	0.35	30
BCB	3	0.34	50
Lamination	0.09	0.41	150
Cu trace, pad	Non-linear	0.3	17.3
Filler polymer	0.05	0.4	167
Chip	129	0.28	2.62
Printed circuit board	18.2	0.3	16
Solder bump (Sn63/Pb37)	Temp. dependent	0.35	23.9

4. FEA-based design for thermo-mechanical reliability assessment

4.1 Macro FE model for COM PLP

The COM PLP is a novel design concept, and its thermo-mechanical behavior in board level is predicted to be different from the conventional WLP during thermal loading. In order to understand the physical behavior of the proposed packaging technology, a three-dimensional FE model (macro model) was established first based on the real sample. The macro model was comprised of the silicon chip, the chip carrier, the filler polymer, the benzocyclobutene (BCB), the lamination material, the copper pad, the eutectic solder, the solder mask and the printed circuit board (PCB). All materials, except for the eutectic solder joint and non-liner copper material, were considered to have elastic properties as listed in Table 1.

In the FE model, the solder joint profile was obtained using the energy-based method, the Surface Evolver [14]-[15]. On the other hand, the FEA technique of multi-point constraint (MPC) was applied in the macro model for mesh transition. As shown in Fig. 4, the MPC interface was selected away from the location of interest, i.e. the solder joint, and effectively reduced the number of elements in the macro model. Due to the symmetry of the package, only a quarter of the COM PLP was modelled in the FE analysis. Symmetric boundary conditions were imposed on the XY and YZ planes of the package. In addition, the Y-direction displacement of the node at the XZ plane origin point was also constrained in order to prevent rigid body motion. Fig. 5 shows the quarter COM PLP finite element model as well as the applied boundary conditions. It consists of 60,000 elements and 67,500 nodes with a total 202,500 degrees of freedom (DOF).

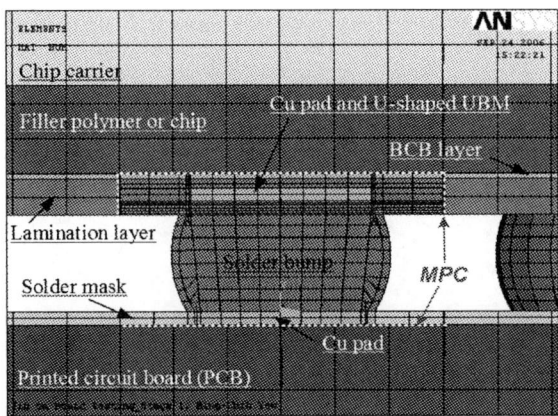

Figure 4. The cross-sectional view of the macro model at the solder bump region.

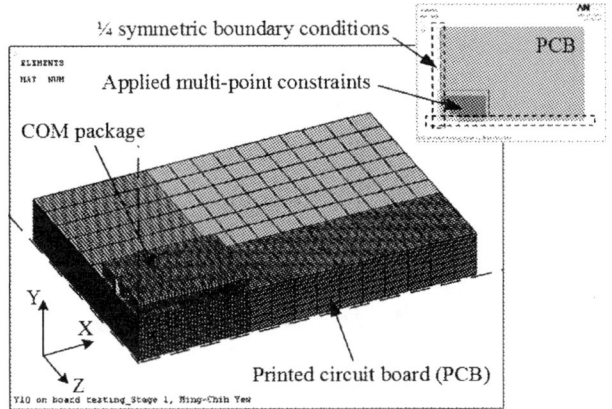

Figure 5. The three-dimensional finite element model (macro model) of the COM packaging structure.

4.2 Numerical results from the macro model

In the board-level FEA, the established FE model was subjected to a thermal cycle loading between -40°C and 125°C (JEDE22-A104-B, condition G). The initial stress-free reference temperature for all materials was 25°C, and the residual stress/strain from the manufacturing process was assumed to be zero. In this research, another three-dimensional FE model of a conventional WLP which has the same packaging dimension, i.e. 4mm × 6mm, was utilized as the comparison to the macro model of COM PLP. The thickness of the chip and solder mask in the FE model are 265μm and 50μm, respectively. Besides, the element mesh density in the interest location is controlled to be the same as the COM PLP FE model. As shown in Fig. 6, the maximum equivalent plastic strain in the conventional WLP happened at the outmost solder joint, and its value after one thermal cycle was 3.19%. From the

Coffin-Manson relationship (Eq. 1), the mean cycles to failure of the conventional structure is 376 cycles. The non-ideal life cycle is due to lack of the improved design, e.g. stress buffer layer, copper post, etc., for releasing the stress from the CTE mismatch.

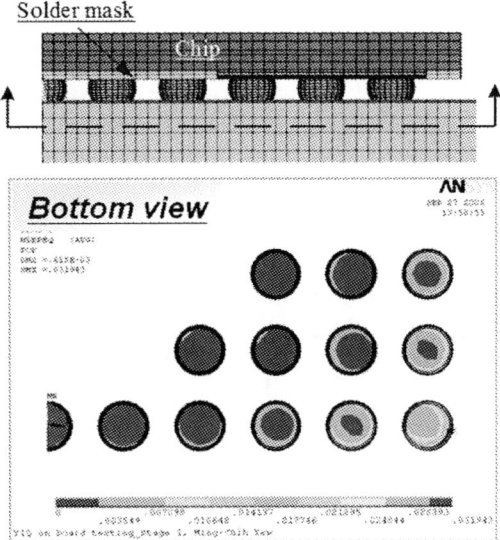

Figure 6. The equivalent plastic strain pattern in the solder joints of the conventional WLP structure. (Thermal loading : -40°C ~ 125°C)

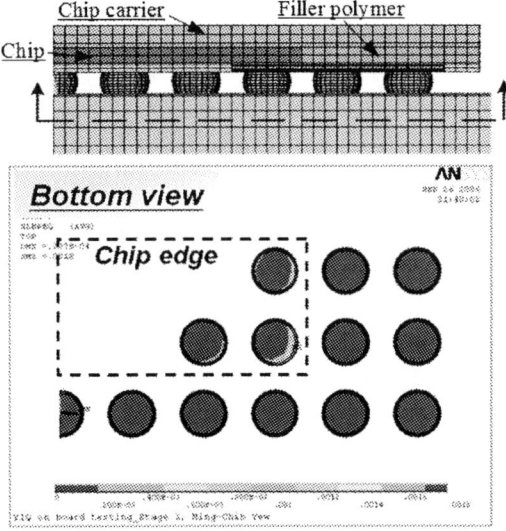

Figure 7. The equivalent plastic strain pattern in the solder joints of the proposed PLP structure. (Thermal loading : -40°C ~ 125°C)

In the FEA of the proposed COM PLP, the maximum equivalent plastic strain after one thermal cycle was effectively dropped to 0.18%. This extremely small strain value leads to a result in which the predicted life cycle for the solder joints is more than 100,000 cycles. It is explained that the major reason of the improvement is the application of soft filler polymer and lamination material. As shown in Fig. 7, the solder bump with the largest distance from the neutral point (DNP) no longer has the maximum strain value because the filler polymer is filled around the chip and works as the stress buffer layer. On the other hand, the lamination material in the COM PLP was used as the substitution of the solder mask in the conventional WLP. The Young's modulus of the lamination material is about one-fortieth to the solder mask; thus, the stress/strain from the CTE mismatch can be easily released by the deformation of the soft pad constraint.

4.3 Investigation of lamination layer and chip/polymer ratio

The effect of the selected lamination material in the PLP was investigated in this study by adjusting the thickness of the first lamination layer. In the real testing sample, the first lamination is used as the bottom insulating layer (BIL) of redistribution metal traces, and its thickness can be controlled by the process of coating. The range of the first lamination thickness is chosen from 3μm to 20μm in the FE analysis, and the corresponding maximum equivalent plastic strain in solder bump is shown in Fig. 8. From the results, the strain value increases obviously as the thickness of BIL approaches zero. It should be noted that although the reliability of solders will get worse as the thickness of the first lamination decreases, the predicted life cycle from the Coffin-Manson relationship could still be more than 3,500 cycles. The material property of the selected lamination is elastic, and it can also be considered as a good buffer layer. It does not matter if the solders are located on the filler polymer or the chip.

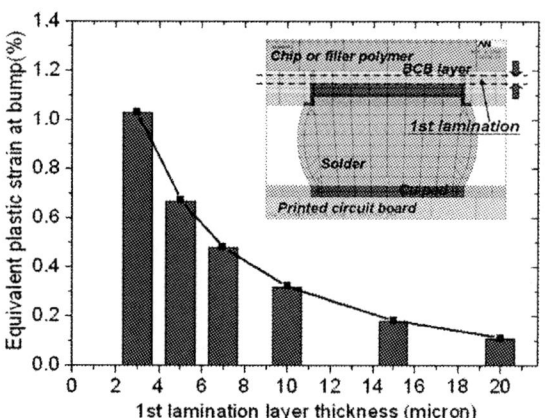

Figure 8. The first lamination layer thickness versus the maximum equivalent plastic strain at bump in the COM packaging structure.

In the developed PLP technology, ICs with different applications and dimensions can be chosen and put on the panel. The filler polymer is used to fill the trenches between the chips and provide a smooth surface for metal redistribution. Based on the primary design concept, a smaller chip/polymer ratio means that the COM PLP has a better fan-out capability. The effect of the chip/polymer

ratio to the packaging reliability is an important issue, and it is a crucial to study this effect through the FEA. In the investigation, six chips with different dimensions of diagonal (1.601mm, 2.016mm, 2.236mm, 2.358mm, 2.574mm and 3.132mm) are selected and put on the same metallic carrier. Fig. 9 shows the maximum equivalent plastic strain in the solder bumps to the diagonal dimension of the chip. Generally speaking, it is found that the focused strain value does not show a clear relation to the chip size.

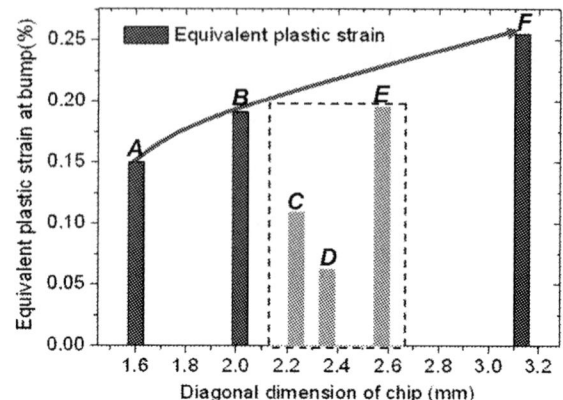

Figure 9. The chip size versus the maximum equivalent plastic strain at the bump. (A, B, F : w/o solders on the chip/polymer edge; C, D, E : with parts of solders on the chip/polymer edge)

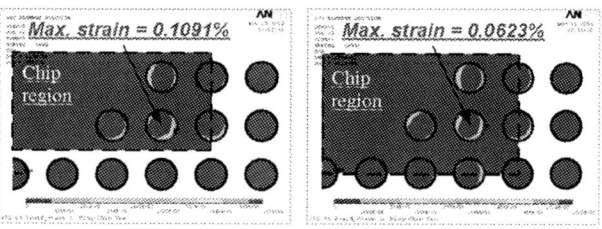

(a) Case C from Fig. 9   (b) Case D from Fig. 9

Figure 10. The equivalent plastic strain pattern in the solder joints of the COM packaging structure under different chip sizes. (a: chip diagonal = 2.236 mm; b: chip diagonal = 2.358 mm)

In fact, two different conditions exist in the six models. In cases A, B and F, there are no solder joints located on the chip/polymer edge, and the predicted plastic strain increases gradually as the dimension of the chip goes up. On the other hand, in cases C, D and E, some solder joints are placed on the chip/polymer edge, and a new physical behavior needs to be considered. By comparing Fig. 7 (case B) and Fig. 10 (cases C and D), it is found that the solders located on the chip/polymer edge can assist in sharing the stress/strain from the CTE mismatch; therefore, the solders in the chip region suffer from a lower strain during thermal cycling. In this study, the reliability assessment of the solder joints in the proposed PLP technology is investigated through the FEA of the macro model. It was found that the predicted life cycles of solder joint are more than 3,500 cycles among all macro level analysis. The outstanding performance is explained by the application of soft filler polymer and lamination material. However, it is also learned that the deformation of the lamination material during thermal loading may affect the reliability of metallic trace in PLP. Therefore, the reliability assessment of copper trace needs to be further studied.

4.4 Macro-micro modeling for COM PLP

In the real sample of PLP, the redistribution lines are embedded in the lamination, and the layout of signal trace is arbitrary. It is difficult to describe the thermal-mechanical behavior of the embedded lines by using a single finite element model. Therefore, the two-level macro-micro modeling methodology in FEA is applied to simulate the details of the interconnection features. Fig. 11 shows the corresponding position between the macro and micro models. There are 12 micro models that included copper trace near the 12 solders in the macro model of PLP. In the micro model, one copper trace with a 40μm width is built close to the pad region, and the nodal displacements of the boundary are interpolated by the data from the macro model. As shown in Fig. 12, the three-dimensional FE model consists of only 4,536 elements, thus effectively reducing the complexity when studying the signal trace reliability using a macro model.

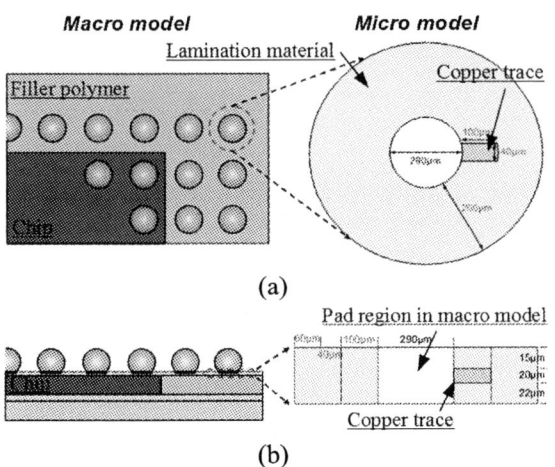

Figure 11. Illustration of two-level macro-micro modeling for COM packaging structure. (a: top view; b: cross-sectional view)

In the micro level FEA investigation, the thermal loading of 25°C to 125°C was chosen. The geometric BC was interpolated from the results of the macro model at 125°C. In addition, the residual stress/strain at reference temperature (25°C) was also assumed to be zero. Due to the material property of the large CTE, it is predicted that the thermal expansion/shrinking of lamination during temperature loading may deform the copper lines and affect their reliability. However, every micro model belonging to each solder joint in the PLP has a different thermal-mechanical behavior because the effects from its

surrounding are complex. Therefore, in this research, the direction of the trace/pad junction is modified in the parametric micro model, and the reliability assessment of the copper trace is performed through this manner. Fig. 12 shows the micro model with copper trace in zero degree. Eight degrees for the direction of the trace/pad junction are chosen from 0° to 360°, with an interval of 45°. On the other hand, in order to describe the phenomenon of cycled deformation, the focused failure criterion is the maximum equivalent plastic strain in the metallic line.

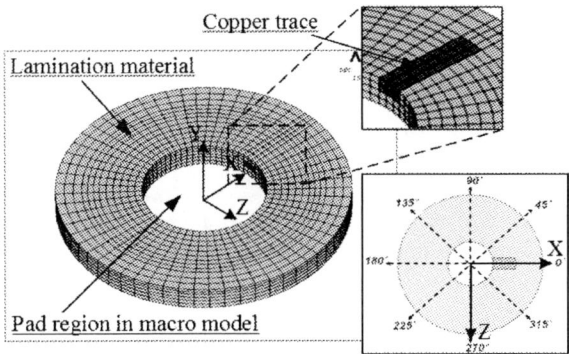

Figure 12. The micro model for the macro-micro analysis, and its layout for the effect of trace/pad junction direction.

In the FEA, all solders in the PLP model are numbered and will be discussed in groups. The solder joints in the polymer region which are located near the edge of polymer/chip can be classified as one group. Considering the solder joints belonging to this group, the copper traces which pass the edge of the polymer/chip are concern. As shown in Fig. 13, the predicted plastic strain in solders 8, 9 and 10 has a relatively high and unstable value as the direction of the trace/pad junction changes from 180° to 360°. Because of the CTE mismatch among the filler polymer, lamination and chip, the edge of the polymer/chip has a large deformation, and this is where the trace/pad junction should not be located.

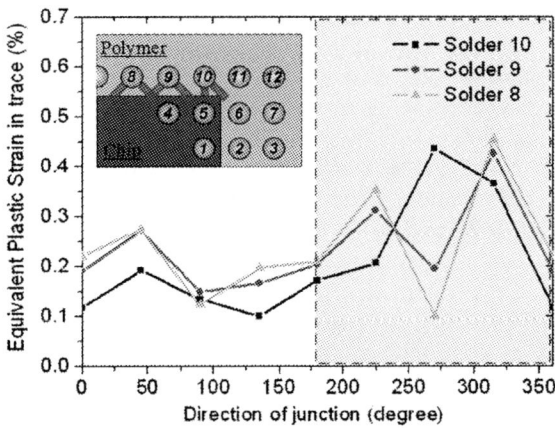

Figure 13. The direction of the trace/pad junction versus the maximum equivalent plastic strain in trace. (temp. = 125°C, solder number = 8, 9, 10)

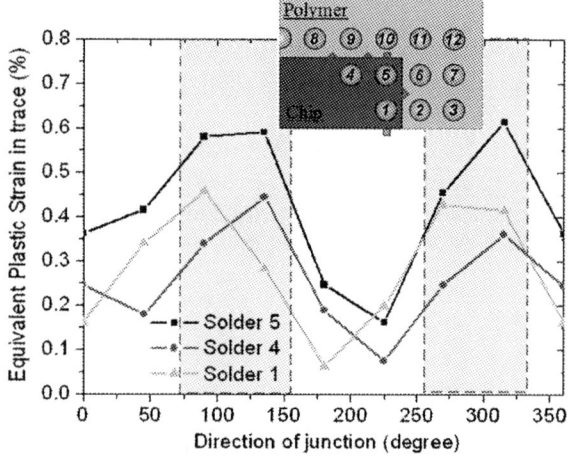

Figure 14. The direction of the trace/pad junction versus the maximum equivalent plastic strain in trace. (temp. = 125°C, solder number = 1, 4, 5)

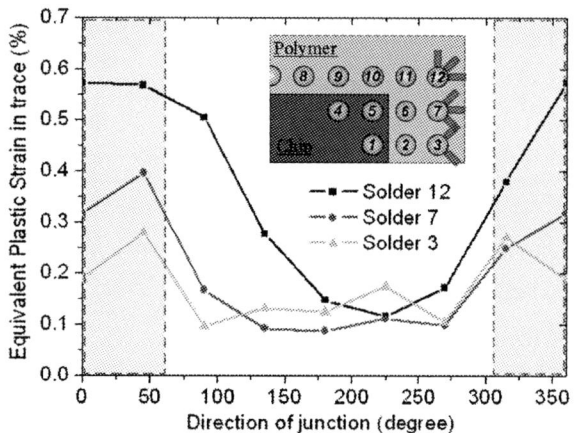

Figure 15. The direction of the trace/pad junction versus the maximum equivalent plastic strain in trace. (temp. = 125°C, solder number = 3, 7, 12)

The solder joints located in the chip region, e.g. solders 1, 4, and 5, are another classified group in the COM PLP. Due to the stiff material of the chip, the in plane expansion (or shrink) of the lamination material is the major reason which affects the reliability of the trace/pad junction. The better design of the junction is its possession of a direction parallel to the CTE mismatch, which can provide a stronger structure to resist the in plane deformation. As shown in Fig. 14, the junction has a higher predicted strain value in the red region, and they are in a direction perpendicular to the CTE mismatch where the phenomenon of bending may easily occur. It should be noted that the effect mentioned may become worse as the DNP of the chip increase; therefore, the design rules are important and need to be established carefully.

Fig. 15 shows the variation of the predicted plastic strain as the junctions are placed around the solder which is close to the edge of the chip carrier. The bathtub-type

curves show that the relatively worse cases in the FEA happen when the direction of the trace/pad junction is close to the edge of the filler polymer. The application of the filler polymer to broaden the chip side area is the design concept of PLP technology. However, a big portion of polymer may induce the warpage in the packaging structure. The warpage effect in PLP depends on the shape of the package, the ratio of the chip/polymer, and the like. Solders 3, 7, and 12 are selected as the representatives to explain the thermo-mechanical behavior near the outer edge of the filler polymer. As shown in Fig. 16, the thermal expansion of the filler polymer curves the chip carrier at 125°C. Besides, the gradient of y-displacement in the lamination material increases near the edge of the package. Therefore, the copper traces embedded in the lamination layer may be deformed, and reliability becomes a concern. One of the possible approaches to improve the reliability of signal traces near the packaging edge is decreasing the expansion (or shrinking) of the polymer material through physical (e.g. advanced structure design) or chemical (e.g. modification of material property) strategies.

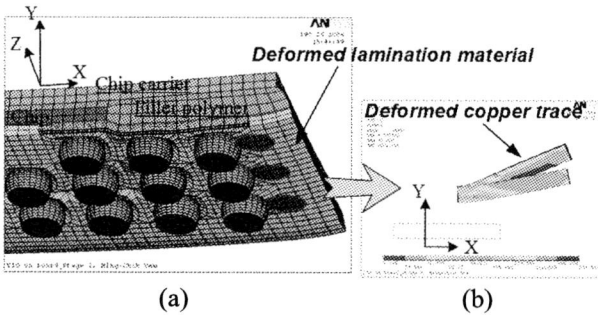

(a)   (b)

Figure 16. (a) The geometric deformation of COM packaging in the macro model at 125°C; (b) The deformation of copper trace in the micro model.
(Scale factor = 15)

5. Conclusions

The panel level package (PLP) technology with fan-out capability was proposed in this study. The signals from an IC with small pitch can be expanded by using the redistribution layers on the applied filler polymer material. From the FEA, the predicted life cycles of solder joints in the PLP were found to have an outstanding performance of more than 3,500 cycles among all macro level analysis. On the other hand, the direction of the trace/pad junction of the redistribution lines was also investigated in this study using the macro-micro modeling methodology. The thermo-mechanical behavior of the proposed COM PLP was different from the conventional WLP, and the reliability assessment was discussed in groups. Generally speaking, the signal traces can achieve better reliability by wisely dealing with the traces near the edge of the polymer/chip and the chip carrier. On the other hand, experiments based on the proposed FE results are on-going, which are planned to validate the design rules of

COM PLP. The PLP technology is being developed and will have a high potential for various applications in the near future.

Acknowledgments

The authors would like to thank the National Science Council (Project NSC94-2212-E-007-017), and the members of the Advanced Chip Engineering Technology (ACET) for supplying the data on the PLP technology.

References

1. Badihi, A., "Ultrathin wafer level chip size package", *IEEE Tranactions on Advanced Packaging*, Vol. 23, No. 2 (2000), pp. 212-214.

2. Lau, J. H. and Lee, R., Chip Scale Package, CSP: Design, Material, Processes, and Applications, McGraw-Hill, (Singapore, 1999).

3. Garrou, P. E., Rogers, W. B., Scheck, D. M., Strandjord, A. J. G., Ida, Y. and Ohba, K., "Stress-Buffer and Passivation Processes for Si and GaAs IC's and Passive Components Using Photosensitive BCB: Process Technology and Reliability Data," *IEEE Tranactions on Advanced Packaging*, Vol. 22, No. 3 (1999), pp. 487-498.

4. Chiu, C. C., Wu, C. J., Peng, C. T., Chou, C. Y. and Chiang, K. N., "Reliability Impact of Highly Temperature-Dependent Underfill Material to the Lead-Free Flip Chip Package," *Proceedings of 7th EuroSimE international conference*, Como, Italy, April 2006, pp.328-335.

5. Kawahara, T., "SuperCSP™", *IEEE Tranactions on Advanced Packaging*, Vol. 23, No. 2 (2000), pp. 215-219.

6. Yew, M. C., Chiu, C. C., Chang, S. M. and Chiang, K. N., "A Novel Crack and Delamination Protection Mechanism for a WLCSP Using Soft Joint Technology," *Soldering and Surface Mount Technology*, Vol. 18, No. 3 (2006), pp. 3-13.

7. Yuan, C. A., Han, C. N., Yew, M. C., Chou, C. Y. and Chiang, K. N., "Design, Analysis and Development of Novel Three-Dimensional Stacking WLCSP," *IEEE Tranactions on Advanced Packaging*, Vol. 28, No. 3 (2005), pp. 387-396.

8. Yew, M. C.,Yuan, C., Han, C. N., Huang, C. S., Yang, W. K. and Chiang, K. N., "Factorial Analysis of Chip-on-Metal WLCSP Technology with Fan-Out Capability," *Proceedings of 20th IEEE IPFA2006 Conference*, Singapore, July 2006, pp.223-228.

9. Yew, M. C., Chou, C. Y., Huang, C. S., Yang, W. K. and Chiang, K. N., "The Solder on Rubber (SOR) Interconnection Design and Its Reliability Assessment Based on Shear Strength Test and Finite Element Analysis," *Journal of Microelectronic Reliability*, Vol. 46 (2006), pp. 1874-1879.

10. Lee, W. W., Nguyen, L. T. and Selvaduray, G. S., "Solder joint fatigue models: review and applicability to chip scale packages," *Microelectronics Reliability*, Vol. 40, No. 2 (2000), pp.231-244.

11. Solomon, H. D., "Fatigue of 60/40 Solder", *IEEE Transactions on Components, Hybrids, and Manufacturing Technology*, Vol. CHMT-9, No. 4 (1986), pp. 91-104.

12. Fan, X., Pei, M. and Bhatti1, P. K., "Effect of Finite Element Modeling Techniques on Solder Joint Fatigue Life Prediction," *Proceedings of 56th Electronic Components and Technology Conference*, San Diego, USA, May 2006, pp. 972-980.

13. Fiori, V. and Orain, S., "A Mutli Scale Methodology to Evaluate Wire Bond Pad Architecture," *Proceedings of 6th EuroSimE international conference*, Berlin, Germany, April 2005.

14. Brakke, K.A., "The Surface Evolver", *Experimental Mathematics*, Vol. 1, No. 2 (1992), pp. 141-165.

15. Chiang, K.N. and Yuan, C.A., "An Overview of Solder Bump Shape Prediction Algorithms with Validations", *IEEE Transactions on Advanced Packaging*, Vol. 24, No. 2 (2001), pp. 158-162.

A Novel Micro/Nano Electromechanical Actuator with Integral Electrodes

Jofre Pallarès[1], Manuel Carmona[2], Marta Duch[3], Marta Gerbolés[3], Lluís Terés[3]

[1]Barcelona International R&D core (BIRD), Centro Nacional de Microelectrónica-CSIC
Bellaterra (Barcelona), Spain
[2]Barcelona R&D Laboratory (BRDL), Epson Europe Electronics GmbH
Sant Cugat del Valles (Barcelona), Spain
[3]Institut de Microelectrònica de Barcelona, Centro Nacional de Microelectrónica-CSIC
Bellaterra (Barcelona), Spain
Manuel.Carmona@epson-electronics.de, tel: +34 5947700 Ext. 1201, Jofre.Pallares@cnm.es, Ext. 1212

Abstract

A novel electromechanical actuator (patent pending) is presented, based on the beam-like deformation of two parallel electrodes, both mounted together in the same device and with a small gap in between. It avoids the need of an actuator electrode placed on the substrate. Simulations, mainly performed with ANSYS, have been performed to obtain the performance of the actuator in static conditions. They show that 2D FEM models can be used with a reasonable accuracy. 3D models has to be more thoroughly developed, although a more realistic structure can be simulated, taking into account the electrical connection of the actuator electrodes . A simple analytical model has also been developed in order to estimate the static displacement of a beam with as many actuator electrodes as possible on top of it. The model is useful for a first estimation of device deformation. The working principle of the device is in this paper demonstrated by simulation, until future measurements are possible.

1. Introduction

Electromechanical actuators are used in a large number of MEMS (like micromirrors, microswitches, resonators, etc.) because of their good properties at small scale compared to other actuation principles [1, 2]. Usually the application of the force is based on placing one of the electrodes at the fixed substrate, while the second electrode is placed onto the movable component. Typical examples are microswitches, accelerometers, micromirrors, etc. [1, 5]. Apart from this typical parallel plate actuation configuration, other different ways to apply the electrostatic force have been developed by different groups [3, 4, 7] for in-plane and out-of-plane actuation.

A way to increase the force and, therefore, displacement for in-plane working devices is based on the array of comb-finger actuators [3]. Usually, these devices require a large area. For out-of-plane displacement, the way to increase the force is usually by increasing the area. Nevertheless, these devices are much more prone to problems like stiction [6]. These devices have usually a very slow response due to the large mass, damping effects and the small force applied at maximum gap conditions.

This work presents a new way to actuate electromechanically MEMS components. This actuator is still patent pending. It is based on building both electrodes on the same device and having a small gap between both electrodes, which can provide good characteristics to the device, like high force and large displacements with low applied voltage.

2. Device Description

The basic actuator is based on the electrostatic force applied between two parallel beams, separated by an electrically non-conductive material. When a voltage difference is applied between both beams and they are placed on the same device, both beams deflect in opposite direction. For a deflection in upward direction, we will choose as beam (to be deflected) the lower one, while the upper one will make the function of actuator electrode (which, in other devices from literature, is usually placed on the substrate). Different actuators can be put "in series" along the beam, allowing an increased maximum displacement of the beam. Figure 1 shows a cross-section of one of the possible implementations of this actuation method, as well as the working principle.

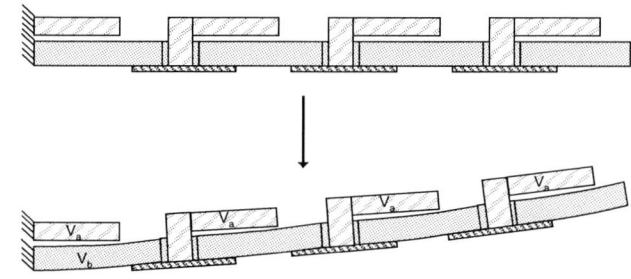

Figure 1. Cross-section of a possible implementation of the actuation principle (upper figure) and working principle (lower figure).

In this example, the lower layer is an electrically non-conductive material (which in this case is made of Si_3N_4; we will called it support material), the upper layer is the actuator electrode and the middle layer is the beam (doing also the function of second electrode). Beam and actuator electrodes are not electrically connected but they are mechanically joined through the support material. The beam itself is anchored to the substrate at the left-most part. Of course, in this figure the electrical connections of the actuator electrodes are not shown. When a voltage difference is applied between the beam and the actuator electrodes, the beam is deformed upwards at the area under the actuator electrodes until pull-in condition is

1-4244-1105-X/07/$25.00 ©2007 IEEE
332

achieved, where the beam and actuator electrodes will collapse. The deformation produced by every actuator electrode is summing up, being able to achieve a large displacement of the beam.

It could be easily imagined a way of interchanging both poly layers to obtain alternatively up or down deflection of the beam.

If the distance between the beam and the actuator electrodes is small (let's say generally under 1 μm, for typical material dimensions), the voltage needed to significantly deform the beam will be small, while achieving a large force. Of course, these values will depend on the specific thicknesses and material properties of every layer.

The expected characteristics of this actuator are the following:

- Low application voltage.
- Large displacement.
- Large force.
- Reduce response time.
- Not very large dimensions and reduced areas.

Different geometries can be used. Through this work, a T-type actuator electrode has been used, which is shown in Figure 2 (Figure 1 shows a cantilever type).

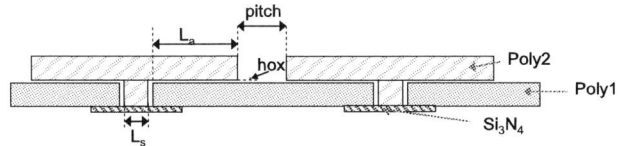

Figure 2. Schematic of the T-type actuator.

This figure also shows the dimension parameters that will be used in the next sections. L_a is considered the length of the actuator, L_s is the length of the support area, h_{ox} is the gap between beam and actuator electrode and pitch is the distance between consecutive actuators (fixed to 2 μm). The thicknesses of the different layers are fixed, with values of 0.18 μm for Si_3N_4, 0.5 μm for Poly1 and 1.0 μm for Poly2. The T-type actuator is expected to perform better than the cantilever type because the support and pitch areas are not contributing to the increase of displacement of the beam and this configuration uses one support area for every two actuator beams.

This device can have a large number of applications, for static and dynamic working regimes. Among them, it could be used for microswitches, microvalves, resonators, etc.

A similar approach to that used in this paper is by using electrodes attached to a mechanical material [7]. The application of a voltage to the electrodes generates a force between the electrodes which forces the bending of the mechanical material. In this case, the problem is to fabricate large enough (or thick enough) electrodes to induce significant bending to the mechanical material and place them close enough. The main difference is that the idea shown in this paper is more easily implemented with standard microelectronic processing and it could be fully implemented with standard CMOS process.

3. Actuator Analytical Model

The relationship between applied voltage and tip deflection of a cantilever beam has already been analysed in different publications [2, 8- 10] for a parallel plate configuration with one fixed electrode. Later we will try to consider deformation of the fixed electrode. In this work, we have made use of the expressions from Lin et al.

We will first focus the attention to a single beam actuator electrode. For DC voltage conditions, the change of angle from the anchor to the tip of the cantilever can be obtained as:

$$\left.\frac{dy}{dx}\right|_{u=1} = 1.375 \cdot \frac{y^*}{L_a}$$

where y is the beam position in the direction of the deformation from the actuator electrode, while y^* is the displacement of the tip. At pull-in voltage conditions, this angle will become:

$$\left.\frac{dy}{dx}\right|_{u=1} = 0.6325 \cdot \frac{y_o}{L_a} \equiv k_a \cdot \frac{y_o}{L_a}$$

where y_o is the initial gap between actuator electrode and beam. Pull-in voltage (or threshold voltage, V_{th}) is given by:

$$V_{th} = \sqrt{\frac{0.285 \cdot E \cdot (h \cdot y_o)^3}{\varepsilon_o \cdot L_a^4}}$$

where E is the Young's Modulus of the beam material. Pull-in is considered to occur at a beam deflection of:

$$y^* = 0.46 \cdot y_o \equiv k \cdot y_o$$

For a cantilever beam of length L_b with multiple actuators (of the T-type), the total deflection of the beam can be obtained by summing up all the displacements by taking into account the beam angle achieved by every single actuator. The expression for the total deflection of the beam (Δy) with multiple actuators is the following:

$$\Delta y = n_{act} \cdot \left[2 \cdot y^* + \left.\frac{dy}{dx}\right|_{u=1} \cdot \left(2 \cdot (L_b - d_s) - 3 \cdot L_a - L_s - L_p \cdot (n_{act} - 1)\right)\right]$$

where n_{act} is the number of T-type actuators over the beam, d_s is an initial offset of the first actuator with respect to the substrate anchor position. L_p is defined as:

$$L_p \equiv 2 \cdot L_a + L_s + pitch$$

At pull-in voltage, the expression can be developed as follows:

$$y_{max} \equiv \Delta y_p = n_{act} \cdot k \cdot y_o \cdot \left[2 + \frac{k_a}{L_a} \cdot \left(2 \cdot (L_b - d_s) - 3 \cdot L_a - L_s - L_p \cdot (n_{act} - 1)\right)\right]$$

If n_{act} is large, then it can be approximated by:

$$y_{max} \equiv \Delta y_p \approx \frac{L_b^2}{L_a \cdot (2 \cdot L_a + L_s + pitch)} \cdot k_a \cdot k \cdot y_o$$

If we fix L_b, n_{act} can be easily calculated and, therefore we can obtain the maximum beam displacement as a function of the other device parameters. Figure 3 shows Δy_p and V_{th} vs. L_a.

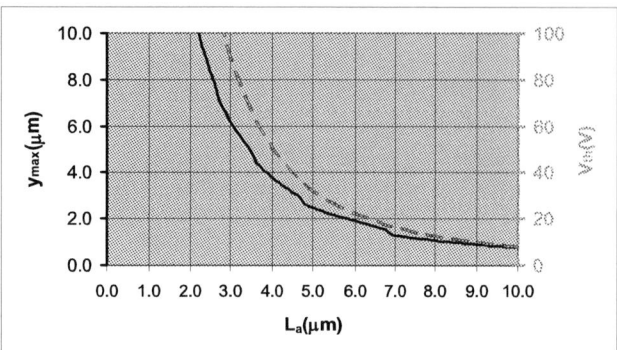

Figure 3. Maximum deflection (y_{max}, continuous line) and threshold voltage (V_{th}, dashed line) vs. actuator length (L_a).

From the maximum displacement, it can be observed some steps at different L_a values. This is due to the discrete change on the value of n_{act} as L_a is changed. As expected, a short L_a value provides higher y_{max} values, although at the price of an increased V_{th}. Another conclusion is that significant y_{max} values can be obtained at reasonable V_{th} values.

Ways to improve the device performance can be:
- Reduction of the beam thickness will provide smaller V_{th}.
- Increase of the gap will provide an increased displacement. If combined with thickness reduction, the displacement will increase for the same voltage.

In Figure 4 it is shown the same curve as previously shown, but with a beam with half the thickness and a double gap.

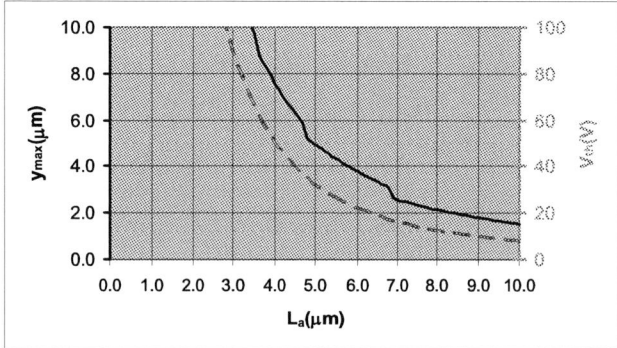

Figure 4. y_{max} and V_{th} vs. L_a for half beam thickness and double gap.

As it can be seen from this figure, the V_{th} curve is kept exactly the same (because they have the same dependency with V_{th}), but y_{max} is increased significantly.

If we consider also the deflection of the actuator electrode, y_{max} will become smaller because the pull-in condition will happen at a reduced voltage. As a first approximation, we can assume that the sum of displacements of both cantilevers will have to be ky_o. Taking into account that voltage is the same for both electrodes and assuming the same material for the beam and the actuator electrode, the displacement will only differ by their thickness ratio. Therefore, it can be obtained the following relationship:

$$\Delta y_p = n_{act} \cdot \frac{k \cdot y_o}{1 + (h_b / h_a)^3} \cdot \left[2 + \frac{k_a}{L_a} \cdot (2 \cdot (L_b - d_s) - 3 \cdot L_a - L_s - L_p \cdot (n_{act} - 1)) \right]$$

V_{th} can also be obtained, taking into account that there is a quadratic relationship between displacement and voltage. Figure 5 shows the same former figures, but for an actuator having double thickness than the beam and taking into account the deflection of the actuator.

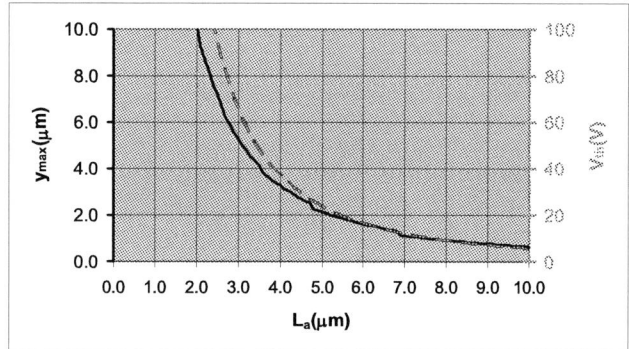

Figure 5. y_{max} and V_{th} vs. L_a taking into account the deflection of an actuator with double thickness than beam.

It can be seen that there is a sensitive reduction of the maximum displacement, accompanied by a reduction of the pull-in voltage. If the actuator has the same thickness, the difference is still more evident, as shown in Figure 6.

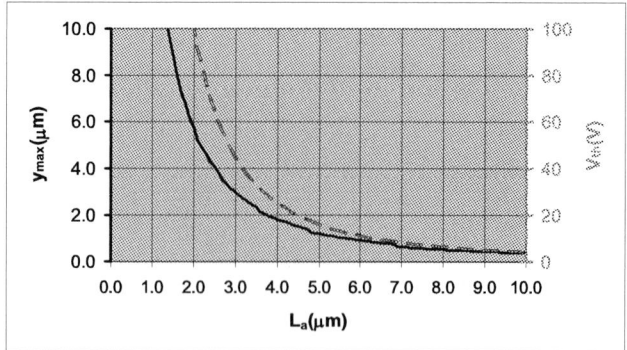

Figure 6. y_{max} and V_{th} vs. L_a taking into account the deflection of an actuator with same thickness than beam.

4. Simulation

Different methods can be used within ANSYS [11] in order to simulate electromechanical problems. The simplest method is to use TRANS126 elements, which allows us to avoid the meshing and re-meshing of the air during solution. TRANS126 is a one-dimensional electromechanical transducer element. The characteristics of this element can be extracted by electrostatic simulations using usual FEM elements or directly by assuming a parallel plate capacitor interaction. In our case, this last method can be used because of the small distance between the actuator electrode and the beam compared to lateral dimensions. Moreover, every connected node maintains a very similar relative position during the beam deformation with small z-displacements and, therefore, the error by using these elements should be very small.

Another option in ANSYS is to use its Multi-Field Solver (MFS). It should be more accurate but also more time-consuming, especially for 3D models.

For a first verification, a simple model has been built in ANSYS in order to simplify its analysis and the comparison with the analytical model. In this way, we avoid the influence of parameters in simulation results that are not taken into account in the model. Later on, the effect of these parameters will be analysed by simulation. This simple model consists of a simple beam with a rigid actuator electrode (by using a high value of its Young's Modulus). Electromechanical actuation is simulated with TRANS126 elements. Plane stress conditions were assumed. Figure 7 shows this 2D model.

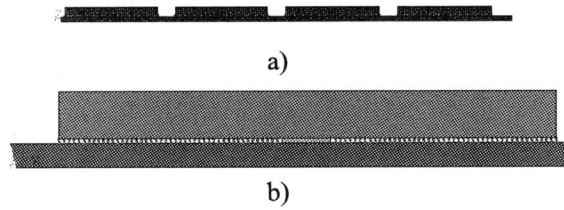

a)

b)

Figure 7. a) 2D simplified model, b) detailed view of T-type actuator, showing anchoring and TRANS126 elements.

Figure 8 shows how a magnified deformation of the structure looks like upon voltage application.

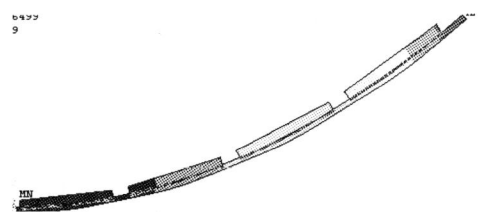

Figure 8. Example of structure deformation obtained by simulation (magnified x10).

Here we can observe for the right-most actuator the combination of z-translation and local deformation of the actuator electrode and beam, which represents one of the main issues when solving this kind of problems. Nevertheless, by using TRANS126 elements, this difficulty is overcome because we do not need to perform any remeshing of the air or any morphing process.

Figure 9 shows the maximum displacement of the beam vs. the applied voltage.

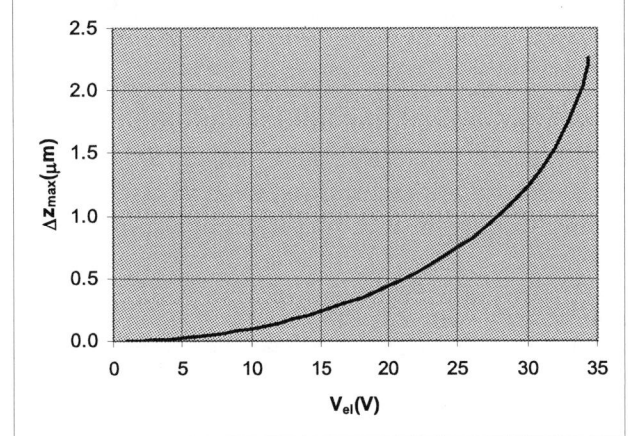

Figure 9. Beam maximum deflection vs. applied voltage for an ideal actuator, with L_a=4.6 μm.

This nearly quadratic dependency is as expected from the theoretical dependency of the electrostatic force with the applied voltage. This simulation can be used to extract the pull-in voltage, although a good resolution with the voltage is needed to extract the correct maximum achievable displacement. Comparing with the analytical model (Figure 3), we can observe that results are similar.

A similar result is obtained for a non-ideal actuator (i.e., without assuming that the actuator electrode is rigid), as shown in Figure 10.

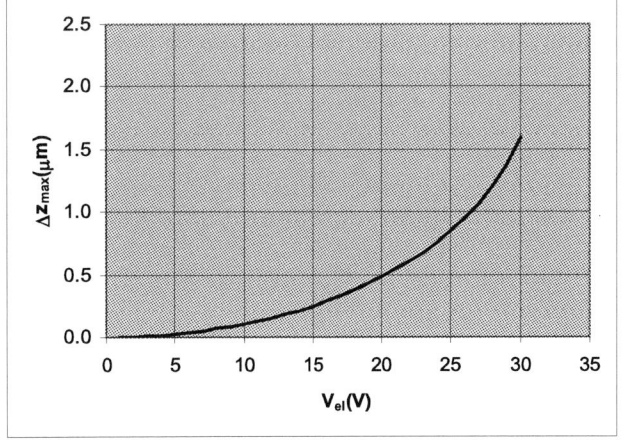

Figure 10. Beam maximum deflection vs. applied voltage for a non-ideal actuator (double thickness of the beam).

For every applied voltage, the maximum displacement is now higher due to the increased forced, as a result of a smaller distance between deformed actuator electrode and beam. Therefore, a smaller voltage is needed for the same beam displacement. On the other hand, pull-in voltage is reduced, as well as displacement of the beam at pull-in voltage. The result is that, as expected, the maximum achievable displacement (at pull-in voltage) is smaller than in the ideal case. But, if the applied voltage is well below pull-in, an increased displacement will be achieved with a non-ideal actuator electrode.

In order to compare these simulation results with the analytical model, the pull-in voltage has been obtained for several actuator dimensions. From literature [8], a value of the pull-in electrode distance should be $0.54y_o$ (2D static analysis) for an ideal actuator electrode. A 1D lumped model provides a value of $0.66y_o$. These ANSYS static simulations provided a value of $0.60y_o$, which represents an error of $\approx 10\%$ from the expected $0.54y_o$ value.

Another approach is to use TRANS109 elements, which uses a morphing process. These simulations have also been run but they crash due to largely deformed air elements. This is due to the combination of z-displacement and local deformation of the elements. It was found only one way to avoid it, by using only one layer of air elements. In this way, all TRANS109 nodes displacements are constraint to the deformation of the mechanical elements. Nevertheless, this could limit the validity of simulation results. In any case, simulation results are slightly nearer to expected results in this case. Now, in the ideal case, the obtained actuator pull-in electrode distance is $0.575y_o$, which is nearer to the theoretical value. Figure 11 shows the beam displacement vs. applied voltage.

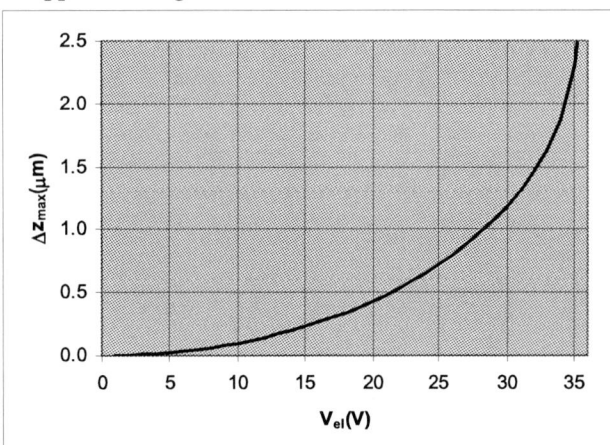

Figure 11. Beam maximum deflection vs. applied voltage for an ideal actuator obtained with TRANS109 elements.

Comparing with Figure 9, the results are quite similar. The differences are that pull-in voltage and, therefore, maximum displacement are slightly higher by using TRANS109 elements, which are even nearer to the expected values.

By using the MFS solver, the situation is similar to using TRANS109 elements. Only one layer of elements can be used in order to avoid non-convergence problems

due to large node deformations of air elements during morphing. Simulation time is increased considerably, but results are even more accurate (especially near pull-in conditions). Pull-in position is now $0.57\,y_o$, which is even nearer the $0.54y_o$ expected value (around 5% error). Figure 12 shows the result obtained for an ideal and a non-ideal actuator electrode.

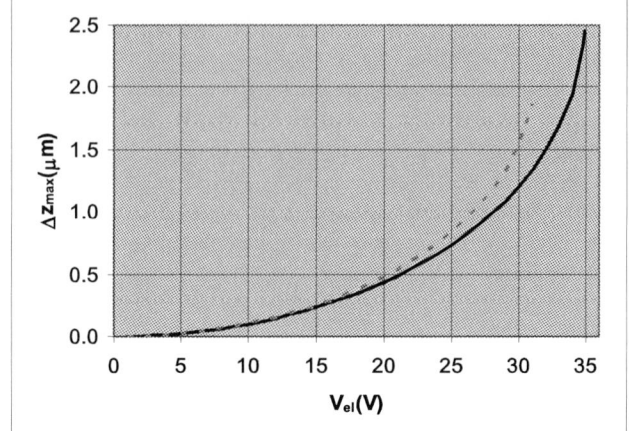

Figure 12. Beam maximum deflection vs. applied voltage for an ideal actuator (continuous line) and non-ideal actuator (dashed line) obtained with MFS solver.

A 2D model was developed, which is geometrically nearer to the real device implementation. In this model, the beam and the actuator electrode are connected via a thin material at the bottom part of device. This model and the device deformation (magnified x10) are shown in Figure 13.

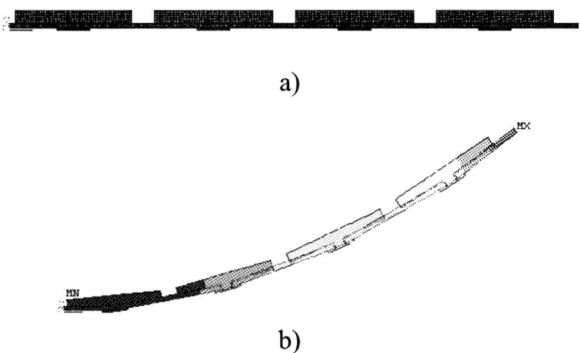

Figure 13. a) Model of a more real device and b) magnified x10 device deformation.

Here, the support is not as rigid as in the previous simpler model, which could impact significantly the device performance. Simulation results are shown in Figure 14.

This result can be explained by the smaller rigidity of the support. As the support material is now only a thin material, the beam can deform easily, allowing a larger displacement with a smaller voltage. This effect will not be present at the real 3D device, as the beam is not discontinued at the support areas.

336

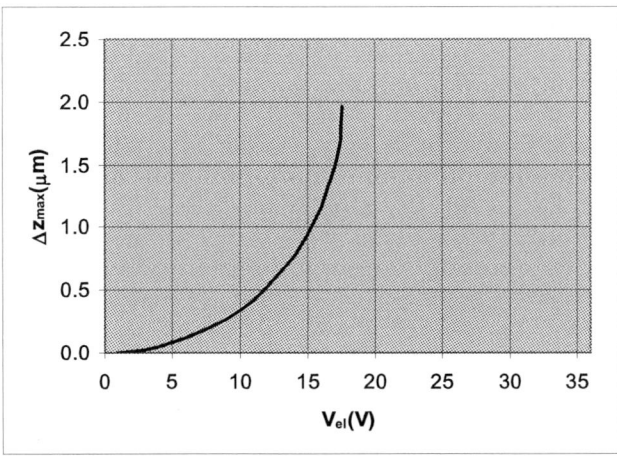

Figure 14. Simulation results obtained for a device more near to real device.

There are many aspects of the performance of this device that the 2D model is not able to take into account. For example, the mechanical effect of the electrical connections (which will contribute to a reduction of the beam displacement) for the actuator electrodes and the width limit of the beam (which could influence, for example, the rigidity of the actuator electrodes). Moreover, and as previously mentioned, the continuity of the beam surrounding the support areas (which will make the beam stiffer).

A 3D model has been developed by using, as before, TRANS126 for simulating the electromechanical coupling. Its use in 3D simulations represents and advantage compared to TRANS109. Also, the ANSYS Multi-Field Solver has been used, although it takes much longer to solve than with TRANS126. Figure 15 shows the model of the structure with TRANS126 elements, including the arm used for electrically connecting the actuator electrodes. It also shows the deformation shape obtained from simulations.

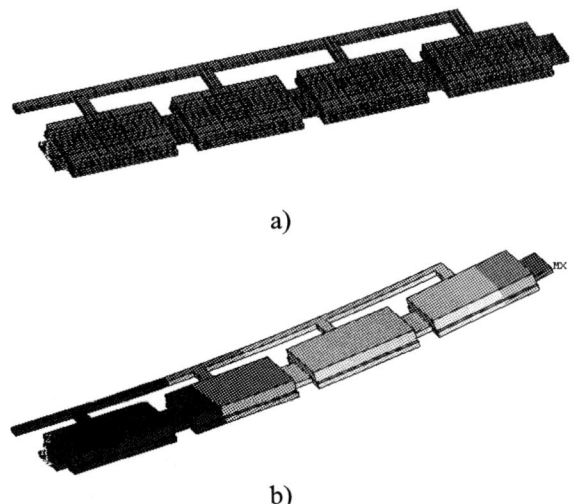

Figure 15. a) 3D model and b) one of the simulation results.

The electrical connection beam is connected to the actuator electrodes at the height of its supports in order to reduce, as long as possible, the influence on the device deformation.

3D simulations with MFS pose more problems. To avoid extremely deformed elements during morphing process, one has to make use of coupling equations to bring the air elements at the translated position. These couplings are made between free air nodes and the nodes of the solid elements (beam or actuator electrodes). The 3D model in appearance is very similar, as well as structure deformation, to that used with TRANS126 elements. Therefore, they are not shown here.

Figure 16 shows the obtained deformation for two cases: without using any electrical connection to the actuation electrodes and with the electrical connection.

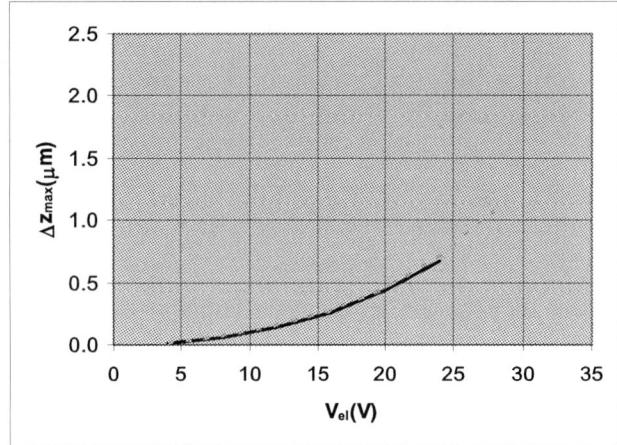

Figure 16. 3D simulation results for two cases: with electrical connection (continuous line) and without considering the electrical connection (dashed line).

The effect of the electrical connection can already be noticed, although not significantly. The reason is that the thickness is half of the actuator electrode and the width is small compared to the device width.

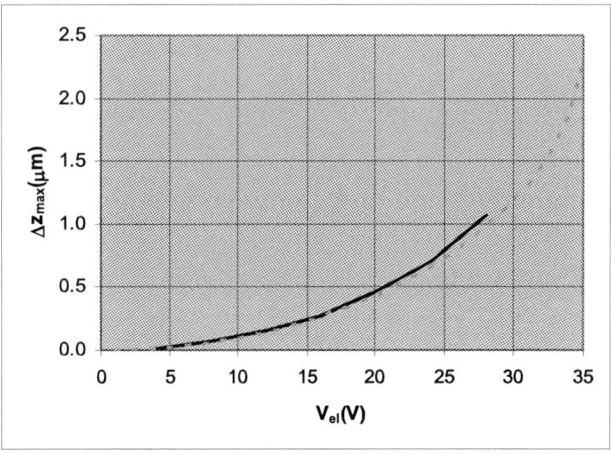

Figure 17. Results comparing deflection of a 3D model without arms and a 2D model.

Comparing the case of 3D with no electrical connection with the results from the 2D model, they provide quite similar results, as shown in Figure 17.

3D model provides larger displacements for the same applied voltage, which can be explained by a slightly higher force (of the region surrounding the anchor part) and a slightly larger deflection of the lateral edges of the actuator.

5. Measurements

No validation of these devices has been possible up to now. They have already been fabricated but they remain stuck down to the substrate, probably due to stiction problems. Also there could be some issues regarding the release process of the beam. Figure 18 shows a top view of one of the fabricated devices.

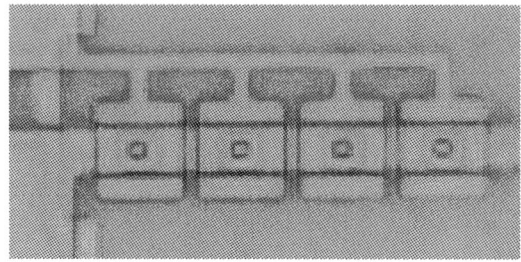

Figure 18. Top view of one of the implemented devices.

Only an electrical test was possible, showing the well known hysteresis effect when excitation is applied and released. In this case, current limitation was applied in order to avoid burning of the devices. If no current limitation is applied, devices are burnt out after pull-in has been reached, as shown in Figure 19.

Figure 19. Burnt out device after pull-in (without limiting current).

The location of the burnt areas is at the highest electrical resistance part: the thin arm for electrical connections.

6. Conclusions

In this work, a novel electromechanical actuator device has been presented. It shows several good performance parameters like low voltage, large displacement, large force and reduced area.

A simple analytical model was developed which can be used to first estimate device parameters.

Simulations have been used to obtain their static characteristics, showing the expected performance. A first conclusion is that 2D models can be near in accuracy to the 3D results, allowing this to use 2D models for approximate estimations of the device performance. 3D models can also be used, although with a significant increase on simulation time. Also, additional node couplings were necessary in order to avoid excessive deformation of air elements, which complicates even more the model generation. Nevertheless, 3D model can be needed when factors, not possible to include in 2D models, can be of relevance to the device deformation.

Due to processing issues, structures were stuck down to the substrate and it was not possible to obtain any measurement of the displacement.

Future work will be the measurement of the devices and comparison with the simulation results.

Acknowledgments

This work has been possible, in part, thanks to the support from the "Departament d'Universitats, Recerca I Societat de la Informació" of the Catalonia's Government.

References

1. Walraven, J.A., "Introduction to Applications and Industries for Microelectromechanical Systems (MEMS)", International Test Conference, Charlotte (USA), October 2003, pp. 647-680.
2. Gaspar, J., Chu, V., Conde, J.P., "Electrostatic actuation of thin-film micro-electro-mechanical structures", Journal of Applied Physics, Vol. 93, No. 12 (2003), pp. 10018-10029.
3. Molfese, A., Nannini, A., Pennelli, G. Pieri, F., "Analysis, testing and optimisation of electrostatic comb-drive levitational actuators", Analog Integrated Circuits and Signal Processing, Vol. 48, No. 1 (2006), pp. 33-40.
4. Rosa, M.A., De Bruyker, D., Völkel, A.R., Peeters, E., Dunec, J., "A novel external electrode configuration for the electrostatic actuation of MEMS based devices", Journal of Micromechanics and Microengineering, 14 (2004), pp. 446-451.
5. Yao, J.J., "Topical Review: RF MEMS from a device perspective" Journal of Micromechanics and Microengineering, 10 (2000), pp. R9-R38.
6. Hariri, A. Zu, J.W., Ben Mrad, R., "Modeling of dry stiction in micro electro-mechanical systems (MEMS)", Journal of Micromechanics and Microengineering, 16 (2006), pp. 1195-1206.
7. Tsou, C., "The design and simulation of a novel out-of-plane miro electrostatic actuator", Microsystem Technologies, No. 12 (2006), pp. 723-729.
8. Lin, H.H., Chu, C.H., Chang, C.L., Chang, P.Z., "Mechanical behavior of RF MEMS", 2nd Int. Symp. on *Acoustic Wave Devices for Future Mobile Communication Systems*, Chiba (Japan), March 2004, pp. 123-128. (www.usl.chiba-u.ac.jp/~ken/Symp2004/PDF/2B2.PDF)
9. I.V. Avdeev, "New Formulation for Finite Element Modelling Electrostatically Driven Microelectro-mechanical Systems", PhD, 2003. (http://etd.library

.pitt.edu/ETD/available/etd-11242003-124652/ unrestricted/avdeev_ilya_phd.pdf).

10. J.B. Muldavin, G.M. Rebeiz, "Nonlinear Electro-Mechanical Modeling of MEMS Switches", IEEE Int. Microwave Theory and Techniques Symp., Phoenix (USA), May 2001, pp. 2119-2122.

11. www.ansys.com

Design of Metal Interconnects for Stretchable Electronic Circuits using Finite Element Analysis

Mario Gonzalez[1], Fabrice Axisa[2], Mathieu Vanden Bulcke[1], Dominique Brosteaux[3],
Bart Vandevelde[1] and Jan Vanfleteren[2,3]

[1] IMEC, Kapeldreef 75, 3001, Leuven, Belgium
[2] TFCG Microsystems, IMEC, Gent-Zwijnaarde, Belgium
[3] ELIS - TFCG Microsystems, University of Ghent, Gent-Zwijnaarde, Belgium

Abstract

In this work, the design of flexible and stretchable interconnections is presented. These interconnections are done by embedding sinuous electroplated metallic wires in a stretchable substrate material. A silicone material was chosen as substrate because of its low stiffness and high elongation before break. Common metal conductors used in the electronic industry have very limited elastic ranges; therefore a metallization design is crucial to allow stretchability of the conductors going up to 100%.

Different configurations were simulated and compared among them and based on these results, a horseshoe like shape was suggested. This design allows a large deformation with the minimum stress concentration. Moreover, the damage in the metal is significantly reduced by applying narrow metallization schemes. In this way, each conductor track has been split in four parallel lines of 15 μm and 15 μm space in order to improve the mechanical performance without limiting the electrical characteristics. Compared with the single copper or gold trace, the calculated stress was reduced up to 10 times.

1. Introduction

Flexible and stretchable electronic circuits is a relatively new concept aiming in a first instance at improving the comfort of consumer's needs. This technology can also be used in many other applications where the ability to deform is an advantage or where the electronics should preferably take the shape of the object in which they are integrated. Some examples of this technology are skin mounted or implantable biomedical devices [1, 2] where the circuit must behave as the tissue itself; textile or wearable electronics [3,4].

Traditional signal and power transmission lines in micro-electronic systems are placed on rigid or at most flexible substrates, thus limiting the commercial applications to relatively small devices for optimal mobility and comfort. Stretchable electronic circuits will integrate different components onto a compliant polymer substrate that may be stretched once or many times depending of the application. One way to make these circuits is to place rigid or flexible components distributed over a polymer surface and then interconnect these components with a stretchable connection. Nevertheless, one of the critical points in this technology is the reliability of the elastic interconnections. This is particularly challenging given the relatively high and complex mechanical loading expected for these applications, i.e. bending, elongation and torsion.

Several technologies have been proposed in recent years such as intrinsic conductive polymers [5,6], pre-stressed metal conductors [7,8,9] or in plane patterned metal conductors [10,11]. Nowadays, metals are the best options to realize these interconnections because of their high electrical performance and relatively low cost. However, in all cases, the main challenge is maintaining the integrity of the circuit during and after flexing or stretching the substrate.

In this work, a description of a Moulded Interconnect Device (MID) technology is given and the shape of metal interconnections is optimized by Finite Element Analysis (FEA) to allow large deformations. Even if simple conductor shapes like triangular or sinusoidal allow higher deformations compared to a straight line, they present a high concentration of stresses in the crest and trough, giving rise to early failures at fairly small deformations. Based on these results, a horseshoe like shape was designed. Stresses are distributed in a wider region instead of being concentrated in a small zone. Common metal conductors (Au, Cu, Ni, Pt) have very limited elastic ranges, therefore a design of an appropriate shape is crucial to allow stretchability of the conductors. For this reason, the interconnection between two points will not be a straight line but a periodic undulating metal track. In this way a sort of two-dimensional spring is obtained. As stretchable polymer we use polydimethylsiloxanes (PDMS) because their low elastic modulus and high stretchability. Moreover, PDMS has been used in many medical implantable devices and also in electronics it is not a new material [12].

As a first step, the 3D FEM simulations were used to compare the performance of different shapes (sinus, "U" shape, half circles, elliptical and "horseshoe") and identify the more promising structures that later will be mechanically tested and optimized. The outcome of thermo-mechanical modeling is presented as a stress or strain distribution in the different parts of the structure. The quantification of the stresses and strains is based on the substrate stiffness, the width of the metal track and the radius of curvature of these conductors.

2. Sample Preparation

Figure 1 outlines the sample fabrication. All processing is done on flexible but non-stretchable substrate which allows using conventional electroplating techniques. In a first instance, a photoresist is spin-coated

1-4244-1105-X/07/$25.00 ©2007 IEEE

on a copper foil and patterned, with the meandering shape by UV-radiation. Then a 4 μm thick gold layer was electroplated, followed by a 2 μm nickel layer to improve solderability. Gold was chosen as metal conductor to facilitate the separation of the sacrificial copper foil. However, other metals such as copper or nickel can be used if an extra thin metal layer is applied on top of the copper foil. Before etching the copper foil, the conductors are encapsulated with 0.25 or 0.5 mm thick elastic polymer. As elastic polymer material Silastic MDX4-4210 from Dow Corning had been chosen because it's a biomedical grade silicone elastomer with a low Young's modulus and high elongation (up to 470% from supplier datasheet). More detailed information about this processing was presented by Brosteaux [13].

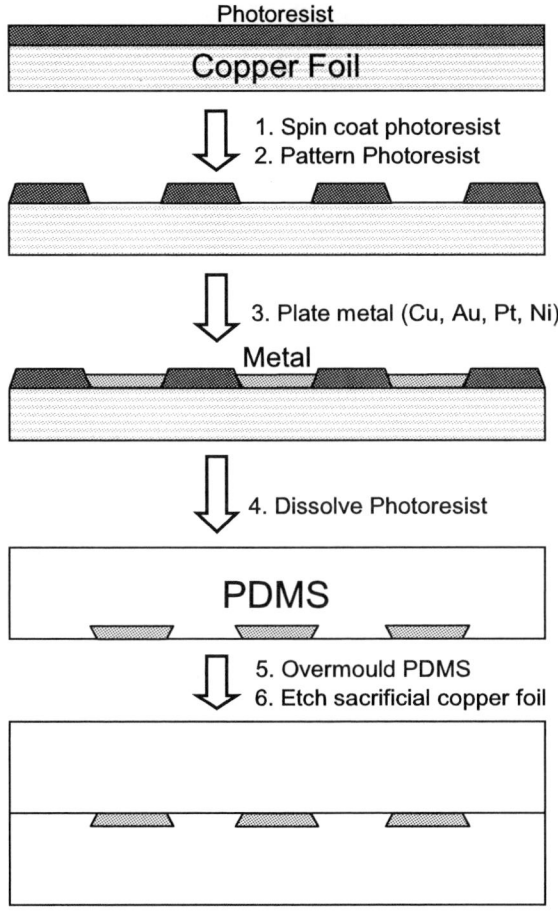

Figure 1. Process sequence for metallic stretchable interconnections embedded in PDMS

3. Preliminary Elastic Model

For this first approach a 2D plain-stress model was done. The objective of this preliminary study is to compare the stresses induced in a copper conductor when a 20% deformation is applied in the axial direction of the meander. The effect of the substrate was neglected in a first instance. The conductor used for these models is

copper with a thickness of 15 μm and a trace width of 90 μm. An amplitude of 700 μm and a period of 500 μm was used for the three proposed configurations.

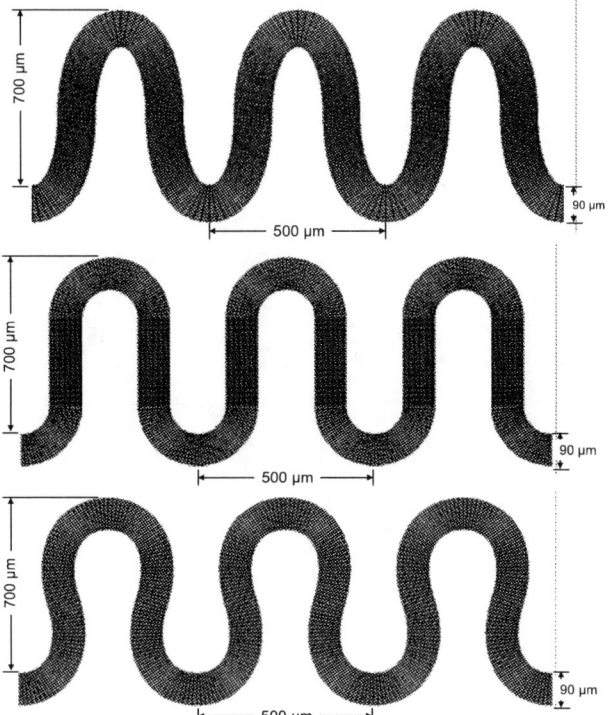

Figure 2. Different conductor shapes

The mechanical behavior of the copper was modeled as being isotropic, linear elastic and temperature independent. In a 3D simulation of these configurations an out of plane deformation is observed when a load or displacement is applied in the axial direction. However, in an actual configuration the metal conductor is embedded into a stretchable substrate and the out of plane deformation is constrained. Therefore, even if a 2D plane stress (in plane deformation) model is not an accurate method because the effects of substrate and metal thickness and the out of plane deformation are not taken into account, it is a good "first view" of the stresses induced in the structure.

Stress comparison of different conductor shapes

Three different conductor shapes were analyzed and compared. In all cases, a total deformation of 20% was applied in the axial direction of the meander. Results of these models are presented graphically in Figure 3. In the case of an elliptical shape (Figure 3A), a high stress concentration is observed in the crest and trough of the line. In order to avoid this concentration of stresses, a rounded design is preferred. The "U" shape (Figure 3B) offers a better stress distribution but is still limited by a reduced radius of curvature. Furthermore the straight vertical lines limit the deformation perpendicular to the axis of the meander when a biaxial deformation is needed. In the optimal shape (Figure 3C), the stress is distributed in an extended part of the conductor. A reduction of 46% in the stress is obtained with this shape vis-à-vis the

341

elliptical one. As a first step, FEM simulations were developed considering linear behavior of the materials, this explains the high values of calculated stress. Nevertheless, as we are just looking for trends, an "A to B" stress comparison is enough for choosing the conductor shape. Predominant failures are expected in regions where the highest concentration of stress is located.

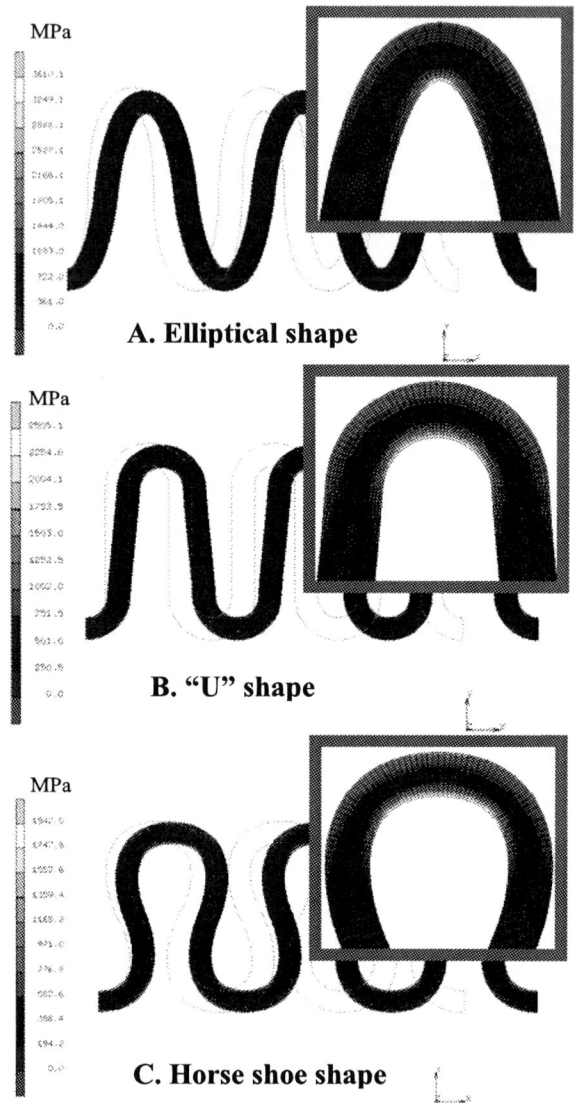

A. Elliptical shape

B. "U" shape

C. Horse shoe shape

Figure 3. Stress distribution in copper conductor line for three different conductor shapes

In order to reduce even more the stresses in the metal conductor, without sacrificing the electrical performance and/or changing the amplitude or period of the design, the line can be subdivided in several lines of smaller width as shown in Figure 4. This "multi-copper trace" improves the stretchability and reduces the induced stresses.

In an idealized case (without substrate) for the geometry proposed here, the stresses are reduced by a factor of 10 if the copper line width is reduced from 90 μm to 15 μm.

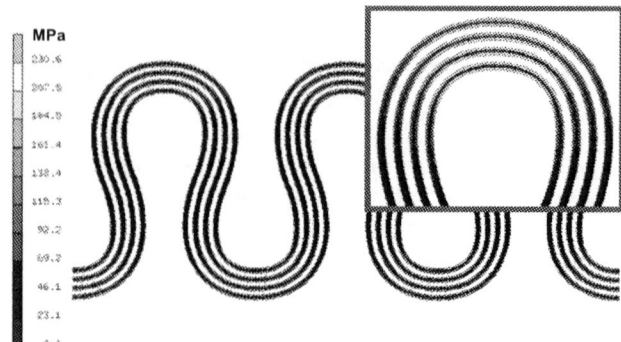

Figure 4. Stress distribution in a multi-track horseshoe conductor design

4. FEM Simulation of the conductor embedded into the stretchable substrate

In order to quantify the strains in the copper meander, the substrate has to be included. When the substrate is stretched in the axial direction, due to the Poisson's effect, the tensile deformation is always accompanied by a lateral contraction as shown in Figure 5. During a uniaxial stretching, metal conductor is in tension in the crest and trough and in compression in the center of the design. Yellow and red colors in the Figure 5 represent a concentration of plastic strains when the structure is stretched 25%. The dashed line indicates the original dimensions of the structure.

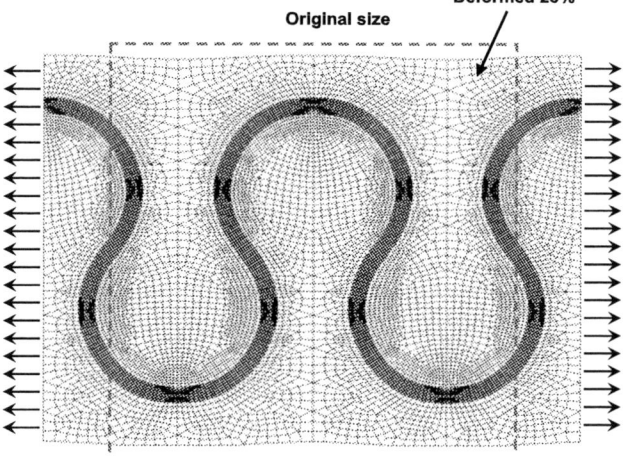

Figure 5. Poisson effect observed during a uniaxial tension test

A qualitative comparison between modeling and experiments shows that calculated regions with high concentration of plastic strain ($\varepsilon^{pl}_{max} \cong 10\%$) correspond to the observed failures (Figure 6). Visible cracks are observed in the crests and troughs of the meander at about 25% of total deformation.

342

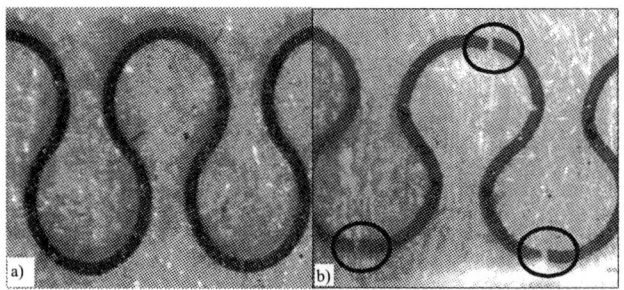

Figure 6. Tensile strain test of horseshoe metal interconnects. (a) Before elongation, (b) After 25% elongation highlighting the failures in the crest and trough

Because the copper conductor is completely embedded into the stretchable matrix, 2D plane stress or plane strain models cannot be used. Those models assume a unique material in the thickness direction. For solving this problem 3D FEM has to be used. However, 3D models require high number of memory and computational time. The same 3D effect can be obtained by using composite shell elements. Those elements are composed of layers of different materials with various layer thicknesses. Furthermore, the material in each layer may be linear or not linear. Therefore, gold or copper can be modelled as a perfect plastic material.

In order to validate the accuracy of the 2D composite elements, a horseshoe meander was modelled by using 3D brick elements and 2D composite shell elements. Results of this comparison are depicted in Figure 7. The continuous line shows the strains calculated with the 3D model and the points represent the plastic strain calculated in the copper meander by using composite 2D elements. As the difference between 2D and 3D elements is negligible and calculation time was divided by 10 it is advisable to use 2D composite shell elements to model these structures.

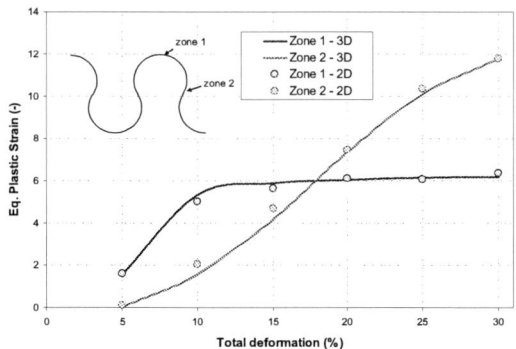

Figure 7. Comparison of the equivalent plastic strain induced in the copper conductor as function of the applied deformation

The deformation of the patterned metal conductor is, in a certain manner, controlled by the deformation of the stretchable substrate. If the conductor is embedded in a stiff substrate, the deformation of the conductor will be the same as the one of the substrate. On the other hand, if a very soft substrate is used, the conductor has some

"freedom" to move inside the substrate, reducing in this way the Poisson's effect and in consequence, the accumulated plastic strain (compressive stresses are reduced).

To quantify the maximum strain in the conductor as a function of the stiffness of the substrate, the meander design presented in Figure 5 was modeled with different Young's modulus of the substrate. In this case, the thickness of copper and substrate were kept constant with values of 17 μm and 100 μm respectively.

Results of this study are presented in Figure 8. For this configuration, when the Young's Modulus of the substrate is about 120 MPa, the plastic strain in the copper is practically the same as the total strain applied to the substrate. This means, that increasing the stiffness of the substrate will decrease the maximum allowable stretchability of the structure without electrical failure.

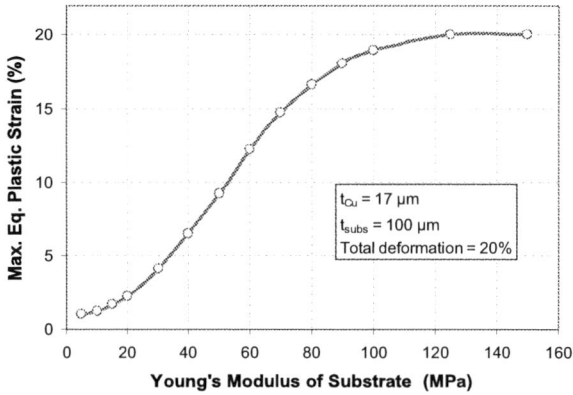

Figure 8. Equivalent plastic strain induced in the copper as function of the Young's Modulus of the substrate

5. Parameters used in the horseshoe design

The horseshoe pattern is created by joining a series of circular arcs as shown in Figure 9, where R is the inner radius, W is the width of the copper trace and theta (θ) is the angle, measured clockwise, where the two arc of circles intersect. When θ=0°, we have a semicircle design, if θ>0°, we obtain the horseshoe design.

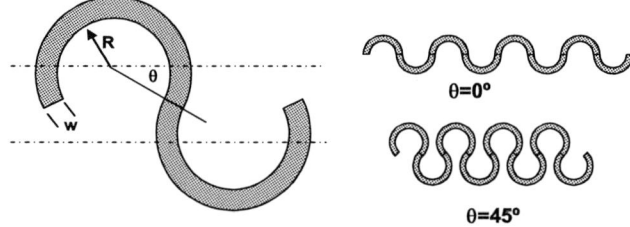

Figure 9. Configuration of the horseshoe design

If we neglect the possible delamination and buckling problems, it is possible to define a scale factor as the ratio R/W. This means that the stress and strain induced in the metal are constant if the ratio R/W is kept constant, independently of the amplitude of the horseshoe design. A series of 70 models were simulated with different R, W and θ in order to find a relation between the damage parameter (plastic strain) and the scale factor R/W.

Figure 10 gives a picture of the relation between plastic strain and the ratio R/W. The different percentages presented in the plots represent the total deformation applied to the structure. These plots show a clear trend: an increase of the scale factor is translated into a reduction of the induced strain. Therefore a narrow copper trace or a large radius of curvature is preferred for these configurations.

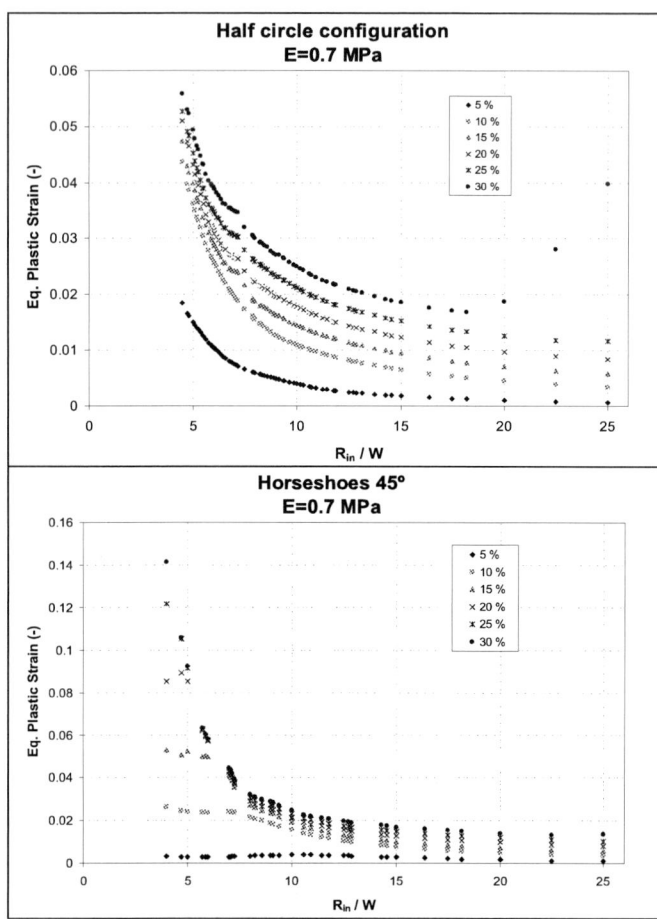

Figure 10. Relation between the equivalent plastic strain and the scale factor (R/W) for a substrate with a Young's Modulus of 0.7 MPa. Top image: semicircle design. Bottom image: Horseshoe design

6. Evaluation of horseshoe design.

Based on the FEM results presented above, the induced stresses in the metal under deformation increase drastically if a wide metal track is used. Therefore the single tracks have been made as narrow as the photolithography process allowed it with sufficient reliability, resulting in a width of 15 μm. The spacing between two single tracks is also 15 μm, making the whole "multitrack" 105 μm wide. On regular points, where the predicted deformation is minimal, neighboring

single tracks are connected to each other in order to compensate single track interruptions caused by process faults or mechanical failure. Samples with three different junction angles (θ) were fabricated and a tensile test was done. Figure 11 depicts the maximal deformation of the different samples in relation to its initial resistivity per unit of length. In the case of semicircles design (θ=0°), we obtain the smaller resistivity but also the smaller elongation. Elongations up to 100% and more were obtained in the case of horseshoe design. The large dispersion observed in these results can be explained by the variability of processing parameters, which are not stable enough (silicone thickness uniformity, crystallization of metal during electroplating, presence of defects in the substrate as air bubbles or voids). In all cases, the variation in the resistivity during elongation was below of 2%.

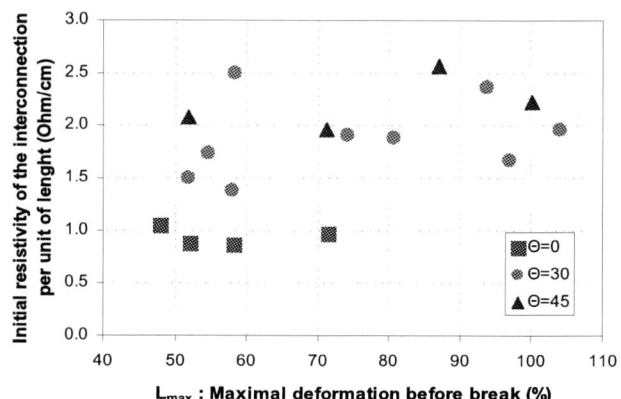

Figure 11. Maximum deformation and resistivity for different patterns of stretchable interconnections

4. Conclusions

Several designs of the electrical conductors satisfying the requirements of stretchability were proposed and compared among them by either FEM or experimental tests. From the modeling results, a horseshoe like shape was found to be the optimal; nevertheless, the magnitude of the stresses is related to the stiffness of the substrate and the geometry of the meander. The damage in the metal is significantly reduced by applying narrow metallization schemes and low elastic modulus of the substrate. In order to increase even more the stretchability of the conductors without limiting the electrical performance, four parallel metal conductors of 15 μm width and 15 μm space were fabricated and tested. Based in experimental results, the maximum stretchability achieved with single and wide gold trace was in the order of 20%, while for the case of multiple and narrow metallizations the stretchability of the circuits above 100% have been demonstrated. These results show the obvious advantage of using an optimized meander shape and multiple conductor lines in order to allow high deformation of the structure.

Acknowledgments

The authors would like to thank the Institute for the Promotion of Innovation by Science and Technology in Flanders (IWT) for the financial support through the SBO-Bioflex project (contract number 04101).

References

1. F. Spelman, "The past, present, and future of cochlear prostheses," *Engineering in Medicine and Biology Magazine, IEEE*, Vol. 18, No. 3, pp. 27-33, May/Jun 1999.
2. G. Clark, "Cochlear implants. Fundamentals and applications", Springer-Verlag, New York, 2003.
3. H. Kudo *et al.*, "A flexible and wearable glucose sensor based on functional polymers with Soft-MEMS techniques". *Biosensors and Bioelectronics* Vol. 22, pp. 558–562, 2006.
4. J. Weber, *et al.*, "Coin-size coiled-up polymer foil thermoelectric power generator for wearable electronics", *Sensors and Actuators A: Physical,* Volume 132, Issue 1, Nov. 2006, Pages 325-330.
5. S. D. Deshpande *et al.*, "Studies on conducting polymer electroactive paper actuators: effect of humidity and electrode thickness" *Smart Mater. Struct.* Vol. 14, pp. 876-880, 2005.
6. S. M. Richardson-Burns, *et al.*, "Polymerization of the conducting polymer poly(3,4-ethylenedioxythiophene) (PEDOT) around living neural cells". *Biomaterials,* Volume 28, Issue 8, March 2007, Pages 1539-1552.
7. S. Wagner *et al.* "Electronic skin: architecture and components". *Physica E.* Vol. 25, pp. 326–334, 2004.
8. S. P. Lacour *et al.* "Stiff subcircuit islands of diamondlike carbon for stretchable electronics", *Journal of Applied Physics,* **100**, 014913, 2006.
9. Y. Sun, *et al.* "Controlled buckling of semiconductor nanoribbons for stretchable electronics", *Nature nanotechnology,* Vol. 1, pp. 201-207, Dec. 2006.
10. D. S. Gray, *et al.*, "High-conductivity elastomeric electronics," *Advanced Materials,* Vol. 16, No. 5, pp. 393-397, Mar. 2004.
11. R. Pelrine, "High-field deformation of elastomeric dielectrics for actuators", *Materials Science and Engineering C,* Vol. 11, pp. 89–100, 2000.
12. J. Brugger *et al.* "Low-cost PDMS seal ring for single-side wet etching of MEMS structures", *Sensors and Actuators A: Physical,* Volume 70, Issues 1-2, Oct. 1998, pp. 191-194
13. D. Brosteaux *et al.* "Elastic Interconnects for Stretchable Electronic Circuits using MID (Moulded Interconnect Device) Technology". *Proc. of the MRS Spring 2006*, Volume 926E.

Numerical Modeling of Electrical Resistance of Interconnections in High-Tech Multilayer PCBs Manufactured by Magnetron Sputtering Deposition of Copper

Janusz Borecki[1], Artur Wymyslowski[2]

[1] Tele and Radio Research Institute, Centre for Advanced Technology of Electronic Interconnections
03-450 Warsaw, Ratuszowa 11, Poland
[2] Wroclaw University of Technology; Faculty of Microsystem Electronics and Photonics
53-439 Wroclaw, Grabiszynska 97, Poland
e-mail: artur.wymyslowski@pwr.wroc.pl

Abstract

In the paper, authors focus on application of numerical simulation methods along with the experimental measurements to access the electrical resistance of interconnections in PCB (Printed Circuit Board). The comparison of experimental and numerical results allow to define the design rules of interconnection properties and specification of interconnection and technology details depending on the selected application. Additionally, it is believed to allow course prediction of deposition process of conductive layer on via walls, without necessity of performing costly and time consuming experiments. In the paper authors present results and investigations which should lead to manufacture interconnections with aspect ratio (relation of via deep to via diameter) higher than 1, and diameter of via in the range of 25 to 150 µm.

1. Introduction

The ongoing progress and intercorrelated demand on miniaturized electronic equipment requires development of new production techniques of High-Tech PCBs. One of the key aspects and most probable research directions are focused on high density of interconnections. This can be achieved in case of Printed Circuit Boards by farther miniaturization of interconnections [1]. On the other hand this requires high quality of manufactured interconnections, which can be defined by the value of electrical resistance. At the development stage, the electrical resistance of interconnections can be predicted by numerical modeling. However, numerical evaluation can be incorrect due to a number of reasons as e.g. electrical resistance of lead tracks or insufficient knowledge of metallized layer thickness of via walls. Nevertheless, thanks to knowledge of electrical properties of deposited conductive layer it is possible to predict leading current of manufactured interconnection.

2. Test samples

To achieve the high aspect ratio miniaturized interconnections in PCB, the microvia technique was used. In our investigations test samples were prepared according to Sequential Build-Up (SBU) technology in which microvias were made with using of Nd:YAG 355 nm LASER µVia DRILL machine model 5200 from Electro Scientific Industries Inc. Microvias were formed directly in RCC material which was pressed on the core (FR-4 laminate with thickness 0.6 mm) in the standard epoxy lamination process. Actually, it is possible to form

blind microvias in RCC material with minimal diameter about 25 µm. However, to investigate the possibilities of copper deposition by magnetron sputtering deposition technique in small diameter vias, the test samples were made with different diameter of microvias (150, 125, 100, 75, 60, 50, 37 and 25 µm). In the result, microvias were formed with aspect ratio 0.6:1, 0.7:1, 1.2:1, 1.5:1, 1.8:1, 2.4:1 and 3.6:1.

The forming of microvias in RCC material was made in few steps. The multi-step process of microvia forming is due to the different level of laser fluency required to ablate different materials (copper layer, dielectric layer).In the first step a window in the top copper layer with using of highest fluency was made. Next, the ablation of dielectric layer to the second (inter) copper land layer with using lower laser fluency was made. At the end, in third step, the bottom of blind microvia was cleaned. The process of microvia formation schematically is presented in figure 1, and in details it was described in paper [6].

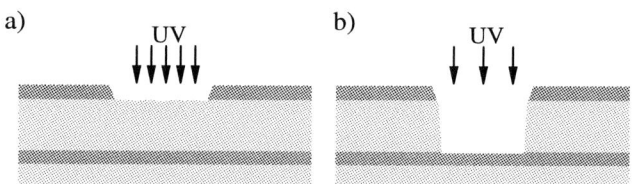

Figure 1: Multi-step process of microvia forming
a)   step 1 – window opening in TOP copper layer
b)   step 2 – dielectric layer ablation

After microvias formation, test samples were specially prepared to allow investigation of the quality of manufactured interconnections. These preparations of test samples were described with details in paper [7, 8]. Test sample which was used during investigations is presented in figure 2. The numbers placed in the rectangles describe the diameter of microvia.

Next, test samples were put to the trials of metallization of microvia walls. It was made by forming an electrical-conductive base layer through magnetron sputtering deposition of copper. The process of sputtering was made in both standard mode (with labour gas, Ar) and new one self-sustaining mode (without labour gas) [2, 3, 4]. Additionally, the deposition of electrical-conductive base layer was made by both static mode (without movement of test samples during process) and by dynamic mode (with movement of test samples during

1-4244-1105-X/07/$25.00 ©2007 IEEE

process). The experiments of microvias walls metallization were presented in detail in paper [7].

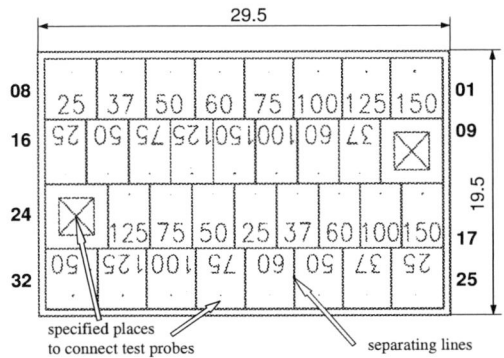

Figure 2: Test sample with different diameter of blind microvias

3. Quality of interconnections

In our investigations the quality of metallized microvias was defined as the electrical resistance value of manufactured interconnections. In order to measure interconnections electrical resistance the laboratory multimeter Agilent model 34401 with four wire option was used. The method of measurements which was used during investigations schematically is shown in figure 3. In order to measure electrical resistance of manufactured interconnections separately, the separating lines around every of blind microvia were made by laser beam ablation (see figure 2 and 3). Additionally, microvias were metallized with applying of steel stencil. It was described with details in paper [8].

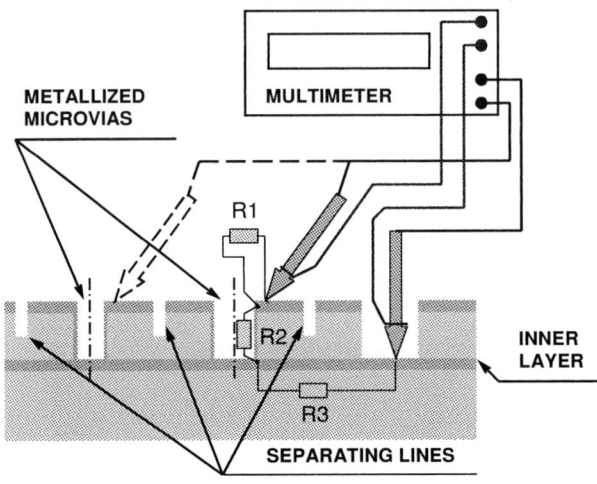

Figure 3: Method of interconnections electrical resistance measurements

As it is shown on the figure 3, the measured resistance consist the resistance of interconnection (R2), but also resistance of connection on TOP layer (R1) and resistance of connection on INNER layer (R3). However, it should be noticed that the thickness of deposited on via walls electrical-conductive base layer is about 1-2 μm, and it is a few times lower than thickness of copper on top layer (17.5 μm thickness of base layer grow by 5-7 μm thickness of deposited layer) and on inner layer (17.5 μm thickness). Additionally, the "width" of manufactured interconnections is about 0.07 to 0.46 mm (for microvia with diameter 150 to 25 μm, respectively), and it is tens times lower than the "width" of connections on the TOP and INNER layer, where the connections are realized by copper plates. In the result, the measured electrical resistance approximately describes the electrical resistance of manufactured interconnections only. However, if the thickness of deposited conductive layer on via walls is higher, of course the resistance of interconnection is smaller, but the influence of connections resistance on TOP and INNER layers is more important. In the result, in this case, the measured resistance is the sum of interconnection resistance (R2) and resistance of connections on TOP and INNER layer (R1 and R3, respectively). It is impossible to simply calculate the electrical resistance of interconnection only. To investigate that effect it is possible to access manufactured interconnections resistance through numerical simulation methods. It allows predicting the resistance of interconnection in function of thickness of electrical-conductive base layer deposited on via walls and also in function of microvia diameter. Besides of that the main advantage from numerical simulation is that it is not necessary to realize all experiments of microvias metallization. It is completely sufficient to make a few experiments to verify the designed model.

Additionally, the quality of manufactured interconnections was examined on the base of its cross-sections. A few of cross-sections are presented in figure 4. According to the figures the deposited conductive layer is uniform and continuous. The thickness of this layer is about few micrometers and it can be thickened by electroplating process.

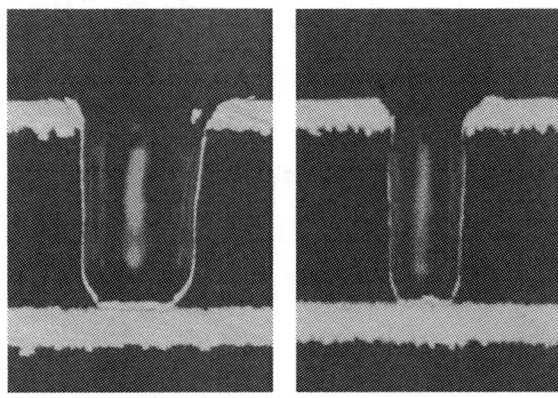

75 μm, aspect ratio 1.2   25 μm, aspect ratio 3.6

Figure 4: Cross-sections of manufactured interconnections

On the basis of achieved cross-sections it was possible to prepare relevant geometrical and numerical model of the interconnections.

4. Model of interconnection resistance

The model of manufactured and investigated interconnections was created on the basis of geometrical dimensions of interconnections, which were obtained from cross-sections. In this way, final model was designed for all interconnections separately (4 models for each diameter of microvia). In consequence, 32 models signed as "model_01", "model_02", ... , "model_32" (the numbers "01", ... ,"32" in figure 2) were made. The general scheme of interconnection structure which was used during designing of models is presented in figure 5.

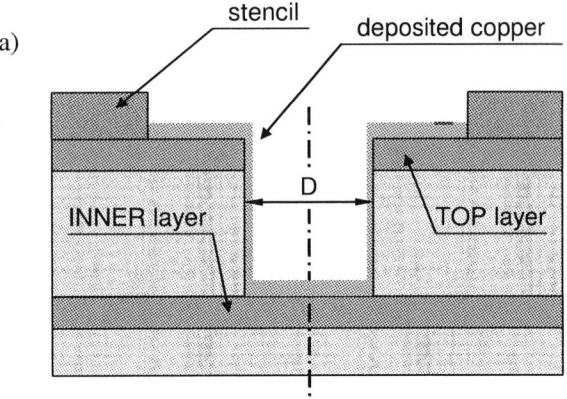

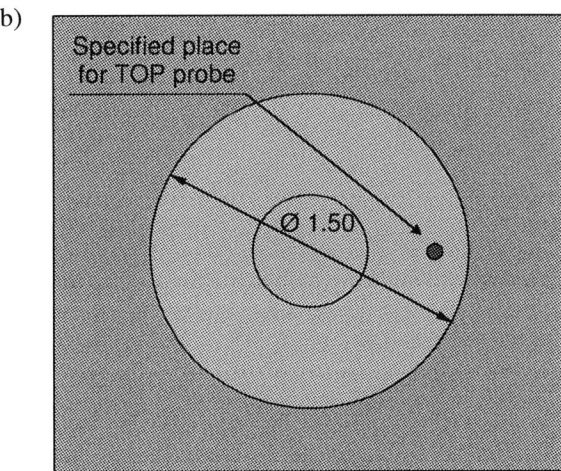

D = 25 ÷ 150 μm

Figure 5: Geometrical configuration and dimensions of the designed model (not in scale)
a)   cross-section of interconnection
b)   TOP view of interconnection

The resistance of interconnection depends on the dimensions of interconnection but mainly on resistivity of conductive layer which covers via wall. It should be underlined that resistivity of the deposited copper and bulk copper are different. Additionally, it depends on the thickness of deposited layer. For example the resistivity of bulk copper is about 1.7·10E-8 Ωm, when the resistivity of deposited copper layer with 3 ÷ 4 μm thickness is about 2.0·10E-8 Ωm [5], and it should be consider during numerical modeling of models.

5. Results and Discussion

Numerical assessment of interconnection electrical resistance was based on defining electrical potential load of 1 Volt and simulation of the relevant density current distribution in test sample. The load is connecting to specified place on INNER layer (see figure 2) and on TOP layer near the microvia with 0.5 mm distance (see figure 5). The distribution of electrical potential in each area of test sample is observed by the different colors (see figures 6-14). On the base of those observations it is possible to describe the influence of each area of test sample (TOP layer – R1, interconnection – R2, INNER layer – R3) in total measured electrical resistance of interconnection. The experiments based on numerical model were conducted for different thicknesses of conductive layer (1, 5 and 10 μm) deposited on microvia walls. Some results of numerical modelling are shown in the figures 6-14.

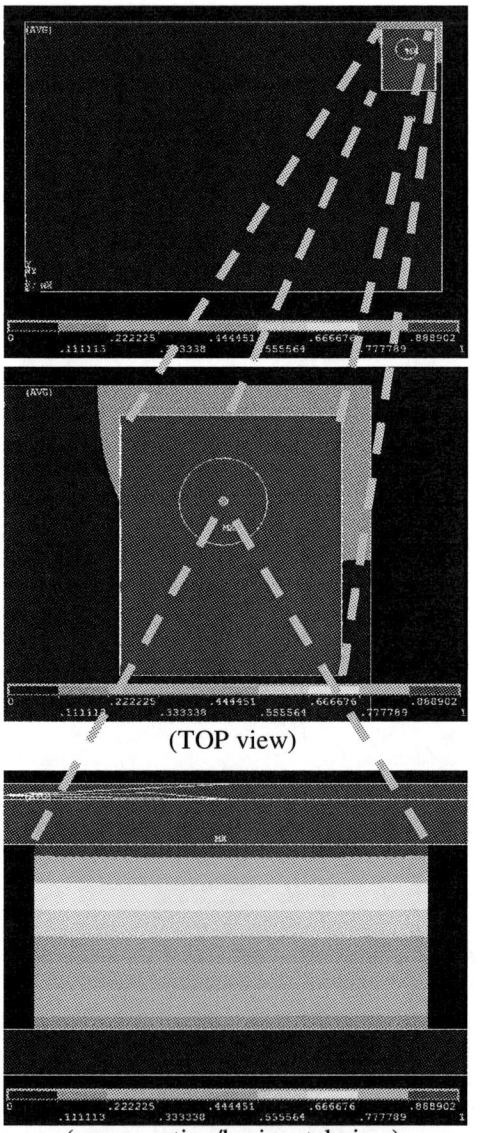

Figure 6: Model of 150 μm microvia (model_01) with 1 μm deposited layer of copper

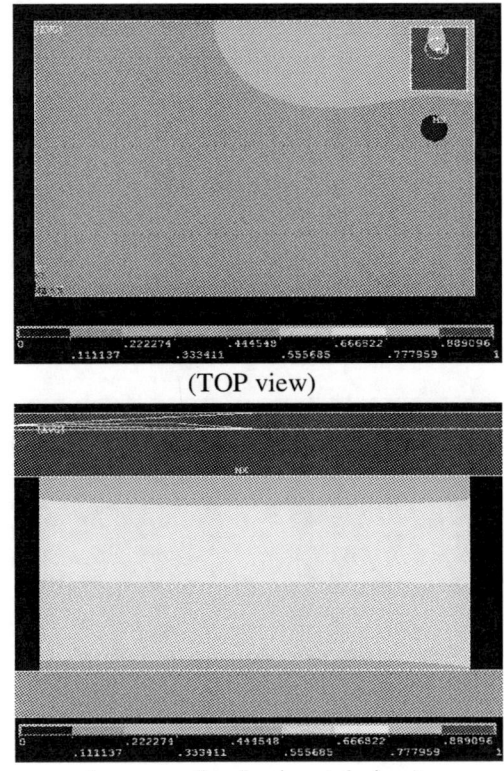

(TOP view)

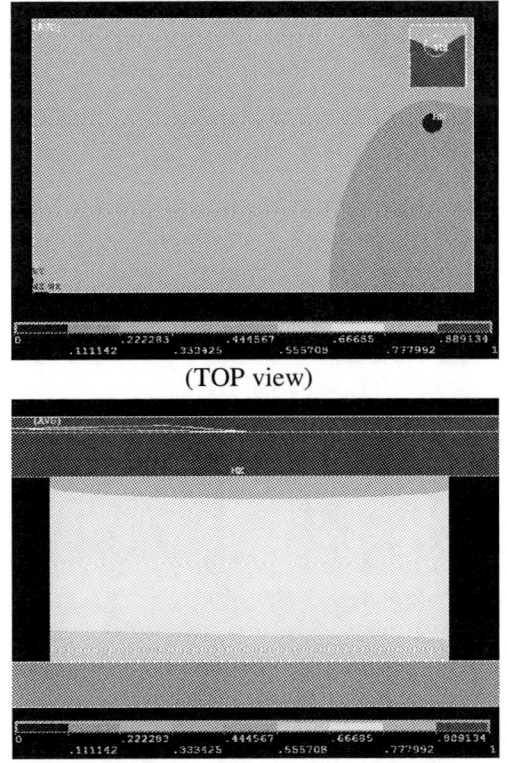

(cross-section/horizontal view)

*Figure 7: Model of 150 μm microvia (model_01)
with 5 μm deposited layer of copper*

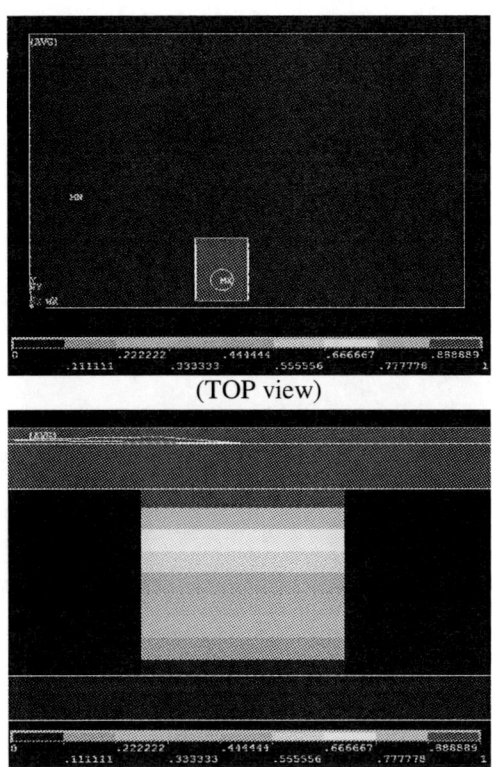

(TOP view)

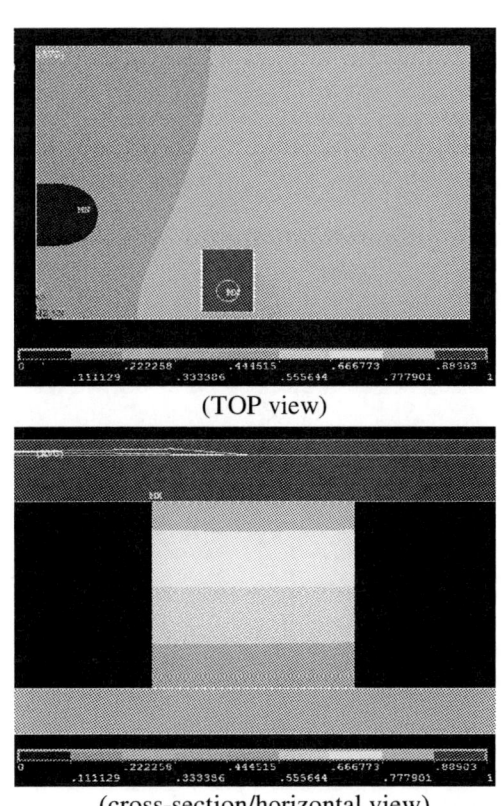

(cross-section/horizontal view)

*Figure 9: Model of 75 μm microvia (model_12)
with 1 μm deposited layer of copper*

*Figure 8: Model of 150 μm microvia (model_01)
with 10 μm deposited layer of copper*

*Figure 10: Model of 75 μm microvia (model_12)
with 5 μm deposited layer of copper*

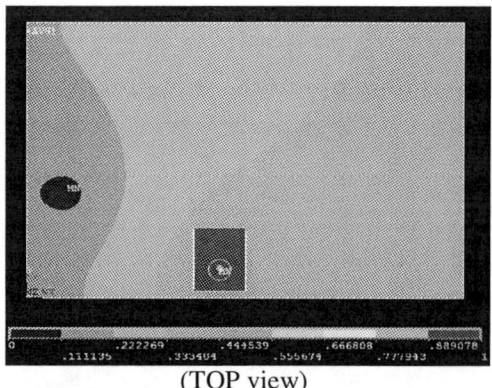

(TOP view)

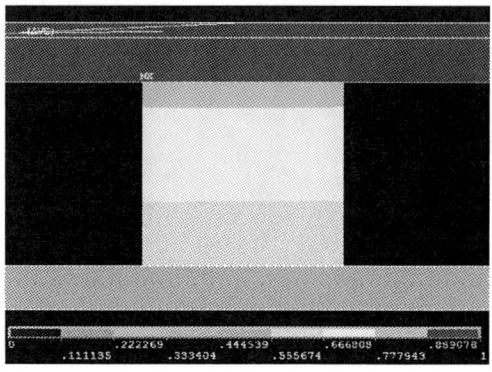

(cross-section/horizontal view)
Figure 11: Model of 75 μm microvia (model_12)
with 10 μm deposited layer of copper

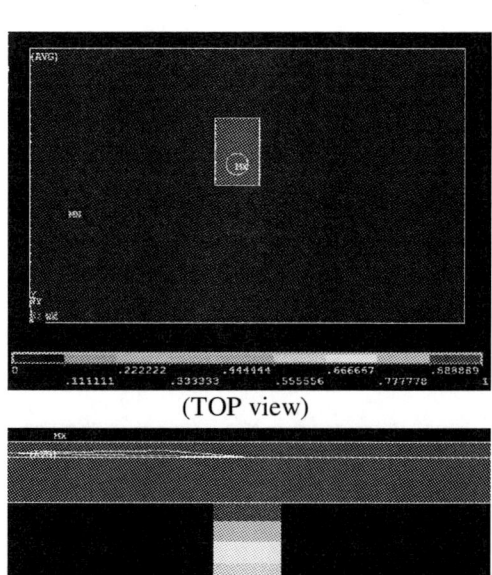

(TOP view)

(cross-section/horizontal view)
Figure 12: Model of 25 μm microvia (model_29)
with 1 μm deposited layer of copper

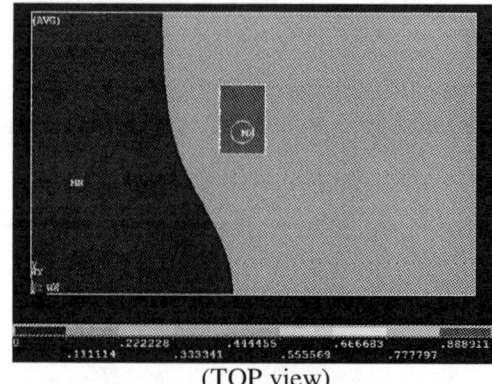

(TOP view)

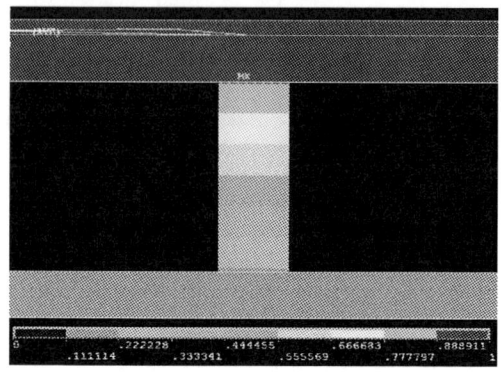

(cross-section/horizontal view)
Figure 13: Model of 25 μm microvia (model_29)
with 5 μm deposited layer of copper

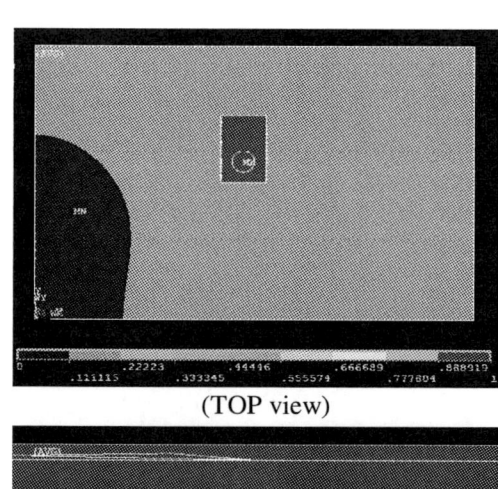

(TOP view)

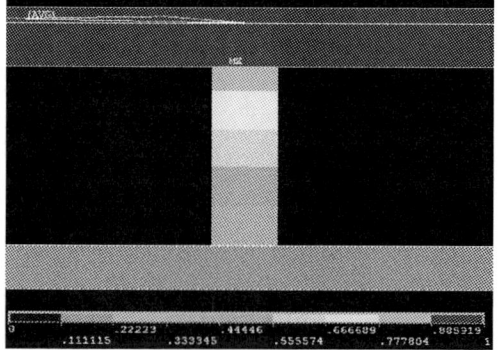

(cross-section/horizontal view)
Figure 14: Model of 25 μm microvia (model_29)
with 10 μm deposited layer of copper

According to the figures 6-8 (about microvia with 150 μm diameter), the influence of thickness of deposited on via walls copper layer (1, 5 and 10 μm) is not so big on electrical resistance of interconnection. However, the influence of resistance of connection on TOP and INNER layer is essential. The contribution of them in total measured resistance is even about 45 %.

A little bit different situation is with 75 μm microvia (see figures 9-11). In this case the influence of thickness of deposited on via walls copper layer is more perceptible, but simultaneously the influence of TOP and INNER connection resistance is smaller.

In reference to smallest microvia (25 μm diameter, figures 12-14) which aspect ratio is about 3.6 (more high than 1) it should be noticed that the resistance of interconnection strong depends on thickness of deposited on via walls electrical-conductive base layer. In consequence, measured resistance is practically equal to resistance of interconnection. The influence of resistance of connection on TOP and INNER layer is negligible.

Application of numerical model allows predicting the electrical resistance of interconnection with set up of whatever microvia diameter and thickness of conductive layer on via wall. The some results of simulations are presented in table 1 and on the graph in figure 15. The intermittent line on the graph marks the minimum measured resistance of manufactured interconnections.

Table 1: Results of simulations of interconnection electrical resistance

Number of model	Microvia diameter	Thickness of deposited base layer		
	[μm]	1 μm	5 μm	10 μm
01	150	27.5 mΩ	2.4 mΩ	2.0 mΩ
12	75	55.1 mΩ	3.6 mΩ	2.8 mΩ
29	25	165.9 mΩ	8.1 mΩ	5.5 mΩ

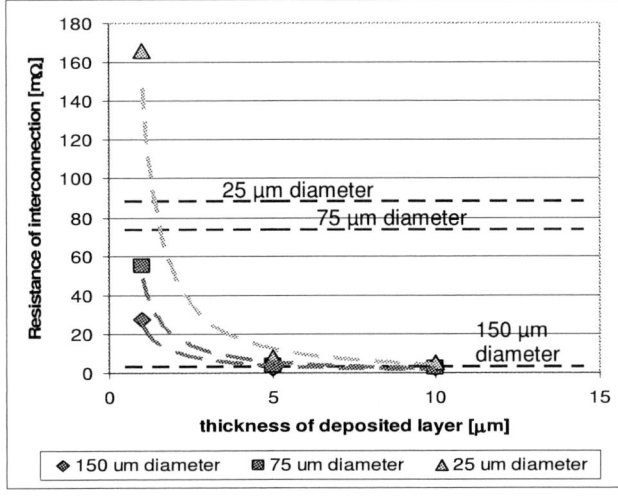

- - - - minimum measured resistance of interconnection

Figure 15: Trend of interconnection electrical resistance in reference to thickness of deposited conductive base layer

6. Conclusions

In the current paper the results of electrical resistance of PCBs interconnections modeling by using numerical simulations method were presented. On the base of manufactured interconnections it was possible to design the model of interconnection. The rightness of construction of designed model was estimated on the base of physical tests. The comparison of results of numerical simulations with results of physical measurements shows that the construction of designed model correctly describes the resistance of manufactured interconnections.

Applying of numerical simulations allows to predict the resistance of interconnection based on whatever microvia diameter and thickness of conductive layer which covers the microvia wall. It is also possible to predict the maximum electrical current which can flow through the interconnection without damage of them.

References

1. Borecki J., "Interconnection for High Density PCB – Ways of Design Optimization", Proceedings of XXVI International Conference of IMAPS Poland Chapter, Warsaw, 25-27 September 2002, pp. 113-117
2. Radzimski Z., Posadowski W.M., Rossnagel S.M., Shingubara S., "Directional copper deposition using dc magnetron self-sputtering", Jurnal Vacuum Science Technology B 16(3), May/Jun 1998 (American Vacuum Society), pp.1102-1106
3. Posadowski W.M., "Plasma Parameters of Very High Target Power Density Magnetron Sputtering", Thin Solid Films 392 (2001), pp. 201-207
4. Posadowski W.M., Brudnik A., "Optical emission spectroscopy of self-sustained magnetron sputtering", Vacuum 53, 1999, pp. 11-15
5. Boo J.-H., Jung M.J., Park H. K., Nam K.H., Han J.G.: „High-rate deposition of copper thin films using newly designed high-power magnetron sputtering source", Surface & Coatings Technology 188-189 (2004), pp. 721-727
6. Borecki J., Felba J., Posadowski W.M.: "Magnetron Sputtering deposition of Copper on Polymers in High Density Interconnection PCB's", Proceedings of 5th International Conference on Polymers and Adhesives in Microelectronics and Photonics, Polytronic 2005, October 23-26, 2005, Wroclaw, Poland, pp. 192-196
7. Borecki J., Felba J., Posadowski W.M.: "Quality of PCB Interconnections Based on Blind Microvias Metallized by Magnetron Sputtering Deposition", Proceedings of 29th International Spring Seminar on Electronics Technology, May 10-14, 2006, St. Marienthal, Germany, paper No. 089.
8. Borecki J., Felba J., Gromek J., Posadowski W.M.: "Interconnections in Multilayer PCBs Based on Microvias Metallized by Magnetron Sputtering Deposition", Proceedings of 1st Electronics Systemintegration Technology Conference, September 5-7, 2006, Dresden, Germany, pp. 525-531

Simulation of Failure Criteria in Dielectric Layers

Khalil Arshak, Ivor Guiney, and Edward Forde
Microelectronic and Semiconductor Research Group,
University of Limerick,
Ireland

Abstract

Breakdown in dielectric layers has previously been assumed to be dependent only on the electric field strength at a certain point in the dielectric lattice [1], [2]. This does not take quantum aspects of the lattice structure into account. The present model takes into account quantum variations of particles in conjunction with different particle characteristics. Dielectric breakdown patterns are characterised by their quantum breakdown probabilities and studies are carried out as a function of the applied electric field, temperature and specific material constants responsible for breakdown, such as the mean free path and the potential barrier present. How these quantities affect the breakdown probability of the dielectric layer is examined and this facilitates the prediction of failure in dielectric layers in semiconductor devices.

1. Introduction

Characterising dielectric breakdown correctly within an insulated material has increased in importance in recent years. This is due to the fact that many semiconductor devices now use dielectric layers of ever diminishing thickness. In many cases these coatings are quite thin [1], thus increasing the likelihood of voltage punch-through and subsequent breakdown.

From the theoretical point of view, dielectric breakdown in homogeneous materials has been described as a stochastic process producing fractal structures that are called electrical trees. One widely used model is the dielectric breakdown model (DBM), first introduced by Niemeyer, Pietronero, and Wiesmann [3], which assumes that the dielectric is homogeneous, i.e., the electrical tree propagates in a dielectric medium without inhomogeneities. The dependence of the breakdown probability on the local electric field in the material is the main feature in the DBM. This is a fact that endeavours to consider the basic mechanism underlying breakdown in real materials. Breakdown channels are produced by stochastic fluctuations that damage the material, thus increasing the local electric field and eventually producing new channels. Since its introduction, the DBM has been broadly studied to describe experimental results [4]-[10], though the physical origin of such stochastic fluctuations is beyond the DBM and cannot be explained by it [11]. It is for this reason that new methods are required.

In this paper we present a new model, which takes quantum probability due to location of charges into account. The existing dielectric breakdown model is outlined in section 2 and the new quantum breakdown model is explained in section 3. We present results in section 4 and finish in section 5 by drawing conclusions from the work.

2. The Dielectric Breakdown Model

In the DBM [3] the dielectric is represented by a rectangular lattice where each lattice point corresponds to a point in the dielectric. The DBM assumes that the tree grows stepwise, starting at a lattice point with electric potential $\phi = 0$ and ending in a counter lattice point where $\phi = 1$. The tree channel growth is governed stochastically by the electric field. The probability P of a tree channel growth at each site of the electrical tree neighbourhood is chosen to be proportional to a power η of the electric field E at such site ($P \propto E^\eta$). The electric field E can be written from ϕ, and therefore

$$P\left(i, k \rightarrow i', k'\right) = \frac{\left(\phi_{i',k'}\right)^\eta}{\sum \left(\phi_{i',k'}\right)^\eta} \qquad (1)$$

where i,k and i',k' represent the discrete lattice positions, ϕ is the electric potential defined for all points of the lattice by the discrete Laplace equation and η is a power of the electric field.

The sum in the denominator refers to all of the possible growth sites (i',k') adjacent to the electrical tree.

The electric field distribution is obtained by solving the Laplace equation considering that the tree structure has the electric potential of the initial lattice point ($\phi = 0$), i.e.

$$\nabla^2 \phi = 0 \qquad (2)$$

As outlined by reference [3], the discrete form of equation 2 on the two-dimensional lattice can be written as:

$$\phi_{i,k} = \frac{1}{4}\left(\phi_{i+1,k} + \phi_{i-1,k} + \phi_{i,k+1} + \phi_{i,k-1}\right) \qquad (3)$$

Iteration of this equation, combined with appropriate boundary conditions, allows for the potential to be obtained.

3. Quantum Breakdown Model

It has been documented that the damage caused by breakdown to the dielectric takes the form of avalanches [12], [13]. Here an electron in the solid is accelerated to a higher kinetic energy, which is sufficient to ionize a molecule on its path thereby producing two electrons. This process is carried out until the electrons produced can no longer acquire sufficient kinetic energy for ionization and are thermalised and trapped. One possible path for the discharge is shown in figure 1. Here a random point in the dielectric lattice of a MOSFET is selected and the potential of each adjacent lattice point is calculated by iteration of the Laplace equation.

An ionization coefficient, $\kappa(\phi, T)$ is defined such that the ionization coefficient is strongly influenced by electric potential, ϕ and absolute temperature, T. A breakdown probability

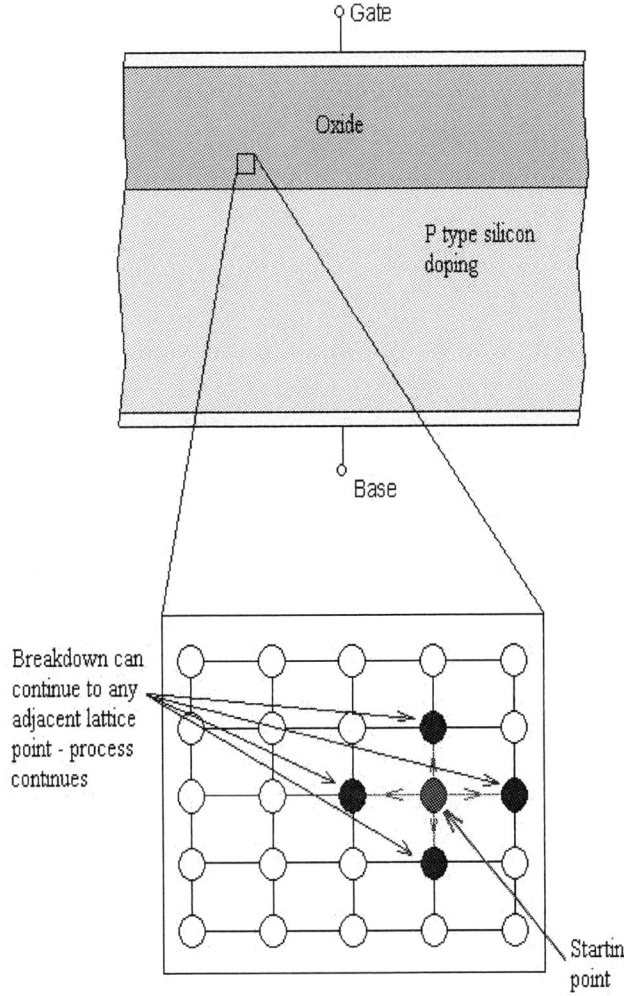

Fig. 1. Cross-section of a MOSFET showing gate oxide layer and lattice breakdown mechanism

is also defined:

$$p_{breakdown}(x'|x)dx = \kappa(\phi,T) \exp\left(-\int_x^{x'} \kappa(\phi,T)dx''\right)dx$$

(4)

$$= \kappa(\phi,T)P_{breakdown}(x'|x)dx$$

where $P_{breakdown}(x'|x)$ is the probability that a charge will survive when traversing between two points x and x' in the lattice within the incremental length dx, valid in the region $0 \leq P_{breakdown}(x'|x) \leq 1$.

Equation 4 represents the probability that breakdown will occur, starting at x, within the incremental length dx. x' in this case is taken to be four different, distinct values representing the four lattice points surrounding each site.

The ionization coefficient $\kappa(\phi,T)$ itself, is represented by the following equation

$$\kappa(\phi,T) = a \exp\left(-\frac{b}{\phi}\right)$$

(5)

where a and b are different for every material under investigation and ϕ is the electric potential obtained by iteration of equation 2. According to [14] and [15], the quantity a is equal to the reciprocal of the mean free path between collisions. In solids, this length is treated as an adjustable parameter [16]. The quantity b is a function of absolute temperature of the dielectric as based on work by Chen *et al.* [17]. In this case

$$b = \frac{3}{2}\int_0^{T_t} \sqrt{\Phi(x)}dx$$

(6)

where T_t is the tunneling distance an electron can travel in the presence of pits and traps and $\Phi(x)$ is the temperature dependent potential barrier of the metal-dielectric interface in electron volts. This is given by

$$\Phi(x) = -k_bT \ln\left[\frac{k(x,T)}{k_0}\right]$$

(7)

where k_b is Boltzmann's constant, T is the absolute temperature, k_0 can be interpreted as an attempt frequency, i.e. the frequency an atom is oscillating around its equilibrium position at a given temperature and is in the order of $10^{12} - 10^{13}s^{-1}$ and $k(x,T)$ is the rate at which an isolated atom will jump to any of the four next local energy minima on a surface. b is defined in electron volts.

This new work introduces inhomogeneity characteristics, assigning different probabilities to the breakdown channel formation, according to the conducting characteristics at each site. The situation can be rationalised by introducing different values for x' in equation 4. Note that this modification affects only the probability assigned to each site adjacent to the electrical tree. Further rationalisation can be achieved by varying the temperature parameter in equation 7. Values for breakdown probability are calculated from equation 4 and the electrical tree follows the path with greatest quantum ionization probability in a stochastic manner similar to the DBM [3].

4. Results and Discussion

We will now present a study of the probability of breakdown within a dielectric with the model developed in the preceding section. Changing the input voltage, temperature of the dielectric or varying the dielectric material via its mean free path and attempt frequency results in different fractal structures following a similar stochastic process to those breakdowns presented in this work.

Graphs of breakdown probability as a function of the temperature of the dielectric material and of the intermolecular spacing within the lattice are presented in figures 2 and 3 respectively. Figure 2 shows the exponential nature of the increase in breakdown probability with increasing dielectric temperature. As the temperature reaches approximately 100K above room temperature, the breakdown probability starts to increase considerably. This can be attributed to the fact that the charged particles within the dielectric now have increasingly greater energy to ionize in an avalanche manner. Figure 3 shows the decrease in breakdown probability as intermolecular distance increases. This is due to the fact that charges will find shorter intermolecular gaps more difficult to traverse, thus diminishing breakdown likelihood. Figure 4 shows the exponential increase in breakdown probability with increased applied voltage to the dielectric layer. A log plot has been used for clarity and the results follow an expected trend, i.e. with increased voltage the molecules in the dielectric lattice will acquire more energy and will more readily ionize and form a breakdown channel. Figure 5 shows the exponential increase in breakdown probability as a function of tunnelling distance within the dielectric lattice. At a small tunnelling distance of 1Å a very small probability exists, however for a relatively large tunnelling distance of 100Å close to unity probability is present. This is due to the fact that dielectric layers in semiconductor devices are usually very thin (in some cases less than 100Å) so that tunnelling distances comparable to these would present an immense probability of breakdown.

The amount of charge build-up in the dielectric can be represented by its continuity equation:

$$\frac{\delta q}{\delta t} - D\nabla \cdot q = q\nu_{ion} - \alpha_{rec}q_iq_e \qquad (8)$$

where D is a diffusion coefficient given by [18], $q_{i,e}$ is the charge generated due to electrons and ions respectively in the dielectric, ν_{ion} is the frequency of ionization within the layer and α_{rec} is the coefficient of recombination of opposite polarity charges. The cross-section of charge build-up in the dielectric layer can be seen in figure 6. This serves to highlight the exponential nature of the overall breakdown.

A more complete description of this process is possible if pits and traps are taken into account. These can be incorporated by an accurate determination of the nature and locations of these discontinuities and allowing for this when determining electric potential at each lattice point. Much work has yet to be undertaken in this area and the dispersion of charge and hence the nature of any breakdown within this area is intended as future study for the authors. Future work is also intended to focus on the modification of this theory to account for a complete picture of three dimensional dielectric breakdown. New solutions of the Laplace equation (see equations 2 and 3) would be necessary, which would involve analysing multiple lattice points in three dimensional Cartesian space. This would require rigorous iterations, which would be considerably more time consuming than those attempted in this work. The lattice grid employed for modeling purposes should additionally be improved upon by incorporating a greater degree of non-uniformity and varying lattice position in order to produce ever increasingly realistic models.

The breakdown probabilities between adjacent sites, as calculated from the new theory in this paper, were found to increase as voltage gradually increases for a given material. The temperature of the dielectric was also found to have a considerable effect on the breakdown probability as can be seen from figure 2. The exponential nature of the ionization coefficient results in a strong dependence on electrical potential and temperature, as stated. The mean free path between collisions, while important, is found to affect the calculations considerably less than electrical potential or temperature. This

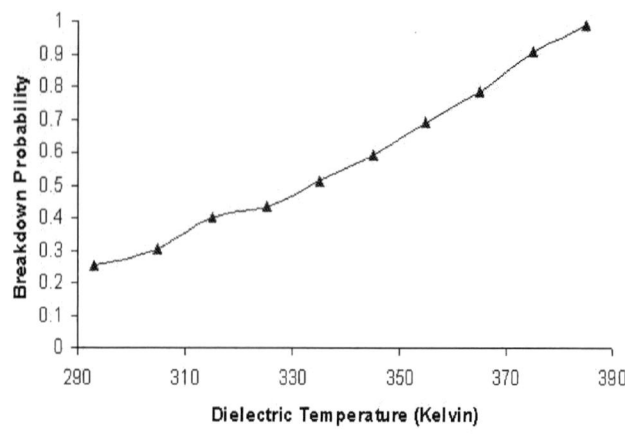

Fig. 2. Dielectric breakdown probability as a function of dielectric temperature

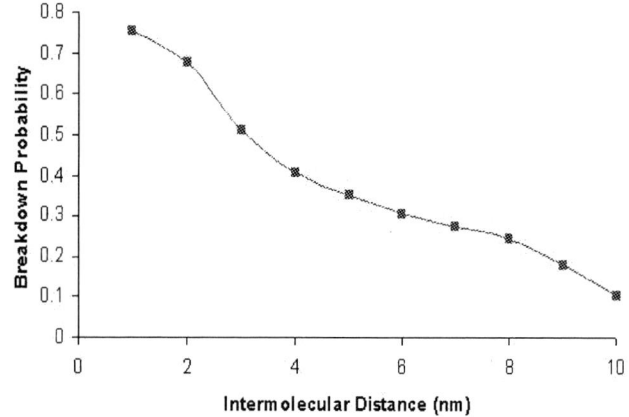

Fig. 3. Dielectric breakdown probability as a function of intermolecular spacing within the dielectric lattice

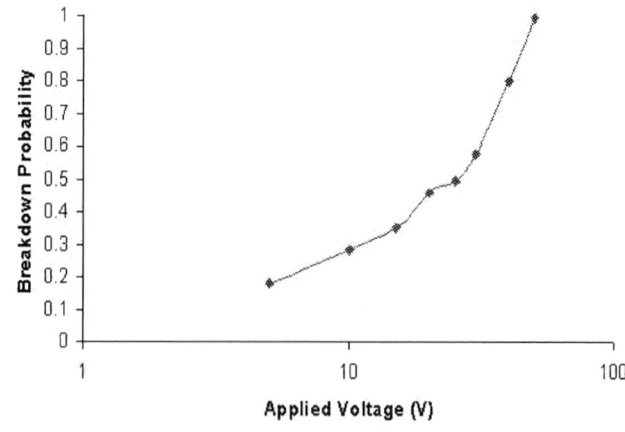

Fig. 4. Dielectric breakdown probability as a function of applied voltage

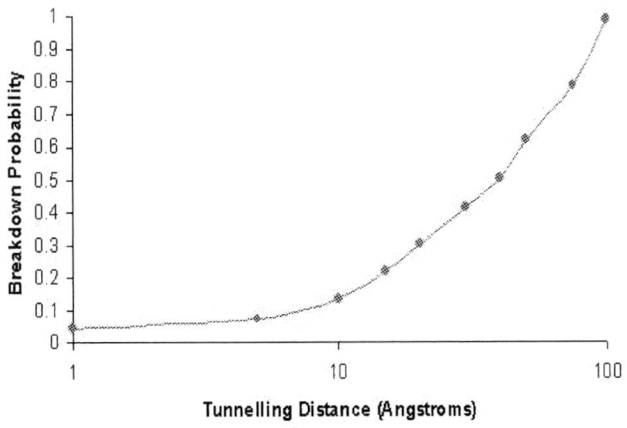

Fig. 5. Dielectric breakdown probability as a function of tunneling distance

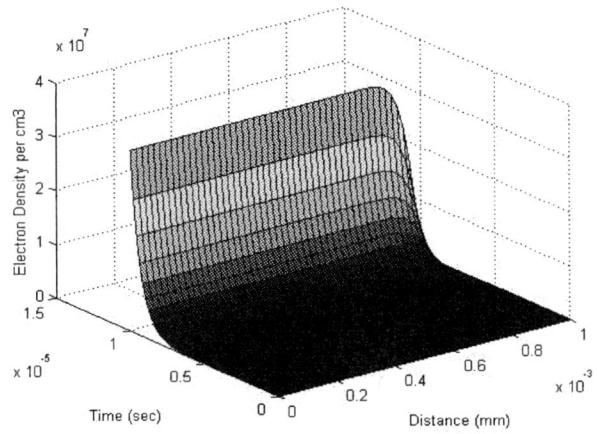

Fig. 6. Charge Evolution in the Dielectric Layer over Time

is not to say that factors other than material properties are more relevant in breakdown probability calculations however. The rate at which an isolated atom will jump to any of the four next local energy minima on a surface is an important material characteristic and is greatly influenced by the insulative nature of the dielectric, i.e. its dielectric constant. Work is currently ongoing to develop a more complete model, which incorporates the dielectric constant of a material into the breakdown probability calculations.

The variable nature of inter-lattice distance has additionally been incorporated in this work, as illustrated by equation 4. Varying of the quantity x' results in a considerable change in the breakdown probability between adjacent sites in the lattice. Thus it is evident that intermolecular distance in conjunction with electric potential, dielectric temperature and the nature of the dielectric affect breakdown probability considerably.

The influence of the potential barrier of the metal-dielectric interface and the tunneling distance on the breakdown probability illustrates the quantum nature of the approach. This assumes the availability of states necessary for tunneling to occur. A particle can tunnel through the potential barrier, but unless there are states available within the barrier, the particle can only tunnel to the other side of the barrier. A more detailed analysis of the tunneling present in this case is desirable in order to form a more complete picture of the breakdown probability. This is intended for future work.

5. Conclusion

In this paper, a new method of predicting breakdown in dielectrics was developed. Conducting particles are distributed at random in the insulating matrix and the dielectric breakdown propagates according to new rules to take into account electrical properties, material properties, particle size and temperature. Intermolecular spacing also becomes important, which is an improvement over existing methods. The theory introduced by the DBM implies that the electrical tree will propagate in the most likely direction from a particular lattice point. This concept has been utilised and expanded upon to form a more complete model of dielectric breakdown.

Acknowledgment

The authors would like to thank John Harris for his invaluable help throughout this work.

References

[1] E. Y. Wu, D. L. Harmon, and L.-K. Han, "Interrelationship of voltage and temperature dependence of oxide breakdown for ultrathin oxides," IEEE Electron Device Lett., vol. 21, pp. 362-364, 2000

[2] B. P. Linder, S. Lombardo, J. H. Stathis, A. Vayshenker, and D. J. Frank, "Voltage Dependence of Hard Breakdown Growth and the Reliability Implication in Thin Dielectrics", IEEE Electron Device Lett., vol. 23, no. 11, 2002.

[3] L. Niemeyer, L. Pietronero, H. J. Wiesmann, "Fractal Dimension of Dielectric Breakdown" Phys. Rev. Lett. **52** 12 (1984)

[4] E. Arian, P. Alstrom, A. Aharony, H.E. Stanley, "Crossover Scaling from Multifractal Theory: Dielectric Breakdown with Cutoffs", Phys. Rev. Lett. **63**, 2005 (1989).

[5] J.M. Cooper, C.G. Stevens, "The Influence of Physical Properties on Electrical Treeing in a Cross-Linked Synthetic Resin", J. Phys. D **23**, 1528 (1990).

[6] A.L. Barclay, P.J. Sweeney, L.A. Dissado, G.C. Stevens, " Stochastic Modelling of Electrical Treeing: Fractal and Statistical Characteristics" J. Phys. D **23**, 1536 (1990).

[7] K. Kudo, "Fractal Analysis of Electrical Trees", IEEE Trans. Electr. Insul. **5**, 713 (1998).

[8] J.L. Vicente, A.C. Razzitte, M.C. Cordero, E.E. Mola, "Thermodynamic Approach to Electrical Tree Formation", Phys. Rev. E **57**, R1 (1998).

[9] I.M. Irurzun, J.L. Vicente, M.C. Cordero, E.E. Mola, "Fractal Analysis of Electrical Trees in a Cross-Linked Synthetic Resin", Phys. Rev. E **63**, 016110 (2001).

[10] I.M. Irurzun, P. Bergero, V. Mola, M.C. Cordero, J.L. Vicente, E.E. Mola, "Dielectric Breakdown in Solids Modeled by DBM and DLA" Chaos, Solitons Fractals **13**, 1333 (2002).

[11] L.A. Dissado, J.C. Fothergill, N. Wise, A. Willby, J. Cooper, "A Deterministic Model for Branched Structures in the Electrical Breakdown of Solid Polymeric Dielectrics", J. Phys. D **33**, L99 (2000).

[12] C. J. Mayoux, "Partial Discharge Phenomena and the Effect of Their Constituents on Polyethylene", IEEE Trans. **EI-11**, 139 (1976)

[13] R. Degraeve, J. L. Ogier, R. Bellens, P. J. Roussel, G. Groeseneken, and H. E. Maes, "A New Model for the Field Dependence of Intrinsic and Extrinsic Time-Dependent Dielectric Breakdown", IEEE Trans. Electron Devices, vol. 45, no. 2, 1998

[14] L.A. Dissado, P.J.J. Sweeney, "Physical Model for Breakdown Structures in Solid Dielectrics", Phys. Rev. B **48**, 16261 (1993).

[15] F. Seitz, "On the Theory of Electron Multiplication in Crystals", Phys. Rev. **76** 1376 (1949)

[16] J. J. O'Dwyer, IEEE Trans. **EI-19**, 1 (1984)

[17] I-C Chen, S. E. Holland, C. Hu, "Electrical Breakdown in Thin Gate and Tunneling Oxides", *IEEE Journal of Solid-State Circuits* **20** 1 (1985)

[18] A. L. Filatov, V. I. Mirgorodsky and V. A. Sablikov, "Photorefractive method of contactless determination of the charge carrier lifetime and diffusion coefficient in semiconductors", *Semicond. Sci. Technoi.* 8 (1993) 694-699.

Dynamic mechanical behavior of SnAgCu BGA solder joints determined by fast shear tests and FEM simulations

Eberhard Kaulfersch[1], Sven Rzepka[2], Vijay Ganeshan[2], Axel Müller[2] and Bernd Michel[1]

1) Fraunhofer Institute for reliability and micro-integration, Micro materials center Berlin and Chemnitz
Gustav Meyer Allee 25, D-13355 Berlin; Otto-Schmerbach-Str. 19, D-09117 Chemnitz, Germany
eberhard.kaulfersch@che.izm.fraunhofer.de, +49 371 866 2029
2) Qimonda Dresden GmbH & Co OHG, Dept. Backend Technologies
Königsbrücker Landstraße 180, D-01099 Dresden, Germany
sven.rzepka@qimonda.com, +49 351 886-1145

Abstract

To allow realistic stress assessments of BGA modules during drop and shock events, the dynamic mechanical behavior of SAC BGA solder joints was determined by a combination of fast shear tests and FEM simulations. The goal has been the development of a validated material model for SnAg1.0Cu0.5 BGA solder joints that accounts for dynamic material hardening and the dependency of the fracture mode on the shear speed. Therefore, fast shear experiments of single joints have been performed applying a DAGE® tester to BGA solder balls being attached to Cu+Ni/Au pads with recording the peak shear forces and fracture surface plots. Dynamic 3-D finite element simulations than have been performed applying both codes, ABAQUS explicit™ and ANSYS/LS-DYNA™, to replicate the shear tests virtually. The simulations involved a rate dependent model for the solder material as well as a cohesive zone approach for the fracture in the IMC layer and included a shear criterion for the damage within the bulk solder. The coefficients of all these models have been calibrated iteratively based on the shear test results. Combining experiments and FEM simulations this way, realistic parameters have been determined for the dynamic hardening of the solder as well as for the solder bulk and the IMC strengths. Applying these parameters, the simulated peak forces match the experimental values at all shear speeds and also the range of speed, in which the transition between bulk damage and IMC fracture takes place, is precisely met. These validated models now allow highly dynamic events of BGA modules to be assessed realistically by means of FEM.

1. Motivation and objective

The high dynamic loads affecting the solder interconnects in cases of mechanical drop and/or shock of BGA modules are important reliability concerns in microelectronics packaging. Design optimization based on direct measurements of the stresses occurring in the solder joints during those events is not feasible. In contrast, finite element simulations are well capable of visualizing the solder joint behavior and of estimating the magnitude of the stress components in all parts of the module. However, the accuracy of such virtual approaches is determined by adequacy, comprehensiveness, and reliability of the models describing the relevant behavior of the materials the structure consists of. In the case of BGA modules,

the data available today for covering the dynamic mechanical hardening of the solder and the fracture toughness of the joints is not sufficient. Hence, effects like the increase in the peek shear force and the change in the fracture site from the bulk of the solder next to the pad/joint interface to the intermetallic compound right at this interface can not be addressed by the simulations, although they are clearly observed in experiments when the shear speed is increased. That means realistic simulations of drop and shock events strongly necessitate the additional dynamic mechanical data.

2. Experimental setup and results

In experimental observations the impact behavior of lead-free solder alloys at both first and second level interconnect (i.e., after solder ball attach and after component assembly, see Fig. 1) has been to be characterized to determine the stress/strain distribution at the substrate/solder/PCB interfaces when subjected to package shear loads at different speeds and parameters. In the present study, the focus was put on first level interconnect changes with the rate of the impact load as reported here.

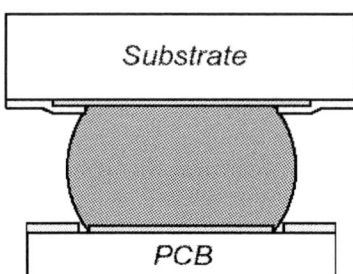

Figure 1: Typical second level BGA interconnect

With reflowed solder balls, that would be of 450 μm in diameter after second level packaging, being attached to Cu+Ni/Au pads, characteristic threshold magnitudes at which the failure mode is changing from fracture in the solder bulk to fracture in the intermetallics were to be obtained. The pad on the substrate side consists of copper with a nickel top layer and a gold flash, covered partially by the solder mask layer. The joint interface is formed by a NiSn intermetallics compound (IMC) of about 1,5 μm thickness, wheras the Ni, IMC and solder layers are usually detached from solder mask (see Fig. 2). A silicon die is adhesively bonded to the substrate top surface and finally overmolded. The BGA joint is formed of

1-4244-1105-X/07/$25.00 ©2007 IEEE   357

SnAg1.0Cu0.5 or SnAg3.5Cu0.75 leadfree solder. In the second level interconnection step, the package gets bonded to an FR-4 PCB.

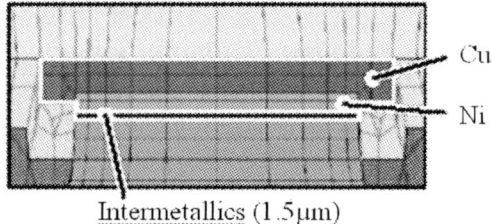

Figure 2: Pad layer sequence

For single ball shear experiments, the package is flipped into a stainless steel frame for support (see Fig. 3). The shear direction is parallel to bond channel. i.e. in the longitudinal package direction. The tool is adjusted to keep a constant shear height of 30 μm from the top of the solder mask.

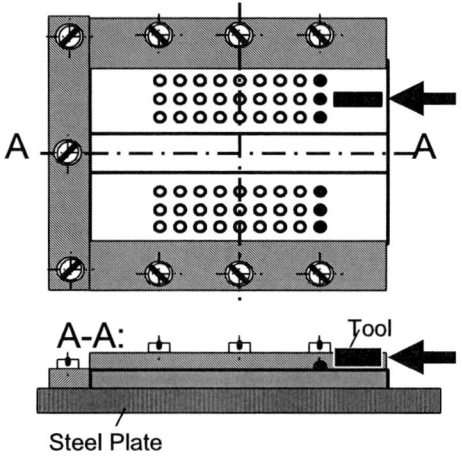

Figure 3: Ball shear setup

In Fig. 4 the actual shape of the ball after reflow and the face area of the tool are visualized. Dimensions of ball shear tool as shown in Fig. 4 below are a shear tool depth of >2.0 mm and a bottom with of 0.75 mm. It is made of ceramic composite material and can be assumed to be a rigid body in comparison to solder bulk material.

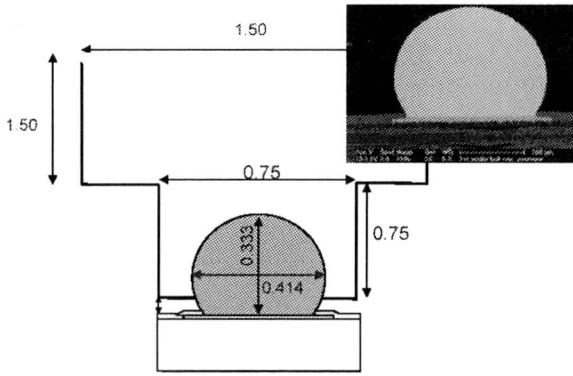

Figure 4: Single ball and shear tool dimensions

During shear experiments, peak shear forces and the fracture surface plots were recorded. The tool speed ranged from 0.35 to 2,000 mm/s and included the following values: 0.35, 10, 500, 1000 and 2000 mm/s. Fig. 5 is showing the observed transition from bulk solder shear via partial solder fracture to so called grey plot fracture with all the damage occurring within the IMC, Fig. 6 covers the corresponding shear strengths. From Fig. 7 it can be experienced, that a transition from bulk fracture to IMC damage takes place at a shear speed range from 1000 to 2000 mm/s.

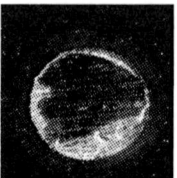  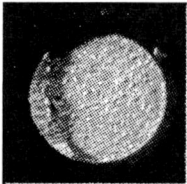

Figure 5: Fracture plots at 10 mm/s (left, bulk fracture), 1000 mm/s (middle, partial bulk fracture) and 2000 mm/s (right, IMC fracture)

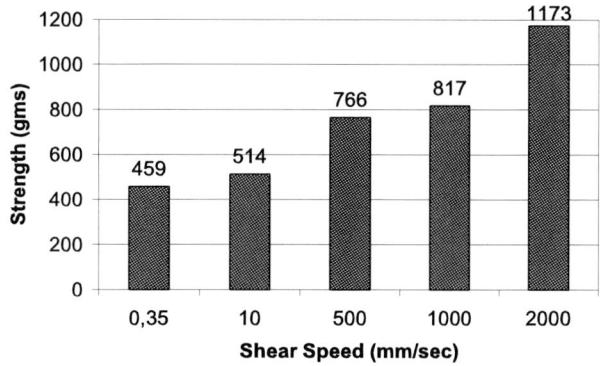

Figure 6: Solder ball shear strength at different shear speeds

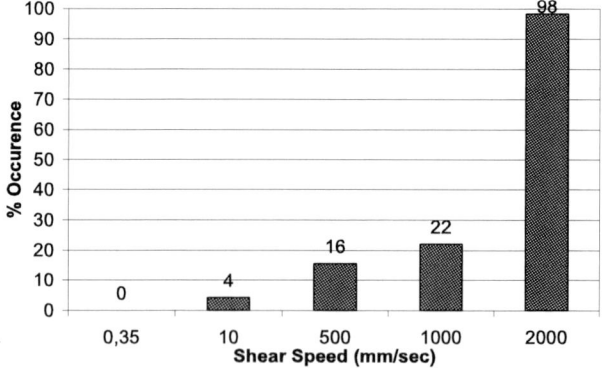

Figure 7: % brittle fractures after solder ball shear test at different shear speeds

3. Finite Element investigations

The response of the solder joint in contact with the shear tool moving at a constant shear speed was calculated in three-dimensional transient analyses using finite element models and appropriate fracture criteria. The simulation task was to characterize the strain rate dependent material behavior of lead-free solder interconnect and to determine threshold magnitudes of changes in the failure mode by meeting the measured peak force values and the transition region between solder bulk damage and fracturing of the SnNi intermetallic layer.

Therefore, FEM techniques were applied to simulate a fast ball shear test under different shear speeds/conditions to validate the strain dependent deformation behavior that correlates with experimental data. Modeling based on the configuration shown in Fig. 3 with only a single ball, surrounded by a solder mask layer, on an underlying structure of Ni on copper pad, organic substrate, Si-chip and mold cap (Fig. 8). The shear tool was assumed to be a rigid body with constant velocity. Bottom plane and adjacent steel frame plane were fixed with displacement boundary conditions. The material model to be developed has to be capable of identifying changes in failure mode locations (bulk solder, IMC, pad lifts, etc.) depending on input loading conditions from the experimental investigations. Pad lift failure was not addressed at the moment.

For solder bulk damage a shear criterion with strain rate dependence of critical plastic strain but with neglected dependence on shear stress ratio has been used: Critical plastic strains for damage initiation were 15% at 0.1 s^{-1}, 5% at 100 s^{-1} and 1% at 1000 s^{-1}.

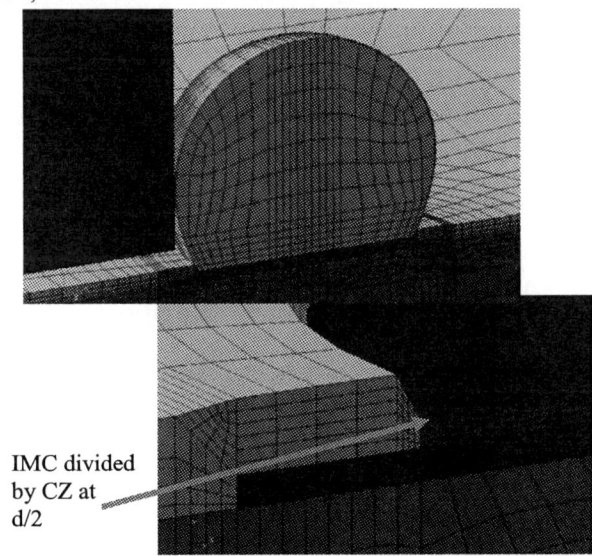

IMC divided by CZ at d/2

Figure 8: Single ball shear symmetric finite element model with zero thickness cohesive zone dividing IMC layer

In recent years, cohesive zone models have been employed to simulate fracture and delamination in solids. The approach is based on the cohesive zone concepts of Dugdale (1960) [1] and Barenblatt (1962) [2]. The cohesive zone model (CZM) originally was suggested by Needleman (1987) [3] for simulation of inclusion debonding from a metal matrix. In various numerical investigations like crack growth in homogeneous ductile materials (Tvergaard & Hutchinson 1992) [4], interface debonding (Tvergaard & Hutchinson 1993) [5], impact damage in brittle materials (Camacho & Ortiz 1996) [6] and the analysis of sandwiched structures (Tvergaard & Hutchin-

son 1994 [7], the cohesive zone approach has been successfully employed. For the modeling of crack growth, zero thickness cohesive zone elements have become an attractive concept, where geometric and constitutive thicknesses are defined independently.

The material data fit procedure performed here included the following IMC damage model: The cohesive zone was based on a cohesive traction separation approach with uncoupled behavior for the normal and shear forces, with a linear traction-separation response (Fig. 9). The characteristics of the cohesive elements are determined by two parameters, cohesive strength and separation energy. The criterion for damage initiation was chosen to be a maximum nominal stress criterion. An assumption for the corresponding shear stress maximum was a factor of 2 times the normal stress. Damage evolution used linear stiffness degrading down to zero stiffness.

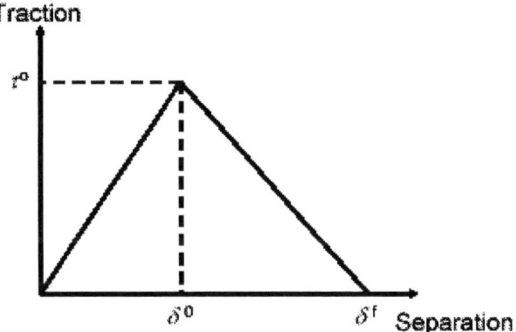

Figure 9: Linear traction-separation response

Failure mode and mechanical behavior of solder alloys are highly dependent on strain rate. In Fig. 10 the computed force-displacement curves are compared to the measured peak force values at varying shear speed.

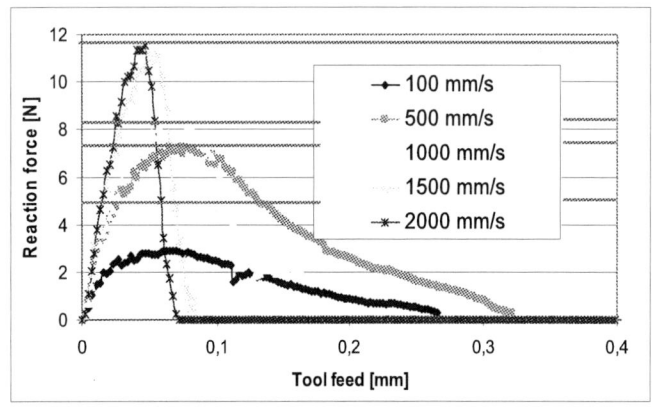

Figure 10: Calculated force-displacement curves for various shear speeds compared to the measured peak forces (see Fig. 6)

If recorded in shear experiments, several parameters can be used to characterize the sheared joint: Fmax, the maximum impact force, which relates to the IMC strength at high shear speeds; the duration and slope of the ascending part of the impact force profile, which represent the ductility of the solder joint; and the area below the ascending part of the impact force profile, which is a measure of the toughness of the solder joint. Recording of the force-displacement relation is difficult especially for high shear speeds, thus, only the peak values are available for tuning the material behavior put into the finite element simulations. The peak force values ranging from the ones in the transition region from bulk damage to IMC fracture towards those at higher velocities are determined solely by the IMC strength, the values below mainly by the solder ductility and bulk damage criterion. The corresponding fracture behavior can be seen from Fig. 11 with a damage path through the ball in the case of 500 mm/s and a completely fractured IMC layer and detached ball at 2000 mm/s. From a transition region between shear speeds of 1000 and 2000 mm/s, respectively, a high normal stress maximum of 350 MPa could be derived for the IMC strength. Fig. 12 depicts the bulk remainder on the pad at a shear speed of 1000 mm/s and a complete fraction of the IMC with no remaining solder at 2000 mm/s.

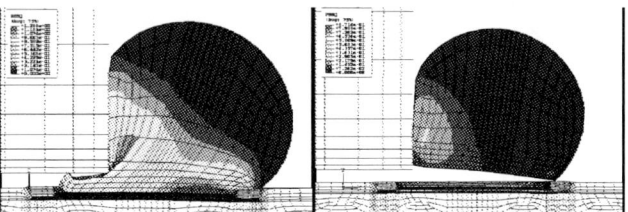

Figure 11: Fracture behavior at 500 mm/s (left) with a damage path through the ball and completely fractured IMC layer and detached ball at 2000 mm/s (right)

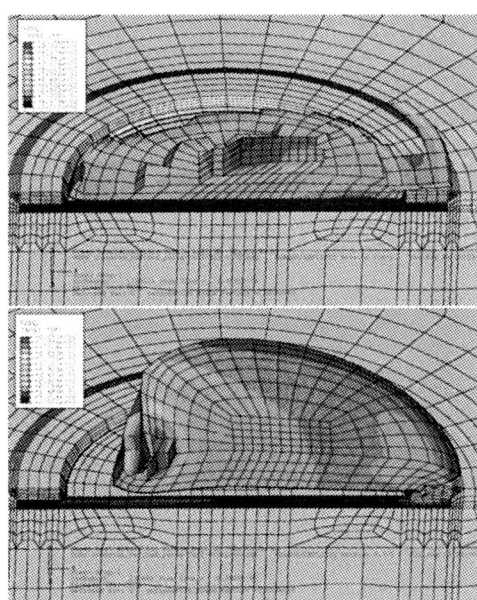

Figure 12: Bulk remainder plots at 1000 mm/s (top) and 2000 mm/s (bottom)

From the material data fit procedure a constitutive law of the following form has been found whereas the actual numbers of the coefficients remain concealed for preservation of company propriety information:

$$\sigma_y = A \, (d\varepsilon/dt)^n + B \, (d\varepsilon/dt)^m$$

with an initial yield strength of 25 MPa. Results seem to be relatively independent on the initial yield strength in the tested interval from 25 to 40 MPa. The IMC strength was stated to be 350 MPa for tension separation. An extension of the methodology to handle pad lift failure in addition to IMC fracture using a second cohesive region beneath the pad is currently under preparation.

4. Conclusions

With the experimental effort of performing high speed single ball shear tests the material behavior including dynamic hardening of the SAC solder alloy has been determined in combination with FEM simulations. Using a cohesive zone formulation for the solder/pad interface, the fracture mode transition from the solder bulk to the IMC could be reproduced numerically. The interface turned out to be of high strength. Applying the constitutive relationship, the simulated and measured peak forces are coinciding well. These validated models now allow highly dynamic events of BGA modules to be assessed realistically by means of FEM. Shear experiments should be enhanced in future with more logging capability to observe the change of shear force versus displacement in more detail. Any information about the shape of the force-displacement curve in terms of force maximum, energy, slope and duration of the ascending part would be valuable.

Acknowledgement

The work for this paper was supported within the scope of technology development by the EFRE fund of the European Community and by funding of the State Saxony of the Federal Republic of Germany.

References

1. Dugdale, D. S., "Yielding of steel sheets containing slits". *J. Mech. Phys. Solids*, No. 8 (1960), pp 100–8
2. Barenblatt, G. I., "The mathematical theory of equilibrium of crack in brittle fracture", *Adv. Appl. Mech.*, No. 7, (1962), pp. 55–129
3. Needleman, A., "A continuum model for void nucleation by inclusion debonding", *ASME J. Appl. Mech.*, No. 54, (1987), pp. 525–531
4. Tvergaard, V., Hutchinson, J. W., "The relation between crack growth resistance and fracture process parameters in elastic-plastic solids", *J. Mech. Phys. Solids*, No. 40, (1992), pp. 1377–1397
5. Tvergaard, V., Hutchinson, J. W., "The influence of plasticity on mixed mode interface toughness", *J. Mech. Phys. Solids*, No. 41, (1993), pp 1119–1135
6. Camacho, G. T., Ortiz, M., "Computational modeling of impact damage in brittle materials", *Int. J. Solids Struct.*, No. 33, (1996), pp. 2899–2938
7. Tvergaard, V., Hutchinson, J. W., "On the toughness of ductile adhesive joints", *J. Mech. Phys. Solids*, No. 44, (1996), pp. 789–800

Mechanical Characterization Analysis of flexible and stretchable Ultra-Thin Substrates by Experiment and FE Simulation

L. Wang[1], T. Zoumpoulidis[2], K.M.B. Jansen[1], M. Bartek[2], G.Q. Zhang[3], L.J. Ernst[1]

[1]Department of Precision and Micro-system Engineering, Delft University of Technology,
Mekelweg 2, 2628 CD Delft, the Netherlands.
Email: l.wang@tudelft.nl ; Phone : +31-15-2786739
[2] Delft Institute of Microelectronics and Submicron Technology/HiTeC, Delft University of Technology, Mekelweg 4, 2628 CD Delft, the Netherlands.
[3]Strategy and Business Development of NXP Semiconductors, Eindhoven, the Netherlands.

Abstract

Initial studies have demonstrated that specially designed and fabricated microelectronics embedded in flexible substrates can maintain functionality when subjected to stretching as well as bending. The acceptable flexibility and stretchability for ultra-thin substrate could be reached by embedding the ultra-thin substrate into flexible polyimide and patterning the silicon into square or hexagon segmentations. In this paper, results of experiments and FE simulations on mechanical issues of poly- and single crystalline silicon on ultra-thin polyimide substrates are presented. Generation of cracks within the silicon and dielectric layers are then studied under controlled bending and tensile tests using bending and tensile tools being specially designed for this purpose. Specimen observation can be done using an optical microscope with possibility of digital recording and evaluation by pattern recognition software. The crack onset and the propagation of cracks are characterized. The results show that the cracks appear first in the middle of the dielectric material in-between the silicon segments. Only at higher loads they propagate or are generated within the silicon itself. The development of first cracks depends significantly on the silicon segmentation size and gaps between segments. The highest flexibility result can be reached such that no cracks are detected under bending tests on cylinders with 2 mm diameter. The segment size and gap size have influence on the tensile deformation. The stiffness of samples with segments is different before and after the first crack occurs, as a consequence of the appearing cracks. Multilevel FEM simulations are performed in order to increase understanding of the major failure processes. The maximum principle strain as found from simulation and from the experiment compares quite well. The maximum local principal strains appear in the gap near the spot where first fracture occurred during testing in the oxide layers.

1. Introduction

Electronic systems in future applications tend to have increasingly higher functional density and will adapt the environment. The demand to miniaturize products especially for mobile applications is continuing to drive the evolution of electronic products and manufacturing methods. One key to miniaturization developed in the past was the use of unpackaged, bare dies [1]. Conventional IC packages form a rigid shell around silicon IC dies. Their purpose is to provide environmental protection, electrical interconnect and heat dissipation. Despite the fact that the majority of current silicon IC's is realized in a very thin top layer of the silicon substrate (<10 μm), the typical thickness of packaged IC dies exceeds 150μm.

During recent years, studies of wearable computer systems, smart clothing, sensitive skin and Radio-Frequency Identification have been executed. Applications of sticking the RFID and smart clothing on non-planar (even high curvature) surfaces of objects issue the challenges to flexible and stretchable substrates. Ultra thin chips (i.e. silicon dies thinned down to ~50 μm total thickness) lend themselves to reach the flexibility to a certain extent [2, 3]. However, sufficient stretchability is not attainable through only thinning the silicon substrates.

The flexible and stretchable substrates for electronic systems were shown with a multitude of material and technologies [4], embedding the components into stretchable non woven substrate and stretchable foil substrate. Previously [5, 6], substrate transfer technology for SOI and non-SOI single-crystalline silicon wafers was demonstrated allowing for high-performance low-power RF applications. One of its very interesting variations is transfer onto flexible ultra-thin (<10 μm) polyimide substrates. Initial studies have demonstrated that active circuitry maintains its functionality even for high bending curvatures, opening new possibilities for embedding electronics in MEMS applications and realization of disposable smart adhesive labels. If next to the vertical thinning also a lateral partitioning of the silicon substrate on sub-millimetre scale is applied, then 3D deformable electronics could be realized. By varying the partition dimensions and the geometry of connecting bridges, the level of acceptable deformations can be controlled. In practical realization such patterned silicon structures have to be embedded into a polymer film to provide local stress relieve and protection. In many applications it will be useful to embed IC's into flexible or even stretchable substrates. The targeted applications of this technology are wireless ID tags and sensor networks.

In this contribution, results of experiments and FE simulations to explore possible mechanical reliability issues of poly- and single crystalline silicon on ultra-thin polyimide substrates are presented. To improve flexibility as well as stretchability, either square or hexagonal

1-4244-1105-X/07/$25.00 ©2007 IEEE

segmentation is applied to the silicon layer before it is transferred onto an ultra-thin polyimide substrate (<10 μm) using a wafer-to-wafer substrate transfer technique based on a temporary glass carrier. Generation of cracks within the silicon and dielectric layers is then studied under controlled bending (on cylinders with diameters of 2 - 10 mm).

The formation of cracks is studied experimentally using bending and tensile tools, specially designed for this purpose. Specimen observation is done using an optical microscope with possibility of digital recording and evaluation by pattern recognition software. The results show that the cracks appear first in the dielectric layers in-between the silicon layer segments and propagate at higher loads only, or are generated within the silicon itself. The development of first cracks depends significantly on the silicon layer segmentation size, which affects both the crack density and the crack width. The crack density increases sharply with strain at the early stage and then increases slightly. The crack width increases steadily. By optimizing the segment structure, segment size and gap between segments, a high flexibility result can be reached, while no cracks are detected under bending tests with cylinder diameters being larger than 2mm

Tensile tests are performed for the samples with differently sized segments and gaps. In the undamaged state, the stiffness of all samples is within a narrow band because of the dominating low stiffness of the polyimide and consequently the minor influence of the high stiffness of the embedded silicon segments. However, the stiffness of the samples diminishes after crack onset. The critical main strain depends on the segment size and the gap in between segments.

In order to achieve more flexible or stretchable ultra-thin substrates, optimization of the geometry and structure is necessary. The large ratio of width (or length) to thickness offers a challenge to reliable FE simulation. Multilevel FEM simulations are performed in order to increase understanding of the major failure processes. Tensile and bending simulations are performed to understand the failure causes. From comparison to the experimental results we learn that the simulations provide good insight into possible places of crack initiation while the ultimate mean strains match well. The consistent simulation results provide the basis for the optimization of the geometry of the samples (including the segment and gap sizes) the materials and the interconnect geometry between the segments.

2. Sample design and preparation

In order to investigate the effect of segment and gap sizes on the flexibility and reliability of the structure, the samples were designed with square and hexagonal partitions (Fig.1) varying in segment length from 150 μm to 2000 μm and in gap size from 20 μm and 250 μm. The test samples were prepared on a 4 inch silicon wafer and consist of a 0.8 μm thick poly-silicon layer, sandwiched between 300 nm thermal and 500 nm PECVD silicon

dioxide layers. The silicon layer was segmented into hexagonal or square partitions varying in size from 150 to 2000 μm. The segmented silicon layer was then transferred onto an ultra-thin polyimide carrier substrate with about 10 μm thickness. The sample preparation process was described in [7]. The vertical structure schematics of the samples are shown in the following figure (Fig. 1).

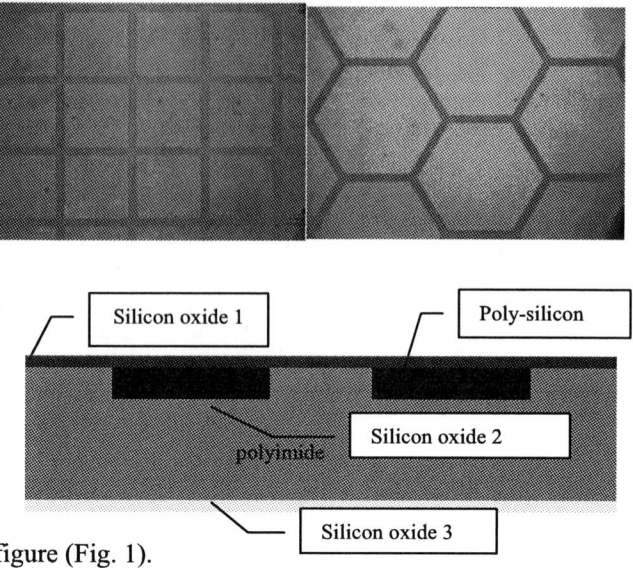

Fig.1: Photos of square and hexagon samples and cross-section schematics

3. Bending test

Special bending test setups [7] were designed and fabricated. On these, the specimen observation can be done by optical microscope during bending tests. The cracks could be detected and recorded by means of an optical microscope with digital recording, the magnification ranging from 25 up to 3000 times and a maximum image resolution of 4800 x 3600 pixels is obtained. The dedicated bending test tool allows the sample to be curved around glass cylinders with different diameters. Due to the ultra-thin sample and residual stress, it is necessary to pre-stress the sample under a small tensile load. The lens can be rotated to observe the cracks around the glass cylinder surfaces.

The samples with square and hexagonal segments varying in size and gap distance were wrapped around the glass cylinders of various diameters. The bending procedure was made up from the following steps: First, the samples were bended around the glass cylinder with the largest diameter (10 mm), and then the Region Of Interest (ROI) was scanned with the microscope to detect cracks. If there were any cracks they will be recorded. Subsequently, the samples were bended around glass cylinders with smaller diameter (8, 6, 4, 2 mm).

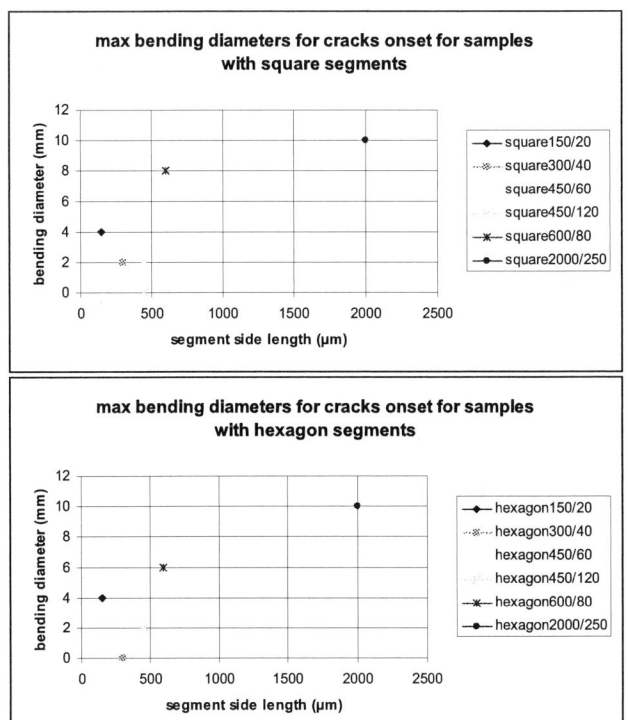

Fig 2. Bending diameter at crack onset versus segment length.

The formation of the first crack depends significantly on the segment sizes and the gaps between the segments for samples with square or hexagonal segments. According to the results in fig. 2, the diameters for the onset of cracking increase with increasing segment size. They decrease with increasing gaps between the segments. For the sample with the square pattern of 450 μm side length and 120 μm gaps and the sample with the hexagonal segment of 300 μm and 40 μm gap, there is no crack, even not for bending around a cylinder with 2 mm diameter. However, for the samples with the 150 μm side length segment we do not observe this trend, the reason probably is the stress concentration due to the small gaps between the segments.

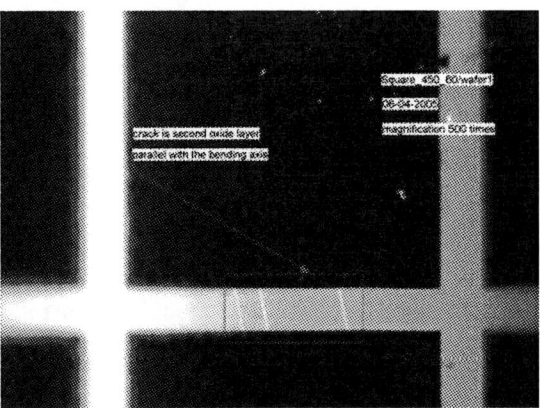

Fig. 3: The cracks under the bending test.

The first crack appears on the second oxide layer (see Fig 1). The cracks are more or less parallel with the axis of the bending cylinder (Fig.3). The silicon segments are ultra thin (about 0.8 μm); therefore the partitions are slightly transparent. Thus possible cracks below segments (in the second oxide layer) would be slightly visible. However such cracks were not detected. Also no cracks were observed in the first layer on top of silicon segments. Actually silicon dioxide cracks only appear to run between silicon segments. Surprisingly such cracks only are present in gaps that stand perpendicular to axis of the bending cylinder and not in gaps that are parallel to this cylinder. This observation holds for all bending tests being performed. So also for the test with the highest curvature (diameter = 2 mm).

4. Tensile test

The tensile tests were performed for samples with hexagonal segments with different sizes and gaps in-between the segments. The tensile test results for the samples with square segments are presented in [7]. The specified tensile setups were also designed with capacity to observe cracks by optical microscope.

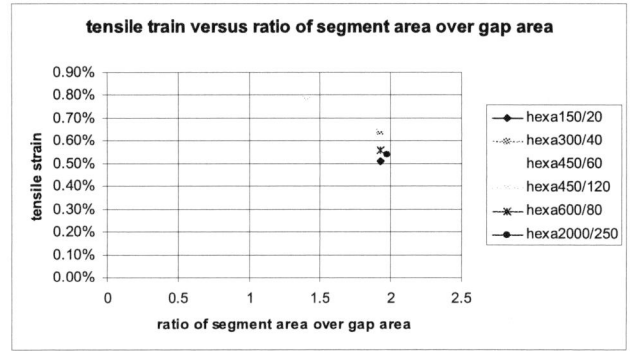

Fig 4. The max strain for onset of cracks versus the ratio "segment area over gap area" (cases with hexagonal segments in tensile testing)

The tensile (failure) strains depend on the ratio of "segment area over gap area". However, the tensile strains not only depend on the ratio of segment area over gap area but also the segment sizes and gaps between the segments. No crack was detected for the samples of ratio 1.4 until the strain is up to 0.78%.

A limiting principal strain value is apparently found from a test of just polyimide with oxide layers subjected to the same tensile test. Here a max. principle strain of 0.8% was found, the standard deviation is 0.08. From this tensile experiment we conclude that the tensile strain of the applied oxide is about 0.8%.

The first crack was detected at the early stage when the sample (mean) strain equals the critical (mean) strain. Later cracks were observed in all silicon dioxide layers and silicon segments. From the above stress-strain curve it can be observed that the initial stiffness for all the samples is almost the same, the slopes change at about 0.8% strain because the cracks than start to affect the samples' stiffness. The result is consistent with the

363

bending test result on just a polyimide layer with oxide layers, where a maximum bending strain in the oxide at first cracking was established also around 0.8%. The stiffness depends on the ratio "segment area over gap area" and the crack density.

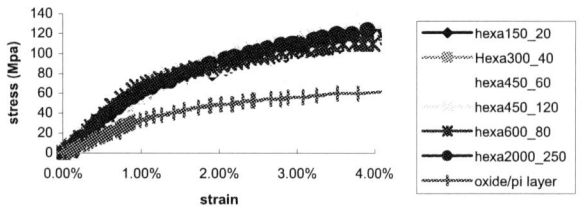

Fig 5. *Mean Stress versus mean Strain curve for samples with and without segments in tensile tests.*

The initial stiffness of a sample only with oxide and polyimide layer is lower than that of a sample including silicon segments. The max principle strain at onset of failure for silicon oxide is about 0.8%.

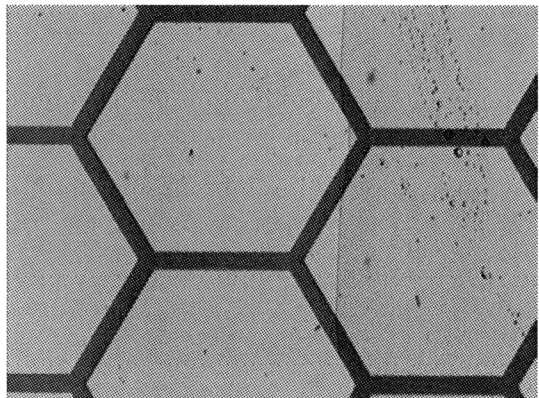

Fig. 6. *photos of crack onset by microscope*

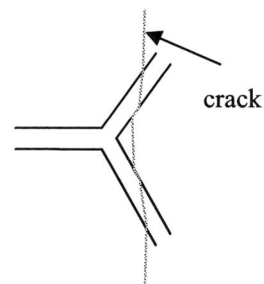

Fig. 7. *schematic of crack onset shape*

Fig.6 and Fig.7 show the shapes and the schematic of the first crack of the samples with hexagon segments under tensile testing. The first crack probably starts in the oxide layers at the gap positions and subsequently propagate into the oxide layers below and on top of the silicon and most probably also within the silicon itself.

5. Simulation

In order to achieve a more flexible or stretchable ultra-thin substrate, optimization of the geometry and structure is necessary. The large ratio of width or length to thickness requires an adequate FE simulation technique. Therefore, multilevel FEM simulations are performed in order to increase understanding of the major failure processes.

All the materials are assumed elastic. Actually, polyimide is a time and temperature dependent (viscoelastic) material. However, the viscoelastic material behaves approximately elastic for the short duration loading at room temperature as applied. Therefore in the present tensile and bending simulations only linear elastic properties are taken into account. The bending simulation results were presented in a previous paper [7].

The large ratio of width or length to thickness offers a challenge for correct FE simulation. Multilevel FE-modelling offers an option to solve this challenge. Representative Unit Cells are selected for different samples to calculate equivalent mechanical properties (Young's modulus, Poisson's ratio). Local models are applied for accurate local stress-strain investigation, while the global simulation results are considered as boundary conditions.

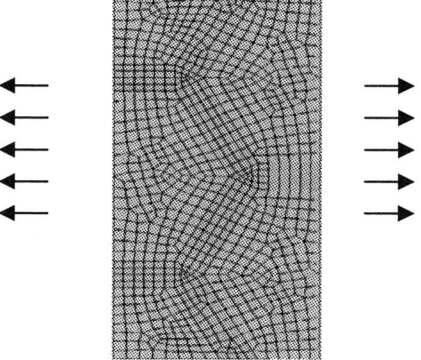

Fig 8. *Local model for the sample with segments*

Since the failure strain of silicon oxide is 0.8% in the previous results [7], the critical local strains of the silicon oxide can be reached for all those samples. Firstly, a mean strain is loaded to the sample with several calculation steps, then the local strains in the oxide layer and silicon segments are being checked. Since the middle oxide layer always reaches its failure critical strain first, the sample's critical mean strain can be obtained when the max. local strain on the silica oxide middle layer reaches the "failure criterion" strain 0.8%.

The simulated critical (mean) strain agrees with the mean tensile strain of samples with segments (See Fig 9). The maximum local principal strains appear in the gap near the spot where first fracture occurred during testing in the oxide layers (see Fig 10). As discussed before, from this spot they subsequently propagate into the oxide layers below and on top of the silicon and most probably

364

also within the silicon itself. The actual propagation of such a crack is no subject of this simulation.

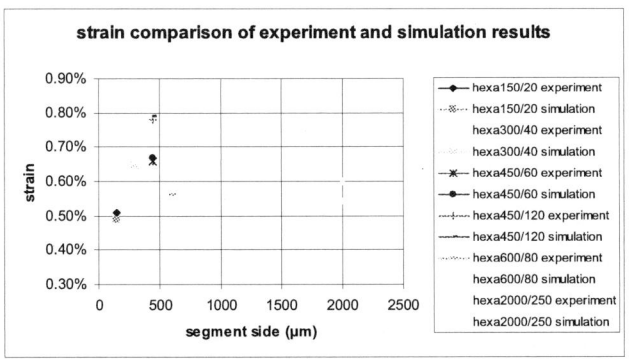

Fig 9. The (mean) strain at crack onset versus segment side (both as obtained from experiments and simulations)

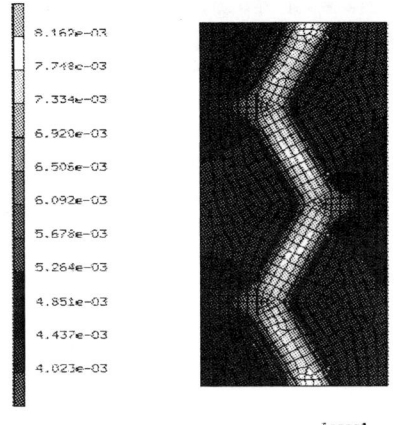

Fig.10. Max principle strain in the middle oxide layer

The simulation results match the experimental results quite well in the mean critical strain and the spot where the maximum local principle strain firstly reaches the failure strain. This provides a good base for optimization of segment size, gap size, interconnecting shape, materials, structures, etc.

6. Conclusions

A promising flexible substrate concept based on vertical thinning and lateral partitioning is proposed. In order to investigate the effect of the segmentation and gap sizes on the mechanical reliability, test samples were designed and prepared on a 4 inch silicon wafer and transferred to a thin polyimide substrate. They consisted of a 0.8 μm thick silicon layer sandwiched between 300 nm thermal and 500 nm PECVD silicon dioxide layers with square partitions varying in size from 150 to 2000 μm. Specific bending tools were fabricated to observe the cracks by optical microscope during the loading with possibility of recording the images and analyzing the crack density and width.

The formation of the first crack depends significantly on the segment sizes and the gaps between the segments for samples with square or hexagonal segments under the bending tests. The tensile (failure) strains depend on the ratio of "segment area over gap area". However, the tensile mean strains also depend on the segment sizes and gaps between the segments. The results show that the flexible ultra thin substrate can be applied on non-planar or even high curvature surfaces. The simulated critical (mean) strain agrees with the mean tensile strain in the tested samples. The maximum local principal strains appear in the gap near the spot where first fracture occurred during testing in the oxide layers. Those correspondences between simulation and measurement results provide a good base of optimization of segment size, gap size, interconnecting shape, materials and structures.

References

1. E.Jung, D. Wojakowski, A. Neumann, C. Landesberger, A. Ostmann, R. Aschenbrenner, and H. Reichl, "Chip-in-Polymer: Volumetric Packaging Solution using PCB technology," *SEMT Technology Symposium: International Electronics Manufacturing Technology (IEMT) Symposium*, pp. 46-49, 2002

2. E. Jung, A. Ostmann, D. Wojakowski, C. Landesberger, R. Aschenbrenner, and H. Reichl, "Ultra thin chips for miniaturized products," *The Conference of Micro System Technologies 2001*, pp. 449-452, March 2001.

3. S. Shkarayev, S. Savastiouk, and O. Siniaguine, "Stress and Reliability Analysis of Electronic Packages With Ultra-Thin Chip," *Transactions of the ASME: Journal of Electronic Packaging*, Vol. 125, pp .98-103, March 2005.

4. T Loeher, D Manessis, R Heinrich, B Schmied, J Vanfleteren, J Debaets,A Ostmann, H Reichl," Stretchable Electronic Systems", *The 8th Electronics Packaging Technology Conference*, 6-8 December 2006, Singapore, pp.271-276.

5. R. Dekker, P.G.M. Baltus, and H.G.R. Maas, "Substrate Transfer for RF Technologies," *IEEE Transactions on Electron Devices*, Vol. 50, No. 3, pp. 747-757, March 2003.

6. R. Dekker; *Substrate Transfer Technology*, PhD thesis, Delft University of Technology, ISBN 90-74445-61-6, June 2004.

7. L. Wang, K.M.B. Jansen, M. Bartek, A. Polyakov, L.J. Ernst, "Mechanical Characterization Analysis of a Segmented Silicon Layer on Ultra-thin Polyimide Substrates by Experiment and FE Simulation", *EuroSimE2006*, Como Italy, April 2006, pp 251-256.

8. Petersen, K.E, "Dynamic micromechanics on silicon: Techniques and devices", *IEEE Transactions on electron devices*, vol. 25, No. 10, pp1241-1250 ,Oct 1978.

Study on the Board-level SMT Assembly and Solder Joint Reliability of Different QFN Packages

Wei Sun, W.H. Zhu, Retuta Danny, F.X. Che, C.K. Wang, Anthony Y.S. Sun and H.B. Tan
United Test & Assembly Center Ltd (UTAC)
Packaging Analysis & Design Center
5 Serangoon North Ave 5, Singapore, 554916
Email: Sun_Wei@sg.utacgroup.com, Tel: +65-65511345

Abstract

The current paper deals with firstly the optimized SMT to assemble various types of QFN (Quad Flat Non-leaded) packages. The important SMT factors such as solder pad and stencil designs will be discussed. Secondly and more importantly, this paper will detail the comprehensive experiment and simulation work done for QFN solder joint reliability modeling. A curve fitted fatigue correlation model together with the use of Schubert's hyperbolic sine lead-free solder constitutive model will be proposed for accurate QFN solder joint reliability prediction.

1. Introduction

The consumer and communication electronics industry demands for lighter, thinner and higher performance packages. QFN package is able to meet those requirements because of its superior thermal and electrical performance. QFN packages can effectively dissipate heat because the exposed pad decreases the thermal resistance. The exposed pad also helps improve electrical performance of QFN by minimizing package ground lead inductance.

Compared to a QFP (Quad Flat Package) package whose leads protrude from the package body, QFN's small body size and footprint are especially favorable in miniaturized electronic devices such as handheld and communication products. Although may meet all package-level reliability requirements, QFN still has board-level concerns. For example, the SMT (surface mount technology) assembly of QFN is one of the board-level challenges. Non-optimized process may lead to low yield of board-level assembly and poor solder joint reliability under temperature cycling on board (TCoB) test. Unlike QFP, whose compliant leads can absorb the strain caused by CTE (coefficient of thermal expansion) mismatch between the package and PCB, the leadless feature of QFN makes it have very rigid connection with the PCB. What's worse, the solder used to connect the package to PCB is solely from paste printing. This makes the solder joint height of QFN is very low, which is another concern under TCoB test.

QFN family includes many members that have different configurations to meet various needs in different applications. Figure 1 describes the schematic pictures of the various QFN packages under the current study. Figure 1-A is a punch-type VQFN which is molded in individual mold cavity. Figure 1-B and 1-C are saw-type VQFNs, which are molded in panel and sawed subsequently in singulation process. The difference between B and C is the lead pullback design. Figure 2 can better illustrate the difference between lead pullback and non-pullback designs. The advantages of using lead pullback design are reduction of lead smearing (which cause short of adjacent leads) and cutting blade wearout in package singulation process (due to the half-etch feature). However, as shown in Figure 3, lead pull back design prevents the solder from wetting the side surface of the exposed lead and thus causes poorer solder joint reliability as compared to lead non-pullback design and punch-type QFN.

HQFN and et-QFN, as shown in Figure 1-D and 1-E, are two QFN packages that are intended for applications requiring enhanced thermal performance. For HQFN, the dummy die attachment possibly increases the board-level solder joint reliability concern as silicon is well-known to be the major source of CTE mismatch. For et-QFN, the die pad is exposed at the package top instead of bottom. Compared with other QFN packages, the lacking of center pad soldering also makes board-level solder joint reliability a concern.

A: VQFN (Punch-type)

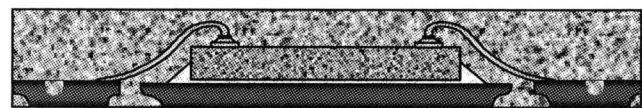

B: VQFN (Saw-type with lead pullback)

C: VQFN (Saw-type without lead pullback)

D: HQFN (Saw-type without lead pullback)

E: et-QFN (Punch-type)

Figure 1: Various QFN package structures

1-4244-1105-X/07/$25.00 ©2007 IEEE

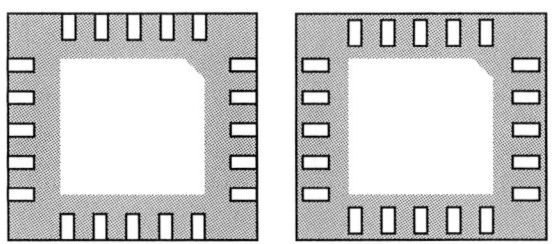

Figure 2: Bottom view of the difference between lead pullback and non-pullback
(left: non-pullback, right: pullback)

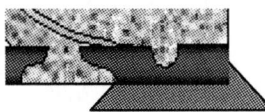

A: Solder joint of saw-type QFN (lead non-pullback)

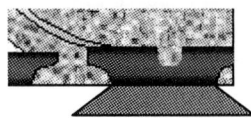

B: Solder joint of saw-type QFN (lead pullback)

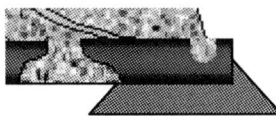

C: Solder joint of punch-type QFN

Figure 3: Solder joint formation for different QFNs

2. SMT assembly of QFN packages

The matrix for current QFN board-level study is listed in Table 1. Some important parameters such as exposed peripheral lead size and center paddle size are given. It is known that the PCB pad design and the stencil design are most critical for successful and quality SMT assembly of QFN packages. Those important design considerations are discussed by Syed and Kang in [1], which is used as the SMT design reference in the current study.

The PCB pad design for all the legs in Table 1 are shown in Figure 4. The solder pad size corresponding to exposed peripheral leads is 0.9x0.3mm for all legs.

Some of the important stencil design parameters are depicted in Figure 5. The stencil used in current study is a 5mil stainless steel stencil by laser cutting. By using those SMT design parameters discussed here, all the QFN legs are assembled successfully with good solder joint fillet formation as is shown in Figure 10.

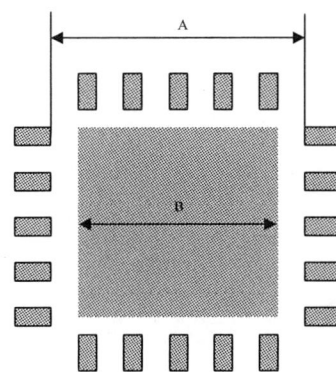

5x5mm et-QFN	A=4mm	B=N.A.
5x5mm VQFN	A=4mm	B=3.6x3.6mm
6x6mm VQFN	A=5mm	B=4.55x4.55mm
8x8mm VQFN/HQFN	A=7mm	B=6.2x6.2mm

Figure 4: PCB pad design for various QFN

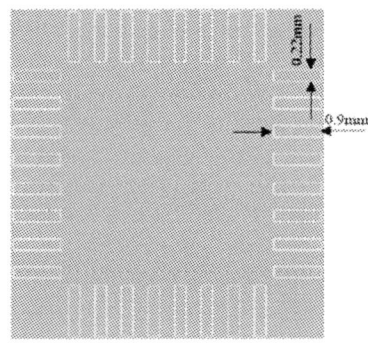

A: Stencil design for 5x5mm et-QFN

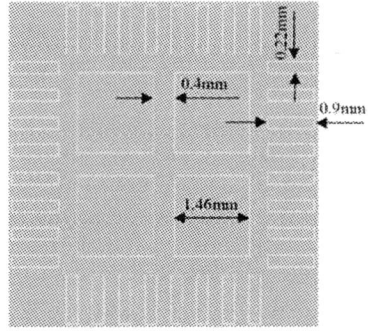

B: Stencil design for 5x5mm VQFN

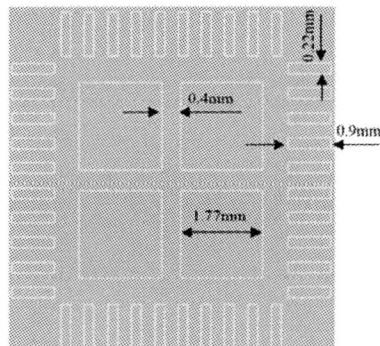

C: Stencil design for 6x6 VQFN

Table 1: QFN matrix under the current study

Leg#	Pkg Type	Size/Pitch (mm)	Exposed Lead Size	Die Paddle (exposed)	I/O	Solder
Leg 1	VQFN-P	5x5x0.90/0.50	0.40x0.25	3.500x3.500	32	SnPb
Leg 2	VQFN-P	5x5x0.90/0.50	0.40x0.25	3.500x3.500	32	Pb-free
Leg 3	et-QFN-P	5x5x0.90/0.50	0.40x0.25	1.920x2.900	32	SnPb
Leg 4	et-QFN-P	5x5x0.90/0.50	0.40x0.25	1.920x2.900	32	Pb-free
Leg 5	et-QFN-P	5x5x0.90/0.50	0.40x0.25	1.920x2.900	32	Pb-free
Leg 6	VQFN-P	6x6x0.90/0.50	0.40x0.25	4.10x4.10	40	Pb-free
Leg 7	VQFN-S	6x6x0.90/0.50	0.35x0.25	4.25x4.25	40	Pb-free
Leg 8	VQFN-S	8x8x0.90/0.50	0.40x0.25	5.900x5.900	56	Pb-free
Leg 9	HQFN-S	8x8x0.90/0.50	0.40x0.25	5.900x5.900	56	Pb-free

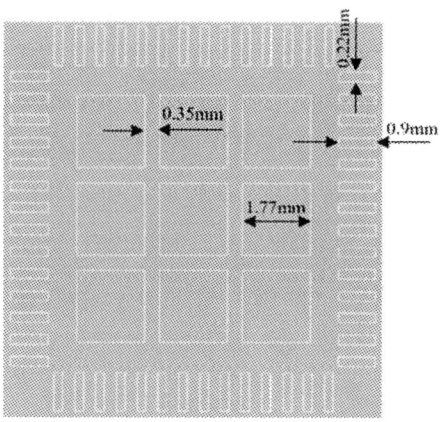

D: Stencil design for 8x8 VQFN/HQFN

Figure 5: Stencil designs for solder paste printing

3. Experiment

3.1 Experimental Matrix Description

There are totally 9 legs of different QFN packages that will go through the TCoB test. Each leg contains a sample size of 33. Table 1 shows the experimental matrix. Leg1 is the control leg. Results of leg2 and leg6 can help to tell the effect of package size on solder joint reliability. Leg3 and leg4 can show the effect of lead-free (SnAg4Cu0.5) and eutectic solder. The difference between leg4 and leg5 is that all et-QFN samples in leg5 are attached with a metal lid to simulate possible application where further thermal dissipation performance is required. In this case, leg5 is to study the effect of such metal lid on solder joint reliability. Comparison of leg6 and leg7 will tell us the effect of punch-type and saw-type (with lead pullback design) on TCoB reliability. Lastly, we can know the effect of dummy die attachment for leg9 on solder joint reliability by comparing its results with leg8.

All the QFN samples use lead-to-lead wirebonding to form the daisy-chain connection on the package side. Together with daisy-chain traces on PCB, the continuity of the whole daisy-chain can be monitored. For current test, the PCB test board used is 6-layer with 2.35mm thickness and OSP pad finish.

3.2 Experimental Results and Discussion

Table 2 listed the test results. Fatigue cycles are 63.2% characteristic lives obtained from weibull plots.

Table 2: Experimental results for all QFN legs

Leg #	Pkg Type	Size (mm)	I/O	Die Size	Solder	Other info	Test Condition	N (63.2%)
Leg 1	VQFN-P	5x5	32	2x2	SnPb			N.A.
Leg 2	VQFN-P	5x5	32	2x2	Pb-free			3170
Leg 3	et-QFN-P	5x5	32	2x2	SnPb			1446
Leg 4	et-QFN-P	5x5	32	2x2	Pb-free		-40C to +125C	1484
Leg 5	et-QFN-P	5x5	32	2x2	Pb-free	with lid	15mins-ramp	1269
Leg 6	VQFN-P	6x6	40	2x2	Pb-free		15mins-dwell	3400
Leg 7	VQFN-S	6x6	40	2x2	Pb-free			2509
Leg 8	VQFN-S	8x8	56	4x4	Pb-free			2324
Leg 9	HQFN-S	8x8	56	4x4	Pb-free			2498

Leg2-VQFN-P VS. Leg6-VQFN-P: Effect of package body size

Leg2 and leg6 have comparable solder joint fatigue lives. Experimental results show that leg 6 is only slightly better than leg2 by 7%. This means that the effect of package body size does not show very clearly from the experiment. Subsequently, simulation will help to identify the better one.

Leg3-et-QFN-P VS. Leg4-et-QFN-P: Effect of lead-free solder

Leg4 with lead-free solder performs only slightly better than leg3 with eutectic solder by 3%. Note that et-QFN has a much lower fatigue life than the rest of QFN packages. This is because of the lacking of center pad soldering for et-QFN. For et-QFN, it center pad is exposed upwards and thus cannot be soldered onto PCB. This is why et-QFN has much lower fatigue life than other QFN packages. Also this may be the cause for the insignificant difference in fatigue life between lead-free and eutectic since both were highly strained in the test and failed early

Leg4-et-QFN-P VS. Leg5-et-QFN-P with Lid: Effect of lid attachment

The purpose of attaching a lid onto et-QFN is to simulate the end user condition where a lid may be used to better dissipate the heat. Its effect on solder joint reliability is seen that lid attachment will reduced the solder joint fatigue life by 15%.

Leg6-VQFN-P VS. Leg7-VQFN-S: Effect of punch- and saw- type (with lead pullback)

Solder joint reliability of leg6 is significantly higher than leg7, showing the strong advantage of punch type over saw type (with pullback) under same package size. Leg6 has no pullback design such that solder can wet the side surface of lead and help to constrain the CTE mismatching.

Leg8-VQFN-S VS. Leg9-HQFN-S: Effect of dummy die

Testing results show that leg 8 and 9 have almost equal reliability performance. Although there is dummy die attachment for leg9, which is expected to have larger CTE mismatch and thus cause poorer reliability, the solder directly under the die edge is the center pad soldering which is strong enough the absorb the extra mismatch caused by the dummy die. In this case, the solder joints at the periphery of package are not affected.

4. Simulation and reliability modeling

4.1 Constitutive and Fatigue Models

For lead-free solder joint reliability modeling, Schubert's hyperbolic sine constitutive model plus his creep strain energy based fatigue correlation model are found to give good prediction accuracy for BGA and CSP packages in study [2-3]. But for QFN solder joint reliability prediction, the accuracy of this approach is unknown. In the current work, Schubert hyperbolic sine solder constitutive model will still be used. The accuracy of his corresponding fatigue model will be examined. Schubert's constitutive model for lead-free solder and his accumulated creep strain energy density based fatigue model are listed in Table 3-5.

Table 3: Constitutive equations for lead-free solders

Solder	Constitutive equation
Sn3.8Ag0.7Cu Sn3.5Ag0.75Cu Sn3.5Ag0.5Cu (Schubert et al. [4])	$\dot{\varepsilon} = 277984[\sinh(0.02447\sigma)]^{6.41} \times \exp\left(\dfrac{-6500}{T\left(^{o}K\right)}\right)$

Table 4: Elastic material properties of lead-free solder [4]

E (MPa)	61251-58.5T (degree K) (Schubert et al. [4])
v	0.36
CTE (ppm/K)	20.0

Table 5: Schubert's fatigue correlation model for his hyperbolic sine constitutive equation [4]

Acc. Creep Energy Density	$N_{cha} = 345w_{acc}^{(-1.02)}$

4.2 FEA model

FE models were set-up using ANSYS FEA software. Figure 6-8 shows the typical meshes for various QFN packages. Due to symmetry, quarter models are used for all the simulations. Solder joint is modeled with 75µm stand-off height and six layers of elements. Three thermal cycles are modeled. As shown in Figure 9, simulation for leg9 predicts that the critical solder is at the package corner and crack is likely to occur between lead and solder. Same observations are found in the simulation results for all of the rest legs. Subsequent failure analyses for randomly selected samples from each leg confirm that solder joint cracks do lie between lead and solder as shown in Figure 10 and solder joints at the package corner register most failures. Based on those findings, accumulated creep strain energy was averaged over the top layer of elements as shown in Figure 11 for subsequent fatigue life predictions.

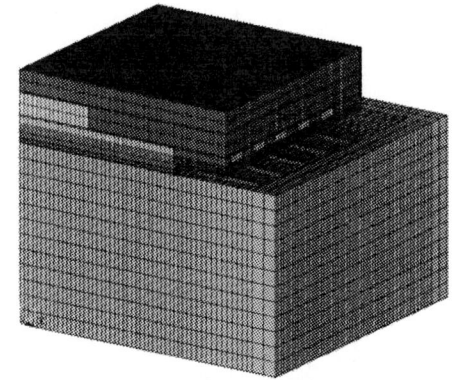

Figure 7: Overall mesh for leg7 (6x6 VQFN-Saw type with lead pull back design)

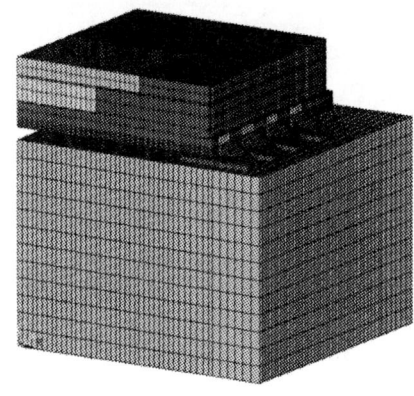

Figure 8: Overall mesh for leg4 (5x5 et-QFN-Punch type without lead pull back design)

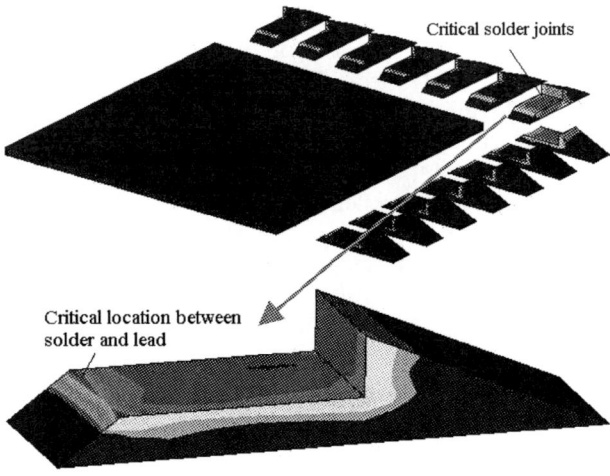

Figure 9: Accumulated creep energy distribution among all solder joints and critical solder joint

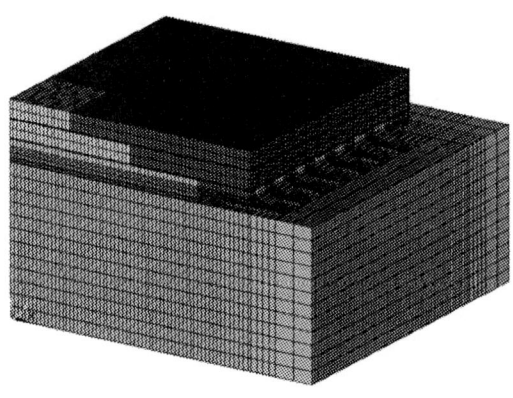

Figure 6: Overall mesh for leg9 (8x8 HQFN-Saw type without lead pull back design)

369

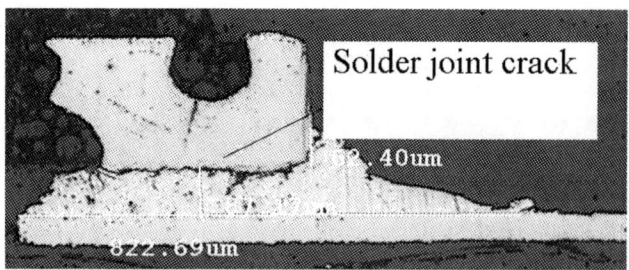

Figure 10: QFN solder joint cracking failure

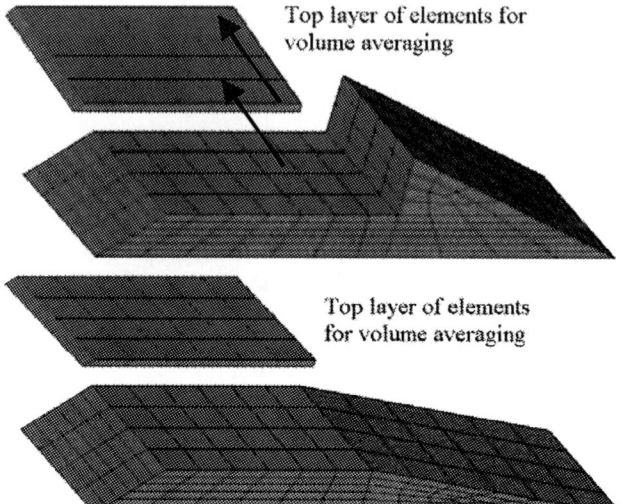

Figure 11: Layer of elements for accumulated creep strain energy averaging

4.3 Simulation results and discussion

The simulated solder joint fatigue life and experimental results are listed in Table 6. It is found that a large difference is there between the two. This probably means that Schubert fatigue prediction model cannot be applied in QFN cases. The constants in the power equation fatigue prediction model need to be adjusted for QFN solder joint reliability prediction. Based on the simulated accumulated creep strain energy density and the corresponding experimental result, a new curve fitted power equation is given in Figure 12. In Table 7, the predicted solder joint fatigue data are listed together with experimental results. It can be seen that the prediction error is within ±20%. Besides, the predicted results generally follow the trends shown in the experimental results. Therefore, the new curve fitted model can predict the QFN solder joint reliability with good accuracy.

Table 6: Solder joint fatigue prediction results using Schubert fatigue model

Leg #	Acc. creep strain energy (for top layer of element) per cycle	Experimental results	Schubert energy-based prediction model
2	0.0193	3170	18380
6	0.0350	3400	10533
8	0.0408	2324	9003
9	0.0409	2498	8998
7	0.0645	2509	5647
4	0.1071	1484	3367
5	0.2772	1269	1276

Table 7: Prediction accuracy of the new curve fitted fatigue prediction model

Leg #	Experimental results	Schubert energy-based prediction model	New curve fitted prediction model	Prediction error
2	3170	18380	3458	9%
6	3400	10533	2742	-19%
8	2324	9003	2582	11%
9	2498	8998	2581	3%
7	2509	5647	2160	-14%
4	1484	3367	1772	19%
5	1269	1276	1223	-4%

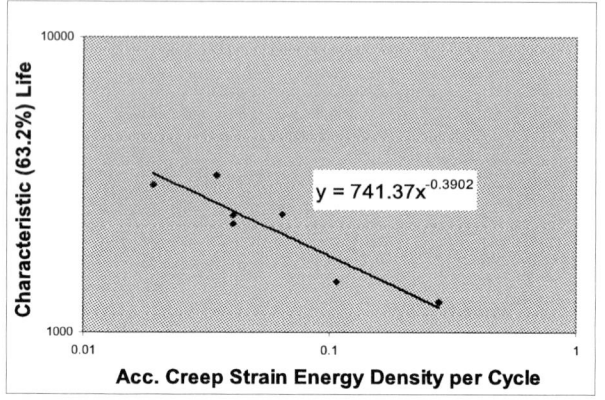

Figure 12: New curve fitted model for QFN solder joint fatigue life prediction

5. Summaries and Conclusions

In the current study, various QFN packages were subject to board-level temperature cycling test under -40C~125C with 15mins dwell/ramp. It is found that et-QFN, which has no center pad soldering to PCB, has much lower fatigue life than other QFN packages whose center large solder joints will significantly help to resist CTE mismatch under TCoB. Also for et-QFN, the attachment of heat spreader on top of the package can reduce the solder joint reliability, as this attachment makes the package rigid and pure shear between package and board is expected to be applied onto the solder joint, causing earlier failure.

Larger body size does not really enable longer board-level solder joint reliability although its center pad soldering is larger and number of peripheral leads is more. This is because larger body size also gives the peripheral solder joint larger distance-to-neutral-point effect.

QFN without lead pull-back design renders better solder joint shape and elongate reliability compared with pull-back design. This is because the solder fillet formation on the exposed side surface of peripheral leads helps to absorb CTE mismatch.

The effect of dummy die attachment for thermal enhancement in HQFN does not affect board-level solder joint reliability, although silicon material is known to be the major source of package/board overall CTE mismatch. This is because under the die influence area there is a large solder joint, thus the additional mismatch is absorbed by this part.

Based on the experimental results, simulation work was firstly performed to examine the applicability of Schubert's hyperbolic since constitutive model plus his fatigue model. However, the large discrepancy between simulation and actual results points out that Schubert's fatigue model does not apply to QFN solder joint reliability prediction. Therefore, a new power equation model is curve fitted. It is seen that the new model gives good accuracy. Thus, a accurate model for QFN solder joint reliability is established for future design and analysis.

6. Acknowledgements

The authors would like to thank UTAC R&D management team for their support. The assembly work done by process group for test vehicles preparation is also greatly appreciated.

References

1. Ahmer Syed and WonJoon Kang, "Board Level Assembly and Reliability Considerations for QFN Type Packages", *2003 SMTA conference*, USA.
2. Sun Wei et el, "Experimental and Numerical Assessment of Board-level Temperature Cycling Performance of SnPb and Pb-free windows-Chip-Scale-Package (wCSP)", *Proceedings of ICEPT2006*, Shanghai, China.
3. Sun Wei et el, "Experimental and Numerical Assessment of Board-level Temperature Cycling Performance for PBGA, FBGA and CSP", *Proceedings of EPTC2006*, Singapore.
4. A. Schubert et al., "Fatigue Life Models for SnAgCu and SnPb Solder Joints Evaluated by Experiments and Simulation", *Proceedings of ECTC2003*, pp.603-610.

Thin Film Interface Fracture Properties at Scales Relevant to Microelectronics

A. Xiao[1,3], L. G. Wang[1], W. D. van Driel[1,2], O. van der Sluis[1,2], D. G. Yang[2], L. J. Ernst[1], G. Q. Zhang[1,2]

[1]Delft University of Technology, Mekelweg 2, 2628 CD Delft, The Netherlands.
[2]NXP Semiconductors, 6534AE Nijmegen, The Netherlands
[3]Email: A.Xiao@tudelft.nl, Phone: +31 15 2786932

Abstract

Nowadays, one of the trends in microelectronic packaging is to integrate multi-functional systems into one package, resulting in more applications of highly dissimilar materials in the form of laminated thin films or composite structures. As a consequence, the number of interfaces increases. Often, the interface between these dissimilar materials is where the failure is most likely to occur especially when the packaged devices are subjected to the thermo-mechanical loading. Prediction of interface delamination is typically done using the critical energy release rate. However, the critical value is dependent on mode mixity. This paper describes our efforts on interface characterization as a function of mode mixity. A new test setup is designed for mixed mode bending testing. It allows for measuring the stable crack growth as the function of mode mixity. The crack length, necessary for calculation of the energy release rate is measured by means of an optical microscope. Finite element simulation is used to interpret the experimental results and thus to establish the critical energy release rates and mode mixities.

1. Introduction

Most micro-electronic packages are composite structures made up from multiple-materials among which thin film coatings. Generally, the interface between two different materials is a weak link due to imperfect adhesion and stress concentrations. At present, interfacial delamination is one of the major concerns in IC packages (figure 1). Failure of these interfaces induces decreased reliability and performance of such packages. Therefore, adequate knowledge of delamination prediction is desirable.

Recently, the extension of linear elastic fracture mechanics in homogenous material to bimaterial interface crack problems has become one of the interests. Many researchers have made important contributions on bi-material interface fracture mechanics. However, the analytical solutions are limited to very simplified cases and can not directly be applied to real engineering applications. General speaking, there are two approaches in fracture analysis: the stress intensity approach and the energy approach. The stress intensity approach regards the crack growth when the stress intensity factor exceeds a critical material specific fracture resistance. Comparing to the stress intensity approach, the energy approach is more attractive [1, 2]. It turns out that the crack propagates as a result of the so-called energy release rate exceeding its critical value. The critical value can be obtained experimentally. However, its measurement is complicated due to the fact that adhesion strength is not only temperature and moisture dependent but also stress state (mode mixity) dependent. In this research, a mixed mode bending method [1,5] is proposed, in which generally, interface delamination growth occurs under combined mode I (opening mode) and mode II (shearing mode) conditions. The mode mixity or mode angle is determined by the ratio from mode I to mode II loading. For an isotropic homogeneous material, a mode angle of 0° describes pure mode I, and mode angles of -90° or 90° describe pure mode II loading (shown in figure 2). In general, critical energy release rate is higher under mode II loading than under mode I loading.

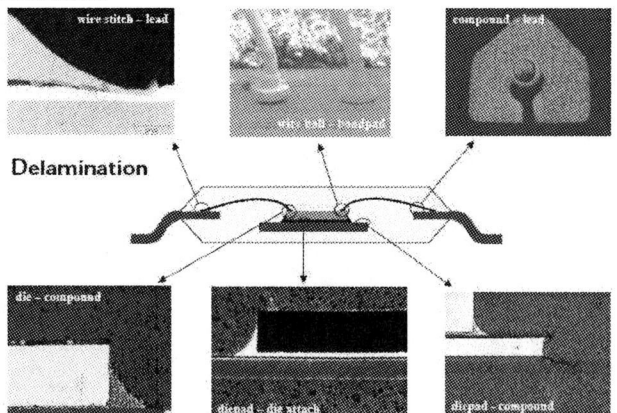

Figure 1. Typical interfacial delamination in lead frame based packages.

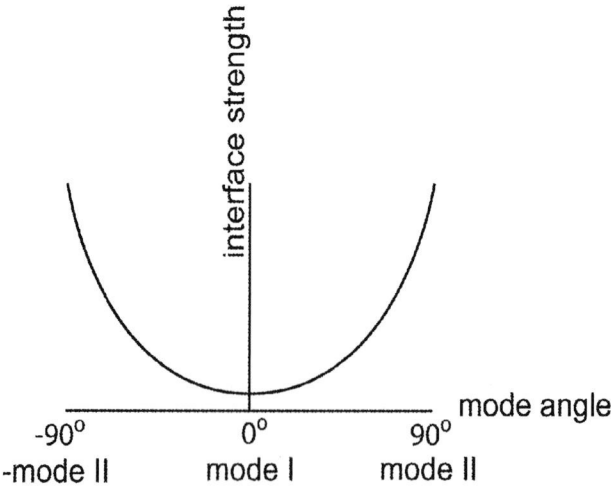

Figure 2. Interface strength versus mode angle.

2. Theory of interfacial fracture mechanics

For an isotropic homogeneous material, usually a crack propagates along the path where pure mode I occurs. For dissimilar laminated thin films, due to material mismatch, the interface cracks propagate under mixed mode combined condition. This means that mode I, mode II and even mode III (3D case) may coexist together.

Linear elastic fracture mechanics is a theory that describes if and how a crack will grow under given loading conditions when assuming an initial crack with given size and location [3]. It assumes the existence of some detectable cracks and predicts the crack propagation during processing and operational cycles. It applies when the nonlinear deformation of the material is confined to a small region near the crack tip compared to the size of the crack. To predict interface delamination, fracture quantities are needed for comparison to the critical data such as fracture toughness. In general, stress intensity factors (SIF) and energy release rate are used to define the loading state at the crack tip.

A criterion for crack growth can be obtained by regarding the energy balance of the material (1), where U represents the energy per unit of time and volume.

$$U_e = U_i + U_a + U_d + U_k \qquad (1)$$

U_e is the total external mechanical energy that is supplied to the material, U_i is the elastic energy that is stored in the material, U_a is the energy dissipated by crack growth, U_d is the energy dissipation caused by other mechanism, and U_k is the change in kinetic energy. It is assumed that U_d is zero, implying that the crack growth is the only cause of energy dissipation. U_k is zero means that crack growth is slow enough for changing in kinetic energy is negligible. The remaining energy balance is know as the Griffith's energy balance (2), which regards energy per unit of newly created fracture surface, or when the material is taken to be constant, per unit of crack length a:

$$\frac{dU_e}{da} - \frac{dU_i}{da} = \frac{dU_a}{da} \qquad (2)$$

Dividing the left hand of equation by the material thickness B, it gives the energy release rate (3).

$$G = \frac{1}{B} \left(\frac{dU_e}{da} - \frac{dU_i}{da} \right) \qquad (3)$$

The energy release rate G is known as Griffith's energy balance, which regards energy released per unit of newly created fracture surface when the crack grows a unit of length. The criterion from Griffith states that crack growth occurs when the energy release rate exceeds a critical value $G > G_c$. The energy release rate appears to be dependent on temperature, moisture and mode mixity so that the criterion for fracture is:

$$G(T, C, \psi) > Gc(T, C, \psi) \ [2] \qquad (4)$$

The mode mixity ψ for a homogeneous material is usually defined as the ratio between mode I to mode II loading and is described by the loading stress state at the crack tip (5).

$$\psi = arctg \frac{K_{II}}{K_I} \qquad (5)$$

Here, K_I and K_{II} represent intensities of mode I (opening) and mode II (shearing) stress states for a crack in a homogeneous material. K_I characterizes the tendency of remote loads to open the crack, while K_{II} characterizes the shear loading.

For an interface crack, due to the elastic mismatch between two materials, the mode mixity can not be simply described by the equation 5. The opening and shearing stresses at the interface ahead of the crack tip, with a distance of r can be calculated from (6).

$$(\sigma_{22} + i\sigma_{12}) = \frac{K}{\sqrt{2\pi r}} r^{i\varepsilon} \qquad (6)$$

Where σ_{12} represents shear stress and σ_{22} represents normal stress. ε is the oscillatory index which is a function of the Young's moduli and the Poisson's ratios. K is the complex stress intensity factor. It is described by:

$$K = K_I + iK_{II} \qquad (7)$$

The mode mixity for an interface crack is described by:

$$\psi = \tan^{-1} \left(\frac{\sigma_{12}}{\sigma_{22}} \right) \qquad (8)$$

According to the basic solution, stress components along the interface are oscillatory [2, 4] and thus can not well be obtained by numerical solutions. Therefore, often an alternative mode mixity definition is used, where the mode mixity is defined by interface stresses (normal and shear) at a chosen length $\hat{L}$ ahead of the crack tip:

$$\psi = \tan^{-1} \left(\frac{\text{Im}(K\hat{L}^{i\varepsilon})}{\text{Re}(K\hat{L}^{i\varepsilon})} \right) \qquad (9)$$

Here the choice of $\hat{L}$ is somewhat arbitrary, but restricted by the dimensions of test samples and the applications within microelectronics.

3. Design of the MMB setup

Measuring interfacial adhesion strength requires loading a sample consisting of two material layers. To determine the interfacial adhesion strength, various test methods (figure 3) have been used. Such as the double cantilever beam (DCB) test, three point bending (TPB) test, and four point bending (FPB) test etc. Note that using the shown test methods to determine the influence of mode mixity on the interfacial fracture toughness, combining normal and shear stresses on the delamination plane, different thickness ratio of material layers has to be generated. However, it is highly impractical as it requires the development of different types of samples for each mode mixity. Also, even when changing the thickness ratio, none of the shown test methods can cover the full range of mode mixity. Moreover, to determine pure mode I, pure mode II, and mixed mode critical values (G_{Ic}, G_{IIc}, and $G_c(\psi)$, respectively), different types of samples need to be subjected to different loading configurations. These configurations can involve different test variables and analysis procedures that can influence test results in ways that are difficult to predict.

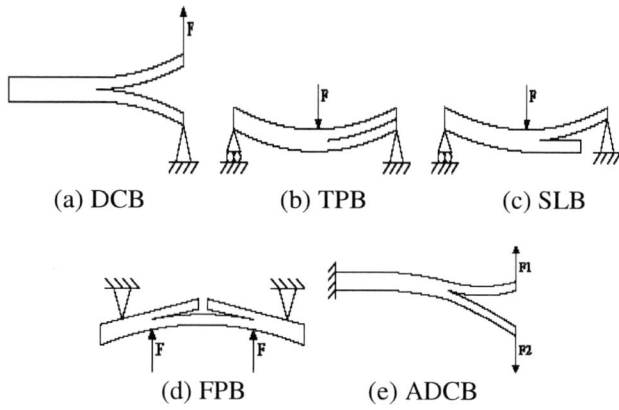

Figure 3. Different test methods for interface strength measurement: (a) Double cantilever beam, (b) Three point bending, (c) Single leg bending, (d) Four point bending, (e) Asymmetric double cantilever beam

The mixed mode bending test method (figure 4), which is used in this research, was first introduced by Reeder and Crews [1990]. This method has been widely used for measuring the interfacial strength experimentally. It provides the stable crack growth over the full range of mode mixities. In their published paper, it had also been proved that the MMB test was rather simple and was believed to offer several advantages over most current mixed mode test methods [5].

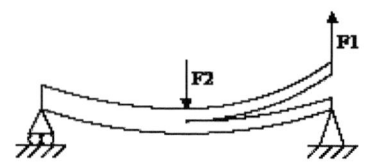

Figure 4. Mixed mode bending (MMB) test method

The test setup is designed and fabricated especially for the mixed mode bending test. It allows transferring two separated loads on a single specimen. A schematic drawing is shown in figure 5.

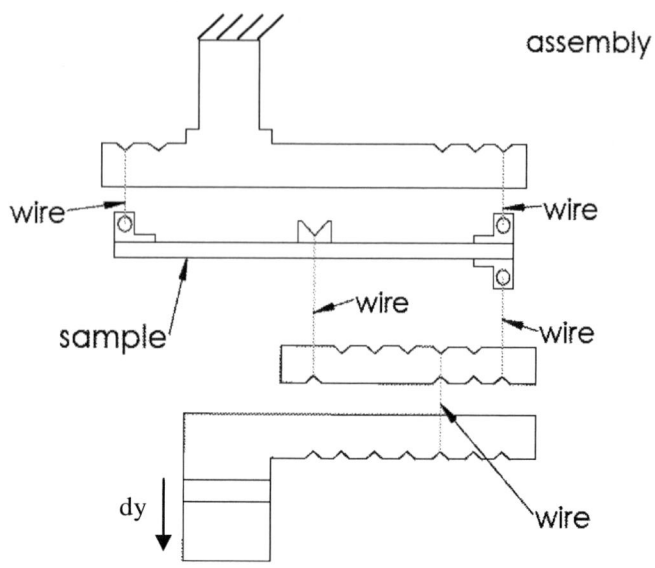

Figure 5. Schematic drawing of the MMB setup

The setup consists of two loading beams and a lever, three attachment hinges, one protecting metal block and several wires. A sample is first glued in between the hinges. The hinges are linked by the wires, and the wires are hitched on the beams and lever. The protecting block is glued on the middle of the sample. This metal block is used to prevent the sample damage and also to prevent wire sliding along the horizontal direction during the experiment. By changing the loading position of the lowest wire, different mode mixities can be controlled. The mode II TPB test occurs when we do not use the lever and directly connect the middle of the sample with the lower loading beam. Mode I DCB test occurs when remove the protecting block, middle beam and left hinge and hitch the lower hinge with wire on the lower beam. The notches on the beams are used to provide the test abilities for different sample length. When attaching a sample in the setup, it seems that the sample is loaded immediately due to the gravity of the middle load transfer beam. However, the weight of this beam is very small. It is not expected that this mass will propagate the initial crack of the sample. The small load of this beam can be simply added to the sample load when interpreting the results of the loading system.

4. Experiment results

In this study, the interface between copper and die attach is investigated. The sample is 28 mm long, 1.2 mm thick and 3 mm wide. It consists of two 5.6 mm thick bonded substrates, two 30 micron thick layers of copper

374

and a 20 micron thick layer of glue. A schematic drawing is shown in figure 6.

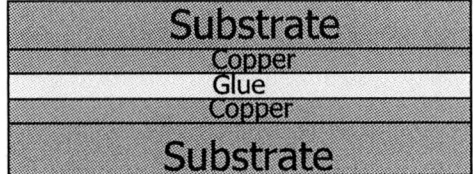

Figure 6. Schematic drawing of test sample

In order to characterize the interface strength more accurately, the test samples are created with identical fabrication processes and materials as used in creating the microelectronic components. The initial pre-stress levels in the test sample are known to play a predominant role in the crack growth behaviour. It is important to note that for large scale samples, high residual stresses may disturb the experiment significantly.

The experiments are performed at room temperature. Firstly, a specially prepared test sample is placed in the load transfer setup. Then, the setup is clamped in a micro tensile tester (actually a sensitive dynamic mechanical analyzer is used), in which various temperature and moisture combinations can be applied. The crack length is monitored and used for calculating the critical energy release rate. It is measured directly using an optical microscope.

Figures 7, 8 and 9 show the force-displacement results from a DCB test, a MMB test (performed at mid loading point), and a TPB test (here the sample is destroyed directly after crack initiation).

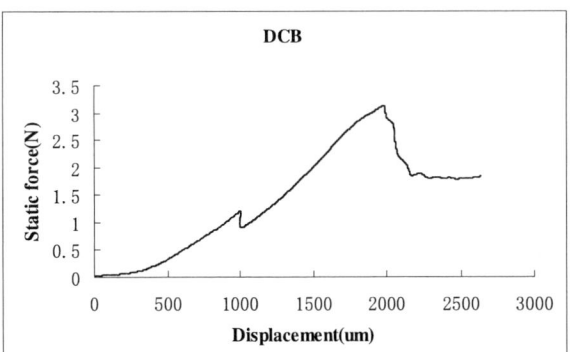

Figure 7. DCB test

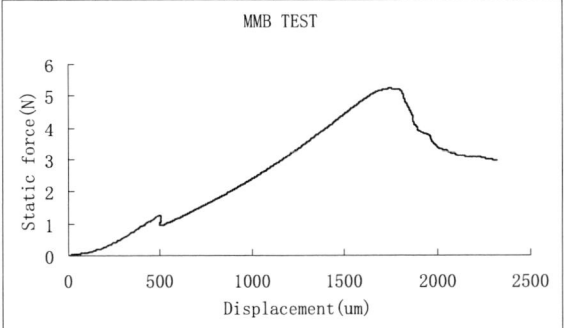

Figure 8. MMB test

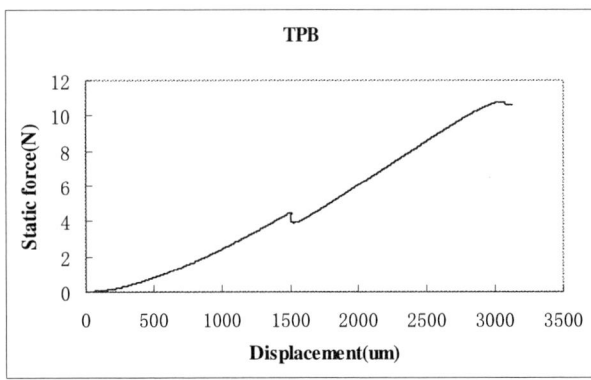

Figure 9. TPB test

Note that in order to speed up the test, at the beginning the system is loaded at high rate. Then, the system stops for 1 minute, and afterwards, it is loaded at low rate continuously. From the test results, initially, the force - displacement curve represents the opening of the pre-crack. When the pre-crack starts to propagate, the force decreases. It is found that the crack growth initially is not stable.

From the graphs, it is found that the tests start with a non-linear response. This is because the test setup consists of wires and these wires provide inelastic deformation. The area under the measured force-displacement curve does not equal the sum of the energy that has been used for a crack growth and the elastic energy stored in the sample. It also contains the energy that is dissipated by the wires. Numerically, it is difficult to include the behaviour of the wires in the finite element simulation. Therefore, the force and crack length relations are measured. A result is shown in figure 10.

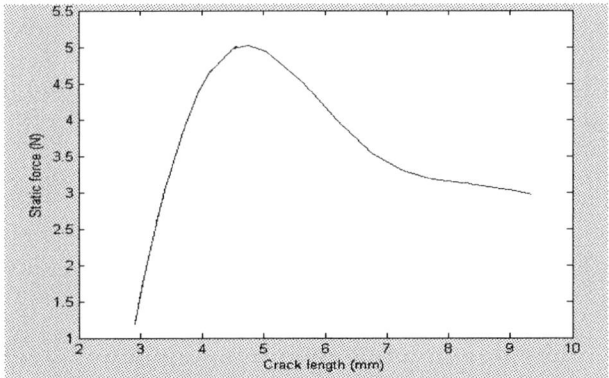

Figure 10. Crack length versus static force

4. Finite element analysis

Experimental data from MMB, DCB, and TPB tests is interpreted through finite element fracture mechanics simulations using a modified J integral concept [6]. The model is shown in figure 11. Quarter-point elements were used around the crack tip to capture the stress singularity.

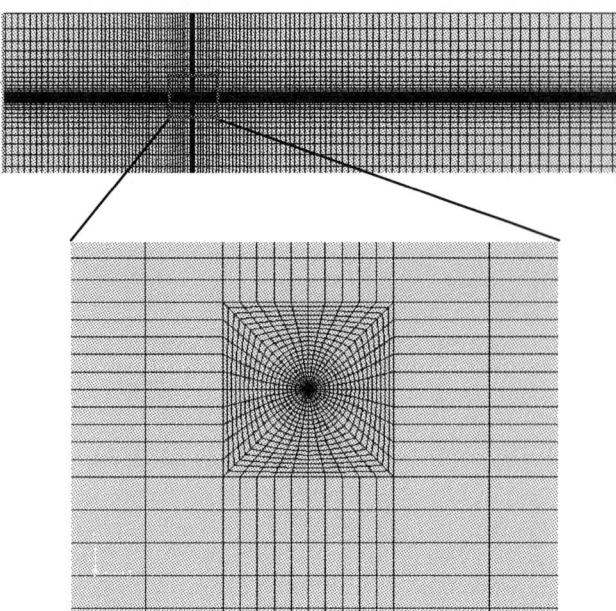

Figure 11. Geometry of 2D FEM model and crack tip mesh

For each crack length the loading is taken from the force-crack length graph. Than a model with the same crack length and loading is used to establish the energy release rate (through J-integral calculation). This established energy release rate is considered as the critical one G_c. The matching mode mixity is calculated with reference length equal to 0.15mm. The result is shown in figure 12.

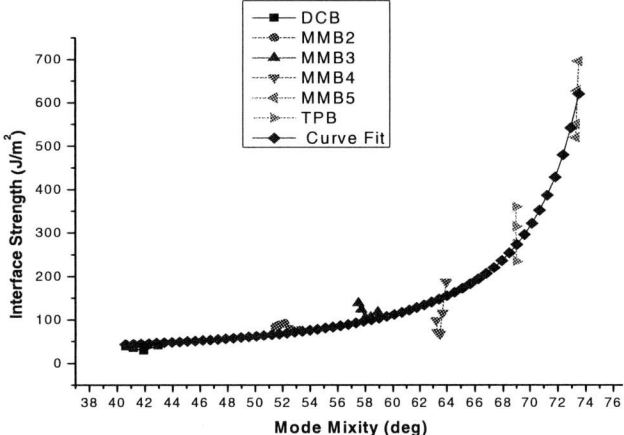

Figure 12. Interface strength versus mode mixity

5. Discussion

The result in figure 12 clearly shows the relationship between interface strength and mode mixity. The interface strength has a minimum mode mixity of 42° (depending on the reference length). One would expect that the minimum interface strength to occur at a mode mixity close to 0°. The definition of mode mixity used to design the test is based on the arbitrarily chosen reference length. In this study, a reference length of 0.15 mm is chosen. Changing the reference length will shift the curve horizontally. According to equation 9, a reference length

of 20 micron (thickness of the glue) would result in a horizontal shift to the right of ~10 degree.

It is also found that glue stiffness plays a dominant role in the mode mixity calculation. Low stiffness of glue could also shift the mode mixity curve. Increasing glue stiffness may indeed decrease the mode mixity. Figure 13 shows the glue young's modulus as function of DCB mode mixity. This result proves that the reason caused mode mixity shifts, is actually the effect of non-linear deformations.

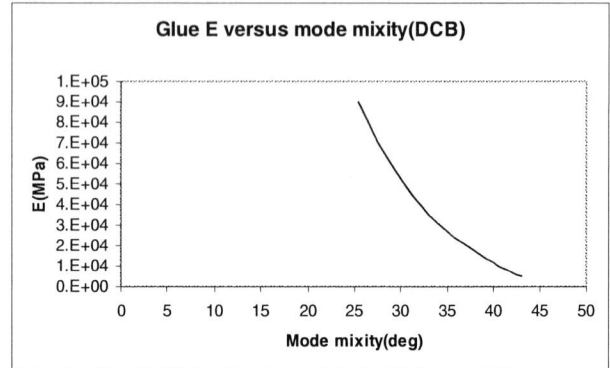

Figure 13. Glue Young's modulus versus DCB mode mixity

6. Conclusions

A newly designed mixed mode bending setup has been used to analyze the interface strength of copper and die attach. The force is measured using a DMA test facility as tensile tester and the crack length is obtained using a microscope. The finite element analysis is used to calculate the critical energy release rate and mode mixity. The results are used to determine the critical energy release rate as a function of mode mixity.

References

1. S. Liu, Y. H. Mei, and Y. Wu *"Bimaterial Interfacial Crack Growth as a Function of Mode-Mixity"* IEEE Transactions on Components, Packaging, and Manufacturing Technology-Part A. Vol 18. NO.3.September 1995.

2. L. J. Ernst et al, *"Fracture and Delamination in Microelectronic Devices"* Proceeding APCFS 2006.

3. M. F. Kanninen and C. H. Popelar *"Advanced fracture mechanics"* Oxford Clarendon Press, 1985. ISBN 0-19-503532-1

4 H. F. Nied, *"Mechanics of Interface Fracture with Applications in Electronic Packaging"* IEEE Transactions on Device and Materials Reliability, Vol 3, No.4, Decemeber 2003.

5 Reeder, J.R., Crews, J.R. *"Mixed-Mode Bending Method for Delamination Testing"* AIAA Journal, Vol. 28, No. 7 (1990), pp.1270-1276.

6 Y. T. He, G. Q. Zhang, W. D. van Driel, *"Cracking Prediction of IC's Passivation Layer Using J-Integral"* Proceeding Electronic Components and Technology conference, IEEE 2003.

7 J. W. Hutchinson and Z. Suo, *"Mixed mode cracking in layered materials"* Advances in Applied mechanics, Vol.29, Academic, New York, 1991.

8 G. Q. Zhang, W, D. Van Driel, and X. J. Fan *"Mechanics of Microelectronic"* solid mechanics and its applications, volume 141 ISBN-10 1-4020-4934-X (HB).

9 C. C. Lee, C. C. Chiu, K. N. Chiang *"Stability of J Integral Calculation in the Crack Growth of Copper/Low-k Stacked Structures"* IEEE Electronic Components and Technology Conf., pp. 885-891, 2006.

10 H. F. Nied, *"Mechanics of Interface Fracture with Applications in Electronic Packaging"* IEEE Transactions on Device and Materials Reliability, Vol 3, No.4, Decemeber 2003

11 A. A. O. Tay, K. Y. Goh, *"A Study of Delamination Growth in the Die-Attach Layer of Plastic IC Packages Under Hygrothermal Loading During Solder Reflow"* IEEE Transactions on Device and Materials Reliability, Vol. 3. no. 4, Dec 2003.

Design for Reliability with AuSn Interconnects

Rainer Dudek, Olaf Wittler, Wolfgang Faust, Birgit Brämer, Matthias Klein[1], Wei Jun[2], and Bernd Michel

Fraunhofer Institute Reliability and Microintegration (IZM)
Micro Materials Center Berlin and Chemnitz
e-mail: rainer.dudek@che.izm.fraunhofer.de

[1] Fraunhofer IZM, Dept. Module Integration & Board Interconnections Technologies
[2] Singapore Institute of Manufacturing Technology, Singapore

Abstract

AuSn is a special purpose interconnect material with advantages concerning its high temperature resistance, used for fluxless soldering for optoelectronic and RF devices as well as in fine pitch flip chip technology. Finite element (FE) analyses were conducted on assemblies with AuSn interconnects. Stress analyses require data on the materials stress-strain behavior, however, in case of AuSn solder joints material characteristics are not easily available. In the paper investigations on material characterization concerning the elastic and plastic behaviour as well as the fatigue properties of electroplated Au and AuSn are reported. It is shown that properties of these materials depend strongly on their composition and their microstructure, the latter being fundamentally affected by processing. By application of thermal lap shear tests the low cycle fatigue behavior of AuSn joints is studied. The results show that inter-phase cracking is the dominant fatigue failure mechanism, which is only indirectly linked to plastic straining calculated under homogeneity assumption. Computational analyses concerning the reliability of AuSn flip chip joints for GaAs pixel detectors are finally reported in the paper.

1. Introduction

Several special purpose applications in high power electronics, photonic interconnect technology, and fine-pitch flip chip technology use alloys of Au and Sn as a soldering material. Because of its relatively low melting temperature of 278 °C the eutectic composition AuSn20 is most appropriate for soft soldering. It is known that AuSn solder joints are mainly composed of the AuSn eutectic (δ phase) and the ζ phase with approximate stoichiometry Au_5Sn, the latter one undergoing a phase transformation at 190°C [1]. In case where thermo-mechanical analyses are attempted for assemblies containing AuSn solder, the definition of the constitutive properties of this heterogeneous material is a serious concern.

In the paper investigations on material characterization concerning the elastic and plastic behaviour as well as the fatigue properties of AuSn are reported. The properties of Au and Sn containing joints depend strongly on their composition and microstructure, the latter being fundamentally affected by processing. The processing effects are particularly observed for heteterogeneous AuSn bumps, see Fig. 1, which form during the reflow process of stacked deposited tin and gold structures, a fine-pitch flip chip technology recently established at IZM [2], [3]. In these joints eutectic AuSn, the gold rich Au_5Sn ζ-phase and pure Au layers occur simultaneously in a relation controlled by the processing conditions.

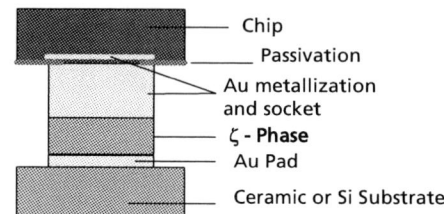

Fig. 1: Schematic of AuSn flip chip bumps.

For the purpose of stress analysis, the heterogeneous structure has to be taken into account, and properties have to be assigned to each of the constituent phases.

Computational analyses concerning the reliability of AuSn flip chip interconnects for GaAs pixel detectors on Si read out chips were one goal of the study. Low cycle fatigue failure of the bumps was originally the concern of the analyses. However, it turned out that for this type of device brittle interface failure is more likely to occur, related to the large GaAs chip sizes (up to 16 mm square), the mismatch of the CTEs between GaAs and Si, and the stiff behaviour of the bumps. In accordance with that finding the material analyses revealed a high yield strength and a low tendency of plastic deformation for the AuSn phases.

2. Materials characterization of AuSn interconnect materials

Elastic-plastic characteristics

The heterogeneous nature of AuSn bumps requires both the characterization of the constituent materials distribution after reflow and the characterization of the different materials, one of them being pure Au. Micro-tensile testing on electroplated Au foil specimens were performed to determine the properties at specimens, which were processed under the same conditions as the bumps to resemble their properties in use as closely as possible. The electroplated foil specimens had a thickness of 20 μm and a dog-bone type outer shape with a parallel length of either 15 mm or 10 mm and a width of either 4 mm or 2 mm at the measuring section.

1-4244-1105-X/07/$25.00 ©2007 IEEE

The measuring results are shown in Fig. 2 for the as-plated and an annealed state. Annealing was performed at 150°C for 10 min. During the loading process, unloading-reloding steps were subsequently included, since no elastic-plastic characteristics were seen at simple loading. A non-linear stress strain response was obtained instead just from the beginning of the loading process. This effect is not yet fully understood, most likely local plastic stretching effects can have caused this behavior. The Young's modulus calculated from the unloading-reloading curves was 62.8 GPa, what is slightly less than the bulk value.

Obviously, annealing changed the hardening behavior of the foil to a great extend. While for the not annealed Au strain hardening leads to high initial yield stresses after the first loading, this strong hardening behavior vanishes after the annealing process and a much lower yield stress is obtained after loading.

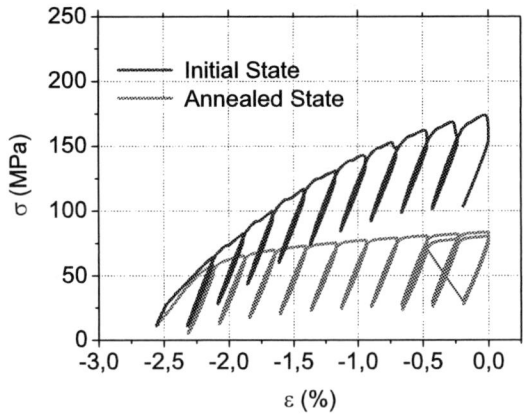

Fig. 2: Stress-strain curves with unloading-reloading steps for electroplated Au foils 20 µm thickness.

Tab. 1: Summary of results from indentation test and comparison to reference data

Material	E [GPa]	E_{IT} [GPa]	H[GPa]
Au Bump, electro-plated, ion etched		84.91)	1.35
Au in AuSn Bump after soldering		72.9	1.25
Au foil electro-plated, annealed	62.8	68.4	1.33
Au foil electro-plated, not annealed		78.5	1.95
ζ phase in AuSn Bump after soldering	57.2	62.1	3.2
Au78Sn22 preform		68	2.4

Local materials characterization experiments were additionally performed [4]. Different specimens, beginning with the Au foil, the lap shear joints, and the heterogeneous bumps were analyzed. On cross-sections of

the flip-chip bumps the location dependent nano-indents were taken in accordance with Fig. 3.

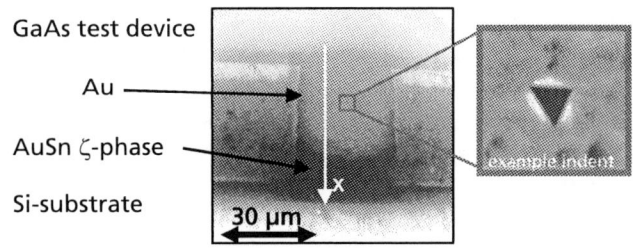

Fig. 3: Location dependent nano-indentation testing on cross-sectioned AuSn flip chip bumps.

The elastic properties of AuSn and of the ζ-phase measured by tensile testing as well those reported in the literature could be confirmed by the nanoindentation tests, see Table 1. It was also confirmed that both the AuSn eutectic and the zeta phase are dominated by elastic-brittle behaviour: They have shown high yield strength and reveal a low tendency of creep [4], especially at temperatures less than 100 °C. Annealing did also lower the yield strength of the AuSn eutectic phase.

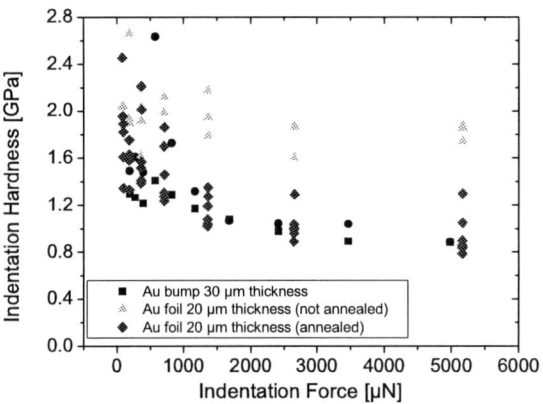

Fig. 4: Hardness as a function of indentation force of electroplated Au, annealed and not annealed, and at bumps.

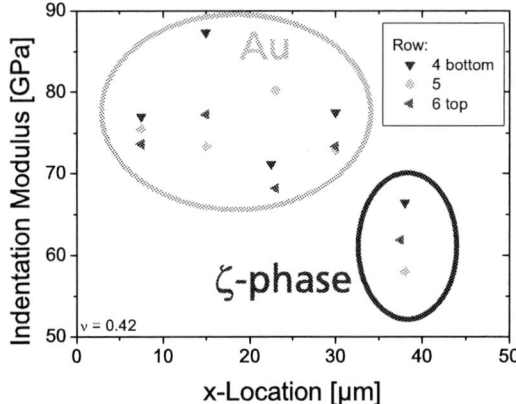

Fig. 5: Indentation modulus in dependence on the position on the bump, x-location starting from the GaAs, at a maximum load of 1 mN.

379

Fatigue characteristics

For the evaluation of the fatigue failure mode the "Thermal Lap Shear Test" [5] was applied, i.e. small lap-shear specimens soldered with AuSn were mounted in a loading frame with slightly different thermal expansion, causing shear loading of the joint material , see Fig. 6. For the tests alumina substrates with Cu-metallization and Ni/Au finish were selected. The pad size was 0.8 mm x 0.8 mm and the thickness of the AuSn solder layer approximately 100 µm. Temperature cycling was performed from RT to approx. 160 °C.

After the first tests it was observed that the AuSn solder behaves fundamentally different from the SnPb or SAC solders studied previously: After a few cycles the ceramic substrates of the lap shear specimens tended to fracture within the ceramic material instead of within the solder joints, for instance in the specimen shown in Fig. 7, which cracked in the ceramic substrate after the 4th cycle. This observation clearly indicated the much higher stiffness, or lower tendency of plastic deformation, of AuSn when compared to other soft solders. Only specimens with a larger amount of voids did not show the ceramic fracture phenomenon and could be studied for higher thermal cycle numbers.

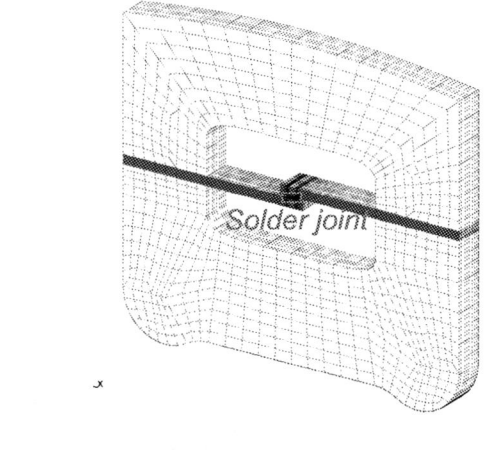

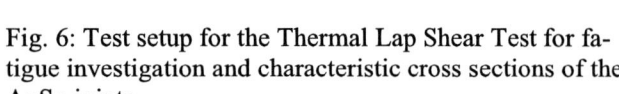

Fig. 6: Test setup for the Thermal Lap Shear Test for fatigue investigation and characteristic cross sections of the AuSn joints.

Fig 8 shows a plane FE-model of the joints with a void distribution pattern taken from a micrograph. Based on the measured materials data given in Table 2, the cyclic accumulated plastic strain distribution pattern was calculated as shown in Fig. 9. It is obvious from the figure that only a maximum of 0.1 % equivalent plastic strain is calculated, even for a high upper dwell temperature of 160 °C. For the same specimen and a similar temperature cyclic amplitude the calculated maximum plastic (creep) strain was approx. 100 times higher for SAC solder [5].

The maximum plastic strains are calculated at the solder-pads interface edges diagonally located at the joint edges. A diagonal path of maximum local plastic straining is calculated for the not voided joint, as it appeared previously for the SAC joint. It is obvious from Fig. 7 that fatigue failure initiation does only partly coincide with maximum plastic straining, since fracture presumably occurs at interphase boudaries (ζ-phase light appearing). In case of the high void-fraction shown in Fig. 9, the diagonal character of maximum local plastic straining is disturbed by the voids. They act as stress concentration locations where local maximum plastic straining occurs.

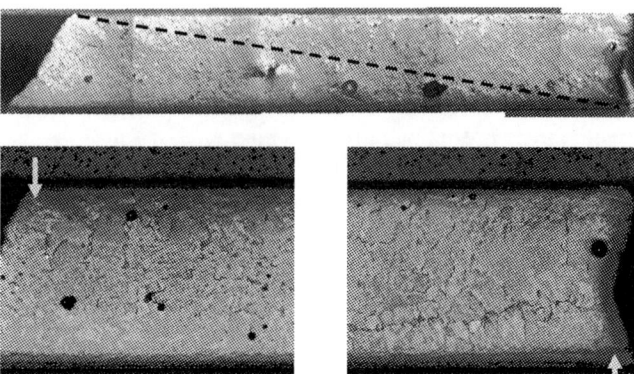

Fig 7: Initiation of fatigue failure in the edge regions of a not voided joint after 4 cycles. Dotted line indicates calculated path of maximum local straining and arrows calculated regions of plastic strain concentrations.

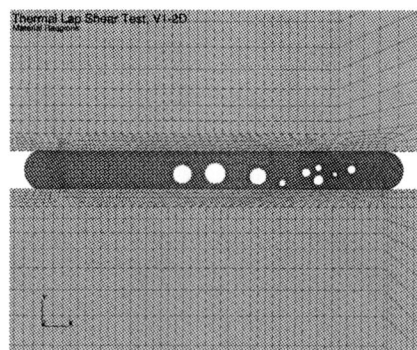

Fig. 8: Plane FE-model for the Thermal Lap Shear Test (joint region shown) with void inclusions in the cross section of the AuSn joint.

In contrast to the low amount of calculated plastic straining, fatigue failure of the joint occurs after a relatively low number of cycles, as was observed by microscopic in-situ analysis of the deforming joint surface. Fig. 10 shows the development of fatigue failure, which was not initiated at the calculated locations of maximum plastic straining, but along the phase boundaries between the constituent phases of the solder. As expected, microcracking starts at the surface of the joints and grows into the volume, to be seen from the FIB analyses shown in Fig 11.

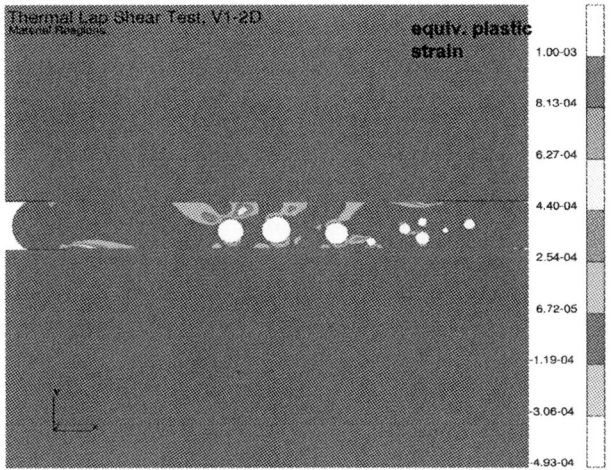

Fig 9: Accumulated cyclic plastic strain distribution pattern after one thermal cycle, RT to 160 °C.

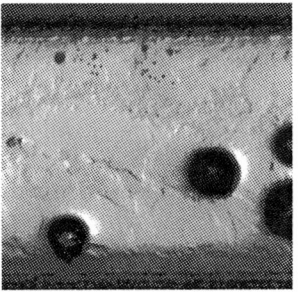 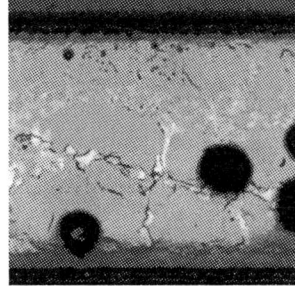

Fig. 10: Development of fatigue failure in the central region of a voided joint, left after 14 cycles, right after 68 cycles.

The experimental observations indicate that the inhomogeneous strain distribution at the different phases have a dominant effect on fatigue failure. High imposed strain imposes local softening, micro-cracking, and relatively early fatigue failure, when plastic deformations occur in the joint. However, due to its high yield stress, plastic straining remains low in many relevant loading situations. Therefore, AuSn joints are more prone to brittle fracture at the interfaces, i.e. the pads, than to fatigue failure within the solder. Theoretical failure predictions based on FEA results which do not account for the intrinsic microstructure of the solder can not easily be given for fatigue failure, since this type of failure depends on the actual microstructure of the joints. But it was already mentioned that critical stress dominated brittle interface fracture is more likely, which can be evaluated by critical stress criteria related to the actual bonding situation.

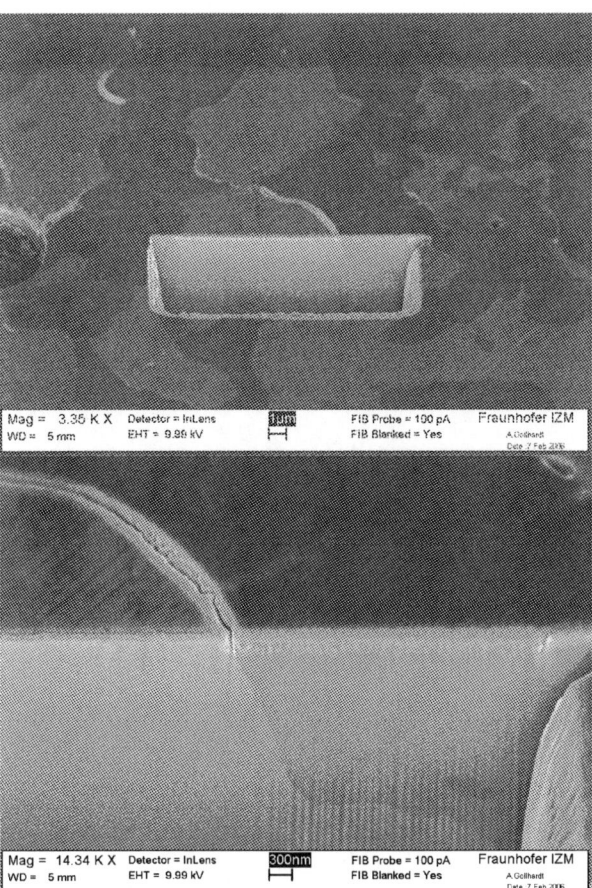

Fig. 11: FIB images on crack initiation at phase boundaries at the lap shear joint surface after 8 thermal cycles.

3. FE based reliability analysis of X-Ray image sensors

The thermo-mechanical reliability of X-ray image sensors consisting of an either 64x64 or 256x256 GaAs pixel detector chip and Si readout chip were analyzed by FEA. Hybrid pixel detectors consist of a sensor chip with an area array configuration of the pixels. Each pixel has to be connected to a readout cell of a chip for local detection of the signal. The pixel detector is flip chip bonded to the IC to read out each pixel diode in parallel with high repetition rate. Pixel detector chip sizes varied from 4 x4 mm ^{2} to 16 x16 mm² and GaAs thickness was 0.5 mm.

The chips were connected by lead-fee solder reflow processes applying both AuSn as well as SnAg electroplated flip chip bumps. An example of the bumped chip, in that case applying SnAg bumps, is shown in Fig. 12. The AuSn bumps were deposited by electroplating of approx. 21 μm Au and 6 μm Sn, forming Au-rich bumps after reflow. On the substrate side 10 μm Au bumps were plated to increase the standoff and the compliance of the joints. An example for the resulting AuSn bump geometry is given in Fig. 12.

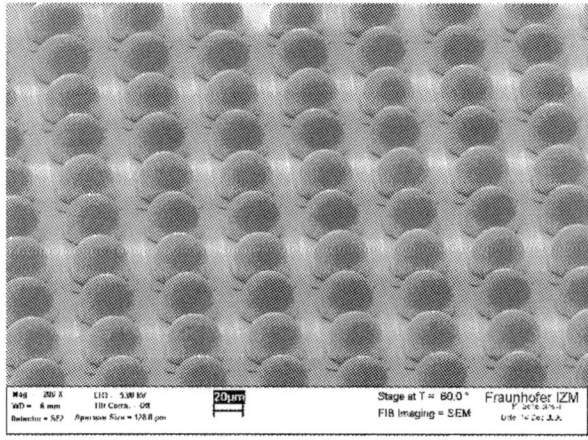

Fig. 12: Bumped wafer with 55 μm pitch, SnAg bumps

As it can be seen from Fig. 13 and Fig. 14, a global-local FE analysis technique was applied, because the 64x64 or 256x256 bump array with 170 or 55 μm pitch could not be modelled in one model in full detail.

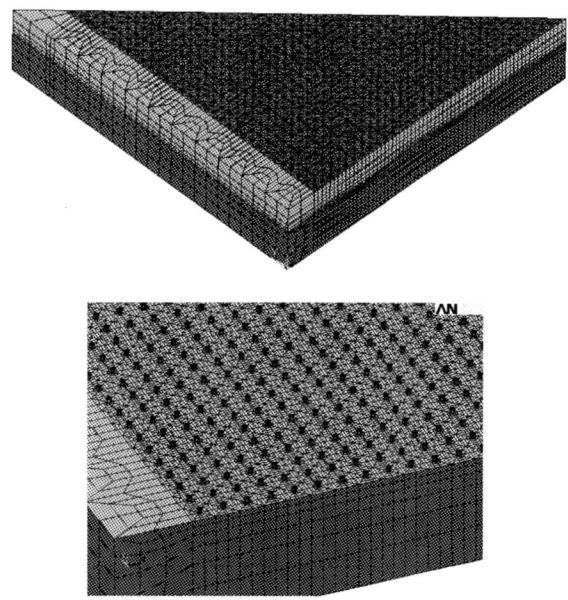

Fig. 13: Global octant FE-model, GaAs on Si chip, 64x64 bump array.

The relatively rigid behaviour of the GaAs chip on the Si chip allowed the simplification of the bumps to be modelled as beams in the global model, while details of the stacked bumps structure could be taken into account in the local model, see Fig. 14.

Temperature dependend elastic-plastic modeling was used for the bump constituent phases, while the creep part was considered negligible. The as-reflowed state without annealing was considered in the analyses, since this state with high yield stresses is the worst case for brittle fracture. During the analyses this type of failure turned out to be the critical one.

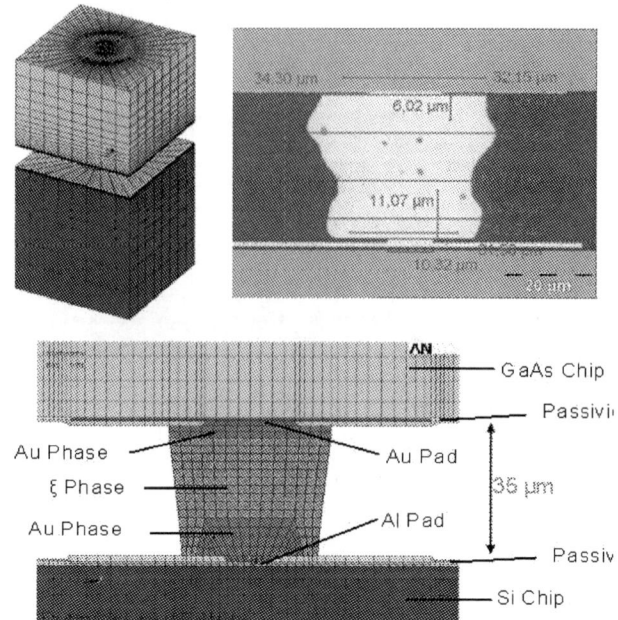

Fig. 14: Local FE-model and cross section of bump vicinity.

Tab. 2: Materials data used for FEA

	Constitut. Model/ Yield stress (MPa)	Young's Modulus (Instantaneous) [GPa]	CTE [10^{-6}/°C]
Si-chip	elastic	168.0	2.8
GaAs chip	elastic-	85.0	6.3
ζ phase in bumps	σ_0: 330 @ 233 K σ_0: 275@ 293 K σ_0: 230 @ 348 K σ_0: 190 @ 398 K σ_0: 140 @ 448 K σ_0: 100 @ 563 K	59.2 @ 233 K 54.0 @ 328 K 48.0 @ 368 K 35.0 @ 423 K 25.0 @ 448 K	14 @ 233 K 14 @ 420 K 17.5 @ 426 K 17.5 @ 550 K
Au – bump	σ_0: 300 @ 200 K σ_0: 270 @ 300 K σ_0: 90 @ 550 K	72 @ 200 K 70 @ 300 K 58 @ 550 K	15
SiO$_2$	elastic	72	1.5
	viscoplastic	48.5 @ 218 K 33. @ 483 K	21
Cu-UBM	σ_0: 280 @ 200 K σ_0: 190 @ 550 K	68	17
Al-pad	σ_0: 180	70	23

The thermal mismatch between Si chip and GaAs chip caused high relative displacements at the bumps, leading to high bump deformations with clear dependence on DNP. Fig. 15 schematically depicts the magnified bump deformation and resulting stresses normal to the chip plane. The calculated bump deformation at the local model of a diagonal edge bump, taken from the global analysis result, is shown in Fig. 16. Extremely high peel stresses up to 1700 MPa acting on the UBMs at the under-bump to chip interface were calculated, see Fig. 17. Accordingly, a very high delamination risk was predicted. In accordance with the prediction, this type of failure was also observed at test assemblies, which failed by UBM delamination or chip cracking either after bonding or after a few thermal cycles. Significant plastic straining occurred only in the gold phase of the bumps. However, fatigue failure was no issue due to earlier brittle failure.

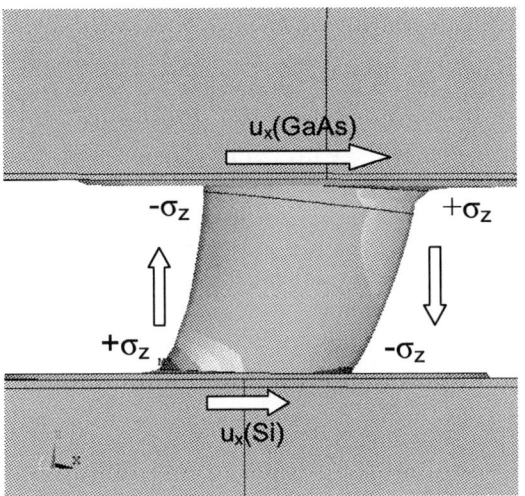

Fig. 15: Effects of thermal mismatch on the bumps.

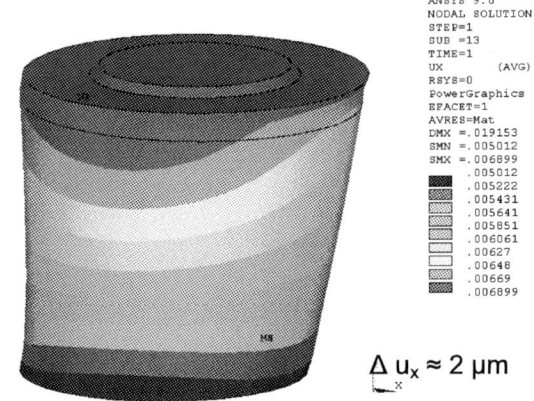

Fig. 16: Thermal mismatch induced bump deformation.

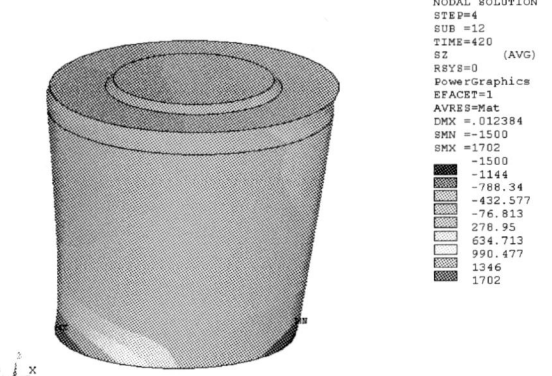

Fig. 17: Bump peel stresses acting on the UBMs.

To circumvent those reliability issues, a SnAg bumping technology was finally chosen, an example shown in Fig. 18. Calculations revealed a reduction of the critical peel stress at the pad-chip interfaces by 300 % to 500 %, dependent on the chip size. The reduction in peel stress is accompanied by a significant increase in plastic (creep) deformation in the solder. Hence, the dominant failure mechanism of AuSn interconnects being pad delamination is now shifted to low cycle fatigue failure of the SnAg bumps. The edge bumps are subjected to mean cyclic creep strains of approximately 10 %, see Fig. 19.

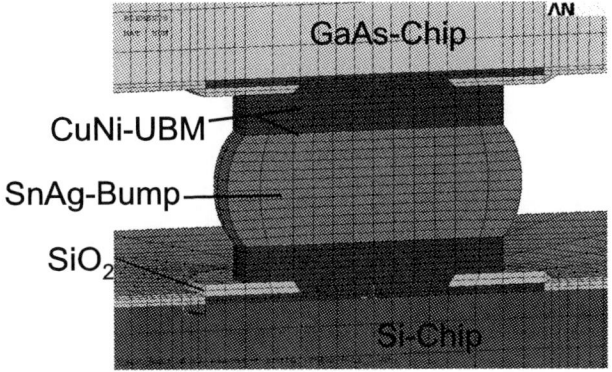

Fig. 18: Bump vicinity of the local FE model with SnAg bumps.

Fig. 19: Accumulated equivalent creep strain in a SnAg edge bump.

By application of a Manson-Coffin type equation reported previously [6], a cyclic life in the range of 100 cycles was expected from these results, which could also be confirmed by testing. Since the assemblies are not subjected to severe environmental temperature loadings, these results were in a range sufficient for use. Enhancement of reliability by underfilling was additionally proven.

Summary

The elastic-plastic characteristics of electrodeposited thin Au and AuSn films (both δ phase and ζ phase) were measured by tensile tests and/or nano-indentation. They fit well into the data range reported in the literature. Testing on electroplated Au foil specimens revealed an elastic-plastic behaviour remarkably away from bulk gold properties. It was also observed that the yield strength of gold is highly affected by annealing. The yield stress of the as-plated material dropped by more than 60% by annealing at 150 °C. Annealing effects were also observed for the ζ phase. However, the process dependence of the properties point to the need of performing measurements for the characteristics of the films actually used.

Thermal lap shear test results and related FEA results revealed that SnAu fatigue failure of the joint occurs along the phase boundaries and does not follow the path of maximum plastic deformation, in contrast to earlier observations for SnPb or SAC soft solders. Accordingly, theoretical fatigue failure predictions based on FEA results, which do not account for the intrinsic microstructure of the solder, cannot easily be given for fatigue failure, since this type of failure depends on the actual microstructure of the joints.

Computational analyses concerning the reliability of AuSn flip chip interconnects for GaAs pixel detectors on Si read-out chips using a global-local FE analysis technique were performed. Details of the layered materials structure of the bumps could be taken into account in the local model. Extremely high peel stresses at the under-bump to chip interface were calculated and, accordingly, a high delamination risk was predicted. In accordance with the prediction, this type of failure was also observed at test assemblies. To circumvent those reliability issues, a SnAg bumping technology was finally chosen. For this type of interconnection technology fatigue failure was well predicted.

Acknowledgement

The authors would like to express their thanks to Dr. H. Oppermann, and Dr. G. Engelmann, IZM Berlin, for the technological developments and processing of samples. They would further like to acknowledge the tensile testing, FIB, and EDX analyses of Mrs. A. Gollhardt and Dr. H. Walter, both are with IZM Micro Materials Center Berlin. Funding of parts of this research work by the Saxon Ministry for Science and the Fine Arts of Germany (project GaAs-Hybrid) and by DFG (German Research Foundation) under contract 242830 is gratefully appreciated.

References

[1] F.G Yost, M.M. Karnowsky, W.D. Drotning and J.H. Gieske: "Thermal expansion and elastic properties of high gold-tin alloys", Metal. Transact. A, vol. 21A, 1990, pp. 1885-1889

[2] M. Hutter, F. Hohnke, H. Oppermann, M. Klein, G. Engelmann, „Assembly and Reliability of Flip Chip Solder Joints Using Miniaturized Au/Sn Bumps", 54th ECTC 2004, Las Vegas, Nevada USA, June 1-4, 2004, proceedings, pp. 49-57

[3] Hutter, M., Thomas, T., Jordan, R., Engelmann, G., Oppermann, H., Reichl, H., Wang, Y., Howlader, M.,Higurashi, E., Suga., "Investigation of Different Flip Chip Assembly Processes Using Au/Sn Microbumps",

MicroSystem Technologies Conference, München, Germany, 5.-6.10.2005, proceedings, 2006, pp. 273-280

[4] O.Wittler, H. Walter, W. Faust, R. Dudek, W. Jun, B. Michel: „Deformation and Fatigue Behaviour of AuSn Interconnects", Proceedings EPTC 2006, Singapore, December 2006, pp. 297-301

[5] R. Dudek, W. Faust, J. Vogel, B. Michel,"A Comparative Study of Solder Fatigue Evaluated by Microscopic In-situ Analysis, On-line Resistance Measurement and FE Calculations," Proceedings EuroSimE 2005, Berlin, Germany, April 2005, pp. 610-618

[6]. A. Schubert, R. Dudek, E. Auerswald, A. Gollhardt, B. Michel, H. Reichl, Fatigue Life Models for SnAgCu and SnPb Solder Joints Evaluated by Experiment and-Simulation", 53rd Electronic Components & Technology Conference 2003, New Orleans, Louisiana USA, May 27-30, 2003, proceedings, pp. 603-610

Creep Analysis of a Lead-free Surface Mount Device

Pradeep Hegde, David Whalley, Vadim. V. Silberschmidt

Wolfson School of Mechanical and Manufacturing Engineering, Loughborough University, Loughborough,
Leicestershire, LE11 3TU, UK

Abstract

In this paper finite element analysis (FEA) is used to understand the effect of a non-uniform temperature distribution on the creep and fatigue behaviour of lead-free solder joints in an electronic assembly comprising of a chip resistor mounted on printed circuit board (PCB). Solder joints in surface mount devices (SMDs) operate over a temperature range as extreme as -55°C to 125°C, which is high compared to the melting temperature of solder alloys. Exposure of solder joints to these temperatures can result in thermo-mechanical fatigue. Eutectic or near- eutectic tin-lead alloys have previously been used as an interconnection material, but the ban imposed on the use of toxic materials in electronic products demands new lead-free solder materials. This paper presents the experiments carried out using a thermal camera to obtain the real temperature distribution in the electronic assembly. These temperature distributions were used in FEA of the chip resistor under temperature cycling conditions. Unlike accelerated tests for obtaining reliability data, FEA is quick and less expensive.

1. Introduction

The environmental impact of lead in electronic products is relatively low, but due to the size of the industry, is becoming a major concern all around the world. The stimulus for the "green movement" is market trend's and customers' perception. Therefore, manufacturers, suppliers and research institutes around the world are investing their efforts into developing lead-free soldering technologies to substitute for tin-lead solder alloys. In addition, researchers are also pondering the pressing need to find a high-performance solder alloy with improved mechanical properties and similar processing characteristics to tin-lead solders [1].

The reliability of lead-free solder joints is still a major concern due to their widespread application in the electronic industry only very recently and therefore there is not a great deal of material data or practical experience available. In this study a near-eutectic lead-free SnAgCu (SAC) solder alloy, with a melting temperature of 217°C, is considered because it is being widely adopted due to its excellent wetting and mechanical properties [2]. When the solder is subjected to a cyclic stress induced by thermal cycling, the reliability of the solder joint depends on its resistance to fatigue. Along with thermo-mechanical fatigue, solder joints are subjected to creep as they operate at high homologous temperatures (T_g, the ratio of absolute operating and melting temperature). In this study a solder joint is subjected to thermal cycles between -55°C and 125°C. This means that they operate between $T_g = 0.44$ and $T_g = 0.81$. It is well documented [3] that creep plays a very important role in deformation behaviour of materials at homologous temperatures close to and above 0.5 if the loading rate is slow enough for creep damage to occur. Since under actual service conditions, the temperature cycle duration is in the order of minutes to days and the homologous temperature is more than 0.5, solder joints formed using SnAgCu alloy are expected to deform primarily due to creep [3]. This is essentially the same as for SnPb solders, but much less is known about the creep fatigue response of Pb free alloys.

Research into the use of finite element analysis (FEA) has been widely carried out to understand the elasto-plastic and creep behaviour of solder joints exposed to uniform (isothermal) temperature cycling conditions [1,3,4,5]. Thermo-mechanical analysis of a chip scale package (CSP) assembled using both lead-free and lead containing solder materials [1] and thermal cycling analysis of flip-chip solder joint reliability [5] are examples. However, experimental studies show that the temperature distribution within an electronic assembly is non-uniform due to different heat dissipation rates in the constituents of the electronic assembly. In addition, the mass distribution within the electronic assembly results in a non-uniform distribution of temperature during rapid changes in ambient temperature or power dissipation. Therefore, this paper focuses on the use of FEA to investigate the effect of a non-uniform temperature distribution on the creep behaviour of SAC solder joints in surface mount devices and a comparison is made with that for an uniform temperature distribution. The finite element analysis is first used to estimate stresses/strains due to cooling from reflow and then three different thermal cycling conditions are applied.

2. Experimental analysis
2.1 Experimental set up

In order to establish an appropriate magnitude for the non-uniform temperature distribution in the electronic assembly, a series of experiments were carried out to acquire the temperature profile in a flip chip assembly under power cycling conditions, using an infrared (IR) camera. Although this flip chip is different to the component (chip resistor) modelled, the general size and interconnection joint distribution makes both flip chip and chip resistor assemblies roughly comparable and means the flip chip experimental results will provide an indication of the temperature gradients to be expected in the chip resistor. The camera measures thermal radiation from the surface, which has a wavelength spectrum and intensity dependent on its temperature, structure and

1-4244-1105-X/07/$25.00 ©2007 IEEE

composition. The higher the temperature, the more the radiation emitted. This infrared radiation, not perceptible with the naked eye, is made visible and measurable by the infrared camera. By analysing this infrared radiation it is possible to measure temperature as well as, indirectly, the thermal conductivity, mechanical stresses, material compositions, defects such as pores and delamination, and various other kinds of inhomogeneities in the materials.

The IR camera used (Fig. 1) has an IR detector head with a focal plane array (FPA) detector sensitive in the range 1 μm – 14 μm. An infrared microscopic lens MWIR 2.5X, with a focus distance of 21-22 mm was used. The camera is interfaced with software to control real-time acquisition and analysis of the infrared data.

The specimens used were identical flip chip assemblies attached to either a copper or FR4 substrate. They were mounted vertically and powered at 1.2 W and cooled by free convection.

The flip chip specimens were silicon-on-silicon multi-chip modules (MCMs) the same as those used in a

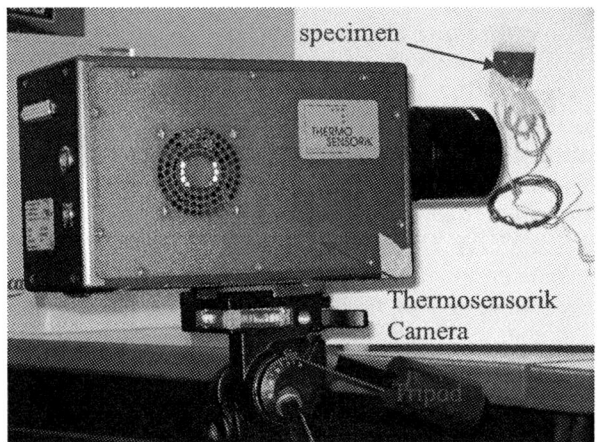

Figure 1 Experimental setup

previously reported experiment [6]. Both MCMs consisted of a 3 mm × 3 mm × 0.5 mm "heater" chip that bore a large central resistive element (the heater) in addition to small aluminium tracks and 36 connection pads. The "carrier" chip was larger at 6 mm × 6 mm × 0.5 mm and included larger ball grid array type pads allowing

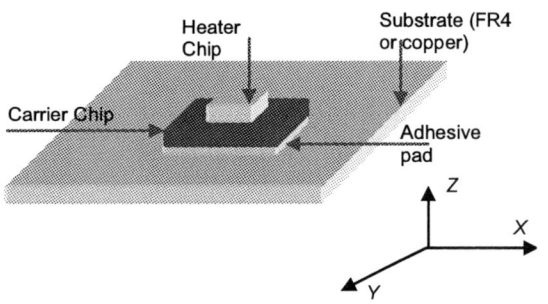

Figure 2 Schematic of flip chip specimens

for external connections to be made, as well as the corresponding pads to match those on the heater chip. The heater chip was attached to the carrier chip so that a standoff height of 35 μm was achieved (without underfill). The MCMs were subsequently attached to the corresponding substrate by a thermally conductive adhesive pad. A schematic of the assembled specimen is shown in Fig. 2.

Figure 1 shows the experimental setup, used for acquiring the temperature data for the flip chip assemblies. The camera, fitted with the micro lens, is mounted on the tripod, the specimen is powered on for few minutes to stabilise the temperature distribution in the specimen, then the lens is focused on the specimen and the temperature profile is captured.

2.2 Experimental results and discussion

The temperature distributions over the chip for a continuous 1.2 W power dissipation are given in Figs. 3 and 4 for free convection. The path used for subsequent temperature distribution analysis, is also shown. These figures demonstrate the effect of the substrate on the chip

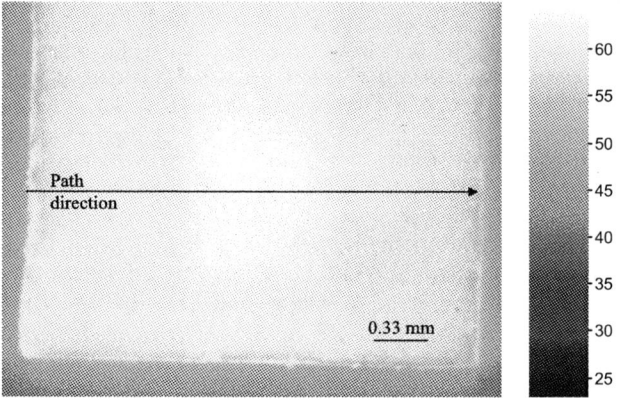

Figure 3 Temperature distributions in a chip mounted on a copper substrate at 1.2W

temperature distribution: the copper substrate results in lower temperatures compared to the FR4 substrate. From comparison of these two temperature profiles, the temperature distribution is symmetric on the chip with a

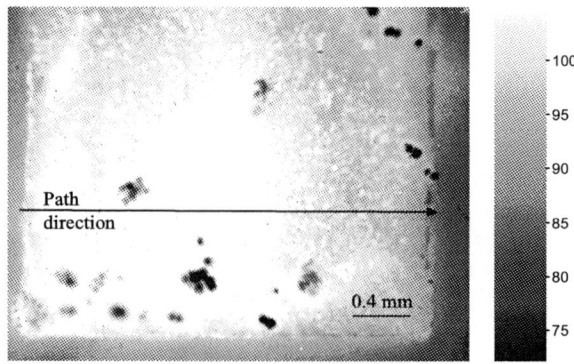

Figure 4 Temperature distributions in a chip mounted on FR4 substrate at 1.2 W

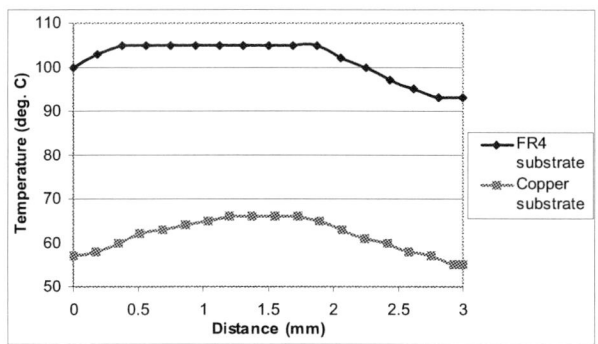

Figure 5 Effect of substrate on temperature distribution in the chip for free convection

copper substrate, while that for the chip with a FR4 substrate is asymmetric. The cool appearing patches may be attributed to non-uniform application of the black paint which is applied to ensure a higher uniform emissivity of the chip surface. Figure 5 shows the temperature distribution across the width of the chip for both types of substrate. It is evident that these temperature distributions are non-uniform: the maximum temperature is observed at the centre of the chip, where the heat is generated, while its boundary is at a lower temperature. Another important observation of this analysis is that the FR4 substrate induces higher thermal gradients in the chip than the copper substrate. This can be explained by the much higher thermal conductivity of copper compared with FR4. The experimental temperature distribution in the chip with a FR4 substrate is used as one of the thermal load cases for creep analysis, as it best represents typical operating conditions of the chip modelled in the finite element analysis.

3. Creep analysis

The geometry of a standard 1206 resistor chip was used for the creep analysis. Figure 6 shows the geometric dimensions of chip resistor modelled for finite element analysis. In the finite element modelling only one half of the geometry was used, due to the symmetry of the structure. The finite element model was created using 2D plane strain elements and a fine mesh pattern is maintained around the interface between component and solder. Figure 7 shows the mesh details.

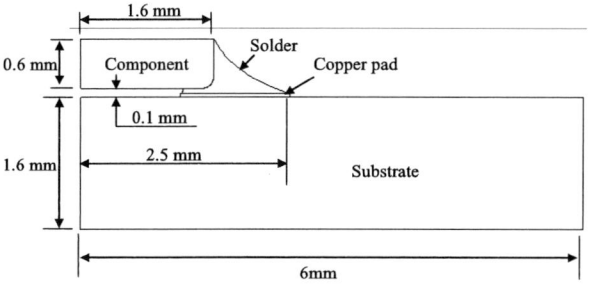

Figure 6 Geometry of 1206 chip resistor

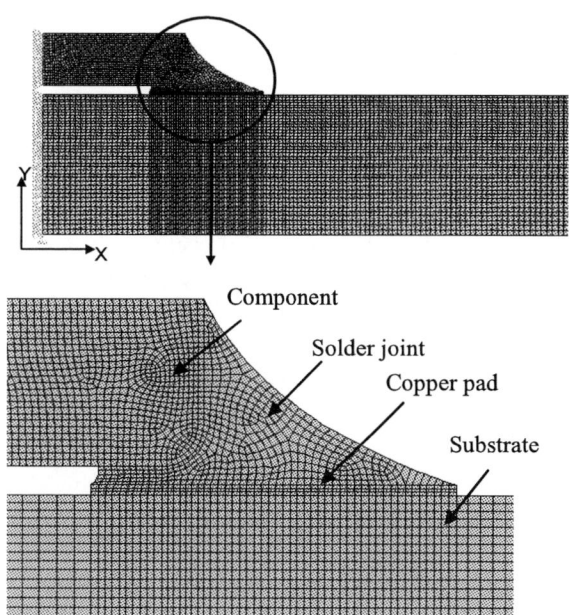

Figure 7 Finite Element Model of 1206 resistor

3.1 Creep constitutive equation

The solder joints of the 1206 resistor were modelled using the temperature-dependent material properties as shown in Table 1. A number of papers have been published [7, 8, 9] on the constitutive equation for creep deformation of SnAgCu alloys and they have identified two mechanisms for steady-state creep deformation. They attributed these to a dislocation climb controlled (low stress) and combined glide/climb (high stress) behaviour and have represented the steady-state creep behaviour using a double power law model. In this paper the creep

Temperature (°K)	Young's Modulus (MPa)	Poisson's ratio	CTE (ppm/°K)	Density (gm/cm³)
218	57300	0.4	12.7	7.5
248	55800	0.4	12.7	7.5
298	52600	0.4	21.2	7.5
248	49300	0.4	21.7	7.5
398	45800	0.4	23.0	7.5

Table 1 Elastic material properties for SnAgCu

model determined by Schubert et al. [7] is used for the steady-state creep behaviour. They also identified two regions for stress-strain rate behaviour, but postulated the high stress region as a power law break-down region, and chose the hyperbolic sine function to fit their creep data:

$$\dot{\varepsilon}_{cr} = A_1[\sinh(\alpha\sigma)]^n \exp\left[\frac{-H_1}{kT}\right] \qquad (1)$$

387

Where $A_1 = 277984$ s^{-1}, $\alpha = 0.02447$ MPa^{-1}, n = 6.41, $\dfrac{H_1}{k} = 6500$, $\dot{\varepsilon}_{cr}$ is steady state creep strain rate, σ is stress, T is absolute temperature.

Plasticity is also included along with creep in the finite element analysis. Plasticity is modelled with bilinear kinematic hardening (BKIN), which includes the Bauschinger effect due to thermal cycling. Table 2 gives the plastic material properties used for SnAgCu [10].

The material properties of 96% alumina (Al$_2$O$_3$) are used for the component body, whilst high-conductivity copper and FR4 material properties [10, 11] are used for pad and substrate respectively.

Temperature (°K)	Yield stress (MPa)	Tangent modulus (MPa)
218	45	5700
248	41	5600
298	32	5260
348	21	4900
398	13	4600

Table 2 Plastic material properties of SnAgCu

3.2 Thermal cycling conditions

In the surface mount assembly process, the components are reflowed in an oven to create the solder joint and the assembly is then returned to room temperature. Therefore, creep analysis is carried out in two steps for three different temperature cycling conditions. In the first step, creep analysis is carried out for the reflow soldering process and relaxation for one hour at room temperature (assuming there is a one hour storage period before the resistor assembly is subjected to thermal cycling). The stress levels at the end of the reflow process give the manufacturing-induced stress in the solder joint, and similarly stress level at the end of relaxation gives the amount of stress after relaxation has taken place in the solder joint due to storage at room temperature. Below are the three different thermal cases used.

Case A: Uniform temperature ranging from 398°K (+125 °C) to 218°K (-55°C), where the entire resistor-substrate assembly is subjected to the same temperature.

Case B: Uniform temperature for the component, solder joint and copper pad ranging from 398°K (+125°C) to 218°K (-55°C), while the substrate's temperature (T$_{sub}$) is also uniform but varies according to the following relation:

$$T_{sub} = 0.18 * T_{comp} + 299.4, \qquad (2)$$

Where T$_{comp}$ is the component's temperature. This equation is deduced from the previously described experimental results.

Case C: This case is more representative of real conditions where the temperature gradient from the experimental results is indirectly used in the thermal cycling. A thermal analysis was first carried out using temperature boundary conditions from the experimental

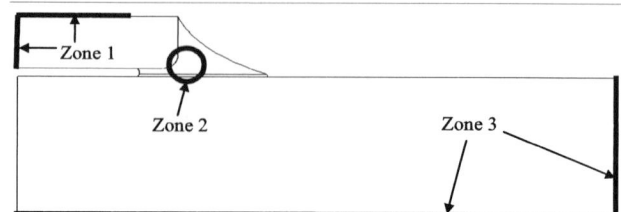

Figure 8 Temperature zones for thermal analysis

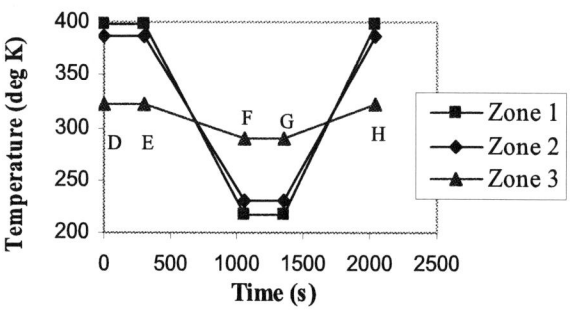

Figure 9 Temperature variations in zone 1, zone 2 and zone 3 in thermal analysis

results to obtain a continuously varying temperature distribution throughout the surface mount assembly. Figure 8 shows the different thermal zones within the resistor assembly used in the thermal analysis. Temperature boundary conditions were applied on the outer surface of the body at zone 1 and zone 3. In zone 2, a set of nodes was selected for temperature boundary condition application. The temperature boundary conditions in zone 2 (solder joint) and zone 3 (substrate) are based on the zone 1 (component) temperature. The

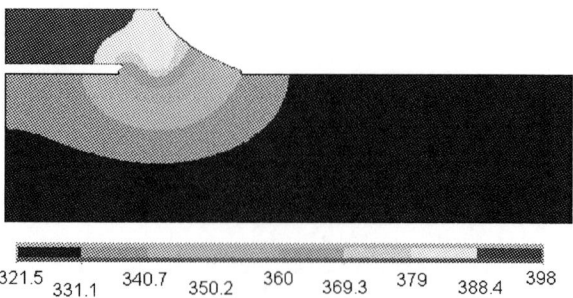

321.5 331.1 340.7 350.2 360 369.3 379 388.4 398

Figure 10 Temperature distribution in resistor assembly when component is at 398°K

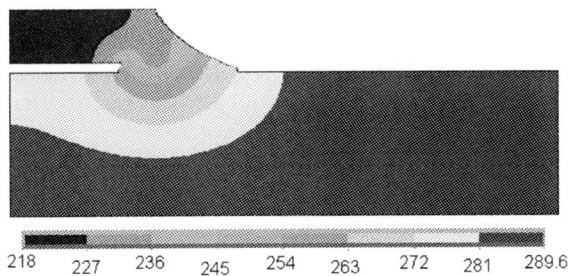

Figure 11 Temperature distribution in resistor assembly when component is at 218°K

relationship between the temperatures in zone1, zone2 and zone3 were deduced from the experimental results. Figure 9 shows the variation of temperature boundary conditions at different zones throughout the thermal cycle. Figures 10 and 11 give the temperature distribution in the 1206 resistor assembly after thermal analysis was carried out when the zone 1 (component) temperatures are 398°K and 218°K respectively. In Case C the resistor assembly is subjected to a thermal cycle between these two extreme temperature profiles.

Figure 12 shows the typical thermal cycle used in creep analysis. Line AB represents the reflow process, where a cooling rate of 4°C/s is used, and line BC represents storage of the resistor at room temperature for an hour. After an hour of storage at room temperature, the component temperature (T_{comp}) is ramped to 398°K (in Case A this is the whole assembly temperature) to start the thermal cycling. A complete thermal cycle starts at D and ends at H. In this thermal cycle there is a ramp of 12 minutes between temperature extremes (398°K and 218°K) and dwells of 5 minutes at the extreme temperatures.

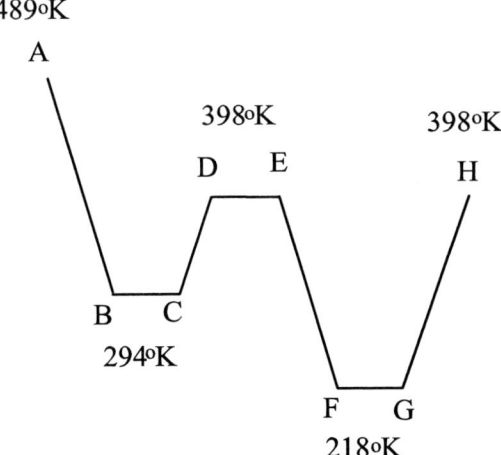

Figure 12 Thermal cycle used for creep analysis based on component temperature

4. Result and discussions

Finite element analysis of the chip resistor assembly was carried out for two thermal cycles and for the three different thermal loading cases described above. Figure 13 shows the shear stress distribution in the solder joint after the reflow process and also the location of maximum shear stress. The maximum shear stress of 25 MPa observed in the solder joint fillet, is mainly due to the mismatch of coefficients of thermal expansion (CTE)

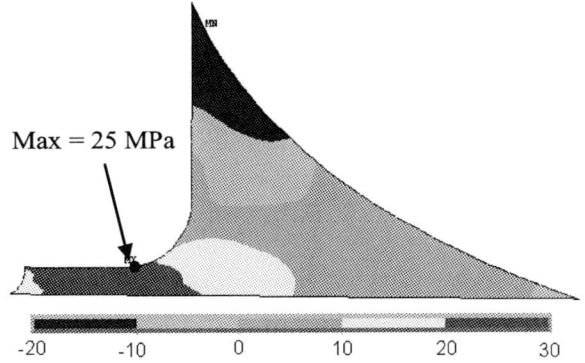

Figure 13 Distribution of shear stress in the solder joint at the end of reflow period (time = 48s)

between component (made from alumina), solder and substrate (made from FR4). This stress is well above the yield stress of SnAgCu solder alloy at room temperature. When the resistor assembly is stored at room temperature, this stress reduces by the solder joints undergoing creep strain. This process is called stress relaxation. The shear stress after stress relaxation for one hour at room temperature reduces to 12 MPa, as can be seen in Fig. 14, which is below the yield stress at room temperature of SnAgCu alloy. The shear stress evolution for the entire creep analysis is shown in Fig. 14, which includes reflow period, relaxation period and 2 thermal cycles, all for the peak stress location in Fig. 13. It is evident from the figure that the shear stress range is 35MPa for Case A and that for Cases B and C is only about 15 MPa. This shows that there is about a 60% reduction in the shear stress

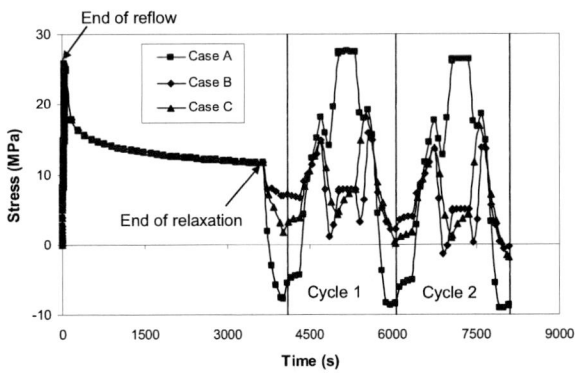

Figure 14 Distribution of shear stress in the solder fillet over time

range for Case B and C.

Accumulated inelastic strain due to thermal cycling is also studied for the solder joint. Figure 15 shows the worst case variation of inelastic strain over time at solder fillet. The total accumulated inelastic strain at the end of two thermal cycles was largest (9.5%) for Case A and smallest for Case C (7.2%). In Case B and Case C the total accumulated creep strain at the end of the thermal cycle is reduced by 13% and 28% respectively compared with case A. It can be observed from Fig. 15 that, even though most of the inelastic strain accumulation has taken place during the reflow and relaxation periods, this depends on parameters such as relaxation time, temperature and number of thermal cycles. In this particular analysis inelastic strain accumulation during the reflow and relaxation period accounts for 50%, 57% and 69% for Case A, Case B and Case C respectively. The

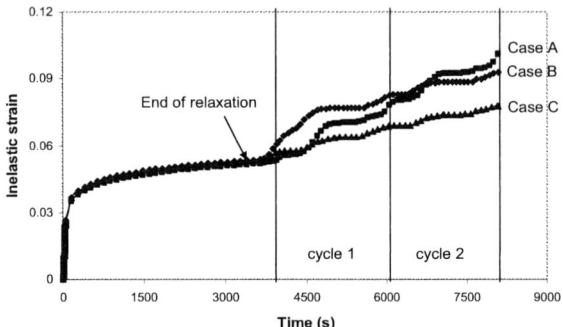

Figure 15 Accumulation of creep strain with time

amount of inelastic strain accumulation is reduced in thermal cycle 2 compared with that in thermal cycle 1. This reduction is only 3% for Case A compared with Cases B and C where it is 23% and 20% respectively. It is however expected that further reductions for subsequent cycles would be smaller.

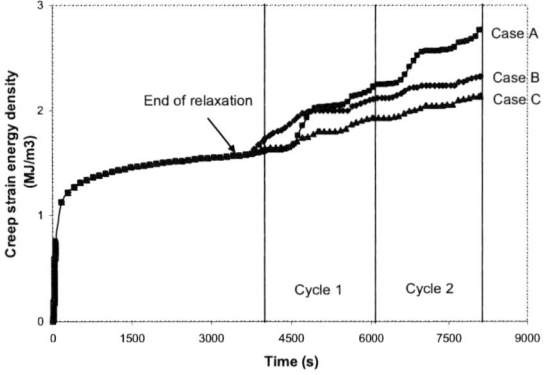

Figure 16 Accumulation of creep strain energy density in solder joint

Figure 16 demonstrates the density of creep strain energy dissipated in the solder joint during the analysis. This variation is quite similar to the variation of

accumulated inelastic strain in the solder joint, with Case A having a higher density of creep strain energy dissipation (2.64 MJ/m^3) due to the higher inelastic strain accumulation and Case C having a lower density of creep strain energy dissipation (1.96 MJ/m^3).

The number of cycles to failure, N_f, for the solder joints were predicted based on the following Coffin-Manson based relationship [3]:

$$N_f = (C'\varepsilon_{acc})^{-1} \qquad (3)$$

Where ε_{acc} = Accumulated inelastic strain per cycle and C' = inverse of creep ductility. The accumulated inelastic strain calculated for the 2nd thermal cycle was considered for these life calculations.

Table 3 gives the predicted lives for the three different thermal loading cases. From comparison of predicted lives for the three different thermal cases, there is more than 100% improvement in the life of the solder joint in case B and case C. Case C is predicted to have the longest life for the solder joint (i.e. 2456 cycles) out of the three cases.

Thermal cycling conditions	C' (inverse of creep ductility)	Acc. inelastic strain in 2nd cycle	Predicted lives (cycles)
Case A	0.0468	0.023	929
Case B	0.0468	0.0104	2054
Case C	0.0468	0.0087	2456

Table 3 Predicted life for chip resistor

5. Conclusions

The experimental results demonstrate a typical non-uniform temperature distribution in an electronic assembly. The finite element study carried out based on the experimental results is a preliminary study to understand the effect of a non-uniform temperature distribution on the fatigue behaviour of lead-free solder joints. Out of the three different thermal loading cases considered, Case C (non-uniform temperature distribution) is predicted to result in lower levels of shear stress, creep strain accumulation and creep strain energy density. However, the accumulation of creep strain and creep strain energy density depend on relaxation time, temperature and number of thermal cycles. Therefore, further creep studies are required considering various relaxation times, temperature and a greater number of thermal cycles. The inelastic strain based estimated lives demonstrate the significant impact of non-uniform temperature distribution (case C) on fatigue life of solder joint in the chip resistor, case C predicting highest number of life cycles. However, the accumulated inelastic strain is from 2nd thermal cycle, which needs further creep analysis to establish the stabilised accumulated inelastic strain per cycle. The capture of thermal data for the actual components studied and a more accurate thermal model, taking into account the thermal mass distribution in the assembly, will be used to assess the interaction of power and thermal cycles on fatigue. Future analysis will also establish the relevant contributions of plastic and creep strains to fatigue damage and the life reduction attributable to post reflow stress relaxation.

References

1. Gonzalez, M., Vandevelde, B., Beyne, E., "Thermo-Mechanical Analysis of a Chip Scale Package (CSP) Using Lead Free and Lead Containing Solder Materials", *Proc 3rd European Microelectronics and Packaging Symposium*, Prague, Czech Republic, June. 2004.

2. Wu, C.M.L., Yu, D.Q., Law, C.M.T., Wang, L., "Properties of Lead-free Solder Alloys with Rare Earth Element Additions", Material Science and Engineering R 44 (2004) pp. 1-44.

3. Syed. A., "Accumulated Creep Strain and Energy Density Based Thermal Fatigue Life Prediction Models for SnAgCu Solder Joints", *Proc 54th Electronic Components and Technology Conf*, Las Vegas, USA, June. 2004, pp. 737-746.

4. Warde, J., Wallach, E.R., "The Prediction of Creep Damage in Surface Mount Components using Finite Element Modelling," NPL Report CMMT(A)218, March. 1999.

5. Pang, J.H.L., Chong, D.Y.R., Low, T.H., "Thermal Cycling Analysis of Flip-Chip Solder Joint Reliability," *IEEE Trans* Compn. and Packag. Technol. Vol. 24, No. 4 (2001), pp. 705-712.

6. Ochana, A.R., Hutt, D.A., Whalley, D.C., Sarvar, F., Al-Habaibeh, A., "Modelling of the Power Cycling Performance of a Si on Si Flip Chip Assembly," *Proc 10th Intersociety Conference on Thermal and Thermochanical Phenomena in Electronics Systems, ITHERM 2006*, San Diego, USA, May. 2006, pp. 243-250.

7. Schubert, A., Dudek, R., Auerswald, E., Gollbardt, A., Michel, B., Reichl, H., "Fatigue Life Models of SnAgCu and SnPb Solder Joints Evaluated by Experiments and Simulations," *Proc 53rd Electronic Components and Technology Conf*, New Orleans, LA, May. 2003, pp. 603-610.

8. Wiese, S., Meusel, E., Wolter, K.-J., "Microstructural Dependence of Constitutive Properties of Eutectic SnAg and SnAgCu Solders," *Proc 53rd Electronic Components and Technology Conf*, New Orleans, LA, May. 2003, pp. 197-206.

9. Morris, J.W., Song, H.G., Hua, F., "Creep Properties of Sn-rich Solder Joints," *Proc 53rd Electronic*

Components and Technology Conf, New Orleans, LA, May. 2003, pp. 54-57.

10. Database for solder properties with emphasis on new lead-free solders http://www.boulder.nist.gov/div853/lead%20free/props01.html

11. Matweb material property database, www.matweb.com.

Author Index

A

Achkar, H.	637
ACHKAR, Hikmat	588
Ahmad, J.	678
AHMAD, Mahmoud AL	588
Albrecht, Arne	560
Ansorge, Frank	447
Arnold, A. K.	555
Arruda, Luciano	545
Arshak, Khalil	352
Auersperg, Jürgen	524
Augustin, Adam	116
Axisa, Fabrice	340

B

Baek, Hyunggil	428
Baelmans, M.	738
Bagdahn, Joerg	672
Bailey, C.	159, 249, 593
Bailey, Chris	243, 653
Bakkers, E. P. A. M.	532
Baldwin-Hendricks, Teri	601
Barbé, J-C	142
Bartek, M.	361
Bauer, J.	7
Belov, Ilja	181
Beng, Lau Teck	150
Bergner, Fredrik	181
Beyne, Eric	491
Bhattacharyya, Bidyut K.	239
Bielen, Jeroen	503, 509
Bohm, C.	319
Boogaart, Marc A.F. van den	62
Borecki, Janusz	346
Bornoff, Robin	181
Bots, Tom	575
Bouarroudj, M.	409
Bousquet, S.	77
Boutaayamou, Mohamed	120
Brämer, Birgit	378
Brizoux, M.	659
Brosteaux, Dominique	340
Brugger, Juergen	62
Bruin, E. de	498
Buiu, Octavian	125
Bulcke, Mathieu Vanden	340
Buttay, Cyril	653

C

Cacchione, Fabrizio	303
Calata, Jesus N.	705
Carles, G.	565
Carmona, Manuel	332
Chae, Seung-Hyun	664
Chaillot, A.	77
Chan, Vincent	51

Chan, Y. S.	631
Chang, Chong-Qing	314
Chao, Brook	664
Chastanet, C.	77
Che, F. X.	283, 366, 416
Chen, Xu	705
Chiang, Kou-Ning	324
Chicharro, J. M.	583
Chin, N.H.	283
Cho, Ji-Man	234
Chunga, Jin Taek	234
Conway, Paul P.	308
Corder, Paul	679
Corfield, Martin	647
Corigliano, Alberto	303, 392
Crécy, F. de	142
Cuddalorepatta, Gayatri	537

D

Damani, M.	678
Danny, Retuta	366
Danto, Y.	659
Dasgupta, Abhijit	537
Dawotola, Alex. W.	532
Deml, Ulrich	207
Dermitzaki, E.	7
Doetzel, Wolfram	257
Dommelen, J.A.W. van	454
Dornel, E.	142
Dowhan, Lukasz	263
Downey, Susan	150
Driel, W. D. van	46, 85, 214, 372, 435, 468, 498, 532
Duch, M.	478
Duch, Marta	332
Dudek, Rainer	263
Dudek, Rainer	378
Dular, Patrick	120
Dumonteil, R.	77
Dupont, L.	409
Dusek, Milos	175

E

Ebert, Matthias	672
Elata, David	291
Eneman, Geert	491
Eng, P. F.	555
Engelen, R. A. B.	214, 468
Erinc, M.	27
Ernst, L. J.	319
Ernst, L. J.	361, 372, 423, 472
Esteve, J.	478
Eymery, J.	142

F

Fachin, Fabio	392
Fan, Haibo	515

Author Index

Faust, Wolfgang 378
Filho, W. C. Maia 659
Fiori, V. 468
Fiori, Vincent 109, 150
Flint, Anthony 745
Flores, Sebastian 314
Fonseca, L. 565
Forde, Edward 352
Frangi, Attilio 623
Freitas, Germano 545
Fremont, H. 659

G

Gallois-Garreignot, Sébastien 109, 150
Galvez, J.L. 403
Ganeshan, Vijay 357
Geers, M.G.D. 27, 454
Gerbach, Ronny 672
Gerbolés, Marta 332
Gessner, Thomas 257
Ghisi, Aldo 392, 623
Gholnejad, Hasan 1
Gille, Thomas 136
Gilles, J. P. 571
Gils, Marcel van 509
Godignon, P. 403
Goebel, J. 549
Goette, Carsten 207
Golinval, Jean-Claude 102
Gómez, E. 478
González, A. 583
Gonzalez, Mario 340, 491
Goux, Ludovic 136
Greer, James C. 62
Grout, Ian 484
Guiney, Ivor 352
GuoZhong, Chai 16
Guy, O. J. 555
Gwyer, D. 593

H

Hainz, Simon 33
Hal, B.A.E. van 454
Hall, Steve 125
HAN, Lei 23
Hang, Sung Woo 234
Hauck, Torsten 131
Hegde, Pradeep 385
Hegde, S. 678
Hering, Siegfried 560
Herkommer, Dominik 537
Hirshberg, Arnon 291
Ho, Paul S. 442, 664
Hoffmann, Martin 560
Hogg, Simon 647
Hölzer, Gisbert 560

Hsieh, Ming-Che 398
Hu, Dyi-Chung 324
Huang, Ching-Shun 324
Hughes, M. 249, 593
Hunt, Christopher 175
Huo, Gang 239

I

Igic, P. 555
Im, Sehyuk 442
Irving, Scott 220, 519, 710
Iwamoto, Nancy 601

J

Janiaud, D. 571
Jansen, K. M. B. 319, 361
Jansen, M. 435
Jiang, Li 56
Johnson, C Mark 647, 653
Jordà, X. 403
Ju, Byeong-Kwon 234
Jun, Wei 378
Jun, Zhou 16
Jungwirth, Mario 33

K

Kalms, A. 549
Kasemset, B. 498
Kaulfersch, Eberhard 357
Kessler, A. 319
Khatir, Z. 409
Kim, Jeongyeol 428
Kim, Kuyoung 428
Kima, Jae Choon 234
Kimb, Gyoung Bum 234
Klein, K. 678
Klein, Matthias 378, 524
Kolchuzhin, Vladimir 257
Korvink, J.G. 549
Kwak, Dongok 428

L

Lall, Pradeep 692
Larson, Amy 601
Lee, Dong Jin 234
Lee, S. W. Ricky 631
Lee, Won Suk 234
Leee, K. J. 678
Lefebvre, S. 409
Lei, Guangyin 705
Leisner, Peter 181
Lepage, Severine 102
Leus, Vitaly 291
Levy, R. 571
Li, Guang 283

Author Index

LI, Junhui 23
Li, Y. P. 723
Liang, Lihua 710
LiHua, Liang 16, 225
Lindgren, Mats 181
Lishchynska, Maryna 62
Lisoni, Judit 136
Liu, Sang 631
Liu, Y. 607
Liu, Yong 220, 519, 710
Liu, Yumin 519
Liu, Z. 549
Lo, G. 678
Loh, Wei-Sun 647
Lombaërt-Valot, I. 77
Lu, Guo-Quan 705
Lu, H. 593
Lu, Hua 51, 243, 647
Luk, Timwah 220, 519, 710

M

Mahalingam, S. 678
Mahmoudi, Jafar 1
Maj, Bartosz 116
Marco, S. 549, 565
Mariani, Stefano 392
Maron, D. 77
Massiot, G. 77
Mathias, H. 571
McDonald, Gavin 509
McManus, K. 249
Megherbi, S. 571
Mehner, Jan 257
Meier, K. 618
Meier, Karsten 93, 207
Meuwissen, Marcel 575
Meyer, Kristin De 136
Michael, Steffen 560
Michel, B. 7
Michel, Bernd 357, 378, 524
Miura, Hideo 686
Morales, A. L. 583
Moreno, R. 583
Moríñigo, José A. 461
Mueller, M. 189
Mueller, Maik 197
Müller, A. 618
Müller, Axel 295, 357
Müller, G. 549
Munier, C. 77
Munier, E. 77
Musallam, Mahera 653

N

Nakamura, Tomoji 442
Nakanishi, Tohru 270, 278

Naumann, Falk 672
Nieto, A. J. 583
Nieuwenhof, Monique van den 575
Niraula, Ratna P. 679
Nithiarasu, P. 555
Nuttall, Keith I. 125

O

Obreja, Vasile V.N. 125
Oh, Joonyoung 428
Ohira, Hiroshi 270, 278
Ohkuma, Hideo 270, 278
Okoro, Chukwudi 491
Oprins, H. 738
Orain, Stéphane 109, 150
O'Shea, Thomas 484

P

Pallarès, Jofre 332
Palmer, Ben 601
Pang, H.L.J. 416
Park, Heung-Woo 234
Parrain, F. 571
Parry, Dr. John 737
Pasion, J. 46
Pecht, Michael 729
Peerlings, R.H.J. 454
Pennec, F. 637
PENNEC, Fabienne 588
Pérez-Castillejos, R 478
Perkins, A. 678
Perpinyà, X. 403
Persaud, Krishna C. 745
Peyrou, D. 637
PEYROU, David 588
Pintado, P. 583
Plana, R. 637
PLANA, Robert 588
Plaza, J. A. 478
Plouseau, D. 77
Polster, Tobias 560
Pons, P. 637
PONS, Patrick 588
Preu, H. 319
Pucha, R. V. 678

Q

Qian, C. 319
Qian, Richard 220
Qiang, Wang 225
Quesada, José Hermida 461

R

Raynal, P. 77
Real, R. A. 46

Author Index

Rebholz, Christian 447
Reichl, H. 7
Requena, Francisco Caballero 461
Rixen, Daniel 69
Rochus, Veronique 69
Rodríguez, Gustavo Adolfo Ardila 167
Roellig, M. 189
Roellig, Mike 197, 207
Rogiers, F. 738
Röllig, M. 618
Röllig, Mike 93
Rongen, R.T.H. 423
Rossi, Carole 167
Ryan, Jeffrey 484
Rzepka, S. 618
Rzepka, Sven 295, 357

S

Sabariego, Ruth V. 120
Salleras, M. 549, 565
Samimi, M. 454
Samitier, J. 549
Santander, J. 565
SARTOR, Marc 588
Schmadlak, Ilko 131
Scholz, D. 618
Schreier-Alt, Thomas 447
Schreurs, P.J.G. 27
Shimokawa, Tomotsugu 641
Shin, Dongkil 428
Silberschmidt, Vadim. V. 385
Silfhout, R. B. R. van 214, 468
Silva, Alexandre Cesar Rodrigues da 484
Sitaraman, S. K. 678
Sluis, O. van der 85, 214, 372, 468, 472
Sluis, Olaf van der 109
Sneath,, Robert W. 745
Soestbergen, M. van 423
Song, Younghee 428
Stecher, M. 319
Steijvers, Henk 575
Stevens, T. 738
Stoyanov, S. 159, 593
Strusevich, N. 159
Stulemeijer, Jiri 503
Sun, Anthony 283
Sun, Anthony Y.S. 366, 416
Sun, F. L. 607
Sun, Wei 283, 366, 416
Suzuki, Ken 686
Swinnen, Bart 491

T

Takami, Kourosh Mousavi 1
Tamakawa, Kinji 686
Tan, H.B. 283, 366, 416

Terés, Lluís 332
Terré, J.Casals- 478
Tilford, Tim 243, 647
Traon, O. Le 571
Tuchband, Brian 729
Tucker, Paul G. 39
Tunga, K. 678
Tyacke, James C. 39

U

Ubachs, R.L.J.M. 85
UCHIBORI, Chihiro J. 442
Udina, S. 565

V

Vallés, E. 478
Vandepitte, Dirk 491
Vandevelde, Bart 340, 491
Vanfleteren, Jan 340
Velandia, Diana Segura 308
Vellvehi, M. 403
Verheyen, Peter 491
Vichare, Nikhil 729
Villard, S. 77
Vries, J. de 435

W

Waal, Adri van der 575
Walter, H. 7
Wang, C.K. 283, 366, 416
WANG, Fuliang 23
Wang, L. 361
Wang, L. F. 607
Wang, L. G. 372
Wang, Shinan 710
Wang, Song-Hao 314
Wang, Wenjie 56
Wang, Z. G. 723
Wei, Hsiu-Ping 324
WeiNa, Hao 16
West, Andrew A. 308
Whalley, David 385
Whalley, David C. 308
Whitehead, Michael 653
Wiese, S. 189, 618
Wiese, Steffen 93, 197, 207
Wilde, Jürgen 611
Wilson, Antony 308
Wittler, Olaf 378
Wolter, K. J. 189, 618
Wolter, Klaus-Juergen 197, 207
Wolter, Klaus-Jürgen 93
Wouters, Dirk J. 136
Wunderle, B. 7
Wymyslowski, Artur 263, 346

Author Index

X

Xiang, Ke 56
Xiao, A. 372
Xiaohong, Weng 225
Xue, Xiangdong 243
Xuefan, Chen 225

Y

Yang, D. G. 46, 372, 498
Yang, Keling 56
Yang, Wen-Kung 324
Yannou, J. M. 159
Ye, Yuming 631
Yellowaga, Deborah 601
Yew, Ming-Chih 324
Yin, C. 593
Ying, Qu Jian 729
Yong, Liu 16
Yu, J.-Hyuk 234
Yuan, C. A. 532
Yuan, Cadmus 472
Yuen, Matthew M.F. 515
Yun, Sang-Kyeong 234

Z

Zbrzezny, Adam Robert 51
Zerbini, Sarah 303, 392
Zhang, G. P. 723
Zhang, G.Q. 46, 85, 214, 361, 372, 423, 472, 532
Zhang, Kai 515
Zhang, Xuefeng 442, 664
Zhao, B.Z. 283
ZHONG, Jue 23
Zhou, Jiang 679
Zhou, Jiemin 56
Zhou, Ming 51
Zhu, W. H. 283, 366, 416
Zhu, X. F. 723
Zoumpoulidis, T. 361
Zukowski, Elena 611

9781424411054

2007 International Conference on Thermal, Mechanical & Multi-Physics Simulation and Experiments in Microelectronics & Micro System

London, United Kingdom
15-18 April 2007

IEEE Catalog Number: CFP07566-POD
ISBN: 978-1-42441-105-4

2007 International Conference on Thermal, Mechanical & Multi-Physics Simulation and Experiments in Microelectronics & Micro System

London, United Kingdom
15-18 April 2007

Volume 2 of 2

IEEE Catalog Number: 07EX1736
ISBN: 1-4244-1105-X

Copyright © 2007 by The Institute of Electrical and Electronics Engineers, Inc.
All Rights Reserved

Copyright and Reprint Permissions: Abstracting is permitted with credit to the source. Libraries are permitted to photocopy beyond the limit of U.S. copyright law for private use of patrons those articles in this volume that carry a code at the bottom of the first page, provided the per-copy fee indicated in the code is paid through Copyright Clearance Center, 222 Rosewood Drive, Danvers, MA 01923.

For other copying, reprint or republications permission, write to IEEE Copyrights Manager, IEEE Operations Center, 445 Hoes Lane, Piscataway, New Jersey USA 08854. All rights reserved.

IEEE Catalog Number:	07EX1736
ISBN:	1-4244-1105-X
LOC:	2007922208

Additional Copies of This Publication Are Available from:

IEEE Service Center
445 Hoes Lane
Piscataway, NJ 08854
IEEE Service Center
445 Hoes Lane
Piscataway, NJ 08854
Phone: (800) 678-IEEE
                 (732) 981-1393
Fax: (732) 981-9667
E-mail: customer-service@ieee.org

Table of Contents

Thermal and hot spot evaluations on oil immersed power Transformers by FEMLAB and MATLAB software's..1
Kourosh Mousavi Takami, Hasan Gholnejad, Jafar Mahmoudi

Molecular Dynamics Simulation for the diffusion of water in amorphous polymers examined at different Temperatures...7
E. Dermitzaki, J. Bauer, H. Walter, B. Wunderle, B. Michel, H. Reichl

Mechanical Characterization and Viscoplastic-Damage Constitutive Model of SnAgCu Solder................16
Zhou Jun, Chai GuoZhong, Liu Yong, Liang LiHua, Hao WeiNa

Atom diffusion mechanism of thermo-sonic flip chip bonding interface..23
Fuliang WANG, Junhui LI, Lei HAN, Jue ZHONG

A coupled numerical and experimental study on thermo-mechanical fatigue failure in SnAgCu solder joints...27
M. Erinc, P.J.G. Schreurs, M.G.D. Geers

Magnetic Field Sensor Using a Physical Model to Pre-Calculate the Magnetic Field and to Remove Systematic Error due to Physical Parameters..33
Simon Hainz, Mario Jungwirth

Unsteady CFD Modelling for Electronics Cooling..39
James C. Tyacke, Paul G. Tucker

Combined Virtual Prototyping and Reliability Testing Based Design Rules for Stacked Die System in Packages...46
W.D. van Driel, R. A. Real, D.G. Yang, G.Q. Zhang, J. Pasion

Evaluation of Thermal Strains in BGA Packages Using Digital Speckle Correlation Technique and FEA........51
Adam Robert Zbrzezny, Vincent Chan, Hua Lu, Ming Zhou

Quantification of creep strain in small lead-free solder joints with the in-situ micro electronic-resistance measurement...56
Li Jiang, Keling Yang, Jiemin Zhou, Ke Xiang, Wenjie Wang

Computational Design and Optimisation of Mechanically Reinforced Masks for Stencil Lithography............62
Maryna Lishchynska, Marc A.F. van den Boogaart, Juergen Brugger, James C. Greer

Extended Finite Element for Electromechanical Coupling..69
Veronique Rochus, Daniel Rixen

Finite Element Modelling (FEM) of Green Electronics in Aeronautical and Military Communication Systems (GEAMCOS)...77
A. Chaillot, G. Massiot, C. Munier, I. Lombaërt-Valot, S. Bousquet, C. Chastanet, D. Plouseau, E. Munier, D. Maron, P. Raynal, S. Villard, R. Dumonteil

Thermomechanical Multiscale Modelling of Substrates..85
R.L.J.M. Ubachs, O. van der Sluis, W.D. van Driel, G.Q. Zhang

Creep Measurements of 200 μm - 400 μm Solder Joints...93
Mike Röllig, Steffen Wiese, Karsten Meier, Klaus-Jürgen Wolter

Finite Element Modeling of Thermoelastic Damping in Filleted Micro-Beams..102
Severine Lepage, Jean-Claude Golinval

Numerical Analysis of the Reliability of Cu/low-k Bond Pad Interconnections Under Wire Pull Test: Application of a 3D Energy Based Failure Criterion..109
Sébastien Gallois-Garreignot, Vincent Fiori, Stéphane Orain, Olaf van der Sluis

Thermal Simulation with Coupled Network Models on System Level..116
Adam Augustin, Bartosz Maj

A PERTURBATION METHOD FOR THE 3D FINITE ELEMENT MODELING OF ELECTROSTATICALLY DRIVEN MEMS...120
Mohamed Boutaayamou, Ruth V. Sabariego, Patrick Dular

Table of Contents

Failure Analysis of Power Silicon Devices at Operation above 200 °C Junction Temperature 125
Vasile V.N. Obreja, Keith I. Nuttall, Octavian Buiu, Steve Hall

Simulation of Wafer Probing Process Considering Probe Needle Dynamics 131
Ilko Schmadlak, Torsten Hauck

Modeling of the mechanical behavior during programming of a non-volatile phase-change memory cell using a coupled electrical-thermal-mechanical finite-element simulator 136
Thomas Gille, Judit Lisoni, Ludovic Goux, Kristin De Meyer, Dirk J. Wouters

Surface Evolution of Strained Thin Solid Films: Stability Analysis and Time Evolution of Local Surface Perturbations 142
E. Dornel, J-C Barbé, J. Eymery, F. de Crécy

Gold Wire Bonding Induced Peeling in Cu/Low-k Interconnects: 3D Simulation and Correlations 150
Vincent Fiori, Lau Teck Beng, Susan Downey, Sebastien Gallois-Garreignotd, Stephane Orain

Reliability Based Design Optimisation for System-in-Package 159
S. Stoyanov, J. M. Yannou, C. Bailey, N. Strusevich

Multiphysic modelling of a microactuator based on the decomposition of an energetic material: application to microfluidics 167
Gustavo Adolfo Ardila Rodríguez, Carole Rossi

Crack Degradation Model Derived From Experimental Strain-Stress Data 175
Milos Dusek, Christopher Hunt

CFD Aided Reflow Oven Profiling for PCB Preheating in a Soldering Process 181
Ilja Belov, Mats Lindgren, Peter Leisner, Fredrik Bergner, Robin Bornoff

The Effect of Downscaling the Dimensions of Solder Interconnects on their Creep Properties 189
S. Wiese, M. Roellig, M. Mueller, K.-J. Wolter

The Dependence of Composition, Cooling Rate and Size on the Solidification Behaviour of SnAgCu Solders 197
Maik Mueller, Steffen Wiese, Mike Roellig, Klaus-Juergen Wolter

Electromigration in Large Volume Solder Joints 207
Karsten Meier, Mike Roellig, Steffen Wiese, Carsten Goette, Ulrich Deml, Klaus-Juergen Wolter

Multi-scale energy-based failure modeling of bond pad structures 214
O. van der Sluis, R.B.R. van Silfhout, R.A.B. Engelen, W.D. van Driel, G.Q. Zhang

Impact of Solder Overflow and ACLV Moisture Absorption of Mold Compound on Package Reliability 220
Richard Qian, Yong Liu, Scott Irving, Timwah Luk

Experimental Determination and Modification of Anand Model Constants for Pb-Free Material 95.5Sn4.0Ag0.5Cu 225
Wang Qiang, Liang Lihua, Chen Xuefan, Weng Xiaohong

Numerical Simulation and Thermal Failure Analysis of SOM Package 234
Jae Choon Kima, Jin Taek Chunga, Won Suk Lee, Gyoung Bum Kimb, Dong Jin Lee, Ji-Man Cho, J.-Hyuk Yu, Byeong-Kwon Ju, Sung Woo Hang, Heung-Woo Park, Sang-Kyeong Yun

A Mathematical Technique to estimate the High Frequency Current Inside the Silicon Die from the Noise Measurements 239
Bidyut K. Bhattacharyya, Gang Huo

Thermal-mechanical Modelling of Power Electronic Module Packaging 243
Hua Lu, Tim Tilford, Xiangdong Xue, Chris Bailey

Multi Physics Modelling of the Electrodeposition Process 249
M. Hughes, C. Bailey, K. McManus

Application of Higher Order Derivatives Method to Parametric Simulation of MEMS 257
Vladimir Kolchuzhin, Jan Mehner, Thomas Gessner, Wolfram Doetzel

Table of Contents

Multi-objective Parametric Approach to Numerical Optimization of Stacked Packages 263
Lukasz Dowhan, Artur Wymyslowski, Rainer Dudek

Research of Stacked VIA's Mechanical Stress 270
Tohru Nakanishi, Hideo Ohkuma, Hiroshi Ohira

Research of Material Properties Reliance for POP with Numerical Analysis 278
Tohru Nakanishi, Hiroshi Ohira, Hideo Ohkuma

Cure shrinkage characterization and its implementation into correlation of warpage between simulation and measurement 283
W.H. Zhu, Guang Li, Wei Sun, F.X. Che, Anthony Sun, C.K. Wang, H.B. Tan, B.Z. Zhao, N.H. Chin

Optimizing the Dynamic Response of RF MEMS Switches using Tailored Voltage Pulses 291
Vitaly Leus, Arnon Hirshberg, David Elata

The Effect of Visco-elasticity on the Result Accuracy of FEM Panel Warpage Simulations Supporting Industrial Microelectronics Packaging 295
Sven Rzepka, Axel Müller

Parametric Study of Fracture Properties in Polycrystalline MEMS 303
Fabrizio Cacchione, Alberto Corigliano, Sarah Zerbini

A Modelling Framework for the Reliability of Safety Critical Electronics 308
Diana Segura Velandia, Paul P. Conway, Antony Wilson, Andrew A. West, David C. Whalley

Development of an Internal Liquid Cooling System for CPU using CAE 314
Sebastian Flores, Song-Hao Wang, Chong-Qing Chang

Kinetic Characterisation of Molding Compounds 319
K. M. B. Jansen, C. Qian, L. J. Ernst, C. Bohm, A. Kessier, H. Preu, M. Stecher

A Study of Failure Mechanism and Reliability Assessment for the Panel Level Package (PLP) Technology 324
Ming-Chih Yew, Hsiu-Ping Wei, Ching-Shun Huang, Dyi-Chung Hu, Wen-Kung Yang, Kou-Ning Chiang

A Novel Micro/Nano Electromechanical Actuator with Integral Electrodes 332
Jofre Pallarès, Manuel Carmona, Marta Duch, Marta Gerbolés, Lluís Terés

Design of Metal Interconnects for Stretchable Electronic Circuits using Finite Element Analysis 340
Mario Gonzalez, Fabrice Axisa, Mathieu Vanden Bulcke, Dominique Brosteaux, Bart Vandevelde, Jan Vanfleteren

Numerical Modeling of Electrical Resistance of Interconnections in High-Tech Multilayer PCBs Manufactured by Magnetron Sputtering Deposition of Copper 346
Janusz Borecki, Artur Wymyslowski

Simulation of Failure Criteria in Dielectric Layers 352
Khalil Arshak, Ivor Guiney, Edward Forde

Dynamic mechanical behavior of SnAgCu BGA solder joints determined by fast shear tests and FEM simulations 357
Eberhard Kaulfersch, Sven Rzepka, Vijay Ganeshan, Axel Müller, Bernd Michel

Mechanical Characterization Analysis of flexible and stretchable Ultra-Thin Substrates by Experiment and FE Simulation 361
L. Wang, T. Zoumpoulidis, K.M.B. Jansen, M. Bartek, G.Q. Zhang, L.J. Ernst

Study on the Board-level SMT Assembly and Solder Joint Reliability of Different QFN Packages 366
Wei Sun, W.H. Zhu, Retuta Danny, F.X. Che, C.K. Wang, Anthony Y.S. Sun, H.B. Tan

Thin Film Interface Fracture Properties at Scales Relevant to Microelectronics 372
A. Xiao, L. G. Wang, W. D. van Driel, O. van der Sluis, D. G. Yang, L. J. Ernst, G. Q. Zhang

Design for Reliability with AuSn Interconnects 378
Rainer Dudek, Olaf Wittler, Wolfgang Faust, Birgit Brämer, Matthias Klein, Wei Jun, Bernd Michel

Table of Contents

Creep Analysis of a Lead-free Surface Mount Device 385
Pradeep Hegde, David Whalley, Vadim. V. Silberschmidt

Multi-Scale Modeling of Shock-Induced Failure of Polysilicon MEMS 392
Aldo Ghisi, Fabio Fachin, Stefano Mariani, Alberto Corigliano, Sarah Zerbini

Packaging Effects of Cu/Low-k Interconnect Structure 398
Ming-Che Hsieh

Validation of Dynamic Thermal Simulations of Power Assemblies Using a Thermal Test Chip 403
X. Jordà, M. Vellvehi, X. Perpinyà, J.L. Galvez, P. Godignon

Thermo-mechanical investigations on the effects of the solder meniscus design in solder joint lifetime for power electronic devices 409
M. Bouarroudj, Z. Khatir, S. Lefebvre, L. Dupont

Development and Assessment of Global-Local Modeling Technique Used in Advanced Microelectronic Packaging 416
F. X. Che, H.L.J. Pang, W. H. Zhu, Wei Sun, Anthony Y.S. Sun, C.K. Wang, H.B. Tan

Transport of Corrosive Constituents in Epoxy Moulding Compounds 423
M. van Soestbergen, L.J. Ernst, G.Q. Zhang, R.T.H. Rongen

Development of Reliability Verification System for Robust Package Design 428
Dongkil Shin, Hyunggil Baek, Joonyoung Oh, Dongok Kwak, Kuyoung Kim, Younghee Song, Jeongyeol Kim

The effect of board stiffness on the solder-joint reliability of HVQFN-packages 435
J. de Vries, M. Jansen, W. van Driel

Investigation of Mechanical Reliability of Cu/low-k Multi-layer Interconnects in Flip Chip Packages 442
Chihiro J. UCHIBORI, Xuefeng Zhang, Sehyuk Im, Paul S. Ho, Tomoji Nakamura

Packaging Design and Testing for High Temperature Applications > 150 °C 447
Thomas Schreier-Alt, Christian Rebholz, Frank Ansorge

An enriched cohesive zone model for numerical simulation of interfacial delamination in microsystems 454
M. Samimi, B.A.E. van Hal, R.H.J. Peerlings, J.A.W. van Dommelen, M.G.D. Geers

Underexpanded Micro-nozzle Flow Simulation with Coupled Thermal-Fluid Modeling 461
José Hermida Quesada, José A. Moriñigo, Francisco Caballero Requena

Optimization of Cu Low-k bond pad designs to improve mechanical robustness using the Area Release Energy method 468
R. A. B. Engelen, O. van der Sluis, R. B. R. van Silfhout, W.D. van Driel, V. Fiori

The chemical-mechanical relationship of the SiOC(H) dielectric film 472
Cadmus Yuan, O. van der Sluis, G. Q. Zhang, L. J. Ernst

Magnetically actuated microvalve for disposable drug infusor 478
M. Duch, J.Casals-Terré, J. A. Plaza, J. Esteve, R Pérez-Castillejos, E. Vallés, E. Gómez

Generating VHDL3AMS Models of Digital-to-Analogue Converters From MATLAB®/SIMULINK® 484
Alexandre Cesar Rodrigues da Silva, Ian Grout, Jeffrey Ryan, Thomas O'Shea

Prediction of the Influence of Induced Stresses in Silicon on CMOS Performance in a Cu-Through-Via Interconnect Technology 491
Chukwudi Okoro, Mario Gonzalez, Bart Vandevelde, Bart Swinnen, Geert Eneman, Peter Verheyen, Eric Beyne, Dirk Vandepitte

Effect of Processing Parameters and Hygro-thermo-mechanical Stresses on the Reliability of Flip Chip Bonding RFID Tags 498
D.G. Yang, E. de Bruin, B. Kasemset, W. D. van Driel

Efficient electrostatic-mechanical modeling of C-V curves of RF-MEMS switches 503
Jeroen Bielen, Jiri Stulemeijer

vi

Table of Contents

Evaluation of Creep in RF MEMS Devices 509
Marcel van Gils, Jeroen Bielen, Gavin McDonald

Investigation of Carbon Nanotube Performance under External Mechanical Stresses and Moisture 515
Haibo Fan, Kai Zhang, Matthew M.F. Yuen

Wafer Probing Simulation for Copper Bond Pad Based BPOA Structure 519
Yumin Liu, Yong Liu, Scott Irving and Timwah Luk

Optimization and Robust Design of Electronics Assemblies under Fracture, Delamination and Fatigue Aspects 524
Jürgen Auersperg, Matthias Klein, Bernd Michel

Mechanical Characterization of III-V Nanowire Using Molecular Dynamics Simulation 532
Alex. W. Dawotola, C. A. Yuan, W. D. van Driel, E. P. A. M. Bakkers, G. Q. Zhang

Stress Relaxation Characterization of Hypoeutectic Sn3.0Ag0.5Cu Pb-free Solder: Experiment and Modeling 537
Gayatri Cuddalorepatta, Dominik Herkommer, Abhijit Dasgupta

Effect of Surrounding Air on Board Level Drop Tests of Flexible Printed Circuit Boards 545
Luciano Arruda, Germano Freitas

Numerical Simulation of Ion Drift within Ion Mobility Spectrometers in High Peclet Conditions using FEM Techniques 549
A. Kalms, M. Salleras, Z. Liu, J.G. Korvink, J. Goebel, G. Müller, J. Samitier, S. Marco

Electro-osmotic Flow Based Cooling System For Microprocessors 555
P. F. Eng, P. Nithiarasu, A. K. Arnold, P. Igic, O. J. Guy

Parameter Identification on Wafer Level of Membrane Structures 560
Steffen Michael, Siegfried Hering, Gisbert Hölzer, Tobias Polster, Martin Hoffmann, Arne Albrecht

A micromachined thermoelectric sensor for natural gas analysis: Thermal model and experimental results 565
G. Carles, S. Udina, M. Salleras, J. Santander, L. Fonseca, S. Marco

Behavioral Modelling of Vibrating Piezoelectric Micro-Gyro Sensor and Detection Electronics 571
S. Megherbi, R. Levy, F. Parrain, H. Mathias, O. Le Traon, D. Janiaud, J. P. Gilles

Validation of constitutive models for electrically conductive adhesives 575
Marcel Meuwissen, Monique van den Nieuwenhof, Henk Steijvers, Adri van der Waal, Tom Bots

Simultaneous Measurement of Young's Modulus and Damping Dependence on Magnetic Fields by Laser Interferometry 583
A. L. Morales, A. J. Nieto, R. Moreno, A. González, J. M. Chicharro, P. Pintado

Validation Of Simulation Platform By Comparing Results And Calculation Time Of Different Softwares. 588
Hikmat ACHKAR, Fabienne PENNEC, David PEYROU, Mahmoud AL AHMAD, Marc SARTOR, Robert PLANA, Patrick PONS

Multi-Physics Modelling for Microelectronics and Microsystems - Current Capabilities and Future Challenges 593
C. Bailey, H. Lu, S. Stoyanov, M. Hughes, C. Yin, D. Gwyer

USING MOLECULAR MODELING TO UNDERSTAND CLEANER EFFICIENCY FOR BARC ("BOTTOM ANTI-REFLECTIVE COATING") AFTER PLASMA ETCH IN DUAL DAMASCENE STRUCTURES 601
Nancy Iwamoto, Deborah Yellowaga, Amy Larson, Ben Palmer, Teri Baldwin-Hendricks

Numerical Simulation of Creep Strain of PBGA Solders under Thermal Cycling 607
F. L. Sun, Y. Liu, L. F. Wang

Comparative Sensitivity Analysis for μBGA and QFN Reliability 611
Jürgen Wilde, Elena Zukowski

Table of Contents

Experimental Determination of Time-Independent Elastic-Plastic Behaviour of Solder Joints at High Strain Rates 618
S. Wiese, K. Meier, D. Scholz, A. Müller, M. Röllig, S. Rzepka, K. J. Wolter

The BGK kinetic model applied to the analysis of gas-structure interactions in MEMS 623
Attilio Frangi, Aldo Ghisi

Development of 2D Modeling Techniques for the Thermal Fatigue Analysis of Solder Joints of a Module Mounted in a 3D Cavity on a Printed Circuit Board 631
Y. S. Chan, S. W. Ricky Lee, Yuming Ye, Sang Liu

A Macro Model Based On Finite Element Method To Investigate Temperature And Residual Stress Effects On RF MEMS Switch Actuation 637
D. Peyrou, H. Achkar, F. Pennec, P. Pons, R. Plana

Atomistic Simulations of Interface Properties in Metals 641
Tomotsugu Shimokawa

Wire Bond Reliability for Power Electronic Modules - Effect of Bonding Temperature 647
Wei-Sun Loh, Martin Corfield, Hua Lu, Simon Hogg, Tim Tilford, C Mark Johnson

Reduced Order Electro-Thermal Models for Real-Time Health Management of Power Electronics 653
Mahera Musallam, Cyril Buttay, C Mark Johnson, Chris Bailey, Michael Whitehead

Reliability Test Method Overiew to Characterize Second Level Interconnects 659
M. Brizoux, H. Fremont, Y. Danto, W. C. Maia Filho

Kinetic Analysis of Electromigration Enhanced Intermetallic Growth and Void Formation in Pb-Free Solders 664
Brook Chao, Seung-Hyun Chae, Xuefeng Zhang, Paul S. Ho

Measurement of Dynamic Properties of MEMS and the Possibilities of Parameter Identification by Simulation 672
Matthias Ebert, Falk Naumann, Ronny Gerbach, Joerg Bagdahn

DESIGN-FOR-RELIABILITY TOOLS FOR HIGHLY INTEGRATED SYSTEM-ONPACKAGE TECHNOLOGY 678
R. V. Pucha, S. Hegde, M. Damani, K. J. Leee, K. Tunga, A. Perkins, S. Mahalingam, G. Lo, K. Klein, J. Ahmad, S. K. Sitaraman

Block-Diagram Based SIMULINK Analysis for the Drop Impact Response of a Mobile Electronic System 679
Jiang Zhou, Paul Corder, Ratna P. Niraula

Fluctuation Mechanism of Mechanical Properties of Electroplated-Copper Thin Films Used for Three Dimensional Electronic Modules 686
Hideo Miura, Ken Suzuki, Kinji Tamakawa

Computational Methods and High Speed Imaging Methodologies for Transient-Shock Reliability of Electronics 692
Pradeep Lall

Low-temperature and Pressureless Sintering Technology for High-performance and High-temperature Interconnection of Semiconductor Devices 705
Guo-Quan Lu, Jesus N. Calata, Guangyin Lei, Xu Chen

3D Modeling of Electromigration Combined with Thermal-Mechanical Effect for IC Device and Package 710
Yong Liu, Scott Irving, Timwah Luk, Lihua Liang, Shinan Wang

Fatigue Strength and Damage Behaviors of Multi-Scale Metallic Films and Multilayers 723
G. P. Zhang, X. F. Zhu, Y. P. Li, Z. G. Wang

Prognostics and Health Monitoring of Electronics 729
Michael Pecht, Brian Tuchband, Nikhil Vichare, Qu Jian Ying

The Changing Role of CFD in Electronics Thermal Design 737
Dr. John Parry

viii

Table of Contents

Digital and Continuous Liquid Cooling for Electronic Systems..738
M. Baelmans, H. Oprins, T. Stevens, F. Rogiers

Solid State Chemical Sensors: Technologies and Applications..745
Krishna C. Persaud, Anthony Flint, Robert W. Sneath,

2007 International Conference on Thermal, Mechanical & Multi-Physics Simulation and Experiments in Microelectronics & Micro System

Volume 2 of 2

Multi-Scale Modeling of Shock-Induced Failure of Polysilicon MEMS

Aldo Ghisi*, Fabio Fachin*, Stefano Mariani*, Alberto Corigliano*, Sarah Zerbini **

* Department of Structural Engineering, Politecnico di Milano. Piazza L. da Vinci, 32. 20133 Milano, Italy
** MEMS Product Division, STMicroelectronics. Via Tolomeo 1. 20010 Cornaredo, (Milano), Italy

Abstract

We investigate the shock-induced stress state and possible failure mechanisms in polysilicon MEMS sensors. In case of accidental drop events, we aim at highlighting the links between drop features, like drop height and impact angles, and the location of the failing detail of the device. Taking into account the small inertial contribution of the sensor to the whole package dynamics, we adopt a decoupled multi-scale numerical approach, where macro-scale analyses (at die length-scale) are used to define the acceleration records to be adopted as loading in meso-scale (at sensor length-scale) simulations.

We show that a commonly adopted indicator to assess drop severity is not able to take in due account the details of the impact event and possible very localized failures at the sensor level.

1. Introduction

Inertial MEMS are often designed to work in mobile devices, and are therefore subject during their life to bumpy handling and accidental drops. In such situations, shock waves impinging upon the sensors can be a major cause of failure.

Some recent researches have attacked this problem, showing links between the drop features and the mechanical response of the packaged device (see, e.g., [1-3]). However, they typically rely on simplified models of the contact conditions between the falling package and the impacted surface (termed target surface henceforth) and of the propagation of shock waves in the package/sensor system. While this approach can quickly furnish results upon calibration, accuracy can get lost in case of very localized failures at sensor level.

In this work we consider a uni-axial polysilicon MEMS accelerometer supported by a naked die and subject to an accidental drop. Within the frame of a finite element approach, we investigate the links between drop height and impact angles (here defined as the Euler angles between the target surface and the bottom surface of the die) and the stress state induced in the sensor. This approach can be straightforwardly extended to fully packaged sensors by simply changing the geometry of the falling body.

Since the ratio between the inertia of the sensor and that of the die-cap assembly is very small, the dynamics of the whole device after the impact is marginally affected by the presence of the MEMS; therefore, we adopt a decoupled multi-scale approach. In the macro-scale analyses, the whole device is modelled while falling and bouncing off the target surface; the effects of the geometry of the device, of the mechanical properties of the device and the target surface, of the drop height and of the impact angles on the acceleration history at the sensor anchor can then be established. In the meso-scale analyses, the formerly obtained acceleration records are adopted as ground motion histories in dynamic simulations at sensor level; possible critical regions, where the stress state exceeds a pre-defined bound on the polysilicon carrying capacity, can thus be identified. In these regions micro-scale analyses accounting for the actual crystal structure of the polysilicon should be adopted, in order to get insights into the details of the failure mechanism, typically linked to intergranular/transgranular dynamic crack growth. This last stage of our approach has been already followed, e.g., in [4-6].

2. Analytical evaluation of the acceleration peak in the post-impact dynamics of die

After a friction-less impact against a flat surface, the falling die repeatedly bounces. An analytical evaluation of the acceleration felt by the whole die can be furnished by Hertz theory. The die is approximated as a compact (spherical-like) body with characteristic dimension R, made of an isotropic elastic material featuring Young's modulus E_d and Poisson's ratio ν_d, whereas the target surface is assumed perfectly flat and made of an isotropic elastic material featuring Young's modulus E_t and Poisson's ratio ν_t. An estimate $\bar{a}$ of the peak acceleration is given by [7]:

$$\bar{a} = \sqrt[5]{\frac{v_{imp}^6 R}{\left[m \left(\frac{1-\nu_t^2}{E_t} + \frac{1-\nu_d^2}{E_d} \right) \right]^2}} \tag{1}$$

where: v_{imp} is the impact velocity; m is the mass of the body.

As for the device here studied, the reference acceleration peak furnished by (1) turns out to be on the order of $\bar{a} = 10^5$ g, being g the gravity acceleration.

3. Analysis of the stress state in a falling polysilicon MEMS

In this Section we discuss the outcomes of the proposed multi-scale finite element approach.

As already mentioned in the Introduction, we do not account for coupling among the different length-scales. Decoupling is allowed in this case by the very small ratio between the weights of the sensor and of the whole die-cap assembly, and by the assumed elastic behavior of the silicon at the macro- and meso-scale. Hence, results are not expected to furnish a detailed description of the possible failure mechanism of the sensor; we are instead interested in localizing possible sites where the stress state exceeds a critical bound on the actual material strength.

3.1 Macro-scale analyses (at die level)

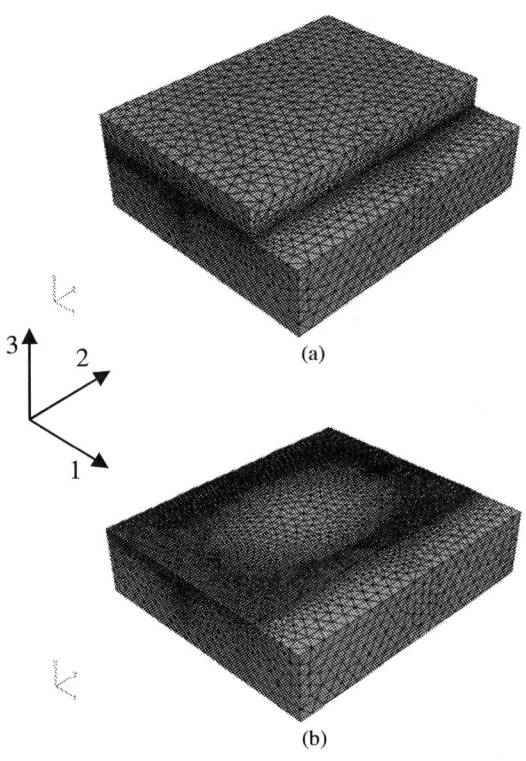

Figure 1: macro-scale analyses (die-cap length scale). View (a) of the whole die-cap mesh and (b) of the die mesh only

At this length-scale we are interested in the propagation of shock-waves in the bulk of the falling die-cap assembly. Therefore, three-dimensional explicit dynamic simulations are run to model the response of the whole body after the impact.

The geometry of the modelled device is shown in Figure 1: a space discretization consisting of about 145,000 tetrahedral, 4-node linear elements and about 30,000 nodes has been adopted. To obtain a high resolution in the proximity of the anchor point, smaller elements with a characteristic size $\ell = 30$ μm have been used on the top surface of the die. The cap has been also discretized, even though it does not appear necessary for impacts on the bottom surface of the die. In fact, in case of impacts on the top surface of the cap, shock-waves propagates after the collision in the direction opposite to the global axis X_3 in Figure 1.

Both die and cap are made of single-crystal silicon, featuring Young's modulus $E_s = 130$ GPa, Poisson's ratio $\nu_s = 0.22$ and mass density $\rho_s = 2330$ Kg/m^3.

Since analyses are aimed at modeling possible drops while handling, two drop heights have been considered: $h_1 = 50$ cm and $h_2 = 150$ cm. The outcomes obtained with the latter one, which obviously induces higher stress peaks in the sensor, are detailed in what follows.

Furthermore, since it is unlikely that the die impinges the target surface with its bottom surface being perfectly horizontal, the effect of tilting has been also addressed. Because of the asymmetric position of the cap with respect to the die center of gravity (see Figure 1), it is expected that while falling the die-cap assembly rotates around axis X_2. Hence, simulations of collisions with the target surface have been run at varying angle of impact about axis X_2, whereas the angle about axis X_1 has been maintained fixed and null.

In Figures 2-6, lateral views of the device bouncing after the impact with a rigid target surface are shown. Because of the asymmetric geometry of the die-cap assembly, rigid body rotations show up in late stage of device response, that is for time $t > 50$ μs.

In cases of small impact angles (within the range 5°-20°), the die experiences repeated collisions with the target surface at opposite edges of the bottom die surface. Possible interactions between the shock waves emanating from the impact locations can take place inside the die; we then conducted all the analyses up to $t_{end} = 100$ μs, which seems a reasonable bound on the time interval to be scanned to detect peak stress states in the sensor at the meso-scale.

During the analyses, the vectorial acceleration at the sensor anchor is continuously monitored. Because rigid body-like motions of the assembly show up for $t > 50$ μs, also the rotational accelerations at the sensor anchor are stored. Rotations are obtained in the analyses by checking the displacements in points neighboring to the anchor location.

Figures 7 and 8 show some acceleration records in the local sensing direction z just after the impact, namely for $0 < t < 2$ μs. In these plots, results are normalized with respect to the analytic estimation furnished by Equation (1), i.e.

$$\alpha_z = \frac{a_z}{\bar{a}} \qquad (2)$$

Due to the multiple reflections of shock-waves at the free surfaces of the body, even in this very small period of time the acceleration histories show sudden variations, characterized by a large amount of peaks.

In Figure 7 it can be noticed that, in case of perfectly flat impacts (impact angle either 0° or 180°) the estimate $\bar{a}$ is not able to account for the actual dynamics of the propagating waves. In fact, while in the top case the acceleration never approaches $\bar{a}$, in the bottom case the value $\bar{a}$ is crossed once at $t \approx 0.1$ μs. On the other hand, Figure 8 shows that tilting affects the acceleration records by strongly reducing peak values.

Figure 2: drop height 150 cm; impact angle 0°.
Snapshots of the bouncing die, taken every 20 μs after the impact event

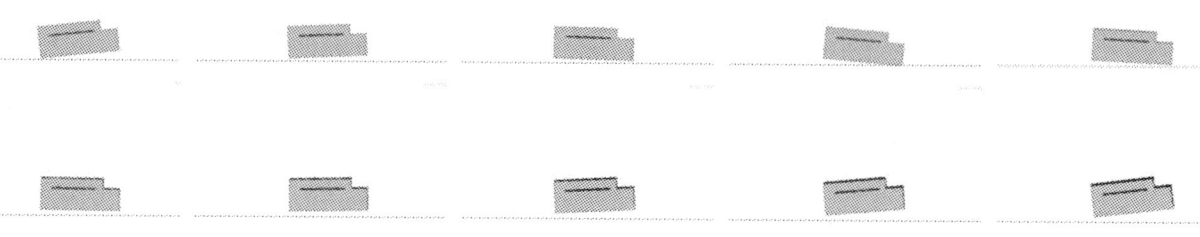

Figure 3: drop height 150 cm; impact angle 10°.
Snapshots of the bouncing die, taken every 10 μs after the impact event

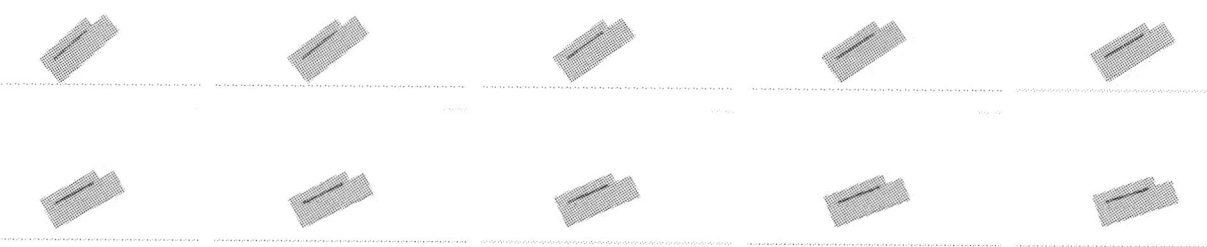

Figure 4: drop height 150 cm; impact angle 45°.
Snapshots of the bouncing die, taken every 10 μs after the impact event

Figure 5: drop height 150 cm; impact angle 90°.
Snapshots of the bouncing die, taken every 20 μs after the impact event

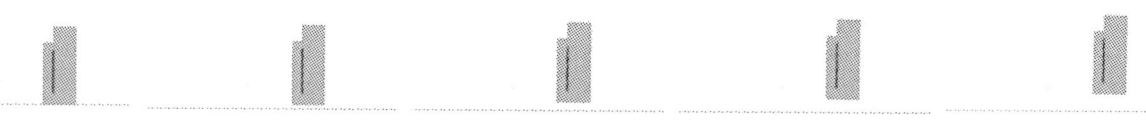

Figure 6: drop height 150 cm; impact angle 180°.
Snapshots of the bouncing die, taken every 20 μs after the impact event

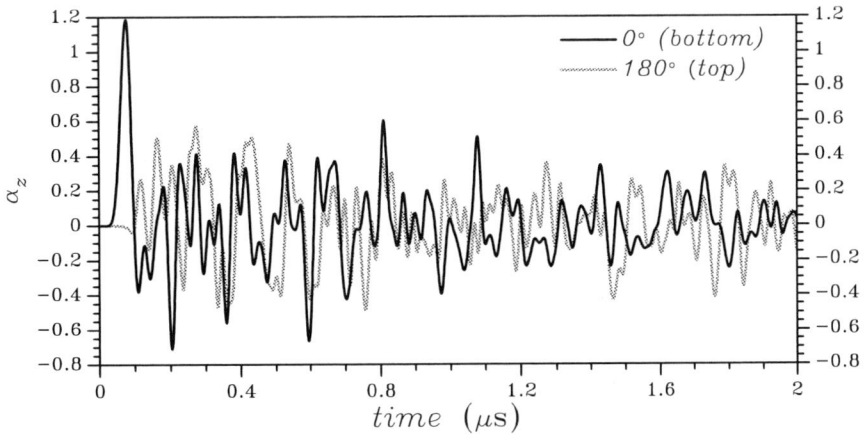

Figure 7: time evolution of the dimension-less acceleration normal to the sensor plane (drop height 150 cm). Comparison between the bottom and top cases

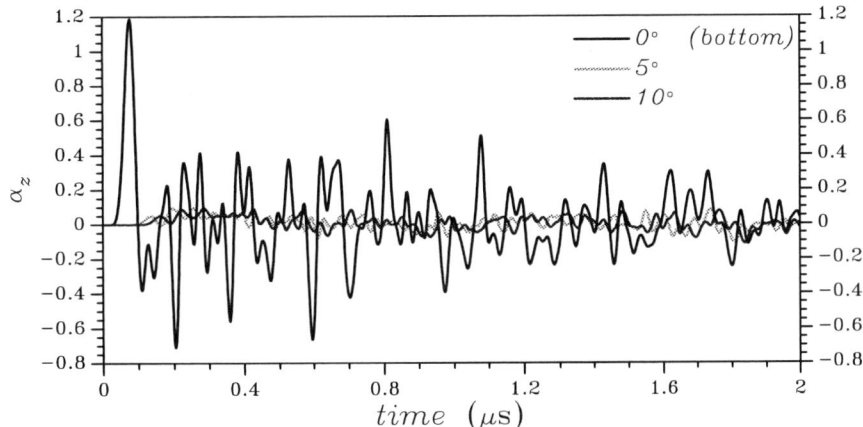

Figure 8: time evolution of the dimension-less acceleration normal to the sensor plane (drop height 150 cm). Effect of tilting

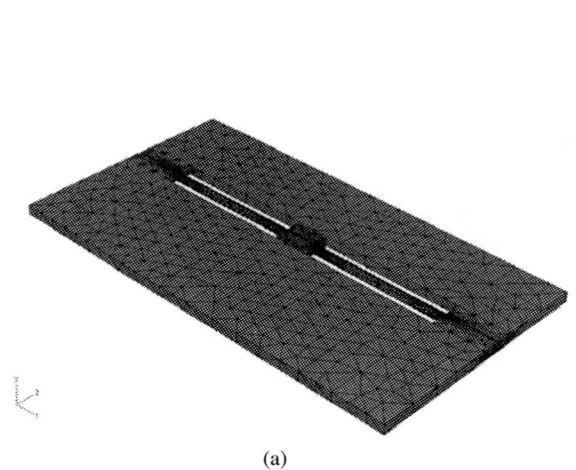

(a)

Figure 9: meso-scale analyses (sensor length scale). View of the whole sensor mesh

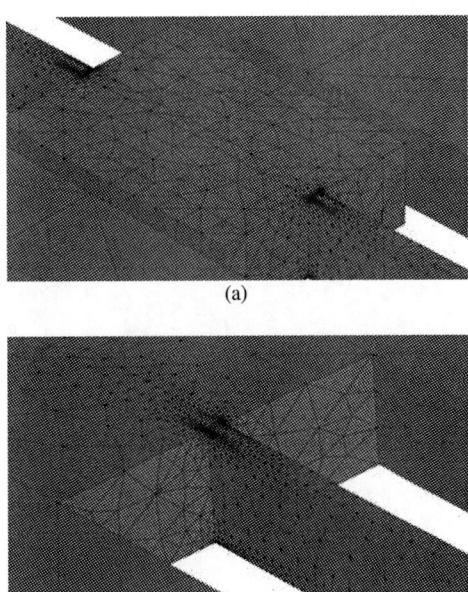

Figure 10: meso-scale analyses (sensor length scale). Details (a) of the anchor-spring and (b) of the spring-plate joint zones

Hence, it can be finally claimed that Equation (1) has to be intended as a rough estimation of the maximum acceleration peak that the anchor experiences after the impact with a rigid target surface. In case of elastic target surfaces, a more detailed campaign of simulations is required to further assess the effects of its mechanical properties.

3.2 Meso-scale analyses (at sensor level)

At this length-scale we are interested in the dynamic response of the uni-axial MEMS accelerometer shown in Figures 9 and 10. Specifically, we aim at detecting where the stress state caused by the impact exceeds or even approaches a pre-defined bound on the actual strength of the silicon.

The sensor is constituted by a seismic mass (or massive plate), connected via two springs (or slender beams) to the anchor point. Both the seismic plate and the beams are made of poly-crystal silicon, featuring Young's modulus $E_p = 145$ GPa, Poisson's ratio $v_p = 0.2$ and mass density $\rho_p = 2330$ Kg/m³. To simply account for the perforated bulk of the seismic plate, reduced

mechanical and mass properties are adopted for the polysilicon in there.

While plate vibrates, in the analyses it has been assumed that the interaction with the surrounding fluid can be disregarded. Furthermore, possible interactions of the movable parts of the sensor (plate and springs) with stoppers and with top/bottom surfaces of the cavity inside the die/cap assembly are disregarded as well.

Outcomes of the simulations clearly show that the bending vibrations of the plate are negligible. On the other hand the springs, because of their small stiffness as compared to the plate one, are subject to bending as well as torsional vibration. The details which are likely to fail after the impact are therefore the spring-anchor and the spring-plate joint sections of both springs.

A detailed resolution of the stress state in these regions appears quite expensive from the computation viewpoint. In fact, the presence of re-entrant corners cause stress concentration, which requires a detailed space discretization to be captured. This requirement motivates the adopted mesh, as shown in the details of Figure 10.

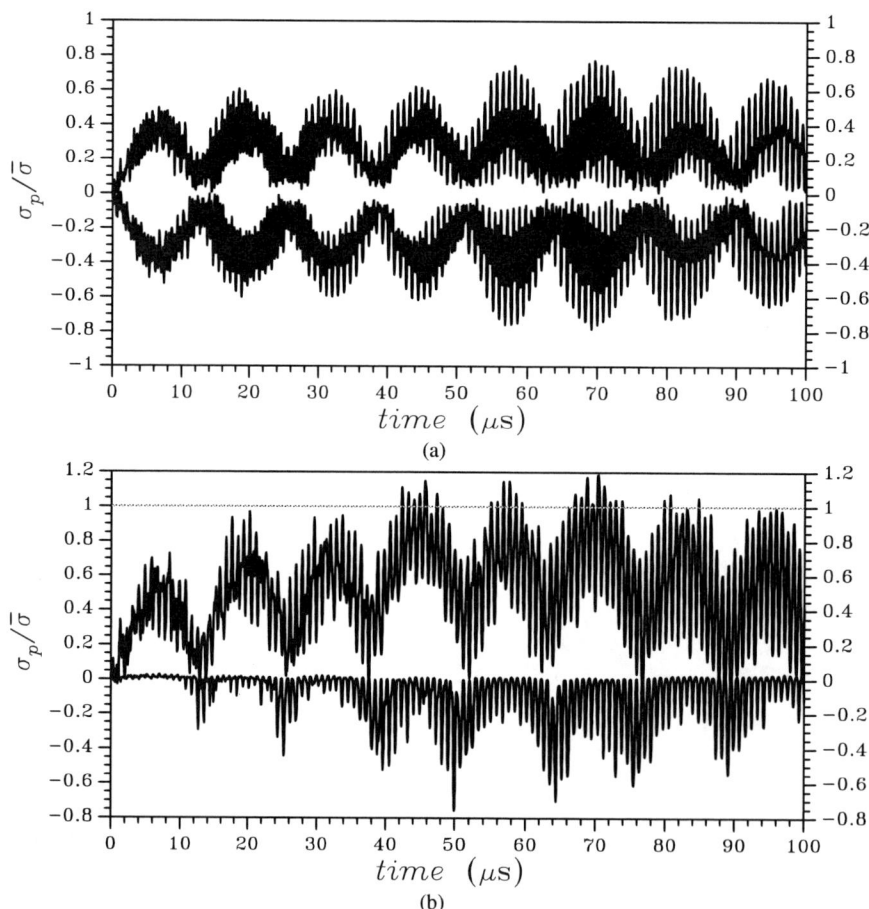

Figure 11: envelope of the dimension-less principal stresses at the spring-anchor joint section (drop height 150 cm). (a) bottom case; (b) top case

As for the joint sections, Figure 11 collects the envelopes of the principal stresses up to $t_{end} = 100$ μs

caused by bottom and top impacts. Here stresses are normalized with respect to an estimate of the polysilicon strength, on the order of $\overline{\sigma} = 3.5$ GPa. It can be noticed that low frequency variations, with period $T_{low} \approx 13$ μs, are superposed to higher frequency ones. These latter ones are linked to higher bending and torsional vibration modes of the beams.

Surprisingly, from Figure 11 it can be concluded that the bottom drop test, characterized by higher acceleration peaks in the sensing direction, does not lead to principal stresses exceeding the material strength. On the contrary the top drop test, characterized by acceleration peaks never approaching $\overline{a}$, leads to maximum tensile principal stresses exceeding $\overline{\sigma}$ in the period $45 < t < 80$ μs.

3.3 Micro-scale analyses (at grain level)

Since the sensor details with high failure likelihood are the spring-anchor and the spring-plate joint sections, the coupled flexural/torsional stress state acting there requires a full three-dimensional analysis of the crystal structure.

Even though this work is not aimed at contributing to a description of the failure mechanisms at the micro-scale, namely at the crystal structure length-scale, two main paths can be envisaged. In case a detailed map of the polycrystalline silicon can be obtained via optical measures, intergranular and/or transgranular dynamic crack growth can be simulated (see, e.g., [4-6] for two-dimensional approximations of the actual problem geometry). In case a detailed map of the silicon crystal structure can not be obtained, Monte Carlo-like approaches have to be followed: the geometry and the local axes of orthotropy of each single crystal, the strength and toughness of each grain and of grain boundaries are allowed to vary stochastically in order to infer an average response and a probability of failure of the sensor detail.

4. Conclusions

A decoupled multi-scale approach to resolve the stress state induced in polysilicon MEMS by drop impacts has been proposed. In the numerics macro-scale calculations (at die level) furnish the acceleration histories at sensor anchor, to be adopted as input loadings for meso-scale analyses (at sensor level).

Main features of the effect of drop height and impact angles on the sensor failure probability have been elucidated.

It has been also shown that, depending on the failure mechanism in the sensor, maximum acceleration peak is not conclusive to assess the safety of the device. In fact, the sensor springs are subject to a flexural/torsional stress state linked to the interplay between shock-waves emanating from the impact locations. Micro-scale analyses should be adopted to obtain an insight into the actual failure mechanism.

Acknowledgments

The financial support of Italian MIUR through PRIN-Cofin2005 programme "*Interfaces in innovative micro and nano structured materials and devices*" is gratefully acknowledeged. Aldo Ghisi wish to thanks Cariplo Foundation for the financial support of the project "*Innovative models for the study of the behaviour of solids and fluids in micro/nano electromechanical systems*".

References

1. G.X. Li, F.A. Shemansky,. "Drop test and analysis on micro machined structures". *Sensor and Actuators A*, Vol. 85 (2000), pp. 280-286.

2. V.T. Srikar, S.D. Senturia, "The reliability of Microelectromechanical systems (MEMS) in shock environments", *Journal of Microelectromechanical Systems*, Vol. 11 (2002), pp. 206-214.

3. H. Shan et al., "Three-Dimensional Modeling and Simulation of a Falling Electronic Device", *Journal of Computational and Nonlinear Dynamics*, Vol. 2 (2007), pp. 22-31.

4. H.D. Espinosa, P.D. Zavattieri, "A grain level model for the study of failure initiation and evolution in polycristalline brittle materials. Part I: theory and numerical implementation. Part II: numerical examples", *Mechanics of Materials,* Vol. 35 (2003), pp. 333–394.

5. A. Corigliano, F. Cacchione, A. Frangi, S. Zerbini, "Simulation of Impact Rupture in Polysilicon Mems", *Proc. Eurosime06*, Como (Italy), April 2006, pp. 197-202.

6. A. Corigliano, F. Cacchione, A. Frangi, S. Zerbini, "Micro-scale simulation of impact rupture in polysilicon Mems" *Proc. ECF16*, Alexandropoulos (Greece), July 2006.

7. E. Falcon, C. Laroche, S. Fauve, C. Coste, "Collision of a 1-D column of beads with a wall", *The European Physical Journal B*, Vol. 5 (1998), pp. 111-131.

Packaging Effects of Cu/Low-k Interconnect Structure

Ming-Che Hsieh

EOL/Industrial Technology Research Institute
Rm.168, Bldg.14, No.195, Sec. 4, Chung Hsing Rd.,
Chutung, Hsinchu, Taiwan 310, R.O.C.
Email: mchsieh@itri.org.tw   Phone: 886-3-5913874   FAX: 886-3-5820374

Abstract

A compressive study of packaging effects of Cu/low-k interconnect structure was represented in this paper by using finite element analysis. The modeling results of 3D-IC inter-wafer Cu/low-k interconnect structure were presented. Since the thermal deformation in 3D-IC package are usually coupled into Cu/low-k interconnect structure and inducing large local deformation to drive interfacial crack formation, the thermal reliability concerns for Cu/low-k chips are becoming the critical reliability issues in the present studies. Moreover, the associated thermal induced equivalent stresses could also result in failures to IC chip and affect its reliability. For the purpose of studying packaging effects of 3D-IC inter-wafer Cu/low-k interconnect structure, three-dimensional finite element analyses were used to analyze the thermal stresses distributions in Cu/low-k interconnect structure with complicated geometries. Furthermore, packaging induced crack driving forces for relevant interfaces in Cu/low-k interconnect structure were deduced. The modified virtual crack closure (MVCC) technique was used to calculate the energy release rate in 3D-IC inter-wafer Cu/low-k interconnect structure. The results not only show the existence of mix fracture mode at these interfaces in Cu/low-k interconnect structure but also indicate the significant packaging effect for 3D-IC inter-wafer Cu/low-k interconnect structure. In addition, the simulated results also show that the corresponding material properties of low-k dielectrics have prominent influences on thermal induced equivalent stresses.

1. Introduction

With the rapid development of IC technologies, the electronic industry now is making its progress to miniaturize the ICs, packages, cards and system levels. In order to develop smaller, lighter, thinner, faster and lower-priced electronic products, 3D-IC package is widely observed and is assuming an important role in electronic industry. In 3D-IC packages, the shortened global wiring length can increase the transistor density in device-level stacking and the shortened chip to chip wiring length can increase the function density in package-level stacking. Due to the shortened wiring length between the devices and chips, the signal delayed effects can be reduced through the shorten interconnections [1-5]. Recently, low-k dielectrics are popularly used to increase the bandwidth, retard the RC signal delayed effects, reduce the inductance and decrease the power consumption in 3D-IC structures. However, by

using these weaker low-k dielectrics, some thermal-mechanical problems accompany, like the problems of thermal-mechanical induced stresses, heat dissipation, induced interfacial delamination/crack and so on. All these induced problems would cause the failures to IC devices as well as 3D-IC structure and would become the critical reliability issues in assembly processes and reliability test procedures. Therefore, the work of thermal stress analysis and the question of concerning packaging effects of driving interfacial delamination for Cu/low-k interconnect structure become more important in the present studies. In this paper, a compressive study of packaging effects of 3D-IC inter-wafer Cu/low-k interconnect structure was represented by using finite element analysis. The thermal induced equivalent stress in 3D-IC inter-wafer Cu/low-k interconnect structure were obtained by FEA simulation with ANSYS software. Moreover, in order to study the information of packaging induced cracks in 3D-IC inter-wafer Cu/low-k structure, the cracks with fixed length were introduced at several relevant interfaces. The modified virtual crack closure (MVCC) technique was used to calculate the energy release rate in Cu/low-k interconnect structure. The crack driving force and three components of energy release rate, G_I, G_{II} and G_{III} that corresponding to fracture modes I, II and III can then be obtained by MVCC technique [5, 6]. The results show that the existence of mix fracture mode at these interfaces in Cu/low-k interconnect structure.

2. Modified Virtual Crack Closure Technique

The modified virtual crack closure (MVCC) technique was used to calculate the energy release rate [6, 7]. Since the critical energy release rate is a function of mix mode, G_I, G_{II} and G_{III} that corresponding to three basic fracture mode I (opening mode), mode II (sliding mode) and mode III (tearing mode) were separately determined. For the eight-node solid elements that shown in Figure.1, the three components of energy release rate G_I, G_{II} and G_{III} can be obtained by the following equations [5-7]

$$G_I = -\frac{1}{2\Delta A} \cdot Z_{Li} \cdot (w_{Ll} - w_{Ll^*}),$$

$$G_{II} = -\frac{1}{2\Delta A} \cdot X_{Li} \cdot (u_{Ll} - u_{Ll^*}), \qquad (1)$$

$$G_{III} = -\frac{1}{2\Delta A} \cdot Y_{Li} \cdot (v_{Ll} - v_{Ll^*}),$$

where $\Delta A = \Delta a \times b$ as shown in Figure 1. ΔA is the area virtually closed, Δa is the length of the elements at the delamination front and b is the width of the elements. In Figure 1, X_{Li}, Y_{Li} and Z_{Li} denote the forces at the delamination front in column L, row i. The corresponding displacement behind the delamination at the upper face node row l are, respectively, denoted as u_{Ll}, v_{Ll} and w_{Ll} and the corresponding displacement at the lower face node row l^* are, respectively, denoted as u_{Ll^*}, v_{Ll^*} and w_{Ll^*}.

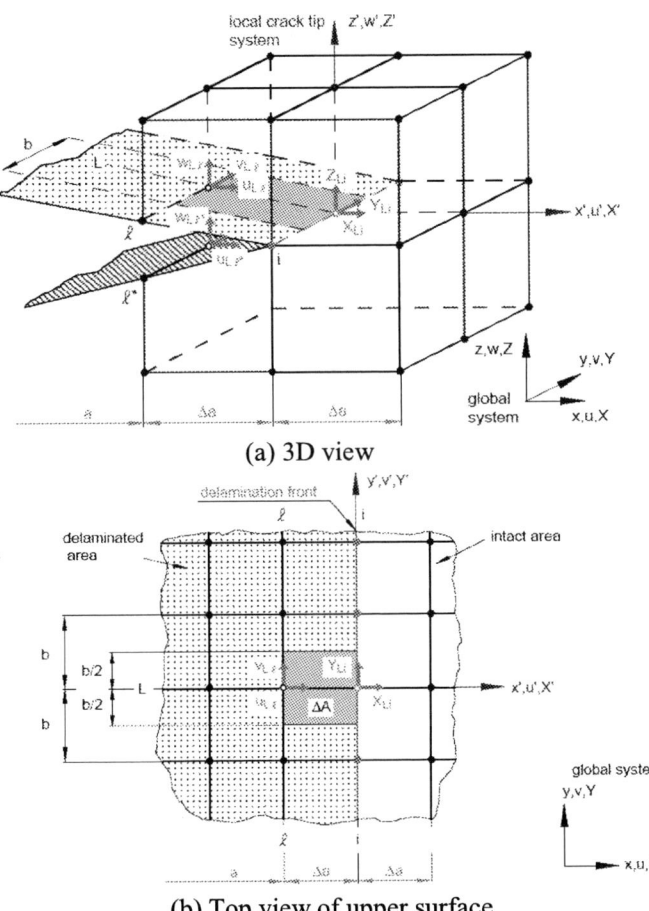

(a) 3D view

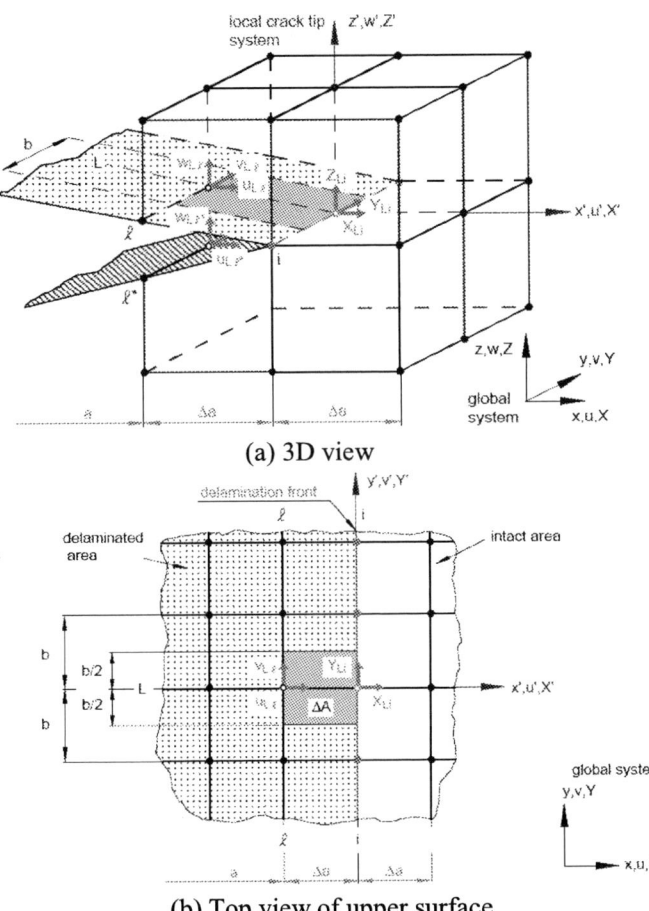

(b) Top view of upper surface
(lower surface terms are omitted for clearity)

Figure 1. Modified virtual crack closure technique [6]

In finite element analysis, the values of G_I / G_T, G_{II} / G_T and G_{III} / G_T may always change with the change of the element size around the crack tip, especially when the element size is very small. (Here, G_T is the total energy release rate and is the summation of three components of energy release rate, i.e., $G_T = G_I + G_{II} + G_{III}$.) However, in our FEA simulation, the element size is relative large, hence the values of G_I / G_T, G_{II} / G_T and G_{III} / G_T are insensitive to the element size.

In general, the interface fracture criterion can be expressed as

$$\left(\frac{G_I}{G_{IC}} \right)^l + \left(\frac{G_{II}}{G_{IIC}} \right)^m + \left(\frac{G_{III}}{G_{IIIC}} \right)^n = 1, \qquad (2)$$

where l, m and n are constants and G_{IC}, G_{IIC} and G_{IIIC} are, respectively, the critical energy release rates for pure fracture mode I, II and III.

3. Thermal Stress Analysis of 3D-IC Inter-wafer Cu/Low-k Interconnect Structure

Since the thermal effects always cause the fatigue or fracture to IC packages, the thermal induced stresses come into existence in IC packages and become a serious reliability concern for Cu/low-k structures. Therefore, the topic of thermal stress analysis and the packaging effects of driving interfacial delaminating in Cu/low-k structure become more attractive in electronic package areas. Consider a 3D-IC inter-wafer Cu/low-k interconnect structure that shown in Figure 2. Its repeated unit consists of three metal levels (M1, M2 and MC) connected by two levels of vias (V1 and V2). MC and M2 are local interconnects in the chain which dimensions are 0.68 μm long, 0.36 μm wide and 0.25 μm height. M1 is a square landing pad with 0.36 $\mu m \times 0.36$ μm dimension and 0.25 μm height. V1 and V2 are the cylinder vias with 0.3 μm in diameter and 0.35 μm tall. M1, M2, V1 and V2 are copper embedded in low-k dielectrics (Parylene-N, SiLK, BCB, SiCOH and PTFE). MC is copper embedded in silicon dioxide. Metal lines are passivated with a Si$_3$N$_4$ cap layer of 20nm thickness. Copper interconnects and vias are covered with Ta-based liners at their bottom and side walls which are taken to be of uniform 20nm thickness [2, 8].

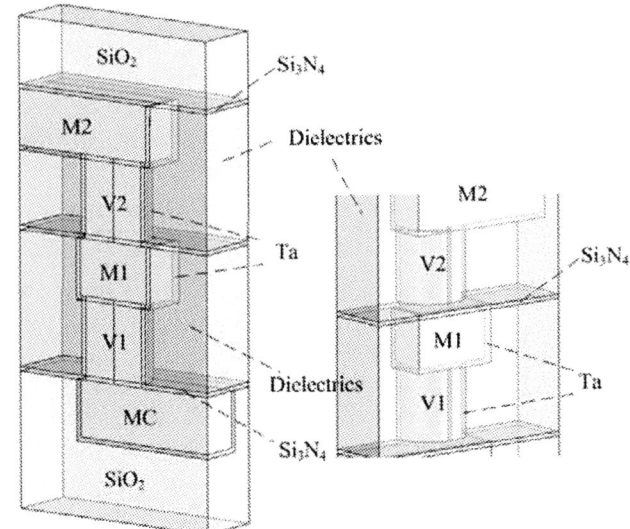

Figure 2. Schematic of 3D-IC inter-wafer structure

For the purpose of obtaining thermal induced equivalent stresses in 3D-IC inter-wafer Cu/low-k interconnect structure, the finite element software package ANSYS was adopted in our FEA modeling. In ANSYS simulation, the finite element model with pure hexahedral elements meshed geometry that contains 83536 eight-node solid elements and 90,861 nodes was used (see Figure 3). All the materials in the 3D-IC inter-wafer structure were assumed to be linear elastic and their material properties were listed in Table 1. In Table 1, ν, E, and CTE are, respectively, the Poisson's ratio, Young's modulus, and the coefficients of thermal expansion.

Table 2 shows the results of global maximum von Mises stress and global maximum displacement in z direction for 3D-IC inter-wafer Cu/low-k interconnect structure when the structure subjected to a uniform temperature change – from -55°C to 150°C. From Table 2 we can see that both the von Mises stress and z-displacement of Cu/SiCOH interconnect structure are smallest than others when copper interconnects and vias were covered with Ta barrier. If the structure was without any Ta barrier, the smallest von Mises stress occurred in Cu/BCB structure while the smallest z-displacement still occurred in Cu/SiCOH structure. However, both the maximum von Mises stress and z-displacement are largest in Cu/Parylene-N structure than other Cu/low-k structures. Compared the Cu/low-k structure with and without Ta barrier, we can find that the structure without Ta barrier has smaller global maximum von Mises stress and larger z-displacement than those with Ta barrier. Alothogh the global maximum von Mises stress in Cu/low-k structure without Ta barrier is smaller than that with Ta barrier, the corresponding von Mises stress distributions in copper interconnects and vias (along z direction in Figure 3) would get larger (except for Cu/SiCOH structure), which is clearly shown in Figure 4 and 5.

From Figure 4, we can see that the value of von Mises stress at the interface Cu/Si$_3$N$_4$ is larger than that at the interface Si$_3$N$_4$/SiO$_2$ and the value of von Mises stress at the interface Cu/Ta is larger than that at the interface Ta/SiO$_2$. These phenomenons may come from the larger CTE mismatch at Cu/Si$_3$N$_4$ interface ($\Delta\alpha = 13.8\,ppm/°C$) than that at Si$_3N_4$/SiO$_2$ interface ($\Delta\alpha = 2.7\,ppm/°C$) and the larger CTE mismatch at Cu/Ta interface ($\Delta\alpha = 10.5\,ppm/°C$) than that at Ta/SiO$_2$ interface ($\Delta\alpha = 6.0\,ppm/°C$), respectively. One should also note that since the material properties in each layer are different, the von Mises stresses shown in Figure 4 are discontinuous at the interfaces (SiO$_2$/Ta/Cu and Cu/Si$_3$N$_4$/SiO$_2$), which seem reasonable in engineering sense. Moreover, we also find that by using SiCOH dielectric instead of using Parylene-N, SiLK, BCB and PTFE would result in smaller von Mises stress in copper interconnects and vias no matter they were covered with or without Ta barrier (see Figure 4 and 5).

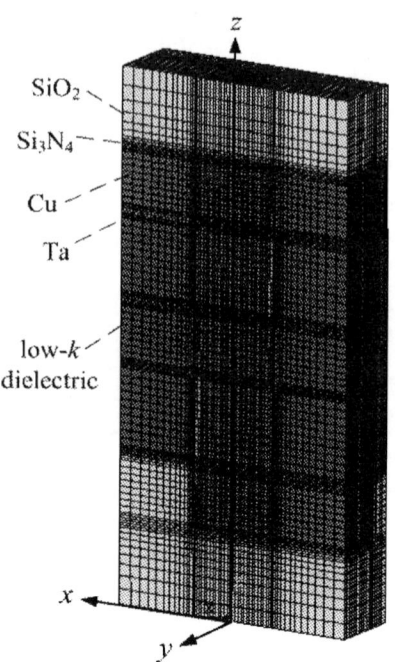

Figure 3. Pure hex-element mesh for 3D-IC inter-wafer Cu/low-k interconnect structure

Table 1. Material properties of 3D-IC inter-wafer structures

Material		E (GPa)	ν	CTE (ppm/°C)	Dielectric constant	Deposition method
SiO$_2$		70	0.22	0.5	N/A	N/A
Cu (MC, M1, M2)		120.5	0.35	17		
Cu (V1, V2)		120.5	0.35	17		
Ta		185	0.30	6.5		
Si$_3$N$_4$		221	0.27	3.2		
Low-k Dielectrics	Parylene-N	2.9	0.4	70	2.58	CVD*
	SiLK	2.5	0.4	66	2.65	SOD**
	BCB	2.9	0.34	52	2.65	SOD
	SiCOH	16.2	0.3	12	2.05	PECVD***
	PTFE	0.5	0.46	135	1.92	CVD

* CVD: Chemical Vapor Deposition

** SOD: Spin-on Dielectric

*** PECVD: Plasma enhanced CVD

Table 2. Values of von Mises stress and z-displacement in Cu/low-k structure with and without Ta barrier

Low-k Dielectrics	With Ta barrier		Without Ta barrier	
	von Mises stress (GPa)	z-displacement (μm)	von Mises stress (GPa)	z-displacement (μm)
Parylene-N	1.7942	0.010546	1.2668	0.012050
SiLK	1.6594	0.009796	1.1532	0.011165
BCB	1.4236	0.008520	1.0852	0.009655
SiCOH	1.0008	0.005459	1.1427	0.005774
PTFE	1.5363	0.010434	1.1028	0.011650

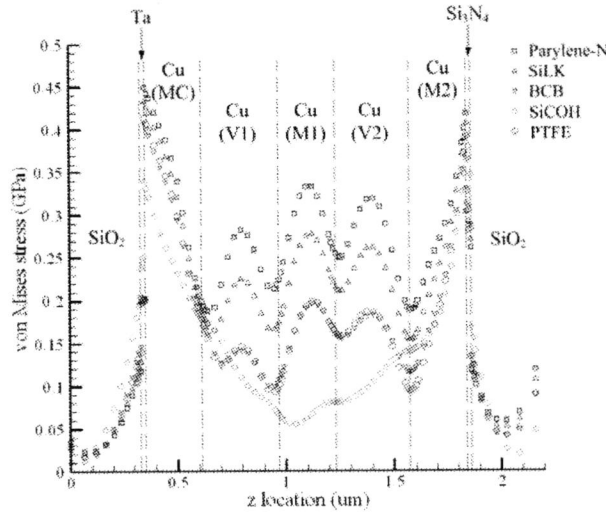

Figure 4. von Mises stress distribution along z direction in Cu/low-*k* interconnect structure (with Ta barrier)

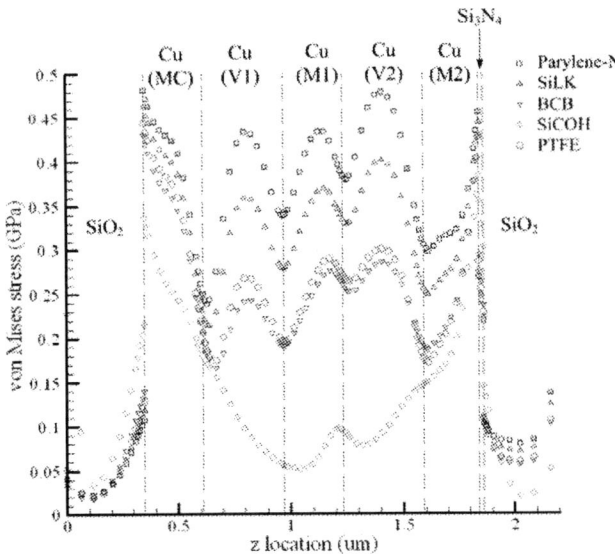

Figure 5. von Mises stress distribution along z direction in Cu/low-*k* interconnect structure (without Ta barrier)

4. Crack Driving Force for 3D-IC Inter-wafer Cu/Low-*k* Structure

In order to obtain the energy release rate for interfacial cracks in 3D-IC inter-wafer Cu/low-*k* interconnect structure, the cracks along five interfaces that illustrated in Figure 6 were introduced. Consider the packaging effect that assumed a stress-free state at -55°C for whole 3D-IC inter-wafer structure and then be heated to 150°C. The crack driving force was obtained at 150°C. Table 3 shows the values of total energy release rate G_T and the ratio of G_I, G_{II}, G_{III} over G_T for five interfacial cracks in Cu/SiCOH interconnect structure that described in Figure 6. From Table 3, we can clearly see that the mix fracture mode existed in Cu/SiCOH interconnect structure. For Cu/SiCOH structure, the value of G_{III} is closed to

zero along interfacial crack 1 (SiO_2/Si_3N_4), which means that the fracture mode III along interfacial crack 1 can be ignored. In addition, the the value of G_{II} is much smaller than G_I and G_{III} along interfacial crack 4 and 5, which also means that the fracture mode along Cu/Ta and Cu/Si$_3$N$_4$ interfaces are almost mode I and mode III dominated.

Moreover, we see that the interfacial crack driving force can be as high as 6.3 J/m^2 along interfacial crack 5 (Cu/Si$_3$N$_4$) (see Table 3). Therefore, the interfacial delamination would be a serious reliability issue for Cu/SiCOH structures. Furthermore, we also see that the interfacial cracks 1, 4 and 5 (along interface SiO$_2$/Si$_3$N$_4$, Cu/Ta and Cu/Si$_3$N$_4$, respectively) are more prone to delamination due to the packaging effect while crack 2 and 3 (along interface SiCOH/ Si$_3$N$_4$ and SiCOH/Ta) have little effect. The results indicate the significant packaging effect for 3D-IC inter-wafer Cu/SiCOH structure.

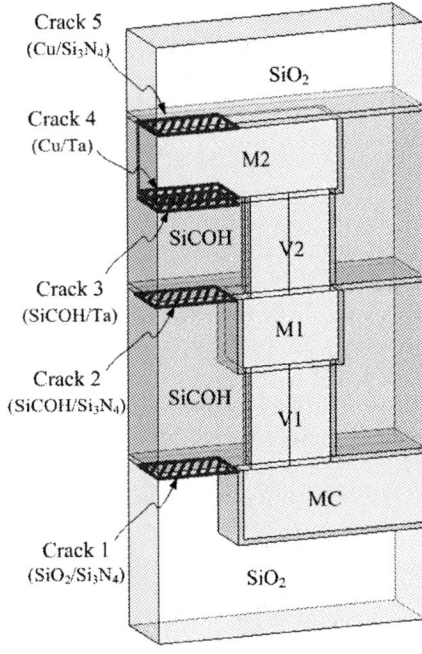

Figure 6. Schematic of Cu/SiCOH interconnect structure embedded interfacial cracks

Table 3. Mode mixty for 3D-IC inter-wafer Cu/SiCOH interconnect structure (from -55°C to 150°C)

	Crack 1	Crack 2	Crack 3	Crack 4	Crack 5
G_I/G_T	0.5093	0.1813	0.7251	0.4311	0.3732
G_{II}/G_T	0.4778	0.4771	0.0506	0.0216	0.0376
G_{III}/G_T	0.0129	0.3416	0.2243	0.5473	0.5892
G_T (J/m^2)	3.2470	0.2927	1.1660	5.2853	6.3055

401

5. Conclusions

In this paper, a compressive study of packaging effects of 3D-IC inter-wafer Cu/low-k interconnect structure was represented by using finite element analysis. The thermal induced equivalent stresses of 3D-IC inter-wafer structure that with copper vias embedded in several low-k dielectrics, Parylene-N, SiLK, BCB, SiCOH and PTFE, respectively, were obtained with the help of three-dimensional FEA simulation. Due to the simulated results, we can find the Cu/SiCOH structure would result in smaller von Mises than other low-k dielectrics. Furthermore, the thermal deformation of package can be directly coupled into the Cu/low-k interconnect structure inducing large local deformation to drive interfacial crack formation. In order to study the information of packaging induced cracks in 3D-IC inter-wafer Cu/low-k structure, five interfacial cracks with fixed lengths were placed at the interfaces of SiO_2/Si_3N_4, SiCOH/ Si_3N_4, SiCOH/Ta, Cu/Ta and Cu/Si_3N_4, respectively. The energy release rate in 3D-IC inter-wafer Cu/low-k interconnect structure can then be obtained by using modified virtual crack closure (MVCC) technique. The results not only show the existence of mix fracture mode at the relevant interfaces but also indicate the significant packaging effect for 3D-IC inter-wafer Cu/low-k interconnect structure.

References

1. J.A. Davis, *et al*, "Interconnect limits on gigascale intergration (GSI) in the 21st century", *Proceedings of the IEEE 89*, Vol. 3 (2001), pp. 305-324.

2. Jing Zhang, *et al*, "Thermal stresses in 3D IC inter-wafer interconnects", *Microelectronic Engineering*, Vol. 82 (2005), pp. 534-547.

3. R.G. Filippi, *et al*, "Thermal cycle reliability of stacked via structures with copper metallization and an organic low-k dielectric", *Proceedings of the 2004 IEEE International Reliability Physics Symposium*, Phoenix, Arizona, April. 2004, pp. 61-67.

4. D. Edelstein, *et al*, "Comprehensive reliability evaluation of a 90 nm CMOS technology with Cu/PECVD low-k BEOL", *Proceedings of the 2004 IEEE International Reliability Physics Symposium*, Phoenix, Arizona, April. 2004, pp. 316-319.

5. Guotao Wang, Ho, P.S., Groothuis, S., **"Packaging effect on reliability of Cu/low k interconnects"**, *IEEE Workshop on Microelectronic and Electro Device*, Boise, Idaho, USA, April. 2004, pp. 28-30.

6. R. Krueger, "The virtual crack closure technique: history, approach and application", *ICASE Report No. 2002-10*, NASA, April. 2002.

7. F. G. Bucholz, R. Sistla, and T. Krishnamurthy, "2D and 3D applications of the improved and generalized modified crack clouse integral method", Computational Mechanics '88, Springer Verlag (New York, 1988).

8. M. C. Hsieh, *et al*, "Thermal stress analysis of Cu/low-k interconnects in 3D-IC structures", *International Microsystems, Packaging, Assembly Conference*, Taipei, Taiwan, October. 2006, pp.27-30.

Validation of Dynamic Thermal Simulations of Power Assemblies Using a Thermal Test Chip

X. Jordà, M. Vellvehi, X. Perpinyà, J.L. Galvez, P. Godignon
Centre Nacional de Microelectrònica CNM - CSIC
Campus UAB Bellaterra, 08193, Cerdanyola del Vallès (Barcelona, Spain)
xavier.jorda@cnm.es, Tel. +34 93 594 77 00, Fax. +34 93 580 14 96

Abstract

Thermal simulation is the main thermal design tool used to predict temperature distributions of complex power electronics assemblies. Nevertheless, the validation of the simulation results remains a complex problem, mainly in dynamic operation, due to the difficulty in measuring semiconductor device temperatures. This paper proposes a methodology for the accurate validation of CFD 3-D thermal simulations of power modules. A test power assembly structure based on two thermal test chips and an insulated metal substrate, has been measured with single power pulse excitations. The corresponding thermal simulations have been performed to reproduce the experimental conditions, close to the real operational ones. A very good agreement between simulation and experience has been obtained. Fine adjustment and analysis of some critical parameters (thermal conductivities, etc.) is possible using this approach.

1. Introduction

Increasing the power density of power modules and assemblies represents a thermal management challenge. Several components dissipating high power levels have to share a small room and, apart from the self-heating phenomenon, the cross coupling effects between devices has also to be understood. The main heat dissipation mechanism in these systems is the conduction through the module substrate, although convection can also take a relevant role depending on each particular application. Thus, a precise knowledge of the devices and substrate characteristics, as well as the particular boundary conditions is crucial to undertake an accurate thermal simulation. In addition, as power electronics is based on switching operation of semiconductor devices, the dynamic aspects can not be avoided, increasing the simulation complexity.

Two main kinds of substrates are used in typical power electronics assemblies. Direct Copper Bonded (DCB) ceramic substrates are the preferred ones for high power applications. They consist basically in a Copper-ceramic layer-Copper sandwich structure. Alumina and Aluminium Nitride ceramic layers show very good thermal properties, although their mechanical properties are not optimal and the AlN cost is relatively high. Insulated Metal Substrates (IMS) represent an interesting alternative for low and medium power applications. Although their thermal properties (i.e., thermal resistance per square cm) are lower than those of the ceramic DCB substrates, they are less expensive, more robust and mechanizeable. The basic structure of an IMS substrate consists in a Copper layer for the circuit layout definition, a metal (usually Aluminium) base-plate, and a thin thermally conductive dielectric layer between the two metals. This dielectric is usually an epoxy filled with ceramic particles, its thermal conductivity and specific heat being very difficult to know.

Concerning the components of power electronics assemblies, the more critical ones are the semiconductor power devices (such as MOSFET, IGBT or fast recovery diode chips), as they show high power densities. Direct measurement of the chip temperature is very difficult (the chip is not always directly accessible, the power circuitry is coupled with the measuring one, etc.) and thermal assessment becomes a complex task.

In the present work, dynamic heating and cooling experiments are performed in IMS-based test power assemblies, using a thermal test chip (TTC) as the main experimentation vehicle. This test chip shows the same thermal behaviour than typical power devices but allows simultaneous and decoupled heat generation and temperature sensing. This fact makes easier the temperature measurements in dynamic operation. Single power pulses up to 52 W and 2.5 s of duration are injected and from the chip temperature evolution, self-heating and coupled thermal effects between chips are deduced. The studied assemblies have been simulated using the CFD software FLOTHERM. The agreement between simulated and experimental results has validated the whole simulation process (modelling, simplifications, boundary conditions, parameter values, etc). The paper also describes the set-up and procedures used to obtain the experimental results.

2. The test power module

A test vehicle based on two thermal test chips (TTC) and an insulated metal substrate (IMS) has been developed to undertake the proposed thermal measurement and simulation work. The method can be extended to other kinds of substrates such as PCB and laminate substrates, or DCBs. These materials show their own specificities and operational characteristics: high power levels for the DCB case and lower power and higher temperatures for the PCB case, etc. The behaviour of the IMS substrates is situated between the PCB and the DBC (medium power applications), constituting a very interesting starting point.

The first (and main) element for the development of the test power module is the TTC. This device is a 6 mm x 6 mm x 0,525 mm Silicon chip developed for thermal tests and assessment of packages and substrates, allowing simultaneous heat dissipation and temperature measurements [1]. It reproduces the thermal behaviour of typical vertical power devices, i.e. a heat generation area on top and a vertical heat flux flowing to the heatsink by conduction mechanisms. The heat is generated by a polysilicon heating resistor distributed on the top, while the temperature is measured with a sensing resistor (an RTD) at the chip centre. Figure 1 shows the top view of a TTC soldered on a Cu pad, with the required interconnections between the resistors terminals and the external tracks, performed with Al wire-bonds.

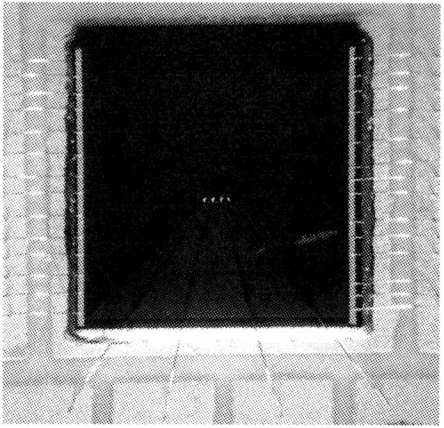

Figure 1: Top view of a thermal test chip showing the wire-bonding connections of both resistors.

Both resistors are electrically isolated from the silicon substrate by a very thin (30 nm) Silicon oxide layer. The thermal resistance of a SiO_2 parallelepiped of 6 mm x 6 mm x 30 nm is $6.4x10^{-4}$ K/W, taking into account a thermal conductivity for this material of 1.3 W/mK. This means that for a dissipated power of 50W, the temperature difference between both oxide surfaces is 0.03 K. Consequently, the thermal influence of this layer can be neglected for all practical operation conditions, where the contribution of the other parts and layers of the assembly show higher contributions. The heating resistor layout consists basically in 130 parallel polysilicon tracks, 20 μm wide and spaced 17.2 μm between them. The heating resistor can be connected with wire-bonds through two long Al pads placed at two edges of the chip (the vertical grey stripes on the right and left of the chip in Fig. 1) and its total equivalent resistance value is 60Ω. It has been demonstrated by simulation (using the FLOTHERM software [2]) that approximately 35 μm below the heating resistor, the heat flux is homogeneous and, consequently, isothermal lines are parallel to the chip backside surface. This structure and this behaviour allow an easy description and modelling of the thermal phenomena inside the chip.

The main advantage of the thermal test chip compared with standard vertical power devices (VDMOS, IGBT, diodes, etc.) used for thermal assessments, is that the chip temperature and power dissipation measurements are decoupled and they can be obtained simultaneously at any time instant. This is not usually the case when the temperature of a power device has to be measured, for example, from a temperature sensitive parameter (TSP).

In this case, after a heating phase, the TSP is measured in a subsequent sensing phase, and the complete device temperature evolution is not completely determined [3]. The TTC used in the present work integrates a centred Platinum resistor on top of the chip. This temperature sense resistor is basically a folded Pt track of 700 – 850 Ω, taking a total area of 700 μm x 700 μm. The resistance value can be accurately measured using the 4-wire technique through the corresponding 4 Pt pads and wire-bonds (see Fig. 1). In the zone of the sense resistor, any heat dissipation is produced and consequently, the temperature at the chip centre is slightly lower than in its surrounding area. This temperature difference has been quantified by simulation (using FLOTHERM) and it is around 2ºC for a 35W steady-state dissipation. The TTC backside is metallized with the same multi-layer used for typical power devices (Ti/Ni/Au in the present case) to allow the same die-attach processes and materials.

The second element making up the test power module is the substrate material. As it has been previously explained, in the present work the used material was a commercial IMS from Denka. Often, the substrate constitutes the most unknown part of the total assembly and the proposed test vehicle is aimed to obtain as much information as possible about the substrate thermal behaviour. For this reason, the test module includes two TTCs attached and connected to the substrate with the same techniques used for true power modules: a soft-solder alloy (SnAgCu) for the die-attach and Al wire-bonds to connect the resistors pads. Figure 2 shows a general view of the fabricated test module, showing also the different pins provided to connect the 4 resistors involved in the measurements to the required external instrumentation. The two TTCs are denoted "20" and "9".

Figure 2: Picture of the test power module based on an IMS and 2 TTCs.

The presented assembly allows the analysis of the self-heating effects as well as the analysis of lateral coupled heating phenomena between chips. In typical applications, power modules are fixed on a heatsink using any kind of thermal interface material (TIM) to improve the thermal contact between the module backside and the heatsink surface. In these conditions, the main heat extraction mechanism is the conduction from the device dissipating area and the heatsink. Thus, the self-heating phenomena will be dominated by the vertical substrate structure. On the other hand, the lateral heat spreading will depend, apart from the substrate materials, on the layout of the Cu pads and tracks. The dimensions of the proposed test module are the same than those of a functional power module developed at CNM implementing a bidirectional switch with 2 IGBTs and 2 diodes [4]. The most unfavourable thermal coupling will occur between the IGBT and the diode placed on the same Cu pad. To analyse this critical case, the 2 TTCs of the test module have been soldered following this configuration, with a 5 mm gap between them.

3. The static measurement method

The first approach followed to obtain information about the thermal behaviour of the test module, was the extension for the multi-chip problem, of a measurement methodology previously established to obtain the thermal resistance (R_{TH}) of packages [5]. The basic idea consists in applying a constant dissipated power in one of the TTCs until reaching the steady-state. In these conditions, the temperature increment between a reference point in the module backside (T_{ref}) and both chips is obtained. Figure 3 shows the basic measurement scheme used to apply the method.

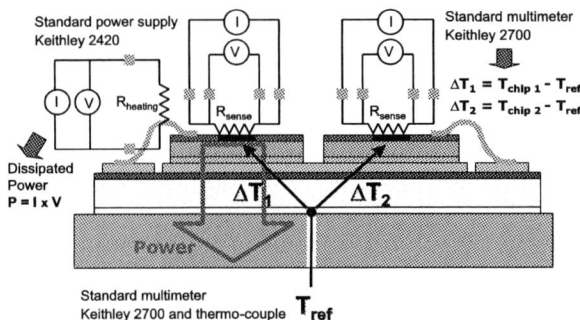

Figure 3: Basic measurement scheme for the static thermal characterisation of the test module.

To perform the chip temperature measurement, a previous calibration step of the sensing resistors is necessary. The test module is placed in an oven and the sensing resistance of both TTCs is measured for temperatures between 20°C and 135°C with 5°C increments. The oven ambient temperature is measured with a PT100 RTD sensor (±0.1°C). Figure 4 shows the corresponding calibration curves for the 2 TTCs used in the present work. As it can be observed, there is a 30Ω resistance shift between both chips, although their temperature variation is almost the same. A linear fit can be well adjusted to the experimental data for each Pt resistor. The test module is attached to a forced convection Al heat-sink and their thermal contact is improved with a silicone based TIM. The heat-sink has a 1.5 mm diameter hole allowing the measurement of the reference temperature at the IMS backside centre, using a small size K-type thermo-couple. One of the advantages of the TTCs is that silicon temperature can be directly evaluated with standard instrumentation. In this sense, a standard multimeter (Keithley 2700) is used to measure

the sensing resistance value in order to deduce the chip temperature from the linear fit derived from the calibration process.

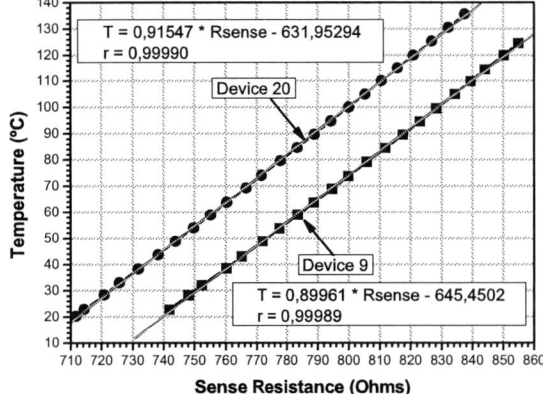

Figure 4: Temperature calibration curves of the Pt sensing resistors of both TTCs.

On the other hand, the heating resistor is excited with a source-measure unit (Keithley 2420) in order to fix the required dissipated power level. To obtain an accurate and reliable characterisation, different dissipated power values are applied to the TTC between 0 and 58W, and the temperature rise between the module backside and both chips is evaluated at each dissipated power. This approach allows obtaining the curves shown in Fig. 5.

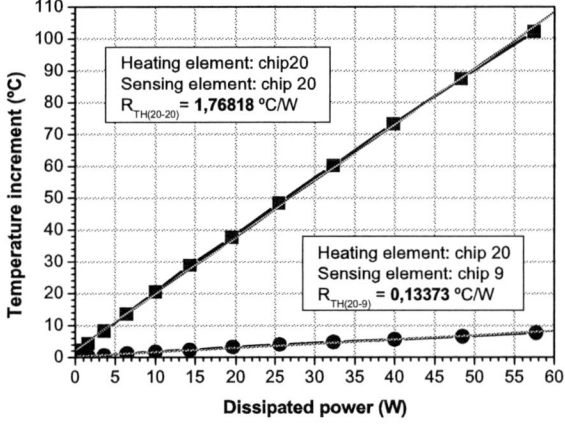

Figure 5: TTCs temperature increments vs dissipated power curves. Coupled and self-heating effects.

The slope of the temperature increment versus power curve for the active chip, gives its thermal resistance value (chip-to-case). This value is mainly related with the thickness and thermal conductivity of the different materials making up the substrate stack. For the IMS case, the predominant parameter in R_{TH} is the thermal conductivity of the dielectric layer, which can be estimated from the present measurements and used in thermal simulators. The coupled thermal effects between chips can also be quantified from the curve of the inactive chip temperature increment versus the active chip power. Although strictly speaking the slope of this curve is not

405

formally a thermal resistance, an R_{TH} value is often also associated to this coupling phenomenon in order to shorten the nomenclature. The presented experimental set-up used for the static measurements, has been taken as the basis for the dynamic measurements.

4. The dynamic measurement method

The approach to obtain dynamic thermal measurements from the test module is the same than for static conditions, although a single power pulse is applied to the TTCs and a multi-channel digital oscilloscope (Tektronix 744A) is used to acquire the time evolution of the different variables. Figure 6 shows the measurement scheme for the dynamic case.

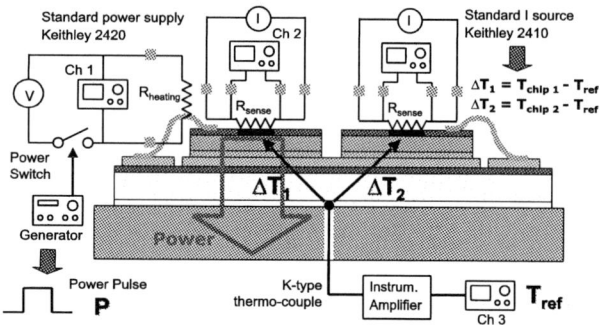

Figure 6: Basic measurement scheme for the dynamic thermal characterisation of the test module.

Concerning the excitation circuit, a power switch implemented with an IGBT is used to apply a voltage pulse to the heating resistor from a DC power supply (Keithley 2420). The pulse duration is determined from a standard waveform generator and the control signal is applied to the IGBT through a gate drive circuit (not represented in Fig. 6). From the TTC sensing resistor point of view, a source-measure unit (Keithley 2410) is used as current source to inject a sensing current to the Pt RTD. The chip temperature is then derived from its voltage drop evolution (recorded in one of the oscilloscope channels) and from the corresponding calibration curve. Finally, the time evolution of the module backside temperature (considered again as reference temperature) has also to be acquired. The K-type thermocouple is connected to an instrumentation amplifier based on the AD595 integrated circuit, which translates the thermocouple signal to a higher voltage level easily measured in the oscilloscope. Although the nominal IC output gives a 10 mV / °C signal, the whole T_{ref} measurement chain is calibrated to obtain more accurate results. In addition, this calibration is performed using the same PT100 sensor used previously for the calibration of the TTC sensing resistors. This allows reducing the error of the temperature differences between the chips and the reference.

Figure 7 shows the different waveforms measured for a 52W and 2.5 seconds power pulse applied to one of the TTCs (chip 20 in this case) with a 100 Hz sampling rate. As it can be observed, the power waveform shows a slight over-shoot at the initial times due to the increase of the

poly-Silicon heating resistance value with temperature. This variation has been approximately evaluated in 0.05 Ω / K. The ambient air temperature being 19°C, the active device reaches 118°C, the inactive one 33°C and the module backside 26°C. Another important aspect that can be observed in Fig. 7 is that after 2.5 seconds the temperatures have not reached their steady state values. In fact, at this moment, the heat-sink is still increasing its temperature under the influence of the TIM between its surface and the module backside. To analyse only the thermal behaviour of the test module, the differences between chip temperatures and the reference one are evaluated and shown in Fig. 8.

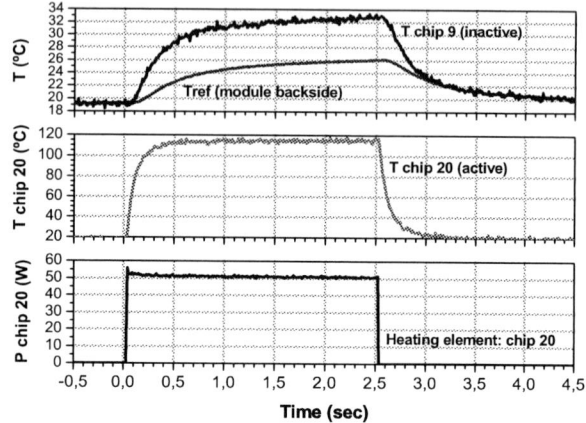

Figure 7: Temperature and power waveforms for a single 52W, 2.5s power pulse excitation.

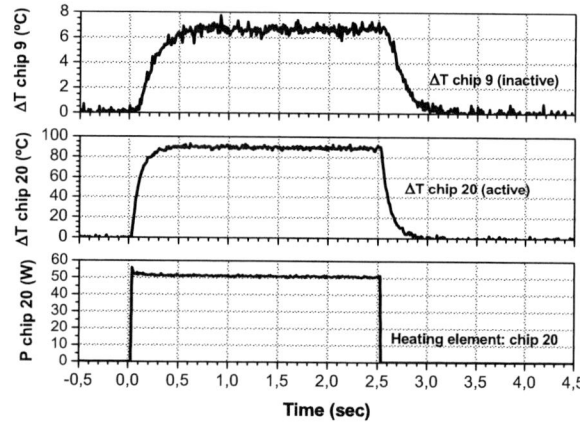

Figure 8: TTCs temperature rise from the module backside for a single 52W, 2.5s power pulse excitation.

The waveforms of Fig. 8 clearly confirm that the chip temperatures and T_{ref} evolve simultaneously and their difference is constant after 0.7 s approximately. The time constant dominating the transient behaviour for the given time scale, corresponds to the Al baseplate of the IMS substrate. Another interesting aspect that can be observed from the temperature increase waveforms is that the inactive device response (number 9 in this case) is clearly delayed from the power excitation. This point can be more clearly observed in the "thermal impedance"

406

representation of Figure 9. This graphical representation corresponds to the log – log plot of the chips temperature increments divided by the dissipated power. In Fig. 9 it can be seen that the temperature rise of the inactive chip starts to react 0.1 s after the application of the power pulse in the active device. The final values of the thermal impedance plots (for t > 0.7 s) match also with the thermal resistance values found from the static curves of Fig. 5. This means that the dynamic measurement method can efficiently replace the static one, because it gives more information with a slightly more complex experimental set-up and, in any case, it involves only standard measurement equipment.

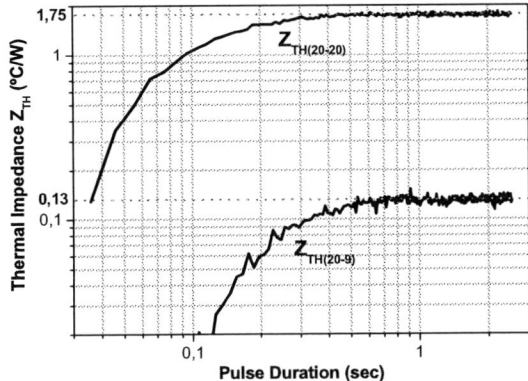

Figure 9: "Thermal impedance" representation of the waveforms of Figure 8.

Other test conditions can be used with the proposed methodology to obtain experimental data for simulation validation purposes (shorter power pulses, chip-to-ambient temperature acquisitions without heat-sink, etc.), but the proposed ones are relatively easy to obtain and they can be easily reproduced in the simulator, as it will be explained in the next section.

5. Thermal simulation validation

The test power module previously presented has been thermally simulated using FLOTHERM. The module model shown in Fig. 10 has been placed over an isothermal surface at 20ºC representing the heat-sink. As the objective is to analyse the thermal behaviour of the module, only the temperature differences between the chips and the substrate backside are analysed with "test points" placed at these locations and the isothermal surface plays only the role of heat extraction device. This approach is valid if the temperature increase of the true module backside remains moderate. If this assumption is not true and the absolute temperatures involved in the measurements are very high, some temperature dependent parameters can change their values. In the first simulations it has been verified that for the present conditions only heat conduction mechanisms are relevant and the air volume over the assembly can be considerably reduced to reduce the number of grid points and the convection phenomena. Another simplification concerns the elimination of the upper Cu tracks and pads that

doesn't play any relevant thermal role. In this sense, only the Cu pad under the chips has been described.

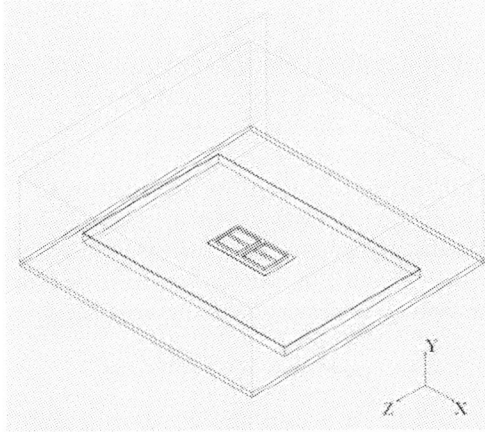

Figure 10: 3-D modelling of the test power module described under the FLOTHERM environment.

Although the module is symmetric, it has been fully described in order to prepare future simulations where any symmetry simplification can not be taken in advantage. The chips have been modelled as 525 μm thick Si slabs with homogeneous heat sources on top, except in the central zone where the Pt sensing resistor is located. This heat source is not critical and its only function is to apply the required dissipated power. Over the Si surface, at the chip centre, the chip temperature is recorded using a first "test point". The typical temperature dependence law of the Si thermal conductivity (K_{TH}) has been considered, as this material shows the stronger temperature dependence and is placed in the main heat flux path. The Si slab is placed over a 30 μm thick SnAgCu die-attach layer with a K_{TH} of 57 W/mK. At its turn, the 70 μm Cu pad lies over the 100 μm thick dielectric layer which constitutes the more critical material of the stack. Concerning its K_{TH} value, this parameter has already been extracted in previous works [5] and 1.25 W/mK has been considered for the present simulations. Nevertheless, if this value is unknown, it can be adjusted in a recursive process to match simulation and experimental results. Concerning the dynamic parameters of the dielectric layer, its density (ρ) and specific heat (c_v) have been only approximately evaluated. The IMS manufacturers don't give many details about the composition of the dielectric layer, although a 60% of Al_2O_3 particles and a 40% of epoxy resin seems to be a realistic approximation. If we consider a proportional contribution (in weight) of these materials to the final ρ and c_v values, we found 3.06×10^{-3} kg/m^3 and 9.55×10^{-2} J/kgK respectively. For the given boundary conditions (heat conduction to the backside) and time scales (a few seconds) the dominating time constant is that of the Al base-plate. Future works will analyse the influence of the dielectric layer parameters in the simulation results. As it has been mentioned, the assembly description is completed with the 2 mm thick Al base-plate (6061 alloy, with K_{TH} = 180 W/mK, ρ = 2.7×10^{-3} kg/m^3 and c_v = 9.63×10^{-2} J/kgK), with the second "test point" placed at its backside centre (T_{ref}). The total number of grid cells is around 900.000, and some grid constraints have been forced along the vertical dimension in the critical layers. Or example, the minimum number of vertical grid cells is 5 in the die-attach, 5 in the Cu pad and 10 in the dielectric layer. Finally, concerning the time steps, 50 ms steps are used during the rising and falling edges of the applied power, while 100 ms steps are used during the slower time evolution phases. In these conditions, the simulation of the test assembly takes 20 minutes in a Pentium 4 / 3GHz processor. Figure 11 shows the simulation results (black squares) and the corresponding experimental results (red lines).

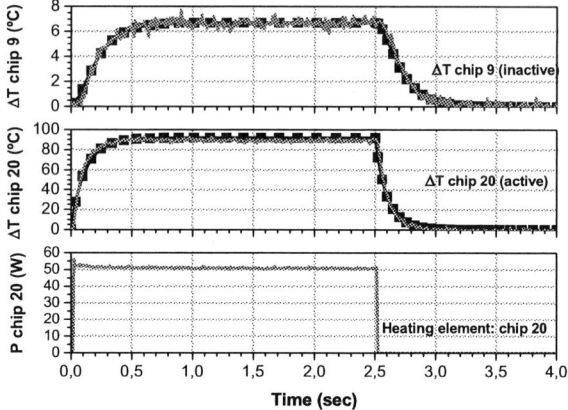

Figure 11: Comparison between experimental (red lines) and simulated (black squares) waveforms.

As it can be observed, a very good agreement has been obtained between the simulated and the measured temperature increments, validating the simulation process of the test power module: structure modelling and description, boundary conditions, thermal parameter values, simplifications, etc.

6. Conclusions

This paper presents a test power assembly based on a thermal test chip, used to validate 3-D CFD simulations of power modules. For the thermal simulation of typical power electronics assemblies it is crucial to determine the correct material parameters values, boundary conditions, etc. but these tasks are not always easy to comply when dealing with complex practical systems. The developed test assembly behaves like a practical or functional module, but it has been designed to give as much thermal information of the structure as possible. The key element of the test assembly is a thermal test chip. This chip thermally behaves like a typical power device but it allows simultaneous power dissipation and temperature measurements using standard measurement instruments. Two of such chips have been used in the test assembly in order to evaluate coupled and self-heating phenomena. The substrate used to develop the test module is an insulated metal substrate, used for medium power applications. The presented measurement set-up is based in a thermal resistance measurement system, but it

introduces the modifications necessary to allow transient temperature acquisition. The basic idea consists in excite one of the thermal test chips with a single power pulse and to measure the temperature rise of both chips from the reference taken at the module backside. The module itself is placed over a forced-convection heatsink (like in true operation conditions) with a small hole to allow the reference temperature measurement. It has been shown that the chips and reference temperatures evolve simultaneously and their difference is constant after 0.7 s approximately. This means that with a 2.5 s power pulse the transient and the static module thermal behaviour can be analysed at once. The same experiment has been reproduced using the FLOTHERM simulator. The main materials involved in the assembly and the main boundary conditions have been analysed. The most critical material of the stack is the dielectric layer of the substrate, which is usually not well described by the manufacturers. The different approaches undertaken to select its thermal parameters have been validated by the good agreement between simulated and experimental results, although in the present conditions, the thermal dynamic behaviour is dominated by the Al baseplate. Other approximations used to simulate the test module have also been validated, such as the elimination of the natural air convection. Future works are oriented to the analysis of shorter heating times and repetitive power pulse excitation.

Acknowledgments

This work was partially supported by the project "Power and Thermal Management of Wide Band Gap Semiconductors" of the European Space Agency (ESA) under contract 17441/NL/CH and by the Spanish Ministerio de Educación y Ciencia under contract TEC2005-087392 (SPACESIC Project).

References

1. Madrid, F., <u>Thermal Conductivity and Specific Heat Measurements for Power Electronics Packaging Materials</u>, PhD Disertation Report, Universitat Autonoma de Barcelona (Bellaterra -Spain, 2005).

2. Flomerics Limited, Flotherm Manual, Issue 1.0 (2003)

3. A. Ammous, B. Allard, H. Morel, "Transient Temperature Measurements and Modeling of IGBT's Under Short Circuit," *IEEE Trans. on Power Electronics*, Vol. 13, No. 1 (1998), pp. 12-25.

4. J.L. Gálvez, X. Jordà, X. Perpiñà, M. Vellvehí, P. Godignon. "Determination of the parasitic inductances in IGBT power modules". *Proc. 11th Electronique de Puissance du Futur (EPF)*, Grenoble (France), July 2006.

5. X. Jordà, M. Vellvehi, F. Madrid, J. L. Gálvez, P. Godignon, J. Millán, "Comparison Between Simulated and Experimental Thermal Resistances of Power Devices Using an Specific Test Chip," *Proc. Mechanical and Multiphysics Simulation and Experiments in Micro-Electronics and Micro-Systems Conference EuroSIME*, Como (Italy), April 2006.

Thermo-mechanical investigations on the effects of the solder meniscus design in solder joint lifetime for power electronic devices

M. Bouarroudj[1], Z. Khatir[1], S. Lefebvre[2], L. Dupont[2]

[1]INRETS-LTN, 2 av. Malleret-Joinville, F94114 Arcueil, France
[2]SATIE-ENS Cachan/CNAM, 61 Av. du prés. Wilson, F94235 Cachan, France

Abstract

The crack and delamination of solder joints between base plates and DCB substrates of power modules is one of the most frequently encountered failure mode and studied in literature. In this paper we present numerical effects of solder meniscus design in solder lifetime prediction. Especially, we show the effect of singular points, which appear in the border edges or corners of DCB or solder joints geometries, in mechanical stress, strain and plastic work evaluations. Furthermore the effect of mesh density on mechanical stress and strains is shown. Finally, we will see the effect of the meniscus design of the solder joint on lifetime estimation using both plastic strain-based and energy-based models.

1. Introduction

Thermal cycling is often responsible for thermo-mechanical damages of power electronics devices [1]. Such constraints lead to crack initiation followed by crack propagation inside solder attach materials. In case of IGBT power modules, the weakest attach layers are the solders between DCB (Direct Copper Bonding) and component base-plates due to their large areas [2].

In previous works [1,3], scan acoustic microscopy (SAM) analyses have shown that DCB substrates are rarely attached to baseplates with homogeneous layer thickness. The most frequent case is an over-thickness in one of the substrate corner and an under-thickness at the opposite one. In such a case, it has been experimentally and numerically shown that the thinnest corner is the weak area for crack initiation when it is subjected to thermal cycles [4]. Nevertheless, too much thick solder layers are not desirable because its results in higher thermal resistances. Generally, after thermal cycles, crack initiation of the solder occurs at the corner of DCB metallization as shown in figure 1 and then propagates from this point. Thus, the geometrical shape and the solder amount at these locations should have some importance for the solder lifetime. As illustration, figure 2 shows a typical solder joint meniscus shape.

Many authors have published results concerning lifetime evaluation of such solder joints for electronic devices. Unfortunately, geometrical singularities like border edges or corners make the simulation results delicate and very sensitive to the mesh density. So, these models are quite good for comparison of different design from each other but are not necessarily well-suited for realistic lifetime prognosis.

In this paper, we present a numerical study of the effect of solder meniscus design on thermo-mechanical results. Especially, we show the effects of singular geometries such as border edges solder joints, in mechanical stress, strain and plastic work evaluations. In addition, we compare lifetimes of such solder joints on the basis of plastic strains and plastic work density.

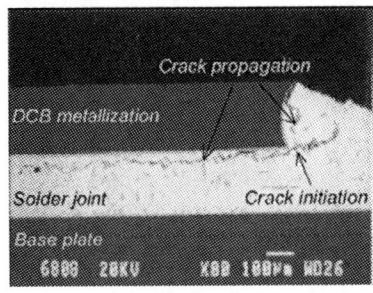

Fig.1 Crack in a solder attach between DCB substrate and base plate of an IGBT power module (from [1]).

Fig.2: Typical meniscus shape of solder joint at the corner edge.

2. Problem statement

Generally, for simplification purpose, the modelled solder layers are geometrically very simple with right angles border edges and corners (Fig.3). Maximum mechanical constraint values are calculated at the geometrical edges where singular points are located. Unfortunately, simulation results are very sensitive with the mesh density, and mathematical procedures must be used in the vicinity of singular locations to avoid singular results. In this aim, some authors focus the mechanical evaluations on elements located in the most stressed region but not at the critical one where numerical singularity occurs [5]. An other frequently used approach is to carry out an average constraint value on selected delamination area [6,5,7].

1-4244-1105-X/07/$25.00 ©2007 IEEE

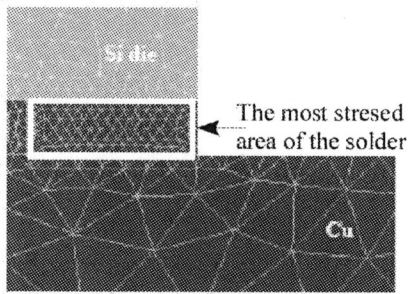

Fig.3 : Typical solder joint model in power electronic packaging [8]

About mesh size effects, authors in [9] have used the second approach and conclude that lifetime prediction of solder decreases with mesh density until the mesh size is small enough to make negligible the difference in the life prediction. In what follows we will verify the criticality of the singular point and in particular the effect of mesh density in different solder designs on mechanical stress and strain evaluations and on plastic work density calculation.

3. Finite Element Models

In order to show solder meniscus effect, we have simulate its thermo-mechanical behaviour with a finite element analysis tool (ANSYS) for an unrealistic model but often used for its simplicity (figure 4 (a)) and three different realistic shape designs (figures 4 (b) to (d)). Material data are given in table 1, DCB ceramic is considered as linear, copper metallization is considered non-linear material (elasto-plastic) and solder is considered visco-plastic and modelled with the Anand model which parameters are given in table 2.

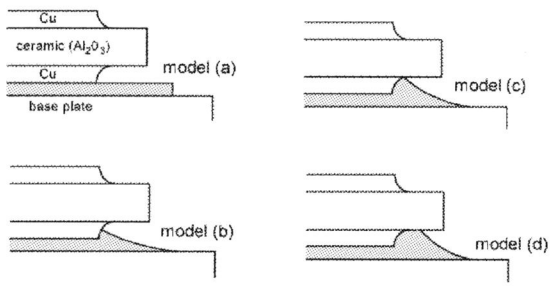

Fig.4: Modelled meniscus shapes of the solder attach.

For comparative purpose, we numerically applied the same thermo-mechanical loading history. First, we simulated the soldering process in order to take into account the initial residual stresses in the assemblies by cooling down from 183°C to 20°C, the highest temperature is the liquidus value of the solder for which all materials of assembly are assumed stress free. Then, we applied ten passive thermal cycles to all designs between -20°C and 120°C with 10 mn dwell times and 10°C/mn temperature variations.

materials	Therm. Cond. (W/m.K)	CTE ($10^{-6}K^{-1}$)	Specific heat (J/Kg.K)	Young modulus (MPa)
Cu	400	17	400	119×10^3
Al_2O_3	30	4.3	765	370×10^3
$Sn_{63}Pb_{37}$	50.6	24.7	180	75842.33-151.68T(°C)

Table 1: Material physical parameters.

s_0 (MPa)	Q/R (K)	A (s^{-1})	ξ	m	h_0 (MPa)	S (MPa)	n	a
12.41	9400	4.10^6	1.5	0.303	1378.95	13.79	0.07	1.3

Table 2: Solder Anand parameters

Due to the symmetry nature of the problem when we consider only the DCB, solder and base plate assembly, we used 2D axisymmetric model. The whole model has been meshed with PLANE42 elements except the solder material which has been meshed with VISCO106 elements for highly non-linear behaviour such as creep. For example, the finite element model of figure 4 (a) shown in figure 5 required 21066 elements and 20205 nodes with 30µm brick elements size at the interesting region of the assembly.

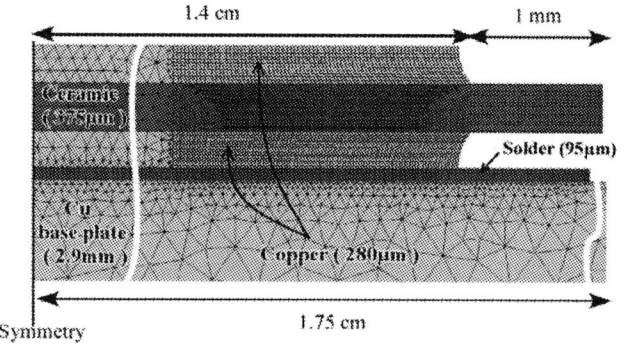

Fig.5: Axisymetric finite element model of the assembly.

4. Results of the FEM simulation

Figure 6 gives the calculated shear stress maps within the solder joint at the end of both temperatures dwells (after stress relaxations) and for all design cases of figure 4. As a result, it can be seen that the maximum shear stresses are located in the same region, at the corner edge of the DCB metallization, both at high and low temperatures, whatever the solder shape design. We can observe that for each temperature and for meniscus shape cases (b) to (d), the stress values are similar. The location and the values of maximum stress are always the same whatever the meniscus shape : around 24.2 MPa near the corner edge of the copper metallization of DCB. But for the commonly modelled shape design (a), stress values are higher and focused in a singular point leading to misestimate the real stress values, especially at low temperature.

410

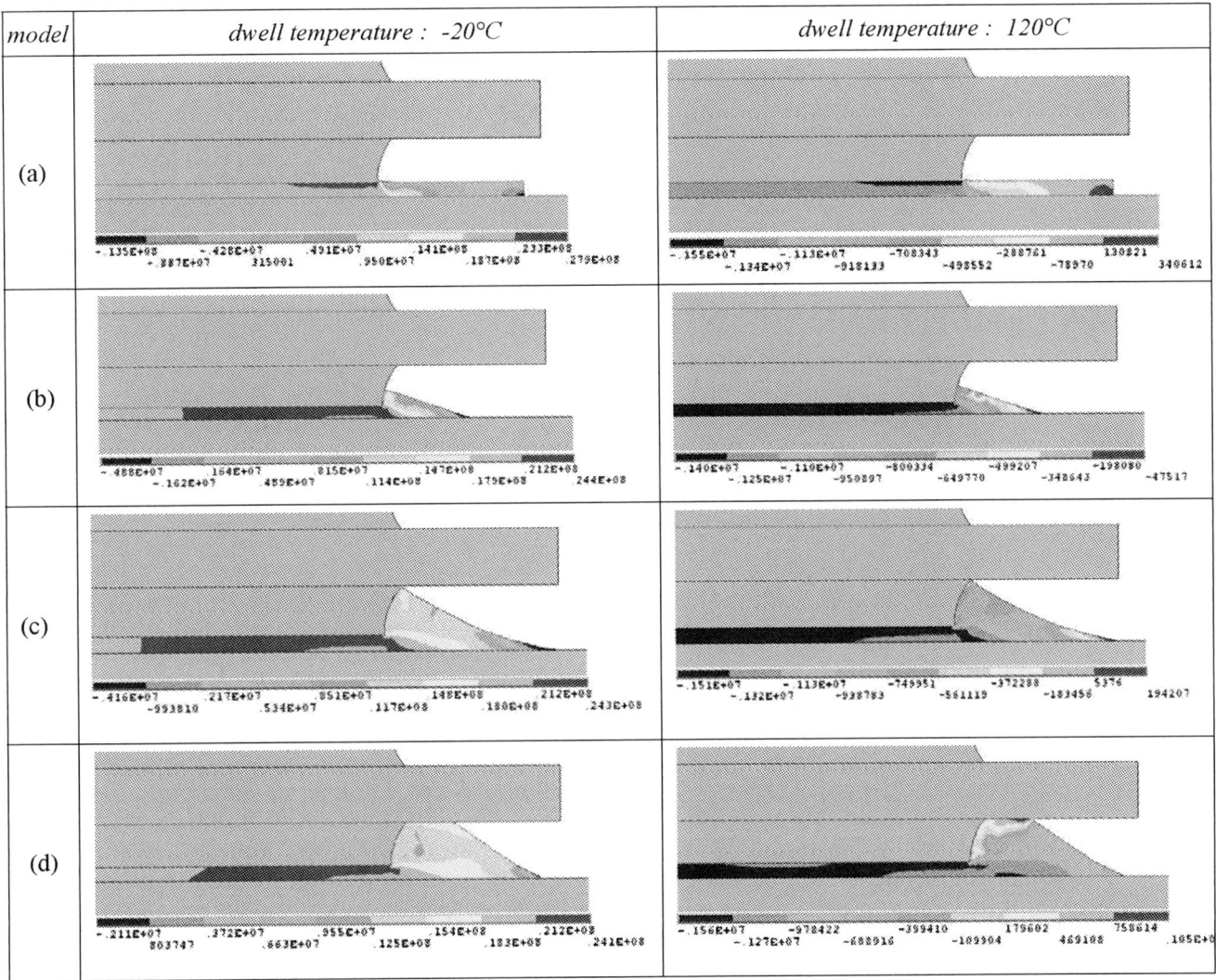

Fig.6: Comparison of shear stress distributions (in Pa) on the solder joint in the different model geometries at the end of both low (-20°C, left column) and high (120°C, right column) temperature dwells for the last cycle.

We will see in the next section that the shear stress, shear strain and cumulated plastic energy density in solder are the main parameters which allow to estimate the solder fatigue lifetime. So, we will compare here the effect of solder meniscus shape, for all models, on both the shear stress distribution and cumulated energy in straight paths along the solder/copper interface as schematically shown in figure 7. In this figure, it can be seen the shear stress distribution along this path for the four models (a) to (d) whereas figure 8 shows the cumulated plastic energy density along the same paths. The initial position (x=0) is the neutral point at the axisymetric axis.

As expected from figure 6, the maximum values of shear stress and cumulated plastic energy are located in the same region for all models. For meniscus shapes (b) to (d), stress and energy values are quite similar but for the common design model (a), results exhibits high peak values at the singular point. Especially, model (a) leads to

high discontinuity on stress values (Fig.7) at the singular point contrary to the other ones. Misestimation, due to model singularity, concerns also the inelastic energy density. As illustration, figure 8 shows the cumulated plastic work, along the same path, after ten thermal cycles.

As discussed above, the mesh density of the FEM model may affect the solder lifetime estimation. Thus, to evaluate this effect, we reduced element size meshes from 30μm to 20μm and then to 6μm for models (a) and (c). Figures 9 and 10 give the effect of mesh size on shear stress in the solder joint along the interface axis, respectively for models (a) and (c), at the vicinity of the corner edge of DCB metallization. For locations distant from the singular point, mesh size has no influence. For model (a), reducing the mesh size amplifies the singularity effect and makes the discontinuity more important (see figure 9). For models (b) to (d), which don't exhibit singularity, mesh size has not significative effect on shear stress. As illustration, figure 10 shows such effect for model (c).

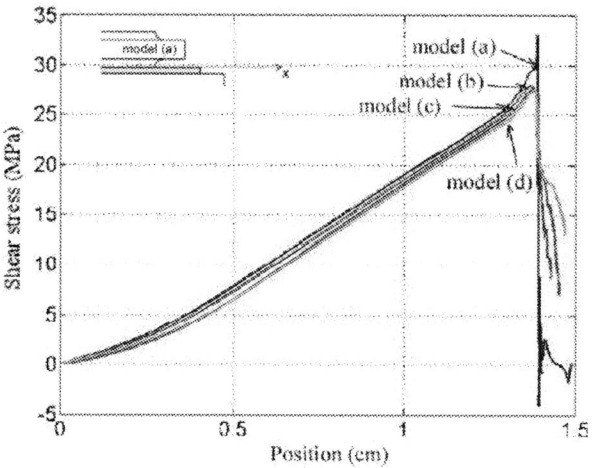

Fig.7: Shear stress along x axis for models (a) to (d).

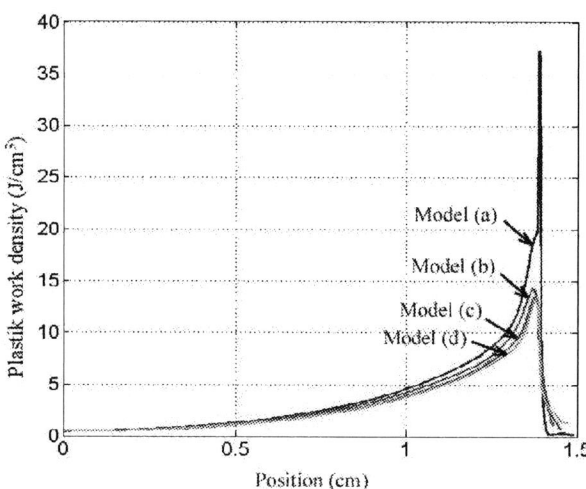

Fig.8: Cumulated plastic work density along the x axis for models (a) to (d) after 10 thermal cycles.

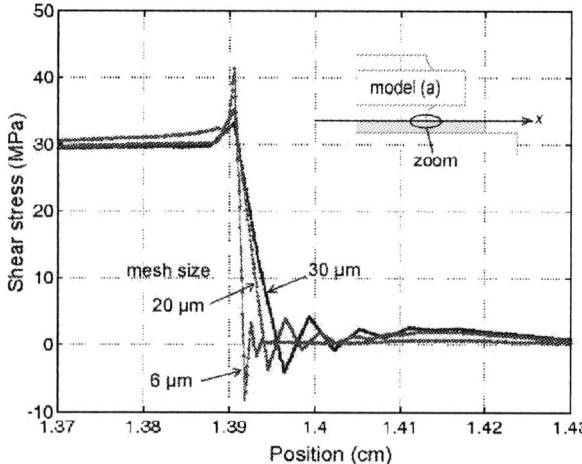

fig 9: Mesh size effect on the shear stress along the x axis for model (a) at the vicinity of singular point

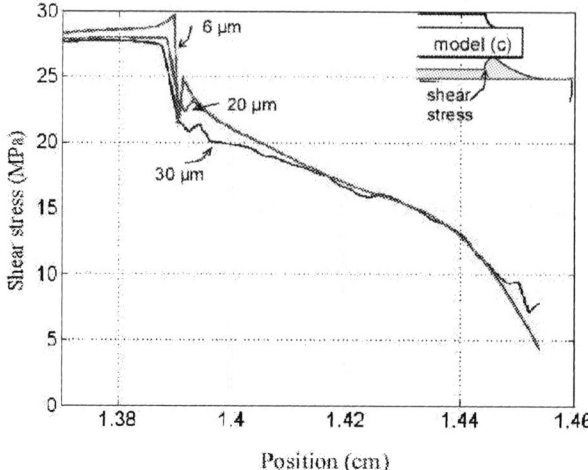

fig 10: Mesh size effect on the shear stress along the x axis for model (c) at the vicinity of singular point.

The variation of shear stress (solid line) versus time is plotted in figure 11 from the model (c) at the location shown in the diagram inserted in the same figure where stresses are maximum. The thermal cycle is also visible in dotted line. Furthermore, the effect of the mesh size (6, 20 and 30µm) on the shear stress is given. We can see that the solder joint exhibits large stress relaxation at the higher temperature dwell and do not at the low level. This is due to the fact that at high temperature, solder becomes less rigid and creep effect is more significant.

As a result, it can be seen in this figure that the mesh density is unsensitive in this model. Only a very light difference (around 1 MPa), visible during the low temperature level, is noticeable between 6µm and 30µm mesh size FE models.

On the contrary, we can see, in figure 12 for model (a), that the mesh size has a higher effect on the shear stress than for model (c). In this figure, results are given at the singular point (SP) for 6µm, 20µm and 30µm mesh size. Because of singular values obtained for the smallest mesh size, we also give results at the nearest node from the singular one (near SP). Globally, results show that whatever the mesh size for model (a) shear stress values are higher than those calculated with model (c). As for model (c) (figure 11), the mesh size has a significant effect only at low temperature level.

As already said, the mesh size has not significant effect in shear stress for model (c). Nevertheless, it has a significant effect on shear strain as visible in figure 13. Figure 14 gives the strain evolution for model (a). For this case, strain levels are strongly higher than those calculated for model (c) leading to a faster plastic work evolution. The cumulated evolution of plastic work density, for ten cycles, is given in figure 15 in the lower area (dotted lines) for model (c) and also for model (a) in the higher area (solid lines). This figure shows that the solder plastic energy which is clearly driven by the shear strain rather than by shear stress depends strongly on the mesh size for both models.

The cyclic variation of plastic strain ($\Delta\gamma$) and plastic work density (ΔW_{pl}), during the last cycle, are given in table 3 both for model (a) and (c) and according to the different mesh sizes. Obtained results give the plastic strain range of model (c) about half than for model (a) whereas the plastic work range for model (c) is about the third of model (a).

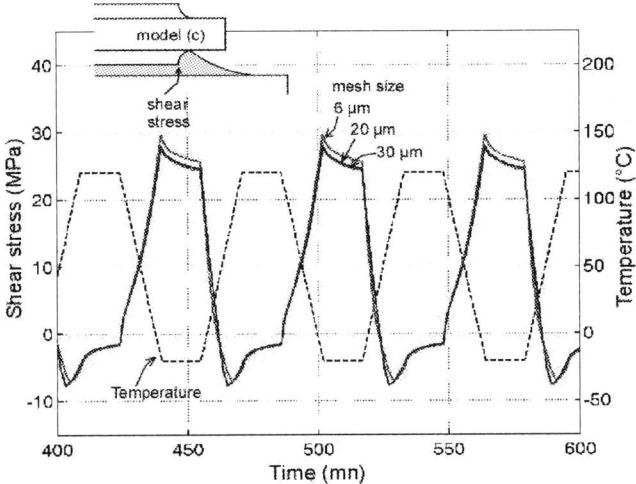

Fig.11: Mesh size effect on the shear stress evolution - model (c).

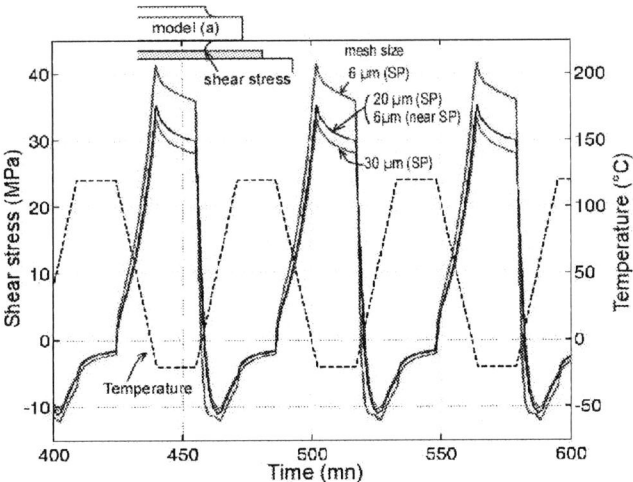

Fig.12: Mesh size effect on the shear stress evolution - model (a)

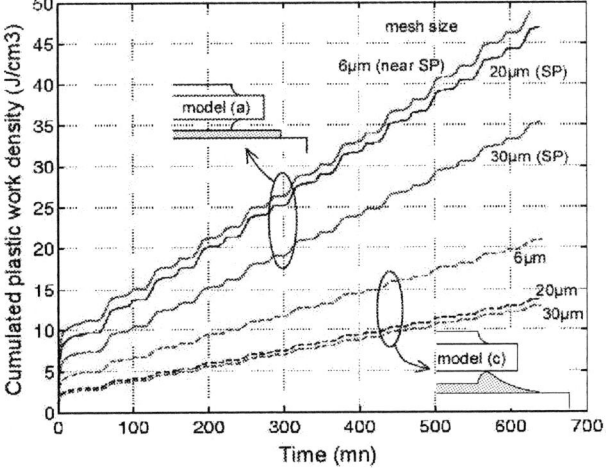

Fig 15: Cumulated plastic work densities in the solder joints

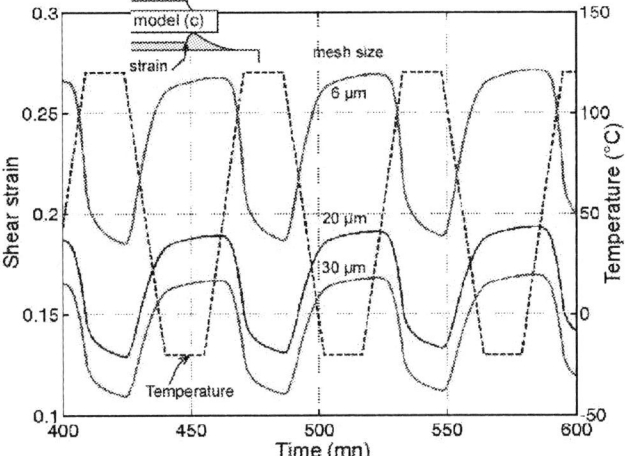

Fig.13: Mesh size effect on the shear strain evolution - model (c).

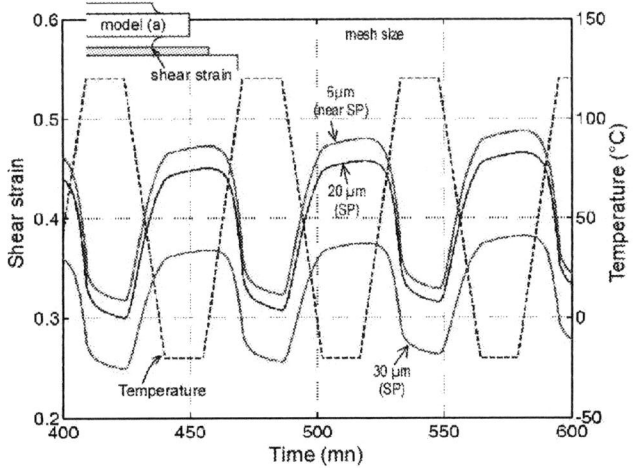

Fig.14: Mesh size effect on the shear strain evolution - model (a).

		6µm	20µm	30µm
Model (a)	$\Delta\gamma$	0.15	0.14	0.11
	ΔWp (J/cm^3)	4.1	3.8	2.8
Model (c)	$\Delta\gamma$	0.08	0.058	0.056
	ΔWp (J/cm^3)	1.63	1.10	1.05

Table 3: Effects of solder geometry and size meshes on plastic strain and plastic energy density variations

In spite of the fact that shear stress is independent of mesh size for model (c), shear strain and then cumulated plastic energy are dependent on the mesh size even if model (c) does not present a singular point. This remark makes

413

delicate the estimation of the lifetime from strain-based or energy-based models.

5. Solder Lifetime evaluation

In the following, we will compare plastic strain-based and plastic energy-based models for the lifetime prognosis of numerically studied solder joints. For this purpose, we used the engelmaier's model (strain-based) [10] and darveaux's model (energy-based) [11,12]. Both models are given in appendix with the parameters corresponding to the solder material (SnPb37).

The obtained results are given in table 4 where we particularly compare the obtained lifetime results for the singular geometry model (a) with the non-singular one (c). The mesh size effect on predicted lifetimes has been also evaluated and given in this table.

For Darveaux's model, we assumed that $N_f(\Delta W_{pl})$ is obtained when the crack propagation has lead to 5% decrease of the solder joint surface area. For the axisymetric models (fig.5), where solder joints are disks of 1.5cm radius, the failure criterion is when the crack length "a" reaches 380 μm.

	Model (a)		Model (c)	
	$N_f(\Delta\gamma)$	$N_f(\Delta W_{pl})$	$N_f(\Delta\gamma)$	$N_f(\Delta W_{pl})$
6 μm	17	50	78	126
20 μm	20	54	169	186
30 μm	36	73	184	197

Table 4: Effects of solder geometry and size meshes on solder joint lifetime estimation.

As expected, results show that model (a), with singular geometry, is more pessimistic with a lower number of cycles to failure compared to (c). In addition, strain-based and energy-based models are quite similar and correlated for design (c) but they are less correlated for design (a). A benefit of using energy-based model, is that it reduces the singular geometry effects. Ratios on lifetimes between design (a) and (c) are around 3 and 5 respectively for energy-based and strain-based models. Finally, an other benefit of energy based model is that it is less sensitive to the mesh size effects. This is all the more true if we use an averaged volumetric plastic work value which reduces the mesh sizing effect.

6. Conclusion

We presented, in this paper, a numerical study of the effect of solder meniscus design on thermo-mechanical results. Especially, it has pointed out the effects of singular points, which appear in the border edges or corners of DCB or solder joints geometries, in mechanical stress, strain and plastic work evaluations. We have shown that singular geometries may lead to over estimation of mechanical stresses as well as plastic strain and plastic work densities inside the solder joints. Nevertheless, for the solder lifetime prognosis, which need such thermo-mechanical values, we have shown that

there is more benefits to use energy-based models rather than strain-based ones. Especially, energy-based models reduce the singular geometry effects, such as corners and border edges, and they are less sensitive to the mesh size of the finite element model.

Appendix

a) Strain based solder lifetime :

We used Engelmaier model [10] to assess the lifetime of the solder joint based on shear strain results:

$$N_f = \frac{1}{2}\left(\Delta\gamma / 2\varepsilon'_f\right)^{1/c}$$

with $2\varepsilon'_f \approx 0.65$ and

$$c = -0.442 - 6\times10^{-4}\,\overline{T}_s + 1.74\times10^{-2}\ln(1+f)$$

Where ε'_f is the fatigue ductility coefficient, N_f the mean cycles to failure and c the fatigue ductility exponent. In our case, the cyclic frequency (f) is equal to 25 cycles/day and the mean cyclic solder joint temperature ($\overline{T}_s$) is 50°C.

a) Energy based solder lifetime :

The Darveaux's model [11,12] has been used to assess the lifetime of the solder joint based on plastic work results:

$$N_0 = 22400\,\Delta W_{pl}^{-1.52} \text{ and } da/dN = 5.86\times10^{-7}\,\Delta W_{pl}^{0.98}$$

In this formulation, N_o is the number of cycles to crack initiation, da/dN the crack growth (in inch/cycle) and ΔW_{pl} is the cyclic plastic work density (in psi).

References

1. J.M. Thebaud *et al.*, "Strategy for designing accelerated ageing tests to evaluate IGBT power modules lifetime in real operation mode", IEEE Trans. On components and packaging technologies, vol.26, n°2, pp.429-438, june 2003.

2. M. Ciappa, "Selected failure mechanisms of modern power modules", Microelectronics Reliability, Vol.42, n°4-5, pp.653-667, April-May 2002.

3. K. Guth, P. Mahnke, "Improving the thermal reliability of large area solder joints in IGBT power modules", CIPS, 2006.

4. J. Yamada et al., "the latest High Performance and High Reliability IGBT Technology in New packages with Conventional Pin Layout", PCIM conference, Nuremberg, 2003.

5. M.Roellig, R. Dudek, S. Wiese, B. Boehme, B. Wunderle, K.J Wolter, B. Michel, "Fatigue analysis of miniaturized lead-free solder contacts based on a novel test concept", Microelectronics Reliability, 2006.

6. J-P Sommer, T.Licht, H. Berg, K. Appelhoff, B. Michel, "Solder Fatigue at High-Power IGBT Modules",CIPS 2006.

7. R. Dudek, H. Walter, R. Doering, B. Michel, "Thermal fatigue modelling for SnAgCu and SnPb solder Joints", 5th International Conference on Thermal and Mechanical Simulation and Experiments in Micro-electronics and Micro-Systems, EuroSim2004.

8. A. Guédon-Gracia *et al.*, "Influence of the thermo-mechanical residual state due to the power assembly process on the lifetime modellization", Micro-electronics Reliability, vol.44, pp.1331-1335, 2004.

9. Q.J. Yang, X.Q. Shi, Z.P. Wang, Z.F. Shi, "Finite-element analysis of a PBGA assembly under isothermal/mechanical twisting loading", Finite Elements in Analysis and Design, Vol.39 pp.819–833, 2003.

10. W. Engelmaier, "Fatigue Life of Leadless Chip Carrier solder joints during power cycling", IEEE Trans. on comp. hybrids and manuf. Technology, vol.6, n°3, sept. 1983.

11. R. Darveaux, "Effect of Simulation Methodology on Solder Joint Crack Growth Correlation", Proceeding of the Electronic Components and Technology Conference, 2000.

12. B.A Zahn, "Impact of ball via configurations on solder joint reliability in tape based chip-scale packages", Proceeding of the 52nd Electronic Components and Technology Conference, 2002.

Development and Assessment of Global-Local Modeling Technique Used in Advanced Microelectronic Packaging

F. X. Che*[1], H.L.J. Pang[2], W. H. Zhu[1], Wei Sun[1], Anthony Y.S. Sun[1] and C.K. Wang[1] and H.B. Tan[1]

[1] United Test & Assembly Center Ltd. (UTAC)
5 Serangoon North Ave 5, Singapore 554916
*Email: FX_Che@sg.utacgroup.com, Tel: 65511465
[2] School of Mechanical and Aerospace Engineering, Nanyang Technological University
50 Nanyang Avenue, Singapore 639798

Abstract

In this study, two types of global-local models are introduced. One is submodeling, in which a coarse global model is used to simulate the whole model and the fine local model is used to simulate the critical region of interest from whole model. The other is global-local-beam (GLB) model, in which the joint is replaced by an equivalent beam with effective stiffness. For submodeling, two different cut boundaries are compared and suitable cut boundary is proposed. In addition, the effective global-local model combining submodeling and GLB modeling technique is also introduced. Case study is presented in this paper. Firstly, PBGA assembly subjected to thermal cycling was investigated using global-local FEA modeling. Secondly, the GLB modeling was used in modal analysis for FCOB assembly. Thirdly, bending simulation was conducted for VQFN assembly using both two-level submodeling and one-level submodeling method.

1. Introduction

Manufacturers of electronic products face with demands for design with high reliability and performance at lower costs. Computational modeling can reduce the product development time to market in a competitive electronic product sector. Commercial finite element analysis (FEA) software, such as ANSYS, has been used extensively to simulate reliability test loads on electronic components, board assemblies and product systems. By using FEA simulations, industry can minimize the requirement for extensive and time-consuming physical testing. This can reduce product development costs, increase reliability and reduce the product time to market. Further miniaturizations of IC component size and higher I/O counts are expected trends in electronic packaging applications. Thus, conventional application of finite element modeling technique will become more difficult as the geometry features become smaller and require higher number of elements so that FEA simulation will require higher speed computing, larger memory size and hard disk storage space. These critical requirements can limit the use of full 3D model applications for finite element reliability analysis of the electronic assemblies. In order to reduce the element size in the FEA simulation for solder joint reliability, some reduced models were used by researchers, including slice model [1-4], one-eighth model [1-2, 4-5], and 2D model [1, 3-4]. Some trade-off in accuracy is expected in these simple models. The comparison details for different FEA models were presented in reference [1]. In this study, a global-local modeling technique was developed. The global-local modeling method with coarse global model and fine submodel was reported by researchers [6-8] for BGA assemblies subjected to thermal or mechanical loadings. Some simplified models for stress-strain analysis of BGA assemblies subjected to thermal cycling or vibration loads, where the solder connections were considered as an effective beam element with similar stiffness, were employed by researchers [9-10].

Case study is presented using developed global-local modeling technique considering different package types and different loading types. Firstly, fatigue life prediction for PBGA assembly subjected to thermal cycling was investigated using submodeling and effective global-local model. Secondly, the GLB modeling was used in modal analysis for FCOB assembly. Thirdly, cyclic bending simulation was conducted for VQFN assembly using both two-level submodeling and one-level submodeling. For two-level submodeling, the first nodal results transfer is done from board-level global model to package-level submodel, then the second nodal results transfer is done from package-level model to solder joint level submodel.

2. Global-Local Modeling Techniques

The global-local model is a good choice for board-level simulation. In this study, two types of global-local models were introduced. One is submodeling, in which a coarse global model was used to simulate the whole model and the fine local model was used to simulate the critical partial area of the whole model. The other is global-local-beam (GLB) model, in which the interconnections were replaced by an equivalent beam with effective stiffness. The results from global-local model simulation were compared to those from fine 3D model simulation for calibration of global-local modeling method.

2.1 Submodeling Method

Submodeling is also known as the cut-boundary displacement method or the specified boundary displacement method. In this study, submodeling was developed using the flip chip on board (FCOB) assembly as FEA calibration study. The geometry size mismatch among FCOB assemblies gives rise to more difficulties in numerical modeling. So the submodeling technique is a good choice to reduce the model size significantly. Before

1-4244-1105-X/07/$25.00 ©2007 IEEE

using submodeling technology, the verification was first conducted to validate the feasibility and accuracy of submodeling as well as the appropriate cut boundary. The silicon chip or die was connected onto PCB by four corner solder joints as shown in Fig. 1 for benchmark study in order to reduce element size. The chip size is 8.5mm×8.5mm×0.65mm. For calibration study, the material of 62Sn36Pb2Ag was selected for solder joint with a standoff height of 0.1mm. The FR-4 PCB has the size of 30mm×10mm×1.13mm. In order to determine the underfill effect on the fatigue life of the electronic product, two cases are studied including FCOB assembly with and without underfill between the die and PCB. For convenience, copper pad and UBM material used in actual FCOB assembly are not considered because of just validation study for submodeling technique. When using the submodel, it is necessary to verify that the cut boundaries are far enough away from the stress concentration region, or interest area.

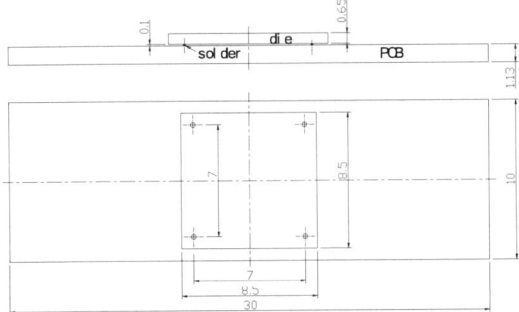

Fig. 1 Geometry size of the FCOB (unit in mm)

Firstly, the main purpose of validation study was to find the reasonable cut boundary and compare the results between the submodel and conventional fine 3D model. The fine 3D model was considered as a reference model, or benchmark model. When the results from submodel agreed with those from reference model, the submodel and corresponding cut boundary were regarded satisfactory. The temperature increment of 20°C was applied to the FCOB assembly as a loading condition. Due to the symmetry, only the quarter part of the assembly was modeled. The 3D models are shown in Fig. 2. Two types of boundary condition as shown in Fig. 3 are conducted to determine the reasonable cut boundary. Fig. 3(a) shows hybrid submodel with a cut boundary far enough away from solder joint interfaces and Fig. 3(b) shows a submodel of solder ball. The grid size of the submodel is same as that of the fine 3D model as shown in Fig. 2(a). The results from submodel were compared with those from the fine 3D model. The Von Mises stress was compared due to its common application in the reliability analysis of the electronic assembly. Figure 4 shows the maximum von Mises stress of solder/die interface comparison between fine 3D model and submodel with two different cut boundaries. It can be seen the first cut boundary gives more accurate result than the second one compared to reference model result.

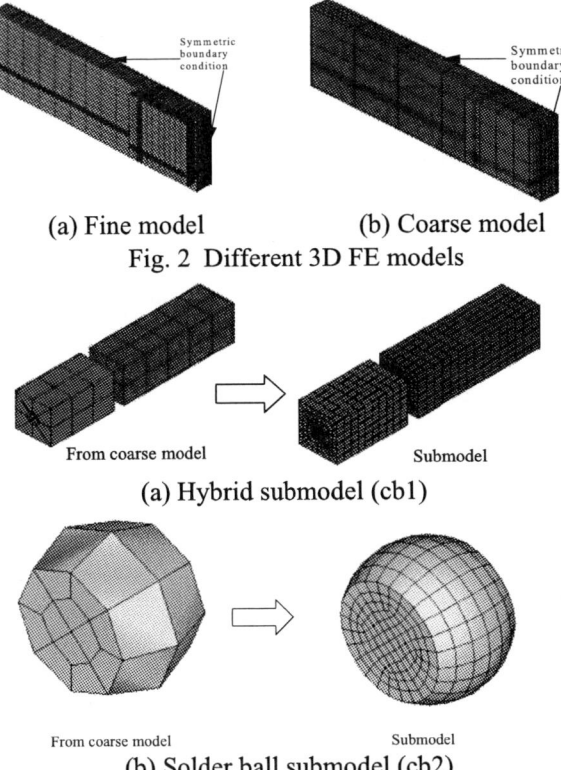

(a) Fine model   (b) Coarse model
Fig. 2 Different 3D FE models

(a) Hybrid submodel (cb1)

(b) Solder ball submodel (cb2)
Fig. 3 Two cut boundaries for different submodels

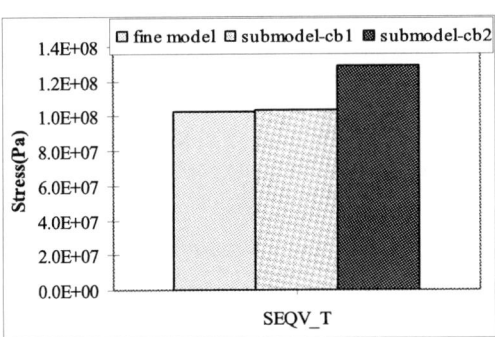

Fig. 4 Von Mises stress for different models

Then, the reasonable cut boundary including die, solder joint and PCB was selected to do the simulation for FCOB specimen when subjected to thermal cycling from –40°C to 125°C with 1hour per cycle and 15mins dwell time. Two cases were considered, that is, FCOB without underfill case and FCOB with underfill case. The elastic-plastic-creep (EPC) model and viscoplastic Anand model are commonly used constitutive models for solder material. These two models can lead to consistent result when thermal cycling loading is simulated [11]. The Anand model was used in this study. The plastic work density is an important parameter in the fatigue life prediction of solder joint when subjected to thermal cycling. Therefore, the plastic work density of center node on die/solder interface was extracted to take comparison and was shown in Fig. 5 for nonunderfill case and Fig. 6 for underfill case.

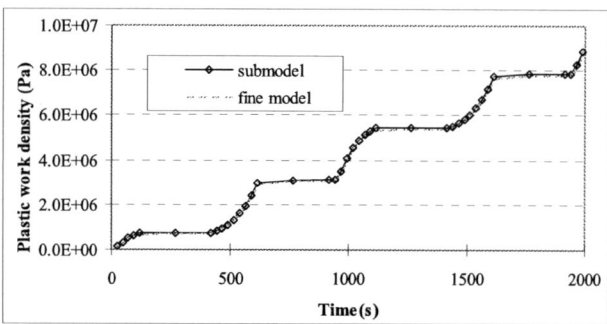

Fig. 5 Plastic work density for nonunderfill case

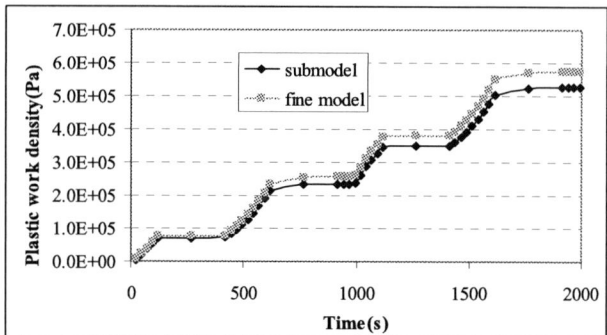

Fig. 6 Plastic work density for underfill case

The error of results obtained from submodeling technology is less than 10% compared with conventional fine 3D model. The advantages of submodeling technique include computational time saving, the hard disc space saving, less memory requirement, and less element size, which can be found from Table 1. When the solid FE model is more complicated, the advantage of submodeling method will become more significant.

Table 1. Advantages of submodel compared to fine model

	Items	Elements	Nodes	Solving time (Mins)	Result file (MB)
Sub-modeling	Coarse model (1)	845	1144	12	113
	Submodel (2)	5580	6176	84	650
Fine 3D model (3)		22905	25408	671	2730
Factor (3)/(1+2)		3.6	3.5	7.0	3.6

2.2 Global-Local Beam (GLB) Model Technique

Vibration and drop analyses need a full model of the assembly because of non-symmetry in out-of-plane displacement. A full detailed model of BGA or FCOB assembly, which includes numerous solder joints and many modules, is difficult to model. The submodeling technique mentioned earlier can reduce the FE mesh complexity, but is still too complex when many components need to be modeled. Therefore, a global-local-beam (GLB) technique was introduced in this study. The key technique in this method is the use of simple elements, such as beam elements with effective stiffness

matrix, to represent solder joints, thus the number of elements needed to model the entire structure can be reduced significantly. This method consists of three steps: stiffness extraction analysis of a single solder joint; deformation analysis of the entire structure and stress strain analysis of solder joint. The specimen selected is same as that shown in Fig. 1. The solder ball is replaced by two-node beam element with effective stiffness and 12 DOFs as shown in Fig. 7. The relation between the generalized nodal force vector $\{F\}$ and the generalized nodal displacement vector $\{u\}$ of beam is expressed in matrix form as follows:

$$\{F\} = \{K\}\{u\} \tag{1}$$

$$\{F\} = \begin{bmatrix} F_{x1} & F_{y1} & F_{z1} & M_{x1} & M_{y1} & M_{z1} & F_{x2} & F_{y2} & F_{z2} & M_{x2} & M_{y2} & M_{z2} \end{bmatrix}^T \tag{2}$$

$$\{u\} = \begin{bmatrix} u_{x1} & u_{y1} & u_{z1} & \theta_{x1} & \theta_{y1} & \theta_{z1} & u_{x2} & u_{y2} & u_{z2} & \theta_{x2} & \theta_{y2} & \theta_{z2} \end{bmatrix}^T \tag{3}$$

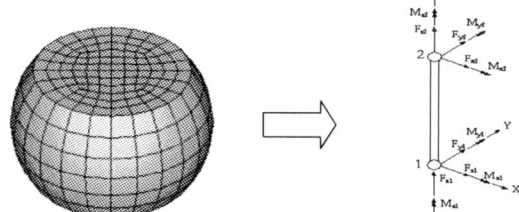

Fig. 7 Solder ball vs. two-node beam element

In the first step, In order to extract the stiffness of the solder joint, unit translational and unit rotational are applied on the surface of the solder joint as a boundary condition. From the solutions the generalized forces imposed at two ends surfaces can be obtained, and these generalized forces form some elements in the stiffness matrix of the solder joint. After 12 different boundary value problems with unit displacement were solved, all the terms in the stiffness matrix can be obtained. When the whole stiffness matrix is obtained, the second step can be simulated. In this step, the solder joints, which connect the chip to the PCB, are modeled as two-node beam elements with the effective stiffness matrix established in the first step. In ANSYS software [12], the element type of MATRIX27 represents an arbitrary two-node element whose geometry is undefined but its elastic kinematic response can be specified by stiffness coefficients. The stiffness matrix constants were input as real constants in the simulation. The chip and PCB can be modeled as solid or shell elements. In this study, these two types of elements are used to model chip and PCB in order to find which one is more accurate. The temperature ranges of $20^{\circ}C$ is selected as loading to validate the feasibility of global-local beam modeling method. From deformation analysis, the weakest solder joint should be selected for next analysis. Usually, the relative deformations between two nodes of the beam element are used as an index to determine the weakest solder joint. After determining the weakest solder joint, stress strain analysis of solder joint can be performed. In this step, the nodal displacements at both ends of the beam element are applied to the detailed solder ball model as a prescribed displacement boundary condition. The translational displacements are applied directly at the nodes located on the two end surfaces of solder joint. The rotations are modified to the translational displacements considering the location of each node on the surface firstly, and then the modified translational displacements are imposed on the corresponding nodes. In FEA simulation, the shell element is usually represented by the middle surface of the structure. In the electronic product, the PCB and chip cannot be assumed as the perfect thin shell element because thickness of PCB and IC chip usually do not satisfy the assumption of the thin shell. Therefore, the thick shell element should be considered in the GLB method. According to the theory of thick plate element, the displacements parallel to the middle surface are given by [13]:

$$u(x, y, z) = z\theta_y(x, y)$$
$$v(x, y, z) = z\theta_x(x, y)$$

(4)

where θ_x, θ_y are the rotations about the x and y axes, z is the coordinate in thickness direction. It is called the modified shell model when Eq. (4) was used in DOF transfer.

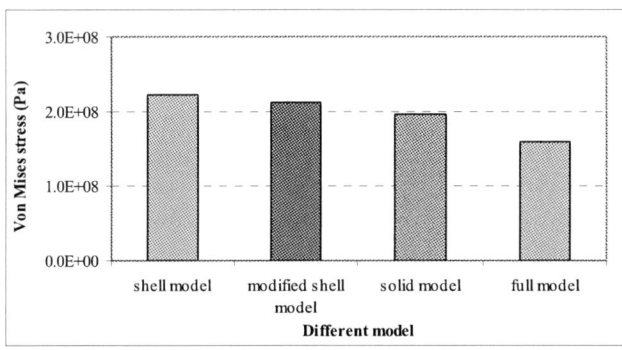

Fig. 8 Von Mise stress comparison for different models

Figure 8 shows the comparison of maximum Von Mises stress in the solder joint from different models. It can be seen that solid model used in GLB model gives rise to more accurate result than shell model. The difference induced by GLB solid model is about 25% compared to the fine 3D model. The difference induced by GLB shell model is larger than GLB solid model, especially for non-modified GLB shell model. The modified shell model is recommended when shell element is used in the entir structure deformation analysis.

2.3 Effective Solder Joint Model

Through analyses for submodeling and GLB model technique mentioned above, it was known that submodeling could lead to more accurate results than GLB model but with more element size and computational resources. It is difficult for GLB model to consider plastic behavior of solder ball because the stiffness of effective beam model for solder joint just considers the elastic deformation of solder ball when

extracting stiffness matrix of solder ball. In order to reduce element size and consider the plastic deformation behavior of solder material, an effective solder joint modeling method was developed by combining submodeling and GLB modeling techniques. In effective solder joint model, the ball shape solder joint was replaced by an effective hexahedron shape solder joint with cubic cross section and the same height as solder ball as shown in Fig. 9. According to GLB modeling technique, the cross section area of effective solder joint can be determined by making solder ball and effective solder joint having the equivalent stiffness in axial tension/compression and shear direction because tension/compression and shear forces are dominant for solder joint when subjected to thermomechanical or mechanical loadings.

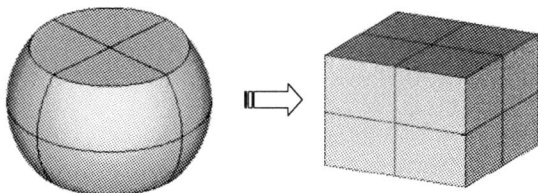

Fig. 9 Real solder ball and its effective solder joint

In the global model, the effective solder joint can be used with fewer elements and plastic deformation can also be considered by an effective solder joint. In order to verify the effective solder joint model, result comparison between effective solder joint model and fine 3D model were conducted considering two cases: FCOB with underfill and without underfill subjected to thermal cycling from $-40^{\circ}C$ to $125^{\circ}C$. Accumulated plastic work density per cycle based on solder/die interface volume averaged method is shown in Fig. 10. It can be seen that effective solder joint model can lead to consistent result with submodel and fine 3D model for both underfill and nonunderfill cases. The practical use of effective solder joint model is that all the ball shape solder joints are modeled as effective solder joints with cubic cross section except the critical solder joint for which the detailed ball shape meshing was used, and then submodel is created and simulated based on the critical solder joint.

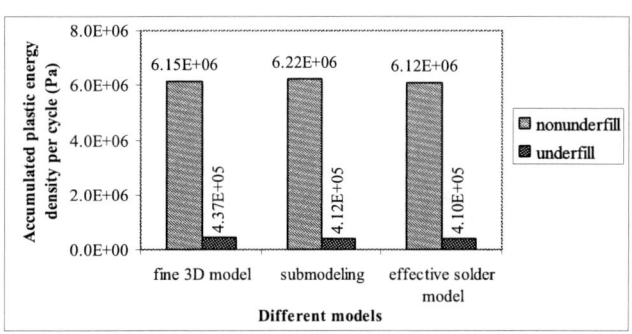

Fig. 10 Accumulated plastic work density per cycle for different models

419

3. Cases Study

3.1 PBGA Assembly Subjected to Thermal Cycling

PBGA specimen was selected for submodeling application case study. The details for thermal reliability test and failure analysis, material properties and loading condition used in FEA modeling and simulation can be referred to the publications [1, 14]. Figure 11 shows FEA models for PBGA assembly including quarter global model and submodel, octant global model and submodel, slice and 2D models, respectively. Quarter model was usually considered as an accurate model. Figure12 shows the equivalent global quarter model and the meshed solder joints. The critical solder joints, for example, solder joints under chip or component corner as shown in Fig. 12, were modeled as real shape solder joint and others were modeled as equivalent cuboid-shape solder joints with the similar effective stiffness as real shape solder joint. The submodel in the effective solder joint model is the same as that used in the quarter model as shown in Fig. 11. The purpose of considering different FE models is to investigate the accuracy of fatigue life prediction when using different FE models. Figure 13 shows the comparison of accumulated volume-averaged strain energy density per cycle of interface layer between solder and component for different models. It can be seen that 2D model gives larger plastic strain energy density than 3D models, thus means that 2D model will underestimate the fatigue life of solder joints compared to 3D models. The consistent FEA results can be obtained for quarter, octant and effective quarter models with using submodeling technique. The comparison between fatigue life prediction and experimental results also verified the accuracy of submodeling application for PBGA assembly [14].

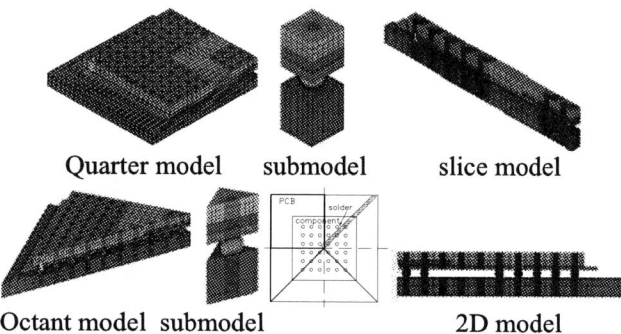

Quarter model   submodel   slice model

Octant model submodel   2D model

Fig. 11 2D and 3D FE models for PBGA

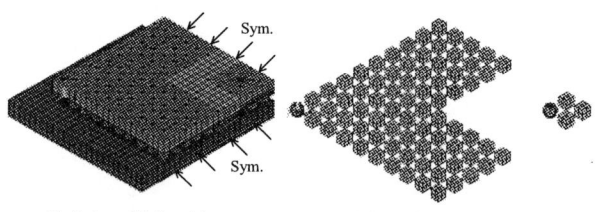

Equivalent global model   Meshed solder joint element

Fig. 12 Effective quarter global FE model

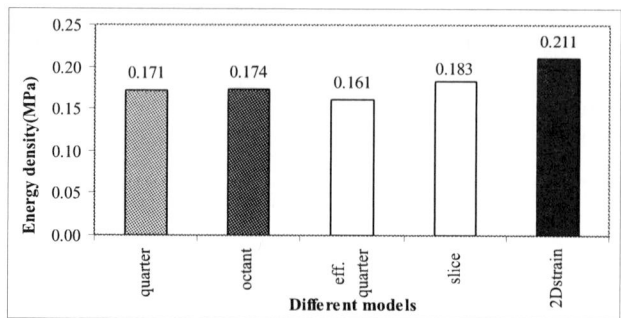

Fig. 13 Plastic work density for different models

3.2 FCOB Assembly Subjected to Vibration

FCOB assembly as shown in Fig. 14 was selected for vibration test and simulation. Six larger chip modules and six smaller chip modules were mounted on the FR-4 PCB. The clamped-clamped along the longer edges boundary condition was used. The details for vibration test and analysis can be referred to publications [15-16]. The modal analysis was conducted to determine the natural frequency. First, the bare PCB was simulated for modal analysis. In addition, each chip will be modeled in order to obtain more accurate results. Because many chip and I/O connectors for FCOB assembly, it is difficult to use the traditional fine 3D model for board level simulation. So, the global-local beam (GLB) model was used in which the IC chips and PCB were modeled as shell and the solder joints were modeled as effective two-node beam elements. The natural frequency of the bare PCB with the clamped-clamped boundary condition can be obtained by [17]:

$$f_n = \frac{3.55}{a^2}\sqrt{\frac{D}{\rho}} \tag{5}$$

where $D = \frac{Eh^3}{12(1-\mu^2)}$, $\rho = \frac{Mass}{Area}$, E is the Young's modulus, μ is the Poisson's ratio, h is the PCB thickness, a is the PCB edge thelength of, ρ is the area density.

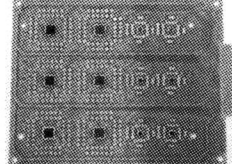

Fig. 14 FCOB assembly layout

Table 2 lists the first natural frequencies obtained from different methods. The frequency obtained from bare PCB has a good agreement with theoretical result. The result of modal analysis using global-local beam (GLB) model has a good agreement with test result.

Table 2. Comparison of the natural frequencies

Model	Bare PCB	vs. Eq. (5)	GLB model	vs. Test
Frequency	208.6	209.4	201.9	194.1

3.3 VQFN Assembly Subjected to Bend

Figure 15 shows the VQFN specimen and schematics of four-point bend test. It is clear that packages between loading span subjected to similar moment load, which increases the sample size in one single four-point bend test. The details for bend test result and analysis can be referred to publications [18-20]. This paper just focuses on the FEA simulation methodology.

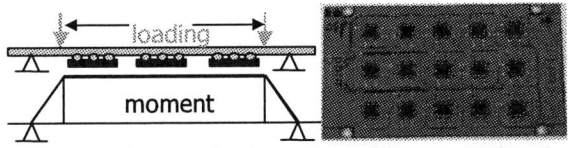

Fig. 15 VQFN and schematics of four-point bend test

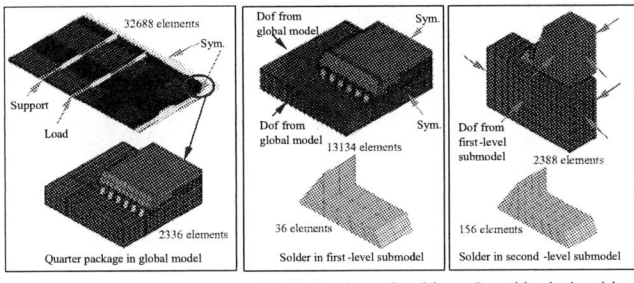

Fig. 16 Two-level submodeling method

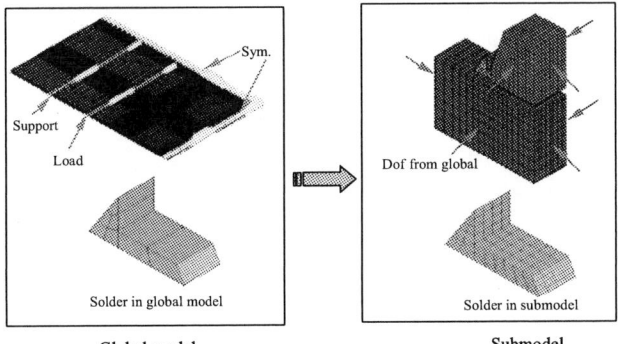

Fig. 17 One-level submodeling method

Two submodeling methods, including two-level submodeling and one-level submodeling, were compared to investigate meshing effect on FEA results. For two-level submodeling method as shown in Fig. 16, the board-level global quarter model was solved firstly. Then DOF results were transferred and interpolated to the cut boundary of the first package-level submodel. Finally, the DOF results from the first-level submodel were transferred and interpolated to the cut boundary of the second-level submodel. The package-level FEA model is referred as a transitional model, which connects the board-level global model to solder joint-level submodel. It was expected that the two-level submodeling method leads to more accurate results but complicated procedure is needed due to twice DOF interpolations. For one-level or traditional submodeling as shown in Fig. 17, the global model and submodel are the same as board-level global

model and solder joint-level submodel used in two-level submodeling method, respectively. The DOF results from global model were transferred to the final submodel directly without using package-level transitional model. However, it maybe gives arise to more error because DOF results were interpolated and transferred from much coarse mesh in the board-level global model to finer mesh in the submodel. In this study, these two submodeling methods were performed for comparison.

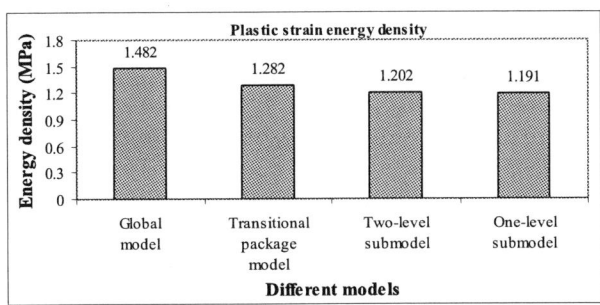

Fig. 18 Solder joint plastic strain energy density for different models

For convenience, the monotonic four-point bending load was used in this validation study. The displacement range from zero to 6mm was added on loading position of global model in 1 second using 6 load steps in FEA simulation. Figure 18 shows the volume-averaged plastic strain energy density at the maximum deflection position for different models considering whole solder joint as an averaging volume. It can be seen that one-level submodel and two-level submodel result in almost the same result. Therefore, one-level submodeling method is effective and sufficient for bending FEA modeling and simulation of VQFN assembly. The FEA simulation accuracy was verified through strain comparison between FEA modeling results and experimental results in references [19-20]. The bending fatigue model for VQFN solder joint has been proposed by combining FEA simulation results and bending reliability testing data [19-20].

4. Conclusions

The global-local modeling method containing submodeling and GLB modeling technique was developed and calibration study was also conducted. Reasonable submodel with hybrid solid, including solder joint, PCB and component, can give more accurate result than submodel with only solder joint. Solid-beam-solid GLB model can lead to more accurate results than shell-beam-shell GLB model. The modified shell model gives less error than non-modified shell elements. GLB model is reasonable for small deformation problem and modal analysis because only elastic deformation behavior is considered for solder joint in the deformation analysis of entire structure. Based on GLB modeling theory, the ball shape solder joint can be replaced by the effective solder joint with cubic cross-section assuming that they have similar stiffness. The effective solder joint model can lead to accurate result with solder joint plastic deformation consideration.

Case studies further validate the feasibility and accuracy of global-local modeling technique used in electronic packaging reliability assessment. Three cases are considered, including PBGA assembly subjected to thermal cycling, FCOB assembly subjected to vibration and VQFN assembly subjected to bend.

References

1. Che, F.X., and Pang, H.L.J., "Thermal Fatigue Reliability Analysis for PBGA with Sn-3.8Ag-0.7Cu Solder Joints", *Proc 6th Electronics Packaging Technology Conf*, Singapore, Dec. 8-10, 2004, pp. 787-792.

2. Pang, H.L.J., Chong, D.Y.R., and Low, T.H, "Thermal Cycling Analysis of Flip-Chip Solder Joint Reliability", *IEEE Transaction on Components and Packaging Technologies*, Vol. 24, No.4 (2001), pp. 705-712.

3. Pang, H.L.J., and Chong, Y.R., "Flip Chip on Board Solder Joint Reliability Analysis Using 2-D and 3-D FEA Models", *IEEE Transaction on Advanced Packaging*, Vol.24, No.4 (2001), pp. 499-506.

4. Akay, H.U., Liu, Y., and Rassaian, M., "Simplification of Finite Element Models for Thermal Fatigue Life Prediction of PBGA Packages", *Journal of Electronic Packaging*, Vol.125 (2003), pp. 347-353.

5. Pang, H.L.J., Seetoh, C.W., and Wang, Z.P., "CBGA Solder Joint Reliability Evaluation Based on Elastic-Plastic-Creep Analysis", *Journal of Electronic Packaging*, Vol. 122 (2000), pp. 255-261.

6. Zhu, J., Quander, S., and Reinikainen, T., "Global/Local Modeling for PWB Mechanical Loading", *Proc 51th Electronic Components and Technology Conf*, Orlando, May 29-June 1 2001, pp. 1164-1169.

7. Gustafsson, G., Guven, I., Kradinov, V., *et al.*, "Finite Element Modeling of BGA Packages for Life Prediction", *Proc 50th Electronic Components and Technology Conf*, Las Vegas, May 21-24, 2000, pp. 1059-1063.

8. Pang, H.L.J., Low, T.H., Xiong, B.S., and Che, F.X., "Design For Reliability (DFR) Methodology For Electronic Packaging Assemblies", *Proc 5th Electronics Packaging Technology Conf*, Singapore, Dec. 10-12, 2003, pp. 470-478.

9. Yang, Q. J., Lim, G.H., Pang, H.L.J., et al., "Vibration Reliability Analysis of a PBGA Assembly under Foundation Excitations", *EEP-Vol. 26-1, Advances in Electronic Packaging, Vol. 1, ASME International Electronic Conf*, 1999, pp. 705-711.

10. Chong, Y.R., Che, F.X., Xu, L.H., Toh, H.J., Pang, J.H.L., Xiong, B.S., and Lim, B.K., "Performance Assessment on Board-level Drop Reliability for Chip Scale Packages (Fine-pitch BGA)", *Proc 56th Electronic Components and Technology Conf*, San Diego, California, May 30-June 2, 2006, pp. 356-363.

11. Che, F. X., Pang, H.L.J., Zhu, W.H., Sun, W., and Sun, Y. S., "Modeling Constitutive Model Effect on Reliability of Lead-Free Solder Joints", *Proc Internatioanl Conf Electronic Packaging Technology*, Shanghai, Aug. 27-29, 2006, pp. 155-160.

12. ANSYS Version 7.0 Manual, 2002, *Ansys Inc*.

13. Petyt, M., Introduction to Finite Element Vibration Analysis, *Cambridge University Press*, 1990.

14. Che, F.X., Pang, H.L.J., Xiong, B.C., Xu, L.H., and Low, T.H., "Lead Free Solder Joint Reliability Characterization for PBGA, PQFP and TSSOP Assemblies", *Proc 55th Electronic Components and Technology Conf*, Florida, May 31-June 3, 2005, pp. 916-921.

15. Pang, H.L.J., Che, F.X., and. Low, T.H., "Vibration Fatigue Analysis For FCOB Solder Joints", *Proc 54th Electronic Components and Technology Conf*, Las Vegas, Nevada, June 1-4, 2004, pp. 1055-1061.

16. Che, F.X., Pang, H.L.J., Wong, F.L., Lim, G.H., and Low, T.H., "Vibration Fatigue Test and Analysis for Flip Chip Solder Joints", *Proc 5th Electronics Packaging Technology Conf*, Singapore, December 10-12, 2003, pp. 107-113.

17. Steinberg, Dave S., Vibration Analysis for Electronic Equipment, John Wiley & Sons (New York, 1998).

18. Che, F.X., and Pang, H.L.J., "Bend Fatigue Reliability Test and Analysis for Pb-free Solder Joint", *Proc 7th Electronics Packaging Technology Conf*, Singapore,, December 2005, pp. 868-872.

19. Pang, H.L.J., and Che, F.X., "Isothermal Cyclic Bend Fatigue Test Method For Lead Free Solder Joints", *Proc 2006 Inter Society Conference on Thermal Phenomena*, San Diego, California, May 30-June 2, 2006, pp. 1011-1017.

20. Che, F.X., and Pang, H.L.J., "Modeling Board-Level Four-Point Bend Fatigue and Impact Drop Tests", *Proc of 56th Electronic Components and Technology Conf*, San Diego, California, May 30-June 2, 2006, pp. 443-448.

Transport of Corrosive Constituents in Epoxy Moulding Compounds

M. van Soestbergen[1,*], L.J. Ernst[2], G.Q. Zhang[2,3], R.T.H. Rongen[4]

[1] Netherlands Institute for Metals Research, [2] Delft University of Technology, Fundamentals of Microsystems
Engineering; [3]Strategy and Business Development of NXP Semiconductors, [4]NXP Semiconductors, Nijmegen
[*]Mekelweg 2, 2628 CD, Delft, the Netherlands
m.vansoestbergen@NIMR.nl

Abstract

Epoxy moulding compounds are the leading encapsulating material in today's microelectronic packaging industry. These compounds are hydrophilic and absorb moisture when exposed to humid environments. As a result of the absorbed moisture, ions in the material will become more mobile. This, in combination with high electrical field strengths due to the continuously decreasing feature sizes of ICs, will result in a large flux of ions toward charged interfaces, such as aluminium bond pads. Hence, ion-related failure mechanisms, such as corrosion, might become more prominent. In this paper a method for obtaining the diffusion coefficients of the different ions at elevated temperatures in saturated moulding compounds is reported. This method is based on determining the total ionic content by ion chromatography after immersing a presaturated sample in a water bath containing the ions and fitting the data to a Fickian diffusion model. The measured diffusion coefficients of NaCl in a commercially available moulding compound at 30 °C and 60 °C are 2.9×10^{-13} m^2/s and 9.5×10^{-13} m^2/s, respectively. These coefficients are used to compare experimental data for bond pad corrosion as function of humidity, time and bias with a multi-physics finite element model.

1. Introduction

Since the introduction of plastic-encapsulated microelectronics, corrosion has been one of the major reliability issues. To qualify the reliability of microelectronic products, lifetime tests are performed at accelerated conditions to reduce testing time. For these reasons, lifetime tests at elevated temperature and humidity are part of the total testing procedure. To extrapolate the lifetime obtained from these test to application conditions, empirical models, such as the well-known Peck model, are commonly used [1,2]. These models all assume that failure is accelerated by the applied temperature and humidity. The effect of electrical bias, however, is not yet clearly specified and is included in these models as an unknown function of bias voltage. The effect of the applied electrical bias is twofold, (1) bias directly influences the corrosion kinetics and (2) bias results in an electrical field that attracts the ions to the bond pads, which will eventually lead to a higher corrosion rate. The latter effect is controlled by the bias as well as the transport properties of the corrosive ions. In literature, however, not much has been reported on the transport properties of corrosive species, such as chloride and sodium in epoxy moulding compounds.

Lantz and Pecht [3] reported the diffusion of NaCl in a commercial biphenyl moulding compound to be slower than the diffusion of moisture. A diffusion cell was used to measure the diffusion coefficient of NaCl, while in addition Time-of-Flight Secondary-Ion-Mass-Spectroscopy (TOF-SIMS) was used to show the concentration profile of a cross-section of the sample. Due to the presence of water the mobility of the ions in a moulding compound will increase, and hence its electrical resistivity decreases. Rauhut [4] showed that after 1000 hours of exposure to steam at 103 kPa pressure, the volume resistivity decreases approximately by a factor of 100 for a highly filled biphenyl compound and by a factor of 300 for a regularly filled epoxy cresol novolac compound.

In this paper we present a method for determining the diffusion coefficients of water-soluble ions in epoxy moulding compounds. This method uses ion chromatography to determine the ion content of the sample after different exposure times to a water bath containing the ions. Next, a finite element model is used to show that there is a large difference between the bulk concentration of ions and their concentration at the bond pads due to the applied electrical field. Finally, the results of the finite element model combined with results from literature are used to show the effect of bias voltage on corrosion.

2. Background

In this section a general scheme for aluminium corrosion in the presence of chloride will be presented. However, the possible sources of corrosive ions that will contaminate the compound will be discussed first.

Small concentration of ions are always present in moulding compounds, examples of these are chlorine (Cl$^-$), sodium (Na$^+$), potassium (K$^+$) and bromine (Br$^-$). Some of these ions are due to the chemical synthesis of the subcomponents of the resins. Epichlorohydrin-based epoxy rings are typical resin components, since they readily react with alcohol and phenols. Functional groups are attached to these epoxy rings in the presence of NaOH. This will finally lead to the formation of NaCl. Although they are purified, a small concentration of NaCl always remains in the resin. Besides, processing steps such as etching of metallizations can introduce corrosive ions. Ions liberated from the die-attach glue at high temperatures might be another source of contamination.

Aluminium in a humid environment corrodes quickly, creating a protective oxide film. In case of an applied electrical potential a distinction can be made between corrosion at the anode and cathode.

1-4244-1105-X/07/$25.00 ©2007 IEEE

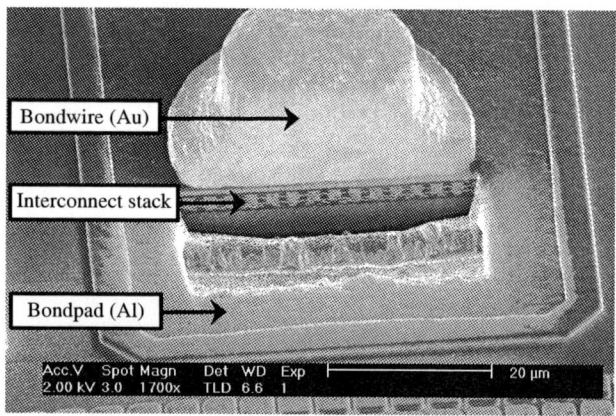

Figure 1: Cross section of the bond ball on an aluminum bond pad.

The chemical reactions occurring at the anode are given by [5,6]:

$$Al \rightarrow Al^{3+} + 3e^-$$
$$2Al^{3+} + 3H_2O \rightarrow 2Al(OH)_3 + 3H^+ \quad (1)$$

Dehydration of $Al(OH)_3$ will lead to alumina (Al_2O_3). The anode attracts negatively charged ions, such as chloride and bromine. These ions will attack the protective oxide layer resulting in pitting corrosion. It is generally accepted that pitting is mainly caused by the attack of chloride, hence [5,7,8]:

$$2Al(OH)_3 + Cl^- \rightarrow 2Al(OH)_2Cl + OH^- \quad (2)$$

After dissolution of the protective oxide film the bare aluminium corrodes according to:

$$Al + 4Cl^- \rightarrow AlCl_4^- + 3e^- \quad (3)$$

The corrosion product now diffuses into the bulk solution and hydrolyzes yielding $Al(OH)_3$ and Cl^- [5-9]. Note that a small concentration of chlorine ions can cause a large amount of corrosion since these ions are liberated during the process.

The corrosion scheme presented above assumes pure aluminium. The aluminium used for bond pads, however, contains amongst other materials a small amount of copper. The copper is present in the aluminium in the form of grains. This in combination with adsorbed electrolyte on the surface will result in a highly active galvanic cell, which will eventually lead to an increase of the corrosion rate [9].

3. Estimating the diffusion coefficient of ions

In this section a technique capable of measuring the mobility of ions will be described. In addition the mobility of sodium (Na^+) and chloride (Cl^-) ions in a commercially available moulding compound will be reported and discussed.

To obtain specimens a biphenyl-based compound has been moulded onto HVQFN lead frames. After moulding, the compound can easily be torn off the lead frames yielding 600 µm thick strips. These strips are cut into smaller parts with dimensions of 20, 10 and 0.6 mm for the length, width and thickness, respectively. These small strips are then soaked in demi-water for 96 hours at 80 °C. Subsequently two sets of five strips are placed in separate beakers containing 200 ml demi-water at 30 °C and 60 °C. After equilibrating the specimens for 24 hours, NaCl was added to obtain a 0.15 M solution. The beakers are covered with aluminium foil to prevent the water from evaporating. The specimens are taken out of the salt solution at different times and were dried to remove any redundant solution from the surface. The specimens are transferred into separate pressure vessels where the ions are extracted from the moulding compound. The vessel consists of a PTFE (Teflon) inner shell with a metal jacket containing 20 ml ultra-pure water. The ultra-pure water extracts are finally analysed by ion chromatography. The results of this analysis are given in Fig. 2.

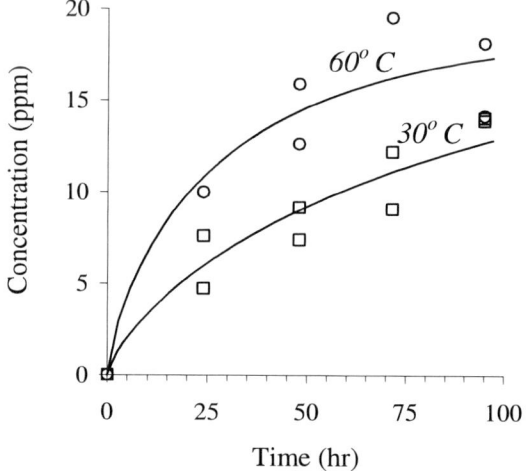

Figure 2: NaCl absorption of a biphenyl moulding compound in 0.15 M solution at 30 °C and 60 °C.

It is assumed that in the absence of an electrical field the diffusion of ions is governed by *Fick's* law. In this case the increase in concentration is approximately given by the expression [10]:

$$c(t) = c_{sat}\left\{ 1 - \exp\left[-7.3\left(\frac{D \cdot t}{h^2} \right)^{0.75} \right] \right\} \quad (4)$$

where c is the concentration of species at the time t, D the diffusion coefficient and h the thickness of the specimen. Subscript sat denotes the saturation level. Fitting Eq. (4) to the results of Fig. 2 yields the following estimations for the diffusion coefficient, 2.9×10^{-13} m^2/s and 9.5×10^{-13} m^2/s for 30 °C and 60 °C, respectively.

The electrical conductivity of dry epoxy moulding compounds at room temperature (25 °C) is typically in the order of picoSiemens per meter. The ion conductivity of a material is given as:

$$\sigma = F\sum_i c_i \cdot z_i \cdot u_i \quad (5)$$

where F is Faraday's constant (96485 C/mol), z the valence number, c the concentration and u the mobility of ionic species i.

The Nernst-Einstein relation can be used to write the ionic mobility in terms of the diffusion coefficient, thus:

$$u = \frac{z \cdot F \cdot D}{R \cdot T} \rightarrow \sigma = \frac{F^2}{RT} \sum_i c_i \cdot z_i^2 \cdot D_i \qquad (6)$$

where R is the universal gas constant (8.3144 J/(K·mol)). The conversion factor from mol/m³ to ppm depends on the mass density of the moulding compound and the atomic mass of the ions. The atomic mass of NaCl is 58.4 g/mol and the mass density of most commercial compounds is in the range of 2 to 3 g/mm², hence the conversion factor is approximately 25. Assuming the total amount of mobile ions in a compound to be 2 mol/m³ with z^2 equal to 1 yields an average diffusion coefficient in the order of 10^{-18} m²/s. This suggests that the presence of water has a significant influence on the transport properties of ions in epoxy moulding compounds.

4. Finite element model

A multi-physics finite element model is used to estimate the effect of an applied electrical bias on the local concentration of ions at the bond pad-moulding compound interface. The flux of ionic species is given by:

$$J_i = -\frac{D_i c_i}{RT} grad(\mu_i) \qquad (7)$$

where μ_i is the chemical potential of species i given by:

$$\mu_i = RT \ln(c_i) + z_i \phi - RT \ln(1 - c_i c*) \qquad (8)$$

where ϕ is the electrical potential, c* the reciprocal of the maximum concentration of species and e is the elementary charge (1.602×10^{-19} C). The chemical potential of the last term of Eq. 8 is in analogy with the potential for hard sphere mixtures reported by Bikerman [11]. Substituting Eq. 8 into 7 yields:

$$J_i = -D_i \left\{ grad(c_i) \left[1 + \frac{c* c_i}{1 - c_i c*} \right] + \frac{Fc_i}{RT} grad(\phi) \right\} \qquad (9)$$

where the concentration of species is given in mol/m³ and grad(ϕ) is the electrical field. Eq. 9 follows from the Nernst-Planck equation for the transport of ions, with an additional term to correct for the maximum concentration of species.

The electrical field is given by the Poisson equation:

$$div(\varepsilon_r \varepsilon_0 E) = F \sum_i z_i c_i \qquad (10)$$

where ε_r is the relative permittivity and ε_0 the permittivity of vacuum (8.854×10^{-12} F/m). At charged interfaces an electrostatic double layer of a few times the Debye length will be formed. In this layer the ions are distributed such that they screen out the electrical field. For monovalent ions the Debye length is given as [12]:

$$L_D = \sqrt{\frac{\varepsilon_r \varepsilon_0 RT}{2F^2 C_\infty}} \qquad (11)$$

where C_∞ denotes the concentration of the bulk solution. It can be readily seen from Fig. 3 that in commercial moulding compounds the Debye length will be in the range of 1 to 100 nm.

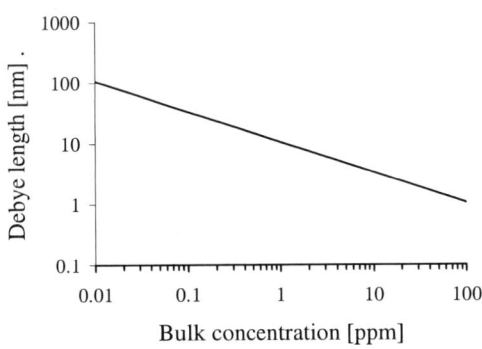

Figure 3: Debye length vs. bulk concentration of ions, ε_r=3.8, T=298 K, ppm to mol/m³ ratio is 25.

Since the length in which the electrical field screens out is in the order of nanometers, an extremely fine grid is necessary at the electrode interface. This will increase the computing time tremendously. Therefore it becomes almost impossible to solve large three-dimensional models. In the absence of concentration gradients, however, the electrical field solely induces the transport of ions. Hence, initially, Eq. 10 remains to be solved without redistribution of ions. Therefore, the results of Eq. 10 will be used to simplify the geometry of the model as much as possible in such a way that the initial electrical field remains equivalent with the original model.

A three-dimensional model according to Fig. 6 is used as a starting point. Two times a quarter of a bond wire including a bond ball on a pad has been modelled using COMSOL Multiphysics 3.3 (fig. 6, left). The bond pads are placed on a 4 µm thick interconnect stack. Although this stack consists of a metal interconnect structure and a dielectric material it is modelled as a solid material with a relative dielectric constant equal to 1. It is assumed that the die underneath the interconnect stack results in a zero potential at the bottom of this stack. The relative dielectric constant of the moulding compound is set to 3.8. An electrical potential of +1 Volt and –1 Volt is set to the left and right bond wire, respectively. The electrical field ratio in z to x–direction of the area between the bond balls and under the dashed line of Fig. 6 is plotted in Fig. 4. It can readily be seen that the electrical field perpendicular to the plane of Fig. 4 is negligible compared to the electrical field parallel to this plane.

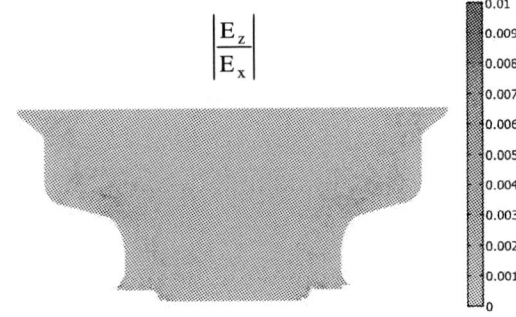

Figure 4: Norm of E_z/E_x for the area between the bond pads (fig. 6); g=15 µm, $\varepsilon_{r,mould}$=3.8, V=1 Volt.

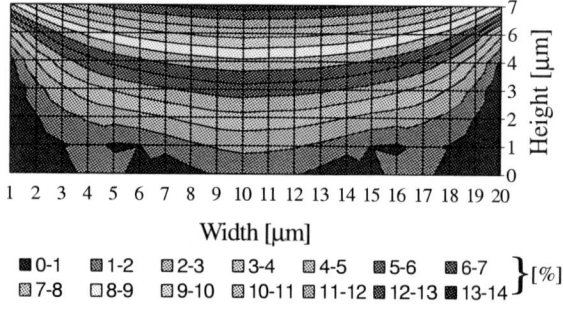

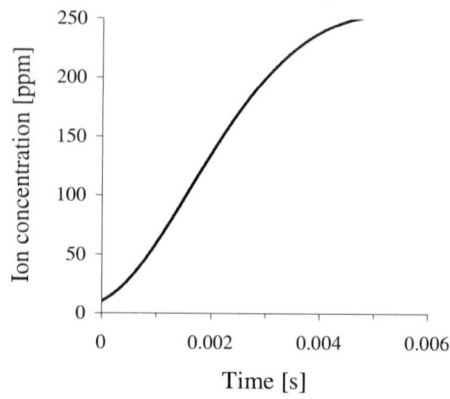

Figure 5: Difference in electrical potential between the original model (fig. 6, middle) and simplified model (fig. 6, right) for the dashed area; V=1 Volt; ε_r=3.8.

Consequently, a two-dimensional model is created (Fig. 6, middle). Since the concentration of ions at the bond pad interface (indicated by P) are of interest in this research, the geometry of the two-dimensional model can still be simplified. Therefore straight lines have replaced all radii of the bond ball (Fig. 6, right). Furthermore, we have decreased the height as much as possible. To show that this model results in a sufficiently accurate estimation of the electrical field between the bond balls, the area indicated by the dashed lines has been analysed (Fig. 6 right). This area has been divided into squares of one by one micrometer. The electrical potential for both the original model (Fig. 6, middle) and the simplified model has been calculated. The difference for every node of the dashed area of the original model and the corresponding nodes of the simplified model is plotted in Fig. 5. It can be seen that around the points of interest the error in electrical potential is never larger than 5%. Therefore, we feel reassured in using the simplified model.

Now Eq. 9 is added to this model to include the transport of ions. All the boundaries of the model are set to insulating. Elements at boundaries where an electrical potential is applied are distributed such that their thickness is in the order of the Debye length (inset Fig.6).

Figure 7: Concentration vs. time at P; V=1 Volt, ε_r=3.8, T=358 K, C_∞=10 ppm, D=10^{-12} m^2/s, C_{max}=250 ppm.

It is assumed that 5 M is the maximum concentration of ions that can be dissolved in water. The weight gain due to water uptake of a moulding compound is typically around 0.2%. Therefore the maximum concentration of ionic species is set to 10 mM ($\approx$ 250 ppm). One species of anions and one species of cations are present in the moulding compound. The diffusion coefficients of both species are assumed to be equal. Next, the temperature is set to 85 °C and the ions are initially randomly distributed with a concentration of 10 ppm.

4. Results

A time-dependent solver has been used to calculate the increase in concentration. A typical plot of the increase in concentration versus time is given in Fig. 7. It can readily be seen that the concentration will reach the maximum concentration of 250 ppm. The time until this saturation level will be reached will decrease with an increase of the applied potential at the electrodes. Fig. 8 shows the decrease in time to saturation as a function of the applied potential. Determining the time to saturation is not completely unambiguous. Therefore not all the points in Fig 9 are on a straight line. Secondly, at low potentials

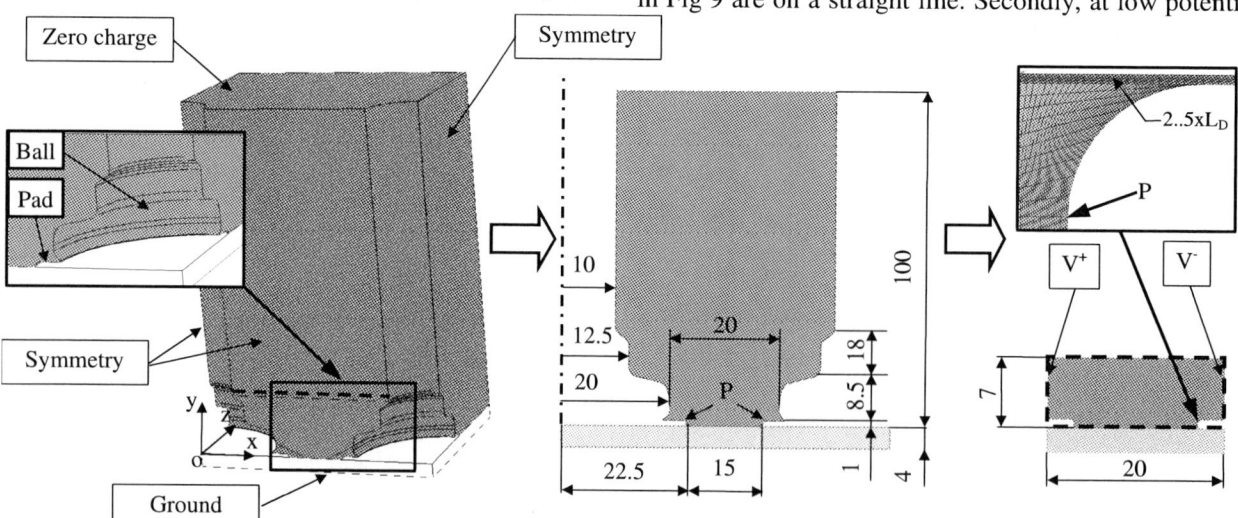

Figure 6: Reduction of the complexity of the FE model; left: 3-D model, middle: intermediate 2-D model, right: final 2-D model; moulding compound indicated by dark grey, interconnect stack by light grey; dimensions in micrometer.

the concentration at the electrodes will not reach the maximum level. Therefore the relation between saturation time and low potentials differs from the relation of Fig. 8.

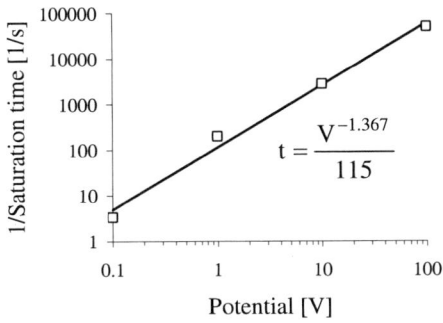

Figure 8: *Reciprocal of the saturation time as a function of the applied voltage; $\varepsilon_r=3.8$, $T=358$ K, $C_\infty=10$ ppm, $D=10^{-12}$ m^2/s, $C_{max}=250$ ppm, $V=0.1..100$ Volt.*

From Fig. 8 it can be seen that the time to saturation is negligible for typical applied potentials used during lifetime tests. However, for less mobile ions the time to saturation will increase, this is shown in Fig. 9. For a high diffusion coefficient the saturation level is reached almost instantaneously, whereas for low diffusion coefficients it will take days until this level is reached.

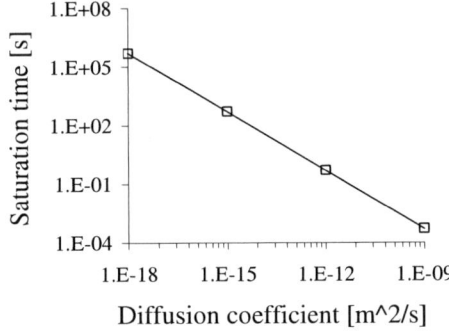

Figure 9: *Saturation time as a function of the diffusion coefficient of the ionic species; $\varepsilon_r=3.8$, $T=358$ K, $C_\infty=10$ ppm, $V=0.1$ Volt $C_{max}=250$ ppm, $D=10^{-18}..10^{-9} m^2/s$.*

5. Conclusions

From literature and the observations reported here, it can be concluded that moisture has a significant influence on the transport properties of ions in epoxy moulding compounds. However, the time scale of the transport of ions in water-saturated compounds is negligible compared to the corrosion time scale. This is in contradiction with the model for the failure rate due to bond pad corrosion reported by Dunn and McPherson [5]. They assume that the corrosion is controlled by the transport of corrosive species. The data reported in this work suggest that there will be an induction time during lifetime test where the corrosion rate is very limited. This induction time finds its origin in the time needed for the moisture to ingress into the moulding compound and thus mobilizing the ions. This is in agreement with the failure rate model proposed by Pecht [6]. Next, in our model the concentration at the

bond pad interface will have a maximum, independent of the applied potential. This maximum concentration might increase significantly when there are flaws in the bonding of the moulding compound such that water can adsorb onto the bond pad. According to Paulson and Lorigan [8] the corrosion rate is proportional with log(c), where c indicates the concentration of corrosive species. So at these flaws the corrosion rate increases. Finally, in the model described above no fluxes of species due to chemical reactions are considered. Since the transport of species through the compound is much faster than the corrosion rate this is assumed to be a good approximation.

Acknowledgments

This research was carried out under project number MC3.05236 in the framework of the Strategic Research programme of the Netherlands Institute for Metals Research in the Netherlands (www.nimr.nl).

References

1. JEDEC standard JEP122C "Failure Mechanisms and Models for Semiconductor Devices" Available at http://www.jedec.org/Catalog/catalog.cfm
2. Peck, D., "Comprehensive Model for Humidity Testing Correlation" *IEEE int. reliability physics symposium proceedings 1986*, pp. 44-50
3. Lantz, L., Pecht, M. G., "Ion Transport in Encapsulants Used in Microcircuit Packaging" IEEE Trans-CPT, Vol. 26, No. 1 (2003) pp. 199-205
4. Rauhut, H. "Dielectric/Electric Epoxy Compound Traits" *Society for Plastic Engineers*, Chicago, IL, March 1996
5. Dunn, C.F., McPherson, J.W "Recent Observations on VLSI Bond Pad Corrosion Kinetics" *J. Electrochem. Soc.*, Vol. 135, No. 3 (1988) pp. 661-665
6. Pecht, M., "A Model for Moisture Induced Corrosion Failures in Microelectronic Packages" *IEEE Trans-CHMT*, Vol. 13, No. 2 (1990) pp. 383-389
7. Iannuzzi, M., "Bias Humidity Performance and Failure Mechanisms of Nonhermetic Aluminum SIC's in an Environment Contaminated with Cl_2", *IEEE Trans- CHMT*, Vol. 6 No. 2 (1983) pp. 191-201
8. Paulson, W.M., Lorigan, R.P. "The Effect of Impurities on the Corrosion of Aluminium Metallization", *International Reliability Physics Symposium*, 1976, pp. 42-47
9. Badawy, W.A. *et al.*, "Electrochemical Behaviour and Corrosion Inhibition of Al, Al-6061 and Al-Cu in Neutral Aqueous solutions," *Corrosion Science*, Vol. 41 (1999) pp. 709-727
10. Shen, C-H, Springer, G. S., "Moisture Absorption and Desorption of Composite Materials", *Journal of Composite Materials*, Vol.10, (1976) pp.2-20
11. J.J. Bikerman, *Philos. Mag.* Vol. 33 (1942) pp. 384
12. P.M. Biesheuvel, "Simplifications of the Poisson-Boltzman Equation for the Electrostatic Interaction of Close Hydrophilic Surfaces in Water", *J. Colloid Interface Sci.*, Vol. 238, (2001) pp. 362-370

Development of Reliability Verification System for Robust Package Design

Dongkil Shin, Hyunggil Baek, Joonyoung Oh, Dongok Kwak,
Kuyoung Kim, Younghee Song, Jeongyeol Kim*
IPT Team, *CAE Team
Memory Division, Samsung Electronics Co.
San#16, Banwol-Dong, Hwasung-City, Gyeonggi-Do, Korea, 445-701
dongkil.shin@samsung.com, 82.31.208.6486

Abstract

To achieve design for reliability (DFR), a reliability verification system (RVS) was developed. Package level and board level reliabilities were predicted by the developed system automatically and the design for six sigma (DFSS) was achieved. Process server and database server were developed. FEM mesh was generated automatically by well-organized builder, simulation conditions were assigned by standardized procedure, and reliabilities were calculated by developed life models. All data related to the simulation were managed by database system. High quality simulation was carried out easily and quickly. Robust package design was achieved by design optimization using the RVS.

1. Introduction

Electronic package needs to be light and thin, and to have small form factor in general. On the other hand, the size of memory package becomes larger because performance-driven market asks high density, high performance, and multi-functional memory device. In addition, new market is emerging so package application becomes various. Therefore, the demand for the development of sophisticated manufacturing process is increasing in the package manufacturing society. Therefore, new process technology and material are required and it needs long time and high cost for the development of a new package [1].

Figure 1 shows a typical example of package development process. Driving force of the development is customer request. Upon receiving with the request, a product is designed. Based on the design, raw materials, subsidiary materials, and infrastructures are prepared and then a prototype is assembled. Feasibility tests are performed based on the design of experiment (DOE) and response surface is found. And then throughout the process qualification and reliability qualification, the product is produced in quantity. Every step happens in the sequential manner. A process cannot move to the next step if the previous process is not clearly finished. In addition, if a problem is found in a certain step, it should go back to the original design process for design modification. So, a lot of resources and times are required.

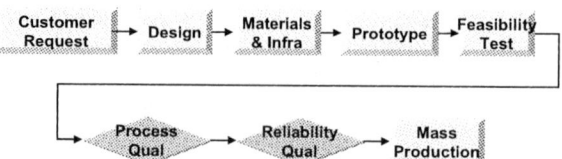

Fig. 1. Sequential procedure of package development

Design modification affects the characteristics of package such as manufacturability, reliability, cost, and so on. Each modified procedure needs extra cost and time. So a smart systems to verify the characteristics of the package not using a real product have been studied for a long time [3-7].

In order to improve the quality of a product with low cost, and to reduce the product introduction time to market, all factors that can affect on the manufacturing should be considered in the design phase. To facilitate manufacturing, a product is designed for manufacturability (DFM). Reliability, one of the critical issues in production, should be considered for design for reliability (DFR). Design for testability (DFT), design for cost (DFC), and any other design for X (DFX) are ways to achieve the best design. Throughout these activities to strengthen the design, design for six sigma (DFSS) is achieved ultimately. Figure 2 depicts the concept of design for six sigma. A rough concept design becomes optimized robust design by DFSS [8].

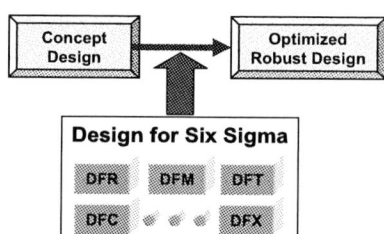

Fig. 2. Design optimization by Design for Six Sigma

2. Design for Reliability by Simulation

Simulation technology is one of the best ways to implement design for reliability (DFR). By simulation, the reliability of a package can be verified in virtual space, without the use of actual package, material, and dimensions. A quality of a product is qualified virtually. An optimized solution can be found by easy parametric study and a new idea can be implemented freely and easily[9,10].

1-4244-1105-X/07/$25.00 ©2007 IEEE   428

Figure 3 shows the general procedures of design for reliability by simulation. Before making prototype, virtual qualification by simulation is performed. By this simulation, a design is optimized until the reliability meets desired limit. The right-hand side of the figure is a magnification of the virtual qualification procedure. By recalculating a certain parameter came from finite element analysis, the life of a package is estimated. Life models are used in calculating the life. The life models are established by analysis of the failure mechanism derived from test data, mass production data, and field data.

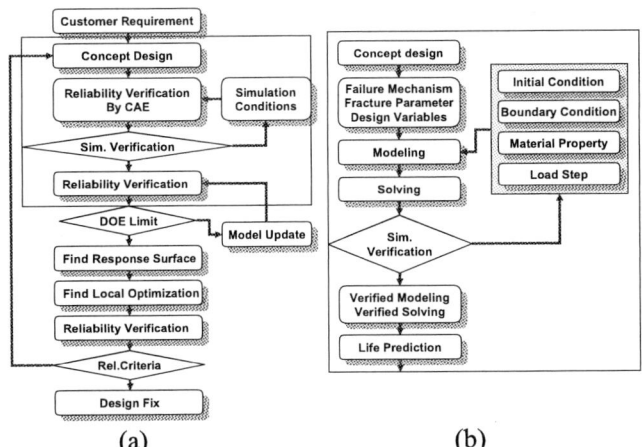

(a)   (b)

Fig. 4. Detail procedures of DFR by simulation

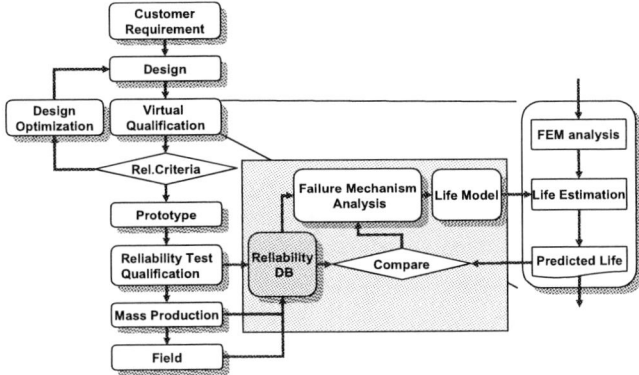

Fig. 3. Procedures of design for reliability

Figure 4 (a) shows the general procedures of the simulation. From the customer requirements, a conceptual design is created. The designed package is then verified by simulation. One important factor to consider in simulation is that the simulation should be verified by a knowledge system, such as previous experience. By using the verified simulation technique, virtual DOE is performed, response surface is drawn, and local optimum design is found. Depending on whether the optimized design meets the required reliability level or not, the final global optimum is found, and if not fount, the concept is redesigned.

Figure 4(b) shows the detail procedures of the simulation. For the virtual qualification, the failure mechanism should be identified clearly. The fracture parameters that can describe the mechanism are defined. Then, design variables are chosen appropriately. These simulation procedures are performed following six-sigma procedures which are the cause and effect diagram (C&E diagram), quality function deployment (QFD), and critical to quality (CTQ). From the result of the simulation verification, simulation conditions, such as initial condition, boundary condition, material property and load step are modified to describe the real system more exactly. By the verified modeling and solving technology, numerical simulation is performed and the life of the package is estimated.

Although the simulation has many advantages, it requires high-level technology to use fluently and needs long time to understand the theoretical factors. Moreover, it needs troublesome computational works. Results are strongly dependent on the user's philosophy. To use the simulation for DFR, the simulation should be simple and fast, and should always guarantee high quality result.

In order to achieve these requirements, we developed a simulation system for DFR, which was reliability verification system (RVS). The RVS simulation is performed simultaneously with the design phase. A designer unfamiliar with simulation can obtain high quality simulation result and can modify one's design according to those results. The simulation procedures are automated and standardized. In general, a fully automated system is sometimes difficult to modify and expand into new creative things. However, this system is flexible for professional computer aided engineering (CAE) users. A CAE engineer can interrupt the procedure, change the route, and implement one's own idea.

Input and out data are manage by a well-organized database system. Actual reliability test data are accumulated in the RVS, and compared with previously predicted life by simulation, as shown in Fig. 3. If there is a large difference between the predicted and actual data, then the life prediction model is modified by the failure mechanism analysis and the physics of failure analysis. Because there is a long time delay between two data, the workflow is controlled by product lifecycle management (PLM) system for efficient updates.

3. Implementation of Reliability Verification System

New workflows were developed for finite element analysis (FEA) automation. Consequently, very fast modeling was established and high quality simulation results were obtained.

The RVS can predict reliability items, which are defined by JEDEC standard[11]. Package level reliability conditions are pre-conditioning, pressure cooker test (PCT), and temperature cycling(TC). Board level reliability conditions are solder joint reliability (SJR), drop, bending, and tip breaking (not standardized yet in JEDEC). In order to predict the life of a package, thermo-mechanical, hygro-mechanical, mechanical, creep, or dynamic analysis is performed. Figure 5 shows the relation between reliability items and simulation technologies. Life is calculated by a predefined life prediction model. The life models correlates the simulated parameters from the numerical analysis with the real life in the specific reliability test.

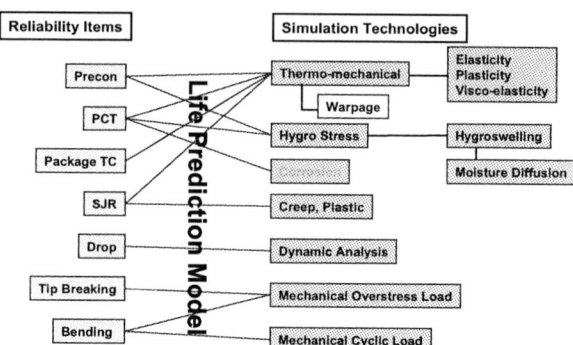

Fig. 5. Relation between reliability items and simulation technology in RVS

Figure 6 shows simulation procedures. A user just inputs the package information and reliability criteria, and RVS makes a finite element method (FEM) model and solves the problem. When the problem is solved, RVS calculates the life by the life prediction model and returns the result to the user.

If the user just wants a simple analysis, only device and package parameters are transferred to the RVS. But if the user wants a detail analysis, PCB design or lead frame design is also transferred to the RVS to consider the effect of copper trace or lead frame structure depending on the package type. For a ball grid array package, the ball layout can also be attached and considered to determine the effect of the ball layout on SJR. A user can interrupt the modeling process and go in a different way from the predefined automatic procedure and can achieve one's own special simulation.

An FEM model is established by the auto mesh generator using the material database(DB) and standard DB. After solving, a report is generated and delivered to the design user. All data generated during the simulation also stored in the database. Whenever the user wants to review the results, the user can search the DB and open the file anytime. Real reliability data are accumulated in the database, also.

All data required for the simulation are supplied by the database, and all data generated by the simulation are stored in the database. The database is managed by the DB manager and closely linked to the simulation process.

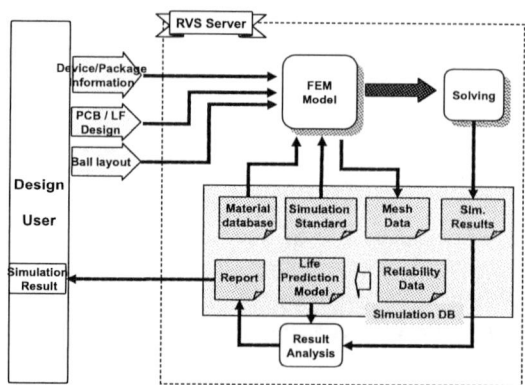

Fig. 6. Procedures of simulation in RVS

Figure 7 shows the working flow of RVS system. The system is composed of the process server and the database server. A user defines a problem and submits the problem to the process server, whish is developed for full automation. In the process server, process scheduler that is a kind of job scheduler gets this problem and creates a new process to solve this problem using process manager.

The database server manages simulation data. The DB has database solution and collaboration solution. They are organized logically with configurable setup.

Detail procedures are as follows.

1) Problem definition : it imports definition file that contains all package information from a user
2) Generating FE-modeling : it generates FE-model for each package
3) Assigning material and section property
4) Applying load
5) Solving : it submits job to the solver automatically.
6) Post-processing : it analyzes result files
7) Reporting : it generates documentation and return the report to the user automatically.

Data are controlled and managed in each tack. These data include FE-models, input files, result files, result values, report files and package information. For this purpose, process manager also automates DB client to store and manage the data in the database server.

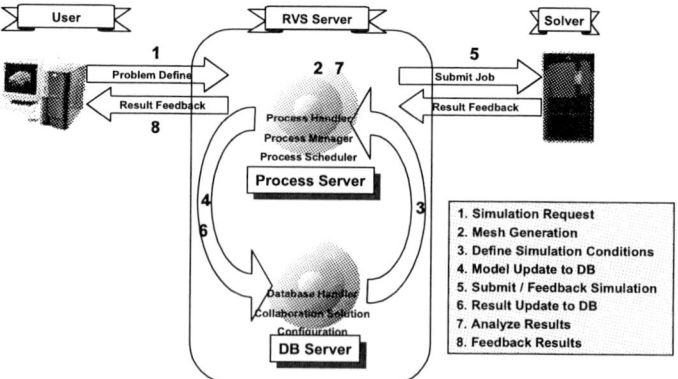

Fig. 7. Architecture of RVS

A simulation needs many input data and it also creates several output files. Various types of files are created such as text, binary, figure and so on, and these files are stored in personal folders. Not-organized or not-managed data are simple and easy to use in short time if they used personally. However, in the long term, these data are only used by a single user who generated the data. In addition, as time goes on, the original data are modified and the history of the data cannot be traced well, and consequently the data are vanished. Because the data are managed by the local system, it is difficult in communicating with coworkers and managers. Data cannot be shared, and the manager is unable to access the data. Another user who is unfamiliar with the previous work will have difficulty and take longer time in reaching the same level of quality as the original, matured analyzer.

To overcome these problems, a total management system was developed by using database systems. By the way, the CAE data have some specific characteristics. It has various file types and large size, and it usually does not need to be kept for long time. When the conventional DB concept is applied to the CAE, it is hard to build, to use, and to keep maintenance. So, we adapted a special DB for CAE users.

In the DB, the simulation database is composed of three groups A, B and C. A RUN is the basic unit of RVS process. Simulation is done RUN-by-RUN, so simulation outputs are stored by at every RUN. A RUN includes input files, simulation results, and a report. FEA inputs and model files are stored in 'Input DB'. FEA results are stored in 'Result DB'. Reports are stored in 'Report DB', respectively.

Data required for each simulation process are stored in group B. These data include material properties, simulation standards, and life model. Material DB includes FEA material input file, which is called by the process manager. Elastic modulus, thermal expansion, density, et al. are systematically managed. The material DB also includes measured raw data and related documents, brochures, and so on. It also contains old history of the property and user can trace it. DB also provides an intuitive comparison between different materials.

Simulation standards provide standardized load steps. Therefore, simulation quality remains at a higher level and input mistakes are prevented. Therefore, a user can trust the simulation result. A report is generated by a predefined template. Life prediction models are implemented in the DB. After simulation, the process calculates the value of the failure parameter, and then calculates the expected life of the package

Reliability is stored in the DB. It takes a long time to obtain the actual reliability data from the design stage. The results are traced by the design code, which is assigned in the designing phase. Actual data are compared with previously predicted data and are merged to the life prediction model to make a better model. The reliability DB also contains any reliability data obtained from the

mass production process and field usage period. Besides the reliability data, it also contains destructive and non-destructive failure analysis data and any related documents to maximize the usage of the database. These integrated database becomes the fundamental of knowledge system [12].

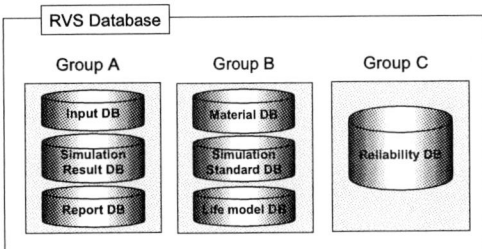

Fig. 8. Database system in RVS

Figure 9 shows the relation between the process manager (PM) and the database. PM and DB pop-up simultaneously on the monitor and the user can easily check the simulation status. The main structure of the PM is divided according to the name of the package, and the substructure is reliability item. The sub categories have their own simulation procedures. Related files are stored in the DB.

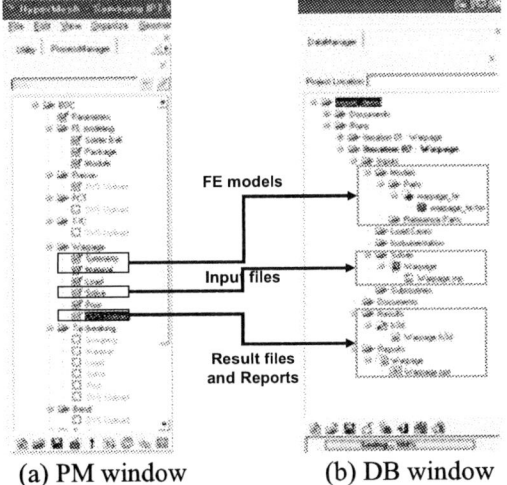

(a) PM window   (b) DB window

Fig. 9. Windows for PM and DB

The lead frame package occupies a large volume in a package and has a higher stiffness than the EMC material, so the full design of the lead frame should be considered in the analysis. However, the complex shape of the lead frame is a time-consuming process. Therefore, we developed an automatic mesh generator, which can model the detailed shape of lead frame in 2D and the down-set / up-set of die pad in 3D. Copper traces in the PCB substrate are converted to FEM mesh automatically. In the design software all layers are transferred to a drawing by a specially developed program. Both the first and last layer are modeled for the numerical simulation, so a failure due to the copper traces can be considered.

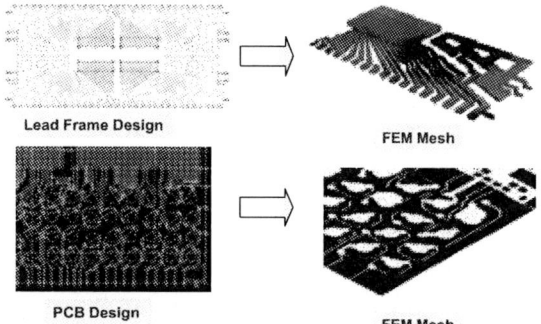

Lead Frame Design

FEM Mesh

PCB Design

FEM Mesh

Fig. 10. Automatic conversion of 2D design data to 3D FEA mesh

In making the FEM mesh, bottom-up method is adopted. From the design data, the package geometry is calculated. In the case of a ball grid array type, the shape of the solder ball is calculated by the surface tension method. Firstly, it builds the unit solder ball model, secondly, constructs the package model, and then finally organizes the board level model. Material properties, boundary conditions, and load steps are assigned from the database as shown in Fig. 11.

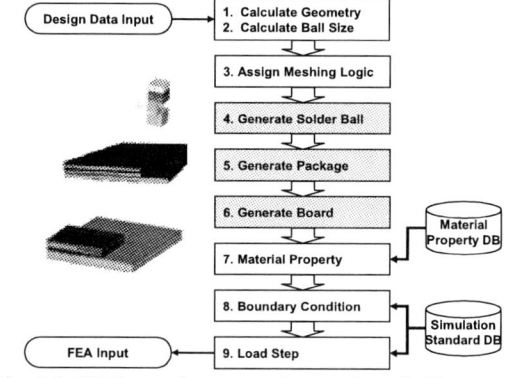

Fig. 11. FEA mesh generation and modeling

4. Package Reliability Simulation

4.1 Hygro-Thermo-Mechanical analysis

Thermo-hygro-mechanical simulation scheme is developed to calculate damage evolution in package caused by temperature change and moisture diffusion. The structure of scheme is shown in Fig. 12. Moisture diffusion and hygro-swelling analysis, thermal analysis, and mechanical deformation & damage evolution analysis are solved in uncoupled way. First, in the moisture diffusion analysis, moisture diffusion rate in package is calculated using Fick's law with the proper boundary condition and material data [13]. The resulting moisture concentration variation in each integration point is stored as a function of time and used in the following thermal analysis and mechanical deformation & damage evolution analysis. Second, in thermal analysis and mechanical deformation & damage evolution analysis, local stress concentration due to thermal strain mismatch and

hygroscopic swelling strain mismatch, and interface strength degradation due to moisture diffusion are calculated. Based on the damage evolution logic which is modified for specific package type, the resulting damage value is used as a mechanical failure criterion.

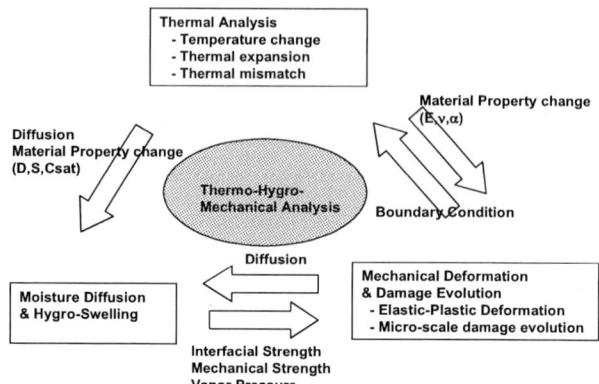

Fig. 12. Thermo-Hygro-Mechanical analysis

4.2 Solder joint reliability analysis

Many design variables affect solder joint reliability, e.g. solder material, pad size, pad shape, ball height, and so on[14]. In addition, there are various failure modes ; bulk crack, inter-metallic layer crack, pad open, and so on. Therefore, numerical estimation is difficult and depends on personal skill. In this study, we standardized the SJR analysis and made a procedure that was independent of user's skill, and established life model for the SJR. Material properties of a solder are modeled using hyperbolic sine rule. The accumulated damage of the solder ball is calculated. The RVS can consider leaded or lead free solder, various pad design, and various types of package. It calculates characteristic life and initial failure life cycle. Complex procedures were simplified by developing user subroutines. Figure 13 shows an automated solder ball and the relation between the calculated dissipated creep energy density per cycle and the real measured characteristic life.

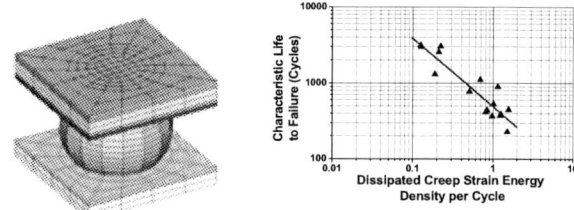

Fig. 13. Unit solder ball for SJR analysis and a relation between strain energy and characteristic life

4.3 Drop analysis

In the drop failure analysis, the effect of inertia force and the stress concentration at a singular point are very important. Therefore, dynamic analysis is necessary [15]. However, the solder ball is small, the structure has complex 3-D shape, and the drop occurs in short time. Failure occurs at a local area of the solder joint, but the size of the board is large and there is no symmetric plane in the board. Therefore, only full model can be used. In order to obtain good results without time loss, global-local modeling technology is implemented, as shown in Fig. 14. The global shape is modeled by a beam-shell element for fast analysis. Local shape is modeled by a solid element for fine analysis. One way to make a local model is by two step procedure : global model → local package model → local ball model. The other way is direct procedure : global model → local ball model. For an efficient analysis, a modal analysis which is based on linear superposition technology is also implemented [16,17].

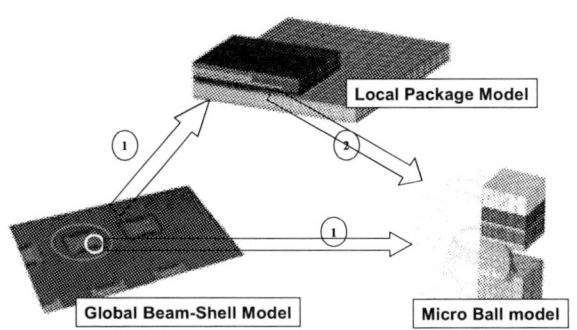

Fig. 14. Methods to predict drop reliability

4.4 Mechanical load analysis

The RVS can simulate bending, twisting and tip breaking failures. These kinds of mechanical loadings are becoming more important with the growth of the market of handheld products. Recently, JEDEC published a new standard for bending reliability [18]. Figure 15 shows two examples of mechanical loading analysis : bending and tip breaking. Three-point bending or four-point bending are available. The standard board of JEDEC and memory module are implemented. For the tip breaking analysis, the package is loaded by a rigid surface resembling the human finger [19].

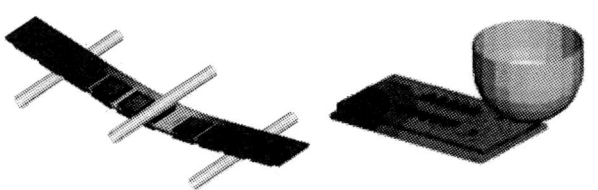

Fig. 15. Three point bending and tip breaking analysis

4.5 Verification of simulation

Simulation conditions are verified by the measured data and correlated with the experimental data. An intuitive method is displacement verification. Out of plane (warpage) by shadow moire and in-plane displacement by moire interferometry are well-developed phase calculation technologies for measuring thermal deformation. Figure 16 shows examples of measured and simulated results. The simulated results well agreed with the experiment results. The verified simulation becomes a master tool of parametric studies and sensitivity analyses. Deflection, load, twisting angle, strain, temperature, and so on are good parameters for the verification of simulation. However, both simulation and experimental measurement cannot explain all the physics of the package. Therefore, calibration and correlation are necessary. Some hidden factors such as process variation and material dependency should be considered carefully [20,21].

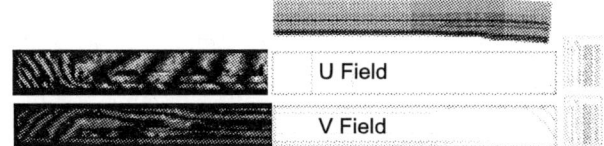

(a) In-plane moire interferometry

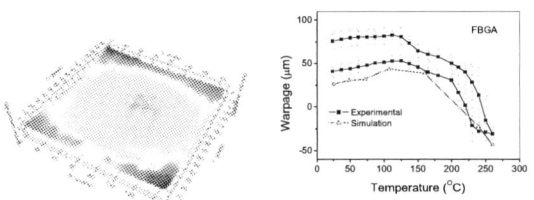

(b) Out-of-plane shadow moire

Fig. 16. Verification of simulation experimental measurement data

5. Summary

In this study, we developed a reliability verification system for robust package design. The reliability of a package was verified and optimized during the design phase. Therefore, the design for reliability was implemented and design for six sigma was achieved.

When a user inputs package information, CAD data, and solder ball layout, the RVS system calculates the reliability, generates a report, and returns the results to the user. For automatic simulation, a process server which controlled every procedures and database server which managed simulation data were developed. By using well-developed commercial programs, easy, fast, and high quality simulation was achieved.

FEM mesh was generated automatically by well-organized builder, simulation conditions were assigned by a standardized procedure, and life were calculated by verified life models. FEA inputs, simulation results, reports, material properties, simulation standards, life models, and reliability data were managed by the RVS database system. Hygro-thermo-mechanical analysis, SJR analysis, drop analysis, and mechanical load analysis are available.

References

1. Song, Y.H. and Lee, D.H., "Technical Challenges in Memory Packaging", *SMTA Pan Pacific Microelectronics Sysposium,* Hawaii, 2004

2. Pecht, M.G. *et al,* "Decreasing the Time-to-Market Through Virtual Risk Assessment and Risk Mitigation," *Proc 21^h Int. Conf. on Microelectronics,* NIS, Yugoslavia, 1997, pp. 25-30.

3. Zhang, G.Q. *et al,* "Virtual thermo-mechanical prototyping of electronic packaging challenges in material characterization and modeling," *Proc Electronic Components and Technology Conference,* 2001

4. Xuan, X. *et al,* "IC Reliability Simulator ARET and Its Application in Design-for-Reliability", *Proc. 12th Asian Test Symposium (STC'03),* 2003.

5. Silfhout R.B.R. *et al,* "Virtual Design and Qualification of IC Backend Structures," *Proc 7th Int. Conf. on EuroSimE,* 2006.

6. Wu,J. *et al,* "Computer Aided Reliability Assessment," *Proc ICEPT,* 2003, pp. 36-39.

7. Snook,I. et al, "Physics of Failure As an Integrated Part of Design for Reliability," *Proc Annual Reliabiliity and Maintainability Symposium,* 2003, pp. 46-54.

8. Brue,G. and Launsby, R.G., Design for Six Sigma, McGraw-Hill (New York, 2003)

9. Tay, A.A.O., "The Role of Simulation in Failure Prediction and Design Optimization in Electronics Packaging," *Proc 7^h Int. Conf. on EuroSimE,* 2006

10. Dasgupta, A., "Computational Challenges for Reliability Assessment of Next-Generation Micro & Nano Systems," *Proc 7^h Int. Conf. on EuroSimE,* 2006

11. JEP70-B, "Quality and Reliability Standards and Publications", JEDEC Solid State Technology Association, (1999)

12. JESD94, "Application Specific Qualification Using Knowledge Based Test Methodology", JEDEC Solid State Technology Association, (2004)

13. Wong, E. H. *et al,* "Advanced Moisture Diffusion Modeling and Characterisation for Electronic Packaging", *ECTC,* 2002, pp. 1297-1303

14. Syed, A., "Accumulated Creep Strain and Energy Density Based Thermal Fatigue Life Prediction Models for SnAgCu Solder Joints", *ECTC,* 2004.

15. Freund,L.B., Dynamic fracture mechanics. Cambridge University Press, (Cambridge, 1990)

16. Shin,D.K., and Lee,J.J., "Fracture parameters of interfacial crack of bimaterial under the impact loading," *Int. J. of Solids and Structures,* Vol. 38 (2001), pp. 5303-5322.

17. Shin, D.K. *et al,* "Development of multi stack package with high drop reliability by experimental and numerical methods," *Proc 56th Electronic Components and Technology Conf,* 2006, pp. 377-382

18. JESD22B113, "Board Level Cyclic Bend Test Method for Interconnect Reliability Characterization of Components for Handheld Electronic Products", JEDEC Solid State Technology Association (2006)

19. Lee,D.Y. *et al,* "Evaluation on Die Cracking due to Handling Damage in FBGA DDP for Memory Device using Handling Impact Test", *IMAPS,* 2004, Long Beach

20. Post, D., Han, B.T., and Ifju, P., High sensitivity Moire, Springer-Verlag, (NewYork, 1994)

21. Shin,D.K. *et al,* "Optimal Structural Design of Multi Chip Packageto Reduce the Failure in Substrate", Key Engineering Materials Vols. 297-300 (2005) pp. 912-917

The effect of board stiffness on the solder-joint reliability of HVQFN-packages

J. de Vries[#], M. Jansen[#], W. van Driel[*,**]

[#]Philips Applied Technologies, High Tech Campus 7, 5656AE Eindhoven, The Netherlands
[*]IMO Back End Innovation, NXP Semiconductors, Gerstweg 2, 6534AE Nijmegen, The Netherlands
[**]Delft University of Technology, Mekelweg 2, 2628CD Delft, The Netherlands
j.w.c.de.vries@philips.com, +31 40 27 48765

Abstract

In this work the results of thermal cycling tests on HVQFN-packages mounted on printed circuit boards are evaluated. The emphasis is on the fatigue life of the soldered interconnections as it is influenced by the printed circuit board, rather than that by the package or the solder.

Data from different experimental set-ups are compared: panel thickness and stand-off height of the packages. New in this respect is inclusion of the base material of the panels as parameter. The test load was set to cycling at -40°C/+125°C and -20°C/+100°C.

The results prove that the essential physical properties governing the fatigue life are the stiffness of the complete assembly and the thermal expansion mismatch between the parts. Numerical simulations support this.

1. Introduction

Over the past few years new types of IC-packages have been developed, among which the quad flat non-leaded (QFN) or micro lead-frame (MLF) packages have become very popular. In this class a variety with a low building height and a heat sink is usually called HVQFN. The advantages of these packages are manifold. They use less board space than other similar packages since they are leadless. Having no solder bumps and leads the electrical properties, such as the self-inductance, are better. The comparatively large heat sink makes them thermally superior.

Still, there are some concerns as to the board level reliability. The low solder joint height of typically a few tens of micrometers leads to a much larger shear strain caused by thermal expansion mismatches. And thus the solder fatigue life is reduced.

Already some, yet not very much, data have been published on the board level endurance of QFN-type of packages. In one of the earliest papers both the mechanical and the thermal solder joint reliability of bump chip carrier- and QFN-packages was compared [1]. An extensive overview was published by Amkor that dealt with design parameters of the QFN-package and of the printed circuit board [2]. These comprise such items as the dimension of the solder lands, the height of the interconnection, die size, board thickness, and of course the test condition. One of the first numerical simulation studies was carried through by STM, and basically covered the same topics as above [3]. Somewhat similar work combining experiments with modeling was issued by Infineon [4] and Intel [5].

Yet in all these valuable studies the basic material properties of the printed circuit board were more or less taken for granted. To date this point has been addressed hardly [6]. After measuring the board material properties the durability of several components – but no QFN – under vibrational or thermal cycling load was simulated.

It is the aim of the present work to show the importance to exactly know the board properties. To this end results from various experiments have been evaluated and completed with additional tests. Numerical simulations serve to support the analyses. Whenever possible, analytical models are formulated to illustrate the problem.

2. Experimental issues & simulation

For the present work HVQFN48-type packages were selected as test carrier. They were made with daisy chains to facilitate on-line monitoring of the integrity of the soldered interconnections. In table I the relevant data of the test packages are listed.

Table I. *Package geometry HVQFN48. The row "experiment" refers to table II.*

Size (mm)	7x7x0.85	
Pad (mm)	5.3x5.3	
Die (mm)	2.2x2.2	3.5x3.5
Pitch (mm)	0.5	
Experiment	A, B	C, D

For all experiments printed circuit boards with four Cu-layers (35 μm) and FR4 were used, the only difference being their layout or thickness. The latter was achieved by adjusting the thickness of the FR4-layer. An example of the boards is shown in figure 1. The panels had NiAu-finish. Each panel could host thirty products.

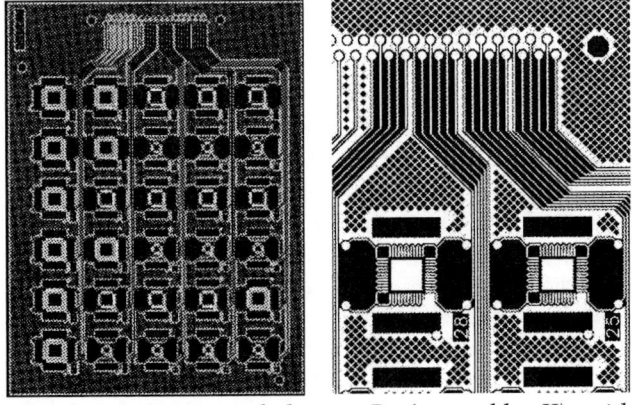

Figure 1. *Test panel layout B (see table II) with connector for on-line monitoring of various HVQFN. Left: entire panel, right: detail of HVQFN48.*

The packages were dried for 24 hours at 125°C prior to reflow soldering. Reflow was done at a peak temperature of 250°C as was determined on the boards. Standard lead-free solder paste (Sn3Ag0.5Cu: SAC305) was applied.

After assembly, the boards were inspected by means of X-ray and cross sectioning. In figure 2 one finds selected pictures of the X-ray inspection of packages immediately after assembling. With some effort even the daisy chains can be discerned. Cross sections of such packages are shown in figure 3.

Figure 2. *X-ray of HVQFN48 soldered on pcb with 3.5 mm die (layout D, left) and 2.2 mm die (layout B, right).*

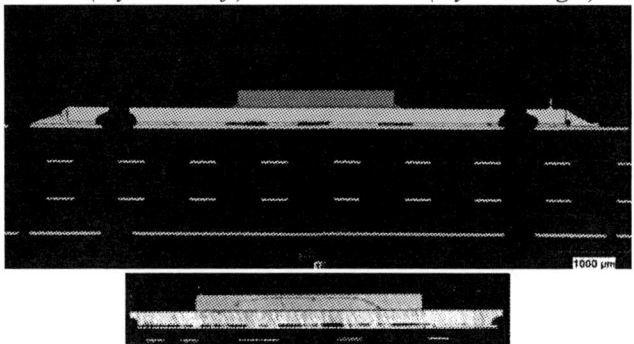

Figure 3. *Cross sections of assembled package. Top die of 2.2 mm, bottom die of 3.5 mm (part of package).*

Then they were subjected to a thermal cycling test in a one chamber system (ESPEC ENX12-7.5 CWL). The test condition was -40°C/+125°C with a cycle time of one hour. In figure 4 typical temperature profiles as recorded on the packages are shown, including the profile that was used for the thermal testing at -20°C/+100°C/1 hr-cycle (one chamber Grenco GTTS 125.20S). Various board level test standards advice to test at least 30 but preferably more products [7]. Table II contains the basic elements of the experimental set-up.

In order to capture the actual moment of failure, on-line monitoring of the daisy-chain resistance is required. Event detectors serve to detect transient interruptions in the chain. In the present case AnaTech 128/256 STD equipment was used.

Commercially available statistical software (Reliasoft, Weibull++ version 6) served to analyze the failure distributions. Weibull or log-normal (LN) statistics are commonly used to evaluate these. Often one takes the latter to describe long term degradation phenomena such as solder fatigue. Such failure distributions can be found

in figures 8, 9, and 10 (all with 95% confidence levels); the results are summarized in the lower part of table II. In some cases early failures were found, which were suspended from the analyses. Together with the surviving samples these data points are marked with separate symbols in the failure distributions.

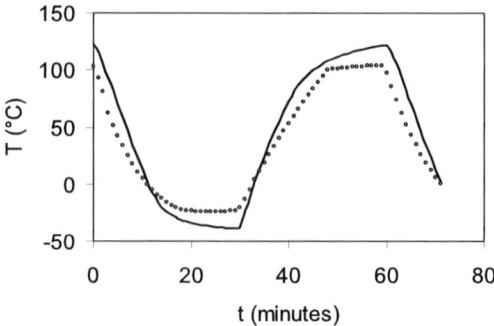

Figure 4. *Temperature profiles -40°C/+125°C (solid) and -20°C/+100°C (dotted) as measured on the packages.*

Numerical simulations were carried out on a quarter model of an assembled package. The moulding compound was simulated with a visco-elastic material model, the solder with a creep model. The finite element suite MARC was used for the modeling. Further details are given in the appropriate sections.

3. Board properties

The elastic modulus and thermal expansion coefficient of complete board stacks including the copper layers were determined. Pieces of the various printed circuit boards were cut out to determine these properties by means of Dynamical Mechanical Analysis (DMA) and Thermal Mechanical Analysis (TMA) respectively. The results of measuring the elastic board properties are collected in figure 5 and, together with the values of the thermal expansion coefficient, in the upper part of table II.

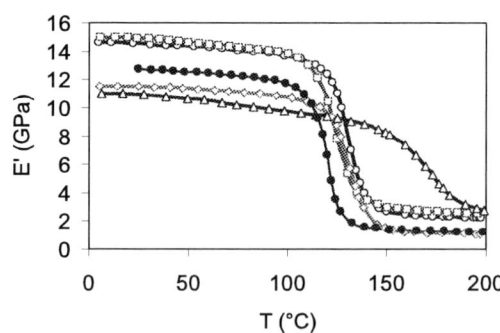

Figure 5. *Storage modulus of pcb's (DMA). See table II: A (◇), B (△), C (●), D1 1.6 mm (○), D2 0.8 mm (□).*

Although only FR4-based panels were used in the experiments, there is an appreciable variation in the mechanical properties of the boards. They are all four-layer panels with 35 μm thick Cu-layers. In the boards made from the same material but of different thickness (types D in table II) the elastic modulus is about 15 GPa below the glass transition temperature (T_g) which lies around 130°C. The stiffness of the 1.6 mm pcb is of course higher than for the corresponding 0.8 mm panel. Figure 5 also shows two other sets of data (A and B): these particular pcb's are of the same thickness, and both have a lower elastic modulus of about 11 GPa below T_g. However, their thermal expansion coefficients differ considerably, as do their respective glass transition points. Finally, the panels of series C have an intermediate value of the elastic modulus. But its glass transition temperature is quite low, such that during thermal cycle testing the temperature may rise to in the glass transition region. To a lesser extent the same holds for the 0.8 mm thick boards.

All of these differences will undoubtedly have effect on the fatigue life of the assmblies. A higher elastic constant or a larger expansion mismatch must induce more stress during thermal cycling.

Table II. Assembly and test parameters: profile, label (ID). Pcb: thickness (t), elastic modulus (E), glass transition temperature (T_g), expansion coefficient (α). Solder joint height (h). Number of samples (N). Log-normal constants: scale (μ) and shape (σ), cycles to x% failures. Open cells: data not determined.

T-profile (°C)	-40/125					-20/100
ID	A	B	C	D1	D2	D3
t (mm)	1.6	1.6	1.2	1.6	0.8	1.6
E (GPa)	11	11	13	15	15	15
T_g (°C)	132	174	120	131	125	131
α (ppm/K)	13.8	25.0		23.8		23.8
h µm	54	34		28		
N	60	40	64	30	30	30
μ	7.94	8.49	8.07	6.57	7.35	7.92
σ	0.34	0.28	0.24	0.31	0.46	0.32
1%	1266	2529	223	349	534	1304
10%	1812	3398	1990	481	863	1824
50%	2813	4881	3080	713	1553	2754

4. Failure analyses

At several moments during the tests samples were taken out for failure analysis. The failure mode was assessed from cross sections made of failed samples. In figures 6 and 7 some examples clearly reveal solder fatigue. Plastic deformation of the soldered joints and cracks within the solder appear.

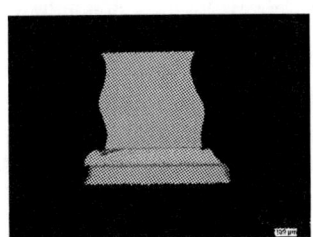 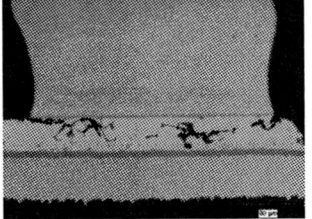

Figure 6. Set B: left 100 µm-bar; right 60 µm-bar.

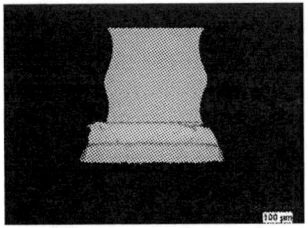 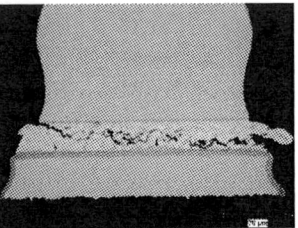

Figure 7. Left: set A 100 µm-bar; right: D1 60 µm-bar.

5. Variation of test condition

In figure 8 the log-normal failure distributions of assemblies subjected to two different test conditions are shown. A summary of the statistical evaluation is listed in the lower half of table II (test runs D1 and D3 with 1.6 mm thick panels). In the first place it must be noted that both distributions run in parallel, as also the shape parameters (σ) show. In the second place from the failure analysis – not shown here – the same failure mode was found. This is a strong indication that both tests invoke the same failure mechanism. The milder temperature cycling condition of -20°/+100°C leads to about a fourfold longer fatigue life (2754 cycles compared to 713 cycles).

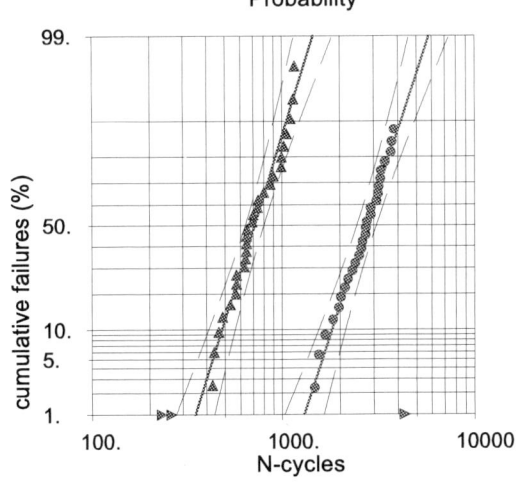

Figure 8. Log-normal distributions. Temperature profile: (▲, D1) -40 °C/+125 °C, (●, D3) -20 °C/+100 °C. Pcb 1.6 mm. Labels see table II.

For comparing the effects of thermal cycling ranges on the lifetime (τ) one usually takes the Coffin-Manson relation:

$$2\tau = \left(\Delta\gamma / 2\varepsilon_f\right)^{1/c}, \tag{1}$$

where $\Delta\gamma$ is the maximum shear strain between the component and the board, ε_f the fatigue ductility coefficient, and c the ductility exponent respectively. For eutectic SnPb-solder the exponent c is about one half, leading to the famous quadratic dependence of the fatigue life on the temperature range. A useful equation to estimate the ductility exponent was given by Engelmaier [8]:

$$c = c_0 + c_1 T_{center} + c_2 \ln\left(1 + 360/t_{dw}\right). \tag{2}$$

The coefficients c_i are specific to the solder material. Further T_{center} is the center temperature of the thermal cycle range, and t_{dw} is the hold time at the extreme temperatures. For three solder alloys the values of c_i and the ductility coefficient ε_f are listed in table III [9].

Table III. *Coefficients c_i for ductility exponent and ductility coefficient ε_f for three solder alloys.*

	SnPb	SAC305	SnAg
c_0	-0.502	-0.347	-0.416
c_1 (1/K)	-7.34E-04	-1.74E-03	-2.10E-03
c_2	1.45E-02	7.83E-03	1.40E-02
ε_f	2.25	3.47	2.25

The acceleration factor can be estimated from the maximum shear strain ($\Delta\gamma$, equation 1), assuming this to be proportional to the temperature range, and the ductility exponent (c, equation 2). For SAC305 the value of c is about 0.4, leading to an exponent of 2.5 in equation (1). The temperature ranges have a ratio of 1.4, which results in an acceleration factor of 2.3. This must be compared to a ratio of 3.8 between the experimental lifetimes. Clearly the relatively simple Coffin-Manson approach does not suffice in the current situation.

Finite element calculations were subsequently done using the creep model proposed by Darveaux [10]:

$$\frac{d\gamma_{cr}}{dt} = C \frac{G(T)}{T} \left[\sinh\left(\theta \frac{s}{G(T)} \right) \right]^n e^{-Q/kT}. \qquad (4)$$

Here $d\gamma/dt$ is the creep strain rate, G is the shear modulus, θ the stress level where the power law breaks down, s the applied stress, n the stress exponent, and Q is the activation energy. For Sn3.5Ag the following values were taken: C = 0.4539, G(T) = 19300-69(T-273) [MPa], θ = 1500, n = 5.5, and Q = 0.5 eV [11, 12]. For more details on the model see the appropriate section 8.

The numerical result leads to a ratio of 3.0 for the fatigue life of the two thermal cycling conditions as compared to the experimental value of 3.8.

6. Effect of solder joint height

The influence of the height of the soldered interconnections on the fatigue life is given in figure 9. All data were obtained on boards of 1.6 mm thickness (see table II, sets A, B, and D1).

Combination of equations (1) and (3) should give the translation factor for the height (h) of the interconnection between the three sets of results. However, it is not possible to obtain a correlation of the results in this manner. We attribute this to the analytical models not taking into account the mechanical properties of the printed circuit boards. And these differ widely for the given sets of data. In fact not one single pair of these three has a combination of comparable values of the elastic modulus and expansion coefficient of the panels.

The mechanism can be readily understood. At a certain temperature the whole assembly experiences no strain. However, when the temperature is changed the various parts will respond by expanding or shrinking, each according to their expansion coefficient. The mismatch which is a result of this action will have to be accommodated by the soldered interconnections, which will be sheared. Usually the following equation is given to describe the maximum shear strain ($\Delta\gamma$):

$$\Delta\gamma = (\Delta L/h)\Delta\alpha\Delta T. \qquad (5)$$

L is the span of the soldered joints under consideration, ΔL their thermal displacement, h is their height, $\Delta\alpha$ is the thermal expansion mismatch between the component and the board, and ΔT is the temperature difference the assembly is subjected to.

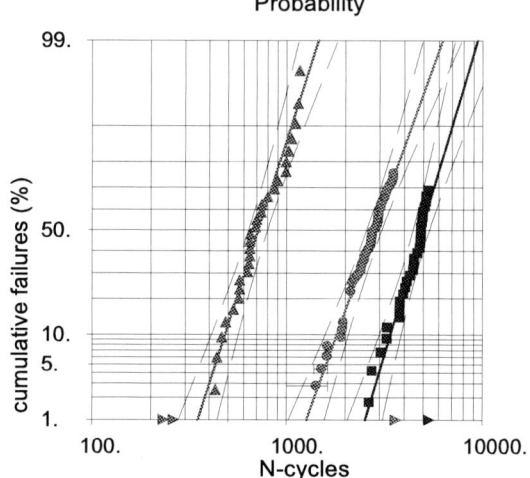

Figure 9. *LN-distributions. Solder joint thickness: (▲, D1) 28 μm, (■, B) 34 μm, (●, A) 55 μm. Pcb 1.6 mm, -40 °C/+125 °C. Labels see table II.*

The strain is thus inversely proportional to the height of the joints. One can use equations (1) and (5) to check the validity of $\tau \sim h^{1/c}$ (τ is the experimentally observed lifetime). Even if one estimates the effective expansion coefficient of the package, which would be about 9 ppm/K, to arrive at the value of the maximum shear strain ($\Delta\gamma$) still no correlation is obtained. Thus, also in this case another type of model is needed.

However, even now no match could be obtained if the properties of the printed circuit boards were held constant.

7. Effect of stiffness of printed circuit board

The stiffness of the printed circuit boards depends on the material constant – the elastic modulus – and the construction – the thickness – of the board. Experimental results for boards of thicknesses ranging from 0.8 to 1.6 mm are shown in figure 10 (data sets C, D1, and D3 of table II). The longer lifetime of the packages mounted onto the thinner boards of 0.8 mm (series D2), which is of the same origin as those of 1.6 mm thickness (series D1), can be readily understood.

Let us use the following simple model to connect the fatigue life of a solder joint to the stiffness of the printed circuit board. Regard the assembly as an one-dimensional spring system. The component is assumed to have infinite stiffness, while the solder and board have a stiffness of k and K respectively. By virtue of the thermal expansion mismatch the solder experiences a displacement (δ) which is partly counteracted by the elastic reaction of the board. From equilibrium of forces one can derive the energy contained in the "solder-spring". Since the lifetime (τ) of the solder joint is inversely proportional to the stored energy, one arrives at an expression for the lifetime as a function of the board stiffness:

$$\tau \propto \left(\frac{1+K/k}{K\delta}\right)^2. \qquad (6)$$

Straightforward mechanics tells us that the stiffness of a slab of thickness t and elastic modulus E is $K \sim Et^3$. The problem in using this basic formula to estimate the ratio of the lifetimes between the various assemblies, is uncertainty about the value of the spring constant of the solder (k). The form of equation (6) already shows that the result depends very strongly on the value of k.

Eventually, the problem could be circumvented by using the same numerical model as described above. The ratio of the lifetimes between 0.8 and 1.6 mm thick panels is 1.7, which must be compared to the experimental value of 2.2.

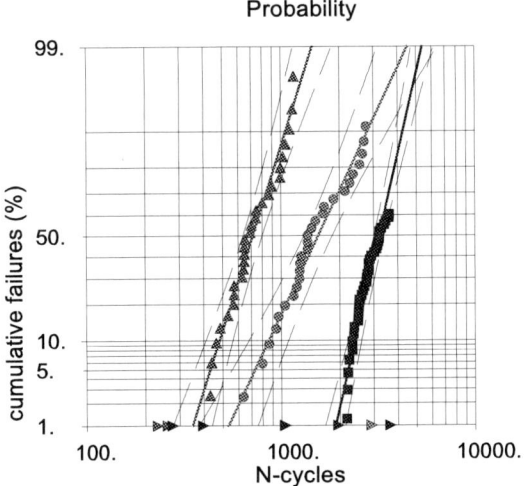

Figure 10. *LN-distributions. Pcb thickness: (▲, D1) 1.6 mm, (■, C) 1.2 mm, (●, D2) 0.8 mm. -40°C/+125°C. Labels see table II.*

As for the assemblies based on the 1.2 mm thick panels (series C in figure 10), they do not fit very well in this argumentation. The lifetime is exceptionally long. However, these very boards have a transition temperature that is low enough (about 120°C, see table II) to enter the temperature cycle range. Thus while stressing the solder joints the elastic modulus of the panels drops strongly, and hence much less stress will build up, leading to a longer fatigue life. Admittedly, the stand-off and the thermal expansion coefficient of the board are not known. At present no quantitative model is available to estimate the size of this effect.

8. Numerical model

The discussions of the experimental results prove the shortcomings of analytical calculations to analyze the results. A few numerical calculations have been presented already which clearly show their advantage over the analytical approach. In this section we will briefly discuss the finite element model and the results of using it to a parameter sensitivity analysis.

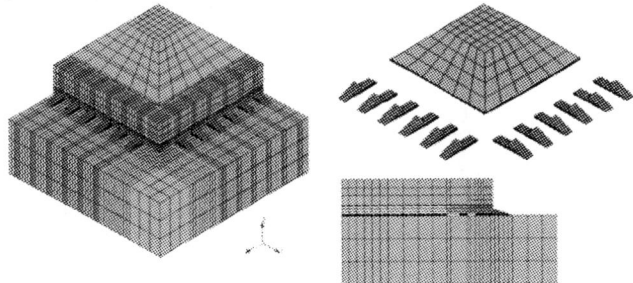

Figure 11. *Quarter FE-model (left), die pad (top right), cross section showing also die and leadframe (bottom right).*

The finite element model consists of a quarter of the complete assembly (see figure 11). Two thermal cycles have been simulated (see figure 12). The strain was determined by averaging over one complete soldered connection at the corner of the package since this is the most critical joint. In figure 12 the plastic strain and the creep strain are shown as they build up during thermal cycling. Plastic strain is only ten percent of the creep strain. As the temperature is at its lower extreme, the creep mechanism is hardly active, while at the higher extreme only little creep occurs because the stress is much lower. Only during the temperature ramps an appreciable amount of creep strain develops.

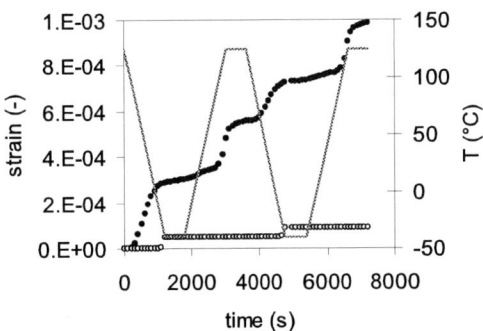

Figure 12. *Temperature profile (right axis, red line), creep strain (left axis, ●), and plastic strain (○).*

Application of the finite element model to various assemblies leads to a satisfactory result as depicted in figure 13. Here, one finds the lifetime normalized to the experimental result of HVQFN48 at -40°C/+125°C on 1.6 mm boards (see table II, series D1). One recognizes the outcome of testing at a different cycling condition which was discussed in section 5 (table II, series D3 and figure 8). Likewise, the tests carried out on packages soldered

439

onto a thinner board are shown (table II, series D2 and figure 10).

To further validate the model, also other HVQFN-packages were modeled and compared to experimental test results. These are a smaller HVQFN24 (4x4 mm²) and a larger HVQFN72 (10x10 mm²). Please note, that these two latter data sets should not be compared to the ones discussed in this work, here they are only put on stage for illustrating the strength of the simulation model.

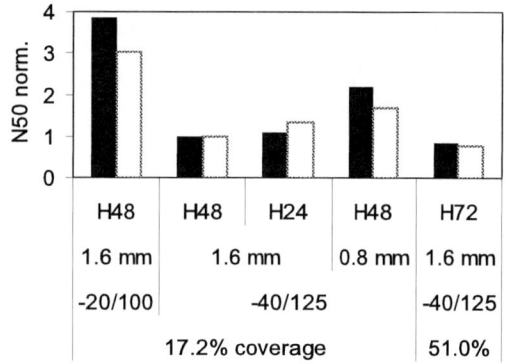

Figure 13. *Experimental lifetime (filled bars) compared to numerical result (open bars), normalized to HVQFN48 at -40°C/+125°C. Along the x-axis from top to bottom: type of package, board thickness, test condition, ratio of die area to die pad area.*

From these results one may safely conclude that the numerical model is sufficiently reliable to proceed with the sensitivity analysis. The analyses of the experiments sofar lead to the notion that the thickness, the elastic modulus, and the thermal expansion coefficient of the printed circuit boards govern the fatigue life of the interconnections and variations therein should not be neglected.

In this case one thermal cycle was modeled. Three board parameters were varied (in brackets the nominal value):

– Thickness 0.5 – 2.5 mm (1.6 mm)
– Elastic modulus 10 – 20 GPa (17.5 GPa)
– Thermal expansion coefficient 10 – 25 ppm/K (17.6 ppm/K)

Again, one of the corner joints was used to average the strain that was translated to fatigue lifetime by means of the Coffin-Manson relation (equation 1). In figure 14 the results are represented in the form of response surfaces of the lifetime for the two pairs of parameters.

Compared to the nominal fatigue life – defined as the lifetime at the following board parameters: thickness 1.6 mm, elastic modulus 17.5 GPa, and expansion coefficient 17.6 ppm/K – the lifetime can vary considerably. If the expansion coefficient of the board increases the lifetime follows, but the more so when thinner panels are used. A factor of 3.5 in lifetime is well possible. With regard to the elastic modulus the effect is to prolong the fatigue life if one would use softer boards, and again: the thinner the board is the better the lifetime will be. As a matter of fact,

the elastic modulus looses its magic power if the printed circuit boards are thicker than about 2 mm.

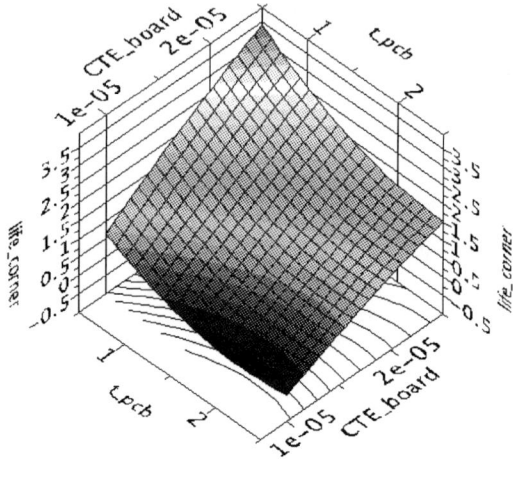

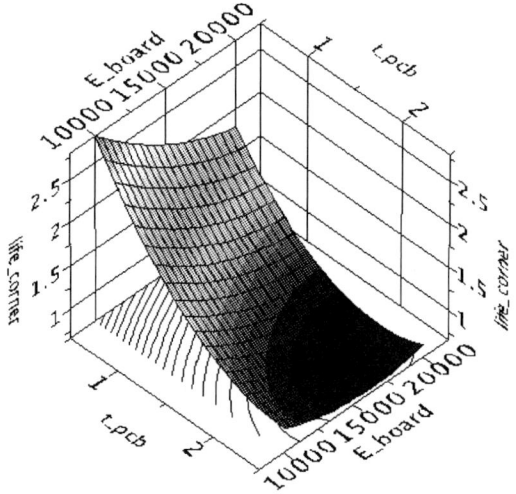

Figure 14. *Response surface of normalized fatigue life against:*
(top) thermal expansion and thickness of board,
(bottom) elastic modulus and thickness of board.

This same analysis carried out on a center joint leads to essentially the same but far less pronounced results.

It will be clear that the numerical model compares better to the experimental results than the analytical attempts. These latter do, however, give a more direct insight in the physics behind the mechanisms that play a role in the fatigue deformation.

9. Conclusions

In this work the authors have discussed experimental endurance results of HVQFN-packages soldered onto printed circuit boards. After analytical and numerical analyses the following conclusions and recommendations can be formulated.

Contrary to the frequently adopted starting point of assuming fixed board properties, they may vary markedly.

These variations have a significant impact on the test result which may mask any variations that have been made by design.

As theoretically expected, the first order effect of board level reliability performance is dominated by the thermal expansion mismatches. However, there is a strong second order effect that is dominated by bending resistance of the printed circuit board, that is in turn characterized by its thickness and/or Young's modulus.

For the investigated package a thin board with a low stiffness can lead to a factor of 2.5 longer board level life time. It is therefore strongly advised to report on the used board materials when discussing board level reliability results.

Also it is likely that the glass transition temperature of the board material affects the solder fatigue life. If T_g enters the range of the temperature cycling test, the mechanism of building up stress changes. To date, this effect can only be addressed qualitatively.

References

1 Hung, S.C., Zheng, P.J., Ho, S.H., Lee, S.C., Wu. J.D., "The comparison of solder joint reliability between BCC++ and QFN," *Proc. Electronic Components and Technology Conf.*, 2001. pp. 1052-1058.

2 Syed, A., Kang, W.J., "Board level assembly and reliability considerations for QFN type packages," *Proc. Surface Mount Technology Association*, 2003, pp. 181-188.

3 Tee, T.Y., Ng, H.S., Yap, D., Zhong, Z., "Comprehensive board-level solder joint reliability modeling and testing of QFN and power QFN packages," *Microelectronics Reliability*, vol. 43 (2003), pp. 1329-1338.

4 Mercado, L.L., Sarihan, V., Fiorenzo, R., "A performance and manufacturability evaluation of bump chip carrier packages,". *IEEE Trans. Adv. Pack.*, Vol. 27 (2004), pp. 151-157.

5 Stoeckl, S., Pape, H. "Improving the solder joint reliability of VQFN-packages," *Proc. Electronics Packaging Technology Conf.*, 2005. pp. 760-767.

6 Qi, H., Ganesan, S., Wu, J., Pecht, M., "Effects of printed circuit board materials on lead-free interconnect durability," *Proc. Int. Conf. On Polymers and Adhesives in Microelectronics and Photonics*, 2005, pp. 140-144.

7 Performance Test Methods and Qualification Requirements for Surface Mount Solder Attachments, IPC9701, 2006.

8 Engelmaier, W., "Fatigue life of leadless chip carrier solder joints during power cycling," *IEEE Trans. Comp. Hybrids & Manuf. Techn.*, Vol. 6 (1983), pp. 232-237.

9 Osterman, M., "Effect of temperature cycle on the durability of lead free interconnects (Sn-Ag-Cu and Sn-Ag)," *CalceEPSC project C05-06*, 2005.

10 Darveaux, R., Banerjee, K., "Constitutive relations for Tin-based solder joints," *IEEE Trans. Comp. Hybrids & Manuf. Techn.*, vol. 15 (1992), pp. 1013-1024.

11 Schubert, A., Dudek, R., Doering, R., Walter, H., Auerswald, E., Gollhardt, A., Schuch, B., Sitzmann, H., Michel, B., "Lead-free solder interconnects: characterization, testing, reliability," *Proc. EuroSimE*, 2002, pp. 62-73.

12 Wiese, S., Schubert, A., Walter, H., Dudek, R., Feustel, F., Meusel, E., Michel, B., "Constitutive behavior of Lead-free solders vs. Lead-containing solders - experiments on bulk specimens and flip-chip joints," *Proc. Electronic Components and Technology Conf.*, 2001, pp. 890-903.

Investigation of Mechanical Reliability of Cu/low-k Multi-layer Interconnects in Flip Chip Packages

Chihiro J. UCHIBORI[1], Xuefeng Zhang[2], Sehyuk Im[2], Paul S. Ho[2], and Tomoji Nakamura[3]

1) Fujitsu Labs. America, Inc., 1240 E. Arques Ave. MS345, Sunnyvale, CA 94085, U.S.A.
Chihiro.UCHIBORI@us.fujitsu.com, +1-408-530-4672
2) Microelectronics Research Center, University of Texas at Austin, Mail Code: R8650, Austin, TX78712, U.S.A.
3) Fujitsu Laboratories Ltd., 50 Fuchigami, Akiruno, Tokyo 197-0833, Japan

Abstract

The impact of Chip-Package Interaction (CPI) on the reliability of Cu/low-k interconnects in a flip-chip package for high performance ULSI was investigated using Finite Element Analysis (FEA). A 3D four-level sub-modeling approach was used to analyze the CPI to link the deformation from the package level to the interconnect level. The energy release rate (ERR) and fracture mode at critical interface were calculated using a A modified virtual crack closure technique (MVCC). The simulation was focused on the die attach process for Pb-free process before underfilling where the maximum CPI effect is expected. First the general characteristics of CPI were analyzed for interfaces in two metal-layer interconnects. The ERR was found to increase rapidly with decreasing modulus of Inter Layer Dielectric (ILD) although the effect of CTE of ILD was found to be small. Next, the CPI for a four metal-layer structure was investigated. Here the ERR for upper M3 and M4 levels were consistently higher than those of lower M1 and M2 levels. If the same low-k ILD is used for all layers, the M4 interfaces show 2.5 times higher ERR than the lower levels. However, when TEOS is used in the M4 level, the ERR at M3 interfaces becomes 35% higher than the M4 level. The wiring dimensions and ILD properties were found to be important in controlling CPI. The CPI impact on ultra low-k reliability and interconnect design rules for the 65nm technology and beyond are discussed.

1. Introduction

For high performance ultra large scale integrated (ULSI) circuit chips, the high I/O density requires the use of flip chip ball grid array (FCBGA) package. During packaging assembly, significant stresses are generated due to the mismatch in thermal expansion coefficient (CTE) between the chip and the package. The chip-package interaction (CPI) will induce large local deformation and stress to drive interfacial crack formation, resulting in interconnect failures. This raises serious reliability concerns beyond the 65nm technology node when the mechanical properties of ultra-low-k ILD materials degrades [1, 2].

Several studies have applied finite element analysis to study the effects of CPI on the mechanical reliability of Cu/low-k structures. Mercado et al. found that the crack driving force at the critical interface increases with increasing number of interconnect layers [3]. They showed that the crack driving force for low-k structures is higher than that for SiO$_2$ interconnects. Subsequent study by Wang et al. found that ILD and packaging materials play an important role in contributing to CPI and interconnect reliability[4]. The crack driving force can be enhanced significantly when spin-on polymer was used as ILD, and became higher for Pb-free solder than eutectic and high lead solders and also for plastic packages compared with ceramic packages. Recently Zhai et al. extended the study to investigate the CPI effect on crack driving force at the packaging as well as the wiring level of multilayered interconnects [5]. In spite of the increasing efforts in developing ultra low-k materials, the CPI impacts are not yet clear but will have to be better understood to ensure the reliablity of the ultra low-k interconnect structures.

In this study, we study the effects of ILD materials and wiring structures on CPI for Cu/low-k interconnects used for high performance ULSI in FCBGA. A FCBGA with a large die (20mm x 20 mm) and high density I/O bumps is investigated using a 3D multi-level sub-modeling technique to calculate the crack driving forces at several interfaces [4, 5]. The analysis is first performed on two metal-layer interconnects to evaluate the effect of material properties on CPI, including a CVD-OSG (k=3.0) [6], MSQ [7], spin-on polymer [4] and porous MSQ based materials with k ~2.3 [8]. Then the study on CPI is extended to wiring structure effects investigating a four metal-layer structure with wiring geometries specified by the design rule. The effects on CPI due to wiring dimensions in combination with different ILD materials are examined. Finally, the implications on ultra low-k reliability and interconnect design rules for 65nm technology and beyond are discussed.

2. Simulations

2. 1. Multi-level Sub-modeling Technique

A 3D multi-level sub-modeling technique was used to calculate CPI, which provided the finite element method linking the mechanical deformation from the package level to the interconnect level. The 4-step submodel bridged the gap in feature sizes between the package and the wafer levels, enabling us to examine in detail the stress-strain distribution in the interconnect structure. At the first step, a quarter of a flip-chip package was modeled using a symmetric boundary condition (Fig. 1a). At the second step, a critical bump area, which was identified by the first step analysis, was modeled (Fig. 1b). At these steps, details of the interconnect structures were

1-4244-1105-X/07/$25.00 ©2007 IEEE

not modeled due to the small dimension of the interconnect. The simulation results at these two steps can be verified by experimental observations based on high–resolution Moiré interferometry [4]. At the third step, a small region with high stress concentration at the interface structures was analyzed (Fig. 1c). Here a portion of solder bump and Si chip and BPSG, ILD and passivation layer were taken into account. Finally, the interconnect region was zoomed in from the third step and the details of interconnect stress distribution were modeled at the fourth step (Fig. 1d). The ERR driving delamination at interconnect interfaces was calculated for several specific interfaces using a modified virtual crack closure (VCC) technique [4].

2. 2. Energy Release Rate

We investigated first the effect of ILD materials on ERR for two metal-layer interconnect structure. The interfaces most prone to delamination chosen for this structure are shown in Fig. 2. The CPI analysis was performed for the die attach process of lead-free solders, where the solder bump was formed by reflow at 250°C before the underfill layer was filled. This induced the maximum thermal stresses and ERR driving force for interfacial delamination in the interconnect structure. We compared first the ERR for CVD-OSG with MSQ and spin-on polymer. Then the calculation was extended to porous MSQ with k of ~2.3 to examine the CPI problem for low-k dielectrics developed for the 65nm node and beyond.

Next the study was extended to investigate the effect of wiring structure on ERR using a four metal-layer interconnect structure. The interfaces of interest for this calculation are shown in Fig. 3. The dimensions of metal line and via layer were specified according to the design rule with 1x for M1 and M2, 2x for M3 and 4x for M4. In this structure, we first calculated ERR for a wiring structure where ultra low-k is used for every level from M1 to M4. Then ERR was calculated with the ILD of M4 changing to SiO2 in order to examine the ILD effect on CPI for multi-level structures. The mechanical properties of the interconnect and ILD used in the calculation are summarized in Table I.

Table. I. Mechanical properties of interconnect and ILD.

Materials	E(GPa)	ν	CTE ($\times 10^{-6}$/K)
Cu	122	0.35	17
Porous MSQ-A	2.0	0.3	10
Porous MSQ-B	5.0	0.3	10
Porous MSQ-C	10.0	0.3	12
Porous MSQ-D	15.0	0.3	10
Porous MSQ-E	10.0	0.3	6
Porous MSQ-F	10.0	0.3	18
CVD-OSG	17.0	0.3	8
Spin-on polymer	2.45	0.35	66
MSQ	6.0	0.35	18

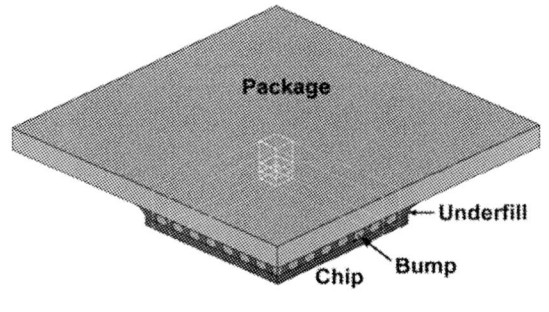

(a) 1st step: Package level

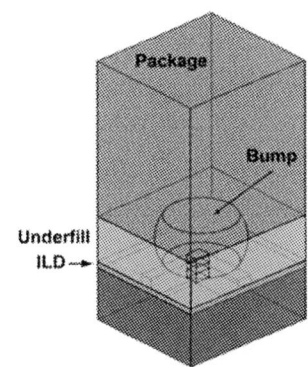

(b) 2nd step: Critical solder region level

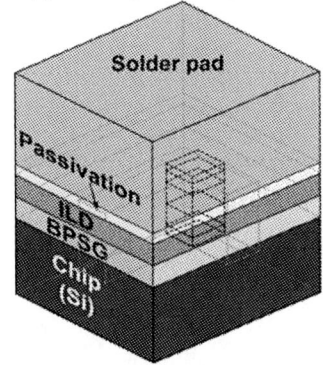

(c) 3rd step: Chip-Bump interface level

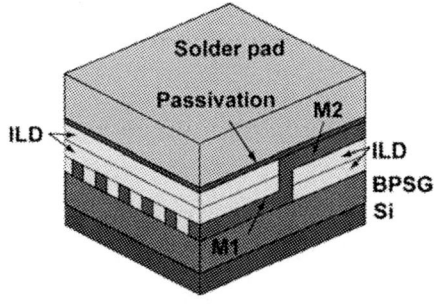

(d) 4th step: Interconnect level

Fig. 1. Schematic illustration of the concept for 4-step sub-modeling.

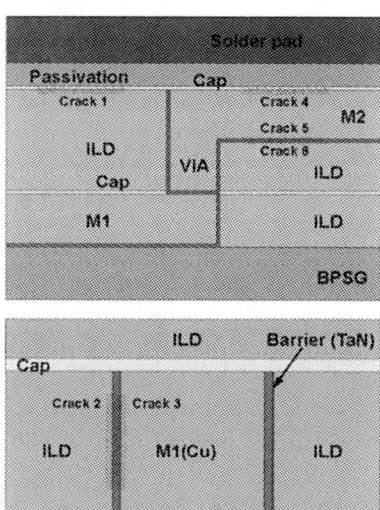

Fig. 2. Interested interfaces in the two metal-layer interconnect structure.

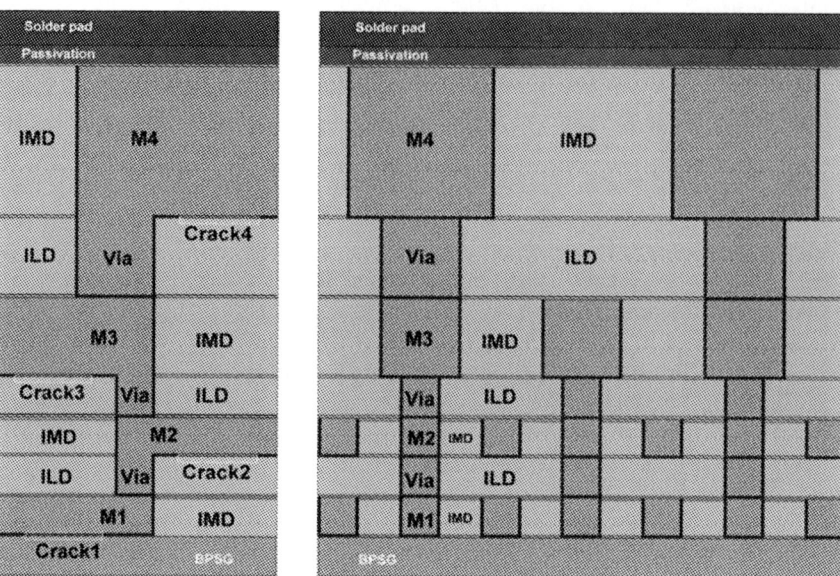

Fig. 3. Interested interfaces in the four metal-layer interconnect structure.

3. Results and Discussion

3. 1. Impact of ILD properties on ERR

Results on ERR obtained by the simulation for cracks 1 to 6 in two-level interconnect structures with different ILD are plotted in Fig.4. The ERR of spin-on polymer [4] and MSQ [7] taken from other papers are included for comparison. The ERR at Crack 2 and 3 which are perpendicular to the chip surface showed the lowest values. This indicates that the crack driving force comes mostly from peeling stresses due to thermally induced warpage of the package to affect mainly the horizontal interfaces: Crack 1 and Cracks 4-6. For these interfaces, the ERR driving forces are the lowest for CVD-OSG, which has the highest Young's modulus (E) compared with the other two materials. The spin-on polymer has the lowest E and thus the largest ERR, reaching 30 J/m^2 for crack 6 at the interface between the ILD and the etch-stop layer (ESL). The ERR for Crack 1 at the ILD/passivation interface is comparable to that of Crack 6 and both are about 6x higher for the organic polymer than CVD-OSG. This can be attributed to the weak elastic modulus of the spin-on polymer, which needs about 6 times higher adhesion strength at those interfaces to yield a mechanical reliability equivalent to interconnects fabricated using CVD-OSG. For these low k materials, only the spin-on polymer has a high CTE. The effects of CTE on ERR will be discussed later.

To examine the mechanical reliability for future interconnects fabricated with porous MSQ with k of less than 2.3, the ERR was calculated and the results are plotted in Fig. 5. Consistent with the results shown in Fig.4, ERR generally increases with decrease in E. It is interesting to compare porous MSQ-C (k=2.3) with fully dense MSQ (k=2.7) since they have similar mechanical

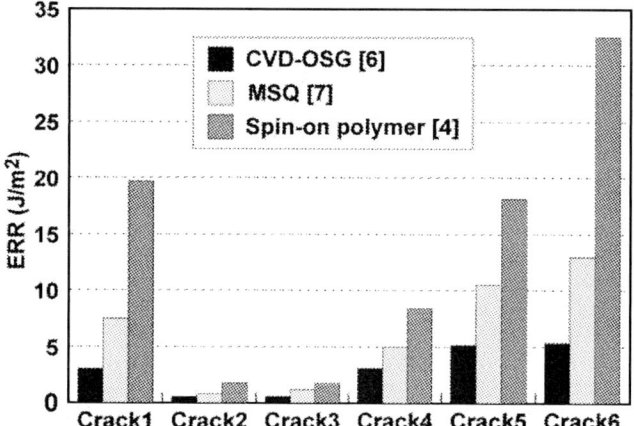

Fig. 4. ERR of CVD-OSG [6], MSQ[7] and Spin-on polymer [4] at Crack1 to 6.

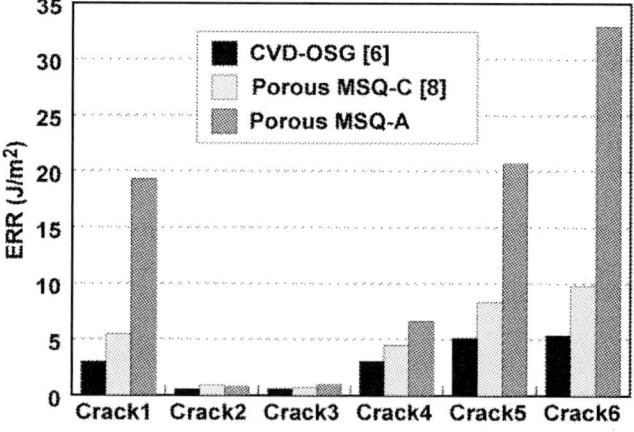

Fig. 5. ERR of CVD-OSG [6], Porous MSQ-A and Porous MSQ-C [8] at Crack1 to 6.

444

properties, their ERR values are similar. In contrast, the porous MSQ-A dielectric due to a low E show almost the same ERR as the spin-on polymer, even the CTE of these materials have 7 times difference. To examine the effect of E, ERR of CVD-OSG, porous MSQ-A and porous MSQ-C dielectric are plotted as a function of E in Fig. 6. Overall, the ERR values at cracks 1, 5 and 6 increase rapidly when the E of the low k dielectric becomes lower than 10 GPa. Since the CTE of these materials are almost the same, about ~10ppm, there is negligible effect due to CTE on ERR for these materials. Therefore, for low k dielectrics with a relatively low CTE, increasing E is effective for improving the mechanical reliability.

To examine the effect of CTE and E on ERR, the ERR of porous MSQ-B to F are plotted in Fig. 7. Note that porous MSQ-C is a low-k ILD developed for 65nm [8], which have material properties different from other low k dielectrics. As already shown in Fig. 6, ERR increase rapidly with decrease in E. However, comparing porous MSQ-C, E and F, their ERR is almost same although their CTE is quite different. This indicates that the ERR is insensitive to CTE.

3. 2. Impact of ILD properties for four metal-layer interconnects

Since the low k interconnect structure for the 65nm technology node will have more than 11 layers and with complex geometry and material combinations [10], we extended this study to a four metal-layer model to investigate the wiring structure effect on CPI. The study was focused on the ILD/ESL interface of Crack 6 in Fig. 2 which showed the highest ERR. The ERR for this interfaces at the four wiring levels are plotted in Fig. 8 where the MSQ type porous ultra low-k material is used for ILD in all the levels. The ERR at the interface of the top M4 layer was found to be 2.5 times higher than that in the lowest M1 layer. This indicates that the top surface layer with the thickest ILD is most prone to delamination. This analysis was extended for the same wiring structure but with SiO_2 replacing the low k ILD in the top layer since ultra low-k ILD is generally not used there. The results in Fig. 9 show that the ERR at the interface in the top M4 layer decreases by 35% and becomes the highest at the M3/M4 interface. This result can be attributed to an elastic mismatching effect where the crack driving force

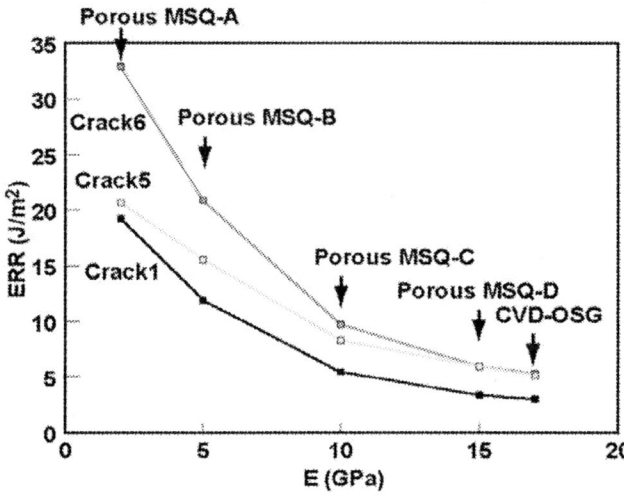

Fig. 6. ERR as a function of Young's modulus where CTE are about 10 ppm.

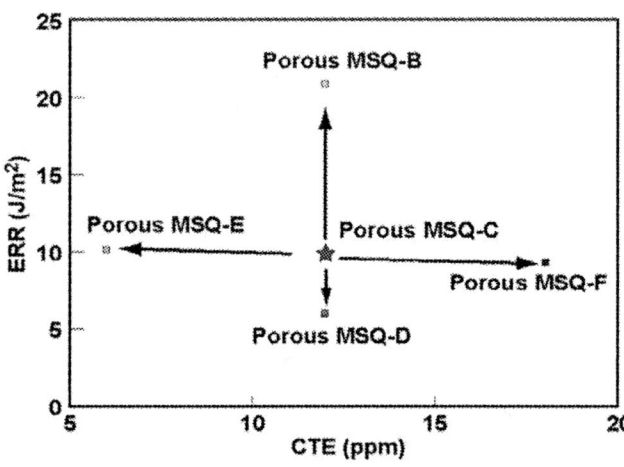

Fig. 7. ERR of Porous MSQ-B to F.

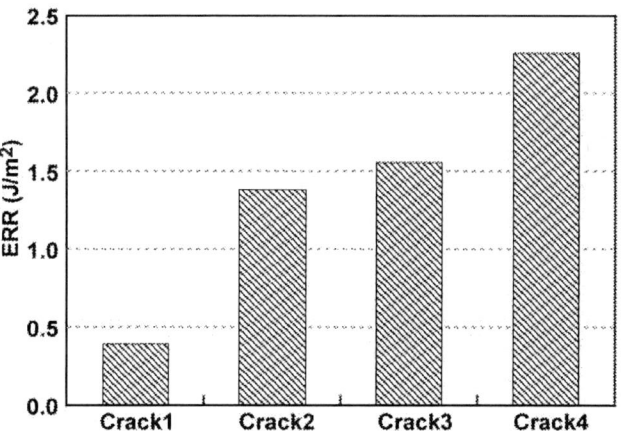

Fig. 8. ERR at ILD/ESL interface in the four metal-layer interconnect structure where the MSQ type porous ultra low-k material is used for ILD in all the levels.

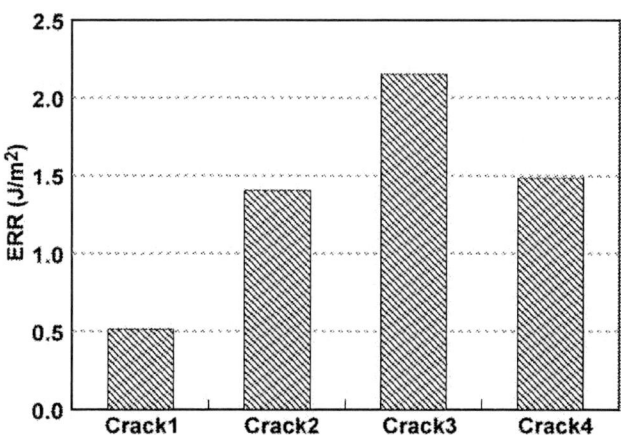

Fig. 9. ERR at ILD/ESL interface in the four metal-layer interconnect structure where ILD of M4 layer is replaced with SiO_2.

445

is enhanced when a stiff top layer is supported by a viscous substrate layer [11]. Such an effect has been demonstrated to enhance channeling cracking in an OSG layer when it was placed on top of a spin-on polymer layer [12]. It is interesting that a similar effect can also enhance the interfacial crack driving force in multilayered interconnects where the stress state can be quite different. It is worth noting that the effect of elastic mismatch depends in general on the thickness and dimension of the wiring structure. The combined effects of ILD materials and wiring dimensions on CPI and crack driving force seem to be interesting and important and will be further studied to help our understanding of the CPI impact for reliability of ULSI interconnect structures.

4. Conclusions

The effect of CPI on Cu/Ultra low-k interconnect reliability for a number of low k materials and multi metal layer structure was investigated to clarify the critical interface of delamination and to obtain the guideline for design rule for 65nm node and beyond. ERR was found to increase rapidly with decreasing modulus but insensitive to CTE. Our results showed that the wiring structure and dimension play an important role in contributing to the crack driving force. The ERR was the highest at the interface between two ILD layers with different elastic moduli. This effect is attributed to elastic mismatch which seems to be important in controlling the ERR for multilayered low k interconnects.

References

1. The International Technology Roadmap for Semiconductors 2003 edition, Semiconductor Industry Association, San Jose, CA., 2003.

2. N. Aoi, T. Fukuda and H. Yanazawa, in *Proc. IEEE 2002 Int. Interconnect Technol. Conf.*, 2002, pp. 72-74.

3. L. L. Mercado, C. Goldberg, S-M. Kuo, T-Y. T. Lee and S. Pozder, "Analysis of Flip-Chip Packaging Challenges on Copper Low-k Interconnects", in *Proc. 53th Electronic Components and Technol. Conf.*, 2003, pp. 1784-1790.

4. G. Wang, C. Merrill, J-H Zhao, S. K. Groothuis and P. S. Ho, "Packaging effects on reliability of Cu/low-k interconnects", *IEEE Trans. Dev. and Mat. Reliability*, vol. 2, pp. 119-128, Dec. 2003.

5. Lei L. Mercado, S-M. Kuo, C. Goldberg and D. Frear, "Impact of Flip-Chip Packaging on Copper/Low-k Structures", *IEEE Trans. Adv. Packaging*, vol. 26, pp. 433-440, November 2003.

6. H. Watatani, T. Owada and S. Fukuyama, unpublished.

7. G.T. Wang, X.F. Zhang and P.S. Ho, *Proc. Inter. Stress Workshop*, AIP Conf. Proc. Series, Vol. 817, 2005, in press.

8. T. Nakamura and A. Nakashima, in *Proc. IEEE 2004 Int. Interconnect Technol. Conf.*, 2004, pp. 175-177.

9. F. G. Bucholz, R. Sistla and T. Krishnamurthy, "2D- and 3D-Applications of the improved and generalized Modified Crack Closure Integral Method", in *Computational Mechanics '88*, S. N. Atluri and G. Yagawa, Eds., Spring Verlag, 1988.

10. I. Sugiura, et. al., in Proc. IEEE 2005 Int. Interconnect Technol. Conf., 2005, pp. 15-17.

11. Liang, Huang, Prevost and Z. Suo, Exp. Mechanics 43, 269 (2003).

12. J. He, G. Xu and Z. Suo, *Proc. 7th Inter. Stress Workshop*, AIP Conf. Proc. Vol. 741, p. 3, 2004; T. Tsui et al., *Proc. MRS Symposium*, 2005.

Packaging Design and Testing for High Temperature Applications > 150 °C

Thomas Schreier-Alt, Christian Rebholz, Frank Ansorge
Fraunhofer Institute Reliability and Microintegration IZM
Micro Mechatronics Centre
Argelsrieder Feld 6, 82234 Oberpaffenhofen
info@mmz.izm.fhg.de, +49(0)8153-90975-00

Abstract

The paper presents final results of a project called "High Temperature Packaging with New Technologies" (HotPaNTs) with a consortium of well known partners from semiconductor and automotive industry. The project investigated system reliability at junction tempertures up to 200°C experimentally and by numerical simulations. Focus is on assembly, reliability testing of packaging technologies and in line electrical testing of the components at high temperatures.

The design phase was accompanied by thermal simulations. We compared several package types (e.g. Flip Chip, QFN paddle up / down, QFP) concerning their performance under static and dynamic thermal loading. In detail two QFN packages have been studied: "paddle down" with main thermal pathways junction / board and "paddle up" with main pathway junction / air. A test chip with internal heat generators and temperature sensors was packaged in both housing types and assembled on different test boards. By numerical simulations and comparison with IR picture recordings, optimized thermal design guidelines (thermal vias, heat spreaders..) could be found.

1. Introduction

By the use of modular, decentralised electronic parts and components within automotive industry an increasing amount of systems are placed "under the hood" or near the powertrain. Mechatronic control units are actually specified for a continuous operation temperature of $T_{amb} \leq 140°C$ when placed within high temperature environments such as exhaust gas, cooling circuit, engine condition monitoring or powertrain. An increasing demand for sensor application within ambient temperatures up to 175°C is emerging with the "X by wire" technology (see table 1).

By combining power electronics with heat generation rates of several watts with IC controllers within one system or even within one package the dissipation loss will increase the junction temperature on the semiconductor chips easily up to $T_{junc} = 200$ °C within these environmental conditions [1]. The trend to combine several chips into miniaturized multi chip modules increases the power density.

Several projects like „MEDEA HiTeC A 306 " have shown the feasibility of integrated circuits at operating temperatures up to $T_{junc} = 200$ °C.

Packaging designers usually avoid classical low cost packaging technologies such as polymer encapsulation when confronted with increased ambient temperatures. The use of metal, ceramic and glass housings is still normative within these applications, but their disadvantages are well known: costs, space consumption and difficulties in high volume batch processing. Unlike frontend progress in miniaturization the package gap is still increasing within high temperature electronics as chip sized package types (CSP) are not in widespread useage yet. Assembly technologies within high-temperature applications suffer from comparable problems: the use of high priced materials for electrical connections (gold, tungsten,…) blocks their useage within low cost and high volume markets.

It is scope of the HotPaNTs project to investigate standard low-cost assambly and packaging technologies for the use at ambient temperaturs up to 175°C. The temperatur rise due to dissipation loss should be limited by innovative cooling concepts to $T_{junc} = 200$ °C.

The project steps within the last five years have been the following:

- Evaluation of thermal boundary conditions for commercial polymer packages due to high-temperature rigidity and reliability.
- Development of high volume assembly technologies with an increased field of application up to $T_{junc} = 200$ °C. Enhanced interconnection technologies have to be applied concerning metallization of electrical interconnects, die attach and molding materials.
- Development of high volume packaging technologies with an increased field of application up to $T_{amb} = 200$ °C (see table 2).
- Evaluation of reliability testing schemes for high-temperature applications. Main focus was on accelerated material ageing near the degradation temperature.
- Development of functionality testing conditions for in-line measurement of electrical performance at high temperatures

Table 1: Requirements and future development for polymer packages for high temperature applications

	today		short term		long term	
	signal	power discretes	signal	SoC	power MCM	signal
T_{amb} [°C]	<140	<150	<150	<140	<150	<200
T_{junc} [°C]	<150	<175	<175	<175	<200	<200
Duration [h]	~	~	<100	~	~	~
Cycling (-40°C / T_{amb})	6000	6000	6000	6000	6000	6000
Chip	+	+	+	+	o/-	o
Contact pad	+	+	+	o	-	-
Bonding system	+	+	+	o	-	-
Moulding compound	+	+	o	-	-	-
Leadframe	+	+	+	+	o	o
Testing WL (@T_{junc})	o	o	-	-	-	-
Testing FP (@T_{amb})	+	+	+	+	+	-

+ technical realization for high volume production available

- technical realization not available

o Limit of actual technical realization

Table 2: Development of packaging concepts for high temperature applications

Subject	Solution	Possible Advantages	Possible Disadvantages
Polymer Packaging	Development of high temperature polymer compounds	Increased T_g and temperature resistance due to degradation and outgassing	Unknown modifications of other physical and chemical properties, processability (esp. flow behaviour)
Die-Attach	Evaluation of high temperature polymers	Increased bond and peel strength, increased T_g	Unknown influence on thermal-mechanical reliability of complete system
	Leadless solder materials	Increased melting point	Unknown influence on thermal-mechanical reliability of complete system, cost increase
Bond system	Single, homogenous metal	No change due to intermetallic phases	Processability decreases
	Design of diffusion barriers	Suppressed change of intermetallic phases	Increased complexity, additional process step
	Development of alternatives to wire bonding (e.g. soldering)	Increased temperature resistance due to degradation	Increased complexity, CTE mismatch, additional process step
Leadframe	Alloy 42	Decreased CTE	Decreased thermal conductivity
	Top-metallization on Cu (Pd,…)	No change due to intermetallic phases	
	Ceramics	High breakthrough voltage, decreased CTE	Decreased bonding strength on polymers after thermal cycling

2. Discussion – High Temperature Reliability Testing

Reaction rates of diffusion driven aging processes (mostly with a chemical origin like oxidation, moisture creep...) are preferably described with a tempearature dependent Arrhenius-type equation with reaction rate v and activation Energy E_0 (Eq. 1).

$$v(T, E_0) = v_0(T) \cdot \exp(E_0 / kT) \qquad (1)$$

For ageing processes with an activation energy E_0 between 0.5 eV and 1.0 eV (typical for most oxidation processes), one can use the well known rule of thumb „every 10K increase of temperature will double the rate of reaction" (Eq. 2).

$$B = \frac{v(T + 10\ K)}{v(T)} \approx e^{\frac{E_0}{k}\left(\frac{10\ K}{T^2}\right)} \approx 2 \qquad (2)$$

according to $E_0 = 0{,}7 \mathrm{eV}, T = 70°C$

Further acceleration by simply increasing temperature becomes questionable near the physical limits at high temperature regions. Due to the nature of the Arrhenius law, the speed up of all degradation mechanisms will decrease by increasing temperatures. The acceleration of reaction depending on the temperature is shown in fig. 1.

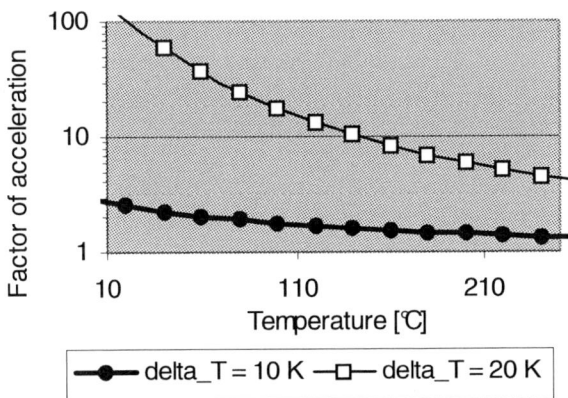

Fig. 1: Acceleration of reaction speed at temperature accelerated test (E_0=0.7eV)

Appart from the decreasing acceleration factor material parameters may change strongly with increasing temperature leading to stress and degradation processes that are originally not related with the quality parameter to be analyzed. Especially when investigating polymers like thermoplastics and thermosets high temperature tests will take place near or even above the glass transition temperature. The changing material parameters as coefficient of thermal expansion CTE, storage and shear modulus as well as thermal conductivity and heat capacity will significantly change mechanical stress and thermal distribution. On the one hand relaxation times get shorter with $T \rightarrow T_g$, on the other hand, otherwise CTE mismatch between different polymers and between polymer and silicon die or leadframe increases.

3. Results – Thermal Reliability Testing

During temperature storage at 220°C (fig. 2) all materials except material "J" show moderate, nearly linear weight loss. Increasing the storage temperature up to 240°C elevates the weight loss of all components (fig. 3). The time behaviour of weight loss is still nearly linear for all components. The molding compounds G, H, I and J show an acceleration factor of approximately 2 for $\Delta T = 20K$. According to equation (1) this corresponds to a typical activation energy of the degradation mechanism of 0,7 eV. Molding compounds B, C, D, E and F show an acceleration factor of 4 which corresponds to a enhanced activation energy of 1,4 eV. The measurement values of material "A" correspond to an activation energy around 2,5 eV, which is a rather untypical value for an EMC.

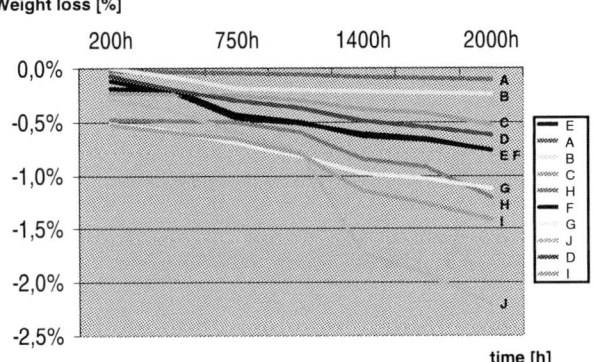

Fig. 2: Weight loss due to temperature storage at 220°C (2000h). Comparison of different EMCs

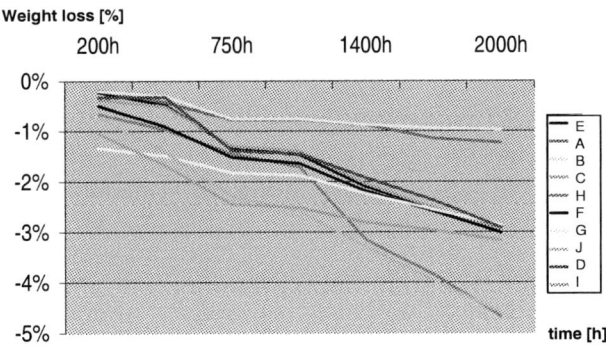

Fig. 3: Weight loss due to temperature storage at 240°C (2000h). Comparison of different EMCs

The material degradation could also be indentified by optical analysis of the parts' cross section. Higher weight loss of the components corresponds with higher thickness of the discolored layer as shown in fig. 4 and 5. The discolored area is porous and brittle with a strong tendency to cracking and flaking.

Material B

Fig. 4: Cross section of material B after temperature storage at 220°C and 240°C for 2000h

Material F

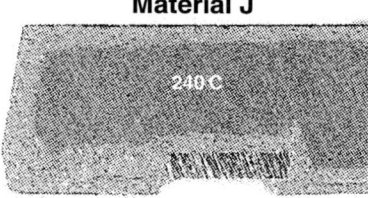

Material J

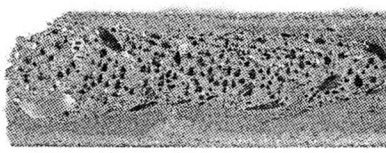

Material H

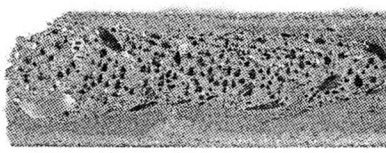

Fig. 5: Cross section of material F, J and H after temperature storage at 240°C for 2000h

The molding compound F and J show a thickness increase of four / two times due to the temperature rise. This is according to an acceleration factor of approximately four / two, dereived from the measurement of weight loss. This implies that these materials already started a "high temperature degradation" mechanism at 220°C.

It has to be stressed, that the thickness of the discolored layer is not increasing according to the weight loss for all components. Material B shows nearly no discoloration at 220°C, only the color intensity changes. According to the high increase of weight loss at 240°C it can be postulated, that the effect of temperature increase from 220°C up to 240°C is not described by an Arrhenius law.

Table 3 shows several degradation mechanisms of EMC materials. As the filler particles itself do not degradate, an overall weight loss of 1% corresponds to a polymer weight loss around 5%, calculated for a filler content of 80 weight-%.

Table 3: Decomposition mechanisms for EMC materials at temperatures > 200°C

Binder (~15%)	Phenolics, epoxies, multi aromatic resins	Cracking process, oxidation of low molecular epoxies
Catalysts (0.5 — 1%)	Peroxides for UP molding compounds	
Curing agents (2—6%)	Hexamethylene tetramine for phenolic novolaks	
Accelerators (0.10 — 5%)	Various, depending on resin system	
Fillers (> 80%)	SiO2, Al2O3	No reaction
Modifiers (3—6%)	Rubber, synthetic resins, graphite, etc.	Outgassing of silans, coating of filler particles, methanol, toluol
Additives (1—3%)	Lubricants, pigments and similar	Outgassing

Bond strength of EMC materials on copper leadframes is another key parameter for component reliability. Especially under humidity storage conditions an insufficient adhesion between polymer and encapsulated material may lead to delamination and popcorn cracking. The bond strenght was determined by molding the EMC onto copper leadframes and storage for 1000h at 220°C.

The results show a decrease in reliability comparable to the previous investigations. Materials with low weight loss show a relativly constand and high bond strength. Materials like component "J" with exceptional weight loss reduce their adhesion ability already after 150h temperature storage.

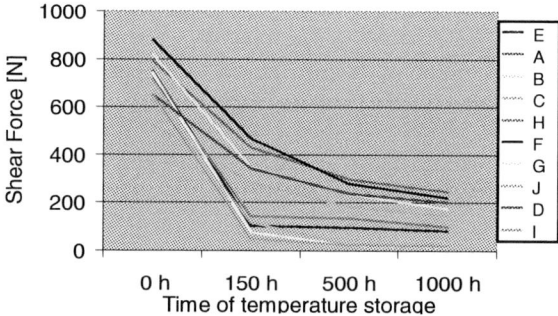

Fig. 6: Shear force resistance due to temperature storage at 220°C for 1000h.

Similar results could be observed for humidity storage at high temperatures. These results are not futher discussed within this paper.

High temperature applications can also weaken wire bonds due to intermetallic formation, especially for Au wire / Al bond pad connections. Early reasearch showed advantages of Al wire / NiAu bond pad systems [6] that also could be approved within this project.

4. Results – Thermal Simulations

Static and transient thermal simulations have been performed in order to optimize thermal cooling paths for dissipated energy at the chip surfaces. The simulations have been complemented by analytical calculations.

There are no fundamental differences in FEA simulations when reaching ambient temperatures $> 175°C$, but it should be kept in mind that the power P

$$P = \sigma \varepsilon A \left(T_1^4 - T_2^4 \right) \qquad (2)$$

dissipated by heat radiation with emissivity ε becomes a significant thermal pathway compared to passive convection by air when entering high temperature regime due to the power of four at the temperature influence (fig. 7).

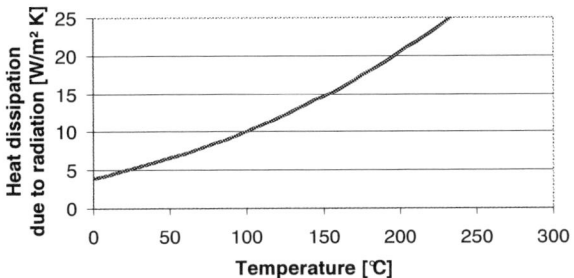

Fig. 7: Head dissipation per Kelvin temperture difference and area due to radiation. At 200°C a ΔT of 1K corresponds to a heat dissipation of 20 W/m² K

The temperature difference ΔT between the parts is nearly linear with the heat dissipated by radiation until $\Delta T = 50K$. Therefore the radiation can simply be implemented by transforming the radiation into a heat convection value α. α is depending on the aspect angle between emitting and absorbing surface.

$$\alpha_{rad} = \sigma \varepsilon \frac{\left(T_1^4 - T_2^4 \right)}{T_1 - T_2} \qquad (3)$$

At an ambient temperatur of 200°C a temperature difference of 1K corresponds to a heat dissipation due to radiation of 20 W/m² K, a value significantly higher then passive air convection.

Within the HotPaNTs project the thermal behaviour of several packaging types (QFN, QSOP, CSP) has been simulated. Within this paper we will focus on the differences between two QFN package types named "paddle up" and "paddle down" indicating the position of the copper leadframe. Fig. 8 shows the insight of a QFN paddle down package with conventional wire bonds and die attach by polymer glue. The leadframe paddle can be attached to a printed circuit board.

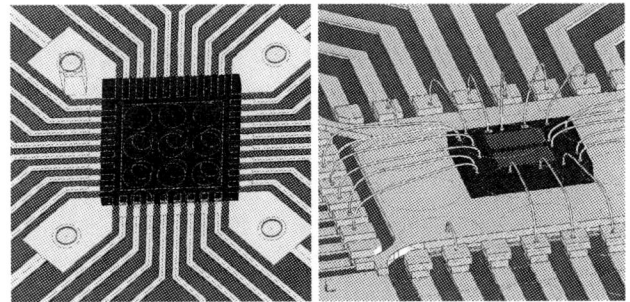

Fig. 8: QFN paddle down package with leadframe mounted onto circuit board.

The thermal performance of attaching the package to the circuit board mainly depends on board design. Without thermal vias and heat sink no significant increase of chip cooling can be obtained. Fig. 9 shows the density of heat conductivity within a QFN paddle down with thermal vias and massive heatsink on the PCB backplane. It can easily be seen that two main thermal pathways exist: 1st directly through the leadframe to the solder pads and 2nd through the metallisation of the thermal vias.

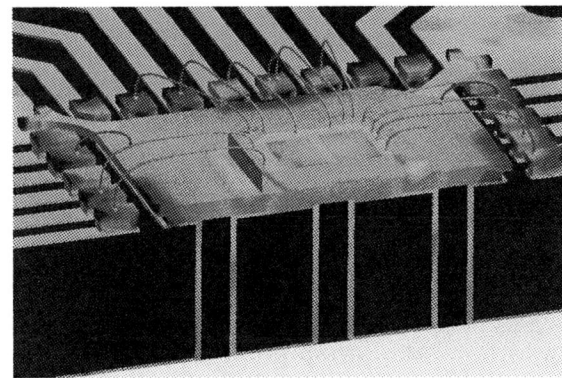

Fig. 9: Density of thermal heat conduction within QFN paddle down package on PCB with thermal vias.

Thermal conductivity of a via region strongly depends on via pitch and plating thickness. Due to the poor thermal conductivity of most polymers ~1 W/m K they can be regarded as thermal insulations as long as direct metal pathways are open.

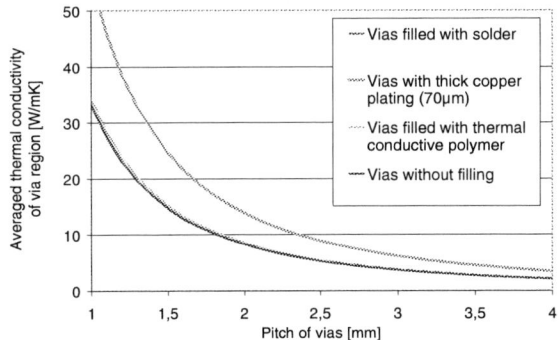

Fig. 10: Average thermal conductivity within via region of a QFN paddle down package on PCB.

The importance of proper chip cooling by backplane connection can be seen in fig. 11. Without attaching the PCB heatsink to any additional heatsink (e.g. by thermopads or bolts) the main thermal pathway will be through the SMD solder pads.

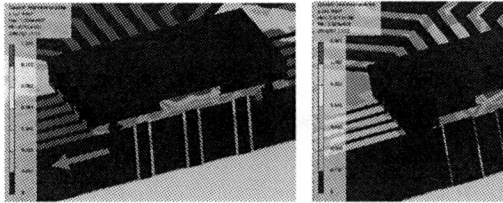

Fig. 11: Density of thermal heat conduction within QFN paddle down package on PCB with thermal vias. left: PCB heatsink not attached to external heatsink right: heatsind attached on cool plate by thermopad

Fig. 12 shows a slight advantage concerning temperature increase of QFN paddle down design compared with paddle up. Both designs permit a temperature increase below 25 K/W, leading to $T_{junc} < 200\,°C$ for $T_{amb} = 175\,°C$.

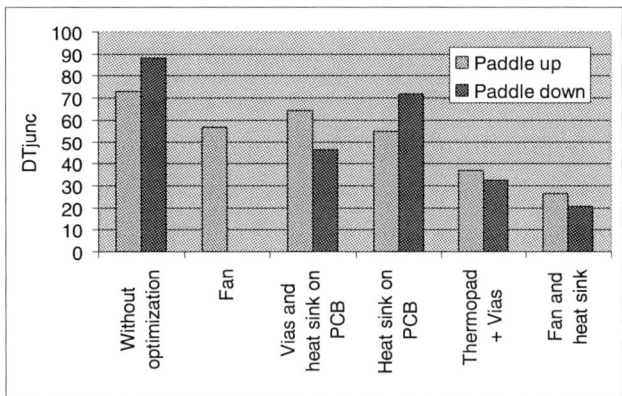

Fig. 12: Temperature increase QFN paddle up / down packages with various thermal boundary conditions

The advantage of paddle down packages might vanish when more than one heat dissipating component is assembled on the PCB. Time consuming case-to-case simulations have generally to manage these tasks, especially whith transient temperature calculations considering a time shifted change of power dissipation at each component.

This problem can be solved with a numerically enhanced analytical calculation. Thereby all thermal resistances of the parts - mainly R (junction to air) and R (junction to board) - area calculated independently by FEA simulations. Afterwards the results are transformed into a schematical arrangement with blocks symbolizing the discrete components. The crosswise thermal relation between the N components is described by a NxN matrix connecting the power dissipation at part n with the correlating temperature increase at all other parts.

$$\Delta \vec{T} = R \cdot \vec{P}$$

$$\begin{pmatrix} \Delta T_1 \\ \Delta T_2 \\ \vdots \\ \Delta T_n \end{pmatrix} = \begin{pmatrix} R_{11} & R_{12} & .. & R_{1n} \\ R_{21} & R_{22} & .. & R_{2n} \\ \vdots & \vdots & \ddots & \vdots \\ R_{n1} & R_{n2} & .. & R_{nn} \end{pmatrix} \cdot \begin{pmatrix} P_1 \\ P_2 \\ \vdots \\ P_n \end{pmatrix} \quad (4)$$

By successive numerical simulations with all all but one $P_i=0$ the coefficients of matrix R can be determined. Doing so it is easy to calculate static temperature distributions for various power scenarios as long as linear superposition is permitted. Transient temperature behaviour can be calculated analogous with three dimensional matrices considering the time dependencies. Calculations with short power peaks < 1 s show the advantage of thermally massive packages compared to highly miniaturized package types (fig. 13). In any case a close thermal contact to a massive circuit board (QFN_down, CSP+) is beneficial for short term heat dissipation. CSP+ denotes a COB design with solder bumps directly attached to thermal vias.

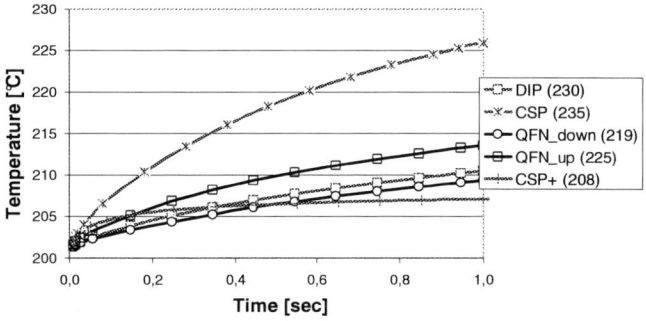

Fig. 13: Transient temperature increase of different packages (static temperature in brackets)

Additional calculations on thermal-mechanical reliability have shown that special attention on CTE mismatch has to be paid when connecting leadless package types to the circuit board. Despite their lower thermal warming the maximum stress is increased due to the strong mechanical coupling to the rigid board. Leaded package types show a higher ability for stress reduction through their flexible copper leads (fig. 14).

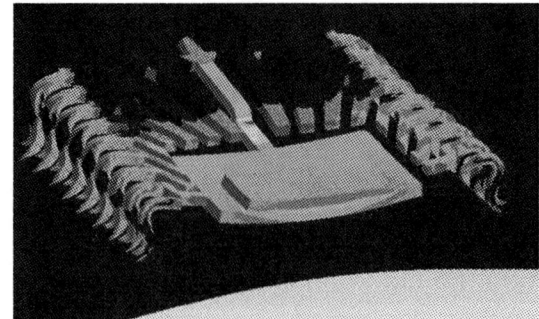

Fig. 14: Von Mises stress within dual inline package (DIP) on circuit board due to 1 Watt heat dissipation

The issue of thermo-mechanical reliability within solder bumps at high temperature environments has been intensively disucussed by Fraunhofer IZM since several years and is discussed in numerous publications (e.g. [2, 8]). Additional research on structural health monitoring at high temperatures up to 200°C was realized by the authors [9].

Experimental correlation of T_{junc} has been performed with specially designed test chips consisting of two heat dissipating areas and five temperature sensors (fig. 15).

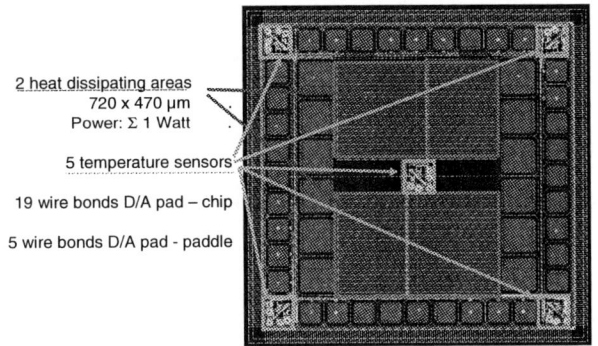

Fig. 15: Atmel test chip for thermal measurements implemented into QFN package types.

4.   Conclusions

In this paper, we have summarized final results of the HotPaNTs project investigating system reliability at junction tempertures up to 200°C. Experimental and numerical results have been presented focussing on reliability testing of packaging technologies and thermal management at high temperatures.

It could be shown that "classical" package types (e.g. leaded packages like QFP) have a better performance under dynamic thermal loading, but show a higher temperature increase when operated under static loading.

New polymer materials have been tested under thermal loading up to $T_{amb} = 200°C$ and could be identified as chemically stable. Nevertheless thermo-mechanical stress can increase significantly for T > Tg restricting the use of thermoset materials within high temperature applications.

We made proposals for several simplificatons of numerical simulatons conserning analysis of radiation and presented a coupled numerical – analytical solution method for temperature prediction of complex electronic boards.

Acknowledgments

This work was developed by support of the German Bundesministerium für Bildung, Wissenschaft und Forschung (BMBF) under project supervision of DLR. We greatly appreciate this support. The authors would like to thank the HotPaNTs consortium, especially Mr. Eitel, Atmel GmbH Germany for valuable cooperation.

References

1. Ansorge, F., Rebholz, C., Schreier-Alt, T., Krumm, R., Reichl, H., "Thermal Management, Characterization of Materials and Packaging. Technologies for High Temperature Electronics", 1st European Advanced Technology Workshop on Micropackaging and Thermal Management, La Rochelle (France), 01.-02.02.2006

2. Kaulfersch, E., Auersperg, J., Michel, B., Schubert, A., "Numerical and Experimental Analysis of Thermo-Mechanical reliability in Interconnection Technology", MST news on "Harsh environment and reliability", 4, 2001

3. Report "Hochtemperaturelektronik für mechatronische Systeme bis 200°C Umgebungstemperatur", Forschungsvereinigung Antriebstechnik e.V., 2001

4. "Grundlagen des Wärmemanagements in Leiterplatten und Baugruppen-Design", J. Adam, Flomerics Ltd., FED-Seminar, 5. Auflage, 2003

5. "Dickkupfer-Leiterplatten: Thermisches Management", Andus Electronic GmbH, Berlin

6. Benoit, J. et al., Proc 4th International High Temperature Electronics conference, p. IX-3, 1994

7. "Wärmeübertragung – Grundlagen und Praxis" P. v. Böckh, Springer-Verlag, 2004

8. Schubert, A., Dudek, R., Michel, B., Reichl, H.: "Package reliability studies by experimental and numerical analysis", *Proceedings of the 3rd International Micro Materials Conference*, B. Michel et al., p. 110, ddp goldenbogen, Dresden, 2000

9. Alt, T., Badstuebner, K., Michel, B., Ansorge, F., "Structural Health Monitoring by embedded Fiber Bragg gratings", Photonics Europe 2004, Strasbourg, France, SPIE Vol. 5459

An enriched cohesive zone model for numerical simulation of interfacial delamination in microsystems

M. Samimi, B.A.E. van Hal, R.H.J. Peerlings, J.A.W. van Dommelen, M.G.D. Geers
Materials Technology Institute, Department of Mechanical Engineering, Eindhoven University of Technology
P.O. Box 513, 5600 MB, Eindhoven, The Netherlands
Email: m.samimi@tue.nl

Abstract

Interfacial failure, mainly in the form of debonding or delamination of brittle interfaces, is one of the major sources of failure in microsystems that consist of multiple thin and stacked layers, manufactured using different materials. A cohesive zone model with a simple traction-separation law is employed to simulate the benchmark test of pure mode I delamination in a double cantilever beam. A local arc-length control procedure is also detailed and its robustness is shown in the case that the quasi-static solution contains limit points due to the brittle nature of the interface considered here. Finally, a bilinear hierarchical extension is proposed to enhance the efficiency and robustness of cohesive zone models by enriching the separation approximation in the process zone of a cohesive crack in brittle interfaces without need for further mesh refinement.

1. Introduction

Due to the goal of increasing functionality and decreasing costs in micro-electronic packages, integrated circuit processing and miniaturized conventional fabrication technologies have been combined to result in microsystems such as Micro Electro Mechanical Systems (MEMS) and System In Packages (SIP). Microsystems consist of multiple thin and stacked layers, manufactured using different materials. The quality, robustness, and reliability of such devices depend, to a large extent, on the adhesion and durability of interfaces. Interfacial failure, mainly in the form of debonding or delamination of brittle interfaces, results in malfunction or failure of integrated microsystems. In order to understand the process of delamination, there is a need for a numerical tool that predicts the interfacial behavior for microelectronic materials in an efficient and robust manner.

Cohesive zone models are widely used to simulate both initiation and propagation of delamination as a result of gradual degradation of the adhesion between two materials when they become separated. In these approaches, decohesion elements are placed at the interfaces between laminae to describe the traction as a function of separation across the process zone in front of the crack tip. Cohesive zone modeling has been applied successfully in interlaminar delamination [1] as well as in quasi-brittle fracture [2].

Application of cohesive zone models in a quasi-static framework is accompanied by some difficulties in the case of relatively brittle interfaces which one often encounters in electronic microsystems. In such cases, the solution of the discretized problem exhibits more limit points where the rate of either displacement or force of a control degree of freedom switches sign. This problem is known as the solution jump problem [3]. Solution jump is related to mesh size; that is, for a given interface, there exists a size of the adjacent elements below which the solution jump does not occur [4]. However, for realistic interface parameters, i.e. a small size of the process zone in brittle interfaces in comparison with other dimensions, the element size has to be extremely small to avoid the solution jump problem, which results in high computational costs.

Arc-length control methods can be used to deal with the mentioned limit points [5]. However, the convergence problems still arise when the limit points become too sharp due to the intrinsic brittle behavior of the system or course meshes. Viscous regularization techniques can be employed to overcome these numerical difficulties at the expense of introducing a non-realistic additional dissipated energy that in turn yields unrealistic results at delamination initiation [3]. Application of local arc-length control methods is useful to deal with problems involving strong deformation localization such as interfacial delamination in brittle interfaces [6,7]. The damage in the cohesive zone elements is used to control the solution procedure without need for line searches. Numerical examples are given showing how the method converges even when the solution contains sharp limit points resulting from course mesh discretization.

Elimination or at least reduction of the oscillations observed in the global load-displacement behavior of systems involving brittle interfaces by local improvement of the description of crack propagation and without further mesh refinement will enhance the efficiency and robustness of cohesive zone models. Therefore, in this paper, a process driven hierarchical extension is proposed to enrich the separation approximation in the process zone of a cohesive crack. In this formulation, the displacement field is enriched such that the crack tip can be located at an arbitrary position within a cohesive zone element. To achieve this, the crack tip position forms an additional degree of freedom. As a result, linear interpolation functions in the cohesive zone element have been modified to incorporate a piecewise linear approximation. The formulation of neighboring bulk elements is also enriched. Enrichment scaling factors are introduced as additional degrees of freedom in both decohesion and bulk elements to ensure compatibility.

2. Cohesive zone model

Using cohesive zone models, the behavior of the structure is split in two parts; the damage free continuum with an arbitrary material law, and the cohesive interfaces between the continuum elements that specify the damage of the material. For the two-dimensional quadrilateral cohesive zone element depicted in figure 1, the field variable of interest is the separation vector $\boldsymbol{\delta}$, which is a measure for the distance between the upper and lower edges in the deformed state. The components of the separation vector in shear and normal direction are δ_s and δ_n, respectively. It should be noted that in the undeformed state, the top and bottom edges coincide; that is, the separation is zero.

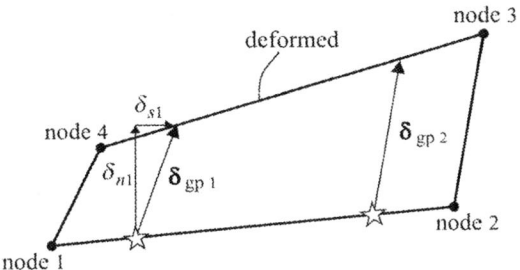

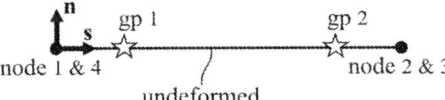

Figure 1: Deformation of a 2D cohesive zone element

In cohesive zone modeling, a traction-separation law (TSL) provides the nonlinear relation between the separation vector $\boldsymbol{\delta}$ and traction vector $\mathbf{t}$. Traction, which is monitored in the integration points of the cohesive zone elements (shown by gp1 & gp2 in figure 1), increases with increasing separation until it reaches a maximum t_{max}. More separation will result in a decrease in element stiffness and an increase in damage until the cohesive zone element fails and the continuum elements become disconnected. Numerous TSLs have been proposed in literature, of which Chandra et. al. [8] have provided an overview.

In this paper, the Smith-Ferrante exponential type of TSL is selected [9] and the nonlinear stiffness relation is considered only in normal direction (see figure 2):

$$t_n = t_{max} \frac{\langle \delta_n \rangle}{\delta_c} \exp\!\left(1 - \frac{\langle \delta_n \rangle}{\delta_c}\right) \tag{1}$$

Note that normal traction t_n cannot take negative values due to the use of McCauley brackets $\langle \delta_n \rangle = \frac{1}{2}\left(\delta_n + |\delta_n|\right)$. The characteristic separation δ_c and the tensile strength t_{max} are the two independent constitutive parameters that describe the above law. However, the work of separation per unit area G_c is frequently used as a constitutive

parameter. In fact, the cohesive zone approach can be related to Griffith's theory of fracture if the area under traction-separation curve is equal to the corresponding fracture toughness G_c:

$$G_c = \int_0^\infty t_n \mathrm{d}\delta_n = \exp(1) t_{max} \delta_c \tag{2}$$

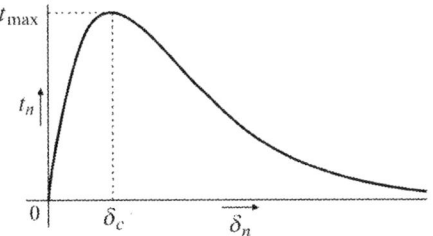

Figure 2: Smith-Ferrante exponential TSL

Irreversible behavior is taken into account by rewriting the as:

$$t_n = K(1-D)\delta_n \tag{3}$$

where K is the initial stiffness defined as:

$$K = \left.\frac{\mathrm{d}t}{\mathrm{d}\delta_n}\right|_{\delta_n=0} = \frac{t_{max}\exp(1)}{\delta_c} \tag{4}$$

and D is a damage variable that monotonically increases from 0 for the undamaged case to 1 for the completely damaged case. The following evolution law governs the damage:

$$D = D(q) = 1 - \exp\!\left(\frac{-q}{\delta_c}\right) \tag{5}$$

where the separation history variable q is the largest effective separation value reached so far and satisfies the Kuhn-Tucker conditions:

$$\left(q - \langle \delta_n \rangle\right) \geq 0 \tag{6.a}$$

$$\dot{q} \geq 0 \tag{6.b}$$

$$\dot{q}\left(q - \langle \delta_n \rangle\right) = 0 \tag{6.c}$$

As a result of the above relations, unloading follows the secant stiffness until a fully unloaded and undeformed state, as can be seen in figure 3. If reloading occurs, the elastic unloading path is followed again until the nonlinear traction-separation curve is reached, which is then followed upon further loading.

The following condition is also introduced to avoid interpenetration of crack faces:

$$t_n = K\delta_n \quad \text{if} \quad \delta_n < 0 \tag{7}$$

According to equation (5), the damage can never reach the value 1; however, the damage in an integration point is set to 1 if the separation history parameter becomes larger than the specified limit $q \geq 6\delta_c$. After passing this limit, the damage growth will be zero and the considered integration point does not contribute to the overall structural stiffness anymore.

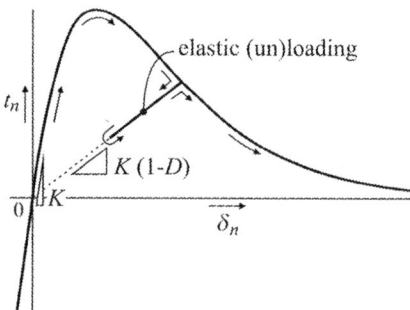

Figure 3: TSL in damage formulation

3. Solution jump problem

Discontinuities in the numerical solutions of rate-independent models are defined as solution jumps. In a simple one-dimensional problem shown in figure 4.a [3], a nonlinear spring with a force-elongation relation similar to the TSL (1) is loaded in series with a linear spring with stiffness k. The total displacement U is considered as the control parameter. When U is increased, after the maximum local response (maximum stress of nonlinear spring t_{max}), two situations may occur depending on k: If k is larger than the absolute value of the tangential stiffness on the whole softening part of the nonlinear force-elongation relation, the local response in nonlinear spring is followed continuously and the global response remains stable (path 1 in figure 4.b). Otherwise, a snap-back situation will arise in which no increase in global displacement is possible if the local response is followed continuously; hence, the global system will jump directly from a to b showing instantaneous failure of the nonlinear spring (path 2 in figure 4.b). In other words, when the local behavior in the nonlinear spring begins to soften, the elastic energy stored in the adjacent linear spring will start to be released by the nonlinear spring, and can be sufficient to break it instantaneously.

Work of separation G_c is determined experimentally; therefore, it is considered to be fixed in this study. On the other hand, the process zone in which damage grows is small compared with the characteristic dimensions of the structure under investigation in the case of brittle fracture. That is, δ_c is very small for brittle interfaces and since G_c has a fixed value, it is likely to face relatively high maximum tractions t_{max} (see equation 2). A high maximum traction results in a high tangential stiffness in the softening part of the nonlinear TSL of the cohesive zone element. In such cases, an unstable solution due to the instantaneous failure of the cohesive zone element can be expected.

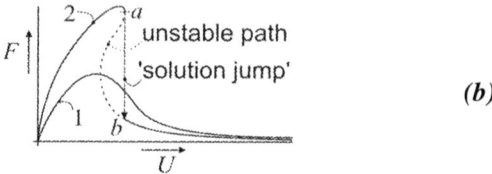

Figure 4: Solution jump problem; a) linear and nonlinear springs in series, b) nonlinear response

The solution jump can be eliminated by refining the mesh sufficiently. In fact, as long as the mesh is fine enough, the elastic energy stored in the region near the interface will be redistributed well when one interface node softens. Tomar et. al. [4] have proposed to use at least 10 elements across the process zone. It means that in brittle interface problems where the size of the process zone is relatively small, the mesh size has to be extremely small, which results in high computational costs.

4. Local arc-length control approach

When a solution jump happens, two solutions exist at the same time, disturbing the performance of the Newton-Raphson scheme. This problem can be observed in the form of limit points such as snap-through and snap-back points in the global load-displacement response of brittle systems. Practically, a Newton-Raphson method under load or displacement control often fails to converge in such cases. Therefore, path following techniques are needed to pass limit points. Crisfield's arc length control method is a global path following approach that controls the iterative update of the external load by constraining the iterative solution update [5]. Since all degrees of freedom contribute with an equal weight, this method cannot lead to convergence if there is a strong localized deformation, such as interfacial delamination. So, it is required that the degrees of freedom contribute to the global solution with different weights in the case of strong deformation localization where limit points become too sharp, as carried out in weighted sub-plane control methods [6,7]. In the case of interfacial delamination simulation using a cohesive zone model, the deformation becomes localized in the cohesive zone elements. The damage in interface elements can be selected as the control function.

The nonlinear discretized equilibrium equation can be stated as:

$$\mathbf{f}_i(\mathbf{u}) = \mathbf{f}_e = \alpha \, \bar{\mathbf{f}}_e \qquad (8)$$

where $\mathbf{u}$ is the solution vector and $\mathbf{f}_i$ and $\mathbf{f}_e$ indicate the internal and external force vectors, respectively. The external force vector can be expressed as the product of a scalar load factor α and the unit external force vector $\bar{\mathbf{f}}_e$.

The local arc-length control method is an incremental-iterative solution procedure. The load variation within a load increment is expressed as:

$$\alpha^{(i)} = \alpha^{(i-1)} + d\alpha^{(i)} \qquad (9)$$

where $d\alpha^{(i)}$ is the unknown load factor update that is solved simultaneously with the iterative solution update $d\mathbf{u}^{(i)}$ by the load control algorithm. $d\mathbf{u}^{(i)}$ is split in two parts:

$$d\mathbf{u}^{(i)} = d\hat{\mathbf{u}}^{(i)} + d\alpha^{(i)} d\overline{\mathbf{u}}^{(i)} \qquad (10)$$

$d\hat{\mathbf{u}}^{(i)}$ and $d\overline{\mathbf{u}}^{(i)}$ are computed using these equations:

$$\mathbf{K}_t^{(i-1)} d\hat{\mathbf{u}}^{(i)} = \mathbf{r}^{(i-1)} \qquad (11.a)$$

$$\mathbf{K}_t^{(i-1)} d\overline{\mathbf{u}}^{(i)} = \overline{\mathbf{f}}_e \qquad (11.b)$$

In the above relations, $\mathbf{K}_t^{(i-1)}$ and $\mathbf{r}^{(i-1)}$ represent the tangential stiffness matrix and the residual vector evaluated at $(\mathbf{u}^{(i-1)}, \alpha^{(i-1)})$, respectively.

A so-called constraint equation is required to govern the iterative load factor update $d\alpha^{(i)}$. This equation is defined on a subplane that involves the introduction of additional variables. As can be seen in figure 5, the load factor α and a new incremental control function ϕ span the subplane.

The control function ϕ consists of a weighted sum of the damage D in the integration points (gp) of all cohesive zone elements [10]:

$$\phi = \sum_{gp} W_{gp} D_{gp} \qquad (12)$$

where W_{gp} represent the weight factors that are computed in the predictor step $(i = 0)$ of each increment:

$$W_{gp} = \left.\frac{\partial D}{\partial \mathbf{u}}\right|_{gp}^{(0)} d\overline{\mathbf{u}}^{(1)} \qquad (13)$$

The control function ϕ is a nonlinear function of the solution vector $\mathbf{u}$ because of the definition of the damage parameter in equations (5,6). Using a Taylor series expansion, its linearization can be performed which results in:

$$D\phi^{(i)} = \sum_{gp} W_{gp} \left.\frac{\partial D}{\partial \mathbf{u}}\right|_{gp}^{(i)} D\mathbf{u}^{(i)} \qquad (14)$$

$$d\hat{\phi}^{(i)} = \sum_{gp} W_{gp} \left.\frac{\partial D}{\partial \mathbf{u}}\right|_{gp}^{(i)} d\hat{\mathbf{u}}^{(i)} \qquad (15)$$

$$d\overline{\phi}^{(i)} = \sum_{gp} W_{gp} \left.\frac{\partial D}{\partial \mathbf{u}}\right|_{gp}^{(i)} d\overline{\mathbf{u}}^{(i)} \qquad (16)$$

where $D\mathbf{u}^{(i)}$ is the incremental solution vector update and evolution law (5,6) governs the damage growth.

The constraint equation is defined in the $\alpha - \phi$ subplane by fixing the incremental control update $D\phi^{(i)}$ to a specified arc-length L:

$$\left| D\phi^{(i)} \right| = L \qquad (17)$$

The iterative load factor update then can be calculated from the above equation. By assuming a monotonically increasing control function ϕ, the considered subplane will not suffer from snap-back points as can be seen in figure 5.

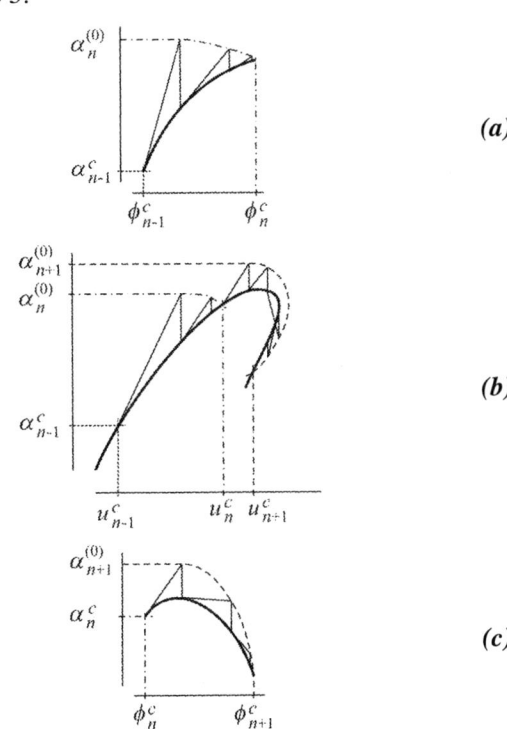

Figure 5: Incremental-iterative procedure of the local arc-length control method; a) $\alpha - \phi$ subplane for increment n, b) load-displacement curve of control degree of freedom u for increments n and $n+1$, c) $\alpha - \phi$ subplane for increment $n+1$

5. Numerical simulation results

The double cantilever beam (DCB) shown in figure 6 is used as a benchmark test for the performance of the cohesive zone model in mode I loading conditions. It consists of two strips of length L, width W, and height H, which are glued together. An initial crack of length A_0 is introduced between the two strips at the left edge of the beam while the right edge is clamped. During the test, the specimen is loaded by the prescribed clamp displacement U at the two moving clamp positions at the left edge. Cohesive zone elements are placed along the interface between the strips and 2D plane strain elements are used to model the bulk material. Table 1 lists the bulk material properties and constitutive parameters of the cohesive zone model.

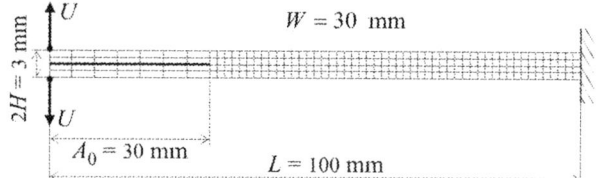

Figure 6: 2D finite element model of a double cantilever beam specimen

Table 1: Bulk material and cohesive zone parameters

Cohesive zone parameters	Bulk material properties
$G_c = 0.36$ N/mm	$E = 130$ GPa
$\delta_c = 0.001$ mm	$\nu = 0.3$
$t_{max} = 132$ MPa	

Figure 7 shows the force-displacement response at the moving clamps using both a local and a global arc-length control method. Before the clamp force F reaches its maximum value, the response is almost linear which is associated with the bending of a cantilever beam of length A_0. The interface starts to delaminate after this point, which causes an overall softening behavior. Moreover, the softening branch shows severe oscillations where both the rate of the force F and the clamp displacement U switch sign. Each oscillation corresponds to the failure of one cohesive zone element.

The global approach converges initially; however, the convergence slows down when the number of failed cohesive zone elements increases and finally breaks down after several oscillations. On the other hand, the weighted subplane control method which is a local approach shows a better performance and simulates the crack growth much further.

The oscillations in the softening branch are caused by a too coarse discretization. Therefore, the number of element in the longitudinal direction is doubled in the initially intact part of the beam. Comparison of the global force-displacement response in both original and refined mesh configurations show that the magnitude of the oscillations in the region of the refined configuration dominated by debonding is reduced substantially (see figure 8).

6. Cohesive zone model enrichment

The separation approximation of cohesive zone elements in the process zone is enriched in order to reduce oscillations such as those observed in figure 7.a without a need for further mesh refinement. The reduction or disappearance of these oscillations is advantageous because the number of increments and iterations that are required for solution convergence will be reduced and the solution becomes less mesh dependent.

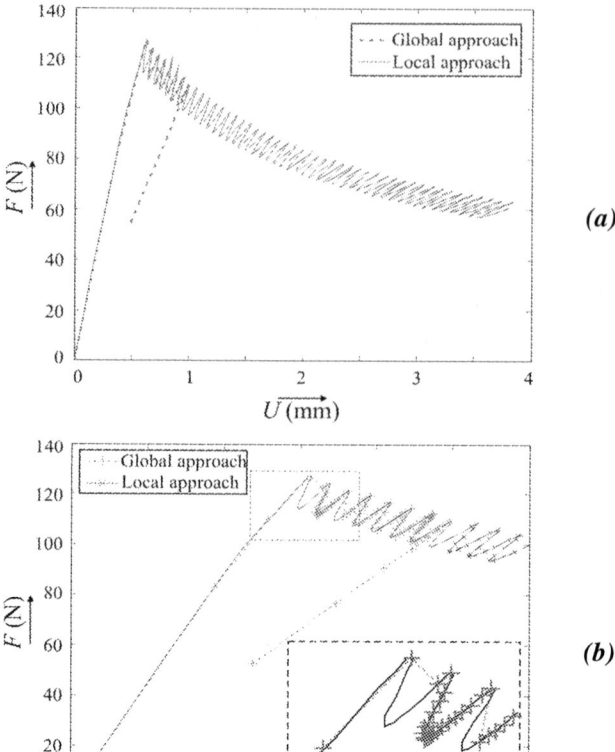

(a)

(b)

Figure 7: Force-displacement response using global and local arc-length control approaches; a) Full range, b) enlarged plot with increment indication

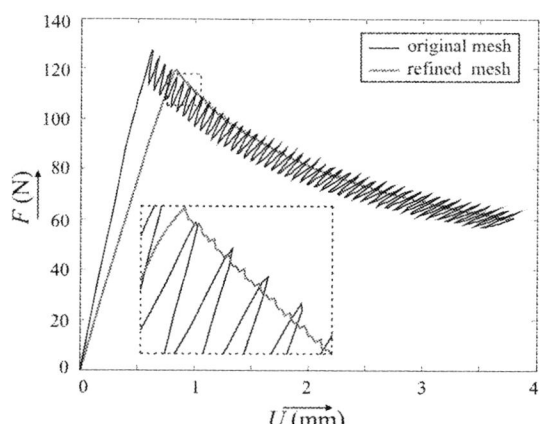

Figure 8: Influence of mesh refinement on the global force-displacement response

As can be seen in figure 9, the linear separation approximation in the cohesive zone element is enriched by adding a piece-wise linear enrichment function ϕ. Here, the crack tip position a forms an additional degree of freedom.

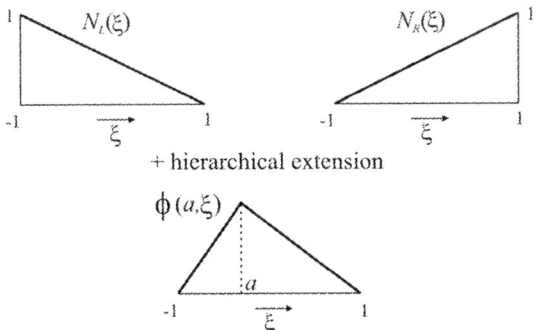

Figure 9: Hierarchical enrichment of linear interpolation functions

The displacement at the top and bottom edges of the cohesive zone element $(\mathbf{u}_t, \mathbf{u}_b)$ is approximated by:

$$\mathbf{u}_t(\xi, a) = \mathbf{u}_4 N_L(\xi) + \mathbf{h}_t \phi(\xi, a) + \mathbf{u}_3 N_R(\xi) \tag{18.a}$$

$$\mathbf{u}_b(\xi, a) = \mathbf{u}_1 N_L(\xi) + \mathbf{h}_b \phi(\xi, a) + \mathbf{u}_2 N_R(\xi) \tag{18.b}$$

where N_L and N_R are linear interpolation functions corresponding to the left and right side of the cohesive zone element, respectively. $\mathbf{h}_t$ and $\mathbf{h}_b$ are unknown enrichment scaling factors that are multiplied by enrichment function and the results are added to displacement approximations at the top and bottom edges of the cohesive zone element, respectively, in order to predict the crack shape more accurately as can be seen in figure 10. The separation ($\boldsymbol{\delta}$) is defined as:

$$\boldsymbol{\delta}(\xi, a) = \mathbf{u}_t(\xi, a) - \mathbf{u}_b(\xi, a) \tag{19}$$

and the piece-wise linear enrichment function is written as:

$$\phi(\xi, a) = \frac{1}{4}(1-a)(1+\xi) - \frac{1}{2}\Re(\xi - a) \tag{20}$$

where $\Re$ is the ramp function defined as:

$$\Re(\xi - a) = \begin{cases} 0 & \text{if } \xi \le a \\ (\xi - a) & \text{if } \xi > a \end{cases} \tag{21}$$

Using the above formulation, which is stated with respect to the normalized coordinate system $(-1 \le \xi \le 1)$, separation can be defined as:

$$\boldsymbol{\delta}(\xi, a) = \mathbf{B}_u(\xi)\mathbf{u} + \mathbf{B}_h(\xi, a)\mathbf{h} \tag{22}$$

where:

$$\boldsymbol{\delta}(\xi, a) = \begin{bmatrix} \delta_s(\xi, a) \\ \delta_n(\xi, a) \end{bmatrix} \tag{23.a}$$

$$\mathbf{u}^T = \begin{bmatrix} u_{1s} & u_{1n} & u_{2s} & u_{2n} & u_{3s} & u_{3n} & u_{4s} & u_{4n} \end{bmatrix} \tag{23.b}$$

$$\mathbf{h}^T = \begin{bmatrix} h_{ts} & h_{tn} & h_{bs} & h_{bn} \end{bmatrix} \tag{23.c}$$

$$\mathbf{B}_u^T(\xi) = \begin{bmatrix} -N_L(\xi) & 0 \\ 0 & -N_L(\xi) \\ -N_R(\xi) & 0 \\ 0 & -N_R(\xi) \\ N_R(\xi) & 0 \\ 0 & N_R(\xi) \\ N_L(\xi) & 0 \\ 0 & N_L(\xi) \end{bmatrix} \tag{23.d}$$

$$\mathbf{B}_h(\xi, a) = \phi(\xi, a) \begin{bmatrix} 1 & 0 & -1 & 0 \\ 0 & 1 & 0 & -1 \end{bmatrix} \tag{23.e}$$

Subscripts s and n denote shear and normal directions in a cohesive zone element, respectively.

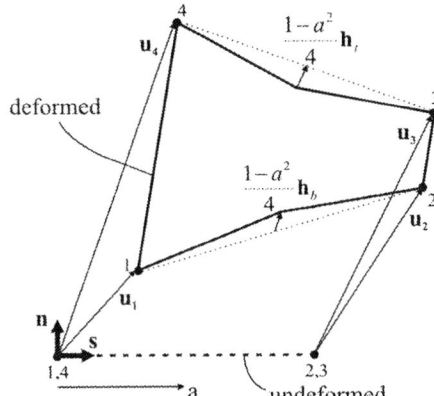

Figure 10: Deformed state of an enriched cohesive zone element

The principle of virtual work that is valid for nonlinear stress-strain (or stress-rate of strain) relations can then be used as the weak form of the equilibrium equations. Moreover, continuity of the displacement field requires that bulk elements adjacent to the enriched cohesive zone elements be enriched as well.

Note that the crack tip position a is considered as a history parameter that is implicitly governed (or driven) by the deformation process. In fact, a can be considered as an element degree of freedom. The force equilibrium, that results from the principle of virtual work, governs its value in this case.

7. Conclusions

Although cohesive zone modeling allows the analysis of both crack nucleation and crack growth, its application to brittle fracture suffers from numerical difficulties because of the small size of the process zone in such interfaces. Mesh refinement is known to be the most trivial remedy. A solution for this problem is the application of local arc-length control approaches. These methods are better suited for this type of physical nonlinear problems where strong deformation localization slows or breaks down the convergence of global methods. In fact, application of a local arc-length control method enables us to pass through limit points in the global load-displacement response.

In this paper, a piece-wise linear enrichment function is proposed to enrich cohesive zone elements in order to enhance their robustness when applied to brittle interfaces by local improvement of the description of crack propagation in such interfaces. In other words, the crack tip can be located at any arbitrary position within a cohesive zone element by enriching the displacement approximations at the top and bottom edges of the cohesive zone element. The location of the crack tip and the magnitude of the enrichment within a cohesive zone element are given by additional degrees of freedom. It will help eliminating oscillations observed in global load-displacement response so that equilibrium conditions can be established without need for further mesh refinement or application of complicated path-following approaches.

References

1. Camanho P, Davila C, "Mixed-mode decohesion finite elements for the simulation of delamination in composite materials", *Technical Report,* NASA Langley Research Center, Report No. TM-2002-211737 (2002).

2. de Borst R, "Numerical aspects of cohesive-zone models", *Engineering Fracture Mechanics*, Vol. 70 (2003), pp. 1743-1757.

3. Chaboche J, Feyel F, Monerie Y, "Interface debonding models: a viscous regularization with a limited rate dependency", *International Journal of solids and Structures,* Vol. 38 (2001), pp. 3127-3160.

4. Tomar T, Zhai J, Zhou M, "Bounds for element size in a variable stiffness cohesive finite element", *International Journal for Numerical Methods in Engineering.* Vol. 61 (2004), pp. 1894-1920.

5. Crisfield M.A, *Non-linear finite element analysis of solids and structures; Vol 1: Essentials*, John Wiley & Sons Ltd. (Chichester, UK, 1997), pp. 266-279.

6. Geers MGD, "Enhanced solution control for physically and geometrically non-linear problems. Part I – the subplane approach", *International Journal for Numerical Methods in Engineering*, Vol. 46 (1999), pp. 177-204.

7. Geers MGD, "Enhanced solution control for physically and geometrically non-linear problems. Part II – comparative performance analysis", *International Journal for Numerical Methods in Engineering*, Vol. 46 (1999), pp. 205-230.

8. Chandra N, Li H, Shet C, Ghonem H, "Some issues in the application of cohesive zone models for metallic-ceramic interfaces", *International Journal of solids and Structures,* Vol. 39 (2002), pp. 2827-2855.

9. Ortiz M, Pandolfi A, "Finite-deformation irreversible cohesive elements for three-dimensional crack-propagation analysis", *International Journal for Numerical Methods in Engineering*, Vol. 44 (1999), pp. 1267-1282.

10. van Hal BAE, Peerlings RHJ, Geers MGD, "Local arc-length control method for cohesive zone modelling", *Submitted*.

Underexpanded Micro-nozzle Flow Simulation with Coupled Thermal-Fluid Modeling

José Hermida Quesada [1], José A. Moríñigo [2], Francisco Caballero Requena [2]

[1] Dept. Aerodynamics and Propulsion, [2] Dept. Space Programmes
National Institute for Aerospace Technology
Ctra. Ajalvir km-4, Torrejón de Ardoz, 28850 Madrid, Spain
email: {hermidaqj, morignigoja, caballerorf}@inta.es

Abstract

Simulation of micro-thrusters intended for space propulsion relies on accurate modeling of those physical processes that occur highly coupled in the micron-sized scale. In fact, micronozzles exhibit significant differences when compared to their large-scale counterparts, so their modeling requires accounting for specific aspects to achieve success in the prediction.

The goal of the present investigation is to simulate the steady and transient operation of a micro-thruster in underexpansion for cold and hot inert gas under a multiphysics, continuum-based Navier-Stokes approach with slip-flow conditions. The modeling incorporates three major effects: gas rarefaction in the slip regime, viscous dissipation and solid-gas heat transfer. Non-equilibrium effects have been addressed with the implementation of a 2nd–order slip-model for velocity and temperature on the walls. The results summarize the attained micro-thruster performance and wafer thermal response. Their dependence on the thermal-fluid coupling is analyzed.

Nomenclature

A : cross-section area, m^2
C_p : specific heat at constant pressure, J·kg^{-1}K^{-1}
E : thrust, N
G : mass flow rate, kg·s^{-1}
h : nozzle radius, m
k : thermal conductivity, W·m^{-1}K^{-1}
Kn : Knudsen number
L_{div} : nozzle divergent length, m
M : Mach number
p : pressure, Pa
Re : Reynolds number
R_{gas} : gas constant, J mol^{-1} K^{-1}
T : temperature, K
U : velocity modulus, m·s^{-1}
x, r : axial, radial coordinates, m
γ : specific heat ratio
λ : molecular mean free path, m
ρ : density, kg·m^{-3}
μ : viscosity, kg·m^{-1}s^{-1}

Subscripts

o : chamber conditions
t : stagnation
th : throat conditions
w : wall

1. Introduction

A variety of micropropulsion devices intended for space applications are being investigated by different groups as they constitutes an enabling technology to support the next space missions, like formation flying, where the orbital geometry of a swarm of spacecrafts should be accurately adjusted by frequent thruster firings. Versatility is also a key issue since maneouvers span a wide range of tasks during the spacecraft lifetime: drag compensation, station-keeping or fine attitude control are examples. All of them demand very small impulse bits and thrust levels (typically within 0.1 to 10mN), at the same time the spacecraft constrains should be met, namely mass, power budget and other aspects as ease of integration and fabrication; in resume, those that reduce the system cost.

In chemical micro-rockets, the nozzle accelerates the flow, hence it plays a fundamental role in the attainable performances. To this respect, two potential concepts based on cold and hot agent gas that expands through DeLaval micro-nozzles have been investigated in recent years. Regarding their hardware complexity, cold-gas thrusters need a storage tank for the fluid and miniaturized piping system with valves, whereas the hot-gas may be produced with a solid-propellant combustion. The later class is simpler as it has no moving parts. Furthermore, its feasibility is demostrated [1-4].

Performance of micro-nozzles operated with cold-gas has been analyzed by several authors, thus solving the standard (non-slip) Navier-Stokes (NS) equations [5,6] and incorporating the rarefaction effects with a slip-flow condition set on the walls [7-8]. Alexeenko *et al.* [9] conducted transient flow simulations coupled with the silicon wafer heating for a 3D micro-thruster under different flow conditions (cold and hot gas). They adopted a hybrid method of Finite Elements for the solid zone and Direct Simulation Monte Carlo for the gas. Additionally, Kujawa *et al.* [10] performed steady-state and transient two-dimensional NS (non-slip) simulations to estimate the thermal losses in a MEMS-based nozzle. Louisos *et al.* [11] extended the previous work to the startup transient. More rencently, Zhang *et al.* [12] carried out steady 2D NS simulations with and without a slip-flow condition for a solid propellant micro-thruster.

In the present paper, steady and transient performance of a micro-thruster is analyzed following a multiphysics approach that fully couples gasdynamics and heat conduction in the wafer. Rarefaction effects are taken into

1-4244-1105-X/07/$25.00 ©2007 IEEE

account with a 2nd-order in *Kn* slip-model. Hence, the Navier-Stokes (NS) and heat conduction equations are solved simultaneously for the gas and solid regions, respectively. A thermal-fluid coupling between them is established at the solid-gas interface.

The paper is organized as follows. Section 2 outlines the micro-device geometry and operation for realistic cold and hot-gas conditions. The modeling of the rarefaction effects, their inclusion into the solver and the solid-fluid coupling, are briefly introduced in section 3. Discussion of the computed steady-state and transients results is summarized in section 4. Some conclusions are provided.

2. Micro-nozzle geometry and flow conditions

A schematic view of the micro-thruster considered for analysis is shown in Fig.1, where the main dimensions are indicated. The axisymmetric conical micro-nozzle has a sharp corner at the throat and smooth walls, 15° half-angle, throat diameter $2h_{th}=300\mu m$, divergent length $L_{div}=5038\mu m$ and an exit-to-throat area ratio of 100. The geometry parameters and operation points with molecular nitrogen are provided in Table 1. Stagnation temperature is prescribed at realistic values: 300°K for cold-gas; and 1500°K for hot-gas (close to the combustion temperature of solid propellants used in gas generators). The chamber stagnation pressure of 0.5bar makes the nozzle to operate in underexpansion for the ambient pressure $p_\infty=10$Pa. Besides, it is noted that common designs can not withstand high combustion pressures.

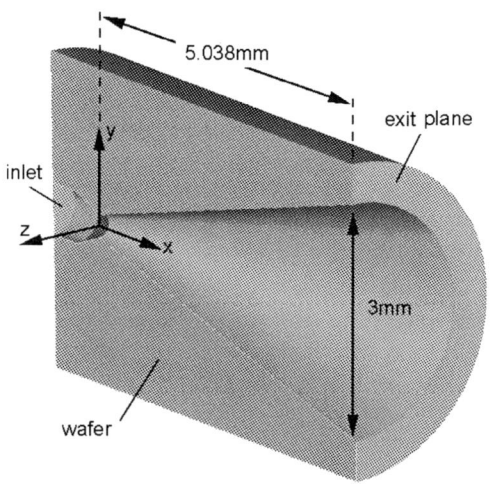

Fig.1: Half-view of the micro-thruster, where the conical nozzle and wall thickness are visible.

According to the flow conditions, the Reynolds number referred to the nozzle throat ($Re_{th}=2\rho Uh_{th}/\mu$) is about ~2040 and ~300 for cold and hot-flow, respectively. Thus, flow is assumed laminar. Micro-thruster material is silicon (Si) of thermal properties $k_{Si}=142$W·m^{-1}K^{-1}, $C_{p,Si}=702$J·kg^{-1}K^{-1}, and $\rho_{Si}=2329$kg·m^{-3}. Sensitivity of the results to the thermal condition imposed on the wafer external wall (back and upper) is explored. Hence, two

boundary conditions (BCs) are considered: insulated wall (modeled as an adiabatic BC); and active cooled surface (modeled isothermal: $T_w=300$°K). Their influence on the flowfield and performance is analyzed in section 4.

Table 1 Micro-nozzle geometry and flow conditions

Agent gas: molecular nitrogen		
Chamber:		
Stagnation pressure	$p_{t,o}$	50000 Pa
Stagnation temperature	$T_{t,o}$	1500 $°K$ (HOT)
		300 $°K$ (COLD)
Freestream static pressure:	p_∞	10 Pa
Geometry (conical):		
Divergent half-angle	α_{div}	15°
Divergent length	L_{div}	5038 μm
Throat section radius	R_{th}	150 μm
Expansion area ratio	A_e/A_{th}	100

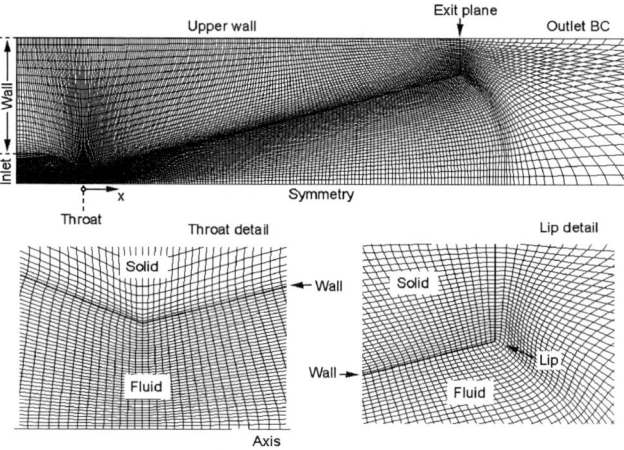

Fig.2: Structured grid in the fluid and wafer zones. The solid-fluid wall is indicated with a thicker line.

3. Continuum-based approach

In the slip-flow regime ($10^{-3}<Kn<\sim0.12$) non equilibrium effects due to rarefaction are confined to a thin region by the wall (Knudsen layer), then the NS equations with appropiate BCs for the velocity slip (U_{slip}) and temperature jump (T_{slip}) at the walls provide an accurate description of the flow.

Hence, the compressible axisymmetric NS equations have been solved using the FLUENT$^©$ code. The set of governing equations is discretized using a 2nd-order upwind algorithm. Ideal gas and Sutherland's law for viscosity are assumed. Time marching for steady and implicit dual-time stepping integration for time-accurate simulations is carried out.

The computational domain and grid are depicted in Fig.2. The domain comprehends the gas zone (nozzle inside and freestream region –outside- where the nozzle discharges) and the wafer zone. Outside the nozzle, an

462

outflow BC states extrapolation of the solution variables from the inner cells at those supersonic cells and imposes $p_\infty=10Pa$ otherwise. The $2L_{div}$-axial extension of the domain downstream the nozzle exit permits to deal with the end-wall effects on the micro-thruster performance. A grid-convergence analysis has shown that the grid (fluid zone: 8853 cells / solid zone: 5742 cells) is adequate to capture the flow features.

The implemented higher-order slip-flow model due to Karniadakis & Beskok [13] states for U_{slip}

$$
\begin{aligned}
U_{slip} - U_w = & \frac{2-\sigma_V}{\sigma_V} \frac{h \cdot Kn}{1-B(Kn)\cdot Kn} \frac{\partial U}{\partial n}\bigg|_w \\
& + \frac{3}{4}\frac{\gamma-1}{\gamma} P_r \frac{k}{\rho R_g T_w} \frac{\partial T}{\partial s}\bigg|_w + O(Kn^3)
\end{aligned}
\tag{1}
$$

where U_w is the wall velocity and $P_r=\mu C_p/k$ the Prandtl number. The Knudsen number $Kn=\lambda/h$ is referred to local properties and the local mean free path reads

$$
\lambda = \mu\sqrt{\frac{\pi}{2\rho p}}
\tag{2}
$$

The function $B(Kn)$ is approximated by the first term b of its Taylor expansion $B(Kn)\approx b+cKn+O(Kn^2)$, then

$$
b = \frac{h}{2}\cdot\frac{d^2U_s/dn^2}{dU_s/dn}\bigg|_w
\tag{3}
$$

being U_s the tangential to wall velocity component and n the normal to wall. The temperature jump is proposed in analogy with eqn. (1)

$$
T_{slip} - T_w = \frac{2-\sigma_T}{\sigma_T}\frac{2\gamma}{\gamma+1}\cdot\frac{h}{P_r}\cdot\frac{Kn}{1-B(Kn)\cdot Kn}\cdot\frac{\partial T}{\partial n}\bigg|_w
\tag{4}
$$

Accommodation coefficients of N_2 in contact with Si are set $\sigma_V\sim0.8$ for momentum (from [14]) and $\sigma_T=1.0$ for energy. The implementation of (1-4) and heat transfer conditions at the solid-gas interface in FLUENT© has been done with the User Defined Functions [15]. The thermal BCs at the solid-gas interface impose the temperature jump defined by eqn. (4) and the heat-flux continuity across it. Thus the heat-flux balance, that determines T_w, comprehends the heat transfer between N_2 and Si, but not the additional supply or removal of energy due to external sources (solar radiation) and cooling or heating from inside.

4. Results and discussion

Steady-state and transient axisymmetric simulations for cold and hot-flow and chamber stagnation pressure $p_{t,o}=0.5bar$ have been conducted for various BCs set at the micro-thruster walls to explore their influence on the flowfield and thruster performances. Hence, the steady-state results comprehend the solutions obtained with a slip-flow BC and adiabatic / isothermal ($T_w=300ºK$) BC at the external (upper-)wall. In the unsteady simulations, two thruster transients have been computed for cold and

hot-gas (with slip-flow BC and isolated external wall). In addition, steady computations with non-slip BC have been carried out to quantify the influence of the slip-flow on the modeling.

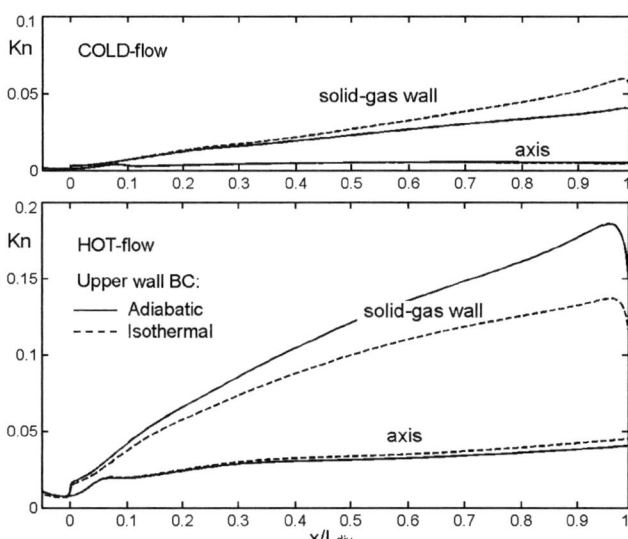

Fig.3: Knudsen number along the solid-gas wall and axis, computed for cold and hot-flow with adiabatic and isothermal ($300ºK$) upper-wall BC.

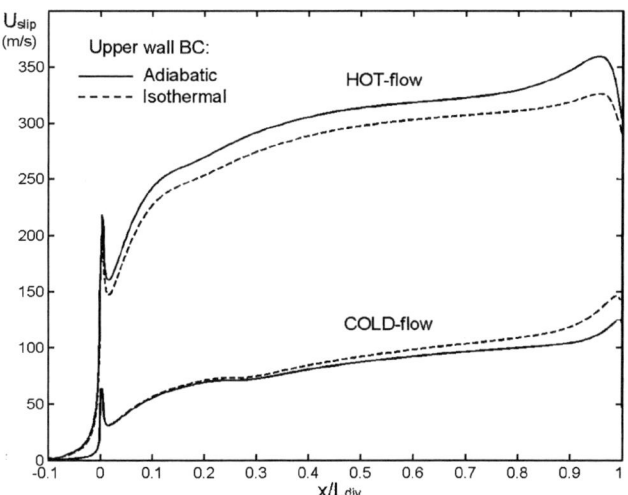

Fig.4: Steady-state slip-velocity U_{slip} on the solid-gas wall for hot ($T_{t,o}=1500ºK$) and cold-flow ($T_{t,o}=300ºK$).

The Knudsen number along the nozzle axis and wall is depicted in Fig.3 for steady simulations. The plots for cold-flow shows that $Kn<0.06$, so the gas stays in the slip-regime. On the contrary, hot-flow simulations reveal a higher degree of rarefaction; in particular, the Kn for the isolated upper-wall simulation shows that the flow close to the wall goes into the transitional regime for almost half the nozzle length. Thus, marginal validity of the slip-model is reached in this case. This issue is revisited later in the discussion of the thrust level. The corresponding slip-velocity and temperature jump are given in Figs. 4 & 5. The U_{slip} peak observed at x=0 in Fig.4 is due to the sharp edge of the throat, that causes a steep local variation of the slip according to the first term in eqn. (1).

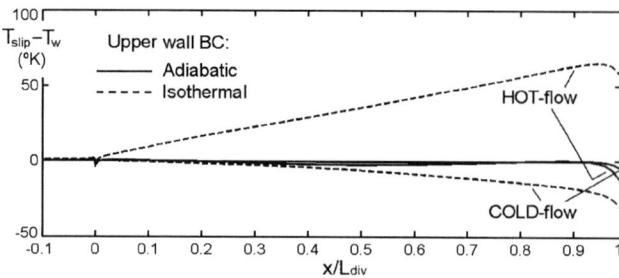

Fig.5: Temperature jump $T_{slip}-T_w$ on the gas-solid wall for hot ($T_{t,o}$=1500°K) and cold-flow ($T_{t,o}$=300°K).

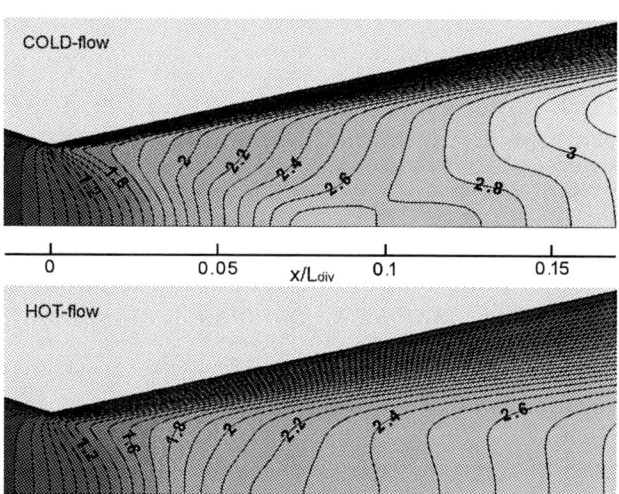

Fig.6: Steady-state Mach field computed for cold and hot-flow simulations (adiabatic upper-wall).

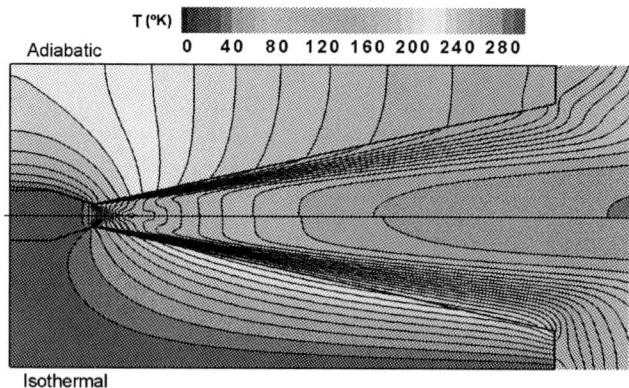

Fig.7: Static temperature field for cold-gas with adiabatic (upper-half) and isothermal (lower-half) upper-wall BC.

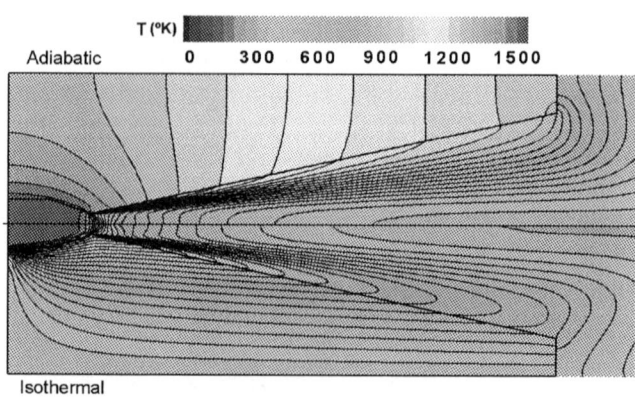

Fig.8: Static temperature field for hot-gas with adiabatic (upper-half) and isothermal (lower-half) upper-wall BC.

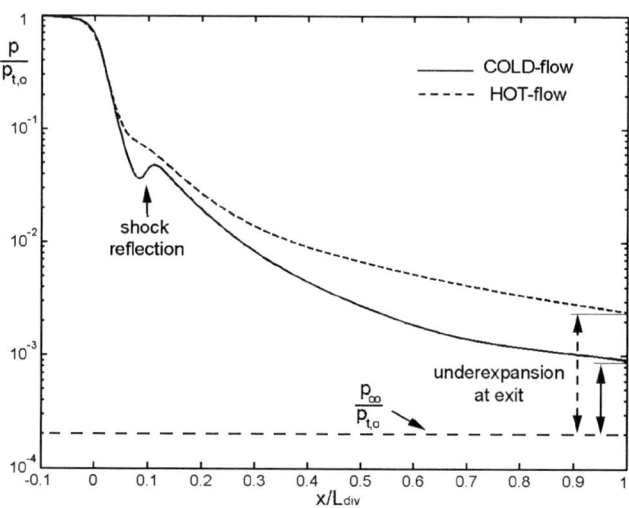

Fig.9: Nondimensional static pressure along the axis for cold and hot-flow simulation with adiabatic upper-wall.

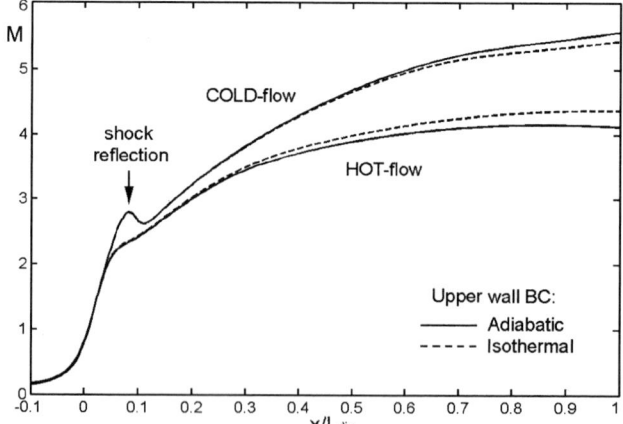

Fig.10: Mach number along the axis for cold and hot-flow simulations.

464

The plots stress that the end-wall effects (nozzle lip expansion) may be significant, above all for the hot-flow, and these become apparent from the abrupt decay of U_{slip} near the exit area. The thick boundary layer at the exit area conditions the gas deflection around the lip since viscous dissipation converts part of the mechanical energy into heat. The growing thickness of the boundary layer is visible in Fig.6 by means of the isoMach contours for cold and hot-flow. Moreover, its behaviour may be inferred from the static temperature fields given in Figs. 7 & 8. It is clear that hot-flow chamber conditions yield a much thicker boundary layer. An obvious consequence is that the gas expands more efficiently in the cold-flow case since the viscous dissipation acts now in a thinner region, then the effective crosswise extension of the core flow is larger. The static pressure (Fig.9) and Mach number (Fig.10) profiles along the nozzle centreline illustrates this behaviour. The visible bump in the cold-flow plot is provoked by an internal shock that originates near the throat edge and reflects on the axis.

Regarding the influence of the external wall BC on the nozzle performance, the hot-flow simulations show that the isothermal upper-wall BC acts lowering the overall temperature in the wafer, as expected (see Fig.8). It is noted that the cooling is particularly effective in the material around the throat region and explains that a larger mass flow rate may pass through, as it is shown in Table 2 (the extra mass flow rate is about ~2%). Besides, the higher acceleration in Mach attained along the axis for the hot-flow with isothermal upper-wall BC, points out to the beneficial effect of cooling the external surface on reducing the viscous losses. Nevertheless, although the temperature maps in Fig. 8 show that cooling makes the boundary layer thinner in the hot-flow field, the situation is complex since thermal loss outweights the lower viscous dissipation and makes the net thrust to drop below the value obtained with the adiabatic upper-wall BC, as it is seen in Table 2. A further insight into the performance may be attained by plotting the deliverable thrust as a function of the axial nozzle length in Fig.11. It shows that the maximum thrust for prescribed chamber conditions is reached at an intermediate cross-section. Thus, from that location on, losses dominate and the thrust drops. As a result, the optimum nozzle length is shorter than L_{div}: 30% and 75%-shorter for the cold and hot-flow cases, respectively. A closer examination of the thrust in hot-flow for both external wall BCs, shows that the isothermal upper-wall BC yields higher thrust in the portion of optimum length, in agreement with the above mentioned effects of viscous losses and cooling. It should be noted that the ineffective divergent portion found in the hot-flow case corresponds to the zone of the early transitional regime shown in Fig.3, where the slip-model accuracy fails.

Table 2 summarizes the micro-thruster performance computed for cold and hot-flow and stresses the interest of developing micro-thrusters based on high temperature gases, since they provide higher specific impulses I_{sp}. The variation of E and I_{sp} computed in hot-flow with non-slip and slip-flow BCs, clearly states the great importance of accounting for gas rarefaction in the modeling to predict the performance. The large overprediction of thrust in hot-flow with non-slip BCs is visible in Fig.11.

Table 2 Nozzle performance for cold and hot-flow

Model / Upper-wall BC	$10^6 \cdot G$ (kg/s)	E(mN)	I_{sp}(s) [1]
COLD-flow			
Non-slip / Adiabatic	7.73	5.28	69.6
Slip / Adiabatic	7.78	5.35	70.1
Slip / IsoThermal	7.76	5.38	70.7
HOT-flow			
Non-slip / Adiabatic	3.41	4.95	148.2
Slip / Adiabatic	3.36	4.30	130.5
Slip / IsoThermal	3.43	4.21	125.1

[1] Specific impulse $I_{sp}=E/(G \cdot g_o)$, where g_o=9.81ms^{-2}.

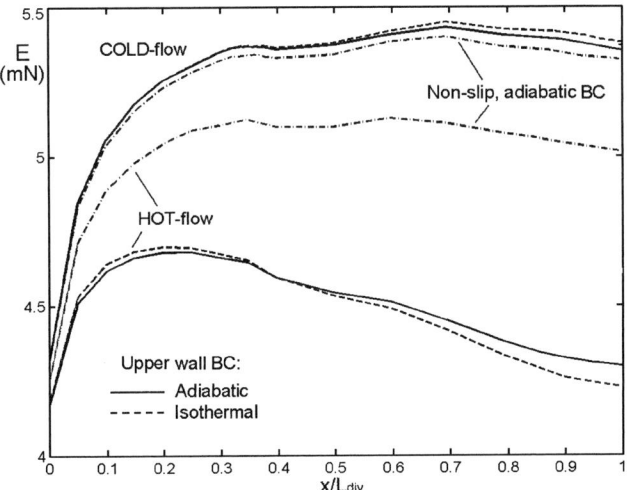

Fig.11: Thrust vs. nozzle length for cold and hot-flow simulations, with upper-wall adiabatic and isothermal BC. Non-slip NS results are also plotted.

To analyze the wafer heating with time, two transient simulations have been accomplished for cold and hot-flow. Both simulations consist of a pseudo-impulsive nozzle startup, where $p_{t,o}$ and $T_{t,o}$ increase from the initial freestream state (t=0s) to the nominal values (t=2ms) defined in Table 1. The time-varying inlet BC follows a sinusoidal law. The different thermal response for cold and hot-flow is illustrated in Fig.12, where the static temperature snapshots sequence is provided for identical time instants. The disparity of the time scales of the gas and solid is evident. Hence, while the thrust reaches almost its definitive steady-state value just after 2.5ms from the start (Fig.13), the temperature field in the wafer needs several tens (in cold) or hundreds (in hot-flow)

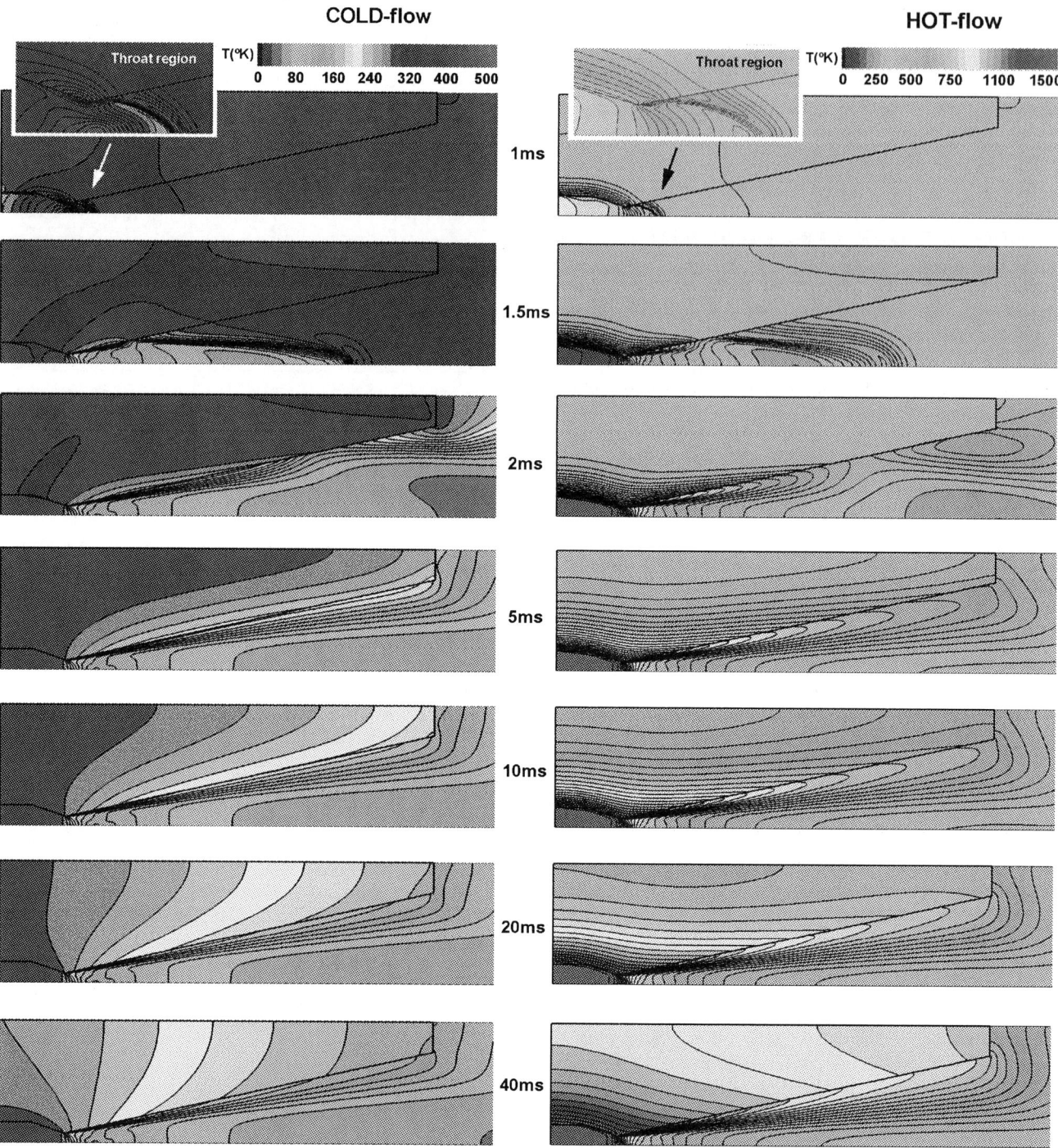

Fig.12: Sequence of static temperature snapshots corresponding to the cold (left-side) and hot-flow (right-side) startup transient. The flow in the throat region at time=1ms is magnified (time instants in milliseconds).

of milliseconds to approach the steady-state. Regarding the wall temperature, the melting temperature of silicon is already surpassed in the entire chamber surface at 40ms for hot-flow. Thus, longer firings would require wall cooling in order to keep the material temperature under acceptable limits.

5. Conclusions

Steady and transient response of a micro-thruster with thermal-fluid coupling has been simulated for two realistic conditions corresponding to underexpanded cold and hot-flow. A 2nd–order in Kn slip-model has been implemented to solve the Navier-Stokes equations in the gas zone coupled with the heat conduction equation in the solid region. The static temperature maps computed for steady-state and transient operation reveal that the growing rate of the boundary layers is higher in the hot-flow case, as expected from the rather different Reynolds numbers that characterize the flow. Therefore, viscous dissipation dominates in the hot-flow. One important consequence is the significant drop of thrust level. However, the decrease of mass flow rate which occurs at higher chamber stagnation temperatures implies a net benefit in the propulsive efficiency (quantified by the I_{sp}), that increases in about ~90% with regard to the cold-flow operation for the same chamber pressure. Results show that an overprediction of the micro-nozzle thrust and specific impulse up to ~18% occurs (hot-flow) when the slip is neglected. This stresses that flow slip at the walls should be taken into account in addition to the solid-fluid coupled modeling.

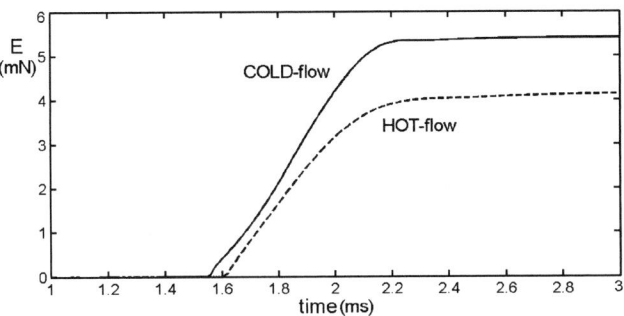

Fig.13: Variation of thrust with time in cold and hot-flow simulations with upper-wall adiabatic BC.

Acknowledgments

This research was supported as part of the micro-propulsion activities in the Small Satellites Programme (INTA), funded by the Spanish Ministry of Defence.

References

1. Lewis D.H., Janson S.W., Cohen R.B., Antonsson E.K.,"Digital micropropulsion", Sensors and Actuators 80 (2000), pp.143-154.
2. Rossi C., Do Conto T., Estève D., Larangôt B. "Design, Fabrication and Modelling of MEMS-based Microthrusters for Space Application", Smart Mater. Struct., 10 (2001), pp.1156-1162.
3. Teasdale D., Milanovic V., Chang P., Pister K.S.J., "Microrockets for Smart Dust", Smart Mater. Struct., 10 (2001), pp.1145-1155.
4. Chaalane A., Larangôt B., Rossi C., Granier H., Estève D., "Main Directions of Solid Propellant Micro-propulsion Activity at LAAS", AIAA Paper 2004-6706 (2004).
5. Ivanov M.S., Markelov G.N., Ketsdever A.D., Wadsworth D.C., "Numerical Study of Cold Gas Micronozzle Flows", AIAA Paper 99-0166 (1999).

6. Markelov G.N., Ivanov M.S., "A Comparative Analysis of 2D/3D Micronozzle Flows by the DSMC Method", AIAA Paper 2001-1009 (2001).
7. Bayt R.L., "Analysis, Fabrication and Testing of a MEMS-based Micropropulsion System", Ph.D. Thesis, MIT (also as report FDRL TR 99-1), 1999.
8. Aleexenko A.A., Gimelshein S.F., Levin D.A., Collins R.J., "Numerical Modeling of Axisymmetric and Three-Dimensional Flows in MEMS nozzles", AIAA Paper 2000-3668 (2000).
9. Alexeenko A.A., Levin D.A., Fedosov D.A., Gimel-shein S.F., "Performance Analysis Micro-thrusters Based on Coupled Thermal-Fluid Modeling and Simulation", J. Prop. and Power Vol.21, Nº1 (2005) pp.95-101.
10. Kujawa J., Hitt D.L., "Transient Shutdown Simulation of a Realistic MEMS Supersonic Nozzle", AIAA Paper 2004-3762 (2004).
11. Louisos W.F., Hitt D.L., "Viscous Effects in Super-sonic Micro-nozzle Flows: Transient Analysis", AIAA Paper 2006-2874 (2006).
12. Zhang K.L., Chou S.K., Ang S.S., "Performance Prediction of a Novel Solid-Propellant Microthruster", J. Prop. and Power, Vol.22, Nº1 (2006), pp.56-63.
13. Karniadakis G., Beskok, A.: Micro Flows: Funda-mentals and Simulation (Springer, 2002) Heidelberg, pp. 54-62.
14. Arkilic E.B.: "Measurement of the Mass Flow and Tangential Momentum Accommodation Coefficient in Silicon Micromachined Channels", Ph.D. Thesis, MIT (also as report FDRL TR 97-1), 1997.
15. User Defined Functions (UDF) Manual, FLUENT 6.2, Fluent Inc., Jan. 2005.

Optimization of Cu Low-*k* bond pad designs to improve mechanical robustness using the Area Release Energy method

R. A. B. Engelen[1,*], O. van der Sluis[1], R. B. R. van Silfhout[1], W.D. van Driel[2], and V. Fiori[3]

[1] Philips Applied Technologies, High Tech Campus 7, 5656 AE Eindhoven, The Netherlands
[2] NXP Semiconductors, Gerstweg 2, 6534 AE Nijmegen, The Netherlands
[3] STMicroelectronics, 850 rue Jean Monnet, F-38926 Crolles Cedex, France
[*] e-mail: Roy.Engelen@Philips.com, phone: +31 (0)40 27 48561

Abstract

In the development of present and future CMOS-technologies (CMOS065 and beyond) for micro-electronic components the combination of state-of-the-art modeling techniques and experimental testing is crucial and provides a challenge to address the thermo-mechanical reliability issues and the demand for shorter time-to-market by the industry[1]. Nowadays these modeling techniques often involve the construction of very detailed Cu Low-*k* IC structures, which is still very time-consuming and computationally demanding.

This paper adresses an alternative modeling strategy that intends to gain fundamental insights and understanding of the mechanisms that have an impact on the thermo-mechanical reliability of a Cu Low-*k* device. As an example the relation between the metal densities of the various layers in a typical Cu Low-*k* IC bond pad structure and the Area Release Energy (ARE) criterion [2] has been investigated.

The modeling and computational effort have been considerably limited by building a simplified two-dimensional bond pad model. This model is constructed such that it is well capable to reveal the impact on the thermo-mechanical reliability and such that it is still representative for the behaviour of advanced Cu Low-*k* structures. When optimizing the bond pad towards lower ARE levels for better mechanical robustness the impact of the metal density within the BE layers is evident. Furthermore, it shows that the metal density in a single layer does not only affect the ARE level in that specific layer, but it also influences the ARE profile within the entire stack.

Clearly, the presented modeling strategy is suitable to identify the design parameters that play an important role when optimizing the thermo-mechanical performance of advanced Cu Low-*k* structures. In combination with experimental tests (e.g. industrial qualification tests, interface strength measurements, etc.) it may provide a robust tool to further improve and ensure the reliability of present and future IC bond pad structures.

1. Introduction

Sophisticated simulation and optimization techniques [3] aim to deal with reduced design margins of future CMOS technology and intend to replace time and money consuming trail-and-error approaches. Such a Virtual Prototyping approach allows to address reliability issues already in the design phase where design changes are more efficient and cheaper. In practice, however, these tools are also addressed when reliability issues have actually been identified. In this case time is very limited and problems can only be solved in a rather ad hoc fashion. The understanding of the underlying failure mechanisms gained from such simulations is limited. As a consequence similar issues may arise in other product and different reliability issues may arise after a design change. In fact, the real challenge is not only to be able to provide input when solving urgent reliability problems but to predict issues in current and future IC technology.

Furthermore, the tendency exists to create extensive and detailed models, which are labor-intensive to construct and are CPU-demanding. Clearly, this may compromise the efficiency of the Virtual Prototyping. For this purpose an alternative approach uses relatively simple models that are still able to capture the most important failure mechanisms. By the reduction of the modeling and computational effort these models can be a versatile tool to ultimately derive design rules for Cu low-*k* IC processes.

In section 2 an example of an IC BE structure within a bond pad model is presented within the context of wire-pull tests. This model is constructed in such a way that the impact of the metal density in each BE layer on the overall mechanical performance can be investigated. The results of these wire-pull simulations are presented in section 3. Finally, the main conclusions are discussed in section 4.

2. Bond pad model with IC BE structure

Within the context of wire-pull qualification testing a 2D model of a wire-bonded bond pad structure is constructed that is capable to study the influence of the metal densities of the individual IC BE layers on the overall mechanical performance.

Geometry and boundary conditions

The 120x120μm bond pad geometry including the golden wire and ball are depicted in Figure 1. The wire is subjected to a pull force of 5g under an angle of 20° with the vertical axis conform common test conditions. In order to mimic a realistic pulling of the wire the top part of the wire is first rotated until it is aligned within the pull force direction. Subsequently, the movement of the wire-end is restricted into this direction. It is important to accurately model the deformation within the wire as this predominantly determines how the bond pad below is loaded. For this purpose the constitutive behavior is modeled using an elasto-plastic model that is able to deal with large deformations in an adequate fashion.

In the out-of-plane direction plane strain conditions are applied. It should be noted that in the out-of-plane direction different thicknesses have to be applied for the pad, wire and ball in order to obtain the correct cross-sectional area and consequently to induce the correct stress levels throughout the model.

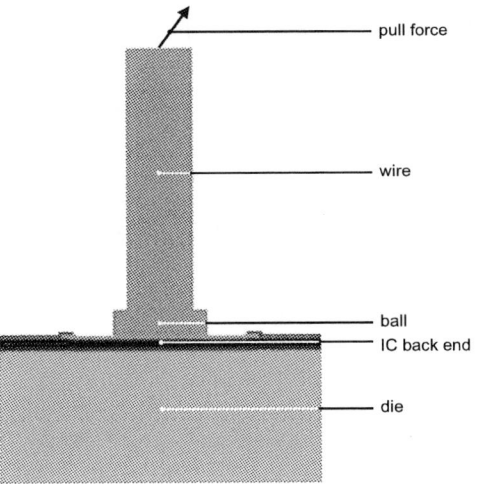

Figure 1 Bond pad geometry (2D) with wire and ball

IC BE layers

In Figure 2 a more detailed view of the IC back end layers is shown. Within the pad region of the BE, which is processed in accordance with the 6M1T metallization option, the different colors represent the various metal and dielectric layers within the BE. In the ring region of the BE only dielectric materials are present. Within each of these regions in the BE effective material properties are calculated in such a fashion that they reflect the variations in the metal density as will be outlined below.

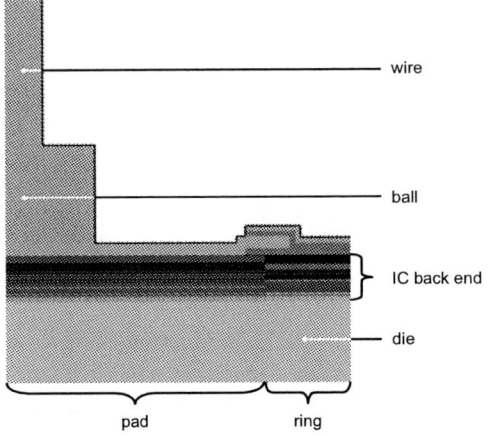

Figure 2 Bond pad model detail: IC back end layers

The effective material properties are calculated as a function of 1) the average metal density in each metal layer and 2) maximum or no vias in each via layer, which

are sufficient to fully determine the effective properties of the BE layers. For a restricted number of densities these effective properties can be determined by homogenization procedures performed on separate local models of repetitive structures within the BE layers. In this case, however, a first-order approximation is chosen instead to determine the effective material properties, which avoids the need to create separate models and which is still sufficiently capable to capture the impact of the varying BE properties at the bond pad model level.

The procedure above is schematically shown in Figure 3 for an arbitrary combination of layers. Part A could be e.g. the amount of metal within a metal layer ρ_{metal}, and part B the dielectric content in the same layer $(1-\rho_{metal})$. Part C is e.g. a full dielectric layer (no vias) $\rho_{dielectric}$.

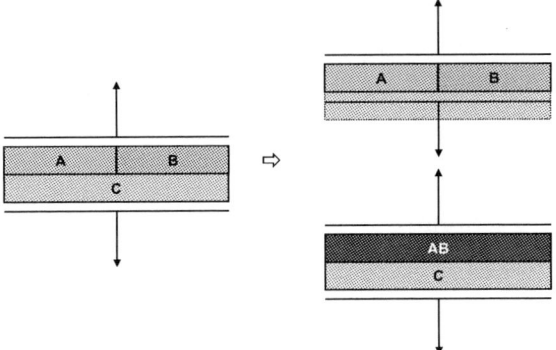

Figure 3 Schematic representation of first-order approximation of effective material properties

The material stiffness of the top layer is calculated using the analogy of a parallel spring:

$$E_{AB} = \rho_A E_A + (1 - \rho_A) E_B \qquad (1)$$

with ρ_A taken as the metal density. The material stiffness of the complete stack is then calculated by considering part AB and C as a serial combination of springs:

$$\frac{1}{E_{ABC}} = \rho_{AB} \frac{1}{E_{AB}} + (1 - \rho_{AB}) \frac{1}{E_C} \qquad (2)$$

with ρ_{AB} the ratio between the thickness of the top and bottom layer, which is known input for a certain CMOS processing technology.

When considering the via levels instead of a metal layer only no or max vias is specified. In this case the corresponding metal density depends on the amount of metal in the metal lines above and below the vias. Bear in mind that vias can only be present in regions that have metal both above and below:

$$\rho_{metal} = \rho_{via} \cdot \min(\rho_{metal,above} ; \rho_{metal,below}) \qquad (3)$$

where ρ_{via} is the maximum local via metal density based on both the via size and minimum spacing conform the applicable design rules.

With the above procedure the effective material properties can be calculated in both the pad and ring regions of the various BE layers as a function of the input parameters (metal density and no/max vias). A typical example of the normalized effective stiffness over the entire stack height is depicted in Figure 4. Whether the effective BE material properties have a significant impact on the overall bond pad behavior and the mechanical performance is investigated in the next section.

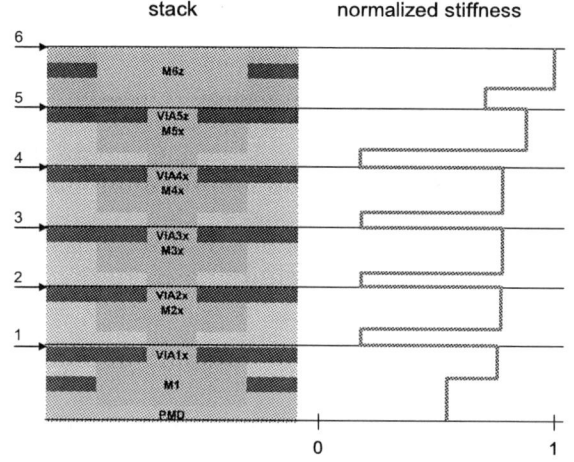

Figure 4 Example of effective material properties in BE layers: normalized stiffness for entire material stack (dimensions not at scale)

3. Wire-pull results

As already mentioned in the previous section it is important to consider a realistic deformation of the wire in the model. The typical wire deformation is depicted in Figure 5. The contour bands in this image represent the amount of equivalent plastic strain that occurs in the wire during the wire pull test. Concentrations of the plastic deformation are found on both sides in the neck of the ball and the wire. Moving away from the neck region a regular profile is found that is typical for bending (as a direct consequence of the rotation during the first steps of the wire pulling test).

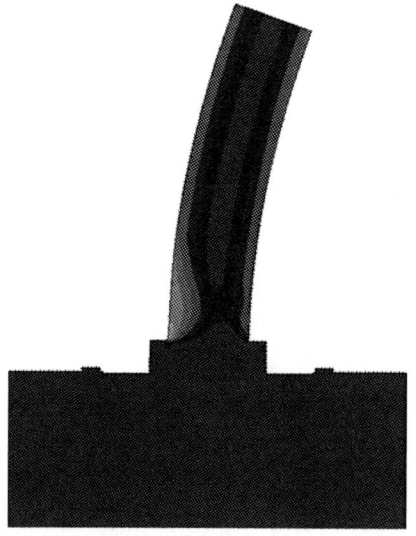

Figure 5 Typical effective plastic deformations within the wire during the wire-pull test

As a consequence the bond pad structure is loaded in accordance with Figure 6. In this image the peel stresses that are induced into the bond pad are visualized. In the left part where the wire is stretched a concentration of tensile peel stresses occurs whereas on the right side where the wire is compressed compressive stresses are induced into the bond pad. In the zoom section the impact of the BE layers is clearly visible.

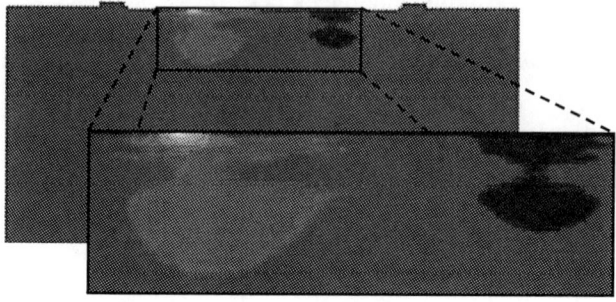

Figure 6 Typical peel stresses as induced into the bond pad during the wire-pull test

The mechanical performance is measured in terms of the Area Release Energy criterion [2-3]. In each interface as indicated by the numbers in Figure 4 the ARE profiles are calculated over the entire width of the bond pad. The normalized maximum ARE values in each level are plotted against the interface number in Figure 7. This graph allows identifying the most critical interface within the IC BE. In this case this would be either interface 4 or 6. It should be noted that interfaces 1-4 and 5-6 have different dielectric materials and hence different interface strengths as well.

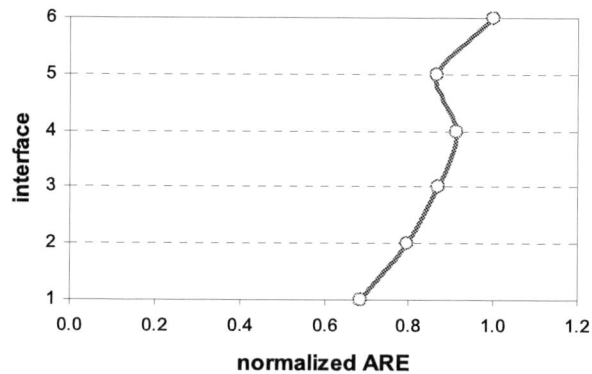

Figure 7 Typical profile of maximum Area Release Energy over the height of the BE layers during the wire-pull test

Besides identifying the most critical interface in the BE layers the ARE criterion is also suitable for comparing various structures with each other. In this case several variations in metal densities and the presence of vias have been analyzed. The result is shown in Figure 8 for the lower 4 BE interfaces only. In this case the left most profile belongs to the optimal BE design as it has overall the lowest ARE values. Similar, the right most profile belongs to the worst BE design.

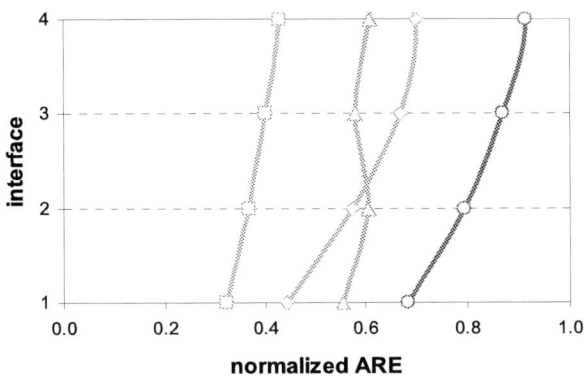

Figure 8 Optimization of various IC BE designs using the ARE criterion

Both the middle profiles are the result of changing the metal density and/or presence of vias in either interface 1-2 of the optimal BE design, or in interface 3-4 of the worst BE design. It is clear that changing the design (in this case the metal density and/or the presence of vias) in a single layer affects the mechanical performance of the entire BE structure. Slight changes to a single layer may improve or worsen the performance of the entire stack.

The need for an energy-based failure criterion like the ARE values in comparison to e.g. a stress-based failure criterion like e.g. the maximum peel stress that occurs over the height of the back end becomes clear when comparing Figure 8 with Figure 9. The latter picture shows the normalized maximum peel stress at each BE interface corresponding to the same BE designs as for the ARE criterion. Evidently, the peel stresses predict the opposite ranking. Moreover, it always predicts the top interface 4 as the most critical one in contrast to the ARE results. As shown in [3] the ARE predictions match the experimentally confirmed ranking of BE structures and critical interface for Cu low-*k* structures.

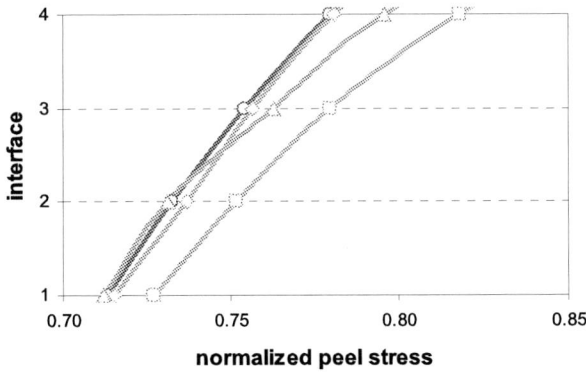

Figure 9 Normalized maximum peel stresses at each interface for the same BE designs as in Figure 8

4. Conclusions and discussion

In this paper an example has been given of a modeling approach that is a versatile tool when optimizing Cu Low-*k* BE design to improve the overall mechanical performance. Key features of such an approach are:
- the model is constructed in such a fashion that there exists a direct link between the input parameters and the criterion that is used to assess the overall performance
- the computational effort to create and simulate alternative designs is very low

Consequently, by performing a smart set of simulations it is possible to gain fundamental insight in how a specific design factor affects the mechanical performance. Of course the difficulty remains that it is usually not known on beforehand what design factors play a significant role. Nevertheless, the same approach could be adopted to assess the impact of a large number of design factors.

In essence the presented modeling approach is suitable for incorporation into an automated optimization procedure. This allows to further optimize the BE design with respect to various design factors at the same time.

With respect to the specific example that has been used to illustrate the approach it can be concluded that the metal density within the IC BE layers indeed is a design factor that affects the mechanical performance.

Moreover, it has been shown that the metal density in a single BE layer does not only locally affects the mechanical performance, but has an impact on the performance of the remaining BE layers as well.

References

1. W. D. van Driel, "Facing the challenge of designing for Cu/Low-k reliability", *Proceedings 7th Inernational Conference on Thermal, Mechanical and Multiphysics Simulation and Experiments in Micro-Electronics and Micro-Systems, EuroSimE 2006*, Como, Italy, 2006, pp. 770-774.

2. M. A. J. van Gils, O. van der Sluis, G. Q. Zhang, J. H. J. Janssen, and R. M. J. Voncken, "Analysis of Cu/low-k bond pad delamination by using a novel failure index", *Microelectronics Reliability*, 47 (2007), pp. 179-186.

3. O. van der Sluis, R. A. B. Engelen, R. B. R. van Silfhout, W. D. van Driel, and M. A. J. van Gils, "Efficient damage sensitivity analysis of advanced Cu Low-k bond pad structures using the Area Release Energy Criterion", *Microelectronics Reliability*, 2007, Accepted.

The chemical-mechanical relationship of the SiOC(H) dielectric film

Cadmus Yuan[*], O. van der Sluis, G. Q. Zhang, L. J. Ernst
Department of Precision and Microsystem Engineering,
Delft University of Technology, The Netherlands

W. D. van Driel
NXP Semiconductors, IMO-BE Innovation BY 1.055, The Netherlands

R. B. R. van Silfhout
Philips Applied Technologies, HTC 7, The Netherlands

B. J. Thijsse
Structure and Change in Materials, Department of Materials Science and Engineering, Delft University of Technology, The Netherlands

Abstract

We propose an atomic simulation techniques to understand the chemical-mechanical relationship of amorphous/porous silica based low-dielectric (low-k) material (SiOC(H)). The mechanical stiffness of the low-k material is a critical issue for the reliability performance of the IC backend structures. Due to the amorphous nature of the low-k material which has till now unknown molecular strucure, a novel algorithm is required to generate the molecular structure. The molecular dynamics (MD) mehtod is used as the simulation tool. Moreover, to understand the variation of the mechanical stiffness and density by the chemical configuration, sensitivity analyses have been performed. A fitting equation based on homogenization theory is established to represent the MD simulation results. The trends which are indicated by the simulation results exhibit good agreements with experiments from literature. Moreover, the simulation results indicate that the slight variation of the chemical configuration can induce significant change of the mechanical stiffness (over 80%) but not the density.

1. Introduction

As feature sizes for the advanced IC continue to shrink, the semiconductor industry is focusing the technology to minimize the intrinsic time delay for signal propagation, quantified by the resistance-capacitance (RC) delay. [1-2] The increasing demands for the electronic performance of the IC wiring have recently driven the replacement from aluminum trace to copper trace, and the alternative materials for SiO_2 film with lower dielectric constant [3]. These new low-k materials can be classified by silsesquioxane based material, silica based material, organic polymers and amorphous carbon. In the silica based matrix material, the attempt to reduce the k value can be obtained by two aspects. One can to replace oxygen by the carbon, hydrogen (organosilicate glass, OSG), or by fluorine (fluorinated silica glass, FSG). Generating the porosity within the material is another efficient approach.

The silicon oxide based low-k materials (SiOC(H), also called black diamond, illustrated in Fig. 1(a)) are preferred by industry because the fabricating processes of this materials exhibits high IC compatibility and high yielding rate. The k value can be reduced in two ways: either chemically by replacing oxygen by the methyl groups or H, OH, or physically by generating porosity within the material [3]. The different Si atoms are indicated with the usual denomination related to the number of O atoms linked to them: mono (M), di (D), tri (T) and quadri (Q) –functional group. The remaining links are of the type Si-R, where R is the –CH_3, O and OH functional group [4]. In addition, when functional group is replaced by a silanol group, it is indicated with OH as superscript. The Fig. 1b illustrates the groups of Q, T, D and M.

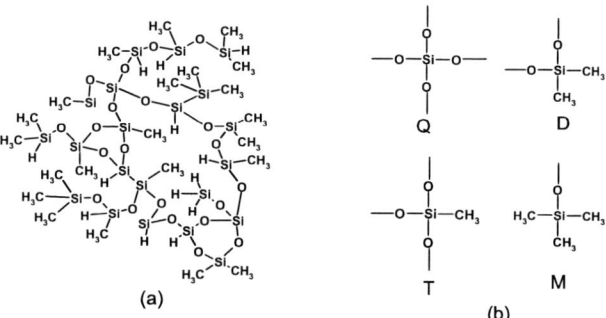

Fig. 1 . Illustration of the chemical structure of SiOC(H). (a) is the illustration of the material. (b) is the illustration of the connection capability of the basic building blocks

Among the materials of advanced IC backend structures, the low-k material has low mechanical stiffness, approximately 5-15 GPa. Experiments [4] show that enhancing the Young's modulus of the low-k material will increase the interfaceial toughness of SiOC(H)/TaN interface, which is known as the most critical interface in these structures. Among all the

---

[*] Corresponding author: c.a.yuan@tudelftt.nl

enhancement methods, the ultraviolet (UV) curing is preferred because the SiOC(H) film can perform the enhancement of the mechanical strength without much loss of the dielectric characteristic. However, the relationship between the chemical composition, porosity and mechanical properties remained unclear, and a trial-and-error design method is still common practice in the design/fabrication of the low-k material in the industry. Therefore, in this study, an atomic modeling method is developed, which is capable to analyze amorphous silica based material with porosity, to systematically study relation between mechanical characteristics of the SiOC(H) low-k film and it's chemical structure.

Theoretically, the amorphous nature of the SiOC(H) film together with the porosity increases the difficulty to directly simulate its nano-scaled mechanical response. Due to the amorphous nature, the atomic structure can be hardly defined. The void in the SiOC(H) molecule occurs randomly, and the size of the void should be also carefully considerred. According the literature, the complicate molecule (like SiOC(H) film) can be modeled when the accurate atomic structures and the potential functions are available. Yuan et al.[5-6] have stated that one can model the long chain complicate dsDNA molecule and metal after the proper atomic structures and the potential functions are obtained, no matter using the analytical solution, finite element method or the molecular dynamics. Falk and Langer [7] have applied the 12-6 Lennard-Jones potential function to describe viscoplastic deformation in amorphous solids.

In this paper, an algorithm which is capable of generating a reasonable molecular structure based on the given concentration of basic building blocks (i.e. Q, T, D, M and void). A series of simulations will be performed to understand the sensitivity of the mechanical stiffness and density with respect to the variation of the concentration of building blocks. Moreover, the fitting function based on homogenization theory is applied to understand the mechanical behavior of SiOC(H). Two sets of experimental results, the SiOC(H) film before and after UV curing, are used to validate the accuracy of the fitting function.

2. Theory

a. Molecular dynamics method

From the quantum mechanics point of view, matters have dual natures: particle and wave. However, while the geometry of the system is large enough, the wave nature of individual components becomes un-apparent and the system becomes determined. The molecular dynamics (MD), which is widely used in IC technology, is a treatment for the many-particle problems, and a determined response is prescribed. This method assumes the atom(s) as solid spheres; their movement is described by coordinate variables. The interactions between the particles are described by the potential functions, also called force fields. When the wave nature of the particle will be ignored or considered implicitly by the potential functions, MD exhibits high efficiency in the simulation

of the nano-scaled molecules. The following paragraphs will introduce the basic theory of MD, potential function, time integration scheme, boundary/initial conditions and limitation of MD.

Theoretically, MD is based on the Newton's second law of motion,

$$\vec{F}_i = m_i \vec{a}_i \qquad (1)$$

for each particle i in a system constituted by N particles. In Eq. (1), m_i is the mass of particle i, $\vec{a}_i = d^2\vec{r}_i / dt^2$ is its acceleration, and $\vec{F}_i$ is the force acting on the particle. Therefore, MD is a deterministic technique: given an initial set of positions and velocities, the subsequent time evolution can be determined.

The interaction force between particles, which is required in Eq. (1), can be defined by the potential functions or force fields:

$$\vec{F}_i = -\frac{\partial}{\partial \vec{r}_i} U(\vec{r}_1,...,\vec{r}_N) \qquad (2)$$

where U is the potential function and $\vec{r}_k, k = 1...N$ is the atomic coordinate.

b. Bar loading method

An atomistic method is established herein to predict the mechanical stiffness parameter, which is represented by the Young's modulus, of the nano-scaled structure. The nano-scaled specimens are simulated by the MD method with an additional energy minimization procedure.

A bar model is established as illustrated in Fig. 2, where one end of the bar is fixed and the opposite end is applied a displacement. The applied displacements and reaction forces which obtained at the fixed end are used to extract the Young's modulus by the elasticity theory. Due to the small deformation assumption of elasticity [9], the total amount of the longitudinal deformation should be less than 1.0% of the total length of the specimen. Moreover, based on Saint-Venant's principle [9], a model with high aspect ratio (L/h) is required to prevent boundary effects, as illustrated in Fig. 2. The loading and boundary conditions are applied at the longitudinal direction. Moreover, due to the linearity assumptions, reaction force outputs are linear with the externally applied displacement. The reaction forces $\vec{F}^i$ (i represent the i-th substeps) at the fixed end) can be extracted either by the force of the pseudo-spring of the anchor point (illustrated in Fig. 3) or the energy gradient of the fixed atoms.

According to linear elasticity theory, the mechanical deformation of the uniaxially loaded bar can be represent as: $\Delta d = FL / EA$ [9], where F, E, L and A represent external mechanical force, Young's modulus, initial length and initial cross section area of the specimen, respectively.

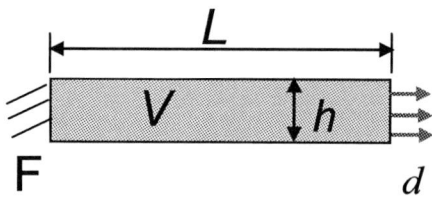

Fig. 2. Illustration to bar loading model

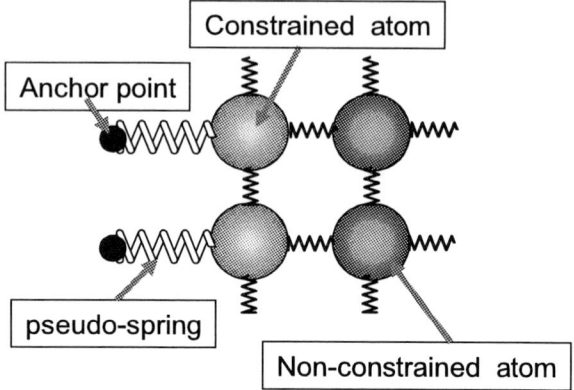

Fig. 3 . Illustration scheme of the constrainted atoms

3. Atomistic Model of SiOC(H)

The building blocks (Fig. 1b), Q, T, D and M, represent Si atoms having four, three, two and one capabilities to connect to other basic blocks, respectively. The size of the void is assumed to be the same as the basic blocks, and no basic group can connect to this. In the molecular modeling of the SiOC(H) film, we further assume that only the single bond would exist between any two basic groups. Moreover, the composition of the low-k film is assumed to follow the four basic blocks (Q, T, D and M). Considering the basic building blocks with silanol group and methyl group, like –OH of T^{OH} and –CH_3 of T, both of them can not provide the connection capability to the other basic building blocks and they have the similar atomic mass. From the mechanical point of view, the transferring of the force will be terminated at the methyl or silanol group; therefore, the blocks with methyl and silanol group will be mechanically similar. Hence, the concentrations of the basic building blocks with silanol group (e.g. T^{OH} and D^{OH}) are merged into the ones with methyl groups (e.g. T, D).

In practice, SiOC(H) films with thickness ranging from 200 to 700 nm were deposited by Chemical Vapour Deposition (CVD) at 350°C. Trimethylsilane and O_2 were used as precursor and gas for film deposition [4]. Due to the similarity of the fabrication process between SiOC(H) and SiO_2, we assume that the connection catalogue of SiOC(H) and SiO_2 are similar. Therefore, these basic blocks are assumed to be distributed onto a three dimensional, where each node has a maximum of 4 connection capabilities. As shown in Fig. 4a, the framework will define where the building block can or can not be located. The building blocks (including the void, Q, T, D and M) will randomly distribute into the framework (Fig. 4a) and the connection between blocks will be established (Fig. 4b). However, most atoms shown in Fig. 4a are not in the equilibrium state because a cubic framework is used. The geometrical optimization procedure [10] is used to minimize the atomic potential energy of the connection catalogue.

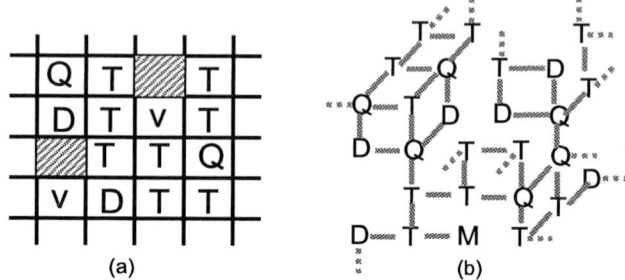

Fig. 4. Illustration of generating algorithm (a) Two-dimensional illustration of the framework and locating of the basic building block. (b) Illustration to the obtained topology of amorphous SiOC(H) molecule

4. MD simulation results and data analysis

4.1. MD simulation parameters

In order to prevent boundary effects, the length and cross section size of both cases are chosen as approximately 10nm and 6.5 nm^2 after the structural relaxation; the number of basic building blocks is 1,224. Both the cases of SiOC(H) molecule before and after UV treatment are simulated by the commercial MD solver Discover (version 2005.2) [10], and the force fields between the atoms are described by COMPASS (definition: cff91, version 2.6) [10]. Both computations are performed on an i686 machine with 2.8GHz CPU and CPU time for each case is approximately 270,500 seconds. In this paper, the canonical ensemble (NVT) ensemble, which conserves the number of atoms (N), the system volume (V) and the temperature (T), is used. Moreover, no periodic boundary condition is applied to any model.

4.2. Parametric analysis on the case of before/after UV treatment

In order to verify the accuracy of the proposed method, two SiOC(H) models, A1 (shown in Fig. 6) and A2, having similar chemical composition as the SiOC(H) before and after UV treatment, have been generated. demonstrates the side view and cross sectional view of model A1, where the dark yellow, red, grey and white spheres represent, respectively, the silicon, oxygen, carbon and hydrogen atom. The simulation results list at the case A1 and A2 of Table I. The simulation shows that the Young's modulus and density of A2 (after UV treatment) is slightly higher than A1 (before UV treatment), and the similar trend is also found in the experiment [4]. Note that the simulated density is defined as the ratio of atomic mass and molecule volume. Note that the molecule volume is defined as the volume which is occupied by the molecular surface. This simple case study demonstrates that the MD simulation has the capability to describe the variation of Young's modulus and density as function of chemical composition.

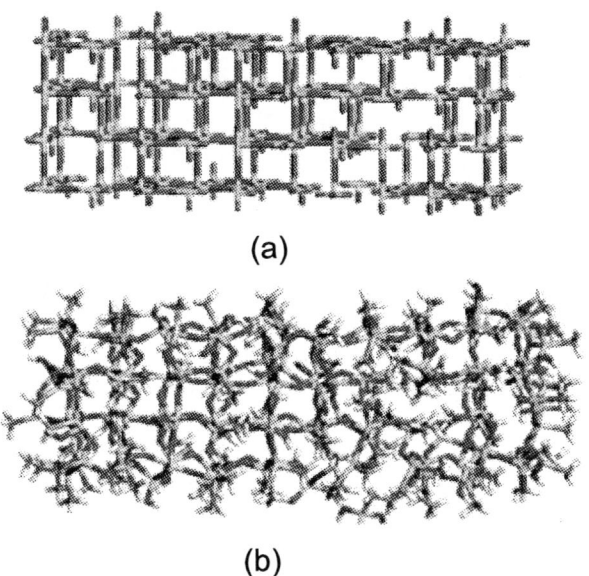

(a)

(b)

Fig. 5. (a) A generated approximate topology of SiOC(H) film. (b) the SiOC(H) film after minimization

4.3. Parametric analysis

Three series of parametric analyses are conducted as listed in Table I: the chemical composition in the B series are similar to A1 and A2; the models C1, C2 and C3 emphasize the effect of Q, T and D, respectively; the D series comprises the extreme cases (e.g. SiO₂ and air). The molecular model generating method, geometrical size and loading/boundary conditions of the B and C series and model D1 follow the same procedure of models A1 and A2. The model D1 (SiO₂) is established by the conventional silicon oxide single lattice rather than the proposed generating algorithm, but the geometry and boundary/loading conditions of model D1 is the same as the other cases. For the model D2(air), both the Young's modulus and density of D2 (air) are assumed as zero, and no computational effort is required. The simulation results of the test cases are listed in Table I, and shown in Fig. 7.

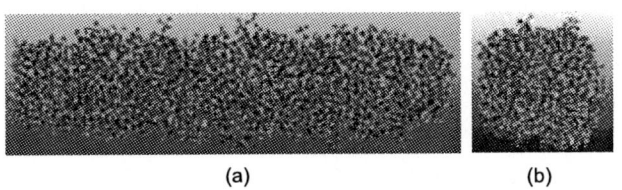

(a)   (b)

Fig. 6. The molecular model A1. (a) side view, (b) cross section view.

Considering the B series, the simulated Young's moduli and densities are similar because the concentration of basic building blocks are similar. The simulated Young's moduli for the C series exhibit large variation, but the density are similar. Therefore, the Young's modulus is highly dependent upon the chemical composition but the density is not.

Table I Parametric analysis of the SiOC(H)

Case	Ratio of basic building blocks			Young's modulus	Density
	Q	T	D	(GPa)	(g/cm³)
A1	16%	44%	29%	13.41	1.91
A2	21%	49%	19%	9.35	1.96
B1	21%	39%	29%	6.92	1.88
B2	31%	29%	29%	11.68	1.97
B3	15%	45%	16%	7.52	1.92
C1	70%	25%	6.0%	26.80	2.69
C2	22%	68%	9.4%	16.39	2.13
C3	11%	19%	69%	3.48	1.69

4.3. Data management

In order to understand how the concentration of Q, T, D and void impact the Young's modulus and density, A response function, $f_{E,density} = c_0 + c_Q r_Q + c_T r_T + c_D r_D + c_{void} r_{void}$, is used to obtain the sensitivity of the parametric analysis For simplification, the ratio of M is merged into D because the ratio of M is relatively small compared to the rest. The coefficients of the response function are normalized by c_Q, and the results are shown in Fig. 8. The sensitivity shows that the basic building blocks of Q and T will positively influence the Young's modulus and density. Increasing the porosity will decrease both Young's modulus and density. Varying the ratio of D will not significantly influence the simulation result.

Moreover, a rather simple fitting function based on homogenization theory is used to describe the numerical results. We denote Young's moduli and densities of 100% Q, T, D are E_Q, E_T, E_D, ρ_Q, ρ_T and ρ_D, respectively. Hence two fitting functions for Young's modulus and density can be written as:

$$E = E_Q r_Q + E_T r_T + E_D r_D \qquad (3a)$$

$$\rho = \rho_Q r_Q + \rho_T r_T + \rho_D r_D \qquad (3b)$$

The coefficients can be obtained by the least square method. Considering the detail experimental data on SiOC(H) molecule before and after UV treatment, the concentrations of basic building blocks are listed in Table II, and following the simulation results (obtained by the fitting equations (3a) and (33b)) and the experimental results. Table II indicated that the MD simulation

475

(represented by the fitting equation) can not provide the quantitative prediction for the SiOC(H) molecule. However, MD can simulate the increasing trend of the Young's modulus and density after the UV treatment within acceptable accuracy. Thus, our work shows that MD provides a tool to perform material design, albeit in a qualitative way. Note that the proposed simulation procedure did not consider the complex fabrication process but includes individual chemical concentration as the input.

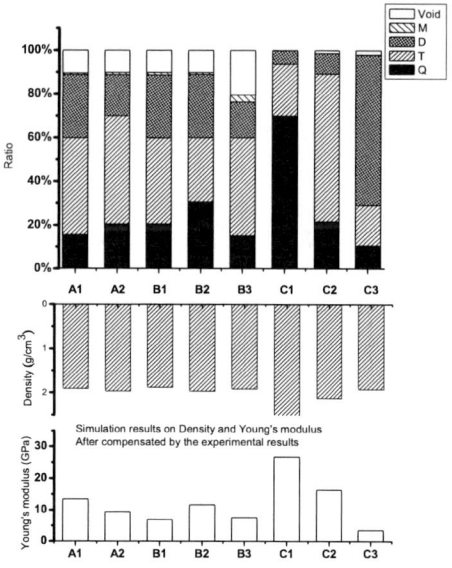

Fig. 7. The plots of concentration of basic blocks, density and Young's modulus (from upper panel to lower panel)

Table II Experimental validation on predicted trend

SiOC(H)		BU*	AU*
Concentration **	Q	15.70%	21.70%
	T	47.40%	49.70%
	D	29.80%	20.50%
By Fitting function	E (GPa)	9.87	12.43
	D (g/cm³)	1.99	2.02
By Experiment	E*** (GPa)	11±1	16±1
	D**** (g/cm³)	1.48	1.52

*: BU and AU represent the SiOC(H) molecule before and after UV treatment

**: obtained by nuclear magnetic resonance (NMR)

***: obtained by nano indentor

****: obtained by X-ray reflectivity (XRR)

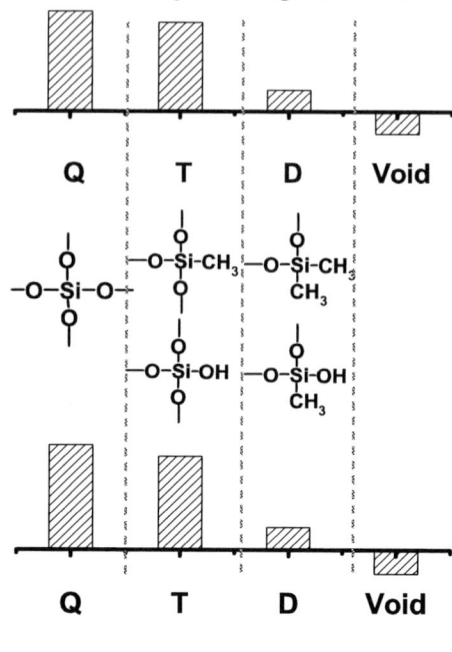

Fig. 8. The sensitivity of Young's modulus and density

Considering the magnitude of the Young's modulus density listed in Table II, the values calculated by MD are several times higher than the experimental results. Possible reasons are:

● No enough defect types are modeled, including dislocations and grain boundaries;

● Size effect: due to the fact that the surface atom may not fulfill the requirement of the octet rule, the surface atoms are often charged. This phenomenon can induce higher mechanical stiffness. Note that the surface charge is considered in the MD simulation ;

● Void collapse: the void presented in the molecular topology might collapse by the structure minimization step. After the minimization step, the void will remain 1/3 to 1/4 compared to the original topology. Therefore, the porosity of the models is much smaller then the one in reality.

However, as the size of each model listed in Table I is controlled and the porosity of SiOC(H) is approximately 10% in the reality, the qualitative trend can be validated by the experimental results.

5. Conclusions

A series of molecular modeling method is presented to simulate the amorphous low-k especially SiOC(H) material. The simulation procedure comprises three steps:

■ generation of amorphous molecular model,

476

- parametric study by molecular dynamics (MD) method and
- data analysis.

Based on the chemical composition of the basic building blocks (Q, T, D, M) and the void, the chemcial topologies are obtained and the structural minimization procedure is then performed to obtain the an approximate molecular structure. A series of parametric studies is performed to understand the sensitivity of Young's modulus and density while varying the chemical concentration. Moreover, a simple fitting function based on the homogenization theory is then applied to acquire the Young's modulus and density as functions of concentration of the basic building blocks. The experimental validation shows that the proposed method can qualitatively represent the trend. Moreover, the simulation results indicate that the slight variation of the chemical configuration can induce significant change of the mechanical stiffness (over 80%) but not the density. However, in order to achieve higher quantitative accuracy, the molecular model should be improved by increasing the geometry size, inluding defects of realistic size and improve method for including the porosity.

Acknowledgments

The authors are grateful to Dr. F. Iacopi for sharing her experimental results and experience of the low-k material. Also, the authors thank Dr. N. Iwamoto for valuable discussions on the simulation technique of molecular dynamics. C. Yuan thanks Dr. C. Menke and Dr. J. Wescott for discussions on numerical simulation technique.

References

1. International Technology Roadmap for Semiconductors, *ITRS*, 2005.
2. Grill, A. and Neumayer, D. A., "Structure of low dielectric constant to extreme low dielectric constant SiCOH films: Fourier transform infrared spectroscopy characterization," *J. Appl. Phys.*, Vol. 94, Issue 10 (2003), pp. 6697-670.
3. Maex, K., Baklanov, M. R., Shamiryan D., Iacopi, F., Brongersma, S. H. and Yanovitskaya, Z. S., "Low dielectric constant materials for microelectronics," *J. Appl. Phys.*, Vol. 93, Issue 11 (2003), pp. 8793-884.
4. Iacopi, F., Travaly, Y., Eyckens, B., Waldfried, C., Abell, T., Guyer, E. P., Gage, D. M., Dauskardt, R. H., Sajavaara, T., Houthoofd, K., Grobet, P., Jacobs, P. and Maex, K., "Short-ranged structural rearrangements and enhancement of mechanical properties of organosilicate glasses induced by ultrabiolet radiation," *J. Appl. Phys.*, Vol. 99 (2006), pp. 053511.
5. Yuan, C. A., Han, C. N., Chiang, K. N., "Investigation of the Sequence- Dependent dsDNA Mechanical Behavior using Clustered Atomistic-Continuum Method," *NSTI Nanotechnology Conference*, Anaheim, California, U.S.A., May 8-12, 2005.
6. Chiang, K. N., Yuan, C.A., Han, C. N., Chou, C. Y. and Cui, Y., "Mechanical characteristic of ssDNA/dsDNA molecule under external loading," *Appl. Phys. Lett.*, Vol. 88, (2006), pp. 023902.
7. Falk, M. L., Langer, J. S., "Dynamics of viscoplastic deformation in amorphous solids," *Phys. Rev. E*, Vol. 57, (1998), pp. 7192- 7205.
8. Rieth, M. , <u>Nano-engineering in science and technology</u>, World Scientific publishing Co. (New Jersey, USA, 2003).
9. Love, A. E. H., <u>A treatise on the mathematical theory of elasticity</u>, Cambridge university press (New York, 1934).
10. Accelrys Inc., <u>Materials Studio™ DISCOVER</u>, Accelrys Inc. (San Diego, 2005).

Magnetically actuated microvalve for disposable drug infusor

M. Duch*, J.Casals-Terré**, J. A. Plaza*, J. Esteve*, R Pérez-Castillejos*, E. Vallés*** and E. Gómez***
*Centro Nacional Microelectronica,
** Technical University of Catalonia,
*** Universitat de Barcelona,
C/ Colom 7-11 Terrassa, Barcelona,Spain
Jasmina.casals@upc.edu, +3493798023

Abstract

A magnetic microfluidic valve has been analyzed to improve its performance. Operation relies on the use of a permanent magnet which interacts with an electrodeposited layer of Co-Ni on a V-shaped cantilever beam. The deflection caused by the magnetic forces opens or closes the fluid flow. The microvalve performance has been optimized by means of finite element analysis. The FEA model has been experimentally validated using confocal microscopy and used to improve the magnetic circuit. Then, a fluidic cell has been built and the microvalve has been demonstrated to work as a flow regulator, when being magnetically actuated, and as a check valve.

Experiments showed that the fabricated prototypes presented a good performance from 0 to 50 mbar, with control on a flow up to 18 sccm.

Index Terms—Microvalve, magnetic actuation, drug infusor, flow regulator.

1. Introduction

PORTABLE drug infusors are used to provide a slow and continuous infusion into the bloodstream, this allows drug treatments to be given in several days using an infusor for home treatments. Current state-of-the-art of portable drug infusor is based on elastomeric latex-free pumps [1,2] which are single passive fluidic resistors. Therefore, each time it is required to modify the drug dose, the fluidic resistor has to be replaced by a new one. As a response to the need of variable passive fluidic resistors for portable drug infusors, "quasi-digital" flow restrictors based on magnetically actuated microvalves are being developed that will allow programming and modifying the dosing flow with only one product [3].

Their main advantages are: portability (reduced size as well as lightweight) and ability to provide continuous drug infusion under precisely controlled flow rates and no power consumption to operate (permanent magnets). These features improve significantly the patient's life quality that can have their treatment at home and carry on their daily activities rather than having to stay overnight at the hospital for the duration of the treatment.

In fluid control purposes, large force is required in order to efficiently interact with the fluid and to generate desired control effects. The significance of applying MEMS in fluid control was already analyzed by Hou [4]. Magnetic actuation can potentially provide both large force and large displacement, [5]. Miller et al. [6] and Judy et al. [4] had both previously reported individually addressable magnetic actuated devices. Barbic et al. [7] designed a micromotor inspired in the macroworld to control the fluid flow.

Recently, Sutanto et al. [8] presented a surface micromachined bistable electromagnetic microvalve, the valve operated up to 50 μl/ min and up to 7.8 kPa, but due to the electromagnetic actuation there was current consumption and the fabrication process was more complex than the one presented in this paper. Similar approach had been previously presented for Bohn et al [9] but at a higher scale (in the orther of mm). A similar device but electrostatically actuated was applied to a Braille display system by Yobas et al [10], the new actuation methodology presented in this paper could reduce the complexity of the system because there is no need of power to maintain the valve closed or opened.

This work proposes a novel, low-power, externally actuated and friendly in use, micromachined flow regulator based on an array of individual cantilever based microvalve. In this paper a detailed description of the behaviour of a single cantilever valve is provided using both FEA simulation and experimental validation. A multiphysics simulation methodology is used to optimize its behaviour. In this model magnetic forces generated by a magnetic field are used as inputs to generate deflection on the mechanical structure.

This paper is organized as follows: the principle of operation of a single cantilever microvalve is described in Section 2. In Section 3, the microfabrication process is described. Section 4, an FEA analysis optimizes of the magnetic circuit and verifies the possibility to use the proposed geometry to achieve the flow control. In Section 5, the proposed microvalve is applied as flow regulator in a fluidic cell, the simulations results are experimentally tested and validated, so it is concluded that this type of simulations are valid to analyze magnetic-structural coupled systems.

2. Principle of actuation

The microvalve has two parts: top beam (Cap part) and bottom membrane (Body part). The top cantilever beam is a V-beam anchored at one end and with a circular plate at its end, see Figure. 1a. The cantilever itself, in order to avoid torsion modes, is designed as a double anchored V- beam producing only out-of-plane linear motion. The V-shaped beam has an electrodeposited Co-Ni layer which causes the top beam deflection in presence of a permanent magnet, see Figure. 1b.

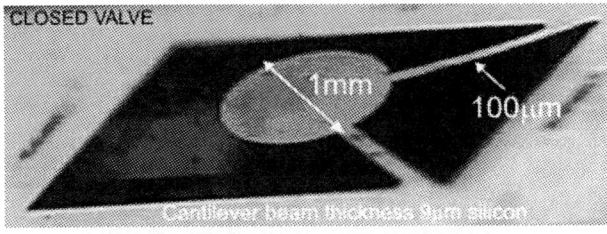

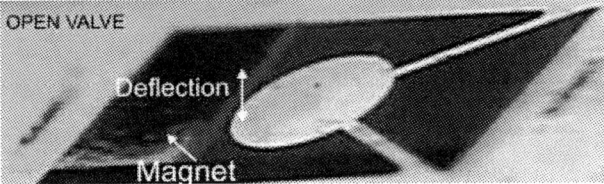

Fig.1. V-shape cantilever beam (a) dimensions (b) deflection of the V- shaped cantilever beam under the presence of a permanent magnet.

Due to the deflection caused by the magnet, the circular part of the top cantilever beam shuts the hole of the bottom part (body part), closing the flow path. The bottom part is a silicon membrane with a central hole. So, depending on the position of the magnet, the cantilever beam opens or closes the fluid path, see Figure 2.

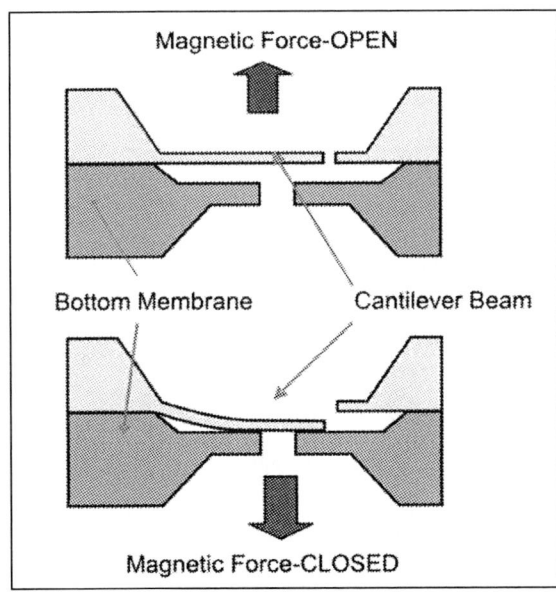

Fig.2. Schematics of the microflow regulator magnetically actuated.

Both parts are joined together using a high ticsotropic synthetic Bruger ® paint, which at high temperatures have low fluidity and join the two silicon parts without blocking the bottom hole or fixing the thin cantilevers.

Once joined the valve is placed in a polycarbonate fluidic cell, which has been machined in order to be able to allocate the magnet and the fluidic inlet and outlet that connect the fluid with the valve. An o-ring is placed between both parts of the fluidic cell which causes a watertight. The fluid is forced into the cavity through the inlet that connects to the watertight cavity and the only way to proceed is to go through the microvalve which is placed in the path of the fluid. See Figure 3 where all the parts that configure the system are plotted.

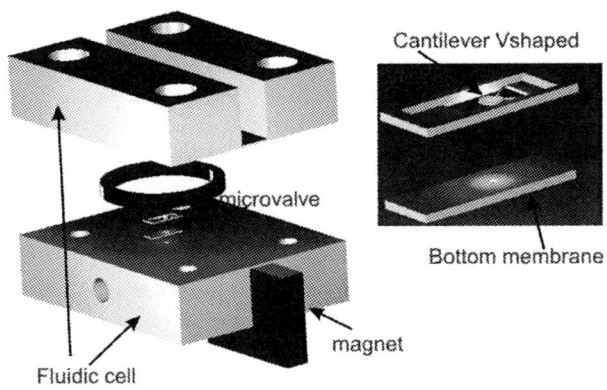

Fig.3. Parts of the magnetic actuated microvalve and polycarbonate fluidic cell.

3. Fabrication

The fabrication process of the disposable parts (silicon parts) combines silicon micromachining technologies and electrodeposition of magnetic CoNi alloy [11,12]. The fabrication flow-chart consists of two parts: the valve caps (Figure 4) and the fluidic body that are ultimately held together.

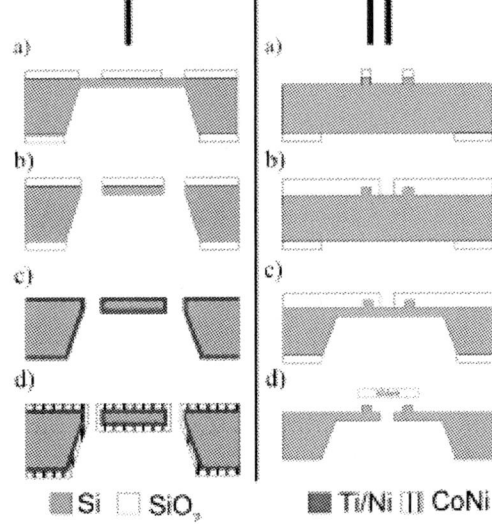

Fig.4. Schematic view of the fabrication process. I Cap part with the valve. II Body part membrane with the hole.

The valve caps are made of microstructured silicon. A silicon oxide layer is used as mask material, figure 4.I.a) for an anisotropic wet etching and RIE etching figure 4.I.b) that define the valves geometry. Next, a Ti/Ni film is deposited as a seed layer, Figure 4.I.c). Finally, a CoNi layer is electrodeposited as a magnetic material, figure 4.d). Meanwhile, silicon oxide layers are also used on the top and backside of the wafer as mask materials for the machining of the second chip Figure 4.II.a). The top silicon oxide layer is used to define a shadow O-ring by a RIE process. Secondly, an additional silicon oxide layer is patterned on the topside to be used as mask material for the last RIE. Following, an anisotropic etching is done to machine the silicon wafer, Figure 4.II.c). And finally, a RIE process is performed to define the hole. Once the two silicon chips are finished, they are attached. The chemical bond of the two parts of the valve is made of a high ticsotropic synthetic paint.

4. FEA Magnetic-Structural Model

Ansys Multiphysics V.10 finite element analysis software is used to perform the required coupled field analysis [13-14]. Magnetic-structural analysis optimized distance between the magnet and the valve, so the microvalve could control the flow path and analyzed the most favorable orientation of the magnet versus the cantilever beam. A direct magnetic/structural analysis has been executed to characterize the motion of the magnetically actuated cantilever beam.

A 3D dimensional model is created. The model contains the permanent magnet, the iron field concentrators and the top part of the valve (cantilever beam + frame) embedded in a square of air, see Figure 5.

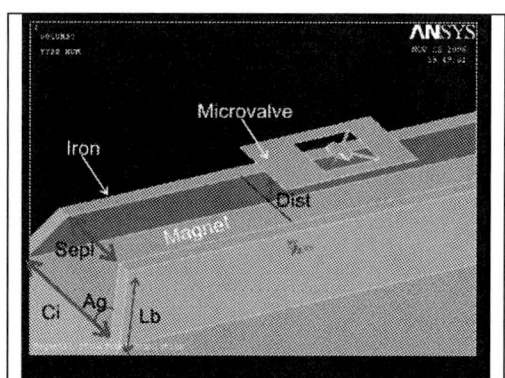

Fig.5. Volumes and variables analyzed using coupled –field structural and magnetic finite element analysis.

The magneto-structural analysis is a two field analysis, so it requires a model that can be used in both fields. In the model of Figure 5 the magnet, the field concentrators (iron), the microvalve (which includes the CoNi layer of magnetic material) and the air surrounding these elements constitute the 3D model of the magnetic part. The magnetic field distribution obtained from this model is applied as input to the structural model formed by the microvalve itself.

The two analysis advance using an iterative coupled field analysis, where physics coupled passes loads across physics field interfaces (in this case the surface of the microvalve). So, two iterations is needed (one for each physics) in sequence is needed to achieve the coupled response.

First the magnetic model, using SOLID96 magnetic element, calculates the magnetic field distribution generated by a NdF-eB235 permanent magnet. This field caused distributed magnetic forces over the cantilever beam (physic interface).

The material properties of the magnet are incorporated in the form of the coercitive force and the relative magnetic permittivity, see Table II. The air domain and the silicon microvalve are also part of the magnetic model and they are magnetically characterized through its relative magnetic permittivity.

The boundary of the model (air) has been parameterized and change till the effect on the model results was irrelevant. The interaction between the two models is accomplished through the FLAG surfaces (interface), which permits the magnetic forces generated through the magnetic field of the magnet to be transferred as mechanical forces in the nodal points of the structural model once meshed using SOLID 45, see Figure 6 where the mesh of the microvalve is plotted .

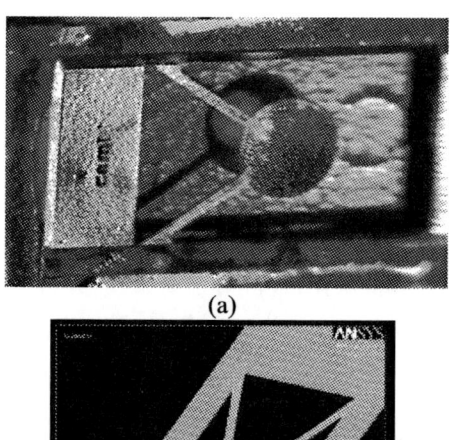

Fig.6. 3-D (a) Photograph or the real microfabricated valve (b) mesh of the top microvalve.

The main dimensions of the volumes in the model are summarized in Table I.

TABLE I
SIMULATED VALVE DIMENSIONS

Dimension		Value
Magnet	Length	25 mm
	Width	5 mm
	Thickness	18 mm
Top cantilever beam	Length	1000 μm
	Width silicon	100 μm
	Thickness silicon	9 μm
	Thickness Co-Ni	1 μm
Frame	Length	7.2 mm
	Thickness	500 μm
Circular plate	Diameter	1 mm
	Thickness silicon	9 μm
	Thickness Co-Ni	1 μm

The magnetic and structural properties of the materials used in the model are summarized in Table II.

TABLE II
MATERIAL PROPERTIES

Property	Value
Silicon Young modulus (E)	1,69. GPa
Iron Young modulus (E)	350 GPa
Magnet Young modulus (E)	150 MPa
Magnet Relative permeability	1,36 mT/{kA/m}
Magnet Coercivity	860 kA/m
Co-Ni Relative permeability	755 mT/{kA/m}
Air Relative permeability	1 mT/{kA/m}
Silicon Relative permeability	1 mT/{kA/m}
Iron Relative permeability	10^4 mT/{kA/m}

3. 1 Magnetic field interaction

The analysis started with a parallel configuration, both the magnet and the microvalve where parallel to each other, see Figure 5. From the magnetic field distribution generated by the magnet, see Figure 7 b, a torsion torque was applied on the circular end of the cantilever which caused a loss of energy that was not used to achieve vertical deflection.

Two different models were done to determine the best orientation of the magnet versus the cantilever beam. When the magnet was placed perpendicular to the valve Figure 7 a, the magnetic field causes bending torque over the beam and the beam deflects downwards, while if the magnet is place parallel to the beam a torsion torque is applied to beam and part of the energy is lost in this torque instead of downwards deflection, see the difference in deflection in Figure 8.

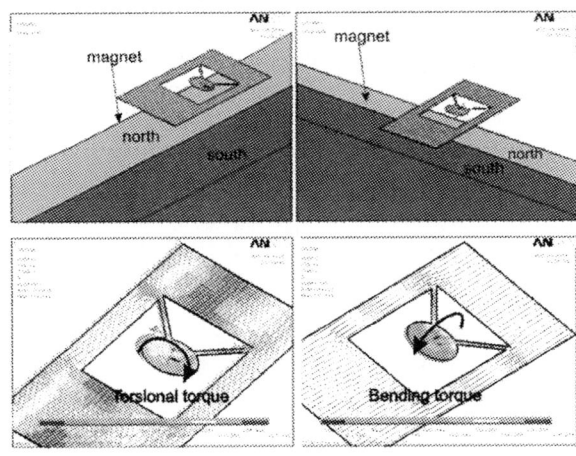

Fig.7. 3D ANSYS model: two different configurations of magnet orientation vs. microvalve and the corresponding magnetic field distribution at the bottom.

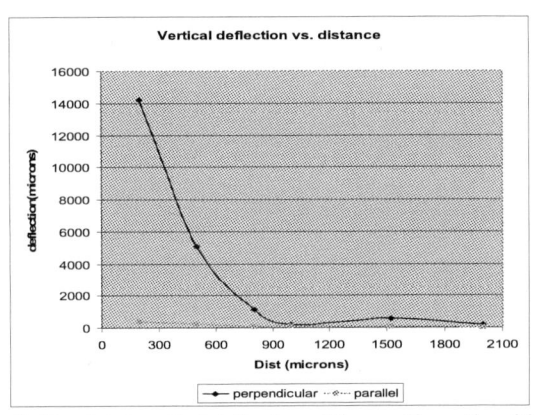

Fig.8. Simulated vertical deflection of the cantilever tip with the microvalve placed perpendicular or parallel to the magnet.

The vertical deflection depends on the number of field lines that cross the cantilever tip, the concentration of lines can be enhanced if we use iron plates as field concentrators, placed at each side of the magnet guiding the field through them to the cantilever. In the Figure 9 we can appreciate that the downwards deflection increases exponentially with iron plates (field concentrators).

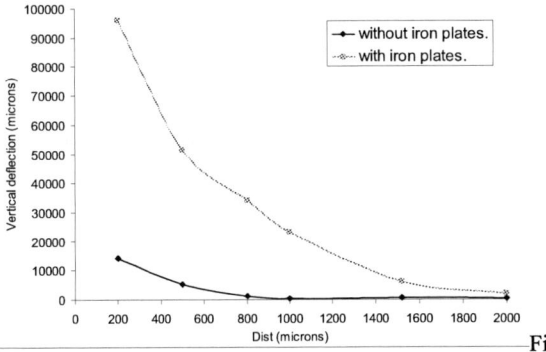

Fig.9. Simulated vertical deflection of the cantilever tip with or without iron plates.

The first purpose of the model is determining the optimal spatial distribution of all the elements that configure the system (distance from the microvalve to the magnet / iron plates, distance between iron plates and angle of the iron plates). In order to start the optimization an initial distance from the iron tips to the Co-Ni layer of the cantilever beam was fixed to 1520 μm (*dist*), due to the wafer thickness of the top and bottom part and the polycarbonate thickness of the fluidic cell.

A run of simulations was performed with these values to determine an initial optimal angle for the deflection of the iron plates. In these simulations, the following relation was deduced: If we use field concentrators the distance between the tips of the concentrators (*Sepl*) has to be larger than the distance between the concentrators and the cantilever beam (*dist*), so the field lines go through the cantilever. And Sepl has to verify that:

$$Sepl = ci - 2 \cdot Lb \cdot \cos(AG) \quad (1)$$

Where:

Sepl is the distance between both tips of the iron plates, *ci* is the width of the magnet, *Lb* is the length of the iron plates from the magnet to the tip, *AG* is the angle of the iron plates, see Figure 5.

To satisfy equation (1) and the initial optimal angle (AG_0) and the initial Lb_0=ci/2, *AG* was optimized considering that the relation between *Lb* and cos(*AG*) in the optimal initial case had to be kept constant:

$$AG = a\cos\left(\frac{Lb_o}{Lb}\cos\left(\frac{AG_o\pi}{180}\right)\right)\frac{180}{\pi} \quad (2)$$

From equation (2) the optimal value of the angle *AG* of the iron plates can be seen in Figure 10 where the vertical deflection of the tip of the cantilever is plotted for different *AG* with the valve perpendicular to the magnet and the circular tip centered in the magnet.

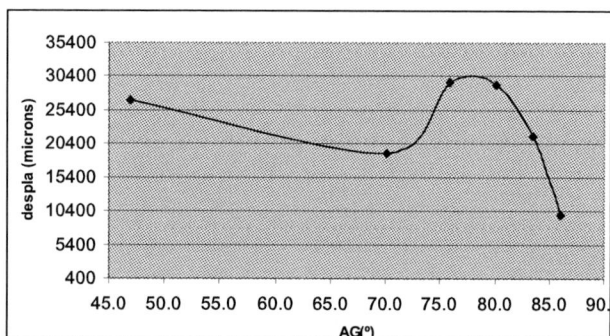

Fig.10. Simulated vertical deflection versus angle of the iron tips.

From the angle obtained of the previous analysis, the distance of the iron plates *Lb* was adjusted. In Figure 11 the optimal *Lb* can be appreciated.

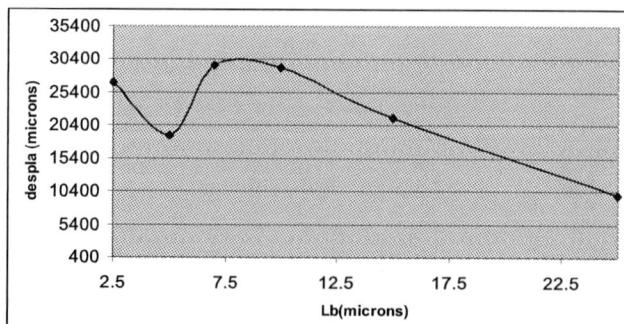

Fig. 11. Simulated vertical deflection versus length of the iron tips.

From these two analyses both *Lb* and *AG* are fixed to its optimal values for the design. The distance from the tip of the iron plates to the magnet is a parameter that has initially been fixed due to the fabrication restrictions (wafers thickness and polycarbonate thickness of the fluidic cell). However, these two parameters can be pushed further if needed. In Figure 12 it is demonstrated the vertical deflection increases exponential with proximity to the magnet, so polycarbonate thinning can easily decrease the force needed to close the membrane hole.

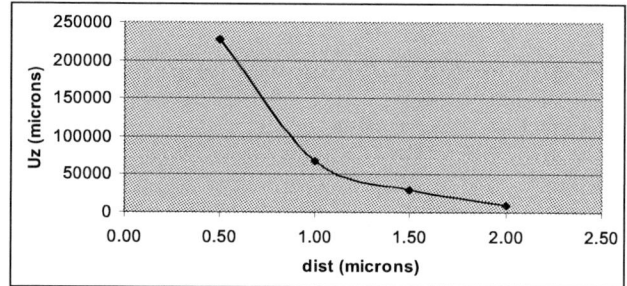

Fig. 12 Simulated vertical deflection versus distance from the tips to the microvalve

3.2 Model Validation

In order to verify the model the results of FEA software where compared with the deflection measured in a confocal microscope model PLμ Confocal Imaging Profiler. Both model and experiment included the top cantilever beam valve (silicon) covered with the electrodeposited CoNi layer and the permanent magnet without iron plates. The configuration was parallel orientation between the magnet and the microvalve. In Table III, the deflection in the model is 100 μm for a distance between the magnet and the microvalve set to 2000 μm compared to 89 μm in the confocal microscope. The deviation is low enough to consider the model valid to proceed with the analysis.

482

TABLE III. MEASURED DISPLACEMENTS OF THE MICRO-VALVE, WITH DIFFERENT MAGNET ORIENTATIONS AND MAGNET OFFSETS.

	without magnet	parallel magnet	perp. magnet	perp. magnet 2 mm offset
Disp. [μm]	0	-59	-5	89

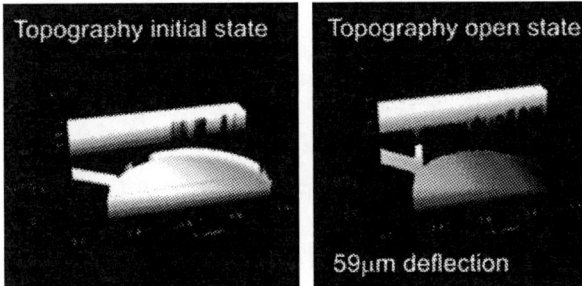

Fig. 13. Micro-valve measured displacements with parallel magnet orientation (Confocal microscope).

4. Conclusions

A new microfluidic flow regulator has been proposed, which uses permanent magnet actuation which does not require any type of power input to control the fluid flow. This provides several advantages in MEMS applications, such portability, light weight and low cost. Its performance has been analyzed and its actuation principle shown, along with a study of its main parameters. An FEA model has been developed to analyze the main design parameters; the performance of the model has been validated comparing the results with the measured deflections of a microfabricated valve using confocal microscopy, with a good agreement between the experimental results and the simulated values. This proved that using direct magnetic-structural coupled field analysis is adequate to simulate this type of devices. So, the model can be used to optimize the performance of the magnetically actuated microvalve for drug infusors. The optimization can be achieved through the magnetic field concentrators that according to the model provide an exponential increase of the deflection achieved by the cantilever valve. The improvement observed through the model will be experimentally proved in a near future.

References

1. DOSI-FUSER www. Leventon.es
2. Department of Health, Bath Institute of Medical Enginering. The Wolfson Centre. "Market survey: Non-electrically powered disposable infusion devices". Bath, UK September 2005.
3. M. Duch, J. Esteve, A. Salas, M.C. Acero, J.A. Plaza, E. Vallés and E. Gómez

"Magnetic "quasi-digital" flow regulator for drug infusor" MicroTas 2005. Boston
4. J. Judy and R. S Muller "Magnetically actuated, addressable microstructures" Journal of Micro electromechanical systems Vol 6 N 3 p 257 1997.
5. J.W. Judy, R.S. Muller and H.H. Zappe, "Magnetic microactuation of polysilicon flexure structures" Journal of Micro electromechanical systems Vol 4 N4 pp 162-169 1995.
6. R.A. Miller, Y.C. Tay, G. Xu, J. Bartha and F. Lin "An electromagnetic MEMS 2x2 fiber optic by pass switch" 1997 International Conference on Solid-State sensors and actuators, Chicago IL, Vol 1. pp 89-92 1997
7. .M. Barbic, J.J. Mock, A.P. Gray, S. Schultz "Electromagnetic micromotor for microfluidics applications" Applied Physics letters v79 n9 27 2001.
8. J.sutanto, P.J. Hesketh, Y.H. Berthlot "Desgin, microfabrication and testing of a CMOS compatible bistable electromagnetic microvalve with latching/unlatching mechanism on a single wafer". Journal of Micromechanis and Engineering v16 n2 Feb 2006.
9. S. Bohn, G.J. Burger, M.T. Korthorst, F. Roseboom "A micromachined silicon valve driven by a miniature bis-stable electro-magnetic actuator" Sensors and Actuators A. v a80 n1 march 2000.
10. L. Yobas, D.M. Durand, G.G: skebe, F.J. Lisy, M.A. Huff "Novel integrable microvalve for refreshable Braille display system" Journal of Microelectromechanical systems v12 n3 June 2003
11. M Duch, J Esteve, E Gomez, R Perez-Castillejos and E Valles. Electrodeposited Co-Ni alloys for MEMS. Journal of Micromechanics and microengineering. Vol 12 2002. pp. 400-405
12. M. Duch, J. Esteve, E. Gómez, R. Pérez-Castillejos and E. Vallés "Development and characterization of Co-Ni Alloys ofr Microsystems Applications" Journal of Electrochemical Society, 149 (4) pp. 201 208 2002.
13. A.khasdia, D.J. Power and J.P. Loughlin "Iterative Magnetic/structural Simulation of a MEMS microshutter" ANSYS Conference in 2004.
14. J.P. Loughlin, R.K. Fettig, S.H. Moseley, A.S. Kutyrev and D.B. Mott "Structural Analysis of a Magnetically Actuated Silicon Nitride Micro-Shutter for Space Applications" ANSYS Conference in Pittsburgh PA,2002.

Generating VHDL-AMS Models of Digital-to-Analogue Converters From MATLAB®/SIMULINK®

Alexandre Cesar Rodrigues da Silva[1], Ian Grout[2], Jeffrey Ryan[3] and Thomas O'Shea[4]

[1]UNESP – School of Engineering at Ilha Solteira – Department of Electrical Engineering, Av. Brasil, 56,
ZIP 15385-000, Ilha Solteira, SP, Brazil

[2,3,4]Department of Electronic and Computer Engineering – University of Limerick, National Technological Park,
Castletroy, Limerick, Co. Limerick, Republic of Ireland

[1]acrsilva@dee.feis.unesp.br, [2]ian.grout@ul.ie, [3]jeffrey.ryan@ul.ie@, [4]thomas.oshea@ul.ie

Abstract

Today, the trend within the electronics industry is for the use of rapid and advanced simulation methodologies in association with synthesis toolsets. This paper presents an approach developed to support mixed-signal circuit design and analysis. The methodology proposed shows a novel approach to the problem of developing behvioural model descriptions of mixed-signal circuit topologies, by construction of a set of subsystems, that supports the automated mapping of MATLAB®/SIMULINK® models to structural VHDL-AMS descriptions. The tool developed, named MS^2SV, reads a SIMULINK® model file and translates it to a structural VHDL-AMS code. It also creates the file structure required to simulate the translated model in the SystemVision™. To validate the methodology and the developed program, the DAC08, AD7524 and AD5450 data converters were studied and initially modelled in MATLAB®/SIMULINK®. The VHDL-AMS code generated automatically by MS^2SV, (MATLAB®/SIMULINK® to SystemVision™), was then simulated in the SystemVision™. The simulation results show that the proposed approach, which is based on VHDL-AMS descriptions of the original model library elements, allows for the behavioural level simulation of complex mixed-signal circuits.

1. Introduction

Top-Down design methodologies have generated a profound impact on electronic system design. They allow for fast and efficient design specification, modelling of design functionality in a suitable description language, verification of design functionality via simulation, and the synthesis of the design model into a target techonology.

Unlike digital designs, which can be implemented using a relatively small number of different primitive logical gates and memory devices, mixed-signal (analogue and digital electronic circuit) and mixed-technology (electrical/electronic and non-electronic components) designs need increasingly sophisticated building blocks to adequately describe and implement the required system operation [1]. Unfortunately, there are not many synthesis tools available within the mixed-signal design domain. The are, for example, tools that use mathematical modelling, such as MATLAB®/SIMULINK®, from Mathworks Inc, to generate the initial design specification via modelling, but there is then the requirement to retarget the model to a form which can be directly implemented in the target technology. Some previous papers that motivated the development of the present work is summarised below.

In the work of Edenfeld [2], the trend of the technology of semiconductors, with the advent of the system with mixed-signal (analogue & digital) parts is presented. He said that projects in the high abstraction level is possible to discover problems in the initial stage of development, which reduces the project time and reduces costs.

In MacMillen's work [3], a complete overview of the technologies, algorithms and methodologies that are used in the EDA (Electronic Design Automation) tools and the economic impact of these technologies is undertaken. Each step of the physical project, simulation, verification, synthesis and test, is discussed. Also discussed are the types of tool sets required to support the project development environment.

The necessity of behavioural modelling in mixed-technology systems (parts with parts electronic, electric and non-electric - mechanical, pneumatic, etc - operating together) is described in the Wilson's work [4]. He presented the interaction between the different domains and the behavioural models using a language such as VHDL-AMS.

In Christen's work [5], a review of the VHDL-AMS language used for analogue and mixed-signal applications is undertaken. The main elements of the language are studied. Pecheaus's work [6], also dealt with the way in which to develop models using the VHDL-AMS language.

In analogue and mixed-signal design, the generation of models is at a high abstraction level. Trofimov [7] wrote that one method to increase the speed of the project design process is to use the top-down design methodology. The use of mixed signal design process models in high abstraction levels permits a complete simulation of the mixed signal device. However, adequate tools to model and simulate the device must be used. It is important to have good interaction between the project tools.

In 2000, Grout [8] developed a prototype of a tool (software toolbox) that analyzed and processed a SIMULINK® block diagram model to produce a VHDL representation of the model. The derived VHDL model is a combination of behavioural, RTL (Register Transfer Logic) and structural definitions mapped directly from the SIMULINK® model. This approach was considered to enable a user to develop and simulate a digital control algorithm using MATLAB® and once developed in MATLAB® it can be converted to VHDL code.

In 2003, Zorzi [9] describe a tools, called I.M.A.Ge-AMS, for writing new spice models using the VHDL-AMS standard. The core of the I.M.A.Ge-AMS tool is the compiler which is derived from the VHDL-AMS compiler, developed for another CAD tool called SANSA.

In 2004, Grout [10] discussed the need to provide suitable provision for, and flexibility in the use of, modelling and simulation languages suitable for supporting the range of mixed-signal microelectronic circuit design, test and test development activities that are encountered in today's complex microelectronic products. In this work, special emphasis was made into developing language support for test development activities. The target area considered is Delta-Sigma (Δ-Σ) modulation for on-chip signal generation in the support of mixed-signal Built-In Self-Test (BIST).

This paper presents an approach to develop and support mixed-signal circuit design and analysis. The methodology proposed shows a novel approach to the problem of developing model descriptions of a mixed-signal circuit topologies, by the construction of a set of subsystems that support the automated mapping of MATLAB®/SIMULINK® models to structural VHDL-AMS descriptions. The tools developed, named MS²SV (MATLAB®/SIMULINK® to SystemVision™), reads a SIMULINK® model file and translates it to a structural VHDL-AMS code. It also creates the file structure required to simulate the translated model in the SystemVison™ environment from Mentor Graphics®. MS²SV was developed in C programming language and it requires a number of predefined library components for the translation process.

At the present time, the library of components has a set of elements such as basic logic gates (AND, OR, NOT, etc), sequential digital circuits such as bistables, shift registers and latches, in addition to control elements created using available subsystems at library. It also includes analogue elements such as summation, subtraction, product, gain and elements for data type conversion and differents signal sources as well.

The reaminder of the paper is presented as follows. Section 2 provides a description of the implementation of the DAC08 Data Converter, which was modeled in both MATLAB®/SIMULINK® and VHDL-AMS using the SystemVision™ environment. The development of the conversion toolbox will be introduced and discussed in section 3. The conversion toolbox will be used in a case study in section 4, which describes the modelling of the AD7524 Data Conveter. Section 5 discusses future work to improve the conversion toolbox and conclusions will be provided in section 6.

2. DAC08 Model in SIMULINK® and SystemVision™

The DAC08 is an 8-bit D/A (Digitial-to-Analogue) converter that uses the R/2R ladder structure in which the output current is a product of a digital number and the input reference current. Each digital input has a weight represented by the gain. The MSB (Most Significant Bit) has a weight of 0.5 and the LSB (Least Significant Bit) has a weight of 0.00390625. In this way, all gains blocks have their own particular weight. In the model used, these gains are added when their inputs has logic "1" and are multiplied by the reference voltage, then generating the output voltage.

With this information, it is possible to create the DAC08 converter models using the components available in SIMULINK® library and using the instances available in SystemVision™ library as well. Figure 1 shows the DAC08 model implemented in SIMULINK® and Figure 2 shows the same model implemented in SystemVision™.

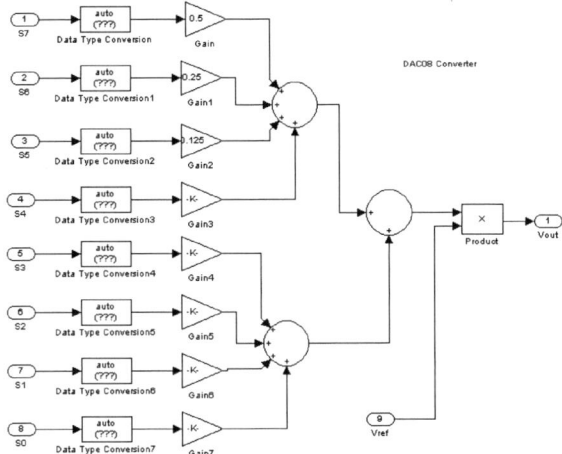

Figure 1. DAC08 model implemented in SIMULINK®.

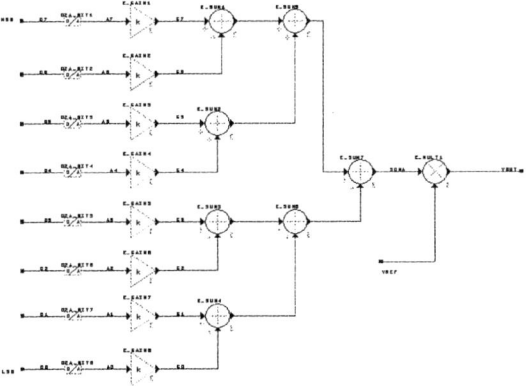

Figure 2. DAC08 model implemented in SystemVision™.

It can be seen that both models consist of a R-2R ladder network structure and a product block. The product block generates the output voltage of the binary value in the digital input. The reference voltage defines the maximum output voltage.

In the model presented in figure 2, it is necessary to use the symbol "d2a_bit" due to the mixed-signal nature of the model. This single model contains analog and digital parts. In SIMULINK®, the "Data Type Conversion" block completes the translation between the

485

analogue and digital parts of the model. The only difference between the two models, presented in Figure 1 and Figure 2, is the number of inputs of the sum blocks. In the SystemVision™ environment, the maximum number of inputs in the in-built SUM blocks is two, where as in SIMULINK® there is no limit to the number of inputs in the SUM blocks.

The VHDL-AMS code of the instances used in the DAC08 Data Converter, generated by SystemVision™ is shown below:

- **MULT**

```
Library IEEE;
Use IEEE.electrical_systems.all;
Entity e_Mult is
   Generic (K: real: = 1.0);   -- Gain
   Port (
      Terminal in1 , in2 : electrical;
      Terminal output   : electrical);
End entity e_Mult;
Architecture behavioral of e_Mult is
   Quantity vin1 across in1 to electrical_ref;
   Quantity vin2 across in2 to electrical_ref;
   Quantity vout across iout through output to electrical_ref;
Begin
   Vout == k * vin1 * vin2;
End architecture behavioral;
```

- **GAIN**

```
Library IEEE;
Use IEEE.MATH_REAL.all;
Use IEEE.ELECTRICAL_SYSTEMS.all;
Entity e_Gain is
   Generic (K: real: = 1.0); -- Gain multiplier
   Port   (Terminal input: electrical;
           Terminal output: electrical);
End entity e_Gain;
Architecture behavioral of e_Gain is
   Quantity vin across input to electrical_ref;
   Quantity vout across iout through output to electrical_ref;
Begin
   Vout == K * vin;
End architecture behavioral;
```

- **SUM**

```
Library IEEE;
Use IEEE.electrical_systems.all;
Entity e_Sum is
   Generic (
      K1: Real: = 1.0;
      K2: Real: = 1.0);
   Port (
      Terminal in1 , in2: electrical;
      Terminal output   : electrical);
End entity e_Sum;
Architecture behavioral of e_Sum is
   Quantity vin1 across in1 to electrical_ref;
   Quantity vin2 across in2 to electrical_ref;
   Quantity vout across iout through output to electrical_ref;
Begin
   Vout == K1*vin1 + K2*vin2;
End architecture behavioral;
```

The motivation for using SIMULINK® as a high-level design model comes from the fact that SIMULINK® is a standard language used for example in area of control. It is also a popular tool used in education and research as well. The choice of VHDL-AMS as the target language is motivated by standard of hardware description language, available in the majority of synthesis environment. It is important to remember that SIMULINK® is purely a simulation tool, therefore, an automatic translation to VHDL-AMS is highly desirable.

3. The MS²SV Toolbox Conversion

The MS²SV toolbox conversion was developed in C language. It reads a file type .mdl (MATLAB® model) describing a model to be implemented and generates all the necessary structure for this model to be simulated on the SystemVision™ environment. The .mdl model is then described in VHDL-AMS.

An overview of methodology and how to use the developed tools is shown in Figure 3.

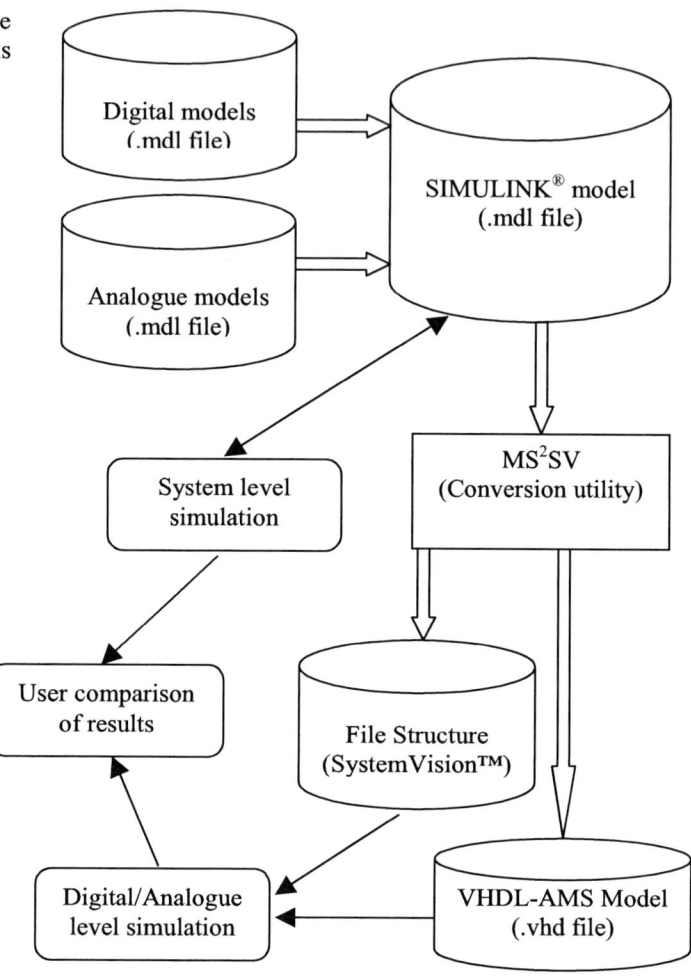

Figure 3. MS²SV and the used methodology.

The steps involved are as below:

(i) Initial specification of the model and simulation using SIMULINK®.

(ii) Conversion of the SIMULINK® model to a correspondent VHDL-AMS model.

(iii) simulation and analysis of the converted model using SystemVision™ environment.

(iv) Comparison of simulation results from both models forms.

(v) Continue the synthesis process (of the VHDL-AMS code parts) if result comparation is acceptable, or return to a new specification.

In the initial phase of specification, the user can only use the components available in the library named LIB_MS2SV. This library has a set of combinational and sequential primitive such as OR, AND, NOT, latches, bistables (flip-flops), counters, shift registers, etc. In addition, it contains analogue primitives such as gain, product and sum. Also available are a number of subsystem created for specific use, along with constant types and pulse sources. Figure 4 shows some of components availabe in the library. Note that the names of components are different to the conventional names, because in the conversion process the conventional names are used as reserved word in the target models.

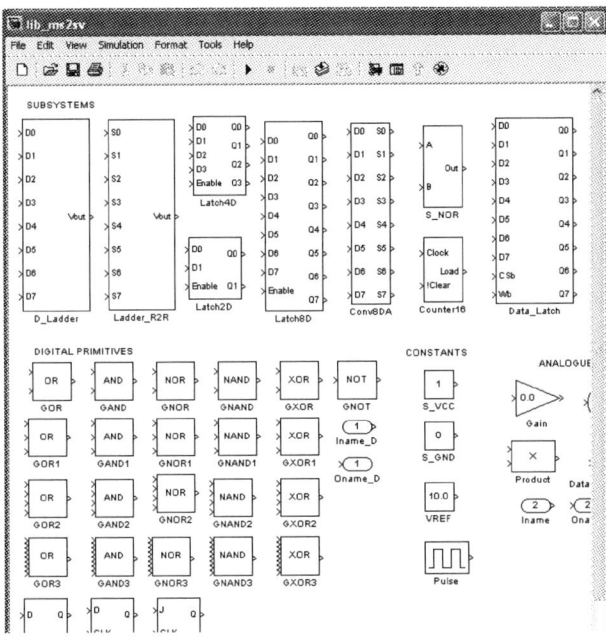

Figure 4. LIB_MS2SV library of components.

One important aspect of the design methodology is related with the names used to specify the input/output ports of the system. The name of all digital signals must finish with "_D" and can not have names which are reserved word in VHDL-AMS. They should also not have names that finish with a number, for example "NOT-1".

When the MS^{2}SV program is executed, the program identifies all the important information inside the .mdl file (SIMULINK® model file) and recognises this information in the library of components LIB_MS2SV. The translation of the structure of the original model into the corresponding VHDL-AMS code structure is then undertaken.

Once the SIMULINK® model has been translated into VHDL-AMS code, a project set is created in SystemVision™ project environment which allows for adequate simulation and analysis of the translated model.

The files structure created to uses the translated model in SystemVision™ environment is shown in Figure 5.

4. Case Study

To validate the methodology and the developed program, the DAC08, AD7524 and AD5450 data converters were studied and initially modelled in MATLAB®/SIMULINK®. The VHDL-AMS code generated automatically by MS^{2}SV was then simulated in the SystemVision™ environment. The top level of the design consists of a structural VHDL-AMS equivalent of the original model block connections.

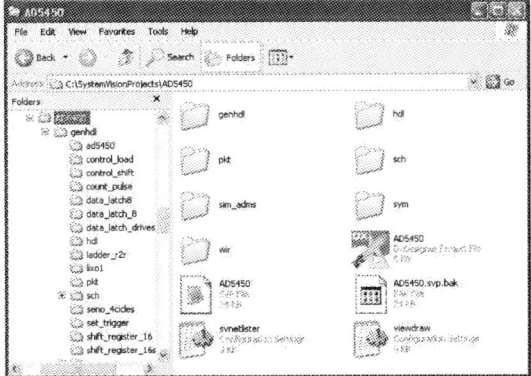

Figure 5. File structure to SystemVision environment.

In this paper, we present the design of AD7524 Data Converter. The AD7524 is a low cost, 8-bit monolithic CMOS DAC designed for direct interface to most microprocessors. Basically, an 8-bit DAC with input latches, the AD7524 load cycle is similar to the "write" cycle of the random access memory.

In Figure 6, the simplified functional block diagram of a AD7524 converter is shown.

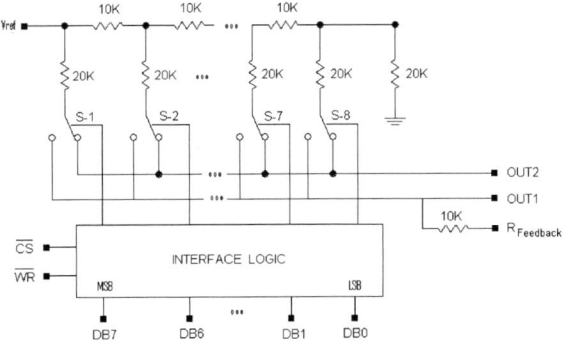

Figure 6. Functional block diagram of a AD7524.

It consists of an inverted R-2R ladder structure and eight channel current switches. The currents that represents the binary weighted values are switched between the OUT1 and OUT2. When all the digital inputs are LOW, the reference current is switched to OUT2.

In the AD7524, the mode selection is controlled by CS_B (chip select) and WR_B (write/read) inputs. In the Write Mode, when CS_B and WR_B are both LOW, the analogue output responds to data activity at the DB_0-DB_7

487

data bus input. When either CS_B or WR_B is HIGH, the AD7524 is in the Hold Mode. In this mode, the output holds the value correpondent to the last digital input present at DB_0-DB_7 prior to CS_B or WR_B assuming the HIGH state.

With the information above, it is sufficient to describe this model in SIMULINK®. Figure 7 shows the SIMULINK® model used to specify the AD7524 data converter. It consists of components only availabe in the Library show in Figure 4. The model consists in the following components:

(i) Pulses to generated the required input waveform.
(ii) Data_Latch to hold the input signal when desired. In this model, the control signals (CS_B and WR_B) are pemanently held at a LOW level.
(iii) Conv8DA, to convert the digital data type to analogue data type.
(iv) Vref and S_GND to define a constant value within the system.
(v) Ladder_R2R network to define the binary weight of the inputs.
(vi) Product to define the output value based on the product of a reference voltage and DAC input code.

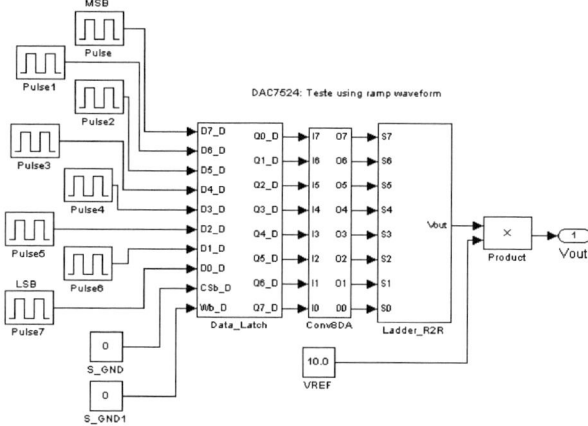

Figure 7. AD7524 Data model on SIMULINK®.

Some of the VHDL-AMS code generated by MS²SV tools is shown below. The signals defined in the architecture part and the instances that is replicated several times are only shown once.

```
-- genhdl/d7425ramp
-- Generated by MS2SV tool version 1.0

LIBRARY ieee;
USE ieee.std_logic_1164.all;
USE ieee.electrical_systems.all;
USE ieee.mechanical_systems.all;
USE ieee.fluidic_systems.all;
USE ieee.thermal_systems.all;
USE ieee.radiant_systems.all;
LIBRARY edulib;
USE work.all;
library fundamentals_vda;
library spice2vhd;
```

```
entity d7425ramp is
Port(
      terminal Vout : electrical);
end entity d7425ramp;

architecture arch_d7425ramp of d7425ramp is
  terminal VREF : electrical;
  signal S_GND : std_logic;
  signal Pulse7 : std_logic;
  terminal Product : electrical;
  terminal Ladder_R2R : electrical;
  signal Data_Latch0 : std_logic;
  terminal Conv8DA0 : electrical;
begin

  V_VREF : entity EDULIB.V_CONSTANT(IDEAL)
    generic map ( LEVEL => 10.0 )
    port map ( POS => VREF,
               NEG => ELECTRICAL_REF );

-- This block is replicate two times
  L_S_GND : entity EDULIB.LEVELSET
    port map ( LEVEL => S_GNDb );

-- This block is replicate eight times
  E_Pulse7 : entity EDULIB.CLOCK_FREQ(IDEAL)
    generic map ( FREQ => 0.0078125 )
    port map ( CLK_OUT => Pulse7 );

  W_Data_Latch                    :        entity
WORK.DATA_LATCH(ARCH_DATA_LATCH)
    port map ( Q7 => Data_Latch0,
               Q6 => Data_Latch1,
               Q5 => Data_Latch2,
               Q4 => Data_Latch3,
               Q3 => Data_Latch4,
               Q2 => Data_Latch5,
               Q1 => Data_Latch6,
               Q0 => Data_Latch7,
               W => S_GND1,
               Cs => S_GND,
               D7 => Pulse7,
               D6 => Pulse6,
               D5 => Pulse5,
               D4 => Pulse4,
               D3 => Pulse3,
               D2 => Pulse2,
               D1 => Pulse1,
               D0 => Pulse);

  W_Conv8DA : entity WORK.CONV8(ARCH_CONV8)
    port map ( OUT0 => Conv8DA0,
               OUT1 => Conv8DA1,
               OUT2 => Conv8DA2,
               OUT3 => Conv8DA3,
               OUT4 => Conv8DA4,
               OUT5 => Conv8DA5,
               OUT6 => Conv8DA6,
               OUT7 => Conv8DA7,
               IN1_D => Data_Latch0,
               IN2_D => Data_Latch1,
               IN3_D => Data_Latch2,
               IN4_D => Data_Latch3,
               IN5_D => Data_Latch4,
               IN6_D => Data_Latch5,
               IN7_D => Data_Latch6,
               IN8_D => Data_Latch7);

  W_Ladder_R2R                    :        entity
WORK.LADDER_R2R(ARCH_LADDER_R2R)
    port map ( Vout => Ladder_R2R,
               S0 => Conv8DA0,
               S1 => Conv8DA1,
               S2 => Conv8DA2,
               S3 => Conv8DA3,
               S4 => Conv8DA4,
               S5 => Conv8DA5,
               S6 => Conv8DA6,
               S7 => Conv8DA7);

  E_Product : entity EDULIB.E_MULT(BEHAVIORAL)
    port map ( IN1 => VREF,
               IN2 => Ladder_R2R,
               OUTPUT => Vout );
end architecture arch_d7425ramp;
```

The corresponding structural VHDL-AMS code that was generated by the MS²SV converter tool can be seen below. Each block presented in Figure 7 has one corresponding description in VHDL-AMS. It is used to simulate an input ramp waveform. The VHDL-AMS description of all instances used in the model are also generated in the adequate position of the directory to be used by the model. All instances that are not a primitive component within SystemVision™ are automatically generated by MS²SV tool. For exemple, the VHDL-AMS code to the instance Ladder_R2R is shows below.

```
-- SubSystem < ladder_r2r >
-- To be used on MS2SV tool version 1.0

LIBRARY ieee;
USE ieee.std_logic_1164.all;
USE ieee.electrical_systems.all;
USE ieee.mechanical_systems.all;
USE ieee.fluidic_systems.all;
USE ieee.thermal_systems.all;
USE ieee.radiant_systems.all;
LIBRARY edulib;
USE work.all;
library fundamentals_vda;
library spice2vhd;

entity ladder_r2r is
Port(
        terminal Vout : electrical;
        terminal S0   : electrical;
        terminal S1   : electrical;
        terminal S2   : electrical;
        terminal S3   : electrical;
        terminal S4   : electrical;
        terminal S5   : electrical;
        terminal S6   : electrical;
        terminal S7   : electrical);
end entity ladder_r2r;

architecture arch_ladder_r2r of ladder_r2r is
    terminal Sum1 : electrical;
    terminal Sum  : electrical;
    terminal Gain7 : electrical;
    terminal Gain6 : electrical;
    terminal Gain5 : electrical;
    terminal Gain4 : electrical;
    terminal Gain3 : electrical;
    terminal Gain2 : electrical;
    terminal Gain1 : electrical;
    terminal Gain  : electrical;

begin
    E_Gain7 : entity EDULIB.E_GAIN(BEHAVIORAL)
      generic map ( K => 0.003906 )
      port map ( INPUT => S0,
                       OUTPUT => Gain );
    E_Gain6 : entity EDULIB.E_GAIN(BEHAVIORAL)
      generic map ( K => 0.007813 )
      port map ( INPUT => S1,
                       OUTPUT => Gain1 );
    E_Gain5 : entity EDULIB.E_GAIN(BEHAVIORAL)
      generic map ( K => 0.015625 )
      port map ( INPUT => S2,
                       OUTPUT => Gain2 );
    E_Gain4 : entity EDULIB.E_GAIN(BEHAVIORAL)
      generic map ( K => 0.03125 )
      port map ( INPUT => S3,
                       OUTPUT => Gain3 );
    E_Gain3 : entity EDULIB.E_GAIN(BEHAVIORAL)
      generic map ( K => 0.0625 )
      port map ( INPUT => S4,
                       OUTPUT => Gain4 );
    E_Gain2 : entity EDULIB.E_GAIN(BEHAVIORAL)
      generic map ( K => 0.125 )
      port map ( INPUT => S5,
                       OUTPUT => Gain5 );
    E_Gain1 : entity EDULIB.E_GAIN(BEHAVIORAL)
      generic map ( K => 0.25 )
      port map ( INPUT => S6,
```

```
                       OUTPUT => Gain6 );
    E_Gain : entity EDULIB.E_GAIN(BEHAVIORAL)
      generic map ( K => 0.5 )
      port map ( INPUT => S7,
                       OUTPUT => Gain7 );
    E_Sum1 : entity WORK.L_SUM4(ARCH_L_SUM4)
      port map ( IN1 => Gain7,
                       IN2 => Gain6,
                       IN3 => Gain5,
                       IN4 => Gain4,
                       OUTPUT => Sum1 );
    E_Sum : entity WORK.L_SUM4(ARCH_L_SUM4)
      port map ( IN1 => Gain3,
                       IN2 => Gain2,
                       IN3 => Gain1,
                       IN4 => Gain,
                       OUTPUT => Sum );
    E_Sum2 : entity EDULIB.E_SUM(BEHAVIORAL)
      port map ( IN1 => Sum1,
                       IN2 => Sum,
                       OUTPUT => Vout );

end architecture arch_ladder_r2r;
```

After the conversion process, the model initially described in SIMULINK®, is prompted to be simulated in the SystemVision™ environment. Figure 8 shows the simulation result when the input is a sine waveform. In this case it was necessary to use two additional instances. One instance is used to read the digitised sine waveform from a specified text file into the model (input stimulus), and another is used to write the result of simulation to a file for further analysis (e.g. using the FFT algorithm within MATLAB®).

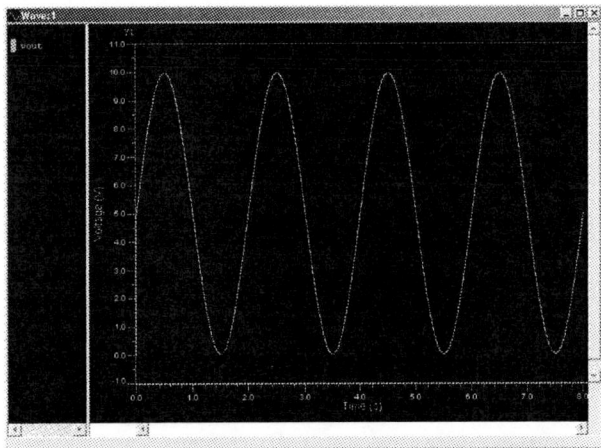

Figure 8. Waveform resulting of D/A conversion.

To evaluate the developed program, the additional data converters ADC08 and the DA5450 (serial input) were also evaluated. The ramp and step waveforms were also used to test the models developed.

5. Future work to improve the MS²SV

The work described here is the first development stage of the translation tool named MS²SV. We presented the approach to develop VHDL-AMS models modelled from the SIMULINK® block diagram. To improve and develop the tool it is necessary to :

(i)   Develop a more consistent interpreter to read the ".mdl" file and to extract the relevant information from the SIMULINK® model in order to increase the flexibility of the tool

(ii) Create additional instances within the library to be used for different data converters models.

(iii) Create the ability for users to create their own custom models for inclusion within the library.

(iv) Create the ability to generate SPICE models in addition to the VHDL-AMS models.

6. Conclusions

This paper has presented a conversion tool developed to support mixed-signal circuit design. The methodology proposed show a novel approach to the problem of developing model descriptions of a mixed-signal circuit topologies, by the construction of a set of subsystems that support the automated mapping to MATLAB®/ SIMULINK® models to structural VHDL-AMS descriptions.

The simulation results show that the proposed approach, which is based on VHDL-AMS descriptions of the original model library elements in SIMULINK®, allows for the behavioural level simulation of complex mixed-signal circuits.

This work considers some of the challenges set by the electronic industry for the further development of simulation methodologies and tools in the field of mixed-signal technology. Although this approach was considered to target data converter modelling and simulation, it has potential to be extended to consider other system designs such as control and mechatronic systems.

Acknowledgments

This work was carried-out with support from CAPES (Proc. 3359-05-0). The authors would like to express their thanks to Sao Paulo State University (UNESP) - Brazil and University of Limerick (UL) – Republic of Ireland.

References

1. Ashenden P.J. *et al*, <u>The System Designer's Guide to VHDL-AMS: Analog, Mixed-signal, and Mixed-technology Modeling,</u> Elsevier Science (San Francisco, 2003).

2. Edenfeld D. *et al*, "2003 Technology Roadmap for Semiconductores," *IEEE Computer Society*, January (2004), pp. 47-56.

3. MacMillen D. *et al*, "An industrial View of Electronic Design Automation," *IEEE Transactions on Computer Aided Design of Integrated Circuits and Systems*, Vol. 19, No.12, December (2000), pp. 1428-1448.

4. Wilson P.R. *et al*, "Multiple Domain Behavioral Modeling Using VHDL-AMS," *Proc of the 2004 International Symposium on Circuits and Systems*, Vol. 5, 23-26, May (2004), pp. V644-647.

5. Christen E., and Bakalar K., "VHDL-AMS – A Hardware Description Language for Analog and Mixed-Signal Applications," *IEEE Trans. on Cicruits and Systems – II: Analog and Digital Signal Processing*, Vol. 46, No. 10, October (1999), pp. 1265-1272.

6. Pecheus F. *et al*, "VHDL-AMS and Verilog-AMS as Alternative Hardware Descriptions Languages for Efficient Modeling of Multidiscipline Systems," *IEEE Trans. on Computer-Aided Design of Integrated Circuits and Systems*, Vol. 24, No. 2, February (2005), pp. 204-225.

7. Trofimov M., Mosin S., "The Realization of Algorithmic Description on VHDL-AMS," *TCSET'2004*, Lviv-Slavsko, Ukraine, February. 2004, pp. 350-352.

8. Grout I., Keane K., "A Matlab to VHDL Conversion Toolbox for Digital Control," *IFAC Symposium on Computer Aided Control System Design (CACSD'2000)*, Salford, UK, September. 2000.

9. Zorzi M. *et al*, "A Tool for Integration of New VHDL-AMS Models in Spice," *Proc of the 2004 International Symposium on Circuits and Systems*, Vol. 4, May. 2004, pp. IV637-640.

10. Grout I, O'Shea T., "MATLAB/VHDL-AMS Modelling and Simulation Support for Microelectronic Circuit Design and Test," *Proc of the 10th International Mixed-Signal Testing Workshop*, 2004, pp. 178-183.

Prediction of the Influence of Induced Stresses in Silicon on CMOS Performance in a Cu-Through-Via Interconnect Technology

Chukwudi Okoro[1,2], Mario Gonzalez[1], Bart Vandevelde[1], Bart Swinnen[1], Geert Eneman[1,2], Peter Verheyen[1],
Eric Beyne[1], Dirk Vandepitte[2]

[1]IMEC

75 Kapeldreef, B-3001 Leuven, Belgium
Email: chukwudi.okoro@imec.be
Tel.: +32(0)16287740

[2]Katholieke Universiteit Leuven, Belgium

Abstract

One approach to 3D chip stacking and integration is to process filled Cu-vias into the Si and to attach them to a next level die by means of thermocompression bonding. This results in induced stresses in the silicon due to the large CTE disparity between copper and silicon, and also from the force applied during thermocompression bonding. These stresses can have an impact on the performance of the transistors and may as well result in die fracture. This paper studies these stresses through Finite Element modeling. We found that the keep-away-zone of the transistors from the copper via where transistor performance is impacted by the through-Si interconnect proximity, is proportional to the via diameter. The bonding temperature is found to be the main cause for the induced stresses during the thermo-compression bonding process. The induced stresses in silicon decrease with decreasing the silicon thickness.

1. Introduction

Three-dimensional stacking of chips has in recent times become a preferred means of achieving higher portability and functionality in electronic devices. This is made possible since 3D-stacked systems offer smaller footprint and volume, and enable shorter signal routing, and reduced wiring density, thereby decreasing delays and capacitance, while increasing speed and/or cutting power requirements [13, 14]. A whole range of interconnection technologies currently exist for the packaging of 3D assemblies [1, 2, 4]. Copper-through-via interconnection technique offers the shortest wiring length between dies, thereby leading to higher chip performance [3, 6]. While many methods of achieving cu-through-via interconnection has been reported, IMEC is developing a 3D-Stacked Integrated Circuit (3D-SIC) in which the connection between different chip stack is achieved by copper to copper bonding. This interconnection method has the advantage of achieving reliable bonds without the formation of intermetallics, which often acts as a site for failure propagation. A schematic of the 3D-SIC architecture is shown in figure 1 and an in depth description of the 3D-SIC process was given in [14].

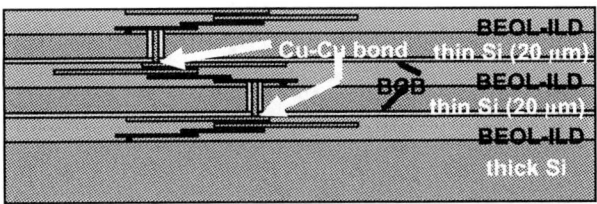

Figure 1: Schematic drawing of the 3D-SIC structure, showing the copper-to-copper connection of different dies

During the thermo-compression bonding process, force and temperature are applied simultaneously, leading to worries about the interconnections realized by this technique. This is because the large mismatch in coefficient of thermal expansion (CTE) between copper (16.7ppm/°C) and silicon (2.3ppm/°C) will induce stresses in the bulk silicon. This might lead to fracture in the silicon or delamination occurrence at the interface of the two materials. These may cause reliability problems such as electrical breakdown [3, 9]. Failure could also result from the applied bonding force.

Moreover, mechanical loads are known to affect transistor performance through piezo-resistive effects: the applied load impacts on the carrier mobility which translates to a change in the performance of the transistors, this effect is known to be dependent on temperature and the doping concentration [10, 11,12]. It then means that copper-through-via interconnections may impact on the performance of transistors that are close to it since the CTE mismatch between copper and silicon induces enormous stresses in the bulk silicon. How far the active region (the transistors) is to be located away from the copper vias, then becomes an important design consideration.

In this work, Finite Element Method (FEM) is used to determine the induced stresses in the silicon resultant after thermo-compression bonding. We describe a parametric study of the influence of geometry, bonding temperature and bonding force on the induced stresses in bulk silicon. Using piezo-resistive coefficients from literature [10, 11, 12] the residual stresses witnessed after the thermo-compression process are translated into stress induced carrier mobility changes for a 3D-SIC architecture.

1-4244-1105-X/07/$25.00 ©2007 IEEE   491

2. Finite Element Model: Description

For this work a 2D axisymmetric modeling technique is used. A unit of the structure is modeled consisting of the copper via, the copper bonding pad, bulk silicon, silicon oxide dielectric layer and a back-end-of-line (BEOL) copper layer. The modeled structure (with the applied boundary conditions and mesh density) is shown in the figure below.

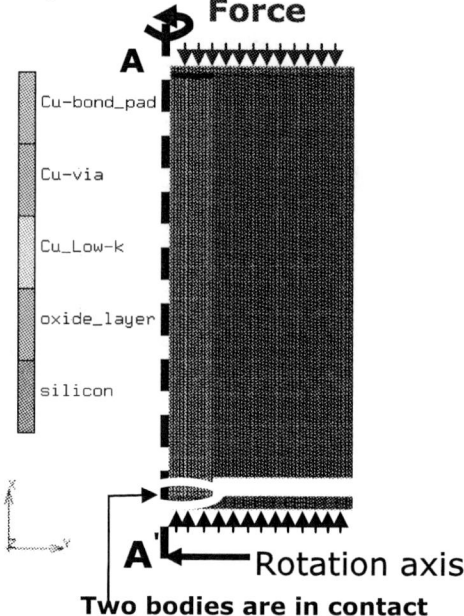

Figure 2: 2D axisymmetric model description

In reality, the actual structure requires a three-dimensional analysis, but the idealized 3D geometry is included in the model by the axisymmetry assumption [9]. The 3D equivalent is shown in figure 3.

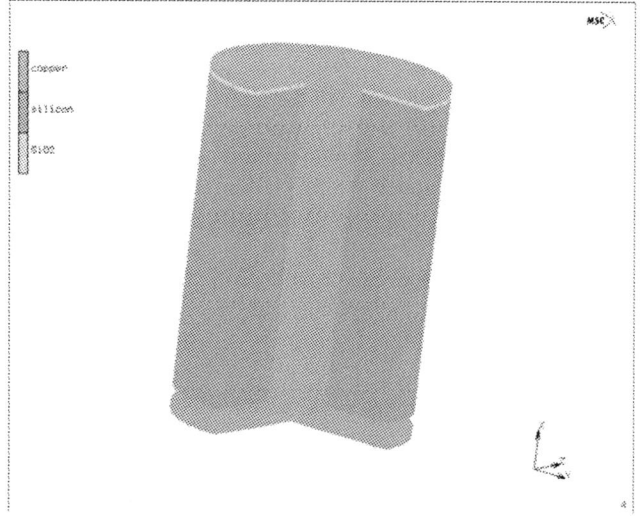

Figure 3: 3D representation of the axisymmetric model

A contact body feature of Msc. Marc software is used to indicate that the copper via is in touch with the copper bonding pad and also used for the application of bonding force to the structure.

The results were obtained at the nodes where the maximum principal stress occurred in the bulk silicon. We also determined the stress components at this node.

Since the active region (being the area where the transistors are integrated) is located within 1µm of the top thickness of silicon, the impact of stress on carrier mobility change is obtained by taking a path plot from the top of the silicon all through across its width at a thickness of 250nm. The obtained residual stress components are correlated with piezo-resistivity tensor calculations to determine the impact of stress on carrier mobility.

It is also worth noting that the piezo-resistivity coefficients tensor calculations are for a doping concentration of $5.10^{17} cm^{-3}$ at room temperature [10]. For this study 3µm to 10µm copper via diameter are considered.

MSc Mentat is the pre- and post processing software program that is used in conjunction with MSC Marc as the analysis software for this simulation

Considered Parameters

For this work, the parameters that are considered are the geometry, the applied bonding temperature and the bonding force.

The via height is varied between aspect ratios of 2 and 10. For a copper via height of 20µm, the via diameter considered is from 3µm to 10µm. The bonding temperature is studied for a temperature range between 250°C and 400°C and the bonding forces, for a range between 150N and 600N uniformly distributed over the die. The nominal values for these parameters are shown in table 1:

Bonding Temperature (°C)	Bonding Force (N)	Via Diameter (µm)	Silicon Thickness (µm)
350	300	5	20

Table 1: Nominal parameter values

For all the reported results (except for the geometry study) the nominal parameter values are used.

The schematic of the considered parameters is shown in figure 4.

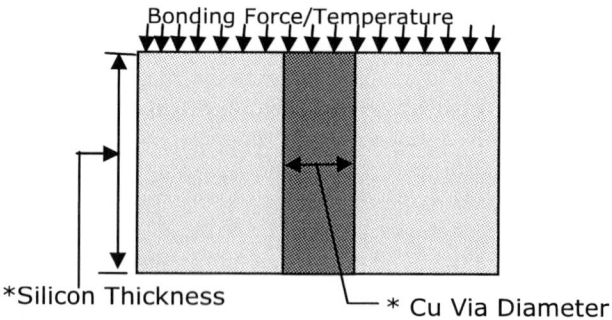

Figure 4: Schematic drawing of the copper through via and the studied parameters

492

Material Properties

Silicon and SiO₂ materials are modeled as elastic materials, while copper is considered as an elastic-plastic material, having yield strength which is temperature dependent. The material properties used are shown in table 2, while figure 5 represents the implemented copper yield strength dependency on temperature.

Material	Young's Modulus (GPa)	Poisson Ratio	CTE (ppm/°C)
Silicon	169	0.26	2.3
SiO₂	75	0.17	0.5
Copper	117	0.3	16.7

Table 2: *Material Properties implemented in finite element modeling*

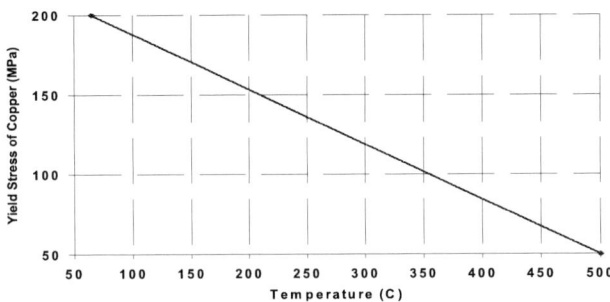

Figure 5: *Copper yield strength vs. temperature graph*

Loading Conditions

As a loading condition, a thermo-compression profile consisting of the bonding force and bonding temperature profiles is used. A pre-bonding heat treatment process accounting for the back-end-of-the-line (BEOL) copper sintering step, done at 420°C is also included in the model. In the model a stress-free state is assumed prior to the sintering process. As of all BEOL process, the sintering step has the highest temperature; we assume that the stress state modeled after sintering (prior to thermocompression bonding) is representative for the stress state in a real case. The force and temperature profiles used for modeling are shown in figure 6.

The stress results are analyzed for increments occurring at the end of the sintering step (referred to as "*before bonding*"), during the bonding process and after the bonding process (called "*after bonding*").
The component stresses are also examined in order to better interpret the results.

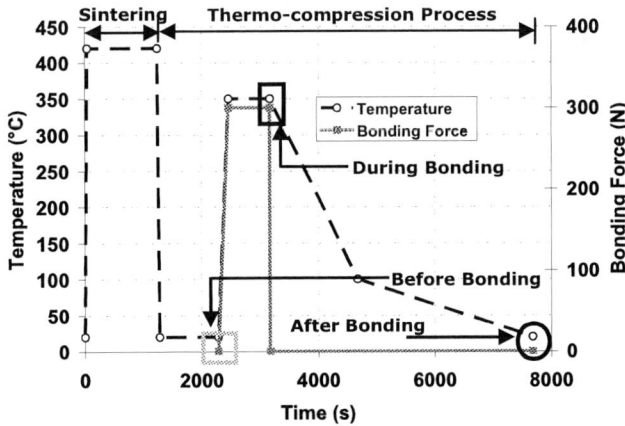

Figure 6: *Sintering and thermo-compression profile used as loading condition in FEM modeling. The regions at which results were analyzed are also indicated.*

Since only a unit of the structure is modeled, it is important to calculate the equivalent amount of force that will be included in the model from that applied to the entire bonded chip.

For flip chip bonding of the 3D-SIC structure as in [14], we use a chip area of 112.4mm² of which 5% of the entire area is assumed to be occupied by copper vias.
Based on this knowledge the force implemented in FEM is calculated for the different force and copper via diameters using the equation below;

$$F_{per}via = \frac{F_{chip}(0.25\pi D_{via}^2)}{(0.05A_{chip})}$$

Where;
F = Force (N)
F_{chip} = Total force (N) applied on the chip
D_{via} = Diameter (μm) of the copper via

3. Results
Influence of Bonding Temperature and Bonding Force

It is observed for the nominal parameter settings (shown in table 2) that *during bonding*, an increase in the bonding temperature or the bonding force led to an increase in the induced stresses in the silicon. Second conclusion is that the bonding temperature parameter showed much more significant influence on the induced stresses than the bonding force. This is evident in figure 7.

A linear relationship is observed as the bonding temperature increases from 250°C to 400°C leading to an increase in stress by about 270MPa at a bonding force of 600N. The influence of the bonding force is found to be a maximum of 90MPa at a bonding temperature of 400°C.

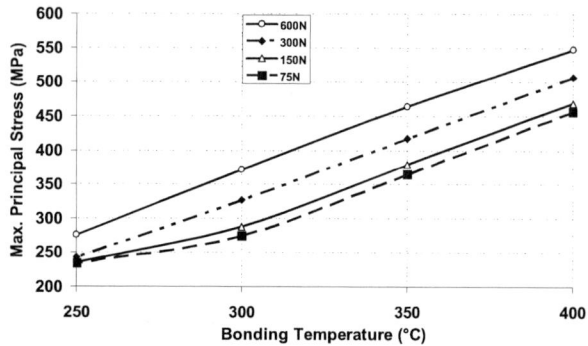

Figure 7: Effect of bonding temperature and force on the induced stresses in silicon

This can be explained by the fact that the large CTE difference between copper and silicon is the major cause of the induced stresses. Because the free expansion of copper is inhibited by the bulk silicon resulting in high induced stresses in the silicon.

The resultant component stresses were studied and it is found that *during bonding* that the maximum principal stress corresponded to the stress in the circumferential direction (σ_{33}). This can be attributed to the expansion of the copper nail as the temperature increases. The maximum compressive stresses occurred in the radial stress direction (σ_{22}), since during bonding the copper via expands outwardly inducing compressive stresses in the bulk silicon. The stress values of the component stresses are presented in table 3.

Bonding Force (N)	Max. Prin. Stress (MPa)	Out-Of-Plane Stress (σ_{11}) (MPa)	Radial Stress (σ_{22}) (MPa)	Circum. Stress (σ_{33}) (MPa)	Shear Stress (σ_{12}) (MPa)
600	464	243	-317	464	-46
300	418	257	-262	418	-49
150	379	246	-219	379	-49
75	365	244	-199	365	-50

Table 3: Values of the different stress components during thermo-compression at 350°C.

After bonding, the resultant residual stresses are found to be independent of the applied force and bonding temperature (as shown in figure 8). This is expected since the stress inducing factors --the bonding temperature and bonding force-- have been taken off. The observed induced residual stresses are found to be lower than the stresses witnessed during bonding. This is because, after the release of the bonding tool, the structure cools down leading to the contraction of the copper via and the relief of induced stresses in the silicon.

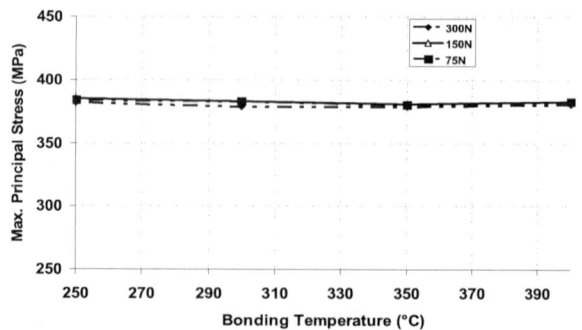

Figure 8: Residual stresses after thermo-compression bonding process

A study of the stress components after bonding reveals that, as a result of the contraction of the copper via, tensile stresses are induced in the silicon in the radial (σ_{22}) direction. And though the circumferential stress (σ_{33}) remains tensile, its stress magnitude is comparatively low as shown in table 4. These effects on the component stresses are anticipated since the expansion and/or the contraction of the copper via is the governing cause of the induced stress in the silicon and is acting in opposite directions.

Bonding Force (N)	Max. Prin. Stress (MPa)	Stress11(σ_{11}) (MPa)	Stress22 (σ_{22}) (MPa)	Stress33 (σ_{33}) (MPa)	Stress12 (σ_{12}) (MPa)
300	379	157	329	12	-105
150	381	142	330	10	-111
75	380	139	329	10	-112

Table 4: Stress components witnessed after the bonding process

A further study shows that the stress distribution and magnitude of the *before bonding* and *after bonding* images are similar (figure 9a and 9c). This therefore implies that the residual stresses witnessed after the bonding process (called *after bonding*) is caused by the pre-bonding heat treatment (during BEOL processing) rather than the thermo-compression bonding process itself.

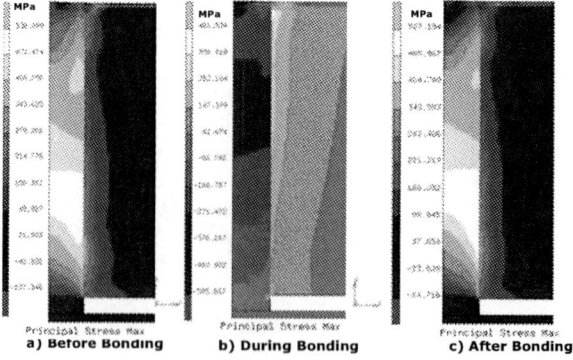

Figure 9: Effect of bonding temperature and force on the induced stresses in silicon.

Effect of Geometry

During bonding, as the silicon thickness is increased from 10μm - 50μm at constant via diameter (resulting to a via aspect ratio from 2 to10). It is found that the induced stress in silicon also increases (see figure 10) as is studied at a bonding temperature and force of 350°C and 300N respectively. This is because increase in the aspect ratio leads to an increased impact of the expanding copper via on silicon walls, causing higher induced stresses in silicon.

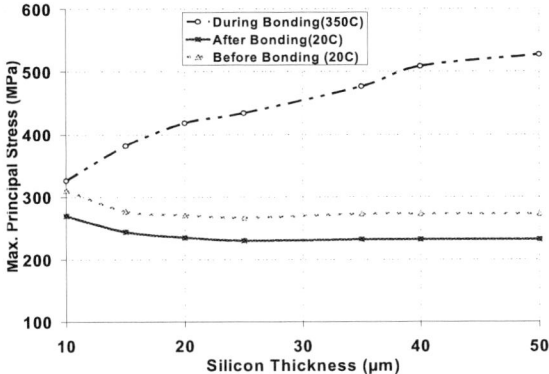

Figure 10: Variation of silicon induced stresses with change in silicon thickness

During bonding, larger via diameters witnessed a decreased induced stresses. This could be explained by the fact that an increase in copper via increases the rigidity of the copper via, and the material has more ability to expand in the out-of-plane direction, thereby resulting to a reduced expansion on the circumferential and radial direction, leading to a reduction of the induced stresses in silicon. This could be seen in figure 11

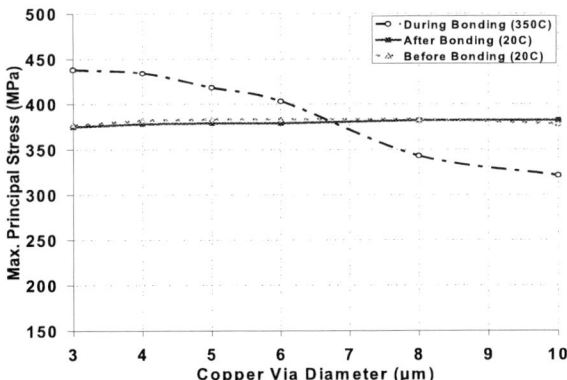

Figure 11: Variation of silicon induced stresses with change in via diameter

For both geometrical studies - via height and diameter -, the maximum residual stresses resultant *before* and *after bond*ing are all found to be independent of the via height and diameter.

An analysis of the stress distribution in the silicon *after bonding*, shows that, though the maximum residual stresses in the silicon is geometry independent, the stress distribution in the silicon increased with increase in the via diameter as shown in figure 12

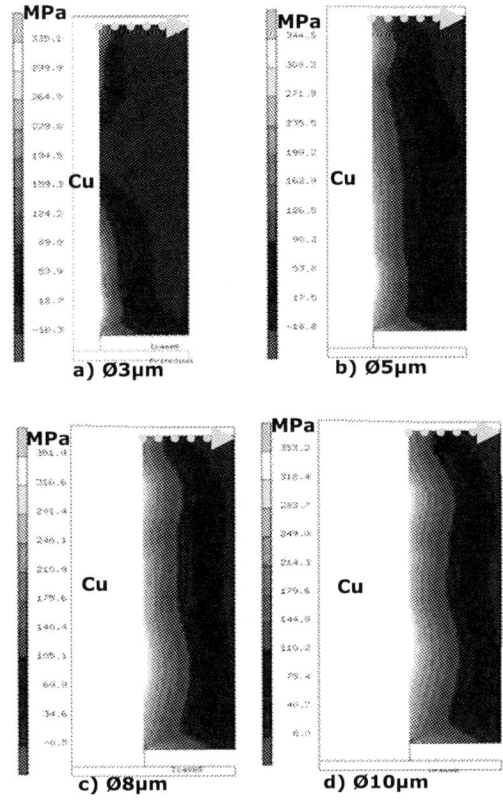

Figure 12: Stress distribution highlighting only the stresses in silicon for different via diameters after bonding at 350°C.

This could be accounted for, by the increase in the via rigidity.

Influence of Via Diameter on CMOS Carrier Performance Change

A path plot taken 250nm from the top of the silicon and across the silicon width (as shown by the arrows in figure 12) is related to the calculated mobility performance change obtained from piezo-resistive coefficient tensor calculations at a doping concentration of 5.10^{17}cm^{-3} (shown in table 5).

	nMOS	pMOS
Parallel Stress	2.6%	-5.4%
Vertical Stress	-3.9%	-0.1%
Perpend. Stress	1.2%	5.9%

Table 5: Relative mobility change on the introduction of 100MPa tensile stress for different orientations of the stress components relative to the transistor channel.

The obtained result is used to determine the keep-away-zone of the transistors from the copper via, while assuming a carrier mobility change tolerance of 5%. This is done separately for the pMOS and nMOS and for transistor current direction positions parallel and perpendicular to the (110) crystal direction.

The keep-away-zone of the active region/transistor is found to be proportional to the diameter of the copper vias, with larger vias resulting to larger via keep-away-zones. The pMOS is revealed to be more sensitive to the copper via induced stresses than the nMOS, and therefore sets the keep-away-zone for a CMOS process. As is shown in figures 13 and 14 for a 5µm via diameter, while the nMOS could be placed at any distance from the copper nail, the keep-away-zone for the pMOS is found to be **1.5µm** for a doping concentration of $5.10^{17} cm^{-3}$. This is because the effect of mechanical loads on piezo-resistivity coefficients of pMOS transistors is much more than that of nMOS transistors, which could also be seen in table 5 [10, 11].

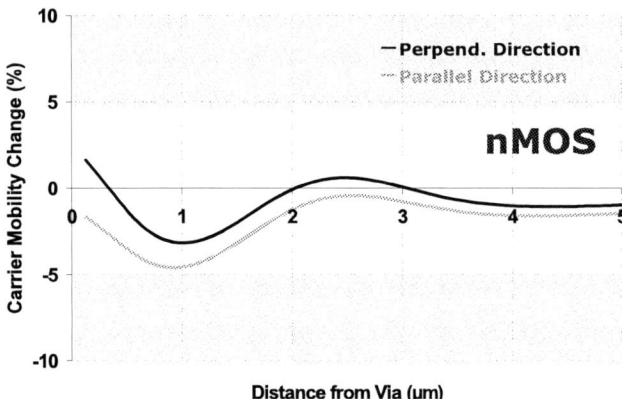

Figure 13: Carrier mobility change variation with distance from copper via for nMOS transistor

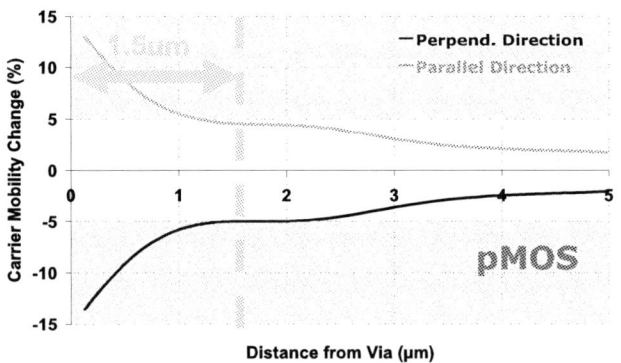

Figure 14: Carrier mobility change variation with distance from copper via for pMOS transistor

A plot of the keep-away-zone with the via diameter in figure 15 for nMOS and pMOS transistors reveals that a decrease in via diameter encourages the location of the active regions closer to the copper via.

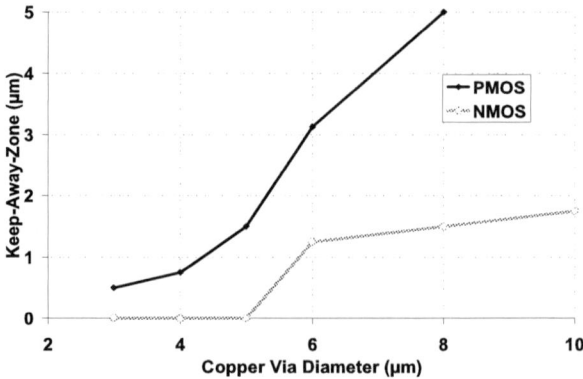

Figure 15: keep-away-zone variation with via diameter for both pMOS and nMOS transistor process.

4. Conclusions

Induced stresses in silicon caused by thermo-compression Cu-to-Cu bonding are analyzed and it is found that

- The temperature component of the thermo-compression process is the main cause of the induced stresses in the silicon rather than the applied thermo-compression bonding force.
- The resultant residual stresses witnessed *after bonding* are mainly caused by the pre-bonding heat treatment undertaken during the sintering step of back-end-of-line (BEOL) copper.
- A decrease in silicon thickness resulted in a decrease in the induced stresses in the silicon during bonding.
- PMOS is found to be more sensitive to stresses induced by copper nails than the NMOS, thereby invariably setting the keep-away-zone.
- The extent of stress distribution around the copper via strongly depends on the copper via diameter, which has an impact on the keep-away-zone distance for CMOS integration.

Acknowledgments

The authors are thankful to the entire 3D-SIC team for all the helpful contributions made in the course of this work. This work was made possible by the support of the IMEC core-partners program.

References

1. Said F. Al-Sarawi *et al*, "Review of 3-D Packaging Technology," *IEEE Trans-CPMT-B*, Vol. 21, No. 1 (1998)
2. Rhett Davis W., *et al*, "Demystifying 3D ICs: The Pros and Cons of Going Vertical" IEEE Design and Test of Computers 2005
3. Miranda P. A., *et al*, "Thermo-mechanical Characterization of Copper-Through-Wafer Interconnects," *Proc. of the 56th Electronic*

Components and Technology Conference, May 2006. pp. 844-848.

4. Karnezos M., "3-D Packaging: Where All Technologies Come Together" SEMI Int'l Electronic manufacturing Technology Symposium, July 2004. pp. 64-67

5. Hara K., *et al* "Optimization for Chip Stack in 3-D Packaging," *IEEE Transactions on Advanced Packaging-B,* Vol. 28, No. 3 August 2005. pp 367-376.

6. Tanida K., *et al*, "Ultra-high-density 3D Chip Stacking Technology," *Proc. of the 53rd Electronic Components and Technology Conference*, May 2003. pp. 1084-1089.

7. Benkart P., "3D Chip Stack Technology Using Through-Chip Interconnects" *IEEE Design and Test of Computers 2005*

8. Takahashi K., *et al* "Development of Advanced 3D Chip Stacking Technology with Ultra-fine Interconnection," *Proc. of the 51st Electronic Components and Technology Conference*, May 2001. pp. 541-546.

9. Gonzalez M., *et al* "Influence of Dielectric Materials and Via Geometry on the Thermomechanical Behavior of Silicon Through Interconnects,"

10. Smith C. S., *"Piezoresistance Effect in Germanium and Silicon" Physical Review* Vol. 94, No.1, April 1954. pp 42-49.

11. Matsuda K., *et al* "Nonlinear piezoresistance effects in Silicon" *J. Appl. Phys.* Vol. 73, No. 4, Feb. 1993.

12. Thompson S. E., *et al* "Uniaxial-Process-Induced Strained-Si: Extending the CMOS Roadmap," *IEEE Transactions on Electron Devices* Vol. 53, No. 5, May 2006.

13. Goldstein H., "Packages" *IEEE Spectrum.* August 2001

14. B. Swinnen, W. Ruythooren, P. De Moor, L. Bogaerts, L. Carbonell, K. De Munck, B. Eyckens, S. Stoukatch, D. Sabuncuoglu Tezcan, Z. Tokei, J. Vaes, J. Van Aelst and E. Beyne, "3D integration by Cu-Cu thermocompression bonding of extremely thinned bulk Si die containing 10μm pitch through Si vias", *Proc. IEDM Conference, December 11-13, 2006, San Fransisco*

Effect of Processing Parameters and Hygro-thermo-mechanical Stresses on the Reliability of Flip Chip Bonding RFID Tags

D.G. Yang, E. de Bruin, B. Kasemset, W. D. van Driel

NXP Semiconductors, P.O. Box 30008, 6534 AE Nijmegen, The Netherlands
Tel: +31(0) 243536353 Fax: +31(0)243533350
E-mail: daoguo.yang@nxp.com

Abstract

The thermal-mechanical reliability of Flip Chip bonding RFID packaging using low cost materials is investigated. The major processing parameters, such die bonding force, curing shrinkage, are discussed. The reliability tests on the new developed packages show that the electrical resistance of some test samples is degraded and unstable after MSL preconditioning and TMCL tests. In order to investigate the roof causes of such failures, an integrated modeling on the bonding process, temperature cycling and moisture absorption were carried out. 3D parametric FE modeling was setup simulate the die bonding process and adhesive curing process, temperature cycling and hygro-swelling. The experimental and numerical results show that the initial deformation in the metal pads caused by the die bonding and ACA curing shrinkage has an important impact on the electrical performance. Moisture-induced swelling reduces the contact pressure between bumps and the metal pads, resulting in electrical instability. It is also found that moisture-induced corrosion between the contact surfaces may also have important influence on the electrical performance, which is under further investigation.

1. Introduction

Radio Frequency Identification (RFID) is a technology that identifies objects using radio frequency technology. Generally, RFID consists of two basic components: a radio signal transponder, or tag, and a reader. The former, RFID tag, consists of a chip that contains identifying information about the attached object and an antenna to communicate that information via radio waves. Because the RFID tags can hold much more information than other data carrier systems such as a bar-code system, RFID is expected to become a dominant alternative for bar-code system and also show great promise for other applications. However, before the huge expansion of the RFID market, there are some crucial issues to be solved, among which cost reduction (large volume production with low cost) and product reliability are the major concerns.

COB (Chip on Board) or COF (chip on flex) technologies have been used in applications such as smart cards as well as smart labels for many years. In COB or COF technology, IC chips are attached to a substrate and electrically connected using wire bonding process. Then, the contacts are protected against environmental influence by encapsulating the chip and the wire bonds. Standard encapsulation compounds are cured by light or heat. The curing process of the encapsulation compounds may take 30 seconds to 2 hours, which depends on the curing temperature. For smart label or ID tag applications, however, this complex and relatively slow process is simply not a good solution.

Flip Chip technology has gained popularity in applications such as smart labels and RFID tags. In the flip-chip technology, IC chips with bumps on its active side is pressed into the substrate metallization with its active side facing the substrate. It has the advantages of low cost and being suitable for large volume production and can meet today's packaging needs of semiconductors and high-speed devices, especially with respect to miniaturizing packages and reducing the interconnection distances for a high standard of performance in RF and high frequency applications. Conductive adhesives, such as isotropic conductive adhesives (ICAs) and anisotropic conductive adhesives (ACAs), are commonly used for providing electrical connections and for bonding purposes. Mostly ICAs have been used for interconnection of bare dies. When they are in use with less noble metals like copper or aluminum as metallization on low cost substrates, ICAs show potential low reliability [1]. Instead, ACAs may be used in such cases to obtain reliable interconnection.

ACA's are generally composed of an adhesive polymer matrix and conductive particles. Filled particles can be metallic particles or metal-coated polymer spheres. Electrical conduction can be established through the conductive particles trapped between the interconnection areas in the compressed direction of the adhesive. The concentration of the conducting particles in the adhesive is well below the percolation threshold, so there is no conduction in the plane of the adhesive. As the contact pads on the two sides of the adhesive are forced together during bonding, they trap a number of conducting particles between the two pads. As the particles are compressed between these pads, the adhesive is squeezed out between the surface of the particle and the pad, and an electrical contact between the particle and the two bond pads is created at points where the particle comes in contact with each bond pad. The polymer matrix (usually epoxy resin) not only serves as a protection of the conducting particles and bumps from environment and external damage, but also provides a bonding to keep contact among the bump, conducting particles, and metal pad. The contact resistance of ACA depends on the processing parameters (bonding pressure, temperature, etc.) and properties of the materials as well [2,3].

In the development stage of the new RFID package, reliability issues aroused in the temperature cycling (TMCL) and moisture absorption tests. Electrical instability and failures were observed. In order to solve the problem, the root causes for such failures and the influence of material properties, package designs and processing conditions on the reliability have to be investigated and understood.

In this paper, firstly, a description of the packaging process is introduced. Secondly, the experimental results and observations are presented. Thirdly, integrated modeling on the bonding process, temperature cycling and moisture absorption are carried out. 3D parametric FE modeling is setup to simulate the die bonding process and adhesive curing process, temperature cycling and hygro-swelling. The impact of the major processing parameters, structure designs, and key material properties on the reliability is investigated.

2. The major processes

Figure 1 shows schematically the steps of the bonding process for the Flip Chip RFID tags. First, the IC chip with the bumps is made. ACA paste is applied onto PET substrate with printed or etched copper and etched aluminum foils. Then, the silicon chip with bumps is bonded to the metal foil and substrate using proper press force. Afterwards, ACA curing provides mechanical bonding between the chip and the metal pad and PET substrate.

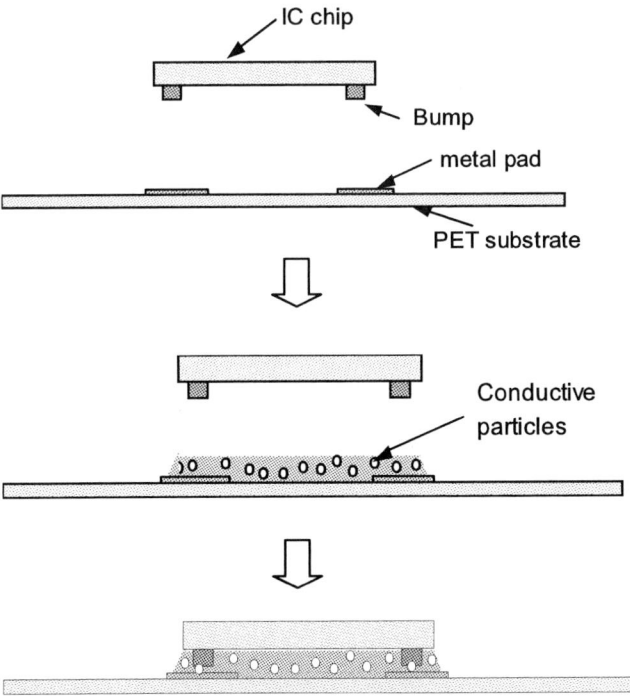

Figure 1: Schematic illustration of the bonding process of Flip Chip RFID tags.

3. Experimental

Cross sectional analysis was conducted to study the positioning of bonding, deformation and the contact status. Figure 2 shows the cross section of a sample with Al pad. The deformation in the metal pad after the bonding process is between 5-10µm. Figure 3 shows a typical microstructure of the bonding interconnection with etched aluminum as metal pad. It is seen that under the press force and pressure stress generated by curing and thermal shrinkage of the adhesive, a conductive particle penetrates into the metal pad. It is believe that such a penetration through the metal pad and/or bump would be favorable to the electrical conduction and may lead to good and long term stable electrical contacts.

MSL preconditioning (85C/85%RH, 168 hours) and TMCL tests were performed. Electrical instability and failures were observed. The analysis and correlation of the MSL and TMCL test data are under way.

Figure 2: Cross section of a sample with Al pad.

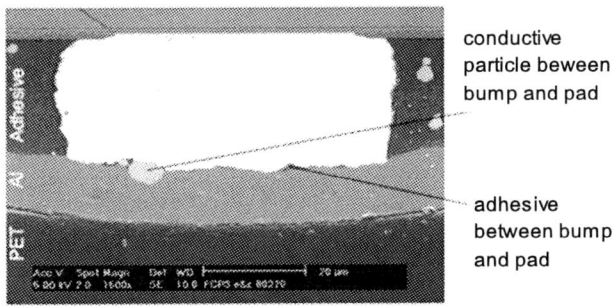

Figure 3 Typical picture of the bonding interconnection with Al pad. Conductive particle penetrates into the Al pad.

4. Finite element modeling

1). Geometry and FE Mesh.

The RFID package consists of silicon die, bumps, PET substrate, metal strap (pad), conductive adhesives and antenna. 3D parametric FE model is built. Figure 4 shows the FE mesh. Antenna is neglected in the simulation, since it is expected that it would have negligible influence on the simulation results. Contact

499

elements are used for both the elements along the interfaces between bump and metal strap.

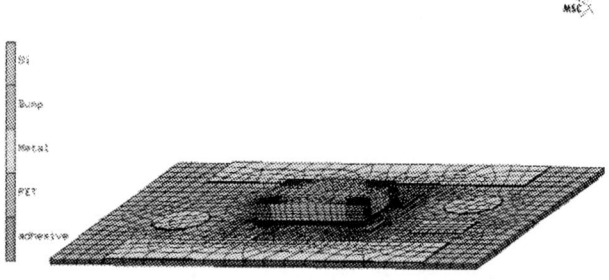

Figure 4: Finite element mesh of the RFID tag.

2). Material Properties

Table 1 lists the material properties used in the modeling. The silicon die is assumed elastic, and the gold bump and metal strap (Al and Cu) are assumed elastic and plastic. The adhesive is a snap curable acrylic-hybrid anisotropically conductive adhesive paste. It is considered to be temperature-dependent elastic. Figure 5 shows the temperature dependency of the Young's modulus.

DMA measurements under 1Hz frequency were performed on the PET films. Figure 6 presents the storage modulus of the two PET films. It can seen that PET B is much stiffer than PET A and it also a higher glass transition temperature (T_g=125°C) than PET A (T_g=75°C).

Table 1 Material properties used in the modeling

Constituent name		E [Mpa]	ν	Yield stress [Mpa]	CTE [ppm/°C]
Silicon die		169000	0.23	-	3
Au bump		75000	0.42	80	14.3
Metal foil	Al	60000	0.35	200	20
	Cu	120000	0.33	220	17.7
ACA		See Fig. 5	0.30	-	50 ($<T_g$) 160 ($>T_g$)
Substrate film	PET A	1Hz DMA (see Fig. 6)	0.40	-	50 ($<T_g$) 90($>T_g$)
	PET B	1Hz DMA (see Fig. 6)	0.40	-	50 ($<T_g$) 90($>T_g$)

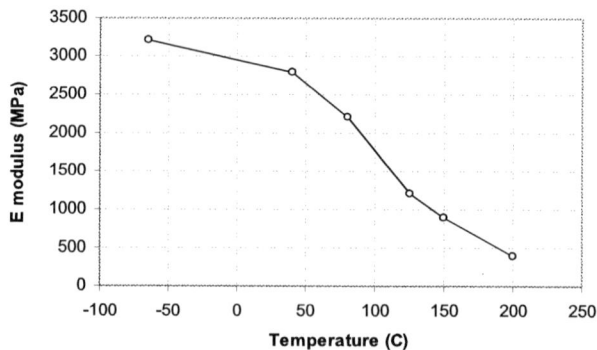

Figure 5: Temperature-dependent E modulus of ACA.

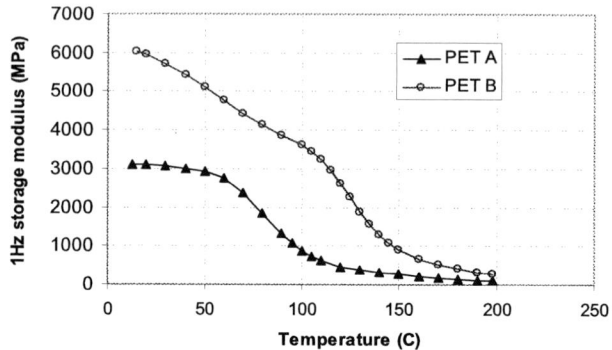

Figure 6: Storage modulus of the PET substrates (PET A and PET B) from Hz DMA measurements

3). Thermal-mechanical loading

The thermal loading profile in the simulation includes die bonding, ACA curing process, cooling down phase, MSL preconditioning, and TMCL. A brief description of the conditions for modeling of the processes and tests are given below:

- Die bonding: the Si die is placed onto the metal pad under the die bonding temperature and the die pressing force is 0.8N. Different bonding temperatures are used.
- ACA is cured under isothermal conditions. The curing temperature is the same as the bonding temperature. During cure, the die pressing force is kept constant at the prescribed value.
- After the curing process, the package is cooled down to room temperature.
- MSL preconditioning (85°C/85%RH, 168 hours) is applied. Moisture diffusion and hygro-swelling is modeled.
- Temperature cycling between –40°C to 125°C is applied.

5. Results and discussion

With the bonding process, the electrical interconnection is established through the contact of bump, conductive particles, and the metal pad. Die press force, curing shrinkage of the adhesive and thermal shrinkage all play certain roles in the contact resistance. When the samples are exposed to moisture conditioning, the polymeric adhesive and the PET substrate will absorb moisture. In addition to the possible chemical reaction such as corrosion, the polymeric materials will swell. It is recognized that moisture swelling is one of the major reasons that cause the increase of the contact resistance [5].

Figure 7 shows the deformation in the Al pad after die bonding process. Figure 8 presents the plot of the vertical displacement along the central surface (A-B shown in Fig. 6). It can be seen that most of the deformation is caused by the curing of the adhesive.

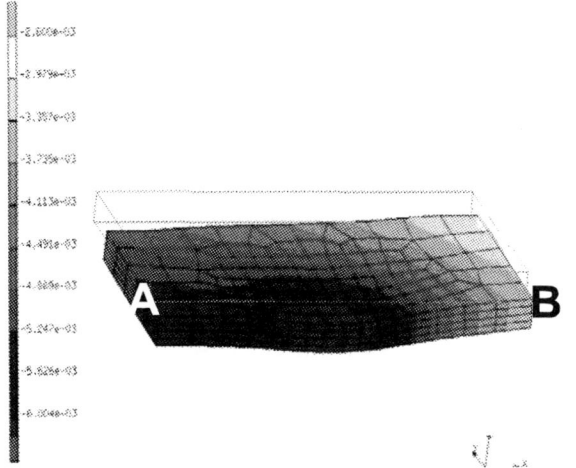

Figure 7: Deformation in Al pad after bonding process.

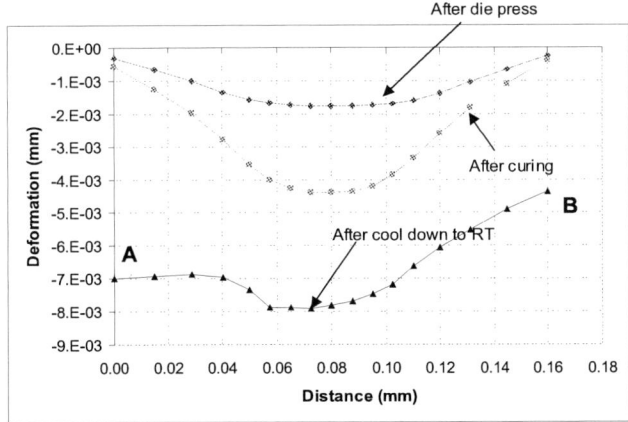

Figure 8: Deformation along the path A-B at different stages of the bonding process.

Figure 9 shows the evolution of the contact stress (normal stress on the bump surface, averaged) during the bonding process and subsequent temperature cycling for different designs and PET materials. D1 and D2 refer to the design of 10μm Al pad and the design of 20μm Al pad, respectively. The upper graph indicates the corresponding temperature history. It can be seen that just after the die press, the levels of the contact stress are virtually the same for all of the cases. At the end of curing, a big difference among these cases can be seen. Thicker metal pad and stiffer PET substrate intend to generate higher compressive stress on the bump. This is due to the curing shrinkage induces higher stress on the bump through stiffer pad. After cooling down to room temperature, both designs give almost the same stress, and stiffer PET shows slightly higher stress. During temperature cycling, stiffer substrate can produce higher compressive during cycling to elevated temperature. PET A also has a higher T_g than PET B.

Figure 10 shows the plot of the compressive stress acting on the bottom surface of the bump for both 150μm and 75μm die. It is shown that die thickness has very little influence on the bump contact stress. The reason is that Si die has much higher stiffness that other materials and the change of this thickness has little impact on the local stress around bump. It should be noted that when die thickness changes from 150μm to 75μm, peak stress in die would increase nearly 60%. However, in our case, the die stress is still below the critical strength value.

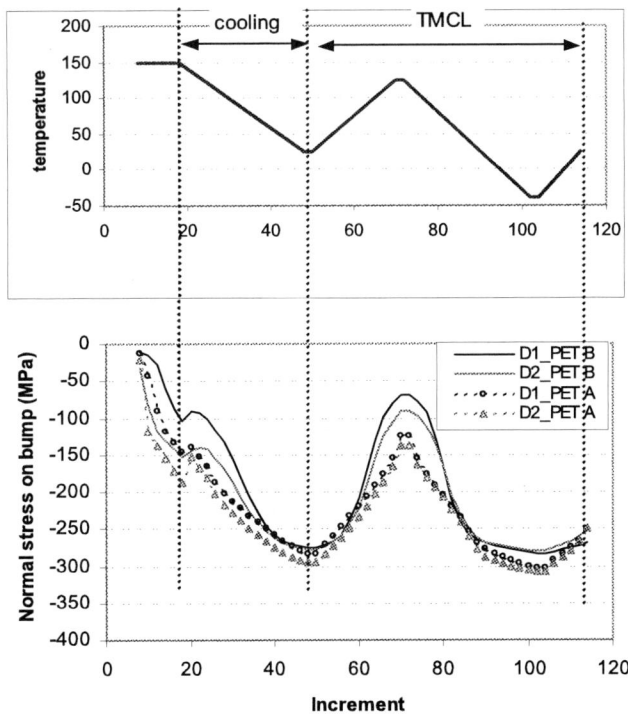

Figure 9: Evolution of the contact stress (normal stress on the bump surface) during bonding process and subsequent temperature cycling. The upper graph shows the corresponding temperature.

501

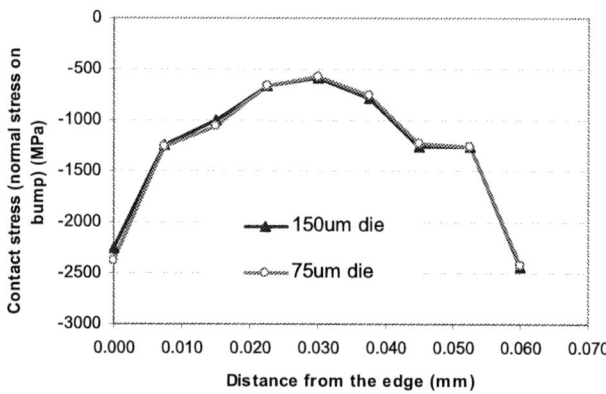

Figure 10: The contact stress (normal stress on the bump surface) along the central line of the bump

6. Conclusions

The thermal-mechanical reliability of Flip Chip bonding RFID packaging using low cost materials is investigated. The reliability tests on the new developed RFID tags show that reliability issues arouse after MSL preconditioning and TMCL tests. 3D parametric FE modeling was setup simulate the die bonding process and adhesive curing process, temperature cycling and hygro-swelling. The experimental and numerical results show that the initial deformation in the metal pads caused by the die bonding and ACA curing shrinkage has an important impact on the electrical performance. Thicker metal pad and stiffer PET substrate intend to generate higher compressive stress on the bump, which is favorable for the electrical conduction. Stiffer substrate with a higher T_g can produce higher compressive stress during cycling to elevated temperature. It is shown also that die thickness has very little influence on the bump contact stress. The effect of moisture-induced swelling is under further investigation.

Acknowledgments

The authors would like to thank Leon Goumans for performing the DMA measurements.

References

1. Wong, C.P., *et al*, "Fundamental study of Electrically Conductive adhesives,", in Proceedings of PEP'97 Conference.
2. Wei, Y. and Sancaktar, E., "A Pressure Dependent conduction Model for Electrically conductive adhesives,", in Proceedings of ISHM'95, 1995, pp: 231-236.
3. Yim, M.J. *et al*, "The Contact Resistance and reliability of Anisotropic Conductive Film (ACF) ", *IEEE Trans-AP*, Vol. 22, No. 2 (1999), pp. 166-173.
4. Kwon W.S. *et al*, "Contraction Stress Development of Anisotropic Conductive Films (ACFs) Flip Chip Interconnection: Prediction and Measurement" , *IEEE Trans-CPT*, Vol. 29, No. 3 (2006), pp. 688-694.
5. Teo, M., et al., "Correlation of Material Properties to Reliability Performance of Anisotropic Conductive Adhesive Flip Chip Packages," *IEEE Trans-CPT*, Vol. 28, No. 1 (2005), pp. 157-164.

Efficient electrostatic-mechanical modeling of C-V curves of RF-MEMS switches

Jeroen Bielen, Jiri Stulemeijer
NXP Semiconductors
Gerstweg 2, 6534 AE Nijmegen, The Netherlands
Tel: +31 24 353 6626, Fax: +31 24 353 6556, E-mail: jeroen.bielen@nxp.com

Abstract

The capacitance versus actuation voltage, the C-V curve, is characteristic for the steady state behavior of a RF-MEMS capacitive switch. It is imperative to have an efficient method to simulate these curves and overcome the convergence problems from pull-in and release instability that is inherent to these electrostatic actuated devices. In this paper we show how the complete CV curve can be calculated in FE code, including conditionally stable parts and zipping regions, which also comprises a non-linear contact model. Efficiency improvement by use of a reduced order model for the electrostatic domain is shown. Validity of the simulation results is shown by comparison to measurements.

1. Introduction

For future ambient intelligent RF-applications, the use of capacitive RF-MEMS switches (Figure 1) is gaining increased attention since these pose a means to e.g. create adaptive antenna matching and reconfigurable RF circuits [1]. Part of this attention is due to the large on-off capacitance ratio that can be realized with this class of MEMS (>1:20), which means the device is in full contact with the bottom dielectric in the activated state. From a modeling point of view means that the open behavior of such a switch must be described accurately, but also the (partially) closed state and the transitions between these states that determine important properties such as release voltage. The release part is not only of interest because of the voltage, but also to identify potential partial release states where sticking can occur when the dielectric gets charged (an undesirable phenomena). This is closely related to the presence of the conditionally stable parts.

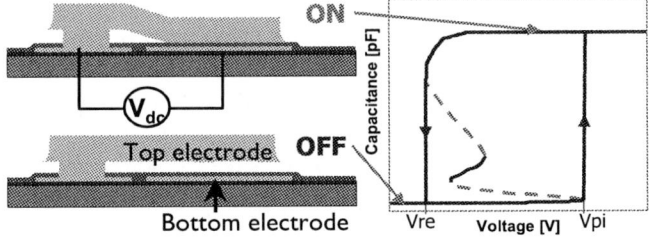

Figure 1: Schematic drawing of electrostatic switch and CV curve with conditionally stable parts in dashed red.

At the pull-in voltage a fold in the C-V curve exists. Trying to find the exact pull-in voltage by applying a voltage sweep and using adaptive stepping until there is no longer convergence requires a large computational effort and is inefficient. A number of methods to overcome this have been reported in literature. Notably, Elata et al. ([2],[3]) introduced the displacement iteration

scheme (DIPIE). The underlying idea of this method is that instead of prescribing a voltage, a displacement constraint is prescribed and a voltage is solved for (Figure 3). This algorithm can be implemented in FE codes by scripting.

This paper presents a DIPIE+ algorithm capable of simulating the entire CV curve of a large variety of devices including stable or conditionally stable parts of the release trajectory. Additional to the DIPIE algorithm a second adaptive algorithm for selecting the location of the control node is introduced in order to allow the iteration scheme to continue after initial contact.

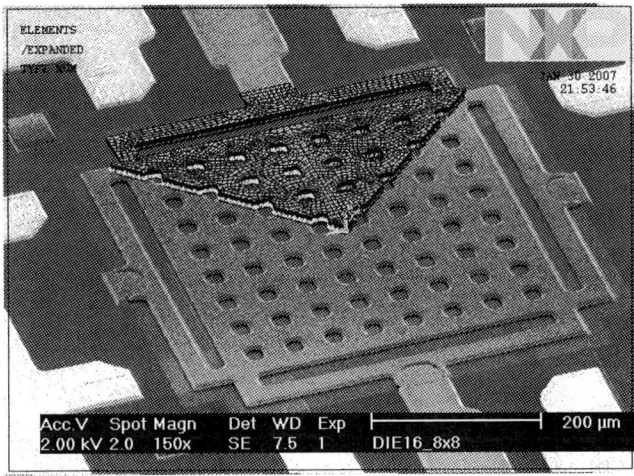

Figure 2: RF-MEMS capacitive switch. FE model overlaid on SEM image of an 8x8 device

Further efficiency can be gained in the implementation of the electrostatic mechanical coupling. Despite the deformation of plate and beams the structure remains to good approximation parallel to the substrate. Based on this observation a reduced order model can be derived. With a detailed electrostatic model that is solved for a small number of gaps, the charge as function of gap is derived for every node in the coarser meshed mechanical model. This reduced order model is subsequently used in the much smaller directly coupled electro-mechanical model for which the DIPIE+ scheme is used. Simulating the entire CV curve requires up to several hundreds of iterations, that are performed on a significantly reduced model. This results in an efficient way to include fringe fields.

A large range of topologies of devices can be solved with this scheme. The simulated results obtained with Ansys have been verified with measurements on various devices. The presence of stable parts in the C-V curve can depend on initial deformations and temperature as will be shown.

1-4244-1105-X/07/$25.00 ©2007 IEEE

2. Method

As shown in the paragraph before, electrostatic switches have an inherent hysteresis, causing the pull-in and release voltage to be different. The states in between the open and close state are instable. For a 1-D problem, a rigid plate suspended by a spring, the steady state equilibrium between mechanical forces F_m and electrostatic force F_{es} is given by:

$$Fm + Fes = k \cdot (z - gap0) + \frac{\varepsilon 0 \cdot Area \cdot V^2}{2 \cdot z^2} = 0 \quad (1)$$

Equation 1 has no unique solution of displacement z when voltage V is prescribed, but it has an unique solution for V(z), as depicted in Figure 3.

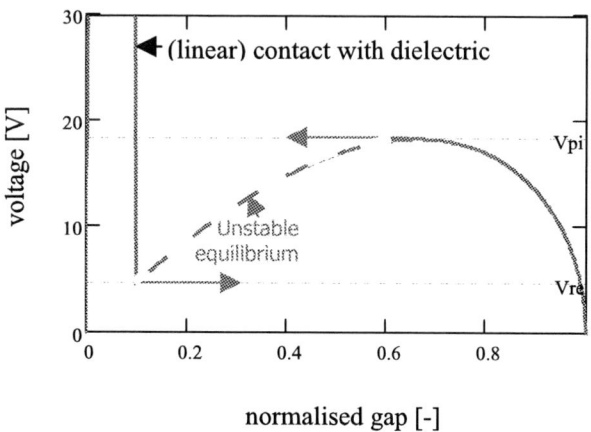

Figure 3: The idea behind the DIPIE algorithm: A prescribed displacement only allows for one equilibrium voltage where a prescribed voltage may have multiple displacements as solution.

One method to calculate the pull-in voltage of an arbitrary shaped electrostatic switch is by applying a voltage sweep, and running the simulator until no convergence is reached. A much more efficient method is introduced by Elata et all [2,3] and termed DIPIE algorithm. Although Figure 3 will not hold for any node in a full 3D structure, the principle of prescribing displacement on one node and searching the voltage that minimizes the reactive force can still be applied in 3D problems, provided a proper control node is used (Figure 5). This graph shows how by first using node 1 to prescribe displacement until it reaches contact and than switching to 2, all equilibrium states can be found.

The DIPIE algorithm is easily implemented in any FE code that has some scripting capability. A displacement is prescribed for the control node and a guess value for the voltage is applied. With a prescribed displacement and in the absence of pull-in instabilities, the FE solver will quickly converge to a solution from which the reaction force on the control node can be retrieved. A simple Newton iteration method then is used to calculate the next voltage estimate to feed to the FE solver. This is repeated until the reaction force is below some error norm.

For an actual 3D FE model however, it is not always guaranteed that the optimal node is (or can) be selected to prescribe the displacement. This may results in pull-in and release hysteresis of a different part of the structure without a nodal displacement prescribed while the Newton scheme attempts to search the voltage that nullifies the reaction force. This means there are potentially two voltages that will have a zero reaction force, the correct solution being the one closest to a previous state. In order to find the correct voltage, a step back in the Newton algorithm needs to be enforced. Since the FE model switches from the 'good' (based on shape or capacitance) to the 'wrong' state when such pull-in is encountered, it needs to be set back to the previous solution of the 'correct' state. This is implemented by using a single frame restart whenever necessary and switching to a regula falsi search algorithm. This approach works quite well, on the expense of convergence speed however.

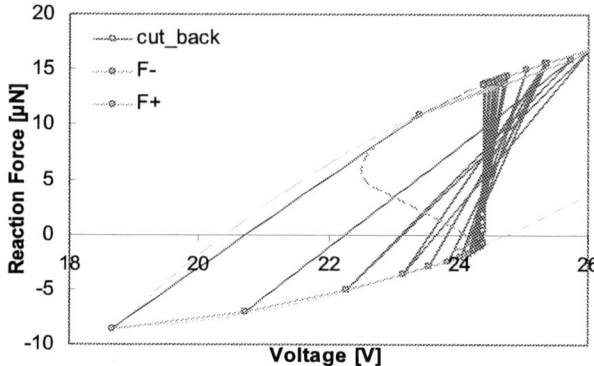

Figure 4: Reaction force versus voltage during Newton iterations with cut-back initiations; the x-axes intersection at 24V is the searched zero and can only be found if the simulator is 'reset' to the lower trace.

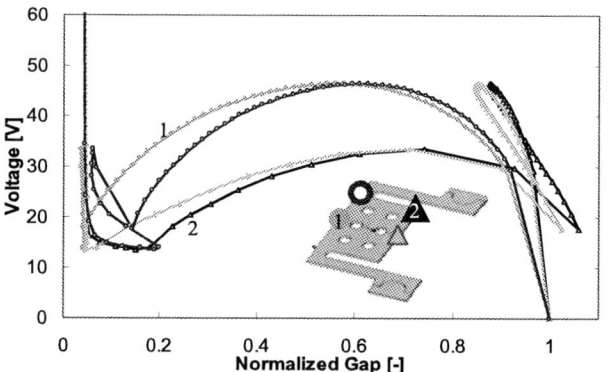

Figure 5: Equilibrium voltage versus gap for four nodes on a device with mode shape changes: Unique voltages can only be found if an appropriate node is used to prescribe displacement.

Control node selection

However, this DIPIE algorithm as described only allows the extraction of pull-in voltage, and not possible additional stable parts of the CV curve at the release part or during zipping. Examples of devices that exhibit such behavior are single cantilevers that can possess a stable state after an initial pull-in when the tip touches the substrate and have a second pull-in resulting in part of the beam in contact with substrate. The beam can zip into further contact with increasing voltage. In order to find these other stable solutions, the entire CV curve, including conditionally stable solutions needs to be calculated.

The DIPIE algorithm as described above is useful for finding the pull-voltage, but cannot find any solutions of the release once the control node has contacted the bottom. Using a node at the side of the plate is not an adequate solution since this will fail to stabilize the centre of a flexible plate. In order to find those solutions, all nodes of the plate must ultimately reach contact so the control node must be adaptively selected.

Three different criteria for selecting the best control node were compared. The most obvious is to take the lowest point of the plate, since this experiences the largest electrostatic pressure. Alternatively, the selection can be based on previous converged displacement iterations. This leads to considering either displacement increment or increment of electrostatic force, the latter possibly being a measure for how quick a location approaches pull-in. Although there were differences in the convergence speed, none of the selection criteria good prevent (i.e. predict) a pull-in hysteresis to occur during the successive displacement iteration, while the voltage was changed to minimize the reaction force. With the topologies that we evaluated, the electrostatic force increment appeared the best criteria.

3. Experimental

Measurements of steady state CV curves presented here have been done on a wafer prober at low frequency under controlled ambient. In the used set-up an LCR meter is used at typically 1MHz to measure the capacitance. A bias voltage is applied externally and slowly stepped with steps of 0.5V or 1V to well above the positive and well below the negative pull-in. Parasitic capacities are characterized by removing the plate of a device and measuring the capacitance again, although de-embedding of S-parameters measured at 1GHz is also used.

During the CV measurements, the dielectric of the device may experience charging, an undesirable phenomenon that is often prevented by applying bipolar actuation [7]. In the CV measurements this will result in both a shift but also 'narrowing' of the voltage axis when one device is measured several times. When characterizing the temperature dependency, CV curves should be corrected for this effect. In the presented measurements this is done by calculating the charging related voltage shift from CV measurements at room temperature before and after the measurement.

4. FE model

The previously described DIPIE+ algorithm has been implemented in Ansys90 by using APDL (Ansys Parametric Design Language). The FE model consists of two FE models that both use the same solid model geometry. One model of the electrostatic domain is used to calculate the capacitance versus gap when the plate receives a uniform vertical displacement. The second model is an (directly coupled) electro-mechanical model that uses the results of the other model for the transducer characteristics. This latter model uses the DIPIE+ algorithm to calculate the CV curve.

Geometry of the models is either fully parametric or imported from CAD drawing. Both physics domains take advantage of any symmetry that applies, either rotational (with help of constraint equations) or $1/8^{th}$ normal symmetry. Local coordinate systems are rotated accordingly and are taken into account for the DIPIE displacement DOF.

Electrostatic model.

When assuming that the plate is to good approximation parallel to the bottom electrode, the electrostatic domain can be effectively captured in a reduced order model that describes the capacitance versus gap of each node on the mechanical model. For this purpose an electrostatic model was build, and looped with a gap ranging from just the effective air gap on closure up to completely open.

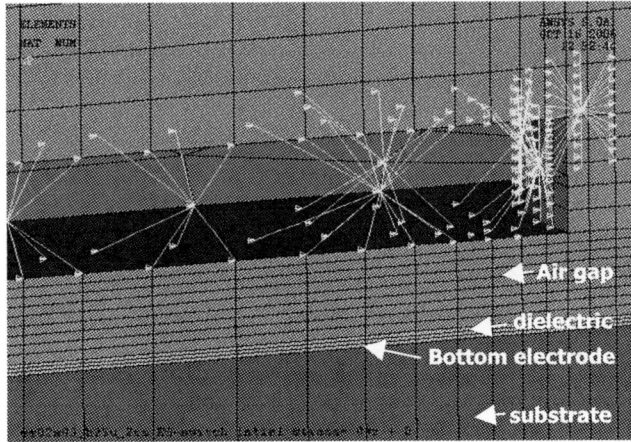

Figure 6: Electrostatic model. Arrows indicate how charge is assigned from the electrostatic nodes to the nearest 'mechanical' node (shell elements at the plate mid-surface).

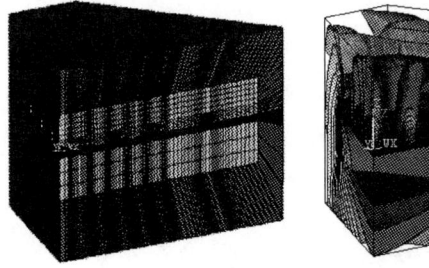 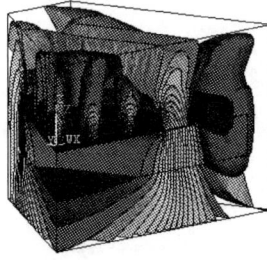

Figure 7: Electrostatic model and iso-potential surfaces

This electrostatic model needs a finer mesh near the edge of the holes than the mechanical model and hence has a different mesh. The charge in each node of the electrostatic calculation is mapped to the closest node on the bottom of the plate of the mechanical mesh. This guarantees that the total capacitance is correct and charge is conserved (this mapping process is schematically depicted in Figure 6). Since there always remains some effective air gap due to microscopic roughness, the electrostatic model does not need to be solved for a completely closed gap. Therefore morphing one and the same mesh can simulate the entire range of the gap.

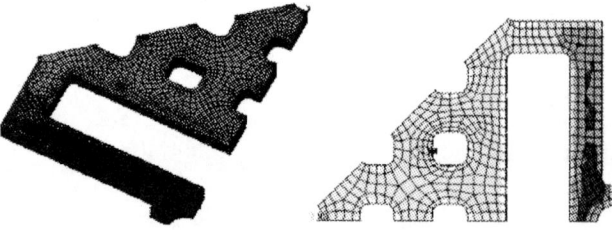

Figure 8 Charge distribution @ 1V (=capacitance) of the ES model (left) and charge density as transferred to the mechanical mesh (right)

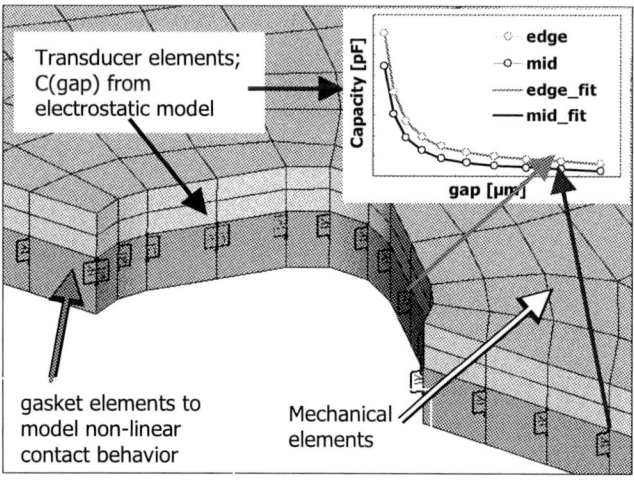

Figure 9: Detail of electromechanical model and an example of the cap(gap) relation used for the transducer elements (inset).

Figure 10: FE model overlaid on SEM image of the 5x5 device

Far field effects are modeled with so called infinite elements that account for the correct boundary conditions of exterior faces of the model. The plate receives zero voltage in order not to add far field capacitance contributions to the plates charge. The only relevant material property in this model is the relative dielectric

permittivity of the dielectric which was measured on some MIM caps.

Electro-mechanical model

The mechanical model is depicted in Figure 9 and pretty straightforward. As long as the primary interest in this simulation is not the stress levels, but the CV curve, the element divisions in the thickness direction of the plate is not very relevant. Element division in x and y do affect the CV curve. Material properties have been measured on special single cantilever structures using a nano-indenter (reported in [4]).

The transducer elements receive their C(gap) from the electrostatic simulation (passed through a table). The table is approximated by Ansys with a function:
$$Cap(gap)=c_0/gap+c_1+c_2 \cdot gap+c_3 \cdot gap^2+c_4 \cdot gap^3 \quad (2)$$
This provides an adequate fit for adding fringe field effects to a parallel plate equation. The gap of the mechanical model also includes the dielectric i.e. gap=gap_air+gap_diel/eps_diel. Since the number of transducer elements is much smaller than the number of elements in the electrostatic model, this approach saves over 100k DOF;s for the model on which the DIPIE algorithm is run.

Non-linear contact model

An important feature of the actually measured CV curves that we also wanted to capture in the model is the dependency of Cclose on voltage. This is caused by bending of the plate around small features that protrude from either of the surfaces and from the non-linear contact deformation. As the effect is equally present in small devices, this deformation behavior of the microscopic roughness is thought to be dominant.

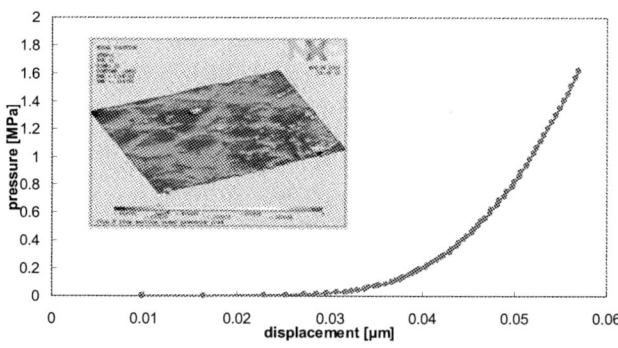

Figure 11: Measured contact pressure versus (electrostatic) gap is caused by microscopic roughness (inset is AFM profile) and can be approximated with an exponential relation.

Therefore, in analogy to [6], a non-linear spring needed to be introduced as contact element since this effect cannot be captured when contact is modeled with linear contact penalty stiffness i.e. the default contact in the transducer element is inadequate. In Ansys a multi-linear gasket element is suited for this purpose although a 1D non-linear spring can be used as well. Properties were extracted from the close part of a CV curve (Figure 11) using equation 1 and paralell plate aproximation for C.

5. Results
Electrostatic results

A typical result of the electrostatic model, capacitance versus gap, is shown in the inset of Figure 9. The values of the capacitance calculated with Ansys is within 1% of values calculated with more dedicated software for HF simulations i.e. Sonnet.

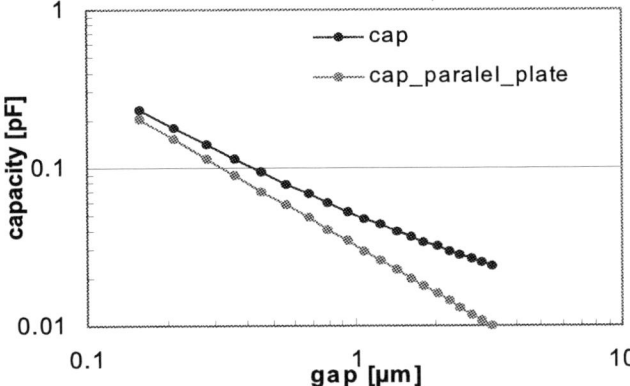

Figure 12: Simulated capacitance versus gap of 2x2 device showing significant contribution of fringe fields to total capacitance

Mesh size sensitivity evaluation was done for both the electrostatic and mechanical model which resulted in the used mesh. It turned out that for the mechanical mesh, if only the CV curve is of interest, and in absence of stress gradients, the plate can be modeled with solid shell elements (1 division in thickness). Lateral element division must suffice to capture roll-off effects caused by the holes i.e. at least 2 elements per side of the holes. Absorbing the holes in an effective Youngs modulus is therefore less accurate.

Measured & simulated CV curves

For validation of the model, simulations were compared to measurements of four different device layouts, ranging in size from 2x2 up to 8x8. From CV measurements on several wafers, a typical device was selected for each layout. In order to have an accurate model the exact realized geometry of these devices was measured in a SEM.

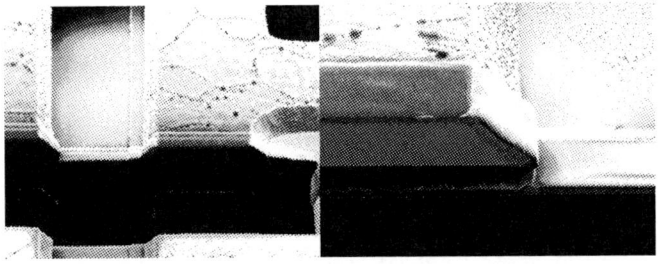

Figure 13:Image of FIB cross-section of a beam (in pull-in due to charging by the FIB) showing the inclined sidewalls.

Layer thickness was measured on FIB cross-sections. The sidewalls of the metal are inclined (Figure 13) but the model has straight walls. To account for this effect, the effective width in terms of second moment of inertia was taken for the beams (=mechanical cross-section). For the

plate, the bottom width was taken since here the electrostatic forces are quantitatively most important.

In addition to these SEM/FIB measurements, the initial bow of the plate was measured with a white light interferometer. These deformations were included in the model by introducing an (uniform) initial stress gradient. In principle it is possible to predict this initial shape as well with FE simulations of the processing steps, but this is considered a different topic and beyond the scope of this paper. The initial deformation of the beams was included in the initial gap by offsetting the entire structure. This approach can be justified by considering that the beams do not undergo large rotations or experience other non-linear effects.

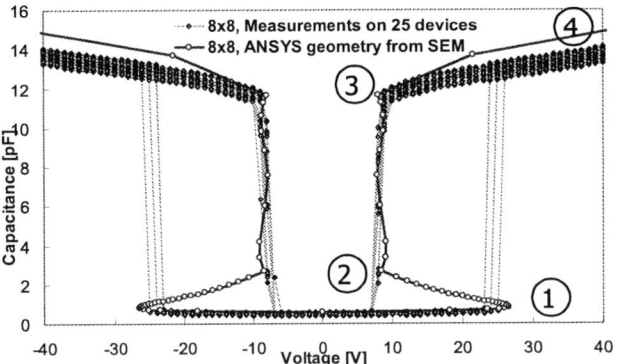

Figure 14: Measured and simulated C-V curves of an 8x8 reference device

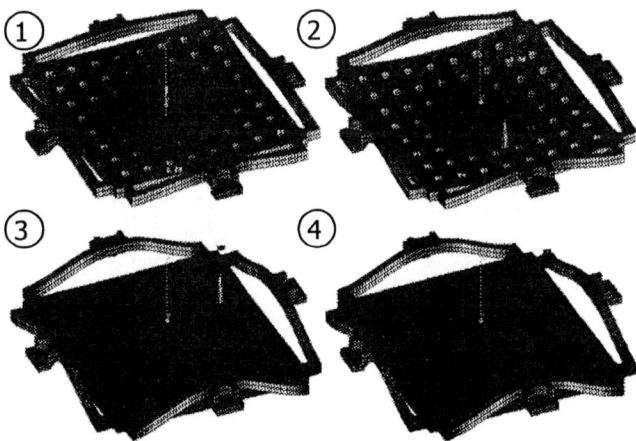

Figure 15: Z-displacement of 8x8 device at four stages depicted in the CV curve of Figure 14

The measured and simulated C-V curves of a 5x5 and an 8x8 device are depicted in Figure 14 and Figure 16. It can be seen that there is good agreement between simulations and measurements. Difference in the close capacitance can be caused by difference in micro roughness of just 10nm (for all simulations the same contact model was used). The pull-in and release voltages can be altered significantly by changing the geometry marginally, possibly less than the accuracy of the geometry measurements. Furthermore, when CV curves show a lot of steps in the release trajectory, the measured

507

release voltage is strongly dependent on the definition and algorithm used to extract it from the measured CV curve.

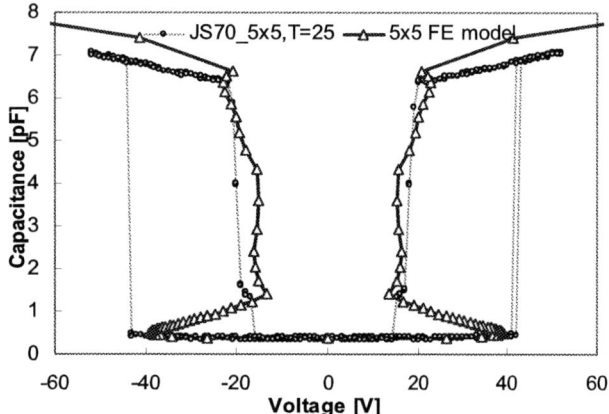

Figure 16: Measured and simulated C-V curves of a 5x5 device

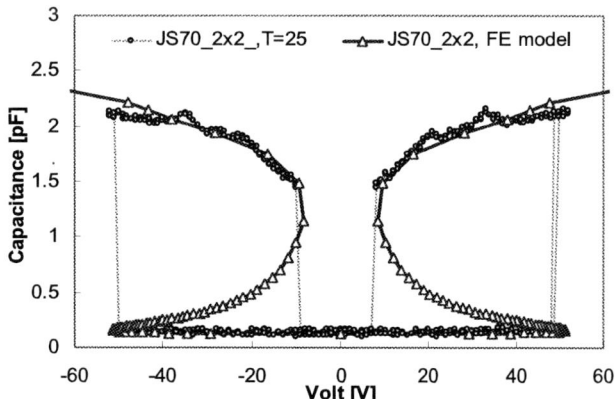

Figure 17: Measured and simulated C-V curves of a 2x2 device

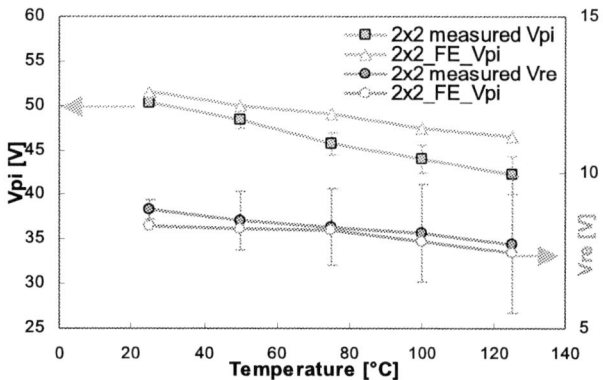

Figure 18: Measured and simulated temperature dependencies of pull-in and release voltage of 2x2 device.

6. Outlook

The presented DIPIE+ algorithm has been applied on various topologies. Although a proper control node selection is essential for finding the equilibrium state, it cannot solve any problem. For example systems that tend to have strong changes in the bending modes or have buckling behavior, like plates tilting around a hinge, still pose problems regardless the control node selection criteria. In those cases several nodes, that are only weakly coupled by the structure itself, may require a prescribed displacement to stabilize it while searching for the voltage to nullify the reactive force on another node. We are currently working on a nesting method of the presented algorithm to stabilize multiple nodes.

7. Conclusions

A displacement iteration scheme with adaptive control node selection that is able to calculate the entire CV curve of real-live topology electro-static switches has been presented. We showed that, with the inclusion of a non-linear contact model, an accurate match with measurements is reached, also in the partial or full contact regimes of the curves. The inclusion of fringe field effects can be achieved by deriving a reduced order model from an electrostatic model that is morphed. Temperature or packaging effects can be taken into account in the mechanical model. This provides a powerful, computationally efficient tool to predict functional parameters of various device types and moreover can certify the design performs well against specifications.

Acknowledgement

Special thanks go to the NXP IC-RF MEMS team, specifically to Ramon Havens for performing the measurements.

References

1. J.T.M. van Beek, P.G. Steeneken, G.J.A.M. Verheijden, J.W. Weekamp, A. den Dekker, M.Giesen, A.J.M. de Graauw, J.J. Koning, F.Theunis, P. van der Wel, B. van Velzen, P. Wessels, "MEMS for wireless communication: application, technology opportunities and issues", *European Microwave week 2006*, Manchester.
2. Elata, D., "Modeling the electromechanical response of RF MEMS", *Eurosime2006*, Como, Italy, 2006.
3. Elata, D., Leus, V., "Switching time, impact velocity and release response of voltage and charge driven electrostatic switches", *IEEE ICMENS'05*.
4. Burg, V., den Toonder, M., van Dijken, A., Hoefnagels, J., Geers, M., "Characterization of free-standing thin film material properties for RF-MEMS", *Eurosime2006*, Como.
5. Ansys™90 theory manual, Ansys Inc.
6. Chan, E.K., Garikipati, K., Dutton, R.W., "Characterization of Contact Electromechanics Through Capacitance-Voltage Measurement and simulations", *J.of MEMS*, Vol.8, No.2, June1999.
7. DelRio, F.W.; Herrmann, C.F.; Hoivik, N.; George, S.M.; Bright, V.M.; Ebel, J.L.; Strawser, R.E.; Cortez, R.; Leedy, K.D, "Atomic layer deposition of Al/sub 2/O/sub 3///ZnO nano-scale films for gold RF MEMS", *IEEE MTT-S International,* Vol.3, June 2004.
8. Greenwood, J.A., Williamson, J.B.P., "Contact of nominally flat surfaces", *Proc.R.Soc.A*, Vol295, 1966

Evaluation of Creep in RF MEMS Devices

Marcel van Gils, Jeroen Bielen, Gavin McDonald
NXP Semiconductors
Gerstweg 2, 6534 AE Nijmegen, The Netherlands
marcel.van.gils@nxp.com

Abstract

RF MEMS are capacitive switches consisting of a suspended aluminum beam that can be pulled down by electrostatic force. At elevated temperatures and high mechanical stresses the aluminum beam can exhibit creep phenomena that result in shifting of device parameters as a function of time. Experimental and numerical methodologies are presented for measuring and predicting the effect of creep on the RF MEMS device performance. A constitutive creep model is implemented in a Finite Element code where the parameters of this constitutive model for creep in thin aluminum layers are determined by wafer curvature experiments. In order to distinguish creep effect from charging effects, special test structures are designed. Simulations on the test with different geometries indicate the effect of creep and can result in design rules for the RF MEMS switches. The numerical predictions and the measured degradation on the RF MEMS switches are compared and conclusions are drawn with respect to the methodology.

1. Introduction

The development and production of MEMS (microelectromechanical systems) devices has shown a rapid growth over the last 15 years. Their unique physical properties through the combination of electrical and mechanical functions have resulted in applications not possible with alternative techniques. However, despite the apparent potential of MEMS many ideas never came to production due to roadblocks. One of the challenging fields in the commercialization of MEMS is reliability. Due to the nature of MEMS the reliability is often unique for the specific MEMS technology and highly multiphysical.

One of the promising MEMS devices is the radio frequency (RF) capacitive switch [1]. The potential use of these switches is in adaptive antenna matching for GSM mobile phones. An SEM photograph of the NXP RF MEMS switch is visible in Figure 1. The switch consists of a 3-5 μm thick aluminum top electrode that is suspended 1-3 μm above a bottom electrode. A silicon nitride layer is deposited on top of the bottom electrode and acts as a dielectric. By applying a voltage difference between the top and bottom electrode an electrostatic force is generated which pulls the top plate down towards the bottom electrode (see Figure 2). At a certain voltage difference (the pull-in voltage) the electrostatic force exceeds the spring force of the suspended aluminum structures and causes the plate to snap down on the silicon nitride layer.

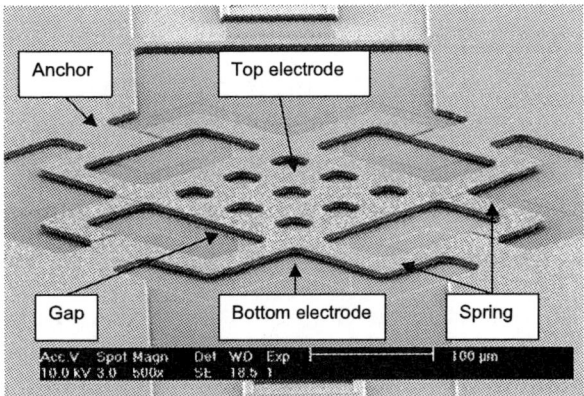

Figure 1: RF MEMS capacitive switch

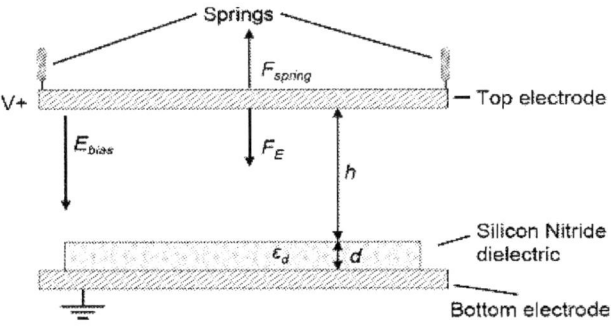

Figure 2: Principle of electrostatic actuation

For a 1-D approximation of the switch the voltage at which this occurs can be derived analytically and yields the expression for the pull-in voltage V_{pi} [14]:

$$V_{pi} = \sqrt{\frac{8kg^3}{27A\varepsilon_0}} \qquad (1)$$

with k the spring stiffness, g the initial gap ($g=h+d/\varepsilon_d$), A the surface of the electrode and ε_0 the dielectric constant of air.

The resulting characteristic of the RF MEMS switch is best visualized using the resulting capacitance versus voltage curve. A typical example of such a C-V curve is visualized in Figure 3 illustrating the hysteresis behavior and the symmetric actuation at both negative and positive voltages.

In the application the RF MEMS device is switched between the low capacitance at the open state (0V) and the high capacitance at a voltage above V_{pi}. The capacitance ratio between the open and closed state is an important characteristic of the switch.

1-4244-1105-X/07/$25.00 ©2007 IEEE

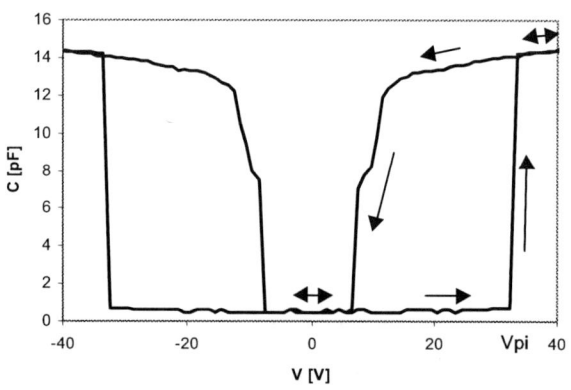

Figure 3: Typical C-V curve of the RF MEMS switch showing a symmetric curve

One of the most important reliability aspects of the RF MEMS switch is the change of the C-V curve that can result in a non-functional device. The two most important failure mechanisms causing a potential change of the C-V curve are charging of the SiN layer [2-4] and mechanical creep of the aluminium top electrode. Charging occurs during prolonged actuation of the switch at high voltages where the resulting electrical field causes trapping of electrons in the SiN layer. The resulting charge in the SiN layer shields the external applied voltage causing a change in the total electrostatic force and thus a shift of the C-V curve. When the build-up charge is high enough it will finally lead to sticking of the device where at 0 V the amount of trapped charge is high enough to keep the switch in the down state. Several groups during the last 7 years have investigated this charging effect and the resulting sticking behaviour. Figure 4 shows a typical result of the shift of the C-V curve due to prolonged actuation and the final sticking of the switch during reliability testing at accelerated conditions.

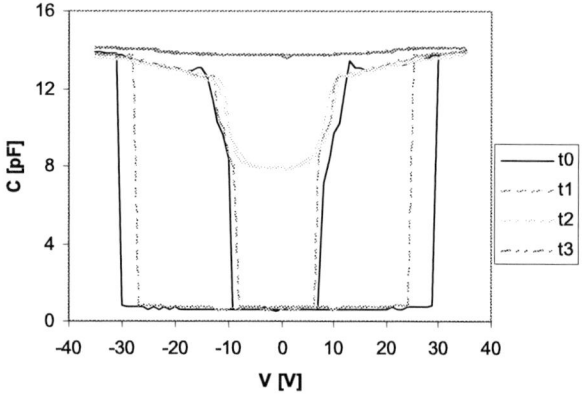

Figure 4: Typical shift of C-V curve due to prolonged actuation of the RF MEMS switch during accelerated testing

The amount of papers investigating creep failure mechanisms is limited [5-8]. However, for MEMS devices based on metal alloys such as aluminum, creep could occur during application of the device when insufficient care is taken. From equation (1) it is clear that

a change in the gap due to creep phenomena will have a large impact in the resulting V_{pi} value. Therefore it is important to investigate the creep properties of the materials used and to optimize the design of the RF MEMS devices or the used materials in order to limit the effect of creep. This paper focuses on experimental and numerical methodologies to evaluate the occurrence of creep in RF MEMS devices. Because during application the charging and creep phenomena occur simultaneously, it is important to be able to distinguish between the two failure mechanisms.

2. Creep properties of thin film aluminum

The material used as the moving top electrode in the RF MEMS devices is an aluminum based alloy which is deposited in commercial PVD equipment. The thickness of the aluminum layer is approximately 3-5 µm. It is well known that the behavior of materials at these small length scales does not necessarily correspond to the behavior of bulk materials. This is due to the fact that the grain sizes are in the same order of magnitude as compared to the characteristic dimensions of the structures, which invalidates the use of properties measured on bulk specimen and may imply that continuum mechanics no longer apply. Studies related to this size effect are numerous during the last 10-15 years [9-11]. Due to this size effect and the small dimension of the devices of interest, novel experimental methods and material models are required. However, at the present date no validated and usable constitutive models based on the microstructure of the material are available in commercial finite element codes. Therefore we stick to the continuum based constitutive equations of elasticity, plasticity and creep.

Creep is usually characterized by measuring the creep strain evolution at a certain constant stress level. There are different stages of creep as can be seen in Figure 5. The first stage is called primary creep and shows a decreasing creep rate. The next stage is secondary or steady state creep: equilibrium is established between the deformation and recovery mechanics. For lifetime predictions this steady state creep is the most important region and most constitutive creep models are developed for this region. The last stage is called tertiary creep and shows an increasing creep finally resulting in failure.

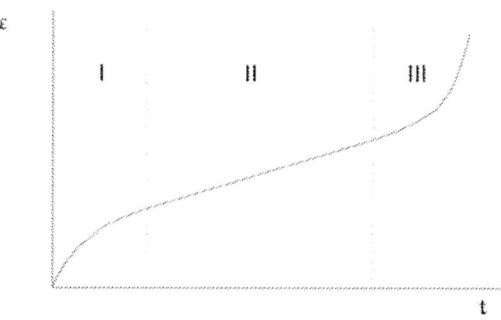

Figure 5: Typical creep strain evolution at a constant stress level showing primary, secondary and tertiary creep regions

The techniques commonly used in literature for measuring the creep properties of thin (aluminum) layers are wafer curvature, nano-indentation or the use of dedicated test structures. These techniques all have their specific pro and cons:

- The wafer curvature technique uses standard (silicon) wafers where on one side a layer of aluminum is deposited. The curvature of the wafer is measured as function of temperature and time. The stress σ_m in the aluminum layer can be calculated from this curvature using the well-known Stoney's equation:

$$\sigma_m = \frac{E_s}{6(1 - v_s)} \frac{t_s^2}{t_f} \left(\frac{1}{R_2} - \frac{1}{R_1} \right) \qquad (2)$$

with E_s and v_s the Young's modulus and Poisson's ratio of the substrate, respectively; t_s the substrate thickness and t_f the film thickness. R_2 is the radius of curvature after processing and R_1 the initial radius of curvature of the wafer before film deposition.

Advantages of the wafer curvature method are its simplicity and accuracy. A disadvantage is the fact that the thin film is not free standing because it is constrained by the silicon substrate. This could influence plasticity and creep behavior because the constraint surface can hinder dislocation movements. Another disadvantage is the fact that the stress and temperature levels can't be controlled independently.

- With nano-indentation a small, sharp diamond tip is indented in the material. From the resulting force-displacement curve the constitutive behavior of the material can be extracted. Advantage of this technique is the ability to control the stress levels. A drawback is the fact that only a very small volume of the material is tested, which is not necessarily representative for the overall behavior. Another drawback is the interpretation of the data, which requires sophisticated finite element simulations to extract the constitutive models.

- The use of dedicated test structures, which resemble the actual device, is usually preferred. However, it is often difficult to accurately control and measure the small forces and displacements.

In this paper both the wafer curvature technique and dedicated test structures are used for characterizing the creep behavior of the aluminum material. Modlinski and co-authors have published data for creep relaxation using the wafer curvature technique [5-8]. In these papers measured data for AlCu is presented together with a fitted constitutive creep model. Because the investigated material and thickness are comparable to our material, we used this data as start of our creep investigation. In [7] the creep relaxation was measured at 72 °C and 101 °C and the following relation, based on dislocation glide, is fitted to the data:

$$\sigma = -\frac{kT}{a} \ln \left[\exp \left(-\frac{a\sigma_0}{kT} \right) + \frac{ac}{kT} t \right] \qquad (3)$$

with

$$a = \frac{\Delta F}{\tau \sqrt{3}} \qquad (4)$$

,

$$c = \left\{ \left[\frac{\dot{\gamma}_0}{2\sqrt{3}} \right] \left(\frac{E}{1 - v} \right) \right\} \exp \left(-\frac{\Delta F}{kT} \right) \qquad (5)$$

σ_0 the biaxial stress at the beginning of the relaxation, k the Boltzmann constant, $\dot{\gamma}_0$ a constant and T the absolute temperature. The fitted values of ΔF and τ for AlCu are 4 eV and 252 MPa respectively.

3. Numerical methodologies

The Finite Element method is used for predicting the behavior of the RF MEMS switches. This is a challenging task due to the multiphysical and nonlinear behavior of the switch:

- The electrostatic force for each lateral position should be predicted as function of the displacement of the top electrode.
- The electrostatic force is non-linearly dependent on the gap. Combined with the linear dependency of the spring force this results in a non-linear response with snap-back behavior and unstable equilibrium.
- Plasticity and creep in the aluminum during actuation should be taken into account resulting in non-linear material behavior.

A detailed description of the simulation methodology with respect to the first two points is presented in these proceedings [15]. In this paper the creep constitutive model for aluminum is added to the presented methodology. This creep equation is implemented in the commercial finite element program ANSYS multiphysics 9.0® [12] using a steady-state creep equation of the exponential form:

$$\dot{\overline{\varepsilon}}_{vm} = C_1 \exp \left(\frac{\overline{\sigma}_{vm}}{C_2} - \frac{C_3}{T} \right) \qquad (6)$$

Equation (3) can be rewritten in the form of equation (6) as shown in [8,13]

With this implementation the creep relaxation wafer curvature measurements are reproduced. The predicted stress relaxation curves are visualized in Figure 6 and indicate the correctness of the FE implementation. The predicted strain rate as function of the Von Mises stress is plotted in Figure 7 for the two different temperatures. It is clear that a large stress dependency exists with significant strain rates starting at 80 MPa.

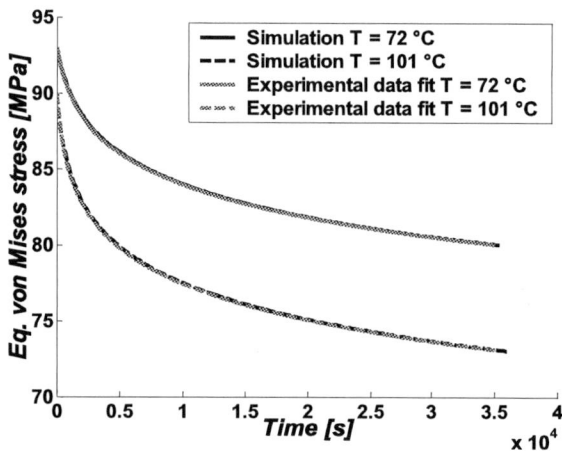

Figure 6: Isothermal stress relaxation experimental data fits compared with the FE predictions

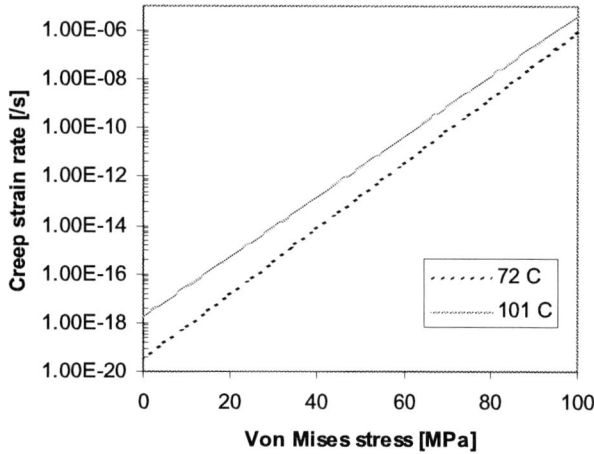

Figure 7: Creep strain rate as function of the Von Mises stress

4. Experimental procedures

4.1 Test structures

In order to evaluate the occurrence of creep on the RF MEMS switches, dedicated test structures are designed and manufactured. The manufacturing process used to fabricate the test structures is identical to the NXP RF MEMS process. This facilitates the transformation of the results to the actual RF MEMS devices.

An important characteristic of the test structures is the ability to distinguish between deformation due to charging and that due to creep. For the actual RF MEMS switches this is difficult because both result in similar shifts of the C-V curve. Figure 8 is a FE model showing the principle of the designed test structure. The top electrode consists of three portions that can be actuated separately by three independent bottom electrodes.

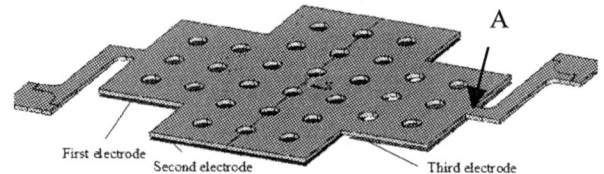

Figure 8: Schematic of the designed test structure showing a central electrode together with two side electrode which can be actuated separately.

For evaluating creep the idea is to actuate the middle electrode above the pull-in voltage. After actuation the middle plate is contacting the SiN surface whereas the side electrodes and the springs will deform but will not touch the SiN layer. Charging at the side electrodes and the springs will be minimized because there is no contact with the SiN layer The ratio in stiffness between the spring and the side electrode will determine the amount of deflections in the spring and side electrodes.

After actuation of the middle plate additional voltages can be placed on the side electrodes in order to create an electrostatic force, which further deflects the side electrodes and springs. In case of creep the springs will show an increased deflection as function of time. This can be monitored using electrical and optical measurement techniques. The principle of the test structure is also visualized in Figure 9.

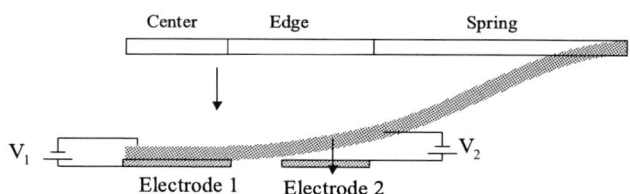

Figure 9: Side view schematic of the designed test structure illustrating the working principle.

Eight different versions with varying spring length and electrode size are fabricated of this test structure.

5. Results

5.1 Experiments

Figure 10 shows photographs of two test structures with different spring length fabricated in the RF MEMS process. The shorter spring length of the left structure will have higher stress levels during actuation and thus a higher risk of creep strain evolution at elevated temperatures.

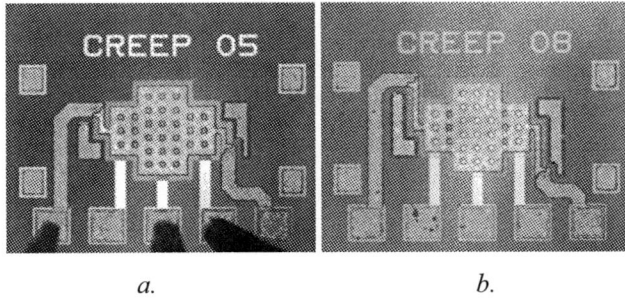

Figure 10: Photographs of two fabricated test structures (creep05 and creep08) with different spring lengths.

For triggering creep the creep05 structure is actuated at different temperatures during a prolonged time. Voltage is applied at the central electrode until the central electrode is pulled down on the SiN layer. While keeping this voltage on the central electrode an additional voltage of 30V is applied on the right electrode causing a further deflection of the side electrode and an increase in mechanical stresses in the spring. The deflection of the electrodes is measured with a white light interferometer at certain time intervals. A typical measured shape of the structure is depicted in Figure 11 showing the deflections of the springs and the slope of the right electrode.

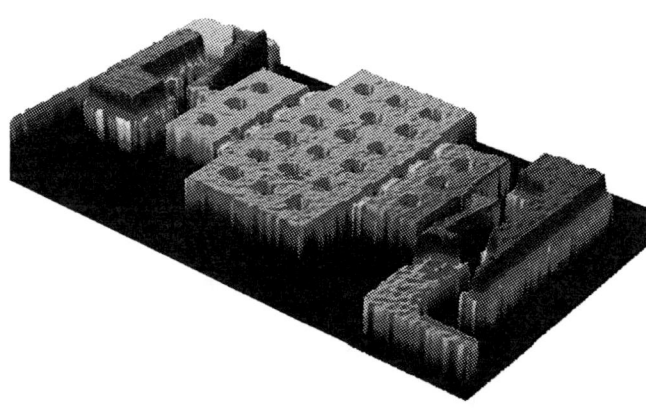

Figure 11: Deflection of the creep05 test structure during actuation measured with white light interferometry.

When creep strains accumulate, it causes the side electrode to further deflect and thus reduce the gap. At a certain critical deflection the side electrode will show a pull-in behavior and snaps down on the SiN layer. After this pull-in the test is stopped and the system is allowed to recover. For short springs it is possible that this pull-in of the side electrodes causes plastic deformation of the springs, which also prohibits recovery.

In Figure 12 the deflection at the end of the spring (point A in Figure 8) is plotted as function of time. From this graph it appears that at 25 °C no creep effects are visible. However, at 75 °C and 100 °C a clear increase in deflection is measured which finally results in a snap down of the side electrode. After relaxation (> 11 days) a permanent deformation is measured which is larger for higher temperatures. This permanent deformation is caused by the developed creep strains and the plastic deformation.

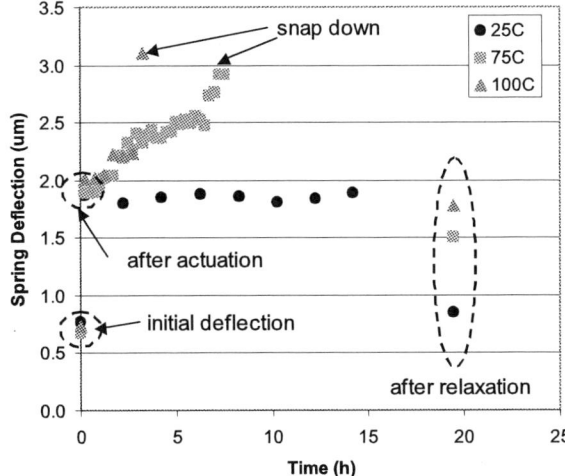

Figure 12: Measured creep deflection for creep05 structure at 3 different temperatures. Relaxation data obtained after 11 days but visualized on arbitrary time in the graph for comparison.

5.2 Simulations

The experiment at 75 °C on the creep05 structures presented in the previous section is also simulated with the FE methodology described in section 3. The Von Mises stresses in the spring of the structure predicted after 8.5 hours of actuation at are visualized in Figure 13. From this picture it is clear that stress concentrations exist near the corners and that the overall stress levels are relative high compared with values in Figure 7.

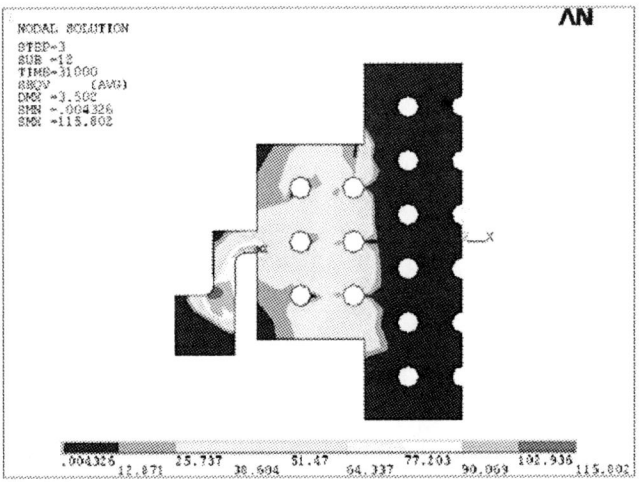

Figure 13: Predicted Von Mises stress levels in the springs of test structure creep05

The predicted equivalent creep strains after 8.5 hours of actuation are visualized in Figure 14. The comparison between the predicted increase of deflection and the measured values is visualized in Figure 15.

513

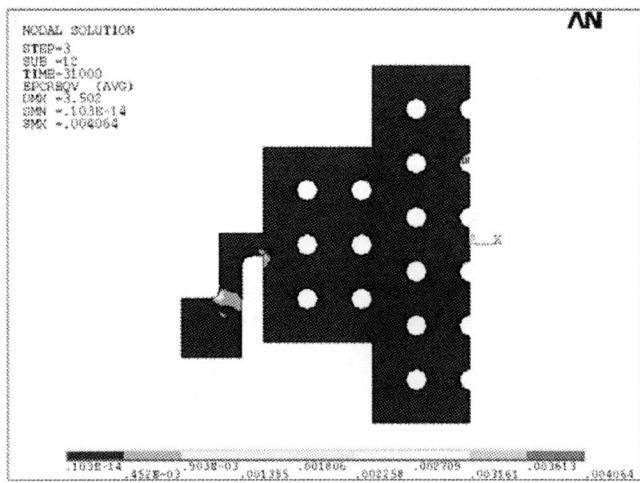

Figure 14: Predicted equivalent creep strain levels in the springs of test structure creep05 after 8.5 hours actuation at 75 °C

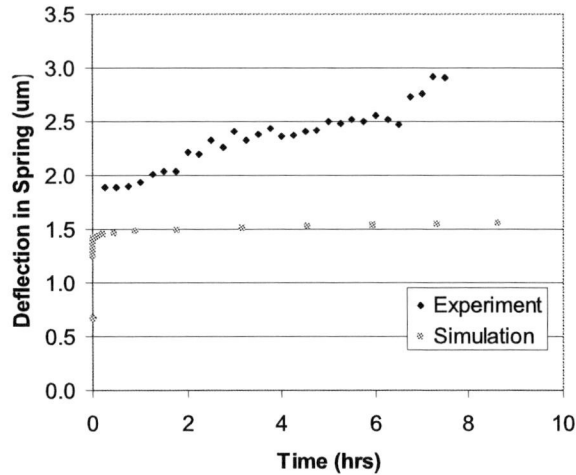

Figure 15: Comparison between measured and predicted deflection due to creep at 75 °C for the creep05 structure.

It is clear that although both the experimental results and the numerical prediction show a creep effect, the difference is large. Apparently the creep constitutive equation from literature for AlCu is not representative for our material.

6. Conclusions

We presented a combined experimental and numerical methodology for the evaluation of creep in the aluminum top electrode of RF MEMS switches. The FE model for the RF MEMS switches including the creep constitutive equation is capable of predicting the creep relaxation measurements. Measurements on the designed test structures indicated the occurrence of creep in the springs. However, the predicted creep levels for the test structures are much lower than the observed values.

Further evaluation of creep degradation using these test structures will be performed in the future. In conjunction to these test we will also perform curvature relaxation measurements on the material used in the RF MEMS switches in order to optimise our constitutive creep equation.

References

1. J.T.M. van Beek *et al*, "MEMS for wireless communication: application, technology opportunities and issues", *European Microwave week 2006*, Manchester.
2. R. W. Herfst, H. G. A. Huizing, P. G. Steeneken, J.Schmitz, "Characterization of dielectric charging in RF MEMS capacitive switches", *Proceedings of the 2006 International Conference on Microelectronic Test Structures*, Austin, TX, USA, IEEE Electron Devices Soc., March 6-9, pp.133-136, 2006.
3. Goldsmith, C. *et al*, "Lifetime characterization of capacitive RF MEMS switches," *Dig.IEEE Int. Microwave Symp.*, 2001, pp. 227-230.
4. Spengen, M. van *et al.*, "A comprehensive model to predict the charging and reliability of capacitive RF MEMS switches", *J. Micromech. Microeng.* Vol. 14 pp. 514-521, 2004.
5. Modlinski, R *et al*, "Creep as a reliability problem in MEMS" *Microelectron. Reliab.* Vol. 44, No 9-11, pp. 1733-1738, 2004.
6. Witvrouw, A. *et al*, "Stress relaxation in Al-Cu and Al-Si-Cu thin films"*J. Mater. Res.* Vol. 14, No. 4, pp. 1246-1254, 1999.
7. Modlinski, R. *et al*, "Creep Characterization of Al alloy thin films for use in MEMS applications", *J. Microelectron. Eng.*, Vol. 76, No 1-4, pp. 272-278, 2004.
8. Modlinski, R. *et al*, "Creep-resistant aluminum alloys for use in MEMS", *J. Micromech. Microeng.* Vol. 15 pp. S165-S170, 2005.
9. Nix, W.D., "Mechanical properties of Thin Films", *Metall. Trans A*, Vol. 20A, No. 11, pp. 2217-2245, 1989.
10. Arzt, E., "Size effects in materials due to microstructural and dimensional constraints: a comparative review", *Acta Materialia*, Vol. 46, No. 16, pp. 5611-5626, 1998
11. Burg, V., Characterization of freestanding thin film properties for RF-MEMS, Ëindhoven University of Technology, Master thesis, Internal Report (2006)
12. Ansys™90 theory manual, Ansys Inc
13. Eijden, R. van, "Predicting the influence of aluminum creep on the performance of RF MEMS switches using FE simulations", Eindhoven University of Technology, Internal Report (2006)
14. Rebeiz G. M., RF MEMS: Theory, Design and Technology (New York: Wiley), 2002.
15. Bielen J. and Stulemeijer, J., "Efficient electrostatic-mechanical modeling of C-V curves of RF-MEMS switches", *Proceedings EuroSimE2007*, London, 15-18 april 2007.

514

Investigation of Carbon Nanotube Performance under External Mechanical Stresses and Moisture

Haibo Fan, Kai Zhang, and Matthew M.F. Yuen
Department of Mechanical Engineering,
Hong Kong University of Science and Technology
Clear Water Bay, Kowloon, Hong Kong SAR, China

Abstract

Due to its remarkable properties, carbon nanotube (CNT) was widely used in different areas, especially in electronic packaging for the improvement of the adhesion and thermal conductivity. CNT as an emerging thermal interface material (TIM) is now widely used to improve thermal dissipation in electronic packaging. CNTs suffer from moisture or high interfacial stresses from the mismatch of the thermal expansion coefficient between different layers in packages during fabrication and assembly. Both of these factors have an effect on the material performance of CNTs. To apply CNTs in electronic packaging, it is important for us to understand these loading effects on the thermal performance of CNTs at an atomic level. Molecular dynamics (MD) simulation is a proper method to study these material properties of CNTs.

In this study MD simulations were conducted to investigate the thermal conduction of CNT under different conditions, including mechanical stresses and moisture. A series of MD models were built using the Materials studio software (Accelrys, Inc). Based on Fourier's law, thermal conductivities of the SWCNT under different conditions were calculated. The MD simulation results showed that the thermal conductivity of SWCNT subjected to axial stress decreased with the stress changing from the compression to tension. Both of moisture and torsion stress degraded the intrinsic thermal conductivity of CNT. This MD simulation gave a basic understanding of the surrounding effect on thermal performance of CNTs and provided information for the assembly of CNT in electronic packaging

Introduction

Thermal dissipation is a key issue in electronic packaging design. Ineffective thermal dissipation inside packages can result in not only delamination failure between different layers but reduce life cycle of the packages. Compared with traditional TIM, carbon nanotube (CNT) has excellent mechanical and thermal properties, and is now attracted more attention of researchers as a TIM to dissipate heat from die to heat sink in electronic packaging.

Investigation of the thermal performance of CNT is rather important for the thermal dissipation in electronic packaging. Due to technically difficulty of experimental measurement of the CNT thermal conductivity, MD simulation of is powerful technique to investigate heat transport in nanostructure. Molecular modeling represents molecular structures numerically and simulates their behavior with the equations of quantum and classical physics and it is one of the fastest growing fields in science.

MD simulation has been successfully applied to predict the thermal conductivity of single walled CNT (SWCNT) or multi-walled CNT (MWCNT) [1-7]. The predicted thermal conductivity of SWCNT or MWCNT varied from several hundreds to thousands w/km. These MD simulations mainly were focused on the CNTs without considering the surrounding effects. However, CNTs are often subjected to moisture and external stresses, which normally are resulted from CNT-TIM assembly, or CTE mismatch of interface materials connected with the two ends of the CNT during the package working environment. These external stresses can not only result in the deformation of CNT but the change the CNT properties. Guo et al. [8] studied the effect of the tensile loading and electronic field on the mechanical and electrostatic properties. They found that the electronic polarization and mechanical deformation induced by an electric field could result in significant change of the electronic properties of a CNT, and the tensile load changed the energy gap of the tube affecting the field emission properties of CNTs. Ding et al. [9] also found that the electronic conductance depended on the strain of metallic armchair CNTs. However, little attention has been focused on the investigation of external stress effect on the thermal conduction of CNT. The mechanism of thermal conduction of CNT under different environmental conditions is not very clear. How do the CNTs perform under these conditions in real packages? Understanding of these issues is rather important for the design of CNT as TIM in electronic packaging. MD simulation can provide fundamental knowledge to predict these effects on the material properties of CNTs at a fundamental level

In this study, MD simulations were conducted to investigate thermal conductivity of CNT under different conditions. Heat flux was applied on the system and temperature in each region of the SWCNT was calculated from the velocities of atoms from MD simulation. Based on Fourier's law, thermal conductivity was obtained from the heat flux and the calculated temperature gradient along the SWCNT. The MD simulation results showed that the surrounding conditions except for compression stress have a negative effect on the thermal conductivity of CNT.

MD simulation of the thermal conductivity of SWCNT under different conditions

Basically, two kinds of MD simulation methods were used for investigation of the heat transfer in CNT. One is equilibrium molecular dynamics (EMD) method based on Green-Kubo relation. The other is non-equilibrium molecular dynamics (NEMD) method based on Fourier's

1-4244-1105-X/07/$25.00 ©2007 IEEE

law. Due to the difficulty of converging the heat flux and the complexity of the autocorrelation function in EMD method, NEMD method was used to predict thermal conductivity by Fan et al. [7]. The same method was also used in this study.

MD simulations were conducted using the Materials Studio (Accelrys) software. The condensed-phase optimized molecular potential for atomistic simulation studies (COMPASS) force field was used in the simulation. COMPASS force field enables accurate and simultaneous prediction of structural, conformational, vibrational, and thermophysical properties for a broad range of molecules in isolation and in condense phases under a wide range of conditions of temperature and pressure. Tersoff-Brenner potential, whose parameters were taken from experiments on diamond and graphite, was often used in some MD simulation of CNT thermal conductivity. However some issues occurred during the MD simulation as mentioned by Bi et al.[3] that the Tersoff-Brenner excluding van der Waals force among the atoms may make longer tube bending which resulted in failure of the thermal conductivity calculation. Moreover, the Tersoff-Brenner potential without nonbonding forces is not suitable to investigate the interaction between CNT and water molecules in this study. With both covalent bonding energy and non-bonding energy COMPASS force filed can reproduce the behavior of CNT reliably in this study.

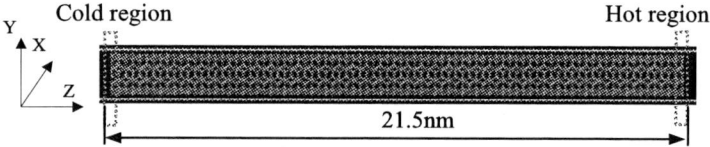

Fig 1. MD model of SWCNT with a finite length under different conditions

As we known, the thermal conductivity of SWCNT is dependent on the length, diameter of SWCNT, defects and temperature. In order to avoid other effects on the CNT thermal conductivity, MD simulation was focused on the defect-free SWCNT with a certain length at room temperature in this study. A (10, 10) SWCNT model with a length of 24.5nm was built with the simulation box periodical in x and y directions, as shown in Fig.1. The axial stresses (including tension and compression stresses) or torsion stress, respectively, were applied to the two ends of the CNT. The stress value varied from 0 to 500Mpa, which are in the range of interfacial stresses occurring in electronic packages.

It is now accepted that open-end CNTs connected with other substrate materials can improve the interfacial adhesion between the CNT and substrates. However, moisture can also penetrate into the tube with larger diameter from the open ends during the assembly. In order to investigate moisture effect on the thermal conductivity of CNTs, MD models with water molecules confined inside the CNT were also built. The water molecules were initially randomly put inside the tube and water molecules were

allowed to move freely in all directions. Energy minimization was performed to find the equilibrated structure of the system. The mass ratio of water molecules inside the tube to the CNT varied from 1.75% to 6.50%.

The MD model was divided into N regions along the z axial direction of the CNT. The two regions at the two ends were defined as a hot and a clod region respectively. Based on the NEMD algorithm proposed by Ikeshoji and hafskjold [10], a constant energy was input to the hot region and the same amount of energy was removed from the cold region. The velocities of atoms in both the hot and cold regions were scaled to accomplish the energy transfer from the hot region to the cold region, in which both the energy and momentum are conserved for the whole system. The heat flux in the system was defined as follows:

$$J = \frac{\Delta E}{S \Delta t} \qquad (1)$$

where J is the heat flux of the CNT, S is the cross-section area of the CNT, and Δt is the simulation time step.

The instantaneous local temperature in each region can be found by:

$$T_k = \frac{1}{3 n_k k_B} \sum_{i=1}^{n_k} m_i v_i^2 \qquad (2)$$

where n_k is the amount of atom in region k, k_B is Boltzmann's constant, m_i and v_i, respectively, are the mass and velocity of atom i.

Based on the temperature gradient along the CNT and the heat flux, the thermal conductivity of the CNT, λ, is obtained by the Fourier law:

$$\lambda = \frac{<J>}{<\partial T / \partial Z>} \qquad (3)$$

where $\partial T / \partial Z$ is the temperature gradient along the CNT and the brackets denote a statistical time average.

All the systems were initially equilibrated for 80ps at the room temperature using the ensemble of the constant number of particles, volume and temperature (NVT). A small energy was then imposed on the system. NEMD simulations were conducted based on the algorithm proposed by Ikeshoji and hafskjold [10] using the ensemble of the constant number of particles, volume and energy (NVE). All the MD simulations run for a long time until a steady state temperature distribution achieved along the SWCNT. The velocity verlet algorithm was used for integration in all MD simulations. The non-bonded interactions included van der Waals and electrostatic forces. The atom based summation with a cutoff distance of 0.95nm was used for the dispersion interactions. All the simulations were performed with an interval of 1 femtosecond (fs) in each MD simulation step. For each kind of MD model, velocities of atoms in each region were averaged over last 1ps period to the temperature calculation along the SWCNT.

Results and discussion

Table 1: Thermal conductivity of CNT under different conditions

	Stress/moisture-free CNT	CNT under axial stresses						Torsion stress (Mpa)				Mass ratio of water molecules to the CNT (%)		
		Compression (Mpa)		Tension (MPa)										
		150	50	50	150	300	500	50	150	300	500	1.75	3.32	6.50
Thermal conductivity (W/mk)	255.7	272.8	265.5	241.8	227.7	174.8	158.9	244.8	208.2	179.6	166.5	193.2	160.4	130.3

Figure 2 showed the temperature profile along the stress-free SWCNT at room temperature. Temperature distributed linearly along the SWCNT and the slope of the fitted straight line showed the temperature gradient. The heat flux was calculated by the energy and the cross section area with a 3.4Å thick annular ring. Based on the above equations, the thermal conductivity of the SWCNT under different conditions was calculated for the SWCNT, and the results were listed in Table 1. It was found that both the external stresses and moisture had a large effect on the thermal conductivity of SWCNT. The thermal conductivity of CNT with water molecules confined inside the tube was smaller that that of the stress-moisture-free SWCNT and decreased with the increase of decreases with increase of moisture inside the tube. The degradation of the thermal conductivity was resulted from the increase of the phonon scattering caused by water molecules. Water molecules inside the tube interact with their neighboring carbon atoms in the tube, which acts as additional scattering centers resulting in the mean free path of phonon and significantly reduced the thermal conductivity.

From the above MD simulations, it was found that the thermal conductivity of SWCNT was dependent on the applied stress. For those CNT subjected to axial stresses, the thermal conductivity monotonically decreased with the stress changing from the compression to tension. The change of the thermal conductivity varied from 4% to 38%. The thermal conductivity of the CNT subjected to torsion stress also decreased with the increase of torsion stress. These results indicated that the stresses had an effect on the transport properties of CNT, not only the electrical conductivity [8-9] but thermal conductivity.

The stress effect on the thermal conductivity can be explained by the kinetic theory. The thermal conductivity of CNT is dominant by phonon scattering and decided by the specific heat capacity, velocity and mean free path of phonon. At the given temperature, the specific heat capacity is the constant and the velocity and mean free path of phonon are affected by the external stresses applied on the CNT. Picu et al [11] investigated the stain and size effect on heat transport in nanostructures and found that the lattice thermal conductivity reduced or enhanced by the tensile or compressive load applied on the nanostructure. They addressed that the phonon velocity was proportional to the square root of the stiffness of the nanostructure which increased in compression and decreased in tension. They also pointed out that the variation of mean free path resulted from the lattice anharmonicity was a dominant factor on the strain dependence of the thermal conductivity. The same conclusions were given by Bhowmick and Shenoy [12] who presented a systematic study of strain effect on the thermal conductivity of an insulating solid.

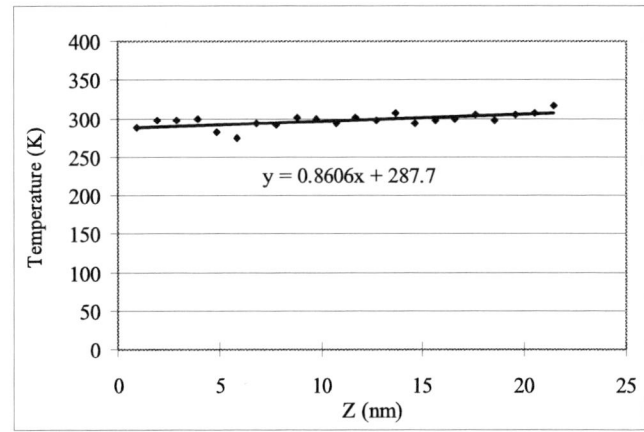

Fig. 2: Temperature distribution along the SWCNT at room temperature

These results are all consistent with the results from MD simulations in this study confirming that the intrinsic thermal conductivity of SWCNT was affected by external stresses. The finding in this study can be also used to explain why the thermal conductivity of a (10,10) CNT is higher than that of a (5,5) CNT. The higher stretching strain along the circumference in a (5,5) CNT made by a rolled grapheme sheet reduced the mean free path of phonon, which results in lower thermal conductivity, which was also mentioned by Che et al [1]. However, attention should be paid to the compression stress effect on the thermal conductivity of CNT. Due to the hollow structure of the CNT, CNT under larger compression stress would buckle or even collapse [13], which should heavily affect the phonon transfer and result in degradation of thermal conductivity.

Except for the compression stress, moisture, tensile and torsion stresses have a negative effect on the thermal conductivity of CNT. That will significantly affect the thermal performance of CNT based assemblies, especially CNT array as TIM in electronic packaging. The induced ineffective thermal dissipation can threaten the reliability of the electronic packages. Therefore, dry environment during the assembly of CNT based TIM is needed to avoid moisture effect, and CTE matched materials connected with CNT should be considered in the design, which is important to enhance the performance of CNT based TIM in electronic packaging.

The results from MD simulations predicted the operational environmental effects on the thermal conductivity of SWCNT. The results indicate that more attention has to be paid to the CNT assembly to prevent the environment-induced degradation of the thermal performance of CNT-array TIM in electronic packages. The present MD simulation approach primarily provides a qualitative prediction of the thermal performance of SWCNT under different conditions. Further studies will be focused on the thermal resistance between CNTs and other electronic materials, such as heat sink and silicon chip.

Conclusions

The study was focused on the operational environmental effects on the thermal conductivity of SWCNT using molecular dynamic simulations. MD results showed that all factor except for compression stress in this study had a negative effect on the thermal conductivity of CNT. The thermal conductivity of SWCNT subjected to axial stress decreased with the stress changing from the compression to tension. While moisture and torsion stress degraded the thermal conductivity of CNT. It is essential to prevent moisture ingress into the thermal interface material. Therefore, attention should be paid to the CNT assembly and materials selection, which is significant for the thermal dissipation in electronic packaging.

References

[1] J. Che, T. Cagin, and W. A. Goddard III, "Thermal conductivity of carbon nanotube," *Nanotechnology* Vol.11, pp.65-69, 2000.

[2] Z. Yao, J. Wang, B. Li, and G. Liu, "Thermal conduction of carbon nanotubes using molecular dynamics," *Phys rev B* Vol.71, 085417, 2005.

[3] B. Bi, Y. Chen, J. Yang, Y. Wang, and M. Chen, "Molecular dynamics simulation of thermal conductivity of single-wall carbon nanotube," *Physics letters* A Vol. 350 pp.150-153, 2006.

[4] M. A. Osman, and D. Srivastava, "Temperature dependence of the thermal conductivity of single-wall carbon nanotube," *Nanotechnology* Vol.12, pp.21-24, 2001.

[5] S. Maruyama, "A molecular dynamics simulation of theat conduction of a finite length single-wallled carbon nanotube," *Microscale Thermophys Eng* Vol.7, pp.41-50, 2003.

[6] K Zhang, H B Fan, M M F Yuen, "Molecular dynamics study of the thermal conductivity of CNT-array-thermal interface material," In: Proc. EMAP, Hong Kong, pp 113-116, 2006.

[7] H B Fan, K Fan, M M F Yuen, "Effect of defects on thermal performance of carbon nanotube investigated by molecular dynamics simulation," In: Proc. EMAP, Hong Kong, pp 451-454, 2006.

[8] Y Guo, and W Guo, "Mechanical and electrostatic properties of carbon nanotubes under tensile loading and electric field," *J. Phys. D: Appl. Phys.* pp.805-811, 2003.

[9] J W Ding, X H Yan, J X Cao, D L Wang , Y Tang and Q B Yang, "Curvature and strain effect on electronic properties of single-wall carbon nanotubes," *J.Phys.: Condes. Mater.* Vol.15, pp.439-445, 2003.

[10] T. Ikeshoji and B. Hafskjold, "Non-equilibrium molecular dynamics calculation of heat conduction in liquid and through liquid-gas interface," *Molecular Physics* Vol.81 pp.251-261, 1994.

[11] R C Picu, T Bocra-Tasciuc, and M C Pavel, "Strain and size effect on heat transport in nanostructure," *J. Appl. Phys.* Vol.93 pp.3535-3539, 2003.

[12] S Bhowmick and V B Shenoy, "Effect of strain on the thermal conductivity of solids," *J.Chem. Phys.* Vol.125, 164513, 2006.

[13] K M Liew, C H Wong, X Q He, M J Tan and S A Meguid, "Nanomechanics of single and multiwalled carbon nanotubes," *Phys. Rev.*, Vol. 69, 115429, 2004.

Wafer Probing Simulation for Copper Bond Pad Based BPOA Structure

Yumin Liu

Fairchild Semiconductor (Suzhou) Corp., 1 Sutong Road, Suzhou, China

Yong Liu, Scott Irving and Timwah Luk
Fairchild Semiconductor Corp., 82 Running Hill Road, South Portland, ME 04106, USA
Email: yliu@fairchildsemi.com; Phone: (207) 761-3155; Fax: (207) 761-6339

Abstract

The bond pad metallization in the bond pad over active (BPOA) structure as well as interconnection lines in the device are shifting from aluminum to copper in recent years, because copper has better mechanical property and improved electrical conductivity compared with Al. In order to prevent copper oxidization, two extra thin layers of Ni and Au are added above the copper pad. With this strucure, reliable wire bonding can be formed. On the other hand, in order to further reduce the die size and fully utilize the space below bond pads, a layout technology, termed Bond Pad Over Active (BPOA), is developed. The BPOA strucure may induce some reliability concern that the wire bonding stress and probing sress may cause failures in the underlying devices. In wafer probing test, both electrical and mechanical contacts are made between the probe tip and the bond pad. This may impact the structures below the bond pad, such as dielectric layers and active devices. It may induce very high stress in dielectric and passivation layers since the contact area of probe tip is very small. In this paper, the impact of passivation thickness, and bond pad metal layer thickness is investigated, as well as the probe tip shape and friction between the probe and bond pad by the finite element simulations. Simulation results show that the trends of different parameters impact on 2-D results match well with the 3-D results.

1. Introduction

The bond pad metallization in the bond pad over active (BPOA) structure as well as interconnection lines in the device are shifting from aluminum to copper in recent years. Fundamentally the Young's modulus and yield strength of copper are much higher than aluminum materials. Another advantage of copper is the improved electrical conductivity compared to Al. However, copper oxide, unlike aluminum oxide, cannot act as a protective film to keep the underlying copper layer from further oxidation while exposed to air at the elevated temperature [1]. Reliable bond wire cannot be formed on such oxidized copper surface. Probing bare copper bond pads also challenges existing wafer-probe technology. In order to prevent copper oxidization, one method currently utilized in the semiconductor industry is to add an extra aluminum layer on the copper bond pad [2-3]. This method will add extra cost in the copper-fab process.

Some research proposed to add an ultra-thin (<30A) inorganic passivation layer on copper pads [4]. One disadvantage of this method is that the passivation layer would need to be deposited after wafer probing test which also infers that oxidation on the wafer would need to be cleaned off before the passivation film is deposited. In this paper, a new method of using copper bond pad with two extra thin layers of Ni and Au is proposed. Reliable bond wire can be formed since the top layer of the bond pad is gold. In order to avoid cracks in interconnect structure and passivation layer, the thickness of Cu, Ni and SiN is optimized by finite element simulations.

Cost competitiveness is a major driving force in the semiconductor industry. Die size reduction is a top priority for cost competitiveness. Advances in processing technology have shrunk the device sizes, resulting in a smaller die core size. However, the space below wire-bond pads has remained relatively underutilized because of the reliability concern that the wire bonding stress and probing stress can cause failures in the underlying devices. Recent studies have examined wire bonding stresses and attempted to improve the use of space below wire-bond pads [5-7]. In this paper, a layout technology, termed Bond Pad Over Active (BPOA), is developed to allow the placement of bond pads over active silicon. The reliability of the copper bond pad based BPOA structure is studied during the wafer probing test.

Wafer probing tests are used to evaluate the electrical performance for most manufactured semiconductor devices before assembly. During the test, both electrical and mechanical contacts are made between the probe tip and the bond pad. This may impact the structures below the bond pad, such as dielectric layers and active devices. It may induce very high stress in dielectric and passivation layers since the contact area of probe tip is very small. Both 2-D and 3-D models are used in the simulations. The impact of passivation thickness and bond pad metal layer thickness is investigated, as well as the probe tip shape and friction between the probe and bond pad. Simulation results show that the trends of different parameters impact of 2-D results match well with the 3-D results.

2. Setting Up the 2-D model

Firstly, a 2-D model of the copper based BPOA strucutre during probing is built. The advantages of such a model are its few meshing constraints, small file size, and fast solution time. It is acceptabel to use the 2-D modeling results for comparative purposes rather than to predict absolute values. A BPOA copper based bond pad structure is shown in Fig. 1. The active circuits are under the bond pad. The bond pad includes an Au film at top, then a Ni film and a thick copper layer and a TiW film at the bottom. Below TiW it is the SiN layer. Then the vias, M1, M2 lines and TEOS layers. The FEA solid model and mesh are shown in Figs. 2-3.

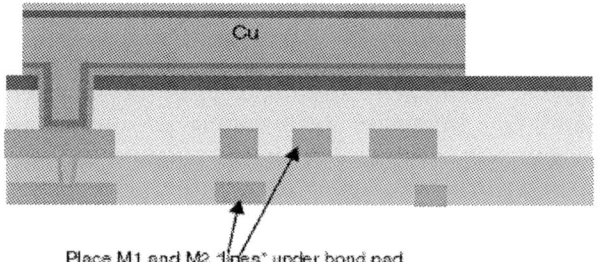

Fig. 1 BPOA structure with copper bond pad

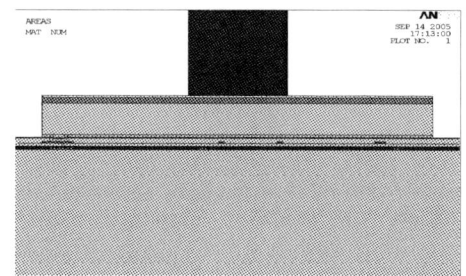

Fig. 2 BPOA FEA 2-D solid model (flat probe

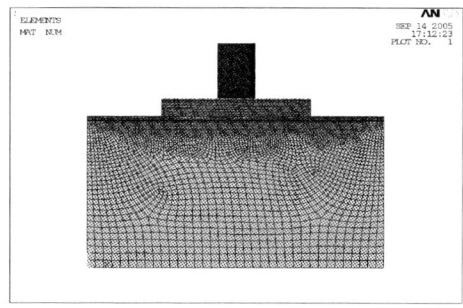

Fig. 3 BPOA contact model with mesh

Table 1. OT vs. Contact Force (BCF=2.15g/mil)

Over Travel (mils)	Contact Force P (gram)	Contact Force P (mN)	Contact pressure p (GPa)
2	4.31	42.2	0.0833
4	8.62	84.4	0.167
6	12.93	126.7	0.25
8	17.2	168.9	0.333
10	21.55	211.1	0.417

The relationship between probe over travel (OT) and probe tip force has been thoroghly studied in [8], which is listed in Table 1. The friction coefficient in the contact model between bond pad and probe tip is 0.3-0.6 (Probe tip vs Au film and then penetrate through Au and Ni films to contact copper). In order to examine the impact of friciton, different friction coefficients are selected for the modeling. The horizontal probe scrub (DX) during probe touch down is about 15 um based on testing. Probe tip is subjected to two loads. One is the probe OT which is converted into probe tip pressure, and the other is the probe scrub.

3. Simulation Results and Discussion of 2-D model
(1) Effect of M2 thickness
The model parameters and simulation results of the metal 2 layer thickness effect are summarized in Table 2. For these two simulations, the probe OT is fixed at 10 mil. It is indicated that the change of metal 2 thickness has little impact on the maximum first principal stress in TEOS and nitride layer

Table 2. Effect of M2 thickness

model parameters						simulation results	
Met 2 (μm)	D3 (μm)	SiN (μm)	Cu (μm)	Ni (μm)	Au (μm)	S1 in TEOS (MPa)	S1 in SiN (MPa)
0.55	1	0.8	10	2	0.5	54	244
0.9	1	0.8	10	2	0.5	54	247

(2) Effect of copper pad thickness
Five different copper pad thickness is investigated, and the other model parameters are listed in Table 3. The probe OT is also fixed at 10 mil. The simulation results show that reducing the thickness of Cu could reduce the first principal stress for both TEOS and SiN, however, it seems that there exists an optimization point.

Table 3 Effect of copper pad thickness

model parameters						simulation results	
Met2 (μm)	D3 (μm)	SiN (μm)	Cu (μm)	Ni (μm)	Au (μm)	S1 in TEOS (MPa)	S1 in SiN (MPa)
0.55	1	0.4	15	2	0.5	86	330
0.55	1	0.4	10	2	0.5	73	293
0.55	1	0.4	5	2	0.5	57	249
0.55	1	0.4	3	2	0.5	51	234
0.55	1	0.4	2	2	0.5	47	284

(3) Effect of Ni layer thickness
Three different Ni layer thicknesses are investigated, and the other model parameters are listed in Table 4. The probe OT is fixed at 6 mil for all the simulations. It can be

520

seen that the maximum first principal stress in TEOS and nitride layer increases with the increase of Ni layer thickness. The dielectric layer 3 has very little impact on the stresses in TEOS and nitride layer.The Ni layer thickness has similar impact on the TEOS and nitride as the copper layer thickness. The max first principal stress of TEOS and nitride locates around the area under the edge of the bond pad. This may explain the effect of copper and Ni layer effect.

Table 4 Effect of Ni layer thickness

model parameters						simulation results	
M2 (µm)	D3 (µm)	SiN (µm)	Cu (µm)	Ni (µm)	Au (µm)	S1 in TEOS (MPa)	S1 in SiN (MPa)
0.9	1	0.4	10	1	0.5	43	178
0.9	1	0.4	10	3	0.5	49	198
0.9	1	0.4	10	5	0.5	54	213
0.9	2	0.4	10	1	0.5	44	184
0.9	2	0.4	10	3	0.5	50	204
0.9	2	0.4	10	5	0.5	56	219
0.9	3	0.4	10	1	0.5	45	189
0.9	3	0.4	10	3	0.5	51	209
0.9	3	0.4	10	5	0.5	57	225

(4) Effect of SiN layer thickness

Three different SiN layer thicknesses are investigated, with the other model parameters listed in Table 5. The probe OT is fixed at 6 mil for all the simulations. It can be seen that the maximum first principal stress in TEOS and nitride layer decreases with the increase of nitride layer thickness. It is also confirmed that the dielectric layer 3 has very little impact on the stresses in TEOS and nitride layer.

Table 5 Effect of SiN layer thickness

model parameters						simulation results	
M2 (µm)	D3 (µm)	SiN (µm)	Cu (µm)	Ni (µm)	Au (µm)	S1 in TEOS (MPa)	S1 in SiN (MPa)
0.9	1	0.4	10	3	0.5	49	198
0.9	1	0.8	10	3	0.5	37	164
0.9	1	1.2	10	3	0.5	31	156
0.9	2	0.4	10	3	0.5	50	204
0.9	2	0.8	10	3	0.5	37	169
0.9	2	1.2	10	3	0.5	32	161
0.9	3	0.4	10	3	0.5	51	209
0.9	3	0.8	10	3	0.5	38	173
0.9	3	1.2	10	3	0.5	32	164

(4) Probe OT effect

In order to study the effect of probe OT, the metal layer 2 is fixed at 0.9µm. The other model parameters are listed in Table 6, and the simulation results are illustrated in Table 6 and Fig. 4. The simulation results show that the maximum first principal stress in TEOS and nitride layer increases with the increase of probe OT, and it has the most impact on the SiN layer. This agrees with the results in [9].

Table 6 Effect of probe OT

model parameters						simulation results	
D3 (µm)	SiN (µm)	Cu (µm)	Ni (µm)	Au (µm)	OT (mil)	S1 in TEOS (MPa)	S1 in SiN (MPa)
1	0.4	5	1	0.5	2	11	49
1	0.4	5	1	0.5	4	22	97
1	0.4	5	1	0.5	6	33	147
1	0.4	5	1	0.5	8	43	193
1	0.4	5	1	0.5	10	52	234

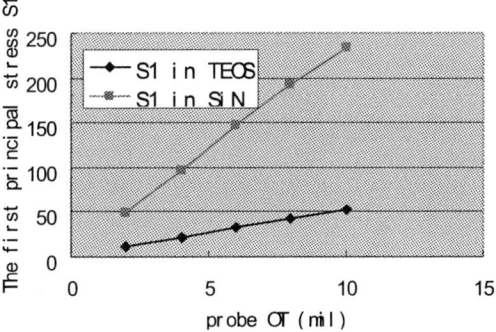

Fig. 4 Effect of probe OT

4. Setting Up of the 3-D model

In order to accurately capture the effects of vias and line layout and further reduce the computation error induced by the assumption of the 2-D model, a 3-D model is built to study the probing test. But the 3-D model takes much longer simulation time. The FEA model and mesh are shown in Figs. 5-6, with half model used because of symmetry.

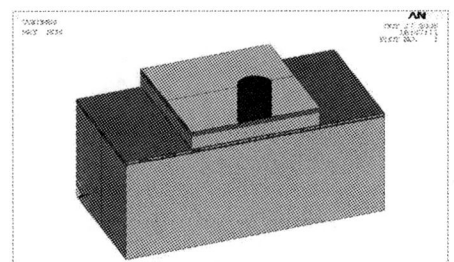

Fig. 5   3-D solid model (flat probe

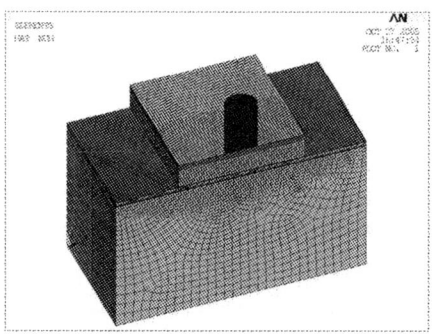

Fig. 6 Mesh of the 3-D solid model (flat probe

Table 7 Model parameters

model parameters (thickness)							
Met2 (μm)	D3 (μm)	SiN (μm)	Cu (μm)	Ni (μm)	Au (μm)	OT (mil)	DX (μm)
0.9	1	0.4	15	5	0.5	6	15

5. Simulation Results and Discussion of 3-D model

(1) Effect of probe tip shape:

The model parameters listed in Table 7 are used for the 3-D simulations. Three types of probe tip shape: flat, flat/round and round are studied to check the effect on the TEOS and nitride layer. The friction coefficient between probe and bond pad is fixed at 0.35. The first principal stress contour in TEOS and nitride layer with the round probe tip is shown in Fig. 7 (a) & 7 (b), respectively. The results comparison for the three types of probe tip is illustrated in Fig. 8. It can be seen that the flat probe tip generates the lowest stress and the round probe tip generates the highest stress. This agrees well with the previous study in [9].

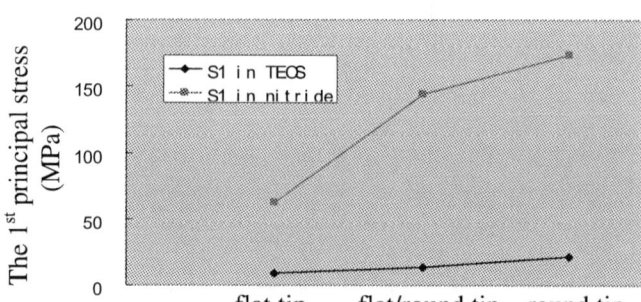

Fig. 8 the 1st principal stress comparison for different probe tip shape

(2) Effect of friction between probe and bond pad

The model parameters listed in Table 7 are adopted. In order to study the effect of friction between the probe and the bond pad, two friction coefficients are used for comparison. One is 0.35, the other is 0.8. The first principal stress contours of TEOS and nitride layer are shown in Fig.9 & Fig.10 with flat/round probe tip. The simulation results of different friction coefficient between probe tip and bond pad are shown. The comparison of different probe tip shape is illustrated in Fig. 11. It can be seen that with the increase of the friction coefficient, the max first principal stress in both TEOS and nitride layer increases, and it has more impact on the TEOS stress than the nitride layer stress. This also agrees with the 2-D case in [9].

By comparing the 3-D simulation results with the 2-D case, it is found that the trends match well for each different parameter's impact. Therefore, the 2-D model can be used if only for comparative purposes to compare each parameter's impact with the benefit of fast simulation time. Another observation is that the absolute result values of the 3-D model are lower than that of the

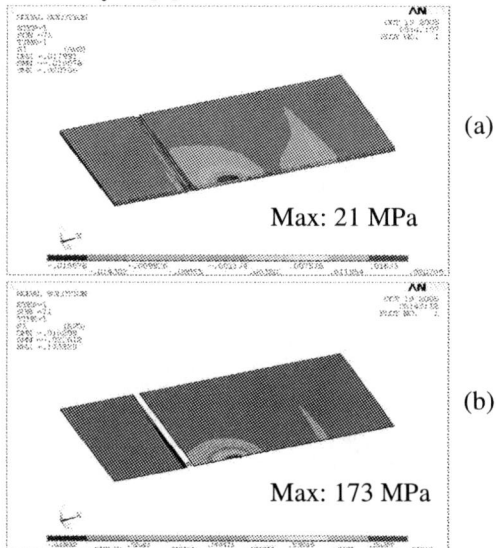

Fig. 7 Contour of the fist principal stress for round probe tip: (a) TEOS, (b) nitride layer

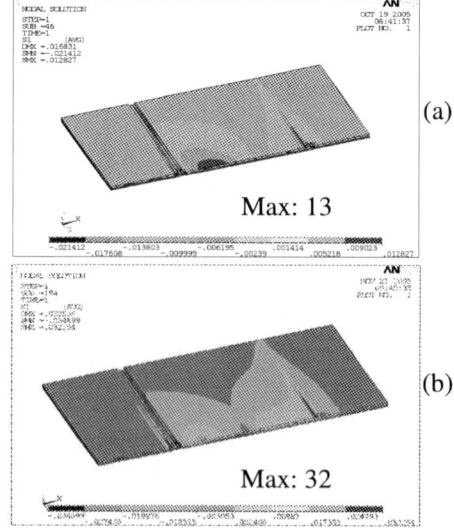

Fig. 9 Contour of the fist principal stress in TEOS for flat/round probe tip: (a) f=0.35, (b) f=0.8

2-D model. This may result from the improper assumptions in the 2-D model. So the 3-D model is recommended if failure criterion is used to judge the models reliability performance.

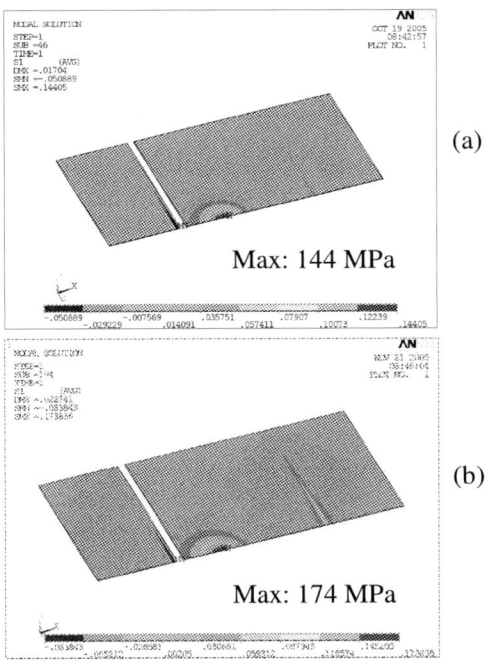

Fig. 10 Contour of the fist principal stress in nitride for flat/round probe tip: (a) *f*=0.35, (b) *f*=0.8

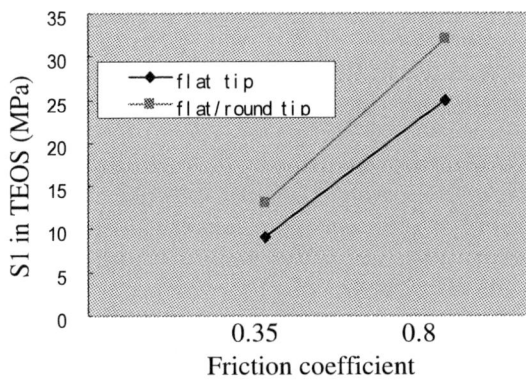

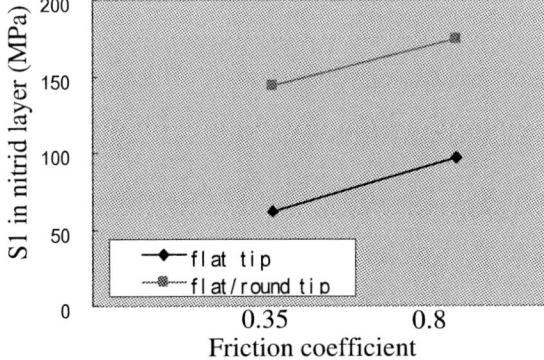

Fig. 11 Effect of friction coefficient

6. Conclusion

Both 2-D and 3-D probing simulations of copper bond pad based BPOA structure are conducted. According to the 2-D simulation results, thinner bond pad, thicker SiN layer and smaller probe OT generate the best result from the viewpoint of the mechanical stress. However, these parameters should be balanced-off according to other factors such as the process feasibility and equipment parameters. Simulation results show that the trends of different parameters impact from 2-D results match well with the 3-D results. So the 2-D model can be used for comparative purposes only. However, the 3-D model is recommended if a failure criterion is used to judge the model reliability performance.

Acknowledgements

The authors wish to thank the support from Automation Development department in Maine and assembly Suzhou, Fairchild Semiconductor Corp.

References

1. Lee, C., Tran, T., Yong, L., Harun, F., and Yong C., "A thermal aging study on both Au-Cu and Au-Al wire-bonded interfaces", Electronic Components & Technology Conference, 2003.

2. Tran, T., Yong, L., Williams, B., Chen, S., and Chen, A., "Fine pitch probing and wirebonding and reliability of aluminum capped copper bond pads", Electronic Components & Technology Conference, 2000.

3. Yong, L., Tran, T., Lee, S., Williams, B., and Ross, J., "Novel Method of Separating Probe and Wire Bond Regions Without Increasing Die Size", Electronic Components & Technology Conference, 2003.

4. Sauter, W., Aoki, T., Hisada, T., Miyai, H., Petrarca, K., Beaulieu, F., Allard, S., Power, J., and Agbesi, M., "Problems with wirebonding on probe marks and possible solutions", Electronic Components & Technology Conference, 2003.

5. Chou, K.Y., et al., "Active devices under CMOS I/O pads," IEEE Transactions on Electron Devices, Vol. 49,No. 12, December 2002, pp. 2279–2287.

6. Hess, K., et al., "Reliability of bond over active pad structures for 0.13-μm CMOS Technology," Electronic Components and Technology Conference 2003.

7. Awad, E., "Active devices and wiring under chip bond pads: stress Simulations and modeling methodology", Electronic Components and Technology Conference 2004.

8. Liu, Y., Desbiens, D., Irving, S., and Luk, T., " Probe Test Failure Analysis of Bond Pad Over Active (BPOA) Structure by Modeling and Experiemnt", Electronic Components and Technology Conference, 2005.

9. Liu, Y., Desbiens, D., Luk, T., and Irving, S., "Parameter optimization for wafer probe using simulation", EuroSimE, 2005.

Optimization and Robust Design of Electronics Assemblies under Fracture, Delamination and Fatigue Aspects

Jürgen Auersperg[1,2], Matthias Klein[3], Bernd Michel[1]

[1] Fraunhofer Institute for Reliability and Microintegration (IZM)
Dept. Micro Materials Center Berlin, Gustav-Meyer-Allee 25, 13355 Berlin, Germany
[2] AMIC Angewandte Micro-Messtechnik GmbH, Volmerstr. 9b, 12489 Berlin, Germany
[3] Fraunhofer Institute for Reliability and Microintegration (IZM)
Dept. Module Integration & Board Interconnection Technologies, Gustav-Meyer-Allee 25, 13355 Berlin, Germany
e-mail: juergen.auersperg@izm.fraunhofer.de

Abstract

Electronics components especially in the field of RF, optoelectronics, high temperature and power applications are often exposed to extreme thermal environmental conditions, mechanical vibrations and shock. Simultaneously, the well known thermal expansion problem of the several materials brought together, residual stresses generated by several steps of the manufacturing process and various kinds of inhomogeneity attribute to interface delamination, chip cracking and fatigue of interconnects. For that reason, design studies on the basis of parameterized FE-models and DOE/RSM-approaches are performed to optimize electronic components at early phases of the device development process. The applied methodologies typically base on classical stress/strain strength evaluations or/and life time estimations of solder interconnects using modified Coffin-Manson approaches. Recent studies show also how the evaluation of mixed mode interface delamination phenomena, classical strength hypotheses along with fracture mechanics approaches and thermal fatigue estimation of solder joints can simultaneously be taken into account.

This contribution is now coupling such an integrated approach with optimization algorithms towards a thermo-mechanical reliable design whereas, the attention is also turned to the robustness against scattering model parameters (scattering of geometry and/or materials properties), in particular.

1. Introduction

Microelectronic packages like Flip-Chip, CSP assemblies or smart, thin electronics devices are typically compounds of materials with quite different Young's modules and thermal expansion coefficients (CTE). Furthermore, various kinds of inhomogeneity, residual stresses from the manufacturing process and extreme thermal environmental conditions are often driving factors for interface delamination, chip cracking and fatigue of solder interconnects. Consequently, numerical investigations by means of nonlinear FEA, fracture mechanics concepts are commonly used for design optimizations using sensitivity analyses – see [1-3]. In doing so, the most important points in simulation are

- the description of the constitutive behavior of all materials present in the model,

- the measurement and preparation of all relevant geometric and materials properties,

- the definition of all necessary boundary conditions, loading conditions as well as data or additional simulation steps required to define the stress free state and residual stresses from manufacturing (this could also include special materials models describing the curing process of polymeric materials – see [4]), and

- (after appropriate simulations) the evaluation of the possibly occurring failure modes.

Especially for design optimizations and sensitivity analyses, most of the publications regarding enhanced microelectronics packages utilize

1. classical strength hypotheses (maximum principal stresses, peel stresses, von Mises stresses, ultimate tensile strength or strains) to estimate the cracking risk of substrates, semiconductors, encapsulations,

2. accumulated equivalent plastic strains to evaluate the fatigue of metals (lead frames, for instance), and

3. Coffin-Manson like approaches based on accumulated equivalent creep strains or during thermal cycling volume weighted dissipated inelastic strain energy to evaluate the thermal fatigue of solder interconnects.

Otherwise, it is common knowledge in mechanical engineering that cracks starting at sharp edges have to be taken into account in order to come to a conservative evaluation of the fracture toughness of the several materials and interfaces present in advanced electronic packages. Also as stress and/or strains fields show singular behavior (and therefore, the evaluation of pure stresses or strains is impossible) the application of fracture mechanics concepts is the recommended procedure. So, coming from K-concept usage, a lot of work was done recently in order to explore mixed mode effects also for the bimaterial interface delamination problem, e.g. [2-3]. Although, in contrast to investigations into bulk material fracture phenomena, integral fracture concepts like the J- or the C*-integral, which have the potential to take into consideration inelastic behavior of the related materials, remain almost unconsidered for use. Major reasons are surely the discontinuities of the stress and strain fields at the interface which do not allow to make use of the J- or C*-Integral implementations as implemented in almost utilized commercial FE-codes (violation of smoothness conditions due to the set of Gauss at the interface leads to loss of path independence).

1-4244-1105-X/07/$25.00 ©2007 IEEE

Therefore, the numerical investigations performed and discussed here are utilizing the energy release rate (ERR) G and the phase angle $\Psi(G)$ between its components as parameters - see also [5].

Hutchinson et al. [6] introduced a complex stress intensity factor (SIF) $K=K_I+iK_{II}$ characterizing the near tip stress field at an interfacial crack with

$$(\sigma_{xx}^{\infty} + i\sigma_{yy}^{\infty})_{\Theta=0} = Kr^{i\varepsilon}/\sqrt{2\pi r} \qquad (1)$$

where the oscillatory exponent ε is

$$\varepsilon = \frac{1}{2\pi}\ln\left[\frac{\left(\frac{\kappa_1}{\mu_1}+\frac{1}{\mu_2}\right)}{\left(\frac{\kappa_2}{\mu_2}+\frac{1}{\mu_1}\right)}\right] \quad \kappa_j = \begin{cases} 3-4\nu_j & plane\,strain \\ (3-\nu_j)/(1+\nu_j) & plane\,stress \end{cases} \qquad (2)$$

Here, μ_j is a shear modulus, ν_j Poisson's ratio and subscripts j symbolize upper and lower materials at the interface, respectively. The most important obstacle for utilizing the well known conceptions from bulk material fracture mechanics is the oscillatory nature of the stress fields under LEFM conditions represented by the $r^{i\varepsilon}$ term. Sun et al. [5] characterized the oscillation zone size for crack opening and shear. The determined radii turn out to be the size of this (possibly overlapping of the crack flanks) zone. Hutchinson [6] suggested for the case $\varepsilon = 0$

$$G(\psi_G) = G_c(\psi_G) \; with$$
$$\psi_G = \tan^{-1}\left(\frac{G_{II}}{G_I}\right) and \; G = G_I + G_{II} \qquad (3)$$

However, the ERR G equals the change in strain energy with crack area expansion (developing new surface area at the crack flanks). As originally proposed by Rybicki and Kanninen [7] and later on enhanced for 3d-cases by Krueger [8] G can be calculated by a virtual crack closure technique (VCCT).

Based on G, Sun and Qian [5] derived accurate (for plane strain and stress linear elastic conditions only) stress intensity factors K_I and K_{II} as well as non-oscillatory energy release rates G_I and G_{II} to be used as a fracture criterion for interfacial cracks. It is to be noted that these models do not take notice of overlapping of the crack flanks in the oscillation zone and do not fully reflect stress/strain fields of really 3D situations with nonlinear material behavior.

In order to obtain the interface toughness over a whole range of mode mixity, Hutchinson and Suo [6] proposed

$$G_c(\psi) = G_{IC}\left[1+(\lambda-1)\sin^2\psi_G\right] \qquad (4)$$

as the critical energy release rate, where λ (a further material/interface parameter) is a function of temperature and moisture concentration, as well.

Note: Experimentally determined results regarding $G_c(\Psi)$ suggest that the exponent is often higher than 2 and the behavior is asymmetrical.

2. Interface Fracture Evaluation – Challenges

In general, mixed mode stress situations naturally found at material interface cracks are a challenge for evaluating the interfacial fracture toughness. The use of mode separation approaches based on linear elasticity assumptions is a theoretical restriction common to nearly all related publications. This is accepted in spite of the well known nonlinear behavior of the materials (sometimes the geometry, too). Alternatives which have the potential to overcome such limitations are the damage modeling using cohesive zone approaches and integral fracture concepts like the M-Integral proposed by Banks-Sills et al. [9].

Regardless of this, one major task prior to delamination investigations is the experimental examination of the interface fracture toughness depending on varying externally applied loading conditions – see also [10-11].

Alternatives to handle bimaterial delamination problems are:

- SIF evaluation by CTOD techniques as outlined by Sun and Qian [5] which is also restricted to LEFM,
- VCCT – restricted to LEFM especially regarding the mode separation of SIF and ERR as described above,
- M-Integral as proposed by Banks-Sills et al. [9] – necessarily to be enhanced for thermal loading conditions, nonlinear and rate dependent materials etc.
- Meso-mechanical (multiscale) approaches incorporating also molecular modeling are subjects of current research,
- Cohesive zone/surface modeling (CZM) approaches as introduced by Xu and Needleman [12] or Alfano and Crisfield [13] which are basically enhancements of the approaches of Barenblatt [14] and Dugdale [15] have the potential of incorporating micromechanical conditions and processes,
- XFEM (extended FEM) utilizing Heaviside enrichment functions additionally to the shape functions of "normal" finite elements in order to allow mesh independent crack propagation simulations are also subjects of current research.

Even if CZM-approaches seem to be THE solution to overcome the drawback of conventional fracture mechanics approaches – to need at least a little pre-crack – their application has also special difficulties:

- CZM-applications often show mesh dependent results,
- Time integration stability controlling algorithms lead to a very fine time stepping with the result of time consuming simulations,
- There are stability problems to handle especially at the moment when an allowed peak-stress is achieved,
- High number or model-parameters have to be measured prior to the simulations – σ_{1max}, σ_{2max}, σ_{3max} (maximum bearable stress levels), G_1, G_2, G_3 (fracture toughness values), β (scheme control parameter), K_a (stiffness until the delamination starts) need to be prepared.

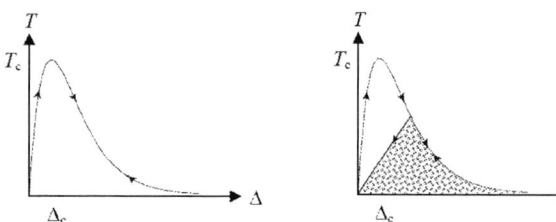

Fig. 1: Reversible (left) and irreversible (right) CZM

Result of simulations is the description of the crack propagation vs. time behavior along the prepared paths. But, necessary for DOE-applications and sensitivity investigations regarding the thermo-mechanical reliability of microelectronics packages are the evaluation and incorporation of the damage process. This means such CZM-elements have to reflect the damage process with the help of inherent parameters that can be utilized by sensitivity of DOE analyses - Fig. 1. On the other hand, the preparation of the necessary model parameters calls for FE assisted experiments utilizing fracture mechanics approaches.

As an example, the interface delamination during and after thermosonic/thermocompression bonding was investigated utilizing the CZM-approach – a detail of the FE-model is shown in Fig. 2. This technology offers a clean, dry bonding process which provides good joining with a short bonding time. The process also uses much less force than a typical thermo compression process, and the bonding temperature is also greatly reduced. Low force and low temperature makes it particularly useful for fragile materials such as low-k dielectric layers.

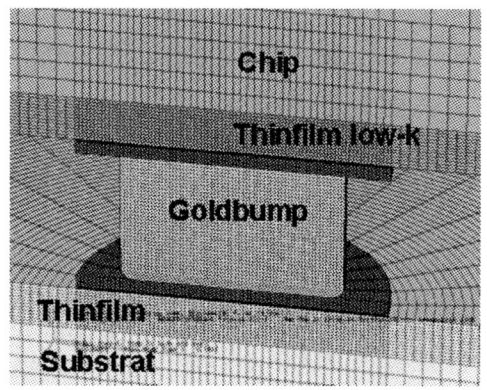

Fig. 2: The FE-Model in the surrounding of a single bump

Without to take into account dynamic aspects here the bonding under 245 °C, the load release, cool down to RT and a subsequent lift-off test were simulated with a FE-model consisting of CZM-elements in all interfaces of the structure.

Fig. 3 demonstrates the function of CZM-elements during the lift-off test after thermosonic/thermo-compression bonding on low-k materials.

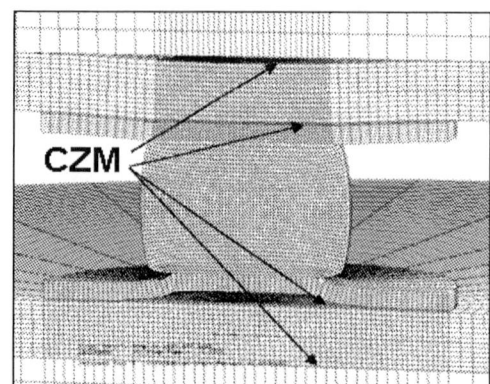

Fig. 3: Delamination underneath and above the bump during the lift-off test

3. Modeling Assumptions

It is the major objective of the FEA-study explained here to contribute to overall thermo-mechanical reliability investigations of electronics assemblies by nonlinear FEA-simulations taking into account several dominating failure modes. That is, thermo-mechanical fatigue of solder interconnects, chip cracking, cracking of substrates or encapsulations as well as the delamination risk have to be taken into consideration, altogether. Jointly with the interface toughness-values determined as shown above, RSM/DOE approaches can then utilize the VCCT-element-based methodology to evaluate the delamination risk of interfaces previously investigated by separate experiments.

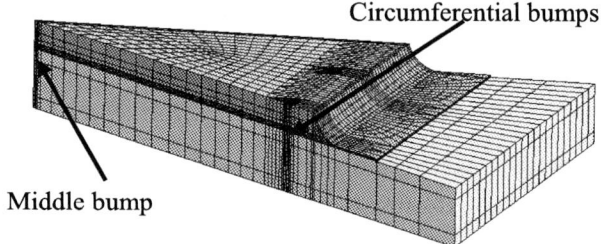

Fig. 4: FE-model of a quarter of an underfilled FC with circumferential bumps and a single middle bump

Exemplarily, a Flip-Chip assembly is investigated here knowing that a lot of knowledge is available which helps to evaluate possible results of these simulations. Fig. 4 shows the FE-model of a quarter of an underfilled FC with circumferential bumps and single middle bump - Fig. 4. More in detail, Fig. 5 explains the several parts/materials partitions in the surrounding of a solder bump.

It is to be noticed that the model consists of two different underfiller regions and a solder mask region opening an underfilled gap to the solder bump – a common design in Flip-Chip technology.

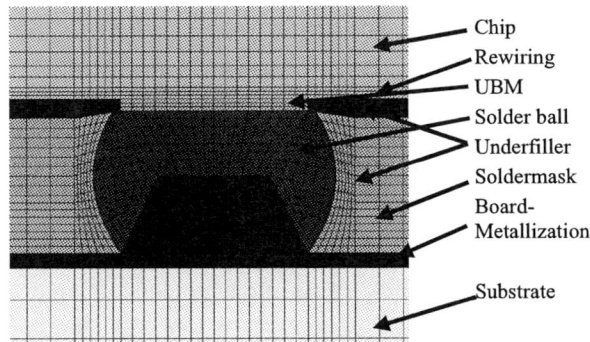

Fig. 5: Detailed FE-model of the bump-region

With the purpose of studying the influence of geometric variations the model has been parameterized so that

e_wide: the distance the chip center to the center of the circumferential solder bumps,

t_chip: the thickness of the chip, and

t_sub: the thickness of the substrate

act as design variables. That means a simulation controlling RSM/DOE or optimization algorithm can handle these variables inside of well defined boundaries – see Fig. 6.

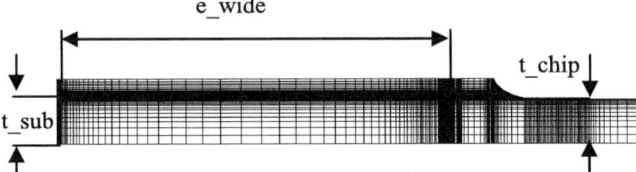

Fig. 6: Geometric parameters for DOE and optimization

As discussed in [19], varying the geometry of the finite element model accounted for keeping the finite elements regular (positive Jacobian) and to avoid element warpage and distortion.

Other parameters like the thickness of the board metallization and chip metallization, the bump diameter and height (80 μm) and material properties are invariable in the model. All metallizations show elastic-plastic materials behavior, the underfiller and solder mask behave as viscoelastic and the solder bumps (Sn96,5Ag3,5) are modeled as viscoplastic material with primary and secondary creep equations.

Besides the symmetry conditions taken into account by the quarter model, the Flip-Chip is assumed to be thermally cycled between -40 °C and 150 °C under ATA conditions and a dwell time of 20 min.

4. Failure Monitoring Preparation

It is the objective of the FEA-studies explained here to take into account failure modes that are essential to the overall thermo-mechanical reliability of the Flip-Chip assembly under investigation. For that reason several failure modes have been assumed to have an effect. It is the purpose of all simulations performed here to show exemplarily the possibility of utilizing different failure hypotheses in parallel rather than to use the best suited one for all failure modes (fracture mechanics approach for

chip cracking or the volume weighted dissipated inelastic strain energy as a measure for solder fatigue for instance).

It is to be noted that the evaluation of delamination risks follows the equation originally proposed by Hutchinson and Suo [6] which was modified similar to a yield function

$$G_{crit} = G(\psi_G) - G_{IC}\left[1 + (\lambda - 1)\sin^{2n}(\psi_G - \psi_0)\right]. \quad (5)$$

G_{crit} is the critical fraction of the phase angle weighted load energy release rate $G(\Psi_G)$. As discussed above the parameters n and Ψ_0 help to better fit measured date. G_{crit} should be less than zero for uncritical delaminations.

1. Risk of chip cracking: The maximum principal stress in all elements that are part of the upper surface of the chip, taken from all simulations → Smax is monitored.

2. Thermal fatigue of solder balls: The mean number of cycles to failure in solder balls → MC2F (minimum of all solder balls) is observed.

3. Risk of delamination between underfiller and chip: A short delamination starting from the chip edge and growing into the interface with a VCCT-element at the crack tip were modeled and taken into account by a special user-subroutine [18] delivering the fraction of the phase angle weighted energy release rate G_edge as defined in Eq. (5) – see Fig. 7.

4. Risk of delamination between underfiller and solder balls: Short delaminations starting from the corner line of the bump (edge with the board metallization) which are growing into the interface between solder and underfiller and VCCT-elements at the crack tips to determine G_BuUF as defined in Eq. (5).

5. Risk of delamination between underfiller and chip-surface: Short delaminations which are growing into the interface towards the bump and in the opposite direction with VCCT-elements at the crack tips were modeled and taken into account by a special user-subroutine [18] delivering G_ChUF as defined in Eq. (5) – see Fig. 8.

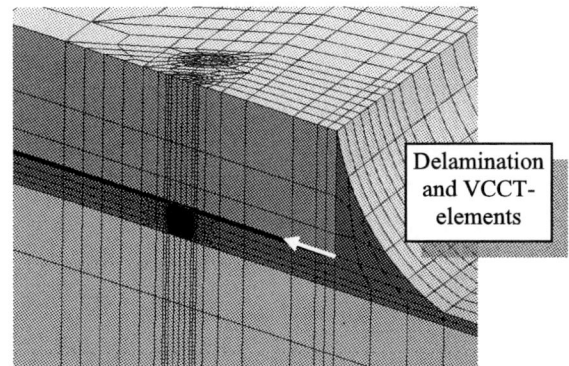

Fig. 7: Assumed delamination between chip and underfiller

The simulation results intended for use in DOE to build a response surface or/and to optimize the structure regarding thermo-mechanical reliability are Smax, MC2F,

G_edge to G_ChUF as described above as well as G_Delam which is max(G_edge to G_ChUF).

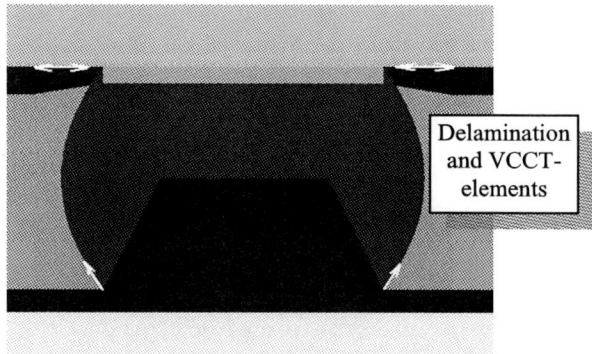

Fig. 8: Assumed delaminations at the solder bump – underfiller and underfill-chip-surface interfaces

5. Numerical Simulation

All simulations were performed with the commercial FEM-code Abaqus™ [16] including special in-house user-subroutines for the description of the creep behavior of solder Sn96,5Ag3,5 and with the in-house VCCT-elements. The software chosen to perform DOE, RSM and optimization was Optimus™ [19]. Fig. 9 graphically illustrates the structure of an Optimus™-project intended for DOE with the aim of delivering geometry related response surfaces of the several failure risk describing output-variables.

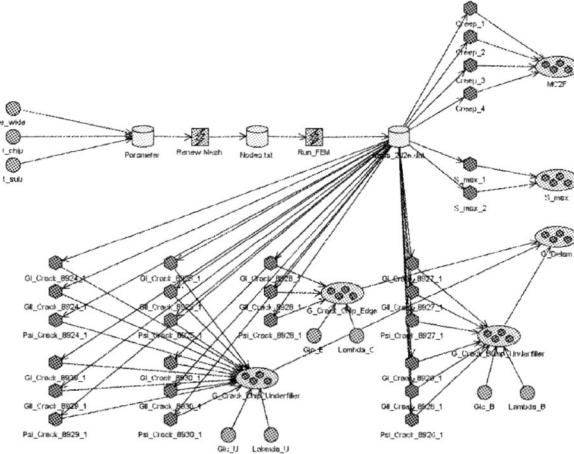

Fig. 9: Structure of an Optimus™-project

The data flow is starting with the geometry parameters. "Renew Mesh" delivers a new set of all nodal coordinates or if necessary a fitted new mesh which is later on used by Abaqus™ for suitable FE-simulations. Optimus™ grabbes then interim data and resulting data Smax, MC2F, G_Delam as described above from the FEA-result files "results.dat".

After loading these results, Optimus™ allows then to extract mathematical models describing the response surfaces of all result data with regard to the variation of all input variables. Several algorithms like the "full factorial model", the "3 level full factorial model" or the "adjustable factorial model" provide such mathematical

representations with a limited number of finite element runs, which is vital especially for highly nonlinear, transient and/or spatial FE-simulations to execute.

6. DOE Results and Discussion

Again to remark, all simulations performed here are exemplarily to show the possibility of utilizing different failure hypotheses together in order to show the influence of geometric or materials properties of finite element models on reliability relevant results. Subsequently such dependences on geometric parameters will be discussed.

- Risk of chip cracking: The maximum principal stress in several elements that are part of the upper surface of the chip was monitored in order to estimate the risk of chip cracking. Fig. 10 points up its dependence on the thickness of chip and substrate for chip diameter chosen as e_wide = 6,214 mm. It indicates that the risk of chip cracking is at lowest for thin chips on thick substrates and vice versa. Chip cracking is most probable for thin chip on thin substrates and vice versa.

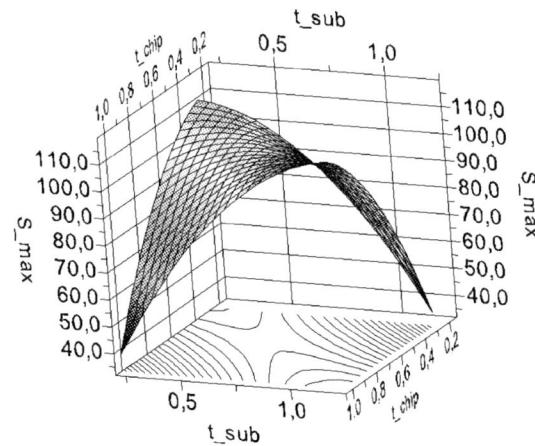

Fig. 10: Dependence of maximum principal stress in the chip surface on thickness of chip and substrate at e_wide = 6,214 mm

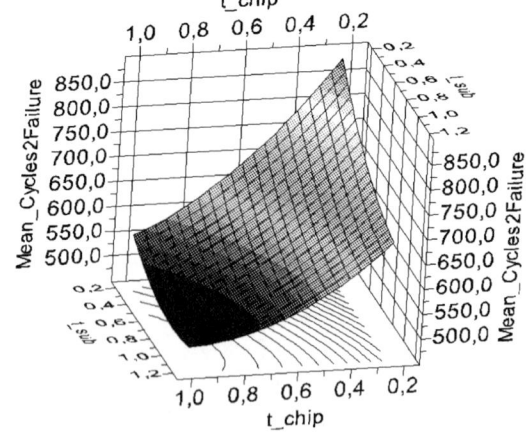

Fig. 11: Dependence of mean number of cycles to failure in solder balls → MC2F on thickness of chip and substrate at e_wide = 6,214 mm

- Thermal fatigue of solder balls: mean number of cycles to failure (MC2F) in solder balls → MC2F for e_wide = 6,214 mm indicates the best values for thin chips on thin substrates and worst values for thick chips on thick or thin substrates.

  These results well correspond with common knowledge – higher flexibility leads to unloading effects for solder interconnects.

- Risk of delamination was observed as mainly dominated by the interface between underfiller and chip near the chip edge: The critical fraction of the phase angle weighted energy release rate (for e_wide = 6,214 mm) - largely independent on the substrate thickness arises with the chip thickness - Fig. 12.

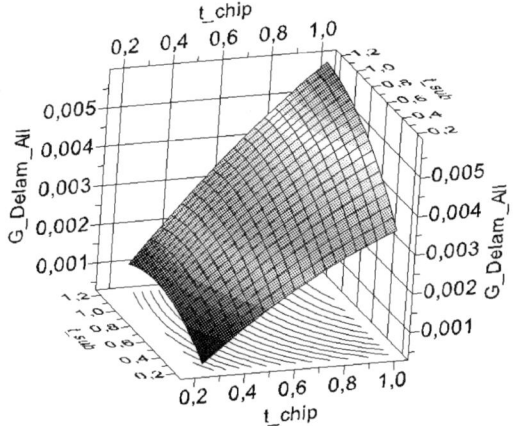

Fig. 12: Dependence of the critical fraction of the phase angle weighted energy release rate of all interfaces on the thickness of chip and substrate

7. Optimization

Even if the results shown here are intended to be exemplary and assuming that the interface toughness values for all interfaces have been measured, these results were utilized for optimization purposes. The several fatigue and failure risk evaluation approaches contribute as follows to the optimization-approach:

- Stress based strength hypotheses ($\sigma_{max} \leq \sigma_c$) act as upper boundaries,
- Cracking risk evaluation by SIF (K), J-Integral of ERR (G) also act as upper boundaries,
- Delamination risk evaluation by VCCT-elements resulting in critical fraction of the phase angle weighted energy release rate as defined in Eq. (5) also act as upper boundaries,
- Thermal fatigue of solder joints evaluations by Coffin-Manson like approaches establishes the response surface.

The application of an appropriate DOE-Software like OPTIMUS™ [19] enables now to optimize the design and the materials selection, to enhance the reliability and also the robustness of found optimal solutions [19].

While the incorporation of fatigue crack growth of interface delaminations and thermal fatigue of solder interconnects directly leads to multi-goal optimization procedures the approach utilized first is a single-goal optimization. The goal is to find the best pair of chip and substrate thicknesses for given chip diagonal e_wide and to avoid at the same time chip cracking and delamination.

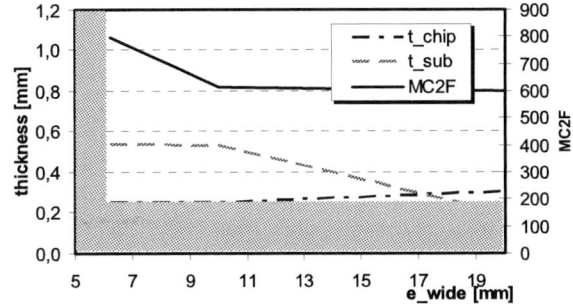

Fig. 13: Optimal thicknesses of chip and substrate for different chip diagonals e_wide and corresponding MC2F

The optimization results show the clear trend to thinner chips and substrates for large dies – see Fig. 13. This was mainly bounded by

- The predefined parameter windows (250 μm for the chip and 200 μm for the substrate thickness as lower boundaries)
- The upper boundary for the maximum tensile stresses in the chip surface predefined with 100 MPa and often by
- The delamination risk monitoring at the chip-underfiller interface near the chip edge G_edge as defined in Eq. (5).

8. Robustness and Design for Six Sigma

The purpose of a reliability analysis is to estimate the probability that a structure will fail to meet a predefined criterion. Naturally some variables of the model are stochastic parameters, as a consequence all results depend on this scattering and show also a stochastic distribution. Assuming the results show also a Gaussian distribution an optimum found for determined model parameters could reside in a region near or directly on a boundary established by the fracture toughness value, for instance. Some of the potential solutions will now reside in the forbidden region of the solution area and will probably faile. In order to reach a robust design, probabilistic design methods like the First Order Reliability Method (FORM) or Second Order Reliability Method (SORM) utilize the so called reliability index which is the distance of the nearest failure point from the mean value. This distance is a direct measure for reliability. It denotes how many variance steps lie between the parameter mean and the nearest failure point. In other words, if we have a distance of six standard deviations – that is 6 Sigma – practical no item exceeds the specifications – see also expressions in [21].

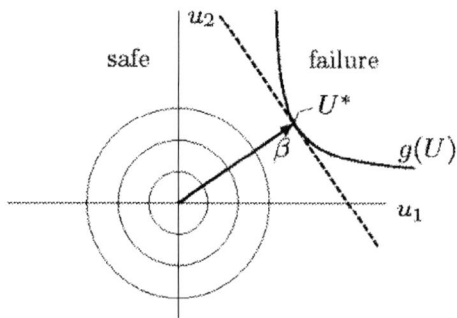

Fig. 14: First Order Reliability Method (FORM) – see Optimus™, theoretical background [19].

In summary, a design for reliability bases on physics of failure and turns out to reach a probabilistic level, such as 6 sigma, which is 3.4 failures in one million parts. Such a design is a "robust design".

Fig. 15 exhibits the dependence of sigma of delamination indicating simulation results on the thickness of chip and substrate – unfortunately far from 6 sigma in this case.

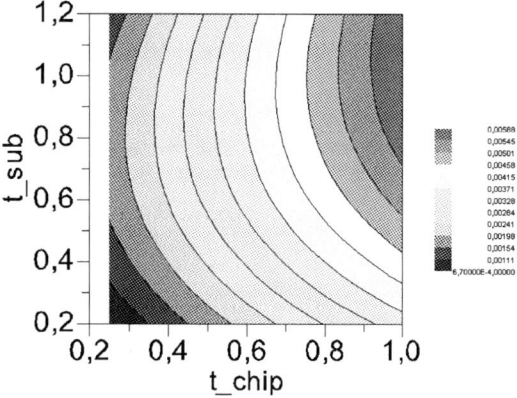

Fig. 15: Dependence of sigma of the critical fraction of the phase angle weighted energy release rate of all interfaces on the thickness of chip and substrate at e_wide = 6,214 mm

As a consequence, the optimal 6 sigma-design deviates from failure regions. The thickness of the chip remains at the parameter boundary while the thickness of the substrate shifts from 540 μm to 770 μm resulting in a reduction of MC2F from 796 to 712 cycles - see Fig. 16.

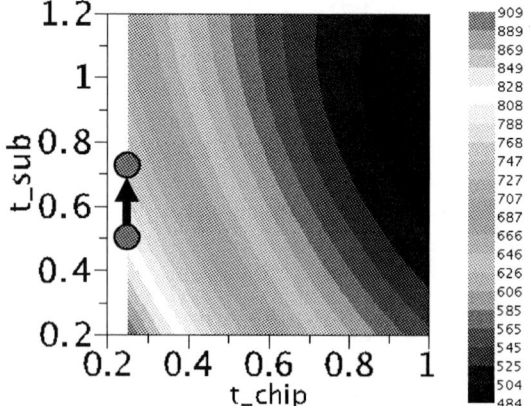

Fig. 16: The position of the optimal 6 sigma-design

9. Interface Cracks under TCT Conditions

As explained above the evaluation of bimaterial interface cracks has not only to look for maximum values of the energy release rate G for instance but also to weight it by the phase angle $\Psi(G)$ and to combine it with $G_c(\Psi)$ curves as determined by several experiments similar to Eq. (5). To show the behavior of $G(\Psi)$ during thermal cycling the underfiller/die interface in the neighborhood of the UBM was observed over 8 thermal cycles.

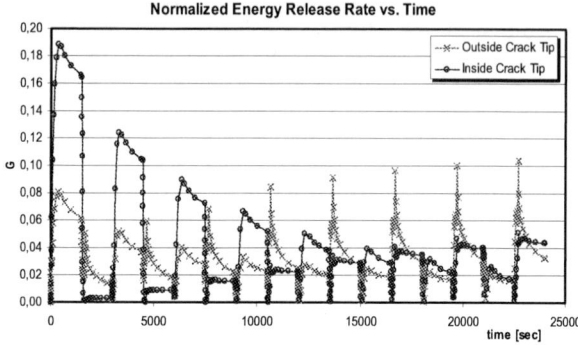

Fig. 17: Normalized energy release rate vs. time during thermal cycling

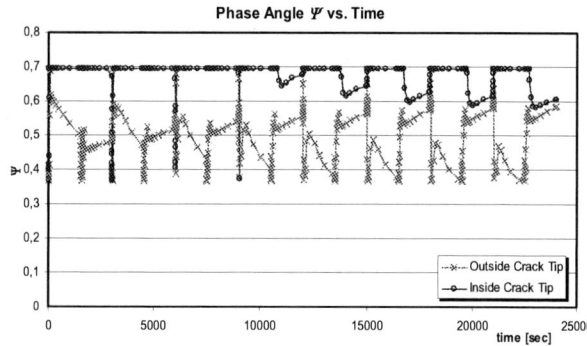

Fig. 18: Phase angle $\Psi(G)$ vs. time during thermal cycling

Fig. 17 and Fig. 18 show the normalized energy release rate and the phase angle $\Psi(G)$ during thermal cycling. Distinct from the well known behavior of measures for solder fatigue (accumulated equivalent creep strains for instance) which tend to be stable after 2-3 thermal cycles the energy release rates as well as the associated phase angles at interface cracks drift over a lot of thermal cycles. It has to be carefully analyzed in the future whether this behavior is combined with a changing fatigue progress in the solder bumps or not.

10. Conclusions

Combined fracture, delamination risk and thermal fatigue evaluations of advanced microelectronics applications have been exemplary carried out on the basis of a parameterized finite element model of a Flip-Chip assembly. The simulation models were prepared to introduce residual stresses, to satisfy the boundary/loading conditions and to determine interface fracture parameters, strength and results that allow to estimate thermal fatigue. The interface delamination investigations outlined and utilized here base on fracture mechanics approaches for bimaterial crack problems. They take into account the special mixed mode situations. Results of DOE-controlled simulations due to the influence of geometric parameters on thermal fatigue of solder balls, the delamination risk on the chip/underfiller interface as well as on the solder-ball/underfiller interface are shown and discussed.

The assumption that some variables of the model are stochastic parameters leads to the consequence that all results show also scattering. First results towards a robust design are to emphasize the potential of the utilized approach. Further developments will introduce fatigue crack problems. This calls for multi-goal optimization procedures.

In summary, these results together with the methodology and software developed by Auersperg et al. [20] show the potential of FEA-based RSM/DOE approaches to evaluate the thermo-mechanical reliability of various electronics assemblies in a more complex way giving at the same time a solid basis for design optimizations.

Acknowledgments

The authors acknowledge the European Community for their financial support of basic investigations, substantially promoting this work (Mevipro project GRD1-2001-40296).

References

1. R. Dudek, R. Doering, B. Michel, G. Petzold, J. Albrecht, C. Wieand, S. Kuhn, "Investigations on the Reliability of Lead-Free CSPs Subjected to Harsh Environments", Proceedings ITherm 2004, Las Vegas, USA, June 2004
2. Liu, Sh., Mei, Y., Wu, T. Y., "Bimaterial Interfacial Crack Growth as a Function of Mode-Mixity", IEEE Trans-CPMT - A, Vol. 18 No. 3, Sept. 1995, pp. 618-626
3. Wang, J., Lu, M., Ren, W., Zou, D., Liu, Sh., "A Study of the Mixed-Mode Interfacial Fracture Toughness of Adhesive Joints Using a Multiaxial Fatigue Tester", IEEE Trans. of Electronics Packaging Manufacturing, Vol. 22 No. 2, April 1999, pp. 166-173
4. K.M.B. Jansen, L. Wang, D.G. Yang, C. van 't Hof, L.J. Ernst, H.J.L. Bressers and G.Q. Zhang, "Constitutive Modeling of Moulding Compounds", In: Proc. of the 54th Electronic Components and Technology Conference (ECTC 2004), Las Vegas, June 1-4, 2004, IEEE Catalog number 04CH37546C, ISSN: 0569-5503, ISBN: 0-7803-8366-4, pp. 890-894
5. Sun, C. T., Qian, W., "The Use of Finite Extension Strain Energy Release Rates in Fracture of Interfacial Cracks", Int. J. Solids Structures, Vol. 34. No. 20, pp. 2595-2609
6. Hutchinson, J. W., Suo, Z., "Mixed Mode Cracking in Layered Materials", Advances in Applied Mechanics, Vol. 29, pp. 63-191
7. Rybicki, E. F., Kanninen, M. F., "A finite element calculation of stress intensity factors by a modified crack closure integral", Eng. Fracture Mechanics, Vol. 9, pp. 931-938
8. R. Krueger, "The Virtual Crack Closure Technique: History, Approach and Applications", NASA/CR-2002-211628, ICASE Report, No. 2002-10
9. L. Banks-Sills and A. Sherer, "A Conservative Integral for Determining Stress Intensity Factors of a Bimaterial Notch", International Journal of Fracture, 115, (2002), 1-26
10. Auersperg, J., Kieselstein, E., Schubert, A., Michel, B., "Mixed Mode Interfacial Fracture Toughness Evaluation for Flip-Chip Assemblies and CSP Based on Fracture Mechanics Approaches", Symp. on Reliability, Stress Analysis, and Failure Prevention of the 2001 ASME IMECE, New York, proc. Vol. 2, IMECE2001/DE-25109
11. Auersperg, J., Kieselstein, E., Schubert, A., Michel, B. „Delamination Risk Evaluation for Plastic Packages Based on Mixed Mode Fracture Mechanics Approaches", Trans. of the ASME, JEP, Vol. 124, Dec. 2002, pp. 318-322
12. Xu, X. and Needleman, A., "Numerical Simulations of Fast Crack Growth in Brittle Solids," J. Mech. Phys. Solids, Vol. 42, 1994, pp. 1397-1434
13. Alfano, G. and Crisfield, M.A. Solution strategies for the delamination analysis based on a combination of local-control arc-length and line searches, Int J Num Meth Engng, 58, pp. 999-1048, 2003
14. Barenblatt, G. I., "The Mathematical Theory of Equilibrium of Cracks in Brittle Fracture," Adv. Appl. Mech., 9, 1962, pp. 55–129
15. Dugdale, D. S., "Yielding of Steel Sheets Containing Clits," J. Mech. Phys. Solids, 8, 1960, pp. 100–104
16. Abaqus User's Manuals, Version 6.5, Hibbitt, Karlsson & Sorensen, Inc. Pawtucket, RI, USA, 2005
17. Auersperg, J.; Dudek, R.; Michel, B, "Investigation of the mechanical-thermal field coupling effect during quasi-static crack growth and application in micro system technology", Proc. of the German ABAQUS User Meeting, Ulm 1995, pp. 138-152
18. Auersperg, J., Vogel, D., Michel, B., "Crack and Delamination Risk Evaluation of Thin Silicon Based Microelectronics Devices", Topic 26 MEMS, Special Session at 11th Int. Conf. on Fracture ICF11, Turin (Italy), March 20-25, 2005, Proc. on CD
19. Noesis Solutions, Optimus 5.0 Software, http://www.noesissolutions.com
20. Auersperg, J., Seiler, B., Cadalen, E., Dudek, R., Michel, B., "Fracture Mechanics Based Crack and Delamination Risk Evaluation and RSM/DOE Concepts for Advanced Microelectronics Applications", Proc. of 6th IEEE EuroSimE conference and exhibition, Berlin (Germany), April 18-20, 2005, pp. 197-200
21. P. Limaye, B. Vandevelde, J. Vd Peer, S. Donders, R. Darveaux, "Probabilistic Design Approach for Package Design and Solder Joint Reliability Optimization for a Lead Free BGA Package", FE-Design Publ., http://www.fe-design.de/ueberblick-titel/abstracts-2005.html#696

Mechanical Characterization of III-V Nanowire Using Molecular Dynamics Simulation

Alex. W. Dawotola[*1], C. A. Yuan[1], W. D. van Driel[1,2], E. P. A. M. Bakkers[3] and G. Q. Zhang[1,2]

1. Department of Precision and Microsystems Engineering, Delft University of Technology, The Netherlands
2. NXP Semiconductors, Nijmegen, The Netherlands
3. Philips Research, Eindhoven, The Netherlands
*a.w.dawotola@tudelft.nl

Abstract

Mechanical stiffness and density of III-V (GaAs) nanowire (NW) are studied by atomistic simulation in the <111>, <110> and <100> directions. Series of molecular models are established and mechanical characteristics of the crystal orientations are considered. The simulation results indicate that the NW exhibits highest structural stiffness in the <111> direction. We also found that GaAs NW exhibits mechanical linearity under 2GPa stress. Moreover, a qualitative comparison of simulation and other calculated results is carried out, and a good agreement is established.

1. Introduction

Nanowires have many interesting properties that are not seen in bulk materials. This is because electrons in nanowires are quantum confined laterally and thus occupy energy levels that are different from the traditional continuum of energy levels or bands found in bulk materials. The mechanic response of nano materials are also found to be different from that of bulk materials [1]. There are different types of nanowires, including metallic (e.g., Ni, Au), semiconducting (e.g., InP, Si), and insulating types(e.g., SiO_2, TiO_2). III-V nanowires are semiconductor nanowires formed from a compound of group III and V elements. They have outstanding electronic and optical properties [2-3] and are considered ideal materials for photonic and electronic nanodevices, such as resonant tunneling diodes, single electron transistors, photoemitters, and photodetectors [4].

In order to guarantee a long-term reliability of the device, the mechanical response of the III-V nanowire should be well understood. However, many mechanical properties of III-V NWs have not been thoroughly elucidated. Important issues in current research include understanding the influence of growth technologies and device processing on the final properties of the nanowire. Experimental techniques such as scanning tunneling microscopy (STM), atomic force microscopy (AFM), transmission electron microscopy (TEM) and nanoindentation have so far been employed in the mechanical characterization of III-V NWs [5-7]. However, in the absence of experimental data, atomistic simulation is expected to give a very good insight into their properties [8]. A simulation approach could aid the understanding of experiments as well as stimulate new experiments through its predictive power [9-10]. Moreover, one way to compliment the understanding gained from experimental methods is through the use of computer simulations, such as molecular dynamics (MD) and Monte Carlo (MC) simulations.

The Young's modulus, E of a rod in tension and the flexural rigidity, EI of a beam undergoing bending are one of the most fundamental mechanical properties of an engineering structure [11]. Through these properties, the stiffness of the material can be estimated. Molecular Dynamics simulation has proved to be effective technique for mechanical characterization of bulk systems on the atomic level [12]. In this work, MD simulation is applied to predict the stiffness (Young's modulus) and density of III-V (GaAs) NW in the <111>, <110> and <100> directions.

Although, GaAs nanowire is anisotropic, it is often desirable to define its Young's moduli as common in isotropic materials. A molecular model which describes the crystal orientation of the nanowire is established and a feasible loading and boundary condition is applied. The Young's modulus of the structure is extracted by analysing the force-displacement response produced in the dynamics of the structure. The density of the structure is also calculated.

2. MD Simulation using ESFF Forcefield

In brief, molecular dynamics simulation method is based on Newton's equation of motion, given by:

$$\vec{f}_i = m_i \vec{a}_i \qquad (1)$$

Where $\vec{f}_i$ is the total force exerted on particle i, m_i and a_i are the mass and acceleration of particle i respectively.

Force, $\vec{f}_i$ can also be expressed as the gradient of potential energy,

$$\vec{f}_i = -\nabla_i V \qquad (2)$$

These two equations are combined to obtain a comprehensive expression:

$$-\frac{dV}{d\vec{r}_i}(i = 1,...,N) = m_i \frac{d^2 \vec{r}_i}{dt^2} \qquad (3)$$

where V is the potential energy of the system.

Based on this expression, Newton's equation of motion can then relate the derivative of potential energy to changes in position as a function of time. Equation (3) is integrated by discretizing time with an interval, Δt and applying a finite-difference integrator that depends on statistical ensemble. In this paper, the microcanonical ensemble (NVE) ensemble, which conserves the number of atoms (N), the system volume and the total energy, is used. Moreover, the velocity-Verlet algorithm is implemented.

For MD simulation, two kinds of information are required. The first is the chemical composition and spatial configuration of the atoms. The second is the forcefield, which defines the mechanical interactions between atoms. The choice of a forcefield determines to a great extent the accuracy of MD simulation. Therefore, an appropriate forcefield should be selected, to predict within a reasonable accuracy the potentials of the atoms. Among the available forcefields in atomistic simulation, the extensible systematic force field (ESFF) is chosen because it is well parameterized for applications to groups III and V elements and their compounds. The ESFF, proposed by Shenghua et al [13], is a rule-based forcefield covering a wide range of atoms. The ESFF is validated for structural characterization of some metallic type elements [14].

In the ESFF, atoms are parameterized using ab initio calculations and fitting of crystal structures. The parameters of the atoms are classified based on the types of interactions involved, which are; bond, angle, torsion, out-of-plane, electrostatic, and van der Waals interaction. The total energy E is expressed as a sum of bond energy E_{bd}, angle energy E_a, torsion energy E_t, out-of-plane energy E_{op}, van der Waals energy E_{vdw} and electrostatic energy E_{es}.

$$E = E_{bd} + E_a + E_t + E_{op} + E_{vdw} + E_s \qquad (4)$$

The bond types are characterized by the bond orders, π lone-pair interactions, and symmetry positions. The bond energy, kcal/mol is expressed in terms of a Morse function:

$$E_{bd} = D_i[1 - e^{-\alpha_i(r-r_i)}]^2 \qquad (5)$$

where r is the distance between the atoms, r_i is the equilibrium bond distance, D_i is the "equilibrium" dissociation energy of the molecule (measured from the potential minimum), and α_i controls the 'width' of the potential and is equal to the square root of half the force constant divided by D_i,

$$\alpha_i = \sqrt{k/2D_i} \qquad (6)$$

The angle energy, unit kcal/mol varies for different types of angles and is classified according to coordination number, symmetry, π-bonding situation, and ring information, if applicable. For a linear angle θ, the angle energy in terms of the force constant, k of the atoms is given by [13]:

$$E_a = k \cos^2 \theta \qquad (7)$$

Torsion types are determined by the central bond order and the torsion energy is calculated only if the central bond involves nonmetal atoms with coordination numbers less than 4. Similarly, the out-of-plane energy is included only when the centre atom is an sp2 hybridized nonmetal atom with more than two bonds or is an atom with D_{3h} or D_{4h} bonding symmetry having less than 5 bonds.

The van der Wals energy expression in ESFF is of Lennard-Jones 6-9 form [13,15-16]:

$$E_{vdw} = \sum_i \sum_{j\neq i}^{j} \varepsilon_{ij}\left(2\frac{r^9_{vij}}{r^9_{ij}} - 3\frac{r^6_{vij}}{r^6_{ij}}\right) \qquad (8)$$

where r_{vij} and ε_{ij} are the van de Wals parameters, and are calculated from the atomic van der Waals radii and well depths of atoms i and j.

The electrostatic energy is calculated by Coulomb's law:

$$F = k_c \frac{|q_i| \cdot |q_j|}{r_{ij}^2} \qquad (9)$$

where: $k_c = 1/4\pi\varepsilon_0$ is the electrostatic constant and ε_0 is the electric constant (permittivity of free space), r_{ij} is the distance between atom i and atom j and q_i and q_j are atomic partial charges, on atoms i and j.

3. Approach

Fig. 1 below gives an overview of modeling and simulation procedures followed in this paper.

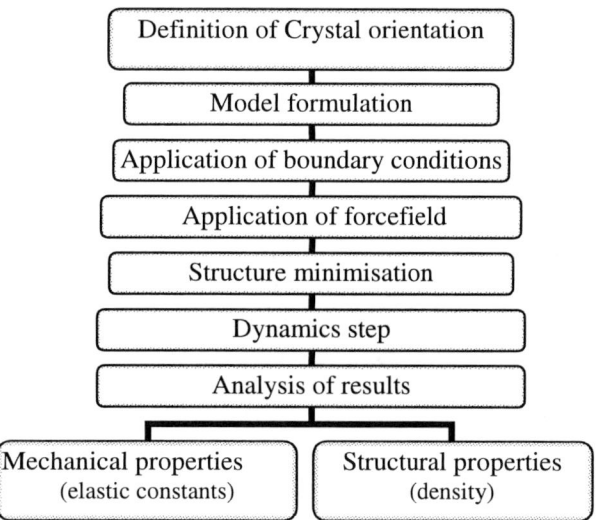

FIG. 1. Simulation steps: a step by step approach to the modelling and simulation process.

With sufficient description of the configuration and potential functions of the molecules, simulation results can quantitatively and qualitatively match experimental results and also interpret the mechanics of a molecular model under external loadings [17-18]. Moreover, the MD simulation results can assist the material scientist to develop a robust material with higher mechanical strength and longer reliability cycles by adjusting the material composition as desired.

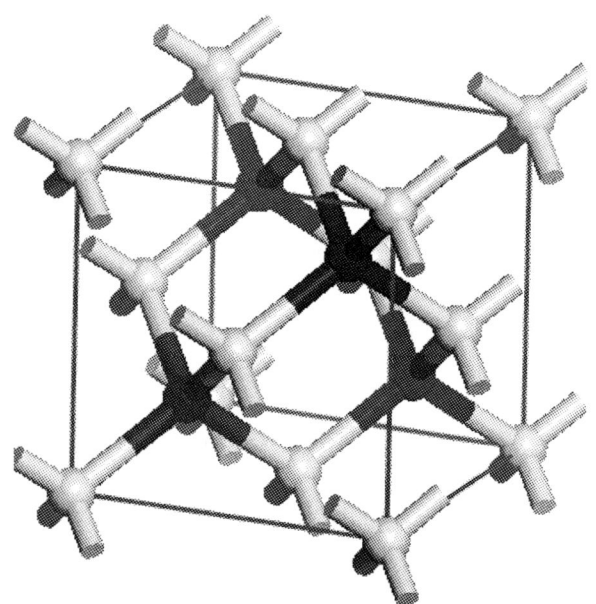

FIG. 2. Crystalline structure of GaAs. *Arsenic and gallium atoms are coloured black and gray, respectively.*

Modelling and simulation are carried out on the commercial software MS modeling 4.0 by Accelrys [19]. The NW model is developed from a sigle crystal of GaAs (which is zinc blende at room temperature, Fig. 2). The

structure is minimized under a minimization step of 15,000 cycles and then stretched in the x-direction. The dynamics time is set at 200ps. The simulation is carried out at room temperature, which is kept constant throughout the simulation with the aid of a Berendsen thermostat.

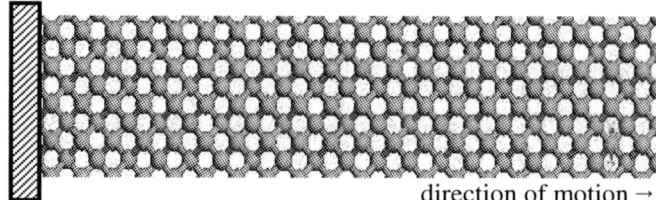

direction of motion →

FIG. 3. The GaAs nanowire model.
In the model, two atom layers are fixed in all directions while the opposite end is in displacement.

The Force-displacement (F-dx) response curves are obtained after the dynamics step of the MD simulation, (Figs. 4, 5 and 6). The Young's modulus is calculated from the elasticity theory:

$$E = \frac{FL}{Adx} \qquad (10)$$

where, A and L are the geometrical properties of the structure, (area and length, respectively). Furthermore, density of the NW is calculated from the ratio of atomic mass to molecular volume.

4. Results

GaAs NW models, which represent three different crystal orientations, all of area $1.5nm^2$ and length 10nm, are established. The III-V nanowires tend to grow in the <111> direction [21], making it the most important growth direction. However, we investigated the <110> and <100> directions to further compare the stiffness of the NW along different directions. The predicted Young's moduli for the NW along directions: <111>, <110> and <100> (using 1,000atoms) are 198.87GPa, 153.54GPa and 147.7 GPa respectively. The result shows that GaAs NW is indeed anisotropic and has highest stiffness in the <111> direction, and least in the <100> direction, due to the condensed packing of the atoms. This inference is further confirmed from the stress-strain plots, which show that a stress of 2GPa applied on the NW strains the material by 1%, 1.25% and 1.5% in the <111>, <110> and <100> directions respectively, confirming the ability of the NW to withstand more stress in the <111> direction. The value of the density obtained, $4.96g/cm^3$ is also comparable, though lesser to the bulk value (5.3g/cm3). Our results show a qualitative agreement, of less than 45% deviation with calculated Young's moduli for GaAs by Brantley, W.A [20], 141.2GPa for <111>, 121.3GPa for <110> and 85.3GPa for <100>, and we observed a similar trend in values.

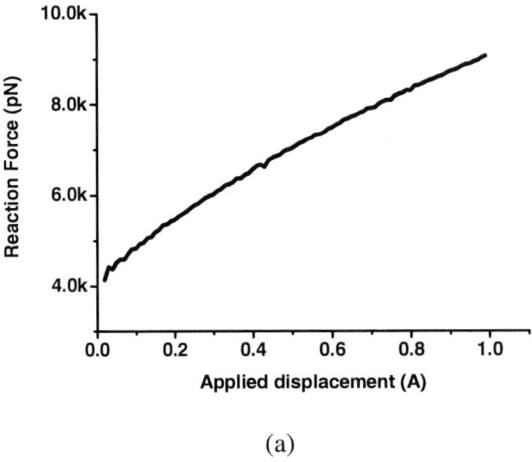

(a)

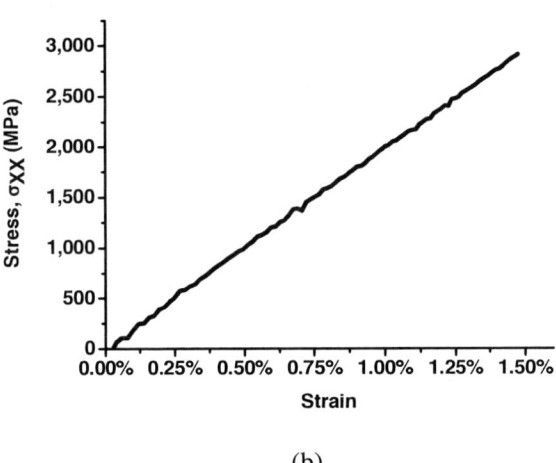

(b)

FIG. 4. Mechanical response of GaAs <111> nanowire (a) Force-displacement response curve. (b) is the stress-strain curve.

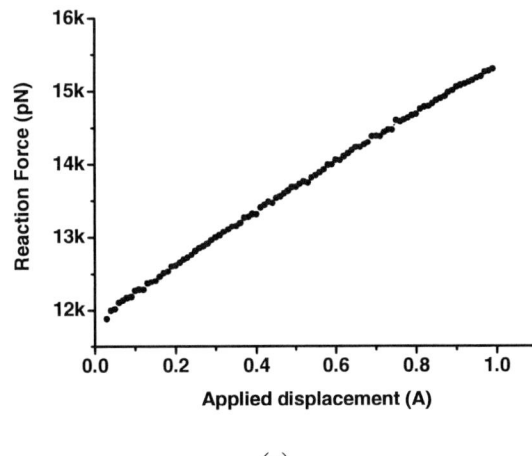

(a)

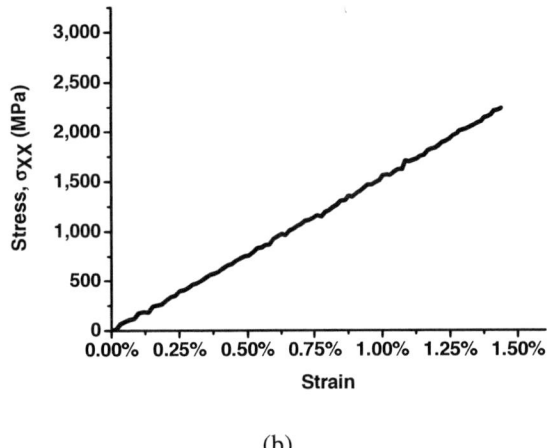

(b)

FIG. 5. Mechanical response of GaAs <110> nanowire (a) Force-displacement response curve. (b) is the stress-strain curve.

5. Conclusions

In this paper, the mechanical and structural properties of III-V nanowire are computed using MD simulation. The simulation predicts the Young's moduli for GaAs nanowire in the three directions, <111>, <110> and <100>. The difference in our results and other numerical values [20] could be due to the size of the material simulated, area 1.5nm^2 and length 10nm, and system error from the forcefield. In our future research, we will investigate the influence of point defects, and size variation on the mechanical properties of III-V nanowires. We also hope to carry out a sensitivity analysis of these effects on the final properties of our NW model.

Acknowledgments

A.W.D and C.AY are grateful to Dr J. Wescott and Dr C. Menke (from Accelrys) for their helpful discussions on III-V MD simulations. A.W.D would also like to appreciate the Netherlands Government for a graduate fellowship.

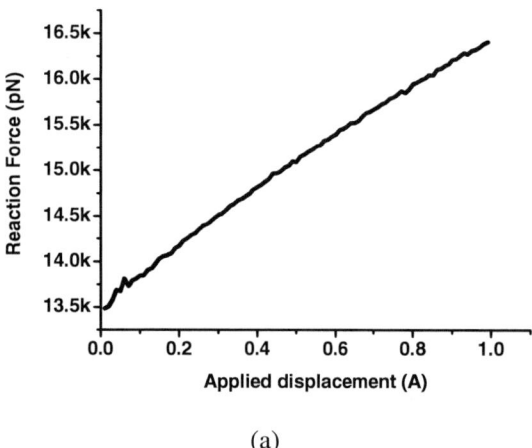

(a)

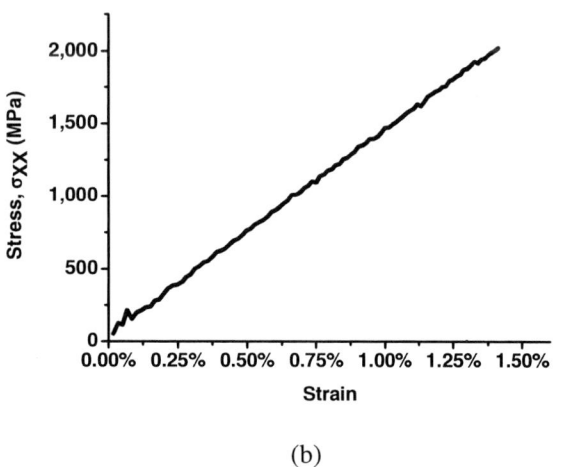

(b)

FIG. 6. Mechanical response of GaAs <100> nanowire. (a) Force-displacement response curve. (b) is the stress-strain curve.

References

1. B. Lee and R. E.Rudd, "First-principles study of the Young's modulus of Si <001> nanowires, eprint arXiv:cond-at/0611073 (2006)

2. X. Duan et al, "Indium phosphide nanowires as building blocks for nanoscale electronic and optoelectronic devices", Nature 409, 66-69 (2001)

3. M.Law et al "semionductor nanowires and nanotubes", Rev. Mater. Res.34, 83 (2004)

4. P. Paiano et al "Size and shape control of GaAs nanowires grown by metalorganic vapor phase epitaxy using tertiarybutylarsine", J. Appl. Phys. 100, 094305 (2006)

5. M. Tabib-Azar et al., "Mechanical properties of self-welded silicon nanobridges", Appl Phy. Ltts., 87 (11), (2005).

6. Z.L. Wang et al., "Measuring the Young's modulus of solid nanowires by in situ TEM", J. of Electron microscopy 51, (2002)

7. Li XD et al, "Direct nanomechanical machining of gold nanowires using a nanoindenter and an atomic force microscope, J. micromech. and microengineering. 15 (3): 551-556 (2005).

8. S. Kodiyalam et al "grain boundaries in gallium arsenide nanocrystals under pressure: a parallel molecular dynamics study" Phys Rev Lett, 86, 55 (2000)

9. E. Z. da Silva et al, "Gold nanowires and the effect of impurities, Nano. Res Lett 1:91–98. (2006)

10. M. P. Allen, "Introduction to molecular dynamics simulation" in Computational Soft Matter: From Synthetic Polymers to Proteins, NIC Series, Vol. 23, pp. 1-28, (2004.)

11. R. Miller and V. Shenoy, "size dependent elastic properties of nano-sized structural elements", Nanotech., vol. 11, No.3, pp. 139-147 (2000)

12. M. Cheng and Y. Lu Structural stability of carbon nanotubes using molecular dynamics and finite-difference time-domain methods , IEEE, Vol 42. Iss 4, pp. 891-894 (2006).

13. S. Shi et al "An extensible and systematic force field, ESFF, for molecular modeling of organic, inorganic, and organometallic systems" J Comp. Chem. 24 No 9 1059-1076 (2003)

14. N. Jager and U. Schilde, "Molecular Mechanics calculations on Chelates of Titanium(IV), Vanadium(IV/V), Copper(II), Nickel(II), Molybdenum(IV/V), Rhenium(IV/V) and Tin(IV) with Di- and Tridentate Ligands Using the New Extensible Systematic Force Field (ESFF) - An Empirical Study", Str. Chemistry. Vol.9. No.2 (1998).

15. D. White "A computationally efficient alternative to the Buckingham potential for molecular mechanics calculations", J. Comp.-aided Mol. Design, Vol. 11, No. 5 (1997)

16. H. Tempel and M.Wolf "Third virial coefficient of ^{4}He for Lennard-Jones (6-9) potential in the temperature range from 18°K to 295°K", Phys Ltts., A, Vol. 24, Iss. 3, pp. 187-188. (1967)

17. K. N. Chiang et al, "Mechanical Characteristic of ssDNA/dsDNA molecule under external loading", Appl. Phy. Lett., Vol. 88, pp. 023902, (2006).

18. C. Yuan et al, "Investigation of the sequence-dependent dsDNA mechanical behavior using clustered atomistic-continuum Method," in proc. NSTI Nano conf., California, U.S.A.(2005).

19. Accelrys Inc., Material Studio™ DISCOVER, San Diego: Accelrys Inc., (2005)

20. W. A. Brantley, "Calculated elastic constants for stress problems associated with semiconductor devices", J. Apply. Phys. Vol. 44, No.1 (1973)

21. M. Verheijen et al, "Growth kinetics of heterostructured GaP-GaAs nanowires", J. Am. Chem. Soc. 128 (4), 1353–1359 (2006).

Stress Relaxation Characterization of Hypoeutectic Sn3.0Ag0.5Cu Pb-free Solder: Experiment and Modeling

Gayatri Cuddalorepatta, Dominik Herkommer, Abhijit Dasgupta
CALCE Electronic Products and Systems Center
Mechanical Engineering Department
University of Maryland, College Park, MD 20742 USA
Phone: 301- 405-5231, Email: gayatric@umd.edu

Abstract

This study investigates the time dependent behavior of lead-free hypoeutectic Sn3.0Ag0.5Cu solder using a combination of experimentation and modeling. Experimental data for hypoeutectic Sn3.0Ag0.5Cu (SAC) is obtained from stress relaxation tests at different temperatures and different initial stress levels, and compared with the benchmark eutectic Sn37Pb, using a modified lap shear test. The test specimen uses a modified Iosipescu specimen with a 180 micron wide solder joint between two notched copper platens. The creep modeling uses both primary creep (generalized exponential model) and secondary creep (Garofalo sine hyperbolic model). The model constants are obtained iteratively by matching the test results with viscoplastic finite element models of the test specimen. As expected, the stress relaxation data confirms that eutectic Sn37Pb relaxes faster than hypoeutectic SAC. For example, at a given stress level of 7 MPa and at a given temperature of 75°C, SAC relaxes only by 68% in the time that Sn37Pb relaxes almost completely. Modeling work suggests that both primary and secondary creep behavior must be included in models to achieve reasonable agreement between stress relaxation tests and simulations. Improvements to existing creep modeling strategies to better capture the relaxation response are discussed.

1. Introduction

Typical field conditions seen in electronic packages are thermal cycling, mechanical cycling, vibration loading and drop, to name a few. Solder joints form critical components of the package since they form not only the electrical interconnection but also the load bearing mechanical interconnection. The most dominant failure mode in solder is fatigue. High temperature and time dependent loading causes viscoplastic material response of solder. Hence in order to understand the material response of solder accurately and assess the package reliability, it is necessary to characterize the viscoplastic constitutive behavior and fatigue durability response.

Owing to the recent ban on the usage of lead due to environmental legislation, the electronics industry is shifting to Pb–free substitutes for solder interconnects. Eutectic SnAg, eutectic SnCu and SnAgCu solders were assessed to be few of the promising alloys for wave soldering and reflow soldering applications. As opposed to the traditionally used eutectic SnPb, the Pb-free solders have not been studied as extensively. The constitutive properties and durability response of these Pb-free substitutes are very important and are currently being studied by several researchers.

The constitutive properties of Sn3.9Ag0.6Cu are considerably different from those of the Sn37Pb solder, owing to the microstructural differences. In particular, the creep resistance of the Sn3.9Ag0.6Cu solder is much higher, especially at lower stresses (Figure 1) [1]. This is attributed to the differences in the microstructure of the SnAgCu and SnPb solders.

Sn3.9Ag0.6Cu consists of small (5-50 nm diameter) intermetallic particles of Cu_6Sn_5 and Ag_3Sn embedded at the grain boundaries and in the matrix. The specifics of the microstructure can vary depending on factors such as solder composition [2], cooling rate [3,4], dimensions of solder joint, to name a few. For example, the microstructure of eutectic SnAgCu undergoes coarsening under slow cooling, and forms a finer network of Ag_3Sn intermetallics at the Sn grain boundaries compared to the solder undergoing fast cooling [3]. Figure 2 shows the difference in microstructure induced due to difference in cooling rates. As opposed to SnAgCu alloys, the Sn37Pb consists of large equiaxed Pb islands embedded in a polycrystalline tin matrix.

Eutectic SnAgCu has been found to be the most viable substitute for Pb-free solders. However, due to cost-benefits, Japan is considering the use of SnAgCu solder with reduced Ag content. Such a solder falls in the hypoeutectic region and due to the reduced silver content would be expected to have different volume fraction of Ag_3Sn intermetallics leading to different material behavior. This study focuses on the stress relaxation of the hypoeutectic Sn3.0Ag0.5Cu solder and provides a comparison with the relaxation seen in traditionally used eutectic Sn37Pb.

Several studies have been conducted on varying compositions of SnAgCu solders to understand the microstructural evolution and their dependence on mechanical constitutive and durability response. Durabiltiy tests conducted include mechanical durability tests [1, 5, 6] as well as thermal cycling durability tests [7] as a function of composition of SnAgCu. Zhang characterized and compared the viscoplastic constitutive behavior of eutectic Sn3.8Ag0.7Cu with eutectic Sn37Pb. Ma et. al.[8] conducted creep tests on SnAgCu and SnAg as a function of varying alloy compositions, and found significant differences in the creep characteristics with increasing Ag and Cu content. The microstructural differences arising from the composition change have been tied to the observed creep characteristics. Hence we

1-4244-1105-X/07/$25.00 ©2007 IEEE

can expect the hypoeutectic Sn3.0Ag0.5Cu to exhibit different creep behavior from the eutectic Sn3.8Ag0.7Cu. It is important that the test results obtained are presented with the microstructural characteristics seen for that specific fabrication profile. It has been seen that the durability [9] and the creep resistance [8] of SnAgCu solders deteriorate with microstructural aging.

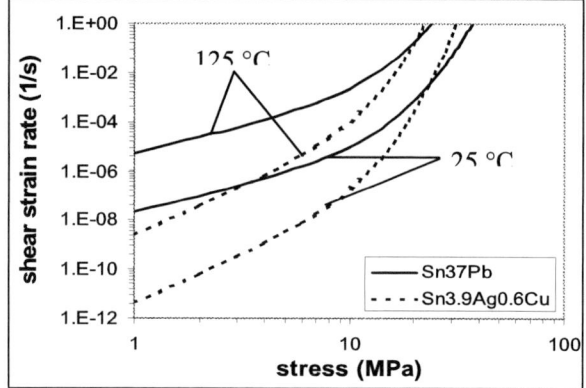

Figure 1: Creep Curves- Sn3.9Ag0.5Cu and Sn37Pb

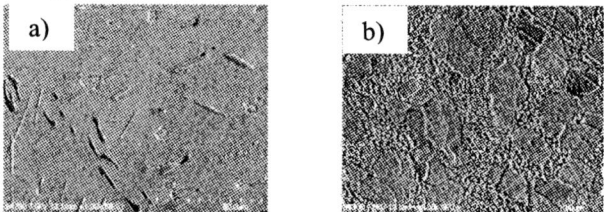

Figure 2: Micro-structure of eutectic SnAgCu Solder for a) fast cooling and b) slow cooling. Microstructural grain coarsening is evident under slow cooling.
(Microstructural Image from Weise et al. [3])

In order to qualify the thermal cycling reliability of electronic products, accelerated testing of electronic packages is conducted. Standard temperature cycling profiles have maximum temperatures of 125^0 C, 100^0 C and 75^0 C while the minimum temperatures are normally 0^0 C or -40^0 C. The package is held at the extreme temperatures from 15 minutes to a few hours. The important parameters in the temperature profile are temperature extremes, amplitude, ramp rate and dwell times.

Under the above described temperature loading, if the hysteresis stress-strain response is evaluated, it is seen that relaxation of stresses occurs during ramping to hot temperatures, although it picks up significantly during the isothermal dwell. The hysteresis response seen in the solder of a BGA 256 I/O package subject to thermal cycling using a 125^0 C to -40^0 C profile, with 15 and 3 hours dwells, is shown for illustration purposes in Figure 3. Two solders have been studied here, namely the Pb-free eutectic SnAgCu solder and the traditionally used eutectic Sn37Pb. It is seen that the stress in the solder relaxes at high temperature and dwells, due to creep in viscoplastic solder. While it is beneficial to reduce the peak stresses through stress relaxation, this comes at the price of creep damage that affects the fatigue life of the solder. As shown in Figure 1, the creep properties of Pb-

free SnAgCu (SAC) solder differ vastly from traditionally used Sn37Pb (SP) solder. Owing to higher creep resistance of the SAC solder, stress relaxation at the hot dwell is expected to be lower for SAC than in SP (Figure 3). Hence the focus of this study is to evaluate and compare the stress relaxation rates of SAC with that of traditionally used eutectic Sn37Pb as a function of initial stress levels and temperature. In this study, virgin solder specimens of hypoeutectic Sn3.0Ag0.5Cu are used.

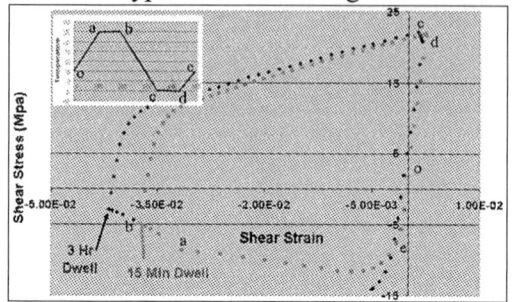

Figure 3: Finite element model hysteresis results showing stress relaxation of solder in BGA 256 under thermal cycling (shown insert)

Arrowhead et al., [10] and Kashyap et al, [11] conducted creep and stress relaxation tests on SnPb solder and compared the creep constants obtained from both tests. Bang et al [12] conducted a survey on the results obtained by the above authors and several others to assess the feasibility of using stress relaxation tests to verify the creep constants. The study concluded that though a reasonable agreement is obtained in most cases, discrepancies in data exist in predicting the exponents, creep regions and transitions. Yang et al.,[13] conducted stress relaxation tests on eutectic Sn3.5Ag using impression tests and compared them with existing creep tests in literature. The activation energies derived were found to agree with those from creep tests only at low stress levels.

In the current study, the creep constants of hypoeutectic Sn3.0Ag0.5Cu are derived iteratively by matching stress relaxation test results and viscoplastic finite element model simulations. The creep constants from the eutectic Sn3.8Ag0.7Cu are used as a baseline for the evaluation of hypoeutectic SAC properties. Primary (generalized exponential) and secondary (Garofalo) creep laws [14] are investigated, for the the solder. The limitations of the methodology used are listed and the future work required to address the issues are described.

2. Approach

The approach is two-fold. First, the stress relaxation testing of Pb-free and eutectic Sn37Pb is conducted as a function of initial stress levels and temperature. Next the creep constants of hypoeutectic SAC are investigated using viscoplastic, non-linear finite element simulations of the stress relaxation test

2.1. Experiments

The stress relaxation tests have been conducted using a thermo-mechanical microscale (TMM) test system, developed at the University of Maryland. This setup has the capability of measuring constitutive and fatigue damage properties of materials. The test setup used in this work has been described in the literature [15] and relevant details are summarized here. For a comprehensive overview of the test setup and the algorithm for the control software, see Haswell [16]. The TMM test system is schematically depicted in Figure 4.

The system primarily comprises of three components; namely a piezoelectric stack actuator, a linear variable displacement transducer (LVDT) and a load cell. The specimen is loaded in the steel grips that are attached to the piezoelectric actuator via a shaft and flexible link configuration. The piezoelectric actuator produces the required cyclic displacements in closed-loop control. The LVDT measures the displacement of the specimen along with the load train while the load cell measures the load. The test is a displacement controlled test. The high resolution, large range of strain-rates, and absence of moving parts or linkages in the load train, yield excellent cyclic tests with zero backlash upon reversal. High temperature isothermal conditions are facilitated with a hot-air, closed-loop heating module.

The specimen configuration used in the current work, is shown schematically in Figure 5. The shear specimen is a simple notched shear specimen similar to that developed by Iosipescu [17]. The advantage of this specimen configuration is that a uniform stress distribution is obtained in the solder region [18]. The specimen fabrication involves soldering two copper platens with the solder of interest. Complete details of the fabrication procedure are presented elsewhere [15, 16].

The fabricated specimens have a solder joint thickness of 180 μm and are deformed in mechanical shear. All specimens are aged for 100 hours at 0.8 T_m (~132 ^{0}C in the case of SnAgCu solders) prior to testing, to obtain a stable microstructure. Unlike in thermal cycling tests, where the cycling stabilizes the microstructure, pre-aging of test specimens is required in room temperature mechanical cycling. The aging also serves to relax any residual stresses developed due to the polishing or reflow of the solder.

The TMM test system has the capability of conducting monotonic constitutive property and durability testing. The test control parameters and raw test data are in terms of the displacement δ and load P. These parameters are converted to average shear stress and strain by

$$\gamma = \frac{\delta_s}{h} \tag{1a}$$

where,
$$\delta_s = \delta - P/K \tag{1b}$$

$$\tau = \frac{P}{A} \tag{2}$$

where γ and τ are the average shear strain and stress, respectively, in the solder; h and A are the solder joint height and cross-sectional area, respectively; and δ_s is the specimen shear deformation obtained from the displacement transducer reading δ, by adjusting for load

train stiffness, K [16]. Since the geometric parameters of specimen vary slightly due to the fabrication process, the stress and strain curves are based on actual measured dimensions to eliminate geometric influences. Isothermal testing at high temperature is conducted using pressurized air heating.

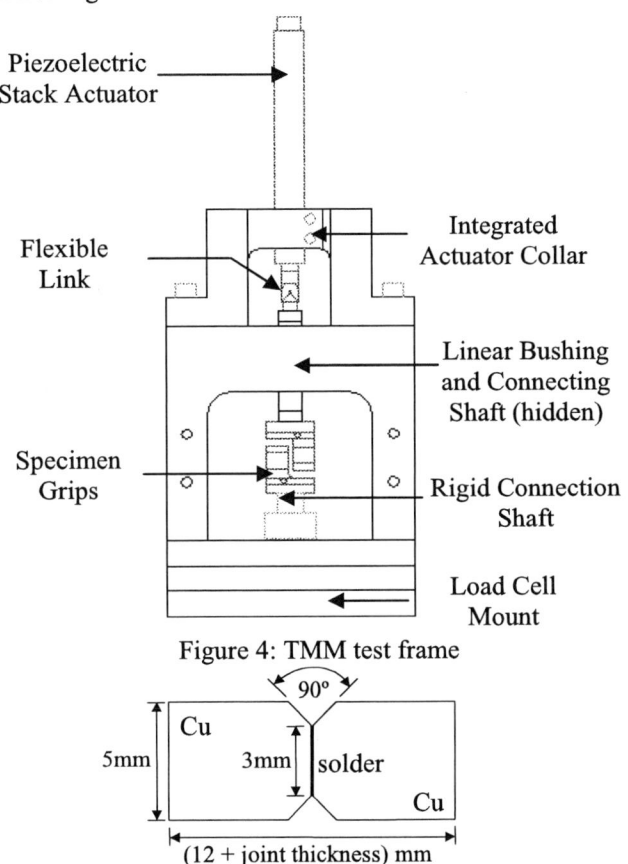

Figure 4: TMM test frame

Figure 5: TMM specimen schematic

The stress relaxation test is a method to evaluate the drop in stress at a constant displacement, owing to the creep behavior of a material. Loading involves ramping to an initial stress level and holding the displacement constant. The stress history is recorded. In contrast, in creep testing a constant load is applied to the test specimen. This is different from creep testing where a constant load is applied to the test specimen. In order to understand the effect of temperature, high temperature isothermal testing is conducted. At high temperature, the maximum stresses are lower due to lower yield stress and due to higher creep deformation encountered during the initial ramp ot the initial deformation level. The input profile and test matrix for the stress relaxation test is displayed in Figure 6 and Table 1. The ramp rate was set to a high value of 10 microns per second to reduce the effect of creep during the loading stage. A minimum of two specimens were tested per load level for repeatability.

Temp. / ^{0}C	25	75
Stress / MPa	8 - 25	4 - 9

Table 1: Test Matrix

2.2. Finite Element Modeling of Stress Relaxation for Determining the Creep Constants

In order to simulate the stress relaxation tests, the portion of the specimen spanning between the two grips is modeled with a simple 2-D finite element model. The model therefore includes the solder layer and the tapered potion of the copper platens, as shown in Figure 7. The viscoplastic behavior of the solder is modeled using partitioned, temperature-dependent elastic, plastic, and creep properties. For comparison purposes, solder creep is modeled using secondary creep alone, as well as combined primary and secondary creep. Garofalo creep law is used to represent the secondary creep, while primary creep is modeled using a generalized exponential law [14]. Copper is modeled using elastic behavior. The material laws and creep properties of eutectic SAC are given in Appendix 1 and 2. The model uses eight-noded, quadratic, plane stress elements.

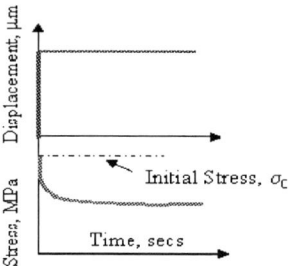

Figure 6: Loading Profile for Stress relaxation

Figure 7: 2D FEA model of Test Specimen

3. Results and Discussion

The results of the stress relaxation tests and the corresponding finite element modeling are presented and discussed here.

3.1. Stress Relaxation Testing Results – Comparison of Hypoeutectic SAC and eutectic Sn37Pb

Stress relaxation results from tests conducted on both solders at various stress levels and at two temperatures levels are presented here. The raw data from multiple tests at a given load level are post-processed to obtain a fit. The fits yielded good correlation factors. The test data presented henceforth represents the post-processed fit data, unless otherwise mentioned. Figures 8 and 9 show the comparison of stress histories for Sn3.0Ag0.5Cu and Sn37Pb at room and high temperature. The 75^0 C high temperature data show the extent to which the stress relaxation rate and creep saturation rate increase at high temperature. The relaxation of SAC solder is seen to be much less and slower than SnPb solder, as expected from the superior creep resistance. Table 2 gives quantitative comparisons of the stress relaxation as a function of the temperature and solder type.

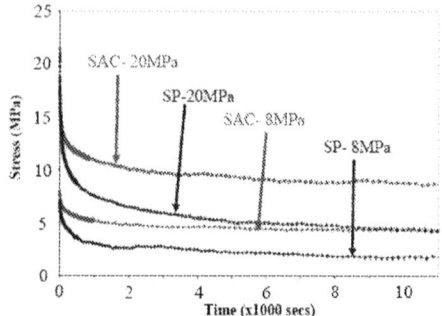

Figure 8: Test results of SAC vs. SP at RT

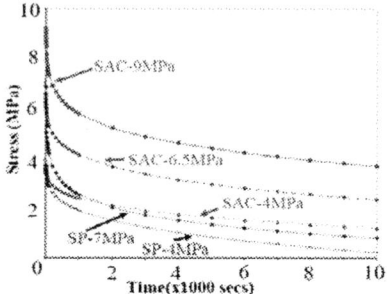

Figure 9: Test results of SAC vs. SP at RT 75^0 C

% Stress Relaxation			
Time, Stress (MPa), Temperature (^{0}C)	SAC	SP	Ratio (SAC/SP)
20 min (7, 75)	40	70	0.57
3 hours (7, 75)	68	100	0.68
20 min (7, 25)	42	58	0.72
3 hours (7, 75)	52	76	0.68

Table 2: Comparison of relaxation in 20 minutes and 3 hours for SAC and SP at RT and 75^0 C

As seen from room temperature comparison in Figure 8, for a given initial stress of 20 MPa, SAC relaxes only by 60% in the time, SP relaxes by 80%. At 75^0 C, a similar comparison at 7 MPa shows that SAC relaxes only 68% in the time that SP relaxes by 80%. The load drop in SAC vs. SP at different stress levels at RT, shown in Figure 10, gives a clear indication of the rate of the stress drop in SP compared to SAC, at different stress levels.

From the relaxation plots obtained from the two solders at the given temperature conditions, it is seen that the response has two distinct portions; an initial steeply descending portion followed by a "steady state" decrease at a much lower slope. It is seen that the majority of the relaxation (~70%) in Sn37Pb occurs in the initial 500 - 1000 seconds. On the other hand, for the Pb-free solder case, the stress drops down to barely 40% in the same time. It is also seen that the steady state slope of eutectic SnPb is higher than that of the Pb-free solder, with the difference being more pronounced with increasing temperatures. Increase in the initial stress increases the relaxation in both case, but more severely in SAC than in SnPbs. Relaxation in the SnPb solder saturates within about within 15 minutes, irrespective of the stress level.

The observed regions of stress relaxation, with an initial rapid stress drop followed by a steady state relaxation at lower rate, is explained in terms of the dominant dislocation mechanisms in Pb-free solders. Pb-free solders are considered to be dispersion strengthened materials, whose superior creep properties arise from the hard intermetallics which act as obstacles (other than grain boundaries) that pin dislocations. In order to pass, the dislocations need to bend and climb past obstacles followed by detachment of the attractive interactions at the trailing end. This requires external drivers in the form of temperature or stress. During the initial stages of the relaxation, the relatively higher stress levels assist in the creep (dislocation creep). As the relaxation process progresses, the available stress driver for overcoming the obstacles is lowered. This causes a reduction in the dislocation climb and detachment, leading to lower rates of relaxation seen in the response. Increase in temperature improves the dislocation climb and detachment, as is evident from the plots at high and room temperature.

Based on the stress relaxation results obtained here, it is seen that for the purposes of accelerated thermal cycling, for profiles with $T_{mac} < = 75^0$ C, dwelling the packages beyond 20 -30 minutes does not seem to have much benefit with respect to the stress saturation obtained (for example, for 7 MPa initial stress, 20 minutes of dwell yields around 50% relaxation, while a 3-hour dwell increases the relaxation only by another 10%)

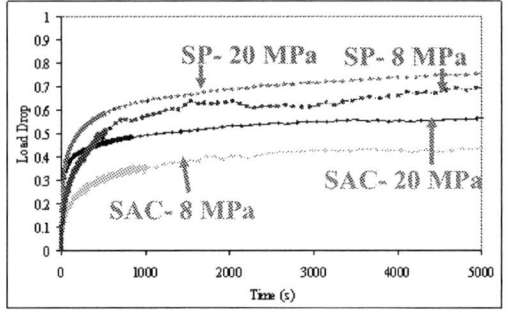

Figure 10: Stress relaxation viewed in terms of load drop.

3.1.1. Post-test microstructural characterization

Post-test microstructural characterization is conducted using the ESEM, to obtain insights on the effect of initial stress and temperature on the damage patterns, if any. It was found that specimens tested at lower stress levels (under 10MPa) showed no visible signs of damage. Specimens tested at higher stress levels however mainly showed two damage mechanisms. On most of the specimens at the higher stress level (20 MPa), small voids occurred along the shear planes (Figure 11 a). It is also possible that these cavities may have been generated by the polishing process dues to highly stressed and highly damaged regions along the shear planes. It was also seen that the material develops cracks along the intermetallics between copper and solder (Figure 11 b).
To contrast the microstructural differences, Figure 11(b) shows the solder joint of a specimen that has been tested under mechanical cycling [9]. The deformation of the solder shows significant microcracking in the bulk of the

solder accompanied with macro cracks close to the interface of solder and copper. Shear bands and striations are observed from the repeated reversal of the applied displacement.

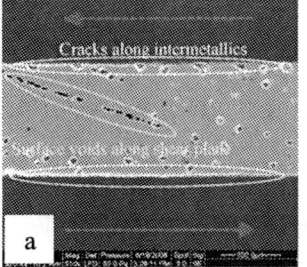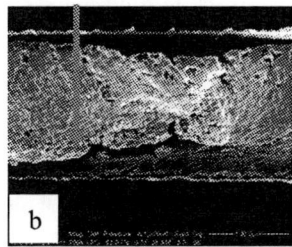

Figure 11: (a) Post-test images of (a) Stress relaxation showing damage along shear plane and interfacial intermetallic cracks, (b) Damage in solder under mechanical cycling [9]

3.2. Viscoplastic Finite Element Modeling of Sn3.0Ag0.5Cu (SAC305) Stress Relaxation for Extracting Creep Model Constants

The stress relaxation test data for the lead-free hypoeutectic SAC 305 material are used to derive creep constitutive properties by iteratively comparing finite element model simulation results with the test results. Creep constants of eutectic SnAgCu solder, developed from prior creep testing [1], are utilized as initial estimates in this iterative process, owing to the similarity in solder composition. The output of the iterations is a set of feasible creep constants, that provide a reasonable fit to the stress relaxation data for SAC305.

Most researchers simulating stress relaxation utilize creep constants obtained from creep tests. Studies have also been conducted to derive creep constants based on stress relaxation tests. Some studies report a reasonable match between the constants obtained from these two approaches. Bang et.al, [12] provide a comprehensive review of such studies for eutectic Sn37Pb. A brief overview of the results is provided here.

While the data from creep and stress relaxation tests were found to be in agreement for most cases surveyed [10, 11], disagreement was seen in the work of other researchers [13, 19]. For example, creep and relaxation tests conducted by Arrowood [10] on SnPb showed reasonable agreement for microstructure with larger grain size, while deviations were seen in the case of microstructure with smaller grain size. Kashyap [11] found good agreement between the results from creep and stress relaxation tests. In general, the stress exponents and the transition stresses between creep regions show good qualitative agreement from both types of tests.

However, Grivas [19] reports discrepancies in the three regions of creep, especially for region I (low stress creep regime) and a much higher stress exponent in region II (moderate stress creep regime), and abrupt transition from region III (high stress creep regime) to I. Some of the possible reasons suggested for the discrepancies observed by different researchers are: microstructural dependence, the inability of the relaxation test to capture superplastic creep region (region II), and the short residence time at a given stress during stress relaxation tests compared to constant stress creep tests.

In the current paper, creep constants from creep tests were not available for simulating stress relaxation. Hence these constants were deduced from the stress relaxation tests, using a heuristic iterative procedure that starts with eutectic SAC creep constants.

3.2.1. Results

Figure 12 shows a comparison for SAC305 solder, between the test results and the corresponding finite element simulation based on only secondary creep at 75^0C. As seen in the figure, modeling the solder with only the secondary creep constants from eutectic SAC does not correctly predict the stress relaxation of hypoeutectic SAC. The relaxation is underpredicted in the model, and the error increases as the initial stress becomes smaller. The time to saturation of stress relaxation is overestimated by the FEA, compared to the experiment. Furthermore, the initial steep drop in the relaxation curve is significantly underpredicted by the secondary creep. Figure 13 shows that there is significant improvement in the agreement, if both primary and creep is included in the FEA simulations. Similar trends were also observed in the room temperature case.

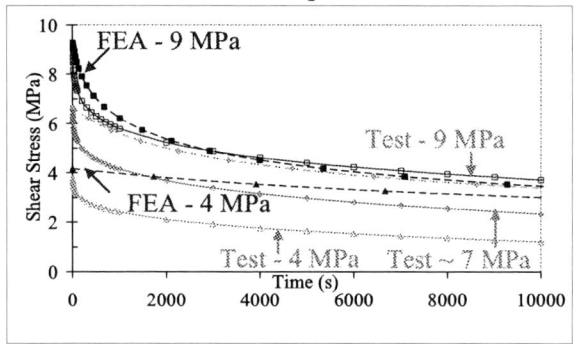

Figure 12: Comparison between FEA results (based on Secondary Creep constants of Eutectic SAC) and Test Results (of Hypo-SAC) at 75^0 C

Since there are a total of nine constants in the primary and secondary creep laws, the iteration process for obtaining an acceptable set of model constants is tedious and a unique solution is not assured. To judge the feasibility of the derived constants, they are compared to baseline eutectic SAC creep constants. For this analysis, the iteration was tailored such that saturation stress past three hours matched reasonably closely with the test results. Based on this procedure, the creep constants obtained for hypoeutectic Sn3.0Ag0.5Cu are given in Appendix 2.

Figures 14 gives the resulting plot of the secondary creep of hypoeutectic SAC in comparison to those for eutectic SAC. The creep resistance of hypoeutectic SAC is slightly inferior to that of eutectic SAC at low stresses, with the trend reversing at high stress. The corresponding relaxation results from the new SAC305 model constants

are plotted in Figures 15-16, along with the test results. Although there is reasonable qualitative agreement with test results, there are quantitative discrepancies. At room temperature the predicted relaxation is less sensitive than the test results, to the initial stress. At high temperature, the model overpredicts the stress relaxation, especially during the early period when the primary creep dominates.

In Figure 17, the relaxation test data are compared to FEA results based on both the eutectic SAC constants (measured from creep tests), as well as the new SAC305 creep constants (iteratively derived from stress relaxation tests). In general, it is seen that eutectic SAC properties underpredict the stress relaxation test data while the SAC305 properties overpredict the relaxation response. However, the steady-state value (after 3 hours) predicted by the SAC305 constants agree well with test results. This information is also summarized by plotting load drop at 10000 seconds in Figures 18-19.

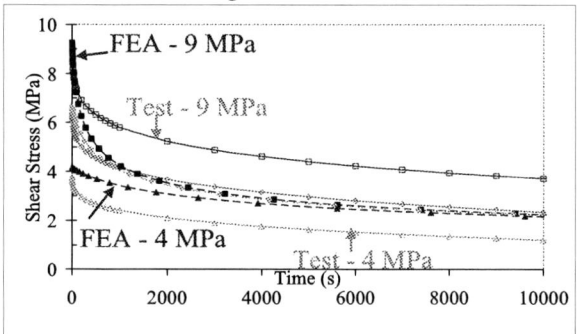

Figure 13: Comparison of FEA with Secondary and Primary Creep of Eutectic SAC and Test Results of Hypo-SAC – 75^0 C test

3.2.2. Discussion of the Creep Constants Obtained from Stress Relaxation

The creep constants obtained from the heuristic iteration in this study do not appear to provide a very close quantitative match to the stress relaxation test results. This set of creep constants needs to be verified from creep tests. For example, the secondary creep rate at high stress suggests that hypoeutectic SAC is more creep resistant than eutectic SAC (Figure 14), which is counter-intuitive. Theoretical considerations suggest that hypoeutectic SAC creep rate should be equivalent, if not inferior, to eutectic SAC, based on the volume fraction of intermetallics pinning dislocations. However, this is the feasible set that provides the best overall fit over the entire range of stress, temperatures and time; while not deviating unreasonably excessively from baseline SAC constants.

This study certainly raises serious questions about the validity of using only secondary creep in FEA models, as is frequently done in the thermal cycling literature. It appears that primary creep plays a very important role whenever the stress is not constant.

Creep testing is currently in progress and will be used to provide a set of creep constants that can be compared to the constants obtained in this study by iteratively matching stress relxation test results.

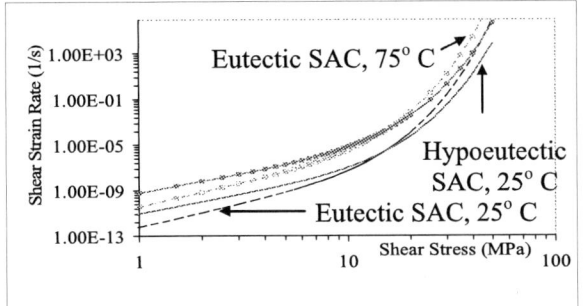

Figure 14: Secondary creep law comparison for hypo-eutectic vs. eutectic SAC

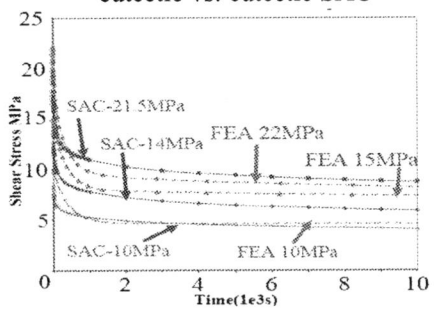

Figure 15: Plot of test results of hypoeutectic SAC with modeling results of the new hypoeutectic SAC constants (from relaxation) – Room Temperature

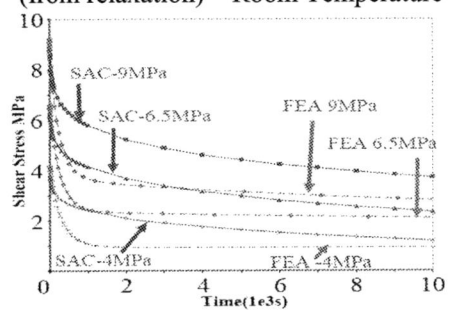

Figure 16: Plot of test results of hypoeutectic SAC with modeling results of the new hypoeutectic SAC constants (from relaxation) – 75 ^{0} C

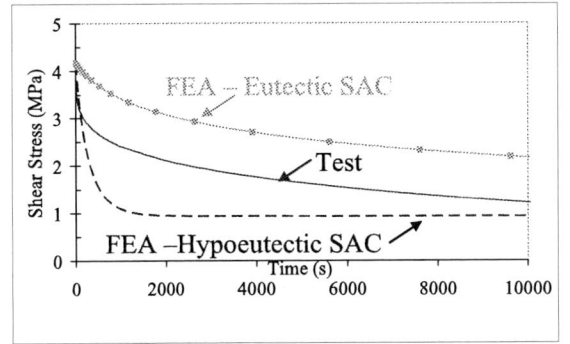

Figure 17: Comparison of test results of hypoeutectic SAC with modeling results of the new hypoeutectic SAC constants (from relaxation) and the eutectic creep constants (from creep testing) -75^0 C

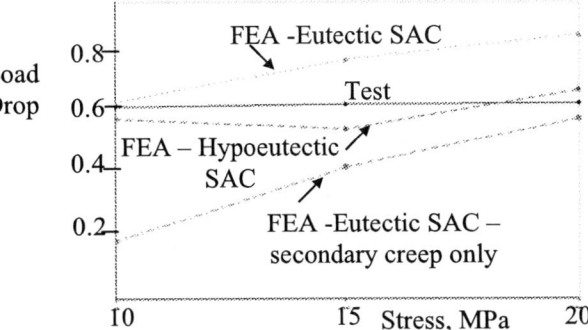

Figure 18: Load Drop at room temperature after 10000 seconds dwell

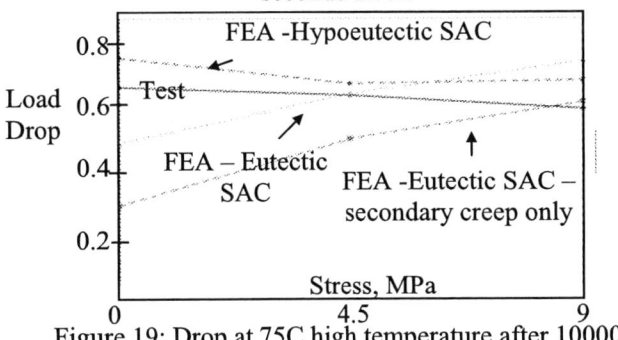

Figure 19: Drop at 75C high temperature after 10000 seconds dwell

4. Summary and Conclusions

Isothermal stress relaxation tests were conducted at different initial stress levels and temperatures, to characterize and compare the relaxation response of hypoeutectic Sn3.0Ag0.5Cu with eutectic Sn37Pb. The relaxation response has two distinct regions, with an initial steep drop followed by a steady state region. The trend is explained in terms of the dominant dislocation mechanisms and their interactions. As expected, the eutectic Sn37Pb showed a higher amount of relaxation compared to hypoeutectic SAC which has superior creep resistance. Based on the test results, it is seen that isothermal dwell past twenty minutes for accelerated thermal cycling profiles with T $_{max}$ < 75 C, does not yield any significant benefit with respect to relaxation and creep deformation. The stress relaxation tests were further utilized to derive the creep constants of hypoeutectic SAC by iteratively matching test and viscoplastic finite element models. The results show that secondary creep alone underpredicts the stress relaxation and does not capture the steep stress drop during the initial time period. Including primary creep with the secondary creep model qualitatively improves the initial steep drop of the stress. Owing to the complexity of the creep laws chosen for modeling, it is necessary to evaluate the creep constants of hypoeutectic SAC using creep tests, to verify the constants derived in this study from stress relaxation tests.

References

1. Zhang, Q., "Isothermal Mechanical and Thermo-mechanical Durability Characterization of Selected Pb-Free Solders" Ph.D. Dissertation, 2004. University of Maryland, College Park, MD, USA,

2. Terashima, S., Kariya, Y., Hosoi, T., and Tanaka, M., ,"Effect of Silver Content on Thermal Fatigue Life of Sn-xAg-0.5Cu Flip-Chip Interconnects," *Journal of Electronic Materials,* Vol. 32, No.12, (2003), pp. 1527-1533

3. Wiese, S., and Wolter, K.-J., "Microstructure and creep behavior of eutectic SnAg and SnAgCu solders", *Microelectronics Reliability*, Vol. 44, (2004),pp. 1923 – 1931.

4. Kang, S.K., Lauro, P., Shih, D., Henderson, D.W., "The Microstructure, Thermal Fatigue, and Failure Analysis of Near-Ternary Eutectic Sn-Ag-Cu Solder Joints", Material *Transactions,* 45(3), (2004),pp. 695-702

5. Lee, K. O., Yu, J., Park, T. S., and Lee, S.B., "Low – Cycle Fatigue Characteristics of Sn-Based Solder Joints", *Journal of. Electronic Materials,* 33, (2004), pp. 249-257

6. Kariya, Y., Morihata, T., Hazawa, E., and Otsuka, M., "Assessment of Low-Cycle Fatigue Life of Sn-3.5mass%Ag-X (X=Bi or Cu) Alloy by Strain Range Partitioning Approach," *Journal of Electronic Materials*, 30(9), (2001),pp. 1184-118

7. Nurmi, S., Sundelin, J., Ristolainen, E., Lepisto, T., "The Effect of Solder Paste Composition on the Reliability of SnAgCu joints", *Microelectronics Reliability,* 44, (2004), pp. 485-494

8. Ma, H., et al, " Reliability of the Aging Lead Free Solder Joint", *IEEE ECTC Proceedings,* 2006

9. Cuddalorepatta, G, et al, "Isothermal Mechanical Durability of Hypoeutectic Sn3.0Ag0.5Cu", *ASME International MechanicalEngineering Congress and Exposition, 2005.*

10. Arrowwood, R., et al, "Stress Relaxation of a Eutectic Alloy in Superplastic Condition",, *Material Science and Engineering,* 92 (1987) pp 23-32 and 33-40

11. Kashyap, P. , Material Science Engineering, (1981) pp. 50, 205.

12. Bang, W.H. et al, " The Correlation Between Stress Relaxation And Steady State Creep Of Eutectic Sn-Pb', *Journal of Electronic Materials,* Vol. 34, No. 10, (2005), pp. 1287-1300

13. Yang, "Impression Stress Relaxation of Sn3.5Ag Eutectic Alloy,"*J. of Mater. Res*Vol. 21, No 10, (2006),pp. 2653-59

14. Skrzypek, J.J., <u>Plasticity and Creep: Theory, Examples and Problems</u>, CRC Press Inc (1993)

15. Haswell P., Dasgupta A., "Viscoplastic Characterization Of Constitutive Behavior Of Two Solder Alloys," *ASME International Mechanical Engineering Congress and Exposition, 1999.*

16. Haswell, P., 2001, "Durability Assessment and Microstructural Observations of Selected Solder Alloys", Ph.D. Dissertation, University of Maryland, College Park, MD, USA

17. Iosipescu, N., "New Accurate Procedure for Single Shear Testing of Metals," *Journal of Materials*, Vol. 2, , September 1967, No. 3, pp. 537-566.

18. Reinkainen, T., Poech, M., Krumm, M., Kivilahti, J., "A finite-element and experimental analysis of stress distribution in various shear tests for solder joints." *Journal of Electronic Packaging, Transactions ASME*, 1998,,. 120 (1), pp. 106-113.

19. Grivas, D., et al, "Deformation of Pb-Sn Eutectic Alloys at Relatively High Strain Rates", *Acta metallurgica*, 27, 731, (1979)

Appendix 1: Constitutive Material Laws used in FEA

1. Solder Properties:

$$\varepsilon = \varepsilon_{el} + \varepsilon_{pl} + \varepsilon_{scr} + \varepsilon_{pcr}$$

ε_{el} = elastic strain, ε_{pl} = plastic strain, ε_{scr} = secondary creep strain, ε_{pcr} = primary creep strain

$$\varepsilon_{el} = \sigma E(T) \qquad \text{\textit{Hooke's Law}}$$

$$\boldsymbol{\varepsilon_{pl}} = \boldsymbol{C_{pl}(T)\sigma^{n(T)}} \quad \textit{Ramberg-Osgood Plastic Law}$$

$$\frac{d\varepsilon_{scr}}{dt} = A'[\sinh(\alpha\sigma)]^n \exp(-\frac{Q}{RT}) \quad \textit{Garofalo Creep Law}$$

$$\frac{\gamma_{pcr-sat} - \gamma_{pcr}}{\gamma_{pcr-sat}} = \exp(-A_{pcr}t) \qquad \gamma_{pcr-sat} = C\tau^n$$

$$\textit{Generalized Exponential Primary Creep}$$

Poison's ratio: 0.35, CTE = 26e-6 1/K

$$E = (-22.1296e6\frac{1}{K} * T + 24.715e9) * Pa$$

2. Copper:

E=117e9 Pa, Poison's Ratio =0.31, CTE=17e-61/K

Appendix 2: Creep Constants of SAC Solders

Primary creep constants for Hypoeutectic Sn3.0Ag0.5Cu (from Stress Relaxation) and Eutectic Sn3.9Ag0.7Cu (from Creep Tests)

Solder	$C_1(Pa^{-1})$	C_2	$C_3(Pa^{-1})$	C_4	$C_5(K^{-1})$
Hypoeut.	3.48E-7	0.5	3.07e-3	0.2	1.10E3
Eutect.	5.65E-11	1.2	1.86E-22	3.3	4.39E3

Secondary creep constants for Hypoeutectic Sn3.0Ag0.5Cu (from Stress Relaxation) and Eutectic Sn3.9Ag0.7Cu (from Creep Tests)

Solder	C_6	$C_7(Pa^{-1})$	C_8	$C_9(K^{-1})$
Hypoeut.	866	1.09e-6	4.02	8574.4
Eutect.	4040	1.09e-6	3.28	8574.4

$$\textit{Where,} \quad \varepsilon_{pcr} = C_1 * \sigma^{C_2} * (1 - e^{-C_3\sigma^{C_4}e^{\frac{C_5}{T}}*t}) \; ;$$

$$\varepsilon_{scr} = C_6(\sinh(C_7 * \sigma))^{C_8} * e^{-C_9/T} * t$$

Effect of Surrounding Air on Board Level Drop Tests of Flexible Printed Circuit Boards

Luciano Arruda, Germano Freitas.
Nokia Technology Institute - INdT
Manaus, Amazonas, Brazil
luciano.arruda@indt.org.br   Phone: +55 92 2126 1034
germano.freitas@indt.org.br   Phone: +55 92 2126 1040

Abstract

Flexible Printed Circuit Boards (FPCBs) are being introduced in portable electronic devices due to their reduced thickness and ability to bend and adapt to various shapes. In order to be applied in mass production products, FPCBs need to be reliable regarding usage conditions. One of the main causes of failure in electronic devices is impact due to drops during transportation and field use. Typically, manufacturers conduct drop tests to quantify the durability of the components. A common tool to verify the reliability of electronic components in early stages of design is the use of the Finite Element Method (FEM).

Because of their reduced thickness and consequent overall stiffness, external effects such as damping as consequence of air resistance may affect structural behavior during impact. This effect might be even more pronounced in confined environments such as the interior of electronic products.

This work proposes to develop a multiphysics simulation model to consider fluid-structure interaction between the FPCB and surrounding air. The structural field will be modeled according to current PCB modeling techniques and the surrounding air will be modeled in such a way to consider pressure distribution on FPCB. Experimental drop tests will also be performed to provide data for comparison with FEM model.

1. Introduction

Electronics industry has pursued miniaturization since its blooming. Huge efforts has been made and the results can be seen in our day-to-day live. New requirements have also appeared since then, such as reduced thickness and non-flat shapes. Flexible Printable Circuit Boards (FPCB) were developed in order to attain these two requirements.

FPCBs are nowadays extensively used in a broad range of products that goes from space shuttles to electronic toys. A common method to evaluate the reliability of FPCBs, in order to verify the suitability for a certain application, is to perform drop tests [3]. Drop tests have been used to evaluate Rigid Printed Circuit Boards and no attention has been given to air damping effect. Since FPCBs have reduced stiffness, this effect should be investigated.

In this paper, a vibration analysis considering surrounding air around Flexible Printed Circuits (FPCBs) in thin film environment was performed. On previous testing, strain gage history data and high-speed video recording had indicated that FPCBs present damping. Since FPCBs have low mass and low stiffness compared to common rigid boards it was decided to investigate the effect of surrounding air on system damping.

One possible solution to evaluate damping due to fluid structure interaction is through Squeeze Film methodology. References have been found on application of this method on MEMS [3, 4]. Since FPCBs will many times be confined inside products, it has been considered that the surrounding air could be considered as a Squeeze Film. Although this method may not bring quantitative results, it may provide information for evaluating if the effort for a more detailed analysis is necessary.

2. Squeeze Film Analyses Assumptions

Squeeze film damping occurs when a plate moves in close proximity to another solid surface, thus alternately stretching and squeezing any fluid that may be present in the space between the moving plate and the solid surface [4].

If the fluid flow in the gap can be assumed quasi-steady and viscous-dominant, and if the gap height is small compared to the plate width, the velocity profile of the fluid between the moving plate and the solid surface can be approximated as parabolic in the thickness direction. In this case, the Navier-Stockes equations governing the fluid flow can be reduced to a scalar equation in terms of the fluid pressure. This squeezed film can be combined with a solid mechanical (finite-element) model in order to perform dynamic fluid-structure interaction analyses.

An assumption required is that the gap height is small compared to the minor dimension of the plate. In general is used the following relationship:

$$h \le \frac{L}{10} \qquad \text{eq. 2.1}$$

where L is the width of the moving plate.

3. Methodology adopted for evaluation

The vibration of a FPCB inside an electronic product is considered to comply with Squeeze Film methodology.

A Squeeze film simulation was performed to evaluate the stiffness and damping ratios of a flex film surrounded by air. Therefore, a matrix of modal squeeze stiffness coefficient K_{ij}, and modal damping coefficients, C_{ij} was calculated, and used to calculate damping parameters.

1-4244-1105-X/07/$25.00 ©2007 IEEE   545

Squeeze Film phenomenon results from movement of the fluid around a plate that generates resistance force because of the fluid viscosity. The consequence is a pressure distribution over the plate, which may act as a spring and/or a damper. Recent studies show that the damping force dominates the spring force at low frequencies, whereas the spring force dominates the damping force at high frequencies.

In order to estimate the frequency range in which the Squeeze Film analysis should be performed, a FPCB has been submitted to a drop test with a strain gage bonded to the central region. Only one gage was used to minimize changes in stiffness. The strain history was then analyzed in frequency domain.

The calculation of the stiffness (K) and damping (C) constants are done by integrating the pressure distribution over the area, then taking these force calculations and feeding them into the equations:

$$C = \frac{F^{Re}}{V_z} \qquad \text{eq. 3.1}$$

$$K = \frac{F^{Im_a}}{V_z} \qquad \text{eq. 3.2}$$

where F^{Re} is the real component and F^{Im_a} is the imaginary component of the pressure force.

4. Test Vehicle and Simulation Model Description

The FPCB test vehicle and drop test setup is shown in figure 4.1. Detail of tri-axial gage used is shown in figure 4.2. The test vehicle used has embedded resistors, capacitors and inductors distributed in two layers. There are six holes with 2.5 mm radius and 36 lateral 1.6x1.6 mm square holes. Test vehicle length is 71 mm, width 48 mm and thickness 1 mm. The FPCB was fixed in 4 points and drop height set to 1000 mm.

The simulation model proposed is shown in figure 4.3. FPCB was modeled in ANSYS [2] using layered solid element SOLID46. Air gap was modeled using fluid element FLUID136. Since holes might affect airflow, all holes present in test vehicle were modeled. Boundary condition used was restraining all degrees of freedom on four corner holes.

The squeeze film element was considered as a 4.8 mm gap between FPCB and a rigid surface. The physical properties used to model the FPCB structure and surrounding air are shown in tables 1 and 2.

Figure 4.1 - FPCB Test Vehicle and Drop Test setup

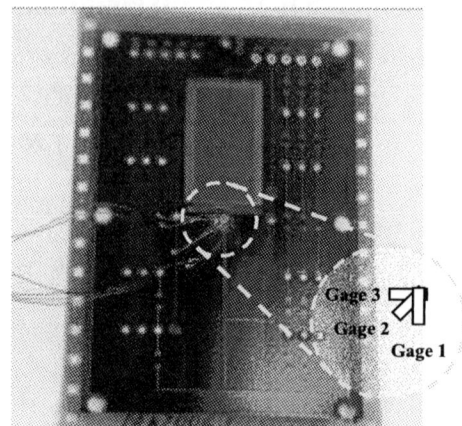

Figure 4.2 - Strain Gage bonded to FPCB.

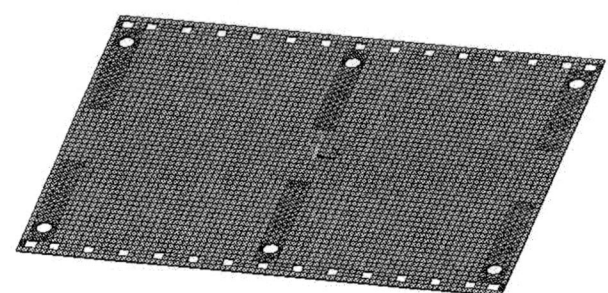

Figure 4.3 - Simulation Model

Table 4.1 – FPCB parameters

Property	Value
Polyimide: Elasticity Modulus, E	4 GPa
Density, γ	1420 Kg/m^2
Poisson Ratio, υ	0.34
Cooper: Elasticity Modulus, E	130 GPa
Density, γ	9520 Kg/m^2
Poisson Ratio, υ	0.35
Board Width, W	48 mm
Board Length, L	98 mm

Table 4.2 – Fluid Parameters

Property	Value
Density of Fluid, ρ	1.2 kg/m^3
Viscosity of Fluid, μ	1.8x10^{-5} N.sec/m^2
Ambient Pressure, $P_{ambient}$	0.1 MPa
PREF[1]	0.1 MPa
MFP[2]	6.4x10^{-8} m
Gap height, h	4.8 mm

1. Reference Pressure for free path
2. Mean free path at reference pressure

5. Experimental Results

The acceleration profile due to impact is shown in Figure 5.1. Strain time history is shown in figure 5.2. Fast Fourier Transform (FFT) was applied to strain history data to evaluate frequency response range, as shown in figure 5.3, where it can be seen that frequency range is from 0 to around 500 Hz.

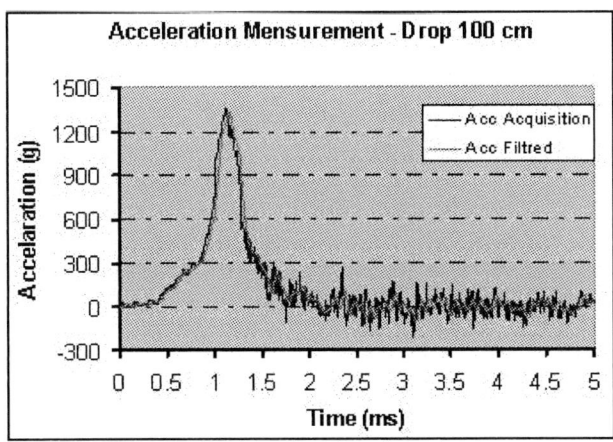

Figure 5.1. Acceleration from impact.

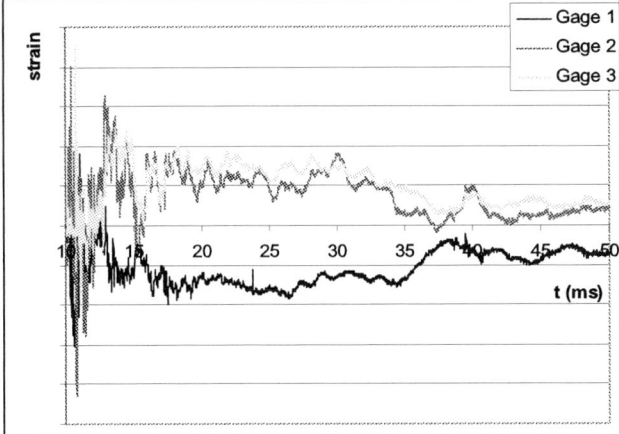

Figure 5.2 - Strain Time History

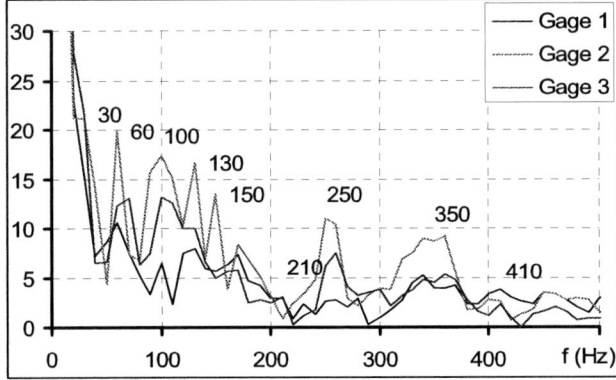

Figure 5.3 – Strain History FFT

6. Fluid-Structural Numerical Analysis

A Squeeze film analysis was performed using ANSYS. The frequency range adopted was from 0 to 500 Hz, based on the results obtained in experimental tests.

The first step to perform a Squeeze Film analysis is to perform a modal analysis. The results from the modal analysis are shown in table 6.1, where the natural frequencies of some of the first modes are shown. It can be seen that the natural frequencies are in the same range as from the FFT performed on experimental tests. An interesting fact observed was that there were many modes found with small difference in natural frequency and similar mode shapes. Due to this fact it was decided to verify higher order modes, and not only the first modes, as is usually considered on structural analysis of PCBs.

Damping ratio obtained from Squeeze Film analysis is shown in figure 6.1. It can be seen that most significant damping occurs on first frequencies, up to 200 Hz range.

Table. 6.1: Numerical Frequency Results

Mode	Frequency, Hz
1	8.9619
2	28.015
3	37.086
10	71.877
13	101.05
22	146.29
30	210.02
36	251.33
55	350.99
63	416.78

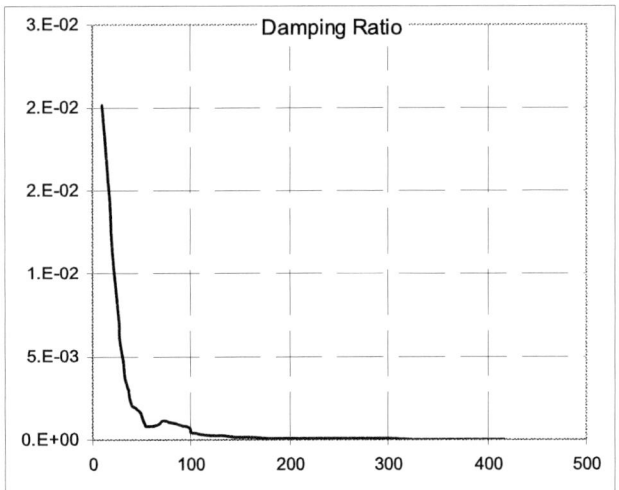

Figure 6.1 – Damping Ratio from Simulation

7. Conclusions

Although surrounding air on FPCBs may not always be considered as Squeeze Films, this first evaluation has provided information on whether damping effect is or is not present, and in which frequency range. Results obtained from Squeeze Film analysis show that most significant damping occurs in the 0 to 200 Hz range. From experimental test FFT analysis it can be seen that this is also the frequency range with highest amplitude during the drop test performed.

547

The results obtained show that it would be worthwhile performing a more detailed fluid structure interaction analysis on an FPCB. A suggestion would be to perform a transient fluid-structure analysis.

If surrounding air has effect on FPCBs during impact, resulting deformations may change, and simulation models without damping may not be able to predict actual FPCB behaviour, thus limiting the use of simulation to predict FPCB reliability.

References

1. Coombs' Jr, Clyde F., Coombs' Printed Circuits Handbook, McGraw Hill (2001), pp. 56.3 – 56.6.
2. ANSYS "Electronics Manual". *Version 10.0* USA.
3. JEDEC Drop Test Standard.
4. M. I. Younis., "Modelling and Simulation of Micromechanical Systems in Multi-physics Fields", *PhD Dissertion, Virginia Polytechnic Institute and State University*, Blacksburg Virginia, 2004.
5. Wang, X.;Liu, Y., Wang, M. & Chen, X. "The Effect of Air-Damping on the Planar MEMs Structures". *Proceeding of HDP'04. IEEE.* 2004,349-352.

Numerical Simulation of Ion Drift within Ion Mobility Spectrometers in High Peclet Conditions using FEM Techniques

A. Kalms[1], M. Salleras[1], Z. Liu[2], J.G. Korvink[2], J. Goebel[3], G. Müller[3], J. Samitier[1], S. Marco[1]

[1]Departament d'Electrònica, Universitat de Barcelona, Spain
[2]University of Freiburg, IMTEK-Institute for Microsystem Technology, Germany
[3]EADS-CRC, Munich, Germany
E-mail: akalms@el.ub.es

Abstract

The goal of this paper is to model and simulate the transport of ions through a miniaturised drift tube of ion mobility spectrometer (IMS). In the first part of this work we test and compare three numerical methods to solve the transient diffusion-advection equation in high Peclet conditions. These methods are the following: Streamline upwind Petrov Galerkin (SUPG), the modified-PDE (MPDE) and the finite calculus method (FIC). In the second part we use the last two numerical methods to study ion transport in a one-dimensional IMS model. Ion peak broadening effects, namely coulombic repulsion, diffusion, transport of two ionic species, have been explored for different initial conditions.

1. Introduction

IMS is an analytical technique used to identify volatile organic substances in air. Nowadays, it is the mainstay of chemical-warfare defense systems [1]. The IMS technique is fundamentally based on ion mobility identification. This means, how rapidly a given ion moves trough a gas under the influence of certain electric field.

The IMS operation principle is relatively simple. Ion mobility is generally measured as the time-of-flight of the ions when this moves against a drift gas. A controlled pulse of ions generated by a ionization source is introduced into a region of uniform electric field (drift tube shown in Fig.1). These field separates different ions based on their different mobilities. Therefore, it is possible to obtain a blueprint (Fig. 1, lower part) of the target compound through their mobilty spectrum at the end of the drift tube.

The IMS technology has fundamental advantages when compared to other sensor techniques: a) high resolution (in the order of ppb), b) fast measurements (milliseconds). Additionally, IMS instruments operate at atmospheric pressure, allowing a smaller analytical unit, lower power requirements, lighter weight, and easier use for field applications. The typical dimensions of these instruments are about tens of centimetres, but, due to the trend towards miniaturization shorter drift tubes are being explored [2-5]. These advantages make IMS a useful technology for trace detector monitoring, usually for the detection of agents, explosives and narcotics, and other environmental purposes, such as safety and industrial hygiene [1-5].

IMS instrument can be miniaturized by using microfabrication techniques [2-6]. However, the reduction of dimensions can generate problems related to the electric field inhomogeneities, effects of the repulsion of Coulomb as well as an increase of intensity of advection effects. Due to the difficulty of design of miniaturized IMS devices that support the level of resolution in the scaled IMS instruments, new micromachined instruments have benn developed based on the IMS technology. These technologies include the Field Asymmetric IMS (FA-IMS) [3,4], the radio frequency IMS (rf-IMS) [5] or Differential IMS (DMS) [6] In each of them the ion mobility alternates between a high-field and a low-field value (differential mobility).

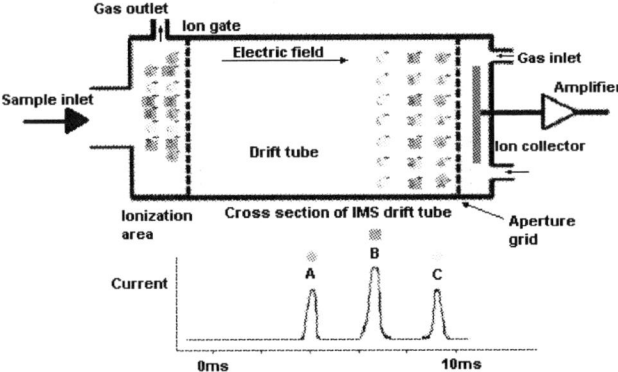

Fig. 1. (Upper figure) IMS scheme. Three different ion mobilities are measured after reaching the ion collector. (Lower figure) The detected signals are the spectra of each ion mobility (A, B or C).

When miniaturizing IMS devices it is of utmost importance to build them with an appropriate geometrical design. Thus, mutual repulsion behavior and electric field homogeneity are key factors on the final system design [2,7]. The movement and behaviour of ions through the drift tube can be modeled by using partial diferential equations (PDEs): the so-called Poisson-Nenst-Plank equations. The governing equations inhere the linear drift-diffusion together with electrostatic coupled problems.

In the field of simulation only few specific IMS CAE solutions can be found. The commercial CAE tools that have been applied in IMS are ion optic simulation software. Those tools are available to simulate and analyse the optics of charged particles through electric and magnetic fields, and to calculate their time of fly [8] although they do not deal with the mutual repulsion of the drift ions.

The purpose of this paper is to search for efficient numerical methods as tools to study the relevant effects that can be important to the miniaturization of the drift tube. In the first section of this paper a description of the physical model and its main chararacteristics are given.

1-4244-1105-X/07/$25.00 ©2007 IEEE

After that, the formulation of the ion drift problem is exposed and a concis theoretical explanation about the numerical methods used as modeling techniques is reported. Included in the 1-D model are the validation of the methods with the analytical solution comparison, and relevant 1-D results. In this last part, we will show the results computed accurately with commercial software and the analysis of the relevant effects. Finally, we give conclusions.

2. Physical system

A typical drift tube is composed of alternate metal and isolator rings (Fig.2). Each ring is set to a fixed electric potential, which value decreases as the rings get closer to the collector so as to have an homogeneous electric field distribution within the drift tube. The number of rings depends on the drift length. The relation between thickness of the drift ring and its inner diameter should be as small as possible to sustain the field quality. Electric fields range from 100 V/cm to 1000 V/cm.

In this paper it is assumed that ions are generated by a UV laser pulse in the ionization area. Commercial IMS devices use radioactive ionization sources (^{63}Ni). To control the entrance of ions in the drift tube, an ion shutter [9] is placed behind the ionization area. It is a conductive grid controlled electrically. However, it is not considered in our simulation . At the end of the drift tube, the aperture grid shields the detector from the induced current flow due to the ion charges moving towards the collector (the so called Faraday plate). This metal plate is set at virtual ground potential. The output signal from the collector is an analog signal that is digitally recorded.

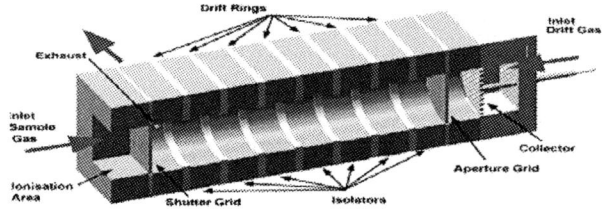

Fig. 2. Cross section of a state of the art IMS drift tube.

The detection levels of interest are traces in the rank of the low concentration range of between 1ppb and 1ppm. The substances to detect include the most sensible explosives, for instances EDGN, NG and DNT, but also samples of average steam pressure like the one of TNT. Their mobilities are resumed in table 1.

Ions move at a speed v (or drift velocity) proportional to the electric field E inside the drift tube with the ion mobility K by means of the expression:

$$\vec{v} = K \cdot \vec{E} \qquad (1)$$

The drift velocity v is determined by two substantial components, diffusion D and the electric field E. If the ion concentration and E are small, then the Nernst Townsend Einstein relationship can be applied, eq.(2)

$$K = \frac{D \cdot q}{k_B \cdot T} \qquad (2)$$

where k_B is the Boltzmann constant, room temperature T=300K and q the elemental charge. Mobility values in IMS technology are between 1 and 3 $cm^2v^{-1}s^{-1}$ [17].

Table 1. Mobilities of interesting chemical compounds.

Explosives (negative ion mobility spectra)	K ($cm^2v^{-1}s^{-1}$)
EDGN	2.70
NG	1.45
DNT	1.84
TNT	1.69
Drugs (positive ion mobility spectra)	
Cocaine	1.27
Cannabinol	1.16

3. Theory

A coupled problem must be solved formulated whith PNP equations. The electrical field must be solved on every point of the drift tube, and the ion distribution must be computed at every time. The bidirectional coupling is made through the electrical field perturbation produced by the ions, and the ion's drift is affected by the electrical field. This highly coupled problem has another numerical problem which is the high advection values present inside the drift tube.

The multi-dimensional transport equation that describes the transient advection-diffusion problem of a single ion species in a drift tube can be formulated as:

$$\frac{d\Phi}{dt} + \vec{v} \cdot \nabla\Phi = D \cdot \Delta\Phi \qquad (3)$$

where Φ is the ion concentration solution, and ∇ is spatial gradient operator and Δ is Laplace operator. The boundary conditions associated with eq.(3) are:

$$\left.\begin{array}{ll} \Phi - \Phi_B = 0 & on \quad \Gamma_\phi \\ n \cdot \nabla\Phi + q_B = 0 & on \quad \Gamma_q \end{array}\right\}$$

where Φ_B is the Dirichlet value condition and q_B is the Neuman value condition.

In turn, electric field comes from the solution of Poisson's equation in which the charge density concentration is defined:

$$\Delta V = -\frac{q \cdot \Phi_0 \cdot \Phi}{\varepsilon} \qquad (4)$$

Here V is the electric potential, and Φ_0 is the initial ion concentration at t=0s and ε is the relative permitivity in air. Boundary conditions for V are constant values at each metal ring.

The intrinsic problem that affects this equation is that for small values of diffusion, small variations on the data produce large variations on the solution Φ. A key factor when considering drift-diffusion problems is the Peclet number (Pe) which relates the amount of drift to the amount of diffusion. The general Pe expression is Pe=v·l/2D, where v is the velocity field and l is a characteristic discretization length. Problems with high advection terms have Peclet numbers much greater than 2.

550

In recent years, great effort has been spent to avoid numerical difficulties resulting from high Peclet numbers within the FEM framework [10]. It is well known that the numerical solution of a advection equation suffers from numerical diffusion and oscillation when a centered type discretization is used for the convection operator. This problem is due to the hyperbolic nature of the convection equation and is shared by the central finite difference and standard Galerkin finite element method, and also by the finite volume methods. This instability has been tried to be solved by several methods using different terminology in different research scopes. For example, FEM people talk about 'stabilization techniques' and CFD people talk about 'modified partial differential equations'. In fact, there is a certain relationship between them. In this work the basic idea of different FEM methods will be reviewed: the first one is based on the standard Galerkin FEM, namely streamline upwind Petrov Galerkin (SUPG, [11]), the second is the modified partial differential equation (MPDE [12,13]) and the last one is the finite calculus-FEM (FIC-FEM [14,15]). Finally, the relationship between these methods for linear convection equation will be shown.

Stabilized finite element methods have been around for more than 20 years. The SUPG method (or streamline diffusion) was introduced by Hughes and Brooks [11]. The formulation of the method starts with the weak form of the discretization based on Galerkin's weak formulation. In this case the weight or test functions have an additional term which contains a stabilization parameter (δ) which can be tuned in order to control the added diffusion. However, parameter values are singular for each different problem condition.

On the one hand, the MPDE method replaces the PDE of the problem by a 2-level 7 points central difference scheme analogue. The differential equation solved by a difference scheme is the modified equation. Afterwards, the modified equation is derived by a two-step process starting with expanding the pure advection term into a Taylor series and then eliminating time derivatives higher than first order and mixed time and space derivatives by Warming, R.F $et\ al.$[12]. From the resulting modified equation with fourth order accuracy of the advection equation, physical dissipation terms and dispersion terms can be identified in order to compensate the numerical oscillations latent in our problem. Come to this point, the modified equation is solved by the FEM.

On the other hand, considering the weak formulation of the advection equation (without the diffusive term) the resulting term is expressed as follows:

$$\int_{\Omega}(w+\delta\cdot(v\cdot\nabla w))\frac{\partial\Phi}{dt}d\Omega + \int_{\Omega}w(v\cdot\nabla\Phi)d\Omega + \\ + \int_{\Omega}\delta(v\cdot\nabla\Phi)(v\cdot\nabla w)d\Omega = 0 \quad (5)$$

where $w+\delta\cdot(v\cdot\Delta w)$ is the test function, used in the SUPG formulation, and δ a scaling parameter. The last term of eq.(5) corresponds to an anisotropic artificial diffusion.

Following the MPDE formulation, we consider the third term from Eq. (3), and multiply it by a test function and integrate over the whole domain to obtain the weak formulation of the advection equation, leading to the formula:

$$\int_{\Omega}\frac{\Delta t}{2}\cdot(v\cdot\nabla)\cdot(v\cdot\nabla\Phi)\cdot w\cdot d\Omega = \int_{\Omega}\frac{\Delta t}{2}\cdot(v\cdot\nabla\Phi)\cdot(v\cdot\nabla w)\cdot d\Omega \quad (6)$$

and considering the last term on the left hand side of eq.(5) and the rigth term of eq.(6) the relation $\delta=\Delta t/2$ equals both terms. This makes the SUPG method coincide with the weak formulation of the MPDE method. However, the last one uses a more strict Taylor expansion to get a modified strong form of initial PDEs. Interestingly, the strong form is more general and compact than the weak form, which is generated for a specific problem.

Recently, a Galerkin FIC-FEM formulation applied to convection-diffusion problems with sharp gradients [14,15] has been published. The stabilised formulation is generated through the application of the FIC to the modified governing differential equations. The FIC method comes from the balance of fluxes in a finite problem domain. In 1-D element of length h, the flux balance of diffusive and convective fluxes (without recombination or generation term) can be formultated as:

$$D\nabla\Phi(x) + v\cdot\Phi(x) - D\nabla\Phi(x-h) + v\cdot\Phi(x-h) = 0 \quad (7)$$

Expanding the last two terms of eq.(7) into a 2nd order Taylor series we find the residual form called FIC governing equation:

$$r - \frac{h}{2}\nabla r = 0 \quad \text{in } \Omega$$

$$r(\Phi) = \frac{d\Phi}{dt} + v\cdot\nabla\Phi + \nabla D\nabla\Phi = 0$$

Boundary conditions

$$\begin{cases} -v\Phi + D\nabla\Phi + q_B - \frac{h}{2}r = 0 \quad \text{on } \Gamma_q \quad \text{and on } \Gamma_p \\ \Phi - \Phi_B = 0 \end{cases}$$

Initial conditions are $\Phi(t=0)=\Phi_0$. In a general case, h is a characteristic length of discretization which acts as a stabilization parameter. It is remarkable that when h tends towards zero, the FIC differential equations remain as the general infinitesimal form. If the balance equation is expressed with the principal curvature directions, the new term represents an orthotropic diffusion. In other words, this formulation comprises the behavior of the so called cross-wind and shock capturing methods. We can select the 2nd order or the 4th order advect stabilization schemes. These two options come from the assumptions of the characteristic length computation. The first one is considered an iterative algorithm where the characteristic length is computed from the last iteration data. For the second one it is assumed that the main curvature direction is approximated by the direction of the gradient vector $\nabla\Phi$ at the element center. In this formulation, the stabilization is carried out by adding an extra nonlinear dissipation term into the discretized equations. The added term is characterized by a single stabilization parameter that is computed by an iterative algorithm.

Although the formulations explained before solve our problem, they are insufficient to realize a general method of calculating the parameters of stabilization for the whole scale of situations of flow.

The cases in which the method SUPG and the MPDE is applied have been implemented by COMSOL Multiphysics™, COMSOL AB, Sweden whereas the method FIC is implemented in Tdyn, COMPASS Ingeniería y Sistemas SA, Spain.

4. One dimensional model

4.1. Model definition

In the following section, the model is defined and the preliminary validations of the numerical methods are presented. This specific 1-D example models the axis of symmetry of a drift tube shaped like a cylinder with the length l=5cm. An electrical potential of V=5kV has been applied on the left end of the cylinder and V=0V on the right end, as boundary conditions, producing an electric field of 10^3V/cm. For simplicity, an initial normalized ion concentration has been considered within a length of 0.2cm. This length represents the depth at which the ions may be distributed when they move at an average drift velocity v =20m/s as in eq.(1), and a time pulse width of t_g=100μs. The drift gas is assumed to be air with a null average velocity.

In order to validate the performance of the different techniques proposed, 1-D convection equation is solved accurately on a reasonably fine mesh, based on the strong form of the MPDE, on the weak form of the SUPG and also with the FIC-FEM. The results are compared to the analytical solution, without considering coupling fields for each one. The analytical solution has been obtained from [16]. A source term which is assumed to be a rectangular pulse is defined at initial time t_0 as (in cylindrical coordinates):

$$S(r,z) = \left[H(z) - H(z - v \cdot t_g) \right]$$

where H is the Heaviside function. Assuming that there is no recombination or generation of ions in the cavity while the ions are drifting, the analytical ion concentration expression is:

$$\Phi(z,t) = \frac{\Phi_0}{2} \left[erf\left(\frac{z - v \cdot t}{2\sqrt{D \cdot t}} \right) - erf\left(\frac{z - v \cdot t - v \cdot t_g}{2\sqrt{D \cdot t}} \right) \right] \quad (8)$$

Note that in this case it is considered that the concentration is small enough to disregard.

4.2. Numerical Results

4.2.1. Stability of the methods

Fig.3 shows the comparison for the analytical result from eq.(8) and results from the numerical simulations. Notable are the numerical oscillations in the SUPG case, where artificial stabilization parameter δ_s has a default value of 0.25. We may observe that not only the

numerical oscillations disappear with both MPDE and FIC methods, but also all numerical results fit very good with the analytical solution. It means that the MPDE and FIC methods are able to solve our model with the same number of elements, and no numerical oscillation appears.

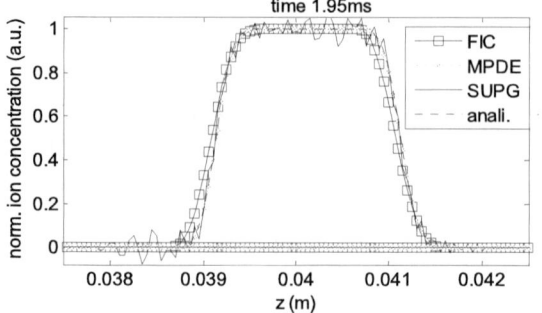

Fig. 3. Comparison of theory (dashed line) with numerical FIC (squared line), MPDE (dashdotted line) and SUPG (solid line), along the z axis. All of the concentrations have been normalized

In a pulsed laser beam, the ion concentration peak shape can be described as Gaussian. It is centered in the first dielectric ring and it is normalized to unity selecting the value of $\Phi_0 = 10^{13}$ ions/m^3 (0.1 ppm).

The next computations has been solved with SUPG, MPDE with the same mesh of 1000 linear elements and FIC with a 2D mesh keeping the same element length of 10μm. No interaction between ions and electrical fields has been assumed (the same case as in Eq. (8)). After solving the electrical field with initial conditions the solution is applied to the advection-diffusion problem. A comparison for different instants are in Fig. 4 for each numerical result. We see, that SUPG stills presents important oscillations compared to the rest of methods. For this reason this method has been disregarded for the rest of the work.

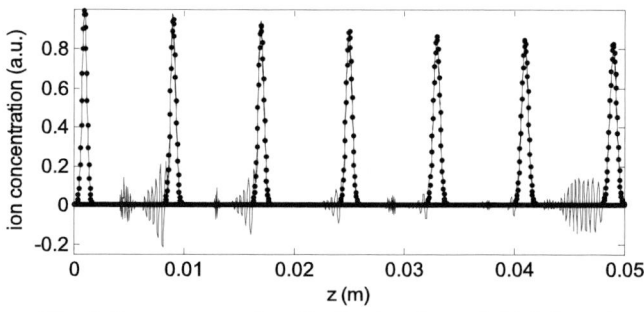

Fig. 4. Ion concentration distribution for an initial Gaussian pulse solved with SUPG, MPDE (dotted line) and FIC each 0.4ms vs. axial position from left to right.

In the FIC-FEM method we can select the 2nd order or the 4th order advect stabilization schemes. In our case, we have found that in our model the 2nd order compute overdiffusive concentrations.

4.2.2. Coulombic repulsion effects

After the validation of the proposed numerical techniques of the 1D model with analytical solution, we compute the ion peak concentration at different initial conditions. In miniaturized instruments with laser ionization ion concentration is higher than in most commercial instruments. We may consider coupled equations on the simulations. The concentration width that corresponds to the half of the maximum height is proportional to the square root of the drift time, and inversely proportional to drift velocity. A comparison for two initial ion concentrations with the same mobility and electrical conditions is made when fields are coupled. The results are plotted for different drift times, see Fig.5.

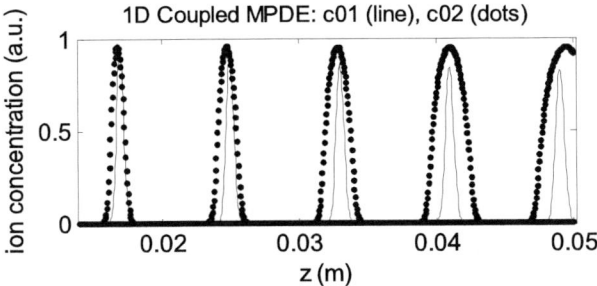

Fig. 5. Ion concentration distribution with initial concentrations $c01=10^{13}m^{-3}$, and $c02=10^{16}m^{-3}$ vs. axial position for $t=400\mu s$, $800\mu s$, 1.2ms, 2.0ms and 2.4ms from left to right. The y-axis has been normalized to 1 in the concentration figures.

As fig.5 shows, peak broadening due to Coulomb repulsion increases for higher ion concentration levels, and also, with the drift tube length. The effects of the coulombic repulsion are a major pulse broadening and a lower maximum of the peaks.

In miniaturized drift tube an homogenous electrostatic field is expected, making possible an exact calculation of the ion mobilities. Two figures of merit are relevant in this analysis. The applied voltage has to be lineal with drift distances and the geometry of the drift tube has to keep as much as possible the homogeneity of the electric field. Considering that the electrical field is not perturbed by geometry, we only analyse the effects of the ions moving inside the drift tube. In Fig.6, the electrical field distribution changes whenever the ion cloud advances, and this change increases for higher initial ion charges. When non coupled equations are considered, electric field remains constant for the whole axial axis.

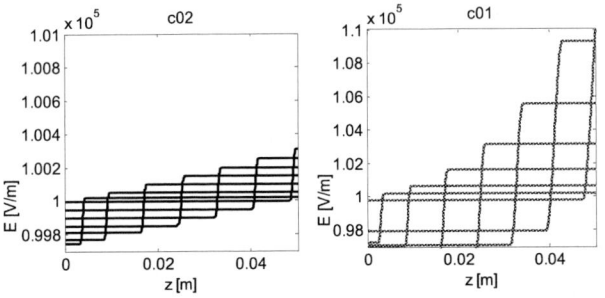

Fig. 6. Electric field distribution vs. axial position (z) for times 0s, 400μs, 800μs, 1.2ms, 2.0ms and 2.4ms. Left figure has an initial concentration of c02 and c01 for right figure (same values as in Fig.5).

In the simulations we have found that MPDE needs more degrees of freedom to evaluate cases with higher

level of concentration. Nevertheless, the FIC method does not raise these limitations.

4.2.3. Diffusion effects

As we have already mentioned, drift velocity changes linearly with electric field and diffusion at fixed temperatures. If we consider an initial single ion concentration of mobility 2 $cm^2v^{-1}s^{-1}$ and we simulate for two different applied voltages (E1, E2), we find both ion peaks at any time. A comparison is made at 0.25ms and 2.5ms in Fig.7

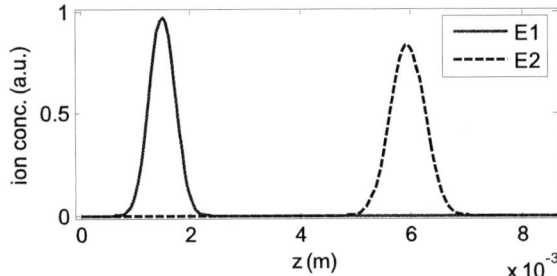

Fig. 7. Peak shapes vs. axial position for E1=1kV/cm, E2=100V/cm, and initial concentration 10^{13} m^{-3}. E1 plots at 0.25ms and E2 plots at 2.5ms.

From Fig.7 we may see how changes in velocity affect peak broadening. Peak widths will stretch, and peak maximums will decrease because of the diffusion effects.

4.2.4. Transport of two ionic species

Fig. 8 shows the electric field distribution when two ion species are considered together in a coupled system. Each concentration has different mobility, which causes the stepping and desplacement of the electric field.

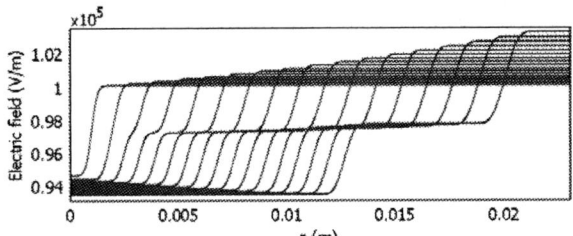

Fig. 8. Electric field distribution for a two ion concentration coupled problem. Each plotted line corresponds with times from 0s until 960μs, each 60 μs.

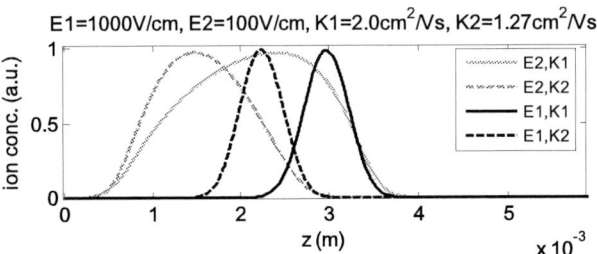

Fig. 9. Peak shapes vs. axial position for two ion mobilities coupled in two different electric field conditions in the drift tube: E1 and E2. The ion species have the same initial ion concentration 10^{14} m^{-3}. Time for the peaks E2 is 1.0ms, time for the peaks E1 is 0.1ms.

553

When two ion clouds are computed coupling their electric charges with the electric field distribution, the ion clouds couple among them. The result (Fig.9) is that the ion peaks are deformed differently while they travel along the drift tube. This effect can be understood with Fig. 8, where the electric field increases due to the coupling of electric charges. Also, the lower the initial ion concentration is, the lower the coupling effect becomes.

On the other side, the Coulomb effects also varies with the bias voltage. For two bias voltages and mobilities (Fig.9). The Coulomb contribution to the broadening, and as consequence, to the resolution has a relation with the initial concentration. The numerical results remark that for higher voltages the initial concentration effect of the broadening of the ion curve is less relevant. For concentrations higher than 10^{15} m^{-3} simulations may need a smaller linear element length or a finer mesh.

5. Conclusions

Different techniques to cope with numerical instabilities due to high Peclet numbers in convection-diffusion have been compared for the simulation of miniaturized IMS devices. Results show that MPDE and FIC formulations provide the required sability while SUPG results depend critically on the right choice of the stabilization parameter. Moreover, MPDE and FIC show excellent agreement with analytical solutions in 1-D problems.

The proposed methods applied to the 1-D model show the relevance of Coulomb and diffusion effects. Coulombic repulsion produces important peak broadening in the range of high ion concentration, causing a degradation of the chemical resolution of the instrument. However these effects are weaker for higher electric fields.

Finally, simulations have revealed that transport of several ion species may couple through the electric field producing deformation in the peak shape. Again those effects weaken for higher electic fields.

Acknowledgments

The authors would like to acknowledge the final support from Network of Excellence GOSPEL (FP6-IST 507610). The authors would also like to thank Dr. J. Garcia-Espinosa for many helpful discussions on the FIC method.

References

1. Ewing, R.D., Atkinson, D.A., Eiceman, G.A and Ewing, G.J., "A critical review of ion mobility spectrometry for the detection of explosives and explosive realted compounds," *Talanta*, Vol. 54 (2001), pp. 515-529.
2. Xu, J., Whitten, W. B., and Ramsey, J. M., "Space charge effects on resolution in a miniature ion mobility spectrometer," *Anal. Chem.*, Vol. 72 (2000), pp 5787 – 5791.
3. Miller, R.A., Eiceman, E.A., Nazarov, E.G. and King, A.T., "A novel micromachined high-field asymmetric

waveform-ion mobility spectrometer," *Sens. Actuators B*, Vol. 67 (2000), pp. 300-306.
4. http://www.owlstonenanotech.com.
5. Miller, R.A., Nazarov, E.G. Eiceman, G.A. and King, A.T.,"A MEMS radio-frequency ion mobility spectrometer for chemical vapor detection," *Sens. Actuators A*, Vol. 91, No. 3 (2001), pp. 301-312.
6. Lambertus, G.R., Fix, C.S., Reidy, S.M., Miller, R.A., Wheeler, D., Nazarov, E. and Sacks, R. "Silicon microfabricated column with microfabricated differential mobility spectrometer for GC analysis of volatile organic compounds," Anal. Chem. Vo. 77 (23) (2005), pp. 7563-7571.
7. Soppart, O. and Baumbach, J.I., "Comparison of electric fields within drift tubes for ion mobility spectrometry," *Meas. Sci. Technol.* Vol. 11 (2000) pp. 1473-1479.
8. Eiceman, G.A., Nazarov, E.G., Rodriguez, J.E., Stone, J.A., "Analysis of a drift tube at ambient pressure: Models and precise measurements in ion mobility spectrometry," *Rev. Sci. Instrum.* Vol. 72, 9 (2001), pp. 3610-3621.
9. Salleras, M., Kalms, A., Kessler, M., Goebel, J., Marco, S., "Electrostatic shutter design for a miniaturized ion mobility spectrometer," *Sens. Actuators B*, Vol. 118 (2006), pp. 338-342.
10. Zienkiewicz, O.C. *et al*, The Finite Element Method Elsevier (Amsterdam 2005), 6th edn, Vol. 3.
11. Brooks, A.N., Hughes, T.J.R., "Streamline upwind/Petrov-Galerkin formulation for convection dominated flows with particular emphasis on the incompressible Navier Stokes equation," *Comput. Methods Appl. Mech. Engrg.*, vol. 32 (1982), pp. 199-259.
12. Warming, R.F., Hyett, B.J., "The modified equation approach to the stability and accuracy analysis of finite difference methods," *J. Comput. Phys.*, vol. 14 (1974), pp. 159-179.
13. Shih, Y.T., Elman, H.C., "Modified streamline diffusion schemes for convection-diffusion problems," *Comput. Methods Appl. Mech. Engrg.*, vol. 174 (1999), pp. 137-151.
14. Oñate, E., Miquel, J., Zárate, F., "Stabilized solution of the multidimensional advection-diffusion-absorption equation using linear finite elements," *Comput. & Fluids*, vol. 36 (2007), pp. 92-112.
15. Oñate, E., Valls, A., García, J., "FIC/FEM formulation with matrix stabilizing terms for incompressible flows at low and high Reynolds numbers" *Comput. Mech.*, vol. 38 (2006), pp. 440-445.
16. Spangler, G.E., Collins, C.I., "Peak Shape Analysis and Plate Theory for Plasma Chromatography" *Anal. Chem.*, vol. 47 (3) (1975), pp. 403-307.
17. Hill Jr., H.H., Siems, W.F., St. Louis, R.H., McMinn, D.G. "Ion Mobility Spectrometry," *Anal. Chem.*, vol. 62 (23) (1990), pp. 1201A-1209A.

Electro-osmotic Flow Based Cooling System For Microprocessors

P.F. Eng, P. Nithiarasu *, A.K. Arnold, P. Igic and O.J. Guy

School of Engineering, University of Wales Swansea, Swansea SA2 8PP, UK

e-mail: P.F.Eng@swansea.ac.uk; P.Nithiarasu@swansea.ac.uk; 175161@swansea.ac.uk

Abstract

An elliptical, electroosmosis device, with many micro channel flow paths has been designed to circulate liquid to cool hot spots on a microprocessor chip. A pair of electrodes deposited 3mm apart, across the channels, was employed to impose external electric potential. A preliminary experimental setup has been fabricated to determine the flow rate and heat transfer of the micro channel device. A pulsed external electric potential signal was applied to reduce bubbles generated as a result of electrolysis. Silicon dioxide has been used as an electrical isolator to avoid current interference between the cooling device and the microprocessor. Preliminary flow rates measured were $0.04\mu l$ min^{-1}, $0.16\mu l$ min^{-1}, $0.24\mu l$ min^{-1} using external potential values of 10V, 20V and 40V, respectively. Temperature measurements have demonstrated that an increased rate of heat transfer is possible, if the liquid is forced to move.

Keywords: Electroosmosis, electronic cooling, micro-pumps, MEMS, multi-disciplinary analysis, numerical, experimental.

1. Introduction

As microprocessor technology continues to progress, electronic systems are expected to generate heat in excess of 200W per square centimeter in the next decade. Transistors packed on to microprocessor chips are sources of hot spots and pose problems for developing efficient cooling methods. Thus, the thermal management of such systems is both challenging and costly. In the past century, the sizes of the heat sinks have had to be increased to maintain the safe device operation of such systems. However, a liquid based cooling mechanism can dissipate more heat due to its high heat capacity, giving better temperature consistency. Forced fluid cooling, using micro-channels has been demonstrated by Tuckerman and Pease [1] and they proved that up to $1000W/cm^2$ can be removed using micro-channel heat sinks. Micro-channels have also been demonstrated to have effectively reduced hot spots in a most cost effective manner. Recently, an electro-kinetic pump was used to force liquid through a sealed loop of micro-channels and micro pipes to transfer heat from a micro heat exchanger to a radiator. The electro-kinetic pump is a non mechanical pump with no moving parts and potentially can give high flow rates. These pumps use an electro-hydrodynamic effect to drive the dielectric liquid. Electro-osmotic pump reported by Yao at el [2] produce up to 1.3 atm of pressure and flow rates greater than 33ml/min, using a porous sintered-glass cylinder with a 40 mm diameter and 1mm thickness to support cooling heat load of 100W.

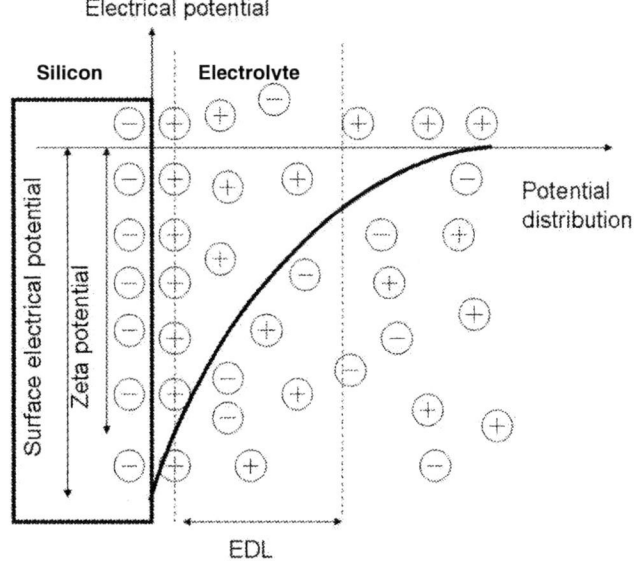

Figure 1: Electric double layer (EDL) by the wall

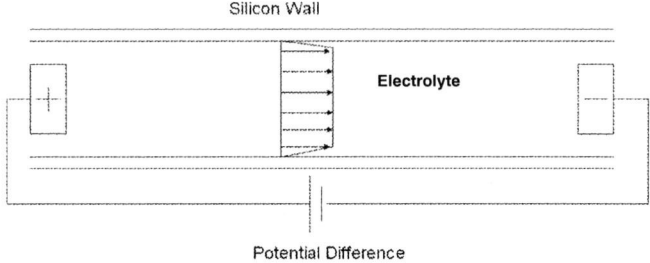

Figure 2: Velocity profile of electro-osmotic flow

Electro-osmotic flow was discovered by Ferdinand Fiodorovuch Reuss in 1809 [3]. Reuss found that application of an electric field across a porous medium filled with a dielectric liquid caused the liquid to flow. This is due to the fact that Silanol (Si-OH) groups at the

---

*Author for correspondence

1-4244-1105-X/07/$25.00 ©2007 IEEE

surface of silicon, when immersed in the liquid such as water, converts Si-OH to Si-O-, resulting in a net negative surface potential (see Figure 1). These negative surface charges attract positive charges in the buffer towards the surface and repel negatives ions from the wall. An electric double layer is formed due to the presence of static charges near the walls.

When an external electric potential difference is applied along the silicon micro-channel, the positive charges in the bulk fluid will be attracted towards the negative electrode. The high fluid velocity near the wall drags along the bulk fluid resulting in the flow of the fluid in the channel. The higher external potential difference, the higher will be the resulting velocity. A typical velocity distribution in a micro-channel is shown in Figure 2.

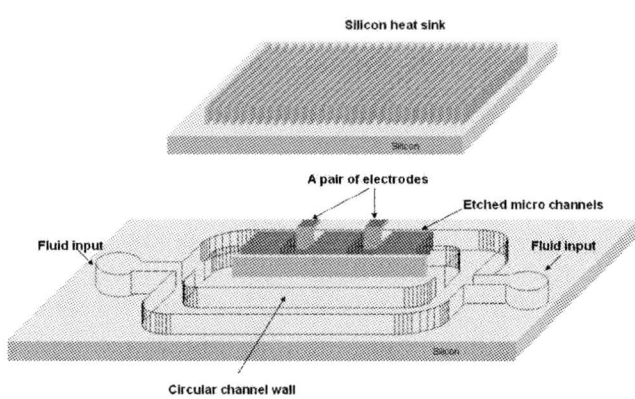

Figure 3: Electroosmotic flow in a heat spreader. Schematic of the model proposed.

Forced liquid circulation, using both mechanical and electro-kinetic pumps have been made in the past [4] but these have not produced adequate cooling for microprocessor applications. Therefore, the present paper proposes a novel way of spreading heat and reducing hot spot temperature. This can be achieved with the support of existing technology such as MEMS and VLSI. Using the existing technology, the proposed liquid based heat spreader can both be fabricated and integrated into a microprocessor. This is expected to reduce the heat sink size and thus the total space requirements for the cooling system.

2. Heat Spread Design

An elliptical structure in Figure 3 and 4 has been designed as a heat spreader for microprocessor chips. Liquid is allowed to flow through micro-channels and elliptical paths on a silicon surface, by subjecting it to a small external electric field via two electrodes. The idea is that by circulating the liquid on the surface of the microprocessor chips, the heat flux from the microprocessor chips becomes uniformly distributed. The design consists of two silicon wafers. The bottom silicon wafer

Figure 4: Elliptical Structure

is assumed to be the microprocessor chip. The elliptical paths and micro-channels are etched at the back of the microprocessor chips. The electrodes are deposited across all the micro-channels as shown in Figure 4. The top silicon wafer acts as both a water tight cover and the heat sink. The plate fin heat sink is expected to be attached to the top surface of the top silicon wafer as shown in Figure 3. With this design if a liquid is continuously circulated within the sealed silicon heat spreader the system is expected to reduce the hot spot temperature, by spreading the heat over the surface of the microprocessor chips.

3. System Fabrication

Photolithography and plasma etching are used to fabricate both micro-channels and elliptical path on the silicon wafer surface [5]. Positive photo-resist was used to etch shallow channels, whilst negative photo-resist was used for deep channel etching. The width and depth of the micro-channels vary between 5 and 50 micro meters (Figure 5). The system consists of a set of more than fifty micro-channels. A pair of thin film electrodes was constructed using pure gold which is evaporated onto two exposed areas of shadow mask ranging from 3mm to 10mm apart. De-ionized water, with a electrical conductivity of 3.0×10^{-4} Sm^{-1}, was used as the working fluid in these micro-channels.

Electrical Isolator

Experiments were carried out with and without fluid movement and with different concentration of fluid. The current flowing through the silicon micro-channels was measured as shown in Figure 7. It shows that current flows through the silicon as a result its lower resistivity than that of the fluid. Current flowing in the silicon affects the performance of microprocessor chips (see Fig-

(a)

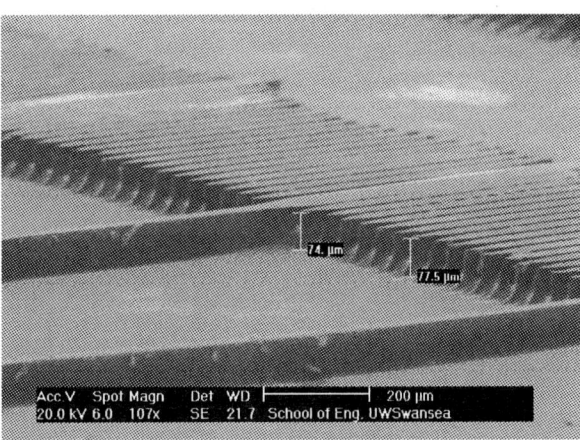

(b)

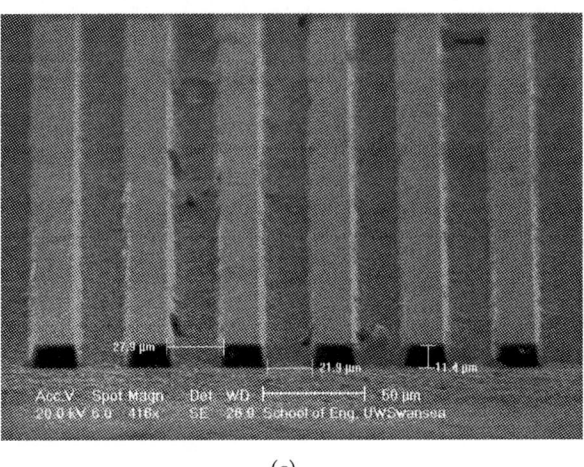

(c)

Figure 5: Electro-osmotic flow through rectangular channels. Typical channels etched on silicon wafer surfaces.

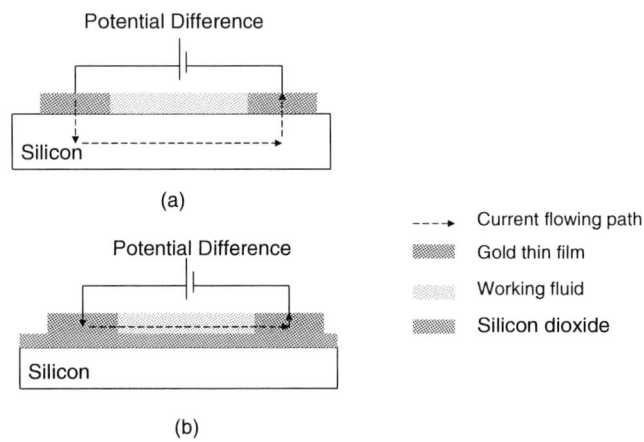

Figure 6: (a)No silicon dioxide (b)Silicon dioxide on silicon sample

ure 6a).

More experiments were carried out with silicon dioxide layer deposited on the inner surface of the micro-channels. A layer of silicon dioxide was thermally grown on the surfaces of micro-channels as shown in Figure 6b. The purpose of silicon dioxide layer was to isolate silicon from the thin gold film used as electrodes[6]. Silicon dioxide is a good electrical isolator and also a moderate thermal conductor. The high resistivity of silicon dioxide results in low or no current flowing through silicon. The results of the current measurement is shown in Figure 8. As seen, current flowing into the silicon is now reduced and current flowing into the liquid is increased.

Bubbles

To improve the flow rates, better electrical conductor such as potassium chloride (KCl) needs to be used. However, the drawback of increasing the electrical conductivity and concentration of the electrolyte is that it can trigger electrolysis. Electrolysis is one of the major problems when running an experiment, especially at high external electrical potentials. In the past, bubbles were generated near the gold electrodes, when a DC voltage of 1.1V or above was applied [7]. The bubbles block the channels and the channel will dry out in a few seconds. Furthermore, the bubbles peel off the films of gold electrode. In order to increase the flow rates and to reduce bubbles formation, experiments were carried out by varying the duty cycle, frequency, fluid electrical conductivity and applied voltage.

At a positive voltage higher than 10.4V and at a negative voltage lower than -10.4V, bubbles start forming near the cathode and anode respectively. To reduce bubble formation, low reversed voltage needs to be applied. We used a pulse like voltage input of 10.4V and -1.8V as shown in Figure 9. This resulted in successful elimination of bubbles. Also, a duty cycle,(D), of less than 30 per cents and a low frequency helped in reduc-

557

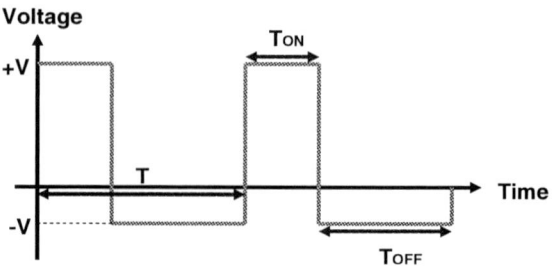

Figure 9: Duty Cycle

Circular Channel No SiO2 layer -2V to 2V

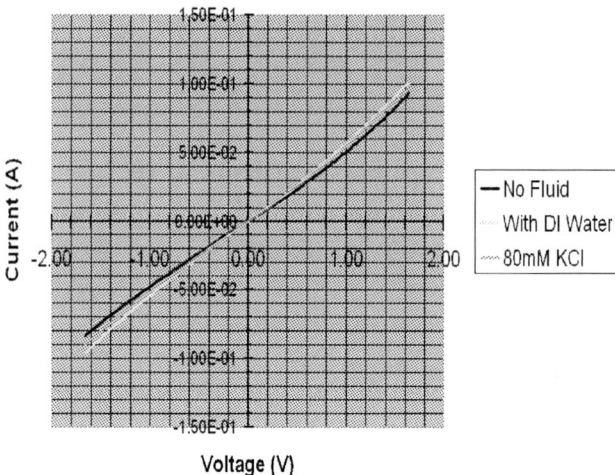

Figure 7: No silicon dioxide

Straight Channel with SiO2 layer -10V to 10V

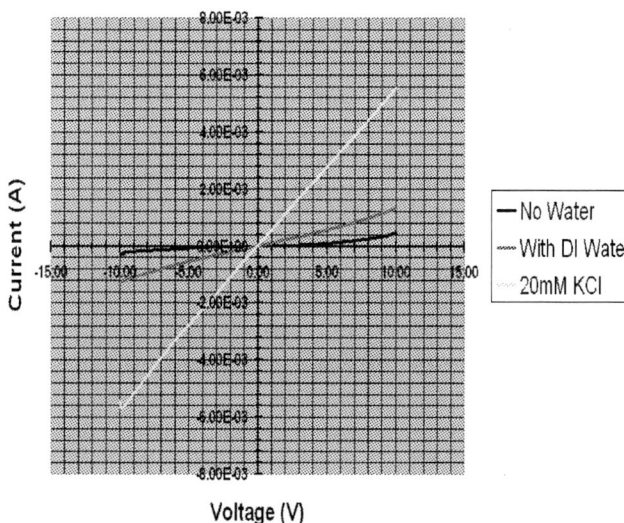

Figure 8: Silicon dioxide on silicon sample

ing the bubble formation near the cathode. Duty cycle is defined as the ratio between the duration of positive voltage(TON) to the total duration (T). The quantities TON and $TOFF$ are defined as (refer to Figure 9).

$$T = TON + TOFF \qquad (1)$$

$$D = \frac{TON}{T} \qquad (2)$$

High concentration of electrolyte has high electrical conductivity and consequently increases the bubble generation. A substantial reduction in bubble generation can be achieved by using pulse cycles of 10.4V and -1.8V AC voltage, with a 30 per cents duty cycle at a frequency of 10Hz.

4. Preliminary Measurements

Flow measurement

Several preliminary experiments were carried out by filling the channels with de-ionized water and connecting the gold electrode pairs with an external voltage source. Flow rate was calculated by digitally capturing the fluid flow via monitoring the motion of a single particle with respect to real time. The preliminary experiments were operated at 10V, 20V and 40V and flow rates were calculated at these voltages. The average velocities for the different voltages are 0.013mm s^{-1}, 0.062mm s^{-1}, 0.078mm s^{-1} respectively. The flow rates calculated are 0.04μl min^{-1} at 10V, 0.16μl min^{-1} for 20V and 0.24μl min^{-1}using 40V (see Table 1).

Temperature measurement

Preliminary experiments show that circulation of water can increase heat transfer rate from the heat source as demonstrated in Figures 10 and 11. The difference in temperature in the vertical axis was calculated with reference to room temperature. As seen, with the water circulation the heat source and electrolyte reach the equilibrium temperature faster. Without the water circulation, the equilibrium was not reached within the observed time period.

External Potential Difference (V)	Velocity (mm/s)	Flow Rate (ul/min)
10	0.013	0.04
20	0.062	0.16
40	0.078	0.24

Table 1: Flow Rate

5. Numerical Modelling

A numerical study of electro-osmotic flow through various geometries has been carried out to complement the experimental work. The fully explicit numerical model [8] uses the Finite Element Method to discretise the modified Navier-Stokes equations. The model has been shown to be accurate, through excellent comparisons to analytical and experimental data [9].

6. Future Work and Conclusion

Bonding is essential to perform the experiments in a real situation. It is, therefore, important to perfect this procedure. Accurate flow measurement is also a challenge, however temperature measurement is sufficient to demonstrate the performance of the proposed heat spreader. Electro-osmotic flow in silicon wafer is an excellent alternative to conventional micro cooling systems. Although the fabrication process needs MEMS technology, the electro-osmotic flow based cooling systems are relatively easy to manufacture. Further research is essential to understand its viability as an alternative cooling system.

Acknowledgements

The financial support of Welsh Assembly Government and Welsh Development Agency (WDA) via Knowledge Exploitation Fund (KEF) number HE 09 COL 1004 is acknowledged.

References

[1]D.B. Tuckerman and R.F.W. Pease, High Performance Heat Sinking for VLSI, *IEEE Electron Device Lett. Vol 2*, 1981, pp 126-129.

[2]S. Yao, D.E. Hertzog, S. Zeng, J.C. Mikkelsen, J.G. Santiago, Porous glass electro-osmotic pumps: design and experiments *Journal of Colloid and Interface Science Vol 268*, 2003, pp 133-142.

[3]R.F. Probstein, *Physicochemical Hydrodynamics,*

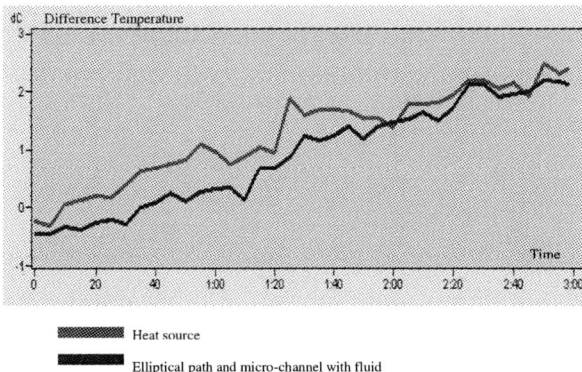

Figure 10: With Flow

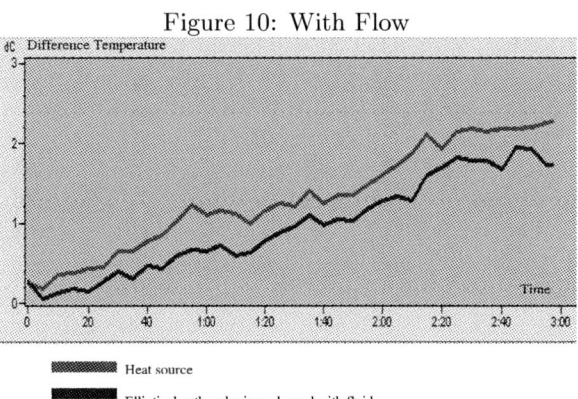

Figure 11: Without Flow

Butterworths, 1989.

[4]D.J. Laser, J.G.Santiago, A Review of Micropumps *Journal of Micromechanics and Microengineering 14,* 2004

[5]S.M. Sze *VLSI Technology, McGraw Hill*, 1983.

[6]S.M. Sze *Physics of Semicondutor Devices*, John Wiley and Sons,1981.

[7]J.O.M. Bockris, A.K.N. Reddy. *Modern Electrochemistry 1 and 2*, Plenum Rosetta, 1973.

[8]A.K. Arnold, P. Nithiarasu and P.G. Tucker, Finite element modelling of electroosmotic flows on unstructured meshes, *International Journal of Numerical Methods for Heat and Fluid Flow*, (to appear, 2006).

[9]P. Dutta, M.J. Kim, K.D. Kihm and A. Beskok, Electro-osmotic flow in a grooved micro-channel configuration: a comparitive study of upiv measurements and numerical simulations. *IMECE2001/MEMS-23895*, 2001.

Parameter Identification on Wafer Level of Membrane Structures

Steffen Michael [1], Siegfried Hering [2], Gisbert Hölzer [2], Tobias Polster [3], Martin Hoffmann [3], Arne Albrecht [3]

[1] Melexis GmbH, Erfurt, Germany
[2] X-FAB Semiconductor Foundries AG, Erfurt, Germany
[3] Technische Universität Ilmenau, Fachgebiet Mikromechanische Systeme, Ilmenau, Germany

Abstract

A fast identification method of membrane structure parameters is investigated for an early stage of the manufacturing process. The approach consists of performing optical measurement of the modal responses of the membrane structures. This information is used in an inverse identification algorithm based on a FE model.

Device characteristics can be determined by measured modal frequencies which are fed into a model based on the FE simulations. The number of parameters to be identified is thereby generally limited only by the number of measurable modal frequencies. A quantitative evaluation of the identification results permits furthermore the detection of defects like cracks which cannot be classified within a FE model.

The approach is validated by first measurements which have shown a good correlation between simulated and measured modal frequencies.

1. Introduction

The development of the two criteria costs and reliability is essential for the further growth of the MEMS market. Efficient test procedures can reduce costs significantly by the detection of faulty sensors before the subsequent packaging and assembly steps. The presented method deals with an approach for a fast and accurate identification of geometrical and material parameters of MEMS with characteristic out-of-plane vibrations.

MEMS devices usually do not permit direct parameter measurement. The indirect parameter identification by modal frequencies is presented in [1], [2]. This approach is taken up for parameter identification of membrane structures. An high precision vibrometer measures optically the electrostatic excited out-of-plane vibrations. The electrostatic excitation is chosen due to the expected modal frequencies in the range of MHz. Conventional excitation methods like ultrasound or piezoelectric stimulation cannot be used in this frequency range.

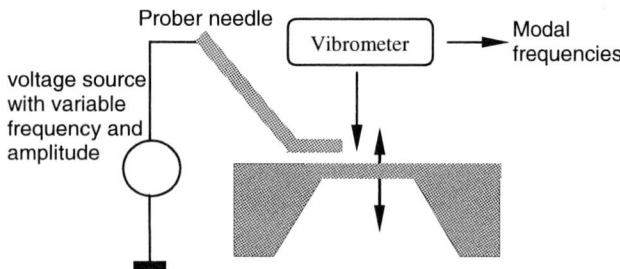

Figure 1: Measurement principle

The mechanical behaviour of the membrane structures is described by a FE model. This model is the base for simulations of modal frequencies versus unknown sensor parameters like the intrinsic stress and process-related variing parameters like the membrane thickness. The time consuming FE simulations have to be done only once for each sensor type and thus do not influence the real identification time.

The simulation results provide the functional dependencies of the modal frequencies versus the sensor parameters which are approximated by polynomials. Finally the sensor parameters are extracted from the measured data by a nonlinear optimization and parameter adaption procedure.

2. Measurement hardware setup

The measurement hardware consists of three components – a Doppler vibrometer integrated in the Micro System Analyzer from Polytec [3], a semi-automatic Suss prober PA200 and an electrostatic excitation unit. The measurement and prober system are coupled and permit hence automated measurements of whole wafers. The specified setup shown in Fig.1 enables the excitation and measurement of out-of-plane vibrations until 2.5MHz at wafer level.

Figure 2: Measurement setup

The fast measurement of vibrations in the interesting MHz frequency range with vertical picometer resolution is realized by a scanning laser Doppler vibrometer. The laser beam of the vibrometer scans automatically over a user defined grid at the surface of the membrane. For each scan point an out-of-plane velocity measurement is performed within some milliseconds.

The excitation of the membrane structures is realized by electrostatic forces. A prober needle is connected to an high voltage (up to 400 volt) excitation signal controlled

1-4244-1105-X/07/$25.00 ©2007 IEEE

by an output signal of the measurement system. The needle is positioned above the membrane surface. With respect to an high excitation force the gap between the needle and the membrane is smaller than 100µm.

3. FE modeling and identification algorithm

The identification system can be subdivided accordingly Fig. 3 into the modules measurement unit, simulation unit and the real identification module with the submodules peak detection, polynomial approximation and optimization.

The automated identification process can be controlled by the parameters

- Polynomial approximation accuracy,
- Sensor parameter range and
- Maximal *Estimated Identification Error (EIE)*.

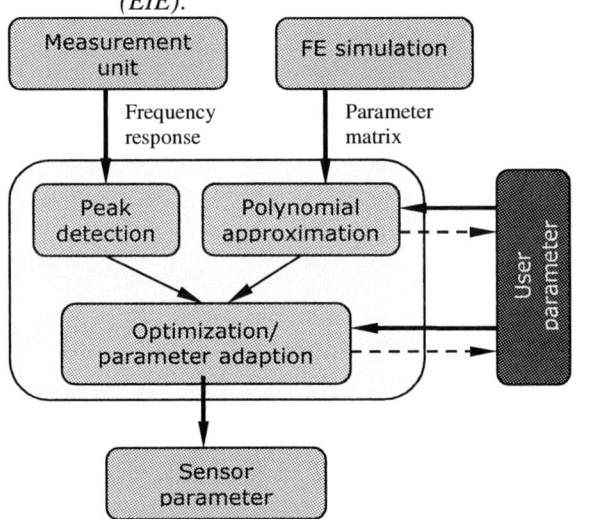

Figure 3: Identification structure

The term EIE is introduced with respect to a quantitative evelution of the identification results. Considering the presented example of an absolute pressure sensor where two parameters (membrane thickness and one membrane edge length) have to be identified. Four modal frequencies are chosen for the identification. Hence six combinations of modal frequencies exist for doing the optimization. For each of the frequency combinations a parameter set p_i is obtained. Then EIE is defined as

$$EIE = \max(p_i) - \min(p_i)$$

and the normed EIE_N correspondingly

$$EIE_n = \frac{\max(p_i) - \min(p_i)}{\operatorname{mean}(p_i)}$$

The mean value of the EIE_N depends on the sensor type as well as the number of parameters. The EIE reflects the measurement and modelling errors. Hence the EIE can be used for the FE model improvement.

The identification of a new sensor type starts within the characterization mode. A fine grid of measurement points permits an unique assignment of frequency peaks to the corresponding modal shapes. Within the

characterization mode suitable modal frequencies are chosen for the identification. Furthermore a maximal EIE_N is defined based on the measurements of some dozen devices. Devices with a bigger EIE have defects not implemented in the FE model like cracks and are declared as faulty dies.

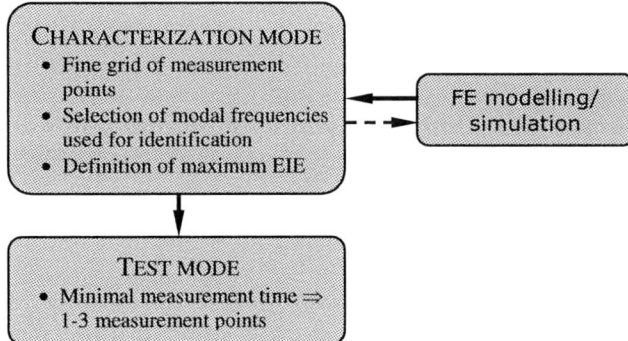

Figure 4: Identification modi

The measurement time corresponds with the number of measurement points. For the identification of whole wafers within the test mode the number of measurement points is reduced to 1-3. Their optimal position is given by the characterization mode. In the current configuration the measurement time per die is about 2-3 seconds. An improvement of the communication between prober and measurement system should reduce the measurement time to 1 second.

3.1 FE modelling

The modelling and simulation is realized in ANSYS. The existing membrane symmetries allow the reduction of the simulation time by using a quarter model. The existence of thermal induced stress due to the passivation layers requires a nonlinear prestressed modal analysis for the parameter sets. The ambient pressure is also considered in the model in case of absolute pressure sensors to obtain accurate identification results.

An additional harmonic analysis is done in order to exclude the influence of the electrode position on the value of the modal frequency.

3.2 Peak picking

The measurement system provides a frequency response of the membrane structures. The characteristic and the SNR of such frequency responses can vary in a wide range due to the different sensor and electrode types. Peaks with very small amplitudes can appear depending on the sensor type and the electrode type and position. Unsymmetries in quadratic membranes can cause closely located peaks.

With respect to a fast and robust algorithm two different approaches were investigated – a fitting by a nonlinear least square algorithm with Gaussian or Lorentzian functions and a conventional local maximum search algorithm. The local maximum search algorithm is chosen due to a faster peak detection.

3.3 Polynomial approximation

Parameter variations are performed by means of the FE model. The resulting parameter matrix is polynomially approximated with respect to a fast and accurate parameter identification.

Based on a user defined accuracy (default value 0.1%) the degree of the polynomial is selected by the approximation module. In case of a twodimensional problem like the presented rectangular membrane the automated approximation results in a polynom of 2nd degree with mixed terms for each frequency f_i

$$f_i = c_{i0} + c_{i1}x_n + c_{i2}z_n + c_{i3}x_nz_n + c_{i4}x_n^2 + c_{i5}z_n^2$$
$$+ c_{i6}x_nz_n^2 + c_{i7}x_n^2z_n + c_{i8}x_n^2z_n^2$$

with x_n as normed membrane edge length and z_n as normed membrane thickness.

If the required accuracy is not reached or an oscillation occurs between the calculated reference points a warning is generated. That indicates a necessary recalculation of the parameter matrix with closer reference points of the sensor parameters.

3.4 Optimization

The polynomial approximation results usually in polynomials of higher orders. Thus a nonlinear least square algorithm solves the multidimensional problems which yields a stable solution for the investigated sensors.

Identified peaks do not correspond necessarily with modal frequencies. The unique assignment of modal frequencies to frequency peaks is done within the optimization module.

4. Applications

4.1 Absolute pressure sensors

Absolute pressure sensors manufactured by grinding bonded wafers are investigated in two different configurations – with quadratic and rectangular membranes.

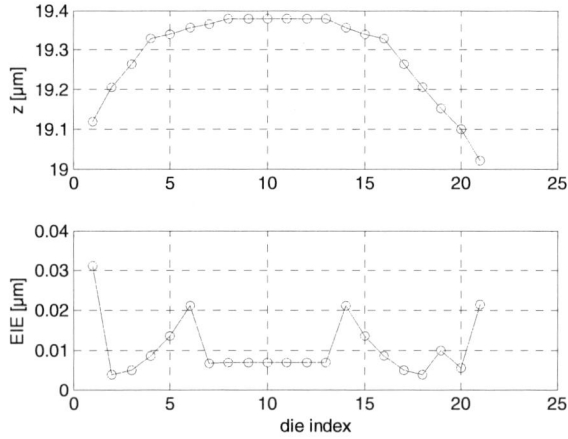

Figure 5: Identified thickness and EIE of a quadratic membrane

The membrane thickness of a testchip with a quadratic membrane (500µm x 500µm) is identified. The characterization mode is used for the adaption of the FE model concerning the stress. The improved model yields a membrane thickness with a very low EIE<100nm. The identification is based on two measurable frequencies ($f_1 \approx 1$MHz, $f_2 \approx 2$MHz).

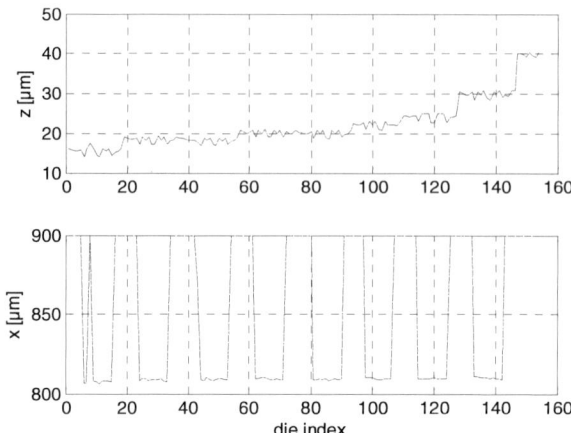

Figure 6: Identified thickness z and edge length x of test chips

Several wafers are processed to get testchips with rectangular membranes and different membrane thicknesses which vary within 15-40µm. Whereas one membrane edge is constant (900µm) the second membrane edge ranges between 800µm and 900µm. For the identification 157 stocasticaly chosen dies from several wafers are selected. The first four modal frequencies can be measured within the whole parameter range and are therefore taken as input for the optimization unit. The optimization unit has to handle in this example the special case of a quadratic membrane where the 2nd and 3rd frequency are identical.

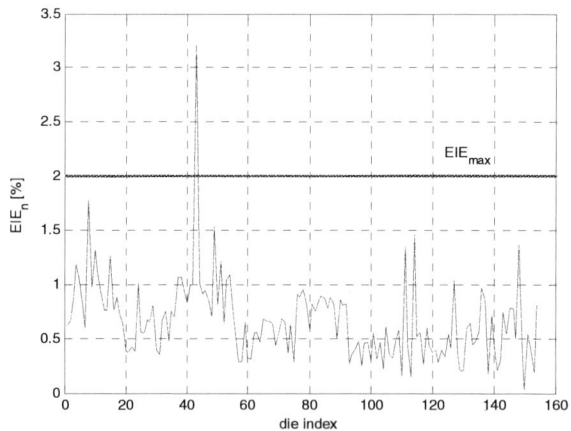

Figure 7: Estimated thickness identification error

Fig.6 shows the identification results where all parameters are within the interval given by the design and process spec. Resulting from the measurements of the characterization mode the upper limit of the EIE_N is defined at 2% for the membrane thickness. The increased number of parameters causes a higher mean EIE_N in comparison to the 0.5% of the presented quadratic membrane due to numerical effects.

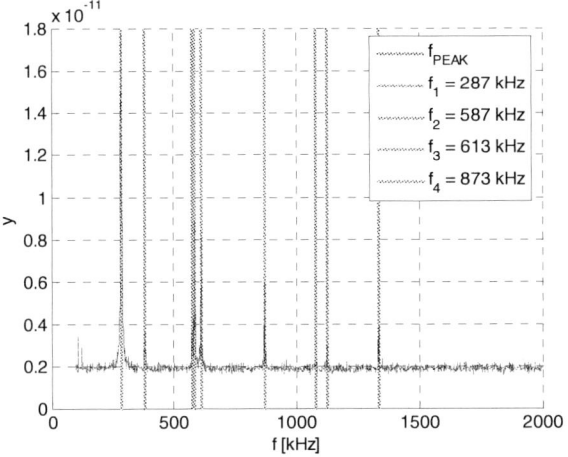

Figure 8: Frequency response of the identified test chip with crack

Considering Fig. 7 chip #43 is out of the defined EIE_N limit. The characteristic of the frequency response does not give a hint on a defect device. A rectangular membrane is indicated by the closely located 2nd and 3rd frequency. Furthermore a parasitic peak occurs at 400kHz. Nevertheless a membrane defect cannot be dissipated by the qualitative validation of the frequency response.

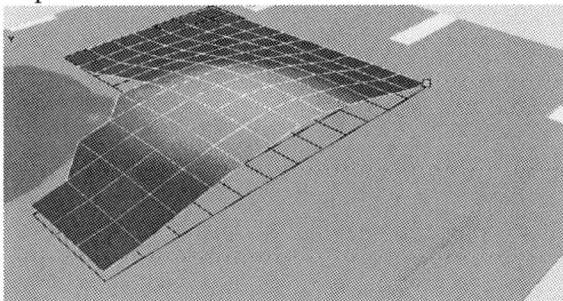

Figure 9: Identified test chip with crack (at the right front)

The visual inspection confirms the identification of die #43 as a wrong one based on the quantitative validation due to a crack at the right front.

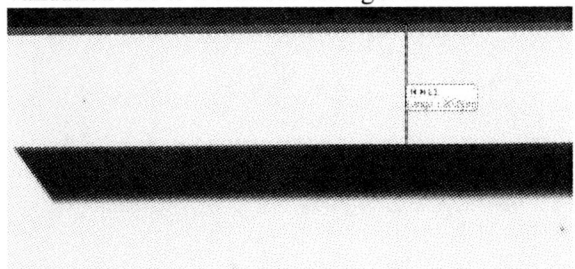

Figure 10: Optical membrane thickness measurement with destructive cross section analysis

The identification results are validated by a destructive cross section analysis at selected dies.

Measured and calculated membrane thicknesses of these dies differ within a range of 1%.

4.2 AlN membrane

At the Technische Universität Ilmenau thin AlN films are processed by reactive sputtering with regard to a usage of these films as piezoelectric layers for micro- and nanosensors [4].

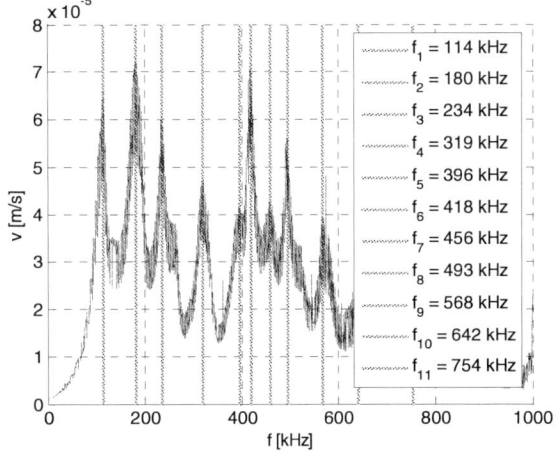

Figure 11: Frequency response of AlN membrane

First investigations are done at a round membrane (diameter 1mm) to identify the expected thickness of 300nm. Depending on the electrode position not all frequency modes are excited. The 4th and 5th as well as the 7th and 8th modal frequency cannot be measured with this electrode. So the 1st, 2nd, 6th and 9th mode is chosen as base for the identification.

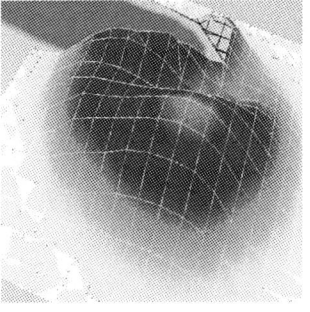

Figure 12: Shape of the 2nd and 6th modal frequency of the AlN membrane

After a model adaption concerning the intrinsic stress a membrane thickness of 304.2nm is calculated. Further characterizations especially of the material properties should decrease the current EIE_N of 4%.

4. Conclusions

The presented approach permits the fast and accurate identification of typical membrane parameters like thickness and lateral dimensions by optical measurements. A quantitative evaluation of the results can detect defects which are not fed into the model. Furthermore the method is well adapted for the integration in the fabrication process of MEMS.

Further works will deal with the application of the approach to other MEMS with characteristic out-of-plane vibrations like IR sensors or beam structures.

Acknowledgments

The works are done within the public research projects PAR-TEST and INNOSENS which are funded by the German Federal Ministry of Education and Research (BMBF).

The measurements were enabled with the gratefully acknowledged support of the Polytec GmbH Waldbronn and the Suss Microtec GmbH Sacka.

References

1. Smith, N. F. et al, "Non-Destructive Resonant Frequency Measurement on MEMS Actuators", *39th Annual International Reliability Physics Symposium*, Orlando, FL, USA, 2001, pp. 99-105
2. Tanner, D. M. et al, "Resonant frequency method for monitoring MEMS fabrication", *Reliability, Testing, and Characterization of MEMS/MOEMS II*, San Jose, CA, USA, 2003, pp. 220-228
3. Rembe, C. et al, "Accurate New 3D-Motion Analyzer for MEMS", *Microsystem Technologies 2003*, Editor H.Reichl, Franzis Verlag, 2003, pp. 435-442
4. Polster, T. et al, "AlN as a piezoelectric material for integrated micro and nano sensors on silicon", *Smart Systems Integration*, Paris, Frankreich, 2007

A micromachined thermoelectric sensor for natural gas analysis: Thermal model and experimental results

G. Carles[1], S. Udina[1], M. Salleras[1], J. Santander[2], L. Fonseca[2], S. Marco[1]

1) Departament d'Electrònica, Universitat de Barcelona, Martí i Franquès, 1, 08028 Barcelona, Spain
2) Centre Nacional de Microelectrònica, Campus UAB, 08193 Bellaterra, Spain

Abstract

In this work, the potential use of a micromachined thermopile based device for analyzing natural gas is explored. Device's response is studied to attain the final goal of detecting variations in natural gas composition through the variation of its thermal conductivity. A FEM thermal model of the device is developed and used to simulate its thermal operation and estimate its sensitivity. Different responses have been recorded experimentally and compared with those obtained from the model developed, showing a good agreement. Results show that small variations in the gas mixture composition can be clearly detected.

1. Introduction

Natural gas property measurement is a field witnessing a remarkable evolution over the past years [1]. Directly related to gas market liberalization within the European Union, gas properties are expected to vary more frequently and more strongly. Precise monitoring of these variations is of great economic and technical importance.

In energy billing, the superior calorific value of the natural gas delivered must be very accurately measured. Other gas properties such as normal density and CO_2 mole fraction are required to convert volume at operating conditions to volume at normal conditions on the basis of ISO12213/3 [2]. Wobbe index and methane number are key parameters in industrial process control, for instance Wobbe index is used to control heat quantity supplied to sensitive combustion processes, as in glass fabrication. Methane number is an important indicator to prevent knocking in gas engines.

The established technologies for such sensible measurements have been calorimeters and Process Gas Chromatographs (PGC) so far, though newer technologies have recently appeared and are progressively being adopted by the gas industry. Some reasons for this are that PGC, though very accurate, are expensive and slow. Where flame calorimeters are bulky, inefficient and costly to maintain. The new methods pursue lower cost, faster responses and are based in methods that correlate particular physical measurements with natural gas properties or even composition.

In this scenario it is apparent that natural gas instrumentation could highly benefit from a Micro-Electro-Mechanical System (MEMS) approach, regarding reliability, ruggedness, maintenance ease and cost. The basic principles in which our device is based have been long used in micromechanized Thermal Conductivity Detectors (TCDs) [3] so widely present in the Gas Chromatography (GC) field. A TCD-like approach has been reported [4] for natural gas evaluation, but some features, mainly time response and resolution, are not optimum for fast and accurate gas analysis, as would be the case with most micromechanized TCDs [5]. From these works the need of accurate simulation tools for the design of this kind of microsystem can be pointed out.

This current work presents simulation model data and experimental measurements data for a promising thermoelectric MEMS for determination of natural gas composition and properties.

2. Thermopile sensor features

Thermopiles are devices which make use of the Seebeck effect to output an electrical signal depending on a temperature difference they are exposed to,

$$V_{out} = \Delta T \cdot Se \cdot n \qquad (1)$$

being Se the Seebeck coefficient and n the number of thermocouples.

The device studied in this work is fabricated with CMOS compatible micromachining processes, leading to many advantages, namely the possibility to integrate electronic signal acquisition and conditioning in the same chip, an enhancement of sensing properties, as well as achieving low power consumption.

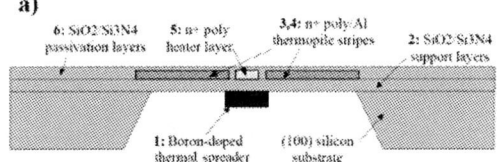

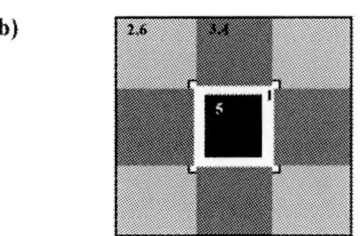

Figure 1: Schematic (a) cross-sectional and (b) top views of the device. 1, silicon thermal spreader; 2, SiO_2/Si_3N_4 support layer; 3,4, n+poly/Al thermopile stripes; 5, polysilicon heater layer; 6, SiO_2/Si_3N_4 passivation layer.

The sensor structure consists of a thin membrane defined on a silicon chip, see figure 1. The membrane is a 1500x1500 μm^2 multilayer sandwich structure of SiO_2/Si_3N_4 which sustains the thermocouple stripes extending from the silicon rim (where their cold junctions

1-4244-1105-X/07/$25.00 ©2007 IEEE   565

stand) to a hotplate in the center of the membrane (where the hot junctions are). The device has 10 thermocouples per side, for a total of 40. A polysilicon heater is located in the hotplate to heat up the hot junctions at the desired temperature and a boron-doped silicon island is located right below as a thermal spreader for better temperature homogeneity across the hotplate. Heater dimensions are $366{\times}310$ μm^2 and the thermal spreader is $450{\times}450$ μm^2. The backside of the dye is attached to a metal casing (TO 8) using a high thermal conductivity epoxy adhesive. The metal casing acts as a heat sink to keep the substrate (and thus rim) temperature approximately constant.

The materials chosen to make the thermocouples are of key importance since they greatly determine the sensor performance. In fact a compromise arises between a high seebeck thermoelectric effect and the goodness of the structure's thermal behaviour [6, 7]. The chosen materials were aluminum and n-doped polysilicon. In table 1 the values of the thermal conductivity of different device materials is shown. These were the values used in the simulations.

Material	Thermal conductivity
Silicon	150 W/(m·K)
Si_3N_4	24 W/(m·K)
SiO_2	1.4 W/(m·K)
Aluminum	235 W/(m·K)
Poly-n+	30 W/(m·K)

Table 1: Material properties

At the beginning of the device fabrication process a 300 μm thick silicon wafer is heavily doped with boron to define the thermal spreader area. After that, the oxide-nitride sandwich layers are deposited (100 nm SiO_2, 300 nm SiN_3 and 50 nm SiO_2). On top of that, the polysilicon resistor is defined on the membrane, by deposition of a 480 nm thick layer and posterior etching to pattern the stripes. Electrical isolation of the resistor is accomplished by deposition of 500 nm oxide layer, after that the 500 nm aluminum thermocouple stripes and electrical contacts are patterned. Then a last layer of passivation (oxide-nitride layer) is deposited. Finally, an anisotropic wet etching process is performed to eliminate the silicon from the backside and thus defining the supporting membrane (the high level boron doping of the silicon island prevents it to be removed). A top view of the final device can be seen in figure 2.

The measurement principle consists of heating up the hotplate by applying a voltage at the heater, thus causing a certain temperature distribution. The heat generated flows through the surrounding gas, as well as through the membrane. Hence the temperature reached at the centre of the membrane will depend on the thermal conductivity of said surrounding gas, as it happens in GC TCDs, but without the presence of a specifically designed fluidic channel. The sensor stationary signal output is related to the thermal conductivity of the gas. The thermal isolation

of the membrane and the high thermal conductivity of the silicon bulk are of great importance in order to enhance the temperature difference between hot and cold junctions. The use of a thermopile to measure the temperature instead of measuring the heater resistance as done in most TCDs, allows for better resolution.

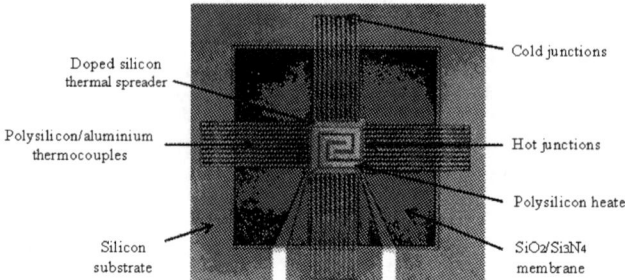

Figure 2: Top view photograph of the device.

3. Sensor characterization

Calibration of our sensor consisted in finding the Seebeck coefficient of the thermocouples and thus knowing the effective curve $V_{out}(\Delta T)$ of the thermopile sensor.

To perform this calibration, the heater's temperature dependence had to be found out, in order to provide a reference to calibrate the thermopile. The resistance versus temperature was recorded using a climatic chamber. A linear variation of the resistance was assumed, as in equation 2, resulting in a value for the temperature coefficient, TCR, of $8.556{\cdot}10^{-4}$ K^{-1} and a value for the resistance, R_0, of 981.1Ω; being $T_0=0$ ºC.

$$ R = R_0\left[1 + TCR(T - T_0)\right] \qquad (2) $$

The resistor placed in the substrate rim is also calibrated in this way, so it can be used to monitor slight changes in the substrate temperature, where the cold junctions are placed.

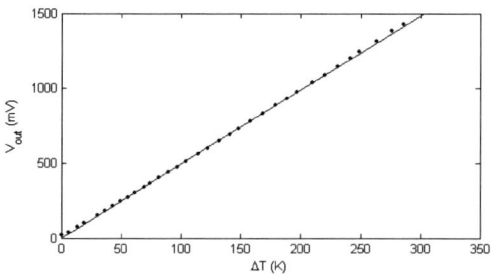

Figure 3: Plot of the thermopile output versus calculated temperature difference.

Dependency of the thermopile signal versus the power applied to the heater was obtained by applying different voltages to it; its resistance was also recorded. Given this, the signal output versus temperature difference between hot and cold junctions could be plotted. Figure 3 shows this voltage signal versus temperature difference. Its slope gives us the Seebeck coefficient of the thermocouples. Assuming it is linear, results in a value for the Seebeck coefficient of 123.9 $\mu V/K$.

4. Thermal conductivity of gas mixtures

The sensor is expected to work immersed in a gas mixture, and its behaviour will be modulated by the thermal conductivity of said gas. It is a must to provide a sufficiently accurate model for the thermal conductivity of a gas mixture in order to perform a useful simulation.

With this in mind, is worth noting that thermal conductivity of a gas mixture is not a linear function of the thermal conductivities of its components. An exact resolution for a mixture of more than 2 gases is yet not known, though many models have been proposed essentially based in empirical measurements, most of which can be reduced to some form of the Wassiljewa equation [8],

$$k_m = \sum_{i=1}^{n} \frac{x_i k_i}{\sum_{j=1}^{n} x_j A_{ij}} \quad (3)$$

being k_m the thermal conductivity of the mixture, k_i the thermal conductivity of component i, x_i the concentration of component i, and A_{ij} some function to be specified.

This model was used, along with the Mason and Saxena [8] proposal for calculating A_{ij},

$$A_{ij} = \frac{\varepsilon \left[1 + \left(\frac{k_{tr,i}}{k_{tr,j}} \right)^{1/2} \left(\frac{M_i}{M_j} \right)^{1/4} \right]^2}{\left[8 \left(1 + \frac{M_i}{M_j} \right) \right]^{1/2}} \quad (4)$$

where M_i is the molecular weight of component i, a value of 1.065 is used for ε and k_{tr} is the coefficient of *frozen* conductivity (A_{ii} is assumed to be unity). These coefficients account for the monatomic value of the thermal conductivity (when no rotational degrees of freedom are considered) and their relations appearing in equation 4 can be computed using,

$$\frac{k_{tr,i}}{k_{tr,j}} = \frac{\Gamma_j}{\Gamma_i} \left[\frac{e^{(0.0464T_{ri})} - e^{(-0.2412T_{ri})}}{e^{(0.0464T_{rj})} - e^{(-0.2412T_{rj})}} \right] \quad (5)$$

where T_r is the reduced temperature ($T_r = T/T_c$) and the values of Γ are calculated using,

$$\Gamma = 210 \left(\frac{T_c M^3}{P_c^4} \right)^{1/6} \quad (6)$$

being T_c and P_c the critical temperature and pressure.

To check the convenience of this chosen model, the values of the thermal conductivity at 20 and 70 °C of several common natural gases [9] were compared with the model's prediction. The Wassiljewa equation provides a better fitting than a simple weighted average as figure 4 shows.

For simulations and laboratory purposes natural gas was approximated as a mixture of its four main components, namely methane, ethane, nitrogen and carbon dioxide. Temperature dependences of these gas thermal conductivities are shown in figure 5 [10, 11].

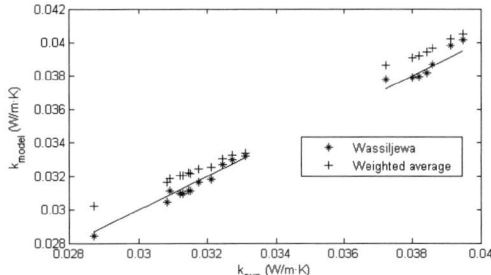

Figure 4: Validation of the exposed model. First group of values corresponds to the calculations at 20 °C while the second ones are at 70 °C.

The different dependences on temperature of the curves in figure 5 are of key importance regarding the goal of distinguishing the different contributions to the mixture.

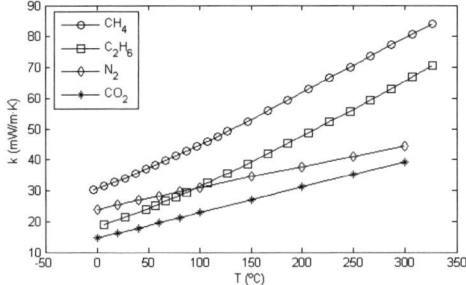

Figure 5: Temperature dependence of the gas thermal conductivities.

5. FEM model

To simulate the operation of the device a Finite Element Method (FEM) model has been developed using ANSYS® (v10.0). The model consists a silicon volume, a gas volume, and the 2D membrane areas (as temperature gradients perpendicular to the membrane can be neglected). In figure 6, the constructed model can be seen. As the device presents symmetry, the model consists of half of it. In addition, boundary conditions are set as to maintain the bottom of the silicon volume (the attaching point to the metal heat sink casing) and the top of the gas volume at ambient temperature, and also to keep the lateral areas in adiabatic conditions. The heat generated at the heater was calculated taking into account its different polysilicon track widths, with heat dissipation being inversely proportional to these.

The properties of each different area in the multilayered membrane were calculated by averaging the properties of each layer across the entire thickness. In the regions containing the thermocouples, a second averaging was performed taking into account the proportion of area corresponding to each material. The effective thermal conductivity can then be calculated as in equation 7.

$$k_{eq} = \frac{\sum_i k_i h_i A_i}{\sum_i h_i A_i} \quad (7)$$

where k_i, h_i and A_i are the thermal conductivities, the thicknesses and area of each material respectively.

The model includes conduction and radiation, but not convection. The Rayleigh number was calculated using equation 8 to check if the system was getting near the critical temperature above which convection becomes significant [13].

$$R_A = \frac{\rho^2 \cdot g \cdot \gamma \cdot L^3 \cdot \Delta T \cdot c_p}{\mu \cdot k} \qquad (8)$$

where ρ is the fluid density, g is the gravity, γ the volumetric coefficient of expansion, L a characteristic length, ΔT the temperature difference reached, c_p the fluid's specific heat, μ its dynamic viscosity and k the thermal conductivity. By using a worst-case estimation of each of these variables, a Rayleigh number of 120 is obtained. The values above which convection is significant vary depending on the system, but the threshold is at some thousands [14], so it was assumed that convection could be neglected.

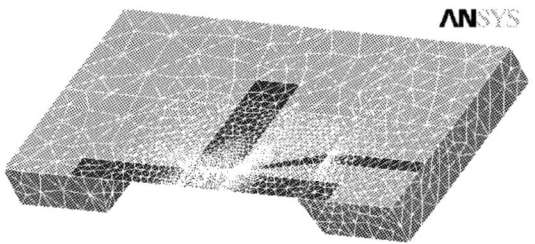

Figure 6: Geometrical view of the model. For clarity the gas volumes have been suppressed.

A stationary temperature distribution calculated with this model beside an IR image of the sensor can be seen in figure 10 for comparison.

6. Model Results and validation

6.1 Preliminary results

Two first sets of simulations were computed in order to explore the sensitivity of the device to gases showing different k at ambient temperature. In one set constant k was assumed, while in the other one k was increased linearly with temperature ($k = A + BT$). For the constant k case, the '*' curve in figure 7 holds, a higher sensitivity can be observed. The other three curves using the linear dependency of k show a decrease in the device sensitivity.

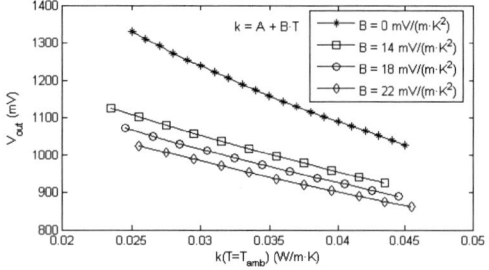

Figure 7: Output signal dependence on the thermal conductivity value (at ambient temperature) of the surrounding gas. 60 mW have been dissipated.

Temperatures reached in the device normal operation are high enough as to consider the temperature dependence of k for obtaining accurate results. However the linear dependency is still an approximation of the true dependency as shown in figure 8.

This preliminary result showed great influence of the dependency $k(T)$ in the sensor response, which is an important result in order to be able to distinguish different natural gas compositions which may have the same k at ambient temperature.

6.2 Final results

After this first two sets of simulations providing qualitative results, two final sets of simulations were computed, one using several synthetic natural gases and calculating the dependency of $k(T)$ using the Wassiljewa model and the data shown in figure 5; and a second set of measurements in air, using tabulated $k(T)$ [12].

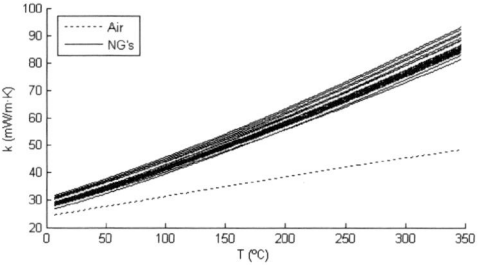

Figure 8: Temperature dependence of the thermal conductivity of the gases considered, according to Wassiljewa equation.

Table 2 shows the considered natural gases based in the composition variation ranges found in Europe [9], while figure 8 shows their thermal conductivity versus temperature when equation 3 is used. For the natural gases 7 different powers were dissipated at the heater, by applying 3 to 9 volts (one volt increments).

CH_4	C_2H_6	N_2	CO_2
100.00	0.00	0.00	0.00
84.58	8.11	6.09	1.22
81.75	8.11	8.92	1.22
80.80	11.89	6.09	1.22
77.97	11.89	8.92	1.22
84.02	8.11	6.09	1.78
81.19	8.11	8.92	1.78
80.24	11.89	6.09	1.78
77.41	11.89	8.92	1.78
82.50	10.00	7.50	0.00
79.50	10.00	7.50	3.00
91.00	0.00	7.50	1.50
71.00	20.00	7.50	1.50
88.50	10.00	0.00	1.50
73.50	10.00	15.00	1.50
81.00	10.00	7.50	1.50
100.00	0.00	0.00	0.00
97.70	2.30	0.00	0.00
87.44	9.38	3.18	0.00
95.24	2.70	2.06	0.00
93.22	4.60	1.08	1.10

Table 2: Composition of the simulated gases, in %.

For the case of air, 3 powers were dissipated corresponding to 21.41, 38.86 and 59.50 mW. These simulations were also computed to allow comparison with radiometric data recorded with the ThermoVision[TM] A40M IR camera from FLIR SYSTEMS (Wilsonville, Oregon, USA).

6.3 Laboratory testbench and validation results

A gas mixing station was specifically configured to output controlled mixtures of 4 of the natural gas main components: ethane, methane, carbon dioxide and nitrogen. Static atmospheres of the synthetic natural gases were supplied to a sensor chamber where measurements were acquired. The gas was supplied at ambient temperature and atmospheric pressure. Expected accuracy of the volumetric gas composition were variations of up to 0.05% in CO_2, 0.3% in C_2H_6, 0.5% in N_2, and 1% in CH_4. A 34970A Agilent multimeter sampling at 25Hz was used to measure the voltage output of the sensor. Measurements in air were taken to allow an IR camera to record images of the sensor operation. A 34401 Agilent multimeter was used in this case.

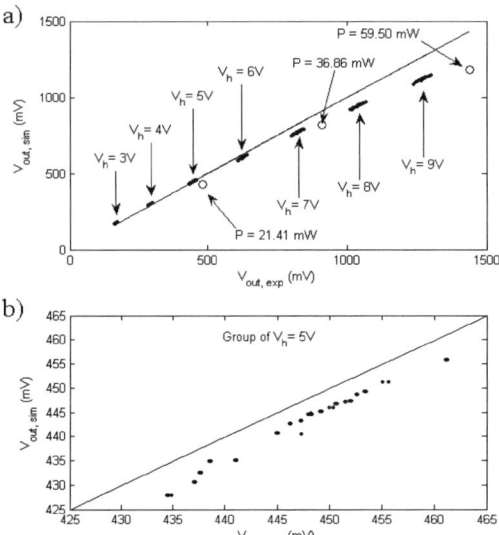

Figure 9: (a) Comparison between measured responses and simulated ones; in every power step (3 to 9 V) are the 42 gas realizations, two series of the compositions shown in table 2 (dots). Three measured and simulated pairs (dissipating 21.41mW, 38.86 mW and 59.50 mW) of outputs using air are also plotted (circles). (b) Zoom to show the variations due to gas mixture composition changes.

Figure 9 compares the results of the simulations with the acquired experimental data. As can be seen in figure 9(a) there is good agreement but for a scale-constant which increases with the applied power. Note that for instance discrepancy in the output at the 5V step is below 0.02%.

6.4 IR imaging and discussion

To better understand what could be happening at the higher powers, an infrared camera was used to record the radiometry map of the sensor in air. Figure 10 shows the IR image when dissipating 59.50 mW. Expected emissivity differences between thermocouple and membrane areas are shown as discontinuous temperature

fields; and dissipation occurring at the heater's legs is also clear, and in good agreement with the simulation. It is in this legs area that the maximum temperatures occur (the influence of this is later discussed). On the other hand the images also confirm that the silicon shows no significant temperature increase with respect to ambient.

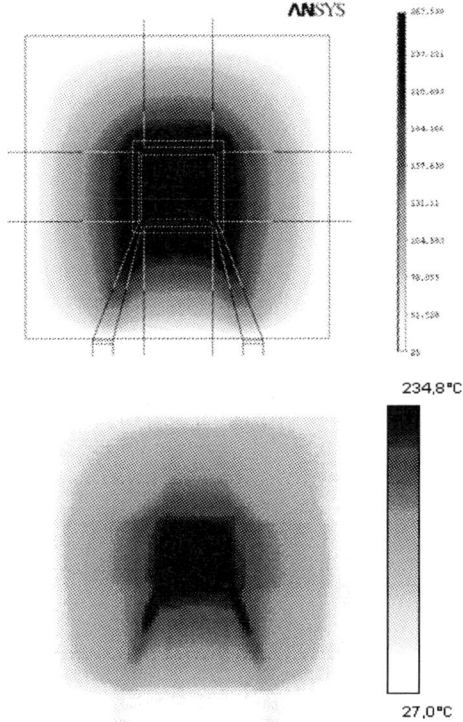

Figure 10: Simulated and recorded temperature map of the device operation with air and 59.50 mW of dissipation.

At the point of publishing this work no emissivity correction had yet been performed due to technical difficulties, so unfortunately no precise numerical temperature information can be extracted from the IR images. Even so, inspection of the images shows an unexpected temperature difference (of at least 15 K at 59.50 mW) between hot junctions in the hotplate.

Note that these temperature differences increase with the applied power. On the other hand this temperature gradient does also affect the polysilicon heater, which we used to calibrate the Seebeck coefficient of the thermocouples. We previously assumed the hot junctions temperature to be the same as the heater's, but while the temperature sensed by the hot junctions is near the edges of the hotplate, temperature given by the heater is an overestimation, since the higher resistivity of the hotter heater areas contribute more to the total resistance. Hence, the hot junctions' average temperature was possibly overestimated, leading to an underestimated Seebeck coefficient. This mismatch increases with applied power suggesting that the Seebeck effect may have an increase with the temperature not considered in the simulation, and thus explaining the discrepancy in figure 9. A closer inspection of the calibration process

569

should be carried out to determine a temperature dependence of the Seebeck coefficient and incorporate it to the simulations for correction of the expected response at the higher powers.

7. Estimation of the sensitivity

The response of our device in front of the concentration of each gas component has three degrees of freedom. Here an estimation of the sensitivity is provided, though it may also depend on the working point and direction because of nonlinear effects.

A number of simulations were performed to explore the output signal variation when small amounts of concentrations are changed. Two sets of concentration variations were used: 100% methane with subsequent addition of 2% of each gas, and some couples previously measured (being possible to compare them). The variation of the signal output over the variation of concentration is assumed to be the device sensitivity to that component, namely,

$$S_i = \frac{\Delta V}{\Delta x_i} \qquad (9)$$

The results are shown in table 3. These values are demonstrating the high sensitivity of the sensor and are encouraging.

Concentration (%)				$S_{i,\text{sim}}$ (mV/%)	$S_{i,\text{mes}}$ (mV/%)
CH_4	C_2H_6	N_2	CO_2		
100	0	0	0	2.45	
98	2	0	0		
100	0	0	0	2.00	
98	0	2	0		
100	0	0	0	2.21	
98	0	0	2		
91	0	7.5	1.5	2.01	2.50
71	20	7.5	1.5		
88.5	10	0	1.5	1.99	1.73
73.5	10	15	1.5		
82.5	10	7.5	0	2.08	3.57
79.5	10	7.5	3		

Table 3: Approach to the sensitivity by simulation using univariate changes in gas mixture composition. Comparison wit experimental results.

8. Conclusions

The use of a thermopile based device for analyzing natural gas was studied. FEM modeling allowed in-depth exploration of the device, obtaining simulations in very good agreement with experimental data. The sensor's performance at detecting small variations in the gas thermal conductivity was explored. The device shows high sensitivity, allowing changes in natural gas composition to be clearly detected. Further signal processing work is to allow the quantification of the natural gas main components.

Acknowledgments

The authors thank to the project TEC2004-07853-C02-01. G.C. and S.U. acknowledge PhD grant from the Spanish Ministry of Education and Science.

References

1. P. Schley, M. Jaeschke, K. Altfeld, "New technologies for gas quality determination", *Proc 22nd World Gas conference*, Tokyo, June 2003.

2. ISO12213: Natural Gas – Calculation of Compression factor. Part 1: Introduction and Guidelines; Part 3: Calculation Using Physical Properties. ISO 12213 International Standard.

3. LGCG online magazine article "The Thermal Conductivity Detector", http://www.lcgcmag.com.

4. D. Puente, J. Gracia, I Ayerdi, "Thermal conductivity microsensor for determining the methane number of natural gas", *Sensors and Actuators B*, Vol. 110, No. 2 (2005), pp. 181-189.

5. Y.E. Wu, K. Chen, C.W. Chen, K.H. Hsu, "Fabrication and characterization of thermal conductivity detectors (TCDs) of different flow channel and heater designs", *Sensors and Actuators A*, Vol. 100, No. 1 (2002), pp. 37-45.

6. M. Salleras, J. Palacín, M. Moreno, L. Fonseca, J. Samitier, S. Marco, "A methodology to extract Dynamic Compact Thermal models under Time-varying boundary conditions: application to a thermopile based IR sensor" *Microsystem Technologies*, Vol. 12, No 1-2 (2005), pp. 21-29.

7. H. Baltes, O. Paul, O. Brand, "Micromachined thermally based CMOS microsensors", *Proceedings of the IEEE*, Vol. 86, No. 8 (1998), pp. 1660-1678.

8. Bruce E. Poling, John M. Prausnitz, John P. O'Connell, The Properties of Gases and Liquids (McGraw-Hill, 2001, fifth edition), pp. 10.1-10.70.

9. Camal Rahmouni, Mohand Tazerout, Olivier Le Corre, "Determination of the combustion properties of natural gases by pseudo-constituents", *Fuel* 82 (2003) 1399-1409.

10. B. A. Younglove, J. F. Ely, "Thermophysical Properties of Fluids. II. Methane, Ethane, Propane, Isobutane and Normal Butane", *J. Phys. Chem. Ref. Data*, Vol. 16, No. 4 (1987), pp. 577-798.

11. F. J. Uribe, "Thermal Conductivity of Nine Polyatomic Gases at Low Density", *J. Phys. Chem. Ref. Data*, Vol. 19, No. 5 (1990), pp. 1123-1136.

12. K. Stephan, A. Laesecke, "The Thermal Conductivity of Fluid Air", *J. Phys. Chem. Ref. Data*, Vol. 14, No. 1 (1985), pp. 227-234.

13. Jin-Qiang Zhong, Jun Zhang, "Thermal convection with a freely moving top boundary", *Physics of Fluids*, Vol. 17, No. 11 (2005), pp. 115105.1-115105.12

14. S. Amiroudine, P. Bontoux, P. Larroude, B. Gilly, B. Zappoli, "Direct numerical simulation of instabilities in a two-dimensional near-critical fluid layer heated from below", *J. Fluid Mechanics* (2001), Vol. 442, pp. 119-140.

Behavioral Modelling of Vibrating Piezoelectric Micro-Gyro Sensor and Detection Electronics

S. Megherbi[1], R. Levy[1], F. Parrain[1], H. Mathias[1], O. Le Traon[2], D. Janiaud[2], J. P. Gilles[1]

[1]Institut d'Electronique Fondamentale, UMR 8622, Univ. of Paris-Sud, 91405 Orsay cedex. France
[2]ONERA-DMPH, 29 av. de la division Leclerc, 92322, Chatillon, France
Email : souhil.megherbi@ief.u-psud.fr

Abstract

This paper describes the development of a model of vibrating piezoelectric micro-gyro sensor using analog hardware description. Our procedure implies several steps with emphasis in model complexity reduction and identification of critical parameters. The proposed macro-model permits multi-physic simulations including mechanical, piezo-electric and electrical analytic descriptions and allows a top-down development approach. As a tool for coding our descriptions, we use analog hardware description language (Verilog-A). For achieving the behavioural computation results, CADENCE simulation environment was used. The critical parameters of the gyro are then studied: the output noise, the output bias and scale factor stability over temperature.

1. Description of the Vibrating Integrated Gyrometer (VIG)

Vibrating micro-gyrometers bodies are machined on silicon or piezoelectric material wafers. The physical phenomenon used for vibrating gyros is the Coriolis force induced by rotation. The VIG used in our study has been developed and integrated by ONERA [2], it's principle is based on a tuning fork which allows two orthogonal modes of vibration; the drive mode along the x-axis which is excited at resonance, and the Coriolis force induced by a rotation along the z-axis that induces a y-axis vibration, the detection mode, whose amplitude is proportional to the angular rate velocity Ω (fig. 1 and fig. 2).

Classically, modelling of such sensors has been focused on the device level, typically using FEM solvers. But, the maturity of the technology needs more and more simulation to achieve precise description. It is our purpose to present a behavioral modelling methodology for our micro-gyro sensor in order to study the limiting effects as thermal instability and parasitic noise upon geometrical and physical parameters.

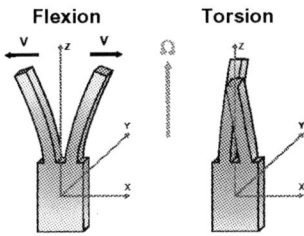

Figure 1: The VIG's tuning fork

Figure 2 : the whole piezo-electric VIG's structure with its electrodes (ONERA-DMPH).

A first model based on the mechanical description is presented on fig.3. It describes the mass and stiffness model of the sensor and its associated equations (eq.1,2).

In this description, the drive and detection modes are respectively the mechanical displacements along x and y axis.

For the drive mode; F_x is the excitation force, V_x the excitation voltage, x the mechanical displacement and m the mass of the mass and stiffness model of the gyrometer.

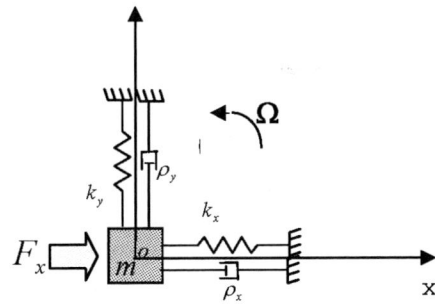

Figure 3: first level mechanical modelling of the VIG

$$F_x = m\ddot{x} + m\frac{\omega_x}{Q_x}\dot{x} + m\omega_x^2 x \qquad (eq. 1)$$

$$m\ddot{y} + m\frac{\omega_y}{Q_y}\dot{y} + m\omega_y^2 y + 2m\Omega\dot{x} = 0 \qquad (eq. 2)$$

To complete this description we add the piezoelectricity parameters describing excitation and detections effects (eq.3, 4).

$$F_x = n_x V_x \qquad \text{(eq. 3)}$$

$$q_y = n_y y \qquad \text{(eq. 4)}$$

In those equations, n_x and n_y are respectively the piezoelectric conversion factors on drive and detection modes.

V_x is the voltage used to control the gyro at drive mode, q_y are the charge detected at the output electrodes.

2. Behavioral modelling of the Vibrating Integrated Gyrometer (VIG) and its electronics

In the case of this HDL model, the mechanical behaviour of the two resonant beams that compose the micro-gyrometer fork (ie the pilot and the detection beam) is described by the way of two distinct serial RLC networks. The well known electromechanical analogy employed consist in considering the voltage across the RLC circuit as the applied or resulting mechanical forces and the current through as the beam tip velocity. According to the mechanical parameters previously presented, the value of R, L and C are as below:

$$R_{xy} = \frac{m.\omega_{xy}}{Q_{xy}},$$

$$L_{xy} = m,$$

$$C_{xy} = \frac{1}{m.\omega_{xy}^2}$$

For each beam, the piezo mechanical coupling is taken into account by a set of equations that describe the behaviour of an ideal transformer assuming the transfer of energy without loss. The ratios of these transformers are equal to the piezoelectric conversion factors n_x and ny. The Coriolis coupling is realized by a current controlled voltage source that links the two RLC resonant circuits. The ratio of conversion of this source is equivalent to a resistance that is proportional to the rate of angular velocity Ω. A second controlled voltage source is employed to take into account parasitic mechanical couplings due to machining defaults. The voltage provided by this source is directly proportional to the position of the pilot resonator. Additionally four parasitic capacitors allow simulating the capacitive coupling between the different electrodes of the system that implies charge injection from the pilot to the detection.

Our behavioral model presented on fig. 4 allows performing DC, AC, but also noise analysis including the mechanical noise due to thermal energy for both resonators. Here the mechanical noise is directly derived from Nyquist-Johnson noise that occurs in resistors in the case of electric circuits. The resistance that we have to consider in our case is the motional resistor R_{xy} that

induced a force white noise generator for each resonator. The magnitude of the mean square force is given by:

$$\overline{f_{nxy}^2} = \frac{4.k.T.m.\omega_{xy}}{Q_{xy}} \qquad \text{(eq. 5)}$$

T is the temperature of the device and k the Boltzman constant.

To analyse the behaviour of the gyrometer over the temperature we have included in the discussed model the temperature dependencies of the quality factors $Q_{x,y}$ and the resonant pulsation $\omega_{x,y}$. The laws of variation of these quantities have been extracted by polynomial fitting from various measurements.

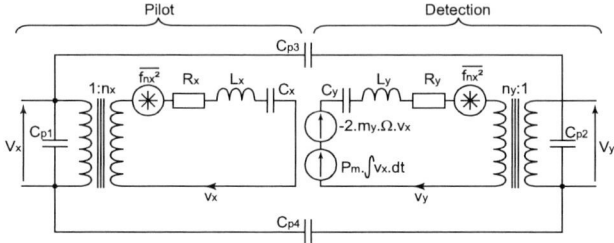

Figure 4: Electric equivalent circuit of the micro-gyrometer.

The associated electronics is presented in fig. 5, it includes a first charge amplifier stage allowing detection of the output signal, a second stage to reduce the common mode signal induced by the capacitive coupling, a third stage to reduce the phase quadrature signal induced by mechanical coupling and to detect the Coriolis signal magnitude. Finally, a low-pass filter stage permits to determine the sensor bandpass [3] [4].

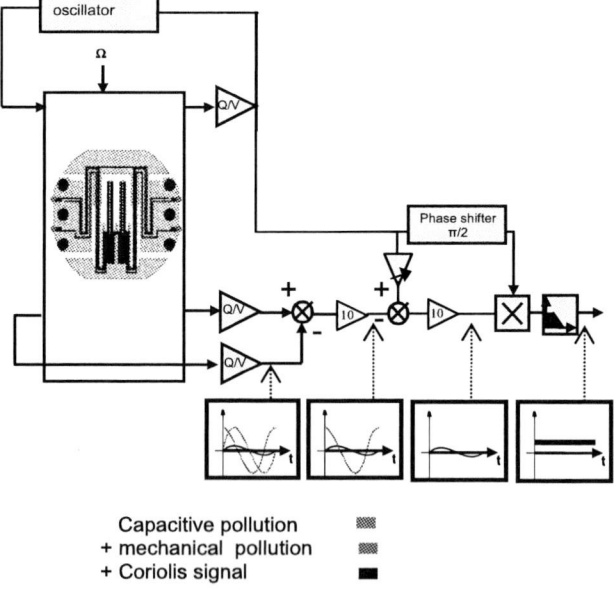

Figure 5: electronic excitation/detection around the VIG.

The complete description of the behavioral Verilog-A description associated to excitation-detection electronics had been developed using CADENCE simulation tools. The results are discussed bellow.

3. Simulation results and discussion

In order to estimate the contribution of the mechanical structure and the associated electronics to the global performances of the gyrometer, different parameters have been studied: thermal output offset and scale factor sensitivities, and output noise.

The following parameters have been deduced by experimental tests and used in our gyrometer description:

	$R_{x,y}$	$L_{x,y}$	$C_{x,y}$	$C_{0x,y}$
drive mode	1,8 MΩ	1,5 MH	1,6 aF	1 pF
detection mode	54 kΩ	69 kH	324 aF	1pF

A first result of the model is the computation of the output charges q_{y+} and q_{y-} by a sweep of the capacitive and mechanical coupling magnitudes. The drive mode is excited at its resonance pulsation ω_x, and the charges on the electrodes of the detection mode q_{y+} and q_{y-} include the Coriolis charges, the charges induced by capacitive coupling, and those induced by mechanical coupling:

$$V_X = X\cos(\omega_x t + \varphi)$$

$$q_y+ = \varepsilon(\Omega).\sin(\omega_x t) + C\sin(\omega_x t) + M\cos(\omega_x t) \quad \text{(eq. 6)}$$

$$q_y- = -\varepsilon(\Omega).\sin(\omega_x t) + C\sin(\omega_x t) + M\cos(\omega_x t) \quad \text{(eq. 7)}$$

$$\underbrace{\qquad\qquad}_{\substack{Coriolis\ signal}} \quad \underbrace{\qquad}_{\substack{Capacitive\\coupling}} \quad \underbrace{\qquad}_{\substack{Mechanical\\coupling}}$$

In the following, we describe the output voltage of the electronics stages by:

$$Vs = SF.\Omega + V_{offset} + V_{noise} \quad \text{(eq. 8)}$$

Vs: is the scale factor, Ω: angular velocity, V_{offset}: output offset voltage, and V_{noise}: output noise voltage.

3.1 Thermal offset sensitivity

For achieving the simulation of this part, we considered $\Omega = 0$ and $V_{noise} = 0$ in the implemented model (cf. eq. 8), the output voltage Vs is then computed as V_{offset} over temperature (T).

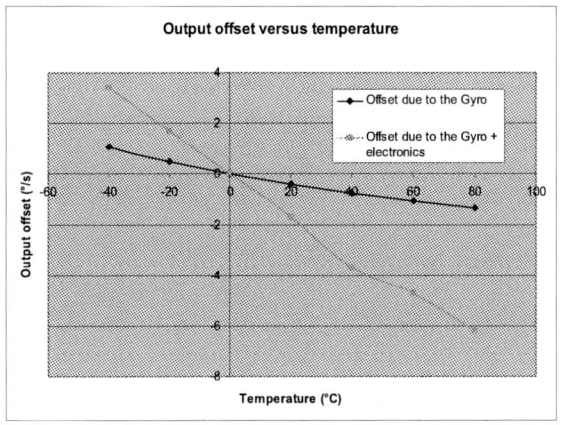

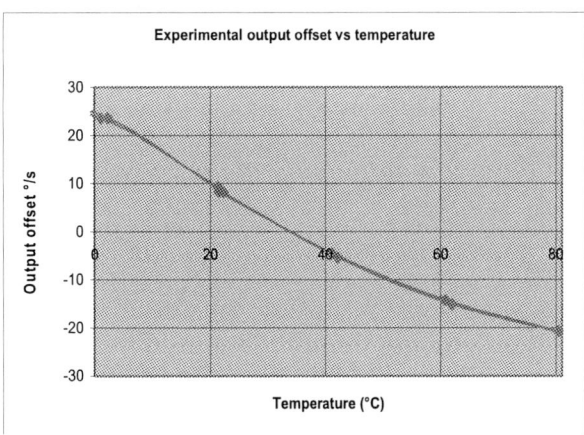

Figures 6a and 6b: Output offset versus temperature, (6a) model simulation, (6b) experimental measurement.

A first analysis of the above curves shows that simulated output offset thermal drift of the gyro and its electronics is smaller than the measured one. In this case, this shows the limit of the model, where other effects must be taken into account [5]. However, we can observe that the contribution of the electronics to offset thermal drift is higher than the gyro's one. For reducing this effect a design of low thermal drift electronics must be improved.

3.2 Thermal Scale Factor sensitivity

For analysing the scale factor sensitivity, we consider in a very high value of omega in the model. This high value permits to reduce the contribution of V_{offset} in the expression of V_s.

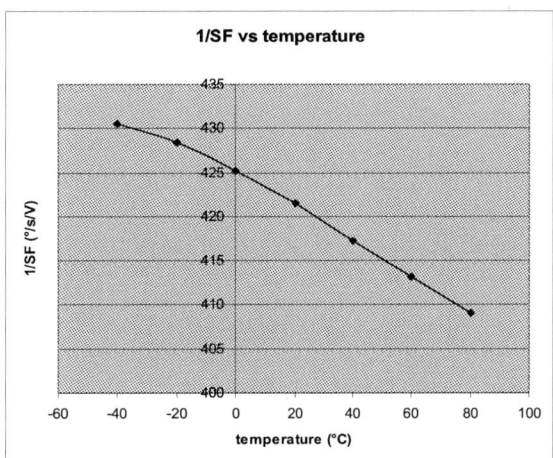

Figure 7a: simulated Scale Factor SF drift versus temperature.

The comparison of the measured and the simulated scale factor variations versus temperature (fig. 7a and 7b) are in good agreement. One can see that the error between the two drifts in the considered temperature range is less that 10%. In our model, we can consider separately the contributions of the frequency, the quality factor, and the electronics drifts versus temperature. We can conclude, in the case of our study, that the drift of the frequency versus temperature, in the considered range, is the main source of scale factor drift.

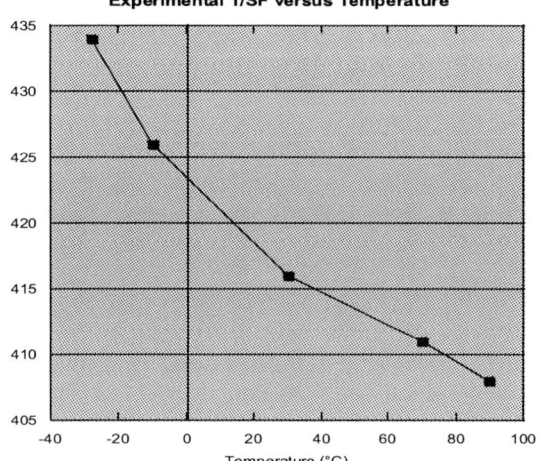

Figure 7b: experimental Scale Factor drift versus temperature.

3.3 Noise analysis

The following figures present the simulated and the measured voltage noise of the gyrometer and the associated electronics.

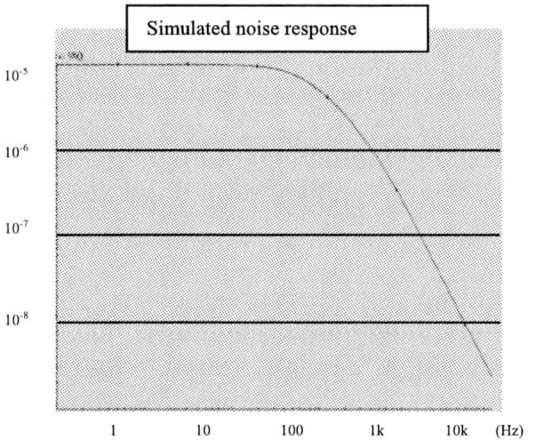

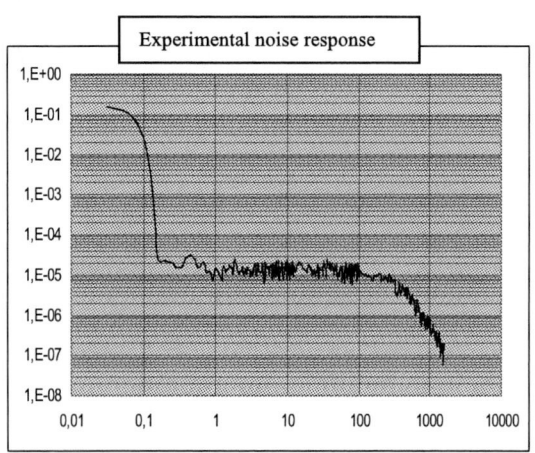

Figures 8a and 8b: Simulated and experimental noise of the gyro and associated electronics.

One can see that the 1/f noise is not considered in the model. Different simulations have been achieved without electronics part; they showed that thermal noise of the mechanical structure can be neglected compared to the electronics noise. The main source of this noise is the charge amplifier stage which must be optimized to reduce global noise.

4. Conclusion

As a principal result, we developed a model of the VIG piezoelectric vibrating gyro using Verilog-A hardware description. This model includes the mechanical and piezoelectric equations describing the gyro's sensor together with the first level surrounding electronics. The main advantage of this description is to permit the study of critical parameters influencing the gyro's performances. Three studies have been presented: output offset thermal drift, scale factor thermal drift and a first order noise analysis.

References

[1] Fedder G K and Jing Q 1999 A hierarchical circuit-level design methodology for microelectromechanical systems *IEEE Trans. Circuits Syst. II* **46** 1309–15.

[2] D. Janiaud, The VIG Vibrating Integrated Gyrometer: a new quartz micromachined sensor, Symposium Gyro Technology 2003, Stuttgart, Germany.

[3] Gabbay L D and Senturia S D 2000 Computer-aided generation of nonlinear reduced-order dynamic macromodels: I. Non-stress-stiffened case *IEEE J. Microelectromech. Syst.* **9** 262–9.

[4] Zhou G Y and Dowd P 2002 A method to include micromechanical components into system level simulation *Sensors Actuators* A **97–8** 386–97

[5] B J Gallacher *et al* 2006 *J. Micromech. Microeng.* **16** 320-331 A control scheme for a MEMS electrostatic resonant gyroscope excited using combined parametric excitation and harmonic forcing.

Validation of constitutive models for electrically conductive adhesives

Marcel Meuwissen, Monique van den Nieuwenhof, Henk Steijvers, Adri van der Waal, Tom Bots
TNO Science and Industry, PO Box 6235, NL 5600 HE Eindhoven, The Netherlands
Telephone +31 40 26 50 482 Fax +31 40 26 50 850 E-mail marcel.meuwissen@tno.nl

Abstract

By means of standard characterisation experiments, the parameters in a viscoelastic model were determined for a commercially available isotropically conductive adhesive. Next, two non-standard tests were conducted to validate the predictions of this model under conditions closer to the practical application of the adhesive. The performance of the viscoelastic model was compared to that of an elastic model. Finally, the model was used to study the thermo-mechanical performance of a photovoltaics laminate during temperature cycling. The numerical simulations predict that the original design of the PV laminate results in excessively high stresses in the adhesive interconnect which are expected to cause failure and therefore a change in design is required.

1. Introduction

Electrically conductive adhesives are frequently being applied in electronics nowadays. They offer potential advantages over solder interconnects in terms of low temperature processing, further miniaturisation, and better environmental compatibility.

When introducing conductive adhesives as solder replacements, their reliability is a key issue. In order to assess the thermo-mechanical reliability of adhesive interconnects and identify possible weaknesses in the design, a combination of physical tests and numerical simulations is commonly carried out. For performing accurate numerical simulations, constitutive models describing the response of the individual materials to thermo-mechanical loadings are required.

The current paper describes the procedure adopted for parameter identification and validation of such a model. The procedure is demonstrated for a commercially available isotropically conductive epoxy adhesive. Within this procedure two non-standard experiments are used for validation of the model. These experiments required only a minimal amount of resources and were simple to conduct, but provided valuable information about the performance of the constitutive models under conditions close to the actual application.

The quantified model was utilised to simulate the thermo-mechanical performance of a photovoltaic (PV) laminate during temperature cycling. In this particular application, solders were originally used as interconnect materials. Because of cost efficiency, adhesive interconnects are being investigated as possible solder replacements.

2. PV Module and Materials

The layout of the studied PV module is schematically shown in Figure 1. This particular module incorporates the back-contacted cell concept [1,2]. The silicon cells in the module are electrically interconnected by a backside foil consisting of conductive tracks supported by a layer of Polyethylene Terephthalate (PET) and Polyvinyl Fluoride (PVF). The PET/PVF layers also act as a barrier layer. The backside foil is connected to the cells using a silver flake filled epoxy adhesive. The cavities between the backside foil and the cells that are not occupied by the adhesive are filled with Ethylene Vinyl Acetate (EVA). The front side of the cells is also covered by a layer of EVA which is attached to a glass plate.

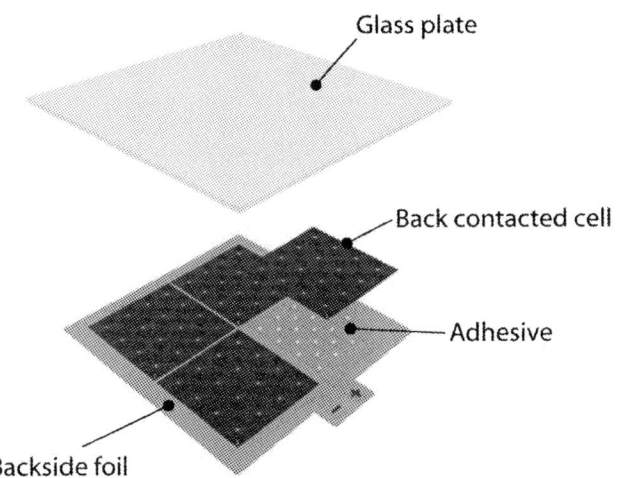

Figure 1: Layout of a PV module incorporating the back-contacted cell concept and electrically conductive adhesive interconnects.

Studies [3] have shown that more than 45% of the field failures of crystalline PVmodules have thermo-mechanical origins such as cell breakage, delamination, and interconnect breakage. Field returns are costly and hamper the acceptance of PV energy. It is therefore necessary to control the thermo-mechanical reliability of new PV module designs.

3. Experiments

From a simulation point of view, the materials used in the PV module can be divided in two categories: those for which the properties are well-known and those for which the properties are not readily available. Materials in the first category are the glass plate, the electrically conductive tracks and the silicon cells. The materials in the latter category exhibit significantly more complex behaviour: the EVA, the adhesive and the PET/PVF foil.

This section illustrates the experiments carried out to characterise the behaviour of the electrically conductive adhesive. The other materials were characterised in a similar manner.

1-4244-1105-X/07/$25.00 ©2007 IEEE

3.1. Characterization Experiments for the Adhesive

Test samples of the adhesive were prepared by pouring uncured material in beam-shaped cavities machined in a Teflon block. The samples were subsequently cured in an oven for 1 hour at approximately 150°C. The curing time was longer than advised on the datasheet to ensure that the material is fully cured. The dimensions of the samples are approximately 30×2×5mm³.

After preparation, the sample was mounted in a TA Instruments DMA 2980 as shown in Figure 2.

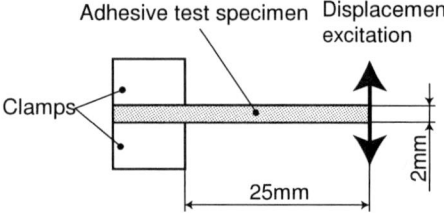

Figure 2: Set-up of the characterisation experiments.

The sample is clamped on one end and a harmonically varying lateral displacement is imposed on the free end. The required force as a function of time is monitored and this information is used to determine the modulus of the material. Many polymeric materials – such as the adhesive tested here – exhibit so-called temperature dependent viscoelastic behaviour, *i.e.* the modulus is a function of time and temperature. For certain classes of materials, this behaviour is conveniently characterised by measuring the modulus using the experiment described above and varying the temperature of the sample and the frequency of the displacement excitation over relevant ranges [4].

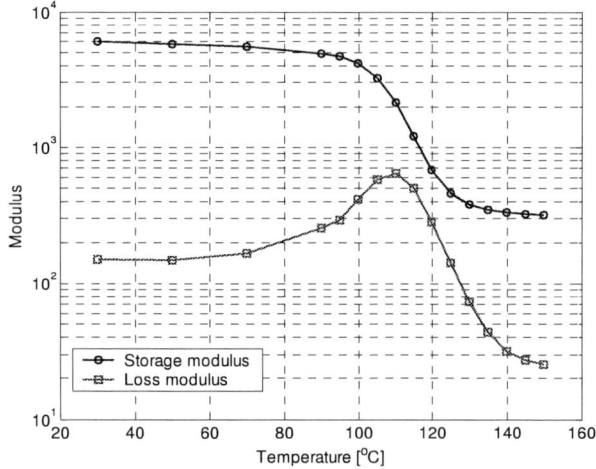

Figure 3: Storage and loss modulus as a function of temperature for a 1 Hz excitation frequency.

The modulus as measured in the experiment can be decomposed into a part associated with the elastic (recoverable) behaviour of the adhesive and a part associated with the viscous (nonrecoverable) behaviour.

These parts of the modulus are termed the storage modulus and loss modulus respectively.

Typical results of the experiment are shown in Figure 3 and Figure 4. Figure 3 shows the storage and loss modulus versus temperature for an excitation carried out at a 1Hz frequency. Using the peak in the loss modulus as a definition of the glass transition temperature [5], this temperature is estimated at 110°C.

Figure 4 shows the mastercurve of the material at 110°C. A linear viscoelastic model is fitted on the data (solid lines in this Figure). A 20 mode Maxwell model [6] is used here. The model fit agrees well with the measurement data.

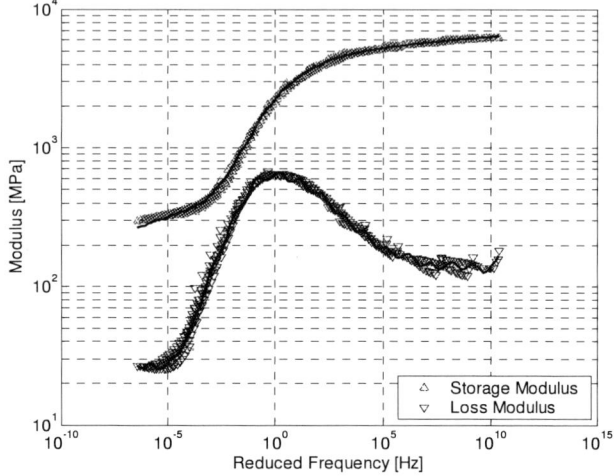

Figure 4: Comparison between model fit (solid lines) and measurements for storage and loss modulus at 110°C for different frequencies.

For the temperature shift factor, the WLF equation [6] is used above a reference temperature and an empirical equation below this reference temperature. A comparison between the measured shift factor and the model fit is shown in Figure 5.

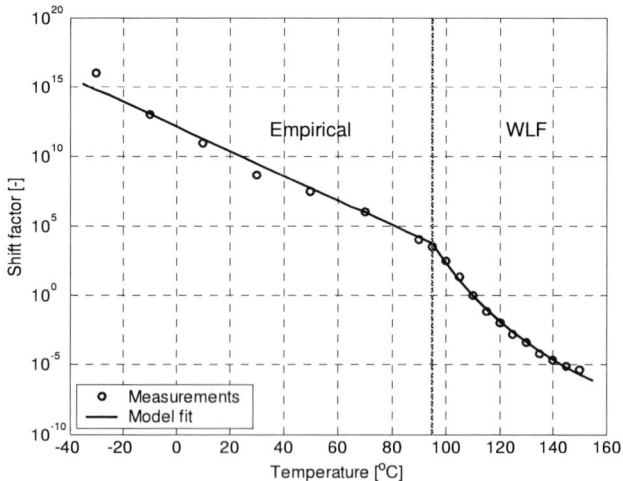

Figure 5: Measured temperature shift factor and model fit.

As a first validation of the model, additional experiments are carried out that differ slightly from the experiments described above. The same test setup is used as before (see Figure 2), but instead of imposing a harmonic excitation, relaxation experiments are carried out at different temperatures. The sample is subjected to a bending strain of 0.1% and this strain is maintained for 1 hour while the development of the stress is monitored. Next, the stress on the sample is released. This situation is maintained for an hour as well. This cycle is repeated several times.

The results are shown in Figure 6 and Figure 7 for two temperatures: -20°C and 120°C. In addition, these figures show the prediction of the viscoelastic model as fitted on the data of the previous experiment.

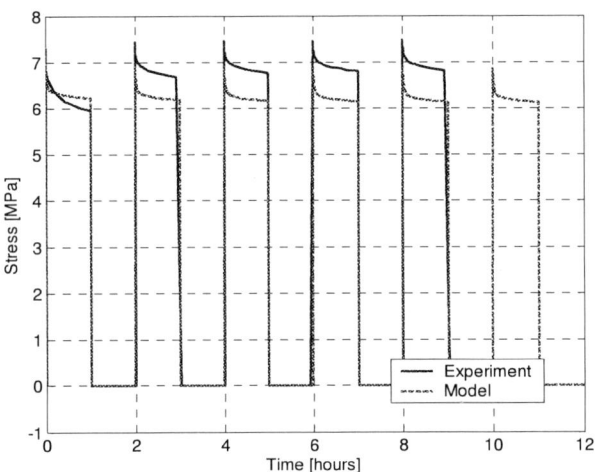

Figure 6: Comparison between model predictions and experiments for a repetitive relaxation experiment at -20°C.

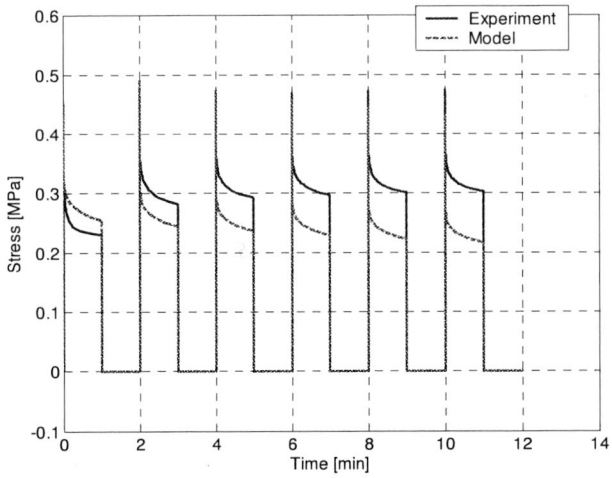

Figure 7: Comparison between model predictions and experiments for a repetitive relaxation experiment at 120°C.

The model predictions show a reasonable agreement with the measurements. There is a difference in absolute stress levels of about 10-15%. Clear stress relaxation is observed in the experiment which is also predicted by the model.

3.2. Validation Experiment 1

The first validation consists of an adhesive-on-strip experiment. In this experiment a thin steel strip is used on which an adhesive layer is applied which is subsequently cured at 150°C. After curing, the sample allowed to cool down in air to room temperature.

The sample is schematically shown in Figure 8 in top and side view. The steel strip has a thickness of 0.1mm and the thickness of the adhesive layer is approximately 0.12mm.

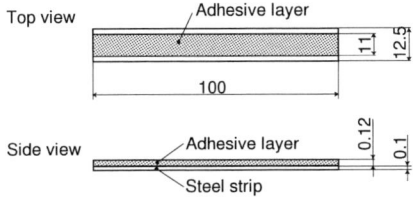

Figure 8: Dimensions of the adhesive on strip sample. All dimensions are in mm.

The sample is mounted in an oven as shown in Figure 9. The temperature in the oven is varied over time. This leads to deflection of the sample due to the difference in coefficient of thermal expansion of the adhesive and the steel strip. The amount of deflection is determined – among other factors – by the properties of the adhesive.

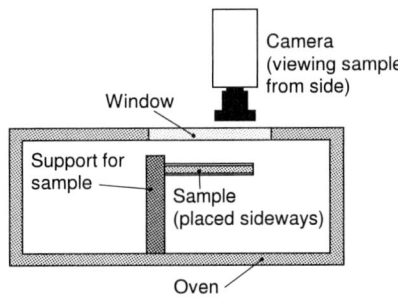

Figure 9: Set-up for measuring the deflection during temperature changes.

One end of the strip is clamped by a support. The strip is mounted sideways in order to cancel out the influence of gravity, *i.e.* deflection of the sample takes place perpendicular to the plane of drawing in Figure 9. The deflection is measured optically using a digital camera placed outside the oven and viewing the side of the sample through a window. The images of the sample are processed digitally to determine changes in deflection of the sample's free end. The temperature in the oven is monitored by a thermo-couple that is placed close to the sample.

Experiments are carried out for two oven temperature profiles as shown in Figure 10. For the first temperature profile, the maximum temperature in the oven is about

100°C and for the second profile, the maximal temperature is nearly 150°C. Additional experiments have been carried out with thermo-couples placed on the samples as well. The sample temperature turns out to differ only a few degrees from the temperature measured in the oven. The temperature variations over de sample were also within a few degrees. In the actual experiments, no thermo couples were placed on the samples, in order to avoid possible disturbances.

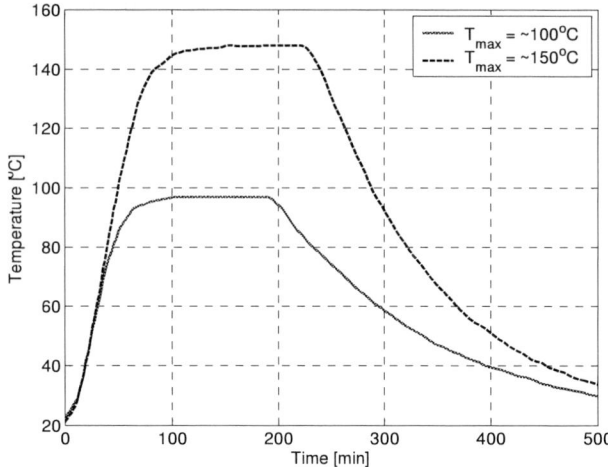

Figure 10: Measured temperature profiles in the oven for the strip bending experiments.

The measured deflection change for the experiment in which the oven temperature attains 150°C is shown in Figure 11. The deflection change is defined as the total deflection of the strip minus the deflection at the start of the experiment at room temperature (approximately -20mm).

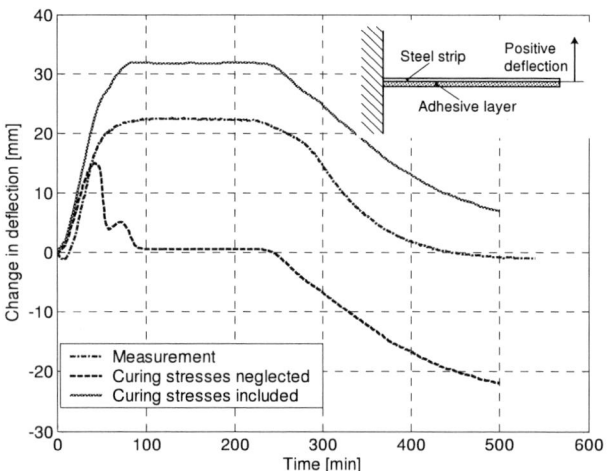

Figure 11: Comparison between measured and calculated deflection change due to temperature changes. The maximum oven temperature is nearly 150°C.

The deflection of the strip was calculated using a generalised plane strain (2D) model implemented in the finite element code MSC.Marc [7]. This model incorporates the viscoelastic model of the adhesive as determined in the previous section. For the steel strip, a linear elastic material model is adopted with parameters taken from literature.

The results of these simulations are also shown in Figure 11. Two different situations are simulated. In the first simulation, the stresses that developed during the initial cool down from curing temperature to room temperature were neglected and in the second prediction these stresses were taken into account. The inclusion of curing stresses clearly results in more accurate predictions of the observed deflections. Nevertheless, even for the simulations with curing stresses included, there are still remarkable differences between the experiments and the models. Possible causes for these deviations are:

- The curing profile used in the characterisation experiments (for determining the model parameters) differs from the profile used for curing the adhesive on the strip material. This will influence the mechanical properties of the cured adhesive.
- In the model, the stress build-up due to curing is only approximately taken into account. No full analysis of the stress-development during adhesive cure is made.
- The sample temperature used as input in the simulations is the measured oven temperature. It has been observed from additional experiments that the sample temperature is a few degrees below the oven temperature and not uniform over the strip.
- The thickness of the adhesive layer is measured at a few discrete points and these values are used in the model for defining the adhesive layer thickness. The deflection of the free strip end is strongly dependent on this parameter.
- The model neglects possible chemical and physical ageing processes that might be developing in the adhesive.

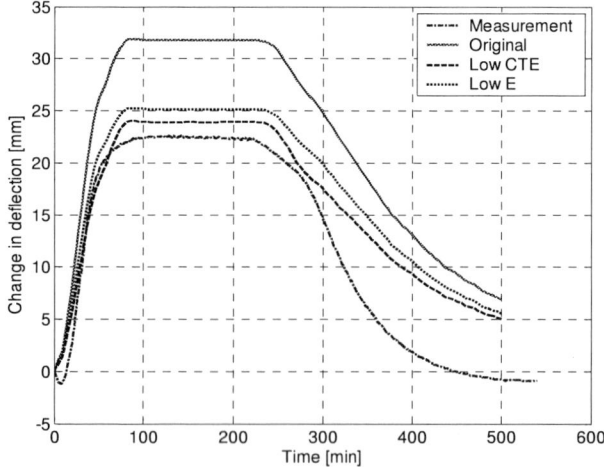

Figure 12: Comparison between measured and calculated deflection change for the original simulation and two additional simulations with varied parameters. The maximum oven temperature is nearly 150°C.

To determine the influence of the parameters in the constitutive model on the deflection of the sample's free end, additional simulations are performed: a simulation in which the modulus of the adhesive is reduced to 75% of its original value, and a simulation in which the coefficient of thermal expansion is halved. The results are shown in Figure 12.

The levels of deflection predictions for the varied parameters are closer to the measurements. It is however unlikely that the coefficient of thermal expansion of the material deviates a factor two from its specified value. A deviation of about 25% in modulus is more likely. Nevertheless, even for the varied simulations, the behaviour during cooldown still differs from the measured behaviour. The reason for this is unclear.

Figure 13 compares a model incorporating linear elastic behaviour for the adhesive to the measurements and the original model incorporating a viscoelastic constitutive model. For the linear elastic model, a temperature dependent modulus is used which is derived from the measurements.

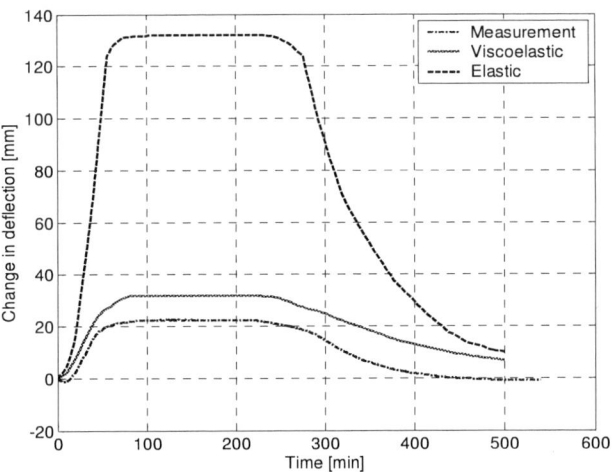

Figure 13: Comparison between measurements, the original model including viscoelastic behaviour for the adhesive and a model incorporating an elastic model. The maximum oven temperature is nearly 150°C.

The elastic model clearly predicts stronger deflections than observed in the experiments. This is due to the assumptions in the elastic model. The model does not account for any relaxation of the strains built up during cooling down from cure temperature to room temperature. The use of elastic models for describing the behaviour in such cases should thus be treated with great care.

3.3. Validation Experiment 2

In the second validation experiment, a sample is used that more closely resembles the actual application of the adhesive in the PV module (see Figure 14).

The sample consists of a 320 μm thick silicon wafer attached to a 330 μm thick backside foil [8]. The backside foil is a layup of aluminium, PET, and PVF. In between the wafer and the backside foil are two penny shaped adhesive interconnects (thickness 200μm, diameter 3mm) surrounded by EVA material [9].

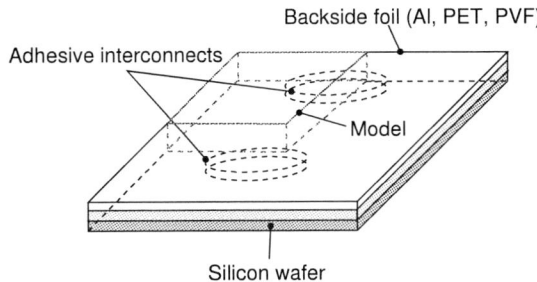

Figure 14: Sample used in the second validation experiment.

The lateral dimensions of the backside foil, wafer adhesive interconnects as well as the location of the interconnects are indicated in Figure 15.

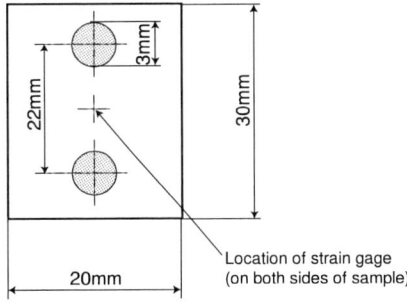

Figure 15: Characteristic lateral dimensions of the sample and location of the adhesive interconnects.

The samples are assembled and the adhesive and the EVA material are cured in an oven at elevated temperature. Strain gages [10] are attached to both the wafer and backside foil at the location shown in Figure 15. The strains are measured in the direction parallel to the long side of the sample. The samples are placed in an oven such that they can bend freely. They are heated from room temperature to about 130°C and cooled down again to room temperature. The complete temperature cycle takes about 5 hours.

The measured strains on the two sides of the sample are shown in Figure 16. Two typical measurements are shown: one in which a multi-crystalline wafer is used and another in which a mono-crystalline wafer is used. The applied strain gages are specified to have a deviation of less than $\pm 1.8 \cdot 10^{-6} \text{°C}^{-1}$ which corresponds to $\pm 1.8 \cdot 10^{-4}$ over the covered temperature range of approximately 100°C.

Some hysteresis appears to occur, but part of this may be caused by the limited accuracy of gages as an apparent hysteresis was also observed in similar experiments carried out on samples consisting of the silicon wafer only.

Only small differences were observed between measurements on the multi-crystalline and mono-crystalline samples.

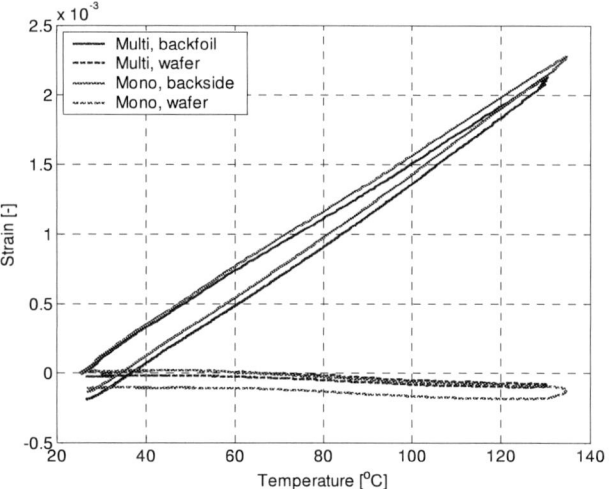

Figure 16: Example of measured total strains at the wafer side and the backside.

A model of the experiment was implemented in the finite element code MSC.Marc. Because of symmetry only a quarter of the sample was modeled (see Figure 14). This model incorporated the viscoelastic model as determined earlier for the adhesive. For the EVA material, a viscoelastic model was adopted as well. The parameters for this model were determined from similar experiments as used for the adhesive characterisation. Measurements showed that the backside foil exhibits anisotropy of about 20%–25%. This anisotropy was neglected and an isotropic model was used for this material. A viscoelastic model was not available and therefore a (less accurate) elastic model with temperature dependent parameters was used. For the silicon wafer and the aluminium layer in the backside foil an isotropic linear elastic model was adopted.

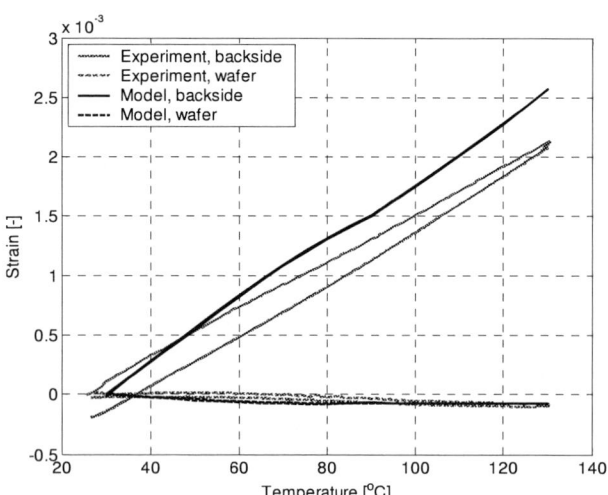

Figure 17: Comparison between measurements on the multi-crystalline sample and finite element model.

A comparison between the model and the experiment is shown in Figure 17. An acceptable agreement between the measurements and the experiments is obtained in terms of attained strain levels for both the wafer and the foil side. The hysteresis as observed in the measurements is much weaker in the predictions.

4. Simulations

A finite element model is implemented to study the behaviour of a PV laminate subjected to temperature cycling. The stresses that develop in the individual components of the laminate are investigated. The single cell laminate considered here is shown in Figure 18.

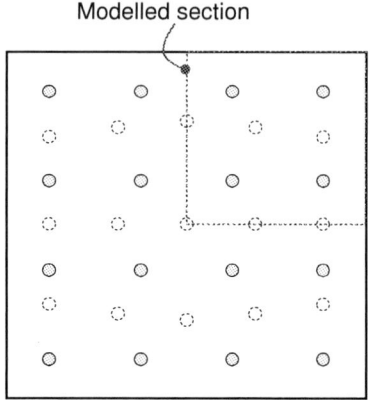

Figure 18: Top view of the single cell laminate as studied by means of numerical simulations. The circles denote adhesive interconnects between the cell and the backside foil. In total, each cell has 31 interconnects: 16 to the frontside of the cell and 15 to the backside.

A single cell has a surface of $150 \times 150 mm^2$. A total of 31 adhesive interconnects are used between the cell and the backside foil: 16 to the frontside of the cell and 15 to the backside of the cell. Each adhesive interconnect has a diameter of 3 mm and a height of 200μm. A schematic cross section of the assembly near an interconnect is shown in Figure 19.

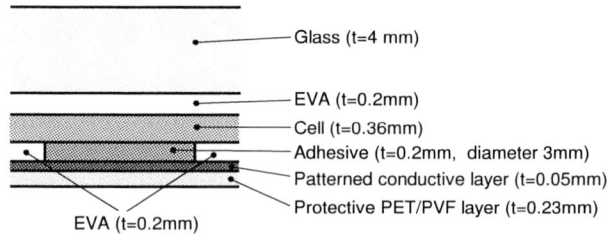

Figure 19: Schematic cross section of the single cell PV laminate. Drawing is not to scale.

Since all materials are assumed to be isotropic and due to the symmetry of the assembly, only a quarter of the single cell PV laminate is modelled as indicated in Figure 18. Because of the large difference in length scales ranging from about 150mm (length and width of the cell) to about 200μm (thickness of the adhesive), a global-local

580

approach is adopted. First a model of the quarter assembly is implemented to calculate the global deformation. Next a detailed local model is used to calculate the stresses and strains in an adhesive interconnect and its vicinity. In this local model, the results of the global model are used as boundary conditions. The approach is schematically shown in Figure 20.

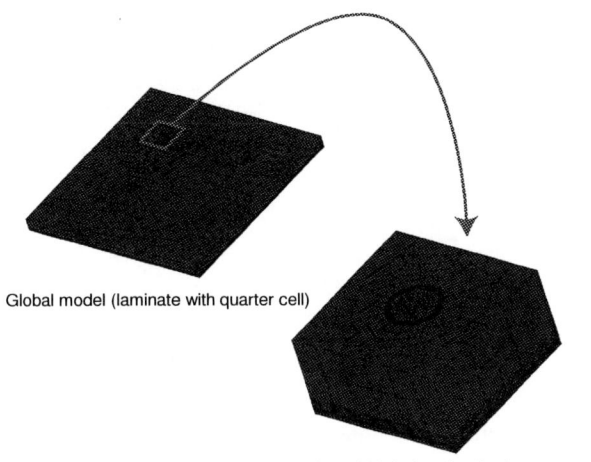

Figure 20: Global-local approach adopted for calculating the stresses in the adhesive interconnects.

The material models for the individual components are summarised in Table 1. The parameters for the glass, aluminium, and wafer are taken from literature. For the other materials, parameters were determined from measurements. For the PET/PVF in the backside foil, an elastic model was adopted, although a viscoelastic model is expected to lead to better results, but the required parameters were not available. The modulus and coefficient of thermal expansion are dependent on temperature. These parameters were determined from measurements.

Table 1: Material models used for the individual components in the model.

Material	Model
Glass	Linear elastic
EVA	Viscoelastic
Silicon	Linear elastic
Adhesive	Viscoelastic
Aluminium	Linear elastic
PET/PVF	Linear elastic, temperature dependent parameters

The imposed temperature cycle is shown in Figure 21. The maximal temperature is 85°C whereas the minimum temperature is -40°C. Prior to the temperature cycle, the cooling down from curing temperature to room temperature was also simulated in order to take into account the residual stress distribution. In all cases, a uniform temperature distribution is used.

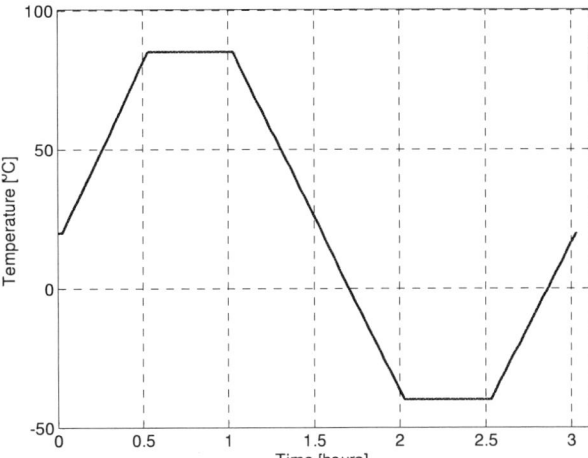

Figure 21: Imposed temperature cycle. A uniform temperature distribution is imposed.

The maximal principal stress in the adhesive interconnects as calculated by the global model is shown in Figure 22. The calculated stresses are very high, in particular in the interconnects far away from the center of the assembly. These stress levels would lead to failure of the interconnects.

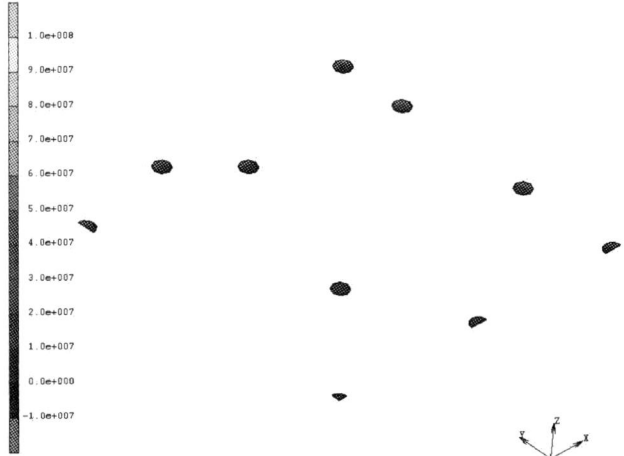

Figure 22: Maximal principal stress in the adhesive interconnect at the highest temperature in the temperature cycle (85°C).

The stress distribution in the adhesive as calculated in more detail by the local model is shown in Figure 23, whereas Figure 24 shows the development of the maximum principal stress in the most critical region of the interconnect as a function of time. The calculated peak stresses are extremely high.

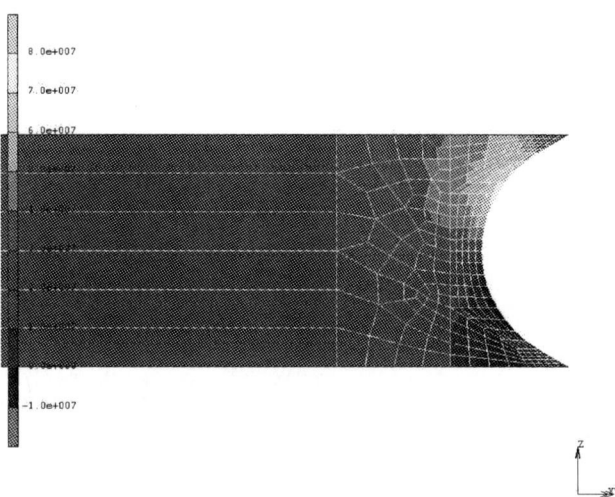

Figure 23: Maximum principal stress in a cross section of the adhesive furthest away from the center of the assembly at the highest temperature in the temperature cycle (85°C).

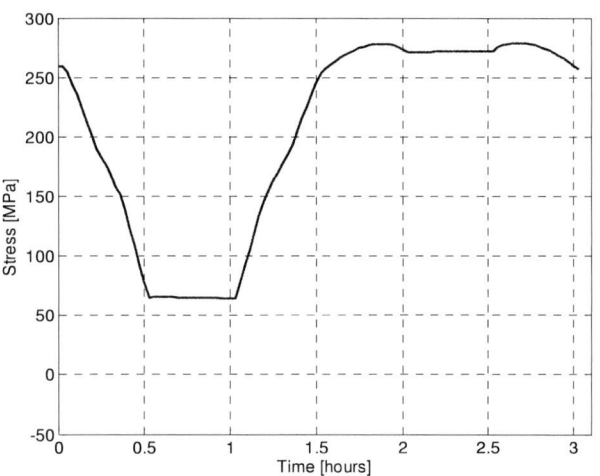

Figure 24: Maximum principal stress in the most critical region of the adhesive as a function of time.

It is expected that a more rigid EVA material would relieve the stresses in the adhesive. Moreover, a more compliant adhesive material is expected to allow for stress reductions. These alternatives are currently being studied.

5. Conclusions

The thermo-mechanical behaviour of a commercially available electrically conductive epoxy adhesive has been determined from standard experiments. The performance of constitutive models was validated in two additional experiments. These two experiments were relatively simple to conduct and provided valuable information about the behaviour of the adhesive under more practical conditions.

It was shown that a viscoelastic model for the adhesive resulted in fair agreement between calculations and measurements. In addition, the influence of the stress

build up during cure had to be taken into account to obtain more accurate predictions of the measurements. Furthermore, a linear elastic model may result in inaccurate results in some cases.

The models were applied to determine the stresses that develop in a PV module with back-contacted cells. The model predicted excessively high stresses, and this configuration would most likely fail. Design changes aimed at reducing the stresses in the adhesive are currently being investigated.

Acknowledgments

The authors acknowledge the financial support of the Dutch programme office for Economy, Ecology, and Technology (EET), which is an initiative of the Ministry of Economic Affairs, the Ministry of Education, Culture and Sciences and the Ministry of Housing, Spatial Planning and Environment, and the financial support of the Point One project MEMSLand (www.point-one.nl).

References

1. Bultman, J.H., et al., "Fast and easy single step module assembly for back-contacted c-Si solar cells with conductive adhesives", 3rd WCPVSEC, Osaka (2003).

2. De Jong, P.C., et al., "Single-step laminated full size PV modules made with back contacted mc-Si cells and conductive adhesives", 19th EPVSEC, Paris (2004).

3. Wohlgemuth, J.H., "Long term photovoltaic module reliability", *NCPV and Solar Program Review Meeting*, NREL/CD-520-33586 (2003), pp. 179–182.

4. Ward, I.M., Sweeney, J., The Mechanical Properties of Solid Polymers, John Wiley & Sons (Chichester, 2004).

5. Cadenato, A., et al., "Determination of gel and vitrification times of thermoset curing process by means of TMA, DMTA and DSC techniques TTT Diagram", J. Thermal Anal. Vol. 49 (1997), pp. 269–279.

6. Macosko, C.W., Rheology, principles, measurements, and applications, Wiley-VCH (New York, 1994).

7. MSC.Software, http://www.mscsoftware.com.

8. Icosolar 3316, supplier: ISOVOLTA AG, Austria, www.isovolta.com.

9. EVA Film VISTASOLAR, supplier: ETIMEX Primary Packaging GmbH, Martin-Adolff-Str. 44, D-89165, Dietenheim, Germany.

10. Kyowa Electronic Instruments Co. Ltd., 1-22-14, Toranomon, Minato-ku, Tokyo, 105-0001, Japan, http://www.kyowo-ei.com

Simultaneous Measurement of Young's Modulus and Damping Dependence on Magnetic Fields by Laser Interferometry

A.L. Morales, A.J. Nieto, R. Moreno, A. González, J.M. Chicharro, P. Pintado
Universidad de Castilla – La Mancha (E.T.S.I. Industriales, Área de Ingeniería Mecánica)
Avda. Camilo José Cela, s/n. Edificio Politécnico. 13071. Ciudad Real (Spain)
AngelLuis.Morales@uclm.es, +34 926 295 300 Ext. 3838

Abstract

The main objective of this work is to characterize the dependence on applied magnetic field of both Young's modulus and damping due to aspects relative to magnetostriction and domain theory. The studied samples are ferromagnetic slender rods longitudinally placed inside a straight solenoid with a pair of Helmholtz coils designed to achieve a homogeneous field along all the longitude of the sample. The longitudinal vibration in the sample is induced by a brief impact perpendicular to the base of the cylinder with a pendulum with a quartz sphere, and its vibration velocity is measured by a compact laser vibrometer based on Doppler Effect. The signal acquired is processed by a software developed in Matlab environment obtaining the vibration frequency (related to the Young's modulus) and the logarithmic decrement. That experimental method has been applied to nickel rods, obtaining maximum variations about 2.5% in Young's modulus and 156% in logarithmic decrement, and offering important advantages over other methods like lack of interaction with the sample, high accuracy, rapidity, no destruction of the sample after the test and possibility of checking small size specimens such as wires.

1. Introduction

When an object is set into vibration due to a stress pulse travelling through the material, Young's modulus represents the linear relation between the perpendicular stress and its elastic deformation whereas damping is referred to the gradual extinction of the oscillation due to the conversion of elastic energy into heat. It is well known that both properties have an intimately dependence on the internal structure of a material, so it is easy to suspect that they will also depend on magnetic interaction externally induced due to the essence of electrons and protons of every element [1].

Magnetoelasticity is the phenomenon in which elastic and magnetic materials change their dimensions and elastic properties depending on their magnetic state and vice versa. In spite of any material presents this effect, only ferromagnetic materials change its properties in an outstanding and noticeable way thanks to the unique existence of magnetic domains inside of them [2,3].

Magnetic materials, and specifically ferromagnetic ones, are widely used in a huge amount of industrial and engineering applications. Elements as different as transformers, electric engines or sensors use this kind of materials in presence of external magnetic fields, so effects like the change in dimensions, elasticity,

resonance frequencies and damping must be characterized and taken into account in order to achieve optimum results.

The dependency of elastic modulus upon magnetic field applied (the so-called ΔE-effect) is mainly measured using resonance-antiresonance techniques, measurement of ultrasound velocity or optical methods like heterodyne speckle interferometry. On the other hand, there are also several experimental methods for measuring damping variations (also called Δδ-effect). Some of them built a resonance curve with forced vibration relating its bandwidth with the internal friction and others are based on the measurement of the attenuation in free vibration of the specimen like torsional pendulum system or heterodyne speckle interferometry that has been also successfully used with this purpose [2,4,5,6,7,8,9].

The proposed method of measurement is developed with the aim of solve the most important drawbacks of the existing methods. First, and maybe the most important advantage, it is a method of simultaneous measurement avoiding the use of one different system for each investigated parameter: Young's modulus and damping coefficient. Second, it is a method based on lack of interaction with the sample, that is, not only it is a non-destructive system but also carries out the measurement without any instrument which modifies the free vibration of the sample. And the last but not the least, it is an optical method with all its inherent advantages but allowing a very fast characterization in opposition to other optical ones in which the samples require previous preparation and frequent calibrations must be done.

2. Experimental set-up description

Figure 1 shows a diagram of the experimental set-up used to carry out the tests, whose main components will be describe in detail next.

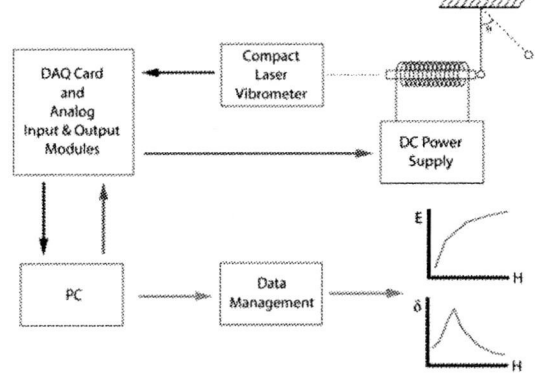

Figure 1. Diagram of the experimental set-up operation

The studied samples are slender rods of nickel, a ferromagnetic material whose magnetostrictive effect is the highest one and presents the lower saturation magnetization (484.1 emu/cm^3 at 20ºC). Their dimensions are 230 mm in length and 10 mm in diameter, but many other different dimensions would be able to test too. Nevertheless it is advisable that the samples present a ratio length-diameter higher than 10 in order to be considered slender and to show a longitudinal vibration frequency within the measure range of the instruments used. The qualifying slender is important to know the demagnetizing field generated by the sample, which is dependent on the material and its shape.

A solenoid is specially designed for this purpose and it is in charge of the generation of a homogeneous magnetic field. This solenoid, showed in Figure 2, presents in its centre a cylindrical free space of 45 mm in diameter in which the ferromagnetic rods are longitudinally placed thanks to a wooden block which fits into it. The samples are also fitted inside the mentioned wooden block in the position where the amplitude of its longitudinal vibration is minimum, that is, its centre.

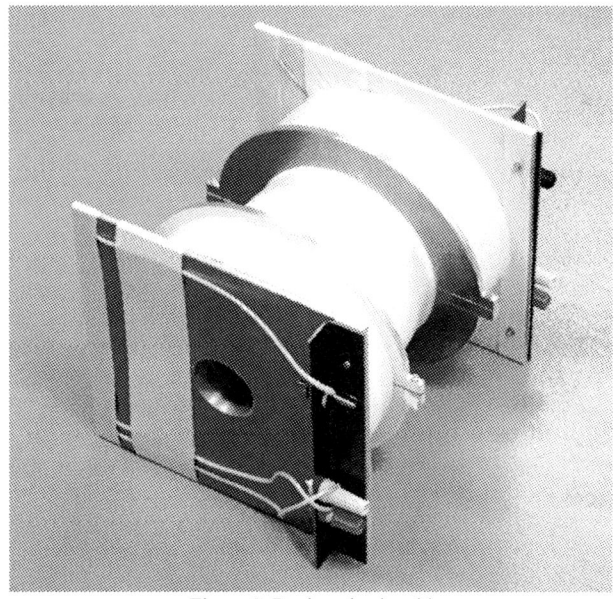

Figure 2. Designed solenoid

Nevertheless, the more important design characteristics are the magnetic ones. The solenoid must be able to generate magnetic fields high enough to saturate the nickel samples with a high homogeneity in length. Both goals are achieved superimposing the effects of two different coils: a straight one which obtains the main component of the field and a couple of Helmholtz coils which contribute to reach the desired homogeneity. The maximum magnitude of each independent source of magnetic field and their full effect are shown in Figure 3. With the full solenoid compound by the straight coil and the Helmholtz ones a maximum value of 1003 Oe is reached, which allows the sample be magnetized in an uniform way along a 60% of its length.

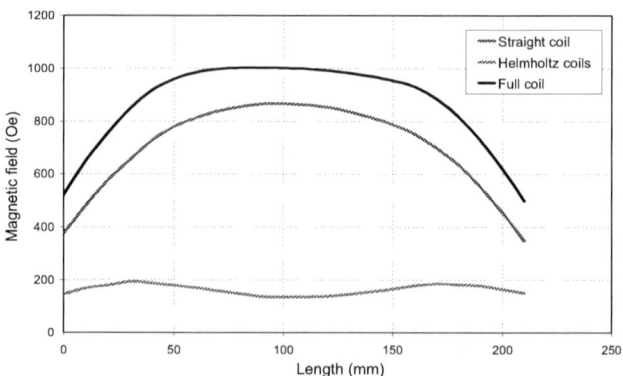

Figure 3. Experimental characterization of each configuration of coils with MG-4D Gaussmeter of Walker Scientific

The maximum value of magnetic field comes limited by a DC supply connected to the solenoid. In this case the system has a Delta SM3004-D programmable power supply of 600 W. Taking into account that the solenoid designed contributes with an electric resistance of almost 80 Ω, the voltage range required is 0-300 V and the maximum current is upper limited by 2 A. Moreover, another limitation appears while the solenoid is being powered: the time of use. Joule Effect losses increase proportionally to the square of the current and for high values of it they can not be dissipated fully by natural convection. So, the time in which the solenoid is working is limited by the time in which the maximum admitted temperature is reached. With preciseness, a dangerous temperature of almost 50ºC is reached after only 13 minutes working at maximum power.

Longitudinal vibration in the sample is induced by a brief impact perpendicular to the base of the cylinder at its centre. A pendulum with a quartz sphere which is 5 mm in diameter is used to apply the impact. This type of excitation allows the sample to vibrate freely in its longitudinal modes.

A compact laser vibrometer, Polytec CLV-1000, is used to measure the velocity of vibration of the sample at the detection point, the opposite to the excitation one, allowing a high bandwidth of 250 kHz. The measure of this instrument must be perpendicular to the vibrating surface and the detection principle is based on Doppler Effect: a laser beam with a constant carrier frequency is sent and, after being reflected by the perpendicular vibrating surface, is modulated with a modulation frequency proportional to the velocity of vibration. Finally, the measured signal, which might be filtered by a low-pass, high-pass or band-pass filter, is converted into an analogical one.

The final step of the experimental set-up is to develop a data acquisition and signal conditioning system which allows the user to control all the measurement process and the subsequent proper management of data. That goal can be divided into two necessary parts: hardware devices and software programation.

From the point of view of hardware requirements to provide the connexion between the PC and both laser and

DC power supply, a National Instruments data acquisition card DAQCard6062E is used. This device presents a higher enough sample rate of 500 Msamples/s in order to detect the highest expected longitudinal vibration frequencies fulfilling the Nyquist theorem. Furthermore, a chassis NI SC-2345 for connecting specialized analog modules of input and output data is used [10].

In order to manage the hardware devices which compound the experimental set-up and synchronize the sending and receipt of data a software developed in Matlab environment has been programmed. The Matlab programming language has been chosen to take advantage of its calculation capacity and the existence of advanced toolboxes in the needed subjects like data acquisition, signal conditioning and signal processing. Furthermore, the useful possibility of developing a user graphical interface was well-spent. In Figure 4 the appearance of the control window with an acquired time response and its Fast Fourier Transform is shown.

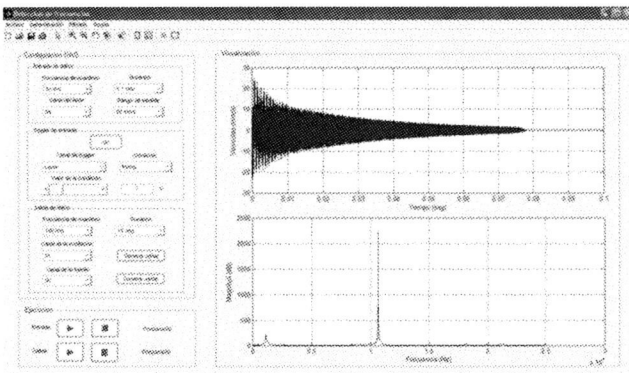

Figure 4. Control window of the software developed

Next, an overview about the whole system operation is done. Thanks to the software programmed in Matlab and the required analog output modules connected to the data acquisition system it is possible to send a suitable analog signal which set the DC power supply in the required level of current. So, the solenoid connected to the source generates a magnetic field which magnetizes the specimens fixed inside. At the same time, the compact laser vibrometer is found pointing the laser beam perpendicularly to the base of the cylindrical sample, but analog input modules connected to the PC does not acquire any piece of information until a trigger has happened. That trigger is prompted thanks to a brief impact, also perpendicular to the base of the cylinder but opposite to the measurement side, which is generated with the pendulum of quartz. When the data is acquired, the signal is processed obtaining the desired parameters, damping coefficient and Young's modulus. Finally the program becomes ready to start again, and the process is repeated as many times as it is necessary.

3. Experimental results

The samples studied were slender nickel rods of 10 mm in diameter and 230 mm in length. Nickel purity and mechanical treatment are the same in every tested sample.

A metallographic analysis gave a nickel purity up to 99.9%, with a density of 8912 kg/m^3, and all the rods were cold worked and heat treated with an annealing process at 900°C in a continuous furnace. Furthermore, taking into account the ratio of the length to the diameter of each kind of sample the demagnetizing effect can not be considered negligible in any case. This circumstance involves that magnetic poles appear in the ends of the magnetized sample generating a field H_d which is opposite in direction to the externally applied one H_{app} and whose magnitude depends on both magnetization M and a factor N_d which characterizes the geometrical dimensions of the sample. So, the effective magnetic field H_{eff} inside the sample is given by the expression

$$H_{eff} = H_{app} - H_d = H_{app} - N_d \cdot M \qquad (1)$$

and the demagnetizing factor can be calculated as a function of the previously mentioned ratio of the specimen r. The ratio of the studied samples is $r = 23$, leading to a demagnetizing factor of $N_d = 0.0674$ [1].

Figure 5 shows a temporal response belonging to an unmagnetized nickel rod in free longitudinal vibration and Figure 6 its corresponding Fast Fourier Transform.

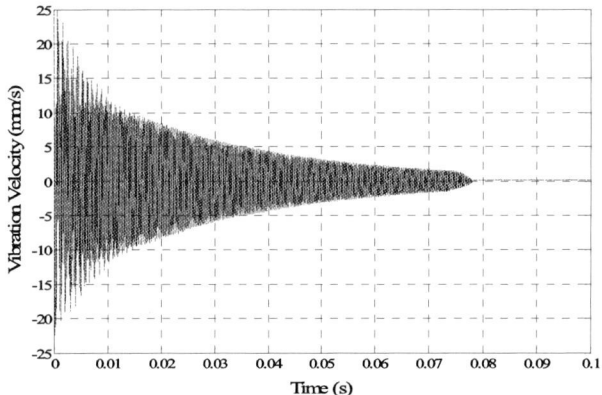

Figure 5. Temporal response of an unmagnetized nickel rod

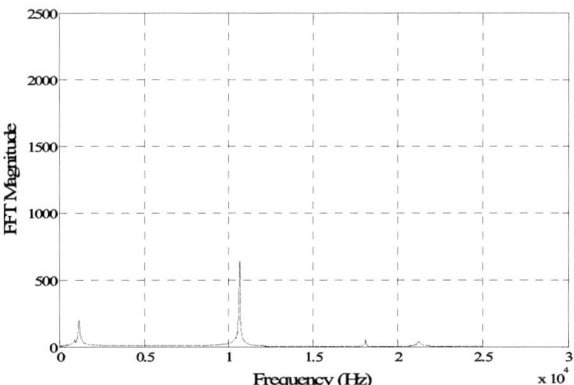

Figure 6. FFT of a rod without magnetic field applied

The previous frequency spectrum can be used to detect the damped frequency corresponding to the longitudinal vibration mode and to recognize other possible undesired oscillation frequencies which appear as a consequence of an imperfect exciting impact. In order to isolate the appropriate compound, a band-pass filter is applied around the corresponding longitudinal frequency. Once the signal has been properly managed, the damping can be obtained taking into account that, from vibration theory, the envelope of the vibration is a decreasing exponential function. Exactly, the expression for theoretical vibration can be written as follows

$$u_z(t) = \left(A \cdot \cos 2\pi f_{dl} t + B \cdot \cos 2\pi f_{dl} t \right) \cdot e^{-\frac{\gamma}{2}t} \quad (2)$$

where u_z is the longitudinal displacement, t is the time, A and B are constants, f_{dl} is the damped longitudinal frequency and γ is the desired attenuation constant. So, relating those theoretical parameters with the mathematical ones, which are calculated as a result of a fitting process, the attenuation constant is obtained. Figure 7 shows one example of a filtered signal and its described decreasing exponential envelope [9].

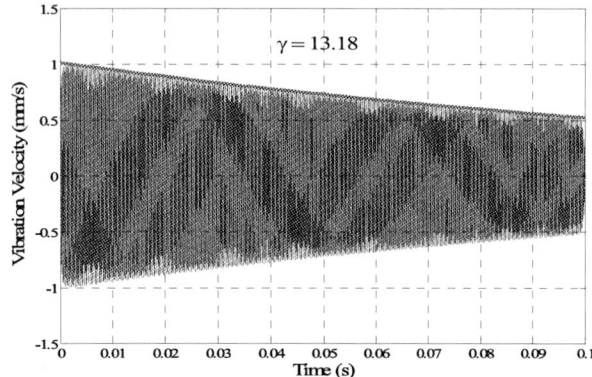

Figure 7. Exponential envelope for an acquired signal

On the other hand, attending to the longitudinal vibration theory of a beam, it is possible to relate the natural vibration frequency f_{nl} with other geometrical and mechanical parameters following the theoretical expression

$$f_{nl} = \sqrt{\frac{n^2}{4l^2}\left(\frac{E}{\rho}\right)} \quad (3)$$

where n is the order of the longitudinal mode used among all the longitudinal ones considered, ρ the material density, l the length of the rod and E the expected Young's modulus. Although the value of the natural frequency is not known, it can be easily obtained thanks to its relation with the damped frequency f_{dl} and the attenuation constant γ, measured from the FFT and the fitting process respectively. Such relation is given by the expression

$$f_{nl} = \sqrt{f_{dl}^2 + \left(\frac{\gamma}{4\pi}\right)^2}. \quad (4)$$

With all those theoretical considerations the desired ΔE-effect and Δδ-effect can be easily achieved. Starting with the damping, the logarithmic decrement δ is required. This parameter is define through the expression

$$\delta = \frac{1}{k}\ln\left(\frac{A_i}{A_{i+k}}\right) \quad (5)$$

where A_i and A_{i+k} are the amplitudes for the oscillation i and $i+k$ respectively, which can be also related to the attenuation constant γ and the longitudinal natural frequency f_{nl} through Eq. (2) as follows

$$\delta = \frac{\gamma}{2f_{nl}}. \quad (6)$$

Below, in Figure 8, the evolution of the logarithmic decrement depending on the magnetic field applied is shown. Such Δδ-effect can be divided into two different stages. In the first one, a fast increase of the logarithmic decrement, about 156%, happens until the magnetic field reaches a value near to the technical saturation field, in which the magnetostriction becomes forced and not spontaneous. The second stage consists of a continuous and gradual reduction of logarithmic decrement, achieving a variation about 91% with regard to the maximum value. The full variation between the unmagnetized state and the magnetized one turns out to be a decrease about 77%.

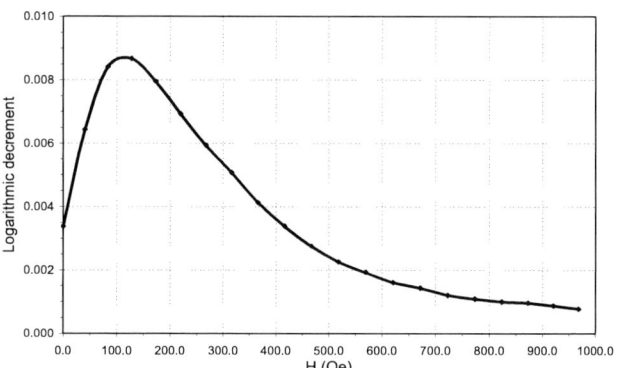

Figure 8. Logarithmic decrement dependence on magnetic field.

The final step of the developed study consists of determining the Young's modulus of the sample from natural frequency. Operating Eq. (3), and taking into account the relation between natural and damped frequencies shown in Eq. (4), the next expression for the Young's modulus can be written

$$E = 4\rho l^2 n^{-2} f_{nl}^2 = 4\rho l^2 n^{-2}\left(f_{dl}^2 + \frac{\gamma^2}{16\pi^2}\right). \quad (7)$$

After making all the needed tests with different applied magnetic fields in any case, the corresponding values of Young's modulus and measured damped frequencies are shown in Figure 9. Such results follows a rising tendency, but two different zones can be again distinguished. The first one is characterized by a fast increase owing to the spontaneous magnetoelasticity whereas the second one happens when the technical saturation has been reached and matches with lower variations. The full ΔE-effect measured was about 2.6% but percentage variations higher than 1.5% can be obtained into a low range of applied magnetic field [11].

586

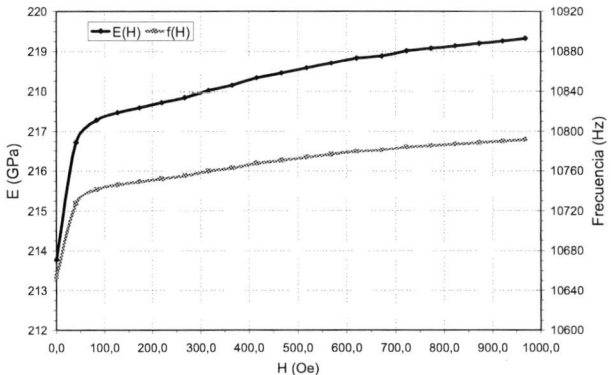

Figure 9. Young's modulus and damped frequency dependence on magnetic field.

Finally, it is important to mention that the structural damping coefficient in nickel, and in general in any metals, is very low. This characteristic means that the attenuation constant can be considered negligible and so the undamped frequency equal to the damped one. Under this assumption the Eq. (7) can be replaced by the next easier and faster expression which can be used with the only information obtained from the experimental frequency spectrum.

$$E \simeq 4\rho l^2 n^{-2} f_{dl}^2 \qquad (8)$$

5. Conclusions

A method for studying both damping and Young's modulus and their dependence upon magnetization is presented. The method is applied to nickel rods of 230 mm in length and 10 mm in diameter, since this material presents the greater magnetostrictive effect and the free vibration is carried out through a brief impact with a simple quartz pendulum. Nevertheless, any other ferromagnetic material can be tested, even with different sample sizes whenever it can be longitudinally excited and measured.

The main advantages of the method applied over others are: lack of interaction with the sample, high accuracy, simple preparation of the sample and rapidity. Furthermore, this method makes possible the use of different exciting systems depending on the studied sample, not only based on free vibration, but also on forced one.

Compared to other interferometric methodologies, this method based on Doppler Effect means a great improvement over those which use speckle interferometry. The main reason lies in the need to aligning frequently the sample with the laser, which leads to long experiments and few points of resolution in the experimental curves. Thanks to the proposed interferometric method lots of points can be tested without requiring the calibration process. As a consequence, unexpected results were found like the existence of a maximum in the Δδ-effect curve which was not detected in other researches [9].

Finally, it is important to notice the giant magnetoelastic effect which nickel is able to undergo: up to 2.6% with respect to ΔE-effect, which is in accordance with the main consulted sources about this effect, and 156% regarding Δδ-effect. Moreover, the main variations take place in the low range of applied magnetic field, so this smart material may take a special significance in numerous technical and scientific applications [13].

References

1. Cullity, B.D., Introduction to Magnetic Materials, Addison-Wesley Publishing Company (Florida, 1994).
2. du Trémolet de Lacheisserie, E., Magnetostriction: Theory and Applications of Magnetoelasticity, CRC Press (Florida, 1993).
3. Bozorth, R.M., Ferromagnetism, D. Van Nostrad Company Inc. (New York, 1951).
4. Hathaway, K.B. et al, "Measurement of high magnetomechanical coupling factor by resonance techniques", Journal of Applied Physics., Vol. 55, No. 6 (1984), p.1765.
5. McSkimin, H.J., Ultrasonic Methods for Measuring the Mechanical Properties of Liquids ans Solids, Physical Acoustics, Vol. 1A, W.P. Mason (Ed.). (New York, 1964).
6. Kakuno, K. et al, "A New Measuring Method of Magnetostrictive Vibration", Journal of Applied Physics, Vol. 50, No. 11 (1979), p.7713.
7. Squire, P.T., "Magnetomechanical Measurements of Magnetically Soft Amorphous Materials", Measurement Science and Technology, Vol. 5, No. 2 (1994), p.67.
8. Chicharro, J.M. et al, "Measurement of Field-Dependence Elastic Modulus and Magnetomechanical Coupling Factor by Heterodyne Interferometry", Journal of Magnetism and Magnetic Materials, Vol. 202, (1999), p.465.
9. Chicharro, J.M. et al, "Measurement of Damping in Magnetic Materials by Optical Heterodyne Interferometry", Journal of Magnetism and Magnetic Materials, Vol. 268, (2004), p.348.
10. Newland, D.E., An introduction to random vibrations and spectral analysis, Longman Group Limited (London, 1975).
11. Seto, W.W., Mechanical vibrations, theory and problems, Schaum (New York, 1964).
12. Rao, S.S., Mechanical vibrations Pearson Prentice Hall (New Jersey, 2004).
13. Ledbetter, H.M. et al, "Elastic Properties of Metals ans Alloys, I. Iron, Nickel, and Iron-Nickel Alloys", Journal of Physical and Chemical Reference Data, Vol. 2, (1973), p.531.

Validation Of Simulation Platform By Comparing Results And Calculation Time Of Different Softwares.

Hikmat ACHKAR[1], Fabienne PENNEC[1], David PEYROU[1], Mahmoud AL AHMAD[1], Marc SARTOR[2], Robert PLANA[1], Patrick PONS[1].
[1]LAAS-CNRS, 7 Avenue du colonel ROCHE, 31077 Toulouse cedex 4
[2]LGMT, 135 avenue de Rangueil, 31077 TOULOUSE
hachkar@laas.fr, Tel: +33 5 61 33 69 30, Fax: +33 5 61 33 62 08

Abstract

The need of a powerful multiphysics software, to simulate different topologies of deformable micro structures, is under investigation. The criteria for choosing the software is the precision of the results with an acceptable time of calculation to simulate many models with different variables. The existance of a new software, COMSOL 3.3, at a time linking all the physics and having a good interactive interface seemed to be a good solution. Being new, and under development, COMSOL 3.3 needed to be valid on different points. As a first step, we validated the numerical platform by comparing results of different softwares and some analytical solutions. The basic advantages and drawbacks of each software were considered for the final decision on the choice of the software that will be used. In terms of time of calculation, we found very intressting results on the speed of COMSOL to solve the problem when compared to ANSYS, while keeping the same precision of calculation.

1. Introduction

In the domain of structural mechanics, the major difficulty in design relies on, modelling MEMS having high aspect ratio, properly defining the material properties used, and well describing the geometry defined by the technological process. Generally, in MEMS simulations, we need to combine with the mechanical simulation other physical behaviours (Electrostatics, Piezoelectricity...), from where comes the need for multi-physics softwares.

For any application in MEMS, deformable micro structures using different kinds of actuation, with repetitive behaviour, are to be simulated in order to study their possible functionning. This functionning is based on designing a mechanical structure deformed piezoelectricaly, thermomechanicaly, or electrostaticaly, mainly to get into contact with another part, in order to do its function.

To facilitate our task, we need a multiphysics software offering at a time, a well developed solver to reduce the time of calculation and a facility to build parametric models (parametric geometry and parametric properties), without loosing too much accuracy on the results.

Following this need, COMSOL 3.3 appeared as a software that is capable to do piezoelectric simulations combined to mechanical structure and contact problems (without forgetting thermal and initial stress effects).

As a first step, we need to validate the results of the numerical platforms existing in our laboratory and specialy COMSOL's results, as being a new software and unable to juge it depending on its backgroung. To do so, results obtained from simulations on COMSOL 3.3 were compared to other numerical results, and to analytical results where a mathematical formulation can be applied. We first simulated the structure shown in figure:1 taken from the literature [2], on a well known and frequently used software, ANSYS 10.0, as well as on COMSOL 3.3, then we compared the results. Because of their imporatance to the functioning of MEMS, the maximum deflexion and the maximum Von Mises stress were chosen as major criterium for the comparision. As a result, the difference between both softwares was acceptable and a very good amelioration in the calculation time was obtained, what makes COMSOL 3.3 an excellent candidate in the world of simulation.

Still in the validation of COMSOL's results in piezoelectric actuation, but this time by analytical results, a simplified structure, shown in the figure:2, for which there exists an analytical solution, was simulated.

A contact model was then simulated on COMSOL 3.3, in a step to validate the results obtained on contact with an analytical model, and we obtained satisfactory results.

In a second step to weigh the usage of COMSOL for our application, we compared the capabilities and the facilities that each software offers in order to make the good decision about the choice of the convenient one. For this purpose, a table listing the advantages and disadvantages for each software was created.

This study will permit us to make the choice of the software the most efficient for our application. Both the accuracy of COMSOL in deflection and stress, and the reduced calculation time makes it very useful for us.

2. Numerical-numerical validation

What we mean by Numerical-numerical validation is the validation of the results obtained on one software by simulating it on another well known and tested one.

In numerical-numerical validation, we compared the results obtained from the simulation on ANSYS v10 and COMSOL 3.3, for the piezoelectricaly actuated structure taken from literature [2], and shown in figure:1.

The approximations applied to this model were, the neglection of the electrodes effect on the mechanical stiffness of the structure, and neglecting the ancorage effect by considering that the structure is perfectly clamped at its ends.

1-4244-1105-X/07/$25.00 ©2007 IEEE

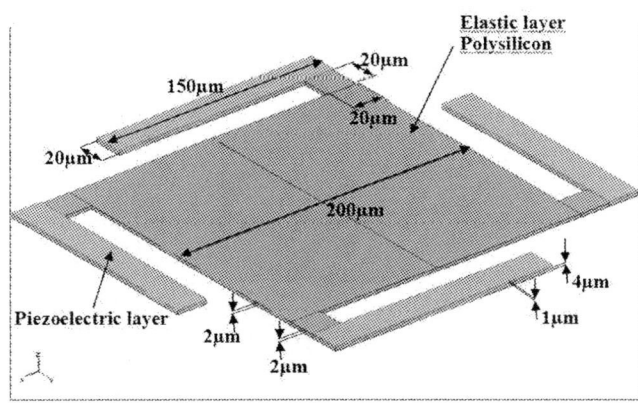

Figure:1. Dimensions of the structure simulated on, ANSYS v10 and COMSOL 3.3, for validation.

The elastic and piezoelectric properties of the material, used in the simulation on both softwares, are listed in table:1 below. The actuation voltage has been taken as 3V.

E	Young's modulus of polysilicon	162 GPa
υ	Poisson's ratio of Polysilicon	0.23
d31	Piezoelectric coupling coefficient of PZT-5H	-274 PC/N

Table:1. Material's elastic and piezoelectric properties.

After studying the effect of the mesh size, or in other words element's size, on the maximum deflection (deflection at the middle of the plate) and on the maximum Von Mises stress, we found it necessary to mesh ANSYS finer than COMSOL in order to have the same results or in other words to have convergence of the results. This refinement of the meshing concludes to a longer time of calculation. Table:2, presents below, a summary of the results and the calculation time, obtained for different mesh size and on different softwares.

Software	COMSOL	ANSYS	ANSYS
Mesh size (μm)	10	5	1.25
Max Von Mises stress (MPa)	129	73	123
Max deflection (μm)	0.8165	0.793	0.821
Calculation time (seconds)	22	33	690

Table:2. Results on deflection and time calculation for different mesh size and different softwares.

We can easily discover the difference of 40% in von Mises stress value between a coarsly and a fine meshed structure. The mesh in ANSYS was refined in order to converge the results and so that the correct stiffness of the structure was represented.

The diffrerence in terms of deflection for the two models is about 0.6% while in terms of Von Mises stress the difference reaches 5% which are acceptable offsets.

We can see clearly that the time of calculation for the two coinciding models are very far, it is mainly due to the necessity to refine the mesh which increase the number of degrees of freedom to solve for. The 30 times diffrence in solving time, makes COMSOL a favourable candidate as finite element software to be used in MEMS simulation, specially in problems where multiple models are to be studied in order to optimize the structure.

This simulation permits us, not only to validate the results of COMSOL 3.3 but also to validate the capacity of COMSOL to do piezoelectric simulations.

3. Numerical-analytical validation

What we mean by Numerical-analytical validation is the validation of the results obtained on COMSOL by analytical results.

Two models were studied for this purpose, a cantilever beam actuated piezoelicaly, and a contact model of a cylinder with a metallic block.

Cantilever model

In this validation, we compared the analytical value of the deformation for the structure shown in Figure:2 below, published in literature [1], obtained from the simplified equation Eq:1 written below and we compared it to the simulated results.

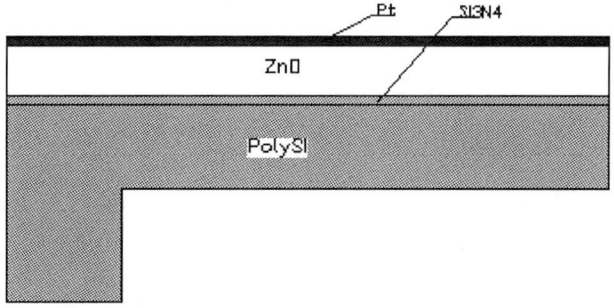

Figure:2. Cantelever structure which has analytical formulation, where the material stack is shown.

The deflection of the cantilever at the abcissa x is given by[1]:

$$\delta(x) = \frac{3t_e\left(t_e + t_p\right)E_e E_p x^2 d_{31}V}{E_e^2 t_e^4 + E_e E_p\left(4t_e^3 t_p + 6t_e^2 t_p^2 + 4t_e t_p^3\right) + E_p^2 t_p^4} \text{ (Eq:1)}$$

Where the symbols signifies:
E_e: Young's modulus of the elastic layer (polySi).
E_p: Young's modulus of the piezoelectric layer (ZnO).
t_e : Elastic layer's thickness.

t_p : Piezoelectric layer's thickness.

V: Actuation voltage between the electrodes.

d_{31}: Piezoelectric layer's coupling coefficient.

Layer	Thickness (µm)	E (GPa)	v	d_{31}(C/N)	Width (µm)
PolySi	1.7	162	0.23	no	30
Si_3N_4	0.2	290	0.28	no	26
ZnO	1	161	0.36	2.3×10^{-12}	26
Pt	0.2	250	0.25	no	26

Table:3. The properties of the multimorphe stack

The properties of the multimorphe stack are listed in Table:3 above. The young's modulus of the isolating layer and the top electrodes (Si_3N_4 and Pt) doesn't enter in equation Eq:1, since their effects are neglected due to their neglected thicknesses but they will be considered in the numerical model. The length of the cantilever was fixed to 500µm in all our analysis.

For the 1µm ZnO thikness, the publication [1] indicates that the analytical deflection is around 0.23µm and the experimental deflection is around 0.25µm. Comparing COMSOL's solution of 0.2349µm, as shown in Figure:3, to the analytical results and the experimental published results [1] we found a difference of 0.2% and 0.6% respectively. This simulation shows accordance in the results between COMSOL's solution from one side and the published experimental data and analytical data from the other side.

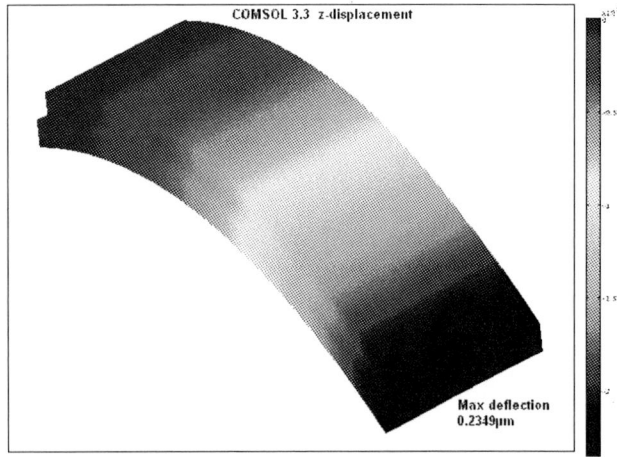

Figure:3. Deflexion result as simulated on COMSOL3.3

Contact model

This model of a metal-to-metal contact simulated with COMSOL Multiphysics had a goal to validate the contact results by analytical values obtained from the formulation of Hertz.

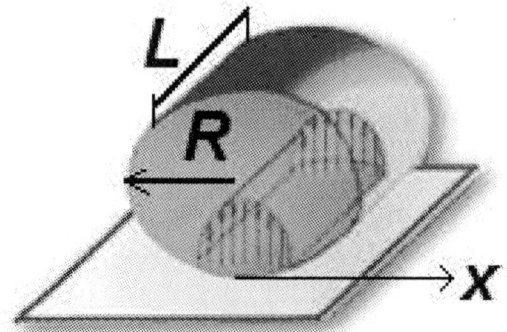

Figure:4. The contact model between a plate and a cylinder formulated by Hertz.

The numerical model consisted of a contact between half a cylinder and a metallic block as shown in Figure:5. It was simulated in order to study the maximum contact pressure and the contact length, then using Hertz's formulation Eq:2, we compared the analytical maximum pressure with the simulation.

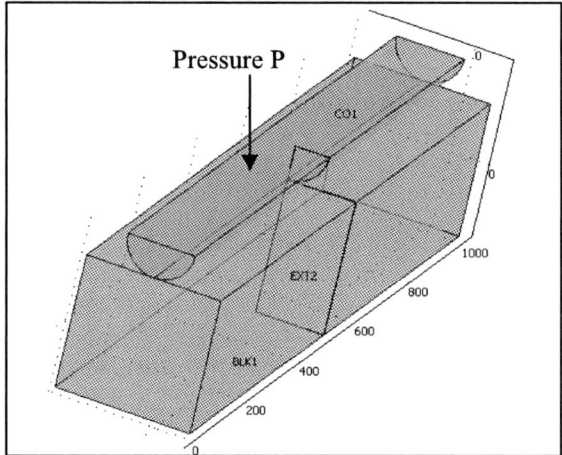

Figure:5. Numerical contact model as built in COMSOL 3.3.

$$P_{max} = \frac{2F}{\Pi a^2} = \sqrt{\frac{FE^*}{\Pi R}} = \sqrt{\frac{2PE^*}{\Pi}} \quad (Eq:2)$$

Where E*, the combined elastic modulus is defined by

$$\frac{1}{E^*} = \frac{1 - v_1^2}{E_1} + \frac{1 - v_2^2}{E_2} \quad (Eq:3)$$

The contact length is given by:

$$a = \sqrt{\frac{4FR}{\Pi E^*}} = \sqrt{\frac{8PR^2}{\Pi E^*}} \quad (Eq:4)$$

a : is the contact length in the x-axis direction.

F=PxA. is the force applied on the cylinder.

P: is the pressure applied to the cylinder.

P_{max}: is the maximum pressure (at x=0).

R: is the radius of the cylinder. Taken 50 mm.

E_1 (MPa)	Young's modulus material 1	70000
E_2 (MPa)	Young's modulus material 2	210000
v_1	Poisson's ratio material 1	0.42
v_2	Poisson's ratio material 2	0.3
P (MPa)	Applied pressure	500

Table:4. Modeling parameters for the contact model.

The assumptions are that both materials, for the cylinder and the metallic block, are homogeneous, elastic and isotropic.

So maximum pressure of 4474 MPa obtained by COMSOL 3.3 is in good agreement with the analytical maximum pressure of 4447MPa within a difference of 0.6%.

As for the contact length, COMSOL evaluated it as 7.22 mm while the analytical solution gave 7.16mm, that is a difference of 0.8% which is a very acceptable.

Now the contact pressure as a function of x can be expressed as:

$$P_c(x) = P_{max} \sqrt{\left(1 - \left(\frac{x}{a}\right)^2\right)} \quad \text{(Eq:5)}$$

The following graph shows the distribution of the contact pressure along the surface of contact (in the x direction), for both COMSOL and analytical solution.

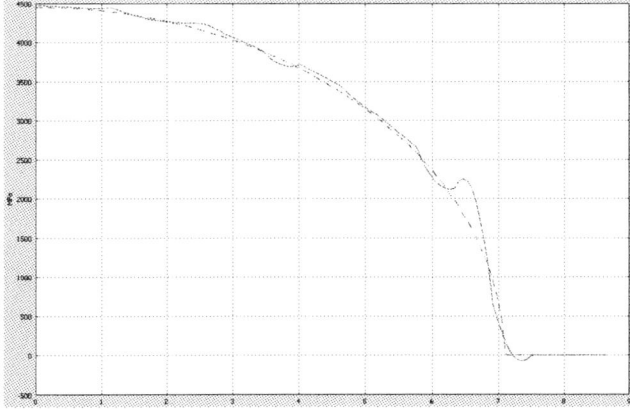

Figure:6. Graph comparing the pressure distribution of analytical and numerical models.

The dotted line showing the analytical values and the continuous one showing the simulated values. Both graphs are superposing and once more validating the contact simulation of COMSOL 3.3.

4. Comparing the capabilities of each software

As a final step, and to conclude to the goal of our study, a brief comparision of the capacities of each software was summarized in Table:5 at the end of the paper.

The major advantages of COMSOL 3.3 is its capacity to couple many different physics at a time as well as the capacity of building a parametric model using COMSOL script. It is to note that routines in COMSOL script are ready for optimization purposes.

One of the drawbacks is its limitation in terms of memory and in solving huge models.

5. Conclusions and perspective

For our application, deformable micro structures are to be simulated. To do so, we need a multiphysics software offering at a time, a well developed solver to reduce the time of calculation and facility to build parametric models, without loosing accuracy on the results.

As a first step, we needed to validate the results of the numerical platforms existing in our laboratory. A numerical-numerical validation was done, where the results of COMSOL 3.3 were compared to those obtained on ANSYS 10. The difference in terms of deflection was 0.6% while in terms of Von Mises stress it was around 5%. It is an acceptable offset, except that the simulation on COMSOL 3.3, was 30 times faster than ANSYS.

Then, a numerical-analytical validation was performed. Two models simulated on COMSOL 3.3 were compared to analytical ones. The first model consisted of a cantilever actuated piezoelectricaly, where the difference was between 0.2% and 0.6%. The second model was about solving a classical contact problem. A cylinder in contact with a metallic block were simulated and while comparing the max pressure, we found 0.6% of difference. The difference in the contact length is about 0.8%, still in acceptable range.

A table, comparing the capacity of each software of our numerical platform, was listed in order to facilitate the choice of software for our application. COMSOL 3.3 seams to fit most our needs, from time point of vue and accuracy point of view.

After this validation, a parametric model built with COMSOL 3.3 will permit us to optimize the geometry in order to fit our needs and to fix the parameters of the technological process for the fabrication.

As a future work to be done, special structures of test are to be done and tested, in order to validate our simulation results by experimental data.

Acknowledgments

Special thanks for the team with whom I'm working.

Software	Application	Analysis type													Initial stress implementation			
		Static	Modal analysis	Time dependent	Harmonic response	Parametric	Transatory quasi-static	Large deformation	Linear bukling	Elastoplasticity	Multiphysics	Electrostatic	Piezoelectric	Thermal	Mean stress σ_0	$\sigma_0(x,y,z)$	Initial deformation ε_0	Thermal deformation ε_{th}
COMSOL	Solid	✓	✓	✓	✓	✓	✓	✓	✓	✓	✓	✓	✓	✓	✓	✓	✓	✓
	Shell	✓	✓	✓	✓	✓	✓	✓		✓	✓	✓	✓	✓	✓		✓	✓
ANSYS	Solid	✓	✓	✓	✓	✓	✓	✓	✓	✓	✓	✓	✓	✓	✓			✓
	Shell	✓	✓	✓	✓	✓	✓	✓	✓		✓	✓	✓	✓	✓			✓
COVENTOR	Solid	✓	✓		✓	✓	✓	✓	✓			✓	✓	✓	✓	✓		✓
	Shell											✓			✓			✓
I-DEAS	Solid	✓	✓		✓		✓	✓	✓	✓		✓	✓	✓	✓			✓
ABAQUS	Shell	✓	✓		✓		✓	✓	✓			✓	✓		✓		✓	✓

Table:5. Summary of the softwares' capacities.

References

1. Don L. DeVoe, Albert P. Pisano, "Modeling and optimal design of piezoelectric cantilever microactuators" JOURNAL OF MICROELECTROMECHANICAL SYSTEMS, VOL. 6, NO. 3, PP. 266-270, SEPTEMBER 1997.

2. X Chen, C H J Fox, S Mc William, "Modeling of a tunable capacitor with piezoelectric actuation" JOURNAL OF MICROMECHANICS AND MICROENGINEERING, 14 (2004) S102-S107.

3. H.A. Rouabah, C.O. Gollasch, M. Kraft, "Design optimisation of an electrostatic MEMS Actuator with low spring constant for an Atom Chip" UNIVERSITY OF SOUTHAMPTON, SCHOOL OF ELECTRONICS AND COMPUTER SCIENCE, Highfield, Southampton, SO17 1BJ, UK.

4. Oliver J. Myers, M. Anjanappa, Carl B. Freidhoff, "Modeling a piezoelectrically Actuated Planar Capacitor Actuator Membrane" Proceeding of the COMSOL users conference.

5. Jae-Hyoung Park, Hee-Chul Lee, Yong-Hee Park, Yong-Dae Kim, Chang-Hyeon Ji, Jonguk Bu, Hyo-Jin Nam, "A fully wafer-level packaged RF MEMS switch with low actuation voltage using a piezoelectric actuator" JOURNAL OF MICROMECHANICS AND MICROENGINEERING, 16 (2006) 2281-2286.

Multi-Physics Modelling for Microelectronics and Microsystems - Current Capabilities and Future Challenges

C Bailey, H Lu, S Stoyanov, M Hughes, C Yin, D Gwyer
Computing and Mathematical Sciences,
University of Greenwich,
London, SE10 9LS, UK
Email: C.Bailey@gre.ac.uk, Tel: +44(0)2083318660, Fax: +44(0)2083318665

Abstract

At present the vast majority of Computer-Aided-Engineering (CAE) analysis calculations for micro-electronic and microsystems technologies are undertaken using software tools that focus on single aspects of the physics taking place. For example, the design engineer may use one code to predict the airflow and thermal behavior of an electronic package, then another code to predict the stress in solder joints, and then yet another code to predict electromagnetic radiation throughout the system.

The reason for this focus of mesh-based codes on separate parts of the governing physics is essentially due to the numerical technologies used to solve the partial differential equations, combined with the subsequent heritage structure in the software codes.

Using different software tools, that each requires model build and meshing, leads to a large investment in time, and hence cost, to undertake each of the simulations.

During the last ten years there has been significant developments in the modelling community around multi-physics analysis. These developments are being followed by many of the code vendors who are now providing multi-physics capabilities in their software tools.

This paper illustrates current capabilities of multi-physics technology and highlights some of the future challenges.

1. Introduction

Increasing global competition is a significant factor impacting the design of modern products. While the product development time in the early 1980s was often years, portable computing and consumer products today have a time-to-market of only a few months. Such rapid times-to-market do not leave room for time-consuming trial and error approaches that have been the normal practice in the past.

Virtual prototyping, or computational modelling, tools that predict thermal, electrical and mechanical phenomena are now playing a key part at the early design stage and impacting delivery of reliable products to market as illustrated in figure 1. Exploitation of these software technologies benefits companies by:

- ✓ minimising the amount of physical prototyping
- ✓ improving quality and performance
- ✓ identifying optimal properties and process conditions

- ✓ generating knowledge of the process
- ✓ getting products to market earlier
- ✓ reducing overall development costs

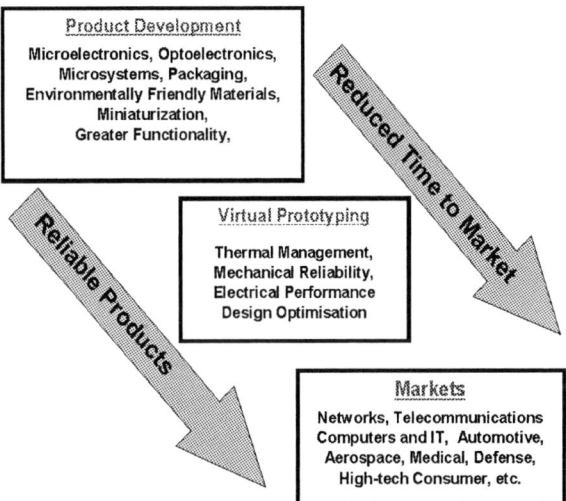

Figure 1 : Virtual Prototyping impact on Product Development

The development of heterogeneous systems that combine digital, analogue, RF, and even fluidic functions into a single piece of silicon has proved difficult and extremely costly. Although major advances are continuing to be made in the semiconductor industry, for highly complex systems containing multi-functional components (i.e. digital, analogue, RF, MEMS, Optics, etc) the System-on-Chip (SoC) option will be very costly if it can be achieved at all.

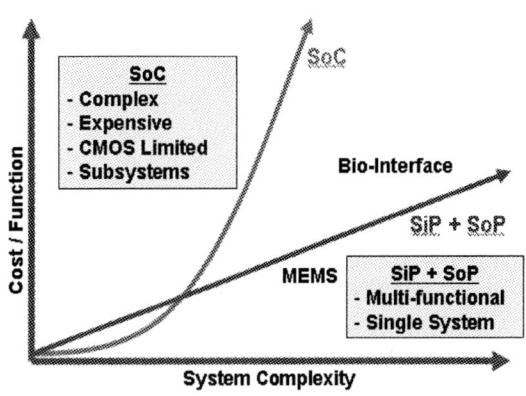

Figure 2: SOC, SIP and SOP

Simulation and Analysis tools have traditionally focused on one aspect of the design requirement; for example: thermal, electrical or mechanical.

System-in-Package and System-on-Package technologies require analysis and simulation tools that can easily capture the complex three dimensional structures and provided integrated fast solutions to issues such as thermal management, reliability, electromagnetic interference, etc.

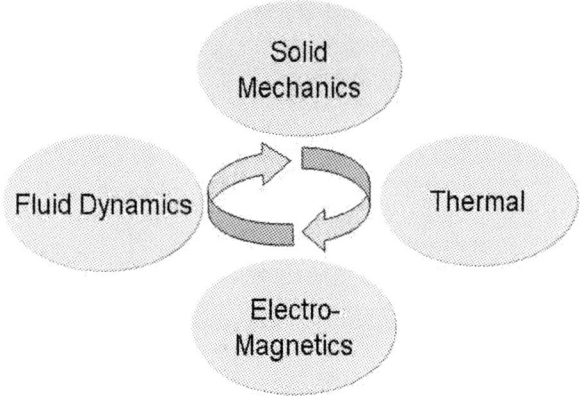

Figure 3: Multi-physics modelling

Technology roadmaps (i.e. ITRS, iNEMI) emphasize the requirement for improved design tools that permit integrated modeling and simulation of materials and processes to accommodate the rapid advancements in technology. Modeling tools can help industry identify potential defects very early in the design cycle and more importantly that can be used to provide optimal process conditions and material properties that will ensure success.

1. Multi-Physics Simulation Strategies

Until recently most of the Computer-Aided Engineerig analysis software tools have been developed in the context of single disciplinary groups such as:

- *Computational Fluid Dynamics (CFD)* solving phenomena such as fluid flow, heat transfer, combustion, solidification, etc

- *Computational Solid Mechanics (CSM)* solving deformation, dynamics, stress, heat transfer, and failures in solid structures

- *Computational Electromagnetics (CEM)* used to solve electromagnectis, electro-statics and magneto-statics.

One reason for this is the underpinning discretisation and solver technolgies used in each discipline. For computational fluid dynamics these have been based on control volume (or finite volme) techniques using segredated iterative solvers. For computational solid mechanics the solver techniques have been finite element based with the resulting matrices solved using direct

solvers. For computational electromagnetics a mixture of finite volume, finite element and even boundary element techniques have been used [1-4].

The distinctive features of the above solver technologies and subsequent software developments over the last thirty years has meant that coupling the physics between the different disciplines has been challenging.

Many of the software vendors now claim a multi-physics or multi-discplinarity capability. In this context the term multi-discplinary means that data generated by one code (i.e. a traditional CFD solver) is transferred to another code (i.e. a traditional CMS solver) to undertake thermal-stress calculations for example. Here data from one code is used as input to the other code either as boundary conditions, loadings or volume sources.

Depending on the class of problem being solved this exchange of data can be classified as one-way or two-way. When the classification is one-way then the calculations from the first solver will influence the calculations in the second solver but not vice-versa. An example of this may be the temperatures calculated from a CFD solver, such as FLOTHERM, which influence the thermal-stress calculations in a solder joint calculated by a CSM solver, such as ABAQUS, MARC or ANSYS, but not vice versa. Figure 4 illustrates this type of approach.

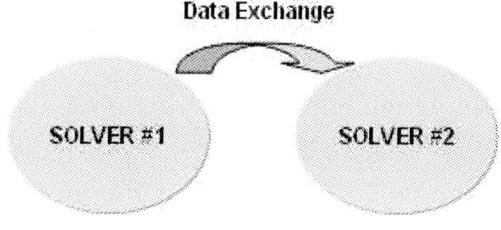

Figure 4: Multi-Physics: One-Way Coupling

Figure 5 highlights this type of one-way coupling where the CFD predictions for airflow and temperature are transferred to a CMS solver in terms of temperature changes, and then the CMS solver will calculate stress due to these changes.

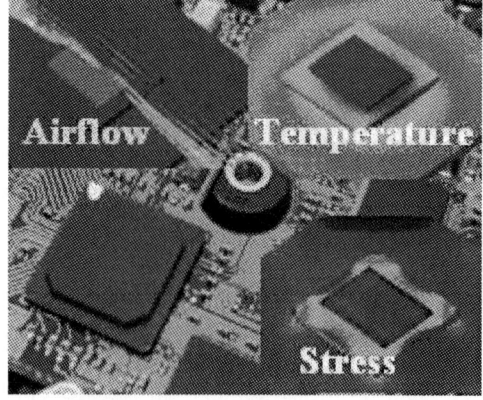

Figure 5: Example of one-way coupling

594

True multi-physics capability is defined by a much tighter integration between the solvers and this in general requires a two-way exchange of data between each as the predictions of one solver influences the other and vice-versa.

Such technology may be emebedded into a single software enviuronement. Multi-physics solvers may require the two-way exchange of data in both time and space. Figure 6 details this type of approach.

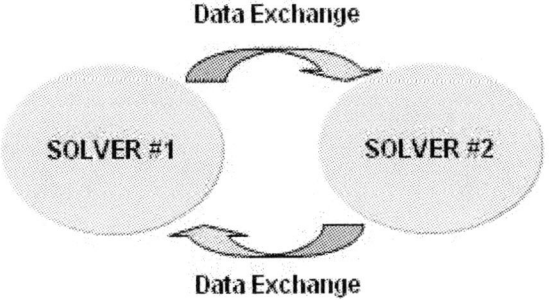

Figure 6: Multi-Physics: Two-Way Coupling

Figure 7 illustrates a process that will require two-way coupling capabilities. This is a simulation of the electrodeposition process, which requires integrated calculations between CFD, CSM and CEM. This work is further explained in the paper by Hughes-et-al in these proceedings.

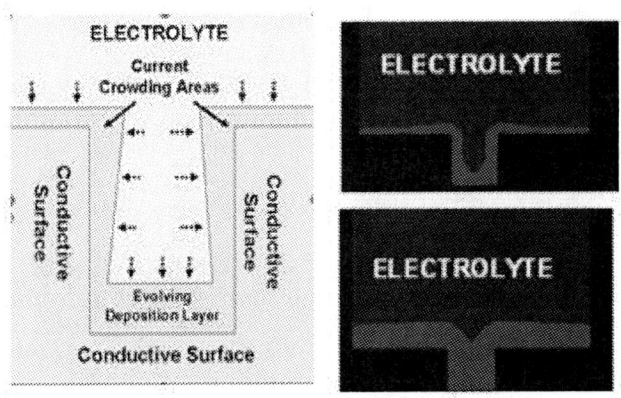

Figure 7: Example of Two-Way Coupling

Simulations that require one-way coupling can be undertaken by some form of file transfer between the solvers. For those that require two-way coupling the complexity increases as each solver may require mesh compatibility and time constraints.

In the last couple of years there have been a number of projects targeted at creating high performance computing tools to facilitate the coupling of distinct mesh-based solvers. For example coupling a CFD code and a CMS code to undertake multi-physics calculations. Some projects include:

- MDICE; a US Airforce funded project to develop a integrated computing environment led by CFDRC [5]

- ICE; a US Army funded project targeted at coupled multi-discplinary simulations across a GRID environment [6]

- MpCCI; an EU funded project to develop a suite of tools to enable the coupling of a wide variety of commercial codes [7]

The other approach is to try and solve all of the physics, and its coupling, within a single high performance computing software framework. This avoids the complexities of coupling different codes as outlined above.

The following table lists a number of software vendors in the microelectronics and microsystems market who have multi-physics or multi-disciplinary capabilities.

Software	Web Address
ANSYS	www.ansys.com
COMSOL	www.comsol.com
ANSOFT	www.ansoft.com
FLOMERICS	www.flomerics.com
PHYSICA	www.physica.co.uk

Table 1: Some Multi-physics codes

3. Fabrication and Assembly

Fabrication and assembly technologies for micrelectronics and microsystems can be complex and governed by interacting physical phenomena. One example is the formation of solder joints [8,9]. This takes place by first printing solder paste onto a printed circuit board and then reflowing the solder in a reflow furnace. The first calculations detailed in figure 8 illustrate CFD calculations for the printing of solder paste across a stencil.

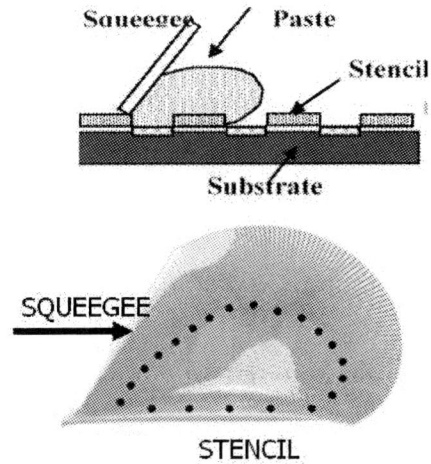

Figure 8: CFD Predictions for Solder Paste Printing

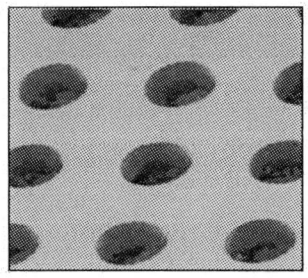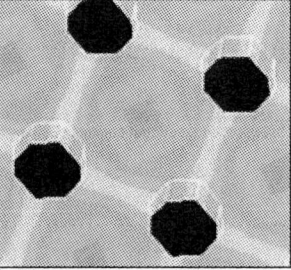

Figure 9: Stress Calculation in the Stencil

Figure 9 illustrates CSM calculations for stress in the stenncil due to the mounting and printing processes. These calculations of provide an insight into stencil behavior over time and its impact on final print quality. Full integration between CFD and CSM calculations for solder print predictions and stencil deformation is a two-way analysis as the fluid dyanmics of the paste places pressure on the stencil and the movement of the stencil impacts the manner in which the paste enters the apertures.

Once the solder is deposited onto a circuit board, and the component is place on the board, then this assembly is passed through a reflow furnace. The shape that the solder takes during this process can be predicted using the code SURFACE EVOLVER. These solder shapes can then be used by a computational mechanics code, such as PHYSICA, to model other interesting physics such as temperature and stress. This is an example of one-way coupling where in this case it is assumed that temperature, and stress does not affect the solder joint shape.

The shapes predicted using Evolver are represented using a surface triangular mesh. This is adequate to capture the evolving surface of the solder. For subsequent PHYSICA simulations of heat flow and stress a volume mesh is required. The interface developed between Evolver and PHYSICA uses PATRAN to generate a three-dimensional mesh from the two-dimensional surface mesh generated by Evolver. The surface to volume mesh procedure is:

1. Equilibrium geometry from Evolver is output into PATRAN format, one file for each body (i.e. solder, board, lead). In addition, a file containing boundary conditions is saved.
2. PATRAN produces a volume mesh from an Evolver surface mesh and the volume mesh files are output in PHYSICA format.
3. PHYSICA input files are created. PHYSICA simulations (temperature, stress, etc) can begin on the predicted solder shape using this volume mesh.

Figure 10 details comparisons between real solder joints and those predicted by Evolver. The top two pictures show the real solder bump and that predicted by Evolver. To illustrate the comparison between the two, the lower plots show the Evolver predictions overlaid onto the picture of the real joint. Clearly, we can see that the Evolver calculations give good agreement with the solder shapes found in reality.

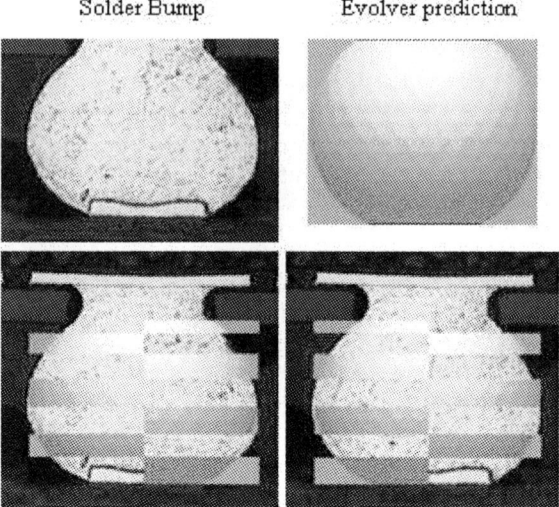

Figure 10: Evolver Calculation for Solder Joint Shape

Once we have the solder joint shape, then we can investigate other important physical phenomena, using in this case the PHYSICA software. Figure 11 shows void formation in solder joints with blind vias. At the very small micro-via dimensions being used, there is a concern that, during the printing process, not all of the micro-via is being filled with solder paste. Therefore, at the start of the reflow process, a void may already be present which, when the solder melts, will rise and form the void observed in the following photograph.

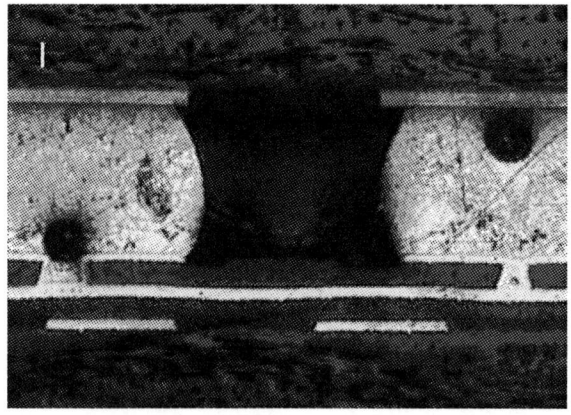

Figure 11: Observed Voids in Solder Joints

To test if the solder printing process is the cause of the final void observed, numerical simulations have been undertaken using the level set method to capture the movement of a void through liqiud solder. Figure 12 details the results from these simulations where solder is assumed to have penetrated the micro-via during printing and wetted its base. Clearly we can see that the void rises over time, due to buoyancy forces. Also presented is the magnitude of marangoni convection which is driven by surface tension gradients along the surface of the solder. .

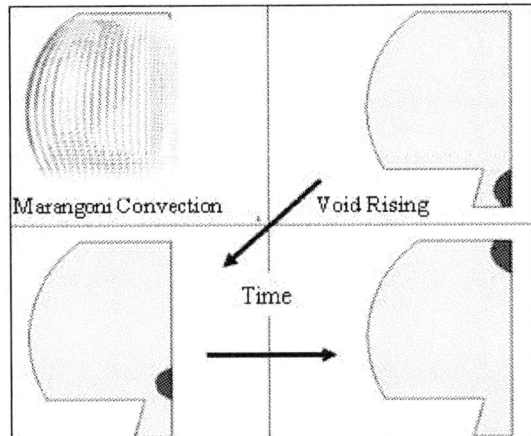

Figure 12: CFD Calculations of Void Movement

These simulations have shown that if the solder material does not wet the base of the micro-via, then it will not be able to rise. This is due to the high surface tension along the solder-void interface at the micro-via exit. Therefore, as long as solder wets the base of the micro-via and the void diameter is smaller than the micro-via diameter then the void will be able to rise into the solder mass from the micro-via. This requires a low initial void volume and a contact angle between the solder and micro-via wall near to zero. Obviously, these initial simulations are ignoring other effects such as gas from the flux that may also result in void formation

The next stage in solder joint formation is its solidification. This can be calculated using CFD techniques. Figure 13 shows the solidification fronts (light region is liquid) of the solder bumps during cool down. The solidification results show the corner (edge) bump solidifying first. The power connections, which are connected to the copper plate in the substrate, solidify locally at a faster rate than the ground connections. These results for solidification time can feed into the previous simulations on void movement to see how long a void has to rise through the solder bump and possibly escape.

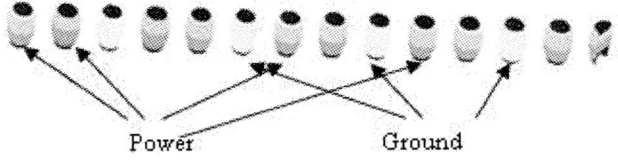

Figure 13: CFD Calculations of Solidification

Figure 14 shows the CMS calculations of stress in the solder joints at the end of the reflow process.

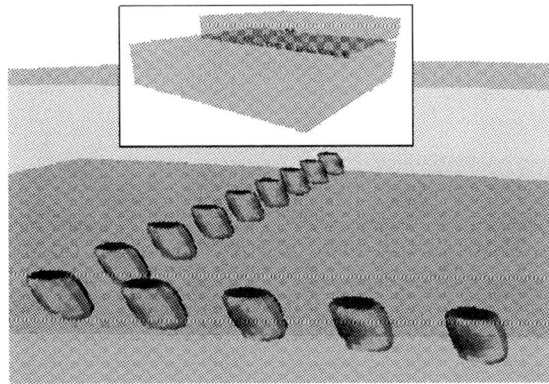

Figure 14: CSM Calculations of Stress after Reflow

4. Test and Reliability

Under test conditions microelectronic and Microsystems components are subjected to extreme environmental conditions, which are meant to promote the failures that would be observed in the field. The type of environmental conditions that a component can be subjected to is changes in temperature, humidity and vibration.

Given that the device may exhibit electrical, optical and fluidic behavior then the performance of the component under accelerated test conditions; the resulting stresses imposed; and the likelihood of failure is a multi-physics process.

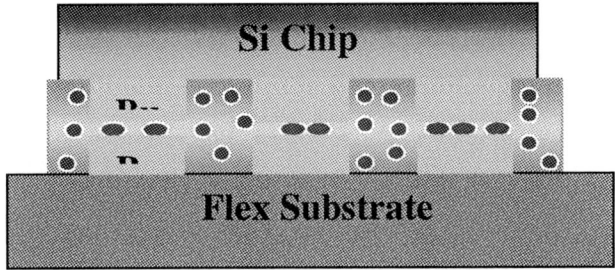

Figure 15: Anisotropic Conductive Film

As an exmaple of a test condition consider the pressure cooker test for anisotropic condictive adhesives [10]. The pressure cooker (or Autoclave) test is one of the most severe tests that polymer or adhesive materials can be subjected to. It involves placing the package into a humid environment (100% RH) at increased pressure (2 atm) and high temperature (120 C) for a period of time and measuring the change in contact resistance between polymer particles in an anisotropic conductive adhesive.

In this simulation the moisture diffusion analysis is coupled with the stress analysis so that the displacement field and the moisture concentration are solved simultaneously. Figure 16 shows the wetness fractions

597

distribution in the Anisotropic Conductive Film (ACF) layer placed between a flexible substrate and die at 1 hour, 3 hour and 12 hours during the autoclave test. The results show that for this flip-chip assembly it is expected that the adhesive will be nearly fully saturated with moisture after 3 hours at 120C, 100%RH and 2atm conditions.

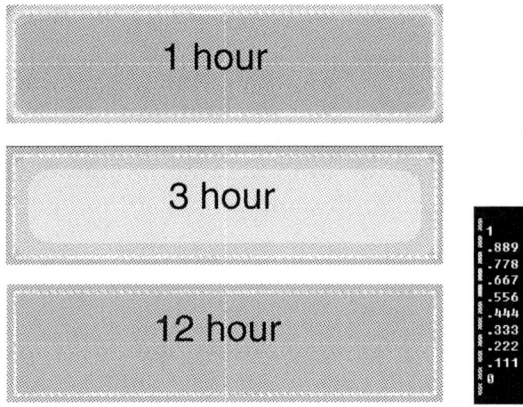

Figure 16: Moisture Ingress during Humidity Test

The above calculations have been undertaken on the whole flip-chip assembly. At the conductive particle scale there is also interest in predicting phenomena taking place. In this case it is the temperature and moisture induced stresses and the effect these can have on the contact resistance between the particle and pad surface. Figure 17 details a finite element model of the bump region which includes conductive particles.

A micro-macro modelling approach is adopted here where moisture and deformation results from the above model are transferred to this local model to predict the stresses around the particle.

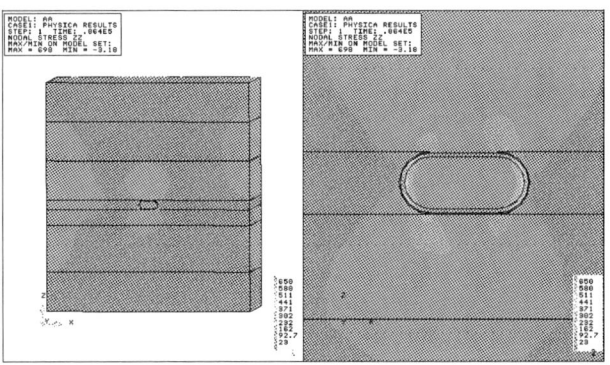

Figure 17: Stress around Conductive Particle due to Temperature and Moisture

The normal stress distribution around the conductive particle is shown in Figure 17. Higher stress was found at the interfaces between the conductive particle and adhesive matrix. This is an example of two-way coupling analysis as the temperature and moisture affect the stress

calculations and the stress calculations, if cracking occurs, will affect the moisture calculations and to some degree the thermal behavior. .

As a second example, consider the following VCSEL device that is flip-chip assembled onto an organic substrate with embedded optical waveguides. The performance of the VCSEL device is governed by the thermal, mechanical and optical characteristics of this assembly. Figure 18 illustrates this package [11, 12].

During operation, the VCSEL device will heat up and the thermal change together with the CTE mismatch in the materials will result in potential misalignment between the VCSEL apertures and the waveguide openings in the substrate. Any degree of misalignment will affect the optical performance of the package.

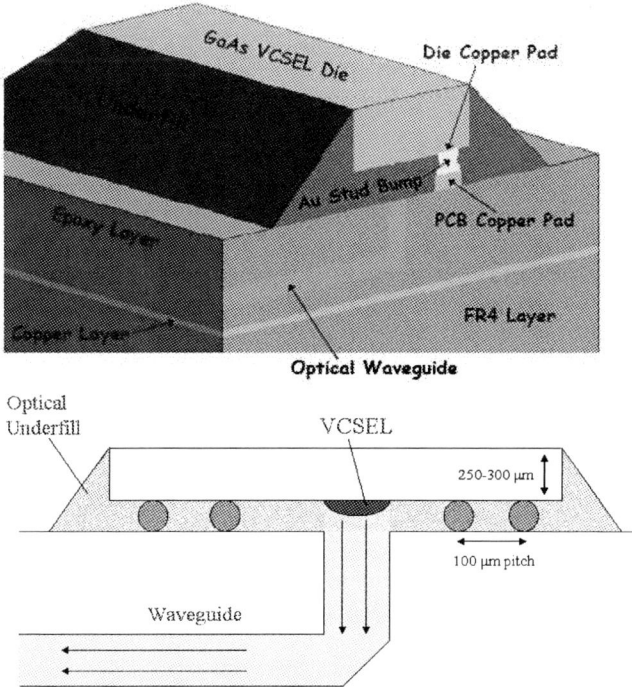

Figure 18: VCSEL Package

The amount of attenuation will depend on the degree of deformation between VCSEL aperture and waveguide entrance. The thermo-mechanical model simulates the VCSEL array heating up due to normal operation. Localised heating in the area surrounding each VCSEL device is seen at around the expected temperature of 85°C.

With the correct heating profile, the resultant stresses can be seen due to the effects of the heating and the CTE mismatch between the various materials present. The thermo-mechanical model showed the greatest general deformation in a single direction, along the horizontal direction as illustrated in figure 19.

598

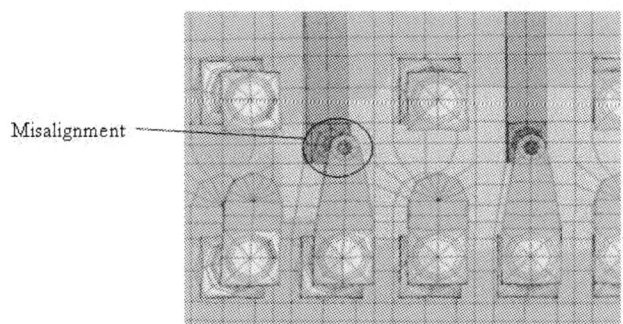

Figure 19: CSM Calculations of VCSEL Opening during Thermal Cycling

From the above simulation results, the misalignment values for these VCSELs were taken into account (i.e. maximum deformation) and used in a subsequent optical model. This is an example of again a one-way coupling simulation where the thermal model will affect the structural behavior of the package but this will not in tern affect the thermal behavior. For the optical calculations, the structural behavior will affect this although the optical calculations will not affect the structural behavior.

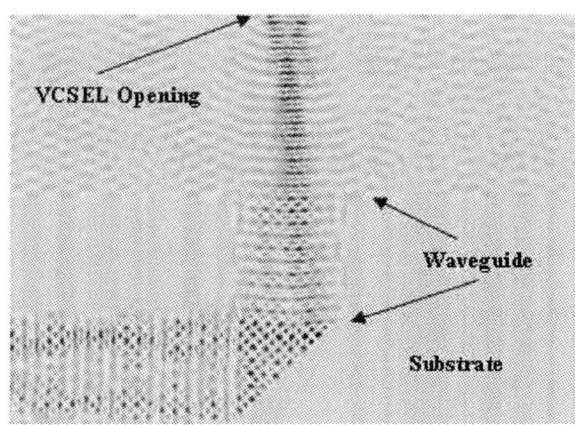

Figure 20: CEM calculations of optical performance of VCSEL Waveguide and Underfill

The objective of the optical simulations is to predict the coupling efficiency of the VCSEL beam to the waveguide entrance. This is characterised by the attenuation value, which is calculated by comparing the level of the optical signal as it leaves the VCSEL aperture, to that leaving the waveguide exit. Any attenuation observed will be because of the misalignment between VCSEL and waveguide; geometry of the waveguide; and the polymer material properties used. Fig. 20 shows a typical propagation contour plot of the optical signal travelling through the waveguide model.

5. Future Trends and Requirements

Although numerical modeling tools, based around CFD, CSM and CEM, are now routinely used in the design of microelectronic and Microsystems devices there are a number of key capability challenges for these tools need to address in the future.

1) Closer Coupling: Microsystems processes are generally governed by close coupling between different physical processes. As seen above numerical modelling tools are now addressing the need for multi-physics calculations, but more work is required to capture the physics accurately and to identify relevant failure models. Applications that involve complex fluid-structure interaction, including large amounts of mesh movement, are particularly challenging.

2) Multi-Discipline Analysis: Codes that can provide ease in data transfer between thermal, electrical, mechanical, environmental, and other designers important. Tools that accomplish this will allow design engineers from different disciplines to trade-off their requirements early in the design process and this will dramatically reduce lead times.

3) Multi-Scale Modelling: Much of the illustrations shown above have used continuum mechanics to solve the governing physics. Some of these calculations can be governed by phenomena taking place at the nano-scale. Multi-physics and multi-scale analysis is a very challenging area and will see a great deal of development in the near future. Modelling techniques that provide seamless coupling between simulation tools across the length scales are required.

4) Faster Calculations: Multi-physics software that solves highly coupled non-linear partial differential equations is compute intensive and slow. There is a need for reduced-order-models (or compact models) to be developed and used at the early stage of design. Although not as accurate as high fidelity models, these provide the design engineer with the ability to quickly eliminate many unattractive designs early in the design process.

To allow fast calculations the porting of multi-physics solvers to high performance computing clusters can result in dramatic speed-up in simulation times and the ability to run very large problems.

5) Life-Cycle Considerations: Life-cycle factors such as reliability, maintenance and end-of-life disposition receive limited visibility in numerical modelling tools. Future multi-physics and multi-scale models will aim to include all life-cycle considerations, such as product greenness, recycling, disassembly and disposal.

6) Variation Risk Mitigation: Microelectronic and microsystem simulations usually ignore process variation, manufacturing tolerances, and uncertainty in the input data. Future models will include these types of parameters to help provide a prediction of manufacturing and reliability risk. This can then be used by the design engineers to enable them to implement a mitigation strategy.

7) Integration with Optimisation Tools: Numerical optimisation techniques bring enormous advantages by offering an automated, logical and time efficient approach to identify the best process/design parameters for reliable components and products. Figure 21 illustrates the link between optimization and process modeling.

Although different design problems have been solved using optimisation procedures, fully integrated or coupled simulation-optimisation software modules are only just appearing and much more is required to fully capture process variation and uncertainty into these optimisation calculations.

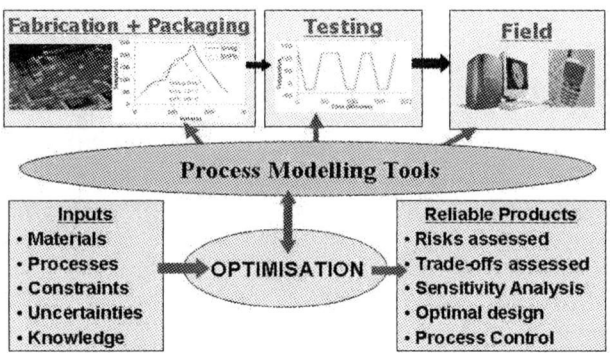

Figure 21: Integrated Process Modelling

8) Modelling through the Supply Chain: Numerical modelling tools require high quality input data in terms of materials data and failure models. Many companies are now using these modelling tools and there is an increasing requirement for companies within each others supply chain to gather and provide relevant modelling data. This is now starting to take place but much more effort is required.

9) Close integration with CAD. There is a trend in the analysis community to closely integrate analysis with Computer-Aided-Design (CAD). Users are demanding this capability with current software and the demand will also be there for new multi-physics software.

6. Conclusions

CAE analysis tools are now being used to underwrite the design of many microelectronic and microsystems components. The demand for greater capability of these tools is increasing dramatically because the user community is faced with the challenge of producing reliable products in ever shorter lead times.

This leads to the requirement for analysis tools to represent the interactions amongst the distinct phenomena and physics at multiple length and time scales. Multi-physics technology is now becoming a reality with many code vendors providing some capability in this area. The strategy in developing a multi-physics framework has been outlined above. Coupling seperate codes together is one approoch. The other is to provide close coupling of the solvers within a single software framework.

But much still needs to be done to satisfy future user requirents. This paper has highlighted some of the current capabilities of Multi-physics technology and the trends and requirements for the future.

Acknowledgements

The authors wish to acknowledge the numerous projects funded by the EPSRC and other funding agencies in the UK for supporting the above developments.

References

1. C. Bailey, H Lu, D Wheeler, "Computational Modelling Techniques for Reliability of Electronic Component on Prinited Circuit Boards", Applied Numerical Mathematics, Vol 40, pp101 -117, 2002, Pub Elsevier

2. M Cross, T Croft, A Slone, A Williams, N Christakis, M Patel, C Bailey, K Pericleous "Computational Modelling of Multi-Physics and Multi-Scale Processes in Parallel", Int Jnl for Computational Methods in Engineering Science and Mechanics, 8, pp1-12, (2007)

3. C Bailey, M Warner, A Agha, K Pericleous, J Parry, C Marroney, H Reeves, I Clark, "Flo/Stress: an Integrated Software Module to Predict Stress in Electronic Products", IEE Journal of Computing and Control Engineering, Vol. 13, No. 3, pp 143 – 148, 2002, Pub IEE. (2002)

4. S Stoyanov, C Bailey, H Lu , M Cross, "Integrated Computational Mechanics and Optimization for Electronic Components, Optimization in Industry", p57 – 71, Pub. Springer Verlag, (2002)

5. MDICE, http://www.cfdrc.com

6. ICE, http://www.arl.mil

7. MpCCI, http://www,mpcci.org

8. G. Glinski C. Bailey, and K Pericleous "Non-Newtonian Computational Fluid Dynamics study of the stencil printing process", IMECHE Journal of Engineering Science, p437-446, 4, V215, 2001

9. C. Bailey, S. Stoyanov, H. Lu, "Reliability predictions for High Density Packaging", Proceedings of High Density Microsystem Design and Packaging and Component Failure Analysis (HDP'04), June 30- July 03, 2004, Shanghai, China, (IEEE, PRC), pp121-127, Pub IEEE

10. C Yin, H Lu, C Bailey and Y C Chan, "Macro-Micro Modelling Analysis for an ACF Flip Chip", Journal of Soldering and Surface Mounting Technology, Vol.18, No.2, pp.27-32. Pub Emerald (2006)

11. P Misselbrook,, D Gwyer, C Bailey, P Conway, K Williams, *Review of the Technology and Reliability Issues Arising as Optical Interconnects Migrate onto the Circuit Board*". Micro- and Opto-Electronic Materials and Structures: Physics, Mechanics, Design, Reliability, Packaging. Chapter 14, pp. 361-382. Pub Springer Verlag, (2007)

12. D Gwyer, P Misselbrook, D Philpott, C Bailey, P Conway, K Williams, "Thermal, Mechanical and Optical Modelling of VCSEL Packaging", Proceedings of the 9th Intersociety Conference on Thermal, Mechanics and Thermomechanical Phenomena in Electronic Systems (ITHERM-2004), Las Vegas, Nevada. Publ. IEEE, Page 405-410. (2004)

USING MOLECULAR MODELING TO UNDERSTAND CLEANER EFFICIENCY FOR BARC ("BOTTOM ANTI-REFLECTIVE COATING") AFTER PLASMA ETCH IN DUAL DAMASCENE STRUCTURES

Nancy Iwamoto[a], Deborah Yellowaga[b], Amy Larson[b], Ben Palmer[b] and Teri Baldwin-Hendricks[b]

[a]Honeywell Specialty Materials
P.O. Box 547
Ramona, CA 92065
nancy.iwamoto@honeywell.com

[b]Honeywell Electronic Chemicals
6760 W. Chicago St.
Chandler, AZ, 85226

Abstract

The invention of inorganic bottom anti-reflective coatings (BARCs) was a promising enabling technology for lithography due to their near unity plasma etch selectivities with the low-k dielectric materials used in advanced via first trench last (VFTL) dual damascene structures. Having these selectivities during plasma etch allowed for the avoidance of via "fence" defects that are often created by the shadowing of the dielectric surrounding the vias by the slower etching organic BARC via fill. The presence of "fence" defects creates an area of high stress in the copper line due to the tight space between the fence and side of the trench that has to be filled, and can cause failure of the device due to shorting of the line from voids caused by copper migration away from the high stress area. In order to achieve a 1:1 selectivity in plasma etch which avoids fence formation, organosiloxane polymers are typically employed in the inorganic BARC materials. However, modification of the organosiloxane film during plasma etching and ashing densifies the material and may also remove organic content from the film, making removal more difficult with traditional cleaning technologies.

This work describes the surface energy and reactivity modeling of the Honeywell cleaner technology that has demonstrated an inorganic BARC removal solution with significantly higher removal rates and etch selectivities compared to currently available products. We will show how molecular modeling influenced the development of the cleaner formulation and the impact on experiments.

1. Introduction and background

BARC ("Bottom Anti-Reflective Coatings") are used as lithography enablers during dual damascene structure build-up. The BARC is added under the photoresist to enhance exposure, and is removed during development and etch/clean. For the BARC to perform properly it must have differential reactivity in order to distinguish it from the interlayer dielectric so that clean and sharp features result from the process [1]. The general steps used for inorganic BARCs is found in Figure 1 showing the location of the BARC, photoresist and interlayer dielectric (ILD). As may be appreciated, the BARC should not develop or etch worse than the ILD to avoid undercutting, but also must be completely removed in order to open features as seen on in the upper processes by ashing or through the use of newly developed Honeywell post ash etchants and cleaners in Figure 1.

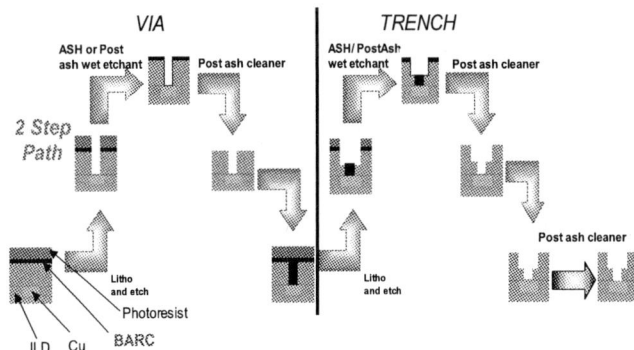

Figure 1. Process flow for inorganic BARCs (iBARC) for dual damascene structures.

One issue that has emerged is the tendency of iBARC layers to densify after the O2 plasma etch step which affects the BARC's reactivity toward further removal by wet etchants and cleaners. Another experimental issue that has emerged is the observation that upon aging, the iBARC layers become more resistant to wet etchants and cleaners. This work will show that the iBARC (iBARC 193, and iBARC 248) developed by Honeywell as well as the Honeywell etchants and cleaners have been tuned together to maximize removal after the O2 plasma, and how the structural nature of the iBARC impacts it's

---

"DISCLAIMER
Although all statements and information contained herein are believed to be accurate and reliable, they are presented without guarantee or warranty of any kind, express or implied. Information provided herein does not relieve the user from the responsibility of carrying out its own tests and experiments, and the user assumes all risks and liability for use of the information and results obtained. Statements or suggestions concerning the use of materials and processes are made without representation or warranty that any such use is free of patent infringement and are not recommendations to infringe any patent. The user should not assume that all toxicity data and safety measures are indicated herein or that other measures may not be required."

1-4244-1105-X/07/$25.00 ©2007 IEEE   601

removal ability and explains both the removal of the O2 plasma densified films and the aging data.

2. Investigations of BARC etching tendencies.

A. Thermodynamic model results

As mentioned previously, after O2 plasma etch, the iBARC tends to densify and change its' reactivity. In order to understand the change in reactivity, a combination of thermodynamic models (using the DFT/QM code DMOL [2]) in conjunction with FTIR analysis was run to understand the underlying changes, and the reasons behind the performance observations.

For the reactivity modeling, general structures that represent network formation were used to understand reactivity of the iBARC as represented by Figure 2 mostly representing cage (middle 4 structures) and ladder (last 2 right stuctures). Most of the work was geared toward the investigation of the reactivity of the cages and ladders, as representatives of network structures. By doing so four, five and six membered ring structures are represented. For the hydrolysis steps (for investigation of etch and cleaning) and condensation steps (for investigation of effects from the O2 plasma as well as aging), usually the last step of the structure formation is calculated to give an indication of tendency of reaction of the different network components.

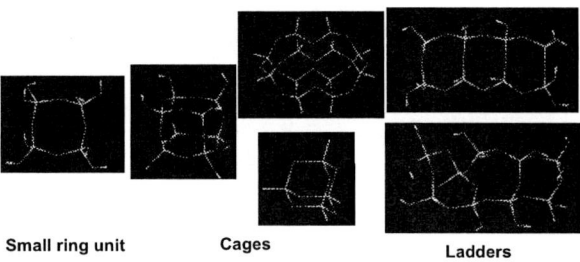

Small ring unit   **Cages**   **Ladders**

Figure 2. General structual units investigated

Figure 3 and 4 summarize the hydrolysis tendencies of the structures as identified by the calculated free energies of reaction. Figure 3 investigated the tendency of hydrolysis due to organic content, as represented by ring cleavage of the small 4 silicon ring unit. It was found that in general, the higher the organic content of the iBARC and the lower the silanol content, the structures are more resistant to hydrolysis as indicated by an increase in the expected endothermicity of the reaction with higher organic content. For the larger structures, it was found that the larger cages and ladders with larger unstrained

multicyclic rings, such as the T8, T12 structure (middle top cage structures in Figure 2, T8 on left and T12 on right) and large ladder (bottom ladder structure in Figure 2) are the most resistant to hydrolysis. For condensation that might occur during aging or plasma etch, the same hydrolsysis models were used by analyzing for the inverse reaction. In doing so, was concluded that the higher the organic content the higher the tendency to condense to network structures; and the more strained the resulting network structure, the lower the tendency to form. However, the small T4 branched (lower left cage structure in Figure 2) structure has low enough free energy at higher temperatures that it might form at higher temperatures (Figure 4, "T4-O-T4"). These general trends were then compared with FTIR analysis, which was applied to look for the structural changes. Together with the thermodynamic tendencies, the structural changes make clear the reasons behind the experimental observations of etch behavior.

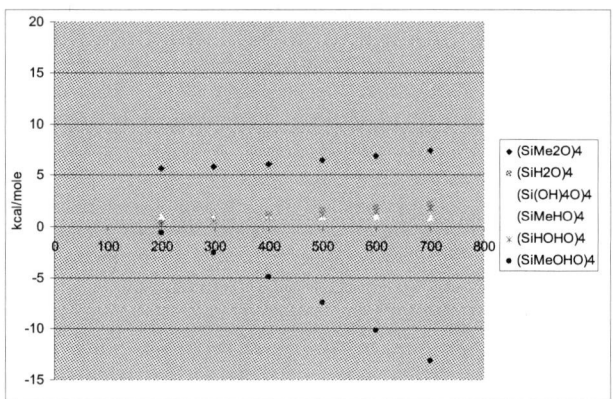

Figure 3. Reactivity demonstrating the effect of organic content in iBARCS on hydrolysis

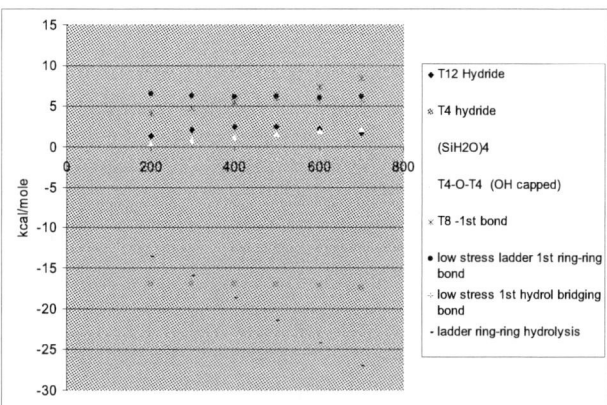

Figure 4. Reactivity demonstrating the effect of structure in iBARCS on hydrolysis

---

"DISCLAIMER
Although all statements and information contained herein are believed to be accurate and reliable, they are presented without guarantee or warranty of any kind, express or implied. Information provided herein does not relieve the user from the responsibility of carrying out its own tests and experiments, and the user assumes all risks and liability for use of the information and results obtained. Statements or suggestions concerning the use of materials and processes are made without representation or warranty that any such use is free of patent infringement and are not recommendations to infringe any patent. The user should not assume that all toxicity data and safety measures are indicated herein or that other measures may not be required."

B. Experimental Film Analysis and Correlation to the Modeled Thermodynamics.

For O2 etched films, the FTIR analysis of iBARC 193 and iBARC 248 films have shown that both 193 and 248 forms of the BARC react to a similar structure (Figure 5). FTIR shows that the main difference before and after O2 plasma treatment is that organic content has been eliminated. This should be the most important effect, as loss of the organic appears to lower the resistance to the wet etch according to the thermodynamic calculations, and so the observed increase in wet etch is mainly due to loss of the organic (Figure 6). However, because plasma etch also involves high local temperatures it there may be more different types of network formation, including the higher strained content that may add to the reactivity of the film. The broadening of the SiOSi peak suggests that network formation is occuring which can include such structures.

For aged films, there are subtle differences, however there is an increase in the OSiO intensity at ~1050cm^{-1} which suggests longer chain and less strained network structures (Figure 7). The lower the ring strain the more stable the structure according to the thermodynamic calculations, so the aged films should be more resistant to etch explaining the experimental observations (Figure 8, top) in which the aged films are more stable. The second observation that can be made is a difference between the iBARC193 and 248 formulations. That is, the iBARC 193 maintains a high quantity of cage structures during aging which should increase it's resistance to hydrolysis over the iBARC248 according to the thermodynamics. This is also observed during wet etch in Figure 8, bottom.

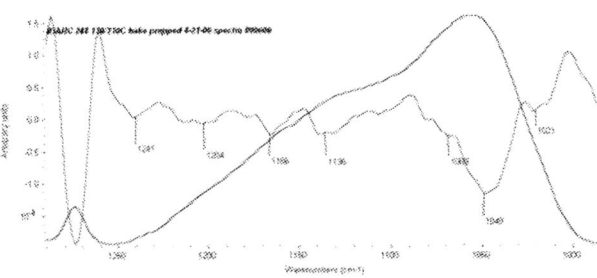

Figure 5. Post O2 Plasma Etch FTIR Example

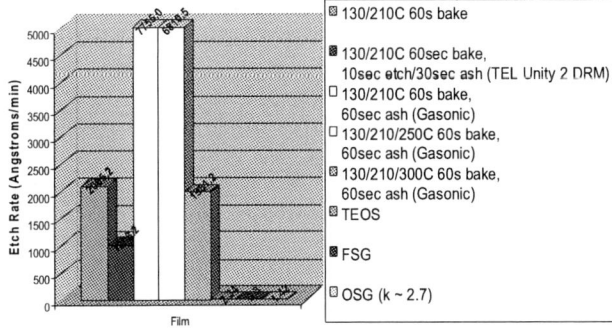

Figure 6. Wet Etch Performance of Post O2 Plasma Films

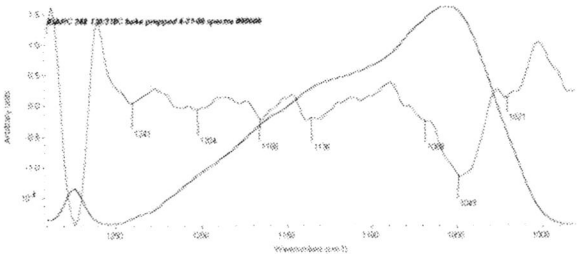

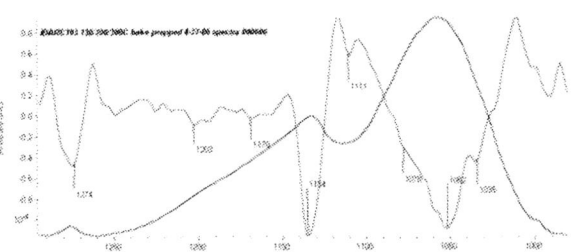

Figure 7. Aged iBARC films (top: iBARC 248; bottom: iBARC 193)

4. Investigations of the Etchant/Cleaner

In addition to structural effects of the iBARC, the other material that was investigated was the effect of the Honeywell etchant/cleaner on efficiency. For the modeling work both the wetting ability and the reactivity of the etchant was investigated. While specifics of the formulation cannot be divulged, the basic trends will be reported here.

For the wetting tendencies, the basic Honeywell etchant/cleaner was modeled on several different surfaces using molecular modeling techniques previously applied to adhesives and underfills [3] in which the surface energy is qualitatively determined by energy drop trends during room temperature dynamics after liquids are

"DISCLAIMER

Although all statements and information contained herein are believed to be accurate and reliable, they are presented without guarantee or warranty of any kind, express or implied. Information provided herein does not relieve the user from the responsibility of carrying out its own tests and experiments, and the user assumes all risks and liability for use of the information and results obtained. Statements or suggestions concerning the use of materials and processes are made without representation or warranty that any such use is free of patent infringement and are not recommendations to infringe any patent. The user should not assume that all toxicity data and safety measures are indicated herein or that other measures may not be required."

introduced to a surface. (Quantitative surface energy determination makes use of calibration that is outside the scope of the current work.) The etchant/cleaner was contrasted for wetting behavior on a simulated iBARC (Figure 9) and also two different ILD structures (Figure 10, 11). Before/after images of the models are shown in Figures 9-11 and indicate that there is better wetting on the iBARC, suggesting that higher intimate contact is achieved using the Honeywell etchant/cleaner, which is a prerequesite to higher reactivity and less residual formation. The reason for the differentially better wetting of the iBARC is due to a balance of charges during the formulation process of the etchant/cleaner leading to the low surface energies required for wetting (Figure 12 middle bar)

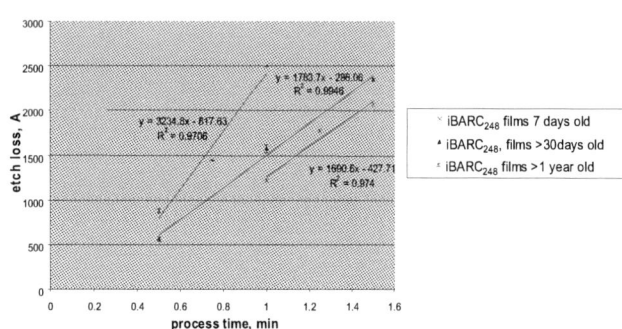

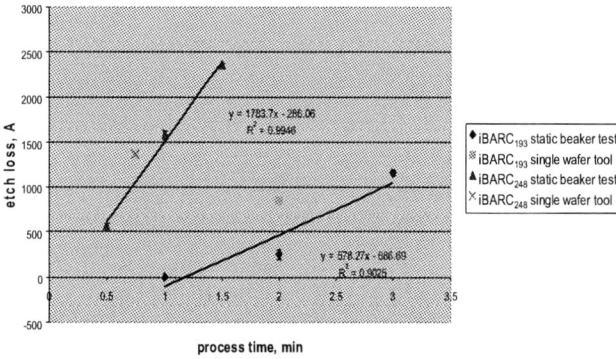

Figure 8. Wet Etch Performance of Aged Films (Top: Example of effect of aging; Bottom: iBARC 248 vs. iBARC 293)

Figure 9. Before/After wetting simulation of the basic Honeywell etchant/cleaner formulation on iBARC 193 showing a larger area interface after wetting between the etchant/cleaner and the iBARC

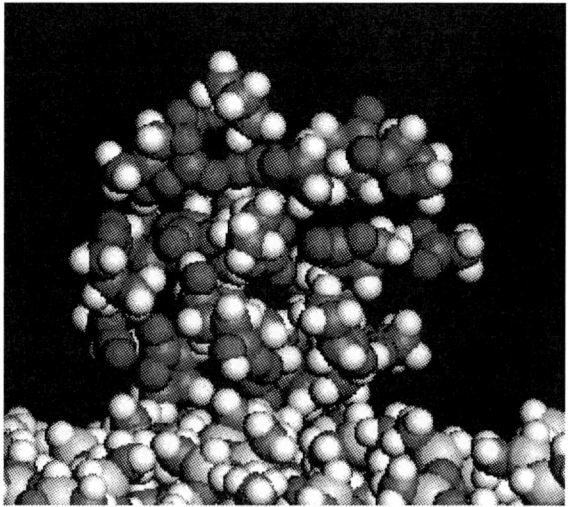

Figure 10 Ater wetting on TMCTS low k

"DISCLAIMER
Although all statements and information contained herein are believed to be accurate and reliable, they are presented without guarantee or warranty of any kind, express or implied. Information provided herein does not relieve the user from the responsibility of carrying out its own tests and experiments, and the user assumes all risks and liability for use of the information and results obtained. Statements or suggestions concerning the use of materials and processes are made without representation or warranty that any such use is free of patent infringement and are not recommendations to infringe any patent. The user should not assume that all toxicity data and safety measures are indicated herein or that other measures may not be required."

Figure 11. After wetting on OMCTS low k.

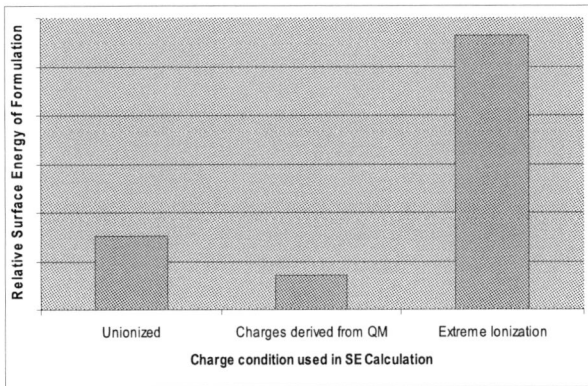

Figure 12. Effect of charge content on surface energy of the Honeywell etchant/cleaner

The interesting consequence of this balance of charges is not only an increase in wetting, but an increase calculated reactivity (Figure 13-14). Figure 13 shows a variety of structures and etchant environments and free energy of hydrolysis. According to the data, in general, increasing the ionic content helps the hydrolysis, but it is not a smooth relationship. That is, identification of the type of ion does not necessarily mean that the reaction is more exothermic, rather the overall effect of the ionization on the formulation must be considered as in Figure 12. Figure 14 shows free energy of hydrolysis of the structural components of a typical ILD (left bars) and the components of the iBARC. Using the Honeywell etchant/cleaner, it appears that the energetics of reaction

are tuned more for the iBARC than for the ILD. So, from comparisons of the modeling results of the surface energies and the reactivities, both the surface energy of the etchant/cleaner as well as the reactivity of the formulation has been tuned in order to wet the iBARC preferentially as well as react better to the iBARC (Figure 15).

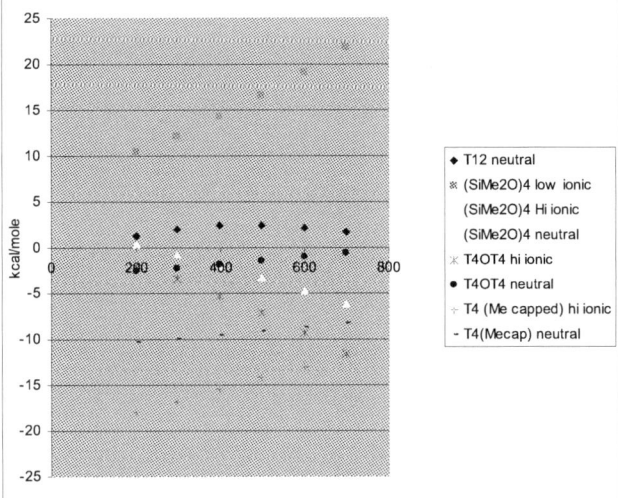

Figure 13. Calculated Reactivity of General Materials Types showing how charge structure may improve free energy.

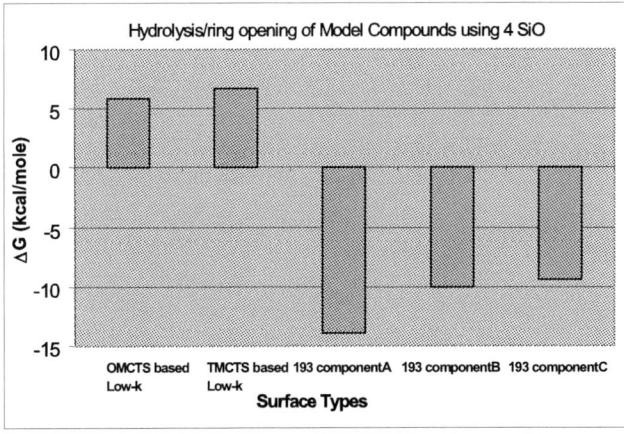

Figure 14. Reactivity differences of ILD materials vs. iBARC 193

"DISCLAIMER
Although all statements and information contained herein are believed to be accurate and reliable, they are presented without guarantee or warranty of any kind, express or implied. Information provided herein does not relieve the user from the responsibility of carrying out its own tests and experiments, and the user assumes all risks and liability for use of the information and results obtained. Statements or suggestions concerning the use of materials and processes are made without representation or warranty that any such use is free of patent infringement and are not recommendations to infringe any patent. The user should not assume that all toxicity data and safety measures are indicated herein or that other measures may not be required."

The tuned response of the Honeywell etchant/cleaner is further demonstrated by a head-to-head experimental comparison of the Honeywell etchant versus different competitor formulations (Figure 15). It was found that the selectivity of the Honeywell formulation exceeds the selectivity of the other vendors, which is in-line with expectations of selectivity from the modeling.

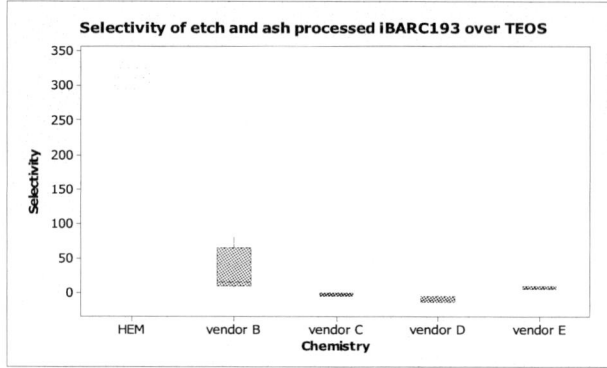

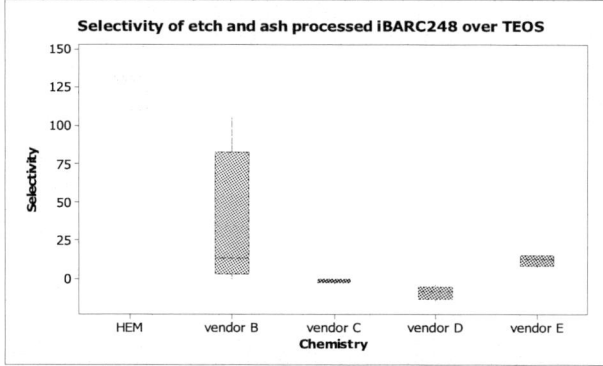

Figure 15. Head-to-head testing of the Honeywell Etchant/Cleaner with other etchant/cleaners showing higher etch rates the Honeywell formulation (top: on iBARC 248; bottom: on iBARC 193).

5. Conclusions

The modeling work described in this paper demonstrates how modeling was used as an aid in understanding the experimental work in developing a new etchant/cleaner for iBARC layers. The modeling demonstrated how the structural changes to both the iBARC and the etchant formulation impacts the performance of the film.

It was found that aged iBARC films are expected to have higher resistance to the wet etch because of the formation of lower strained networks/rings and that differences in etching between types of iBARCs upon aging are predictable based upon the types of ring

structure, with the iBARC 193 more resistant to etchants than iBARC 248 based upon its' higher cage composition. In addition it was found that higher wet etch ability after plasma etch can be attributed directly to lower organic content. By modeling the etchant itself, it was shown that the Honeywell etchant has been tuned to maximize its' wetting properties to the surface of the iBARC, as well as maximizing its' chemistry to react preferentially to the iBARC rather than the ILD.

In order to gain a complete picture, a combination of quantum mechanic (thermodynamic) models which rely on electron distribution as underlying principles as well as classical mechanics (for the surface energy modeling) which rely on classical force field definition, were used demonstrating that molecular methods spanning multiple scales are a better way at understanding the performance properties.

Acknowledgments

We would like to thank the following people: Aaron Bicknell, SEZ America and Gale Hansen, SEZ America

References

1. "Highly Selective Removal of Plasma Etched and Ashed Inorganic BARC Materials"; Deborah Yellowaga, Amy Larson, Ben Palmer, Teri Baldwin-Hendricks and Nancy Iwamoto; Center for Microcontamination Control, 4th International Surface Cleaning Workshop; Nov 8th, 2006; Boston, MA

2. Accelyrs, Inc., San Diego, CA. For this work the DNP basis set employing the GGA/BP functionals were used.

3. a. "Molecular Dynamics and Discrete Element Modeling Studies of Underfill", N. Iwamoto, M. Li, S.J. McCaffrey, M. Nakagawa, G. Mustoe, 1998 International Journal of Microcircuits and Electronic Packaging, v. 21 No. 4, Fourth Quarter 1998, pp. 322-328.

   b. "Simulation Underfill flow for Microelectronics Packaging" N. Iwamoto, M. Nakagawa, G.G.W Mustoe, Proceedings of the 49th Electronic Components and Technology Conference, June 1-4, 1999, San Diego, CA, pp. 294-301.

   c. "Predicting Material Trends Using Discrete Newtonian Modeling Techniques", N.E. Iwamoto, M. Nakagawa, G. Mustoe, Nepcon West '99 Proceedings V. III, pp. 1689-1698

   d. "Molecular Modeling and Discrete Element Modeling Applied to the Microelectronics Packaging Industry"; N.E. Iwamoto, Masami Nakagawa; Micro Materials 2000 Conference; Berlin, Germany, April 17-19,2000

"DISCLAIMER
Although all statements and information contained herein are believed to be accurate and reliable, they are presented without guarantee or warranty of any kind, express or implied. Information provided herein does not relieve the user from the responsibility of carrying out its own tests and experiments, and the user assumes all risks and liability for use of the information and results obtained. Statements or suggestions concerning the use of materials and processes are made without representation or warranty that any such use is free of patent infringement and are not recommendations to infringe any patent. The user should not assume that all toxicity data and safety measures are indicated herein or that other measures may not be required."

Numerical Simulation of Creep Strain of PBGA Solders under Thermal Cycling

F.L. Sun, Y. Liu, L.F. Wang

Material Science & Engineering College, Harbin University of Science and Technology.
52# Xuefu road, Harbin, China, 150080
sunfengl@yahoo.com

Abstract

An octant 3D Finite Element(FE) model of Plastic Ball Grid Array (PBGA) assembly system was developed according to actual package product. The FE method was used to study the creep strain and reliabilities of solder joints under four thermal cycles. The distribution of the creep strain of solder joints under temperature loading was analyzed, and the reason of which was investigated. 95.5Sn-3.8Ag-0.7Cu and 63Sn-37Pb were taken as two kinds of solder to be compared in this study. The creep strain, low cycle fatigue life and reliability of both solders were evaluated. Results showed that for the whole joints array, the creep strain was much larger around the die, the reason of which might be attributed to the larger thermal expansion mismatches between the die and other components in this area. Comparing the reliability of the two kinds joints in this paper: 63Sn-37Pb joints creep more seriously in the cycling process. And 95.5Sn-3.8Ag-0.7Cu joints had longer fatigue life, which showed a higher reliability under thermal cycling.

1. Introduction

With the drastic advances that have taken place in microelectronics over the past three decades, ball grid array (BGA) packages are increasingly being used in microsystems applications. BGA package can meet many requirements such as high I/O density, small profile, high performance and low cost[1]. However, ceramic ball grid array (CBGA) packages have been used extensively in the microsystems industry, and the use of plastic ball grid array (PBGA) packages is relatively new, especially for automotive and aerospace applications where harsh thermal conditions prevail.

A PBGA assembly system usually consists of a package, solder joints and a printed circuit board (PCB). Solder joints offer mechanical support, electrical conduction and thermal dispersion for the microelectronic devices. The CTE and stiffness mismatch among the components results in thermal stresses in solder joints during temperature and power cycling. The damage caused by these stresses accumulates as the electronic assembly is subjected to multiple cycles, cause the ultimate failure of solder joints[2]. So solder joints reliability is a major concern for the whole electronic assembly system.

The melting points of most solder alloys are usually above 180°C. This means that the homologous temperature(the ratio of operating and melting temperature in absolute scale) of a solder alloy is usually above 0.57 at room temperature[2]. It is well documented that creep plays a very important role in deformation

behavior of materials at homologous temperatures close to or above 0.5 if the loading rate is slow enough for creep deformations to occur. Under most existing conditions, the temperature cycle duration is in the order of minutes to days and the homologous temperature is usually above 0.5, So solder joints formed by most solder alloys are expected to deform primarily due to creep.

To investigate the solder joints reliability, thermal cycling, which use a thermal-time load to obtain the function situation of the assembly system, has been used extensively. Because it can reflect the mismatch of Coefficient of Thermal Expansion (CTE) among the constituent parts of the assembly system. This kind of reliability test usually needs a long time. To save testing time, computational simulation has been commonly employed to analyze the thermal cycling process. It can offer a approximate predictiction of the joints reliabiity before extensive testing.

Pb free solder is fast becoming a reality in electronic manufacturing due to marketing and legislative pressures. All kinds of solder alloys have been put forward to replace the commonly used SnPb solder. Amongst all the candidates, Sn-Ag-Cu (with diferent composition) has been regarded as one of the most promising replacements [3-4].

In this paper, a 3D finite elements model of PBGA assembly system was created to analyze the creep behavior of solder joints, which is subjected to cyclic stress generated during the thermal cycling. The cycling number of failure was predicted with the creep strain energy density from simulating results and an empirical relationship found in literature. Sn-3.8Ag-0.7Cu and 63Sn-37Pb are considered as time-and temperature-dependent material in the whole analysis. The performances of these two kinds of solder are compared as well.

2. FEA modeling and simulation

The geometries of the finite element model is based on a real PBGA assembly system. The outline of components is showed in figure 1. Detailed specifications are described in table 1.

Due to its square symmetry, only one-eighth of the structure needs to be modeled. The finite element mesh was obtained by cutting from diagonal and centerline planes according to geometry symmetric property. This modeling method can save considerable computer calculating time. Moreover it will not reduce the accuracy of the result. So it has been employed a lot in studies, which performed well in analyzing work. The finite element mesh is shown in figure 2.

1-4244-1105-X/07/$25.00 ©2007 IEEE

Figure 1: Outline of geometric model

Table 1: Summary of component spicifications

Item	spicification
No. of package I/O	272
Ball size (diameter)	0.76mm
Ball pitch	1.27mm
Die size	$5.08 \times 5.08 \times 0.5 mm^3$
BTsubstrate thickness	0.56mm
PCB substrate thickness	1.57mm

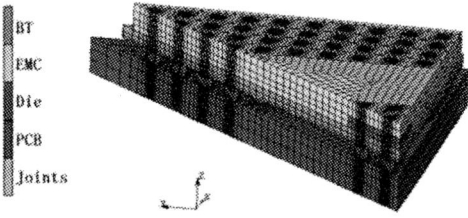

(a) The whole FEA model

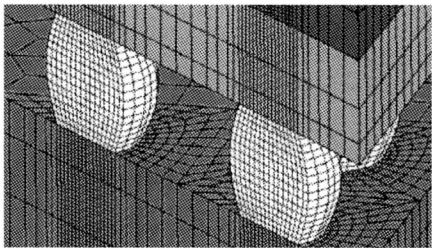

(b) magnified image of central joints

Figure 2:   FEA model of PBGA assembly

It is well known that the thermal fatigue failure of solder joints is dominated by the secondary creep[5]. Therefore, the subsequent analysis concentrates on the creep strain generated in the process of the thermal cycling. The creep behavior of solders is modeled using Garofalo-Arrhenius steady-state creep constitutive equation:

$$\frac{\partial \varepsilon}{\partial t} = C_1 \frac{G}{T} \left[\sinh(\frac{\alpha}{G}\delta) \right]^n \exp\left(-\frac{Q}{kT} \right) \qquad (1)$$

Where C_1 is a material constant $(Ks^{-1}/N/mm^2)$. G is the temperature-dependent shear modulus. T is the absolute temperature (K), α defines the stress level at which the power law stress dependence breaks down. n is the stress exponent. Q is creep activation energy (ev). k is the Boltzmann's constant $(8.617 \times 10^{-5}$ ev/k). Table 2 shows the values of constants for Sn63Pb37 in equation (1).

Table 2: Constants for Sn63Pb37 in euqation (1)

$G=(28388-56T)N/mm^2$			
$C_1(k/s/N/mm^2)$	α	n	Q(eV)
16.7	866	3.3	0.548

Equation (1) can be re-written into equation (2) as following, and the constants C_1, C_2, C_3 and C_4 for Sn-3.8Ag-0.7Cu are given in Table 3.

$$\frac{\partial \varepsilon}{\partial t} = C_1 \left[\sinh(C_2 \delta) \right]^{C_3} \exp\left(-\frac{C_4}{T} \right) \qquad (2)$$

Table3: Constants for Sn-3.8Ag-0.7Cu solder in equation(2)[6]

$G=(27360-40.5T)N/mm^2$			
$C_1(1/s)$	$C_2(mm^2/ N)$	C_3	$C_4(K)$
32000	0.037	5.1	6524.7

The relevant elastic properties of the package components are listed in Table4 which include the elastic properties of the solder alloys as a function of temperature[7-8]. BT is short for bismaleimide triazine; EMC is short for Epoxy Molding Compound.

Table4: Material Properties

Material	Young's Modulus E(GPa)	Pois-son's Ratio μ	CTE α (ppm/℃)
EMC	14	0.23	16.0
Die	169	0.26	2.3
PCB	18.2	0.25	15
BT	22	0.30	$\alpha_x \, \alpha_y$ 15.5 α_z 52.5
Sn-3.8Ag-0.7Cu	48.5 (233k) 37.9 (398k)	0.36	20
63Sn-37Pb	40.5 (233k) 15.4 (398k)	0.35	21

Temperature load applied to the model specifies the range of -40°C to +125°C, with 15 min ramp, 15 min hold at hot, and 15 min hold at cold. Four temperature cycles are executed in order to confirm the stabilization of the creep responses.

2.   Simulation Results analysis

Figure 4a shows the distribution of equivalent creep strain in the solder joints array. It shows that the highest creep strain was found to be located at the solder joints near the die for both 63Sn-37Pb and Sn-3.8Ag-0.7Cu solder alloys. The cause of this distribution is analyzed as following:

Factors which result in greater creep strain are greater assembly stiffness, greater temperature excursions, higher temperature range, greater CTE mismatch, greater distance from neutral point(DNP), smaller joint heights, and lower cyclic frequency[9]. For solder joints in the same array , which experience the same thermal cycles, determinants for different creep strain are CTE mismatch and distance from neutral point(DNP), which result in higher deformation and greater internal stress. This can be seen from euqation (1), in which stress plays a deciding role and the key factor for the creep strain differences in the solder joints array. High stress areas will creep more rapidly. As the highest creep strain located at the central areas, rather than the edge, the effect of CTE mismatch on creep is more remarkable than that of DNP. That means CTE mismatch is more prominent for creep strain in this assembly model. Because the CTE of the die is much smaller than that of other components in neighborhood, so the expansivity mismatch in this area is much more seriously than other areas. Figure 5 depicts the deformed situation of the components near the die. It shows that the die intensively restrains the thermal deformation of other components around it. And this will generate a great deal of stress. So the equivalent stress of solder joints array in this area is also the highest (see figure 4b).

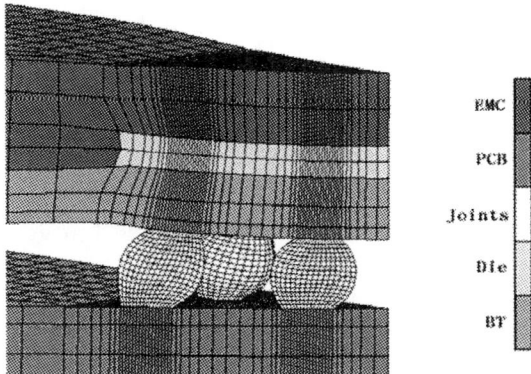

Figure 4: the deformation of the components near the die

Low cycle fatigue model is frequently used to predict fatigue life of solder joints subjected to thermal cycling loading. Coffin-Manson strain-based model is one of the most commonly used model. And besides strain-based model, strain energy-based methods have also been applied to the analysis of solder joints. The life prediction model used in this work was taken From Bert Zahn for 63Sn37Pb and 95.5Sn-4.0Ag-0.5Cu soldler[10]. The life prediction model is shown in equation (3), Parameters of the model are shown in table 5 :

Table5: Parameters of life prediction equation

Solder alloys	C_1	C_2
63Sn-37Pb	902.57	-0.5791
95.5Sn-3.8Ag-0.7Cu	2054.6	-0.1081

$$N_f = C_1(\Delta W_{in})^{C_2} \qquad (3)$$

ΔW_{in} is inelastic strain energy density range, which was extracted from numerical results as a damage parameter for comparison between different solder joints. For most of thermal cycle experiments, strain rates are on the order of $10^{-5} sec^{-1}$, very little time-independent plastic deformation take place. Hence, fatigue life can be correlated solely in terms of the creep strain per cycle[9]. So in this work, ΔW_{in} is approximately considered to be equal to ΔW_{cr}, which can be determined from the area of the stable shear stress and shear creep strain hysteresis loop(see figure6). Since the hystersis looops have become stabilized after the first 2 cycles, the subsequent cycles can represent a steady state of secondary creep, which was assumed in the solder consititutive. The results of computional caculationg are sumerized in table 6.

Table6: Parameters of life prediction equation

solder	$\Delta\varepsilon(\%)$	ΔW_{in} (MPa)	N_f (cycle)
63Sn-37Pb	1.53	0.336	1698
95.5Sn-3.8Ag -0.7Cu	1.05	0.225	2413

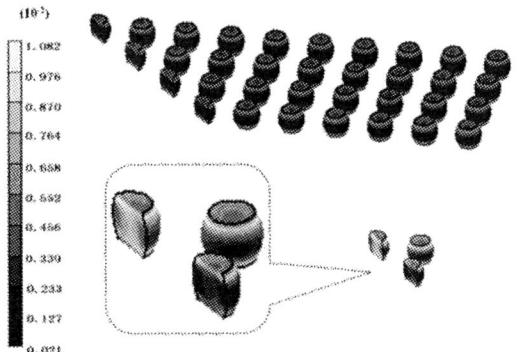

(a) The distribution of equivalent creep strain

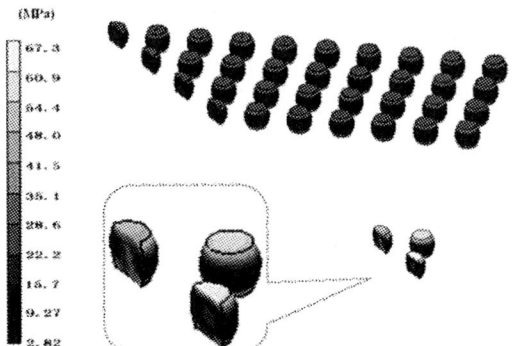

(b)The distribution of equivalent stress

Figure 3 : The distribution of equivalent creep Strain and equivalent stress in the solder array

It can be seen that the creep strain range and shear creep strain energy density range of eutectic Sn-Pb solder are all larger than that of 95.5Sn-3.8Ag-0.7Cu solder. Analyzing the shapes and sizes of loops at the location with the maximum creep strain range for both solder joints, it reveals that the eutectic Sn-Pb joints creep more seariously in the process of thermal cycling. And the joints made of 95.5Sn-3.8Ag-0.7Cu solder with longer fatigue cycles exhibited a higher reliability.

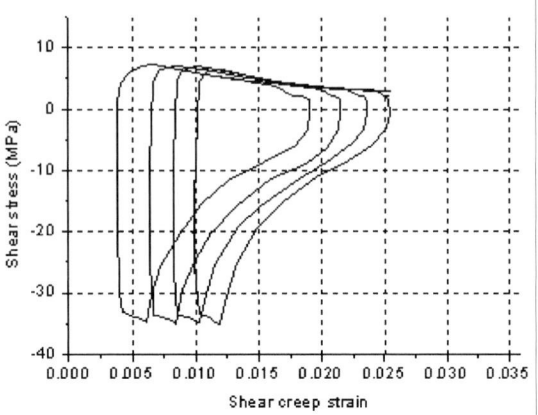

(a) 63Sn-37Pb solders

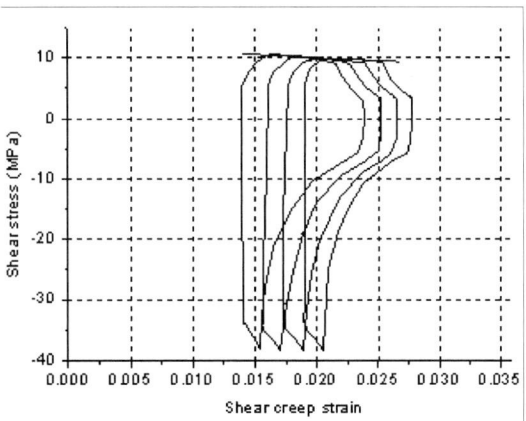

(b) 95.5Sn-3.8Ag-0.7Cu solders

Figure 5: Hysteresis Loops of Solder Joints Under Thermal Cycling

3. Conclusions

The highest creep strain were found to be located at the solder joints near the die for both 63Sn-37Pb and Sn-3.8Ag-0.7Cu solder alloys. This might be attributed to the severe expansivity mismatch between the die and the other components. For the model developed in this study, the effect of CTE mismatch on creep is more remarkable than that of DNP.

Eutectic Sn-Pb solder Joints creeped more seriously during thermal cycling. And the 95.5Sn-3.8Ag-0.7Cu solder joints show a higher reliability. PBGA components assembled with 95.5Sn-3.8Ag-0.7Cu solder improved

about 42% with respect to the lifetime of the packages assmbled with eutectic Sn-Pb solder.

References

1. F.X. Che ,John H.L. Pang, "Thermal Fatigue Reliability Analysis for PBGA with Sn-3.8Ag-0.7Cu Solder Joints," *Proc. Electronics Packaging Technology Conference*, 2004, pp. 787-792

2. Ahmer Syed, "Accumulated Creep Strain and Energy Density Based Thermal Fatigue Life Prediction Models for SnAgCu Solder Joints," *Proc. Electronic Components and Technology Conference*, 2004, pp. 737-746

3. Abtew M., Selvaduray G., "Lead free solders in microelectronics," Mater Sci Eng, vol. 27 (2000), pp. 98-102.

4. Zeng K, Tu K. N., Mater Sci. Eng. Rep. vol.38 (2003), pp. 55

5. Darveaux, R., "Solder Joint Fatigue Model," *Proc. Syniposium on Design & Reliability of Solders and Solder Interconnections*, Orlando, FL, February. 1997, pp. 213-218.

6. Pang, H.L.J., Xiong, B.S. and Low, T.H., "Creep and Fatigue Characterization of Lead Free 95.5Sn-3.8Ag-0.7Cu Solder",*Proc. 54th ECTC*, Las Vegas, Nevada, June. 2004, pp.1333-1337.

7. Jiang Tingbiao, "Finite Element Simulation Analysis on Cure-residual Stress in PBGA Device", *Electronic Components & Materials*, Vol. 24, No. 2 (2005), pp. 3-6

8. Rainer Dudek, Hans Walter, Ralf Doering, and Bernd Michel. "Thermal Fatigue Modeling for SnAgCu and SnPb Solder Joints", *Proc. 5th. Int. Conf. on Thermal and Mechanical Simulation and Experiments in Microelectronics and Micro-Systems*, EuroSimE 2004,pp. 557-564

9. Robert Darveaux, Kingshuk Banerji, "Fatigue analysis of flip chip assemblies using thermal stress simulations and a Coffin-Manson relation," *Proc. 41st Electronic Components and Technology Conference*, 1991, pp. 797-805

10. Bert A.Zahn. "Solder Joint Fatigue Life Model Methodology for 63Sn37Pb and 95.5Sn4Ag0.5Cu Materials," *Proc.Electronic Components and Technology Conference*, 2003, pp. 83-94

Acknowledgment

The financial support work from the Natural Science Foundation of China (No. 50575060) is gratefully acknowledged.

Comparative Sensitivity Analysis for µBGA and QFN Reliability

Jürgen Wilde, Elena Zukowski
University of Freiburg,
Department of Microsystems Engineering
Laboratory for Assembly and Packaging
Georges-Koehler-Allee 103,
79110 Freiburg, Germany

Abstract

The estimation of product reliability during the design is one of the key questions in microelectronics. The assembly lifetime is a function of such parameters as geometry, material properties and loads. All these influences exhibit systematic and stochastic variations. The effect of variability can be analysed by a probabilistic FE-simulation and statistical methods. This paper presents an approach for the prediction of thermal fatigue life of two CSP types, the µBGA and the QFN. Besides geometry parameters, also material properties and cycle temperatures are used as variable inputs. Based on a preliminary study the input parameters were defined as normal distributed. Sensitivities of the lifetime to the design parameters were computed and ranked after FE-simulations for both µBGA and QFN packages parameters had been performed. The fatigue life prediction of solder joints used in this work is based on a Coffin-Manson model and it was performed using the stress-strain data extracted from FE-simulations. As there exists a dependency on the solder deformation behaviour, the correct choice of the deformation model of lead-free solder alloys is an important aspect of this work. Summarising, in this work a probabilistic simulation method was developed to compute realistic failure distributions of two CSP types.

1. Introduction

Under the current rapid change of electronic technology, the need to develop assemblies that are increasingly reliable is indispensable. FE analysis can be used to predict performance, lifetime, and reliability already in the design phase. For lifetime calculations, analytical models no longer play an essential role in the technology qualification phase.

Differences in the CTE of the assembly materials result in large stresses and strains during thermal cycles. This is the dominant mechanism leading to failure of soldered assemblies. It is important to predict not only thermal stresses and strains within the system, but also to analyse how different design parameters influence the stress. Furthermore, it is necessary to determine which material properties, geometry parameters and manu-facturing process conditions are critical for reliability.

Reliability of soldered joints under temperature loads is critical not only for BGA but also for QFN (Quad Flat Non-leaded) packages, one of the most advanced package types. For complementing previous studies, which focused mostly on BGAs, this paper will compare the two packages µBGA and VQFN. Here the solder fatigue life of the assemblies is predicted using the Coffin-Manson equation in combination with parametric FE-models.

2. FE-Model

The investigation of design parameters like material properties, or geometry was carried out using the parametric study approach. A configurable model was constructed, so that a parametric analysis could be performed with variable geometry and material properties without having to rebuild the model for each change. The capability to update finite elements models rapidly allows one to perform Monte Carlo simulations of assemblies with multiple random variables. The existing parameterised deterministic model of µBGA and QFN [1] was extended to a probabilistic model by treating the geometry, dimensions and material properties as random variables.

2.1 Geometric Details

A quarter symmetry 3D non-linear finite-elements model was generated using a modular modelling method [1].

Fig.1 A quarter symmetry 3D FE-µBGA model

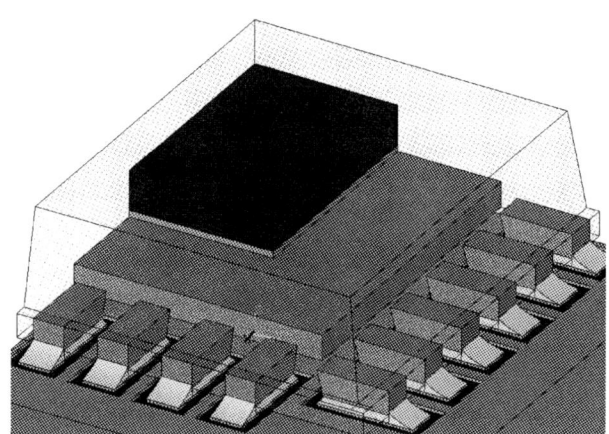

Fig.2 A quarter symmetry 3D FE-QFN model

Packages chosen for this study included a QFN and μBGA packages with 0.5 mm and 0.8 mm pitch size respectively. The connection of the chip to the substrate is accomplished using 49 I/O in a 7×7 matrix for the μBGA and 40 I/O (8×12) in the case of QFN solder joints. Both packages were mounted to non-solder mask defined 1.5 mm thick FR4 board.

Figures 1-3 illustrate the constructed FE-models and provide a detailed view of the copper lands and the solder joint region.

Fig.3a Detailed view of the copper lands-solder joint region of a metallographic QFN cross-section

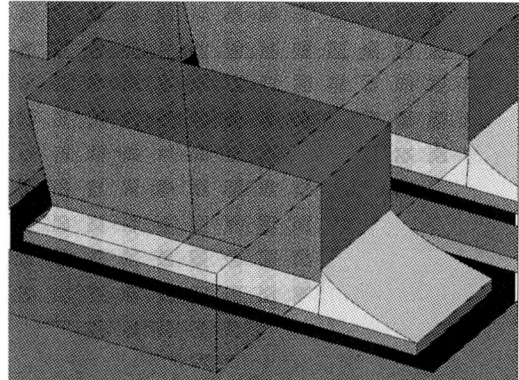

Fig.3b Detailed view of the copper lands-solder joint region of QFN model

2.2 Material Model

Most of the materials properties were assumed to be linear elastic.

Table 1. Material properties used in simulations, (all data taken at 25°C)

Material	CTE, ppm/K	E Modulus, MPa
Silicon	2.6	Orthotropic $c_{11}=165640$ $c_{12}=63940$ $c_{44}=79510$
FR4 (Interposer)	15(X); 16(Y); 50(Z)	23000
Die att. adhesive	65	1500
Solder mask	60	3500
Cu	16,7	130000
FR4 (PCB)	12(X); 17(Y); 45(Z)	23000

Two variants of moulding compounds, which are identified as LS#1 (Low Stress) and LS#2, are chosen for the investigations. Moulding compound LS#1 has a low T_g and as consequence the CTE difference is high at high temperatures (Fig. 4a).

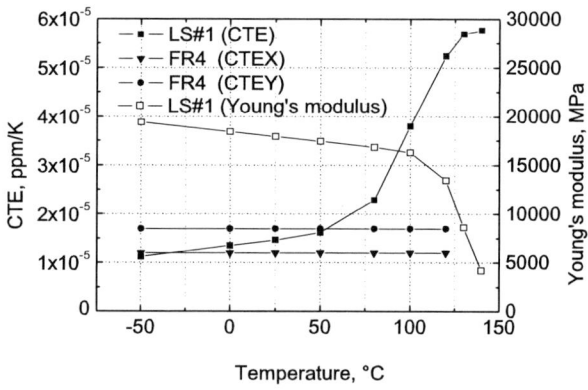

Fig.4a CTE of moulding compound LS#1 [2]

The second moulding compound has lower CTE than the substrate at all simulated temperatures (Fig4.b).

Temperature dependent values of CTE (22ppm/K@25°C) and Young's modulus (50GPa@25°C) for SnAgCu solder were found in literature [3].

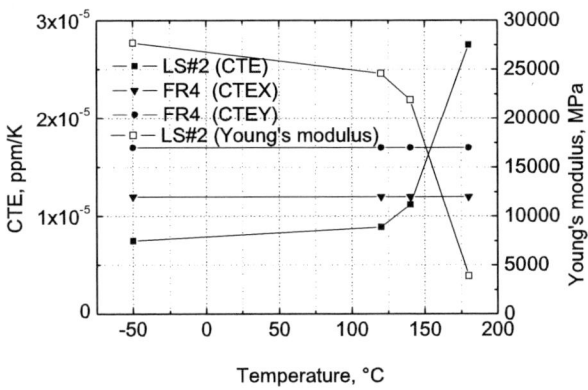

Fig.4b CTE of moulding compound LS#2

The mechanical properties of solder alloys are non-linear. For modelling the creep deformation of solder materials in FE-simulations, the well-known generalized Garofaldo equation [4] was used. The inelastic strain rate is given as

$$\frac{d\varepsilon_{cr}}{dt} = C_1 \left[\sinh(C_2\sigma)\right]^{C_3} \exp\left(-\frac{C_4}{T}\right) \quad (1)$$

σ is von-Mises stress.

During the last years, several constitutive models for Sn95.5Ag3.8Cu0.7 were published, but there is yet no material model, which is accepted as general standard. The analysis of published creep data was performed with five test simulations. The corresponding constants of steady-state creep deformation are shown in the Table 2.

Table 2. Deformation constants of steady-state creep for SnAg3.8Cu0.7 and SnAg3.8Cu1 [10] alloy

C_1[1/s]	C_2 [1/MPa]	C_3	C_4 [K]	Reference
3.20E+04	0.037	5.10	6525	[5]
2.78E+05	0.024	6.41	6500	[6]
5.01E+02	0.032	4.96	5434	[7]
9.81E+03	0.045	8.03	8547	[8]
8.78E+08	0.052	5.13	11210	[9]
2.60E-03	0.185	3.00	4655	[10]

For comparison, the isothermal lines of the different data sets from Table 2 are plotted in Figs. 5a and 5b. The plotted data cover a wide range of stress levels and creep rates [6]. The graph corresponding to [8] shows higher creep resistance than the other models. The largest differences between the data sets are observed at low stresses and low strains.

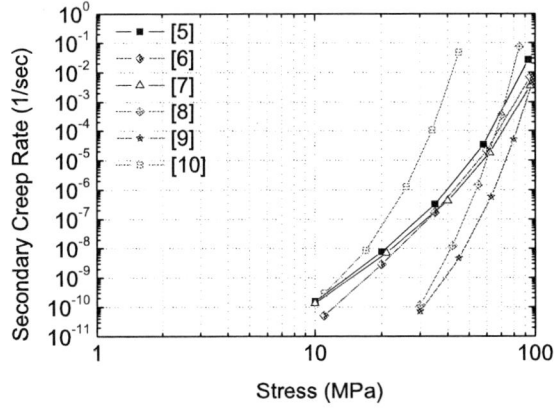

Fig.5a Creep strain rate versus von-Mises stress for Sn95.5Ag3.8Cu0.7 at -40°C

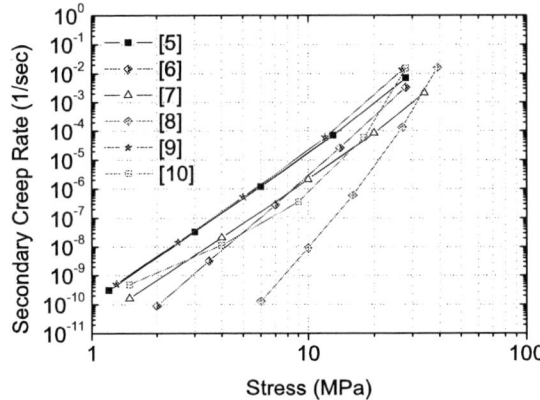

Fig.5b Creep strain rate versus von-Mises stress for Sn95.5Ag3.8Cu0.7 at 125°C

The corresponding hysteresis loops for simulated shear stress versus creep shear strain in Sn95.5Ag3.8Cu0.7 solder joints of BGAs are plotted in Fig.6 for thermal cycles between -40°C and 125°C.

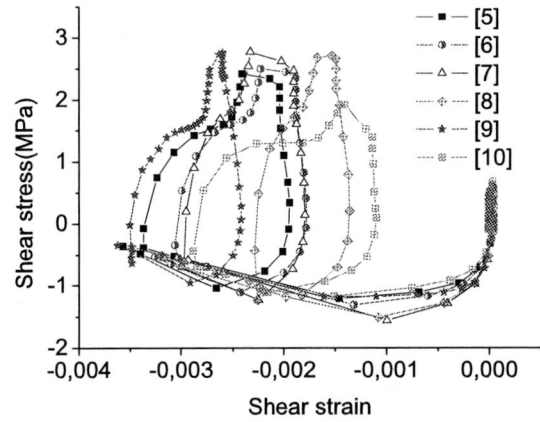

Fig. 6 Simulation of shear stress vs. shear strain creep hystereses for Micro-BGA packages

The simulated equivalent creep strain amplitude and the predicted life cycles for each creep models for are given in Table 3.

Table 3. Equivalent creep strain amplitude and number of cycles to failure computed with Equation 3

$\Delta\varepsilon_{cr}$	N_f	Reference
1.33E-2	1215	[5]
1.26E-2	1300	[6]
1.21E-2	1365	[7]
1.02E-2	1700	[8]
1.02E-2	1698	[9]
1.54E-2	998	[10]

From the life prediction results, it becomes evident that the data have some scattering. The creep-fatigue lives of µBGA solder joints predicted with [8] and [9] are higher than those predicted with [5], [6], [7] and [10]. In the authors′ opinion the data from [6] represent average values for the considered sinh-model parameters and therefore these data are used for our FE-simulations.

3. Input parameters and loads

Material properties and geometry data are used as analysis inputs. Geometric features exhibit variations, which are caused by manufacturing and assembly processes. Among these are the component placement accuracy on the substrate (diagonal distance from corner joint to corner joint), screen-printing process variations and solder paste parameters like the standoff height. Some geometry parameters and material properties influence the solder joint reliability significantly. For example, the materials and geometry, that define the package structure, have influence on the effective CTE of the package. Therefore, CTE is a function of the coefficients of thermal expansion of the materials, the thickness and the moduli of elasticity [11, 12].

As shown in Fig. 7 a cyclic thermal load is defined by a temperature transient from –40°C to 125°C, with 30 *min* dwell time and 10 *s* ramp time.

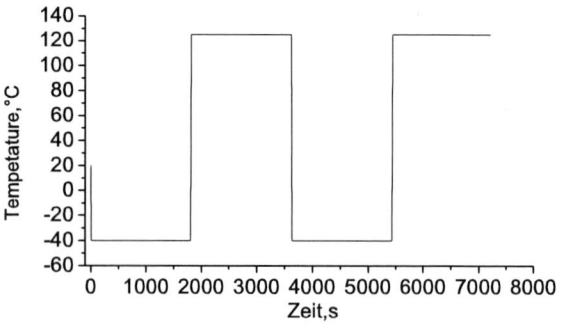

Fig.7 Passive temperature loading cycle used in simulations

For creep analysis, it is important to study the responses for multiple thermal cycles until the hysteresis loops become stabilized. In our simulations creep shear strain is stabilized after the second cycle. Therefore only two thermal cycles were simulated.

4. Results and Discussion
4.1 Fatigue model

For strain-based analysis, Coffin-Manson fatigue models [13-15] are widely used for low cycle fatigue analysis.

$$\Delta\varepsilon_{pl} = C \cdot N_a^{\alpha} \qquad (2)$$

Here $\Delta\varepsilon_{pl}$ is the effective von-Mises viscoplastic strain range and α and C are temperature-dependent parameters, which can be obtained experimentally from strain-cycling tests. In practice, for experimental or analytical determination of the creep strain amplitude $\Delta\varepsilon_{cr}$ can be used at high homologous temperatures in the Coffin-Manson equations for the prediction of the number of cycles to failure N_f.

Inelastic strain and stress in µBGAs and VQFNs solder joints are taken from FEA. For the evaluation of assembly lifetime, the following damage model [16] was applied:

$$N_f = 4.5(\Delta\varepsilon_{cr})^{-1.295} . \qquad (3)$$

In thermal cycling tests on many components the times or numbers of cycles to failure will be distributed over rather large intervals. The two-parameter Weibull function is considered as a suitable distribution function for failure times or cycles N_f of the assemblies:

$$F(t) = 1 - e^{-\left(\frac{t}{\alpha}\right)^{\beta}} \quad or \quad F(N_f) = 1 - e^{-\left(\frac{N_f}{\alpha}\right)^{\beta}} \qquad (4)$$

Cycles to failure distributions for Weibull plots can be obtained from thermal cycling simulations only when distributed input parameters are used.

4.2 Probabilistic sensitivity analysis

Probabilistic sensitivities are important in order to improve a design toward a more reliable version. From Monte Carlo simulations, sensitivities can be derived by calculation of correlation coefficients between input variables and output parameters. A frequently used non-parametric correlation tests for sensitivity calculation is the Spearman test based on rank-order correlation [4].

$$r_s = \frac{\sum_i^n (R_i - \overline{R})(S_i - \overline{S})}{\sqrt{\sum_i^n (R_i - \overline{R})^2} \sqrt{\sum_i^n (S_i - \overline{S})^2}} \qquad (5)$$

There R_i is the rank of x_i within the set of observations $[x_1, x_2, ... x_n]$, S_i is the rank of y_i within the set of

$[y_1, y_2, ... y_n]$, and $\overline{S}, \overline{R}$ are average ranks of a R_i and S_i respectively. Values of sensitivity r_S can be distributed in a range from +1 (perfect correlation), through 0 (no correlation), to -1 (perfect negative correlation).

4.3 Effect of solder pad opening size and area ratio

The geometric characteristics can have a significant effect on the thermo-mechanical reliability. Two variants of ball geometry were regarded: NSMD pads on boards and SMD pads on package substrates. For both types the ball radius (225 µm) and the connection radius on the package side (170 µm) are the same and pad radii on the substrate side are different (150 µm and 200 µm respectively). So for variant #1 (small substrate pad) $R_s > R_p$, while for variant #2 (large substrate pad) $R_s < R_p$.

Fig.8 Solder ball models with different copper pad designs.

The two-parameter Weibull distribution was used to characterize the lifetime distribution of assembled packages. Smaller pads lead to higher reliability while larger pads have lower lifetime and a higher failure probability.

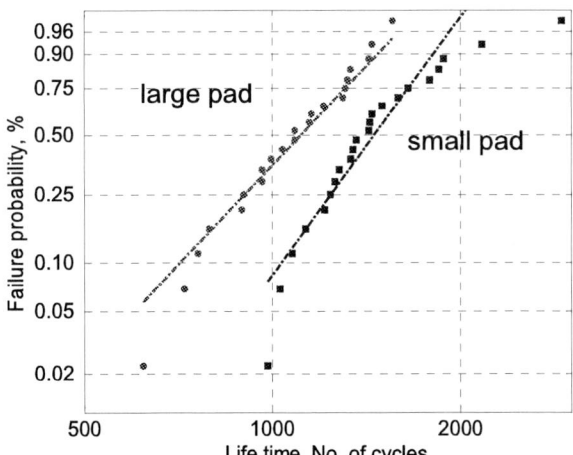

Fig.9 Computed cycles to failure distributions for two different solder pad opening size (µBGA49)

Sensitivities of lifetime to the change of such input variables as package pad opening radius (R_p) and substrate pad radius (R_s) were calculated. Comparison of these coefficients shows that the pad opening radius is a powerful parameter for package reliability optimisation.

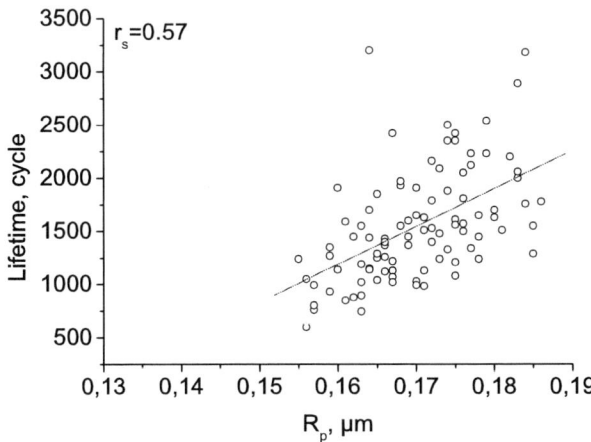

Fig.10a Sensitivity of lifetime to package pad opening radius (µBGA49)

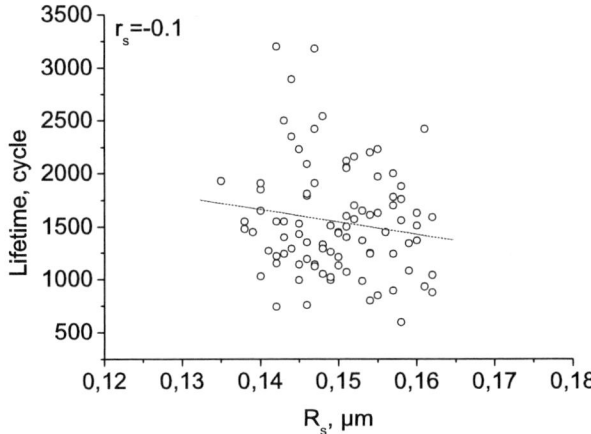

Fig.10b Sensitivity of lifetime to substrate pad radius (µBGA49)

For comparison, we have investigated the influence of QFN solder joint geometry parameters on package lifetime.

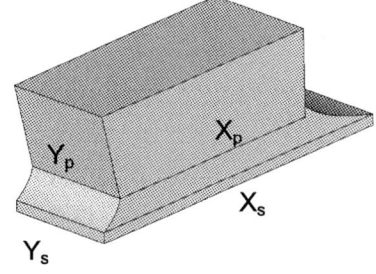

Fig.11 Interconnect geometry parameters on VQFN

For VQFN the fatigue life of soldered joints shows a positive sensitivity to increasing Cu-pad dimensions, which determine solder joint geometry. However, PCB pad width Y_s has the strongest influence ($r_s = 0.33$) on the lifetime. Therefore, the increase of solder pads and solder land areas leads to increasing solder joint reliability.

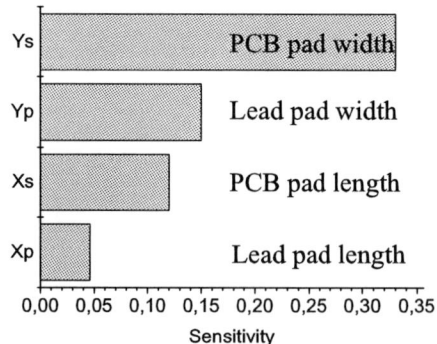

Fig.12 Lifetime sensitivity plot of interconnect geometry variables for the FE-model of an QFN

It is shown, that µBGAs with larger solder mask opening size and QFNs with longer peripheral lead length have longer fatigue lives.

4.4 Effect of moulding compound CTE on fatigue life

The material data of the moulding compound can be important for the assembly reliability. Both properties Young's modulus and CTE values change dramatically around the glass transition temperature. Therefore, the influence of the variance of the moulding compounds´ CTE on the package lifetime was investigated.

A higher CTE of a moulding compound increases the effective package CTE and reduces the CTE mismatch with PCB. This factor which tends to increase the fatigue life. We have investigated two variants: In one case, the CTE of the moulding compound is lower than that of the PCB (LS#2). In the second case (LS#1) the difference in CTEs between moulding compound und PCB was inverse in a high temperature range, $\alpha_{MC} > \alpha_{PCB}$.

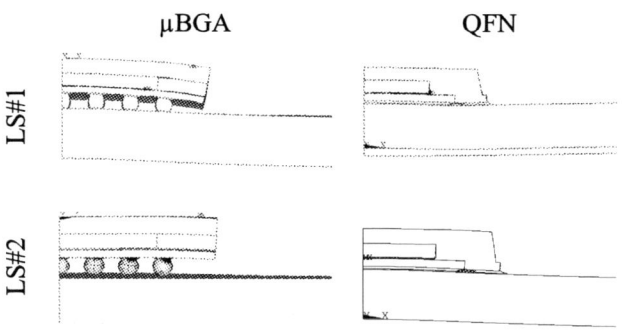

Fig.13a Package deformations for two different moulding compounds at -40°C (displacement scale factor=50)

A qualitative comparison of the thermal-mechanical behaviour of packages with different moulding compound CTEs is shown in Fig. 13. It can be seen, that at –40°C

both variants of mould compound show the same modes of package deformation.

At 125°C, the deformation behaviour of assemblies with both moulding compounds is different. For packages with inverse $\Delta\alpha$ ($\alpha_{MC} > \alpha_{PCB}$), the stress of moulding compound in the die overhang area increases due to an increase of the thermal expansion of the moulding compound in the die overhang area.

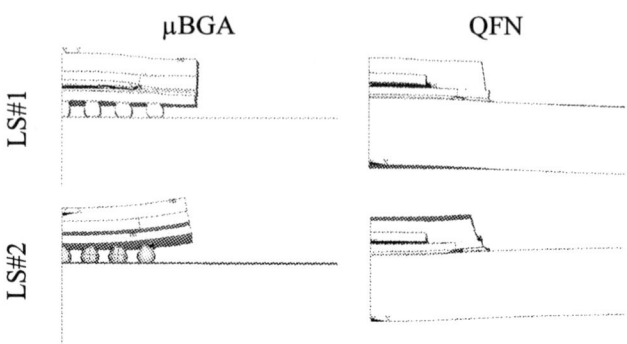

Fig.13b Package deformations for two different moulding compound CTEs at 125°C (displacement scale factor=50)

Finite elements modelling indicated that LS#1 would improve fatigue life compared with LS#2 (Fig. 14).

Fig. 14 Cycles to failure for two different moulding compounds (temperature cycling -40°C to +125°C for VQFN-40)

Another interesting fact is that for LS #1 the location of the critical zone is changed due to the influence of the CTE of the moulding compound. In order to investigate the effect of the variability of geometry and material properties a Monte Carlo simulation was conducted. All input values were changed according to their distributions and strain-based fatigue life predictions were made. In the case when LS#2 was used for the µBGA, the critical zone with the highest strain range was found in the corner balls for all simulations. For LS#1, the critical zone was displaced to the next ball in the diagonal inward direction. Nevertheless for some geometry value sets the zone of maximum strain appeared in the corner balls.

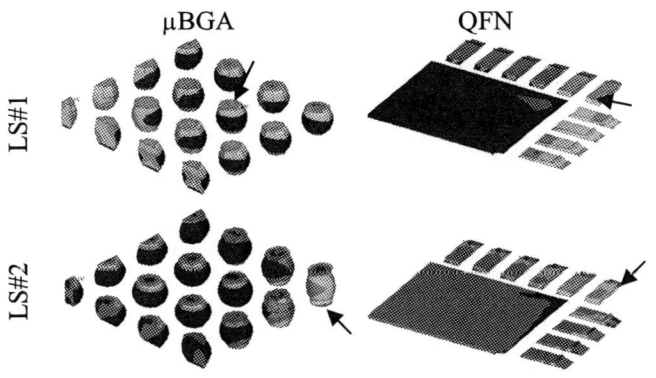

Fig.15   Location of the critical zone for two different moulding compound CTEs (displacement scale factor=50)

5. Conclusions

A comparative analysis of μBGA and QFN is undertaken regarding geometric and material parameters. Stress and creep analyses were utilized to predict the reliability of lead free solder joints during thermal cycling. Simulations results coupled with sensitivity analyses show the same behaviour for both types of packages.

- Increase of solder mask opening sizes for μBGA and peripheral lead length for QFN increased also package fatigue life.
- Values of the CTE difference between moulding compounds and substrates can change the location of the critical zones
- It is important use optimal moulding compound CTE, in order to prevent inverse CTE difference.

Acknowledgments

The described investigations were funded by the Federal Ministry of Economics and Labour BMWA under the AIF research program (AIF project number 13.821 N) and supported by the Research Association on Welding and Allied Processes of the German Welding Society DVS (DVS-No. 10.040). The authors are very grateful for this support.

References

[1] Zukowski, E., Deier, E., Wilde, J., "Correct modelling of geometry and materials properties in the thermo-mechanical finite-elements-simulation of chip scale packages", *Proc. Eurosime 2005*, pp. 545-552.

[2] S. Fischer, "Einfluss der Aufbau- und Verbindungs-technik auf die funktionalen Eigenschaften thermomechanisch belasteter Sensoren",

Dissertation, University of Freiburg, 2006, ISBN-10:3-89959-527-0

[3] Database for solder properties with emphasis on new lead-free solders, Release 4.0, *National Institute of Standards&Technology and Colorado School of Mines*, www.boulder.nist.gov/div853/lead%20free/solders.html

[4] ANSYS 10.0 Documentation

[5] John, Pang, H. L., Xiong, B. S., Low, T. H., "Creep and fatigue characterization of lead free 95.5Sn-3.8Ag-0.7Cu solder", *Electronic Components and Technology Conference*, 2004, Vol.2, pp.1333- 1337

[6] Schubert, A., Dudek, R., Auerswald, E., Gollhardt, A., Michel, B., Reichl, H., "Fatigue life models of SnAgCu and SnPb solder joints evaluated by experiments and simulations," *53rd ECTC Conference Proc.*, 2003, pp. 603-610

[7] Pang, J., Xiong, B., "Mechanical Properties for 95.5Sn-3.8Ag-0.7Cu Lead-Free Solder Alloy," *IEEE Transactions on Components and Packaging Technologies.* 2005, Vol. 28, no. 4, pp. 830-840.

[8] Dušek, M. and Hunt, C., "Measurement of materials properties of lead-free solders for modelling requirements", *Proc. SMTA Int. Conf.*, 2003, pp.753-757

[9] Pang, J.H.L., Xiong, B.S., Neo, C.C., Zhang, X.R. and Low, T.H., "Bulk solder and solder joint properties for lead-free 95.5Sn-3.8Ag-0.7Cu solder alloy", *Proc. 53rd ECTC*, 2003, pp.673-679

[10] Wiese, S., Schubert, A., Walter, H., Dudek, R.,Feustel, F., Meusel, E., Michel, B. , "Constitutive behaviour of lead-free solders vs. lead-containing solders-experiments on bulk specimens and flip-chip joints", *Proc. 51rd ECTC*, 2001, pp.890-902

[11] Chaparala, S.C.; Roggeman, B.D.; Pitarresi, J.M.; Sammakia, B.G.; Jackson, J.; Griffin, G.; McHugh, T., "Effect of geometry and temperature cycle on the reliability of WLCSP solder joints", *IEEE Trans. CPMT*, Vol. 28, No 3, 2005, pp. 441- 448.

[12] Lall, P.; Singh, N.; Suhling, J.C.; Strickland, M.; Blanche, J. "Thermo-mechanical reliability tradeoffs for deployment of area array packages in harsh environments joints", *IEEE Trans. CPMT*, Vol. 28, No. 3, 2005, pp. 457- 466.

[13] IPC-SM-785, Guidelines for accelerated testing of surface mount solder attachments. *IPC, Lincolnwood, IL, USA*, 1992.

[14] Schmidt, C. G; Simons, J. W. and Kanazawa, C. H., "Thermal fatigue behaviour of J-lead solder joints", *IEEE Trans. CPMT-Part A*, 1995,Vol. 18, No. 3

[15] Solomon, J.D;. "Fatigue of 60/40 solder", *IEEE Trans. CHMT*, (1986). Vol. CHMT-9, pp 423-433.

[16] Dudek, R., Rzepka, S., Dobritz, S., Doring, R., Keybig, K., Wiese, S., Michel, B.: "Fatigue life prediction and analysis of wafer level packages with SnAgCu solder balls", *Electronics Systemintegration Technology Conference*, 2006, Vol. 2, pp.903-911

Experimental Determination of Time-Independent Elastic-Plastic Behaviour of Solder Joints at High Strain Rates

S. Wiese[1], K. Meier[1], D. Scholz[2], A. Müller[2], M. Röllig[1], S. Rzepka[2], K.-J. Wolter[1]

1) Dresden University of Technology, Electronic Packaging Laboratory, Germany
2) Qimonda Dresden GmbH & Co. OHG, Dresden, Germany
TU Dresden, IAVT, D-01062 Dresden, Germany
wiese@avt.et.tu-dresden.de, +49 351 33172

Abstract

In order to investigate the time-independent behaviour of solder joints at high strain rates, a test setup was created capable of measuring the force vs. time reaction of small solder joint in shear loading at strain rates between $10^{-1} - 10^4$ [1/s]. The paper describes the design of this high strain rate tester. Details providing for the fast and high resolution force measurement are given. The results of analytic and FEM analyses applied to optimize the force sensor are presented. The paper also points out the challenges of high strain rate measurements of small solder joints.

The validation tests applied alumina chips with five SnAgCu solder joints, which had barrel shape with about 200 μm diameter and stand-off height. The test results show good reproducibility of the force vs. time reactions. They match the benchmark results obtained at moderate shear speeds with existing equipment. In addition, fractographic analysis clearly found the two failure modes, bulk and brittle, that had been expected. Hence, the new test setup seems well suited for the tasks it has been designed for.

1. Introduction

Although the majority of electronic products fail by thermomechanical fatigue, specific devices, e.g. mobile phones, smart cards, notebooks etc., experience a substantial amount of pure mechanical stress in addition. These purely mechanical loads in form of bending and twisting of the substrate are caused by vibration and drop impacts. In contrast to thermomechanical fatigue, the pure mechanicals loadings are characterised by quick stress changes. When the stress conditions change fast, the solder material has no time to creep. Therefore, it deforms plastically (high strain rate plasticity). However, most of the current constitutive models for FEM simulation account for elastic and creep deformation terms only. This can lead to significant errors when purely mechanical loadings of electronic assemblies are simulated based on these models. Hence, the existing constitutive models need to be upgraded by a term for high strain rate plasticity.

As pointed out by Frost and Ashby [1], conditions of explosive or shock loading may lead to strain rates that are very large ($> 10^2 \text{ s}^{-1}$). In such case the interaction of moving dislocations with phonons or electrons can limit

its velocity. At low strain rates the velocity of moving dislocations is usually limited by obstacles such as the lattice, hard particles or the dislocation itself. The rate equation for drag limited glide can be expressed as [1]:

$$\dot{\gamma} = \frac{c \cdot \rho \cdot G}{B} \cdot \left(\frac{\tau}{G} \right) \quad (1)$$

where $\dot{\gamma}$ is the shear rate, ρ is the density of mobile dislocations, G is the shear modulus, B is the drag coefficient, τ is the shear stress and c is a constant which includes the appropriate Taylor factor.

At high strain rates at which phonon drag dominates, ρ approaches a constant value [2, 3]. In that case the strain rate depends linearly on stress, and can be expressed by [1]:

$$\dot{\gamma} = A \cdot \left(\frac{\tau}{G} \right) \quad (2)$$

where A is a constant and can be estimated to be $A \approx 5 \cdot 10^6 \text{ s}^{-1}$ using the data from [2, 3].

According to the considerations in [1] about the deformation behaviour at high strain rates there will be strong hardening effect at high strain rates. This hardening effect will be responsible for the transition from ductile fracture modes within solder joints to brittle fracture modes. In order to be able to calculate this transition, the values of the constant A should be determined experimentally for various solders and metallization systems.

2. Concept of test setup

Deformation behaviour at high strain rates is usually characterized by a Split Hoppkins pressure bar test. In this test, strain gages were applied on a bulk sample, in order to measure the dynamic response, the incident, transmitted and reflected stress wave that propagates through the sample are recorded. The strain-stress-response of the material can be calculated based on stress wave theory.

Tests on SnAg3.8Cu0.7 solder samples were conducted by Pang and co-workers [4]. Bulk solder samples were tested having a total length of 50 mm, a gage length of 10 mm and a diameter of 4 mm. The tests were carried out at the three different strain rates $7 \cdot 10^2 \, s^{-1}$, $9 \cdot 10^2 \, s^{-1}$ and $1.3 \cdot 10^3 \, s^{-1}$, respectively.

Based on the results of these tests, the deformation behaviour of SnAg3.8Cu0.7 solder at high strain rates ($> 10^2 \, s^{-1}$) was described by the following rate equation [4]:

$$\sigma = 64.1 MPa \cdot \dot{\varepsilon}^{0.082} + 1.44 \cdot 10^{-5} MPa \cdot \dot{\varepsilon}^{2.22} \quad (3)$$

Equations (1) and (2) are very similar but they differ in the order of the strain rate dependency. The theoretical considerations [1] led to a linear relation while the experiments [4] rather suggest a quadratic dependency. Thus, further investigations are needed.

Using the results from [4], $10^4 \, s^{-1}$ has been deduced as maximum strain rate typically occurring in solder joints. Accordingly, the test setup has been designed to allow direct measurements of the yield stress at strain rates up to this level.

Figure 1: Test fixture for shear tests at high strain rate – slider with recirculating ball bearing guidance is mounted on 3 m long rail (above), specimen is hold in a rotational manipulator on the slider (below).

Figure 2: Test fixture for shear tests at high strain rate – the force sensor is mounted opposite to slider on the frame.

The test setup is depicted in Fig. 1. It contains a slider that is guided on 3 m long rail by a recirculating ball bearing. This bearing provides optimum stability of the slider during the impact with the sample. The weight of the slider is approximately 5 kg, in order to optimize the ratio of impact energy to kinetic energy. This way a constant speed during the impact should be achieved. The sample is attached to the slider, i.e., the moving part of the equipment, while the unit of shear tool and force sensor is mounted on a x-y-table firmly connected to the frame of the test fixture (Fig. 1, 2). In particular, the sample is mounted on a rotational table at the slider so that it can be adjusted to the shear tool precisely.

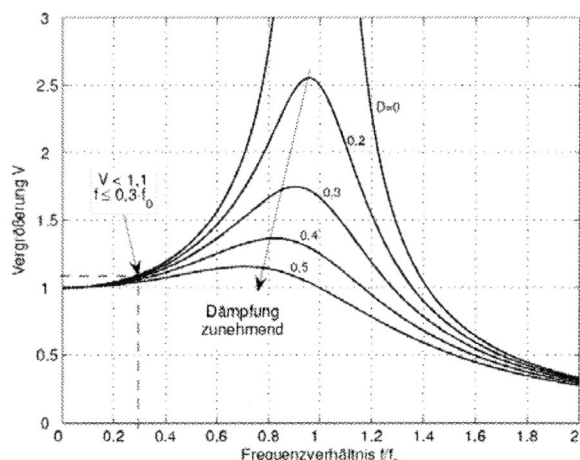

Figure 3: Relation between resonance frequency and amplification V of a mass-spring-system in dependence of the damping D.

The force sensor is the most critical part of the test fixture because it forms a mass-spring-system. If the incident stress wave has a frequency close to the resonance frequency of the force sensor, the force signal will be amplified strongly, which causes errors in force measurement. It can be seen from Fig. 3 that if the frequency of the incident stress wave is below one third of the resonance frequency of the force sensor, the error in force measurement will be less then 10 %.

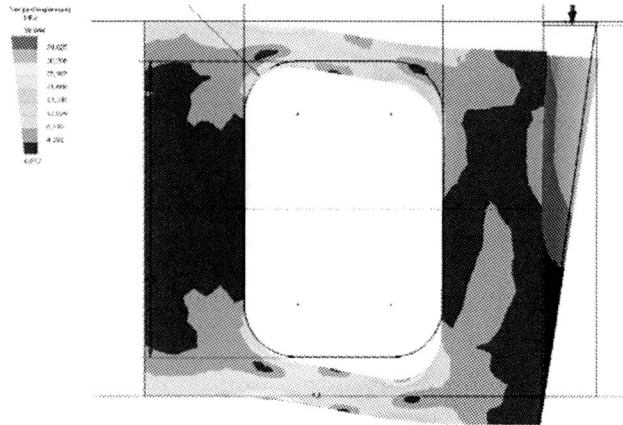

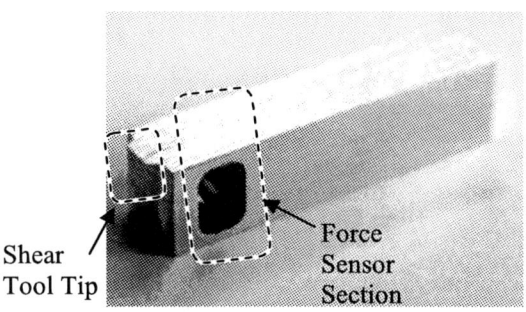

Figure 4: Unit of shear tool and force sensor: Design optimization was carried out by FE-Analysis (Top), shear-chisel-design was manufactured by precision mechanical tooling (centre), strain gages were applied on the top and bottom side of the sensor section (bottom).

In order to create a suitable force sensor for the test fixture, a design optimization was carried out by FEM-simulation. The final design is depicted in Fig. 4. A resonance frequency of $f = 29$ kHz could be achieved by series of design and material optimizations.

3. Experimental

The specimen consists on two alumina chips (3.2 mm X 1.2 mm) soldered against each other by 4 corner balls and one centre ball (Fig. 5). The solder balls were manually made in the lab by micro-punching circular platelets (diameter = 300 μm, thickness = 140 μm) out of thin solder sheets. The sheets were rolled from solder ingots. The composition of the original ingots was SnAg2.5Cu0.5. The balls were soldered on a AgPd-metallization (thick film paste), that was printed on the alumina chips, to provide circular landing pads with a diameter of 200 μm. Bumped specimens were cut of the alumina sheet and soldered against each other. Further details of the specimen are discussed in [5].

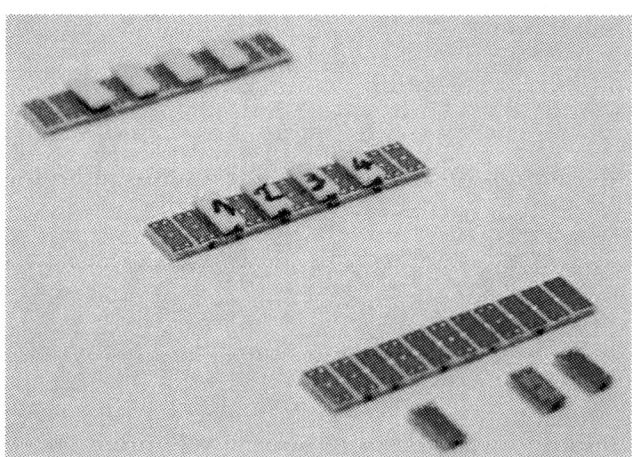

Figure 5: Specimen design: The specimen consists of two alumina chips (3.2 mm X 1.2 mm) soldered against each other by 4 corner balls and one centre ball (above). The balls were soldered on a Ag thick film metallization. Only the top chips were separated by cutting off the alumina sheet while the bottom chips were kept together in one row. After assembly this row was mounted into the test setup (below).

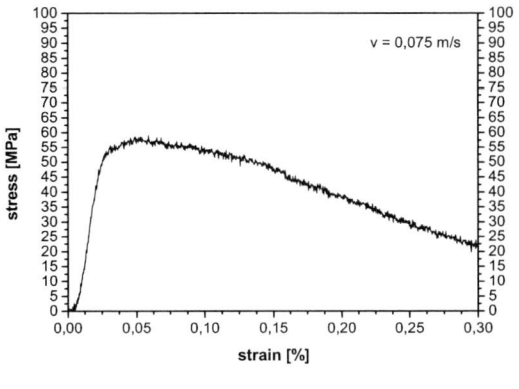

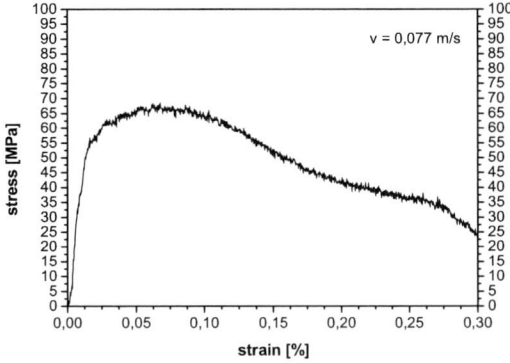

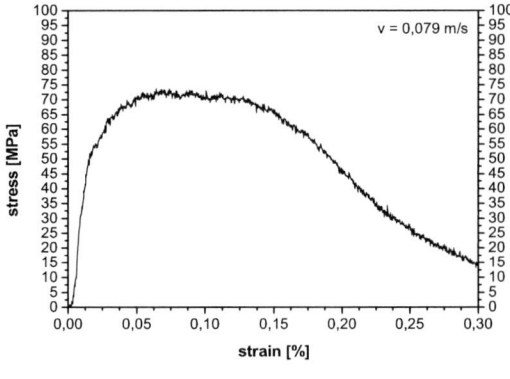

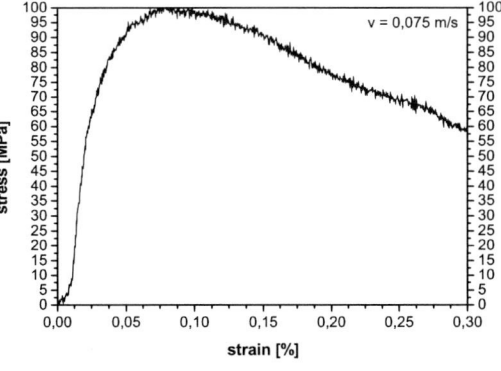

The first validation tests yielded encouraging results (Fig. 6). The moderate shear speed chosen corresponds to a shear strain rate of $4 \cdot 10^2 \, s^{-1}$, which is about the upper limit of the existing test equipment ([7]). At this loading, a yield stress at the order of 50 MPa was determined for the samples studied here. This magnitude matches the values obtained in earlier tests on flip chip solder joints [7] very well. As further seen in Fig. 6, the reproducibility of the readings is fairly good despite the dynamic nature of the test and the usual differences in joint size and shape among the individual samples. Thus, the new test setup is found well suited to the tasks it is designed for.

In addition, a fractographic analysis has been carried out (Fig. 7) after the shear test in order to interpret the strain stress results. Dominantly, the joints failed with fractures through the bulk of the solder next to the bottom chip. Again, this is in perfect agreement to the earlier tests at similar shear strain rates [7]. Nevertheless, first indications of brittle fracture mode have also been found in the current tests denoted as '2' in Fig. 7. A larger share of brittle mode failures is expected when the shear speed is increased.

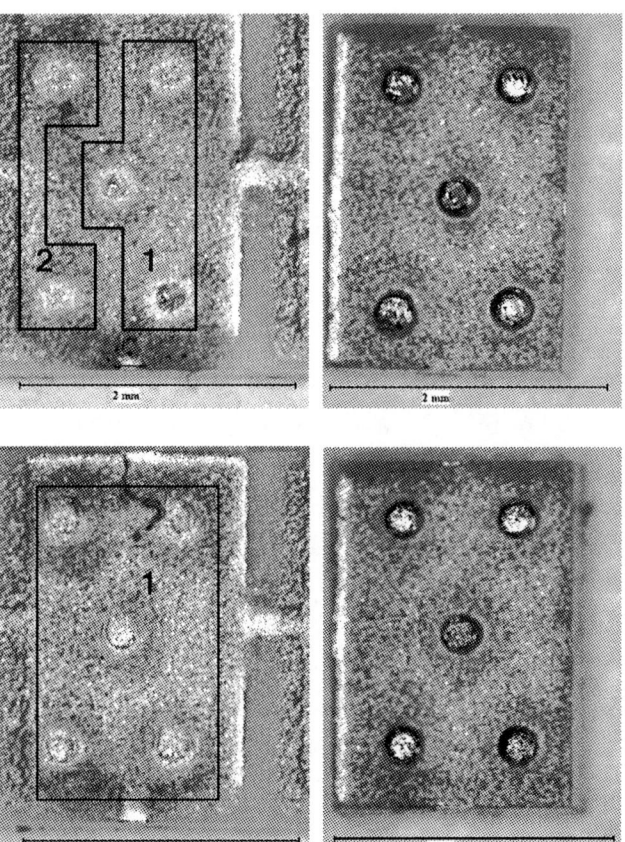

Figure 6: Results of impact shear test at speed of approximately 0.08 m/s. Stress-strain-plots were calculated using the method described in [6].

Figure 7: Fractographic analysis of the tested specimen. The Fracture occurs within the bulk of the solder (field 1) or on the solder/metallization interface (field 2).

4. Conclusions

A test setup was created that is capable to measure the force vs. time reaction of solder joints in shear loading at strain rates between $10^{-1} - 10^4$ [1/s]. The purpose of such tests is to investigate the deformation behaviour of solder joints at high strain rates. It is believed that at high strain rates ($> 10^2$ s^{-1}) solder joints undergo a significant strengthening, which is probably responsible for the transition from ductile fracture modes within solder joints to brittle fracture modes, that are observed at drop tests of electronic assemblies. The first results, that were gained on the new test setup match well to the results gained on previous setups for solder joint testing.

Acknowledgement

The work for this paper was supported within the scope of technology development by the EFRE fund of the European Community and by funding of the State Saxony of the Federal Republic of Germany.

References

1. Frost, H. J.; Ashby M. F.: Deformation-Mechanism Maps: The Plasticity and Creep of Metals and Ceramics, Pergamon Press (Oxford, 1982), pp. 9-10

2. Kumar, A.; Hauser, F. E.; Dorn, J. E. (1968) Acta Metallica, vol. 16 , pp. 1189

3. Kumar, A.; Krumble, R. G. (1969) Journal of Applied Physics, Vol. 40, pp. 3475

4. Pang, J.H.L.; Xiong, B. S.; Low, T.H.: Comprehensive Mechanics Characterization of Lead-Free 95.5Sn-3.8Ag-0.7Cu Solder. Micromaterials and Nanomaterials, Vol. 3 (2004), pp. 156–161

5. Röllig, M.; Dudek, R.; Wiese, S.; Wunderle, B.; Wolter, K.-J.; Michel, B.: Novell test concept for experimental lifetime prediction of miniaturized lead-free solder contacts. Proceedings of EuroSIME 2005, Berlin, April 18 -20 (2005), pp. 86 – 90

6. Röllig, M.; Wiese, S.; Wolter, K.-J.: Extraction of material parameters for creep experiments on real solder joints by FE-analysis. Proceedings of the Eurosime Confernce 2006, Como (I), April 24 -26 (2006), pp. 281-289

7. Wiese, S.; Rzepka, S.: Time-independent elastic-plastic behaviour of solder materials. Microelectronics Reliability. Vol. 44 (2004), pp. 1893-1900

The BGK kinetic model applied to the analysis of gas-structure interactions in MEMS

Attilio Frangi, Aldo Ghisi
Politecnico di Milano, Italy, `attilio.frangi@polimi.it`

Abstract

The BGK model of the Boltzmann equation is applied to the analysis of damping in silicon inertial MEMS working at low-moderate frequencies. Assuming small perturbations, the linearised steady-state 2D equation is implemented in a deterministic manner in order to avoid noise intrinsic in statistical approaches. Implementation details are discussed and the comparison with available experimental data in terms of forces exerted on the suspended shuttle is presented.

1 Introduction

While many features of inertial MEMS are nowadays fully dominated, the evaluation of fluid damping is still an intriguing and partially unresolved topic and strongly affects the structural response. Damping is due to gas flow in very small gaps between the movable and fixed elements of the MEMS. Since the gaps are typically only a few micrometers wide, the molecular mean free path is not negligible compared to the gap width and the gas cannot be treated as a continuous phase. The parameter employed to estimate the degree of rarefaction in a gas is the Knudsen number $Kn = \lambda/d$, where λ is the mean free molecular path and d is a typical flow dimension, e.g. the gap between electrodes [12, 18]. At ambient pressure and for several inertial MEMS like the one analyzed in Section 4, the Knudsen number is of the order of 10^{-2}, which means that the flow mainly develops in the slip flow regime. The evaluation of damping in this regime has been thoroughly investigated in the literature [18, 21, 22] and more recently in [14, 15, 33] using the Stokes model and Boundary Element techniques. According to the latter contributions, the use of integral equations and a series of simplifying hypotheses suited for moderate working frequencies permits the full scale 3D simulation of MEMS in the slip regime and guarantees excellent agreement with experimental results.

When environmental pressure or MEMS typical dimensions are further reduced, the flow enters the transition regime in which the regions where kinetic effects are important have the same size of the flowfield. Several authors have proposed to compute via analytical

or semi-analytical approaches corrected parameters to be employed in classical continuum numerical tools. A correction has been obtained by replacing the static viscosity coefficient with an effective viscosity whose computation is based on the solution of the linearized Boltzmann equation for the one-dimensional Poiseuille and Couette flow problems [28, 29, 30]. Alternatively, other authors (e.g. [4]) suggest to employ slip boundary conditions which depend on the Knudsen number and should prove accurate in the whole pressure range. These formulas yield accurate results in specific situations but have been obtained under several simplifying hypotheses which are not always met by real 3D MEMS.

In principle, a correct theoretical description of gas flow in the transition regime can be obtained by solving the Boltzmann Equation (BE) [8, 12], a complex non-linear integro-differential equation providing the distribution function of molecular velocities at any flowfield location. Mathematical difficulties prevent from obtaining closed form solutions of BE in cases of practical interest, but efficient Monte Carlo methods have been developed for its numerical treatment[5]. Unfortunately, in most MEMS flows reference Mach numbers and deviations from local equilibrium are small and difficult to capture by traditional statistical Monte Carlo methods. A number of possible modifications to statistical particle schemes have been proposed [7, 13] but research in this direction is still very active. Small deviations from equilibrium could be computed more accurately by noise-free deterministic methods. However, their huge memory demand has limited the adoption of such methods to space homogeneous or one-dimensional problems.

A deterministic approach to the numerical solution of kinetic equations becomes viable if the complicated collision integral in the BE is replaced by a simpler expression. As described below, in the BGK model kinetic equation [6] the term giving the collisional rate of change of the distribution function is simply proportional to the departure from local equilibrium. In its simpler and more usefull form, the model contains a single disposable function (the collision frequency) which depends on local density and temperature and assigns the same decay rate to all kinetic modes. Hence, the hydrodynamic limit of the model is only partially correct since the collision frequency can be tuned to obtain either the correct fluid viscosity or thermal conductivity, but not both. In spite of its shortcomings, the BGK model is often more accurate than expected, particularly in problems where momentum and heath transport are not equally important and the collision frequency can be adjusted to match the most important transport coefficient. Its applications to rarefied gases date back to the first semi-analytical solutions of [11, 27] for Poiseuille and Couette flow. Subsequently, many different numerical applications have been presented in the literature [1, 2, 19, 20, 23, 32, 34] focusing especially on high speed applications. More recently, a number of papers have appeared where low-speed flows of various complexity have been studied by deterministic numerical solutions of the BGK model kinetic equation [25, 26]. However, a complete validation of working hypotheses is still lacking since no applications to real MEMS and comparisons with experimental data are presented. Hence, in this paper a simple and straightforward discretization technique is adopted to solve numerically the BGK model equation associated to a rarefied monatomic gas flowing in a two-dimensional domain. The flow geometry is derived from a MEMS for which experimental values of damping forces are available. It is shown that excellent agreement with experimental results can be obtained in a wide range of pressures. The paper content is organized as follows: the theoretical background is presented in Section 2, the numerical method is discussed in Section 3, and an example is finally presented together with the comparison with available experimental results.

2 Formulation

Let $f(\mathbf{x}, \boldsymbol{\xi})$ denote the velocity distribution function of molecules, where $\mathbf{x}$ are space coordinates and $\boldsymbol{\xi}$ is molecular velocity. If ∇ denotes the gradient with respect to $\mathbf{x}$, the BGK model of the Boltzmann equation [2, 12, 32] reads:

$$\frac{\partial f}{\partial t} + \boldsymbol{\xi} \cdot \nabla f = \nu(\rho, T)(f_M - f) \qquad (1)$$

where the right hand side relaxation term replaces the collision operator of the Boltzmann equation which accounts for binary collisions between particles; $\nu(\rho, T)$ is the collision frequency which is assumed independent of $\boldsymbol{\xi}$ and f_M is the local equilibrium Maxwellian:

$$f_M = \frac{\rho}{(2\pi \mathcal{R} T)^{3/2}} \exp\left(-\frac{|\boldsymbol{\xi} - \mathbf{v}|^2}{2\mathcal{R} T}\right) \qquad (2)$$

It can be shown [12] that the correct fluid viscosity $\mu(T)$ in the hydrodynamic limit can be obtained from

eqn. (1) by setting

$$\nu(\rho, T) = \frac{\rho \mathcal{R} T}{\mu(T)} \qquad (3)$$

Macroscopic velocity $\mathbf{v}$, density ρ and temperature T are moments of f in the velocity space:

$$\rho = \int_{R^3} f d\boldsymbol{\xi} \qquad \rho \mathbf{v} = \int_{R^3} f \boldsymbol{\xi} d\boldsymbol{\xi}$$

$$T = \frac{1}{3\mathcal{R}\rho} \int_{R^3} f |\boldsymbol{\xi} - \mathbf{v}|^2 d\boldsymbol{\xi} \qquad (4)$$

Finally $\mathcal{R}$ is the specific gas constant (the universal constant divided by the molar mass). The BGK model eqn. (1) can thus be interpreted as a relaxation towards a local Maxwellian equilibrium state. The gas interacts with both fixed and movable surfaces immersed in a virtually unconfined domain which is often truncated at a sufficient distance from the structures. To simplify the gas-wall interaction, it is assumed that the scattering from the wall is either diffused or specular or a combination of the two. As a matter of fact silicon surfaces originating from etching procedures are very rough so that, as a first approximation, diffuse reflection will be assumed in the sequel. If $\mathbf{n}$ denotes the unit normal vector to the solid surface S_R pointing inside the fluid domain, diffuse reflection [12] means that molecules are re-emitted from the surface according to the wall distribution function

$$f = \frac{\rho_w}{(2\pi \mathcal{R} T_w)^{3/2}} \exp\left(-\frac{|\boldsymbol{\xi} - \mathbf{v}_w|^2}{2\mathcal{R} T_w}\right) \qquad (5)$$

for $(\boldsymbol{\xi} - \mathbf{v}_w) \cdot \mathbf{n} > 0$ on S_R, where T_w is the wall temperature and $\mathbf{v}_w$ is the wall velocity and ρ_w is fixed so as to impose zero net flux across the surface:

$$\rho_w = \left(\frac{2\pi}{\mathcal{R} T_w}\right)^{1/2} \int_{R^3, (\boldsymbol{\xi} - \mathbf{v}_w) \cdot \mathbf{n} < 0} (|\boldsymbol{\xi} - \mathbf{v}_w| \cdot \mathbf{n}) f d\boldsymbol{\xi} \qquad (6)$$

On the outer fictitious boundary S_F, on the contrary, the inflow is forced to have the same features of the far region assumed at equilibrium with $f = f_0$, where f_0 is the Maxwellian at rest defined by temperature T_0, density ρ_0 and zero macroscopic velocity:

$$f = f_0 = \frac{\rho_0}{(2\pi \mathcal{R} T_0)^{3/2}} \exp\left(-\frac{|\boldsymbol{\xi}|^2}{2\mathcal{R} T_0}\right) \qquad (7)$$

for $\boldsymbol{\xi} \cdot \mathbf{n} > 0$ on S_F. Equations (1),(5),(7) represent a non-linear large-scale problem, since the velocity distribution f generally depends on the two 3D vector variables $\mathbf{x}, \boldsymbol{\xi}$. However, in this investigation we choose to analyse 2D situations in which the geometry is invariant with respect to translations along x_3 and $v_3 = 0$.

In order to take advantage of these simplifications, following [2] we introduce two new distribution functions

$$\chi = \int_R f \mathrm{d}\xi_3 \qquad \psi = \int_R \xi_3^2 f \mathrm{d}\xi_3 \qquad (8)$$

associated to the equations:

$$\frac{\partial \chi}{\partial t} + \boldsymbol{\xi} \cdot \nabla \chi = \frac{\rho}{\sigma}(\chi_M - \chi)$$

$$\frac{\partial \psi}{\partial t} + \boldsymbol{\xi} \cdot \nabla \psi = \frac{\rho}{\sigma}(\psi_M - \psi) \qquad (9)$$

which can be obtained from eqn. (1) integrating over R w.r.t. ξ_3 (after multiplying by ξ_2^3 in the latter case). In eqn. (2):

$$\chi_M = \int_R f_M \mathrm{d}\xi_3 = \frac{\rho}{2\pi \mathcal{R}T} \exp\left(-\frac{|\boldsymbol{\xi} - \mathbf{v}|^2}{2\mathcal{R}T}\right)$$

$$\psi_M = \int_R \xi_3^2 f_M \mathrm{d}\xi_3 = \frac{\rho}{2\pi} \exp\left(-\frac{|\boldsymbol{\xi} - \mathbf{v}|^2}{2\mathcal{R}T}\right)$$

In eqn. (2) and in the sequel bold letters denote 2D vectors, e.g. $\boldsymbol{\xi} = \xi_1 \mathbf{e}_1 + \xi_2 \mathbf{e}_2$ and integrations are limited to the 2D space. Density, mean velocity and temperature can be expressed in terms of the χ and ψ as:

$$\rho = \int_{R^2} \chi \mathrm{d}\boldsymbol{\xi} \qquad \rho \mathbf{v} = \int_{R^2} \boldsymbol{\xi} \chi \mathrm{d}\boldsymbol{\xi}$$

$$T = \frac{1}{3\mathcal{R}\rho} \int_{R^2} |\boldsymbol{\xi} - \mathbf{v}|^2 \chi \mathrm{d}\boldsymbol{\xi} + \frac{1}{3\mathcal{R}\rho} \int_{R^2} \psi \mathrm{d}\boldsymbol{\xi}$$

Far field conditions (normal $\mathbf{n}$ pointing towards fluid domain) become $\chi = \chi_0$, $\psi = \psi_0$ for $\boldsymbol{\xi} \cdot \mathbf{n} > 0$ on S_F, where

$$\chi_0 = \frac{\rho_0}{2\pi \mathcal{R}T_0} \exp(-\frac{|\boldsymbol{\xi}|^2}{2\mathcal{R}T_0}) \qquad \psi_0 = \frac{\rho_0}{2\pi} \exp(-\frac{|\boldsymbol{\xi}|^2}{2\mathcal{R}T_0})$$

while diffused reflection at rigid walls S_R reads:

$$\chi = \frac{\rho_w}{2\pi \mathcal{R}T_w} \exp\left(-\frac{|\boldsymbol{\xi} - \mathbf{v}_w|^2}{2\mathcal{R}T_w}\right) \qquad (10)$$

$$\psi = \frac{\rho_w}{2\pi} \exp\left(-\frac{|\boldsymbol{\xi} - \mathbf{v}_w|^2}{2\mathcal{R}T_w}\right) \qquad (11)$$

for $(\boldsymbol{\xi} - \mathbf{v}_w) \cdot \mathbf{n} > 0$, with:

$$\rho_w = \frac{(2\pi)^{1/2}}{(\mathcal{R}T_w)^{1/2}} \int_{R^2, (\boldsymbol{\xi} - \mathbf{v}_w) \cdot \mathbf{n} < 0} |(\boldsymbol{\xi} - \mathbf{v}_w) \cdot \mathbf{n}| \chi \mathrm{d}\boldsymbol{\xi}$$

2.1 Linearization of the BGK model

Despite simplifications, the numerical solution of the above equations is still a very demanding task which can be greatly simplified by the following key hypotheses. Perturbations in inertial MEMS are generally small enough so as to apply a linear expansion to the distribution functions

$$\chi \simeq \chi_0(1 + \chi_1) \qquad \psi \simeq \psi_0(1 + \psi_1) \qquad (12)$$

Moreover, if a MEMS is working at a relatively low frequency ω such that the non-dimensional Stokes number $\mathrm{St} = \omega \rho D^2 / \eta \ll 1$, a quasi static approach applies [15] (η denotes the dynamic viscosity of the gas, ρ is air density and D is the typical flux length). The velocity of rigid walls is hence enforced as a boundary condition and the time derivatives in eqn. (2) can be dropped. It is worth stressing that these hypotheses permit a straightforward numerical solution of the model and are fully respected by the real structure analyzed in Section 4. However, a general solution without these restrictive assumptions is always possible at a greater computational cost.

Setting $\beta_0 = 1/(2\mathcal{R}T_0)$, $\tilde{\boldsymbol{\xi}} = \beta_0^{1/2} \boldsymbol{\xi}$ and:

$$M^{\chi_0} = \frac{1}{\pi} \int_{R^2} \exp(-\left|\tilde{\boldsymbol{\xi}}\right|^2) \chi_1 \mathrm{d}\tilde{\boldsymbol{\xi}} \qquad (13)$$

$$\mathbf{M}^{\chi_1} = \frac{1}{\pi} \int_{R^2} \tilde{\boldsymbol{\xi}} \exp(-\left|\tilde{\boldsymbol{\xi}}\right|^2) \chi_1 \mathrm{d}\tilde{\boldsymbol{\xi}} \qquad (14)$$

$$M^{\chi_2} = \frac{1}{\pi} \int_{R^2} \left|\tilde{\boldsymbol{\xi}}\right|^2 \exp(-\left|\tilde{\boldsymbol{\xi}}\right|^2) \chi_1 \mathrm{d}\tilde{\boldsymbol{\xi}} \qquad (15)$$

$$M^{\psi_0} = \frac{1}{\pi} \int_{R^2} \exp(-\left|\tilde{\boldsymbol{\xi}}\right|^2) \psi_1 \mathrm{d}\tilde{\boldsymbol{\xi}} \qquad (16)$$

first order expansions for the macroscopic quantities can be easily developed:

$$\rho \simeq \rho_0(1 + \rho_1) \simeq \rho_0(1 + M^{\chi_0})$$

$$\mathbf{v} \simeq \frac{1}{\beta_0^{1/2}} \mathbf{M}^{\chi_1}$$

$$\beta \simeq \beta_0 \left(1 + (M^{\chi_0} - \frac{2}{3}M^{\chi_2} - \frac{1}{3}M^{\psi_0})\right)$$

hence, with $\tilde{\nu}_0 = \beta_0^{1/2} \rho_0 \mathcal{R}T_0 / \mu_0$:

$$\tilde{\boldsymbol{\xi}} \cdot \nabla \chi_1 + \tilde{\nu}_0 \chi_1 = \tilde{\nu}_0 \left[M^{\chi_0} + 2\tilde{\boldsymbol{\xi}} \cdot \mathbf{M}^{\chi_1} + \right.$$

$$\left. (1 - \left|\tilde{\boldsymbol{\xi}}\right|^2)(M^{\chi_0} - \frac{2}{3}M^{\chi_2} - \frac{1}{3}M^{\psi_0})\right] \qquad (17)$$

Similarly, for the second equation:

$$\tilde{\boldsymbol{\xi}} \cdot \nabla \psi_1 + \tilde{\nu}_0 \psi_1 = \tilde{\nu}_0 \left[M^{\chi_0} + 2\tilde{\boldsymbol{\xi}} \cdot \mathbf{M}^{\chi_1} - \right.$$

$$\left. \left|\tilde{\boldsymbol{\xi}}\right|^2 (M^{\chi_0} - \frac{2}{3}M^{\chi_2} - \frac{1}{3}M^{\psi_0})\right] \qquad (18)$$

Moreover, since all the walls are assumed to remain at $T_w = T_0$, the linear term in the expansion for β can be dropped out with the consequence that eqns. (17) and (18) decouple and can be solved independently. If the aim of our analyses is the evaluation of forces exerted on the structures by the fluid, only χ is required, as shown in the sequel, and eqn. (18) does not need to be solved for and focus is set, henceforth, on the single 2D equation:

$$\tilde{\boldsymbol{\xi}} \cdot \nabla \chi_1 + \tilde{\nu}_0 \chi_1 = \tilde{\nu}_0 \left[M^{\chi_0} + 2\tilde{\boldsymbol{\xi}} \cdot \mathbf{M}^{\chi_1} \right] \qquad (19)$$

Accordingly, also boundary conditions are linearised. On the far field boundary S_F we have:

$$\chi_1 = 0 \quad \text{for} \quad \boldsymbol{\xi} \cdot \mathbf{n} > 0 \qquad (20)$$

If we assume that $T_w = T_0$ and that:

$$\tilde{\mathbf{v}}_w = \frac{\mathbf{v}_w}{\sqrt{2\mathcal{R}T}} \ll 1$$

the linearised equation of diffuse reflection on S_R reads:

$$\chi_1 = \rho_{w1} + 2\tilde{\boldsymbol{\xi}} \cdot \tilde{\mathbf{v}}_w \quad \text{for} \quad \boldsymbol{\xi} \cdot \mathbf{n} > 0 \qquad (21)$$

with

$$\rho_{w1} = \sqrt{\pi} \tilde{\mathbf{v}}_w \cdot \mathbf{n} - \frac{2}{\sqrt{\pi}} \int_{R^2, \tilde{\boldsymbol{\xi}} \cdot \mathbf{n} < 0} (\tilde{\boldsymbol{\xi}} \cdot \mathbf{n}) \exp(-\left| \tilde{\boldsymbol{\xi}} \right|^2) \chi_1 \mathrm{d}\tilde{\boldsymbol{\xi}}$$

Once the distribution function has been solved for, the global force acting on the solid walls can be computed by integrating on the wall the stress tensor $\boldsymbol{\sigma}$ [8]. After some algebraic manipulations:

$$\begin{aligned}
\boldsymbol{\sigma} &= \frac{\rho_0}{2\beta_0} \mathbb{1} + \frac{\rho_0}{\pi \beta_0} \int_{R^2} \exp\left(-\left| \tilde{\boldsymbol{\xi}} \right|^2 \right) \tilde{\boldsymbol{\xi}} \otimes \tilde{\boldsymbol{\xi}} \chi_1 \mathrm{d}\tilde{\boldsymbol{\xi}} \\
&= \rho_0 \mathcal{R} T_0 \left[\mathbb{1} + \frac{2}{\pi} \int_{R^2} \exp\left(-\left| \tilde{\boldsymbol{\xi}} \right|^2 \right) \tilde{\boldsymbol{\xi}} \otimes \tilde{\boldsymbol{\xi}} \chi_1 \mathrm{d}\tilde{\boldsymbol{\xi}} \right]
\end{aligned} \qquad (22)$$

3 Numerical implementation

As discussed in the Introduction, a deterministic technique is preferred for the solution of eqn. (19) in order to avoid statistical noise. The usual approach for addressing steady state problems with the non-linear BGK model (e.g. [23, 24]) is to solve an unsteady problem with steady boundary conditions and letting time evolve to infinity. This guarantees in general robust convergence properties. However the linearized quasi steady BGK model eqn. (19) can be solved directly by means of a semi-implicit iterative approach with no recourse to time evolution. Moreover, since MEMS layouts are intrinsically very regular due to process requirements, the finite volumes (or finite differences) technique can be applied without restrictions and is here preferred to other domain methods like Finite Elements. In order to privilege simplicity, the region of interest is meshed with a structured cartesian grid of N_C rectangular cells and χ_1 has been assumed constant over each cell. The 2D velocity space is "truncated" and only the square surface $|\tilde{\xi}_1| \leq \tilde{U}, |\tilde{\xi}_2| \leq \tilde{U}$ is considered, and $\tilde{U}$ is set such that the contributions from

larger velocities are negligible in the r.h.s. of eqn. (19) $M^{\chi_0} + 2\tilde{\boldsymbol{\xi}} \cdot \mathbf{M}^{\chi_1}$. The truncated velocity space is partitioned into N_U^2 equal square velocity cells of size $\Delta \tilde{U}$ and χ_1 is assumed to be constant over each velocity cell and equal to the value in the cell center. The global number of unknowns is thus $N_C N_U^2$.

Let $x^{(i)}, y^{(j)}$ be the coordinates of the center of a generic space cell (in row i an column j of the cartesian grid) and $\tilde{\xi}^{(k)}, \tilde{\xi}^{(\ell)}$ the coordinates of the center of a generic velocity cell. Let also $\Phi_{i,j,k,\ell}$ denote $\chi_1(x^{(i)}, y^{(j)}, \tilde{\xi}^{(k)}, \tilde{\xi}^{(\ell)})$. Employing a classical first order upwind scheme, the discretized version of eqn. (19) reads:

$$2s_k \tilde{\xi}^{(k)} \frac{\Phi_{i,j,k,\ell} - \Phi_{i-s_k,j,k,\ell}}{\Delta H_i + \Delta H_{i-s_k}} + 2s_\ell \tilde{\xi}^{(\ell)} \frac{\Phi_{i,j,k,\ell} - \Phi_{i,j-s_\ell,k,\ell}}{\Delta V_j + \Delta V_{j-s_\ell}} \qquad (23)$$

$$+ \tilde{\nu}_0 \Phi_{i,j,k,\ell} = \tilde{\nu}_0 \left[M^{\chi_0} + 2\tilde{\xi}^{(k)} M_1^{\chi_1} + 2\tilde{\xi}^{(\ell)} M_2^{\chi_1} \right]$$

where $s_m = \text{sign}(\tilde{\xi}^{(m)})$ and $\Delta H_i, \Delta V_j$ are the dimension of the rectangular cell of center $x_1^{(i)}, x_2^{(j)}$. It should be remarked that, in order to introduce numerical diffusion only in flow direction, the upwind scheme depends on the sign of the molecular velocity (see [2]). These equations, supplemented by the boundary conditions eqns. (20) and (21) yield a linear system of large dimensions which is solved with the following semi-implicit iterative scheme. The list of the unknowns is initialised to zero: $\boldsymbol{\Phi}^{(0)} = 0$. Using the current estimate $\boldsymbol{\Phi}^{(n)}$ we enforce the boundary conditions eqns. (20) and (21) yielding the new estimate $\boldsymbol{\Phi}^{(n+1)}$ for $\boldsymbol{\xi} \cdot \mathbf{n} > 0$ on S_F and S_R. Thanks to the upwind discretization utilized, the numerical solution of the system of these equations can be recast by a simple renumbering of the unknowns into a triangular sparse form. This can be solved at a cost which is proportional to the number of unknowns $N_C N_U^2$. The procedure converges rapidly, as reported in the following section, and this steady-state semi-implicit approach turns out to be much faster than an explicit time-dependent technique.

Unfortunately the space cell size depends on the free molecular path λ. Good results can be obtained if in the regions of high gradients the mesh size is some fraction of λ. The proposed numerical tool is thus ideal for analyses at low pressures, as evidenced in the sequel, but in view of the microdimensions of MEMS, it still maintains a practical applicability up to the lower limits of the slip flow region, above which where alternative "continuum-like" approaches should be employed. Before describing and discussing numerical results, it is worth observing that the adoption of a first order upwind scheme for the discretization of the advective term has been mainly dictated by the simplicity of the resulting algorithm. The effects of numerical diffusion have been controlled through the choice of the proper grid size. Moreover, employing a more sophisticated (and complicated) scheme would have not necessarily granted higher accuracy. Actually, a number of discontinuity lines for each molecular velocity propagate in the flowfield starting from boundaries. Since the number of discrete molecular velocity is rather high (a few hundreds), discontinuity lines appear everywhere in the flowfield thus spoiling the performances of a higher order scheme.

4 Application to MEMS

The biaxial accelerometer of Figure 1, produced by STMicroelectronics with a surface micro-machining process, has been selected to test the proposed approach. The same structure has already been analysed in [15] to validate, in the slip-flow regime, an integral equation approach allowing for the analysis of the full-scale 3D structure. The accelerometer consists of a central "shuttle" suspended by means of suitable "springs" and four series of external "stators" attached to the substrate. The shuttle is essentially free to move parallel to the xy plane and is otherwise constrained. Both parts are endowed with a series of long and thin plates interdigited into capacitors serving both as actuators and sensors (see also Figure 2). The length of the longest plates is $277\,\mu$m, the in-plane width is $3.9\,\mu$m, the height is $15\,\mu$m and the air gaps between plates are $2.6\,\mu$m. The gap between the shuttle and the substrate is $4.2\,\mu$m.

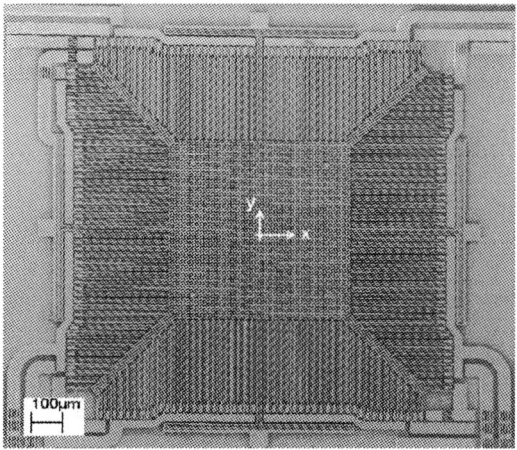

Figure 1: Biaxial accelerometer

During the experimental tests, performed at different pressures, the accelerometer has been set in oscillation along the y direction by means of electrostatic actuation in a wide range of frequencies centered at the undamped resonating frequency $f_0 = 4400\,$Hz. The actuating forces are such that the maximum amplitude of oscillation of the plates is much smaller than the air gaps. A set of plates, i.e. those parallel to the x direction, generate a Poiseuille-like flow, while those parallel to the y direction induce a Couette-like flux. In [15] it has been shown that quantitatively correct results can be obtained by analyzing and scaling the results obtained with a single "unit" like the one depicted in Figure 2, in which the motion of the shuttle occurs in the direction orthogonal to the long plates and induces a Poiseuille flow. Indeed, the contribution to damping from the central mass and from the other sets of interdigited plates can be reasonably neglected, as a first approximation. The advantage of this simplification is that the plates in Figure 2 are sufficiently long to ignore 3D effects and permit a 2D analysis by focusing the attention on section A, where the structure is depicted in dark grey and the light-grey region is, on the contrary, the fluid domain analysed.

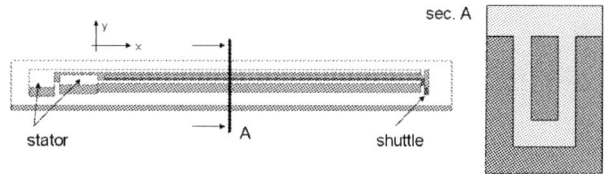

Figure 2: Single unit employed for the analysis of the structure

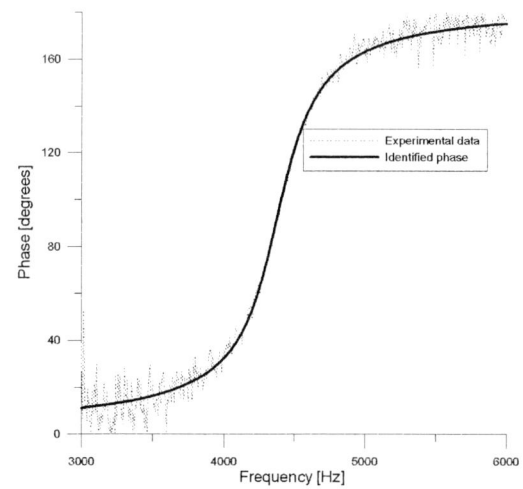

Figure 3: Plot of phase lag versus frequency at a given pressure

The typical output employed for the validation of the proposed approach is presented in Figure 3. The light-grey line corresponds to the measurements of the phase lag (between the sinusoidal electrical input and the shuttle vibration) performed for different frequencies at a given pressure. The black line, on the contrary, is the best fit of the same experimental data obtained by considering the structure as a mass-spring-damper 1D model with given mass and stiffness and unknown damper coefficient which is actually provided by the fitting procedure. In the context of the present quasi-static approach, the coefficient identified can thus be interpreted as the "experimentally measured" force exerted by the gas on the shuttle when the shuttle has a unit velocity along y at the specific pressure considered. The same procedure can be repeated at different far field pressures p_0 yielding the evolution of the measured force as a function of pressure.

	1 bar	10^{-1} bar	10^{-2} bar	10^{-3} bar
M_1	$0.44\,10^{-1}$	$0.68\,10^{-2}$	$0.10\,10^{-2}$	$0.11\,10^{-3}$
M_2	$0.28\,10^{-1}$	$0.63\,10^{-2}$	$0.10\,10^{-2}$	$0.11\,10^{-3}$
M_3	$0.19\,10^{-1}$	$0.61\,10^{-2}$	$0.10\,10^{-2}$	$0.11\,10^{-3}$
M_4	$0.14\,10^{-1}$	$0.60\,10^{-2}$	$0.10\,10^{-2}$	$0.11\,10^{-3}$
M_5	$0.12\,10^{-1}$	$0.60\,10^{-2}$	$0.10\,10^{-2}$	$0.11\,10^{-3}$

Table 1: Forces [μN] acting on the a cross section of the shuttle of unit depth

Section A in Figure 2 has been discretized with several structured mesh; the air region above the section is rather limited in extent, but different analyses employing "larger" domains did not put in evidence significant variations of the forces on the shuttle. The force acting on the shuttle cross section (of depth 1 μm), when the shuttle moves with unit velocity towards the stator, is considered as the primary output.

In Table 1 the numerical results for different meshes and pressures are collected. Far field and wall temperature has been set to $T_0 = 293\,$K and the dynamic viscosity to $\mu_0 = 1.8\,10^{-5}$ in SI units. The meshes utilized are: $M1, M2, M3, M4, M5$, with a typical cell size of $0.4\,\mu$m, $0.2\,\mu$m, $0.1\,\mu$m, $0.05\,\mu$m and $0.025\,\mu$m, respectively. The analysis stops when the relative norm of the increment over one iteration: $\|\mathbf{\Phi}^{(n+1)} - \mathbf{\Phi}^{(n)}\|/\|\mathbf{\Phi}^{(n)}\|$ becomes smaller than the fixed tolerance 10^{-5}. The speed of convergence depends, as largely expected, on the value of the far-field pressure p_0 imposed. For mesh M3, the number of iterations and the global computing time at different pressures are collected in Table 2. All the analyses have been run on a Dell Precision 490 with a Xeon 3.2GHz Dual Core processor employing a serial Fortran90 code.

As already recalled in the previous section, the size of the cells to be employed in kinetic approaches depends on the mean free molecular path in the sense that at least 4-5 cells per molecular path should be utilized. The cost of the approach thus rapidly increases with p_0. The width of the channel between the stator and the shuttle is 2.6 μm which explains the rapid convergence for pressures below 0.1 bar and the coarse prediction at ambient pressure, where $\lambda = 0.064\,\mu$m. However, is worth stressing once more that the approach proposed is intended for medium-low pressure applications, i.e. for $p < 0.1$ bar, since other simpler approaches can

be employed in the slip flow regime. Among different possible choices, Boundary Element (BE) techniques have been employed in [15] in the slip regime and perfectly compare with experimental results in this pressure range so that, at the level of detail of Figure 4, the difference between experiments and BE results could not be appreciated.

pressure	1 bar	10^{-1} bar	10^{-2} bar
Iterations for mesh M3	2269	211	106
CPU time	521s	49s	24s

Table 2: Mesh M3: number of iterations and CPU time

Hence, the analyses have been here performed up to 1 bar only for the purpose of comparing results with experimental data, as presented in Figure 1 for M5. This motivated the use of extremely fine meshes which would be otherwise unnecessary. The finest one contains 321152 cells of almost uniform size and $\tilde{U} = 4$ with 20^2 cells in the velocity space. While at high pressures an even coarser discretization of the velocity space should be enough, at high Knudsen numbers the use of a larger number of velocities is generally recommended. However, according to the numerical tests performed in this work, the damping forces (the only of interest for the analysis at hand) all differ by few percents when increasing the number of velocities. This is possibly due to the fact that the output of interest, the overall viscous force, is a global measure. It should also be recalled that the maximum speeds of MEMS at these frequencies are still very low, and always below 1m/s.

This issue stimulated further investigations. The numerical procedure was tested against classical results for 1D flows (Couette and Poiseuille) showing very good accuracy. Moreover, also alternative numerical schemes have been employed, i.e. the method of moments in which the velocity is discretized by means of Hermite polynomials, and the method of Half-Range-Gauss-Hermite quadrature which is strictly related to the former and was prosed in [19, 34]. These approches, which are very efficient especially at high pressure, provide damping forces which compare very well with the method adopted herein. The issue of obtaining an estimate of damping forces at lower pressures has also been addressed in [16] using a corrected viscosity approach applied to the BE Stokes solver with similar results.

5 Conclusions

The application of the BGK model of the Boltzmann equation to the analysis of damping in MEMS has been investigated, focusing on the specific class of silicon

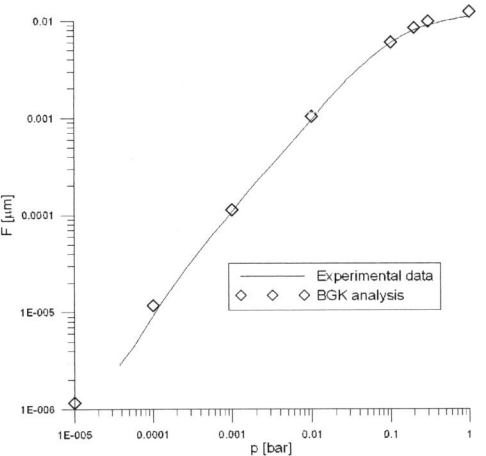

Figure 4: Biaxial accelerometer: comparison between experimental and numerical forces [μN] for the finest mesh M5

inertial MEMS working at low-moderate frequencies. The linearised steady-state 2D equation has been implemented in a deterministic manner in order to avoid noise intrinsic in statistical approaches. Numerical results in terms of forces exerted on the suspended shuttle have been compared with available experimental results showing encouraging predictive capabilities, especially at low pressure where the extension to 3D seems straightforward. Since the accuracy of the approach is directly linked to the ratio between the molecular mean free path and the cell size, the computational burden grows considerably at pressures near the lower bound of the slip regime where very fine meshes are required. In this case, the extension to 3D problems requires the recourse to large computing facilities and parallel calculus. Possibly, the implementation of higher order methods should improve convergence and is currently under investigation as well as further testing with different experimental data.

Acknowledgements

The authors wish to thank Dr. Benedetto Vigna, head on MEMS business unit - STMicrolectronics, for providing experimental data on the accelerometer. Financial support from "Fondazione Cariplo" is gratefully acknowledged.

References

[1] Andries P., Bourgat J.F., le Tallec P., Perthame B. : Numerical comparisons between the Boltzmann and ES-BGK models for rarefied gases, *Comp. Meth. Appl. Mech. Engng.*, 191, 3369–3390 (2002)

[2] Aoki K., Kanba K., Takata S. : Numerical analysis of a supersonic rarefied gas flow past a flat plate, *Phys. Fluids*, 9, 1144–1161 (1997)

[3] Aoki K., Takata S., Nakanishi T. : Poiseuille-type flow of a rarefied gas between two parallel plates driven by a uniform external force, *Phys. Rev.*, 65, 026315-1–22 (2002).

[4] Beskok A., Karniadakis G.E., Trimmer W. : Rarefaction and compressibility effects in gas microflows, *J. Fluids Engng.*, 118, 448–456 (1996)

[5] Bird G.A. : *Molecular gas dynamics and the direct simulation of gas flows*, Clarendon Press, , 1994

[6] Bhatnagar P.L., Gross E.P., Krook M. : A model for collision processes in gases. I. Small amplitude processes in charged and neutral one-component systems, *Phys. Rev.*, 94, 511–525 (1954)

[7] Cai C., Boyd I., Fan J., Candler G. : Direct simulation methods for low-speed microchannel flows, *J. Thermophys. Heat Transfer*, 14, 368–380 (2000)

[8] Chapman S., Cowling T.G. : *The mathematical theory of non-uniform gases*, Cambridge University Press, Cambridge, 1960

[9] Cho Y., Kwak B., Pisano A.P., Howe R.T. : Slide film damping in laterally driven microstructures, *Sensors and Actuators A*, 40, 31– (1993)

[10] Cho Y., Pisano A.P., Howe R.T. : Viscous damping model for laterally oscillating microstructures, *Journal of Microelectromechanical Systems*, 3, 81– (1994)

[11] Cercignani C., Daneri A. : Flow of a rarefied gas between two parallel plates, *Journal of Applied Physics*, 34, 3509–3513 (1963)

[12] Cercignani C. : *The Boltzmann equation and its applications*, Springer, New York, 1988

[13] Fan J., Shen C : Statistical simulation of low speed unidirectional flows in transition regime, *Rarefied gas dynamics, Brun ed.*, 2, 245–253 (1999)

[14] Frangi A. : A fast multipole implementation of the qualocation mixed-velocity-traction approach for exterior Stokes flows, *Engng. Analysis with Boundary Elem.*, 29, 1039–1046 (2005)

[15] Frangi A, Spinola G., Vigna B. : On the evaluation of damping in MEMS in the slip-flow regime, *Int. J. Num. Meth. Engng.*, 68, 1031–1051 (2006)

[16] Cercignani C.,Frangi A.,Lorenzani S., Vigna B. : BEM approaches and simplified kinetic models for the analysis of damping in deformable MEMS, *Engng. Analysis with Boundary Elem.*, accepted for publication (2006)

[17] Fukui S., Kaneko R. : Analysis of ultra-thin gas film lubrication based on the linearized Boltzmann equation, *JSME International Journal*, 30, 1660–1666 (1987)

[18] Gad-el-Hak M. : The fluid mechanics of microdevices - the Freeman scholar lecture, *J. Fluids Eng.*, 121, 5–33 (1999)

[19] Li Zhi-Hui., Zhang Han-Xin : Numerical investigation from rarefied flow to continuum by solving the Boltzmann model equation, *Int. J. Numer. Methods in Fluids*, 42, 361–382 (2003)

[20] Loyalka S.K., Petrellis N., Storvick S.T. : Some exact numerical results for the BGK model: Couette, Poiseuille and thermal creep flow between plates, *Z. Angew. Math. Phys.*, 30, 514–521 (1979)

[21] Gad-el-Hak M., ed. : *The MEMS handbook*, CRC Press, , 2002

[22] Karniadakis G.E. and Beskok A. : *Micro flows, fundamentals and simulation*, Springer, New York, 2002

[23] Mieussens L. : Discrete Velocity Model and Implicit Scheme for the BGK Equation of Rarefied Gas Dynamics, *Math. Models and Meth. Appl. Sci.*, 8, 1121–1149 (2000)

[24] Mieussens L. : Discrete velocity models and numerical schemes for the Boltzmann-BGK equation in plane and axisymmetric geometries, *J. Comput. Phys.*, 162, 429–466 (2000)

[25] Naris S. and Valougeorgis D. : The driven cavity flow over the whole range of the Knudsen number, *Phys. Fluids*, 17, 097106–12 (2005)

[26] Sharipov F. : Rarefied gas flow through a slit. Influence of the boundary condition, *Phys. Fluids*, 8, 262–268 (1995)

[27] Sone Y. : Flow induced by thermal stress in rarefied gases, *Physics of fluids*, 15, 1418–1423 (1971)

[28] Veijola T, Kuisma H., Lahdempera J., Ryhanen T. : Equivalent circuit model of the squeezed gas film in a silicon accelerometer, *Sensors & Actuators*, 48, 239–248 (1995)

[29] Veijola T, Kuisma H., Lahdempera J. : The influence of gas-surface interaction on gas-film damping in a silicon accelerometer, *Sensors & Actuators*, 66, 83–92 (1998)

[30] Veijola T, Turowski M. : Compact damping models for laterally moving microstructures with gas-rarefaction effects, *Journal of Microelectromechanical Systems*, 10, 263–273 (2001)

[31] Williams M.M.R. : A review of the rarefied gas dynamics theory associated with some classical problems in flow and heat transfer, *Z. Angew. Math. Phys.*, 52, 500–516 (2001)

[32] Yang J.Y., Huang J.C. : Rarefied flow computation using nonlinear model Boltzmann equations, *J. of Computat. Physics*, 120, 323–339 (1995)

[33] Ye W., Wang X., Hemmert W., Freeman D., White J. : Air damping in lateral oscillating micro-resonators: a numerical and experimental study, *Journal of MEMS*, 12, 557–566 (2003)

[34] Zhang X., Tang W.C. : Viscous air damping in laterally driven micro-resonators, *Sensors and Materials*, 7, 415– (1995)

Development of 2D Modeling Techniques for the Thermal Fatigue Analysis of Solder Joints of a Module Mounted in a 3D Cavity on a Printed Circuit Board

Y. S. Chan*, S. W. Ricky Lee
Electronic Packaging Laboratory, Center for Advanced Microsystems Packaging
Hong Kong University of Science & Technology, Clear Water Bay, Kowloon, Hong Kong
*Tel: +852-2358-8356, Fax: +852-2358-8357, E-mail:epsing@ust.hk

Yuming Ye and Sang Liu
Huawei Technologies Co. Ltd., Bantian, Shenzhen, P. R. China

Abstract

Unlike the common types of SMT packages such as BGA and QFP, the component under investigation is a module mounted in a cavity on a PCB with their lead-fingers hanged over the edges of the cavity. This PCB assembly has a 3D configuration in nature such that regular 2D modeling is not capable to solve the problem. However, it is impractical to perform a 3D thermal fatigue analysis for this structure due to the very limited time to market requirement in the industry. In order to solve this problem, a 2D finite element modeling methodology with the use of artifical elements (effective block) to supplement the necessary boundary conditions is proposed. The focus of this paper is put on the calculation of the material properties of the effective block, and also on the application of it to solve the current 3D thermal fatigue problem. The good matching between the 2D modeling results and those from experiments suggests that the proposed methodology is an effective one, in addition to its high efficiency inborn.

As 3D modeling comprises larger complexity and requires much heavier computational effort, it is usual to perform 2D modeling before any 3D simulation work for preliminary results. Yet, the 3D condition can be largely degenerated when it comes down to the 2D level. The present study demonstrates that the use of effective blocks is a way of enhancing the applicability of 2D models for solving more complicated problems.

1. Introduction

The thermal fatigue analysis of solder joint is a very important field in the electronic industry and has been well studied for years. There are extensive studies regarding this problem on common SMT packages like BGA and flip chip [1-8]. Taking the advantage of symmetry, simplification of 3D models to strip models or even 2D models of these packages can be achieved for quick solution evaluation without much loss in accuracy [9-10]. This is benificial and actually crucial to the competitive electronic industry as the product time to market is highly limited.

However, different from those traditional SMT packages, the component under investigation is a module mounted in a cavity on a PCB with their lead-fingers hanged over the edges of the cavity (Figure 1). The configuration of this assembly is a 3D one and regular 2D modeling across the cross-section A-A (Figure 2) is not sufficient to represent the complete situation as the hatched PCB regions which govern the deformation of the cavity are omitted.

In order to solve this 3D thermal fatigue problem efficiently, a 2D modeling approach using an effective block to supplement the effect of the hatched PCB regions is proposed in this paper (Figure 3). This deliberately added effective block is designed such that it can provide very similar mechanical influence as the hatched PCB regions on the entire system under the accelerated thermal cycling (ATC) condition. The calculation of the material properties of this effective block is discussed in detail in the coming sessions. A validation work regarding the applicability of it then follows and finally, the use of it in the 2D model to solve the 3D thermal fatigue problem is presented and the computational results are compared with those from the experiment.

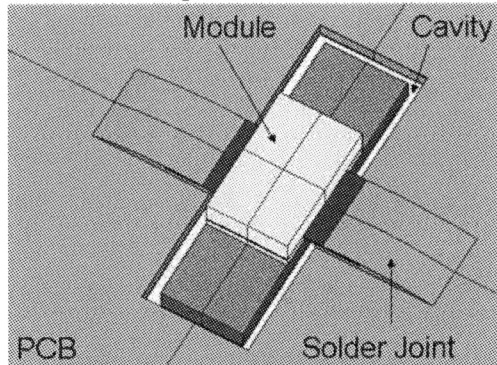

Figure 1. Module Mounted in a Cavity on a PCB

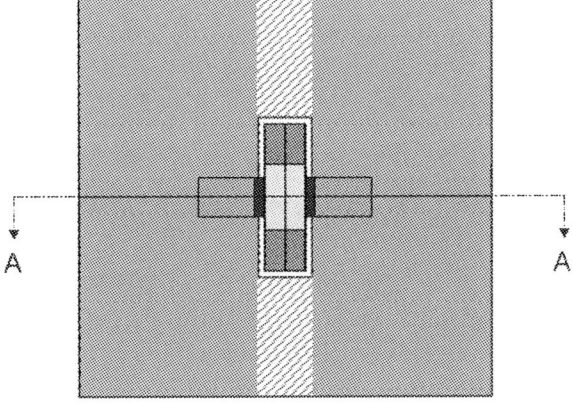

Figure 2. PCB Regions Omitted in Regular 2D Model

1-4244-1105-X/07/$25.00 ©2007 IEEE

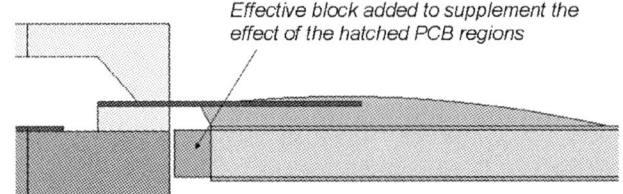

Figure 3. Effective Block Added in the 2D Model

2. Effective Material Properties of a Block Composed of Two Isotropic Materials

Since the PCB composes of layers of copper and FR4, the effective block needs to consists both the material properties of copper and FR4. Consider a laminate composes of two isotropic materials as shown in Figure 4. This laminate can be represented by a new block of material of the same dimension with the same axial mechanical responds. The effective material properties of this block are summarized in equation set (1).

$$
\left\{
\begin{aligned}
E_x &= E_1 \cdot \frac{H_1}{H} + E_2 \cdot \frac{H_2}{H} \\
\frac{1}{E_y} &= \frac{1}{E_1} \cdot \frac{H_1}{H} + \frac{1}{E_2} \cdot \frac{H_2}{H} \\
E_z &= E_x \\
\nu_{xy} &= \nu_1 \cdot \frac{H_1}{H} + \nu_2 \cdot \frac{H_2}{H} \\
\nu_{yz} &= \nu_{yx} = \frac{E_y}{E_x} \nu_{xy} \\
\nu_{xz} &= \frac{\nu_1 \cdot H_1 E_1 + \nu_2 \cdot H_2 E_2}{H_1 E_1 + H_2 E_2} \\
G_{xy} &= G_1 \cdot \frac{H_1}{H} + G_2 \cdot \frac{H_2}{H} \\
G_{yz} &= G_{xy} \\
G_{xz} &= G_{xy} \\
\alpha_x &= \frac{\alpha_1 \cdot H_1 E_1 + \alpha_2 \cdot H_2 E_2}{H_1 E_1 + H_2 E_2} \\
\alpha_y &= \alpha_1 \cdot \frac{H_1}{H} + \alpha_2 \cdot \frac{H_2}{H} \\
\alpha_z &= \alpha_x
\end{aligned}
\right.
\tag{1}
$$

For the sake of simplicity without loss of the purpose of the current study, equation set (1) is used to evaluate the effective material properties of the PCB under the assumption that the FR4 in our study is isotropic.

Equation set (1) can be derived easily by preserving the axial mechanical loading capability of the block [11]. As a result of this consideration in the derivation, the new block carries the same axial mechanical responds as the original one, but there could be errors regarding the bending and shear responds. Yet in our case of which the problem is driven mainly by the axial loading, the use of equation set (1) remains valid.

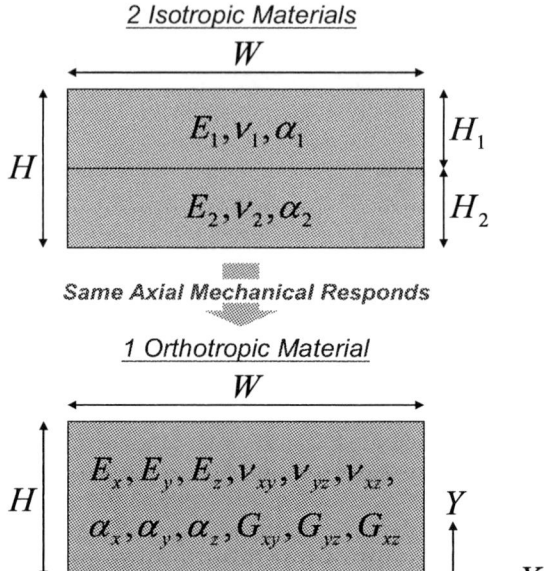

Figure 4. Transformation of a Block of Two Isotropic Materials into One Orthotropic Material

3. Preservation of Mechanical Responds of a Block after Dimension Change

An essential step to calculate the material properties of the effective block includes the computation of the effective material properties of a block after dimension change.

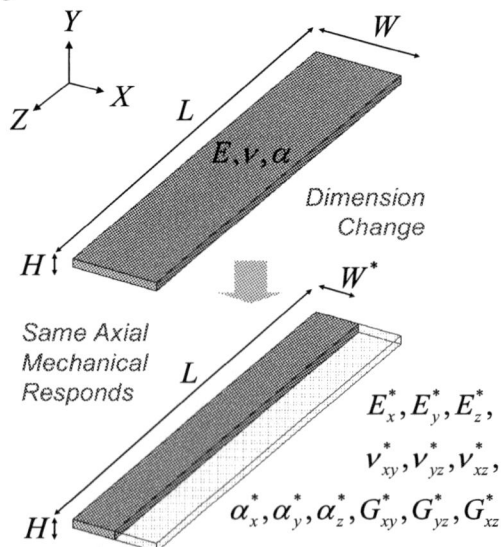

Figure 5. Dimension Change of a Block with Axial Mechanical Responds Preserved

632

Consider an isotropic block with material properties E, v and α and dimensions $W \times L \times H$. If its dimensions are now changed to $W^* \times L \times H$ while having its axial mechanical responds preserved (Figure 5), its new material properties can be calculated by using equation set (2).

$$
\begin{cases}
E_x^* = E_x \cdot C_x \\[6pt]
E_y^* = E_y \cdot \dfrac{1}{C_x} \\[6pt]
E_z^* = E_z \cdot \dfrac{1}{C_x} \\[6pt]
v_{xy}^* = v_{xy} \cdot C_x \\[6pt]
v_{yz}^* = v_{yz} \\[6pt]
v_{xz}^* = v_{xz} \cdot C_x \\[6pt]
G_{xy}^* = G_{xy} \cdot C_x \\[6pt]
G_{yz}^* = G_{yz} \cdot \dfrac{1}{C_x} \\[6pt]
G_{xz}^* = G_{xz} \cdot C_x \\[6pt]
\alpha_x^* = \alpha_x \cdot \dfrac{1}{C_x} \\[6pt]
\alpha_y^* = \alpha_y \\[6pt]
\alpha_z^* = \alpha_z
\end{cases}
\tag{2}
$$

where $C_x = \dfrac{W^*}{W}$ is a factor for the dimension change in the x-direction. The derivation for equation set (2) is similar to that of equation set (1) and hence similar assumptions are boren in it.

4. Material Properties of the Effective Block

The calculation of the material properties of the effective block includes successive transformations of the hatched PCB regions. The 1st transformation is to transform the hatched PCB laminate into a block of the same dimensions ($W \times L \times H$) with effective material properties by using equation set (1). The 2nd transformation refers to a dimension change of the effective block from W to W^* by using equation set (2). W^* depends on the space available between the package and the PCB and it is this room allows the use of the effective block a feasible way in solving the 3D problem. The 3rd transformation includes a dimension change of the effective block from L to L^*. After this transformation, the effective block has the same length as the PCB domain. The 2nd and 3rd transformation of the effective

block is shown in Figure 6. The 4th transformation requires a dimension change from L^* to L^{**} as shown in Figure 7. This is required as we are comparing the stiffness of the leadframe and that of the PCB which are of different domain size. In order to accommodate this effect in the 2D model, this last transformation is critical. As the 3rd and 4th transformation are along the same direction, they can be combined so that the net effect is a dimension change from L to L^{**}.

In short, the calculation of the material properties of the effective block requires first the use of equation set (1) to transform the isotropic copper and FR4 materials into one orthotropic material block. Then followed by using equation set (2) two times to have the dimensions of the effective block changed from $W \times L \times H$ to $W^* \times L^{**} \times H$.

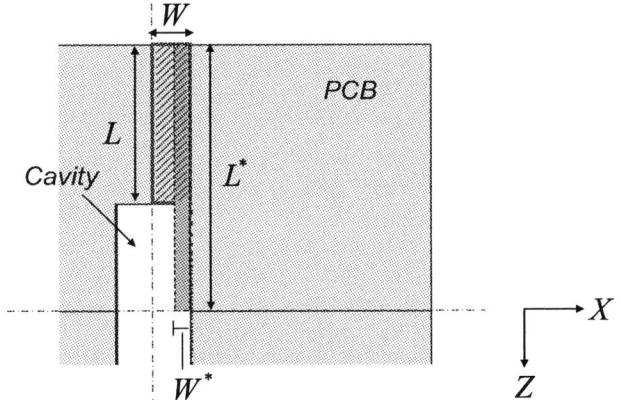

Figure 6. The 2nd and 3rd Transformation of the Effective Block

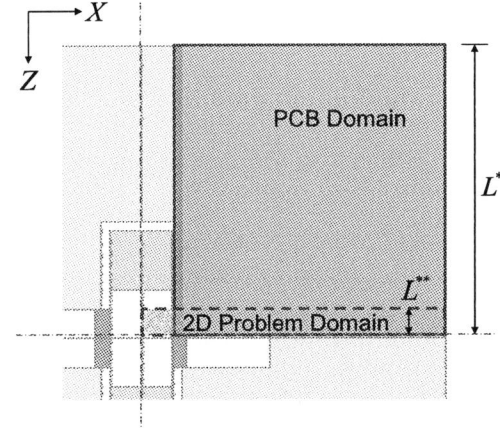

Figure 7. The 4th Transformation of the Effective Block

5. Validation of the Use of the Effective Block

To validate the capability of the effective block for supplementing the effect of the hatched PCB regions on the deformation of the cavity, two finite element models are developed. One of them is a 3D model with geometry simplified from the detailed model. The other one is a 2D model of the simplified 3D model with an effective block supplemented. The simplified models comprise only the lead frame and the PCB with solder as the interconnect in between. The simplified 3D model is shown in Figure 8 and the simplified 2D model is shown in Figure 9. The material properties used are summarized in Table 1.

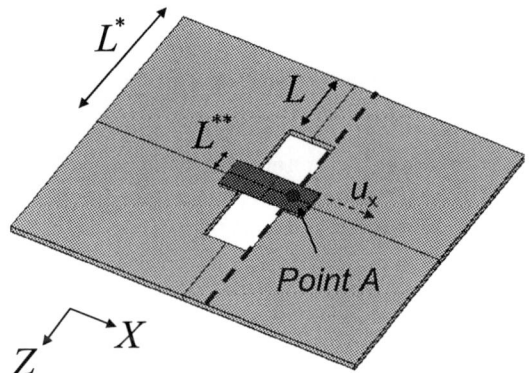

Figure 8. Simplified 3D Model for Validation

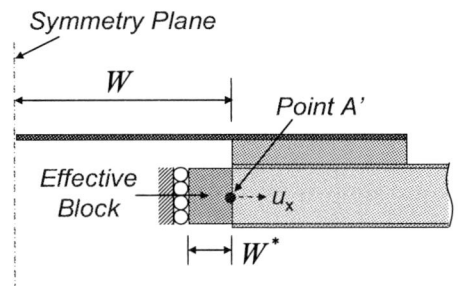

Figure 9. Simplified 2D Model for Validation

Table 1. Material Properties for Simplified Models

Materials	E (MPa)	v (non-dim)	α (ppm/°C)
Solder	19.2	0.35	24.5
FR4	26.9	0.2	11
Copper	76	0.35	17
Leadframe	145	0.3	4.45

The models are assigned with a thermal loading of 100°C with reference temperature set to 25°C. The deformation of the PCB at the edge of the cavity in the x-direction u_x determined by the two models are compared with each other. There are two representative results from the simplified 3D model. One is the local deformation obtained at point A and the other one is the averaged deformation acquired along the dash line as shown in Figure 8. For the simplified 2D model, only the local result obtained at point A' is used for comparison (Figure 9). Since the 2D model can be solved based on the plane strain or plane stress assumptions, results from both cases are taken for the study.

The deformations of the PCB predicted by the two models for different size of the cavity (by adjusting L) are summarized in Figures 10 and 11. Results from the 3D model are the same for both Figures. The difference

comes with the 2D modeling results such that Figure 10 is obtained by using an effective FR4 block while Figure 11 is acquired by using an effective PCB block. The effective FR4 block takes the properties of FR4 into account only while the effective PCB block consists both the material properties of copper and FR4.

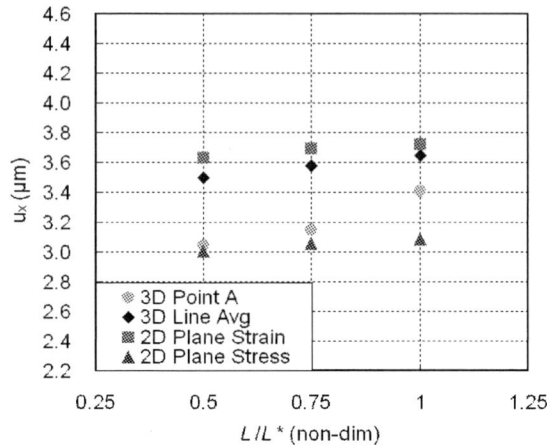

Figure 10. Results Comparison Between 2D and 3D simplified Models - Use of Effective FR4 Block

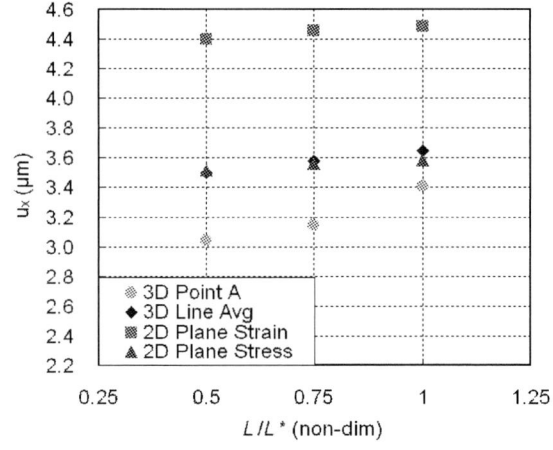

Figure 11. Results Comparison Between 2D and 3D simplified Models - Use of Effective PCB Block

Investigating the 3D modeling results, it is observed that the line averaged deformations are larger than the local deformations at point A. This is natural since point A is just right below the leadframe where its constraint on the deformation of the cavity is the most significant. Since this local influence is in the z-direction which is not included in the 2D model, only the line averaged results from the 3D models are meaningful for comparing with the 2D ones.

Since the elastic modulus and coefficient of thermal expansion (CTE) of copper are both larger than FR4, the 2D modeling results with the use of effective PCB block are larger than those with using the effective FR4 block. Having taken most information into account, the 2D modeling results with using the effective PCB block are

634

supposed to be the closest to the 3D line averaged results. But there are large overshooting comes with the corresponding plane strain results. The reason lies on the much larger PCB domain size (L^*) than the 2D problem domain size (L^{**}). As the effective PCB block deform as the same extent as the PCB domain under thermal loading, its large deformation in the z-direction actually boost the deformation in the x-direction under the plane strain condition. With the deformation in the z-direction neglected for the plane stress case, good matching between the 2D results and the 3D line averaged results is observed. Therefore, the use of the effective block in 2D model for supplementing the mechanical influence of the hatched PCB regions on our 3D problem is valid and in the next session, the effective PCB block will be used in the 2D model for solving the 3D thermal fatigue problem.

6. Solving the 3D Thermal Fatigue Problem

Figure 12 shows the 2D finite element model used for this study which carries an effective PCB block. The material properties for various materials are summarized in Table 2.

Table 2. Material Properties of the Real Problem

Materials	E (MPa)	v (non-dim)	α (ppm/°C)
63Sn-37Pb	34.4-19.2 (0-100°C)	0.35	24.5
Silicon	131	0.3	2.8
FR4	26.9	0.2	11
Copper	76	0.35	17
FeNi	145	0.3	4.45
W85Cu15	248	0.33	6.5
Aluminum	68	0.33	24
Ceramic	357	0.3	7.7

The implicit generalized Garofalo creep model is employed for solving this thermal fatigue problem:

$$\dot{\varepsilon}_{Cr} = C_1 \left[\sinh\left(C_2 \sigma\right) \right]^{C_3} e^{-\frac{C_3}{T}} \tag{3}$$

where $\dot{\varepsilon}_{Cr}$ = Creep strain rate (1/s)

σ = Stress (MPa)
T = Temperature (K)
$C_1 = 463(508\text{-}T)/T$ (1/s)
$C_2 = 1/(37.78 \times 10^6 \text{-} 74414T)$ (1/Pa)
$C_3 = 3.3$ (non-dim)
$C_4 = 6360$ (K)

The thermal cycle profile has a range from 0 to 100°C. The dwell time at each extreme is 8 mins and the ramp time is 7 mins. The stress free temperature is assumed to be 25°C. In the real situation, the package and the PCB are screwed on an aluminum base plate. So in the 2D model, the package is fixed on the base plate. A frictionless contact is assigned between the PCB and the base plate. And the PCB is fixed to the base plate at the far end. The boundary conditions for the 2D model is given in Figure 13.

After solving the model for 5 temperature cycles, it is found that the maximum creep strain energy density is located at the critical position C (Figure 14), which means a crack will be initiated there. Using the crack growth correlation equation:

$$N = N_0 + \frac{a}{\dfrac{da}{dN}} = \alpha_1 (\Lambda W)^{\beta_1} + a \alpha_2^{-1} (\Lambda W)^{-\beta_2} \tag{4}$$

where N = Total number of cycles (cycles)
N_0 = Cycles for crack initiation (cycles)
a = Crack length (mm)
da/dN = Crack propagation rate (mm/cycle)
ΔW = Creep strain energy density per cycle (MPa)
$\alpha_1 = 54.2$ (cycle-MPa)
$\beta_1 = -1$ (non-dim)
$\alpha_2 = 0.000349$ (mm/cycle-MPa$^{1.13}$)
$\beta_2 = 1.13$ (non-dim)

a crack length of 0.627 mm will be initiated from the critical position C after 1373 cycles, which shows very close prediction as the experimental results which found that a crack of 0.627 mm was initiated from the critical position C after 1270 cycles.

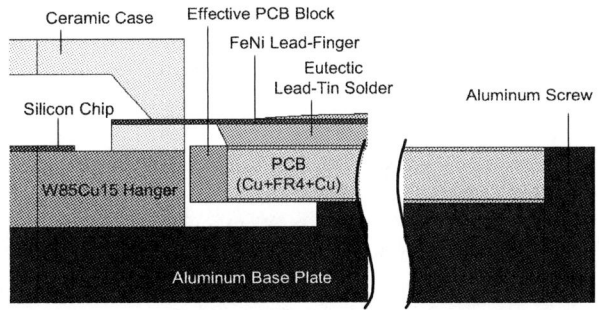

Figure 12. 2D Model with Effective PCB Block for Solving the Real Problem

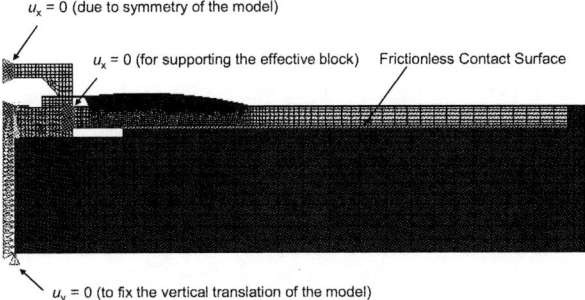

Figure 13. Boundary Conditions for the Real Problem

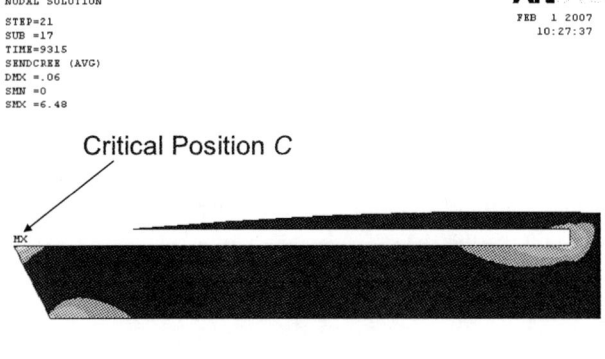

Figure 14. Creep Strain Energy Density of the Solder Joint at the End of the 5th Cycle

7. Conclusions

The current 3D thermal fatigue problem encountered with a module mounted in a cavity on a PCB has been solved effectively and efficiently with the proposed 2D modeling methodology. The essense of this proposed 2D approach lies on the use of an effective block to supplement the necessary boundary conditions which is hidden in the regular 2D model so that the fortified 2D model is sufficient to represent the 3D problem. The effective block is derived and acquired scrupulously by sucessive transformations of those hidden parts and therefore, it responds mechanically as if that of the hidden parts which have been demonstrated in this paper.

Based on the large adaptability of the use of the effective block, the proposed 2D methodology is not limited to the current 3D thermal fatigue problem, but also valid for other complex problems on which the results of interest is 2D while the problem domain is 3D in nature. Inspite of the fact that the computating capabity is getting more and more advanced, it is still too expensive to carry out a comprehensive 3D analysis regarding the problems which are complicated in geometry and requires iterative solution solving (transient and dynamic problems). The propsed modeling approach with the use of effective block provides a way which enhances the applicability of 2D models for solving more complicated 3D problems, which can be a very powerful tool during the product design stage.

References

1. Darveaux, R. *et al*, <u>Reliability of Plastic Ball Grid Array Assembly</u>, McGraw-Hill (New York, 1995).
2. Lau, H. J. and Pao, Y. H., <u>Solder Joint Reliability of BGA, CSP, Flip Chip and Fine Pitch SMT Assemblies</u>, McGraw-Hill (New York, 1997).
3. Pang, H. L. J., Chong, D. Y. R. and Low, T. H., "Thermal Cycling Analysis of Flip-Chip Solder Joint Reliability," *IEEE Transaction on Components and Packaging Technologies*, Vol. 24, No.4 (2001), pp. 705-712.
4. Zhai, C. J. and Sidharth, B. R., "Board Level Solder Reliability vs. Ramp Rate & Dwell Time During Temperature Cycling," *Proc. 41st IEEE International Reliability Physics Symposium,* Mar 30-Apr 4, 2003, pp. 447- 451.
5. Lee, S. W. R. and Lau, C. Y. D., "Computational Model Validation with Experimental Data from Temperature Cycling Tests of PBGA Assemblies for the Analysis of Board Level Solder Joint Reliability," *Proc 5th EuroSimE Conference,* Brussels, Belgium, May 10-12, 2004, pp. 115-120.
6. Lau, C. Y. D. and Lee, S. W. R., "Computational Analyses on the Effects of Irregular Conditions During Accelerated Thermal Cycling Tests on Board Level Solder Joint Reliability," *Proc 6th Electronic Packaging Technology Conference*, Singapore, Dec 8-10, 2004, pp. 516-521.
7. Che, F. X., Pang, H. L. J., Xiong, B. S., Xu, L. and Low, T. H., "Lead Free Solder Joint Reliability Characterization for PBGA, PQFP and TSSOP Assemblies," *Proc 55th Electronic Components and Technology Conference*, Orlando, Florida, CA, May 31-Jun 3, 2005, pp. 916-921.
8. Fan, X. *et al*, "Effect of Finite Element Modeling Techniques on Solder Joint Fatigue Life Prediction of Flip-Chip BGA Packages," *Proc 56th Electronic Components and Technology Conference*, San Diego, CA, May 30-Jun 2, 2006, pp. 972-980.
9. Yao, Q. and Qu, J., "Three-Dimensional Versus Two-Dimensional Finite Element Modeling of Flip-Chip Packages," *Journal of Electronic Packaging*, Vol. 121, 1999, pp. 196-201.
10. Pang, H. L. J. and Chong, D. Y. R., "Flip Chip on Board Solder Joint Reliability Analysis Using 2-D and 3-D FEA Models," *IEEE Transactions on Advanced Packaging*, Vol. 24, No. 4 (2001), pp. 499-506.
11. Staab, H. G., <u>Laminar Composites</u>, Butterworth-Heinemann (1999), pp. 71-76.

A Macro Model Based On Finite Element Method To Investigate Temperature And Residual Stress Effects On RF MEMS Switch Actuation

D. Peyrou, H. Achkar, F. Pennec, P. Pons, R. Plana
LAAS-CNRS, MINC-M2D Group
7, Av. Colonel Roche, 31077 Toulouse cedex 4, France
Email : dpeyrou@laas.fr Phone : +33-561 33 69 30, Fax: +33-561 33 62 08

Abstract

Till nowadays, MEMS design suffers from the lake of efficient and easy-to-use simulation tools considering the complete MEMS design procedure, from individual MEMS component design to complete system simulation. Finite element analysis (FEA) methods offer high efficiency and are widely used to model and simulate the behaviour of MEMS components.

However, as MEMS are subject to multiple coupled physical phenomena at process level, such as initial stress, mechanical contact, temperature, thermoelastic, electromagnetic effects, which highly affects the component, we need to integrate these effects in our model. Doing so, finite element models may involve large numbers of degrees of freedom so that full simulation can be prohibitively time consuming. As a consequence, designers must simplify models or concentrate on interesting results in order to obtain accurate but fast solution.

Some multiphysics' softwares, such as COMSOL [3], allow Reduced Order Modeling (ROM) or macro models which considers the global behaviour of the device. Thus designers can create automatically, for example, their own Simulink (Matlab ©) library from a multiphysics finite element modelization, in order to develop a global behavioural model of the whole component.

This work deals with a Simulink macro model, generated from a three-dimensional multiphysics finite element analysis (FEA) using COMSOL, aiming to investigate the pull-in and pull-out voltage of microswitches.

1. Introduction

RF Capacitive microswitches are widely studied during the last decade, which demonstrated high isolation, low insertion, large bandwidth and unparalleled signal linearity from dc to 100 GHz frequencies range. Most of these switches uses electrostatic actuation to perform the capacitive shunt for the RF signal (figure 1) because of their low power consumption and their simple manufacturing processes in the clean room techniques. Despite these benefits, RF Mems capacitive switches haven't been yet seen in commercial products for reliability causes, limits in signal power handling and packaging problems. The long term reliability is limited by stiction between the freestanding metal membrane and the dielectric layer coat on the bottom electrode. This stiction is due to both dielectric charging and roughness effects, which will impact the isolation. As low actuation voltage can improve significantly the lifetime, it is essential to design a microswitch able to achieve pull in contact at low voltage (typically 5 V).

Therefore, modeling and simulation of electrostatic actuation taking into account the physical phenomena at process level (residual stress [1-2], thermo-mechanicals effects, buckling...), play an important role during the design phase to predict the device's characteristics.

2. Static modeling of electrostatically actuated microswitches

We have used the hypothesis of a thin elactic isotropic beam under large deflection with an electrostatic actuation which is modeled on Simulink as a "rigid plate" attached to a spring above a fixed electrode (figure 1).

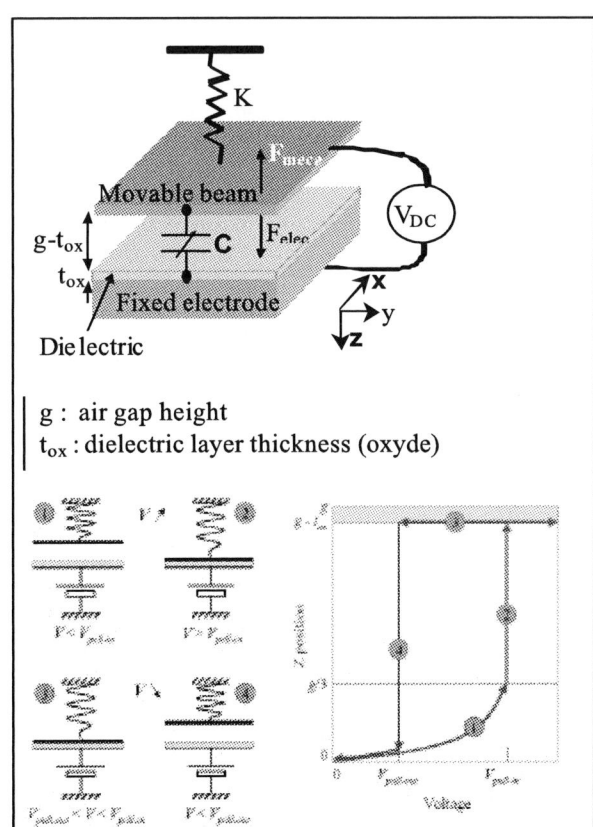

g : air gap height
t_{ox} : dielectric layer thickness (oxyde)

Figure 1 Electrostatic pull-in and pull-out scheme from an elementary model with a spring and rigid plate

In this case, pull-in voltage can be easily obtained from the equilibrium equation between the spring force and the electrostatic force [5], using the following expression:

$$V_p = \sqrt{\frac{8\,k\,g_0^3}{27\,e_0\,A}}$$

With $\begin{cases} A & \text{Electrode's area} \\ k & \text{Beam's stiffness} \\ \varepsilon_0 = 8{,}854.10^{-12} \text{ (vacuum's permittivity)} \\ g_0 & \text{initial gap between the beam and the electrode)} \end{cases}$

To evaluate the stiffness of the beam, we first extracted the constant stiffness, of an analytical campled-campled beam model as shown in (figure 2), by Castigliano's formulae using the energetic method, as described below.

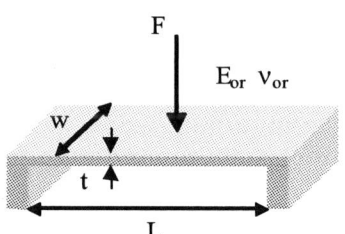

- $w = 40\ \mu m$
- $L = 400\ \mu m$
- $t = 2.7\ \mu m$
- F : Force
- E_{or} : Young modulus of gold = 80 GPa
- ν_{or} : poisson's ratio of gold = 0.42

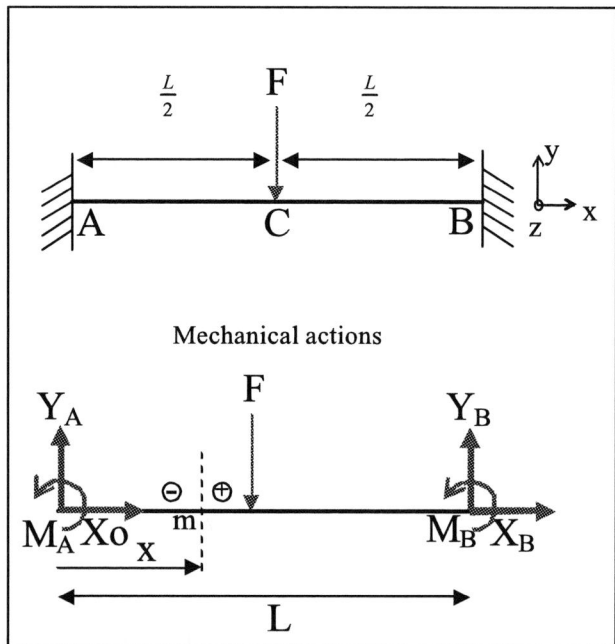

Figure 2 Analytical model definition of a fixed-fixed beam with a punctual load

The adopted sign convention to define the internal forces for the theoretical model represented in (figure 2) is the right convention:

The total energy density in the beam is formulated as:

$$[W_{def}] = \frac{1}{2}\int_0^L \frac{X_A^2}{ES}dx + \frac{1}{2}\int_0^L \frac{Y_A^2}{GS}dx$$
$$+ \frac{1}{2EI}\int_{L/2}^L \left(-M_A + \frac{F}{2}(L-x)\right)^2 dx$$
$$+ \frac{1}{2EI}\int_0^{L/2} \left(M_A - \frac{F}{2}x\right)^2 dx$$

So by deriving this energy, we can find all the hyperstatic forces and of course the stiffness of the beam:

$$\frac{\partial W_{def}}{\partial X_A} = 0 \Rightarrow X_A = X_B = 0$$

$$\frac{\partial W_{def}}{\partial M_A} = 0 \Rightarrow \dots M_A = -M_B = \frac{FL}{8}$$

$$\frac{\partial W_{def}}{\partial F} = d = \frac{FL^3}{192EI} \quad \text{deflection at the center (point C)}$$

$$K = \frac{F}{d} = \frac{192EI}{L^3} \quad \text{stiffness of the beam}$$

with $I = \frac{wt^3}{12}$ inertia of the crosssection

The numerical application give us a stiffness: **K=15,75 N/m**. Unfortunately, this theoretical calculation is far from the real stiffness extracted from experimental data, as shown in figure 3.

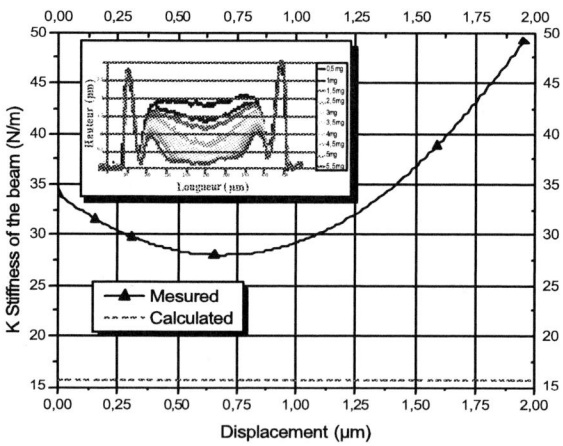

Figure 3 Graph comparing Experimental stiffness to theoretical stiffness

These differences result from the non linear behaviour of the beam's stiffness which was not taken into account in the analytical expression. To solve this problem, a finite element analysis of the beam, loaded with an equivalent electrostatic pressure applied to the entire beam's surface, was performed in order to study the deflection. This electrostatic pressure was not computed with a multiphysic simulation but it was introduced as an analytical pressure depending on the deflection of the considered point.

The COMSOL macro model consisted of the structural behaviour taking into account the thermal and residual stress effects [1-2,4] without considering the electrostatic actuation as described in figure 4.

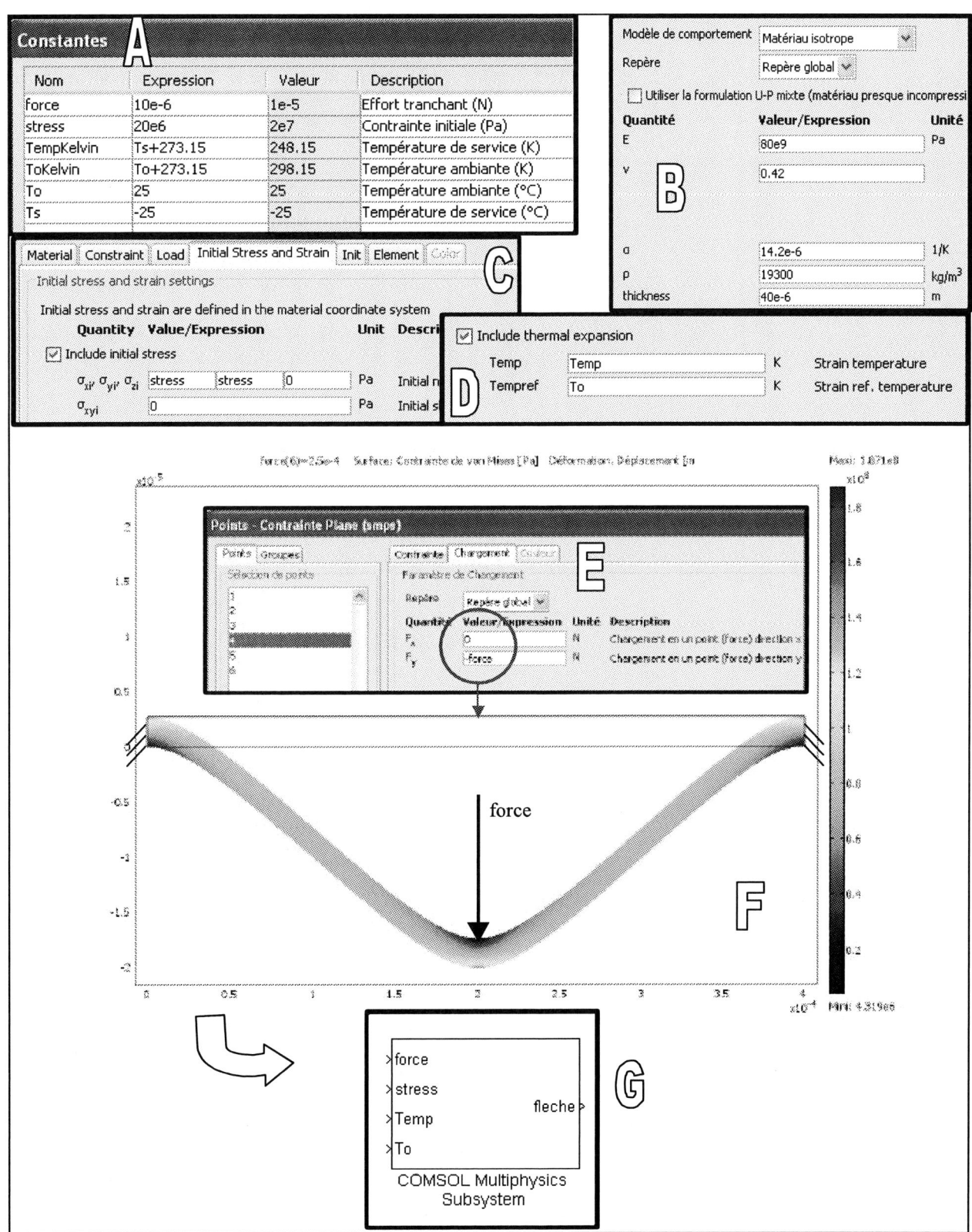

Figure 4 COMSOL Model: (a) Constants' definition - (b) Material's properties – (c) Initial stress – (d) Including thermal expansion– (e) Boundary conditions – (f) Results and postprocessing – (g) Simulink macro model exportation

3. Simulation and results

So, our approach consisted of seperating thermo mechanical and electrostatic domains into sequential analysis. A macro model representing only the thermo mechanical behaviour was integrated in a Simulink scheme to perform the electrostatic actuation (figure 5).

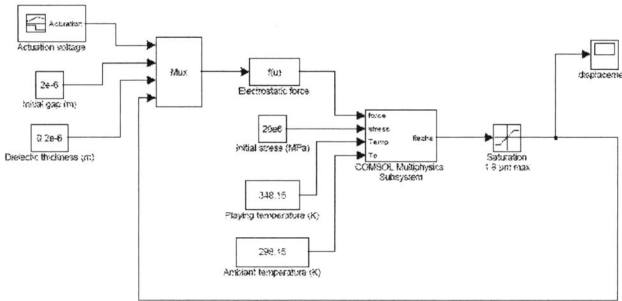

Figure 5 Simulink's model scheme

The main difficulty in setting up the pressure was to take into account the variation of the surface's normal vector during the deformation.

By using NANSON's formula which allows to connect an element of surface oriented in the non deformed configuration, in an element of area in the deformed configuration (with the convention vector in bold):

$$ds = JF^{-T}d\mathbf{S}$$

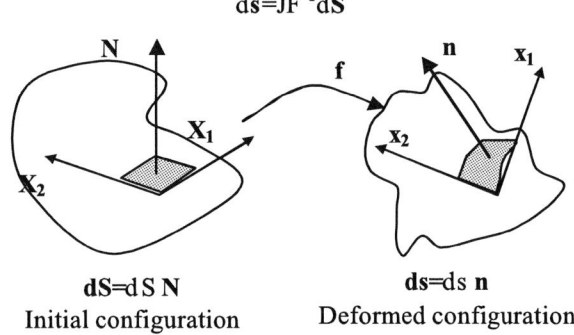

dS=d S **N**
Initial configuration

ds=ds **n**
Deformed configuration

With J, the jacobian of the vectorial transformation f to cross from the initial state to the deformed state and F the strain gradient, we have:

$$F = \overline{\overline{grad}}\,\mathbf{f} = \frac{\partial f_i}{\partial X_j} = \frac{\partial x_i}{\partial X_j} = \begin{pmatrix} F_{11} & F_{12} \\ F_{21} & F_{22} \end{pmatrix}$$

So, for our application, it is enough to implement the value of the surface load according to the initial configuration in COMSOL. According to the formula of NANSON, this pressure is obtained by:

$$\begin{Bmatrix} F_x \\ F_y \end{Bmatrix} = pJF^{-T}\begin{Bmatrix} N_x \\ N_y \end{Bmatrix}$$

Contact pressure along x and y directions in the initial configuration

Contact pressure

Normal vector in the deformed state

This method allows us to easily implement residual stress and various operational temperatures in the macro model in order to study the performance of microswitches.

The simulations showed that both residual stress and temperature significantly affects the pull-in and pull-out voltages. In fact, the analysis shows that the change in the operating climat from room temperature to -50 °C or an increase in the residual stress state, has to be overcomed by higher pull-in and pull-out voltages (figure 6).

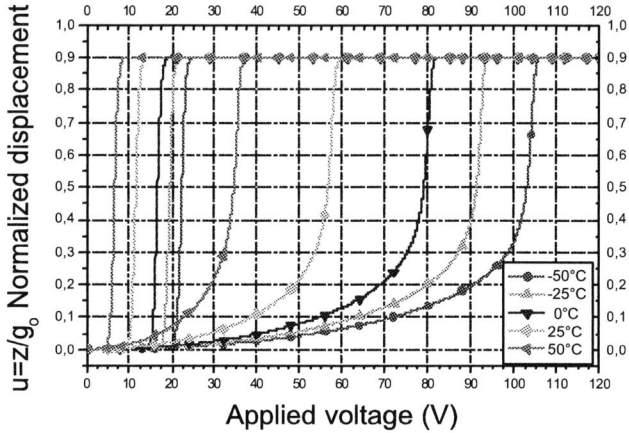

Figure 6 Rigid-plate displacements versus actuation voltage taking into account the non-linear stiffness of the beam (hysteresis).

4. Conclusions

This original approach permits us to reduce the calculation time and by consequence opens for us a new perspective to improve the structural model, by taking into account the real profile of the microswitches as discussed in a previous work, where we validated a reverse engineering technique.

Acknowledgments

The authors would like to acknowledge the Délégation Générale de l'Armement (DGA) for its support through the contract 0134032.

References

1. W. Fang and J.A. Wickert, "Determining mean and gradient residual stresses in thin films using micromachined cantilevers", J. Micromech. Microeng. 6 (1996) p301-309
2. Youn-Hoon Min, Yong-Kweon Kim, "In situ measurement of residual stress in micromachined thin films using a specimen with composite layered cantilever", J. Micromech. Microeng. 10 (2000) p314-321
3. http://www.comsol.com/
4. D. Peyrou et Al. "Multiphysics Softwares Benchmark on Ansys / Comsol Applied For RF MEMS Switches Packaging Simulations", Eurosime 2006, Côme (Italie), 24-26 april 2006, 8p
5. G. M. Rebeiz « RF MEMS Theory, Design and Technology » ,édition Wiley, 2003, p38

Atomistic Simulations of Interface Properties in Metals

Tomotsugu Shimokawa

Division of Mechanical Science and Engineering, Graduate School of Natural Science and Technology,
Kanazawa University
Kakuma-machi, Kanazawa, Ishikawa 920-1192, JAPAN
simokawa@t.kanazawa-u.ac.jp

Abstract

The effects of grain boundary structure on the mechanical properties of polycrystalline aluminum are investigated by using molecular dynamics and quasicontinuum simulations. Three problems are simulated: (a) tensile deformation of nanocrystalline aluminum with different grain boundary misorientation distributions, (b) interaction between edge dislocations and tilt grain boundaries, and (c) grain boundary motion under shear deformation. It is found that the arrangements of grain boundary dislocations, which are determined not only by the misorientation angle but also by the deviation angle, strongly control the grain boundary motion; therefore, it is concluded that individual grain boundary structures influence the macroscopic mechanical properties in nanostructured materials.

1. Introduction

The mechanical properties of polycrystalline metals are strongly governed by their internal structures. In coarse-grained metals, the plot of the yield stress σ_y against $d^{-1/2}$ gives a line with a constant slope k_y; this is due to the Hall-Petch relationship [1, 2] according to which grain boundaries act as barriers for dislocation movements. Generally, grain boundary structures depend on five macroscopic and three microscopic degrees of freedom; therefore, it is important to investigate the effects of individual grain boundary structures and the distribution of grain boundary characteristics on the macroscopic mechanical properties of polycrystalline materials [3].

For examining grain boundary properties in detail, atomic simulation is a powerful tool because it can directly depict the defect structures at the atomic level within the limits of time and space. Thus, in this study, we adopt the molecular dynamics (MD) and quasicontinuum (QC) methods to explore the dynamic properties of grain boundaries and to express the long-range strain field due to the dislocation interaction with a grain boundary, respectively. The QC method [4, 5] is a concurrent multiscale method that couples atomistic and continuum descriptions. Therefore, this method can efficiently simulate the interaction between defect structures by expressing regions distant from the lattice defects as continuum fields.

In this study, asymmetrical tilt grain boundaries are adopted. In order to examine the effect of the grain boundary misorientation angle and the deviation angle from the symmetrical plane on the macroscopic mechanical properties of polycrystalline metals, three different problems are simulated: tensile deformation tests of nanocrystalline metals by the MD method and two kinds of shear deformation tests of bicrystals by the QC and MD methods.

2. Models and Methodologies

Three different types of models are produced in order to investigate the influence of grain boundary structures on the mechanical properties of polycrystalline aluminum. All the models use the embedded atom method of Mishin et al. [6]. This method accurately reproduces the energy values of stacking faults.

2.1. Dependence of mechanical properties of nanocrystalline metals on grain boundary characteristics

We consider quasi two-dimensional models with a grain sizes d ranging from 5 nm to 80 nm (see Fig. 1(a, b)). The length of all the models along the X direction is approximately 1.1 nm and each model comprises eight (110) atomic planes. The analysis models are composed of unit structures that consist of eight hexagonal grains A~H in which the crystal orientation along the X direction is fixed at $\langle 110 \rangle$. Two types of grain arrangements are considered in order to produce different distributions of the grain boundary misorientation angle θ; we term these two arrangements as model 1a and model 1b, respectively. All the models have an identical texture because they consist of the same eight grains A~H. Figure 1(c) shows the distributions of θ in both the models, and each of the values is shown in Fig. 1(a, b); the deviation angles ϕ of the actual grain boundary plane from the symmetrical grain boundary plane are also provided in parentheses.

There are between 1.1 and 3.2 million atoms in each model. A periodic boundary condition is adopted for all directions. A tensile deformation is caused in the Z direction at a strain rate of 8×10^8 1/s, whereas a deformation in the Y direction is produced by maintaining the stress σ_{yy} at zero by the Parrinello-Rahman method [7]. The temperature is maintained at 300 K during all the simulations. The local face-centered-cubic (fcc) and hexagonal-close-packed (hcp) crystal structures in addition to defect atomic structures are classified using common neighbor analysis (CNA) [8].

2.2. Mechanism of interaction between dislocations and grain boundaries

Figure 1(d) shows the schematic of the analysis model used for investigating the interaction mechanism between edge dislocations and the $\langle 112 \rangle$ asymmetrical tilt grain boundaries through quasicontinuum simulations. The analysis model comprises two crystal grains A and B, and the dislocation source in grain A is a crack. In this study, we intend to express the movement of dislocations from the crack to the grain boundary and the interaction between the dislocations and the grain boundary. For this reason, the region near the dislocation slip plane in grain A and that near the grain boundary between grains A and B have full atomistic resolution. On the other hand, the regions distant from

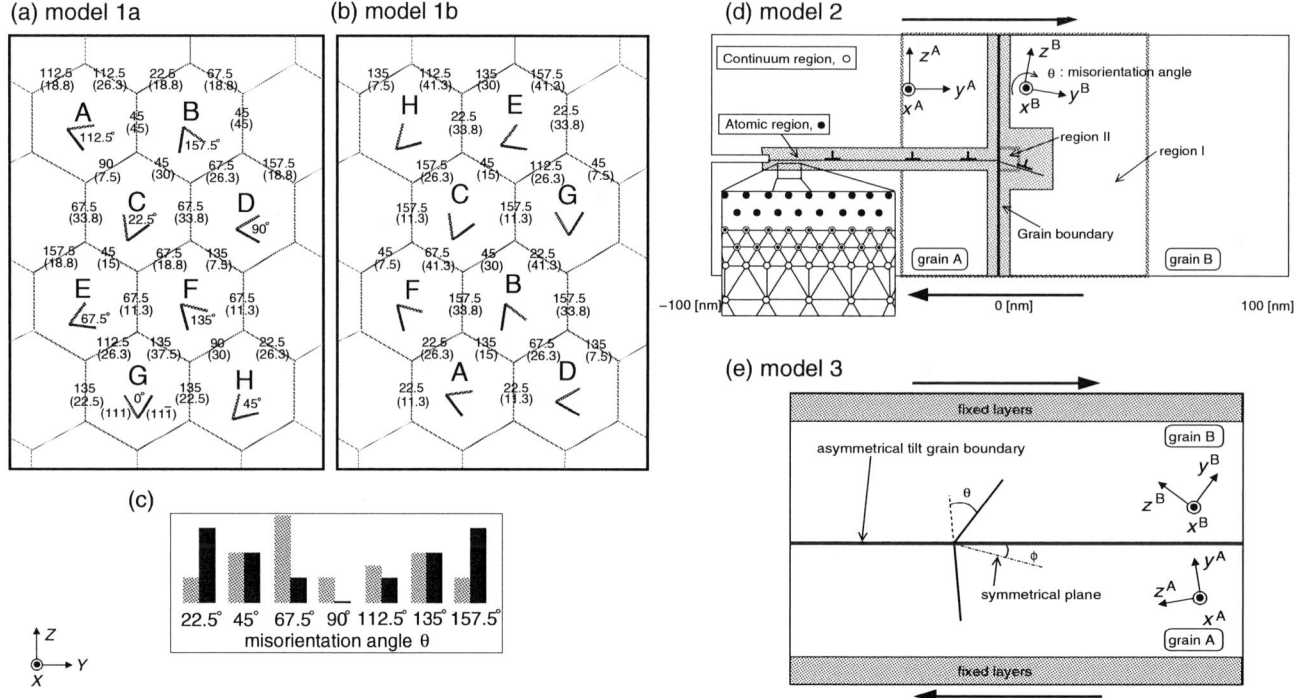

Figure 1: Analysis models. (a)(b) Arrangements of eight crystal grains in the nanocrystalline models: model 1a and model 1b. (c) Distributions of grain boundary misorientation angle in model 1a (gray) and model 1b (black). (d) Coupling model to investigate the interaction between edge dislocations and tilt grain boundaries in quasicontinuum method. (e) Grain boundary model.

the slip plane or the grain boundary are divided into finite elements. Therefore, the mechanics of the atoms in an element are determined by the positions of the node atoms. The enlarged picture in Fig. 1(d) shows the atomic arrangements in the vicinity of the coupling interface between the atomistic and the continuum regions. The solid, open, and double circles represent the *nonlocal* atoms corresponding to the atomistic region, *local* atoms corresponding to the continuum regions, and *quasi-nonlocal* atoms corresponding to the buffer layers that seamlessly couple the atomistic and continuum regions, respectively [9]. Periodic boundary conditions are adopted along the X and Z directions. In order to generate edge dislocations from the crack in grain A under shear deformation, as shown in Fig. 1(d), the crystal orientations along the x^A, y^A, and z^A directions of grain A are set to $[11\bar{2}]$, $[\bar{1}10]$, and $[111]$, respectively. The crystal orientation along the x^B direction of grain B is also set to $[11\bar{2}]$; therefore, the grain boundary structure in this model is controlled by the rotational angle θ about the X direction. In this study, two small-angle tilt grain boundaries with misorientation angles of 13° and 30° are considered; these two boundaries characterize model 2a and model 2b, respectively. For each model, the dimensions along the X, Y, and Z directions are approximately 24, 200, and 55 nm, respectively. The distance between the crack tip and the grain boundary is approximately 40 nm. The sum of the nonlocal, local, and quasi-nonlocal atoms in each model is approximately 1.5 million. If the analysis model is expressed by only nonlocal atoms, the total number of atoms is approximately 16 million. Consequently, the degree of

freedom in this QC model is one-tenth of that in a full atomistic model.

In order to simulate the interactions between the incoming dislocations and the tilt grain boundaries, the shear strain increment $\Delta\gamma_{ZY}$ is repeatedly applied to the analysis models. The energy of the analysis model for each $\Delta\gamma_{ZY}$ is minimized by the conjugate gradient method; no thermally activated process is considered in these simulations. The value of $\Delta\gamma_{ZY}$ is 0.002. Under the shear deformation, the Z-directional displacement of the surface nodes along the Y direction is fixed.

2.3. Deformation mechanism of the asymmetrical grain boundaries

In order to examine the influence of ϕ on the grain boundary motion, we produce the $\langle 112 \rangle$ asymmetrical tilt grain boundary model, as shown in Fig. 1(e). The crystal orientations of grains A and B are the same as those in model 2. Periodic boundary conditions are adopted along the X and Y directions. Four deviation angles are adopted in this study: $\phi = 0°$, 3.6°, 15.7°, and 42.9°; we refer to the models corresponding to these angles as model 3a, model 3b, model 3c, and model 3d, and the θ value for all the models is 21°. The dimensions along the X and Z directions are approximately 1.5 nm and 25 nm.

In order to apply shear deformation, atoms in the gray regions with a width of 1.5 nm at the surfaces are moved at a velocity of 5 m/s. We consider two different temperatures—300 K and 800 K—for all the simulations.

642

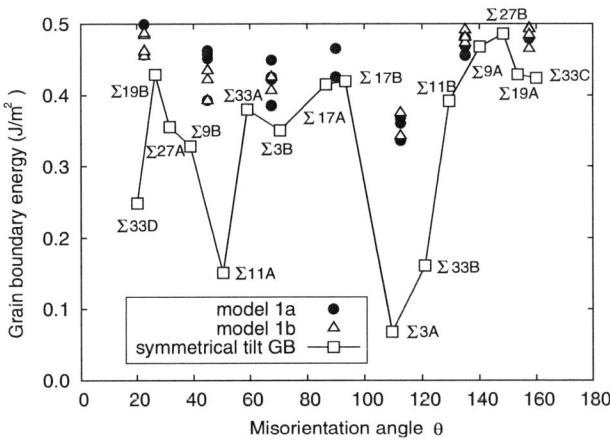

Figure 2: Grain boundary energy versus θ.

3. Results and Discussion

3.1 Relationship between grain boundary characteristics and mechanical properties of nanocrystalline aluminum

Figure 2 depicts the relationship between the grain boundary energy and θ in model 1a and model 1b. The grain boundary energy is calculated by performing simulated annealing at 0.1 K. The values of the ⟨110⟩ symmetrical grain boundary are also shown in Fig. 2. All grain boundaries in model 1a and model 1b are asymmetrical boundaries, as shown in Fig. 1(a, b); therefore, the grain boundary energies show higher values than the energies of symmetrical boundaries. It is noteworthy that grain boundaries with the same θ show different energy values; therefore, it can be confirmed that the grain boundary energy depends not only on θ but also on φ. The influence of φ on the grain boundary motion is discussed in detail in the following sections.

Figure 3 shows the relative proportion of the grain boundary region in simulated nanocrystalline metals (f_{gb}^{MD}) versus d. Light, medium, and dark gray atoms in the inset represent the local fcc, defect, and hcp structures, respectively. The broken curves represent f_{gb}^{ideal}, which is esti-

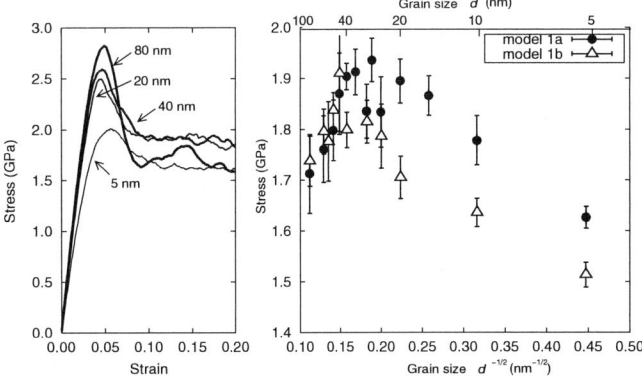

Figure 4: (a) Stress-strain curves of model 1a. (b) Relationship between σ_f and d^{-1/2}.

mated by an ideal model with a grain boundary thickness of 0.54 nm [11]. The values of f_{gb}^{MD} and f_{gb}^{ideal} are almost identical; therefore, it can be inferred that the grain boundary thickness in both the models does not depend on d and that the grain boundary thickness b cannot be neglected when d decreases. Consequently, we can deduce that the influence of the grain boundary characteristic on the mechanical properties of polycrystalline metals increases as the grain size decreases to the order of nanometers.

Figure 4(a) shows the stress-strain curves in the case of model 1a. In this study, the average flow stress σ_f estimated by averaging the tensile stress after a strain ε = 0.1 is regarded as the material strength. Figure 4(b) shows the relationship between the flow stress σ_f and the inverse square root of the grain size $d^{-1/2}$. It can be observed that a transition from grain-size hardening to grain-size softening occurs in both the models; the flow stress increases as the grain size decreases to 30~40 nm, which illustrates the Hall-Petch effect. On the other hand, grain-size softening can be observed as the grain size decreases below 30~40 nm, which illustrates the inverse Hall-Petch effect. It is noteworthy that the deformation resistance in the grain-size hardening region in model 1b is larger than that in model 1a, but this tendency is reversed in the grain-size softening region. In the former region, intragranular deformation is dominant, as shown in Fig. 5(a); therefore, the

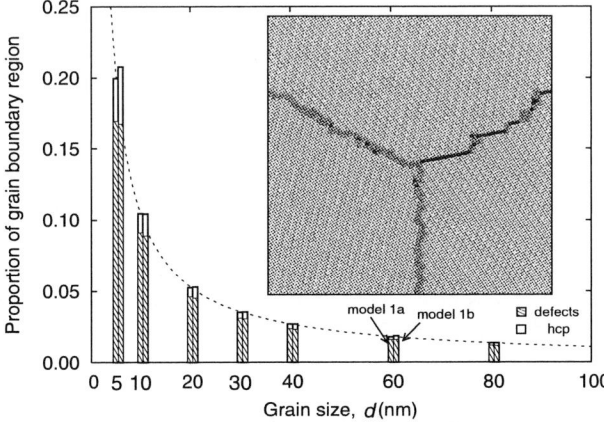

Figure 3: Proportion of grain boundary regions in model 1. Broken line represents the analytical proportion of grain boundary region estimated by an ideal model with a grain boundary thickness of 0.54 nm.

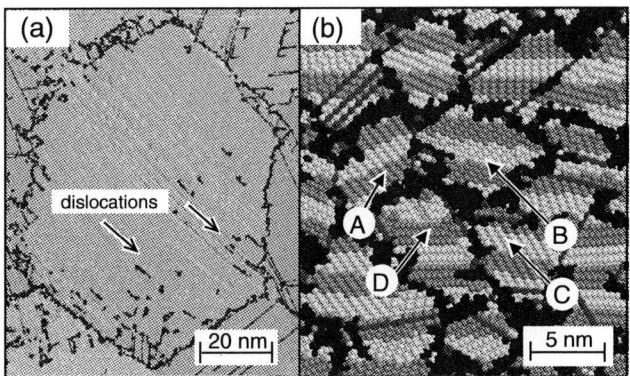

Figure 5: (a) Intragranular deformation in the 80 nm grain when ε = 0.2 in model 1a. (b) Intergranular deformation in the 5 nm grain when ε = 0.12 in model 1a. Grain boundary sliding and grain rotation occur.

643

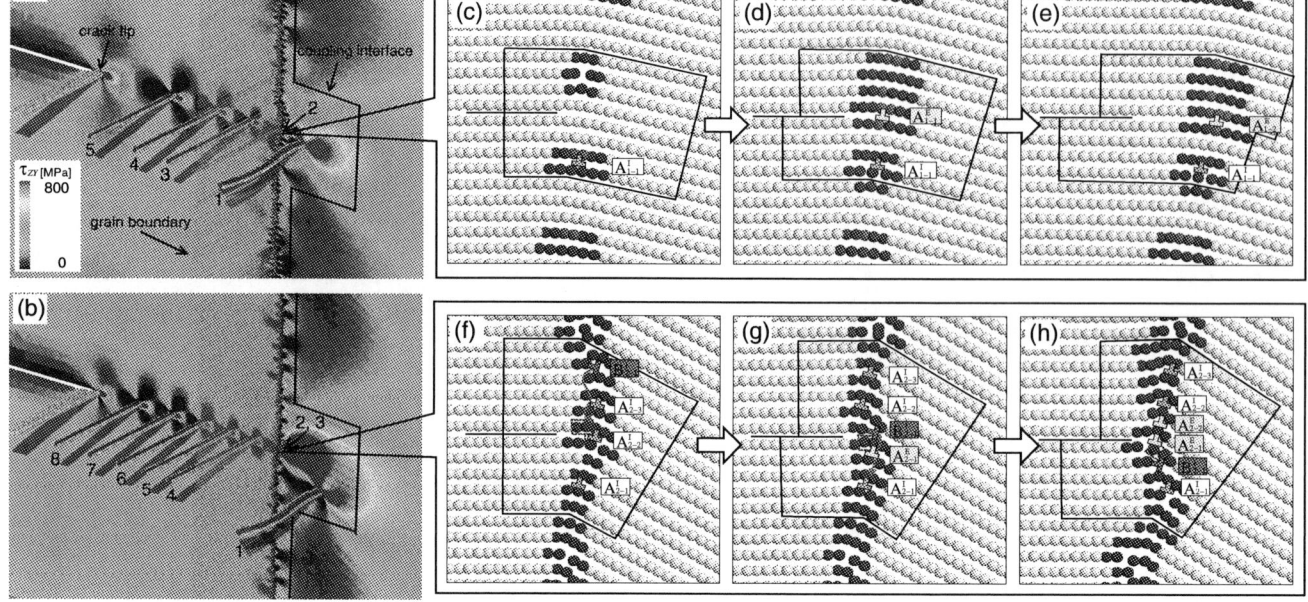

Figure 6: Interactions between lattice edge dislocations from the crack tip and asymmetrical tilt grain boundaries under shear deformation. (a) model 2a: $\theta = 13.0°$, $\phi = 7°$, $\gamma_{ZY} = 0.024$; (b) model 2b: $\theta = 29.5°$, $\phi = 14°$, $\gamma_{ZY} = 0.034$. Atomic configurations of the tilt grain boundaries in model 2a and model 2b under shear deformation: (c) $\gamma_{ZY} = 0.002$, (d) $\gamma_{ZY} = 0.012$, (e) $\gamma_{ZY} = 0.024$, (f) $\gamma_{ZY} = 0.002$, (g) $\gamma_{ZY} = 0.012$, and (h) $\gamma_{ZY} = 0.028$.

grain boundary acts as an obstacle in the path of the dislocation motion. On the other hand, in the grain-size softening region, intergranular deformation, i.e., grain boundary sliding, is dominant (as shown in Fig. 5(b)). Consequently, the grain boundary role changes as the grain size decreases, and the intergranular and intragranular deformations could also strongly depend on the grain boundary structure defined by θ and ϕ. In the following sections, the influences of the grain boundary structure on the interaction between dislocations and grain boundaries observed in the grain-size hardening region and on the grain boundary motion observed in the grain-size softening region are investigated in detail.

3.2 The mechanism of interaction between edge dislocations and tilt grain boundaries

Figure 6(a, b) shows the final atomic configurations in each model under the shear deformation applied in this study. In order to observe the defect structures easily, the atoms in the local fcc structure are not shown in Fig. 6(a, b). The atoms in the local hcp and defect structures are shown in brown and gray, respectively. The defect atoms in the grain boundaries are shown as transparent circles. The dislocations resulting from the crack tip are numbered according to their order of appearance. The distributions of τ_{ZY} are also shown in each background. The black lines show the coupling interface between the atomistic and continuum regions in grain B, and the seamless distribution of τ_{ZY} can be observed. The dislocation transmission in model 2a occurs when γ_{ZY} attains the value of 0.024 and the grain boundary absorbs the second incoming dislocation, as shown in Fig. 6(a). Similarly, the dislocation transmission in model 2b occurs when γ_{ZY} attains the value of 0.034 and two incoming dislocations are absorbed by the grain boundary, as shown in Fig. 6(b).

Figure 7 shows the relationship between the macroscopic and microscopic shear stresses (τ_{ZY}^{Mac} and τ_{ZY}^{Mic}) and the shear strain γ_{ZY} in order to investigate the influence of dislocation pile-up on the stress concentration. The macroscopic and microscopic stresses are evaluated in region I: -30 nm $\leq Y \leq 50$ nm, and in region II: 0 nm $\leq Y \leq 5$ nm and -2.5 nm $\leq Z \leq 2.5$ nm, respectively (as shown in Fig. 1(d)).

The slopes of τ_{ZY}^{Mac} and τ_{ZY}^{Mic} are almost identical for both the models before the first dislocation emission from the crack tip because the anisotropy factor—$2C_{44}/(C_{11} - C_{12})$—is very close to one (approximately 1.25) for the adopted atomic potential. After the first dislocation emission from the crack tip when $\gamma_{zy} = 0.012$, the slopes of τ_{ZY}^{Mac} decrease in comparison with those under elastic deformation. On the other hand, the slopes of τ_{ZY}^{Mic} increase with the number of dislocations in the pile-up [10]. The number of dislocations in model 2b is larger than that in model 2a; hence, the critical force on the edge dislocation for it to be ejected from the tilt grain boundaries in model 2b is larger than that in model 2a. Consequently, the transmission of the plastic deformation in model 2a occurs under a macroscopic shear stress that is lower than that of model 2b due to the difference in the grain boundary structures.

In order to elucidate the accommodation of incoming dislocations in the grain boundaries, Figs. 6(c) to 6(h) show a series of atomic configurations in the vicinity of the grain boundaries where the incoming dislocations are absorbed. Analytically, two sets of uniformly spaced edge dislocations are required to construct an asymmetrical low-angle tilt grain boundary [12]. The Burgers vectors of these grain boundary dislocations—b_1 and b_2—are perpendicular to each other. In model 2a, one vacancy and one primary intrinsic grain boundary dislocation (IGBD) are found in the

644

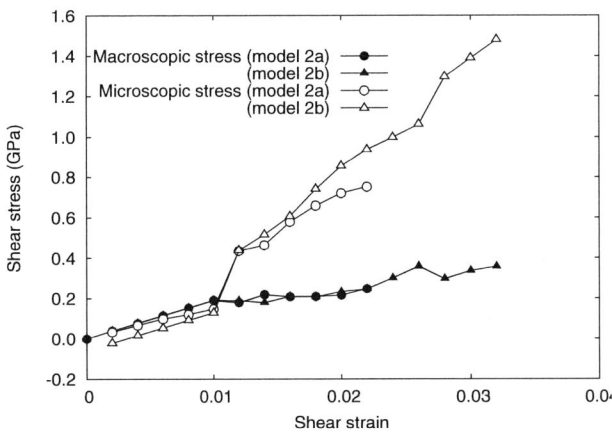

Figure 7: Relationship between shear stress and γ_{ZY}.

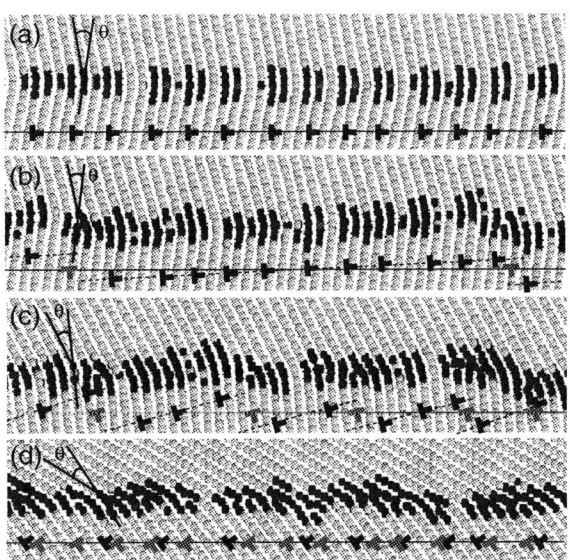

Figure 8: Initial atomic configurations of asymmetrical tilt grain boundaries with $\theta = 21°$. (a) $\phi = 0°$ (b) $\phi = 3.6°$ (c) $\phi = 15.7°$ (d) $\phi = 42.9°$

Burgers circuit, as shown in Fig. 6(c). No secondary IGBD is observed near the slip plane; hence, the grain boundary around the slip plane can be considered to be symmetrical. In model 2b, as shown in Fig. 6(f), three primary IGBDs and one secondary IGBD are found in the Burgers circuit. Hence, the effect of the secondary IGBD on the accommodation of extrinsic grain boundary dislocations (EGBDs) can be investigated by comparing the results of these two models.

When γ_{ZY} attains the value of 0.012 in model 2a, the incoming dislocation is absorbed at the site between the vacancy and the IGBD A^I_{1-1} as the EGBD A^E_{1-1}, as shown in Fig. 6(d). Subsequent dislocations from the crack tip pile up behind A^E_{1-1} with an increase in γ_{ZY}, as shown in Fig. 6(a). Finally, A^E_{1-1} is ejected from the grain boundary when $\gamma_{ZY} = 0.024$ and the next incoming dislocation is absorbed as A^E_{1-2} at the same site, as shown in Fig. 6(e). Consequently, this grain boundary can accommodate only one EGBD in the slip plane of the incoming dislocations.

In the equilibrium state in model 2b, as shown in Fig. 6(f), the IGBD A^I_{2-2} already exists in the slip plane of the incoming dislocation. Therefore, it can be easily inferred that the first incoming dislocation will pile up behind A^I_{2-2}, as in the case of model 2a. However, the first incoming dislocation does not pile up but is absorbed by the grain boundary as A^E_{2-1}, as shown in Fig. 6(g). This is due to the sliding of the secondary IGBD B^I_{2-1} along the grain boundary and the ejection of the atomic group in the red (dark gray) box, which corresponds to the magnitude of the Burgers vector of the incoming edge dislocation, from its atomic plane. Consequently, dislocations A^I_{2-2}, A^I_{2-3}, and A^E_{2-1} change their atomic planes; the dislocations climb up and climb down occur without the diffusion process. As γ_{ZY} increases, the incoming dislocations pile up behind B^I_{2-1}. When γ_{ZY} attains the value of 0.028, B^I_{2-1} slides once again along the direction of the arrow, and the second incoming dislocation is simultaneously absorbed as A^E_{2-2} with a climb up due to the sliding of B^I_{2-1}. There is no secondary IGBD in the upper region of the slip plane, as observed in Fig. 6(h); therefore, the EGBDs are not accommodated at the grain boundary and the incoming dislocations pile up behind A^E_{2-1}. Finally, A^E_{2-1} is ejected from the grain boundary by the dislocation pile-

up when γ_{ZY} attains the value of 0.034. It is noteworthy that the secondary IGBD B^I_{2-1} does not slide alone under a macroscopic shear stress τ_{ZY} of 200~400 MPa in this study. Hence, the stress-assisted sliding along the grain boundary plane of the secondary IGBD with the absorption of the incoming dislocations could strongly affect the accommodation of the EGBDs.

3.3 Grain boundary motion under shear deformation

In order to investigate the ϕ dependence of the grain boundary motion, we construct four $\langle 112 \rangle$ asymmetrical tilt grain boundary models as shown in Fig. 1(e). Figure 8 shows the atomic configurations of asymmetrical tilt grain boundaries with $\theta = 21°$. The models have different deviation angles—$\phi = 0°$, $3.6°$, $15.7°$, and $42.9°$. We label these models as model 3a, model 3b, model 3c, and model 3d. Two sets of grain boundary dislocations are also shown in Fig. 8. The spacings of two dislocation sets along the boundary correspond well to that calculated from the expressions $b_1/\theta \cos \phi$ and $b_2/\theta \sin \phi$ [12], e.g., in the case of $\phi = 15.7°$, the spacings of the primary and secondary dislocations (shown in black and gray) are obtained as 0.84 nm and 2.16 nm in model 3c, while they are analytically calculated as 0.81 nm and 2.36 nm, respectively. Therefore, the grain boundaries in our atomic model comprise only IGBDs.

Figure 9 shows the grain boundary displacements from the initial positions under shear strain γ_{ZY} at 300 K and 800 K, respectively. In the case of the symmetrical grain boundary in model 3a, grain boundary migration occurs by the movement of primary dislocations toward grain B. In the case of small deviation angles ($\phi = 3.6°$ and $15.7°$) it can be observed that grain boundary migrations also occur under shear deformation by the movement of primary and secondary dislocations; the primary dislocations move toward grain B and the secondary dislocations move along the grain boundary. Although it can be observed that the rate of the grain boundary migration depends on ϕ, no tem-

645

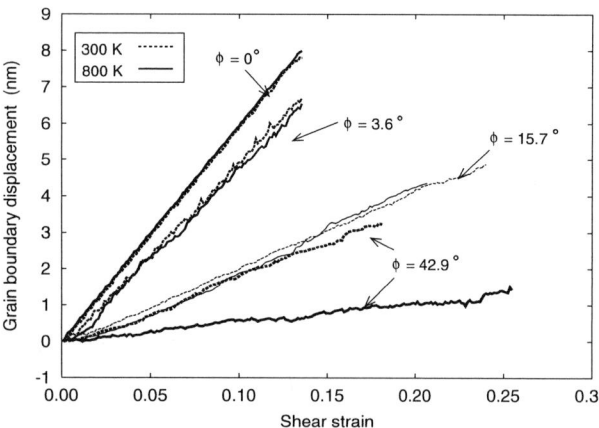

Figure 9: Grain boundary positions under shear deformation.

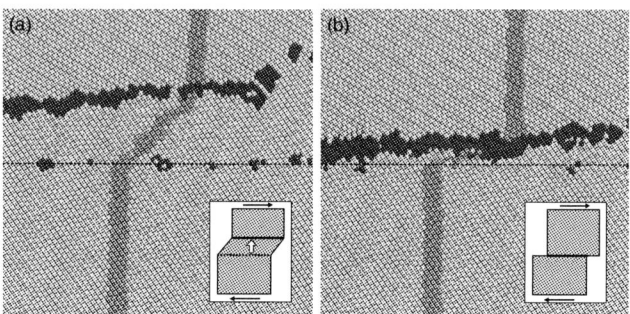

Figure 10: Atomic configurations for (a) 300 K and (b) 800 K around the grain boundary with θ = 21° and φ = 42.9° when γ = 0.188. Broken lines represent the initial grain boundary positions.

perature dependence can be detected. However, in the case of model 3d with a large ϕ, the influence of temperature on the grain boundary motion is observed to be remarkable. Figure 10 shows the atomic configurations around the grain boundary when $\gamma_{ZY} = 0.188$ in model 3d. Gray atoms along the Z direction are colored at the initial configuration in order to help understand the grain boundary motion. For 300 K, the grain boundary migration is similar to that in the other models; however, for 800 K, grain boundary sliding can be observed. Consequently, it can be observed that the grain boundary motion strongly depends on ϕ at the tilt grain boundary with the same θ; thus, the structure of grain boundary dislocations controls the grain boundary motions.

4. Conclusions

The relationship between grain boundary structures and the mechanical properties of polycrystalline aluminum is investigated by simulating the following problems through molecular dynamics and quasicontinuum methods: (1) the influence of the grain boundary characteristic distribution on the macroscopic mechanical properties of the nanocrystalline metals, (2) the accommodation mechanism of incoming dislocations at the tilt grain boundary, and (3) the dependence of the deviation angle on the grain boundary motion. It is observed that the arrangements of grain boundary dislocations control the grain boundary motion;

therefore, from the point of view of the grain boundary engineering, it could be worth considering the deviation angle of grain boundaries in the design of nanostructured materials.

Acknowledgments

The author acknowledges support from the Ministry of Education, Science, Sports and Culture (Grant-in-Aid for Young Scientists (B), 2004, 16760063). This study was supported in part by a Grant-in-Aid for Scientific Research from the Ministry of Education, Culture, Sports, Science and Technology (MEXT), Japan on Priority Areas "Giant Straining Process for Advanced Materials Containing Ultra-High Density Lattice Defects". The author also thanks Mr. Wataru Katayama and Mr. Kengo Suzuki for their help in the computations.

References

1. Hall, E. O., "The Deformation and Ageing of Mild Steel: III Discussion of Results", *Proc. Phys. Soc. B*, Vol. 64, (1951), pp. 747-753.
2. Petch, N. J., "The Cleavage Strength of Polycrystals", *J. Iron Steel Inst.*, Vol. 174, (1953), pp. 25-28.
3. Watanabe, T. *et al*, " The Control of Brittleness and Development of Desirable Mechanical Properties in Polycrystalline Systems by Grain Boundary Engineering", *Acta Mater.*, Vol. 47, (1999), pp. 4171-4185.
4. Tadmor, E. B. *et al*, "Mixed Atomistic and Continuum Models of Deformation in Solids", *Langmuir*, Vol. 12, (1996), pp. 4529-4534.
5. Shenoy, V. B, *et al*, "An Adaptive Finite Element Approach to Atomic-scale Mechanics - the Quasicontinuum Method", *J. Mech. Phys. Solids*, Vol. 47, (1999), pp. 611-642.
6. Mishin, Y. *et al*, "Interatomic Potentials for Monoatomic Metals from Experimental Data and *ab initio* Calculations", *Phys. Rev. B*, Vol. 59, (1999), pp.3393-3407.
7. Parrinello, M. *et al*, "Polymorphic Transitions in Single Crystals: A New Molecular Dynamics Method", *J. Appl. Phys.*, Vol. 52, (1981), pp. 7182-7190.
8. Jónsson, H. *et al*, "Icosahedral Ordering in the Lennard-Jones Liquid and Glass", *Phys. Rev. Lett.*, Vol. 60, (1998), pp. 2295-2298.
9. Shimokawa, T. *et al*, "Matching Conditions in the Quasicontinuum Method: Removal of the Error Introduced at the Interface between the Coarse-grained and Fully Atomistic Region", *Phys. Rev. B*, Vol. 69, (2004), pp. 214104(1-10).
10. Hirth, J. P. *et al*, <u>Theory of Dislocations: Second edition</u>, McGraw-Hill (New York 1982).
11. Shimokawa, T. *et al*, "Grain-size Dependence of Relationship between Intergranular and Intragranular Deformation of Nanocrystalline Al by Molecular Dynamics Simulations", *Phys. Rev. B*, Vol. 71, (2005), pp. 224110(1-8).
12. Read, W. T. *et al*, "Dislocation Models of Crystal Grain Boundaries", *Phys. Rev.*, Vol. 78, (1950), pp. 275-289.

Wire Bond Reliability for Power Electronic Modules - Effect of Bonding Temperature

Wei-Sun Loh[1], Martin Corfield[1], Hua Lu[2], Simon Hogg[3], Tim Tilford[2], C Mark Johnson[4]

[1] Department of Electronic and Electrical Engineering, University of Sheffield, Mappin Street, Sheffield, S1 3JD.

[2] School of Computing and Mathematical Sciences, University Of Greenwich, Park Row, London SE10 9LS

[3] The Institute of Polymer Technology and Materials Engineering (IPTME), Loughborough University, Loughborough, LE11 3TU, U.K.

[4] School of Electrical and Electronic Engineering, University of Nottingham, University Park, Nottingham, NG7 2RD, U.K.

Tel.: +44 (0) 114 222 5890. Email: w.s.loh@sheffield.ac.uk.

Abstract

In this paper, thermal cycling reliability along with ANSYS analaysis of the residual stress generated in heavy-gauge Al bond wires at different bonding temperatures is reported. 99.999% pure Al wires of 375 μm in diameter, were ultrasonically bonded to silicon dies coated with a 5μm thick Al metallisation at 25°C (room temperature), 100°C and 200°C, respectively (with the same bonding parameters). The wire bonded samples were then subjected to thermal cycling in air from -60°C to +150°C. The degradation rate of the wire bonds was assessed by means of bond shear test and via microstructural characterisation. Prior to thermal cycling, the shear strength of all of the wire bonds was approximately equal to the shear strength of pure aluminum and independent of bonding temperature. During thermal cycling, however, the shear strength of room temperature bonded samples was observed to decrease more rapidly (as compared to bonds formed at 100°C and 200°C) as a result of a high crack propagation rate across the bonding area. In addition, modification of the grain structure at the bonding interface was also observed with bonding temperature, leading to changes in the mechanical properties of the wire. The heat and pressure induced by the high temperature bonding is believed to promote grain recovery and recrystallisation, softening the wires through removal of the dislocations and plastic strain energy. Coarse grains formed at the bonding interface after bonding at elevated temperatures may also contribute to greater resistance for crack propagation, thus lowering the wire bond degradation rate.

1. Introduction

Thick aluminum wire bonding is the most commonly employed interconnect technology in power electronic modules. The reliability of Al wire bonds depends on the bond strength between the Al wire and the IGBT chip [1]. However, the wire bonds are susceptible to heel crack [2] where failure arises from flexing due to thermal expansion or overworked bond heel during ultrasonic bonding [3]. Additional fatigue failures are caused by thermo-mechanical damage mechanisms caused by the mismatch of thermal expansion coefficients (CTE) between the aluminum wire and silicon die at the contact interface. This failure mode is aggravated by wide thermal cycling ranges.

A recent study by Jin Onuki [4] has shown that the deterioration rate of the Al wire bonds may possibly be reduced by increasing the grain size with heat treatment. Furthermore, work by Komiyan et al. [5] has also shown that bonds formed at elevated temperatures with low ultrasonic energy could also exhibit high bonding strength. This was attributed to the ease of deformation at the bonding interface, and enhancement of the actual bonded area, resulting in a reduction of voids. However, at present there is little information on the deformation mechanism and Al grain structure changes after the bonding process, and their significance in affecting the crack propagation rate at the bonding interface. This paper will focus on the thermal cycling reliability to predict how grain structure, and thus the material properties change with bonding temperature, and how material property changes affect the residual stress in the bond heel using finite element analysis. Conclusions are drawn concerning the application of high temperature bonding as a mechanism for enhancing the thermal cycling reliability of wire bonds.

2. Wire bonding process

High purity (99.999%) aluminum wires of diameter 375μm were ultrasonically bonded to silicon dies coated with a 5μm thick aluminium top metal at 25°C (room temperature), 100°C and 200°C, using a ultrasonic power of 1.6W for 150ms. The temperature of the bonding samples were monitored throughout the bonding process. Figure 1 shows a schematic diagram of the high temperature heavy-gauge aluminum wire bonding experimental setup.

The wire bonded samples were then subjected to repetitive passive thermal cycling in air from -60°C to +150°C. The degradation behaviour of the wire bonds were evaluated by measuring the bond shear strength at regular intervals and via micro structural analysis by using optical microscopy and Electron Backscatter Diffraction (EBSD).

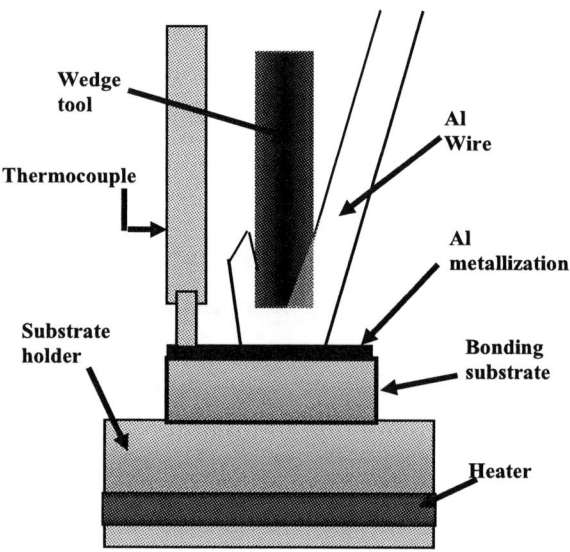

Figure 1: Schematic representation of the high temperature ultrasonic Al wire bonding experimental setup.

3. Reliability of high temperature bonded Al wire bonds

The mean shear force required to shear the wire bonds increases steadily with bonding temperature, however, this increase is directly porportional to the average increase in bond foot area, as listed in Table 1. Therfore, the mean shear strength for the Al wire bonds bonded at RT, 100°C and 200°C remains almost constant and approximately equal to the shear strength of 99.999% pure aluminum (50MPa) [6]. This implies that good wire bonding has been achieved over the full range of bonding temperatures used in the present study.

Table 1: Average size, mean shear force and normalised shear stress of the wire bonds bonded at RT, 100°C and 200°C respectively.

Bonding Temp. (°C)	Average bond foot area (mm^2)	Mean shear force (N)	Normalised shear stress (MPa)
25 (RT)	0.32	16.1	50.3
100	0.38	18.5	48.4
200	0.40	19.4	47.5

The normalised shear force drops steadily with increasing thermal cycles irrespective of bonding temperature, as shown in figure 2, due to a reduction in bonded area resulting from fatigue crack propagation. The shear strength of room temperature bonded samples was observed to decrease more rapidly with thermal cycling, suggesting a higher crack propagation rate across the bonding area.

Wire bond lift-off for the room temperature bonded Al wires occurs after 1500 thermal cycles. In contrast, bond lift-off was observed for 100°C bonded wires at 2100 thermal cycles and 200°C bonded wires at 2400 cycles. As the shear strength of the wire bonds is sensitive to the length of the fatigue crack [7], it can be assumed that the degradation rate of the wire bonds depends strongly on the crack propagation rate at the bonding interface. This may indicate that, high temperature bonding inhibits crack propagation, giving rise to improved reliability.

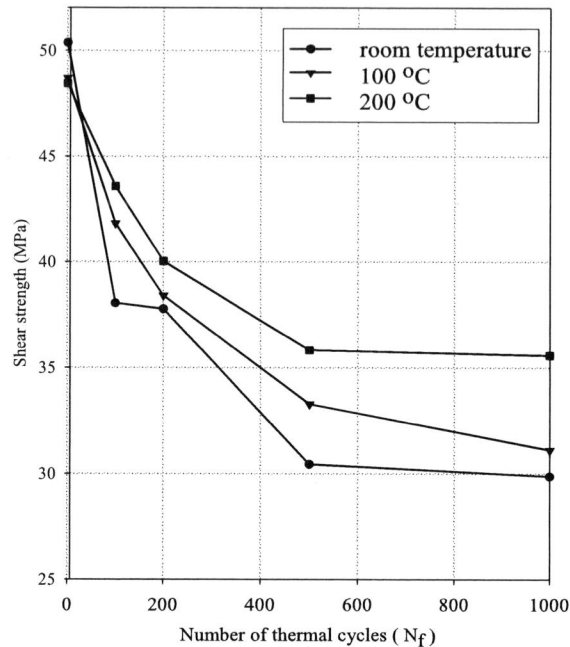

Figure 2: Mean shear strength for Al wire bonds, bonded at (a) RT, (b) 100°C and (c) 200°C as a function of thermal cycles

4. Grain structure analysis

Cross section analysis

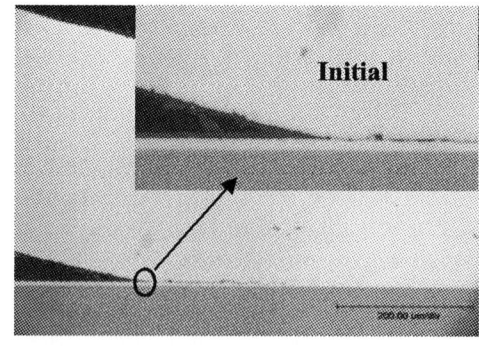

Figure 3: Cross section of an aluminum wedge bond bonded at RT showing crack initiation after bonding. The inset shows a magnified view of the initial crack.

During ultrasonic bonding, Al wire is plastically deformed due to the applid bonding force and ultrasonic energy, increasing the dislocation density and strain energy at the bonding interface. Residual stress is produced by heterogeneous plastic deformation and thermal contraction resulting from CTE mismatch between the Al wire and silicon die after removal of the heat gradient. Pre-cracks are often found at the bond heel, a consequence of the mechanical deformation and unavoidable flexing of the wire when forming the wire loop during the bonding process, as shown in Figure 3.

A significant amount of thermo-mechanical stress is induced along the bonding interface especially at the bond heels due to the mismatch of CTE between the Al wire and the silicon substrate [8] during cyclic loading. The fatigue crack propagates from the bond heel towards the bond centre, along the plane just above the bonding interface, between the Al wire and metallization, leading to relaxation of the residual stress. The crack area widens by continued shear deformation.

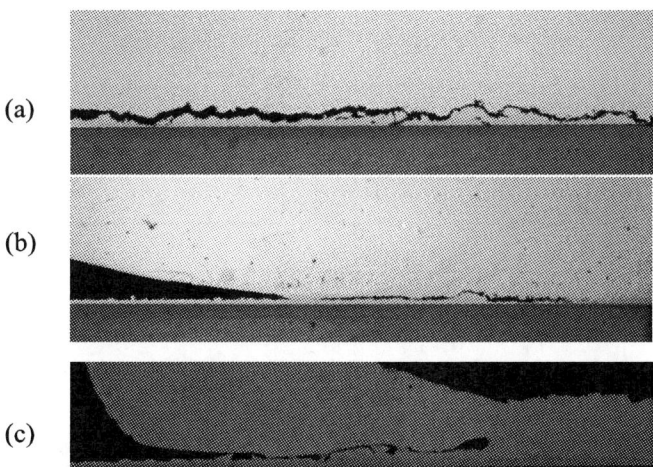

Figure 4: Fatigue crack (magnification x100) after 1500 thermal cycles for wire bonds joined at (a) 25°C, (b) 100°C and (c) 200°C, respectively.

As shown in Figure 4, for room temperature bonded samples, the fatigue crack propagates along the bonded area and eventually shears off the wire bonds after 1500 thermal cycles. The fatigue crack in 200°C bonded samples is much shorter than in 100°C bonded samples suggesting that high temperature bonding has suppressed crack propagation in the wire bonds.

EBSD pattern analysis

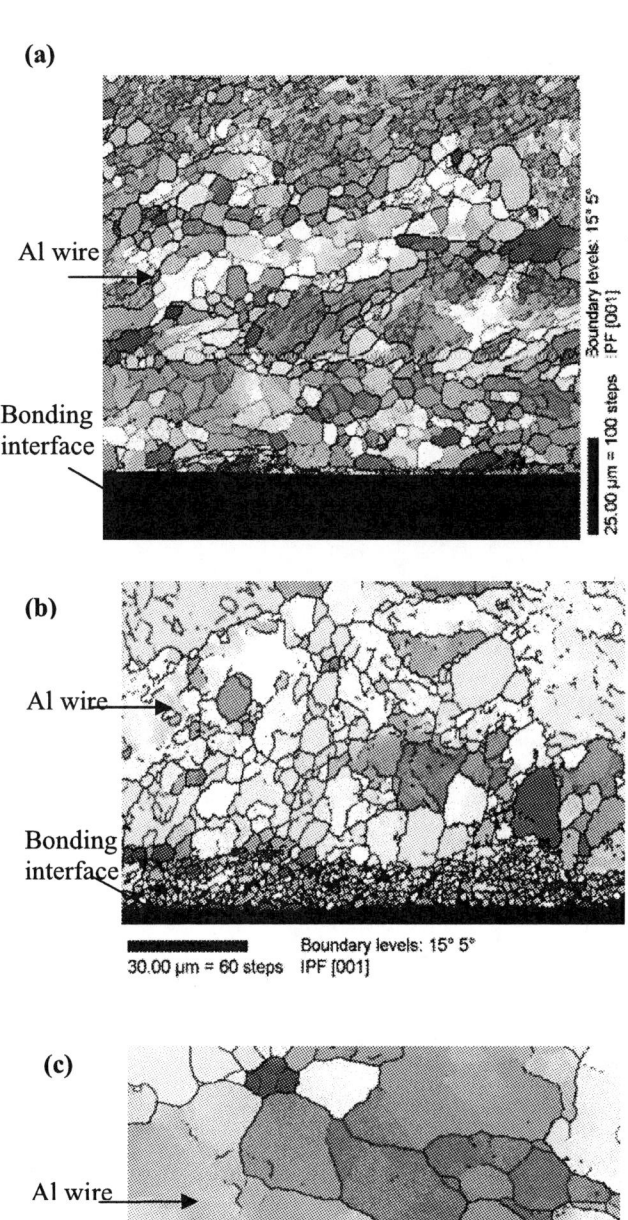

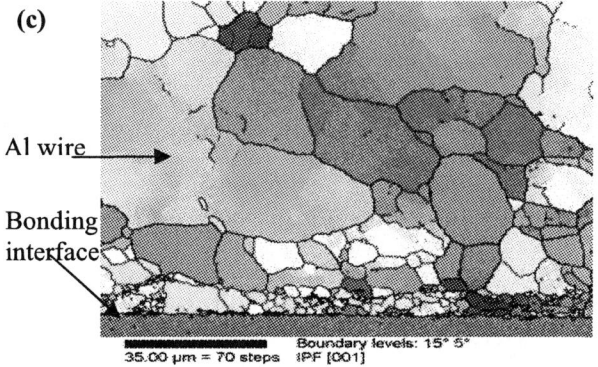

Figure 5: EBSD analysis for heavy-gauge Al wire bonds, bonded at (a) RT, (b) 100°C and (c) 200°C

The EBSD pattern represents crystallographic orientation of the selected bonding area as shown in Figure 5(a-c). The black shaded areas found near the bonding interface are "non-index" areas where accurate information can not be obtained [9].

As shown in Figure 5(a), closely packed fine grains are formed near the bonding interface resulting from strain hardening due to accumulation of dislocations after plastic deformation of the Al wire bonds. The recovery and subsequent recrystallisation of the grains within the deformed area are promoted by the high temperature bonding process, as shown in Figure 5(b) and 5(c), leading to relaxation of residual stress along with re-arrangement/reduction in the density of dislocations near the bonding interface [10]. As mentioned in section 4, crack propagation is found to be much slower for samples bonded at high temperature which can be attributed to the grain structure changes at the interface. The presence of coarser grains along the bonding interface not only reduces the residual stress and dislocation density but also may give rise to much greater resistance to crack propagation. Consequently, it may be concluded that strong and reliable bonds can be achieved by high temperature bonding.

5. Computer modeling of the stress in wirebond

As mentioned in section 4, when the wire bonds are cooled down after bonding, residual stress is generated at the bonding interface. This stress inevitably affects the reliability of the wirebond structure.

The residual stress in wire bonds is usually expected to be higher for higher temperature bonding temperatures due to the greater temperature excursion. However the Al grain structures may vary with bonding temperature, as shown in Figure 5, and materials properties such as the yield stress will also be temperature dependent [2,11]. A lack of detailed information about the temperatures attained, deformation mechanisms and grain structure changes during the bonding process makes accurate quantitative analysis impossible. However, finite element analysis can be employed to investigate the effect of varying thermal load and material properties on the residual stress in the Al wire. ANSYS [12] has been used for this analysis.

Figure 6 illustrates a 3D model of the wire bond. This model contains a slice of the device along the wire and uses periodic boundary conditions to represent the effect of the array of wires. In order to further reduce the model size, a mirror plane symmetry of the structure is taken so that only half of the wire and the surrounding structure need to be included. The final model contains approximately 20,000 elements.

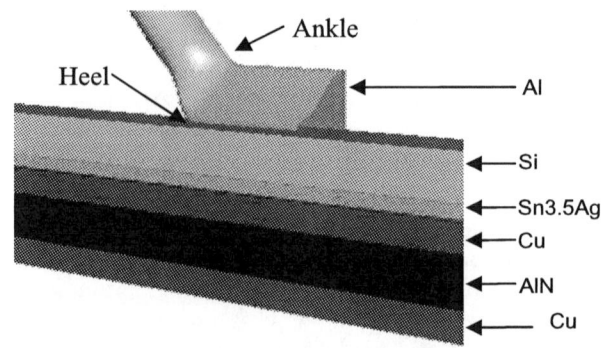

Figure 7: The layered structure of the wirebond heel.

Figure 7 shows the layered structure of the materials used in the model. Elastic-plastic material properties are used for Cu and Al. A creep law is used for SnAg solder and the material parameters can be found in [13]. Elastic material properties are listed in Table 2.

Table 2: Elastic properties of the materials used in the model. The unit of temperature is Celsius.

	E(GPa)	ν	CTE ppm/°C
AlN	310	0.24	5.6
Al	70	0.33	24.5
Si	113	0.29	3
Sn3.5Ag	54.05-0.193T	0.4	21.85+0.02039T
Cu	103.42	0.3	17

As the precise temperature-time conditions of the bondng process are unkown the investigation is limited to a constant thermal load as a driver for the residual stress calculation. It is recognised that this is a relatively crude approximation to the true temperature-time profile resulting from the ultrasonic bonding process, however it serves to illustrate the effects of variations in load and material properties. The deformation and stress in the model under a thermal load of $\Delta T= -190°C$ is shown in Figure 8. For the Al wire, the stress concentrates at the wire-die interface and at the ankle of the Al wire bond.

Figure 6: 3D Finite element model of the wire bond.

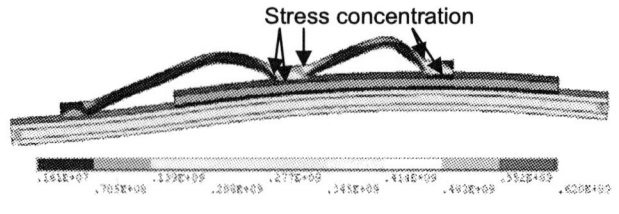

Figure 8: Stress distribution in wirebond under a thermal load of -190°C.

650

Pre-cracks are usually found at the wire-die interface after bonding and these will act as subsequent stress concentrators. The numerical model will focus on the detailed stress values at the wire-die interface, and will require a refined mesh of the interface region.

It was found that removing the wire loops above the wire-die interface, as illustrated in Figure 9, only affected the stress at the wire-die interface by just over 10%. This is relaitively small compared to the overall stress level. Therefore, in the following analysis, the wire loops were removed and the mesh at the interface has been refined to allow detailed paramteric analysis.

Consequently, a section of the bonded wire without ceramic substrate has been used in the following modelling analysis to predict the wire-die interfacial stress at the heel as shown in Figure 10.

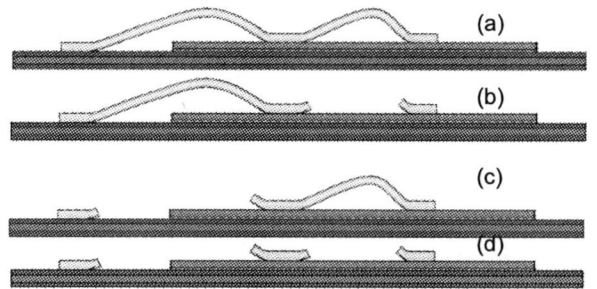

Figure 9: Wire bond models to evaluate of the Al wire loop on the stress at the wire-die interface.

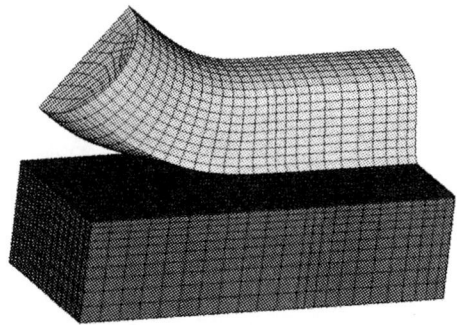

Figure 10: A simplified wire bond model which takes into account only the local CTE mismatch between the Al wire and the die.

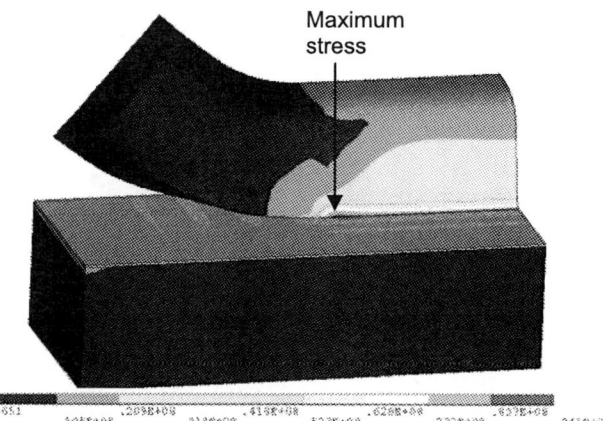

Figure 11: Typical residual von Mises stress distribution of the wirebond model at the room temperature.

Figure 11 shows the residual Von-Mises (V-M) stress when the temperature cools down from 200°C to 25°C, and the yield stress of Al is assumed to be 20 MPa. The maximum predicted V-M stress is about 94 MPa at the heel of the Al wirebond. In this model, there is no stress concentration at the ankle as the wire loop which contributes to the stress is excluded.

A total of 12 simulations have been carried out to analyse the effect of changes in the material properties and thermal load, on the predicted maximum stress at the heel of the Al wire.

The material properties which have been studied are the tangent modulus (E_t), the Young's modulus (E) and the yield stress (σ_y) of the Al wire. The material properties and the modeling results are listed in Table 3.

Table 3: Simulation results for the maximum stress in wire bond heel. T_{load} and σ_{max} are the temperature load and the maximum V-M stress respectively.

RUNS	E_t (GPa)	E(GPa)	σ_y(MPa)	T_{load}	σ_{max}
1	6.2	62	10	-75	49
2	6.2	62	20	-75	61
3	6.2	62	10	-175	94
4	6.2	62	20	-175	110
5	3.1	62	10	-75	35
6	3.1	62	20	-75	45
7	3.1	62	10	-175	58
8	3.1	62	20	-175	73
9	3.1	55.8	10	-75	35
10	3.1	55.8	20	-75	45
11	3.1	55.8	10	-175	58
12	3.1	55.8	20	-175	72

In Table 3, the material properties for RUNS#1 and #2, (modulus (E) and yield stress (σ_y)) are recognised values for standard Aluminium. The material properties and the tangent modulus will change with respect to the grain sizes. The results show that the predicted maximum

V-M stress at the wire-die interface is sensitive to changes in these properties.

It is clear that both the bonding process thermal load and the material properties govern the predicted residual stress level. Higher peak bonding temperatures may result in a much greater predicted residual stress if the material properties do not alter with bonding temperature. However, if the tangent modulus and the yield stress decrease with increasing bonding temperature, then a smaller increase or even a decrease in the residual stress would be expected. It is also important to note that CTE mismatch has been assumed to be the only cause of the residual stress. Other effects such as annealing i.e. recovery, recrystallisation and grain growth, which may occur during and after the bonding process are not accounted for in this analysis. Further experimental work is planned in order to establish the precise nature of the residual stress resulting from the bonding process.

6. Conclusions

The initial shear strength of the wire bonds prior to thermal cycling remains virtually constant with respect to bonding temperature, with no appreciable differences observed in the shear strength, with bonding temperature.

During thermal cycling, the mean shear strength of the wire bonds (for all bonding temperatures) drops significantly, due to the reduction in bonded area, resulting from fatigue crack propagation from the bond heel to the centre of the bond.

The shear strength of the room temperature bonded samples was observed to decrease more rapidly with thermal cycling than for those bonded at the higher temperatures; a result of a higher crack propagation rate across the bond. This suggests that high temperature bonding is a good candidate for enhancing the reliability of heavy-gauge aluminum wire bonds.

The combination of both heat and pressure induced by the high temperature wire bonding technology may assist in relaxation of the residual stress and re-arrangement/reduction in the density of dislocations at the bonding interface by means of recovery, recrystallisation and grain growth, thus lower the crack propagation rate.

Current computer modeling results have shown that the residual stress at the Al wire-die interface can be influenced by changes to the Al material properties resulting from changes in the grain structure. The stress at the wire-die interface is most sensitive to changes in the tangent modulus and yield stress. Further work is required to identify the conditions present close to the bond foot during the bonding process and the resulting changes in the wire material properties.

Acknowledgments

The authors wish to acknowledge the support of the Innovative Electronics Manufacturing Research Centre (IeMRC) and the United Kingdom Department of Trade and Industry for their support of the project 'Modelling of power modules for lifetime, accelerated testing, reliability and risk'. The authors would also like to thank the project partners Semelab plc., Dynex Semiconductor Ltd., Goodrich Engine Control, Raytheon Systems Ltd., SR Drives Ltd., and Areva T&D Ltd for their valuable collaboration.

References

1. Lee. R. Levine, "Wire Bonding in Optoelectronics", *Advancing Microelectronics*, Vol. 29(1), pp. 17-19, Jan. 2002.
2. S. Ramminger, N. Seliger and G. Wachutka, "Reliability Model for Al Wire Bonds Subjected to Heel Crack Failures", *Microelectronics Reliability*, Vol. 40, pp. 1521-1525, 2000.
3. K.C. Joshi, "The Formation of Ultrasonic Bonds Between Metals", *Welding Journal*, Vol. 50, pp.840-848, 1971.
4. Jin Onuki, Masahiro Koizumi and Masateru Suwa, "Reliability of Thick Al Wire Bonds in IGBT Modules for traction Motor Drives", *IEEE. Trans. on Adv. Packaging*, Vol.23 (1), pp.108-112, Feb. 2000
5. Takao Komiyama, Yasunori Chonan, Jin Onuki, Masahiko Koizumi and Tatsuya Shigemura, "High Temperature Thick Al Wire Bonding Technology for High Power Modules", Jpn. *J. Appl. Phys.*, vol. 41, pp. 5030-5033, 2002.
6. E. A. Brandes, G. B. Brook, " Smithells Metals Reference Book", 7th edition, Butterworth-Heinemann, 1992.
7. M. Gonzalez, B. Vandevade, R. Van Hoof and E. Beyne, "Characterisation and FE analysis on the Shear Test of Electronic Materials", *Microelectronics Reliability*, Vol. 44(12), pp. 1915-192, 2004.
8. G. Lefranc, B. Weiss, C. Klos, J. Dick, G. Khatibi and H. Berg, "Aluminium Bond-Wire Properties after 1 Billion Mechanical Cycles", *Microelectronics Reliability*, vol. 43, pp.1833-1838, 2003.
9. Adam J, "Electron Backscatter Diffraction in Materials Science", Kluwer Academic, New York, 2000.
10. Cotterill P, "Recrystallisation and Grain Growth in Metals", Surrey Univeristy Press, 1976.
11. Held, M., Jacob, P., Nicoletti, G., Scacco, P., and Poech, M.H., *Proc International Conference On Power Electronics And Drive Systems*, Vol. 1 and 2 (1997), pp.425-430
12. ANSYS is a product of ANSYS, inc., http://www.ansys.com
13. Lau, J.H. (editor), "Ball Grid Array Technology", McGraw-Hill (1995), pp396

Reduced Order Electro-Thermal Models for Real-Time Health Management of Power Electronics

[1]Mahera Musallam,[2]Cyril Buttay, [3]C Mark Johnson,[4]Chris Bailey,[5]Michael Whitehead

[1, 2, 3]Department of Electrical and Electronic Engineering, University of Nottingham, UK

[4]School of Computing and Mathematical Sciences, University of Greenwich, UK

[5]Department of Electronic & Electrical Engineering, University of Sheffield, UK

e-mail: Mahera.Musallam@nottingham.ac.uk, Cyril.Buttay@ nottingham.ac.uk,

Mark.Johnson@nottingham.ac.uk, c.bailey@gre.ac.uk, Michael.Whitehead@sheffield.ac.uk

Abstract

Real-time health management of power electronic devices is emerging as a key tool in the drive for cost, weight and volume in achieving efficient systems with high reliability and performance. Implementation of such schemes requires real-time electro-thermal models that can be used to predict the temperatures of device junctions, interfaces etc. that cannot ordinarily be measured during service. This paper presents a real-time reduced-order thermal model that includes a multi-rate computational method to improve efficiency. Results from the developed reduced-order model are presented along with experimental measurements that validate the model. The model is used as a tool to investigate the effects of module temperature variations for health management purposes.

1. Introduction

High temperatures and temperature cycles can dramatically reduce the lifetime and the reliability of power electronic modules. Thus, real-time monitoring and active management of thermal cycles is an important component of any power electronic health management system [1]. An essential part of such systems is an efficient real-time thermal model [2] that can be used to predict the temperatures of inaccessible locations within the module e.g. device junctions.

Thermal calculations for power modules are usually based on manufacturers' data which give the static junction to case thermal resistances for the main heat transfer paths. To accurately represent the module's dynamic thermal behavior, a multiterm exponential representation of the transient thermal impedance is typically employed. For real-time applications the number of terms in such a model must be chosen to give adequate accuracy without incurring excessive computational overhead. Furthermore, it is important to note that the heat flux from any device dissipating power to the heatsink will affect the temperature of other devices within the module as shown in figure (1). Thus it may be necessary to account for thermal cross-coupling effects.

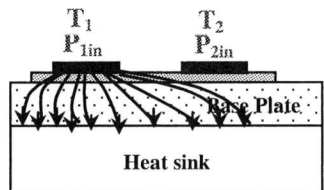

Figure (1) Cross section of the module and heatsink

2. Thermal Model Representation

To accurately predict the effect of thermal cycling under arbitrary input conditions an efficient electro-thermal model of the system (components and thermal management) is developed. This model considers multiple heat sources applied to the system. Thus, the total junction temperature for any particular device within the module will result from a combination of its own self-heating and the contributions from any other heated devices. A convenient approach to combining the effects from a number of heat sources is to represent the total increase in temperature (e.g. junction temperature) by a linear superposition of all the heat sources contributions within the module. This will be reasonable provided the non-linearities in the system are small e.g. change of thermal conductivity with temperature.

As an example of such an approach, consider a power module containing two semiconductor elements (dies). The self-heating and cross-coupling effects of the thermal behavior can be represented by a 2x2 transfer function matrix. This matrix is represented in equation (1) where $a_{11}....a_{22}$ represents the transfer function of each heat transfer path including the self-heating and cross-coupling effects within the module, T_A is the ambient temperature, $P_{1in}....P_{2in}$ are the heat sources and $T_1....T_2$ are the junction temperatures of both dies.

$$\begin{bmatrix} T_1 \\ T_2 \end{bmatrix} = \begin{bmatrix} a_{11} a_{12} \\ a_{21} a_{22} \end{bmatrix} \begin{bmatrix} P_{1in} \\ P_{2in} \end{bmatrix} + T_A \qquad (1)$$

For convenience, each heat transfer path within the module is represented by a Foster network with RC elements. For this reduced-order model, a maximum number of four terms was used as shown in figure (2).

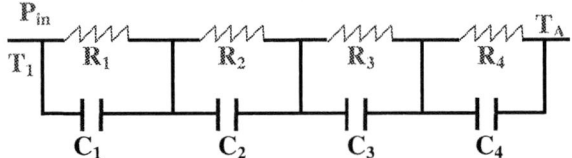

Figure (2) An example of heat transfer path modelling using Foster RC network [4]

For each path, the amount of the electrical power depends on both the temperature difference and the time rate of change of the temperature [3]. The thermal parameters of each term in the model can be determined experimentally as described elsewhere [4].

3. Real-Time Model Implementation

Pulse width modulation (PWM) is widely employed in the control of power electronic systems and thus the model assumes that PWM is used as a basis for controlling the modelled power electronic devices. PWM generates a fixed frequency pulse train whose duty cycle is modulated, resulting in the variation of the average value of the modulated waveform [5]. Figure (3) shows a typical PWM signal for a sinusoidal modulation reference signal. The current through the device and thus the heating power consist of a pulse train at the PWM frequency, with a variation in average power at the modulation frequency.

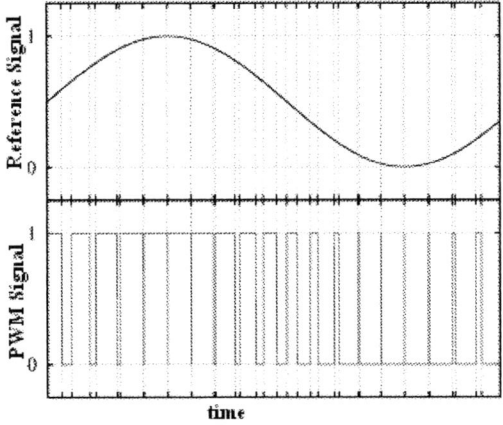

Figure (3) PWM modulation signal using sinosoidal reference wave form

The real time thermal model uses three real-time input variables, namely the current, the real-time estimated temperature as a feedback control and the duty cycle of a PWM system controller. The current is sampled sysnchronously with the PWM so that each sample interrupt occurs in the middle of the on-state cycle as shown in the figure (4).

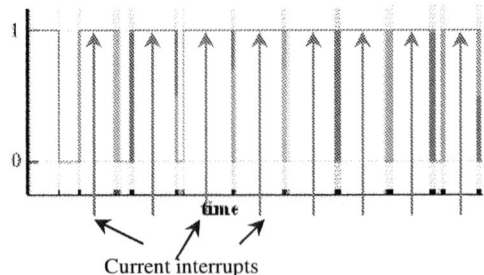

Figure (4) Occurance of current interrupts through the PWM modulation

Real-time implementation on a PWM-based system constrains many of the processes to run at the PWM frequency. This is generally much faster than is strictly necessary to resolve the thermal behaviour with sufficient accuracy. The thermal model thus uses multi-rate computational blocks where the power dissipated in each device is found at each PWM cycle (the high-rate process) and the reduced-order thermal model is updated every N PWM cycles (the low-rate process). Figure (5) represents the real time implementation of the model. The power dissipation in each device is obtained using lookup tables which are functions of the module current, duty ratio and the estimated temperatures. The dissipated power is averaged over a number (N) of PWM cycles.

In order to implement properly the thermal model, the mathematical relationships between temperature, the input power and time constant for each element of each heat path (figure (2)) are derived.

For each n element:

$$P_{in} = \frac{T_n}{R_n} + C_n \frac{dT_n}{dt} \quad \text{or}$$

$$T_n = \frac{1}{C_n} \int (P_{in} - \frac{T_n}{R_n}) dt + T_A$$

Numerical integration of the equations is achieved using a discrete form with explicit time propagation of the variables (the explicit form is well-suited for fast execution with small time-steps).

The thermal time constants of the individual devices within the module determine the maximum time step for the reduced-order thermal model, hence allowing a suitable value for N to be chosen. Thus the thermal model running frequency is synchronised to the PWM frequency. In practice, both the thermal model and PWM processes are triggered by timer controlled interrupts, the thermal model effectively running as a slower background process.

4. Real-Time Modelling Results

The model of figure (5) was developed in MATLAB/Simulink [6] using the real time software toolbox and implemented on a dSPACE [7] real-time system incorporating a hardware PWM card (DS5101). In practice, the dSPACE system generates the PWM signals and uses the PWM hardware counter to generate the

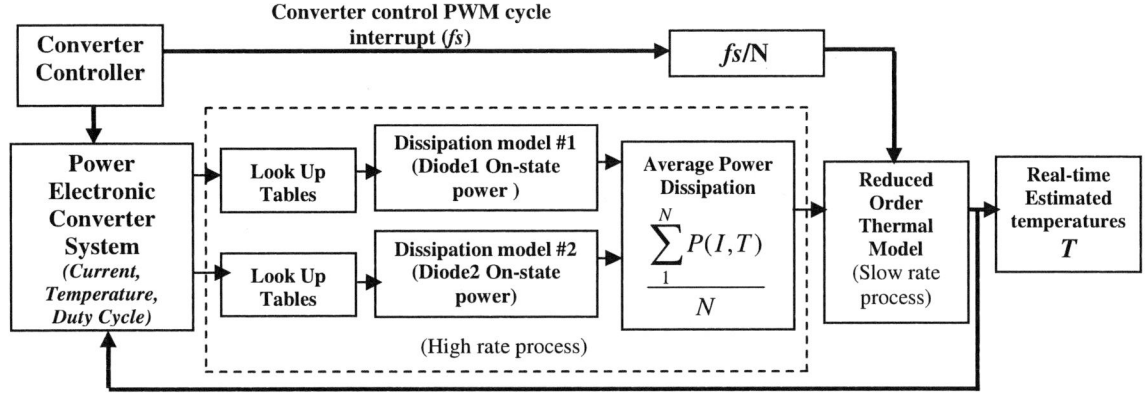

Figure (5) Real-time implementation of the thermal model

interrupts for the current sampling process and the execution of the thermal model.

As a test case, a model of a diode mounted on a heatsink was implemented and the effect of self-heating with a PWM heating current studied. The PWM frequency was set to 2 kHz and a sinusoidal modulation signal of 10 Hz applied. Using the experimentally determined model parameters, the maximum time step for the thermal model was found to be just over 3 ms. A timestep of 1 ms and a value of N equal to 2 was thus selected for implementation of the slow-rate thermal model process. The diode current was set to a fixed value of 25 A and an ambient temperature of 45°C was assumed.

Figure (6) shows the output from the model under the stated conditions. Note that the dissipated power and temperature response are both dominated by the modulation frequency of 10 Hz. The PWM carrier of 2 kHz is not evident on this figure since the model values are based on the average of 2 PWM cycles.

5. Experimental Verification

Making non-invasive, accurate experimental measurements of semiconductor junction temperature when the device is operating in a power electronic converter is a significant challenge. One method is to measure the diode forward voltage drop under constant current conditions and use the inherent temperature coefficient of the junction voltage to infer the instantaneous temperature [8]. In general, it is not possible to achieve the conditions to implement such a scheme since i) it is rarely possible to make the low noise differential voltage measurement required and ii) the circuit topology and control of the system rarely allows a constant current bias to be applied. To get around these difficulties a specific test circuit was designed in which the current through a diode (the heated element) was controlled by a PWM-controlled MOSFET switch as shown in figure (7).

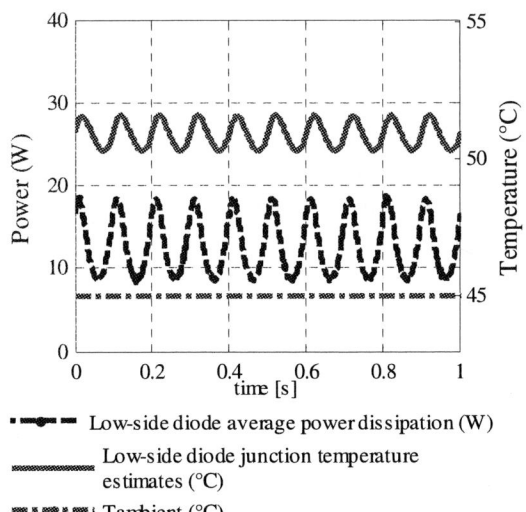

Figure (6) Real time low-side diode junction temperature estimates at (10 Hz) modulation frequency

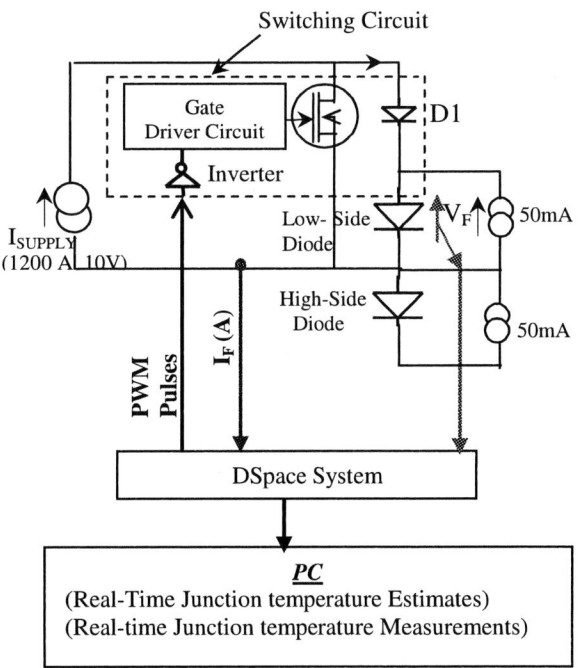

Figure (7) Circuit arrangement for measuring the low-side junction temperature

When the switch is closed the current is diverted away from the diode. Diode D1 isolates the low-side diode from the switch and permits the use of a bias current, in this case 50 mA, to be used to make the forward voltage measurement. When the switch opens the full heating current flows through the diode thus emulating the diode operation in a typical power circuit. Two diodes (low-side and high-side) were mounted on the same substrate tile and attached to a common heatsink, thus allowing the thermal cross-coupling effects to be studied.

6. Real-Time Module Measurements

For this experiment, the low-side diode was self heated with a PWM current pulse train of amplitude 25 A. The ambient temperature was set to a stable value of 45°C by attaching the power module to a temperature controlled water circuit. dSPACE was used to generate the PWM converter control pulses with a sinusoidal modulation signal at 10 Hz. The PWM pulses are fed to the physical module and the generated real time forward current (dSPACE ADC_1) across the diode is fed synchronously with the PWM signal as input to the thermal model as mentioned above. To measure the instantaneous diode temperature, the diode forward voltage V_f is fed to the dSPACE ADC_2 synchronously with the PWM frequency and is sampled in the middle of the off-state PWM cycle i.e. when the diode is under a constant bias current of 50 mA. Calibration between the diode forward voltage (as output of the dSPACE ADC_2) and the junction temperature was achieved by making a separate experimental test based on heating up the diode over a range of temperatures and measuring the forward voltage at each value. The calibration equation for the low-side diode is represented as follows:

$$T(^\circ C) = -555.56 V_f + 470.34$$

The actual measurements were very noisy and so they were treated by averaging the continuous signal over a number of modulation cycles within dSPACE. The real time low-side diode junction temperature results are shown in figure (8).

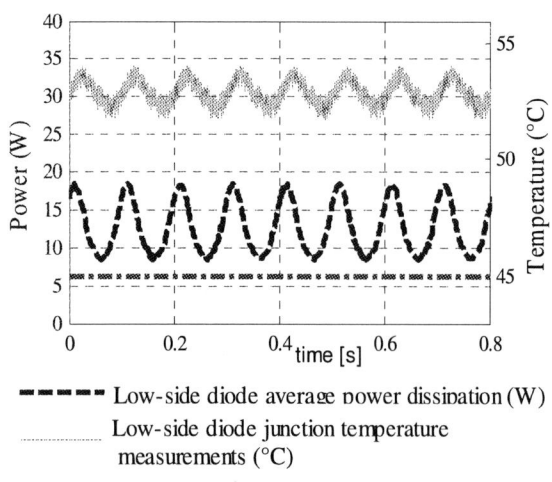

- - - - Low-side diode average power dissipation (W)

.... Low-side diode junction temperature measurements (°C)

▪▪ ▪ ▪▪ ▪ ▪ Tambient (°C)

Figure (8) Real time low-side diode junction temperature measurements at (10 Hz) modulation frequency

6. Validation of the results

Figure (9) shows a comparison between the real-time juntion temperature estimations of the low-side diode obtained using the real time thermal model with the real time junction measurements under the same input conditions.

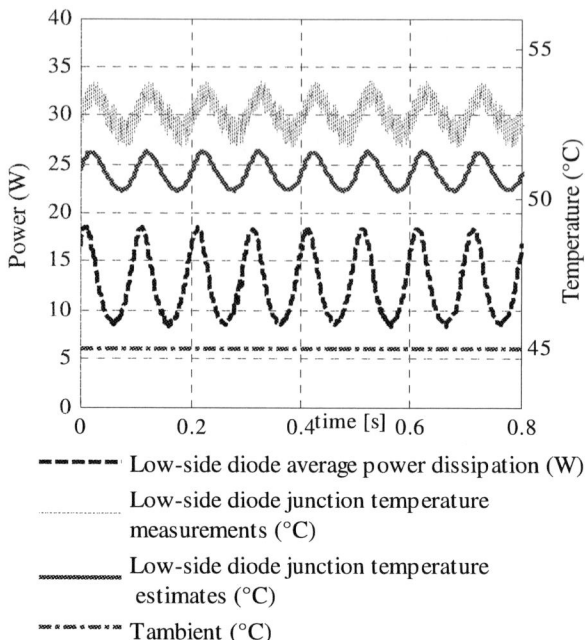

- - - - Low-side diode average power dissipation (W)

.... Low-side diode junction temperature measurements (°C)

▧▧▧▧▧ Low-side diode junction temperature estimates (°C)

▪▪ ▪ ▪▪ ▪ ▪ Tambient (°C)

Figure (9) Comparison between the real time junction temperature estimates and the real time junction temperature measurements for the low-side diode at (10 Hz) modulation frequency

These results show that there is a slight difference between the diode junction temperature measurements and the estimated ones of about 2°C. This difference is due to accuracy errors within the equipments used in these experiments. The dSPACE input ADC card (DS 2001) has offset error of ±1.5mV which contributes to an error of approximately 1°C besides other considered errors as the card gain error ±0.2%...etc. The accuracy of the measured data could also be affected by the errors within the used testing equipments, such as variations of the water cooler temperature of ±1°C. In addition, a slight delay appears between the real time junction temperature estimates and the measurements, possibly as a consequence of small errors in the thermal model parameters.

This test was repeated with a modulation signal of 1 Hz with the same input conditions. Results are shown in figure (10). Comparing the results in figures (9) and (10) shows that the variations in the real time diode junction temperature estimates for modulation frequency 10 Hz are less than those at 1 Hz through the modulation cycle. The same applies for the variations of the junction temperature measurements.

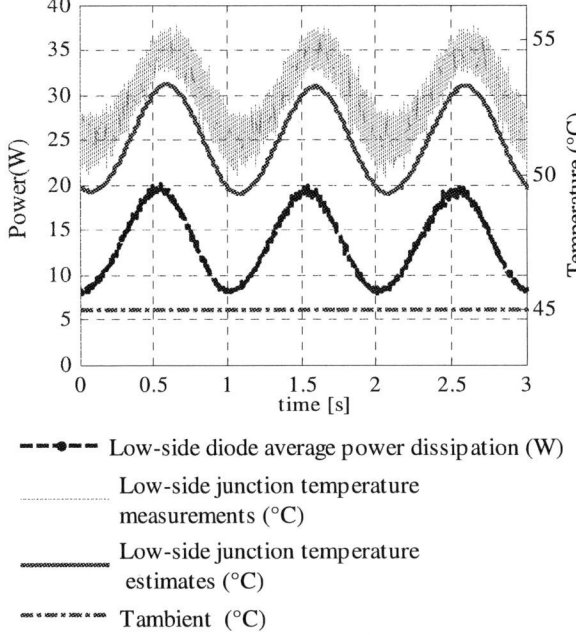

━ ● ━ Low-side diode average power dissipation (W)

.... Low-side junction temperature
measurements (°C)

▬▬▬▬ Low-side junction temperature
estimates (°C)

━ ⋅ ━ ⋅ ━ Tambient (°C)

Figure (10) the real time junction temperature estimates and the real time junction temperature measurements for the low-side diode at modulation frequency (1Hz).

The thermal model was used to obtain the low-side diode junction temperature estimates over a range of modulation frequencies from 1 Hz to 100 Hz. For each frequency the estimated and measured temperature variation over one modulation cycle were obtained. Figure (11) illustrates the thermal response of the diode and heatsink as a function of the PWM modulation frequency. As might be expected the thermal response is a strong function of frequency. The agreement between the estimated and experimental values of temperature increase is excellent, emphasising the suitability of the modelling technique.

7. Use of the developed thermal model for health management of power devices.

Repetitive power cycling can badly affect the performance and reliability of power electronic devices [9,10]. A typical power module consists of several different materials (semiconductor, metal, ceramic) with different physical parameters, such as different coefficients of thermal expansion. When these materials are bonded together mechanical stresses are generated and these stresses change as the temperature changes. During thermal cycling of a device, mechanical stresses on the wire bonds and on the interface bond between adjacent layers which cause failure of these bonds or of the materials themselves. Knowledge of the thermal cycling range seen by a device can be coupled with a mechanical stress model and a physics of failure model of the appropriate wear-out mechanism to generate information on life consumption. For many wear-out processes [11] such a technique may also be a practical way of predicting the end-of-life processes of devices since many of the physical wear-out mechanisms (e.g. propagation of cracks in substrate tiles, wire bond lift...etc.) give little external indication of impending failure. Equally, for other wear-out processes, differences between experimental and estimated parameters can be used as a prognostic tool for early indication of approaching end of life.

A validated real-time model, such as that presented above, can be used to estimate the effect of operational conditions (e.g. variations in load cycle) on the degree of thermal cycling seen by the devices. As an example of extreme power cycling the low-side diode temperature measurements and estimates were obtained with square wave modulation signal at 0.2 Hz and a supply current of 35 A. The results in figure (12) show that significant temperature cycling occurs within the device as a result of variations in the device power losses.

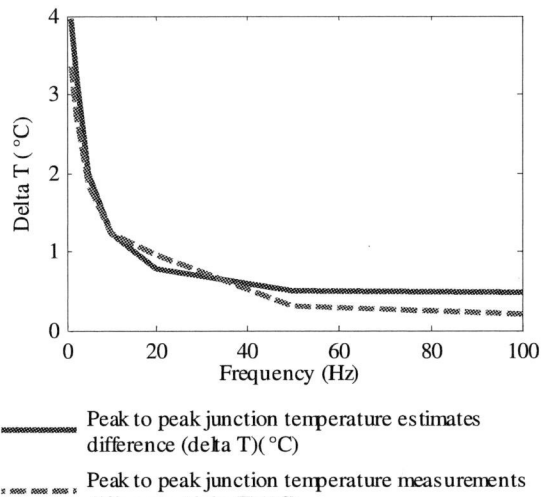

▬▬▬▬ Peak to peak junction temperature estimates
difference (delta T)(°C)

━ ⋅ ━ ⋅ ━ Peak to peak junction temperature measurements
difference (delta T)(°C)

Figure (11) Low-side diode junction temperature estimates and measurement variations over a range of modulation frequency.

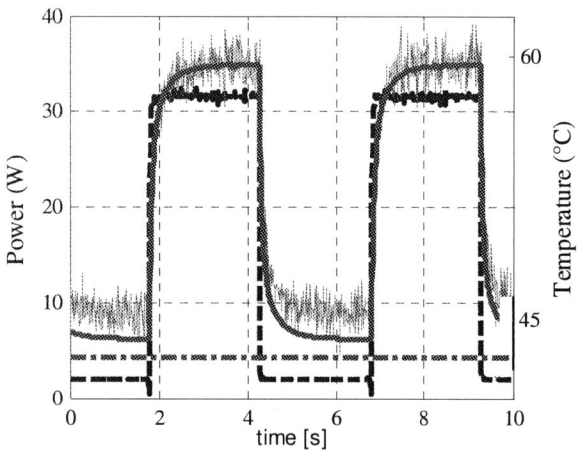

-- Low-side diode average power dissipation (W)

.... Low-side diode junction temperature
measurents (°C)

——— Low-side diode junction temperature
estimates (°C)

· ▬ · ▬ Tambient (°C)

Figure (12) Low-side diode temperature variations through
power cycling with modulation frequency (0.2) Hz.

8. Conclusions

A reduced-order real-time implementation of a thermal model suited to health monitoring applications in power electronics has been presented. Real-time thermal modelling of the junction temperatures was accomplished using simplified circuit representations and a multi-rate process in which the data acquisition for the model occurs at a high rate (the PWM frequency) whilst the model itself runs at a fraction of this frequency. The complete model was implemented in dSPACE using experimentally determined model parameters and validated using a PWM controlled diode bridge embedded within a dedicated test circuit. This enabled direct comparisons to be made with the equivalent predicted system variables (e.g. voltage, temperature). Comparison of the real time temperature estimates with the measured values shows that the reduced order model was effective in estimating the diode junction temperature with an error of less than 2°C over a wide range of modulation conditions. It can be concluded that this real-time model is well suited to the continuous monitoring of the internal behavior of the electro-thermal effects within power electronic modules and thus be used as part of a health management tool.

Acknowledgments

This work is funded by the Innovative Electronics Manufacturing Research Center.

References

[1] S. Stoyanov, W. Mackay, C. Bailey, D Jibb, C Cregson, "Lifetime Assessment of Electronic Components for High Reliability Aerospace Applications", *6th Electronics Packaging Technology Conference - EPTC 2004*, p324 – 329, ISBN 0-7803-8821-6, Pub IEEE, (2004).

[2] Musallam, M.; Acarnley, P.P.; Johnson, C.M.; Pritchard, L.; Pickert, V.; "Open loop real-time power electronic device junction temperature estimation" Industrial Electronics, 2004 IEEE International Symposium Volume 2, 4-7 May 2004 Page(s):1041 - 1046 vol. 2

[3] L.S. Pritchard, P.P. Acarnley and C.M. Johnson, 'An investigation of the structural and thermal transfer characteristics of commercial power MOSFET devices using experimental and modelling techniques', *IEE International Conference on Power Electronics, Machines and Drives*, Conference Publication No.487, 2002, pp562-567.

[4] Michael Whitehead, C Mark Johnson, "Determination of Thermal Cross-Coupling Effects in Multi-Device Power Electronic Modules", in *Proc. PEMD 2006 International Conference on Power Electronics Machines and Drives*, 2006.

[5] Holmes D. G., 'Pulse Width Modulation for Power Converters: Principles and Practice', John Wiley & Sons, INC., Publication, 2003, USA, ISBN 0-471-20814-0.

[6] www.mathworks.com

[7] www.dspace.ltd.uk

[8] J.Reimann, U.Franke, J.Petzolt, R.Krümmer and L.Lorenz, 'System integration-thermal aspects of chip utilization of power devices and control', ISPSD,Germany, Conference proceedings.

[9] Ohring M, 'Reliability and Failure of Electronic Materials and Devices', Academic Press, 1998, USA, ISBN 0-12-524985-3.

[10] Green A.E., Bourne A.J., 'Reliability Technology', John Wiley and Sons, ISBN 047132480 9, 1972, UK.

[11] Bosch, E.G.T. Components and Packaging Technologies, IEEE Transactions on Components, Packaging and Manufacturing Technology, Part A: Packaging Technologies, Volume 26, Issue 1, Date: March 2003, Pages: 173 – 178.

Reliability Test Method Overview to Characterize Second Level Interconnects

M. Brizoux [a], H. Frémont [b], Y. Danto [b], W. C. Maia Filho [a], [b]

[a] Engineering & Process Management, Thales Services SAS, Orsay 91404, France
[b] IMS, ENSEIRB, UMR 5818 Université Bordeaux 1, Talence 33405 cedex, France

Abstract

This paper deals with electronic board assembly reliability studies, where most of the failures in the field are known to be caused by electrical discontinuities, and often intermittent ones. To choose the test method and to define the testing conditions, the mission profile has to be evaluated. It is necessary to verify and consider all the possible failure mechanisms in order to avoid any violation of acceleration factor hypothesis for the reliability prediction. The test vehicle has to be representative of the real products and to take into account all the relevant parameters which can change the test results, like raw materials, subparts dimensions, etc. The last point, and one of the most important, is the failure detection. This paper illustrates a methodology applicable to the second level interconnect reliability evaluation, and proves that the early detection of intermittent failures is necessary.

1. Introduction

The current trends in electronics are towards high frequency signal operation, higher interconnects density and continuous reliability improvement interconnects. They are used in non protected environments, where they are submitted to severe constraints. It is the case for instance in automotive or space applications, where electronic boards have to face harsh mission profiles, where thermal cycling, shocks, vibrations, moisture and other chemical aggression can be simultaneously present. Table 1 is an example of a civil aviation mission profile.

As electronic is now involved in nearly all human activities, the reliability concern has become crucial, linked to personal safety, economic efficiency and product marketing. Very low failure rates and precise prediction of the lifetime are now expected from electronic system manufacturers. Reliability testing strategy must then evolve from the classical statistical median life testing to the use of more sophisticated degradation laws, based on the physics of the failure, and reliability predictions need computer assistance.

2. Reliability Testing Method Overview

In order to evaluate the reliability, accelerated aging tests are still applied to evaluate lifetime and to determine acceptable reliability levels. The test vehicle is a key parameter for the whole reliability test validity. Two main items must be defined: the object to be tested, and the stresses it must be submitted to.

The life time expectation of a structure can be estimated following its mission profile by the acceleration of the failure mechanisms. The accelerated tests are used in laboratory to reproduce in reduced period of time the accumulated damage similar to that observed in operation. There are several manners to accelerate a cyclic failure mechanism: by the increase of the stress intensity level or by the increase of the frequency of the stress application or also by the combination of both.

It is imperative to understand the relation between the accelerated test and the operational conditions to allow the acceleration factor estimation, which is the relation between the number of cycles or the period of time necessary to reproduce the failure or the accumulated damage. Furthermore, it is important to avoid the introduction of additional failure mechanisms,

For the electronic component solder joints, several different tests can be used: the mechanical accelerated tests (vibration, torsion), the thermo-mechanical accelerated tests (thermal cycling and thermal shock), or still the concomitant or sequential combination of both. The accelerated test to use depends on the failure mechanism, which corresponds to the mission profile.

In most cases, in the development phase for instance, it is impossible to experimentally test all the foreseeable configurations. So, the test must be carried out on the probable most critical ones. Analytical simple models, and more sophisticated finite element models may be helpful to extend the test results to different configurations of a same family [2,3].

Table 1: Example of a mission profile in civil avionic application [1]

Mission phase			Moisture and Temperature		Thermal Cycling				Mechanical stress
Description	State	Duration [h]	T [°C]	Relative humidity [%]	ΔT [°C]	Number of cycles per year	Cycle duration [h]	T_{max} [°C]	Vibration [g_{RMS}]
At ground after power on	On	797	47	30	33	330	2	47	-
At ground between 2 flights	On	1193	55	30	15	647	1,5	55	-
Take off / Landing	On	84	47	5	-	-	-	-	2,5
Flight	On	4083	40	5	-	-	-	-	0,3
Garage	Off	2603	14	70	10	108	24	19	-

1-4244-1105-X/07/$25.00 ©2007 IEEE

Another key point is the criterion used for the comparison between the accelerated test result and the failure in the field. The failure criterion or the early failure indicator must be representative of the operating failure: failure induced by the same mechanisms must be identically identified during test and operation. Based on this fact, the definition of failure criteria is fundamental to ensure repeatability and high confidence level of test results.

Electronic component solder joints have fundamentally three functions: mechanical support, thermal and electrical conduction. The operational failure of a solder joint is generally detected by the interruption of the electric conduction because the failure does not happen at the same time for all the joints of a component and because the electric function in high frequency is affected by intermittent defects.

Once this parameter is defined, the monitoring procedure used during the accelerated test must be able to detect the failure as soon as it appears. Otherwise, if the failure is detected only long time after its first occurrence, the acceleration calculation and hence the reliability results may be strongly affected. This can be the case in solder joint assembly reliability testing, where different electrical continuity measurement techniques can be used to evaluate the crack propagation leading at the end to open circuits.

3 Experiments

In this paper we describe the methodology we used to apply torsion test to estimate BGA assembly reliability.

3.1 Test vehicle definition [4]

In order to replace long accelerated thermal cycling tests for second level interconnect fatigue evaluation, torsion tests were developed some years ago [5,6]. This kind of test is expected to reproduce the same failure within a few days. The principle is to apply to a board equipped with electronic packages cyclic torsion efforts at low torsion angles from the neutral position. The efforts applied on the board simulate a relative displacement between the board and the package solder joint pads. But today, no standard torsion test is used in industry because the loading conditions on package solder joints are unknown and torsion test results coming from different test vehicle designs are not reproducible. In fact, the stresses are not homogeneous all over the board and in consequence not evenly distributed on all the solder joints. Tests are only reproducible if all stresses are known.

In order to improve this situation, and to make usable this kind of testing, printed circuit board using industrial representative materials and fabrication processes board was studied under torsion in order to determine the strain type and distribution. Three factors were found to be critical: the edge effects, the homogeneous strain area and the strain linearity as a function of torsion angle. Measurements with strain gauges and ANSYS simulations were complementarily used to point out a

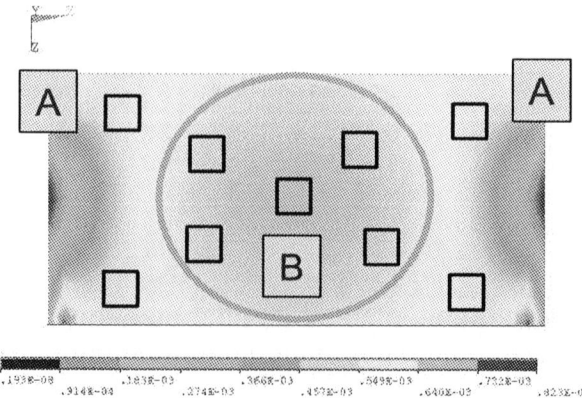

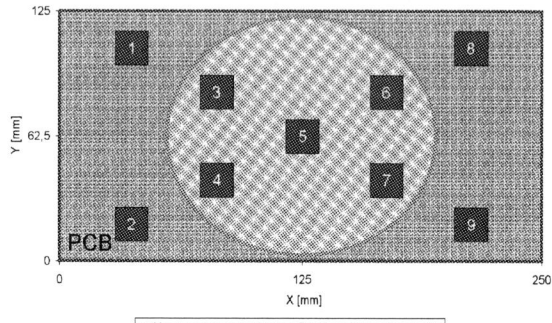

Figure 1: Definition of a homogenous area for torsion test: simulation results (top), validation through measurements (bottom)

homogeneous area of strain, and in consequence to determine the area correctly exploitable for the torsion test (figure 1). Samples located out of the homogenous area (N°1,2,8,9 on the figure) are not submitted to the same stress as the other ones; results coming from these components are not significant, and hence cannot be taken into account for reliability predictions. Consequently, the test vehicle did not contain components assembled out of this area, but all the components assembled inside the homogenous area were assumed to be stressed with the same acceleration factor.

Edge effects and homogeneous strain area define the rules to develop dedicated vehicles for torsion tests. These vehicles should have long PCB and the packages should preferentially be assembled on center area. Stress distribution in solder joint matrix was also analyzed. Simulations focused on the most stressed solder joint indicated the failure mechanisms as a function of torsion angle values. As a result, torsion test must be applied under low angles in order to induce the same failure mechanism as those found on the field. The actual angle values were then fixed after a step stress first evaluation, assuming that this condition was filled, and so as the test duration was not prohibitive. Two dwell times and temperatures were also chosen.

3.2 Monitoring

Different kinds of assembled BGA packages (PBGA 256 and 1156 CBGA 996...) were submitted to the torsion tests defined above, in order to analyze their

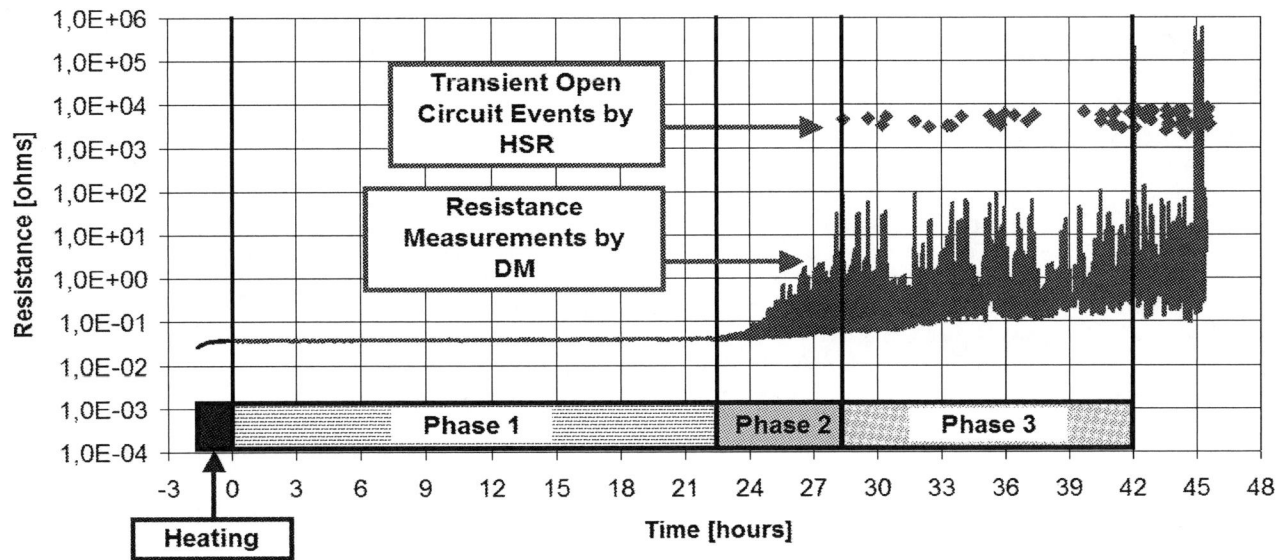

Figure 2 - Results overview of two monitoring test methods

second level interconnect reliability. Special daisy chain components and dedicated PCB allowed in-situ electrical continuity measurements: electrical continuity is usually used as an indicator of second level interconnect signal integrity and can be monitored by several test bench configurations, basically classified in 3 categories.

- Periodic checkout stops: Sometimes followed by cross sectional analysis, this kind of test only provide approximate reliability data because a failure can be identified only a long time after the first failure event.
- Low sampling rate: These methods normally use data recorders. These equipments are relatively cheap, can monitor thousands of channels in pooling mode, have a good resistance resolution (normally sub-ohmic) and can withstand electrical perturbations. Their principal disadvantages are the blindness to transient signals and the increase of the number of cables for 4-wires measurements.
- High sampling rate: Data recorders or event detectors operating at a high sampling rate, more than 1MS/s (Mega sample per second), allow to identify and record transient events in tens or hundreds of channels at the same time (using common-ground measurements). But they are costly, have low resistance resolution, are subject to electromagnetic interferences and require dedicated software to compile all the data recorded.

Two monitoring equipments simultaneously analyzed the daisy chain circuit. The first one was a digital 6 ½ digit multimeter HP34401 (here named DM). 4-wires measurement was applied; the second one was a high sampling rate multi-channel data-recorder (here named HSR).

Torsion tests were used to apply stresses and to induce failures on solder joints. Optimal test conditions set so as to induce creep and relaxation equivalent to those seen in accelerated tests applied for harsh environment mission profile (about 20 MPa).

The result overview of the whole test period is plotted in figure 2. Results were classified in 3 phases according to the resistance value fluctuations. The first one has a stable resistance value and indicates the solder joint integrity. The second phase corresponds to a very low resistance fluctuation due to a complete or nearly complete crack in one solder joint of the daisy chain circuit. The last phase begins when the first transient event occurs. These transient events are the first electrical signal linked to the failure that is representative of the failure in the field. The same transient events were found under thermal-cycling or thermal-shock tests.

The question to be answered is hence to determine the appropriate failure criterion for continuous monitoring test: is it the first event, the IPC criterion (10 events in 10% of cycle period), or the permanent failure?

3.3 Failure criterion

To answer this question, it was necessary to improve the understanding of intermittent failure, by physical approach, using electrical measurements and modeling, and observation with SEM. A zoom on the first event shows that the resistance value drops during a dwell time

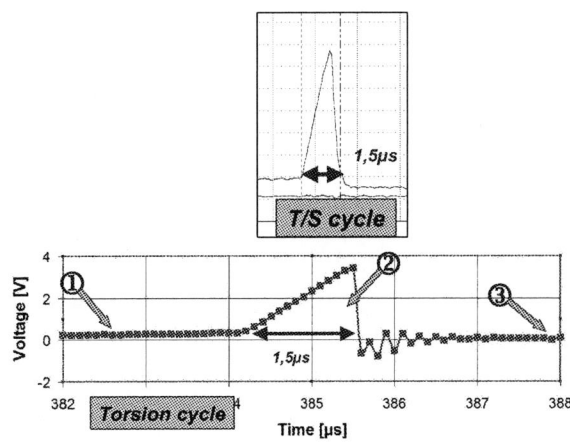

Figure 3: First transient event during thermal shock or torsion test

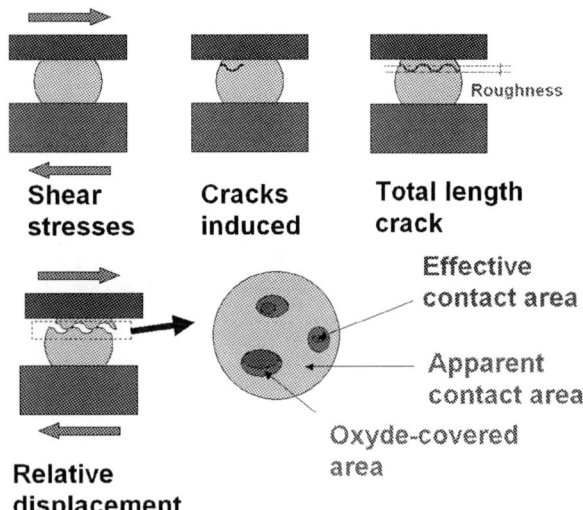

Figure 4: - Failure mechanism schematic

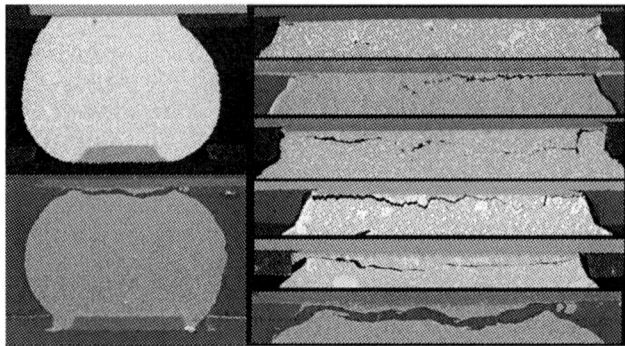

Figure 5: - Example of gap increase

and that the first event is detected at this time by HSR. This phenomenon was observed for every detected single event, and is similar to those found under thermal-cycling or thermal-shock tests. The transient open circuit signal can be divided in 3 different stages (figure 3):

- Relatively slow increase of resistance during some milliseconds (1);
- High increase of voltage during about one microsecond (2);
- Voltage drop and oscillation (3) similar to electrical discharge. The resistance level after the transient event is lower than 1 Ω.

Electrical modeling using SPICE based software was useful to extract the part of the signal due to the measurement system from the contribution of the ball failure.

The resistance fluctuations can be explained by electrical contact theory, as developed in [7]. A schematic view of the failure mechanism is shown in figure 4, and figure 5 illustrates the gap increase. Shear stresses are applied over solder joints by thermal expansion mismatch between board and package or by board deformation. These forces induce cracks in the solder joints; and the crack propagation takes time to reach all the solder joint length.

To summarize, the results obtained by physical, electrical measurements and electrical modeling clearly demonstrated that:

- A failed solder joint can be considered as a simple open switch in series with a low value variable resistor;
- The first event is an indicator of the full length crack;
- The resistance fluctuation phase begins with the first transient event, and ends when the first open circuit is larger than 1 second;
- This fluctuation is created by the gap growth between two cracked surfaces by micro-erosion and microfusion effects (see figure 4).

The duration of the resistance fluctuation phase depends on the type of stress applied to the assembly. For instance, for thermal shock tests, it can reach several hundreds of cycles. It also depends on solder alloy hardness, package type (BGA, QFN...), I/O number, stress level, etc. This demonstrates that transient events can be observed only after that total-length crack is produced. Important delay can be found if appropriate monitoring method is not used. Sampling rate has to be higher than 1MS/s or resistance resolution lower than 0.1Ω in order to early identify failure and to obtain an adequate failure criterion more representative of failures on the field.

4. Conclusions

In order to properly characterize second level interconnects reliability taking into account intermittent failure which happen in the field, several continuous monitoring test method have been assessed and discussed.

This study clearly demonstrates that :

- Unrealistic time to failure can be found if inappropriate monitoring method is used. Sampling rate has to be higher than 1MS/s or Resistance resolution has to be lower than 0.1Ω. Therefore from a pragmatic standpoint, to correctly manage hundreds of channels (one channel per daisy chain) with high speed clock rate applications High Sampling Rate method is the most useful and highly recommended.
- The first transient event is "the mirror" of a total length crack of a solder joint. A simple electrical SPICE model gives simulation results in good accordance with the experiment. Moreover electrical contact theory can be applied to calculate the contact resistance value and to explain the resistance fluctuations

This HSR continuous monitoring test method becomes mandatory to predict with realism the reliability of solder joint in fatigue of very dense and high speed digital electronic boards. Indeed with the first transient events typically in the range of 1µs opening time for solder joint interconnects, it is more than enough to produce a functional board failure taking into account the speed of microprocessor, memories...

Nevertheless reliability prediction of second level interconnects will be realistic only if the design of the test

662

vehicle is representative of the functional boards and if the process used for assembly is mature and under control.

Taking into account the cost and the duration of reliability tests, it is more and more necessary to use FEM simulations tools and analytical models to optimize the reliability test program.

References

1. "Guide FIDES 2004 Edition A - Méthodologie de fiabilité pour les systèmes électroniques DGA - DM/STTC/CO/477-A," Direction Générale de l'Armement, Ministère Français de la Défense, 2004.

2. Y. Danto, J.-Y. Delétage, F. Verdier, H. Frémont Evolution of Reliability Assessment In PCB Assemblies: *Proceeding of the SBMicroelectronics* (2001)

3. J.Y. Delétage, F. J. M. Verdier, B. Plano, Y. Deshayes, L. Béchou and Y. Danto, Reliability estimation of BGA and CSP assemblies using degradation law model and technological parameters deviations, *Microelectronics Reliability*, 43(7) 2003, pp 1137-1144.

4. W. C. Maia Filho, M. Brizoux, H. Frémont, Y. Danto. Solder Joint Loading Conditions under Torsion Test. *EUROSIME 2006.*

5. Dickinson, Gerard T. et al., "Circuit board assembly torsion tester and method," United States Patent, Patent number 5,567,884, Oct.22, 1996.

6. Pang, John H. L. et al, "Mechanical Deflection System (MDS) Test and Methodology for PBGA Solder Joint Reliability," *IEEE Transactions on Advanced Packaging,* Vol. 24, No. 4 (2001), pp. 507-514.

7. W. C. Maia Filho, M. Brizoux, H. Frémont, Y. Danto. Improving Understanding of Intermittent Failures under Continuous Monitoring Methods. *ESREF 2006.*

Kinetic Analysis of Electromigration Enhanced Intermetallic Growth and Void Formation in Pb-Free Solders

Brook Chao, Seung-Hyun Chae, Xuefeng Zhang and Paul S. Ho
Microelectronics Research Center, The University of Texas at Austin
10100 Burnet Rd., Bldg 160
Austin TX 78758-4445
Mail Code: R8650

Abstract

The kinetics for electromigration (EM) enhanced intermetallic growth and void formation in Sn-based Pb-free solder joints with Cu under bump metallization (UBM) was analyzed. The simulated diffusion couple comprised the two terminal phases, Cu and Sn, as well as the two intermetallic phases, Cu_3Sn and Cu_6Sn_5, formed between them. The diffusion and EM parameters were obtained by solving the inverse problem of the EM-enhanced intermetallic growth and found to be consistent with the literature values. Finite difference method was used to solve the mass transport kinetics within the intermetallic phases and across each interface of interest. Simulation showed that with zero current and no EM effects, intermetallic growth follows a parabolic law, suggesting a diffusion controlled mechanism for thermal aging. However, under significant current stressing $(4.12 \times 10^4$ and 5.16×10^4 Amp/cm^2), the growth of the dominant intermetallic Cu_6Sn_5 clearly follows a linear law, suggesting a reaction controlled mechanism for electromigration. The kinetic results were consistent with the experimental observations. The vacancy transport under EM was also analyzed and the results showed substantial increase in vacancy concentration in the Cu_6Sn_5 phase near Cu_3Sn/Cu_6Sn_5 interface. The peaking of the vacancy concentration predicts EM-induced Kirkendall void formation at this region and its effect on solder reliability is discussed.

1. Introduction

The reliability of flip chip packages has emerged as a critical concern because of the demand for increasing current density with smaller solder bump size. A common failure mode for flip chip packages is an electrical open due to void formation induced by intermetallic compound (IMC) growth at the interface between the solder and under bump metallurgy (UBM). Failures of this type have been reported after prolonged current stressing at an elevated temperature and have been identified as a result of electromigration (EM) [1]. Under EM, IMC growth can be significantly enhanced and accompanied by Kirkendall void formation which plays an important role in controlling the EM lifetime of solder joints in flip-chip packages [2]. Gan and Tu reported distinct characteristics for IMC growth at both anode and cathode and formulated a kinetic model to account for the current polarity effect on IMC growth [3]. In their model, the two intermetallic phases, Cu_3Sn and Cu_6Sn_5, formed at the solder/UBM interface were not distinguished but

treated as a single phase for simplicity. Following this study, Orchard and Greer analyzed the EM effect on compound growth at interfaces, taking into account the effect of interfacial reaction barriers but still treated only a single intermetallic phase [4].

Gurov and Gusak were the first to consider the formation of dual intermetallics under an electric field in the kinetic analysis and presented asymptotic solutions at steady state [5]. These authors found that the growth kinetics of the dual intermetallics can follow distinct growth modes depending on the balance of the interdiffusion and electromigration fluxes in individual compound layers. This was expressed in terms of a parameter designated as interdiffusion electromigration coefficient in the present paper.

This paper reports a kinetic analysis for EM-enhanced growth of dual intermetallic layers in Pb-free solders. The analysis was applied to evaluate experimental results on IMC formation observed under EM between a Cu UBM and a Sn-based Pb-free solder. The interfacial reaction was treated as growth of dual compound layers in a Cu-Sn diffusion couple taking into account the current-driven mass transport of Cu and Sn atoms. The complexity of the dual IMC formation under EM necessitates the use of a finite difference method (FDM) to analyze the IMC growth kinetics at each discretized time step and to extract the interdiffusion EM coefficients from the experiments. The numerical approach enabled detailed examination of the effect of current in enhancing the growth of individual compounds. In addition, the rate of vacancy transport can be deduced from the flux balance in the layered structure. The result provides insight to void formation induced by EM during IMC growth, which controls EM damage formation.

2. Kinetic Formulation

A Cu-Sn diffusion couple was considered in which two intermetallic phases Cu_3Sn and Cu_6Sn_5 formed between pure Cu and pure Sn phases. The two intermetallic phases have very narrow composition ranges as indicated by the phase diagram, Fig. 1(a), and the values of the boundary composition that are listed in Table 1. Fig 1(b) shows the schematic composition profile of the diffusion couple. Within each phase, Cu and Sn atoms diffuse simultaneously and, in addition, they are subjected to an EM driving force due to the electron current from the under bump metallization (UBM) toward the top surface metallurgy (TSM). Darken's equation for interdiffusion [6] can be modified to apply to this case, taking into account the atomic flux induced by both chemical potential and

1-4244-1105-X/07/$25.00 ©2007 IEEE

external electric field. The atomic flux due to chemical diffusion can be expressed as

$$J_{Cu,i}^{Chem} = -D_{Cu,i}\frac{\partial C_i}{\partial x}$$

$$J_{Sn,i}^{Chem} = -D_{Sn,i}\frac{\partial C_{Sn,i}}{\partial x} = D_{Sn,i}\frac{\partial C_i}{\partial x} \quad (1)$$

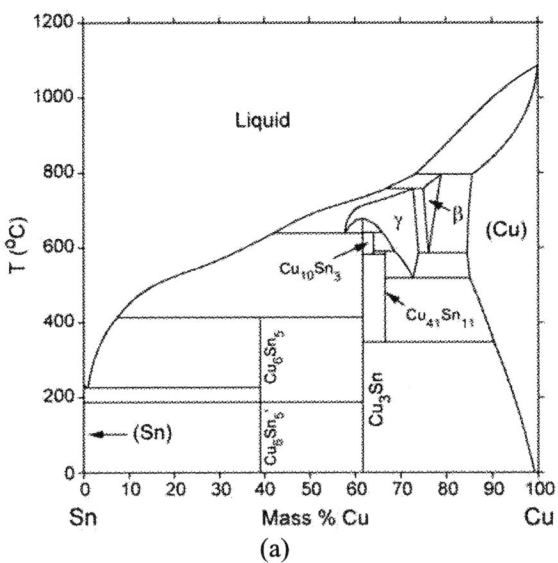

(a)

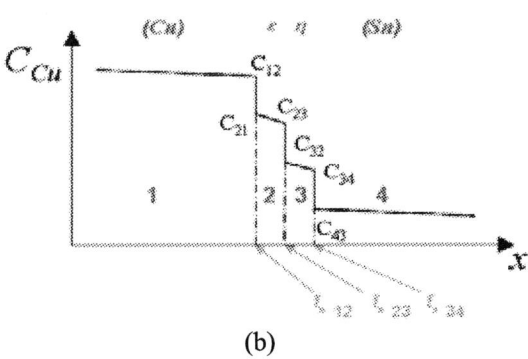

(b)

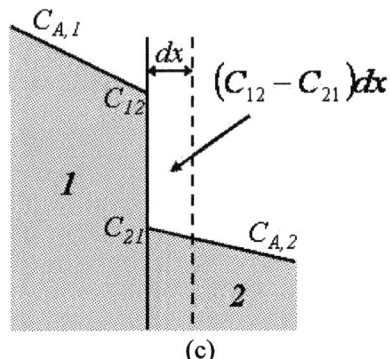

(c)

Fig 1 (a) Sn-Cu binary phase diagram, (b) Cu composition profile of Cu-Sn diffusion couple, (c) Cu composition profile near an interface

Table 1 Atomic fraction of Cu at interfaces [4]

C_{12}	C_{21}	C_{23}	C_{32}	C_{34}	C_{43}
0.993	0.765	0.755	0.549	0.541	0.00006

In this paper, all compositions C are expressed as mole fraction of Cu and the subscript Cu is omitted for simplification unless otherwise denoted. The running index i denotes the phase in which interdiffusion takes place. It is assumed that the vacancy concentration is low and therefore the density of the phase is $C_0 = C_{Cu} + C_{Sn}$.

The current induced atomic flux is expressed as

$$J_{Cu,i}^{EM} = C_{Cu,i}\frac{D_{Cu,i}}{kT}Z_{Cu,i}^* e\rho_i j = C_i D_{Cu,i}\phi_{Cu}j$$

$$J_{Sn,i}^{EM} = C_{Sn,i}\frac{D_{Sn,i}}{kT}Z_{Sn,i}^* e\rho_i j = (C_0 - C_{Sn,i})D_{Sn,i}\phi_{Sn}j \quad (2)$$

where $\phi = \frac{Z^*}{kT}e\rho$ is the electromigration factor

Combining the chemical diffusion flux and current induced flux into Fick's 2nd law gives the governing equation of current enhanced interdiffusion within each phase.

$$\frac{\partial C_i}{\partial t} = \frac{\partial}{\partial x}\left[\widetilde{D}(C_i)\frac{\partial C_i}{\partial x} + \widetilde{\phi}_i C_i(1-C_i)j\right] \quad (3)$$

$$\widetilde{\phi}_i = D_{Sn,i}\phi_{Sn,i} - D_{Cu,i}\phi_{Cu,i} \quad (4)$$

where $\widetilde{\phi}_i$ is the effective interdiffusion-electromigration coefficient of phase i as defined in Ref [5]

Fig 1(c) shows the composition profile across the 1/2 interface and its associated movement assuming chemical diffusion and electromigration are the only contributions to atomic flux. The velocity of interface migration can be described based on the mass conservation principle:

$$v = \frac{dx}{dt} = v_{Chem} + v_{EM}$$

$$= \frac{1}{C_{ij} - C_{ji}}\left\{\left(\widetilde{D}_j\frac{\partial C_{Cu,j}}{\partial x} - \widetilde{D}_i\frac{\partial C_{Cu,i}}{\partial x}\right) + j\left[C_{ji}(C_0 - C_{ji})\widetilde{\phi}_j - C_{ij}(C_0 - C_{ij})\widetilde{\phi}_i\right]\right\}$$

(5)

Both chemical diffusion and current induced diffusion cause the interfaces to move. Each compound phase grows or shrinks as a result of the migration of its interface boundaries.

3. Numerical Simulation

The numerical simulation was programmed with a Matlab 6.5 program using the finite difference method. The simulated domain was meshed into 1000 fixed elements (1001 fixed nodes) and each with a size of 100nm. At the two boundaries of each phase, an extra cladding node was attached to the end of the nodal array in order to represent the moving boundary, as shown in Fig 2. Overall, the simulation domain contained 1001 fixed nodes and 6 movable nodes addressing the moving interfaces. The fixed nodes had constant positions but variable composition, while the moveable nodes had variable positions but constant composition.

665

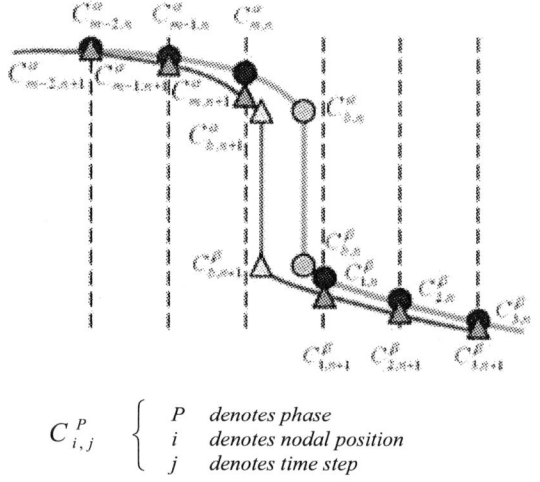

$$C_{i,j}^{P} \left\{ \begin{array}{ll} P & \text{denotes phase} \\ i & \text{denotes nodal position} \\ j & \text{denotes time step} \end{array} \right.$$

Fig 2 Nodal arrays and moving interfacial nodes

Eq (3) was applied in each phase and solved for the boundary conditions listed in Table 1. In dealing with the cladding interfacial nodes and their first and second adjacent nodes, partial differences were adopted according to the temporal position of the movable interfacial nodes. Once the composition profile of each phase was obtained, migration of interfaces was determined by Eq (5) based on the local composition gradient at the interfaces. Forward or backward differences were used to calculate the composition gradient at the interfaces. Care was taken to ensure that interfaces moved by a distance less than 1/100 of the element size in each iterative step in order to maintain the stability of the iteration. Current density is an adjustable parameter in this program and can be set to zero to investigate the intermetallic growth kinetics by thermal aging only.

4. Measurements of Intermetallic Growth

EM experiments were performed on Sn-3.5Ag lead-free solder bumps with a Cu-based UBM stack as shown in Fig 3(a) to measure current enhancement effects on IMC growth kinetics and Kirkendall voids formation [7]. The test structures were subjected to a high current stressing (4.12×10^4 and 5.16×10^4 Amp/cm^2) at elevated temperature (~140°C) for a prolonged period of time (>400 hr). The joule heating effect was found to cause the actual solder temperature to rise about 10-15°C over the nominal temperature 140°C. Therefore the actual solder temperature in this experiment was around 150-155°C.

In the EM tests, electron current flowed from the under bump metallurgy (UBM) to the top surface metallurgy (TSM). Cross-sectional images were taken using SEM to investigate the evolution of IMC phases and formation of Kirkendall voids. Each phase was identified by energy-dispersive x-ray analysis (EDX) performed in a scanning electron microscope. The growth rates of the intermetallics along the electron flow direction were evaluated by averaging the area of each phase in the SEM images over the diameter of the bumps.

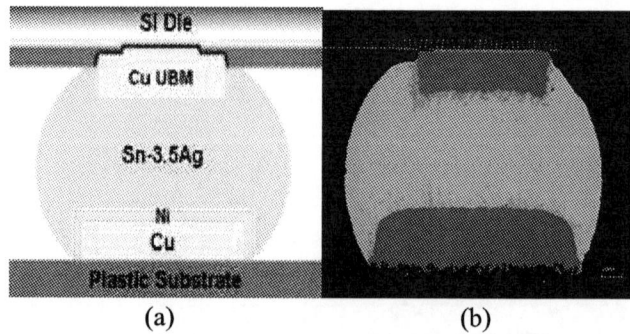

(a)	(b)

Fig 3 (a) Schematic of a solder bump prior to current stressing, (b) cross section of a control sample subject to 140°C annealing but not current stressing for 300 hours

In the EM experiments, the control samples, Fig 3(b), were subjected to only thermal aging at test temperature of 140°C. They showed minimal thickness change compared with the current stressed samples after >300 hour. Fig 4 compares the cross-sectional micrographs of a Sn-3.5Ag lead free solder bump after current stressing in (a) and prior to current stressing in (b). At the cathode, electron current flowed from UBM to TSM causing the Cu-UBM to dissolve into the Sn solder. The material reaction led to the growth of the Cu$_6$Sn$_5$ phase extending to the TSM end, which was identified by EDX. The other intermetallic phase, Cu$_3$Sn, however, remained as a thin layer between the Cu$_6$Sn$_5$ and the remaining Cu UBM.

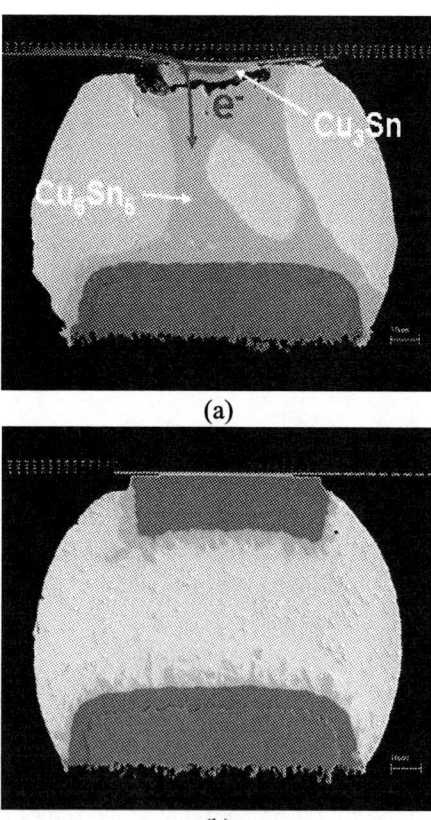

Fig 4 Cross section of a solder bump (a) after current stressing, and (b) prior to current stressing

5. Results and Discussion

A. Diffusivity and EM Effective Charge

The analysis of the intermetallic growth kinetics requires information on the diffusion coefficients of both Cu and Sn atoms in all four phases (pure Cu and pure Sn as the terminal phases and the two intermetallic phases). The coefficient of self-diffusion in pure Cu and pure Sn phases are readily available in ASM and diffusion handbooks [8,9,10]. Sn has been identified to have a tetragonal crystal structure [11] and anisotropic diffusion coefficients were reported [8]. Since the interdiffusion model in this paper is one dimensional, the anisotropy is neglected and only an average self-diffusion coefficient is used [9]. Dyson, Anthony and Turnbull reported the extremely rapid and anisotropic mobility of Cu in the pure Sn phase due to its fast diffusion by an interstitial mechanism.[12]

The diffusion coefficients of Cu and Sn atoms in the intermetallics are not well documented although the interdiffusion coefficients of the Cu-Sn intermetallics have been reported. Both Mei et al [13] and Onishi et al [14] have deduced the interdiffusion coefficients based on intermetallic growth in annealing experiments. Although the intrinsic diffusivities and activation energies they deduced were different, their interdiffusion coefficients were reasonably consistent at the temperature of interest. Bader et al investigated the formation and growth of IMC in thin Cu-Sn-Cu tri-layers and reported that the interdiffusion coefficient for Cu_3Sn in thin films [15] and Tu derived the diffusion coefficients of Cu-Sn thin films by investigating the aging effect of Cu-Sn bi-layer [16]

The EM driving force is measured by a dimensionless parameter, the effective charge number $Z*$ as defined below [17]:

$$F_{eff} \equiv |e| Z * E \qquad (6)$$

Self electromigration has been reported for both Cu and Sn and their effective charge numbers have been reported [18,19,20,21]. However, this parameter has not been reported for Cu and Sn as dilute solute in each other. Hsieh and Huntington reported the effective charge number for Cu as dilute solute in pure Pb in the range of 0.6~3.25 [22]. The effective charge number Z of Cu in Pb

and in Sn is expected to be similar because Pb and Sn are both quadrivalent metals in group IV with similar electronic configuration and Cu atoms diffuse interstitially in both host metals. The effective charge number of Cu in pure Pb is herein taken as that in pure Sn with reservation.

In addition to the interdiffusion coefficients of the intermetallic phases, the interdiffusion electromigration factors and information of the diffusion coefficients of both diffusants Cu and Sn are required to calculate EM enhanced IMC growth according to Eq (4). The individual diffusion coefficients of Cu and Sn in the phases of interest are not readily available in literature. However, the diffusion coefficients can be obtained together with effective charge numbers by virtue of the solution of the inverse problem of the experimental intermetallic growth [23]. The diffusion coefficients so derived can be verified by calculating the interdiffusion coefficients based on the composition of each phase and comparing with the literature reported values.

The $Z*$ and diffusion coefficients used in the simulation are listed in Table 2. The diffusion coefficients of the diffusants Cu and Sn were obtained individually using the above mentioned inverse optimization method. The interdiffusion coefficients calculated accordingly were also listed and compared to the literature reported values. They were found to be very consistent and this suggests that these are reasonable values.

B. Kinetics of Intermetallic Growth

To investigate the current enhancement effect on IMC formation, the growth rates under EM were compared to the growth rates under thermal aging. Under thermal aging, the rate of diffusion-controlled intermetallic growth can be expressed as:

$$v_d = \frac{1}{C_{ij} - C_{ji}} \left(\tilde{D}_j \frac{\partial C_{Cu,j}}{\partial x} - \tilde{D}_i \frac{\partial C_{Cu,i}}{\partial x} \right) \qquad (7)$$

It is clear that diffusion controlled growth rate is proportional to the composition gradient. As the intermetallic thickens the composition gradient across the phase decreases, so the growth rate decays over time following a parabolic law in general.

Table 2 Derived diffusion coefficients and effective charge number at 150°C

Phase	Diffusant	$Z*$	D_{150} (m²/s)	Interdiffusion coefficient (m²/s)	Interdiffusion coefficient *literature* (m²/s)
Cu_3Sn	Cu	26.5	3.67×10^{-17}	1.87×10^{-16}	2.87×10^{-17} [14] ~ 3.81×10^{-16} [15]
	Sn	23.6	2.35×10^{-16}		
Cu_6Sn_5	Cu	26.0	7.04×10^{-16}	6.74×10^{-16}	1.61×10^{-16} [14] ~ 4.19×10^{-16} [24]
	Sn	36.0	6.49×10^{-16}		

The contribution of electromigration to the growth rate, however, is obtained from a different equation

$$v_c = \frac{j}{C_{ij} - C_{ji}} \left[C_{ji} \left(C_0 - C_{ji} \right) \widetilde{\phi}_j - C_{ij} \left(C_0 - C_{ij} \right) \widetilde{\phi}_i \right] \quad (8)$$

In contrast to Eq (7), the growth rate in Eq (8) is a function of the current density and the composition across the interface. Hence, under a steady current density, EM-induced growth remains at a constant rate even when the intermetallic thickens. The fact that this growth is only dependent on interfacial compositions not only leads to the conclusion that the time dependency is expected to be linear but also suggests that this mechanism is controlled by interfacial reactions.

The growth behavior for the two limiting cases can be readily deduced. At high current density, current induced growth dominates and the overall rate follows a linear law. At low or zero current density, however, current induced growth is absent and therefore the IMC growth should follow a parabolic law. In general, the growth behavior will be more complicated, depending on the balance of the mass fluxes driven by interdiffusion and EM.

Kinetic analysis was first conducted for intermetallic growth due to thermal aging ($j=0$), as shown in Fig 6.

Fig 6 shows the variation of intermetallic thickness due to thermal aging at 150°C along with the experimental data from a Cu-Sn thin film aging experiment [25]. The simulation results are in good agreement with the empirical data and the IMC growth kinetics apparently follows parabolic law.

Simulation was then conducted for intermetallic growth under EM at a nominal temperature 140°C (joule heating corrected temperature ~150°C) and two current stressing conditions: $4.12 \times 10^4 A/cm^2$ and $5.16 \times 10^4 A/cm^2$. Fig 6(a) and 6(b) show the simulated time dependency of the thickness of Cu_3Sn and Cu_6Sn_5 under the two EM conditions, where the experimental data points are from Ref [7].

In Figs 5 and 6, Cu_3Sn thickened with time both under electromigration and thermal aging. It seems that Cu_3Sn exhibits more significant thickening under thermal aging than under electromigration. However, this in part results from the fact that the initial thicknesses of the aging sample and the EM sample are different since the aging experiment was done with thin film samples whereas the EM experiment was done with solder joints in a package.

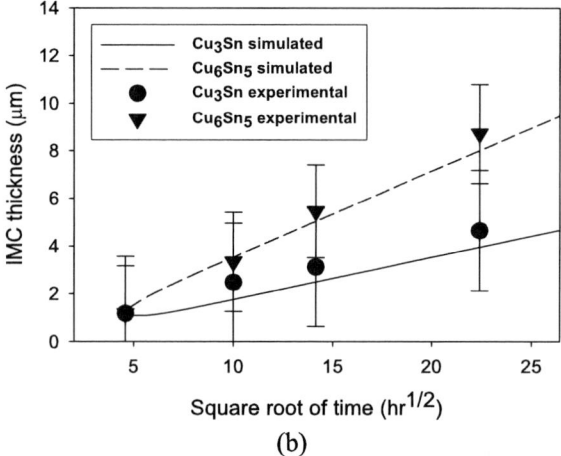

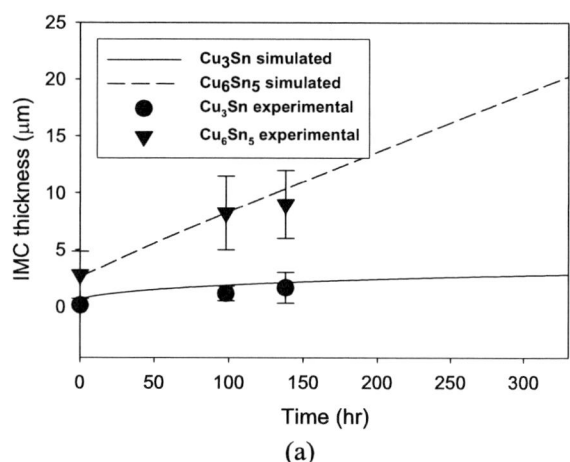

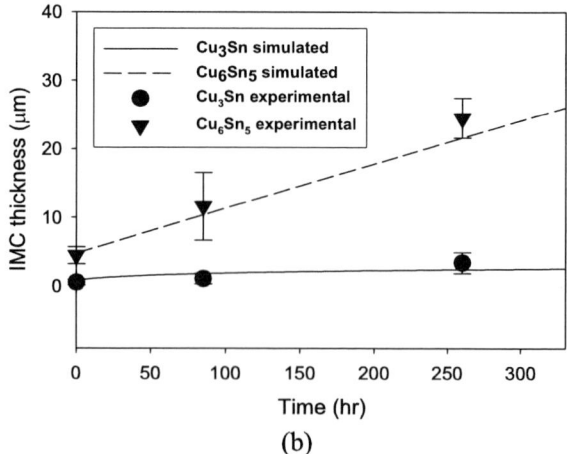

Fig 5 IMC growth for thermal aging (a) plotted against time and (b) plotted against square root of time

Fig 6 Electromigration enhanced IMC growth under (a) $4.12 \times 10^4 A/cm^2$ A/cm2 and (b) $5.16 \times 10^4 A/cm^2$

The overall intermetallic thickness of the EM sample was much thicker than that of the thermal aging sample because the initial intermetallic thickening of the EM sample was done during solder reflow. The fact that Cu_3Sn remains a very thin conformal layer between Cu UBM and the adjacent phase adds to the difficulty of experimental determination of Cu_3Sn thickness. However, in general, the simulation still predicted the Cu_3Sn thickening within the error bars of the experimental data.

While the EM effect on the thickening of Cu_3Sn is not yet clear, the thickening of Cu_6Sn_5 was significantly enhanced by EM, as shown in Fig 7(b). Simulation predicted the thickening of Cu_6Sn_5 of both EM and aging experiments. Most important of all, it shows that the thickening of Cu_6Sn_5 not only is significantly enhanced by EM but has a linear time dependency.

This kinetic model successfully rendered an accurate prediction of intermetallic growth and movement, which was consistent with the experimental observation. However, there was still some discrepancy between simulation and experimental results. First, in the simulation, the Cu_6Sn_5 phase did not extend itself as far into the solder and reach the TSM as was evidently observed in Fig 4(a). The discrepancy arose from the fact that the model was 1-D while the solder bumps under test were three dimensional structures. In a 3-D solder bump, Cu_6Sn_5 not only grows along the electron current path due to current crowding [1] but also shrinks along lateral direction due to dissolution of the Cu UBM.

It is also important to extend some discussion regarding the intermetallic morphology observed in samples subjected to current stressing, Fig 4(a), and those subjected only to thermal aging, Fig 3(b). The distinct morphologies of the two intermetallic phases of reflowed solders are well known: a layer-type Cu_3Sn phase and a scallop-type Cu_6Sn_5, as shown in Fig 4(b). However, under high current stressing, the morphological development of the intermetallic phases noticeably followed the distribution of current flux divergence, as shown in Fig 4(a). This indicated that kinetics under EM dominated the evolution of the intermetallic morphology.

In order to understand the non-planarity of IMC, it is critical to distinguish the reflow reaction, one that occurs at solid/liquid interfaces, and the subsequent aging and EM reactions, ones that occur at solid/solid interfaces.

For reflow reaction, Kim, *et al.* indicated that the size of Cu_6Sn_5 scallops approximately follows $t^{1/3}$ dependence growth kinetics, t being the reflow time, and the scallops form by Ostwald ripening reactions[26]. Görlich, Schmitz, and Tu indicated that the wetting angle between adjacent scallops is an equilibrium feature at the reflow temperature and is related to the surface tension balance between molten solder/Cu_6Sn_5 interface and the grain boundary [27]. Based on these studies, it became clear that during reflow reactions, the IMC morphology and its growth kinetics is different from the phase growth in the solid state.

For solid state aging, Tu et al. found that, for post-reflow eutectic SnPb solder, the Cu_6Sn_5 morphology can gradually change from scallop type to layer type at a very slow rate [28]. They also indicated that the solid state aging kinetics follow a parabolic law. It is confirmed in the present paper that the parabolic kinetics can be attributed to the diffusion driven IMC growth model. In EM, however, we show that Cu_6Sn_5 growth can be significantly enhanced by current along the electron path and that the growth kinetics follows a linear law, suggesting a reaction controlled mechanism.

The comparisons between the Cu/Sn reactions during reflow, aging, and EM are summarized in Table 3. Although the equilibrium IMC morphologies of these reactions are different, their kinetic behaviors can be explained by flux-driven approaches. Gusak and Tu devised a flux-driven ripening theory to explain the growth kinetics and size distribution of Cu_6Sn_5 scallops during solder reflow reactions [29]. We presented in this paper the kinetic model that takes into account both diffusion and EM induced flux. This model can therefore explain the kinetic behaviors of solid state aging and EM based on the current density given.

It is worth noting that the effect of the reaction history has in the analysis of the solder/UBM IMC structure. As indicated by Tu et al., the kinetics of reflow reactions can be four orders of magnitude faster than the solid state aging [28]. Therefore, Cu_6Sn_5 phase of reflowed solders exists in scallop shapes even after a prolonged solid state aging. However, high current stressing can substantially increase the solid state kinetics and completely alter the morphology within the same reaction time.

Table 3 Reactions at Solder/IMC interface

Reaction	Critical interface	IMC Morphology		Mechanism	Kinetics	Ref
		Cu_3Sn	Cu_6Sn_5			
Reflow	Solid/liquid	Layer	Scallop	Ostwald ripening	$t^{1/3}$	26
Aging	Solid/solid	Layer	Layer	Diffusion controlled	$t^{1/2}$	7, 29
EM	Solid/solid	Layer	Electron path	Reaction controlled	T	7

C. EM Enhanced Void Formation

As shown in Fig 4(a), crack formation in the intermetallics was the dominant failure mode of solder joints under EM [2, 7, 30]. This failure mode has been attributed as the result of Kirkendall void formation accompanying the IMC growth due to reaction of Cu and Sn [2]. To understand EM-induced void formation due to intermetallic growth, we incorporate in the kinetic analysis the evolution of vacancy concentration in the kinetic analysis. For this purpose, vacancy transport was considered in the following continuity equation:

$$\frac{\partial C_V}{\partial t} = -\frac{\partial J_V}{\partial x} \qquad (9)$$

$$J_V = (D_{Cu} - D_{Sn})\frac{\partial C}{\partial x} - J_{Cu}^{EM} - J_{Sn}^{EM} \qquad (10)$$

The vacancy concentration increases at the infinitesimal distance when

$$(D_{Cu} - D_{Sn})\frac{\partial^2 C}{\partial x^2} - j(D_{Cu}\phi_{Cu} + D_{Sn}\phi_{Sn})\frac{\partial C}{\partial x} < 0 \quad (11)$$

It is clear that both the first and second derivative of the composition play a role in terms of the time variation of the vacancy concentration. Results from this analysis are shown in Fig 7. Fig 7(a) shows the simulated composition profile of each phase after a prolonged and significant current stressing ($5.16 \times 10^4 A/cm^2$ and >300hrs) while Fig 7(b) shows the concentration of vacancy accruing over time. It is observed in this figure that under EM, a high concentration of vacancy can exist trailing the advancing Cu/Cu_3Sn interface. Given the high concentration of vacancy in this region, Kirkendall voids are likely to form as a result. The predicted high vacancy concentration and subsequent void formation are consistent with experimental observations. Under EM tests at the nominal temperature 140°C, the dominant failure mode of the solder joints tested was due to void formation at the Cu_6Sn_5 side of the Cu_6Sn_5/Cu_3Sn interface.

This analysis of vacancy transport does not include any vacancy annihilation effects, therefore vacancy concentration in the Cu_6Sn_5 was overestimated. Provided vacancy annihilation at the sinks was considered, the peak vacancy concentration should be reduced and it should concentrate more locally at a location behind but very close to the receding Cu_3Sn/Cu_6Sn_5 interface, as observed in experimental images Fig 4(a) and 7(c).

The IMC growth is usually accompanied by an atomic volume change which reaches -4.8% for Cu_6Sn_5 and can generate significant tensile and shear stresses upon cooling to room temperature. Following the kinetic analysis, a finite element analysis was performed and yielded indeed a local stress state with significant concentration due to EM enhanced IMC formation. Such a stress state when combined with Kirkendall void formation can induce crack formation at the UBM/solder interface to cause EM failure of the solder joint. More details of the EM damage analysis and observation will be published separately.

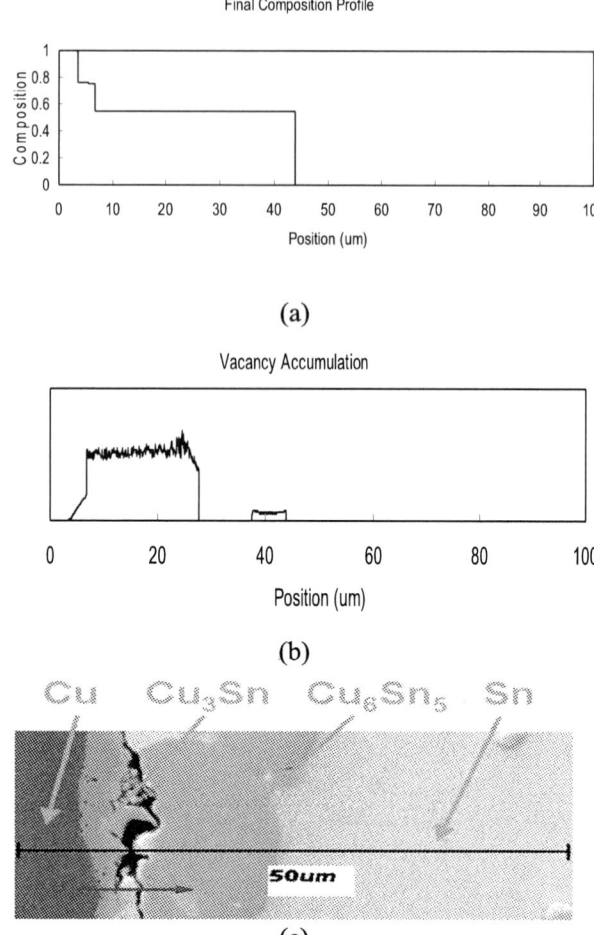

(a)

(b)

(c)

Fig 7 (a) Composition profile after prolonged high curreint stressing ($5.16 \times 10^4 A/cm^2$ and >300 hr) (b) vacancy accumulation derived from vacancy transport model (c) expanded SEM image of a failing solder joint showing the interfaces of intermetallic phases

6. Conclusion

A kinetic model of EM enhanced intermetallic growth in a Pb-free solder joint with Cu-UBM was formulated taking into account the interdiffusion and current driving forces. Simulation showed that the EM enhanced IMC growth follows a linear law whereas the IMC growth follows a parabolic law under thermal aging. This result leads to the conclusion that IMC growth occurs in a reaction controlled mechanism under EM while it occurs under diffusion control during thermal aging. This was verified by EM and thermal aging experiments. A kinetic analysis of vacancy transport was also incorporated and the results showed substantial increase of vacancy concentration at the Cu_6Sn_5 phase near Cu_3Sn/Cu_6Sn_5 interface. Experimental observation verified that significant void formation occurred as predicted. This suggests that the local escalation of vacancy concentration played an integral role in the void formation and growth in the intermetallics under current stressing.

Acknowledgments

This research was supported in part by the Semiconductor Research Corporation. The authors would like to thank Dr. Peng Su, Trent Uehling, and Lakshmi N. Ramanathan of Freescale Semiconductor for funding support and assistance in experiments and also like to thank Professor Venkat Ganesan at the University of Texas at Austin for useful discussion of numerical techniques.

References

1. Yeh, *et. al.*, *Applied Physics Letters*, Vol. 80 (2003), p. 580

2. Zeng, *et. al.*, *Journal of Applied Physics*, Vol. 97, 024508 (2005)

3. Gan and Tu, *Journal of Applied Physics*, Vol. 97, 063514 (2005)

4. Orchard and Greer, *Applied Physics Letters*, Vol. 86, 231906 (2005)

5. Gurov and Gusak, Physics of Metals and Metallography, Vol. 52, No. 4 (1981), p. 75

6. Darken, *Trans. Met. Soc AIME*, Vol. 175 (1948), p. 184

7. Chae *et al.*, *Journal of Materials Science: Materials in Electronics*, Vol. 18, No. 1-3 (2007), pp. 247-258

8. Peterson, <u>Solid State Physics</u>, 22, Turnbull and Ehrenreich (Eds.), Academic Press (New York, 1968)

9. Seith and Heumann: Diffusion of Metals: Exchange Reactions, Springer Verlag Press, Berlin, p 65 and 68 (1962)

10. Davis (Ed), <u>ASM Specialty Handbook: Copper and Copper Alloys</u>, ASM International (2001), pp. 235

11. Deshpande and Sirdeshmukh, *Acta Crystallography*, Vol. 14 (1961), p. 355

12. Dyson, Anthony, and Turnbull, *Journal of Applied Physics*, Vol. 38 (1967), p. 3408

13. Mei, Sunwoo, and Morris Jr., *Metallurgical Transactions A*, Vol. 23A (1992), p. 857

14. Onishi and Fujibuchi, *Trans. Japan Institute of Metals*, Vol. 16, No. 9 (1975), p. 539

15. Bader, Gust and Hieber, *Acta Metallurgica et Materiala*, Vol. 43, No. 1 (1995), p. 329

16. Tu, *Acta Metallurgica*, Vol. 21 (1973), p. 347

17. Huntington (Ed), <u>Diffusion in Solids: Recent Developments</u>, A S Nowick and J J Burton (1974), pp. 303-352 (1974)

18. Grone, *Journal of Physics and Chemistry of Solids*, Vol. 20 (1961), p. 88

19. Sullivan, *Journal of Physics and Chemistry of Solids*, Vol. 28 (1967), p. 347

20. Grimme, <u>Atomic Transport in Solids and Liquids</u> (1971)

21. Liu, Chen and Tu, *Journal of Applied Physics*, Vol. 88, No. 10 (2000), p. 5703

22. Hsieh and Huntington, *Journal of Physics and Chemistry of Solids*, Vol. 39 (1978), p. 867

23. Chao, et. al., *Acta Materialia* (2007)

24. Hoshino, Iijima, and Hirano, Transactions of Japan Institute of Metals, Vol. 21 (1980), p. 674

25. Siewert, Madeni and Liu, *Proc. of APEX conference on Electronics Manufacturing* (2003)

26. Kim, Liou, and Tu, Applied Physics Letters, Vol. 66, No. 18 (1995), p. 2337

27. Görlich, Schmitz, and Tu, Applied Physics Letters, Vol. 86 (2005), 053106

28. Tu et al., Journal of Applied Physics, Vol. 89, No. 9 (2001), p. 4843

29. Gusak and Tu, Physics Review B, Vol. 66 (2002), 115403

30. Ding et al., Journal of Applied Physics, Vol. 99 (2006), 04906

Measurement of Dynamic Properties of MEMS and the Possibilities of Parameter Identification by Simulation

Matthias Ebert, Falk Naumann, Ronny Gerbach, Joerg Bagdahn
Fraunhofer Institute for Mechanics of Materials (IWMH), Heideallee 19, D-06120 Halle, Germany
matthias.ebert@iwmh.fraunhofer.de
Tel.: +49 (0) 345 / 5589 117
Fax: +49 (0) 345 / 5589 101

Abstract

In the paper basic investigations for the nondestructive quality testing methods for MEMS (Micro-Electro-Mechanical Systems) are presented that can be applied on wafer level in early stage of the manufacturing process. The dynamic measurements of test specimen are performed by laser-Doppler-vibrometry (LDV). Finite Element (FE) models with different specifications are created to identify parameters from the measured eigenfrequency values. Different aspects of the numerical results and of experimental investigations of dynamic properties are presented for pressure sensor test structures. Results of identification of one and more parameters are shown.

1. Introduction

The nondestructive determination of geometrical and material parameters of MEMS is a key element for the monitoring of manufacturing processes. Therefore, optical measurement techniques have gained great importance. They allow a fast and non-destructive detection of mechanical parameters of MEMS. Different methods were already developed to solve these tasks. A first method for a nondestructive measuring system for testing MEMS or MOEMS at chip level was already presented in Gorecki [1]. This system is based on an interferometric system. The out-of-plane deflection and dynamic properties of silicon membranes and electrostatic actuators were investigated with this measuring system.

The combination of laser-Doppler-vibrometry and analytical or numerical calculations is another approach for this task. This method is characterized by a fast measurement and a wide field of application. First investigations to determine material parameters were done on both side clamped cantilever beams to characterize the Young's modulus and residual stresses of the beam materials by Zhang et al. [2] and on MEMS optical scanners to identify geometrical parameters and residual stress by Kurth & Doetzel [3].

Thin films are used in many technical applications like semiconductor devices and MEMS. The knowledge of the material properties of thin films is required for the design and fabrication. Often the parameter can not be determined from separated test structures of this material of thin films. Sometimes it is not possible to produce the test structures or the behavior is specific as thin film. Today tests like the wafer curvature tests are used to determine residual stresses of thin films after deposit processes during a manufacturing process.

This paper presents a further developed method to evaluate MEMS structure parameters with laser-Doppler vibrometry and numerical calculations by FE models. First results were published in Ebert et al. [6] and Gerbach et al. [7]. The measured dynamic properties are used for optimization of FE- models and the identification of unknown material and especially in this paper geometrical parameters. A lot of aspects of experimental measurements and numeric results are shown. Two aims should be achieved, at first to characterize reliable the parameter of membrane test structures and in the next step to use the membranes to characterize thin films on it.

The investigated classes of MEMS are membranes of pressure sensor test structures, delivered by the partner X-Fab Semiconductor Foundries AG inside the project "Par-Test" [8].

2. Types and Manufacturing of Silicon Membranes

Membrane structures are a functional important part of microsystems and are used e.g. in micro-pressure-sensors and microphones. Different materials are used for such structures. Silicon volume micro mechanics is one possible and very important process to manufacture membranes. Hereby the membranes are etched into a silicon wafer. The form and the clamping conditions of these structures are dependent on the used etchant. Typical clamping conditions of silicon membranes are shown in Figure 1.

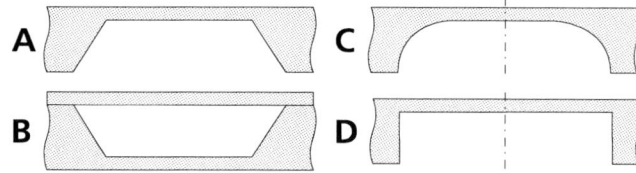

Figure 1 Typical shapes and clamping conditions of silicon membranes

Anisotropic etching (e.g. with KOH, Figure 1A) is used to manufacture quadratic or rectangular shapes. The etch stop is either created by a different doping level, by an intermediate oxide layer or by time control. Figure 1B shows another possibility to produce rectangular shaped membranes. Hereby a cavity is etched in a wafer. The membrane is generated by the bonding and grinding of another wafer. Circular membranes (Figure 1C and D) can be manufactured by isotropic etching processes to an etching stop layer.

In all cases membranes or plates in the mechanical sense are created with specific static and dynamic behavior independence of geometry, material, support conditions and other influences like prestress.. A membrane of case B (see Figure 2) is investigated. The membranes have relatively constant length in both directions and stable support conditions. A varying parameter is the grinded membrane. Fluctuations of thickness are possible inside the area of one membrane or about the average thickness of a wafer. The known thickness of the membrane in an early stage is important information for the next steps of production.

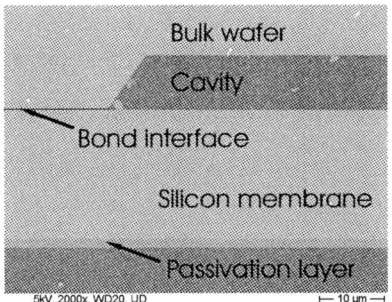

Figure 2 Cross section of the pressure sensor test structure (SEM, 2000x)

3. Measuring techniques of dynamic properties of microsystems

3D-structural-vibration measurements have to reveal the information about the 3D-geometry and the 3D-vibration pattern with a complete frequency spectrum to provide an experimental basis for dynamic model identification and modal analysis verification. This task is solved in the macroscopic world by using 3D-scanning vibrometry combined with scanning light pulse delay measurements. However, 3D-scanning vibrometry cannot be used to analyze the modal behavior of MEMS because light is usually not scattered on a deposited surface and light-pulse-delay measurements have not sufficient resolution for micrometer-sized structures. Two techniques are known nowadays for motion measurements in MEMS: the scanning laser-Doppler-vibrometry with strobed videomicroscopy [7] and strobed white-light interferometry (WLI) [8]. LDV cannot be used to measure 3D geometry and WLI cannot measure a high-resolution vibration spectrum (e.g. 10 Hz resolution) in a reasonable time (less than 15 seconds).

The Micro System Analyzer (MSA) from Polytec [9], see Figure 3 was used for the investigations in this paper. Some investigations were done by partners in the "Par-Test" project [6], some with the MSA of the Institute.

4. Measuring setup for out-of-plane vibrations

The measuring setup shown in Figure 4 is used to characterize the dynamic properties of the membrane structures.

A probe tip is positioned above the membrane by a micro positioner. The mechanical system is stimulated to vibrations by an electrical voltage with different frequencies.

Typical mode shapes of the membrane will be stimulated in case of resonance between the excitation and an eigenfrequency of the mechanical structure. To perform investigations on wafer level the used measuring setup is integrated on a manual probe station. This enables an exact and repeatable realization of the measurements.

Figure 3 Micro System Analyser from Polytec at IWMH

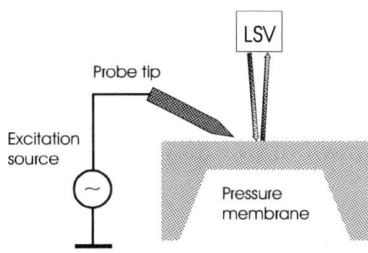

Figure 4 Schematically measuring setup

5. Characterization of dynamic properties

5.1. Numerical characterization of dynamic properties

The results of the identification process of structural parameters should be reliable and accurate as good as possible. So it is necessary to investigate all possible influences on the result. This is shown in this chapter for some of the investigation mainly by numerical models.

Finite Element Model

In this work FE- models with three-dimensional brick elements were used, see a quarter of a model in Figure 5. Only in this way it is possible to characterize the real support conditions of such sensor. The symmetry conditions are considered. Thin layers like the passivation layer of 1.5μm are a challenge of FE- modeling of MEMS. In this work no sub modeling techniques or other techniques to relate fine and wide meshes areas were necessary.

Definitely, the membrane is a clamped plate, so that analytical solutions can be used. Some properties of the membrane like anisotropic behavior and possible prestress is analytically complicated to describe. A comparison of an isotropic analytical model and an isotropic FE-model already has shown differences in the results up to 5%. The analytical solution is stiffer, because it assumes a

673

completely clamped plate. Investigations to the convergence of eigenvalues in dependence of mesh were done and considered to reduce the deviation of numerical results

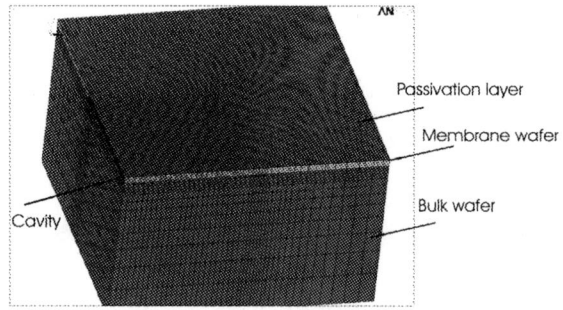

Figure 5 FE- model (quarter), view on symmetry areas

Variation of cross section of the membrane

Mainly constant membrane thicknesses are assumed in numerical models. In the experimental investigation the variation of thickness was investigated by destroying cross section analysis. It is shown that a fluctuation of membrane thickness exists. This can be a half wave (membrane 2 and 3) or a double wave (membrane 1 and 4), as shown for membrane 1 and 2 in Figure 6. The mean values are also shown.

A function for the membrane thickness is interpolated and considered in the FE- model. There are relative differences in the first frequencies smaller than 1% between constant mean and variable thickness. Some improvements to the experimental values are observed. Figure 7 shows the error in percent from the measured value.

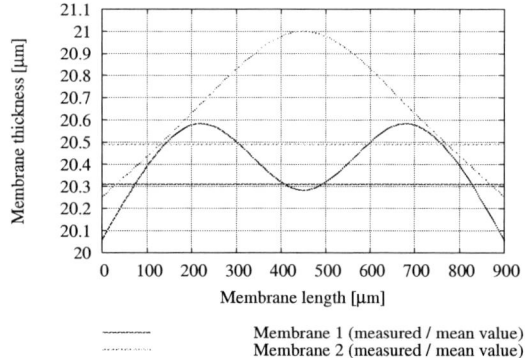

Figure 6 Variation of membrane thickness form a SEM cross section analysis

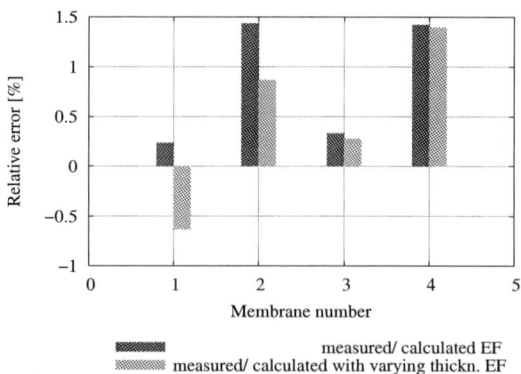

Figure 7 Relative deviations of first eigenfrequency with fluctuating membrane cross section

5.2 Experimental characterization of dynamic properties

In this chapter some aspects of the determination of dynamic properties are described, which are important for a reliable and fast identification process.

Determination of resonance frequencies

A tool for the automatic determination of resonance frequencies is developed. It is possible to use single point measurements, averaged measurements or the better conditioned coherence function to determine reliable the "peak" to the same mode shape of comparable structures. Figure 8 shows an example for a coherence function and measured averaged spectra. It is done for the velocities. Some parameters have to chosen, this includes the "knowledge" about the behavior of the structure. The detection without visualization of spectra must be very reliable and has to advice deviations from the expected behavior.

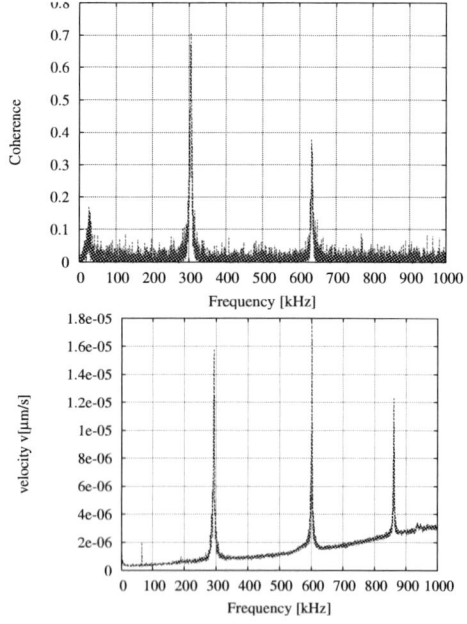

Figure 8 Coherence function (up) and averaged velocity amplitude spectra

Statistical evaluation of a repeated single measurement

The dynamic out-of- plane measurement was repeated with the same structure for 50 times. Aim is to investigate the stability of the measurement result. One side is the repeatability of the measurement in dependence of the LDV. This seems to be very accurately. The other side is the repeatability of the support of the test structures and the slightly different location of the excitation. The support could change by some contamination. Different locations of excitation should not influence the frequencies. So the excitation point has no influence on the mechanical structure in the sense o f a weak spring.

The result of the measurement is a distribution function. Figure 9 shows the distribution for the first eigenfrequency. The mean value is 306.4 kHz with a coefficient of variation of 1.22%. That value is similary for the higher frequencies. The value is small, but the value is absolutely important for accurate identification. The reason for the runaways is not absolutely clear. The conclusion of this result is that such an investigation should be done before the dynamic measurement and the following identification with new types of test structures.

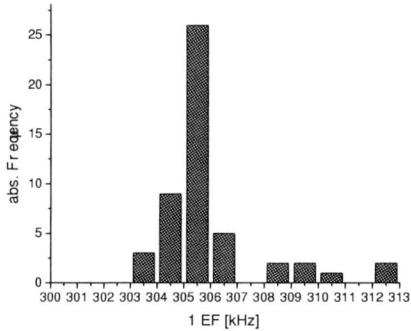

Figure 9 Distribution function for the 1ˢᵗ eigenfrequency of a repeated measurement with the same test structure

Pre-deformation of a membrane

The cavity is evacuated during the fabrication of the membrane. So a pressure difference between environment and cavity exist. This should lead to a deformation of the membrane. The height profile of the membrane was measured by the white light interferometry function of the MSA.

An example of such deformations is shown Figure 10. Measured values are app. between 600 and 900 μm. These values variations are possible because of different process dependent pressure and different membrane thickness. The calculated pressures in the FE- model are between 0.07 and 0.1 bar for the measured displacements. This corresponds to differences near air pressure to vacuum.

Mechanical pressure on the membrane with high slenderness ratio leads to stiffening of the membrane. The ratio is for the investigated structure app. 50. Higher ratios up to 300 exits. This stiffening indicates higher eigenfre-

quencies. This was investigated only for the describe pressure differences. The change of frequencies is very small, for the first eigenfrequency of about 0.35% from 319.3 to 320.4 kHz. But this geometric nonlinear behavior should be considered as part of the demanded accuracy. The changing of frequencies under higher pressure will be further investigated as possibility to get more than one set of frequencies of a structure without of changing the support of structure.

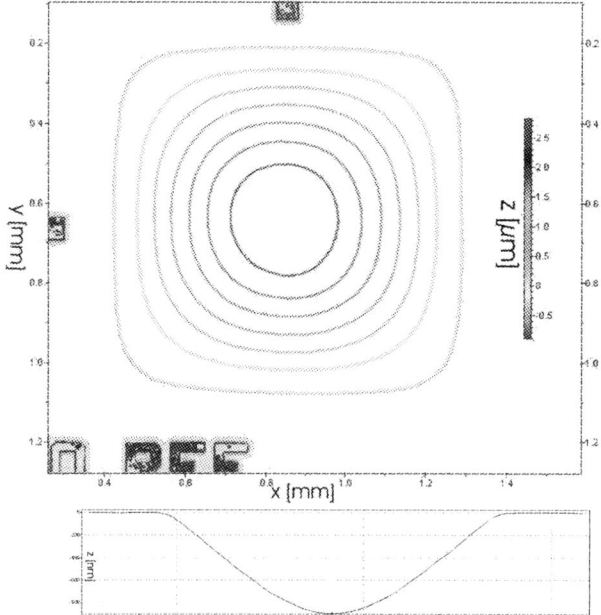

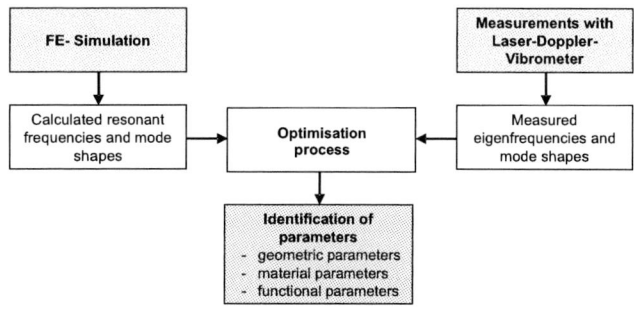

Figure 10 Contour lines profile and cut through the membrane at initial state

1 Parameter Identification

The methodology of parameter identification based on the combination of dynamic measurements and the FE analyses and is shown in Figure 11.

The result of this optimization process between experimental and numerical results is the identification of material, geometrical or functional parameters under given constraints.

Figure 11 Schematic methodology of parameter identification

Important for the identification process are the sensitivities of the dynamic properties to variations of the identification parameters. In the case of the investigated membrane the sensitivities of the membrane length, width and thickness on the first and higher natural frequencies have to be known. Further the sensitivity to variations of the passivation layer and the pre-stress of temperature processes have to be estimated.

The sensitivity of mode shapes to parameter variations is also investigated by using the Modal Assurance Criterion (MAC) value. The value calculates the affinity of two eigenvectors of the same dimension and from the same geometric positions of the structure. A Mac value of 100% means complete correlation of the vectors and 0% means no correlation between the vectors. It can be shown that different thicknesses of the membrane inside the complex structure not change the MAC value. That was naturally shown in the FE- model. The interesting comparison of the MAC ten measured membranes with known thicknesses after cross section analysis has shown the same result, see an example for a grid in Figure 12 . All MAC values have nearly equal values and no tendency in dependence of thickness can be shown. The MAC value is a useful indicator in the case of significant defects of the membrane or to compare "unsafe" peaks of different structures by mode shapes.

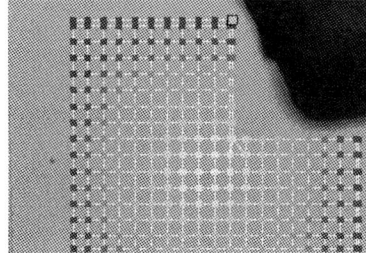

Figure 12 Grid for experimental measurement of mode shapes

2   Results of parameter identification

In the following parameters of the pressure sensor test structures are identified. Figure 13 shows the one parameter case. The thickness is identified. One and three frequencies (overdetermined system) were used for identification. There is no significant difference between the two identified thicknesses. The deviation of the identified values to the experimental values is smaller than 1.8%. This is equivalent to app. 400nm accuracy

Further the identification was successful done for two parameters. The width of the membrane is the second parameter. Three frequencies are used for the optimization. The identified thicknesses are very similar to the first results. The optimization delivered no better results for the membrane thickness in comparison with the experimental results. Some deviations up o 1.25% occur from the width of 900μm in the optimization. But no experimental results exist for the width. So these are not a verifiable value.

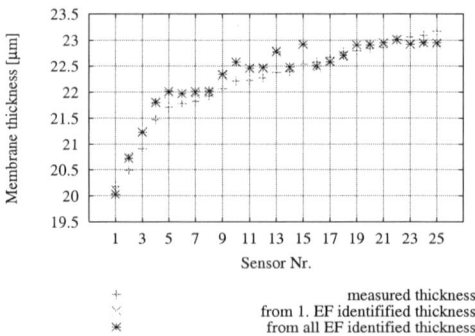

Figure 14 Identified membrane thickness by use of one and up to three frequencies

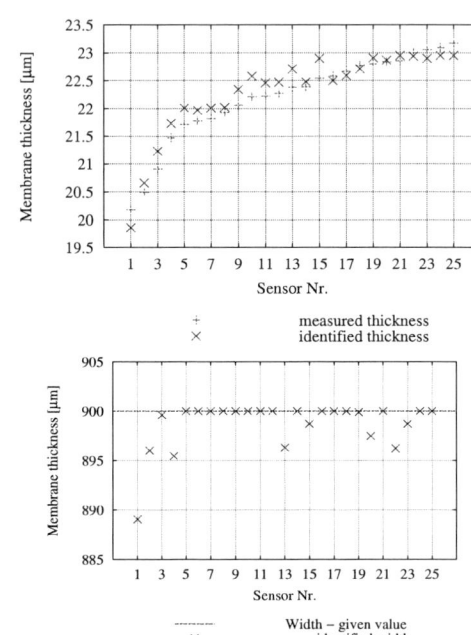

Figure 15 Identified membrane thickness and width by use of three frequencies

3   Conclusions and outlook

Basic investigations for the dynamic measurment, the simulation of dynamic properties and the subsequent optimzation process for parameter identfication of MEMS are introduced. The work is part of the method develpoment to characterize the status of MEMS at early stage of production on wafer level. The out-of-plane vibrations of pressure membranes were measured by LDV. Results of parameter identifcatio are introduced.

All knowledges about the influences of structrue properties on eigenfrequencies, the accuracy of the simulation models and the possibilities and sensitivities of the optimization procedure can be transfered to other types of MEMS. Further it is possible to use the change of dynamic properties to estimate the material parameters of thin films on membranes or cantilevers by this nondestructive method.

Acknowledgments

The financial support of the project "PAR-TEST" by the German Federal Ministry of Education and Research (BMBF) (contract 16SV1940, [6]) is gratefully acknowledged.

Furthermore the authors like to thank Siegfried Hering from X-Fab Semiconductor Foundries AG and Steffen Michael from the Melexis GmbH for the preparation of test structures and the provided MSA measurement results.

References

[1] C. Gorecki et al.: "Multifunctional Interferometric Platform for On-chip Testing of the Micromechanical Properties of MEMS/MOEMS", Proc. SPIE, 5543, San Jose, USA, 2628 Januar 2004, pp. 63-69, 2004

[2] L. M. Zhang et. al.: "Measurement of Youngs's modulus and Internal Stresss in Silicon Micro-resonators Using a Resonant Frequency Technique", Meas. Sci. Technol. 1, pp.1343-1346, 1990

[3] S. Kurth, W. Doetzel: "Experimental Adaption of Model Parameters for Microelectromechanical Systems (MEMS)", Sensors and Actuators A, vol. 62, no. 1-3, pp. 760-764, 1997

[4] M. Ebert et. al.: "Numerical Identification of Geometric Parameters from Dynamic Measurements of Grinded Membranes on Wafer Level", Proc. 7th EuroSimE, Como, Italy, pp. 208-213, 2006

[5] Gerbach, R. et al.: "Identification of geometrical parameters of MEMS from measured resonant frequencies", in Proc. of 17th MicroMechanics Europe (MME2006), University of Southampton (Ed.), Southampton, UK (2006) 49-52

[6] http://www.memunity.com/par-test.htm

[7] C. Rembe et al.: "Measuring MEMS in Motion by Laser Doppler Vibrometry" in Optical Inspection of Microsystems, 1st ed., W. Osten, Ed., New York, Taylor & Francis Group, 2007

[8] S. Petitgrand et. al.: "3D Measurement of Micromechanical Devices Vibration Mode Shapes with a Stroboscopic Interferometric Microscope", Optics and Lasers in Engineering, vol. 36, no. 2, pp. 77-101, 2001

[9] http://www.polytec.com/ger/158_6392.asp

DESIGN-FOR-RELIABILITY TOOLS FOR HIGHLY INTEGRATED SYSTEM-ON-PACKAGE TECHNOLOGY

R.V. Pucha, S. Hegde, M. Damani, K. J. Leee, K. Tunga, A. Perkins,

S. Mahalingam, G. Lo, K. Klein, J. Ahmad and S.K. Sitaraman*

Packaging Research Center

Georgia Institute of Technology, Atlanta GA 30332-0405

*Voice: 404-894-3405, Fax: 404-894-9342, E-mail: suresh.sitaraman@me.gatech.edu

With the dramatic advances made in Microsystems industry, System-on-a-Package (SOP) technology holds promise in terms of reduction in size, cost, and improved performance. To be able to achieve such benefits in an integrated system, it is necessary not to compromise the overall reliability of the system. Therefore, the SOP technology will require up-front system-level design-for-reliability approaches and appropriate reliability assessment methodologies to ensure the reliability of digital, optical, and RF functions as well as their interfaces. Design-for-reliability requires (i) Mechanics-based reliability prediction models for various failure mechanisms associated with Digital, Optical, and RF Functions, and their interfaces in the system (ii) Design optimization models for the selection of suitable materials and processing conditions, for reliability as well as functionality and (iii) System-level reliability models understanding the component and functional interaction. This presentation will focus on the reliability assessment of digital, optical, and RF functions in SOP-based microsystems [1]. Upfront physics-based design-for-reliability models for various functional failure mechanisms are presented to evaluate various design options and material selection even before the prototypes are made. Advanced modeling methodologies and algorithms to accommodate material length scale effects, due to enhanced system integration and miniaturization are presented. System-level mixed-signal reliability is discussed thorough system-level reliability metrics relating component level failure mechanisms to system-level signal integrity as well as statistical aspects.

Block-Diagram Based SIMULINK Analysis for the Drop Impact Response of a Mobile Electronic System

Jiang Zhou, Paul Corder, and Ratna P. Niraula
Department of Mechanical Engineering
Lamar University, Beaumont TX 77710, USA
jenny.zhou@lamar.edu

Abstract

In this paper, the block diagram based SIMULINK models were developed to evaluate the dynamic response of a portable electronic product with various impact configurations. The visual interfaces of the developed models present results in a way that people can immediately identify the effects of changing system parameters. It was found that time durations of the input profiles play an important role in the dynamic response. The system response can be designed by carefully choosing the impact time. Certain input pulse time results in the response with very low ringing after first or second peaks.

1. Introduction

Dynamic performance during drop impact is a great concern to semiconductor and electronic product manufacturers, especially for portable devices such as mobile phones. Recently more attention and effort have been devoted toward upon mechanical modelling of microelectronic packages under drop impact loads [1-8].

In this paper, the block diagram based SIMULINK analysis was introduced to determine the dynamic response of the portable electronic systems with various impact configurations. SIMULINK is an interactive, block-diagram-based tool for modelling and analyzing dynamic systems, and it is tightly coupled with MATLAB and supported by blocksets and extensions. Using such a tool, the relationship between input and output can be obtained and visualized easily and quickly with selected system parameters. As an example, a board level predictive dynamic model was established and the SIMULINK models were developed and used in building block diagrams, performing simulations, as well as analyzing results. The visual interface of the developed models present results in a way that one can immediately identify the effects of changing system parameters.

In the section which follows, a simplified 2 DOF model is developed for a mobile phone subjected to an impact load during accidental drop or during drop test. Four different initial conditions and input load profiles used in the drop test were also identified. In Section 3, the block diagram based SIMULINK models were developed for both the free vibration with initial conditions and the forced vibration with various impact profiles. In Section 4, the developed models were verified by correlating the model results to the analytical solution of the initial conditions problem for a specific mobile phone. Then, forced vibration responses are discussed in Sections 5, 6 and 7, for the half-sine, rectangular, and input signals. Finally, the major conclusions are summarized in Section 8.

2. Analytical Model

Many mobile phones are composed of a PCB board with packages amounted on it, and a plastic housing to hold the PCB board. The mobile phone can be simplified as a two-degree-of-freedom system, as shown in Fig. 1. m_1, b_1 and k_1 are mass, damping ratio, and spring constant of the PCB board; m_2, b_2 and k_2 are mass, damping ratio, and spring constant of the housing; f_1 and f_2 are external forces for the PCB and housing, respectively; x_1 and x_2 are displacements for PCB and housing, respectively.

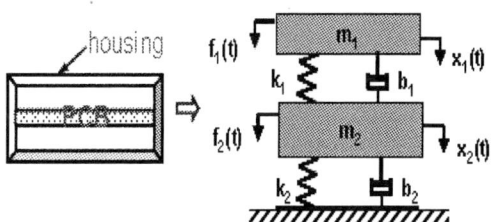

Fig. 1 Simplified model of the mobile phone

The equation of moton for the system in question is, thus,

$$\begin{bmatrix} m_1 & 0 \\ 0 & m_2 \end{bmatrix}\begin{Bmatrix} \ddot{x}_1(t) \\ \ddot{x}_2(t) \end{Bmatrix} + \begin{bmatrix} b_1 & -b_1 \\ -b_1 & b_1+b_2 \end{bmatrix}\begin{Bmatrix} \dot{x}_1(t) \\ \dot{x}_2(t) \end{Bmatrix} + \begin{bmatrix} k_1 & -k_1 \\ -k_1 & k_1+k_2 \end{bmatrix}\begin{Bmatrix} x_1(t) \\ x_2(t) \end{Bmatrix} = \begin{Bmatrix} f_1(t) \\ f_2(t) \end{Bmatrix}$$

(1)

The deflection due to PCB bending is the primary driver of solder joint failure during drop impact. The acceleration of PCB board causes the dynamic forces. Thus, the dynamic response of the PCB assembly under impact are important variables to be investigated.

Different assumptions for the initial conditions and external excitations in the current research literature on system and board level drop impact analysis on the potable electronic devices [1 - 3] are listed in the following.

Case 1: Free vibration with initial conditions:

$$\begin{Bmatrix} x_1(t) \\ x_2(t) \end{Bmatrix} = 0, \text{ and } \begin{Bmatrix} \dot{x}_1(t) \\ \dot{x}_2(t) \end{Bmatrix} = \begin{Bmatrix} \sqrt{2gh} \\ \sqrt{2gh} \end{Bmatrix} \qquad (2)$$

where h is the drop height. There is no external excitations in this case, i.e.,

$$\begin{Bmatrix} f_1(t) \\ f_2(t) \end{Bmatrix} = \begin{Bmatrix} 0 \\ 0 \end{Bmatrix} \qquad (3)$$

Case 2: Forced vibration - half-sine input forced vibration with zero initial conditions, as shown in Fig 2(a):

$$f_1(t) = f_2(t) = \begin{cases} G_p \sin(\dfrac{\pi t}{\tau}), & 0 \leq t \leq \tau \\ 0, & t > \tau \end{cases} \qquad (4)$$

where G_p and τ are amplitude and duration of the half-sine curve. The initial conditions are zero.

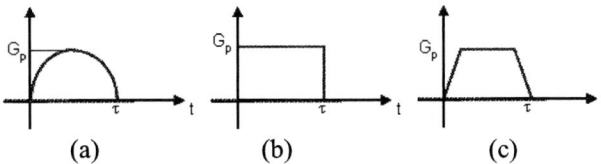

(a)   (b)   (c)

Fig. 2 Forcing functions: (a) sine curve, (b) rectangular curve, (c) trapezoid curve

Case 3: Forced vibration – rectangle pulse input forced vibration with zero initial conditions, as shown in Fig 2(b):

$$f_1(t) = f_2(t) = \begin{cases} G_p, & 0 \leq t \leq \tau \\ 0, & t > \tau \end{cases} \qquad (5)$$

where G_p and τ are amplitude and duration of the rectangle pulse curve. The initial conditions are also zero.

Case 4: Forced vibration – trapezoid pulse input forced vibration with zero initial conditions, as in Fig. 2 (c). Here, G_p and τ are amplitude and duration of the trapezoid pulse curve. The initial conditions are zero. Free vibration with initial conditions in Case 1 is used to simulate the drop impact of a real-world drop scenario. Forced excitations assumptions are used in the component level shock tests in electronic industries. A half-sine forcing function input profile is usually used to simulate

the end-user impact conditions, and the trapezoidial input profile simulates the shipping shock environment.

3. SIMULINK Model Development

The closed-form analytical solutions for Eq. (1) with initial conditions and external excitations in Cases 1-4 are complicated. In this paper, a set of SIMULINK models were developed to determine the system's dynamic response with different input profiles. SIMULINK is block-diagram-based model to analyze and design dynamic systems. SIMULINK models can be obtained from the differential equations.

Rewriting the Eq. (1)

$$\begin{cases} \ddot{x}_1 = \dfrac{1}{m_1}\left[f_1(t) + b_1(\dot{x}_2 - \dot{x}_1) + k_1(x_2 - x_1)\right] \\ \ddot{x}_2 = \dfrac{1}{m_2}\left[f_2(t) + b_1(\dot{x}_1 - \dot{x}_2) - b_2\dot{x}_2 + k_1(x_1 - x_2) - k_2 x_2\right] \end{cases}$$

$$(6)$$

The block diagram of the system represented by Eq. (6) is sketched and the corresponding SIMULINK models are shown in Fig. 3 and Fig. 4.

Fig. 3 is the block diagram and SIMULINK model for the free vibration with initial conditions in Case 1. Initial conditions are allowed to input in the velocity integrators and displacement integrators. System parameters can be controlled by changing the corresponding constants, labelled in the model figures as 'mass1', 'stiffness1', 'damping1', and 'mass2', 'stiffness2', 'damping2'. Four scopes are used to monitor the two displacements and two accelerations of the two masses which represent the PCB and housing of a mobile phone.

Fig. 4 is the block diagram and SIMULINK model for the forced vibration with Cases 2 – 4 different input profiles and zero initial conditions. The input profiles are coded separately in MATLAB and need to be executed before running the SIMULINK model. Each case requires different individual MATLAB code. MATLAB results for the input functions are then connected to the SIMULINK model. Similar to the model in Fig. 3, system parameters are allowed to vary by changing the corresponding constants. The input signal, two displacements and two accelerations for the two masses are monitored by the use of five scopes.

It is very convenient to conduct the parametric studies with these SIMULINK models. Dynamic responses of the system can also be visualized and quickly extracted.

680

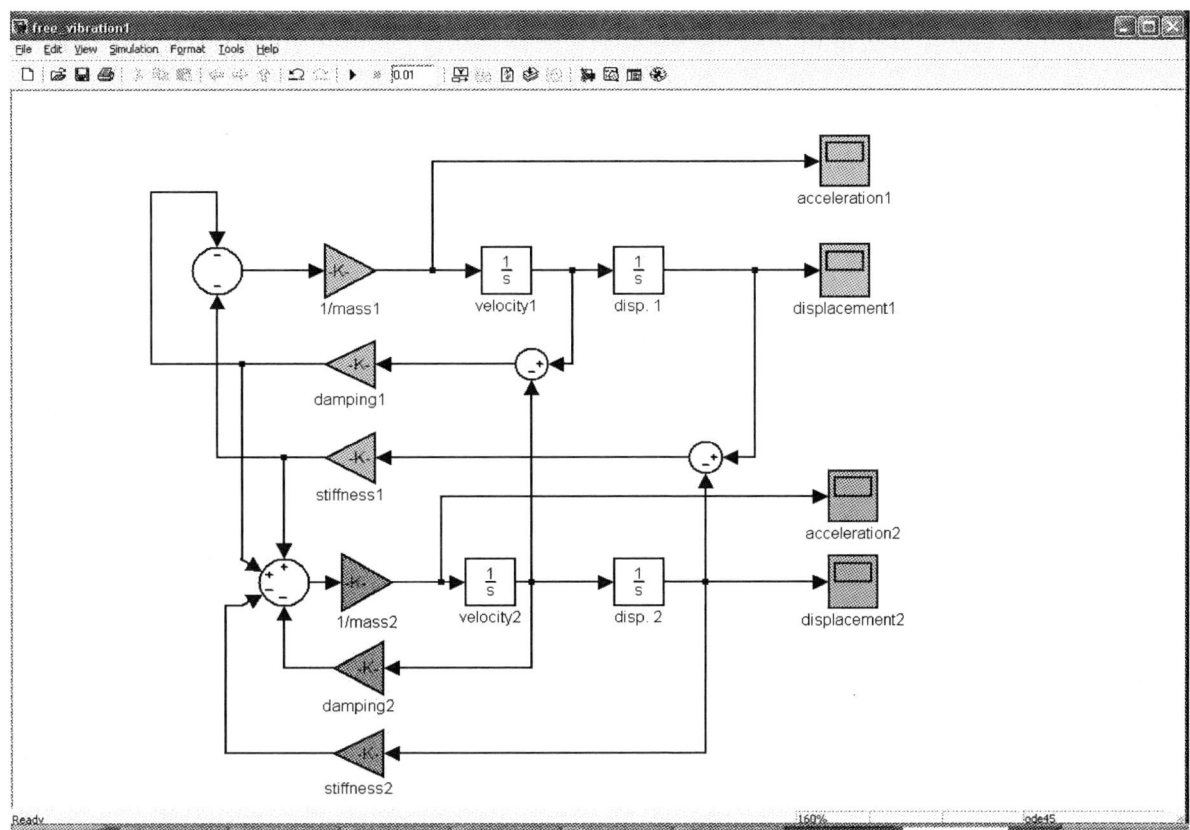

Fig. 3 Block diagram and SIMULINK model for free vibration with initial conditions in Case 1

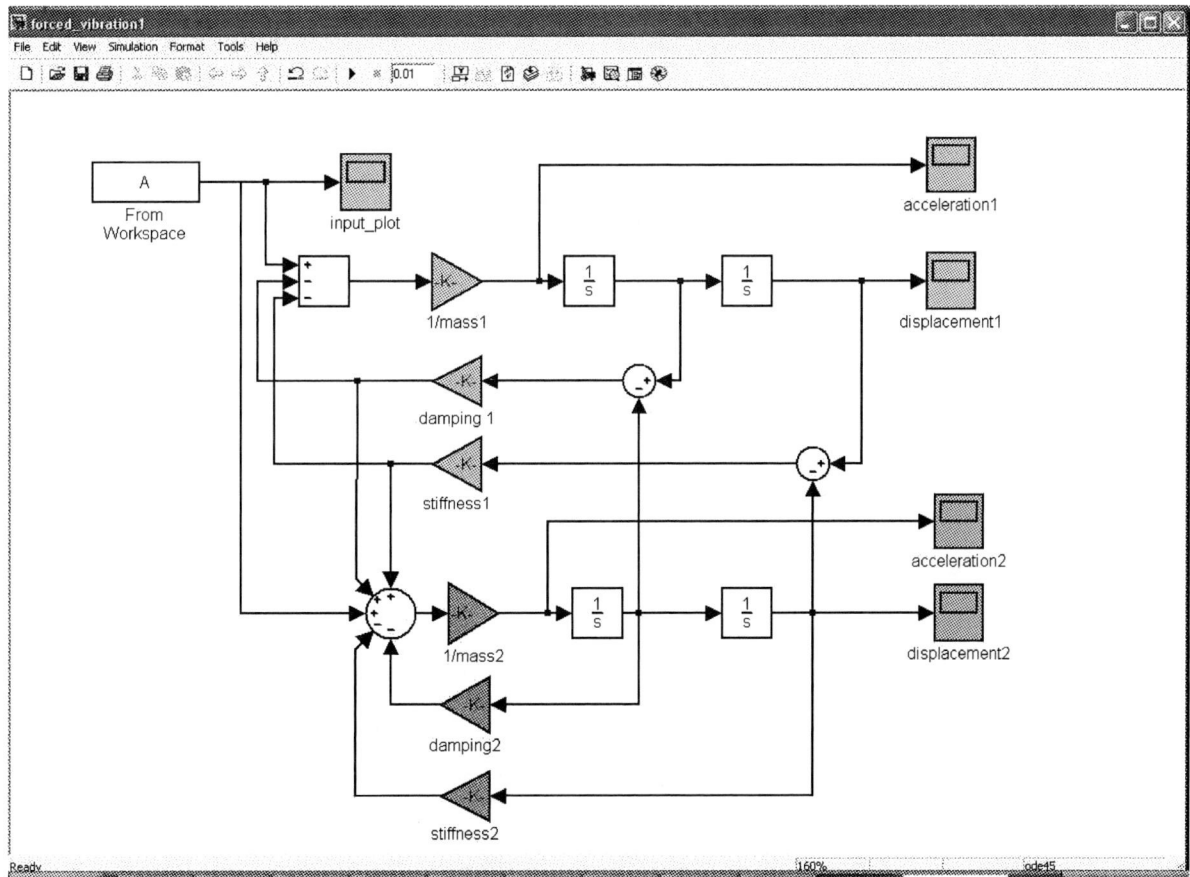

Fig. 4 Block diagram and SIMULINK model for forced vibration with differenct input profiles

4. Model Verification and Results of Free Vibration with Initial Conditions

A specific public used cellular phone is used to conduct the model evaluation. The masses of the PCB board and housing of the cellular phone are 30 gram and 35 gram, respectively. The spring constants for th PCB and housing are calculated as 1465 N/mm and 590 N/mm, respectively, using FEA analysis shown in our previous paper [1]. Assume the drop height is 1.5 m. For the first step analysis, the dampings of the system are neglected. The two resonant frequencies are calculated using MATLAB to be $\omega_1 = 456$ Hz, and $\omega_2 = 1582$ Hz. Fig. 5 and Fig. 6 show the displacement and acceleration associated with PCB boards directly from two scopes in the SIMULINK model shown in Fig. 3. The maximum displacement is ~0.21 mm, and the maximum acceleration is ~2e4 m/s^2, as plotted in Figures 5 and 6. These results are coincident with our previous theoretical analysis [1], which validates the present SIMULINK models.

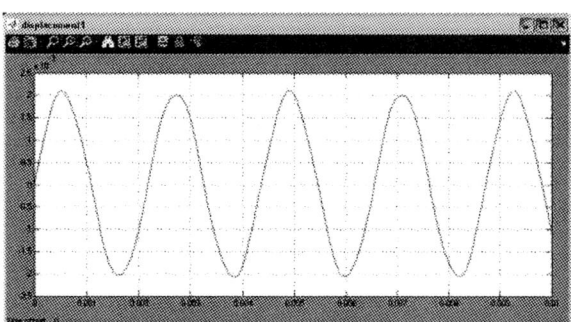

Fig. 5 Displacement of the PCB board with initial conditions and without input signals

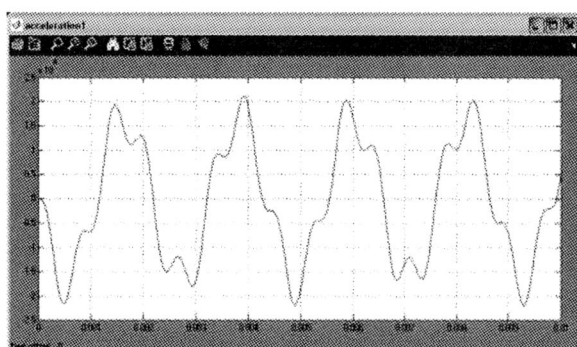

Fig.6 Acceleration of the PCB board with initial conditions and without input signals

5. Forced Vibration with Half-Sine Forcing Inputs

The SIMULINK model in Fig. 4 was used to find the dynamic response of the system to a half-sine input forcing function. The first resonant frequency ω_n of the system is assumed to be the same as what is in the previous section of the case study. Input duration is taking as 1x, 1.5x, and 2x of the period calculated from ω_n. Fig. 7 is the input plot displayed in the input scope in SIMULINK, when $\tau = 2/\omega_n = 2T$. Scopes' results

obtained were sent to MATLAB to facilitate the plots presented in this paper. Displacement and acceleration of the PCB board with half-sine input signals for various impact time duration are plotted in Fig. 8 and Fig. 9.

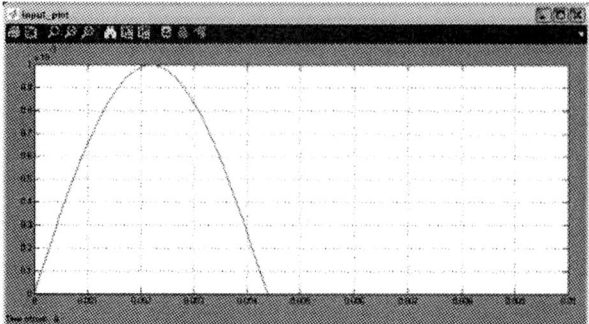

Fig. 7 Half-sine input signal displayed in the scope, when $\tau = 2/\omega_n = 2T$

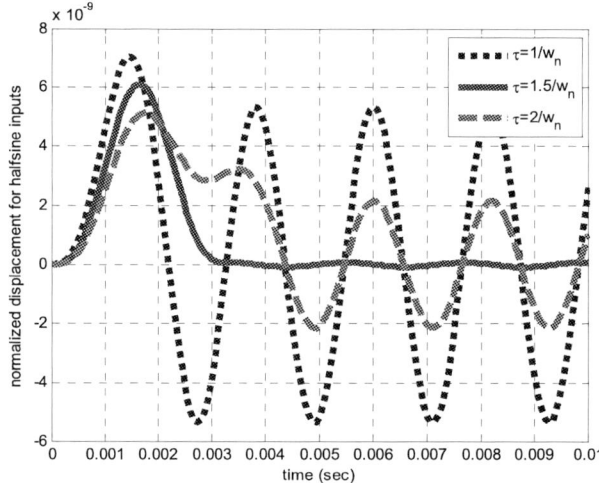

Fig. 8 Displacement of the PCB board with half-sine input signals

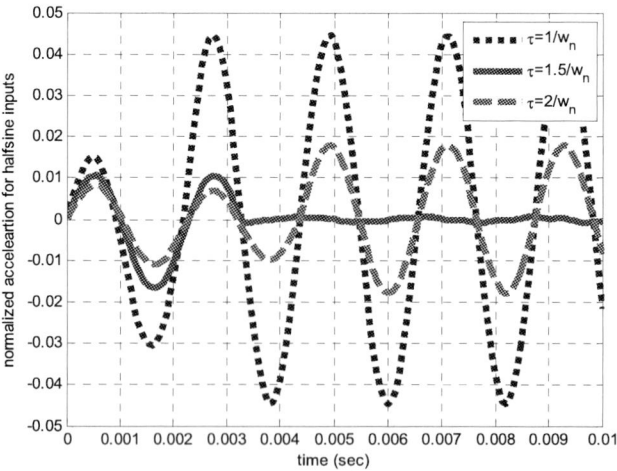

Fig. 9 Acceleration of the PCB board with half-sine input signals

It is very interesting to note that when $\tau = 1.5/\omega_n$, the responses have a predominant mode. The amplitudes of second and after peaks are very small, compared with the dominant first peak, for both displacement and acceleration. SIMULINK simulation tests showed that, for all $\tau = (i + \frac{1}{2})/\omega_n$, when $i = 1, 2, 3, \ldots$, there exist predominant modes, and very low residual ringing occurs. It is very important to control the input time duration such that the second peak is minimized in the componant level shock test in the electronic industries. Inspired by this findings based on the SIMULINK results for the 2 degree of freedom (2-DOF) system, theoretical study was performed for the 1-DOF system. A different publication is being prepared to introduce the theoretical study of the dynamic response control. Here, only the 1-DOF closed-form theoretical solutions are plotted in Figure 10. It is found that the same conditions applied for the 1DOF system to produce the minimum residual ringing, that is when $\tau = (i + \frac{1}{2})/\omega_n$, when $i = 1, 2, 3, \ldots$.

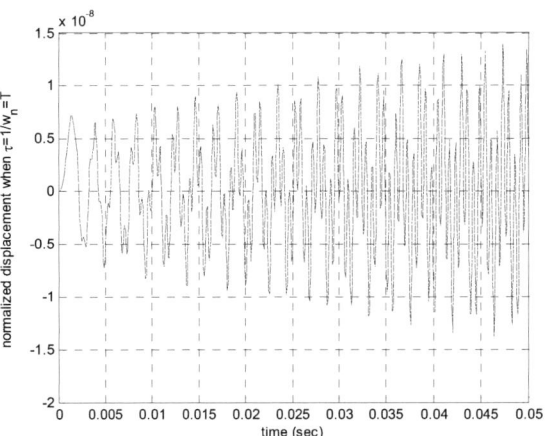

Fig. 11 Displacement of the PCB board with half-sine input signals when $\tau = 1/\omega_n$ or $\tau = T$

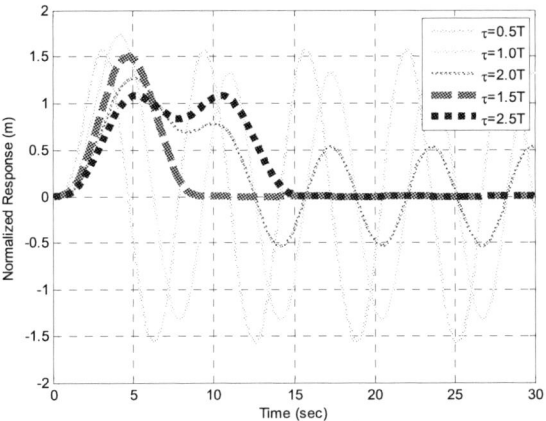

Fig. 10 Theoretical solution of the displacement with half-sine input signals for 1DOF system

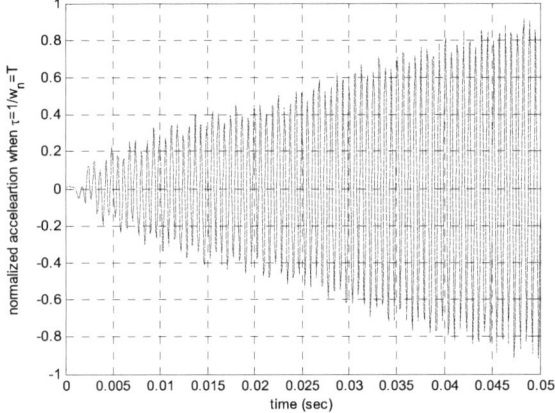

Fig. 12 Acceleration of the PCB board with half-sine input signals when $\tau = 1/\omega_n$ or $\tau = T$

When the results were plotted for extended time ranges, it was surprising to see that the evaluated mobile phone is actually a dynamically unstable system. Fig. 11 and Fig. 12 show the displacement and acceleration with half-sine input signals when $\tau = 1/\omega_n$ or $\tau = T$, and Fig. 13 and Fig. 14 plot the displacement and acceleration with half-sine input signals when $\tau = 1.5/\omega_n$ or $\tau = 1.5T$. In Fig. 13, the predominant first peak is remarkable. With both input durations, the displacements and accelerations increase unboundly with time. By carefully choose the system parameters, such as, masses and spring constants, the instability can be avoided. These SIMULINK models are a very useful tool in design to quickly check the instability of the system. Of course, in the real world, system damping takes effect so that it is possible that the real system is still stable.

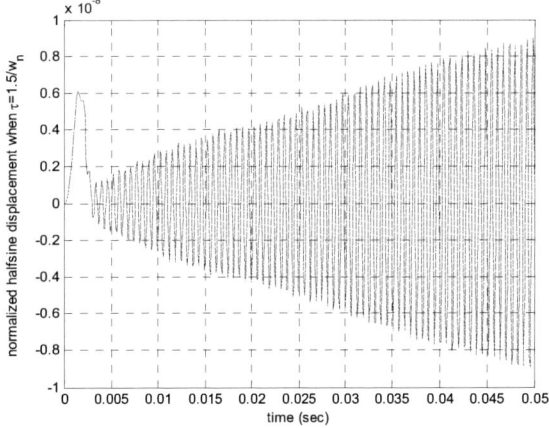

Fig. 13 Displacement of the PCB board with half-sine input signals when $\tau = 1.5/\omega_n$ or $\tau = 1.5T$

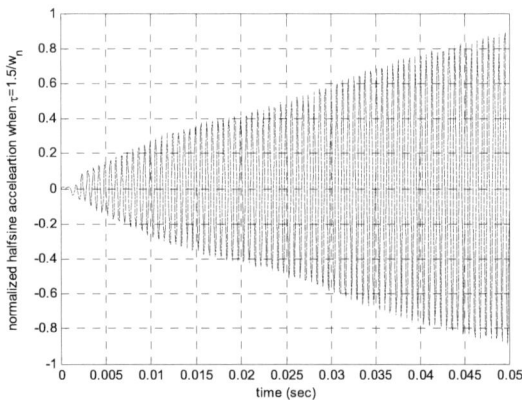

Fig. 14 Acceleration of the PCB board with half-sine input signals when $\tau = 1.5/\omega_n$ or $\tau = 1.5T$

6. Forced Vibration with Rectangular Pulse Inputs

Similar to the previous half-sine inputs, the SIMULINK model in Fig. 4 was also used to find the dynamic response of the system with rectangular input signals. Input durations are also taking as one time, 1.5 times, and 2 times of the period calculated from the ω_n. Displacement and acceleration of the PCB with rectangular input signals for various input time durations are plotted in Figures 15 and 16.

Note that when $\tau = 1.0/\omega_n$ or $\tau = 2.0/\omega_n$, the responses have the predominant mode. Tests in SIMULINK simulation showed that, for all $\tau = i/\omega_n$ (i = 1,2,3,...), there exists predominant modes, and little residual ringing occur for the rectangular impact load inputs. Theoretical study for the 1-DOF system showed the same no residual ringing condition as in the 2-DOF system.

Similar unstable results with unbound increasing displacement and acceleration can be found as in Figures 11 – 14 for the rectangular input profiles.

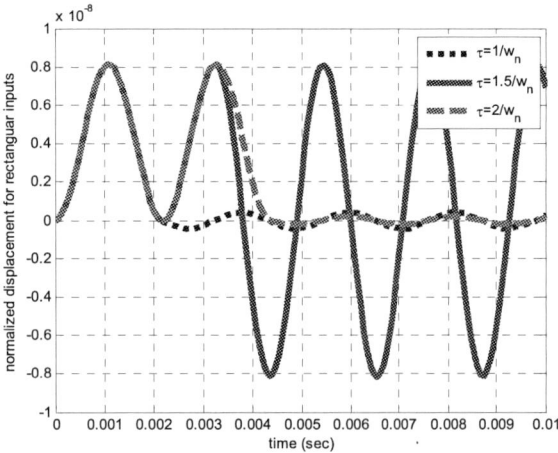

Fig. 15 Displacement of the PCB board with rectangular input signals for various input time durations

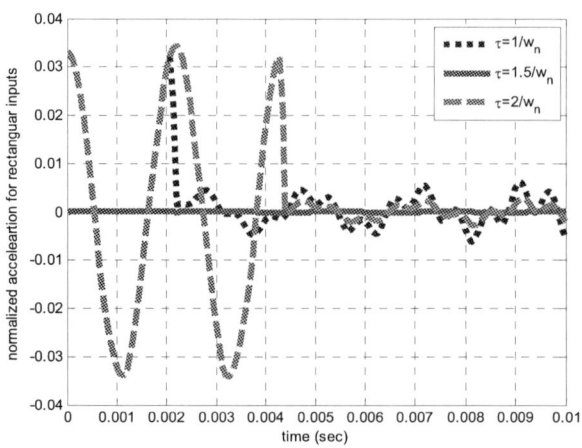

Fig. 16 Acceleration of the PCB board with rectangular input signals for various input time durations

7. Forced Vibration with Trapezoidal Pulse Inputs

Fig. 17 is the trapezoidal input profile displayed in the input scope in SIMULINK, when $\tau = 2/\omega_n = 2T$. Displacement and acceleration of the PCB board with trapezoid input signals for different input time durations are plotted in Fig. 18 and Fig. 19.

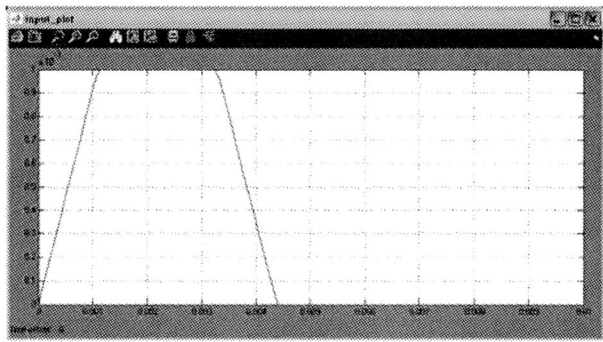

Fig. 17 Trapezoidal input profile displayed in the input scope in SIMULINK model, when $\tau = 2/\omega_n$

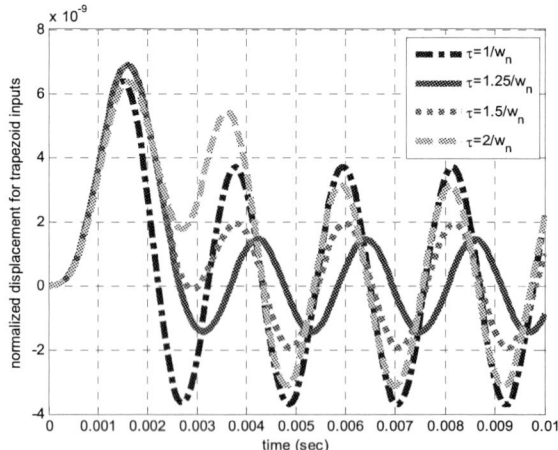

Fig. 18 Displacement of the PCB board with trapezoidal input signals for various input time durations

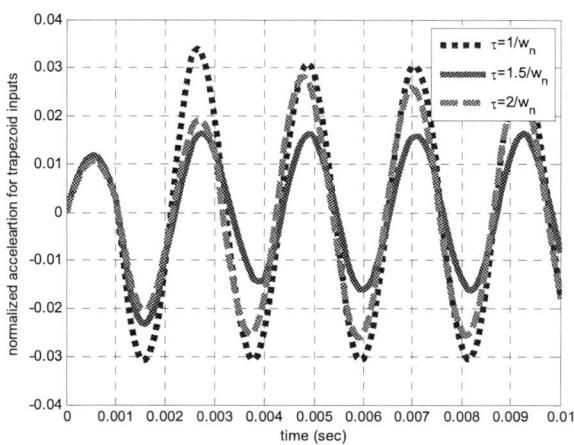

Fig. 19 Acceleration of the PCB board with trapezoidal input signals for different input time durations

Notice that the predominant modes occur when τ is somewhere between $\tau = 1.0/\omega_n$ and $\tau = 1.5/\omega_n$. In Fig. 18, there is a much smaller 2nd peak response for $\tau = 1.25/\omega_n$. Similar unstable results with unbound increasing displacement and acceleration can be found as in Figures 11 – 14 for the trapezoidal impact load profiles.

8. Conclusion

The following conclusions are drawn from the analyses of this paper.

- A mobile phone experiencing elevated accelerations during accidental drop or during drop testing was simplified as a two-degree-of-freedom model.

- Four different initial conditions and input profiles used in the drop test by manufacturers were identified. They are free vibration with initial conditions, and forced vibration with input profiles of half-sine, rectangular pulse, and trapezoidal pulses.

- Block-diagram based SIMULINK models were developed to evaluate the impact response with any input profiles and initial conditions. The developed models were verified by correlating the model results to the analytical solution of the initial conditions problem.

- Dynamic responses of a mobile phone were evaluated with three different input signals. It is found that that the evaluated system is instable when the damping effect is neglected.

- Time duration of the input profiles play an important role in the dynamic response. In the other words, the system response can be designed by carefully choosing the impact time. It is very meaningful in the design of drop test in the electronic industries.

- It is found that certain input pulse time results in low ringing response. For the half-sine input, the

no ringing conditions are $\tau = (i + \frac{1}{2})/\omega_n$, where $i = 1, 2, 3, \ldots$. For the rectangular input, the no ringing conditions are $\tau = i/\omega_n$, where $i = 1, 2, 3, \ldots$. For trapezoid input, the no ringing conditions are $i/\omega_n < \tau < (i + \frac{1}{2})/\omega_n$, where, $i = 1, 2, 3, \ldots$. The conclusions are held for both 1-DOF and 2-DOF systems.

- The developed SIMULINK models include the damping constants, while the results here were obtained by ignoring the damping. Further research can be performed by the models and theoretical analysis for the effect of damping.

- SIMULINK is a very powerful and useful tool in the design of the portable electronic products to quickly extract the system response and analyze the stability. The visual interface of the developed models presents results in a way that one can immediately identify the effects of changing system parameters.

- The developed models can be applied not only to the design of the portable electronic products, but to other engineering applications where dynamic response analysis is needed.

References

1. Zhou, J., Kallolimath K., and Lahoti, S., "Analytical and numerical analysis of drop impact behavior for a portable electronic device," *IEEE International Conference on Thermal, Mechanical and Multiphysics Simulation and Experiments in Micro-Electronics and Micro-Systems (EuroSimE)*, April 17-20, 2006, Milan, Italy, pp 138 – 144
2. Suhir E., "Dynamic Response of a Rectangular Plate to a Shock Load, with application to Portable Electronic Product," *IEEE Transactions on Components, Packaging and Manufacturing Technology*, Vol.17, No. 3, 1994.
3. Suhir E., "Is The Maximum Acceleration An Adequate Criterion Of The Dynamic Strength of a Structural Element in an Electronic Product?" *IEEE Transactions on Components, Packaging, and Manufacturing Technology*, Vol. 20, No. 4, December 1997.
4. Suhir E., "Could Shock Tests Adequately Replace Drop Tests", *8th International Symposium of Advance Packaging*, September 2002.
5. Lall, P., Panchagade, D., Liu, Y., Johnson, W., Suhling, J., "Models for Reliability Prediction of Fine-Pitch BGAs and CSPs in Shock and Drop-Impact", *54th Electronic Components and Technology Conference*, June 2004.
6. Luan J., and Tee T. Y., "Effect of Impact pulse parameters on Consistency of the Board level Drop Test and Dynamic Responses," *Electronic Components Technology Conference*, 2005.
7. Tee T.Y., Ng H.S., Lim C.T., Pek E., and Zhong Z., "Board Level Drop Test and Simulation of TFBGA Packages for Telecommunication Applications," *The Proc. of the 53rd IEEE/EIA Electronic Components and Technology Conference*, 2003.
8. Tee, T. Y., Ng, H. S., Lim, C. T., Pet, E., and Zhong, Z. W., "Impact Life Prediction Modeling of TFBGA Packages under Board Level Drop Test", *Microelectronics Reliability Journal*, Vol. 44(7), pp.1131-1142. 2004.

Fluctuation Mechanism of Mechanical Properties of Electroplated-Copper Thin Films Used for Three Dimensional Electronic Modules

Hideo Miura, Ken Suzuki, and Kinji Tamakawa

Fracture and Reliability Research Institute, Graduate School of Engineering, Tohoku University

6-6-11-712 Aoba, Aramaki, Aoba-ku, Sendai, Miyagi 980-8579, Japan

hmiura@rift.mech.tohoku.ac.jp, Tel./Fax. +81-22-795-6986/4311

Abstract

The mechanical properties of copper thin films formed by cold-rolling and electroplating were compared using tensile test and nano-indentation. Both the Young's modulus and tensile strength of the films were found to vary drastically depending on the microstructure in the films. The Young's modulus of the cold-rolled film was almost same as that of bulk material. However, the Young's modulus of the electroplated thin film was about a fourth of that of bulk material. The micro structure of the electroplated film was polycrystalline and a columnar structure with a diameter of a few hundred-micron. The strength of the grain boundaries of the columnar grains seemed to be rather week. Such a columnar structure causes the cooperative grain boundary sliding and the film shows low elasticity and superplastic deformation. In addition, there was a sharp distribution of Young's modulus along the thickness direction of the film. Though the modulus near the surface of the film was close to that of bulk material, it decreased drastically to about a half within the depth of about 1 μm. There was also a plane distribution of Young's modulus near the surface of the film.

1. Introduction

Electronic products such as mobile phones and PCs have been miniaturized continuously and their functions have been improved drastically [1]. Three dimensionally stacked structures such as multi-chip modules and multi-chip packages are indispensable for increasing the assembly density. Cu thin-film interconnections and small bumps are going to be applied to high speed devices as shown in Fig. 1 because of their low electronic conductivity. However, mechanical properties such as Young's modulus and tensile strength of thin films vary significantly depending on the micro structure of the films and the surface energy of the films. In particular, the thickness of the film has become less than 1 μm in semiconductor applications. Since the micro crystallographic structure of the films varies drastically depending on the conditions of film deposition methods such as sputtering and electroplating, their mechanical properties may change due to the change of the microstructure of the deposited films. The microstrucure of the copper films strongly depends on their formation process. Conventionally, copper films has been formed by cold-rolling. Recently, sputtering and electroplating have been applied to the formation of the copper thin films thinner than 1 μm. The density, average diameter of grains, and crystallographic orientation of the fims also vary depending on the formation process. In a thin film deposition process, an intrinsic stress higher than 1 GPa ocuurs in the deposited film easily due to the mismatch of lattice constant between the film and a substare or thermodynamical inequiribrium atomic bonding caused by low temperature depsotion [2][3]. Such a high residual stress also changes the effective mechanical properies of the deposited film such as hardness and elasticity.

In addition, it has been found that there is a size effect on the physical and chemical properties of thin film materials [4][5]. It is well known that both the mechanical and electrical properties of carbon nanotubes are quite different from those of bulk carbon materials [6]-[8]. As a matter of fact, a silicon thin film of about 100 nm thick shows stable plastic deformation even at room temperature [9], though it is believed to be a brittle material. Thin films with columnar grain structures were found to show strong anisotropic mechanical properties between their thickness direction and the plane direction [10][11]. The strength of their grain boundaries plays an important role on the anisotropy. The intrinsic stress of thin films is another important issue to be considered for reliability of thin film products [12].

Therefore, the effect of the microstructure of copper thin films on the mechanical properties such as Young's modulus and the strength of the film formed by cold-rolling and electroplating were measured using nano-indentation technique and a tensile test of the films. The distribution of these properties was measured three-dimensionally.

2. Variation of micro texture of copper thin films

Electroplated copper thin films were formed on a stainless steel substrate (SUS304) using copper sulfate solution. Direct current of about 25 mA/cm^2 was applied to the substrate during electroplating. The average thickness of the electroplated film was about 10 μm.

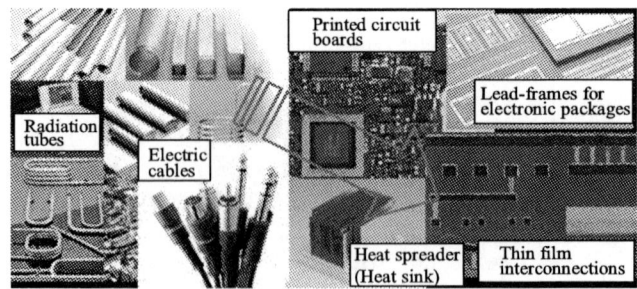

Fig. 1 Example applications of bulk materials and thin films of copper

1-4244-1105-X/07/$25.00 ©2007 IEEE

Fig. 2 scanning electron micrograph of cross-sectional texture of an electroplated copper thin film

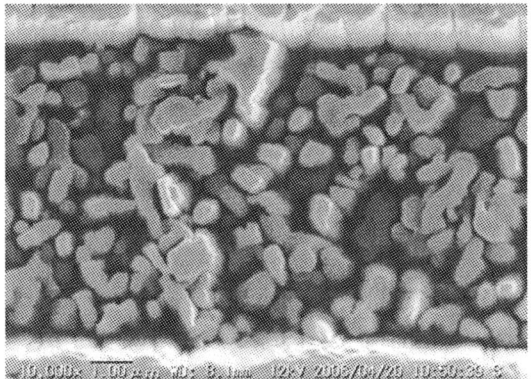

Fig. 3 Columnar grains of an electroplated copper thin film

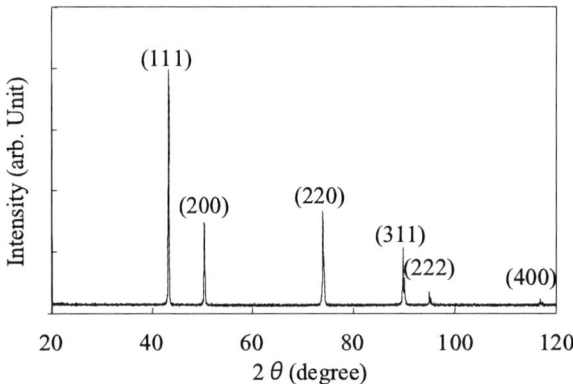

Fig. 4 X-ray diffraction pattern of an electroplated copper thin film

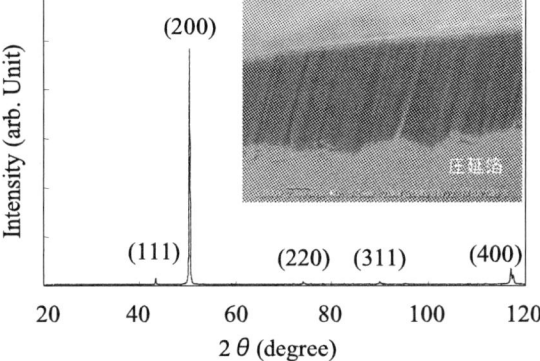

Fig. 5 X-ray diffraction pattern of a cold-rolled copper thin film

It was easy to delaminate the electroplated film from the substrate. A scanning electron micrograph of the cross-sectional view of the electroplated film is shown in Fig. 2. The surface morphology of the film was rather rough. The average surface roughness was about 200 nm. The film was found to consist of thin columnar grains. The average diameter of the columnar was about a few hundreds nm. This columnar structure was caused by direct current perpendicular to the substrate during electroplating.

To observe the grain structure clearly, a focused ion beam was irradiated to the film. A scanning electron micrograph of a plane view of the partially etched film is shown in Fig. 3. This photograph clearly indicates that the film is composed of thin columnar grains of about 500 nm in diameter. In addition, grain boundaries are observed sharply. This means that the grain boundaries were etched off easily because they were rather porous comparing with conventioanl grain boundaries of poly-crystalline material. Thus, the micro texture of this electroplated copper film is anisotropic. There is no grain boundaries along thickness direction of the film, while there are a lot of grain boundaries along plainar direction of the film.

Figure 4 shows a typical x-ray diffraction pattern of an electroplated copper film. The crystalographic direction of the columnar grains were mainly (111). This is because the (111) plane is the most dense crystallographic plane of a fcc (face centered cubic) crystal. However, (110) plane

was the main orientation of some electroplated films. Thus, the crystallographic direction of electroplated films are not uniform.

Figure 5 shows a typical x-ray diffraction pattern and cross-sectional structure of a cold-rolled copper thin film. It is well known that the cold-rolled film is highly oriented to (100) direction due to the secondary recrystallization. The cross-sectinal structure of the film is rather dense and it is hard to observe grain boundaries directly.

The micro texture of copper thin films, therefore, vary drastically depending on their formation process. In particular, the thin columnar structure with porus grain boundaries observed in electroplated films may give rise to strong anisotropy of mechanical properties of the film.

3. Mechanical properties of copper thin films

A simple tensile test for thin films and nano-indentation technique were applied to the films to measure the mechanical properties of the copper films thinner than 10 μm. Examples of the measured stress-strain curves along the plane direction of the films are shown in Fig. 6. The stress-strain curve of a cold-rolled copper film is similar to that of a bulk material. However, tensile strength of the film is about 780 MPa and this value is about twice of that of a bulk material. On the other hand, the fracture elongation of the electroplated copper film is about six times higher than that of the

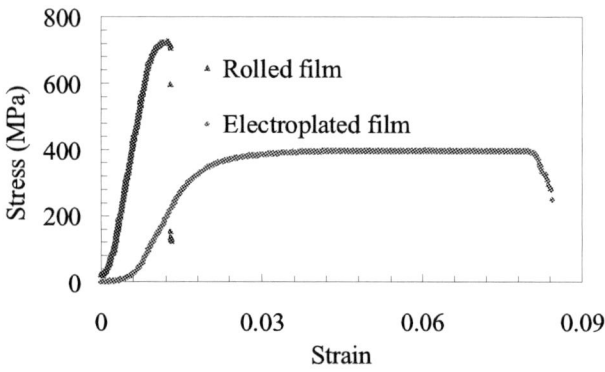

Fig.6 Example of stress-strain curves of copper thin films

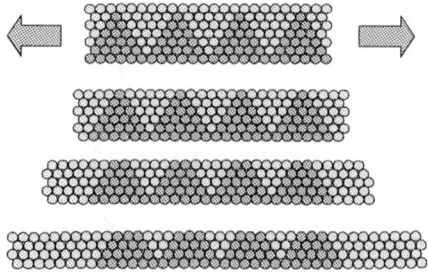

Fig. 7 Basic idea of deformation of columnar structure based on cooperative grain boundary sliding

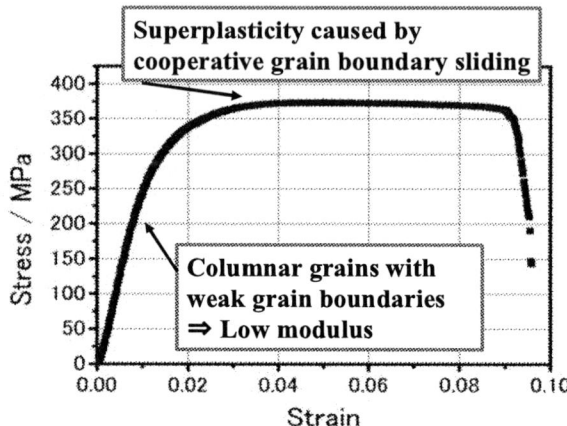

Fig. 8 Dominant factors of stress-strain curve of an electroplated copper thin film

sputtered copper film, though the tensile strength of the films is same as that of bulk copper. The Young's modulus of the electroplated copper film is about 30 GPa and this value is about one fourth of that of bulk copper. The fracture strain of the electroplated copper films varied widely among the measured films from about 0.05 to 0.3.

In addition, no stain hardening is observed in the stress-strain curve of the electroplated copper films. It shows superplasticity during platic deformation. Such a deformation process can not be observed in the conventional bulk copper materials. Therefore, it is very important to understand the mechanism of the change of these mechanical properties of electroplated thin films to assure their reliability.

Since there was no difference in the chemical composition between the cold-rolled films and electroplated films, this difference in the mechanical properties should be attributed to the change of the micro texture of the films. It is well known that the mechanical properties of materials are function of the average diameter of grains and crystallographic orientation. But the reported change of Young's modulus, for example, was within 10%. Thus, the observed columnar grain structure should be the main reason for the drastic change of the mechanical properties of the electroplated copper films.

Recently, it was reported that the superplastic deformation can occur even in brittle ceramic material when the material consists of columnar grain structure with weak grain boundaries [13][14]. The cooperative grain boundary sliding shown in Fig. 7 enables the superplastic deformation of brittel materials. In this deformation process, each thin columnar grain does not deform substantially. The columnar grains only slide easily along grain boundaries as dislocations move on crystallographic slip planes.This sliding occurs only when the strength of grain boundaries is rather weak and its driving force is a function of square of average diameter of grains. Thus, this cooperative grain boundary sliding is activated when the diameter of the columnar grain becomes thinner and thinner. As is shown in Fig. 3, the

diameter of the electroplated copper film is much smaller than 1 μm and the grain boundaries seems to be rather porous. Thus, this coopaeratuve grain boundary sliding is the main reason for the superplastic deformation of the film.

This thin columnar grain structure with weak (rather porous) grain boundaries can explain the unique stress-strain curve of the electroplated copper thin films as shown in Fig. 8. The drastic low elastic constant of the electroplated copper thin films is caused by the weak grain boundaries. The superplastic deformation of the film is due to the cooperative grain boundary sliding of the columnar grains.

When the electroplated film wa fractured, the fractured surface was almost the same as that of a brittle material as shown in Fig. 9. There was no clear shape change of the film around the fractured edge due to plastic deformation. No apparent ductile fracture morphology was obserbed on the fractured surface either. In addition, some columanar grain boundaries were observed on the fractured surface. Thus, the fracture of this film occurred mainly at grain boundaries. The fractured surface was seemed to be a brittle material.

Thus, the measured unique stress-strain curve of the electroplated copper thin film is not caused by the change of mechanical properies of copper crystal, but the

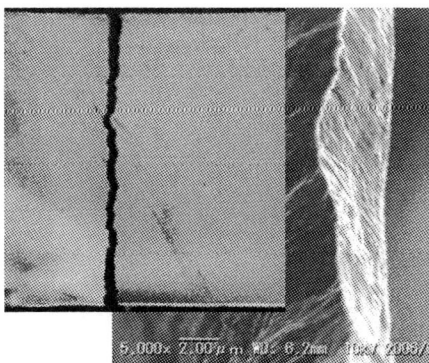

Fig. 9 Fractured Surface of an electroplated copper thin film

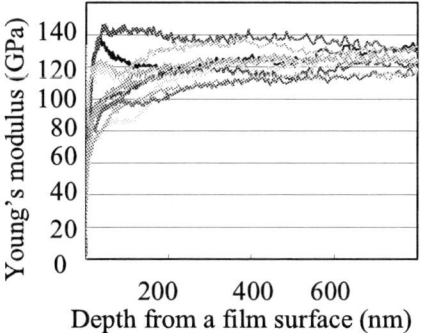

Fig.10 Depth profile of Young's modulus of cold-rolled copper thin films

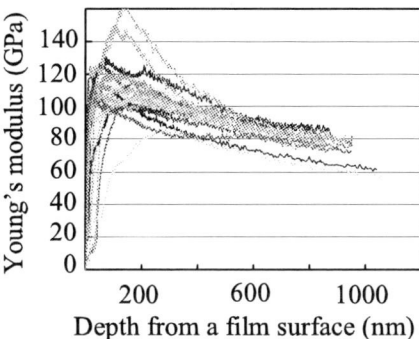

Fig.11 Depth profile of Young's modulus of electroplated copper thin films

unique micro texture of the film, i.e., thin columnar garain structure with weak grain boundaries.

4. Three-dimensional distribution of Young's modulus in copper thin films

The mechanical properties of the electroplated copper thin films along their thickness direction was measured using Nano Indenter DCM-SA2 made by MTS Corp. The resolution of the displacement measuremnet was 0.2 pm and that of the applied force was 1 nN. The strain rate during the measurement was fixed at 0.05/s. In addition, the continuous stiffness measurement method was applied to measure the thickness distribution of Young's modulus of the film. An alternative current was applied to a load cell coil to make small vibration at a tip of indenter. The Young's modulus at a certain point was measured during each small unloading process of this small vibration. No change of the mechanical properties caused by strrain hardening occurs by minimizing the amplitude of this vibration.

Figure 10 shows the distributions of Young's modulus of the cold-rolled thin films along their thickness direction. The Young's modulus of the film was almost constant of about 120 GPa and this value agrees well with that of bulk material. However, there is a sharp gradient of Young's modulus within 1 μm from the surface of the electroplated film. It decreases drastically to less than 80 GPa at about 1 μm from the surface as shown in Fig. 11. The Young's modulus of the inner part of the film corresponds well to that along plane direction of the film. One of the reasons of this decrease of Young's modulus can be explained as follows. When a tip of the indenter starts to contact with the film, only one columnar grain starts to deform. Thus, the measured properties are close to those of bulk material. The total number of the columnar grains under the indenter tip increases monotonically with the increase of the applied stress. This means that the weak grain boundaries are also included in the contact area. Thus, the measured properties are averaged values of the columnar grains and weak grain boundaries. The averaged value should be samaller than those of bulk material. The shape of this distribution also varies drastically depending on the microstructure of the film. Thus, there is a complicated three-dimensional

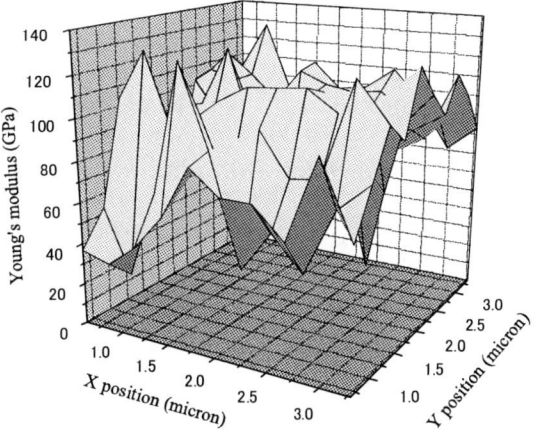

Fig.12 Planar distribution of Young's modulus of electroplated copper thin films at a depth of 100 nm

distribution of Young's modulus in the electroplated copper films.

Figure 12 shows the planar distribution of the Young's modulus of the electroplated copper thin films at about 100 nm in depth. The indenter tip was scanned two-dimensionally by a pitch of 300 nm. The plotted area is 3.0-micron square. The Young's modulus of the film near the surface of about 100 nm varies from about 30 GPa to 130 GPa. This remarkable fluctuation of Young' modulus

689

strongly depends on the position where the indenter was contacted. When the indenter was pressed to a center of a columnar grain, the measured Young's modulus agreed with that of bulk material. On the other hand, the measured value decreased drastically when the indenter was pressed to a grain boundary of the columnar structured film.

The micro texture of the electroplated film also varies spatially as shown in Fig. 13. There were large grains with the thin columnar structure. The diameter of the large grain reached about a few μm, while that of the thin columnar was about 200 nm. The Young's modulus of the large grain was alomost the same as that of bulk, and there was no drastic planar distribution of the modulus. Therefore, it is hard to discuss the mechanical reliability of the film because the weak points exist randomly in the film.

The mechanical properties of the electroplated copper film change further depending on the thermal treatment at temperatures higher than 200°C after the film formation. This is because that the microstructure of the film can change easily by such low temperature annealing. Recrystallization can occur in the film even at such low temperature. Therefore, it is very important to control the mechanical properties of the electroplated copper films by optimizing the film formation process in order to assure the reliability of the products.

5. Fluctuation of Young's modulus in electroplated copper interconnection films

The electroplated copper thin films have been used in semiconductor devices for thin film interconnections. Since the micro texture of the electroplated film is changed by the current density during electroplating, the mechanical properties of the film vary in one semiconductor device if there is a distribution of the current density in the device during its fabrication process. Thus, various interconnection patterns were formed by electroplating on a 2-inch Si wafer as shown in Fig. 14. The width of the interconnection was varied from 10 μm to 150 μm. The electrodes for electroplating were arranged only on the outskirts of the wafer. Thus, there must be the gradient of current density in the wafer.

The distribution of the Young's modulus of the electroplated thin film interconnection was measured using the Nano Indenter as shown in Fig. 15. It is easy to determine the indentation point by using precisely controlled piezo-actuators.

It was confirmed that the quality of the formed interconnection film varied on the wafer surface locally because there was the distribution of the current density on the Si wafer. In particular, the surface roughness of the film showed drastic change on the wafer depending on both the width and period of the interconnections. Fig. 16 shows an example of the distribution of Young's modulus along the thickness direction of the film which consists of fine grains. The Young's modulus of the film is almost constant of about 110 GPa as was observed in a cold-rolled film shown in Fig. 10. On the other hand, when the

Fig. 13 Example of micro texture of an electroplated copper thin film (Mixture of a large grain and thin columns)

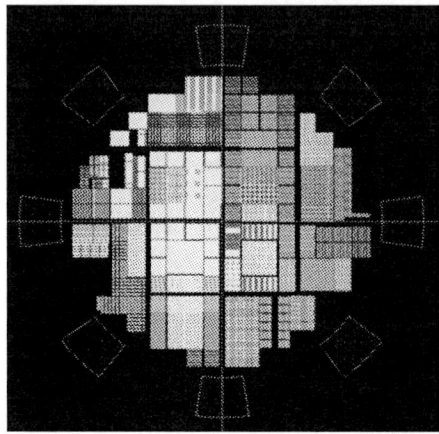

Fig. 14 Test interconnection patterns on a 2-inch Si wafer

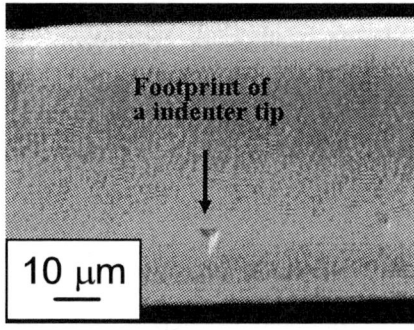

Fig.15 Example of a footprint of an indenter tip on an interconnection film

film is composed of larger grains, its Young's modulus is rather low as shown in Fig. 17. In this area, the Young's modulus of the film near the surafce is very low. It increases monotonically with depth and reaches about 70 GPa at a depth of 1 μm. This low modulus was caused by the decrease of the density of the film because of high speed growth of the film. It was confirmed that the density of the film is a function of the local current density. The density decareses drastivcally when the current density during electroplating exceeds the certain critical value. Therefore, the quality of the electroplated film sometimes varies significantly when there is a large distribution of the current density during electroplating in a substrate.

4. Conclusions

The mechanical properties of copper thin films vary drastically depending on their micro crystallographic structure. A complicated and anisotropic three-dimensional distribution of Young's modulus appears in the electroplated film with columnar grain structures. The microstructure changes depending on the conditions of electroplating such as current density, temperature, and the composition of solution. Therefore, it is very important to control the microstructure of the electroplated copper film to keep its mechanical properties constant.

Acknowledgments

This research work is partly supported by the Grants-in-Aid for Scientific Research and the 21COE program of Tohoku University.

References

1. International Technology Roadmap for Semiconductors 2005, (2005)
2. H. Ohta, et. al., Proc. of Asian Pacific Conf. on Fracture and Strength '93, Tsuchiura, Japan, (1993), pp. 735-738.
3. H Miura, "Residual Stress of Thin Films", J. of Japanese Society of Tribologists, Vol. 40, No. 3 （1995.3）pp. 228-233.
4. A. Inoue, "Stabilization of metallic supercooled liquid and bulk amorphous alloys", ACTA MATERIALIA, Vol. 48 (1), (2000), pp. 279-306.
5. J. Carsley, A. Fisher, W. Milligan, et. al., "Mechanical behavior of a bulk nanostructured iron alloy", Metallurgical and Materials Trans. A-Physical Metallurgy and Materials Science, Vol. 29 (9), (1998), pp. 2261-2271.
6. M. Yu, B. Files, S. Arepalli, and R. Ruoff, "Tensile loading of ropes of single wall carbon nanotubes and their mechanical properties", PHYSICAL REVIEW LETTERS Vol. 84 (24), (2000), pp. 5552-5555.
7. J. Salvetat, J. Bonard, N. Thomson, A. Kulik, et. al., "Mechanical properties of carbon nanotubes, Applied Physics A-Materials Science & Processing", Vol. 69 (3), (1999), pp. 255-260.
8. M. Kiuchi, Y. Isono, S. Sugiyama, et. al., "Development of On-chip Micro Tensile Testing Device for Mechanical-Electrical Characteristics of Carbon Nanowires", Proc. of Mechanical Engineering Congress, 2004 Japan. Vol. 1, (2004), pp. 367-368.
9. H. Kito, T. Kikuchi, and Y. Isono, "Development of Micro Fatigue Tester Based on AFM Technique for MEMS Materials", Proc. of Mechanical Engineering Congress, 2004 Japan, Vol. 1, (2004), pp.371-372.

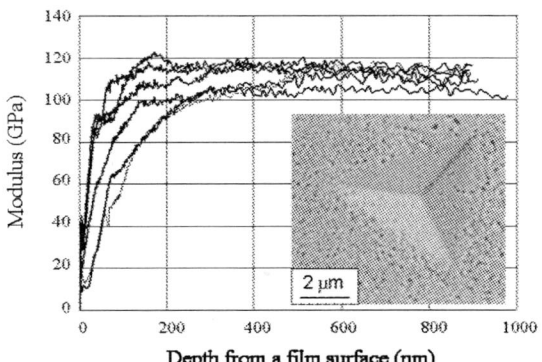

Fig. 16 Depth profile of Young's modulus of electroplated copper thin films with fine grains

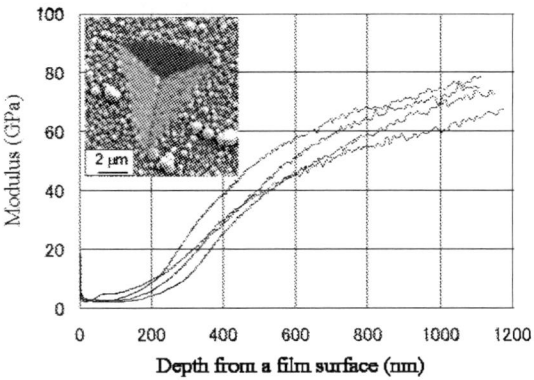

Fig. 17 Depth profile of Young's modulus of electroplated copper thin films with large grains

10. H. Ogi, N. Nakamura, and M. Hirao, "Advanced resonant ultrasound spectroscopy for measuring anisotropic elastic constants of thin films", Fatigue Fract. Engng. Mater. Struct., Vol. 28, (2005), pp. 657-663.
11. N. Nakamura, H. Ogi, and M. Hirao, "Resonance Ultrasound Spectroscopy for Measuring Elastic Constants of Thin Films," Japanese Journal of Applied Physics, Vol. 43, No. 5B (2004), pp. 3115-3118.
12. H. Miura and S. Ikeda, IEICE Trans. on Electronics, "Mechanical Stress Simulation for Highly Reliable Deep-Submicron Devices", Vol. E82-C, No. 6, (1999), pp. 830-838.
13. H. Muto and M. Sakai, "A Novel Deformation Mechanism for Superplastic Deformation," Key Engineering Materials, Vol. 166 (1999), pp. 103-108.
14. H. Muto and M. Sakai, "The Large-Scale Deformation of Polycrystalline Aggregates: Cooperative Grain-Boundary Sliding," Acta Materialia, Vol. 48 (2000), pp. 4161-4167.

Keynote Presentation

Computational Methods and High Speed Imaging Methodologies
for Transient-Shock Reliability of Electronics

Pradeep Lall
Auburn University
Department of Mechanical Engineering
and Center for Advanced Vehicle Electronics
Auburn, AL 36849, USA
Tele: (334) 844-3424
E-mail: lall@eng.auburn.edu

Abstract

Product level assessment of drop and shock reliability relies heavily on experimental test methods. Prediction of drop and shock survivability is largely beyond the state-of-art. However, the use of experimental approach to test out every possible design variation, and identify the one that gives the maximum design margin is often not feasible because of product development cycle time and cost constraints.

In this paper, the modeling approaches and high-speed experimental techniques for first-level solder interconnects in shock and drop of electronics assemblies have been discussed. The shock and vibration reliability prediction of electronic interconnects involves multiple scales from macro-scale transient-dynamics of electronic assembly to micro-structural damage history of interconnects. Previous modeling approaches include, solid-to-solid sub-modeling using a half test PCB board, shell-to-solid sub-modeling technique using a quarter-symmetry model. Inclusion of model symmetry in state-of-art models saves computational time, but targets primarily symmetric mode shapes. Explicit modeling approaches has been presented, which enable prediction of both symmetric and anti-symmetric modes, which may dominate an actual drop-event. Approaches presented include, smeared property models, Timoshenko-beam element models, explicit sub-models, and continuum-shell models.

A failure-envelope approach based on wavelet transforms and damage proxies has been discussed to model drop and shock survivability of electronic packaging. Data on damage progression under transient-shock and vibration in both 95.5Sn4.0Ag0.5Cu and 63Sn37Pb ball-grid arrays has been presented. The concept of relative damage index has been used to both evaluate and predict damage progression during transient shock. The failure-envelope provides a fundamental basis for development of component integration guidelines to ensure survivability in shock and vibration environments at a user-specified confidence level.

Transient dynamic behavior of the board assemblies in free and JEDEC-drop has been measured using high-speed strain and displacement measurements. Correlation of model predictions with dynamic measurements has been presented for acceleration, strain and resistance using high-speed data acquisition systems capable of capturing in-situ strain, continuity and acceleration data in excess of 5 million samples per second. Ultra high-speed video at 150,000 fps per second has been used to capture the deformation kinematics. Life Prediction have been correlated with experimental data for both leaded and leadfree ball-grid arrays.

1. Introduction

Emerging trends of portable computing and communication applications towards smaller, lighter, form-factors has driven the need for robust-designs under overlapping environments of shock and vibration. Electronic products may be subjected to drop and shock due to mishandling during transportation or during normal usage. Test methods for drop reliability can be broadly classified into board-level and product-level tests, under constrained and unconstrained or free drop. Examples of board-level constrained drop include the JEDEC test method. The JEDEC test standard [2003] is often used to evaluate and compare the drop performance of surface mount electronic components for handheld electronic product applications. Correlation of the board-level tests to product-level performance is often challenging. Product-level failures are often influenced by housing design, in addition to drop-orientation, which may not always be perpendicular to the board surface [Lim 2002]. Factors such as drop height, mass of the product, impact orientation and the properties of the impacting surface affect the forces and the accelerations that are experienced by the product during impact. Design changes encompass an iterative process for improving the impact resistance of the electronic product. Use of experimental approach to test out every possible design variation, and identify the one that gives the maximum design margin is often not feasible because of product development cycle time and cost constraints.

Previous researchers have investigated the effect of board design and packaging technology [Saha 2004, Chai 2005], solder alloy composition, intermetallic growth [Chiu 2004, Lall 2005], surface finishes [Chong 2005] on solder joint reliability. Supplementary restraint mechanisms investigated include the use of corner-bonded underfills [Lall 2004, Tian 2005] and pre-applied underfills [Hannan 2004, Morganelli 2005]. Magnitude of acceleration, effect of drop orientation on the product

1-4244-1105-X/07/$25.00 ©2007 IEEE

performance of fine-pitch electronics has been studied. [Wu 1998, Lim 2002, Tan 2005]. Failure modes typically include, cracking of the printed circuit board dielectric, copper trace fractures, copper-delamination, first-level and second-level solder-interconnects failure, silicon and dielectric failures in Cu Low-K devices. Transition to wafer-level chip scale package (CSP) and flip chips with smaller finer pitches and smaller solder joint interconnects requires the need for better predictive capabilities for determination of design margins.

Modeling efforts have focused on prediction of the transient dynamic behavior, and included the use of explicit finite-elements with and without smeared properties [Lall 2004, 2005, Xie 2002, 2003, Wu 1998, 2000], implicit finite elements [Irving 2004, Pitaressi 2004, Syed 2005], and submodeling [Tee 2003, Wong 2003, Zhu 2001, 2003, 2004]. Previously, prediction of failure has been investigated using fracture mechanics [Shah 2004], von-mises stress [Tee 2004], and board-strain based damage index [Lall 2005].

Strain behavior of electronic assemblies is not elastic and the large transient deformation are often not accurately represented by the small strain theory. In addition, the final or the peak strain does not capture the final strain or damage state of the assembly. The rate-of-change of deformation and the latent damage in previous drops impacts the susceptibility to failure during successive drops. Dynamic responses at board level [Sogo 2001, Mishiro 2002] at product level [Lim 2002, 2003] have been measured by previous researchers. Failure may not happen in the first drop, and damage may be cumulative. There is need for techniques and damage proxies which enable the determination of damage equivalency and cumulative damage during overstress and repetitive loading for various packaging architectures.

2. Modeling Approaches

One of the challenges in modeling shock response of electronic products, is the multiple-scale differences between the dimensions of the individual layers, such as solder interconnects, copper pad, chip-interconnects and the dimensions of the electronic assembly, which makes the computational effort needed to attain fine mesh to model chip interconnects while capturing the system-level dynamic behavior very challenging. Various modeling approaches have been pursued to reduce the computational time required for simulation. Zhu [2001] applied solid-to-solid sub-modeling technique to analyze BGA reliability for free board level drop using half PCB board. Shell-to-solid sub-modeling using beam-shell based quarter symmetry global model was employed by Ren, et. al. [2003, 2004] to further reduce the computational time. Previously, the JEDEC JESD22-B111 has been modeled using the input-G method [Tee 2004]. Symmetry of load and boundary conditions has been used to attain computational efficiency and decrease the model size. The assumption of symmetry in state-of-art models targets symmetric modes predominantly. The explicit time-integration is most suitable for solving wave

propagation problems such as drop impact [Lall 2004, 2005]. The simulation time is determined by the size of the time step which is directly proportional to the length of the smallest element in the model. Board level drop simulations using smeared property approach have been carried out and validated with experimental data. Transient dynamic responses of board assemblies have been predicted fairly accurately while achieving computational efficiency.

Explicit Modeling Approach
The transient dynamic response of a printed circuit board under drop impact can be represented in the finite element domain with step-by-step direct integration in time for the explicit formulations. The governing differential equation of motion for a dynamic system can be expressed as, [Lall 2006[a, b, c, d]]

$$[M]\{\ddot{D}\}_n + [C]\{\dot{D}\}_n + \{R^{int}\}_n = \{R^{ext}\}_n \qquad (1)$$

For a linear problem, $\{R^{int}\}_n = [K]\{D\}_n$, where [M], [C] and [K] are the mass, damping and stiffness matrices respectively and $\{D\}_n$ is the nodal displacement vector at various instants of time. Methods of explicit direct integration calculate the dynamic response at time step n+1 from the equation of motion, the central difference formulation and known conditions at one or more preceding time steps [Lall 2006[a, b, c, d]],

$$\left[\frac{1}{\Delta t^2}M + \frac{1}{2\Delta t}C\right]\{D\}_{n+1} = \{R^{ext}\}_n - \{R^{int}\}_n$$
$$+ \frac{2}{\Delta t^2}[M]\{D\}_n - \left[\frac{1}{\Delta t^2}M - \frac{1}{2\Delta t}C\right]\{D\}_{n-1} \qquad (2)$$

Equation (2) has been combined with equation (1) at time step n. In the implicit algorithm, the dynamic response at time step n+1 has been calculated from known conditions at present time-step, in addition to one or more preceding time-steps. Using Newmark relations and the average acceleration method, the equation of motion can be written as follows [Lall 2006[a, b, c, d]],

$$[K^{eff}]\{D\}_{n+1} = \{R^{ext}\}_{n+1}$$
$$+ [M]\left\{\frac{1}{\beta \Delta t^2}\{D\}_n + \frac{1}{\beta \Delta t}\{\dot{D}\}_n + \left(\frac{1}{2\beta}-1\right)\{\ddot{D}\}_n\right\}$$
$$+ [C]\left\{\frac{\gamma}{\beta \Delta t}\{D\}_n + \left(\frac{\gamma}{\beta}-1\right)\{D\}_n + \Delta t\left(\frac{\gamma}{2\beta}-1\right)\{\ddot{D}\}_n\right\}$$

$$(3)$$

Where,
$$[K^{eff}] = \frac{1}{\beta \Delta t^2}[M] + \frac{\gamma}{\beta \Delta t}[C] + [K] \qquad (4)$$

and γ and β are numerical factors that control the characteristics of the algorithm such accuracy, numerical stability and amount of algorithmic damping. All the terms on the right hand side of Equation (2) are known and have already been calculated at earlier time steps, however the same is not true of Equation (3). The mass matrix, [M] has been diagonalized, using the lumped approach, improving computational efficiency, because time step is executed very quickly without solution of simultaneous equations. Use of the lumped mass approach increases the allowable step time but is limited to the explicit formulation. For the implicit formulation, the effective stiffness matrix, [K^{eff}] is not a diagonal matrix, even if the mass and damping matrices are diagonalized, since it contains the stiffness matrix, [K]. The diagonal mass matrix in implicit formulation therefore provides very little computational economy. Furthermore, the implicit method is usually more accurate when [M] is the consistent mass matrix, thus increasing the computational time and storage space.

Element size in the explicit model has been limited due to the conditional stability of the explicit time-integration, which influences the critical value for the time step, to avoid instability and error accumulation in the time integration process. This limiting criterion increases the number of time steps required to span the time duration of an analysis. Explicit time-integration is well suited to wave propagation problems including drop impact, because the dynamic response of the board decays within a few multiples of the longest period. Most implicit formulations, are unconditionally stable, which means that the process is stable regardless of the size of the time step, thus allowing a fewer number of time steps as compared to the explicit method. However, high deformation rates involved in impact, using the implicit formulation with a large time step might introduce too much strain increase in a single time-step, causing divergence in large deformation analysis. A large time-step may cause the contact force, which is proportional to the penetration of the contact bodies, to be very large at the contact causing local distortion and failure. Advantage of being able to use a larger time step with implicit methods can only be used in a limited manner for impact analysis. The explicit formulation is better suited to accommodate material and geometric non-linearity without any global matrix manipulation.

Explicit-Models and Element Formulations
Reduced integration elements have been used in the analysis because they use fewer integration points to form the element stiffness matrices, thus reducing the computational-time for simulation of transient dynamic events. First-order elements perform better, when large strains or very high-strain gradients are expected as in the case of impact. Higher order elements have higher frequencies than lower order elements and tend to produce noise when stress waves move across an FE mesh. Therefore, lower order elements are better than higher order elements at modeling a shock wave front.

Two types of shell elements are available in Abaqus™ including,: conventional shell elements and continuum shell elements. The use of both elements has been investigated for modeling transient-dynamic events. The conventional shell elements discretize the surface by defining the element's planar dimensions, its surface-normal, and its initial curvature. Surface thickness is defined through section properties. Quadrilateral elements have been used with linear interpolation. The conventional shell-element is a reduced integration element which accounts for large strains and large rotations

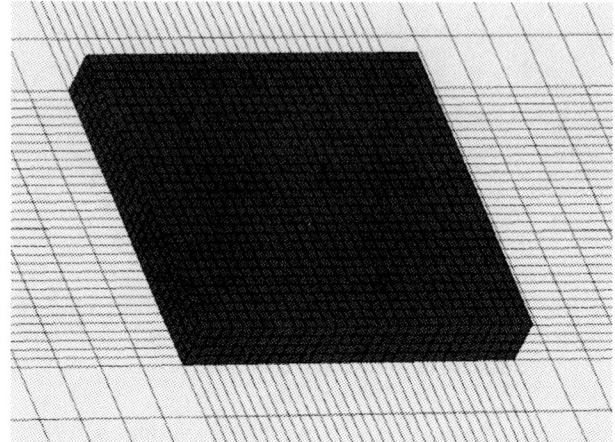

Figure 1: Printed Circuit Assembly has been modeled using Smeared Properties with Conventional Shell Elements. [Lall 2004, 2005]

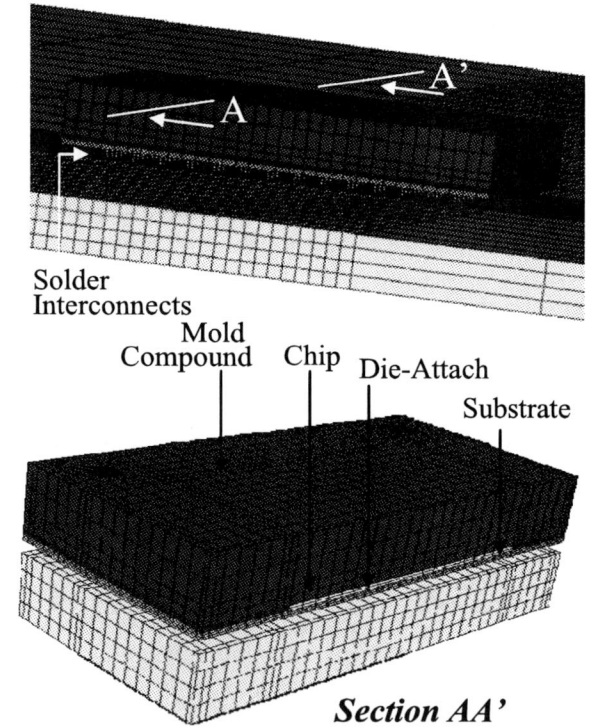

Figure 2: Printed-Circuit Assembly with Timoshenko-Beam Element Interconnects and CONTINUUM Shell-Elements. [Lall 2006[a, b]]

Continuum shell elements resemble three-dimensional solid elements and discretize the entire three-dimensional body. The continuum shell elements are formulated such that their kinematic and constitutive behavior is similar to conventional shell elements. The continuum shell element (SC8R), has three-translational degrees of freedom at each node and the element accounts for finite membrane strains and arbitrarily large rotations [Abaqus 2005[a]]. Shell elements are used to model the printed circuit board, since the thickness dimension is significantly smaller than the other dimensions and the stresses in the thickness direction are smaller than in the in-plane directions. First order tetrahedral elements (C3D4) have not been used for analysis, since they have a simple, constant-strain formulation and very fine meshes are required for an accurate solution.

that the strain in the beam's cross-section is the same in any direction in the cross-section and throughout the section. For fine pitch solder interconnects, with very low stand-off heights, the constant cross-section assumption is a fairly good approximation.

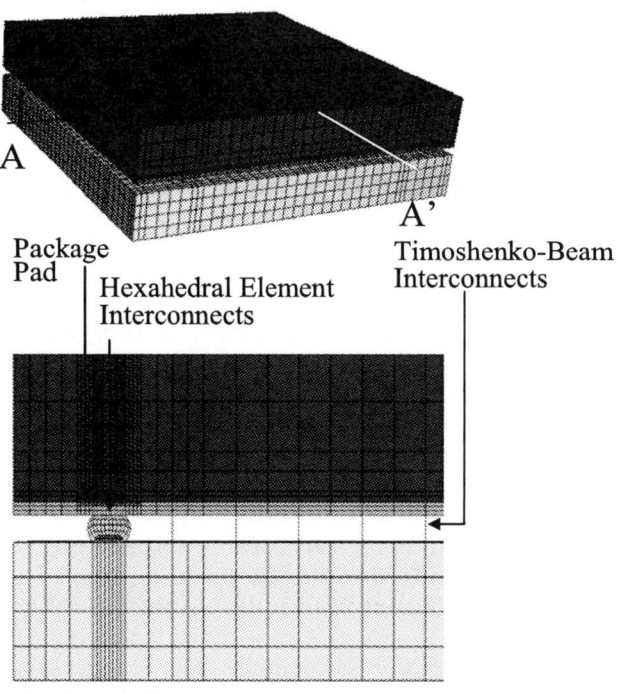

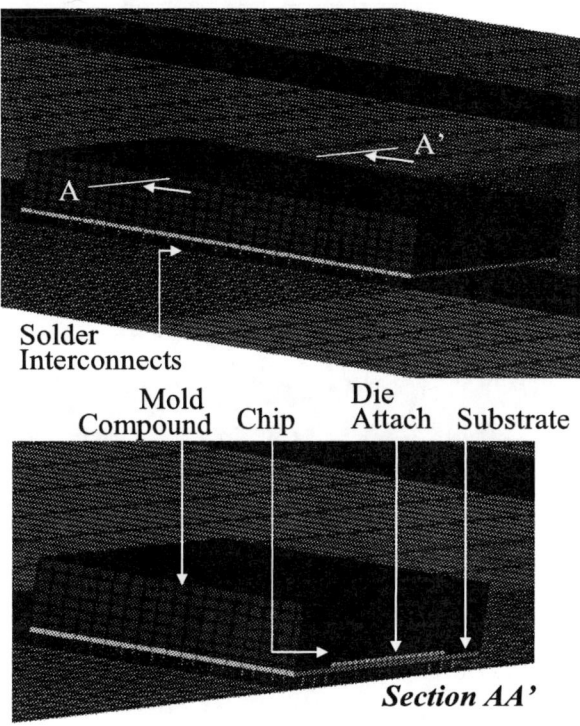

Section AA'

Figure 3: Printed-Circuit Assembly with Timoshenko-Beam Element Interconnects and CONVENTIONAL Shell-Elements. [Lall 2006[a, b]]

Section AA'

Figure 4: Global-Local Explicit Sub-Modeling with Hexadedral-Element Corner Interconnects, Timoshenko-Beam Element Interconnects and PCB meshed with Hexahedral Reduced Integration-Elements. [Lall 2006[a, b]]

Interconnects modeling has been investigated using two element types including the three-dimensional, linear, Timoshenko-beam element (B31) and the eight-node hexahedral reduced integration elements. Three-dimensional beams have six degrees of freedom at each node including, three translational degrees of freedom (1–3) and three rotational degrees of freedom (4–6). The rotational degrees-of-freedom have been constrained to model interconnect behavior. The B31 elements allow for shear deformation, i.e., the cross-section may not necessarily remain normal to the beam axis. [Abaqus 2005[b]]. Shear deformation is useful for first-level interconnects, since it is anticipated that the shear flexibility may be important. It is assumed throughout the simulation that, the radius of curvature of the beam is large compared to distances in the cross-section and that the beam cannot fold into a tight hinge. It is also assumed

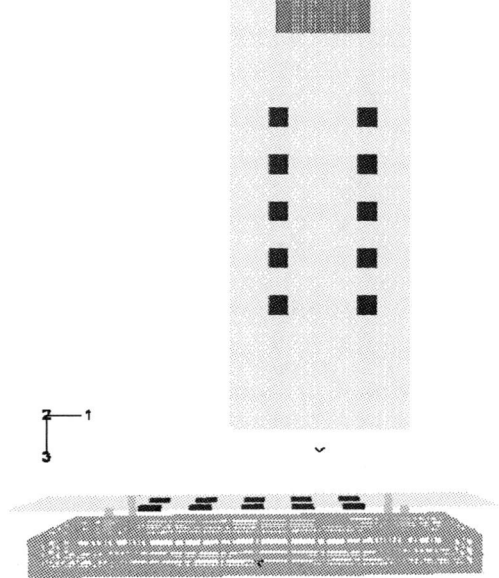

Figure 5: Drop-orientation has been varied from 0° JEDEC-drop to 90° free-drop. [Lall 2006[a, b]]

Table 1: Comparison of Actual and Simulated Component Masses. [Lall 2006[a, b]]

Component	Actual Mass (gm)	Simulated Weight (gm)
PCB	28.15	28.25
CSP	0.140	0.142
Weight	31.8	31.8

Four explicit model approaches have been investigated including, smeared property models (Figure 1), Timoshenko-beam element interconnect models with continuum shell-element (Figure 2), Timoshenko-beam element interconnect models with conventional shell-element (Figure 3), and the explicit sub-models with combination of Timoshenko-beam elements and reduced integration hexahedral element corner interconnects (Figure 4). For each different type of element used for the PCB, the various component layers such as the substrate, die-attach, silicon die, mold-compound have been modeled with C3D8R elements. The solder interconnections have been modeled with two-node beam elements (B31). Smeared properties have been derived all the individual components based on volumetric averaging [Lall 2004, 2005]. The simulated weight of the model for the PCB and all the components closely approximates the actual weights as shown in Table 1. The concrete floor has been modeled using rigid R3D4 elements. In case of free vertical drop, a weight has been attached on the top edge of the board. Node to surface contact has been employed between a reference node on the rigid floor and the impacting surface of the test assembly. The drop orientation has been varied from 0° JEDEC-drop to 90° free-drop (Figure 5).

Explicit sub-modeling has been accomplished using a local model, in addition to the global model. The local model is finely meshed and includes all the individual layers of the CSP and the corresponding PCB portion. The four corner solder interconnections are created using solid elements while the remaining solder joints are modeled using beam elements. Shell-to-solid sub-modeling technique has been employed to transfer the time history response of the global model to the local model. Displacement degrees of freedom from the global model are interpolated to the local model and applied as boundary conditions. The corresponding initial velocities for the respective drop orientation were assigned to all the components of the sub-model.

3. Damage Detection

Repeatability of drop orientation is critical to measuring a repeatable response. Small variations in the drop orientation can produce vastly varying transient-dynamic board responses. Significant effort was invested in developing a repeatable drop set-up. The drop height is varied to obtain shock pulses of various magnitudes, in addition to the 1500 Gs, 0.5 millisecond duration, half-sine input pulse (Figure 6). In addition to JEDEC drop-test, free-drop test of the printed circuit assemblies can be studied with motion control setup of the drop-tower.

Figure 7 shows component locations on the test boards instrumented with strain sensors. Figure 8 shows strain and continuity data acquired during the drop event using a high-speed data acquisition system at 2.5 to 5 million samples per second. Figure 9 shows the drop-event simultaneously monitored with ultra high-speed video camera operating at 40,000 frames per second. Targets are mounted on the edge of the board to allow high-speed measurement of relative displacement during drop. [Lall 2004, 2005, 2006[a, b, c, d]]

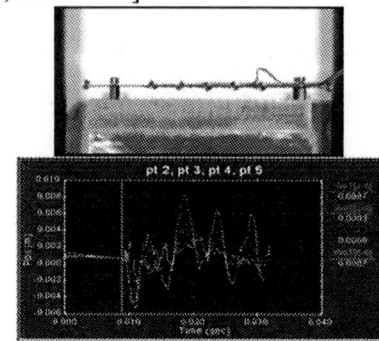

Figure 6: Measurement Relative Displacement During Impact. [Lall 2005, 2006[b]]

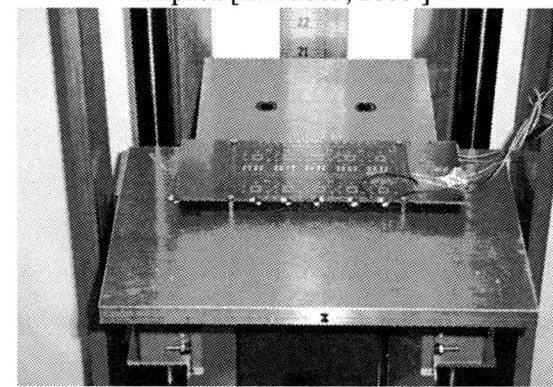

Figure 7: Experimental Set-up for Controlled Drop. [Lall 2005, 2006[b]]

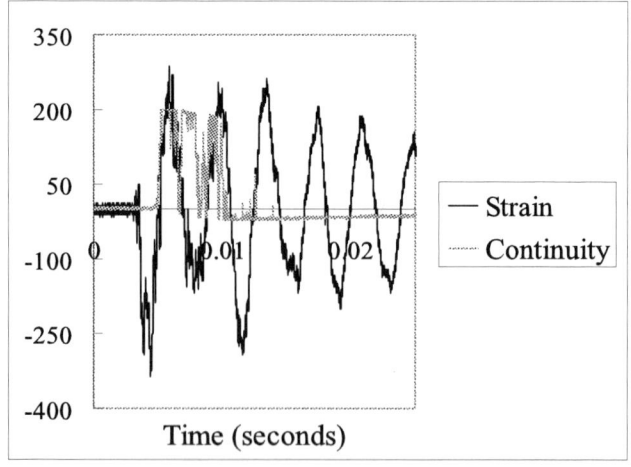

Figure 8: Package Strain and continuity transient history in JEDEC drop-shock, for the 8mm, 95.5Sn4Ag0.5Cu ball-grid array. [Lall 2005, 2006[b]]

Figure 8 shows strain, displacement, orientation angle, velocity, acceleration, and continuity data acquired simultaneously. Figure 6 shows a typical relative displacement plot measured during the drop event using an image tracking software to quantitatively measure displacements. The position of the vertical line in the plot represents the present location of the board (i.e. just prior to impact in this case) in the plot with "pos (m)" as the ordinate axis. The plot trace subsequent to the white scan is the relative displacement of the board targets w.r.t. to the specified reference. Figure 7 shows the board instrumentation for strain and relative displacement during horizontal JEDEC drop. In addition to relative displacement, velocity, and acceleration of the board prior to impact was measured. This additional step was necessary since, the boards were subjected to a controlled drop, in order to reduce variability in drop orientation.

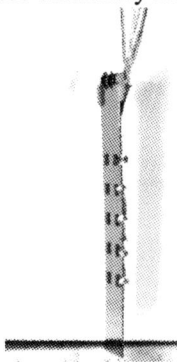

Figure 9: Test Set-Up for Free-Drop of Printed Circuit Assemblies. [Lall 2005, 2006[b]]

Figure 8 shows the strain history during JEDEC drop-shock for a 8mm 95.5Sn4Ag0.5Cu ball-grid array package. Failure in the device has been identified as an increase in voltage drop. Different locations on the test board exhibit different strain histories during the same drop and different number of drops to failure. However, the strain histories are very consistent and repeatable at the same component location on the test board for various drops. The strain history is also very repeatable for the same component location across various test boards.

4. Model Correlation

In this section, the field quantities and derivatives of field-quantities have been compared from both various explicit finite element models and experimental data in free-drop and JEDEC-drop. In addition, the transient strain histories of the solder interconnects have been correlated to location of failure. Susceptibility of the CSPs to chip-fracture in shock or drop has been investigated using the modeling approach.

Correlation of Transient Mode-Shape, Peak Relative Displacement, and Peak Strain

Figure 10 shows the transient mode shapes of the printed circuit board from high-speed video and explicit finite-element simulation at 2.4 ms and 4.8 ms after impact. The model prediction shows good correlation with the experimentally observed mode shapes. The peak relative-

displacements (Table 2) have been correlated between smeared property and conventional-shell models at the various locations along the board length. The error in the predicted values is in the neighborhood of 8-25% for both models.

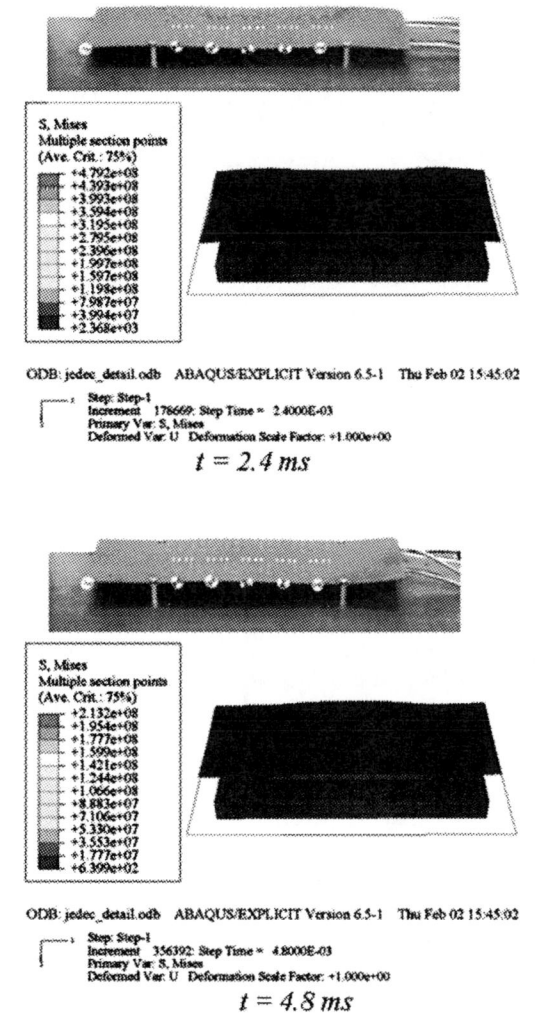

Figure 10: Correlation of Transient Mode-Shapes during JEDEC-Drop. [Lall 2006[a, b]]

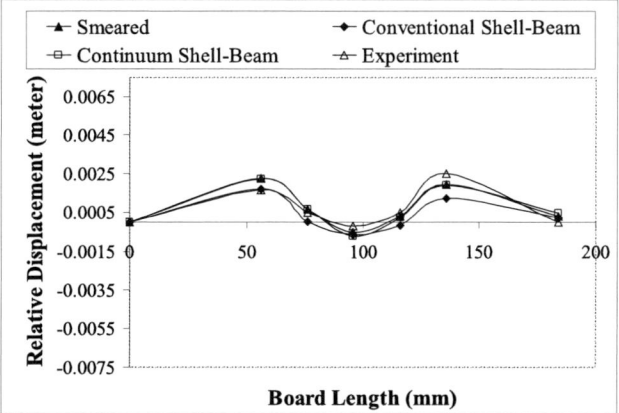

Figure 11: Correlation Between Experimental Relative Displacement of Board Assembly at 2.4 ms with Model Predictions under zero-degree JEDEC drop-test. [Lall 2006[a, b]]

Table 2: Correlation of Peak Relative-Displacement Values with high-speed experimental data in zero-degree JEDEC Drop (mm). [Lall 2006[a, b]]

	Loc E1	Loc E3	Loc E5
Experiment (mm)	3.61	4.47	4.58
Smeared Property	3.86	3.35	3.39
Error (%)	-7.03	24.93	25.86
Timoshenko-Beam, Continuum Shell	3.80	4.16	3.26
Error (%)	-5.17	6.97	28.73
Timoshenko-Beam, Conventional Shell	4.43	4.85	4.15
Error (%)	-22.8	-8.62	9.21

Figure 11 shows the correlation of the board relative displacement 2.4 ms after impact, from high-speed image analysis with the model predictions from smeared, continuum-shell with Timoshenko-beam, conventional-shell with Timoshenko-beam models. The peak strain values have also been correlated for the models versus experimental data in 90-degree orientation free-drop. The peak strain values (Table 3) exhibit error in the rage of 10-30%. All the three-modeling approaches including smeared properties, conventional-shell with beam elements, and continuum-shell with beam elements exhibit similar results. The wires on the right-side of the board are for strain and continuity measurement during the shock-event.

Table 3: Correlation of Peak-Strain Values from Model Predictions Versus Experiments for 90-degree Free-Drop. [Lall 2006[a, b]]

	Loc C1	Loc C3	Loc C5
Experiment (microstrain)	1417	2248	1667
Smeared Property	1603	1563	1424
Error (%)	-13.15	30.48	14.56
Timoshenko-Beam Continuum Shell	1820	1990	1960
Error (%)	-28.47	11.49	-17.60
Timoshenko-Beam Conventional Shell	1760	1630	2070
Error (%)	-24.24	27.50	-24.20

Solder Interconnect Strain Histories
Figure 12 and Figure 13 show the strain plots from the Timoshenko-Beam Element with Conventional-Shell model prediction for the solder interconnection located at the outermost corner of the package and in the solder interconnect located at the corner of the fourth-row from the outside, during a 0° JEDEC-drop. Plots indicate that the transient strain history is very different at the four-corners of the chip-scale package. Therefore, the susceptibility of the solder interconnects to failure may be different in different corners.

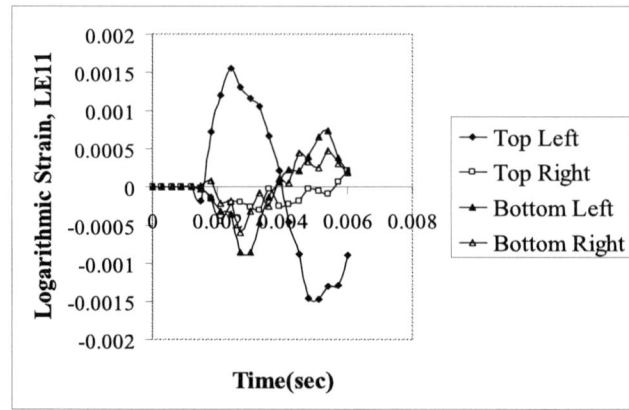

Figure 12: Timoshenko-Beam Element with Conventional-Shell Model Prediction of Transient Strain History at the Package Corner Solder Interconnect during 0° JEDEC-Drop. [Lall 2006[a, b]]

A comparison of transient strain histories in Figure 12 and Figure 13 reveals that, a large portion of the strain is carried by the outside row of the solder interconnects. For this reason, the explicit sub-model includes reduced-integration hexahedral elements for the corner solder interconnect. The beam elements allow output of the axial strain and the transverse shear strain only. The hexahedral element mesh solder interconnects provide insight into the logarithmic strain, LE23, distribution in the solder interconnects. Model results indicate that the strains are maximum at the solder-joint to package interface and the solder-joint to printed circuit board interface, indicating a high probability of failure at these interfaces. Failure analysis of the samples reveals that the observed failure modes correlate well with the model predictions. (Figure 15)

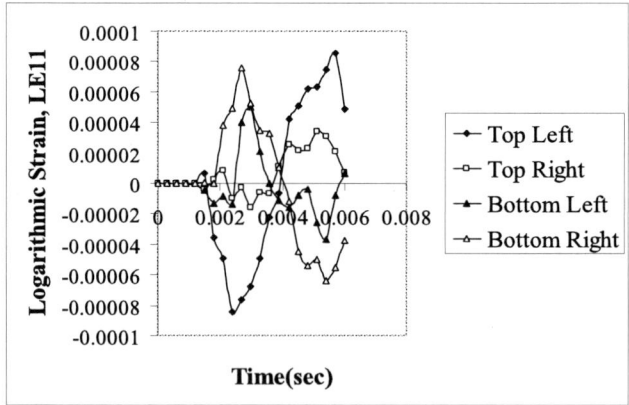

Figure 13: Timoshenko-Beam Element with Conventional-Shell Model Prediction of Transient Strain History in the Solder Interconnect Located at the Corner of the 4-Inner Row from Outside during 0° JEDEC-Drop. [Lall 2006[a, b]]

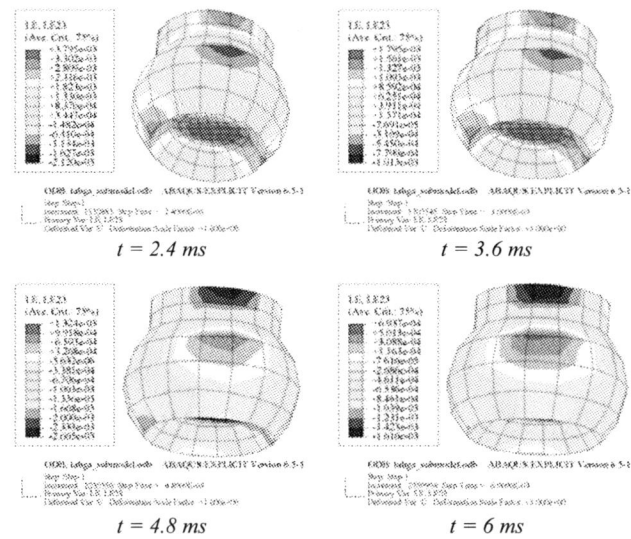

t = 2.4 ms   *t = 3.6 ms*

t = 4.8 ms   *t = 6 ms*

Figure 14: Global-Local Explicit Sub-Model Predictions of Transient Logarithmic Shear Strain, LE23, in the Solder Interconnect of one of the Chip-Scale Packages on the Printed Circuit Board Assembly during a Zero-Degree JEDEC-Drop. [Lall 2006[a, b]]

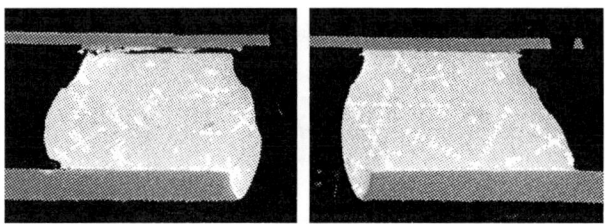

Figure 15: Cross-section of corner solder interconnect in the failed samples showing higher susceptibility of the samples to fail at the package-to-solder interconnect interface or the solder-to-printed circuit board interface. [Lall 2006[a, b]]

5. Wavelet Decomposition for Transient Analysis

Wavelets have been used in several areas including data and image processing [Martin 2001], geophysics [Kumar 1994], power signal studies [Santoso 1996], meteorological studies [Lau 1995], speech recognition [Favero 1994], medicine [Akay 1997], and motor vibration [Fu 2003, Yen 1999]. However, the application of wavelets to analysis of transient-response of electronics under shock and vibration is new. In this paper, wavelets and wavelet transforms have been used to analyze transient signals acquired during drop-impact of printed circuit board assemblies. (Figure 16)

Wavelets based time-frequency analysis is specifically useful to analyze non-stationary signals. A time-frequency representation describes simultaneously when a signal component occurs and how its frequency spectrum develops with time, so as to extract the transients or sudden spikes in the signal. Wavelet transform is more suitable to determine the frequency spectrum of the transient strain, acceleration or displacement signals as the Fourier transform extracts frequency information for the complete duration of the stress signal, using sine and cosine functions that are uniform in time. The Fourier Transform does not contain any time dependence of the signal and therefore cannot provide any local information regarding time evolution of its spectral characteristics. Transient impulses often occur as discontinuities in the signal. Representation of the local characteristics of signal in Fourier Transform is very inefficient and requires large number of Fourier Components.

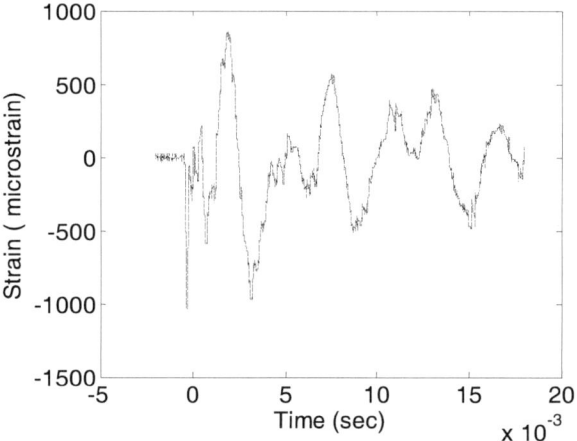

Figure 16: Transient Strain During Drop-Impact of Printed Circuit Assembly From 6ft. [Lall 2005, 2006[b]]

One candidate for extracting the local frequency information from a transient strain signal could have been the Windowed Fourier Transform. However, in a Windowed-Fourier Transform, a time series is examined under a fixed time-frequency window, i.e. the resolution or interval is constant in both time and frequency domains. In transient dynamics, a wide range of frequencies are involved and a fixed time window of the Windowed-Fourier Transform tends to have a large number of high-frequency cycles and a few low-frequency cycles or parts of cycles. This causes over-representation of the high-frequency components and an under-representation of the low-frequency components. Therefore, different resolutions are required to analyze a variety of signal components of different duration. The accuracy of extracting frequency information is limited by the length of the window relative to the duration of the singularity in the signal.

The use of wavelet transforms enables the examining of the transient signal shown in Figure 16, at different time windows and frequency bands, i.e. at different time resolutions and frequency resolutions, by controlling translation and dilation of wavelets and achieve optimal resolution with the least number of base functions. The size of the time window is controlled by the translation or positioning of the wavelet while the width of the frequency band is controlled by the dilations or scaling of the wavelet. In this case, wavelets therefore enable higher frequency-resolution and lower time-resolution in low frequency part, and at the same time enable lower frequency-resolution and higher time resolution in high frequency part. The wavelet transform is defined by [Lall 2005, 2006[b]]

$$Wf(u,s) = \langle f, \psi_{u,s} \rangle = \frac{1}{\sqrt{s}} \int_{-\infty}^{+\infty} f(t) \, \psi^* \left(\frac{t-u}{s} \right) dt \qquad (4)$$

where the base atom ψ^* is the complex conjugate of the wavelet function which is a zero average function, centered around zero with a finite energy. The function f(t) is decomposed into a set of basis functions called the wavelets with the variables s and u, representing the scale and translation factors respectively.

Variable Amplitude Transient-Load Analysis

In transient-shock and drop, the loads vary in both amplitude and frequency. Electronic structures very rarely experience constant amplitude loading. To analyze the structures in operating conditions, strain measurements and relative displacement measurements taken at specific locations, have been analyzed using Daubechies, D_{10} wavelet with 12-level decomposition.

For cycle-counting analysis the second approximation, A_2 has been used as input to the rainflow algorithms. In rainflow analysis, the strain time history data is drawn with time axis vertical with increasing time downwards. Rainflow cycles are then defined analogous to rain falling down the roof. Detailed rules for cycle counting are described in [ASME 1997, Bannantine 1990, and Downing 1982]. A flow of rain is begun at each strain reversal in the history and is allowed to continue to flow unless, (a) the rain began at a local maximum point (peak) and falls opposite a local maximum point greater than that from which it came. (b) the rain began at a local minimum point (valley) and falls opposite a local minimum point greater (in magnitude) than that from which it came. (c) it encounters a previous rainflow.

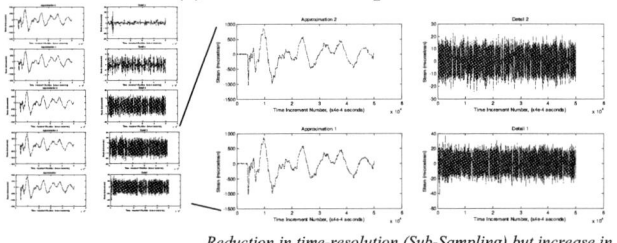

Reduction in time-resolution (Sub-Sampling) but increase in frequency resolution. (Spans ½ Frequency Band)

Figure 17: Approximation and Details 1-6 for a 12-Level Decomposition for Transient Strain Signal During Drop, Based on Daubechies 10 Wavelet.

The transient-dynamic data in time-domain has been reduced into histograms of load cycle amplitudes and number of cycles. Number of cycles is calculated using cycle counting algorithm for the transient-strain history. Figure 18 is the histogram for one of the drops to failure showing the number of cycles for a specific strain amplitude during the transient signal. The transient strain signal has a high number of very small strain amplitude cycles and very few number of large strain amplitude cycles. Damage from both repeatable and non-repeatable drops has been analyzed. For repeatable drops, the strain

histograms are very similar. Damage has been computed for each component, till failure.

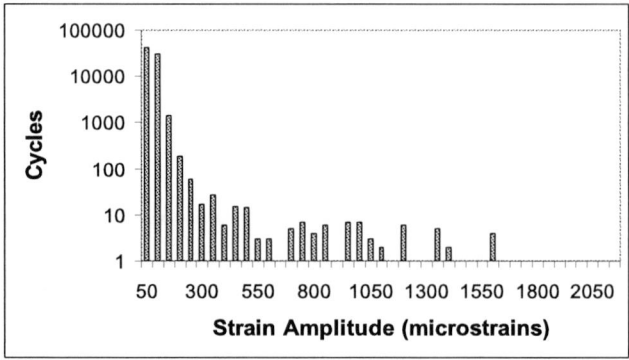

Figure 18: Cycles Versus Strain Amplitude for Transient Strain History from Rainflow Algorithm. Histogram is for 95.5Sn4Ag0.5Cu Solder Joint Failures during the Drop for 132 I/O, 8 mm Ball Grid Array. [Lall 2005, 2006[b]]

All information about the sequence of the individual strain variation is lost during counting. The resulting cumulative frequency distribution histogram, gives the overall number of load cycles for each load amplitude. A relative damage index has been defined such that the damage magnitude at failure is "1". Linear superposition of damage has been assumed in this study. The equation can be re-written as follows based on the assumed logarithmic relationship between strain and number of cycles (Coffin-Manson Relationship), [Lall 2005, 2006[b]]

$$\sum_{k=1}^{M} \frac{D_k}{D} = 1 \qquad \sum_{k=1}^{M} \frac{N_k}{A \left(\frac{\Delta \varepsilon_k}{2} \right)^n} = 1 \qquad (5a, b)$$

where, "k" is the bin-index for the histogram, $\Delta \varepsilon / 2$ is the printed circuit board strain amplitude, M is the total number of bins in the histogram, N is the number of cycles subjected on the sample in the kth histogram bin during all the drops until-failure of the device, and D is the damage index. Data from data sets has been solved as follows, [Lall 2005, 2006[b]]

$$\left(\sum_{k=1}^{N} \frac{N_k}{A \left(\frac{\Delta \varepsilon_k}{2} \right)^n} \right)_i = 1, \quad \left(\sum_{k=1}^{N} \frac{N_k}{A \left(\frac{\Delta \varepsilon_k}{2} \right)^n} \right)_{i+1} = 1 \qquad (6a, b)$$

where the index "i" indicates the drop number. Each data-set includes test-vehicles which have been dropped-to-failure, therefore the cumulative relative damage index is 1. The load histories have been varied by varying the angle of impact by a small magnitude. Subtracting the two equations, we get,

$$\left(\sum_{k=1}^{N} \frac{N_k}{\left(\frac{\Delta \varepsilon_k}{2} \right)^n} \right)_i - \left(\sum_{k=1}^{N} \frac{N_k}{\left(\frac{\Delta \varepsilon_k}{2} \right)^n} \right)_{i+1} = 0 \qquad (7)$$

700

Average values of the exponent, "n" and coefficient "A" have been computed over the complete data-set. Since, the values have been computed for board strain or package strain, they are specific to the test structure analyzed. The relative damage in any particular drop is computed based on total damage at failure (Figure 19, Figure 20). The advantage of the proposed approach is that it can be used to calculate damage in the test structures of interest, instead of an idealized test specimen. The test samples are cross-sectioned after failure, and the test data was sorted based on failure modes.

6. Definition of the Failure Envelope and Model Correlation

The relative damage index has been used to predict the number of drops to failure for the 8 mm Ball Grid Array, 95.5Sn4Ag0.5Cu, 132 I/O. The location for the predictions is different from location at which the experimental data was acquired. Therefore, the transient-strain history and the damage progression is also different.

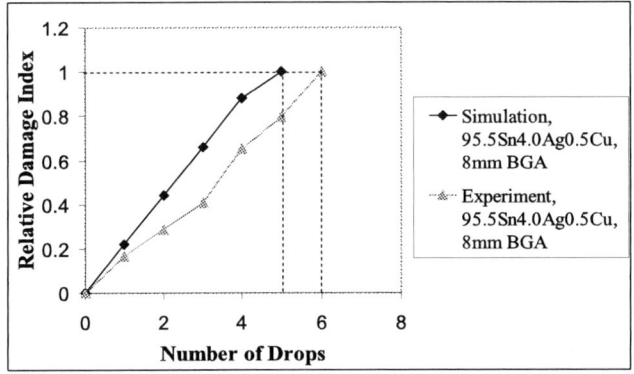

Figure 19: Correlation of Damage Progression and Number of Drops to Failure Between Experiment and Simulation for the 8 mm, 95.5Sn4Ag0.5Cu, 132 I/O Ball Grid Array. [Lall 2005, 2006[b]]

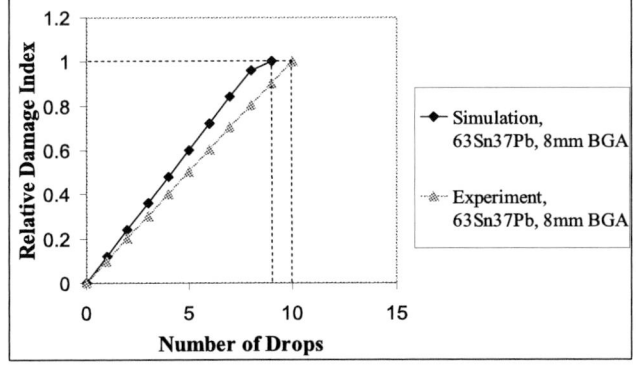

Figure 20: Correlation of Damage Progression and Number of Drops to Failure Between Experiment and Simulation for the 8 mm, 63Sn37Pb, 132 I/O Ball Grid Array. [Lall 2005, 2006[b]]

Model prediction indicates 5 drops-to-failure and correlates well with the experimental data, which indicates that the ball-grid arrays at the location of

measured transient strain failed after 6 drops. The present data-set is one has been chosen for correlation, because it is a representative average of the damage progression at this location. Number of drops-to-failure at this location ranged between 5-7 drops. The correlation is for a non-repeatable drop, indicated by the non-uniform damage progression in experimental data-set (Figure 19).

Model predictions and correlation with experimental data for the 63Sn37Pb, 8mm BGA is shown in Figure 20. The model predictions indicate failure after 9 drops-to-failure at the location of the transient strain trace. Representative average experimental data for the failure location exhibits failure after 10 drops. The relative damage index approach outlined in this paper provides a method to define the failure envelope for packaging architectures. The solder interconnect strain is not as easily measurable as board strain. The proposed methodology enables evaluation of the failure envelope in the application of interest and for the packaging architecture in question. Damage during the life of the product should not exceed "1" for the design to have good survivability in drop and shock applications. The constants used for damage progression are specific to the package architecture and boundary conditions. The proposed methodology is amenable to implementation not just at board-level but also at system-level.

7. Summary and Conclusions

Four explicit modeling techniques have been investigated for modeling shock loading of printed circuit board assemblies. The focus of the paper is on modeling multiple scales from first-level interconnects to assembly-level transient dynamics. Modeling techniques investigated include, smeared properties, Timoshenko-beam with Conventional Shell Elements, Timoshenko-Beam with Continuum Shell Elements, and Explicit Sub-modeling. The paper extends the state-of-art, which presently focuses on prediction of interconnect stresses based on assumptions of symmetry of geometry and boundary conditions. In this paper, modeling techniques have been developed to capture system-level dynamics in addition to interconnect transient-stress and transient-strain histories, without any assumption of assembly symmetry, has been demonstrated. The ability to eliminate symmetry assumptions enables the modeling of asymmetric modes in addition to symmetric modes. The model predictions have been correlated with experimental data from high-speed video, high-speed image analysis and high-speed strain acquisition.

Methodology for development of the failure envelope for area-array packaging architectures has been presented. Wavelet transforms which have been used extensively in several areas including data and image processing, geophysics, power signal studies, meteorological studies, speech recognition, medicine, and motor vibration, have been applied to analysis of transient-response of electronics under shock and vibration. Wavelets have been used to avoid the problem of fixed time-frequency window with windowed Fourier transforms, which causes over-representation of the high-frequency components and an under-representation of the low-frequency components of a transient drop-impact signal. The need of different resolutions required to analyze a variety of signal components of different duration has been addressed by using wavelet transforms. The Daubechies D_{10} wavelet with 12-level decomposition has been used to analyze non-stationary transient dynamic signals.

A relative damage index has been developed for prediction of the number of drops-to-failure under transient loads. The research presented attempts to address the need for techniques and damage proxies which enable the determination of failure-envelopes and cumulative damage during overstress and repetitive loading for various packaging architectures. The approach is based on assembly strains, since there are experimental limitations of measuring field-quantities and their derivatives at the board-solder joint interface, primarily because of the small size of interconnects in fine-pitch ball-grid array packages. Explicit finite element models in conjunction with the proposed approach have been used to predict survivability of fine-pitch ball-grid arrays in transient-shock and vibration. The approach has been applied to both leadfree (95.5Sn4.0Ag0.5Cu) and leaded (63Sn37Pb) ball-grid array architectures. The validation has been presented for both repeatable and non-repeatable drops.

Acknowledgments

The work presented here in this paper has been supported by a research grant from the Semiconductor Research Corporation (SRC), Research ID 1283.

References

Abaqus Documentation, Finite Elements and Rigid Bodies, Getting Started with Abaqus Explicit: Keywords Version, Version 6.5, Section 4.1, 2005[a]

Abaqus Documentation, Beam Element Overview, Abaqus Theory Manual, Version 6.5, Section 3.5.1, 2005[b]

Akay, M.; Wavelet Applications in Medicine; IEEE Spectrum, Vol: 34 , Issue: 5, pp. 50 – 56, 1997.

American Society of Mechanical Engineers (ASME), Standard Practices for Cycle Counting in Fatigue Analysis, ASTM E1049-85, 1997.

Bannantine, J. A., Comer, J.J., Handrock, J.L., "Fundamentals of Metal Fatigue Analysis", Prentice Hall, 1990.

Downing, S. D., and Socie, D. F., Simple Rainflow Counting Algorithms, International Journal of Fatigue, Vol. 4, No. 1, pp. 31-40, 1982.

Chai, T.C., Quek, S., Hnin, W.Y., Wong, E.H., Board-Level Drop Test Reliability of IC Packages, Electronic Components and technology Conference, Orlando, Florida, pp. 630-636, May 31-June 3, 2005

Chiu, T.C., Zeng, K., Stierman, R., Edwards, D., Ano, K., Effect of Aging on Board Level Drop Reliability for Pb-Free BGA Packages, Electronic Components and

technology Conference, Las Vegas, Nevada, pp. 1256-1262, June 1-4, 2004

Chong, D. Y. R., Ng, K., Tan, J. Y. N., Low, P. T. H., Pang, J. H. L., Che, F. X., Xiong, B. S., Xu, L. Drop Impact Reliability Testing for Lead Free and Leaded Soldered IC Packages, Electronic Components and technology Conference, Orlando, Florida, pp. 622-629, May 31-June 3, 2005.

Favero, R.F.; Compound Wavelets: Wavelets For Speech Recognition, Proceedings of the IEEE-SP International Symposium on Time-Frequency and Time-Scale Analysis, pp. 600 – 603; 1994.

Fu, Y., Wen-Sheng Li; Guo-Hua Xu; Wavelets and singularities in bearing vibration signals; International Conference on Machine Learning and Cybernetics, Vol. 4, Pages:2433 - 2436, 2003.

Hannan, N., Kujala, A., Mohan, V., Morganelli, P., Shah, J., Investigation of Different Options of Pre-Applied CSP Underfill for Mechanical Reliability Enhancements in Mobile Phones, Electronic Components and technology Conference, Orlando, Florida, pp. 770-774, May 31-June 3, 2005

Irving, S., Liu, Y., Free Drop Test Simulation for Portable IC Package by Implicit Transient Dynamics FEM, 54th Electronic Components and Technology Conference, pp. 1062 – 1066, 2004.

JEDEC Solid State Technology Association, Board-Level Drop Test Method of Components for Handheld Electronic Products, No. JESD22-B111, 2003.

Kumar, P, Georgiou, E. F. , Wavelets in Geophysics, Wavelets and its Applications, Academia Press, 1994.

Lall, P., Gupte, S., Choudhary, P., Suhling, J., Solder-Joint Reliability in Electronics Under Shock and Vibration using Explicit Finite Element Sub-modeling, Proceedings of the 56th IEEE Electronic Components and Technology Conference, San Diego, California, pp.428-435, May 30-June 2, 2006[a].

Lall, P., Panchagade, D., Iyengar, D., Suhling, J., Life Prediction and Damage Equivalency for Shock Survivability of Electronic Components, Proceedings of the ITherm 2006, 10th Intersociety Conference on Thermal and Thermo-mechanical Phenomena, San Diego, California, pp.804-816, May 30-June 2, 2006[b].

Lall, P., Choudhary, P., Gupte, S., Health Monitoring for Damage Initiation & Progression during Mechanical Shock in Electronic Assemblies, Proceedings of the 56th IEEE Electronic Components and Technology Conference, San Diego, California, pp.85-94, May 30-June 2, 2006[c].

Lall, P., D. Panchagade, Y. Liu, R. W. Johnson, and J. C. Suhling, Models for Reliability Prediction of Fine-Pitch BGAs and CSPs in Shock and Drop-Impact, *IEEE Transactions on Components and Packaging Technologies*, Volume 29, Number 3, pp. 464-474, September 2006[d].

Lall, P., Panchagade, D., Choudhary, P., Suhling, J., Gupte, S., Failure-Envelope Approach to Modeling Shock and Vibration Survivability of Electronic and MEMS Packaging, Electronic Components and

Technology Conference, Orlando, FL, pp. 480-490, June 1-3, 2005.

Lall, P., Panchagade, D., Liu, Y., Johnson, W., Suhling, J., "Models for Reliability Prediction of Fine-Pitch BGAs and CSPs in Shock and Drop-Impact", 54th Electronic Components and Technology Conference, pp. 1296 – 1303, 2004.

Lau, K.-M., and H.-Y. Weng, Climate Signal Detection Using Wavelet Transform: How to Make a Time Series Sing, Bulletin of the American Meteorological Society, No. 76, pp. 2391–2402, 1995.

Lim, C.T. and Low, Y.J., "Drop Impact Survey of Portable Electronic Products," 53rd Electronic Components and Technology Conference, pp. 113-120, 2003.

Lim, C.T. and Low, Y.J., "Investigating the Drop Impact of Portable Electronic Products," 52nd Electronic Components and Technology Conference, pp. 1270-1274, 2002.

Martin, M.B.; Bell, A.E.; New Image Compression Techniques Using Multiwavelets and Multiwavelet Packets; IEEE Transactions on Image Processing, Vol: 10, Issue: 4, Pages: 500 – 510; 2001

Mishiro, K., "Effect of the Drop Impact on BGA/CSP Package Reliability," Microelectronics Reliability Journal, Vol. 42(1), pp. 77-82, 2002.

Morganelli, P., Shah, J., Wheelock, B., Mohan, V., Laffey, M., Partial Pre-Applied Underfill for Assembly and Reliability of Pb-Free CSPs, Electronic Components and technology Conference, Orlando, Florida, pp. 223-227, May 31-June 3, 2005.

Pitaressi, J., Roggeman, B., Chaparala, S., Mechanical Shock Testing and Modeling of PC Motherboards, Electronics Components and Technology Conference, Las Vegas, Nevada, pp. 1047 – 1054, June 1-4, 2004.

Ren, W., Wang, J., Reinikainen, T., Application of ABAQUS/Explicit Submodeling Technique in Drop Simulation of System Assembly, Electronic Packaging Technology Conference, pp. 541-546, 2004.

Ren, W., Wang, J., Shell-Based Simplified Electronic Package Model Development and its Application for Reliability Analysis , Electronic Packaging Technology Conference, pp. 217-222, 2003.

Santoso, S.; Powers, E.J.; Grady, W.M.; Hofmann, P.; Power quality assessment via Wavelet transform analysis; IEEE Transactions on Power Delivery, Volume: 11 , Issue: 2, Pages: 924 – 930; 1996.

Saha, S. K., Mathew, S., Canumalla, S., Effect of Intermetallic Phasis on Performance in a Mechanical Drop Environment: 96.5Sn3.5Ag Solder on Cu and Ni/Au Pad Finishes, Electronic Components and technology Conference, Las Vegas, Nevada, pp. 1288-1295, June 1-4, 2004.

Shah, K., Mello, M., Ball Grid Array Solder Joint Failure Envelope Development for Dynamic Loading, Electronic Components and technology Conference, Las Vegas, Nevada, pp. 1067-1074, June 1-4, 2004

Sogo, T. and Hara, S., "Estimation of Fall Impact Strength for BGA Solder Joints," ICEP Conference Proc., Japan, pp. 369-373, 2001.

Syed, A., Kim, S. M., Lin, W., Khim, J.Y., Song, E.K., Shin, J.H., Panczak, T., An Methodology for Drop Performance Prediction and Application for Design Optimization of Chip Scale Packages, Electronic Components and technology Conference, Orlando, Florida, pp. 472-479, May 31-June 3, 2005

Tan, L.B., Ang, C.W., Tan, V.B.C., Zhang, X., Modal and Impact Analysis of Modern Portable Electronic Products, Electronic Components and technology Conference, Orlando, Florida, pp. 645-653, May 31-June 3, 2005

Tee, T. Y., Luan, J., Pek, E., Lim, C. T., Zhing, Z., Advanced Experimental and Simulation Techniques for Analysis of Dynamic Responses during Drop Impact, Electronic Components and technology Conference, Las Vegas, Nevada, pp. 1088-1094, June 1-4, 2004

Tee, T. Y. , Hun Shen Ng, Chwee Teck Lim, Eric Pek , Zhaowei Zhong , "Board Level Drop Test and Simulation of TFBGA Packages for Telecommunication Applications" 53rd Electronic Components and Technology Conference, pp. 121-129, 2003.

Tian, G., Liu, Y., Johnson, W., Lall, P., Palmer, M., Islam, M. N., Corner Bonding of CSPs: Processing and Reliability, IEEE Transactions on Electronic Packaging Manufacturing, Volume 28, Number 3, pp. 231-240, July 2005.

Xie, D., Minna Arra, Dongkai Shangkai, Hoang Phan, David Geiger and Sammy Yi, "Life Prediction of Lead free Solder Joints for Handheld Products", Presented and published at Telecom Hardware Solutions Conference, Doubletree Hotel, Plano, Texas, USA, May 15-16, 2002.

Xie, D., Minna Arra,, Yi, S., Rooney, D., Solder Joint Behavior of Area Array Packages in Board-Level Drop for Handheld Devices, 53rd Electronic Components and Technology Conference, pp. 130 – 135, 2003

Wong, E. H., Lim, C. T., Field, J. E., Tan, V. B. C., Shim, V. P. M., Lim, K. T., Seah, S. K. W., Tackling the Drop Impact Reliability of Electronic Packaging, ASME International Electronic Packaging Technical Conference and Exhibition, July 6 -11, Maui, pp. 1 – 9, 2003.

Wu J., "Global and Local Coupling Analysis for Small Components in Drop Simulation," 6th International LSDYNA Users Conference, pp. 11:17 - 11:26, 2000.

Wu, Jason, Goushu Song, Chao-pin Yeh, Karl Wyatt, " Drop/Impact Simulation and Test Validation of Telecommunication Products", Intersociety Conference on Thermal Phenomena, pp. 330-336, 1998.

Yen, G.Y.; Kuo-Chung Lin; Wavelet packet feature extraction for vibration monitoring; Proceedings of the 1999 IEEE International Conference on Control Applications, Vol.2, Pages:1573 - 1578, 1999.

Zhu, L., "Submodeling Technique for BGA Reliability Analysis of CSP Packaging Subjected to an Impact Loading," InterPACK Conference Proceedings, 2001.

Zhu, L., "Modeling Technique for Reliability Assessment of Portable Electronic Product Subjected to Drop Impact Loads," 53rd Electronic Components and Technology Conference, pp. 100-104, 2003.

Zhu, L., Marccinkiewicz, W., Drop Impact Reliablity Analysis of CSP Packages at Board and Product System Levels Through Modeling Approaches, Inter Society Conference on Thermal and Thermo-mechanical Phenomena, pp. 296 – 303, 2004.

Low-temperature and Pressureless Sintering Technology for High-performance and High-temperature Interconnection of Semiconductor Devices

Guo-Quan Lu (gqlu@vt.edu)[a], Jesus N. Calata[a], Guangyin Lei[a], and Xu Chen[b]

[a]Department of Materials Science and Engineering, Virginia Tech, Blacksburg, VA, USA
[b]School of Chemical Engineering, Tianjin University, Tianjin, the People's Republic of China

Abstract

We present an interconnect technology based on low-temperature and pressureless sintering of a nanoscale metal paste to achieve high-performance and high-temperature packaging of semiconductor devices. The nanoscale metal paste, consisting of nanoparticles of silver mixed in an organic binder/solvent vehicle, can be sintered at temperatures close to 275°C. Measurements on electrical and thermal properties of the sintered die interconnect gave it at least five times better than the soldered or epoxied attachment. Die-shear tests of the sintered joints showed a bonding strength of about 25 MPa. The sintered joints exhibited excellent reliability in aging and temperature-cycling tests. Since silver melts at 961°C, the sintered interconnect can be used for wide band gap semiconductor devices (SiC or GaN), which are operable over 300°C where none of the existing solder alloys or epoxies can be used. In summary, the low-temperature sintering of nanoscale metal paste is shown to be a reliable, lead-free interconnect solution for high-temperature and high-performance packaging needs.

1. Introduction

Steady progresses made in the growth of wide bandgap semiconductor materials have led to the emergence of SiC devices for switching power electronics and GaN devices for solid-state lighting. These wide bandgap devices were shown[1-3] to have excellent properties with an added advantage of functioning at elevated temperatures in the 300°C – 500°C range. The latter opens the possibility of using these devices under extreme conditions since stringent heat sinking and cooling requirements can be relaxed. Specifically, the ability to use SiC devices at high temperatures would enable integration of power electronics circuits inside a motor, thus reducing weight and size of the motor system for automotive or aerospace applications. Moreover, the ability to support the junction temperature of a GaN LED at 350°C would allow much higher current density flowing through the device, resulting in a super bright but still reliable solid-state light source.

However, for these devices to be used under extreme conditions, high-temperature materials for interconnection and packaging have to be developed. Today, solder alloys—lead-tin and lead-free—are still the workhorse materials for interconnecting power semiconductor devices and GaN LEDs. Very few studies have addressed the need for high-performance and high-temperature interconnect solutions. Johnson *et. al.*[4] have extensively evaluated performance and reliability of

several hard solders, like gold-tin, gold-germanium, and gold-silicon eutectic alloys, as high-temperature die attachment. These solders can be processed using a typical solder-reflow oven or furnace with a maximum temperature higher than the solders' melting points. Currently, these eutectic alloys are the only materials of choice for high-end applications where devices have to be operated at a junction temperature above 125°C. An example of these applications is high-power diode laser module for telecommunications systems. However, these hard solders have much lower thermal and electrical conductivities compared to those of pure noble metals, like silver.

Chuan and Lee[5] reported a high-temperature interconnect solution based on alloying of silver and indium at low temperatures. In their study, pure silver and indium thin films were separately vacuum-deposited on a device and/or substrate. Then, the joining process took place by heating the device/substrate assembly to a temperature above the melting point of indium (157°C) to allow the formation of Ag-In alloy via atomic interdiffusion. From the Ag-In binary phase diagram [6], Ag-In solid solution can be made to melt at temperatures exceeding 400°C, thus making the technique a potential candidate for high-temperature die attachment that can be processed at a lower temperature. Unfortunately, the reported low-temperature alloying process took over 20 hours to complete, a process that is too costly to be practical. Like the gold-based solder alloys, the Ag-In alloy also has poor thermal and electrical properties. Finally, the existence of several brittle intermetallic phases in the binary system poses a serious reliability threat to the long-term performance of the joints.

To take advantage of the excellent properties and high melting temperatures of noble metals for attaching devices, Scheuermann [7] and Schwarzbauer [8] explored sintering of commercial silver pastes for joining devices to a substrate. Silver, silver-palladium, and copper pastes are widely used in the fabrication of hybrid and co-fired microelectronic packages. Unfortunately, these metal pastes, consisting of micron-size metal particles in organic binder/dispersant vehicles, have to be densified at temperatures over 650°C to be useful as a conductor. To lower the sintering temperature of silver pastes below 300°C, Schwarzbauer applied a quasi-hydrostatic pressure on the device/substrate assembly to speed up the sintering kinetics of silver paste at low temperatures. It was reported that with a pressure of about 40 MPa, the silver die-attach layer underwent significant densification at 250°C and had excellent

1-4244-1105-X/07/$25.00 ©2007 IEEE

thermal and electrical properties. The sintered silver joints were also found to be highly reliable. The technique had been successfully used in the fabrication of some experimental power electronic modules at Semikron.

Recently, we reported our study [9] on the pressure-assisted sintering of commercial silver pastes for high-temperature interconnect. Our results confirm most of Schwarzbauer's findings. The sintered silver attachment had a density of 80% and thermal and electrical conductivities about half of the bulk values. However, our experience with the technique shows that the need for pressure significantly complicates the die-attach process and subjects brittle semiconductor materials to possible microscopic cracking. These are the likely reasons why manufacturers are still concerned about using the pressure-assisted sintering technique in production.

In this article, we present a low-temperature (as low as 275°C), pressureless (i.e. no externally applied pressure) sintering technology that utilizes nanosilver paste to provide superior electrical, thermal and mechanical properties, and high-temperature capability for device interconnection. Some of this work has been published earlier in references [10-12]. Here, we present recent advances in the development.

2. Experimental

Details on the making of the nanoscale silver paste used in this study have been presented elsewhere[12, 13]. The material is being commercialized by NBE Technologies, LLC, 2200 Kraft Drive, Suite 1425, Blacksburg, VA 24060, USA and can be obtained by contacting nbetech@verizon.net. In this section, we describe the making of samples used for measuring the electrical conductivity of sintered silver, the bonding strength of sintered die-to-substrate joint, and the joint reliability.

2.1 Sintering time-temperature profile

Figure 1 shows an exemplary time-temperature profile used to sinter the nanosilver paste. The entire process has to be done in air to burn-out organics in the paste. The four drying steps at 50°C, 75 °C, 100°C, and 125°C are necessary to gradually remove all the solvents in the paste prior to sintering. These incremental heating steps were accomplished with a Sikama solder reflow belt furnace. The rapid ramp-up to sintering temperatures is recommended to minimize coallesce of the nanoparticles prior to densification, which is initiated [13] by the removal of organic binder molecules in the material. We achieved rapid ramp-ups by transferring samples at 125°C directly onto the hot plate set at one of the two sintering temperatures. For substrates and devices whose bonding surfaces are metallized with silver, sintering takes place at 275°C for 10 mins,while 325°C is needed for gold metallized surfaces to enhance interdiffusion between silver and gold.

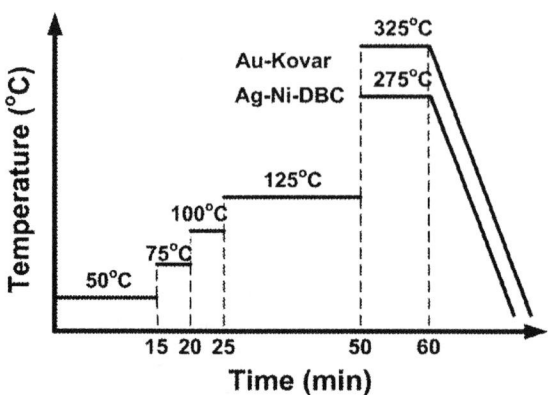

Figure 1: Recommended sintering time-temperature profile for applying the nanosilver paste to die-attach devices.

2.2 Sample Preparation

Figure 2 shows pictures of samples prepared as resistor patterns for measuring the electrical conductity of the sintered silver. The resistor pattern was formed by stencil-printing the paste on a bare alumina substrate. The print thickness was about 20 μm. After sintering, the film shrank mainly through its thickness to roughly 5~8 μm. Four-point probe test was used to get accurate measurement of the resistance. Then, from mesurements of the film thickness and width, we calculated the resistivity of the sintered silver. Finally, using the Wiedemann-Franz law [14], we estimated its thermal conductivity.

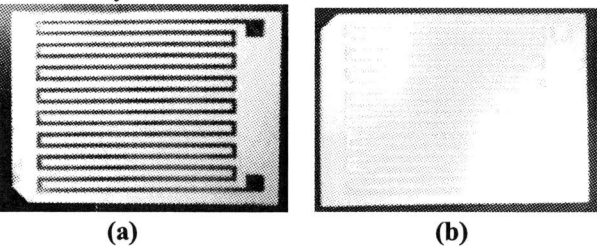

(a)   **(b)**

Figure 2: (a) Stencil-printed nanosilver paste resistor pattern on a bare alumina substrate (1.7 inch x 2.3 inch); (b) after sintering with final thickness of about 40 μm.

Figure 3 shows representative pictures of devices attached on substrates. Silicon and silicon carbide devices ranging from 4 mm² to 16 mm² were used. Both had silver as backside metal for attchment. Four types of substrate were used: (1) alumina direct-bond-copper (DBC) electroplated with about 10 μm nickel and then 10 μm silver; (2) alumina DBC electroplated with about 10 μm nickle and then 1 μm gold; (3) Kovar metallized with a few μm-thick gold; and (4) copper electroplated with 10 μm silver. To attach the devices, we first printed the paste onto a substrate using a stencil or a spacer to a thickness between 50 μm and 100 μm, and then pressed the devices, one at a time, on the wet silver print with a little wiggling/sliding motion to ensure good wetting on device backside. Here, too much pressing pressure (> 400

706

kPa) should be avoided to prevent squeezing out the paste and burying or shorting the device.

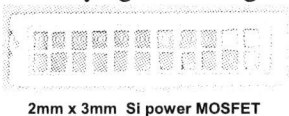

2mm x 3mm Si power MOSFET
on Ag-coated DBC

2mm x 2mm SiC devices on
Au-coated DBC

4mm x 4mm SiC devices on
Ag-coated Copper

2mm x 3mm Si MOSFET on Kovar

Figure 3: Examples of attached silicon and SiC devices on four types of substrates using the nanosilver paste and the sintering process.

As described in the previous section, devices attached on silver metallized substrates were sintered at 275°C for 10 mins, while those on gold metallized substrates were sintered at 325°C for 10 mins. After sintering, some attached devices were sheared off, using an in-house built tester, to determine their bonding strengths. Some of the devices attached on silver-metallized DBC were subjected to aging at 300°C in an oven, then at different times were taken out and sheared off to determine the effect of aging on bonding strength. We decided on 300°C for aging because it is high enough—close to 2/3 of silver melting point (ratio of the absolute temperatures)—for microstructure annealing and is low enough to avoid extensive interdiffusion between electroplated metals and underlying copper on the DBC substrate.

Finally, devices mounted on Ag-metallized DBC and Au-coated Kovar substrates were subjected to temperature cycling between -40°C and 150°C to evaluate the reliability of the sintered joints. The temperature cycling was carried out in a Tenney Jr. temperature cycling chamber. A soak time at the maximum and minimum temperatures of about 5 minutes each was included in the cycle. This corresponds to test condition M and soak mode 2 in the JEDEC standard temperature cycling specifications (JEDEC Standard No. 22-A104-B). The chamber is limited to less than 1 cycle per hour (condition G) such that we are unable to test for higher cycle rates.

3. Results and Discussion

Electrical resistance measurements of the sintered resistor patterns gave an average electrical conductivity of 4×10^7 $(\Omega\text{-m})^{-1}$, compared to 6.3×10^7 $(\Omega\text{-m})^{-1}$ for bulk silver [15]. The corresponding thermal conductivity calculated by the Wiedemann-Franz law is 290 W/m-K versus 459 W/m-K for the bulk [16]. The nearly 40% lower conductivity values found in the sintered films are attributed to a substantial porosity (about 20%) remaining in the sintered microstructure. Figure 4 is a typical scanning electron micrograph of the sintered microstructure showing a uniformly distributed micron-size pores. It is expected that denser microstructures can be achieved by going to higher sintering temperatures and

longer times. But, we argue that the porous microstructure is advantageous for reliability because it lowers the elastic modulus of the sintered attachment. Attachment with lower-modulus material is more compliant to a large difference in coefficients of thermal expansion (CTEs) between the device and substrate, thus resulting in lower thermo-mechanical stresses at the joint. In our earlier work [12], we found an elastic modulus of 10 GPa for the porous material versus 67 GPa for the bulk silver.

Figure 5 shows the typical values obtained on the die-shear strengths from Ag- and Au-coated substrate surfaces. In both cases, the die-shear strengths are generally greater than 20 MPa and averaged at around 25 MPa, which is comparable to the published die-shear strengths obtained in our earlier work [11] and those found in soldered [17] and epoxied joints [18]. The larger scatter in the data is believed to come from lack of precise control over the device placement done by hand and feel. Examination of sheared-off surfaces revealed attachment silver left on the device and substrate suggesting a cohesive failure in the sintered material.

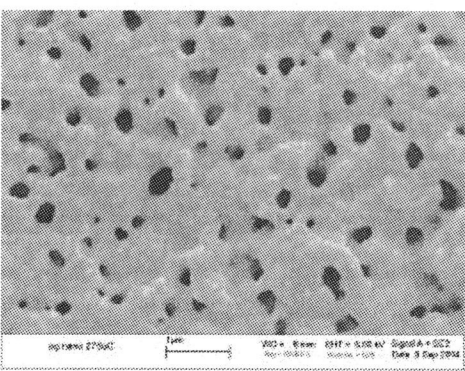

Figure 4: A typical SEM image of the sintered silver film showing a uniform distribution of about 20% micro-size posority.

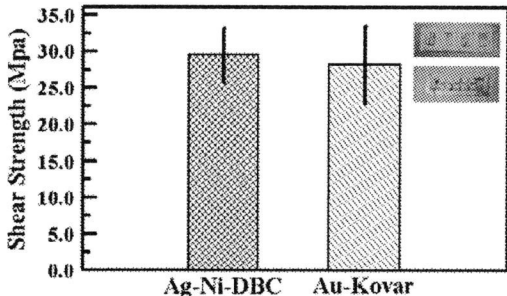

Figure 5: Measurements of die-shear strengths on devices mounted on Ag- and Au-coated substrates.

Figure 6 is a plot of die-shear strength versus aging time from devices attached on Ag-coated DBC substrates that were heated at 300°C in an oven. Taking into consideration the scatter in die-shear strengths, we believe that the aging had no significant effect on the joint strength. In a recent study on using lead-free solder for die-attach [19], the authors reported a drastic decrease in bonding strength after aging. The decrease was attributed to microstructural changes in the solder alloy at the aging temperature of 150°C. In our case, since the aging temperature of 300°C is much lower than the 961°C melting point of silver and the sintering process had spent much of the surface/interfacial free energies, we do not expect the once sintered microstructure to undergo significant microstructural evolution. The sintered microstruture is likely to densify a little further at the aging temperature. This, we believe, will strengthen the joint by increasing the cohesion of the attachment as well as adhesion between the attachment and its adherents (i.e. device and substrate).

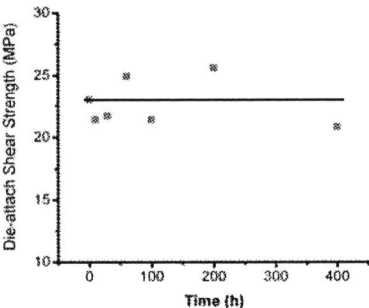

Figure 6: Measurements of die-shear strengths on devices mounted on Ag- and Au-coated substrates.

Figure 7 (a) and (b) are plots of die-shear strength vs. number of cycles from the temperature-cyling tests on devices attached to Au-coated Kovar and silver-coated DBC substrates, respectively. In both, the die-shear strength decreased slowly with number of cycles, but still retained about 20 MPa after 600 cycles. The cycling experiment was stopped because we ran out of samples to test. New batches of samples have been made to allow us to extend the temperature-cycling experiemnt.

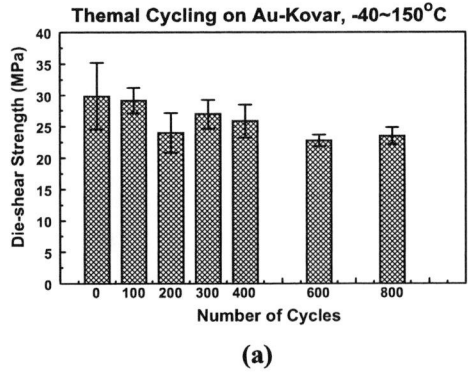

(a)

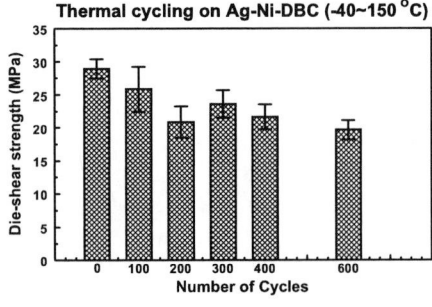

(b)

Figure 7: Plots of die-shear strength versus number of cycles from -40°C-150°C temperature-cycling tests

of (a) devices mounted on Au-coated Kovar substrate and (b) Ag-coated DBC.

To demonstrate that the low-temperature sintering technique can be used for die-attach as well as for interconnecting other terminals, we made packages of IGBT devices shown in figure 8. First, the devices were attached on Ag-coated DBC by the sintering process at 275°C, and then copper straps that were electroplated with μm-thick silver were interconnected to the other terminals by the second sintering process. In the package shown on the left, the gate terminal was wire-bonded; while on the right the gate interconnection was formed by sintering along with the source interconnection. A benefit of using the sintering technique is that once the material is sintered it will not melt unless it is heated to 961°C (the melting point of silver). Thus, the same sintering process can be repeatedly used in a packaging process that involves multiple joining steps. This eliminates the need to design a hierarchy of process temperatures necessary to accommodate different reflow schedules when solder alloys are used.

Wire bond

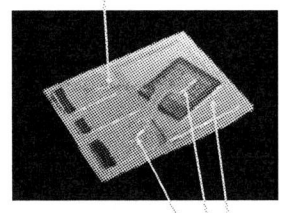 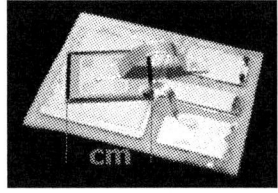

Sintered Ag joints   **All Sintered-Ag Connected IGBT**

Figure 8: Examples of IGBT devices interconnected on both attached on using the low-temperature sintering process.

4. Conclusions

Table 1 summarizes the characteristics of widely used die-attach materials together with those obtained in this study on the low-temperature and pressureless sintered nanoscale silver paste. Since the sintered joints are made of pure silver, they have substantially higher (at least five times) electrical and thermal conductivities than those of the other materials. The relative properties are expected to be even better for the sintered silver joints when devices are powered up since the other materials would be operating much closer to their melting/decomposition temperatures.

Because of its relative density of roughly 80%, the sintered silver has a low elastic modulus of about 10 GPa, which lowers thermo-mechanical stresses at the joints, thus improves joint reliability. Results of our temperature-cycling tests showed high reliability of the sintered joints. As large voids are known to form in a soldered die-attach layer, in the sintered attachment residual pores are small (micron-size) and uniformly distributed, thus eliminating hot spots underneath the device. Furthermore, formation of brittle intermetallic phases at soldered joints is always a concern for joint

708

reliability. This, however, is not a concern with the sintered joint since elemental metal is used.

Table 1: Summary of typical properties of commerical die-attach materials and nanosilver paste of this work.

	Processing temperature	Max. use temperature	Electrical conductivity 10⁷ (O-m)⁻¹	Thermal conductivity (W/K-m)	Die-shear Strength (MPa)
Lead-tin solder	217°C	< 183°C	0.69	51	35
Lead-free solder	260°C	< 225°C	0.75	70	35
Gold-tin solder	340°C	< 280°C	0.029	58	30 - 60
Silver epoxy	100 ~ 200°C	< 200°C	0.1	10	10 - 40
P-less Sintered nano-Ag	~ 275°C	< 961°C*	4.0	290	20 ~ 35

As the low-temperature sintering technology for interconnecting devices does not require external pressure, and the nanosilver paste can be readily printed or dispensed, the nanosilver paste can be a one-to-one replacement for solders and epoxies that are commonly used in today's manufacturing processes. Finally, by properly selecting diffusion-barrier and encapsulation materials to prevent silver electromigration, the sintering technology with the nanosilver paste can be a viable interconnection solution for high-temperature packages.

Acknowledgments

We are grateful for the financial support from the US National Science Foundation under Award Number EEC-9731677, the American Competitiveness Institute, US Army Research Laboratory, and the National Natural Science Foundation of China (No.50528506) and its Program of Introducing Talents of Discipline to Universities (No:B06006).

References

1. Wang, X.W., W. J. Zhu, X. Guo, T. P. Ma, J. B. Tucker, and M.V. Rao, *High temperature (450°C) reliable NMISFET's on p-type 6H-SiC.* IEEE IEDM Technical Digest, 1999: p. 209 - 212.
2. Wondrak, W., R. Held, E. Niemann, and U. Schmid, *SiC devices for advanced power and high-temperature applications.* IEEE Transactions on Industrial Electronics, 2001. **48**(2): p. 307-308.
3. Singh, R., S. H. Ryu, D. C. Capell, and J. W. Palmour, *High temperature SiC trench gate p-IGBTs.* IEEE Transactions on Electron Devices, 2003. **50**: p. 774 - 784.
4. Johnson, R.W., M. Palmer, C. Wang, and Y. Liu, *Packaging Materials and Approaches for High Temperature SiC Power Devices.* Advancing Microelectronics, 2004. **31**(1): p. 8-11.
5. Chuang, R.W., and C. C. Lee, *Silver-Indium Joints Produced at Low Temperature for High Temperature Devices.* IEEE Transactions on Components and Packaging Technologies, 2002. **25**(3): p. 453-458.

6. Massalski, T.B., ed. *Binary Alloy Phase Diagrams.* 2nd Edition ed. 1990, ASM International: Materials Park, OH. 34.
7. Scheuermann, U., *Low Temperature Joining Technology - A High Reliability Alternative to Solder Contacts,* in *Workshop on Metal Ceramic Composites for Functional Application.* 1997: Vienna.
8. Schwarzbauer, H., *Method of Securing Electronic Components to a Substrate.* 1998: US.
9. Zhang, Z., and G-Q. Lu, *Pressure-Assisted Low-Temperature Sintering of Silver Paste as an Alternative Die-Attach Solution to Solder Reflow.* IEEE Transactions on Electronics Packaging Manufacturing, 2002. **25**(4): p. 279 - 283.
10. Zhang, Z., J. N. Calata, J. G. Bai, and G-Q. Lu. *Nanoscale Silver Sintering for High-Temperature Packaging of Semiconductor Devices.* in *Proc. of 2004 TMS Annual Meeting & Exhibition.* 2004. Charlotte, NC.
11. Bai, J.G., and G-Q. Lu, *Thermomechanical Reliability of Low-Temperature Sintered Silver Die-Attached SiC Power Device Assembly.* IEEE Trans. on Device and Materials Reliability, 2006. **6**(3): p. 436 - 441.
12. Bai, J.G., Z. Z. Zhang, J. N. Calata, and G-Q. Lu, *Low-Temperature Sintered Nanoscale Silver as a Novel Semiconductor Device-Metallized Substrate Interconnect Material.* IEEE Transactions on Components and Packaging Technologies 2006. **29**(3): p. 589 - 593.
13. Bai, J.G., *Ph.D. Dissertation: Low-Temperature Sintering of Nanoscale Silver Paste for Semiconductor Device Interconnection,* in *Materials Science and Engineering.* 2005, Virginia Polytechnic Institute and State University: Blacksburg.
14. Ashcroft, N.W., and N.D. Mermin, *Solid State Physics.* 1976: Holt, Rinehart, and Winston.
15. Matula, R.A., J. Phys. Chem. Ref. Data, 1979. **8**(4): p. 1147.
16. Ho, C.Y., R.W. Powell, and P.E. Liley, J. Phys. Chem. Ref. Data, 1972. **1**: p. 279.
17. Hwang, J.S., ed. *Chapter Six: Solder technologies for electronic packaging and assembly.* 3rd Edition ed. Electronic Packaging and Interconnection Handbook, ed. C.A. Harper. 2000, McGraw-Hill: New York.
18. Nguyen, G.P., et al.,, *Conductive adhesives.* Circuits Assembly, 1993: p. 36.
19. Anderson, I.E., and J.L. Harringa, *Elevated temperature aging of solder joints based on Sn-Ag-Cu-Effects on joint microstructure and shear strength.* J. Electron. Mater. , 2004. **33**: p. 1485.

3D Modeling of Electromigration Combined with Thermal-Mechanical Effect for IC Device and Package

Yong Liu, Scott Irving and Timwah Luk, Fairchild Semiconductor Corp., South Portland, Maine, USA
Lihua Liang and Shinan Wang, Fairchild-ZJUT Joint Lab,
Zhejiang University of Technology, China

Abstract

This paper studies the numerical simulation method for electromigration in IC device and solder joint in a package under the combination of high current density, thermal load and mechanical load. The three dimensional electromigration finite element model for IC device/interconnects and solder joint reliability are developed and tested. Numerical experiment is carried out to obtain the electrical, thermal and stress fields with the migration failure under high current density loads. The direct coupled analysis and in-direct coupled analysis that include electrical, thermal and stress fields are investigated and discussed. The viscoplastic Anand constitutive material model with both SnPb and SnAgCu lead-free solder materials is considered in the paper. An IC device is studied to show the modeling methodology and the comparison with previous test data. A global CSP package with PCB is modeled using relative coarse elements. In order to reduce the computational costs and to improve the calculation accuracy, a refined mesh sub-model is constructed. The sub-model technique is studied in a direct and indirect coupled multiple fields. The comparison of voids generation through numerical example in this paper and previous experimental result is given.

1. Introduction

It is well known that as the electronics industry continues to push for high performance and miniaturization, the demand higher current densities, which may cause electromigration failures, not only in a IC interconnect but also in solder bumps of a IC package[1]. Early studies of electromigration mostly concentrated on interconnects of integrated circuit packages such as Al-Cu alloy interconnect wire or pure Cu interconnect wire [2-4]. In classical electromigration studies, Black's equation [2] has been successful in characterizing the operation life of aluminum and copper traces on a chip, but a lot of experimental work and modeling have shown that the Black's theory is not enough accurate to evaluate the reliability or unable to reasonably simulate the failure due to void formation[5-7]. Black's equation assumes that the current density in aluminum is constant [2], which is not true in solder bumps in which current density varies in different location. In recent years the researchers have found the phenomenon of electromigration occurs inside the solder which is adjacent to the under bump metallization (UBM) layer. The void propagates along the interface between solder and the UBM, which would induce open failures of the joints. The electromigration failure mechanism in solder balls is distinctly different from that in Al or Cu interconnects [8-11].

There are a few papers that proposed an algorithm for 3D void simulation with consideration of the electromigration, the thermomigration and the stress migration [12-14]. Although these works can simulate the void formation in the interconnect lines of integrated circuits, the limit of their algorithms makes it unable to provide an accurate solution to the life prediction of IC interconnect lines. Further, their work has not yet been applied to solder joint reliability analysis. We have done some initial work to consider both the thermal and electromigration in power cycling [15]. Afterward, we presented a numerical study of electromigration in solder joints under the combination of high direct current density, thermal loads and mechanical loads [16-17], but only a simple traces-bump model is studied and the solder bump material is assumed to be elastic. In fact, the simple traces-and-bump model doesn't get an appropriate thermal field and the thermal mismatch in a whole package with PCB system. Because the solder bumps have viscoplastic deformation at high temperature, the simple trace-bump model will induce the errors in multi-physics electromigration analysis.

This paper further develops the fully 3D electromigration model with consideration of both IC interconnects in a device and solder joints in a package by FEA modeling. It considers the combination analysis of electric-thermal-structural coupled fields based on ANSYS multi-physics simulation platform. The atomic flux divergence calculation method considers three mechanisms which includes the electromigration, the thermomigration and the stress migration. It also considers the void generation, and location. Evolution of the parameters corresponding to void growth and electricity is recorded during simulation. Meanwhile, the corresponding time to failure life is studied. The direct coupled and in-direct coupled analysis that include electrical, thermal and structure fields are presented and discussed. A sub-model technique is introduced in the 3D multiple physics analysis. A global chip scale package (CSP) with PCB is modeled using relative coarse elements, and a refined mesh sub-model is constructed for the detailed electromigration, thermomigration and stress-migration coupled analysis. Finally, the numerical examples for IC device and solder joints in a flip chip CSP are presented, and the comparison with the measurement result is discussed.

2. Basic migration formulation and Algorithm

Based on Black's equation [2], the atomic flux due to electromigration, thermomigration and stress migration in a conductor can be expressed as following [12]:

1-4244-1105-X/07/$25.00 ©2007 IEEE

$$\vec{J}_{Tol} = \vec{J}_{Em} + \vec{J}_{Th} + \vec{J}_S$$

$$= \frac{ND}{kT} Z^* e\rho\vec{j} - \frac{ND}{kT^2} Q^* \frac{\nabla T}{T} \qquad (1)$$

$$- \frac{ND}{kT} \Omega \nabla \sigma_m$$

where N is the atomic concentration; k is Boltzmann's constant; e is the electronic charge; Z^* is the effective charge which is determined experimentally; T is the absolute temperature, ρ is the resistivity which is calculated as $\rho = \rho_0 (1 + \alpha(T - T_0))$, α is the temperature coefficient of the metallic material; D is the diffusivity, $D = D_0 \exp\left(-\frac{E_a}{kT}\right)$, E_a is the activation energy, D_0 is the thermally activated diffusion coefficient; $\vec{j}$ is the current density vector; Q^* is the heat of transport; Ω is the atomic volume; σ_m is the local hydrostatic stress.

Without considering the volumetric amount of displaced material, the divergences of atomic flux for electronic migration, thermal migration and stress migration can be expressed as below:

$$div(\vec{J}_{Tol}) = \left(\frac{E_a}{kT^2} - \frac{1}{T} + \alpha\frac{\rho_0}{\rho}\right) \cdot \vec{J}_{Em} \cdot \nabla T$$

$$+ \left(\frac{E_a}{kT^2} - \frac{3}{T} + \alpha\frac{\rho_0}{\rho}\right) \cdot \vec{J}_{Th} \cdot \nabla T$$

$$+ \frac{NQ^* D_0}{3k^3 T^3} j^2 \rho^2 e^2 \exp\left(-\frac{E_a}{kT}\right)$$

$$+ \left(\frac{E_a}{kT^2} - \frac{1}{T}\right) \cdot \vec{J}_S \cdot \nabla T \qquad (2)$$

$$- \frac{2EN\Omega D_0 \alpha_l}{3(1-\nu)kT} \exp\left(-\frac{E_a}{kT}\right)\left(\frac{1}{T} - \alpha\frac{\rho_0}{\rho}\right)\nabla^2 T$$

$$- \frac{2EN\Omega D_0 \alpha_l}{3(1-\nu)kT} \exp\left(-\frac{E_a}{kT}\right)\frac{j^2 \rho^2 e^2}{3k^2 T}$$

where E is Young modulus, ν is the Poisson ratio, α_l is the expansion coefficient of the metallization. The divergences of atomic flux can be expressed further

$$div(\vec{J}_{Tol}) = N \cdot F(\vec{j}, T, \sigma_m, E_a, D_0, E, \cdots) \qquad (3)$$

The time dependent evolution of the local atomic concentration is given as:

$$div(\vec{J}_{Tol}) + \frac{\partial N}{\partial t} = 0 \qquad (4)$$

From Eq. (2-3), the divergence value is in proportion to the atomic concentration N and to the function F in which different physical parameters are included. So Eq. (4) is equivalent to the following equation (5):

$$NF + \frac{\partial N}{\partial t} = 0 \qquad (5)$$

Integrating Eq. (5), it can be obtained that

$$\int_{N_i}^{N_{i+1}} \frac{dN}{N} = \int_{t}^{t_{i+1}} -F dt \qquad (6)$$

Due to the explicit Euler algorithm, Eq. (6) can be simplified as

$$N_{i+1} = N_i e^{-F_i \Delta t_i} \qquad (7)$$

where N_i and F_i are the atomic concentration and value of F in ith step, respectively. Eq. (7) is the incremental formula of the atomic concentration. From Eqs. (1)-(3) and (7), the local time-dependent iterative scheme has been developed, which is different from previous work.

In this paper, both the direct coupled analysis and in-direct coupled analysis are studied. The directly coupled method and its flow chart have been proposed in our previous papers [16-17]. Fig. 1 shows the algorithm of in-directly coupled analysis. In structure analysis, the solder bump is meshed with visco107 element for ANAND model. But the visco107 element in ANSYS® dose not have the element birth and death feature. To achieve the "element death" effect, ANSYS® does not actually remove "killed" elements. Instead, it deactivates them by changing the element material attribute from ANAND model to elastic model with 1.0E-6 of elastic module and density of source model. This actually multiplies the stiffness by a severe reduction factor. Here, the criterion is that the material of a element in the model is considered to be empty when its atomic concentration reaches 10% of the initial concentration. So, once the atomic concentration value of a element has reached the above criterion, it will be killed.

In the in-direct coupled algorithm flow chart, finding the solution of divergence of electric, thermal and stress migration is a core part of the methodology. Following is the algorithm to find the divergence.

Let i_{step} be the iterative steps from initial time T_0 to failure time T_f.

1). Resume electric-thermal couple sub-model and solve;

2). Compute the divergences for electronic migration, thermal migration:

2.1) Create element table for temperature, current density, temperature gradient.

2.2) Element table operation: $\vec{j} \cdot \nabla T, \nabla^2 T, j^2$;

2.3) For elem$\in$ [1, emax], where emax is the size of element table

(a) Get T, $\vec{j} \cdot \nabla T$, $\nabla^2 T$ and j^2 of every element

(b) AT(elem)=T, AJDT(elem)=$\vec{j} \cdot \nabla T$, ADT2(elem)= $\nabla^2 T$, AJS2(elem)= j^2, ATG(elem)=∇T

(c) Compute $div(\vec{J}_{Em})$ and $div(\vec{J}_{th})$

2.4) Output all the parameters and array to a file

2.5) Save and exit

3). Resume structural sub-model and solve;

4). Compute the divergences for stress migration:

4.1) Resume all the parameters and array

4.2) For elem$\in$ [1, emax]

(a) Get T, $\vec{j} \cdot \nabla T$, $\nabla^2 T$ and j^2 of every element

(b) Get the local hydrostatic stress σ_m^i of every node inside current element

(c) Create the shape function array ψ_i

(d) $\sigma_m = \sum_{i=1}^{n} \sigma_m^i \psi_i$, $\dfrac{\partial \sigma_m}{\partial x_i} = \sum_{i=1}^{n} \sigma_m^i \dfrac{\partial \psi_i}{\partial x_i}$

(e) Compute $div\left(\vec{J}_S\right)$

(f) $div\left(\vec{J}_{Tol}\right) = div\left(\vec{J}_{EM}\right) + div\left(\vec{J}_{Th}\right) + div\left(\vec{J}_S\right)$

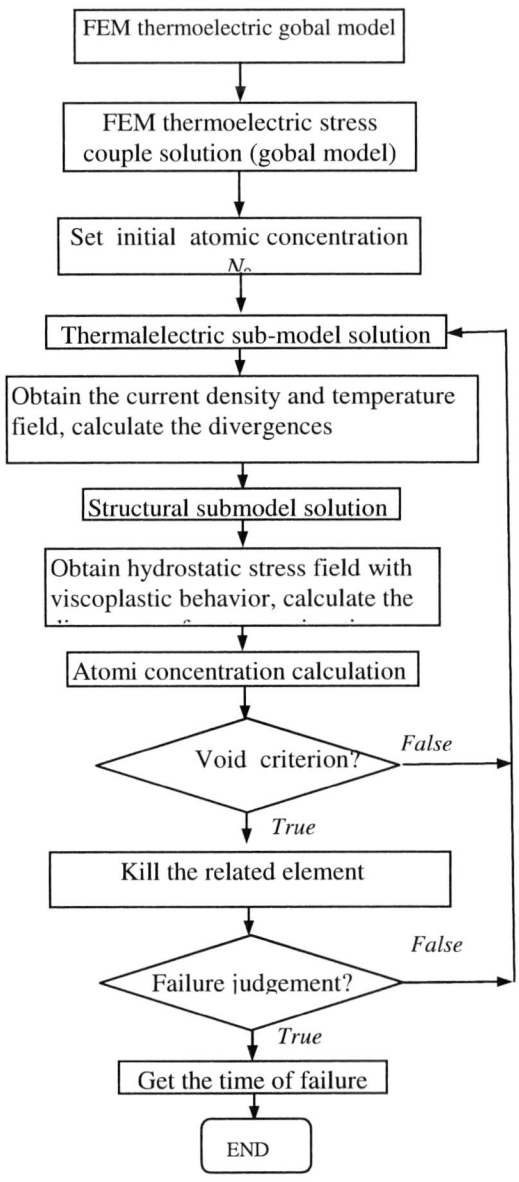

Figure 1 Flow chart of in-direct coupled analysis

3. Electromigration Examples from IC device and Package

This section will give two application examples, one is an IC device and another one is a flip chip CSP package. Direct couple analysis methodology is applied to IC device and both direct and in-direct couple methodologies are applied to solder joint connection of a package.

3.1 A Sweat structure

The three-dimensional finite element model for a sweat structure (from [13]) is shown in Fig.2. Here due to symmetry, only a quarter of the structure was modeled. The symmetry from this simplification can reduce the number of elements and save computing time. The linear regular hexahedral elements were map-meshed, which can also save computing time and improve the accuracy.

This paper focuses only on high load conditions from 12e6A/cm² up to 24e6A/cm². Different values of loading are taken within this range in order to study the influence of the electrical charge on the lifetime of the metallization structure. The structure is considered to be stress-free at room temperature for the simulations

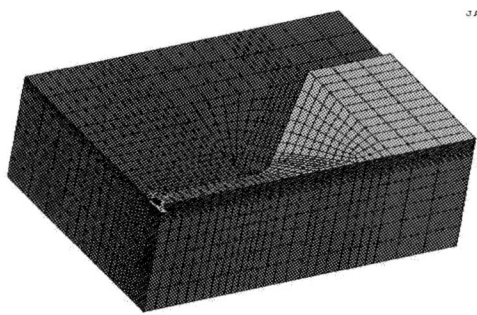

Figure 2 A sweat structure and mesh

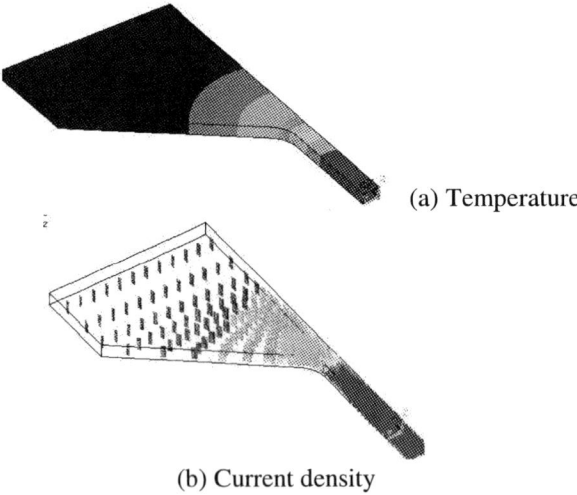

(a) Temperature

(b) Current density

Figure 3 Temperature and current density distributions with current load 0.123A

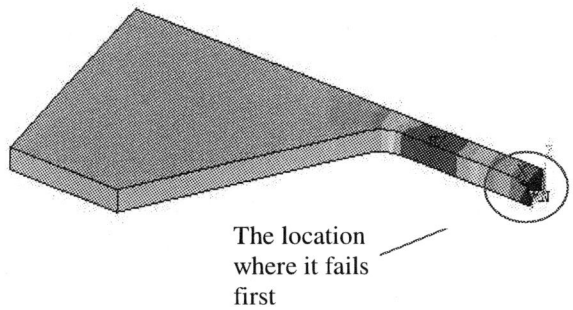

The location where it fails first

(a) Atomic density distribution (dark blue indicate very small value: void)

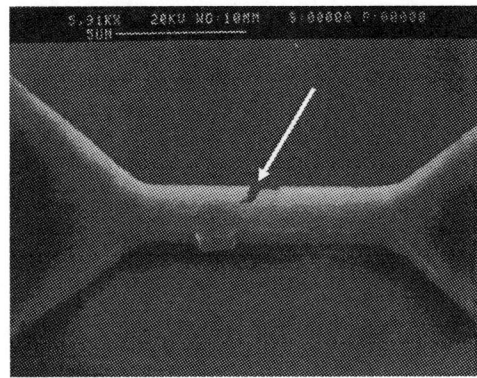

(b) Void generation [13]

Figure 4 Void comparison between and simulation and test

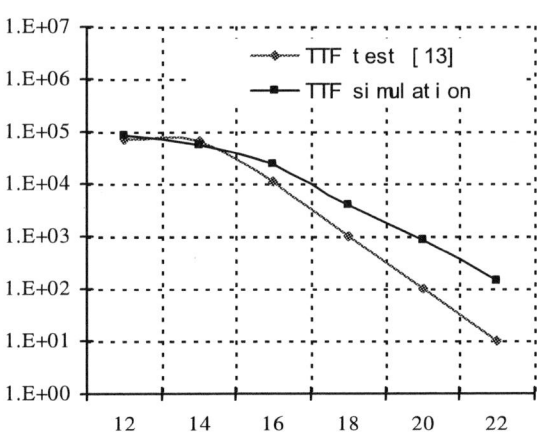

Figure 5 Time to failure (TTF) comparison

Fig. 3 shows the temprature distribution and current density distribution under current load 0.123A. Fig. 4 shows the atomic density distribution of the sweat structure, which indicates that failure firstly happens near the middle of sweat structure. The result was in good agreement with the picture observed in the experiment [13], as shown in Fig. 4(b). Furthermore, according to atomic density distribution, the hillock formation can be simulated, such as the red location in Fig. 4(a).

A comparison of the simulation results for TTF and the previous experimental test results obtained in [13] is also presented in Fig.5. The simulated TTF life data are very consistent with the experimental data. In considering all the assumptions (Al diffusivity, effective charge number, and vacancy formation criterion, etc.) that were taken and the simplifications of the models made, the simulation results seem quite reasonable.

3.2 A flip chip CSP with viscoplastic solder bumps

Gee et al [18] have done the electromigration test for Lead-free and PbSn solder joint in a CSP package. Components used in the system include: silicon chip, under bump metallurgy UBM), aluminum trace, copper trace and solder bumps. To correlate our simulation methodology with the experimental data. The CSP package in reference [18] is modeled, which has 36 bumps with 500 μm pitch. The exterior 20 solder bumps are assumed to connect with each other in a daisy chain as shown in Fig. 6. Sub-model technique in ANSYS is introduced to get the better response of the electronic migration. The global thermal-electric coupled field model uses Solid-69 element and the global stress model uses Visco-107 element for solder bumps and Solid-45 element for the remaining parts of the model. The global structure is modeled using relative coarse elements in the first step. In the second step, a refined thermal-electric coupled field sub-model and a refined stress sub-model with UBM (Al/NI(V)/Cu) layer are then constructed as shown in Fig. 7. Gee, et al [18] confirmed that the UBM has greater resistance to the electromigration, and the UBM/Solder interface (solder side) seems more likely to have electromigration voids. So in this paper, a very fine mesh is set along the interface.

Two solder materials (a SnPb and a SnAgCu) are introduced in this paper. Their related thermal mechanical and electrical constants used in the simulation are selected from references [15-16, 19-20], while the Anand parameters are listed in Tab.1.

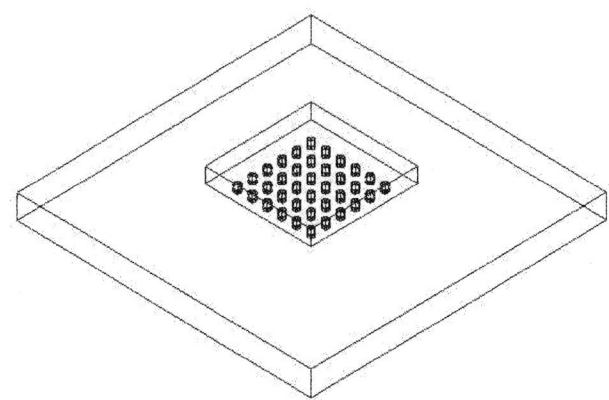

(a) A CSP package structure

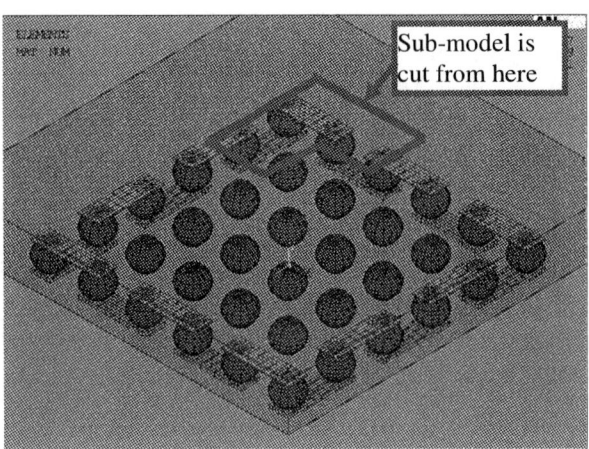

(b) Local view

Figure 6 A CSP package model

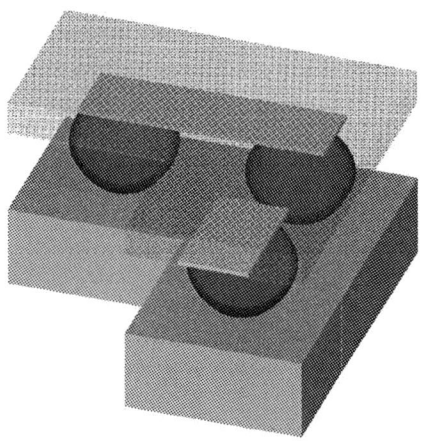

(a) Solid sub-model

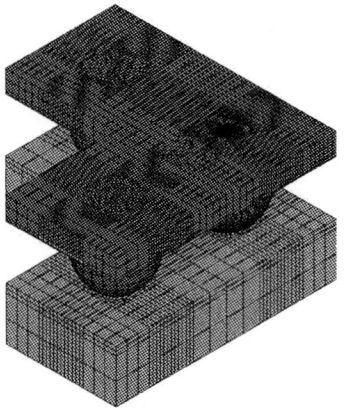

(b) Sub-model mesh

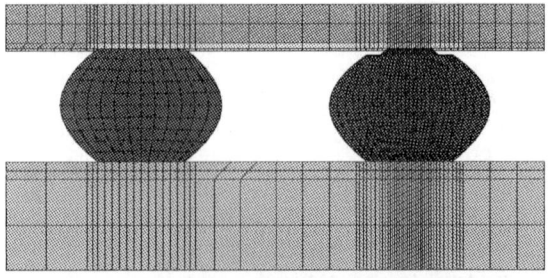

(c) Front view of the sub-model with fine mesh on corner solder bump

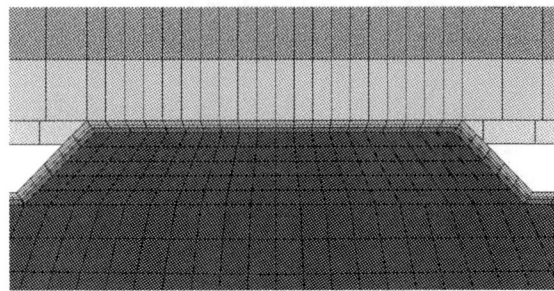

(d) Local view of the UBM structure

Figure 7 Thermal-electric sub-model and stress sub-model

Table1 ANAND model parameters for SnAgCu and SnPb

Description	Symbol	63Sn37Pb [20]	95.5Sn4.0Ag0.5Cu [21]
Pre-exponential factor	A(1/s)	4e6	325
Activation energy	Q/ R(K)	9400	10561
Stress multiplier	Ξ	1.5	10
Strain rate sensitivity of stress	M	0.303	0.32
Coef. For deformation resistance saturation value	$\hat{s}$ (MPa)	13.79	42.1
Strain rate sensitivity of saturation value	N	0.07	0.02
Hardening coefficient	h0(MPa)	1378.95	800000
Strain rate sensitivity of hardening coeff.	A	1.3	2.57
Initial value of s	s0(MPa)	12.41	20

The electromigration parameters of both SnPb and SnAgCu solder bumps are selected from previous references [1,5-6,8-11,18]. For SnPb material: activation energy E_a=0.8eV; effective charge number Z^*=-33; self-diffusion-coefficient D_0=0.016 m^2/s; heat of transport Q^*=0.0094eV; initial electrical resistivity R_0=15.5e-8Ω·m; temperature coefficient resistance α=3.0e-3 /k; atomic volume Ω=2.48e-29 m^3/atom. For SnAgCu material: activation energy E_a=0.8eV; effective charge number Z^*=-23; self-diffusion-coefficient D_0=0.027m^2/s; heat of transport Q^*=0.0094eV; initial electrical resistivity R_0=13.3e-8Ω·m; temperature coefficient resistance $α$=2.8e-3/k; atom volume $Ω$=2.71e-29 m³/atom.

Fig. 8 shows the current flow direction in a global model. The free convection boundary condition is applied with 20 W/m²·°C film coefficient and 50°C bulk temperature.

Simulation results and discussions

In this section, four topics are discussed: (1) The impact of direct coupling and in-direct coupling analysis; (2) Impact of different current loadings; (3) Impact of different solder material (SnPb vs SnAgCu); (4) Void generation.

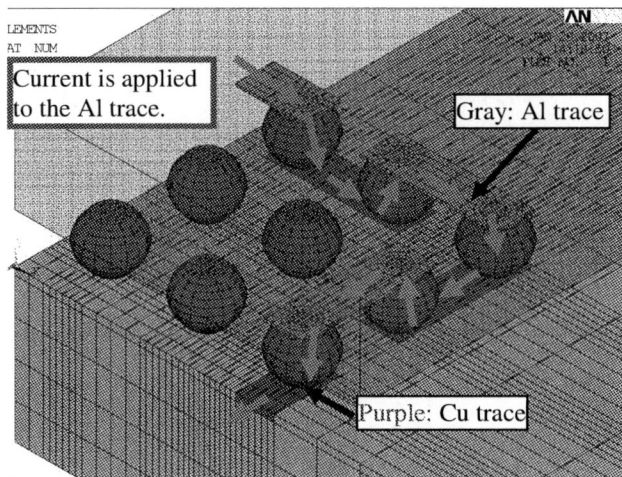

Figure 8 Current flow direction in a global model

3.2.1 The impact of direct couple analysis and in-direct couple analysis

In the simulation the solder bump model/element is considered to be elastic in the direct couple analysis while it is considered to be viscoplastic in the in-direct coupling analysis, because there is no directly coupled element which can be used for non-linear analysis.

Fig. 9 shows the current density distribution in sub-model and solder bump with 1.7A. From Fig. 9 it can be seen that the current density at the corner is approximately one order of magnitude higher than the average current density. It agrees the conclusion of reference [2]. Fig. 10 shows the temperature gradient distribution in sub-model and bump with 1.7A. The temperature distribution agrees with the test results very well [18]. Due to conductivity performance, the maximum temperature gradient value is 17046 K/m (relative small), which would induce small divergences of thermomigration.

Fig. 11 shows the hydrostatic stress distribution in the sub-model with direct coupling analysis (elastic) and in-direct coupling analysis (viscoplastic solder) under the current loading 1.7A. From Fig. 11, it can be observed that the maximum hydrostatic stress obtained by in-direct coupling analysis with viscoplastic solder bump is less than one third value obtained by the direct coupling method with elastic solder bump.

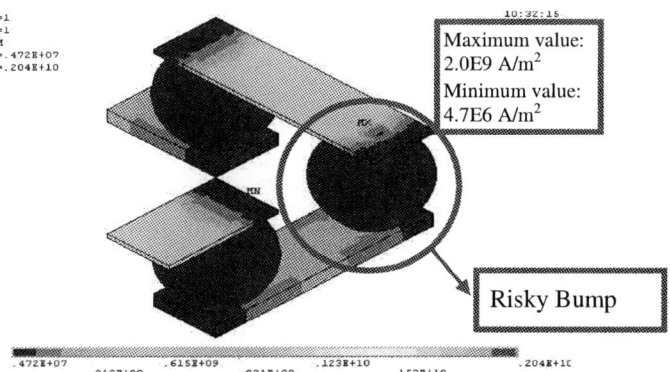

(a) Current density distribution for trace-and-bump

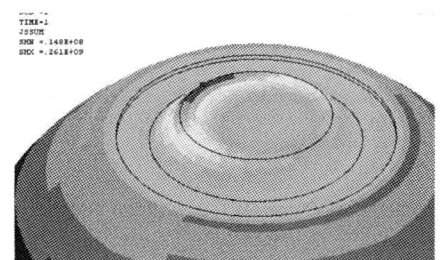

(b) Local current density in corner risky bump

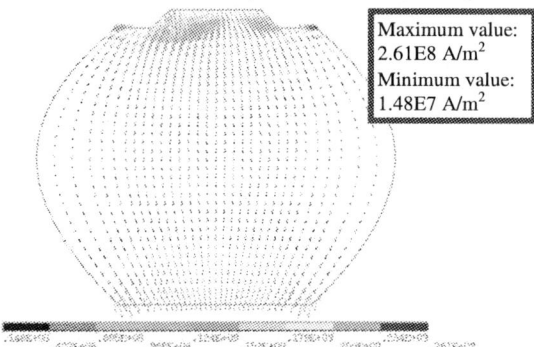

(c) Current density vector contour for risky bump

Figure 9 Current density distribution with current load 1.7A

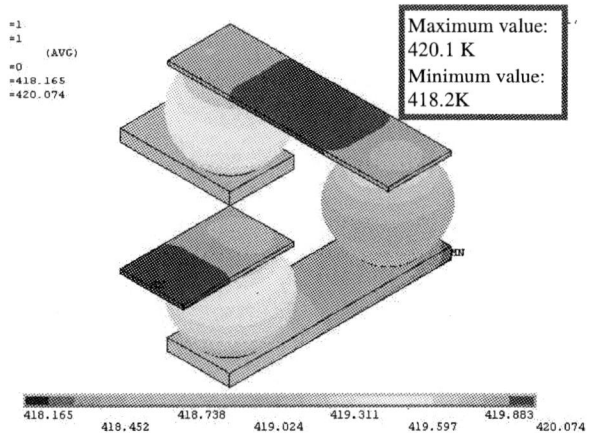

(a) Temperature distribution for trace-and-bump

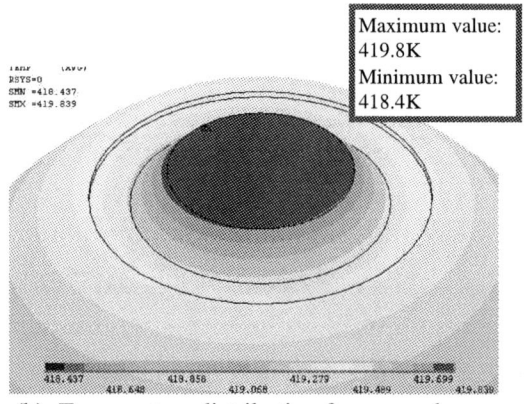

Maximum value:
419.8K

Minimum value:
418.4K

(b) Temperature distribution for corner bump

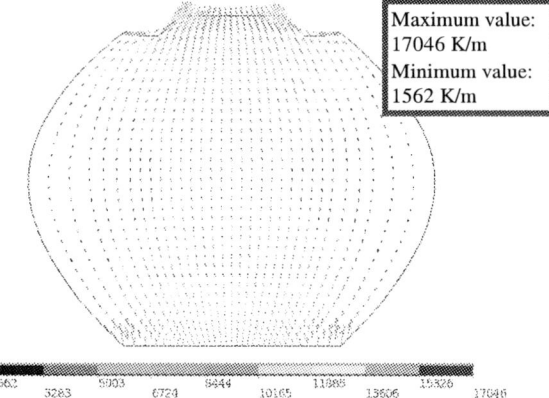

Maximum value:
17046 K/m

Minimum value:
1562 K/m

(d) Temperature gradient vector contour for risky bump

Figure 10 Temperature and its gradient distribution in sub-mode and solder bump

Fig. 12 shows the total atomic flux divergence contour due to electromigration, thermomigration and stress migration in both elastic solder bump (direct couple method) and viscoplastic SnPb solder bump (in-direct couple method) with current loading 1.7A. From Fig.12, it can be seen that the significantly different divergence distributions from the direct coupling method and the in-direct coupling method, and the maximum total divergence value of direct method is larger than that of the in-direct method.

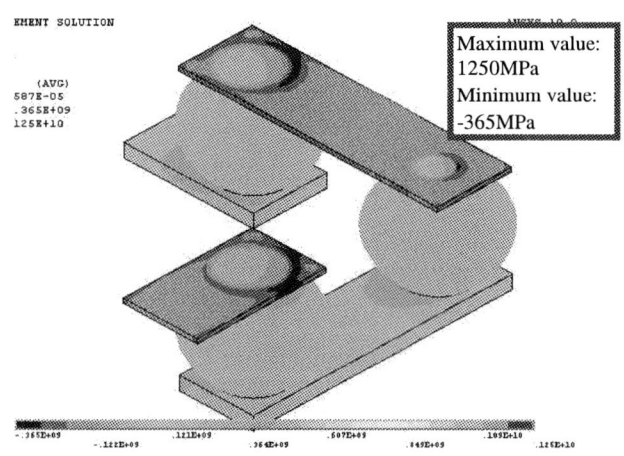

Maximum value:
1250MPa

Minimum value:
-365MPa

(a) Elastic stress of sub-model by direct couple method

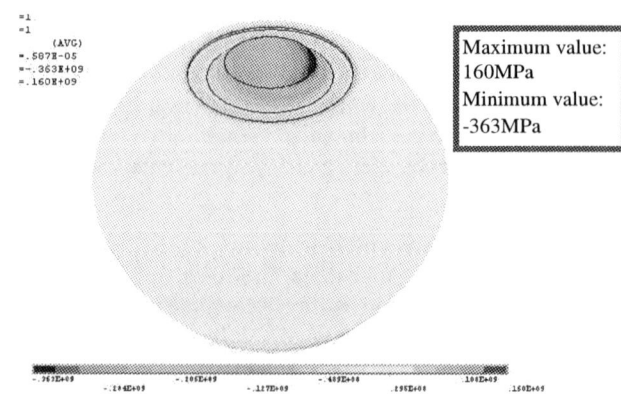

Maximum value:
160MPa

Minimum value:
-363MPa

(b) Elastic stress in the corner solder bump

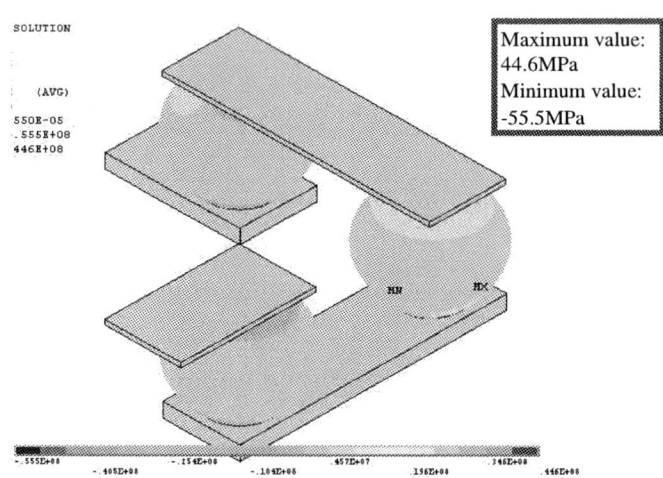

Maximum value:
44.6MPa

Minimum value:
-55.5MPa

(c) viscoplastic stress of sub-model with in-direct couple analysis

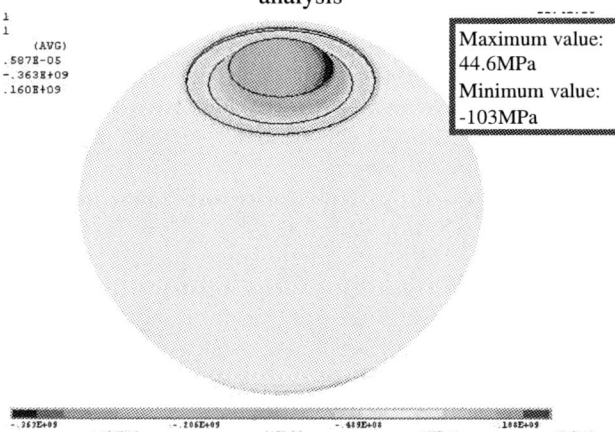

Maximum value:
44.6MPa

Minimum value:
-103MPa

(e) Viscoplastic stress in the corner solder bump

Figure 11 Hydrostatic stress (pressure) distribution

716

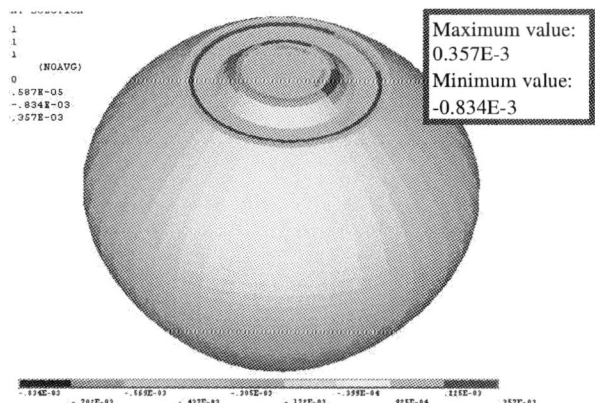

Maximum value:
0.357E-3
Minimum value:
-0.834E-3

(a) Elastic solder bump with direct couple method

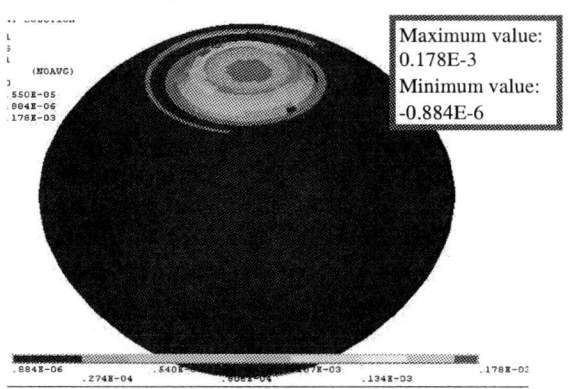

Maximum value:
0.178E-3
Minimum value:
-0.884E-6

(b) Viscoplastic solder bump with in-direct coupled method

Figure 12 Total atomic flux divergence due to electromigration, thermomigration and stress migration with direct and in-direct couple methods

3.2.2 Impact of different current loadings

The results of figs.13-16 give the solution of a corner SnPb joint under the in-direct method with different current loads.

Fig. 13 shows the maximum current density of the corner solder bump with different current loading 1.65 A, 1.7A and 1.8A. Fig. 14 shows the maximum temperature of the corner solder bump subjected to current loads 1.65 A, 1.7A and 1.8A. Fig. 15 shows the maximum hydrostatic stress in the corner solder bump with current 1.65 A, 1.7A and 1.8A. From these figures, it can be observed that the current density, the temperature, the hydrostatic pressure linearly increase as the current load increases.

Fig. 16 shows the total atomic flux divergence in the corner solder joint varies with different current 1.65 A, 1.7A and 1.8A. The result indicates that as the current increases the total divergence increases rapidly.

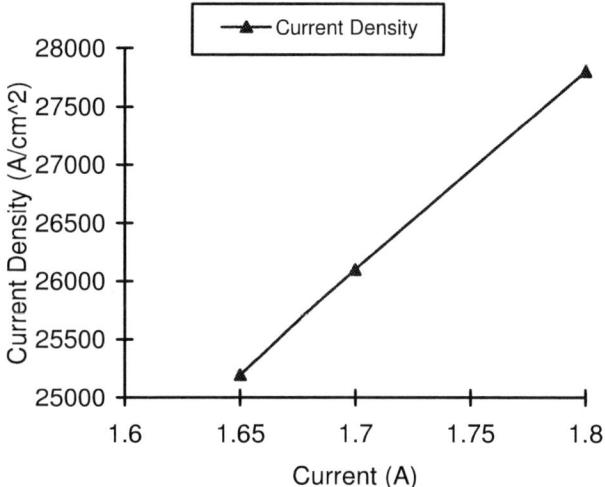

Figure 13 Maximum current density of SnPb solder bump with different current loads

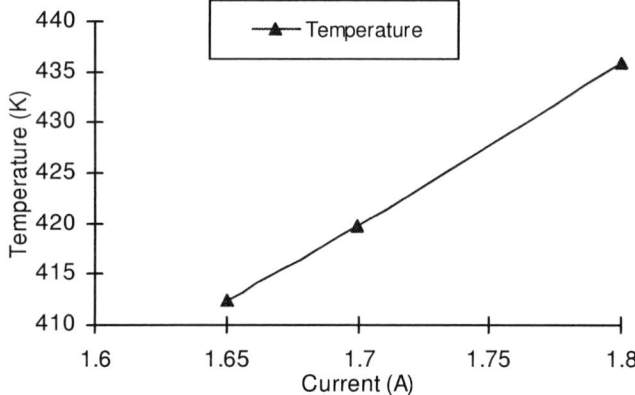

Figure 14 Maximum temperature of viscoplastic SnPb solder bump with different current

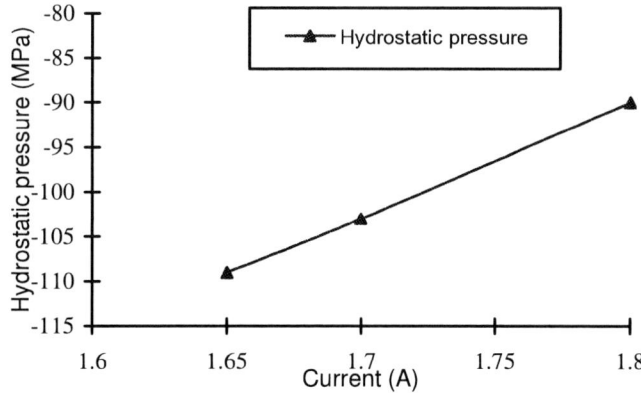

Figure 15 Maximum hydrostatic stress of viscoplastic SnPb solder bump with different current

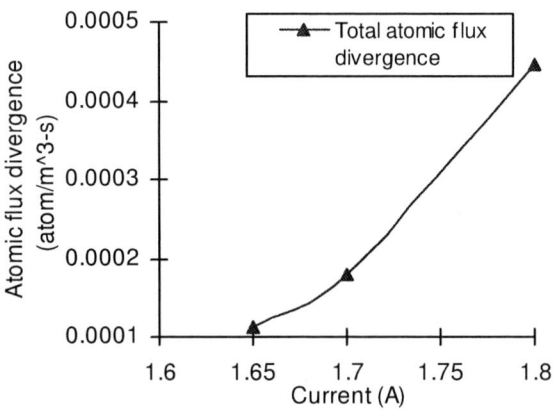

Figure 16 Maximum total atomic divergence of viscoplastic SnPb solder bump with different current

3.2.3 Impact of different materials (SnPb vs SnAgCu)

Fig.17 shows the current density distribution in viscoplastic SnPb and SnAgCu corner solder bump with 1.7A. Fig.18 shows the temperature distribution in viscoplastic SnPb and SnAgCu solder bump with 1.7A. Fig.19 shows the hydrostatic stress distribution in viscoplastic SnPb and SnAgCu corner solder bump with 1.7A. Fig. 20 shows the total atomic flux divergence contour due to electromigration, thermomigration and stress migration in viscoplastic SnPb and SnAgCu corner solder bump with 1.7A. Tab.2 lists the max. divergences due to electromigration, thermomigration and stress migration in viscoplastic SnPb and SnAgCu corner solder bump. It can be seen that for the same temperature and current densities, the SnAgCu solder seems is slightly more resistant to electromigration failure than the PbSn solder.

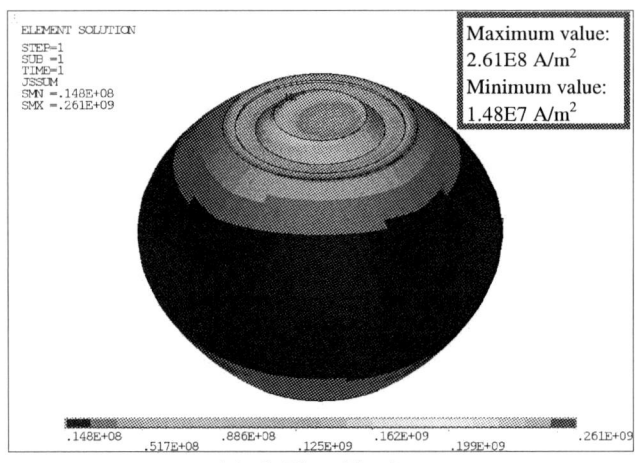

(a) SnPb solder bump

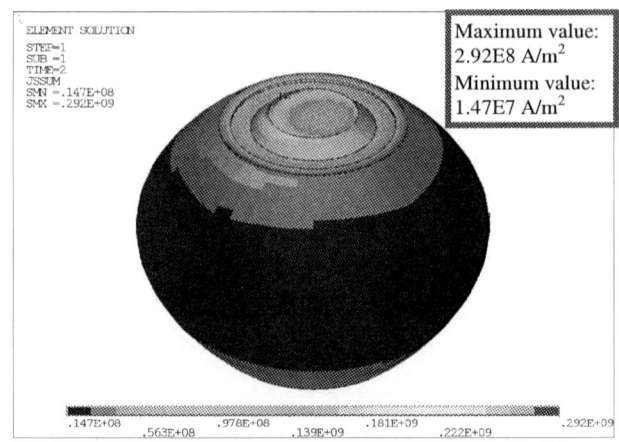

(b) SnAgCu solder bump

Figure 17 Current density distribution with two different materials

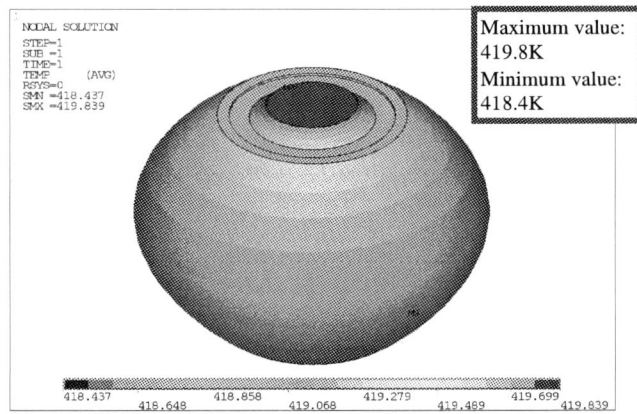

(a) Viscoplastic SnPb solder bump

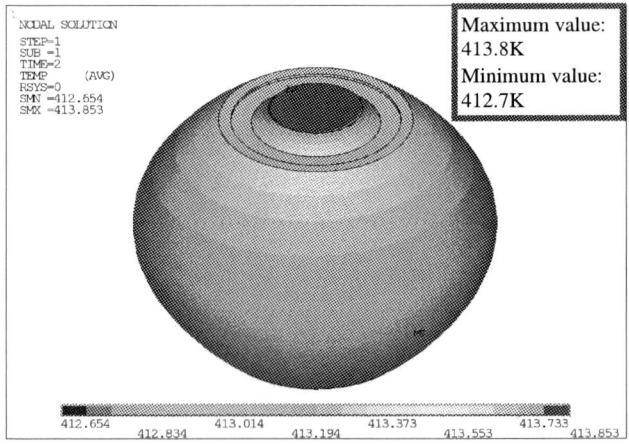

(b) Viscoplastic SnAgCu solder bump

Figure 18 Temperature distribution with two different materials

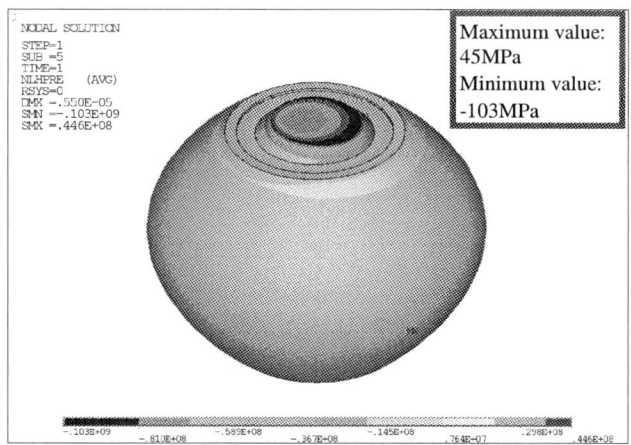

(a) Viscoplastic SnPb solder bump

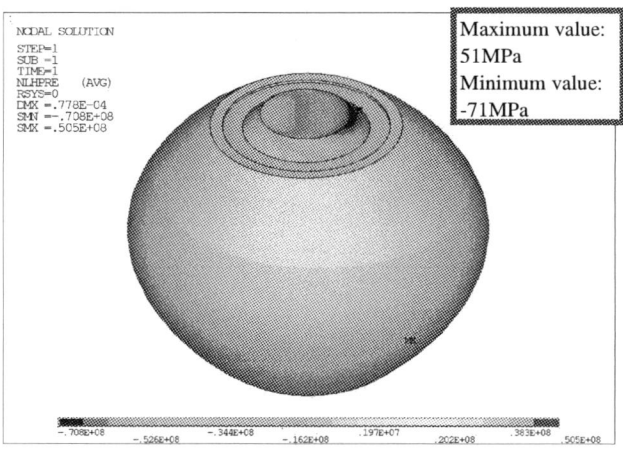

(b) Viscoplastic SnAgCu solder bump

Figure 19 Hydrostatic stress distribution with two different materials

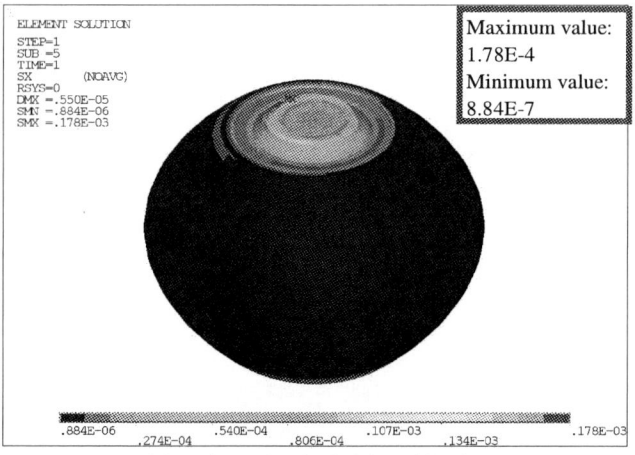

(a) Viscoplastic SnPb solder bump

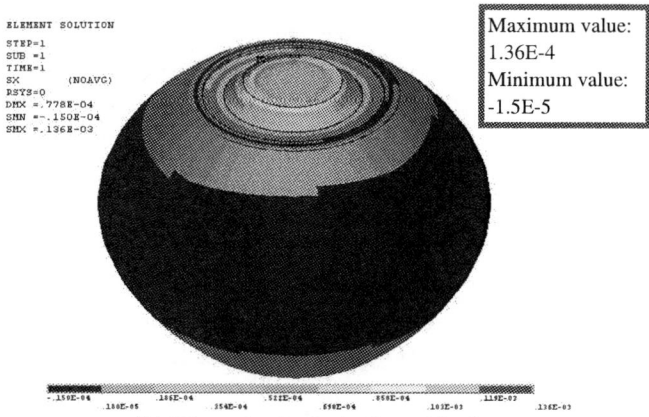

(b) Viscoplastic SnAgCu solder bump

Figure 20 Total atomic flux divergence with different materials

Table. 2 The maximum atomic flux divergence due to electromigration, thermomigration and stress migration with two different solder materials.

Material	Div Jth	Div Jem	Div Js	Div Jtot
PbSn	0.758E-06	0.154E-03	0.237E-04	0.178E-03
SnAgCu	0.841E-06	0.119E-03	0.162E-04	0.136E-03

3.2.4 Void formation

Fig.21 shows the void formation along the interface of solder and UBM in a SnPb corner solder bump at 414 hours under 1.7A. As the void grows, we define the void time to failure criteruion is the ratio of the void area and the area of top solder contact to UBM 50%. Fig.22 shows the void has reached the 50% criterion at 817.5 hours. It can be seen that the void evolves along the interface of UBM and solder bump which agrees the test results of reference [18] quite well (see Fig. 23). The failure time by simulation is 817.5h, while the time to failure 15% dR/R in PbSn solder bump of the experimental result is 500-1500h in [18].

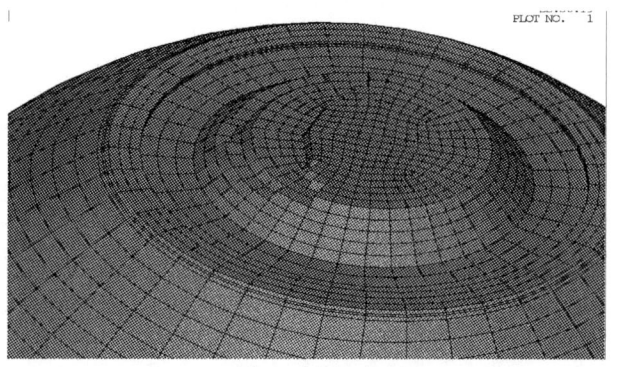

(a) void formation

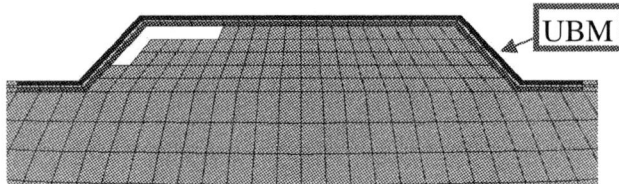

(b) Cross-section of void formation with UBM

(c) Current density contour

Figure 21 Void formation of SnPb solder bump at 414 hours

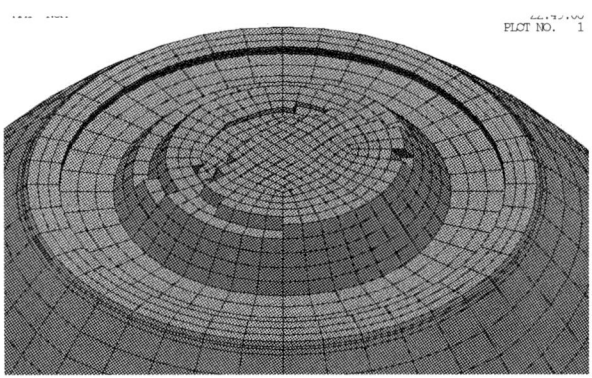

(a) void formation

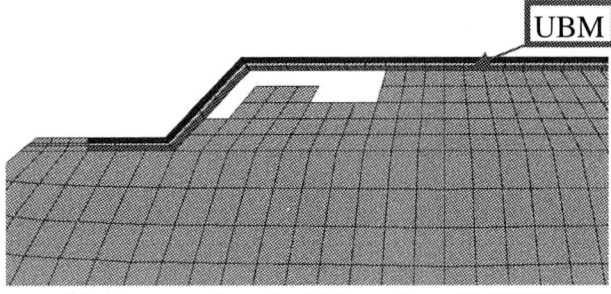

(b) Cross-section of void formation with UBM

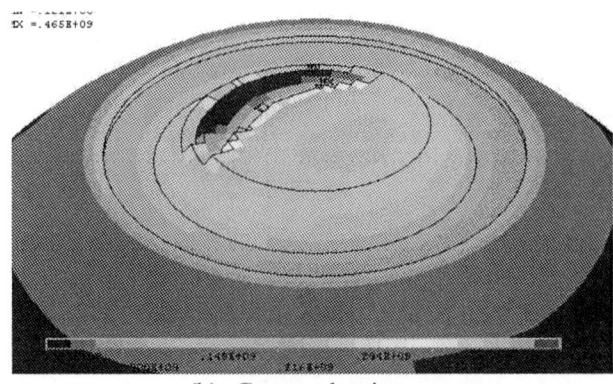

(b) Current density contour

Figure 22 Void formation of SnPb solder bump at 817.5 hours

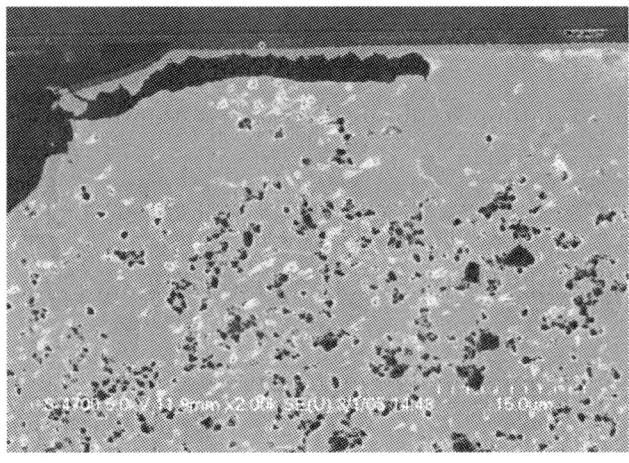

Figure 23 SEM of void of SnPb solder bump [18]

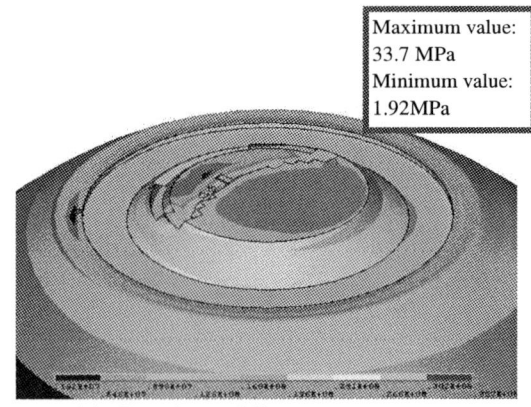

(a) Local von-Mises stress

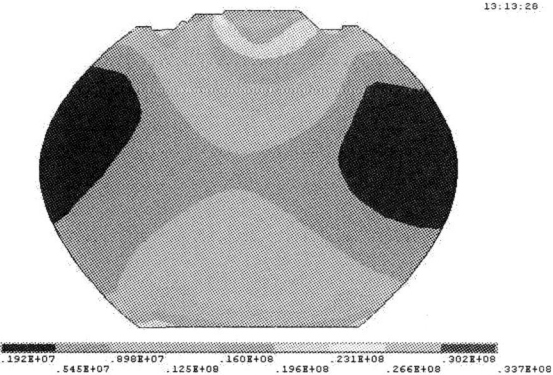

(b) Cross-section of the corner solder bump

Figure 24 von Mises stress of SnPb solder at 414 hour

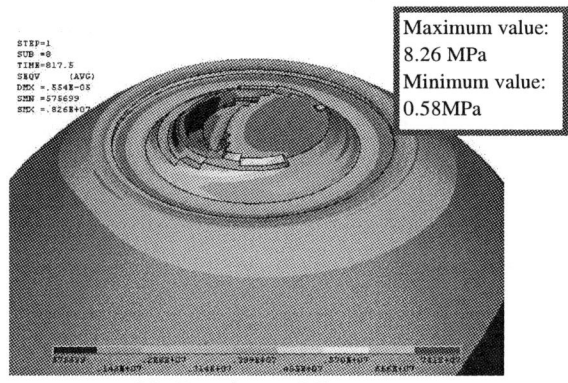

Maximum value:
8.26 MPa
Minimum value:
0.58MPa

(a) Local von-Mises stress

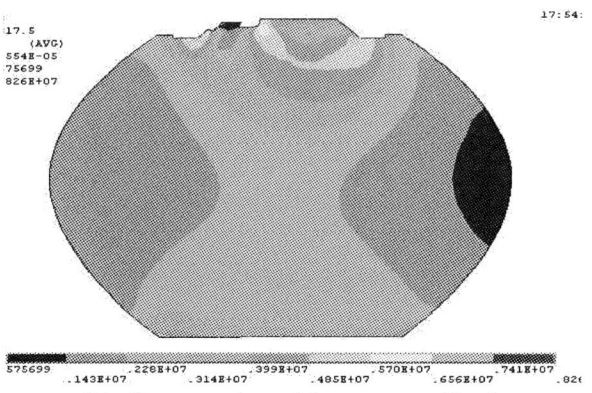

(b) Cross-section of the corner solder bump

Figure 25 von-Mises stress of SnPb solder at 817.5h

Figs. 24-25 show the von-Mises stress contour pictures of viscoplastic SnPb solder bump at 414 hours and 817.5 hours. The results show as the void grows, the von-Mises stress relaxes.

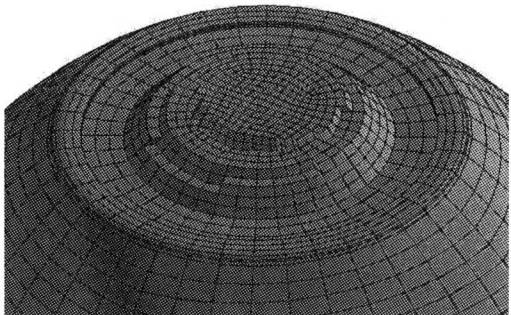

(a) Void at time to failure of SnAgCu solder material

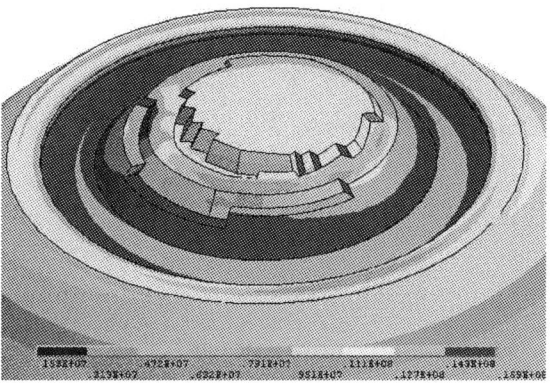

(b) von-Mises stress at time to failure

Figure 26 Pb-free material SnAgCu time to failure
at 1074.6 hours

Fig.26 gives the time to failure of void generation at 1074.6 hours and its von-Mises stress distribution. It is clear that the pb-free material SnAgCu has longer TTF life than SnPb material.

5 Conclusions

The three-dimensional electrical, thermal and stress direct and in-direct coupled analyses for electromigration simulation are presented to examine and quantify the effects of current crowding, joule heating and stress in an IC device and solder joint of a package. A sub-modeling technique is developed to better simulate the electromigration, thermal migration and stress migration. Comparison between our numerical simulation results and the experimental results from previous work has shown the good agreement.

The simulation has disclosed that direct coupling method seems to give higher stress and atomic flux divergence than in-direct couple method, this is because the element in ANSYS direct coupling analysis only has the linear properties, while the indirect method may make full use of the materials non-linear properties. The simulated results also found that Pb-free material SnAgCu seems to have less migration failure as compared to SnPb material.

Since the ultimate goal of this research project is to develop a damage mechanics model for both IC interconnects and solder joints under high current density stressing, it is important to investigate the mechanical property degradation of solder alloy during the migration process. Also, to get the right electromigration parameters for a Pb-free material requires a large amount of experimental work. These would be our future research projects.

Acknowledgments

The authors thank Fairchild Semiconductor and Chinese National Science Foundation (10372093) for support.

References

1. Choi, W. J., Yeh, E. C. C., Tu, K. N., "Electromigration of Flip Chip Solder Bump on Cu/Ni(V)/Al Thin Film under Bump Metallization," Electronic Components and Technology Conference. 2002, pp.1201-1205.

2. Black,J.R, Proc. 6th Annual Reliability Physics Symp., (IEEE, New York, 1967) pp.148-159.

3. James, R. B.,"Electromigration Failure Modes in Aluminum Metallization for Semiconductor Devices," Roceedings of the IEEE, Vol. 57, No. 9(1969), pp.1587-1594

4. Blech,I.A., "Electromigration in thin aluminum films on tianium nitride", Journal of Applied Physics, vol. 47, No.4 (1976), pp 1203-1208.

5. Tu, K. N., "Recent Advances on Electromigration in every-large-scale-integration of Interconnects," Journal of Applied physics, Vol. 94, No. 9 (2003), pp. 5451-5473.

6. Liang, S. W., Shao, T. L., Chen, C., "3-D Simulation on Current Density Distribution in Flip-Chip Solder Joints with Thick Cu UBM under Current Stressing," Electronic Components and Technology Conference, 2005, pp. 1416-1420.

7. Rinne, G., "Electro`migration in SnPb and Pb-Free Solder Bumps," IEEE ECTC2004, pp. 974-978.

8. Liang, S. W., Chang, Y. W., Shao, T. L., Chen, Chih., Tu, K. N., "Effct of three-dimensional current and temperature distributions on void formation and propagation in filip-chip solder joints," Applied physics Letter, Vol. 89, No.2 (2006), pp.021117

9. Nah, J. W., Ren, Fei, Tu, K. N. et al, "Electro migration in Pb-free filp chip solder joints on flexible substrates," Applied physics Letter, Vol. 99, No.2 (2006), pp. 023520

10. Lai, Y. S,, Chen, K. M., Kao, C. L. et al., "Electromigration of Sn-37Pb and Sn-3Ag-1.5Cu/Sn-3Ag-0.5Cu composite flip-chip solder bumps with Ti/Ni(V)/Cu under bump metallurgy." Microelectronics Reliability, 2006, .

11. Yue, H., Basaran, C., Hopkins, D., Lin, M.,"Modeling Deformation in microelectronics BGA solder joints under high current density, part I: simulation and testing," IEEE ECTC2005, 2005:1437-1444.

12. Dalleau, D. and Weide-Zaage, K., "Three-dimensional Voids Simulation in Chip-level Metallization Structures: A Contribution to Reliability Evaluation. Microelectronics Reliability," Vol. 41 (2001), pp. 1625–1630.

13. Dalleau, D., Weide-Zaage, K., "3-D Time-Depending Simulation of Void Formation in a SWEAT Metallization Structure," Proceedings of EuroSimE, April 2002.

14. Weide-Zaage, K, Dalleau, D., Yu, X. B.,"Static and Dynamic Analysis of Failure Locations and Void Formation in Interconnects due to Various Migration Mechanisms," Materials Science in Semiconductor Prcessing, No. 6 (2003), pp. 85-92.

15. Liu, Yong and Scott, Irving, "Power Cycling Simulation of an IC Package: Considering Electro migration,and Thermal-Mechanical Failure," ECTC2003, pp. 415-421

16. Liang, L.H. Xu, Y. J. and Liu Y., "Electro-Migration Study in Solder Joint and Interconnects of IC packages," Proceedings of EuroSIME2006, Apr. 2006, Como/Italy, pp. 464-470.

17. Liang, L.H. and Liu Y., "Reliability study in solder joint under electromigration thermal-mechanical load," International Conference on electronics packaging tchnology ,ICEPT2006, pp.861

18. Gee S, Kelkar N, Huang J, Tu K N. Lead-free and PbSn bump electromigration testing. Proceedings of InterPACK 2005, IPACK2005-73417, July 17-22.

19. Joseph Alison King, Material Handbook for Hybrid Microelectronics, Artech House, boston, 1998.

20. Darveaux,R., Effect of simulation methodology on solder joint crack growth correlation and fatigue life prediction, ASME J. of Electronic Package, Vol.124, 2002, pp147-152.

21. Qiang Wang, et al, Experimental Determination and Modification of the Anand Model Constants for 95.5Sn4.0Ag0.5Cu, Eurosime 2007, London, UK, April, 2007.

Fatigue Strength and Damage Behaviors of Multi-Scale Metallic Films and Multilayers

G. P. Zhang[*], X. F. Zhu, Y. P. Li and Z.G. Wang
Shenyang National Laboratory for Materials Science
Institute of Metal Research, Chinese Academy of Sciences
72 Wenhua Road, Shenyang 110016, P. R. China
* Corresponding author, E-mail: gpzhang@imr.ac.cn; Tel. & Fax: +86-24-23971938

Abstract

An understanding of fatigue reliability of thin metal films and multilayers with layer thickness ranging from micrometers to nanometers is becoming more and more important not only due to the rapid development of micro and nano technology, but also because of the demends of the fundamental research interests in small scale materials. In this paper, we will firstly present a couple of new testing methods for fatigue of metallic mutlilayers. Then, the current state in studies on fatigue strength and damage behaviors of thin metal films and multilayers are reviewed. Fatigue strength of the thin metal films investigated previously and that of the Cu-X multilayers measured by us recently were presented. Microscopic characterization of fatigue crack initiation in the thin metal films and the multilayers shows that there exists a significant variation in fatigue damage behaviors with decreasing layer thickness. The relationship between fatigue properties and microstructures of the materials, especially effects of length scale is discussed. The possible research direcitons about fatigue of metallic multilayers are suggested.

1. Introduction

Thin metal films and multilayered structures stacked alternatively by two or multiple components on a substrate are widely used in microelectronic devices and microsystems [1,2]. For example, metallization film and low-k dielectric layer or diffusion resistant layer are stacked alternatively on a Si substrate in a chip. Today, the miniaturization of the micro-devices leads to a tendency that individual layer thickness of the multilayer is now shrinking from submicrometer to nanometer scale. It is expected that the decrease in individual layer thickness would give rise to the variation in mechanical properties of the thin films and multilayers [3-5]. That would have an influence on yield strength and fatigue strength of the materials. As a result, the reliability of the thin metal films and multilayers under cyclic loading becomes a key issue.

To evaluate the fatigue properties of the multilayered composites, it is necessary to consider the following two important questions corresponding to fundamental and experimental aspects. One is a suitable fatigue testing method, which could be available to evaluate the fatigue properties of multilayers with different individual layer thickness from submicrometers to nanometers [6]. The other is size and interface effects on the fatigue properties of the thin films and multilayers. In this paper, we will first describe several new experimental methods employed currently for fatigue tests of the multilayers, and then the results from the fatigue tests of the metallic metal films and multilayers will be presented and discussed.

2. Testing Methods

Like thin metal films, fatigue testing of the metallic multilayers with a thickness around micrometers or less would be very difficult due to their small thickness. To obtain accurate and convincible fatigue data, firstly the fatigue machine should have an ultra low force load cell and a high resolution of displacement, secondly, the multilayered sample should be easily clamped and/or mounted in the machine, and thirdly, a simple and relatively homogenous stress state in the multilayered sample should be obtained. That would facilitate the extraction of fatigue data and make a reasonable comparison with bulk materials easier.

In view of these factors, we adopted the similar testing method as that used by Hommel et al. [7] to conduct fatigue tests of metallic multilayers which were deposited on a polyimide (PI) substrate with a high elasticity and mechanical stability. Through applying a uniaxial load to the metallic multilayer/a compliant PI substrate composite shown in Fig. 1, the multilayer/substrate composite is strained cyclically by tensile loading and unloading while the film is deformed elastically and plastically in tension and in compression on loading and unloading, respectively [8]. Using this method, a cyclic tension-compression load could be applied to the multilayers with

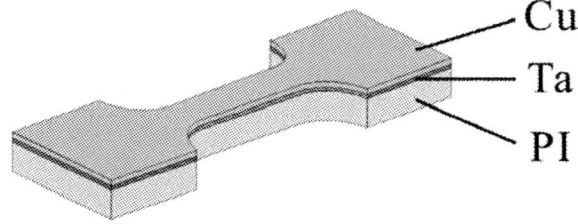

Fig. 1 Schematic illustration of tensile fatigue sample of metallic multilayer on a polyimide substrate.

a thickness even around tens of nanometers. Especially, the bulk-like fatigue sample of metallic multilayered composite can be fabricated easily and fixed by relatively large clamps on the fatigue machine. Importantly, the samples for transmission electron microscopy (TEM) can be also easily obtained from the fatigued specimens by conventional preparation methods.

Using this method, we have tested Cu-Ta and Cu-Ni multilayers with a total thickness of 1 μm and a individual layer thickness ranging from 10 nm to 500 nm. Experimental results demonstrate that such a method seems to be more suitable for fatigue investigations of the multilayers at room temperature [9].

In addition, a device based on the resonant frequency method was constructed by Wang et al. [10,11], as shown schametically in Fig. 2. By this method, fatigue properties of self-supported Cu/Nb multilayers with a total thickness of 40 μm and individual layer thickness of 40 nm were measured. Their results show that the resonant frequency method is efficient for fatigue testing of self-supported thin film materials. This method provides a new fatigue testing technique to investigate fatigue damage behaviors of metallic multilayers and/or thin films under a very high frequency and a relatively easy way to monitor fatigue failure lifetime compared with other methods.

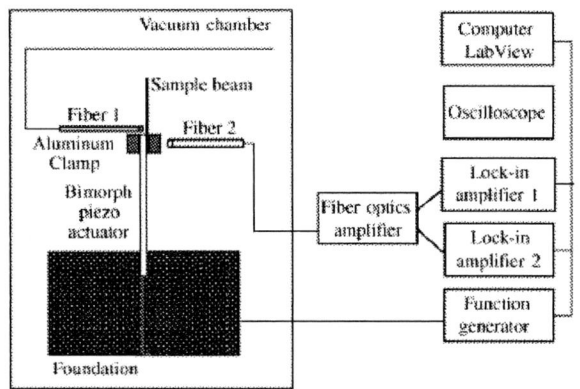

Fig. 2 Schematic of the resonant frequency device to perform fatigue tests with fully reversed stress amplitudes on thin films [10].

3. Fatigue Strength

Transition from the traditional engineering to micro/nano technology drives the research interests in fatigue toward small-scale materials, such as thin films, ultrafine wires and belts, multilayers. On the small scale, the design and reliability of the materials may not follow the conventional theories since the fatigue behavior of small-scale materials has been proven to be different from that of bulk materials due to size effects [8-30]. A couple of examples shown below can illustrate the variation in fatigue strength with the change of film thickness.

Judelewicz et al. [13] reported that Cu foils 100 μm thick always failed after a number of cycles which was 10-30 times lower than that for the thinner foils. Hong and Weil [16] found that fatigue strengthening coefficient of 25 μm and 33 μm thick Cu films is higher than that of bulk Cu. Such a size effect has recently been examined by Kraft et al. [17] through systematic fatigue testing of thin Cu films with a thickness from ~0.4 μm to ~3.0 μm. They

found that the fatigue strength of the 0.4 μm thick films is higher than that of micrometer-thick films, as shown in Fig. 3. Furthermore, Zhang et al. [18] measured fatigue strength of nanometer-thick Cu films, and found that the fatigue strength can be further increased even at the nanometer scale, but it is somewhat lower than that of submicrometer-thick films reported by Kraft. A comparison between the data reported by Zhang et al. [18] and Kraft et al. [17] is presented in Fig. 3. In addition to the Cu films [13-22], other investigations of thin metal foils and films, such as Ag films [15,23], Ni foils [24], steel foils [25,26] and even amorphous metal films [27] were also conducted. In general, the enhanced fatigue strength of the thin metal fims is attributed to the confinement of film thickness and small grain size to dislocation activities [21].

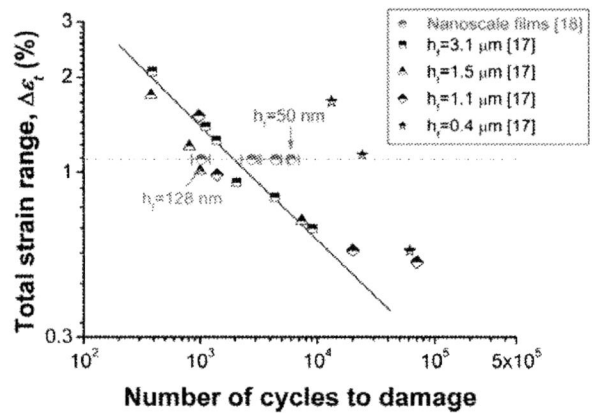

Fig. 3 A comparison of fatigue strength between nanometer-thick Cu films and micrometer/submicrometer-thick films.

Another comparison of fatigue strength between SUS 304 stainless steel thin film used in MEMS device and its bulk counterpart was conducted by Zhang et al. [26]. As can be seen in Fig. 4, the fatigue strength of SUS304 stainless steel films 25 μm thick is about 250 MPa and significantly higher than that of the bulk SUS304 stainless steel with the similar grain size.

Even though a number of studies on fatigue properties of thin metal films have been carried out, up till now there are very limited literature reporting fatigue properties and damage behaviors of metallic multilayers [10,11,28,29]. Stoudt et al. [28] reported that a 5 μm thick Cu/Ni multilayer with equal layer thickness of 20 nm on a Cu substrate can effectively suppress fatigue crack initation through decreasing surface roughening and enhance fatigue strength of the Cu substrate. Direct evaluation of fatigue strength of metallic multilayers was conducted recently by Wang et al. [10], who used the resonant frequeny method as mentioned above. Fig. 5 presents the normalized fatigue strength of the Cu/Nd multilayers. It was found that the ratio of the fatigue endurance limit to the ultimate tensile strength was about 0.35.

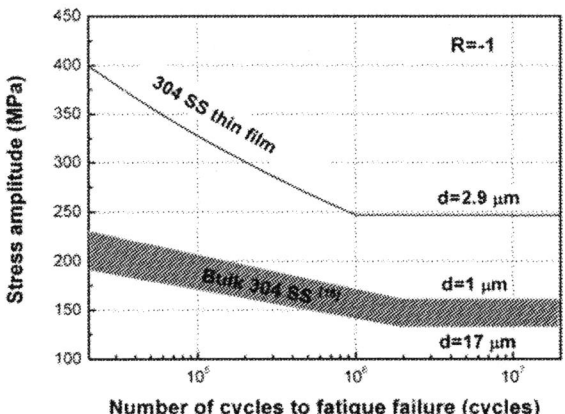

Fig. 4 A comparison of fatigue strength between SUS304 stainless steel film 25 μm thick and its bulk counterpart.

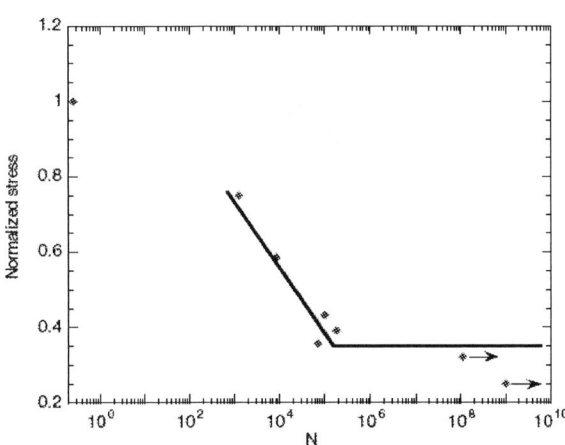

Fig. 5 Normalized S–N curve of the 40 nm Cu/Nb multilayers. The normalization is defined as ratio of maximum stress to ultimate tensile stress. [10]

Fatigue tests of Cu/Ta and Cu/Ni multilayers with individual layer thickness from 10 nm to 500 nm have been performed recently by Zhu and Zhang [9]. The multilayers were deposited on a polyimide substate by radio-frequency magnetron sputtering system. The multilayered composites were cyclically tensiled, and that resulted in the cyclic tension-compresion load on the mutlilayers. Preliminary results of the Cu/Ta multilayers show that the fatigue resistance to cracking would increase with decreasing the individual layer thickness down to 20 nm, while it begins to decrease when the individual layer thickness is 10 nm. The dependence of the fatigue strength on layer thickness is consistent with the variation in yield strength of the multilayers with layer thickness. Similarly, the Cu-Ni multilayers were also found to have the layer thickness dependent fatigue strength [9]. It is suggested that the enhanced fatigue stregnth of the metallic multilayers may orginate from the strengthening effects of fine scale layer thickness, nanometer scale grain size as well as interface.

4. Fatigue Damage Behaviors

For a thin metal film, there exist two length scales corresponding to film thickness and grain size. If the in-plane grain size of the film is larger than the film thickness, the film should have a columnar grain structure, while if it is less than the film thickness, the film may have an equal-axial grain structure. Fig. 6 presents an illustration of the relationship between film thickness and grain size of the thin Cu films investigated previously [30].

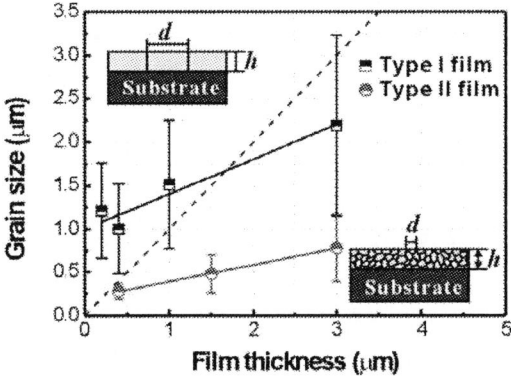

Fig. 6 Schematic illustration of the relationship between film thickness and grain size in the thin films studied by Zhang et al. [30]

Fatigue damage behaviors of thin Ag films [15,23] and Cu films [17,19-22,30] have been studied by means of focused ion beam (FIB) microscopy and transmission electron microscopy (TEM). These systematic observations established the relationship between fatigue damage behaviors and microstructures associated with film thickness and grain size. As can be seen from Fig. 7 [30], fatigue damage of the micrometer-thick Cu films results from typical "bulk-like" extrusions (Fig. 7(a)), which can be correlated with dislocation wall/cell structures (Fig. 7 (b)), while fatigue damage of the submicrometer-thick Cu films orginates from interface-induced cracking (Fig. 7(c)) without the formation of the extrusions and the dislocation structures (Fig. 7(d)). Combined with the experimental results, a theoretical model [30] related to length scales in the films further provides an in-depth understanding of length-scale dependent fatigue damage behaviors in the thin metal films.

Fig. 8 presents a mechanism map of fatigue of thin metal films at different length scales (both film thickness and grain size) based on the extensive experimental results in the literature and the prediction of the proposed length-scale-controlled model [30]. In the map, the filled-symbols indicate the occurrence of fatigue extrusions and typical dislocation structures, while the open-symbols mean the interface-induced damage and individual dislocations. These experimental results can be well separated into three *Regimes* by the proposed model. *Regime I* is the "bulk-like fatigue behavior", which is characterized by dislocation structures and surface

725

extrusions. *Regime II* is the "transition behavior" characterized by dislocation tangles and small surface extrusions or intense slip bands. While *Regime III* indicates the "small volume" behavior, which is characterized by individual dislocations and interface-mediated damage behavior such as cracking along grain boundaries (GBs) and twin boundaries (TBs), voiding at GBs, and grain boundary diffusion. In addition, Fig. 8 can also be divided into two large *Regimes* by a 45° line. The upper *regime* corresponds to length scales for which fatigue behavior is predominately controlled by the film thickness, whereas the lower *regime* corresponds to grain size controlled fatigue behavior. However, it is in fact that a wide distribution of grain sizes in a given film means that there will be competitive and/or coupled fatigue mechanisms.

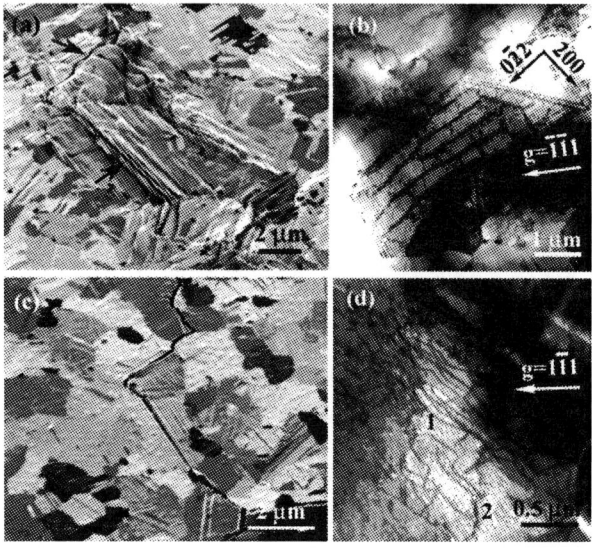

Fig. 7 Fatigue extrusions forming at the surfaces of the 3.0 μm thick film (a) and the corresponding dislocation walls (b); Interface-induced fatigue cracking forming at the surfaces of the 0.2 μm thick film (c) and the individual dislocations in the 0.4 μm thick film (d).

Based on a number of the experimental observations of the fatigue damage behaviors in submicrometer-thick metal films, it is expected that grain boudary-mediated mechanism will dominate fatigue behavior of nanometer-thick films. Our recent study [8, 9] on fatigue properties of the Cu/Ta multilayers seems to provide direct evidence of the GB-mediated fatigue damage behavior. In the 500 nm thick Cu/250 nm thick Ta bilayered films, it was found that GBs can be aligned in the localized deformation region at the film surface, as shown in Fig. 9, and that was attributed to grain boundary sliding (GBS) combined with a small amount of grain rotation due to the high stress concentration at the crack tip [8]. Furthermore, the cross-sectional view of the multilayers with differnet individual layer thickness reveals that GBS and/or grain rotation seem to become more prevalent in the multilayer with nanometer-scale

individual layer thickness [9]. However, a lot of work are still needed to be conducted by means of TEM and other microscopical analytical methods.

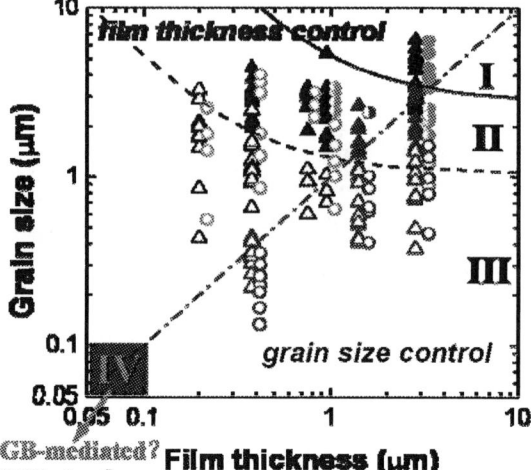

Fig. 8 Fatigue damage mechanism map for thin Cu films. Experimental observations are summarized with data symbols. The lines separating the map into regime I, II and III were obtained by the theoretical model. The further prediction of fatigue damage mechanisms at the length scale less than 100 nm is proposed.

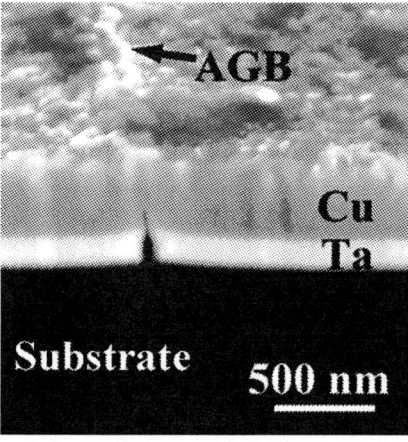

Fig. 9 SEM cross-sectional view of the damage behavior Cu/Ta bilayered film subjected to cyclic loading showing aligned grain boundaries (AGB) ahead of the crack tip.

Thin films and multilayers, as the basic structural components in microsystem and MEMS, have multi-physical scales and heterogenous interfaces, which strongly affect mechanical properties of the materials. Especially, fatigue properties of the materials may not simply be predicted and analyzed by the conventional theories developed from bulk materials. The previous investigations of fatigue behaviors of the thin metal films and the metallic multialyers have shown the obvious differences in fatigue properties between the film and the bulk counterpart. Thus, further systematic and in-depth studies should be conducted for all kinds of thin film systems and multilayers with different length scales and interface structures. In particular, the decrease in the length scale makes the interfaces and surfaces of the materials more active to be involved in fatigue properties. In view of these aspects, the following research directions may be taken into account in the furture as:

(1) Roles of interfaces, including GBs, TBs and the interfaces between two constituents of the multilayers and between the film and the substrate, in the origin of fatigue damage.

(2) Effects of individual layer thickness and grain size on fatigue resistance of metal thin films and multilayers.

(3) Unified understanding and more suitable theoretical models for the fatigue behaviors of the materials with small length scales.

(4) Data base of fatigue strength of all kinds of thin metal films and multilayers with multiscale obtained through newly-developed testing methods for small-scale materials.

5. Summary

Both the extensive studies of fatigue properties of the thin metal films and the limited investigations of fatigue of metallic multilayers reveal the general tendency that fatigue behaviors of the thin metal films and multilayers with different length scales strongly depend on layer thickness, grain size as well as interface, which are expected to have an influence on dislocation activities. Interface-mediated mechanism would become more prevalent at nanoscale, and should be considered as one of the most important factors in the evaluation of fatigue properties and the reliable design of metal film/multilayer-based structures, which are servicing in micro and nano systems now.

Acknowledgments

This work was supported by "The Hundred Talent Plan" of Chinese Academy of Sciences, the National Basic Research Program of China (Grant No. 2004CB619303) and the National Natural Science Foundation of China (Grant Nos. 50471081 and 50571103).

References

1. M. Ohring: Reliability and Failure of Electronic Materials and Devices, Academic Press (San Diego, 1998) pp. 91-99.
2. Menz W, Mohr J, Paul O, Microsystem Technology, Wiley-VCH, (Weinheim, 2001) pp. 289-368.
3. Arzt E., "Size Effects in Materials due to Microstructural and Dimensional Constraints: A Comparative Review" *Acta Mater.* Vol. 46 (1998), pp. 5611-5626.
4. Was G. S, Foecke T, "Deformation and fracture in microlaminates" *Thin Solid Films*, 286 (1996) pp. 1-31.
5. Misra A, Verdier M, Lu Y.C, Kung H, Mitchell T. E, Nastasi M, Embury J. D, "Structure and mechanical properties of Cu-X (X=Nb, Cr, Ni) nanolayered composites" *Scripta. Mater.* 39 (1998) pp. 555-560.
6. Zhang G. P, Volkert C. A, Schwaiger R, Mönig R, Kraft O, "Fatigue and Thermal Fatigue Damage Analysis of Thin Metal Films" *Proceedings of EuroSIME 2006*, Como, Italy, April. 2006, pp. 549-554.
7. Hommel M, Kraft O, Arzt E., "A new method to study cyclic deformation of thin films in tension and compression" *J. Mater. Res.* Vol. 14, No. 6 (1999), pp. 2373-2376.
8. Zhu X. F, Zhang G. P, Tan J, Zhu S. J, "Cracking behavior of Cu/Ta bilayered films under cyclic loading" Submitted to *J. Mater. Res.* (2007).
9. Zhu X. F, Zhang G. P, "Fatigue of metallic multilayers" *Unpublished Research.*
10. Wang Y. C, Misra A, Hoagland R. G, "Fatigue properties of nanoscale Cu/Nb multilayers" *Scripta Mater.* Vol. 54 (2006) pp. 1593-1598.
11. Wang Y. C, Hoechbauer T, Swadener J. G, Misra A. Hoagland R. G, Nastasi M, "Mechanical Fatigue Measurement via a Vibrating Cantilever Beam for Self-Supported Thin Solid Films" *Experi. Mech.* Vol. 46 (2006) pp. 503–517.
12. Hofbeck R, Hausmann K, Ilschner B, Künzi H. U., "Fatigue of Very Thin Copper and Gold Wires" *Scripta Metall.* Vol. 20 (1986), pp. 1601-1605.
13. Judelewicz M, Künzi H. U, Merk N, Ilschner B., "Microstructural Development during Fatigue of Copper Foils 20-100 μm Thick" *Mater. Sci. Eng.* A Vol. 186 (1994), pp.1-142.
14. Read T. D, "Tension-tension Fatigue of Cu Thin Films" *Int. J. Fatigue*, Vol. 20 (1998), pp. 203-209.
15. Schwaiger R, Kraft O., "High Cycle Fatigue of Thin Silver Films Investigated by Dynamic Microbeam Deflection" *Scripta Mater.*, Vol. 41 (1999), pp. 823-829.
16. Hong S, Weil R, "Low Cycle Fatigue of Thin Copper Foils" *Thin Solid Films*, Vol. 283 (1996), pp. 175-181.
17. Kraft O, Wellner P, Hommel M, Schwaiger R, Arzt E, "Fatigue Behavior of Polycrystalline Thin Copper Films" *Z. Metall.* Vol. 93 (2002), pp. 392-400.
18. Zhang G. P, Sun K. H, Zhang B, Gong J, Sun C, Wang Z. G, "Tensile and fatigue strength of ultrathin copper films" To be published in *Mater. Sci. Eng.* A (2007).
19. Kraft O, Schwaiger R, Wellner P, "Fatigue in thin films: lifetime and damage formation" *Mater Sci. Eng.* A Vol. 319-321 (2001), pp. 919-923.
20. Schwaiger R, Dehm G, Kraft O, "Cyclic Deformation of Polycrystalline Cu Films" *Phil Mag.*, Vol. 83 (2003), pp. 693-710.
21. Zhang G. P, Schwaiger R, Volkert C. A, Kraft O. "Effect of Film Thickness and Grain Size on Fatigue-

Induced Dislocations Structures in Cu Films" *Phil Mag Lett.*, Vol. 83 (2003), pp. 477-483.

22. Zhang G. P, Volkert C. A, Schwaiger R, Arzt E, Kraft O, "Damage Behavior of 200-nm Thin Copper Films under Cyclic Loading" *J Mater Res.*, Vol. 20 (2005), pp. 201-207.

23. Schwaiger R, Kraft O, "Size effects in the fatigue behavior of thin Ag films" *Acta Mater.*, Vol. 51 (2003), pp. 195-206.

24. Allameh S. M, Lou J, Kavishe F, Buchheit T, Soboyejo W. O, "An investigation of fatigue in LIGA Ni MEMS thin films" *Mater. Sci. Eng.* A Vol. 371 (2004) 256-266.

25. Zhang G. P, Takashima K, Shimojo M, Higo Y. "Fatigue Behavior of Microsized Austenitic Stainless Steel Specimens" *Mater Lett.*, Vol. 57 (2003), pp.1555-1560.

26. Zhang G. P, Takashima K, Higo Y, "Fatigue strength of small-scale type 304 stainless steel thin films" *Mater. Sci. Eng.* A Vol. 426 (2006) 95-100.

27. Maekawa S, Takashima K, Shimojo M, Higo Y, "Fatigue tests of Ni-P amorphous alloy microcantilever beams" *Jpn. J. Appl. Phys.* Vol. 38 (1999) pp. 7194-7198.

28. Stoudt M. R, Cammarata R. C, Ricke R. E, "Suppression of fatigue cracking with nanometer-scale multilayered coatings" *Scripta Mater.* Vol. 43 (2000) pp. 491-496.

29. Stoudt M. R, Ricke R. E, Cammarata R. C, "The influence of a multilayered metallic coating on fatigue crack nucleation" *Int. J. of Fatigue* Vol. 23 (2001) pp. 215-223.

30. Zhang G. P, Volkert C. A, Schwaiger R, Wellner P, Arzt E, Kraft O, "Length-scale-controlled Fatigue Mechanisms in Thin Copper Films" *Acta Mater.* Vol. 54 (2006) pp. 3127-3139.

Prognostics and Health Monitoring of Electronics

Michael Pecht[1], Brian Tuchband[1], Nikhil Vichare[2], and Qu Jian Ying[3]

[1]CALCE – University of Maryland, College Park, MD 20742 USA
[2]Dell Inc., Austin, TX 78753 USA
[3]Tangshan Tianyi Chinese Medical, Western Medical Hospital

Abstract

As a result of intense global competition, companies are considering novel approaches to enhance the operational efficiency of their products. In addition, competitive market requirements as well as demands for increased warranties and the severe liability of product failures have forced manufacturers to improve the reliability of their products. Higher field reliability requires knowledge of in-service use and life cycle environmental and operational conditions. Thus, interest has been growing in monitoring the ongoing health of products and systems in order to predict failures and provide warnings in advance of catastrophic failure. This paper provides a basic understanding of prognostics and health monitoring of products and systems and the techniques being developed to enable prognostics for electronic systems.

1. Reliability and Prognostics

Reliability is the ability of a product or system to perform as intended (i.e., without failure and within specified performance limits) for a specified time, in its life-cycle environment. Prognostics is the prediction of the future state of health based on current and historical health conditions [1].

Commonly used electronics reliability prediction methods generally do not accurately account for the life-cycle environment of electronic equipment. This arises from fundamental flaws in the reliability assessment methodologies used, and uncertainties about the product life-cycle loads [2]. In fact, traditional reliability prediction methods based on the use of handbooks have been shown to be misleading and to provide erroneous life predictions, a fact that led the U.S. military to abandon their electronics reliability prediction methods [2]. Although the use of stress and damage models permits a more accurate account of the physics-of-failure [3], their application to long-term reliability predictions based on extrapolated short-term life testing data or field data is typically constrained by insufficient knowledge of the actual operating and environmental application conditions of the product.

Prognostics and health monitoring (PHM) techniques combine sensing, recording, and interpretation of environmental, operational, and performance-related parameters indicative of a system's health. Product health monitoring can be implemented through the use of various techniques to sense and interpret the parameters indicative of (i) performance degradation, such as deviation of operating parameters from their expected values; (ii) physical or electrical degradation, such as

material cracking, corrosion, interfacial delamination, or increases in electrical resistance or threshold voltage; or (iii) changes in a life-cycle environment, such as usage duration and frequency, ambient temperature and humidity, vibration, and shock. Based on the product's health, determined by its monitored life-cycle conditions, maintenance procedures can be developed. Health monitoring therefore permits new products to be concurrently designed for a life-cycle environment known through monitoring.

The framework for prognostics is shown in Figure 1. Sensor data from various levels of an electronic product or system will be monitored in-situ and analyzed using prognostic algorithms. Different implementation approaches can be adopted individually or in combination. These approaches will be discussed in detail in the next section. Ultimately, the objective is to predict the advent of failure in terms of a distribution of remaining life, level of degradation, or probability of mission survival.

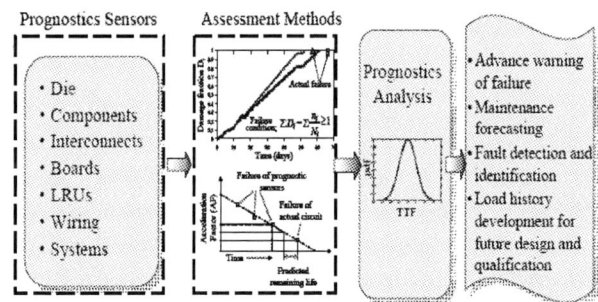

Figure 1. Framework for prognostics and health monitoring

2. PHM for Electronics

PHM has emerged as one of the key enablers for achieving efficient system-level maintenance and lowering life-cycle costs. In November 2002, the U.S. Deputy under Secretary of Defense for Logistics and Materiel Readiness released a policy called condition-based maintenance plus (CBM+) [4].

CBM+ represents an effort to shift unscheduled corrective equipment maintenance of new and legacy systems to preventive and predictive approaches that schedule maintenance based upon the evidence of need. The importance of PHM implementation was explicitly stated in the DoD 5000.2 policy document on defense acquisition [5], which states that "program managers shall optimize operational readiness through affordable, integrated, embedded diagnostics and prognostics, and embedded training and testing, serialized item management, automatic identification technology (AIT),

1-4244-1105-X/07/$25.00 ©2007 IEEE

and iterative technology refreshment." Thus, PHM has become a requirement for any system sold to the DoD. A 2005 survey of eleven CBM programs highlighted "electronics prognostics" as one of the most needed maintenance-related features or applications, without regard for cost [6], a view also shared by the avionics industry [7].

If one can assess the extent of deviation or degradation from an expected normal operating condition for electronics, this information can be used to meet several powerful goals, which include (1) advanced warning of failures; (2) minimizing unscheduled maintenance, extending maintenance cycles, and maintaining effectiveness through timely repair actions; (3) reducing the life-cycle cost of equipment by decreasing inspection costs, downtime, and inventory; and (4) improving qualification and assisting in the design and logistical support of fielded and future systems. In other words, since electronics are playing an increasingly large role in providing operational capabilities for today's systems, prognostic techniques have become highly desirable.

3. PHM Concepts and Methods

The first efforts in diagnostic health monitoring of electronics involved the use of built-in test (BIT), defined as an on-board hardware-software diagnostic means to identify and locate faults. A BIT can consist of error detection and correction circuits, totally self-checking circuits, and self-verification circuits.

Several studies conducted on the use of BIT for fault identification and diagnostics showed that BIT can be prone to false alarms and can result in unnecessary costly replacement, re-qualification, delayed shipping, and loss of system availability. However, there is also reason to believe that many of the failures were "real" but intermittent in nature [8]. The persistence of such issues over the years is perhaps because the use of BIT has been restricted to low-volume systems. Thus, BIT has generally not been designed to provide prognostics or remaining useful life due to accumulated damage or progression of faults. Rather, it has served primarily as a diagnostic tool.

The different approaches to prognostics and the state of research in electronics PHM are presented here. Three current approaches include (1) the use of fuses and canary devices; (2) monitoring and reasoning of failure precursors; and (3) monitoring environmental and usage loads for damage modeling.

3.1 Fuses and Canaries

Expendable devices, such as fuses and canaries, have been a traditional method of protection for structures and electrical power systems. Fuses and circuit breakers are examples of elements used in electronic products to sense excessive current drain and to disconnect power from the concerned part. Fuses within circuits safeguard parts against voltage transients or excessive power dissipation, and protect power supplies from shorted parts. For

example, thermostats can be used to sense critical temperature limiting conditions, and to shut down the product, or a part of the system, until the temperature returns to normal. In some products, self-checking circuitry can also be incorporated to sense abnormal conditions and to make adjustments to restore normal conditions, or to activate switching means to compensate for the malfunction [9].

The word "canary" is derived from one of coal mining's earliest systems for warning of the presence of hazardous gas using the canary bird. Because the canary is more sensitive to hazardous gases than humans, the death or sickening of the canary was an indication to the miners to get out of the shaft. The canary thus provided an effective early warning of catastrophic failure that was easy to interpret. The same approach has been employed in prognostic health monitoring. Canary devices mounted on the actual product can also be used to provide advance warning of failure due to specific wearout failure mechanisms.

Mishra et al. [10] studied the applicability of semiconductor-level health monitors by using pre-calibrated cells (circuits) located on the same chip with the actual circuitry. The prognostics cell approach, known as Sentinel Semiconductor™ technology, has been commercialized by Ridgetop Group to provide an early warning sentinel for upcoming device failures [11]. The prognostic cells are available for 0.35, 0.25, and 0.18 micron CMOS processes; the power consumption is approximately 600 microwatts. The cell size is typically 800 μm^2 at the 0.25 micron process size. Currently, prognostic cells are available for semiconductor failure mechanisms such as electrostatic discharge (ESD), hot carrier, metal migration, dielectric breakdown, and radiation effects.

The time to failure of these prognostic cells can be pre-calibrated with respect to the time to failure of the actual product. Because of their location, these cells contain and experience substantially similar dependencies as does the actual product. The stresses that contribute to degradation of the circuit include voltage, current, temperature, humidity, and radiation. Since the operational stresses are the same, the damage rate is expected to be the same for both the circuits. However, the prognostic cell is designed to fail faster through increased stress on the cell structure by means of scaling.

Scaling can be achieved by controlled increase of the current density inside the cells. With the same amount of current passing through both circuits, if the cross-sectional area of the current-carrying paths in the cells is decreased, a higher current density is achieved. Further control in current density can be achieved by increasing the voltage level applied to the cells. A combination of both of these techniques can also be used. Higher current density leads to higher internal (joule) heating, causing greater stress on the cells. When a current of higher density passes through the cells, they are expected to fail faster than the actual circuit.

Figure 2 shows the failure distribution of the actual product and the canary health monitors. Under the same environmental and operational loading conditions, the canary health monitors wear out faster to indicate the impending failure of the actual product. Canaries can be calibrated to provide sufficient advance warning of failure (prognostic distance) to enable appropriate maintenance and replacement activities. This point can be adjusted to some other early indication level. Multiple trigger points can also be provided, using multiple cells evenly spaced over the bathtub curve.

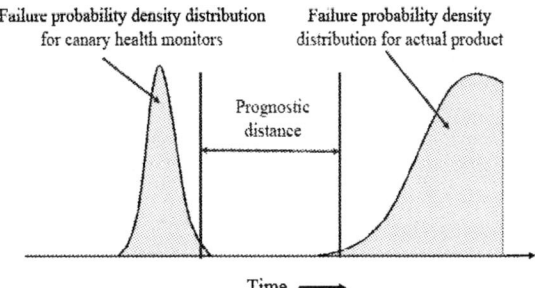

Figure 2. Advanced warning of failure using canary structures

The extension of this approach to board-level failures was proposed by Anderson et al. [12], who created canary components (located on the same printed circuit board) that include the same mechanisms that lead to failure in actual components. Anderson et al. identified two prospective failure mechanisms: (1) low cycle fatigue of solder joints, assessed by monitoring solder joints on and within the canary package; and (2) corrosion monitoring, using circuits that are susceptible to corrosion. The environmental degradation of these canaries was assessed using accelerated testing, and degradation levels were calibrated and correlated to actual failure levels of the main system. The corrosion test device included an electrical circuitry susceptible to various corrosion-induced mechanisms. Impedance spectroscopy was proposed for identifying changes in the circuits by measuring the magnitude and phase angle of impedance as a function of frequency. The change in impedance characteristics can be correlated to indicate specific degradation mechanisms.

Still, there remain unanswered questions with the use of fuses and canaries for PHM. For example, if a canary monitoring a circuit is replaced, what is the impact when the product is re-energized? What protective architectures are appropriate for post-repair operations? What maintenance guidance must be documented and followed when fail-safe protective architectures have or have not been included? This approach is difficult to implement in legacy systems, because it may require re-qualification of the entire system with the canary module. Also, the integration of fuses and canaries with the host electronic system could be an issue with respect to real estate on semiconductors and boards. Finally, the company must ensure that the additional cost of implementing PHM can

be recovered through increased operational and maintenance efficiencies.

3.2 Monitoring and Reasoning of Failure Precursors

A failure precursor is an event that signifies impending failure. A precursor indication is usually a change in a measurable variable that can be associated with subsequent failure. For example, a shift in the output voltage of a power supply would suggest impending failure due to a damaged feedback regulator and opto-isolator circuitry. Failures can then be predicted by using a causal relationship between a measured variable that can be correlated with subsequent failure.

A first step in PHM is to select the life-cycle parameters to be monitored. Parameters can be identified based on factors that are crucial for safety, that are likely to cause catastrophic failures, that are essential for mission completeness, or that can result in long downtimes. Selection can also be based on knowledge of the critical parameters established by past experience and field failure data on similar products and on qualification testing. More systematic methods, such as failure mode mechanisms and effects analysis (FMMEA) [13], can be used to determine parameters that need to be monitored.

Pecht et al. [14] proposed several measurable parameters that can be used as failure precursors for electronic components including switching power supplies, cables and connectors, CMOS integrated circuits, and voltage-controlled high-frequency oscillators. Testing was conducted to demonstrate the potential of select parameters for viably detecting incipient failures in electronic systems.

Supply current monitoring is routinely performed for testing of CMOS ICs. This method is based upon the notion that defective circuits produce an abnormal or at least significantly different amount of current than the current produced by fault-free circuits. This excess current can be sensed to detect faults. The power supply current (Idd) can be defined by two elements: the Iddq-quiescent current and the Iddt-transient or dynamic current. Iddq is the leakage current drawn by the CMOS circuit when it is in a stable (quiescent) state. Iddt is the supply current produced by circuits under test during a transition period after the input has been applied. Iddq has been reported to have the potential for detecting defects such as bridging, opens, and parasitic transistor defects. Operational and environmental stresses, such as temperature, voltage, and radiation, can quickly degrade previously undetected faults and increase the leakage current (Iddq). There is extensive literature on Iddq testing, but little has been done on using Iddq for in-situ PHM. Monitoring Iddq has been more popular than monitoring Iddt [15].

Smith et al. [15] developed a quiescent current monitor (QCM) that can detect elevated Iddq current in real time during operation. The QCM performed leakage current measurements on every transition of the system clock to get maximum coverage of the IC in real time. Pecuh et al. [16] and Xue et al. [17] proposed a low-power built-in current monitor for CMOS devices. In the Pecuh et al. study, the current monitor was developed and tested on a series of inverters for simulating open and short faults. Both fault types were successfully detected and operational speeds of up to 100 MHz were achieved with negligible effect on the performance of the circuit under test. The current sensor developed by Xue and Walker enabled Iddq monitoring at a resolution level of 10 pA. The system translated the current level into a digital signal with scan chain readout. This concept was verified by fabrication on a test chip.

Kanniche et al. [18] developed an algorithm for health monitoring of voltage source inverters with pulse width modulation. The algorithm was designed to detect and identify transistor open circuit faults and intermittent misfiring faults occurring in electronic drives. The mathematical foundations of the algorithm were based on discrete wavelet transform (DWT) and fuzzy logic (FL). Current waveforms were monitored and continuously analyzed using DWT to identify faults that may occur due to constant stress, voltage swings, rapid speed variations, frequent stop/start-ups, and constant overloads. After fault detection, "if-then" fuzzy rules were used for VLSI fault diagnosis to pinpoint the fault device. The algorithm was demonstrated to detect certain intermittent faults under laboratory experimental conditions.

Self-monitoring analysis and reporting technology (SMART), currently employed in select computing equipment for hard disk drives (HDD), is another example of precursor monitoring [19]. HDD operating parameters, including the flying height of the head, error counts, variations in spin time, temperature, and data transfer rates, are monitored to provide advance warning of failures (see Table 1). This is achieved through an interface between the computer's start-up program (BIOS) and the hard disk drive.

Systems for early fault detection and failure prediction are being developed using variables such as current, voltage, and temperature, continuously monitored at various locations inside the system. Sun Microsystems refers to this approach as continuous system telemetry harnesses [20]. Along with sensor information, soft performance parameters such as loads, throughputs, queue lengths, and bit error rates are tracked. Prior to PHM implementation, characterization is conducted by monitoring the signals of different variables to learn a multivariate state estimation technique (MSET) model. Once the model is established using this data, it is used to predict the signal of a particular variable based on learned correlations among all variables [21]. Based on the expected variability in the value of a particular variable during application, a sequential probability ratio test (SPRT) is constructed. During actual monitoring the SPRT will be used to detect the deviations of the actual signal from the expected signal based on distributions (and not on single threshold value) [22].

During implementation, the performance variables are continuously monitored using sensors already existing in Sun Microsystems' servers and recorded in a circular file structure. The file retains data collected at high sampling rates for seventy-two hours and data collected at a lower sampling rate for thirty days. For each signal being monitored, an expected signal is generated using the MSET model.

Table 1. Monitoring parameters based on reliability concerns in hard drives

Reliability Issues	Parameters Monitored

• Heads/head assembly
  – crack on head
  – head contamination or resonance
  – bad connection to electronics module
• Motors/bearings
  – motor failure
  – worn bearing
  – excessive run-out
  – no spin
• Electronic module
  – circuit/chip failure
  – interconnection/solder joint failure
  – bad connection to drive or bus
• Media
  – scratch/defects
  – bad servo
  – ECC corrections
	• Head flying height: A downward trend in flying height will often precede a head crash.
	• Error Checking and Correction (ECC) use and error counts: The number of errors encountered by the drive, even if corrected internally, often signals problems developing with the drive.
	• Spin-up time: Changes in spin-up time can reflect problems with the spindle motor.
	• Temperature: Increases in drive temperature often signal spindle motor problems.
	• Data throughput: Reduction in the transfer rate of data can signal various internal problems.

This signal is generated in real time based on learned correlations during characterization (see Figure 3). A new signal of residuals is generated, which is the arithmetic difference of the actual and expected time-series signal values. These differences are used as input to the SPRT model, which continuously analyzes the deviations and provides an alarm if the deviations are of concern. The monitored data is analyzed to (1) provide alarms based on leading indicators of failure, and (2) enable use of monitored signals for fault diagnosis, root-cause analysis of no-fault-founds (NFF), and analysis of faults due to software aging [23].

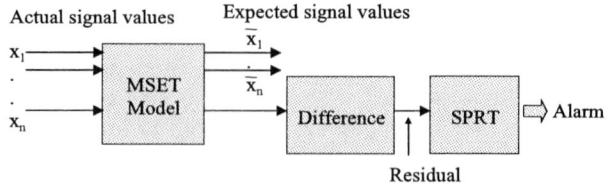

Figure 3. Sun Microsystems' approach to PHM

In general, to implement a precursor reasoning-based PHM system, it is necessary to identify the precursor variables for monitoring, and then develop a reasoning algorithm to correlate the change in the precursor variable with the impending failure. This characterization is typically performed by measuring the precursor variable under an expected or accelerated usage profile. Based on the characterization, a model is developed--typically a parametric curve-fit, neural-network, Bayesian network, or a time-series trending of a precursor signal. This approach assumes that there is one or more expected usage profiles that are predictable and can be simulated in a laboratory setup. In some products the usage profiles are predictable, but this is not always true.

For a fielded product with highly varying usage profiles, an unexpected change in the usage profile could result in a different (non-characterized) change in the precursor signal. If the precursor reasoning model is not characterized to factor in the uncertainty in life-cycle usage and environmental profiles, it may provide false alarms. Additionally, it may not always be possible to characterize the precursor signals under all possible usage scenarios (assuming they are known and can be simulated). Thus, the characterization and model development process can often be time-consuming and costly and may not always work.

3.3 Monitoring Environmental and Usage Loads for Damage Modelling

The life-cycle environment of a product consists of manufacturing, storage, handling, operating and non-operating conditions. The life-cycle loads (Table 2), either individually or in various combinations, may lead to performance or physical degradation of the product and reduce its service life [24].

Table 2. Examples of life-cycle loads

Load	Load Conditions
Thermal	Steady-state temperature, temperature ranges, temperature cycles, temperature gradients, ramp rates, heat dissipation
Mechanical	Pressure magnitude, pressure gradient, vibration, shock load, acoustic level, strain, stress
Chemical	Aggressive versus inert environment, humidity level, contamination, ozone, pollution, fuel spills
Physical	Radiation, electromagnetic interference, altitude
Electrical	Current, voltage, power, resistance

The extent and rate of product degradation depends upon the magnitude and duration of exposure (usage rate, frequency, and severity) to such loads. If one can measure these loads in-situ, the load profiles can be used in conjunction with damage models to assess the degradation due to cumulative load exposures.

The assessment of the impact of life-cycle usage and environmental loads on electronic structures and components was studied by Ramakrishnan et al. [24]. This study introduced the life-consumption monitoring (LCM) methodology (Figure 4), which combined in-situ measured loads with physics-based stress and damage models for assessing the life consumed.

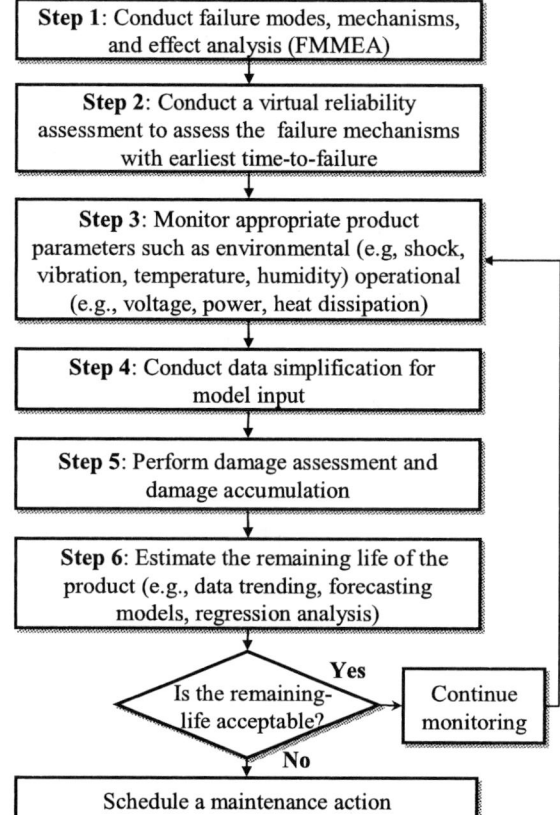

Figure 4. CALCE life-consumption monitoring methodology

The application of the LCM methodology to electronics PHM was illustrated with two case studies. The test vehicle consisted of an electronic component-board assembly placed under the hood of an automobile and subjected to normal driving conditions in the Washington, DC, area. The test board incorporated eight surface-mount leadless inductors soldered onto an FR-4 substrate using eutectic tin-lead solder. Solder joint fatigue was identified as the dominant failure mechanism. Temperature and vibrations were measured in-situ on the board in the application environment. Using the monitored environmental data, stress and damage models were developed and used to estimate consumed life.

The remaining life of the test board, estimated by LCM, is compared in Figure 5 with estimates obtained using similarity analysis, SAE handbook data, and the actual measured life. As shown in Figure 5, the remaining life estimated by either similarity analysis or using data from SAE handbook differs significantly from the actual life of the board, whereas the remaining life estimated by

733

LCM is in excellent agreement with actual life. The discrepancies between either similarity analysis or SAE estimates and actual life are attributed to the fact that neither approach accounts for the accident that the car experienced on day 22. Only LCM accounted for this unforeseen event because the operating environment was being monitored in-situ.

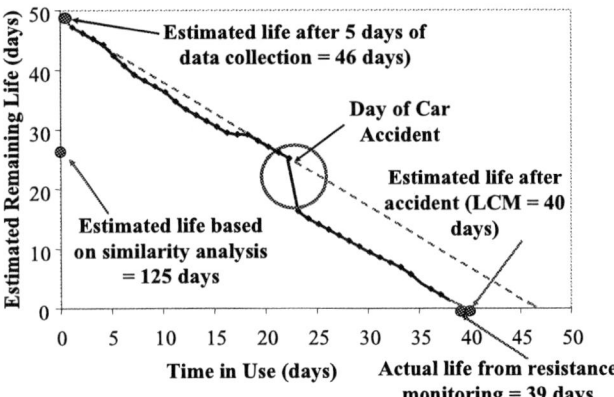

Figure 5. Remaining-life estimation of test board

Mathew et al. [25] applied the LCM methodology in conducting a prognostic remaining-life assessment of circuit cards inside a space shuttle solid rocket booster (SRB). Vibration-time history, recorded on the SRB from the pre-launch stage to splashdown, was used in conjunction with physics-based models to assess the damage caused due to vibration and shock loads. Using the entire life-cycle loading profile of the SRBs, the remaining life of the components and structures on the circuit cards were predicted. It was determined that an electrical failure was not expected within another forty missions. However, vibration and shock analysis exposed an unexpected failure of the circuit card due to a broken aluminum bracket mounted on the circuit card. Damage accumulation analysis determined that the aluminum brackets had lost significant life due to shock loading.

Shetty et al. [26] applied the LCM methodology for conducting a prognostic remaining-life assessment of the end effector electronics unit (EEEU) inside the robotic arm of the space shuttle remote manipulator system (SMRS). A life-cycle loading profile for thermal and vibrational loads was developed for the EEEU boards. Damage assessment was conducted using physics-based mechanical and thermomechanical damage models. A prognostic estimate using a combination of damage models, inspection, and accelerated testing showed that there was little degradation in the electronics and they could be expected to last another twenty years.

Vichare et al. [27] outlined generic strategies for in-situ load monitoring, including selecting appropriate parameters to monitor and designing an effective monitoring plan. Methods for processing the raw sensor data during in-situ monitoring to reduce the memory requirements and power consumption of the monitoring device were presented. Approaches were also presented

for embedding intelligent front-end data processing capabilities in monitoring systems to enable data reduction and simplification (without sacrificing relevant load information) prior to input in damage models for health assessment and prognostics.

Embedding the data reduction and load parameter extraction algorithms into the sensor modules as suggested by Vichare et al. can lead to reduction in on-board storage space, low power consumption, and uninterrupted data collection over longer durations. A time-load signal can be monitored in-situ using sensors, and further processed to extract cyclic range (Δs), cyclic mean load (Smean), and rate of change of load (ds/dt), using embedded load extraction algorithms. The extracted load parameters can be stored in appropriately binned histograms to achieve further data reduction. After the binned data is downloaded, it can be used to estimate the distributions of the load parameters. The usage history is used for damage accumulation and remaining life prediction.

Efforts to monitor life-cycle load data on avionics modules can be found in time-stress measurement device (TSMD) studies. Over the years TSMD designs have been upgraded using advanced sensors, [28] and miniaturized TSMDs are being developed due to advances in microprocessor and non-volatile memory technologies [29].

Searls et al. [30] undertook in-situ temperature measurements in both notebook and desktop computers used in different parts of the world. In terms of the commercial applications of this approach, IBM has installed temperature sensors on hard drives (Drive-TIP) [31] to mitigate risks due to severe temperature conditions, such as thermal tilt of the disk stack and actuator arm, off-track writing, data corruptions on adjacent cylinders, and outgassing of lubricants on the spindle motor. The sensor is controlled using a dedicated algorithm to generate errors and control fan speeds.

Strategies for efficient in-situ health monitoring of notebook computers were provided by Vichare et al. [32] In this study, the authors monitored and statistically analyzed the temperatures inside a notebook computer, including those experienced during usage, storage, and transportation, and discussed the need to collect such data both to improve the thermal design of the product and to monitor prognostic health. The temperature data was processed using two algorithms: (1) ordered overall range (OOR) to convert an irregular time-temperature history into peaks and valleys and also to remove noise due to small cycles and sensor variations, and (2) a three-parameter Rainflow algorithm to process the OOR results to extract full and half cycles with cyclic range, mean, and ramp rates. The effects of power cycles, usage history, CPU computing resources usage, and external thermal environment on peak transient thermal loads were characterized.

Skormin et al. [33] developed a data-mining model for failure prognostics of avionics units. The model provides a means of efficiently clustering data on parameters measured during operation, such as vibration, temperature, power supply, functional overload, and air pressure. These parameters are monitored in-situ on the flight using time-stress measurement devices. The objectives of the model are (1) to investigate the role of measured environmental factors in the development of particular failure; (2) to investigate the role of combined effects of several factors; and (3) to re-evaluate the probability of failure on the basis of known exposure to particular adverse conditions. Unlike the physics-based assessments made by Ramakrishnan and Pecht, the data-mining model relies on the statistical data available from the records of a time-stress measurement device (TSMD) on cumulative exposure to environmental factors and operational conditions. The TSMD records, along with calculations of probability of failure of avionics units, are used for developing the prognostic model. The data mining enables an understanding of the usage history and allows tracing the cause of failure to individual operational and environmental conditions.

4. Implementation of PHM in an Electronic System

Implementing an effective PHM strategy for a complete product or system may require integrating different prognostic health monitoring approaches. The first step is an analysis to determine the weak link(s) in the system based on the potential failure modes and mechanisms to enable a more focused monitoring process. Once the potential failure modes, mechanisms, and effects (FMMEA) have been identified, a combination of canaries, precursor reasoning, and life-cycle damage modeling may be necessary, depending on the failure attributes. In fact, different approaches can be implemented based on the same sensor data.

For example, operational loads of computer system electronics such as temperature, voltage, current, and acceleration, can be used with damage models to calculate the susceptibility to electromigration between metallizations and thermal fatigue of interconnects, plated-through holes, die attach, etc. Also, the processor usage, current, and CPU temperature data can be used to build a statistical model that is based on the correlations between these parameters. This model can be appropriately trained to detect thermal anomalies and identify signs for certain transistor degradation.

Future electronic system designs will integrate sensing and processing modules that will enable in-situ PHM. Advances in sensors, microprocessors, compact non-volatile memory, battery technologies, and wireless telemetry have already enabled the implementation of sensor modules and autonomous data loggers. For in-situ health monitoring, integrated, miniaturized, low-power, reliable sensor systems operated using portable power supplies (such as batteries) are being developed. These sensor systems have a self-contained architecture requiring minimum or no intrusion into the host product, in addition to specialized sensors for monitoring localized parameters. Sensors with embedded algorithms will enable fault detection, diagnostics, and remaining life prognostics, which will ultimately drive the supply chain. The prognostic information will be linked via wireless communications to relay needs to maintenance officers. Automatic identification techniques such as radio frequency identification (RFID) will be used to locate parts in the supply chain, all integrated through a secure web portal to acquire and deliver replacement parts quickly on an as-needed basis.

References

1. Vichare, N. *et al*, "Prognostics and Health Management of Electronics", *IEEE Transactions on Components and Packaging Technologies*, Vol. 29, No. 1, March 2006, pp. 222-229.

2. Pecht, M. *et al*, "The IEEE Standards on Reliability Program and Reliability Prediction Methods for Electronic Equipment," *Microelectronics Reliability*, Vol. 42, 2002, pp. 1259-1266.

3. Dasgupta, A., "The Physics-of-Failure Approach at the University of Maryland for the Development of Reliable Electronics," *Proceedings of 3rd International Conference on Thermal and Mechanical Simulation in (Micro) Electronics (EuroSimE)*, pp. 10-17, 2002.

4. Condition Based Maintenance Plus, http://www.acq.osd.mil/log/mppr/CBM%2B.htm.

5. DoD 5000.2 Policy Document, Defense Acquisition Guidebook, Chapter 5.3 – Performance Based Logistics, December 2004.

6. Cutter, D. *et al*, "Condition-Based Maintenance Plus Select Program Survey," Report LG301T6, January 2005, viewed from http://www.acq.osd.mil/log/mppr/CBM%2B.htm, October 2005.

7. Kirkland, L. *et al*, "Avionics Health Management: Searching for the Prognostics Grail," *Proceedings of the IEEE Aerospace Conference*, Vol. 5, 6-13 March 2004, pp. 3448 – 3454.

8. Williams, R. *et al*, "An Investigation of 'Cannot Duplicate' Failure," *Quality and Reliability Engineering International*, Vol. 14, 1998, pp. 331-337.

9. Ramakrishnan, A. *et al*, Electronic Hardware Reliability, Avionics Handbook, CRC Press, Boca Raton, Florida, December 2000, pp. 22-1 - 22-21.

10. Mishra, S. *et al*, "In-situ Sensors for Product Reliability Monitoring," *Proceedings of the SPIE*, Vol. 4755, 2002, pp. 10-19.

11. Ridgetop Semiconductor-Sentinel Silicon™ Library, "Hot Carrier (HC) Prognostic Cell," August 2004.

12. Anderson, N. *et al*, "Framework for Prognostics of Electronic Systems," *Proceedings of International Military and Aerospace/Avionics COTS Conference*, Seattle, WA, August 3-5, 2004.

13. Ganesan, S. *et al*, "Identification and Utilization of Failure Mechanisms to Enhance FMEA and FMECA," *Proceedings of the IEEE Workshop on*

Accelerated Stress Testing and Reliability (ASTR), Austin, Texas, October 3-5, 2005.

14. Pecht, M. *et al*, <u>Guidebook for Managing Silicon Chip Reliability</u>, CRC Press, Boca Raton, FL, 1999.

15. Smith, P. *et al*, "Practical Implementation of BICS for Safety-Critical Applications," *Proceedings of IEEE International Workshop on Current and Defect Based Testing-DBT*, 30 April 2000 pp. 51 – 56.

16. Pecuh, I. *et al*, "1.5 Volts Iddq/Iddt Current Monitor," *Proceedings of the IEEE Canadian Conference on Electrical and Computer Engineering*, Vol. 1, 9-12 May 1999, pp. 472 – 476.

17. Xue, B. *et al*, "Built-In Current Sensor for IDDQ Test," *Proceedings of the IEEE International Workshop on Current and Defect Based Testing-DBT*, 25 April 2004, pp. 3 – 9.

18. Kanniche, M. *et al*, "Wavelet based Fuzzy Algorithm for Condition Monitoring of Voltage Source Inverters," *Electronic Letters*, Vol. 40, No. 4, February 2004.

19. Self-Monitoring Analysis and Reporting Technology (SMART), PC Guide, http://www.pcguide.com/ref/ hdd/perf/qual/featuresSMART-c.html, viewed on August 30, 2005.

20. Gross, K., "Continuous System Telemetry Harness," Sun Labs Open House, 2004, http://research.sun.com/sunlabsday/docs.2004/talks/1 .03_Gross.pdf, viewed in August 2005.

21. Whisnant, K. *et al*, "Proactive Fault Monitoring in Enterprise Servers," *2005 IEEE International Multiconference in Computer Science and Computer Engineering*, Las Vegas, NV, June 2005.

22. Mishra, K. *et al*, "Dynamic Stimulation Tool for Improved Performance Modeling and Resource Provisioning of Enterprise Servers," *Proceedings of the 14th IEEE International Symposium on Software Reliability Engineering (ISSRE'03)*, Denver, CO, November 2003.

23. Vaidyanathan, K. *et al*, "MSET Performance Optimization for Detection of Software Aging," *Proceedings of the 14th IEEE International Symposium on Software Reliability Engineering (ISSRE'03)*, Denver, CO, November 2003.

24. Ramakrishnan, A. *et al*, "A Life Consumption Monitoring Methodology for Electronic Systems," *IEEE Transactions on Components and Packaging Technologies*, Vol. 26, No. 3, September 2003, pp. 625-634.

25. Mathew, S. *et al*, "Prognostic Assessment of Aluminum Support Structure on a Printed Circuit Board," *ASME Journal of Electronic Packaging*, Vol. 128, Issue 4, pp. 339-345, December 2006.

26. Shetty, V. *et al*, "Remaining Life Assessment of Shuttle Remote Manipulator System End Effector," *Proceedings of the 22nd Space Simulation Conference*, Ellicott City, MD, October 21-23, 2002.

27. Vichare, N. *et al*, "Monitoring Environment and Usage of Electronic Products for Health Assessment and Product Design", *IEEE Workshop on Accelerated Stress Testing and Reliability (ASTR)*, Austin, TX, USA, October 2-5, 2005.

28. Harchani, N. *et al*, "Time Stress Measurement Device: System Design and Synthesis," *Proceedings of the SPIE*, Vol. 4051, 2000, pp. 337-48.

29. Rouet, V. *et al*, "Development and Use of a Miniaturized Health Monitoring Device," *Proceedings of the IEEE International Reliability Physics Symposium*, 2004, pp. 645-646.

30. Searls, D. *et al*, "A Strategy for Enabling Data Driven Product Decisions Through a Comprehensive Understanding of the Usage Environment," *Proceedings of IPACK'01*, Kauai, Hawaii, USA, July 8-13, 2001.

31. Herbst, G., "IBM's Drive Temperature Indicator Processor (Drive-TIP) Helps Ensure High Drive Reliability," IBM White Paper, http://www.hc.kz/pdf/drivetemp.pdf, viewed in September 2005.

32. Vichare N. *et al* "In-Situ Temperature Measurement of a Notebook Computer - A Case Study in Health and Usage Monitoring of Electronics," *IEEE Transactions on Device and Materials Reliability*, Vol. 4., No. 4, December 2004, pp. 658-663.

33. Skormin, V. *et al*, "Data Mining Technology for Failure Prognostic of Avionics," *IEEE Transactions on Aerospace and Electronic Systems*, Vol. 38, No. 2, April 2002, pp. 388 – 403.

The Changing Role of CFD in Electronics Thermal Design

Dr. John Parry, CEng
Flomerics Ltd
81 Bridge Road, East Molesey, Surrey KT8 9HH, UK
john.parry@flomericsgroup.com

Invited Keynote Talk

Abstract

The presentation will discuss how the use of computational fluid dynamics (CFD) technology in the thermal design of electronics systems has changed since its earliest use in electronics design in the late 1980s.

1. Rationale for Using CFD

The use of CFD largely post-dates the very high module power densities seen at the end of the bipolar technology period, as shown in Figure 1. Perhaps a relevant question to ask is "why was CFD used?" A decade earlier engineers in the computer industry were able to cool much higher heat fluxes, as shown in Figure 1.

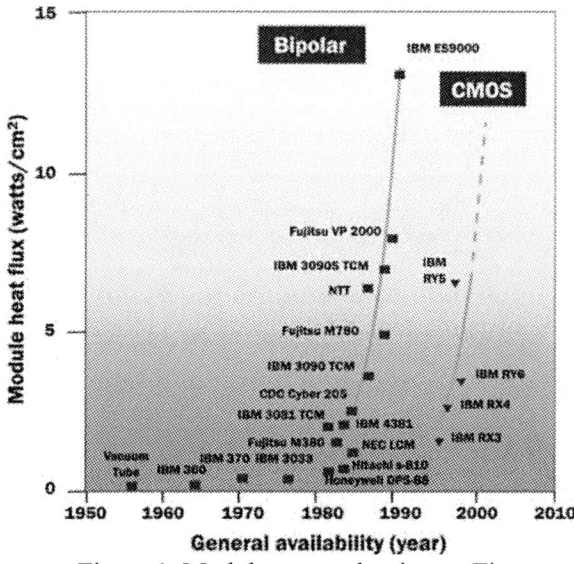

Figure 1: Module power density vs. Time

What had changed? What problems were engineers now trying to address, and where did CFD fit in the overall design flow? The presentation will attempt to address these questions by walking through the changes that have occurred since the late 1980s through to the present day.

2. Walkthrough of the use of CFD

Since the late 1980s the challenges facing designers have changed considerably. This has driven efforts in research, supply chain and design flow integration, and even international standardisation. Over the last two decades the types of problem being solved have changed; cooling solutions have changed, and the market and user demographics have changed. Consequently the tools have had to change considerably - in terms of feature set, scalability, and ease-of-use. A key factor has been improved integration with the mechanical and electronics design flows, with new software tools emerging to provide integrated toolsets that can facilitate significant time savings in today's evolving electronics design flows. The presentation will track these changes over time and against the changing fortunes of the electronics sector over this time period, which was severely affected by the "dot-com bubble" bursting and the ensuing recession in the telecom, computing and networking sectors.

3. Viewpoints on the changing use of CFD

To help understand the changes that have taken place the presentation will look at the changes from a number of 'viewpoints' such as changes in user demographics; design practices; data sharing; thermal challenges at the different packaging levels; and cooling technologies.

4. Conclusions

To conclude, the presentation will discuss what changes are likely to be required in the future, as electronics and mechanical design flows continue to merge, with thermal management an increasingly important and integral part of the overall design process.

Will companies favour simple, robust CFD tools with tight integration with electronics design automation (EDA) tools for use at the concept design stage by electronics engineers and designers with little or no expertise in heat transfer?

Driven by the product lifecycle management (PLM) paradigm, will MCAD tools find use earlier in electronics product design, so CFD tools with tight MCAD integration find preference in the digital design world?

To some extent the answers will depend on how EDA and MCAD suites evolve. Some quotes from management guru Peter Drucker are relevant: "Trying to predict the future is like trying to drive down a country road at night with no lights while looking out the back window", and "The only thing we know about the future is that it will be different.". A final quote is perhaps the most encouraging: "The best way to predict the future is to create it".

1-4244-1105-X/07/$25.00 ©2007 IEEE   737

Digital and Continuous Liquid Cooling for Electronic Systems

M. Baelmans[1], H. Oprins[1,2], T. Stevens[1], F. Rogiers[1]

[1]Katholieke Universiteit Leuven, Department of Mechanical Engineering
Celestijnenlaan 300A, 3001Leuven-Heverlee, Belgium
martine.baelmans@mech.kuleuven.be
[2]IMEC, Kapeldreef 75, 3001 Leuven-Heverlee, Belgium

Abstract

In this paper both digital and liquid cooling concepts are proposed. Microchannel cooling techniques either actuated by a separate feed pump or actuated by electrowetting are investigated and their cooling capabilities and pumping requirements are assessed. Though the continuous liquid microchannel cooling clearly outperforms the other systems with respect to global thermal resistances, the low energy consumption and the flexibility as a reconfigurable cooling system, make the electro-actuated cooling systems attractive.

1. Introduction

Control of component temperatures and temperature gradients is essential for successful operation and reliability of electronics products [1]. However, increasing heat fluxes combined with smaller dimensions imply high local temperatures. Moreover, recent IC's typically have a non-uniform spatial power distribution with regions of hot spots. At the location of these hot spots, large temperature gradients are expected and may cause unacceptably high mechanical stress. As conventional cooling methods such as natural convection or fan-induced air-cooling cannot cope with the increasing demand for electronics cooling, more sophisticated cooling techniques are required. As such, novel cooling methods with improved performance are crucial to face the growing need of the electronics market.

In this context, forced convection microchannel liquid cooling is often envisaged as a promising technique [1]. As the downscaling of cooling channels has an important influence on pressure drops and heat transfer in the laminar fluid flows through these microchannels, a lot of attention is paid to these aspects in literature [2]. Pressure losses on micro-scale are typically studied in order to check the validity of classical laminar correlations and to study the discrepancy of the friction factor, the Nusselt number and the critical Reynolds number between micro-scale and conventionally sized channels.

In micro-channel heat sinks, optimization of channel geometry is necessary and should be oriented towards various types of heat sources (homogeneous, hot-spots). Liu and Garimella [3] investigated the optimization of micro-channel heat sink geometries with respect to minimal thermal resistance and concluded that even the simple analytical models offer sufficiently accurate predictions for practical design of water-cooled micro-channel heat sinks. In addition they found that entrance effects have a negligible effect on the calculation of heat transfer and fluid flow and therefore on the optimization

results. However, Ryu et al. stated that entrance effects might have contribution up to 15 % on the thermal resistance and therefore it might be important to take it into account when computing the cooling capacity of the heat sink [4]. Stevens et al. [5] investigated the effect of microchannel aspect ratios in combination with entrance effects. In this paper it was concluded that optimal heat sink design strongly depend on the choice of optimization constraints. Further, it turned out that in practice production techniques put a limit on the maximal channel heights and minimal wall widths. They concluded that the entrance effect is important in the determination of the optimal channel geometry; in their cases this effect reduces the thermal resistance with 3% only, while the required number of channels decreases with approximately 40%.

Though improved heat transfer can be achieved by the use of microchannels, a critical issue in applying microchannels for electronics cooling is the substantial pressure drop encountered over the heat exchanger. Thereby the necessary pumping device is most often quoted as a weak component with respect to reliable operation. In view of further miniaturization and integration, novel pumping technologies have been studied. In recent reviews on micropumps [6,7], several techniques that have been proposed so far rely on electrohydrodynamic, electrowetting, electro-osmotic or piezoelectric actuation.

In this paper we will explore microchannel cooling techniques actuated either by a separate feed pump that can provide the required pressure drop and mass flow for a global cooling system or by electrowetting at IC level. Electrowetting is a well-known technique to manipulate fluids on a millimeter or micrometer scale [8-10]. As such it is already used in miniaturized chemical and biological reactors [8,10]. This principle is recently investigated for its use in electronics cooling applications [11,12]. In the next section the working principles of the techniques under investigation are discussed. Subsequently, their performance is assessed and their characteristic features compared.

2. Different microchannel cooling configurations

The features of three microchannel cooling systems are explored in this paper: 1) a continuous liquid flow is circulated through the microchannels by means of a separate feed pump; 2) a digital droplet flow is pulled through the small channels using electrowetting actuation, which is controlled by numerous electrodes that are sequentially activated; 3) electrowetting is used to periodically fill and empty microchannels from a larger reservoir. Each of the above mentioned systems has its own characteristics and constraints: the continuous liquid flow is governed by the allowable pressure drop, whereas the electro-actuated droplet flow is limited by the maximal achievable droplet velocity. The third system on the other hand is limited by the filling frequency and the corresponding filling lenghts that can be achieved by the electrowetting action.

In order to assess these features, the three cooling approaches are elaborated for a 10x10mm chip with uniform heat flux distribution. The heat sink is made of silicon (k = 148 W/mK). Water is used as coolant with constant thermal properties evaluated at room temperature (ρ = 997 kg/m^3, μ = 8.55 10^{-4} Ns/m^2, c_p = 4179 J/kgK and k = 0.613 W/mK). The flow is incompressible and enters the microchannels with a prescribed temperature (T_{in} = 20°C). For convenience all system performances are compared based on the achieved thermal resistance.

3. Continuous liquid cooling

Based on experimental evidence [3,13,14] the assessment of continuous liquid cooling in microchannels can be studied by means of classical correlations for conventionally-sized channels as long as the Reynolds number Re_{Dh} (based on the hydraulic diameter D_h of the channels) is well below the critical Reynolds number of 2300. Indeed, experimental data shows that this number can significantly reduce in microchannels to values for the critical Reynolds number around 1600 [14].

In order to assess the overall performance of the continuous liquid cooling case, the main counteracting heat transfer phenomena are evaluated leading to a rough estimate for the optimal channel hydraulic diameter, the total thermal resistance and the required pumping power. The maximal heat transfer removal is on the one hand determined by the caloric capacity of the coolant given by

$$Q_{cal} = \dot{M}c_p(T_{out} - T_{in})$$

where $\dot{M}$ is the total mass flow through the heat sink, c_p the specific heat of the coolant and T_{in} and T_{out} the inlet and outlet coolant temperature respectively. On the other hand, assuming a constant heat flux to the channel, the convective heat transfer to the coolant

$$Q_{conv} = hA(T_{max} - T_{out})$$

is determining the total cooling capacity of the system. Here, $h = Nu.k/D_h$ is the convective heat transfer coefficient, Nu the Nusselt number and k the thermal conductivity of the coolant. T_{out} and T_{max} are the coolant outlet temperature and the maximal chip temperature respectively. A is the available area for heat transfer. The caloric capacity will become the limiting factor at small hydraulic diameters, whereas the convective heat transfer will govern the heat transfer at large channels. It was shown in [5] that conduction effects in most practical cases, due to the relative thick wall measures are negligible.

Based on correlations for pressure drop the total mass flow can be further elaborated in terms of geometrical parameters of the heat sink (see Figure 1):

$$\Delta p = f \frac{L}{D_h} \frac{\rho u^2}{2} = \frac{u\mu}{2} \frac{L}{D_h^2} f Re_D,$$

with f the Darcy friction factor, L the length of the channel, the mean coolant velocity u and $Re_D = \rho V D_h/\mu$ the Reynolds number based on the hydraulic diameter.

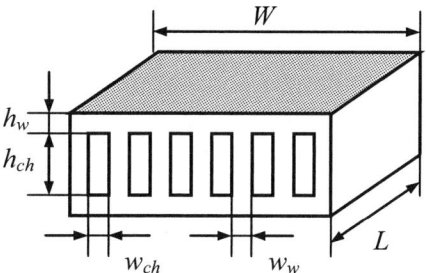

Figure 1: Heat sink geometrical design parameters

With the definition of the channel hydraulic diameter $D_h = 2w_{ch}h_{ch}/(w_{ch}+h_{ch})$ and the aspect ratio $\alpha = h_{ch}/w_{ch}$, and taking account that for fully developed laminar flow $f \cdot Re_D$ equals the Darcy friction factor constant C_{fRe}, the heat transfer removal by the coolant can be written as

$$Q_{cal} = \rho n \frac{(1+\alpha)^2}{2\alpha} \frac{\Delta p D_h^4}{\mu L C_{fRe}} c_p (T_{out} - T_{in})$$

with n the number of parallel channels in the heat sink. The caloric thermal resistance is then

$$R_{cal} = \frac{(T_{out} - T_{in})}{Q_{cal}} = \frac{2\alpha}{\rho n (1+\alpha)^2} \frac{\mu L C_{fRe}}{\Delta p D_h^4 c_p}.$$

Similarly, based on the geometrical relations, the convective thermal resistance becomes:

$$R_{conv} = \frac{(T_{max} - T_{out})}{Q_{conv}} = \frac{\alpha}{Nu \ k \ L \ n \ (1+\alpha)^2}$$

When for simplicity no wall thickness between the channels is taken into account, thus $n = W/w_{ch} = 2\alpha W/[(1+\alpha)D_h]$, it can be seen that for a given aspect ratio R_{cal} evolves with D_h^{-3} and R_{conv} with D_h. This results in an optimal hydraulic diameter, as can be seen in Figure 2, where the thermal resistance is plotted as a function of D_h for a heat sink with square channels (α = 1). The total thermal resistance of the heat sink becomes the series connection of the caloric and convective thermal resistances as obtained by equations (2) and (4). The optimal hydraulic diameter can now be calculated analytically by minimizing the total thermal resistance. This leads to:

$$D_h^4 = \frac{6Nu \ C_{fRe}\mu \ k \ L^2}{\rho \ \Delta p \ c_p}$$

It can be seen that the aspect ratio α is not explicitly present in this equation. However, it should be noted that, for the assumption of fully developed laminar flow, both the Darcy friction factor constant C_{fRe} and the Nusselt number Nu depend on the channel aspect ratio. For heat transfer, the Nusselt correlation for constant applied heat flux at 4 walls of the channel is used [15].

739

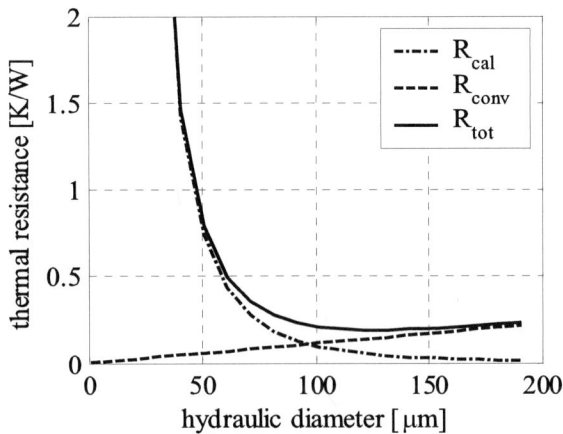

Figure 2: Optimal hydraulic diameter for $\alpha = 1$ and $\Delta p = 0.6$ bar and no wall thickness

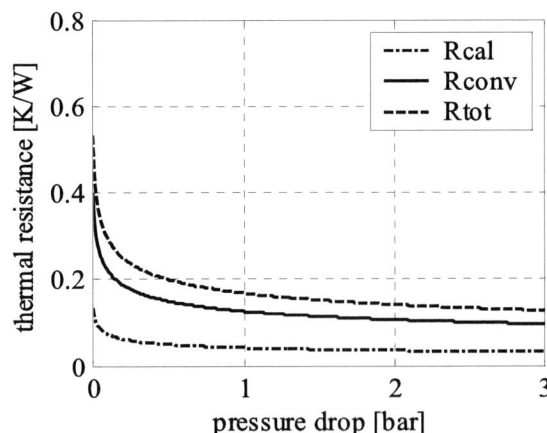

Figure 3: Total thermal resistance as a function of aspect ratio $\alpha = 1$ for $\Delta p = 0.6$; 1; 1.5; 2; 2.5 and 3 bar

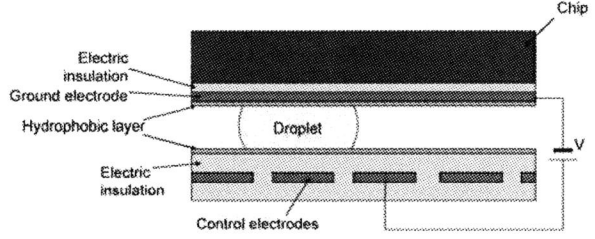

Figure 4: Total thermal resistance as a function of pressure drop for $\alpha = 1$

In [5] it is shown that at one hand the presence of finite walls will increase the overall thermal resistance (e.g. at $\Delta p = 0.6$ the thermal resistance was doubled by a

factor 2 for wall thicknesses of 100µm). On the other hand, as shown in Figure 3, this can be compensated by an increase in aspect ratio (the thermal resistance is approximately inverse proportional to α for $\alpha > 2$). In Figures 4 and 5 it is shown that an increase in pressure drop can further enhance the cooling capacities, with the penalty of an increased energy consumption. The power dissipation at 0.6 bar pressure difference amounts to 300 mW at a total volumetric flow of 5 ml/s. This high power consumption and relatively high pressure drops remain critical issues especially with respect to reliability and control of the cooling system, despite the promise of achieving very low thermal resistances.

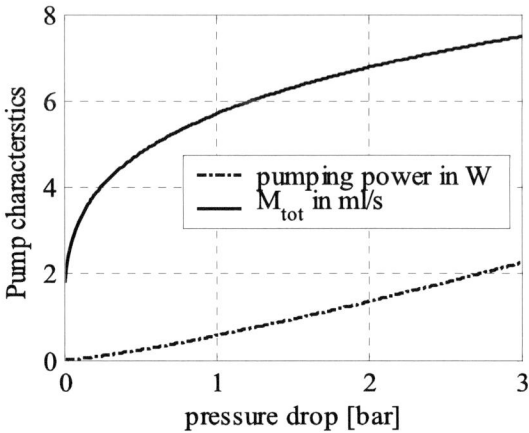

Figure 5: Pumping power and total mass flow through the cooler as a function of pressure drop for $\alpha = 1$

4. Electro-actuated liquid droplets

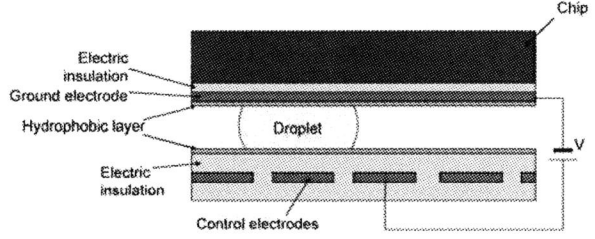

Figure 6: Schematic of an electrostatic actuated system used for chip cooling

The principle of electrostatic actuation of droplets in small channels is shown in Figure 6. In this system, a so-called spacer introduces a small gap between a chip and the substrate. On the chip side, a full metallization layer is deposited between two dielectric materials. Thereby, the dielectric layer at the gap side is chosen to be hydrophobic. The metal layer at the chip side is grounded. At substrate side, separate, individually addressable electrodes, separated by an insulator are mounted. By connecting one or several electrodes to a voltage source while keeping the others grounded, an

740

electric field will be generated which induces charges in the solid-liquid and liquid-vapour interfaces of the droplet. If the voltage between the electrodes is sufficiently high the droplet moves towards the powered electrodes due to the electrostatic force originating from the attraction between induced charges and charged electrodes. By switching the electrodes sequentially the droplet can be transported along the electrode path. With a good choice of the powering frequency for the different succeeding pads, the droplets can have a continuous movement through the gap removing the heat of the chip. The actuated droplet will take away the heat produced in the chip by passing through the gap. When the droplets pass the end of the gap, they should be gathered and cooled at another location of the system, before they are re-entering the chip cooling system. This system has important advantages over other cooling systems: the droplets are transported over the electrodes without the need for mechanical pumps and valves; low power consumption [8]; compact and light structure; the thermal performance is independent of the orientation of the device. However, the cooling capacity of this system is expected to be well below the one of forced continuous liquid cooling.

In order to assess the cooling capabilities of this system and its possible improvements the electro-actuating phenomenon should be investigated in detail. To this end, the droplet geometry in a static situation was combined with an electric field finite element simulation in [12]. For a droplet radius of 1.5 mm, a substrate thickness of 5 mm with a 2 μm insulating coating, a channel height of 1 mm and electrodes of 0.9 mm spaced with 0.1 mm the achieved electrostatic force is shown in Figure 7.

liquid flow, i.e. for the droplet flow an actuation power per volumetric flow of 26 mJ/l is required whereas for the continuous flow this ratio amounts to 60 J/l. The first number should of course be further increased to take account for the efficiency of the electrical switching system. On the other hand, also for continuous liquid cooling the electrical and mechanical pump efficiencies were not taken into account. The cooling capacity of the droplet on the other hand should be investigated by looking to the detailed convection patterns in the droplet. This can be done by performing CFD computations on a droplet [16]. These CFD computations reveal the rolling feature of the droplet, leading to a thermal barrier in between the upper and lower convection cell (see Figure 8) at one hand and an increased heat transfer over the conduction effect due to convection on the other. Transient simulations finally lead to global thermal resistances for the chip as shown in Figure 9. These thermal resistances are crucially dependent on the droplet velocity.

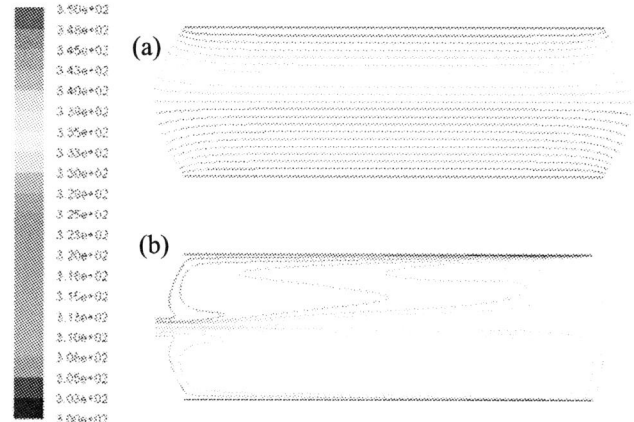

Figure 8: Temperature distribution (K) (a) Stationary droplet. (b) Moving droplet 10cm/s

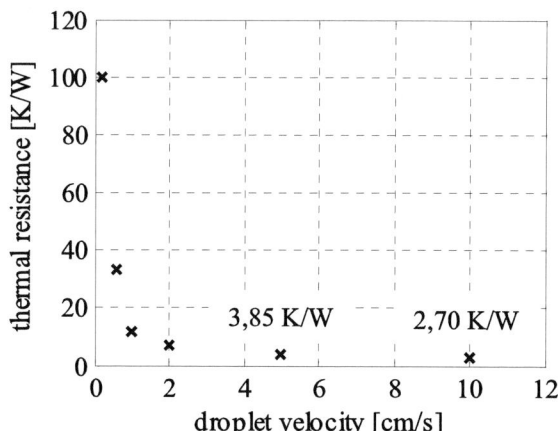

Figure 9: Total thermal resistance of the droplet flow for different droplet velocities.

Figure 7: Horizontal component of the electrostatic force as a function of the droplet position

Using this information for simulation of the droplet motion led to experimentally observed values of the order of 10 cm/s. This number leads to a typical power excerted on the droplets in the order of 2 μW per droplet or for the global system under investigation to 10 μW at a volumetric flow of 392 μl/s. This means that the rolling movement of the droplet outperforms the continuous

The above analysis typically leads to a cooling rate of 18.5 W for the 10x10 mm² chip at 50K temperature

741

difference. In this context, also the low pumping requirements should be re-assessed. Indeed, for a global thermal resistance of 2.7 K/W, the continous flow through microchannels would achieve values for the pressure drop around 1.5 Pa and mass flows of 0.35 g/s, which also would result in a power consumption of 0.5 μW only. Though the cooling rate is significantly reduced in comparison with the continuous flow configuration, the droplet flow might be favourable when reconfigurable systems are needed. Furthermore, the controllable function of the electrodes make this cooling option attractive to produce efficient and programmable cooling systems. In addition to the possible use for efficient and uniform cooling of the chip, this system yields also a high potential to preferentially cool hot spots. Indeed, in this systems droplets can be directed towards specific locations. However, there are still many questions to be solved with respect to reliability in general, wetting of the surface by previous liquid droplets, bio-fouling, etc.

5. Micro channel electrowetting

The electrostatic actuation, used in the previous sections to induce droplet flow, is often referred to as electrowetting as it can be used to alter the wetting properties of a droplet [17,18]. Indeed, the shape of a droplet on an insulated substrate can be altered by applying a voltage difference U between substrate and liquid. The principle of electrowetting is illustrated in Figure 10. On a flat substrate, this effect is governed by Lippmann's equation:

$$\cos \theta = \cos \theta_Y + \left(\frac{U}{U_L} \right)^2$$

where $U_L = (2D\gamma_{LV}/\varepsilon_0\varepsilon_r)^{1/2}$ is a characteristic of the system, with D the silicon oxide layer thickness, ε_0 the dielectric permittivity of vacuum, ε_r the dielectric constant, γ_{LV} the liquid-vapor surface tension and θ_Y the Young's contact angle of a droplet on a flat surface. It is this change in contact angle that can be used to move liquid over solid walls. This phenomenon can also be used to wet microchannels. Indeed, a liquid droplet deposited on an open microchannel with rectangular cross-section has multiple equilibrium states. There are typically two prominent states in which the drop can reside on the channels: depending on channel geometry and liquid properties, the drop either sits on top of the channels or it wets the channels.

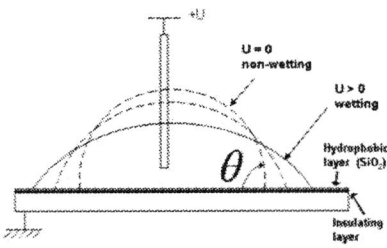

Figure 10: Principle of electrowetting

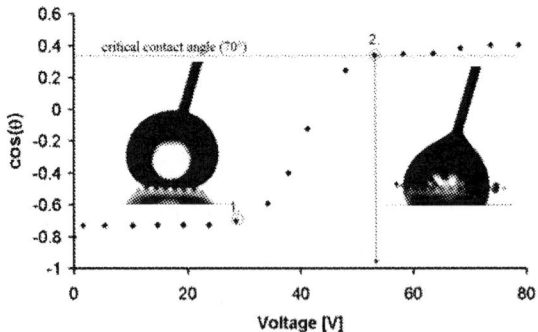

Figure 11: Contact angle measurements for a water-glycerol-salt mixture on a microchannel structure as a function of the applied voltage

As an example, a 5μl water droplet deposited on the channels will not wet the microchannels as shown in Figure 11, left. It will sit on top of the channels. However, by applying a voltage, the liquid wants to wet the surface and the liquid will penetrate in the channels (Figure 11, right). If the voltage is even more increased, the liquid will spread by filling the microchannels. The filling length of the liquid in the microchannel is finite and determined by the applied voltage at one hand and by the frequency of the voltage source on the other. The transition of a 5 μl droplet of the water-salt-glycerol mixture on the microchannels is monitored by the contact angle in Figure 11. Transition point '1' indicates the start of the transition. At point '2' the liquid wets all channels. From that point on, the filling of the microchannels grows with increased voltage. The filling and emptying of the channels in the presented experiments was established with a filling frequency of 0.1 Hz. A filling length of 1750μm is realized for an actuation voltage of 53.6V.

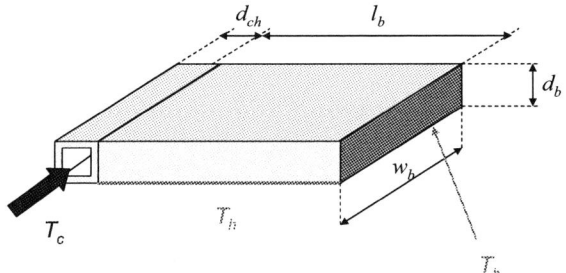

Figure 12: Schematic of the reference cooling system

To assess the enhancement in liquid cooling that can be achieved by electrowetting, cooling capabilities of a system with electrowetted microchannels are compared to those of a system without microchannels. The reference system is a silicon block with dimensions $l_b \times w_b \times d_b$ as presented in Figure 12. At the left surface, liquid flowing through an adjacent channel extracts the heat from the system. In general most of the heat is dissipated at the bottom side of the chip. Here we consider a worst case cooling scenario where the heat dissipation is concentrated at the right side of the block. For this worst case situation, the heat dissipation is assumed to be concentrated at the right side of the block. Therefore, the right surface and the bottom surface of the block are assumed to be heated up to a maximum allowed temperature T_h. In first approximation, the heat that can be extracted by the reference system is given by

$$\dot{Q} = \frac{\Delta T}{R_{th,cond} + R_{th,conv}}$$

For the 10x10 mm^2 chip under investigation and with $d_{ch} \cong 1$ mm and $d_b \cong 1$ mm, the maximum heat transfer that can be extracted is limited both by the heat transfer coefficient in the channel and by the heat conduction through the Si-block. For this numerical example a thermal resistance of 55,6 K/W would be achieved. The cooling capabilities of the system can be increased by the addition of electrowetted system microchannels in the silicon block, which are connected to the main channel. In these microchannels a pulsating flow will be generated by supplying an electric voltage. A fraction v of the cross section of the silicon is replaced by microchannels. The total heat extracted by the system consists of a convective part by the pulsating liquid and a conductive part through the silicon block. The heat extracted by the liquid filling one channel during one filling period can be analytically calculated using the time dependent filling function of the liquid in the channel. (See [17] for details on the calculations). The filling length L of the liquid in the microchannels can be described by following equation:

$$L = K\sqrt{t}$$

where K is a parameter depending on the wetting properties of the liquid, the geometry of the channels and the applied voltage. From experimental data, we can conclude that for the case at 0.1 Hz filling frequencies and an actuation voltage of 53.6 V, K=0.0015.

For a certain microchannel fraction v the conductive heat transfer through the silicon can be added to calculate the total heat extracted by the system. The effect of the microchannel fraction is linear between the reference case of convection of the main channel only and the hypothetic value for v =1. In practice the value of v is limited to 0.5. The dependence of the heat transfer with the filling period and the fraction v is shown in Figure 13.

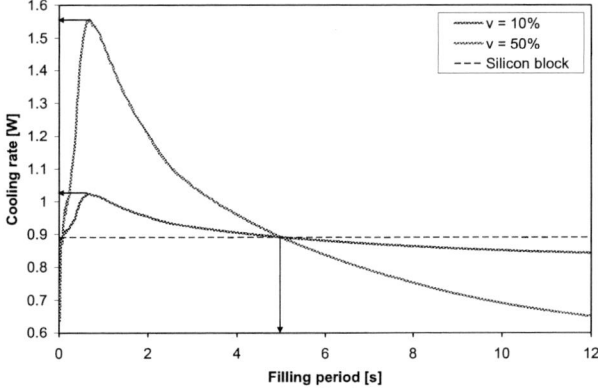

Figure 13: Cooling rate of the electrowetted channels as function of the filling period and for different values of the microchannel fraction

From this figure it is clear that the different curves do intersect at a critical filling period. For shorter filling periods, the heat transfer will be improved by inserting microchannels, for higher filling periods the electrowetting deteriorates the cooling. For the example under consideration, this critical filling period amounts to 4.4s. A maximal cooling rate is achieved for an optimal filling frequency. For v=0.5 the rate is nearly twice as high as for conduction cooling only. For the 100µm square microchannels this optimal frequency amounts to 4 Hz. This is still a feasible frequency that can be achieved in the system using the electrowetting actuation. At this frequency the cooling capacity can be increased with 55% in comparison with a pure conductive system. This lead to a total thermal resistance of 35.9 K/W.

Though much higher thermal resistances are induced than in the cooling techniques elaborated in the previous sections, this system might be promising for use as an integrated enhanced cooling technique. Furthermore, it is expected that significantly better results can be achieved by further optimizing the channel layout and by the use of other fluids, such as liquid metals.

6. Conclusions and challenges for further research

In this paper digital and continous liquid cooling techniques were explored and their cooling capabilities assessed. Though the continuous liquid flow requires a separate feed pump to force the liquid flow though the channels, it clearly surpasses the electrowetted systems with respect to thermal resistance as low as 0.2 K/W. At a channel pressure drop of approximately 0.5 bar, the microchannel seems to realize a good compromise between pumping power and cooling performance. The electro-actuated droplet flow on the other hand is expected to realise lower cooling rates. However, its pumping power per mass flow rate is relatively low, and its flexible manipulation opens new possibilities for integrated and controllable cooling systems. Finally the electrowetted microchannel filling only slightly improves the cooling rate compared to conduction. Though not completely elaborated yet, it might find its applicability in small scale and integrated cooling systems.

Some of the systems proposed in this paper might take advantage by using other liquids or even by exploiting two-phase heat transfer. In addition, further optimization of the electrowetted systems, is expected to improve their performance. Furthermore, challenges for the proposed systems are their technolgical implementation and reliability (the long term effect of coatings, bio-fouling, etc.).

Acknowledgments

This research is financed by a scholarship for H. Oprins and the SBO-project 030288 "PowerMEMS" of the IWT, the Institute for the Promotion of Innovation by Science and Technology in Flanders, Belgium. Part of the work was performed in co-operation C. Nicole, L. van der Tempel and C. Lasance of Philips Research, the Netherlands.

References

1. Tuckerman, D. B. and Pease, R. F. W., "High-Performance Heat Sinking for VLSI", *IEEE Electron Device Letters*, Vol. 2 (1981), pp. 126-129.
2. Morini G.L., "Single-phase convective heat transfer in microchannels: a review of experimental results", *Int. J. Thermal Sciences*, Vol. 43 (2004), pp. 631-651.
3. Liu, D. and Garimella, S. V., "Analysis and optimization of the thermal performance of micro-channel heat sinks", *Int. J. Numerical Methods Heat Fluid Flow*, Vol. 15 (2005), pp.7-26.
4. Ryu, J. H., Choi, D. H. and Kim, S. J., "Numerical optimization of the thermal performance of a microchannel heat sink", *Int. J. Heat Mass Transfer*, Vol. 45 (2002), pp.2823-2827.
5. Stevens, T., Rogiers, F., Baelmans, M., "Optimization of micro-channel heat sink geometry", *Proc. of the 13th International Heat Transfer Conference (IHTC-13)*, Sidney, Australië, August 2006.
6. Laser D.J., Santiago J.G., "A review of micropumps", *J. of Micromech. and Microeng.*, Vol. 14 (2004), R35-R64.
7. Singhal V., Garimella S.V. and Raman A., "Microscale pumping technologies for microchannel cooling systems", *Applied Mechanics Review*, Vol. 57 (2004), pp. 191-221.
8. Pollack M.G. et al, "Electrowetting-based actuation of droplets for integrated microfluidics", *The Royal Society of Chemistry, Lab Chip*, Vol. 2 (2002), pp. 96-101.
9. Washizu M., "Electrostatic actuation of liquid droplets for microreactor applications", IEEE Trans. Industry Appl., Vol. 34(1998), No. 4, pp.732-737.
10. Cho S., Moon H., Kim C.-J., "Creating, transporting, cutting and merging liquid droplets by electrowetting-based actuation for digital microfluidic circuits", *Journal of microelectromechanical systems,* Vol 12 (2003), No. 1, pp. 70-80.
11. Pamula V.K. and Chakrabarty K., "Cooling of integrated circuits using droplet-based microfluidics", *Proc. ACM Great Lakes Symposium on VLSI*, Washington, DC, USA, April 2003, pp. 84-87.
12. Oprins H., Vandevelde B., Beyne E., Borghs G., Baelmans M., "Selective Cooling of microelectronics using electrostatc actuated liquid droplets, Proc. *THERMINIC*, Sophia Antipolis, France, September 2004, pp. 207-212.
13. Judy, J., Maynes, D., Webb, B.W., "Characterization of frictional pressure drop for liquid through microchannels", *Int. J. Heat Mass Transfer*, Vol. 45 (2002), pp. 3477-3489.
14. Stevens, T., Rogiers, F., Delport, S., Vleugels, P., Peirs, J., Baelmans, M., "Collector pressure losses in micro heat exchangers", *Proc. of the Conference on Thermal Issues in Emerging Technologies, ThETA 1*, Cairo, Egypt, January 2007, pp. 11-17.
15. Shah, R. K. and London, A. L., "Laminar Flow Forced Convection in Ducts", Advances in Heat Transfer, Supplement 1, Academic Press (New York, 1978).
16. Oprins, H., Danneels, J., Van Ham, B., Vandevelde, B., Baelmans, M., "Convection heat transfer in electrostatic actuated liquid droplets for electronics cooling, *Proc. of THERMES II 2007 (Thermal challenges in next generation electronic systems),* Santa Fe, New Mexico, January, 2007, pp. 223-230, ed. S.V. Garimella and A.S. Fleischer, Millpress, Rotterdam, 2007, ISBN 978-90-5966-051-9.
17. Oprins, H., Nicole, C.C.S., Baret, J.C., Van der Veken, G., Lasance, C., Baelmans, M., "On-Chip Liquid Cooling with Integrated Pump Technology", *Proc. of SEMITHERM (Semiconductor thermal measurement and management symposium)*, San Jose, California, USA, March 2005,pp. 347-353.
18. Oprins, H., Van der Veken, G., Lasance, C., Baelmans, M., "On-Chip Liquid Cooling with Integrated Pump Technology", accepted for publication in *IEEE Transactions on Components and Packaging Technologies*, 2007.

Solid State Chemical Sensors: Technologies and Applications

Krishna C. Persaud, Anthony Flint, Robert W. Sneath[1],
The University of Manchester
SCEAS, PO Box 88, Manchester M60 1QD
[1]Silsoe Odours Ltd, Silsoe, UK.
Email: krishna.persaud@manchester.ac.uk

Abstract

The agricultural industry needs to be more accountable with regard to environmental pollution. This has arisen because the adoption of increasingly intensive farming methods in agriculture, with the associated production of significant quantities of solid and gaseous waste emissions that has, in recent years, been linked to adverse environmental change. Solid state chemical sensors are highly desirable, but available technology still lacks robustness for use in the agricultural field. This paper reviews sensing, sampling and measurement technology applicable to real-time monitoring of gaseous fluxes.

1. Introduction

Examples of agricultural emissions of interest in recent years have included ammonia (NH_3), carbon dioxide (CO_2), hydrogen sulphide (H_2S) and methane (CH_4). Of these CO_2 and CH_4 are both odourless and colourless and yet they can be present in potentially dangerous concentrations within poorly ventilated livestock buildings or underfloor waste slurry pits. CO_2, though not highly toxic, can cause asphyxiation in enclosed spaces if there is insufficient oxygen (O_2) available. Hydrogen sulphide (H_2S) is considered to be one of the most toxic gases and is commonly evolved from porcine excreta. It has a distinctive odour of rotten eggs, recognizable at concentrations of 1ppm or less, yet at 100ppm or greater, H_2S will deaden the sense of smell (olfactory fatigue) and no odour may be obvious. Ammonia also has a distinctive odour and is an irritant to the eyes at 20-25ppm with a risk of damage to the corneas at higher concentrations and/or prolonged exposure. Ammonia exposure in the workplace has also been linked to inflammatory reactions of the lung tissue.

Medical symptoms arising from acute exposure to toxic gases are dependent upon concentration and exposure time. However, empirical research has shown that there is not usually a simple reciprocal relationship between these two factors. Typically concentration is considered more important than the exposure time when assessing the harm potential for any given incident . Ammonia will generally cause irritation of the upper respiratory tract at <200ppm, breathing difficulties at <500ppm for a few minutes, whilst at 5000 – 10,000ppm it will result in lower respiratory tract damage and death may occur at concentrations >10,000ppm (1-3)

In contrast to acute incidents of toxic gas release, in agriculture it is long-term exposure that is of primary concern. Currently, for workers, the long-term exposure limit (LTEL) for ammonia is a time-weighted average (TWA) of 25ppm ($17mg.m^{-3}$) over an 8 hour period with a short-term exposure limit (STEL) of 35ppm ($24mg.m^{-3}$) over 10 minutes. However, for livestock, depending on species, exposure times and country of origin, the maximum acceptable concentrations (MAC) are more stringent than for humans because they are permanently kept in these atmospheres and their long-term comfort and safety is considered paramount. An ammonia-flux measurement system that operates in real-time is highly desirable.

2. Instrumentation
2.1. Gas chromatography

Gas chromatography (GC) is a sample component separation technique which uses a heated column containing a well characterized immobilized layer upon the inner wall known as the stationary phase. A carrier-gas eg. nitrogen (N_2) or helium (He), the mobile phase, is fed through the column. The sample is injected into the gas stream and components of the sample elute from the column where they may be detected by a suitable sensing device. Three commonly used detectors are the thermal conductivity detector (TCD), the flame ionisation detector (FID) and the electron capture detector (ECD). The column retention time for the sample components depends in part upon the strength of the adsorption interaction (affinity) with the stationary phase. The stronger the interaction is, the longer it takes for a component to elute from the column.

Retention times are also influenced by the species molecular size and shape. Further, the thermal regime will also have an influence, thus the column is normally contained within a programmable oven.

Although originally developed as a laboratory instrument, portable systems are now available, and GC's are commonly used in the field, in vehicle-based mobile laboratories. Their power supply requirements, relatively costly columns and the need for a carrier-gas make them unsuitable for adaptation to distributed multi-point sampling. Further, by virtue of their mode of operation, they do not generally offer near real-time performance. Electronic Sensor Technology (USA) has developed a breakthrough chemical vapor analysis technology which is able to analyze vapours in just 10 seconds. This new technology based upon ultra-high speed gas chromatography, has resulted in a revolutionary instrument called the zNose®. This incorporates a surface acoustic wave detector with high sensitivity (4).

1-4244-1105-X/07/$25.00 ©2007 IEEE

2.2. Mass spectrometry

The mass spectrometer (MS) typically uses a high-energy electron beam impact to convert atomic and molecular species, within a high-vacuum chamber, into ionic fragments. A series of detector plates (instead of the CRT phosphor) are located increasingly down-range of the electron gun. The mass/charge ratio of ionic fragments at the detector plates are then analysed. Many such systems often use a quadrupole ion-trap configuration where four parallel poles are charged with both direct current (DC) and an oscillating radio-frequency (RF) voltage. Ions follow a highly specific oscillatory path to the detector and thus by scanning both DC and AC potentials, the range of ions produced from the sample may be detected. The MS may be used as the detector for a GC and the GC-MS configuration is commonly considered to provide a "gold standard" in terms of many sample analyses.

Their complexity and cost may rule them out even for many single-point sampling applications in the field and on this basis are less likely to be used for regular monitoring of agricultural environments than a stand-alone GC.

2.3 Ion mobility spectrometry

Ion mobility spectrometry (IMS) is an electrophoretic technique that allows ionised analyte molecules to be separated on the basis of their mobilities in the gas phase. The technique has found widespread application as a detector, most noticeably for chemical warfare agents on the battlefield and for explosives and narcotics at ports and airports. This technique is rapidly being applied to environmental monitoring in specific situations (4;5).

2.4 Infra-red spectrophotometry

Monochromatic light energy is fed into a sample, heat energy is radiated and the absorption spectrum is recorded. Typically, a contemporary spectrophotometer is of the dual beam type where monochromatic light travels via a splitter through both a reference and sample cell, to two photo-detectors. The output signals from the detectors are fed into a difference amplifier from where the final signal is taken. The requirement for the gas sample to be placed in a sample cell presents a fundamental problem when designing multi-point gas sensing systems as fewer instruments require a greater number of longer sample delivery lines with the need for line switches and pumps. Further, separate provision for airflow monitoring would also be needed.

Fourier-transform infra-red (FTIR) spectroscopy operates in the frequency domain monitoring a range of frequencies concurrently such that several target analytes may be searched for simultaneously (6;7). The sample data from the frequency domain is converted to the time domain using an inverse transform function. This method would be useful for monitoring complex atmospheres as found within livestock housing. There have been attempts

to miniaturize these instruments and to produce devices with no moving parts that can be used in the field for monitoring airborne pollutants. However, implementation of this technique still requires complex and costly instrumentation, which is contrary to the requirement for low-cost flux monitoring. Tuneable diode-laser IR absorption spectrometry uses an extremely sensitive, narrow line-width diode to provide detailed spectral information. Tuning is by adjustment of diode temperature or current and the laser output may also be modulated to enhance signal processing. As with the FTIR spectrometer, instrumentation is complex and costly.

Current methods of measuring atmospheric ammonia include IR gas-correlation spectroscopy using a modulating reference cell containing an ammonia sample at a high enough concentration to produce strong absorption and a second cell containing a gas that does not absorb at the same wavelengths as the target analyte. The chopped light beam source is further modulated by the rotating dual-cell filter wheel. This technique is popular because it is not affected by humidity. Again, this type of already complex instrument would require the installation of sample lines, switching circuits and pumps together with additional airflow monitoring equipment.

2.5 Laser-based reflectometry

This uses the same basic principles as the spectrophotometer. Light detection and ranging (LIDAR) reflectometric systems measure back-scattering of pulsed laser light from atmospheric aerosols (Mie scattering) or molecules (Rayleigh scattering). Some of the laser light is reflected back along the transmittance path and captured by a photo-detector. Wavelengths used are typically in the ultra-violet or visible (UV-Vis) band. Absorption spectra provide detail regarding the absorbing species and the pulse-return time delay together with the angle of inclination provides spatial information. A variant of this system is differential absorption LIDAR or DIAL which is typically used to monitor for specific pollutants, often exhaust plumes from industrial chimney-stacks. DIAL uses two simultaneous light-pulses at adjacent wavelengths (8;9). Scattering effects are assumed to be wavelength independent. One of the wavelengths chosen is strongly absorbed by the target species whilst the other is not. If the pollutant is not present then the signals are identical, otherwise signal difference is used to determine concentration. These systems may be located in fixed positions for monitoring industrial plants but can also be operated from a moving platform such as a ground-based vehicle or aircraft. Although this technique can provide local remote sensing at the margins of structures, it requires uninterrupted line of sight.

2.6 Chemiluminescent nitric oxide analyzer

Chemiluminescence is the emission of light where the energy is provided by a chemical reaction rather than by some external source. An example of the application of this type of reaction is one where nitric oxide (NO) may be measured indirectly by its reaction with ozone (O_3) which produces nitrogen dioxide (NO_2) via a radical (reactive uncharged molecular fragment having an unpaired electron) species together with the emission of light energy. The light is emitted as a consequence of changes to the bonding energy between molecules.

About 10% of the NO_2 is produced (inside a light-tight reaction chamber) in an electronically excited state and emits visible light on its return to the ground-state where light intensity, with excess ozone present, is proportional to the mass-flow rate of nitric oxide. The detector is a photo-multiplier tube which provides an amplified signal proportional to the light emitted by the reaction (10).

Ammonia may be converted to nitric oxide by passing the gas through a stainless steel tube heated to 600-800°C. Thus ammonia may be measured indirectly by placing a thermal ammonia converter upstream of the nitric oxide analyzers' chemiluminescence reaction chamber.

2.7 Array Based Gas Sensor Instrumentation

The array-based gas analysis instrument typically comprises an array of sensors where each sensor has partial specificity to molecules or atoms within a broad chemical class. Each sensors' affinity, for specific analytes within a given class, may be slightly or considerably different from that of the other sensors within the same array. The sensor array output is used to produce a feature-rich map that may be interpreted by pattern recognition software in order to classify the sample.

The hardware, in addition to the sensor array, signal conditioning and data acquisition, broadly comprises an array header through which the sample is pumped, array heater for thermal stability and/or improved response and valve switching to purge the lines and header with an inert gas. Each sensor might, for example, be employed as a chemo-resistor in a constant-current, amplifier feedback circuit. Surface and bulk electronic changes of the sensor due to adsorbing and desorbing species, produces a change in the amplifier output voltage that is proportional to analyte concentration over a characteristic dynamic range.

Constraints may be placed upon the choice of sensor technology suitable for e-Nose instrumentation due to the multi-channel interfacing requirements and the need to acquire data as rapidly as possible. Technologies that have been successfully employed include metal-oxide semi-conductors (MOS) and conducting polymers. Although both MOS and conducting polymer sensors can be reversibly responsive to ammonia, the former require elevated temperatures of around 400-550 °C to function correctly, detecting adsorbed decomposition by-products, whilst the latter have a direct response to the analyte of interest, at or near ambient temperature.

Recent advances in low power, large scale integrated (LSI) micro-electronics enables "smart-sensor" modules to have a greater degree of autonomy via local signal conditioning and increased data storage capacity (11-13). Thus, it is now a practical proposition to construct small multi-sensor, data acquisition modules for use in distributed, large area gas sensing systems. The combination of presently available, low-cost, low-power, high specification digital and linear circuitry with conducting polymer sensor arrays offers the basis for development of a multi-point, distributed ammonia flux measurement system.

2.8 Ion-selective electrode

The movement of electrical charge due to chemical processes occurring within liquid media is typically due either to the passage of electrons during oxidation-reduction or the diffusion of ions from a region of high ionic concentration to one of lower concentration. Oxidation-reduction reactions may be observed using a simple potentiometric electrochemical cell comprising an indicator electrode, a reference electrode and a liquid or solid-phase electrolyte. For dissimilar electrode materials oxidation occurs at one electrode and the electrons pass to the reducing electrode, thus a potential difference may be measured (often via an amplification stage) across the electrodes depending upon the nature of the electrode materials and the electrolyte. It should be noted that standard potentials and slope factors (sensitivity) are temperature dependent and typically some form of temperature compensation circuitry may also be required. Electrochemical measurements may be made either with the electrochemical cell in equilibrium, when no current flows (potentiometric where "static" potential is measured) or away from equilibrium conditions (amperometric or voltammetric) where current flows. Techniques based upon these two modes are commonly used to determine the concentration of ionic species within aqueous samples. For example, an electro-chemical cell may be used as the detector for field-based liquid-phase atmospheric denuders.

Where ion selectivity is conferred upon an electrochemical cell by the addition of a semi-permeable membrane, the cell is often referred to as an ion selective electrode (ISE). The device may be used as the basis for a gas sensor as changes to the population of ionic species within an electrolyte due to absorption from a headspace gas sample, can change the ion concentration gradient across the membrane, thus altering the potential observed. This type of device may be calibrated to provide either a pH reading or, for the general case, a p[ion] reading..

Because of the direct link that electro-chemical measurements have between chemical processes and the measurement circuitry and also because the ammonium (NH^{4+}) ISE has been successfully employed elsewhere for the detection of ammonia in aqueous samples, this relatively simple instrument is a strong contender for adaptation to a distributed monitoring system (14-17).

The ammonium ISE can be adapted for use as an ammonia gas sensor, detecting dissolved ammonia in aqueous samples. This ISE uses a hydrophobic gas permeable membrane to separate the sample solution from an ammonium chloride (NH_4Cl) internal solution. Dissolved ammonia diffuses through the membrane, changing the internal solution pH sensed by a pH electrode. A chloride ion (Cl^-) selective electrode acts as the reference having constant activity (a_i), where a_i, unlike concentration, takes account of ionic interactions. The ISE exhibits a potential difference that is dependent upon the activity of a specific ion in the electrolyte. This is because electrolyte ions interact with the solvent, becoming solvated (solvent sheathed) and the interactions between solvated ions are such that their activity is different from their concentration (c) in solution by an activity coefficient factor.

An ISE is ideally constructed in such a way that there is selective passage of a single ionic species from one phase, the sample solution to the membrane phase. The membrane may be solid glass or liquid in a support matrix. Charge separation takes place at the interface producing a potential difference.

Typically, selectivity for ionic species is a distinct advantage for the ISE, however, positive interferences may arise from the presence of amine species, whilst mercury (Hg) and silver (Ag) complex with ammonia to give a lower reading than would otherwise occur. Further, despite the basic cell having a relatively simple construction, these devices become more complex once a suitable sample delivery system is included.

2.2.8 Wet chemistry denuders

A denuder scavenges gaseous species from the atmosphere where the reactive media may take the form of either a permeable solid matrix, liquid reservoir or alternatively a thin adsorbent solid or liquid layer, the sample gas passing over a coated surface. These devices integrate flux over time providing a time-averaged value for concentration where the sample flow rate and exposure time are known.

A reservoir-type denuder or bubbler, has the sample pumped through a reactive liquid, trapping the analyte. One application is the determination of sulphur dioxide (SO_2) in the atmosphere where air is pumped through water, trapping SO_2 as sulphuric acid (H_2SO_4). As atmospheric ammonia is also absorbed, the solution is analysed for the presence of sulphate (18;19). The disadvantage of these devices is the need to pump the sample air through permeable solid or liquid media at a constant rate.

A denuder that has the sample passing over the surface of a thin liquid or solid coating is often referred to as an impinger. This type of denuder does not impede the sample airflow in the same way as a bubbler and it is possible to construct an impinger-type denuder that passively integrates the analyte over time without the need for a pump. The adsorptive passive samplers are a good example of this type of device, employing a solid reactive coating.

An example of the liquid-coated absorptive impinger is the annular-design (two concentric tubes) ammonia analyzer from Anasys©, Albergen, The Netherlands. This active device has the air sample drawn between two rotating glass tubes which have the facing inner surfaces constantly re-coated with a thin film of water. Ammonia dissolves into the film and this aqueous solution is pumped away from the denuder and mixed with an alkali to liberate the ammonia which then passes via a permeable membrane into a flow of pure water where changes in conductivity are constantly recorded. Although this type of device requires a pump, it has the advantage of combining operation in near real-time, relative simplicity and relatively low power consumption. The disadvantages include the need for each device to have a water-source and drain.

3. Discussion

No ideal solid state sensor technology yet exists for monitoring specific gases in the harsh context of agriculture. In the context of distributed multi-point environmental measurements in an agricultural setting, the favoured gas sensing technologies may fall into one of the following areas –

- Laser-based (DIAL) reflectometry with airflow detection

- The ammonium ISE with airflow detection

- Liquid or solid-phase irreversible impinger-type denuders

- Near-ambient temperature reversible conducting polymers with airflow detection

Cost effective, distributed instrumentation arrays have been investigated to realise a robust automated ammonia flux measurement system suitable for use at the periphery of naturally ventilated livestock buildings (20). The system is designed to permit the calculation or ammonia flux indirectly from measurements of ammonia concentration, using conducting polymer sensors, and wind speed with direction, via a commercially available airflow sensor. It is necessary to be able to compensate for anomalous flow regimes within the sampler system, which can arise from changes to the external wind direction. Wind conditions, across the site of interest, should be faithfully recorded throughout, otherwise models based upon the acquired data may be unrealistic.

4. Conclusions

The key to the identification and final selection of appropriate technologies suitable for incorporation into an ammonia flux instrument lies in understanding the demanding nature of the application. Factors that strongly influence instrument design are strongly influenced by the operational environment and the likely atmospheric conditions. A broad sampling methodology for non-invasive distributed multi-point sampling for the entire site in near real-time is desirable. The choice of sensor technology that will work in such environments is seriously limited and presents a challenge to sensor researchers.

Acknowledgments

We acknowledge the immense contribution to the field of agricultural emission measurements made by our late friend and colleague Dr. V.R. Phillips of Silsoe Research Institute. Anthony Flint carried out a PhD in the area of ammonia flux measurement at the University of Manchester in collaboration with Silsoe Research Institute.

References

1. Phillips VR, Holden MR, Sneath RW, Short JL, White RP, Hartung J, et al. The development of robust methods for measuring concentrations and emission rates of gaseous and particulate air pollutants in livestock buildings. Journal of Agricultural Engineering Research 1998;70(1):11-24.

2. Phillips VR, Cowell DA, Sneath RW, Cumby TR, Williams AG, Demmers TGM, et al. An assessment of ways to abate ammonia emissions from UK livestock buildings and waste stores. Part 1: ranking exercise. Bioresource Technology 1999;70(2):143-55.

3. Phillips VR, Lee DS, Scholtens R, Garland JA, Sneath RW. A review of methods for measuring emission rates of ammonia from livestock buildings and slurry or manure stores, Part 2: Monitoring flux rates, concentrations and airflow rates. Journal of Agricultural Engineering Research 2001;78(1):1-14.

4. Staples EJ. First quantitatively validated electronic nose for environmental testing of air, water, and soil. Abstracts of Papers of the American Chemical Society 2000;219:U120.

5. Haiduc I, Bocos-Bintintan V. Modern techniques in harmful pollutants detection. Revista de Chimie 2006;57(9):973-7.

6. Gao MG, Liu WQ, Zhang TS, Liu JG, Lu YH, Wang YP, et al. Remote sensing of atmospheric trace gas by airborne passive FTIR. Spectroscopy and Spectral Analysis 2006;26(12):2203-6.

7. Xu LH, Feng YQ, Chen JQ. Study on simultaneous analysis of indoor air multi-component VOCs with FTIR. Spectroscopy and Spectral Analysis 2006;26(12):2197-9.

8. Clemitshaw KC. A review of instrumentation and measurement techniques for ground-based and airborne field studies of gas-phase tropospheric chemistry. Critical Reviews in Environmental Science and Technology 2004;34(1):1-108.

9. Vasil'ev BI, Mannoun OM. IR differential-absorption lidars for ecological monitoring of the environment. Quantum Electronics 2006;36(9):801-20.

10. Navas MJ, Jimenez AM, Galan G. Air analysis: Determination of nitrogen compounds by chemiluminescence. Atmospheric Environment 1997;31(21):3603-8.

11. Hauptmann PR. Selected examples of intelligent (micro) sensor systems: state-of-the-art and tendencies. Measurement Science & Technology 2006;17(3):459-66.

12. Persaud KC, Wareham P, Pisanelli AM, Scorsone E. "Electronic nose" - A new monitoring device for environmental applications. Sensors and Materials 2005;17(7):355-64.

13. Ulivieri N, Distante C, Luca T, Rocchi S, Siciliano P. IEEE1451.4: A way to standardize gas sensor. Sensors and Actuators B-Chemical 2006;114(1):141-51.

14. Hart JP, Serban S, Jones LJ, Biddle N, Pittson R, Drago GA. Selective and rapid biosensor integrated into a commercial hand-held instrument for the measurement of ammonium ion in sewage effluent. Analytical Letters 2006;39(8):1657-67.

15. Pivarnik L, Ellis P, Wang X, Reilly T. Standardization of the ammonia electrode method for evaluating seafood quality by correlation to sensory analysis. Journal of Food Science 2001;66(7):945-52.

16. Saleh MA, Ewane E, Wilson BL. Monitoring the Houston Ship Channel for inorganic pollutants by ion selective electrodes, ion chromatograpy, and inductively coupled plasma spectroscopy. Chemosphere 1999;39(13):2357-64.

17. Pivarnik LF, Thiam N, Ellis PC. Rapid determination of volatile bases in fish by using an ammonia ion-selective electrode. Journal of Aoac International 1998;81(5):1011-22.

18. Phillips VR, Lee DS, Scholtens R, Garland JA, Sneath RW. A review of methods for measuring emission rates of ammonia from livestock buildings and slurry or manure stores, Part 2:

Monitoring flux rates, concentrations and airflow rates. Journal of Agricultural Engineering Research 2001;78(1):1-14.

19. Flechard CR, Fowler D. Atmospheric ammonia at a moorland site. II: Long-term surface-atmosphere micrometeorological flux measurements. Quarterly Journal of the Royal Meteorological Society 1998;124(547):759-91.

20. Flint TA, Persaud KC, Sneath RW. Automated indirect method of ammonia flux measurement for agriculture: effect of incident wind angle on airflow measurements. Sensors and Actuators B-Chemical 2000;69(3):389-96.

Author Index

A

Achkar, H. 637
ACHKAR, Hikmat 588
Ahmad, J. 678
AHMAD, Mahmoud AL 588
Albrecht, Arne 560
Ansorge, Frank 447
Arnold, A. K. 555
Arruda, Luciano 545
Arshak, Khalil 352
Auersperg, Jürgen 524
Augustin, Adam 116
Axisa, Fabrice 340

B

Baek, Hyunggil 428
Baelmans, M. 738
Bagdahn, Joerg 672
Bailey, C. 159, 249, 593
Bailey, Chris 243, 653
Bakkers, E. P. A. M. 532
Baldwin-Hendricks, Teri 601
Barbé, J-C 142
Bartek, M. 361
Bauer, J. 7
Belov, Ilja 181
Beng, Lau Teck 150
Bergner, Fredrik 181
Beyne, Eric 491
Bhattacharyya, Bidyut K. 239
Bielen, Jeroen 503, 509
Bohm, C. 319
Boogaart, Marc A.F. van den 62
Borecki, Janusz 346
Bornoff, Robin 181
Bots, Tom 575
Bouarroudj, M. 409
Bousquet, S. 77
Boutaayamou, Mohamed 120
Brämer, Birgit 378
Brizoux, M. 659
Brosteaux, Dominique 340
Brugger, Juergen 62
Bruin, E. de 498
Buiu, Octavian 125
Bulcke, Mathieu Vanden 340
Buttay, Cyril 653

C

Cacchione, Fabrizio 303
Calata, Jesus N. 705
Carles, G. 565
Carmona, Manuel 332
Chae, Seung-Hyun 664
Chaillot, A. 77
Chan, Vincent 51

D

Chan, Y. S. 631
Chang, Chong-Qing 314
Chao, Brook 664
Chastanet, C. 77
Che, F. X. 283, 366, 416
Chen, Xu 705
Chiang, Kou-Ning 324
Chicharro, J. M. 583
Chin, N.H. 283
Cho, Ji-Man 234
Chunga, Jin Taek 234
Conway, Paul P. 308
Corder, Paul 679
Corfield, Martin 647
Corigliano, Alberto 303, 392
Crécy, F. de 142
Cuddalorepatta, Gayatri 537

D

Damani, M. 678
Danny, Retuta 366
Danto, Y. 659
Dasgupta, Abhijit 537
Dawotola, Alex. W. 532
Deml, Ulrich 207
Dermitzaki, E. 7
Doetzel, Wolfram 257
Dommelen, J.A.W. van 454
Dornel, E. 142
Dowhan, Lukasz 263
Downey, Susan 150
Driel, W. D. van 46, 85, 214, 372, 435, 468, 498, 532
Duch, M. 478
Duch, Marta 332
Dudek, Rainer 263
Dudek, Rainer 378
Dular, Patrick 120
Dumonteil, R. 77
Dupont, L. 409
Dusek, Milos 175

E

Ebert, Matthias 672
Elata, David 291
Eneman, Geert 491
Eng, P. F. 555
Engelen, R. A. B. 214, 468
Erinc, M. 27
Ernst, L. J. 319
Ernst, L. J. 361, 372, 423, 472
Esteve, J. 478
Eymery, J. 142

F

Fachin, Fabio 392
Fan, Haibo 515

Author Index

Faust, Wolfgang 378
Filho, W. C. Maia 659
Fiori, V. 468
Fiori, Vincent 109, 150
Flint, Anthony 745
Flores, Sebastian 314
Fonseca, L. 565
Forde, Edward 352
Frangi, Attilio 623
Freitas, Germano 545
Fremont, H. 659

G

Gallois-Garreignot, Sébastien 109, 150
Galvez, J.L. 403
Ganeshan, Vijay 357
Geers, M.G.D. 27, 454
Gerbach, Ronny 672
Gerbolés, Marta 332
Gessner, Thomas 257
Ghisi, Aldo 392, 623
Gholnejad, Hasan 1
Gille, Thomas 136
Gilles, J. P. 571
Gils, Marcel van 509
Godignon, P. 403
Goebel, J. 549
Goette, Carsten 207
Golinval, Jean-Claude 102
Gómez, E. 478
González, A. 583
Gonzalez, Mario 340, 491
Goux, Ludovic 136
Greer, James C. 62
Grout, Ian 484
Guiney, Ivor 352
GuoZhong, Chai 16
Guy, O. J. 555
Gwyer, D. 593

H

Hainz, Simon 33
Hal, B.A.E. van 454
Hall, Steve 125
HAN, Lei 23
Hang, Sung Woo 234
Hauck, Torsten 131
Hegde, Pradeep 385
Hegde, S. 678
Hering, Siegfried 560
Herkommer, Dominik 537
Hirshberg, Arnon 291
Ho, Paul S. 442, 664
Hoffmann, Martin 560
Hogg, Simon 647
Hölzer, Gisbert 560

Hsieh, Ming-Che 398
Hu, Dyi-Chung 324
Huang, Ching-Shun 324
Hughes, M. 249, 593
Hunt, Christopher 175
Huo, Gang 239

I

Igic, P. 555
Im, Sehyuk 442
Irving, Scott 220, 519, 710
Iwamoto, Nancy 601

J

Janiaud, D. 571
Jansen, K. M. B. 319, 361
Jansen, M. 435
Jiang, Li 56
Johnson, C Mark 647, 653
Jordà, X. 403
Ju, Byeong-Kwon 234
Jun, Wei 378
Jun, Zhou 16
Jungwirth, Mario 33

K

Kalms, A. 549
Kasemset, B. 498
Kaulfersch, Eberhard 357
Kessler, A. 319
Khatir, Z. 409
Kim, Jeongyeol 428
Kim, Kuyoung 428
Kima, Jae Choon 234
Kimb, Gyoung Bum 234
Klein, K. 678
Klein, Matthias 378, 524
Kolchuzhin, Vladimir 257
Korvink, J.G. 549
Kwak, Dongok 428

L

Lall, Pradeep 692
Larson, Amy 601
Lee, Dong Jin 234
Lee, S. W. Ricky 631
Lee, Won Suk 234
Leee, K. J. 678
Lefebvre, S. 409
Lei, Guangyin 705
Leisner, Peter 181
Lepage, Severine 102
Leus, Vitaly 291
Levy, R. 571
Li, Guang 283

Author Index

LI, Junhui 23
Li, Y. P. 723
Liang, Lihua 710
LiHua, Liang 16, 225
Lindgren, Mats 181
Lishchynska, Maryna 62
Lisoni, Judit 136
Liu, Sang 631
Liu, Y. 607
Liu, Yong 220, 519, 710
Liu, Yumin 519
Liu, Z. 549
Lo, G. 678
Loh, Wei-Sun 647
Lombaërt-Valot, I. 77
Lu, Guo-Quan 705
Lu, H. 593
Lu, Hua 51, 243, 647
Luk, Timwah 220, 519, 710

M

Mahalingam, S. 678
Mahmoudi, Jafar 1
Maj, Bartosz 116
Marco, S. 549, 565
Mariani, Stefano 392
Maron, D. 77
Massiot, G. 77
Mathias, H. 571
McDonald, Gavin 509
McManus, K. 249
Megherbi, S. 571
Mehner, Jan 257
Meier, K. 618
Meier, Karsten 93, 207
Meuwissen, Marcel 575
Meyer, Kristin De 136
Michael, Steffen 560
Michel, B. 7
Michel, Bernd 357, 378, 524
Miura, Hideo 686
Morales, A. L. 583
Moreno, R. 583
Moríñigo, José A. 461
Mueller, M. 189
Mueller, Maik 197
Müller, A. 618
Müller, Axel 295, 357
Müller, G. 549
Munier, C. 77
Munier, E. 77
Musallam, Mahera 653

N

Nakamura, Tomoji 442
Nakanishi, Tohru 270, 278

Naumann, Falk 672
Nieto, A. J. 583
Nieuwenhof, Monique van den 575
Niraula, Ratna P. 679
Nithiarasu, P. 555
Nuttall, Keith I. 125

O

Obreja, Vasile V.N. 125
Oh, Joonyoung 428
Ohira, Hiroshi 270, 278
Ohkuma, Hideo 270, 278
Okoro, Chukwudi 491
Oprins, H. 738
Orain, Stéphane 109, 150
O'Shea, Thomas 484

P

Pallarès, Jofre 332
Palmer, Ben 601
Pang, H.L.J. 416
Park, Heung-Woo 234
Parrain, F. 571
Parry, Dr. John 737
Pasion, J. 46
Pecht, Michael 729
Peerlings, R.H.J. 454
Pennec, F. 637
PENNEC, Fabienne 588
Pérez-Castillejos, R 478
Perkins, A. 678
Perpinyà, X. 403
Persaud, Krishna C. 745
Peyrou, D. 637
PEYROU, David 588
Pintado, P. 583
Plana, R. 637
PLANA, Robert 588
Plaza, J. A. 478
Plouseau, D. 77
Polster, Tobias 560
Pons, P. 637
PONS, Patrick 588
Preu, H. 319
Pucha, R. V. 678

Q

Qian, C. 319
Qian, Richard 220
Qiang, Wang 225
Quesada, José Hermida 461

R

Raynal, P. 77
Real, R. A. 46

Author Index

Rebholz, Christian ... 447
Reichl, H. ... 7
Requena, Francisco Caballero ... 461
Rixen, Daniel ... 69
Rochus, Veronique ... 69
Rodríguez, Gustavo Adolfo Ardila ... 167
Roellig, M. ... 189
Roellig, Mike ... 197, 207
Rogiers, F. ... 738
Röllig, M. ... 618
Röllig, Mike ... 93
Rongen, R.T.H. ... 423
Rossi, Carole ... 167
Ryan, Jeffrey ... 484
Rzepka, S. ... 618
Rzepka, Sven ... 295, 357

S

Sabariego, Ruth V. ... 120
Salleras, M. ... 549, 565
Samimi, M. ... 454
Samitier, J. ... 549
Santander, J. ... 565
SARTOR, Marc ... 588
Schmadlak, Ilko ... 131
Scholz, D. ... 618
Schreier-Alt, Thomas ... 447
Schreurs, P.J.G. ... 27
Shimokawa, Tomotsugu ... 641
Shin, Dongkil ... 428
Silberschmidt, Vadim. V. ... 385
Silfhout, R. B. R. van ... 214, 468
Silva, Alexandre Cesar Rodrigues da ... 484
Sitaraman, S. K. ... 678
Sluis, O. van der ... 85, 214, 372, 468, 472
Sluis, Olaf van der ... 109
Sneath,, Robert W. ... 745
Soestbergen, M. van ... 423
Song, Younghee ... 428
Stecher, M. ... 319
Steijvers, Henk ... 575
Stevens, T. ... 738
Stoyanov, S. ... 159, 593
Strusevich, N. ... 159
Stulemeijer, Jiri ... 503
Sun, Anthony ... 283
Sun, Anthony Y.S. ... 366, 416
Sun, F. L. ... 607
Sun, Wei ... 283, 366, 416
Suzuki, Ken ... 686
Swinnen, Bart ... 491

T

Takami, Kourosh Mousavi ... 1
Tamakawa, Kinji ... 686
Tan, H.B. ... 283, 366, 416

Terés, Lluís ... 332
Terré, J.Casals- ... 478
Tilford, Tim ... 243, 647
Traon, O. Le ... 571
Tuchband, Brian ... 729
Tucker, Paul G. ... 39
Tunga, K. ... 678
Tyacke, James C. ... 39

U

Ubachs, R.L.J.M. ... 85
UCHIBORI, Chihiro J. ... 442
Udina, S. ... 565

V

Vallés, E. ... 478
Vandepitte, Dirk ... 491
Vandevelde, Bart ... 340, 491
Vanfleteren, Jan ... 340
Velandia, Diana Segura ... 308
Vellvehi, M. ... 403
Verheyen, Peter ... 491
Vichare, Nikhil ... 729
Villard, S. ... 77
Vries, J. de ... 435

W

Waal, Adri van der ... 575
Walter, H. ... 7
Wang, C.K. ... 283, 366, 416
WANG, Fuliang ... 23
Wang, L. ... 361
Wang, L. F. ... 607
Wang, L. G. ... 372
Wang, Shinan ... 710
Wang, Song-Hao ... 314
Wang, Wenjie ... 56
Wang, Z. G. ... 723
Wei, Hsiu-Ping ... 324
WeiNa, Hao ... 16
West, Andrew A. ... 308
Whalley, David ... 385
Whalley, David C. ... 308
Whitehead, Michael ... 653
Wiese, S. ... 189, 618
Wiese, Steffen ... 93, 197, 207
Wilde, Jürgen ... 611
Wilson, Antony ... 308
Wittler, Olaf ... 378
Wolter, K. J. ... 189, 618
Wolter, Klaus-Juergen ... 197, 207
Wolter, Klaus-Jürgen ... 93
Wouters, Dirk J. ... 136
Wunderle, B. ... 7
Wymyslowski, Artur ... 263, 346

Author Index

X

Xiang, Ke 56
Xiao, A. 372
Xiaohong, Weng 225
Xue, Xiangdong 243
Xuefan, Chen 225

Y

Yang, D. G. 46, 372, 498
Yang, Keling 56
Yang, Wen-Kung 324
Yannou, J. M. 159
Ye, Yuming 631
Yellowaga, Deborah 601
Yew, Ming-Chih 324
Yin, C. 593
Ying, Qu Jian 729
Yong, Liu 16
Yu, J.-Hyuk 234
Yuan, C. A. 532
Yuan, Cadmus 472
Yuen, Matthew M.F. 515
Yun, Sang-Kyeong 234

Z

Zbrzezny, Adam Robert 51
Zerbini, Sarah 303, 392
Zhang, G. P. 723
Zhang, G.Q. 46, 85, 214, 361, 372, 423, 472, 532
Zhang, Kai 515
Zhang, Xuefeng 442, 664
Zhao, B.Z. 283
ZHONG, Jue 23
Zhou, Jiang 679
Zhou, Jiemin 56
Zhou, Ming 51
Zhu, W. H. 283, 366, 416
Zhu, X. F. 723
Zoumpoulidis, T. 361
Zukowski, Elena 611

9781424411054